Edited by Detlef Stolten and Bernd Emonts

Fuel Cell Science and Engineering

Related Titles

Li, X.

Polymer Electrolyte Membrane Fuel Cells

ISBN: 978-0-470-87110-2

Stolten, D., Scherer, V. (eds.)

Efficient Carbon Capture for Coal Power Plants

2011
ISBN: 978-3-527-33002-7

Stolten, D. (ed.)

Hydrogen and Fuel Cells
Fundamentals, Technologies and Applications

2010
ISBN: 978-3-527-32711-9

Hirscher, M. (ed.)

Handbook of Hydrogen Storage
New Materials for Future Energy Storage

2010
ISBN: 978-3-527-32273-2

Barbaro, P., Bianchini, C. (eds.)

Catalysis for Sustainable Energy Production

2009
ISBN: 978-3-527-32095-0

Mitsos, A., Barton, P. I. (eds.)

Microfabricated Power Generation Devices
Design and Technology

2009
ISBN: 978-3-527-32081-3

Vielstich, W.

Handbook of Fuel Cells

6 Volume Set
2009
ISBN: 978-0-470-74151-1

Edited by Detlef Stolten and Bernd Emonts

Fuel Cell Science and Engineering

Materials, Processes, Systems and Technology

Volume 1

WILEY-VCH Verlag GmbH & Co. KGaA

The Editors

Prof. Detlef Stolten
Forschungszentrum Jülich GmbH
IEF-3: Fuel Cells
Leo-Brandt-Straße
52425 Jülich
Germany

Dr. Bernd Emonts
Forschungszentrum Jülich GmbH
IEF-3: Fuel Cells
Leo-Brandt-Straße
52425 Jülich
Germany

We would like to thank the following institutions for providing us with the photographic material used in the cover illustration: IdaTech Fuel Cells GmbH, EnergieAgentur.NRW, and Forschungszentrum Jülich GmbH.

All books published by **Wiley-VCH** are carefully produced. Nevertheless, authors, editors, and publisher do not warrant the information contained in these books, including this book, to be free of errors. Readers are advised to keep in mind that statements, data, illustrations, procedural details or other items may inadvertently be inaccurate.

Library of Congress Card No.: applied for

British Library Cataloguing-in-Publication Data
A catalogue record for this book is available from the British Library.

Bibliographic information published by the Deutsche Nationalbibliothek
The Deutsche Nationalbibliothek lists this publication in the Deutsche Nationalbibliografie; detailed bibliographic data are available on the Internet at <http://dnb.d-nb.de>.

© 2012 Wiley-VCH Verlag & Co. KGaA, Boschstr. 12, 69469 Weinheim, Germany

All rights reserved (including those of translation into other languages). No part of this book may be reproduced in any form – by photoprinting, microfilm, or any other means – nor transmitted or translated into a machine language without written permission from the publishers. Registered names, trademarks, etc. used in this book, even when not specifically marked as such, are not to be considered unprotected by law.

Print ISBN: 978-3-527-33012-6
ePDF ISBN: 978-3-527-65027-9
ePub ISBN: 978-3-527-65026-2
mobi ISBN: 978-3-527-65025-5
oBook ISBN: 978-3-527-65024-8

Cover Design Formgeber, Eppelheim
Typesetting Laserwords Private Limited, Chennai, India
Printing and Binding betz-druck GmbH, Darmstadt

Printed in the Federal Republic of Germany
Printed on acid-free paper

Contents to Volume 1

List of Contributors XIX

Part I Technology 1

1 **Technical Advancement of Fuel-Cell Research and Development** 3
   Bernd Emonts, Ludger Blum, Thomas Grube, Werner Lehnert, Jürgen Mergel, Martin Müller, and Ralf Peters
1.1 Introduction 3
1.2 Representative Research Findings for SOFCs 4
1.2.1 Tubular Concepts 4
1.2.2 Planar Designs 6
1.2.3 Actors and Major Areas of Development 8
1.2.4 State of Cell and Stack Developments 10
1.3 Representative Research Findings for HT-PEFCs 11
1.3.1 Actors and Major Areas of Development 11
1.3.2 Characteristic Data for Cells and Stacks 12
1.4 Representative Research Findings for DMFCs 12
1.4.1 DMFCs for Portable Applications 13
1.4.2 DMFCs for Light Traction 14
1.5 Application and Demonstration in Transportation 17
1.5.1 Fuel Cells and Batteries for Propulsion 17
1.5.2 On-Board Power Supply with Fuel Cells 22
1.6 Fuel Cells for Stationary Applications 24
1.6.1 Stationary Applications in Building Technology 24
1.6.2 Stationary Industrial Applications 26
1.7 Special Markets for Fuel Cells 26
1.8 Marketable Development Results 27
1.8.1 Submarine 27
1.8.2 DMFC Battery Chargers 27
1.8.3 Uninterruptable Power Supply/Backup Power 29
1.8.4 Light Traction 30

1.9	Conclusion 30	
	References 32	
2	**Single-Chamber Fuel Cells** 43	
	Têko W. Napporn and Melanie Kuhn	
2.1	Introduction 43	
2.2	SC-SOFCs 44	
2.2.1	Basic Principles of Single-Chamber Fuel Cell Operation 44	
2.2.2	Catalysis in SC-SOFCs 46	
2.2.3	Heat Production and Real Cell Temperature 47	
2.2.4	Current Collection 48	
2.2.5	Electrode and Electrolyte Materials 48	
2.2.6	Anode Materials 48	
2.2.7	Cathode Materials 49	
2.2.8	Electrolyte Materials 50	
2.3	SC-SOFC Systems 50	
2.3.1	Electrolyte-Supported SC-SOFCs 50	
2.3.2	Anode-Supported SC-SOFCs 51	
2.3.3	SC-SOFCs with Coplanar Electrodes 52	
2.3.3.1	Cell Performance 52	
2.3.3.2	Miniaturization 56	
2.3.3.3	Limitations and Challenges 57	
2.3.4	Fully Porous SC-SOFCs 59	
2.3.5	Tubular SC-SOFCs 60	
2.4	Applications of SC-SOFCs Systems 60	
2.5	Conclusion 61	
	References 61	
3	**Technology and Applications of Molten Carbonate Fuel Cells** 67	
	Barbara Bosio, Elisabetta Arato, and Paolo Greppi	
3.1	Molten Carbonate Fuel Cells overview 67	
3.1.1	Operating Principle 67	
3.1.2	Operating Conditions 69	
3.1.3	Geometry and Materials 70	
3.1.4	Reforming 71	
3.1.5	Balance of Plant 73	
3.1.6	Vendors 75	
3.1.7	State of the Art 75	
3.2	Analysis of MCFC Technology 76	
3.2.1	Approach 76	
3.2.2	Technology Optimization 79	
3.2.3	Scientific Knowledge 81	
3.3	Conventional and Innovative Applications 86	
3.3.1	Distributed Generation 86	
3.3.2	Carbon Capture, Storage, and Transportation 87	

3.3.3	Hydrogen Co-generation	89
3.3.4	Renewable Fuels	89
3.3.5	Other Applications	90
3.4	Conclusion	90
	List of Symbols	91
	References	92

4 Alkaline Fuel Cells 97

Erich Gülzow

4.1	Historical Introduction and Principle	97
4.2	Concepts of Alkaline Fuel-Cell Design Concepts	99
4.2.1	Traditional Stacks	100
4.2.2	Eloflux Cell Design	100
4.2.3	Falling Film Cell	101
4.2.4	Bipolar Stack Concept by DLR	101
4.2.5	Hydrocell Concept	102
4.2.6	Ovonics Concept	103
4.2.7	Stack Design with Anion-Exchange Membranes	104
4.2.8	Alkaline Direct Ethanol Fuel Cells Assembled with a Non-Platinum Catalyst	104
4.2.8.1	Electrode Types	105
4.2.9	PTFE-Bonded Gas Diffusion Electrodes	105
4.2.10	Double-Skeleton Electrodes	106
4.2.10.1	Preparation and Electrode Materials	106
4.2.10.2	Dry Preparation of PTFE-Bonded Gas Diffusion Electrodes	108
4.2.11	Reduction of NiO	111
4.2.12	Production of Cathode Gas Diffusion Electrodes	113
4.3	Electrolytes and Separators	113
4.4	Degradation	114
4.4.1	Gas Diffusion Electrodes with Raney Nickel Catalysts	114
4.4.2	Gas Diffusion Electrodes with Silver Catalysts	121
4.5	Carbon Dioxide Behavior	123
4.6	Conclusion	126
	References	126

5 Micro Fuel Cells 131

Ulf Groos and Dietmar Gerteisen

5.1	Introduction	131
5.2	Physical Principles of Polymer Electrolyte Membrane Fuel Cells (PEMFCs)	132
5.3	Types of Micro Fuel Cells	134
5.3.1	Hydrogen-Fed Micro Fuel Cell	134
5.3.2	Micro-Reformed Hydrogen Fuel Cell	135
5.3.3	Direct Methanol Fuel Cell (DMFC)	135
5.3.4	Direct Ethanol Fuel Cell (DEFC)	136

5.4	Materials and Manufacturing 137
5.4.1	Miniaturization 137
5.5	GDL Optimization 138
5.5.1	Flow-Field Design 139
5.5.2	Miniaturized DMFC 141
5.5.3	Discharge of Carbon Dioxide 142
5.5.4	Passively Operating DMFC 142
5.6	Conclusion 142
	References 143
6	**Principles and Technology of Microbial Fuel Cells** **147**
	Jan B.A. Arends, Joachim Desloover, Sebastià Puig, and Willy Verstraete
6.1	Introduction 147
6.2	Materials and Methods 149
6.2.1	Electrode Materials 149
6.2.2	Membrane 151
6.2.3	Configurations and Design 151
6.2.4	Measurements, Techniques, and Reporting Values 152
6.2.4.1	Biological Measurements 152
6.2.4.2	Electrochemical Measurements 152
6.2.4.3	Reporting Performance 156
6.3	Microbial Catalysts 157
6.3.1	Anode Reactions 157
6.3.1.1	Electron Donors 158
6.3.1.2	Biocatalysis 158
6.3.1.3	Electron-Transfer Mechanisms 159
6.3.2	Cathode Reactions 160
6.3.2.1	Biocatalysts 161
6.3.2.2	Electron-Transfer Mechanisms 161
6.3.2.3	Electron Acceptors 162
6.3.3	Pure Cultures and Mixed Microbial Communities 162
6.3.4	Photosynthetic Biocatalysts 163
6.3.5	Biological Limitations 163
6.4	Applications and Proof of Concepts 164
6.4.1	Energy and Wastewater Concept 164
6.4.1.1	Wastewater Treatment 164
6.4.1.2	Sediments, Plants, and Photosynthesis in a BES 168
6.4.1.3	Electro-Assisted Anaerobic Digestion 168
6.4.2	Product Concept 169
6.4.2.1	Desalination 169
6.4.2.2	Caustic Soda and Hydrogen Peroxide Production 170
6.4.2.3	Organic Alcohols and Acids 170
6.4.3	Providing Environmental Services 171
6.4.3.1	Recalcitrant Compounds 171
6.4.3.2	Greenhouse Gas Mitigation 171

6.4.3.3	Heavy Metal Recovery/Removal	172
6.4.3.4	Biosensors and Environmental Monitoring	172
6.5	Modeling	173
6.6	Outlook and Conclusions	173
	Acknowledgments	173
	References	174

7	**Micro-Reactors for Fuel Processing**	**185**
	Gunther Kolb	
7.1	Introduction	185
7.2	Heat and Mass Transfer in Micro-Reactors	185
7.3	Specific Features Required from Catalyst Formulations for Microchannel Plate Heat-Exchanger Reactors	188
7.4	Heat Management of Microchannel Plate Heat-Exchanger Reactors	190
7.4.1	Reforming	191
7.4.2	Water Gas Shift Reaction	195
7.4.3	Preferential Oxidation of Carbon Monoxide	197
7.4.4	Selective Methanation of Carbon Monoxide	200
7.5	Examples of Complete Microchannel Fuel Processors	201
7.6	Fabrication of Microchannel Plate Heat-Exchanger Reactors	206
7.6.1	Choice of Construction Material	206
7.6.2	Micromachining Techniques	207
7.6.3	Sealing Techniques	209
7.6.4	Reactor–Heat Exchanger Assembly	210
7.6.5	Catalyst Coating Techniques	210
	References	212

8	**Regenerative Fuel Cells**	**219**
	Martin Müller	
8.1	Introduction	219
8.2	Principles	220
8.3	History	222
8.4	Thermodynamics	223
8.5	Electrodes	226
8.5.1	Electrodes for Alkaline Electrolytes	226
8.5.1.1	Alkaline Fuel Cells (AFCs)	227
8.5.1.2	Alkaline Electrolysis	227
8.5.1.3	Alkaline URFCs	228
8.5.2	Polymer Electrolyte Membrane (PEM)	229
8.5.2.1	PEM Electrolyzers	230
8.5.2.2	PEMFCs	231
8.5.2.3	PEM URFC	231
8.6	Solid Oxide Electrolyte (SOE)	233
8.7	System Design and Components	234

8.8	Applications and Systems *236*	
8.8.1	Stationary Systems for Seasonal Energy Storage *237*	
8.8.2	RFC Systems for Aviation Applications *239*	
8.9	Conclusion and Prospects *240*	
	References *241*	

Part II Materials and Production Processes *247*

9 Advances in Solid Oxide Fuel Cell Development Between 1995 and 2010 at Forschungszentrum Jülich GmbH, Germany *249*
Vincent Haanappel

9.1	Introduction *249*
9.2	Advances in Research, Development, and Testing of Single Cells *250*
9.2.1	SOFCs with an LSM Cathode *250*
9.2.1.1	1995–1998 *250*
9.2.1.2	1998–2002 *252*
9.2.1.3	2002–2005 *254*
9.2.1.4	2005–2010 *259*
9.2.2	SOFCs with an LSC(F) Cathode *259*
9.2.2.1	2000–2006 *259*
9.2.2.2	2006–2010 *266*
9.2.3	Advances in Testing of SOFCs *268*
9.2.3.1	Testing Housing *269*
9.2.3.2	SOFC Specifications *270*
9.2.3.3	SOFC Testing Procedure *270*
9.3	Conclusions *272*
	Acknowledgments *272*
	References *272*

10 Solid Oxide Fuel Cell Electrode Fabrication by Infiltration *275*
Evren Gunen

10.1	Introduction *275*
10.2	SOFC and Electrochemical Fundamentals *275*
10.3	Current Status of Electrodes; Fabrication Methods of Electrodes *276*
10.3.1	Methods for Coating Electrode Materials *276*
10.4	Electrode Materials *278*
10.4.1	Anode Materials *280*
10.4.2	Cathode Materials *281*
10.5	Infiltration *281*
10.5.1	Motivation for Infiltration *281*
10.5.2	Infiltration Applications *282*
10.5.2.1	Anodes Produced by Infiltration *284*
10.5.2.2	Cathodes Produced by Infiltration *290*
10.6	Conclusion *295*
	References *297*

11	**Sealing Technology for Solid Oxide Fuel Cells** *301*
	K. Scott Weil
11.1	Introduction *301*
11.1.1	Solid Oxide Fuel Cells (SOFCs) *301*
11.1.2	Functional Requirements for pSOFC Seals *304*
11.2	Sealing Techniques *306*
11.2.1	Rigid Bonded Seals *308*
11.2.1.1	Glass and Glass–Ceramic Sealants *309*
11.2.1.2	Ceramic Seals *318*
11.2.2	Compressive Seals *319*
11.2.2.1	Metal Gaskets *320*
11.2.2.2	Mica-Based Seals *320*
11.2.2.3	Hybrid Mica Seals *321*
11.2.3	Bonded Compliant Seals *323*
11.2.3.1	Brazing *324*
11.2.3.2	Bonded Compliant Seal Concept *327*
11.3	Conclusion *328*
	References *329*

12	**Phosphoric Acid, an Electrolyte for Fuel Cells – Temperature and Composition Dependence of Vapor Pressure and Proton Conductivity** *335*
	Carsten Korte
12.1	Introduction *335*
12.2	Short Overview of Basic Properties and Formal Considerations *337*
12.2.1	Anhydride and Condensation Reactions *337*
12.2.2	Acidity and Protolytic Equilibria *337*
12.2.3	Composition Specifications and Condensation Equilibria *338*
12.3	Vapor Pressure of Water as a Function of Composition and Temperature *339*
12.3.1	Number of Independent Variables, Gibb's Phase Rule *339*
12.3.2	Evaluated Literature Data for the Vapor Pressure of Phosphoric Acid in the Temperature Range between 25 and 170 °C *340*
12.4	Proton Conductivity as a Function of Composition and Temperature *344*
12.4.1	Mechanism of the Electrical Conductivity in Phosphoric Acid *344*
12.4.2	Evaluated Literature Data for the (Proton) Conductivity of (Aqueous) Phosphoric Acid in the Temperature Range Between 0 and 170 °C *344*
12.4.3	Non-Arrhenius Behavior for the Ionic Transport *346*
12.4.4	Enthalpy of Activation for the Ionic Transport *350*
12.4.5	Evaluated Data for the Dynamic Viscosity of Aqueous Phosphoric in the Temperature Range from 23 to 170 °C *352*
12.5	Equilibria between the Polyphosphoric Acid Species and "Composition" of Concentrated Phosphoric Acid *353*

12.5.1	Evaluated Literature Data for the Polyphosphoric Acid Equilibria	354
12.6	Conclusion	356
	References	357

13	**Materials and Coatings for Metallic Bipolar Plates in Polymer Electrolyte Membrane Fuel Cells**	**361**
	Heli Wang and John A. Turner	
13.1	Introduction	361
13.2	Metallic Bipolar Plates	363
13.2.1	Bare Metallic Bipolar Plates	363
13.2.2	Light Alloys	366
13.2.3	Coated Stainless-Steel Bipolar Plates	368
13.3	Discussion and Perspective	370
13.3.1	Substrate Selection	371
13.3.2	Coatings and Surface Modification	372
	Acknowledgments	374
	References	374

14	**Nanostructured Materials for Fuel Cells**	**379**
	John F. Elter	
14.1	Introduction	379
14.2	The Fuel Cell and Its System	380
14.3	Triple Phase Boundary	382
14.4	Electrodes to Oxidize Hydrogen	384
14.5	Membranes to Transport Ions	388
14.6	Electrocatalysts to Reduce Oxygen	393
14.7	Catalyst Supports to Conduct Electrons	397
14.8	Future Directions	402
	References	403

15	**Catalysis in Low-Temperature Fuel Cells – an Overview**	**407**
	Sabine Schimpf and Michael Bron	
15.1	Introduction	407
15.2	Electrocatalysis in Fuel Cells	408
15.2.1	Oxygen Reduction in PEMFCs	410
15.2.1.1	Platinum-Based Catalysts	411
15.2.1.2	Non-Platinum Catalysts	415
15.2.1.3	Platinum-Free Noble Metal Catalysts	415
15.2.1.4	Metal/N/C Catalysts	415
15.2.1.5	Transition Metal Chalcogenides	417
15.2.2	Oxygen Reduction in Other Low-Temperature Fuel Cells	417
15.2.2.1	Direct Fuel Cells	417
15.2.2.2	Alkaline Fuel Cells	418
15.2.3	Hydrogen Oxidation and CO Poisoning	418
15.2.4	Catalysis in Direct Fuel Cells	420

15.3	Electrocatalyst Degradation *421*	
15.4	Novel Support Materials *422*	
15.5	Catalyst Development, Characterization, and *In Situ* Studies in Fuel Cells *423*	
15.6	Catalysis in Hydrogen Production for Fuel Cells *424*	
15.6.1	Hydrogen Production from Methanol to Heavy Hydrocarbons *425*	
15.6.1.1	Introduction *425*	
15.6.1.2	Catalytic Steam Reforming (SR) *426*	
15.6.1.3	Catalytic Partial Oxidation (CPO) *427*	
15.6.1.4	Autothermal Reforming (ATR) *428*	
15.6.2	Carbon Monoxide Removal *429*	
15.6.3	Catalysis in the Production of Hydrogen from Biomass *430*	
15.7	Perspective *431*	
	References *431*	

Part III Analytics and Diagnostics *439*

16 Impedance Spectroscopy for High-Temperature Fuel Cells *441*
Ellen Ivers-Tiffée, André Leonide, Helge Schichlein, Volker Sonn, and André Weber

16.1	Introduction *441*
16.2	Fundamentals *443*
16.2.1	Principle of Electrochemical Impedance Spectroscopy *443*
16.2.1.1	Operating Principle of Frequency Response Analyzers *445*
16.2.2	Impedance Data Analysis *446*
16.2.2.1	Evaluation of Data Quality *446*
16.2.2.2	Complex Nonlinear Least-Squares (CNLS) Fit *447*
16.2.2.3	Distribution Function of Relaxation Times (DRT) *450*
16.3	Experimental Examples *452*
16.3.1	Process Identification *453*
16.3.1.1	Variation of Temperature *454*
16.3.1.2	Variation of Anodic Water Partial Pressure *455*
16.3.1.3	Variation of Cathodic Oxygen Partial Pressure *456*
16.3.1.4	Conclusions *457*
16.3.2	Equivalent Circuit Model Definition and Validation *458*
16.3.2.1	Cathodic Oxygen Partial Pressure Dependence *460*
16.3.2.2	Anodic Water Partial Pressure Dependence *461*
16.3.2.3	Thermal Activation *462*
16.3.2.4	Conclusions *464*
16.4	Conclusion *465*
	References *466*

17	**Post-Test Characterization of Solid Oxide Fuel-Cell Stacks** *469*	
	Norbert H. Menzler and Peter Batfalsky	
17.1	Introduction *469*	
17.1.1	Reasons for Post-Test Analysis *470*	
17.1.2	Methods of Post-Test Analysis *471*	
17.2	Stack Dissection *472*	
17.2.1	Thermography *473*	
17.2.2	Stack Embedding *474*	
17.2.3	Photography and Distance Measurements *475*	
17.2.4	Optical Microscopy *477*	
17.2.5	Topography *482*	
17.2.6	Scanning Electron Microscopy (SEM) and Energy-Dispersive X-Ray (EDX) Analysis *482*	
17.2.7	X-Ray Diffraction (XRD) *484*	
17.2.8	Wet Chemical Analysis *486*	
17.2.9	Other Characterization Techniques *488*	
17.2.10	Lessons Learned from Post-Test Stack Dissection and Analysis *488*	
17.3	Conclusion and Outlook *489*	
	Acknowledgments *490*	
	References *491*	
18	***In Situ* Imaging at Large-Scale Facilities** *493*	
	Christian Tötzke, Ingo Manke, and Werner Lehnert	
18.1	Introduction *493*	
18.2	X-Rays and Neutrons *494*	
18.2.1	Complementarity of X-Rays and Neutrons *494*	
18.2.2	Principles of Radiography and Tomography *496*	
18.2.2.1	Transmission and Attenuation *496*	
18.2.2.2	Synchrotron X-Ray Sources and X-Ray Tubes *496*	
18.2.2.3	Tomography and Tomographic Reconstruction *497*	
18.2.2.4	Artifacts *498*	
18.2.2.5	Image Normalization Procedure *499*	
18.3	Application of *In Situ* 2D Methods *500*	
18.3.1	PEFCs *500*	
18.3.1.1	X-Rays *500*	
18.3.1.2	Neutron Radiography *504*	
18.3.2	DMFCs *507*	
18.3.2.1	CO_2 Evolution Visualized by Means of Synchrotron X-Ray Radiography *508*	
18.3.2.2	Combined Approach of Neutron Radiography and Local Current Density Measurements *509*	
18.3.3	HT-PEFCs *511*	
18.4	Application of 3D Methods *513*	

18.4.1	Neutron Tomography	513
18.4.2	Synchrotron X-Ray Tomography	514
18.5	Conclusion	517
	References	518

19 Analytics of Physical Properties of Low-Temperature Fuel Cells 521
Jürgen Wackerl

19.1	Introduction	521
19.2	Gravimetric Properties	524
19.3	Caloric Properties	527
19.4	Structural Information: Porosity	530
19.5	Mechanical Properties	531
19.6	Conclusion	535
	References	536

20 Degradation Caused by Dynamic Operation and Starvation Conditions 543
Jan Hendrik Ohs, Ulrich S. Sauter, and Sebastian Maass

20.1	Introduction	543
20.2	Measurement Techniques	546
20.2.1	Reference Electrode	546
20.2.2	Current Density Distribution	548
20.2.3	Cyclic Voltammetry	549
20.3	Dynamic Operation at Standard Conditions	550
20.4	Starvation Conditions	553
20.4.1	Overall Hydrogen Starvation	553
20.4.2	Hydrogen Starvation During Start-up/Shut-down	555
20.4.3	Local Hydrogen Starvation	558
20.4.4	Oxygen Starvation	561
20.5	Mitigation	562
20.5.1	Materials and Design	563
20.5.2	Operation Strategies	563
20.6	Conclusion	565
	References	565

Part IV Quality Assurance 571

21 Quality Assurance for Characterizing Low-Temperature Fuel Cells 573
Viktor Hacker, Eva Wallnöfer-Ogris, Georgios Tsotridis, and Thomas Malkow

21.1	Introduction	573
21.2	Test Procedures/Standardized Measurements	574
21.2.1	Preconditioning of the Fuel Cell	574
21.2.2	Humidification Sensitivity Test	574
21.2.2.1	Setting the Test Conditions (Test Inputs)	574

21.2.2.2	Measuring the Test Outputs 577
21.2.2.3	Data Post Processing 578
21.2.3	On/Off Aging Test 578
21.2.3.1	Setting the Test Conditions (Test Inputs) 578
21.2.3.2	Measuring the Test Outputs 578
21.2.3.3	Data Post-Processing 580
21.2.4	Performance Test 581
21.2.4.1	Setting the Test Conditions (Test Inputs) 581
21.2.4.2	Measuring the Test Outputs 582
21.2.4.3	Data Post-Processing 583
21.2.5	Long-Term Durability Test 583
21.2.5.1	Setting the Test Conditions (Test Inputs) 583
21.2.5.2	Measuring the Test Outputs 584
21.2.5.3	Data Post-Processing 585
21.2.6	Dynamic Load Cycling Aging Test 585
21.2.6.1	Setting the Test Conditions (Test Inputs) 585
21.2.6.2	Measuring the Test Outputs 585
21.2.6.3	Data Post-Processing 586
21.3	Standardized Test Cells 587
21.4	Degradation and Lifetime Investigations 587
21.4.1	Analysis of MEA Aging Phenomena 587
21.4.2	Load Cycling 588
21.5	Design of Experiments in the Field of Fuel-Cell Research 592
	References 593
22	**Methodologies for Fuel Cell Process Engineering** 597
	Remzi Can Samsun and Ralf Peters
22.1	Introduction 597
22.2	Verification Methods in Fuel-Cell Process Engineering 597
22.2.1	Design of Experiments 598
22.2.1.1	2^2 Factorial Design 599
22.2.1.2	3^2 Factorial Design 601
22.2.1.3	2^3 Factorial Design 604
22.2.1.4	2^{n-k} Fractional Factorial Designs 609
22.2.2	Evaluation of Measurement Uncertainty 610
22.2.2.1	Summary of Procedure to Evaluate and Express Uncertainty 611
22.2.2.2	The Use of the Monte Carlo Method to Evaluate Uncertainty 612
22.2.2.3	Practical Example of the Use of the Monte Carlo Method to Evaluate Uncertainty 613
22.2.3	Determination of Conversion in Reforming Processes 616
22.3	Analysis Methods in Fuel-Cell Process Engineering 628
22.3.1	Systems Analysis via Statistical Methods 628
22.3.2	Predictive Method to Determine Vapor–Liquid and Liquid–Liquid Equilibria 630
22.3.2.1	Residual Hydrocarbons in the Reformer Product Gas 632

22.3.2.2	Evaporation of Model Fuels *634*	
22.3.3	Model Evaluation for Nonlinear Systems of Equations *637*	
22.3.4	Pinch-Point Analysis *639*	
22.4	Conclusion *641*	
	Acknowledgments *642*	
	References *642*	

Contents to Volume 2

List of Contributors *XIX*

Part V   Modeling and Simulation *645*

23	**Messages from Analytical Modeling of Fuel Cells** *647* *Andrei Kulikovsky*
24	**Stochastic Modeling of Fuel-Cell Components** *669* *Ralf Thiedmann, Gerd Gaiselmann, Werner Lehnert, and Volker Schmidt*
25	**Computational Fluid Dynamic Simulation Using Supercomputer Calculation Capacity** *703* *Ralf Peters and Florian Scharf*
26	**Modeling Solid Oxide Fuel Cells from the Macroscale to the Nanoscale** *733* *Emily M. Ryan and Mohammad A. Khaleel*
27	**Numerical Modeling of the Thermomechanically Induced Stress in Solid Oxide Fuel Cells** *767* *Murat Peksen*
28	**Modeling of Molten Carbonate Fuel Cells** *791* *Peter Heidebrecht, Silvia Piewek, and Kai Sundmacher*
29	**High-Temperature Polymer Electrolyte Fuel-Cell Modeling** *819* *Uwe Reimer*
30	**Modeling of Polymer Electrolyte Membrane Fuel-Cell Components** *839* *Yun Wang and Ken S. Chen*
31	**Modeling of Polymer Electrolyte Membrane Fuel Cells and Stacks** *879* *Yun Wang and Ken S. Chen*

Part VI Balance of Plant Design and Components *917*

32 Principles of Systems Engineering *919*
Ludger Blum, Ralf Peters, and Remzi Can Samsun

33 System Technology for Solid Oxide Fuel Cells *963*
Nguyen Q. Minh

34 Desulfurization for Fuel-Cell Systems *1011*
Joachim Pasel and Ralf Peters

35 Design Criteria and Components for Fuel Cell Powertrains *1045*
Lutz Eckstein and Bruno Gnörich

36 Hybridization for Fuel Cells *1075*
Jörg Wilhelm

Part VII Systems Verification and Market Introduction *1105*

37 Off-Grid Power Supply and Premium Power Generation *1107*
Kerry-Ann Adamson

38 Demonstration Projects and Market Introduction *1119*
Kristin Deason

Part VIII Knowledge Distribution and Public Awareness *1151*

39 A Sustainable Framework for International Collaboration: the IEA HIA and Its Strategic Plan for 2009–2015 *1153*
Mary-Rose de Valladares

40 Overview of Fuel Cell and Hydrogen Organizations and Initiatives Worldwide *1181*
Bernd Emonts

41 Contributions for Education and Public Awareness *1211*
Thorsteinn I. Sigfusson and Bernd Emonts

Index *1223*

List of Contributors

Kerry-Ann Adamson
Pike Research – Cleantech
Market Intelligence
180–186 Kings Cross Road
London WC1X 9DE
UK

Elisabetta Arato
University of Genoa
PERT, Process Engineering
Research Team
Via Opera Pia 15
16145 Genoa
Italy

Jan B.A. Arends
Ghent University
Faculty of Bioscience Engineering
Laboratory of Microbial Ecology
and Technology (LabMET)
Coupure Links 653
9000 Ghent
Belgium

Peter Batfalsky
Forschungszentrum Jülich
GmbH
ZAT
Leo-Brandt-Straße
52425 Jülich
Germany

Ludger Blum
Forschungszentrum Jülich
GmbH, IEK-3
Leo-Brandt-Straße
52425 Jülich
Germany

Barbara Bosio
University of Genoa
PERT, Process Engineering
Research Team
Via Opera Pia 15
16145 Genoa
Italy

Michael Bron
Martin-Luther-Universität
Halle-Wittenberg
Naturwissenschaftliche Fakultät
II – Chemie, Physik, und
Mathematik
Institut für Chemie – Technische
Chemie I
von-Danckelmann-Platz 4
06120 Halle
Germany

List of Contributors

Ken S. Chen
Sandia National Laboratories
7011 East Avenue
MS9154, Livermore
CA 94550
USA

Kristin Deason
NOW GmbH
Nationale Organisation
Wasserstoff- und
Brennstoffzellentechnologie
Fasanenstraße 5
10623 Berlin
Germany

Joachim Desloover
Ghent University
Faculty of Bioscience Engineering
Laboratory of Microbial Ecology
and Technology (LabMET)
Coupure Links 653
9000 Ghent
Belgium

Mary-Rose de Valladares
International Energy Agency
Hydrogen Implementing
Agreement (IEA HIA)
9650 Rockville Pike
Bethesda
MD 20814
USA

Lutz Eckstein
RWTH Aachen University
Institut für Kraftfahrzeuge (IKA)
Steinbachstraße 7
52074 Aachen
Germany

John F. Elter
Sustainable Systems LLC, 874
Old Albany Shaker Road, Latham
NY 12110
USA

and

University of Albany, State
University of New York
College of Nanoscale Science and
Engineering
NanoFab 300 East, 257 Fuller
Road, Albany
NY 12222
USA

Bernd Emonts
Forschungszentrum Jülich
GmbH, IEK-3
Leo-Brandt-Straße
52425 Jülich
Germany

Gerd Gaiselmann
Universität Ulm
Institut für Stochastik
HelmholtzStraße 18
89069 Ulm
Germany

Dietmar Gerteisen
Fraunhofer Institute for Solar
Energy Systems ISE
Department of Fuel Cell Systems
Heidenhofstraße 2
79110 Freiburg
Germany

Bruno Gnörich
RWTH Aachen
Institut für Kraftfahrzeuge (IKA)
SteinbachStraße 7
52074 Aachen
Germany

Paolo Greppi
University of Genoa
PERT, Process Engineering
Research Team
Via Opera Pia 15
16145 Genoa
Italy

Ulf Groos
Fraunhofer Institute for Solar
Energy Systems ISE
Department of Fuel Cell Systems
Heidenhofstrasse 2
79110 Freiburg
Germany

Thomas Grube
Forschungszentrum Jülich
GmbH, IEK-3
Leo-Brandt-Straße
52425 Jülich
Germany

Erich Gülzow
Deutsches Zentrum für Luft- und
Raumfahrt eV (DLR)
Institut für Technische
Thermodynamik
Pfaffenwaldring 38–40
70569 Stuttgart
Germany

Evren Gunen
TUBITAK Marmara Research
Center
Energy Institute
Dr. Zeki Acar Cad.
Baris mah. No: 1
Gebze
Kocaeli 41470
Turkey

Vincent Haanappel
Forschungszentrum Jülich
GmbH, IEK-3
Leo-Brandt-Straße
52425 Jülich
Germany

Viktor Hacker
Graz University of Technology
Institute of Chemical Engineering
and Environmental Technology
Inffeldgasse 25/C/II
8010 Graz
Austria

Peter Heidebrecht
Max Planck Institut
Dynamics of Complex Technical
Systems
Sandtorstraße 1
39106 Magdeburg
Germany

Ellen Ivers-Tiffée
Karlsruher Institut für
Technologie (KIT)
Institut für Werkstoffe der
Elektrotechnik (IWE)
Adenauerring 20b
Gebäude 50.40
76131 Karlsruhe
Germany

Mohammad A. Khaleel
Boston University
Department of Mechanical
Engineering
110 Cummington Street
Boston
MA 02215
USA

List of Contributors

Gunther Kolb
Institut für Mikrotechnik Mainz GmbH
Energietechnik und Katalyse
Carl-Zeiss-Straße 18–20
55129 Mainz
Germany

Carsten Korte
Forschungszentrum Jülich GmbH, IEK-3
Leo-Brandt-Straße
52425 Jülich
Germany

Melanie Kuhn
Massachusetts Institute of Technology
Department of Materials Science and Engineering
77 Massachusetts Avenue
Cambridge
MA 02139
USA

Andrei Kulikovsky
Forschungszentrum Jülich GmbH, IEK-3
Leo-Brandt-Straße
52425 Jülich
Germany

Werner Lehnert
Forschungszentrum Jülich GmbH, IEK-3
Leo-Brandt-Straße
52425 Jülich
Germany

Werner Lehnert
Forschungszentrum Jülich GmbH, IEK-3
Leo-Brandt-Straße
52425 Jülich
Germany

André Leonide
Karlsruher Institut für Technologie (KIT)
Institut für Werkstoffe der Elektrotechnik (IWE)
Adenauerring 20b
Gebäude 50.40
76131 Karlsruhe
Germany

Sebastian Maass
Robert Bosch GmbH
Corporate Sector Research and Advance Engineering
CR/ARC1 – Energy Storage and Conversion
Robert-Bosch-Platz 1
70839 Gerlingen-Schillerhöhe
Germany

Thomas Malkow
European Commission
Directorate-General Joint Research Centre
Institute for Energy and Transport
Westerduinweg 3
1755 LE Petten
The Netherlands

Ingo Manke
Helmholtz-Zentrum Berlin
Hahn-Meitner-Platz 1
D-14109 Berlin
Germany

Norbert H. Menzler
Forschungszentrum Jülich
GmbH, IEK-1
Leo-Brandt-Straße
52425 Jülich
Germany

Jürgen Mergel
Forschungszentrum Jülich
GmbH, IEK-3
Leo-Brandt-Straße
52425 Jülich
Germany

Nguyen Q. Minh
University of California,
San Diego
Center for Energy Research
9500 Gilman Drive
La Jolla
CA 92093-0417
USA

Martin Müller
Forschungszentrum Jülich
GmbH, IEK-3
Leo-Brandt-Straße
52425 Jülich
Germany

Téko W. Napporn
Université de Poitiers
Electrocatalysis Group (e-lyse),
IC2MP UMR 7285 CNRS
4 rue Michel Brunet
86022, Poitiers
France

Jan Hendrik Ohs
Robert Bosch GmbH
Corporate Sector Research and
Advance Engineering
CR/ARC1 – Energy Storage and
Conversion
Robert-Bosch-Platz 1
70839 Gerlingen-Schillerhöhe
Germany

Joachim Pasel
Forschungszentrum Jülich
GmbH, IEK-3
Leo-Brandt-Straße
52425 Jülich
Germany

Murat Peksen
Forschungszentrum Jülich
GmbH, IEK-3
Leo-Brandt-Straße
52425 Jülich
Germany

Ralf Peters
Forschungszentrum Jülich
GmbH, IEK-3
Leo-Brandt-Straße
52425 Jülich
Germany

Silvia Piewek
Max Planck Institute
Dynamics of Complex Technical
Systems
Sandtorstraße 1
39106 Magdeburg
Germany

Sebastiá Puig
University of Girona
Faculty of Sciences
Institute of the Environment
Laboratory of Chemical and
Environmental Engineering
(LEQUIA-UdG)
Campus Montilivi s/n
17071 Girona
Spain

Uwe Reimer
Forschungszentrum Jülich
GmbH, IEK-3
Leo-Brandt-Straße
52425 Jülich
Germany

Emily M. Ryan
Boston University
Department of Mechanical
Engineering
110 Cummington Street
Boston
MA 02215
USA

and

Pacific Northwest National
Laboratory
902 Battelle Boulevard
Richland
WA 99352
USA

Remzi Can Samsun
Forschungszentrum Jülich
GmbH, IEK-3
Leo-Brandt-Straße
52425 Jülich
Germany

Ulrich S. Sauter
Robert Bosch GmbH
Corporate Sector Research and
Advance Engineering
CR/ARC1 – Energy Storage and
Conversion
Robert-Bosch-Platz 1
70839 Gerlingen-Schillerhöhe
Germany

Florian Scharf
Forschungszentrum Jülich
GmbH, IEK-3
Leo-Brandt-Straße
52425 Jülich
Germany

Helge Schichlein
Karlsruher Institut für
Technologie (KIT)
Institut für Werkstoffe der
Elektrotechnik (IWE)
Adenauerring 20b
Gebäude 50.40
76131 Karlsruhe
Germany

Sabine Schimpf
Martin-Luther-Universität
Halle-Wittenberg
Naturwissenschaftliche Fakultät
II – Chemie, Physik, und
Mathematik
Institut für Chemie – Technische
Chemie I
von-Danckelmann-Platz 4
06120 Halle
Germany

Volker Schmidt
Universität Ulm
Institut für Stochastik
HelmholtzStraße 18
89069 Ulm
Germany

Thorsteinn I. Sigfusson
University of Iceland
Sæmundargötu 2
101 Reykjavík and Innovation
Centre Iceland
Keldnaholt
112 Reykjavik
Iceland

Volker Sonn
Karlsruher Institut für
Technologie (KIT)
Institut für Werkstoffe der
Elektrotechnik (IWE)
Adenauerring 20b
Gebäude 50.40
76131 Karlsruhe
Germany

Kai Sundmacher
Max Planck Institute
Dynamics of Complex Technical
Systems
Sandtorstraße 1
39106 Magdeburg
Germany

Ralf Thiedmann
Universität Ulm
Institut für Stochastik
HelmholtzStraße 18
89069 Ulm
Germany

Christian Tötzke
Technische Universität Berlin
StraBe des 17. Juni 135
D-10623 Berlin
Germany

and

Helmholtz-Zentrum Berlin
Hahn-Meitner-Platz 1
D-14109 Berlin
Germany

Georgios Tsotridis
European Commission
Directorate-General Joint
Research Centre
Institute for Energy and
Transport
Westerduinweg 3
1755 LE Petten
The Netherlands

John A. Turner
National Renewable Energy
Laboratory
1617 Cole Boulevard
Golden
CO 80401
USA

Willy Verstraete
Ghent University
Faculty of Bioscience Engineering
Laboratory of Microbial Ecology
and Technology (LabMET)
Coupure Links 653
9000 Ghent
Belgium

Jürgen Wackerl
Forschungszentrum Jülich
GmbH, IEK-3
Leo-Brandt-Straße
52425 Jülich
Germany

Eva Wallnöfer-Ogris
Graz University of Technology
Institute of Chemical Engineering
and Environmental Technology
Inffeldgasse 25/C/II
8010 Graz
Austria

Heli Wang
National Renewable Energy
Laboratory
1617 Cole Boulevard
Golden
CO 80401
USA

Yun Wang
University of California
Department of Mechanical and
Aerospace Engineering
4231 Engineering Gateway
Irvine
CA 92697
USA

André Weber
Karlsruher Institut für
Technologie (KIT)
Institut für Werkstoffe der
Elektrotechnik (IWE)
Adenauerring 20b
Gebäude 50.40
76131 Karlsruhe
Germany

K. Scott Weil
Pacific Northwest National
Laboratory
902 Battelle Boulevard
Richland
WA 99352
USA

Jörg Wilhelm
Forschungszentrum Jülich
GmbH, IEK-3
Leo-Brandt-Straße
52425 Jülich
Germany

Part I
Technology

1
Technical Advancement of Fuel-Cell Research and Development

*Bernd Emonts, Ludger Blum, Thomas Grube, Werner Lehnert,
Jürgen Mergel, Martin Müller, and Ralf Peters*

1.1
Introduction

The world energy demand is growing at a rate of 1.8% per year. As a consequence of increasing industrialization, it has now shifted to today's developing countries. Since the higher demand is largely met with the fossil fuel reserves that are also responsible for emissions of greenhouse gases (GHGs) and other pollutants, emissions from developing countries may account for more than half of the global CO_2 emissions by 2030. The industrialized countries should therefore take the challenge to lead the way towards the development of new energy systems. This requires a comprehensive energy strategy that takes into account the entire cycle from development to supply, distribution, and storage in addition to conversion. It also includes considering the impact on the producers and users of energy systems. Short- and long-term goals to be addressed are greater energy efficiency and better integration of renewable energy sources. On this path characterized by technical developments, as an efficient and clean technology, fuel cells can make a substantial contribution. In the long term, alongside electricity, hydrogen will be a major energy vector.

A sustainable energy supply that is largely CO_2 free and based on electricity and hydrogen will be supplemented by fuel cells, which convert energy very efficiently. Since fuel-cell systems run very quietly and deliver high-quality electricity, they are particularly suitable for application in sensitive and sophisticated applications, such as in hospitals, IT centers, and vehicles. The efficiency of fuel cells, which rises with decreasing load, is nearly independent of system size and has proven to reduce energy consumption and regulated emissions significantly when used for vehicle propulsion. Even if conventional fuels such as diesel or natural gas are used, energy can be saved and emissions reduced in combination with reformers for mobile on-board power supply and decentralized energy supply. Fuel cells have the potential to convert hydrogen and other fuels into electricity very efficiently, producing negligible pollution. Furthermore, they are sufficiently flexible to be adapted to the different intermittent renewable energy sources that will enrich the

Fuel Cell Science and Engineering: Materials, Processes, Systems and Technology,
First Edition. Edited by Detlef Stolten and Bernd Emonts.
© 2012 Wiley-VCH Verlag GmbH & Co. KGaA. Published 2012 by Wiley-VCH Verlag GmbH & Co. KGaA.

1 Technical Advancement of Fuel-Cell Research and Development

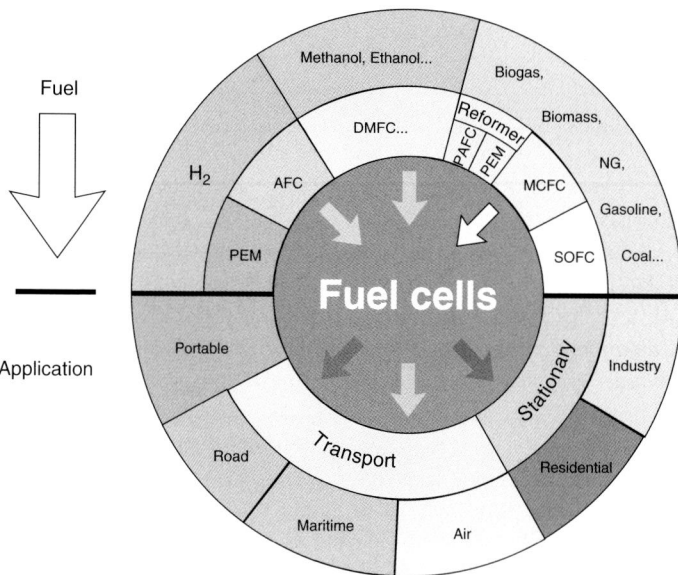

Figure 1.1 Fuel-cell technologies, possible fuels, and applications [1].

energy mix in the future. The numerous possible energy carriers, from solids (e.g., coal and biomass) and liquids (diesel, methanol, and ethanol) to gases (e.g., natural gas, biogas, and hydrogen) in combination with proven fuel-cell technologies shown in Figure 1.1 can be used in all those fields of application requiring a stable power supply. Fuel-cell systems conditioned in different ways satisfy power requirements from a few watts for portable 4C applications to the megawatt range for stationary applications such as decentralized combined heat and power (CHP) generation.

Global funding initiatives for research, development, and demonstration accompany the already great efforts of industry, and support fuel-cell technology with regard to the complex replacement processes required for capturing future markets.

1.2
Representative Research Findings for SOFCs

Two main concepts for solid oxide fuel cells (SOFCs) are currently under development: the tubular and the planar designs. In terms of long-term stability, the tubular concept has demonstrated the best results, while the planar design promises higher power densities.

1.2.1
Tubular Concepts

The standard tubular design is based on a porous cathode tube, of which a part is coated with a ceramic interconnect as a vertical stripe along the tube. The remaining surface of the cathode tube is coated with a thin electrolyte, where the overlapping with the interconnect is the critical part concerning gas tightness. The electrolyte is coated with the anode material. The Japanese company TOTO started to use this standard tubular design in 1989. TOTO invented cheap manufacturing technologies, called the TOTO wet process, based on slurry coating and sintering [2]. It uses tubes with a length of 0.66 m and an external diameter of 16.5 mm. Fuel gas is supplied to the outside of the cell while air is supplied to the inside via a thin ceramic tube, the so-called air supply tube. The cathode consists of lanthanum–strontium–manganese, the interconnect is made of lanthanum–calcium–chromate, the electrolyte of ScSZ, and the anode of Ni/YSZ. These cells can attain power densities of up to 330 mW cm^{-2} [3]. Twelve tubes are connected with nickel materials in a 2 × 6 arrangement (2 in parallel, 6 in series) to form bundles or stacks. The current path along the circumference of the tubes causes a high internal resistance, which limits the power density. As a result of high cathode polarization, an operating temperature of 900–1000 °C is necessary in order to achieve high power density (HPD).

The tubular concept of Siemens (derived from the activities of Westinghouse, already started in the 1970s) was based on a porous lanthanum–calcium–manganese cathode tube with a wall thickness of 2.2 mm and a length of 1.8 m, of which 1.5 m can be utilized electrically. A lanthanum–calcium–chromate interconnect, which serves to carry power away from the cathode, is deposited as a stripe on this tube by atmospheric plasma spraying (APS). A YSZ electrolyte layer is then sprayed on to the rest of the tube by means of APS and sintered until it is gas-tight. In a final step, the anode (Ni/YSZ) is also applied by means of APS [4]. The tubes are connected to form bundles using nickel felt. The operating temperature is in the range 950–1000 °C in order to achieve the required power density of ∼200 mW cm^{-2}. In order to overcome the problem of high ohmic resistance of the tubular design, Siemens developed a modified concept using flattened tubes with internal ribs for reduced internal resistance (HPD tubes). A similar design, albeit anode supported, is being developed by the Japanese company Kyocera [5] and the Korean research institution KIER [6]. Siemens was also working on another design variant known as the Delta 9 design, which makes further increases in power density possible. Based on in-house analyses of the cost reduction potential of the tubular design and derived designs, Siemens abandoned this development work in late 2010 [7].

Another type of tubular cells uses the anode as the tube material. The US company Acumentrics develops anode-supported tubes with a length of 45 cm and an external diameter of 15 mm [8].

A different tubular design is being pursued in Japan by Mitsubishi Heavy Industries (MHI). The single cells are positioned on a central porous support tube and electrically connected in series using ceramic interconnect rings. This leads to an increased voltage at the terminals of the individual tubes. The fuel is fed into the inside of the tube and air is supplied to the exterior [2, 9]. The maximum tube length is 1.5 m with an external diameter of 28 mm. With these specifications, power densities of up to 325 mW cm^{-2} at 900 °C have been reported [10].

1.2.2
Planar Designs

Planar designs can be broken down into electrolyte-supported and electrode-supported designs. The former uses the electrolyte to stabilize the cell mechanically. The electrolyte is 100–200 µm thick for a cell area in the range 10 × 10 cm^2. Owing to the comparatively high ohmic resistance of the thick electrolyte, typical operating temperatures of this design are 850–1000 °C. For operation at very high temperatures, ceramic interconnects made of lanthanum–chromate are preferentially used. There is an obvious trend towards metallic interconnects, as these ceramic plates are restricted in size, require high sintering temperatures, have different thermal expansion behavior in oxidizing and reducing atmospheres, and have comparatively low electrical and thermal conductivities. The advantage of ceramic plates is the low level of corrosion and therefore low degradation of the contacts, which sustains the interest in this material. The metallic interconnects allow (and also demand) a reduction in operating temperature and make the manufacture of larger interconnect plates possible. The high thermal conductivity reduces the temperature gradients in the stack and allows greater temperature differences between the gas inlet and outlet, reducing the amount of air required for cooling. As the thermal expansion coefficient of conventional high-temperature alloys is much higher than that of zirconia, a special alloy referred to as CFY (chromium with 5% iron and 1% yttrium oxide) was jointly developed by the Austrian company Plansee and Siemens. This alloy is used by different companies throughout the world for their stacks, including Hexis (formerly Sulzer Hexis) and Fraunhofer IKTS in Dresden, Germany, and also Bloom Energy.

When Siemens discontinued its planar activities, Fraunhofer IKTS took over a large proportion of the existing know-how and has been systematically refining the technology. Cells are being developed in close cooperation with Kerafol, a company which has also been working closely together with H.C. Starck – another cell manufacturer in Germany – in the area of electrolyte–substrate cell production since 2009.

In the Hexis design, fuel is supplied to the center of the electrolyte-supported circular cell (diameter 120 mm), from where it flows to the outer rim of the cell. Here, the fuel that has not reacted within the cell is burned. Air is supplied from the outside and heats up as it flows towards the center of the cell. It then flows back outside the cell in parallel with the fuel. The stack is typically operated at 900 °C. Between 50 and 70 cells are stacked together, generating a power of 1.1 kW [11]. In order to reduce manufacturing costs, Hexis has since altered the two-layer interconnect design to a one-plate concept [12].

Similar designs are also used by the Japanese companies Kyocera, Mitsubishi Materials Corporation (MMC), Nippon Telegraph and Telephone. (NTT), and Toho Gas. Fraunhofer IKTS and Bloom Energy both use conventional cross-flow

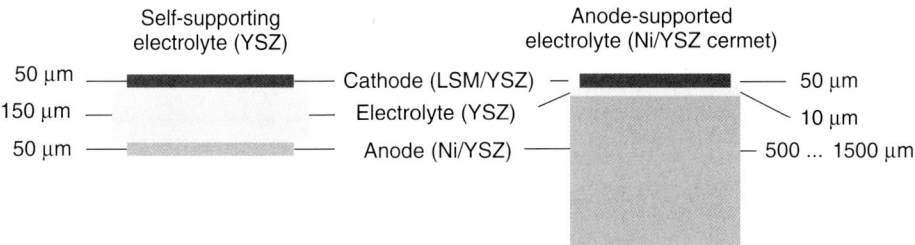

Figure 1.2 Anode substrate (right) in comparison with electrolyte substrate (left).

designs with electrolyte-supported cells soldered on to CFY interconnects. A joint development by MHI and Chubu Electric Power is the MOLB (mono-block layer built) design. Cells up to a size of 20 × 20 cm² are manufactured. They are based on a corrugated electrolyte layer. The electrolyte thus also contains the gas channels. This simplifies the design of the interconnect, allowing planar ceramic plates to be used. The largest stack of this type was built from 40 layers and delivered 2.5 kW at 1000 °C [13]. In 2005, MHI began testing cells measuring 40 × 40 cm² in 10-layer stacks as a basis for increased system output [14].

Since electrolyte resistance is the most significant obstacle to further decreasing the operating temperature, manufacturing thinner electrolytes constitutes a major challenge. This challenge can be overcome by shifting the function of mechanical stabilization from the electrolyte to one of the electrodes. For this concept, the anode tends to be preferred because it exhibits much better electrical conductivity. Therefore, no increase in ohmic resistance occurs when the electrode thickness is increased (see Figure 1.2). Nickel cermet also has good mechanical stability, which allows larger cells to be produced.

When Forschungszentrum Jülich began working on the development of this concept in 1993, it was one of the first institutions to do so. Since then, many developers throughout the world have come to regard this concept as the next generation of SOFCs. It allows the operating temperature to be reduced to between 650 and 800 °C while retaining and even surpassing the power density of electrolyte-supported cells operated at 900 °C. At the same time, this design allows cheaper ferritic chromium alloys to be used for the interconnects because their thermal expansion coefficient corresponds to that of the anode substrate.

At Forschungszentrum Jülich, anode substrates with a thickness of between 1 and 1.5 mm are manufactured by warm pressing. The electrolyte with a thickness of 5–10 μm is deposited on the substrate by means of vacuum slip casting. The stack design is based on a co-flow or counter-flow arrangement. The latter is favored for operation on natural gas with internal reforming. A 60-layer stack delivered 11.9 kW at a maximum temperature of 800 °C (average temperature in the stack ~700 °C) when operated on methane with internal reforming [15].

Similar concepts have been developed, for example, by Versa Power Systems (VPS) in Canada, Delphi and PNNL in the USA, and Topsøe Fuel Cells and Risø

National Laboratory in Denmark. In Germany, the companies H.C. Starck and CeramTec manufacture these anode-supported cells. Most of these institutions have also developed concepts using pure metal substrates instead of the anode cermet, to improve mechanical and redox stability.

A completely different design has been developed by Rolls Royce. Short electrode and electrolyte strips are applied to a porous, flat ceramic substrate. These single cells are connected electrically in series using ceramic interconnect strips, which leads to a high voltage output of one unit at a low current. Fuel gas is supplied to the inside of the supporting substrate and air to the outside. The operating temperature is about 950 °C [16]. Kyocera together with Tokyo Gas are developing a similar concept [17].

At DLR in Stuttgart, Germany, a concept in which all cell layers are produced by means of plasma spraying processes was developed in the mid-1990s. The cells are based on a metal substrate which promises to be more resistant to oxidation than the nickel-based anode substrate. Even though the power densities were increased in the last few years, they are still considerably below the values achieved for the anode substrates [18, 19].

1.2.3
Actors and Major Areas of Development

In the late 1990s, some of the most important developers in Europe, Daimler-Benz/Dornier and Siemens, discontinued their activities on planar SOFCs. After an interim phase, the number of companies and research facilities involved in SOFC development has increased again (Table 1.1). The planar technology is being developed further at research institutions such as Forschungszentrum Jülich, DLR in Stuttgart, and Fraunhofer IKTS in Dresden (all of them had already cooperated with Siemens and Dornier in individual fields in the 1990s) and at companies such as Staxera in Germany and Topsøe Fuel Cells in Denmark.

During the last two decades of the last century, Westinghouse (since 1998 Siemens) dominated developments in the USA. Since the Solid State Energy Conversion Alliance (SECA) program started, the situation has changed. Various activities in the field of planar SOFCs have been restarted or expanded, and some new consortia were founded. In its second phase, the SECA program is focusing on the development of power generation technology for a cost-effective, highly efficient central power station (>100 MW$_{el}$). The industry teams involved are Fuel Cell Energy (FCE) and VPS, UTC Power and Delphi, and Rolls-Royce, assisted by numerous research institutions [7]. A tremendous development took place at a new company, Bloom Energy, whose activities are partly based on those of Ion America, taken over by Bloom Energy. As of the end of 2011, Bloom Energy has sold more than 80 systems with a nominal power of 100 kW. They employ more people than all other developers in North America together.

During the 1990s in Japan, more than 10 companies were engaged in planar SOFC development. Because the goals of the NEDO "Sunshine" project could not

Table 1.1 The most important SOFC developments worldwide.

Continent	Facilities/employees	Designs	Development focus[a]	
Europe	Industrial enterprises 17	Planar design	(1) Systems	12
		Anode substrate	Stacks	12
		Electrolyte substrate	Cells	11
	Research facilities 6	Metallic substrate	(2) Materials	9
		Porous ceramic substrate	(3) Fabrication	3
			Powders	2
	Employees 750–850	CGO electrolyte for 550 °C	System components	2
			Fuel processing	2
		Metallic interconnect	Interconnects	1
			Reformers	1
			Cell and stack testing	1
			Stack and system testing	1
			Modeling	1
North America	Industrial enterprises 12	Planar design	(1) Cells	16
		Tubular design	Stacks	14
		Microtubes	(2) Systems	9
	Research facilities 5	Anode substrate	Materials	6
		Electrolyte substrate	(3) Systems (low power)	3
		Metallic substrate	Modeling	2
	Employees >2000	Metallic interconnect	System testing	1
		Ceramic interconnect		
Asia and Australia	Industrial enterprises 14	Planar design	(1) Cells	17
		Tubular design	Stacks	16
		Microtubes	Systems	13
	Research facilities 4	Anode substrate	Materials	13
		Electrolyte substrate	(2) Systems (low power)	3
		Flat tubes	Systems (pressurized)	1
	Employees 600–750	Metallic interconnect		
		Ceramic interconnect		

[a] From most to least frequent.

be achieved completely, a reorientation took place and other companies started SOFC development. A demonstrative research project on small systems was started in 2007. By the end of 2011, more than 130 units in the range 0.7–8 kW had been installed based on different stack concepts, developed by TOTO, MHI, MMC, Kansai Electric Power Company (KEPCO), Kyocera, and Tokyo Gas [9].

1.2.4
State of Cell and Stack Developments

The field of cell and stack development comprises numerous activities. This makes it difficult to provide an overview, particularly one that is based on comparable operating conditions. There are three different types of cells:

- anode-supported cells at an operating temperature of 750 °C
- electrolyte-supported cells at 800–900 °C
- tubular cells at 900–1000 °C.

Table 1.2 lists the results achieved in terms of cell power density (at a cell voltage of 0.7 V), active cell area, degradation rates, duration of relevant long-term measurements carried out, and the power of the constructed stacks.

Although the different operating conditions and fuels used prevent direct comparisons, it is obvious that the highest energy densities are achieved with anode-supported cells, preferably with lanthanum strontium cobalt ferrite (LSCF) cathodes, although a number of tubular designs have clearly improved over the last few years. In addition to energy density, the manufacturable cell size is an important factor in characterizing the potential of the technology. In the meantime, the degradation values of planar cells are in the same range as those of the tubular cells produced by Siemens. At the same time, the demonstrated operating times have increased significantly (tubular 40 000 h, planar 26 000 h). Both properties are shown in Table 1.2.

With respect to the development status of system technology and long-term stability, the best results have been achieved with the tubular design by Siemens. However, Siemens has ceased work in this area. The majority of developers see a clear advantage in the cost reduction potential of planar technology. This is due on the one hand to more cost-effective manufacturing technologies and on the

Table 1.2 Results achieved for different SOFC concepts.

Parameter	Anode-supported cells, 750 °C	Electrolyte-supported cells, 800–900 °C	Tubular cells, 900–1000 °C
Power density at 0.7 V (W cm^{-2})	0.46–2.0	0.03–0.63	0.11–0.53
Active cell area (cm^2)	20–960	80–840	30–990
Cell degradation rate (% per 1000 h)	1.4–0.2	1.0–0.5	2.0–0.1
Cell operating time (h)	≤26 000	≤10 000	≤4 000
Stack power (kW)	0.1–25	0.4–5.4	–
Ref.	[2, 8, 9, 15, 18–21, 23, 24, 26, 29–33, 36]	[2, 8, 9, 11–14, 20, 22, 25, 27, 34, 35]	[3–10, 16, 17, 20, 28]

other to the higher power density. In this context, there is a clear trend towards an anode-supported design using ferritic chromium steel as an interconnect material. In addition to a higher power density, this concept also allows the operating temperature to be reduced to below 800 °C.

1.3
Representative Research Findings for HT-PEFCs

One of the objectives of high-temperature polymer electrolyte fuel cell (HT-PEFC) development is to increase operating temperatures to between 150 and 180 °C. Higher temperatures make heat removal easier with a smaller cooling surface than in low-temperature polymer electrolyte fuel cells. In addition, the temperature level of the heat removed is higher and can therefore be easily utilized. Due to the higher operating temperature, HT-PEFCs also tolerate a higher proportion of carbon monoxide in the fuel. As a consequence, gas purification is simpler and therefore cheaper. As the membranes do not need to be wetted, costly water management is unnecessary.

A combination of phosphoric acid and polybenzimidazole (PBI) is currently the most interesting material for HT-PEFC membranes. PBI membranes doped with phosphoric acid can be manufactured in a synthesis process using different methods. The basic difference between them lies in whether doping with phosphoric acid is part of polycondensation, that is, whether it takes place *in situ*, or whether doping takes place by soaking the PBI foil in phosphoric acid, or whether it is affected via the gas diffusion layer (GDL) or the catalyst. The polycondensation method was developed and patented by BASF, which is currently the only company manufacturing membranes in this way.

1.3.1
Actors and Major Areas of Development

A number of companies and research institutions are responsible for advances in development. Industry contributions to R&D have been made by BASF, for example, which took over Pemeas in 2006. Pemeas was established by Celanese and a consortium of investors in 2004. Two years earlier, Celanese had begun its launch of a pilot production unit for high-temperature polymer membrane electrode assemblies membrane electrode assembly (MEAs). Another company contributing to R&D is Sartorius, which first became involved in the development of HT-PEFC MEAs and stacks of up to 2 kW in 2001. In 2009, Elcomax took over MEA activities from Sartorius and has marketed MEAs for HT-PEFCs since then. In addition, Fumatech has produced membranes based on AB-PBI (polybenzimidazole) since 2005. Danish Power Systems, a research-based development company which was founded in 1994, is distributing MEAs under the tradename Dapozol [38]. Recently, Advent Technologies started the production of MEAs. The membranes are not based on PBI but are also doped with phosphoric acid. Samsung Advanced Institutes of Technology published data on their own MEAs with excellent performance. An overview of actual HT-PEFC membranes can be found in the literature [39–41] Serenergy, a Danish company, is at present the only supplier of commercial HT-PEFC stacks in the kilowatt range [42]. Plug Power developed HT-PEFC stacks for the use in stationary applications and Volkswagen developed HT-PEFC stacks for automotive application, but stopped these activities recently.

In addition to industrial companies, several research and university institutes are working worldwide in the field of HT-PEFC. Forschungszentrum Jülich is developing HT-PEFC stacks in the power range up to 5 kW for on-board power supply running on diesel and kerosene. In addition, MEAs based on AB-PBI membranes provided by Fumatech are being developed. The Centre for Solar Energy and Hydrogen Research Baden-Württemberg, the Fuel Cell Research Center in Duisburg and the Fraunhofer Institute for Solar Energy Systems (FhG-ISE) also have activities in the field of HT-PEFCs. Their main focus is on stack and system development. Aalborg University in Denmark mainly investigates systems, stacks, and cells [43] whereas the Technical University of Denmark is well known for its research in the field of membranes and MEAs [44]. The key aspect of Spanish groups is on membrane and electrode development (e.g., [45, 46]) and the key aspect of the group at Newcastle University, UK, is on modeling, membrane, and electrode development [47]. Well-known groups performing membrane-related science can be found in the USA at the University of South Carolina and Case Western Reserve University [48, 49]. In recent years, increasing interest in HT-PEFCs can also be observed in China at the Dalian Institute of Chemical Physics [50]. In Korea, the Korea Institute of Science and Technology [51] and the Korean Institute of Energy Research [52] have reported relevant results in this field. Other academic and research institutions, for example, in Russia, are also active in R&D. An overview of the major findings in the HT-PEFC area can be found elsewhere [53–55].

1.3.2
Characteristic Data for Cells and Stacks

The power density of MEAs has reached a high level. BASF and Sartorius have the longest experience with their development, which is reflected in the high power densities and low degradation rates they achieve. MEAs based on the membrane materials produced by Fumatech, which embarked on the technology later, are developing rapidly. Their power densities are now on a par with those of Sartorius MEAs. Figure 1.3 gives an overview of the development of performance data.

The area-specific power densities of HT-PEFC stacks measured at 0.5 V, 160 °C, and with H_2 as a fuel increased from ~180 mW cm^{-2} in 2006 (Sartorius) and 2008 (Forschungszentrum Jülich) to 500 mW cm^{-2} in 2010 (Forschungszentrum Jülich and Fraunhofer ISE) [72–85].

1.4
Representative Research Findings for DMFCs

The development of direct methanol fuel cells (DMFCs) was reactivated all around the world around 1990 thanks to the use of membranes made of sulfonated

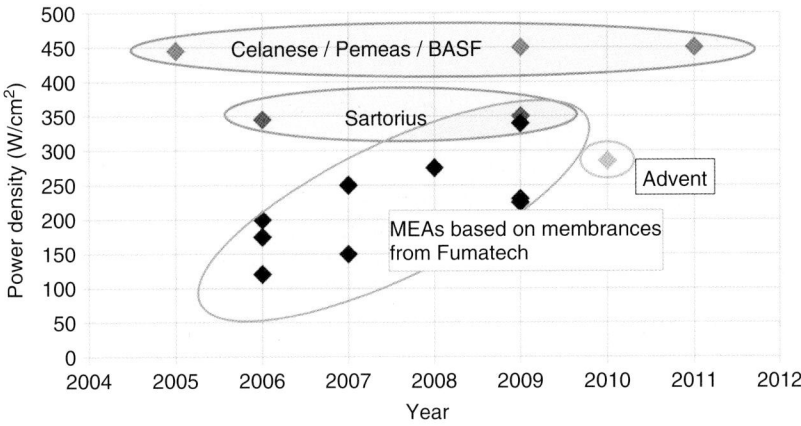

Figure 1.3 Power densities of MEAs manufactured by different developers [56–71] for comparable operating conditions. Temperature, 160 °C; fuel, pure hydrogen, pressure; 1 bar; cell voltage, 0.5 V.

fluoropolymers (Nafion) instead of electrolytes containing sulfuric acid. Development initially focused on mobile applications and then mainly on the area of "portables" as a possible replacement for batteries, since the increasing energy demand, in particular for modern cell phones (greater functionality, larger displays, etc.) also increases the discharge rate of batteries.

1.4.1
DMFCs for Portable Applications

More energy can be provided with the available volume and therefore longer lifetimes can be achieved for portable applications <50 W, since methanol or the DMFC system has a higher energy density than Li batteries. A significant share of global research and development work on DMFCs designed for portable applications is being carried out in China, South Korea, Japan, and Taiwan. This is illustrated by the fact that about two-thirds of publications on DMFCs are by Asian research organizations or companies [86]. Therefore, the first products focusing the power supply of small electronically devices (up to 5 W) were also developed in Asia. One example is the commercial available system from Toshiba for the charging of small electronic devices [87].

Nevertheless, the results obtained in Europe and North America are not insignificant. The Federal Ministry of Education and Research [Bundesministerium für Bildung und Forschung (BMBF)] intensively promoted this development in Germany with the "Micro Fuel Cell" lead innovation initiative in the last few years. The Fraunhofer Institute for Solar Energy Systems (FhG-ISE) in Freiburg has also been involved in the development of DMFCs since the mid-1990s, but only in the power range <50 W. In addition, it also develops direct ethanol fuel cells (DEFCs)

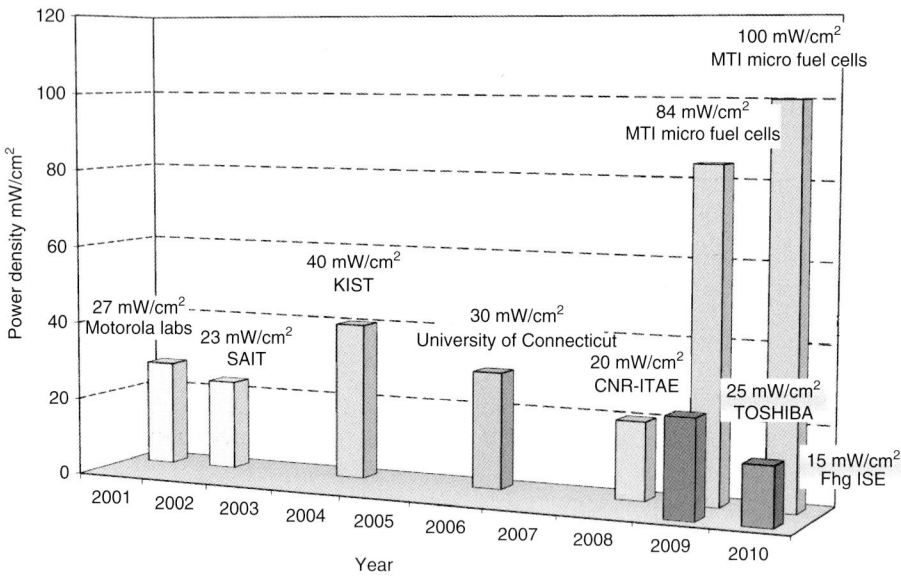

Figure 1.4 Development of the power density of passive DMFC systems at room temperature (20–30 °C).

in the power range from a few milliwatts to several tens of watts. In 2009, a planar micro fuel cell system produced by means of injection molding was developed with a consortium of small- and medium-sized businesses and in close cooperation with FWB Kunststofftechnik. Four injection-molded micro fuel cell modules were integrated into this 2 W DMFC system. The system supplies power for a locating system for container tracking for several weeks. In addition, a passively operated DMFC for an electrocardiograph (ECG) device was also set up [88].

In an international comparison, the US company MTI Micro Fuel Cells developed a passive system (see Figure 1.4) with a power density of 100 mW cm^{-2} [89]. However, no information is available on the catalyst loading of the MEA. Catalyst loadings in the range 6–16 mg cm^{-2} per cell can be found in the literature for passive DMFC MEAs.

1.4.2
DMFCs for Light Traction

Within the framework of a European project from 1996 to 1999, Siemens developed a DMFC stack with a single electrode area of 550 cm^2 together with IRD Fuel Cells (Denmark) and Johnson Matthey (UK) [90]. At a temperature of 110 °C, an oxygen pressure of 3 bar and a cell voltage of 500 mV, a power density of 200 mW cm^{-2} was achieved. Only 50 mW cm^{-2} was achieved at 500 mV in air operation at 80 °C and 1.5 bar. Unsupported Pt and PtRu catalysts with a loading of 1–4 mg cm^{-2} per electrode were used. Over the years, IRD Fuel Cells has improved their systems and they are now providing DMFC systems with 800 and 500 W power output. The current systems are operated with pure methanol or with a methanol–water mixture and with air [91].

In 2000, DaimlerChrysler presented the first go-cart powered by a DMFC system with a net power of 2 kW$_{el}$. The DMFC system, which was operated with pure oxygen on the cathode side, was developed in cooperation with Ballard Power Systems [92]. The 3 kW stack had an operating temperature of 100 °C.

The Center for Solar Energy and Hydrogen Research [Zentrum für Sonnenenergie- und Wasserstoff-Forschung (ZSW)] in Ulm, Germany, has developed and studied MEAs with nonfluorinated homopolymers together with the Institute of Chemical Process Engineering [Institut für Chemische Verfahrenstechnik (ICVT)] in Stuttgart. With these materials, power densities of up to 240 mW cm^{-2} were achieved in 2002, but with a high catalyst loading of >11 mg cm^{-2} per cell, a cell temperature of 110 °C, an anode pressure of 2.5 bar and a cathode pressure of 4 bar [93]. The first stack consisting of 12 cells was manufactured in 2004 with these MEAs using conventional Nafion 105 membranes. An electrical power of almost 120 W was achieved in air operation at an operating temperature of 70–90 °C, which corresponds to a power density of 55–60 mW cm^{-2} at a single cell voltage of 500 mV [94].

Forschungszentrum Jülich has been intensively involved in the development of DMFCs since 1996. As part of a dissertation on DMFC systems analysis, it was demonstrated that a pressure of more than 3 bar and an air:methanol ratio of 1.75 is required for an operating temperature of 110 °C [95]. Since this ratio is normally insufficient for the stable operation of a DMFC stack, higher air ratios require a higher cathode pressure, which has a negative impact on system efficiency. DMFC systems in the temperature range 80–100 °C are therefore not practical owing to the high losses for cathode air compression and the unfavorable heat management [96]. Even though a specific power density of 100 mW cm^{-2} at 500 mV was achieved at 110 °C after a short development time, further work concentrated on optimizing DMFC stacks under normal pressure. For example, a specific power density of 50 mW cm^{-2} in a compact 500 W DMFC stack under normal pressure at 70 °C was first achieved in 2002 [97]. A first 2 kW demonstration system was set up in 2003. Other systems followed in electric scooters in which the lead acid batteries were replaced with DMFC hybrid systems. In this way, the specific power density was increased to 90 mW cm^{-2} at 80 °C, and the catalyst loading was reduced from 6 to 4 mg cm^{-2} per cell [98].

As in Germany, DMFC development in Korea started in the mid-1990s with the construction of small stacks in the range of several watts. Subsequently the developed DMFC systems became more powerful and more efficient. In 2004, the Korea Institute of Energy Research (KIER) presented a 400 W stack with 42 cells and a catalyst loading of 10 mg cm^{-2}. In tests the stack was operated without additional heating and a power density of ~45 mW cm^{-2} at 500 mV was measured [99]. In 2009, an optimized DMFC system for a scooter consisting of two sub-stacks each with ~700 W peak power was realized. The maximum power density of the MEA in stack tests is above 90 mW cm^{-2} [100].

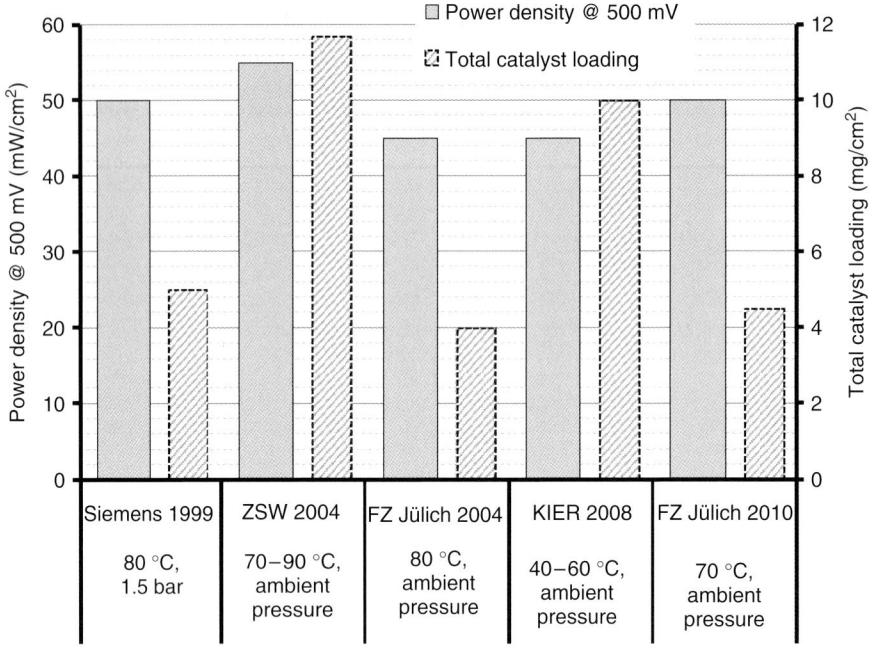

Figure 1.5 Specific power densities of membrane electrode assemblies in DMFC stacks in the power class >100 W.

Also in Japan, companies and research institutions developing DMFC systems. Yamaha Motor, for example, developed several DMFC systems for a two-wheel scooters. In 2003, the first scooter system with an output of 500 W was presented. In recent years, this system was improved and in 2007 a highly efficient 1 kW DMFC system reached a power density of 146 mW cm^{-2}. The fuel cell system achieves 30% system efficiency [100–102]. In the USA, Oorja Protonics has also developed DMFC systems with a power output of ∼800 W for small forklifts. However, there is not much information available about the system setup and the system components [103]. Figure 1.5 provides an overview of the power densities of MEAs achieved at an average individual voltage of 500 mV and a temperature range of 70–80 °C.

The best fuel-cell stacks and systems today currently achieve lifetimes of at least 3000 h. SFC Energy, for example, guarantees an operating life of 3000 h within 36 months for its commercial DMFC systems. However, these systems only have a maximum power of 65 W, and the guaranteed 3000 h can also involve replacement of a stack [104]. The Danish company IRD Fuel Cell Technology markets an 800 W DMFC system which also has a lifetime of 3000 h [105].

In terms of reducing the degradation of the electrochemically active DMFC components, Forschungszentrum Jülich was able to increase the long-term stability from less than 50 h to more than 9000 h under real operating conditions from 2005 to 2011 by clarifying degradation mechanisms of MEAs in DMFC systems. For this

Table 1.3 Development of significant membrane modifications.

Approach	Impact	Ref.
Ionically or covalently bonded materials	Stabilized membrane structure	[109]
Sulfonated or nonsulfonated block copolymers	Limited swelling, methanol permeation, and long-term stability High proton conductivity	[110, 111]
Additives in Nafion and sPEEK	Improved conductivity/methanol permeability behavior	[112, 113]
Fully sulfonated polysulfones	Clearly suppressed methanol uptake	[114]
Barrier coatings/layers, for example, via plasma treatment	Reduced methanol permeation Increased membrane resistance	[115, 116]
Polymers with quaternary ammonium groups	Limited long-term stability ($>60\,°C$), conductivity, and availability of ionomer solution	[117–119]

purpose, new corrosion-resistant and carbon-supported PtRu catalysts were used, in particular for the V3.3-2 DMFC system [106].

For both DMFC systems for light traction and for DMFC systems for portable applications, Nafion is still the standard membrane material. A general overview of the polymer electrolyte membrane materials, their modifications, and their function can be found in. [107] and with the focus on the DMFC operation in [108].

In the late 1990s and early 2000s, nonfluorinated homopolymers were studied as promising alternatives. In simplified terms, however, reduced methanol permeation and reduced conductivity are combined in these materials to achieve a DMFC performance comparable to that of Nafion-based MEAs, and the membranes had to be so thin that it was not possible to reduce substantially the absolute value for fuel loss by permeation. Table 1.3 provides an overview of the most significant membrane modifications.

1.5
Application and Demonstration in Transportation

1.5.1
Fuel Cells and Batteries for Propulsion

The major objectives of a reorientation in the energy supply sector, which is generally considered necessary, are the reduction of global and local impacts on the environment, the reduction of dependence on imported energy raw materials, and economic policy-related aspects. The German government has published numerous strategic documents on the role of mobility for reaching these objectives and has established programs for funding the associated energy conversion and storage technologies [120–124]. The technological reorientation within the transport sector focuses on the development of vehicle drives with batteries and fuel cells and also hybrid drives including plug-in hybrid electric vehicles (PHEVs). The goal of the German government calls for 1 million electric vehicles with batteries to be sold by 2020 and 6 million by 2030 [121]. If hydrogen is used in highly efficient fuel cells, renewable power that cannot be utilized due to grid stability can be stored temporarily in the form of hydrogen. Liquid fuels with a high energy density will still be required in the long term, predominantly for heavy trucks, aircraft, and ships. In the future, systems with fuel cells in the power range from ~5 kW to over 1 MW could be used for the on-board power supply in such vehicles, providing a suitable fuel by reforming the fuel at hand. A renewable basis for liquid fuels is biomass, which can be converted into suitable fuels using biochemical or thermochemical processes. The following section deals with the assessment of the primary energy demand, GHG emissions and the cost of electric vehicle concepts with fuel cells and batteries.

Vehicle concepts whose drive structure and storage dimensions facilitate all-electric drive operation include the following:

- plug-in hybrids with a combustion engine or fuel cell system (PHEVs) with a range of up to 50 km in battery operation
- electric vehicles with fuel cell and battery (fuel-cell hybrid electric vehicles, FCHVs) with a range of over 400 km
- electric vehicles with a battery (battery electric vehicles, BEVs) with a range of up to 200 km.

PHEVs will not be considered in greater depth here, since their low battery capacity requires frequent operation of the combustion engine. The main factors for the comparability of these novel concepts to current vehicles are costs, but also the performance of the drive in terms of top speed and acceleration, and the range that can be achieved between refuelings or battery rechargings. Electric drives are considered to be easy to drive owing to the torque curve of the motor. No gear transmission is required at moderate top speeds, thus streamlining the system.

The fifth to sixth generation of some concept cars is already under development for FCHVs in hydrogen operation and is produced in processes nearing series production, with a total of nearly 800 units built since 1994. Today, only polymer electrolyte fuel cells (PEFCs) are used with operating temperatures between 80 and 95 °C. The preferred storage type is compressed gas storage at 700 bar.

Progress made in the development of fuel cell drives can be shown by referring to performance data for passenger cars (Figure 1.6). The figure shows that in an early phase of development, the stack performances were comparatively low, certainly due to considerably lower power densities or specific performances of fuel-cell systems. This allowed only a limited driving performance. The range was

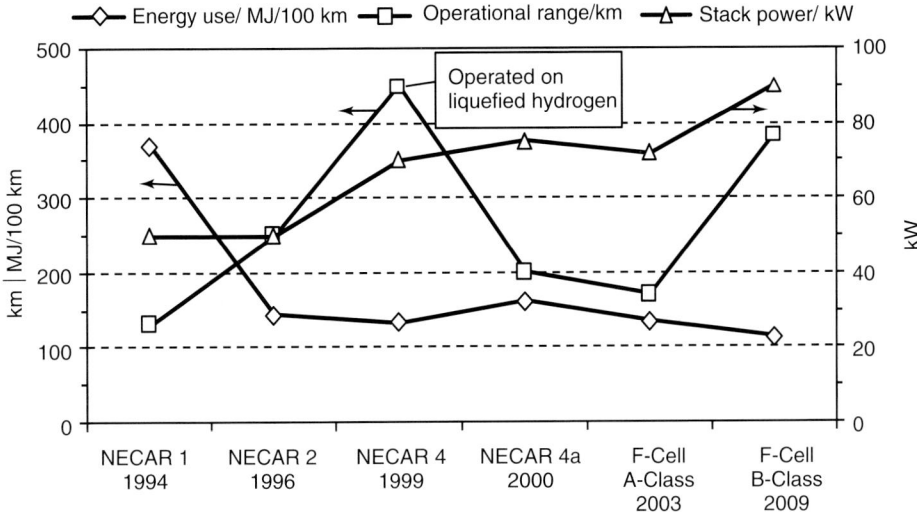

Figure 1.6 Development of range, stack performance, and energy demand of FCHVs.

also initially low when compressed hydrogen (CH2) was used, but was increased by improving the systems and applications of the 700 bar technology to ~400 km and by reducing the energy demand also shown in Figure 1.6, so that there are few restrictions today [125]. NECAR 4 with liquid hydrogen (LH2) already had a range of 450 km back in 1999 – in an unspecified driving profile – and therefore greater than values achieved with the 700 bar technology today. However, the LH2 technology is rarely used in passenger cars today. Other values that document the increase in range are available for the Toyota FCHV-adv, for example, which has a range of up to 830 km [126], whereas its predecessor's range was only 330 km.

Challenges for the further development of fuel-cell vehicles involve increasing power density, specific power and lifetime, improving storage technologies, and achieving competitive costs. Market success will be determined by the availability of the supply infrastructure. Progress in the development of fuel-cell systems can be measured using parameters such as specific power and power density, precious metal requirements and cold start properties. Louie [127] stated that current fuel-cell stacks successfully cold start at $-30\,°C$ and have a specific power of $1–1.5\,kW_{el}\,kg^{-1}$, corresponding to $2–2.25\,kW_{el}\,l^{-1}$.

Few data are available for the precious metal requirements, which are one of the most important cost drivers in fuel-cell technology. The progress achieved in the reduction of precious metal requirements for PEFCs in automotive applications is shown in Table 1.4.

It is difficult to document the progress made in the development of H_2 storage owing to insufficient data. Broad ranges with large time overlaps are usually stated for the important parameters – specific energy, energy density, and cost. Consistent data for specific lines of developments are not available. For Figure 1.7, the storage densities and specific energies according to the US Department of

Table 1.4 Precious metal requirements for PEFCs in automobile applications.

Company	Precious metal requirement per FCHV (g)	Year	Ref.
Daimler	30	2009	[128]
General Motors	112	2000	[129]
	80	2007	[130]
	30	2013	[130]

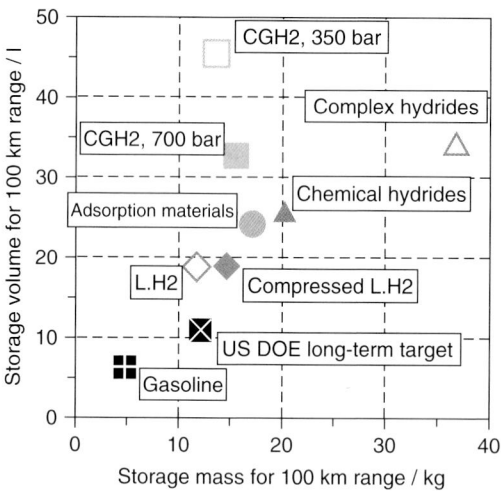

Figure 1.7 Mass and volume of different storage systems relative to a range of 100 km.

Energy (DOE) [131] for 2009 were therefore assessed using the specific energy requirements of FCHVs per 100 km in order to take into account the much more efficient utilization of energy in FCHVs in comparison with gasoline-powered passenger cars used as a reference here. The values in the figure represent the mass or volume of storage systems required for a range of 100 km, assuming that the FCHVs are designed for a range of 400 km. The data are taken from simulations based on the New European Driving Cycle [or Motor Vehicle Emissions Group (MVEG)] for determining the mass-dependent mechanical energy requirement at the wheels of small to mid-sized passenger cars [132, 133].

Some BEVs are produced in series today [134]. Lithium ion technology is used for this purpose, because it is best suited to requirements such as high specific energy and energy density, but also provides long lifetime and low self-discharge rates. Table 1.5 contains information on these batteries in comparison with lead

Table 1.5 Performance data for batteries.

Parameter	Lead acid	Ni–metal hydride	Li ions
Theoretic specific energy (Wh kg^{-1})	167	214	420
Effective specific energy (Wh kg^{-1})	35–49	45–75	65–150
Specific power (W kg^{-1})	227–310	250–1000	600–1500
Energy density (Wh l^{-1})	70–96	125–182	130–300
Power density (W l^{-1})	445–620	600–2800	1200–3000
Lifetime (years)	2–6	12	7–10
Lifetime (cycles × 1000)	0.1–0.3	2.5–300	2–300
Self-discharge (% per month)	2–3	20–30	2–10
Temperature range (°C)	−30 to 70	−10 to 60	−25 to 50

acid and nickel metal hydride batteries. On the cell level, a specific energy of up to 185 Wh$_{el}$ kg^{-1} is achieved today [135]. In automotive applications, allowances must be made in this respect for spare capacity and system integration into an overall module including cooling, control electronics, and housing. For example, the figure for the Opel Ampera based on the data sheet in [136] is ∼90 Wh$_{el}$ kg^{-1}.

Grube and Stolten [132] reported detailed simulations based on the MVEG that were carried out for a comparison of BEVs with a range of 200 km and FCHVs with a range of 400 km. Table 1.6 gives an overview of primary energy demand, GHG emissions, and system costs, comparing vehicles with fuel cells and vehicles with batteries. It shows that, if the relevant cost targets can be met, the costs of the power supply system with fuel cells may be considerably lower at a range that is twice as high. With respect to the well-to-wheel (WtW) balance of primary energy input and GHG emissions, FCHVs are at a slight disadvantage if natural gas is used to produce hydrogen. These values can be further reduced by switching to primary energies for generating power and hydrogen with lower GHG emissions. The recharging of BEVs and refueling of FCHVs pose special challenges. In principle,

Table 1.6 Comparison of electric cars with batteries and with fuel cells [132].

Parameter	Electric car	
	BEV 200	FCHV 400
Cost of power supply system (€)	9733	7401
Stored energy (MJ)	102	396
Vehicle mass (kg)	1325	1313
Specific primary energy, well-to-wheel (MJ km^{-1})	1.61	1.65
Specific GHG emissions (g km^{-1})	88	97

refueling times, fueling procedures, and refueling intervals for fuel-cell vehicles are comparable today.

From the present point of view, grid services provided by BEVs are possible in principle; however, the influence of the depth of discharge and also charge and discharge capacities on battery lifetime will have to be considered. Long battery lifetimes can be achieved if the depth of discharge is in the range of a small percentage and if temperatures are considerably lower than 60 °C [137]. It is important for BEV balances to take into account the time of day at which batteries are typically recharged, since the energy must be fed in at the same time. In the short and medium term, hydrogen provision may benefit from the integration of residual hydrogen from industry, until new production capacities can be built in the long term, preferably on the basis of renewable energies and with the option of energy storage.

1.5.2
On-Board Power Supply with Fuel Cells

On-board power supply is required by almost all mobile applications. In the past few years, studies were conducted on on-board power supplied by auxiliary power units (APUs) for the transport of goods and passengers by sea and air. This includes aircraft, ships, passenger cars and, above all, trucks. Numerous US studies have considered the use of fuel cells in "line haul sleeper trucks" [138–141]. Targets for different APU applications are compiled in Table 1.7. The targets for applications in aircraft and ships are defined by considerably fewer values than for combined heat and power generation. The cost targets are most ambitious for passenger car applications. The power density target for aircraft applications is roughly the

Table 1.7 Targets for different APU applications and for stationary systems based on natural gas.

	Aircraft	Passenger car	Truck	CHP
Ref.	[142–144]	[145]	DOE targets 2015/2020 [146, 147]	
Power range (kW)	100–400	10	1–10	1–10
Efficiency (%)	40	<35	35/40	42.5/45
Specific cost	€1500 kW^{-1}	€40 kW^{-1}	$600 kW^{-1}	$450 kW^{-1}
Durability (h)	20 000/40 000	5000	15 000/20 000	40 000/60 000
Dynamic aging (% per 1000 h)	–	–	1.3/1	0.5/0.3
Power density (W l^{-1})	750	333	35–40	–
Mass-specific power (W kg^{-1})	0.5–1	250	40–45	–
System availability (%)	–	–	98/99	98/99
Cold starting	–	<1 s	10/5 min	30/20 min
Load cycle 10–90%	–	<1 s	3/2 min	3/2 min
Partial load	–	1:50	–	–

Table 1.8 Power densities achieved and targeted for APU systems.

Power density ($W_{el}\,l^{-1}$)	2000	2003	2005	2006	2008	2010	2011	2012	2015	2020
DOE 2006 targets	–	–	25	70	–	100	–	–	100	–
DOE target review	–	–	–	–	–	–	–	30	35	40
Delphi 3.5 kW$_{el}$	10	20	25	–	17	–	–	–	–	–
Cummins 1.5 kW$_{el}$	–	–	–	–	–	5	–	–	–	–
Webasto 1 kW$_{el}$	–	–	–	8	–	–	–	–	–	–
Truma 0.25 kW$_{el}$	–	–	–	–	–	–	2.7	–	–	–

same as for passenger cars, whereas the high long-term stability requirement corresponds more closely to the targets for stationary applications.

Table 1.8 shows the power densities achieved for different fuel-cell systems for on-board power supply in the power range from 250 W_{el} to 3.5 kW$_{el}$. The US company Delphi obviously made great progress in making systems more compact from 2000 to 2005. Based on these status data, the DOE established targets for the period up to 2020, first in 2006 and again in 2009. However, it became clear in 2008 that the power density decreased to 17 $W_{el}\,l^{-1}$ with increasing system autonomy and technical maturity. The power density to be achieved by 2020 is now 40 $W_{el}\,l^{-1}$. Many of the targets set by the DOE are tailored specifically to SOFCs for application in trucks. Other system providers have been and still are the companies Webasto (Germany) and Cummins (USA). A 250 W_{el} device for on-board power supply in campers manufactured by Truma (Germany) is also shown. The device works with liquefied petroleum gas (LPG) as an energy carrier; a micro steam reformer from IMM (Germany) produces a hydrogen-rich gas.

How can the power density of a system for on-board power supply be significantly increased? Figure 1.8 shows the power density of catalysts for partial oxidation, autothermal reforming, and steam reforming. Depending on the process and the fuel used, the power density is between 10 and 50 kW$_{el}\,l^{-1}$. Mixing zones and heat exchangers are part of the design, resulting in power densities of 1–5 kW$_{el}\,l^{-1}$ for reformers as a core component for fuel production. PEFC stacks for the automobile sector achieve up to 1.5 kW$_{el}\,l^{-1}$. SOFC stacks manufactured by Delphi (USA) have a residual value of at least 720 $W_{el}\,l^{-1}$. Less compact stacks have a power density in the same range as HT-PEFC stacks of between 50 and 200 $W_{el}\,l^{-1}$ today. If the different components for fuel production are interconnected as a package, 120 $W_{el}\,l^{-1}$ can be achieved with current designs. For example, if the best SOFC stack is combined with a compact partial oxidation reactor (3.3 kW$_{el}\,l^{-1}$), an HPD of 600 $W_{el}\,l^{-1}$ could be achieved at best. However, only 17 $W_{el}\,l^{-1}$ can be obtained today since further system components such as heat exchangers, pumps, blowers, and compressors and also suitable electronics are required. Making these systems more compact will require intensive efforts regarding component development and innovative system design.

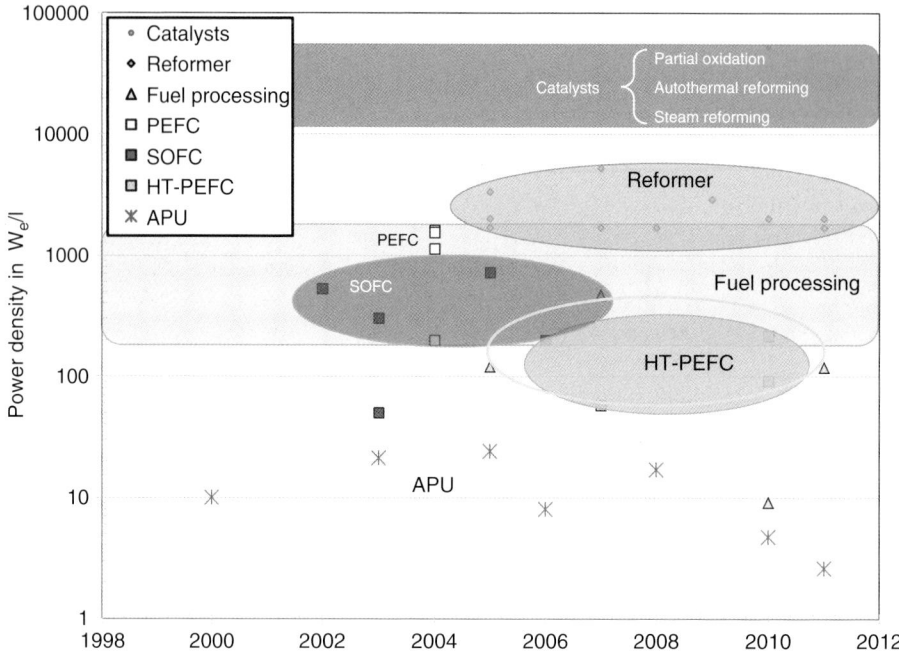

Figure 1.8 Power densities achieved for catalysts, reformers, fuel processing systems, fuel-cell stacks, and APUs.

1.6
Fuel Cells for Stationary Applications

1.6.1
Stationary Applications in Building Technology

In 1996, Vaillant started its 13 year development work on fuel-cell heaters based on natural gas as an energy carrier. It took 6.5 years to develop a prototype. Another 6.5 years were planned until the product was launched on the market. Fifteen second-generation demonstration systems were installed in 2002. The reliability of the systems tripled. From 2004 to 2006, the focus was on field testing. Fifty-six third-generation demonstration systems were tested in Germany, The Netherlands, Spain, and Portugal. The volume of the natural gas reformer was cut by half, the electrical efficiency was increased from 18 to 30%, and the stacks achieved lifetimes of up to 6000 h. It was found that for the third generation, the minimum power could be reduced from 2.5 to 1 kW. A lifetime of up to 12 000 h seemed possible. There were, however, problems with the reliability of the electronics, with controls, and with conventional system components. Work on conventional PEFCs ceased at the end of 2005. In the end, the cost objectives could not be achieved with Nafion membranes and the relatively complex process of fuel production. Work on HT-PEFCs followed, and also a cooperation with Staxera and Fraunhofer IKTS for the development of an SOFC building energy system which started in 2006. Field tests were planned starting in 2011 [148, 149].

In 1997, the first devices for building energy supply based on a PEFC prototype manufactured by Dais Analytic Corporation (DAC) in Boston were installed at HGC Hamburg Gas Consult. Nine "alpha units" were installed and tested until 1999. European Fuel Cell GmbH (EFC), a 1999 spin-off, started to build an improved "beta unit" in 2001. This company was then taken over in 2002 by the Baxi Group (the third-largest heating technology group in Europe with Brötje as one of its well-known brands). A preproduction series was planned starting in 2009. When the field tests with a "gamma" version are successfully completed, 40 000 devices of the marketable aggregate are to be produced per year in the period from 2010 to 2013 [150–152].

The company inhouse engineering GmbH was founded in 2005 as a spin-off from Schalt- und Regeltechnik GmbH. The company is developing 5 kW fuel-cell systems based on PEFCs and natural gas reforming. Field tests were planned to start in 2009. HT-PEFCs are being considered in a development project. The development partners are TU Bergakademie Freiberg (responsible for the reformer), DBI Freiberg (burners, cooling device), and the University of Magdeburg (control system). The prototype with four cells had an electric power of 150 W in 2009. A 30-cell system with 1.5 kW was subsequently to be developed for operation with a reformate. When the project ended in August 2009, the subsystem for fuel production in particular had been successfully developed. No information was made available on the overall system [153].

All the R&D work presented so far uses natural gas as an energy carrier. In the MÖWE collaborative project, the partners, that is, S&R Schalt- und Regeltechnik GmbH, Öl-Wärme-Institut GmbH (OWI), Behr GmbH, and Umicore AG, studied the development of a steam reformer based on light heating oil for decentralized hydrogen production and system testing using a stationary PEFC system. Recent research findings with diesel steam reformers show that degradation already becomes apparent after a few operating hours [154]. Regenerating the catalyst by oxidation of the carbonaceous deposits is possible [155]. However, there is still a considerable need for development.

Since 2006, a relatively large number of companies have constructed complete SOFC systems. Most of these systems are initial laboratory test systems. However, they still highlight the impressive progress made during the last few years. Hexis has the most experience with small building energy systems and has made great progress in terms of reliability and long-term stability with Galileo, a system that it has developed further [12]. Ceramic Fuel Cell Limited (CFCL) has already demonstrated a net electrical efficiency of 60% with a 1.5 kW system [156]. SOFCs received an extra boost by the work of Bloom Energy, which operated behind closed doors for several years and has been selling 100 kW systems since 2009.

Current work on stationary applications of fuel cells for building technology are being carried out within the German Callux demonstration project by the heater manufacturers Baxi Innotech, Hexis, inhouse engineering GmbH, and Vaillant, together with the utility companies E.ON Ruhrgas, MVV, EnBW, EWW, and VNG, and with the research institution ZSW. Up to 800 field test systems are to be installed by 2012, preferably in private households; the 100th device was built and installed in November 2010. The systems are based on SOFCs and PEFCs, can be fueled with natural gas or bio natural gas, and have an electrical power of 1 kW$_{el}$ and a thermal power of 2 kW$_{th}$ in the combined heat and power system.

1.6.2
Stationary Industrial Applications

Based on Ballard's development in the 1990s of PEFC stacks for mobile applications, stationary applications were also developed. The first 210 kW combined heat and power system was tested by Ballard in Burnaby (Canada). The system was developed together with Alstom. A field test program with improved 250 kW systems (BEWAG, EBM-Schweiz, etc.) was launched in 2000 [157]. In the test period from 2001 to 2004, seven 250 kW systems were delivered worldwide. Component reliability was one of the main problems. In addition, the low electrical efficiency of 35% was a great disadvantage in competition with conventional technology. The longest operating time was 6500 h.

Development work at MTU resulted in a 250 kW system with molten carbonate fuel cells (MCFCs) operated at Ruhrgas in Dorsten in 1997. The system had a short lifetime, but provided valuable information for improvements. In total, 35 HotModul systems had been delivered (together with FCE) by the end of 2006. Several systems had a lifetime of >30 000 h until the stack had to be replaced. The overall electrical efficiency is between 45 and 46% at 50–100% of the rated power [158].

Based on the good results achieved with eight SOFC systems of the 25 kW class (cell tubes with a length of 50 cm) up to 1995 (accumulated operating time 35 000 h; more than 100 thermo cycles; degradation ~0.2% per 1000 h), Siemens started to develop a system in the 100 kW class with cell tubes 1.5 m long. The first 100 kW SOFC system was put into operation in Arnhem (The Netherlands) in 1997. Extensive maintenance was required after 3500 operating hours, since several tubes were damaged. Subsequently, the system was operated for more than 20 000 h with negligible degradation and an overall electrical efficiency of 46% [159]. In 2000, a hybrid system operated under pressure (coupled with a gas turbine) was put into operation in the USA. Despite numerous breakdowns, this system achieved ~2000 operating hours and an overall electrical efficiency of 52% [160].

1.7
Special Markets for Fuel Cells

In the past few years, a number of applications for fuel cells have emerged that may enter the market early [161, 162]. This is not a matter of niche applications, but as these markets are less price sensitive they are well suited to act as door openers for the broad application of fuel cells. These include:

- on-board power supply for campers and sailing yachts
- electric vehicles such as e-bikes, scooters, and vehicles for tourists
- battery replacement for emission-free drives in electric warehouse trucks
- uninterruptable power supply (UPS) and emergency power supply, for example, for telecommunications, hospitals, control centers, and IT centers
- micro applications (4C applications).

The power required for these applications ranges from 100 mW to ~50 kW. Many applications use methanol in DMFCs, methanol or LPG in connection with a reformer in PEFCs, or hydrogen in PEFCs, as energy carriers. Table 1.9 shows the most important developments for this segment in the last 20 years with a special focus on the German situation.

1.8 Marketable Development Results

1.8.1 Submarine

Siemens tested on-board power supply for submarines with a 100 kW alkaline fuel cell on the U1 submarine of the German Navy in 1988. The 6 month test runs were very successful and constituted the basis of the German Navy's decision to equip the new 212 generation of submarines with fuel cells. However, the more recent PEFC development was used. The fuel cell developed by Siemens is based on a Nafion membrane with a high platinum loading and a metallic bipolar plate with an elastomer seal, all developed specifically for the utilization of H_2/O_2.

Nine modules of the 34 kW type were connected in a 300 kW system and used in the German Navy submarines. The further development led to 120 kW modules that are currently being used in the class 214 submarines intended for export [178]. Initial tests were carried out in submarines in 2003 [179]. These fuel-cell systems have been available as a commercial product for this particular application for a few years.

1.8.2 DMFC Battery Chargers

Toshiba launched its DMFC-driven Dynario battery charger system in 2009. Owing to the use of concentrated methanol solution, the system has a size of only 150 × 21 × 74.5 mm^2 and a low weight of 280 g. The system has a hybrid structure, which uses a lithium ion battery charged by the DMFC to store electricity. The Dynario system and its fuel cartridge fully comply with the International Electrotechnical Commission's safety standards; it was sold in Japan, together with a special fuel cartridge for simple and fast refueling, in a limited edition of 3000 units [87]. With a DMFC device that is called Mobion, the US company MTI has developed a product for a similar market to Toshiba's Dynario systems. The strategy

Table 1.9 Development status of fuel-cell systems for special markets.

Market	Developer	Country	Fuel cell	Fuel	Max. continuous system power	Ref.
On-board power supply	SFC Energy	Germany	DMFC	CH_3OH	<250 W_{el}	[163]
	Truma	Germany	HT-PEFC	LPG	210–250 W_{el}	–
	EnyMotion	Germany	PEFC	LPG	250 W_{el}	[163]
Electric vehicles	Clean Mobile/SFC Energy	Germany	DMFC	CH_3OH	250–500 W_{el}	[163]
	Masterflex	Germany	PEFC	H_2	200 W_{el}	[163]
Material handling	Siemens	Germany	PEFC	H_2	10 kW_{el}	[164]
	Still/Linde/Proton Motor	Germany	PEFC	H_2	18 kW_{el}	
	Still/Hydrogenics	Germany/Canada	PEFC	H_2	10 kW_{el}	[165]
	Still/Hoppecke/Linde/Nuvera	Germany/Italy	PEFC	H_2	5 kW_{el}	–
	Still/Hoppecke/Linde/Hydrogenics	Canada/Germany	PEFC	H_2	12 kW_{el}	[165]
	Gruma/Linde/Hydrogenics	Germany/Canada	PEFC	H_2	–	[166]
	Forschungszentrum Jülich	Germany	DMFC	CH_3OH	1.3 kW_{el}	[167]
	Nuvera	Italy	PEFC	H_2	10 kW_{el}	[168–170]
	Proton Motor	Germany	PEFC	H_2	2, 4, or 8 kW_{el}	[171]
	Plug Power/Ballard	USA/Canada	PEFC	H_2	1.8 or 10 kW_{el}	[172–174]
	Oorja Protonics	USA	DMFC	CH_3OH	0.8 kW_{el}	[103]
UPS	Future E/Ballard Power Systems	Germany/Canada	PEFC	H_2	0.5, 1, or 2 kW_{el}	[163]
	IdaTech	USA	PEFC	CH_3OH	2.5 or 5 kW_{el}	[175]
	IdaTech	USA	PEFC	H_2	2.5 or 5 kW_{el}	[176]
	Dantherm/Ballard	Denmark/Canada	PEFC	H_2	1.6 or 5 kW_{el}	[177]
4C applications	FhG-ISE	Germany	DMFC	CH_3OH	–	–
	Friwo/Solvicor/FhG-ISE	Germany	DMFC	CH_3OH	~100 mW_{el}	[163]

of the Mobion DMFC is the use of highly concentrated methanol in a simple system where the water management is done within the MEA by back-diffusion of water from the cathode to the anode [180]. At various exhibitions, Samsung SDI has shown their DMFC systems that can be used for charging mobile phone batteries or for powering laptops. In 2009 they presented a DMFC system with a power output of 25 W that was developed for military application [181, 182].

Commercial DMFC systems are currently being sold by SFC Energy in Brunnthal-Nord (Germany). It has sold more than 20 000 DMFC systems in the power range up to 250 W from when it was founded in 2000 until 2011 [183]. These fuel cells were developed as battery chargers for a voltage range from 12 to 24 V specifically for the reliable provision of power for mobile and portable applications in the power range up to 250 W. They are mainly used in the recreational sector, that is, for supplying power to electricity consumers in mobile homes, caravans, and sailboats. A lead acid battery is usually charged by a DMFC with a continuous output of 25–90 W, which means that the systems have a charging capacity of 600–2200 Wh per day. Other areas of application are transportation and safety engineering and also environmental sensor technology.

A frequently cited disadvantage of methanol is its toxicity. What is often forgotten, however, is that methanol is much less toxic than the conventional energy carrier gasoline, which owing to its benzene content is, moreover, also carcinogenic and teratogenic. Methanol, in contrast, although acutely toxic, is neither teratogenic nor carcinogenic and is much more readily biodegradable than gasoline or diesel. It is therefore not harmful to handle methanol if hermetically sealed containers are used that permit safe handling, as in the case of SFC's fuel cartridges. These cartridges, for example, have the seal of approval from the German Technical Control Board (TÜV-GS Siegel) and are licensed for transportation purposes on-board land vehicles, ships, and aircraft.

1.8.3
Uninterruptable Power Supply/Backup Power

In addition to the portable battery charger systems, the uninterruptable power supply of critical systems for telecommunication infrastructure and hospitals is an interesting market for fuel-cell systems [184]. The US company Hydrogenics supplies the HyPM XR system that is available with a power output of 4, 8, and 12 kW. These systems can be scaled up to a power output of 100 kW and they are easy to integrate into an electrical cabinet [185]. The companies Dantherm and Ballard have a cooperation where Ballard is the stack supplier and Dantherm the system integrator. The Dantherm power systems have a power output of 1.6 or 5 kW [177, 186–188]. The US company IdaTech offers systems that are constructed for outdoor use. These ElectraGen systems are available for operation with hydrogen or for use with a methanol–water mixture in combination with a reformer. The reformer can provide gaseous hydrogen from the supplied liquid mixture of methanol and water; this makes storing hydrogen cylinders on-site unnecessary and reduces the overall costs of the fuel supply. The systems are available with a power output of 2.5 or 5 kW [175, 176].

German system integration companies that already have products on the market are Rittal, P21, b + w Electronics and FutureE Fuel Cell Solutions. They all use PEFC systems, with either pure hydrogen or methanol as an energy carrier. The methanol-based systems are fuel-cell backup power systems from IdaTech. All other system providers use fuel-cell systems based on pure hydrogen. DMFC systems manufactured by SFC Energy (see Section 1.8.2) are also used for emergency power supply in the low-power range.

Among the system integration providers mentioned above, only P21 produces its own fuel cell stacks. Rittal uses stacks from Ballard for high-power UPS systems (RiCell Flex fuel-cell system up to 20 kW) and stacks from Schunk for less powerful systems (RiCell 300, 600, 900 W). A strategic partnership established between Rittal and IdaTech in 2003 for the integration of fuel-cell systems for emergency power systems has since been terminated. Rittal now purchases stacks from Ballard via FutureE Fuel Cell Solutions and from Schunk, who developed their own stacks. FutureE Fuel Cell Solutions is a start-up company founded in 2006 which sold its first fuel-cell emergency power systems of the Jupiter family in 2008. These are modular hydrogen systems with a maximum power of 40 kW.

1.8.4
Light Traction

Owing to financial support from the government, fuel-cell users in the USA are entitled to a federal tax credit up to $3000\,kW^{-1}$, so the market for light traction applications especially in the USA is booming [189]. Different companies have developed fuel-cell systems for material handling applications and numerous such systems are now running in different distribution centers around the USA. The companies Plug Power, Nuvera, Hydrogenics, and Oorja Protonics are the main actors in the market. In the year 2011 alone for Plug Power it was possible to double the number of systems sold annually to 1456 modules up to October [190].

Whereas Plug Power, Nuvera, and Hydrogenics using pressurized hydrogen as fuel (350 bar) the Oorja Protonic systems use methanol. The Oorja DMFC system has a lower power output than the hydrogen systems and it does not replace the complete battery, but in combination with a conventional battery it is used as a range extender.

In the light traction sector, the only marketable products in Germany are also manufactured by SFC Energy. They are based on the DMFC EFOY Pro Series. These devices are referred to as "range extenders," internal chargers for the batteries of electric scooters or small electric cars with a maximum electric power of 90 W, which make the range of these vehicles greater.

1.9
Conclusion

Intensive global research and development efforts together with groundbreaking demonstration activities have conclusively demonstrated that in selected fields of application, fuel cells have a high potential for effectively addressing the challenge of reducing CO_2 emissions and conserving resources by replacing conventional energy technology in the future. Combined with conventional energy carriers, fuel cells will then contribute substantially to sustainability in the energy supply sector.

Laboratory and field tests have demonstrated the suitability of SOFCs, particularly with anode-supported cells, for applications in the field of decentralized combined heat and power generation and large on-board power supply systems. These cells have a power density of up to $2\,\text{W}\,\text{cm}^{-2}$ at $0.7\,\text{V}$, a size of up to $960\,\text{cm}^2$, a stack power of up to $25\,\text{kW}$, and technically relevant aging rates of 0.2% per 1000 h. However, coupling SOFCs with a combined cycle process for highly efficient power generation with systems efficiencies of more than 80% will require considerable efforts to improve SOFC reliability and complex innovations in terms of systems and plant technology. HT-PEFCs with PBI membranes doped with phosphoric acid have been developed for about 15 years and are therefore still relatively new. Nevertheless, there is conclusive evidence that they are suitable for electrochemically converting fuels containing CO and that fuels available on the market can therefore be used, for example, diesel and kerosene. Reducing the cell overpotential on the cathode side is considered to be a means for greatly enhancing the electrochemical performance of the cells. Today, power densities of $0.5\,\text{W}\,\text{cm}^{-2}$ can be achieved in HT-PEFC stacks with an operating temperature of $160\,^\circ\text{C}$ and a cell voltage of $0.5\,\text{V}$. DMFCs are of great interest for all those applications that are powered by batteries today and struggle with limitations in terms of functionality and operating life. In spite of high catalyst loadings of $\sim 4\,\text{mg}\,\text{cm}^{-2}$ per cell, the power densities of DMFCs are hardly above $50\,\text{mW}\,\text{cm}^{-2}$, depending on the pressure ($<4\,\text{bar}$), the operating temperature ($70\text{--}100\,^\circ\text{C}$) and the air:methanol ratio. This is essentially a consequence of methanol permeation to the cathode, which promising R&D approaches such as the production of barrier layers by means of plasma treatment intend to address.

The transportation sector aims to use fuel cells to increase the efficiency of passenger vehicle and bus drivetrains and also for on-board power supply used in heavy goods vehicles. Hybridization with a suitable battery dimensioned for the intended purpose is crucial for the performance and efficiency of the overall system. Mid-sized cars field tested today have a specific energy requirement of slightly over $100\,\text{MJ}$ per $100\,\text{km}$. With an electric stack power of $90\,\text{kW}$ and a tank system for gaseous hydrogen at $700\,\text{bar}$, vehicles have a range of almost $400\,\text{km}$. Cold-start ability at up to $-30\,^\circ\text{C}$, a noble metal loading of $30\,\text{g}$ per vehicle, and a stack power density of $\sim 2\,\text{kW}_{\text{el}}\,\text{l}^{-1}$ – values achieved during development – show that current passenger cars with a fuel-cell drive are now technically mature enough to be launched on the market as soon as ongoing field tests are completed. On-board power supply (APUs) for aircraft, ships, railroads, and trucks is targeted at compact, efficient, and low-emission energy systems that use the fuel available on-board, for example, kerosene, diesel, or LPG. For power generation by means of fuel cells, this requires an adapted fuel production system and also further system components. Current APUs with an average power density of $\sim 17\,\text{W}_{\text{el}}\,\text{l}^{-1}$ are 2–44 times below the estimated target values.

The strategy for stationary applications of fuel cells for decentralized energy supply in the power range from 1 to $5\,\text{kW}_{\text{el}}$ for private households and from

100 to X000 kW$_{el}$ for industrial and local facilities has changed numerous times. Development is currently focused on small- and medium-sized SOFC systems, which are undergoing intensive field testing all over the world. The benchmark for the electrical efficiency of a 1.5 kW$_{el}$ system produced by CFCL is 60%.

Fuel-cell applications in the special markets segment, for example, on-board power supply, lightweight electric vehicles, emergency power supply, and very small electronic systems, are also undergoing in-depth field testing. The fuel-cell types used are DMFCs, HT-PEFCs, and PEFCs with upstream LPG reforming and PEFCs with hydrogen. The electrical system powers are between 100 mW$_{el}$ for 4C applications and 18 kW$_{el}$ for forklifts. Special fuel-cell systems for the power supply of submarines with PEFCs, for small consumers in the recreational sector with DMFCs, uninterruptible power supply with PEFCs, and DMFCs, and also for driving microcars with DMFCs, have achieved a position in the market.

References

1. European Commission (2003) *Wasserstoffenergie und Brennstoffzellen – Eine Zukunftsvision*, EUR 20719 DE, European Commission, Brussels.
2. Fujii, H. (2002) Status of national project for SOFC development in Japan, presented at the Solid Oxide Fuel Cells Meeting, Palm Springs, CA, November 18, 2002.
3. Hiwatashi, K., Murakami, H., Shiono, M., Saito, T., A. Abe, T., and Ueno, A. (2007) Development Status of Tubular type SOFC in TOTO in Proceedings of the Fuel Cell Seminar, Palm Springs, CA.
4. Kabs, H. (2002) Advanced SOFC technology and its realization at siemens westinghouse, in Proceedings Bilateral Seminars 33, Materials and Processes for Advanced Technology, Egyptian–German Workshop, Cairo, Egypt, April 7–9, 2002 (ed. D. Stöver and M. Bram), pp. 91–101.
5. Yoshida, H., Seyama, T., Sobue, T., and Yamashita, S. (2011) Development of residential SOFC CHP system with flatten tubular segmented-in-series cells stack. *ECS Trans.*, **35** (1), 97–103.
6. Lim, T.-H., Kim, G.-Y., Park, J.-L., Song, R.-H., Lee, S.-B., and Shin, D.-R., (2007) Operating Characteristics of Advanced 500W class Anode-supported Flat Tubular SOFC Stack in KIER, *ECS Trans.*, **7** (1), 193–197, ISBN 978-1-56677-554-0.
7. Pierre, J., (2010) Stationary Fuel Cells in Proceedings of the 11th Annual SECA Workshop, Pittsburg, PA.
8. Bossel, U. (ed.) (2004) Proceedings of The Fuel Cell World, Lucerne, Switzerland.
9. Hosoi, K., Ito, M., and Fukae, M. (2011) Status of national project for SOFC development in Japan. *ECS Trans.*, **35** (1), 11–18.
10. Nishiura, M., Koga, S., Kabata, T., Hisatome, N., Kosaka, K., Ando, Y., and Kobayashi, Y., (2007) Development of SOFC-micro gas turbine combined cycle system., *ECS Trans.*, **7** (1), 155–160, ISBN 978-1-56677-554-0.
11. Schmidt, M. (1998) The Hexis project: decentralised electricity generation with waste heat utilisation in the household. *Fuel Cells Bull.*, **1** (1), 9–11.
12. Schuler, A. and Mai, A. (2011) Stationäre SOFC im Praxistest, presented at the DACH Workshop, Reutte, Austria, February 23, 2011.
13. Nakanishi, A., Hattori, M., Sakaki, Y., Miyamoto, H., Aiki, H., Takenobu, K., and Nishiura, M. (2002) Development of MOLB type SOFC, in Proceedings of the Fifth European SOFC Forum, Lucerne, Switzerland, July 1–5, 2002, vol. 2 (ed. J. Huijsmans), pp. 708–715.

14. Onodera, T., Kanehira, S., Takenobu, K., Nishiura, M., Nakanishi, A., Hattori, M., Sakaki, Y., Kimura, K., Kobayashi, Y., Ando, Y., and Miyamoto, H., (2005) Development of MOLB Type SOFC in Proceedings of the Fuel Cell Seminar, Palm Springs, CA.
15. Steinberger-Wilckens, R., Blum, L., Buchkremer, H.-P., Gross, S., Haart, L.G.J., de Hilpert, K., Nabielek, H., Quadakkers, W.J., Reisgen, U., Steinbrech, R.W., and Tietz, F. (2006) Overview of the development of solid oxide fuel cells at Forschungszentrum Juelich. *Int. J. Appl. Ceram. Technol.*, **3** (6), 470–476.
16. Gardner, F.J., Day, M.J., Brandon, N.P., Pashley, M.N., and Cassidy, M. (2000) SOFC technology development at Rolls Royce. *J. Power Sources*, **86**, 122–129.
17. Kayahara, Y. and Yoshida, M. (2008) Residential CHP Program by Osaka Gas Company and KYOCERA Corporation in Proceedings of the Eighth European SOFC Forum, Lucerne, Switzerland.
18. Schiller, G., Henne, R., Lang, M., and Müller, M. (2004) Development of solid oxide fuel cells by applying DC and RF plasma deposition technologies. *Fuel Cells*, **4** (1–2), 56–61.
19. Szabo, P., Arnold, J., Franco, T., Gindrat, M., Refke, A., Zagst, A., and Ansar, A. (2009) Progress in the metal supported solid oxide fuel cells and stacks for APU. *ECS Trans.*, **25** (2), 175–185.
20. Ponthieu, E. (1999) Status of SOFC Development in Europe, in Proceedings of the 6th International Symposium Solid Oxide Fuel Cells - SOFC VI, (eds. S.C. Singhal and M. Dokiya), Electrochemical Society, Pennington, NJ. pp. 19–27.
21. Larsen, P.H., Bagger, C., Linderoth, S., Mogensen, M., Primdahl, S., Joergensen, M.J., Hendriksen, P.V., Kindl, B., Bonanos, N., Poulsen, F.W, and Maegaard, K.A, (2001). Status of the Danish SOFC Program, in Proceedings of the 7th International Symposium Solid Oxide Fuel Cells – SOFC VII, (eds. H. Yokokawa and S.C. Singhal), Electrochemical Society, Pennington, NJ, PV 2001–16 pp. 28–37.
22. Nguyen, M. (2003) Solid Oxide Fuel Cell Systems for Power Generation Applications, in Proceedings of the 8th International Symposium Solid Oxide Fuel Cells – SOFC VIII, (eds. M. Dokiya and S.C. Singhal), Electrochemical Society, Pennington, NJ. PV 2003–07 pp. 43–47.
23. Steinberger-Wilckens, R., de Haart, B., Buchkremer, H.-P., Nabielek, H., Quadakkers, J., Steinbrech, R., Tietz, F., and Vinke, I. (2004) Recent Results of Solid Oxide Fuel Cell Development at Forschungszentrum Juelich, in Proceedings of the Fuel Cell Seminar, San Antonio, TX. pp. 120–123.
24. Christiansen, N., Hansen, J., Kristensen, S., Holm-Larsen, H., Linderoth, S., Hendriksen, P., Larsen, P., and Mogensen, M. (2004) SOFC Development Program at Haldor Topsøe/Risø National Laboratory – Progress Presentation in Proceedings of the Fuel Cell Seminar, San Antonio, TX. pp. 124–127.
25. Mai, A., Iwanschitz, B., Weissen, U., Denzler, R., Haberstock, D., Nerlich, V., and Schuler, A. (2011) Status of Hexis' SOFC Stack Development and the Galileo 1000 N Micro-CHP System, *ECS Transactions*, **35** (1), 87–96.
26. Matsuzaki, Y., Baba, Y., Ogiwara, T., and Yakabe, H. (2002) Development of Anode-Supported SOFC with Metallic Interconnectors for Reduced-Temperature Operation, in Proceedings of the Fifth European Solid Oxide Fuel Cell Forum, Vol. 2, (eds. J. Huijsmans and U. Bossel), Lucerne Publishing, Lucerne, pp. 776–783.
27. Dinsdale, J., and Föger, K. (2004) Ceramic Fuel Cells Product Demonstration Program in Proceedings of the Fuel Cell World, European Fuel Cell Forum Lucerne, pp. 298–305.
28. (2007) Progress in Acumentrics' Fuel Cell Program, in Proceedings of the 8th Annual SECA Workshop, San Antonio, TX

29. Steinberger-Wilckens, R., Buchkremer, H.P., Malzbender, J., Blum, L., de Haart, L.G.J., and Pap, M. (2010) Recent developments in SOFC research at Forschungszentrum Jülich, im Proceedings of the 9th European SOFC Forum, Lucerne, Switzerland, pp. 2-31–2-39.
30. Tang, E., Pastula, M., Wood, T., and Borglum, B. (2009) 10 kW SOFC Cell and Stack Development at Versa Power Systems in Proceedings of the 33rd International Conference on Advanced Ceramics and Composites, Daytona Beach, FL.
31. Inagaki, T., Nishiwaki, F., Kanou, J., Miki, Y., Hosoi, K., Miyazawa, T., and Komada, N. (2004) Intermediate Temperature SOFC Modules and System in *Proceedings of the Fuel Cell Seminar*, San Antonio, TX. pp. 113–116.
32. Hayashi, K., Miyasaka, A., Katou, N., Yoshida, Y., Arai, H., Hirakawa, M., Uwani, H., Kashima, S., Orisima, H., Kurachi, S., Matsui, A., Katsurayama, K., and Tohma, E. (2011) Progress on SOFC power generation module and system developed by NTT, SPP and THG, *ECS Trans.*, **35** (1), 121–126.
33. Hwang, S.C., Kim, S.G., Kim, D., and Jun, J.H. (2010) Development of kW-class anode-supported planar SOFC stacks at RIST, in Proceedings of the 9th European SOFC Forum, Lucerne, Switzerland, pp. 2-29–2-30.
34. Mai, A., Haanappel, V.A.C., Tietz, F., and Stöver D. (2006) Ferrite-based perovskites as cathode materials for anode-supported solid oxide fuel cells, Part II: influence of the CGO interlayer. *Solid State Ionics*, **177**, 2103–2107.
35. Wunderlich, C. and Lawrence, J. (2010) Status of system integration and robustness of Staxera ISM products, in Proceedings of the 9th European SOFC Forum, Lucerne, Switzerland, pp. 2-116–2-123.
36. Love, J., Amarasinghe, S., Selvey, D., Zheng, X., and Christiansen, L. (2009) Development of SOFC Stacks at Ceramic Fuel Cells Limited. *ECS Trans.*, **25** (2), 115–124.
37. US DOE (2004) Office of Fossil Energy Fuel Cell Program Annual Report., US Department of Energy, Washington, DC.
38. Danish Power Systems (2012) Danish Power Systems, http://www.daposy.com/ (last accessed 25 January 2012).
39. Bose, S., Kuila, T., Nguyen, T., Kim, N., Lau, K., and Lee, J. (2011) Polymer membranes for high temperature proton exchange membrane fuel cell: recent advances and challenges. *Prog. Polym. Sci.*, **36**, 813–843.
40. Asensio, J., Sánchez, M., and Gómez-Romero, P. (2010) Proton-conducting membranes based on benzimidazole polymers for high-temperature PEM fuel cells. A chemical quest. *Chem. Soc. Rev.*, **39**, 3210–3219.
41. Daletou, M., Kallistis, J., and Neophytides; S. (2011) Materials, proton conductivity and electrocatalysis in high-temperature PEM fuel cells, in *Interfacial Phenomena in Electrocatalysis*, Modern Aspects of Electrochemistry, vol. 51 (ed. C.G. Vayenas), Springer, Berlin, pp. 301–368.
42. Serenergy (2012) The Power of Simplicity, http://www.serenergy.dk/ (last accessed 25 January 2012).
43. Aalborg University, Department of Energy Technology (2012) Introduction on Fuel Cell and Battery Systems, http://www.et.aau.dk/research-prog rammes/fuel-cell-systems/ (last accessed 25 January 2012).
44. DTU Energy Conversion (2011) Proton Conductors, http://www.energi.kemi.dtu.dk/Forskning/ Publications.aspx (last accessed November 2011).
45. Lobato, J., Cañizares, P., Rodrigo, M., Úbeda, D., and Pinar, F. (2011) A novel titanium PBI-based composite membrane for high temperature PEMFCs. *J. Membr. Sci.*, **369**, 105–111.
46. Lobato, J., Cañizares, P., Rodrigo, M., Linares, J., and Pinar, F. (2010) Study of the influence of the amount of $PBI-H_3PO_4$ in the catalytic layer of a high temperature PEMFC. *J. Hydrogen Energy*, **35**, 1347–1355.

47. Newcastle University, School of Chemical Engineering and Advanced Materials (2011) Professor Keith Scott, http://www.ncl.ac.uk/ceam/staff/profile/keith.scott#tab_publications (last accessed December 2011).
48. University of South Carolina, Department of Chemistry and Biochemistry (2011) Brian C. Benicewicz, http://www.chem.sc.edu/people/faculty StaffDetails.asp?SID=872 (last accessed December 2011)
49. Case Western Reserve University (2012) Robert Savinell, http://engineering.case.edu/profiles/rfs2 (last accessed 25 January 2012).
50. Dalian Institute of Chemical Physics (2011) Laboratory of Fuel Cells, http://www.english.dicp.ac.cn/04rese/08/01.htm (last accessed 25 January 2012).
51. Korea Institute of Science and Technology (2012) Fuel Cell Research Center, http://www.kist.re.kr/en/iv/en_fu_in.jsp (last accessed 25 January 2012).
52. Korea Institute of Energy Research (2012) Energy Innovation for a Better Life, http://www.kier.re.kr/open_content/eng/main_page.jsp (last accessed Deember 2012).
53. Lehnert, W., Wannek, C., and Zeis, R. (2010) in *Innovations in Fuel Cell Technology* (eds. R. Steinberger-Wilckens and W. Lehnert), RSC Publishing, Cambridge, pp. 45–75.
54. Wannek, C. (2010) in *High-Temperature PEM Fuel Cells: Electrolysis, Cells, and Stacks Hydrogen and Fuel Cells, Fundamentals, Technologies and Applications* (ed. D. Stolten), Wiley-VCH Verlag GmbH, Weinheim, pp. 17–40.
55. Li, Q., Jensen, J.O., Savinell, R.F., and Bjerrum, N. (2009) High temperature proton exchange membranes based on polybenzimidazoles for fuel cells. *Prog. Polym. Sci.*, **34**, 449–477.
56. Kallitsis, J.K., Geormezi, M., Gourdoupi, N., Paloukis, F., Andreopoulou, A.K., Morfopoulou, C., and Neophytides, S.G. (2010) Influence of the molecular structure on the properties and fuel cell performance of high temperature polymer electrolyte membranes. *ECS Trans.*, **33** (1), 811–822.
57. Reiche, A. (2006) Presented at the Sartorius Fuel Cell Seminar, Honolulu, HI.
58. Henschel, C. (2005) Hochtemperatur-PEM-Brennstoffzellen: auf dem Weg zum Konkurrenzfähigen CHP-System. Presented at the f-cell Symposium, Stuttgart, Germany, September 27, 2005.
59. Schmidt, T.J. (2006) Durability and degradation in high-temperature polymer electrolyte fuel cells. *ECS Trans.*, **1** (8), 19–31.
60. Schmidt, T.J. and Baurmeister, J. (2008) Properties of high-temperature PEFC Celtec®-P 1000 MEAs in start/stop operation mode. *J. Power Sources*, **176**, 428–434.
61. Hartnig, C., and Schmidt, T.J. (2011) Simulated start–stop as a rapid aging tool for polymer electrolyte fuel cell electrodes. *J. Power Sources*, **196**, 5564–5572.
62. Hofmann, C. (2009) Untersuchungen zur Elektrokatalyse von Hochtemperatur-Polymerelektrolyt membran-Brennstoffzellen (HT-PEMFCs). Dissertation, Universität Göttingen.
63. Bauer, B., Schuster, M., Klicpera, T., Zhang, W., and Jeske, M. (2009) Fumion® – Fumaperm® – Fumasep®, presented at Statusseminar HT-PEMFC und SOFC, Stand der Forschung und Handlungsbedarf für die Zukunft, Berlin, January 21–22, 2009.
64. Siegel, C., Bandlamudi, G., and Heinzel, A. (2011) Systematic characterization of a PBI/H_3PO_4 sol–gel membrane – modeling and simulation. *J. Power Sources*, **196**, 2735–2749.
65. Agert, C., Aicher, T., Hutzenlaub, T., Kurz, T., and Zobel, M. (2007) Aktuelle entwicklungen in der reformer- und HT-PEM technik am Fraunhofer ISE, presented at the 5. Riesaer Brennstoffzellen-Workshop, Riesa, Germany, February 27, 2007.
66. Wannek, C., Dohle, H., Mergel, J., and Stolten, D. (2008) Novel VHT-PEFC MEAs based on ABPBI membranes for

APU applications. *ECS Trans.*, **12** (1), 29–39.
67. Wannek, C., Kohnen, B., Oetjen, H.-F., Lippert, H., and Mergel, J. (2008) Durability of ABPBI-based MEAs for high-temperature PEMFCs at different operating conditions. *Fuel Cells*, **8** (2), 87–95.
68. Konradi, I. (2009) Einsatz von Porenbildner in der kathodischen Katalysatorschicht von Hochtemperatur-Polymerelektrolyt-Brennstoffzellen, Fachhochschule Aachen, Jülich.
69. Wannek, C., Lehnert, W., and Mergel, J. (2009) Membrane electrode assemblies for high-temperature polymer electrolyte fuel cells based on poly(2,5-benzimidazole) membranes with phosphoric acid impregnation via the catalyst layers. *J. Power Sources*, **192**, 258–266.
70. Wannek, C., Konradi, I., Mergel, J., and Lehnert, W. (2009) Redistribution of phosphoric acid in membrane electrode assemblies for high temperature polymer electrolyte fuel cells. *Int. J. Hydrogen Energy*, **23**, 9479.
71. Huslage, J. (2008) VW high-temperature PEM MEA – status and prospects, presented at the Carisma Workshop.
72. Sartorius (2009) HT-PEM-Stack: Entwicklungsvorhaben 0326893, presented at the BMBF Statusseminar, 23 January 2009.
73. Scholta, J., Zhang, W., Jörissen, L., and Lehnert, W. (2008) Conceptual design for an externally cooled HT-PEMFC stack. *ECS Trans.*, **12** (1), 113–118.
74. Scholta, J., Messerschmidt, M., Jorissen, L., and Hartnig, C. (2009) Externally cooled high temperature polymer electrolyte membrane fuel cell stack. *J. Power Sources*, **190**, 83–85.
75. Pasupathi, S., Ulleberg, ø., Bujlo, P., and Scholta, J. (2010) in *Proceedings of the 18th World Hydrogen Energy Conference 2010 – WHEC 2010: Parallel Sessions Book 1*, Schriften des Forschungszentrums Jülich, Energy and Environment, vol. 78-1 (eds. D. Stolten and T. Grube), Forschungszentrums Jülich GmbH, Jülich. pp. 131–136, ISBN 978-3-89336-651-4.
76. Ogrzewalla, J. (2008) Verbundprojekt PEM-APU: Bordstromversorgung für gasbetriebene Fahrzeuge auf Basis einer PEM Brennstoffzelle, presented at the BMBF Statusseminar, Berlin, Germany, January 22, 2008.
77. Abschlussbericht zum Verbundvorhaben (2010) Bordstromversorgung für Gasbetriebene Fahrzeuge auf Basis einer PEM Brennstoffzelle, Förderkennzeichen FKZ 0327728A.
78. Scholta, J., Messerschmidt, M., and Jörissen, L. (2010) Design and operation of a 5 kW HT-PEMFC stack for an onboard power supply, presented at the Fuel Cell Seminar 2010, San Antonio, TX, October 18–22, 2010.
79. Löhn, H. (2010) Leistungsvergleich von Nieder- und Hochtemperatur-polymermembran-Brennstoffzellen – experimentelle Untersuchungen, Modellierung und numerische Simulation, Dissertation, TU Darmstadt.
80. Heinzel, A., Burfeind, J., Bandlamudi, G., Kreuz, C., Derieth, T., and Scharr, D. (2007) HT-PEM-Stackentwicklung am ZBT, presented at the Riesaer Brennstoffzellen Workshop, Technologieorientiertes Gründerzentrum, Glaubitz, Germany, February 27, 2007.
81. Beckhaus, P., Burfeind, J., Bandlamudi, G., and Heinzel, A. (2010) in *Proceedings 18th World Hydrogen Energy Conference 2010 – WHEC 2010: Parallel Sessions Book 1*, Schriften des Forschungszentrums Jülich, Energy and Environment, vol. 78-1 (eds. D. Stolten and T. Grube), Forschungszentrums Jülich GmbH, Jülich. pp. 125–130, ISBN 978-3-89336-651-4.
82. Kolb, G., Schelhaas, K.-P., Wichert, M., Burfeind, J., Hesske, C., and Bandlamudi, G. (2009) Entwicklung eines mikrostrukturierten Methanolreformers gekoppelt mit einer Hochtemperatur-PEM-Brennstoffzelle. *Chem. Ing. Tech.*, **81** (5), 619–628.
83. Zentrum für Brennstoffzellentechnik (2010) Statusbericht 2010. http://www.zbt-duisburg.de/_downloads/

zbt-statusbericht-2010-web.pdf (last accessed 25 January 2012).

84. Jungmann, T., Kurz, T., and Groos, U. (2010) HT- and LT-PEM Fuel Cell Stacks for Portable Applications, presented at the Hannover Trade Fair 2010, 23 April 2010, Hannover, Germany.

85. Bendzulla, A. (2010) Von der Komponente zum Stack: Entwicklung und Auslegung von HT-PEFC-Stacks der 5 kW-Klasse, Dissertation, Schriften des Forschungszentrums Jülich, Reihe Energie und Umwelt, Band 69, ISBN 978-3-89336-634-7.

86. Suominen, A., and Tuominen, A. (2010) Analyzing the direct methanol fuel cell technology in portable applications by a historical and bibliometric analysis. *J. Bus. Chem.*, **7** (3), 14.

87. Toshiba (2009) Toshiba Launches Direct Methanol Fuel Cell in Japan as External Power Source for Mobile Electronic Devices. News Release, 22 October 2009, *http://www.toshiba.co.jp/about/press/2009_10/pr2201.htm* (last accessed 5 January 2012).

88. Tian, X., Sandris, T.J.G., and Hebling, C. (2010) in *Proceedings of the 18th World Hydrogen Energy Conference 2010 – WHEC 2010, Parallel Sessions Book 1*, Schriften des Forschungszentrums Jülich, Energy and Environment, vol. 78-1 (eds. D. Stolten and T. Grube), Forschungszentrums Jülich GmbH, Jülich. pp. 227–232, ISBN 978-3-89336-651-4.

89. MTI Micro (2011) The Mobion® Chip Breakthrough, *http://www.mtimicrofuelcells.com/technology/breakthrough.asp* (last accessed 22 February 2012).

90. Baldauf, M. and Preidel, J. (1999) Status of the development of a direct methanol fuel cell. *J. Power Sources*, **84** (2), 161–166.

91. IRD Fuel Cell Technology (2011) DMFC Generator 800 W, *http://www.ird.dk/pdf/IRDdmfcgenerator%20800W.pdf* (last accessed 5 January 2012).

92. Lamm, A. and Müller, M. (2010) *System Design for Transport Applications, Handbook of Fuel Cells*, John Wiley & Sons, Ltd., Chichester.

93. Jörissen, L., Gogel, V., Kerres, J., and Garche, J. (2002) New membranes for direct methanol fuel cells. *J. Power Sources*, **105** (2), 267–273.

94. Gogel, V., Frey, T., Zhu, Y.-S., Friedrich, K.A., Jörissen, L., and Garche, J. (2004) Performance and methanol permeation of direct methanol fuel cells: dependence on operating conditions and on electrode structure. *J. Power Sources*, **127** (1–2), 172–180.

95. von Andrian, S. and Meusinger, J. (2000) Process analysis of a liquid-feed direct methanol fuel cell system. *J. Power Sources*, **91** (2), 193–201.

96. Dohle, H., Mergel, J., and Stolten, D. (2002) Heat and power management of a direct-methanol-fuel-cell (DMFC) system. *J. Power Sources*, **111** (2), 268–282.

97. Dohle, H. et al. (2002) Development of a compact 500 W class direct methanol fuel cell stack. *J. Power Sources*, **106** (1–2), 313–322.

98. Mergel, J., Gülzow, E., and Bogdanoff, P. (2004) Stand und Perspektiven der DMFC, Fachtagung Forschungsverbund Sonnenenergie, Wasserstoff und Brennstoffzellen, Berlin, Germany.

99. Joh, H.-I., Hwang, S.Y., Cho, J.H., Ha, T.J., Kim, S.K., Moon, S.H., and Ha, H.Y. (2008) Development and characteristics of a 400 W-class direct methanol fuel cell stack. *Int. J. Hydrogen Energy*, **33**, 7153–7162.

100. International Energy Agency (2009) International Energy Agency – Advanced Fuel Cells Implementing Agreement. Annual Report 2009, p. 40, *http://www.ieafuelcell.com/documents/AnnualReport2009_v3.pdf* (last accessed 5 January 2012).

101. Yamaha Motor (2007) Yamaha Motor models to be displayed at the 40th Tokyo Motor Show 2007 (Part 2). FC-Dii (Special Exhibition Model/Prototype), *http://www.yamaha-motor.co.jp/global/news/2007/10/05/tms-mc.html* (last accessed 5 January 2012).

102. Olah, G.A., Goeppert, A., and Prakash, G.K.S. (2009) *Beyond Oil and Gas. The Methanol Economy*, Wiley-VCH Verlag GmbH, Weinheim, pp. 209–212.
103. Oorja Protonics (2010) OorjaPac_Model_III, http://www.oorjaprotonics.com/PDF/OorjaPac_Model_III_Product_Sheet.pdf (last accessed 5 January 2012).
104. SFC Energy (2011) EFOY Brennstoffzellen – Garantiebedingungen, http://www.sfc.com/de/sfc-service-garantiebedingungen.html (last accessed 18 January 2012).
105. IRD (2011) Fuel Cell Technology for Today and Tomorrow. DMFC Generator 850 W, http://www.ird.dk (last accessed 12 April 2011).
106. Cabello-Moreno, N., Crabb, E.M., Fisher, J.M., Russell, A.E., and Thompsett, D. (2009) Improving the stability of PtRu catalysts for DMFC, Meeting Abstracts, 216th ECS Meeting, Abstract 983, The Electrochemical Society Pennington (NJ).
107. Peighambardoust, S.J., Rowshanzamir, S., and Amjadi, M. (2010) Review of the proton exchange membranes for fuel cell applications. *Int. J. Hydrogen Energy*, **35**, 9349–9384.
108. Ahmad, H., Kamarudin, S.K., Hasran, U.A., and Daud, W.R.W. (2010) Overview of hybrid membranes for direct-methanol fuel-cell applications. *Int. J. Hydrogen Energy*, **35**, 2160–2175.
109. Kerres, J.A. (2001) Development of ionomer membranes for fuel cells. *J. Membr. Sci.*, **185** (1), 3–27.
110. Gogel, V., Jörissen, L., ChromikA., SchonbergerF., LeeJ., SchaferM., KrajinovicK., and KerresJ. (2008) Ionomer membrane and MEA development for DMFC. *Sep. Sci. Technol.*, **43** (16), 3955–3980.
111. Meier-Haack, J. TaegerA., VogelC., SchlenstedtK., LenkW., and LehmannD. (2005) Membranes from sulfonated block copolymers for use in fuel cells. *Sep. Purif. Technol.*, **41** (3), 207–220.
112. Mathuraiveeran, T., Roelofs, K., Senftleben, D., and Schiestel, T. (2006) Proton conducting composite membranes with low ethanol crossover for DEFC. *Desalination*, **200** (1–3), 662–663.
113. Jung, D.H., Cho, S.Y., Peck, D.H., Shin, D.R., and Kim, J.S. (2002) Performance evaluation of a Nafion/silicon oxide hybrid membrane for direct methanol fuel cell. *J. Power Sources*, **106**, 173–177.
114. Schuster, M., Kreuer, K.-D., Andersen, H.T., and Maier, J. (2007) Sulfonated poly(phenylene sulfone) polymers as hydrolytically and thermooxidatively stable proton conducting ionomers. *Macromolecules*, **40** (3), 598–607.
115. Walker, M., Baumgärtner, K.M., Feichtinger, J., Kaiser, N., Räuchle, E., and Kerres, J. (1999) Barrier properties of plasma-polymerized thin films. *Surf. Coat. Technol.*, **119**, 996–1000.
116. Abdelkareem, M.A., Morohashi, N., and Nakagawa, N. (2007) Factors affecting methanol transport in a passive DMFC employing a porous carbon plate. *J. Power Sources*, **172**, 659–665.
117. Scott, K., Yu, E., Vlachogiannopoulos, G., Shivar, M., and Duteanu, N. (2008) Performance of a direct methanol alkaline membrane fuel cell. *J. Power Sources*, **175** (1), 452–457.
118. Hou, H.Y., Sun, G., He, R., Sun, B., Jin, W., Liu, H., and Xin, Q. (2008) Alkali doped polybenzimidazole membrane for alkaline direct methanol fuel cell. *Int. J. Hydrogen Energy*, **33** (23), 7172–7176.
119. Bianchini, C., Bambagioni, V., Filippi, J., Marchionni, A., Vizza, F., Bert, P., and Tampucci, A. (2009) Selective oxidation of ethanol to acetic acid in highly efficient polymer electrolyte membrane-direct ethanol fuel cells. *Electrochem. Commun.*, **11** (5), 1077–1080.
120. BMU (2007) Eckpunkte für ein integriertes Energie- und Klimaprogramm.
121. BMWi and BMU (2010) Energiekonzept für eine Umweltschonende, zuverlässige und Bezahlbare Energieversorgung.
122. BMWi (2008) Mobilität und Verkehrstechnologien – Das 3. Verkehrsforschungsprogramm der Bundesregierung.

123. BMBF (2009) Nationaler Entwicklungsplan Elektromobilität.
124. BMBF (2010) Ideen. Innovation. Wachstum. Hightech-Strategie 2020 für Deutschland.
125. Daimler (2010) Innovative Ideen im Luxussegment, http://www.daimler.de (last accessed 18 April 2011).
126. Toyota (2011) What We Care About – What You Care About, http://www.toyota.com (last accessed 18 April 2011).
127. Louie, C. (2009) Automotive fuel cells stack technology – status, plans, outlook, in Proceedings, ZEV Technology Symposium 2009, Sacramento, CA, September 21–22, 2009.
128. von Helmolt, R. (2009) Fuel cell or battery vehicles? Similar technology, different infrastructure, in Proceedings, f-cell, Stuttgart, Germany, September 28–29, 2009.
129. Thiesen, L.P., von Helmolt, R., and Berger, S. (2010) in *Proceedings of the 18th World Hydrogen Energy Conference 2010 – WHEC 2010, Parallel Sessions Book 6*, Schriften des Forschungszentrums Jülich, Energy and Environment, vol. 78-1 (eds. D. Stolten and T. Grube), Forschungszentrums Jülich GmbH, Jülich. pp. 91–96, ISBN 978-3-89336-656-9.
130. Wirtschafts Woche (2010) Brennstoffzelle Reloaded, http://www.wiwo.de/technologie/auto/wasserstoffautos-brennstoffzelle-reloaded/5666152.html (last accessed 18 January 2012).
131. DOE (2010) Multi-Year Research, Development and Demonstration Plan: Planned Program Activities for 2005–2015, http://www.energy.gov/ (last accessed 18 April 2011).
132. Grube, T., and Stolten, D. (2010) *Bewertung von Fahrzeugkonzepten mit Brennstoffzellen und Batterien*, Innovative Fahrzeugantriebe, VDI Verlag, Düsseldorf.
133. JRC (2008) Well-to-Wheels Analysis of Future Automotive Fuels and Powertrains in the European Context. TANK-TO-WHEELS Report, Appendix 1, Version 3, JRC Joint Research Centre of the European Commission, Brussels.
134. Mitsubishi (2010) Product Information. Mitsubishi i-MiEV, Mitsubishi.
135. Saft America (2010) Datenblatt zu VL 52 E – High Energy Cell, http://www.saftbatteries.com (last accessed 15 September 2010).
136. General Motors (2010) Technische Daten zur Batterie, http://ampera.opel.info (last accessed 22 September 2010).
137. Rosenkranz (2009) Li-Ion Batterien: Schlüsseltechnologie für das Elektroauto, in Proceedings, Erster Deutscher Elektro-Mobil Kongress, Bonn, Germany, June 16, 2009.
138. Lutsey, N., Brodrick, C., and Lipman, T. (2007) Analysis of potential fuel consumption and emissions reductions from fuel cell auxiliary power units (APUs) in long-haul trucks. *Energy*, **32** (12), 2428–2438.
139. Baratto, F. and Diwekar, U. (2005) Life cycle assessment of fuel cell-based APUs. *J. Power Sources*, **139** (1–2), 188–196.
140. Contestabile, M. (2010) Analysis of the market for diesel PEM fuel cell auxiliary power units onboard long-haul trucks and of its implications for the large-scale adoption of PEM FCs. *Energy Policy*, **38** (10), 5320–5334.
141. Jain, S., Chen, H., and Schwank, J. (2006) Techno-economic analysis of fuel cell auxiliary power units as alternative to idling. *J. Power Sources*, **160** (1), 474–484.
142. Heinrich, H.-J. (2007) Brennstoffzellensysteme für Anwendungen in der Luftfahrt, in Proceedings, Symposium der Wasserstoffgesellschaft, Hamburg, Germany.
143. Hari Srinivasan, J.Y., Welch, R., Tulyani, S., and Hardin, L. (2006) *Solid Oxide Fuel Cell APU Feasibility Study for a Long Range Commercial Aircraft Using UTC ITAPS Approach*, Proceedings, Volume I: Aircraft Propulsion and Subsystems Integration Evaluation, NASA Glenn Research Center, Cleveland, OH.

144. Mallika Gummalla, A.P., Braun, R., Carriere, T., Yamanis, J., Vanderspurt, L.H.T., and Welch, R. (2006) *Fuel Cell Airframe Integration Study for Short-Range Aircraft.* Proceedings, Volume I: Aircraft Propulsion and Subsystems Integration Evaluation, NASA Glenn Research Center, Cleveland, OH.
145. Docter, A., Konrad, G., and Lamm, A. (2000) Reformer für Benzin und benzinähnliche Kraftstoffe. *VDI-Ber.*, **1565**, 399–411.
146. DOE (2009) Distributed/Stationary Fuel Cell Systems, http://www1.eere.energy.gov/hydrogenandfuelcells/fuelcells/systems.html (last accessed 16 November 2009).
147. Peters, R. (2010) in *Large APUs for Vessels and Airplanes, Innovations in Fuel Cell Technology* (ed. R. Steinberger-Wilckens), Royal Society of Chemistry, Cambridge.
148. Vaillant (2005) Das Vaillant Brennstoffzellen-heizgerät: Stand der Felderprobung. *VDI-Ber.*, **1874**, 147–158.
149. Dauensteiner, A. (2008) Brennstoffzellenheizgeräte für die Hausenergieversorgung: Die Zukunft der Kraft-Wärme-Kopplung. *VDI-Ber.*, **2036**, 197–204.
150. Winkelmann, T. (2005) Das Brennnstoffzellenheizgerät für das Einfamilienhaus. *VDI-Ber.*, **1874**, 159–164.
151. Braun, M. (2008) Das Brennstoffzellenheizgerm Feldtest: Erfahrungen und Ausblick. *VDI-Ber.* **2036**, 179–187.
152. Klose, P. (2009) Pre-series fuel-cell-based μ-CHP units in their field test phase, presented at the 11th Grove Fuel Cell Symposium, London.
153. Kraue, H., Grosser, K., Gerber, J., Nitschke, J., Arnold, J., and Giesel, S. (2005) Entwicklung und Erprobung von Brennstoffzellen-Mini-BHKW im Bereich bis 12 kW elektrische Leistung. *VDI-Ber.*, **1874**, 301–303.
154. Maximini, M., Huck, T., Wruck, R., and Lucka, K. (2010) Investigations on steam reforming catalysts for diesel fuel, in *Proceedings of the 18th World Hydrogen Energy Conference 2010 – WHEC 2010, Parallel Sessions Book 3*, Schriften des Forschungszentrums Jülich, Energy and Environment, vol. 78-3 (eds. D. Stolten and T. Grube), Forschungszentrums Jülich GmbH, Jülich. pp. 283–290. ISBN 978-3-89336-653-8.
155. Mengel, C., Konrad, M., Wruck, R., Lucka, K., and Köhne, H. (2008) Diesel steam reforming for PEM fuel cells. *J. Fuel Cell Sci. Technol.*, **5** (2), 021005-1–021005-5.
156. Föger, K. (2010) Commercialisation of CFCL's residential power station – BlueGen, in Proceedings of the 9th European SOFC Forum, Lucerne, Switzerland, pp. 2 / 22–2/28.
157. Pokojski, M. (2004) Betriebserfahrungen mit einer 250-kW-PEM-Brennstoffzelle. *BWK*, **56** (7–8), 54–58.
158. Bischoff, M. (2006) Large stationary fuel cell systems: status and dynamic requirements. *J. Power Sources*, **154** (2), 461–466.
159. George, R. (2000) Status of tubular SOFC field unit demonstrations. *J. Power Sources*, **86** (1), 134–139.
160. George, R. and Casanova, A. (2003) Developments in Siemens Westinghouse SOFC program, in Proceedings of the Fuel Cell Seminar, San Diego, CA.
161. Bundesministerium für Verkehr, Bau und Stadtentwicklung (2007) Nationaler Entwicklungsplan – Version 2.1. Nationales Innovationsprogramm Wasserstoff- und Brennstoffzellentechnologie (NIP). Strategierat Wasserstoff Brennstoffzellen, BMVBS, Berlin.
162. US Department of Energy (2011) The Department of Energy Hydrogen and Fuel Cells Program Plan, http://www.hydrogen.energy.gov/pdfs/program_plan2011.pdf, p. 54 (last accessed 5 January 2012).
163. Zweck, A. (2010) *Studie zur Entwicklung eines Markteinführungsprogramms für Brennstoffzellen in Speziellen Märkten*, VDI Technologiezentrum, Düsseldorf.
164. Szyszka, A. (1999) *Schritte zu einer (Solar-) Wasserstoff-Energiewirtschaft, 13 Erfolgreiche Jahre Solar-Wasserstoff-Demonstrationsprojekt der SWB in Neunburg vorm Wald, Oberpfalz,*

Schriftenreihe Solarer Wasserstoff, Solar - Wasserstoffanlage, vol. 26, SWB.
165. McConnell, V.P. (2010) Fuel cells in forklifts extend commercial reach. *Fuel Cells Bull.*, **9**, 12–19.
166. McConnell, V.P. (2010) Rapid refill, high uptime: running forklifts with fuel cells. *Fuel Cells Bull.*, **10**, 12–19.
167. Mergel, J., Glüsen, A., and Wannek, C. (2010) in *Hydrogen and Fuel Cells – Fundamentals, Technologies and Applications* (ed. D. Stolten), Wiley-VCH Verlag GmbH, Weinheim, pp. 41–57.
168. Nuvera Fuel Cells (2011) PowerEdge WL 10 Specifications, http://www.nuvera.com/pdf/Nuvera_PowerEdge_W.pdf (last accessed 5 January 2012).
169. Nuvera Fuel Cells (2011) PowerEdge RL 40-F Specifications, http://www.nuvera.com/pdf/Nuvera_PowerEdge_R.pdf (last accessed 5 January 2012).
170. Nuvera Fuel Cells (2011) PowerEdge CM 48 Specifications, http://www.nuvera.com/pdf/Nuvera_PowerEdge_C.pdf (last accessed 5 January 2012).
171. Proton Motor (2010) PM 200 Wasserstoff Brennstoffzelle, http://www.proton-motor.de/fileadmin/Dokumente_2011/PM_200_Datenblatt_deutsch.pdf (last accessed 5 January 2012).
172. Plug Power (2011) Fuel Cell Power for Today's Material Handling Equipment. http://www.plugpower.com/Libraries/Documentation_and_Literature/Series_1000_Product_Spec_Sheet.sflb.ashx (last accessed 5 January 2012).
173. Plug Power (2011) Fuel Cell Power for Today's Material Handling Equipment. http://www.plugpower.com/Libraries/Documentation_and_Literature/Series_2000_Product_Spec_Sheet.sflb.ashx (last accessed 5 January 2012).
174. Plug Power (2011) Fuel Cell Power for Today's Material Handling Equipment. http://www.plugpower.com/Libraries/Documentation_and_Literature/Series_3000_Product_Spec_Sheet.sflb.ashx (last accessed 5 January 2012).
175. IdaTech (2011) ElectraGen™ ME System. http://www.idatech.com/uploadDocs/ElecGenMEDS_110919.pdf (last accessed 5 January 2012).
176. IdaTech (2010) ElectraGen™ H2-I System. http://www.idatech.comuploadDocs/ElecGenH2-I_021210.pdf (last accessed 5 January 2012).
177. Dantherm Power (2011) Data Sheets. http://www.dantherm-power.com/Documentation/Data_sheets.aspx (last accessed 18 January 2012).
178. Siemens, AG (2001) Industrial Solutions and Services – Marine Solutions, PEM Fuel Cells for Submarines, Order No. E10001-A930-V2-7600. Hamburg
179. Bornemann, E. (2004) U31 – The boat is back again, Magazin Loyal, 22–24.
180. MTI Micro (2011) Mobion® vs. Traditional DMFCs, http://www.mtimicrofuelcells.com/technology/differentiation.asp (last accessed 5 January 2012).
181. Samsung (2011) Fuel Cell Battery, http://samsungsdi.com/nextenergy/fuel-cell-battery.jsp (last accessed 5 January 2012).
182. Yoon, S.K., Na, Y., Joung, Y., Park, J., Kim, Y.L., Hu, L., Song, I., and Cho, H. (2009) Direct methanol fuel cell systems for portable applications, presented at the Fuel Cell Seminar, Palm Springs, CA.
183. SFC Energy (2011) Meilenstein: SFC Energy AG verkauft 20.000 EFOY-Brennstoffzell. Press Release, 18 January 2011, http://www.investor-sfc.de/de/pm.php?type=pm&id=224&year=2011&lang=de (last accessed 18 January 2012).
184. Fuel Cell and Hydrogen Energy Association (2010) Fuel Cells and Hydrogen Energy 2011. Recap and Highlights: Backup Power, http://fchea.posterous.com/fuel-cells-and-hydrogen-energy-2011-recap-and-56715 (last accessed 5 January 2012).
185. Hydrogenics (2010) HyPM® Fuel Cell Power Modules, http://www.hydrogenics.com/assets/pdfs/HyPM-%20Fuel%20Cell%20Power%20Module%20Brochure%202010.pdf (last accessed 5 January 2012).

186. Fuel Cells Bulletin (2007) Ballard secures materials handling contract, backup power order. *Fuel Cells Bull.*, **2007** (5), 4–5.
187. Fuel Cells Bulletin (2008) Dantherm, Ballard support hydroelectric in BC project. *Fuel Cells Bull.*, **2008** (8), 4.
188. Fuel Cells Bulletin (2009) Dantherm backup power for McKesson. *Fuel Cells Bull.*, **2009** (4), 4.
189. US Department of Energy (2011) Early Markets: Fuel Cells for Material Handling Equipment, *http://www1.eere.energy.gov/hydrogenandfuelcells/education/pdfs/early_markets_forklifts.pdf* (last accessed 5 January 2012).
190. Fuel Cell and Hydrogen Energy Association (2011) Fuel Cells and Hydrogen Energy 2011. Recap and Highlights: Materials Handling, *http://fchea.posterous.com/fuel-cells-and-hydrogen-energy-2011-recap-and* (last accessed 5 January 2012).

2
Single-Chamber Fuel Cells

Têko W. Napporn and Melanie Kuhn

2.1
Introduction

Increasing energy demands, environmental and availability issues related to the use of fossil fuels, and technical progress and downsizing of high-performance electronic systems are challenging us to develop clean, efficient, easily accessible, and rechargeable power-generating systems. Fuel cells have emerged as promising candidates for efficiently producing clean energy for various applications. High efficiencies are expected as these systems are not Carnot limited, and the chemical energy of the fuel and oxidant is directly converted into electrical energy. However, the costs of such systems are still high and are largely due to the nature of the materials used (catalysts, electrolytes, bipolar plates, and their accessories). Conventional fuel cells use two sealed, separate compartments for the fuel and the oxidant. The fuel and the oxidant can then be supplied separately to the respective electrodes, permitting controlled reactions at each electrode necessary for high cell performance. Gas management and manifolding, however, add additional costs to the system and become especially challenging when several single cells are to be assembled in stacks, and for miniaturized fuel cell systems. Moreover, the two gas compartments need to be gas-tight and adequate sealing is required that is chemically and thermomechanically compatible with the cell component materials and can withstand operation at elevated temperatures and during heating–cooling cycles. In solid oxide fuel cells (SOFCs), a high-temperature fuel cell system, common sealants are glass and ceramic materials that face long-term stability and durability issues and can crack during operation of the fuel cell.

The single-chamber fuel cell (SC-FC) is a sealing-free fuel cell in which both electrodes are situated in a single compartment and a mixture of fuel and oxidant is supplied directly to the cell (Figure 2.1). SC-FCs promise higher thermomechanical strength in addition to a compact and simplified design. However, single-chamber operation requires selective electrode materials that catalyze the fuel reaction only at the anode and the oxidant reduction only at the cathode. The SC-FC concept dates back to fuel cell research for applications in space in the 1960s [1]. During the last 30 years, the performance of SC-FCs has increased considerably, mainly due to the

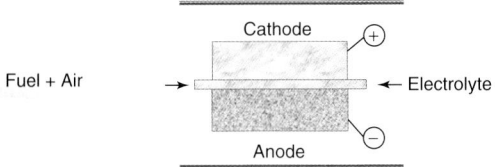

Figure 2.1 Schematic principle of a single-chamber FC.

use of different electrode materials, improvements in cell design, and stack assembly. SC-FCs can be operated on both gaseous and liquid fuels, and the concept has been applied to alkaline fuel cells [1], mixed-reactant direct methanol fuel cells [2], microbial fuel cells [3] and SOFCs [4]. This chapter covers progress in and challenges for developing efficient single-chamber solid oxide fuel cells (SC-SOFCs).

2.2
SC-SOFCs

The first all-solid-state electrochemical device operated in a fuel–oxidant gas mixture was developed by Eyraud *et al.* [5]. A N|Al$_2$O$_3$|Pd cell showed an open-circuit voltage (OCV) of −0.35 V in humidified air, which increased further when hydrogen was added. The pioneering work of Hibino *et al.*, who introduced the term *single chamber* [6], led from the proof-of-concept to SC-SOFCs delivering a significant current [7]. SC-SOFCs are commonly built either as planar electrolyte- or anode-supported cells [7, 8] but, with ongoing efforts to miniaturize cells, SC-SOFCs with coplanar microelectrodes [9, 10] and microtubular SC-SOFCs have been emerging [11]. Gold, silver, platinum, palladium, and nickel were used as electrode materials at the beginning, which were then replaced by conventional SOFC materials such as Ni-based cermet anodes and perovskite-type cathodes. Catalytic activity, selectivity, and stability in the gas mixture are still the main issues for suitable electrode materials for use in SC-SOFCs. Additional parameters affecting the cell performance include (i) operating conditions such as flow rate, temperature, fuel-to-oxygen ratio (R_{mix}), and choice of fuel [4], (ii) the design of the gas chamber [12, 13], since the cell support should promote an efficient interaction between the gas mixture and the electrode surface, (iii) inert, stable current collector materials [13], and (iv) heat management with respect to the exothermal fuel reactions [4, 14].

2.2.1
Basic Principles of Single-Chamber Fuel Cell Operation

The SC-FC uses one compartment to convert directly a mixture of fuel and oxidant into electrical energy based on the difference in catalytic activity and selectivity of the electrode materials for the fuel reaction. Considering the CH$_4$–air system, the complete reaction is

$$CH_4 + 2O_2 \longrightarrow CO_2 + 2H_2O \tag{2.1}$$

This overall reaction is the sum of two electrochemical reactions at the respective electrodes:

$$\text{Anode: } CH_4 + 4O^{2-} \longrightarrow CO_2 + 2H_2O + 8e^- \quad (2.2)$$

$$\text{Cathode: } 4 \times \left(\frac{1}{2}O_2 + 2e^- \longrightarrow O^{2-}\right) \quad (2.3)$$

The cell voltage, E_r, can be calculated by the Nernst equation:

$$E_r = E° + \frac{RT}{nF}\ln\left(\frac{a_{CH_4}a_{O_2}^2}{a_{CO_2}a_{H_2O}^2}\right) = E° + \frac{RT}{8F}\ln\left(\frac{p_{CH_4}p_{O_2}^2}{p_{CO_2}p_{H_2O}^2}\right) \quad (2.4)$$

where n is the number of electrons ($n = 8$ in this case) involved in the reaction per mole of reactant, a_x is the electrochemical activity of the species x, $E°$ is the standard reversible cell potential, F is the Faraday constant (96 485 C mol^{-1}), R is the ideal gas constant (8.314 J mol^{-1} K^{-1}), and T is the temperature (K). Assuming that the gases are ideal, the electrochemical activity of each gas can be replaced by its partial pressure p.

In any fuel cell system, the chemical energy of the reactants is converted directly into electrical energy. Thereby, the Gibbs free energy ΔG is related to the reversible cell potential:

$$E_r = -\frac{\Delta G}{nF} \quad (2.5)$$

The driving force for the migration of the oxygen ions from the cathode to the anode through the electrolyte results from the oxygen chemical potential or oxygen partial pressure gradient between the anode (10^{-15}–10^{-30} atm) and cathode (0.21 atm) [15]. Therefore, the cell voltage is also given by

$$E_r = \frac{RT}{nF}\ln\left(\frac{p_{O_2,\text{cathode}}}{p_{O_2,\text{anode}}}\right) \quad (2.6)$$

where p_{O_2} is the oxygen partial pressure at cathode and anode [16].

The fuel cell efficiency ε can be expressed as the product of thermodynamic efficiency ε_{th}, potential efficiency ε_p, and current efficiency ε_f [4, 16]:

$$\varepsilon = \varepsilon_{th}\varepsilon_p\varepsilon_f = \varepsilon_{th}\varepsilon_p\varepsilon_{FU} = \frac{\Delta G}{\Delta H}\frac{E_{cell}}{E_r}\frac{I}{I_f} = \frac{\Delta G}{\Delta H}\left(-\frac{nFE_{cell}}{\Delta G}\right)\frac{I}{I_f} \quad (2.7)$$

where ΔG is the Gibbs free energy converted into electrical energy, ΔH the reaction enthalpy, E_{cell} the operating cell voltage, I the current produced by the cell at the maximum cell power output, and I_f the theoretical current generated for 100% electrochemical conversion of the fuel given by Faraday's law [16]. The difference between the operating and reversible cell voltages results from ohmic, activation, and concentration polarization losses during cell operation. The current or Faradaic efficiency corresponds to the losses that occur due to parasitic side reactions where reactants are consumed instead of in the electrochemical reactions, and can therefore also be referred to as the *fuel utilization*, ε_{FU}.

In reality, currently used SOFC anode materials do not promote the direct electrochemical conversion of the fuel [Eq. (2.1)]. The partial oxidation [Eq. (2.8)] of the fuel is mostly assumed to be the primary reaction at the anode in SC-SOFCs, leading to the required difference in oxygen partial pressure and the generation of an OCV between the two electrodes [17]:

$$CH_4 + \frac{1}{2}O_2 \longrightarrow CO + 2H_2 \qquad (2.8)$$

This reaction is followed by the electrochemical conversion of the produced syngas. The different reactions at the electrodes can be written as follows:

$$\text{Anode: } 2H_2 + 2O^{2-} \longrightarrow 2H_2O + 4e^- \qquad (2.9)$$

$$CO + O^{2-} \longrightarrow CO_2 + 2e^- \qquad (2.10)$$

$$\text{Total anode reaction: } 2H_2 + CO + 3O^{2-} \longrightarrow CO_2 + 2H_2O + 6e^- \qquad (2.11)$$

$$\text{Cathode: } \frac{3}{2}O_2 + 6e^- \longrightarrow 3O^{2-} \qquad (2.12)$$

The overall reaction from the electrochemical processes is

$$2H_2 + CO + \frac{3}{2}O_2 \longrightarrow CO_2 + 2H_2O \qquad (2.13)$$

The Nernst equation for this system is

$$E_r = E° + \frac{RT}{6F} \ln \left(\frac{p_{H_2}^2 p_{CO} p_{O_2}^{\frac{3}{2}}}{p_{CO_2} p_{H_2O}^2} \right) \qquad (2.14)$$

In this case, compared with the direct electrochemical oxidation of the fuel [Eq. (2.2)], the Faraday efficiency is only 75% since the production of syngas (H_2 + CO) is not an electrochemical process, and electrons (six instead of eight) are generated only by the electrochemical oxidation of the syngas. In addition to this intrinsically reduced efficiency, current SC-SOFC systems show very low fuel utilization (1–8%) and thus efficiencies [4, 18, 19]. While gas intermixing, small-scale electrodes, and high flow rates contribute to the low efficiency, non-ideally selective electrode materials and parasitic reactions are the primary reason. The further development of SC-SOFCs therefore requires active and selective materials for optimized performance.

A fuel-to-oxygen ratio corresponding to the partial oxidation of the fuel (i.e., $R_{mix} = 2$ in the CH_4–O_2 system) was found appropriate for SC-SOFC operation [13, 15, 20, 21]. This optimized R_{mix} value also depends on the electrode structures and thickness. Fuel-rich gas mixtures can cause carbon deposition whereas oxygen-rich gas mixtures can favor complete fuel oxidation and increase the risk of explosions.

2.2.2
Catalysis in SC-SOFCs

Despite the numerous investigations on understanding SC-SOFCs over the past 20 years, little attention has been paid to the crucial role of catalysis in such systems [6, 19, 22–24]. The operation of SC-SOFCs requires high selectivity of the electrode materials for the fuel reaction. Considering a CH_4–air gas mixture, optimal electrochemical conversion of the fuel at the anode corresponds to Eq. (2.2). However, in real SC-SOFC systems, partial fuel oxidation occurs over the anode catalyst [Eq. (2.8)]. Moreover, the use of a gas mixture in the same compartment at elevated operating temperatures also triggers several other catalytic, but parasitic, reactions such as the direct, non-electrochemical combustion of the fuel. Hibino et al. [6] showed that a Ni or Ni–$Ce_{0.8}Gd_{0.2}O_{1.9}$ [gadolinia-doped ceria (GDC)] anode exposed to a methane–air mixture ($R_{mix} = 1$) produced not only H_2 and CO but also CO_2. They additionally observed that an $La_{1-x}Sr_xMnO_{3-\delta}$ [lanthanum strontium manganite (LSM)] cathode could also promote the oxidation of the fuel. The cathode activity for methane oxidation was confirmed by Morel et al. [23].

Recently, Savoie et al. [24] reported an extensive investigation on catalysis in SC-SOFCs with Ni–YSZ (yttria-stabilized zirconia) anodes. Their investigations were performed on three different half-cells exposed at various temperatures to a methane-air gas mixture. The detected outlet gases contained CO and H_2, but also CO_2, suggesting that the complete oxidation reaction also occurred. Furthermore, at a fixed gas flow rate, the outlet gas composition was found to depend on both R_{mix} and the thickness of the anode. At 600 °C and $R_{mix} = 1.2$, 33% of methane was catalytically converted on a thin anode (0.05 mm), and more than double on a thick anode (1.52 mm). An increase in temperature led to an increase in methane conversion. At 800 °C and $R_{mix} = 1.2$, the yield of H_2 was 14 and 38% for thin and thick anodes, respectively. The production of syngas was significantly reduced at lower temperatures. From this study, it is obvious that the optimization of SC-SOFC systems requires extensive catalytic studies on the electrode materials.

2.2.3
Heat Production and Real Cell Temperature

The direct interaction between the fuel–air mixture and the electrode materials leads to partial or complete oxidation reactions. These reactions are exothermic processes [24] and the resulting heat release increases the cell temperature. The control of the overheating and the measurement of the true cell temperature are essential to efficient operation of the system.

The exothermic effect of the fuel–air mixture on the local electrode temperature can be observed by applying thermocouples to the anode and cathode surface [25] or the electrolyte [13, 26]. Temperature gradients of up to 100–160 °C with respect to the furnace temperature were monitored in different gas mixtures. With platinum used either as electrode or current collector, the overheating effect is further enhanced whereas blank YSZ electrolyte substrates with no catalyst material showed negligible overheating [13]. Local temperature gradients between the fuel inlet and the outlet side of an electrode were also observed [27–29]. When the temperature increase can be sufficiently monitored and controlled to prevent localized and overall damage to the cell, the heat generated could be used for cogeneration [27] or operation at very low furnace temperatures [30], or to thermally self-sustain the SC-SOFCs [19].

2.2.4
Current Collection

Current collection is a crucial issue for SC-SOFCs. The current-collector material should be inert to the fuel–air mixture and have sufficient thermal stability to withstand the temperature increase during SC-SOFC operation. Therefore, the operating conditions impose strict constraints on the choice of the current-collector material. Most research on SC-SOFCs thus far has implied the use of platinum for current collection even though this material has shown catalytic activity in the fuel–air mixture. The presence of platinum led to a temperature rise of 160 °C compared with the furnace temperature [13, 26]. In order to reduce the overheating, platinum can be embedded inside each electrode layer [31]. Alternatively, gold current collectors were found to be inert towards the fuel and cause no overheating in a methane–air mixture [26]. However, gold reacts with nickel to form a gold–nickel eutectic at temperatures above 800 °C, imposing lower operating temperatures. In microscale SC-SOFCs with coplanar electrodes, a major challenge is to find suitable and efficient current-collector designs in addition to material constraints.

2.2.5
Electrode and Electrolyte Materials

The electrode and electrolyte materials used in SC-SOFCs are similar to the ones being used in conventional SOFCs operated with separate fuel and air compartments. However, selectivity and catalytic activity requirements and overheating impose different constraints on the cell component materials under single-chamber operating conditions.

2.2.6
Anode Materials

Ni-based materials are found to present reasonable selectivity and activity for use as anodes in SC-SOFCs. Specifically, composite anodes of Ni and samarium, gadolinium, or cerium oxides are being investigated, and the addition of metal catalysts can further enhance the cell performance [31–34]. An increase of 36% in performance was observed, for instance, when a simple Ni anode was replaced by a mixed conductive $NiO-ZrO_2-CeO_2$ anode in a methane–air mixture at 950 °C [35]. At temperatures lower than 550 °C, a significant increase in the OCV was achieved by adding PdO to a Ni-SDC (samaria-doped ceria, $Ce_{0.8}Sm_{0.2}O_{1.9}$) anode [32]. In addition to the composition, the anode microstructure also affects the cell performance [36].

Despite the good performance with Ni-based anodes, their degradation with time and nickel volatilization are critical issues for stable, long-term cell operation [21]. Loss of nickel through the formation of $Ni(OH)_2$ is especially favored in oxygen-rich mixtures where a larger quantity of water vapor can be generated. Moreover, in the gas mixture, the Ni-based anode undergoes a reduction–oxidation process (Ni $\rightleftharpoons$ NiO) which engenders volume changes between Ni and NiO and ultimately causes damage to the anode, especially in nickel-rich and thick anodes. According to Zhang et al. [37], the reduction–oxidation cycles are accompanied by oscillations in reactant and product concentrations and also temperature fluctuations over the catalyst. Oxygen reacts with Ni on the catalyst surface and forms NiO, thereby decreasing the partial oxidation reactions of methane. Methane in turn reacts with NiO in an endothermic reduction reaction, resulting in a decrease in temperature. Reoxidation of the newly formed Ni follows, and so on. Voltage fluctuations during cell operation are a result of these oxidation–reduction cycles [21] and microstructural changes such as the formation of Ni aggregates are indications of the redox cycling of Ni–YSZ [38]. Additionally, the use of Ni-based electrodes requires cell initialization through NiO reduction under reducing atmospheres and carbon deposition needs to be avoided [4].

2.2.7
Cathode Materials

LSM is the most commonly used cathode material in SC-SOFCs. However, the sintering process and the cell operating temperature strongly affect its catalytic activity and selectivity [23, 39, 40]. Under single-chamber operation, the cathode is required only to promote the oxygen reduction [Eq. (2.3)]. However, when sintered at low temperature, LSM shows catalytic activity towards hydrocarbon oxidation [23, 39, 40]. When sintered at higher temperatures, an increased density leads to negligible catalytic activity for methane conversion [23].

Most investigations on cathode materials focus on finding new materials that are selective only for the oxygen reduction. In a propane–air mixture, $La_{0.2}Sr_{0.8}Co_{0.8}Fe_{0.2}O_{3-\delta}$ [lanthanum strontium cobalt ferrite (LSCF)], which was less affected by the sintering temperature, presented a better performance than LSM [40]. Mixed conductive materials such as $Ln_{0.7}Sr_{0.3}Fe_{0.8}Co_{0.2}O_{3-\delta}$ (Ln = Pr, La, Gd) [41] are considered for improved oxygen conductivity and electrocatalytic activity. For operation at lower temperatures, $La_{0.5}Sr_{0.5}CoO_3$ [lanthanum strontium cobaltite (LSC)] was investigated [42], and $La_{0.8}Sr_{0.2}Sc_{0.1}Mn_{0.9}O_3$ [lanthanum strontium scandium manganite (LSSM)] [43] was specifically proposed for anode-supported SC-SOFCs.

A promising new cathode material with a high activity for oxygen reduction is $Ba_{0.5}Sr_{0.5}Co_{0.8}Fe_{0.2}O_3$ [barium strontium cobalt ferrite (BSCF)] [14]. At reduced temperatures, BSCF shows very low activity for the fuel oxidation; hydrocarbon conversion only becomes activated at higher operating temperatures. Owing to its low electrode resistance, high oxygen ion conductivity, and fast oxygen exchange kinetics, high power densities can be obtained with BSCF-based SC-SOFCs.

2.2.8
Electrolyte Materials

The most common electrolyte material is YSZ, which, however, cannot be used at temperatures below 700 °C because of its reduced conductivity [16]. Similar to conventional dual-chamber SOFCs, GDC- and SDC-based electrolytes are being investigated for intermediate- and low-temperature operation [44, 45]. Compared with YSZ, cells with a SDC electrolyte showed higher current densities and lower ohmic resistance [44]. Cells with GDC electrolytes were operated at temperatures as low as 300 °C [45]. Reducing the electrolyte thickness additionally contributes to a reduced ohmic cell resistance [46]. Barium-based electrolyte materials, $BaLaIn_2O_{5.5}$ [47], proton-conducting electrolytes, $CaZr_{0.9}In_{0.1}O_{3-\alpha}$ [33], and multi-layer electrolyte structures [48] have also been investigated.

2.3
SC-SOFC Systems

Whereas electrolyte-supported cells are composed of a thin layer of anode and cathode materials on an electrolyte substrate, a thick anode support (400–2000 μm) is used with a thin electrolyte layer (a few microns) in the anode-supported cell configuration. The latter has the advantage of increased stability and mechanical resistance while at the same time the ohmic loss is decreased with the decrease in electrolyte thickness. Fuel cells with coplanar electrodes and porous electrolytes are specific to single-chamber operating conditions as the use of a fuel–air gas mixture relieves constraints on sealing, cell design, and electrolyte processing. SC-SOFCs with coplanar electrodes have great potential for miniaturization for small-scale power applications. Microtubular SC-SOFCs promise high fuel utilization.

2.3.1
Electrolyte-Supported SC-SOFCs

The electrolyte-supported SC-SOFC was the first cell configuration to be investigated for single-chamber operation. In 1993, significant current densities were obtained by Hibino and Iwahara for the first time for an SOFC operated in a methane–air mixture ($R_{mix} = 2$) at 950 °C [7]. The Ni|YSZ|Au cell delivered an OCV of 0.35 V and a maximum power of about 2.36 mW cm^{-2}. Further improvement of the system was obtained with a Pt|$SrCe_{0.95}Yb_{0.05}O_{3-\alpha}$|Au cell which delivered an OCV of 0.66 V [49]. For a Pt cathode and Au anode, the performance could be enhanced with various electrolyte materials, and a maximum power of 170 mW cm^{-2} at 950 °C was obtained with a Pt|$BaCe_{0.8}Y_{0.2}O_{3-\alpha}$|Au cell [50]. YSZ electrolytes doped with TiO_2, Tb_4O_7, and Pr_6O_{11} [49, 51], MnO_2-doped Pt, and Au electrodes [52] were also investigated before turning to cells composed of conventional SOFC materials [53]. OCVs of about 1 V and power densities of several hundred milliwatts per square centimeter can be obtained with these cells at high and intermediate operating temperatures [4], and operation at temperatures below 450 °C becomes possible by using the heat released from the fuel reactions [25].

The first SC-SOFC stack with a serial connection of electrolyte-supported single cells delivered an OCV of 3 V [54]. An electrolyte-supported SC-SOFC miniature cell module consisting of two cells connected in series delivered an OCV of 1.6 V at 500 °C and a power density of 4.5 mW at 575 °C in a propane–air gas mixture ($R_{mix} = 0.56$) [55].

2.3.2
Anode-Supported SC-SOFCs

In SOFCs, one of the major sources of polarization losses is the ohmic resistance [16]. A possibility for reducing the ohmic resistance is to decrease the electrolyte thickness to a few tens of micrometers [56], which in turn also permits a decrease in the operating temperature. In order to maintain mechanical stability, a thick anode is used as support for the thin electrolyte. Anode-supported SC-SOFCs use the same electrode and electrolyte materials as the electrolyte-supported SC-SOFCs.

The first investigation on anode-supported cells operated under single-chamber conditions was reported by Jasinski et al. [39]. With 760 mW cm^{-2} measured for a Ni–SDC|SDC|BSCF–SDC cell at 750 °C in a methane–air mixture, anode-supported SC-SOFCs delivered the highest power output for SC-SOFCs measured thus far [57]. They could also be successfully operated with liquid fuels [45] and at very low temperatures (200–250 °C) [30].

With the exothermic fuel reactions generating sufficient heat, anode-supported SC-SOFCs were operated as thermally self-sustaining systems. After cell start-up in a preheated furnace at 500 °C, a cell temperature of around 580 °C could be maintained without any external heating system [19]. Using a spiral Swiss roll heat exchanger and combustor, the appropriate operating temperatures could also be maintained through the combustion of the cell exhaust gases without any external heat input [58].

Series connection of anode-supported cells can provide OCVs over 1 V [19, 59] and power densities sufficient to power an MP3 player [19] or small fan [59]. However, gas intermixing between electrodes of adjacent cells imposes several constraints on stack assembly. When the distance between two cells became smaller than 4 mm, the oxygen partial pressure at the cathode of the cell close to the gap was affected by the reactions occurring at the anode of the adjacent cell [60]. Similarly, gas intermixing caused a near-zero OCV for an anode-facing-cathode configuration of a Ni–SDC|SDC|BSCF–SDC two-cell stack, whereas the anode-facing-anode stack configuration delivered a high enough OCV and power to operate a 1.5 V MP3 player [19]. A symmetrical star-shaped anode-facing-cathode stack configuration was proposed to allow a uniform gas distribution over the single cells and to minimize gas intermixing [59].

Compared with electrolyte-supported cells, anode-supported cells can deliver higher power outputs and be operated at lower temperatures [30, 39, 57]. Due to the use of a thick anode, the cell stability can be enhanced but the reduction of NiO during cell initialization is more critical [13].

2.3.3
SC-SOFCs with Coplanar Electrodes

The concept of fuel cells with coplanar electrodes was first proposed in 1965 by van Gool [61]. In his "surface-migration" cell, a comb-like anode and cathode form an interdigitated electrode array on the electrolyte substrate. The gap between the electrode catalysts is bridged by the electrolyte across which the ions are transported via surface migration. The first high-temperature SC-SOFC with coplanar electrodes was reported by Hibino et al. in 1995 [9]. A cell with a Pd electrode and an Au electrode arranged side-by-side on a $BaCe_{0.8}Gd_{0.2}O_{3-\alpha}$ electrolyte generated an OCV of 0.7 V and a maximum current of 24 mA at 950 °C in a methane–oxygen mixture ($R_{mix} = 2$).

Compared with the conventional anode–electrolyte–cathode three-layer SOFC cell structure, the coplanar electrode configuration does not require a very thin or ultra-thin electrolyte. The ohmic cell resistance is mainly governed by the electrode width and the gap between adjacent anode and cathode as the oxygen ions are transported across the gap through the electrolyte. The cells can be constructed on thick, mechanically strong electrolyte substrates, resulting in higher thermomechanical cell stability. As several single cells can be assembled in stacks on the same electrolyte substrate, the coplanar electrode design permits compact fuel cell stacks.

2.3.3.1 Cell Performance
In the simplest geometry, an SC-SOFC with coplanar electrodes consists of a single pair of anode and cathode arranged side-by-side on the same side of the electrolyte substrate (Figure 2.2a). In the fuel–air gas mixture, a sufficient difference in oxygen partial pressure can be established between the two adjacent electrodes, leading to reasonable OCVs [4, 62]. The oxygen ions are transported from the cathode to the anode through the electrolyte bridging the gap between the two electrodes (Figure 2.2a). According to performance simulations [63, 64] and impedance analysis [65], the ohmic loss is the major source of polarization in SC-SOFCs with coplanar electrodes while the concentration polarization is minimal [64]. The electrode reaction resistance can be lowered by using mixed conductive thin-film cathodes and microstructured, nanoporous thin-film nickel anodes [66]. As demonstrated in experimental studies [9, 62], the ohmic loss is mainly governed by the resistance to the ionic conduction, and therefore depends on the length of the ionic conduction path, that is, the inter-electrode gap and electrode width (Figure 2.2b), in addition to the electrolyte surface morphology. Several other cell geometry-specific parameters also influence the cell performance.

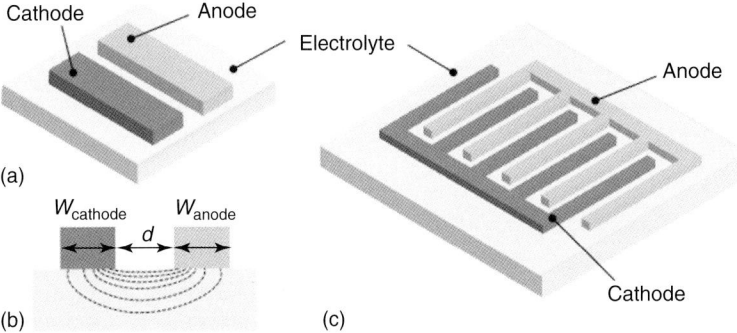

Figure 2.2 Schematics of SC-SOFC with coplanar electrodes: side-by-side arrangement for a single electrode pair: (a) top view and (b) side view; w is the electrode width and d the inter-electrode distance. The dashed lines are a representative illustration of the oxygen ion pathway [67]. (c) Interdigitated electrode pattern.

2.3.3.1.1 Inter-Electrode Gap

As the oxygen ions are transported across the electrolyte between adjacent anode and cathode, the ionic conduction path is mainly determined by the size of the inter-electrode gap. Several experimental studies performed under different operating conditions (gas composition, temperature) and for cells constructed from different cell component materials showed that a decrease in the inter-electrode gap leads to a decrease in the ohmic cell resistance and an increase in power output [9, 20, 62, 68]. These observations were also confirmed by a computational analysis where maximum cell performance was obtained for the smallest gap size studied [64]. The OCV, however, was found to be independent of changes of the inter-electrode gap [20]. When the anode and cathode are located in grooves in the electrolyte, ions can also be transferred through the electrolyte material filling the inter-electrode gap, and a reduction in ohmic cell resistance can be achieved [65].

2.3.3.1.2 Electrode Width

When comparing a conventional three-layer cell structure with a coplanar electrode configuration, Hibino and co-workers [46, 62] observed a lower cell performance for the coplanar cell although the electrolyte thickness of the three-layer cell was chosen to be identical with the inter-electrode gap of the coplanar cell. The reason for the performance difference was attributed to a longer ionic conduction path in the coplanar electrode configuration where the electrode widths also contribute to the conduction path. Similar cell performance could be obtained for the two different electrode configurations when both the electrode widths and inter-electrode gap were designed to match the electrolyte thickness of the three-layer cell. Cell performance measurements as a function of electrode width in SC-SOFCs with coplanar electrodes confirmed the increase in maximum cell power with decrease in electrode width at a fixed inter-electrode gap [62, 68].

2.3.3.1.3 Electrolyte Surface Roughness

The ohmic resistance of SC-SOFCs with coplanar electrodes also decreases with decrease in electrolyte surface roughness, providing an indication that the ion conduction mainly occurs along the electrolyte surface or at a-near-surface depth of the electrolyte surface (Figure 2.2b) [62, 67, 69]. Additionally, the decrease in surface roughness was found to lower the electrode reaction resistance [62].

2.3.3.1.4 Electrode Thickness

Although an increase in electrode thickness from 10 to 80 µm did not significantly affect the OCV and maximum power output according to finite element modeling [70], the voltage stability of SC-SOFCs with coplanar electrodes was found to depend on the thickness of a nickel-rich conduction layer on top of a 20 µm thick contact layer [68]. For 40 and 200 µm thick conduction layers, the OCV decreased after reduction of the anode and exhibited oscillatory fluctuations, resulting from nickel oxidation–reduction cycles, blocked reaction sites, and/or nickel volatilization. The use of a 300 µm thick Ni-rich layer compensated for the nickel instability issues and the cell yielded a stable OCV close to 1 V without any major fluctuations during 80 h of testing.

2.3.3.1.5 Electrolyte Thickness

One of the advantages of the coplanar electrode design is the possibility of using a thick, mechanically strong electrolyte substrate for enhanced robustness of the cell without any deterioration of the cell performance. Finite element modeling showed that the cell performance improved with an increase in electrolyte thickness from 10 to 80 µm with the cell power output converging for a thickness above 40 µm [64]. As the electrolyte resistance is inversely proportional to the area [66], higher ohmic loss is expected for a very thin electrolyte where the oxygen ion flow is limited [67].

2.3.3.1.6 Electrode Shape

Instead of a single pair of rectangular anode and cathode, SC-SOFCs with coplanar electrodes can be constructed from comb-like electrodes which can be arranged in an interdigitated array on the electrolyte substrate (Figure 2.2c) [61, 69]. The comb-like electrode can be considered as a parallel connection of several single side-by-side cells [71]. The use of interdigitated electrodes maximizes the effective surface area while minimizing the conduction path [69]. Ahn et al. [20] observed a decrease in the OCV with increasing number of electrode pairs, so that the OCV could only be maintained when the cells were divided into two-pair cells separated by large gaps. Although the low OCVs were attributed to a higher system complexity favoring intermixing of reaction gases and a decrease in the cathode oxygen partial pressure [72], the feasibility of SC-SOFCs with coplanar interdigitated electrodes has been demonstrated by measuring OCVs over 0.7 V and maximum power densities of several tens of milliwatts per square centimeter [34, 38, 71–73]. Moreover, the short-term stability of the OCV was shown to improve with increasing number of electrode pairs [10]. Other proposed electrode designs include U-shaped electrodes [74] and curvilinear electrodes of arbitrary complex geometry [75]. Stable

OCVs of 0.7 V [74] and 0.9 V [75] confirmed that the complexity of the electrode geometry did not impede the establishment of a reasonable OCV. Additionally, SC-SOFCs with curvilinear microelectrodes of different geometries but similar electrode dimensions (electrode width, inter-electrode gap, and electrode surface area) exhibited similar OCVs and power densities, indicating that the electrode shape did not significantly affect the cell performance.

2.3.3.1.7 Cell Stacks

Series connection of two single cells with side-by-side electrodes on the same electrolyte substrate produced double the voltage of the single cell [69]. Similarly, the power output was doubled for two cells connected in parallel and further increased for a three-cell stack [68, 69].

In addition to cell design and geometry, the cell performance also depends on the testing conditions. Especially the gas composition and orientation of the electrodes with respect to the incoming gas stream were found to affect the OCV and cell power.

2.3.3.1.8 Gas Flow Direction

There are three possibilities for the electrode alignment of SC-SOFCs with coplanar electrodes with respect to the gas flow direction: (i) parallel gas flow where the gas mixture reaches both electrodes simultaneously, (ii) perpendicular gas flow with anode first (the gas mixture encounters the anode first), and (iii) perpendicular gas flow with cathode first [68]. Perpendicular gas flow with cathode first [64, 68, 74] and parallel gas flow [20], especially for a cell stack [68], provide the best cell performance, whereas the lowest cell power was measured for a cell position with the anode being exposed first to the gas mixture [20, 68]. When the anode is placed ahead in the gas stream, oxygen is primarily consumed in the partial fuel oxidation reaction and little oxygen reaches the cathode.

2.3.3.1.9 Gas Mixture Composition (Fuel-to-Oxygen Ratio)

An optimum gas mixture composition cannot be easily identified from the literature as there were significant differences in cell component materials, chamber design, operating temperature, and so on between the various studies. However, as a general observation, the highest power output was measured for a fuel-to-oxygen ratio close to the stoichiometric ratio for the partial oxidation of the fuel, that is, $R_{mix} = 2$ in the case of methane–air mixtures [20, 68, 71]. While higher R_{mix} values generally favor carbon deposition and are unsuitable for cell operation [15], R_{mix} values >3 result in the best cell performance in wet gas mixtures where methane is also consumed by steam reforming [20, 76]. The lowest OCVs were observed in oxygen-rich mixtures [68, 71].

2.3.3.1.10 Choice of Fuel and Lower Operating Temperatures

Methane is the standard fuel used for SC-SOFCs with coplanar electrodes at intermediate and high SOFC operating temperatures [4]. At lower temperatures, the reduced catalytic activity of the anode for methane partial oxidation leads to

Table 2.1 Summary of SC-SOFCs with coplanar electrodes: smallest electrode dimensions and highest power output for different fabrication approaches

Fabrication technique	d (mm)	w (mm)	Electrode pairs	$T_{furnace}$ (°C)	OCV (V)	P_{max} (mW cm^{-2})	Ref.
Painting	0.5	0.5	1	600	0.8	245	[62]
Sputtering	0.005	0.015	1	400	0.38	–	[77]
Screen-printing	0.2	0.5	1	800	0.9	40	[68]
Screen-printing	0.5	–	26	700	0.8	40	[34]
Micromolding	0.014	0.1	19	650	0.77	17	[71]
Photoresist molding	0.02	0.02	100	500	0.2	67	[78]
Microfluidic lithography	0.05	0.1	16	800	0.67	105	[76]
Direct-writing	0.255	~0.6	1	900	0.8	101	[20]
Direct-writing	0.5	~0.2	5	700	0.8	1.3	[38]

poor performance in methane–air mixtures [62]. At 500–600 °C, Ni-based anodes can still catalyze the partial oxidation of higher hydrocarbon fuels, and similar OCVs and power output were obtained for ethane, propane, and butane. The addition of metal catalysts to the Ni-based anodes further enhances the catalytic activity for hydrocarbon partial oxidation even at reduced temperatures. The highest power output for an SC-SOFC with coplanar electrodes reported to date was 245 mW cm^{-2} and was obtained in a butane–air mixture at 600 °C when PdO was added to the anode (Table 2.1) [62].

2.3.3.2 Miniaturization

Based on equivalent circuit models of the cell component resistances [66], maximum cell performance of SC-SOFCs with coplanar electrodes is predicted for very small electrode widths (6–10 µm) and gap sizes (2–12 µm). Performance comparisons of macro-, milli-, and microcells [71] revealed a 10 times higher power density for the micro SC-SOFC which had smaller inter-electrode gaps and electrode widths. As closely spaced small-scale electrodes lower the ohmic resistance and the interdigitated electrode pattern maximizes the electrode surface area, miniaturization of SC-SOFCs with coplanar, interdigitated electrodes is expected to yield suitable cell performance for small- and microscale power applications. The fabrication of microcells (Figure 2.3) presents many challenges and requires the manufacturing of coplanar microscale electrode patterns from multicomponent ceramic materials.

The first SC-SOFCs with coplanar electrodes were fabricated by painting electrodes with a brush onto the electrolyte, yielding inter-electrode gaps and electrode widths larger than 0.5 mm [9]. Interdigitated electrode patterns were not fabricated by this method. Cells with side-by-side anode and cathode and cells with interdigitated electrode structures were also prepared by screen printing

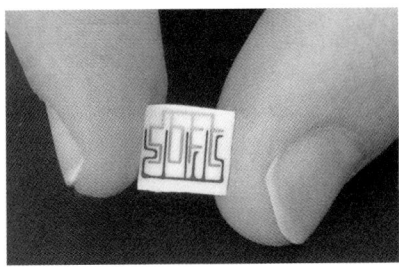

Figure 2.3 SC-SOFC with coplanar microelectrodes [63].

[34, 68, 71]. Minimum gap sizes and electrode widths were in the order of 0.2 and 0.5 mm, respectively. Using photo-masks, coplanar microscale Pt and Au electrode structures were fabricated by sputtering [77]. Although the cells delivered very low OCVs, the smallest inter-electrode gaps (5 μm) and electrode widths (15 μm) reported so far were realized by this fabrication technique.

Combining microfabrication techniques with wet ceramic processing, microscale electrodes can be fabricated from conventional SOFC materials and electrode porosity and microstructure can be tailored by subsequent sintering. Robot-controlled direct-write microfabrication and micromolding or microfluidic lithography have been employed so far. Interdigitated electrode patterns of ∼20–100 anode–cathode pairs with microscale dimensions were successfully fabricated by the latter technique [71, 72, 78]. Although the OCVs were relatively low, these cells yielded the highest maximum power densities for SC-SOFCs with coplanar interdigitated electrodes [72, 78]. Microfluidic lithography [72] is based on the vacuum-assisted infiltration of polymer mold channels with electrode suspensions. Similarly, mold capillaries are filled with electrode suspensions in micromolding in capillaries [71]. The filling process, however, was found to yield incomplete filling and low reproducibility. In multi-step photoresist molding with thermosetting polymer [78], coplanar microelectrode arrays are fabricated based on a modified photolithography approach where the patterned photoresist mold is filled with electrode suspensions. Robot-controlled direct-write microfabrication consists in extruding electrode suspensions through micronozzles and depositing electrode structures on the electrolyte substrate using a robot-controlled deposition apparatus. Side-by-side anode and cathode and also interdigitated electrode structures were fabricated from conventional SOFC materials with inter-electrode distances and electrode widths as small as ∼100 μm [10, 20, 38, 74, 75, 79]. Without the need for masks or molds, the direct-write technique allows the fabrication of a wide variety of electrode shapes and sizes [10, 74, 75]. The electrode width and gap sizes depend on suspension composition and rheological properties in addition to the process parameters [20, 38, 80].

Table 2.1 presents the coplanar SC-SOFC with the smallest electrode dimensions and highest cell performance for each fabrication technique.

2.3.3.3 Limitations and Challenges

With reasonable OCVs and power outputs in the milliwatts range, SC-SOFCs with coplanar electrodes are a promising approach for compact, robust, and small-scale power generation. However, the development of such fuel cells faces several challenges, as outlined below.

2.3.3.3.1 Miniaturization Limits

Although microscale, closely spaced electrodes were found to yield lower cell resistance and higher power, a limiting gap size of ∼1 µm is imposed by the electrical field approaching electrical breakdown conditions (3×10^6 V m^{-1} in air) for a potential difference of 1 V between the anode and cathode [61]. Additionally, very small inter-electrode gaps (<50 µm [81]) favor gas intermixing and turbulent gas flow, thus affecting the establishment of a sufficiently high oxygen partial pressure gradient between adjacent electrodes and resulting in very low OCVs [72, 77, 78]. For cells with a single coplanar electrode pair, a minimum or critical electrode width was observed below which the generation of an OCV was impeded by the small electrode surface area [10]. A non-zero, but fluctuating OCV was obtained for electrode widths near the critical size and a larger electrode with a larger area allowed higher and stable OCVs even during aging [10, 68]. An upper size limit is imposed, however, by an increase in ohmic resistance due to a long ion conduction path for the electrode regions furthest from the inter-electrode gap. Microstructural changes induced by the partial oxidation of methane over a nickel catalyst revealed that the whole anode surface area does not participate equally in the electrochemical reactions but that the reactions principally occur close to the inter-electrode gap [10, 68, 71]. The parallel connection of multiple electrode pairs with dimensions below the critical width in closely spaced interdigitated electrode structures is a compromise for the trade-off between a larger electrode area for better OCV stability and small electrode widths for reduced ohmic resistance [10]. Finally, the practical feasibility of microscale electrode patterns is limited by the available fabrication techniques.

2.3.3.3.2 Gas Intermixing

Reactant gases and reaction products from one electrode can also easily diffuse to the adjacent electrode and affect the oxygen partial pressure. While reasonable OCVs can be achieved with microscale interdigitated electrode patterns [71], very low OCVs are attributed to turbulent flow and gas intermixing between the closely spaced microelectrodes [72, 78], especially under wet gas conditions [20, 76].

2.3.3.3.3 Chemical Interaction between Coplanar Electrodes

In combination with a LSM-based cathode, a Ni–YSZ anode and YSZ electrolyte appeared blackened after sintering whereas cells with a LSCF cathode did not exhibit such a color change [73]. The black coloration was attributed to the diffusion of manganese from the cathode to the anode through the electrolyte. Chemical interaction between closely spaced coplanar anodes and cathodes could affect the cell performance and material compatibility studies could facilitate the selection of suitable cell component materials.

2.3.3.3.4 Cell Efficiency

The cell efficiency of SC-SOFCs with interdigitated electrodes was estimated to be below 1%, with a fuel utilization of less than 0.1% [63]. In addition to insufficient catalytic selectivity of the electrode materials, the small size of SC-SOFCs with coplanar electrodes limits the electrochemical conversion of fuel.

2.3.3.3.5 Current Collection

Efficient current collection is a major challenge in SC-SOFCs with coplanar, interdigitated microelectrodes. The current collectors are generally applied at the beginning of an electrode array where the single electrode lines are connected together, thus avoiding short-circuits and the need for precisely machined current collectors to fit on the electrode surface. However, this current-collection method results in long conduction paths, which can cause excessive ohmic losses [71]. Current collection at the beginning of the electrode array, which represented about 20% of the total electrode area, was responsible for a performance drop of over 50% in power density compared with current collection over the whole electrode surface [79].

2.3.3.3.6 Nickel Instability

The reactivity of nickel in methane–gas mixtures can cause a decrease in nickel content, observed in the form of a whitened region in the anode close to the inter-electrode gap [10, 68, 71]. Nickel is lost by forming volatile compounds during the partial oxidation of methane, which thus seems to be primarily catalyzed in the region close to the gap. The current density is highest in this part of the electrode where microstructural changes and chemical instability are aggravated by localized heating [18]. Large, thick anodes were found to be less subject to degradation [68].

2.3.3.3.7 Microscale Electrodes

The small electrode size not only imposes fabrication challenges and limits the feasibility of SC-SOFCs with coplanar microscale electrodes, it also limits detailed analyses of these cells. SC-SOFCs with coplanar microscale electrodes yield very low conversion of the reactant gases, so that differences between input and output gases cannot be easily detected by mass spectrometry and information about the reactions that occur cannot be obtained. Similarly, the small electrode size makes impedance analysis difficult. The lack of fundamental studies and appropriate characterization and fabrication techniques leaves the working principles of SC-SOFCs with coplanar electrodes to a great extent unexplored.

2.3.4 Fully Porous SC-SOFCs

SC-SOFCs do not require a fully dense, gas-tight electrolyte. Respecting a minimum thickness to avoid contact between the electrodes and short-circuits [18], SC-SOFCs can be fabricated with porous electrolytes at lower manufacturing temperatures and costs. An oxygen partial pressure gradient can be established across a porous electrolyte and OCVs only 0.1 V lower as for cells with a dense electrolyte can be obtained [82]. Compared with a fully dense layer, a porous electrolyte exhibits increased conductivity [82], probably due to the contribution of surface ionic conduction [61]. Depending on the gas-flow configuration, fully porous SC-SOFCs can be operated in a flow-by or flow-through mode with the latter permitting enhanced mass transport, shorter residence time of the gas mixture, and high fuel utilization [18]. For both flow regimes, the cell performance is significantly affected by the gas flow rate, and the OCV and cell power increase with increase in flow rate [82, 83]. Higher flow rates promote the partial oxidation reaction over the anode, leading to higher cell temperatures (of over 100 °C) and lower cell resistance [82]. Moreover, higher flow rates impede diffusion of reaction products from the anode to the cathode through the electrolyte [83].

2.3.5
Tubular SC-SOFCs

As tubular SOFCs are an intrinsically sealing-free cell design, the main advantage of operating tubular cells under single-chamber conditions is the simplification of gas manifolding in cell stacks and miniaturized cells. Compared with other SC-SOFC cell configurations, the tubular cell design might improve mass transport and prevent gas intermixing, which in return allows higher fuel utilization and cell efficiencies [11]. However, when microtubes are used, mass transport limitations at the anode result from a low gas velocity inside the tube and can lower the cell performance [84]. Introducing the gases at higher velocities can improve the cell performance to the detriment of fuel utilization [11]. Anode-supported microtubular SC-SOFCs operated in methane–air mixtures exhibited the highest maximum power densities at $R_{mix} = 1$ because of a higher cell temperature from methane combustion in oxygen-rich gas mixtures [11]. Conversely, the exothermic reactions cause temperature gradients along the cell length together with a gradient in OCV and power output [29, 84]. The overheating from the exothermic reactions also causes mechanical failure of silver current collectors during long-term cell operation [85] and electrode degradation [29, 86].

2.4
Applications of SC-SOFCs Systems

The proof of concept that SC-SOFCs can be considered for portable power generation was provided by Shao *et al.* [19], who successfully powered a 1.5 V MP3 player with a miniature two-cell SC-SOFC stack. The heat generated from the exothermic reactions allowed rapid start-up from cold start to stable power in less than 1 min and thermally self-sustained the system. However, the low cell efficiency of current SC-SOFCs questions their practical implementation, and energy-harvesting applications where energy is generated from waste gases and efficiency plays a secondary role seem more adequate. Hibino and co-workers installed an SC-SOFC stack in the exhaust system of a motorcycle and a stable power of over 1 W could be generated from the exhaust gases [87]. Thermal cycling and mechanical damage did not significantly degrade the cell performance, indicating the potential of SC-SOFCs for long-term use in harsh environments. Such energy-harvesting applications could extend from on-board automotive exhaust conversion to energy generation from industrial and domestic exhaust gases.

With regard to low cell efficiency, other potential applications could benefit from the intrinsic properties of SC-SOFCs, such as the seal-free cell design, cell operation in fuel–air gas mixtures at elevated temperatures, possible rapid start-up and their capacity for heat generation and hydrocarbon reforming. Within this scope, SC-SOFCs are considered as gas, temperature, or pressure sensors under mixed gas conditions [88]. Through the hydrocarbon conversion, electrical power and chemicals could be cogenerated [50]. Finally, miniaturized SC-SOFCs have been proposed for integration with microelectromechanical systems (MEMS) [89]. With more and more possible applications of SC-SOFCs emerging, patents filed by industrial corporations address methods of fabrication, cell design, and gas management [4].

2.5
Conclusion

Since the beginning of SC-SOFC research in the 1990s, significant progress in this technology has been made in terms of cell design, fabrication, materials, and generated power output. Due to fuel cell operation in a fuel–oxidant gas mixture, the main advantages of single-chamber operation include a compact, sealing-free structure, simplified gas manifolding, fabrication, and stack assembly, higher thermomechanical strength, thermal self-sustainability and a great potential for miniaturization. In addition to the conventional planar and tubular cell configurations, new cell designs become possible where fully porous cells promise lower fabrication costs and coplanar cells are especially suited for miniaturization. Future research on SC-SOFCs needs to target the imminent challenges of this technology such as low fuel utilization, cell efficiency, and selectivity of electrode materials to expedite the use of SC-SOFCs in portable power generation, sensors, and energy-harvesting applications.

References

1. Grueneberg, G. (1965) Electrochemical Conversion of Nuclear Energy, in *Fuel Cells-Modern processes for the electrochemical production of energy* (eds. W. Vielstich and D.J.G. Ives), Wiley-Interscience, New York, pp. 374–376.
2. Shukla, A.K., Raman, R.K., and Scott, K. (2005) Advances in mixed-reactant fuel cells. *Fuel Cells*, 4, 436–447.
3. Cheng, S., Liu, H., and Logan, B.E. (2006) Power densities using different cathode catalysts (Pt and CoTMPP) and polymer binders (Nafion and PTFE)

in single chamber microbial fuel cells. *Environ. Sci. Technol.*, **40** (1), 364–369.

4. Kuhn, M. and Napporn, T. (2010) Single-chamber solid oxide fuel cell technology – from its origins to today's state of the art. *Energies*, **3** (1), 57–134.

5. Eyraud, C., Lenoir, J., and Gery, M. (1961) Fuel cells using electrochemical properties of adsorbents – Piles à combustibles utilisant les propriétés electrochimiques des adsorbats. *C. R. Acad. Sci.*, **252** (11), 1599–1600.

6. Hibino, T., Wang, S., Kakimoto, S., and Sano, M. (1999) Single chamber solid oxide fuel cell constructed from an yttria-stabilized zirconia electrolyte. *Electrochem. Solid-State Lett.*, **2** (7), 317–319.

7. Hibino, T. and Iwahara, H. (1993) Simplification of solid oxide fuel cell systems using partial oxidation of methane. *Chem. Lett.*, **7**, 1131–1134.

8. Shao, Z., Kwak, C., and Haile, S.M. (2004) Anode-supported thin-film fuel cells operated in a single chamber configuration 2T-I-12. *Solid State Ionics*, **175** (1–4), 39–46.

9. Hibino, T., Ushiki, K., Sato, T., and Kuwahara, Y. (1995) A novel design for simplifying SOFC system. *Solid State Ionics*, **81**, 1–3.

10. Kuhn, M., Napporn, T.W., Meunier, M., Vengallatore, S., and Therriault, D. (2009) Miniaturization limits for single-chamber micro solid oxide fuel cells with coplanar electrodes. *J. Power Sources*, **194** (2), 941–949.

11. Akhtar, N., Decent, S.P., Loghin, D., and Kendall, K. (2009) Mixed-reactant, micro-tubular solid oxide fuel cells: an experimental study. *J. Power Sources*, **193** (1), 39–48.

12. Stefan, I.C., Jacobson, C.P., Visco, S.J., and De Jonghe, L.C. (2004) Single chamber fuel cells: flow geometry, rate and composition considerations. *Electrochem. Solid-State Lett.*, **7** (7), A198–A200.

13. Napporn, T.W., Jacques-Bedard, X., Morin, F., and Meunier, M. (2004) Operating conditions of a single-chamber SOFC. *J. Electrochem. Soc.*, **151** (12), A2088–A2094.

14. Shao, Z. and Haile, S.M. (2004) A high-performance cathode for the next generation of solid-oxide fuel cells. *Nature*, **431**, 170–173.

15. Buergler, B.E., Grundy, A.N., and Gauckler, L.J. (2006) Thermodynamic equilibrium of single-chamber SOFC relevant methane–air mixtures. *J. Electrochem. Soc.*, **153** (7), A1378–A1385.

16. Minh, N.Q. and Takahashi, T. (1995) *Science and Technology of Ceramic Fuel Cells*, Elsevier, Amsterdam.

17. Yano, M., Tomita, A., Sano, M., and Hibino, T. (2007) Recent advances in single-chamber solid oxide fuel cells: a review. *Solid State Ionics*, **177** (39–40), 3351–3359.

18. Riess, I. (2008) On the single chamber solid oxide fuel cells. *J. Power Sources*, **175** (1), 325–337.

19. Shao, Z., Haile, S.M., Ahn, J., Ronney, P.D., Zhan, Z., and Barnett, S.A. (2005) A thermally self-sustained micro solid-oxide fuel-cell stack with high power density. *Nature*, **435**, 795–798.

20. Ahn, S.-J., Kim, Y.-B., Moon, J., Lee, J.-H., and Kim, J. (2007) Influence of patterned electrode geometry on performance of co-planar type single chamber SOFC. *J. Power Sources*, **171**, 511–516.

21. Jacques-Bedard, X., Napporn, T.W., Roberge, R., and Meunier, M. (2006) Performance and ageing of an anode-supported SOFC operated in single-chamber conditions. *J. Power Sources*, **153** (1), 108–113.

22. Hao, Y., Shao, Z., Mederos, J., Lai, W., Goodwin, D.G., and Haile, S.M. (2006) Recent advances in single-chamber fuel-cells: experiment and modeling. *Solid State Ionics*, **177** (19–25), 2013–2021.

23. Morel, B., Roberge, R., Savoie, S., Napporn, T.W., and Meunier, M. (2007) Catalytic activity and performance of LSM cathode materials in single chamber SOFC. *Appl. Catal. A*, **323**, 181–187.

24. Savoie, S., Napporn, T.W., Morel, B., Meunier, M., and Roberge, R. (2011) Catalytic activity of Ni-YSZ anodes in a single-chamber solid oxide fuel cell reactor. *J. Power Sources*, **196** (8), 3713–3721.

25. Hibino, T., Hashimoto, A., Inoue, T., Tokuno, J.-I., Yoshida, S.-I., and

Sano, M. (2001) A solid oxide fuel cell using an exothermic reaction as the heat source. *J. Electrochem. Soc.*, **148** (6), A544–A549.
26. Napporn, T., Morin, F., and Meunier, M. (2004) Evaluation of the actual working temperature of a single-chamber SOFC. *Electrochem. Solid-State Lett.*, **7** (3), A60–A62.
27. Morel, B., Roberge, R., Savoie, S., Napporn, T.W., and Meunier, M. (2007) An experimental evaluation of the temperature gradient in solid oxide fuel cells. *Electrochem. Solid-State Lett.*, **10** (2), B31–B33.
28. Morel, B., Roberge, R., Savoie, S., Napporn, T.W., and Meunier, M. (2009) Temperature and performance variations along single chamber solid oxide fuel cells. *J. Power Sources*, **186** (1), 89–95.
29. Akhtar, N., Decent, S.P., and Kendall, K. (2010) Cell temperature measurements in micro-tubular, single-chamber, solid oxide fuel cells (MT-SC-SOFCs). *J. Power Sources*, **195** (23), 7818–7824.
30. Tomita, A., Hirabayashi, D., Hibino, T., Nagao, M., and Sano, M. (2005) Single-chamber SOFCs with a $Ce_{0.9}Gd_{0.1}O_{1.95}$ electrolyte film for low-temperature operation. *Electrochem. Solid-State Lett.*, **8** (1), A63–A65.
31. Buergler, B.E., Siegrist, M.E., and Gauckler, L.J. (2005) Single chamber solid oxide fuel cells with integrated current-collectors. *Solid State Ionics*, **176** (19–22), 1717–1722.
32. Hibino, T., Hashimoto, A., Yano, M., Suzuki, M., Yoshida, S.-I., and Sano, M. (2002) High performance anodes for SOFCs operating in methane–air mixture at reduced temperatures. *J. Electrochem. Soc.*, **149** (2), A133–A136.
33. Van Rij, L.N., Le, J., Van Landschoot, R.C., and Schoonman, J. (2001) A novel Ni-CERMET electrode based on a proton conducting electrolyte. *J. Mater. Sci.*, **36** (5), 1069–1076.
34. Yoon, S.P., Kim, H.J., Park, B.T., Nam, S.W., Han, J., Lim, T.H., and Hong, S.A. (2006) Mixed-fuels fuel cell running on methane–air mixture. *J. Fuel Cell Sci. Technol.*, **3** (1), 83–86.
35. Lamas, D.G., Cabezas, M.D., Fabregas, I.O., De Reca, N.E.W., Lascalea, G.E., Kodjaian, A., Vidal, M.A., Amadeo, N.E., and Larrondo, S.A. (2007) NiO/ZrO_2-CeO_2 anodes for single-chamber solid-oxide fuel cells operating on methane/air mixtures. *ECS Trans.*, **7** (1), 961–970.
36. Jou, S. and Wu, T.H. (2008) Thin porous Ni-YSZ films as anodes for a solid oxide fuel cell. *J. Phys. Chem. Solids*, **69** (11), 2804–2812.
37. Zhang, X., Hayward, D.O., and Mingos, D.M.P. (2003) Further studies on oscillations over nickel wires during the partial oxidation of methane. *Catal. Lett.*, **86** (4), 235–243.
38. Kuhn, M., Napporn, T., Meunier, M., Vengallatore, S., and Therriault, D. (2008) Direct-write microfabrication of single-chamber micro solid oxide fuel cells. *J. Micromech. Microeng.*, **18** (1), 015005.
39. Jasinski, P., Suzuki, T., Byars, Z., Dogan, F., and Anderson, H.U. (2003) Comparison of anode and electrolyte support configuration of single-chamber SOFC. In ECS Transactions – 8th International Symposium on Solid Oxide Fuel Cells, SOFC-VIII, Paris, France, 27 April–2 May 2003, No. 2003-7, pp. 1101–1108.
40. Jasinski, P., Suzuki, T., Zhou, X.D., Dogan, F., and Anderson, H.U. (2003) Single chamber solid oxide fuel cell – investigation of cathodes. *Ceram. Eng. Sci. Proc.*, **24** (3), 293–298.
41. Ruiz de Larramendi, I., Lamas, D.G., Cabezas, M.D., Ruiz de Larramendi, J.I., Walsöe de Reca, N.E., and Rojo, T. (2009) Development of electrolyte-supported intermediate-temperature single-chamber SOFCs using $Ln_{0.7}Sr_{0.3}Fe_{0.8}Co_{0.2}O_{3-\delta}$ (Ln = Pr, La, Gd) cathodes. *J. Power Sources*, **193** (2), 774–778.
42. Pinol, S. (2006) Stable single-chamber solid oxide fuel cells based on doped ceria electrolytes and $La_{0.5}Sr_{0.5}CoO_3$ as a new cathode. *J. Fuel Cell Sci. Technol.*, **3**, 434–437.
43. Zhang, C., Zheng, Y., Lin, Y., Ran, R., Shao, Z., and Farrusseng, D. (2009) A comparative study of $La_{0.8}Sr_{0.2}MnO_3$

and $La_{0.8}Sr_{0.2}Sc_{0.1}Mn_{0.9}O_3$ as cathode materials of single-chamber SOFCs operating on a methane–air mixture. *J. Power Sources*, **191** (2), 225–232.

44. Hibino, T., Hashimoto, A., Inoue, T., Tokuno, J.-I., Yoshida, S.-I., and Sano, M. (2000) A low-operating-temperature solid oxide fuel cell in hydrocarbon–air mixtures. *Science*, **288** (5473), 2031–2033.

45. Yano, M., Kawai, T., Okamoto, K., Nagao, M., Sano, M., Tomita, A., and Hibino, T. (2007) Single-chamber SOFCs using dimethyl ether and ethanol. *J. Electrochem. Soc.*, **154** (8), B865–B870.

46. Hibino, T., Tsunekawa, H., Tanimoto, S., and Sano, M. (2000) Improvement of a single-chamber solid-oxide fuel cell and evaluation of new cell designs. *J. Electrochem. Soc.*, **147** (4), 1338–1343.

47. Asahara, S., Michiba, D., Hibino, M., and Yao, T. (2005) Single chamber SOFC using $BaLaIn_2O_{5.5}$ solid electrolyte. *Electrochem. Solid-State Lett.*, **8** (9), 449–451.

48. Tomita, A., Teranishi, S., Nagao, M., Hibino, T., and Sano, M. (2006) Comparative performance of anode-supported SOFCs using a thin $Ce_{0.9}Gd_{0.1}O_{1.95}$ electrolyte with an incorporated $BaCe_{0.8}Y_{0.2}O_{3-\alpha}$ layer in hydrogen and methane. *J. Electrochem. Soc.*, **153** (6), A956–A960.

49. Hibino, T., Asano, K., and Iwahara, H. (1994) Solid oxide fuel-cell which can work in uniform-gas phase using partial oxidation of methane. *Nippon Kagaku Kaishi*, **7**, 600–604.

50. Iwahara, H. (1998) A one-chamber solid electrolyte fuel cell for chemical cogeneration. *Ionics*, **4** (5), 409–414.

51. Asano, K. and Iwahara, H. (1997) Performance of a one-chamber solid oxide fuel cell with a surface-modified zirconia electrolyte. *J. Electrochem. Soc.*, **144** (9), 3125–3130.

52. Hibino, T., Kuwahara, Y., and Wang, S. (1999) Effect of electrode and electrolyte modification on the performance of one-chamber solid oxide fuel cell. *J. Electrochem. Soc.*, **146** (8), 2821–2826.

53. Hibino, T., Wang, S., Kakimoto, S., and Sano, M. (2000) One-chamber solid oxide fuel cell constructed from a YSZ electrolyte with a Ni anode and LSM cathode. *Solid State Ionics*, **127** (1,2), 89–98.

54. Goedickemeier, M., Nussbaum, D., Kleinlogel, C., and Gauckler, L.J. (1997) Solid oxide fuel cells with reaction-selective electrodes. 192nd Meeting of the Electrochemical Society, Abstract No. 2191, p. 2562.

55. Suzuki, T., Jasinski, P., Anderson, H.U., and Dogan, F. (2004) Single chamber electrolyte supported SOFC module. *Electrochem. Solid-State Lett.*, **7** (11), A391–A393.

56. Suzuki, T., Jasinski, P., Petrovsky, V., Anderson, H.U., and Dogan, F. (2004) Anode supported single chamber solid oxide fuel cell in CH_4–air mixture. *J. Electrochem. Soc.*, **151** (9), A1473–A1476.

57. Shao, Z., Mederos, J., Chueh, W.C., and Haile, S.M. (2006) High power-density single-chamber fuel cells operated on methane. *J. Power Sources*, **162** (1), 589–596.

58. Ahn, J., Ronney, P.D., Shao, Z., and Haile, S.M. (2009) A thermally self-sustaining miniature solid oxide fuel cell. *J. Fuel Cell Sci. Technol.*, **6** (4), 041004.

59. Wei, B., Lü, Z., Huang, X., Liu, M., Jia, D., and Su, W. (2009) A novel design of single-chamber SOFC micro-stack operated in methane–oxygen mixture. *Electrochem. Commun.*, **11** (2), 347–350.

60. Liu, M., Lu, Z., Wei, B., Huang, X., Chen, K., and Su, W. (2009) Effect of the cell distance on the cathode in single chamber SOFC short stack. *J. Electrochem. Soc.*, **156** (10), B1253–B1256.

61. van Gool, W. (1965) The possible use of surface migration in fuel cells and heterogeneous catalysis. *Philips Res. Rep.*, **20**, 81–93.

62. Hibino, T., Hashimoto, A., Suzuki, M., Yano, M., Yoshida, S.-I., and Sano, M. (2002) A solid oxide fuel cell with a novel geometry that eliminates the need for preparing a thin electrolyte film. *J. Electrochem. Soc.*, **149** (2), A195–A200.

63. Kuhn, M. (2009) *Direct-Write Microfabrication and Characterization of Single-Chamber Micro Solid Oxide Fuel Cells with Coplanar Electrodes*, PhD thesis, École Polytechnique de Montréal.
64. Chung, C.-Y. and Chung, Y.-C. (2006) Performance characteristics of micro single-chamber solid oxide fuel cell: computational analysis. *J. Power Sources*, **154** (1), 35–41.
65. Wang, Z., Lu, Z., Wei, B., Huang, X., Chen, K., Liu, M., Pan, W., and Su, W. (2010) A configuration for improving the performance of coplanar single-chamber solid oxide fuel cell. *Electrochem. Solid-State Lett.*, **13** (3), B14–B16.
66. Fleig, J., Tuller, H.L., and Maier, J. (2004) Electrodes and electrolytes in micro-SOFCs: a discussion of geometrical constraints. *Solid State Ionics*, **174** (1–4), 261–270.
67. Crumlin, E. (2007) *Architectures for Individual and Stacked Micro Single-Chamber Solid Oxide Fuel Cells*, MSc thesis, Massachusetts Institute of Technology, pp. 33–34.
68. Jacques-Bedard, X., Napporn, T.W., Roberge, R., and Meunier, M. (2007) Coplanar electrodes design for a single-chamber SOFC: assessment of the operating parameters. *J. Electrochem. Soc.*, **154** (3), B305–B309.
69. Hibino, T., Ushiki, K., and Kuwahara, Y. (1996) New concept for simplifying SOFC system. *Solid State Ionics*, **91** (1–2), 69–74.
70. Chung, C.Y., Chung, Y.C., Kim, J., Lee, J., and Lee, H.W. (2006) Numerical modeling of micro single-chamber ceria-based SOFC. *J. Electroceram.*, **17** (2–4), 959–964.
71. Buergler, B.E., Ochsner, M., Vuillemin, S., and Gauckler, L.J. (2007) From macro- to micro-single chamber solid oxide fuel cells. *J. Power Sources*, **171**, 310–320.
72. Ahn, S.-J., Lee, J.-H., Kim, J., and Moon, J. (2006) Single-chamber solid oxide fuel cell with micropatterned interdigitated electrodes. *Electrochem. Solid-State Lett.*, **9** (5), A228–A231.
73. Kuhn, M., Napporn, T.W., Meunier, M., and Therriault, D. (2010) Single-chamber micro solid oxide fuel cells: study of anode and cathode materials in coplanar electrode design. *Solid State Ionics*, **181** (5–7), 332–337.
74. Son, J.-W., Ahn, S.-J., Kim, S.M., Kim, H., Choi, S.-H., Moon, J., Kim, H.-R., Kim, S.E., Lee, J.-H., Lee, H.-W., and Kim, J. (2006) Fabrication and operation of co-planar type single chamber solid oxide fuel cells. Presented at the 7th European SOFC Forum, Lucerne, Switzerland.
75. Kuhn, M., Napporn, T., Meunier, M., Therriault, D., and Vengallatore, S. (2008) Fabrication and testing of coplanar single-chamber micro solid oxide fuel cells with geometrically complex electrodes. *J. Power Sources*, **177** (1), 148–153.
76. Lee, D., Ahn, S.-J., Kim, J., and Moon, J. (2010) Influence of water vapor on performance of co-planar single chamber solid oxide fuel cells. *J. Power Sources*, **195** (19), 6504–6509.
77. Crumlin, E.J., La O, G.J., and Shao-Horn, Y. (2007) Architectures and performance of high-voltage, microscale single-chamber solid oxide fuel cell stacks, *ECS Trans.*, **7** (1), 981–986.
78. Kim, H., Choi, S.-H., Kim, J., Lee, H.-W., Song, H., and Lee, J.-H. (2010) Microfabrication of single chamber SOFC with co-planar electrodes via multi-step photoresist molding with thermosetting polymer. *J. Mater. Process. Technol.*, **210** (9), 1243–1248.
79. Kuhn, M., Napporn, T.W., Meunier, M., and Therriault, D. (2008) Experimental study of current collection in single-chamber micro solid oxide fuel cells with comblike electrodes. *J. Electrochem. Soc.*, **155** (10), B994–B1000.
80. Kuhn, M., Rao, R.B., and Therriault, D. (2009) Viscoelastic inks for direct-write microfabrication of single-chamber micro solid oxide fuel cells with coplanar thick electrodes. *Mater. Res. Soc. Symp. Proc.*, **1179**, 111–116.
81. Jasinski, P. (2008) Micro solid oxide fuel cells and their fabrication methods. *Microelectron. Int.*, **25** (2), 42–48.

82. Suzuki, T., Jasinski, P., Petrovsky, V., Anderson, H.U., and Dogan, F. (2005) Performance of a porous electrolyte in single-chamber SOFCs. *J. Electrochem. Soc.*, **152** (3), A527–A531.
83. Buergler, B.E. Ph.D. thesis (2006) *Single Chamber Solid Oxide Fuel Cells*, ETH, Zurich.
84. Akhtar, N., Decent, S.P., and Kendall, K. (2010) Numerical modelling of methane-powered micro-tubular, single-chamber solid oxide fuel cell (MT-SC-SOFC). *J. Power Sources*, **195** (23), 7796–7807.
85. Akhtar, N., Decent, S.P., and Kendall, K. (2009) Structural stability of silver under single-chamber solid oxide fuel cell conditions. *Int. J. Hydrogen Energy*, **34** (18), 7807–7810.
86. Akhtar, N. and Kendall, K. (2011) Silver modified cathode for a micro-tubular, single-chamber solid oxide fuel cell. *Int. J. Hydrogen Energy*, **36** (1), 773–778.
87. Nagao, M., Yano, M., Okamoto, K., Tomita, A., Uchiyama, Y., Uchiyama, N., and Hibino, T. (2008) A single-chamber SOFC stack: energy recovery from engine exhaust. *Fuel Cells*, **8** (5), 322–329.
88. Tomita, A., Namekata, Y., Nagao, M., and Hibino, T. (2007) Room-temperature hydrogen sensors based on an In^{3+}-doped SnP_2O_7 proton conductor. *J. Electrochem. Soc.*, **154** (5), J172–J176.
89. Tuller, H.L. (2004) Integration of solid state ionics and electronics: sensors and power sources. *J. Ceram. Soc. Jpn.*, **112** (5), S1093–S1098.

3
Technology and Applications of Molten Carbonate Fuel Cells

Barbara Bosio, Elisabetta Arato, and Paolo Greppi

3.1
Molten Carbonate Fuel Cells overview

3.1.1
Operating Principle

Fuel cells (FCs) are electrochemical devices able to convert the chemical energy of reactions directly into electric energy [1]. The electrochemical conversion is possible in molten carbonate fuel cells (MCFCs) thanks to a ceramic matrix consisting of $LiAlO_2$ filled with a combination of molten alkali metal carbonates, which acts as an electrolyte and allows the transfer of carbonate ions from the cathode to the anode [2]. The electrochemical reactions occurring in MCFCs are as follows:

$$\text{Cathode}: CO_2 + \frac{1}{2}O_2 + 2e^- \longrightarrow CO_3^{2-} \tag{3.1}$$

$$\text{Anode}: H_2 + CO_3^{2-} \longrightarrow H_2O + CO_2 + 2e^- \tag{3.2}$$

$$\text{Overall reaction}: H_2 + \frac{1}{2}O_2 + CO_2 \text{ (cathode)}$$
$$\longrightarrow H_2O + CO_2 \text{ (anode)} \tag{3.3}$$

The net result of these reactions, as shown in Figure 3.1, is the production of water and electric power.

In addition to the reaction involving H_2 and O_2 to produce H_2O, Eq. (3.3) shows a transfer of CO_2 from the cathode gas stream to the anode gas stream via the CO_3^{2-} ion, with 1 mol of CO_2 transferred with 2 F of charge, that is, 2 mol of electrons.

Hence MCFCs require the presence of CO_2 at the cathode; this can be obtained either from an external supply of CO_2 from a different source, or by an external transfer of CO_2 from the anode exit gas to the cathode inlet gas [exhaust gas recirculation (EGR)], or, alternatively, the production of CO_2 by burning the anode exhaust gas, which is mixed directly with the cathode inlet gas. In a stand-alone

Fuel Cell Science and Engineering: Materials, Processes, Systems and Technology,
First Edition. Edited by Detlef Stolten and Bernd Emonts.
© 2012 Wiley-VCH Verlag GmbH & Co. KGaA. Published 2012 by Wiley-VCH Verlag GmbH & Co. KGaA.

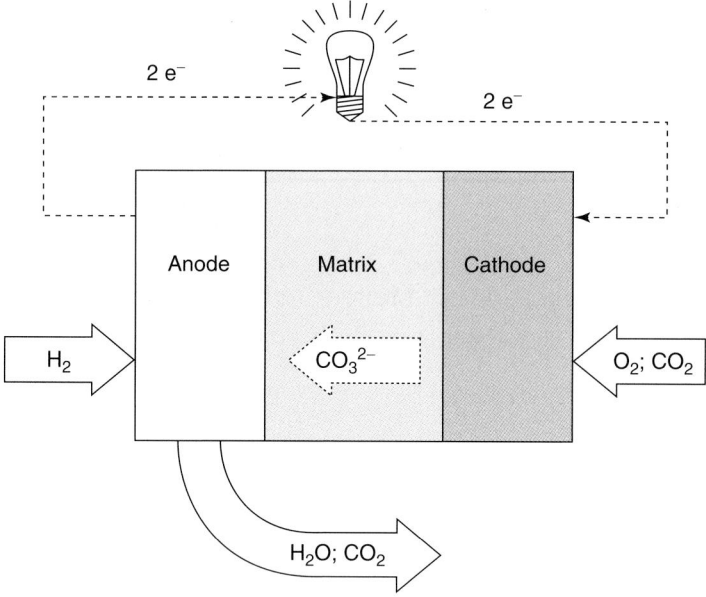

Figure 3.1 Net result of electrochemical reactions occurring in MCFCs.

MCFC system, the CO_2 generated at the anode is usually transferred to the cathode externally.

MCFCs also benefit from the presence of CO_2 at the anode as in sour fuels or biofuel-derived gases; in this case, the benefit is a decrease in the tendency of the electrolyte to evaporate.

In fact, other chemical reactions are also taking place in the system at the same time:

$$CO + H_2O \longrightarrow H_2 + CO_2 \quad \text{water gas shift (WGS)} \quad (3.4)$$

$$CO + \frac{1}{2}O_2 \longrightarrow CO_2 \quad \text{CO cross-over and combustion} \quad (3.5)$$

$$H_2 + \frac{1}{2}O_2 \longrightarrow H_2O \quad \text{H}_2 \text{ cross-over and combustion} \quad (3.6)$$

$$CH_4 + H_2O \longrightarrow 3H_2 + CO \quad \text{methane steam reforming} \quad (3.7)$$

That are the water gas shift (WGS) reaction [Eq. (3.4)], which is very important when fuel rich in CO is fed; the combustion reactions [Eqs. (3.5) and (3.6)] due to an undesired gas cross-over from one electrode side to the other, and finally the methane steam reforming [Eq. (3.7)], which can provide the hydrogen when the fuel contains methane.

Note that CO, generally fed with hydrogen to the anode, is not used directly in the electrochemical oxidation, but is beneficial because it produces additional H_2 when combined with water in the WGS reaction.

Finally, as indicated above, MCFCs have the specific characteristic of concentrating CO_2 on the anodic side. This peculiar property is related to the electrochemical reactions that take place inside MCFCs.

In fact, when an MCFC is working, the reactants on the cathodic side are CO_2 and O_2 (oxidant gas), which diffuse in the nickel oxide electrode to reach the electrolyte and react, following Eq. (3.1). Carbonate ions (CO_3^{2-}) diffuse in the electrolyte to reach the nickel anodic electrode and react with H_2 fed to the anode, following Eq. (3.2). The products of Eq. (3.2) (H_2O, CO_2) diffuse in the anode before leaving.

The capacity of MCFCs to concentrate CO_2 and to produce highly efficient energy will also be discussed in Section 3.3.2.

3.1.2
Operating Conditions

MCFCs usually work at atmospheric pressure or under a slight overpressure (about 0.3–0.4 MPa), but some tests have been run at up to 1.2 MPa [3–5].

The FC operates in the temperature range 855–960 K, where the alkali metal carbonates form a highly conductive molten salt, with carbonate ions providing ionic conduction. A lower temperature causes problems with electrolyte solidification, and higher temperatures have drawbacks for the materials and corrosion. At the high operating temperatures employed, noble metal catalysts are not required; in fact, nickel (anode) and nickel oxide (cathode), that is, low-cost materials, are adequate to promote the reaction.

The conversion of the electrochemical reactions in the FCs is never complete owing to concentration polarization. Typically, the H_2 conversion is in the range 65–80% and the CO_2 conversion is in the range 50–60%. O_2 is normally in excess.

MCFCs can supply current densities in the range 0–1500 A m^{-2} and cell voltages in the range 0.6–1 V as a function of the operating conditions (e.g., fuel composition and flow rate).

One of the more critical features in the operation of MCFCs is the local thermal management on the cell plane [6]. The temperature at each point of the cell plane must first be high enough to prevent solidification of the electrolyte; this thermodynamic lower bound depends on the alkali metal carbonate mixture chosen; for example, the phase diagram in Figure 3.2 shows the stability region of the liquid phase for the commonly employed Li_2CO_3–K_2CO_3 system. Furthermore the local temperature should be everywhere higher than 855 K to guarantee good ionic conduction [7], but must also be lower than 960 K for the prevention of cell-component damage [8, 9].

The maximum local temperature on the cell plane T^{max}, is usually controlled by manipulating the inlet temperature and the total flow of the cathodic gas, which is used to remove heat.

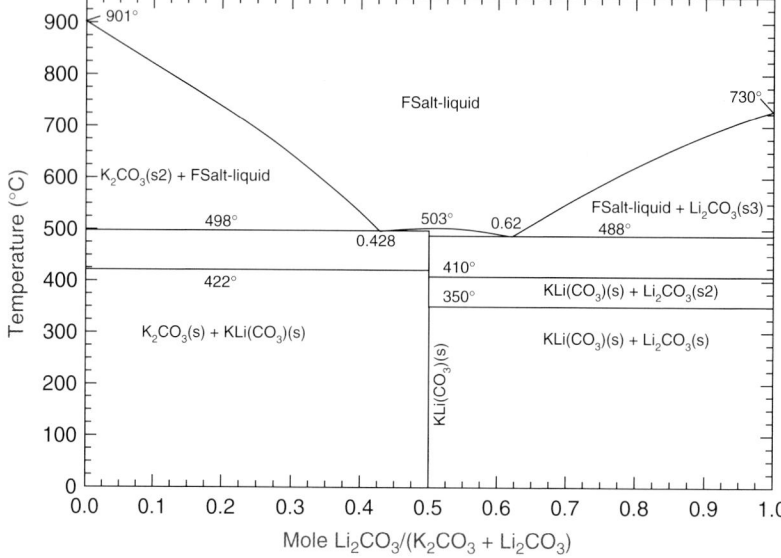

Figure 3.2 Phase diagram showing the stability region of the liquid phase for the commonly employed Li_2CO_3–K_2CO_3 system.

3.1.3
Geometry and Materials

MCFCs typically have a planar cell geometry, with areas ranging from bench scale 0.01 m² to full scale up to 1 m² [10]. The cell form is normally square or rectangular [11], but recently circular cells have also been proposed [12]. The flow pattern is cross-flow in most cases, but according to Kim et al. [13] the counter-flow pattern is associated with lower irreversible losses due to ohmic resistance and anode and cathode activation. Other workers [14] have found by numerical simulation that the co-flow configuration has a higher net output power in some cases.

The main characteristics of the components are summarized in Table 3.1.

Table 3.1 Characteristics of the MCFC components.

Component	Material	Surface area (m² g⁻¹)	Thickness (mm)
Anode	Ni–Cr or Ni–Al	0.1–1	0.2–0.5
Cathode	Lithiated NiO–MgO	0.5	0.5–1
Electrolyte support	α- or γ-$LiAlO_2$	0.1–12	0.5–1
Electrolyte	Li_2CO_3–K_2CO_3/Na_2CO_3	–	0.5–1
Bipolar plate	Metal-coated alloy: Incoloy 825, 310S, or 316 steel	–	0.2–0.5

The electrolyte management, that is, the optimum distribution of molten carbonate electrolyte in the different cell components, is critical for achieving high MCFC performance: various processes, and also corrosion reactions, creepage of the matrix, and salt vaporization, can occur and contribute to the redistribution of the molten carbonate in MCFCs.

Finally, the stainless-steel bipolar plate consists of a separator and current collectors. The plate is exposed to the anodic environment on one side and the cathodic environment on the other. The low oxygen partial pressure on the anodic side of the bipolar plate prevents the formation of a protective oxide coating and, on the cathode side, the contact electrical resistance increases as an oxide scale builds up. Active research is focused on finding alloys for bipolar current-collector materials that function well in both anodic and cathodic environments, have a low cost and ohmic resistance, and have good corrosion resistance [15].

3.1.4
Reforming

If, rather than with hydrogen or syngas, the MCFC system is fed with hydrocarbons or oxygenated compounds such as glycerol [16] or other alcohols [17–20], the fuel must typically undergo water steam reforming to be converted to mainly H_2 prior to being introduced into the FC and electrochemically oxidized.

In some cases, the fuel is supplied as-is to the FC, for example, the bioethanol–water mixture described by Devianto et al. [17]; however, several negative effects, such as a decrease in performance due to low H_2 content, electrolyte loss due to evaporation, and corrosion of the anode cell frame, have been reported.

Schaefer et al. [18] presented a review of conventional, compact, and quick-reacting reactor designs and catalysts for the production of hydrogen from hydrocarbon- and oxygen-containing feedstocks such as (bio-)ethanol or methanol. Bioethanol reforming and feeding to an MCFC has also been studied theoretically [19] and reviewed [20].

Moreover, the sorption-enhanced steam methane reforming (SE-SMR) process has been proposed to reduce CO_2 and CO to a low level and produce almost pure hydrogen [21], which is used as a fuel feed to an MCFC anode. The novel electricity-generation system proposed, which can operate at lower energy consumption and with an almost pure hydrogen feed, is helpful for the performance and lifetime of the MCFC.

Finally, other nonhydrocarbon fuels have been proposed that can be converted to hydrogen through decomposition rather than reforming [22]. As expected, steam reforming provides greater hydrogen production; however, the hydrogen is less pure and the carbon capture is lower than in the decomposition process. In considering primary fuels, decomposition of alcohols could be applied to MCFCs; of these, decomposition of ethanol is preferable because it gives the highest H_2/CO ratio.

Thus, in the case of the conventional external reforming MCFC shown in the scheme in Figure 3.3a, the fuel is fed to an external fuel processor. The external

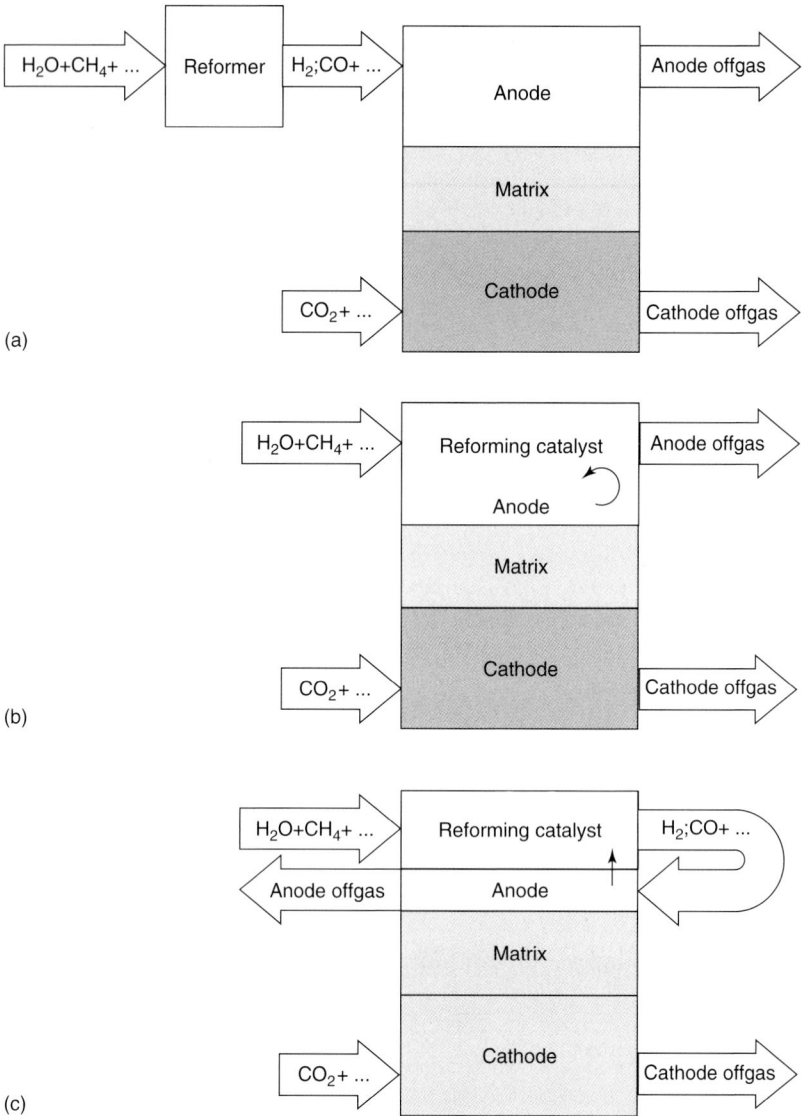

Figure 3.3 Schemes of (a) conventional external reforming MCFC, (b) direct internal reforming, and (c) indirect internal reforming

reforming MCFC approach used in Italy today and previously used by Korean manufacturers allows for greater control of the reforming process of more difficult fuels.

However, the internal reforming in an MCFC is a process-intensification technique that eliminates the need for a separate fuel processor for reforming fuels;

it provides a highly efficient, simple, reliable, and cost-effective alternative to the conventional external reforming MCFC system and is currently being tested in the United States, Germany, Japan, and Korea. This concept is practical in high-temperature FCs where the steam reforming reaction can be sustained with catalysts, and is generally suited to natural gas. The internal reforming MCFC can be realized by closely coupling the reforming reaction and the electrochemical oxidation reaction within the FC.

There are two alternative approaches to internal reforming MCFC: indirect internal reforming (IIR) (Figure 3.3c) and direct internal reforming (DIR) (Figure 3.3b). In the first approach, the reformer section is separate, but adjacent to the FC anode. This cell takes advantage of the thermal integration of the exothermic heat of the cell reaction with the endothermic reforming reaction, but a disadvantage is that the conversion of methane to hydrogen is not promoted as efficiently as in the direct approach. In the DIR cell, hydrogen consumption in the reactive anode compartment due to Eq. (3.2) reduces its partial pressure, thus enhancing the conversion for the methane steam reforming reaction [Eq. (3.7)].

3.1.5
Balance of Plant

The balance of plant (BoP) is required to integrate an MCFC to form a complete, stand-alone power plant.

Depending on the fuel and the configuration, the BoP may include one or more of the following components:

- fuel clean-up, to condition the fuel and remove the undesired pollutants and also sulfur or halogen compounds;
- external reformer;
- CO_2 separation equipment; for carbon capture, storage, and transportation applications, see Section 3.3.2;
- H_2 separation equipment; for hydrogen cogeneration applications, see Section 3.3.3;
- catalytic burner, to remove the partially converted fuel;
- blower for configurations with cathode outlet EGR.

Similarly to turbocharged internal combustion engines and to large-scale steam turbines within combined cycle power plants, MCFCs also can be integrated with turbines.

Different types of MCFC–gas turbine (GT) integration are possible. The configuration proposed by Ansaldo Fuel Cells (AFCo) uses the compressor to pressurize the air fed to the FC; the FC exhaust is burnt with additional methane in the turbine burner and expanded. This configuration couples the Brayton cycle top pressure and the MCFC pressure, and is most suited for integration with microturbines or in any case with low compression ratios.

An example of such a system configuration is shown schematically in Figure 3.4, the performance of which was analyzed by Greppi *et al.* [23]. The fuel-cell

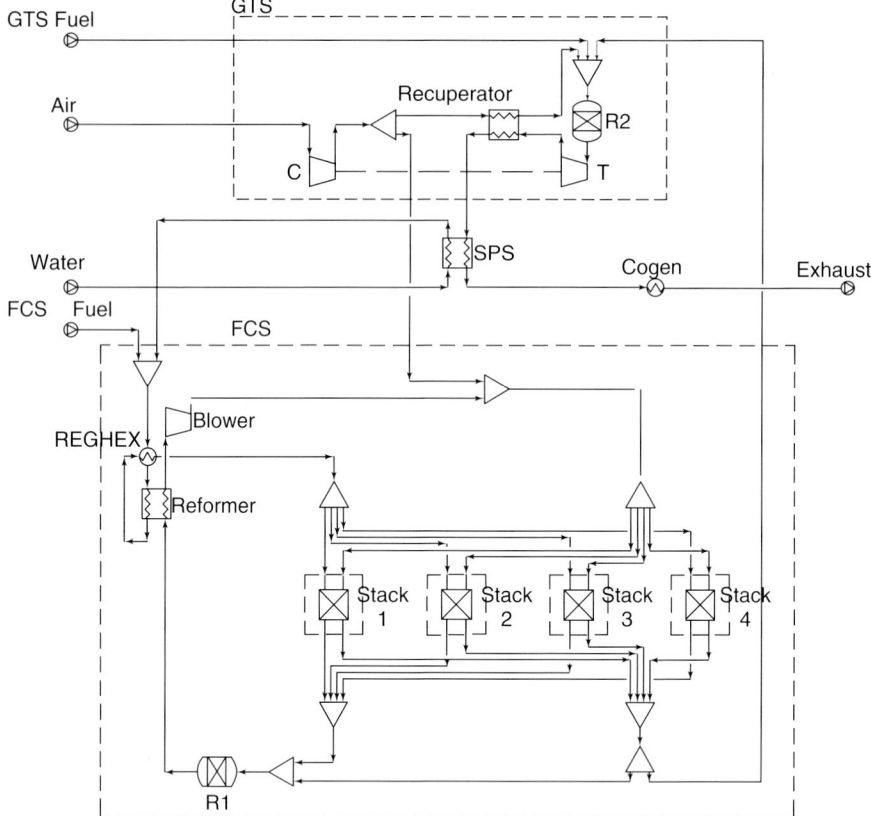

Figure 3.4 Example of integrated MCFC–GT system.

subsystem (FCS) consists of four separate vessels containing planar, rectangular, and cross-flow stacks together with the common BoP which is placed outside the vessels: the external reformer, catalytic burner (R1), preheater (REGHEX), and recycle blower. The gas turbine subsystem (GTS) is a recuperated gas turbine-driven air compressor with a steam production subsystem (SPS) and a co-generative heat exchanger (COGEN) on the turbine exhaust. A net electric power output of 1160 kW corresponding to a plant efficiency of about 53% was computed by neglecting the AC/DC conversion losses and the consumption of energy by such accessories as the control system and fuel booster compressor. Under such conditions, 19% of the net electrical power was provided by the GTS. In addition, a net thermal power output of about 700 kW was available.

Fuel cell energy (FCE) proposes integration with an externally "fired" turbine, in the sense that the heat is supplied by the MCFC. In this way, the Brayton cycle and the MCFC pressure are decoupled.

Another possibility is to exploit the high thermal level of the MCFC exhaust in an organic Rankine cycle.

3.1.6
Vendors

MCFC technology is still under development, but it is commercially available today from at least five international vendors [24].

Currently, in the United States, FuelCell Energy (FCE) (Danbury, CT) is actively pursuing the commercialization of MCFCs, while progress at Gencell (Southbury, CT) is in the development phase. Over the last decade, FCE has notably reduced FC system manufacturing costs and is now approaching its target of making FC systems competitive with traditional power plants.

Three companies are known to be pursuing MCFC technology in the Far East: Ishikawajima-Harima Heavy Industries (IHI) (Japan), Posco (Korea), and Doosan Heavy Industries and Construction (DHI) (Korea). Although there has been some recent academic activity [21, 25, 26], and a development program reportedly reached the kilowatts scale at the Fuel Cell Institute of Shanghai Jiaotong University [27, 28], no Taiwanese or Chinese industrial vendors are known to the authors.

In Europe, AFCo (Italy) and MTU CFC Solutions (Germany) have been playing important roles in the development of external and internal reforming MCFC systems, respectively, but both of them have currently discontinued their MCFC activities.

3.1.7
State of the Art

The development of MCFC technology has been more intensive for stationary, military, and marine applications, where their relatively large size and weight and slow start-up do not represent limitations. MCFCs are available for use with a wide range of conventional fuels, such as natural gas and coal-based power plants, and renewable fuels, such as biogas, landfill gas, and syngas. The current promising sizes are from 250 kW up to a few megawatts.

Moreover, the relatively high operating temperature of the MCFC results in several benefits: no expensive electro-catalysts are needed as the nickel electrodes provide sufficient activity, and both CO and certain hydrocarbons can serve as fuels for the MCFC, as they are converted to hydrogen within the stack on special reformer plates, simplifying the BoP and improving the system electric efficiency by 40–50%. In addition, the high temperature of the waste heat allows the use of a co-generation cycle to increase the system efficiency further by 50–60%, and up to 70% in the case of GT integration.

However, the requirement of a certain quantity of CO_2 at the cathode to form the carbonate ion is a drawback in the plant management. Furthermore, the high contact resistances and the cathode resistance increase overpotentials and decrease efficiency, and concentration polarization limits the electric power densities. Finally, the main challenge for MCFC developers derives from the very corrosive and mobile electrolyte, which requires the use of nickel and high-grade stainless steel for the cell hardware. Higher temperatures promote material problems, and also cathode dissolution into lithium oxide, anode sintering, and metallic component corrosion, impacting on costs, mechanical stability, and stack life. These challenges have resulted in high costs of installation, in the range €3000–4000 kW^{-1} depending on the scale, and a stack life limited to about 25 000 h; for this technology to become competitive with more mature technologies, the installation costs should be lower than €1100 kW^{-1} and consecutive stack operation in excess of 40 000 h.

3.2
Analysis of MCFC Technology

As stressed in the previous sections, there are still some technical issues to be resolved before MCFCs can be placed on the market in competition with traditional energy systems.

In this scenario, research activity plays an important role, and MCFC technology is at present the object of intensive study in order to improve performance.

3.2.1
Approach

One of the more reliable study approaches consists in effective integration of theory and practice: the results can be interesting from an academic point of view for the understanding of phenomena and from an industrial point of view in terms of technology and optimization of operating conditions.

An example of such an approach is illustrated in Figure 3.5, where modeling and experimental activities were at the basis of the authors' cooperation with AFCo to study the scale-up process from the electrode to the plant scale [29].

In terms of simulation, the choice of either a simplified or detailed modeling approach or of either *ad hoc*-developed or commercial tools depends on the analytical aim. Generally, the best solution is to couple the accuracy of detailed *ad hoc* blocks for MCFC stacks or other innovative components with the flexibility of commercial packages simulating the behavior and the interactions of standard components.

In this way, the modeling approach conforms to different simulation aims from the investigation of micro-phenomena to global plant design. The different degrees of detail could be defined as follows:

- Electrode micro-modeling, based on analysis of transport phenomena, locally accounting for voltage losses by means of semiempirical kinetic relationships.
- Cell two-dimensional macro-modeling, based on local mass, energy, charge, and momentum balances; this gives maps of the main electrical, chemical, and physical parameters on the cell plane as simulation results.
- Stack three-dimensional macro-modeling, also including heat transfer along the vertical coordinate of the stack.
- System macro-modeling, where the overall system is simulated by taking account of the auxiliary components and the plant balance.

3.2 Analysis of MCFC Technology | 77

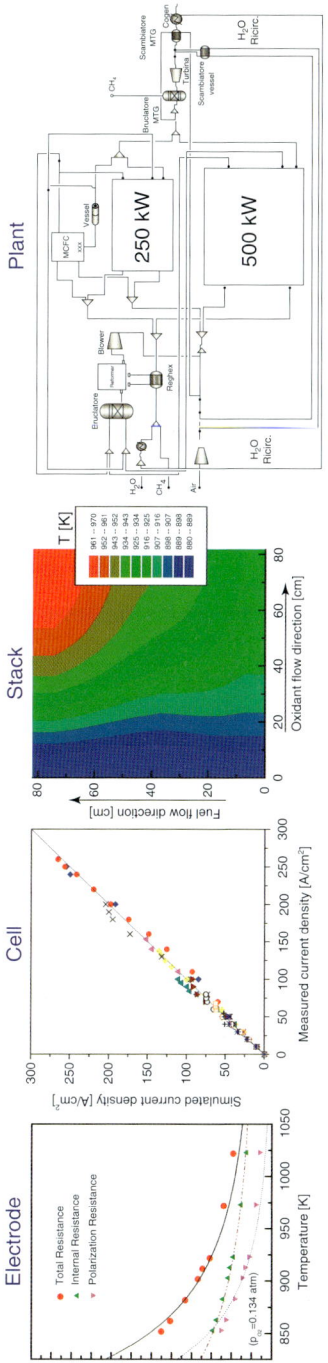

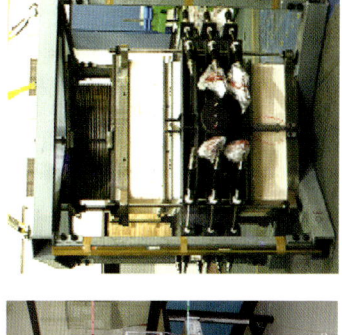

Figure 3.5 Illustration of modeling and experimental activities to study the scale-up process from the electrode to the plant scale.

The authors' experience has led them to develop two codes for the stack simulation, namely MCFC-D3S (Molten Carbonate Fuel Cell Dynamic and Steady-State Simulation [30]) and SIMulation of Fuel Cells (SIMFCs [6]). The former is specific to AFCo technology and the basis of a patent [31] and the latter is more general and applicable to different technologies.

The codes, based on the integration of the electrode, cell, and stack models, allow the calculation of the maps of the current density, Nernst voltage, polarizations, internal resistance, gas and solid temperatures, pressure drops, and the compositions and flow rates of the gaseous streams.

These tools can be used for several applications: planning, optimization, diagnosis, and control. Their integration in overall system simulation allows reliable plant design and the setting up of the operating conditions while checking the local constraints, for example, the maximum temperature on the cell plane or the maximum difference between the anode and cathode pressures.

For the study of complex systems or of transient plant behavior, more simplified models can also be used, provided that proper experimental validation is obtained. For macro-modeling of the system, an *ad hoc* steady-state code was developed on the basis of LIBrary for Process Flowsheeting (LIBPF) technology [31], capable either of integrating the detailed codes for stack simulation of the MCFC-D3S and SIMFC, or of using an intermediate-fidelity model for the stack to improve the calculation performance when required.

Experimental activity can also be carried out at different scales. When the MCFC apparatus is considered, and not the overall plant, the laboratory cell represents the classical bench scale, characterized by dimensions of the order of a postcard, intermediate between the laboratory scale (a postage stamp) and the commercial scale (a poster).

When the cell area is increased, the uniformity of the operating conditions on the cell plane decreases and so the electrode complexity increases. It is already important to take into account this aspect at the bench scale, where electrode complexity is often underestimated, in order to interpret correctly the data and the phenomena occurring.

Usually the experimentation follows a test procedure, which involves:

- Test facility controls and instrument calibration.
- Pressurization: accurate control of operating pressure and differential pressure between vessel and electrodes.
- Start-up phase: accurate cell warming during which carbonates melt and the conditioning of cell components ends.
- Preliminary tests: estimation of possible gas cross-over across the tile and of gas leakages between cell and vessel, conduction of reference characteristic curves to check cell performance, analysis of the test repeatability and of the times required to reach steady state, and so on.
- Repetitive tests: gas chromatographic analysis, internal resistance evaluation with current interruption or impedance measurement methods, reference characteristic curves to control aging, and so on.

- Specific tests: long-life tests, poisoning tests, performance evaluation under different operating conditions, and so on.
- Post-test analysis: porous component stability estimation, electrolyte content measurement, corrosion estimation, and so on.

Recently, standard procedures have been set up within the "Fuel Cell Systems Testing, Safety and Quality Assurance (FCTES^{QA})" project, which is a specific targeted research project co-financed by the European Commission within the Sixth Framework Program.

The main aim of FCTES^{QA} is to address the aspects of pre-normative research, benchmarking, and validation through round-robin testing of harmonized, industry-wide test protocols and testing methodologies for FCs.

In conclusion, the proposed study approach consists in an accurate simulation supported by suitable experimental investigation, resulting in the best guide for developing new system solutions.

3.2.2
Technology Optimization

Simulation and experimental results can be used to investigate the phenomena occurring in FCs, to optimize the operating conditions, to evaluate the effects of fuel choice on FC performance, to control scale-up processes, to suggest new technological solutions, and to guide experimental tests carried out in cooperation with many industrial partners.

Many examples of technology optimization could be presented with regard to:

- matrix thickness, sufficiently thin to allow proper ionic conduction and sufficiently resistant to avoid gas cross-over between the anodic and cathodic sides
- current-collector material, able to withstand mechanical stress and corrosion
- electrode wettability, to optimize chemical and physical kinetics, and so on.

Here a specific example is reported based on the authors' experience and concerning a key problem in MCFC operation, namely thermal management.

In general, the preferred operating temperature range of MCFCs is 925–955 K, but gradients up to about 100 K m^{-1} can be present in the cell during operation. In addition, an input gas temperature of at least 855 K is necessary to guarantee good ionic conduction inside the cells, while maximum local temperatures higher than 960 K should be avoided because they can cause problems such as electrolyte loss and corrosion.

The first limit can easily be managed, whereas the second can only be managed through taking many local measurements or, more properly, reliable detailed simulation models.

As the temperature maps of an MCFC plane can be very irregular, some parts of the cell can work under critical conditions even if the average temperature is not too high, and this aspect is critical for industrial optimization of MCFC performance.

Different techniques have been tried to obtain a uniform temperature distribution on the cell plane.

In particular, given the predominant role of fluid dynamics in the problem, different solutions have been proposed based on the use of nonuniform inlet gas flows [6] or on double-side feeding and adjustment of the mass flow rate of the cathodic gas supplied to the various cells [12].

Considering a specific geometry of the current collectors, which also work as gas distributors, a detailed fluid dynamic model has been set up and integrated in the SIMFC code mentioned above in order to take account not only of the gas flow along straight paths parallel to each other (common in the literature), but also of the real behavior of the gases, which can also have transversal paths.

It was observed that the main component of the velocity of the anodic and cathodic gases was along the principal direction, but a transversal component appeared: the percentage ratio of its maximum value to the maximum value of the longitudinal component was about 3.3% for the anodic gas and 3.8% for the cathodic gas.

Our investigations considered, in line with an ENEA report [24], uniform inlet gas pressure profiles, progressively linearly decreasing profiles, and progressively linearly increasing profiles combined to build eight non-uniform inlet flow patterns.

Figure 3.6 shows the temperature maps for the eight different patterns for $d = 0.005\%$, where d represents the unilateral deviation of the nonuniform profile, which is the ratio between the variation in the inlet gas pressure on one side (18 Pa) and the mean inlet gas pressure (0.36 MPa).

Comparing the different temperature distributions, it is apparent that the temperature maps of patterns (a) and (b) are similar to the reference case; the maps of patterns (c), (f), and (h) are worse; and the maps of patterns (d), (e), and (g) are more uniform with a lower maximum temperature. The main variations are given by the cathodic inlet gas distribution, whereas the anodic one does not affect the maps: in fact, the same temperature distributions are given by patterns (a) and (b), patterns (c), (f), and (h), and patterns (d), (e), and (g). This suggests that only the cathodic gas influences the temperature distribution on the cell plane because its flow rate exceeds that of the anodic gas.

In particular, patterns (d), (e), and (g) give a T^{max} reduction of about 0.7% and a ΔT reduction of about 8%; this means that the temperature map is more uniform with a lower peak. On the other hand, patterns (c), (f), and (h) produce less uniform temperature maps with higher peaks: T^{max} increases by about 0.7% and ΔT increases by about 7%. Moreover, as discussed above, the pattern (d), (e), and (g) results are about the same, so that pattern (d) is chosen as the best one for the temperature management, considering that with this configuration it is possible to maintain a uniform constant anodic gas inlet pressure and vary only the cathodic one.

Studying different pressure profiles, Figure 3.7 demonstrates that the variation of ΔT decreases rapidly when d is increased up to 0.03%, whereas it starts to increase again for higher values of d. For this reason, the optimum configuration for the temperature distribution on the cell plane could be identified in pattern (d) with $d = 0.03\%$, which would seem to be the best compromise between the uniformity of the map and the lower T^{max}.

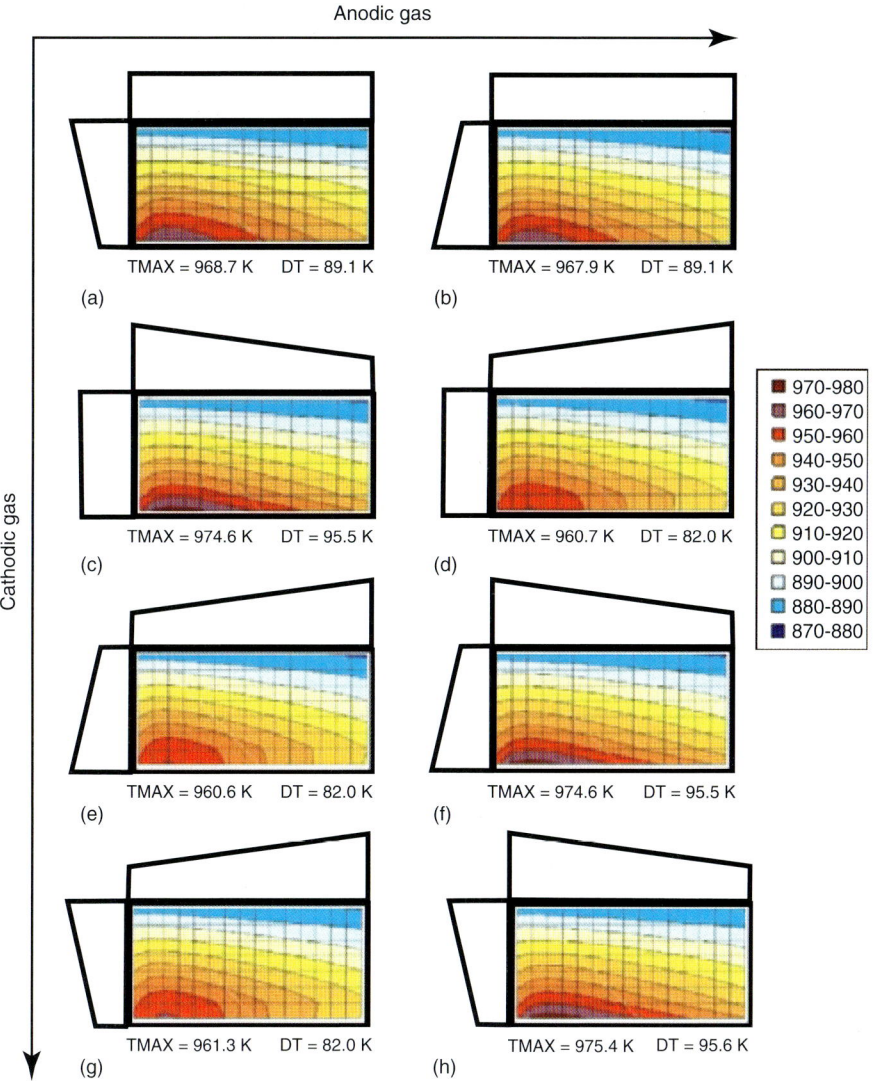

Figure 3.6 Temperature maps for the eight non-uniform inlet flow patterns for $d = 0.005\%$.

A similar approach could be applied to different current-collector geometries and lead to new technological solutions aimed at optimizing MCFC performance.

3.2.3
Scientific Knowledge

At the basis of each technological solution is a detailed scientific knowledge of the underlying physical phenomena. Currently, the area of fundamental research

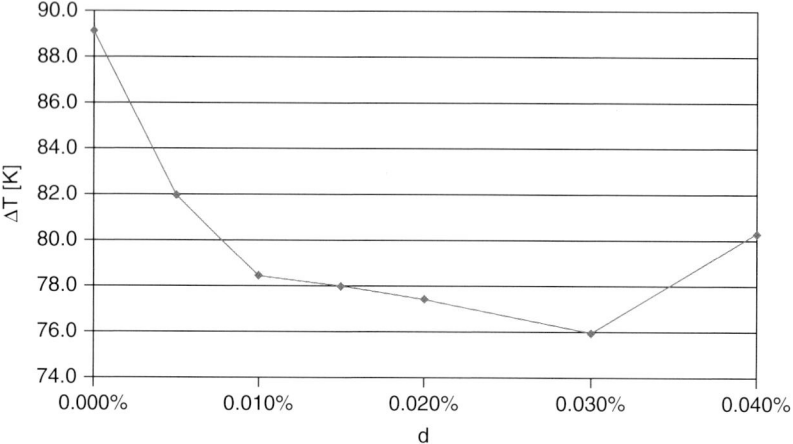

Figure 3.7 Variation of $\Delta T = T_{max} - T_{min}$ on the cell plane with d.

where the most effort is vested is the study of kinetic and mass transport, in order to improve them from both an electrochemical and a physical point of view. MCFC behavior needs to be studied for the particular MCFC application field, so that for plants based on CO_2 concentration (see Section 3.3.2) a detailed evaluation of CO_2 mass transport mechanisms is needed to increase the related utilization factor. On the other hand, for plants based on biogas feeding, a detailed knowledge of poisoning phenomena is needed to increase the related tolerance limits.

In general, reactant diffusion in the electrode structure, that is, physical kinetics, plays a fundamental role in MCFC performance, limiting efficiency to a maximum utilization factor of about 75 and 56% for H_2 and CO_2, respectively.

When the circuit is open and no current flows, the activity of each reactant on the electrode surface is equal to its activity in the bulk. The Nernst thermodynamic voltage expression can be written in terms of the bulk activity, or the concentrations in the ideal solution simplification, and yields the open-circuit voltage (OCV) for the overall electrochemical reaction [Eq. (3.3)] of hydrogen combustion.

The cell voltage is always lower than the OCV because when the electrochemical reaction occurs, the voltage decays by a term ΔE_{TC} due to thermodynamic concentration polarization:

$$\Delta E_{TC} = \frac{RT}{v_e F} \Sigma v_i \ln\left(\frac{c_{i,s}}{c_{i,b}}\right) \tag{3.8}$$

The fraction inside the parentheses can be expressed [32] in terms of the local limit current density J_L, that is, the maximum current reachable when the limiting reactant concentration on the electrode surface approaches zero, and further simplified to

$$\Delta E_{TC} = \frac{v_L}{v_e} \frac{RT}{F} \ln\left(1 - \frac{J}{J_L}\right) \tag{3.9}$$

The limit current density is usually interpreted by means of these thermodynamic statements. However, as is well known, only the ideal or maximum efficiency depends on electrochemical thermodynamics, whereas the effective efficiency depends on electrode kinetics.

By taking thermodynamic constraints into account in the framework of a kinetic analysis, a slightly different and more consistent concentration polarization expression has been proposed [32]:

$$\Delta E_C = -\frac{\alpha_L}{\alpha_e}\frac{RT}{F}\ln\left(1 - \frac{J}{J_L}\right) \qquad (3.10)$$

where the voltage decay due to concentration polarization is a function not only of J/J_L, that is, the ratio between current density and limit current density, but also of the reaction rate orders α_e and α_L, related to the electrons (e) and the limiting reactant (L).

The reaction rate orders are not known *a priori* like the stoichiometric coefficients ν_e and ν_L, but on the condition that the limiting current is well determined, the differences between the thermodynamic and kinetic equations:

$$\frac{\eta - \eta_T}{\eta_T} = \frac{\Delta E_C - \Delta E_{TC}}{\Delta E_{TC}} = -\frac{\alpha_L \nu_e}{\alpha_e \nu_L} - 1 = -\frac{\gamma_L \nu_e}{\alpha_e \nu_L} \qquad (3.11)$$

are not very important. In particular, the operating conditions where γ_L, the thermodynamic consistency parameter [32], is significantly different from zero, kinetic polarization, and thermodynamic polarization show differences in the V/J performance curve concerning the shape of the "knee" in the transition from linear behavior to limit current, as shown in Figure 3.8.

Obviously, a better knowledge of the kinetics provides more optimization possibilities.

In addition, details can also be important when, on the basis of validated kinetics of an MCFC running under standard conditions, operation with poisoned feeds is investigated. This is another significant example of applied research activity.

As already mentioned, MCFCs can be used with many different gases, such as natural gas, coal gas, biogas, and landfill gas, but certain contaminants present in these fuels can be harmful to MCFC operation and cause degradation of performance and a reduction in cell life.

An anaerobic digestion gas typically contains up to 200 ppm of H_2S and other pollutants such as halogens, organics, and other hydrocarbons. In landfill gases, hydrocarbons, up to 200 ppm of H_2S, and a range of chlorine/fluoride-containing compounds can be found. In pipeline natural gas, the total sulfur content is of the order of 10–80 ppm, with about 4 ppm of H_2S and 4 ppm of mercaptans (thiols). Coal typically has a higher sulfur content than biomass and, as a result, the syngas produced from coal can contain 0.1–1% or more of H_2S. Depending on the type of gasifier, some of the sulfur in the product gases may be oxidized in the form of COS or SO_2 [33–36].

A number of studies on the MCFC tolerance towards sulfur compounds can be found in the literature [36–46], but it is not completely clear what the maximum tolerance limit is, as it depends on many factors such as temperature, pressure,

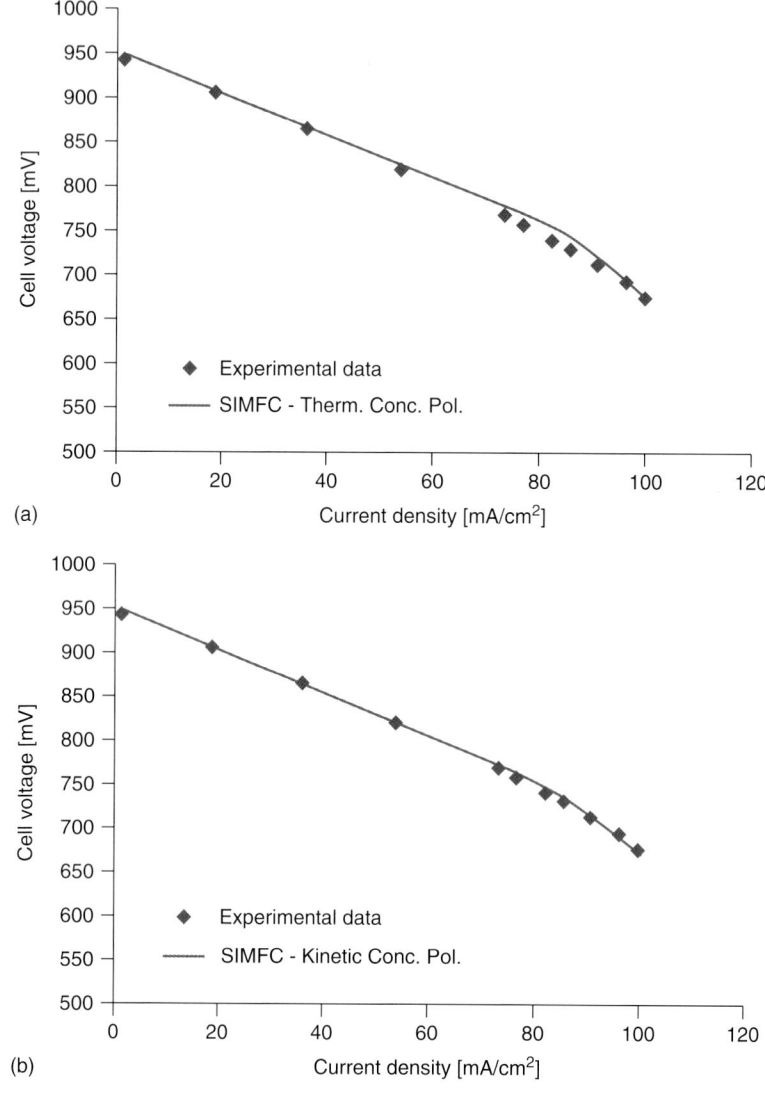

Figure 3.8 Cell voltage versus current density by considering thermodynamic (a) and kinetic (b) polarization.

gas composition, current density, cell components, and system configuration (e.g., recycle and clean-up). The main experimental results are summarized in Table 3.2.

The major cause of MCFC poisoning is believed to be H_2S, although compounds such as COS and CS_2 seem to have a similar effect [33]. The effects and mechanisms involved are as follows:

- reaction with nickel, deactivation of the electrocatalyst and consequently voltage drop

Table 3.2 Experimental results for operation in the presence of sulfur compounds.

Conditions	Anode	Cathode	Ref.
$P = 1$ atm, gas utilization 75%	$H_2S < 10$ ppm	$H_2S < 0.5$ ppm $SO_2 < 1$ ppm	[1]
	H_2S, COS < 1 ppm	–	[47]
	$H_2S < 0.1$ ppm	–	[35, 48]
	H_2S, COS, $CS_2 < 0.5–1$ ppm	–	[40]
40 000 h without removal systems	$H_2S < 0.01$ ppm	–	[1]
40 000 h with removal from burner	$H_2S < 0.5$ ppm	–	[1]

- oxidation to SO_2 and consequent reaction with carbonates
- poisoning of catalytic sites for the WGS reaction.

An experimental campaign was recently carried out at the Fuel Cell Research Center laboratories of the Korea Institute of Science and Technology (KIST) using its $0.01\,m^2$ single-cell facilities. The investigation was run on synthetic gases simulating the conditions of MCFCs integrated with a combined cycle power plant and allowed a preliminary analysis of the effect of H_2S on MCFC performance, highlighting how the main operating parameters affect poisoning phenomena.

The performance decay due to H_2S during endurance tests for up to 600 h was fairly constant under constant operating conditions (see Figure 3.9). The sulfur adsorption reaction effect on the OCV was analyzed, to isolate it from the electrochemical reaction contribution, demonstrating how the WGS reaction is inhibited at Ni–Al anodes. An exponential dependence of voltage decay due to

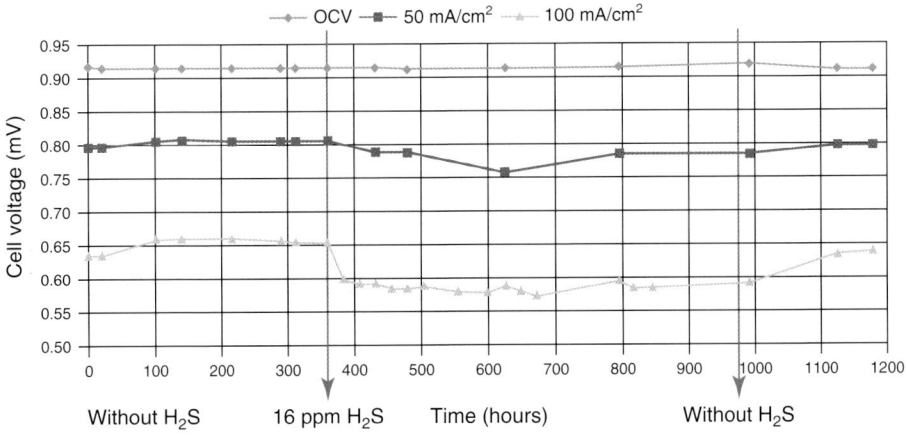

Figure 3.9 Performance decay due to H_2S during endurance tests for up to 1200 h.

H_2S on the current density was observed, even when varying the total flow rate to maintain constant utilization factors and compositions, in spite of the contrary influence of the mass transport phenomena. The temperature effect is such that low temperatures emphasize poisoning damage.

The results obtained, confirming the possibility of using MCFCs when up to 10 ppm of H_2S is present in the feed, have provided suggestions for approaching the interpretation of phenomena for setting up a kinetic equation which also takes account of the H_2S effects and allows reliable predictive calculations.

3.3
Conventional and Innovative Applications

MCFCs are currently in operation in more than a 100 installations around the world. Although the number of MCFC developers and the investment levels are much reduced compared with a decade ago, development and demonstrations continue.

The most significant conventional application of MCFCs is in the distributed generation of electric energy in small to medium sized stationary power plants. This application will be described first and then the innovative applications proposed or already developed will be outlined.

3.3.1
Distributed Generation

MCFC energy plants fit with a new philosophy for the production and distribution of electricity: instead of huge plants of the order of gigawatts that supply entire cities or regions, such as traditional power plants, smaller plants that supply small areas or single structures.

For this reason, as already mentioned, plants of megawatt size are typically considered on the basis of the worldwide experience in distributed generation [5, 49–54]. The high thermal level of the FC exhaust can also be exploited for co-generation in combined heat and power (CHP) applications [55].

The applications of distributed generation include wastewater treatment, plant manufacturing, large hotels, hospitals, prisons, computer data centers, colleges and universities, and so on.

FCE supplies self-contained electric power generation systems with an external manifold, with DIR MCFCs, and with nominal electric power output in the range 0.3–3 MW. GenCell is developing distributed power systems in the range 40–125 kW. IHI has focused on the commercialization of a pressurized 300 kW MCFC co-generation system, with internal manifold and external reformer, that can be fed by methane or biogas, and also in a hybrid configuration (MCFC–Micro Gas Turbine).

The South Korean government has been supporting the growth of the FC industry in the framework of the country's economic policy; POSCO Power and the

Korean Electric Power Research Institute (KEPRI) are the MCFC main contractors for power plant construction in collaboration with the KIST. MCFC modules of 100 and 250 kW with internal manifolds and external reforming of natural gas have already been delivered, for a total installed power of about 30 MW [56]. DHI is developing a 300 kW module with internal reforming.

POSCO Power is the exclusive supplier of MCFC systems in Korea under a technology transfer agreement with FCE; it has just completed the construction of its FC stack manufacturing plant in the city of Pohang. The facility now has the capacity to produce 100 MW worth of MCFC stacks annually and the complex has installed a 2.4 MW FC power plant.

AFCo has developed its own technology, adopting an external reforming solution thermally integrated in the plant for modules ranging from 100 to 500 kW.

MTU has developed and installed a number of modules belonging to the 250 kW power series, using stacks supplied by FCE and running with different gaseous feeds.

3.3.2
Carbon Capture, Storage, and Transportation

The carbon capture, storage, and transportation (CCS&T) is an important option for reducing the impact of greenhouse gas (GHG) emissions and mitigate climate change [57], allowing the use of fossil fuels to produce energy in an environmentally friendly way.

As an innovative application of MCFC, the possibility of using them in CCS&T has been proposed by several groups. This application involves MCFC plant sizes that can be significantly larger than those actually used in the distributed generation strategy; in this case modularity can be played upon.

For reducing carbon emissions, two MCFC features are relevant: the high efficiency and the capability to segregate CO_2 while at the same time producing energy, in contrast with other capture technologies that are heavy net energy consumers.

The high electrical efficiency of MCFCs, up to 50%, means that at a predefined power level, lower fuel consumption is possible in comparison with conventional electric generators. For example, less natural gas or other hydrocarbons are used to produce the primary MCFC fuel, that is, H_2, and consequently less CO_2 is sent into the atmosphere, demonstrating that high efficiency makes MCFCs a natural candidate for CO_2 abatement.

Furthermore, MCFCs are the only FC technology that allows CO_2 concentration, meaning an innovative way of separating CO_2 for its final abatement.

This peculiar property is related to the reactions that take place inside MCFCs [Eqs. (3.1)–(3.3)], whereby the CO_2 is concentrated on the anodic side. The CO_2 in the concentrated stream still needs to be separated with a conventional capture technology, but the mass flow is considerably reduced.

These important and innovative characteristics can be used to reduce the CO_2 emissions of a fossil fuel-fed power plant in a post-combustion configuration, by continuously feeding the power-plant exhaust to the MCFC cathodic inlet. The concentrated CO_2 on the anodic side can then easily be separated from the richer stream with a reduced size separation plant thanks to the lower flow. Therefore, the use of MCFCs to concentrate CO_2 could be applied in combination with the conventional CO_2 separation methods, such as absorption in solvents, absorption by a solid, separation with a membrane, and cryogenic separation. The CO_2 energy cycle is reported in Figure 3.10, coupling the CO_2 removal system with the production of electric power.

It is also feasible to integrate the MCFC thermally and physically with a natural gas combined cycle (NGCC) with the objective of carbon capture. From the thermal point of view, the endothermic natural gas reforming for the syngas production can be sustained with an energy side-draw from the Brayton cycle natural gas burner, which can, at least thermodynamically, easily be adapted to deliver the heat at the required high temperature. The material integration consists in supplying the medium-pressure water steam to the natural gas reforming with a side-draw from the heat recovery steam generator. This opens the steam cycle but avoids equipment duplication.

MCFC integration with an NGCC power plant was investigated using LIBPF, a process flow-sheeting and modeling tool arranged as a C++ library. A very simple "black box" model was used for the carbon capture and an intermediate fidelity model for the FC stack; the latter makes it possible to take the specific constraints of the electrochemical reactor into consideration to a certain extent. The required CO_2 concentration to 95% purity specification has been realized with a black box separation unit with specific energy consumption in the range of the typical applicable technologies (Monoethanolamine, gas separation membranes). The results highlighted an increase in the electric energy production of the MCFC +

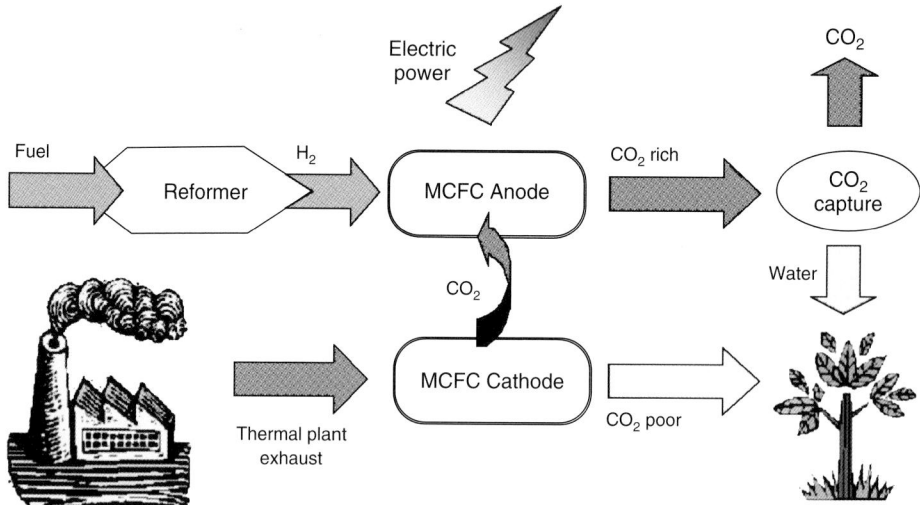

Figure 3.10 The CO_2 energy cycle in a MCFC system.

NGCC plant with a slight energy efficiency decrease and an average CO_2 emission abatement.

3.3.3
Hydrogen Co-generation

MCFC, like other FC types, can not completely convert the H_2 in the feed. Therefore, there is the possibility of recovering and purifying the unconverted hydrogen, obtaining a combined heat, hydrogen, and power (CHHP) generation plant.

Some groups have investigated this option specifically for MCFCs. A biogas-fueled MCFC-GT has been coupled to a pressure swing absorption (PSA) system for hydrogen production [58]. CHP and CHHP with a phosphoric acid fuel cell (PAFC) and MCFC has been compared with conventional combustion technologies [59]. The authors reported energy and GHG emission benefits for CHHP systems when applied to actual facilities in two different climatic regions of the United States under variable electricity-to-heat load ratios.

Direct carbon fuel cells (DCFCs) are based on the same electrolyte and carbonate ion transport mechanism as MCFCs, but are fed with solid carbon rather than hydrogen. It is an interesting technology from the thermodynamic point of view, but still in the initial research phase. The DCFC would not be worth mentioning in an MCFC review in the context of hydrogen as an energy vector, because hydrogen is not involved in their chemistry and electrochemistry. However, a recent contribution [60] suggested the possibility of producing the carbon required as reactant for the DCFC from methane, which is thermally decomposed to solid carbon in an external reactor while at the same time hydrogen is formed as a by-product. In this configuration, the methane decomposition and the DCFC work together as a unit simultaneously producing electric energy, hydrogen, and pure CO_2 at the anode.

3.3.4
Renewable Fuels

The efficiency and pollution advantages of MCFCs can be leveraged to obtain an even greater environmental benefit if the hydrogen is produced from renewable sources.

Commercially, MCFCs are proposed by several vendors for biogas from anaerobic digestion processes of organic residues [37] and "opportunity fuels" such as landfill gas. These gases raise the issue of hydrogen sulfide poisoning, which is discussed in Section 3.2.2.

Various solid and liquid renewable sources have also been proposed for producing syngas for MCFCs: wood [61, 62], chestnut coppice [63], bioethanol [17, 19, 20], and the glycerol obtained as a by-product in biodiesel production with the transesterification process [16]. Oxygenated liquid fuels are challenging for the reforming step (see Section 3.1.4).

3.3.5
Other Applications

Other applications mentioned in the literature include hydrogen waste streams, syngas from integrated gasifier combined cycle (IGCC), nonstationary, or military applications, and natural gas depressurization.

Waste or by-product streams rich in hydrogen occur in several petrochemical and electrochemical processes, such as chlor-alkali electrolysis and electroplating. When no chemical use for the hydrogen is possible, its conversion to electricity can be performed in an MCFC; the good co-generation opportunities typically found in industrial environments should make MCFCs preferable to other FC technologies.

Syngas from an IGCC can be used as fuel for an MCFC; the integration with such a complex plant offers several options for optimization. A pilot-scale MCFC power plant that can be installed side-by-side with an existing IGCC plant has been proposed [64].

Mobile applications of MCFCs have only been envisaged in the naval sector for ships and military submarines. Possible maritime fuels are liquefied natural gas (LNG), diesel, and kerosene; the desulfurization step is critical as some of these fuels have high sulfur contents. An MCFC can provide a ship-board electric energy (auxiliary power unit, APU), while the conventional internal combustion engine provides mechanical energy; the challenge of feeding the MCFC system with available fuels (marine diesel and kerosene) has been the topic of the EU-funded MC-WAP FP6 Project (MCFCs for Waterborne APplication) [65].

Military applications envisage the deployment of MCFCs to supply the energy required to recharge portable electronic appliances in the battlefield; in this context the low-noise and highly efficient operation of MCFC power generation is invaluable. The easier logistics of liquid fuels has oriented research towards adapting MCFCs to operation on commercial propane [66]. MCFCs have also been considered for air-independent propulsion (AIP) in non-nuclear military submarines because of the low noise.

When natural gas is depressurized at the end of a transportation pipeline, a large amount of energy is released. Part of this energy can be recovered and additional "ultra-clean" energy can be produced with very high efficiency in the hybrid, multi-megawatt plant proposed by FCE for this application.

3.4
Conclusion

MCFC technology can produce electricity with efficiencies as high as 50% while maintaining low pollutant levels in the flue gases, making it the most environmentally friendly device commercially available today for producing distributed electric energy or CHP from natural gas and from renewable or opportunity fuels. Thanks to the modular construction, these attractive characteristics hold across the wide 0.2–20 MW nominal power range [67].

On the other hand, the installation costs of MCFC plants are still too high and their durability and reliability are no match for more mature technologies such as internal combustion engines or turbo-machinery. For these reasons, MCFCs are currently economically sustainable only in the presence of:

- feed-in tariff (FIT) schemes
- renewable portfolio standards (RPSs), that is, policies that assimilate electric energy produced from MCFCs to renewable sources
- stringent air pollution standards

as have been put in place in Korea [56] and in some states in the United States [67].

The mature technologies are approaching the limits of their thermal cycles, their material operating conditions, and their down- or up-scalability, whereas for FCs there is considerable room for technological improvement, which can only be achieved with a sustained, systematic research and development effort.

For example, the efficiency of an oxy-fuel, coal-fed IGCC–MCFC with CO_2 capture has been projected [68] to a theoretical limiting efficiency of 58% higher heating value, whereas the technical limit for an integrated NGCC–MCFC has been evaluated at 77% electrical efficiency, lower heating value base (F. Yoshiba, personal communication).

The MCFC is potentially one of the key technologies to fulfill the "20–20–20" objectives [57] and it is strategic to invest in a sustained, systematic research and development effort to harvest its benefits.

List of Symbols

c_i	=	reactant concentration of the component i, mol m^{-3}
e^-	=	electron
E	=	voltage, V
ΔE	=	polarization, V
F	=	Faraday's constant, C mol^{-1}
J	=	current density, A m^{-2}
R	=	gas constant, J mol^{-1} K^{-1}
T	=	temperature, K

Greek Letters

α_i and β_i	=	reaction rate orders related to the component
α_e and β_e	=	reaction rate orders related to e$^-$ = transfer coefficients
γ	=	thermodynamic consistency parameter in Eq. (3.11)
η	=	dimensionless polarization
ν_i	=	stoichiometric coefficient of component i
ν_e	=	stoichiometric coefficient of e$^-$

Subscripts

b = bulk
C = related to concentration polarization
e = electron
L = limiting value
s = electrode surface
T = thermodynamic

References

1. EG&G Technical Services (2004) *Fuel Cell Handbook*, 7th edn, US Department of Energy, Office of Fossil Energy, National Energy Technology Laboratory, Morganstown, WV.
2. Sundmacher, K., Kienle, A., Pesch, H.J., Berndt, J.F., and Huppmann, G. (eds.) (2007) *Molten Carbonate Fuel Cells: Modeling, Analysis, Simulation, and Control*, Wiley-VCH Verlag GmbH, Weinheim.
3. Yoshiba, F., Izaki, Y., and Watanabe, T. (2004) Wide range load controllable MCFC cycle with pressure swing operation. *J. Power Sources*, **137** (2), 196–205.
4. Baranak, M. and Atakül, H. (2007) A basic model for analysis of molten carbonate fuel cell behavior. *J. Power Sources*, **172** (2), 831–839.
5. Yoshiba, F. (2008) Test results and efficiency estimation of 1.2 MPa pressurized MCFC module. *J. Fuel Cell Sci. Technol.*, **5** (2), 021011/1–021011/12.
6. Bosio, B., Marra, D., and Arato, E. (2010) Thermal management of the molten carbonate fuel cell plane. *J. Power Sources*, **195** (15), 4826–4834.
7. Arato, E., Bosio, B., Massa, R., and Parodi, F. (2000) Optimisation of the cell shape for industrial MCFC stacks. *J. Power Sources*, **86** (1–2), 302–308.
8. Perez, F.J., Hierro, M.P., Duday, D., Gomez, C., Romero, M., and Daza, L. (2000) Hot-corrosion studies of separator plates of AISI-310 stainless steels in molten-carbonate fuel cells. *Oxid. Met.*, **53** (3–4), 376–398.
9. Huijsmans, J.P.P., Kraaij, G.J., Makkus, R.C., Rietveld, G., Sitters, E.F., and Reijers, H.T.J. (2000) An analysis of endurance issues for MCFC. *J. Power Sources*, **86** (1–2), 117–121.
10. Bischoff, M., Farooque, M., Satou, S., and Torazza, A. (2010) MCFC fuel cell systems, in *Handbook of Fuel Cells*, vol. 4, Part 12, 1260–1275 (eds. W. Vielstich, A. Lamm, and H.A. Gasteiger), John Wiley & Sons, Ltd., Chichester.
11. Arato, E., Bosio, B., Massa, R., and Parodi, F. (2000) Optimisation of the cell shape for industrial MCFC stacks. *J. Power Sources*, **86** (1–2), 302–308.
12. Verda, V. and Sciacovelli, A., (2011) Design improvement of circular molten carbonate fuel cell stack through CFD analysis. *Appl. Thermal Eng.*, **31** (14–15), 2740–2748.
13. Kim, Y.J., Chang, I.G., Lee, T.W., and Chung, M.K. (2010) Effects of relative gas flow direction in the anode and cathode on the performance characteristics of a molten carbonate fuel cell. *Fuel*, **89** (5), 1019–1028.
14. Yoshiba, F., Ono, N., Izaki, Y., Watanabe, T., and Abe, T. (1998) Numerical analyses of the internal conditions of a molten carbonate fuel cell stack: comparison of stack performances for various gas flow types. *J. Power Sources*, **71** (1–2), 328–336.
15. Boden, A., Yoshikawa, M., and Lindbergh, G. (2007) The solubility of Ni in molten $Li_2CO_3-Na_2CO_3$ (52/48) in $H_2/H_2O/CO_2$ atmosphere. *J. Power Sources*, **166** (1), 59–63.
16. Urbani, F., Freni, S., Galvagno, A., and Chiodo, V. (2011) MCFC integrated system in a biodiesel production process. *J. Power Sources*, **196** (5), 2691–2698.
17. Devianto, H., Li, Z., Yoon, S.P., Han, J., Nam, S.-W., Lim, T.-H., and Lee, H.-I. (2009) The effect of water on direct ethanol molten carbonate fuel cell, *Catal. Today*, **146** (1–2), 2–8.

18. Schaefer, A., Farrauto, R., Schwab, E., and Urtel, H. (2010) Hydrocarbon reforming catalysts and new reactor designs for compact hydrogen generators. *DGMK Tagungsber.*, **3**, 89–95.
19. Silveira, J.L., Caetano de Souza, A.C., and Evaristo da Silva, M. (2006) Thermodynamic analysis of direct steam reforming of ethanol in molten carbonate fuel cell. *J. Fuel Cell Sci. Technol.*, **3**, 346–350.
20. Frusteri, F. and Freni, S. (2007) Bio-ethanol, a suitable fuel to produce hydrogen for a molten carbonate fuel cell. *J. Power Sources*, **173** (1), 200–209.
21. Yan, W., Sui, S., Xiu, G., and He, Y. (2009) Combination of sorption-enhanced steam methane reforming and electricity generation by MCFC: concept and numerical simulation analysis. *Sep. Sci. Technol.*, **44** (13), 3013–3044.
22. Khaodee, W., Wongsakulphasatch, S., Kiatkittipong, W., Arpornwichanop, A., Laosiripojana, N., and Assabumrungrat, S. (2011) Selection of appropriate primary fuel for hydrogen production for different fuel cell types: comparison between decomposition and steam reforming. *Int. J. Hydrogen Energy*, **36** (13), 7696–7706.
23. Greppi, P., Bosio, B., and Arato, E. (2008) A steady-state simulation tool for MCFC systems suitable for on-line applications. *Int. J. Hydrogen Energy*, **33** (21), 6327–6338.
24. McPhail, S., Moreno, A., and Bove, R. (2009) International Status of Molten Carbonate Fuel Cell (MCFC) Technology. ENEA Report RSE/2009/181, ENEA, Rome.
25. Liu, S.-F., Chu, H.-S., and Yuan, P. (2006) Effect of inlet flow maldistribution on the thermal and electrical performance of a molten carbonate fuel cell unit. *J. Power Sources*, **161** (2), 1030–1040.
26. Zhang, H., Lin, G., and Chen, J. (2011) Performance analysis and multi-objective optimization of a new molten carbonate fuel cell system. *Int. J. Hydrogen Energy*, **36** (6), 4015–4021.
27. Fuel Cell Institute, Shanghai Jiaotong University (2010), *http://fuelcell.sjtu.edu.cn/CN/showzy.asp?id=30* (last accessed 18 November 2011).
28. Yu, L.J., Ren, G.P., and Jiang, X.M. (2008) Experimental and analytical investigation of molten carbonate fuel cell stack. *Energy Convers. Manage.*, **49** (4), 873–879.
29. Bosio, B., Costamagna, P., Parodi, F., and Passalacqua, B. (1998) Industrial experience on the development of the molten carbonate fuel cell technology. *J. Power Sources*, **74** (2), 175–187.
30. Bosio, B. and Arato, E. (2005) Molten carbonate fuel cell system simulation tools, in *Advances in Fuel Cells* (ed. X.W. Zhang), Research Signpost, Trivandrum, pp. 277–290.
31. Parodi, F., Bosio, B., and Arato, E. (2005) Method and system of operating molten carbonate fuel cells. European Patent EP 1834371, filed 1 April 2005, granted 28 July 2010.
32. Bosio, B., Arato, E., and Costa, P. (2003) Concentration polarisation in heterogeneous electrochemical reactions: a consistent kinetic evaluation and its application to molten carbonate fuel cells. *J. Power Sources*, **115** (2), 189–193.
33. Watanabe, T., Izaki, Y., Mugikura, Y., Morita, H., Yoshikawa, M., Kawase, M., Fumihiko Yoshiba, F., and Asano, K. (2006) Applicability of molten carbonate fuel cells to various fuels. *J. Power Sources*, **160** (2), 868–871.
34. Bove, R. and Lunghi, P. (2005) Experimental comparison of MCFC performance using three different biogas types and methane. *J. Power Sources*, **145** (2), 588–593.
35. Iaquaniello, G. and Mangiapane, A., (2006) Integration of biomass gasification with MCFC. *Int. J. Hydrogen Energy*, **31** (2), 399–404.
36. Zaza, F., Paoletti, C., LoPresti, R., Simonetti, E., and Pasquali, M. (2011) Multiple regression analysis of hydrogen sulphide poisoning in molten carbonate fuel cells used for waste-to-energy conversions. *Int. J. Hydrogen Energy* **36** (13), 8119–8125.
37. Ciccoli, R., Cigolotti, V., Lo Presti, R., Massi, E., McPhail, S.J., Monteleone, G., Moreno, A. Naticchioni, V., Paoletti, C.,

Simonetti, E., and Zaza, F. (2010) Molten carbonate fuel cells fed with biogas: combating H_2S, *Waste Manage.*, **30** (6), 1018–1024.

38. Zaza F., Paoletti C., Lo Presti R., Simonetti E., and Pasquali M. (2010) Studies on sulfur poisoning and development of advanced anodic materials for waste-to-energy fuel cells applications. *J. Power Sources*, **195** (13), 4043–4050.

39. Urban, W., Lohmann, H., and Salazar Gomez, J.I. (2009) Catalytically upgraded landfill gas as a cost-effective alternative for fuel cells. *J. Power Sources*, **193** (1), 359–366.

40. Cigolotti, V., Massi, E., Moreno, A., Polettini, A., and Reale, F. (2008) Biofuels as opportunity for MCFC niche market application. *Int. J. Hydrogen Energy*, **33** (12), 2999–3003.

41. Weaver, D. and Winnik, J., (1989) Sulfation of molten carbonate fuel cell anode. *J. Electrochem. Soc.*, **136** (6), 1679–1686.

42. Vogel, W.M. and Smith, S.W. (1982) The effect of sulfur on the anodic H_2 (Ni) electrode in fused $Li_2CO_3–K_2CO_3$ at 650 °C. *J. Electrochem. Soc.*, **129** (7), 1441–1445.

43. Townley, D., Winnick, J., and Huang, H.S. (1980) Mixed potential analysis of sulphuration of molten carbonate fuel cells. *J. Electrochem. Soc.*, **127** (5), 1104–1106.

44. Kawase, M., Mugikura, Y., and Watanabe, T. (2000) The effects of H_2S on electrolyte distribution and cell performance in the molten carbonate fuel cell. *J. Electrochem. Soc.*, **147** (4), 1240–1244.

45. Uchida, I., Ohuchi, S., and Nishina T. (1994) Kinetic studies of the effects of H_2S impurity on hydrogen oxidation in molten $(Li + K)CO_3$. *J. Electroanal. Chem.*, **369** (1–2), 161–168.

46. Sammells, A.F., Nicholson, S.B., and Ang, P.G.P. (1980) Development of sulphur-tolerant components for the molten carbonate fuel cell. *J. Electrochem. Soc.*, **127** (2), 350–357.

47. Ronchetti, M. and Iacobazzi, A. (2002) Celle a Combustibile-Stato Dello Sviluppo e Prospettive Della Tecnologia, ENEA, Rome, p. 87.

48. Dayton, D.C. (2001) Fuel Cell Integration: a Study of the Impacts of Gas Quality and Impurities. Milestone Completion Report, National Renewable Energy Laboratory, Golden, CO.

49. Campanari, S., Iora, P., Silva, P., and Macchi, E. (2007) Thermodynamic analysis of integrated molten carbon fuel cell–gas turbine cycles source for sub-MW and multi-MW scale power generation. *J. Fuel Cell Sci. Technol.*, **4** (3), 308–316.

50. Mugitani, N. (2003) Development of 300 kW class MCFC compact system. Presented at Fuel Cell Seminar, Miami Beach, FL.

51. Hishinuma, Y. and Kunikata, M. (1997) Molten carbonate fuel cell power generation systems. *Energy Convers. Manage.* **38** (10–13), 1237–1247.

52. Ken'ichi, M., Masaaki, T., Toshio, I., and Yoshihiro, A. (1999) Development of 1000 kW molten carbonate fuel cell (MCFC) pilot plant and 250 kW stack. *IHI Eng. Rev.*, 32. (2), 53–61.

53. Ishikawa, T. and Yasue, H. (2000) Start-up, testing and operation of 1000 kW class MCFC power plant. *J. Power Sources*, **86** (1–2), 45–150.

54. He, W. (1998) Dynamic model for molten carbonate fuel-cell power-generation systems. *Energy Convers. Manage.*, **39** (8), 775–783.

55. Elisangela, M. and Silveira, J.L. (2002) Study of fuel cell cogeneration system applied to a dairy industry. *J. Power Sources*, **106** (1–2), 102–108.

56. Ryu, B.H., Moon, H., Chang, I.G., Lee, G.P., and Lee, T.W. (2011) MCFC development status in Korea. Presented at International Workshop on Molten Carbonates and Related Topics, Chimie-ParisTech, Paris, 21–22 March 2011.

57. European Commission (2010) European Union Climate and Energy Package: Renewable Energy Directive (2009/28/EC), Emissions Trading Directive (2009/29/EC) and Directive on the Geological Storage of Carbon Dioxide (2009/31/EC), http://ec.europa.eu/clima/policies/package/index_en.htm (last accessed 20 November 2011).

58. Verda, V. and Nicolin, F. (2010) Thermodynamic and economic optimization of a MCFC-based hybrid system for the combined production of electricity and hydrogen. *Int. J. Hydrogen Energy*, **35** (2), 794–806.
59. Han, J., Elgowainy, A., and Wang, M. (2010) Fuel-cycle analysis of fuel cells for combined heat, hydrogen, and power generation. In Proceedings of the ASME International Mechanical Engineering Congress and Exposition, Lake Buena Vista, FL, 13–19 November 2009, Vol. 6, pp. 367–375.
60. Hemmes, K., Houwing, M., and Woudstra, N. (2010) Modeling of a methane fuelled direct carbon fuel cell system. *J. Fuel Cell Sci. Technol.*, **7** (6), 061008 / 1–061008/6.
61. Kivisaari, Y., Björnbom, P., and Sylwan, C. (2002) Studies of natural gas and biomass fuelled MCFC systems. *J. Power Sources*, **104** (1), 115–124.
62. Morita, H., Yoshiba, F., Woudstra, N., Hemmes, K., and Spliethoff, H. (2004) Feasibility study of wood biomass gasification/molten carbonate fuel cell power system – comparative characterization of fuel cell and gas turbine systems. *J. Power Sources*, **138** (1–2), 31–40.
63. Orecchini, F., Bocci, E., and di Carlo, A. (2008) Process simulation of a neutral emission plant using chestnut's coppice gasification and molten carbonate fuel cells. *J. Fuel Cell Sci. Technol.*, **5** (2), 021015 / 1–021015/9.
64. Greppi, P., Bosio, B., and Arato, E. (2009) Feasibility of the integration of a molten carbonate fuel-cell system and an integrated gasification combined cycle. *Int. J. Hydrogen Energy*, **34** (20), 8664–8669.
65. Bensaid, S., Specchia, S., Federici, F., Saracco, G., and Specchia, V. (2009) MCFC-based marine APU: comparison between conventional ATR and cracking coupled with SR integrated inside the stack pressurized vessel. *Int. J. Hydrogen Energy*, **34** (4), 2026–2042.
66. Daly, J., Steinfeld, G., Moyer, D.K., and Holcomb, F.H. (2007) Molten carbonate fuel cell operation with dual fuel flexibility. *J. Power Sources*, **173** (2), 925–934.
67. Farooque, M., Leo, A., and Venkataraman, R. (2011) Emergence of the stationary DFC power plants. Presented at International Workshop on Molten Carbonates and Related Topics, Chimie-ParisTech, Paris, 21–22 March 2011.
68. Yoshiba, F. (2007) Thermal efficiency of a high performance CO_2 recovery IG/MCFC system (in Japanese). In The Japan Society of Mechanical Engineers (JSME) Annual Meeting, (3), pp. 75–76, *http://ci.nii.ac.jp/naid/110007085578* (last accessed 20 November 2011).

4
Alkaline Fuel Cells
Erich Gülzow

4.1
Historical Introduction and Principle

The history of alkaline fuel cells (AFCs) is a story of discoveries in basic science posing a challenge to applied research and technology development, a story of 100 years of experiments and false starts that finally led to the first practical fuel cell in the 1930s [1].

> The alkaline fuel cell (AFC), also known as the Bacon fuel cell after its British inventor, is one of the most developed fuel-cell technologies and is the cell that flew Man to the Moon. NASA has used alkaline fuel cells since the mid-1960s, in Apollo-series missions and on the Space Shuttle. [2]

The main hurdle for the commercialization of fuel cells is still their high cost. Tremendous progress and innovations have been realized through the decades, but full mass-market commercialization has not been realized.

In 1839, the English lawyer, judge, and physical scientist William R. Grove (1811–1896) [3, 4] described the operation of the fuel cell. His invention is closely related to the work of the German chemist Christian Friedrich Schönbein (1799–1868) [5], who described the principle of fuel cells for the first time and who was in contact with Grove. Because of their close interaction, both scientists should be recognized as the discoverers of the full cell.

These scientists investigated the reaction in acidic electrolytes, and the first AFCs were described by Reid in 1902 [6] and by Noel [7] in 1904.

The inspiration for the development of fuel cells was the aim to enter the market. Therefore, it was necessary to increase the performance of the early fuel cells to reach technically relevant values. Consequently, the challenge was to improve the properties of the materials or find new alternative materials that performed better, had longer lifetimes, and had a less negative environmental impact. The materials, for example, the catalysts and the electrode structures, were developed intensively, and a few important milestones are described below.

Fuel Cell Science and Engineering: Materials, Processes, Systems and Technology,
First Edition. Edited by Detlef Stolten and Bernd Emonts.
© 2012 Wiley-VCH Verlag GmbH & Co. KGaA. Published 2012 by Wiley-VCH Verlag GmbH & Co. KGaA.

In 1923, Schmid [8, 9] improved fuel-cells electrode, and Scharf [10] developed the hydrogen diffusion electrode. A major advance in alkaline electrochemistry was achieved by Raney [11], who developed the well-known Raney nickel (high surface-area Ni) catalyst. With this high surface-area electrode, it was possible to achieve a technically relevant performance with a simple catalyst material.

In the subsequent decades, scientists integrated fuel cells into larger systems. Foremost among them was Bacon [12–14] at Cambridge, who developed the Allis Chalmers tractor (1959, USA) with a 15 kW fuel cell.

In Germany, Justi and Winsel advanced the AFC technology during the 1950s and 1960s [15]. They invented the so-called double-skeleton catalyst electrode (DSK) (with a bimodal pore structure), an important milestone for terrestrial fuel-cell development in Europe. The subsequent development by Siemens was based on these technological advancements. Winsel and Sauer introduced bifunctional rolled electrodes 20 years later. In the 1960s, Kordesch (in Austria) [16] developed AFCs and built demonstration vehicles, including bikes and cars, powered by fuel cells. With these early demonstration projects, in particular with the fuel-cell-powered car, Kordesch showed the potential of this technology for demanding applications.

In the 1960s and 1970s, space technology became important in Russia and in the USA. A Belgian company, Elenco, developed a city bus with a 78 kW fuel cell in 1994. Additionally, Siemens developed a military submarine in the 1990s with a fuel cell of ~ 100 kW.

The AFCs are realized by different designs and applications that have various requirements. The materials used for AFCs depend strongly on the AFC design type. Therefore, the materials required for a classic AFC are significantly different from those employed for anion-exchange membrane (AEM)-containing fuel cells or direct methanol AFCs. Because of the function of the components, defined structural and chemical properties have to be realized; for example, the electrodes have to be highly electrocatalytically active and need to exhibit superior transport properties. Consequently, some of the components consist of different materials, including composites of two or more components, required to fulfill all of the functional requirements.

During the years, many problems have been solved, and many technical issues have been described. In general, AFCs are fairly easy to operate, have very high electrical efficiencies, and are extremely suitable for dynamic operating modes. They can be built into small, compact systems and large power plants. They can be designed without noble metals and thus at a low production cost. However, many European researchers have stopped working in the field because of the higher performance obtained with polymer electrolyte fuel cells (PEFCs) and solid oxide fuel cells (SOFCs).

This chapter provides an overview of the techniques and activities and also some of the misconceptions of AFCs. In particular, the differences between AFCs and PEFCs are discussed.

AFCs have the advantage of exhibiting high theoretical electrical efficiencies. These fuel cells are mostly used with highly pure gases and are focused on space and military applications that can support the additional costs of pure fuels and

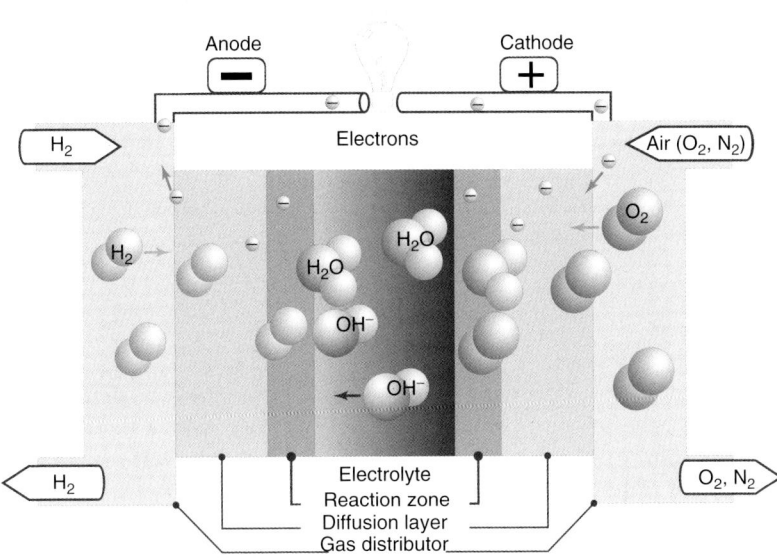

Figure 4.1 Scheme of an alkaline fuel cell (AFC).

pure oxidizers. The KOH electrolyte, used in AFCs with a standard concentration of ~30–45 wt%, has an advantage over acidic electrolytes because the oxygen reduction kinetics are much faster with alkaline than with acidic electrolytes, making the AFC a very attractive system for specific applications. The AFC was one of the first fuel cells to be used in space; it was used in the Apollo missions and the Space Shuttle program, and it was planned for use in the European Hermes Project.

In alkaline media, the relevant reactions are based on OH^- ions. The OH^- species formed by the cathodic reduction of oxygen travel though the electrolyte solution to the anode, where they recombine with hydrogen to produce water (Figure 4.1). The overall main reactions at the electrodes are the following:

$$\text{Cathode:} \quad \frac{1}{2}O_2 + H_2O + 2e^- \rightarrow 2OH^- \quad (4.1)$$

$$\text{Anode:} \quad H_2 + 2OH^- \rightarrow 2H_2O + 2e^- \quad (4.2)$$

4.2 Concepts of Alkaline Fuel-Cell Design Concepts

The stack design of AFCs is different from the most common designs of PEFCs or SOFCs. The stack design is similar to that of standard primary batteries. AFCs often have a liquid alkaline electrolyte in two cell chambers that are divided by electrodes or separators. Sometimes, the AFC also has three chamber cells. However, two compartments are required for the hydrogen and oxygen (air) reactants.

The operating temperature of the fuel cell depends on the application and the load dynamics. AFCs can be operated at up to 230 °C depending on the design and the heat management of the cell. New developments in AFCs are mainly based on AEMs that can be used to immobilize the electrolyte such as in the proton exchange membrane of a PEFC. The design of these new cells is similar to the PEFC stack design; however, as described later, there are some disadvantages, including the limited stability and lifetime.

4.2.1
Traditional Stacks

The traditional fuel-cell stack design has three chambers. The first compartment is for the fuel, the second is for the oxidizing reactant, and the third is filled with the electrolyte and sometimes contains additional materials that prevent short-circuiting of the cell. The electrochemical energy carrier can be pure hydrogen or a gas with a high hydrogen content, including reformers or liquids such as ammonia and hydrazine. The barriers between the cell chambers are typically electrodes with a bubble point pressure such as gas diffusion electrodes (GDEs) with small pores.

The cooling system is combined with the electrolyte loop. Hence there is no need for a separate cooling system such as in the PEFC. However, the direct contact between the electrolyte and the gas chambers allows the water (the reaction product) to dilute the electrolyte. Therefore, to prevent a low ion conductivity of the electrolyte, re-concentration of the electrolyte is necessary to maintain a constant KOH concentration (because the conductivity is decreased by increasing the water flow to the electrolyte).

4.2.2
Eloflux Cell Design

Standard galvanic cells are independently sealed, whereas the Eloflux cell is completely different. The main difference is that one electrode is used for transporting both the gas and the electrolyte. Therefore, this cell design reduces costs by simplification of the cell components. There are fewer parts with less degradation and corrosion. Eloflux was described in 1965 by Winsel and Wendtland [17–20].

The Eloflux principle is based on flexible rolled electrodes and a liquid electrolyte. The electrodes are extremely porous, containing bifunctional transport systems with hydrophobic and hydrophilic pores that allow the transport of gases and electrolytes through a thin separator. In contrast to other cell designs, the electrolyte is transported through the separator and the electrodes in a direction perpendicular to the electrode plane. Therefore, the electrodes and the diaphragms contain an interconnecting system of narrow pores in which the electrolyte is dispersed. The larger pores are more hydrophobic because of the polytetrafluoroethylene (PTFE) coating on the electrodes and are filled with hydrogen or oxygen. Because of the different pore sizes, the gases do not enter the narrow pores that are filled with electrolyte; for gas to enter narrow hydrophilic pores, significantly higher pressures are required. Therefore, the gas transport in these pores is independent of the electrolyte and ion transport. However, the electrolyte is diluted by the water produced, and the water is therefore removed from the circulated electrolyte using an external water vaporizer or a dialysis concentrator. The electrolyte loop is similar to that of other AFCs and functions as a cooling system.

The Eloflux system contains resins instead of plates, with integrated seals to seal the stack. The KOH electrolyte is aggressive and therefore it is difficult to seal the stack effectively. However, the application of resins is an efficient method for sealing the system. The only disadvantage of this resin-embedded stack is that the contacts for the current must be inserted in the resin.

4.2.3
Falling Film Cell

An unusual type of fuel-cell stack design is the falling film fuel cell. The first falling film process based on chlor-alkali electrolysis was developed by Höchst in 1981 to save electrical power [21]. The goal was to be less limited by the active area of the electrode. Traditional electrodes are limited in size because of the hydrostatic pressure that becomes heterogeneous for large sizes. For automotive or portable applications, the size of the traditional design is acceptable and suitable because a high volumetric power density is an important requirement. However, for stationary applications, larger cells have a significant advantage because they cost less.

In the completely differently designed falling film fuel cell, the electrolyte flows from the top to the bottom of the cell. The hydrostatic pressure is compensated by an equally large but oppositely oriented hydrodynamic pressure gradient [22]. The pressure difference between the electrolyte on the front side of the electrode and gas on the back side of the electrode remains constant over the entire area of the vertical electrode. Hence it is possible to construct a large-area cell, which is important for cells without a KOH-tight diaphragm. The gap between the electrodes is 0.5 mm, and the cell therefore operates at very high current densities (>25 kA m^{-2}). The cost of this new type of stack is fairly low, and the power density is very high [23].

The largest falling film cell is 0.25 m wide and 1 m tall and has been tested as an electrolyzer. Since 1990, GDEs have been successfully used in falling film cells.

This example demonstrates that the careful design of the stack for a specific application leads to significant improvements of AFCs and all other fuel-cell technologies.

4.2.4
Bipolar Stack Concept by DLR

AFC stacks are frequently designed with a monopolar concept for which single cells are externally connected. For PEFCs, a bipolar configuration is commonly employed, and the monopolar concept is only used for specific low-power applications or for special applications that require a single-cell *in operando* replacement.

Bipolar designs have also been used for AFCs. For the Apollo missions, General Electric developed a bipolar AFC operated at 200–230 °C with 85 wt% KOH as the electrolyte, and KOH was heated above 100 °C to achieve a sufficiently high ion conductivity. However, in the Apollo cells, the electrolyte was immobilized in a matrix material because of the high vibrational stability that was required. Therefore, this system could not include an electrolyte loop as a cooling system. Bipolar AFC stacks were also used in other space missions, for example, the Space Shuttle program and the Russian Buran program (as part of the Foton system). All of the bipolar AFC systems developed for space applications contain an immobilized electrolyte to reduce the influence of micro-gravitation and orientation.

A bipolar configuration similar to the PEFC design but for liquid electrolytes was developed by the German Aerospace Center (Deutsches Zentrum für Luft- und Raumfahrt, DLR) [24, 25].

The DLR bipolar stack consists of two endplates, the bipolar plates, the electrodes, the diaphragms, and the sealing. Figure 4.2 shows a scheme of the DLR AFC stack. The endplates integrate the gas supply and the electrolyte circulation, and also the connectors for the electrical wires. Additionally, the stack endplates apply a homogeneous contact pressure on the entire stack to obtain low a resistance between each component of the cell. The bipolar plates distribute the gases in each cell with appropriate flow fields and integrate an internal manifold for the individual cells.

Because of the greater thickness of the typically used AFC electrodes and the resulting higher gas diffusion layers inside the electrodes, the complexity of optimizing the flow fields for AFCs is significantly lower than that for PEFCs. The only function of the diaphragms between the electrodes is the avoidance of the direct contact that leads to a short-circuit. This new kind of bipolar stack was developed and tested at DLR and showed significant volumetric and gravimetric improvements compared with monopolar configurations. The parasitic current between each single cell must be monitored to reduce power loss and to prevent stack damage. However, the parasitic current can be reduced significantly with optimized designs.

4.2.5
Hydrocell Concept

An example of a very large-area electrode is the Hydrocell. Hydrocell is a company [26] in Finland that develops portable fuel cells. This type of fuel cell has a large cylindrical electrode outside the stack for air breathing; therefore, this system consists of a single cell with a large active area and an adapted DC/DC conversion. To increase the power output, various small systems can be connected. The fuel cells developed by Hydrocell are based on a cylindrical electrode. This electrolyte is a gel (developed by Hydrocell) that prevents leakage of the alkaline solution. The advantage of gel electrolytes over liquid electrolytes is that the fuel cell can be

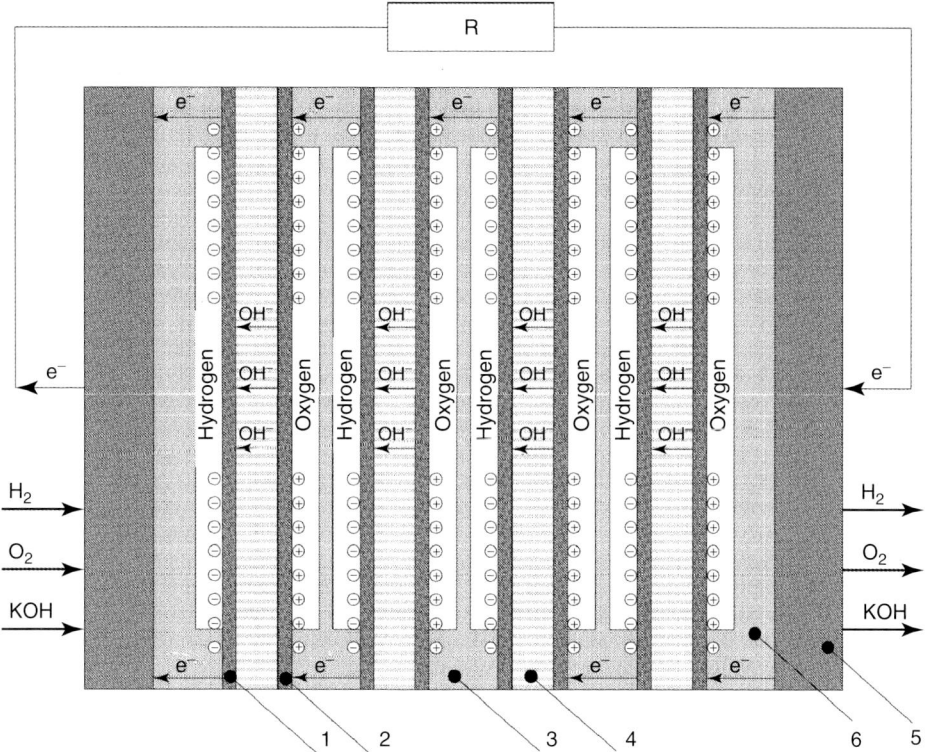

Figure 4.2 Scheme of a bipolar-configured AFC: 1, anode; 2, cathode; 3, bipolar plate; 4, electrolyte with spacer (diaphragm); 5, endplate; 6, monopolar plate. Source: DLR.

operated independently of its orientation, thus rendering it highly portable. For this fuel cell, hydride storage with 40 l of hydrogen can be integrated into the design [27].

4.2.6
Ovonics Concept

To obtain a rapid start-up device, Ovonics (USA) developed an AFC design that combines the fuel cell and the fuel storage material. The hydride for hydrogen storage is locally combined directly with the catalysts on the anodic side, and a non-noble metal oxide is located on the cathode side [28–30]. The hydride storage material is integrated into the anode; therefore, this fuel cell can be also operated like a nickel hydride accumulator that allows ultrafast start-up at temperatures below $-20\,°C$. In addition, these cells can store electrical energy by loading the anode with hydrogen produced during the electrolysis. These AFCs may be useful for different applications such as stationary, portable, and transport power supply units. The applications of these fuel cells range from military, telecom, and automotive equipment to auxiliary and uninterruptible power supplies.

4.2.7
Stack Design with Anion-Exchange Membranes

The most recent AFC design uses AEMs. Because of this solid-state electrolyte membrane, stack designs based on AEMs are similar to the stack designs for PEFCs. The electrolyte loop is eliminated, which simplifies the cell design from a systems perspective, and the cooling is instead similar to that of PEFCs. The AFC can still be credited with a reduced need for expensive noble metals.

At present, only a few materials are available for AEMs (stability requirements are demanding); therefore, this type of stack has been used in laboratories but is not yet commercially available.

One recent example is notable for its innovative application and has already been mentioned here. A new stack concept from the University of Applied Science in Offenburg (Germany), included a new catalyst from the Italian company ACTA and an AEM membrane from the Japanese membrane producer Tokuyama. This consortium developed an AEM-based ethanol-powered AFC and successfully inserted it into an electric vehicle. This stack was not able to work for an extended time because the system was built without an ethanol loop and without a KOH re-concentrator, but it is an impressive demonstration of AEM technology and the direct use of ethanol. DLR together with the University of Düsseldorf are working with laboratory-scale fuel cells to understand the ethanol oxidation mechanism in detail.

4.2.8
Alkaline Direct Ethanol Fuel Cells Assembled with a Non-Platinum Catalyst

To combine the advantages of AFCs with the benefits of PEFCs, AEMs are developed and used specifically for direct ethanol fuel cells. Direct ethanol fuel cells are considered attractive electrical power sources because they promise clean energy from non-hazardous, widely available, and inexpensive biofuels using an existing infrastructure (e.g., in Brazil).

However, poor performances and reduced lifetimes need to be overcome for the commercialization of adapted PEFCs. Alkaline direct ethanol fuel cells are favored because they feature an increased performance and allow the use of cheaper platinum-free catalysts.

Acetate is often the main product of the ethanol oxidation reaction. However, for concentrated alkaline solutions, side products are formed, including polymerized acetaldehyde and carbonate. Polyacetaldehyde is regarded as an unwanted by-product because it coats and blocks the catalyst layer. The complete oxidation of ethanol to carbonate is highly desired because it doubles the number of electrons per reagent molecule. The catalytic cleavage of the C–C bond in aqueous media is a challenging task in catalyst research. At DLR, this is a research topic in collaboration with the University of Düsseldorf [31, 32].

The setup consists of a technically oriented single cell with an AEM (Tokuyama) between two Hypermec electrodes (Acta SpA, Italy) with an active area of 50 cm^2.

The fuel cell was characterized for different alkali concentrations and temperatures by current–voltage measurements [$V(i)$ or $E(i)$ curves], short-term operations under load, electrochemical impedance spectroscopy (EIS), and pH monitoring. The passage of ethanol through the membrane was determined by redox titration of the cathodic exhaust.

Power densities up to 70 mW cm^{-2} could be achieved, but insufficient long-term stability under load was observed. Furthermore, the $V(i)$ or $E(i)$ characteristics showed pronounced hysteresis at room temperature that indicated the formation of poisoning intermediates during the electrode reactions. The pH declined during the operation, and the EIS measurements showed an increased membrane resistance over time.

The reaction product outcome was investigated by Raman spectroscopy. Raman spectroscopy is a vibrational spectroscopic technique suitable to distinguish qualitatively and quantitatively between reaction species in aqueous media. This kind of investigation is only possible with a closed loop for the electrolyte and ethanol. Additionally, a flow-through cell was inserted into the anodic electrolyte cycle using a micro-Raman spectrometer to understand the reaction mechanism better.

The measurements showed that the passage of CO_2 from the cathodic air through the membrane leads to a detectable amount of carbonate. With the use of CO_2-free synthetic air, acetate was unambiguously identified as the main product of the reaction. Complete oxidation to carbonates could be excluded with certainty. Nevertheless, trace amounts of precursors of polymerized acetaldehyde, for example, crotonaldehyde, were detected.

Another main objective is to understand the effects of using CO_2-containing air with AEMs. Many groups worldwide are testing AEMs, but the role of carbonate inside the membrane during long-term operation is still not sufficiently understood.

For stationary applications, CO_2 should be avoided because the replacement of membranes is expensive or technically impossible.

4.2.8.1 Electrode Types

To develop a highly efficient AFC, an electrode structure with more than one pore system or with a bimodal pore distribution is required. The liquid and gaseous reactants must reach the electrolyte that has direct contact with the anode and cathode catalysts. For AFCs, different electrode types have been used. The main types of electrodes are the PTFE-bonded GDE and the DSK.

4.2.9
PTFE-Bonded Gas Diffusion Electrodes

PTFE-bonded GDEs can be manufactured by different methods. One method is preparation under wet conditions (using inks or paste materials) followed by drying and tempering steps. Another method is dry preparation, which is described below. The dry method avoids the use of solvents and lowers the energy consumption because the drying steps are not required. The layers are often thicker when using the dry procedure than with the wet procedure [33].

4.2.10
Double-Skeleton Electrodes

DSKs [34] contain various types of pores. Within the Raney nickel granules, there are pores with diameters in the 20–80 Å range that open large inner surfaces with high catalytic activities. Raney nickel consists of a hydrophilic inner pore system that is filled with the electrolyte even at very high gas pressures. Between the particles of the supporting skeleton structure there are pores with diameters of up to a few micrometers that are also filled with the electrolyte under the operating conditions. Only the coarse pores, preferably formed near the Raney nickel granules, are filled with gas.

It is important that the electrolyte-containing pores and the gas-containing pores are interconnected for the production of a three-phase reaction zone.

In hydrogen fuel cells, DSKs are used in their optimized form and are called Janus electrodes [35]. These electrodes consist of three layers. The middle layer is the working layer that contains the catalyst and the coarse, gas-filled pores. The outer layers are covering layers that contain fine electrolyte-filled pores that prevent the gas from escaping from the electrode. This construction is easier to handle because pressure variations do not strongly influence the stability of the three-phase reaction zone.

4.2.10.1 Preparation and Electrode Materials

Today, the most commonly used electrode is the PTFE-bonded GDE, and various procedures have been described for their production. The preparation procedure for rolled electrodes will be described in detail to illustrate the composition of PTFE-bonded gas electrodes. Rolled electrodes are composites of a metal catalyst and PTFE supported by a metal grid on the back side of the electrode. The metal grid is responsible for high electrical conductivity and also for mechanical support and stability.

PTFE is used as an organic binder to link the catalyst particles and to produce a multifunctional electrode. For GDEs, hydrophobic and hydrophilic pore systems are required. For porous catalysts, for example, the Raney nickel catalyst, hydrophilic materials allow the penetration of the electrolyte into the electrode and the transport of the ions between the reaction zones; the hydrophobic pore system – outside the catalyst – is required for the transport of gases to the reaction zones. In addition, to provide mechanical stability, the PTFE in the electrodes forms a hydrophobic pore network.

The structure can be compared to a composite of particles that are linked by PTFE fibers that cover each particle. PTFE yields a hydrophobic pore network that is not flooded by the alkaline electrolyte, and the electrode can be supplied with the fuel gas from the back side of the electrode.

Different catalysts are used for AFCs. For the hydrogen oxidation reaction, carbon-supported platinum and platinum–palladium catalysts (e.g., noble metal catalysts) are suitable. However, one of the advantages of the AFC compared with acid electrolyte fuel cells, including the phosphoric acid fuel cell (PAFC) and the

PEFC, is that non-noble metals are also stable in an alkaline environment. The greatest advantages of AFCs are that non-noble catalysts can be used and that an expensive and lifetime-reducing ion-conducting membrane can be avoided, thus drastically reducing the material cost.

Porous nickel is a frequently used catalyst for hydrogen oxidation with AFCs. Porous nickel is prepared from Raney nickel to obtain an inexpensive material with a large active surface area. The initial material is a nickel–aluminum alloy containing 50 wt% of each metal. In contact with the potassium hydroxide solution, the catalyst is activated by the dissolution of the aluminum at 350 K. After this procedure, the catalyst contains less than 5% of aluminum. During the activation process, the nickel is loaded with 0.5–1.2 hydrogen atoms per nickel atom [36], forming a pyrophoric nickel hydride. After the catalyst activation process, the powder is washed and dried to give an AFC catalyst with a high porosity and a large active surface. Because hydrogen is present in the metal and because of the high reactivity of the metallic nickel surface (pyrophoric character), contact between the metal and oxygen must be avoided.

Therefore, after drying, the catalyst powder has to be passivated by controlled oxygen exposure under temperature control. Vacuum chambers or special chemical reactors are required for this purpose. During the preparation step, the absorbed hydrogen is removed, and the catalyst surface is subsequently oxidized and will be reduced before electrochemical use.

Many cathode catalyst materials have been used. For noble metal catalysts, platinum was mainly used in fuel cells for space applications. For terrestrial use, one has to use less expensive materials, and non-noble metal catalysts are therefore mainly employed. Bacon used lithium-doped nickel oxide as a cathode catalyst for high-temperature AFCs. Lithium-doped nickel oxide has a sufficient electrical conductivity at temperatures above 150 °C. Currently, mainly Raney silver and pure silver catalysts are favored. Developments of silver-supported materials containing PTFE are sometimes successful. Silver catalysts are usually prepared from silver oxide, Raney silver, and supported silver. Typically, the catalysts on the cathode are supported by PTFE because it is highly stable under basic and acidic conditions. In contrast, carbon is oxidized at the cathode in contact with oxygen, when carbon is used as an inexpensive support material. In the past, the silver catalysts frequently contained mercury as part of an amalgam to increase the stability and the lifetime of the cathode. Because mercury is partially dissolved during the activation procedure (see below) and during the fuel-cell operation, some electrolyte contamination can be observed. Because of the environmental hazard of mercury, this metal is currently not used in silver catalysts.

In addition, oxide catalysts are common in cathodes, for example, the spinel $MnCo_2O_4$ and the perovskite $La_{0.1}Ca_{0.9}MnO_3$ catalysts [37] that are used in AFCs (including the Hydrocell AFC), the perovskite $La_{1-x}Sr_xMnO_3$ catalysts [36] that are useful for SOFC applications [38], and the MnO catalysts [39, 40] that are suitable for AFC cathodes. In addition, W_2C has been employed in AFC cathodes [41].

To achieve a high efficiency and high performance, it is necessary to obtain large three-phase reaction zones. As previously described, the use of additives to obtain a hydrophobic pore system for gas transport is needed. The three-phase reaction zone is formed by the co-occurrence of the electrolyte, the catalyst, and the gaseous reactant phases. PTFE is commonly selected as a hydrophobic agent because it is stable under extreme conditions. In addition, PTFE functions as an organic binder in the electrodes because it is elastic under mechanical stress. Therefore, PTFE fibers that cross the electrode will be formed during the rolling process of the electrode production. A PTFE powder is typically used in the dry preparation of PTFE-bonded electrodes in combination with reactive mixing, whereas a PTFE suspension is used in the wet preparation.

To adjust the pore system during the preparation process, a pore former is sometimes added. These pore formers must be eliminated after the electrode preparation, for example, by their dissolution in water or by their decomposition into gaseous components such as sodium or ammonium bicarbonate.

4.2.10.2 Dry Preparation of PTFE-Bonded Gas Diffusion Electrodes

The production methods for multifunctional GDEs based on Sauer [42] and Winsel [43] have been further developed for electrodes in AFCs [44–49]. The electrode preparation procedure starts with the reactive mixing of the raw materials in a high-speed powder mill under water cooling. Usually, the catalyst mixture is then inserted vertically with a calender (with a line funnel under the influence of gravity) and rolled to form a tape. Finally, the catalyst tape is pressed on to a metal grid from one or both sides. The metal grid improves the mechanical stability of the GDE and acts as a current collector to increase lateral conductivity. Monopolar AFCs are often designed with current collector at one point on the electrode, where it contacts the outside of different cells. A schematic of the preparation technique is given in Figure 4.3. This preparation technique can be fully automated, and it circumvents environmentally harmful steps and waste emission, thus allowing the continuous, environmentally friendly manufacture of low-cost GDEs. This preparation technique is partly used in the preparation of batteries, and the cost structure is well known.

The anodes consist mainly of a mixture of the Raney nickel catalyst, copper powder, and PTFE powder on a copper grid as described by Rühling [50]. The Raney nickel catalyst is composed of highly porous metal particles and is formed from an aluminum–nickel alloy by dissolving the aluminum. The advantages of using the Raney material are described above.

After reactive mixing of the different materials for the electrodes, the PTFE powder, the copper powder, and the oxidized nickel catalyst in a knife mill, the powder is rolled into a self-carrying strip using a calender. This self-carrying strip is then rolled on to a metal grid to form the electrode. The rolled electrode has a thickness of $\sim$100–350 μm.

The PTFE is smeared during this process and forms ultrathin PTFE fibers and a PTFE film that cover the oxidized catalyst (Figure 4.4).

Because of the PTFE film, only a low porosity can be measured at the surface of the rolled electrode, and the active surface is therefore too small for acceptable performance. The pore system and the specific surface area of the AFC anodes

4.2 Concepts of Alkaline Fuel-Cell Design Concepts

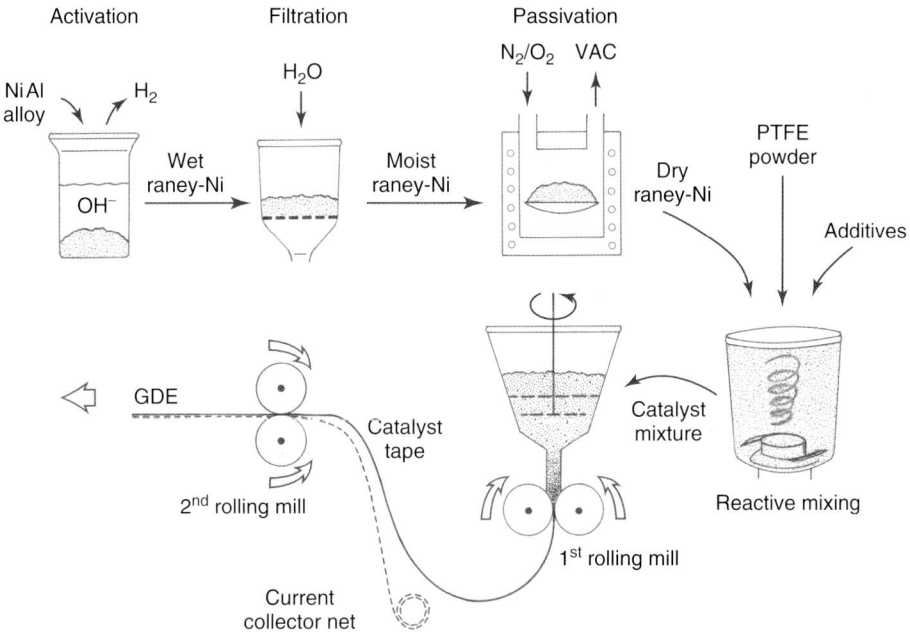

Figure 4.3 Preparation of gas diffusion electrodes (GDEs) with a Raney nickel catalyst VAC = vacuum. Source: DLR.

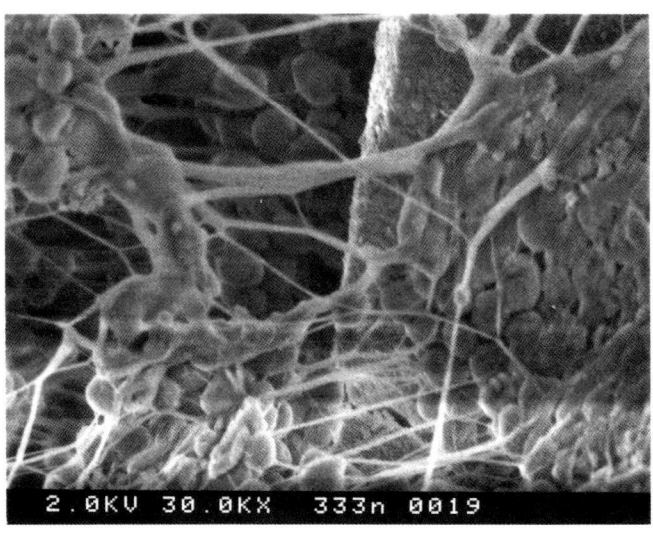

Figure 4.4 SEM image of a non-activated electrode recorded with an electron beam voltage of 2 kV. Source: DLR.

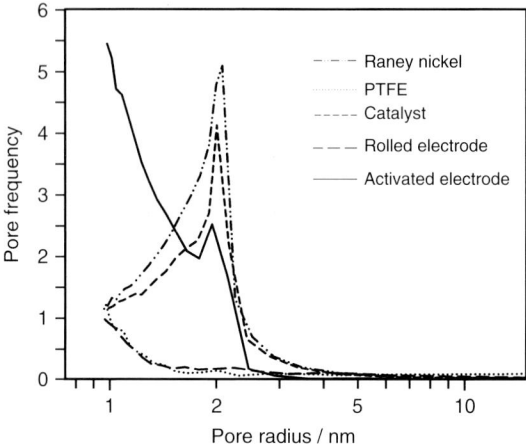

Figure 4.5 Pore frequency of the electrode components for different electrode preparation steps. Source: DLR.

are determined by the uncovered surface of the metallic catalyst. For the nickel catalyst, a pore system with a pore radius of 2 nm is recorded. PTFE has no surface according to nitrogen adsorption measurements. Because of the rolling process during the electrode preparation, the specific surface area of the electrode is only ~7 m^2 g^{-1} before the activation procedure, which is a significantly smaller area than that of the nickel catalyst. The surface of the catalyst is covered by a PTFE film that is smeared out when the electrode is rolled (Figure 4.5).

To prevent electrode burning during the production process, the Raney nickel is passivated by oxidation at a controlled temperature and pressure. To recover the catalytic activity and fuel-cell performance, the catalyst has to be reactivated.

The standard activation procedure is to reduce the nickel oxide electrochemically in 30% KOH for 18 h with the hydrogen generating at a current density of 5 mA cm^{-2} [50]. In addition to the electrochemical reduction conditions, hydrogen is exposed to the electrodes from the back side.

During the reactivation process, different alterations to the electrode composition and structure occur: the polymer changes, the nickel oxide is reduced to metallic nickel, and the residual aluminum from the Raney nickel is dissolved.

The specific surface area increases (measured by nitrogen adsorption) and the pore size increases (to 2 nm) for longer activation times (Figure 4.5). The pore system of the nickel catalyst is re-exposed by the activation process. Only a pore system with a pore radius of 2 nm can be observed. The specific surface area does not increase linearly with time. For activation times between 4.5 and 8.5 h at 5 mA cm^{-2}, a significant increase in the pore frequency and in the specific surface area from 21 to 53 m^2 g^{-1} are observed (Figure 4.6). During prolonged activation times, the specific surface area increases gradually from a low value of 64 m^2 g^{-1} after 18 h to 72 m^2 g^{-1} after 115 h.

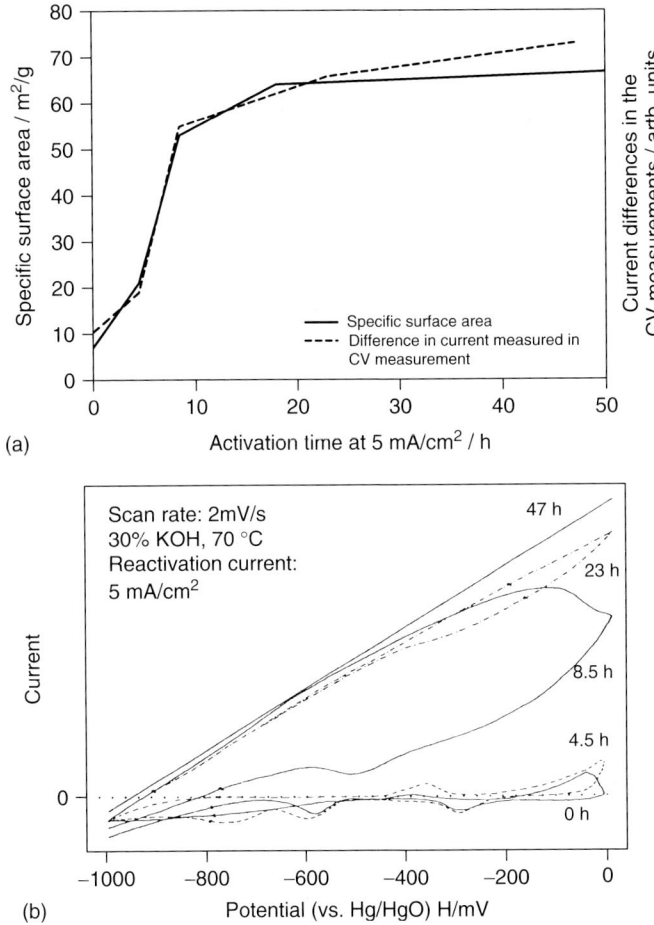

Figure 4.6 (a) Specific surface area measured by nitrogen adsorption (solid line) and (b) the current differences observed by cyclic voltammetry (CV) for different activation times at a current density of 5 mA cm^{-2} (dashed line). Source: DLR.

The increased specific surface area clearly indicates that the PTFE film covering the catalyst is removed during the activation process. The removal of the PTFE film at the anode is the rate-determining step in the activation procedure.

4.2.11
Reduction of NiO

To investigate the electrode activation procedure, X-ray photoelectron spectroscopy (XPS) is typically used. A controlled transfer from the solution to the ultrahigh vacuum chamber has to be realized with an appropriate setup to prevent the surface behavior of the samples from changing. Depth profiling by sputtering (e.g., with argon) was performed to obtain information about the surface composition. Nickel oxide is observed on the surface of the catalyst in the XPS spectra at the start of the depth profiling. This nickel oxide layer on the surface may be formed during the transfer process from the KOH electrolyte to the vacuum system. The interaction between nickel surfaces and water vapor has been described in detail [51, 52]. The oxidation state of the electrodes can be determined from the contributions of the metallic nickel and the nickel oxide in the XPS spectra recorded during the depth profiling.

The nickel catalyst surface is reduced after 4.5 h at a current density of 5 mA m^{-2}. The electrodes are not expected to be reduced after a short activation time because the catalysts are still significantly covered by the PTFE film, observed with XPS and specific surface area measurements. Perforations in the PTFE film are sufficient to allow contact between the electrolyte and the nickel catalyst and therefore to reduce the nickel oxide. During the activation process, the perforations in the PTFE film are increased or a new contact between the electrolyte and the metal catalyst is formed. The contact area may be created by the hydrogen that is formed on the nickel surface and that separates the PTFE film from the catalyst surface. Therefore, the initial nickel signal of the activated electrode (observed by XPS) is increased. The specific surface area increases without the complete removal of the PTFE film.

The hydrogen loading can be measured after various activation times using cyclic voltammetry (CV). No variance is observed in the electrochemical measurements using CV, indicating a nonlinear increased specific surface area during the activation procedure. The surface area measurements by gas adsorption and CV correlate well (Figure 4.6).

From the specific surface area measurements, it can be assumed that a large area of the catalyst is still covered by the PTFE film after short activation times. This film obstructs the contact area for the gas and also the access of the electrolyte to the catalyst. In conclusion, the activation process commences at the surface of the electrode, and the activation zone increases in depth with increasing activation time. Two effects cause this behavior. The electrolyte penetration into the electrode is hindered by the hydrophobic character of PTFE. With increasing PTFE damage and time, the electrolyte and the activation zone penetrate further into the electrode. The perforation in the PTFE film can enhance the electrolyte penetration into the electrode.

Energy-dispersive X-ray analysis at DLR did not reveal the oxidation state of the nickel because the oxygen signal overlaps with other signals from the electrode. Scanning electron microscopy (SEM) images of the activated electrodes show slight formation of cracks on the nickel catalyst, and these cracks are more significant after long-term operation. The cracks on the nickel may be due to the loading of the nickel with hydrogen during the activation and the electrochemical operation, a process known as hydrogen embrittlement of nickel [53, 54].

4.2.12
Production of Cathode Gas Diffusion Electrodes

Silver electrocatalysts are suitable high-performance catalysts for oxygen reduction in alkaline solutions. However, these catalysts are Raney like, meaning that a large amount of material is used. To reduce the amount of material, a supported catalyst concept employed for PEFC catalysts is suitable. To achieve high surface areas for porous GDEs, silver particles have to be dispersed in a porous matrix. Therefore, a preparation technique that encompasses reactive powder–polymer mixing and rolling (RMR) is often used, similar to the technique employed in the production of anode GDEs [55]. Porous PTFE-bonded GDEs consist of an electrocatalytic powder (e.g., silver oxide) and the organic binding agent PTFE within a metal-wire gauze that mechanically stabilizes the electrode and collects the current.

To show high performance, the silver oxide catalyst has to be activated by reducing it to the metallic form. The silver oxide reduction can be performed galvanostatically simultaneously with the oxygen reduction reaction (ORR) before use or *in situ* in the fuel cell. The following equations can be assumed for the reactions that occur:

$$Ag_2O + H_2O + 2e^- \rightarrow 2Ag + 2OH^-$$
$$E_0 = 247 \text{ mV vs. Hg/HgO} \tag{4.3}$$

$$O_2 + 2H_2O + 4e^- \rightarrow 4OH^-$$
$$E_0 = 303 \text{ mV vs. Hg/HgO} \tag{4.4}$$

4.3
Electrolytes and Separators

In principle, three different electrolyte types can be distinguished: liquid flow electrolytes (e.g., potassium hydroxide, sodium hydroxide, or their mixtures), immobilized electrolytes (i.e., the liquid electrolyte is fixed in a matrix or a gel), and anion-exchange polymer electrolytes (e.g., solid polymer electrolytes in PEFCs as described above).

For liquid electrolytes, a diaphragm or separator material is needed to separate the anode from the cathode to avoid a short-circuit. A common diaphragm material before the 1980s was asbestos, which was highly stable under alkaline conditions; however, the restrictions on the use of asbestos in industrialized countries led to its abandonment, and the development of alternative separator materials became necessary.

Examples of alternative materials are presented below, including sintered diaphragms, porous polyethylene (PE) plates, nonwoven polypropylene (PP) materials, and PP plates with micropores (Celgard). The characteristics of these diaphragm materials in a 30% KOH electrolyte solution are listed in Table 4.1.

Table 4.1 Diaphragm materials.

Name	Material	Thickness	Pore size (μm)
PE plates	Polyethylene, sintered	1–3 mm	125–175
PP nonwoven material	Polypropylene	2–3 mm	–
Celgard 3401	Polypropylene plate	25 μm	0.117

Table 4.2 Resistance of electrolyte-filled diaphragms with an active area of 50 cm^2 at 70 °C.

Material	Measured resistance (mΩ)	Specific resistance (Ω cm^2)	Specific resistance of the diaphragm (Ω cm^2)[a]	Specific resistance of the diaphragm thickness (Ω cm)[b]
PE plates, 3 mm	20.47	1.02	0.78	2.6
Celgard 3401 (on the anode surface and the cathode surface), 1.5 mm distance between the electrodes	7.00	0.35	0.23	11.5
Nonwoven PP, 3 mm distance between the electrodes	5.15	0.26	0.09	0.3

[a] Additional electrolyte resistance induced by the diaphragm.
[b] Specific resistance of the diaphragm material.

However, there are many additional materials, and their application depends strongly on the designed system.

The internal resistance of the fuel cell is influenced by the thickness of the electrolyte film between the electrodes. Other factors that are important for conductivity include the position of the diaphragms that determine the distance between the electrodes, the concentration of the electrolyte and the average length of the electrolyte paths. The resistances of the various diaphragms to electrolyte solutions are listed in Table 4.2. The resistance of the electrolyte decreases with increasing temperature.

4.4
Degradation

4.4.1
Gas Diffusion Electrodes with Raney Nickel Catalysts

The durability and lifetime are important factors for the commercialization of fuel cells. Therefore, the degradation of fuel-cell components has been subjected to an increasing number of studies, mainly focused on PEFCs [56–67]. For AFCs, several degradation processes have been studied [68–75], and all were focused on the electrodes because the electrolyte in AFCs can easily be exchanged.

Currently used electrodes are flexible, and the most commonly used electrodes are therefore PTFE-bonded electrodes. The degradation of PTFE-bonded electrodes is described below for different types of nickel anodes and one silver cathode.

Three different electrode types have been investigated at DLR:

- type 1 electrodes consisting of a mixture of pure Raney nickel and PTFE powder on a nickel grid
- type 2 electrodes consisting of a mixture of a Raney nickel catalyst, a copper powder additive, and PTFE powder on a copper grid
- type 3 electrodes consisting of nickel catalysts that have been size-selected before the electrode preparation.

The Raney nickel catalyst is a highly porous metallic catalyst that is formed from an aluminum–nickel alloy after dissolving the aluminum. The basic alloy composition of the catalyst and the PTFE powder are identical for all electrodes. Investigations of the degradation of type 2 electrodes have been described in detail. In addition to the decreased electrochemical performance [76], the electrode properties and their components change during the fuel-cell operation. Physical characterization of the electrodes after operation shows that the pore-size distribution changes (Figure 4.7), although the specific surface area is almost unchanged.

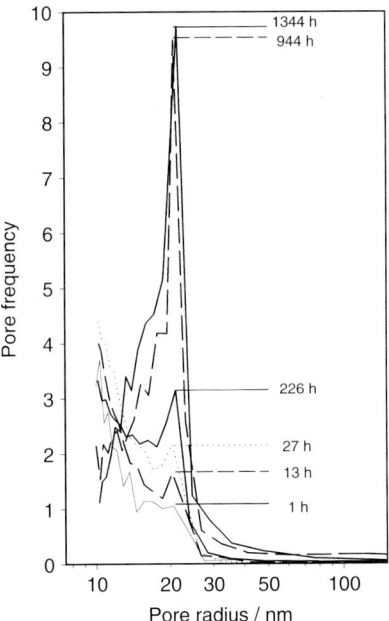

Figure 4.7 Pore-size distribution measured by nitrogen adsorption of electrodes after various operating times at a current density of 100 mA cm^{-2}. Source: DLR.

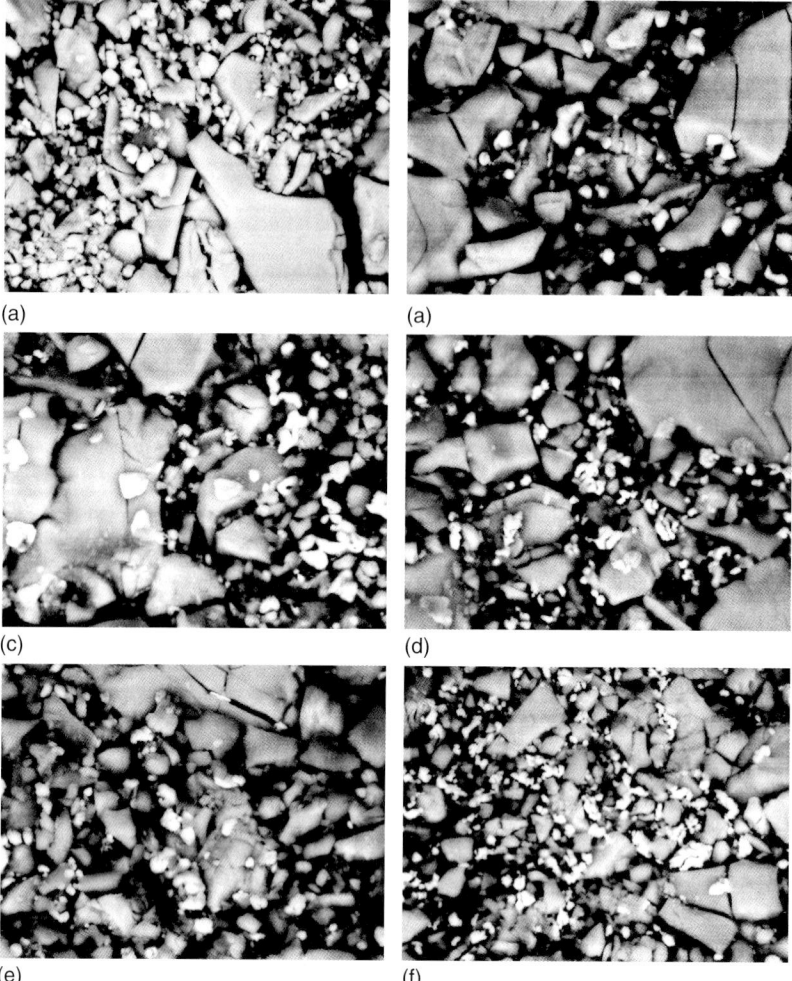

Figure 4.8 SEM images of new anodes and electrochemically stressed anodes at various times at a current density of 100 mA cm^{-2}. The non-activated (a) and the activated, operated electrodes (b) are shown at the top. Below, the same electrodes are shown after (c) 13, (d) 226, (e) 944, and (f) 1344 h of operation. Source: DLR.

The nickel catalyst particles are disintegrated during the electrochemical operation (Figure 4.8). The hydrogen-induced embrittlement is caused by the hydrogen loading and discharging of the nickel catalyst; therefore, the electrochemical conditions influence the hydrogen loading and discharging. Finally, the disintegration of the nickel catalysts can also depend on the operating conditions, including the operating current density. The pore network of the electrodes (measured by nitrogen adsorption) is characterized by the open pores in the nickel catalyst; therefore, the disintegration of the nickel particles is influenced by the pore-size distribution.

For high anode overpotentials, the nickel is oxidized, the copper is partially dissolved in the electrolyte and the residual aluminum in the electrode is further dissolved during the fuel-cell operation. Additionally, the decomposition or disappearance of PTFE is observed during the electrochemical operation. This result is very surprising because PTFE should be stable during all electrochemical conditions. PTFE is also used in PEFCs as a hydrophobic material in gas diffusion layers and electrodes and is at present a major topic of fuel-cell research.

During the electrode activation process, similar changes are also observed, including the dissolution of the copper and the subsequent transportation of copper to the electrolyte. A changed copper concentration on the electrode surface, depending on the time and the current density, was observed by XPS.

The type 1 and 2 electrodes have a similar initial electrochemical performance. The initial performance of the type 3 electrodes is slightly improved. The same initial electrochemical performances of the type 1 and 2 electrodes stem from the identical nickel catalyst used for both electrode types. The improvement of the electrical conductivity by copper does not improve the electrochemical performance. In contrast, the size-selected nickel catalysts in the type 3 electrodes lead to an improved electrochemical performance. During long-term electrochemical experiments, a decreased performance is observed for all three electrode types. The decreased electrochemical performance (Perf) indicator, measured by the conductance, current density, and voltage, is nonlinear for the investigated electrodes, and the Perf indicator is approximated by the following function:

$$\text{Perf}_1(t) = a \exp(-t/t_0) + c \tag{4.5}$$

If the polarization induced by the decreasing electrochemical performance during the electrochemical loading increases significantly above 100 mV, the electrode rapidly loses its electrochemical performance, shown for the lifetime curve of a type 1 electrode with an operation time of 1200 h.

The type 2 electrodes were operated with current densities from 25 to 300 mA cm^{-2}. The type 3 electrodes were used with current densities between 100 and 150 mA cm^{-2} during the long-term experiments. Figure 4.9 shows the decreased electrochemical performance of a type 2 electrode operating at different current densities. To compare different samples, the time curves in Figure 4.9c have been normalized by their different initial electrochemical performance values.

The decreased electrochemical performance of the type 2 electrodes cannot be described by a single exponential function of the type in Eq. (4.5). The time-dependent electrochemical performance can instead be described by the following combination of two exponential functions:

$$\text{Perf}_2(t) = a \exp(-t/t_0) + b \exp(-t/t_1(i)) + c \tag{4.6}$$

The first exponential function depends only on the time, whereas the second exponential function depends on the current density (i) and the time.

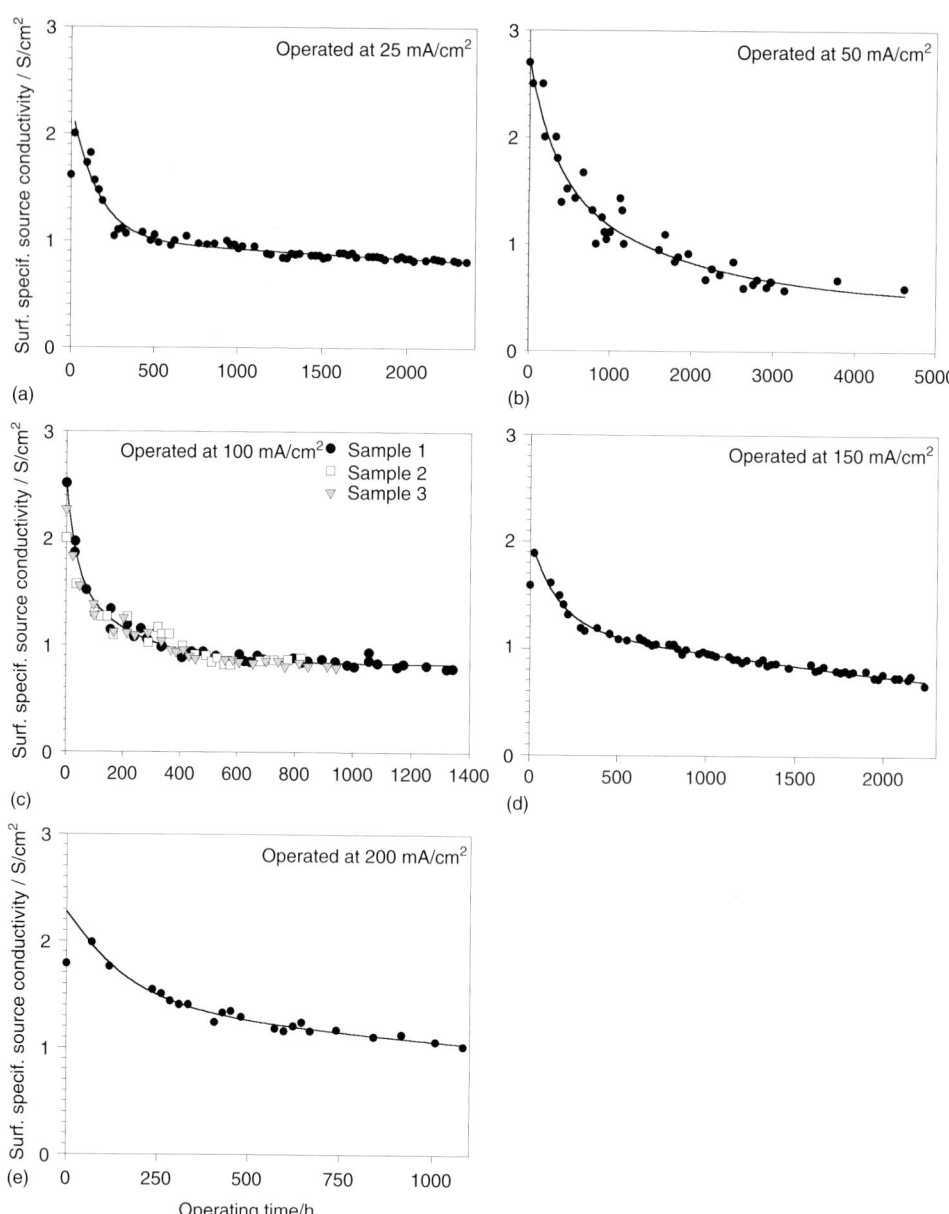

Figure 4.9 Time-dependent surface-specific conductivity (inverse of the area-specific resistance) of a type 2 nickel anode recorded at current densities of (a) 25, (b) 50, (c) 100, (d) 150, and (e) 200 mA cm^{-2}. Source: DLR.

In contrast, the time-dependent electrochemical performance of the type 3 electrodes can be described by the following single exponential function that depends on the charge density:

$$\text{Perf}_3(t) = a\,\exp\left[-t/t_0(i)\right] + c \tag{4.7}$$

The decreased electrochemical performance of the three electrode types is not significantly different. For the type 1 and 3 electrodes, the time constant (t_0) was ~250 h under the described operating conditions and at a current density of 100 mA cm^{-2}. This time constant was calculated from the long-term behavior of surface properties, specifically the source conductivity according to Eqs. (4.5) and (4.7). For the type 2 electrodes (at a current density of 100 mA cm^{-2}), the time constant t_0 of Eq. (4.6) was also 250 h. Therefore, the time constants are nearly equal for all three electrode types.

The disintegration of the nickel particles should be comparable for the type 1 and 2 electrodes because the same catalyst is used in both electrodes. Because of the size selection of the nickel catalyst in the type 3 electrodes, the nickel particles disintegrate differently from those of the other electrode types. For all of the electrode types, the concentration of small nickel particles increases; however, because of the disintegration of large nickel particles in the type 1 and 2 electrodes, nickel particles of different sizes are formed. Therefore, the change in the size distribution is not as significant as that for the nickel catalyst in the type 3 electrodes; only the large nickel catalyst particles are removed from the particle-size distribution in the type 1 and 2 electrodes. In the type 3 electrodes, the initial particle-size distribution is shifted to smaller particle sizes.

A decreased electrochemical performance, induced by electrochemical stress, was observed at DLR for GDEs with nickel catalysts in the anodes of AFCs. For all electrodes, a time constant of ~250 h is observed at a current density of 100 mA cm^{-2}. This time constant depends on the current density, a dependence that is stronger for the type 3 electrodes than for the type 2 electrodes. The second time constant for describing the degradation of the electrochemical performance of the type 2 electrodes depends strongly on the current density.

Electrodes with size-selected nickel catalysts are not expected to have nearly the same characteristic degradation times as electrodes with standard nickel catalysts. The disintegration of the nickel particles that affects the electrochemical activity is different. For the type 2 and 3 electrodes, the different time dependence on the pore frequency (for pores with a 2 nm radius) reflects the different nickel particle disintegration (Figure 4.10). The strong correlation between the electrochemical performance (measured by the surface-specific conductivity) and the pore frequency observed with the type 2 electrodes could not be observed with the type 3 electrodes.

During high loads, the electrode potential shifted to positive values, nickel was oxidized, the electrode was passivated, and the reaction stopped. This loss of electrochemical performance is not correlated with the alteration of the pore-size distribution or the pore frequency.

However, the kind of nickel degradation pathway depends strongly on the activation procedure and the morphology of the electrodes. Nickel oxidation is not

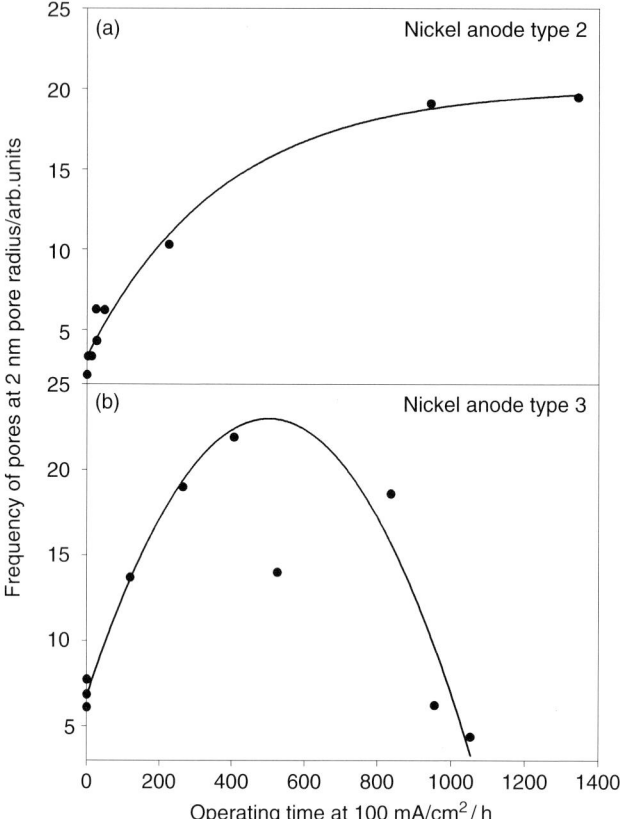

Figure 4.10 Pore frequency with a pore radius of 2 nm as a function of time (measured by nitrogen adsorption) of a type 2 electrode (a) and a type 3 electrode (b). Source: DLR.

generally the dominant degradation process. Hence the lifetime of the electrode is expected to depend on the current density caused by the higher anodic polarization at higher current densities.

For the type 2 and 3 electrodes, the copper distribution changes because the copper is partially dissolved and transported during the operation, and the copper is partly redeposited.

More grain boundaries are formed because of the disintegration of the nickel particles, and poor electrical contacts between the particles should therefore yield an increased electrical resistance that results in higher losses. However, the most significant effect is that the potential difference in the electrode increases, the three-phase zone can be extended less, and the electrochemical performance therefore decreases.

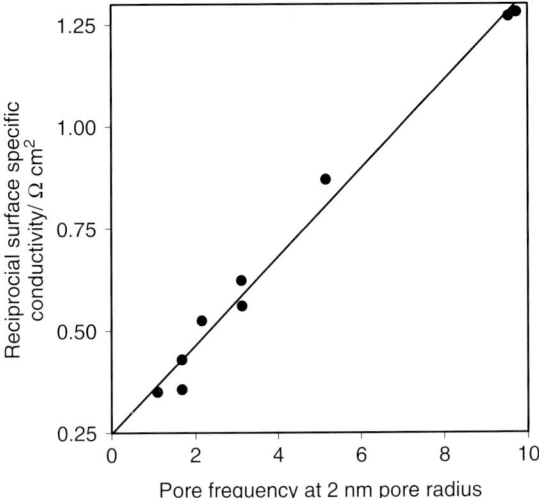

Figure 4.11 Correlation between the final reciprocal surface-specific conductivity and the pore frequency (with a pore size of 2 nm) of the electrodes after various operating times at a current density of 100 mA cm^{-2}.

Figure 4.11 shows the correlation between the final reciprocal surface-specific conductivity and the pore frequency, that is, the reciprocal value of the final experimentally determined surface-specific conductivity is plotted against the pore frequency for the stressed electrodes. Both parameters are strongly correlated.

4.4.2
Gas Diffusion Electrodes with Silver Catalysts

Non-platinum materials such as silver are often used as cathode catalysts. For a silver GDE immersed in a complete AFC, the solubility of the silver can sometimes cause a transfer of the silver to the anode, and the resulting anode plating may damage the anodic catalyst (usually Raney nickel). During open-circuit operation (idle period), oxidation of silver may occur. Especially carbon-containing cathodes suffer from oxidation at higher temperatures. Shutdown and restarting procedures can also cause damage.

During oxygen reduction (at 70 °C in 30 wt% KOH) with a silver GDE, a decreased electrochemical performance is observed. This decreased electrochemical performance shows a linear progression with a maximum gradient of ~20 μV h^{-1}. After 5000 h of operation, the voltage loss is ~100 mV. Considering the high cell voltages of AFCs, this decrease in the electrochemical performance of the cathode is ~3% per 1000 h of operation (e.g., based on a 5000 h operating time).

For some applications, including automotive operation with an estimated operating time of ~5000 h, these kinds of catalysts and electrodes are sufficient. For stationary applications with up to 40 000 h of operation, the long-term efficiency must be improved or the cost of stack exchanges must be accounted for.

To understand the mechanism of the degradation of the physical and electrochemical components, two different pore systems must be distinguished: first, the pore system in the electrode formed by the interparticle space between the catalysts (that allows for the gas transport), and second, the pore system in the silver catalyst. In the first type of pore system, the hydrophobic character is dominant.

Physical characterization of the electrodes has demonstrated that the roughness of the silver catalyst decreases, hence the surface area also decreases [77]. The surface roughness is correlated with the second type of pore system. For Pt catalysts in PEFCs, the electric field gradient was sometimes the driving force for the mobility of the platinum. For a silver GDE, the electric field gradients are expected to be enhanced in the pores. The decreased surface roughness of the silver catalyst is because of the known high exchange-current density of silver. The electrochemical characterization shows that the electrochemical active surface area, which is related to the double-layer capacity, decreases, as is expected from the decreased catalyst roughness.

The decreased surface roughness does not significantly influence the electrochemical performance. This observation is corroborated by the nearly constant transfer resistance as measured by EIS (Figure 4.12) [78].

The second degradation pathway of the electrodes (observed by physical characterization) is the alteration of the PTFE. The decomposition of PTFE is typically

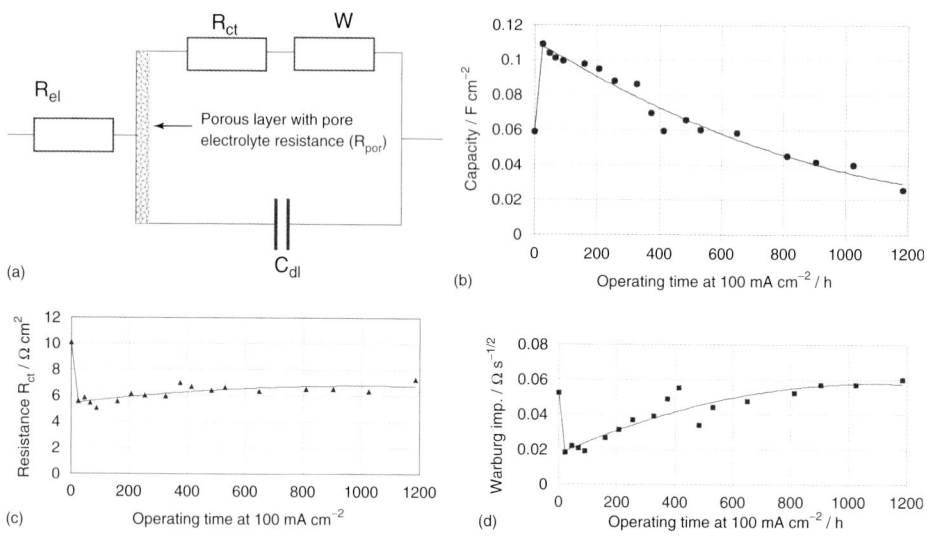

Figure 4.12 Equivalent circuit (porous electrode model) for the evaluation of the impedance spectra of silver gas diffusion electrodes (a). Time dependence of the double layer capacity (b), of the charge-transfer resistance (c), and of the Warburg impedance (d) obtained with a least-squares fit of the measured impedance spectra of the open-circuit potential at the silver GDE after different operating times.

related to a decreased hydrophobicity. The EIS revealed that the increased Warburg impedance is related to the increased transport resistance.

Hydrophobic properties are important for gas transport in the electrode. The decreased hydrophobicity allows the electrolyte to flood more pores, thus hindering the gas transport.

4.5 Carbon Dioxide Behavior

Most reports describe the formation of carbonates in the electrodes as the main problem with AFCs that use air rather than pure oxygen. Studies using half-cells, single cells, and short stacks have also been performed at DLR. Electrodes (anodes and cathodes) have been investigated after being operated with pure fuel gases and with CO_2-contaminated fuel gases. The tests were always terminated without problems after 3500 h of operation without a significant indication of this degradation mechanism.

In systematic investigations, anodes and cathodes were electrochemically stressed and characterized, for example, in an electrochemical half-cell configuration consisting of an electrode in a holder manufactured from acrylic glass, a nickel counter electrode and an Hg/HgO reference electrode. With this experimental setup, the electrodes were supplied with gases without overpressure from the back side (the side with the metallic web). The electrode holder was inserted in a temperature-controlled vessel (353 K) containing 30% KOH. The vessel contained ~1.5 l of the electrolyte. The electrode had an active area of 6 cm². The GDEs were operated for up to several thousand hours with a constant loading of 100 or 150 mA cm^{-2}. The advantage of the half-cell tests compared with long-term experiments with a full-cell configurations is that the electrodes and their degradation can be investigated individually.

To study the mechanism of CO_2 poisoning, the electrodes were operated using gases containing 5% CO_2 (and pure gases as references). Before operation with CO_2-containing gases, all electrodes were operated with pure gases for 100 h. The CO_2–O_2 gas mixture had a 150 times higher CO_2 concentration than air. This high concentration was used to enhance the potential effect of CO_2.

For the electrochemical characterization, $V(i)$ curves were recorded and an *IR* correction was used to compensate for the ohmic losses in the measured potential (the ohmic resistance was measured by current interruption). The slope of the $V(i)$ curve (the surface-specific conductivity) was calculated as a characteristic parameter for the electrochemical performance of the anodes and the cathodes.

Additionally, the CO_2 contents of the reaction gases were quantitated by gas chromatography, and the carbonate concentration in the electrolyte was measured after various operating times.

Figure 4.13 shows the electrochemical performance during long-term experiments performed with silver cathodes. The decreased electrochemical performances of all of the cathodes were similar, and no significant difference was

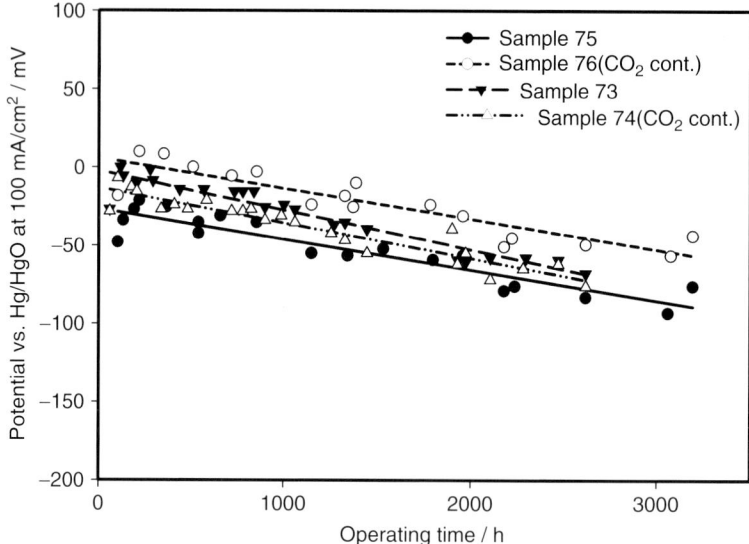

Figure 4.13 Change in the electrochemical performance of the silver cathode during operation with pure oxygen and oxygen containing 5% CO_2. Source: DLR.

observed whether the experiments were performed with pure oxygen or with CO_2-containing oxygen. These results indicate that CO_2 has a minor influence on the electrochemical performance. The slope of the time-dependent electrochemical performance showed a decrease of 17 $\mu V\,h^{-1}$ at the cathode. EIS measurements recorded during a long-term experiment with silver cathodes in pure oxygen showed that the decreased electrochemical performance was induced by a decreased active surface area and increased transport blockage. The decreased electrochemical performance could be explained by the alteration or decomposition of the PTFE. PTFE is required for the gas transport and influences the extension to the three-phase zone.

XPS measurements were performed after operation of the silver anodes for more than 2600 h. The element distributions of the electrodes operated with pure oxygen and with CO_2-containing oxygen were equal in the depth profile measurements (Figure 4.14). In particular, the carbon signal in the XPS spectra is of interest because CO_2 can form carbonates that may be deposited on or in the electrodes. The XPS spectra of the electrodes operated with pure gases and CO_2-containing gases did not differ significantly, indicating that no carbonate was deposited on the electrode surface or in the pores of the electrode.

The carbonate in the alkaline electrolyte forms CO_3^{2-} ions that should migrate with the electrical field towards the anode. Therefore, the anodes also need to be investigated to assess the influence of CO_2 on AFCs. If carbonates are deposited on the anode, they could block the electrode pores. Hence the pore system and the specific surface area of the nickel anodes were investigated by nitrogen adsorption

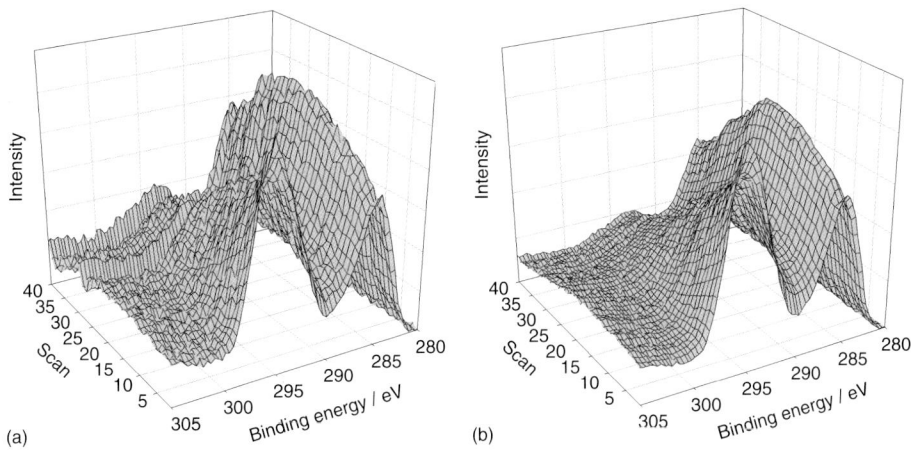

Figure 4.14 XPS spectra of the carbon C1s signal recorded during depth profile measurements of a cathode operating for 2633 h with pure oxygen (a) and of a cathode operating for 2682 h with oxygen containing 5% CO_2 (b).

techniques. The specific surface area was nearly constant for anodes operated with pure hydrogen and with CO_2-containing hydrogen for various times, and no change induced by the carbon dioxide could be observed. The investigations showed no influence of CO_2.

In addition, anodes were investigated by XPS after operation for over 2600 h in the presence of pure hydrogen and in the presence of CO_2-containing hydrogen. No carbonate deposition could be observed. In addition, SEM measurements were also performed with electrodes operated in the presence of either pure gases or CO_2-containing gases. Consistent with the porosimetry and XPS measurements, no effects of the CO_2 were observed by SEM.

The electrochemical performance and the physical characterization of the anodes and the cathodes showed that CO_2 has no significant influence on the electrodes.

The electrolyte is also important for changes observed in the conductivity. Measurements of the CO_2 content in the residual gas (by gas chromatography at the back side of the fuel-cell electrodes) showed that the residual gas of the anode contained ~10% of the CO_2 content of the feed gas; the cathode contained ~20%. These measurements indicate that most of the CO_2 is dissolved in the alkaline solution by forming K_2CO_3. Therefore, the carbonate concentration in the alkaline electrolyte increases. As expected, the carbonate concentration increases linearly during the operation time for CO_2-containing gases; however, no difference between the anode and cathode could be observed. During long-term experiments, the electrodes are unaffected, but increased K_2CO_3 concentrations (up to 450 g l^{-1}) lead to lower conductivities. The solubility limit of K_2CO_3 is between 50 and 60 wt% for the temperature rang used e [79]. Therefore, for the experiments shown, the solubility limit had not been reached.

In contrast to the electrodes, the electrolyte is influenced by the CO_2-containing gas supply. During operation with CO_2-containing gases, the carbonate concentration increases linearly with the operating time.

It is acceptable to replace the electrolyte after a few thousand hours because electrolyte exchanges are very simple.

The electrochemical performance is not affected by CO_2. In addition, physical examination of the electrodes showed that CO_2 has no effect on the electrode components. In summary, the CO_2 content in the feed gases has no influence on the electrodes with a porous structure.

4.6
Conclusion

Two main problems associated with AFCs are CO_2 exposure and the liquid electrolyte. The first problem can be solved with a cleaning step or with an optimized porous structure. The second problem causes leaks and requires expensive sealing techniques or the use of AEMs. Many companies are working on AEMs and not only for fuel-cell applications. Electrochemical electrolyzer reactors would also benefit from AEMs. However, there is currently no long-term stable membrane that functions without a liquid electrolyte phase, and it is unclear whether carbonate formation affects the interior of the membrane when the fuel cell is operated with air.

For liquid AFCs, there are some on-going research activities.

Within the last 15 years, some corporate activities have been under way.: In 2000, Zevco received significant publicity for its goal to start the large-scale production of AFCs in Cologne, Germany. However, they stopped their operation in 2001. This stoppage caused major drawbacks in activities worldwide because many system integrators wanted to use the products from Zevco. Another company with significant development efforts was Astris in Canada. Astris developed a proprietary stack and constructed stationary and mobile application demonstrations. However, Astris recently halted their development efforts, sold their intellectual property to investors and stopped their alkaline activities a short time thereafter. Today, only AFC Energy has significant AFC research activities in Europe. They focus on systems for stationary applications, but commercial products are not yet available.

In conclusion, it is my conviction that AFCs have a future for battery-range extender applications because of their cost advantage.

References

1. Guelzow, E., Nor, J., Nor, P., and Schulze, M. (2006) *Fuel Cell Rev.* 3, 19–25.
2. Wikipedia.(2012) Alkaline Fuel Cell, http://en.wikipedia.org/wiki/Alkaline_fuel_cell (last accessed 2 February 2012).
3. Gülzow, E. and Schulze, M. (2008) in *Alkaline Fuel Cells, Materials for Fuel Cells* (ed. H.M Gasik), Woodhead Publishing, Cambridge. 64–100.
4. Wikipedia (2012) William Robert Grove, http://en.wikipedia.org/wiki/William_

Robert_Grove (last accessed 2 February 2012).
5. Wikipedia (2012) Christian Friedrich Schönbein, *http://en.wikipedia.org/wiki/Christian_Schoenbein* (last accessed 2 February 2012).
6. Reid, J.H. (1902) US Patents 736,016,736,017; Reid, J.H. (1904) US Patent 757,637.
7. Noel, P.C.L. (1904) French Patent 350,111.
8. Schmid, A. (1923) *Die Diffusions-Gas-Elektrode*, Verlag F. Enke, Stuttgart.
9. Schmid, A. (1933) *Helv. Chim. Acta*, **7**, 169.
10. Scharf, P., DRP 46466 (Patent) (1888).
11. Wikipedia (2012) Murray Raney, *http://en.wikipedia.org/wiki/Murray_Raney* (last accessed 2 February 2012).
12. Wikipedia (2012) Francis Thomas Bacon, *http://en.wikipedia.org/wiki/Francis_Thomas_Bacon* (last accessed 2 February 2012).
13. Bacon, F.T. (1950) British Patent 667,298; Bacon, F.T. (1954) DAS 1,025,025; Bacon, F.T. (1955) British Patent 725,661.
14. Doehren, H. and Euler, J. (1965) *Der Heutige Stand der Brennstoffelemente*, Akademische Verlagsgesellschaft, Frankfurt.
15. (a) Justi, E. and Winsel, A. (1962) *Kalte Verbrennung*, Franz Steiner Verlag, Wiesbaden. (b) Winsel, A. and Wendtland, R. (1965) DE Patent DBP No. 1546719.
16. Kordesch, K. (1999) *Fuel Cells and Their Applications*, Wiley-VCH Verlag GmbH, Weinheim.
17. Winsel, A. and Wendtland, R. (1965) DE Patent DBP No. 1496214.
18. Winsel, A. (1983) *DECHEMA Monogr.*, **92**, 1885–1913.
19. Winsel, A. and Wendtland, R. (1971) US Patent 3,597,275.
20. Gaskatel GmbH (2011) Gaskatel, *http://www.gaskatel.de* (last accessed 2 February 2012).
21. Staab, R. (1987) *Chem.-Hzg.-Tech.*, **59**, 316–319.
22. Tetzlaff, K.-H. and Wendel, W. (1988) *DECHEMA Monogr.*, **112**, 325–337.
23. Tetzlaff, K.-H., Walz, R., and Gossen, C.A. (1994) *J. Power Sources*, **50**, 311–319.
24. Gülzow, E., Schulze, M., and Gerke, U. (2006) *J. Power Sources*, **156**, 1.
25. Gülzow, E., Schulze, M., and Gerke, U. (2004) Development of an alkaline fuel cell stack with bipolar configuration, presented at the Fuel Cell Seminar, San Antonio, TX.
26. *http://www.hydrocell.fi/en/fuel-cells/generalobservations/* (last accessed 15 February 2012).
27. *http://www.hydrocell.fi/en/fuel_cells/hc-100.html*.
28. Ovshinsky, S.R., Menjak, Z., Venkatesan, S., Gradinarova, L., Menjak, A., and Wang, H. (2004) Alkaline fuel cell pack with gravity fed electrolyte circulation and water management system. US Patent Application 2004-0161652.
29. Reichman, B., Fetcenko, M.A., Ovshinsky, S.R., Young, K., Mays, W., and Strebe, J. (2005) Low temperature alkaline fuel cell. US Patent Application 2005-0064274.
30. Ovonics (2003) *Fuel Cells Bulletin.*, **2003** (8), 3–4.
31. Gülzow, E., Beyer, M., Friedrich, K.A., Pengel, S., Fischer, P., and Bettermann, H. (2011) *ECS Trans.*, **30** (1), 345–351.
32. Gülzow, E., Schulze, M., Friedrich, K.A., Fischer, P., and Bettermann, H. (2011) *ECS Trans.*, **30** (1), 65–76.
33. Kiros, Y. and Schwartz, S. (2000) *J Power Sources*, **87**, 101.
34. Winsel, A. (1969) *Electrochim. Acta*, **14**, 961.
35. Sauer, H. and Spahrbier, D. (1976) US Patent 3,977,902.
36. Miyazaki, K., Sugimura, N., Matsuoka, K., Iriyma, Y., Abe, T., Matsuoka, M., and Ogumi, Z. (2008) *J. Power Sources*, **178**, 683.
37. Oy Hydrocell (2012) Materials, *http://www.hydrocell.fi/en/materials/* (last accessed 2 February 2012).
38. Singhal, S.C. and Kendall, K. (2003) *High Temperature Solid Oxide Fuel Cells: Fundamentals, Design and Applications*, Elsevier, Amsterdam.
39. Wang, Y. and Xia, Y. (2006) *Electrochem. Commun.*, **8**, 1775.

40. Verma, A., Jha, A.K., and Basu, S. (2005) *J. Power Sources*, **141**, 30.
41. W2C: Meng, H., Wu, M., Hu, X.X., Nie, M., Wie, Z.D., and Shen, P.K. (2006) *Fuel Cells*, **6**, 447.
42. Sauer, H. German Patent DE 2,941,774.
43. Winsel, A. (1985) German Patent DE 3,342,969.
44. Gülzow, E. and Schnurnberger, W. (1991) *DECHEMA Monogr.*, **124**, 675.
45. Gülzow, E., Holzwarth, B., Schulze, M., Wagner, N., and Schnurnberger, W. (1992) *DECHEMA Monogr.*, **125**, 561.
46. Gülzow, E., Bolwin, K., and Schnurnberger, W. (1990) *DECHEMA Monogr.*, **121**, 483.
47. Gülzow, E., Bolwin, K., and Schnurnberger, W. (1989) *DECHEMA Monogr.*, **117**, 365.
48. Rahman-Ur, S., Al-Saleh, M.A., Al-Zakri, A.S., and Gultekin, S. (1997) *J. Appl. Electrochem.*, **27**, 215.
49. Gülzow, E., Holzwarth, B., Schnurnberger, W., Schulze, M., Steinhilber, G., and Wagner, N. (1992) PTFE bonded gas diffusion electrodes for alkaline fuel cells, presented at the 9th World Hydrogen Energy Conference, 1992, Paris.
50. Rühling, K. (1986) Untersuchungen an neuartigen PTFE-gebundenen Raney-Nickel- und Silberelektroden, Diploma Thesis. University of Kassel.
51. Lorenz, M. and Schulze, M. (1999) *Fresenius' J. Anal. Chem.*, **365**, 154.
52. Schulze, M., Reissner, R., Lorenz, M., and Schnurnberger, W. (1999) *Electrochim. Acta*, **44**, 3969.
53. Kitagawa, S. (1979) in Proceedings of JIMIS-2, Hydrogen in Metals, Minikami, Japan, 1979, p. 497.
54. Schulze, M. and Gülzow, E. (2004) *J. Power Sources*, **127**, 252.
55. Gülzow, E., Wagner, N., and Schulze, M. (2003) *Fuel Cells*, **3**, 67.
56. Davies, D.P., Adcock, P.L., Turpin, M., and Rowen, S.J. (2000) *J. Appl. Electrochem.*, **30**, 101.
57. Davies, D.P., Adcock, P.L., Turpin, M., and Rowen, S.J. (2000) *J. Power Sources*, **86**, 237.
58. Mattsson, B., Ericson, H., Torell, L.M., and Sundfolm, F. (2000) *Electrochim. Acta*, **45**, 1405.
59. Hübner, G. and Roduner, E. (1999) *J. Mater. Chem.*, **9**, 409.
60. Schulze, M., Lorenz, M., Wagner, N., and Gülzow, E. (1999) *Fresenius' J. Anal. Chem.*, **365**, 106.
61. Büchi, F.N., Gupta, B., Haas, O., and Scherer, G.G. (1995) *Electrochim. Acta*, **40**, 345.
62. Gülzow, E., Helmbold, A., Kaz, T., Reissner, R., Schulze, M., Wagner, N., and Steinhilber, G. (2000) *J. Power Sources*, **86**, 352.
63. Gupta, S., Tryk, D., Zecevic, S.K., Aldred, W., Guo, D., and Savinell, R.F. (1998) *J. Appl. Electrochem.*, **28**, 673.
64. Gülzow, E., Fischer, M., Helmbold, A., Reissner, R., Schulze, M., Wagner, N., Lorenz, M., Müller, B., and Kaz, T. (1998) Innovative production technique for PEFC and DMFC electrodes and degradation of MEA-components, in Fuel Cell Seminar, Palm Springs, November 16–19, 1998, p. 469.
65. Gülzow, E., Kaz, T., Lorenz, M., Schneider, A., and Schulze, M. (2000) Degradation of PEFC components, in Fuel Cell Seminar 2000, Portland, October30– November 2, 2000, p. 156.
66. Wilson, M.S., Valerio, J.A., and Gottesfeld, S. (1995) *Electrochim. Acta*, **40**, 355.
67. Dirven, P.G., Engelen, W.J., and Van Der Poorten, C.J.M. (1995) *J. Appl. Electrochem.*, **25**, 122.
68. Gülzow, E., Schulze, M., Steinhilber, G., and Bolwin, K. (1994) Carbon dioxide tolerance of gas diffusion electrodes for alkaline fuel cells, in Proceedings of Fuel Cell Seminar, San Diego, CA, 1994, p. 319.
69. Gülzow, E., Schulze, M., and Steinhilber, G. (2002) *J. Power Sources*, **106**, 126.
70. Gultekin, S., Al-Saheh, M.A., Al-Zakri, A.S., and Abbas, K.A.A. (1996) *Int. J. Hydrogen Energy*, **21**, 485.
71. Kiros, Y. and Schwartz, S. (2000) *J. Power Sources*, **87**, 101.
72. Schulze, M., Bolwin, K., Gülzow, E., and Schnurnberger, W. (1995) *Fresenius' J. Anal. Chem.*, **353**, 778.
73. Khalidi, A., Lafage, B., Taxil, P., Gave, G., Clifton, M.J., and Cezac, P. (1996) *Int. J. Hydrogen Energy*, **21**, 25.

74. Schulze, M., Gülzow, E., and Steinhilber, G. (2001) *Appl. Surf. Sci.*, **179**, 252.
75. Wagner, N., Gülzow, E., Schulze, M., and Schnurnberger, W. (1996) Gas diffusion electrodes for alkaline fuel cells, in *"Solar Hydrogen Energy"*, *German–Saudi Joint Program on Solar Hydrogen Production and Utilization Phase II 1992–1995* (eds. H. Steeb and H. Aba Oud), DLR, Stuttgart, ISBN 3-89100-028-6.
76. Schulze, M., Schneider, A., and Gülzow, E. (2004) *J. Power Sources*, **127**, 213.
77. Wagner, N., Gülzow, E., and Schulze, M. (2004) *J. Power Sources*, **127**, 264.
78. Gülzow, E. and Schulze, M. (2004) *J. Power Sources*, **127**, 243.
79. Lide, D.R. (ed.) (1996) *Handbook of Chemistry and Physics*, 77th edn, CRC Press, Boca Raton, FL.

5
Micro Fuel Cells

Ulf Groos and Dietmar Gerteisen

5.1
Introduction

For ecological reasons, the demand for more efficient, non-polluting energy technologies is increasing rapidly. Fuel-cell systems show promise in meeting the future energy demands and have a substantial range of applications, such as in stationary power generation systems, fuel cell-powered vehicles, and battery replacement devices. Battery hybrid systems in the range from milliwatts to several hundred watts for use in portable electronic devices or even as a battery alternative are often seen as the first niche market for fuel-cell systems due to the less stringent demands on systems costs in this area. The early market for micro fuel cells can be categorized in three different sectors: (i) portable electronic devices such as mobile phones, laptops, emergency lights, and power toys [1]; (ii) military applications to power navigation devices and mobile phones; and (iii) the healthcare sector, for example, micro-biofuel cells to power implants such as heart pacemakers and glucose sensors [2]. The topic of microbial fuel cells (MFCs) is addressed in Chapter 6.

In fuel cells, the highest power density is achieved by using pure hydrogen as fuel, due to the nearly nonpolarizable hydrogen electrode (anode). Because the hydrogen storage does not scale with the miniaturization of polymer electrolyte membrane fuel cells (PEMFCs), hydrogen-fed fuel cells are not suitable for small system-integrated battery replacement devices [2]. Nevertheless, this technology is very interesting for battery recharging devices. Passive direct methanol fuel cell (DMFC) systems that use highly concentrated methanol and do not need external pumps are best suited for the mobile phone market [1]. For meeting the high power requirements of laptops, however, an active DMFC system is preferable [3]. PEMFCs use noble metals such as platinum to accelerate the redox reactions. They are not yet cost competitive with primary batteries and face competition with rechargeable batteries such as lithium ion and lithium polymer batteries [2].

To meet the needs of portable electronics, research has focused on miniaturizing the fuel cell itself in addition to its peripheral components, so-called "balance of plant" (BoP). Further research is needed on improving the heat dissipation,

Fuel Cell Science and Engineering: Materials, Processes, Systems and Technology,
First Edition. Edited by Detlef Stolten and Bernd Emonts.
© 2012 Wiley-VCH Verlag GmbH & Co. KGaA. Published 2012 by Wiley-VCH Verlag GmbH & Co. KGaA.

water management, and fuel supply to the point where the system can operate reliably under ambient conditions and be sufficiently rugged for everyday use [4]. The relatively low power demand of portable devices, compared with automotive applications or stationary power generators, permits a variety of new design concepts, material selection options, and fabrication techniques.

The global production of portable fuel cells is continually increasing. About 70% of these units are based on hydrogen-fed PEMFCs, 24% consist of DMFC units, and the remaining 6% are related to other technologies [3].

In addition to making fuel cells compact and lightweight, technical challenges such as durability, improved reliability, and increased power density have to be solved. Depending on the fuel used (methanol or hydrogen), the system concept (active or passive), and the operating conditions (temperature, humidification, and stoichiometry), the reported power densities range from 30 to 500 mW cm^{-2}. On the cathode, the oxygen reduction reaction (ORR) and reactant delivery to the active sites are still an issue. Chan *et al.* [5] showed that the power output of a pure oxygen-driven micro fuel cell is double that of an air-driven cell. Cha *et al.* [6] analyzed the influence of the channel size and gas diffusion layer (GDL) thickness on the mass transport phenomena in the micro-scale flow field. They found improved air convection into the GDL by having a larger pressure drop in small channels. Operational parameters such as backpressure and cell temperature were investigated with impedance spectroscopy by Hsieh *et al.* [7]. They showed that the influence of these parameters on a single-cell setup is similar to that on a micro fuel-cell stack [8].

The results clearly show an improved performance when using humidified gases, because they keep the membrane well hydrated. Therefore, to increase the power density without increasing the size and complexity of the system, proper water management without an external humidifier is an important objective.

5.2
Physical Principles of Polymer Electrolyte Membrane Fuel Cells (PEMFCs)

PEMFCs use a proton-conducting polymer membrane as electrolyte. The membrane is squeezed between two porous electrodes [catalyst layers (CLs)]. The electrodes consist of a network of carbon-supported catalyst for the electron transport (solid matrix), partly filled with ionomer for the proton transport. This network, together with the reactants, forms a three-phase boundary where the reaction takes place. The unit of anode catalyst layer (ACL), membrane, and cathode catalyst layer (CCL) is called the membrane–electrode assembly (MEA). The MEA is sandwiched between porous, electrically conductive GDLs, typically made of carbon cloth or carbon paper. The GDL provides a good lateral delivery of the reactants to the CL and removal of products towards the channel of the flow plates, which form the outer layers of a single cell. Single cells are connected in series to form a fuel-cell stack. The anode flow plate with structured channels is on one side and the cathode flow plate with structured channels is on the other side. This so-called bipolar plate

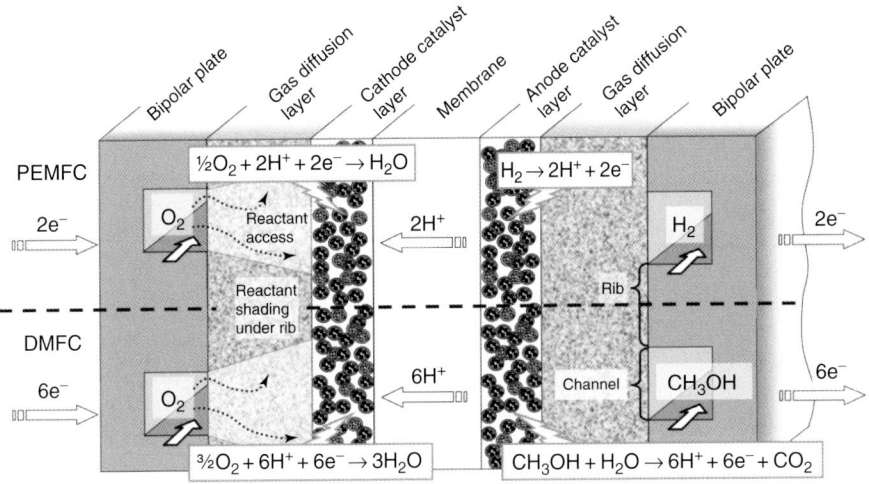

Figure 5.1 Schematic of a PEMFC. The cathode and anode reactions are given for a hydrogen PEMFC and a DMFC. At high current density or liquid water saturation of the GDL, the shading of the catalyst layer by the rib area causes mass transport limitation.

(BP) acts, on the one hand, as a positive cathode for one cell and, on the other, as a negative anode for the next cell. A schematic diagram of the fuel cell configuration and basic operating principles of a hydrogen-fed PEMFC and a DMFC are shown in Figure 5.1.

At the anode of a PEMFC, hydrogen is oxidized, creating protons and electrons. The polymer membrane provides proton-conducting pathways, whereas the electrons are forced through an external circuit by a potential difference between the anode and cathode. Within the CCL, oxygen is reduced to water in the presence of protons and electrons. The respective half-cell reactions, typically catalyzed by cost-intensive platinum, are

Anode: $H_2 \longrightarrow 2H^+ + 2e^-$

Cathode: $O_2 + 2H^+ + 2e^- \longrightarrow H_2O$

In the case of a DMFC, the anodic oxidation reaction, mostly catalyzed by a Pt–Ru alloy, is

Anode: $CH_3OH + H_2O \longrightarrow 6H^+ + 6e^- + CO_2$

where the reduction reaction on the cathode is, in principle, the same as in a PEMFC:

Cathode: $\frac{3}{2}O_2 + 6H^+ + 6e^- \longrightarrow 3H_2O$

The maximum reversible cell voltage is determined by the thermodynamics of the overall reaction. For both fuel-cell types, PEMFC and DMFC, an open-circuit voltage greater than 1 V should theoretically be reached. Since a DMFC suffers strongly from an incomplete methanol oxidation reaction and from mixed potential formation, caused by methanol crossover through the membrane to the cathode side, only the PEMFC reaches values of around 1 V. By drawing current, internal losses such as activation, ohmic, and mass transport losses further reduce the theoretical cell voltage for both fuel cell types, albeit to different extents.

5.3
Types of Micro Fuel Cells

Depending on the application, different scenarios can be considered for scaling down fuel-cell systems to millimeter size.

In general, future micro fuel cell design should forego heavy bolts and thick BPs in order to miniaturize the system. The other components, such as the GDL and MEA, are negligible in volume and weight. As a result, the material selection for the BPs becomes very important. Thin sheets that are stiff enough to achieve uniform low-contact resistance between MEA, flow field and current collector are needed. The bolts for the stack compression can be replaced by, for example, clamps, straps, or coiling techniques [9].

5.3.1
Hydrogen-Fed Micro Fuel Cell

The simplest and most effective way to produce electricity in a PEMFC is to use pure hydrogen as fuel. Owing to the fast kinetics of the hydrogen oxidation reaction, a high power density with high efficiency is possible, leading to a compact and lightweight fuel-cell stack. So far, the hydrogen storage in compressed gas cylinders is still too bulky and in metal hydride containers too heavy. Research concentrated on developing efficient hydrogen storage with high volumetric and gravimetric energy density is needed in order to realize viable storage sizes and weights that can compete with batteries. Recent studies indicate that carbon nanotubes show potential in this area [10–12].

With respect to the cell itself, water management is one of the most critical topics for reliable PEMFC operation [13–17].

Water management involves several phenomena. First, water is produced by the ORR in the CL. Owing to the difference in water concentration, water diffuses from the cathode (high water concentration) to the anode (low water concentration). Because of the migration of protons, water molecules are dragged from the anode to the cathode (electroosmotic drag). To mitigate membrane dehydration, the gas streams may become humidified, so water is introduced into the fuel cell by an external humidifier. The water distribution depends strongly on the local temperature distribution. The main effects of the heat management are condensation of water vapor in the gas streams resulting in flooding or having undersaturated gases which in turn leads to dehydration of the membrane. The complexity of the water management of fuel cells becomes evident if the dependency of all the phenomena is understood: high current densities lead to high water production, resulting in high water diffusion from the cathode to the anode, but also to a high electroosmotic drag from the anode to the cathode. The water production might lead to pore flooding in the CL or GDL, which might stop the gas transport and thus the reaction. On the other hand, the heat production will rise due to the low electrical efficiency of about 50%. This might cause dry zones of low humidity, which could enforce the dehydration of the membrane and thus minimize the proton conductivity and the reaction.

A further effect to be considered, making the description even more complex, is that the membrane tends to swell with high water content and shrink with low water content. This means that the porous layers, especially the GDL, are compressed and the free pore space is reduced by the swelling membrane. Additionally, the wetting properties of the surfaces, especially of the GDL, depend on the surface treatment and might change during the lifetime of the cell due to degradation.

Some approaches to solving the challenges in water management of PEMFCs are discussed in Section 5.4.

5.3.2
Micro-Reformed Hydrogen Fuel Cell

To overcome the problem of the complex hydrogen storage, an on-board fuel reformer unit, reforming either hydrocarbons or chemical hydrides, is possible [18] (Figure 5.2). The challenge is to down-scale the reformer, since increasing the surface-to-volume ratio leads to higher heat losses to the environment. Additionally, gas purification units are necessary to avoid catalyst poisoning since the hydrogen purity after the first reforming step is not high enough for a conventional MEA. Here, the use of a polybenzimidazole (PBI)-based high temperature-tolerant proton-exchange membrane may facilitate the design of the micro fuel processor by direct feeding of reformed gas to the fuel cell without pretreatment [2]. The required BoP for combining reformer and fuel cell did not scale with the same magnitude. Here, microelectromechanical system (MEMS) technology has high potential to minimize a reformer/fuel cell unit.

5.3.3
Direct Methanol Fuel Cell (DMFC)

A DMFC system is an attractive power source because of the high energy density of methanol and its availability and rapid refilling capability. The last advantage results from the fact that methanol is a liquid fuel. This makes it relatively simple to install an appropriate infrastructure, based on existing infrastructures. To realize a real competitor to a battery or even to a methanol-reforming system combined with a PEMFC, the power density and fuel utilization have to be increased. For this, a solution has to be found to minimizing the fuel crossover through the membrane, which is one of the main performance losses. If a membrane is found

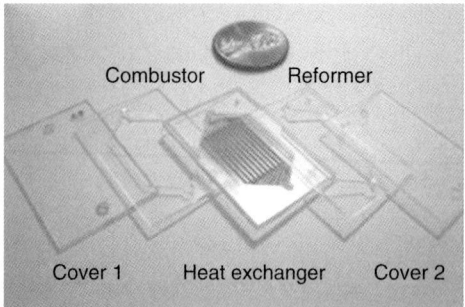 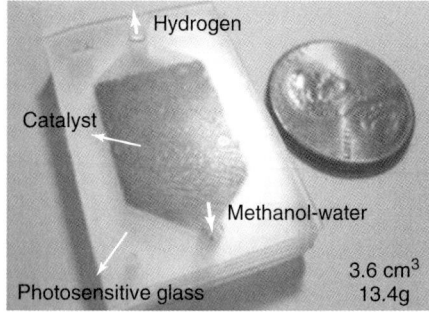
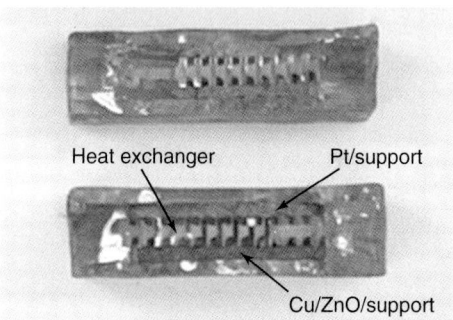 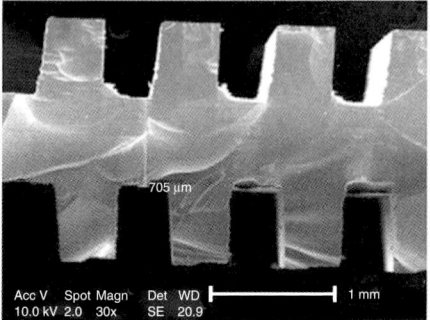

Figure 5.2 Complete micro methanol reformer. From Kim [19], with permission of *International Journal of Hydrogen Energy*.

that is nonpermeable for methanol, pure methanol can be used instead of a highly diluted aqueous methanol solution, which is the method used so far. Investigating various types of materials of interesting for BPs is the focus of research activities aimed at miniaturization. These materials include silicon [20], printed-circuit boards [21, 22], photopolymers [23], and flexible polymer substrates [24].

5.3.4
Direct Ethanol Fuel Cell (DEFC)

Direct ethanol fuel cells (DEFCs) produce power directly from ethanol without prior reforming. Compared with methanol used as the fuel for DMFCs, ethanol is nontoxic, environmentally friendly, and universally available, and making the handling easy. Since ethanol is also a liquid alcohol like methanol, the technological issues of crossover, discharge of carbon dioxide, and passive operation are comparable.

The main challenge is the electrocatalytic cleavage of the carbon–carbon bond and the complete reaction from ethanol to carbon dioxide. This reaction is enhanced in alkaline compared with acidic chemistry. Additionally, an alkaline environment permits the use of non-precious metals as catalysts. As in alkaline fuel cells, hydroxyl ions are transported from the cathode to the anode, water is produced at the anode, and no electroosmotic drag occurs. Drying out is very unlikely owing to the use of ethanol–water solutions at the anode. In addition to these advantages over proton-conducting fuel cells, the main challenge is to inhibit the reaction of KOH and CO_2 from the cathode air to give K_2CO_3, which will precipitate and block the electrodes. Also, state-of-the-art membranes have to be washed with alkaline solutions (KOH). KOH is not fixed within the membrane, and will diffuse out of the latter. This means that the user has to be aware of the presence of liquid with pH 14.

5.4
Materials and Manufacturing

5.4.1
Miniaturization

For micro fuel cells, miniaturization of the fuel-cell system is a key issue. Research in this area includes developing passively operating fuel cells in order to reduce the peripheral spends and the parasitic energy losses and minimize the system complexity. Development must also focus on low-cost production materials such as silicon or thermoplastics, and fabrication techniques such as MEMS technology or injection molding.

Most developers focus on a diffusion-driven air supply with air breathing cathodes. This can be realized by planar fuel cell designs, where the cells are not stacked, as in high-power fuel cells, but rather are oriented side-by-side in one plane (Figure 5.3).

In order to minimize the distance between the cells and to reduce the efforts for assembly, a monoelectrolytic design is preferred. In such a design, only one

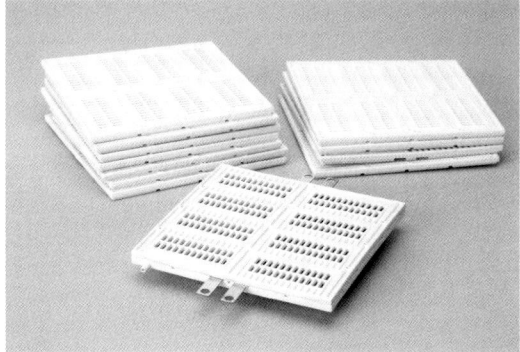

Figure 5.3 Planar fuel cells (here DMFC), which are manufactured by injection molding. In front are modules with eight cells. One can clearly see the open cathodes, which allow a diffusion-driven air supply. Developed by FWB Kunststofftechnik and Fraunhofer ISE (Germany).

Figure 5.4 Segmentation of an MEA by laser ablation. The cross-section shows the ablation of the electrode by a laser beam and the corresponding and well-defined segmentation.

membrane is used for all cells and no sealing between the cells is necessary. To realize this design, the electrodes of the MEA have to be segmented in order to avoid a short-circuit between the cells. This can be done by laser ablation (Figure 5.4): a laser beam evaporates the electrodes immediately on both sides. For the PEMFC with very thin membranes, a cell interval of 40 μm is sufficient to avoid a protonic short-circuit, which corresponds to the width of several laser beams [25]. Laser ablation further guarantees that the segments of both electrodes are positioned exactly with respect to each other.

By using a monoelectrolytic design, the electrical connection from one cell to the next has to be realized externally to the membrane. This can easily be done with electrically isolated printed circuit boards (PCBs) made of a polymer or resin material and also with ceramic multilayer technologies. Of course, other approaches such as using micro-system technology, thin-film technology, or wafer technology are also appropriate [4].

An alternative way to miniaturize fuel cells is found in microfluidic fuel cells. In these cells, oxidants and fuel co-flow without a physical barrier such as a membrane [26, 27]. Some researchers have developed selective electrodes, which avoid mixed potentials, and can thus operate with a mixed flow of oxidants and fuel in one flow field [28].

5.5
GDL Optimization

Avoidance of electrode flooding requires that water is transported from the electrodes out of the cell. A critical component is the GDL, which is a multifunctional component. Its microstructure assures a fine distribution of the reactants over the electrodes. It is a spatially finely distributed current collector. Further, it reduces the mechanical stress on the membrane by its compressibility, and the pores of the GDL allow the diffusion and storage of water. Many investigations have been carried out on the surface treatment of GDLs, including the application of a microporous layer (MPL) and a hydrophilic and also a hydrophobic coating [29–32]. The optimum design depends largely on the overall boundary conditions such as membrane material and thickness, electrode configuration, gas flows, operation and control strategies, and environmental conditions. As the physical understanding of two-phase flow in porous media with mixed wettability and its treatment is not yet fully elucidated, experimental analysis is still necessary [33].

As the lateral transport of water in a GDL is faster than the vertical transport, creating vertical water transport channels (WTCs) can optimize the transport properties in the GDL. Here again, experimental data show the applicability of laser techniques, where tailored WTCs significantly enhance the water transport in the vertical direction by the use of capillary forces [33].

5.5.1
Flow-Field Design

The outer layer of a fuel cell is the current collector, or BP, for the case of a cell in a stack. As the name implies, the BP carries the electrons from the anode of one cell to the cathode of the adjacent cell. Additionally, a flow channel structure is embedded in the BP to deliver oxygen and fuel over the cell area. Further, the liquid water that is produced by the reaction is removed by the gas stream in the channels. By means of an elaborate flow-field design, the appearance of inhomogeneties in reactant concentrations, humidity, and temperature within the cell area can be reduced to a certain extent. The gas velocity and the pressure loss in the flow channel strongly impact liquid water removal in the channel itself and in the subjacent GDL by gas shortening. The BPs give the fuel cell mechanical stability and therefore assure a homogeneous contact pressure within the cell area. For large fuel cells, the BPs are conventionally made of metal or graphite. For micro fuel cells, well-known materials from micro- and nanotechnology were considered, since the fabrication processes in this area are well established. Silicon-based separators were identified by Kim *et al.* [34] as a successful alternative to conventional BPs for a micro PEMFC. Typical MEMS technologies such as deep silicon etching, photomasked electroplating, physical vapor deposition (PVD), anodic bonding, and spin coating can be applied to create a flow pattern in silicon wafers [35]. Lee and Chuang [36] replaced the conventional GDL by a porous silicon substrate. The platinum catalyst was deposited on the surface and inside the porous silicon by PVD. Zhang *et al.* [37] developed a novel porous gas diffusion medium with improved thermal and electrical conductivities and controllable porosity fabricated from a copper foil using micro-/nanotechnology in a PEMFC. In addition to silicon, low-temperature co-fired ceramic (LTCC) material can be used as an alternative for the BP of micro fuel cells. For series-connected planar fuel cells, PCB offer an interesting platform, since a multilayer design provides high flexibility for the current collector [2]. Madou and co-workers used carbon obtained by pyrolyzing polymer precursors (called

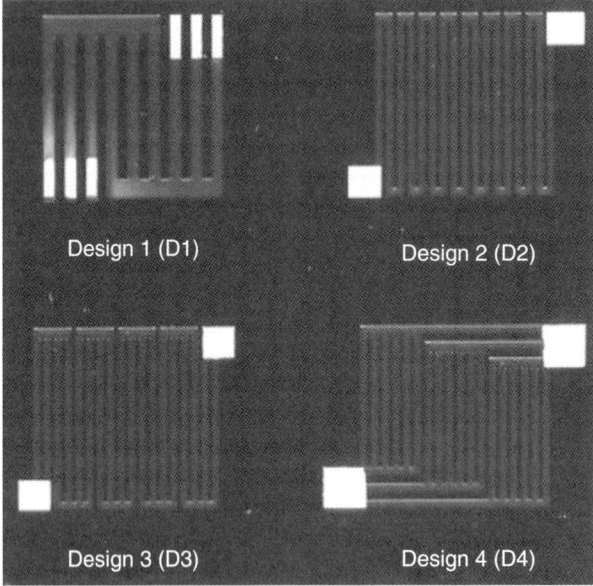

Figure 5.5 Four designs of flow fields for micro DMFCs. From Lu and Reddy [44], with permission of *International Journal of Hydrogen Energy*.

C-MEMS process) for the bipolar fluidic plates [38–40]. Chan et al. [5] developed a micro fuel cell using polymeric micromachining. The MEA was embedded in a gold-coated poly(methyl methacrylate) substrate. Hsieh et al. [41] proposed an SU-8 photoresist microfabrication process for the fuel-cell flow structure. Hahn et al. [42] used reactive ion etching (RIE) to machine microchannels in stainless-steel plates. Seyfang et al. [43] developed a novel, simplified concept for micro PEMFCs, omitting the GDL by using micro flow channels in glassy carbon sheets.

Four designs of flow fields for micro DMFCs are shown in Figure 5.5.

When water droplets reach the surface of the GDL and come into contact with the flow field, the flow-field design dictates the water management. If the gases are forced with active components through the flow-field channels (e.g., by a compressor or a fan), a one-channel meander design allows maximum water transport. On the other hand, a one-channel meander leads to a high pressure drop, which needs strong force. Parallel channels have the lowest pressure drop, but if one channel is blocked, for example by a water droplet, no active flow will free the channel. An interdigitated flow field, which forces the gas flow over the channel rib through the GDL, has many disadvantages for the water management, as the pressure drop is high and water droplets are collected at the ends of the channels without the possibility of them being forced out of the flow field. Often a meander design of several channels is chosen as an optimum [45]. In a PEMFC, to force water droplets out of the channels purging is often used at the anode flow field. Here a valve at the channel end opens to push out the anode gases and water.

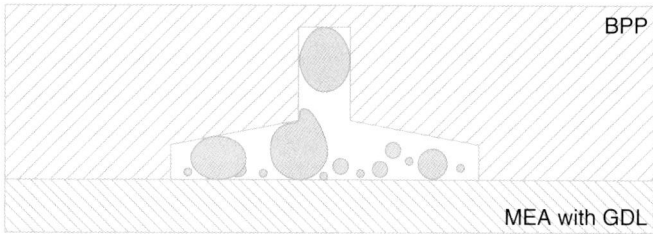

Figure 5.6 Example of a geometry that uses capillary forces to drive out water droplets passively [46].

In addition to hydrogen, inert gases such as nitrogen, which pass through the membrane from the cathode side, are also found at the anode. Because hydrogen is released by the purging, developers try to prolong the time between purgings.

To reduce the necessary gas flow and minimize the need for purging, the use of capillary forces may be helpful. On the one hand, the flow-field channel design could be arranged to make use of capillary forces to drive water droplets out of the channel (Figure 5.6). This could be realized by a T-shaped channel geometry with a tapered ceiling of the bottom channel, which leads into a small top channel [45].

Another approach is to integrate capillaries into the separator plate of the flow field [47–49]. The capillaries end within the channel and suck water droplets to the other side of the separator plate, where they evaporate. Using a porous material, which is laid over the separator plate, the distribution of the water droplets over the outer surface of the separator plate and the evaporation are enhanced. If done so, a gas-impermeable membrane should be integrated in order to prevent the loss of hydrogen on the anode side.

5.5.2
Miniaturized DMFC

DMFCs convert methanol directly into power, heat, and carbon dioxide. There are several issues particular to this type of cell that must be addressed.

As the methanol molecule is chemically very similar to the water molecule, methanol also diffuses into the typically used Nafion membrane, which leads to a loss of the fuel. Even more critical, however, is the mixed potential, due to the presence of methanol on both the anode and cathode. Additionally, methanol is toxic and release of methanol into the environment should be minimized. There are two conventional, well-established methods to treat this problem. First, thick membranes are used. This means that the ionic resistance is also relatively high, which leads to a lower power density compared with a PEMFC. Second, methanol is used in a very low-concentration solution at the anode (usually $1–2\,\mathrm{mol\,l^{-1}}$). To take advantage of highly concentrated methanol as a fuel in the cartridge and the high energy density, the fuel within the system must be diluted with product water.

5.5.3
Discharge of Carbon Dioxide

If not discharged, the CO_2 gas produced in the anode reaction would fill and block the flow field [50]. Therefore, the gas flow has to be separated from the liquid flow of the methanol–water solution in the anode. One approach is to use microstructured flow fields with a T-shaped channel design. The methanol is distributed in a wide channel on the bottom. If CO_2 bubbles evolve, they will grow until they reach the ceiling of the channel. If the channel ceiling is tapered and leads to a small channel, set on top of the wide channel, the bubbles will be driven into the small channel by capillary forces [51]. The same principle is used for separating water droplets from a gas stream, as described above.

As in micro fuel cells, the use of a valve is not desirable owing to size, reliability, costs, and parasitic energy losses. The integration of a CO_2-permeable membrane can be used to separate the gas bubbles from the methanol solution. Additionally, the buoyancy of the bubbles in the aqueous solution can be used to force the methanol convection [52, 53].

5.5.4
Passively Operating DMFC

Passive DMFCs usually operate with air-breathing open cathodes in a planar design. The main challenge is the passive operation of the anode. If liquid methanol or a methanol–water solution is used, capillary forces can be made use of by appropriately designing the microchannels. Unfortunately, most capillary flow-field designs need clearly defined structures so that the bubbles experience continuously growing forces along their way through the flow-field channel. If a GDL with its undefined surface structure is used above the channel, the driving force for the gas bubbles may be disturbed. Also, the two-phase problem of both liquid methanol and gaseous CO_2 is an issue.

An interesting solution might be to operate with methanol vapor so that only a one-phase gas flow appears at the anode [54]. Different methods for evaporating methanol passively or actively in micro DMFCs have been described [55, 56].

With vapor-fed micro DMFCs, a higher fuel efficiency, which also means a higher energy density and a better performance in addition to a higher power density, could be achieved [56, 57].

5.6
Conclusion

To realize micro fuel cells, there is a wide diversity of technological approaches beginning with a variety of primary fuels, different fuel cell technologies, and a range of manufacturing possibilities. Much work has already been done in the field of micro fuel cells, which have contributed to the understanding of the processes, effects, and interdependencies in fuel-cell operation. Commercialization of micro fuel cell on a large scale s is not foreseeable in the near future. Micro fuel-cell systems still face challenges. The advantages of long operation times with no or little maintenance and the high energy content of the primary fuels have already contributed to market-ready micro fuel-cell products in niche markets within the industrial sector, such as container tracking and sensors. Micro fuel-cell products for the consumer mass market, such as mobile phones and laptops, are still in the development phase, where miniaturization remains a key focus.

References

1. Wang, Y., Chen, K.S., Mishler, J., Cho, S.C., and Adroher, X.C. (2011) A review of polymer electrolyte membrane fuel cells: technology, applications, and needs on fundamental research. *Appl. Energy*, **88** (4), 981–1007.
2. Kundu, A., Jang, J.H., Gil, J.H., Jung, C.R., Lee, H.R., Kim, S.H., Ku, B., and Oh, Y.S. (2007) Micro-fuel cells – current development and applications. *J. Power Sources*, **170** (1), 67–78.
3. Butler, J. (2009) Portable Fuel Cell Survey 2009, Fuel Cell Today, Royston, available at *http://www.fuelcelltoday.com* (last accessed 29 November 2011).
4. Cowey, K., Green, K.J., Mepsted, G.O., and Reeve, R. (2004) Portable and military fuel cells. *Curr. Opin. Solid State Mater. Sci.*, **8** (5), 367–371.
5. Chan, S.H., Nguyen, N.-T., Xia, Z., and Wu, Z. (2005) Development of a polymeric micro fuel cell containing laser-micromachined flow channels. *J. Micromech. Microeng.*, **15** (1), 231–236
6. Cha, S.W., O'Hayre, R., Saito, Y., and Prinz, F.B. (2004) The scaling behavior of flow patterns: a model investigation. *J. Power Sources*, **134** (1), 57–71.
7. Hsieh, S.-S., Yang, S.-H., and Feng, C.-L. (2006) Characterization of the operational parameters of a H_2/air micro PEMFC with different flow fields by impedance spectroscopy. *J. Power Sources*, **162** (1), 262–270.
8. Hsieh, S.-S., Feng, C.-L., and Huang, C.-F. (2006) Development and performance analysis of a H_2/air micro PEM fuel cell stack. *J. Power Sources*, **163** (1), 440–449.
9. Kurz, T. and Keller, J. (2011) Heat management in a portable high temperature PEM fuel cell module with open cathode. *Fuel Cells*, **11** (4), 518–525.
10. Chang, J.-K., Tsai, H.-Y., and Tsai, W.-T. (2008) Effects of post-treatments on microstructure and hydrogen storage performance of the carbon nano-tubes prepared via a metal dusting process. *J. Power Sources*, **182** (1), 317–322.
11. Dimitrakakis, G.K., Tylianakis, E., and Froudakis, G.E. (2008) Pillared graphene: a new 3-D network nanostructure for enhanced hydrogen storage. *Nano Lett.*, **8** (10), 3166–3170.
12. Anon. (2009) Graphene-CNT superstructure holds promise for hydrogen storage. *Nano Today*, **4** (1), 6.
13. Li, H., Tang, Y., Wang, Z., Shi, Z., Wu, S., Song, D., Zhang, J., Fatih, K., Zhang, J., Wang, H., Liu, Z., Abouatallah, R., and Mazza, A. (2008) A review of water flooding issues in the proton exchange membrane fuel cell. *J. Power Sources*, **178** (1), 103–117.
14. Bazylak, A. (2009) Liquid water visualization in PEM fuel cells: a review. *Int. J. Hydrogen Energy*, **34** (9), 3845–3857.
15. Owejan, J.P., Gagliardo, J.J., Sergi, J.M., Kandlikar, S.G., and Trabold, T.A.. (2009) Water management studies in PEM fuel cells. Part I: fuel cell design and *in situ* water distributions. *Int. J. Hydrogen Energy*, **34** (8), 3436–3444.
16. Lu, Z., Kandlikar, S.G., Rath, C., Grimm, M., Domigan, W., White, A.D., Hardbarger, M., Owejan, J.P., and Trabold, T.A. (2009) Water management studies in PEM fuel cells. Part II: *ex situ* investigation of flow maldistribution, pressure drop and two-phase flow pattern in gas channels. *Int. J. Hydrogen Energy*, **34** (8), 3445–3456.
17. Yousfi-Steiner, N., Moçotéguy, P., Candusso, D., Hissel, D., Hernandez, A., and Aslanides, A. (2008) A review on PEM voltage degradation associated with

water management: impacts, influent factors and characterization. *J. Power Sources*, **183** (1), 260–274.

18. Qi, A., Peppley, B., and Karan, K. (2007) Integrated fuel processors for fuel cell application: a review. *Fuel Process. Technol.*, **88** (1), 3–22.

19. Kim, T. (2009) Micro methanol reformer combined with a catalytic combustor for a PEM fuel cell. *Int. J. Hydrogen Energy*, **34** (16), 6790–6798.

20. Jiang, Y.Q., Wang, X.H., Zhong, L., and Liu, L. (2006) Design, fabrication and testing of a silicon-based air-breathing micro-direct methanol fuel cell. *J. Micromech. Microeng.*, **16**, 233–239.

21. O'Hayre, R., Braithwaite, D., Hermann, W., Lee, S.-J., Fabian, T., Cha, S.-W., Saito, Y., and Prinz, F.B. (2003) Development of portable fuel cell arrays with printed-circuit technology. *J. Power Sources*, **124** (2), 459–472.

22. Lim, S.W., Kim, S.W., Kim, H.J., Ahn, J.E., Han, H.S., and Shul, Y.G. (2006) Effect of operation parameters on performance of micro direct methanol fuel cell fabricated on printed circuit board. *J. Power Sources*, **161** (1), 27–33.

23. Cha, S.W., O'Hayre, R., Park, Y.-I., and Prinz, F.B. (2006) Electrochemical impedance investigation of flooding in micro-flow channels for proton exchange membrane fuel cells. *J. Power Sources*, **161** (1), 138–142.

24. Ito, T., Kimura, K., and Kunimatsu, M. (2006) Characteristics of micro DMFCs array fabricated on flexible polymeric substrate. *Electrochem. Commun.*, **8** (6), 973–976.

25. Schmitz A., Wagner S., Hahn R., Weil A., Tranitz M., and Hebling C. (2003) Segmentation of MEA by laser ablation. In Proceedings of 2nd European Polymer Electrolyte Fuel Cell Forum, Lucerne, Switzerland, pp. 323–339.

26. Kjeang, E., Djilali, N., and Sinton, D. (2009) Microfluidic fuel cells: a review. *J. Power Sources*, **186** (2), 353–369.

27. Mousavi Shaegh, S.A., Nguyen, N.-T., and Chan, S.H. (2011) A review on membraneless laminar flow-based fuel cells. *Int. J. Hydrogen Energy*, **36** (9), 5675–5694.

28. Priestnall, M.A., Kotzeva, V.P., Fish, D.J., and Nilsson, E.M. (2002) Compact mixed-reactant fuel cells. *J. Power Sources*, **106** (1–2), 21–30.

29. Park, G.-G., Sohn, Y.-J., Yang, T.-H., Yoon, Y.-G., Lee, W.-Y., and Kim, C.-S. (2004) Effect of PTFE contents in the gas diffusion media on the performance of PEMFC. *J. Power Sources*, **131** (1–2), 182–187.

30. Atiyeh, H.K., Karan, K., Peppley, B., Phoenix, A., Halliop, E., and Pharoah, J. (2007) Experimental investigation of the role of a microporous layer on the water transport and performance of a PEM fuel cell. *J. Power Sources*, **170** (1), 111–121.

31. Nakajima, H., Konomi, T., and Kitahara, T. (2007) Direct water balance analysis on a polymer electrolyte fuel cell (PEFC): effects of hydrophobic treatment and micro-porous layer addition to the gas diffusion layer of a PEFC on its performance during a simulated start-up operation. *J. Power Sources*, **171** (2), 457–463.

32. Kitahara, T., Konomi, T., and Nakajima, H. (2010) Microporous layer coated gas diffusion layers for enhanced performance of polymer electrolyte fuel cells. *J. Power Sources*, **195** (8), 2202–2211.

33. Gerteisen, D., Heilmann, T., and Ziegler, C. (2008) Enhancing liquid water transport by laser perforation of a GDL in a PEM fuel cell. *J. Power Sources*, **177** (2), 348–354.

34. Kim, J.-Y., Kwon, O.J., Hwang, S.-M., Kang, M.S., and Kim, J.J. (2006) Development of a miniaturized polymer electrolyte membrane fuel cell with silicon separators. *J. Power Sources*, **161** (1), 432–436.

35. Lee, S.-J., Cha, S.W., Liu, Y.C., O'Hayre, R., and Prinz, F.B. (2000) High power-density polymer–electrolyte fuel cells by microfabrication, *Micro Power Sources*. **2000-3**, 121–136.

36. Lee, C.-Y. and Chuang, C.-W. (2007) A novel integration approach for combining the components to minimize a micro-fuel cell. *J. Power Sources*, **172** (1), 115–120.

37. Zhang, F.-Y., Advani, S.G., and Prasad, A.K. (2008) Performance of a metallic

38. Lin, P.-C., Park, B.Y., and Madou, M.J. (2008) Development and characterization of a miniature PEM fuel cell stack with carbon bipolar plates. *J. Power Sources*, **176** (1), 207–214.
39. Wang, Y., Pham, L., Vasconcellos, G.P.S.d., and Madou, M. (2010) Fabrication and characterization of micro PEM fuel cells using pyrolyzed carbon current collector plates. *J. Power Sources*, **195** (15), 4796–4803.
40. Park, B.Y. and Madou, M.J. (2006) Design, fabrication, and initial testing of a miniature PEM fuel cell with micro-scale pyrolyzed carbon fluidic plates. *J. Power Sources*, **162** (1), 369–379.
41. Hsieh, S.-S., Huang, C.-F., Kuo, J.-K., Tsai, H.-H., and Yang, S.-H. (2005) SU-8 flow field plates for a micro PEMFC. *J. Solid State Electrochem.*, **9** (3), 121–131.
42. Hahn, R., Wagner, S., Schmitz, A., and Reichl, H. (2004) Development of a planar micro fuel cell with thin film and micro patterning technologies. *J. Power Sources*, **131** (1–2), 73–78.
43. Seyfang, B.C., Kuhnke, M., Lippert, T., Scherer, G.G., and Wokaun, A. (2007) A novel, simplified micro-PEFC concept employing glassy carbon micro-structures. *Electrochem. Commun.*, **9** (8), 1958–1962.
44. Lu, Y. and Reddy, R.G. (2011) Effect of flow fields on the performance of micro-direct methanol fuel cells. *Int. J. Hydrogen Energy*, **36** (1), 822–829.
45. Metz, T., Paust, N., Mfiller, C., Zengerle, R., and Koltay, P. (2008) Passive water removal in fuel cells by capillary droplet actuation. *Sens. Actuators A Phys.*, **143** (1), 49–57.
46. Eccarius, S., Litterst, C., and Koltay, P. (2006) Offenlegungsschrift DE 10 2006 002 926 A1.
47. Strickland, D.G. and Santiago, J.G. (2010) *In situ*-polymerized wicks for passive water management in proton exchange membrane fuel cells. *J. Power Sources*, **195** (6), 1667–1675.
48. Ge, S., Li, X., and Hsing, I.M. (2005) Internally humidified polymer electrolyte fuel cells using water absorbing sponge. *Electrochim. Acta*, **50** (9), 1909–1916.
49. Litster, S. and Santiago, J.G. (2009) Dry gas operation of proton exchange membrane fuel cells with parallel channels: non-porous versus porous plates. *J. Power Sources*, **188** (1), 82–88.
50. Argyropoulos, P., Scott, K., and Taama, W.M. (1999) Carbon dioxide evolution patterns in direct methanol fuel cells. *Electrochim. Acta*, **44** (20), 3575–3584.
51. Litterst, C., Eccarius, S., Hebling, C., Zengerle, R., and Koltay, P. (2006) Increasing mu DMFC efficiency by passive CO_2 bubble removal and discontinuous operation. *J. Micromech. Microeng.*, **16** (9), S248–S253.
52. Prakash, S., Mustain, W., and Kohl, P.A. (2008) Carbon dioxide vent for direct methanol fuel cells. *J. Power Sources*, **185** (1), 392–400.
53. Meng, D.D. and Kim, C.J. (2009) An active micro-direct methanol fuel cell with self-circulation of fuel and built-in removal of CO_2 bubbles. *J. Power Sources*, **194** (1), 445–450.
54. Chang, I., Ha, S., Kim, S., Kang, S., Kim, J., Choi, K., and Cha, S.W. (2009) Operational condition analysis for vapor-fed direct methanol fuel cells. *J. Power Sources*, **188** (1), 205–212.
55. Guo, Z. and Faghri, A. (2007) Vapor feed direct methanol fuel cells with passive thermal-fluids management system. *J. Power Sources*, **167** (2), 378–390.
56. Eccarius, S., Krause, F., Beard, K., and Agert, C. (2008) Passively operated vapor-fed direct methanol fuel cells for portable applications. *J. Power Sources*, **182** (2), 565–579.
57. Kim, H. (2006) Passive direct methanol fuel cells fed with methanol vapor. *J. Power Sources*, **162** (2), 1232–1235.

6
Principles and Technology of Microbial Fuel Cells

Jan B.A. Arends, Joachim Desloover, Sebastià Puig, and Willy Verstraete

6.1
Introduction

Microbial fuel cells (MFCs), or as they are called nowadays bioelectrochemical systems (BESs), were first described about 100 years ago. Ever since Potter [1], inspired by the findings of other biologists in the late nineteenth and early twentieth centuries, devised the first MFC, these devices have undergone substantial research and development. An MFC can be distinguished from other fuel cells by the fact that at least one of the two reactions is catalyzed by a biological component (Figure 6.1) [2]. Generally this is the anode reaction, where organic carbon or another electron donor is oxidized by microorganisms while a solid-state electron acceptor (i.e., the electrode) is subsequently reduced.

Since a few years ago, microbial-catalyzed cathode reactions have also been investigated and developed. The microorganisms present on the cathodic electrode are able to receive electrons from a solid material and subsequently reduce a final electron acceptor. Nutrients for growth and maintenance are obtained from the liquid or gas phase. Not only whole microorganisms are able to catalyze redox reactions, but also enzymes can have a catalytic function in a fuel cell [3]. The focus of this chapter is on microbial-catalyzed reactions although the basics also hold true for enzymatic fuel cells. In contrast to chemical fuel cells (CFCs), biological fuel cells (BFCs) are more delicate as extreme operating temperatures ($4\,°C < T_{optimal} > 40\,°C$) and pressures ($>1$ bar) have an adverse effect on the biological catalysts. Microbiological life has been observed above these thresholds, up to $121\,°C$ and $110\,MPa$ [4, 5], but these microorganisms are not found in abundance. Although no BES has yet been described under these conditions, some thermophilic organisms are able to respire with metal oxides at $100\,°C$ [6]. Some of the same materials and even the same reactants can be applied in a BES as in a CFC. For the electrodes, materials such as carbon, stainless steel, palladium, and platinum are often used. Electron donors and acceptors used in a CFC can also be used in a BES, that is, hydrogen and methanol as anodic reactants, while oxygen is most frequently used in the cathode. Hydrogen and small organic compounds can also be used in a biologically catalyzed anode, whereas oxygen is a suitable

Fuel Cell Science and Engineering: Materials, Processes, Systems and Technology,
First Edition. Edited by Detlef Stolten and Bernd Emonts.
© 2012 Wiley-VCH Verlag GmbH & Co. KGaA. Published 2012 by Wiley-VCH Verlag GmbH & Co. KGaA.

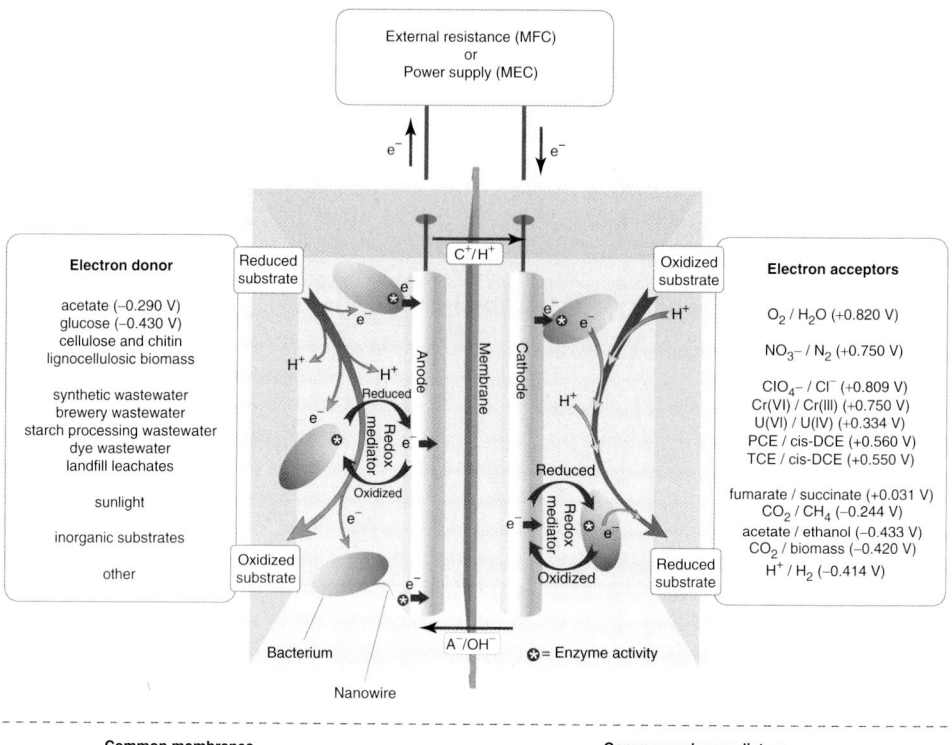

Figure 6.1 Overview of a BES with examples of electron donors and acceptors, membranes and redox mediators. A^-: anion; C^+: cation; e^-: electron.

reactant for a biologically catalyzed cathode. In addition to these small molecules, heterogeneous mixtures of larger molecules or recalcitrant compounds can also be applied in a BES.

Regarding the naming of these devices, many different terms have been coined. An MFC designates a fuel cell where one or both reactions are microbial catalyzed. A BFC has the same meaning but can also include the use of purified enzymes to catalyze a reaction. A microbial electrolysis cell (MEC) is a system where no power is harvested. Instead, some extra potential is applied to electrons coming from the anode to perform a reaction that under ambient conditions, based on thermodynamics, is endergonic. It is counterintuitive to supply potential to a fuel cell but it is not about producing electrical power, but about creating a product at the cathode such as hydrogen (H_2) or methane (CH_4) gas, caustic soda (NaOH), hydrogen peroxide (H_2O_2), or small organic acids (e.g., acetate and propionate). MxC is the term used for research that can be applied to either an MFC or an MEC. In a microbial desalination cell (MDC), a third, middle, compartment is installed that can be filled with salt water to be desalinated by the electric field [7]. S-MFC and P-MFC denote a sediment MFC and a plant MFC [8], respectively. Here, organic material in a waterlogged soil and/or organic material provided by means of rhizodeposition from living plants in a waterlogged soil are the main anodic substrates. An MEC in a sediment setting has not yet been described. As the emphasis of MFC research has shifted recently from power output to other functions or applications, the general term BES will be used henceforth in this chapter.

In the following sections, a general overview of the microbial aspects of a BES is presented. An attempt is made to clarify the possibilities and challenges of BES research compared with conventional fuel cells discussed elsewhere in this book. This chapter should be considered as a starting point for delving further into the concept of BESs.

6.2
Materials and Methods

As BESs have not yet outgrown the laboratory, a wide variety of materials have been investigated for electrode purposes. However, the most commonly used material, for both the anode and cathode, is (modified) carbon. Materials used for the construction of the system are usually glass reactor flasks or plastic frames, both of which are, of course, nonconductive.

6.2.1
Electrode Materials

Electrode materials for a BES require a high specific surface area in order to create a high volumetric current density; $1.3 \text{ m}^2 \text{ g}^{-1}$ for granules (Mersen, Wemmel, Belgium) and $0.6 \text{ m}^2 \text{ g}^{-1}$ for felt materials (Alfa Aesar, Ward Hill, MA, USA) are commonly found in a BES. When reporting specific electrode surfaces, one has to keep in mind that most surface areas are determined by means of N_2 gas adsorption. With three-dimensional electrodes, it is essential to determine the surface available for biological interaction instead of the surface available for N_2 adsorption. Different types of carbon and stainless steel have mostly been used for both anode and cathode electrodes. Carbon materials that are frequently applied are granular, felt, cloth, brushes, or solid blocks [9–12].

As carbon itself has a low catalytic activity, various attempts have been made to increase the performance of carbon electrodes. The use and effectiveness of catalysts on an electrode depend on the catalyst loading and stability and, of course, the operating conditions of the electrode compartment [13, 14]. Especially the use of undefined (waste) streams in combination with a microbial catalyst can lead to fouling of the chemical coating on the electrode. Treatment of activated carbon felt with ethylenediamine and with nitric acid increased the power density by 25% and 58%, respectively, and decreased the start-up time of the bioanode. This increase was attributed to modified surface characteristics of the original material [15]. Tungsten carbide-modified anodes in combination with a pyrolyzed iron(II)phthalocyanine-modified cathode has also been shown to improve both anodic and cathodic reactions [16, 17]. This is remarkable since tungsten is not a noble metal and thus cheaper than platinum. Also, Park and Zeikus investigated the modification of electrode materials, and obtained a 1000-fold increase in power density when using a graphite anode doped with Mn^{4+} and a graphite cathode doped with Fe^{3+} atoms compared with woven graphite control electrodes [18].

The modification of the cathode with either a soluble iron-chelated compound or a fixed iron complex was investigated by Aelterman *et al.* [19]. They concluded that iron-EDTA coated on the cathode is a sustainable alternative. However, the power and current densities were the highest with the nonsustainable hexacyanoferrate iron complex. Unmodified carbon and stainless-steel felt were investigated for use as a biological oxygen-reducing cathode. Both are suitable materials but operational parameters such as salinity and presence of a biofilm have a strong impact on the performance of the material [14]. Freguia *et al.* showed that granular graphite can also efficiently catalyze the oxygen reduction reaction by itself without any (biological) catalyst [20]. Current densities (17 mA m^{-2} projected cathode area) were attributed to the high specific surface area of the material. A range of carbon materials were tested by Scott *et al.* [21] for application as a cathode in marine sediment fuel cells. These included unmodified carbon cloth, paper and sponge, unmodified graphite, and reticulated vitreous carbon (RVC).

In addition to unmodified materials, also materials modified with, for example, carbon paper coated with Fe and Fe–Co tetramethoxyphenylporphyrin (TMPP), platinized carbon, and platinized titanium electrodes were used [22]. From the results, it could be seen that the cathode catalyzed with Fe–CoTMPP resulted in a maximum power output of 60 mW m^{-2} whereas the unmodified carbon sponge gave about 38 mW m^{-2}. This showed that the choice of material is crucial, as modified materials do not always outperform unmodified electrode materials [21].

For MECs, research has been focused on finding materials for efficient hydrogen gas generation at the cathode. Selembo *et al.* studied metal-based electrodes such as stainless steel, nickel, and platinum. Their results showed good performance for a metal coated with an NiO$_x$ catalyst [23]. They noted, however, that long-term stability of the catalyst needs improvement.

Recent advances in nanofabrication provide a unique opportunity to develop efficient materials for BESs. Nanoparticles and carbon nanotubes have been used as materials for electrodes in BESs [9]. Nickel powders and both chemically and biologically produced palladium nanoparticles have been used for hydrogen production in an MEC [23–25]. Liu *et al.* demonstrated that nanostructured manganese oxide could also be used for oxygen reduction in an MFC [26]. Finally, the current density was improved 20-fold compared with a plain graphite anode using graphite disks with gold and palladium nanoparticles [27].

6.2.2
Membrane

As in a chemical fuel cell, a membrane is also of importance for a BES. Membraneless BESs also showed promising results and can be a reasonable alternative [28, 29]. The main purposes of a membrane are (i) to separate reactants in order to prevent internal short-circuiting and (ii) to provide a means of internal charge balancing between the anode and cathode reactions. Membranes used so far in BES research include dialysis membranes, cation-exchange membranes (CEMs), proton-exchange membranes (PEMs), and anion-exchange membranes (AEMs) [30, 31] (Figure 6.1). In addition to the conventional membranes, non-ion-exchange separators have also been investigated. These include cloth [32], filtration membranes [33], bipolar plates [34], and baked earth [35]. AEMs seem to be the best choice for decreasing internal resistance and increasing internal charge transport [30, 31]. The use of a membrane of any kind has the drawback that it promotes a pH gradient between the two compartments, which interferes with transport of OH^- and H^+ ions. A difference of one pH unit between the two compartments can lead to a voltage loss of 59 mV. This can be calculated with the Nernst equation [36].

6.2.3
Configurations and Design

Most BESs have been designed as conventional fuel cells, that is, two flat electrode compartments separated by a membrane. Design parameters can be adjusted from this traditional design. These parameters include electrode spacing, flow patterns, reactor volumes and electrode surface areas. The original H-type fuel cells are mostly used for studies of physiological parameters of the biocatalyst and not for optimizing output [37, 38].

In addition to the conventional design, tubular upflow designs have been developed to cope with high internal resistances to charge transfer [39, 40]. Some research has been focused on miniaturizing a BES, based on the concept that a smaller BES, having lower internal resistance, while electrically connected can deliver larger volumetric power outputs [33, 41]. Miniaturization has also led to systems that are easier to set up and can be studied in large numbers [42–44]. Another advantage of small systems is the possibility of studying pure or defined cultures in detail [42, 45, 46]. Most of this work has so far been focused on the anode compartment.

Small milliliter-scale fuel cells are mostly used with a membrane electrode assembly (MEA), which means that the conductive electrode material is coated on a membrane. This has led to a one-chamber design that contains the anode compartment. The cathode electrode is open to the air [47, 48].

Large-scale systems have also been used to study performance. However, results regarding cubic meter-scale BESs have not yet become available. Large-scale BESs are mostly stacked systems [49]. Owing to the biological nature, the phenomenon of stack reversal can occur, which leads to a decrease in usable output [50, 51]. So far, BESs with sizes up to 20 l have been characterized [52–55].

6.2.4
Measurements, Techniques, and Reporting Values

6.2.4.1 Biological Measurements

In a BES, the main catalysts on the electrode are of biological origin. A distinction can be made between living and nonliving catalysts. Living catalysts are microorganisms; the other type of catalysts of biological origin are enzymes and metabolites. In most BESs studied to date, mixed microbial communities were used to perform the electrochemical reactions at the anode and cathode. To determine the composition of these microbial communities, various techniques can be applied. All these techniques have their own (dis)advantages and the reader is referred to the various publications cited in this section for detailed information. These include (i) clone libraries followed by sequence analysis of the 16S rRNA gene [42, 56, 57] and (ii) denaturing gradient gel electrophoresis (DGGE), which separates the 16srRNA gene based on its GC content [58, 59]. The resulting "fingerprint" can be analyzed by means of the microbial resource management (MRM) toolbox [60, 61]. Another option is to perform cloning and sequencing on fragments of interest (in)directly from the gel. (iii) A third measurement technique to determine the composition is terminal restriction fragment length polymorphism (T-RFLP). The result is again a "fingerprint" of the microbial community, which can be studied by means of sequencing or the MRM toolbox.

In addition to these fingerprint techniques based on the molecular composition of the microbial catalyst, imaging techniques are also frequently applied. The two most frequently used methods are fluorescent *in situ* hybridization (FISH) and scanning electron microscopy (SEM). FISH enables a researcher to hybridize colored probes tagged with a specific DNA-sequence to certain (groups of) microorganisms. A biofilm can then be visualized by means of microscopy and the distribution of microorganisms in the biofilm observed. SEM is a technique that enables a researcher to see directly the microorganisms on the electrode and visualize biofilm structures [38, 62–64]. A drawback of these imaging techniques is that the biofilm does not remain completely intact during sample preparation, but also the limited availability of FISH probes hampers a visual inspection of an electroactive biofilm. A drawback of all of the techniques described above is that no direct link can be made between the metabolism of the microorganisms and their presence.

6.2.4.2 Electrochemical Measurements

Several electrochemical techniques have been used to study BESs. A fairly simple technique is to poise the potential of the electrode or cell voltage of the system

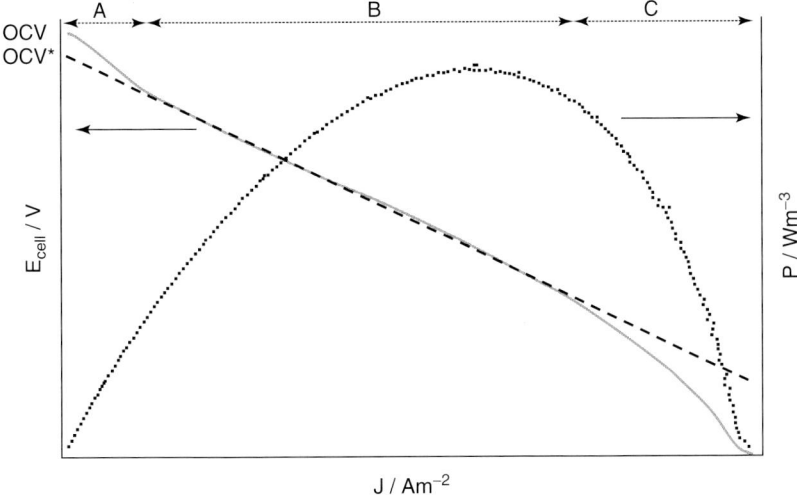

Figure 6.2 Typical polarization curves that can be recorded from a BES. OCV, open cell voltage; OCV*, open cell voltage used to determine current-dependent internal resistance. Sections: A, activation overpotentials are the dominant voltage loss in this region; B, current-dependent voltage loss is the dominant voltage loss in this region; C, transport and concentration polarization are the dominant voltage losses in this region. Solid line, cell voltage[E_{cell}(V)]; dotted line, power density [P(W m^{-3})]; dashed line, linearized cell voltage to determine R_{int}.

and record the resulting current output, also known as *potentiostatic control*. When the potential is varied over time and the resulting current is recorded, this is called a *potentiodynamic controlled experiment*. By varying the cell potential, either by potentiodynamic control or by stepwise lowering of the external resistance, one can record a polarization curve and subsequently calculate a power curve. A polarization and a power curve are a good approach to (readily) verify the capabilities of the system under study (Figure 6.2).

The internal resistances, or energy losses, of a fuel cell can be characterized using different methods (Figures 6.2 and 6.3). The four main methods that have been used in BES research are (i) current interrupt, (ii) calculation from a polarization curve, (iii) calculation from a power curve, and (iv) electrochemical impedance spectroscopy (EIS). Interpretation of the results from these calculations is a challenge since the biological nature of the catalyst and the polymorphous spacing of the catalyst on the electrode can severely influence the outcome of these measurements. The total internal resistance of a BES consists of various components (Figure 6.3) [2, 36, 65, 66]. A brief introduction to these various components based on a polarization curve is given here. However, owing to the intricate nature of a BES, the reader is referred to the literature cited for more detailed information.

The difference between the theoretical cell potential (calculated based on the redox reactions that occur at the electrodes and the Nernst equation) and the cell

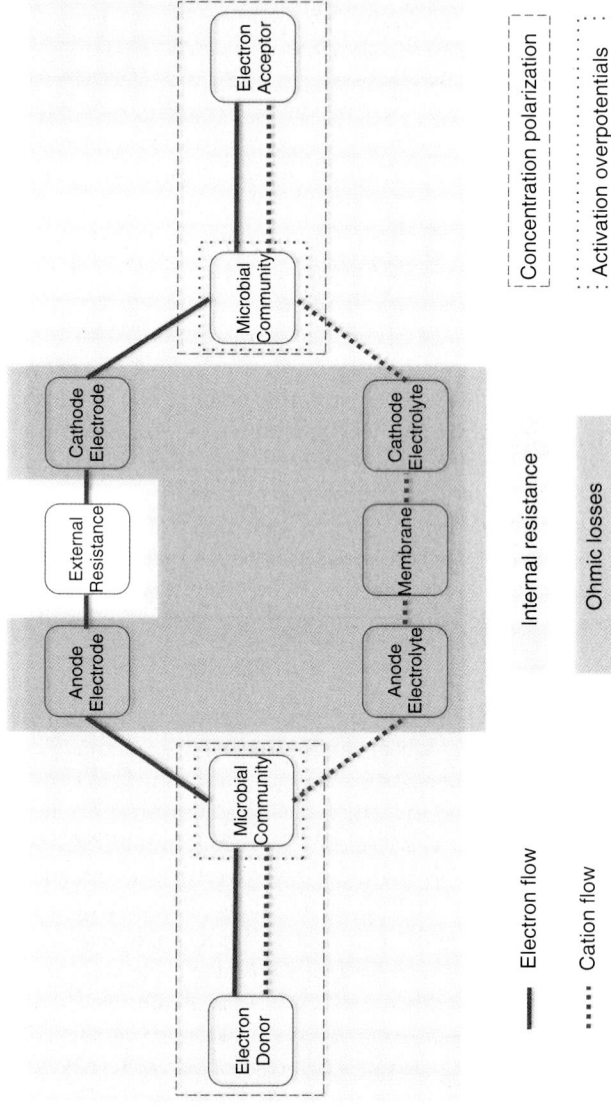

Figure 6.3 Schematic overview of internal resistances occurring in a BES. In an ideal situation, internal charge balancing is done by means of protons. However, cations but also anions can perform this task.

electromotive force E_{emf} or open-circuit voltage (OCV) arise due to the enzymes or terminal electron transport molecules that are involved in establishing the transfer of electrons to the electrode. This is dependent on the metabolism of the microorganism(s) catalyzing the reaction [2, 36, 65].

On closing the circuit and allowing a small current to run through the system, the activation overpotentials are the greatest source of voltage loss. Activation polarization losses can be determined by the Tafel equation [20] when concentration polarization is not taken into account (Figure 6.2, section A, and Figure 6.3) [2, 36, 65].

Allowing more current to flow through the system gives rise to ohmic losses (Figure 6.2, section B, and Figure 6.3). These losses are current dependent and become more influential compared with the overpotentials as more current is allowed to flow [2, 65]. These losses are due to electrical contacts between the electrode and current collector [2, 36, 65], anolyte and catholye conductivity, resistance to ion transport through the membrane/separator [31, 67], and the pH difference between the two electrodes [66]. Although the loss that occurs due to the increasing pH difference is a thermodynamic parameter (i.e., it alters the theoretical cell potential), the change in pH is caused by an increase in current and imperfect transport in the BES. The total sum of all internal losses during current generation can be determined from the linear part of the polarization curve with $R_{int} = (E^{*}_{OCV} - E_{cell})/I$ [66]. To determine the influence of the different components in the total internal resistance, separate experiments need to be conducted, such as conductivity and pH measurements. The voltage drop over the membrane can be measured with standard reference electrodes on both sides of the membrane, corrected for conductivity and pH of the electrolyte [31, 67]. The internal resistance calculated by Ohm's law can also be expressed as a parameter of the reactor under specific operating conditions. This is done by replacing current (I) by current density (J). The resulting resistance is consequently expressed in $\Omega\,m^2$ or $\Omega\,m^3$.

Increasing the current even more leads to a nonlinear voltage drop. The dominant factor for potential loss can now be related to mass transport limitation and concentration polarization (Figure 6.2, section C, and Figure 6.3). Examples are lack of reactant supply to the electrode and a build-up of product at the electrode. These voltage losses cannot be expressed as a resistance any longer owing to their independence of the current. Mass transport and concentration polarization losses can be amended by adopting different cell designs or flow regimes [68, 69].

The total internal resistance can also be obtained from a power curve from the notion that at maximum power the internal resistance is equal to the external resistance. Consequently, the internal resistance can be calculated with $R_{int} = P/I^2$. Recording a polarization curve in a BES needs to be performed with some care. Researchers have recorded curves with multiple fed-batch cycles and one external resistor per cycle, during one fed-batch cycle with multiple resistors, and during continuous-mode operation [70]. Therefore, it is advisable that before comparing results, the experimental conditions should also be taken into account. This was excellently shown by Winfield *et al.*, who analyzed the effect of scan rate, biofilm maturity and feedstock concentration on the polarization curve [71, 72].

A polarization curve can be recorded with simple means such as a variable resistor box. A potentiostat equipped with a three-electrode setup and appropriate software is used for polarization curves, EIS, and current interrupt.

The current interrupt method can be used to determine the ohmic resistance of a fuel cell. A resistor is used to close the circuit, which enables the cell to give a stable potential and current output. Subsequently, the external resistor is removed and the instantaneous potential change is used for the calculation of the ohmic resistance by means of $R_\Omega = \Delta V/I$ [36]. EIS is a more sophisticated technique to characterize BES [40]. It entails applying an alternating potential with set amplitude on a set cell potential. The results are analyzed by fitting the data to an equivalent circuit with the potentiostat software and internal resistance is determined from a Nyquist plot. However, the use of EIS has not been widely applied in BES research and therefore no consensus yet exists on frequency range, amplitude, and interpretation of the data with the equivalent circuit [10, 12, 40, 73–75].

Another technique used to study electron transfer mechanisms in more detail is cyclic voltammetry (CV) or linear sweep voltammetry (LSV) [36]. As with the techniques mentioned above, owing to the microbiological nature of the catalyst, the results must be interpreted carefully. Voltammetric techniques are used to characterize redox-active species in a three-electrode setup. This characterization can be carried out on liquid samples, either as such, centrifuged or filtered, or on electrodes with the biofilm attached. By scanning all these fractions, one can obtain an idea of the nature (soluble, membrane bound, etc.) and the mid-peak potentials of the redox-active compounds present. These data can help in interpreting the electron-transfer mechanisms that are being employed at the specific electrode. As most liquids used in BES research are fairly heterogeneous and sometimes undefined, mid-peak potentials and current responses do not always coincide with theoretical patterns or patterns obtained from the pure compound in defined media. Examples of differences in the physiological state of the microorganism and a detailed discussion on the use of CV in BES research were clearly presented by Fricke *et al.* [76]. CV was successfully used to demonstrate the presence of various indigenously produced redox-active mediators in the anolytes of various BESs [46, 77]. Biological catalysis of the cathodic oxygen reduction reaction has also been verified by CV [69, 78]. To perform a voltammetric technique, in the past a potentiostat was required; however, Cheng *et al.* recently devised a computer-controlled switchboard that performs the same function as a potentiostat [79]. In pure electrochemical research, the peak current of a redox signal can be related to the concentration of the active compound when the electrode surface area is known. So far, this analysis has not yet been performed on a BES. This is mainly due to the large and sometimes unknown surface area of the electrodes and to the heterogeneous mixture of redox-active compounds.

6.2.4.3 Reporting Performance

Various different means of reporting power, current, and potential have been described in the BES literature. It is essential, however, that outputs and results are reported in a standardized way. In fuel-cell research, current and power are reported as density per unit electrode surface area, which is a good standard measure for CFCs. However, in BES research, current densities have been reported in different ways. From an environmental engineering perspective, cubic meters are the preferred measure, whereas from a (bio)electrochemical perspective, square meters per unit electrode surface area are the preferred measure. This resulted in expressions such as current and power output per square meter of electrode surface, per square meter of projected electrode surface, per square meter of membrane area, per cubic meter of total anode, cathode, or reactor volume, and per cubic meter of net anode, cathode, or reactor volume. One can easily appreciate now that this plethora of reporting methods is prone to confusion. Since no standard has so far been set up, we recommend that workers should provide clearly the data needed to convert between various measurements in the experimental section of published papers or book chapters.

The same holds true for potentials; there is a standardized method, which is to report them versus the standard hydrogen electrode (SHE) or normal hydrogen electrode (NHE) [2, 36]. Furthermore, cell potentials are only relevant when the external resistance is also noted.

6.3
Microbial Catalysts

Microbial electrocatalysis relies on microorganisms as catalysts for reactions occurring at electrodes. The microorganisms involved are able to transport electrons in and out of the cell, a process known as extracellular electron transfer (EET), and can catalyze both oxidation and reduction reactions [80, 81]. Their catalytic properties have been confirmed by the fact that they are able to lower the overpotentials (lower energy loss) at both anodes [82] and cathodes [56, 69], giving an increased performance of the system. Nevertheless, they cannot be considered as true catalysts since part of the substrate/electron donor is consumed for growth.

So far, most research has been done on biological anodes, cathodes being mainly abiotic and of minor interest. In the last few years, researchers have realized that the (bio)cathode still remains one of the weakest points of this technology; this has led to a steep increase in the scope of possible electrode materials and cathode reactions, both chemical and biological.

6.3.1
Anode Reactions

In bioanodes, an (in)organic electron donor is oxidized by microorganisms with concomitant liberation of electrons and protons (Figure 6.1). The electrons produced are shuttled through the internal electron transport chain of the microorganisms and are deposited on the anode. The energy level of the electrons deposited on the electrode is dependent on the terminal electron transfer molecule.

This also determines the amount of energy that an organism can use for its own metabolism and the amount that is available for work. Two strategies for energy conservation by microorganisms are also used in bioanodes: *Shewanella* spp. make use of substrate-level phosphorylation whereas *Geobacter* spp. make use of oxidative phosphorylation for energy conservation. Electrons are transferred through a cascade of cytochromes, quinones, and other electron-transfer molecules from the electron donor through the inner cell membrane and periplasmic space until they finally reach a terminal electron-transfer molecule in the outer cell membrane [83–85]. This is the case for Gram-negative microorganisms, that is, organisms containing an outer cell membrane and a small amount of peptidoglycan. The exact electron-transfer mechanisms have not yet been elucidated for Gram-positive bacteria, that is, containing a large peptidoglycan layer in their cell wall.

6.3.1.1 Electron Donors

Recently, Pant *et al.* summarized the large variety of substrates that can be handled by bioanodes (see also Figure 6.1) [86]. The potential practical applicability of this technology can be appreciated since different types of wastewaters have been found suitable to drive electron-donating reactions. Furthermore, in addition to organic electron donors, also inorganic waste streams, for example, containing sulfides, can be applied as a substrate for the anode [62]. More recently, even sunlight has been reported as an energy source for BESs. This requires photosynthetically active plants or microorganisms, and has the additional advantage that CO_2 is fixed from the atmosphere [87].

Attention has to be paid to the type of substrate used and the loading rate as they influence the composition of the microbial community, but also influence the performance, including power density and coulombic efficiency [11, 57].

6.3.1.2 Biocatalysis

The term biocatalyst relates to the microorganisms present that play a catalytic role in the liberation and shuttling of electrons conserved in the substrate to the electrode. As mentioned before, they cannot be considered as true catalysts, since they consume part of the substrate for growth and maintenance purposes. Substantial research has been carried out to try to establish the key players in bioanode processes, that is, electrochemically active microorganisms capable of respiring with insoluble materials [also called electricigens or anode respiring bacteria (ARB)]. Although some specialist species have frequently been found in bioanodes, such as *Geobacter* and *Shewanella* spp., microbial community analysis has revealed a broad spectrum of microorganisms capable of living in anode biofilms and (possibly) respiring with electrodes, comprising both Gram-positive and Gram-negative bacteria [46, 59, 88, 89]. The thickness of biofilms on anodes is typically in the range 10–50 μm, although cases of even thicker biofilms have been reported (Figure 6.4) [90]. Bacteria densities within biofilms can be difficult to quantify owing to the amorphous nature of the various electrodes materials.

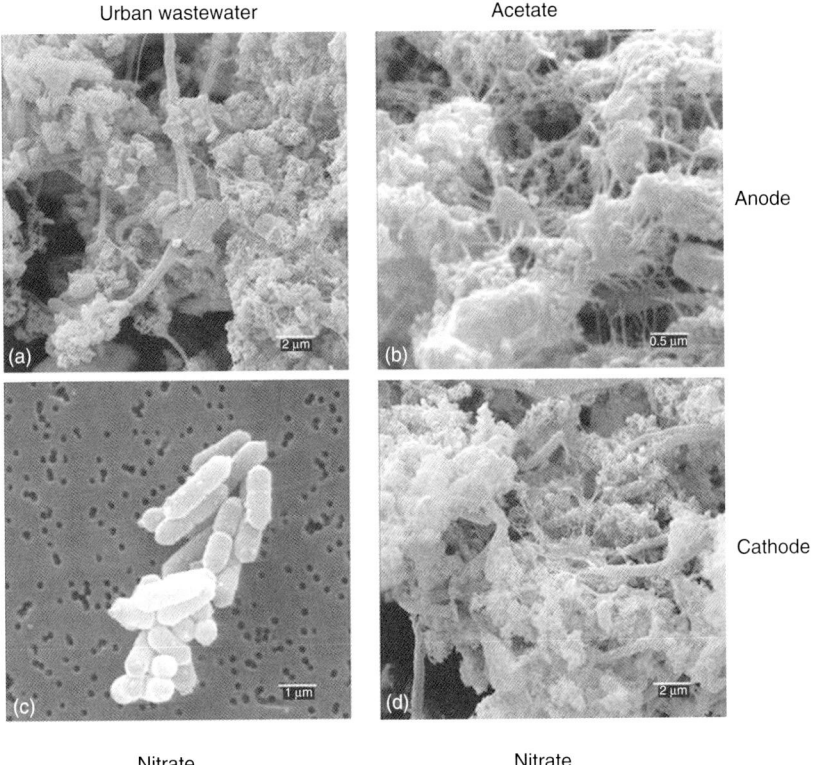

Figure 6.4 SEM images of biofilms on electrodes. (a, b) Two anodes, operated with different electron donors. Putative nanowires can clearly be seen in (b). (c, d) Two cathodes with nitrate as the electron acceptor.

Values of 50 g VS l^{-1} biofilm have been reported [90], where VS (volatile solids) is a general measure for biomass as used in environmental engineering.

6.3.1.3 Electron-Transfer Mechanisms

One of the most intriguing aspects of these biocatalysts is the way in which these microorganisms shuttle electrons to the anode. Over the years, it has come to light that several electron-transfer mechanisms can take place, either directly or indirectly. Furthermore, all of these mechanisms seem to be occurring simultaneously within the microbial community in an attempt to maximize the use of the substrate present for microbial benefit. Further analysis of the microbial community revealed that Gram-negative bacteria have much stronger EET capacities than Gram-positive bacteria [91–93]. As a result, all the proposed transfer mechanisms are based on studies with Gram-negative isolates that have frequently been found in the microbial community of BESs.

6.3.1.3.1 Direct Electron-Transfer Mechanisms

Direct EET can be defined as the transfer of electrons to the electrode without the need for a mobile component able to diffuse to and from the cell for electron transport [80]. This therefore requires physical contact between the bacterial cell membrane and the electrode surface. Current models for direct EET are based on pure culture studies of Gram-negative bacteria such as *Geobacter sulfurreducens* [94, 95] and *Shewanella oneidensis* [37, 41, 96], which are both dissimilatory metal-reducing organisms. Several putative pathways have been proposed and reviewed [89, 97, 98]. They involve at least a series of periplasmatic and outer membrane complexes, of which cytochromes seem to play a pivotal role. Regarding *Geobacter* spp., OmcS and OmcZ seem to be the most important electron-transferring cytochromes in the final electron-transfer step [99–101]. Additionally, pili, which are appendages of the cell wall, also seem to be involved in the transport of electrons [63]. These so-called nanowires have also been proposed as possible electron-transport mechanisms between different microorganisms [96, 102] and as a means for oxic metabolism in anoxic sediments [103]. From a practical point of view, direct electron transport eliminates to a certain extent diffusion limitations inherent in indirect transport, and simplifies solid–liquid separations as the biocatalyst is immobilized in the reactor [80]. For more detailed explanations regarding electron-transfer mechanisms the reader is referred to reviews by Richardson [85], Shi *et al.* [84], and Bird *et al.* [83].

6.3.1.3.2 Indirect Electron-Transfer Mechanisms

Indirect EET involves the use of so-called electron shuttles which physically transfer electrons from the cell to the electrode [80]. Commonly applied mediators include humic substances such as anthraquinone 2,6-disulfonate (AQDS) [104]. Furthermore, Thrash and Coates reviewed the electron shuttles used in BESs, and reported that the addition of a chemical shuttle can be expensive, toxic, and prone to wash-out of the system [81]. In addition to artificial redox mediators, some microorganisms are able to produce their own mediators such as secondary metabolites like phenazines [46, 92] and flavins [77]. Finally, primary metabolites such as sulfur species [105] and hydrogen gas [106] are also able to convey electrons toward electrodes.

Overall, mediators are able to enhance the electrical interconnectivity between electrochemically active microorganisms and the electrode, and hence increase the active range of the electrode beyond the biofilm into the bulk liquid. Nevertheless, care should be taken when applying them artificially.

6.3.2
Cathode Reactions

In cathodes, the electrons and protons delivered by the anodic oxidation reaction are used to drive reduction reactions. Cathode reactions in BESs can be either purely chemical or biologically catalyzed. Typical chemical cathode reactions used in BESs are oxygen [13, 107] and proton reduction [29, 108] to water and hydrogen gas, respectively. Hexacyanoferrate is a chemical electron acceptor commonly used in cathodes. It has frequently been applied as it allows bioanode processes to be investigated without the need to pay much attention to the cathode. However, the need for replenishment and possible toxic effects limits its application to laboratory-scale experiments.

In addition to chemical cathodes, biocathodes have also been developed, and actually fit much better with the sustainable nature of BESs than chemical cathodes. Therefore, there is increasing interest in the development of biologically driven cathode reactions, as has been demonstrated by the increasing number of biocathode studies in recent years, from two publications in 2004 to 41 in 2010 (Web of Knowledge, keyword "biocathode").

As the scope of this chapter is focused on the biological aspects of electrochemical systems, only biocathodes and their electron-transfer mechanisms will be discussed further.

6.3.2.1 Biocatalysts

Also in biocathodes microorganisms play a pivotal role in catalyzing reduction reactions. So far, most of the electrochemically active bacteria in biocathodes have been reported to be Gram-negative, although Gram-positive bacteria are also able to play a role in the transfer of electrons. This indicates that, similarly to anodic communities, there is a potentially wide capability of bacteria to catalyze electrode reactions [38, 78, 80, 109, 110].

6.3.2.2 Electron-Transfer Mechanisms

Whereas numerous studies have reported on microbial-assisted electron transfer towards electrodes (bioanodes), only limited information is available on the reverse process. Especially the microbial electron uptake from cathodes by microorganisms needs thorough investigation.

Recent biocathode studies have shown that electrochemically active bioanodes may be turned into biocathodes by changing the environmental and operating conditions [111, 112]. Hence Rosenbaum *et al.* revisited some known anodic EET mechanisms and evaluated their potential role in cathodic EET mechanisms [110]. Their main conclusion was that also in cathodes both direct and indirect mechanisms take place, and are very similar to processes at the bioanode. However, the main difference is that the redox-active components operate at higher redox potentials. Furthermore, it has not yet been demonstrated that biocatalyzed electron transfer is a respiratory process yielding energy for electrochemically active microorganisms [110, 113]. For instance, it has been shown that biological oxygen reduction at the cathode can occur passively without active involvement of the bacterium [114]. Nevertheless, Clauwaert *et al.* hypothesized an electron-transfer mechanism with concomitant generation of a proton motive force in the microorganisms present in denitrifying biocathodes [115].

6.3.2.2.1 Direct Electron-Transfer Mechanisms

It has already been discussed that cytochromes play an important role in direct anodic EET mechanisms (Section 5.3.2.2.1). However, as these compounds cover a broad redox-active range, they were investigated successfully for their involvement in direct cathodic EET [110, 113].

Furthermore, it is becoming apparent that hydrogenase-containing microorganisms are also able to accept directly electrons from polarized electrodes [110].

6.3.2.2.2 Indirect Electron-Transfer Mechanisms

During indirect cathodic EET, the same type of mediators (artificial or naturally produced) as reported in indirect anodic EET studies can be applied or involved [81]. Consequently, the same advantages and possible negative effects apply to the cathode side. However, manganese oxides have also been applid to shuttle electrons for oxygen reduction, and have the advantage that, owing to their solid nature, wash-out is slowed [116].

6.3.2.3 Electron Acceptors

A large variety of possible reactions have been established in both MFC and MEC operating modes for biocathodes. Operating the system as an electrolysis cell broadens the range of possible cathode reactions, as in this case large cathodic overpotentials or thermodynamically unfavorable conditions can be overcome (e.g., hydrogen production). This is done by giving an extra energy boost to the electrons originating from the anode oxidation reaction by means of a power supply. Recently, Rosenbaum *et al.* [110] and Huang *et al.* [113] reviewed the possible specific reduction reactions that can take place in biocathodes. A summary of these reactions is given in Figure 6.1, and indicates that these biocathodes show potential for use in a wide range of niche applications.

6.3.3
Pure Cultures and Mixed Microbial Communities

Both mixed and pure cultures can be used for the establishment of biological anodes or cathodes. From a practical point of view, mixed-culture electrodes seem to be more robust and resilient. Several groups have observed a higher power output when comparing mixed and pure culture systems [56]. This could be explained by either synergistic effects or the presence of a more productive exoelectrotroph in the mixed culture biofilms [56, 117]. However, several other scenarios could have occurred, such as pH effects and underdeveloped biofilms of the pure cultures [56].

Pure cultures are ideal for studying fundamental aspects or for the catalysis of species-specific reactions. Thrash and Coates gave an overview of pure cultures used in BESs and suggested that these kinds of studies will help us understand the means by which we can stimulate microbial communities to use electric current [81]. Furthermore, a completely new application would be the use of pure isolates for high-quality cathodic processes aimed at the specific generation of products, and has been reviewed by Rabaey and Rozendal [80].

6.3.4
Photosynthetic Biocatalysts

The use of photosynthetic microorganisms or plants in combination with a BES is a concept that has existed for several years, and is discussed in detail in Section 6.4.1.2. However, the direct involvement of photosynthetic bacteria, meaning direct electron transport with electrodes, remains to be shown. Nevertheless, some authors have already suggested this phenomenon. For instance, the cyanobacterium *Synechocystis* sp. PCC 6803 was reported to possess potentially conductive nanowires, and therefore could be involved in direct electron transfer with electrodes [96]. Stronger indications have been reported by Cao *et al.*, who carried out bicarbonate reduction with light in a photo-biocathode. Direct electron transport was assumed because these autotrophic microorganisms did not generate oxygen, electrocatalysts were absent, and flushing of the cathode to wash out soluble mediators did not affect the current production [118]. Further research is needed with a pure photosynthetic bacterial culture to verify direct involvement with electrodes. From then on, one could really talk about a photosynthetic microbial catalyst.

6.3.5
Biological Limitations

Applying microorganisms in a technology brings some biological limitations. First, proper physical parameters such as pH (around neutral) and temperature (15–30 °C) need to be maintained for good microbial metabolism. However, as always, some bacteria are able to function under slightly different conditions. For instance, the microbial reductive dechlorination of perchlorate was the highest at pH 8.5 [119]. Shear forces also seem to be an important physical parameter as they influence the biofilm thickness and density, and also the power output [64].

Second, biofilm thickness, structure, composition, and density affect the flux of substrates and products within the biofilm. The latter can result in large overpotentials, which have a negative impact on the performance of the system. In the case of bioanodes, higher power production was observed from thicker anodic biofilms [120]. Strikingly, the reverse effect has been observed in cathodic biofilms [35].

Finally, even if the microorganisms are not limited by their environment, there are some intrinsic limitations associated with the microbial metabolism such as their growth rate, uptake rate of nutrients, and electron transfer capacities. Certainly cathodic communities seem to encounter harsh conditions as they have to invest energy for CO_2 fixation in their biomass (autotrophic microorganisms), and have to be able to take up electrons from the cathode at a high potential.

6.4
Applications and Proof of Concepts

Applications for BESs have widened tremendously over the last decade. To provide a framework for placing BESs, three concepts are used. The first is energy production from a dilute (waste) liquid substrate. For various reasons but mainly the low power density, BESs are at present not yet competitive with current large-scale energy-producing or water-treatment technologies. Second, apart from only energy production, BESs are also considered as a useful concept for product formation from waste streams. Third, a BES is able to facilitate a reaction or control a microbial community in such a way that it provides a sustainable alternative for (non)existing technologies.

To underpin the usefulness of a BES for a certain treatment or application, one has to consider the total cost and benefits, also in terms of sustainability, of applying the BES option compared with another option. Recently, several groups have laid the basis for an analysis of the economic viability of implementing a BES system [121–123]. A study on the lifecycle assessment (LCA) of high-rate anaerobic treatment, MFCs and MECs by Foley *et al.* revealed that an MFC does not provide a significant environmental benefit to the conventional anaerobic treatment [124]. However, an MEC can provide environmental benefits through the displacement of chemical products (e.g., NaOH, H_2O_2). This study suggests that anaerobic digestion, which is furthermore empowered by the recovery of, for example, phosphate or nitrogen [125], is at present the most sustainable full-scale technology and that the overall BES technology still has to advance strongly towards a target conversion of some 1000 A per cubic meter of reactor at the cubic meter scale in order to become competitive. The major drawback of LCA and economic analysis is that a large pool of real-life data is not available; however, several start-up companies have been founded, so more knowledge and experience are expected in the near future. A (non-exhaustive) literature review of representative labscale MFC and MEC outputs is given in Table 6.1.

6.4.1
Energy and Wastewater Concept

6.4.1.1 Wastewater Treatment
Wastewater has been regarded as a cost in all respects. Owing to its negative value and low organic material content, wastewater is viewed as the preferred substrate for energy-producing BESs. Domestic wastewater is usually treated using different energy-demanding technologies with high efficiency levels [126]. Aelterman *et al.* demonstrated that an MFC might occupy a market niche in terms of a stand-alone source of electricity for use in wastewater treatment [127]. A maximum power density of 25 mW m^{-2} was reached by oxidizing the organic matter from urban wastewater [128]. Puig *et al.* demonstrated the pH effect on electricity production with short- and long-term treatment of wastewater, achieving a power density up to 1.8 W m^{-3} and an organic removal rate of

1.9 kg COD m^{-3} day^{-1} [77% chemical oxygen demand (COD) removal efficiency] [129]; COD is a general term to describe the amount of organic material or "energy" present in a solution. Ahn and Logan treated domestic wastewater at an organic loading rate of 54 kg COD m^{-3} day^{-1} (25.8% COD removal efficiency) and a maximum power generation of 0.42 W m^{-2} (12.8 W m^{-3}) [130].

However, wastewater also contains nitrogen, which must be treated before discharge into the environment. Moreover, the denitrification process (nitrate/nitrite reduction to dinitrogen gas) is highly dependent on the influent carbon-to-nitrogen ratio. At present this ratio is unbalanced, limiting the removal efficiency. Autotrophic nitrate/nitrite removal in the biocathode of a BES has been successfully demonstrated by Clauwaert and co-workers [115, 152] and Puig et al. [132]. Autotrophic nitrogen removal in a BES permits denitrification at really low carbon-to-nitrogen ratios in comparison with conventional heterotrophic denitrification.

However, wastewater usually contains ammonium instead of nitrate, which should be previously oxidized. Ammonium oxidation to nitrate could be done *in situ* in a BES or in an external compartment. This concept was demonstrated by coupling a double-chamber MFC with a nitrifying bioreactor [131, 153] or by introducing a low dissolved oxygen concentration in the cathode [154, 155].

Industrial wastewaters are highly contaminated with a wide range of organic and inorganic compounds, nutrients, and heavy metals. The treatment of some of these has been assessed using BESs. One of the best examples is brewery wastewater because of the type of organic matter and the lack of high concentrations of inhibitory substances [156, 157]. Brewery wastewater was treated in an air-cathode MFC with maximum power densities of 264 and 528 mW m^{-2}. Bakery, paper, and dairy wastewater have also been evaluated in an air-cathode MFC for current production [135]. Paper wastewater was shown to produce the highest current density in comparison with the other influents, independent of substrate biodegradability. Heavily polluted wastewaters such as landfill effluents have been treated using MFCs. Electricity could be produced from those effluents with organic removal efficiencies similar to those obtained by a biological aerated filter under similar controlled conditions [133]. Landfill leachate treatment and electricity production in an MFC at high nitrogen concentration (6.0 g l^{-1}) and conductivity (73.6 mS cm^{-1}) have also been demonstrated [134]. Up to 8.5 kg COD m^{-3} day^{-1} of biodegradable organic matter was removed at the same time as electricity (344 mW m^{-3}) was produced, even at free ammonia concentrations around 900 mg N-NH$_3$ l^{-1}.

An MEC has also been applied to treat domestic and industrial (winery and swine) wastewaters. Swine wastewater with a COD of 12–17 g l^{-1} was treated with COD removal efficiencies ranging from 19 ± 15 to 72 ± 4%, with hydrogen recoveries of 17 ± 7 to 28 ± 6% based on COD removal [147]. Organic matter removal efficiency and energy recovery were higher for MFC- than MEC-treated winery and domestic wastewaters. The energy recovery for winery wastewater in MFCs was 0.26 kWh kg^{-1} COD compared with −0.32 kWh kg^{-1} COD for MECs. Hydrogen production costs of winery and domestic wastewaters with as MEC were determined as $4.51 kg^{-1} H$_2$ for winery wastewater and $3.01 kg^{-1} H$_2$ for

Table 6.1 Literature review of representative MFC and MEC outputs.

Cell type	Application	Volume (ml)	Operational mode	Current (power) density	Output	Ref.
Microbial fuel cells (MFCs)	Domestic wastewater (organic matter)	28	Air-cathode MFC	12.8 W m^{-3} (0.42 W m^{-2})	—	[130]
	Domestic wastewater (nitrogen removal)	242		1.8 W m^{-3}	—	[129]
		716	Two-chamber MFC	8 W m^{-3} (58 A m^{-3})	—	[115]
		250		—	—	[131]
		650		3.2 W m^{-3} (20.8 A m^{-3})	—	[132]
	Landfill leachate	900	Two-chamber MFC	0.002 mW m^{-2} (0.003 A m^{-2})	—	[133]
	Industrial wastewaters (bakery, brewery, paper, and dairy)	167	Air-cathode MFC	0.34 W m^{-3}	—	[134]
		45	Air-cathode MFC	0.06–0.005 W m^{-2} (0.12–0.01 A m^{-2})	—	[135]
	Refractory contaminants (furfural)	200	Two-chamber MFC	15.9 W m^{-3} (39.9 A m^{-3})	—	[136]
	Refractory contaminants (ceftriaxone sodium)	100	Air-cathode MFC	11 W m^{-3}	—	[137]
	Refractory contaminants (glucose and azo dye)	220	Two-chamber MFC	0.4 W m^{-2} (2.7 W m^{-3})	—	[138]
	Refractory contaminants (azo dye: Methyl Orange)	—	Two-chamber MFC	0.006 W m^{-2}	—	[139]
	Refractory contaminants (nitrobenzene)	364	Two-chamber MFC	17.06 W m^{-3}	—	[140]
	Water desalination	57	Three-chamber microbial desalination cells	31 W m^{-3} (2 W m^{-2})	90% salt removal	[7, 141]
	Copper recovery	800	Two-chamber MFC with bipolar membrane	0.8 W m^{-2} (3.2 A m^{-2})	—	[142]

Table 6.1 (Continued)

Cell type	Application	Volume (ml)	Operational mode	Input voltage	Output	Ref.
Plant/sediment MFCs	Bioremediation of organic contaminants (acetate and benzoate)	—	Sediment batteries	0.016 W m^{-2}	—	[143]
	Bioelectricity and biomass production	—	Cylindrical plant MFC	0.222 W m^{-2}	—	[144]
	Solar energy-powered MFC	3300	Solar energy-powered MFC	0.041 W m^{-2}	—	[145]
Microbial electrolysis cell (MECs)	Winery and domestic wastewater	28	Two-chamber MEC	−0.90 V	0.17 and 0.28 m^3 H$_2$ m^{-3} day^{-1}	[146]
	Swine wastewater	28	Two-chamber MEC	−0.50 V	0.9–1.0 m^3 H$_2$ m^{-3} day^{-1}	[147]
	Water desalination	57	Three-chamber microbial desalination cells	−0.55 V	0.16 m^3 H$_2$ m^{-3} day^{-1}	[148]
	Hydrogen peroxide generation	518	Two-chamber MEC	−0.50 V	1.9 ± 0.2 kg H$_2$O$_2$ m^{-3} day^{-1}	[149]
	Caustic soda production	3313	Lamellar-type reactor	−1.77 V (1015 A m^{-3})	NaOH (3.4 wt%)	[150]
	Ethanol production	1672	Two-chamber MEC	−0.55 V	1.82 mM EtOH; 0.012 m^3 H$_2$ m^{-3} day^{-1}	[151]

domestic wastewater [146], which are lower than the estimated market value of hydrogen ($6.00 kg^{-1} H$_2$).

6.4.1.2 Sediments, Plants, and Photosynthesis in a BES

A niche in the energy concept is occupied by MFCs in natural environments. Sediment MFCs have been described by the pioneering research of Reimers, Tender, and Bond and co-workers [143, 158, 159]. These types of MFCs are the only designs that have been shown to have a commercial application. These benthic systems are used to power remote sensors that require long-term remote low-power supplies [160–162]. The main advantage of these systems is that on-site battery replacements are no longer required. Dewan *et al.* made a step forward by developing a method to evaluate and optimize energy harvesting when an MFC was used with a capacitor to power sensors monitoring the environment [162].

The main limiting factor for high power output of a sediment MFC is the low flux of reduced compounds towards the anode electrode. These are not actively provided since transport relies on passive mechanisms such as natural water currents and settling of (dead) organic material. A solution to this problem was found by adding living photosynthesizing plants to the system. A living plant excretes a range of oxidizable organic material from its roots, up to 40% of the plants' photosynthetic productivity, and can thus be a continuous supplier of fuel for a sediment system [87, 163]. The plant-powered sediment MFC concept was demonstrated by three groups in 2008 [8, 164, 165]. Since then, research on this topic has expanded to include plants in more engineered systems [144].

Photosynthesis by microorganisms and algae as a driven force for a BES has been researched in both the anode and the cathode. Recently, several reviews of BESs based on photosynthesis have been published [87, 166, 167]. In the anode, various strategies have been explored to convert sunlight into oxidizable compounds or into direct electricity by means of biology. In the cathode, options for photo-driven reactions are limited. The most viable option is to use algae for *in situ* generation of oxygen to drive the cathodic reduction reaction. Strik *et al.* developed a bio-electrode with reversible properties, that is, during illumination it produced oxygen and acted as a cathode whereas during dark periods it acted as an anode [145].

6.4.1.3 Electro-Assisted Anaerobic Digestion

Anaerobic digestion is applied worldwide to recover energy from organic wastes. The organics are treated either in completely stirred tank reactors or in so-called upflow anaerobic sludge bed (UASB) reactors [168]. However, the microbiology involved is complex: the organics have to be depolymerized to higher fatty acids; the latter are then converted by syntrophic acetogenic bacteria (SAB) to acetate and hydrogen. The acetate is cleaved by acetoclastic methanogens (AMs), but it can also be oxidized to hydrogen and CO$_2$ by the so-called syntrophic acetate oxidizers (SAOs). The latter molecules are subsequently converted to methane by the hydrogenotrophic methanogens (HMs). At present, it is not exactly clear how the metabolite fluxes occur and how one can stimulate these different key players in the biomethanation process.

By implementing the MEC configuration, organics, such as various fatty acids, are converted to hydrogen, which can then function as a booster for the HM [169]. This conversion only requires ~30% of the energy otherwise required to generate H_2 from water by electrolysis [170]. Several studies have indicated that, even when a few percent of the overall COD flow is used by an electro-assisted process, the biomethane formation is significantly enhanced in terms of rate and efficiency [169, 171]. This suggests that the electro-assistance particularly helps to alleviate the low-rate processes otherwise taken care of by the fragile SAB and SAO. At present, no full-scale implementations of BES in anaerobic digestion processes have been realized. This may be because the overall rates of conversion of BESs remain relatively low (in the order of a few kilogram COD equivalents per cubic meter of reactor per day [68]) and that the capital expenses of BESs are still a factor 10 too high relative to those of plain aerobic digester systems [122].

6.4.2
Product Concept

In addition to power production, other opportunities to apply microbial-catalyzed electrochemical reactions have been sought. This has led to two paths of research. The first makes use of the electrochemical properties of a BES and entails desalination of water in a three-compartment BES [7] and the formation of caustic soda [150] or peroxide [149]. The formation of these products is based on the better migration of certain cations compared with others. The second path is based on the bioelectrochemical properties of a BES. Here specialized microbial (mixed) cultures are supplied with a current which they are able to convert into a pure compound [38, 151]. Thus far, these applications have been proven at the laboratory scale; their commercial applicability is strongly dependent on the end-user product specifications and the costs of competing processes.

6.4.2.1 Desalination
Current water desalination techniques (i.e., reverse osmosis, electrodialysis, and distillation) are energy intensive (3–5 kWh m^{-3} for reverse osmosis) and some use membranes operated at high pressures (70 kPa) [172]. Water desalination was accomplished in a three-chamber MFC with two membranes (an AEM placed adjacent to the anode and a CEM positioned next to the cathode). When current was produced by bacteria at the anode, ionic species in the middle chamber were transferred into the two electrode chambers, thereby desalinating the water in the middle chamber [7]. This new method of desalination was able to remove 90% of the salt content without pressurizing or use of an external power source. The MDC design was optimized using oxygen at the cathode, rather than hexacyanoferrate, and combined with new ion-exchange membranes [141]. In this way, 50% of the solution conductivity was reduced. However, the desalinated liquid needed further polishing with reverse osmosis as the voltage was not stable and the osmotic pressure was reduced between the anode and desalination chamber over time.

To cope with this, the process was further optimized by operating the system as a microbial electrodialysis cell (MEDC). The energy efficiencies obtained (η_E) in the MEDCs reached $231 \pm 59\%$ (5 g l^{-1} NaCl) and $213 \pm 38\%$ (20 g l^{-1} NaCl), suggesting that sufficient hydrogen was produced to power the MEDC [148].

6.4.2.2 Caustic Soda and Hydrogen Peroxide Production

Rabaey et al. produced caustic soda from brewery wastewater while producing currents up to 1015 A m^{-3} [150]. The lamellar-type reactor with a total liquid volume of 1.63 l (66% anodic) produced caustic soda in the cathodic compartment taking advantage of the sodium diffusion from the anode and the proton consumption in the cathode compartments. The energy input as the ratio of electricity to caustic soda produced was 1.6 kWh kg^{-1} NaOH, which corresponds to an operational cost of around $0.1 kg^{-1} low-strength caustic soda. This price is lower than the market price of above $0.5 kg^{-1} [150]. This impressive result should be confirmed after long-term performance to assess future applications.

Hydrogen peroxide (H_2O_2) was produced in an MEC at an applied voltage of 0.5 V [149]. Cathodic reduction of oxygen to H_2O_2 was catalyzed by inexpensive carbon materials. Under these conditions, the MEC was capable of producing around 1.9 ± 0.2 kg H_2O_2 m^{-3} day^{-1} from acetate with energy requirements of around 0.9 kWh kg^{-1} H_2O_2. However, the low concentration of H_2O_2 produced (0.13 ± 0.01 wt%) is at present not suitable for sale. Both products are envisioned to be used for cleaning in place (CIP) as a waste stream is converted on-site into a cleaning solution, but still need to be thoroughly evaluated in comparison to competing processes and long-term operation.

6.4.2.3 Organic Alcohols and Acids

Formation of organic acids and alcohols has also been demonstrated using a BES. The cathode compartment of a BES was inoculated with acetogenic bacteria in order to reduce carbon dioxide to acetate and other multicarbon extracellular products (e.g., 2-oxobutyrate) [38]. The electron recovery in acetate and 2-oxobutyrate was 86% of the electrons transferred at the cathodes. This process was termed *microbial electrosynthesis* for its capability to convert carbon dioxide and water to multicarbon extracellular organic compounds. The authors suggested that although acetate has economic value, a more important consideration is that acetate is formed from acetyl coenzyme A (acetyl-CoA), which is the central intermediate for the genetically engineered production of a wide range of chemical commodities and also potential liquid transportation fuels. However, further research is still necessary to define the potential applications.

Ethanol can also be produced in a BES. Steinbusch et al. reported on the biological reduction of acetate with hydrogen to ethanol [151]. To stimulate acetate reduction at the cathode with mixed microbial cultures, they applied a cathode potential of -0.55 V. Four major products were formed at the cathode: ethanol, hydrogen, n-butyrate, and the non-reversible reduced methylviologen. Ethanol production had a coulombic efficiency of 49%. The highest rate of ethanol production was 60 mg m^{-3} cathodic compartment day^{-1}. Hydrogen was produced at 0.012 Nm^3

cathodic compartment day^{-1} or 0.0035 Nm^3 H$_2$ m^{-2} day^{-1}. The study concluded that the product concentrations and rates were still very low in the BES compared with other waste conversion processes, such as anaerobic digestion.

6.4.3
Providing Environmental Services

6.4.3.1 Recalcitrant Compounds

The production of different types of products and processes such as dyes, explosives, pesticides, paper printing, color photography, pharmaceuticals, cosmetics, and leather goods leads to large amounts of recalcitrant compounds that are not readily degradable under natural conditions and are typically not removed from wastewater by conventional wastewater treatment systems. Physicochemical methods, such as adsorption, coagulation–flocculation, and especially advanced oxidation (ozonation, photocatalysis, Fenton oxidation, etc.) have been used to treat them. These processes are energy and cost intensive. BES technology can be considered as a reliable alternative to these processes. It has been demonstrated that microorganisms can degrade many types of toxic and refractory compounds, such as phenols, indoles, azo dyes, and halogenated compounds, and generate electricity simultaneously.

Furfural and Acid Orange are toxic and refractory pollutants that are used in oil refineries and petrochemical refining units to extract dienes from other hydrocarbons. Luo *et al.* [136] demonstrated that electrons and protons were released from the biodegradation of furfural in the anode chamber and transported to the cathode chamber to drive the Fenton-like reaction catalyzed by FeVO$_4$ towards degrading Acid Orange 7. Li *et al.* [138] studied the azo dye removal mechanism in a double-chambered MFC. The azo bond was biologically cleaved through azo dye decolorization under anaerobic conditions in the anode chamber and abiotically in the cathode chamber. The aromatic amines were removed in the aerobic chamber.

Nitrobenzene originates from numerous industrial and agricultural activities, and can be abiotically removed at a cathode coupled to microbial oxidation of acetate at an anode, as shown by Mu *et al.* [140]. Finally, a model antibiotic (ceftriaxone) was shown to be oxidatively removed and to boost current production from glucose in the anode in an air-cathode MFC [137].

The chlorinated solvent 1,2-dichloroethane was removed in the anode of an MFC without the production of any toxic intermediate at a maximum removal rate of 40 g m^{-2} anode surface day^{-1} [173]. Biogenic palladium nanoparticles produced by the precipitation of Pd on the surface of bacteria have been used as a catalyst for the dehalogenation of trichloroethylene and diatrizoate in the cathode of an MEC [25].

6.4.3.2 Greenhouse Gas Mitigation

Root residues (mainly organic compounds) from rice plants produce methane. It is estimated that worldwide rice agriculture contributes 7–20% of the total methane emissions [174]. De Schamphelaire *et al.* showed that a plant MFC could offer the prospect to steer and control the sediment redox potential and thus abate undesirable processes such as methylation of metals and emissions of methane. Moreover, the plant MFC produced sustainable power up to 330 W ha^{-1} from the oxidation of the plant-derived compounds [8]. The greenhouse gas mitigation potential was identified as a potentially strong parameter in an LCA of a BES [123].

6.4.3.3 Heavy Metal Recovery/Removal

Mining and metallurgical wastewaters and leachates contain heavy metals such as copper, nickel, cobalt, and zinc. A potentially new application of BESs is in the removal and recovery of metals. Mining and metallurgical industries are the main contributors of anthropogenic copper emissions to the environment [142]. Ter Heijne *et al.* showed that Cu^{2+}, added as $CuCl_2$ at pH 3, could be reductively removed in the cathode compartment of an MFC with concomitant energy generation [142]. Copper was deposited under anaerobic and aerobic conditions on the graphite electrode. The authors proposed that Cu^{2+} reduction as a cathodic reaction is an interesting option to consider for the improvement of MFC performance as the oxygen reduction reaction was also improved by the presence of copper.

Uranium and mercury have also been removed using BESs. Lovley and co-workers demonstrated that *Geobacter sulfurreducens* can use electrons derived from electrodes to reduce U(VI) to the less soluble U(IV), and that it may be possible to remove and recover uranium with poised electrodes [175, 176]. Power generation (433.1 mW m^{-2}) coupled with mercury removal (240 mg Hg^{2+} l^{-1} day^{-1}, 99% removal efficiency in batch) in the cathode of an MFC was shown to be possible. Mercury was recovered as elemental Hg (on the cathode surface) and Hg_2Cl_2 (deposited on the bottom of the chamber) [177].

6.4.3.4 Biosensors and Environmental Monitoring

The use of microbial techniques to monitor the natural environment has mainly been done by means of toxicity assays. A wide array of research has already been carried out on enzymatic biosensors with glucose oxidase being the focal point [178]. With the discovery of ARB, researchers have seen the opportunity to create biosensors that directly link environmental parameters to a small current. So far, most research has been focused on biological oxygen demand (BOD) sensors [133, 179]. The applicability of a BES biosensor in the field remains to be proven. Several issues need to be addressed, including calibration, functional stability of the active unit (i.e., biofilm or immobilized enzyme) under different operating and storage conditions, and competitiveness with existing technologies [178]. Small steps have been made towards understanding the behavior of MFC-based biosensors by looking at toxicity responses [180] and stable baseline outputs [181]. Williams *et al.* used a BES sensor from a different perspective. Their approach was to monitor *in situ* microbial activity in anoxic soils.

They were able to link an increase in MFC current to an enhanced removal of U(VI) [176].

6.5
Modeling

The large amount of data currently available and the urge to implement BES in real applications have led to the first attempts to model a BES. The focus thus far has been on the anode, where Picioreanu and co-workers have modeled the transport behavior of substrates and products related to the anode configuration and flow regimes [182, 183]. Anode respiration and kinetics were modeled by using a Butler–Volmer–Monod model, which described experimental data better than the Nernst–Monod model [184]. The modeling of the relationship between the competing processes of methanogenesis and electricigenesis showed that periodically adjusting the external resistance can give an advantage to ARB over methanogenic archaea [185].

6.6
Outlook and Conclusions

In this chapter, an attempt has been made to give an overview of the great possibilities and challenges of BES. Owing to their biological nature and inherent versatility of biocatalysts, tremendous opportunities lie ahead if the focus is shifted from power production towards sustainable engineering. For the future, we expect to see more and more proofs of concepts regarding "difficult" transformations of recalcitrant compounds and product formation with added value. In addition to broadening of applications, BES research will yield new insights into the (eco)physiology of biocatalysts. To move from the bench to the field, research has to focus on applying the knowledge gained to niches where BES has a competitive edge over known technologies. It can already be said that only large-scale power generation does not hold a viable future for BESs. A successful future lies in the application of BESs to product formation and applied green technology.

Acknowledgments

J.B.A.A. is supported by the European Community's Seventh Framework Programme (FP7/2007-2013) under Grant Agreement No. 226532). J.D. is supported by the Institute for the Promotion and Innovation through Science and Technology in Flanders (IWT-Vlaanderen, SB-091144). S.P. acknowledges the Spanish Government (MCYT-CTQ2008-06865-C02-01/PPQ and CONSOLIDER-CSD2007-00055) for financial support. The authors thank Peter Clauwaert and Tim Lacoere for help with the graphical representations.

References

1. Potter, M.C. (1911) Electrical effects accompanying the decomposition of organic compounds. *Proc. R. Soc. Lond. B*, **84** (571), 260–276.
2. Logan, B.E., Hamelers, B., Rozendal, R., Schröder, U., Keller, J., Freguia, S., Aelterman, P., Verstraete, W., and Rabaey, K. (2006) Microbial fuel cells: methodology and technology. *Environ. Sci. Technol.*, **40** (17), 5181–5192.
3. Ivanov, I., Vidakovic-Koch, T., and Sundmacher, K. (2010) Recent advances in enzymatic fuel cells: experiments and modeling. *Energies*, **3** (4), 803–846.
4. Lauro, F.M. and Bartlett, D.H. (2008) Prokaryotic lifestyles in deep sea habitats. *Extremophiles*, **12** (1), 15–25.
5. Kashefi, K. and Lovley, D.R. (2003) Extending the upper temperature limit for life. *Science*, **301** (5635), 934–934.
6. Kashefi, K., Moskowitz, B.M., and Lovley, D.R. (2008) Characterization of extracellular minerals produced during dissimilatory Fe(III) and U(VI) reduction at 100 degrees C by *Pyrobaculum islandicum*. *Geobiology*, **6** (2), 147–154.
7. Cao, X., Huang, X., Liang, P., Xiao, K., Zhou, Y., Zhang, X., and Logan, B.E. (2009) A new method for water desalination using microbial desalination cells. *Environ. Sci. Technol.*, **43** (16), 7148–7152.
8. De Schamphelaire, L., Van Den Bossche, L., Dang, H.S., Hofte, M., Boon, N., Rabaey, K., and Verstraete, W. (2008) Microbial fuel cells generating electricity from rhizodeposits of rice plants. *Environ. Sci. Technol.*, **42** (8), 3053–3058.
9. Zhou, M., Chi, M., Luo, J., He, H., and Jin, T. (2011) An overview of electrode materials in microbial fuel cells. *J. Power Sources*, **196** (10), 4427–4435.
10. Logan, B., Cheng, S., Watson, V., and Estadt, G. (2007) Graphite fiber brush anodes for increased power production in air-cathode microbial fuel cells. *Environ. Sci. Technol.*, **41** (9), 3341–3346.
11. Aelterman, P., Versichele, M., Marzorati, M., Boon, N., and Verstraete, W. (2008) Loading rate and external resistance control the electricity generation of microbial fuel cells with different three-dimensional anodes. *Bioresour. Technol.*, **99** (18), 8895–8902.
12. Ter Heijne, A., Hamelers, H.V.M., Saakes, M., and Buisman, C.J.N. (2008) Performance of non-porous graphite and titanium-based anodes in microbial fuel cells. *Electrochim. Acta*, **53** (18), 5697–5703.
13. Zhao, F., Harnisch, F., Schröder, U., Scholz, F., Bogdanoff, P., and Herrmann, I. (2006) Challenges and constraints of using oxygen cathodes in microbial fuel cells. *Environ. Sci. Technol.*, **40** (17), 5193–5199.
14. De Schamphelaire, L., Boeckx, P., and Verstraete, W. (2010) Evaluation of biocathodes in freshwater and brackish sediment microbial fuel cells. *Appl. Microbiol. Biotechnol.*, **87** (5), 1675–1687.
15. Zhu, N., Chen, X., Zhang, T., Wu, P., Li, P., and Wu, J. (2011) Improved performance of membrane free single-chamber air-cathode microbial fuel cells with nitric acid and ethylenediamine surface modified activated carbon fiber felt anodes. *Bioresour. Technol.*, **102** (1), 422–426.
16. Rosenbaum, M., Zhao, F., Schröder, U., and Scholz, F. (2006) Interfacing electrocatalysis and biocatalysis with tungsten carbide: a high-performance, noble-metal-free microbial fuel cell. *Angew. Chem. Int. Ed.*, **45** (40), 6658–6661.
17. Zhao, F., Harnisch, F., Schröder, U., Scholz, F., Bogdanoff, P., and Herrmann, I. (2005) Application of pyrolysed iron(II) phthalocyanine and CoTMPP based oxygen reduction catalysts as cathode materials in microbial fuel cells. *Electrochem. Commun.*, **7** (12), 1405–1410.
18. Park, D.H. and Zeikus, J.G. (2003) Improved fuel cell and electrode designs for producing electricity from microbial

degradation. *Biotechnol. Bioeng.*, **81** (3), 348–355.

19. Aelterman, P., Versichele, M., Genettello, E., Verbeken, K., and Verstraete, W. (2009) Microbial fuel cells operated with iron-chelated air cathodes. *Electrochim. Acta*, **54** (24), 5754–5760.
20. Freguia, S., Rabaey, K., Yuan, Z., and Keller, J. (2007) Non-catalyzed cathodic oxygen reduction at graphite granules in microbial fuel cells. *Electrochim. Acta*, **53** (2), 598–603.
21. Scott, K., Cotlarciuc, I., Head, I., Katuri, K.P., Hall, D., Lakeman, J.B., and Browning, D. (2008) Fuel cell power generation from marine sediments: investigation of cathode materials. *J. Chem. Technol. Biotechnol.*, **83** (9), 1244–1254.
22. Cheng, S. and Logan, B.E. (2007) Ammonia treatment of carbon cloth anodes to enhance power generation of microbial fuel cells. *Electrochem. Commun.*, **9** (3), 492–496.
23. Selembo, P.A., Merrill, M.D., and Logan, B.E. (2010) Hydrogen production with nickel powder cathode catalysts in microbial electrolysis cells. *Int. J. Hydrogen Energy*, **35** (2), 428–437.
24. Huang, Y.-X., Liu, X.-W., Sun, X.-F., Sheng, G.-P., Zhang, Y.-Y., Yan, G.-M., Wang, S.-G., Xu, A.-W., and Yu, H.-Q. (2011) A new cathodic electrode deposit with palladium nanoparticles for cost-effective hydrogen production in a microbial electrolysis cell. *Int. J. Hydrogen Energy*, **36** (4), 2773–2776.
25. Hennebel, T., Benner, J., Clauwaert, P., Vanhaecke, L., Aelterman, P., Callebaut, R., Boon, N., and Verstraete, W. (2011) Dehalogenation of environmental pollutants in microbial electrolysis cells with biogenic palladium nanoparticles. *Biotechnol. Lett.*, **33** (1), 89–95.
26. Liu, X.-W., Sun, X.-F., Huang, Y.-X., Sheng, G.-P., Zhou, K., Zeng, R.J., Dong, F., Wang, S.-G., Xu, A.-W., Tong, Z.-H., and Yu, H.-Q. (2010) Nano-structured manganese oxide as a cathodic catalyst for enhanced oxygen reduction in a microbial fuel cell fed with a synthetic wastewater. *Water Res.*, **44** (18), 5298–5305.
27. Fan, Y., Xu, S., Schaller, R., Jiao, J., Chaplen, F., and Liu, H. (2011) Nanoparticle decorated anodes for enhanced current generation in microbial electrochemical cells. *Biosens. Bioelectron.*, **26** (5), 1908–1912.
28. Clauwaert, P. and Verstraete, W. (2009) Methanogenesis in membraneless microbial electrolysis cells. *Appl. Microbiol. Biotechnol.*, **82** (5), 829–836.
29. Call, D. and Logan, B.E. (2008) Hydrogen production in a single chamber microbial electrolysis cell lacking a membrane. *Environ. Sci. Technol.*, **42** (9), 3401–3406.
30. Rozendal, R.A., Hamelers, H.V.M., Molenkmp, R.J., and Buisman, J.N. (2007) Performance of single chamber biocatalyzed electrolysis with different types of ion exchange membranes. *Water Res.*, **41** (9), 1984–1994.
31. Sleutels, T.H.J.A., Hamelers, H.V.M., Rozendal, R.A., and Buisman, C.J.N. (2009) Ion transport resistance in microbial electrolysis cells with anion and cation exchange membranes. *Int. J. Hydrogen Energy*, **34** (9), 3612–3620.
32. Zhang, X.Y., Cheng, S.A., Wang, X., Huang, X., and Logan, B.E. (2009) Separator characteristics for increasing performance of microbial fuel cells. *Environ. Sci. Technol.*, **43** (21), 8456–8461.
33. Biffinger, J.C., Ray, R., Little, B., and Ringeisen, B.R. (2007) Diversifying biological fuel cell designs by use of nanoporous filters. *Environ. Sci. Technol.*, **41**, 1444–1449.
34. Harnisch, F., Schröder, U., and Scholz, F. (2008) The suitability of monopolar and bipolar ion exchange membranes as separators for biological fuel cells. *Environ. Sci. Technol.*, **42** (5), 1740–1746.
35. Behera, M., Jana, P.S., and Ghangrekar, M.M. (2010) Performance evaluation of low cost microbial fuel cell fabricated using earthen pot with biotic and abiotic cathode. *Bioresour. Technol.*, **101** (4), 1183–1189.

36. Logan, B.E. (2008) *Microbial Fuel Cells*, John Wiley & Sons, Inc., Hoboken, NJ.
37. Rosenbaum, M., Cotta, M.A., and Angenent, L.T. (2010) Aerated *Shewanella oneidensis* in continuously fed bioelectrochemical systems for power and hydrogen production. *Biotechnol. Bioeng.*, **105** (5), 880–888.
38. Nevin, K.P., Woodard, T.L., Franks, A.E., Summers, Z.M., and Lovley, D.R. (2010) Microbial electrosynthesis: feeding microbes electricity to convert carbon dioxide and water to multicarbon extracellular organic compounds. *mBio*, **1** (2), e00103-10.
39. Rabaey, K., Clauwaert, P., Aelterman, P., and Verstraete, W. (2005) Tubular microbial fuel cells for efficient electricity generation. *Environ. Sci. Technol.*, **39** (20), 8077–8082.
40. He, Z., Wagner, N., Minteer, S.D., and Angenent, L.T. (2006) An upflow microbial fuel cell with an interior cathode: assessment of the internal resistance by impedance spectroscopy. *Environ. Sci. Technol.*, **40** (17), 5212–5217.
41. Ringeisen, B.R., Henderson, E., Wu, P.K., Pietron, J., Ray, R., Little, B., Biffinger, J.C., and Jones-Meehan, J.M. (2006) High power density from a miniature microbial fuel cell using *Shewanella oneidensis* DSP10. *Environ. Sci. Technol.*, **40** (8), 2629–2634.
42. Hou, H., Li, L., Cho, Y., De Figueiredo, P., and Han, A. (2009) Microfabricated microbial fuel cell arrays reveal electrochemically active microbes. *PLoS ONE*, **4** (8), e6570.
43. Chen, Y.-P., Zhao, Y., Qiu, K.-Q., Chu, J., Lu, R., Sun, M., Liu, X.-W., Sheng, G.-P., Yu, H.-Q., Chen, J., Li, W.-J., Liu, G., Tian, Y.-C., and Xiong, Y. (2011) An innovative miniature microbial fuel cell fabricated using photolithography. *Biosens. Bioelectron.*, **26** (6), 2841–2846.
44. Moriuchi, T., Sumida, S., Furuya, A., Morishima, K., and Furukawa, Y. (2009) Development of a flexible direct photosynthetic/metabolic biofuel cell for mobile use. *Int. J. Precis. Eng. Manuf.*, **10** (1), 75–78.
45. Zuo, Y., Xing, D.F., Regan, J.M., and Logan, B.E. (2008) Isolation of the exoelectrogenic bacterium *Ochrobactrum anthropi* YZ-1 by using a U-tube microbial fuel cell. *Appl. Environ. Microbiol.*, **74** (10), 3130–3137.
46. Pham, T., Boon, N., Aelterman, P., Clauwaert, P., De Schamphelaire, L., Vanhaecke, L., De Maeyer, K., Höfte, M., Verstraete, W., and Rabaey, K. (2008) Metabolites produced by *Pseudomonas* sp. enable a Gram-positive bacterium to achieve extracellular electron transfer. *Appl. Microbiol. Biotechnol.*, **77** (5), 1119–1129.
47. Liu, H. and Logan, B.E. (2004) Electricity generation using an air-cathode single chamber microbial fuel cell in the presence and absence of a proton exchange membrane. *Environ. Sci. Technol.*, **38** (14), 4040–4046.
48. Zhuang, L., Feng, C.H., Zhou, S.G., Li, Y.T., and Wang, Y.Q. (2010) Comparison of membrane- and cloth-cathode assembly for scalable microbial fuel cells: construction, performance and cost. *Process Biochem.*, **45** (6), 929–934.
49. Ieropoulos, I., Greenman, J., and Melhuish, C. (2008) Microbial fuel cells based on carbon veil electrodes: stack configuration and scalability. *Int. J. Energy Res.*, **32** (13), 1228–1240.
50. Aelterman, P., Rabaey, K., Pham, H.T., Boon, N., and Verstraete, W. (2006) Continuous electricity generation at high voltages and currents using stacked microbial fuel cells. *Environ. Sci. Technol.*, **40** (10), 3388–3394.
51. Logan, B.E. and Oh, S.E. (2007) Voltage reversal during microbial fuel cell stack operation. *J. Power Sources*, **167** (1), 11–17.
52. Dekker, A., Ter Heijne, A., Saakes, M., Hamelers, H.V.M., and Buisman, C.J.N. (2009) Analysis and improvement of a scaled-up and stacked microbial fuel cell. *Environ. Sci. Technol.*, **43** (23), 9038–9042.
53. Clauwaert, P., Mulenga, S., Aelterman, P., and Verstraete, W. (2009) Litre-scale microbial fuel cells operated in a complete loop. *Appl. Microbiol. Biotechnol.*, **83** (2), 241–247.

54. Logan, B.E. (2010) Scaling up microbial fuel cells and other bioelectrochemical systems. *Appl. Microbiol. Biotechnol.*, **85** (6), 1665–1671.
55. Zhang, F., Jacobson, K.S., Torres, P., and He, Z. (2010) Effects of anolyte recirculation rates and catholytes on electricity generation in a litre-scale upflow microbial fuel cell. *Energy Environ. Sci.*, **3** (9), 1347–1352.
56. Rabaey, K., Read, S.T., Clauwaert, P., Freguia, S., Bond, P.L., Blackall, L.L., and Keller, J. (2008) Cathodic oxygen reduction catalyzed by bacteria in microbial fuel cells. *ISME J.*, **2** (5), 519–527.
57. Chae, K.J., Choi, M.J., Lee, J.W., Kim, K.Y., and Kim, I.S. (2009) Effect of different substrates on the performance, bacterial diversity, and bacterial viability in microbial fuel cells. *Bioresour. Technol.*, **100** (14), 3518–3525.
58. Erable, B., Roncato, M.A., Achouak, W., and Bergel, A. (2009) Sampling natural biofilms: a new route to build efficient microbial anodes. *Environ. Sci. Technol.*, **43** (9), 3194–3199.
59. De Schamphelaire, L., Cabezas, A., Marzorati, M., Friedrich, M.W., Boon, N., and Verstraete, W. (2010) Microbial community analysis of anodes from sediment microbial fuel cells powered by rhizodeposits of living rice plants. *Appl. Environ. Microbiol.*, **76** (6), 2002–2008.
60. Marzorati, M., Wittebolle, L., Boon, N., Daffonchio, D., and Verstraete, W. (2008) How to get more out of molecular fingerprints: practical tools for microbial ecology. *Environ. Microbiol.*, **10** (6), 1571–1581.
61. Wittebolle, L., Marzorati, M., Clement, L., Balloi, A., Daffonchio, D., Heylen, K., De Vos, P., Verstraete, W., and Boon, N. (2009) Initial community evenness favours functionality under selective stress. *Nature*, **458** (7238), 623–626.
62. Rabaey, K., Van De Sompel, K., Maignien, L., Boon, N., Aelterman, P., Clauwaert, P., De Schamphelaire, L., Pham, H.T., Vermeulen, J., Verhaege, M., Lens, P., and Verstraete, W. (2006) Microbial fuel cells for sulfide removal. *Environ. Sci. Technol.*, **40** (17), 5218–5224.
63. Reguera, G., Mccarthy, K.D., Mehta, T., Nicoll, J.S., Tuominen, M.T., and Lovley, D.R. (2005) Extracellular electron transfer via microbial nanowires. *Nature*, **435** (7045), 1098–1101.
64. Pham, H.T., Boon, N., Aelterman, P., Clauwaert, P., De Schamphelaire, L., Van Oostveldt, P., Verbeken, K., Rabaey, K., and Verstraete, W. (2008) High shear rate enrichment improves the performance of the anodophilic microbial consortium in a microbial fuel cell. *Microb. Biotechnol.*, **1**, 487–496.
65. Clauwaert, P., Aelterman, P., Pham, T.H., De Schamphelaire, L., Carballa, M., Rabaey, K., and Verstraete, W. (2008) Minimizing losses in bio-electrochemical systems: the road to applications. *Appl. Microbiol. Biotechnol.*, **79** (6), 901–913.
66. Fan, Y., Sharbrough, E., and Liu, H. (2008) Quantification of the internal resistance distribution of microbial fuel cells. *Environ. Sci. Technol.*, **42** (21), 8101–8107.
67. Ter Heijne, A., Hamelers, H.V.M., De Wilde, V., Rozendal, R.A., and Buisman, C.J.N. (2006) A bipolar membrane combined with ferric iron reduction as an efficient cathode system in microbial fuel cells. *Environ. Sci. Technol.*, **40** (17), 5200–5205.
68. Sleutels, T.H.J.A., Lodder, R., Hamelers, H.V.M., and Buisman, C.J.N. (2009) Improved performance of porous bio-anodes in microbial electrolysis cells by enhancing mass and charge transport. *Int. J. Hydrogen Energy*, **34** (24), 9655–9661.
69. Ter Heijne, A., Strik, D.P.B.T.B., Hamelers, H.V.M., and Buisman, C.J.N. (2010) Cathode potential and mass transfer determine performance of oxygen reducing biocathodes in microbial fuel cells. *Environ. Sci. Technol.*, **44** (18), 7151–7156.
70. Watson, V.J. and Logan, B.E. (2011) Analysis of polarization methods for elimination of power overshoot

in microbial fuel cells. *Electrochem. Commun.*, **13** (1), 54–56.
71. Winfield, J., Ieropoulos, I., and Greenman, J. (2010) The overshoot phenomenon as a function of internal resistance in microbial fuel cells. *J. Biotechnol.*, **150** (Suppl. 1), 23–23.
72. Ieropoulos, I., Winfield, J., and Greenman, J. (2010) Effects of flow-rate, inoculum and time on the internal resistance of microbial fuel cells. *Bioresour. Technol.*, **101** (10), 3520–3525.
73. He, Z., Shao, H., and Angenent, L.T. (2007) Increased power production from a sediment microbial fuel cell with a rotating cathode. *Biosens. Bioelectron.*, **22** (12), 3252–3255.
74. Manohar, A.K., Bretschger, O., Nealson, K.H., and Mansfeld, F. (2008) The use of electrochemical impedance spectroscopy (EIS) in the evaluation of the electrochemical properties of a microbial fuel cell. *Bioelectrochemistry*, **72** (2), 149–154.
75. Borole, A.P., Aaron, D., Hamilton, C.Y., and Tsouris, C. (2010) Understanding long-term changes in microbial fuel cell performance using electrochemical impedance spectroscopy. *Environ. Sci. Technol.*, **44** (7), 2740–2744.
76. Fricke, K., Harnisch, F., and Schröder, U. (2008) On the use of cyclic voltammetry for the study of anodic electron transfer in microbial fuel cells. *Energy Environ. Sci.*, **1** (1), 144–147.
77. Marsili, E., Baron, D.B., Shikhare, I.D., Coursolle, D., Gralnick, J.A., and Bond, D.R. (2008) *Shewanella* secretes flavins that mediate extracellular electron transfer. *Proc. Natl. Acad. Sci. U. S. A.*, **105** (10), 3968–3973.
78. Cournet, A., Délia, M.-L., Bergel, A., Roques, C., and Bergé, M. (2010) Electrochemical reduction of oxygen catalyzed by a wide range of bacteria including Gram-positive. *Electrochem. Commun.*, **12** (4), 505–508.
79. Cheng, K.Y., Cord-Ruwisch, R., and Ho, G. (2009) A new approach for *in situ* cyclic voltammetry of a microbial fuel cell biofilm without using a potentiostat. *Bioelectrochemistry*, **74** (2), 227–231.
80. Rabaey, K. and Rozendal, R.A. (2010) Microbial electrosynthesis – revisiting the electrical route for microbial production. *Nat. Rev. Microbiol.*, **8** (10), 706–716.
81. Thrash, J.C. and Coates, J.D. (2008) Review: direct and indirect electrical stimulation of microbial metabolism. *Environ. Sci. Technol.*, **42** (11), 3921–3931.
82. Lowy, D.A., Tender, L.M., Zeikus, J.G., Park, D.H., and Lovley, D.R. (2006) Harvesting energy from the marine sediment–water interface II. – Kinetic activity of anode materials. *Biosens. Bioelectron.*, **21** (11), 2058–2063.
83. Bird, L.J., Bonnefoy, V., and Newman, D.K. (2011) Bioenergetic challenges of microbial iron metabolisms. *Trends Microbiol.*, **19** (7), 330–340.
84. Shi, L.A., Richardson, D.J., Wang, Z.M., Kerisit, S.N., Rosso, K.M., Zachara, J.M., and Fredrickson, J.K. (2009) The roles of outer membrane cytochromes of *Shewanella* and *Geobacter* in extracellular electron transfer. *Environ. Microbiol. Rep.*, **1** (4), 220–227.
85. Richardson, D.J. (2000) Bacterial respiration: a flexible process for a changing environment. *Microbiology*, **146** (3), 551–571.
86. Pant, D., Van Bogaert, G., Diels, L., and Vanbroekhoven, K. (2010) A review of the substrates used in microbial fuel cells (MFCs) for sustainable energy production. *Bioresour. Technol.*, **101** (6), 1533–1543.
87. Strik, D.P.B.T.B., Timmers, R.A., Helder, M., Steinbusch, K.J.J., Hamelers, H.V.M., and Buisman, C.J.N. (2011) Microbial solar cells: applying photosynthetic and electrochemically active organisms. *Trends Biotechnol.*, **29** (1), 41–49.
88. Logan, B.E. (2009) Exoelectrogenic bacteria that power microbial fuel cells. *Nat. Rev. Microbiol.*, **7** (5), 375–381.
89. Lovley, D.R. (2006) Bug juice: harvesting electricity with microorganisms. *Nat. Rev. Microbiol.*, **4** (7), 497–508.

90. Lee, H.S., Torres, C.I., and Rittmann, B.E. (2009) Effects of substrate diffusion and anode potential on kinetic parameters for anode-respiring bacteria. *Environ. Sci. Technol.*, **43** (19), 7571–7577.
91. Milliken, C.E. and May, H.D. (2007) Sustained generation of electricity by the spore-forming, Gram-positive, *Desulfitobacterium hafniense* strain DCB2. *Appl. Microbiol. Biotechnol.*, **73**, 1180–1189.
92. Rabaey, K., Boon, N., Hofte, M., and Verstraete, W. (2005) Microbial phenazine production enhances electron transfer in biofuel cells. *Environ. Sci. Technol.*, **39** (9), 3401–3408.
93. Wrighton, K.C., Agbo, P., Warnecke, F., Weber, K.A., Brodie, E.L., Desantis, T.Z., Hugenholtz, P., Andersen, G.L., and Coates, J.D. (2008) A novel ecological role of the firmicutes identified in thermophilic microbial fuel cells. *ISME J.*, **2** (11), 1146–1156.
94. Bond, D.R. and Lovley, D.R. (2003) Electricity production by *Geobacter sulfurreducens* attached to electrodes. *Appl. Environ. Microbiol.*, **69** (3), 1548–1555.
95. Caccavo, F. Jr., Lonergan, D.J., Lovley, D.R., Davis, M., Stolz, J.F., and Mcinerney, M.J. (1994) *Geobacter sulfurreducens* sp. nov., a hydrogen- and acetate-oxidizing dissimilatory metal-reducing microorganism. *Appl. Environ. Microbiol.*, **60** (10), 3752–3759.
96. Gorby, Y.A., Yanina, S., Mclean, J.S., Rosso, K.M., Moyles, D., Dohnalkova, A., Beveridge, T.J., Chang, I.S., Kim, B.H., Kim, K.S., Culley, D.E., Reed, S.B., Romine, M.F., Saffarini, D.A., Hill, E.A., Shi, L., Elias, D.A., Kennedy, D.W., Pinchuk, G., Watanabe, K., Ishii, S., Logan, B., Nealson, K.H., and Fredrickson, J.K. (2006) Electrically conductive bacterial nanowires produced by *Shewanella oneidensis* strain MR-1 and other microorganisms. *Proc. Natl. Acad. Sci. U. S. A.*, **103** (30), 11358–11363.
97. Lovley, D.R. (2008) The microbe electric: conversion of organic matter to electricity. *Curr. Opin. Biotechnol.*, **19** (6), 564–571.
98. Torres, C.I., Marcus, A.K., Lee, H.S., Parameswaran, P., Krajmalnik-Brown, R., and Rittmann, B.E. (2010) A kinetic perspective on extracellular electron transfer by anode-respiring bacteria. *FEMS Microbiol. Rev.*, **34** (1), 3–17.
99. Leang, C., Qian, X.L., Mester, T., and Lovley, D.R. (2010) Alignment of the c-type cytochrome OmcS along pili of *Geobacter sulfurreducens*. *Appl. Environ. Microbiol.*, **76** (12), 4080–4084.
100. Inoue, K., Leang, C., Franks, A.E., Woodard, T.L., Nevin, K.P., and Lovley, D.R. (2011) Specific localization of the c-type cytochrome OmcZ at the anode surface in current-producing biofilms of *Geobacter sulfurreducens*. *Environ. Microbiol. Rep.*, **3** (2), 211–217.
101. Richter, H., Nevin, K.P., Jia, H.F., Lowy, D.A., Lovley, D.R., and Tender, L.M. (2009) Cyclic voltammetry of biofilms of wild type and mutant *Geobacter sulfurreducens* on fuel cell anodes indicates possible roles of OmcB, OmcZ type IV pili, and protons in extracellular electron transfer. *Energy Environ. Sci.*, **2** (5), 506–516.
102. Morita, M., Malvankar, N.S., Franks, A.E., Summers, Z.M., Giloteaux, L., Rotaru, A.E., Rotaru, C., and Lovley, D.R. (2011) Potential for direct interspecies electron transfer in methanogenic wastewater digester aggregates. *mBio*, **2** (4), e00159-11.
103. Nielsen, L.P., Risgaard-Petersen, N., Fossing, H., Christensen, P.B., and Sayama, M. (2010) Electric currents couple spatially separated biogeochemical processes in marine sediment. *Nature*, **463** (7284), 1071–1074.
104. Holmes, D.E., Bond, D.R., and Lovley, D.R. (2004) Electron transfer by *Desulfobulbus propionicus* to Fe(III) and graphite electrodes. *Appl. Environ. Microbiol.*, **70** (2), 1234–1237.
105. Dutta, P.K., Keller, J., Yuan, Z.G., Rozendal, R.A., and Rabaey, K. (2009) Role of sulfur during acetate oxidation in biological anodes. *Environ. Sci. Technol.*, **43** (10), 3839–3845.
106. Schröder, U., Niessen, J., and Scholz, F. (2003) A generation of

microbial fuel cells with current outputs boosted by more than one order of magnitude. *Angew. Chem. Int. Ed.*, **42** (25), 2880–2883.
107. Cheng, S., Liu, H., and Logan, B.E. (2006) Power densities using different cathode catalysts (Pt and CoTMPP) and polymer binders (Nafion and PTFE) in single chamber microbial fuel cells. *Environ. Sci. Technol.*, **40** (1), 364–369.
108. Rozendal, R.A., Hamelers, H.V.M., Euverink, G.J.W., Metz, S.J., and Buisman, C.J.N. (2006) Principle and perspectives of hydrogen production through biocatalyzed electrolysis. *Int. J. Hydrogen Energy*, **31** (12), 1632–1640.
109. Gregory, K.B., Bond, D.R., and Lovley, D.R. (2004) Graphite electrodes as electron donors for anaerobic respiration. *Environ. Microbiol.*, **6** (6), 596–604.
110. Rosenbaum, M., Aulenta, F., Villano, M., and Angenent, L.T. (2011) Cathodes as electron donors for microbial metabolism: which extracellular electron transfer mechanisms are involved? *Bioresour. Technol.*, **102** (1), 324–333.
111. Cheng, K.Y., Ho, G., and Cord-Ruwisch, R. (2010) Anodophilic biofilm catalyzes cathodic oxygen reduction. *Environ. Sci. Technol.*, **44** (1), 518–525.
112. Rozendal, R.A., Jeremiasse, A.W., Hamelers, H.V.M., and Buisman, C.J.N. (2008) Hydrogen production with a microbial biocathode. *Environ. Sci. Technol.*, **42** (2), 629–634.
113. Huang, L.P., Regan, J.M., and Quan, X. (2011) Electron transfer mechanisms, new applications, and performance of biocathode microbial fuel cells. *Bioresour. Technol.*, **102** (1), 316–323.
114. Freguia, S., Tsujimura, S., and Kano, K. (2010) Electron transfer pathways in microbial oxygen biocathodes. *Electrochim. Acta*, **55** (3), 813–818.
115. Clauwaert, P., Rabaey, K., Aelterman, P., De Schamphelaire, L., Ham, T.H., Boeckx, P., Boon, N., and Verstraete, W. (2007) Biological denitrification in microbial fuel cells. *Environ. Sci. Technol.*, **41** (9), 3354–3360.
116. Rhoads, A., Beyenal, H., and Lewandowski, Z. (2005) Microbial fuel cell using anaerobic respiration as an anodic reaction and biomineralized manganese as a cathodic reactant. *Environ. Sci. Technol.*, **39** (12), 4666–4671.
117. Erable, B., Vandecandelaere, I., Faimali, M., Delia, M.L., Etcheverry, L., Vandamme, P., and Bergel, A. (2010) Marine aerobic biofilm as biocathode catalyst. *Bioelectrochemistry*, **78** (1), 51–56.
118. Cao, X.X., Huang, X., Liang, P., Boon, N., Fan, M.Z., Zhang, L., and Zhang, X.Y. (2009) A completely anoxic microbial fuel cell using a photo-biocathode for cathodic carbon dioxide reduction. *Energy Environ. Sci.*, **2** (5), 498–501.
119. Butler, C.S., Clauwaert, P., Green, S.J., Verstraete, W., and Nerenberg, R. (2010) Bioelectrochemical perchlorate reduction in a microbial fuel cell. *Environ. Sci. Technol.*, **44** (12), 4685–4691.
120. Nevin, K.P., Richter, H., Covalla, S.F., Johnson, J.P., Woodard, T.L., Orloff, A.L., Jia, H., Zhang, M., and Lovley, D.R. (2008) Power output and coulombic efficiencies from biofilms of *Geobacter sulfurreducens* comparable to mixed community microbial fuel cells. *Environ. Microbiol.*, **10** (10), 2505–2514.
121. Fornero, J.J., Rosenbaum, M., and Angenent, L.T. (2010) Electric power generation from municipal, food, and animal wastewaters using microbial fuel cells. *Electroanalysis*, **22** (7–8), 832–843.
122. Rozendal, R.A., Hamelers, H.V.M., Rabaey, K., Keller, J., and Buisman, C.J.N. (2008) Towards practical implementation of bioelectrochemical wastewater treatment. *Trends Biotechnol.*, **26** (8), 450–459.
123. Pant, D., Singh, A., Van Bogaert, G., Gallego, Y.A., Diels, L., and Vanbroekhoven, K. (2011) An introduction to the life cycle assessment (LCA) of bioelectrochemical systems (BES) for sustainable energy and product generation: relevance and key

aspects. *Renew. Sustainable Energy Rev.*, **15** (2), 1305–1313.
124. Foley, J.M., Rozendal, R.A., Hertle, C.K., Lant, P.A., and Rabaey, K. (2010) Life cycle assessment of high-rate anaerobic treatment, microbial fuel cells, and microbial electrolysis cells. *Environ. Sci. Technol.*, **44** (9), 3629–3637.
125. Carballa, M., Moerman, W., De Windt, W., Grotaerd, H., and Verstraete, W. (2009) Strategies to optimize phosphate removal from industrial anaerobic effluents by magnesium ammonium phosphate (MAP) production. *J. Chem. Technol. Biotechnol.*, **84** (1), 63–68.
126. Abegglen, C., Ospelt, M., and Siegrist, H. (2008) Biological nutrient removal in a small-scale MBR treating household wastewater. *Water Res.*, **42** (1–2), 338–346.
127. Aelterman, P., Rabaey, K., Clauwaert, P., and Verstraete, W. (2006) Microbial fuel cells for wastewater treatment. *Water Sci. Technol.*, **54** (8), 9–15.
128. Rodrigo, M.A., Canizares, P., Lobato, J., Paz, R., Saez, C., and Linares, J.J. (2007) Production of electricity from the treatment of urban waste water using a microbial fuel cell. *J. Power Sources*, **169** (1), 198–204.
129. Puig, S., Serra, M., Coma, M., Cabré, M., Balaguer, M.D., and Colprim, J. (2010) Effect of pH on nutrient dynamics and electricity production using microbial fuel cells. *Bioresour. Technol.*, **101** (24), 9594–9599.
130. Ahn, Y. and Logan, B.E. (2010) Effectiveness of domestic wastewater treatment using microbial fuel cells at ambient and mesophilic temperatures. *Bioresour. Technol.*, **101** (2), 469–475.
131. Virdis, B., Rabaey, K., Yuan, Z.G., Rozendal, R.A., and Keller, J. (2009) Electron fluxes in a microbial fuel cell performing carbon and nitrogen removal. *Environ. Sci. Technol.*, **43** (13), 5144–5149.
132. Puig, S., Serra, M., Vilar-Sanz, A., Cabré, M., Bañeras, L., Colprim, J., and Balaguer, M.D. (2011) Autotrophic nitrite removal in the cathode of microbial fuel cells. *Bioresour. Technol.*, **102** (6), 4462–4467.
133. Greenman, J., Gálvez, A., Giusti, L., and Ieropoulos, I. (2009) Electricity from landfill leachate using microbial fuel cells: comparison with a biological aerated filter. *Enzyme Microb. Technol.*, **44** (2), 112–119.
134. Puig, S., Serra, M., Coma, M., Cabré, M., Balaguer, M.D., and Colprim, J. (2011) Microbial fuel cell application in landfill leachate treatment. *J. Hazard. Mater.*, **185** (2-3), 763–767.
135. Velasquez-Orta, S.B., Head, I.M., Curtis, T.P., and Scott, K. (2011) Factors affecting current production in microbial fuel cells using different industrial wastewaters. *Bioresour. Technol.*, **102** (8), 5105–5112.
136. Luo, Y., Zhang, R., Liu, G., Li, J., Qin, B., Li, M., and Chen, S. (2011) Simultaneous degradation of refractory contaminants in both the anode and cathode chambers of the microbial fuel cell. *Bioresour. Technol.*, **102** (4), 3827–3832.
137. Wen, Q., Kong, F., Zheng, H., Yin, J., Cao, D., Ren, Y., and Wang, G. (2011) Simultaneous processes of electricity generation and ceftriaxone sodium degradation in an air-cathode single chamber microbial fuel cell. *J. Power Sources*, **196** (5), 2567–2572.
138. Li, Z.J., Zhang, X.W., Lin, J., Han, S., and Lei, L.C. (2010) Azo dye treatment with simultaneous electricity production in an anaerobic–aerobic sequential reactor and microbial fuel cell coupled system. *Bioresour. Technol.*, **101** (12), 4440–4445.
139. Liu, R.H., Sheng, G.P., Sun, M., Zang, G.L., Li, W.W., Tong, Z.H., Dong, F., Lam, M.H.W., and Yu, H.Q. (2011) Enhanced reductive degradation of methyl orange in a microbial fuel cell through cathode modification with redox mediators. *Appl. Microbiol. Biotechnol.*, **89** (1), 201–208.
140. Mu, Y., Rozendal, R.A., Rabaey, K., and Keller, J. (2010) Electrochemically active bacteria assisted nitrobenzene removal

from wastewater. *J. Biotechnol.*, **150** (Suppl. 1), 147–147.
141. Mehanna, M., Saito, T., Yan, J.L., Hickner, M., Cao, X.X., Huang, X., and Logan, B.E. (2010) Using microbial desalination cells to reduce water salinity prior to reverse osmosis. *Energy Environ. Sci.*, **3** (8), 1114–1120.
142. Ter Heijne, A., Liu, F., Van Der Weijden, R., Weijma, J., Buisman, C.J.N., and Hamelers, H.V.M. (2010) Copper recovery combined with electricity production in a microbial fuel cell. *Environ. Sci. Technol.*, **44** (11), 4376–4381.
143. Bond, D.R., Holmes, D.E., Tender, L.M., and Lovley, D.R. (2002) Electrode-reducing microorganisms that harvest energy from marine sediments. *Science*, **295** (5554), 483–485.
144. Helder, M., Strik, D.P.B.T.B., Hamelers, H.V.M., Kuhn, A.J., Blok, C., and Buisman, C.J.N. (2010) Concurrent bio-electricity and biomass production in three plant-microbial fuel cells using *Spartina anglica*, *Arundinella anomala* and *Arundo donax*. *Bioresour. Technol.*, **101** (10), 3541–3547.
145. Strik, D.P.B.T.B., Hamelers, H.V.M., and Buisman, C.J.N. (2010) Solar energy powered microbial fuel cell with a reversible bioelectrode. *Environ. Sci. Technol.*, **44** (1), 532–537.
146. Cusick, R.D., Kiely, P.D., and Logan, B.E. (2010) A monetary comparison of energy recovered from microbial fuel cells and microbial electrolysis cells fed winery or domestic wastewaters. *Int. J. Hydrogen Energy*, **35** (17), 8855–8861.
147. Wagner, R.C., Regan, J.M., Oh, S.-E., Zuo, Y., and Logan, B.E. (2009) Hydrogen and methane production from swine wastewater using microbial electrolysis cells. *Water Res.*, **43** (5), 1480–1488.
148. Mehanna, M., Kiely, P.D., Call, D.F., and Logan, B.E. (2010) Microbial electrodialysis cell for simultaneous water desalination and hydrogen gas production. *Environ. Sci. Technol.*, **44** (24), 9578–9583.
149. Rozendal, R.A., Leone, E., Keller, J., and Rabaey, K. (2009) Efficient hydrogen peroxide generation from organic matter in a bioelectrochemical system. *Electrochem. Commun.*, **11** (9), 1752–1755.
150. Rabaey, K., Bützer, S., Brown, S., Keller, J., and Rozendal, R.A. (2010) High current generation coupled to caustic production using a lamellar bioelectrochemical system. *Environ. Sci. Technol.*, **44** (11), 4315–4321.
151. Steinbusch, K.J.J., Hamelers, H.V.M., Schaap, J.D., Kampman, C., and Buisman, C.J.N. (2010) Bioelectrochemical ethanol production through mediated acetate reduction by mixed cultures. *Environ. Sci. Technol.*, **44** (1), 513–517.
152. Clauwaert, P., Desloover, J., Shea, C., Nerenberg, R., Boon, N., and Verstraete, W. (2009) Enhanced nitrogen removal in bio-electrochemical systems by Ph control. *Biotechnol. Lett.*, **31** (10), 1537–1543.
153. Virdis, B., Rabaey, K., Yuan, Z., and Keller, J. (2008) Microbial fuel cells for simultaneous carbon and nitrogen removal. *Water Res.*, **42** (12), 3013–3024.
154. Virdis, B., Rabaey, K., Rozendal, R.A., Yuan, Z.G., and Keller, J. (2010) Simultaneous nitrification, denitrification and carbon removal in microbial fuel cells. *Water Res.*, **44** (9), 2970–2980.
155. Xie, S., Liang, P., Chen, Y., Xia, X., and Huang, X. (2011) Simultaneous carbon and nitrogen removal using an oxic/anoxic-biocathode microbial fuel cells coupled system. *Bioresour. Technol.*, **102** (1), 348–354.
156. Feng, Y., Wang, X., Logan, B.E., and Lee, H. (2008) Brewery wastewater treatment using air-cathode microbial fuel cells. *Appl. Microbiol. Biotechnol.*, **78** (5), 873–880.
157. Wen, Q., Wu, Y., Cao, D.X., Zhao, L.X., and Sun, Q. (2009) Electricity generation and modeling of microbial fuel cell from continuous beer brewery wastewater. *Bioresour. Technol.*, **100** (18), 4171–4175.
158. Reimers, C.E., Tender, L.M., Fertig, S., and Wang, W. (2001) Harvesting energy from the marine sediment-water interface. *Environ. Sci. Technol.*, **35** (1), 192–195.

159. Tender, L.M., Reimers, C.E., Stecher, H.A., Holmes, D.E., Bond, D.R., Lowy, D.A., Pilobello, K., Fertig, S.J., and Lovley, D.R. (2002) Harnessing microbially generated power on the seafloor. *Nat. Biotechnol.*, **20** (8), 821–825.
160. Tender, L.M., Gray, S.A., Groveman, E., Lowy, D.A., Kauffman, P., Melhado, J., Tyce, R.C., Flynn, D., Petrecca, R., and Dobarro, J. (2008) The first demonstration of a microbial fuel cell as a viable power supply: powering a meteorological buoy. *J. Power Sources*, **179** (2), 571–575.
161. Donovan, C., Dewan, A., Heo, D., and Beyenal, H. (2008) Batteryless, wireless sensor powered by a sediment microbial fuel cell. *Environ. Sci. Technol.*, **42** (22), 8591–8596.
162. Dewan, A., Donovan, C., Heo, D., and Beyenal, H. (2010) Evaluating the performance of microbial fuel cells powering electronic devices. *J. Power Sources*, **195** (1), 90–96.
163. Günter, N. and Volker, R. (2007) in *The Rhizosphere; Books in Soils, Plants, and the Environment* (eds P. Roberto, V. Zeno, and N. Paolo), CRC Press, pp. 23–72.
164. Kaku, N., Yonezawa, N., Kodama, Y., and Watanabe, K. (2008) Plant/microbe cooperation for electricity generation in a rice paddy field. *Appl. Microbiol. Biotechnol.*, **79** (1), 43–49.
165. Strik, D.P.B.T.B., Hamelers, H.V.M., Snel, J.F.H., and Buisman, C.J.N. (2008) Green electricity production with living plants and bacteria in a fuel cell. *Int. J. Energy Res.*, **32** (9), 870–876.
166. Rosenbaum, M. and Schröder, U. (2010) Photomicrobial solar and fuel cells. *Electroanalysis*, **22** (7–8), 844–855.
167. Rosenbaum, M., He, Z., and Angenent, L.T. (2010) Light energy to bioelectricity: photosynthetic microbial fuel cells. *Curr. Opin. Biotechnol.*, **21** (3), 259–264.
168. Pham, T.H., Rabaey, K., Aelterman, P., Clauwaert, P., De Schamphelaire, L., Boon, N., and Verstraete, W. (2006) Microbial fuel cells in relation to conventional anaerobic digestion technology. *Eng. Life Sci.*, **6** (3), 285–292.
169. Sasaki, K., Sasaki, D., Morita, M., Hirano, S., Matsumoto, N., Ohmura, N., and Igarashi, Y. (2010) Bioelectrochemical system stabilizes methane fermentation from garbage slurry. *Bioresour. Technol.*, **101** (10), 3415–3422.
170. Ditzig, J., Liu, H., and Logan, B.E. (2007) Production of hydrogen from domestic wastewater using a bioelectrochemically assisted microbial reactor (BEAMR). *Int. J. Hydrogen Energy*, **32** (13), 2296–2304.
171. Weld, R.J. and Singh, R. (2011) Functional stability of a hybrid anaerobic digester/microbial fuel cell system treating municipal wastewater. *Bioresour. Technol.*, **102** (2), 842–847.
172. Semiat, R. (2008) Energy issues in desalination processes. *Environ. Sci. Technol.*, **42** (22), 8193–8201.
173. Pham, H., Boon, N., Marzorati, M., and Verstraete, W. (2009) Enhanced removal of 1,2-dichloroethane by anodophilic microbial consortia. *Water Res.*, **43** (11), 2936–2946.
174. Sass, R.L., Andrews, J.A., Ding, A., and Fisher, F.M. (2002) Spatial and temporal variability in methane emissions from rice paddies: implications for assessing regional methane budgets. *Nutr. Cycling Agroecosyst.*, **64** (1), 3–7.
175. Gregory, K.B. and Lovley, D.R. (2005) Remediation and recovery of uranium from contaminated subsurface environments with electrodes. *Environ. Sci. Technol.*, **39** (22), 8943–8947.
176. Williams, K.H., Nevin, K.P., Franks, A., Englert, A., Long, P.E., and Lovley, D.R. (2010) Electrode-based approach for monitoring *in situ* microbial activity during subsurface bioremediation. *Environ. Sci. Technol.*, **44** (1), 47–54.
177. Wang, Z., Lim, B., and Choi, C. (2011) Removal of Hg^{2+} as an electron acceptor coupled with power generation using a microbial fuel cell. *Bioresour. Technol.*, **102** (10), 6304–6307.

178. Kissinger, P.T. (2005) Biosensors – a perspective. *Biosens. Bioelectron.*, **20** (12), 2512–2516.
179. Di Lorenzo, M., Curtis, T.P., Head, I.M., Velasquez-Orta, S.B., and Scott, K. (2009) A single chamber packed bed microbial fuel cell biosensor for measuring organic content of wastewater. *Water Sci. Technol.*, **60** (11), 2879–2887.
180. Patil, S., Harnisch, F., and Schroder, U. (2010) Toxicity response of electroactive microbial biofilms – a decisive feature for potential biosensor and power source applications. *ChemPhysChem*, **11** (13), 2834–2837.
181. Stein, N.E., Hamelers, H.V.M., and Buisman, C.N.J. (2010) Stabilizing the baseline current of a microbial fuel cell-based biosensor through overpotential control under non-toxic conditions. *Bioelectrochemistry*, **78** (1), 87–91.
182. Picioreanu, C., Head, I.M., Katuri, K.P., Van Loosdrecht, M.C.M., and Scott, K. (2007) A computational model for biofilm-based microbial fuel cells. *Water Res.*, **41** (13), 2921–2940.
183. Picioreanu, C., Van Loosdrecht, M.C.M., Curtis, T.P., and Scott, K. (2010) Model based evaluation of the effect of pH and electrode geometry on microbial fuel cell performance. *Bioelectrochemistry*, **78** (1), 8–24.
184. Hamelers, H.V.M., Ter Heijne, A., Stein, N., Rozendal, R.A., and Buisman, C.J.N. (2010) Butler–Volmer–Monod model for describing bio-anode polarization curves. *Bioresour. Technol.*, **102** (1), 381–387.
185. Pinto, R.P., Srinivasan, B., Manuel, M.F., and Tartakovsky, B. (2010) A two-population bio-electrochemical model of a microbial fuel cell. *Bioresour. Technol.*, **101** (14), 5256–5265.

7
Micro-Reactors for Fuel Processing

Gunther Kolb

7.1
Introduction

Within the last two decades, micro-reactors have not only become established in chemical research but increasingly also in production processes, especially in the case of fine chemicals. In parallel with this development in the field of liquid-phase chemistry, gas-phase reactions have been the subject of numerous investigations. This quickly led to the introduction of heterogeneous catalysts in the reactors, which are normally present in the form of small fixed beds or coated on the walls of the microchannels. The potential of microstructured reactors for process intensification attracted much attention in the field of energy technology, wherever compact solutions were required, covering the entire field from portable and mobile power generation to stationary and decentralized production of fuels and power. Fuel-cell technology requires hydrogen or at least synthesis gas as fuel and therefore the step towards microstructured fuel processor technology was an obvious one.

In contrast to ceramic and metallic monoliths developed for automotive exhaust treatment purposes, which nowadays carry channels on the microscale and are actually "micro-reactors" by definition, the micro-reactors discussed in this chapter rather cover plate heat-exchanger technology with channels on the microscale. An overview of the fundamentals, practical applications, and production issues of micro-reactors for fuel-processing purposes is provided.

7.2
Heat and Mass Transfer in Micro-Reactors

Micro-reactors by definition are chemical reactors that carry one or a multitude of channels at least one dimension of which does not exceed 1 mm. Micro-reactors have some specific features compared with conventional technology. The flow regime is usually laminar, diffusion paths for heat and mass transfer are very small, and they have a small surface-to-volume ratio, which leads to domination of surface effects compared with volumetric effects. The share of wall material is higher than in macroscopic devices and therefore the heat transfer in the wall material contributes significantly to the overall heat transfer and needs to be considered when designing micro-reactors.

An analytical solution exists for the velocity profile under conditions of laminar, incompressible flow in a rectangular channel [1]. An approximate solution was derived by Purday [2] for a rectangular channel with a width $2a$ and a height $2b$:

$$u(x, y) = u_{max} \left[1 - \left(\frac{x}{a}\right)^s\right]\left[1 - \left(\frac{y}{b}\right)^r\right] \tag{7.1}$$

for a maximum flow velocity u_{max}. Correlations for s and r were given by Hartnet and Kostic [3]. To determine the pressure drop over a channel, the following equation can be used:

$$\frac{dp}{dz} = -f\frac{2\rho U^2}{D_h} \tag{7.2}$$

where dp/dz is the pressure gradient along the channel, ρ the density of the fluid, U the mean flow velocity, and D_h the hydraulic diameter, which is the ratio $4A/P$ of the channel cross-sectional area A and its perimeter P. A simple correlation for the friction factor was derived by Shah and London [4]:

$$f = \frac{24}{Re} - (1 - 1.3553\,\alpha + 1.9467\,\alpha^2 - 1.7012\,\alpha^3 + 0.9564\,\alpha^4 - 0.2537\,\alpha^5) \tag{7.3}$$

where α is the channel aspect ratio, that is, the ratio of channel width to channel depth.

In practical micro-reactors designed for fuel processor systems the channel dimensions are in the range 250–1000 μm in height and from 250 μm to several millimeters in width. Practical micro-reactors are comprised of a multitude, normally thousands, of channels of identical dimensions which are operated in parallel. Owing to the high number and relatively large dimensions of the channels, the pressure drop is in the order of a few millibars to a few tens of millibars.

Owing to the high degree of parallelization, the scale-up of the micro-reactor is relatively simple provided that equipartition of the fluid flow through the channels is achieved. It was shown by Walter et al. that even at low Reynolds numbers Re of around 30, vortices can be formed in distribution chambers upstream of multichannel geometries, which lead to deviations from the mean flow rate inside the channels of about 20% [5].

Different approaches can be followed to achieve flow equipartition. If the channels of one flow path of the reactor cover the entire length of the plate stack (similar to monolithic reactors), all channels of all plates of this flow path have a common distribution chamber. Alternatively, the flow between the channels of an individual plate is distributed in a relatively wide inlet region and a second flow chamber ensures the distribution over the different plates. In practical systems, the introduction of distribution grids is frequently required to ensure flow equipartition, as shown in Figure 7.1a [6]. Alternatively, a trumpet-shaped inlet geometry could be used as described by O'Connell et al. [7] (see Figure 7.1b).

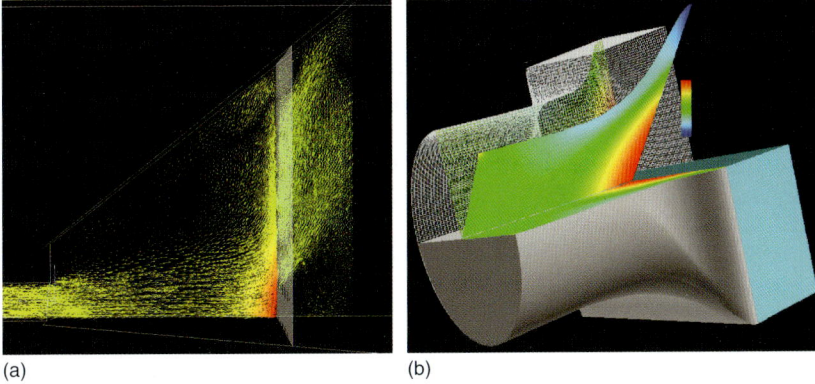

(a) (b)

Figure 7.1 (a) Results of numerical calculations of a flow distribution grid for a microchannel reactor for autothermal reforming of isooctane [6]. Source: IMM. (b) Pressure drop distribution over the channel inlet of a microstructured diesel steam reformer [7]. Source: IMM.

In practical microchannel systems, the Nusselt number Nu, which is the dimensionless ratio of the heat transfer coefficient k times the hydraulic diameter D_h to the heat conductivity of the material λ:

$$Nu = \frac{kD_h}{\lambda} \qquad (7.4)$$

is in many cases in the range 3–7 for straight channels and a gaseous medium at low pressure drop. Despite these relatively low numbers, the small diffusion paths allow for fast heat transfer. For liquids, much higher Nu values are achievable owing to the higher heat conductivity of the medium.

It is obvious that straight microchannels are the simplest design feasible for micro-reactors and that other geometries such as sinusoidal channels or fin geometries exist, which have the potential for increasing the heat transfer considerably. Hardt et al. [8] demonstrated that Nu values exceeding 100 can be achieved for air as medium in microfin arrangements, as shown in Figure 7.2.

However, an increase in the heat transfer at constant pressure drop, which can be expressed as the ratio of the Nusselt number and the Poiseuille number Po, is not achieved for most channel geometries. Po is the ratio of the friction factor f according to Eq. (7.2) to Re. Drese and Hardt showed that diamond-shaped fins have the potential to increasing the Nu/Po ratio by a factor of more than 2, if the ratio of fin length to fin width is increased by a factor of 4, because recirculation zones are suppressed by the longer fins [9].

Keeping the system pressure drop low is a critical issue for practical systems because compression of gases requires energy. The parasitic power losses originating from compressors are known to reduce the efficiency of fuel-cell systems considerably [10]. Steam reforming of liquid fuels is an exception here because only liquid pumps for both fuel and water and no compressor are required to supply

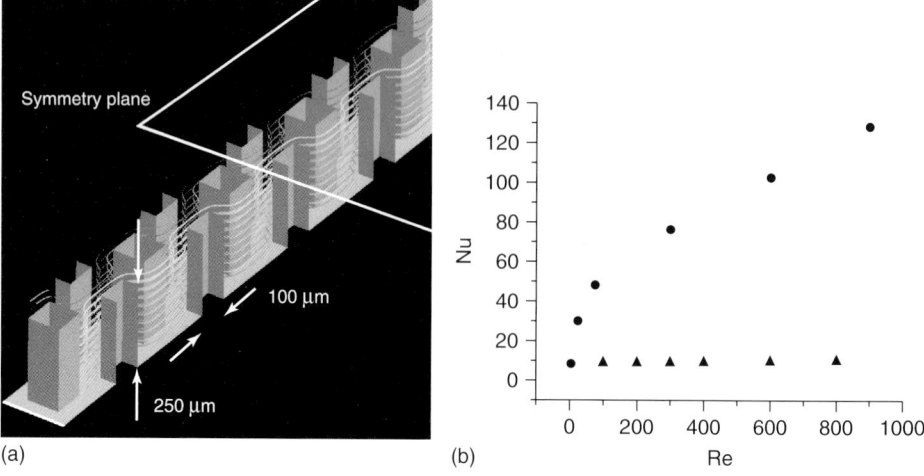

Figure 7.2 (a) Microfin arrangement applied by Hardt et al. [8]. Source: IMM. (b) Nusselt number as function of Reynolds number obtained for the fin arrangement (circles) and parallel plates (triangles). Source: IMM.

the feed to the fuel processor. Fazeli and Behnam demonstrated the advantages of zig-zag shaped channels over straight channels for methanol steam reforming [11].

The heat conductivity of the wall material decreases the efficiency of plate heat exchangers [12], especially if their length to-width ratio is in the region of 1 because the devices tend to show isothermal behavior, especially when a highly heat conductive material such as aluminum is used. It is therefore necessary to increase this ratio, to minimize the wall material and to maximize the porosity of the heat exchanger. To achieve high efficiency of gas/gas heat exchangers, the selection of wall materials with low heat conductivity such as polymers or even glass should be considered whenever feasible from the viewpoints of operating conditions and ease of fabrication.

On the other hand, heat-exchanger reactors for reactions such as steam reforming should be as isothermal as possible, which is discussed in Section 7.3.

7.3
Specific Features Required from Catalyst Formulations for Microchannel Plate Heat-Exchanger Reactors

To make a fuel-processing reactor out of a microstructured plate heat exchanger, heterogeneous catalysts need to be introduced into the microchannels, usually by wash-coating, similar to the procedure established for automotive exhaust clean-up.

Because environmental conditions and gas compositions are less well defined in small-scale power generation devices compared with plants on an industrial scale, especially during start-up and shut-down, suitable catalysts for small fuel processors need to be more robust against exposure to air and moisture compared with catalysts designed for large-scale industrial use which run under constant conditions for several years. For example, during start-up of a small system, reduction of the catalyst with hydrogen, which is frequently applied in industrial systems, is not possible at all, because no hydrogen is available. In contrast, it might be required to heat up the reactor with hot air or combustion gases before the reforming process is started and the catalyst activity must not suffer from this treatment in a practical system. All these aspects require consideration when selecting a catalyst formulation for a steam reformer of smaller than industrial scale.

Another aspect is the low catalyst mass per unit reactor volume that can be introduced into the plate heat-exchanger reactor. This drawback is counterbalanced by much better utilization of the catalyst owing to the improved heat and mass transfer. Consequently, more active catalysts are required which compensate the lower mass. At the same time, the cost of the catalyst is less of an issue compared with, for example, fixed-bed reactor technology. These statements will be confirmed by a practical example below.

Methanol is an attractive fuel for low-power applications, because the reaction temperature required for steam reforming is limited to values below 300–400 °C, which in turn minimizes heat losses from a small-scale system. Hence numerous research groups working on microchannel steam reforming are focusing on methanol as fuel. The carbon monoxide content present in reformate produced by methanol steam reforming is the lowest of all fuels compared at the same molar steam-to-carbon (S/C) ratio. Assuming an S/C ratio of >2 and a reaction temperature of 300 °C, not more than 1.2% of CO will be present in the feed [13]. This is related to the water gas shift (WGS) equilibrium and reduces the workload of the subsequent gas-purification steps.

Catalyst coatings under development for methanol steam reforming may be divided into copper- and precious metal-based systems.

The main advantage of commercial Cu/ZnO catalysts is their relatively high activity at operating temperatures below 300 °C. The catalysts are usually sensitive to temperatures exceeding 300 °C, which is a drawback when start-up procedures with hot combustion gases may result in temporary temperature excursions above 300 °C. In other words, it is difficult to heat a reactor to an operating temperature close to 300 °C within a few minutes without locally exceeding this temperature. Furthermore, Cu/ZnO catalysts are pyrophoric, which means that they show temperature excursions when exposed to air.

Bravo et al. [14] coated commercial $CuO/ZnO/Al_2O_3$ catalyst into capillaries and achieved 97% conversion at 97% carbon dioxide selectivity at a volume hourly space velocity (VHSV) of 3.9 l (h g_{cat})$^{-1}$, which is a typical value and is comparable to results obtained by many research groups for this type of catalyst [15–18]. The VHSV obtained by Bravo et al. is a rather low value compared with hydrocarbon reforming, where the VHSV is usually in the order of several hundred l (h g_{cat})$^{-1}$. The low value originates from the low reaction temperature of methanol steam reforming and therefore the advantages of Cu/ZnO catalysts mentioned above have another drawback, which is obviously not remotely compensated for by the improved mass transfer in microchannels.

Several CuO/ZnO catalysts were investigated by Pfeifer et al. [19]. They compared the performance of these catalysts with a Pd/ZnO catalyst, with the noble metal catalyst showing more activity. The formation of a Pd/Zn alloy at higher reduction temperatures was identified as crucial to achieving lower CO selectivity. It was postulated that metallic Pd particles promoted the formation of excess CO. Similar results were obtained by Chin et al. [20] and Hu et al. [21]. The PdZn alloy was assumed to be formed not only during reduction by pure hydrogen but also *in situ* in the hydrogen-rich reaction mixture of methanol steam reforming [22]. Later, Pfeifer et al. [23] prepared Pd/Zn catalysts by both pre- and post-impregnation of zinc oxide wash-coats with palladium. These samples showed CO concentrations below the WGS equilibrium. For both preparation routes, the highest activity was determined for samples containing 10 wt% palladium, which were also the most stable against deactivation. The VHSV amounted to $18 \, l \, (h \, g_{cat})^{-1}$ for the activity tests, which is more than four times higher than the results achieved by Bravo et al. [14], which were discussed above.

However, the danger of metallic Pd(0) formation remains high for Pd/ZnO catalysts. The search for an alternative catalyst formulation to overcome this drawback has been the focus of recent studies by Men et al. [24]. It was found that $Pd/In_2O_3/Al_2O_3$ catalysts are promising candidates. Further investigations by the same group revealed that $Pt/In_2O_3/Al_2O_3$ catalysts are even more active and show at least 10 times higher activity than Cu-based systems [25]. VHSV values of more than $100 \, l \, (h \, g_{cat})^{-1}$ were achieved. This higher activity was achieved at much higher reaction temperatures (400 °C) compared with Cu-based systems. Under these conditions, the catalysts show extremely low selectivity towards CO, resulting in a concentration of less than 1 vol.% in the reformate at Pd:In and Pt:In ratios of 1:2, which was determined as the optimum. The novel catalyst contains high concentration of noble metals (35 wt%), but about $300 \, l \, h^{-1}$ of hydrogen can be produced over 1 g of catalyst (noble metal and carrier). Hence about 1 g of noble metal is required for an electric power equivalent of 1 kW.

7.4
Heat Management of Microchannel Plate Heat-Exchanger Reactors

The majority of micro-reactors reported in the literature are still dedicated to catalyst evaluation. These reactors are usually monolith-type laboratory devices without heat-exchange functions, which allow for the removal of the microstructured plates after testing [26–35]. These are supplied by electrical power for heating and are still far away from a practical application. Therefore, the design of these reactors will not be discussed in detail below, bearing in mind that they are useful tools for catalyst screening and characterization.

A micro-reactor as part of a practical fuel processor is generally designed as a plate heat exchanger, a conventional heater being shown in Figure 7.3a. If only one

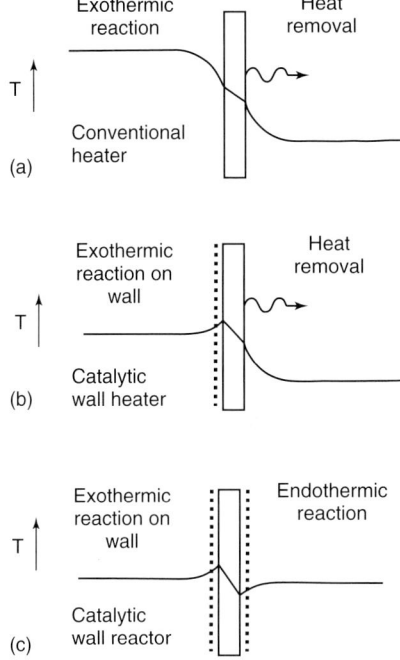

Figure 7.3 Three different reactor concepts according to Redenius et al. [36]: (a) conventional heater; (b) catalytic wall heater; (c) catalytic wall reactor.

flow path of the heat exchanger is coated with catalyst, it turns into a catalytic wall heater, as shown in Figure 7.3b. The application of catalytic wall heaters improves the temperature management of exothermic reactions such as partial oxidation of fuels and, even more prominently, the CO clean-up reactions, namely water-gas shift (WGS) and preferential oxidation of CO.

When catalyst is introduced on to the walls of the second flow path, a catalytic wall reactor is formed, as shown in Figure 7.3c. This design has enormous potential for directly coupling exothermic reactions (such as steam reforming) and exothermic reactions (such as catalytic combustion of fuel cell anode off-gas), which are then only separated by the few hundred micrometer metal foils between the two coatings.

For the various reactions of fuel processing, different flow arrangements in plate heat exchangers have been proven to be the optimum, which is discussed below from a theoretical point of view and illustrated by practical applications.

7.4.1
Reforming

It was shown by Frauhammer et al. that a counter-flow arrangement is not suitable for coupling endothermic reactions such as steam reforming with exothermic reactions [37]. Rather, a co-current flow arrangement should be chosen, which has been proven to be effective for practical applications such as steam reforming of methanol, liquefied petroleum gas (LPG), and diesel by Kolb and co-workers [38, 39], Wichert et al. [40], and O'Connell et al. [7, 41]. Recent work by Hsueh et al., which contradicts this conclusion, is misleading [42]. Catalytic combustion of hydrogen and residual CO contained in the fuel cell anode off-gas is well suited as the heat source for steam reforming of all kinds of fuels. The temperature profile in micro-reactor plate heat exchangers, which are operated as a coupled steam reformer–catalytic afterburner, depending on the reactor design and size, either is isothermal [39] or declines slightly over the length of the reactor axis. This has been explained by simulations [43] and proven experimentally [7]. However, a declining temperature profile is not a practical problem, especially when higher hydrocarbons are the feedstock, because the feedstock is more or less completely converted to lighter hydrocarbons, carbon oxides, and hydrogen at the inlet section of the reactor. Lower temperatures do not impair the catalyst stability at the reactor center and outlet section, because conversion of light hydrocarbons takes place there. Lower temperatures at the exit are even beneficial because they reduce the carbon monoxide concentration of the product owing to a shift of the equilibrium of the WGS reaction. Too low temperatures, however, favor methane formation owing to the equilibrium of the methanation reaction. Zhai et al. demonstrated that higher wall conductivity is beneficial for the isothermality of the reactor and therefore metallic construction material is preferred over ceramics [44].

Below, some examples of development work aimed at integrated steam reformer/afterburner reactors are briefly described.

A combined evaporator and methanol reformer was developed by Park et al. [29], as the hydrogen source of a 5 W fuel cell. The device was heated by electrical heating cartridges. Prior to coating the channels with a commercial $CuO/ZnO/Al_2O_3$ catalyst (Synetix 33-5 from ICI), an alumina sol was coated as interface on the channel surface. The catalyst was reduced in 10% hydrogen in nitrogen at 280 °C prior to exposing it to the reaction mixture. At a reaction temperature of 260 °C, 90% methanol conversion was achieved and the CO concentration in the reformate was lower than 2 vol.%. Later, Park et al. [45] developed a combined afterburner–methanol reformer with an electric power equivalent of 28 W, which was sealed by brazing; 99% methanol conversion could be achieved at a reaction temperature of 240 °C.

Reuse et al. [46] combined endothermic methanol steam reforming with exothermic methanol combustion in a plate heat-exchanger reactor, which was composed of a stack of 40 foils (see Figure 7.4). Each foil carried 34 S-shaped channels. Cu/ZnO catalyst from SüdChemie (G-66MR) was coated in the channel system for the steam reforming reaction. Cobalt oxide catalyst served for the combustion reaction. The reactor was operated in co-current mode. The steam reformer was operated at a S/C ratio of 1.2. At reaction temperatures between 250 and 260 °C, more than 95% conversion and more than 95% carbon dioxide selectivity were achieved.

Wichert et al. reported about long-term experiments performed at a microstructured coupled steam reformer/catalytic afterburner for LPG. The reactor was

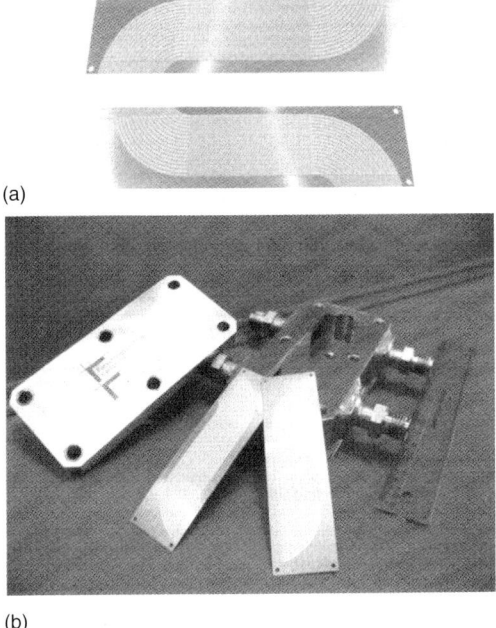

Figure 7.4 Integrated reformer–combustor for methanol steam reforming [46].

operated in a co-current flow arrangement. For 1060 h test duration the reactor was operated at a S/C ratio of 4.0 and temperatures around 750 °C with 29 start-up and shut-down cycles in total. Complete conversion of propane was observed for all experimental conditions as shown in Figure 7.5. Selectivity towards carbon species was determined. The selectivity for the by-product methane was very low. The methane content in the reformate amounted to 1.9 vol.% on average. The gas composition of the reformate determined from the thermodynamic equilibrium calculations of propane reforming, WGS, and methanation reactions at the reactor outlet temperature of 675 °C agreed well with the experimental tests. The reactor, which was constructed of stainless steel 1.4841 (German classification), was not affected. Additionally, the welding seams were not damaged during the long-term test (as proved by repeated leakage tests at overpressure). No effects of the elevated temperature and start–stop cycles on the mechanical integrity of the catalyst coatings were observed, which could be demonstrated after dismantling the reactor.

Cremers *et al.* [47] and Pfeifer *et al.* [48] presented a reactor combining endothermic methane steam reforming with the exothermic combustion of hydrogen stemming from the fuel-cell anode off-gas (see Figure 7.6a). NiCroFer 3220H was applied as the reactor material. The reactor was designed to power a fuel cell with 500 W electrical power output. The steam reforming side of the reactor was operated at a S/C ratio of 3 and temperatures exceeding 750 °C.

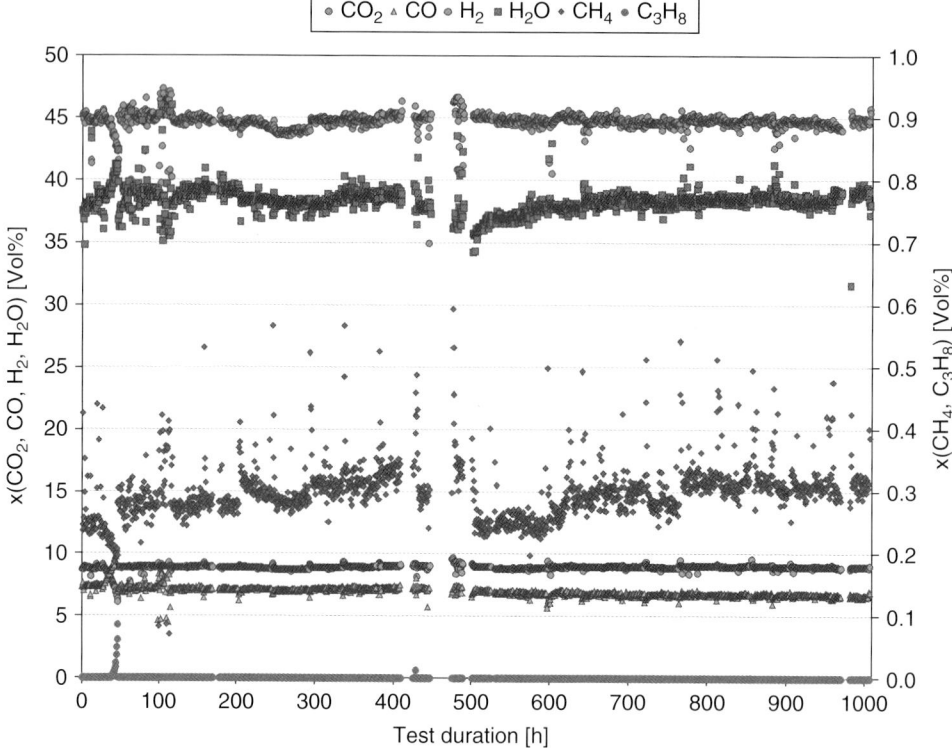

Figure 7.5 Gas composition of the reformer product as determined for the integrated propane steam reformer–catalytic burner during 1000 h test duration [40].

The development and evaluation of a reactor based on microchannel technology for the reforming of diesel fuel for a 5 kW fuel cell was reported by O'Connell et al. [7]. The reactor itself was based on an integrated reformer–burner heat-exchange reactor concept (see Figure 7.6b). It had a co-current flow arrangement, which combined diesel steam reforming with combustion of fuel cell anode and cathode off-gas surrogate. Diesel oxidative steam reforming was performed at temperatures above 750 °C and at various S/C ratios, down to a minimum of 3.17, and up to an electrical power equivalent of 5 kW for a duration of 38 h. Over 98% total diesel conversion was observed at all times over the testing period. O'Connell et al. calculated from Aspen modeling software the following scheme for endothermic oxidative diesel steam reforming, performed at S/C = 3, O/C = 0.3, and 800 °C, taking into consideration the equilibrium of the WGS reaction:

$$C_{11.3}H_{24.6} + 1.7O_2 + 33.9H_2O \longrightarrow$$
$$6.0CO + 5.3CO_2 + 25.5H_2 + 20.7H_2O \quad \Delta H_{1073\,K} = 853\,\text{kJ mol}^{-1} \quad (7.5)$$

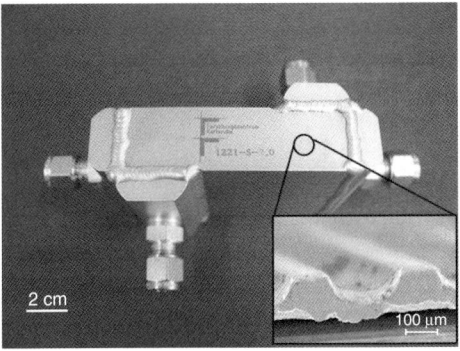

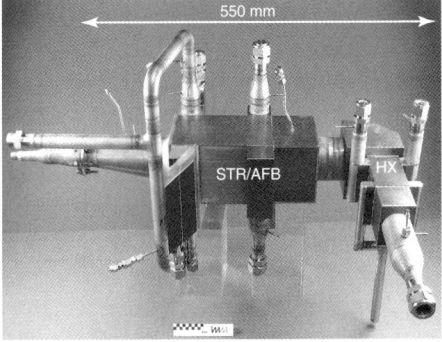

Figure 7.6 (a) Combined methane reformer–combustor designed for 500 W electrical power output [49]. (b) 5 kW$_{net,el.}$ microstructured diesel steam reformer–catalytic afterburner coupled to a heat exchanger [7]. Source: IMM.

7.4.2
Water Gas Shift Reaction

The slightly exothermic WGS reaction requires a countercurrent cooling concept to achieve an optimum temperature profile, which allows a high reaction rate at high temperature at the reactor inlet and a decreasing temperature towards the reactor outlet. The latter shifts the equilibrium of the reaction in a favorable direction. Such a temperature profile could be adjusted in a plate heat exchanger as proposed for the first time by Zalc and Löwffler [50]. They showed that an optimum temperature profile exists, which significantly improves the CO conversion compared with isothermal and adiabatic (uncooled) operation of the reactor according to the calculations as shown in Figure 7.7a and b. The low catalyst utilization in fixed-bed WGS reactors is much improved by the wall-coated microchannel reactors. Recent developments in the field of noble metal-based catalysts make possible the operation of these WGS catalysts in wall-coated monoliths or plate heat exchangers at a high VHSV of 180 l (h g$_{cat}$)$^{-1}$. The application of microchannel plate heat-exchanger

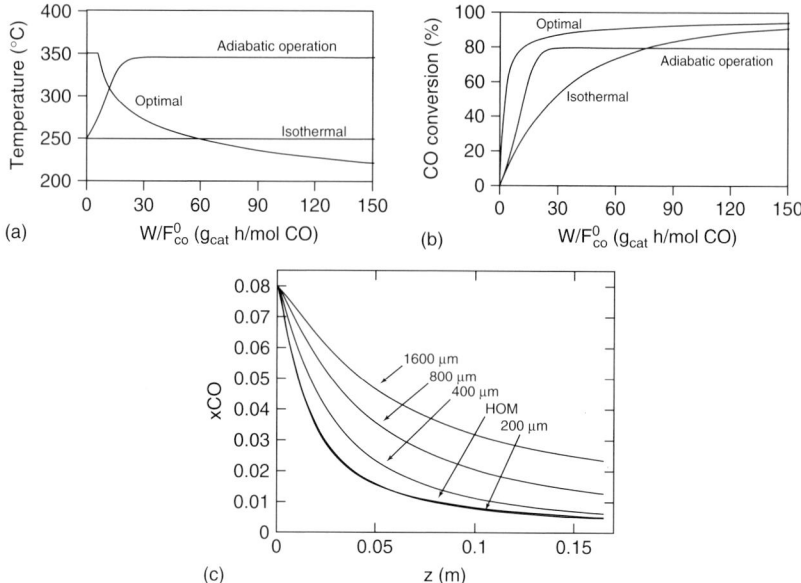

Figure 7.7 (a) Isothermal, adiabatic, and optimum temperature profiles and (b) carbon monoxide concentration versus modified residence time W/F_{CO}^0 [50]. (c) Carbon monoxide molar fraction versus channel length for different channel heights in a WGS heat-exchanger reactor. HOM corresponds to the results from a homogeneous model (no diffusion limitation) [54]. Source: IMM.

technology saves more than 50% of the catalyst compared with conventional or monolithic two-stage adiabatic designs [51–53]. Hence significantly less catalyst, only a single reactor, and no intermediate heat exchanger or water injection are required when applying an integrated heat exchanger.

Baier and Kolb [54] showed that reducing the flow rates (turn-down) improves the performance of a WGS heat-exchanger reactor owing to the longer residence time. They also investigated the effect of diffusion limitations in channels of different sizes. For a channel height of 200 µm, the absence of diffusion limitations was observed (see Figure 7.7c). Severe diffusion limitations were observed for a channel height of 800 µm, which is in line with earlier experimental work performed by Pasel et al. [55]. Hence the channel height in practical applications should remain below this value. However, a trade-off is required between catalyst utilization and pressure drop in a practical system [44]. Additionally, increasing the channel height from 200 to 800 µm doubles the reactor length required to achieve the same degree of CO conversion.

The results of the work of Baier and Kolb were successfully applied by the same group to the construction of WGS heat-exchanger reactors on the kilowatt scale [56–58].

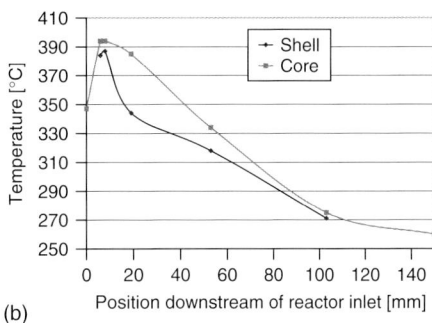

Figure 7.8 (a) Integrated WGS reactor/heat exchanger designed for a 5 kW fuel cell system [57]. Source: IMM. (b) Internal temperature profile achieved during operation of a similar reactor [56]. (Source: IMM.)

A typical temperature profile, which was determined experimentally in a plate heat-exchanger reactor, is shown in Figure 7.8a. After a slight temperature peak at the reactor inlet, which originated from the high initial heat of reaction, the reactor temperature decreased towards the outlet of the reactor (see Figure 7.8b). The content of CO in the reformate surrogate could be reduced from 10.6 to 1.05 vol.%, which corresponded to 91% conversion. Somewhat lower temperatures were observed at the shell of the reactor, especially in the inlet section, which was attributed to heat losses to the environment.

7.4.3
Preferential Oxidation of Carbon Monoxide

Both the preferential oxidation reaction and the hydrogen oxidation taking place in parallel are highly exothermic, which can lead to local overheating of fixed catalyst beds [59]. The CO in the reformate needs to be reduced to levels below 100 ppm, which are regarded as acceptable for state-of-the-art reformate-tolerant polymer electrolyte membrane fuel cells (PEMFCs).

Kahlich *et al.* studied the preferential oxidation of CO over platinum/alumina and gold/iron oxide catalysts [60]. Because their platinum catalyst was operated at a very high reaction temperature of 200 °C, incomplete conversion was obtained at lower space velocities. The reverse WGS reaction occurred as soon as oxygen was consumed over the surplus of catalyst. These conditions correspond to a partial load of a fuel processor. Reverse WGS reaction is favored at higher temperatures. Hence noble metal catalysts need to be optimized for operation at low temperatures as close to 100 °C as possible to minimize the reverse WGS. Kahlich *et al.* solved this issue by switching to a gold/α-alumina catalyst, less reverse WGS being observed.

Giroux *et al.* proposed a two-stage approach with inter-stage cooling for the preferential oxidation of CO for monolithic reactors. They found that the addition of the small air flows required for the second oxidation stage was difficult. Therefore, the first stage was operated at low O/CO ratio of 1.2 and incomplete conversion of CO. This also limited the undesired oxidation of hydrogen to a minimum [52]. Kim *et al.* proposed a two-stage preferential oxidation for a microreactor at the smallest scale even without oxygen addition between the stages [61].

As a simpler alternative to the two-stage adiabatic reactor concept, Zalc and Löwffler proposed the application of plate heat-exchanger technology for the preferential oxidation of CO to improve the heat management of the highly exothermic process [50]. Because the optimum operating temperature of state-of-the-art preferential oxidation catalysts is in the region of 100 °C, evaporation cooling with water preferably in a co-current flow arrangement is obviously the best strategy for the heat management of plate heat-exchanger reactors. Some examples of successful operation of such reactors are discussed below.

Lopez *et al.* described the operation of a folded-plate reactor which was operated with water cooling in a co-current flow arrangement, and multistage (four stages) air addition was used. The Au catalyst formulation was operated well below 100 °C at an O/CO ratio (λ value) of 3. The CO could be reduced to values below 100 ppm and the reactor had 0.4–0.6 kW electric power equivalent [62].

Ouyang *et al.* studied the preferential oxidation of CO in a silicon reactor [63] which was fabricated by photolithography and deep reactive ion etching (see also Section 7.5). Each reactor had two gas inlets for reformate and air, a premixer, a single reaction channel, and an outlet zone where the product flow was cooled. The single channel was 500 µm wide, 470 µm deep, and 45 mm long. Platinum/alumina catalyst containing 2 wt% platinum was deposited on the channel walls at a thickness of 2–5 µm. The channel was then sealed by anodic bonding with a Pyrex glass plate. The surface area of the catalyst was very high (400 m^2 g^{-1}). The catalyst was reduced in undiluted hydrogen at 400 °C for 4 h before the tests. A low flow rate of 5 ml min^{-1} of reformate surrogate, containing 1.7 vol.% of carbon monoxide, 21 vol.% of carbon dioxide, 68 vol.% of hydrogen, and 9 vol.% of nitrogen, and 0.5 ml min^{-1} of air were fed into the reactor, which corresponded to an O/CO ratio of 2.5. The experiments were performed by applying a temperature ramp program from ambient to 300 °C. Full conversion of CO was achieved at a reaction temperature of 170 °C. Up to a reaction temperature of 300 °C, full conversion could be maintained, which is surprising. This result was attributed to the improved heat transfer in the microchannel, which avoided hot spots leading to a reverse WGS reaction. Similar work was also recently reported by Hwang *et al.* [64].

Kolb *et al.* reported on the design and testing of a preferential oxidation (PrOx) reactor on the kilowatts scale (see Figure 7.9a) [38]. Evaporation cooling applying patented technology [65] was chosen in a co-current flow arrangement. The reformate entered the preferential oxidation reactor from both sides, while the water was distributed in a front distribution chamber at the reactor inlet. Similarly to the inlet, the purified reformate left the reactor at both sides, while the superheated steam was gathered in a single outlet manifold. Surrogate of reformate equivalent to WGS reactor off-gas was prepared by dosing H_2, CO, CO_2, N_2, O_2, and steam

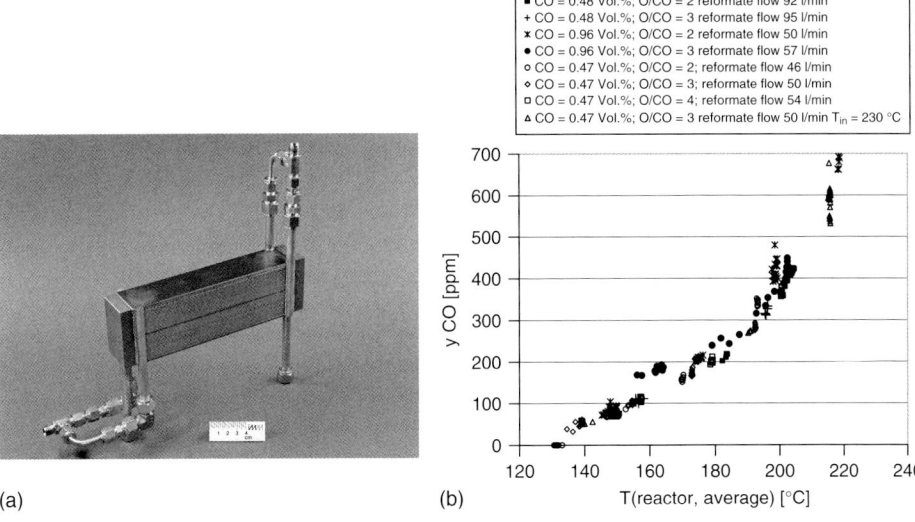

Figure 7.9 (a) Co-currently operated reactor for preferential oxidation; cooling was performed by evaporation of water [56]. Source: IMM. (b) CO concentration in the off-gas of the preferential oxidation reactor at different total reformate flow rates, CO inlet concentrations, and O/CO values. Source: IMM.

as reactor feed. The flow rate of the reformate surrogate fed to the PrOx reactor was varied from 42 to 99 l min^{-1}, the CO concentration in the feed was set to either 0.47 or 0.96 vol.% and the O/CO ratio was increased from 2 to 4 so as to assess the performance of the reactor. Under these conditions, the prototype reactor worked as a 1.44 kW PrOx reactor. A typical gas feed composition setting was the following: 44.40 vol.% H_2, 0.47 vol.% CO, 15.25 vol.% CO_2, 7.47 vol.% N_2, 0.49 vol.% O_2, and 31.92 vol.% H_2O.

Stable operation and a narrow reactor temperature range of 10 °C could be adjusted inside the reactor at least for lower λ values, that is, O/CO = 2. Water at 122 g h^{-1} was fed at a pressure of 1.6 bar into the integrated evaporation cooler under these conditions. The steam left the superheating channels at a temperature of ~175 °C, which was about 10 °C below the temperature of the reformate outlet. The heat removal by water evaporation and superheating amounted to 94 W under these conditions, which correlated well with the heat generation of 88 W by hydrogen and carbon monoxide combustion. The remaining cooling power was required to cool the reformate to the off-gas temperature of the reactor. Figure 7.9b shows a plot of the CO concentration as determined in the PrOx reactor off-gas against the reactor temperature (average values were chosen) for different experimental conditions. Surprisingly, it was clearly demonstrated that the off-gas concentration did not decrease with increasing O/CO values but was dependent only on the reactor temperature. Hence a higher surplus of air (higher O/CO value) did not improve the reactor performance. On the contrary, the lowest values

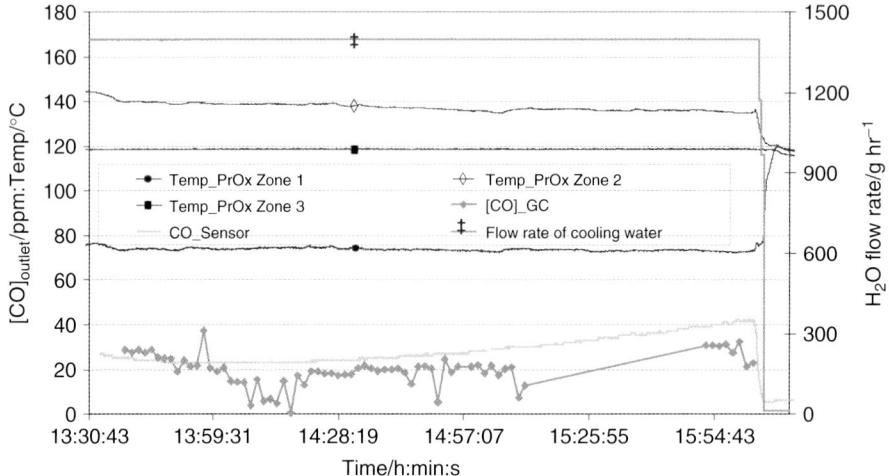

Figure 7.10 Typical steady-state operation of the microstructured 5 kW$_{el}$ preferential oxidation reactor at 100% load level; the reactor was cooled by evaporation cooling in a co-current flow arrangement.

(below the detection limit, which was 5 ppm for the analytical equipment applied) were found for an O/CO value of 2, because a lower reactor temperature could be achieved owing to the lower heat generation. Higher temperatures might favor the reverse WGS reaction leading to a higher CO content of the purified reformate.

O'Connell et al. reported of a 5 kW$_{el}$ one-stage preferential oxidation reactor, which was designed and evaluated for the clean-up of surrogate diesel reformate [57]. Both partial load operation and load changes could be carried out without significant overshoots of CO. Figure 7.10 shows an example of stable operation for 2.5 h of the reactor at 100% load. Having started with a CO content in the feed of 1.0 vol.%, the CO content was reduced to well below 50 ppm, the lowest value being 23 ppm. Also in the graph are the results obtained with an infrared sensor which agree well with the gas chromatographic (GC) analyses. The reactor was then integrated with a 5 kW$_{el}$ WGS reactor upstream for the purposes of reducing the CO levels in the reformate exit stream to levels below 100 ppm [58]. Load changes for both reactors could also be carried out without significant overshoots of CO.

7.4.4
Selective Methanation of Carbon Monoxide

Selective methanation of CO, although basically simpler than preferential oxidation, because no air addition to the reformate is required, suffers not only from the critical issue of competing CO_2-methanation [66] but also from temperature management problems [67], which can be solved by multistaged operation when conventional adiabatic fixed beds or monoliths are applied.

Similarly to many other cases described above, the microchannel plate heat-exchanger technology offers advantages for simplification of the reactor.

7.5
Examples of Complete Microchannel Fuel Processors

Some authors have described methanol steam reformers in the low-power range of a few watts and less, which were dedicated to power supply to electronic devices such as mobile phones. Consequently, production techniques were applied that are known from microelectromechanical systems (MEMS) and silicon was the construction material chosen in many cases. Applying these techniques, electrical heaters for start-up and temperature sensors may be integrated into the devices. In some cases, catalysts were introduced into the microchannels by sputtering, which usually leads to a low surface area and activity [68]. In case fixed catalyst beds were applied [69], the pressure drops were consequently high, which would increase the power demand of the dosing equipment of future systems. In the event that electrical heating was chosen as energy source for steam reforming [70], about 30% of the electrical energy produced by the fuel cell would be required for operating the fuel processor [71]. This is rather a conservative value and such simplified systems suffer from poor efficiency.

Yoshida et al. [72] designed an integrated methanol fuel processor from silicon and Pyrex glass substrates for a power equivalent of 10 W. It was composed of an integrated device, which contained functional layers for steam reforming, evaporation, and combustion (see Figure 7.11). Whereas a commercial Cu/ZnO catalyst served for reforming, the Pt/TiO_2 combustion catalyst was prepared by a sol–gel method. A high power density of 2.1 W cm^{-3} was determined for the device. Kim et al. reported an even smaller methanol fuel processor made from Foturan glass with a power equivalent of 1.5 W [73].

Little information is available on the status of microstructured fuel processor developments in industry. Terazaki et al. [74] at Casio developed a fuel processor made of 13 glass plates containing evaporators, a steam reformer, a hydrogen off-gas burner, and CO clean-up functionalities. The device was insulated by vacuum packaging and radiation losses were minimized by a thin Au layer. The finished device was sealed by anodic bonding. The fuel processor was tested in connection with a miniaturized fuel cell [75]. A methanol conversion of 98% was observed with the production of 2.5 W of electrical energy. Later, a similar methanol reformer with integrated heating functionalities was presented by Kawamura et al. [76] at Casio. This device was developed in cooperation with the University of Tokyo. The reformer carried only a single meandering channel, which was 600 μm wide, 400 μm deep, and 333 mm long. The Cu/ZnO catalyst required reduction under hydrogen. Full methanol conversion could be achieved at 250 °C and the thermal power equivalent of the hydrogen product was in the region of 3.3 W. Subsequently, a complete fuel processor was developed [77] that included an anode off-gas burner, which supplied the methanol reformer with energy. A palladium-based catalyst

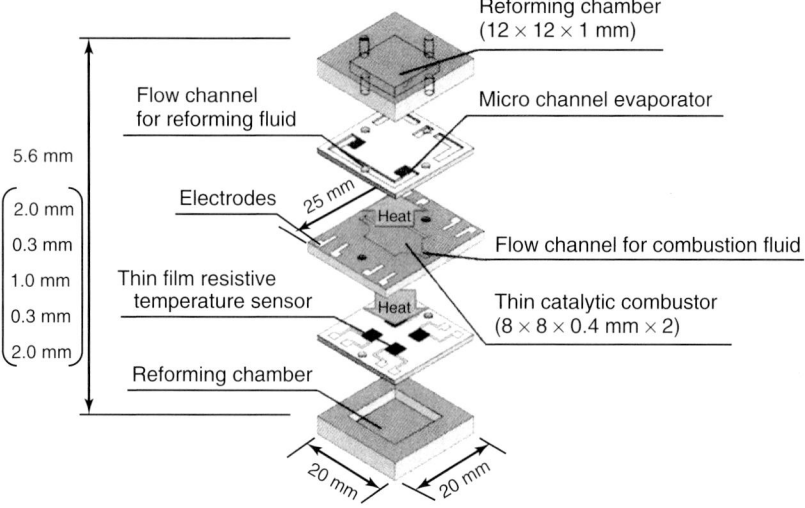

Figure 7.11 Small-scale methanol reformer–evaporator–burner system with 4 W electrical power equivalent [72].

was used for reforming. The preferential oxidation reactor was operated between 110 and 130 °C. The fuel processor had a volume of 19 cm^3 and a weight of 30 g including vacuum layer insulation and radiation shields, which reduced the heat losses of the system to 1.2 W. The electrical power consumption was in the region of 70 mW. The fuel processor was combined with a fuel cell and balance-of-plant as shown in Figure 7.12. The system was operated at the Fuel Cell Seminar 2006 for demonstration purposes.

A complete methanol fuel processor for the electrical power equivalent range from 60 to 170 W was presented by Holladay *et al.* [78]. The device, which is shown

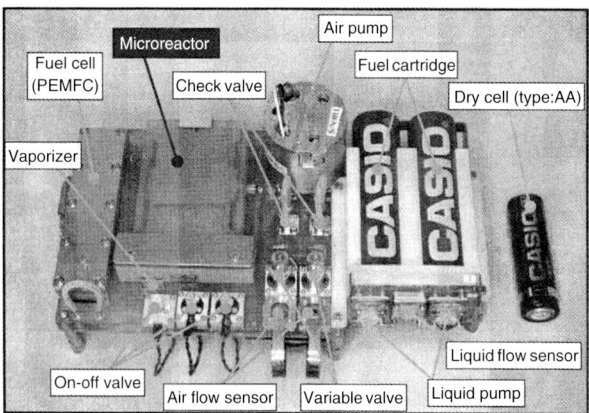

Figure 7.12 A 2.5 W methanol fuel processor–fuel cell system developed by Casio [77].

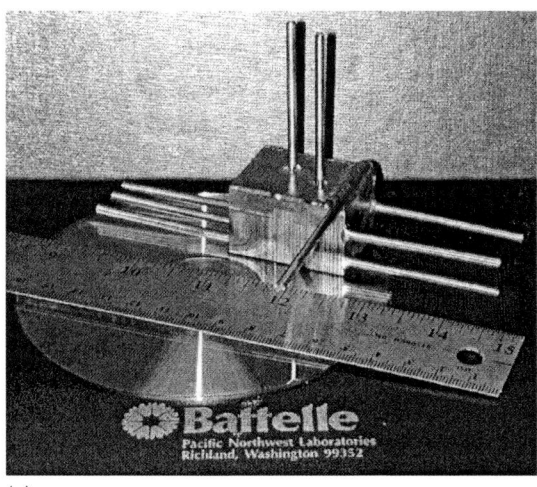 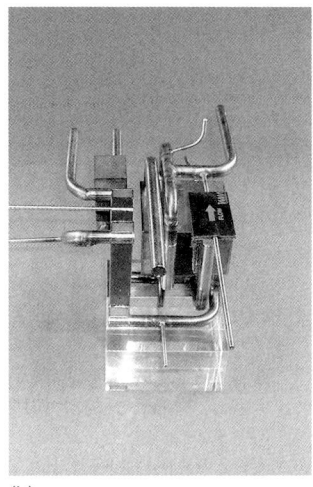

(a) (b)

Figure 7.13 (a) Integrated methanol fuel processor with 100 W power equivalent developed by Holladay et al. [78]. (b) Integrated methanol fuel processor with 100 W power equivalent developed by Kolb et al. [39]. Source: IMM.

in Figure 7.13a, had a volume of <30 cm^3, a mass <200 g and a thermal efficiency of >80%. The reformer was operated at a reaction temperature of 350 °C.

Kolb et al. reported the development of an integrated microstructured fuel processor with a electrical net power output of 100 W [39]. The fuel processor (see Figure 7.13b) worked very stably under normal operating conditions under both full and partial load. A very narrow temperature profile of only 3 K was achieved in the reformer, which was crucial for the Pd/ZnO catalyst technology applied. The heat demand of oxidative steam reforming created a slightly decreasing temperature profile in this reactor in the case of full load, whereas a maximum temperature was formed in the reactor center under conditions of partial load. Methanol conversion was always complete (>99.9%). At O/C = 0.25, S/C = 1.7, and 100% load the carbon monoxide content of the reformate amounted to 1.8 vol.% and the hydrogen content to about 50%. This gas composition did not change significantly at 50% load, merely the carbon monoxide content decreased to 1.6 vol.%. The fuel processor was tested separately and then coupled to a high-temperature fuel cell, which had been developed in parallel. The high-temperature PEMFC stack generated an electrical power output of 103 W (14.7 V at 7 A) when operated with the reformate from the fuel processor.

Because of the existing distribution grid for LPG and its widespread application in caravans and trailers, it is an attractive fuel for the electrical power supply of such vehicles. Truma (Putzbrunn, Germany), Europe's largest manufacturer of heating systems for caravans and trailers, has developed a fuel processor/fuel cell system (brand name VeGA) together with IMM (Institut für Mikrotechnik Mainz GmbH).

The power output of the systems is about 300 W, and 50 W are consumed by the balance-of-plant components, leaving a net electric power output of 250 W for the consumer.

The utilization of microstructured plate heat-exchanger technology made a compact design of the fuel processor possible. A fully integrated and automated system and the fuel processor, at an earlier stage of the development, are shown in Figure 7.14a and b, respectively. The fuel processor consisted of an integrated microstructured evaporator and a microstructured reformer, both integrated with microstructured catalytic burners, heat exchangers, and microstructured WGS. Wichert et al. reported performance data of one of these complete LPG fuel processors which had been operated for up to 3500 h in combination with high-temperature PEMFC stacks [40]. Additionally, the reformate gas composition of a complete miniaturized self-sustaining LPG fuel processor including a microstructured evaporator, a microstructured propane steam reformer, a WGS stage, and several heat exchangers that was running for 1350 h at Truma was evaluated (see Table 7.1). Fifty of these VeGA systems had been fabricated in 2008 and tested in field trials until 2009. Some of the systems had been operated by dedicated end-users in caravans for between 6 and 15 months, one of them in Africa.

Irving et al. [79] presented a microreactor with fixed catalyst beds that was capable of reforming gasoline, diesel, methanol, and natural gas by steam reforming at temperatures up to 800 °C. Not only reforming but also mixing of fuel and steam, heat exchange, evaporation, and feed preheating were covered by the integrated device, which was made from both stainless steel and ceramics. Gasoline steam reforming at a flow rate of 0.1 g min^{-1} gasoline feed was performed at S/C ratios between 5 and 8. The dry reformate contained 70 vol.% hydrogen. Methane formation and catalyst deactivation were negligible owing to the high S/C ratio

(a) (b)

Figure 7.14 (a) The 250 W_{el} fuel cell–fuel processor system VeGA developed jointly by Truma and IMM. Photograph courtesy of Truma. (b) Fuel processor of the VeGA (early stage of development). Source: IMM.

Table 7.1 Gas composition of the WGS product as determined for the fully integrated propane fuel processor at two different load levels; the fuel processor had been operated before together with a fuel cell for 1350 h; fuel quality: C_3H_8 2.5; S/C = 4.0; the propane concentration was always below the detection limit (100 ppm).

$x(CH_4)$ (vol.%)	$x(CO_2)$ (vol.%)	$x(CO)$ (vol.%)	$x(H_2)$ (vol.%)	$x(H_2O)$ (vol.%)	Load (%)
0.44	14.02	1.04	47.80	36.69	50
0.40	13.87	1.09	47.32	37.32	50
0.43	13.80	1.77	48.73	35.28	100
0.42	13.85	1.67	48.66	35.40	100

applied. Sulfur contained in the gasoline led to hydrogen sulfide formation. Lowering the S/C ratio in the case of isooctane steam reforming increased the methane concentration in the reformate to 5 vol.%. At a feed rate of 0.3 g min^{-1} and a S/C ratio of 4, 100% conversion and a methane content well below 5 vol.% were achieved for a mixture of 60% isooctane, 20% toluene, and 20% dodecane that also contained 476 ppm sulfur. However, the VHSV was rather low at about 30 l (h g_{cat})$^{-1}$. Later, a multi-fuel processor developed by Innovatec was presented by Irving and Pickles [80], which is shown in Figure 7.15. It was sized for a 1 kW$_{el}$ PEMFC and again microstructured components were used to build the reformer, while the gas purification relied on membrane technology. The fuel processor has been operated with wide variety of fuels from methane to biofuels and JP-8 fuel. When operated with methane it had a power output of 1 kW, but it was lower for diesel.

Figure 7.15 Multi-fuel processor developed by Irving and Pickles for a 1 kW PEM fuel cell [80]; the device contains microstructured components and membrane separation.

7.6
Fabrication of Microchannel Plate Heat-Exchanger Reactors

In this section, the fabrication of microstructured reactors and fuel processors from metals is discussed, and also their bonding, sealing, and packaging. Ceramic and polymer fabrication techniques are not considered because the application of such materials is less advanced for micro-reactors and it would exceed the scope of this chapter. Whereas ceramics offer potential in the highest temperature range exceeding 1000 °C, polymers are promising materials for low-temperature plate heat exchangers and especially for the extraction of the last portion of energy out of the fuel processor off-gases before releasing them to the environment. Detailed information about the fabrication of ceramic and polymer micro-reactors and heat exchangers can be found elsewhere [81–84], as can details on the fabrication of chip-like microdevices, which are similar to the fabrication of MEMS [64, 69, 85].

The choice of materials for microstructured fuel processors is mainly determined by the operating temperature and pressure. Because fuel processors are future mass products, only cheap fabrication techniques can be considered in the longer term. Hence only techniques suited for future mass production of microstructured fuel processors are briefly discussed below; other techniques have been discussed in detail in other books [86, 87].

The fabrication technologies are divided into two different groups, namely erosive and generative techniques. Technologies such as embossing and molding belong to the class of generative fabrication, whereas laser micromachining, for example, is an erosive technique. From the viewpoint of fabrication costs, it is obvious that generative techniques should be the preferred option for mass production, because they are faster in most cases and no material is lost.

7.6.1
Choice of Construction Material

Plate heat exchangers are stacked arrangements with a multitude of parallel minichannels and high surface-to-volume ratios in the range 200 m^2 m^{-3} [88]. The preferred construction material is stainless steel. For applications at temperatures exceeding 900 °C, as required for diesel reforming, nickel-based alloys or FeCr alloys are an alternative. A disadvantage of the latter material is its brittleness. For low-temperature processes such as methanol reforming and carbon monoxide purification, copper and especially aluminum are suitable. Their orders of magnitude higher heat conductivity compared with stainless steel provide other options for thermal management, for example, to achieve isothermal conditions, as needed for evaporators or reactors operating within narrow temperature windows. Since the higher heat transport is also given in the axial direction (i.e., through the wall material), a thermal engineering evaluation is required before the decision on the final choice of material is made. However, aluminum is a less preferred material when corrosion issues are considered.

The choice of materials for plate heat exchanger/reactors also depends on the desired dynamic properties of the microsystem. One important parameter is the energy demand for fuel processor start-up, which results from the product of specific heat capacity and density of the construction material. For a given geometry and volume of the device, aluminum is favored over copper and stainless steel.

7.6.2
Micromachining Techniques

Precision micromachining of plates by milling, drilling, slotting, and planning is comparable to the techniques well known in conventional dimension machining. Especially micromilling is a useful fabrication technique readily at hand for experimental work and rapid prototyping, but does not satisfy future mass production issues.

Microelectrodischarge machining (μEDM) is controlled spark micromachining between a conductive electrode and a conductive working piece under a dielectric fluid [89]. μEDM techniques are suitable for rapid prototyping, but not (directly) for the mass production of microstructured devices. However, since mold inserts can be fabricated in that way, μEDM opens up indirectly the latter route by permitting injection molding, most prominently leading to polymeric microstructures made in high numbers, but also to metallic or ceramic microstructures.

Etching techniques were initially developed for silicon micromachining and are suited to mass production. In silicon, very small channels can be fabricated [90]. For many metals, etching is a cheap and well-established technology, but not as widespread and readily available as reported above [91]. It is competitive for mass production and covers a wide range of channel depths from about 100 up to 600 μm, which is the channel size usually applied in microstructured reactors for fuel processing applications. However, for applications in the kilowatt range, large microstructured foils are required and costs become a critical issue when etching is applied. The fabrication sequence starts with coating a photosensitive polymer mask material on to the steel substrate. The mask is exposed to light via a structure primary mask. After development of the polymer, etching is performed to remove material, typically with iron trichloride solution. Wet chemical etching yields semielliptical or semicircular microstructures with fairly high surface roughness (in the range of some microns), again due to the isotropic etching. Figure 7.16 shows a stainless-steel microchannel structure manufactured by wet chemical etching. The semielliptical shape of the etched microstructure is also shown in Figure 7.16.

Punching is an inexpensive, but less frequently applied, technique for making microchannels and is also suitable for mass production [92]. Punching usually does not allow grooves or channels to be made on substrates; rather, breakthroughs are opened by punching action due to complete material removal at the exposed sites. Accordingly, unstructured plates need to be inserted between the punched plates, as kinds of top and bottom plates for each channel array, in order to achieve a sealed microchannel system. The ability for holes and breakthrough

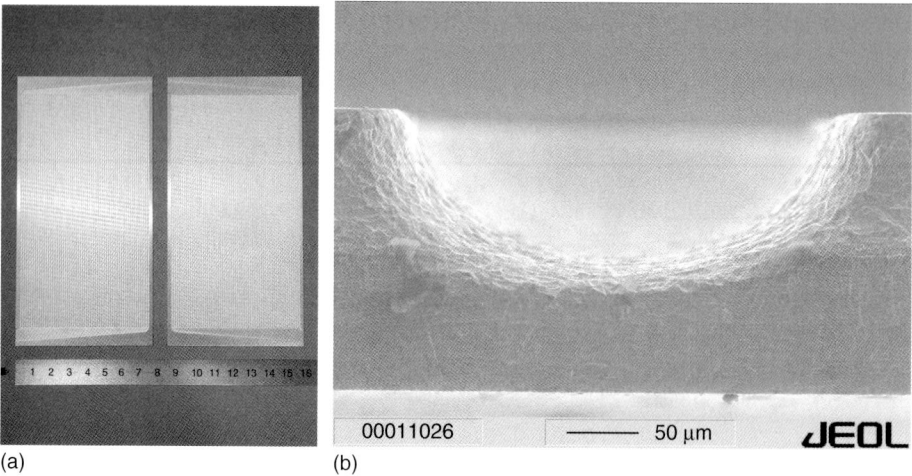

Figure 7.16 (a) Microchannels introduced into a stainless-steel plate by wet chemical etching. (b) Detailed view of a cut etched microchannel in stainless steel with a semielliptical shape. Source: IMM.

generation allows for an easy means of fluid distribution to connect the various plates, besides the pure channel manufacture. For punching, the manufacture of a precise negative model is needed, using, for example, precision machining or spark erosion techniques. Structures of different heights allow the formation of geometrically complex microstructures with holes, slits, and openings combined with more conventional structures such as channels and voids within a single working step. Particular care has to be taken during the punching process to avoid any bends around the recesses created, otherwise tight sealing will be difficult to achieve. Coating of punched microdevices with catalyst has hardly been explored and remains a challenging task.

Embossing is another inexpensive technique for manufacturing metal foils, highly suitable for mass production (see Figure 7.17) [93, 94]. Even microstructures with a structure size down to a few micrometers can be achieved by embossing [95]. In Figure 7.17, details of an embossed microchannel metal foil are shown.

Currently, corrugated metal foils fabricated by rolling are applied in the field of automotive exhaust gas systems [96]. The metallic monoliths fabricated in this way actually have channel dimensions in the sub-millimeter range and can be truly termed *"microtechnical devices."*

Laser ablation is a frequently applied fabrication technique of proven industrial suitability [41, 42]. However, fabrication of microchannels of several hundred micrometers depth, as typically required for many applications using microstructured reactors, will take too long and therefore the method is not cost competitive. For smaller channel dimensions, laser ablation is a viable option, especially for applications on the smaller scale.

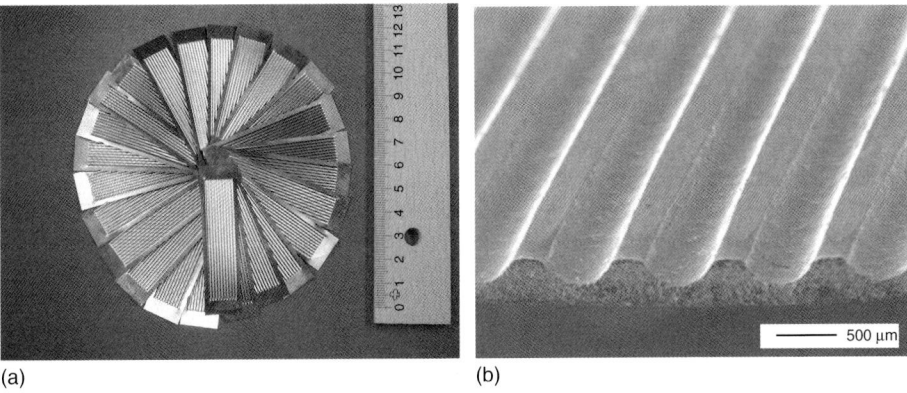

(a) (b)

Figure 7.17 (a) Microstructured and embossed heat exchanger. Source: IMM. (b) Microchannels introduced in a stainless-steel foil by embossing. Source: Karlsruhe Institute of Technology.

7.6.3
Sealing Techniques

Sealing by gaskets is the method of choice for interconnection and stacking of the plates and also facilitates catalyst coating and cleaning due to easy dismantling. Frequently applied gasket materials such as nitrile, neoprene, and Viton limit the operating temperature to 200 °C [88]. Above 200 °C, metallic gaskets are the choice. A disadvantage of the concept of sealing by gaskets is the need to insert screws and thus to have a bulky housing. Especially for smaller devices, the thermal mass is increased considerably, which, for example, increases the start-up time.

Irreversible sealing techniques typically make use of elevated temperatures, for which compatibility with the plate material and its coatings has be tested. This is not critical for thermal process engineering devices such as heat exchangers and evaporators, especially when made out of stainless steel. For chemical reactors, the main issue is the thermal stability of the catalyst coating, if being filled in or attached before the sealing (which is the current mainly applied method). Catalyst deactivation may arise from about 300 °C to more than 800 °C depending on the catalyst formulation, which can pose a serious limit to the applicability of the sealing method. When the technique is not compatible, the catalyst or the catalyst coating has to be inserted into the device after the sealing procedure.

Laser welding is a viable technique for the sealing of plate heat exchanger/reactors. The cost of the welding procedure is mainly dictated by the power of the laser applied. The spatially limited energy input protects incorporated catalyst from damage. Examples of reactors that were sealed by laser welding are shown in Figures 7.6, 7.8, 7.9, 7.13, and 7.14.

Electron beam welding, well established in the automotive industry, is an alternative with similar locally limited energy input.

In contrast, diffusion bonding requires high temperatures with no spatial restriction and also high vacuum. The material is compressed and heated to temperatures close to the melting point, which generates a quasi-boundary-free single workpiece. The Karlsruhe Research Center (KTI) has frequently applied diffusion bonding for almost two decades for their microstructured heat exchangers and reactors (see Figure 7.4) [97, 98].

Brazing techniques also find use for the sealing of compact plate heat exchangers. The brazing lots often contain heavy metals such as cadmium and tin, which are poisonous to catalysts [52]. Therefore, the brazing step should be followed by a thorough cleaning step before applying the catalyst coating [52]. If brazing is done with catalyst coated plates, however, the same temperature considerations as outlined above with regard to the catalyst coatings hold for the melting temperature of brazing lots.

7.6.4
Reactor–Heat Exchanger Assembly

Correct positioning of the microstructured plates to form stacks or other arrangements is crucial, since only small distortions during bonding may lead to severe deviations from the ideal microchannel shape. Alignment techniques can be based on simple mechanical methods (e.g., the use of alignment pins), edge-catches in a specially designed assembling device, or optical methods. Most of these methods stem from silicon micromanufacturing, with precise alignment of multiple mask layers. At the microscale, especially burr formation from mechanical micromachining or laser machining has to be avoided or minimized, as otherwise misalignment of the microstructures will occur.

7.6.5
Catalyst Coating Techniques

To coat metallic surfaces with a catalyst, a pretreatment to improve the adherence is required [99]. In addition to mechanical roughening, chemical and thermal pretreatment are also frequently applied. Fecralloy, the construction material for metallic monoliths, is usually pretreated at temperatures between 900 and 1000 °C. An alumina layer of about 1 μm thickness is formed on the Fecralloy surface under these conditions, which is an ideal basis for catalyst coatings. However, metal oxide layers are formed on stainless steel and may also serve as an adhesion layer.

Aluminum substrates are frequently pretreated by anodic oxidation to generate a porous surface, which may serve as a catalyst support itself or as an adhesion layer for a catalyst support [99].

Once the surface has been pretreated, the coating slurry needs to be prepared. The most common method is to prepare a dispersion of finished catalyst, sometimes including gelation steps. Ceramic monoliths are usually wash-coated by these means. The catalyst carrier or the catalyst itself [100] is mixed with a binder such as poly(vinyl alcohol) or methylhydroxyethylcellulose [101], acid, and solvent, usually water. A smaller particle size improves adhesion [102, 103].

It has been demonstrated that the slurry viscosity determines the thickness of the coating. The viscosity itself is determined by the concentration of particles, pH value, and surfactant addition [104].

Sol–gel methods include a gelation procedure, also known as *peptization* of the sol. The time demand for this procedure may vary considerably from hours to weeks. A sol is prepared by polycondensation of alkoxides. Alumina sol may be prepared from aluminum alkoxide or pseudo-boehmite [AlO(OH) · xH_2O]. Addition of additives such as urea provides porosity in the catalyst layer by thermal composition during calcination [105]. The sol then serves as a binder for the particles, which form the coating. Active metals can be incorporated into the sol. Usually sol–gel methods produce coatings of lesser thickness, in the range of a few micrometers. Therefore, hybrid methods of sol–gel and wash-coating are sometimes applied, which make greater coating thicknesses possible.

The amount of catalyst material that can be coated on to a monolith ranges between 20 and 40 g m^{-2}, and plate heat exchangers may even take up more catalyst when coated prior to the sealing procedure, because the access to the channels is better. A screen-printing method has been developed by the present author's group to introduce the catalyst suspension into the microchannels as shown in Figure 7.18.

Alternative but less commonly applied techniques are spray coating, which requires a decrease in the viscosity of the slurry or sol [99], flame spray deposition [106], and electrophoretic deposition [107].

After the deposition, drying and calcinations steps usually follow, the latter being a temperature treatment in air or other gases for a defined duration. Normally the dried samples are not immediately put into a hot furnace but rather heated up gradually. The final temperature of calcination needs to ensure that organic materials such as binders are completely removed.

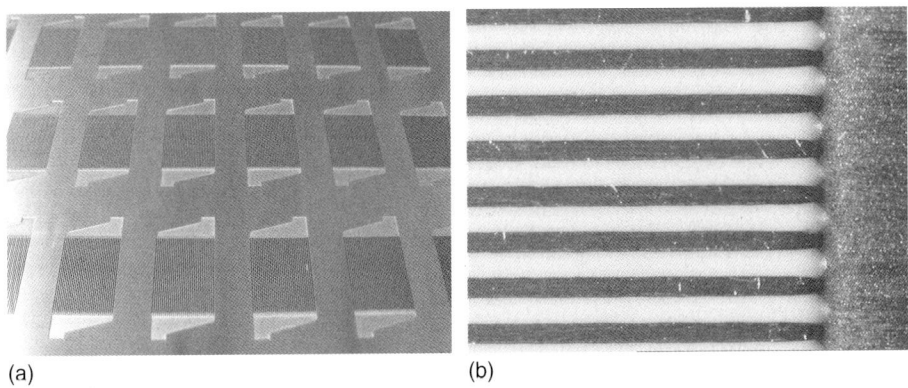

(a)   (b)

Figure 7.18 (a) Semiautomated catalyst coating by screen-printing. Source: IMM. (b) Microchannels filled with catalyst suspension after screen-printing. Source: IMM.

References

1. Hardt, S. (2004) *Chemical Micro Process Engineering – Fundamentals, Modelling and Reactions*, Wiley-VCH Verlag GmbH, Weinheim, pp. 125–256.
2. Purday, H.F.P. (1949) *An Introduction to the Mechanics of Viscous Flow*, Dover, New York.
3. Hartnet, J.P. and Kostic, M. (1989) Heat transfer to Newtonian and non-Newtonian fluids in rectangular ducts. *Adv. Heat Transfer*, **19**, 247–356.
4. Shah, R.K. and London, A.L. (1978) Laminar flow forced convection in ducts. *Adv. Heat Transfer*, **102** (2), 1–256.
5. Walter, S., Frischmann, G., Broucek, R., Bergfeld, M., and Liauw, M. (1999) Fluiddynamische Aspekte in Mikroreaktoren. *Chem. Ing. Tech.*, **71** (5), 447–455.
6. Kolb, G., Baier, T., Schürer, J., Tiemann, D., Ziogas, A., Ehwald, H., and Alphonse, P. (2008) A micro-structured 5 kW complete fuel processor for iso-octane as hydrogen supply system for mobile auxiliary power units. Part I – development of the autothermal catalyst and reactor, *Chem. Eng. J.*, **137** (3), 653–663.
7. O'Connell, M., Kolb, G., Schelhaas, K.P., Schürer, J., Tiemann, D., Ziogas, A., and Hessel, V. (2009) Development and evaluation of a microreactor for the reforming of diesel fuel in the kW range. *Int. J. Hydrogen Energy*, **34**, 6290–6303.
8. Hardt, S., Ehrfeld, W., Hessel, V., and Vanden Bussche, K.M. (2003) Strategies for size reduction of microreactors by heat transfer enhancement effects. *Chem. Eng. Commun.*, **190** (4), 540–559.
9. Drese, K.S. and Hardt, S. (2000) In *Proceedings of MICRO.Tec 2000*, VDE Verlag, Berlin, pp. 371–374.
10. Kolb, G. (2008) *Fuel Processing for Fuel Cells*, 1st edn, Wiley-VCH Verlag GmbH, Weinheim.
11. Fazeli, A. and Behnam, M. (2010) Hydrogen production in a zigzag and straight catalytic wall coated micro channel reactor by CFD modelling. *Int. J. Hydrogen Energy*, **35**, 9496–9508.
12. Stief, T., Langer, O.U., and Schubert, K. (1999) Numerical investigations of optimal heat conductivity in micro heat exchangers. *Chem. Eng. Technol.*, **21** (4), 297–302.
13. Larson, A.T. (1947) Preparation of hydrogen, US Patent 2,425,625.
14. Bravo, J., Karim, A., Conant, T., Lopez, G.P., and Datye, A. (2004) Wall coating of a $CuO/ZnO/Al_2O_3$ methanol steam reforming catalyst for micro-channel reformers. *Chem. Eng. J.*, **101** (1–3), 113–121.
15. Men, Y., Gnaser, H., Zapf, R., Kolb, G., and Ziegler, C. (2004) Steam reforming of methanol over $Cu/CeO_2/\gamma\text{-}Al_2O_3$ catalysts in a microchannel reactor. *Appl. Catal. A*, **277**, 83–90.
16. Men, Y., Gnaser, H., Zapf, R., Kolb, G., Hessel, V., and Ziegler, C. (2004) Parallel screening of $Cu/CeO_2/\gamma\text{-}Al_2O_3$ for steam reforming of methanol in a 10-channel micro-structured reactor. *Catal. Commun.*, **5**, 671–675.
17. Men, Y., Gnaser, H., Ziegler, C., Zapf, R., Hessel, V., and Kolb, G. (2005) Characterization of $Cu/CeO_2/\gamma\text{-}Al_2O_3$ thin film catalysts by thermal desorption spectroscopy. *Catal. Lett.*, **105** (1–2), 35–40.
18. Men, Y., Kolb, G., Zapf, R., Tiemann, D., Wichert, M., Hessel, V., and Löwe, H. (2008) A complete miniaturised microstructured methanol fuel processor/fuel cell system for low power applications. *Int. J. Hydrogen Energy*, **33** (4), 1374–1382.
19. Pfeifer, P., Fichtner, M., Schubert, K., Liauw, M.A., and Emig, G. (2000) In *Proceedings of Microreaction Technology: 3rd International Conference on Microreaction Technology, Proceeding of IMRET 3*, Springer, Berlin, pp. 372–382.
20. Chin, Y.H., Dagle, R., Hu, J., Dohnalkova, A.C., and Wang, Y. (2002) Steam reforming of methanol over highly active Pd/ZnO catalyst. *Catal. Today*, **77**, 79–88.

21. Hu, J., Wang, W., VanderWiel, D., Chin, C., Palo, D., Rozmiarek, R., Dagle, R., Cao, J., Holladay, J., and Baker, E. (2003) Fuel processing for portable power applications. *Chem. Eng. J.*, **93**, 55–60.
22. Chin, Y.H., Wang, Y., Dagle, R.A., and Li, X.S. (2003) Methanol steam reforming over Pd/ZnO: catalyst preparation and pretreatment studies. *Fuel Process. Technol.*, **83**, 193–201.
23. Pfeifer, P., Schubert, K., Liauw, M.A., and Emig, G. (2004) PdZn catalysts prepared by washcoating microstructured reactors. *Appl. Catal. A*, **270**, 165–175.
24. Men, Y., Kolb, G., Zapf, R., O'Connell, M., and Ziogas, A. (2010) Methanol steam reforming over bimetallic Pd–In/Al_2O_3 catalysts in a microstructured reactor. *Appl. Catal. A*, **380**, 15–20.
25. Kolb, G., Keller, S., Pecov, S., Pennemann, H., and Zapf, R. (2011) Development of micro-structured catalytic wall reactors for hydrogen production by methanol steam reforming over novel Pt/In_2O_3/Al_2O_3 catalysts. *Chem. Eng. Trans.*, **24**, 133–138.
26. Cominos, V., Hardt, S., Hessel, V., Kolb, G., Löwe, H., Wichert, M., and Zapf, R. (2005) A methanol steam micro-reformer for low power fuel cell applications. *Chem. Eng. Commun.*, **192** (5), 685–698.
27. Kolb, G., Cominos, V., Drese, K., Hessel, V., Hofmann, C., Löwe, H., Wörz, O., and Zapf, R. (2002) A novel catalyst testing microreactor for heterogeneous gas phase reactions. In *Proceedings of the 6th International Conference on Microreaction Technology, IMRET 6, New Orleans, 11–14 March 2002*, AIChE Publication No. 164, American Institute of Chemical Engineers, New York, pp. 61–72.
28. Pfeifer, P., Schubert, K., Liauw, M.A., and Emig, G. (2003) Electrically heated microreactors for methanol steam reforming. *Chem. Eng. Res. Des.*, **81** (A7), 711–720.
29. Park, G.-G., Seo, D.J., Park, S.-H., Yoon, Y.-G., Kim, C.-S., and Yoon, W.-L. (2004) Development of microchannel methanol steam reformer. *Chem. Eng. J.*, **101** (1–3), 87–92.
30. Seo, D.J., Yoon, W.-L., Yoon, Y.-G., Park, S.-H., Park, G.-G., and Kim, C.-S. (2004) Development of a micro fuel processor for PEMFCs. *Electrochim. Acta*, **50**, 719–723.
31. Yu, X., Tu, S.T., Wang, Z., and Qi, Y. (2006) Development of a microchannel reactor concerning steam reforming of methanol. *Chem. Eng. J.*, **116**, 123–132.
32. Kundu, A., Ahn, J.E., Park, S.S., Shul, Y.G., and Han, H.S. (2008) Process intensification by micro-channel reactor for steam reforming of methanol. *Chem. Eng. J.*, **135**, 113–119.
33. Thormann, J., Pfeifer, P., Kunz, U., and Schubert, K. (2008) Reforming of diesel fuel in a microreactor. *Int. J. Chem. Reactor Eng.*, **6**, P1.
34. Thormann, J., Maier, L., Pfeifer, P., Kunz, U., Deutschmann, O., and Schubert, K. (2009) Steam reforming of hexadecane over a Rh/CeO_2 catalyst in microchannels: experimental and numerical investigation. *Int. J. Hydrogen Energy*, **34**, 5108–5120.
35. Peela, N.R., Mubayi, A., and Kunzru, D. (2011) Steam reforming of ethanol over Rh/CeO_2/Al_2O_3 catalysts. *Chem. Eng. J.*, **167**, 578–587.
36. Redenius, J.M., Schmidt, L.D., and Deutschmann, O. (2001) Millisecond catalytic wall reactors: I. Radiant burner. *AIChE J.*, **47** (5), 1177–1184.
37. Frauhammer, J., Eigenberger, G., von Hippel, L., and Arntz, D. (1999) A new reactor concept for endothermic high-temperature reactions. *Chem. Eng. Sci.*, **54** (15–16), 3661–3670.
38. Kolb, G., Hofmann, C., O'Connell, M., and Schürer, J. (2009) Micro-structured reactors for diesel steam reforming, water-gas shift and preferential oxidation in the kilowatt power range. *Catal. Today*, **147**, 176–184.
39. Kolb, G., Schelhaas, K.P., Wichert, M., Burfeind, J., Hesske, C., and Bandlamudi, G. (2009) Development of a microstructured methanol fuel processor coupled to a high temperature

PEM fuel cell. *Chem. Eng. Technol.*, **32** (11), 1739–1747.

40. Wichert, M., Men, Y., O'Connell, M., Tiemann, D., Zapf, R., Kolb, G., Butschek, S., Frank, R., and Schiegl, A. (2011) Self-sustained operation and durability test of a 300 W-class micro-structured LPG fuel processor. *Int. J. Hydrogen Energy*, **36**, 3496–3504.

41. O'Connell, M., Kolb, G., Schelhaas, K.P., Schuerer, J., Tiemann, D., Ziogas, A., and Hessel, V. (2009) An investigation into the transient behaviour of a microreactor system for the reforming of diesel fuel in the kW range. *Chem. Eng. Technol.*, **32** (11), 1790–1798.

42. Hsueh, C.-Y., Chu, H.-S., Yan, W.-M., and Chen, C.-H. (2010) Numerical study of heat and mass transfer in a plate methanol steam micro reformer with methanol catalytic combustor. *Int. J. Hydrogen Energy*, **35**, 6227–6238.

43. Petrachi, G.A., Negro, G., Specchia, S., Saracco, G., Maffetone, P.L., and Specchia, V. (2005) Combining catalytic combustion and steam reforming in a novel multifunctional reactor for on-board hydrogen production from middle distillates. *Ind. Eng. Chem. Res.*, **44**, 9422–9430.

44. Zhai, X., Ding, S., Cheng, Y., Jin, Y., and Cheng, Y. (2010) CFD simulation with detailed chemistry of steam reforming of methane for hydrogen production in an integrated micro-reactor. *Int. J. Hydrogen Energy*, **35**, 5383–5392.

45. Park, G.-G., Yim, S.-D., Yoon, Y.-G., Lee, W.-Y., Kim, C.-S., Seo, D.-J., and Eguchi, K. (2005) Hydrogen production with integrated microchannel fuel processor for portable fuel cell systems. *J. Power Sources*, **145**, 702–706.

46. Reuse, P., Renken, A., Haas-Santo, K., Görke, O., and Schubert, K. (2004) Hydrogen production for fuel cell application in an autothermal microchannel reactor. *Chem. Eng. J.*, **101** (1–3), 133–141.

47. Cremers, C., Stummer, M., Stimming, U., Find, J., Lercher, J.A., Kurtz, O., Crämer, K., Haas-Santo, K., Görke, O., and Schubert, K. (2003) Micro-structured reactors for coupled steam-reforming and catalytic combustion of methane. In *Proceedings of the 7th International Conference on Microreaction Technology, IMRET 7, Lausanne, 7–10 September 2003*, DECHEMA, Frankfurt, p. 100.

48. Pfeifer, P., Bohn, L., Görke, O., Haas-Santo, K., and Schubert, K. (2005) Microstructured components for hydrogen production from various hydrocarbons. *Chem. Eng. Technol.*, **28** (4), 474–476.

49. Pfeifer, P., Bohn, L., Görke, O., Haas-Santo, K., and Schubert, K. (2004) Mikrostrukturkomponenten für die Wasserstofferzeugung aus unterschiedlichen Kohlenwasserstoffen. *Chem. Ing. Tech.*, **76** (5), 618–620.

50. Zalc, J.M. and Löwffler, D.G. (2002) Fuel processing for PEM fuel cells: transport and kinetic issues of system design. *J. Power Sources*, **111**, 58–64.

51. TeGrotenhuis, W.E., King, D.L., Brooks, K.P., Holladay, B.J., and Wegeng, R.S. (2002) Optimizing microchannel reactors by trading-off equilibrium and reaction kinetics through temperature management. In *Proceedings of the 6th International Conference on Microreaction Technology, IMRET 6, New Orleans, 11–14 March 2002*, AIChE Publication No. 164, American Institute of Chemical Engineers, New York, pp. 18–28.

52. Giroux, T., Hwang, S., Liu, Y., Ruettinger, W., and Shore, L. (2005) Monolithic structures as alternatives to particulate catalysts for the reforming of hydrocarbons for hydrogen generation. *Appl. Catal. B*, **56**, 95–110.

53. Kim, G.Y., Mayor, J.R., and Ni, J. (2005) Parametric study of microreactor design for water gas shift reactor using an integrated reaction and heat exchange model. *Chem. Eng. J.*, **110**, 1–10.

54. Baier, T. and Kolb, G. (2007) Temperature control of the water-gas shift reaction in microstructured reactors. *Chem. Eng. Sci.*, **62** (17), 4602–4611.

55. Pasel, J., Cremer, P., Stalling, J., Wegner, B., Peters, R., and Stolten, D. (2002) Comparison of two different reactor concepts for the water-gas

shift reaction. In Proceedings of the Fuel Cell Seminar, Palm Springs, CA, 18–21 November 2002, pp. 607–610.

56. Kolb, G., Schürer, J., Tiemann, D., Wichert, M., Zapf, R., Hessel, V., and Löwe, H. (2007) Fuel processing in integrated microstructured heat-exchanger reactors. *J. Power Sources*, **171** (1), 198–204.

57. O'Connell, M., Kolb, G., Schelhaas, K.P., Schuerer, J., Tiemann, D., Ziogas, A., and Hessel, V. (2010) The development and evaluation of microstructured reactors for the water-gas shift and preferential oxidation reactions in the 5 kW range. *Int. J. Hydrogen Energy*, **35**, 2317–2327.

58. O'Connell, M., Kolb, G., Schelhaas, K.P., Schuerer, J., Tiemann, D., Keller, S., Reinhard, D., and Hessel, V. (2010) Investigation on the combined operation of water gas shift and preferential oxidation reactor system on the kW scale. *Ind. Eng. Chem. Res.*, **49** (21), 10917–10923.

59. Ouyang, X. and Besser, R.S. (2005) Effect of reactor heat transfer limitations on CO preferential oxidation. *J. Power Sources*, **141**, 39–46.

60. Kahlich, M.J., Gasteiger, H.A., and Behm, R.J. (1998) Preferential oxidation of CO over Pt/γ-Al_2O_3 and Au/α-Fe_2O_3: reactor design calculations and experimental results. *J. New Mater. Electrochem. Syst.*, **1**, 39–46.

61. Kim, K.-Y., Han, J., Nam, S.W., Lim, T.-H., and Lee, H.-I. (2008) Preferential oxidation of CO over CuO/CeO_2 and $Pt-Co/Al_2O_3$ catalysts in micro-channel reactors. *Catal. Today*, **131**, 431–436.

62. Lopez, E., Kolios, G., and Eigenberger, G. (2007) Preferential oxidation in a folded-plate reactor. *Chem. Eng. Sci.*, **62**, 5598–5601.

63. Ouyang, X., Bednarova, L., Besser, R.S., and Ho, P. (2005) Preferential oxidation (PrOx) in a thin-film catalytic microreactor: advantages and limitations. *AIChE J.*, **51** (6), 1758–1771.

64. Hwang, S.M., Kwon, O.J., Ahn, S.H., and Kim, J.J. (2009) Silicon based micro-reactor for preferential CO oxidation. *Chem. Eng. J.*, **146**, 105–111.

65. Kolb, G. and Tiemann, D. (2005) Mikroverdampfer, German Patent DE 10 2005 017 452.

66. Galletti, C., Specchia, S., and Specchia, V. (2011) CO selective methanation in H_2-rich gas for fuel cell applications: microchannel reactor performance with Ru-based catalysts. *Chem. Eng. J.*, **167**, 616–621.

67. Xu, G., Chen, X., and Zhang, Z.G. (2006) Temperature-staged methanation: an alternative method to purify hydrogen-rich fuel gas for PEFC. *Chem. Eng. J.*, **121**, 97–107.

68. Pattekar, A.V., Kothare, M.V., Karnik, S.V., and Hatalis, M.K. (2001) A microreactor for *in-situ* hydrogen production by catalytic methanol reforming. In Proceeding of the 5th International Conference on Microreaction Technology – IMRET 5, Strasbourg, 27–30 May 2001, pp. 332–342.

69. Pattekar, A.V. and Kothare, M.V. (2004) A microreactor for hydrogen production in micro fuel cells. *J. Microelectromech. Syst.*, **13** (1), 7–18.

70. Shah, K., Ouyang, X., and Besser, R.S. (2005) Microreaction for microfuel processing: challenges and prospects. *Chem. Eng. Tech.*, **28** (3), 303–313.

71. Pattekar, A.V. and Kothare, M.V. (2005) A radial microfluidic fuel processor. *J. Power Sources*, **147**, 116–127.

72. Yoshida, K., Tanaka, S., Hiraki, H., and Esashi, M. (2006) A micro fuel reformer integrated with a combustor and a microchannel evaporator. *J. Micromech. Microeng.*, **16**, 191–197.

73. Kim, T., Hwang, J.S., and Kwon, S. (2007) A MEMS methanol reformer heated by decomposition of hydrogen peroxide. *Lab Chip*, **7**, 835–841.

74. Terazaki, T., Nomura, M., Takeyama, K., Nakamura, O., and Yamamoto, T. (2005) Development of multi-layered microreactor with methanol reformer for small PEMFC. *J. Power Sources*, **145**, 691–696.

75. Ogura, N., Kawamura, Y., Yahata, T., Yamamoto, K., Terazaki, T., Nomura, M., Takeyama, K., Nakamura, O., Namai, T., Terada, T., Bitoh, H., Yamamoto, T., and Igarashi, A. (2004)

Small PEMFC system with multi-layered microreactor. In Proceedings of PowerMEMS, Kyoto, 28–30 November 2004, pp. 142–145.

76. Kawamura, Y., Ogura, N., Yamamoto, T., and Igarashi, A. (2006) A miniaturised methanol reformer with Si-based microreactor for a small PEMFC. *Chem. Eng. Sci.*, **61**, 1092–1101.

77. Kawamura, Y. (2007) A micro fuel processor with microreactor for a small fuel cell system. Presented at Small Fuel Cells Conference, Miami, FL, 8–9 March 2007.

78. Holladay, J.D., Wang, Y., and Jones, E. (2004) Review of developments in portable hydrogen production using microreactor technology. *Chem. Rev.*, **104**, 4767–4790.

79. Irving, P.M., Lloyd Allen, W., Healey, T., and Thomson, W.J. (2001) in *Microreaction Technology – IMRET 5: Proceedings of the 5th International Conference on Microreaction Technology* (eds. M. Matloszedssnm, W. Ehrfeld, and J.P. Baselt), Springer, Berlin, pp. 286–294.

80. Irving, P.M. and Pickles, J.S. (2006) Operational requirements for a multi-fuel processor that generates hydrogen from bio- and petroleum based fuels. Presented at Fuel Cell Seminar, Honolulu, 15–17 November 2006.

81. Knitter, R., Göhring, D., Risthaus, P., and Haussell, J. (2001) Microfabrication of ceramic microreactors. *Microsyst. Technol.*, **7**, 85–90.

82. Alm, B., Knitter, R., and Haussell, J. (2005) Development of a ceramic micro heat exchanger – design, construction, and testing. *Chem. Eng. Technol.*, **28** (12), 1554–1560.

83. Knitter, R., Dietrich, T., Baltes, H., Brand, O., Fedder, G.K., Hierold, C., Korvink, J., and Tabata, O. (eds.) (2006) *Micro Process Engineering*, vol. **5**, Wiley-VCH Verlag GmbH, Weinheim.

84. Brandner, J.J., Gietzelt, T.H., Henning, T., Kraut, M., Moritz, H., and Pfleging, W. (2006) in *Advanced Micro and Nanosystems*, Micro Process Engineering, Vol. **5** (eds. H. Baltes, O. Brand, G.K. Fedder, C. Hierold, J. Korvink, and O. Tabata), Wiley-VCH Verlag GmbH, Weinheim pp. 267–321

85. Kim, T. and Kwon, S. (2006) Design, fabrication and testing of a catalytic microreactor for hydrogen production. *J. Micromech. Microeng*, **16**, 1760–1768.

86. Ehrfeld, W., Hessel, V., and Löwe, H. (2000) *Microreactors*, Wiley-VCH Verlag GmbH, Weinheim.

87. Hessel, V., Kolb, G., and Brandner, J. (2009) in *Microfabricated Power Generation Devices* (eds. A. Mitsos and P.I. Barton), Wiley-VCH Verlag GmbH, Weinheim, pp. 7–37.

88. Dudfield, C.D., Chen, R., and Adcock, P.L. (2000) A compact CO selective oxidation reactor for solid polymer fuel cells powered vehicle applications. *J. Power Sources*, **86**, 214–222.

89. Ho, K.H. and Newman, S.T. (2003) State of the art electrical discharge machining (EDM). *Int. J. Mach. Tools Manuf.*, **43** (13), 1287–1300.

90. Lopez, E., Irigoyen, A., Trifonou, T., Rodriguez, A., and Llorca, J. (2010) A million-channel reformer on a fingertip: moving down the scale in hydrogen production. *Int. J. Hydrogen Energy*, **35**, 3472–3479.

91. Brandner, J., Bohn, L., Schygulla, U., Wenka, A., and Schubert, K. (2003) in *Microreactors, Epoch-Making Technology for Synthesis – High Technology Information* (ed. J.I. Yoshida) CMC, Tokyo, pp. 75–87.

92. Boljanovic, V. (2004) *Sheet Metal Forming Processes and Die Design*, Industrial Press, New York.

93. Shiu, P.P., Knopf, G.K., Ostojic, M., and Nikumb, S. (2008) Rapid fabrication of tooling for microfluidic devices via laser micromachining and hot embossing. *J. Micromech. Microeng.*, **18**, 025012.

94. Theis, H.E. (1999) *Handbook of Metalforming Processes*, CRC Press, Boca Raton, FL.

95. DeVries, W.R. (1992) *Analysis of Material Removal Processes*, Springer, New York.

96. Reed-Hill, R. (1991) *Physical Metallurgy Principles*, 3rd edn, PWS Publishing, Boston.

97. Bier, W., Keller, W., Linder, G., Seidel, D., Schubert, K., and Martin, H. (1993) Gas-to-gas heat transfer in micro heat exchangers. *Chem. Eng. Process.*, **32** (1), 33–43.
98. Schubert, K., Brandner, J., Fichtner, M., Linder, G., Schygulla, U., and Wenka, A. (2001) Microstructure devices for applications in thermal and chemical process engineering. *Microscale Thermal Eng.*, **5**, 17–39.
99. Meille, V. (2006) Review of methods to deposit catalysts on structured surfaces. *Appl. Catal. A*, **315**, 1–17.
100. Zapf, R., Becker-Willinger, C., Berresheim, K., Holz, H., Gnaser, H., Hessel, V., Kolb, G., Löb, P., Pannwitt, A.K., and Ziogas, A. (2003) Detailed characterization of various porous alumina based catalyst coatings within microchannels and their testing for methanol steam reforming. *Chem. Eng. Res. Des. A*, **81**, 721–729.
101. Germani, G., Stefanescu, A., Schuurman, Y., and van Veen, A.C. (2007) Preparation and characterization of porous alumina-based catalyst coatings in microchannels. *Chem. Eng. Sci.*, **62** (18–20), 5084–5091.
102. Agrafiotis, C. and Tsetsekou, A. (2000) The effect of powder characteristics on washcoat quality. Part I: alumina washcoats. *J. Eur. Ceram. Soc.*, **20**, 815–824.
103. Avila, P., Montes, M., and Miro, E.E. (2005) Monolithic reactors for environmental applications. A review on preparation techniques. *Chem. Eng. J.*, **109**, 11–36.
104. Cristiani, C., Valentini, M., Merazzi, M., Neglia, S., and Forzatti, P. (2005) Effect of aging time on chemical and rheological evolution in γ-Al_2O_3 slurries for dip-coating. *Catal. Today*, **105**, 492–498.
105. Tomasic, V. and Jovic, F. (2006) State-of-the-art in the monolithic catalysts/reactors. *Appl. Catal. A*, **311**, 112–121.
106. Thybo, S., Jensen, S., Johansen, J., Johannesen, T., Hansen, O., and Quaade, U.J. (2004) Flame spray deposition of porous catalysts on surfaces and in microsystems. *J. Catal.*, **223**, 271–277.
107. Wunsch, R., Fichtner, M., Görke, O., Haas-Santo, K., and Schubert, K. (2002) Process of applying Al_2O_3 coatings in microchannels of completely manufactured microstructured reactors. *Chem. Eng. Technol.*, **25** (7), 700–703.

8
Regenerative Fuel Cells
Martin Müller

8.1
Introduction

Fuel cells (FCs) are usually defined as electrochemical devices that convert the chemical energy of a reaction directly into electrical energy and heat. An FC consists of an electrolyte layer with a porous anode and cathode electrode on both sides of the electrolyte.

A regenerative fuel cell or reverse fuel cell (RFC) is an electrochemical device that can be operated like a standard FC and also like an electrolyzer. The RFC consumes electricity and a substance to produce another substance. This latter substance can be stored and on demand the RFC can convert the stored substance back to the base substance and electricity. The typical medium used in the RFC is water that is separated into hydrogen and oxygen during charging and then the oxygen and hydrogen are reconverted to water during discharging:

$$H_2 + \frac{1}{2}O_2 \rightleftharpoons H_2O \tag{8.1}$$

In an RFC system, attention is focused on the storage of electrical energy, which is the reason why such systems are compared with other electrochemical storage techniques such as batteries. Beck and Rüetschi compared different electrochemical storage techniques with aqueous electrolytes [1]. In their study, batteries were benchmarked by their energy density and their cycle life. The highest theoretically achievable specific energy in this benchmark is 1550 Wh kg^{-1}, reached by an inorganic lithium battery. The practical maximum energy density of this battery type is given as 220 Wh kg^{-1}.

Under standard conditions with gaseous educts and products, hydrogen has an energy content of 241.82 kJ mol$_{LHV}^{-1}$ or 33.3 kWh kg^{-1} and in terms of volume 2.75 kWh m^{-3}. The theoretical specific energy of the hydrogen is about 20 times higher than that of the best battery. Values achievable in practice are not so high because the tank and the device for converting chemical energy into electrical energy (FC) and the device for converting electrical energy into chemical energy (electrolyzer) must also be taken into account. By using lightweight pressure

Fuel Cell Science and Engineering: Materials, Processes, Systems and Technology,
First Edition. Edited by Detlef Stolten and Bernd Emonts.
© 2012 Wiley-VCH Verlag GmbH & Co. KGaA. Published 2012 by Wiley-VCH Verlag GmbH & Co. KGaA.

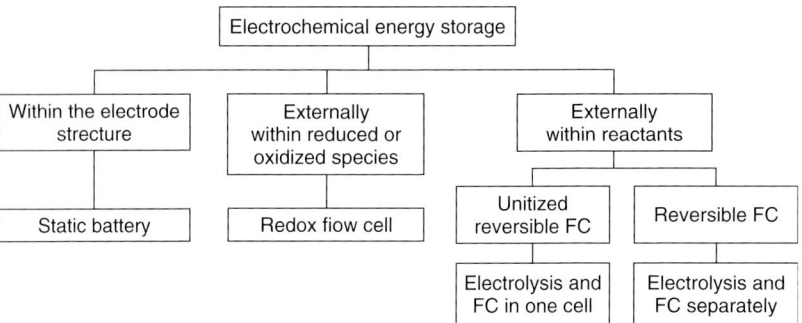

Figure 8.1 Types of energy storage in electrochemical cells.

vessels, a specific energy of 0.4–1 kWh kg^{-1} can be achieved in RFC systems [2]. Compared with conventional secondary batteries, a further advantage of the RFC and the redox flow cell is that the energy carrier is stored externally because in this way the stored energy content and the maximum system power are independent of each other [3]. Compared with the redox flow cell, the specific energy content of the RFC is definitely higher. A rough categorization of electrochemical storage systems is given in Figure 8.1.

Other problems associated with batteries are self-discharging when not in use and the influence of the ambient temperature on performance [4]. In an RFC system, hydrogen and in some cases oxygen are stored in pressure tanks and the rate of hydrogen or oxygen losses, depending on the diffusion of the hydrogen through the walls of the tank, is very low. When the RFC system is started up, it heats up to the operating temperature and has the same performance at low or high ambient temperatures. In spite of all the advances in relation to batteries, the round-trip efficiency of RFC systems is lower. However, it is not affected by storage duration.

A special form of RFC system is the unitized regenerative fuel cell (URFC). Such systems have only one device for both the conversion of chemical into electrical energy and the reconversion of electrical into chemical energy. The advantage of such systems is the reduced weight due to the reduced number of components. The disadvantage is that the efficiency and durability are not as high as in normal RFC systems since the electrodes have to operate in both directions. These systems are probably more suitable for space and aviation applications where weight is a major concern. The costs of URFCs may, however, be lower because of the reduced number of components. Systems with separate devices for electrolysis and FC operation are more interesting for applications that focus on durability and efficiency.

8.2
Principles

Regeneratively produced electrical energy from wind and photovoltaics is usually supplied discontinuously into the electricity grid. If the grid is also fed with electrical

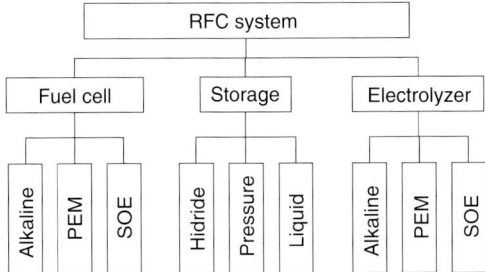

Figure 8.2 Core components of an RFC system.

energy converted from fossil fuels the output of these conventional power plants can be reduced when considerable amounts of regenerative energy are available. If the amount of regeneratively produced electricity available exceeds the amount of conventionally produced electricity, or if no electric grid is available, it is useful to store the excess energy. Apart from other technologies, the conversion of electrical energy into chemical energy is very attractive due to the high energy densities.

The most important components of such RFC systems are an electrolyzer, an FC, and a storage tank (see Figure 8.2). Water is usually converted into hydrogen and oxygen in the electrolyzer by the use of electrical energy. If the system is used in an application where the weight or the costs of the storage system are a critical issue and air is available, then oxygen will not be stored. However, if no air is available, for example in space, oxygen must be stored. When electric power is required, hydrogen and oxygen can be reconverted into water with the production of electricity. If oxygen is used instead of air, it is possible to operate the FC more efficiently.

Different technologies can be used for RFC systems. Electrolyzers can be classified by the electrolyte they use: alkaline, polymer electrolyte membrane (PEM) and solid oxide electrolyte (SOE) [5]. Alkaline, PEM, and SOE electrolyzers can also be used for FCs. The electrolytes differ according to the ion that is conducted and their operating temperatures (Table 8.1).

A special form of RFC systems is the URFC. In such systems, the electrochemical device operates as an electrolyzer and an FC. The difficulty is to identify and develop catalyst materials that are durable in an oxidizing and reducing environment [6, 7].

Hydrogen storage techniques can be roughly classified into liquid, pressurized, and metal hydride [8]. The development of storage technologies is a broad research field that will not be considered in this chapter.

Table 8.1 Types of electrolytes.

Electrolyte?	Ion	Operating temperature (°C)
Alkaline electrolyte	OH^-	60–120
Polymer electrolyte	H^+	20–100
Solid oxide electrolyte	O^{2-}	700–1000

8.3
History

This chapter focuses on the electrochemical components. The development of alkaline electrolyzers started during the 1920s and soon these electrochemical converters attained practical relevance in industrial processes [9, 10]. In the 1930s and 1940s, work with alkaline fuel cells (AFCs) started and the first ideas concerning RFC systems were put forward [11]. The idea was to use the excess electrical energy produced during the night for the conversion of water into hydrogen and oxygen. In times with a high demand for electric power, hydrogen and oxygen would be reconverted into electricity. This technique was also intended to reduce investments for the installation of the electric grid [12, 13].

From the 1950s to the mid-1970s, the development of FCs and RFCs was driven by the space exploration programs of the US National Aeronautics and Space Administration (NASA). During the Gemini program, polymer electrolyte membrane fuel cells (PEMFCs) were developed and used. In the Apollo program, AFCs were used and the focus was on reliable and reproducible stacks and systems [14].

Table 8.2 Time line of RFC developments.

1917	Patent: "Apparatus for the electrolytic decomposition of water," using filter press plates to separate oxygen from hydrogen during evolution in the electrodes [9]
1920s	Development of alkaline electrolysis started
1930	Patent for pressure electrolyzers [10]
1930s	Development of alkaline fuel cell started
1932	Patent for reversible fuel cell system to compensate peak load in an electric grid. "Verfahren zum Speichern und Verteilen elektrischer Energie" [12, 13]
1939	Operation of a reversible fuel cell at 100 °C and at a pressure of 2300 psi with 27% KOH and an overall efficiency of 47% by Bacon [11]
1945	Gunn and Hall are granted a very broad patent for a fuel cell with aqueous electrolytes. This patent also includes an RFC system [22]
1950s	Development of PEFC (Gemini program) [23–25]
1960	Patent concerning the alkaline URFC (Apollo program) [26]
1970s	Alkaline fuel cell for spacecraft, NASA and Giner, Inc. [27]
1970s	Various NASA patents concerning RFC systems [28–30]
1970s	Development of PEM electrolysis at General Electric
1970s	SOE electrolysis and reverse operation with tubular cells at Dornier, Lurgi, Westinghouse [31–34]
1980s	Unitized alkaline reversible fuel cell developed by NASA and Giner, Inc.
1990s	Development of unitized PEM fuel cells started
1990s	Large stationary systems in kW-power range for seasonal energy storage were built and tested. A combination of alkaline electrolysis and PEMFC was frequently used
1996	50 W prototype PEM URFC which operated for 1700 10-min charge–discharge cycles. Degradation was only a small percentage at maximum current densities [2]
2003	18.5 kW URFC was installed for the propulsion of the unmanned Helios aircraft and was tested during test flights [35]

From the mid-1970s to the end of the 1980s, the existing FC concepts were further improved. The development of electrodes and cells for alkaline URFCs started and powerful and durable electrodes were identified [15, 16]. Research was also focused on the development of SOE electrolyzers. In Europe, the USA, and Japan, different programs were implemented focusing on the improvement of this technology [17, pp. 227–242].

In the 1990s, the development of PEM URFCs started [18]. The advantages of the PEM compared with the AFC are its simplicity and safe operation without a liquid electrolyte. At the same time, RFCs were tested for the seasonal storage of regeneratively produced electricity in various test plants [19, 20]. In 2003, an RFC system powered the autonomous unmanned solar aircraft Helios [21, pp. 175–176].

The time line of RFC developments is summarized in Table 8.2.

8.4
Thermodynamics

In order to understand the function of an RFC system, it is necessary to consider the media and energy flow in the system components. Figure 8.3 shows a flow chart of a possible RFC configuration. The system is supplied with water and electrical power.

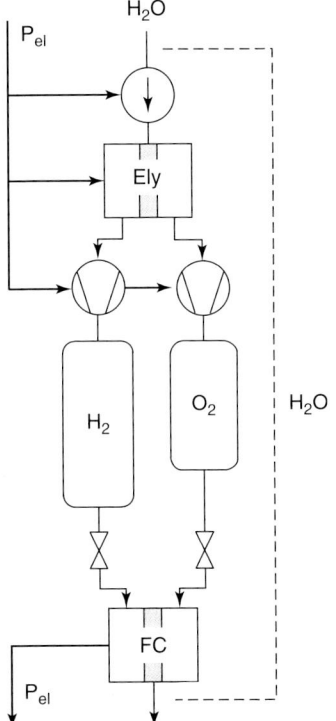

Figure 8.3 Flow chart of an RFC system.

The water is pumped into the electrolyzer (Ely) and it is converted to hydrogen and oxygen with the consumption of electrical energy (P_{el}). After compression, hydrogen and oxygen can be stored in pressure tanks. When electrical energy is needed, hydrogen and oxygen can be reconverted into water and electrical energy in an FC.

For users of RFC systems, it is important to know the overall efficiency, or the so-called *round-trip efficiency*. A measure of this efficiency is the relation between the electrical energy that is applied to the system during the charging period and the electrical energy that can be recovered during the discharging period:

$$\eta_{RFC} = \frac{\int P_{el\_discharging} \, dt}{\int P_{el\_charging} \, dt} \tag{8.2}$$

This overall or round-trip efficiency is the product of the efficiencies of the system components. These components are the electrolyzer (Ely), the FC, the storage pressure vessels, and the compressors for the compression of hydrogen and oxygen and the electrical converters. If a URFC is used instead of an RFC, there is only one electrochemical device in the system. However, this device operates at different efficiencies in the electrolyzer mode and the FC mode:

$$\eta_{RFC} = \eta_{Ely} \, \eta_{FC} \, \eta_{storage} \, \eta_{converter} \tag{8.3}$$

In practical tests, it was shown that the converters have a notable influence on the whole system, since the entire current flow takes place via these converters and is therefore affected by the converter efficiency. The focus here is on electrochemical processes and the efficiency of converters, and their development will not be considered.

The electrolyzer efficiency can be calculated in relation to the higher heating value or to the lower heating value. If the electrolyzer is fed with liquid water, it is reasonable to use the higher heating value to evaluate the quality of the electrolyzer design, but here the focus is on the system's efficiency and the comparability of all components with other systems. Usually in technical devices the efficiency of the conversion of chemical energy into another form of energy is given in relation to the lower heating value. Therefore, in the following equations and calculations the lower heating value is used. The related lower heating value voltage can be calculated using the enthalpy of water in Table 8.3, the number of ions z and the Faraday constant F:

$$U_{LHV} = \frac{-H^f_{H_2O}}{zF} \tag{8.4}$$

The number of ions is two, the enthalpy $H^f_{H_2O}$ is -241.82 kJ mol^{-1} and the lower heating value voltage U_{LHV} is 1.253 V. The efficiency of the electrolyzer is represented by the lower heating value voltage in relation to the applied voltage U_{Ely}:

$$\eta_{Ely} = \frac{U_{LHV}}{U_{Ely}} \tag{8.5}$$

Table 8.3 Thermodynamic properties of hydrogen, oxygen, and water[a] [36, p. 422].

Property	O_2	H_2	H_2O	
M (g mol)	32	2.02	18.02	
R (kJ kg^{-1} K^{-1})	0.26	4.13	0.46	
c_p (kJ/(kg K))	0.92	14.3	1.86	4.18
H^f (kJ mol^{-1})	0	0	−241.82	−285.83
S (J mol^{-1} K^{-1})	205.14	130.68	188.83	69.91
State of matter	g	g	g	l

[a] Molar mass M, gas constant R, heat capacity c_p, enthalpy H^f, entropy S; g, gaseous; l, liquid; $T_0 = 298.15$ K; $p_0 = 100$ kPa.

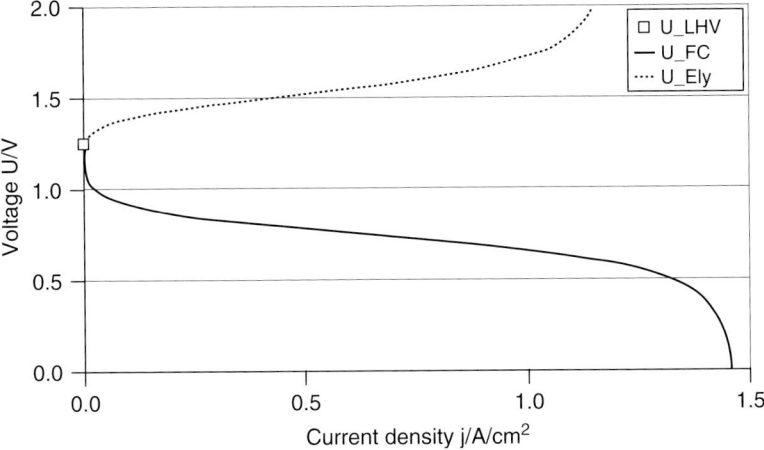

Figure 8.4 Typical pattern of RFC voltages in FC and Ely operation.

The efficiency of the energy conversion in the FC can be calculated by dividing the cell voltage U_{FC} by the heating value voltage U_{LHV}:

$$\eta_{FC} = \frac{U_{FC}}{U_{LHV}} \tag{8.6}$$

Figure 8.4 shows a typical pattern of the cell voltages as a function of the current density in an RFC. The plots are calculated and not measured, but the values are in accordance with measurements. U_FC is the polarization curve during FC operation and U_Ely is the polarization curve in electrolyzer mode. The square (U_LHV) indicates the lower heating value voltage. A difference in the maximum current density achievable in FC or electrolyzer operation is typical of real systems and can be affected by the use of different electrolytes in RFCs or in URFCs by different diffusion behaviors of the operating media.

The efficiency of the whole energy storage system is given by the energy output in relation to the energy input:

$$\eta_{storage} = \frac{H_{output}}{H_{input}} \tag{8.7}$$

The energy input is not only the chemical energy content of the hydrogen, but also the energy that is needed to compress the hydrogen and the oxygen and the energy that is needed to operate the feed pump. The total input of energy is given by the chemical energy content of the hydrogen added to the result of multiplying the compression and the pumping power by the operating time:

$$H_{input} = H_{H_2} + \left(P_{comp.} + P_{pump}\right) t_{comp} \tag{8.8}$$

A more detailed thermodynamic description of the relevant components and a simulation study of an RFC system were given by Ghosh [37]. To reduce the power consumption for compressing the product gases, it is possible to apply the pressure by the feed pump before entering the electrolyzer. However, it has to be taken into account that this increases the voltage during electrolysis, according to the Nernst equation. This decrease in electrolyzer efficiency therefore has to be compared with the increase in compressor efficiency.

8.5
Electrodes

If an RFC system has a separate FC and electrolyzer, it is possible to choose different materials for the electrolyzer and for the FC electrodes. The materials can be adapted in terms of ideal performance and high durability. If a URFC system is considered, the electrodes have to be suitable for both hydrogen consumption and evolution or oxygen consumption and evolution. This bifunctional operation subjects the electrodes to considerable stress and can reduce their performance and durability. In addition to the type of operation (conventional or bifunctional), the composition of the electrodes is greatly dependent on the electrolytes used. In the following subsections, the different electrolytes (alkaline, PEM, SOE) are described and also suitable electrodes for each type of operation.

8.5.1
Electrodes for Alkaline Electrolytes

In RFCs with alkaline electrolytes, the conducted ion is hydroxide (OH^-). The typical electrolyte is a mixture of potassium hydroxide (KOH) and water. The electrolyte is circulated in the cells by a pump and fixed by microporous structures. Another option is to fix the electrolyte in a matrix [38, p. 4-4] or to use an anion-exchange membrane. For electrolysis, polymeric diaphragms have been developed with a high porosity and small pore size in order to achieve low gas permeability [39]. The typical operating temperatures are in the range 60–120 °C, but higher temperatures

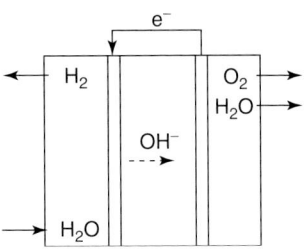

Figure 8.5 Alkaline electrolysis.

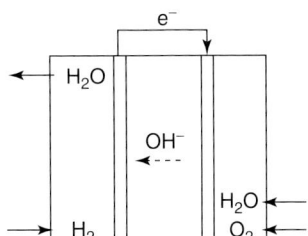

Figure 8.6 Alkaline fuel cell.

up to 200 °C are also possible. Figure 8.5 shows the flow of hydrogen, oxygen, and water, and the electronic and ionic flow during electrolysis. Figure 8.6 shows the flow of an alkaline cell in FC operation.

8.5.1.1 Alkaline Fuel Cells (AFCs)

The oxygen reduction in alkaline electrolytes is very attractive owing to the low overpotentials. A wide variety of catalyst materials, including non-noble metals, can be used. Suitable catalyst materials are platinum (Pt), gold (Au), cobalt (Co), nickel (Ni), and manganese (Mn). The catalyst material for FC operation differs depending on whether the cell is fed with pure oxygen or with air. When AFCs are fed with media containing carbon dioxide such as air, CO_2 reacts with the electrolyte. Usually the electrodes of an AFC consist of several PTFE-bonded layers. The backing material is in contact with the reactant gas followed by the diffusion layer and the active layer contacts the electrolyte [40]. In a broad study of AFC oxygen electrodes operated with pure oxygen, different porous carbon electrodes with various amounts of platinum and gold were examined, and also an unsupported gold platinum alloy. It was found that the unsupported gold catalyst performed best. The particles were Teflon bonded and the load was 20 mg cm^{-2}. The heavy gold load contributes to the electronic conductivities and avoids the corrosion problems of carbon [41]. Typical voltages and the associated efficiency that can be achieved in operation with air or oxygen are given in Table 8.4.

8.5.1.2 Alkaline Electrolysis

Cathodic catalyst materials are typically nickel based [17, pp. 257–260]. A study of the electrocatalytic properties of nickel-based electrocatalysts showed that increasing the molybdenum content on the electrode surface reduces the overpotentials for hydrogen [42]. An investigation of materials for hydrogen evolution under real

Table 8.4 Voltage and efficiency in AFC at 75 °C and ambient pressure [38, pp. 4–8].

Current density (A cm^{-2})	Voltage (air) (V)	Voltage (oxygen) (V)	Efficiency (air LHV)	Efficiency (oxygen LHV)
0.1	0.85	0.88	0.68	0.70
0.5	–	0.7	–	0.56

Table 8.5 Voltage and efficiency in alkaline electrolyzer, 120 °C, (Ni, Co)S/Co$_3$O$_4$ [46].

Current density (A cm^{-2})	Voltage (V)	Efficiency (LHV)
0.3	1.65	0.76
1.0	1.8	0.70

working conditions showed that local potential distribution, temperature, and electrolyte concentration significantly influence performance [43].

In addition to the mechanisms of oxygen evolution at nickel anodes [44], the surface properties of Co$_3$O$_4$ electrodes have also been investigated [45]. Another possibility is the use of mixed oxides of a spinel or perovskite type for anode oxygen evolution; examples are La$_{0.5}$Sr$_{0.5}$CoO$_3$, NiCo$_2$O$_4$–PTFE, and Co$_3$O$_4$ [46]. Representative results of electrochemical measurements in an alkaline electrolyzer are given in Table 8.5.

8.5.1.3 Alkaline URFCs

In addition to low overpotentials during operation, the electrodes in a URFC must have high stability with respect to oxygen evolution and oxygen reduction and, accordingly, for hydrogen evolution and oxidation. In a broad study, mixed metal oxides with a high surface area that demonstrated improved electrochemical activity were examined. The materials evaluated are candidates for electrocatalysts and for supports for electrocatalysts in electrolyzers and FCs. The material screening included La, Ti, Ca, Mg, Sr, Zr, Pb, Ru, Co, Mn, and Ni alloys. The highest surface area and an adequate electrocatalytic oxygen activity was found for the perovskite LaCoO$_3$ [47].

Support and catalyst materials suitable for use in oxygen electrodes operated with 30% KOH at 80 °C are investigated in terms of

- manufacturing
- electrical conductivity
- resistance to chemical corrosion
- catalytic activity (O$_2$ evolution and reduction).

Table 8.6 Performance of alkaline URFC with 30% KOH at 80 °C [15].

Current (A cm^{-2})	Voltage Ely (V)	Voltage FC (V)	Efficiency Ely	Efficiency FC	Efficiency round trip
0.1	1.50	0.88	0.84	0.70	0.59
0.5	1.70	0.70	0.74	0.56	0.41

The catalyst loading is in the range of 10–22 mg cm^{-2}. The most favorable catalysts in these measurements are $Na_xPt_3O_4$ and $PbIrO_x$. At a current density of 0.1 A cm^{-2}, the potential versus the reversible hydrogen electrode (RHE) in the oxygen reduction mode is 0.9 V, which is comparable to the Pt/Au catalyst mentioned above, and in the oxygen evolution mode 1.5 V [15]. $LiNiO_x$ and ZrN with lower catalytic activity are identified as suitable materials for the support of a bifunctional alkaline oxygen electrode. The most promising catalyst is $Na_xPt_3O_4$ and three other catalysts have been successfully integrated into dual-character electrodes: Pt/IrO_2, Pt/RhO_2, and Rh/RhO_2. The measured cell potentials are comparable to the first results, but during oxygen evolution the potential is reduced to 1.4 V [16].

The mechanisms of oxygen evolution and reduction on Teflon-bonded $NiCo_2O_4$ spinel in alkaline solutions were also examined. The influence on the electrode performance of KOH concentration from 0.1 to 1 M and electrolyte temperature from 25 to 75 °C was also studied [48]. A selection of values measured in real cell operation at an electrolyte temperature of 80 °C and 30% KOH is given in Table 8.6. The column "round-trip efficiency" is the product of the efficiency in electrolyzer (Ely) and FC modes.

8.5.2
Polymer Electrolyte Membrane (PEM)

In a PEMFC, the ion conducted is the proton (H$^+$). A typical electrolyte material is the perfluorosulfonic acid (PFSA) membrane Nafion from DuPont. In electrolysis mode, water is oxidized at the anode and oxygen is generated, and hydrogen is generated at the cathode (see Figure 8.7). In Figure 8.8, the cell is operated in the reverse mode. Hydrogen is consumed at the anode and oxygen is consumed at the cathode, where hydrogen is also generated.

Advantages of PEM-based RFC systems compared with alkaline RFCs are

- higher power densities
- simpler system design due to the use of a polymer foil instead of an alkaline substance
- excellent partial-load range and rapid response to fluctuating power input
- compact design allowing high-pressure operation.

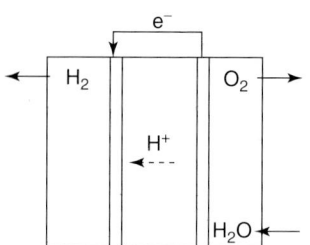

Figure 8.7 PEM electrolysis.

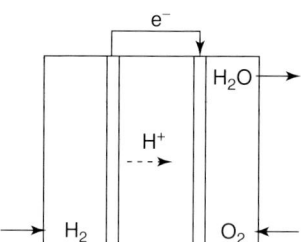

Figure 8.8 PEM fuel cell.

The challenges are to reduce the costs and to improve the durability [17, pp. 271–289]. RFCs have a separate electrolysis and FC module or are constructed as URFCs where both reactions take place in the same cell. A brief overview of the electrodes for separate electrolyzer and FC is given first, followed by a more detailed description of the bifunctional electrodes for URFCs.

8.5.2.1 PEM Electrolyzers

The best catalyst material for hydrogen evolution is platinum; typical loadings are in the range of 1–2 mg cm^{-2}. Suitable catalyst materials for the evolution of oxygen with acidic electrolytes are, in order of their catalytic activities, Ir/Ru > Ir > Rh > Pt. It is also possible to use oxides and mixtures of the metals [49, 50].

The best performance was achieved with platinum for hydrogen evolution and iridium for oxygen evolution. The catalyst loading was between 0.5 and 3 mg cm^{-2}. Cobalt clathrochelates were also tested for oxygen evolution, but their performance is much lower [51]. Typical voltages at 80–90 °C and 0.1 MPa are given in Table 8.7.

As mentioned in Section 8.4, it is useful to improve the round-trip efficiency by applying the pressure for hydrogen and oxygen storage on to the liquid water. This

Table 8.7 Voltage and efficiency of PEM electrolyzers 80–90 °C, 0.1 MPa [51].

Current density (A cm^{-2})	Voltage (V)	Efficiency (LHV)
0.1	1.45	0.86
1.0	1.7	0.74
2.0	1.9	0.65

Table 8.8 Voltage and efficiency in PEM fuel cell [54].

Current density (A cm^{-2})	Voltage (V)	Efficiency (LHV)
0.1	0.84	0.67
1.0	0.7	0.56
1.5	0.6	0.48

means that the electrolyzer must be operated under pressure. The increase in cell voltage is shown by the measurements, and calculations must be made to determine which pressure level is useful for operation with the highest efficiency [52].

8.5.2.2 PEMFCs

The PEMFC is a well-established technology. Its advantage is the simple setup and the high-power densities [53]. The typical catalyst is platinum. The catalyst loading at each electrode is less than 1 mg cm^{-2}. As a diffusion layer, carbon paper or carbon felt is used and is impregnated to adapt the wettability. Typical voltages at ~80 °C and atmospheric pressure are given in Table 8.8.

8.5.2.3 PEM URFC

In a unitized regenerative PEMFC, the catalysts must be suitable for hydrogen and oxygen consumption and production. This is especially difficult for oxygen: Platinum is ideal for oxygen reduction but is not suitable for oxygen evolution. The same is true of iridium oxide, which is ideal for oxygen evolution but not for reduction. One approach for designing electrodes is to combine an appropriate ratio of these monofunctional catalysts in one electrode. The other approach is to develop a catalyst with a bifunctional activity. $Na_xPt_3O_4$, for example, is considered to be a suitable catalyst material [18]. However, not only is the catalyst material important for stable and durable operation, the electrode structure must also be able to perform in both modes. Therefore, it is necessary to have areas with hydrophobic and hydrophilic impregnation. Titanium is often used as current-collector material to protect the oxygen electrode from carbon corrosion [55, 56].

A good catalyst for the bifunctional hydrogen electrode is platinum, and for the bifunctional oxygen electrode the most promising catalyst material is a mixture of platinum and iridium oxide. The use of thin catalyst layers in the electrode helps to minimize mass transport and ohmic limitations [57]. In addition to the catalyst composition, the ionomer content, the catalyst layer thickness, and the PTFE content are varied and the influence of these variations on performance has been described [58–63]. The highest efficiency can be achieved using a catalyst with a high amount of platinum and a low amount of iridium [58, 60].

Another approach is the investigation of catalyst deposition. It is possible to apply different catalysts to the active area of the cell in single layers or as a mixture or in segmented areas [64]. A new technique is the chemical deposition of platinum nanoparticles on iridium oxide for oxygen electrodes [65].

Table 8.9 Performance with Pt:Ir 85:15 catalyst at ~75 °C.

Current (A cm^{-2})	Voltage Ely (V)	Voltage FC (V)	Efficiency Ely	Efficiency FC	Efficiency round trip
0.1	1.45	0.90	0.86	0.72	0.62
0.5	1.56	0.77	0.80	0.61	0.49
1.0	1.66	0.68	0.75	0.54	0.41

The scale-up from laboratory cells to a 10-cell URFC stack with 250 cm^{-2} per cell was demonstrated by Kato et al. [66]. In this stack, the performance is as high as in the laboratory cells and at the same level as that of cells optimized for electrolysis or FC operation. The manufacture and design of electrodes (Pt:Ir) and the test of a seven-cell URFC stack with 256 cm^{-2} active area for each cell was described by Grigoriev and co-workers [67, 68]. In this stack, the maximum power in electrolysis mode is 1.5 kW and the efficiency is 72% at 0.5 A cm^{-2}. An efficiency of 44% can be achieved in FC operation.

Some values for the voltage and efficiency with Pt : Ir $\approx$ 85 : 15 for the electrolyzer and FC modes are given as an example in Table 8.9.

Another operational concept employs a unidirectional current and proton flow in electrolyzer and FC modes. This means that during electrolysis hydrogen is generated at the cathode of the electrolyzer and in FC mode the electrode must be suitable for oxygen consumption (Figures 8.9 and 8.10). Oxidation and reduction reactions are assigned to the electrodes. Using this concept, it is easier to identify electrode materials that are suitable for both electrodes [69–71]. Ledjeff et al. characterized the development of membrane electrode assemblies for this type of

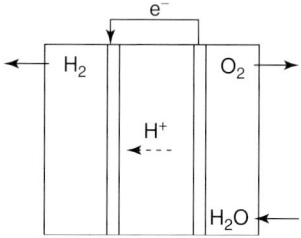

Figure 8.9 Alternative PEM electrolysis operation concept.

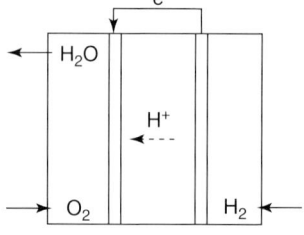

Figure 8.10 Alternative PEM fuel cell operation concept.

reversible PEMFC. The best catalyst material for reduction is platinum and the best material for oxidation is a mixture of iridium and platinum [72]. During the 1990s, attention was focused on the development of this type of PEM URFC, but owing to the complex media supply research is now concentrated on conventional PEM URFCs.

8.6
Solid Oxide Electrolyte (SOE)

The SOE is a ceramic material that is conductive for oxygen ions (O^{2-}). A popular electrolyte material is yttria-stabilized zirconia (YSZ). This material has good ionic conductivity at temperatures above 700 °C. The advantages of the SOE RFC are its good efficiency due to the negligible kinetic overpotentials and the use of nonprecious metals as catalyst. The disadvantage is the complex system setup due to the need for heat exchangers to achieve the high operating temperatures. SOE electrolysis has been described in several papers [73–75] and the solid oxide fuel cell (SOFC) has also been described in numerous articles [17, pp. 227–242, 76]. Figures 8.11 and 8.12 show the flow of electrons and ions in SOE electrolysis and in an SOFC.

To achieve a high electrochemical performance in an SOE RFC, the electrodes must have a low concentration polarization, low ohmic losses, and low activation polarization. For adequate durability, the electrolyte must be stable in oxidizing and reducing environments, and the mechanical properties must also be appropriate. The hydrogen electrode must be resistant to redox operation and have low grain-coarsening rates. The oxygen electrode must have sufficient oxygen evolution kinetics to avoid electrode delamination.

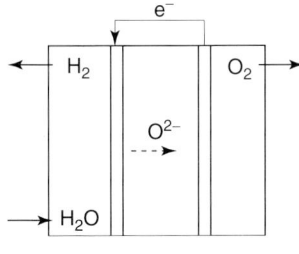

Figure 8.11 SOE electrolysis.

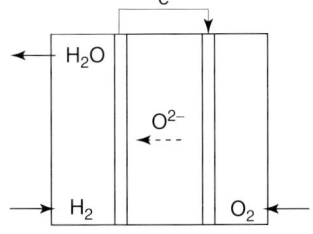

Figure 8.12 SOE fuel cell (SOFC).

Table 8.10 Performance data achieved with an SOE URFC at ~850 °C [79].

Current (A cm^{-2})	Voltage Ely (V)	Voltage FC (V)	Efficiency Ely	Efficiency FC	Efficiency round trip
0.5	1.03	0.83	1.22	0.66	0.81
1.0	1.17	0.76	1.07	0.61	0.65
1.5	1.32	0.68	0.95	0.54	0.51

In FC and electrolysis operating modes, the performance of the oxygen electrode decreases in the order lanthanum strontium cobalt ferrites (LSCFs) > lanthanum strontium ferrites (LSFs) > lanthanum strontium manganites (LSMs). The degradation stability of LSCFs and LSFs in electrolysis mode is better than that of LSMs and YSZ.

The materials used for hydrogen electrodes are Ni- and Cu-based or conducting ceramic electrodes. Owing to possible polarization losses for hydrogen electrodes in electrolysis mode (different diffusion behaviors of H_2 and H_2O), thinner electrodes and smaller particles are preferred [78]. Typical performance data for an SOE URFC are given in Table 8.10.

8.7
System Design and Components

In addition to the development of electrodes and electrode structures, there are many options for the system and stack setup. Some patents and papers focus on the basic principles of the system, whereas other patents describe advantageous designs in detail.

1) Systems consisting of separate cells or electrodes for electrolysis and FC operation often have a simple setup. The advantage of such systems is that electrolysis and FC operation are completely decoupled and can take place at the same time. The voltage levels and the currents can be adapted to the demands of the user by an appropriate design of the electrolyzer and the FC. Another advantage is that no pumps or other system components are needed. Figure 8.13 shows a fully encapsulated RFC system with solid electrolyte consisting of a chamber for oxygen, hydrogen, and a liquid water absorption material [30]. A passive RFC system that operates at variable pressures where the transport of liquid water is achieved by gravity has been patented by Sprouse [80]. A special system with liquid electrolyte is shown in Figure 8.14. The key feature of this passive system is the FC electrodes, which float on the surface of the electrolyte [77].

2) An example of a more complex setup is a system that operates as a power booster. In this system, a conventional FC is supplied with hydrogen by a

8.7 System Design and Components | 235

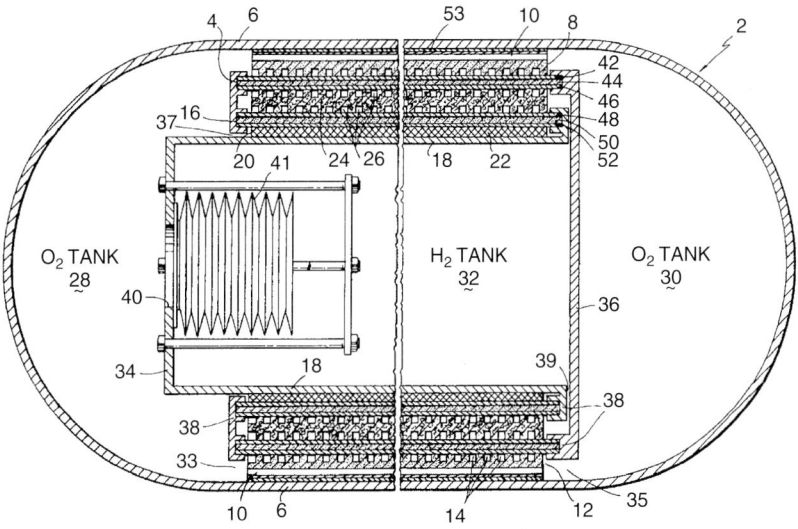

Figure 8.13 Fully encapsulated RFC [30].

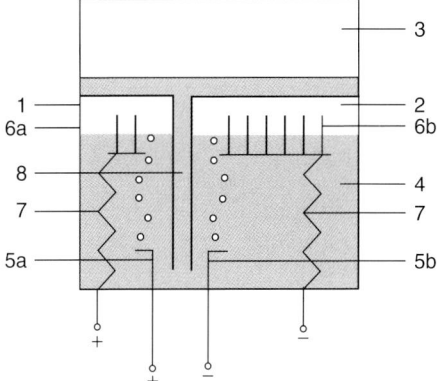

Figure 8.14 RFC with liquid electrolyte and floating electrodes [77].

reformer. During normal operation, part of the electricity produced is converted into hydrogen and oxygen by an electrolyzer. If more than the average amount of power is necessary, the stored hydrogen and oxygen can be fed into the FC instead of the reformate and air, which increases the power of the FC. The system is very complex and from today's point of view it would be better to use a battery or a supercapacitor to provide peak power [81].

3) The idea of the short-time storage of electrical energy in a reverse-operated FC within the structure of the electrodes has been described [82, 83]. The storage volume is not large, but it is big enough for the storage of braking energy in a vehicle.

4) Systems with one oxidation and one reduction electrode were developed in the 1990s. The advantage of this technique is the more durable electrodes. The

disadvantage is the complex system required to supply the media to the cell [49, 69].

In addition to the stability of the catalysts and the electrodes of a URFC, the durability of the stack and system components has also been investigated. Titanium plates are the best candidates as bipolar material owing to their sturdiness in highly acidic and humid conditions with no corrosion at high positive overpotentials [84, 85].

Many papers have dealt with the modeling of hybrid RFC energy-storage systems coupled with photovoltaic or wind electrical power generation for residential applications [86]. All these simulations suffer from the numerous degrees of freedom. For residential applications, the RFC systems have to cope with other storage techniques and an important factor is the cost of the stored energy. The costs of different energy conversion techniques have been compared [87, 88].

For space, submarine, or military applications, the focus is not so much on the cost but on the function of the system. For space and aviation, the proportion of efficiency and the specific weight (Wh kg^{-1}) is preeminent. It was found that in spite of the reduced number of system components, the UFC does not have any advantage in comparison with a discrete RFC in terms of function. The round-trip efficiency is given as 34%, and the energy density is 500 Wh kg^{-1} [89]. For this reason, the minimization of the mass of RFC systems for on-board energy storage is an interesting topic, which is discussed in several reports. According to the circumstances of use in the Earth's atmosphere or outer space, different requirements are placed on the electrochemical components and the storage system. For example, at the low temperatures in space, the condensation and evaporation of oxygen and hydrogen are an issue, but this topic does not need be considered in terrestrial applications with simple pressure vessels. Also, the lack of gravity can influence mass transport inside the stacks and system components [90–93].

8.8
Applications and Systems

The systems already implemented are stationary systems for the seasonal energy storage of regeneratively produced electrical energy and systems for space and aviation applications, where the focus is on low weight per unit stored energy content.

Stationary systems will be introduced first. In many applications, the functionality, durability, and efficiency of the system components were evaluated. The key components of such systems are the FC, the electrolyzer, and the hydrogen/oxygen storage system. The focus of the development of stationary systems is not on the reduction of space demand or weight, but on improving the efficiency of the whole cycle and reducing the costs. These systems are often coupled with a photovoltaic array for the production of regenerative electricity.

8.8.1
Stationary Systems for Seasonal Energy Storage

During the 1990s, many RFC systems for the storage of regeneratively produced electricity were tested at different locations. One of the first systems was tested in California and consisted of an alkaline bipolar electrolyzer operated at 790 kPa. The oxygen that was produced during electrolysis was not used in the system. A PEFC was used for the reconversion of the hydrogen into water. The system was constructed in June 1991 and went into full-time automatic operation in August 1993, and was subsequently operated for more than 3900 h [94].

Many systems have also been developed and tested in Europe. The setup of all the systems was very similar. As an example of all stationary systems implemented, the regenerative PHOEBUS system at Forschungszentrum Jülich will be described in detail. The idea was to store the surplus electrical energy produced by a solar field on the library building in the summer months for use in winter when radiation rates are lower (Figures 8.15 and 8.16).

Lead acid batteries were used for short-time energy storage. For the seasonal storage, the electrical energy is converted in an alkaline electrolyzer with 30% KOH solution operated at 80 °C and a pressure of 700 kPa. After electrolysis, the hydrogen thus produced is compressed to a pressure of 15 MPa and oxygen is compressed

Figure 8.15 Solar field on library building at Forschungszentrum Jülich [19].

Figure 8.16 Electrolyzer, fuel cell, and hybridization batteries at Forschungszentrum Jülich [19].

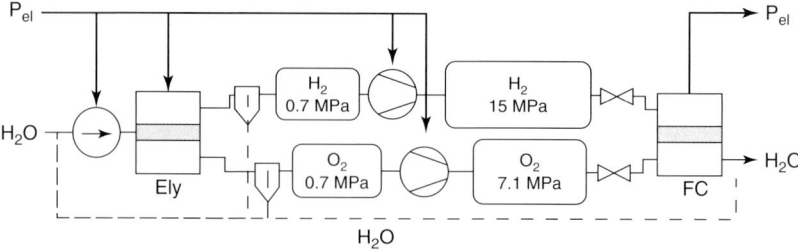

Figure 8.17 Regenerative PEM fuel cell system with separate electrolyzer and fuel cell.

Table 8.11 Specifications of PHOEBUS.

Photovoltaic	43 kW$_{peak}$
Electrolyzer	26 kW at 0.7 MPa
PEMFC	5.6 kW
Batteries	300 kWh
H$_2$ tank	25 m^3 at 15 MPa
O$_2$ tank	20 m^3 at 7.1 MPa

to a pressure of 7.1 MPa. Pneumatically driven piston compressors perform this compression. No figures are available for the energy demand of the compressors, but Meurer gave the proportion of compressor power consumption in relation to the energy content of the hydrogen as 15% [95]. A flow chart of the plant is shown in Figure 8.17. Technical data for the key components are shown in Table 8.11.

Initially, an AFC with a maximum power of 6.5 kW was used in the system. However, owing to problems with automatic operation, the complex design and losses of the liquid electrolyte, a 5 kW PEMFC stack was later integrated into the system [19, 96].

Starting in 1993, the system was operated for more than 6 years. Some advanced concepts with the aim of increasing the efficiency were developed on the basis of the operating experience gained with the PHOEBUS system. In order to reduce electrical losses, the system configuration was modified so that it could be operated with fewer converters. Owing to the high power consumption of the compressors, work was started on developing a high-pressure electrolyzer for 12 MPa. The advantage is that the pressure can be applied to the liquid feed water, which consumes only a fraction of the energy needed by mechanical compression [97–100]. A study of the cost optimization of the PHOEBUS system was made by Ghosh [37].

Another regenerative system was also tested in Neunburg vorm Wald, Germany [20]. Similar components were tested in this project, but the focus was on the supply of hydrogen for different applications and not on the seasonal storage of energy.

Similar autonomous energy supply systems with renewable energy have been implemented in Saudi Arabia as a part of the German–Saudi Hysolar Program. Near Riyadh, a 350 kW solar hydrogen production demonstration plant was tested

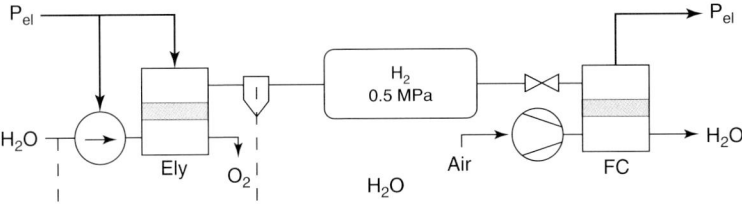

Figure 8.18 Simple RFC system.

[101]. A system with a bipolar alkaline 5 kW$_{peak}$ electrolyzer and a 3 kW PEMFC was also tested in Italy [102].

Alkaline electrolyzers are used in most of the systems described above. Vanhanen et al. [103] described a simple RFC system based on a PEM electrolyzer and a PEMFC (Figure 8.18). The power of the electrolyzer was only 30 W. Furthermore, a metal hydride hydrogen store is used instead of high-pressure tanks. The store can be charged with a pressure of 5–10 bar and discharged at ambient pressure. The round-trip efficiency of this RFC system is 30%. An efficiency of 40% should be possible if the components are further improved.

Proton Energy Systems developed the UNIGEN RFC for uninterruptible power supply with a storage capacity of 50 h [104]. The system consists of four power-generating modules of the Ballard NEXA type of 1.2 kW each and one low-pressure (1.72 MPa) and one high-pressure (13.8 MPa) hydrogen-generating module. In 2007, Bergen et al. described the development of an RFC system based on commercially available components. The system's setup is similar to that of the PHOEBUS module. A Stuart Energy 6 kW electrolyzer is used for hydrogen generation. The FC is a Ballard NEXA PEM of 1.2 kW. Hydrogen can be stored in a compressed form in Dynatec composite tanks or in metal hydrides manufactured by Ovionics and Palcan [105].

8.8.2
RFC Systems for Aviation Applications

A great deal of research effort has been devoted by NASA to developing FCs and reversible FCs for space applications. One of the first operational areas for FC systems was the supply of electrical energy to spacecraft. As mentioned in Section 8.3, all US spacecraft (Gemini, Apollo, Space Shuttle) are powered by PEMFCs or AFCs [106]. The need for and the advantages of RFCs for space applications can be summarized by light weight and compactness, very long cycle life, and in-orbit maintainability. The combination of FC and electrolyzer is seen as a long-term solution to the problem of energy storage in space [107].

The PEM offers major advantages compared with the alkaline technologies. The advantages are enhanced safety, longer life, lower weight, improved reliability and maintainability, higher peak to nominal power capability, and compatibility with propulsion-grade reactants [108]. Compared with batteries, RFCs have an advantage in weight but no advantage in the system volume [14].

Owing to the ongoing development of RFC technology, it was possible to improve the FC system so that such a system was integrated into the solar-powered lightweight Helios aircraft. A record flight height of 30 km was achieved with this unmanned aircraft and it is hoped to replace, for example, telecommunication satellites by aircraft of this type. The RFC has a power of 18 kW and the idea is to store the surplus electrical energy generated by the photovoltaic system by day for operation during the night. Owing to problems with the remote controls, the plane crashed during a test flight [35]. In spite of the crash, the solar electric aircraft from NASA represents a development for the future, so that similar aircraft can, for example, fly over Mars powered by RFCs [109, 110].

8.9
Conclusion and Prospects

RFC technology has proven its functionality and durability in different demonstration projects. For stationary systems, the advantage of the RFC is its ability to provide long-term energy storage with low losses in the stored media. For mobile applications, especially in avionics, the low mass per kWh of stored energy is advantageous.

The efficiencies of alkaline- and PEM-based systems are similar. The advantages of the PEM systems are the higher power output and improved safety due to the solid electrolyte. The SOE-based systems have shown the highest efficiency (65% at 1 A cm^{-2}), but the technology suffers from low dynamic operation and material problems at high temperatures. These systems are also complex owing to the heat exchangers needed to evaporate the water and to preheat the supplies.

The URFC represents a special configuration of regenerative systems. In such systems, the conversion of electrical energy into chemical energy and the reverse conversion occur in the same electrochemical device. This reduces the number of core components, but the additional expense for system components such as valves and inertization has to be taken into account. The stress on the electrodes in URFCs is higher, which reduces their durability. On the other hand, saving one electrochemical device could be advantageous in terms of investment costs.

The highest power densities can be achieved in SOE RFCs with more than 1 W cm^{-2}. For PEM RFCs, the power densities achieved are also in the region of 1 W cm^{-2}, but the efficiency is only about 40%.

The future energy supply must become more and more independent of fossil fuels and it will be based on regenerative energy sources with discontinuous operation. Therefore, the need to store electrical energy will grow. RFCs are especially attractive for off-grid applications owing to their high energy densities. If further space exploration programs are started, this could be a great opportunity to develop new RFC systems with improved efficiency and with mass-reduced storage technology. If costs can additionally be decreased, the RFCs will also be an interesting option for future terrestrial applications.

References

1. Beck, F. and Rüetschi, P. (2000) Rechargeable batteries with aqueous electrolytes. *Electrochim. Acta*, **45**, 2467–2482.
2. Mitlitsky, F., Myers, B., Weisberg, A.H., Molter, T.M., and Smith, W.F. (1999) Reversible (unitized) PEM fuel cell devices. *Fuel Cell Bull.*, **11**, 6–11.
3. Smith, W. (2000) The role of fuel cells in energy storage. *J. Power Sources*, **86**, 74–83.
4. Perrin, M., Malbranche, P., Lemaire-Potteau, E., Willer, B., Soria, M.L., Jossen, A., Dahlen, M., Ruddell, A., Cyphelly, I., Semrau, G., Sauer, D.U., and Sarre, G. (2006) Comparison for nine storage technologies results from the INVESTIRE network. *J. Power Sources*, **154**, 545–549.
5. Dutta, S. (1990) Technology assessment of advanced electric hydrogen production. *Int. J. Hydrogen Energy*, **6**, 379–386.
6. Jörissen, L. (2006) Bifunctional oxygen/air electrodes. *J. Power Sources*, **155**, 23–32.
7. Yim, S.-D., Park, G.-G., Sohn, Y.-J., Lee, W.-Y., Yoon, Y.-G., Yang, T.-H., Um, S., Yu, S.-P., and Kim, C.-S. (2005) Optimization of PtIr electrocatalyst for PEM URFC. *Int. J. Hydrogen Energy*, **30**, 1345–1350.
8. Sherif, S.A., Barbir, F., and Veziroglu, T.N. (2005) Wind energy and the hydrogen economy – review of the technology. *Solar Energy*, **78**, 647–660.
9. Dohmen, A. (1917) Apparatus for the electrolytic decomposition of water. US Patent 1,211,687.
10. Noeggerath, J.E. (1930) Elektrolyseur, insbesondere Druckelektrolyseur. German Patent DE 508480.
11. Bacon, F.T. (1969) Fuel cells, past, present and future. *Electrochim. Acta*, **14**, 569–585.
12. Niederreither, H. (1932) Verfahren zum Speichern und Verteilen elektrischer Energie. German Patent DE 648941.
13. Niederreither, H. (1933) Method of producing, storing, and distributing electrical energy by operating gas batteries, particulary oxy-hydrogen gas batteries and electrolyzers. US Patent 2,070,612.
14. Burke, K.A. (2003) Fuel Cells for Space Science Applications. Technical Memorandum NASA/TM-2003-212730, NASA.
15. Swette, L. and Giner, J. (1988) Oxygen electrodes for rechargeable alkaline fuel cells. *J. Power Sources*, **22**, 399–408.
16. Swette, L., Kackley, N., and McCatty, S.A. (1991) Oxygen electrodes for rechargeable alkaline fuel cells. *J. Power Sources*, **36**, 323–339.
17. Stolten, D. (ed.) (2010) *Hydrogen Energy*, Wiley-VCH Verlag GmbH, Weinheim.
18. Swette, L., LaConti, A.B., and McCatty, S.A. (1994) Proton-exchange membrane regenerative fuel cells. *J. Power Sources*, **47**, 343–351.
19. Schucan, T. (2000) Case Studies of Integrated Hydrogen Energy Systems. Final Report of Subtask A: Case Studies of Integrated Hydrogen Energy Systems, International Energy Agency, Paris.
20. Szyszka, A. (1998) Ten years of solar hydrogen demonstration project at Neunburg vorm Wald, Germany. *Int. J. Hydrogen Energy*, **10**, 849–860.
21. Olah, G.A., Goeppert, A., and Prakash, G.K.S. (2005) *Beyond Oil and Gas: the Methanol Economy*, Wiley-VCH Verlag GmbH, Weinheim.
22. Gunn, R. and Hall, W.C. (1938) Fuel cell. US Patent 2,384,463.
23. Grubb, W.T. (1959) Fuel cell. US Patent 2,913,511.
24. Niedrach, L.W. (1967) Electrode structure and fuel cell incorporating the same. US Patent 3,297,484.
25. Appleby, A.J. and Yeager, E.B. (1986) Solid polymer electrolyte fuel cells (SPEFCs). *Energy*, **11**, 137–152.
26. Ludwig, F.A. (1960) Energy conversion cell. US Patent 3,132,972.
27. Kordesch, K.V. (1978) 25 years of fuel cell development (1951–1976). *J. Electrochem. Soc.*, **125**, 77C–91C.

28. Webb, J.E., Wilner, M., Frank, H.A., and Klein, M.G. (1970) Electrochemically regenerative hydrogen–oxygen fuel cell. US Patent 3,507,704.
29. Bloomfield, D.P., Hassett, N.A., and Stedman, J.K. (1974) Regenerative fuel cell. US Patent 3,839,091.
30. Stedman, J.K. (1976) Regenerative fuel cell. US Patent 3,981,745.
31. Erdle, E., Dönitz, W., Schamm, R., and Koch, A. (1992) Reversibility and polarization behaviour of high temperature solid oxide electrochemical cells. *Int. J. Hydrogen Energy*, **10**, 817–819.
32. Dönitz, W., Dietrich, G., Erdle, E., and Streicher, R. (1988) Electrochemical high temperature technology for hydrogen production or direct electricity generation. *Int. J. Hydrogen Energy*, **5**, 283–287.
33. Dönitz, W. and Erdle, E. (1985) High-temperature electrolysis of water vapor – status of development and perspectives for application. *Int. J. Hydrogen Energy*, **5**, 291–295.
34. Isenberg, A.O. (1981) Energy conversion via solid oxide electrolyte electrochemical cells at high temperatures. *Solid State Ionics*, **3/4**, 431–437.
35. Noll, T.E., Brown, J.M., Perez-Davis, M.E., Ishmael, S.D., Tiffany, G.C., and Gaier, M. (2004) Investigation of the Helios Prototype Aircraft Mishap. Volume I. Mishap Report, NASA.
36. Baehr, H.D. (1988) *Thermodynamik*, Springer, Berlin.
37. Ghosh, P.C. (2003) Cost optimisation of a self-sufficient hydrogen based energy supply system. PhD thesis, Aachen University. Berichte des Forschungszentrums Jülich 4049.
38. Williams, M.C. (2004) *Fuel Cell Handbook*, US Department of Energy, Morgantown, WV.
39. Kerres, J., Eigenberger, G., Reichle, S., Schramm, V., Hetzel, K., Schnurnberger, W., and Seybold, I. (1996) Advanced alkaline electrolysis with porous polymeric diaphragms. *Desalination*, **104**, 47–57.
40. Bidault, F., Brett, D.J.L., Middleton, P.H., and Brandon, N.P. (2009) Review of gas diffusion cathodes for alkaline fuel cells. *J. Power Sources*, **187**, 39–48.
41. Singer, J. and Srinivasan, V. (1985) Evaluation Parameters for the Alkaline Fuel Cell Oxygen Electrode, Technical Memorandum NASA/TM-87155, NASA.
42. Hu, W. (2000) Electrocatalytic properties of new electrocatalysts for hydrogen evolution in alkaline water electrolysis. *Int. J. Hydrogen Energy*, **25**, 111–118.
43. Divisek, J., Schmitz, H., and Steffen, B. (1994) Electrocatalyst materials for hydrogen evolution. *Electrochim. Acta*, **11/12**, 1723–1731.
44. Cappadonia, M., Divisek, J., von der Heyden, T., and Stimming, U. (1994) Oxygen evolution at nickel anodes in concentrated alkaline solutions. *Electrochim. Acta*, **11/12**, 1559–1564.
45. Boggio, R., Carugati, A., and Trasatti, S. (1987) Electrochemical surface properties of Co_3O_4 electrodes. *J. Appl. Electrochem.*, **17**, 828–840.
46. Wendt, H. and Imarisio, G. (1988) Nine years of research and development on advanced water electrolysis. A review of the research programme of the Commission of the European Communities. *J. Appl. Electrochem.*, **18**, 1–14.
47. Ham, D.O., Moniz, G., and Taylor, E.J. (1988) High surface area dual function oxygen electrocatalysts for space power applications. *J. Power Sources*, **22**, 409–420.
48. Angelo, A.C.D., Gonzalez, E.R., and Avaca, L.A. (1991) Mechanistc studies of the oxygen reactions on $NiCo_2O_4$ spinel and the hydrogen evolution reaction on amorphous Ni–Co sulphide. *Int. J. Hydrogen Energy*, **16**, 1–7.
49. Ahn, J. (1991) Ein neues Konzept bifunktioneller Elektroden für eine integrierte Wasserelektrolyse- und H_2/O_2−Brennstoffzelle mit Polymerelektrolyt. PhD thesis, Oldenburg University.
50. Lu, P.W.T. and Srinivasan, S. (1979) Advances in water electrolysis technology with emphasis on use of the solide polymer electrolyte. *J. Appl. Electrochem.*, **9**, 269–283.
51. Millet, P., Ngameni, R., Grigoriev, S.A., Mbemba, N., Brisset, F., Ranjbari, A.,

and Etiévant, C. (2010) PEM water electrolyzers: from electrocatalysis to stack development. *Int. J. Hydrogen Energy*, **35**, 5043–5052.
52. Ledjeff, K., Heinzel, A., Peinecke, V., and Mahlendorf, F. (1994) Development of pressure electrolyser and fuel cell with polymer electrolyte. *Int. J. Hydrogen Energy*, **19**, 453–455.
53. Ledjeff-Hey, K. and Heinzel, A. (1996) Critical issues and future prospects for solid polymer fuel cells. *J. Power Sources*, **61**, 125–127.
54. Gasteiger, H., Baker, D.R., Carter, R.N., Gu, W., Liu, Y., Wagner, F.T., and Yu, P.T. (2010) *Electrocatalysis and Catalyst Degradation Challenges in Proton Exchange Membrane Fuel Cells*, Wiley-VCH Verlag GmbH, Weinheim.
55. Wittstadt, U., Wagner, E., and Jungmann, T. (2005) Membrane electrode assemblies for unitised regenerative polymer electrolyte fuel cells. *J. Power Sources*, **145**, 555–562.
56. Grootjes, A.J. and Makkus, R.C. (2004) Development of an oxygen electrode for a unitized regenerative PEM fuel cell. Poster presentation at Fuel Cell Seminar 2004, San Antonio, TX.
57. Shao, Z.-G., Yi, B.-L., and Han, M. (1999) Bifunctional electrodes with a thin catalyst layer for 'unitized' proton exchange membrane regenerative fuel cell. *J. Power Sources*, **79**, 82–85.
58. Jung, H.-Y., Park, S., and Popov, B.N. (2009) Electrochemical studies of an unsupported PtIr electrocatalyst as a bifunctional oxygen electrode in a unitized regenerative fuel cell. *J. Power Sources*, **191**, 3 57–361.
59. Ioroi, T., Kitazawa, N., Yasuda, K., Yamamoto, Y., and Takenaka, H. (2001) IrO_2–deposited Pt electrocatalysts for unitized regenerative polymer electrolyte fuel cells. *J. Appl. Electrochem.*, **31**, 1179–1183.
60. Ioroi, T., Kitazawa, N., Yasuda, K., Yamamoto, Y., and Takenaka, H. (2000) Iridium oxide/platinum electrocatalysts for unitized regenerative polymer electrolyte fuel cells. *J. Electrochem. Soc.*, **147**, 2018–2022.
61. Ioroi, T., Oku, T., Yasuda, K., Kumagai, N., and Miyazaki, Y. (2003) Influence of PTFE coating on gas diffusion backing for unitized regenerative polymer electrolyte fuel cells. *J. Power Sources*, **124**, 385–389.
62. Ioroi, T., Yasuda, K., Siroma, Z., Fujiwara, N., and Miyazaki, Y. (2002) Thin film electrocatalyst layer for unitized regenerative polymer electrolyte fuel cells. *J. Power Sources*, **112**, 583–587.
63. Doddathimmaiah, A. and Andrews, J. (2009) Theory, modelling and performance measurement of unitised regenerative fuel cells. *Int. J. Hydrogen Energy*, **34**, 8157–8170.
64. Altmann, S., Kaz, T., and Friedrich, K.A. (2009) Development of bifunctional electrodes for closed-loop fuel cell applications. Presented at the 216th ECS Meeting, Vienna.
65. Yao, W., Yang, J., Wang, J., and Nuli, Y. (2007) Chemical deposition of platinum nanoparticles on iridium oxide for oxygen electrode of unitized regenerative fuel cell. *Electrochem. Commun.*, **9**, 1029–1034.
66. Kato, A., Masuda, M., Takahashi, A., Ioroi, T., Yoshida, T., Ito, H., and Tetsuhiko, M. (2008) Development of an unitized reversible cell. Abstracts of the 17th World Hydrogen Conference, Brisbane.
67. Grigoriev, S.A., Millet, P., Porembsky, V.I., and Fateev, V.N. (2011) Development and preliminary testing of a unitized regenerative fuel cell based on PEM technology. *Int. J. Hydrogen Energy*, **36**, 4164–4168.
68. Grigoriev, S.A., Millet, P., Dzhus, K.A., Middleton, H., Saetre, T.O., and Fateev, V.N. (2010) Design and characterization of bi-functional electrocatalytic layers for application in PEM unitized regenerative fuel cells. *Int. J. Hydrogen Energy*, **35**, 5070–5076.
69. Ahn, J. and Holze, R. (1992) Bifunctional electrodes for an integrated water-electrolysis and hydrogen–oxygen fuel cell with a solid polymer electrolyte. *J. Appl. Electrochem.*, **22**, 1167–1174.
70. Ahn, J. and Ledjeff, K. (1991) Verfahren zur Energiespeicherung und

Energiewandlung. German Patent DE 4027655.

71. Holze, R. and Ahn, J. (1992) Advances in the use of perfluorinated cation exchange membranes in integrated water electrolysis and hydrogen/oxygen fuel cell systems. *J. Membr. Sci.*, **73**, 87–97.

72. Ledjeff, K., Mahlendorf, F., Peinecke, V., and Heinzel, A. (1995) Development of electrode/membrane units for the reversible solid polymer fuel cell (RSPFC). *Electrochim. Acta*, **3**, 315–319.

73. Zhang, H., Lin, G., and Chen, J. (2010) Evaluation and calculation on the efficiency of a water electrolysis system for hydrogen production. *Int. J. Hydrogen Energy*, **35**, 10851–10858.

74. Yu, B., Zhang, W.-Q., Xu, J.-M., and Chen, J. (2010) Status and research of highly efficient hydrogen production through high temperature steam electrolysis at INET. *Int. J. Hydrogen Energy*, **35**, 2829–2835.

75. Shin, Y., Park, W., Chang, J., and Park, J. (2007) Evaluation of high temperature electrolysis of steam to produce hydrogen. *Int. J. Hydrogen Energy*, **32**, 1486–1491.

76. de Haart, L.G.J., Mayer, K., Stimming, U., and Vinke, I.C. (1998) Operation of anode-supported thin electrolyte film solid oxide fuel cells at 800 °C and below. *J. Power Sources*, **71**, 302–305.

77. Gebert, A. and Krieg, T. (1998) Regenerative elektrochemische Brennstoffzelle. German Patent DE 19731096.

78. Guan, J., Minh, N., Ramamurthi, B., Ruud, J., Hong, J.-K., Riley, P., and Weng, D. (2007) High Performance Flexible Reversible Solid Oxide Fuel Cell. Final Technical Report, DE-FC36-04GO14351, GE Global Research Center.

79. Mogensen, M., Jensen, S.H., Hauch, A., Chorkendorff, I., and Jacobsen, T. (2006) Performance of reversible solid oxide cells: a review. Proceedings, 7th European Fuel Cell Forum 2006, Lucerne.

80. Sprouse, K.M. (1994) Regenerative fuel cell system. US Patent 5,306,577.

81. McElroy, J.F., Chludzinski, P.J., and Dantowitz, P. (1987) Fuel cell power supply with oxidant and fuel gas switching. US Patent 4,657,829.

82. Ovshinsky, S.R., Venkatesan, S., and Corrigan, D.A. (2004) The ovionic regenerative fuel cell, a fundamentally new approach. Abstracts, Hydrogen and Fuel Cells, Toronto.

83. Menjak, Z., Venkatesan, S., Aladjov, B., Gradinarova, L.M., Wang, H.; Ovshinsky, S.R., and Dhar, S.K. (2006) Regenerative bipolar fuel cell. US Patent 7,014,953.

84. Jung, H.-Y., Huang, S.-Y., and Popov, B.N. (2010) High-durability titanium bipolar plate modified by electrochemical deposition of platinum for unitized regenerative fuel cell (URFC). *J. Power Sources*, **195**, 1950–1956.

85. Jung, H.-Y., Huang, S.-Y., Ganesan, P., and Popov, B.N. (2009) Performance of gold-coated titanium bipolar plates in unitized regenerative fuel cell operation. *J. Power Sources*, **194**, 972–975.

86. Maclay, J.D., Brouwer, J., and Samuelsen, G.S. (2007) Dynamic modeling of hybrid energy storage systems coupled to photovoltaic generation in residential applications. *J. Power Sources*, **163**, 916–925.

87. Levene, J., Kroposki, B., and Sverdrup, G. (2006) Wind energy and production of hydrogen and electricity – opportunities for renewable hydrogen. Presented at POWER-GEN Renewable Energy and Fuels Technical Conference, Las Vegas.

88. Floch, P.-H., Gabriel, S., Mansilla, C., and Werkoff, F. (2007) On the production of hydrogen via alkaline electrolysis during off-peak periods. *Int. J. Hydrogen Energy*, **32**, 4641–4647.

89. Barbir, F., Molter, T., and Dalton, L. (2005) Efficiency and weight trade-off analysis of regenerative fuel cells as energy storage for aerospace applications. *Int. J. Hydrogen Energy*, **30**, 351–357.

90. Li, X.-J., Yu, X., Shao, Z.-G., and Yi, B.-L. (2010) Mass minimization of discrete regenerative fuel cell (RFC) system for on-board energy storage. *J. Power Sources*, **195**, 4811–4815.

91. Burke, K.A. (2003) Unitized Regenerative Fuel Cell System Development. Technical Memorandum NASA/TM-2003-212739, NASA.
92. Burke, K.A. (2005) Unitized Regenerative Fuel Cell System Gas Storage-Radiator Development. Technical Memorandum NASA/TM2005-213442, NASA.
93. Baldwin, R., Pham, M., Leonida, A., McElroy, J., and Nalette, T. (1990) Hydrogen–oxygen proton-exchange membrane fuel cells and electrolyzers. *J. Power Sources*, **29**, 399–412.
94. Lehman, P.A., Chamberlin, C.E., Pauletto, G., and Rocheleau, M.A. (1997) Operating experience with a photovoltaic–hydrogen energy system. *Int. J. Hydrogen Energy*, **22**, 465–470.
95. Meurer, C., Barthels, H., Brocke, W.A., Emonts, B., and Groehn, H.G. (1999) PHOEBUS – an autonomous supply system with renewable energy: six years of operational experience and advanced concepts. *Solar Energy*, **67**, 131–138.
96. Barthels, H., Brocke, W.A., Bonhoff, K., Groehn, H.G., Heuts, G., Lenartz, M., Mai, H., Mergel, J., Schmid, L., and Ritzenhoff, P. (1998) PHOEBUS-Jülich: an autonomous energy supply system comprising photovoltaics, electrolytic hydrogen, fuel cell. *Int. J. Hydrogen Energy*, **23**, 295–301.
97. Schnurnberger, W., Hug, W., and Peinecke, V. (1995) Fortgeschrittene Elektrolysetechniken zur Wasserstoffherstellung im intermittierenden Betrieb. *Chem. Ing. Tech.*, **67**, 1320–1323.
98. Emonts, B. (2001) PHOEBUS. Wissenschftlicher Ergebnisbericht/Scientific Report, Forschungszentrum Jülich.
99. Emonts, B., Janssen, H., and Stolten, D. (2008) 10 Jahre PHOEBUS-Projekt. Presentation, Forschungszentrum Jülich GmbH.
100. Barbir, F. (2005) PEM electrolysis for production of hydrogen from renewable energy sources. *Solar Energy*, **78**, 661–669.
101. Abaoud, H. and Steeb, H. (1998) The German–Saudi Hysolar Program. *Int. J. Hydrogen Energy*, **23**, 445–449.
102. Galli, S. and Stefanoni, M. (1997) Development of a solar-hydrogen cycle in Italy. *Int. J. Hydrogen Energy*, **5**, 453–458.
103. Vanhanen, J.P., Lund, P.D., and Tolonen, J.S. (1998) Electrolyser–metal hydride-fuel cell system for seasonal energy storage. *Int. J. Hydrogen Energy*, **4**, 267–271.
104. Porter, S. (2004) Unigen regenerative fuel cell for uninterruptible power supply. Proceedings, Annual Merit Review, Philadelphia.
105. Bergen, A., Schmeister, T., Pitt, L., Rowe, A., Djilali, N., and Wild, P. (2007) Development of dynamic regenerative fuel cell system. *J. Power Sources*, **164**, 624–630.
106. Warshay, M. and Prokopius, P.R. (1989) The Fuel Cell in Space: Yesterday, Today and Tomorrow, NASA Technical Memorandum 102366, NASA.
107. Appleby, A.J. (1988) Regenerative fuel cells for space applications. *J. Power Sources*, **22**, 377–385.
108. Pérez, M.E., Loyselle, P.L., Hoberecht, M.A., Manzo, M.A., Kohout, L.L., Burke, K.A., and Cabrera, C.R. (2001) Energy Storage for Aerospace Applications. Technical Memorandum NASA/TM-2001-211068, NASA.
109. Bents, D., Scullin, V., Chang, B., Johnson, D., and Garcia, C. (2005) Hydrogen–Oxygen PEM Regenerative Fuel Cell Energy Storage System. Technical Memorandum NASA/TM-2005-213381, NASA.
110. McElroy, J.F. (1992) SPE Water Electrolyzers in Support of the Lunar Outpost. Technical Report 19930018788, NASA.

Part II
Materials and Production Processes

9
Advances in Solid Oxide Fuel Cell Development Between 1995 and 2010 at Forschungszentrum Jülich GmbH, Germany

Vincent Haanappel

9.1
Introduction

This chapter presents an overview of the main advances in solid oxide fuel cells (SOFCs) regarding research and development (R&D), measurement standards, and quality assurance (QA) in SOFC testing at the Forschungszentrum Jülich (FZJ). Starting from the mid-1990s, significant improvements in electrochemical performance were achieved, fulfilling targets such as lowering the operating temperature, increasing durability and reliability, increasing power density, and lowering production costs. Development work at FZJ was mainly focused on two types of anode-supported cells (ASCs), one with an $La_xSr_yMn_zO_3$ [lanthanum strontium manganite (LSM)] cathode and the other with an $La_xSr_yCo_z(Fe_{1-z})O_3$ [lanthanum strontium cobaltite ferrite (LSC(F))] cathode. In 2010, the performance of cells with an LSM cathode was about five times higher than that achieved in 1995, and the increase in performance of cells with an LSC(F) cathode was even higher. Details of the improvements in the production techniques, processing conditions, and material and microstructural parameters for cells with an LSM or LSC(F) cathode are outlined in Sections 9.2.1 and 9.2.2, respectively. In this period, other types of cathode materials were also developed and tested. Some of them will also be discussed in the following sections; however, the performance of these cells was not always satisfactory.

During this period, FZJ also started to develop its own standardization and QA in SOFC testing with the objective of obtaining performance measurement data of good accuracy and precision in a consistent, reliable, and repeatable manner. All aspects were systematically examined, including the physical experimental set-up, SOFC cell performance, and SOFC testing procedure. Parts of this standardization and QA are described in Section 9.2.3. All these aspects of R&D, standardization and QA finally led to excellent performance data with one type of cell resulting with a theoretical output of more than 4.4 A cm^{-2} at 800 °C and 700 mV, hence more than 3 W cm^{-2}. All experimental data from the tested cells were compared with each other, with the main key parameters to characterize the measured performance, the current density, and/or area specific resistance (ASR) at a defined cell output

---

Fuel Cell Science and Engineering: Materials, Processes, Systems and Technology,
First Edition. Edited by Detlef Stolten and Bernd Emonts.
© 2012 Wiley-VCH Verlag GmbH & Co. KGaA. Published 2012 by Wiley-VCH Verlag GmbH & Co. KGaA.

voltage. Only during the early years was the overpotential η the main parameter to compare experimental results from cells and half-cells with each other: the lower the overpotential, the better the performance. Nowadays in the literature, only current densities and ASRs are reported to represent electrochemical performance of tested cells. In the following, the key parameter current density is utilized for the electrochemical characterization of different types of cells.

9.2
Advances in Research, Development, and Testing of Single Cells

9.2.1
SOFCs with an LSM Cathode

From the beginning in the mid-1990s, FZJ intensified the production and testing of SOFCs. Since the operating temperatures of electrolyte-supported cells (ESCs) were about 1000 °C, R&D targets were to lower them to below 900 °C. Regarding these ECSs, a systematic approach was initiated to develop electrodes with high catalytic activity for the electrochemical reactions. In other words, the main objective was to reduce the overpotentials and thus to improve performance. During that period, experiments were based on testing small half-cells with a diameter of about 2 cm and with a cathode surface area of around 0.75 cm² [1]. The most common materials used for SOFCs were lanthanum–strontium–manganese cathodes and yttria-stabilized zirconia electrolyte. However, this combination led to various undesirable interactions between the electrode material and the underlying electrolyte, detrimentally influencing the overpotential. The first experiments dealt with ESCs with the following cathode materials: $La_{0.84}Sr_{0.16}MnO_3$, $La_{0.79}Sr_{0.16}Mn_{1-\mu}Co_\mu O_3$, and addition of noble metals to these cathode materials. A shift from electrolyte-supported to anode substrate-supported cells (ASCs) during the end of the 1990s resulted in a further improvement in the performance of SOFCs produced with thinner electrolytes, and as a result lower overpotentials. The first ASC-type single cells being measured were as large as 10×10 cm² and showed a current density of 0.3 A cm^{-2}. Results from the first single-cell measurements performed at FZJ were published in 1998 [2]. Nowadays, ASCs with $La_{0.65}Sr_{0.30}MnO_3$ are being used as the standard-type cells with an LSM cathode. Regarding these LSM-type perovskites, significant progress was made regarding cell performance. Figure 9.1 shows the status of the development of this type of cell with an LSM cathode at FZJ from 1995 onwards.

9.2.1.1 1995–1998
The first FZJ measurements at relatively high temperatures, that is, 950 °C, involved ESCs. During this period, optimization of the cathode material was started. However, the use of relatively thick electrolytes required a minimum realistic operating temperature of ∼900 °C as the electrolyte resistance to ion transport is then the performance bottleneck.

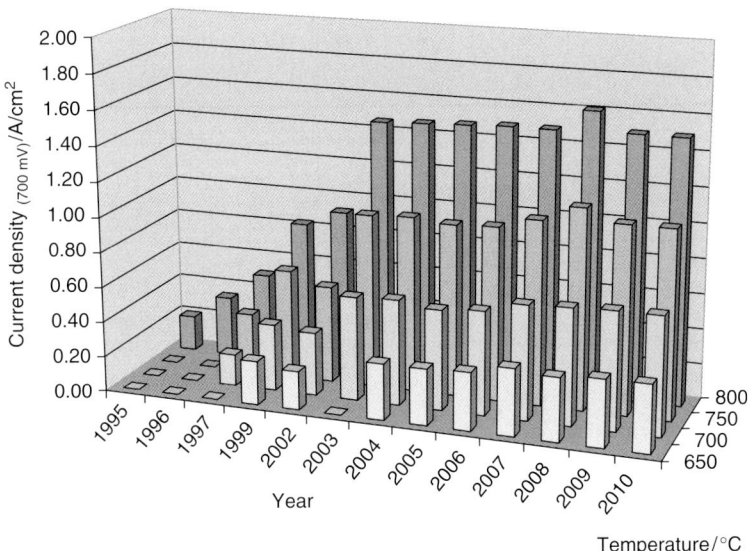

Figure 9.1 Status of development of SOFCs with LSM cathode at FZJ.

These first experiments were based on symmetric cells with an electrode diameter of 10 mm, so-called *button cells*. The electrolyte was of the yttria-stabilized zirconia (YSZ) type, 8YSZ based on 8 mol% yttria-stabilized zirconia, with a thickness between 130 and 150 μm. On both sides, the electrode was applied by screen-printing or wet powder spraying (WPS) with a thickness of 50 μm, and the porosity was about 30%. A schematic view of a cross-section of the cell geometry is depicted in Figure 9.2. Various types of cathodes were screened by potentiodynamic current–potential measurements. Comparison of the electrochemical behavior in relation to material composition was based on the measured current density at an overpotential η of -0.1 V.

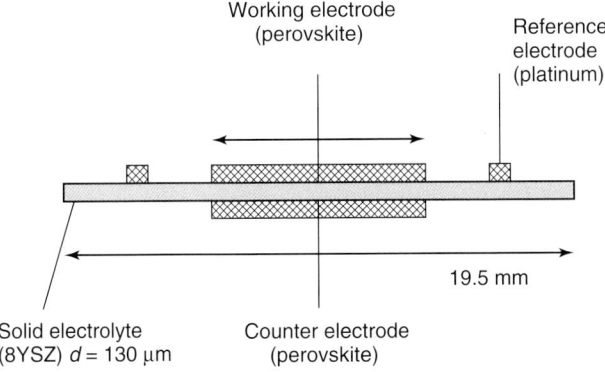

Figure 9.2 Schematic view of the cross-section of the measuring cell [1].

Table 9.1 Current density and apparent activation energy of various types of cathode materials at $\eta = -0.1$ V, $T = 950\ °C$, $p_{O_2} = 20$ kPa [3].

Cathode material	Current density (A cm^{-2})	Activation energy, E_a (kJ mol^{-1})
$La_{0.84}Sr_{0.16}MnO_3$	0.028	214
$La_{0.84}Sr_{0.16}MnO_3 + Pd$	0.130	138
$La_{0.84}Sr_{0.16}MnO_3 + Pt$	0.040	173
$La_{0.84}Sr_{0.16}MnO_3 + Pd/Pt$	0.320	140
$La_{0.79}Sr_{0.16}MnO_3$	0.075	223
$La_{0.79}Sr_{0.16}Co_{0.1}Mn_{0.8}O_3$	0.079	138
$La_{0.79}Sr_{0.16}Co_{0.2}Mn_{0.8}O_3$	0.115	111
$La_{0.79}Sr_{0.16}Co_{0.2}Mn_{0.8}O_3 + Pd$	0.156	143
$La_{0.79}Sr_{0.16}Co_{0.2}Mn_{0.8}O_3 + Pd/Pt$	0.282	140
$La_{0.65}Sr_{0.30}MnO_3$	0.088	165
$La_{0.65}Sr_{0.30}MnO_3$–8YSZ composite	0.360	195
$La_{0.65}Sr_{0.30}MnO_3$–8YSZ composite + Pd	0.420	182

These current densities and the apparent activation energies as a function of the tested materials are given in Table 9.1 [3]. As shown in this table and in the case of undoped LSM-type cathode material, the lowest apparent activation energy and the highest current density were obtained with $La_{0.65}Sr_{0.30}MnO_3$. This composition was the result of a systematic approach at FZJ regarding optimization of electrical conductivity and reactivity with the adjacent electrolyte layer [4–6]. In addition to these compositions, other types of electrodes were also tested, namely $Pr_{1-x}Sr_xMn_{0.80}Co_{0.20}O_3$ [praseodymium strontium manganite cobaltite (PSMC) with $x = 0.2, 0.25$, and 0.3] and $La_{0.65}Sr_{0.30}Fe_{0.80}Co_{0.20}O_3$ [lanthanum strontium ferrite cobaltite (LSFC)]. Characterization of half-cells with PSMC revealed low performance due to insufficient adhesion between the electrolyte and the cathode. With the PSMC and LSFC cathodes, the formation of low-conductive $La_2Zr_2O_7$ and $SrZrO_3$ layers between the electrolyte and cathode during sintering resulted in lower performance. In the case of $La_{0.65}Sr_{0.30}MnO_3$, the thermal expansion coefficient (TEC) was the lowest and approached that of the 8YSZ electrolyte [7]. Hence $La_{0.65}Sr_{0.30}MnO_3$ was chosen as the standard cathode material for SOFCs. A composition with 8YSZ increased the performance further, owing to an increase in the so-called three-phase boundaries (TPBs). It is suggested that the oxidation reactions take place mainly at these TPBs.

9.2.1.2 1998–2002

Whereas in the mid-1990s cells were based on a relatively thick electrolyte foil with a current density at $800\ °C$ and 700 mV of about 0.20 A cm^{-2} (see also Figure 9.1), an increase in performance was obtained with the introduction of the ASC, also called the *Jülich substrate concept* [8] (see Figure 9.3). Owing to the introduction of this

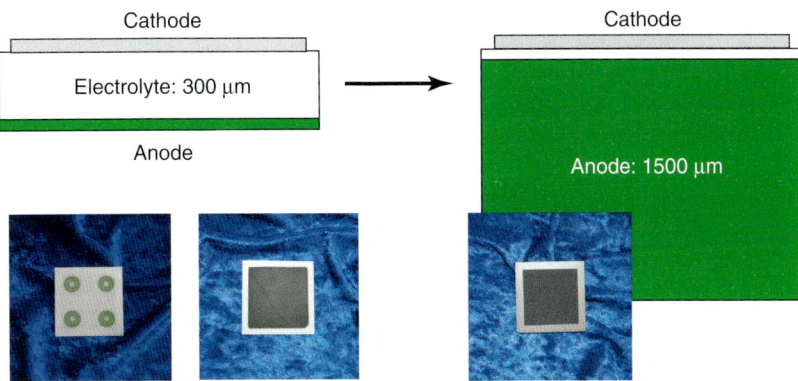

Figure 9.3 Schematic view of cross-sections of an electrolyte-supported cell (ECS) and an anode-supported cell (ASC). Also are depicted the ESCs used as button cells and as a cell with dimensions 40 × 40 mm². A 40 × 40 mm² ASC is shown on the right.

type of SOFC, the cell resistance was obviously lowered. This was explained by the relatively thin electrolyte layer of about 15 μm. As such, the operating temperature could be lowered, which involved several advantages for SOFC stacks, such as the use of cheaper materials for interconnects and manifolds, reducing sealing and corrosion problems, and increased lifetime and reliability. The coarse-pored support of 40 × 40 mm² was about 1.7 mm thick. Under pure hydrogen with 3 vol.% water vapor as the fuel gas and air as the oxidant, the measured current density at 800 °C and 700 mV was obviously improved, up to about 0.6 A cm^{-2}. The $La_{0.65}Sr_{0.30}MnO_3$ cathode layer (LSM) for this ASC type of cell was applied by WPS. The continuous improvements in manufacturing techniques and the microstructure of materials led to an increase in electrochemical performance. Thus, during the period 1998–2002, the electrochemical performance increased from 0.6 to 0.9 A cm^{-2} (800 °C and 700 mV). LSM powder with $d_{50} = 0.5$ μm was used as the cathode current-collector layer (CCCL).

By this time, it had been discovered that the chosen reduction temperature of the NiO in the cermet also influenced the cell performance [9]. It was found that the heating and reduction processes have to be defined well, since wrongly chosen conditions resulted in lower and significant scatter of performance results, because the process of "electrode activation" is not yet fully completed. This "electrode activation" of SOFCs refers to one or more processes that generally occur during the first period of performance measurements; see also Section 9.2.3.

An ASC-type cell with an LSM cathode heated and reduced at 800 °C showed a decrease in the ASR during the first 70 h of exposure at 800 °C under constant current load. The ASR of a similar cell heated and reduced at 900 °C was comparable to that of a cell reduced at 800 °C and exposed for 70 h at 800 °C under constant current load. This means that "electrode activation" is much faster at 900 °C than at 800 °C. Figure 9.4 shows the cell voltage of an ASC-type cell which was reduced

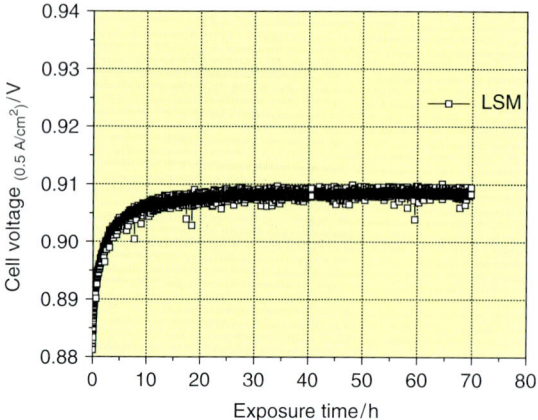

Figure 9.4 Cell voltage of an ASC-type cell (reduced at 800 °C) with an LSM cathode at 800 °C and 0.5 A cm^{-2} as a function of the exposure time [9].

at 800 °C with an LSM cathode at 800 °C and 0.5 A cm^{-2} as a function of the exposure time. It shows clearly the increase in cell voltage at constant current load during the first 30 h of exposure.

9.2.1.3 2002–2005

In 2002, R&D of ASCs with an LSM cathode was further intensified. Optimization of processing and microstructural parameters of LSM cathodes to improve the electrochemical performance of ASCs was started and was focused on various parameters, such as the LSM:8YSZ mass ratio of the cathode functional layer (CFL), the grain size (ground or unground) of the LSM powder for the CCCL, the thickness of the CFL and the CCCL, the influence of calcination of 8YSZ powder being used for the CFL, the presence of a noble metal in the cathode, and the use of tape casting versus warm pressing as the production process for anode substrates [10–12]. This study finally resulted in a performance improvement from 0.9 to about 1.5 A cm^{-2} (800 °C and 700 mV). One parameter that had a significant impact on cell performance was the 8YSZ:LSM ratio of the CFL, located adjacent to the electrolyte. A series of cells were produced with mass ratios ranging between 30:70 and 70:30. Figure 9.5 shows the current density at 700 mV of SOFCs with an LSM cathode between 700 and 800 °C as function of the 8YSZ:LSM mass ratio of the CFL.

From Figure 9.5, it is obvious that the highest performance was obtained with an YSZ:LSM mass ratio of 50:50. This was explained as the maximum of the total reactive area; the maximum length of the TPB is reached when the two components are mixed in equal amounts. However, this is only true when similar grain sizes are employed for both materials and the percolation of ions and electrons in this CFL is not significantly affected. The maximum performance obtained was 1.0 A cm^{-2} when a mass ratio of 50:50 was used. In this case, the d_{90} of the CCCL was around 2.2 µm. A further improvement in performance could be achieved when the grain

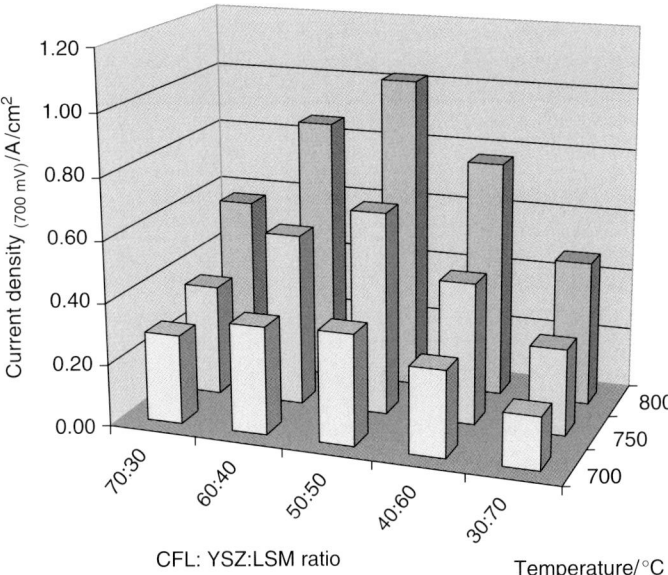

Figure 9.5 Current density at 700 mV of ASCs with an LSM cathode between 700 and 800 °C as function of the 8YSZ:LSM ratio of the cathode functional layer (CFL).

size of the CCCL was taken into consideration. A systematic variation of the main grain size of the LSM powder used for the CCCL layer showed that the highest performance was achieved with LSM powder with $d_{90} = 26$ μm (unground). This improvement was correlated with a higher porosity allowing oxygen diffusion through the LSM layer to the TPBs located in the inner CFL. Figure 9.6 shows the current density as a function of temperature and the grain size (d_{90}) of the CCCL.

In addition, the sintering temperature and the thickness of the CFL and CCCL were also optimized. As an example, thickness variations of the CFL can significantly influence the electrochemical performance of ASCs. From the literature, it was concluded that above a critical thickness of the CFL, the electrochemical performance was more or less maximized ([10] and references therein). On the other hand, it can be expected that a CFL with a much larger thickness also might adversely influence the performance due to longer diffusion paths for the gas. Results have indeed shown that the performance was decreased with increasing thickness of the CFL [10]. To be realistic, not every approach will finally result in a success story. Experimental results, as listed in Table 9.1, and performed with symmetrical button cells showed an improvement in the current density and a decrease in the apparent activation energy when doped with palladium or with platinum. However, this was only valid for the materials $La_{0.84}Sr_{0.16}MnO_3$ and an $La_{0.65}Sr_{0.30}MnO_3 - 8YSZ$ composite. Based on this, a study was initiated aimed at increasing the power output of the ASC-type cells manufactured at FZJ by small additions of Pd, Ag, or Pt to the cathode [12]. Four routes were used to add these

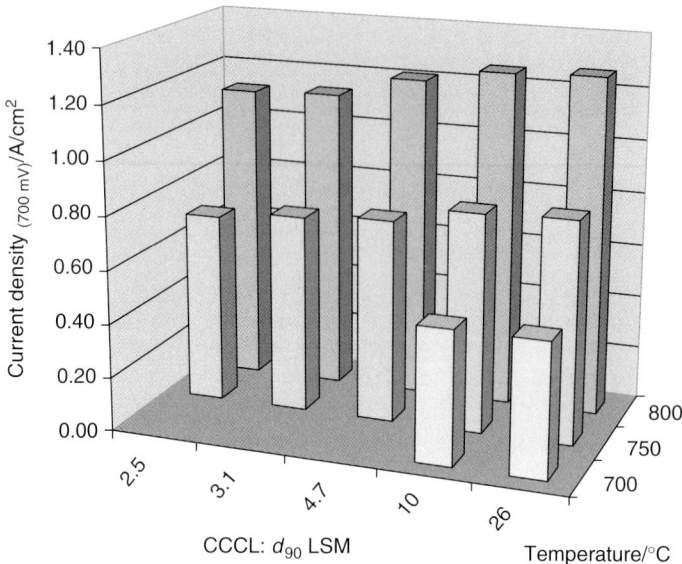

Figure 9.6 Current density at 700 mV of ASCs with an LSM cathode between 700 and 800 °C as a function of the grain size (d_{90}) of the cathode current-collector layer (CCCL).

noble metals: infiltration of the cathode layer, deposition on the electrolyte surface, mixing LSM powder, and synthesis of LSM powder with the addition of $AgNO_3$.

The results showed that between 750 and 900 °C, no electrocatalytic effect occurred with respect to the presence of Pt, added either by deposition on the electrolyte or by mixing with cathode powders. Infiltration of the cathode with a Pd solution or mixing with Pd black also did not result in a positive effect. A small catalytic effect was only found with Pd on activated carbon and in particular at lower operating temperatures. Cells prepared with Ag powder and Ag_2O showed an improved electrochemical performance compared with Ag-free cells sintered at the same temperature of 920 °C. However, in comparison with Ag-free cells sintered at the standard temperature of 1100 °C, lower current densities were measured. All these efforts finally led to an optimized SOFC which is nowadays used and denoted the standard-type ASC, manufactured at FZJ, and described in Table 9.2. A scanning electron microscopy (SEM) image of the fracture surface of a single cell with optimized microstructural parameters is shown in Figure 9.7 with, from top to bottom, CCCL, CFL, electrolyte (EL), anode functional layer (AFL), and anode substrate.

In addition to targeting a general increase in specific power output, significant attention was paid to improving long-term stability and durability under various operating conditions, simulating those expected in real stack systems. A decrease in the power output over time is also described as degradation of cell performance and could be due to a number of different mechanisms. Yokokawa et al. [13] and Tu and Stimming [14] presented an overview of the literature on possible degradation

Table 9.2 Characteristics of the "state-of-the-art" cells with LSM cathode.

Functionality	Production technique(s)	Type	Sintering temperature (°C)	Thickness
Anode substrate	Warm pressing/ tape casting	NiO/8YSZ	1230	0.7–1.5 mm
Anode functional layer	Vacuum slip casting	NiO/8YSZ	1400	~10 μm
Electrolyte	Vacuum slip casting	8YSZ	1400	7–15 μm
Cathode functional layer	Screen-printing	8YSZ/LSM	1100	~10 μm ($d_{90} = 1$ μm)
Cathode current collector layer	Screening-printing	LSM	1100	70–80 μm ($d_{90} = 26$ μm)

mechanisms limiting long-term performance. From these papers, it was obvious that degradation of the cell could occur on both the anode and cathode sides of the cell under various experimental conditions. Degradation rates from long-term endurance tests can be described as a change in the cell output voltage under constant current load over some defined time interval, generally presented in the dimension $\Delta V \, kh^{-1}$. In this case, the values obtained in that ΔV interval are

Figure 9.7 SEM image of the fracture surface of a "state-of-the-art" single cell with optimized processing and microstructural parameters.

dependent on the chosen constant current loading. Thus, if reported in terms of ΔV, the degradation rate is seen to be completely dependent on the choice of the operating specific current density–lower values will necessarily be reported for those operating at lower constant specific current densities. The actual physical system property that is changing over time is the electrical resistance of the cell. This means that a change in the ASR with respect to time ($\Delta ASR/\Delta t$) is a more reliable measure of performance degradation upon which to make comparisons and hence commercial or research decisions. To extract ASR values, however, requires periodic collection of current–voltage data. More information on long-term endurance tests of single cells under various operating conditions can be found elsewhere [15].

The cell voltage of a "state-of-the-art" SOFC at 800 °C with an LSM cathode versus operating time and under a constant current load of 0.5 A cm^{-2} is shown in Figure 9.8. The fuel gas was pure hydrogen, saturated with 3 vol.% water vapor. Related to the output voltage change versus time, it was found that the degradation rate was negligible over the whole operating period; the cell voltage even increased slightly. After regular intervals of about 500 h, current–voltage measurements were made in the temperature range 600–950 °C. Calculations revealed that the change in ASR with time was also negligible. From these results, it can be concluded that this type of cell showed excellent performance behavior with respect to both the power output and to the long-term stability.

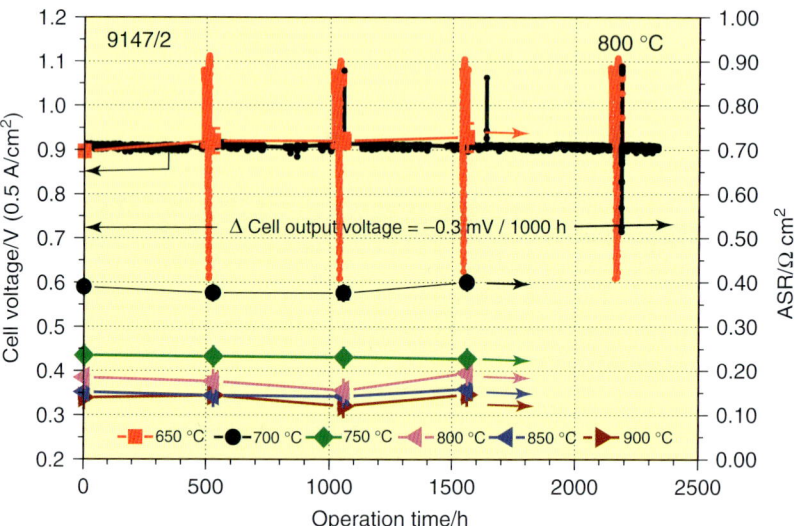

Figure 9.8 Cell output voltage and area specific resistance at various temperatures versus operating time of a "state-of-the-art" cell with an LSM cathode under constant current load [800 °C; 0.5 A cm^{-2}; fuel gas, H$_2$ (3%H$_2$O), 1000 ml min^{-1}; oxidant, air, 1000 ml min^{-1}].

9.2.1.4 2005–2010

During this period, "state-of-the-art" cells with an LSM cathode were produced for testing SOFC stacks and even for complete SOFC systems. At regular intervals, a quality control was needed to assure the high-performance density of about 1.5 A cm^{-2} at 800 °C and 700 mV. Figure 9.1 shows that the current density at various temperatures did not change significantly over the final 5 years. It is worth mentioning that these measured current densities are based not only on the development of SOFCs itself, but also on the introduction of QA in SOFC testing routines. This latter aspect is discussed later in Section 9.2.3.

9.2.2
SOFCs with an LSC(F) Cathode

In addition to the LSM-type cathode cells, LSC(F)-type cells have also been developed and improved. Cells with an LSC(F)-type cathode can be used at temperatures lower than 750 °C, in contrast to the LSM-type cathode cells, which perform best above 750 °C. The decrease in the power density at lower temperatures, in particular for cells with an LSM cathode, is mainly due to overpotentials at the cathode side. Therefore, research was focused on new cathode materials with higher electrocatalytic activity than the "state-of-the-art" LSM perovskites. As a result, LSC(F) was chosen as the material for operating temperatures lower than 800 °C, since it possesses high electronic and ionic conductivity, which make this material an excellent candidate as an MIEC (mixed ionic and electronic conductive) cathode.

9.2.2.1 2000–2006

Figure 9.9 shows the evolution of performance of LSC(F)-type cells since 2000. Again, significant progress has been made over the past 10 years. The first trials with cells with a size of 50 × 50 mm^2 were carried out in 2000, achieving a current density of about 0.7 A cm^{-2} at 800 °C and 700 mV. During that period, progress was made by optimizing the stoichiometry of the LSC(F) cathode, since the La^{3+} : Sr^{2+} ratio obviously influences the deficiency concentration of this type of perovskite. As a result, work to improve their performance was successful, and by 2002 a current density of 1.3 A cm^{-2} had been attained with La$_{0.58}$Sr$_{0.40}$Co$_{0.2}$Fe$_{0.8}$O$_{3-\delta}$ as the composition of the perovskite. Much as in the previous section, a systematic approach to optimizing performance variables was undertaken [16, 17].

In spite of the fact that these iron- and cobalt-containing perovskites have excellent electrochemical properties, care has to be taken since they have a significantly higher TEC than the commonly used 8YSZ electrolyte. Furthermore, this type of perovskite might form SrZrO$_3$ with the 8YSZ electrolyte at high temperatures [18]. As a consequence, Ce$_{0.8}$Gd$_{0.2}$O$_2$ [gadolinium-doped cerium oxide (CGO)] has to be employed as a diffusion barrier to prevent the formation of strontium zirconate. Strontium is less strongly bound in the LSC(F) than in LSM.

A more systematic study of the variation of the composition of the ferrite-based perovskites was started in 2003. The materials examined were La$_{1-x-y}$Sr$_x$Co$_{0.2}$Fe$_{0.8}$O$_{3-\delta}$ ($x = 0.2$ and 0.4; $y = 0$–0.05), La$_{0.8}$Sr$_{0.2}$FeO$_{3-\delta}$ (LSF), La$_{0.7}$ Ba$_{0.3}$

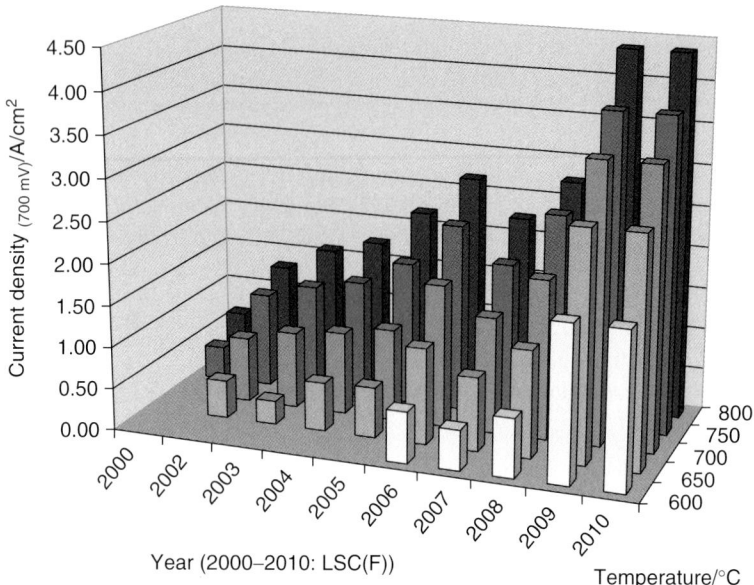

Figure 9.9 Status of development of SOFCs with an LSC(F) cathode at FZJ.

$Co_{0.2}Fe_{0.8}O_{3-\delta}$ (LBCF), and $Ce_{0.05}Sr_{0.95}Co_{0.2}Fe_{0.8}O_{3-\delta}$ (CSCF). In all cases, an interlayer of the composition $Ce_{0.8}Gd_{0.2}O_2$ was applied between the electrolyte and the cathode to prevent undesired chemical reactions. It was found that after sintering, the cathodes made of CSCF and L60SCF showed cracks and were partially spalled off from the underlying CGO layer. This was explained by the presence of thermal stresses due to the high TEC of the materials. Regarding the other types of perovskites, adhesion to the underlying layer was sufficient, but there was a small zone of the Sr-rich phase detectable for all the materials with 40% Sr on the A-site (L55SCF, L58SCF, L60SCF). In comparison with the perovskites with 20% Sr substitution, that is, L78SCF and L80SCF, the performance was much better, since it is known that a higher amount of Sr atoms instead of the trivalent lanthanum on the A-site increased the ionic and electronic conductivity and the surface exchange of oxygen, which can be explained by the larger number of oxygen vacancies and electronic holes. Together with a screen-printed CGO layer, the highest current density was obtained with an L58SCF cathode of 1.8 A cm^{-2} at 800 °C and 700 mV. With this type of perovskite, a systematic study was started on the influence of sintering temperature on the electrochemical performance. In this respect, cells with an L58SCF cathode were sintered at 1040, 1080, and 1120 °C and tested to find the optimum electrochemical performance. Figure 9.10 shows the current–voltage curves at 750 °C of cells with L58SCF cathodes sintered at various temperatures.

From Figure 9.10, it is clear that the electrochemical performance is significantly lower after sintering at 1120 °C than after 1080 °C, which can be explained by the smaller intrinsic surface area of the cathode due to particle growth. The higher

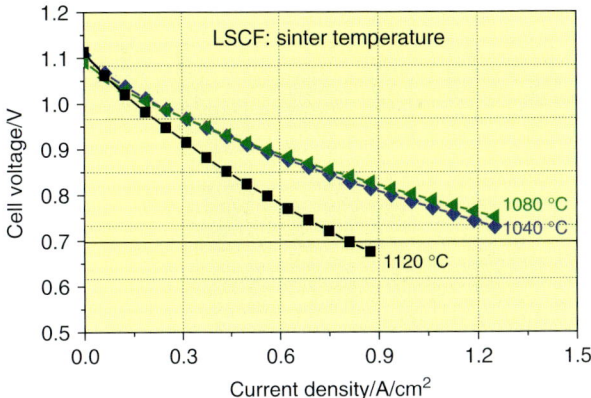

Figure 9.10 Current–voltage curves at 750 °C of cells with an L58SCF cathode sintered at various temperatures including a screen-printed CGO layer.

sintering temperature also increases the $SrZrO_3$ formation, which lowers the performance of the cathode. A much lower sintering temperature also leads to a lower current density, which can be explained by poorer adhesion between the cathode and the diffusion barrier layer.

Not only was the L58SCF cathode optimized, but also optimization of the CGO diffusion barrier layer, applied by screen-printing, was considered [17]. A series of cells were produced with variation of the sintering temperature of the CGO layer between 1200 and 1325 °C. Electrochemical measurements under standard conditions, with pure hydrogen saturated with 3 vol.% water vapor, showed the highest current density with cells when the diffusion barrier layer had been sintered at 1250 °C prior to deposition and sintering of the cathode at lower temperature. Figure 9.11 shows the current density at 700 mV of cells with an LSC(F) cathode as a function of the sintering temperature of the CGO diffusion barrier layer.

A decrease in the current density for sintering temperatures above 1250 °C could be explained by the increased formation of a solid solution between the 8YSZ electrolyte and the CGO layer, whereas a decrease in power densities at sintering temperatures lower than the optimum could be explained by lower adherence of the diffusion barrier layer. Finally, cells with an optimized CGO diffusion barrier and LSC(F) cathode layer achieved current densities of about 2.3 A cm^{-2} at 800 °C and 700 mV. This type of cell is denoted at FZJ as the "state-of-the-art" cell with an LSC(F) cathode. Table 9.3 shows the characteristics of these "state-of-the-art" cells with a screen-printed CGO diffusion barrier layer and with an L58SCF-type cathode.

Initially, improvement in electrochemical performance was the main objective; however, long-term stability or durability is probably even more important for marked success. An acceptable performance degradation rate would be a loss of <1% per 1000 h. Figure 9.12 shows the results of a long-term endurance test performed with this "state-of-the-art" cell, operated over a period of about 2000 h

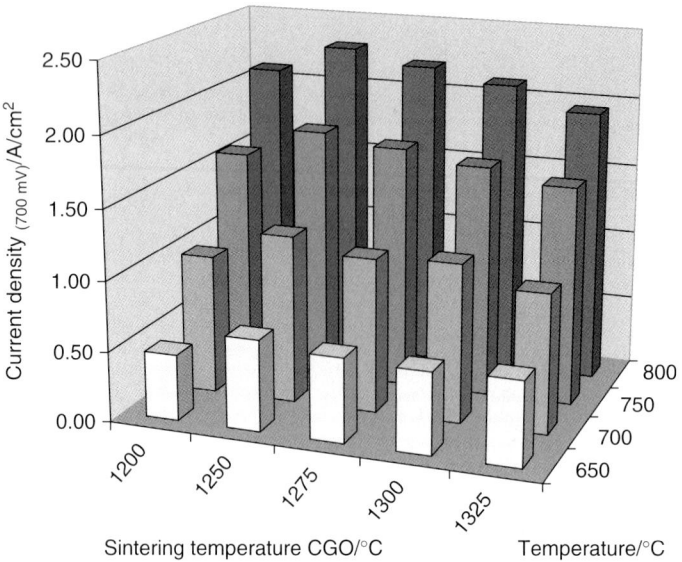

Figure 9.11 Current density at 700 mV of anode-supported cells with an LSC(F) cathode as a function of the sintering temperature of the screen-printed CGO diffusion barrier layer.

Table 9.3 Characteristics of the "state-of-the-art" cells with LSC(F) cathode.

Functionality	Production technique(s)	Type	Sintering temperature (C)	Thickness
Anode substrate	Warm pressing/ tape casting	NiO/8YSZ	1230	0.7–1.5 mm
Anode functional layer	Vacuum slip casting	NiO/8YSZ	1400	~10 µm
Electrolyte	Vacuum slip casting	8YSZ	1400	7–15 µm
Diffusion barrier layer	Screen-printing	CGO ($d_{50} = 0.2$ µm)	1250	7–10 µm
Cathode	Screen-printing	L58SCF	1080	40–50 µm

at 800 °C. The applied current density was 0.5 A cm^{-2}. The aging or degradation rate per 1000 h was about 1.3% or about 10 mV kh^{-1}. Calculations of the change in the ASR were obtained from current–voltage measurements carried out at regular intervals. As can be observed in Figure 9.12, the calculations revealed a so-called aging rate of 28 mΩ cm^2 kh^{-1}, thus demonstrating an acceptable aging rate. However, R&D is still ongoing to reduce the aging rate below the threshold value of 1% per 1000 h.

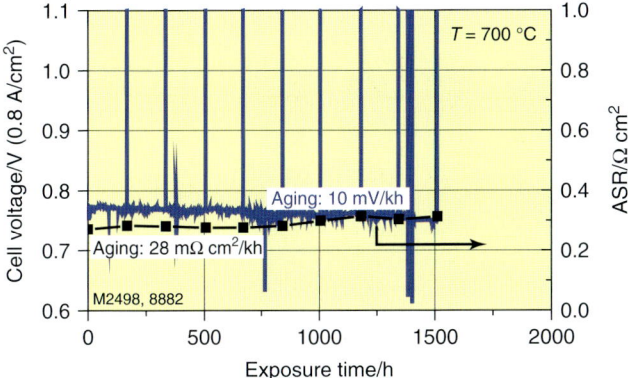

Figure 9.12 Cell output voltage versus operating time of a "state-of-the-art" cell with an LSC(F) cathode under constant current load [800 °C; 0.5 A cm^{-2}; fuel gas, H$_2$(3%H$_2$O), 1000 ml min^{-1}; oxidant, air, 1000 ml min^{-1}].

In addition to cells with an LSC(F) cathode, research at FZJ was also focused on the development of new or the modification of existing cathodes with improved performance and stability. Since increased attention is being paid to a new type of material, that is, a mixed conducting K$_2$NiF$_4$-type cathode, which can be considered as a combination of AMO$_3$ perovskite and AO rock-salt layers, a first attempt was made by the preparation of an Nd$_2$NiO$_4$-type cathode. The influence of the presence of a screen-printed CGO layer between the electrolyte and cathode, the sintering conditions, and the grain size of the powder used for applying the cathode on the electrochemical properties was investigated [19]. Based on results from current–voltage characteristics and permeation and diffusion measurements, it could be concluded that electrochemical performance was promising, and comparable to the "state-of-the-art" cells with an LSM cathode. However, it was also clear that for maximizing the performance of this type of SOFC, more research is needed to optimize the processing, material, and microstructural parameters. More promising results were obtained with a Pr$_{0.58}$Sr$_{0.4}$Co$_{0.2}$Fe$_{0.8}$O$_{3-\delta}$(PSCF) cathode [20]. This type of material was chosen since catalytic measurements have shown that replacing La by Pr resulted in higher oxygen deficiencies [21]. Since the mixed valence of Fe and Co and the resulting oxygen vacancies seem to play an important role in the catalytic process, an additional contribution of the Pr^{3+}/Pr^{4+} valence change was expected to improve the performance further. Furthermore, this composition should also suppress the formation of by-products along the cathode–electrolyte interface, such as Ln$_2$Zr$_2$O$_7$ and SrZrO$_3$, where Ln is La or Pr. The microstructure of the cathode layer was considered to be similar to that of the LSC(F) cathode. With respect to the various sintering temperatures of the cathode, it was clear that the highest performance was obtained with 1020 or 1040 °C (see Figure 9.13).

Two cells with a CGO layer applied by screen-printing and a PSCF cathode were exposed for about 1000 h at 750 °C with a constant current load of 0.5 A cm^{-2}.

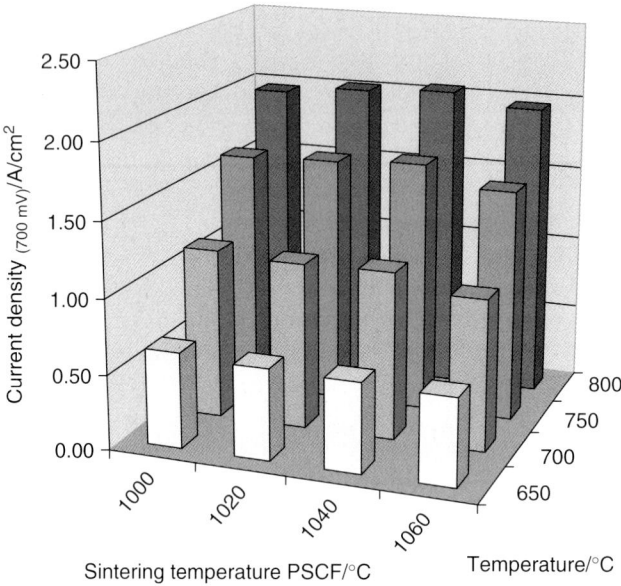

Figure 9.13 Current density at 700 mV of anode-supported cells with a PSCF cathode as a function of the sintering temperature.

The aging rates were calculated and normalized to an operating time of 1000 h. The average aging rate over the final 500 h of exposure was about 9 mV kh^{-1}. This makes this type of cathode an excellent candidate for cathode applications at relatively low operating temperatures.

Not all the attempts with new types of materials to be used for cathode applications were successful [22]. In particular, the use of Cu-containing lanthanum ferrite-based cathodes resulted in deterioration of the performance, since Cu exhibited severe chemical interaction with the 8YSZ electrolyte. Various compositions were systematically investigated, namely, $La_{0.58}Sr_{0.4}Fe_{0.8}Cu_{0.2}O_{3-\delta}$, $La_{0.58}Sr_{0.4}Fe_{0.6}Cu_{0.2}Co_{0.2}O_{3-\delta}$, $La_{0.58}Sr_{0.4}Fe_{0.7}Cu_{0.1}Co_{0.2}O_{3-\delta}$, and $La_2Ni_{0.6}Cu_{0.4}O_4$. Cu was added since it has been suggested that its presence enhanced the sintering process of the cathode and also the electrocatalytic activity of the cathode [23].

Figure 9.14 shows a cross-section of a cell with a cathode consisting of $La_{0.58}Sr_{0.4}Fe_{0.8}Cu_{0.2}O_{3-\delta}$ sintered at 1050 °C. Severe changes in microstructure and composition occurred. On the top of the cross-section, the cathode layer was coarse grained with Cu- and Fe-rich phases near the cathode/CGO interface. The CGO layer was transformed into a dense Sr-, Zr-, and La-rich layer with below it a relatively thick oxide layer rich in Sr, La, and Zr. This was followed by a thin reaction zone, containing Sr, La, Zr, Ce, and Gd. Only a very small layer of the YSZ electrolyte was left. These results showed that not only are the electrochemical characteristics of the individual layers always a critical issue for application, but

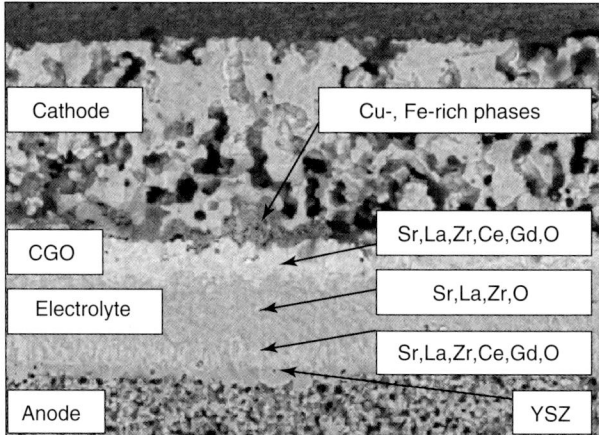

Figure 9.14 SOFC with an $La_{0.58}Sr_{0.4}Fe_{0.8}Cu_{0.2}O_{3-\delta}$ cathode (sintering temperature: 1050 °C) after performance test.

also the compatibility between the various layers. Some experimental results with other types of perovskites, and which can be considered as a prescreening test, can be found elsewhere [24].

So far, only various types of cathode layers with only one type of CGO diffusion barrier layer being deposited by screen-printing were described. The thickness of this layer was always 7–10 μm and showed a porous structure. Since the CGO diffusion barrier layer shows such a structure, Sr and La surface diffusion along the CGO grain cannot be totally excluded. Finally, this will lead to the formation of insulating compounds. In comparison with cells without such a layer, the presence of this CGO layer will strongly reduce the formation of Sr-containing insulating phases. In addition, it also allows better electrical contacting.

A change of the deposition technique of the CGO layer from screen-printing to reactive sputtering or electron beam physical vapor deposition (PVD) led to an improvement in electrochemical performance [17]. This improvement was due to a thinner and denser interlayer improving contacting and decreasing Ohmic resistance. Two options were investigated: reactive sputtering and electron beam PVD. Reactive sputtering resulted in a thin (<1 μm) and dense CGO layer and yielded a current density of 2.7 and 1.7 A cm^{-2} at 800 and 700 °C, respectively (700 mV). Deposition of the CGO layer by electron beam PVD also resulted in excellent performance; however, the power output was slightly lower. SEM images of the fracture of two types of the CGO diffusion barrier layer are shown in Figure 9.15.

It can be concluded that with the change of the deposition technique from screen-printing to reactive sputtering PVD for applying the CGO layer, significant progress was made. Thus, in combination with the optimized LSC(F) cathode, a power output of almost 2 W cm^{-2} (2.7 A cm^{-2}) was achieved at 800 °C and 700 mV.

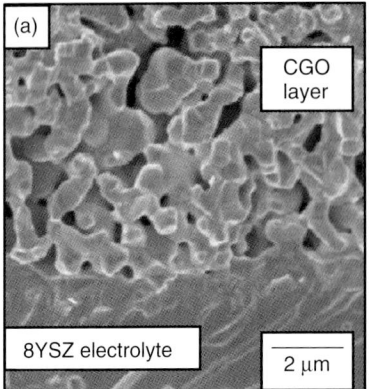 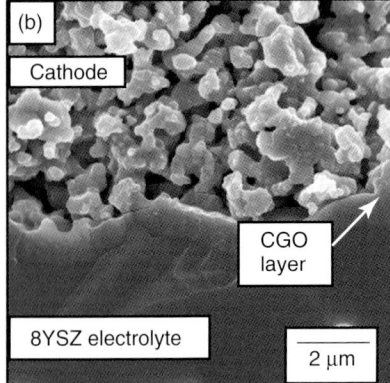

Figure 9.15 SEM images of the fracture surface of two types of CGO diffusion barrier layer, (a) applied by screen-printing and (b) applied by RS-PVD.

9.2.2.2 2006–2010

Since it was known that the use of a PVD-applied CGO layer resulted in a significant improvement in the electrochemical performance, the next step was to optimize further the electrochemical properties of the cathode layer. A first screening test was based on the use of $La_{0.58}Sr_{0.40}Co_{1.0}O_{3-\delta}$ [lanthanum strontium cobaltite (LSC)] as the cathode. Regarding this type of perovskite, it should be taken into account that the TEC of this material (about 23×10^{-6} K^{-1}) is much higher than that of 8YSZ, which lies around $11-12 \times 10^{-6}$ K^{-1} [25]. Such a TEC mismatch will probably result in detachment of the cathode layer. Probably with an intermediate layer, such as a CGO diffusion barrier layer, the mechanical stability can be improved. The first results were obtained with a series of measurements including variations in the sintering temperature and are depicted in Figure 9.16.

From these experiments, it can be concluded that the combination of a PVD-deposited CGO interlayer and a screen-printed unsintered LSC cathode resulted in high electrochemical performance. At 600 °C, the current density achieved was 0.7 A cm^{-2}, corresponding to an ASR of 0.35 Ω cm^2 at 700 mV. Since the TEC of the LSC cathode is significantly higher than that of the electrolyte, as reported above, this aspect will be of minor importance when these are applied for stationary applications. After testing, it was found that no delamination of the cathode occurred. Therefore, circumvention of detachment of the cathode was probably achieved by the application of an interlayer based on ceria.

At FZJ, R&D was still ongoing, aimed at improving the electrochemical characteristics not only of the diffusion barrier layer and the cathode layer, but also of the electrolyte. The standard deposition technique of the 8YSZ electrolyte was vacuum slip casting, which resulted in an electrolyte with a thickness of about 10 μm. In order to prepare cells with a thinner electrolyte, resulting in a lower overall ASR, other deposition techniques were considered, such as PVD and sol–gel deposition. With these techniques, very thin layers can be applied. Unfortunately, cells with

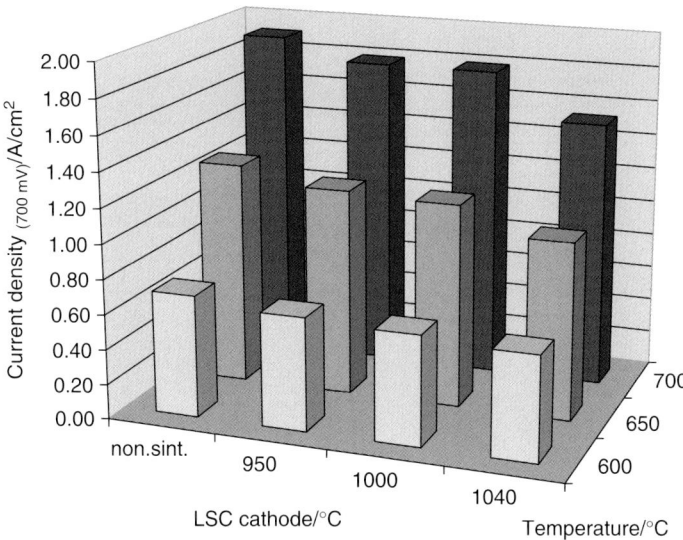

Figure 9.16 Current density (700 mV) at 600, 650, and 700 °C of tested anode-supported single cells with an LSC cathode as function of sintering temperature.

an electrolyte and CGO diffusion barrier layer, both applied by PVD and both with a thickness of 1 µm, did not show the expected outcome. Owing to the relatively low deposition temperature, much lower than the standard sintering temperature of the vacuum slip casting-deposited 8YSZ layer, the open-circuit voltage of the cells was obviously lower than that expected from the theoretical Nernst voltage, probably due to a relatively high leakage of the electrolyte. The current density of this type of cell was similar to that of cells with an electrolyte over vacuum slip casting and a CGO layer over screen-printing.

More promising results were obtained with a sol–gel-deposited 8YSZ electrolyte. The thickness of the electrolyte was about 2 µm and the sintering temperature was set at 1400 °C. Based on the electrochemical data, it was concluded that the tested cells showed excellent performance. Even at the lowest testing temperature of 600 °C, the current density was 1.9 A cm^{-2}; see also Figure 9.9 (year 2009, 2010).

At 800 °C and 700 mV, the power output was more than 3 W cm^{-2}, corresponding to a theoretical current density of 4.4 A cm^{-2}. However, not only is the performance itself an important cell characteristic, but also the long-term stability of such cells and their behavior when they are used in real SOFC stack systems. These tests still have to be performed in order to obtain more knowledge about the long-term behavior under simulated experimental conditions. Characteristics of the cell with the maximum achieved performance are given in Table 9.4. Figure 9.17 shows SEM images of the fracture surface of the "state-of-the-art" and the "high-performance" cells. The much thinner CGO and electrolyte layer of the "high-performance" cell is clear.

Table 9.4 Characteristics of the "high-performance" cells with an LSC cathode.

Functionality	Production technique(s)	Type	Sintering temperature (C)	Thickness
Anode substrate	Warm pressing/tape casting	NiO/8YSZ	1230	0.7–1.5 mm
Anode functional layer	Vacuum slip casting	NiO/8YSZ	1400	~10 μm
Electrolyte	Sol–gel	8YSZ	1400	2 μm
Diffusion barrier layer	PVD	CGO	800 (deposition temperature)	0.5 μm
Cathode	Screen-printing	L58SC	800	60 μm

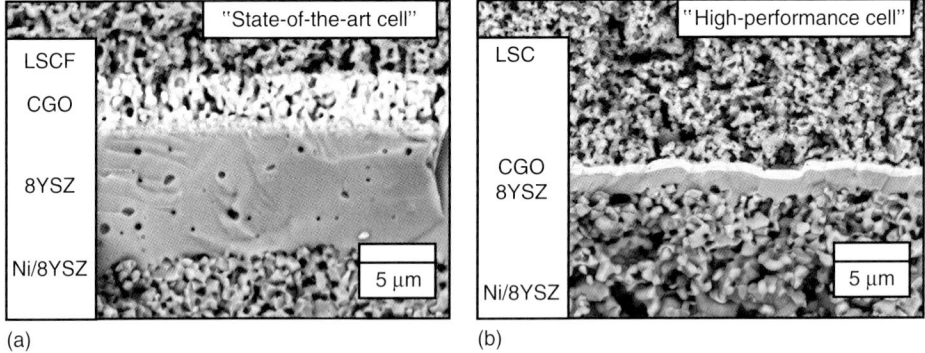

Figure 9.17 SEM images of the fracture surface of (a) the "state-of-the-art" cell and (b) the "high-performance" cell.

9.2.3
Advances in Testing of SOFCs

To achieve performance measurements including data with good accuracy and precision in a consistent, reliable, and repeatable manner, FZJ started to develop its own standardization and QA in SOFC testing. Such a requirement becomes increasingly important in making decisions as the technology approaches precommercial or commercial implementation. This included all aspects of the SOFC testing process being considered and systematically examined. It is not the aim here to discuss every step and every testing parameter in the FZJ SOFC testing routine, but some aspects are described in more detail. More details on standardization and QA are available elsewhere [26, 27].

The completed standardized testing procedure is based on generic QA standards. It is clear that the ISO 9000 series are the most widely accepted generic quality standards; it was not the aim to adopt fully certified ISO 9000 quality systems, but

FZJ followed the principles of this quality system. By adopting the principles of the ISO 9000 system, it can be assured that any system that was implemented will be fully compatible with other existing and possible future SOFC-specific standards. In any case, a first step was made in this direction.

9.2.3.1 Testing Housing

The first single-cell tests at FZJ were carried out on ESCs. For this purpose, an alumina housing was constructed that would fit into a tubular furnace. During the design of the housings, special attention was paid to the gas and air supply. Sealing was accomplished using gold wire. Platinum and nickel meshes were used as contacts on the cathode and anode side, respectively. After changing from ESCs to ASCs, the housing had to be extended with a hood and a bottom plate. As a result of the porous anode used, the sealing had to be on the cathode side, leaving the anode open to the fuel atmosphere. By applying the hood and bottom plate, a closed compartment for the anode was created. The housings could now no longer be used in a tubular furnace so a change to "top hat" furnaces was made. The extra space available in the new type of furnaces prompted a design change for the cell housings. The first housings of the new type were constructed from MACOR, which is a machinable alumina silicate-type ceramic. Problems with the stability of this material under high temperatures and low oxygen partial pressures together with the fact that the material contains alkaline earth metals that can be detrimental to the cell performance led to the construction of cell housings from aluminum oxide (ALSINT 99). In the adapted design, the gas tightness of the hood was improved by increasing the wall thickness and polishing both the surface of the bottom plate and the rim of the hood. Figure 9.18a shows the first type of housing testing, made from alumina and used at FZJ.

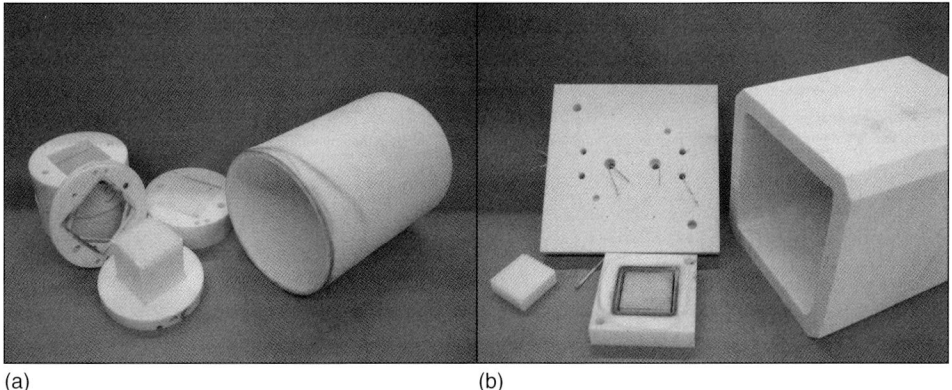

(a)   (b)

Figure 9.18 Two types of test housing for ASC single-cell testing at FZJ: (a) first type of test housing; (b) present test housing.

Currently, this relatively robust and optimized type of cell testing housing is used for all single-cell measurements (Figure 9.18b). In the bottom plate of the measuring housing, aluminum oxide tubes have been cemented in place to provide connections to the gas delivery system.

9.2.3.2 SOFC Specifications

To avoid undesirable variations in performance data, the SOFCs themselves have to meet requirements before information extracted from electrochemical measurements can be considered reliable. The list of requirements to be met includes the maximum deviation from nominal dimensions, maximum curvature of the cell, maximum gas leakage (through the electrolyte), absolute cell thickness, chemical composition of the SOFC (to avoid chemical interactions with the cell housing), and the effective active area and thickness of the applied cathode. Since sealing of the gas compartments is via a gold gasket, the cell dimensions are critical. Of course, some variations in the cathode thickness exist that cannot be ignored. To accommodate this, the client has to supply, in addition to the cell itself, the accurately measured cathode thickness. Based on this, the gold seals are manufactured in varying thickness with single micron tolerance. Experience and leak-testing experiments then allow, from retained records, the ideal gold seal thickness for each cell tested, thus ensuring minimum gas leakage and maximum electrical contacting. If these critical points are well considered before starting the SOFC testing procedure, only then will performance measurements yield repeatable and reliable results. Nevertheless, the testing procedure itself is also of major importance, which will also be part of standardization and QA in single-cell testing.

9.2.3.3 SOFC Testing Procedure

The SOFC testing procedure at FZJ is broken down into a number of consecutive events, activities, decisions, and outputs. It is not the aim here to explain every critical measurement or experimental parameter in detail, but some selected examples are illustrated, that is, reduction temperature and dwell time between individual current–voltage points. Since the need for a high degree of reliability and repeatability of performance values is still growing, a high level of standardization of the operating conditions in the actual measurement process is indispensable. Hence every step in the testing procedure was critically analyzed, optimized, and laid down in the standardized testing procedure making part of the established quality assurance system in the frame of single-cell testing.

9.2.3.3.1 Reduction Temperature

A previous study [9] showed that reduction of the anode at 900 °C or the application of a constant current to ASCs manufactured at FZJ resulted in a high steady-state electrochemical performance, including complete cell conditioning. In the case of reduction at 800 °C, the performance was lower and needed a longer time to reach steady-state performance. The performance was less consistent and significant scatter between data sets from individual cells occurred. This makes comparison between individual cells difficult. Therefore, it is recommended to carry out prenormative research to optimize their preconditioning and stabilization procedures so that the measured electrochemical performance reflects complete cell conditioning. Only then will the data obtained show high reliability and consistency and very often higher performance than might otherwise be seen.

9.2.3.3.2 Dwell Time between Individual Current–Voltage Points

Another important factor in the electrochemical measurements is the chosen dwell time between changing voltage and taking the next measurement. Figure 9.19 shows the current–voltage plots collected at 800 °C leaving different time periods between changing the current density and measuring the voltage (current density step 0.0625 A cm^{-2} in all cases). It is clear that the data points start to converge at around a 30 s dwell time. Smaller intervals resulted in a measurably lower performance, indicating that the electrochemical equilibrium had not yet been reached. Based on this, a dwell time of 30 s was defined. However, this is only valid for FZJ cells; other cells may show faster or slower response times. Hence defining the parametes for each individual setup and/or cell is of major importance.

As already pointed out, the need for standardizing SOFC measurements and QA is strongly recommended for approaching extant problems with consistency and repeatability and for reliability of data. The cell testing procedure and the QA system used at FZJ were developed starting in 2004. All the aspects were entered in the QA system following the outlines of the ISO 9000 series standards, from SOFC specifications until reporting of the experimental data. It was found that only measurements performed under such specifications and with optimized experimental conditions resulted in consistent and reliable tests and measurement results.

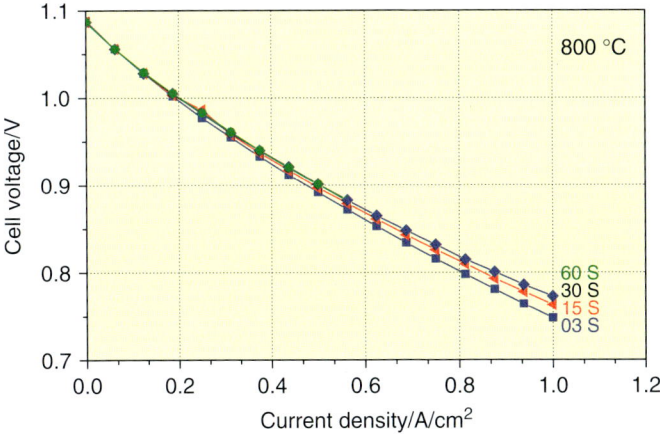

Figure 9.19 Current–voltage plots as a function of the dwell time between individual points.

9.3
Conclusions

Starting from the mid-1990s, FZJ intensified the R&D of various SOFCs, initially based on electrolyte-supported and later on the anode-supported concept. During these 15 years of SOFC production and testing, all activities have finally resulted in significant improvements in the electrochemical performance and a better understanding of the chemical and electrochemical behavior of the SOFCs. Whereas the first SOFCs achieved a current density at 800 °C and 700 mV of about 0.2 A cm^{-2}, measurements with improved SOFCs tested in 2010 resulted in a current density of about 4.4 A cm^{-2}, representing a 20-fold increase in performance. Such an improvement was due to a systematic improvements of individual processing and material and microstructural parameters. Finally, the introduction of standardization and QA also contributed substantially to optimized testing conditions, resulting in SOFCs with excellent performance and long-term stability, based on consistent and reliable test and measurement results.

Acknowledgments

The author thanks all members of the Jülich SOFC development team, the cell testing team, the FZJ metallurgical and mechanical testing group, and the FZJ central chemical laboratories, and also others from the overall FZJ SOFC team including visiting scientists, all of whom contributed in some way or another.

References

1. Erning, J.W., Hauber, T., Stimming, U., and Wippermann, K. (1996) Catalysis of the electrochemical processes on solid oxide fuel cell cathodes. *J. Power Sources*, **61**, 205–211.
2. de Haart, L.G.J., Mayer, K., Stimming, U., and Vinke, I.C. (1998) Operation of anode-supported thin electrolyte film solid oxide fuel cells at 800 °C and below. *J. Power Sources*, **71**, 302–305.
3. Wippermann, K. (1997) Abschluss-bericht des BMFT-Projekts ET 9515 A/1 Aufklärung der Kinetik in der Hochtemperatur-Brennstoffzelle vom 1.1.93-31.12.96. Teil "Kinetic der Sauerstoffreduktion an der SOFC-Kathode". Internal Report, Forschungszentrum Jülich, Jülich.
4. Stochniol, G., Syskakis, E., and Naoumidis, A. (1995) Chemical compatibility between strontium-doped lanthanum manganite and yttria-stabilized zirconia. *J. Am. Ceram. Soc.*, **78** (4), 929–932.
5. Syskakis, E., Jungen, W., and Naoumidis, A. (1993) Properties of perovskite powders and ceramics for solid oxide fuel cell application. In *International Conference on Materials by Powder Technology: PTM 93, Dresden, 23–26 March 1993* (ed. F. Aldinger), DGM Informationsgesellschaft, Oberursel, pp. 707–715.
6. Syskakis, E., Robens, W., and Naoumidis, A. (1993) Structural and high temperature electrical properties of $La_{y-x}Sr_xMnO_3$ perovskite materials. *J. Phys. IV, Colloq. C7, Suppl. J. Phys. III*, **3**, 1429–1434.
7. Wilkenhöner, R. (1997) Herstellung von Elektroden für die Festelektrolyt-Brennstoffzelle durch

Nasspulverspritzen. PhD thesis, Forschungszentrum Jülich.
8. Buchkremer, H.P., Dieckmann, U., and Stöver, D. (1996) Component manufacturing and stack integration of anode-supported planar SOFC system. In *Proceedings of the Second European Solid Oxide Fuel Cell Forum, Oslo, Norway, 6–10 May 1996* (ed. B. Thorstensen), European SOFC Forum, Oberrohrdorf, pp. 221–228.
9. Haanappel, V.A.C., Mai, A., and Mertens, J. (2006) Electrode activation of anode-supported SOFCs with LSM- and LSCF-type cathodes. *Solid State Ionics*, **177**, 2033–2037.
10. Haanappel, V.A.C., Mertens, J., Rutenbeck, D., Tropartz, C., Herhof, W., Sebold, D., and Tietz, F. (2005) Optimisation of processing and microstructural parameters of LSM cathodes to improve the electrochemical performance of anode-supported SOFCs. *J. Power Sources*, **141**, 216–226.
11. Mertens, J., Haanappel, V.A.C., Tropartz, C., Herzhof, W., and Buchkremer, H.P. (2006) The electrochemical performance of anode-supported SOFCs with LSM-type cathodes produced by alternative processing routes. *J. Fuel Cell Sci. Technol.*, **3**, 125–130.
12. Haanappel, V.A.C., Rutenbeck, D., Mai, A., Uhlenbruck, S., Sebold, D., Wesemeyer, H., Röwekamp, B., Tropartz, C., and Tietz, F. (2004) The influence of noble-metal-containing cathodes on the electrochemical performance of anode-supported SOFCs. *J. Power Sources*, **130**, 119–128.
13. Yokokawa, H., Tu, H., Iwanschitz, B., and Mai, A. (2008) Fundamental mechanism limiting solid oxide fuel cell durability. *J. Power Sources*, **182**, 400–412.
14. Tu, H. and Stimming, U. (2004) Advances, aging mechanisms and lifetime in solid oxide fuel cells. *J. Power Sources*, **127**, 284–293.
15. Haanappel, V.A.C., Röwekamp, B., Tropartz, C., Wesemeyer, H., Smith, M.J., and de Haart, L.G.J. (2008) Long-term endurance tests of single cells and defining degradation. In *Proceedings of the 8th European Solid Oxide Fuel Cell Forum, Lucerne, Switzerland, 30 June–4 July, 2008.* (ed. R. Steinberger-Wilckens), European SOFC Forum, Oberrohrdorf, CD, B 0906.
16. Mai, A., Haanappel, V.A.C., Uhlenbruck, S., Tietz, F., and Stöver, D. (2005) Ferrite-based perovskites as cathode materials for anode-supported solid oxide fuel cells – Part I. Variation of composition. *Solid State Ionics*, **176**, 1341–1350.
17. Mai, A., Haanappel, V.A.C., Uhlenbruck, S., Tietz, F., and Stöver, D. (2006) Ferrite-based perovskites as cathode materials for anode-supported solid oxide fuel cells – Part II. Influence of the CGO interlayer. *Solid State Ionics*, **177**, 2103–2107.
18. Tu, H.Y., Takeda, Y., Imanishi, N., and Yamamoto, O. (1999) $Ln_{0.4}Sr_{0.6}Co_{0.8}Fe_{0.2}O_{3-\delta}$ (Ln = La, Pr, Nd, Sm, Gd) for the electrode in solid oxide fuel cells. *Solid State Ionics*, **117**, 277–281.
19. Haanappel, V.A.C., Lalanne, C., Mai, A., and Tietz, F. (2009) Characterization of anode-supported solid oxide fuel cells with Nd_2NiO_4 cathodes. *J. Fuel Cell Sci. Technol.*, **6**, 041016 - 1–041016-6.
20. Haanappel, V.A.C., Mai, A., Uhlenbruck, S., and Tietz, F. (2009) Characterization of anode-supported solid oxide fuel cells with PSCF cathode. *J. Fuel Cell Sci. Technol.*, **6**, 011007 - 1–011007-6.
21. Mai, A. (2004) Katalytische und elektrochemische Eigenschaften von eisen und kobalthaltigen Perowskiten als Kathoden für die oxidkeramische Brennstoffzelle (SOFC). PhD thesis, Ruhr-Universität Bochum.
22. Haanappel, V.A.C., Bär, B., Tropartz, C., Mertens, J., and Tietz, F. (2010) Various lanthanum ferrite-based cathode materials with Ni and Cu substitution for anode-supported solid oxide fuel cells. *J. Fuel Cell Sci. Technol.*, **7**, 061017-1–061017-4.
23. Park, C.Y., Azzarello, F.V., and Jacobson, A.J. (2006) The oxygen non-stoichiometry and electrical conductivity of $La_{0.7}Sr_{0.3}Cu_{0.2}Fe_{0.8}O_{3-\delta}$. *J. Mater. Chem.*, **16**, 3624–3628.
24. Serra, J.M., Vert, V.B., Betz, M., Haanappel, V.A.C., Meulenberg,

W.A., and Tietz, F. (2008) Screening of A-substitution in the system $A_{0.68}Sr_{0.3}Fe_{0.8}Co_{0.2}O_{3-\delta}$ for SOFC cathodes. *J. Electrochem. Soc.*, **155** (2), B204–B214.

25. Zhao, F., Peng, R., and Xia, C. (2008) LSC-based electrode with high durability for IT-SOFCs. *Fuel Cells Bull.*, **2**, 12–16.

26. Haanappel, V.A.C. and Smith, M. (2007) Quality assurance and solid oxide fuel cell testing at Forschungszentrum Jülich. *J. Fuel Cell Sci. Technol.*, **4**, 194–202.

27. Haanappel, V.A.C. and Smith, M. (2007) A review of standardising SOFC measurement and quality assurance at FZJ. *J. Power Sources*, **171**, 169–178.

10
Solid Oxide Fuel Cell Electrode Fabrication by Infiltration
Evren Gunen

10.1
Introduction

Solid oxide fuel cells (SOFCs) are devices that convert electrochemical energy using a fuel gas into electrical energy. SOFCs are based on an oxide ion-conducting electrolyte and they offer some advantages over other fuel cell types such as lower materials costs, lower sensitivity to fuel impurities, and higher efficiency. These advantages result from the high operating temperature of 600–900 °C, which permits the use of nonprecious metals as electrocatalysts, heat recovery, and high ionic conductivity of components. Currently there is a need to reduce the high manufacturing costs that lead to expensive production of electricity, to reduce the operating temperature to 500–800 °C, and to enhance the tolerance for direct hydrocarbon fuels in SOFC technology [1, 2]. Research is being carried out to modify electrode structures in order to achieve these goals worldwide.

Infiltration is a well-known method in heterogeneous catalyst applications. Infiltration, also known as *impregnation*, is commonly used for the preparation of supported catalyst systems. It involves the dispersion of a metal salt component into a support. It has recently been applied to produce enhanced SOFC electrodes in order to solve the problems mentioned above. This chapter is intended to provide an overview of the application of infiltration in manufacturing SOFC electrodes, the procedures that are implemented to impregnate the electrode materials and the improvements achieved with infiltrated electrodes.

10.2
SOFC and Electrochemical Fundamentals

SOFC is composed of a dense oxygen ion-conducting electrolyte separating a porous anode and a porous cathode as a single cell. Electrical interconnections can be used to combine the individual cells to produce a stack. The cell is operated by supplying oxygen at the cathode which is reduced at the cathode–electrolyte interface. At the same time, fuel gas is fed to the anode and it is oxidized at the electrolyte–anode–fuel triple phase boundary (TPB) to release electrons. The resulting cell voltage can be calculated using the Nernst equation:

$$V_{\text{Nernst}} = V_0 + \frac{RT}{2F} \ln \left[\frac{P_{H_2 \text{anode}} \left(P_{O_2 \text{cathode}} \right)^{\frac{1}{2}}}{P_{H_2O \text{anode}}} \right] \quad (10.1)$$

where V_0 is the equilibrium potential for the overall SOFC reaction at its standard state, R is the gas constant, F is the Faraday constant, T is absolute temperature, P_{O_2} is the partial pressure of O_2 at the cathode and P_{H_2} and P_{H_2O} are the partial pressure of H_2 and H_2O, respectively. When current is applied, the voltage decreases mainly due to three kinds of polarization: ohmic polarization caused by the electrode and electrolyte resistance, concentration polarization caused by mass-transfer limitations between the gas phase and TPB, and activation polarization due to slow charge-transfer reactions at the electrodes. Reduction of these polarizations can be achieved by application of improved materials, cell design, and electrode microstructure [3].

10.3
Current Status of Electrodes; Fabrication Methods of Electrodes

Recent R&D efforts on SOFCs have been aimed at lowering the operating temperature from 1000 to 500–800 °C owing to the physical and chemical degradation of SOFC component materials [4, 5]. Intermediate operating temperatures also provide the opportunity to select the interconnect materials from inexpensive metals. The disadvantage of a reduced operating temperature, however, is that with lower temperatures the overall electrochemical performance of SOFCs will also be reduced owing to increased electrode polarization resistances and decreased electrolyte conductivity. The problem associated with the electrolyte conductivity has been solved by employing thin electrolytes on supported electrodes [6]. Generally, the thin electrolyte is supported on the anode; however, there are also applications of cathode-supported SOFCs. Conventionally, the electrode substrate is first produced by a variety of methods, as will be explained below, then the electrode functional layer is coated on the substrate by the method of choice. The substrate production methods are summarized in Table 10.1 [7].

10.3.1
Methods for Coating Electrode Materials

Electrode materials can be applied to SOFC substrates using a number of different methods. Menzler *et al.* [7] categorized the coating methods into three groups:

- wet chemical or ceramic technologies such as screen-printing, slip casting, and spraying
- thermal spraying methods such as atmospheric and vacuum plasma spraying (VPS)

Table 10.1 Substrate manufacturing technologies [7].

Substrate manufacturing technique	Geometry	Thickness (μm)	Slip/paste/suspension characteristics
Tape casting	Planar	100–800	Low viscosity (Pa s), azeotropic solvents, various organics
Pressing	Planar, microtubular	Planar 500–2000, microtubular wall thickness 100–500	High viscosity, pressing agent, fewer organics
Extrusion	Planar, tubular, quasi-tubular	Wall thickness 500–2000	High viscosity (1000 Pa s), water-soluble binders, additional organics
Calendering	Planar	100–2000	Low viscosity (Pa s), azeotropic solvents, various organics

- gas-phase deposition methods such as chemical vapor deposition (CVD), electrochemical vapor deposition (EVD), and physical vapor deposition (PVD).

A summary of the coating methods is given in Table 10.2 [7].

Beckel et al. [8] summarized thin-film deposition techniques in a slightly different grouping; clustering the techniques mainly under vacuum deposition and liquid precursor-based thin-film deposition, and mentioned some interesting methods that were not included by Menzler et al. [7], such as polymeric precursors and impregnation. Table 10.3 gives their deposition method groupings.

Each deposition technique has advantages and disadvantages in terms of microstructure and the quality of the deposited film. The technique to be applied is chosen according to the desired thickness of the resulting layer, layer microstructure, substrate area and type, material, process complexity, and process cost. For thicknesses of more than 30 μm thermal spraying, for 5–100 μm ceramic technologies, and for less than 2 μm PVD or CVD are chosen. If the substrate material is ceramic, then ceramic technologies are preferred; for metal substrates, thermal spraying methods are generally used. Screen-printing, an inexpensive, easily scalable method, has been the most commonly used coating method for planar substrates. Sputtering is another easy to apply process for coating large areas, which makes it suitable for industrial manufacturing; however, controlling the growth of complex compounds can be problematic.

Pulsed laser deposition is an expensive method owing to the necessary equipment and is still applied only on the laboratory scale. The film quality is similar to that with sputtering, but the disadvantage of this method is the poor uniformity of surface coverage [7, 8].

Table 10.2 Coating methods [7].

Coating technology grouping	Examples	Film thickness (μm)	Industrialization status
Wet chemical or ceramic	Screen-printing: most common, industrially established	5–100	Established
	Spraying; wet powder spraying, electrophoresis and electrostatic spraying	5–100	Established
	Slip casting	1–20	Laboratory scale
	Roller/curtain coating	5–20	Laboratory scale to commercialized
	Dip, spin coating	2–10, 10 nm^{-2} sol–gel	Laboratory scale
Thermal spraying methods	Atmospheric plasma spraying	50–300	Laboratory scale
	Vacuum plasma spraying	30–150	Laboratory scale
	Low pressure plasma spraying	20–100	Laboratory scale
	Suspension plasma spraying, high-velocity oxy fuel spraying	5–50	Laboratory scale
Gas-phase deposition	Physical vapor deposition	Thermal evaporation	–
		Sputtering	
	Chemical vapor deposition	–	–

Tape casting, tape calendering, slurry coating, and EVD are methods that have been widely used to manufacture anode functional layers or anode supports [9]. EVD is an expensive method compared with the other methods. Tape casting is generally the preferred method for the substrates of anode-supported cells [10]. The electrode which is produced by conventional methods such as tape casting and screen-printing have to be processed by high-temperature sintering. In order to lower the cost and prevent the defects caused by the sintering of larger area cells, atmospheric plasma spraying (APS), which differs from VPS in the pressure conditions, is applied. The criteria for selecting the method of fabrication of electrodes are basically the cost, the possibility of automation, reproducibility, and precision [11].

10.4
Electrode Materials

Electrode materials for SOFCs are fairly limited since they need to fulfill a number of requirements simultaneously, as summarized in Table 10.4 [12].

Table 10.3 Summary of Beckel et al.'s thin-film deposition techniques [8].

Deposition method grouping	Technique	
Vacuum deposition techniques	Physical vapor deposition	Sputtering
		Pulsed laser deposition
	Chemical vapor deposition	–
Liquid precursor-based thin-film deposition	Spray deposition methods	Electrostatic spray deposition
		Flame spray deposition
		Pressurized gas spray deposition
		Ultrasonic spray pyrolysis
		Mist spray pyrolysis
	Electrophoretic deposition	–
	Spin and dip coating	Metal–organic decomposition
		Polymeric precursors/sol–gel method; alkoxide precursor, halide precursors, nitrate and acetate precursors
		Impregnation/infiltration
		Slurry coating

Table 10.4 Requirements for the SOFC components [12].

Property	Anode	Cathode
Ionic conductivity	If possible	If possible
Electronic conductivity	High (>10^3 S cm^{-1} at 1000 °C)	High
Catalytic properties	High rates for fuel reduction Internal reforming capabilities	High rates for oxygen reduction
Chemical stability in oxidizing environments	–	–
Chemical stability in reducing environments	–	–
Sulfur tolerance	–	–
Gas tightness	Porosity 20–40%	Porosity 20–40%
Cyclability	–	–
Mechanical strength	If anode-supported cell configuration	If cathode-supported cell configuration
Thermal conductivity	–	–

The anode is exposed to hydrogen or hydrocarbons as fuel. It should be a mixed conductor with dominantly electronic conducting to allow the transportation of the electrons produced as a result of the chemical reaction at the anode surface. The electrode composition, powder particle size, and the manufacturing method are important parameters that affect the electronic and ionic conductivity and the activity for the electrochemical reactions. The electrical resistance is composed of internal resistance and electrode thickness. The electrodes should be porous enough to allow the reactant gases to make contact with the TPB over an extended area. The electrodes should be chemically stable not only in the atmosphere of the electrode but also with respect to the electrolyte and the current collector. The electrodes also need to have good thermal compatibility with the electrolyte and the current collector. The anode needs to be fuel flexible, hence it must have good carburization and sulfidation resistance. Since the cathode is exposed to air, it has to have good oxidation resistance [12, 13].

10.4.1
Anode Materials

The standard anode material for SOFC is a ceramic–metallic (cermet) composite of NiO and yttria-stabilized zirconia (YSZ) with YSZ electrolyte. If gadolinium-doped ceria (GDC) or samarium-doped ceria (SDC) is used as electrolyte, then Ni is coupled with GDC or SDC. Ni provides mechanical support for the anode, electronic conductivity, and catalytic activity whereas YSZ provides the thermal expansion match with the YSZ electrolyte along with ionic conductivity to extend the TPB [14]. The amount of Ni in the electrode must be at least 30 vol.% to provide electronic conductivity [15]. Ni–YSZ anodes are commonly produced by sintering an NiO and YSZ powder mixture at 1300 °C. The mixture is applied as a slurry on top of a YSZ layer and the two layers are co-fired. The NiO is reduced to Ni under H_2 at 800 °C and above to form a porous Ni–YSZ cermet. The cermet microstructure, namely particle size, surface area, Ni particle connectivity, porosity, and the TPB, is critical for Ni–YSZ performance and long-term stability. Ni cermet anodes are easy to produce and work efficiently; however; they also have disadvantages such as poor redox stability, low sulfur tolerance, carbon deposition in case of hydrocarbon fuel usage, and agglomeration of nickel particles [16]. There have been studies to enhance the performance of Ni–YSZ cermets using various materials or applications to replace Ni [17–19]. Fluorite anode materials are one group of these alternative materials. CeO_2-based ceramics reveal mixed ionic and electronic conductivity (MIEC) in reducing atmospheres and they have good catalytic activity and high carbon deposition resistance. The disadvantages of Ce, however, are the degradation and cracks formed due to the lattice expansion as a result of the reduction of Ce^{4+} to Ce^{3+}. GDC and SDC have also been used to enhance ceria anodes. Doped ceria anodes have been shown to be resistant to carbon deposition [20, 21]. Perovskites have also been used as anodes, among which $LaCrO_3$-based anodes have shown good catalytic activity and sulfur tolerance and low carbon deposition [22]. Other perovskite materials under study for anodes include $La_{0.75}Sr_{0.25}Cr_{0.5}Mn_{0.5}O_3$ [lanthanum strontium chromite manganite (LSCM)] [23], $La_{0.6}Sr_{0.4}Co_{0.2}Fe_{0.8}O_3$ – GDC [lanthanum strontium cobaltite ferrite (LSCF)–gadolinium-doped cerium oxide (CGO)] [24], and $La_{0.8}Sr_{0.2}Cr_{0.97}V_{0.03}O_3$ [lanthanum strontium chromite vanadite (LSCV)]–YSZ [25], but their long-term stability has not yet been demonstrated.

10.4.2
Cathode Materials

The most common SOFC cathode material has been perovskites, particularly strontium-doped LaMnO$_3$ [lanthanum strontium manganite (LSM)] because of its high electrochemical activity for oxygen reduction at high temperatures, good thermal stability, and good thermal expansion match with YSZ. However, at lower operating temperatures it exhibits significantly high polarization resistance. Mixing LSM with YSZ or GDC to produce a cathode solves this problem to a certain extent; however, at temperatures lower than 750 °C it has relatively high impedance [26]. Cathodes produced by replacing Mn with Co and/or Fe have also been used. Examples are lanthanum strontium ferrite (LSF), lanthanum strontium cobaltite ferrite (LSCF), and lanthanum strontium cobaltite (LSCo). The major problem with these cathode materials is that they react with YSZ at the sintering temperatures [12]. More detailed information on SOFC cathodes can be found in a review paper [27].

10.5
Infiltration

10.5.1
Motivation for Infiltration

Infiltration has been widely applied as a catalyst preparation method; however, the application of infiltration to SOFCs does not have a long history. The term is used interchangeably with impregnation and in the case of SOFCs it implies the deposition of nanoparticles into a porous presintered scaffold. The scaffold is sintered at high temperatures, which provides good bonding of the electrode and the electrolyte, forming a good bridge between the particles for effective electron and ion conduction. A liquid solution of the chosen electrode material which generally contains metal nitrates is applied to the presintered porous structure and calcined to much lower temperatures than in conventional electrode fabrication methods. Low-temperature calcination of the electrode materials prevents reactions with electrolyte material producing low-conductivity phases. An example is perovskite oxide cathode materials such as LSM, LSC, and LSF reacting with YSZ to produce SrZrO$_2$ [28]. LSM–YSZ composite electrodes are generally calcined above 1100 °C; however, by infiltration of LSM into YSZ, it is possible to deposit LSM particles into YSZ at 800 °C [29, 30]. This low-temperature deposition also enhances the TPB by extending the reaction area. As mentioned briefly above, the electrochemical reaction can only occur at the TPB. In case of connection loss of the phases to one another, the reaction fails to occur. Ions from the electrolyte and gas-phase molecules from fuel or air need to be present at the site and electrons need to be transported from the site in order for the site to be active for the reaction. When an MIEC electrode is used, the electrode includes electrolyte material through

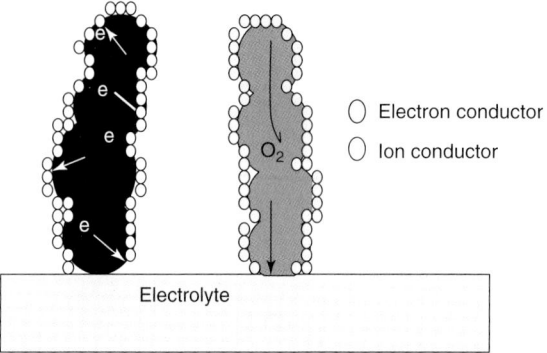

Figure 10.1 Schematic representation of two types of infiltration: left, electronically conducting backbone with ionically conducting particles; right, ionically conducting backbone with electronically conducting particles [31].

the ion-conducting channels of which the ions migrate to the electrode. When the electrode is infiltrated into a presintered backbone, there will be an extended area in the TPB for reaction to occur [32]. The microstructure of the electrode–electrolyte formed by infiltration is depicted schematically in Figure 10.1 [31]. Here, an electronically conducting electrode is infiltrated into an ionic conducting backbone. Basically, this application is a way of producing an MIEC for this specific example. The result of the infiltration is an extended area for charge transport in addition to the TPB. The infiltrated particles are nanosized and, because the application temperature is lower than the sintering temperature, they preserve their nanosize. Several electrodes produced by infiltration have shown good stability; for example, porous ionically conducting material that was infiltrated with LSM exhibited good continued stability over 500 h of operation at 650 °C [33].

Infiltration processes can be applied to enhance the electrode performance, such as in application of additional catalyst one more time after the infiltrated electrode production or once more after producing electrodes by any method. This means that infiltration can be used to produce an electrode or to improve an already produced electrode by additional catalyst application.

Different types of backbones can be used for infiltrating particles. Table 10.5 summarizes the presintered backbone types and presents some examples of infiltrated materials [34].

10.5.2
Infiltration Applications

Studies of infiltration applied to SOFC electrodes go back to 1999 with the work of Gorte's group [35]. They developed a methodology to apply electrode and catalysis materials into porous electrolyte scaffolds to obtain better performing anodes and cathodes. The methodology can be described as first producing a three-layer backbone (e.g., porous YSZ–dense YSZ–porous YSZ) by tape casting, laminating

Table 10.5 Backbones and infiltration applications [34].

Backbone type				
	Electronic	Ionic	MIEC	Composite
Examples	Cathode: LSM Anode: Ni, Cu	Cathode: YSZ, GDC Anode: YSZ, CGO	Cathode: LSCF, BSCF Anode: LCrM, SrTiO$_3$	Cathode: LSM, YSZ Anode: Ni–YSZ
Functions of as-sintered electrode	Electronic pathway, catalysis	Ionic pathway	Ionic and electronic pathways, catalysis	Ionic and electronic pathways, catalysis
Functions of infiltrated material	Ionic pathway, catalysis	Electronic pathway, catalysis	Secondary ionic pathway, catalysis	Local ionic and electronic pathways, catalysis

Abbreviation: LCrM, lanthanum chromium manganite.

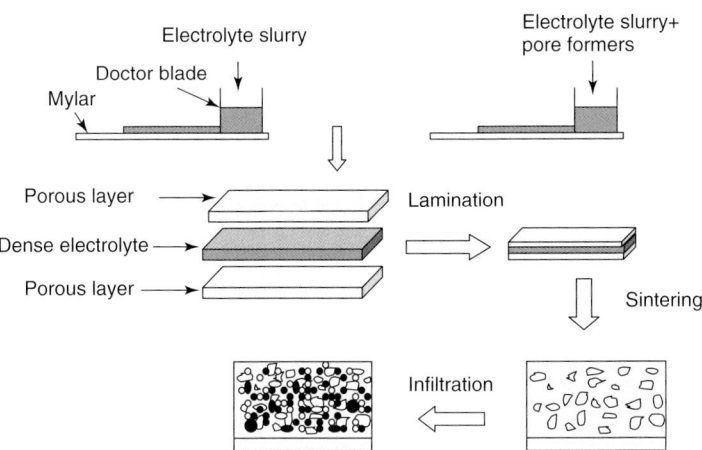

Figure 10.2 General procedure of backbone production and infiltration.

the YSZ layers, and sintering followed by infiltrating the electrode materials into the porous YSZ as depicted by Figure 10.2 [36]. The motivation for investigating a new method for electrode fabrication was mainly sintering the YSZ phase to high temperatures before adding the electrode materials to prevent damage due to high temperatures to the electrode materials and also to replace Ni with another material that would allow dry methane operation [37]. It was then shown that replacing Ni with Cu and adding Cu into the porous YSZ backbone resulted in electrochemical performance similar to that with Ni, with the additional advantage of not forming carbon layers under dry methane operation. Following Gorte and co-workers' studies, other researchers throughout the world have started to become interested in the infiltration approach and have extended the subject to cathodes and to different types of backbones as shown in Table 10.5.

10.5.2.1 Anodes Produced by Infiltration

The most commonly used SOFC anode is the Ni-based cermet anode. However, nickel cermet anodes have the major drawback of carbon fiber formation on the Ni surface when hydrocarbons are used, eliminating the direct oxidation of methane [38, 39]. Conventional Ni–YSZ anodes also undergo irreversible expansion during a redox cycle, resulting in electrolyte cracking [40]. Possible solutions to these problems could either involve systems engineering or materials and microstructural changes and also their combination. Gorte *et al.* developed the approach of infiltrating different catalysts into porous tape-cast YSZ scaffolds. They started with Cu to replace Ni and produced Cu–YSZ composite anodes which were found to be stable in hydrocarbon fuels [41]. The infiltration approach was developed because the melting temperature of copper oxide was not suitable for calcining copper oxide and YSZ together. This problem was solved by first producing the porous YSZ scaffold, which was later infiltrated by Cu salts. In the first applications of infiltration, porous YSZ was prepared using a mixture of zircon fibers and <0.3% Si and pasted on to the dense tape-cast YSZ using glycerol and sintered at 1550 °C for 2 h. Cathode material was added and an aqueous solution of $Cu(NO_3)_2$ was then infiltrated into the pores of the YSZ and calcined at 950 °C [35]. In addition to Cu, CeO_2 was added as catalyst for hydrocarbon oxidation. The cell produced with Cu infiltration resulted in a power density equal to that of nickel cermet anodes operated under the same conditions.

This first application of infiltrated anodes was by its nature electrolyte supported. In the following years, Gorte's group developed a procedure that allowed the cell to be electrode supported. This improved application permitted a much thinner electrolyte to be produced than in the previous application. The YSZ electrolyte was tape cast first, and a second layer of YSZ containing graphite as pore former was cast on top of the electrolyte green tape and they were sintered together at 1550 °C for 4 h. The graphite was pyrolyzed after sintering, resulting in a porous YSZ layer. The cathode was a $YSZ/Sr-LaMnO_2$ paste applied on the other side of the dense YSZ. The nitrate salts of copper and ceria were infiltrated on the porous YSZ and the cell was calcined at 950 °C. The SOFC produced was tested using H_2, CH_4, and C_4H_{10}. The application of copper and ceria together resulted in an increase in the effective TPB due to ceria's MIEC, thus increasing the performance of the SOFC [42, 43]. Following this bilayer tape casting development, the same group continued to work on applications of Cu, combining it with Ni and applying different of Cu : Ni ratios using infiltration, and tested the SOFC cell with CH_4. The combination of Cu and Ni resulted in negligible carbon deposition after 500 h of operation [44].

Porosity is one of the most important parameters which affects the anode (and also the cathode) structure to be infiltrated, hence it needs to be controlled for an effective infiltration application. One of the methods of forming porosity in ceramic structures is to use powders of different particle sizes. Sintering of these powders creates a porous network, but this method can only produce porosities up to 40–50% [45]. Addition of pore formers to the tape-casting slurry has resulted in higher porosities than using powders of different particle sizes [46]. Examples of the pore-former agents that are added to the tape-casting slurry are graphite, polyethylene, rice or potato starch, and poly(methyl methacrylate) (PMMA). A study on the amount and type of pore formers used and the porosity obtained showed that the porosity of YSZ increased almost linearly with increasing amount of pore former [47]. Aside from the above-mentioned pyrolyzable pore formers, NiO has also been tried as pore former for creating a porous YSZ backbone [48]. This method involves first producing Ni–YSZ cermet and then removing the nickel by boiling it in nitric acid. The advantage of using NiO for pore forming is that it allows the determination of the pore size distribution; however, the extra step of acid leaching makes it a less preferred method compared with addition of a pyrolyzable pore former. Ye et al. used ammonium oxalate to create pores to be infiltrated [49]. The scaffold was composed of a thin, dense scandia-stabilized zirconia (ScSZ) electrolyte, supported by a thick, porous ScSZ layer, with thicknesses of 20 and 600 μm, respectively. The porosity obtained by using 40 wt% ammonium oxalate was 55% after sintering. The $Cu-CeO_2-YSZ$ infiltration system which was explained above was replaced with a $Cu-CeO_2-ScSZ$ system in their work and tested using ethanol as fuel. For the cathode, $(Pr_{0.7}Ca_{0.3})_{0.9}MnO_3$ [praseodymium calcium manganite (PCM)] was screen-printed on to dense ScSZ with a thickness of 30 μm after sintering. The total amount of Cu and CeO_2 was 30 wt% and the ratio of Cu to CeO_2 was varied between 2.5 : 1 and 1 : 1 with four different cells at temperatures of 800, 750, and 700 °C. The cell with a 2.5 : 1 ratio of Cu to CeO_2 showed the best power density performance and it was the first $Cu-CeO_2-ScSZ$ anode that had stable performance after 50 h of operation under ethanol.

The thickness of the porous scaffold that will be infiltrated to produce the anode is an important parameter. The electrochemical reaction occurs at the TPB, as already mentioned above. It has been reported by various researchers that the maximum extension distance of the TPB from the electrolyte into the electrode is 10 μm [50–52]. This observation shows that if the porous scaffold produced is thicker than 10 μm, it will be difficult for the infiltrated anode catalyst to be in the TPB zone, hence the power density to be obtained from the SOFC will be lower than it should be.

Another solution to overcome the thickness issue is the addition of materials with MIEC to the anode to extend the TPB area [53]. A related issue to the thickness of the porous backbone is the mechanical strength. If the backbone is required to be of a thickness of around 10 μm and to have pores, then the mechanical strength will be very different than that of conventionally produced electrodes. Especially if the SOFC is anode supported then the anode should have a thickness in the region of 100 μm.

Gorte and co-workers added a support layer on a porous electrode to produce mechanically resistive SOFCs with infiltrated electrodes [54]. The support layer can be produced using the same material as the electrolyte. The support layer

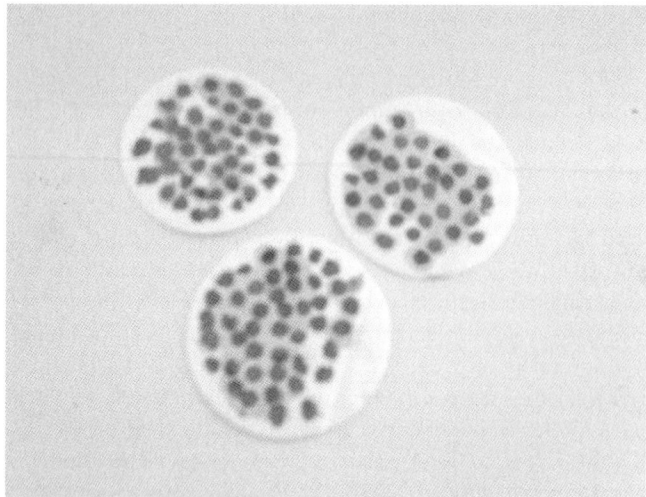

Figure 10.3 Infiltration application through a support layer; cells produced at TUBITAK Marmara Research Center Fuel Cell Laboratory.

is produced using the same method as the electrolyte layer and small holes are produced on the support layer using a punch, then the layers are laminated and sintered together. The electrode solution is infiltrated through the holes of the functional layer of the anode. An application of this procedure is shown in Figure 10.3, made in TUBITAK Marmara Research Center's Fuel Cell Laboratory. The support layer can also be made of a composite of the electrolyte material and an electronically conducting material, for example, YSZ–$La_{0.3}Sr_{0.7}TiO_3$ [lanthanum strontium titanate (LST)]. LST is preferred with YSZ because it does not undergo a solid-state reaction with YSZ and it increases the electronic conductivity of the outer layer of the electrode. The mechanical strength is higher when pure YSZ is used but adding LST to the YSZ provides better electrical contact. A similar approach can also be applied to the cathode.

The starting motivation of the infiltration approach in SOFCs had been to replace Ni for direct hydrocarbon usage. However, Ni has been a widely used anode material and research using Ni has continued. There have been efforts to solve problems associated with Ni, such as particle agglomeration, using infiltration. One of the methods tried was first to produce an Ni–YSZ anode in a conventional way, for example, by preparing the YSZ electrolyte by tape casting and sintering it, followed by screen-printing Ni–YSZ made from NiO and YSZ powders [55]. After screen-printing the LSM cathode, the Ni–YSZ cermet anode was infiltrated with nitrate solutions of $Y(NO_3)_3 + ZrO(NO_3)_2$, $Sm(NO_3)_3 + Ce(NO_3)_3$, and $Ce(NO_3)_3$. The amount of the oxide material infiltrated was determined by weighing the anode after infiltration. Electrochemical performance tests were carried out using H_2 and current interruption transient curves yielded the overpotential and the electrode ohmic resistance. The tests showed that addition of the above-mentioned nitrate salts lowered the polarization, especially the nitrate salts containing ceria, by coating on the Ni surface and inhibiting the agglomeration of Ni particles.

Another way of improving Ni–YSZ anodes using infiltration is to produce an Ni–YSZ anode by applying $Ni(NO_3)_2 \cdot 6H_2O$ solution to a porous layer of YSZ tape cast on dense YSZ electrolyte [56]. The porous YSZ was reported to be 800 μm thick and the dense YSZ was 30 μm. Two anodes were prepared by applying $Ni(NO_3)_2 \cdot 6H_2O$ and $Ni(NO_3)_2 \cdot 6H_2O + Ce(NO_3) \cdot 6H_2O$ solution to the porous YSZ then calcining to 800 °C to form NiO and $NiO-CeO_2$. Pt paste was applied on the cathode side to complete the single cell. The infiltrated anode performance was compared with that of an NiO–YSZ/YSZ composite anode containing 56 wt% NiO. The increased Ni content from 20 to 25 wt% in the infiltrated anode proved to have a significant effect on the power density; however, it was reported that when the Ni content was increased to 30 wt%, although the ohmic resistance decreased with increase in Ni content, the power density decreased. This observation was attributed to the possibility of agglomeration of the reduced Ni on the YSZ surface. The addition of CeO_2 to the Ni anode further increased the power density of the single cell. It has been shown that a minimum Ni loading of 30 vol.% of is necessary for sufficient electrical conductivity [57]. Singh and Krishnan studied the effects of lower Ni loadings on electrical conductivity by infiltration into a porous scaffold, with the aim of attaining better redox stability [58]. They reported that the electrical conductivity of the anode was improved significantly from 10^{-3} S cm^{-1} for porous YSZ without any Ni to 15 S cm^{-1} for 5 vol.% Ni and 155 S cm^{-1} for 10.5 vol.% Ni in 37% porous YSZ, which was reported to be comparable to the results with the conventionally produced NiO–YSZ anode with a 30 vol.% Ni content.

The infiltration technique is also applied to enhance anodes produced conventionally. Jiang et al. [59] investigated the effects of submicron YSZ and GDC infiltrated on an NiO anode. The anode was prepared by coating a slurry of NiO mixed with poly(ethylene glycol), which was then painted on a YSZ electrolyte. The YSZ suspension and $Gd_{0.2}Ce_{0.8}(NO_3)_x$ were applied on Ni anodes using a dropper and heated at 850 °C for 1 h. The electrochemical performance test results for the anode revealed that the electrode polarization resistance of the YSZ infiltrated anode (21 vol.% YSZ) was decreased ∼3.5-fold and that of the GDC infiltrated anode (8.5 vol.%) GDC was decreased ∼7.0-fold. Ni is an electronic conductor and because of that the TPB is limited to the electrode–electrolyte interface. The infiltration of Ni with ionic conducting YSZ and GDC improves the TPB region, resulting in lower electrode polarization. In another study, this work was extended to test a GDC-impregnated Ni anode under humidified methane [60]. The tests were performed under open-circuit conditions at 800 °C as a function of exposure time and the GDC-infiltrated Ni anode was compared with the pure Ni anode and Ni–GDC cermet anode. The pure Ni anode proved to have very poor electrocatalytic activity for the reaction in methane and the impedance arcs increased rapidly in time, showing unstable behavior, whereas the Ni–GDC cermet anode showed much better activity and stability. However, the best performance was obtained with the GDC-impregnated Ni anode: the electrode polarization was ∼25 times

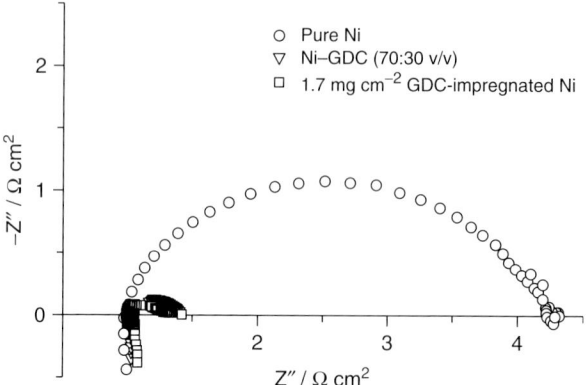

Figure 10.4 Impedance curves of Ni anode, Ni–GDC cermet anode, and GDC-impregnated Ni anode in 97% $H_2 - 3\%H_2O$ at 800 °C under open-circuit conditions.

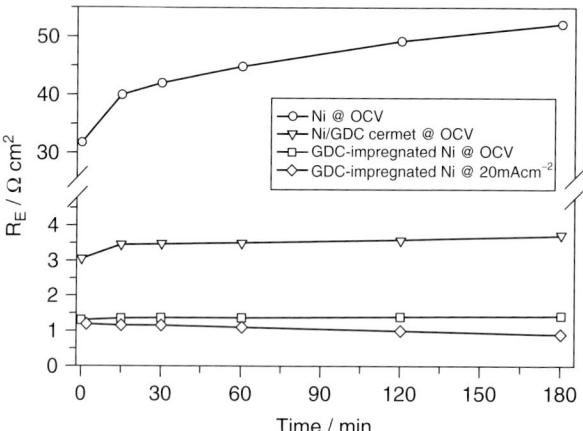

Figure 10.5 Comparison of electrode polarization resistance of Ni anode, Ni–GDC cermet anode, and GDC-impregnated Ni anode in 97%CH_4–!3%H_2O at 800 °C. Reprinted from *Journal of Power Sources*, Wang, W., Jiang, S.P., Tok, A.I., and Luo, L, GDC-impregnated Ni anodes for direct utilization of methane in solid oxide fuel cells, **159**, 68–72. Copyright 2006, with permission from Elsevier.

smaller than that of the pure Ni anode and ~2 times smaller than that of the Ni–GDC cermet anode under the same conditions (Figures 10.4 and 10.5).

Another method of infiltration used in the production of electrodes is to add catalyst to pre-made electrodes. Gorte's group further investigated LST–YSZ anodes by infiltrating Pd and ceria to add catalytic activity to the porous LST-YSZ anodes produced by mixing the LST and YSZ powders and tape casting [61]. In the same study, the replacement of porous LST–YSZ with $Y_{0.04}Ce_{0.48}Zr_{0.48}O_2$ was also examined. However, the reported anode performances were poor compared with the performance achieved with Pd-doped ceria coating the YSZ scaffold

[62]. In another study by Gorte's group, infiltrated LST anode performance was examined, this time adding it to porous YSZ by infiltration, and the performance was compared with that of LSCM–YSZ composites produced by the same group [63]. Both of the infiltrated oxides were 45 wt% and both cells had the same cathode. The studies showed that LSCM–YSZ anodes had better power densities than LST–YSZ-based anodes. This result was attributed to the much higher ionic conductivity of LSCM compared with LST under reducing conditions. Savaniu and Irvine produced an La-doped $SrTiO_3$ anode with A-site deficiency infiltrated with GDC and Cu to obtain more conductive phases than the stoichiometric ones at the same oxygen partial pressure [64]. The dense YSZ and porous LST_{A-} anode and the porous YSZ scaffold were prepared by tape casting. The porous LST_{A-} anode was infiltrated with 10–15 mg cm^{-2} CGO and 8–10 mg cm^{-2} Cu. The conductivity measurements of LST_{A-} with and without CGO + Cu infiltration showed that the electrical conductivity of LST_{A-} was enhanced significantly after the infiltration. In efforts to produce Ni-free anodes, LSCM was studied in an attempt to increase the electrical conductivity [65]. Addition of Cu and Cu–Pd was considered to offer a solution to this problem. Infiltration was preferred owing to the low melting temperature of CuO. The effects of Cu and Cu–Pd infiltrations were studied on LSCM anodes.

A different electrolyte, $La_{0.8}Sr_{0.2}Ga_{0.83}Mg_{0.17}O_{2.815}$ [strontium- and magnesium-doped lanthanum gallate (LSGM)], was used and an interlayer of La_2O_3-doped CeO_3 (LDC) was screen-printed on the electrolyte in order to decrease the interaction of LSGM and LDC. An LSCM anode was produced by screen-printing on the LDC layer. Cu was applied on the LSCM as an aqueous solution of $Cu(NO_3)_2 \cdot 5H_2O$ and calcined at 850 °C to form CuO, and Pd was inserted into the LSCM + Cu using an aqueous solution of $(NH_3)_4PdCl_2 \cdot H_2O$. Addition of 20 wt% Cu to the LSCM anode reduced the overpotential significantly. The LSCM with Cu anode yielded a current density five times greater than that without Cu for hydrogen oxidation, and a similar result was obtained for methane oxidation. The addition of Pd to the LSCM + Cu anode further increased the performance. LSCM was also produced by infiltrating porous YSZ with LSCM nitrate salts [66].

This work introduced an interesting way of pore forming using an Ni leaching method and a pore-former agent together. The porous YSZ backbone was produced by mixing NiO and YSZ powders and tapioca pore former, pressed into a pellet and fired at 1000 °C for 2 h followed by coating the YSZ electrolyte and sintering at 1400 °C for 4 h. Then the bilayer was placed in nitric acid to soak out the Ni. As a result of this procedure, a porous YSZ layer with 70% porosity was obtained. The anode solution of LSCM to be impregnated was prepared by dissolving prefired La_2O_3 in dilute nitric acid and adding $Cr(NO_3)_3 \cdot 9H_2O$, $Sr(NO_3)_2$, and $Mn(NO_3)_2$ followed by addition of urea as complexing and dispersing agent. The YSZ scaffold was dipped into the LSCM solution and heated to 300 °C, and the infiltration procedure was repeated to reach the desired loading. After the infiltration of the anode, the $La_{0.8}Sr_{0.2}MnO_3$ slurry was coated on the YSZ electrolyte. The authors noted that heating at 300 °C was not high enough for single-phase LSCM production. In addition to infiltrated LSCM, Ni and Ag were infiltrated to the anode to increase the catalytic activity and electronic conductivity.

So far, the studies on infiltration described above have all been applied to a tape-cast presintered porous backbone and on planar substrates. A different approach was followed by Zhang et al. [67], where the NiO–YSZ porous substrate for a tubular SOFC was prepared by gel casting, which was chosen mainly because of the cost-effectiveness and ease of scaling up. In order to prepare the gel casting mixture, first NiO and YSZ powders were mixed. Acrylamide monomer, N,N'-methylenebisacrylamide cross-linker, and poly(acrylic acid) were added and dissolved in distilled water. The NiO–YSZ powder mix was added and the mixture was ball-milled for 3 h. Ammonium persulfate was added as initiator and the mixture was ball-milled for a further 30 min. Then the slurry was poured into a mold made of two glass tubes fixed coaxially on a bottom plate. The slurry was in a gel form owing to the polymerization and the gel which was formed was removed from the mold, dried for 3–5 days at room temperature and fired at 1100 °C. After the preparation of an NiO–YSZ porous support and a porous functional NiO–YSZ and firing them at 1150 °C, a thin layer of YSZ electrolyte was applied by dip coating. The anode was then infiltrated with an aqueous solution of $Sm(NO_3)_3$ and $Ce(NO_3)_3$. The authors preferred the gel-casting method to produce the anode substrate due to cost; however, compared with tape casting this procedure is extremely long with the 3–5 days drying time.

10.5.2.2 Cathodes Produced by Infiltration

The conventional cathode material for SOFCs has been Sr-doped $LaMnO_3$(LSM), mainly because it has relatively high electrochemical activity for the oxygen reduction reaction and it has good stability and compatibility with YSZ [68]. LSM cathodes have been improved by adding YSZ or GDC for ionic conductivity. The high temperature necessary for sintering YSZ causes solid–state reactions between LSM and YSZ [69] and also some other conducting oxides such as Sr-doped $LaFeO_3$ (LSF) and $LaSrCoO_3$ (LSCo). Gorte and co-workers [69] applied the infiltration method they had developed and applied to anodes to LSM cathodes to prevent this solid-state reaction problem by producing porous YSZ that was sintered together with the dense YSZ electrolyte. The infiltration method to produce LSM–YSZ (or the other conduction oxides mentioned above) composite cathode provides the advantage of eliminating the necessary barrier layer between LSM and YSZ which needs to be added to prevent the solid-state reactions. Gorte and co-workers [69] prepared the porous YSZ backbone as described earlier. The impregnation solution of LSF was prepared by dissolving $La(NO_3)_3 \cdot 6H_2O$, $Fe(NO_3)_3 \cdot 9H_2O$ and $Sr(NO_3)_2$ in water with an La:Sr:Fe molar ratio of 0.8:0.2:1 and citric acid was added. The solution was stirred for 24 h at 333 K. The solution was applied to the porous YSZ multiple times and after each application calcined at 1123 K to remove the citrate and nitrate ions. The LSF–YSZ cathode that was produced by infiltration was compared with the conventional LSM–YSZ cathode and the impedance results showed that the former cathode showed better performance. In another study the same group produced an LSM–YSZ cathode by infiltration [70]. The aqueous solution was prepared by using the nitrate salts of La, Sr, and Mn with an La : Sr : Mn molar ratio of 0.8 : 0.2 : 1. It was noted that the amount of LSM in YSZ should be 40 wt% in order to have sufficient electronic conductivity. In order to achieve that, multiple impregnations had to be conducted, and after each infiltration a calcination procedure was performed at various temperatures to observe the effect of the calcination temperature on the electrical conductivity. The highest electrical conductivity was found to be that of the LSM–YSZ cathode calcined at 1523 K. The SOFC single cell was completed by producing the $Co - CeO_2 - YSZ$ anode. The performance of the infiltrated LSM–YSZ cathode was compared with that of the same cathode produced by the conventional method with LSM–YSZ ink applied to YSZ electrolyte. The electrochemical performances of the cell with the infiltrated cathode and the conventional cell were similar; however, the infiltrated cell had attractive mechanical properties. Conventional LSM–YSZ cathodes are sintered at 1523 K so that the electrode can adhere to the electrolyte. For infiltrated cathodes, the YSZ is sintered at 1823 K before the addition of the perovskite, which makes the cathodes very strong. This advantage can also be used in producing cathode-supported cells.

Another infiltration application of LSM cathodes was investigated by Jiang and Wang [71, 72]. They chose to enhance the LSM cathode by infiltrating GDC into the cathode. The impregnation solution was prepared using a 3 M solution composed of 10 mol% $Gd(NO_3)_3$ and 90 mol% $Ce(NO_3)_3$. The solution was applied on top of the LSM cathode by placing a drop on the cathode followed by calcination at 850 °C. After six applications of the infiltration of the GDC solution, the GDC loading was 5.8 mg cm^{-2}. The cathode performance was measured by painting Pt paste as reference electrode and comparison with the LSM cathode without infiltration. It was shown that the application of GDC increased the performance significantly, taking advantage of the low-temperature process of infiltration to circumvent the negative effect of the interfacial reactions between the YSZ electrolyte and MIEC cathodes.

Following the above two-step manufacturing process of porous LSM application on a dense electrolyte and infiltrating GDC into the pores of LSM, Xu et al. [73] produced LSM–$Sm_{0.2}Ce_{0.8}O_{1.9}$(SDC) cathodes by infiltrating SDC into the porous LSM. The same group later produced an LSM–SDC cathode by first producing porous SDC on SDC electrolyte and infiltrating LSM into the porous SDC [74]. The resulting cell performance was compared with that of the cell produced with SDC-infiltrated LSM and the conventionally produced LSM–SDC cathode. The results showed that the interfacial polarization resistance of the infiltrated LSM into SDC cathode was much lower than that of both the conventional LSM–SDC and the SDC-infiltrated LSM cathodes.

De Jonghe's group studied the effects of nanoparticle infiltration on LSM cathodes to improve the catalytic activity of LSM at reduced temperatures [75]. $Sm_{0.6}Sr_{0.4}CoO_{3-x}$ [samarium strontium cobaltite (SSC)] was chosen to infiltrate into LSM because it is an effective electrocatalyst for oxygen reduction reactions owing to its MIEC properties. The SSC solution was prepared using the nitrate salts and urea and LSM–YSZ cathodes were saturated into the SSC nitrate–urea solution and heated at 800 °C. Compared with the conventional infiltration applications the firing time is very long. It was shown that after the SSC infiltration, 20–80 nm SSC particles were added to the LMS pores. The cell performance after the addition of the SSC particles was shown to be significantly improved.

Recently, Hansen et al. attempted to address the question of whether the effect of impregnated particles was dependent on the electrode structure [76]. They prepared two LSM–YSZ cathodes, one impregnated with LSM and the other with Al_2O_3, to study the effect of the scaffold structure on the activity of the infiltrated electrodes. The cells prepared were symmetrical LSM–YSZ produced by slurry spraying and screen-printing of size 4 × 4 mm. The solutions to be impregnated were prepared using the nitrates of La, Sr, and Mn for LSM nanoparticles and of Al for Al_2O_3 nanoparticles. The LSM-impregnated cells showed a decrease in the area specific resistance (ASR) and slurry-sprayed cells had better performance. The infiltration of Al_2O_3 had a negative effect on the ASR of the cells.

Another recent study to enhance the LSM cathode was conducted by Ai et al. [77] in order to decrease the polarization resistance of the LSM cathode at temperatures of 600–800 °C. Whereas $Ba_{0.5}Sr_{0.5}Co_{0.8}Fe_{0.2}O_3$ [barium strontium cobalt ferrite (BSCF)] is a good MIEC and has a high electrocatalytic activity, its thermal expansion coefficient is significantly higher than that of YSZ and it reacts with YSZ at 800 °C and higher temperatures. In order to eliminate these disadvantages of BSCF, Ai et al. infiltrated the material as a solution of nitrate salts into the presintered backbone of porous LSM–YSZ/dense YSZ of area 0.5 cm^2 and fired it at 800 °C. The BSCF-infiltrated cathode had a polarization resistance ∼12 times smaller than that of the pure LSM cathode at 800 °C. The infiltrated LSM cathode also resulted in a power density about four times higher than that of the pure LSM cathode at 650 °C.

Some infiltration applications of precious metals to SOFC cathodes to improve the performance have been reported. An example is a Pd-infiltrated YSZ cathode stabilized with Mn [78]. It was shown that addition of Pd nanoparticles enhanced the oxygen reduction reaction and Pd infiltration to YSZ reduced the electrode polarization resistance to 0.22 Ωcm^2. Liang et al. [78] compared the electrochemical performance of Pd-infiltrated LSM-based cathodes with that of the regular LSM–YSZ cathode and showed that the LSM with Pd cathode was significantly superior, with 0.9 Ωcm^2 at 600 °C. One disadvantage of the Pd-infiltrated cathode was that at 800 °C the Pd particles agglomerated. In order to improve the stability of the Pd-infiltrated cathode, Liang et al. added Mn to the Pd. Solutions of Pd and $Pd_{0.95}Mn_{0.05}$ were infiltrated into porous YSZ. Half-cells were prepared by pasting Pt on the other side of the cell. Electrochemical impedance spectroscopic analysis showed that the electrode polarization resistance of the $Pd_{0.95}Mn_{0.05}$–YSZ cathode (0.18 Ωcm^2) was lower than that of the Pd–YSZ cathode (0.23 Ωcm^2) after 30 h of operation under a 200 mA cm^{-2} current at 750 °C. Scanning electron microscopy (SEM) images of the two cathodes showed that after 30 h of operation the Pd–YSZ cathode had a continuous layer of Pd instead of Pd particles whereas the $Pd_{0.95}Mn_{0.05}$–YSZ cathode retained the individual particles, confirming the stability of the $Pd_{0.95}Mn_{0.05}$–YSZ cathode.

Xiong et al. [79] produced a PCM cathode in an anode-supported cell infiltrated with SDC. The Ni–YSZ anode and Sc-stabilized ZrO_2 were multilayer cast and sintered together. The PCM cathode was screen-printed and, following the sintering, $Sm_{0.2}Ce_{0.8}(NO_3)_x$ solution was infiltrated into the cathode. The electrochemical performances of three different cells were measured: a cell with no SDC, a cell with a 1.3 mg cm^{-2} and a cell with a 2.6 mg cm^{-2} SDC loading. The maximum power density was obtained with the cell with 2.6 mg cm^{-2} SDC.

Nicholas and Barnett [80] conducted a study to evaluate the effect of infiltration solution additives, especially Triton X-100 and citric acid, which could influence the infiltrated nanoparticle morphology along with the phase purity on an $Sm_{0.5}Sr_{0.5}CoO_{3-x}$(SSC)–$Ce_{0.9}Gd_{0.1}O_{1.95}$(GDC) cathode. For that purpose, a pure 0.5 M nitrate solution, a solution of 3 wt% Triton X-100 (weight of Triton X-100/weight of nitrates) containing 0.5 M nitrate, and a citric acid solution containing 0.5 M nitrate with a 1 : 1 citric acid : nitrate ratio were prepared. GDC was screen-printed on to both sides of dense GDC disks and $La_{0.6}Sr_{0.4}Co_{0.2}Fe_{0.8}O_{3-x}$ was screen-printed as current collector. X-ray diffraction (XRD) scans of the SSC solutions after heating at 800 °C for 1 h indicated that all of the samples mostly contained SSC phases and pure nitrate, and Triton X-100 solutions contained Co_3O_4, $SmCoO_3$, and $SrCoO_3$ phases, while XRD scans of the citric acid-containing SSC sample showed no impurities. It was concluded that the presence of the additives did not have a significant effect on the infiltrated SSC particles and shape. The infiltrate loading, infiltrate solution concentration, and the number of infiltrations were shown to have a great impact on cathode polarization.

LSCF is another promising cathode material for intermediate-temperature SOFCs [81, 82]. Owing to its much higher oxygen ionic conductivity than LSM, LSCF is highly active for oxygen reduction reactions at lower temperatures. However, LSCF has a tendency to react chemically and is thermally incompatible with YSZ. This disadvantage can be eliminated if LSCF is infiltrated into YSZ as a solution. Chen et al. investigated the preparation of LSCF infiltration solution and its application into porous YSZ [83]. The LSCF solution was prepared using $La(NO_3)_3 \cdot 6H_2O$, $Sr(NO_3)_2$, $Co(NO_3)_2 \cdot 6H_2O$, $Fe(NO_3)_3 \cdot 9H_2O$, fluorocarbon surfactant, isopropyl alcohol, and deionized water. The solution was infiltrated into porous YSZ disks via ultrasonic treatment and fired at 700 °C for 1 h. The cell produced had an active area of 0.5 cm^2. XRD results for the infiltrated LSCF-YSZ composite showed that there were no reaction products of LSCF and YSZ. The electrode polarization resistance of the LSCF-infiltrated YSZ cathode was 0.539, 0.218, 0.089, and 0.047 Ωcm^2 at 600, 650, 700, and 750 °C, respectively, which are fairly low compared with the LSCF cathode produced conventionally, for which a value of 0.35 Ωcm^2 at 700 °C was reported. The advantages of the LSCF-infiltrated YSZ cathode over LSCF–YSZ produced conventionally is that the infiltrated LSCF particles are nanosized and well dispersed in the YSZ pores, thus creating a larger TPB area for the oxygen reduction reaction and the good contact of LSCF particles and YSZ to allow fast ionic migration and transport. The same group recently produced the same type of cell with a larger active area of 81 cm^2 with the goal of evaluating the performance of the infiltrated cathode in a large-scale setting as opposed to the laboratory-scale button cell [84]. The SOFC cell was composed of an NiO–YSZ anode substrate produced by tape casting, an anode functional layer screen-printed on top of the substrate, a dense YSZ screen-printed on the functional anode layer, porous YSZ screen-printed on the dense YSZ, and LSCF infiltrated in porous YSZ. Two cells with 17 and 37 wt% LSCF were produced and tested. The power density of the cell with the 37 wt% LSCF was twice as high as that of the cell with 17 wt% LSCF at 700 °C (Figure 10.6).

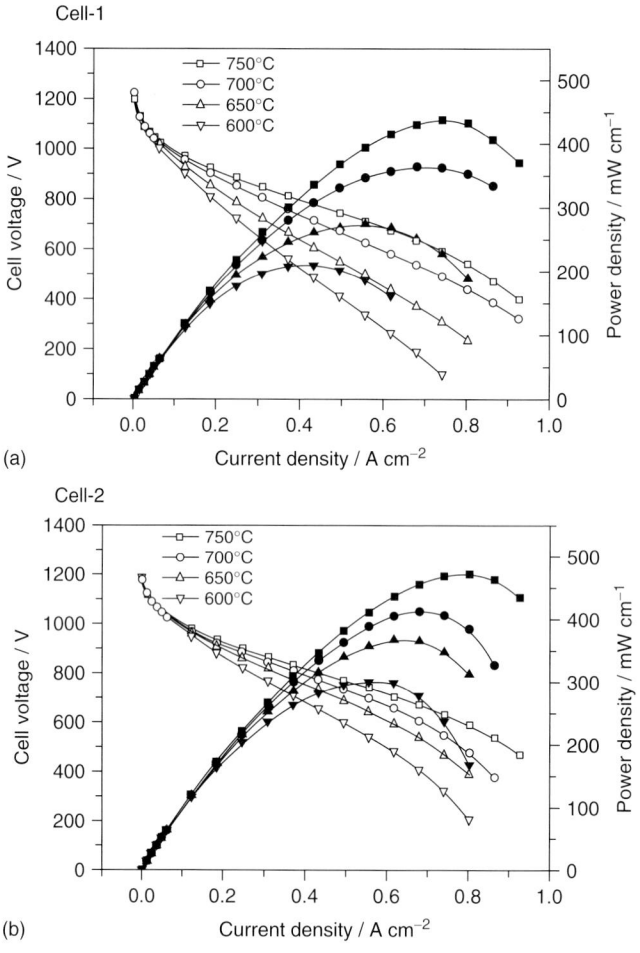

Figure 10.6 Cell performance of the anode-supported planar cells with impregnated LSCF + YSZ composite cathode measured at different temperatures under 4 l min^{-1} of H$_2$ and 4 l min^{-1} of air. (a) Cell-1, LSCF loading ~ 0.6 mg cm^{-2}; (b) Cell-2, LSCF loading ~ 1.3 mg cm^{-2}. Reprinted from *Journal of Power Sources*, Chen, J., Liang, F, Yan, D., Pu, J., Chi, B., Jiang, S.P., and Jian, L., Performance of large-scale anode supported solid oxide fuel cells with impregnated La$_{0.6}$Sr$_{0.4}$Co$_{0.2}$Fe$_{0.8}$O$_{3-\delta}$ + Y$_2$O$_3$ stabilized ZrO$_2$ composite cathodes, **195**, 5201–5205. Copyright 2010, with permission from Elsevier.

A different approach was taken by Nie et al. to produce an LSCF cathode using infiltration [85]. The porous LSCF was produced by tape casting. Both sides of YSZ were covered with SDC slurry to prevent the reaction between LSCF and YSZ, and as a result symmetrical LSCF–SDC–YSZ–SDC–LSCF cells were produced. Nitrate solutions of SDC were prepared using $Sm(NO_3)_2 \cdot 6H_2O$ and $Ce(NO_3)_3 \cdot 6H_2O$ in water. The solution was infiltrated into both sides of the LSCF. The cell with an active area of $0.3\,cm^2$ was then fired at $900\,°C$ for $1\,h$. The SEM images of the SDC infiltration after each infiltration showed the particle size and the film growth on LSCF backbone (Figure 10.7). The blank LSCF cathode had an ASR of 1.09, 0.40, 0.15, and $0.064\,\Omega cm^2$ at 650, 700, 750, and $800\,°C$, respectively. Addition of $0.25\,mol\,l^{-1}$ SDC improved the ASRs by more than 50% compared with the blank LSCF while increasing the SDC content to $0.35\,mol\,l^{-1}$ caused an increase in the ASR. This result was attributed to the reduced length of the SDC–LSCF phase boundaries due to the reduced porosity of the SDC coating.

There are some unclear issues associated with LSCF, such as the effect of the ceria interlayer on the cathode performance and the effect of the Co : Fe ratio. Adijanto et al. investigated the effect of various Co : Fe ratios in the infiltrated LSCF, $La_{0.8}Sr_{0.2}Co_xFe_{1-x}$ with $x = 0.2$–0.8 [86]. The LSCF solutions were prepared using the nitrate salts and the solution was infiltrated into the porous YSZ backbone. Infiltration steps were repeated until a $40\,wt\%$ loading was reached. After the infiltration was complete, the samples were calcined at either 1123 or $1373\,K$ for $4\,h$. The LSCF samples calcined at $1123\,K$ did not show any evidence of reaction at the interface of LSCF and YSZ for all the Co : Fe ratios; however; when calcined at $1373\,K$, samples containing Co mole fractions higher than 0.2 formed insulating $La_2Zr_2O_7$ and $SrZrO_3$ phases.

10.6
Conclusion

Infiltration can provide some advantages over conventionally produced SOFC electrodes. Thermal incompatibilities between the electrolyte and electrode material can be reduced owing to the low-temperature application of the electrode material into the electrolyte after sintering the electrolyte and the porous backbone first. The high-temperature sintering of the porous backbone also allows the application of materials at temperatures lower than that necessary for the electrolyte sintering, hence solid-state reactions with electrode materials and electrolyte materials are prevented when they are applied into the porous backbone via infiltration. The method was first applied to SOFCs in searching for Ni-free anodes for direct oxidation of hydrocarbons. Having proven its positive effect in that application, researchers worldwide have carried out investigations of SOFC electrode production or electrode improvements using the infiltration technique.

Despite the advantages of the methods mentioned above, it also introduces some complexities into the production process. In order to achieve the necessary

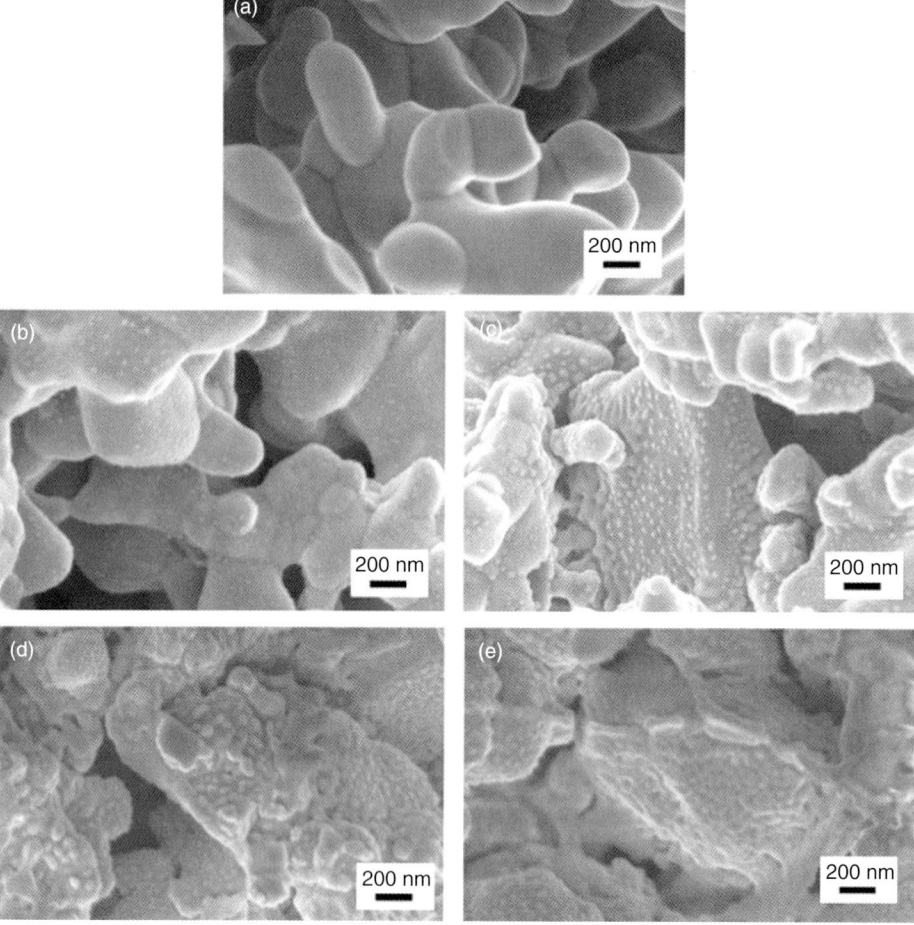

Figure 10.7 Cross-sectional views (SEM images) of a blank LSCF cathode (a) and SDC-infiltrated LSCF cathodes with different concentrations of SDC solutions: (b) 0.05, (c) 0.10, (d) 0.25, and (e) 0.35 mol l^{-1}. Reprinted from *Journal of Power Sources*, Nie, L., Liu, M., Zhang, Y., and Liu, M., La$_{0.6}$Sr$_{0.4}$Co$_{0.2}$Fe$_{0.8}$O$_{3-\delta}$ cathodes infiltrated with samarium-doped cerium oxide for solid oxide fuel cells, **195**, 4704–4708. Copyright 2010, with permission from Elsevier.

amounts of catalytic material in the electrodes, multiple infiltration steps need to be executed. Also, the infiltrated materials have to be calcined after each application, which results in additional process time and cost. Automation and optimization work need to be carried out on the infiltration method to reduce the multiple applications and thus process costs. As reported above, the research on and application of infiltration on SOFC electrodes have mostly been performed on laboratory-scale cells. The method should now be investigated for larger area cells in order to be applicable to SOFC technology.

References

1. Steele, B.C.H. and Heinzel, A. (2002) *Nature*, **414**, 345–352.
2. Serra, J.M. and Buchkremer, H.P. (2007) *J. Power Sources*, **172**, 768–774.
3. Liu, B. and Zhang, Y. (2008) *J. Univ. Sci. Technol. (Beijing)*, **15** (1), 84–90.
4. He, T., Lu, Z., Huang, Y., Guan, P., Liu, J., and Su, W. (2002) *J. Alloys Compd.*, **337**, 231–236.
5. Kim, S.D., Hyun, S.H., Moon, J., Kim, H.H., and Song, R.H. (2005) *J. Power Sources*, **139**, 67–72.
6. Leng, Y.J. and Chan, S.H. (2004) *J. Mater. Sci.*, **39**, 4405–4439.
7. Menzler, N.H., Tietz, F., Uhlenbruck, S., Buchkremer, H.P., and Stöver, D. (2010) *J. Mater. Sci.*, **45**, 3109–3135.
8. Beckel, D., Bieberle-Hütter, A., Harvey, A., Infortuna, A., Muecke, U.P., Prestat, M., Rupp, J.L.M., and Gauckler, L.J. (2007) *J. Power Sources*, **173**, 325–345.
9. Singh, P. and Minh, N.Q. (2004) *Int. J. Appl. Ceram. Technol.*, **1**, 5–15.
10. Simnovis, D., Thulen, H., Dias, F.J., Naoumidis, A., and Stöver, D. (1999) *J. Mater. Process. Technol.*, **92–93**, 107–111.
11. Tietz, F., Buchkremer, H.P., and Stöver, D. (2002) *Solid State Ionics*, **152/153**, 373–381.
12. Antoni, L. (2004) *Mater. Sci. Forum*, **461–464**, 1073–1090.
13. Goodenough, J.B. and Huang, Y.-H. (2007) *J. Power Sources*, **173**, 1–10.
14. Singhal, S.C. and Kendall, K. (2003) *High Temperature Solid Oxide Fuel Cells: Fundamentals, Design, and Applications*, Elsevier, Amsterdam.
15. Dees, D.W., Claar, T.D., Easler, T.E., Fee, D.C., and Mrazek, F.C. (1987) *J. Electrochem. Soc.*, **134**, 2141–2146.
16. Matsuzaki, Y. and Yasuda, I. (2000) *Solid State Ionics*, **132**, 261–269.
17. Murray, E.P., Tsai, T., and Barnett, S.A. (1999) *Nature*, **400**, 649–651.
18. McIntosh, S., Vohs, J.M., and Gorte, R.J. (2002) *Electrochem. Acta*, **47**, 3815–3821.
19. Zhan, Z.I. and Barnett, S.A. (2005) *Science*, **308**, 844–848.
20. Marina, O.A. and Mogensen, M. (1999) *Appl. Catal. A*, **189**, 117–126.
21. Ramirez-Cabrera, E., Atkinson, A., and Chadwick, D. (2000) *Solid State Ionics*, **136**, 825–831.
22. Sfeir, J. (2003) *J. Power Sources*, **118**, 276–285.
23. Tao, S.W. and Irvine, J.T.S. (2003) *Nat. Mater.*, **2**, 320–323.
24. Sin, A., Kopnin, E., Dubitsky, Y., Zaopo, A., Arico, A.S., Gullo, L.R., Russo, D.L., and Antonucci, V. (2005) *J. Power Sources*, **145**, 68–73.
25. Vernoux, P., Guillodo, M., Fouletier, J., and Hammou, A. (2000) *Solid State Ionics*, **135**, 425–431.
26. Minh, N.Q. (1993) *J. Am. Ceram. Soc.*, **76**, 563–588.
27. Adler, S.B. (2004) *Chem. Rev.*, **104**, 4791–4843.
28. Shah, M. and Barnett, S.A. (2008) *Solid State Ionics*, **179**, 2059–2064.
29. Jorgensen, M.J., Primdahl, S., Bagger, C., and Mogensen, M. (2000) *Solid State Ionics*, **139**, 1–11.
30. Sholklapper, T.Z., Lu, C., Jacobson, C.P., Visco, S.J., and De Jonghe, L.C. (2006) *Electrochem. Solid-State Lett.*, **9**, A376.
31. Jiang, Z., Xia, C., and Chen, F. (2010) *Electrochim. Acta*, **55**, 3595–3605.
32. Jiang, S. (2006) *Mater. Sci. Eng. A*, **418** (1–2), 199–210.
33. Sholklapper, T.Z., Radmilovic, V., Jacobson, C.P., Visco, S.J., and De Jonghe, L.C. (2007) *Electrochem. Solid-State Lett.*, **10**, 135.
34. Sholklapper, T.Z., Jacobson, C.P., Visco, S.J., and De Jonghe, L.C. (2007) *Electrochem. Solid-State Lett.*, (2008) *Fuel Cells*, **5**, 303–312.
35. Craciun, R., Park, S., Gorte, R.J., Vohs, J.M., Wang, C., and Worrel, W.L. (1999) *J. Electrochem. Soc.*, **146**, 4019–4022.
36. Vohs, J.M. and Gorte, R.J. (2009) *Adv. Mater.*, **21**, 943–956.
37. Gorte, R.J. and Vohs, J.M. (2009) *Curr. Opin. Colloid Interface Sci.*, **14**, 236–244.
38. Gross, M.D., Vohs, J.M., and Gorte, R.J. (2007) *J. Mater. Chem.*, **17**, 3071–3077.
39. Atkinson, A., Barnett, S., Gorte, R.J., Irvine, J.S., and McEvoy, A.J. (2004) *Nat. Mater.*, **3**, 17–27.

40. Busawon, A.N., Sarantaridis, D., and Atkinson, A. (2008) *Electrochem. Solid-State Lett.*, **11** (10), B186–B189.
41. Gorte, R.J., Vohs, J.M., and McIntosh, S. (2004) *Solid State Ionics*, **175**, 1–6.
42. Park, S., Gorte, R.J., and Vohs, J.M. (2001) *J. Electrochem. Soc.*, **148**, A443–A447.
43. Gorte, R.J., Park, S., Vohs, J.M., and Wang, C. (2000) *Adv. Mater.*, **12**, 1465–1469.
44. Kim, H., Lu, C., Worrell, W.L., Vohs, J.M., and Gorte, R.J. (2002) *J. Electrochem. Soc.*, **149**, A247–A250.
45. Mortensen, A. and Suresh, S. (1995) *Int. Mater. Rev.*, **40**, 239–265.
46. Corbin, S.F. and Apte, P.S. (1999) *J. Am. Ceram. Soc.*, **82** (7), 1693–1701.
47. Boaro, M., Vohs, J., and Gorte, R.J. (2003) *J. Am. Ceram. Soc.*, **86** (3), 295–300.
48. Kim, H., da Rosa, C., Boaro, M., Lu, C., Vohs, J.M., and Gorte, R.J. (2002) *J. Am. Ceram. Soc.*, **85** (6), 1473–1476.
49. Ye, X.F., Huang, B., Wang, S.R., Wang, Z.R., Xiong, L., and Wen, T.L. (2007) *J. Power Sources*, **164**, 203–209.
50. Wang, X., Nakagawa, N., and Kato K. (2001) *J. Electrochem. Soc.*, **148** (6), A565–A569.
51. Tanner, C.W., Fung, K.Z., and Virkar, A.V. (1997) *J. Electrochem. Soc.*, **144** (1), 21–30.
52. Gorte, R.J. and Vohs, J.M. (2003) *J. Catal.*, **216**, 477–486.
53. Fleig, J. (2003) *J. Power Sources*, **118**, 276–285.
54. Ahn, K., Jung, S., Vohs, J.M., and Gorte, R.J. (2007) *Ceram. Int.*, **33**, 1065–1070.
55. Jiang, S.P., Duan, Y., and Love, J. (2002) *J. Electrochem. Soc.*, **149** (9), A1175–A1183.
56. Qiao, J., Sun, K., Zhang, N., Sun, B., Kong, J., and Zhou, D. (2007) *J. Power Sources*, **169**, 253–258.
57. Minh, N.Q. (1993) *J. Am. Ceram. Soc.*, **76**, 563–588.
58. Singh, C.A. and Krishnan, V. (2008) *ECS Trans.*, **6** (21), 25–32.
59. Jiang, S.P., Zhang, S., Zhen, D.Y., and Wang, W. (2005) *J. Am. Ceram. Soc.*, **88** (7), 1779–1785.
60. Wang, W., Jiang, S.P., Tok, A.I.Y., and Luo, L. (2006) *J. Power Sources*, **159**, 68–72.
61. Kim, G., Vohs, J.M., and Gorte, R.J. (2008) *J. Electrochem. Soc.*, **155** (4), B360–B366.
62. Gross, M.D., Vohs, J.M., and Gorte, R.J. (2007) *J. Electrochem. Soc.*, **154** (7), B694–B699.
63. Lee, S., Kim, G., Vohs, J.M., and Gorte, R.J. (2008) *J. Electrochem. Soc.*, **155** (11), B1179–B1183.
64. Savaniu, C.D. and Irvine, J.T.S. (2011) *Solid State Ionics*, **192** (1), 6–8.
65. Lu, X.C. and Zhu, J.H. (2007) *Solid State Ionics*, **178**, 1467–1475.
66. Zhu, X., Lu, Z., Wei, B., Chen, K., Liu, M., Huang, X., and Su, W. (2010) *J. Power Sources*, **195**, 1793–1798.
67. Zhang, L., Gao, J., Liu, M., and Xia, C.J. (2008) *Alloys Compd.*, **482**, 168–172.
68. Jiang, S.P. (2003) *J. Power Sources*, **124**, 390–402.
69. Huang Y., Vohs J.M., and Gorte, R.J. (2004) *J. Electrochem. Soc.*, **151** (4), A646–A651.
70. Huang Y., Vohs J.M., and Gorte, R.J. (2005) *J. Electrochem. Soc.* **152** (7), A1347–A1353.
71. Jiang, S.P. and Wang, W. (2005) *Solid State Ionics*, **176**, 1351–1357.
72. Jiang, S.P. and Wang, W. (2005) *J. Electrochem. Soc.*, **152** (7), A1398–A1408.
73. Xu, X.Y., Jiang, Z.Y., Fan, X., and Cia, C.R. (2006) *Solid State Ionics*, **177**, 2113–2117.
74. Xu, X.Y., Cao, C., Xia, C., and Peng, D. (2009) *Ceram. Int.*, **35**, 2213–2218.
75. Lu, C., Sholklapper, T.Z., Jacobson, C.P., Visco, S.J., and De Jonghe, C. (2005) *J. Electrochem. Soc.*, **153** (6), A1115–A1119.
76. Hansen, K.K., Wandel, M., Liu, Y.L., and Mogensen, M. (2010) *Electrochim. Acta*, **55**, 4606–4609.
77. Ai, N., Jiang, S.P., Lü, Z., Chen, K., and Su, W. (2010) *J. Electrochem. Soc.*, **157** (7), B1033–B1039.
78. Liang, F.L., Chen, J., Jiang, S.P., Wang, F.Z., Chi, B., Pu, J., and Jian, L. (2009) *Fuel Cells*, **5**, 636–642.
79. Xiong, L., Wang, S., Wang, Z., and Wen, T. (2010) *J. Rare Earths*, **28** (1), 96–99.

80. Nicholas, J.D. and Barnett, S.A. (2010) *J. Electrochem. Soc.*, **157** (4), B536–B541.
81. Mai, A., Haanappel, V.A.C., Uhlenbruck, S., Tietz, F., and Stöver, D. (2005) *Solid State Ionics*, **176**, 1341–1350.
82. Tietz, F., Mai, A., and Stöver, D. (2008) *Solid State Ionics*, **179**, 1509–1515.
83. Chen, J., Liang, F., Liu, L., Jiang, S., Chi, B., Pu, J., and Jian, L. (2008) *J. Power Sources*, **183**, 586–589.
84. Chen, J., Liang, F., Yan, D., Pu, J., Chi, B., Jiang, S., and Jian, L. (2010) *J. Power Sources*, **195**, 5201–5205.
85. Nie, L., Liu, M., Zhang, Y., and Liu, M. (2010) *J. Power Sources*, **195**, 4704–4708.
86. Adijanto, L., Küngas, R., Bidrawn, F., Gorte, R.J., and Vohs, J.M. (2011) *J. Power Sources*, **196**, 5797–5802.

11
Sealing Technology for Solid Oxide Fuel Cells
K. Scott Weil

11.1
Introduction

Over the past decade, global energy usage worldwide has expanded by more than 25% [1]. As much as 85% of this consumption in various developed countries, including the United States, is currently sustained by fossil fuels: crude oil, coal, and natural gas [2]. The economic, environmental, and geopolitical risks of continuing on this path are well known [3, 4] and serve as powerful motivations to enhance energy conservation, improve power generation efficiency, and increase the usage of domestic (preferably renewable and sustainable) energy sources. However, no simple alternative has emerged. While heat engine technology (e.g. steam turbines and internal combustion engines) currently meets much of the world's vast and varied power generation needs, efficiency in these systems is bounded by the Carnot limit, which in turn is fixed by the maximum temperature to which the constituent materials in this type of equipment can be exposed. Electrochemical devices, such as fuel cells, are constrained by no such limit and routinely exhibit system efficiencies greater than twice that of a comparably sized heat engine. Hence they offer a means of more effectively utilizing fossil fuel resources in the short term, while cost-effective renewable fuels are being developed. Fuel cells also display other attributes, such as high power density, utilization of a number of fuel feedstocks from various sources, low pollution emissions, quiet operation, few moving parts (and therefore potentially low maintenance), and modularity/scalability from sub- to mega-watt size. The devices are beginning to find commercial markets for both stationary and portable applications, including waste-to-energy conversion, auxiliary power units (APUs), and materials handling equipment [5–7].

11.1.1
Solid Oxide Fuel Cells (SOFCs)

The key feature of any fuel cell is that it converts the chemical energy of an incoming fuel stream directly into electrical energy via an electrochemical reaction [6]. As with other electrochemical devices, a fuel cell relies on an

Fuel Cell Science and Engineering: Materials, Processes, Systems and Technology,
First Edition. Edited by Detlef Stolten and Bernd Emonts.
© 2012 Wiley-VCH Verlag GmbH & Co. KGaA. Published 2012 by Wiley-VCH Verlag GmbH & Co. KGaA.

electrolyte to facilitate ionic conduction while maintaining electrical charge separation. The type of electrolyte employed generally defines the conditions under which the fuel-cell stack will operate optimally and has implications on the required fuel purity, the type of electrocatalysts needed for peak performance, the potential for internal reformation, and the rate at which the system is capable of generating full power from a cold start. While solid oxide fuel cells (SOFCs) are the least mature of the various fuel cell technologies developed to date with respect to commercialization, they offer a number of distinct advantages.

Specifically, SOFCs are the only type of fuel cell constructed entirely from solid materials. All others employ a liquid electrolyte and require a means of maintaining that liquid under proper balance so that the individual cells will neither dry out nor flood, in either case losing function. Direct hydrocarbon fuel reformation is possible within a SOFC system because it operates at high temperature, on the order of 650 °C or higher, and can utilize both the resulting hydrogen and carbon monoxide product gases in subsequent electrochemical reactions. This potentially eliminates the need for an external fuel reformer and enables a wide array of commercially available hydrocarbon fuels to be used in powering the stack, including natural gas, methanol, and coal gas or syngas [8]. The latter capability may turn out to be a significant commercial advantage that allows the technology to transition adeptly from the current nonrenewable fossil fuel-based energy economy to one eventually predicated on renewable fuels. The large amount heat generated during SOFC operation can not only sustain internal fuel reformation and keep the stack at its proper operating temperature, but can also be utilized to increase the overall efficiency of the system [8], making these devices among the most efficient yet invented for the conversion of chemical fuels into electrical power. High-temperature operation also mitigates the need for expensive noble metal electrocatalysts, currently an economic barrier for low-temperature polymer electrolyte membrane fuel cell technology.

There are two primary cell geometries that are employed in SOFC stacks, planar and tubular. Planar stacks are constructed from an alternating series of flat ceramic cells and metal interconnect plates. Each cell consists of a solid electrolyte material sandwiched between a ceramic cathode and anode. During operation, gaseous fuel flows across and into the porous anode layer where it oxidizes and in the process gives up electrons to an external circuit. Electrons returning from the external circuit to the cathode reduce the oxidant (oxygen in air) flowing across that side of the cell. Most SOFC designs rely on oxygen anion transport from the cathode through the electrolyte to the anode, where combination with an incoming fuel produces water and/or CO_2 that exit the system as exhaust vapor. Voltage is built up by electrically connecting the anode of one cell to the cathode of the next using a conductive interconnect, thereby forming a stacked array of cell–interconnect repeat units. Tubular designs employ tubes of solid electrolyte coated on the inside with an anode or cathode layer and on the outside with the opposite. Depending on the coating arrangement, either air or fuel is passed through the inside of the tube while the other gas is passed along the outside. It is generally easier to achieve robust sealing in tubular stacks than in planar SOFC designs because tubular seals tend to be exposed to lower temperatures [8].

Over the past several years, commercial interest in the planar stack concept (i.e. planar solid oxide fuel cell or pSOFC) has accelerated, primarily because these devices operate over a wider range of temperatures and can be produced at lower cost on a dollars per kilowatt basis [9]. In addition, the compact nature of pSOFCs affords significantly higher volumetric power densities and smaller footprints than those of tubular devices. The former design feature is of particular importance for mobile and portable applications, whereas the latter is valuable in stationary applications. Because of the need to maintain separation between the fuel and oxidant gases in any fuel cell design, hermeticity across the electrolyte membrane is paramount. The presence of leaks, due either to flaws that originate during stack manufacture or that form because of cell or seal degradation during stack operation, lead to reduced system performance, lower power generation efficiency, and substandard fuel utilization [10]. A simple measure of the power loss associated with a modest leak can be estimated with the Nernst equation [8]:

$$E = \frac{RT}{4F} \ln \left[\frac{P_{O_2}(c)}{P_{O_2}(a)} \right] \qquad (11.1)$$

where E is the voltage generated, T is the operating temperature, R and F are the universal gas and Faraday constants, respectively, and $P_{O_2}(c)$ and $P_{O_2}(a)$ are the partial pressures of oxygen at the cathode and anode, respectively. In "leaky" cells, for example, where $P_{O_2}(a)$ increases from 10^{-18} to 10^{-10} atm due to the leak, the voltage (and proportionally the maximum amount of power) is 46% less than that of a hermetic cell. Leaks can also cause local hot spots or, worse, widespread internal combustion in the stack, both of which induce accelerated degradation of the cells [10]. In a planar stack design, this means that the electrolyte layer must be dense and connected to the rest of the device with a high-temperature, gas-tight seal. One of the fundamental challenges in fabricating pSOFCs is how to join the thin electrochemically active ceramic cell effectively to the metallic body of the device and thereby create a rugged, hermetic, and chemically stable seal. Typical conditions under which these devices are expected to operate, and to which the accompanying seals will be exposed, include (i) an average operating temperature of 750 °C, (ii) continuous exposure to an oxidizing atmosphere on the cathode side and a wet reducing gas on the anode side, and (iii) an anticipated device lifetime as high as 50 000 + h.

Although sometimes underappreciated during stack and system development, proper seal design and sealant selection are critical to the long-term performance of these devices. The progress made in SOFC sealing technology over the past few years has been extensive enough that several companies are in the initial stages of commercializing kilowatt-size pSOFC systems [9, 11]. With this in mind, this chapter outlines the general requirements for sealing pSOFC devices, reviews the current state-of-the-art, and comments on future development needs for pSOFC sealing technology. It should be noted that the seals discussed here can also be employed in tubular SOFC stack fabrication, although the functional requirements for sealing in these stacks are generally less rigorous than those demanded by planar designs.

11.1.2
Functional Requirements for pSOFC Seals

The starting point for seal selection is consideration of the specific stack design and the conditions under which the stack will operate. Numerous pSOFC stack designs are currently under development, many of which have been reviewed in References 9, 12, and 13. The two generic examples shown in Figure 11.1 illustrate the impact of cell geometry on seal considerations. In the cell-to-edge design in Figure 11.1a, the footprint of the cell matches that of the interconnect plate, each with the same pattern of gas manifold holes that upon stacking form chimneys used in transporting fuel and air. Sealing is required along the peripheral interface between each cell and adjacent interconnect plate. Most pSOFC stacks are constructed with cells that are anode supported, that is, a thick *porous* anode layer serves as the structural foundation for the rest of the cell. When the footprint of this type of cell contains manifold holes for fuel and air flow, both the outer edges of the cell and the inner edges of each manifold hole must also be sealed to prevent fuel from mixing and combusting with the air outside the stack or with that being transported internally.

In the cell-to-frame or window frame design shown in Figure 11.1b, the cell is smaller than the interconnect plate, contains no holes, and is joined to an intermediary component (a metallic window frame) that incorporates the necessary gas

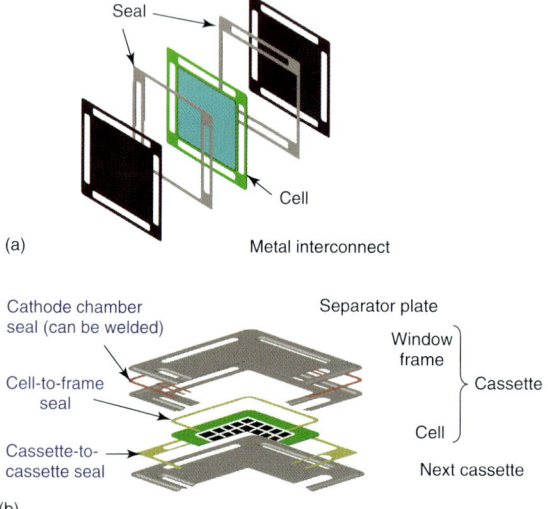

Figure 11.1 Examples of two general cell geometries: (a) the cell-to-edge design and (b) the cell-to-frame design.

porting. Two seals are employed, one between the cell and window frame–separator plate assembly to form a cassette repeat unit and the other between each cassette in the stack. In both designs, the seal between each repeat unit must be electrically insulating to prevent internal shorting. In addition, a third seal, not shown, is required between the stack and the manifold system that supplies fresh fuel and air and removes the exhaust gases. The latter is typically a base plate equipped with the necessary connections or is a set of headers that transport the gases to and from the stack.

A factor closely related to stack design that can have a tremendous impact on seal selection is the stack assembly procedure. A common mistake in pSOFC sealant selection and/or development is to fail to consider how the individual cell or stack manufacturing steps that precede or follow each of the sealing processes can affect either the long-term integrity of the seal or overall performance of the cells and stack. Cells are typically produced by traditional tape-casting, screen-printing, and sintering processes (or, more recently by large-scale coating techniques such as plasma spraying) to form the required anode–electrolyte–cathode laminate architecture. To mitigate interdiffusion and interfacial reactions between these layers, mechanisms that can produce deleterious phases and/or microstructures [14], any processing steps to which the cell is subsequently exposed (e.g., joining/sealing) must be carefully controlled with respect to maximum soak temperature, soak time, and process atmosphere. It has been well established, for example, that high-temperature vacuum or inert-gas exposure can cause chemical reduction and associated performance degradation in many of the new generation of high power density cathode materials now employed in commercially available cells [15]. In some cases, the use of tailored intermediate layers can minimize these problems [14]. Additional considerations include the cost of the sealant materials, the cost of the sealing process, and the potential for mass manufacture. For these reasons, air-fired sealing processes are often favored because they maintain the proper oxidation state in the cell materials and are readily scalable to low-cost, automated or semiautomated, high-volume stack production.

Seal selection is also based on the type of application in which the pSOFC system will be used. Mobile pSOFC systems, such as APUs, require seals that can be subjected to repeated thermal cycling, thermal shock, and dynamic mechanical loading. Conversely, stationary pSOFC stacks are generally exposed to less aggressive thermal and mechanical stress conditions, but are expected to operate at least an order of magnitude longer than their portable counterparts. In both cases, the sealant material must exhibit minimal reactivity with the adjacent components and display high-temperature chemical stability in both air and wet fuel gas environments. Summarized in Table 11.1 is a generic set of requirements for SOFC seals, broken down by functional category. Because seal selection is closely tied to pSOFC stack design and system application, it is dependent on a number of design factors, including: individual cell and stack materials and geometries, stack assembly sequence, thermal gradients expected across the seal and other stack components, maximum weight and/or volume of the power plant, anticipated external forces, and required system heating and cooling rates.

Table 11.1 Functional requirements for pSOFC seals [16, 17].

Thermomechanical	Chemical
• Stable, hermetic sealing, or marginal, nonlocalized leak rate; >5000 h for mobile applications and >50 000 h for stationary applications at 650–900 °C	• Long-term chemical stability under simultaneous oxidizing/wet fuel environments at 650–900 °C
• Mitigation of CTE mismatch stresses or CTE matching, typically $9.5–12 \times 10^{-6}$ K^{-1} for YSZ[a]-based electrolytes and ferritic stainless-steel window frame or interconnect components	• Long-term chemical compatibility with the adjacent sealing surfaces
• Acceptable bond strength or the use of compressive loading to maintain the seal (i.e., load frame design)	• Resistant to hydrogen embrittlement
• Resistant to degradation due to thermal cycling/thermal shock during stack start-up and shut-down	
• Robust under external static and dynamic forces[b]	

Design/fabrication	Electrical
• Low cost	• Non-conductive (non-shorting configuration),[b] typically $>10^4$ Ω cm at operating temperature
• Facile application/processing	• Electrical resistivity >500 Ω cm between cells and stacks at nominal operating conditions (0.7 V at 500 – 700 mA cm^{-2})
• High reliability with respect to achieving initial hermeticity (seal conforms to the substrate surfaces)	
• Acceptable sealing environment/temperature (i.e., has little effect on the subsequent performance of the stack)	
• Design flexibility – for example, allows use of Ni-based alloys in the interconnect[b]	

[a] Yttria-stabilized zirconia.
[b] These factors are stack-design and application specific.

11.2
Sealing Techniques

The various options for sealing and joining the ceramic and metal components in pSOFCs can be broadly classified into one of three categories: rigid bonded seals, compressive seals, and bonded compliant seals. As outlined in Table 11.2, rigid

Table 11.2 Categories of pSOFC sealing.

Type of seal	Advantages	Limitations
Rigid bonded seals • Sealant bonds to adjacent pSOFC components and is nondeformable at room temperature Examples: silicate and barium aluminosilicate glasses	• Leak-tight when properly designed Generally inexpensive Electrically insulating Tend to display acceptable stability in the reducing and oxidizing atmospheres of the stack Can be engineered to exhibit a coefficient of thermal expansion (CTE) matching those of the adjacent pSOFC components in the final joint, thereby mitigating the generation of thermally induced stresses Typically good wetting/adhesive behavior on both sealing surfaces [generally yttria stabilized zirconia (YSZ) and stainless steel] Can be readily applied to the sealing surfaces as a powder dispersed in a paste or as a tape cast sheet	• Susceptible to brittle failure under tensile stresses that can arise during heating and cooling or under externally applied loads Susceptible to changes in CTE that occur during microstructural evolution as the sealant ages under high-temperature stack operation; generally leads to seal degradation under thermal cycling Require a careful balance of properties that in turn depends on a detailed understanding of compositional and microstructural development in the sealant; substantial property knowledge gaps remain
Compressive seals • Employ deformable materials that do not bond to the adjacent pSOFC components Examples: mica, mica–glass composite, and metal gaskets	• No need for CTE matching; offers designers greater freedom in stack materials selection Inexpensive and easy to produce and apply Allow for mid-term stack repair Electrically insulating Acceptable stability in reducing and oxidizing atmospheres Largely mitigates the potential for cell bowing and accompanying nonuniform gas distribution in the stack	• Require a corrosion (oxidation)- and creep-resistant load frame to maintain sufficient load on the seal; this leads to higher system weight and thermal mass Not leak-tight; even under ideal conditions, some amount of cross-leakage between the fuel and oxidant gas streams occurs

(continued overleaf)

Table 11.2 (Continued)

Type of seal	Advantages	Limitations
Bonded compliant seals		
• Sealant bonds to adjacent pSOFC components and plastically deforms at room temperature and above Example: air brazes	• Leak-tight when properly designed Plastic deformation lowers the requirement for CTE matching and expands the choice of stack materials Easy to produce and apply Typically good wetting/adhesive behavior on both sealing surfaces Excellent thermal cycling characteristics Can be inexpensive, depending on the composition employed	• Generally metal based and therefore requires electrically insulating layers to prevent shorting Some compositions are susceptible to degradation under simultaneous exposure to hydrogen on one side and oxygen on the other and therefore require diffusion barrier layers to prevent this

bonded seals are hermetic joints that are nondeformable at room temperature, whereas compressive seals employ deformable materials that do not bond to the adjacent pSOFC components. BCSs are hermetic joints that can plastically deform at or above room temperature. Each option has inherent advantages and limitations that are also briefly described in Table 11.2. A more detailed review of the design considerations, properties, manufacture, and general performance of each seal type is provided below.

11.2.1
Rigid Bonded Seals

Because rigid bonded seals are nondeformable at room temperature (i.e., brittle), they are susceptible to fracture and probabilistic failure when exposed to tensile stresses, such as those generated by thermal expansion mismatches between the sealant and adjacent substrates or by thermal gradients that arise in the stack [18]. Failure typically originates at small flaws or discontinuities in the microstructure of the sealant. Under a tensile stress field, these can grow into a crack that propagates in a rapid and unstable fashion. The failure strength of these seals is defined by those defects that are larger than a critical size, as related through the Griffith crack model:

$$\sigma_f = \frac{K_{IC}}{\sqrt{\pi a_c}} \tag{11.2}$$

where σ_f is the tensile stress at failure, K_{IC} is the inherent resistance of the sealing material to crack growth, and a_c is the critical flaw size that leads to crack propagation [19]. The magnitude of tensile stress across the seal due to residual thermal mismatch can be estimated by

$$\sigma_1 = E_1 \Phi \int_{T_0}^{T_b} (\alpha_1 - \alpha_2) dT \qquad (11.3)$$

where E is the Young's modulus of the material for which the residual stresses are being calculated, α_1 and α_2 are the coefficients of thermal expansions (CTEs) of each material, T_b is the temperature at which bonding takes place (i.e., where the joint is initially stress free), T_0 is the temperature to which the joint is cooled (typically room temperature), and Φ is a geometric factor equal to $(1 + S_1 E_1 / S_2 E_2)^{-1}$ for a planar joint configuration [19], S_1 and S_2 being the cross-sectional areas of each material. According to Eq. (11.3), thermal expansion mismatch can be alleviated to a large extent by proper materials selection and/or tailoring of the sealant to accommodate the CTE for the adjacent substrates. As has been predicted through finite element analysis and shown experimentally both in individual cells and in full stacks, even a relatively small amount of thermal expansion mismatch can cause bowing in the cells that leads to fuel and air maldistribution in the stack and degrades system performance [20, 21]. For these reasons, the metal stack components (frames, separators, and spacers) are typically fabricated from ferritic stainless steel (CTE of $12-13 \times 10^{-6}$ K^{-1}) to match approximately the composite CTE of the cell ($10.5-12.5 \times 10^{-6}$ K^{-1}, depending on whether the cell is electrolyte or anode supported). Rigid bonded seals made from glasses, glass–ceramics, or ceramic composites have been developed with CTEs in this range.

Residual thermal stresses can also arise due to nonuniform heating or cooling of the stack (i.e., thermal shock), nonuniform heat generation across the cells attributable to the natural variation in fuel stream composition (i.e., water content) from inlet to outlet, and steady state heat loss through the faces and edges of the stack. Although the underlying thermal gradients in the stack that are responsible for this cannot be avoided entirely, their severity, and therefore the accompanying thermomechanical stresses, can be mitigated through careful design of the stack, including the appropriate choice and placement of insulation, the combined use of thermochemical and thermomechanical modeling to design flow fields and determine temperature distributions in the stack, and the development of a proper operational protocol for heating and cooling the system [22, 23]. Residual stress can also be counteracted to some extent by the proper choice of rigid bonded sealant, for example, one that retains viscoelasticity at operating temperature over the lifetime of the device.

11.2.1.1 Glass and Glass–Ceramic Sealants

The most common sealants currently used in joining pSOFC cell assemblies and stacks are high-temperature glasses and glass–ceramics. These materials are typically designed to soften and flow at a temperature above that required for stack operation, then "set" to form a hermetic seal by chemically bonding with the adjoining substrates. During the sealing operation, typically some fraction of the original glass devitrifies, or begins to undergo crystallization, forming a composite of nano- and microscale ceramic crystalline phases embedded within a glassy matrix. This time-dependent phenomenon raises the material's viscosity and sets the seal. Although the resulting composite can be stronger than the starting glass, extensive devitrification often leads to the formation of pores and cracks (i.e., gas leak paths and mechanical degradation within the seal) and should be avoided. Hence the primary challenge in selecting or developing a glass–ceramic sealant is to achieve a proper balance of material properties such that the sealing process is consistent and repeatable during stack assembly and the resulting seal is chemically, mechanically, and electrically stable over the lifetime of the stack. After sealing, the average CTE through the sealant should ideally remain constant as a function of time at the operating temperature and the sealant should not exhibit excessive reaction with the adjoining components, most notably the metallic window frame or interconnect.

In evaluating a glass–ceramic material for viability as a pSOFC sealant, there are three key thermal properties to consider initially: (i) the glass softening temperature (T_s; the temperature above which the glass first begins to soften), (ii) the glass transition temperature (T_g; the temperature above which the material behaves viscoelastically and below which it becomes brittle), and (iii) the temperature-dependent CTE. Note that all three properties change as the sealant continues to devitrify as a function of time at temperature. Although the three parameters play a role in defining the sealing process and the stability of the resulting seal, T_s and T_g are generally more important to the former whereas CTE is more relevant to the latter. For example, during cell and stack fabrication, the starting glass must be soft enough at the temperature of sealing to spread across and wet the sealing surfaces (usually under a temporary compressive load), yet not so fluid that it flows out from between the substrates and results in open gaps. Through proper control of crystallization, it is possible to raise the viscosity of the sealant during the sealing operation so that it attains the proper stiffness after initial wetting to minimize excessive flow, as illustrated in Figure 11.2.

Although T_s should be greater than the operating temperature for the stack, typically between 650 and 800 °C, it should also be low enough to avoid excessive oxidation of the metallic stack components during sealing, generally less than 950–1000 °C for the types of ferritic stainless steel commonly employed in pSOFCs. Similarly, the proper T_g can be identified by recognizing that if the nominal operating temperature of the stack is greater than T_g, the seal will exhibit viscoelastic behavior and be capable of relieving thermal stresses and self-healing cracks that might be generated during transient operation. Therefore, without consideration of other factors, T_g should be ideally as low as possible. However, in reality, sealant viscosity is inversely related to the difference between the stack operating temperature (T_{oper}) and T_g. That is, the larger is $T_{oper} - T_g$, the lower is the viscosity of the sealant at T_{oper} and therefore the more susceptible the sealing material is to viscous creep and possible loss of hermeticity during stack sealing and/or operation [24]. The acceptable CTE range for the sealant is established by

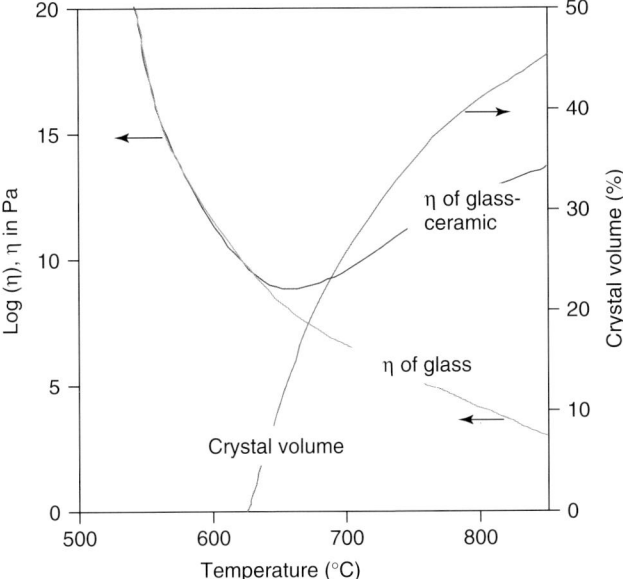

Figure 11.2 Plot of the general relationship between viscosity and temperature in a nascent glass and one undergoing incipient crystallization [17]. Crystallization causes an increase in viscosity of the resulting glass–ceramic relative to the nascent glass.

considering the CTEs of the substrates to be joined. For example, the desired CTE for a glass used to seal a ferritic stainless-steel frame and an Ni-supported cell with an yttria-stabilized zirconia (YSZ) electrolyte is between 10.0 and 12.0×10^{-6} K^{-1}. Long-term CTE stability depends both on the composition and microstructure of the bulk glass–ceramic and on the composition and microstructure at the interfaces with the substrates. Examples of this are discussed below.

Prior to devitrification, high-temperature sealing glasses consist of an amorphous network of oxide molecules that can be categorized as network formers, network modifiers, and intermediate oxides. Network formers such as SiO_2, B_2O_3, GeO_2, and P_2O_5 generally constitute the bulk of the glass structure and form a highly cross-linked, polyhedral network of covalent bonds. Network modifiers such as the alkali and alkaline earth metal oxides (e.g., Li_2O, Na_2O, K_2O, MgO, CaO, SrO, and BaO) display a greater degree of ionicity and locally disrupt the covalently bonded glass structure through the presence of nonbridging oxygen ions. These species lower the relative number of strong bonds in the material and thereby decrease the viscosity of the melt and lower both T_s and T_g [25]. Intermediate oxides such as Al_2O_3, TiO_2, ZrO_2, and Ga_2O_3 can serve as both network formers (e.g., by replacing Si^{4+} within the polyhedral network) and/or modifiers (i.e., by not participating directly in the network structure, but instead disrupting it via ionic bonding). In addition, other oxides that do not directly participate in the

Table 11.3 Common compositional modifiers for silicate-based glass–ceramic sealants.

Glass constituent	Example	Function
Network formers	SiO_2	Base constituent in silicate glasses
	B_2O_3	Reduces T_g, T_s, and viscosity and improves wetting
Network modifiers	BaO, CaO, MgO, SrO, Na_2O, K_2O	Reduce T_g and T_s and raise CTE
Intermediate oxides	Al_2O_3, TiO_2, ZrO_2, Ga_2O_3	Allow control over viscosity and wetting through the rate of crystallization
Other additives	V_2O_5	Reduces surface tension
	Cr_2O_3	Minimizes Cr depletion in adjacent metal parts and improves surface adherence
	La_2O_3, Nd_2O_3, Y_2O_3	Used as viscosity modifier and long-term CTE stabilizer
	CuO, NiO, CoO, MnO	Improve surface adherence

formation or disruption of the oxide network, and therefore are not considered glass constituents, can be included. Oxides such as La_2O_3, Nd_2O_3, ZnO, NiO, and Y_2O_3 are often added to modify the glass viscosity, surface adherence, and long-term CTE stability.

A list of common constituents often employed in high-temperature glass–ceramic sealants is provided in Table 11.3, along with their functional characteristics and the corresponding effects that they can have on T_s, T_g, and initial CTE. In silicate glasses, for example, both T_s and T_g increase with increasing SiO_2 content up to ~55 mol% and generally range from 725–750 °C and 675–725 °C, respectively. Because B_2O_3 additions disrupt cross-linking between silica tetrahedra by forming planar trigonal units, borosilicate glasses exhibit less molecular network rigidity and therefore display comparatively lower viscosities and reduced glass softening and glass transition temperatures, the last two by as much as 50–125 °C for $B_2O_3 : SiO_2$ ratios up to 0.5 [26]. Similarly, because network modifiers disrupt the glass network molecular structure, they also decrease T_s and T_g. On a mass basis, the addition of a network modifier may have a greater impact on T_s and T_g than a network former [27]. The extent of the effect depends on the ionic strength of the modifying cation species, that is, its field strength, ionic radius, polarizability, and coordination number [25]. The role that an intermediate oxide plays in modifying T_s and T_g is less straightforward to predict because of its dual function as network former or modifier [25]. The interested reader is referred to several studies that detail property changes in specific glass–ceramic compositions [28, 29].

The bulk CTE of a sealing glass depends on its glass structure symmetry, bond bending, and molar free volume [29]. Pure SiO_2 glass, which exhibits a high degree of symmetry, has a low CTE (0.6×10^{-6} K^{-1}), whereas lower symmetry

B_2O_3-based glasses display CTEs that are an order of magnitude higher. The addition of a network modifier to a silicate or borosilicate glass typically leads to an increase in initial CTE relative to the unmodified material [25]. As with T_s and T_g, the extent of this effect depends on the degree of ionicity that the modifying cation displays. Generally, the lower the ionic strength or the higher the concentration of the modifying agent, the greater is the increase in initial CTE observed. For example, the addition of 2.6 mol% K_2O to an MgO–BaO–silicate glass increases CTE by 3×10^{-6} K^{-1} over that of the baseline composition [24, 30]. In the case of adding an intermediate oxide, the effect on CTE depends on whether the oxide acts as a network former or modifier. For example, Al_2O, often a constituent in high-temperature sealing glasses, causes a substantial decrease in CTE when it participates directly in the glass network.

Although resistance to excessive devitrification is a critical requirement for a pSOFC sealing glass, this characteristic is difficult to evaluate fully in multicomponent glasses and consequently has not been as comprehensively studied as other material properties. Devitrification may occur during sealing and/or high-temperature use and subsequently impacts T_s, T_g, and CTE, and also sealant strength [27, 28, 31]. Three sources of devitrification are observed in pSOFC sealing glasses: (i) at the interface with either substrate, (ii) along the inner and outer free surfaces of the seals, and (iii) within the bulk of the sealing material. In all three cases, the phenomenon involves localized phase separation accompanied by the nucleation and growth of secondary crystalline phase(s) from the original vitreous matrix, the extents of which depend on both thermodynamic and kinetic considerations. For example, the activation energy required for interfacial devitrification is typically reduced by the presence of structural heterogeneities, such as scratches or roughness, on the surfaces of either substrate that the sealant contacts. Partial dissolution of or reaction with the substrate, such as the inherent, thin iron–chromium oxide scale on the sealing surfaces of stainless-steel components, can also initiate local compositional changes in the glass that lead to devitrification. In addition, interfacial stresses (e.g., due to CTE mismatch) can enhance the nucleation of crystalline phases in the sealant [31].

Similarly, surface devitrification is enhanced by flaws, microheterogeneities, such as dust particles and other impurities picked up during processing, and stresses, in this case on the free surfaces of the sealant [32]. It should be noted that all primary glass network formers (SiO_2, B_2O_3, and P_2O_5) are susceptible to volatilization in wet gases at least to some extent; that is, the environment on the anode side of the cell, which becomes progressively more humid from the fuel inlet to the exhaust manifold. Alkali metal oxides are also susceptible to vaporization at typical pSOFC operating temperatures, whereas alkaline earth metal oxides, Al_2O_3, and rare earth oxides are not [33]. Over time, the loss of these constituents from the anode-side surface of the sealant generates a depletion zone that can also induce surface devitrification.

Bulk devitrification tends to originate at small inclusions and particulates (e.g., insoluble crystalline nucleating agents) inside the glass. Whereas interfacial and surface devitrification are typically difficult to control, bulk devitrification can be managed through proper design of the heat-treatment schedule used in the sealing process, and by manipulating the starting glass composition and/or incorporating filler additions [24–29, 34]. The temperature associated with the onset of devitrification and the highest temperature at which devitrification occurs can be used to quantify the extent to which bulk devitrification is expected during nominal stack operation. Both parameters are measured via combined thermomechanical analysis and differential scanning calorimetry. The difference between the two temperatures is directly proportional to the resistance of the sealant to bulk devitrification [27].

With respect to compositional effects, it is generally found that the greater the amount of network former of a given polyhedral structure type present, such as SiO_2, which forms tetrahedral units, or B_2O_3, which forms planar trigonal networks, the greater is the resistance of the glass to devitrification. Conversely, mixing network formers that form different polyhedral networks reduces this resistance. In the same way, adding constituents that reduce network connectivity, such as network modifiers that increase the concentration of nonbridging oxygen, also makes the glass more susceptible to crystallization [25]. The effect of adding an intermediate oxide again depends on whether the constituent acts as a network former or modifier. Al_2O_3, for example, generally improves glass stability in silicate and borosilicate sealants by participating in network formation and thereby mitigating the tendency for phase separation [35]. The addition of other oxides also has varying effects. For example, La_2O_3 additions often enhance crystallization by serving as nucleation sites for various lanthanum silicate-based crystalline phases such as $La_2Si_2O_7$, $LaBO_3$, and $Ba_4La_6O(SiO_4)_4$. Conversely, small amounts of Cr_2O_3 increase the activation energy for devitrification in borosilicate glasses, thereby enhancing its resistance to crystallization.

Classes of glass and glass–ceramic sealants that have been designed using the above guidelines and tested for use in pSOFCs can be categorized based on their network former and modifying constituents. Phosphate-, boron-, and silica-based sealants and combinations of these have all been evaluated [33, 36, 37]. Work conducted by Larsen et al. [36] revealed a number of challenging problems with glasses containing phosphate as the primary network former. The phosphate component in the glass both reduces the overall CTE, typically well below the desired $11–12.5 \times 10^{-6}$ K^{-1} target range, and tends to be volatile at the stack operating temperature, particularly in the wet environment of the anode. The resulting gas species react with the Ni–YSZ-based anode to form nickel phosphide and zirconium oxyphosphate and thereby degrade the cell's electrochemical properties [36]. Borate- and borosilicate-based glasses and glass–ceramics have also been shown to be compositionally unstable in humid high-temperature environments. Boron oxide is readily hydrolyzed under the typical operating conditions of the anode and will form several volatile products, including $B_2(OH)_2$ and $B_2(OH)_3$ [37, 38]. This corrosion phenomenon is most apparent for sealant compositions containing >10 mol% B_2O_3 [37]. However smaller concentrations of boria can usually be tolerated and used to enhance the wettability or modify the initial T_s, T_g, and CTE of the glass [37]. Silica-based glasses and glass–ceramics offer greater promise as high-temperature pSOFC sealants and are further classified based on the type of network modifier(s) employed (i.e. alkali and/or alkaline earth oxide) and whether Al_2O_3 is incorporated as a key.

A number of silicate and aluminosilicate glass compositions have been developed that meet the T_s, T_g, and initial CTE targets for pSOFC applications [34, 39–47]. A truncated list is provided in Table 11.4. In general, the use of alkaline earth rather than alkali metal network modifiers pose fewer problems with respect to volatilization, interfacial reaction with adjacent substrates, and electrical conductivity at high temperatures, although the alkali metal additions tend to mitigate devitrification [48]. However, several of the alkaline earth metal-containing glasses, including G-18 and YSO46, exhibit good resistance to bulk devitrification. These sealants have undergone further testing in contact with several substrates under steady-state pSOFC operating conditions to determine the potential for long-term interfacial reaction and/or devitrification. Substrates tested included: YSZ, chromia-forming ferritic stainless-steel alloys 446SS and Crofer22APU, and alumina-forming ferritic stainless-steel alloys FeCrAlY and aluminized Crofer22APU. The barium aluminosilicate-based G-18 glass adheres well to YSZ with little chemical interaction, but tends to form interfacial $BaCrO_4$ in contact with the chromia-forming alloy (uncoated 446SS). Over several thousand hours of exposure at operating temperature, this product phase thickens and become porous, yielding interfaces that are weak and susceptible to thermomechanically induced cracking, for example, when the stack undergoes thermal cycling during typical operation [49]. In some cases, an interfacial corrosion product forms due to the depletion of chromium in the adjacent stainless steel, eventually causing deterioration in stack performance [50]. Analogous observations were made with the SrO-bearing YSO-1 glass, which showed no discernible reaction with YSZ, but formed $SrCrO4$ in contact with Crofer22APU, a chromia former. In contrast, the interface between G-18 and an alumina former exhibits a submicron-thick reaction zone composed of $BaAl_2SiO_8$.

As shown in Figure 11.3, devitrification has a measurable effect on the mechanical behavior of the bulk glass. Plotted in each graph are data for the bulk G-18 material in the nascent condition (i.e., after undergoing a simulated bonding heat treatment of 1 h in ambient air at 850 °C) and after aging in air at 750 °C for 1000 h beyond the as-joined condition. The two sets of samples tested at room temperature both display typical brittle behavior, shown in Figure 11.3a, with elastic failure occurring at maximum load/stress. The nascent material exhibits higher strength than the aged sample, ~85 versus ~45 MPa, respectively. Devitrification often causes weakening at temperatures below T_g because localized regions of residual stress that arise as a result of the evolving inhomogeneous microstructure serve as sites for crack initiation. At temperatures above T_g, much of the localized residual stress can be accommodated if the viscosity of the glassy matrix is low or if sufficient time is allowed for plastic flow (i.e., annealing). The latter is likely the reason why the difference in strength between the nascent and aged materials is relatively small at 750 °C, ~48 versus 38 MPa, respectively. Note that the curvature apparent in the $\sigma - \varepsilon$ curves for both materials at high temperature is an indication of residual glass present.

Table 11.4 Selected SOFC sealing glass compositions and properties.

	Composition (mass%)					Properties			Ref.
Name	BaO	B$_2$O$_3$	Al$_2$O$_3$	SiO$_2$	Other	T_g(°C)	T_s(°C)	CTE (10^{-6} K^{-1})	
Alkaline earth metal silicates									
BCAS4	55.6	–	5.5	25.3	9.2 CaO	609	672	11.9	[39]
G-18	56.4	7.3	5.4	22.1	8.8 CaO	630	685	10.8	[34]
Mg1.5-40-15B-8Zn	47.9	12.1	–	27.8	7.5 ZnO, 4.7 MgO	616	653	10.5	[40]
YSO-1	–	5.9	–	23.0	49.6 SrO, 15.2 Y$_2$O$_3$, 6.3 CaO	695	733	11.5–12.1	[41]
YSO46	–	9.3	–	27.2	42.2 SrO, 14.9 Y$_2$O$_3$, 3.7 CaO, 2.7 ZnO	645	693	11.68–11.76	[42]
YSO-7	–	6.6	–	23.0	49.5 SrO, 15.2 Y$_2$O$_3$, 5.7 CaO	685	730	11.4–11.6	[41]
Alkali and alkaline earth metal silicates									
SACNZn	–	–	19.4	38.4	31.6 CaO, 1.2 Na$_2$O, 0.27 MgO	740	NR^a	10.0	[43]
SCN-1	8.2	–	2.8	66.9	10 K$_2$O, 7.2 NaO, 3.3 CaO, 0.5 TiO$_2$, 0.2 Fe$_2$O$_3$	468–494	540–600	11.7	[44]
Alkaline earth metal borosilicate									
L2	47.4	10.3	9.0	17.7	15.6 La$_2$O$_3$	656	710	11.1	[45]
Alkaline earth metal borates									
MA1	–	33.9	9.4	–	8.1 MgO, 48.6 La$_2$O$_3$	623	671	9.0–10.3	[46]
BM1	45.6	43.0	–	–	10.5 MgO	570	616	9.36–10.77	[46]
BMA1 (GC)	6.7	32.2	8.9	–	6.0 MgO, 46.2 La$_2$O$_3$	620	656	9.2–10.0	[46]
BM2 (GC)	50.1	40.2	–	–	9.7 MgO	582	593	11.18–12.13	[46]
Aluminosilicate									
A06S70	–	0.8	5.2	3.3	81.1 Ce$_{0.8}$Sm$_{0.2}$O$_2$, 9.6 La$_2$O$_3$	NR	921	9.4	[47]

^{a}NR, not reported.

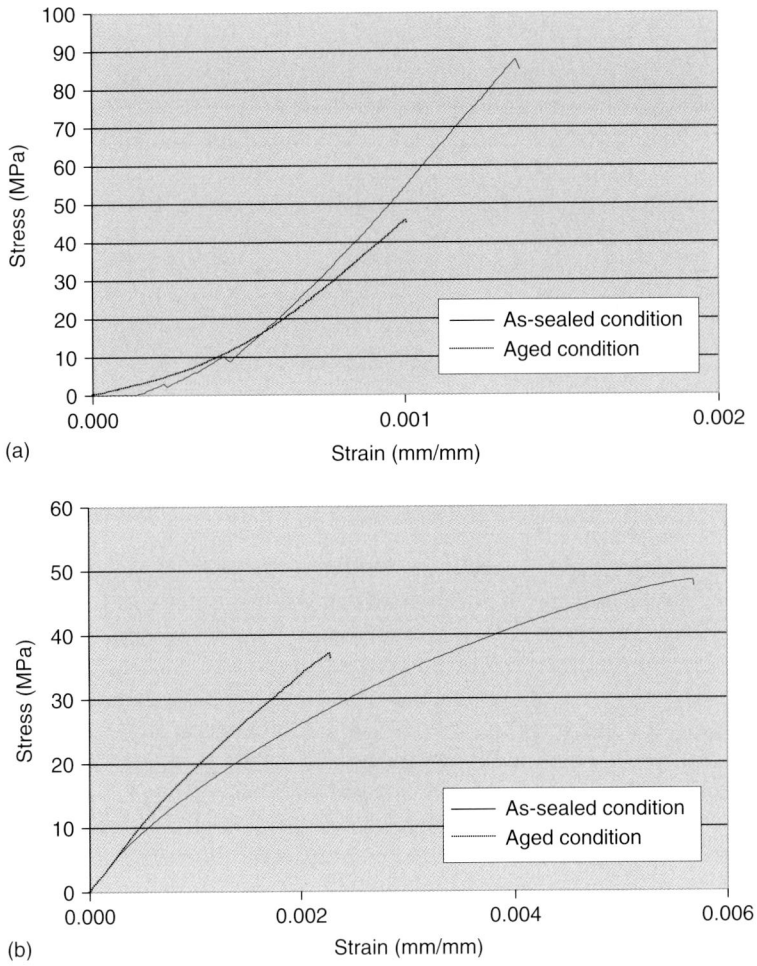

Figure 11.3 Stress–strain curves for the bulk G-18 sealant, as determined from flexural testing conducted at (a) room temperature and (b) 750 °C. Plotted in each case are curves for the material in the "as-sealed" condition (after 4 h at 850 °C) and after aging the as-sealed material for 1000 h at 750 °C [34].

Seals also need to withstand hundreds to thousands of thermal cycles during operation over the lifetime of the stack [17]. During each cycle, the interfaces of the seal glass/cell component(s) are subjected to thermomechanical stress, typically in the order of 20–150 MPa [51]. Both cyclic and long-term bond strength can be improved through several approaches, including (i) further refinements in glass chemistry, (ii) novel methods of crack deflection, blunting, and bridging of the type reported in several recent patents [52, 53], (iii) the use of component coatings (e.g., aluminization) that modify both the composition and morphology of the sealing surface [54], and (iv) the addition of insoluble reinforcements such as tetragonal ZrO_2, which under stress can undergo a phase transition that toughens the sealant material. Each of these enhancements is currently under development.

An additional consideration in glass and glass–ceramic seal design is the application of the sealant to the components to be joined. Glass and glass–ceramic seals can be readily produced by a number of standard ceramic fabrication techniques, including tape casting, automated paste dispensing, screen-printing, and spray deposition. In nearly all cases, glass powder (or frit) is first prepared by mixing the constituent oxides at high temperature to form a melt that is subsequently quenched and solidified, then crushed and milled to an average particle size of 10–50 µm. The frit is mixed with an appropriate organic binder system to form a paste that can be applied to the sealing surfaces of the cell and interconnect or window frame via one of the above approaches. The keys to practical paste formulation and application include:

- Defining the most appropriate frit particle size (average size and size distribution) through thermal and thermomechanical characterization. Particle size can influence the initiation and subsequent rate of glass devitrification [25].
- Determining the appropriate as-dispensed thickness and width of the sealant and developing an application process that consistently achieves this target. Thermal stress analysis can be used to determine optimal as-formed seal thickness for transient and steady-state operation. To retain this seal thickness over the lifetime of the stack, a small concentration of insoluble, monosize particulate can be incorporated into the sealant to achieve a minimum stand-off distance, for example, between the cell and window frame.
- Identifying the desired paste viscosity required for application and subsequent joining. For example, a low-viscosity paste may be desired in order to apply the sealant by screening-printing. To achieve this, a volatile solvent is typically employed so that upon drying the paste reaches the proper viscosity and tackiness require for stack lay-up. Paste viscosity also defines the maximum rates at which the seals can be applied to various components and the stacks fabricated.
- Establishing the maximum paste shelf life required for the cell and stack fabrication process. Like other slurry systems, the powder and binder can undergo settling, solvent loss, and/or slow, long-term reactions (e.g., hydrolysis) that will change the dispensing properties of the paste, such as its viscosity.
- Accounting for material losses during sealant application and incorporating recycling if needed. For example, while tape cast sheets of sealant material are a convenient means of ensure proper glass layer thickness, they can lead to substantial losses unless the sealing footprint for which the tape will be cut is properly designed or the tape can be resolubilized and recycled.

11.2.1.2 Ceramic Seals

Other pSOFC joining techniques that have been considered include the use of high-temperature cements and sealants formed by reaction bonding. Although commercial ceramic adhesives such as Duco and Sauereisen cements have been exceedingly useful in small-scale cell testing, they do not display the degree of CTE matching required for stack fabrication and often crack when cooled to room temperature. Yamamoto *et al.* evaluated a composite composed of mica particulate in a glass–ceramic matrix and found acceptable CTE and chemical stability with YSZ, but reaction with the lanthanum chromite separator used at the time [55]. As will be discussed below, this early attempt at a mica-containing seal has morphed into the hybrid glass–mica compressive seals that are beginning to find use in stack fabrication. Ceramic sealants formed by *in situ* reaction have also been investigated as an alternative method of rigid bonded sealing. Generally, reaction-based approaches require heat treatment at high temperatures. However, the use of pre-ceramic polymer precursors significantly lowers the temperatures required for joining. These precursors include organosilane polymers that convert to SiC or SiO_xC_y when heated to temperatures ranging from 800 to 1400 °C [56]. Lewinsohn *et al.* reviewed the merits of this approach and investigated its potential use for pSOFC joining [57]. In general, they found that the use of polymeric precursors is straightforward and no more difficult than applying a glass seal and that the resulting joining material is microstructurally and compositionally stable up to temperatures beyond that required for stack operation. However, the pyrolysis of these polymers is accompanied by the formation of gaseous reaction products and high-volume shrinkage, which often causes pores and cracks to develop in the joint during processing [58] and leads to a reduction in joint strength. These problems can be overcome to some extent by incorporating suitable filler materials, which also allows the CTE properties of the sealant to be modified. However, the technique requires further development, as well as a demonstration of long-term seal stability and acceptable thermal cycling properties.

11.2.2
Compressive Seals

Compressive seals employ deformable materials that do not bond to the pSOFC components but instead serve as gaskets. Thus sealing is achieved when the entire stack is compressively loaded. Because the sealing material conforms to the adjacent surfaces and is under constant compression during use, it forms a dynamic seal, that is, the sealing surfaces can slide past one another without disruption of the level of hermeticity and the individual stack components are free to expand and contract during thermal cycling with no need to consider CTE matching. This offers stack designers the freedom to consider other than ferritic stainless steels in designing metal componentry. The gaskets are readily produced and easy to apply. Additionally, they offer the potential for mid-term stack repair; by releasing the compressive load, disassembling the stack, replacing the damaged cell or separator components, and installing new gasket seals in stack reassembly. However, in order to employ compressive seals in a pSOFC stack, a load frame is required to maintain the desired level of compression on the stack over the entire period of operation and the stack components must be capable of withstanding the sealing load. The load frame introduces several complexities in stack design, including oxidation of the frame material, load relaxation due to creep, and increased weight and thermal mass, and therefore reduced specific power and thermal response of the overall system. In addition to increased system cost, these factors limit the use of compressive seals in mobile applications.

11.2.2.1 Metal Gaskets

The use of metal gaskets has been investigated for compressive sealing. Small-scale coupon testing indicates that nonoxidizing noble metals such as gold and silver may be viable in forming hermetic seals at pressures of ~25 MPa and higher due to sufficient deformation at stack operating temperatures [59]. The key question is how durable these gasket materials are under prototypic long-term isothermal and thermal cycle conditions and whether they can be made cost-effective for eventual commercial use. Additional concepts include using stamped metal gaskets of the type employed in sealing pressure vessels. In this case, oxidation-resistant alloys such as stainless steel and nickel-based superalloys are fabricated into gaskets with deformable C-shaped, corrugated, or hollow-tube cross-sections. A soft noble metal coating can be applied to improve hermeticity, particularly against a rough or uneven sealing surface. However, little information is available as to the effectiveness of these seals. An obvious disadvantage of this type of seal is that it is electrically conductive and therefore are subject to potential problems with electrical shorting.

11.2.2.2 Mica-Based Seals

An alternative to metal gaskets is the use of mica-based materials. Micas belong to a class of layered minerals known as *phyllosilicates* and are composed of cleavable silicate sheets. These materials are well known for their high electrical resistivity and uniform dielectric constant and in principle can be exposed to high temperatures under both reducing and oxidizing conditions. In practice, they often contain waters of hydration that will be lost during heating, making the resulting dehydrated material more friable and flexible than the original. Various forms have been investigated for use in compressive pSOFC seals, including muscovite [$KAl_2(AlSi_3O_{10})(F,OH)_2$] paper, muscovite single-crystal sheets, and phlogopite [$KMg_3(AlSi_3O_{10})(OH)_2$] paper, each ranging in thickness from 100 to 200 µm. Of these, the cleaved muscovite sheet exhibits the lowest leak rates [60]. The commercial mica papers are composed of small, discrete mica flakes that are oriented with their cleavage planes approximately parallel to each other and held together with an organic binder. Small samples of cleaved mica sheets were found to exhibit leak rates of 0.33–0.65 sccm·cm^{-1} (standard cubic centimeters per centimeter of seal length) at 800 °C and under 100 psi, whereas leak rates for the muscovite and phlogopite papers were approximately an order of magnitude higher [60]. Although hermeticity is desired, leak rates below ~0.04 sccm·cm^{-1} are likely acceptable in a number of applications, representing a decrease in fuel flow rate of ~0.2% for a 60-cell stack comprised of cells with an active area of 14 cm × 14 cm [61]. Two basic leak pathways that exist in mica-based seals are (i) along the interfaces between the mica and the cell components and (ii) through the open spaces (or slits) that form between the sublayers when the material is heated and the chemically bound water (and binder, in the case of the mica flake paper) is lost. To counter these and reduce leakage below the maximum acceptable level, several different hybrid compressive seal concepts have been conceived that combine mica with various secondary materials [61–64].

11.2.2.3 Hybrid Mica Seals

In a single-crystal, compressed mica seal, the dominant leak paths are along the interfaces with the ceramic and metal sealing surfaces, as shown schematically in Figure 11.4a. The leak rate can be reduced substantially by inserting a compliant interlayer in these regions, such as a deformable metal or glass as indicated in Figure 11.4b [61–64]. To date, two hybrid designs have been examined in detail, mica–silver and mica–glass. Both exhibit leak rates that are several orders of magnitude lower than that of cleaved muscovite alone, when compressed and tested under equivalent loading conditions. Table 11.5 gives the measured leak rates for several hybrid mica seal types in comparison with single-crystal muscovite and phlogopite paper, as a function of time at 800 °C and under thermal cycling. Seals constructed from phlogopite paper and 25 μm thick silver foil interlayers exhibit excellent thermal stability and very low leak rates (0.01–0.02 sccm · cm^{-1}) when tested isothermally at temperatures up to 800 °C under flowing reducing gas on one side and air on the other. No substantial physical degradation of the phlogopite was observed in these test specimens, although energy-dispersive spectroscopy indicated a small loss of fluorine [62]. Although some migration of silver into the mica paper was reported after >28 000 h of high-temperature exposure, the amount was small enough that there was no concern about electrical shorting [62]. Corresponding thermogravimetric analysis of the silver interlayers indicated that any loss of silver via evaporation would be minimal; ~1 wt% over 40 000 h of operation at 800 °C [62]. Leak testing conducted under a prototypic dual atmosphere environment (air on one side and 30% H$_2$O in H$_2$ on the other) at 800 °C also indicated good isothermal and thermal cycle stability up to the test limits, over 28 000 h and 119 cycles, respectively [62, 63]. Leak rates ranged from 0.02 to 0.03 sccm · cm^{-1}, again with no discernible changes in the microstructures or compositions of either sealing constituent over the conditions tested [62, 63].

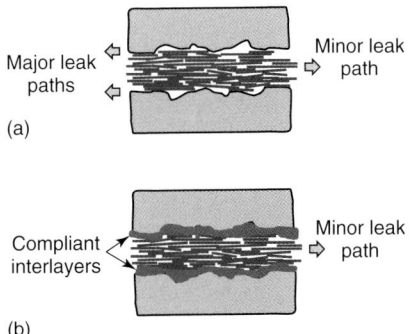

Figure 11.4 Schematic diagrams of (a) a conventional and (b) a hybrid mica seal.

Table 11.5 Sealing properties of selected mica-based seals.

Seal type (compressive pressure in parentheses)	Leak rate (sccm cm^{-1} length of seal)		Ref.
	Time at 800 °C	Thermal cycles after aging	
Single-crystal muscovite (100 psi)^a	0.20 after <2 h of aging	–	[60]
Phlogopite paper (100 psi)^a	2.28 after <2 h of aging	–	[61]
Ag–mica (12 psi)^b	~0.03 after 1000 h	~0.04 after 50 cycles (aged 1000 h)	[62]
	<0.04 after 2000 h	<0.03 after 30 cycles (aged 200 h)	
	<0.03 after 4000 h	~0.03 after 42 cycles (aged 4000 h)	
Ag–mica (12 psi)^c	~0.03 after 1000 h	<0.03 after ~40 cycles (aged 1000 h)	[63]
G-18–phlogopite (6 psi)^b	~0.05 after ~1000 h	~0.32 after 21 cycles (aged ~1000 h)	[64]
G-18m–phologopite^d (6 psi)^b	<0.03 after 1000 h	~0.16 after 6 cycles (aged ~1000 h)	[64]
G-6–phlogopite^d (6 psi)^b	<0.005 after 500 h	<0.01 after 55 cycles (aged 500 h)	[64]

^{a}Tested under He. Seals were compressed between Inconel and LaCrO$_3$.
^{b}Tested under an internal flow of 2.64% H$_2$ – Ar + ~3% H$_2$O. Seals were compressed between Inconel and 8YSZ.
^{c}Tested under an internal flow of 70% H$_2$ + 30%H$_2$O. Seals were compressed between Inconel and 8YSZ.
^{d}G-18m: G-18 glass with 5 wt% TiO$_2$ added. G-6: commercial low-melting borosilicate glass [64].

Test results have been reported for three different glass–mica hybrid variants, each composed of phlogopite paper sandwiched between interlayers prepared from either: (i) a standard barium aluminosilicate glass (the G-18 glass discussed above), (ii) the same glass with 5 wt% of TiO$_2$ added to enhance crystallization (referred to as G-18m), or (iii) a vitreous borosilicate glass (G-6) [64]. The three variants give insight into the effects of glass composition and devitrification on seal quality. As with the silver–mica seals, the glass hybrids were isothermally aged at 800 °C in a simulated SOFC environment prior to short-term thermal cycling. The seals formed with the G-18 interlayers displayed extensive reaction with the phlogopite and, as seen in Table 11.5, exhibit poor thermal cycle stability. The use of TiO$_2$ to force more rapid crystallization in this glass mitigates subsequent reaction with the mica paper. However, because the resulting glass–ceramic is substantially more viscous, the seal exhibits poorer surface wetting and adherence than the base glass sealant and thus poor thermal cycle stability, with leak rates as high as 0.16 sccm · cm^{-1}. The borosilicate glass displays limited reaction with the phlogopite, yielding a hybrid seal that exhibits low leak rates (<0.01 sccm·cm^{-1}) upon isothermal aging and thermal cycling. Although this glass has a high alkali metal content, and therefore may not be suitable for long-term SOFC stack sealing application, it indicates the interlayer characteristics needed to form a successful glass–mica hybrid seal.

The sealing characteristics of mica paper can also be enhanced by infiltrating the mica particulate with a wetting or melt-forming agent such as $Bi(NO_3)_3$ or H_3BO_3 [65], to block the leakage pathways through the mica gasket. However, attention must given to the reactivity of such infiltrates with the adjacent pSOFC components and to the long-term stability of the infiltrate under the operating conditions of the stack. Efforts have also considered infiltration of ceramic fiber papers, in one case using fumed silica [66]. The silica-infiltrated alumina paper seals displayed approximately an order of magnitude decrease in leak rate at 800 °C (0.03 sccm·cm^{-1}, when compressed at 0.1 MPa or 14.5 psi) in comparison with phlogopite and muscovite mica.

A hydromechanics model has been developed recently to understand the underlying factors that control leak rate in mica-based seals [67]. Among the findings from this work were that: (i) when the leak paths are simulated as slits of various height, length, and width, the leak rate is proportional to slit height to the third power and slight width to the first power and inversely proportional to slit length, (ii) the leak rate increases with the total number of slits and is strongly related to the slit size distribution, (iii) the leak rate is proportional to the pressure difference across the seal, and (iv) increasing the level of seal compression and spreading the applied compressive load more uniformly across the seal length by using a compliant interlayer reduce the leak rate by reducing the average slit size, and also the height and length of the leakage pathways within the mica material and along the interfaces with adjacent components [67]. With the latter in mind, future development of this sealing approach should focus on improving load uniformity in full-scale seals, both over the entire footprint of a given individual cell seal and over all of the seals from one end of the stack to the other. To some extent this work has already started and hybrid seals are beginning to find use as "detachable joints" at the interface between the stack and gas manifold in full-scale test and demonstration units [68].

11.2.3
Bonded Compliant Seals

Unlike rigid bonded sealing, the sealant used in BCS forms a joint that plastically deforms at or above room temperature. This can mitigate the effects of thermal expansion mismatch stresses to some extent and lessen the requirements of CTE matching between stack components. However, there are still potential issues with cell bowing and the accompanying nonuniformities in gas distribution. In addition, all of the sealing concepts in this category are currently metal based and therefore electrically conductive. Therefore, in order to use a BCS concept as the sole sealing solution for a given pSOFC stack design, each seal must be accompanied by an electrically insulative layer to prevent internal shorting, for example, a thin ceramic coating or an impenetrable thermally grown oxide on the sealing surfaces of the metal frames [69]. BCS can also be used to complement a second sealing technique in a multi-seal stack design, for example, the cell-to-frame seal in Figure 11.1b with glass employed as the cassette-to-cassette seals used in building the stack. Although the literature abounds with comments implying that glass–ceramic seals are the only viable means of sealing SOFC components and stacks for long-term use, both compressive and metal-based compliant seals have found their way into full-scale demonstration stacks and the latter are finding use in emerging commercial applications [68]. In addition, efforts to improve stack manufacturability have led to laser-welded cassette and cassette-to-cassette sealing [70].

11.2.3.1 Brazing

One of the most reliable methods of joining dissimilar materials is brazing. In this technique, a filler metal with a liquidus well below that of the materials to be joined is heated to a point at which it becomes molten and under capillary action fills the gap between the sealing surfaces. When cooled, a solid joint forms. Most filler metals developed for brazing will only properly wet metal and not ceramic surfaces. However, active metal brazing is a specialized version designed to overcome this problem. Active braze alloys (ABAs) incorporate into the filler metal a reactive element, such as titanium, zirconium, or molybdenum. These constituents chemically transform the ceramic surface at the ceramic–filler metal interface via a reduction or displacement reaction, creating an intermediate layer that is in chemical equilibrium with both the underlying ceramic and the molten braze filler metal. Typically more metallic in nature than the ceramic substrate, this reacted surface is readily wetted by the remaining filler material. Owing to the high oxidation potential of the added reactive species, ABA filler metals typically require a stringent firing atmosphere, with $p_{O_2} < 10^{-5}$ atm. The process is scalable to high-volume production and can be used to join a wide range of ceramics, including alumina and various carbides, nitrides, and silicides, as well as stainless steel and nickel-based alloys [71].

However, two problems are commonly encountered when using ABAs for pSOFC sealing:

1) The resulting joint is not sufficiently oxidation resistant under the target operating conditions because ABA filler metals are typically Ni- or Cu-based. It has been found that even noble metal ABAs suffer from this problem [72].
2) The brazing process causes irreversible chemical reduction and structural/property degradation in the electrochemically active ceramic components used in pSOFC stacks. This is because of the need to carry out brazing under nonoxidizing conditions (i.e., in vacuum, inert gas, or reducing gas) [73].

To overcome these issues, an alternative brazing technique has been developed specifically for use in fabricating solid-state electrochemical devices [74]. Referred to as *air brazing*, the technique employs a molten oxide that is at least partially soluble in a noble metal solvent to promote wetting of the ceramic sealing surface. Due to its ability to dissolve in the noble metal liquid, the oxide compound serves as an *in situ* oxygen buffer, raises the chemical activity of dissolved oxygen, and enhances the metal's wetting characteristics on various oxide substrates. Like glass sealing, air brazing is carried out solely in air. Thus when brazing to a metal substrate, bonding will occur between the filler metal and an oxide scale that grows on the structural metallic component during the joining operation. The presence of this scale prevents substrate dissolution in the filler metal and this, combined with the nature of the two key constituents in the filler metal, conveys oxidation resistance to the joint at moderate to high temperatures. Those metal alloy substrates that exhibit a high degree of metal-to-scale adhesion afford the highest strength joints.

Although there are potentially a number of metal oxide–noble metal systems that can be considered for air brazing, Ag–CuO has been explored in greatest detail. As little as 1.4 mol% of copper oxide in silver results in a good balance of wettability and adhesion on ceramic sealing surfaces, thereby producing high-strength joints of the type shown in Figure 11.5. In addition, work has shown that alloying agents such as Pd, Al, and TiO_2 and inert particulates such as YSZ powder or fiber can be added to Ag–CuO to modify its use temperature, oxygen solubility characteristics, wetting behavior, and strength [75–77]. The effects have recently been reviewed in detail [78].

In designing an air braze filler metal, it is critical that there is some degree of mutual solubility between the metal and oxide constituent(s) in the liquid state. Experimental evidence suggests that the likely indicators for this behavior are measurable oxygen solubility (i.e., oxygen activity) in the liquid metal constituent, compatible melting points between the metal and oxide to be alloyed, and multiple valence states in the cation species of the oxide constituent that allow it to serve as an efficient oxygen buffer in the noble metal. Phase formation and phase equilibria in the filler metal system must be appropriate for the application of interest. For

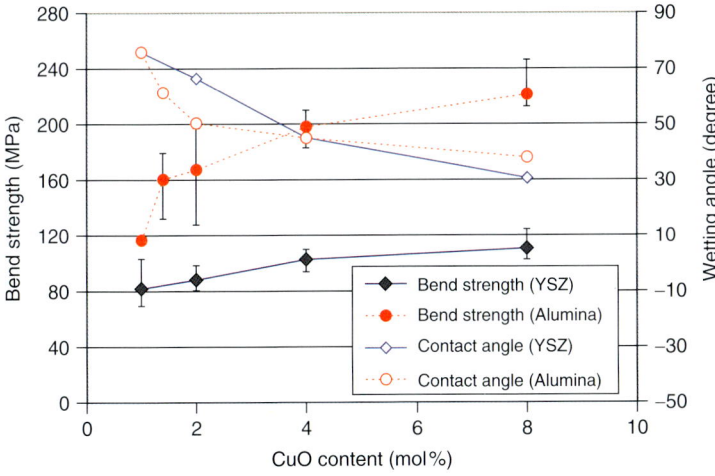

Figure 11.5 Comparison of room-temperature four-point bend strength and contact angle between alumina and YSZ joints brazed at 1000 °C using CuO–Ag braze [79].

example, the filler metal liquidus temperature should be well below the solidus temperatures of the materials to be joined, in fact, low enough to avoid property degradation (e.g., creep) in these materials during the joining operation. Conversely, the solidus temperature must be high enough to avoid remelting or substantial softening of the filler metal during secondary joining operations and during device operation. In addition, the degree of solidification shrinkage, the potential for secondary phase formation, and the possibility of excessive substrate erosion during joining should be understood so that the joint can be designed in a way that avoids through-seal void formation and undue stress concentration. The filler metal alloy must also properly wet the adjoining substrates to ensure gap filling and hermeticity within the joint and promote adhesion between the joined components. As with glass–ceramic sealant design, wetting may depend on dissolution and/or reaction of the substrate with the filler metal. Filler metal composition, joining temperature, and time at temperature all play important roles in defining wettability.

Ag–CuO base filler metals can be applied to the substrates in a number of forms, including as an Ag–Cu metal foil (the copper will oxidize *in situ* during joining), as a powder-based cast tape, or as a paste using an automated dispenser or via screen-printing. Commonly, a 1.4 mol% CuO–Ag filler metal is used to air braze pSOFC components. Typical brazing conditions include fixturing the joint under a loading pressure of ~0.2 psi, heating in air at 5 °C min^{-1} to a soak temperature of 1000 °C and holding for 30 min or less, and cooling at 5 °C min^{-1} [78]. However, recent work has shown that both soak temperature and soak time can be reduced substantially by optimizing the heating schedule. The resulting joint is highly ductile and can be thermally cycled at a rapid rate numerous times with no measurable degradation of either hermeticity or joint strength [80, 81]. Figure 11.6 shows a comparison between the thermal cycling properties of

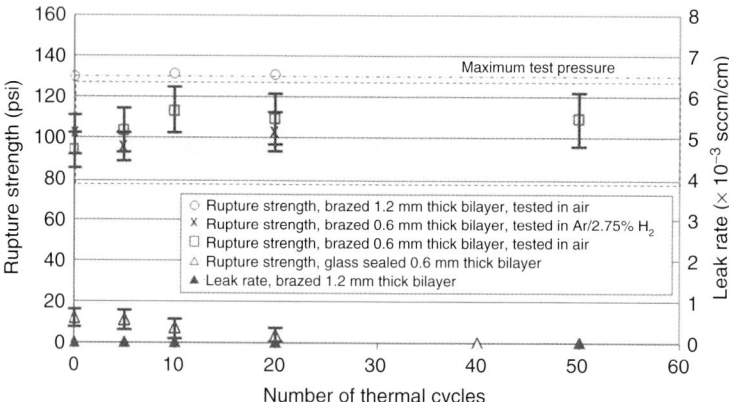

Figure 11.6 Rupture strength and leak rate of Ag1CuO and G-18 sealed specimens as a function of thermal cycling (thermally cycled between room temperature and 750 °C at 75 °C min^{-1}). The specimens consist of bilayer anode-supported discs (550 μm NiO–YSZ/50 μm YSZ or 1150 μm NiO–YSZ/50 μm YSZ) sealed to a 100 μm thick 446SS washer [80].

YSZ–Fecralloy joint strength specimens joined with Ag$_4$CuO and with G-18, the barium aluminosilicate glass–ceramic sealant discussed previously. Note that the glass joints experience a substantial loss in strength beyond 10 thermal cycles at 75 °C min^{-1}, whereas the rupture strength of the air-brazed specimens remains constant as a function of thermal cycling.

Over the course of their development, silver-based sealants have faced two challenges in addition to design considerations regarding internal shorting: (i) long-term silver volatility and (ii) dual-atmosphere degradation. Testing of silver under prototypic anode and cathode conditions indicated that less than 0.161 µg cm^2 h^{-1} will be lost from a typical sealant profile, which at present does not appear to be a significant long-term effect on cell or stack performance [82]. Dual-atmosphere degradation results from silver's solubility for both hydrogen and oxygen. When the sealant is simultaneously exposed to both gases during operation (H$_2$ on the anode side and O$_2$ on the cathode), these species can react within the silver matrix to form internal steam bubbles [83]. The rate of pore formation depends on the temperature of exposure, the pressures and flow rates of both gases across the seal, and the seal height and width (i.e., exposure area). Studies of ceramic–metal joints brazed with the Ag–CuO filler metal indicated that the phenomenon takes place relatively slowly. After 1000 h of dual-atmosphere exposure at 800 °C, some microstructural change does occur, but in general the joints remain mechanically sound [84]. In full-scale testing, stacks composed of air brazed cells that operated nominally at 750 °C beyond 2000 h exhibited no loss in hermeticity [85]. Additional dual-atmosphere testing is needed to determine the long-term degradation effects in silver-based air-brazed seals. However, concerns about dual-atmosphere degradation past 20 000 h may be preventable by incorporating a thin diffusion barrier (e.g., an aluminosilcate or aluminizing layer) inside the braze seal. Short-term testing has shown this concept to be potentially viable [86].

11.2.3.2 Bonded Foil Seal Concept

An alternative compliant sealing concept being developed utilizes a thin stamped metal foil that is bonded to both sealing surfaces, as shown in Figure 11.7. Unlike a metal or mica gasket, this seal is non-sliding. When properly designed, the foil yields or deforms under modest thermomechanical loading and limits the transfer of these stresses to the adjacent ceramic and metal components. Because the metal foil offers a greater degree of geometric deformation than the air-brazed seal, this sealing concept can accommodate a wider array of alloys for use in the pSOFC interconnect and/or frames [20, 87]. If high-CTE nickel-based alloys could be used, for example, the mechanical, oxidation, and through-scale electrical properties of the interconnect would be significantly improved relative to those fabricated from ferritic stainless steel [88]. Recent modeling studies indicated that the sealing concept is scalable, but that the design must account for a small amount of cell bowing through the use of stiffened interconnects. In addition, refinements in foil geometry to mitigate the bending effect should be evaluated [87].

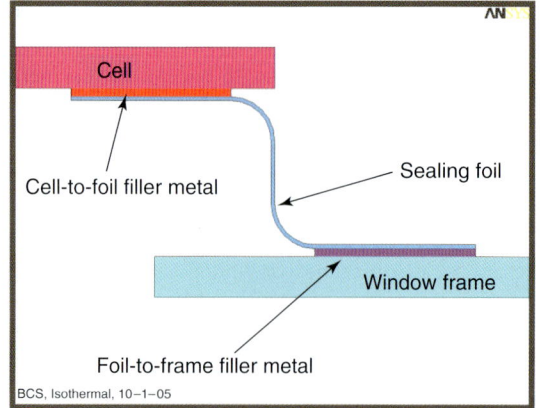

Figure 11.7 (a) Schematic diagram of the BCS design in cross-section and (b) a one-quarter symmetry model of the BCS design.

11.3
Conclusion

Planar SOFCs offer the promise of energy-efficient, high-density power generation, to utilize more effectively both limited fossil fuel resources in the immediate future and renewable resources as they develop in the longer term. To fulfill this promise, robust sealing technologies must be developed that can meet the functional requirements of stack manufacturers. However, it is likely that no single sealing technique will satisfy all stack designs and system applications.

Glass and glass–ceramic seals have been proven effective for stacks that undergo few thermal cycles, of which cycling is carried out slowly. However, there remain some concerns regarding long-term durability due to slow, but inevitable, changes in sealant composition and microstructure as a result of diffusion from adjacent components. New efforts to modify joint geometry, rather than simply the initial glass or glass-ceramic chemistry, may help overcome these problems. Significant progress has been made in developing and demonstrating compressive hybrid mica seal designs, in particular silver–mica seals, and these seals are beginning to find niche use in several near-commercial stacks. The use of air brazing has also shown promise in full-scale stack testing and modified versions that mitigate dual atmosphere degradation are likely to find use in early commercial stacks manufactured for transportation and other rapid thermal cycle applications.

References

1. International Energy Agency (2010) 2010 Key World Energy Statistics, http://www.iea.org/textbase/nppdf/free/2010/key_stats_2010.pdf (last accessed 24 November 2011).
2. US Energy Information Administration (2009) International Energy Outlook 2009. Report DOE/EIA-0484, Energy Information Administration, Washington, DC.
3. Salameh, M.G. (2003) The new frontiers for the United States energy security in the 21st century. *Appl. Energy*, **76**, 135–144.
4. Zervos, A., Lins, C., and Muth, J. (2010) Fuel Energy Policy–2050, European Renewable Energy Council, Brussels.
5. Haynes, C. and Wepfer, W.J. (2000) "Design for power" of a commercial grade tubular solid oxide fuel cell. *Energy Convers. Manage.*, **41**, 1123–1149.
6. EG&G Technical Service (2004) *Fuel Cell Handbook*, 7th edn, US Department of Energy, Office of Fossil Energy, National Energy Technology Laboratory, Washington, DC.
7. US Department of Energy, Office of Energy Efficiency and Renewable Energy (2011) 2010 Fuel Cell Technologies Market Report, http://www1.eere.energy.gov/hydrogenandfuelcells/pdfs/2010_market_report.pdf (last accessed 24 November 2011).
8. Boudghene Stambouli, A. and Traversa, E. (2002) Solid oxide fuel cells (SOFCs): a review of an environmentally clean and efficient source of energy. *Renew. Sust. Energy Rev.*, **6**, 433–455.
9. Fontell, E., Phan, T., Kivisaari, T., and Keränen, K. (2006) Solid oxide fuel cell system and the economical feasibility. *J. Fuel Cell Sci. Technol.*, **3**, 242–253.
10. Rasmussen, J.F.B., Hendriksen, P.V., and Hagen, A. (2008) Study of internal and external leaks in tests of anode-supported SOFC. *Fuel Cells*, **8**, 385–393.
11. Mulot, J., Niethammer, M., Mukerjee, S., Haltiner, K., and Shaffer, S. (2008) Development update on Delphi's solid oxide fuel cell systems. Presented at Fundamental Development of Fuel Cells Conference 2008, Nancy, France, 10–12 December 2008.
12. Singhal, S.C. and Kendall, K. (2003) *High Temperature Solid Oxide Fuel Cells*, Elsevier, New York.
13. Bujalski, W., Dikwal, C.M., and Kendall, K. (2007) Cycling of three solid oxide fuel cell types. *J. Power Sources*, **171**, 96–100.
14. Sholklapper, T.Z., Chun Lu, C., Jacobson, C.P., Steven, J., Visco, S.J., and De Jonghe, L.C. (2006) LSM-infiltrated solid oxide fuel cell cathode. *Electrochem. Solid-State Lett.*, **9**, A376–A378.
15. Basu, R.N. (ed.) (2007) *Materials for Solid Oxide Fuel Cells. Recent Trends in Fuel*

16. Fergus, J. (2005) Sealants for solid oxide fuel cells. *J. Power Sources*, **147**, 46–57.
17. Weil, K.S. (2006) The state-of-the-art in sealing technology for solid oxide fuel cells. *JOM*, **58**, 37–44.
18. Selverian, J.H., O'Neil, D.A., and Kang, S. (1994) Performance testing and strength prediction of ceramic-to-metal joints. *J. Eng. Gas Turbines Power*, **116**, 622–628.
19. Howe, J.M. (1993) Bonding, structure, and properties of metal/ceramic interfaces. Part 2: interface fracture behaviour and property measurement. *Int. Mater. Rev.*, **38**, 257–271.
20. Weil, K.S. and Koeppel, B.J. (2008) Comparative finite element analysis of the stress–strain states in three different bonded solid oxide fuel cell designs. *J. Power Sources*, **180**, 343–353.
21. Yuan, P. (2008) Effect of inlet flow maldistribution in the stacking direction on the performance of a solid oxide fuel cell stack. *J. Power Sources*, **185**, 381–391.
22. Recknagle, K.P., Williford, R.E., Chick, L.A., Rector, D.R., and Khaleel, M.A. (2003) Three-dimensional thermo-fluid electrochemical modeling of planar SOFC stacks. *J. Power Sources*, **113**, 109–114.
23. Lin, C.K., Huang, L.H., Chiang, L.K., and Chyou, Y.P. (2009) Thermal stress analysis of planar solid oxide fuel cell stacks: effects of sealing design. *J. Power Sources*, **192**, 515–524.
24. Mahapatra, M.K. and Lu, K. (2010) Glass-based seals for solid oxide fuel and electrolyzer cells – a review. *Mater. Sci. Eng. R*, **67**, 65–85.
25. Scholze, H. (1991) *Glass: Nature, Structure, and Properties*, Springer, New York.
26. Lim, E.S., Kim, B.S., Lee, J.H. and Kim, J.J. (2006) Effect of BaO content on the sintering and physical properties of $BaO - B_2O_3 - SiO_2$ glasses. *J. Non-Cryst. Solids*, **352**, 821–826.
27. Mahapatra, M.K., Lu, K., and Reynolds, W.T. Jr. (2008) Thermophysical properties and devitrification of $SrO - La_2O_3 - Al_2O_3 - B_2O_3 - SiO_2$–based glass sealant for solid oxide fuel/electrolyzer cells. *J. Power Sources*, **179**, 106–112.
28. Donald, I.W., Mallison, P.M., Metcalfe, B.L., Gerrard, L.A., and Fernie, J.A. (2011) Recent developments in the preparation, characterization, and applications of glass- and glass–ceramic-to-metal seals and coatings. *J. Mater. Sci.*, **46**, 1975–2000.
29. Kumar, V., Arora, A., Pandey, O.P., and Singh, K. (2008) Studies on thermal and structural properties of glasses as sealants for solid oxide fuel cells. *Int. J. Hydrogen Energy*, **33**, 434–438.
30. Budd, M. (2002) Method of forming a glass ceramic material. US Patent 6,475,938.
31. Bansal, N.P. and Gamble, E.A. (2006) Crystallization kinetics of a solid oxide fuel cell seal glass by differential thermal analysis. *J. Power Sources*, **147**, 107–115.
32. Yinnon, H. and Uhlmann, D.R. (1982) A kinetic treatment of glass formation VII: transient nucleation in non-isothermal crystallization during cooling. *J. Non-Cryst. Solids*, **50**, 189–202.
33. Zhang, T., Fahrenholtz, W.G., Reis, S.T., and Brow, R.K. (2008) Borate volatility from SOFC sealing glasses. *J. Am. Ceram. Soc.*, **91**, 2564–2569.
34. Meinhardt, K.D., Kim, D.S., Chou, Y.S., and Weil, K.S. (2008) Synthesis and properties of a barium aluminosilicate solid oxide fuel cell glass–ceramic sealant. *J. Power Sources*, **182**, 188–196.
35. Sun, T., Xiao, H., Guo, W., and Hong, X. (2010) Effect of Al_2O_3 content on $BaO - Al_2O_3 - B_2O_3 - SiO_2$ glass sealant for solid oxide fuel cell. *Ceram. Int.*, **36**, 821–826.
36. Larsen, P.H., Primdahl, S., and Mogensen, M. (1996) Influence of sealing material on nickel/YSZ solid oxide fuel cell anodes. In *High Temperature Electrochemistry: Ceramics and Metals. Proceedings of the 17th Risø International Symposium on Materials Science* (eds. F.W. Poulsen, N. Bonanos, S. Linderoth, M. Mogensen and B. Zachau-Christiansen), Risø National Laboratory, Roskilde, pp. 331–338.

37. Ghosh, S., Das Sharma, A., Kundu, P., and Basu, R.N. (2008) Glass-based sealants for application in planar solid oxide fuel cell stack. *Trans. Indian Ceram. Soc.*, **67**, 5748–5754.
38. Zhou, X.D., Templeton, J.W., Zhu, Z., Chou, Y.S., Maupin, G.D., Lu, Z., Brow, R.K., and Stevenson, J.W. (2010) Electrochemical performance and stability of the cathode for solid oxide fuel cells. *J. Electrochem. Soc.*, **157**, B1019–B1023.
39. Ghosh, S., Das Sharma, A., Kundu, P., Mahanty, S., and Basu, R.N. (2008) Development and characterizations of $BaO-CaO-Al_2O_3-SiO_2$ glass–ceramic sealants for intermediate temperature solid oxide fuel cell application. *J. Non-Cryst. Solids*, **354**, 4081–4088.
40. Pascual, M.J., Guillet, A., and Durán, A. (2007) Optimization of glass–ceramic sealant compositions in the system $MgO-BaO-SiO_2$ for solid oxide fuel cells (SOFC). *J. Power Sources*, **169**, 40–46.
41. Chou, Y.S., Stevenson, J.W., and Singh, P. (2007) Novel refractory alkaline earth silicate sealing glasses for planar solid oxide fuel cells. *J. Electrochem. Soc.*, **154**, B644–B651.
42. Chou, Y.S., Stevenson, J.W., Xia, G.G., and Yang, Z.G. (2010) Electrical stability of a novel sealing glass with (Mn,Co)-spinel coated crofer22APU in a simulated SOFC dual environment. *J. Power Sources*, **195**, 5666–5673.
43. Smeacetto, F., Salvo, M., D'Hérin Bytner, F.D., Leone, P., and Ferraris, M. (2010) New glass and glass–ceramic sealants for planar solid oxide fuel cells. *J. Eur. Ceram. Soc.*, **30**, 933–940.
44. Liu, W.N., Sun, X., and Khaleel, M.A. (2011) Study of geometric stability and structural integrity of self-healing glass seal system used in solid oxide fuel cells. *J. Power Sources*, **196**, 1750–1761.
45. Sohn, S.B., Choi, S.Y., Kim, G.H., Song, H.S., and Kim, G.D. (2004) *J. Am. Ceram. Soc.*, **87**, 254–254.
46. Ghosh, S., Das Sharma, A., Mukhopadhyay, A.K., Kundu, P., and Basu, R.N. (2010) Effect of BaO addition on magnesium lanthanum aluminoborosilicate-based glass–ceramic sealant for anode-supported solid oxide fuel cell. *Int. J. Hydrogen Energy*, **35**, 272–283.
47. Wang, S.F., Lu, C.M., Wu, Y.C., Yang, Y.C., and Chiu, T.W. (2011) $La_2O_3-Al_2O_3-B_2O_3-SiO_2$ glasses for solid oxide fuel cell applications. *Int. J. Hydrogen Energy*, **36**, 3666–3672.
48. Chou, Y.S., Thomsen, E.C., Williams, R.T., Choi, J.P., Canfield, N.L., Bonnett, J.F., Stevenson, J.W., Shyam, A., and Lara-Curzio, E. (2011) Compliant alkali silicate sealing glass for solid oxide fuel cell applications: thermal cycle stability and chemical compatibility. *J. Power Sources*, **196**, 2709–2716.
49. Yang, Z., Meinhardt, K.D., and Stevenson, J.W. (2003) Chemical compatibility of barium–calcium–aluminosilicate-based sealing glasses with the ferritic stainless steel interconnect in SOFCs. *J. Electrochem. Soc.*, **150**, A1095–A1101.
50. Batfalsky, P., Haanappel, V.A.C., Malzbender, J., Menzler, N.H., Shemet, V., Vinke, I.C., and Steinbrech, R.W. (2006) Chemical interaction between glass–ceramic sealants and interconnect steels in SOFC stacks. *J. Power Sources*, **155**, 128–137.
51. Weil, K.S., Deibler, J.E., Hardy, J.S., Kim, D.S., Xia, G.G., Chick, L.A., and Coyle, C.A. (2004) Rupture testing as a tool for developing planar solid oxide fuel cell seals. *J. Mater. Eng. Perform.*, **13**, 316–326.
52. Yang, Z., Coyle, C.A., Baskaran, S., and Chick, L.A. (2005) Gas-tight metal/ceramic or metal/metal seals for applications in high temperature electrochemical devices and method of making. US Patent 6,843,406.
53. Weil, K.S. Chick, L.A., Coyle, C.A., Hardy, J.S., Xia, G., Meinhardt, K.D., Sprenkle, V.L., and Paxton, D.M. (2007) High strength insulating joints for solid oxide fuel cells and other high temperature applications. European Patent EP 1836138.
54. Choi, J.P. and Weil, K.S. (2010) Aluminization of metal substrate faces. US Patent Application 20100297341.
55. Yamamoto, T., Ito, H., Mori, M., Mori, N., and Watanabe, T. (1996)

Compatibility of mica glass–ceramics as gas-sealing materials for SOFC. *Denki Kagaku*, **64**, 575–581.

56. Lewinsohn, C.A., Colombo, P., Reimanis, I., and Ünal, Ö. (2001) Stresses occurring during joining of ceramics using preceramic polymers. *J. Am. Ceram. Soc.*, **84**, 2240–2244.

57. Lewinsohn, C.A. and Elangovan S. (2003) Development of amorphous, non-oxide seals for solid oxide fuel cells, in *27th International Cocoa Beach Conference on Advanced Ceramics and Composites* (eds. W.M. Kriven and H.-T. Lin), American Ceramic Society, Westerville, OH, pp. 317–322.

58. Pippel, E., Woltersdorf, J., Colombo, P., and Donato, A. (1997) Structure and composition of interlayers in joints between SiC bodies. *J. Eur. Ceram. Soc.*, **17**, 1259–1265.

59. Duquette, J. and Petric, A. (2004) Silver wire seal design for planar solid oxide fuel cell stack. *J. Power Sources*, **137**, 71–75.

60. Simner, S.P. and Stevenson, J.W. (2001) Compressive mica seals for SOFC applications. *J. Power Sources*, **102**, 310–316.

61. Chou, Y.S., Stevenson, J.W., and Chick, L.A. (2002) Ultra low leak rate of hybrid compressive mica seals for solid oxide fuel cells. *J. Power Sources*, **112**, 130–136.

62. Chou, Y.S., Stevenson, J.W., Hardy, J., and Singh, P. (2006) Material degradation during isothermal aging and thermal cycling of hybrid mica seal with Ag interlayer under SOFC exposure conditions. *J. Electrochem. Soc.*, **153**, A1591–A1598.

63. Chou, Y.S. and Stevenson, J.W. (2009) Long-term ageing and materials degradation of hybrid mica compressive seals for solid oxide fuel cells. *J. Power Sources*, **191**, 384–388.

64. Chou, Y.S., Stevenson, J.W., Hardy, J., and Singh, P. (2006) Material degradation during isothermal ageing and thermal cycling of hybrid mica seals under solid oxide fuel cell exposure conditions. *J. Power Sources*, **157**, 260–270.

65. Chou, Y.S. and Stevenson, J.W. (2004) Novel infiltrated phlogopite mica compressive seals for solid oxide fuel cells. *J. Power Sources*, **135**, 72–78.

66. Le, S., Sun, K., Zhang, N., and Shao, Y. (2007) Comparison of infiltrated ceramic fiber paper and mica base compressive seals for planar solid oxide fuel cells. *J. Power Sources*, **168**, 447–452.

67. Sang, S., Pu, J., Jiang, S., and Jian, L. (2008) Prediction of H_2 leak rate in mica-based seals of planar solid oxide fuel cells. *J. Power Sources*, **182**, 141–144.

68. Baron, S. (2004) Opening doors to fuel cell commercialisation: intermediate temperature (500–850 °C) solid oxide fuel cells (IT-SOFCs) explained. *Fuel Cell Today*, www.fuelcelltoday.com (last accessed 24 November 2011).

69. Reisdorf, G.F., Keller, J.M., Haltiner K.J. Jr., Mukerjee, S., Weil, K.S., and Hardy, J.S. (2005) Ceramic coatings for insulating modular fuel cell cassettes in a solid oxide fuel cell stack. US Patent 7,422,819.

70. Mukerjee, S., Shaffer, S., Zizelman, J., Chick, L., Meinhardt, K., Sprenkle, V., Weil, S., Paxton, D., and Deibler, J. (2003) Solid oxide fuel cell stack for auxiliary power units: a development update, in *2003 Fuel Cell Seminar* (eds. J. Mizusaki and S. Singhal), Courtesy Associates, Miami, Fl, pp. 852–855.

71. do Nascimento, R.M., Martinelli, A.E., and Buschinelli, A.J.A. (2003) Review article: recent advances in metal-ceramic brazing. *Cerâmica*, **49**, 178–198.

72. Weil, K.S. and Rice, J.P. (2004) Substrate effects on the high-temperature oxidation behavior of a gold-based braze filler metal. *Scr. Mater.*, **52**, 1081–1085.

73. Sun, C., Hui, R., and Roller, J. (2010) Cathode materials for solid oxide fuel cells: a review. *J. Solid State Electrochem.*, **14**, 1125–1144.

74. Weil, K.S., Kim, J.Y., and Hardy, J.S. (2005) Reactive air brazing: a novel method of sealing SOFCs and other solid-state electrochemical devices. *Electrochem. Solid-State Lett.*, **8**, A133–A136.

75. Kim, J.Y., Hardy, J.S., and Weil, K.S. (2007) Ag–Al based air braze for high

temperature electrochemical devices. *Int. J. Hydrogen Energy*, **32**, 3754–3762.
76. Darsell, J.T. and Weil, K.S. (2008) The effect of filler metal composition on the strength of yttria stabilized zirconia joints brazed with Pd–Ag–CuO$_x$. *Metall. Mater. Trans. A*, **39**, 2095–2105.
77. Kim, J.Y., Weil, K.S., and Choi, J.P. (2008) Metal–ceramic composite air braze with ceramic particulate. US Patent Application 20080217382.
78. Weil, K.S., Darsell, J.T., and Kim, J.Y. (2011) Air Brazing: a New Method of Ceramic–Ceramic and Ceramic–Metal Joining. In Ceramic Integration and Joining Technologies (eds. M. Singh, T. Ohji, R. Asthana, and S. Mathur), John Wiley & Sons, Inc., New York, Chapter 4.
79. Kim, J.Y., Hardy, J.S., and Weil, K.S. (2005) Silver–copper oxide based reactive air braze (RAB) for joining yttria-stabilized zirconia. *J. Mater. Res.*, **20**, 636–643.
80. Weil, K.S., Coyle, C.A., Darsell, J.T., Xia, G.G., and Hardy, J.S. (2005) Effects of thermal cycling and thermal aging on the hermeticity and strength of silver–copper oxide air-brazed seals. *J. Power Sources*, **152**, 97–104.
81. Kuhn, B., Wessel, E., Malzbender, J., Steinbrech, R.W., and Singheiser, L. (2010) Effect of isothermal aging on the mechanical performance of brazed ceramic/metal joints for planar SOFC stacks. *Int. J. Hydrogen Energy*, **35**, 9158–9165.
82. Meulenberg, W.A., Teller, O., Flesch, U., Buchkremer, H.P., and Stover, D. (2001) Improved contacting by the use of silver in solid oxide fuel cells up to an operating temperature of 800 °C. *J. Mater. Sci.*, **36**, 3189–3195.
83. Kleuh, R.L. and Mullins, W.W. (1968) Some observations on hydrogen embrittlement of silver. *Trans. Metall. Soc. AIME*, **242**, 237–243.
84. Kim, J.Y., Hardy, J.S., and Weil, K.S. (2007) Dual atmosphere tolerance of Ag–CuO based air brazes. *Int. J. Hydrogen Energy*, **32**, 3655–3663.
85. Southern Company Services (2006) Power Systems Development Facility, Topical Report Gasification Test Run TC12. DOE Report DE-FC21-90MC25140, *http://www.netl.doe.gov/technologies/ coalpower/gasification/projects/adv- gas/25140/FinalRpt.pdf*.
86. Weil, K.S., Hardy, J.S., Kim, J.Y., and Choi, J.P. (2010) Diffusion barriers in modified air brazes. US Patent 7,691,488.
87. Weil, K.S. and Koeppel, B.J. (2008) Thermal stress analysis of the planar SOFC bonded compliant seal design. *Int. J. Hydrogen Energy*, **33**, 3976–3990.
88. Yang, Z.G., Weil, K.S., Paxton, D.M., and Stevenson, J.W. (2003) Evaluation and selection of heat resistant alloys for SOFC interconnect applications. *J. Electrochem. Soc.*, **150**, A1188–A1201.

12
Phosphoric Acid, an Electrolyte for Fuel Cells – Temperature and Composition Dependence of Vapor Pressure and Proton Conductivity

Carsten Korte

12.1
Introduction

In addition to its use in phosphoric acid fuel cells (PAFC) and more recently in phosphoric acid-based high temperature polymer electrolyte membrane fuel cells (HT-PEMFC), phosphoric acid is a widely used compound in the chemical industry. Phosphoric acid is the second most important mineral acid in terms of volume and value, being exceeded only by sulfuric acid. It is mainly used for the production of fertilizers and industrial phosphates, metal surface treatment, and the acidulation of beverages [1].

An overview of the basic properties of phosphoric acid with emphasis on technical processing was given by Schrödter *et al.* [1]. Two phase diagrams for the system H_2O–P_2O_5 were given and are shown in Figure 12.1. Part (a) shows the liquid–solid equilibrium in the temperature range from −100 to 80 °C and in the composition range from 0 to 80 wt% P_2O_5. Part (b) describes the liquid–vapor equilibrium at ambient pressure (1 bar) in the temperature range from 100 to 900 °C and in the composition range from 0 to 100 wt% P_2O_5. In addition to the phase diagrams, the density, the specific heat, and the kinematic viscosity are given as a function of composition and temperature. No evaluated information is provided on the proton conductivity. More detailed information on the vapor pressure for the temperature range from 120 to 200 °C, which is important for the operation of HT-PEMFCs, can be found in the original data of Brown and Whitt [2].

Because of this lack of evaluated data, in this chapter a literature survey of vapor pressure and conductivity data with emphasis on the temperature and composition range relevant for fuel cell applications is presented.

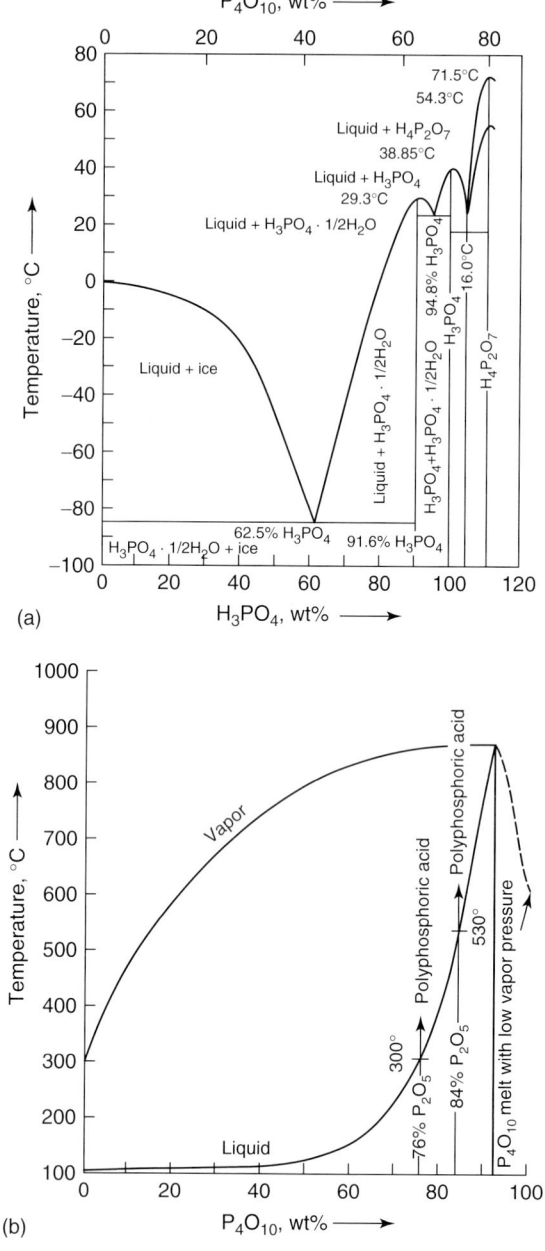

Figure 12.1 (a) Liquid–solid and (b) vapor–liquid phase diagrams of the H_2O–P_2O_5 system [1]. (Reproduced from *Ullmann's Encyclopedia of Industrial Chemistry* by courtesy of Wiley-VCH Verlag GmbH.)

12.2
Short Overview of Basic Properties and Formal Considerations

12.2.1
Anhydride and Condensation Reactions

Phosphoric acid is formed by the reaction of phosphorus pentoxide ("P_2O_5" = P_4O_{10}) with water:

$$P_4O_{10} + 6H_2O \rightleftharpoons 4H_3PO_4 \tag{12.1}$$

The equilibrium of this reaction is virtually to the right-hand side. In a sequence of condensation equilibria, the monomer orthophosphoric acid (H_3PO_4) forms polyphosphoric acids, for example, (pyro-) diphosphoric acid ($H_4P_2O_7$), triphosphoric acids ($H_5P_3O_{10}$), and so on:

$$2H_3PO_4 \rightleftharpoons H_4P_2O_7 + H_2O \quad K_1 \tag{12.2a}$$
$$H_3PO_4 + H_4P_2O_7 \rightleftharpoons H_5P_3O_{10} + H_2O \quad K_2 \tag{12.2b}$$
$$H_3PO_4 + H_5P_3O_{10} \rightleftharpoons H_6P_4O_{13} + H_2O \quad K_3 \tag{12.2c}$$
$$H_3PO_4 + H_6P_4O_{13} \rightleftharpoons H_7P_5O_{16} + H_2O \quad K_4 \tag{12.2d}$$
$$\ldots \quad K_i \tag{12.2e}$$

where

$$K_i = \frac{c_{\text{polyphosphoric acid},i+1} c_{H_2O}}{c_{H_3PO_4} c_{\text{polyphosphoric acid},i}} \tag{12.3}$$

Metaphosphoric acid, $(HPO_3)_n$, is the formal limiting composition if only unbranched chains are formed. In addition to $H[-O-PO(OH)-]_nOH$ chains there are also $[-O-PO(OH)-]_n$ rings, for example

$$H_5P_3O_{10} \rightleftharpoons H_3P_3O_9 + H_2O \quad K_2' \tag{12.4}$$

The equilibrium constants K_i of each equilibrium are functions of temperature and pressure:

$$K_i = f(T, p) \tag{12.5}$$

Thus, if the thermodynamic equilibrium is established, the concentration of each polyphosphoric acid species is essentially a function of only temperature T and the initial concentration of P_2O_5.

12.2.2
Acidity and Protolytic Equilibria

Orthophosphoric acid is a medium-strength proton donor. On a large scale in a dilute aqueous solution, only the first proton is dissociated [3–5]:

$$H_3PO_4 + H_2O \rightleftharpoons H_2PO_4^- + H_3O^+ \tag{12.6}$$

Table 12.1 Acidity coefficients pK_A for ortho- and diphosphoric acid at 25 °C [6, 7].

H_3PO_4, equilibria	$pK_A^a = -\log K_A$	$H_4P_2O_7$, equilibria	pK_A^a
$H_4PO_4^+ - H_3PO_4$	~−3	$H_4P_2O_7 - H_3P_2O_7^-$	0.91
$H_3PO_4 - H_2PO_4^-$	2.16	$H_3P_2O_7^- - H_2P_2O_7^{2-}$	2.10
$H_2PO_4^- - HPO_4^{2-}$	7.21	$H_2P_2O_7^{2-} - HP_2O_7^{3-}$	6.70
$HPO_4^{2-} - PO_4^{3-}$	12.32	$HP_2O_7^{3-} - P_2O_7^{4-}$	9.32

^{a}All pK_A values for 25 °C.

At a composition of 7 wt% P_2O_5 (10 wt% H_3PO_4), a degree of dissociation of about 16% is reached. In contrast, diphosphoric acid is a strong proton donor. It exceeds orthophosphoric acid by one order of magnitude when comparing the equilibrium constants K_A (see Table 12.1):

$$K_A = \frac{c_{A^-} c_{H_3O^+}}{c_{HA}} \qquad (12.7)$$

In concentrated aqueous solutions, above 61.6 wt% P_2O_5 (85 wt% H_3PO_4), the ratio between water and phosphoric acid molecules is <1. In such highly concentrated solutions autoprotolysis also takes place [8]:

$$2H_3PO_4 \rightleftharpoons H_2PO_4^- + H_4PO_4^+ \qquad K_{AP} = \frac{c_{A^-} c_{H_2A^+}}{(c_{HA})^2} \qquad (12.8)$$

or

$$3H_3PO_4 \rightleftharpoons H_3O^+ + H_2P_2O_7^{2-} + H_4PO_4^+ \qquad (12.9)$$

The protonation of an H_3PO_4 molecule forms a phosphazidium ion, $H_4PO_4^+$. The autoprotolysis constant K_{AP} for orthophosphoric acid is relatively high: at 25 °C, for 100 wt% H_3PO_4 a value of about 0.16 is measured [9]. According to Munsen, more than 50% of all molecules are dissociated [10]. The autoprotolysis constant is about three orders of magnitude higher than that for anhydrous H_2SO_4 ($K_{AP} = 2.5 \times 10^{-4}$) and about 13 orders of magnitude higher than that for H_2O ($K_{AP} \approx 10^{-14}$).

12.2.3
Composition Specifications and Condensation Equilibria

The composition of an (aqueous) phosphoric acid solution can be given as a molar concentration, a molar fraction, or as weight percentage of a distinct component. Because of the condensation equilibria forming polyphosphoric acids, discussed above, this can be ambiguous. One has to distinguish between the actual concentrations of each possible species in the thermodynamic equilibrium and the "initial" concentrations of the components H_2O and H_3PO_4, assuming that all P_2O_5 only forms H_3PO_4. Thus, concentrations specified in this chapter are

Table 12.2 Some typical compositions of (aqueous) phosphoric acid, given as weight percent and molar fractions of P_2O_5 and H_3PO_4.

wt% P_2O_5	wt% H_3PO_4	$x(P_2O_5)$	$x'(H_3PO_4)$	Designation
61.56	85	0.17	0.52	"Concentrated" H_3PO_4
72.42	100	0.25	1	"100%" H_3PO_4
79.76	109.79	0.33	–	"100%" $H_4P_2O_7$
88.74	122.52	0.50	–	"100%" $(HPO_3)_x$

preferentially given in wt% P_2O_5:

$$\text{wt\% } H_3PO_4 = 2\frac{M_{H_3PO_4}}{M_{P_2O_5}} \times \text{wt\%} P_2O_5 \qquad (12.10a)$$

$$x(P_2O_5) = 1 - x(H_2O) = \frac{1}{1 - \frac{M_{P_2O_5}}{M_{H_2O}}\left(1 - \frac{100\%}{\text{wt\%} P_2O_5}\right)} \qquad (12.10b)$$

$$x'(H_3PO_4) = 1 - x'(H_2O) = \frac{1}{1 - \left(\frac{M_{H_3PO_4}}{M_{H_2O}} - \frac{M_{P_2O_5}}{M_{H_2O}}\frac{50\%}{\text{wt\%} P_2O_5}\right)} \qquad (12.10c)$$

Accordingly, for the molar fractions $x(P_2O_5)$, $x(H_2O)$, $x'(H_3PO_4)$, and $x'(H_2O)$ and for wt% H_3PO_4 in Table 12.2, the formation of polyphosphoric acids is not taken into account. An overview is given for the most commonly used and characteristic composition of aqueous phosphoric acid.

12.3
Vapor Pressure of Water as a Function of Composition and Temperature

12.3.1
Number of Independent Variables, Gibb's Phase Rule

If we want to treat the H_2O–H_3PO_4 and the H_2O–P_2O_5 systems, it is important to know the number of independent thermodynamic parameters that are necessary to describe the state of the system in a unique way. This can be done by using Gibb's phase rule:

$$F + P = K + 2 \qquad (12.11)$$

where K is the number of independent components, P the number of phases, and F the degrees of freedom for a given state.

In the case of a system with two coexisting phases ($P = 2$, i.e., liquid and vapor) and two independent components ($K = 2$, i.e., H_2O and H_3PO_4), we have to take four independent thermodynamic parameters into account to describe the state of the system uniquely. As a set of parameters, the molar fraction of H_2O in the liquid phase, $x_{H_2O,l}$, and in the vapor phase, $x_{H_2O,g}$, the total pressure, p_{tot}, and the temperature, T, can be chosen. In the equilibrium, the temperature and the pressure in each phase have to identical: $T = T_l = T_g$ and $p_{tot} = p_l = p_g$. The molar fraction of the other component H_3PO_4 is not independent: $x_{H_2O,l} + x_{H_3PO_4,l} = 1$ and $x_{H_2O,g} + x_{H_3PO_4,g} = 1$.

According to Eq. (12.11), the number of degrees of freedom F in a system with two coexisting phases is two. If we assume that there is essentially no phosphoric acid or P_4O_{10} in the vapor phase (e.g., $x_{H_3PO_4,g} = 0$), the partial pressure of water, $p_{H_2O,g}$, in the vapor phase is identical with the total pressure, $p_{H_2O,g} = p_{tot} \, x_{H_2O,g} \approx p_{tot}$, that is, reducing the number of variables by one. Therefore, for two coexisting phases the system can be unequivocally described in the equilibrium state by only temperature, T, vapor pressure of water, $p_{H_2O,g}$, and one composition variable in the liquid, for example, $x_{H_2O,l}$ or $x_{H_3PO_4,l}$. The following cases can be considered:

T fixed ⟶ $p_{H_2O,g}$ and $x_{H_2O,l}$ free   that is, $p_{H_2O,g} = f(x_{H_2O,l})$
p–x diagram for a fixed temperature T

$p_{H_2O,g}$ fixed ⟶ $x_{H_2O,l}$ and T free   that is, $x_{H_2O,l} = f(T)$
T–x diagram for a fixed pressure p, see Figure 12.1

$x_{H_2O,l}$ fixed ⟶ $p_{H_2O,g}$ and T free   that is, $p_{H_2O,g} = f(T)$
p–T diagram for a fixed composition x

Each condensation equilibrium for a polyphosphoric acid species that is additionally taken into account adds one additional component (see Section 12.2.1). Via the laws of mass action, each equilibrium also adds one additional relation between the molar fractions of the components. Accordingly, these additional composition variables are not independent. Therefore, for a phosphoric acid solution containing H_2O, H_3PO_4, $H_4P_2O_7$, $H_5P_3O_{10}$, and so on, the number of independent components is still only two.

Instead of $x_{H_2O,l}$ or $x_{H_3PO_4,l}$, the initial concentration of P_2O_5 can also be used. Because of the condensation reactions in the equilibrium state, all other composition variables are unique functions of the initial concentration of P_2O_5 and temperature T. The temperature dependence of the equilibrium constants K_i is described by Eq. (12.5).

12.3.2
Evaluated Literature Data for the Vapor Pressure of Phosphoric Acid in the Temperature Range between 25 and 170 °C

On this topic, only a limited number of experimental studies are available in the literature, mainly from the period 1930–1970. Data on the vapor pressure of water in a phosphoric acid solution in the important temperature and composition range can be found in several papers [2, 11–14]. Using all the data, it is possible to evaluate a more detailed diagram for the isotherms in the composition range from 0 to 65–75 wt% P_2O_5 and a pressure range up to 1013 mbar (see Figure 12.2).

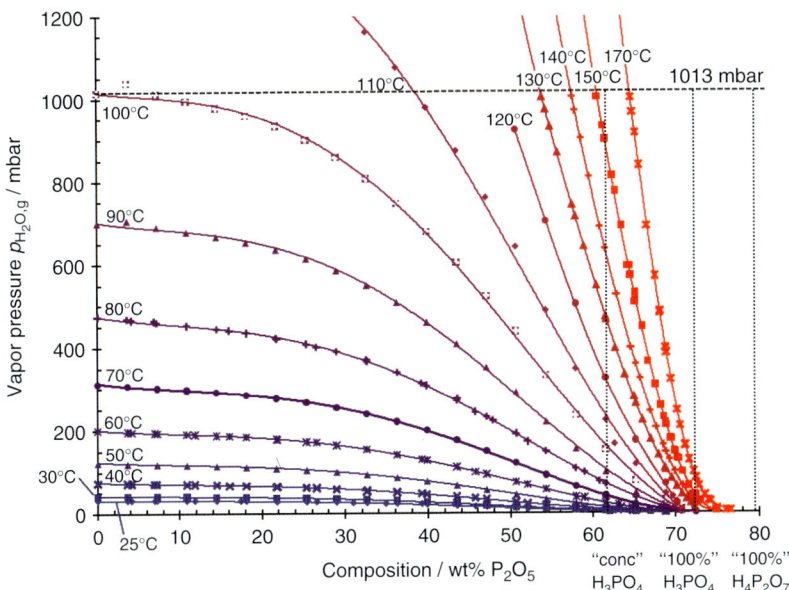

Figure 12.2 Vapor pressure $p_{H_2O,g}$ of the H_2O–P_2O_5 system as a function of composition and temperature. Evaluated data from [11–14], isotherms with interpolated compositions. Isotherms for 100 and 110 °C partially extrapolated for low concentrations. The solid lines connecting the data points are guides for the eye.

The data from the given references fit fairly well together without discontinuities. Intermediate compositions were interpolated by polynomial fitting. Intermediate temperatures were linearly interpolated by using a $\ln p_{H_2O,g}$ versus $1/T$ plot according to the Clausius–Clapeyron equation. The isotherms for 100 and 110 °C at low concentrations are extrapolated. The extrapolated data points fit sufficiently well to the vapor pressure data of pure water [7].

The isotherms in Figure 12.2 represent contour curves of a $p_{H_2O,g}$–T–wt% P_2O_5 plot. The resulting surface represents the boundary between the phase field of liquid phosphoric acid and the phase field of coexisting liquid phosphoric acid and vapor ("bubble point plane"). Cross-sections for a fixed composition in a $p_{H_2O,g}$–T–wt% P_2O_5 plot will result in $p_{H_2O,g}$–T diagrams (see Figure 12.3).

Using the Clausius-Clapeyron equation, the enthalpy of evaporation of water, $\Delta_{vap} H_{H_2O}$, from aqueous phosphoric acid can be calculated from the $p_{H_2O,g}$–T data in Figure 12.3:

$$\frac{\partial \ln p_{H_2O,g}}{\partial (1/T)} = -\frac{\Delta_{vap} H_{H_2O,g}}{R} \tag{12.12}$$

As depicted in Figure 12.4, for low concentrations below 40 wt% P_2O_5, the enthalpy of evaporation is almost constant and amounts about 43 kJ mol^{-1}. This does not differ significantly from the value found for pure water [7]. At higher

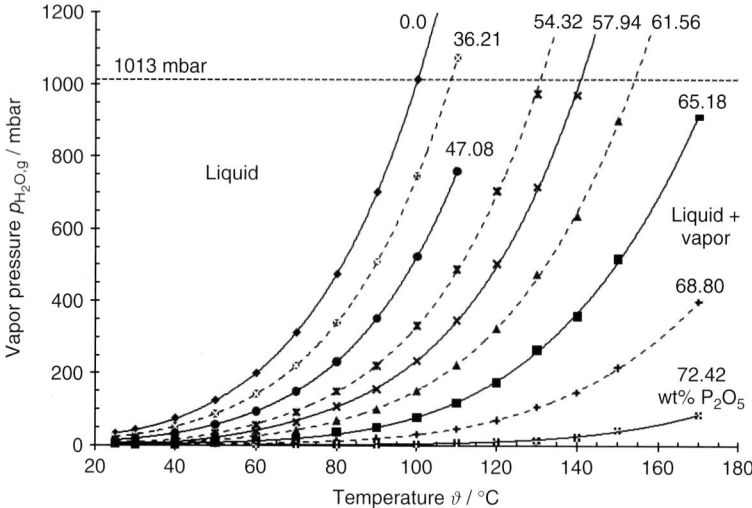

Figure 12.3 $p_{H_2O,g}-T$ phase diagram of the $H_2O-P_2O_5$ system constructed from the data in Figure 12.2 for constant compositions in the liquid phase. Bubble point curves for compositions from 0 to 72.4 wt% P_2O_5. The dashed and solid lines are according to the Clausius–Clapeyron equation.

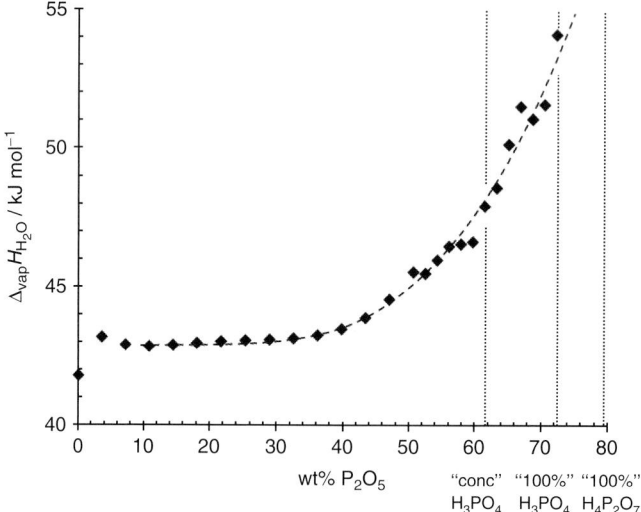

Figure 12.4 Enthalpy of evaporation of water, $\Delta_{vap}H_{H_2O}$, from aqueous phosphoric acid evaluated from $p_{H_2O,g}-T$ data in Figure 12.3 using the Clausius–Clapeyron equation. The dashed line connecting the data points serves as a guide for the eye.

concentrations, it increases strongly, reaching a value of about 53 kJ mol^{-1} at 72.4 wt% P_2O_5 (100 wt% H_3PO_4). The enthalpies of evaporation in Figure 12.4 are averaged values. For low concentrations, the data used were measured in the range from 25 to 50 °C, and for higher concentrations the data were measured in the range between 40–50 and 90 °C. Thus, the average temperature for the data is not the same.

Cross-sections for a fixed vapor pressure of H_2O in the $p_{H_2O,g}$–T–wt% P_2O_5 plot in Figure 12.2 will result in T–wt% P_2O_5 diagrams as depicted in Figure 12.5. In this diagram, the missing dew point curves were constructed according to the study by Brown and Whitt [2]. For temperatures below 300 °C, the partial pressures of phosphoric acid and P_4O_{10} are negligible (see also Figure 12.1b). For a distinct partial pressure of water, the phosphoric acid–water system exits as a single phase only in a field below the corresponding bubble point curve. Above this curve and below the dew point curve, liquid aqueous phosphoric acid and vapor coexist. Entering the two-phase field, for example, by increasing the temperature, will result in preferential evaporation of water until the thermodynamic equilibrium is reached; see Figure 12.5, dashed arrows. At 160 °C, an aqueous solution of 64 wt% P_2O_5 is in equilibrium with water vapor with a partial pressure of about 1 bar.

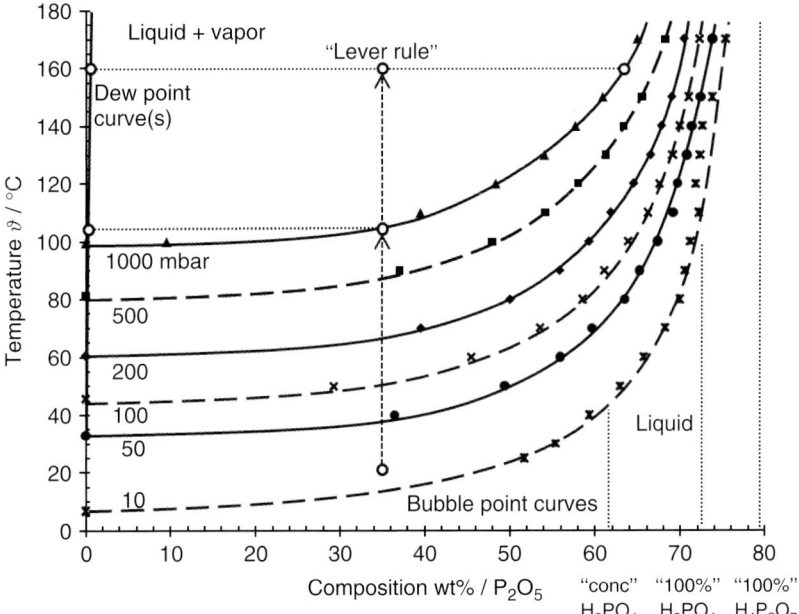

Figure 12.5 T–wt% P_2O_5 phase diagram of the H_2O–P_2O_5 system constructed from the data in Figure 12.2 for constant pressures $p_{H_2O,g}$. Bubble point curves for pressures $p_{H_2O,g}$ from 10 to 1000 mbar. Dew point curve extrapolated according data from Brown and Whitt [2].

12.4
Proton Conductivity as a Function of Composition and Temperature

12.4.1
Mechanism of the Electrical Conductivity in Phosphoric Acid

In (aqueous) phosphoric acid, charge can be transferred by proton migration or by migration of anions such as $H_2PO_4^-$, $H_3P_2O_7^-$, or $H_2P_2O_7^{2-}$ [8, 15]. Depending on the concentration, protons can be transported by hopping mechanisms via H_2O molecules or H_3PO_4 molecules, called *structural diffusion* or *Grotthus mechanism*, or by a *vehicle mechanism* through migration of H_3O^+ or $H_4PO_4^+$ ions.

Generally, proton migration via a hopping mechanism is much faster than the migration of anions. Hence the measured conductivity of phosphoric acid is mainly protonic with a transference number close to one. NMR studies demonstrated that in the case of low concentrations of phosphoric acid, protons will be transferred mainly between H_2O molecules (see Figure 12.6a) [16–18]. In the case of high phosphoric acid concentrations, if the fraction of H_2O molecules is only very small compared with the fraction of H_3PO_4 molecules, protons will be transferred directly between H_3PO_4 molecules (see Figure 12.6b). The existence of a hopping mechanism for protons in anhydrous phosphoric acid is also supported by a relatively small activation volume [19].

12.4.2
Evaluated Literature Data for the (Proton) Conductivity of (Aqueous) Phosphoric Acid in the Temperature Range Between 0 and 170 °C

Also for the electrical conductivity there are only a limited number of comprehensive experimental studies available in the literature. Most of these studies were published before 1970. Data on aqueous phosphoric acid solution in the important temperature and composition range can be found in a number of papers [8, 13, 15, 20–25].

In Figures 12.7 and 12.8, the conductivity data are plotted as a function of composition and temperature in the range from 0 to 86 wt% P_2O_5 and from

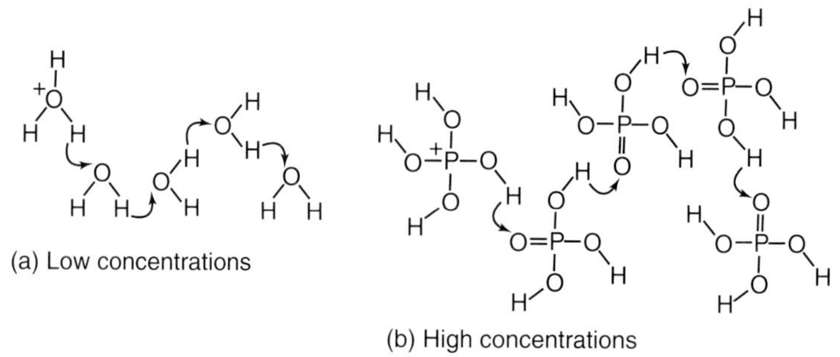

Figure 12.6 Proton conduction via hopping mechanisms in the H_2O–H_3PO_4 system for different concentration regions.

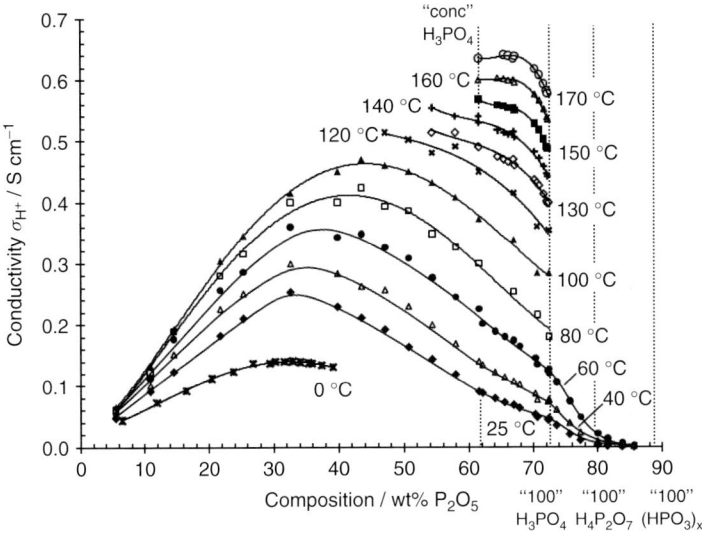

Figure 12.7 Conductivity of the H_2O–P_2O_5 system as a function of composition and temperature, using evaluated data from [13, 15, 24]. The solid lines connecting the data points serve as a guide for the eye.

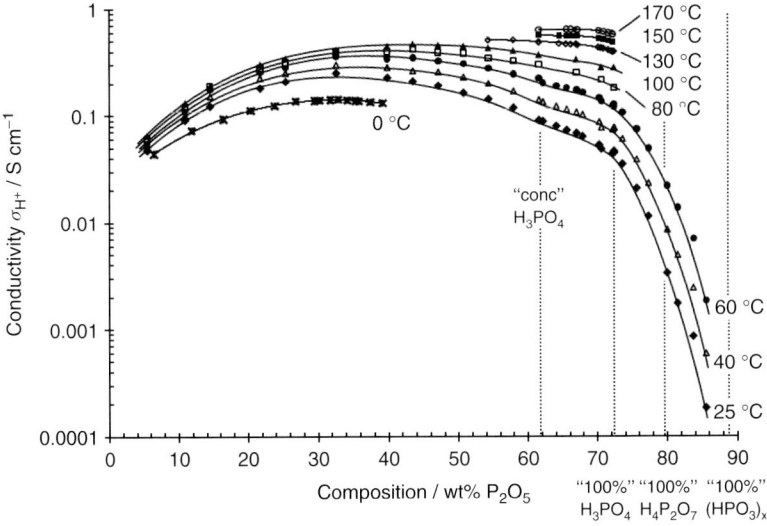

Figure 12.8 Conductivity of the H_2O–P_2O_5 system as a function of composition and temperature in a semilogarithmic plot, using evaluated data from [13, 15, 24]. The solid lines connecting the data points serve as a guide for the eye.

25 to 170 °C. The data from most of the papers fit fairly well together without discontinuities [8, 13, 15, 24, 25], but the data from two papers [22, 23] differ considerably from the others; the measured conductivities are about a factor of 1.5 smaller.

In the composition range from 0 to 90 wt% P_2O_5, the conductivity of phosphoric acid varies by three orders of magnitude. At 0 °C, it reaches a maximum at a composition of about 35 wt% P_2O_5. At 100 °C, the maximum shifts to a composition of 45 wt% P_2O_5. Because there is no data available for higher temperatures and low P_2O_5 contents, the maximum cannot be detected for temperatures above 100 °C. For concentrations above the maximum, the conductivity decreases slightly, and for concentrations above 72 wt% P_2O_5, that is, "100 wt%" H_3PO_4, it decreases strongly (see Figure 12.8).

12.4.3
Non-Arrhenius Behavior for the Ionic Transport

The activation enthalpy $\Delta H^{\#}_{H+}$ for the ionic (proton) transport in phosphoric acid can be calculated from the slope in a $\ln(T\sigma_{H+})$ versus $1/T$ plot:

$$\frac{\partial \ln(T\sigma_{H+})}{\partial\left(\frac{1}{T}\right)} = -\frac{\Delta H^{\#}_{H+}}{R} \qquad (12.13)$$

In Figures 12.9–12.11, Arrhenius plots were constructed for three composition ranges by using the literature data [8, 13, 15, 20, 21, 24] and by linear interpolating some missing compositions. Depending on composition, (aqueous) phosphoric acid is a viscous liquid or a glassy solid. For higher concentrations above 10 wt% P_2O_5, the conductivity shows considerably non-Arrhenius behavior (see Figure 12.9).

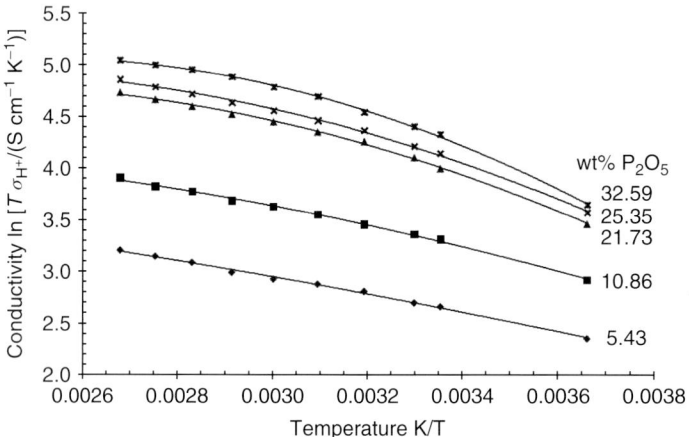

Figure 12.9 Conductivity of the $H_2O–P_2O_5$ system in the composition range from 5.43 to 32.59 wt% P_2O_5 in an Arrhenius plot [15, 21].

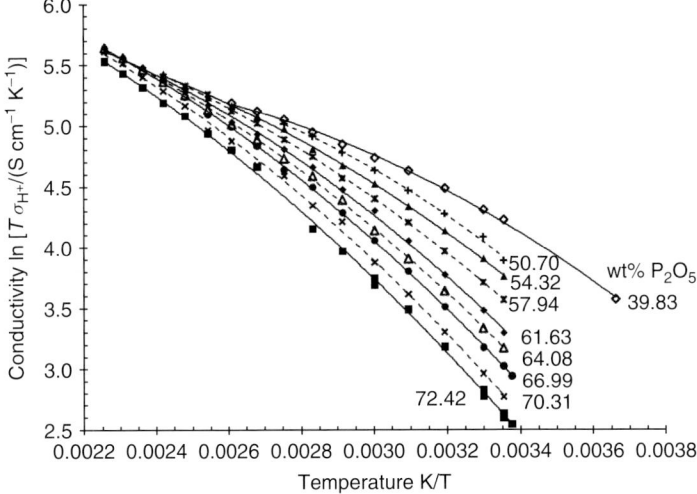

Figure 12.10 Conductivity of the $H_2O-P_2O_5$ system in the composition range from 57.94 to 72.42 wt% P_2O_5 in an Arrhenius plot [8, 13, 15, 20, 24].

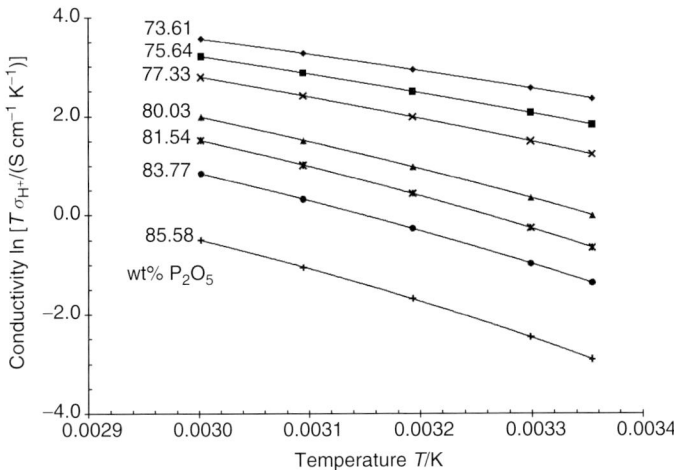

Figure 12.11 Conductivity of the $H_2O-P_2O_5$ system in the composition range from 73.61 to 85.58 wt% P_2O_5 in an Arrhenius plot [24].

This can be observed for many ionic conductors when measuring the ionic conductivity in the vicinity of the glass transition temperature T_g, where the dynamic viscosity η is strongly increasing, for example, in inorganic glasses and polymers. The rate-determining step for the molecular, atomic, or ionic transport in such highly viscous systems is the formation of free volume in the neighborhood of a mobile particle rather than an activated jump over a potential barrier in a free neighbor position. Empirically from experimental studies such as the Vogel–Tammann and Hesse–Fulcher equations [26–28] or from theoretical models such as the "theory of free volume" [29–31], an exponential relation is found between dynamic viscosity η and temperature T:

$$\eta = \eta_0 e^{\frac{B_\eta}{T-T_0}} \tag{12.14}$$

where B_η, η_0, and T_0 are material-dependent parameters. In particular, T_0 is a critical temperature related to the glass transition temperature T_g. In a simple model, the dynamic viscosity increases up to infinity if the temperature of the system approaches the critical temperature T_0. At the critical temperature T_0, the free volume between the particles is assumed to be zero.

In a simple hydrodynamic approach, transport parameters such as the ionic conductivity σ_i, the diffusion coefficient D_i, and the electrochemical mobility u_i of ionic/atomic species i in liquid/glassy systems are linked to the dynamic viscosity η by the Stokes–Einstein equation:

$$\eta D_i = \frac{kT}{6\pi R_i} \tag{12.15a}$$

or

$$\eta u_i = \frac{z_i e_0}{6\pi R_i} \tag{12.15b}$$

or

$$\eta \sigma_i = \frac{(z_i e)^2 N_A c_i}{6\pi R_i} = \text{constant} \tag{12.15c}$$

where R_i is the hydrodynamic radius of the species i and c_i its molar concentration. On introducing the molar ionic conductivity $\lambda_i = \sigma_i/c_i = z_i F u_i$, Eq. (12.15c) can be rewritten in such a way that the product of the dynamic viscosity and the molar ionic conductivity depends only on the material parameter ionic charge z_i and hydrodynamic radius R_i:

$$\eta \lambda_i = \frac{e^2 N_A}{6\pi} \frac{z_i^2}{R_i} = \text{constant} \tag{12.16}$$

In particular, Eq. (12.16) is also called the *Walden product* [32]. In this approach, we assume that every microscopic ion motion is coupled with a viscous flow event. If there is not only one mobile charge carrier, Eq. (12.16) can be rewritten using the total molar conductivity $\Lambda = \sum_i v_i \lambda_i$ as

$$\eta \Lambda = \frac{e^2 N_A}{6\pi} \sum_i \frac{z_i^2 v_i}{R_i} = \text{constant} \tag{12.17}$$

Equations (12.16) and (12.17) are valid for strong electrolytes in case of complete dissociation. Combining Eqs. (12.14) and (12.15c) results in an "Arrhenius-like" exponential relation between temperature T and conductivity σ_i, but also involving a critical temperature T_0:

$$\sigma_i = \frac{(z_i e)^2 N_A c_i}{6\pi \eta_0 R_i} e^{-\frac{B_\eta}{T-T_0}} = A e^{-\frac{B_\eta}{T-T_0}} \tag{12.18a}$$

or

$$\lambda_i = \frac{(z_i e)^2 N_A}{6\pi \eta_0 R_i} e^{-\frac{B_\eta}{T-T_0}} = A' e^{-\frac{B_\eta}{T-T_0}} \tag{12.18b}$$

The pre-exponential factor A often exhibits a slight temperature dependence ($A = A' T^{-1/2}$), hence Eq. (12.18a) can be extended as follows:

$$\sigma_i = A'' T^{-1/2} e^{-\frac{B_\eta}{T-T_0}} \tag{12.19}$$

For all known systems, the material-dependent parameter B_η in Eq. (12.14) does not necessarily have the same value as the parameter B_η in Eq. (12.18a). In a system with a higher probability of ionic charge motions than for viscous flow events, the strict coupling between dynamic viscosity and conductivity via the Stokes–Einstein equation does not hold [33, 34]. By introducing another independent material parameter, $B_\sigma \neq B_\eta$, Eq. (12.18a) can be rewritten as

$$\sigma_i = A e^{-\frac{B_\sigma}{T-T_0}} \tag{12.20}$$

This decoupling leads to a modification of Eq. (12.17):

$$\eta^\alpha \lambda_i = \text{constant with } \alpha = \frac{B_\sigma}{B_\eta} \tag{12.21}$$

The exponent α in Eq. (12.21) is a constant proportional to the extent of decoupling and can be determined as a slope in a log λ_i versus log $(1/\eta)$ plot, the Walden plot.

In Figure 12.12, a $\log \lambda_{H^+}$ versus $\log(1/\eta)$ plot for aqueous phosphoric acid is depicted, using literature viscosity data [15, 35, 36] and conductivity data [15, 24] for the temperature range from 25 to 100 °C and for the composition range from 5.43 to 72.42 wt% P_2O_5. Data for the density to calculate the concentration of phosphoric acid were taken from [7]. All slopes are considerably less than 1.

The molar conductivity λ_{H^+} of aqueous phosphoric acid is only a function of the dynamic viscosity η within the parameter field of the data used. Hence the general relation according to Eq. (12.21) holds independently of temperature and composition, but with a varying exponent α. For low concentrations (5.43 wt% P_2O_5) an exponent α of about 0.29 can be detected. The exponent α increases with increasing concentration. For pure phosphoric acid (72.42 wt% P_2O_5) it will reach a value of about 0.65. Equations (12.16), (12.17), and (12.21) are strictly valid only in the case of complete dissociation, using the limiting molar conductivity. For aqueous phosphoric acid, the degree of dissociation depends on the concentration. Concerning the mobile protonic charge carriers, the degree of dissociation varies between 0.16 (about 7 wt% P_2O_5) and about 0.38 (about 70 wt% P_2O_5) at a temperature of 25 °C [3–5]. This will shift the data in Figure 12.12 by a value between −0.4 and −0.8 because of the logarithmic plot.

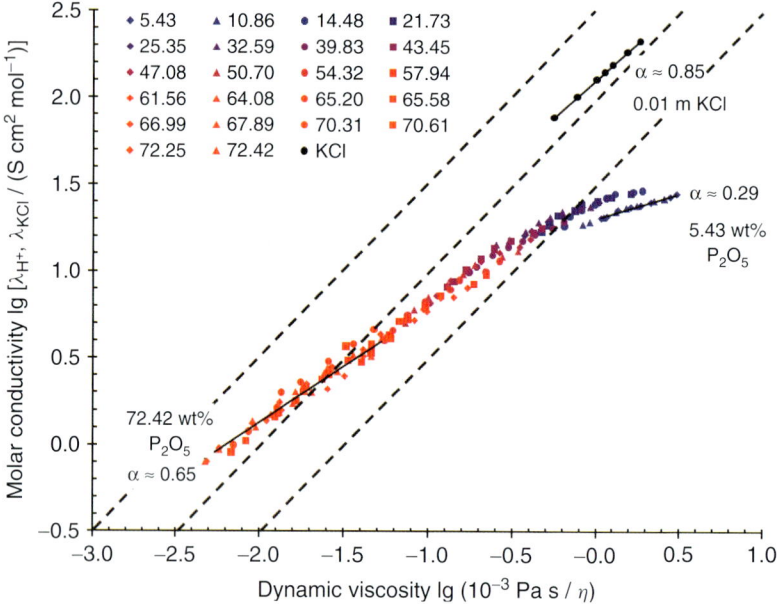

Figure 12.12 Log λ_i versus log$(1/\eta)$ plot for the H_2O–P_2O_5 system in the composition range from 5.43 to 72.42 wt% P_2O_5 and in the temperature range from 25 to 100 °C [7, 15, 24, 35, 36]. Reference data for aqueous 0.01 m KCl in the same temperature range have been added [7]. The slope of 1 is marked by the dashed lines.

Decoupling is found in phosphoric acid for the whole concentration range. This means that the activation energy for proton transport ($\sim B_\sigma$) is always lower than that for viscous flow ($\sim B_\eta$), and demonstrates again that the proton transport takes place by a proton-hopping mechanism independent of viscous motion [37]. In a system without decoupling, ideally an exponent α of 1 is expected. In Figure 12.12, data for aqueous 0.01 m KCl solution for the same temperature range is added as a reference. In this case, an exponent α with a value of 0.85 is found.

12.4.4
Enthalpy of Activation for the Ionic Transport

Using the conductivity data from the literature [8, 13, 15, 20, 21, 24] in Figures 12.9–12.11, the activation enthalpy $\Delta H^{\#}_{H+}$ can be determined according to Eq. (12.13) from the slopes. A relation between the parameter B_σ in Eq. (12.20) and the activation enthalpy $\Delta H^{\#}_i$ according to the Arrhenius law can be obtained by inserting Eq. (12.20) in Eq. (12.13):

$$\Delta H^{\#}_i = RT + R\left(\frac{T}{T-T_0}\right)^2 B_\sigma \tag{12.22}$$

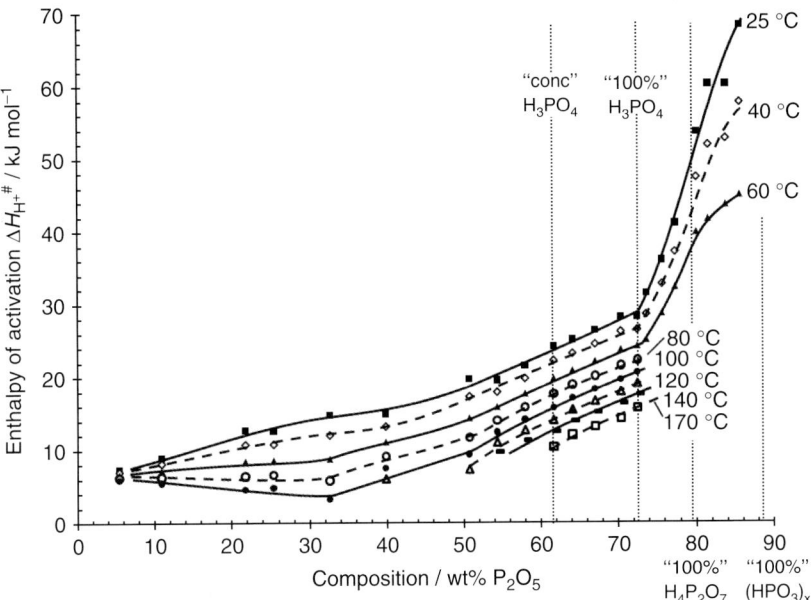

Figure 12.13 Enthalpy of activation $\Delta H^{\#}_{H+}$ for proton transport for the $H_2O-P_2O_5$ a system s a function of the composition and temperature. Enthalpy of activation evaluated from the slopes in Figures 12.9–12.11. Conductivity data from [8, 13, 15, 20, 21, 24]. The dashed and solid lines are guides for the eye. The considerable deviations for the composition of 39.8 wt% P_2O_5 and temperatures above 80 °C from the continuous curve progression may be artifacts because of the limited precision of the data.

Thus, alternatively, the data can be fitted by using Eq. (12.20). The enthalpy of activation $\Delta H^{\#}_i$ can be calculated from B_σ according Eq. (12.22). In Figure 12.13, the activation enthalpy $\Delta H^{\#}_{H+}$ for proton transport is plotted in the composition range from 5.43 to 85.58 wt% P_2O_5 and in the temperature range from 25 to 170 °C.

For low-concentration phosphoric acid, the temperature dependence of the activation enthalpy $\Delta H^{\#}_{H+}$ in this temperature range is only small. For 7.5% H_3PO_4 (5.43 wt% P_2O_5), $\Delta H^{\#}_{H+}$ has a value of about 6–7.5 kJ mol^{-1}. For higher concentration phosphoric acid, the temperature dependence is more pronounced. This can also be deduced from Figures 12.9–12.11 just from the increasing non-Arrhenius behavior, that is, from the increased "bending" of the plotted curves. For 100 wt% H_3PO_4 (72.42 wt% P_2O_5), $\Delta H^{\#}_{H+}$ has a value of about 28 kJ mol^{-1} at 25 °C and 16 kJ mol^{-1} at 170 °C. The activation energy increases with decreasing H_2O content.

For compositions above 72.42 wt% P_2O_5, the increase in the activation enthalpy is much stronger than at lower concentrations. For a composition of 85.58 wt% P_2O_5, $\Delta H^{\#}_{H+}$ reaches values of about 68 kJ mol^{-1} at 25° and of 45 kJ mol^{-1} at 60°.

The values for $\Delta H^{\#}_{H+}$ at low concentrations are comparable to data in the literature for proton conduction in water or in other diluted aqueous solutions.

For pure water, a value of about 10.9 kJ mol^{-1} and for 0.2 m hydrochloric acid (0.7 wt% HCl) a value of about 12 kJ mol^{-1} were found [38–41]. The observed increase in the activation enthalpy for proton transfer can be explained by a change in the dominating molecule carrying the proton. At low concentrations, protons are mainly transferred between H_2O molecules and at high concentrations mainly between H_3PO_4 molecules (see Section 12.4.1, Figure 12.6). Additionally, the dynamic viscosity increases. This is also consistent with the observed change in the exponent α in the log λ_i versus log $(1/\eta)$ plot in Figure 12.12. The proton transfer via the H_2O molecules is faster compared with the transport via the H_3PO_4 molecules because of a smaller activation enthalpy. This results in stronger decoupling for the more dilute phosphoric acid.

Because of the combination of a relatively fast proton transport (high mobility) with an extraordinarily high autoprotolysis constant, resulting in a high intrinsic concentration of charge carriers, 100 wt% H_3PO_4 (72.42 wt% P_2O_5) exhibits the highest proton conductivity compared with other proton-conducting liquids such as H_2O and H_2SO_4. The proton transport in pure H_2O is faster than that in H_3PO_4, but the autoprotolysis constant is about 13 orders of magnitude lower, resulting in a very low concentration of charge carriers (H_3O^+ and OH^-) and thus a much lower conductivity.

12.4.5
Evaluated Data for the Dynamic Viscosity of Aqueous Phosphoric in the Temperature Range from 23 to 170 °C

Data for the dynamic viscosity of aqueous phosphoric acid for the same temperature and composition ranges as given for the conductivity data can be found in the literature [8, 15, 35, 36]. In Figure 12.14, the dynamic viscosity is plotted as ln η versus $1/T$.

For compositions above 60 wt% P_2O_5, considerable deviations from linearity can be detected, that is, according to the Vogel–Tammann–Fulcher equation Eq. (12.14) T_0 is not negligibly small compared with T. Hence the critical temperature T_0 increases with increasing concentration and approaches the temperature range of the viscosity data.

For concentrated phosphoric acids above 62 wt% P_2O_5 (85 wt% H_3PO_4), the critical temperature T_0 can be determined from the data with acceptable errors by fitting with Eq. (12.14). For lower concentrations, T_0 becomes too small compared with the temperature T, resulting in large fitting errors because of poor convergence. The critical temperatures determined by fitting for compositions above 62 wt% P_2O_5 are in good agreement with the findings for other glassy and highly viscous systems, where T_0 is usually about 50 K lower than the glass transition temperature T_g (see Figure 12.15). The glass transition temperatures used as reference were determined by measuring the line broadening in ^1H NMR spectra [42].

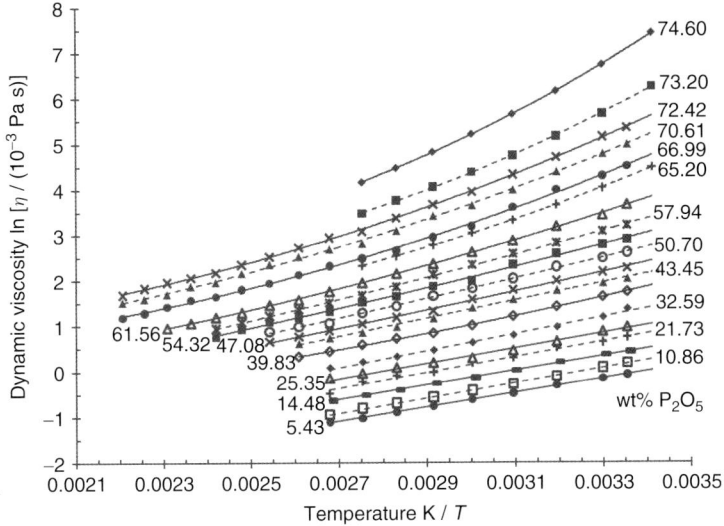

Figure 12.14 Dynamic viscosity η of the H_2O–P_2O_5 system in a $\ln \eta$ versus $1/T$ plot as a function of the composition. Viscosity data from [15, 35, 36].

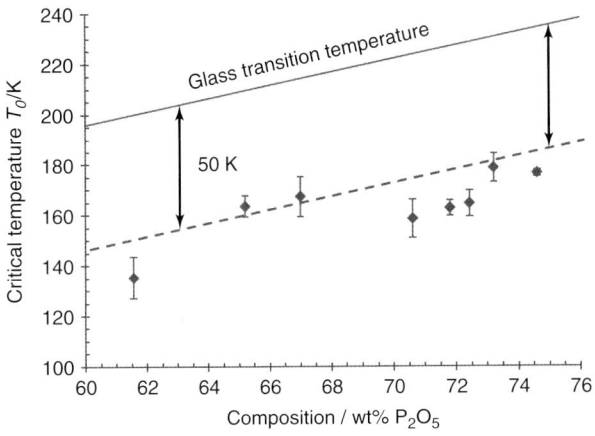

Figure 12.15 Critical temperature T_0 determined by fitting the viscosity data in Figure 12.14 from [8, 15] with the Vogel–Tammann–Fulcher equation. Glass transition temperatures taken from [42].

12.5
Equilibria between the Polyphosphoric Acid Species and "Composition" of Concentrated Phosphoric Acid

As stated in Section 12.2.1, the equilibrium constants K_i of each condensation reaction, leading to a distinct polyphosphoric acid species i, are essentially functions of temperature T. Considering the laws of mass action according to Eq. (12.3), the equilibrium concentrations of each polyphosphoric acid species i are connected to the equilibrium constants K_i and the initial concentration of P_2O_5. Hence the equilibrium concentration of each polyphosphoric acid species in concentrated phosphoric acid is (essentially) only a function of temperature T and of the initial concentration of P_2O_5.

12.5.1
Evaluated Literature Data for the Polyphosphoric Acid Equilibria

The number of experimental studies in the literature on this topic is even smaller than for conductivity and vapor pressure data. All studies found are from the period between 1955 and 1959. Studies using chromatographic methods to separate the polyphosphoric acid species were performed by Higgins and Baldwin, Jameson, and Huhti and Gartaganis [43–45]. A formal approach using models from polymer science was presented by Westman and Beatty [46].

In the studies of Jameson [44] and Huhti and Gartaganis [45], phosphoric acid with a composition of 67–89 wt% P_2O_5 was investigated using paper chromatography for analysis of the polyphosphoric acid contents. The samples were prepared by mixing the corresponding amounts of (concentrated) 85 wt% phosphoric acid and P_2O_5 and subsequent melting of the mixture in a sealed container at temperatures between 250 and 400 °C. Another way is to concentrate phosphoric acid solutions by evaporating water at a sufficiently high temperature (400 °C). In the study of Higgins and Baldwin [43], concentrated phosphoric acid was mixed with aqueous ^{32}P-labeled phosphoric acid and dried at room temperature at a pressure of 3–4 mbar. The phosphoric acid samples were heated to 100 and 176 °C in a container covered with a filter cone. Small amounts were taken in intervals of hours and analyzed using ion chromatography with an anion-exchange resin. The activity of the ^{32}P labeling was used to identify the fractions containing polyphosphoric acids.

In all of these experimental studies, the equilibrium was established at an elevated temperature (≥ 100 °C) and the subsequent analysis was carried out at room temperature. This can only be done if the kinetics of all condensation equilibria are very slow at room temperature and fast enough to establish a thermodynamic equilibrium at the experimental temperature. It was found that the reaction kinetics of the hydrolysis and condensation reactions were very slow at room temperature and a high pH. Using the rate constants for the hydrolysis of diphosphate and triphosphate from other literature references, a half-life of about 30–100 days can be estimated for this species in a dilute solution at room temperature and pH 9–11 [47, 48].

As depicted in Figure 12.16, the results of Jameson [44] and Huhti and Gartaganis [45] are in very good agreement with each other. According to Huhti and Gartaganis, the equilibrium concentrations of the polyphosphoric acid species were reached within 5 min for an equilibration temperature above 250 °C. They did not found

12.5 Polyphosphoric Acid Species and "Composition" of Concentrated Phosphoric Acid

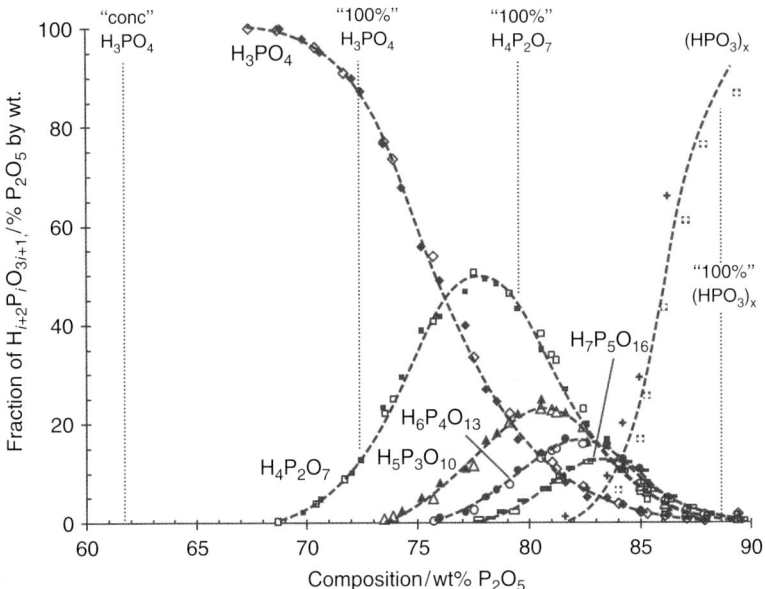

Figure 12.16 Fractions of the different polyphosphoric acids in the H_2O–P_2O_5 system as a function of P_2O_5 content. Data points taken from Jameson [44] are marked with solid symbols and data points taken from Huhti and Gartaganis [45] are marked with open symbols.

any significant temperature dependence of the equilibrium concentrations in the investigated temperature interval between 250 and 400 °C.

The results of Higgins and Baldwin [43], as shown in Table 12.3, are difficult to compare with the other studies, because the exact P_2O_5 content of the phosphoric acid samples after equilibration was not indicated. One can assume that compositions according to the phase diagram in Figure 12.5 were reached. For a phosphoric acid sample dried in vacuum with a pressure of 3–4 mbar at room temperature, a composition of about 60 wt% P_2O_5 should be reached. For samples heated to temperatures of 100 and 176 °C in quasi-open containers (i.e., about 1 bar total pressure), the compositions are more difficult to estimate.

At a temperature of 176 °C and an initial composition of 60 wt% P_2O_5, the H_2O partial pressure of aqueous phosphoric acid is higher than 1 bar. Thus, on heating in a quasi-open container, the concentration of phosphoric acid increases by preferential evaporation of water. A new equilibrium between the gas and liquid phases at an H_2O partial pressure of 1 bar is attained, if a composition of 66 wt% is reached. At a temperature of 100 °C and an initial composition of 60 wt% P_2O_5, the H_2O partial pressure of aqueous phosphoric acid is only about 200 mbar. The H_2O partial pressure at room temperature in air is usually smaller (air humidity). For a container with only a small volume for the gas phase, an equilibrium between the gas and liquid phases for an H_2O partial pressure of 200 mbar can be attained only

Table 12.3 Fractions of different polyphosphoric acids in the $H_2O–P_2O_5$ system at temperatures of 25, 100, and 176 °C according to Higgins and Baldwin [43].

	Fraction (%)		
	25 °C	100 °C	176 °C
H_3PO_4	99.7	96	49
$H_4P_2O_7$	0.3	4	40
$H_5P_3O_{10}$	–		10
$H_6P_4O_{13}$	–		2
$H_7P_5O_{16}$	–		0.2
wt% P_2O_5	60	60?	66

by negligible evaporation of H_2O from the liquid phase. If there are only minor vapor losses to the surroundings through the filter cone, it can be assumed that the composition of the phosphoric acid sample should not change significantly.

The data of Higgins and Baldwin indicate that the fraction of polyphosphoric acids will increase with increasing equilibration temperature. The fractions given for temperatures of 25, 100, and 176 °C are not at all in agreement with the data of Jameson and of Huhti and Gartaganis. According to the latter data, aqueous phosphoric acid should consist almost only of orthophosphoric acid for a composition of 60–66 wt% P_2O_5 (see Figure 12.16).

The data found in the literature are not completely consistent. Especially the temperature dependence of the polyphosphoric acid equilibria below 200 °C is not clear. In addition to the thermodynamic states of the equilibria, there is also no extensive information on the kinetics of these condensation reactions as a function of temperature. According to other studies in the literature, the kinetics at room temperature are presumably slow, that is, the half-life of the polyphosphoric acid species might be in the order of hours. At temperatures above 200 °C, the equilibria will be established within minutes. There is a lack of information on kinetic data for the temperature range from 150 to 180 °C, which is important for HT-PEMFC operating regimes.

12.6
Conclusion

By compiling the available literature data on the vapor pressure of aqueous phosphoric acid as a function of temperature and of composition, a fairly comprehensive description of the liquid–gas-phase field can be made for the temperature range and the total pressure range, which is important for HT-PEMFC operation. The same applies to the literature data on the proton conductivity and the dynamic viscosity as a function of temperature and composition. There is a lack of conductivity data for high temperatures above 120 °C and high concentrations above 100 wt% H_3PO_4 (72.42 wt% P_2O_5).

From the compiled vapor pressure and conductivity data, the evaporation enthalpy and the activation enthalpy for proton conduction were calculated as a function of composition. The critical temperature according the Vogel–Tammann–Fulcher law was determined from the viscosity data and compared with glass transition temperatures from other studies using NMR spectroscopy. A correlation between dynamic viscosity and molar conductivity was found. As expected, a considerable decoupling between ionic conduction and viscous flow can be determined from a Walden plot, which is based on proton-hopping mechanisms in phosphoric acid.

The activation enthalpy for proton conduction in high-concentration phosphoric acid is remarkably higher than that for aqueous proton conductors. In dilute phosphoric acid, the activation enthalpy is comparable to those of other aqueous proton conductors. In dilute phosphoric acid, very fast proton transport takes place via water molecules. In high-concentration phosphoric acid, the proton transport takes place mainly via phosphoric acid molecules, but slower than the transport via water molecules. However, because of the combination of relatively high proton mobility with an extraordinarily high autoprotolysis constant, resulting in a very high intrinsic concentration of charge carriers, 100 wt% H_3PO_4 exhibits the highest proton conductivity compared with other liquid proton conductors.

Only sparse information could be found on the equilibria between the different species of polyphosphoric acids in an aqueous solution of phosphoric acid. The literature data found for equilibration temperatures above 200 °C are largely consistent. The data found for equilibration temperatures below 200 °C do not fit the data above 200 °C. Moreover, detailed information on the kinetics of the equilibration still awaits further research.

References

1. Schrödter, K., Bettermann, G., Staffel, T., Wahl, F., Klein, T., and Hofmann, T. (2008) Phosphoric acid and phosphates, *Ullmann's Encyclopedia of Industrial Chemistry*, Wiley-VCH Verlag GmbH, Weinheim. DOI: 10.1002/14356007.a19 465.pub3.
2. Brown, E.H. and Whitt, C.D. (1952) Vapor pressure of phosphoric acids. *Ind. Eng. Chem.*, **44** (3), 615–618.
3. Rudolph, W.W. (2010) Raman- and infrared-spectroscopic investigations of dilute aqueous phosphoric acid solutions. *Dalton Trans.*, **39**, 9642–9653.
4. Elmore, K.L., Hatfield, J.D., Dunn, R.L., and Jones, A.D. (1966) Dissociation of phosphoric acid solutions at 25°. *J. Phys. Chem.*, **69** (10), 3520–3525.
5. Rudolph, W. and Steger, W. (1991) Dissociation, structure, and rapid proton exchange of phosphoric acid in dilute aqueous solutions. V. Vibrational spectra of phosphoric acid. *Z. Phys. Chem.*, **172** (1), 49–59.
6. Holleman, A.F., Wiberg, E., and Wiberg, N. (1995) *Lehrbuch der anorganischen Chemie*, Walter de Gruyter, Berlin, p. 101.
7. Haynes, W.M. (ed.) (2010) *CRC Handbook of Chemistry and Physics*, 91st edn., CRC Press, Taylor & Francis, Boca Raton, FL, p. 91.

8. Greenwood, N.N. and Thompson, A. (1959) The mechanism of electrical conduction in fused phosphoric and trideuterophosphoric acids. *J. Chem. Soc.*, 3485–3492.
9. Rondini, S., Longhi, P., Mussini, P.R., and Mussini, T. (1987) Autoprotolysis constants in nonaqueous solvents and aqueous organic solvent mixtures. *Pure Appl. Chem.*, **59** (12), 1693–1702.
10. Munson, R.A. (1964) Self-dissociative equilibria in molten phosphoric acid. *J. Phys. Chem.*, **68** (11), 3374–3377.
11. Fontana, B.J. (1951) The vapor pressure of water over phosphoric acids. *J. Am. Chem. Soc.*, **73** (7), 3348–3350.
12. Kablukov, I.A. and Zagwosdkin, K.I. (1935) Die dampfspannung der phosphorsäurelösungen. *Z. Anorg. Allg. Chem.*, **224** (3), 315–321.
13. McDonald, D.I. and Boyack, J.R. (1969) Density, electrical conductivity, and vapor pressure of concentrated phosphoric acid. *J. Chem. Eng. Data*, **14** (3), 380–384.
14. Schmalz, E.O. (1970) Bestimmung der Dampfdruckkurven von Wasser über Phosphorsäure. *Z. Phys. Chem. (Leipzig)*, **245** (5–6), 344–350.
15. Chin, D.T. and Chang, H.H. (1989) On the conductivity of phosphoric acid electrolyte. *J. Appl. Electrochem.*, **19** (1), 95–99.
16. Dippel, T., Kreuer, K., Lassègues, J., and Rodriguez, D. (1993) Proton conductivity in fused phosphoric acid; a $^1H/^{31}P$ PFG-NMR and QNS study. *Solid State Ionics*, **61** (1–3), 41–46.
17. Aihara, Y., Sonai, A., Hattori, M., and Hayamizu, K. (2006) Ion conduction mechanisms and thermal properties of hydrated and anhydrous phosphoric acids studied with $^1H, ^2H,$ and ^{31}P NMR. *J. Phys. Chem. B*, **110**, 24999–25006.
18. Chung, S.H., Bajue, S., and Greenbaum, S.G. (2000) Mass transport of phosphoric acid in water: a ^1H and ^{31}P pulsed gradient spin-echo nuclear magnetic resonance study. *J. Chem. Phys.*, **112** (19), 8515–8521.
19. Fontanella, J.J., Wintersgill, M.C., Wainright, J.S., Savinell, R.F., and Litt, M. (1998) High pressure electrical conductivity studies of acid doped polybenzimidazole. *Electrochim. Acta*, **43** (10–11), 1289–1294.
20. Smith, A. and Menzies, A.W.C. (1909) The electrical conductivity and viscosity of concentrated solutions of orthophosphoric acid. *J. Am. Chem. Soc.*, **31** (11), 1191–1194
21. Campbell, A.N. (1926) The conductivity of phosphoric acid solutions at 0°. *J. Chem. Soc.*, 3021–3022.
22. Kakulin, G.P. and Fedorchenko, I.G. (1962) Electric conductance of concentrated phosphoric acid. *Zh. Neorg. Khim.*, 7, 2485–2486.
23. Fedorchenko, I.G., Kakulin, G., and Kondratenko, Z.V. (1965) Electric conductance of concentrated phosphoric acid at 100–200 °C. *Zh. Neorg. Khim.*, **10** (8), 1945–1946.
24. Wydeven, T. (1966) Electrical conductivity of concentrated phosphoric acid from 25 °C to 60 °C. *J. Chem. Eng. Data*, **11** (2), 174–176.
25. Tsurko, E.N., Neueder, R., Barthel, J., and Apelblat, A. (1999) Conductivity of phosphoric acid, sodium, potassium, and ammonium phosphates in dilute aqueous solutions from 278.15 K to 308.15 K. *J. Solution Chem.*, **28** (8), 973–999.
26. Vogel, H. (1921) Das temperaturabhängigkeitsgesetz der Viskosität von Flüssigkeiten. *Phys. Z.*, **22**, 645–646.
27. Fulcher, G.S. (1925) Analysis of recent measurements of the viscosity of glasses. *J. Am. Ceram. Soc.*, **8** (6), 339–355.
28. Tammann, G. and Hesse, W. (1926) Die Abhängigkeit der Viscosität von der Temperatur bei unterkühlten Flüssigkeiten. *Z. Anorg. Allg. Chem.*, **156** (1), 245–257.
29. Doolittle, A.K. (1951) Studies in Newtonian flow. II. The dependence of the viscosity of liquids on free-space. *J. Appl. Phys.*, **22** (12), 1471–1475.
30. Turnbull, D. and Cohen, M.H. (1970) On the free-volume model of the liquid–glass transition. *J. Chem. Phys.*, **52** (6), 3038–3041.
31. Cohen, M.H. and Turnbull, D. (1959) Molecular transport in liquids

and glasses. *J. Chem. Phys.*, **31** (5), 1164–1169.
32. Walden, P. (1906) Über organische Lösungs- und Ionisierungsmittel. III. Teil: innere Reibung und deren Zusammenhang mit dem Leitvermögen. *Z. Phys. Chem.*, **55**, 207–246.
33. McLin, M.G. and Angell, C.A. (1991) Ion-pairing effects on viscosity/conductance relations in Raman-characterized polymer electrolytes: lithium perchlorate and sodium triflate in PPG(4000). *J. Phys. Chem.*, **95** (23), 9464–9469.
34. Xu, W., Cooper, E.I., and Angell, C.A. (2003) Ionic liquids: ion mobilities, glass temperatures, and fragilities. *J. Phys. Chem., B*, **107** (25), 6170–6178.
35. Kondratenko, Z.V. and Fedorchenko, I.G. (1959) *Zh. Neorg. Khim.*, **4**, 985.
36. Kondratenko, Z.V., Fedorchenko, I.G., and Kovalev, I.A. (1967) Mathematical calculation of the density and viscosity of concentrated phosphoric acid solutions. *Zh. Prikl. Khim.*, **40** (9), 1947–1951.
37. Herath, M.B., Creager, S.E., Kitaygorodskiy, A., and DesMarteau, D.D. (2010) Perfluoroalkyl, phosphonic and phosphinic acids as proton conductors for anhydrous proton-exchange membranes. *Chem. Phys. Chem.*, **11**, 2871–2878.
38. Duecker, H.C. and Haller, W. (1962) Determination of the dissociation equilibriums of water by a conductance method. *J. Phys. Chem.*, **66**, 225–229.
39. Perrault, G. (1961) Sur la conductibilité protonique dans l'eau pure. *C. R. Acad. Sci.*, **252**, 4145–4147.
40. Horne, R.A., Courant, R.A., and Johnson, D.S. (1966) The dependence of ion-, proton-, water and electron transport processes on solvent structure in aqueous electrolyte solutions. *Electrochim. Acta*, **11** (8), 987–996.
41. Loewenstein, A. and Szöke, A. (1961) The activation energies of proton transfer reactions in water. *J. Am. Chem. Soc.*, **84** (7), 1151–1154.
42. Ellis, B. (1976) The glass transition temperature of phosphoric acids. *Nature*, **263**, 674–676.
43. Higgins, C.E. and Baldwin, W.H. (1955) Dehydration of orthophosphoric acid. *Anal. Chem.*, **27** (11), 1780–1783.
44. Jameson, R.F. (1959) The composition of 'strong' phosphoric acids. *J. Chem. Soc.*, 752–759.
45. Huhti, A.L. and Gartaganis, P.A. (1956) The composition of the strong phosphoric acids. *Can. J. Chem.*, **34** (6), 785–797.
46. Westman, A.E.R. and Beatty, R. (1966) Equations for calculating chain length distributions in polyphosphoric acids and polyphosphate glasses. *J. Am. Ceram. Soc.*, **49** (2), 63–67.
47. Crowther, J.P. and Westman, A.E. (1954) Hydrolysis of the condensed phosphates. I. Sodium pyrophosphate and sodium triphosphate. *Can. J. Chem.*, **32**, 42–48.
48. Campbell, D.O. and Kilpatrick, M.L. (1954) A kinetic study of the hydrolysis of pyrophosphates. *J. Am. Chem. Soc.*, **76** (3), 893–901.

13
Materials and Coatings for Metallic Bipolar Plates in Polymer Electrolyte Membrane Fuel Cells

Heli Wang and John A. Turner

13.1
Introduction

A polymer electrolyte membrane fuel cell (PEMFC) is a device that converts hydrogen and oxygen gases into electricity with water as the only by-product. Owing to its high efficiency and near-zero emission, PEMFCs are of great interest as a clean energy device [1]. Cost reduction of the components in a PEMFC stack is the major barrier in its civilian applications. One of the important components is the bipolar plate. This multifunctional component provides the electrical connectivity from cell to cell. Its functions also include distributing gases and removing heat from the active areas. These require that the bipolar plate material should be chemically stable, highly electrically and thermally conductive; have low contact resistance with the backing, good mechanical strength, low gas permeability, low cost, and allow uniform reactant gas distribution and product removal. For transportation purposes, a light weight and small volume are also required. Certainly, not a single material could satisfy all the above requirements. Graphite bipolar plates dominated the application in the early stages. For civilian applications, however, the large volume associated with the gas permeability for graphite, the high cost of machining gas channels, and its brittleness are the largest hurdles for the material. Therefore, the bipolar plate is one of the significant challenges in the cost reduction of PEMFCs because it accounts for a significant amount of cost and weight [2–8]. In 2002, bipolar plates accounted for about 90% total stack weight and about 67% of total stack material costs [4]. This decreased to about 78% total weight and 37% total cost in 2004 [2], and 75% total weight and about 11–45% of total stack cost with graphite bipolar plates in 2006 [8]. Further cost reductions are required for a successful application of PEMFCs in the civilian market, and the US Department of Energy (DOE)'s 2015 technical targets for bipolar plates [9] are listed in Table 13.1.

General reviews and design analyses concerning bipolar plates for PEMFCs have been carried out periodically [10–14], and the bipolar plate materials are commonly divided into composite and metallic materials. Efforts with composite materials have been carried out in research organizations and industry worldwide [15–23]. Wilson and co-workers developed a composite bipolar plate with a vinyl

Table 13.1 US DOE technical targets: bipolar plates [9].

Characteristics	2005 status[a]	2015
Cost[b] ($ kW^{-1})	10[c]	3
Weight (kg kW^{-1})	0.36	<0.4
H_2 permeation rate [cm^3 s^{-1} cm^{-2} @80 °C, 3 atm (equivalent to <0.1 mA cm^{-2})]	<2 × 10^{-6}	<2 × 10^{-6}
Corrosion (μA cm^{-2})	<1[d]	<1[d]
Electrical conductivity (S cm^{-1})	>600	>100
Resistivity[e] (Ω cm)	<0.02	0.01
Flexural strength[f] (MPa)	>34	>25
Flexibility (% deflection at mid-span)	1.5–3.5	3–5

[a] First year for which status was available. 2005 status is for carbon plates, except for corrosion status, which is based on metal plates.
[b] Based on 2002 dollars and costs projected to high-volume production (500 000 stacks per year).
[c] Status is from 2005 TIAX study and will be periodically updated.
[d] May have to be as low as 1 nA cm^{-2} if all corrosion product ions remain in ionomer.
[e] Includes contact resistance.
[f] Developers have used ASTM C-651-91: Standard Test Method for Flexural Strength of Manufactured Carbon and Graphite Articles Using Four-Point Loading at Room Temperature.

ester and graphite powder over a decade ago [15–17]. Some interesting progress in graphite–polymer composite bipolar plate work has been demonstrated more recently, including a conductive composite bipolar plate with a graphite-filled polymer [20], a carbon composite bipolar plate composed of 90% graphite powder and 10% unsaturated polymer [21], a vinyl ester–graphite composite bipolar plate [22], and injection-molded plates based on thermoplastic and carbon compounds [23] and also on graphite–polymer composites [24]. These approaches have improved the conductivity of the composite bipolar plates and reduced the volume needed to satisfy the gas permeability requirement. However, composite bipolar plates still have challenges in operating at a moderate temperature range (e.g., 100–140 °C) due to the material's nature, and in cold-starting of the fuel cell stack. The automobile industry requires a moderate operating temperature range so the present heat-processing system in a car can be reused to decrease the cost of redesigning and testing. The cold-start is required in practice so that a fuel-cell car can be operated in cold winters. On the other hand, metallic bipolar plates do not have such problems. One advantage of using metal bipolar plates is the cold start of a fuel-cell vehicle. Metallic bipolar plates, particularly made of stainless steels (SSs), have been attracting increasing attention owing to their excellent bulk conductivity, high strength, high chemical stability, wide range of choice, ability for mass production, and low cost. The relevant DOE 2015 targets (Table 13.1) for metallic bipolar plates are a cost of $3 kW^{-1}, weight less than 0.4 kg kW^{-1}, corrosion rate less than 1 μA cm^{-2}, and a resistivity of 0.01 Ω cm^2. Owing to the research trends in metallic bipolar plates, a detailed review is needed to clarify the aspects. In this chapter, we focus on this topic only.

13.2
Metallic Bipolar Plates

SSs, Ni-based alloys, Ti-based alloys, and aluminum-based alloys have been investigated widely and considered as candidates for PEMFC bipolar plate materials [10–12]. In 1995, Vanderborgh and co-workers attempted to use metallic Al and Ti alloys as bipolar plates [5] and subsequently developed a corrosion test cell for evaluating the metallic bipolar plates [25]. Hornung and Kappelt in 1998 used Fe-based alloys as bipolar plates [26]. However, the possible corrosion products from the materials when serving in PEMFC environments might be a problem. The dissolved metal ions (Fe^{3+}, Cr^{3+}, etc.) may either poison the membrane and lower its conductivity, or pollute the active catalyst layers [1, 5, 27–30]. Both will result in degradation of the membrane electrode assembly and eventually reduce the lifetime of the fuel-cell stack. Moreover, the surface layers with corrosion products may result in an increase in the surface resistance, thus reducing the output of the cell [10]. In this context, two types of metallic bipolar plates have been studied: bare alloys and coated alloys.

13.2.1
Bare Metallic Bipolar Plates

The most investigated metallic material type for bipolar plates is SS owing to its high chemical stability, good mechanical strength, wide choice of alloys, available mass production pathways, and low cost. The relative high density of the material can be compensated for by the application of thin sheet or by using SS screen flow fields [31]. Therefore, the greatest challenges for the material are corrosion resistance in PEMFC environments and contact resistance.

Type 316/316L SS (with a composition of 16–18% Cr, 10–14% Ni, 2% Mo, and Fe as the remainder) is the most widely studied SS for bipolar plate applications. Two different results account for the adoption of 316/316L SS as PEMFC bipolar plates. On the one hand, 316/316L SS is considered a good candidate. After 72 h of testing in 0.1 M K_2SO_4 (pH 1) at 80 °C, Ma et al. [27] recommended 316 SS owing to its good corrosion resistance in a simulated environment and low cost. The contact resistance for the tested 316 SS samples was reported as 11 mΩ cm^2 on the simulated PEMFC anode side and 41 mΩ cm^2 on the cathode side, compared with 10 mΩ cm^2 for a bare sample [27]. The authors did not specify the compaction force used for the contact resistance measurements, possibly because of the difficulty with the method applied. Davies et al. [32, 33] ran a stack with 316 SS bipolar plates for 3000 h and observed no significant degradation of the stack performance. However, they recommended using steels of higher alloying content such as 310 and 904L SS owing to the thinner oxide films formed with higher Cr and Ni contents in these SSs. A thicker oxide film on 316 SS resulted in less power output than with those SSs with higher Cr and Ni contents [32, 33]. A long-term PEMFC stability test with SS bipolar plates resulted in a stable stack output, indicating that the contact resistance was not significantly changed during the test period and that the corrosion of the steel is not a significant issue. Kumar and Reddy fabricated a foamed 316 SS plate for gas distribution, which gave a much higher output than the conventional 316 SS with machined channels [34]. Lee and co-workers [35, 36] electrochemically treated 316L SS that gave good corrosion resistance and low contact resistance compared with the untreated material. The stack output decreased with untreated 316L SS bipolar plates whereas the stack with treated 316L SS bipolar plates gave a stable output. This difference was related to the lower level of poison on the membrane in the treated 316L SS compared with the untreated SS [35, 36].

On the other hand, Scholta et al. [37] reported high currents, at a level of 3 mA cm^{-2} at 0.9 V (versus RHE), during polarization of 316, 316L, and 316Ti SSs in 1 M H_2SO_4 solution. An interfacial contact resistance (ICR) of 100–120 mΩ cm^2 was reported for 316L SS treated for 24 h in steam [37]. When made into bipolar plates and assembled in a stack, a decrease in cell performance was recorded for both 316 and 316L SS bipolar plates in a 150 h operating period; hence an intermediate layer (carbon paper) was needed [37]. Although the results of corrosion cell testing at the Los Alamos National Laboratory (LANL) indicated that 316 SS may be acceptable for short-term application [25], a high dissolution rate and a large voltage drop in long-term tests were observed. Hence a more "noble" SS with greater corrosion resistance than 316 SS was recommended [17]. Moreover, General Motors (GM) Research Laboratories reported [38] unacceptable corrosion levels of bare 316L and 317L SS on the simulated PEMFC anode side. On the simulated PEMFC cathode side, the steels experienced passivation, hence the contact resistance increased. The corrosion at the simulated PEMFC anode side and the high contact resistance at the simulated PEMFC cathode side made these bare steels unacceptable for PEMFC bipolar plate application [38].

A high dissolution rate of 316L in a single-cell test was also observed by Wind et al. [28]. After 100 h of operation in the fuel cell, around 76 µg cm^{-2} of nickel and 5–10 µg cm^{-2} of iron and chromium were detected on the membrane [28]. Owing to the passive film formation and higher contact resistance, the stack output with 316L SS bipolar plates showed a decreasing power output [28]. In contrast, Pozio et al. found that the dissolution of the 316L SS end plate was mainly of iron, with rates of ~94 µg h^{-1} at the cathode side and ~142 µg h^{-1} at the anode side; hence they preferred 310 and 904L SS plates [39]. It was reported that 316L SS gives a 1–10 µA cm^{-2} current under simulated PEMFC conditions [40–42], higher than 904L and 254SMO SSs in simulated PEMFC environments [43]. The contact resistance of bare 316L was reported to be fairly high [42], significantly higher than that of 310S, 904L SS, and Ni-based alloys [44, 45].

Makkus et al. [46, 47] used 316L SS as a basis for their alloy development. Although 316L SS provided a reasonable stable stack output, they claimed that it is not an optimum choice for PEMFC bipolar plates owing to the interfacial contact resistance [46, 47]. High-contact resistance could be due to a surface oxide film [28]. The dissolution of 316L SS bipolar plates per 100 h of operation at 4 bar compaction force was reported as 200 ppm Fe, 60 ppm Cr, and 35 ppm Ni, mainly on the anode side [47]. The contamination of metal ions was considered the reason for the decreasing performance of the PEMFC stacks [46, 47]. Similarly, Lindbergh and co-workers [48] mentioned the high interfacial contact resistance with bare 316L SS, especially at the PEMFC cathode side. Instead, they used Pt-coated 316L SS plates for their PEMFC tests. Cao and co-workers [49] found that 316 SS was in an active state in H_2-purged dilute HCl (pH 2) solution. A porous corrosion product layer was formed on 316 SS in this environment [49]. In a comparative study, Cunningham *et al.* reported that a PEMFC stack with 316L SS bipolar plates showed much lower output but a rapidly degrading performance compared with those with polymer coating-protected 316L SS bipolar plates [50].

The above somewhat conflicting results are considered to be due to the differences in the surface preparation conditions of the SS and the difference in test conditions. On the other hand, the results also suggest the complicated environment encountered for the application. Therefore, from the performance point of view, 316/316L SS might not be the best choice for PEMFC bipolar plate application, although some groups considered that this material could be the most feasible choice in a multiple selection model [51]. The corrosion resistance of 316/316L SS in PEMFC environments and the high interfacial contact resistance are still major challenges [10–12, 51]. Therefore, other groups suggested using 310 and 904L SSs instead of 316/316L SS [32, 33, 39, 43]. This suggestion is also supported by our investigation of austenite SSs in aggressive, simulated PEMFC environments [52], where 316L SS showed a higher corrosion current and higher interfacial contact resistance than the other SSs tested such as types 904L and 349 SS. From a performance point of view, higher alloying SSs such as types 349, 446, and 2205 are better choices for PEMFC bipolar plate application [52–54], similarly to previous work. This conclusion is based on their superior corrosion resistance.

The second most tested SS is 304/304L SS based on its superior economic advantage over 316/316L SS. Compared with 316/316L SS, much greater dissolution was encountered with bare 304 SS in simulated PEMFC environments [27, 37, 55–59], especially on the anode side [27]. A 304 SS bipolar plate showed a faster voltage decay (46 mV) than an 310S plate (22 mV) in 1000 h of stack operation, and corrosion of the 304 SS bipolar plate was observed on the cathode side of the stack [60]. Moreover, 304/304L SS showed higher contact resistance than 316/316L SS [33, 37, 44, 45, 56–58], which could be related to the lower alloying content [33]. For these reasons, except for use in some small-scale demonstration fuel cells [61] or for cladding with another functional layer [62], 304/304L SS has normally acted as a substrate for coating deposition [44, 45, 55–59, 63–65].

Further, some high N-bearing SSs have been evaluated for bipolar plate application [66–68] and showed promising results both in simulated PEMFC environments and in stack testing. A high N-bearing SS is more economical than the conventional SSs. More importantly, the N alloying could provide a higher corrosion resistance in many environments. The interfacial contact resistance of SSs is 100–200 mΩ cm^2

in the compaction force region of 140 N cm^{-2} [54]. A higher compaction force gives a lower interfacial contact resistance. Even so, approaching the DOE target of 10 mΩ cm^2 with a bare high-N bearing SS is still a great challenge [9]. Therefore, although those alloys have superior corrosion resistance in PEMFC environments, their surfaces need to be modified to decrease the interfacial contact resistance.

Other investigated SSs include austenite 310 SS [27, 32, 33, 40, 45, 47, 60], 317/317L SS [27, 38, 47], 321 SS [33, 47], 347 SS [33, 69], 904/904L [32, 33, 43–45, 47, 69, 70], ferrite 400 series SSs [27, 53, 71–73], and duplex SSs [54, 69]. W- and Ta-modified 316 SS were studied [74]. High Cr–Ni-bearing SS [69] and some other Fe-based alloys [26, 46, 47] were also investigated. Ni alloys [25–27, 33, 44, 45] and Ni–Cr alloy [33, 34] have also been evaluated for bipolar plate application. Ni alloy (In-825) experienced heavy corrosion in a simulated PEMFC anode environment after a 72 h test, although it behaved excellently in a simulated cathode environment [27]. The contact resistance of Ni alloy is comparable to that of type 316 SS [27]. Cu–(2%)Be alloy [75] and Cu–5.3Cr alloy [76] were tested in simulated PEMFC environments and showed reasonably low currents. In addition, some amorphous alloys [77, 78] were investigated. Fe-based $Fe_{50}Cr_{18}Mo_8Al_2Y_2C_{14}B_6$ and $Zr_{75}Ti_{25}$ showed better corrosion resistance than 316L SS in simulated PEMFC environments [77, 78]. In terms of cost, however, these alloys could not play a major role in PEMFC bipolar plate application.

13.2.2
Light Alloys

Ti-based alloys and Al-based alloys are attractive to the automobile industry due to their light weight. An early study showed a low corrosion current for Ti alloy in a pH 4 solution [5]. The excellent corrosion resistance of the Ti was also confirmed in a corrosion cell test in a pH 3 solution for over 1500 h [25]. No detectable Ti ions could be observed in 25 days of tests. N-implanted Ti should improve the performance of Ti alloys further. In 1 M H_2SO_4 solution, Ti was found to have a corrosion rate comparable to that of graphite under the same conditions [37]. Ti alloy also showed comparable contact resistance after 24 h of boiling treatment in water [37]. Although Ti has a better corrosion resistance than 316 SS in PEMFC environments, comparable or higher interfacial contact resistance was reported with Ti [33, 79], resulting in significant output degradation in a 400 h fuel cell test [33]. This is in agreement with the change of ICR at 220 N cm^{-2} from 32 mΩ cm^{-2} before the test to 250 mΩ cm^{-2} after the test [33]. On the other hand, a simulative test for Ti in a 0.1 M K_2SO_4 solution of pH 1 at 80 °C at −0.2 V (vs. Ag/AgCl) and 0.4 V for 72 h revealed a parts per million level of Ti ions in the electrolytes [27]. The contact resistance of the Ti increased significantly after the test with about a 50% increase on the anode side (H_2 side) and about five times that of a freshly abraded SS on the cathode side (air side) [27]. For this reason, Ti needs a conductive coating to perform well as a bipolar plate. Nitrogen ion implantation seems to be an effective method to reduce the contact resistance of Ti-based alloys [25, 79]. A surface-treated Ti showed an almost identical low ICR compared with graphite with no degradation in a durability test of about 8000 h [80]. Au-plated, Pt-coated, InO$_2$-sintered, and other surface-modified Ti-based bipolar plates have been developed [79, 81–83].

Al and Al-based alloys have cost advantages over Ti-based alloys. In early studies on Al alloys as PEMFC bipolar plates, Borup and Vanderborgh found that a high corrosion rate was associated with bare Al, and carbon-coated Al could reduce the corrosion rate by more than two orders of magnitude in pH 4 solutions [5]. A dissolution rate of about 0.1–0.3 mA cm^{-2} was reported in a 5 h corrosion cell test with Al foil to simulate PEMFC anode and cathode sides [25]. The contact resistances increased after the test. Degradation of Al both in cell output and solubility was clearly seen from a 90 h test [25]. The pH of the electrolytes after the test increased to 7.4 for the anode chamber and 8.9 for the cathode chamber [25]. Similar pH increases on both the anode and cathode sides was confirmed in a stack test using Al end plates [39]. Moreover, the low corrosion resistance of bare Al alloy was again confirmed in a simulated PEMFC environment [84]. Therefore, for Al-based alloys serving as PEMFC bipolar plates, surface modification or coating is essential. It was reported that Au-coated Al bipolar plates showed a similar behavior to graphite bipolar plates in the initial fuel cell stack test while the performance with the coated Al bipolar plates decayed very rapidly because the gold lifted from the plate and became embedded in the membrane with high reactivity of Al [79]. A diamond-like coating was successfully deposited on Al-based 5052 alloy by means of physical vapor deposition (PVD), and the coated Al alloy bipolar plate showed a lower contact resistance than 316L SS and good single-cell performance [85, 86]. However, owing to the higher corrosion rate of the substrate Al alloy than the passive film formed on SSs, the lifetime of such coated Al bipolar plates still needs further testing. For cost reduction, the same group treated the 5052 alloy chemically and found that the corrosion resistance and conductivity of the chemically treated alloy could not reach the same levels as the PVD coating [87]. Electrical and electroless depositions of Ni–P, Ni–Mo–P, and other Ni-based alloys were also attempted on Al alloys [88–90]. It turned out that the corrosion resistance of such surfaces with Ni-based alloys needs further improvement in PEMFC environments. Similarly to the application on SSs, some conductive polymers were also tried for coating Al alloys for PEMFC bipolar plate application [91].

Tawfik and co-workers developed a carbide-based amorphous alloy coating for Al alloy [92–95]. In their 1500 h stack operation, the initial performance of the fuel cell stack with the treated Al bipolar plate was better than that with a graphite plate, although a slight degradation of output was seen with the former [92]. The 1000 h fuel cell testing showed no significant power degradation due to corrosion, and the modified coated Al bipolar plates behaved better than graphite plates in power output [93–95]. The carbide-based coating showed good durability in a fuel-cell stack environment, confirmed by the polarization curves before and after a 1000 h stack test [93–95]. Moreover, their recent cost diagram illustrated that the cost of the modified Al plates could be lower than that of graphite plates after a few years of operation [94].

13.2.3
Coated Stainless-Steel Bipolar Plates

Coatings have been used to improve the corrosion resistance and contact resistance of SSs. The coating should be conductive, have good adhesion with the substrate, and have a compatible thermal expansion coefficient with that of the substrate. Coatings are normally divided into two groups [10, 11]: carbon-based and metal-based. The former includes graphite, conductive polymers, diamond-like carbon, and organic self-assembled monopolymers. The latter includes noble metals, metal nitrides, and metal carbides. Some recent reviews have summarized different coatings in detail, based either on substrate materials or on coating compositions [12, 94].

Pt-coated 316 SS bipolar plates behaved excellently in short-term laboratory fuel-cell stack tests; the contact resistance was reduced significantly due to the Pt coating [48]. However, visible evidence of corrosion of the bipolar plates was observed at the anode side of the stack in longer tests [48]. Wind et al. [28] showed that gold-coated 316L SS bipolar plates performed similarly to graphite plates during 1000 h of fuel-cell operation. Similar results were reported by Hentall et al. with gold-coated 316L SS bipolar plates [79] and 316L SS with other low-cost coatings. In an attempt to coat 304, 310, and 316 SSs with different Zr-bearing and Au coatings, it was found that only Zr-coated SSs satisfied the DOE goal of corrosion resistance in simulated PEMFC environments in the short term [40]. An Au coating more than 10 nm thick was needed to achieve the ICR goal [40]. For a similar reason, an Au foil about 10 nm thick was rolled on the SS surface to develop bipolar plates [96]. To reduce the cost, less expensive metals have been tried. An Nb coating on 430 SS [72] showed low ICR and promising corrosion resistance in simulated PEMFC environments. A process of cladding Nb on 430 SS was also developed by Weil et al. [97]. Further, 316L SS was implanted with Ni ions at a dose of $\sim 10^{17}$ cm^{-2} and showed lower ICR and good corrosion resistance in simulated PEMFC environments [42].

Cunningham et al. [50] developed a three-layer carbon coating to protect the 316L SS bipolar plates. Conductive polymer-coated 317L SS also showed excellent behavior in a PEMFC environment for a couple of days [38]. Fukutsuka et al. [57] used a plasma-assisted chemical vapor deposition (CVD) process to obtain a thin carbon coating less than 1 μm thick. Coated 304 SS showed a significant decrease in ICR and improved corrosion resistance in simulated PEMFC environments [57]. Carbon-coated 304 SS was obtained out by a CVD process, and the bipolar plates produced showed excellent performance and stability, even better ICR and performance than graphite in a stack test [64, 65]. Conductive composite coatings on SS for PEMFC bipolar plates have also been actively investigated, and were even considered as a feasible trend [98–100]. Carbon-based film was developed by pulsed bias arc ion plating on 316L SS, and it showed lower ICR and better corrosion resistance than bare 316L in simulated PEMFC environments [100]. Attempts were made to produce a carbon–resin composite-coated 304 SS bipolar plate and it showed promising corrosion resistance in simulated PEMFC environments, although the ICR was slightly higher than that of graphite [63]. Polypyrrole and polyaniline coatings were electrodeposited successfully on 304 SS, and the coated steel showed a significant ICR reduction [56]. In simulated PEMFC environments, 304 SS coated with these polymers showed improved corrosion resistance over bare 304 SS [56, 58, 59]. Polypyrrole-coated 316L also showed a promising improvement in simulated PEMFC anode environments [101]. On the cathode side, the corrosion current remained the same as that of bare 316L [101]. A new approach with a nylon 6–316L SS fiber composite bipolar plate was tested and showed a worse fuel-cell stack performance than the graphite equivalent [102]. These somewhat conflicting results indicate that carbon and conductive polymers may be not be reliable candidates for coating SSs for PEMFC bipolar plates.

A surface-treatment process developed by Lee and co-workers [35, 36] showed the beneficial effect of treated 316L in simulated PEMFC environments and low ICR. Stack tests of 300 h showed no power degradation with the treated 316L SS bipolar plates [35, 36]. A low-temperature carburization-treated 316 SS gave much better corrosion resistance in simulated PEMFC environments [103].

A conductive SnO_2 : F coating has been also investigated at the US National Renewable Energy Laboratory (NREL) [104–107] as a feasible coating to protect metallic substrates from corrosion and also to provide better surface conductivity, which is due to the stability of the coating in most environments, the expertise of NREL in the application of the coating, and low-cost considerations. It was observed that the behavior of SnO_2 : F-coated steels depends strongly on the substrate steel. In view of the corrosion resistance in simulated PEMFC environments, SnO_2 : F behaved excellently on 349, 446, and 2205 SSs [104–106]. The corrosion resistance of the other SSs (such as 317L and 444) was also improved. Lower dissolution rates of the SSs were recorded due to the coating [106]. However, the interfacial contact resistance was not improved much, as to the existing surface oxide film underneath was not modified. A pre-etching process before the SnO_2 : F coating deposition can alter this surface oxide film and thus significantly reduce the contact resistance of the coated SSs [107]. However, the trade-off with the modified process is that the corrosion resistance of the coated SSs in simulated PEMFC environments was decreased [107]. More work is needed to improve the surface resistance and maintain the corrosion resistance by the conductive SnO_2 : F coatings.

Owing to its high conductivity, excellent stability in a wide range of environments, wide industrial applicability, and cost considerations, TiN is naturally one of the best coating candidates for SSs for PEMFC bipolar plate application. Ma et al. [27] illustrated that a commercially available TiN coating on 316L SS was promising in only a short period of time, mostly dissolved after a 1 h test. The TiN coating showed a beneficial effect in simulated PEMFC cathode environments, resulting in much lower dissolution of the substrate. In simulated PEMFC anode environments, however, the TiN coating actually increased dissolution of the substrate 316L SS, which was related to the instability of substrate and possible a quality issue (pinholes) with the coating [27]. Pitting corrosion of TiN-coated 316L, with a 15 μm thick layer obtained by means of PVD, in simulated PEMFC cathode environments was also reported [41]. The same authors also observed a high corrosion rate of 15 μm TiN-coated 410 SS when tested in simulated PEMFC cathode environments

[71]. Again, these were related to pinholes in the TiN coating. On the other hand, Hentall et al. reported that a TiN coating on 316L SS reduced the interfacial contact resistance almost to the level of graphite [79], and TiN on Ti behaved excellently in the real fuel-cell stack operation. Cao and co-workers deposited 2–4 μm TiN on 316L SS by means of a hollow-cathode discharge ion-plating coater [108]. In a 1000 h oxygen (with 2 μm TiN) and 240 h hydrogen (with 4 μm TiN)-simulated environment tests, they reported that the TiN coating offered higher corrosion resistance and higher electrical conductivity than the substrate SS alone [108]. On the other hand, prolonged tests also revealed some dissolution of the coating [108]. Actually, it was reported that a 1 kW PEMFC stack with 1 μm TiN-coated 316 SS bipolar plates worked well for over 1000 h, although the output was slightly lower than that with the control stack with graphite bipolar plates [109]. Coating 310S bipolar plates with nanosized TiN resulted in almost the same stack output performance as that with graphite bipolar plates [110]. These results indicate that TiN is an excellent coating candidate for SSs. AS in the case of other bare SSs, bare 316/316L SS does not seem like not the best choice as the PEMFC bipolar plate substrate based on its corrosion resistance in PEMFC environments; in some aggressive situations, the corrosion of the coating and the substrate would result in problems. However, if a more corrosion-resistant substrate is selected, a TiN coating is definitely an excellent choice, both practically and economically.

Brady and co-workers developed a thermal nitridation process to form pinhole-free CrN/Cr_2N coatings on a model Ni–50Cr alloy [111, 112]. The nitrided Ni–50Cr alloy showed promising corrosion resistance and low contact resistance. After a 4000 h test in simulated anode and cathode conditions in a LANL corrosion cell, the ICR of the nitrided Ni–50Cr plate showed almost no significant change [113]. Moreover, a 1000 h fuel-cell test with the nitrided Ni-50Cr bipolar plate gave negligible membrane contamination and no increase in cell resistance [113]. This method was applied to SSs, and a discontinuous layer formed on austenite 349 SS resulted in heavy corrosion in simulated PEMFC environments, although the contact resistance was significantly reduced due to the nitridation [114]. On the other hand, the same procedure produced a modified surface for a ferric 446 SS [115]. Although it is not a pinhole-free surface, the modified surface with nitrided 446 SS is a mixture of metal oxides and nitrides, which showed excellent low interfacial contact resistance while maintaining the excellent corrosion resistance of the substrate [115].

13.3
Discussion and Perspective

From energy-density and volume-production points of view, the use of metallic bipolar plates in PEMFC stacks is important for the automobile industry. Metallic bipolar plates could provide thinner and lighter choices and improved thermal and bulk electrical conductivity. The use of SSs makes PEMFCs cost-effective. Whereas 316/316L SS may not be the optimum choice for the PEMFC bipolar plate material, it is generally accepted that 310, 904L, 349, 446, and 2205 bare SSs are excellent candidates for this application [32, 33, 39, 52–54]. Some high N-bearing SSs such as type AL219 also behaved better than 316L SS in simulated PEMFC environments [66]. However, SSs form passive films in PEMFC environments, composed mainly of chromium oxide, in the region of a few nanometers thick [53, 54, 116]. On the one hand, such a thin passive film prevents the SS from further corrosion. On the other hand, it also increases the interfacial contact resistance due to the high electrical resistance of chromium oxide. Therefore, from the contact-resistance point of view, it is difficult for bare SSs to achieve the DOE targets for bipolar plates (Table 13.1). Other bare alloys either have similar difficulties (such as Al and Ti alloys) or are not economically available (such as Ni alloys). Therefore, a feasible consideration would be to apply a conductive coating or surface modification process to a cost-effective substrate. Therefore, our approach is a combination of a corrosion-resistant and cost-effective substrate with a coating or surface modification. With this combination, even if the modified surface/coating develops pinholes in the application, the substrate will passivate immediately in PEMFC environments. In other words, the corrosion resistance of the substrate will be maintained, and the conductivity of the modified surface/coating will be utilized. This concept is supported by thermally nitrided 446 SS, in which the surface layer is not continuous [115] but still provides low contact resistance and excellent corrosion resistance in simulated PEMFC environments.

13.3.1
Substrate Selection

Because the corrosion resistance of the coating in PEMFC environments depends strongly on the substrate, it is natural that the selection of the substrate material is the first important step. Considering the above results with SSs, the corrosion resistance of the bare substrate SS in PEMFC environments should be at or above that of 316/316L SS under identical conditions. These are numerous SSs that have much better stability than 316L SS in simulated PEMFC environments. Table 13.2 summarizes the estimated cost of the substrate material, ICR values, and corrosion results for the substrate SSs in simulative PEMFC environments compared with DOE 2015 targets for metal bipolar plates. Three tested nitrided alloys are also listed to show that the technical part of the DOE goal can be met by means of a thermal nitridation process. From the cost point of view, however, high N-bearing SSs (types 201 and AL219 SSs) should be more advantageous.

It is not surprising that high-alloying SSs (349, AISI446, and 2205) showed excellent corrosion resistance in PEMFC environments, much better than 316L SS did. Table 13.2 clearly indicates that the high-nitrogen-bearing steels have a significant cost advantage. Therefore, it is worthwhile investigating such steels in more detail.

Table 13.2 DOE targets and the substrate stainless-steel materials.

Goal/alloy	ICR at 140 N cm^{-2} (mΩ cm^2)^a	Current at −0.1 V vs. SCE(H$_2$) (μA cm^{-2})^b	Current at 0.6 V vs. SCE (air) (μA cm^{-2})	Cost^c ($ kW)$^{-1}$
DOE2015 Goal	10	<1	<1	3
316L	154	+2.5 to +12	0.7–11	3.41
349	110	−4.5 to −2.0	0.5–0.8	3.61
AISI446	190	−2.0 to −1.0	0.3–1.0	4.08
2205	130	−0.5 to +0.5	0.3–1.0	3.53
201	158	−0.5 to +8.5	0.8–2.0	2.18
AL219	730	−3.3 to −1.5	1.0–3.0	2.65
Nitrided 446	6.0	−1.7 to −0.2	0.7–1.5	N/A^d
Nitrided FeCrV alloy	4.8	−9.0 to −0.2	1.5–4.5	N/A
Nitrided AL29-4C	6.0	−6.5 to −3.0	0.3–0.5	N/A

^{a}ICR values were taken with bare samples. If the SS is polarized in simulated PEMFC environments, a passive film forms; therefore, the ICR values with passive films are normally higher than those with bare samples.
^{b}Negative currents at −0.1 V mean a cathodic current. The surface is considered cathodically protected with a cathodic current; therefore, the metallic ions' dissolution reaction is in equilibrium, and the anodic corrosion of the steel is reduced to almost zero.
^{c}Cost data were based on the average trading price of cold-rolled coil 316 SS at the London Metals Exchange for a whole year, 2005 ($2.150 lb^{-1}) [117]. The market price of electrolytic chromium metal was $4.50 lb^{-1}, the price at end of 2004 [118]. Market prices of Ni and Mo were $6.70 lb^{-1} and $23.5 lb^{-1} in February 2006 [119]. Mn was priced at 0.85 lb^{-1} in the market at the end of 2005 [120]. The prices of 349, 446, and 2205 SSs were calculated according to the differences in the alloying composition. Thus, we obtained $2.27 lb^{-1} for 349 SS, $2.57 lb^{-1} for 446 SS, and $2.22 lb^{-1} for 2205 SS. In the same way, we also obtained $2.61 lb^{-1} for 317L SS (not listed), $2.83 lb^{-1} for 904L SS (not listed), $1.38 lb^{-1} for type 201 SS, and $1.67 lb^{-1} for type AL219 SS. By assuming six cells kW^{-1} for a PEMFC stack and the dimensions of a bipolar plate are 24 × 24 × 0.0254 cm (which gives a 400 cm^2 utilization surface area in a 0.01 in-thick sheet), a single plate should weigh 120 g, and six plates would weigh 720 g (density 8 g cm^{-3} assumed). If we use the same method as above, the costs are $4.14 kW^{-1} for 317L SS and $4.49 kW^{-1} for 904L SS. Both are too high for further consideration as substrate materials.
^{d}Not applicable.

13.3.2
Coatings and Surface Modification

From the cost point of view, precious metals (such as Au and Pt) are surely out of contention for practical coatings on SS substrates, although they might be used for short laboratory tests. In fact, electrochemical corrosion cells will be generated from the possible pinholes in the coatings due to the electrochemical dissimilarity of the precious metals and SSs in a PEMFC environment. Difficulties encountered with the carbon-based and conductive polymer-based coatings are application at intermediate temperatures, the cold-start issue, and the differences in thermal expansion coefficients between the coating itself and the substrate SS. The risk of the coating peeling off is fairly high in automobile applications. On the other hand, the above cost calculation for the candidate substrate SSs indicates that the cost of the surface modification or applying a coating should be limited. Considering the cost of the surface modification or coating, we propose the following methods:

- **TiN**: The application of TiN is a very common and cost-effective process in many industries. Some different (and somewhat conflicting) results were encountered with this coating when applied on SSs for PEMFC bipolar plates. However, most investigations were performed with TiN-coated 316/316L SSs. As stated above, bare 316/316L SS might be not an optimum choice for PEMFC bipolar plates. Because bare 316/316L SS is not very stable in this environment, any substrate defects will affect the final application of this coating. On the other hand, TiN-coated 316L SS had a very low interfacial contact resistance and had much better corrosion resistance than the bare steel in PEMFC environments. Therefore, if using substrate steels that have proved corrosion resistance in PEMFC environments, the combination of the substrate steel and TiN coating will provide an economical approach.
- **Cr_xN**: Work on thermal nitridation with a model Ni–50Cr alloy led to the development of pinhole-free nitride layers on the alloy [112–114]. This nitride layer, consisting of CrN and Cr_2N [111–113], offered excellent low interfacial contact resistance and excellent corrosion resistance even in stack tests. Applying the same nitridation procedure to type 446 SS showed that the nitrided material developed a modified surface layer that provided low interfacial contact but still maintained the excellent corrosion resistance of the substrate [115]. Instead of forming a continuous nitride layer on model Ni–50Cr alloy, the surface chemistry of nitrided 446 SS consisted of nitrogen-modified oxides [115]. A short pre-oxidation step greatly modified the process and the performance of thermally nitrided alloys [73, 121–124].

Pozio and co-workers applied a CrN coating on 304 SS and tested in simulated PEMFC environments [44, 45, 72]. Although much improved performance in contact resistance and corrosion resistance was achieved in simulated environments with the coated steel, the long-term stability of the coating under fuel-cell operating conditions is still not clear. A similar coating applied on 316L SS showed low surface resistance and improved corrosion resistance in simulated PEMFC cathode environments [125]. Coating a Cr_xN film on 316L SS by pulsed bias arc ion plating (PBAIP) showed high interfacial conductivity and good corrosion resistance in simulated PEMFC environments [126]. The surface of Cr-electroplated 316L SS developed a Cr_2N layer after thermal nitridation, with much better electrical conductivity than CrN, showed significantly lowered interfacial contact resistance, and demonstrated better corrosion resistance [127]. However, these processes are not economic because either a high temperature was encountered or an expensive coating process was applied.

- **Surface plasma modification**: Plasma treatment is a conventional and economical method in industry. It was shown that thermal nitridation altered the surface chemistry of 446 SS. This process can be performed at a much lower temperature and much lower cost by means of plasma nitridation. By the surface plasma modification, the steel's surface can also be covered with a layer of modified oxides, which is conducting and corrosion resistant. Plasma nitridation was applied to 304L and 316L SSs for PEMFC bipolar plates [128–131] and showed that plasma nitridation at a low temperature significantly altered the surface of the SSs. A nitrogen-rich layer was formed on the surface, resulting in lower contact resistance and better corrosion resistance in simulated PEMFC environments [128–130]. A similar procedure was applied to 349TM SS at NREL [132, 133]. It turned out that 349TM SS treated in a nitrogen plasma developed a more insulating layer on the surface [132]. The steel treated in a nitrogen plasma showed excellent corrosion resistance in simulated PEMFC environments but high interfacial contact resistance [132]. On the other hand, steel treated in an ammonia plasma gave lowered interfacial contact resistance [133]. This difference was solely due to the higher plasma/energy density associated with the ammonia plasma. On the other hand, a bias gave better contact resistance [131]. Further experiments are in progress at NREL.

In summary, both TiN coating and surface plasma modification require suitable substrates with good corrosion resistance in PEMFC environments. These combinations provide economical ways to overcome the technical and cost barriers associated with PEMFC bipolar plates. Both substrate SSs and surface modification methods are cost-effective. It is considered that the DOE targets will be matched with such methods.

Acknowledgments

This work was supported by the Fuel Cells Technologies Program of the US Department of Energy.

References

1. Steele, B.C.H. and Heinzel, A. (2001) *Nature*, **414**, 345.
2. Tsuchiya, H. and Kobayashi, O. (2004) *Int. J. Hydrogen Energy*, **29**, 985.
3. Brandon, N.P., Skinner, S., and Steele, B.C.H. (2003) *Annu. Rev. Mater. Res.*, **33**, 183.
4. Jeong, K.S. and Oh, B.S. (2002) *J. Power Sources*, **105**, 58.
5. Borup, R.L. and Vanderborgh, N.E. (1995) *Mater. Res. Soc. Symp. Proc.*, **393**, 151.
6. Garland, N.L. (2008) Fuel cells sub-program sessions. Presented at the 2008 DOE Hydrogen Program Merit Review and Peer Evaluation Meeting; available at *http://www.hydrogen.energy.gov/pdfs/review08/fc_0_garland.pdf* (last accessed 28 November 2011).
7. Papageorgopoulos, D. (2009) Fuel cell technologies. Presented at the 2009 DOE Hydrogen Program and Vehicle Technologies Program Merit Review and Peer Evaluation Meeting; available at *http://www.hydrogen.energy.gov/pdfs/review09/fc_0_papageorgopoulos.pdf* (last accessed 28 November 2011).

8. Jayukumar, K., Pandiyan, S., Rajalakshmi, N., and Dhathathreyan, K.S. (2006) *J. Power Sources*, **161**, 454.
9. US DOE Office of Energy Efficiency and Renewable Energy (2007) Hydrogen, Fuel Cells and Infrastructure Technologies Program: Multi-Year Research, Development and Demonstration Plan, pp. 3.4–3.26; available at http://www1.eere.energy.gov/hydrogenandfuelcells/mypp/pdfs/fuel_cells.pdf (last accessed 28 November 2011).
10. Hermann, A., Chaudhuri, T., and Spagnol, P. (2005) *Int. J. Hydrogen Energy*, **30**, 1297.
11. Mehta, V. and Cooper, J.S. (2003) *J. Power Sources*, **114**, 32.
12. Brett, D.J.L. and Brandon, N.P. (2007) *J. Fuel Cell Sci. Technol.*, **4**, 29.
13. Cooper, J.S. (2004) *J. Power Sources*, **129**, 152.
14. Wang, Y. and Northwood, D.O. (2009) in *Polymer Electrolyte Membrane Fuel Cells and Electrocatalysts* (eds. R. Esposito and A. Conti), Nova Science Publishers, New York, p. 33.
15. Busick, D.N. and Wilson, M.S. (1999) *Fuel Cells Bull.*, **2**, 6.
16. Busick, D. and Wilson, M. (2000) *Mater. Res. Soc. Symp. Proc.*, **575**, 247.
17. Wilson, M.S., Zawodzinski, C., Bender, G., Zawodzinski, T.A., and Busick, D.N. (2000) in Proceedings of the 2000 Hydrogen Program Annual Review; available at http://www1.eere.energy.gov/hydrogenandfuelcells/annual_review2000.html (last accessed 28 November 2011).
18. Besmann, T.M., Klett, J.W., Henry, J.J. Jr., and Lara-Curzio, E. (2000) *J. Electrochem. Soc.*, **147**, 4083.
19. Scholta, J., Rohland, B., Trapp, V., and Focken, U. (1999) *J. Power Sources*, **84**, 231.
20. Middleman, E., Kout, W., Vogelaar, B., Lenssen, J., and de Waal, E. (2003) *J. Power Sources*, **118**, 44.
21. Cho, E.A., Jeon, U.-S., Ha, H.Y., Hong, S.-A., and Oh, I.-H. (2004) *J. Power Sources*, **125**, 178.
22. Kuan, H.-C., Ma, C.-C.M., Chen, K.H., and Chen, S.-M. (2004) *J. Power Sources*, **134**, 7.
23. Heinzel, A., Mahlendorf, F., Neimzig, O., and Kreuz, C. (2004) *J. Power Sources*, **131**, 35.
24. Müller, A., Kauranen, P., von Ganski, A., and Hill, B. (2006) *J. Power Sources*, **154**, 467.
25. Weisbrod, K.R., Prier, D. II, and Vanderbrogh, N.E. (1999) in 1999 Progress Report, US DOE Hydrogen and Fuel Cell Program, US Department of Energy, Washington, DC, p. 117.
26. Hornung, R. and Kappelt, G. (1998) *J. Power Sources*, **72**, 20.
27. Ma, L., Warthesen, S., and Shores, D.A. (2000) *J. New Mater. Electrochem. Syst.*, **3**, 221.
28. Wind, J., Späh, R., Kaiser, W., and Böhm, G. (2002) *J. Power Sources*, **105**, 256.
29. Okada, T. (2003) in *Handbook of Fuel Cells–Fundamentals, Technology and Applications*, vol. **3**, (eds W. Vielstich, H.A. Gasteiger, and A. Lamm), John Wiley & Sons, Ltd., Chichester, p. 627.
30. Okada, T., Ayato, Y., Dale, J., Yuasa, M., Sekine, I., and Asbjornsen, O.A. (2000) *Phys. Chem. Chem. Phys.*, **2**, 3255.
31. Cleghorn, S.J.C., Ren, X., Springer, T.E., Wilson, M.S., Zawodzinski, C., Zawodzinski, T.A., and Gottesfeld, S. (1997) *Int. J. Hydrogen Energy*, **22**, 1137.
32. Davies, D.P., Adcock, P.L., Turpin, M., and Rowen, S.J. (2000) *J. Power Sources*, **86**, 237.
33. Davies, D.P., Adcock, P.L., Turpin, M., and Rowen, S.J. (2000) *J. Appl. Electrochem.*, **30**, 101.
34. Kumar, A. and Reddy, R.G. (2004) *J. Power Sources*, **129**, 62.
35. Lee, S.-J., Huang, C.-H., Lai, J.-J., and Chen, Y.-P. (2004) *J. Power Sources*, **131**, 162.
36. Lee, S.-J., Lai, J.-J., and Huang, C.-H. (2005) *J. Power Sources*, **145**, 362.
37. Scholta, J., Rohland, B., and Garche, J. (1997) in *Proceedings of the 2nd International Symposium on New Materials for Fuel Cell and Modern Battery Systems* (eds. O. Savadogo and P.R. Roberge), Ecole Polytechnique de Montreal, Montreal, p. 330.

38. Abd Elhamid, M.H. and Mikhail, Y. (2004) in *Fuel Cell Technology: Opportunities and Challenges, Topical Conference Proceedings of the 2002 AIChE Spring National Meeting* (eds. G.J. Igweand and D. Mah), American Institute of Chemical Engineers, New York, p. 460.
39. Pozio, A., Silva, R.F., De Francesco, M., and Giorgi, L. (2003) *Electrochim. Acta*, **48**, 1543.
40. Yoon, W., Huang, X., Fazzino, P., Reifsnider, K.L., and Akkaoui, M.A. (2008) *J. Power Sources*, **179**, 265.
41. Wang, Y. and Northwood, D.O. (2007) *J. Power Sources*, **165**, 293.
42. Feng, K., Shen, Y., Mai, J., Liu, D., and Cai, X. (2008) *J. Power Sources*, **182**, 145.
43. Agneaux, A., Plouzennec, M.H., Antoni, L., and Granier, J. (2006) *Fuel Cells*, **6**, 47.
44. Silva, R.F. and Pozio, A. (2007) *J. Fuel Cell Sci. Technol.*, **4**, 116.
45. Silva, R.F., Franchi, D., Leone, A., Pilloni, L., Masci, A., and Pozio, A. (2006) *Electrochim. Acta*, **51**, 3592.
46. Makkus, R.C., Janssen, A.H.H., de Bruijn, F.A., and Mallant, R.K.A.M. (2000) *Fuel Cells Bull.*, **3**, 5.
47. Makkus, R.C., Janssen, A.H.H., de Bruijn, F.A., and Mallant, R.K.A.M. (2000) *J. Power Sources*, **86**, 274.
48. Ihonen, J., Jaouen, F., Lindbergh, G., and Sundholm, G. (2001) *Electrochim. Acta*, **46**, 2899.
49. Li, M.C., Zeng, C.L., Luo, S.Z., Shen, J.N., Lin, H.C., and Cao, C.N. (2003) *Electrochim. Acta*, **48**, 1735.
50. Cunningham, N., Guay, D., Dodelet, J.P., Meng, Y., Hlil, A.R., and Hay, A.S. (2002) *J. Electrochem. Soc.*, **149**, A905.
51. Shanian, A. and Savadogo, O. (2006) *J. Power Sources*, **159**, 1095.
52. Wang, H., Sweikart, M.A., and Turner, J.A. (2003) *J. Power Sources*, **115**, 243.
53. Wang, H. and Turner, J.A. (2004) *J. Power Sources*, **128**, 193.
54. Wang, H., Teeter, G., and Turner, J.A. (2005) *J. Electrochem. Soc.*, **152**, B99.
55. Ren, Y.J. and Zeng, C.L. (2007) *J. Power Sources*, **171**, 778.
56. Ren, Y.J. and Zeng, C.L. (2008) *J. Power Sources*, **182**, 524.
57. Fukutsuka, T., Yamaguchi, T., Miyano, S.-I., Matsuo, Y., Sugie, Y., and Ogumi, Z. (2007) *J. Power Sources*, **174**, 199.
58. Joseph, S., McClure, J.C., Chianelli, R., Pich, P., and Sebastian, P.J. (2005) *Int. J. Hydrogen Energy*, **30**, 1339.
59. Lucio García, M.A. and Smit, M.A. (2006) *J. Power Sources*, **158**, 397.
60. Kumagai, M., Myung, S.-T., Kuwata, S., Asaishi, R., and Yashiro, H. (2008) *Electrochim. Acta*, **53**, 4205.
61. Lee, S.-J., Chen, Y.-P., and Huang, C.-H. (2005) *J. Power Sources*, **145**, 369.
62. Hong, S.-T. and Weil, K.S. (2007) *J. Power Sources*, **168**, 408.
63. Kitta, S., Uchida, H., and Watanabe, M. (2007) *Electrochim. Acta*, **53**, 2025.
64. Chung, C.-Y., Chen, S.-K., Chiu, P.-J., Chang, M.-H., Hung, T.-T., and Ko, T.-H. (2008) *J. Power Sources*, **176**, 276.
65. Chung, C.-Y., Chen, S.-K., Chin, T.-S., Ko, T.-H., Lin, S.-W., Chang, W.-M., and Hsiao, S.-N. (2009) *J. Power Sources*, **186**, 393.
66. Wang, H. and Turner, J.A. (2008) *J. Power Sources*, **180**, 791.
67. Kumagai, M., Myung, S.-T., Asaishi, R., Katada, Y., and Yashiro, H. (2008) *J. Power Sources*, **185**, 815.
68. Kumagai, M., Myung, S.-T., Kuwato, S., Asaishi, R., Katada, Y., and Yashiro, H. (2009) *Electrochim. Acta*, **54**, 1127.
69. Iversen, A.K. (2006) *Corros. Sci.*, **48**, 1036.
70. André, J., Antoni, L., Petit, J.-P., De Vito, E., and Montani, A. (2009) *Int. J. Hydrogen Energy*, **34**, 3125.
71. Wang, Y. and Northwood, D.O. (2007) *Int. J. Hydrogen Energy*, **32**, 895.
72. Pozio, A., Silva, R.F., and Masci, A. (2008) *Int. J. Hydrogen Energy*, **33**, 5697.
73. Lee, K.-H., Lee, S.-H., Kim, J.-H., Lee, Y.-Y., Kim, Y.-H., Kim, M.-C., and Wee, D.-M. (2009) *Int. J. Hydrogen Energy*, **34**, 1515.
74. Kim, K.M. and Kim, K.Y. (2007) *J. Power Sources*, **173**, 917.
75. Nikam, V.V. and Reddy, R.G. (2005) *J. Power Sources*, **152**, 146.

76. Lee, H.-Y., Lee, S.-H., Kim, J.-H., Kim, M.-C., and Wee, D.-M. (2008) *Int. J. Hydrogen Energy*, **33**, 4171.
77. Jayaraj, J., Kim, Y.C., Kim, K.B., Seok, H.K., and Fleury, E. (2005) *Sci. Technol. Adv. Mater.*, **6**, 282.
78. Jin, S., Ghali, E., and Morales, A.T. (2006) *J. Power Sources*, **162**, 294.
79. Hentall, P.L., Lakeman, J.B., Mepsted, G.O., Adcock, P.L., and Moore, J.M. (1999) *J. Power Sources*, **80**, 235.
80. Hodgson, D.R., May, B., Adcock, P.L., and Davies, D.P. (2001) *J. Power Sources*, **96**, 233.
81. Murphy, O.J., Cisar, A., and Clarke, E. (1998) *Electrochim. Acta*, **43**, 3829.
82. Wang, S.-H., Peng, J., Lui, W.-B., and Zhang, J.-S. (2006) *J. Power Sources*, **162**, 486.
83. Wang, S.-H., Peng, J., and Lui, W.-B. (2006) *J. Power Sources*, **160**, 485.
84. Fu, Y., Hou, M., Yan, X., Hou, J., Luo, X., Shao, Z., and Yi, B. (2007) *J. Power Sources*, **166**, 435.
85. Lee, S.-J., Huang, C.-H., Chen, Y.-P., and Hsu, C.-T. (2005) *J. Fuel Cell Sci. Technol.*, **2**, 290.
86. Lee, S.-J., Huang, C.-H., and Chen, Y.-P. (2003) *J. Mater. Process. Technol.*, **140**, 688.
87. Lee, S.-J., Huang, C.-H., Chen, Y.-P., and Chen, Y.-M. (2005) *J. Fuel Cell Sci. Technol.*, **2**, 208.
88. Bai, C.-Y., Chou, Y.-H., Chao, C.-L., Lee, S.-J., and Ger, M.-D. (2008) *J. Power Sources*, **183**, 174.
89. Abo El-Erin, S.A., Abdel-Salam, O.E., El-Abd, H., and Amin, A.M. (2008) *J. Power Sources*, **177**, 131.
90. Nasralah, M.M. and Adel, A.M. (2007) *ECS Trans.*, **5**, 107.
91. Joseph, S., McClure, J.C., Sebastian, P.J., Moreira, J., and Valenzuela, E. (2008) *J. Power Sources*, **177**, 161.
92. Hung, Y., El-Khatib, K.M., and Tawfik, H. (2005) *J. Appl. Electrochem.*, **35**, 445.
93. Hung, Y., El-Khatib, K.M., and Tawfik, H. (2006) *J. Power Sources*, **163**, 509.
94. Tawfik, H., Hung, Y., and Mahajan, D. (2007) *J. Power Sources*, **163**, 755.
95. Hung, Y., Tawfik, H., and Mahajan, D. (2009) *J. Power Sources*, **186**, 123.
96. Matsuura, T., Kato, M., and Hori, M. (2006) *J. Power Sources*, **161**, 74.
97. Weil, K.S., Xia, G., Yang, Z.G., and Kim, J.Y. (2007) *Int. J. Hydrogen Energy*, **32**, 3724.
98. Zhang, H., Yi, B., Hou, M., Qiao, F., and Zhang, H. (2003) *Chin. J. Power Sources*, **27**, 129.
99. Huang, N., Yi, B., Hou, M., and Ming, P. (2005) *Prog. Chem.*, **17**, 963.
100. Fu, Y., Lin, G., Hou, M., Wu, B., Shao, Z., and Yi, B. (2009) *Int. J. Hydrogen Energy*, **34**, 405.
101. Wang, Y. and Northwood, D.O. (2006) *J. Power Sources*, **163**, 500.
102. Kuo, J.-K. and Chen, C.-K. (2006) *J. Power Sources*, **162**, 207.
103. Nikam, V.V., Reddy, R.G., Collins, S.R., Williams, P.C., Schiroky, G.H., and Henrich, G.W. (2008) *Electrochim. Acta*, **53**, 2743.
104. Wang, H., Turner, J.A., Li, X., and Bhattacharya, R. (2007) *J. Power Sources*, **171**, 567.
105. Wang, H. and Turner, J.A. (2007) *J. Power Sources*, **170**, 387.
106. Wang, H. and Turner, J.A. (2008) *ECS Trans.*, **16**, 1879.
107. Wang, H., Turner, J.A., Li, X., and Teeter, G. (2008) *J. Power Sources*, **178**, 238.
108. Li, M., Luo, S., Zeng, C., Shen, J., Lin, H., and Cao, C. (2004) *Corros. Sci.*, **46**, 1369.
109. Cho, E.A., Jeon, U.-S., Hong, A.-A., Oh, I.-H., and Kang, S.-G. (2005) *J. Power Sources*, **142**, 177.
110. Kumagai, M., Myung, S.-T., Asaishi, R., Sun, Y.-K., and Yashiro, H. (2008) *Electrochim. Acta*, **54**, 574.
111. Brady, M.P., Weisbrod, K., Zawodzinski, C., Paulauskas, I., Buchanan, R.A., and Walker, L.R. (2002) *Electrochem. Solid-State Lett.*, **5**, A245.
112. Brady, M.P., Paulauskas, I., Buchanan, R.A., Weisbrod, K., Wang, H., Walker, L.R., and Miller, L.S. (2003) Presented at the 2nd European PEFC Forum, Lucerne, Switzerland.
113. Brady, M.P., Weisbrod, K., Paulauskas, I., Buchanan, R.A., More, K.L., Wang, H., Wilson, M., Garzon, F., and Walker, L.R. (2004) *Scr. Mater*, **50**, 1017.

114. Wang, H., Brady, M.P., More, K.L., Meyer, H.M. III, John A., and Turner, J.A. (2004) *J. Power Sources*, **138**, 86.
115. Wang, H., Brady, M.P., Teeter, G., and Turner, J.A. (2004) *J. Power Sources*, **138**, 79.
116. Wang, H. and Turner, J.A. (2006) *ECS Trans.*, **1**, 263.
117. MEPS International (2010) World Stainless Steel Prices, http://www.meps.co.uk/Stainless%20Prices.htm (last accessed 28 November 2011).
118. Papp, J.F. (2004) Chromium, in US Geological Survey Minerals Yearbook 2004, http://minerals.er.usgs.gov/minerals/pubs/commodity/chromium/chrommyb04.pdf (last accessed 28 November 2011).
119. Stox Network (2011) MineralSTOX.com, http://www.mineralstox.com/ (last accessed 28 November 2011).
120. US Geological Survey (2005) Mineral Industry Surveys: Manganese in October 2005, http://minerals.usgs.gov/minerals/pubs/commodity/manganese/mnmis1005.pdf (last accessed 28 November 2011).
121. Paulauskas, I.E., Brady, M.P., Meyer, H.M. III, Buchanan, R.A., and Walker, L.R. (2006) *Corros. Sci.*, **48**, 3157.
122. Brady, M.P., Yang, B., Wang, H., Turner, J.A., More, K.L., Wilson, M., and Garzon, F. (2006) *JOM*, **58**, 50.
123. Yang, B., Brady, M.P., Wang, H., Turner, J.A., More, K.L., Young, D.J., Tortorelli, P.F., Payzant, E.A., and Walker, L.R. (2007) *J. Power Sources*, **174**, 228.
124. Brady, M.P., Wang, H., Yang, B., Turner, J.A., Bordignon, M., Molins, R., Abd Elhamid, M., Lipp, L., and Walker, L.R. (2007) *Int. J. Hydrogen Energy*, **32**, 3778.
125. Nam, N.D. and Kim, J.-G. (2008) *Jpn. J. Appl. Phys.*, **47**, 6887.
126. Fu, Y., Hou, M., Lin, G., Hou, J., Shao, Z., and Yi, B. (2008) *J. Power Sources*, **176**, 282.
127. Nam, D.-G. and Lee, H.-C. (2007) *J. Power Sources*, **170**, 268.
128. Tian, R.J., Sun, J.C., and Wang, L. (2007) *J. Power Sources*, **163**, 719.
129. Tian, R., Sun, J., and Wang, L. (2006) *Int. J. Hydrogen Energy*, **31**, 1874.
130. Tian, R., Sun, J., and Wang, J. (2008) *Int. J. Hydrogen Energy*, **33**, 7507.
131. Han, D.-H., Hong, W.-H., Choi, H.S., and Lee, J.J. (2009) *Int. J. Hydrogen Energy*, **34**, 2387.
132. Wang, H., Teeter, G., and Turner, J.A. (2010) *J. Fuel Cell Sci. Technol.*, **7**, 021018.
133. Wang, H., Teeter, G., and Turner, J.A. (2010) *J. Fuel Cell Sci. Technol.*, **7**, 021019.

14
Nanostructured Materials for Fuel Cells
John F. Elter

14.1
Introduction

There is a revolution going on in science and engineering that is literally changing our way of life. We are witnessing, in real time, the application of an exponentially expanding field of knowledge known as *nanotechnology*, the science, engineering, and economics of what happens when the things that we design, control, and manufacture get really small.

Examples of this technological revolution are driven by advances in physics, chemistry, materials science, biology, and medicine. Indeed, one of the key attributes of nanotechnology is the interdisciplinary approach that it demands when engineering new materials, structures, and devices that find application in products that span just about every possible field of application. As an example, because of nanotechnology, today's multifunctional communication devices such as cell phones contain integrated circuits that have nearly 10^{10} transistors per chip, operationally executing 10^{10} instructions per second. Computers that incorporate these same nanostructures will be used to design devices with even smaller characteristic dimensions. This cybernetic process of computers designing even faster and more powerful computers is made possible because of this rapidly expanding body of knowledge – nanotechnology. The accelerating computational power that is the result of this nanotechnological revolution is being used to invent, from first principles, whole classes of new materials, structures, and devices that can find application in communication, energy, health, and defense.

The term nanotechnology refers to the science and engineering of new materials, structures, and devices and processes at the so-called "nanoscale," which covers the range of sizes typically from 10^{-9} to 10^{-7} m. When particles become very small, their properties change for two reasons. Whereas a large particle of a material will have properties characteristic of the "bulk" material of which it is made, for very small particles, the interaction of the particle with its surroundings becomes dominated by the characteristics of its surfaces and interfaces, which are in turn determined by the structural and electronic properties of the atoms at or near the surface. This is because as a particle becomes smaller and smaller, a larger

Fuel Cell Science and Engineering: Materials, Processes, Systems and Technology,
First Edition. Edited by Detlef Stolten and Bernd Emonts.
© 2012 Wiley-VCH Verlag GmbH & Co. KGaA. Published 2012 by Wiley-VCH Verlag GmbH & Co. KGaA.

and larger fraction of its atoms reside at its surface. The particle's interaction with its surroundings is now dominated by forces related to its surface, rather than by those related to its mass. Second, the concept of the "continuum" as a reliable description of the particle also breaks down, and other methods need to be adopted, including those of quantum mechanics. These two reasons underlie both the "geometric" and "electronic" effects, which are different but related attributes of nanoparticles. With greater and greater computational power brought on by the nanotechnological revolution, these effects can now be studied theoretically and, coupled with more fundamental measurements, can provide tremendous insight into the nature of the relevant processes at the nanoscale.

It is from the nanoscale perspective that one needs to consider the capture, storage, and conversion of energy. Indeed, electrochemical processes at the nanoscale are fundamental to the capture of energy in photoelectric devices, the storage of energy in batteries and supercapacitors, and the conversion of energy with fuel cells and electrolyzers. Indeed, mimicking the most fundamental energetic process of Nature, photosynthesis, involves electrochemical processes at the nanoscale. This chapter specifically focuses on the application of nanotechnology to the electrochemical conversion of energy through the use of fuel cells, devices that control the reaction of hydrogen and oxygen to produce electricity, water, and heat.

14.2
The Fuel Cell and Its System

Like many energy conversion devices, fuel cells are not new. Many attempts to design, develop, and commercialize fuel cells have been undertaken since their invention by Sir Walter Grove in 1838. One of the best-known applications of fuel cells was that made by General Electric for use in the Gemini space program in the 1960s. That application was particularly important because it began the serious examination of the use of a solid polymer membrane as the electrolyte for exchanging electrons for protons at the electrode/membrane interfaces, and transporting the proton from the hydrogen oxidizing anode electrode to the oxygen reducing cathode electrode. This so-called *polymer electrolyte membrane* (PEM) became and has remained as the dominant design paradigm of the modern day PEM fuel cell (PEMFC).

It is worth noting that the PEMFC itself, although the focus of this chapter, needs to be considered within the context of an overall fuel-cell system. The system provides for the delivery of the reactants to and from the fuel cell, the conditioning of the power generated by the fuel cell, and the management of the reaction by-products, water, and heat. A greatly simplified fuel-cell system intended for use in transportation is shown schematically in Figure 14.1. This particular system is used by the US Department of Energy (DOE) for establishing performance metrics and assessing manufacturing costs for fuel cells for transportation. Here is shown a hydrogen–air PEMFC system, in which hydrogen is stored on-board in a tank as a compressed gas, oxygen is delivered with air supplied by a turbo compressor, waste

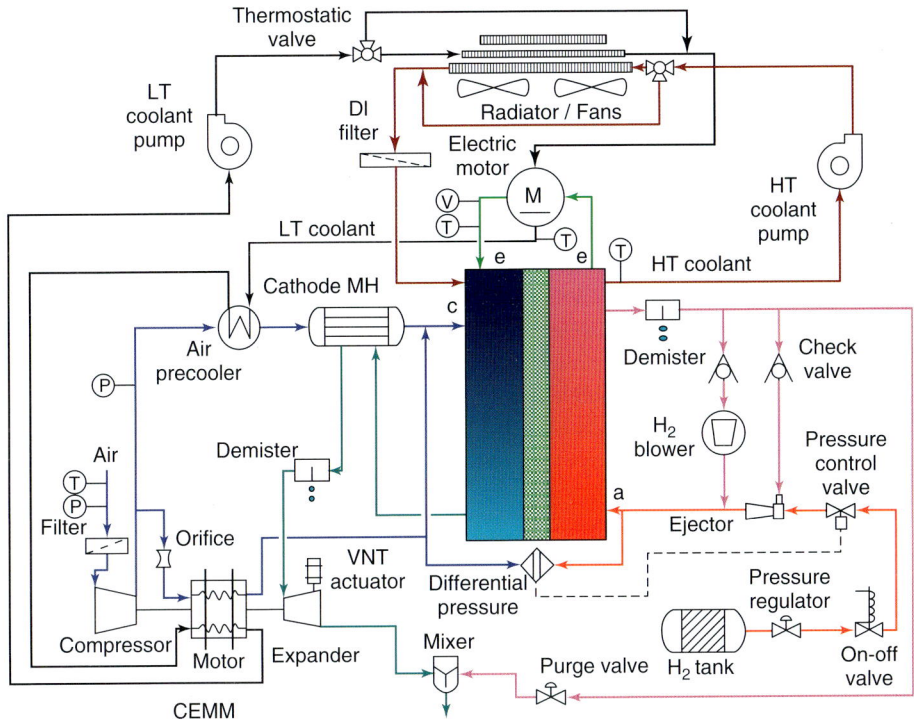

Figure 14.1 DOE hydrogen–air fuel cell system.

heat is absorbed by circulating liquid which is rejected to the atmosphere through a radiator, and water vapor is recirculated back through a humidifier to provide a moist supply of air to the fuel cell to maintain stack health. Not shown explicitly is the power electronics subsystem, which conditions the generated electrical power. Also not shown is the actual construction of the "stack" of individual fuel cells that are connected in series to produce the desired level of voltage and current. This is accomplished through the use of multifunctional bipolar plates which act as current collectors and channel the reactant streams to the anode and cathode electrodes, the gas diffusion layer (GDL) that evenly distributes the gaseous reactants to the electrode, and the proton-conducting membrane which prevents the mixing of the reactant gases while electrochemically connecting the anode to cathode. Both the anode and cathode electrodes consist of a porous and conductive structure allowing protons, electrons, and reactants to meet, combine, and react on the surface of nanoscale catalysts. The GDL, electrodes, and membrane are packaged as a unit called the membrane electrode assembly (MEA).

The fuel-cell system in turn needs to be considered a part of a larger meta-system, one that includes the delivery of the hydrogen to the compressed tank. Although the most abundant element on Earth, hydrogen does not occur naturally. It needs to be generated by some means, and then delivered on demand to the fuel cell. Sustainable methods for the delivery of hydrogen are through the on-site electrolysis of water, preferably with renewably supplied electricity, or by creating a hydrogen-rich fuel stream by the reforming of a renewable fuel. In either case, the cost of hydrogen generation and delivery needs to be included in the operational cost of the fuel-cell system producing electricity and useful heat. Both of these approaches have been and continue to be studied in great detail, especially for fuel-cell applications in transportation, where the need for a ubiquitous supply of hydrogen is of paramount importance.

Nanostructured materials play an important role all along the hydrogen value chain, including in the reforming and the electrolysis methods of hydrogen generation, and these will be briefly discussed further below. However, the emphasis in this chapter is on the role that nanotechnology is playing in the advancement of PEMFC systems for mobile and stationary energy conversion applications. Particular focus is placed on the role of nanostructures in providing the pathways and reaction sites for hydrogen and oxygen to produce electricity, water, and heat at the highest possible efficiency. Particular attention is placed upon those materials that utilize polymer electrolytes as membranes, electrocatalysts and their support structures, and advanced diffusion media that improve performance while reducing cost and broadening operational latitude.

14.3
Triple Phase Boundary

The generally accepted mental model of how a fuel cell works is based upon the concept the so-called "triple phase boundary" (TPB), the site on a catalyst surface where gaseous fuel, protons, and electrons meet and react either to oxidize hydrogen or to reduce oxygen. It is at the TPB that fuel-cell reactants must come into close proximity to enable charge to be transferred. This charge transfer takes place at the nanoscale, and involves nanoscale materials that are required to catalyze the basic fuel-cell reactions involving hydrogen oxidation and oxygen reduction. Hence the number and nature these reaction sites on the surface of the catalyst nanoparticles play a fundamental role in defining the efficiency and effectiveness of the fuel cell. Maintaining them over time as the fuel cell starts up, is cycled, and is shut down, perhaps thousands of times during the system's operation and in the presence of real-world contaminants, is a fundamental challenge in meeting the durability requirements of the particular application.

At the anode of a hydrogen–oxygen fuel cell, for example, hydrogen gas must come into contact with a catalyst "active site" on which the hydrogen adsorbs and is electrochemically oxidized into protons, which are conducted away through the PEM to the cathode, and electrons which pass through an electrical load comprising the external circuit connecting the anode to the cathode. Hence the hydrogen must arrive at the solid TPB, generally in the gas phase, while the proton must leave the reactant site through a proton-conducting pathway, and the electron through an electron-conducting pathway. The composition and structure of the anode electrode must provide these pathways for the reactants to arrive at and the reaction products to leave from the TPB. Intuitively, the electrode must be structured so that it does not impede the flow of gas, while at the same time offering a high internal surface area for the density of TPBs needed to meet the power demand.

A similar situation occurs at the cathode electrode. In this case, the TPB is the reaction site where the gaseous oxygen meets and adsorbs on an active site, and electrochemically reacts with the arriving proton and electron, releasing the products of the reaction: water, and heat. The functional requirements demanded by the TPB for the cathode electrode are therefore complex. The electrode must be porous to enable gases to permeate and reach the TPB at a rate sufficient to keep up with the electricity demand. It also needs to be able to expel the product water at the same rate. To do this, it needs to provide pores of the proper size, distribution, and hydrophilicity to prevent the so-called "flooding" of the electrode. The electrode needs to be not only electrically and ionically conductive to provide the pathways for the electrons and protons, but also thermally conductive in order to transport heat from the reaction site to prevent the cell from overheating.

The functional requirements, particularly of the cathode electrode, are conceptually easy to imagine, but in fact are difficult to achieve and maintain. A significant characteristic of the hydrogen–air fuel cell is the inherent difficulty in reducing oxygen to water. In contrast to the oxidation of hydrogen at the anode, which is quite facile, the reduction of oxygen at the cathode is very "sluggish," and to enable the reaction to proceed at reasonable rates, a high internal surface area and non-trivial amounts of expensive catalyst, typically platinum, are required to reduce the activation barrier to the reduction reaction.

Since the early days of the Gemini program, which used porous electrodes made of large amounts of Pt black bonded with Teflon, significant advances have been made to reduce the amount of Pt used in the MEA. Petrow and Allen, for example, patented the idea of using carbon-supported Pt (Pt/C) as a means to achieve the higher dispersions and lower Pt loadings while maintaining electronic pathways to the catalyst sites [1]. Likewise, the use of ion-conducting polymers, or ionomers, to provide proton-conducting pathways to the Pt/C reaction sites while acting as a bonding agent further reduced the amount of Pt needed for reasonable operation [2].

A highly simplified schematic of the current conventional electrode structure for a PEM fuel cell is shown in Figure 14.2. Here we see Pt or Pt alloy nanoparticles supported on high surface area conductive carbon particles, and bonded together by a proton-conducting hydrophobic ionomer that extends from the PEM into the catalyst layer. The overall electrode structure is highly porous and random. Consequently, its properties are difficult to describe from a microscopic perspective, and in general macroscopic "effective" properties are used when attempting to model the structure [3]. The electrode is manufactured as an "ink" which is then sprayed or doctored on to the PEM, creating a "catalyst-coated membrane" (CCM). Sometimes the ink is applied to the gas diffusion media, which may or may not have a microporous interfacial layer. The amounts of Pt, C, and ionomer vary by application, and the processes of ink formulation and MEA manufacture are

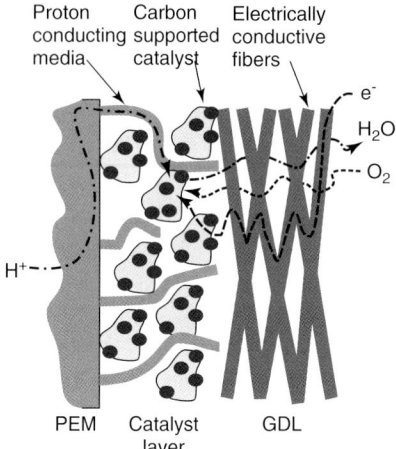

Figure 14.2 Schematic diagram of a conventional fuel-cell electrode.

generally held as trade secrets. Typical reported values are around 0.4 mg$_{Pt}$ cm^{-2} and 0.8 mg$_{Nafion}$ cm^{-2} when using 20 wt% Pt/C, with at least a 50% variation around these nominal values, depending on the method of fabrication [4].

The electrode structure depicted schematically in Figure 14.2 is the vantage point from which this discussion of the role of nanostructured materials in fuel cells begins. While discussing the functional properties of fuel cell components such as the electrodes, and the nanoscale technological approaches to achieving them, it is necessary to keep the cost and durability constraints constantly in perspective. These constraints obviously vary by application. For light-duty vehicle transportation, the current goal is to achieve parity with the existing gasoline engine in terms of cost and performance. As such, the DOE, working hand-in-hand with industry, has issued a set of detailed performance targets. These targets place demands on gravimetric and volumetric power density, as limits on the amount of precious metal group (PMG) catalyst content and on degradation rates to achieve durability targets. These targets are discussed in greater detail below.

14.4
Electrodes to Oxidize Hydrogen

The hydrogen oxidation reaction (HOR) [and its reverse, the hydrogen evolution reaction (HER)], occurs at the anode, is expressed by the electrochemical equation

$$H_2 \rightleftharpoons 2H^+ + 2e^- \tag{14.1}$$

This reaction is considered electrochemically reversible, and serves as a standard reference electrode against which to measure the thermodynamic potentials of all other reduction/oxidation reactions. By convention, the HOR is assigned a value of 0.0 V for the equilibrium potential E_0 under standard (normal) conditions (25 °C, 1 atm). As a reference electrode, it is denoted as the standard hydrogen electrode (SHE) or normal hydrogen electrode (NHE). When referenced to vacuum, so as to permit comparison with an electrode material's Fermi level, the value has been shown to be ~4.44 ±0.02 V, the 20 mV error being due to the uncertainty of the work function of the metal used in the measurement [5].

Being the standard electrochemical reactions against which all other electrochemical reactions are measured, the HOR and HER have attracted considerable attention, and early theories of the underlying mechanism have suggested several possible reaction routes, depending on whether the electrolyte is basic or acidic. The postulated equations for the HER/HOR in an alkaline solution are analogous to those in an acidic solution except that in alkaline solution hydrogen is formed from water, and in acid solution from hydronium (H_3O^+). Here, for the sake of discussion, we will consider acidic electrolytes. In this case, the generally accepted pathways for the HOR are [6]

Tafel reaction: $H_2 + 2M \rightarrow 2MH_{ads}$ (14.2)

Heyrovsky reaction: $H_2 + H_2O + M \rightarrow MH_{ads} + H_3O^+ + e^-$ (14.3)

Volmer reaction: $MH_{ads} + H_2O \rightarrow H_3O + e^- + M$ (14.4)

where term M refers to an active site on the catalyst's metal surface on which the hydrogen adsorbs, forming an adatom H_{ads} on the surface. The rate-determining step of this process varies, depending on the reaction conditions and the specific catalyst sites. In fact, an understanding of the HOR and HER reactions on Pt involves consideration of the "structure sensitivity" of the reactions to the particular crystal planes Pt(*hkl*) that are exposed to the H_2. Such studies have been conducted in ultra-high vacuum (UHV) conditions on well-defined Pt crystal planes, and have revealed much about the sensitivity of the reactions to crystal structure, and also to the important effect of adsorbed species in blocking sites from the reaction [7].

Expressions for the rates of these mechanistic reactions can be derived by assuming a reasonable adsorption isotherm, such as that proposed by Langmuir. For a PEMFC with Pt as the catalyst, it is generally agreed that the HOR process involves the Tafel and Volmer reactions, with the Tafel reaction being the rate-determining step. In this case, the rate of the reaction can be expressed in the form of the well-known Butler–Volmer (B–V) equation [8]:

$$i = i_0 \left[e^{\frac{\alpha F \eta}{RT}} - e^{-\frac{(1-\alpha) F \eta}{RT}} \right] \quad (14.5)$$

where it can be seen that the magnitude of the current density is determined by the overpotential, $\eta = E - E_0$, the transfer coefficient α, which measures the symmetry of the redox reaction, and the exchange current density, i_0, which is a measure of the intrinsic activity of the reaction at equilibrium. Experimentally, it has been found that the exchange current density exhibits a strong sensitivity to the state of the catalyst surface, the nature of the electrolyte, and the presence of any contaminants. For the HOR, a wide range of exchange current densities have been reported in the literature. Additionally, there is a subtle but important effect associated with the state of the MH_{ads} adatom, as there are both underpotentially developed (UPD) and overpotentially developed (OPD) states of the MH_{ads} adatom that influence the HER and HOR processes on Pt. Excellent reviews of the HER/HOR mechanisms, with different perspectives, have been given by Markovic [9] and Conway and Tilak [10].

In effect, the HOR/HER reactions, as archetypal processes, have proven to be an excellent model for studying the subject of electrocatalysis. In turn, these basic experimental studies have guided theoretical work that, together, have significantly advanced the ability to explain the basic interactions. Nørskov et al. [11], for example, examined the trends in the exchange current density for the HER on different catalyst materials. Using density functional theory (DFT), which is a quantum mechanical modeling method for investigating the electronic structure of a many-body system, the theoretically calculated hydrogen chemisorption energies on close-packed (111) surfaces of a number of transition and noble metals were correlated with the experimentally measured exchange current densities. This correlation is shown in Figure 14.3.

What is interesting about this "volcano" plot is that it shows that of all the metals examined, the activity appears to be the maximum on Pt, for which the evolution reaction is thermoneutral at the equilibrium potential ($\Delta G_{H+} \approx 0$). This fact also helps in interpreting the volcano plot. For metals which lie on the left-hand leg

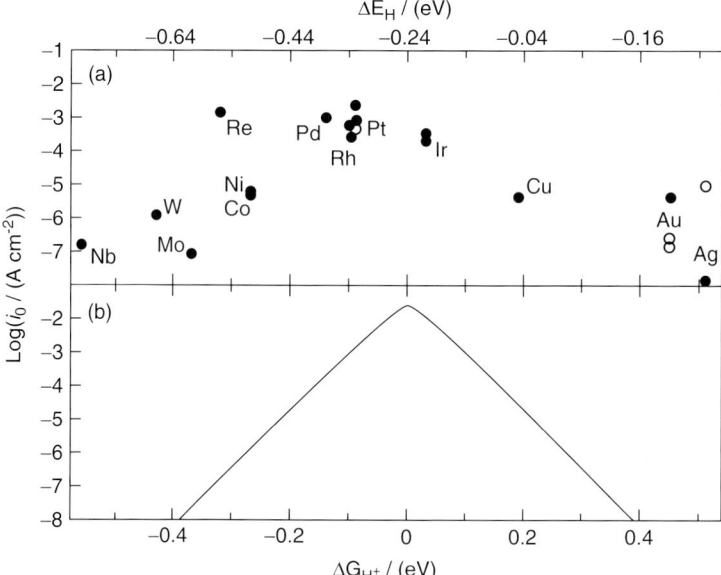

Figure 14.3 Trends in exchange current for HER. (a) Experimentally measured exchange current, $\log(i_0)$ for HER over different metal surfaces plotted as a function of the calculated H chemisorption energy per atom, ΔE_H. Single crystal data are indicated by open symbols. (b) The result of the simple kinetic model plotted as a function of the free energy for hydrogen adsorption, $\Delta G_{H*} = \Delta E_H + 0.24$ eV. Reproduced with permission from [11].

of the plot, metals with a decreasing adsorption energy have a decreasing reaction rate, indicative of the lack of available sites because hydrogen bonds more strongly. Likewise, for the less reactive metals on the right-hand leg of the plot, the rate decreases with increasing ΔE_H because proton transfer becomes more difficult as the hydrogen bonds are weakly adsorbed. Hence the volcano plot is a good representation of the well-known Sabatier principle. This same interpretation will also apply to the oxygen reduction reaction (ORR) to be considered below. The point here is that the "experiment" and the "theory" are mutually supportive. Confidence in the theory can then help guide advances in material optimization.

Based upon what has been said above, the neat hydrogen–air fuel cell, in an acidic environment, requires the use of Pt as the catalyst for promoting the HOR reaction. Owing to the high cost of Pt, however, it is a requirement that the amount of Pt used for the HOR be kept to a minimum. Fortunately, since the HOR reaction is facile, the overpotential is generally small, and small amounts of Pt nanoparticles, when well dispersed on a conductive high surface area support, can achieve the reaction rates necessary for the HOR. The HOR is not a problem with pure hydrogen.

However, when the hydrogen is produced by the catalytic "reforming" of a carbon-based fuel (using nanoscale catalyst materials), there are traces of carbon monoxide (CO) left in the "reformate," and the CO acts as a poison by adsorbing on the Pt surface and reducing the available number catalyst sites that would serve as a TPB. The influence of CO on the HOR has been studied in depth by surface scientists, as CO adsorption is frequently used as a diagnostic to understand reaction mechanisms and determine electrochemically active surface areas. Experimentally, it has been found that the level of CO in the reformate stream need not be high to have a significant impact on cell performance. Indeed, CO levels in the 10 ppm range can easily cause a voltage loss of 200 mV at a current density of 0.6 A cm^{-2} [12]. Hence reformate systems require an effective mitigation strategy for overcoming this poisoning effect. One component of such a strategy is to employ a "CO-tolerant" anode catalyst.

The most notable CO-tolerant catalyst is an alloy mixture of platinum and ruthenium. The basic mechanism of this "bifunctional catalyst" has the Ru interact first with water to produce an activated Ru(H$_2$O) or Ru(OH) site, adjacent to a CO-covered Pt site, which then interact to oxidize CO to CO$_2$ [13]. Here again, there exists a wealth of experimental data relating adsorption to surface structure, and these data are in need of a theory to help understand the design rules for designing nanomaterials for this particular fuel-cell application. This theory has, to a large extent, been provided by Hammer and Nørskov [14] in their d-band theory, which has been developed using DFT. The d-band theory relates bond strength to the d-band center, ε_d (relative to the Fermi level) by predicting the interaction of the d-band in influencing the donation of metal electrons with the bonding and antibonding orbitals. The theory enables one to explain trends in the interaction between adsorbent and the transition metals because variations in the coupling can be attributed to the interaction with the metal's d-states, given that the sp-states are virtually the same for the transition and noble metals [15].

Other Pt alloys in addition to Pt–Ru have been examined as CO-tolerant catalysts for use in fuel cells. These include both Pt–Sn and Pt–Mo, which were studied by Shubina and Koper [16], and Pt_3Sn, which was examined by Gasteiger and co-workers [17, 18]. In most cases, however, the CO-tolerant catalyst of choice is typically based upon Pt–Ru.

In addition to catalytic approaches to dealing with the poisoning effects of reformate, there are system-level approaches that sometimes employ nanostructured materials for dealing with the purity requirement of PEMFCs. These include the use of nanostructured separation membranes that can purify the reactant streams not only of CO, but also of other poisons such as hydrogen sulfide [19].

14.5
Membranes to Transport Ions

The most important functional requirement of the PEM is to provide for the transport of protons through the polymer electrolyte from anode to cathode. However, it also has to perform other critical functions, including preventing the mixing of the reactants, accommodating any pressure differential between the anode and cathode, and maintaining stable electrical, chemical, and mechanical properties over the expected range under operating conditions and over the lifespan of the fuel cell's operation. The PEM must also be chemically compatible with the other components used in the fuel-cell system, including the materials used for the electrodes, gas diffusion media, seals, coolants, and bipolar plates. It must also be robust in the oxidizing and reducing environments of the fuel cell, and against any electrochemical side reactions that may occur during the operation of the system. Finally, it must be easily manufactured at low cost. It is extremely important to appreciate how the properties of the polymer electrolyte, particularly those at the nanoscale, dictate the requirements on the fuel cell, stack, and system design.

Of all of the potential materials which can meet these requirements, only perfluorinated sulfonic acid (PFSA) membranes have so far found widespread acceptance in low-temperature (65–80 °C) hydrogen fuel-cell applications. The most widely used form of these PFSA-type membranes is produced by DuPont under the trade name Nafion. Other PFSA types of PEMs are sold by Asahi Chemical (Flemion and Aciplex) and Solvay Solexis (Hyflon). Gore sells its reinforced PFSA membrane as an integral part of an MEA as its PRIMEA series. Likewise, 3M also sells its own PFSA-type membrane as part of its MEA. One of the best known and basic properties of the current PEMs is that their proton conductivity is a strong function of their level of hydration.

Sulfonated fluoropolymer membranes are based on a polytetrafluoroethylene (PTFE) backbone that is modified by adding a side chain ending in a sulfonic acid group ($-SO_3H$) to the backbone. The resulting macromolecule contains both hydrophilic regions associated with the sulfonic acid group and hydrophobic regions associated with the backbone. A hydrated PFSA membrane forms a nanoscale two-phase system consisting of a water–ion phase that is distributed throughout a partially crystallized perfluorinated matrix phase. The current level of understanding is that as the membrane adsorbs water, the first water molecules cause the sulfonated group to dissociate, forming hydronium (H_3O^+) ions. The water that hydrates the membrane forms counterions that are localized on the sulfonated end groups, which act as nucleation sites. As more water is added, the counterion clusters coalesce to form even larger clusters, until a continuous phase is formed with properties that approach those of bulk water. The hydration level is measured in terms of a parameter λ, which is the equal to the number of absorbed waters per sulfonated group. For $\lambda = 0$, by definition, there is no water. This anhydrous form of the membrane rarely exists, since complete removal of water requires temperatures near the decomposition of the polymer. Researchers have performed molecular dynamics simulations which indicate that the primary hydration shell around the sulfonated group grows to a level of λ equal to about five waters [20]. As more water is added, for $\lambda > 6$, the added water is screened by the more strongly bound water of the primary hydration shell, and it can be considered a free phase. Saturation occurs in Nafion at $\lambda = 22$. One of the consequences of this strong coupling is that proton transport within the membrane is accompanied by the transport of water. This effect is known as "*electroosmotic drag*," and it complicates the required management of the fuel cell by-products: water and heat.

Kreuer *et al.* [21] provided an in-depth review of the basic mechanisms of transport in proton conductors. Transport of the proton can occur by two mechanisms: structural diffusion and vehicular diffusion. Vehicular diffusion is the classical Einstein diffusive motion. The structural diffusion is associated with "hopping" of the proton along water molecules (the so-called "Grotthuss" mechanism). In the nanosized confined hydrophilic spaces within the membrane, both mechanisms are operative. What is important here is that the underlying mechanism of transport in PEMs changes as a function the level of hydration. Understanding the nature of these mechanisms and their dependence on the level of hydration and molecular structure is important in the development of advanced PEM materials that are more tolerant of higher temperatures and lower levels of saturation.

Given this fundamental and strong dependence of proton conductivity with water content, it is necessary to maintain the hydration level locally within the membrane. Even though the product of the fuel-cell reaction is water, and conceptually it should be possible for the local hydration requirement to be met by the natural diffusion of water within the membrane, it turns out that at the flux levels demanded in most applications, the level of hydration must be met through *active management* of water distribution. Various methods of "water management" have been implemented. One design method, employed by Plug Power, is to configure the system so that the relative humidity (RH) of the reactant streams coming into the stack is nearly saturated. Another approach, used by Intelligent Energy, is to provide for water injection. United Technologies employs porous plates to distribute the product water generated during operation. Whatever the design approach taken, it needs to avoid the situation known as "flooding," where too much water results in blockage of pathways to the TPB, or fuel starvation, and permanent performance degradation.

Therefore, the requirement for active management of the PEM water content adds considerable system cost, complexity, and unreliability. Because of this, there is considerable research under way addressing the development of alternative classes of acid-based polymer electrolytes for fuel cells. There are excellent reviews of the progress being made in the development of these alternative polymer electrolytes [22].

The cost and complexity of the thermal and water management systems could be significantly reduced, and the reliability improved, if the fuel cell operated at higher temperatures (up to 120 °C) and at lower RH. Accordingly, the DOE has initiated a major effort to develop new membranes that can operate at temperatures up to 120 °C without the need for *external* humidification, and a variety of strategies are being pursued. Since, as discussed above, the membrane microstructure has been observed to have a substantial effect on conductivity, various approaches are attempting to improve the temperature and humidification latitude by altering the nanostructure within the membrane. These include the use of block copolymers with hydrophilic and hydrophobic properties that create nanostructured domains for proton conduction [23], and also composite membranes with functionalized additives for water retention at higher temperatures [24].

An important consideration in the development of proton-conducting ionomers for intermediate-temperature (90–120 °C) PEMFCs is the role of the polymer equivalent weight and the length of the pendant side chain. Early studies comparing Nafion, which has a long side chain (LSC), ($OCF_2CFCF_3-OCF_2CF_2SO_3H$), with a PFSA membrane from Dow with a shorter side chain (SSC), ($OCF_2CF_2SO_3H$), indicated that the Dow material had superior proton conductivity, better water uptake, and a higher glass transition temperature than its LSC counterpart from DuPont, making it a likely candidate for higher temperature applications [25]. These findings have been duplicated in more recent work comparing the SSC Hyflon ion membranes from Solvay Solexis [26]. Again, the underlying assumption is that the SSC ionomer membranes should be better suited to higher temperature fuel cell operation.

Motivated by these findings, Liu *et al.* [27], using molecular dynamics, recently attempted to examine the relationship between the molecular structure of PFSA ionomers and the nanoscale morphology of the hydrated membranes. They investigated the effect of the length of the polymer side chain to which the sulfonic acid group is attached, the equivalent weight (EW) of the ionomer, and the molecular weight (MW) of the polymer electrolyte. The effect of hydration level on the membrane morphologies was investigated from $\lambda = 3$ to 12. The authors used pair correlation functions, histograms of hydronium ion (H_3O^+) hydration, probability distributions for water cluster size, and self-diffusivities to describe the morphology of the hydrated membrane. Interestingly, their simulations (which were of necessity limited to using 40 polymer electrolyte molecules) indicated that with a longer side chain, there tends to be more clustering of the sulfonate groups and more local water–water clustering, but a more poorly connected aqueous domain. Decreasing the EW in either the SSC PFSA or Nafion resulted in more clustering of the sulfonate groups and more local water–water clustering and a better connected aqueous domain. Intuitively, connectivity enhances and confinement reduces water mobility, and so a decrease in EW, which enhances connectivity and reduces confinement, results in an increase in diffusivity. An increase in side chain length was found to diminish connectivity but reduces confinement, which together resulted in little change in the observed water diffusivity. There is a lower limit on the EW that can be used, as the loss of membrane crystallinity results in increased water solubility and degraded mechanical properties required for adequate durability.

Most recently, 3M has been investigating a class of membranes based on multi-acid side chain (MASC) ionomers, which also allow a lower EW with more crystallinity. A particular ionomer with MASC, a perfluoroimide acid (PFIA), with an EW of 625, has been selected for further optimization, as this new ionomer provides for lower resistance and improved performance under hot, dry conditions. A PFIA membrane, combined with nanofiber supports and chemically stabilizing additives, has been reported to meet almost all of the DOE performance requirements. The most recent status was reported at the 2011 DOE Annual Merit Review [28]. Based on the rate of progress being made with this approach, it appears that it is likely that membranes that can tolerate higher temperatures and lower hydration levels will be incorporated into future MEAs, allowing lower system costs and improved reliability.

Other proton-conducting PEMs have been developed for use in high-temperature (160–200 °C) fuel-cell systems that are ideal for combined heat and power stationary applications [29], and are based on the use of phosphoric acid-doped high-temperature membranes. The first detailed study of the effect of doping a high-temperature polymer membrane such as polybenzimidazole (PBI) with phosphoric acid was conducted at Case Western Reserve University [30], although earlier work on the properties of proton-conducting acid polymer blends was carried out by Lassegues *et al.* [31]. PEMs that do not rely on water as the basis of their proton conductivity offer the potential advantage of simplifying the fuel-cell system, since these systems can be operated without complex external humidification schemes. Furthermore, because these systems run at high temperature, they can tolerate higher levels of CO in the reformate. This allows for further simplification of the system in terms of reforming [elimination of preferential oxidation (PROX)] and control. Additionally, the higher operating temperature simplifies thermal management and provides a source of high-quality heat for combined heat and power applications. Consequently, BASF [32] and other companies are developing PBI-based PEMs for high-temperature stationary fuel-cell applications.

Advent Technologies is developing new high-temperature PEMs based on aromatic polyether polymers and copolymers containing polar pyridine moieties in the main chain. These products, called Advent TPS, are based on the idea of creating acid–base interactions in order to obtain high proton conductivity [33]. Advent's treatment of the membrane with phosphoric acid can be controlled by varying the pyridine based monomer content. This results in highly conducting PEMs in which the polar pyridine groups strongly retain the phosphoric acid molecules, due to their protonation. This serves to inhibit the leaching out of the phosphoric acid [34].

In comparison with low-temperature PFSA-based fuel cells, there is a sizeable increase in the polarization losses with phosphoric acid-based fuel cells. This is known to be caused by the presence of phosphoric acid and/or its anions that adsorb on the surface of the catalyst [35]. Because of this, high-temperature stacks based on phosphoric acid-based PEMs are required to be larger in order to achieve the same power output, and also have the requirement for higher Pt loadings. These two effects, larger stacks and higher loadings, have a sizeable negative impact on the system costs. Reducing or eliminating the effect of the adsorbed anion species would therefore have significant benefit at both the stack and system levels. Such a strategy would again require nanoscale structures and material considerations, and will be discussed further below.

So far, the discussion has been limited to *proton*-conducting PEMs typical of acidic electrolytes. There are also efforts under way to develop alkaline-based polymer electrolytes that have nanoscale backbone and side-chain structures that permit the conduction of the hydroxyl *anion*, OH^-. In this case, the anion-exchange membrane (AEM) allows the replacement of precious metal nanoparticle catalysts, which are required in acidic electrolytes, with those based on inexpensive and abundant metals, such as Co, Fe, and Ni. In particular, Cellera Technologies is pursuing the development of a new alkaline technology by transitioning from the proton-conducting PEM and ionomer to an OH^- ion-conducting membrane and ionomer.

Tokuyama, a Japanese company specializing in membrane technology for electrodialysis and desalination, has undertaken development of AEMs in OH^- form, targeting fuel-cell applications. Tokuyama's 901 membrane anion conductivity, $30\,mS\,cm^{-2}$, at roughly half that of the proton conductivity of the perfluorinated membranes, is at an acceptable level for fuel-cell development. Other material properties, such as dimensional stability due to the swelling as a result of the uptake of water, are also reasonable and are, in fact, better than those of typical PFSA membranes [36].

Historically, there has always been concern about the effect of CO_2 from the air feed on the OH^--conducting ionomer. Since conversion of the OH^- ion in the alkaline ionomer to bicarbonate (and/or carbonate) ion is a very likely process, the cell performance could suffer from the effects on both lower ionomer conductivity and, particularly, the kinetics of electrode processes. This is an issue that has plagued most alkaline-based systems which obtain one of their reactants from the air, including metal–air batteries. The approach developed by Cellera for handling CO_2 upstream of the stack supposedly totally avoids any need for routine replacement of system components, yet this remains a significant concern. Cellera has demonstrated power densities of $200\,mW\,cm^{-2}$ at a cell voltage of 0.4 V, with no platinum catalysts for a hydrogen–air stack, with no addition of liquid electrolyte, and with no humidification from an external source (S. Gottesfeld, personal communication). For all these positive reasons, alkaline-based PEM systems are beginning to attract significant attention.

14.6
Electrocatalysts to Reduce Oxygen

The most energetically lossy reaction in a PEM fuel cell is associated with the ORR. In acidic electrolytes, the ORR can proceed along several pathways. In one, the so-called "direct" four-electron pathway, four electrons are transferred to the oxygen, and the process is represented overall by the reaction

$$O_2 + 4H^+ + 4e^- \rightarrow 2H_2O \quad (U = 1.23\text{V vs. SHE}) \quad (14.6)$$

In the so-called "series" two-electron or peroxide pathway, the reaction is represented by

$$O_2 + 2H^+ + 2e^- \rightarrow H_2O_2 \quad (U = 0.695\text{V vs. SHE}) \quad (14.7)$$

In fact, these representations of the two pathways are overly simplistic and do not describe the underlying mechanisms by which the overall reactions actually take place. Indeed, the ORR is, next to the HOR/HER, one of the most widely studied electrochemical reactions. In spite of this, there remain a number of important and fundamental aspects of the ORR that are still not thoroughly understood or agreed upon. This is because the ORR is, in fact, fairly complex, and many alternative reaction pathways have been proposed [37].

Experience has shown that of the pure elements, the most effective catalyst material for this reaction is Pt. As such, there has been a concerted effort first to understand the role of Pt in the electrocatalytic processes at the atomic and nanoscale, and then to apply this understanding to the design and fabrication of nanostructured materials with significantly lower Pt content. As in the case of the HOR, several proposed mechanisms for the ORR have been put forth. One is the so-called *dissociative* mechanism, in which oxygen in the gas phase reacts with an available reaction site to dissociate and form adsorbed atomic oxygen, to which both a proton and electron react to form an adsorbed hydroxyl (OH) group; the hydroxyl then subsequently reacts with another proton and electron to form water. In the *associative* mechanism, molecular oxygen adsorbs on the reaction site first to form the intermediate O_2H, which then reacts with a proton and electron to form water and an adsorbed atomic oxygen (or hydrogen peroxide). To provide insight into the nature of these two mechanisms, Nørskov et al. examined both mechanisms theoretically, using DFT, with a view to examining the stability of the reaction intermediates on a Pt(111) surface [38]. They found that, depending on the metal and the potential, both reaction pathways may contribute, but in either case the transfer of the proton and electron to the associated intermediate is the rate-determining step, which in turn is strongly dependent on the binding energy. In general, if the binding energy is too strong, the transfer is not stable, and if it is too weak, the electron and proton transfer cannot effectively take place. Introducing an appropriate measure of activity, their analysis shows a dependence on the binding energy of both O and OH as a function of the metal type, as shown in Figure 14.4.

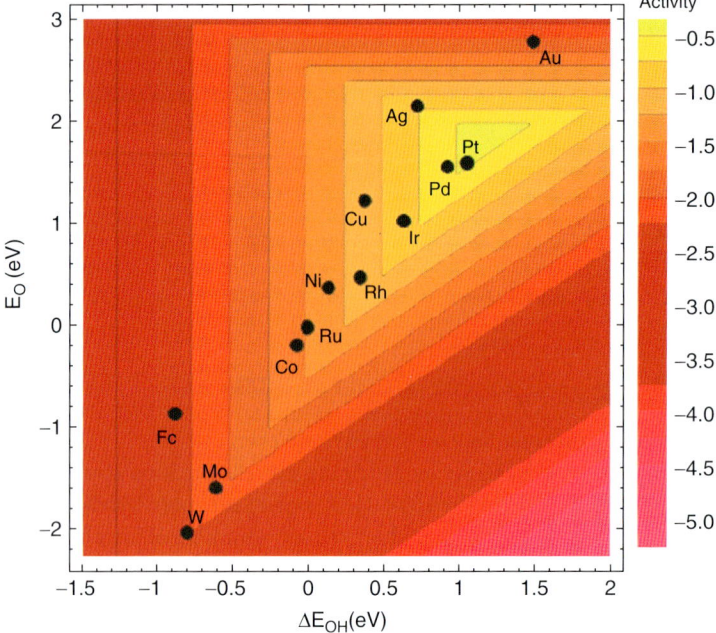

Figure 14.4 Trends in the ORR activity as a function of both the O and OH binding energy. Reproduced with permission from [38].

Here, as in the case of the HOR, a "volcano"-type plot emerges that indicates that of the elementary metals, Pt is best. However, it is evident from this figure that there is room for improvement if a nanostructured material can be found with the optimum binding energies for the reaction intermediates. Several approaches that attempt to achieve this balance in binding energy are possible.

One of the well-known approaches both to improve the activity and to reduce the amount of Pt is to use alloys of Pt and other transition metals. Markovic and co-workers focused on the catalytic properties and the stability of Pt alloyed with 3d-transition metals [39]. They performed in-depth studies of electrochemistry on well-defined bimetallic surfaces, in particular those with nanosegregated profiles associated with polycrystalline Pt_3M alloys (M = Ni, Co, Fe, V, Ti). They found that upon annealing these alloy particles in a UHV environment, the Pt segregates on the topmost surface, forming a "skin" of Pt atoms. After mild sputtering under UHV, the surface composition corresponds to the bulk ratio of alloying components. When exposed to the acidic electrochemical environment, the sputtered surface formed a "skeleton" surface profile, as the non-Pt atoms dissolved out of the surface, leaving a Pt-rich, but "non-smooth" surface, whereas the Pt-skin surfaces remained unchanged. It was concluded that the Pt outer surface layer protects the subsurface transition metal atoms from further dissolution.

The important findings in their study were that the skin and skeleton surfaces behaved differently from a catalytic perspective, suggesting that the underlying layers influence the electronic properties of the topmost surface. However, each type of surface (skin and skeleton) was found to have its own "volcano"-type behavior in terms of the specific activity (SA) dependence on the position of the d-band center (relative to the Fermi level), with maximum activity occurring for Pt_3Co for both types of surfaces. Here again, the Sabatier type of principle was found to be at play. An active catalyst is a balance between two opposing effects, namely the adsorption energy of O_2 and reaction intermediates against the adsorption strength of "spectator" oxygenated species that block reaction sites. If the d-band is too close to the Fermi level, then the oxygen is bound too strongly and the ORR is limited by the availability of spectator free sites, and if the d-band is too far from the Fermi level, then the surface binds O_2 and the intermediates too weakly to promote the ORR. In fact, the computational screening of the same alloy materials by DFT produced very similar results, proving the power of the d-band theory in describing electrocatalytic behavior [40].

An alternative, but related, approach to reduce Pt content and yet increase catalytic activity is the nanostructured "core–shell" approach taken by Adzic and co-workers [41]. As an example, a cartoon of a core–shell structure is illustrated in Figure 14.5, in which a nanostructured catalyst particle is shown that has a Pt monolayer surface with a Pd sublayer on top of a core material of metal M. This core–shell nanostructure is architected to take into account the surface contraction effects that shift the d-band center and reduce the oxygen binding energy, the natural surface segregation effects, and the stability offered by a contiguous monolayer. Clearly, the amount of Pt is reduced because the core can be made of less costly materials.

A specific example of the fabrication and characterization of a core–shell catalyst system is given in the recent work of Adzic *et al.* in which the effects of the thickness of the Pt shell, the mismatch in the lattice parameters, and the particle size on specific and mass activities were systematically studied for the case of a Pt shell on both Pd and Pd_3Co cores [41]. The number of Pt shell layers on the Pd (or Pd_3Co) core was controlled by utilizing a copper underpotential development (Cu-UPD)-mediated electrodeposition method. The composition of the nanostructures was determined using atomic resolved images from scanning transmission electron microscopy (STEM) coupled with element-sensitive electron energy loss spectroscopy (EELS). It was found that with a 4 nm Pd core and with a 4.6 nm Pd_3Co core, the Pt shells ranged from one to four atomic layers thick. When evaluated electrochemically in a rotating disk electrode (RDE), the ORR activities showed enhancement factors of up to 3 in SA and up to 9 in

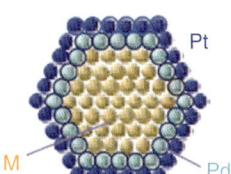

Figure 14.5 Cartoon of a second-generation core–shell nanostructured catalyst particle.

mass activity (MA). Finally, the authors used DFT to calculate the binding energy for oxygen, and attributed the enhancement in SA to the contraction in the Pt (111) facets as a result of the lattice mismatch. As in the case of the Pt-skin approach, this is another excellent example of the impact that computational and experimental nanotechnology is having on the ability to offer pathways to improve the ORR. Indeed, the ability to conceptualize, fabricate, characterize, and model nanostructures which test mental models of the mechanisms underlying the ORR is accelerating the rate of knowledge accumulation.

It is important at this point to note that potential catalyst materials are evaluated in terms of several critical performance related parameters, Of these, three are very important, namely the SA, which is equal to the current measured per unit of actual Pt catalyst surface area, the MA, which is defined in terms of the measured current per unit mass of Pt, and the *total Pt group metal loading* per unit of *geometric* area. All three are critical to achieving the overall fuel-cell cost objectives. The DOE transportation target for SA is $720\,\mu\text{A cm}_{Pt}^{-2}$, measured at 900 mV (*iR*-free) at 150 kPa pressure, 80 °C, and 100% RH. Similarly, the DOE target for MA is $0.44\,\text{A mg}_{Pt}^{-1}$, measured after 1050 s under the same conditions as for SA. The target for total platinum group metal (PGM) loading has recently been set at $0.125\,\text{mg cm}^{-2}$. DOE *durability targets* are defined in terms of cell voltage loss (<30 mV), percentage electrochemical surface area (ECSA) loss (<40%), and percentage MA loss (<40%) under specific stress test conditions. Of course, targets are expected to be met in the production MEA, but for development purposes, 50 cm^2 single-cell experiments are performed. For screening purposes, catalyst activity is typically measured by using an RDE with an aqueous electrolyte that is meant to serve as a surrogate for the polymer electrolyte. The techniques involved have been described in the literature [42] and a high degree of correlation has been achieved between the characterization and evaluation of Pt catalysts in the RDE and in an actual MEA. By using the RDE, or rotating ring disk electrode (RRDE), which is used also to measure the peroxide generated in the ORR, it is possible to rank qualitatively different nanostructured catalysts with respect to the baseline Pt performance, and also to study the effects of the electrolyte type, surface structure of the catalyst, and catalyst support and particle size effects (PSEs).

There is an important relationship between the MA and SA for the ORR, which has been the subject of considerable investigation and controversy, and that relates to the PSE of the catalyst nanoparticle. Catalyst particle size influences the electrochemical reactions in various ways. It is known that low-index crystallographic planes (e.g., <111>, <100>) have higher catalytic activities for the ORR in comparison with low-coordinated edges and corners, and that the number of these low-coordinated sites changes dramatically with particle size. As the size of the particle becomes smaller, there is an exponentially greater proportion of these low-coordination atoms. Interestingly, as shown by Kinoshita [43], both the *mass*-averaged distribution for the Pt(111) facet as a function of particle size, and the *surface*-averaged distribution for the same face as a function of particle size, correlated with the observed behavior of the MA and SA, respectively. The mass-averaged distribution and the MA both showed a broad maximum at particle sizes between 3 and 6 nm, whereas the surface-averaged distribution and the SA showed a decrease with decreasing particle size. However, the explanation of the PSE is not as simple as that proposed by Kinoshita. As demonstrated by Mukerjee and McBreen [44] in experiments using *in situ* X-ray absorption spectroscopy, the particle size itself is affected by its electrochemical environment, and the strength of adsorbed intermediates such as OH_{ads} is particle size dependent, which can also explain the reduced SA on smaller particles. Likewise, the chemical potential also changes with particle size [45]. Most data on the PSE are based on Pt/C materials; it is only recently that there have been investigations of PSE on alternative supports, and these are mentioned briefly below.

It is important to understand that the surface upon which the ORR is taking place is in fact *dynamic* and is constantly changing in response to the electrochemical environment. Surface oxides form and surface atoms undergo place exchange and otherwise reconfigure as a function of the changing potential of the "real" fuel-cell environment, which is far removed from the UHV environment in which fundamental mechanisms are probed for understanding. Furthermore, it is not just a balance in the adsorption energy of oxygen and its intermediates that determines the ORR rate, but also the availability of reaction sites on the catalyst surface, and clearly these two effects are inter-related phenomenon.

Anion adsorption can also have a significant effect on the ORR kinetics, since these specifically adsorbed species occupy reaction sites. In phosphoric acid, for example, the ORR is strongly influenced by the detrimental effects of adsorption of specifically adsorbed anions [46]. The surface coverage of phosphoric acid anions is the highest at the (111)-oriented facets because of the matching structure of tetrahedral anions and the threefold symmetry of the (111) planes. The catalytic activity for the ORR follows the sequence Pt(110) > Pt(100) > Pt(111). Structural dependence is predominantly determined by the adsorption of phosphoric acid anions, which is strongly structure dependent on Pt surfaces [47]. These facts suggest that methods for designing improved catalysts for the ORR clearly have to be based on decreasing the coverage of phosphoric acid anions by means of nanostructured materials. One such approach, for example, would be to fabricate nanostructures that have predominately oriented (110) surfaces.

14.7
Catalyst Supports to Conduct Electrons

The current dominant paradigm in electrode design is to support catalyst nanoparticles on a high surface area carbon support. Carbon has many attractive features. In addition to its high surface area, it has excellent conductivity, and is very inexpensive. There are a large number of types of high surface area carbons, and these are discussed from the aspect of their electrochemical and physicochemical properties, at length, in an excellent treatise by Kinoshita [48]. One of the most often used carbon blacks for PEMFC applications is the furnace black Vulcan XC72R from Cabot. This material is has been used extensively as the support in catalyst development. As such, it is well characterized in terms of its use in acidic and alkaline fuel cells, with much of what has been learned coming from studies of its behavior in phosphoric acid fuel cells. As a point of reference, the average XC72R carbon particle size is ~30 nm and the nominal surface area is 250 m^2 g^{-1}.

The properties and characteristics of carbon black particles can be determined by a variety of techniques. Adsorption of an inert gas such as nitrogen, with an appropriate theory, can be used to determine the specific surface area of the porous carbon materials. Other characterization techniques include the use of small-angle X-ray scattering (SAXS) to determine the particle size distribution and the fractal nature of the particle surface, and the use of scanning electron microscopy (SEM) to determine the characteristic particle size. Watt-Smith *et al.* used these techniques and advanced absorption models that attempt to account for the "fractal" nature of the carbon particle surface in characterizing porous carbon electrode materials used in PEMFCs [49]. Interestingly, the data obtained for the XC72R sample they used did not fit the usual Brunauer–Emmett–Teller (BET) adsorption model, nor did its fractal dimension, obtained independently by SAXS, match the surface adsorption model which incorporated fractal characteristics. The authors suggested that the XC72R sample's surface, whose specific surface area derived from a distribution of model isotherms using DFT calculations was determined as 126 m^2 g^{-1}, was composed of a broad range of nanoscale and mesoscale pore sizes.

Of course, a picture is worth a thousand words, and high-resolution transmission electron microscopy (HRTEM) is a very useful tool for this purpose. An example of the HRTEM of Vulcan XC72, with and without Pt deposited, is shown in Figure 14.6, obtained by More [50]. Most carbon black surfaces are comprised of small (4–5 nm), highly textured crystallites of graphitic domains with many surface steps and defects exhibiting a meso-graphitic structure. It has been found that Pt nanoparticles nucleate uniformly across the highly defective surfaces of Vulcan. Once the Vulcan has been graphitized so that there are fewer surface defects, the

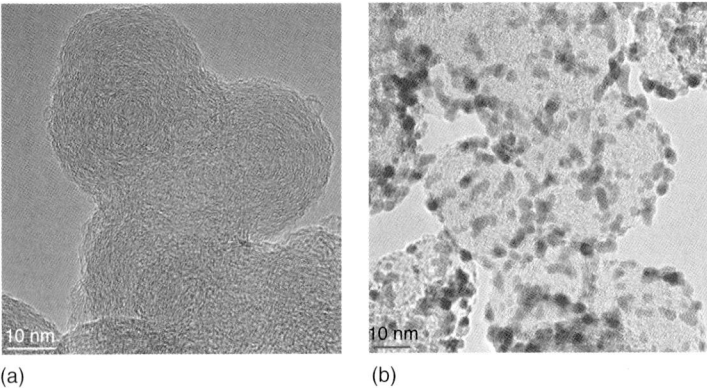

(a) (b)

Figure 14.6 (a) Structure of Vulcan carbon black particle and (b) the distribution of Pt particles nucleated on the surface [50].

Pt nucleates predominantly at regions "between" carbon particles or at graphite folds rather than on flat graphite surfaces.

Finally, it should be mentioned that the use of STEM with three-dimensional tomography is providing even more detail into the internal structure of the carbon support materials, and in the distribution of the ionomer within the catalyst layer [51].

One of the issues associated with the use of carbon materials for catalyst supports is that they can corrode. Although the rate of electrochemically driven carbon corrosion (with an equilibrium potential of 0.207 V vs. SHE) is very low under "normal" fuel-cell operating conditions (the exchange current density is of the order of 10^{-19} A cm^{-2} [52]), significant rates of corrosion can occur under certain higher potential conditions relevant to fuel-cell operation, such as during start-up and shut-down [53] and anode fuel starvation. A comparison of carbon corrosion rates of various commercial carbons has been reported by Yu et al. [54], which corroborates that this is a significant problem need resolution.

A recent study by Cherstiouk et al. [55] attempted to correlate the rate of carbon corrosion with the microstructural characteristics of the carbon particles. They characterized the microstructure of commercial furnace blacks, including XC72, along with proprietary carbon supports from the Sibunit family, a carbon black–pyrolytic carbon synthetic composite. Also included were samples of carbon nanofilaments of various structures, including nanotubes. Their study indicated that correlations could be established between the rate of corrosion (normalized to the specific surface areas) to a generalized microstructural characteristic that is roughly proportional to the fraction of grain boundaries between quasi-graphitic crystallites. This correlation supported their hypothesis that the corrosion of carbon particles occurs inwards along grain boundaries, specifically at defect sites. Their results also showed that carbon particles that supported Pt nanoparticles had, in general, higher rates of corrosion, suggesting that Pt helps to catalyze the corrosion reaction. Of primary interest was their observation that carbon support materials of the Subunit family showed a lower rate of corrosion than XC72, and better stabilization of Pt nanoparticles.

The fact that amorphous carbons corrode has led to led to the search for alternative, carbonless support materials. One of the most significant developments in this regard has been 3M's development of their so-called nanostructured thin-film (NSTF) electrodes, where nanoscale thin films of Pt (and its alloys) are vacuum deposited on highly structured crystalline "whiskers" of an organic pigment, grown via annealing of vacuum-deposited films of the organic pigment Perylene Red [56]. An SEM image showing the unique nanostructure as first reported by 3M is shown in Figure 14.7. These nanostructures have controllable high aspect ratios in the range 20–50, area number densities in the range 3–5 billion whiskers cm^{-2}, and cross-sections of $\sim$50 nm. The Pt loading for the image shown is $\sim$0.25 mg cm^{-2}. This nanostructure of catalyst-coated organic whiskers is transferred to the PEM using the "decal" method. All processes are highly compatible with high-volume roll-to-roll manufacturing.

Figure 14.7 SEM image of 3M NSTF Pt-coated "whiskers" of crystalline organic Perylene Red.

There are several remarkable features associated with this electrode structure. First, of course, it has no carbon support to corrode. Consequently, the NSTF electrodes are very stable in high-voltage conditions. Electronic pathways to and from the reaction sites are obviously provided by the "contiguous" film of Pt on the organic whiskers. This pathway needs to extend from the reaction site on the Pt-coated whisker to the GDL. Second, there is also no proton-conducting ionomer in the electrode. This again goes against the conventional wisdom of what is required to establish and maintain the pathways to the TPB. Obviously, the transport of protons to the reaction site occurs on the surface of the Pt catalyst. It has been hypothesized that there must exist hydroxyl or other groups on the Pt surface which provide some mechanism for proton transport over the distances characteristic of the electrode thickness. In the case of NSTF, the electrode thickness is typically less than 1 μm, in comparison with the conventional Pt/C electrode which is typically 10 μm thick. One of the issues with the 3M electrode is that because it is very thin, it is more prone to flooding, especially under the more stressful conditions of a wet and cold environment.

Performance-wise, another interesting feature of the 3M NSTF electrode is that in spite of the significantly improved SA achieved as a result of the unique catalyst surface structure (SAs in the range 2100 $\mu A\ cm_{Pt}^{-2}$ have been reported [57]), the MA is limited by the geometric surface area. Extensive measurements of the ECSA by hydrogen adsorption/desorption with cyclic voltammetry indicates values in the range of $\sim 10\ cm^2$ per cm^2 of planar electrode surface area, whereas the ECSA of conventional Pt/C electrode structure is about 7.5 times larger (77 cm^2 per cm^2 of planar electrode surface area). Furthermore, the Pt coating can result in a "rough" surface as shown in Figure 14.7, where the "whiskerettes" expose Pt(111) facets, which have grain structures of ~10 nm, about 3–5 times the sizes of catalyst particles supported on carbon. Accordingly, one of the efforts undertaken by 3M has been directed at increasing the effective surface area in order to capture the benefit of the much larger intrinsic SA of the NSTF.

Since the introduction of this nanostructured electrode, 3M has made steady progress in developing, characterizing, and improving it and the MEA manufacturing process. Efforts have been directed towards improving the issues related to MA and water management. At the most recent 2011 DOE review, 3M reported progress across all fronts [58]. A PGM total content of less than 0.18 mg_{Pt} kW^{-1}, a PGM loading in the range 0.15–0.20 mg_{PGM} cm_{MEA}^{-2} (including both anode and cathode) with a PtCoMn alloy, an MA of 0.24 A mg_{Pt}^{-1} (with PtCoMn alloy), an SA of 2100 μA cm_{Pt}^{-2}, and durability metrics exceeding most of the DOE targets. Hence 3M is meeting nearly all of the DOE targets except the MA target. In this regard, it is worthy noting that 3M has uncovered an anomalous but apparently real significant improvement in MA associated with a Pt_3Ni_7 alloy. STEM Z-contrast images taken with high-angle annular dark field (HAADF) STEM reveal that the tips of the "whiskerettes" are Pt rich (typically 57%Pt/43%Ni) in comparison with the gravimetric composition (30%Pt/70%Ni) of the whole whisker. It is tempting to speculate that this nanostructure at the whisker tip is somehow unique in its ability to enhance the ORR reaction through modification of the binding energy of the oxygen intermediates. However, post-processing of the structure with a "surface energy treatment," which no doubt modified this nanostructure, showed a substantial further increase in activity. Performance with the Pt_3Ni_7 material was reported to have essentially met the DOE MA target of 0.44 A mg_{Pt}^{-1}. Limiting current values for the Pt_3Ni_7 were reduced, however, most likely due to leaching out of the nickel, suggesting that further optimization is required.

The unique lessons taught by the 3M nanostructures, and the performance levels being achieved in SA, MA, total loading, and durability, have inspired others to examine similar approaches. Accordingly, efforts are currently under way to mimic various aspects of this approach, namely eliminating the usual carbon support and using contiguous thin films on more stable, low-cost supports. For example, Kim *et al.* are examining Pt-covered amine-functionalized multiwall carbon nanotubes [59]. Their preliminary results for acid- and heat-treated samples, measured in the thin-film mode with the RDE and normalized to the ECSA, are impressive. The intrinsic ORR activity (*iR*-corrected) of ~ 0.95 mA cm_{Pt}^{-2} at 0.9 V vs. SHE is comparable to that in other so-called 1D approaches such as those reported by 3M and others [60]. Perhaps more significantly, the reported MA of ~ 0.48 A mg_{Pt}^{-1} at 0.9 V is better than that of other 1D nanostructured catalysts and commercial supported Pt catalysts [61].

Other researchers are also examining the use of templates to provide a support for contiguous Pt films. The DOE National Renewable Energy Laboratory (NREL), for example, is examining extended, continuous Pt nanostructures for use in thick (in comparison with the NSTF approach), dispersed electrodes by using spontaneous galvanic displacement (SGD) with silver nanowires as the template [62]. The Ag nanowires are 50–250 nm in diameter and microns in length. Very encouraging preliminary results were obtained with limited effort. SGD has many independent synthesis parameters that can be controlled, and the structures demonstrated at this time were reported as far from optimal. SGD, being a solution-based synthesis route, is scalable to increased production volumes using economically viable routes that have precedence in commercial nanoparticle electrocatalysts for fuel cells.

There is also considerable interest in the use of *metal oxides* as catalyst support materials, for several reasons. First, of course, the metal oxides, most notably titanium-based oxides, are known to be electrochemically "inert" over a wider potential range than most carbon supports. Second, these materials are widely produced and cost-effective, finding wide-ranging applications, including those related to energy and the environment. Third, they have the benefit of assuming a wide variety of nanoscale morphologies. Finally, they have the potential to impart a "metal–support interaction" that can enhance electrocatalytic behavior. For these reasons, a number of studies have examined the use of metal oxides (and metal carbides) for use in fuel cells, and an in-depth review of many of these works was given by Antolini and Gonzalez [63].

TiO_2 is a wide-bandgap semiconductor, with unique properties that will dictate how it is used as a support. If a "continuous" or contiguous layer of metal catalyst covers the metal oxide, then that layer can provide for the necessary electronic and protonic pathways to and from the reaction site, in much the same way as in the 3M electrode. If, however, the TiO_2 nanostructures provide support to low loadings of catalyst nanoparticles, then any insulating properties that exist at the nanoscale will be problematic. In this case, one can include conductive carbon materials in the electrode [64], in much the same manner as that in which inks are made, and the electrode should have better stability against corrosion since the corrosion-catalyzing Pt will be supported on the oxide material rather than the carbon Alternatively, one can dope the TiO_2 support with niobium or other atoms to create n-type defect states in the bandgap [65], or use sub-stoichiometric Magneli phases (Ti_nO_{2n-1}) of the oxide, such as Ti_4O_7, which possesses electrical conductivity comparable to that of graphitized carbon [66]. In addition, the well-known interaction between TiO_2 and gold nanoparticles in the electrooxidation of CO [67] suggests that there are likely catalyst particle size and nanostructured support interactions that can be exploited in such a way that the support can be used to promote and enhance further the ORR reaction. Research into the use as supports of metal oxides, not just TiO_2, but also other oxides, is an interesting, exciting, and important area of research applicable to the use of nanostructured materials for fuel cells [68].

14.8
Future Directions

This chapter provides a snapshot of some of the advances being made in the application of nanostructures to fuel cells. Space limitations prevented an in-depth discussion of the directions highlighted here, but it is hoped that the selection of topics was sufficient to convey the message that nanotechnology is the catalyst for the twenty-first century's achievement of a sustainable way of living. Going forward, we can look forward to the continued and exponential accumulation of knowledge made possible by the convergence of info and nano, and, now more than ever, bio. Indeed, it is by examining how Nature goes about its business of fabricating structures for the efficient capture, generation, and storage of energy, with bottom-up manufacturing in aqueous-based environments using Earth-abundant materials, providing pathways and enzymes for proton and electron transfer, that the next generation will more than likely finally realize the objective of truly sustainable development.

References

1. Petrow, H.G. and Allen, R.J. (1977) US Patent 4,044,193.
2. Wilson, M.S. and Gottesfeld, S. (1992) High performance catalyzed membranes of ultra low Pt loadings for polymer electrolyte fuel cells. *J. Electrochem. Soc.*, **139**, L28–L30.
3. Weber, A.Z. and Newman, J. (2004) Modeling transport in polymer electrolyte fuel cells. *Chem. Rev.*, **104**, 4697–4726.
4. Litster, S. and McLean, G. (2004) PEM fuel cell electrodes. *J. Power Sources*, **130**, 61–76.
5. Trasatti, S. (1986) The absolute electrode potential: an explanatory note (recommendations 1986). *Pure Appl. Chem.*, **58** (7), 955–966.
6. Bogatsky, V.S. (2006) *Fundamentals of Electrochemistry*, 2nd edn., John Wiley & Sons, Ltd., Chichester.
7. Markovic, N.M. and Ross, P.N. (2002) Surface science studies of model fuel cell catalysts. *Surf. Sci. Rep.*, **45**, 121.
8. Breiter, M.W. (2003) Reaction mechanism of H_2 oxidation/evolution reaction, in *Handbook of Fuel Cells: Fundamentals; Technology and Applications*, Electrocatalysis, Vol. 2 (eds. W. Vielstich, H.A. Gasteiger, and A. Lamm), John Wiley & Sons, Ltd., Chichester, 361–367.
9. Markovic, N.M. (2003) in *Handbook of Fuel Cells*, Vol. 2, (eds. W. Vielstich, A. Lamm, and H.A. Gasteiger), John Wiley & Sons, Ltd., Chichester, pp. 367–393.
10. Conway, B.E. and Tilak, B.V. (2002) Interfacial processes involving electrocatalytic evolution and oxidation of H_2 and the role of chemisorbed H. *Electrochim. Acta*, **47**, 3571–3594.
11. Nørskov, J.K., Bligaard, T., Logadottir, A., Kitchin, J.R., Chen, J.G., Pandelov, S., and Stimming, U. (2005) Trends in the exchange current density for hydrogen evolution. *J. Electrochem. Soc.*, **152** (3), J23–J26.
12. Ralph, T.R. and Hogarth, M.P. (2002) Catalysis for low temperature fuel cells. Part II, the anode challenges. *Platinum Met. Rev.*, **46**, 117–135.
13. Gasteiger, H.A., Markovic, N., Ross, P.N., and Cairns, E.J. (1994) CO electrooxidation on well-characterized Pt–Ru alloys. *J. Phys. Chem.*, **98**, 617–625.
14. Hammer, B. and Nørskov, J.K. (2000) Theoretical surface science and catalysis – calculations and concepts. *Adv. Catal.*, **45**, 71–129.
15. Chorkendorff, I. and Niemanstverdriet, J.W. (2007) *Concepts of Modern Catalysis and Kinetics*, 2nd edn., Wiley-VCH Verlag GmbH, Weinheim.
16. Shubina, T.E. and Koper, M.T. (2002) Quantum-chemical calculations of CO and OH interacting with bimetallic surfaces. *Electrochim. Acta*, **47**, 22–23.
17. Gasteiger, H.A., Markovic, N.M., and Ross, P.N. (1995) Electrooxidation of CO and H_2/CO mixtures on a well-characterized Pt_3Sn electrode surface. *J. Phys. Chem.*, **99**, 16757–16767.
18. Wang, K., Gasteiger, H.A., Markovic, N.M., and Ross, P.M. (1996) On the reaction pathway for methanol and carbon monoxide electrooxidation on Pt-Sn alloy versus Pt–Ru alloy surface. *Electrochim. Acta*, **41**, 2587–2593.
19. Du, B., Pollard, R., Elter, J.F., and Ramani, M. (2009) Performance and durability of a polymer electrolyte fuel cell operating with reformate:

effects of CO, CO$_2$, and other trace impurities in *Polymer Electrolyte Fuel Cell Durability* (eds. F.N. Buchi, M. Inaba, and T.J. Schmidt), Springer, New York, 341–366.

20. Vishnyakov, A. and Niemark, A.V. (2000) Molecular study of Nafion membrane solvation in water and methanol. *J. Phys. Chem. B*, **104**, 4471–4478.

21. Kreuer, D.K., Paddison, S., Spohr, E., and Schuster, M. (2004) Transport in proton conductors for fuel cell applications: simulations, elementary reactions, and phenomenology. *Chem. Rev.*, **104**, 4637–4678.

22. Scherer, G.G., (2008) *Polymers for Fuel Cells II (Advances in Polymer Science)*, Vol. **26**, Springer

23. Abhishek, R., Yu, X., Dunn, S., and McGrath, J.E. (2009) Influence of microstructure and chemical composition on proton exchange membrane properties of sulfonated–fluorinated hydrophilic–hydrophobic multiblock copolymers. *J. Membr. Sci.*, **327** (1–2), 118–124.

24. Lipp, L. (2011) High temperature membrane with humidification independent cluster structure. Presented at the DOE Hydrogen Program Merit Review and Peer Evaluation, Washington, DC, 10 May 2011.

25. Eisman, G.A. (1990) The application of Dow Chemical's perfluorinated membranes in proton-exchange fuel cells. *J. Power Sources*, **29**, 389–398.

26. Ghielmi, A., Vaccarono, P., Troglia, C., and Arcella, V. (2005) Proton exchange membranes based upon the short-side-chain perfluorinated ionomer. *J. Power Sources*, **145**, 108–115.

27. Liu, J., Suraweera, N., Keefer, D.J., Cui, S., and Paddison, S.J. (2010) On the relationship between polymer electrolyte structure and hydrated morphology of perfluorosulfonic acid membranes. *J. Phys. Chem. C*, **114**, 11279–11292.

28. Hamrock, S. (2011) Membranes and MEAs for dry, hot operating conditions. Presented at the DOE Hydrogen Program Merit Review and Peer Evaluation, Washington, DC, 10 May 2011.

29. Elter, J.F. (2010) Fuel cells for buildings, in *Hydrogen and Fuel Cells* (ed. D. Stolten), Wiley-VCH Verlag GmbH, Weinheim, 755–789.

30. Savinell, R., Yeager, E., Tryk, D., Landau, U., Wainright, J., Weng, D., Lux, K., Litt, M., and Rogers, C. (1994) A polymer electrolyte for operation at temperatures up to 200 °C. *J. Electrochem. Soc.*, **141** (4), L46–L48.

31. Lassegues, J.C., Schoolmann, D., and Trinquet, O. (1992) Proton conducting acid polymer blends. *Solid State Ionics*, **51**, 443–448.

32. Schmidt, T.J. (2009) High-temperature polymer electrolyte fuel cells: durability insights, in *Polymer Electrolyte Fuel Cell Durability* (eds. F.N. Buchi, M. Inaba, and T.J. Schmidt), Springer, New York, 199–221.

33. Pefkianakis, E.K., Deimede, V., Daletou, M.K., Gourdoupi, N., and Kallitsis, J.K. (2005) Novel polymer electrolyte membrane, based on pyridine containing poly(ether sulfone), for application in high temperature fuel cells. *Macromol. Rapid Commun.*, **26**, 1724–1728.

34. Kallitsis, J.K., Geormezi, M., and Neophytides, S. (2009) Polymer electrolyte membranes for high-temperature fuel cells based on aromatic polyethers bearing pyridine units. *Polym. Int.*, **58**, 1226–1233.

35. Neyerlin, K.C., Singh, A., and Chu, D. (2008) Kinetic characterization of a Pt–Ni/C catalyst with a phosphoric acid doped PBI membrane in a proton exchange membrane fuel cell. *J. Power Sources*, **176**, 112–117.

36. Fukuta, K. (2011) Electrolyte materials for AMFCs and AMFC performance. Presented at the DOE AMFC Workshop, Washington, DC, 8 May 2011.

37. Kinoshita, K. (1992) *Electrochemical Oxygen Technology*, John Wiley & Sons, Inc., New York.

38. Nørskov, J.K., Rossmeisl, J., Logadottir, A., Lindqvist, L., Kitchin, J.R., Bligaard, T., and Jónsson, H. (2004) Origin of the overpotential for oxygen reduction at a fuel-cell cathode. *J. Phys. Chem. B*, **108**, 17886–17892.

39. Stamenkovic, V.R., Mun, B.S., Mayrhofer, K.J.J., Ross, P.N., and Markovic, N.M. (2006) Effect of surface composition on electronic structure,

stability, and electrocatalytic properties of Pt–transition metal alloys: Pt-skin versus Pt-skeleton surfaces. *J. Am. Chem. Soc.*, **128**, 8813–8819.

40. Stamenkovic, V., Mun, B.S., Mayrhofer, K.J.J., Markovic, N.M., Rossmeisl, J., Greeley, J., and Nørskov, J.K. (2006) Changing the activity of electrocatalysts for oxygen reduction by tuning the surface electronic structure. *Angew. Chem. Int. Ed.*, **45**, 2897–2901.

41. Wang, J.X., Inada, H., Wu, L., Zhu, Y., Choi, Y., Liu, P., Zhou, W.-P., and Adzic, R. (2009) Oxygen reduction on well defined core–shell nanocatalysts: particle size, facet and Pt shell thickness effects. *J. Am. Chem. Soc.*, **131**, 17298–17302.

42. Garsany, Y., Baturina, O.A., Swider-Lyons, K.E., and Kocha, S. (2010) Experimental methods for quantifying the activity of platinum electrocatalysts for the oxygen reduction reaction. *Anal. Chem.*, **82**, 6321–6328.

43. Kinoshita, K. (1990) Particle size effects for oxygen reduction on highly dispersed platinum in acid electrolytes. *J. Electrochem. Soc.*, **137**, 845–848.

44. Mukerjee, S. and McBreen, J. (1998) Effect of particle size on the electrocatalysis by carbon supported Pt electrocatalysts: an *in-situ* XAS investigation. *J. Electroanal. Chem.*, **448**, 163–171.

45. Nagaev, E.L. (1992) Equilibrium and quasi equilibrium properties of small particles. *Phys. Rep.*, **222**, 199–307.

46. He, Q., Yang, X., Chen, W., Mukerjee, S., Koel, B., and Chen, S. (2010) Influence of phosphate adsorption on the kinetics of oxygen electroreduction on low index Pt (*hkl*) single crystals. *Phys. Chem. Chem. Phys.*, **12**, 12544–12555.

47. Tanaka, A., Adzic, R., and Nikolic, B. (1999) Oxygen reduction on single crystal platinum electrodes in phosphoric acid solutions. *J. Serb. Chem. Soc.*, **64** (11), 695–705.

48. Kinoshita, K. (1988) *Carbon*, John Wiley & Sons, Inc., New York.

49. Watt-Smith, M.J., Rigby, S.P., Ralph, T.R., and Walsh, F.C. (2008) Characterization of porous carbon electrode materials used in proton exchange membrane fuel cells via gas adsorption. *J. Power Sources*, **184**, 29–37.

50. More, K. (2011) Characterization of fuel cell materials. Presented at the DOE Hydrogen Program Merit Review and Peer Evaluation, Washington, DC, 10 May 2011.

51. Uchida, H., Song, J.M., Suzuki, S., Nakazawa, E., Baba, N., and Watanabe, M. (2006) Electron tomography of Nafion ionomer coated on Pt/carbon black in high utilization electrode for PEFCs. *J. Phys. Chem. B*, **110**, 13319–13321.

52. Meyers, J.P. and Darling, R.M. (2005) Model of carbon corrosion in PEM fuel cells. *J. Electrochem. Soc.*, **153**, A1423.

53. Tang, H., Qi, Z., Ramani, R., and Elter, J.F. (2006) PEM fuel cell cathode carbon corrosion due to the formation of air/fuel boundary at the anode. *J. Power Sources*, **158**, 1306–1312.

54. Yu, P.T., Gu, W., Zhang, J., Makharia, R., Wagner, F.T., and Gasteiger, H.A. (2009) Carbon support requirements for highly durable fuel cell operation, in *Polymer Electrolyte Fuel Cell Durability* (eds. F.N. Buchi, M. Inaba, and T.J. Schmidt), Springer, New York, 29–53.

55. Cherstiouk, O.V., Simonov, A.N., Moseva, N.S., Cherepanova, S.V., Simonov, P.A., Zaikovskii, V.I., and Savinova, E.R. (2010) Microstructure effects on the electrochemical corrosion of carbon materials and carbon-supported Pt catalysts. *Electrochim. Acta*, **55**, 8453–8460.

56. Debe, M.K. (2003) Novel catalysts, catalysts support and catalysts coated membrane methods, in *Handbook of Fuel Cells – Fundamentals, Technology and Applications*, Vol. **3** (eds. M.W. Vielstich, A. Lamm, and H.A. Gasteiger), John Wiley & Sons, Ltd., Chichester, 576–589.

57. Debe, M.K., Schmoeckel, A.K., Vernstrom, G.D., and Atanasoki, R. (2006) High voltage stability of nanostructured thin film catalysts for PEM fuel cells. *J. Power Sources*, **161**, 1002–1011.

58. Debe, M. (2011) Advanced catalysts and supports for fuel cells. Presented at the

Annual Merit Review: DOE Hydrogen and Fuel Cells and Vehicle Technologies Programs, Washington, DC, 10 May 2011.

59. Kim, J., Lee, S.W., Carlton, C., and Shao-Horn, Y. (2011) Pt-covered multiwall carbon nanotubes for oxygen reduction in fuel cell applications. *J. Phys. Chem. Lett.*, **2**, 1332–1336.

60. Sun, S., Zhang, G., Geng, D., Chen, Y., Li, R., Cai, M., and Sun, X. (2011) A highly durable platinum nanocatalyst for proton exchange membrane fuel cells: multi-armed starlike nanowire single crystal. *Angew. Chem. Int. Ed.*, **50**, 422–426.

61. Gasteiger, H.A., Kocha, S.S., Sompalli, B., and Wagner, F.T. (2005) Activity benchmarks and requirements for Pt, Pt-alloy, and non-Pt oxygen reduction catalysts for PEMFCs. *Appl. Catal. B., Environ.*, **56**, 9–35.

62. Pivovar, B. (2011) Extended continuous Pt nanostructures in thick, dispersed electrodes. Presented at the DOE Annual Merit and Peer Evaluation Meeting, Washington, DC, 10 May 2011.

63. Antolini, E. and Gonzalez, E.R. (2009) Ceramic materials as supports for low-temperature fuel cell catalysts. *Solid State Ionics*, **180**, 746–763.

64. Liu, X., Chen, J., Liu, G., Zhang, L. Zhang, H., and Yi, B. (2010) Enhanced long-term durability of proton exchange fuel cell cathode by employing Pt/TiO_2/C catalysts. *J. Power Sources*, **195**, 4098–4103.

65. Elezovic, N.R., Babic, B.M., Gajic-Kristajic, L., Radmilovic, V., Kristajic, N.V., and Vracar, L.J. (2010) Synthesis, characterization and electrocatalytic behavior of Nb − TiO_2/Pt nanocatalyst for oxygen reduction reaction. *J. Power Sources*, **195**, 3961–3968.

66. Ioroi, T., Senoh, H., Yamazaki, S., Siroma, Z., Fujiwara, N., and Yasuda, K. (2008) Stability of corrosion-resistant Magneli-phase Ti_4O_7-supported PEMFC catalysts at high potentials. *J. Electrochem. Soc.*, **155**, B321–B326.

67. Haruta, M. (1997) Size- and support-dependency in the catalysis of gold. *Catal. Today*, **36**, 153–166.

68. Hayden, B.E. and Suchsland, J.-P. (2009) Support and particle size effects in electrocatalysis, in *Fuel Cell Catalysis* (ed. M.T. Koper), John Wiley & Sons, Inc., Hoboken, NJ, 567–592.

15
Catalysis in Low-Temperature Fuel Cells – an Overview

Sabine Schimpf and Michael Bron

15.1
Introduction

Fuel cells are devices that convert the chemical energy stored in a fuel and an oxidant into electrical energy. This conversion is associated with an electrocatalytic reaction occurring at both the anode (oxidation of the fuel) and the cathode (reduction of the oxidant), during which electrons are released or consumed, respectively. The electrons may, in an electrical circuit connected to the electrodes, drive electrical devices, while an ion-conducting separator (e.g., a membrane) between the electrodes closes the electrical circuit. The driving force for the electrons is the potential difference between the (in this case) negative anode and the positive cathode.[1] The electrode potentials are, in turn, defined by the thermodynamics of the individual electrode reactions, as described by the Nernst equation [1]. However, as for any chemical process, while the equilibrium state is defined by thermodynamics, for the nonequilibrium case (i.e., in the fuel cell the situation where electricity is withdrawn from the system) very often kinetics, that is, the rates of the individual processes, determine the overall situation. Hence the kinetics of both the anode and the cathode reaction may play a major role in determining the overall fuel cell performance. Both reactions are associated with electrocatalysts, which have to be optimized for the given reaction, type of fuel cell, conditions, and so on. This chapter describes the state-of-the-art in the field of the development of electrocatalysts for low-temperature fuel cells together with the challenges and problems with the individual reactions. Recent trends are highlighted.

Although for a long period the focus of research and development in low-temperature fuel cells was on polymer electrolyte membrane (PEM)-based systems working with a solid acid electrolyte (at least for the intended daily-life applications such as cars and portable electronics), recently there has been a clear tendency towards the development of alkaline fuel cells (AFCs) [2–4], which was

---

1) During electrolysis, the terms negative and positive refer to the cathode and anode, respectively. By definition, the anode is the electrode, where negative charge carriers transfer from solution to the electrode (or positive charge carriers from the electrode into the solution), and the cathode vice versa.

Fuel Cell Science and Engineering: Materials, Processes, Systems and Technology,
First Edition. Edited by Detlef Stolten and Bernd Emonts.
© 2012 Wiley-VCH Verlag GmbH & Co. KGaA. Published 2012 by Wiley-VCH Verlag GmbH & Co. KGaA.

triggered by, among others, the development of anion-conducting membranes that avoid the so-far used liquid electrolytes, for example aqueous KOH, which are prone to reactions such as carbonate formation and need high-purity gases at both the cathode and the anode. From the viewpoint of electrocatalysis, AFCs have the clear advantage that several electrocatalytic reactions appear to be catalyzed much more easily, that is, with much lower kinetic barriers. Furthermore, stability issues make the use of less expensive catalyst materials feasible in alkaline fuel cells. Given this, a short overview of recent developments in catalysts for AFCs is also presented.

However, electrocatalytic reactions are not the only process in fuel cells, where catalysis plays a role. Equally important, at least as long as hydrogen is obtained via traditional means employing fossil fuels, is fuel processing. Although the principles of the underlying reactions have been investigated for many years, research in this field is still ongoing, especially in relation to fuel cells. Catalytic aspects of these reactions are also covered.

The aim of this chapter is not to present a comprehensive literature survey, which, with such a broad range of topics as discussed here, would be a too laborious task. Rather, the intention is to summarize general ideas and principles and refer the interested reader to recent reviews or papers, where available, to allow them to deepen their knowledge further in the respective field.

15.2
Electrocatalysis in Fuel Cells

As described above, fuel cells convert chemical energy into electrical energy by oxidizing a fuel at the anode of a fuel cell and reducing an oxidant at the cathode. Ions [e.g., protons in the case of a polymer electrolyte membrane fuel cell (PEMFC)] migrate through the membrane separating the electrodes, while these electrodes are electrically connected through an external circuit including the consumer load. The reactions for a typical hydrogen–oxygen fuel cell are given by the following equations:

$$\text{Anode: } H_2 \xrightarrow{\text{electrocatalyst}} 2H^+ + 2e^- \tag{15.1}$$

$$\text{Cathode: } O_2 + 4e^- + 4H^+ \xrightarrow{\text{electrocatalyst}} 2H_2O \tag{15.2}$$

Both oxidation and reduction require an electrocatalyst to proceed at a satisfactory rate. This electrocatalyst, as in classical heterogeneous catalysis, interacts with the reactants to allow favorable reaction pathways with low activation barriers; however, a charge transfer must also occur. Hence it becomes clear that in addition to interacting with the reactants, the electrocatalyst must be in electrical contact with the electrodes on which the catalyst is deposited. Furthermore, ions (H^+, OH^-) which are released at one electrode and consumed at the other must diffuse/migrate to and from the active sites, hence the presence of the electrolyte is also required.

15.2 Electrocatalysis in Fuel Cells

This so-called three-phase boundary between the electrode, mostly solid electrolyte, and gas phase is rather complex.

Many processes occur in this three-phase boundary or in the overall fuel cell which may limit the overall performance. From a simple electrochemical point of view, the cell voltage E is associated with the fuel cell efficiency η as described by the following very simple equations; for a more detailed treatment of this topic and a discussion of other efficiency issues, the reader is referred to the literature [5]:

$$\Delta_R G = -nFE_0 \qquad (15.3)$$

$$\eta_{th} = \frac{\Delta_R G}{\Delta_R H} = -\frac{nFE_0}{\Delta_R H} \qquad (15.4)$$

where $\Delta_R G$ is the Gibbs free energy, n the number of electrons transferred, and $\Delta_R H$ the reaction enthalpy. The subscript "th" denotes the theoretical value. E_0 denotes the equilibrium cell voltage, and thus the difference between two electrode potentials which are in equilibrium, that is, in the case when no current is flowing. As soon as electricity is withdrawn from the system, the electrode potentials shift and the cell voltage and consequently the electrical efficiency of the system become smaller:

$$\eta_{el} = -\frac{nFE}{\Delta_R H} \qquad (15.5)$$

where the subscript "el" denotes the electrical efficiency, and E without a subscript is the actual measured cell voltage.

Thus, any decrease in cell voltage accounts for the efficiency loss of the overall system. There are various issues that might lead to a decrease in cell voltage, including mass transfer, that is, limited transport of fuel to or water from the active sites, and ohmic losses within the electrode [5]. However, another important cause of a decrease in cell voltage is the so-called overpotential, which denotes the difference between the equilibrium potential of the electrode and its potential during electrocatalytic reaction. It is important to note, however, that the overpotential is not a fixed quantity that can be used to compare different catalysts in different studies, but depends on many factors; the most important one being the specific current density. The cause of the overpotential is a kinetic limitation. There are various processes that lead to a kinetic limitation of the electrode reaction and thus to an unfavorable shift in electrode potential. These include electrocatalyst poisoning, too weak or too strong adsorption of reaction intermediates, and, in particular, charge-transfer limitations, that is, a kinetic limitation of the charge transfer through the phase boundary, which is characteristic of electrocatalytic reactions and is typically described by the Butler–Volmer equation [1]. The various limiting factors are characteristic of the individual electrode reactions and will be discussed there. Figure 15.1 summarizes the individual losses occurring from the viewpoint of electrochemistry in a fuel cell.

Figure 15.1 indicates that mass transfer, the three-phase boundary, ohmic losses, and so on, are of relevance for the overall fuel cell performance and have to be optimized. However, it also becomes clear that the most demanding challenge

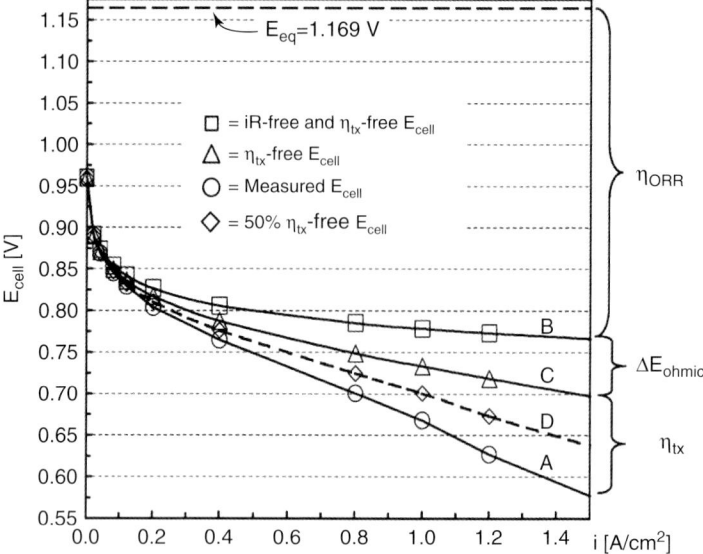

Figure 15.1 Representation of the individual voltage losses occurring in a PEMFC. (A) 50 m² single-cell H₂–air performance at $T_{cell} = 80\,°C$. (B) E_{cell} versus i for the mass transport-free and ohmically corrected (i.e., iR-free) $E_{cell}-i$ curve shown in A. (C) Addition of the ohmic losses, ΔE_{ohmic}, to the polarization curve shown in B. (D) $E_{cell}-i$ curve shown in A corrected for 50% of the mass transport losses. Reprinted from [6], where more experimental details can be found, with permission from Elsevier. © Elsevier 2005.

in electrocatalysis is still to develop catalysts that catalyze the respective reaction at a high rate at an electrode potential which is close to the thermodynamic (equilibrium) potential of the electrode reaction, leading to a high cell voltage and thus a high electrical efficiency. This is in particular true for the oxygen reduction reaction (ORR) (see Figure 15.1), but is also of relevance for, for example, methanol oxidation and others, as discussed below. In the following, the relevant fuel cell reactions are discussed in the light of these issues.

15.2.1
Oxygen Reduction in PEMFCs

The ORR [Eq. (15.2)] is considered one of the major challenges in electrocatalysis, from both fundamental and applied points of view. Some recent reviews have summarized the state-of-the-art and actual developments in this field [6, 7]. Despite extensive research, the ORR is not well understood, and practical oxygen reduction catalysts and electrodes for fuel cells experience a large overpotential for this reaction, that is, a kinetic barrier, which contributes a voltage loss of ∼25% (see above). One of the reasons for the sluggishness of this reaction is that four electrons have to be transferred and also four protons added for the complete reduction of oxygen to water in an acidic electrolyte. Hence the ORR exhibits

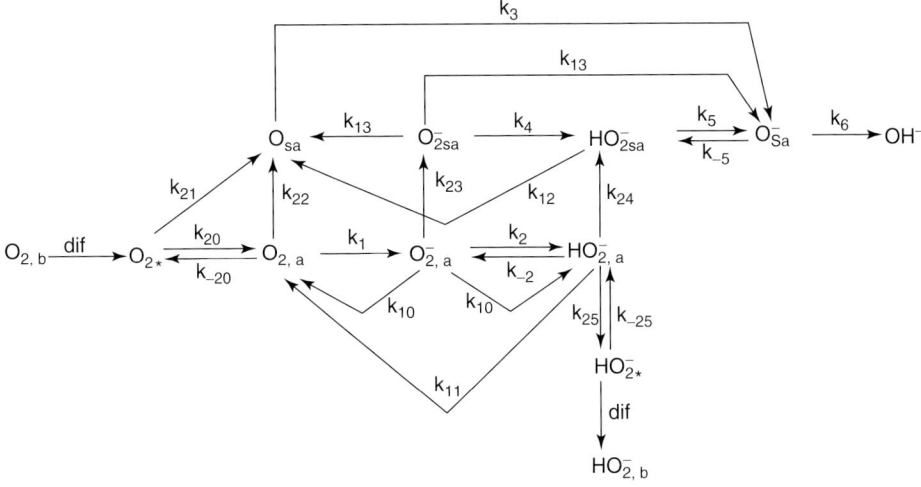

Figure 15.2 General scheme of oxygen reduction. Reprinted from [8] with permission from Elsevier. © Elsevier 1987.

a complex reaction network with various possible pathways and a significant number of elementary steps [8], several of which might form the rate-determining reaction. Furthermore, the incomplete reduction of oxygen to hydrogen peroxide in a two-electron, two-proton reaction is possible, which, from the viewpoint of fuel cells, is undesired since it exhibits a much less positive electrode potential of 0.68 V vs. SHE (standard hydrogen electrode) under standard conditions, leading to further efficiency losses. Furthermore, H_2O_2 has been claimed to be responsible for fuel cell catalyst corrosion (see below) and membrane degradation [9]. Mechanistic studies on oxygen reduction at metal surfaces have been reported and the mechanism depends on the pH and electrode material [10, 11]. Figure 15.2 demonstrates the possible steps during oxygen reaction in the case of an alkaline electrolyte.

15.2.1.1 Platinum-Based Catalysts

The state-of-the-art in oxygen reduction catalysts for PEMFCs and the goals to be achieved for fuel cell commercialization were summarized by Gasteiger et al. in 2005 [6]. The standard catalyst for this application is Pt supported on carbon black, which catalyzes the ORR at a reasonable rate; however, despite being the most active single metal for the ORR in acidic solutions, it still exhibits large overpotentials. If a catalytic reaction is slow, a simple idea to enhance the overall kinetics is to use more catalyst, that is, to provide more active sites to obtain a higher turnover. However, Pt is expensive and rare. The worldwide estimated Pt resources are 40 000 t, and the amount of Pt necessary for an 85 kW stack is about 72–94 g [6], that is, $\sim 1\,g\,kW^{-1}$, with the target of $<0.2\,g\,kW^{-1}$. Ignoring the fact that there are many other applications in which Pt plays an important role, such as catalytic conversions in the chemical industry (e.g., during nitric acid production, cancer treatment pharmaceuticals, and others, leading to further limited availability and enhanced price of Pt), the estimated number of cars to be equipped with PEMFCs containing 1 g kW^{-1} of Pt is about 400 million, containing Pt worth about US$5400, assuming a price per troy ounce (31.1 g) as 1800 US$ as of August 2011, according to Johnson Matthey (*http://www.platinum.matthey.com/*). From these numbers, it becomes clear that it is essential to reduce the amount of Pt in a fuel cell stack at least down to the demanded 0.2 g kW^{-1} [6] or even further.

A decrease in the Pt loading could in principle be achieved by decreasing the particle size of the Pt in the cathode. Although this would lead to an increase in specific surface area, the situation is complicated by the fact that the ORR on Pt exhibits a particle size effect: The surface specific activity of Pt decreases as the particles become smaller [12–14]. This effect is probably related to the increased adsorption of oxygen species or OH blocking surface sites [13, 15]; however, it has the consequence that an increase in the active surface area by reducing the particle size is overcompensated by the decrease in specific activity with particle sizes smaller than ~4 nm. Furthermore, it is well known that structural changes in Pt particles may occur under conditions relevant to fuel cells (see below). Therefore, finding and preparing the optimum particle size also requires them to be stabilized in an operating fuel cell [16].

Various strategies have been put forward in recent years to overcome the kinetic limitations discussed above and at the same time reduce the amount of Pt in a fuel cell stack. Two principle ideas can be identified: the use of less expensive catalyst materials, where probably a larger amount of catalyst could be used, or the development of more active materials, which catalyze oxygen reduction at lower overpotentials, thus holding the promise that high performance can be achieved with less catalyst. Several approaches combine these ideas: Pt alloy catalysts and core–shell systems.

Alloying of Pt to prepare catalysts with favorable properties was an early issue during phosphoric acid fuel cell development, as reviewed in [6], and continues to be an active field of research in PEMFCs [6, 7, 17–19]. Pt alloy catalysts have been shown to exhibit higher mass specific activities for the ORR than Pt-only catalysts, with PtCo-based systems being the most active [20, 21]. Leaching of the non-noble metal from the Pt alloy has been considered an important issue [22]. Leached ions may poison the membrane by blocking proton-conducting sites and leaching may change/reduce the activity of the overall fuel cell system. However, it has been shown that structures form where a thin Pt layer surrounds an alloy nanoparticle in core–shell particles [23]. In addition to others, two main reasons for enhanced activity of alloys and core–shell nanoparticles have been put forward: modification of the electronic structure of Pt by the alloying metal, namely a shift in the d-band center, and also geometric effects, that is, differences in interatomic distances, in the alloy compared with the pure metal, with the former being the favored explanation. Figure 15.3 demonstrates so-called volcano plots correlating the catalytic activity of several Pt alloys with their d-band center position for so-called "Pt-skin" and "Pt-skeleton" catalysts, which differ in the surface atomic arrangement [21].

Electrochemical leaching of Pt alloy catalysts has been deliberately applied to form core–shell materials with enhanced activities, either in half-cell tests [24, 25] or even

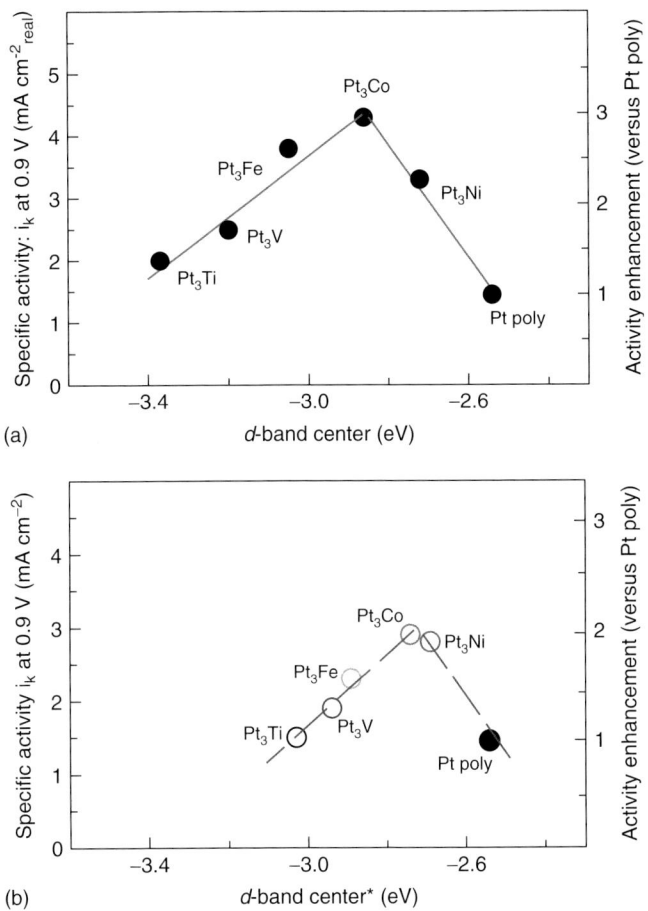

Figure 15.3 Relationships between the catalytic properties and electronic structure of Pt$_3$M alloys. Relationships between experimentally measured specific activity for the ORR on Pt$_3$M surfaces in 0.1 M HClO$_4$ at 333 K versus the d-band center position for (a) the Pt-skin and (b) the Pt-skeleton surfaces. Reprinted by permission from Macmillan Publishers Ltd: Nature Mat. [21] copyright 2007.

employing a membrane electrode assembly (MEA) [26]. An enhancement of activity of up to fivefold, depending on the starting alloy, has been observed [26]. Figure 15.4 demonstrates schematically the process of dealloying and core–shell formation. Recent preparation strategies have been aimed at the deliberate formation of core–shell structures by electrochemical preparation strategies. The latter include the underpotential deposition of a monolayer of copper on a core nanoparticle and the subsequent redox exchange of Cu by Pt [27], and the direct electrochemical deposition of Pt around metal nanoparticles [28, 29]. Pt monolayer catalysts seem to be a fairly promising alternative in terms of cost issues, since they should lead to the optimum Pt utilization. Furthermore, very promising activity data

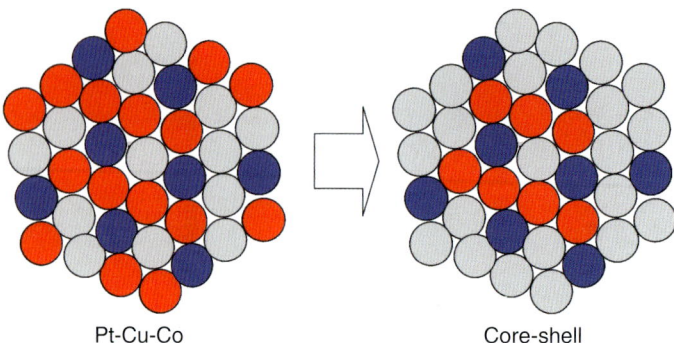

Figure 15.4 Schematic representation of the formation of a Pt-enriched core–shell alloy nanoparticle by voltammetric dealloying of a Cu-rich alloy precursor (Pt, gray; Cu, red; Co, blue). Reproduced with permission from [25] copyright Wiley-VCH Verlag GmbH & Co. KGaA.

have been reported. On the other hand, some doubts remain regarding stability, in particular given the already mentioned structural changes of catalysts under working conditions.

There are several parameters for Pt-based oxygen reduction catalysts that need be tuned in order to obtain optimum activity, including particle size, alloy formation/degree of alloying, and core–shell effects. Although these seem to be rather divergent issues, recently one common explanation for the various dependencies has emerged. It appears that electronic/d-band effects, which influence the strength of OH adsorption, are the cause of the variation in ORR activity. Strong adsorption of OH on Pt leads to site blocking [13, 17, 30] and thus reduced ORR activity. This Pt–OH formation and reduction is potential dependent and the less positive the potential for formation/reduction, the less active the catalyst is. It has been shown that OH adsorption is reduced on alloys and skin-layer catalysts [20, 21, 31] and also larger Pt particles [15]. A too weak OH adsorption, on the other hand, leads to weak oxygen activation, leading to volcano plots as presented in Figure 15.3. However, it should be kept in mind that several other explanations for ORR activity dependence on Pt-based materials have been put forward and it is outside the scope of this chapter to discuss these critically.

Another parameter to be tuned is the shape of the nanoparticles. Since the activity of Pt nanoparticles in the ORR depends on particle size, it is obvious that different single-crystal surfaces [32] and also corners, edges, and defects in the nanoparticle behave differently in the ORR, owing to the different strengths of OH adsorption at these different sites. Hence it should be possible to tune the ORR activity by manifesting certain particle shapes with preferential orientation of the individual exposed single-crystal planes together with the optimum presence of edges and corners. Indeed, such efforts have been reported and preferential particle shapes have been found [33]. Although these studies are important from a fundamental point of view, doubts remain that it ever will be possible to stabilize these nanoparticle shapes under conditions relevant to fuels cell (see below).

15.2.1.2 Non-Platinum Catalysts

Given the limited resources and the high cost of Pt, it is no surprise that many other materials have been tested as prospective oxygen reduction catalysts for fuel cells, which should provide high activity and low overvoltage for the ORR combined with long term-stability, as recently reviewed [34]. A wealth of papers in this area discuss transition metal chalcogenides [35–38], as first proposed by Alonso-Vante and Tributsch [39], and also carbon materials modified with nitrogen or transition metal–nitrogen complexes ("Me/N/C catalysts," with Me being mainly Fe and Co; see below, however), which initially were derived from macrocyclic transition metal complexes, but have since developed in various directions [40]. For both kinds of catalysts, there is an ongoing debate in the literature about the active sites and the mechanism for the ORR on these materials, and it is outside the scope of this chapter to discuss the various points of views. However, a short overview of recent developments in this field is given in the following. In addition, other noble metals have been also investigated as ORR catalysts in acidic electrolytes.

15.2.1.3 Platinum-Free Noble Metal Catalysts

Among the noble metals tested for oxygen reduction in acidic electrolytes are Ag, Au, Pd, Rh, Ir, and Ru [34, 41, 42]. However, none of them was able to compete with Pt in terms of activity, and it is still a matter of debate whether Au nanoparticles reduce oxygen in a two- or a four-electron reduction. However, recently some Pd alloys have been shown to exhibit activities similar to those of Pt. Led by thermodynamic guidelines, Bard and co-workers prepared arrays of catalyst spots of bi- and multimetallic catalysts consisting of Pd, Au, Ag, Co, Ti, V, and Mn and screened their ORR activity by scanning electrochemical microscopy (SECM) [43–45]. PdCo and PdV proved to be nearly as active as Pt.

15.2.1.4 Metal/N/C Catalysts

Jasinski discovered in 1964 that Co phthalocyanines are highly active for the electrocatalytic reduction of oxygen [46]. In subsequent studies, it was found that other macrocyclic N_4–chelates with other central metal ions such as Fe are also active oxygen reduction catalysts and that heat treatment of carbon-supported macrocycles had a positive influence on stability and also activity [47, 48]. It was shown that the heat-treatment temperature and the gas atmosphere, among others, are important conditions in obtaining high activity and high stability, and the large amount of literature in this field has been reviewed [38, 40, 49]. However, the positive effect of heat-treatment temperatures up to 950 °C, where decomposition of the macrocycle is expected, led to the suspicion that it should be possible to prepare similarly active catalysts starting from much simpler precursors than macrocyclic transition metal complexes. Indeed, active catalysts can be prepared that combine a carbon support (which might be formed also *in situ*), a transition metal salt, and a nitrogen source in a heat-treatment step [50], and such diverse nitrogen precursors as nitrogen-containing polymers [51, 52], ammonia [53], acetonitrile [54], and phenanthroline [55] have been employed. This active field of research has been reviewed recently by Jaouen *et al.* [56]. For a long time, it appeared that the activity of such catalysts is insufficient to replace Pt in fuel cell applications and was only about one-tenth of that of the latter. As pointed out by Gasteiger *et al.*, it is not an option simply to prepare thicker gas diffusion electrodes (GDEs) with 10 times higher catalyst mass, since mass transfer losses and conductivity problems within the electrodes then become an issue [6]. Recent reports in this field now indicate that activities high enough to be competitive for fuel cell applications can indeed be achieved [51, 57]. However, obviously stability and activity are not independent of each other and high activity often comes along with low stability [56]. The future will show if these interesting materials will find their way into commercial fuel cell applications.

The active site of the above materials is still not resolved beyond doubt, but there are strong indications that an iron ion coordinated to nitrogen ions is the structure responsible for oxygen reduction [56, 57]. Figure 15.5 displays a sketch of the active site as it is also claimed to be located in micropores [57]. It has been demonstrated that at least two different kinds of active sites having Fe in different coordination spheres exist, and that these sites perform differently with regard to activity and stability. The uncertainty regarding the active sites had its cause, amongst other reasons, in several reports that claimed that metal-free nitrogen–carbon compounds are also active electrocatalysts, especially in alkaline solution [58, 59]. Active nanostructures consisting of carbon and nitrogen [carbon nanofibers and carbon nanotubes (CNTs)] have been prepared and found to be active for the ORR [60–62]. Activity has been correlated with edge plane exposure, the presence of pyridinic N centers, and the basicity of the nitrogen atoms. The transition metal in this view acts mainly as a catalyst for nanostructure formation, including enhanced defect density. These materials, although sometimes discussed in connection with the above metal/N/C materials, should be better considered

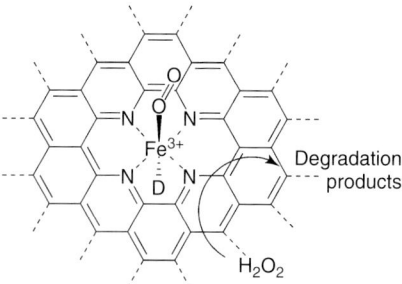

Figure 15.5 Proposed model for the structure of the active site of Fe/N/C catalysts. Reprinted with permission from Schulenburg *et al.*, J. Phys. Chem. B 2003, 107, 9034–9041. © 2003 American Chemical Society.

as their own particular type of catalysts and should be denoted carbons [7] or nitrogen-modified carbons.

15.2.1.5 Transition Metal Chalcogenides

Transition metal chalcogenides for the ORR initially stem from research on Chevrel phases for photoelectrocatalysis and have been found to reduce oxygen readily under conditions relevant to fuel cells [35–38]. The most often used transition metal for such catalysts is Ru; however, other metals such as W, Os, and Rh have also been employed. The early syntheses of the novel catalyst materials were based on the thermal decomposition of molecular carbonyl clusters of the respective transition metals in the presence of other catalyst constituents by refluxing in an appropriate organic solvent such as xylene [63]. However, other synthesis methods have also been employed, including colloidal synthesis [64, 65]. Although also in this field several different views on the reasons for activity and the nature of the active site have been published, there are strong indications that the main action of the chalcogenide compound is to provide electronic stabilization for the active transition metal cluster or nanoparticle, which would otherwise be oxidized, as demonstrated for Ru_xSe_y catalysts [66]. Furthermore, an electronic modification of the active Ru centers to form narrow, highly occupied d-states responsible for high catalytic activity has also been claimed. The different views were discussed by Lee and Popov [36]. The activity obtained with transition metal chalcogenides is generally lower than that of state-of-the-art Pt-based catalysts with similar characteristics. However, it has been shown that transition metal chalcogenides are inert towards methanol and therefore may outperform Pt-based materials in direct methanol fuel cells (DMFCs) (see below).

15.2.2
Oxygen Reduction in Other Low-Temperature Fuel Cells

15.2.2.1 Direct Fuel Cells

As discussed above, a major driving force to replace or modify Pt as a catalyst for PEMFCs is high cost and low abundance. However, there is one other important reason why Pt may be not the material of choice, which comes into play in DMFCs. Here, methanol may cross over from the anode side to the cathode, with the result that a mixed potential is formed, leading to a lowering of the cell voltage and efficiency losses. Furthermore, methanol may form CO at the cathode, leading to catalyst poisoning. Strategies to overcome these issues include the development of novel membranes that are impermeable to methanol. However, another strategy is to develop catalysts that are insensitive or tolerant towards methanol, that is, which do not catalyze methanol oxidation and therefore do not suffer from mixed potentials or from poisoning effects. Particularly the above-discussed transition metal chalcogenides have been shown to perform well in DMFCs, with a performance superior to that of Pt [37]. Similar results have been obtained for metal/N/C catalysts [49] and with nanostructured nitrogen-modified carbons [67]. However, noble metal-based catalysts were also investigated with regard to methanol tolerance, including Pt-based materials [68].

15.2.2.2 Alkaline Fuel Cells

In the literature, it is often claimed that the ORR in alkaline media proceeds at a relatively high rate at reasonably low overpotentials compared with acidic solutions. However, as detailed by Spendelow and Wieckowski, a detailed analysis of the data presented in the literature places doubt on these statements and the ORR on Pt in alkaline solution seems to be similar to that on Pt in acidic solution [69]. Other Pt group metals, including Pd, Ir, and Ru, have also been tested but none of them was able to compete with Pt in terms of activity [2, 69]; however, Pd was claimed to be fairly active [41]. As with acid fuel cells, alloying of Pt with a second or third metal has also been applied in an attempt to enhance activity.

However, one great advantage of less concentrated alkaline solutions is the enhanced stability of many metals. This leads to the favorable situation that other, much less expensive, materials than Pt can be employed, including Ag and Ni [2, 69].

Again, the situation is different if organic fuels instead of hydrogen are used at the anode. As with PEMFCs, methanol or other organic fuels may penetrate through the membrane to the cathode side leading to catalyst poisoning and mixed potential formation. Here, Pd-based catalysts showed higher methanol tolerance than Pt-based catalysts [2]. High methanol tolerance combined with good activity has also been found for Ag-based materials [2].

As with PEMFCs, efforts have been done to replace metal catalysts completely with non-metal catalysts or at least by non-precious metal catalysts. Amongst these are, as for PEMFCs, macrocyclic complexes of porphyrins and phthalocyanines, and materials derived from these via heat-treatment steps. Very promising results, however, have recently been obtained using nitrogen-modified CNTs and carbon nanofibers [58, 59]. Here, activities close or similar to that of Pt were observed. However, the stability of these materials in a long-term durability study has still to be demonstrated.

15.2.3
Hydrogen Oxidation and CO Poisoning

Hydrogen is considered as the most promising fuel for low-temperature fuel cells. The electrocatalytic oxidation of H_2 [Eq. (15.1)] at the state-of-the-art anode catalyst, Pt, is rather facile and the overvoltage and hence the efficiency loss introduced by this reaction are negligible even at high current densities. Therefore, the amount of platinum needed for a PEMFC employing pure hydrogen, which might be produced, for example, by water electrolysis using renewable energy, is low ($\sim 0.05\,\mathrm{g\,cm^{-2}}$ [6]). Indeed, over potentials between 5 and 30 mV have been estimated for such low catalyst loadings, with the former value being the more likely [70]. However, a hydrogen economy is facing several challenges, including hydrogen storage, distribution, and transportation. Furthermore, as discussed in

Section 15.6, hydrogen, rather than being produced by electrolysis, is mainly produced from fossil fuels and therefore contains fairly high concentrations of CO and other gases, which may act as severe catalyst poisons, reducing the cell efficiency drastically. Therefore, it is necessary to reduce the CO concentration to values far below 50 ppm, adding further shift and preferential oxidation of carbon monoxide (PROX) stages to the hydrogen production facility, which increases the system complexity and lowers efficiency, and this effect is more severe in the case of on-board hydrogen generation. Although new developments in the field of high-temperature PEMFCs [71] may help to reduce the adverse effect of CO poisoning, stability issues are amplified in high-temperature PEMFCs. Hence there is a demand for novel electrocatalyst that are less prone to poisoning. Typically, bi- or multimetallic catalysts are discussed for these applications [18, 19]. The most established combination for hydrogen oxidation in the presence of CO is PtRu, but other combinations have also been tested, including PtSn, PtMo, PtW, and PtNi in varying atomic ratios, and ternary combinations [18, 19]. Two main mechanisms have been put forward to help to explain the enhanced CO tolerance of the catalysts: ligand effects, also referred to as electronic effects [72, 73], and a bifunctional mechanism [74, 75]. According to the bifunctional mechanism, CO is strongly adsorbed at Pt surface sites of the catalyst. To oxidize this poisoning CO to CO_2, oxygen-containing species are needed, which form at neighboring sites of the second metal (e.g., Ru) at much less positive potentials than at Pt, thus helping to liberate the poisoned sites and making them available for H_2 oxidation again. In the electronic effect, on the other hand, the electronic influence of the second metal on the electronic (d-band) structure of Pt is considered. Recent density functional theory (DFT) calculations point to the ligand effects as the main mechanism responsible for enhanced CO tolerance [76].

Furthermore, for Pt-only catalysts, particle size effects in CO oxidation have been observed with CO monolayers being more readily oxidized using larger particles [77, 78], whereas smaller particles favor CO bulk oxidation [13, 77]. These particle size effects indicate the structure sensitivity of the CO oxidation, as also found in surface science studies [32, 79]. The enhanced CO bulk oxidation on smaller than larger particles has been ascribed to a higher energy of adsorption of OH, which enhances the catalytic CO oxidation [13, 77]. On the other hand, with CO monolayer formation the strong CO adsorption on smaller particles may hinder CO_{ad} oxidation and thus reduce activity. The influence of particle shapes on CO oxidation has also been investigated [33].

In addition to Pt, recently Pd has been discussed as an anode catalyst for hydrogen and direct fuel cells in addition to oxygen reduction [80] (see also above). Pd, being chemically similar to Pt, is cheaper and much more abundant. The activity of Pd for hydrogen oxidation is relatively low, but the addition of small amounts of Pt enhances the Pd activity to a value similar to that of Pt, and PtPd catalysts with larger proportions of Pt are even less prone to CO poisoning than PtRu.

15.2.4
Catalysis in Direct Fuel Cells

Direct fuel cells are devices that directly convert the chemical energy stored in carbon-containing chemicals into electrical energy by oxidizing them at the anode of the fuel cell without prior conversion into hydrogen, as will be discussed in Section 15.6. Fuels suitable for direct conversion in low-temperature fuel cells typically contain oxygen atoms and include C_1–C_3 mono- and polyhydric alcohols and formic acid.

The archetypical direct fuel cell is the DMFC. Like the PEMFC, the DMFC uses a proton-conducting membrane to separate the anode and cathode, and protons liberated during electrocatalytic methanol oxidation [Eq. (15.6)] at the anode are involved in oxygen reduction at the cathode. However, whereas in the hydrogen fuel cell the anode reaction is straightforward, the methanol oxidation is comparably sluggish, which is mainly attributed to poisoning effects.

$$CH_3OH + H_2O \longrightarrow CO_2 + 6H^+ + 6e^- \tag{15.6}$$

Efficient catalysts for methanol oxidation, similarly to CO oxidation, therefore have to provide oxygenated species at the electrode surface to help oxidize the poisoning species [74]. Hence it is no surprise that similar catalyst compositions are chosen for both CO and methanol oxidation, with PtRu being the most promising and most frequently investigated combination [81].

Another frequently discussed fuel for direct fuel cells is ethanol [82], which has the advantage of being readily available via fermentation from biomass, such as sugar cane and sugar beet or more recently also from lignocelluloses. In contrast to methanol, complete ethanol oxidation requires the cleavage of a C–C-bond. Hence different catalyst compositions turned out to be active for this reaction and the most promising combination to date is PtSn [82]. However, as with methanol, the slow oxidation kinetics will impede the application of direct ethanol fuel cells, if ever commercialized, in high-power applications such as cars. The influence of particle shapes on the oxidation of methanol and ethanol has also been investigated [33, 83].

Alcohol oxidation is much more facile in alkaline media. Although H_2/O_2 fuel cells using a liquid electrolyte such as highly concentrated KOH solution were employed successfully during the Apollo and Gemini space programs, they were not well suited for daily life applications owing to the corrosive electrolyte, the necessity to use very pure gases and the resulting high costs. However, the recent development of anion-exchange membranes for alkaline fuel cells has revived interest in the latter [2–4]. The great advantage of alkaline fuel cells is, as already mentioned, the much more facile electrode kinetics at the anode, leading to relatively low overpotentials, reduced poisoning and hence smaller efficiency losses. Additionally, there is the hope that much less expensive catalysts can be employed in the future. Recent developments in this field have been reviewed [2]. Currently, a variety of materials are being considered for oxidation (for the ORR see above). As above, bimetallic catalysts of Pt in combination with Ru, Mo, Sn, Re, Os, Rh, Pb, and Bi have been tested for alcohol oxidation in alkaline media [2], showing less poisoning. Additionally, non-Pt materials have also been investigated, with much activity in the field of Pd-based monometallic and multimetallic materials [80].

15.3
Electrocatalyst Degradation

The suitability of an electrocatalyst for application in PEMFCs is not only defined by its activity, but also by its stability under practical conditions. These conditions include the highly corrosive environment of strong acids and the strongly oxidizing conditions due to pressurized oxygen or air at elevated temperatures of up to or higher than 100 °C in a humid atmosphere. Catalyst degradation might become an even more urgent issue for future high-temperature PEMFCs working at 150 °C or even higher [71]. Furthermore, during fuel cell operation, additional effects such as fuel starvation at the anode [84] may enhance irreversible catalyst damage and degradation. Hence it is important to conduct fuel cell degradation studies employing realistic MEA arrangements, preferably using appropriate *in situ* techniques (see below). However, basic principles on catalyst aging can also be unraveled using half-cell measurements with aqueous electrolytes.

Various mechanisms have been identified that might lead to fuel cell catalyst degradation and concomitant loss of activity [9]. These include particle growth [16] (loss of electrochemically active area) by agglomeration [85] and Ostwald ripening [86], and also loss of Pt which might occur via Pt dissolution [87] and subsequent deposition in the membrane (chemical reduction by crossover hydrogen) [86, 88], precipitation in the cathode ionomer phase [86], or detachment of whole Pt particles from the support as observed in half-cell measurements employing liquid electrolytes [89]. It is not surprising that these degradation effects depend on the conditions, for example, humidity and oxygen partial pressure [90].

Nanostructural changes in alloy particles also play a role [23, 91]. Interestingly particle growth is not necessarily associated with losses in activity [85]. However, from these findings, it becomes very clear that preparing a catalyst with high activity is only part of the story. Equally important is to find measures, if available, to stabilize these particles under reaction conditions. In addition to these effects leading to physical changes of the catalyst, chemical effects of degradation also have to be considered, including irreversible poisoning by fuel contaminants. Furthermore, support corrosion plays a major role [92, 93]. Indeed, carbon is thermodynamically unstable at the typical cathode potentials that occur in fuel cells. Carbon corrosion is inevitably correlated with changes in the active material.

Various parameters have been identified that influence carbon corrosion, including humidity, with higher humidity enhancing corrosion [94], reactant partial pressure, fuel starvation, and also the carbon characteristics, including surface area and degree of graphitization [94]. As expected, both potential and variations in potential have also a strong influence on carbon corrosion [95].

15.4
Novel Support Materials

Support materials in electrocatalysis, as in heterogeneous catalysis, first serve as substrates for the deposition and stabilization of the active material (in most cases metallic nanoparticles). Furthermore, the supports have to be conductive to provide electronic pathways to and from the active sites. The latter also implies that the catalyst deposition during electrode preparation must be carried out in such a way that a conductive but highly porous network may form (see above, triple-phase boundary).

For a long time, carbon black has been considered the support material of choice for fuel cell electrocatalysts. However, as detailed above, one of the mechanisms considered responsible for fuel cell catalyst degradation is the corrosion of the support material. Consequently, alternative support materials are sought that must fulfill the following criteria: they must be highly conductive to avoid ohmic losses within the electrode, they must withstand the conditions experienced in PEMFCs such as a strongly acidic environment and strongly oxidizing or reducing conditions at temperatures of 100 °C or higher, and they must provide and maintain high surface area and anchoring sites for catalyst particles.

Two classes of materials are currently under consideration for this application: transition metal oxides such as TiO_2 and Nb_2O_5, and CNTs and carbon nanofibers [93, 96]. Especially the latter have been shown to fulfill the above requirements and to be suitable support materials. In particular, the high conductivity and chemical stability are often put forward for the use of CNTs. However, surface functionalization is necessary: functional groups at the CNT surface may act as anchoring sites for metal nanoparticles and may further help to stabilize them under the conditions of a working fuel cell. Such functional groups are often created by simply oxidizing the CNT by refluxing in concentrated nitric acid.

In addition to oxidized CNTs, nitrogen-modified CNTs have also been discussed as support material [97]. In these, the nitrogen functional groups fulfill the same task as the oxygen-containing groups. Furthermore, it is claimed that nitrogen-modified CNTs might have different electronic properties than pure ones, leading to superior metal–support interactions that might have a positive influence on the catalytic activity [97].

As mentioned above, any of the new support materials have to be available in a form that allows the fabrication of GDEs using standard industrial techniques. However, although electrode preparation has been optimized in an empirical approach, doubts still remain about whether the utilization of the electrocatalyst is high enough and the porosity optimum. Recently, various papers have put forward the ideas of designing the GDEs by deliberately preparing nanostructured electrodes [98]. Recent ideas include the growth of aligned CNTs [58, 99, 100], which can also be branched [101], as shown schematically in Figure 15.6. It has been shown that such electrodes based on aligned CNTs indeed can be processed into a GDE for fuel cells [99].

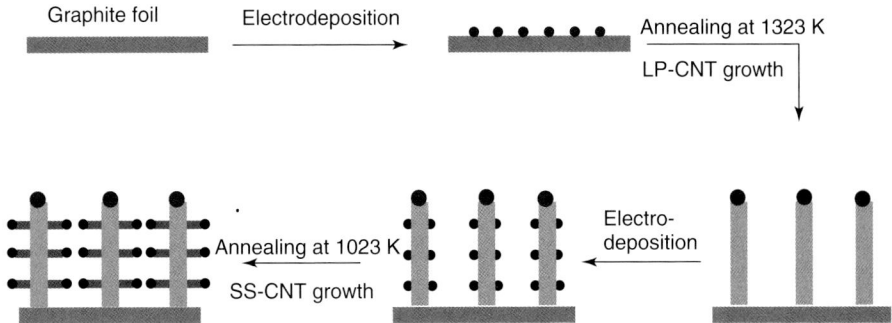

(LP-CNT: long primary carbon nanotube; SS-CNT: short secondary carbon nanotube)

Figure 15.6 Schematic representation of the synthesis of hierarchical carbon nanocomposites. After Fe electrodeposition and subsequent iron-catalyzed growth of CNTs from cyclohexane at 1323 K, again Fe was deposited and secondary short CNTs were grown at 1023 K. Reproduced with permission from [101]. Copyright Wiley-VCH Verlag GmbH & Co. KGaA.

15.5
Catalyst Development, Characterization, and *In Situ* Studies in Fuel Cells

Various strategies have been pursued in the development of new electrocatalysts. The commonly used approach is the experimental one, where new catalyst materials, often an active metal on a carbon or oxide support, are prepared and tested with respect to their electrocatalytic activity in half-cell measurements with thin-film electrodes in an aqueous electrolyte using classical electrochemical techniques [1, 102] such as cyclic voltammetry, rotating disc electrode, chronoamperometry, and impedance spectroscopy. Combinatorial approaches have also been published, for example, those investigating an array of catalyst spots with SECM [43–45, 103]. Although these approaches allow for the testing of a large number of catalyst compositions in a short time under very well-defined conditions, the applied experimental parameters are often different from those encountered in a fuel cell (e.g., aqueous instead of solid electrolytes, room temperature, dissolved instead of gaseous reactants, temperatures $>80\,°C$). Therefore, while active catalysts can be identified with these approaches, their suitability for fuel cell application and especially the evaluation of their long-term stability require fuel cell tests employing at least single MEA arrangements.

In addition to these purely experimental approaches, strategies have also been put forward where possible candidate catalysts have been selected based on theoretical considerations, for example, based on thermodynamic guidelines [43] or even on DFT calculations [17, 104]. Nevertheless, progress in fuel cell catalysis requires the understanding of structure–activity relationships, which is usually attempted by a detailed characterization of the prepared materials with state-of-the-art physicochemical methods, including transmission electron microscopy (TEM), X-ray diffraction (XRD), X-ray photoelectron spectroscopy (XPS), and X-ray absorption spectroscopy (XAS).

A detailed understanding of the individual processes occurring during electrocatalysis, reaction pathways, influence of adatoms and alloys poisoning effects, and so on, is often rather demanding, given the complexity of real-world fuel cell catalysts, including distribution in nanoparticle size, structural heterogeneity, and different exposed surface structures [105], and hence is often restricted to surface-science (often single-crystal) studies using electrochemical and also suitable *ex situ* and *in situ* spectroscopic techniques [32, 106]. With these approaches, a wealth of information at the atomic level can be obtained and such studies have deepened our understanding of electrocatalysis and of surface processes in general. However, given the above-described dynamics of catalysts under fuel cell test conditions, it becomes clear that a real understanding of electrocatalytic processes *in a fuel cell* can hardly be obtained by simply extrapolating single-crystal results to nanoparticles and requires *additionally* the application of *operando* techniques, that is, techniques which are capable of monitoring the structure and structural changes of the catalyst during fuel cell operation and correlate these with performance data. Only by combining both the surface-science and the *operando* approaches does a comprehensive understanding of fuel cell electrocatalysis seem to be possible. Additionally, so-called *postmortem* studies, that is, studies on the catalyst structure after being used in the cell, could prove to be useful.

Owing to the complex architecture of the fuel cell and especially the presence of a flow field and gas diffusion layer, only a few spectroscopic *operando* techniques are available. The most powerful ones include the use of X-rays, which to a certain extent may penetrate through the above-mentioned carbon materials. One of these techniques, XAS, is very useful in this respect since it provides information about catalyst structure, electronic properties [107–110], and, in certain cases, surface species on these catalyst nanoparticles [30]. Several spectroscopic test cells have been proposed [109, 110], one of which was demonstrated recently to allow for *in situ* investigation of fuel cell catalysts without any compromise regarding cell design [111]. Another useful X-ray technique is XRD, which has recently been applied successfully to monitor oscillations in particle growth on Pt/C catalysts [112].

15.6
Catalysis in Hydrogen Production for Fuel Cells

Hydrogen is an ideal fuel for fuel cells owing to its high reactivity and zero emission. Currently, however, hydrogen is not readily available to the public, and neither its production nor distribution infrastructure are widespread, although a recent study by the European Union ("A portfolio of power-trains for Europe: a fact-based analysis", www.zeroemissionvehicles.eu) revealed that the costs for the implementation of a hydrogen infrastructure will be lower than the costs for the implementation of a charging infrastructure for battery electric vehicles (BEVs) and plug-in hybrid electric vehicles (PHEVs). Nevertheless, sustainable hydrogen production is a key target in the development of future alternative energy systems for providing a clean and affordable energy supply [113]. Hydrogen production for fuel cells differs from large-scale industrial hydrogen production in a number of ways, such as size, pressure, product quality, and system integration [114]. Irrespective of the fuels used, for example, methanol, logistic fuels, or biomass, the hydrogen generation is mostly performed by catalytic reactions, as discussed in the following for different feedstocks. As the issue of hydrogen production for fuel cells is very complex and cannot be discussed here in depth, the reader will be referred to literature.

15.6.1
Hydrogen Production from Methanol to Heavy Hydrocarbons

15.6.1.1 Introduction

Methanol reformer systems were the first systems to be developed for on-board fuel processing in the mid-1980s, hence methanol is one of the most studied hydrogen precursors for fuel cells with the advantage of being an easily storable and transportable liquid and being readily available. On the other hand, methanol is highly toxic and highly flammable. It is a secondary derivative fuel and requires new infrastructure. In addition to its use in DMFCs, it can be converted into hydrogen in a reformer by steam reforming (SR):

$$CH_3OH_{(g)} + H_2O_{(g)} \longrightarrow CO_2 + 3H_2 \quad \Delta H_R^{298} = 49.2 \text{ kJ mol}^{-1} \quad (15.7)$$

The SR of methanol is typically performed in the temperature range 523–573 K. Cu/ZnO catalysts are most frequently used, often supported on alumina. The use of palladium- and rhodium-containing catalysts has also been reported [115], in addition to some promoters [116]. The main by-product is carbon monoxide. To minimize the concentration of CO in the product stream, the reaction is performed using a stoichiometry of steam to methanol of >1. The conversion of CO with H_2O is known as the water gas shift (WGS) reaction and is addressed later [Eq. (15.9)]. Detailed information about the reaction path of methanol SR on Cu/ZnO catalysts and the active sites of Cu/ZnO catalysts together with details of the Pd/ZnO system can be found elsewhere [116].

Compared with other hydrocarbon fuels such as ethanol, methane, and gasoline, SR of methanol exhibits a relatively high theoretical conversion efficiency, low conversion temperature, and low by-product formation [117]. Partial oxidation and autothermal reforming (ATR) of methanol are also possible, but play a minor role.

In general, for comparably simple fuels (methanol, ethanol, and methane), processes on catalytic surfaces are much easier to understand than for fossil-based fuels, consisting of hundreds of different hydrocarbon species with a seemingly unlimited number of chemical reactions.

The advantage of using conventional fuels such as gasoline and diesel or similar hydrocarbons is that no extra capital cost for developing the infrastructure is needed. On the other hand, there is an increasingly active debate about whether using these fuels to make hydrogen is beneficial while global warming is an issue. The medium-term objective should therefore be the production of hydrogen from other liquid fuels that contain heavy hydrocarbons produced from biomass. Nevertheless, the developments in the field of hydrogen generation from conventional fuels could provide a bridge for the transition of hydrogen production from fossil fuels to renewable production from biomass [113].

The main reforming processes for heavy hydrocarbons, as already mentioned for the case of methanol, are SR, partial oxidation, and ATR. Regarding the catalysts used, mainly four potential challenges have to be met [113]:

- activity
- carbon formation
- sulfur poisoning
- thermal stability.

15.6.1.2 Catalytic Steam Reforming (SR)

For higher hydrocarbons the SR involves the splitting of hydrocarbons with steam [Eq. (15.8)], the already mentioned WGS reaction [Eq. (15.9)], and the formation of methane [Eq. (15.10)]:

$$C_n H_m + n H_2O \rightleftharpoons nCO + \left(\frac{m}{2} + n\right) H_2 \quad \text{SR}, \Delta H_R > 0 \tag{15.8}$$

$$CO + H_2O \rightleftharpoons CO_2 + H_2 \quad \text{WGS reaction}, \Delta H_R < 0 \tag{15.9}$$

$$CO + 3H_2 \rightleftharpoons CH_4 + H_2O \quad \text{formation of methane}, \Delta H_R < 0 \tag{15.10}$$

The product composition is controlled mainly by thermodynamics, which favor the formation of methane at lower temperatures of about 623 K, and of hydrogen at higher temperatures of about 1223 K [113]. The required heat for endothermic SR can be provided by the total oxidation of hydrocarbons. An increase in the molar ratio of water vapor to carbon content also causes a decrease in carbon monoxide content following Eq. (15.9), a decrease in methane content, and an increase in carbon monoxide content according to Eq. (15.10). More details about thermodynamic considerations [116] and more information about the elementary steps of the reaction of aliphatic hydrocarbons [113] are available in the literature. The SR process has been well examined, since the SR of natural gas is the dominant process for industrial hydrogen production with technical availability for large-scale application and cost-effectiveness [118].

15.6.1.2.1 Desulfurization

Most fuels contain sulfur components, which poison non-precious metal catalysts such as Ni, Co, Fe, and Cu. Particularly logistic fuels contain various sulfur compounds [thiols (mercaptans), thiophenes, benzothiophenes, and dibenzothiophenes] in various concentrations, which can lead to active-site poisoning. The group 8 metals used in reforming catalysts strongly adsorb sulfur, which prevents or affects the further adsorption of hydrocarbons. It is assumed that metal poisoning by sulfur components involves hydrogenolysis of the S-containing molecule

[Eq. (15.11)]. The resulting H_2S would finally lead to the formation of a stable and inactive M–S species on the catalyst [Eq. (15.12)]:

$$M + S - R \longrightarrow M + R' + H_2S \; (M = \text{metal}) \tag{15.11}$$

$$H_2S + M \longrightarrow M - S + H_2 \tag{15.12}$$

The rate of metal poisoning by sulfur, and hence the sulfur resistance, depend on the nature of the metal and the reaction conditions [113].

In particular, conventional nickel catalysts (low sulfur tolerance) do not allow the treatment of high-sulfur heavy hydrocarbon feedstocks, especially in the case of pre-reforming, where the low operating temperatures demand the highest sulfur tolerance [113].

As a consequence, catalysts with improved sulfur tolerance are needed. Alternatively, gas cleaning in a ZnO bed is necessary if H_2S or thiols have to be removed, whereas thiophenes require reduction to H_2S by hydrogen on Co–Mo catalysts [116] before adsorption in the ZnO bed:

$$H_2S + ZnO \longrightarrow ZnS + H_2O \quad \Delta H_R^{298} = -75 \text{ kJ mol}^{-1} \tag{15.13}$$

More details about desulfurization can be found elsewhere [115].

15.6.1.2.2 Carbon Formation

The formation of carbon is a frequent side reaction, which has to be avoided, however, if possible. In particular, SR of heavy hydrocarbons involves a higher risk of carbon formation due to the presence of aromatic structures and the low thermal stability of some components leading to two different kinds of carbonaceous deposits: elemental carbon formed directly from heavy hydrocarbons:

$$C_nH_m \longrightarrow nC + \frac{m}{2}H_2 \tag{15.14}$$

or coke formed by heavy polycyclic structures from condensation/dehydrogenation reactions of the hydrocarbons at elevated temperatures [113].

15.6.1.3 Catalytic Partial Oxidation (CPO)

Partial oxidation is the conversion of fuels under an oxygen-deficient atmosphere:

$$C_nH_m + \frac{n}{2}O_2 \rightleftharpoons nCO + \frac{m}{2}H_2 \tag{15.15}$$

Partial oxidation is characterized by short reaction times (milliseconds – roughly two orders of magnitude faster than SR) and high temperatures (1123–1273 K) with the advantage that in contrast to SR no expensive superheated steam is necessary.

Catalysts for partial oxidation consist mostly of a noble metal supported on porous ceramic monoliths [113]. Two mechanisms have been proposed: the direct partial oxidation mechanism, where hydrocarbon fuel and oxygen are adsorbed and react directly on the catalyst surface to give H_2 and CO, and the combustion reforming mechanism, which assumes that the reaction is initiated near the entrance of the catalyst bed by complete dissociation of the hydrocarbon through multiple dehydrogenation and C–C bond cleavage, followed by the reaction of adsorbed O_2 with carbon and hydrogen to form CO, CO_2, and H_2O [113]. Although the system seems simple, as only fuel and air feed are needed without evaporation processes as required for SR, it is a complicated process. Thus, catalytic partial oxidation (CPO) of heavy hydrocarbons is poorly understood. One reason is that experiments showed that the individual behavior of the components present in the mixture is not simply an average of all the constituent molecules. Changes in the adsorption equilibria of the mixtures compared with the pure components may cause kinetic inhibitions and changes in the reaction pathway, for example, the most reactive components consume all oxygen leading to pyrolysis of the less active components. Around 1073 K, a number of catalytic and autothermal reactions occur: total oxidation, SR, hydrocarbon cracking, methanation, and WGS reaction. Hence the results obtained for individual components do not easily allow the prediction of the behavior of mixtures [113]. Often Ni-based catalysts are used, which are often promoted to reduce the extent of carbon formation, sulfur poisoning, and sintering. The catalysts are typically modified with alkali and alkaline earth metals to neutralize the acidic sites on supports responsible for carbon formation [113].

Carbon formation is a major complication during CPO of higher hydrocarbons, whereas *sulfur poisoning* plays a minor role (the applied high temperatures are less favorable for the adsorption of H_2S, which is the thermodynamically stable form under typical CPO conditions). Regarding deactivation of catalysts, especially at these high temperatures, the following phenomena concerning the *thermal stability* have to be considered: formation of solid metal–support components (oxides, aluminates, e.g., Co, Ni), sintering of metal (Rh, Ir, Pt) and/or support, and metal loss via the formation of volatile metal oxides (Pd, Ru, Pt) or carbonyl compounds (Ni, Co) [113].

15.6.1.4 Autothermal Reforming (ATR)

ATR is the combination of SR and CPO. The necessary heat for SR [Eq. (15.8)] is delivered by partial oxidation CPO:

$$C_nH_m + aO_2 + (n-2a)H_2O \longrightarrow nCO + \left(n - 2a + \frac{m}{2}\right)H_2 \quad (15.16)$$

The advantages of ATR are:

- no expensive superheated steam necessary, in contrast to SR
- higher thermal efficiency
- better control of carbon formation than in CPO.

For liquid fuels, reformers are used that incorporate a fast oxidation zone and a separate slow SR zone, where the O/C and the steam/C ratios are optimized to yield thermoneutral ATR. Unsurprisingly ATR is a complex combination of homogeneous partial oxidation, thermal cracking, dehydrogenation, SR, WGS reaction, methanation, and others. Owing to this complexity, the reaction mechanism occurring in liquid fuel ATR is not well understood. Catalysts used in the ATR are the same as those used for SR and CPO: group VIII, IX, and X metals, especially Ni, Pt, Pd, Rh, and Ru.

The presence of steam during fuel heating decreases the rate of pyrolysis and cracking product formation and thus *carbon formation* as compared with CPO, which however is still significant. On the other hand, *sulfur poisoning* is increased by the lower temperatures used in ATR, which enhances sulfur adsorption on the catalyst surface. Sulfur poisoning of active sites for SR in turn increases gas-phase chemistry and consequently carbon formation. Regarding the *thermal stability*, sintering also causes deactivation of the less noble metals (Ni, Co) used in ATR catalysts and is accelerated in the presence of steam. The problem of volatilization is the same as already described for CPO [113].

An overview of the research strategies in catalyst development for hydrogen production from heavy hydrocarbons regarding the selection of the active phase, promoters, supports, and control of the synthesis of materials for customizing the crystallinity, electronic structure, and morphology of catalysts at the nanoscale was given in detail by Navarro Yerga et al. [113]. In particular, carbon formation (ensemble size control, modification of electronic properties, modified supports by addition of promoters, for example, alkali metals, lanthanoids, to enhance steam adsorption), sulfur poisoning (modification of metal atoms by alloying, changing the particle size to suppress sulfur adsorption), and finally thermal stability (tailoring metal–support interactions for SR and highly stable oxides for ATR and CPO catalysts) were discussed [113]. Peters presented an overview regarding the state of the art in ATR of gasoline and diesel [116].

15.6.2
Carbon Monoxide Removal

Whereas in industrial hydrogen production purification of hydrogen is performed by pressure swing adsorption, the low pressure used in hydrogen production for fuel cells and the low CO tolerance of PEMFCs (<50 ppm) require a complex, multi-step catalytic purification. This includes low-temperature shift reaction and methanation reaction or PROX [114]. The shift reaction can reduce the CO fraction from about 10 to 1 vol.% CO, but cannot achieve the low level of CO necessary to prevent significant degradation of the PEMFC stack performance by CO poisoning. Two catalytic reactions are used in most fuel processor designs to reduce the CO level to <50 ppm: PROX and selective CO methanation (SELMETH).

Most designs use PROX, where low amounts of CO are oxidized in the presence of high hydrogen concentrations [Eq. (15.17)] and the undesired side-reaction [Eq. (15.18)], which is regarded as the most economic and efficient approach:

$$CO + \frac{1}{2}O_2 \longrightarrow CO_2 \quad \Delta H = -280 \text{ kJ mol}^{-1} \quad (15.17)$$

$$H_2 + \frac{1}{2}O_2 \longrightarrow H_2O \quad \Delta H = -240 \text{ kJ mol}^{-1} \quad (15.18)$$

Preferential CO oxidation has been studied using different noble metals (Pt, Pd, Ru, Rh, and Au) and different supports (silica, alumina, and titania) or over ceria-

or ceria–zirconia-supported transition metals (Cu, Co, Ni) [119]. In ceria supports the Ce(III)–Ce(II) redox cycle is fast and the oxygen mobility is facilitated, hence these supports are able to adsorb oxygen reversibly. A more detailed overview is given elsewhere [116].

Less common is the use of SELMETH. An advantage of this approach is that no additional component feed is needed. The most serious drawback, however, is that valuable hydrogen is consumed, not only for the highly exothermic reaction itself [Eq. (15.19)], but also for the slightly less exothermic major side-reaction, the CO_2 hydrogenation [Eq. (15.20)]:

$$CO + 3H_2 \rightleftharpoons CH_4 + 2H_2O \quad \Delta H_R^{473} = -214.1 \text{ kJ mol}^{-1} \quad (15.19)$$

$$CO_2 + 4H_2 \rightleftharpoons CH_4 + 2H_2O \quad \Delta H_R^{473} = -174.7 \text{ kJ mol}^{-1} \quad (15.20)$$

Hence temperature control is of particular significance. In addition, the well-known shift reaction [Eq. (15.9)] plays an important role in this reaction system since owing to insufficient temperature control the shift equilibrium is pushed towards CO_2. The particular challenge for the catalyst is to hydrogenate CO selectively in the presence of high contents of CO_2. Suitable catalysts can be found among the group VIII transition metals and the noble metals (Ru, Ni, Co, Fe, and Mo). More details can be found elsewhere [116].

15.6.3
Catalysis in the Production of Hydrogen from Biomass

Notable advances have been made in hydrogen production from biomass, but the process is still in the initial stages of development [120–122]. A catalyzed process under mild conditions is aqueous-phase reforming (APR), first proposed by Dumesic's group [123]. This process features some advantages: no energy for water vaporization is needed, bioproducts that cannot be vaporized could still be used, wet or water-soluble feedstocks can be used without a dehydration step, and the process is operated at relatively low temperatures (around 500 K) where the WGS reaction occurs, leading to low CO concentrations.

For glycerol, the stoichiometric reaction equation is

$$C_3H_8O_3 + 3H_2O \longrightarrow 7H_2 + 3CO_2 \quad (15.21)$$

Under the applied conditions, C–C bond cleavage and the WGS reaction [Eq. (15.2)] are also possible [124]:

$$C_3H_8O_3 \longrightarrow 4H_2 + 3CO \quad (15.22)$$

Further reaction of CO and/or CO_2 with H_2 would lead to methanation or Fischer–Tropsch reactions [125]. The selectivity towards hydrogen or alkanes can be adjusted by the choice of the active metal and the supports of the catalysts. In this way, higher selectivities towards hydrogen could be achieved if Pt, Ni, and Ni–Sn are used on alumina and titania as support material, whereas Ru, Rh, and Ni and also SiO_2–Al_2O_3 as support revealed higher alkane selectivities [124, 126].

15.7
Perspective

Fuel cell catalysis and electrocatalysis have been a very active field of research for the last two decades or longer. Progress has been made in such diverse fields as mechanistic studies, development of novel catalyst materials and their characterization, and testing in real fuel cells, and all these research areas are important to gain an in-depth understanding of fuel cell catalysis, which in turn may help to identify further avenues of research. However, despite the tremendous achievements that we have witnessed, demanding challenges still remain. These include first a reduction of the Pt loading in PEMFCs. As discussed above, promising candidate catalysts have been developed, but none of them has been shown to be active and stable in the long term under fuel cell conditions. This includes also alloy and core–shell nanoparticles, which have to withstand the dynamic changes to which they are subjected during fuel cell operation to avoid leaching of the second/third metal and to maintain structures of optimum activity. Research in this field must focus on long-term stability and reduced degradation, and this includes real fuel cell studies. Researchers should also be more careful when announcing the next "breakthrough," which has very often been obtained only under idealized conditions.

References

1. Bard, A.J. and Faulkner, L. (2001) Electrochemical Methods–Fundamentals and Applications, 2nd edn., John Wiley & Sons, Inc., New York.
2. Yu, E.H., Krewer, U., and Scott, K. (2010) Principles and materials aspects of direct alkaline alcohol fuel cells. Energies, **3**, 1499–1528.
3. Arges, C.G., Ramani, V., and Pintauro, P.N. (2010) Anion exchange membrane fuel cells. Interface, **19** (2), 31–35.
4. Gülzow, E. (2003) Alkaline fuel cells. Fuel Cells, **4** (4), 251–255.
5. Chen, E. (2003) Thermodynamics and electrochemical kinetics, in Fuel Cell Technology Handbook (ed. G. Hoogers), CRC Press LLC, Boca Raton, FL, Chapter 3.
6. Gasteiger, H.A., Kocha, S.S., Sompalli, B., and Wagner, F.T. (2005) Activity benchmarks and requirements for Pt, Pt-alloy, and non-Pt oxygen reduction catalysts for PEMFCs. Appl. Catal. B: Environ., **56**, 9–35.
7. Gewirth, A.A. and Thorum, M.S. (2010) Electroreduction of dioxygen for fuel-cell applications: materials and challenges. Inorg. Chem., **49**, 3557–3566.
8. Anastasijevic, N.A., Vesovic, V., and Adzic, R.R. (1987) Determination of the kinetic parameters of the oxygen reduction reaction using the rotating-disk electrode. Part I. Theory. J. Electroanal. Chem., **229**, 305–316.
9. Borup, R., Meyers, J., Pivovar, B., Kim, Y.S., Mukundan, R., Garland, N., Myers, D., Wilson, M., Garzon, F., Wood, D., Zelenay, P., More, K., Stroh, K., Zawodzinski, T., Boncella, J., McGrath, J.E., Inaba, O.M., Miyatake, K., Hori, M., Ota, K., Ogumi, Z., Miyata, S., Nishikata, A., Siroma, Z., Uchimoto, Y., Yasuda, K., Kimijima, K.-I., and Iwashita, N. (2007) Scientific aspects of polymer electrolyte fuel cell durability and degradation. Chem. Rev., **107**, 3904–3951.
10. Adzic, R. (1998) Recent advances in the kinetics of oxygen reduction, in Electrocatalysis (eds. J. Lipkowski and

11. Appleby, A.J. (1993) Electrocatalysis of aqueous dioxygen reduction. *J. Electroanal. Chem.*, **357**, 117–179.
12. Kinoshita, K. (1990) Particle size effects for oxygen reduction on highly dispersed platinum in acid electrolytes. *J. Electrochem. Soc.*, **137** (3), 845–848.
13. Mayrhofer, K.J.J., Blizanac, B.B., Arenz, M., Stamencovic, V.R., Ross, P.N., and Markovic, N.M. (2005) The impact of geometric and surface electronic properties of Pt-catalysts on the particle size effect in electrocatalysis. *J. Phys. Chem. B*, **109**, 14433–14440.
14. Mayrhofer, K.J.J., Strmcnik, D., Blizanac, B.B., Stamenkovic, V., Arenz, M., and Markovic, N.M. (2008) Measurement of oxygen reduction activities via the rotating disc electrode method: from Pt model surfaces to carbon-supported high surface area catalysts. *Electrochim. Acta*, **53**, 3181–3188.
15. Mukerjee, S. and McBreen, J. (1998) Effect of particle size on the electrocatalysis by carbon-supported Pt electrocatalysts: an *in situ* XAS investigation. *J. Electroanal. Chem.*, **448**, 163–171.
16. Wikander, K., Ekström, H., Palmqvist, A.E.C., and Lindbergh, G. (2007) On the influence of Pt particle size in the PEMFC cathode performance. *Electrochim. Acta*, **52**, 6848–6855.
17. Nilekar, A.U., Xu, Y., Zhang, J., Vukmirovic, M.B., Sasaki, K., Adzic, R.R., and Mavrikakis, M. (2007) Bimetallic and ternary alloys for improved oxygen reduction catalysis. *Top. Catal.*, **46** (3–4), 276–284.
18. Antolini, E. (2007) Platinum-based ternary catalysts for low temperature fuel cells. Part I. Preparation methods and structural characteristics. *Appl. Catal. B*, **74**, 324–336.
19. Antolini, E. (2007) Platinum-based ternary catalysts for low temperature fuel cells. Part II. Electrochemical properties. *Appl. Catal. B*, **74**, 337–350.
20. Stamenkovic, V.R., Mun, B.S., Mayrhofer, K.J.J., Poss, P.N., and Markovic, N.M. (2006) Effect of surface composition on electronic structure, stability, and electrocatalytic properties of Pt–transition metal alloys: Pt-skin versus Pt-skeleton surfaces. *J. Am. Chem. Soc.*, **128**, 8813–8819.
21. Stamenkovic, V.R., Mun, B.S., Arenz, M., Mayrhofer, K.J.J., Lucas, C.A., Wang, G. et al. (2007) Trends in electrocatalysis on extended and nanoscale Pt–bimetallic alloy surfaces. *Nat. Mater.*, **6**, 241–247.
22. Toda, T., Igarashi, H., Uchida, H., and Watanabe, M. (1999) Enhancement of the electroreduction of oxygen on Pt alloys with Fe, Ni, and Co. *J. Electrochem. Soc.*, **146** (10), 3750–3756.
23. Chen, S., Gasteiger, H.A., Hayakawa, K., Tada, T., and Shao-Horn, Y. (2010) Platinum-alloy cathode catalyst degradation in proton exchange membrane fuel cells: nanometer-scale compositional and morphological changes. *J. Electrochem. Soc.*, **157**, A82–A97.
24. Koh, S. and Strasser, P. (2007) Electrocatalysis on bimetallic surfaces: modifying catalytic reactivity for oxygen reduction by voltammetric surface dealloying. *J. Am. Chem. Soc.*, **129**, 12624–12625.
25. Srivastava, R., Mani, P., Hahn, N., and Strasser, P. (2007) Efficient oxygen reduction fuel cell electrocatalysis on voltammetrically dealloyed Pt–Cu–Co nanoparticles. *Angew. Chem. Int. Ed.*, **46**, 8988–8991.
26. Mani, P., Srivastava, R., and Strasser, P. (2011) Dealloyed binary PtM3 (M = Cu, Co, Ni) and ternary PtNi3M (M = Cu, Co, Fe, Cr) electrocatalysts for the oxygen reduction reaction: performance in polymer electrolyte membrane fuel cells. *J. Power Sources*, **196** (2), 666–673.
27. Adzic, R.R., Zhang, J., Sasaki, K., Vukmirovic, M.B., Shao, M., Wang, J.X., Nilekar, A.U., Mavrikakis, M., Valerio, J.A., and Uribe, F. (2007) Platinum monolayer fuel cell electrocatalysts. *Top. Catal.*, **46**, 249–262.
28. Ruvinsky, P.S., Pronkin, S.N., Zaikovskii, V.I., Bernhardt, P., and Savinova, E.R. (2008) On the enhanced electrocatalytic activity of Pd overlayers

on carbon-supported gold particles in hydrogen electrooxidation. *Phys. Chem. Chem. Phys.*, **10**, 6665–6676.

29. Kulp, C., Chen, X., Puschhof, A., Schwamborn, S., Schuhmann, W., and Bron, M. (2010) Electrochemical synthesis of core–shell catalysts for electrocatalytic applications. *ChemPhysChem*, **11**, 2854–2861.

30. Teliska, M., Murthi, V.S., Mukerjee, S., and Ramaker, D.E. (2005) Correlation of water activation, surface properties, and oxygen reduction reactivity of supported Pt–M/C bimetallic electrocatalysts using XAS. *J. Electrochem. Soc.*, **152** (11), A2159–A2169.

31. Stamenkovic, V., Schmidt, T.J., Ross, P.N., and Markovic, N.M. (2002) Surface composition effects in electrocatalysis: kinetics of oxygen reduction on well-defined Pt_3Ni and Pt_3Co alloy surfaces. *J. Phys. Chem. B*, **106**, 11970–11979.

32. Markovic, N.M. and Ross, P.N. Jr. (2002) Surface science studies of model fuel cell electrocatalysts. *Surf. Sci. Rep.*, **45**, 117–229.

33. Tian, N., Zhou, Z.-Y., and Sun, S.-G. (2008) Platinum metal catalysts of high-index surfaces: from single-crystal planes to electrochemically shape-controlled nanoparticles. *J. Phys. Chem. C*, **112**, 19801–19817.

34. Wang, B. (2005) Recent development of non-platinum catalysts or oxygen reduction reaction. *J. Power Sources*, **152**, 1–15.

35. Feng, Y., Gago, A., Timperman, L., and Alonso-Vante, N. (2011) Chalcogenide metal centers for oxygen reduction reaction: activity and tolerance. *Electrochim. Acta*, **56** (3), 1009–1022.

36. Lee, J.W. and Popov, B.N. (2007) Ruthenium-based electrocatalysts for oxygen reduction reaction – a review. *J. Solid-State Electrochem.*, **11**, 1355–1364.

37. Alonso-Vante, N. (2006) Carbonyl tailored electrocatalysts. *Fuel Cells*, **6**, 182–189.

38. Zhang, L., Zhang, J., Wilkinson, D.P., and Wang, H. (2006) Progress in preparation of non-noble electrocatalysts for PEM fuel cell reactions. *J. Power Sources*, **156**, 171–182.

39. Alonso-Vante, N. and Tributsch, H. (1986) Energy conversion catalysis using semiconducting transition metal cluster compounds. *Nature*, **323**, 431–432.

40. Dodelet, J.P. (2006) Oxygen reduction in PEM fuel cell conditions: heat-treated non-precious metal-N_4 macrocycles and beyond, in N_4-*Macrocyclic Metal Complexes* (eds. J.H. Zagal, F. Bedioui, and J.-P. Dodelet), Springer, New York, pp. 83–147.

41. Lima, F.H.B., Zhang, J., Shao, M.H., Sasaki, K., Vukmirovic, M.B., Ticianelli, E.A., and Adzic, R.R. (2007) Catalytic activity–d-band center correlation for the O_2 reduction reaction on platinum in alkaline solution. *J. Phys. Chem. C*, **111**, 404–410.

42. Bron, M. (2008) Oxygen reduction at carbon black supported gold catalysts. *J. Electroanal. Chem.*, **624**, 64–68.

43. Fernández, J.L., Walsh, D.A., and Bard, A.J. (2005a) Thermodynamic guidelines for the design of bimetallic catalysts for oxygen electroreduction and rapid screening by scanning electrochemical microscopy. M–Co (M: Pd, Ag, Au). *J. Am. Chem. Soc.*, **127**, 357–365.

44. Fernández, J.L., Raghuveer, V., Manthiram, A., and Bard, A.J. (2005c) Pd–Ti and Pd–Co–Au electrocatalysts as a replacement for platinum for oxygen reduction in proton exchange membrane fuel cells. *J. Am. Chem. Soc.*, **127**, 13100–13101.

45. Fernández, J.L., White, J.M., Sun, Y., Tang, W., Henkelman, G., and Bard, A.J. (2006) Characterization and theory of electrocatalysts based on scanning electrochemical microscopy screening methods. *Langmuir*, **22**, 10426–10431.

46. Jasinski, R. (1964) A new fuel cell cathode catalyst. *Nature*, **201**, 1212–1213.

47. Jahnke, H., Schönborn, M., and Zimmermann, G. (1976) Organic dyestuffs as catalysts for fuel cells. *Top. Curr. Chem.*, **61**, 133–181.

48. Wiesener, K. (1986) N_4-chelates as electrocatalysts for cathodic oxygen reduction. *Electrochim. Acta*, **31** (8), 1073–1078.

49. Bezerra, C.W.B., Zhang, L., Lee, K., Liu, H., Marques, A.L.B., Marques, E.O., Wang, H., and Zhang, J. (2008) A review of Fe–N/C and Co–N/C catalysts for the oxygen reduction reaction. *Electrochim. Acta*, **53**, 4937–4951.
50. Lalande, G., Côté, R., Guay, D., Dodelet, J.P., Weng, L.T., and Bertrand, P. (1997) Is nitrogen important in the formulation of Fe-based catalysts for oxygen reduction in solid polymer fuel cells? *Electrochim. Acta*, **42**, 1379–1388.
51. Wu, G., More, K.L., Johnston, C.M., and Zelenay, P. (2011) High-performance electrocatalysts for oxygen reduction derived from polyaniline, iron, and cobalt. *Science*, **332**(6028), 443–447.
52. Jin, C., Nagaiah, T.C., Xia, W., Spliethoff, B., Wang, S., Bron, M., Schuhmann, W., and Muhler, M. (2010) Metal-free and electrocatalytically active nitrogen-doped carbon nanotubes synthesized by coating with polyaniline. *Nanoscale*, **2**, 981–987.
53. Faubert, G., Côté, R., Dodelet, J.P., Lefèvre, M., and Bertrand, P. (1999) Oxygen reduction catalysts for polymer electrolyte fuel cells from the pyrolysis of Fe^{II} acetate adsorbed on 3,4,9,10-perylenetetracarboxylic dianhydride. *Electrochim. Acta*, **44**, 2589–2603.
54. Faubert, G., Côté, R., Guay, D., Dodelet, J.P., Dénès, G., Poleunis, C., and Bertrand, P. (1998) Activity and characterization of Fe-based catalysts for the reduction of oxygen in polymer electrolyte fuel cells. *Electrochim. Acta*, **43**, 1969–1984.
55. Bron, M., Fiechter, S., Hilgendorff, M., and Bogdanoff, P. (2002) Catalysts for oxygen reduction from heat treated, carbon supported iron phenanthroline complexes. *J. Appl. Electrochem.*, **32**, 211–216.
56. Jaouen, F., Proietti, E., Lefevre, M., Chenitz, R., Dodelet, J.-P., Wu, G., Chung, H.T., Johnston, C.M., and Zelenay, P. (2011) Recent advances in non-precious metal catalysis for oxygen-reduction reaction in polymer electrolyte fuel cells. *Energy Environ. Sci.*, **4** (1), 114–130.
57. Lefèvre, M., Proietti, E., Jaouen, F., and Dodelet, J.-P. (2009) Iron-based catalysts with improved oxygen reduction activity in polymer electrolyte fuel cells. *Science*, **324**, 71–74.
58. Gong, K., Du, F., Xia, Z., Durstock, M., and Dai, L. (2009) Nitrogen-doped carbon nanotube arrays with high electrocatalytic activity for oxygen reduction. *Science*, **323**, 760–764.
59. Nagaiah, T.C., Kundu, S., Bron, M., Muhler, M., and Schuhmann, W. (2010) Nitrogen-doped carbon nanotubes as a highly efficient cathode catalyst for the oxygen reduction reaction in alkaline medium. *Electrochem. Commun.*, **12**, 338–341.
60. Biddinger, E.J., von Deak, D., and Ozkan, U.S. (2009) Nitrogen-containing carbon nanostructures as oxygen-reduction catalysts. *Top. Catal.*, **52**, 1566–1574.
61. Matter, P.H., Zhang, L., and Ozkan, U.S. (2006) The role of nanostructure in nitrogen-containing carbon catalysts for the oxygen reduction reaction. *J. Catal.*, **239**, 83–96.
62. Kundu, S., Nagaiah, T., Xia, W., Wang, Y., Dommele, S., Bitter, J., Santa, M., Grundmeier, G., Bron, M., Schuhmann, W., and Muhler, M. (2009) Electrocatalytic activity and stability of nitrogen-containing carbon nanotubes in the oxygen reduction reaction. *J. Phys. Chem. C*, **113**, 14302–14310.
63. Solorza-Feria, O., Ellmer, K., Giersig, M., and Alonso-Vante, N. (1994) Novel low-temperature synthesis of semiconducting transition metal chalcogenide electrocatalysts for multielectron charge transfer: molecular oxygen reduction. *Electrochim. Acta*, **39** (11/12), 1647–1653.
64. Hilgendorff, M., Diesner, K., Schulenburg, H., Bogdanoff, P., Bron, M., and Fiechter, S. (2002) Preparation strategies towards selective Ru-based oxygen reduction catalysts for direct methanol fuel cells. *J. New Mater. Electrochem. Syst.*, **5** (2), 71–81.

65. Zaikovskii, V.I., Nagabhushana, K.S., Kriventsov, V.V., Loponov, K.N., Cherepanova, S.V., Kvon, R.I., Bönnemann, H., Kochubey, D.I., and Savinova, E.R. (2006) Synthesis and structural characterization of Se-modified carbon-supported Ru nanoparticles for the oxygen reduction reaction. *J. Phys. Chem. B*, **110**, 6881–6890.

66. Schulenburg, H., Hilgendorff, M., Dorbandt, I., Radnik, J., Bogdanoff, P., Fiechter, S., Bron, M., and Tributsch, H. (2006) Oxygen reduction at carbon supported ruthenium–selenium catalysts: selenium as promoter and stabilizer of catalytic activity. *J. Power Sources*, **155** (1), 47–51.

67. Biddinger, E.J. and Ozkan, U.S. (2007) Methanol tolerance of CN_x oxygen reduction catalysts. *Top. Catal.*, **46**, 339–348.

68. Salgado, J.R.C., Antolini, E., and Gonzalez, E.R. (2005) Carbon supported Pt–Co alloy as methanol-resistant oxygen-reduction electrocatalysts for direct methanol fuel cells. *Appl. Catal. B*, **57**, 283–290.

69. Spendelow, J.S. and Wieckowski, A. (2007) Electrocatalysis of oxygen reduction and small alcohol oxidation in alkaline media. *Phys. Chem. Chem. Phys.*, **9**, 2654–2675.

70. Gasteiger, H.A. and Garche, J. (2006) in *Handbook of Heterogeneous Catalysis*, Vol. 8 (eds. G. Ertl, H. Knözinger, F. Schüth, and J. Weitkamp), Wiley-VCH Verlag GmbH, Weinheim, pp. 3081–3121.

71. Zhang, J., Xie, Z., Tang, Y., Song, C., Navessin, T., Shi, Z., Song, D., Wang, H., Wilkinson, D.P., Liu, Z.-S., and Holdcroft, S. (2006) High temperature PEM fuel cells. *J. Power Sources*, **160**, 872–891.

72. McBreen, J. and Mukerjee, S. (1995) In situ X-ray absorption studies of a Pt–Ru electrocatalyst. *J. Electrochem. Soc.*, **142**, 3399–3404.

73. Watanabe, M., Zhu, Y., Igarashi, H., and Uchida, H. (2000) Mechanism of CO tolerance at Pt-alloy anode catalysts for polymer electrolyte fuel cells. *Electrochemistry*, **68**, 244–251.

74. Watanabe, M. and Motoo, S. (1975) Electrocatalysis by ad-atoms. Part II. Enhancement of the oxidation of methanol on platinum by ruthenium ad-atoms. *Electroanal. Chem. Interfacial Electrochem.*, **60**, 267–273.

75. Gasteiger, H.A., Markovic, N.M., and Ross, P.N. (1995) H_2 and CO electrooxidation on well-characterized Pt, Ru, and Pt–Ru. 2. Rotating disk electrode studies of CO/H_2 mixtures at 62 °C. *J. Phys. Chem.*, **99**, 16757–16767.

76. Liu, P., Logadottir, A., and Norskov, J.K. (2003) Modeling the electro-oxidation of CO and H_2/CO on Pt, Ru, PtRu and Pt_3Sn. *Electrochim. Acta*, **48**, 3731–3742.

77. Arenz, M., Mayrhofer, K.J.J., Stamenkovic, V., Blizanac, B.B., Tomoyuki, T., Ross, P.N., and Markovic, N.M. (2005) The effect of the particle size on the kinetics of CO electrooxidation on high surface area Pt catalysts. *J. Am. Chem. Soc.*, **127**, 6819–6829.

78. Maillard, F., Schreier, S., Hanzlik, M., Savinova, E.R., Weinkauf, S., and Stimming, U. (2005) Influence of particle agglomeration on the catalytic activity of carbon-supported Pt nanoparticles in CO monolayer oxidation. *Phys. Chem. Chem. Phys.*, **7**, 385–393.

79. Markovic, N.M., Lucas, C.A., Grgur, B.N., and Ross, P.N. (1999) Surface electrochemistry of CO and H_2/CO mixtures at Pt(100) interface: electrode kinetics and interfacial structures. *J. Phys. Chem. B*, **103**, 9616–9623.

80. Antolini, E. (2009) Palladium in fuel cell catalysis. *Energy Environ. Sci.*, **2** (9), 915–931.

81. Antolini, E. (2003) Formation of carbon-supported PtM alloys for low temperature fuel cells: a review. *Mater. Chem. Phys.*, **78** (3), 563–573.

82. Lamy, C., Rousseau, S., Belgsir, E.M., Coutanceau, C., and Léger, J.-M. (2004) Recent progress in the direct ethanol fuel cell: development of

new platinum–tin electrocatalysts. *Electrochim. Acta*, **49**, 3901–3908.
83. Subhramannia, M. and Pillai, V.K. (2008) Shape-dependent electrocatalytic activity of platinum nanostructures. *J. Mater. Chem.*, **18**, 5858–5870.
84. Patterson, T.W. and Darling, R.M. (2006) Damage to the cathode catalyst of a PEM fuel cell caused by localized fuel starvation. *Electrochem. Solid-State Lett.*, **9**, A183–A185.
85. Wilson, M.S., Garzon, F.H., Sickafus, K.E., and Gottesfeld, S. (1993) Surface area loss of supported platinum in polymer electrolyte fuel cells. *J. Electrochem. Soc.*, **140** (10), 2872–2877.
86. Ferreira, P.J., la O', G.J., Shao-Horn, Y., Morgan, D., Makharia, R., Kocha, S., and Gasteiger, H.A. (2005) Instability of Pt/C electrocatalysts in proton exchange membrane fuel cells – a mechanistic investigation. *J. Electrochem. Soc.*, **152** (11), A2256–A2271.
87. Darling, R.M. and Meyers, J.P. (2003) Kinetic model of platinum dissolution in PEMFCs. *J. Electrochem. Soc.*, **150** (11), A1523–A1527.
88. Yasuda, K., Taniguchi, A., Akita, T., Ioroi, T., and Siroma, Z. (2006) Platinum dissolution and deposition in the polymer electrolyte membrane of a PEM fuel cell as studied by potential cycling. *Phys. Chem. Chem. Phys.*, **8**, 746–752.
89. Mayrhofer, K.J.J., Meier, J.C., Ashton, S.J., Wiberg, G.K.H., Kraus, F., Hanzlik, M., and Arenz, M. (2008) Fuel cell catalyst degradation on the nanoscale. *Electrochem. Commun.*, **10**, 1144–1147.
90. Bi, W., Sun, Q., Deng, Y., and Fuller, T.F. (2009) The effect of humidity and oxygen partial pressure on degradation of Pt/C catalyst in PEM fuel cell. *Electrochim. Acta*, **54**, 1826–1833.
91. Kobayashi, M., Hidai, S., Niwa, H., Harada, Y., Oshima, M., Horikawa, Y., Tokushima, T., Shin, S., Nakamori, Y., and Aoki, T. (2009) Co oxidation accompanied by degradation of Pt–Co alloy cathode catalysts in polymer electrolyte fuel cells. *Phys. Chem. Chem. Phys.*, **11**, 8226–8230.
92. Maass, S., Finsterwalder, F., Frank, G., Hartmann, R., and Merten, C. (2008) Carbon support oxidation in PEM fuel cell cathodes. *J. Power Sources*, **176**, 444–451.
93. Yu, X. and Ye, S. (2007) Recent advances in activity and durability enhancement of Pt/C catalytic cathode in PEMFC. Part II: degradation mechanism and durability enhancement of carbon supported platinum catalyst. *J. Power Sources*, **172**, 145–154.
94. Stevens, D.A., Hicks, M.T., Haugen, G.M., and Dahn, J.R. (2005) *Ex situ* and *in situ* stability studies of PEMFC catalysts: effect of carbon type and humidification on degradation of the carbon. *J. Electrochem. Soc.*, **152** (12), A2309–A2315.
95. Shao, Y., Wang, J., Kou, R., Engelhard, M., Liu, J., Wang, Y., and Lin, Y. (2009) The corrosion of PEM fuel cell catalyst supports and its implications for developing durable catalysts. *Electrochim. Acta*, **54**, 3109–3114.
96. Shao, Y., Liu, J., Wang, Y., and Lin, Y. (2009) Novel catalyst support materials for PEM fuel cells: current status and future prospects. *J. Mater. Chem.*, **19**, 46–59.
97. Shao, Y., Sui, J., Yin, G., and Gao, Y. (2008) Nitrogen-doped carbon nanostructures and their composites as catalytic materials for proton exchange membrane fuel cell. *Appl. Catal.B*, **79**, 89–99.
98. Centi, G. and Perathoner, S. (2010) Problems and perspectives in nanostructured carbon-based electrodes for clean and sustainable energy. *Catal. Today*, **150**, 151–162.
99. Prehn, K., Warburg, A., Schilling, T., Bron, M., and Schulte, K. (2009) Towards nitrogen-containing CNTs for fuel cell electrodes. *Compos. Sci. Technol.*, **69**, 1570–1579.
100. Alexeyeva, N., Shulga, E., Kisand, V., Kink, I., and Tammeveski, K. (2010) Electroreduction of oxygen on nitrogen-doped carbon nanotube modified glassy carbon electrodes in acid and alkaline solutions. *J. Electroanal. Chem.*, **648**, 169–175.

101. Li, N., Chen, X., Stoica, L., Xia, W., Qian, J., Assmann, J., Schuhmann, W., and Muhler, M. (2007) The catalytic synthesis of three-dimensional hierarchical carbon nanotube composites with high electrical conductivity based on electrochemical iron deposition. *Adv. Mater.*, **19** (19), 2957–2960.
102. Pletcher, D., Greff, R., Peat, R., Peter, L.M., and Robinson, J. (2010) *Instrumental Methods in Electrochemistry*, Woodhead Publishing, Cambridge.
103. Walsh, D.A., Fernández, J.L., and Bard, A.J. (2006) Rapid screening of bimetallic electrocatalysts for oxygen reduction in acidic media by scanning electrochemical microscopy. *J. Electrochem. Soc.*, **153**, E99–E103.
104. Greeley, J., Stephens, I.E.L., Bondarenko, A.S., Johansson, T.P., Hansen, H.A., Jaramillo, T.F., Rossmeisl, J., Chorkendorff, I., and Nørskov, J.K. (2009) Alloys of platinum and early transition metals as oxygen reduction electrocatalysts. *Nat. Chem.*, **1**, 552–556.
105. van Hardeveld, R. and Hartog, F. (1969) The statistics of surface atoms and surface sites on metal crystals. *Surf. Sci.*, **15**, 189–230.
106. Markovic, N.M. and Ross, P.N. (2000) Electrocatalysts by design: from the tailored surface to a commercial catalyst. *Electrochim. Acta*, **45**, 4101–4115.
107. Niemantsverdriet, J.W. (2007) *Spectroscopy in Catalysis*, 3rd edn., Wiley-VCH Verlag GmbH, Weinheim.
108. Mathew, R.J. and Russell, A.E. (2000) XAS of carbon supported platinum fuel cell electrocatalysts: advances towards real time investigations. *Top. Catal.*, **10**, 231–239.
109. Russell, A.E. and Rose, A. (2004) X-ray absorption spectroscopy of low temperature fuel cell catalysts. *Chem. Rev.*, **104**, 4613–4635.
110. Roth, C. and Ramaker, D.E. (2010) in *Fuel Cell Science: Theory, Fundamentals, and Biocatalysis* (eds. A. Wieckowski and J.K. Norskov), John Wiley & Sons, Inc., Hoboken, NJ, pp. 511–544.
111. Petrova, O., Kulp, C., van den Berg, M.W.E., Klementiev, K.V., Otto, B., Otto, H., Lopez, M., Bron, M., and Grünert, W. (2011) A spectroscopic proton-exchange membrane fuel cell test setup allowing fluorescence X-ray absorption spectroscopy measurements during state-of-the-art cell tests. *Rev. Sci. Instrum.*, **82**, 044101.
112. Smith, M.C., Gilbert, J.A., Mawdsley, J.R., Seifert, S., and Myers, D.J. (2008) In situ small-angle X-ray scattering observation of Pt catalyst particle growth during potential cycling. *J. Am. Chem. Soc.*, **130**, 8112–8113.
113. Navarro Yerga, R.M., Álvarez-Galván, M.C., Mota, N., Villoria de la Mano, J.A., Al-Zahrani, S.M., and Fierro, J.L.G. (2011) Catalysts for hydrogen production from heavy hydrocarbons. *ChemCatChem*, **3** (3), 440–457.
114. Rostrup-Nielsen, J. (2010) Reforming and gasification – fossil energy carriers, in *Hydrogen and Fuel Cells* (ed. D. Stolten), Wiley-VCH Verlag GmbH, Weinheim, pp. 291–305.
115. Kolb, G. (2008), *Fuel Processing for Fuel Cells*, Wiley-VCH Verlag GmbH, Weinheim, pp. 57–128.
116. Peters, R. (2006) in *Handbook of Heterogeneous Catalysis*, vol. **8** (eds. G. Ertl, H. Knözinger, F. Schüth, and J. Weitkamp), Wiley-VCH Verlag GmbH, Weinheim, pp. 3045–3080.
117. Lee, M.T., Werhan, M.D., Hwang, J., Hotz, N., Greif, R., Poulikakos, D., and Grigoropoulos, C.P. (2010) Hydrogen production with a solar steam-reformer and colloid nanocatalyst. *Int. J. Hydrogen Energy*, **35**, 118–126.
118. Häussinger, P. Lohmüller, R., and Watson, A.M., (2000) Hydrogen, 2. Production. in *Ullmann's Encyclopedia of Industrial Chemistry*, pp. 251–269.
119. Schmal, M., Scheunemann, R., Ribeiro, N.F.P., Bengoa, J.F., and Marchetti, S.G. (2011) Synthesis and characterisation of Pt/Fe–Zr catalysts for the CO selective oxidation. *Appl. Catal. A: Gen.*, **392**, 1–10.
120. Balat, H. and Kirtay, E. (2010) Hydrogen from biomass – present and future scenario and future prospects. *Int. J. Hydrogen Energy*, **35**, 7416–7426.
121. Ni, M., Leung, Y.C., Leung, M.K.H., and Sumathy, K. (2006) An overview

of hydrogen production from biomass. *Fuel Process. Technol.*, **87**, 461–472.
122. Nath, K. and Das, D. (2003) Hydrogen from biomass. *Curr. Sci.*, **85** (3), 265–271.
123. Cortright, R.D., Davda, R.R., and Dumesic, J.A. (2002) Hydrogen from catalytic reforming of biomass-derived hydrocarbons in liquid water. *Nature*, **418**, 964–967.
124. Wen, G., Xu, Y., Ma, H., Xu, Z., and Tian, Z. (2008) Production of hydrogen by aqueous-phase reforming of glycerol. *Int. J. Hydrogen Energy*, **33**, 6657–6666.
125. Davda, R.R., Shabaker, J.W., Huber, G.W., Cortright, R.D., and Dumesic, J.A. (2005) Hydrogen from catalytic reforming of biomass-derived hydrocarbons in liquid water. *Appl. Catal. B: Environ.*, **56**, 171–186.
126. Chheda, J.N., Huber, G.W., and Dumesic, J.A. (2007) Liquid-phase catalytic processing of biomass-derived oxygenated hydrocarbons to fuels and chemicals. *Angew. Chem. Int. Ed.*, **46**, 7164–7183.

Part III
Analytics and Diagnostics

16
Impedance Spectroscopy for High-Temperature Fuel Cells
Ellen Ivers-Tiffée, André Leonide, Helge Schichlein, Volker Sonn, and André Weber

16.1
Introduction

Impedance spectroscopy has been established for many years as a powerful measurement technique for the electrical characterization of electrochemical systems. The power of the method lies in the fact that by small-signal perturbation it reveals both the relaxation times and relaxation amplitudes of the various processes present in a dynamic system over a wide range of frequencies.

Electrochemical impedance spectroscopy (EIS) is especially useful if the system performance is governed by a number of coupled processes, each proceeding at a different rate. In these systems, steady-state polarization curves are not useful because static measurements can only identify relatively simple reaction mechanisms that are known to be clearly dominated by a single rate-determining step. However, fuel cells are prominent examples of complex dynamic materials systems that generally do not fulfill this requirement. The physical and chemical processes contributing to the internal resistance of the cell determine their dynamic behavior over a wide range of frequencies. The relaxation times typically span more than 15 orders of magnitude, extending from fast processes that sustain cell operation, for example, current flow, gas conversion, and charge transfer, to long-term degradation processes limiting the lifetime of the cell (Figure 16.1). Owing to practical factors limiting the frequency range of impedance measurements at fuel cells, the method is feasible for processes with relaxation times ranging from microseconds up to tens of seconds. Slower processes exhibiting time constants from several minutes to hundreds of hours are favorably observed in the time domain, for example, by analyzing the response of the cell on a step function of the current, or the potential, respectively.

Processes in fuel-cell systems typically involve complex multi-step reactions that can proceed along several parallel reaction pathways, for example, adsorption, heterogeneous reactions at interfaces, and bulk and surface transport. In the case of advanced electrodes, neither of these reaction steps dominates the overall cell characteristics. The extent to which the various steps contribute to cell performance is governed by their complex dependence on each other and also on the particular

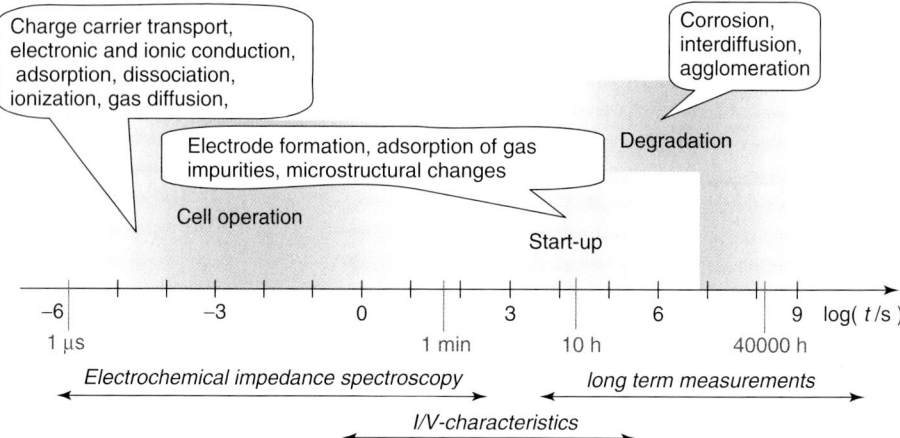

Figure 16.1 Relaxation times of physical processes present in fuel cell operation and corresponding electrical measurement techniques. The dynamic range spans over 15 orders of magnitude. Fast processes are covered by electrochemical impedance spectroscopy.

operating conditions and design of the cell. In order to clarify their contributions to the internal resistance of the cell, EIS has to be used. The technique has therefore become widely used for characterizing the intrinsic loss factors of fuel cell operation.

However, the relaxation times of the physical processes themselves cannot be observed directly from the measurement data if their impedance contributions overlap in the spectrum. Therefore, the impedance data have to be analyzed with respect to the underlying dynamic processes. The physical interpretation of this kinetic information is the key to predicting fuel-cell properties under different operating conditions and different material configurations and thus to permitting a well-directed improvement of fuel-cell performance.

Computer-controlled impedance analysis devices and data acquisition software enable the materials scientist to acquire large amounts of impedance data in a short time. The difficulty in successfully interpreting these data by relating relaxation processes to the underlying physical mechanisms is probably due to the interdisciplinarity of the subject, where a whole range of problems have to be addressed simultaneously. Apart from the preparation and processing of materials and composite structures, one must be able to obtain mathematical expressions describing the frequency response of the relevant processes for the system under investigation, that is, to simulate the impedance from physical models for reactions and transport processes. On the other hand, the researcher also has to be familiar with the concept of impedance and the corresponding measurement technique, which is rather complex, especially in the case of high-temperature conditions. Furthermore, the subsequent data analysis and interpretation of electrochemical models require knowledge in the fields of signal processing and system identification.

Building meaningful models is often experienced to be the most difficult step of the process and is therefore conducted at the end of the investigation. The importance of modeling as an integral part of the investigation is emphasized. Finding, adjusting, and identifying appropriate models should be carried out prior to and during the experimental work.

This chapter provides the fundamentals of impedance spectroscopy. Special emphasis is placed on the issues of fuel-cell development, where impedance spectra are particularly difficult to measure and evaluate owing to the large number and great complexity of concurring physical and chemical processes that contribute to the overall electrical response of the cell.

16.2
Fundamentals

16.2.1
Principle of Electrochemical Impedance Spectroscopy

Whenever an electrical current I_{load} passes through an electrochemical cell, the cell voltage U_{cell} deviates from its equilibrium value U_{OCV} under open-circuit conditions. When this deviation, the so-called overvoltage $U_{Pol} = U_{OCV} - U_{cell}$, and the corresponding cell current I_{load} are known, a polarization resistance R_{Pol} can be assigned to the system. In general, the multitude of chemical and physical processes in the system leads to a complicated, nonlinear relationship between U_{cell} and I_{load}. Therefore, even in the DC case, the definition of the polarization resistance is difficult. Moreover, polarization resistances that are determined from AC measurements are fundamentally different from those determined from DC measurements. When the system is perturbed by an AC input current signal, the AC voltage signal observed at the cell's terminals is phase shifted with respect to the perturbation input and a complex impedance value $\underline{Z}_{cell}$ can be assigned to the system from the phase angle and amplitude gain of the response [1–4]. At this point, it is worth mentioning that in fuel-cell science generally the area specific resistance (ASR), that is, the measured resistance normalized by the (active electrode) area, should be analyzed. In this way, different materials and/or electrode structures can be compared independently with respect to their active surface. In other words, in contrast to the resistance R, the ASR is a material-specific value.

However, owing to the highly nonlinear behavior of an electrochemical system, the system response generally depends not only on the frequency of the input signal but also on its amplitude. The output measurement signal then contains higher harmonics of the AC perturbation and the impedance depends on the amplitude of the perturbation. The quantitative analysis of these effects is not feasible; therefore, it is crucial to keep the perturbation amplitude sufficiently small in order to be able to neglect these nonlinear effects. The most practical way to check this requirement is to increase the input signal to the maximum value at which the response is still independent of the perturbation amplitude. A theoretical linearity check is the

Kramers–Kronig consistency of the data [5]. In this regard, extensive experimental work [6, 7] has shown that a voltage response of 10–12 mV (rms) should represent a good compromise between a high signal-to-noise ratio, corresponding to a high data quality, and the linearity constraint.

For frequencies that are low compared with the relaxation frequencies in the system, the impedance obtained approaches the differential resistance as determined from the DC measurement, that is, the slope of the current–voltage $(I - U)$ characteristics for the particular point of operation (Figure 16.2a).

The basic experimental arrangement for impedance measurement is shown in Figure 16.2b. A sinusoidal current of small amplitude $i(t) = i_0 \sin(\omega t)$ is superposed on a defined bias current I_{load} and the sinusoidal voltage response $u(t) = u_0(\omega) \sin[(\omega t + \varphi(\omega)]$ is measured (Figure 16.2a). From the ratio between the complex variables of voltage and current, the impedance is calculated as follows [8]:

$$\underline{Z}(\omega) = \frac{\underline{u}(t)}{\underline{i}(t)} = \frac{u_0(\omega)}{i_0} e^{j\varphi(\omega)} = \left|\underline{Z}(\omega)\right| e^{j\varphi(\omega)} = \mathrm{Re}\left\{\underline{Z}(\omega)\right\} + j\,\mathrm{Im}\left\{\underline{Z}(\omega)\right\}$$

(16.1)

where $\omega = 2\pi f\,(\mathrm{s}^{-1})$ represents the angular frequency and $\varphi(\omega)$ the frequency-dependent phase shift between voltage and current.

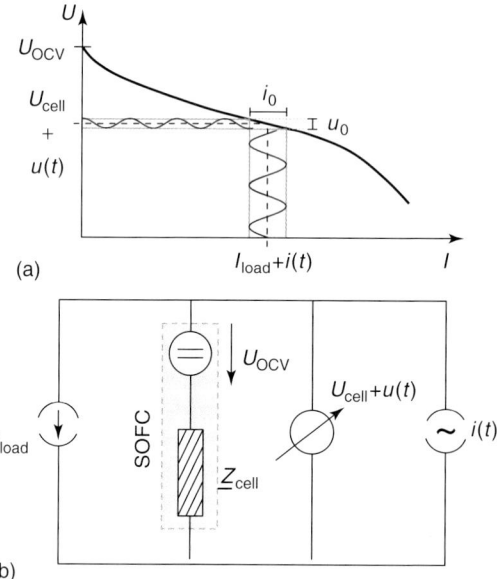

Figure 16.2 (a) Schematic plot of voltage versus current density of a solid oxide fuel cell (SOFC) (I–U curve). A sinusoidal current of small amplitude $i(t)$ is superposed to a defined bias current I_{load} and the voltage response $u(t)$ is measured. (b) Basic experimental setup for the impedance measurement of a real SOFC with an internal impedance $\underline{Z}_{\text{cell}}$ [8].

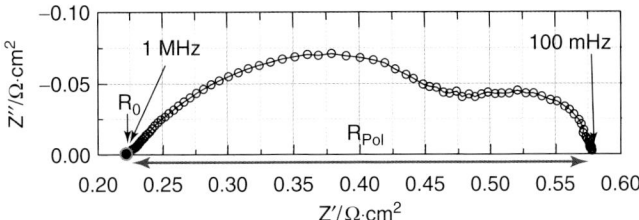

Figure 16.3 Typical Nyquist plot recorded on a real anode-supported SOFC single cell. The high-frequency intercept (for $\omega \to \infty$) with the real axis corresponds to the purely ohmic resistance R_0. The difference between the low- and high-frequency intercept is the so-called polarization resistance R_{Pol} of the cell (cell Z1_153).

In general, this measurement is performed for a discrete quantity of frequency values in a defined frequency range and the recorded impedance values are usually plotted in the complex plane, denoted a Nyquist plot. Figure 16.3 gives an example of a Nyquist plot measured on an anode supported SOFC single cell. The high-frequency intercept (for $\omega \to \infty$) with the real axis corresponds to the purely ohmic resistance R_0 of the cell, whereas the intercept at the lower frequency (for $\omega \to 0$) is identical with the differential resistance, obtained from the corresponding $I - U$ characteristic at the given operating point. The difference between the low- and high-frequency intercepts is the so-called polarization resistance R_{pol} of the cell. R_{pol} is the sum of each single polarization resistance caused by the individual loss mechanisms.

16.2.1.1 Operating Principle of Frequency Response Analyzers

In the past, bridge methods relying on the null detection of a balance condition have been used for impedance measurements. The frequency range was usually restricted to 20 Hz–20 kHz. The determination of each impedance value involved cumbersome analog signal analysis in the frequency domain. A review of these basic measurement arrangements is given in [1]. In the last three decades, automated impedance measurement equipment, termed *"frequency response analyzer"* has become available to the researcher and is used almost exclusively except for special custom-made bridge setups when very high accuracy is required. Frequency response analyzers are digitally demodulated, computer-controlled, stepped-frequency impedance meters. Typical representatives of these devices are the Solartron 1260 and Zahner IM6, both costing US$20 000–40 000 [9, 10]. Commercial suppliers typically provide a full-scale suite of instruments including potentiostats, galvanostats, amplifiers, high-current boosters, and data acquisition and analysis software. These packages allow the convenient and high-precision measurement of impedance spectra over a wide range of frequencies.

In frequency response analyzers, the impedance is determined from the correlation of the cell response $u(t)$ with two reference signals that are generated from the input signal. One reference signal is in-phase with the input whereas the other

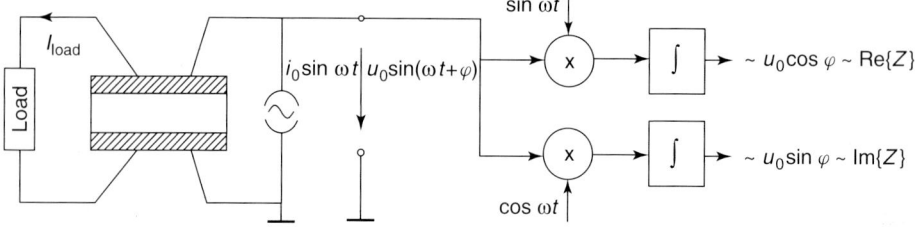

Figure 16.4 Orthogonal signal correlation principle of a frequency response analyzer for impedance determination.

is phase shifted by $\pi/2$. As depicted in Figure 16.4, the output signal is correlated with each of the reference signals and the mean of the resulting signals is taken by integration over several cycle periods. $u(t)$ consists of the fundamental harmonic component, of possible higher harmonic components due to nonlinearity of the system, and of measurement noise $n(t)$:

$$u(t) = u_0 \sin(\omega t + \varphi) + \sum_k A_k \sin(k\omega t + \varphi_k) + n(t) \tag{16.2}$$

If $n(t)$ is stochastic, its mean value vanishes for a sufficiently long measurement interval T_{int}. When T_{int} is a full number of cycles, the higher harmonics are also cancelled by the integration. We therefore obtain at the outputs of the frequency response analyzer a signal that is proportional to the real and the imaginary part of the impedance (Figure 16.4):

$$\frac{u_0}{T_{int}} \int_0^{T_{int}} \sin(\omega t + \varphi) \sin \omega t \, dt = \frac{u_0}{2} \left[\cos \varphi - \frac{1}{T_{int}} \int_0^{T_{int}} \cos(2\omega t + \varphi) \, dt \right]$$

$$= \frac{1}{2i_0} |Z(\omega)| \cos \varphi(\omega) = \frac{1}{2i_0} \text{Re}\{Z(\omega)\}$$

$$\frac{u_0}{T_{int}} \int_0^{T_{int}} \sin(\omega t + \varphi) \cos \omega t \, dt = \frac{u_0}{2} \left[\sin \varphi - \frac{1}{T_{int}} \int_0^{T_{int}} \sin(2\omega t + \varphi) \, dt \right]$$

$$= \frac{1}{2i_0} |Z(\omega)| \sin \varphi(\omega) = \frac{1}{2i_0} \text{Im}\{Z(\omega)\} \tag{16.3}$$

16.2.2
Impedance Data Analysis

16.2.2.1 Evaluation of Data Quality

The measurement data quality is of crucial importance for a reliable interpretation of impedance curves. Both the quality and quantity of information that can be extracted from impedance data are implicitly connected to the noise level and the compliance of the measured curve with the principles of causality, linearity, and stability.

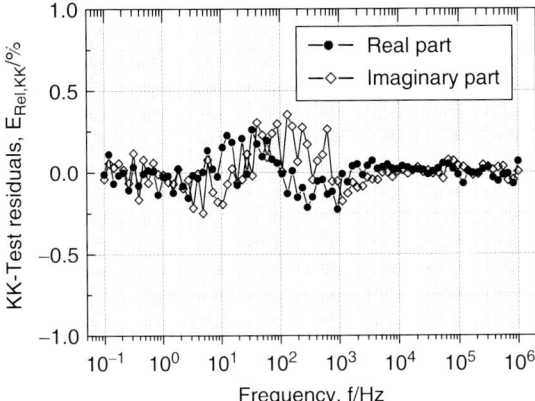

Figure 16.5 Kramers–Kronig test residuals of a typical impedance spectrum, calculated with the "KK Test for Windows" software [12, 13].

The Kramers–Kronig validation [11] is a well-established method to assess consistency and quality of measured impedance spectra. The Kramers–Kronig relations are integral equations, which constrain the real and imaginary components of the impedance for systems that satisfy the conditions of causality, linearity, and stability [11].

Figure 16.5 shows a Kramers–Kronig validation for a typical impedance spectrum calculated with the "KK Test for Windows" software [12, 13]. For most of the spectrum, the relative errors of both real and imaginary data are <0.4 %, confirming an excellent data quality.

16.2.2.2 Complex Nonlinear Least-Squares (CNLS) Fit

As the next step, an equivalent circuit model (ECM), which suitably describes the measured impedance response, may be proposed.

The unknown values of the equivalent circuit elements are generally evaluated by a non-linear least-squares (NLS) fitting algorithm. The least-squares fitting procedure aims at finding a set of parameters which minimizes the sum [1]:

$$Q\left(\underline{p}\right) = \sum_{k=0}^{N-1} \left\{ w_{k,r} \left[\Phi'\left(\omega_k, \underline{p}\right) - Z'(\omega_k) \right]^2 + w_{k,i} \left[\Phi''\left(\omega_k, \underline{p}\right) - Z''(\omega_k) \right]^2 \right\} \to \text{Min}$$

(16.4)

with

$$\underline{\Phi}\left(\omega_k, \underline{p}\right) = \Phi'\left(\omega_k, \underline{p}\right) + j\Phi''\left(\omega_k, \underline{p}\right) \text{ and } \underline{Z}(\omega_k) = Z'(\omega_k) + jZ''(\omega_k)$$

(16.5)

and

$$\underline{Z}(\omega_k) = Z'(\omega_k) + jZ''(\omega_k)$$

(16.6)

where $\underline{\Phi}(\omega_k, \underline{p})$ represents the model fit function and $\underline{p} = (R_0, R_1, R_2, C_1, Q_2, ..., R_N, Q_N)$ is the fit parameter vector [1]. $Z'(\omega_k)$ and $Z''(\omega_k)$ are the real and imaginary part of the measured impedance curve $\underline{Z}(\omega_k)$, respectively. The factors $w_{k,r}$ and $w_{k,i}$ are the weighting factors associated with the kth data point.

Figure 16.6 shows a typical complex nonlinear least-squares CNLS fit procedure that has been carried out on a measured impedance curve [14]. The applied

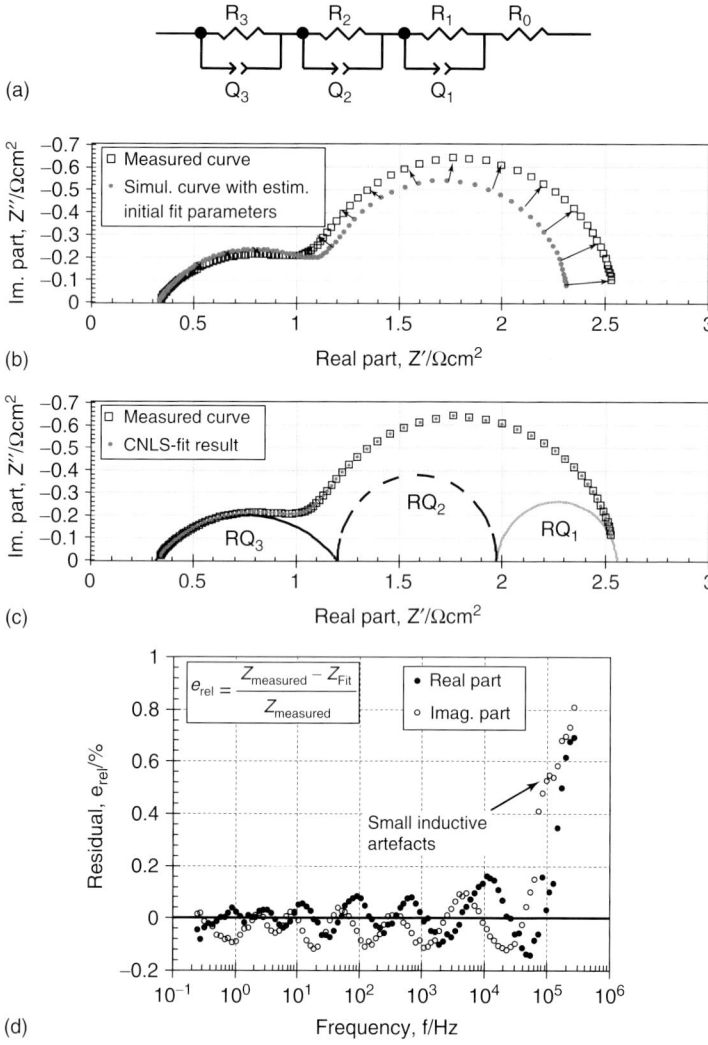

Figure 16.6 (a) A priori defined equivalent circuit model consisting of a series connection of three RQ elements (RQ_1, RQ_2, RQ_3) and an ohmic resistance (R_0). (b) Measured impedance curve along with the simulated curve obtained from the equivalent circuit with the estimated initial parameters. (c) CNLS fit result. (d) Residual pattern of the CNLS fit [14].

equivalent circuit consists of an ohmic resistor R_0 and three RQ elements connected in series (Figure 16.6a). Before starting the actual fitting procedure, appropriate initial fit parameter values of the respective impedance elements need to be defined (i.e., estimated). Figure 16.6b shows the measured impedance curve along with the simulated curve obtained from the equivalent circuit with the estimated initial parameters. Finally, the ECM is fitted to the impedance spectra by the CNLS fitting algorithm (Figure 16.6c).

In general, the fit quality is evaluated by analyzing the relative errors (residuals) between the fit result and the measured curve at each single data point. Figure 16.6d shows the residuals of the CNLS fit performed above. In this case, the residuals are distributed uniformly around the frequency axis yielding errors of <0.2%. Only at higher frequencies (>200 kHz) do inductive artifacts caused by the wiring become noticeable. Nevertheless, the error remains <1%, which is sufficiently small. This analysis clearly indicates that the ECM is well chosen and describes the measured impedance spectrum over the recorded frequency range. However, no conclusion about the physical correctness of an ECM can be drawn from the analysis of a single impedance spectrum. The residuals are "only" a mathematical quantity that describes the level of equivalency between the measured curve and the theoretical impedance spectrum of the proposed model. This discussion leads us directly to the weaknesses of the CNLS fitting approach, which can be summarized as follows [8]:

1) Poor resolution in the frequency domain.
2) An *a priori* defined electrical equivalent circuit is needed.
3) Ambiguity of the equivalent circuit.

The poor resolution of Nyquist plots is problematic because the individual impedance-related processes of complex electrochemical systems, such as SOFC single cells, are numerous and their contributions to the impedance curve overlap. This problem is aggravated by the fact that not every process contributes in the same way to the total polarization loss. Therefore, it is very difficult to discover processes with a small contribution, because they are almost totally covered by the processes that show a large polarization loss.

Another weakness of the CNLS analysis method is the ambiguity of the equivalent circuits, meaning that identical impedance spectra may be obtained from different circuits [1, 8]. Therefore, it is questionable to propose an *a priori* model without any knowledge about the real number and physical origin of the polarization processes contributing to the impedance response of the cell.

To overcome these disadvantages, an alternative approach for analyzing impedance spectra is presented here. The ECM and the optimal starting parameters for the CNLS algorithm are obtained by a preidentification of the impedance response by calculating and analyzing the corresponding distribution function of relaxation times (DRT). The DRT approach is particularly advantageous for the analysis of anode-supported SOFC single cells coupled to thin electrolytes (thickness <20 µm), where reference electrodes are not applicable for the separation of anode and cathode losses [15].

However, in all cases, the ultimate evaluation of the physical correctness of a proposed equivalent circuit should be performed by analyzing the quantitative dependence of the model parameters for a wide range of operating conditions, as is shown later.

16.2.2.3 Distribution Function of Relaxation Times (DRT)

The DRT method uses the fact that every impedance function that obeys the Kramers–Kronig relations can be represented as a differential sum of infinitesimally small RC elements (Figure 16.7). This sum goes from 0 to ∞ [16].

Let us consider a serial connection of RC elements (Figure 16.7). Let $R_{\text{pol},n} = \gamma_n R_{\text{pol}}$ be the real part resistance and τ_n the relaxation time of the nth RC element. R_{pol} represents the total polarization resistance of the circuit. Hence for the total impedance the following equation holds [16]:

$$Z_{\text{pol}}(\omega) = \sum_{n=1}^{N} \frac{R_{\text{pol},n}}{1+j\omega\tau_n} = R_{\text{pol}} \sum_{n=1}^{N} \frac{\gamma_n}{1+j\omega\tau_n} \qquad (16.7)$$

with

$$\sum_{n=1}^{N} \gamma_n = 1 \qquad (16.8)$$

where $\omega = 2\pi f$ is the angular frequency, N is the number of RC elements, and j is the imaginary unit. Here γ_n weights the contribution of the nth polarization process to the total polarization loss.

Instead of a finite number of RC elements, one can assume now an infinite number with relaxation times ranging continuously from 0 to ∞ and obtain the

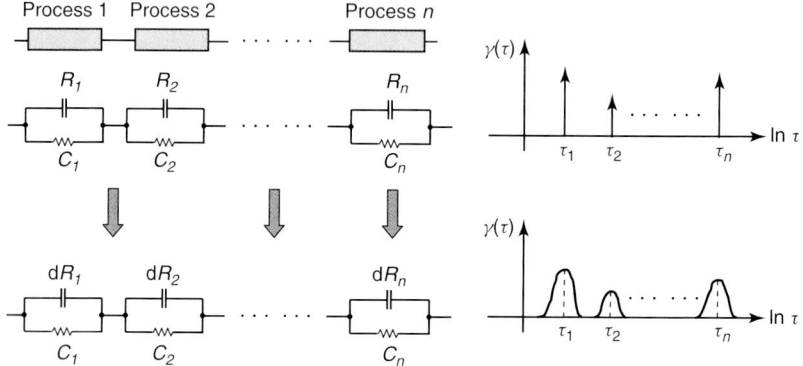

Figure 16.7 Interpretation of EIS data in terms of equivalent circuit models and distribution of relaxation times. Dynamic processes are represented in the distribution by peaks: in the case of ideal RC processes by Dirac impulse functions at the corresponding relaxation time $[\delta(\tau - \tau_n)]$, whereas real processes exhibit peaks distributed around a main relaxation time τ_n [16].

following integral equation [16]:

$$\underline{Z}_{pol}(\omega) = R_{pol} \int_0^\infty \frac{\gamma(\tau)}{1+j\omega\tau} d\tau \qquad (16.9)$$

with

$$\int_0^\infty \gamma(\tau) d\tau = 1 \qquad (16.10)$$

The integral Eq. (16.9) links the distribution function $\gamma(\tau)$ with the frequency-dependent part of the measured impedance $\underline{Z}_{pol}(\omega)$. In Eq. (16.9) $[\gamma(\tau)/(1+j\omega\tau)] d\tau$ represents the fraction of the overall polarization resistance with relaxation times between τ and $\tau + d\tau$. This implies that the area comprised by a peak equals the total polarization resistance of the respective dynamic process.

The problem now is to obtain $\gamma(\tau)$ from $\underline{Z}_{pol}(\omega)$. The real and imaginary parts of the impedance data of a linear, time-invariant system are connected by the Kramers–Kronig transformations. Therefore, it is sufficient to consider the imaginary part of the impedance only [16]:

$$\text{Im}\{\underline{Z}_{pol}(\omega)\} = Z''(\omega) = R_{pol} \int_0^\infty \frac{\omega\tau}{1+(\omega\tau)^2} \gamma(\tau) d\tau \qquad (16.11)$$

After the substitution of the frequency variables with the following expressions:

$$x = \ln\frac{\omega}{\omega_0};\ y = \ln(\omega\tau);\ dy = \frac{1}{\tau} d\tau \qquad (16.12)$$

Equation (16.11) adopts the following form:

$$Z''(\omega) = -R_{pol} \int_0^\infty \frac{e^y}{1+e^{2y}} \gamma(\tau) d\tau \qquad (16.13)$$

with

$$\text{sech}(x) = \frac{2}{e^y + e^{-y}} \text{ and } \hat{g}(y-x) = \gamma(\tau)\tau \qquad (16.14)$$

Equation (16.13) is recognizable as the following convolution product:

$$Z''(x) = -\frac{R_{pol}}{2} \int_{-\infty}^\infty \text{sech}(y)\hat{g}(y-x) dy = -\frac{R_{pol}}{2} \text{sech}(x)^*\hat{g}(x) = -\frac{1}{2}\text{sech}(x)^*g(x) \qquad (16.15)$$

with $g(x) = R_{pol}\hat{g}(x)$.

Equation (16.13) can be converted into an algebraic product by Fourier transformation and, finally, the distribution function $g(x)$ is obtained by back-transformation. This inversion problem is controlled by application of digital filters in the Fourier space and an extrapolation technique to enlarge the frequency range of the data. The detailed procedure is described elsewhere [8].

However, in the following the inversion problem shown in Eq. (16.11) was not solved by the above-mentioned Fourier space transformation and subsequent digital filtering. Alternatively, the convolution product, cf. Eq. (16.13), can be regarded as a Fredholm integral equation of the first kind which can be described by the following equation [8, 17]:

$$\int_a^b K(x,y)g(y)dy = z(x) \tag{16.16}$$

with K representing the Kernel of a given function $z(x)$ and $g(y)$ representing the unknown solution. For the solution of this inverse problem, also called an "ill-posed problem," the Tikhonov regularization [18–21] method is applied and leads, principally, to the same result as the Fourier transformation method.

16.3
Experimental Examples

In the following, a physicochemical ECM for anode-supported single cells (ASCs) with thin-film electrolyte, developed at Forschungszentrum Jülich (FZJ), is presented.

Widely known is the approximation of the measured impedance spectra by a CNLS fit procedure, which results in a model function represented by an equivalent circuit [22]. However, this approach necessitates an *a priori* definition of the ECM. Consequently, the ECM is applied without knowing the number and nature of ohmic and polarization processes contributing to the total cell impedance. This leads to a severe ambiguity of the adopted model [1].

This problem can be avoided, for example, by the following alternative approach for analyzing impedance spectra. First, the starting parameters for the CNLS algorithm are obtained by a preidentification of the impedance response by calculating and analyzing the corresponding DRT. The DRT method, presented in detail in [16], is of particular relevance for the EIS analysis of SOFC single cells with thin-film electrolyte and a supporting structure such as anodes, cathodes, and others where reference electrodes are not applicable [15].

The measurements presented were performed on ASCs with identical material composition, but with nickel/8 mol % yttria-doped zirconia (8YSZ) anode substrates with a thickness of 1000–1500 µm [23]. The consecutive layers consisted of an 8YSZ thin-film electrolyte (10 µm), an interlayer made of $Ce_{0.8}Gd_{0.2}O_{2-\delta}$ (GCO, 8 µm) and a cathode of $La_{0.58}Sr_{0.4}Co_{0.2}Fe_{0.8}O_{3-\delta}$ (L58SCF, 45 µm). Details on the preparation and properties of these cells can be found elsewhere [24–26].

16.3.1
Process Identification

The higher resolution of the DRT identifies polarization processes with characteristic frequencies separated by only half a decade. This superiority is proven by comparing a measured EIS spectrum (Figure 16.8a) and the corresponding DRT calculation (Figure 16.8b) of an ASC. Whereas the impedance curve shows only two depressed semicircles, the calculated DRT reveals at least five individual polarization processes (P_{1C}, P_{1A}, P_{2C}, P_{2A}, and P_{3A}).

However, the physical nature of all processes contributing to the overall polarization loss of the cell still has to be identified. As a consequence, a series of ~300 impedance measurements were performed, in which only one parameter at a time was varied: cathodic oxygen partial pressure, anodic hydrogen/water partial

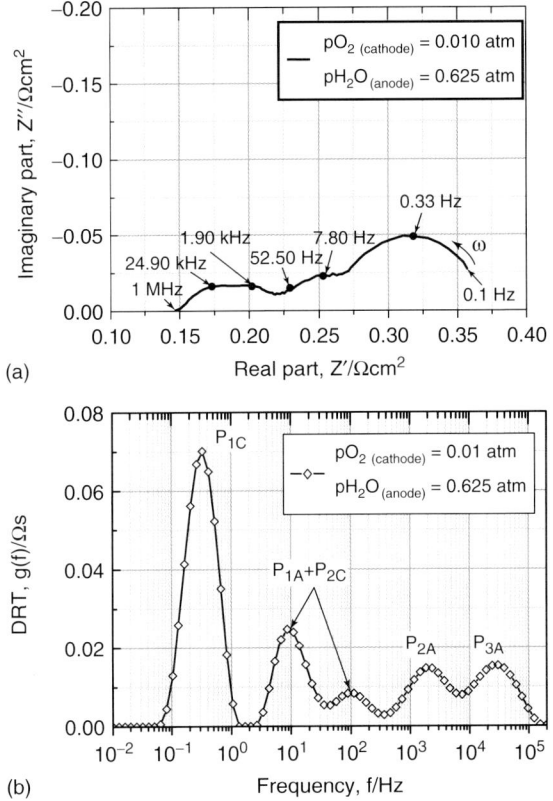

Figure 16.8 (a) Typical impedance spectra of an anode-supported cell with L58SCF cathode recorded at $T = 800\,°C$, $pO_{2(cathode)} = 0.01$ atm, $pH_2O_{(anode)} = 0.625$ atm and (b) corresponding distribution function of relaxation times (DRT). Unlike the Nyquist plot, at least five processes are visible in the distribution curve (cell Z1_153).

pressure (fuel utilization), or temperature. Subsequently, the DRTs were calculated and carefully interpreted with respect to their operating parameter dependency.

16.3.1.1 Variation of Temperature

Typical impedance spectra and corresponding DRT calculations at four temperatures, 730, 750, 800, and 860 °C, are shown in Figure 16.9. The impedance spectra were measured at a fuel gas composition of 62.5 %H_2O and 37.5 %H_2. Air was used as cathode gas.

By optical inspection of the EIS curves (Figure 16.9a), up to three depressed semicircles can be identified, which increase in length and area with decrease in temperature. No further conclusions can be drawn from the data.

In contrast, the DRT calculation reveals four individual processes, and it is evident that the processes P_{2C}, P_{2A}, and P_{3A} are characterized by a pronounced thermal activation, as their peak area increases with decrease in temperature. Process P_{1A} shows, by contrast, a negligible dependence on the operating temperature.

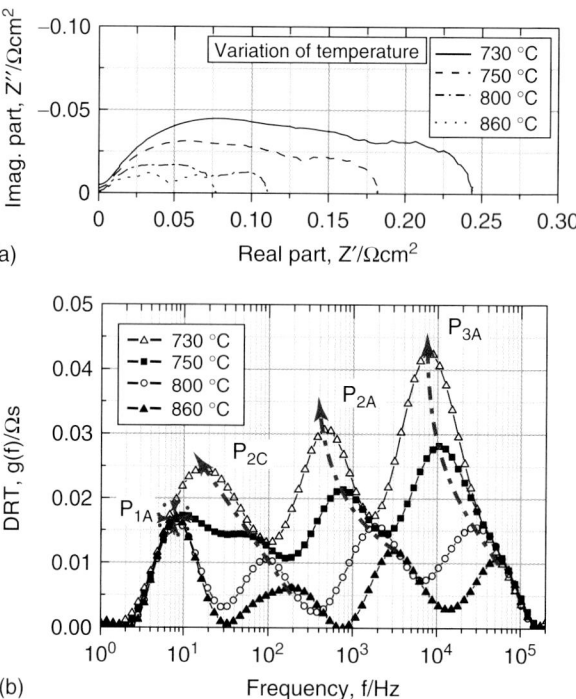

Figure 16.9 (a) Series of impedance spectra (ohmic part subtracted) and (b) corresponding distribution curves at four temperatures [$pH_2O_{(anode)}$ = 0.625 atm (balance H_2), $pO_{2(cathode)}$ = 0.21 atm (air)] (cell Z1_153).

16.3.1.2 Variation of Anodic Water Partial Pressure

In order to identify the anodic processes and their dependence on the partial pressure of water, $pH_2O_{(anode)}$, in the fuel gas, the H_2O content was varied stepwise between 4.88 and 62.5% (balance H_2). Air was used as cathode gas.

Figure 16.10b shows the DRTs calculated from impedance spectra (cf., Figure 16.10a), recorded at four different $pH_2O_{(anode)}$. Both processes P_{1A} and P_{2A} show a significant dependence on the water content in the fuel gas, whereas process P_{3A} is characterized by a minor dependence.

Particular attention should be paid to process P_{1A}. As will be shown later, in the equivalent circuit this process is modeled by a generalized finite length Warburg (G-FLW) element accounting for the diffusion loss within the anode substrate. The DRT of a G-FLW element exhibits a large peak located at the characteristic frequency followed by smaller peaks at higher frequencies. Depending on the operating conditions, this makes it difficult to identify the process P_{2C} as this has a maximum in a frequency range similar to that of the two major P_{1A} peaks between 10 and 100 Hz, thus leading to a possible overlap of the peaks characteristic of both P_{1A} and P_{2C} (Figure 16.10b).

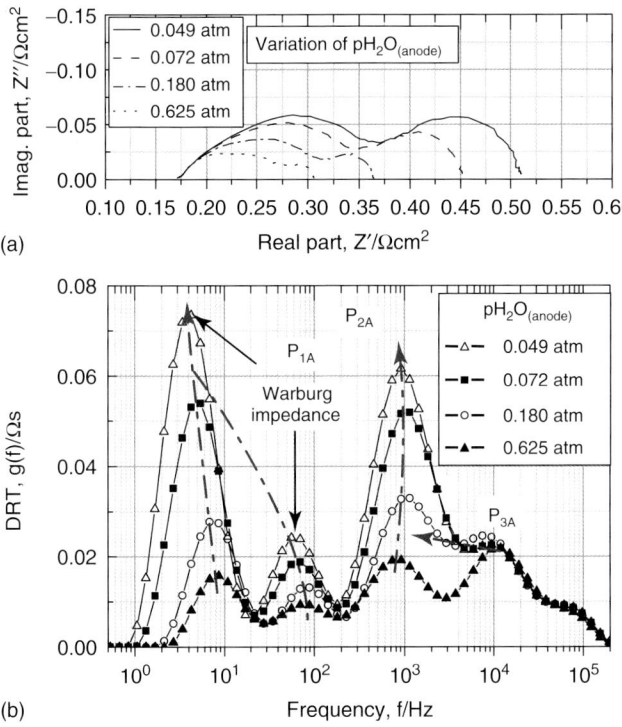

Figure 16.10 (a) Series of impedance spectra and (b) corresponding distribution curves at four different $pH_2O_{(anode)}$ [$pO_{2(cathode)}$ = 0.21 atm (air), $T = 757\ °C$] (cell Z1_153).

16.3.1.3 Variation of Cathodic Oxygen Partial Pressure

The oxygen content of the gas mixture supplied to the cathode was varied between 21% in case of air and 1% balance nitrogen, whereas the composition of the fuel gas was kept constant at a ratio 62.5%H_2O to 37.5%H_2. The high water content was used to keep the anodic polarization losses at a minimum (Figure 16.10b), which is advantageous for the deconvolution of the cathodic contributions.

Figure 16.11b shows the influence of the oxygen partial pressure, $pO_{2(cathode)}$, on the distribution of relaxation times (the corresponding impedance spectra are shown in Figure 16.11a). It is clearly visible that a "new" process, denoted P_{1C}, evolves in the frequency range below 10 Hz at oxygen contents ≤0.05 atm. At the same time, process P_{2C}, already identified in Figure 16.9, is shifted towards lower frequencies, thus strongly overlapping with the two major peaks between 7 and 100 Hz related to the Warburg-type process P_{1A}; see arrows on P_{1A} in Figure 16.11b.

In order to analyze the temperature dependence of the newly identified process P_{1C}, additional impedance measurements were conducted at three temperatures (Figure 16.12a), and at the lowest oxygen partial pressure, $pO_{2(cathode)} = 0.01$ atm.

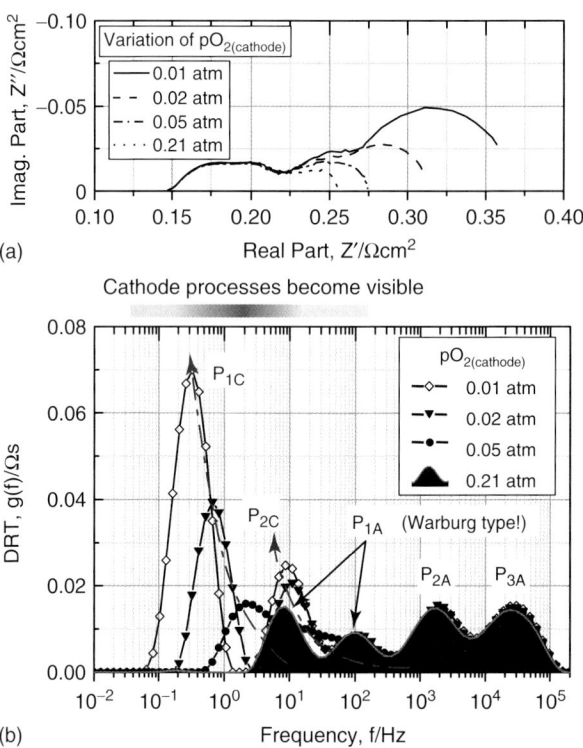

Figure 16.11 (a) Series of impedance spectra and (b) corresponding distribution curves at four different $pO_{2(cathode)}$ [$pH_2O_{(anode)} = 0.625$ atm (balance H_2), $T = 800$ °C] (cell Z1_153).

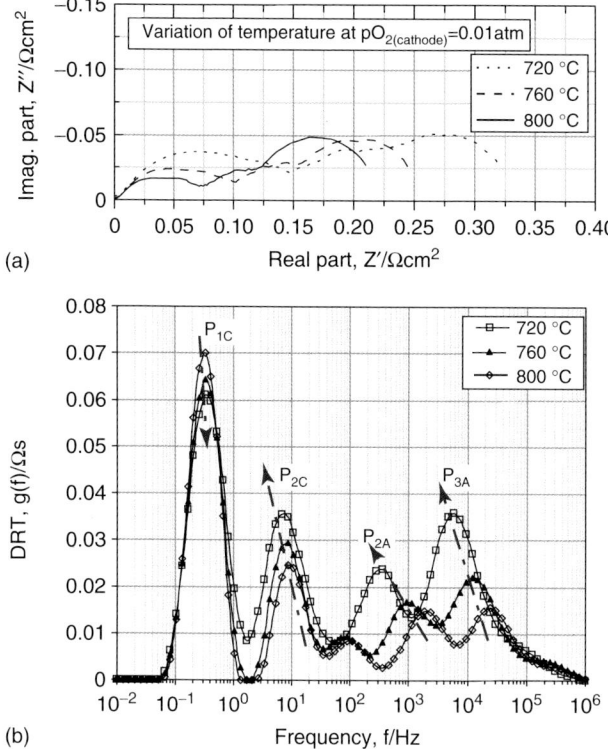

Figure 16.12 (a) Series of impedance spectra (ohmic part subtracted) and (b) corresponding distribution curves for three different temperatures at very low cathodic oxygen partial pressure [$pH_2O_{(anode)}$ = 0.625 atm (balance H_2), $pO_{2(cathode)}$ = 0.01 atm] (cell Z1_153).

Figure 16.12b shows the DRTs calculated from the impedance spectra. P_{1C} is characterized by a minor dependence on the operating temperature, showing a slight decrease with decrease in temperature.

16.3.1.4 Conclusions

The observations made above suggest that the overall polarization loss caused by the cathode is strongly related to the two processes P_{1C} and P_{2C}. On the other hand, P_{1A}, P_{2A}, and P_{3A} show a more or less pronounced dependence on $pH_2O_{(anode)}$ but no dependence on $pO_{2(cathode)}$, demonstrating that these three processes are ascribable to the anode.

Based on the analyzed characteristic dependences of the identified processes, a first hypothesis about their physical origin can be made:

1) Processes P_{1A} and P_{1C} exhibit a very low thermal activation behavior (a negligibly small, even negative, activation energy). Moreover, both processes

Table 16.1 List of processes identified from DRT analysis, together with their characteristic frequency range, gas partial pressure, and temperature dependence

Process	f_{max} (Hz)	Dependences	Assigned electrode
P_{1C}	0.3–10	$pO_{2(cathode)}$, temperature (low)	Cathode
P_{2C}	2–500	$pO_{2(cathode)}$, temperature	Cathode
P_{1A}	4–10	$pH_{2(anode)}$, $pH_2O_{(anode)}$, temperature (low)	Anode
P_{2A}	200–3000	$pH_{2(anode)}$, $pH_2O_{(anode)}$, temperature	Anode
P_{3A}	3000–50000	$pH_{2(anode)}$, $pH_2O_{(anode)}$, temperature	Anode

are strongly affected by gas composition changes. All of these are telltale signs that P_{1C} and P_{1A} are related to gas-phase diffusion [27, 28].

2) Processes P_{2C}, P_{2A}, and P_{3A} are characterized by a pronounced thermal activation, therefore these losses are most likely related to the activation polarization in the anode and cathode, respectively.

Table 16.1 gives an overview of the processes identified by the DRT analysis together with their characteristic frequency range, gas partial pressure, and temperature dependence.

16.3.2
Equivalent Circuit Model Definition and Validation

Based on the extensive DRT analysis shown above, the ECM depicted in Figure 16.13 is proposed. The equivalent circuit consists of six serial impedance elements: the processes P_{1C}, P_{2A}, and P_{3A} are modeled by an RQ element, whereas P_{1A} and P_{2C} are modeled by a G-FLW element and a Gerischer element, respectively. The ohmic losses are accounted by an ohmic resistor R_0.

Figure 16.14a shows a typical CNLS fit result applied to the imaginary part of the impedance curve depicted in Figure 16.8a. The residuals (relative errors) are distributed uniformly around the frequency axis, showing no systematic deviation

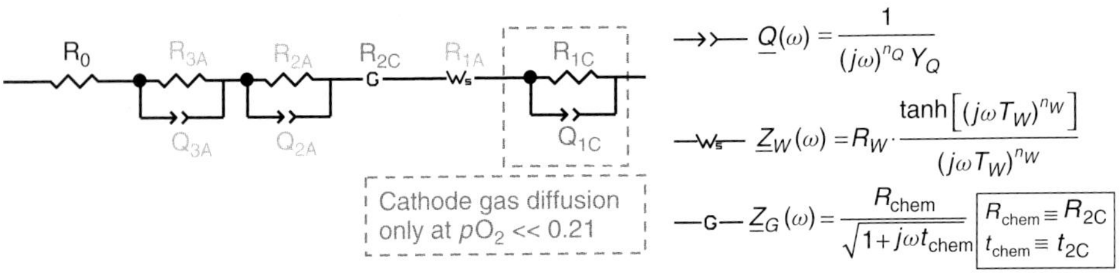

R0
R3A
R2A
R2C
R1A
R1C
Q3A
Q2A
Q1C
G
Ws
Cathode gas diffusion only at pO2 << 0.21
$Q(\omega) = \frac{1}{(j\omega)^{n_Q} Y_Q}$
$\underline{Z}_W(\omega) = R_W \cdot \frac{\tanh\left[(j\omega T_W)^{n_W}\right]}{(j\omega T_W)^{n_W}}$
$\underline{Z}_G(\omega) = \frac{R_{chem}}{\sqrt{1 + j\omega t_{chem}}}$
$R_{chem} \equiv R_{2C}$
$t_{chem} \equiv t_{2C}$

Figure 16.13 Proposed equivalent circuit model for the CNLS fit analysis of impedance spectra.

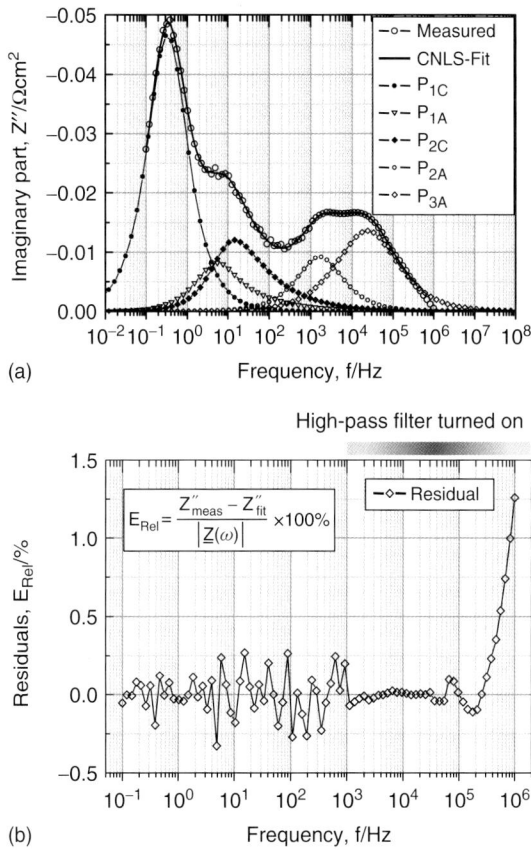

Figure 16.14 (a) CNLS fit of the imaginary part of the impedance spectrum shown in Figure 16.8a; (b) residual pattern of the fit (cell Z1_153).

(Figure 16.14b). For most of the spectrum the relative errors lie below an absolute value of 0.25%. Only from ~300 kHz upwards do inductive artifacts caused by the electrical wiring becomes noticeable. The high quality of this CNLS fit confirms the plausibility of the proposed equivalent circuit. However, as already stated in the previous section, the residuals are "only" a mathematical quantity that describes the level of equivalence between the measured curve and the theoretical impedance spectrum of the proposed model. Therefore, no ultimate conclusion about the physical correctness of the ECM is allowed from just evaluating the residuals. Hence, to prove the physical validity of a proposed ECM, impedance spectra recorded over a wide range of operating conditions, such as various gas compositions and temperatures, were analyzed by the CNLS approximation. In this way the equivalent circuit parameters could be evaluated with regard to physical considerations.

16.3.2.1 Cathodic Oxygen Partial Pressure Dependence

In Figure 16.15 the polarization resistances obtained by the CNLS fit (using the equivalent circuit depicted in Figure 16.13) are plotted against the partial pressure of oxygen. It should be mentioned that in this case the resistance R_{1A} was kept fixed during the entire fit procedures. This approach was essential because otherwise the similar summit frequencies of P_{1A} and P_{2C}, especially at lower $pO_{2(cathode)}$, would have destabilized the fit algorithm.

In accordance with the DRT analysis (Figure 16.11), R_{2A} and R_{3A} are independent of the change in O_2, whereas the two resistances R_{1C} and R_{2C} both show an almost linear trend in the double logarithmic plane with a slope of -1.08 and -0.26, respectively. The slope of R_{2C} is in accordance with literature values obtained on symmetrical L58SCF cells. For instance, Esquirol et al. [29] reported a slope of -0.21 for the resistance associated with the activation polarization of an $La_{0.6}Sr_{0.4}Co_{0.2}Fe_{0.8}O_{3-\delta}$ cathode.

In the following, we focus our attention on process P_{1C}. It can easily be shown that the cathodic gas diffusion resistance can be described by the following equation [30]:

$$R_{D(cathode)} = \left(\frac{RT}{4F}\right)^2 L_{cathode} \frac{1}{D^{eff}_{O_2,N_2}} \left(\frac{1}{pO_{2(cathode)}} - 1\right) \left(1.0133 \times 10^5 \frac{Pa}{atm}\right)^{-1}$$

(16.17)

where $L_{cathode}$ is the cathode thickness and $D^{eff}_{O_2,N_2}$ is the effective diffusion coefficient for a mixture of oxygen and nitrogen; R, T, and F have their usual meanings.

The model Eq. (16.17) was fitted to the experimental data with the effective diffusion coefficient $D^{eff}_{O_2,N_2}$ representing the unknown variable ($L_{cathode}$ was set to

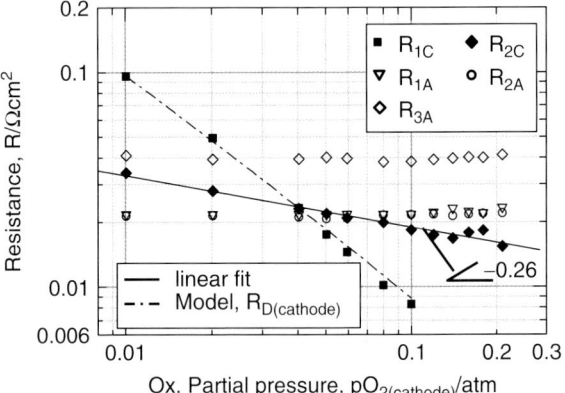

Figure 16.15 Characteristic dependence of fitted equivalent circuit elements on the cathodic oxygen partial pressure. The dashed line indicates the model prediction according to Eq. (16.17) [$pH_2O_{(anode)} = 0.625$ atm (balance H_2), $T = 800\ °C$] (cell Z1_153).

45 μm). The fit result is shown in Figure 16.15 (dashed line). As can be seen, the agreement between the model and the experimentally obtained resistance R_{1C} is fairly good. From the fit, $D^{\mathrm{eff}}_{O_2,N_2} = 2.5 \times 10^{-6}$ m^2 s^{-1} is estimated. This value is in accordance with values reported in the literature for this type of cathode structure [31]. This result strongly supports the hypothesis that the low-frequency process P_{1C} reflects the mass transfer resistance caused by the gas-phase diffusion in the pores of the L58SCF electrode.

16.3.2.2 Anodic Water Partial Pressure Dependence

In Figure 16.16, the polarization resistances obtained from the CNLS fit are plotted against the water partial pressure (balance H_2). During this fit procedure, the resistance R_{2C} was kept fixed in order to ensure a stable fit. The polarization contribution caused by the gas-phase diffusion in the pores of the LSCF electrode can be neglected when air is used as cathode gas, hence the resistance R_{1C} was set fixed to zero. The two resistances R_{2A} and R_{3A} both show an almost linear trend in the double-logarithmic plane with slopes of -0.44 and -0.20, respectively. R_{1A} shows the highest dependence on changes in the water content.

In the following, we focus our attention on process P_{1A}. As discussed in the previous section, P_{1A} is characterized by a negligible thermal activation and shows attributes that are expected to be seen by gas diffusion processes [27]. To support this hypothesis, the following widely used equation, describing the resistance caused by diffusion limitations in a porous anode structure, was employed [27]:

$$R_{D(\mathrm{anode})} = \left(\frac{RT}{2F}\right)^2 L_{\mathrm{anode}} \frac{1}{D^{\mathrm{eff}}_{H_2O,H_2}} \left(\frac{1}{pH_{2(\mathrm{anode})}} + \frac{1}{pH_2O_{(\mathrm{anode})}}\right)$$
$$\times \left(1.0133 \times 10^5 \frac{\mathrm{Pa}}{\mathrm{atm}}\right)^{-1} \qquad (16.18)$$

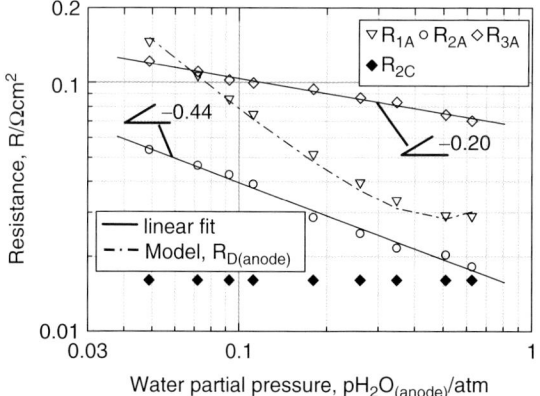

Figure 16.16 Characteristic dependence of fitted equivalent circuit elements on the anodic partial pressure of water (balance H_2) [$pO_{2(\mathrm{cathode})} = 0.21$ atm (air), $T = 757$ °C] (cell Z1_153).

where L_{anode} is the anode thickness (in this case 1.5 mm). $D_{H_2O,H_2}^{\text{eff}}$ represents the effective diffusion coefficient, which may depend on the porosity and tortuosity of the electrode in addition to the effects of Knudsen diffusion.

Equation (16.18) was fitted to the experimental data with $D_{H_2O,H_2}^{\text{eff}}$ representing the unknown variable. The fit result is shown in Figure 16.16 (dashed line). As can be seen, the model approximates very well the trend of the experimentally obtained resistance R_{1A}. From the fit, a reasonable effective diffusion coefficient $D_{H_2O,H_2}^{\text{eff}} = 4.12 \times 10^{-5}$ m^2 s^{-1} is estimated. This result strongly supports the hypothesis that the low-frequency process P_{1A} reflects the mass transfer resistance caused by the gas-phase diffusion in the pores of the Ni/YSZ anode substrate.

16.3.2.3 Thermal Activation

It should be briefly noted that in earlier work [6] (cf., Figures 16.15 and 16.16) the temperature-dependent resistances R_{2A} and R_{3A} were treated separately. However, later extensive work [7, 32] has demonstrated that the two identified processes P_{2A} and P_{3A} are related to three coupled processes, all of them taking place in the anode functional layer (AFL):

1) the oxygen ion transport within the YSZ matrix of the anode
2) the charge-transfer resistance at the TPB Ni/YSZ/gas phase
3) the gas diffusion loss inside the AFL.

For this reason, the resistances R_{2A} and R_{3A} should not be considered separately but only as a sum, describing the overall polarization resistance caused by the interaction of the three processes listed above. However, the gas diffusion loss inside the 7 μm thick AFL can be neglected after appositely activating the cell: the AFL is a dense layer after initial reduction. It takes ~150 h of cell operation until a certain open porosity in the AFL structure is reached. From then on, gas diffusion in the AFL can be neglected for cells with an AFL thickness (t_{AFL}) <8 μm, that is, for the cell type analyzed in this study [32].

The resistances $R_{2C} \equiv R_{\text{act,cat}}$ and $R_{2A} + R_{3A} \equiv R_{\text{act,an}}$ obtained by the CNLS fit of the impedance curves recorded on an FZJ ASC cell are plotted against the temperature in Figure 16.17. In order to facilitate the fit procedure, the resistance $R_{1A} = 38$ mΩ cm^2 was kept constant.

As can be seen, the resistances $R_{\text{act,cat}}$ and $R_{\text{act,an}}$ can be approximated very well with a linear fit, demonstrating an almost perfect Arrhenius behavior. From the slope of the fitted lines, the activation energies $E_{\text{act,cat}} = 1.45$ eV and $E_{\text{act,an}} = 1.09$ eV are obtained.

In the literature, activation energies between 0.7 and 1.1 eV are reported for technical Ni/YSZ anodes ([33] gives an overview). The value determined in this work ($E_{\text{act,an}} = 1.09$ eV) is in good agreement with results by de Boer [34], Geyer et al. [35], and Brown et al. [36], who found activation energies ranging from 1 to 1.1 eV.

It should be noted, however, that all these values are *apparent* activation energies. As Sonn et al. [7] showed in great detail for symmetrical Ni/8YSZ anodes of

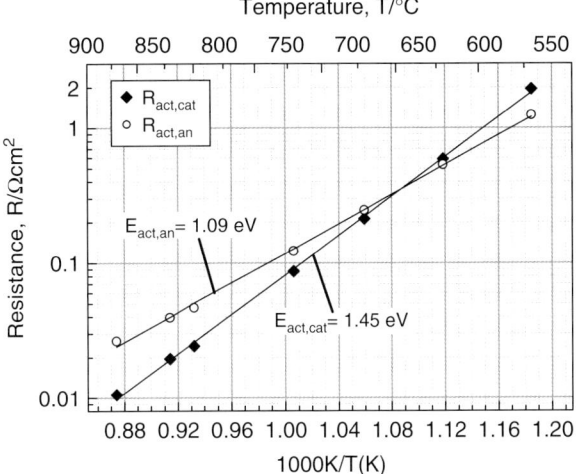

Figure 16.17 Characteristic dependence of fitted equivalent circuit elements on the operating temperature [$pH_2O_{(anode)} = 0.20$ atm (balance H_2), $pO_{2(cathode)} = 0.21$ atm (air)] (cell 1_188).

electrolyte-supported cells, the thermal behavior of the activation polarization in fact depends on two coupled processes: on the one hand, the oxygen ion transport within the YSZ matrix of the anode, and on the other, the charge-transfer resistance at the TPB Ni/YSZ/gas phase. By using a transmission line model [37–39] (Figure 16.18), instead of two serial RQ elements (RQ_{2A}, RQ_{3A}), Sonn et al. [7] were able to determine separate values for the activation energy of the ionic transport losses $R_{ion} = Lr_{ion}$ (0.904 eV) and of the charge-transfer resistance $R_{ct} = r_{ct}/L$ (1.35 eV) (Figure 16.19), where L is the electrode thickness.

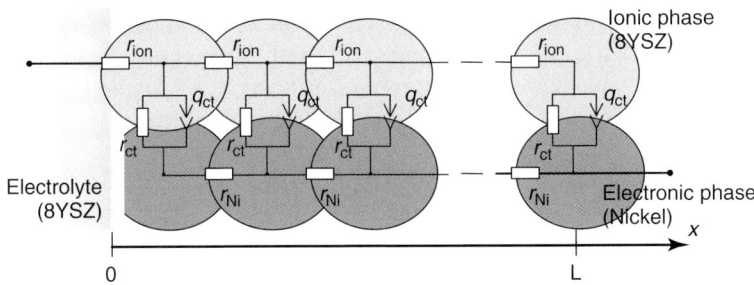

Figure 16.18 Modeling scheme for a porous two-phase Ni/8YSZ composite electrode (transmission line model). r_{ion} and r_{Ni} describe the resistance per unit length (Ωm^{-1}) of the effective ionic and electronic conductivity of the 8YSZ and Ni phase, respectively. The $r_{ct}q_{ct}$ element stands for the charge-transfer resistance-length r_{ct} (Ωm) and double-layer polarization (constant phase element q_{ct}).

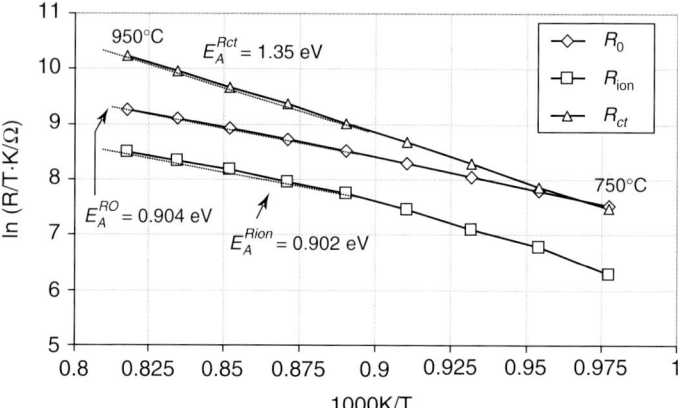

Figure 16.19 Fitted values of the transmission line model. R_{ion} denotes the ionic transport resistance, R_{ct} the charge-transfer resistance, and R_0 the ohmic resistance of the 8YSZ electrolyte substrate. r_{Ni} was set fixed to zero.

In comparison with the anode, the activation energy of the L58SCF cathode shows a significantly higher value. Moreover, the values reported in the literature vary very strongly from 1.2 up to 1.6 eV [29, 40–43]. The best agreement with the value determined in this work is with Esquirol et al. [29], who reported an activation energy of 1.39 ± 0.09 eV. It should be briefly noted that there may be a strong influence of the cathode microstructure in this respect. As shown recently, for example, by Peters et al. [44], the mean porosity and the mean grain size of chemically similar cathodes can differ strongly, thus leading to a scatter of the activation energy.

16.3.2.4 Conclusions

In summary, the presented CNLS analysis confirms all hypotheses made in the previous section. P_{1A} occurs owing to inhibited gas diffusion within the Ni/YSZ anode substrate. The high-frequency processes P_{2A} und P_{3A} are related to gas diffusion coupled with charge-transfer reaction and ionic transport in the Ni/YSZ AFL structure. The two processes P_{1C} and P_{2C} are cathodic processes: P_{1C} characterizes the gas diffusion losses within the pores of the cathode and is, therefore, negligible in air, whereas the faster process P_{2C} is inherently electrochemical, accounting for the losses resulting from oxygen incorporation and oxygen ion transport within the cathode (LSCF).

In Figure 16.20, a schematic fit result of a typical measured impedance curve along with the simulated Nyquist plots of each single impedance element is shown.

Table 16.2 gives a final overview of the processes identified by the DRT analysis together with their characteristic frequency (f_{max}) range, gas partial pressure, temperature dependence and physical origin.

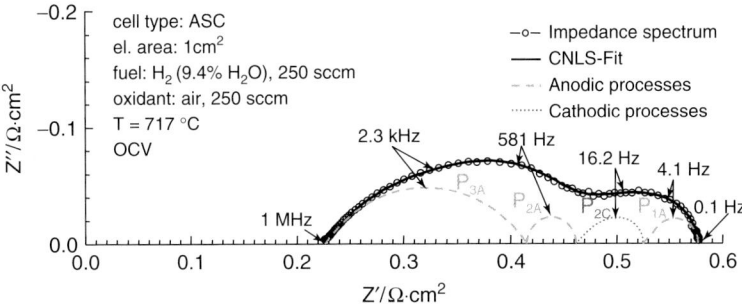

Figure 16.20 Schematic fit result of a typical measured impedance curve along with the simulated Nyquist plots of each single impedance element (cf., Figure 16.13) (cell Z1_153).

Table 16.2 List of processes known to take place in an ASC cell, together with their temperature, gas partial pressure, and frequency dependences and also the magnitude of the real part of the corresponding resistance ($T = 570\text{--}870\ ^\circ\text{C}$, $H_2O \approx 5.5\text{--}65\%$ (balance H_2) at the anode and $O_2 \approx 1.0\text{--}21\%$ (balance N_2) at the cathode)

Process	Element	f_{max}, ASR	Dependences	Physical origin
P_{1C}	RQ $n_{Q,1C} \approx 0.97$	0.3–10 Hz 2–100 mΩ cm²	$pO_{2(cathode)}$ T (very low)	Gas diffusion in the cathode structure
P_{2C}	Gerischer	2–500 Hz 10 mΩ cm² to 2 Ω cm²	$pO_{2(cathode)}$, T	Oxygen surface exchange kinetics and O^{2-} diffusivity in the bulk of the cathode
P_{1A}	G-FLW $n_{W,1A} = 0.45\text{--}0.49$	4–10 Hz 20–150 mΩ cm²	$pH_{2(anode)}$, $pH_2O_{(anode)}$ T (very low)	Gas diffusion in the anode substrate
P_{2A} + P_{3A}	RQ $n_{Q,2A} = 0.89\text{--}0.95$ RQ $n_{Q,3A} = 0.65\text{--}0.75$	200 Hz to 3 kHz 3–50 kHz $R_{2A} + R_{3A}$ 10 mΩ cm² to 2 Ω cm²	$pH_{2(anode)}$, $pH_2O_{(anode)}$ T $pH_{2(anode)}$, $pH_2O_{(anode)}$ T	($P_{2A} + P_{3A}$) gas diffusion coupled with charge-transfer reaction and ionic transport (AFL: anode functional layer)

16.4 Conclusion

This chapter provides the fundamentals of impedance spectroscopy. Special emphasis is placed on the issues of SOFCs, where impedance spectra are particularly difficult to evaluate owing to the number and complexity of concurring physical and chemical processes contributing to the overall electrical response of the cell.

ASCs with a thin-layer electrolyte are characterized at open-circuit voltage conditions over a broad range of operating conditions, including different temperatures and various cathode and anode gas compositions. More than 300 impedance spectra have been recorded and subsequently analyzed by calculating the DRT. The high resolution of the DRT combined with the numerical accuracy of the CNLS fit allowed the separation of cathodic and anodic processes, the former including oxygen surface exchange kinetics, diffusivity of oxygen ions, and gas-phase diffusion, and the latter including gas diffusion, charge-transfer reaction and ionic transport in the Ni/YSZ anode.

An ECM is proposed that consists of six serial impedance elements: three RQ elements, a G-FLW element, a Gerischer element, and an ohmic resistor.

References

1. Macdonald, J.R. (1987) *Impedance Spectroscopy. Emphasizing Solid Materials and Systems*, John Wiley & Sons, Inc., New York.
2. Hamann, C.H., Hamnett, A., and Vielstich, W. (1998) *Electrochemistry*, John Wiley & Sons, Inc., New York.
3. Bard, A.J. and Faulkner, L.R. (2001) *Electrochemical Methods. Fundamentals and Applications*, John Wiley & Sons, Inc., New York.
4. Gerischer, H. (1954) *Z. Elektrochem.*, **58**, 278.
5. Boukamp, B.A. (1993) *Solid State Ionics*, **62**, 131.
6. Leonide, A., Sonn, V., Weber, A., and Ivers-Tiffée, E. (2008) *J. Electrochem. Soc.*, **155**, B36.
7. Sonn, V., Leonide, A., and Ivers-Tiffée, E. (2008) *J. Electrochem. Soc.*, **155**, B675.
8. Schichlein, H. (2003) Experimentelle Modellbildung für die hochtemperatur-Brennstoffzelle SOFC, PhD thesis, Universität Karlsruhe (TH), Verlag Mainz, Aachen.
9. Gabrielli, C. and Keddam, M. (1996) *Electrochim. Acta*, **41**, 957.
10. Agarwal, P. and Orazem, M.E. (1992) *J. Electrochem. Soc.*, **139**, 1917.
11. Orazem, M.E. and Tribollet, B. (2008) *Electrochemical Impedance Spectroscopy*, John Wiley & Sons, Inc., New York.
12. Boukamp, B.A. (1995) *J. Electrochem. Soc.*, **142**, 1885.
13. Boukamp, B.A. (2004) *Solid State Ionics*, **169**, 65.
14. Leonide, A. (2005) *Impedanzanalyse von Ni/CerMet-Anoden*, Diplomarbeit (Master thesis, in German), Institut für Werkstoffe der Elektrotechnik, Universität Karlsruhe (TH).
15. Mogensen, M. and Hendriksen, P.V. (2003) in *High Temperature Solid Oxide Fuel Cells. Fundamentals, Design and Applications* (eds. S.C. Singhal and K. Kendall), Elsevier, Amsterdam, pp. 261–289.
16. Schichlein, H., Müller, A.C., Voigts, M., Krügel, A., and Ivers-Tiffée, E. (2002) *J. Appl. Electrochem.*, **32**, 875.
17. Schäfer, H. and Sternin, E. (1997) *Phys. Can.*, **3/4**, 77.
18. Louis, A.K. (1989) *Invers und Schlecht Gestellte Probleme*, Teubner, Stuttgart.
19. Tikhonov, A.N. and Arsenin, V.Y. (1977) *Solution of Ill-Posed Problems*, John Wiley & Sons, Inc., New York.
20. Tikhonov, A.N., Goncharsky, A.V., Stepanov, V.V., and Yagola, A.G. (1995) *Numerical Methods for the Solution of Ill-Posed Problems*, Kluwer, Dordrecht.
21. Weese, J. (1992) *Comput. Phys. Commun.*, **69**, 99.
22. Boukamp, B.A. (1986) *Solid State Ionics*, **20**, 31.
23. Meulenberg, W.A., Menzler, N.H., Buchkremer, H.P., and Stöver, D. (2002) *Ceram. Trans.*, **127**, 99.
24. Mai, A., Haanappel, V.A.C., Tietz, F., Vinke, I.C., and Stöver, D. (2003) in *Proceedings of the 8th International Symposium on SOFCs (SOFC-VIII)* (eds. S.C.

Singhal and M. Dokiya), Electrochemical Society, Pennington, NJ, p. 525.
25. Buchkremer, H.P., Diekmann, U., and Stöver, D. (1996) in *Proceedings of the 2nd European Solid Oxide Fuel Cell Forum* (ed. B. Thorstensen), U. Bossel, Oberrohrdorf, p. 221.
26. Becker, M., Mai, A., Ivers-Tiffée, E., and Tietz, F. (2005) in *Proceedings of the 9th International Symposium on SOFC (SOFC-IX)* (eds. S.C. Singhal and J. Mizusaki), Electrochemical Society, Pennington, NJ, p. 514.
27. Primdahl, S. and Mogensen, M. (1999) *J. Electrochem. Soc.*, **146**, 2827.
28. Adler, S.B. (2004) *Chem. Rev.*, **104**, 4791.
29. Esquirol, A., Brandon, N.P., Kilner, J.A., and Mogensen, M. (2004) *J. Electrochem. Soc.*, **151**, A1847.
30. Kim, J., Virkar, A.V., Fung, K., Mehta, K., and Singhal, S.C. (1999) *J. Electrochem. Soc.*, **146**, 69.
31. Ackmann, T., de Haart, L.G.J., Lehnert, W., and Stolten, D. (2003) *J. Electrochem. Soc.*, **150**, 783.
32. Leonide, A., Ngo Dinh, S., Weber, A., and Ivers-Tiffée, E. (2008) in *Proceedings of the 8th European Solid Oxide Fuel Cell Forum* (ed. R. Steinberger-Wilckens), European Fuel Cell Forum, Lucerne, p. A0501.
33. Primdahl, S. and Mogensen, M. (1997) *J. Electrochem. Soc.*, **144**, 3409.
34. de Boer, B. (1998) SOFC anode hydrogen oxidation at porous nickel and nickel/yttria stabilised zirconia cermet electrodes, PhD thesis, University of Twente, Enschede, NL.
35. Geyer, J., Kohlmuller, H., Landes, H., and Stubner, R. (1997) in *Solid Oxide Fuel Cells V*, Electrochemical Society Proceedings Series (eds. U. Stimming, S.C. Singhal, H. Tagawa, and W. Lehnert), PV 97-18, Electrochemical Society, Pennington, NJ, pp. 585–593.
36. Brown, M., Primdahl, S., and Mogensen, M. (2000) *J. Electrochem. Soc.*, **147**, 475.
37. Paasch, G. (2000) *Electrochem. Commun.*, **2**, 371.
38. Paasch, G., Micka, K., and Gersdorf, P. (1993) *Electrochim. Acta*, **38**, 2653.
39. Bisquert, J., Belmonte, G.G., Santiago, F.F., Ferriols, N.S., Yamashita, M., and Pereira, E.C. (2000) *Electrochem. Commun.*, **2**, 601.
40. Steele, B.C.H. and Bae, J.M. (1998) *Solid State Ionics*, **106**, 255.
41. Sahibzada, M., Benson, S.J., Rudkin, R.A., and Kilner, J.A. (1998) *Solid State Ionics*, **115**, 285.
42. Bae, J.M. and Steele, B.C.H. (1998) *Solid State Ionics*, **106**, 247.
43. Jiang, S.P. (2002) *Solid State Ionics*, **146**, 1.
44. Peters, C., Weber, A., and Ivers-Tiffée, E. (2008) *J. Electrochem. Soc.*, **155**, B730.

17
Post-Test Characterization of Solid Oxide Fuel-Cell Stacks

Norbert H. Menzler and Peter Batfalsky

17.1
Introduction

The Forschungszentrum Jülich has been working on the development of solid oxide fuel cells (SOFCs) for ~20 years and more than 400 stacks of various sizes and designs have been operated during this period. Roughly half of the operated stacks were characterized after operation. A careful post-test characterization is necessary to understand, for example, failures, electrical characteristics during stack operation, and interaction phenomena depending on the operational conditions. Additionally, postmortem analysis is helpful for future materials and material combination development to ensure high power output, long-term stability, and low stack degradation. Questions concerning geometrical contacting, manufacturing accuracy, and tolerances are also addressed in these analyses.

Various characterization methods have been developed in the last few years to optimize the dissection procedure, to maximize scientific output with respect to future developments and failures, to understand degradation, and to ensure reproducible methods in order to minimize effects caused by the dissection procedure itself.

Unfortunately, up to now only limited information has been published about post-test analysis, methods, and results, and this is true of our own research center in addition to any other developer, be it companies or R&D centers [1–4]. Only single effects after stack operation, single-cell testing, or model experiments and theoretical considerations have been published, such as metal corrosion, Cr poisoning, and thermodynamics [5–15]. Therefore, this chapter summarizes for the first time the stack dissection method, the underlying reasons for stack post-test analysis, the individual characterization techniques and methods, and the results obtained so far.

17.1.1
Reasons for Post-Test Analysis

The reasons for special post-test characterizations are numerous; a few examples are given below, but this list does not claim to be exhaustive:

- Failure during operation
  - short circuiting
  - noncontacted planes
  - leakage
  - low open-circuit voltage (OCV) of cells
  - spontaneous voltage drop
  - falling below a certain voltage level, such as in the range of the Ni reoxidation voltage
  - external cause, such as failure of control unit, measurement unit, or furnace, lightning strike, voltage breakdown.
- Stack objective achieved and subsequent component characterization
  - material changes, such as materials chemistry, crystallographic phases, microstructural changes, development of oxide scales
  - material interactions, such as sealant–metal, sealant–electrolyte, cell materials, contact materials–metal
  - geometric aspects, such as component thickness, creep, distances, tolerances, contact situation, comparison with blueprint.
- Search for specific phenomena depending on operational conditions
  - degradation phenomena on the anode side, such as coke formation, metal dusting, sulfur poisoning, reoxidation damage, chromium poisoning on the cathode side, on metal (corrosion), or in the sealant (devitrification, secondary phase formation)
  - material interactions or changes based on different operational conditions such as the influence of the operating temperature, the applied current density, the fuel gas used, the fuel and air utilization, and during thermo-, load, or redox cycling tests.

If a stack is to be analyzed after operation, it is necessary to define in advance which of the above-mentioned reasons are the basis for the post-test characterization or whether the stack is characterized with respect to special questions. On this basis, a first important decision has to be made: will the stack be dissected plane-by-plane and subsequently the components characterized individually, or should the stack or stack parts be embedded. In particular, in order to retain the geometry and the contact situation, it might be useful to embed the stack completely or partly in a polymeric resin. After embedding, a water jet is used to cut samples in various positions and at various cutting lines, and these samples can then be extracted for post-test analysis. If the stack is not to be embedded, it is possible after dissection to extract single samples from the individual parts such as the frame with cell, the interconnect, or the bottom and top plate by laser cutting for further investigations.

17.1.2
Methods of Post-Test Analysis

There are various methods of characterizing the stack, stack components, or single samples after operation. A few examples are given below:

- Whole stack
  - *leak testing* in assembled condition and *leak detection* with color intrusion method
  - *thermography* to localize short-circuiting
  - *photography* of all stack sides
  - *optical microscopy*, including distance measurements
  - *distance* and *thickness measurements* with micrometer screw and feeler gauge.
- Stack planes
  - *photography* of every plane (anode and cathode side) in the inverse sequence of stack assembly
  - *optical microscopy* at, for example, defects, optically visual interactions, cracks, corrosion
  - *topography* to characterize bending or surfaces.
- Stack components (cell, frame, interconnect)
  - photography
  - optical microscopy
  - topography to characterize bending, surfaces, or roughness
  - scanning electron microscopy (SEM), surfaces, fractured, or polished cross-sections) + EDX (energy dispersive X-ray) analysis
  - transmission electron microscopy (TEM)
  - X-ray diffraction (XRD) for crystal phase analysis
  - wet chemical analysis for chemical composition, incorporation of foreign elements, stoichiometries, and so on.
- Stack embedding
  - whole stack
  - part stack
  - cutting by sawing, water jet, or laser.

Especially, in order to characterize and compare the stack plane-by-plane, it is important to visualize the planes before assembling and after dissection. Therefore, to ensure good and long-term stack operation, the postmortem characterization starts with the quality assurance system for the single components, for example, the cells [16], the sealant, and the metal parts such as interconnect and frame, continues during stack assembly, stack operation, and stack dissection, and ends with the special post-test analysis. Figure 17.1 illustrates this schematically. For simplification, the diagram only includes the cells, the sealant, and the metal parts. In real stack assemblies, contact materials such as the anodic nickel mesh or contact layers on the cathode side and, for example, evaporation protection layers or electronic insulating layers are applied and also analyzed.

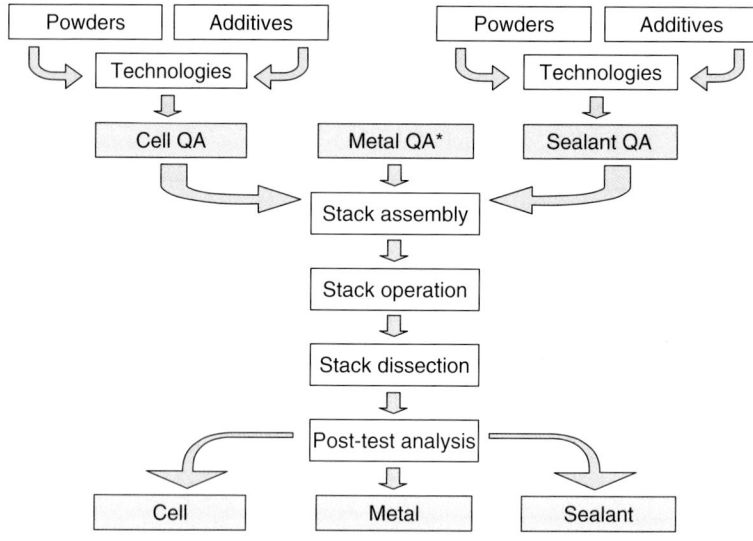

Figure 17.1 Scheme of stack life including necessary quality assurance to ensure high quality of post-test analysis.

The quality and reproducibility of the materials and processes used are ensured by consecutive numbering of all stack parts (cells, interconnects, frames, glass–ceramic batches), handling instructions for all processing steps, standardized stack operation, and stack dissection protocols. All information is collected on a central server system. Hence after stack testing, dissection, and post-test analysis, conclusions can be drawn with respect to the materials used, processes, and testing. There is a complete chain from the starting materials to the post-test analysis and this is also necessary to develop the SOFC system for marketability.

In the following section, typical steps in the stack dissection are presented, followed by sections introducing special technologies for analysis that are illustrated with examples.

17.2
Stack Dissection

A special laboratory has been set up for stack dissection. The laboratory contains all the necessary tools for stack measurement, stack dissection, stack documentation, and basic characterization methods. The laboratory includes an extractor hood in which the plane-by-plane dissection is carried out, a camera that is mounted above the stack to photograph the planes on-line, a PC in which the photographs are copied directly equipped with collector software to label special photographs or add comments and findings, an optical microscope for documentation of optically visible areas, which is also equipped with a camera (the pictures taken are included in the stack dissection protocol), all necessary tools for stack dismantling and measuring, and cabinets for storage of stack parts. All stacks are stored automatically for several years so that samples can be extracted for further investigations in the future.

17.2.1
Thermography

The sequence of stack dissection is governed by the questions that arise during stack operation. For example, if short-circuiting was measured, stacks with fewer planes (if the stack has too many planes it heats up more uniformly and cannot be characterized) were characterized thermographically before stack dismantling. Thermographic analysis is carried out by feeding electric current to the stack. Electrically conducting paths transport the heat, induced by the current earlier than in insulating areas. Thus, by taking heat images during current feeding, the stack is visualized in different colors and the short-circuiting areas became visible. In Figure 17.2, two examples are presented. If the stack contains more than, for example, five planes plus the bottom and top plate, but shows short-circuiting between two individual planes, the stack is partly dissected in advance and subsequently the remaining part with the short-circuit is characterized by thermography. Therefore, this tool can be used for small stacks with only a few repeating units but also for power stacks.

After taking the heat images, the stack can be dismantled and the area of interest characterized in detail. As an example, in one of the two stacks presented in Figure 17.2 a metallic flitter in the glass–ceramic sealing was found to be causing the short-circuiting. The metallic flitter might result from the interconnect or the frame conditioning or from the furnace and was introduced during stack assembly. Such stack failure can be avoided by assembling the stack in a low-dust laboratory and by photographing every plane during stack assembly.

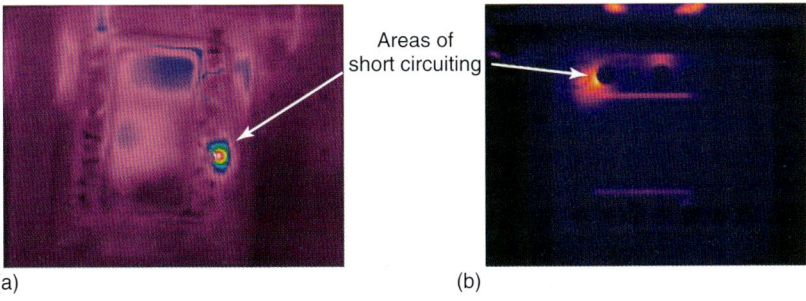

Figure 17.2 Photographs taken with a heating image camera showing areas of short-circuiting of two different stacks.

17.2.2
Stack Embedding

To characterize the contacting and the interfacial reactions, in some cases, stacks are embedded partly or completely in a polymeric resin. The embedding must be done carefully to ensure venting of the stack to fill all cavities with the liquid resin. If not all of the cavities and spaces are filled, there is a risk of artifacts due to water jet cutting. The stack is cut with a water jet since this type of cutting tool has various advantages. First, the thickness of the stack and the adjacent bottom and top plate are in the centimeter range, which is a challenge for other techniques such as laser cutting. Second, the stack consists of various different materials and material combinations and all of them should be cut in a comparable manner. Third, water-jet cutting does not melt the cut edges as laser cutting does. If sawing is applied, the material differences between the metal or ceramic and the polymeric resin are a key problem because due to sawing the polymer may become liquid and subsequently it may smear the other parts and again artifacts are produced from the cutting procedure. The contamination from the cutting parts in the water jet is negligible because all samples are post-processed by grinding and polishing and thus any remaining contaminants are removed.

In Figure 17.3, a partly embedded stack is shown. Whether the stack is embedded completely or partly depends on whether the questions to be answered relate to operation or construction. Additionally, if the stack is partly embedded a decision has to be made concerning which area is to be embedded and in which direction (parallel or perpendicular to the gas flow) it will be cut.

If the stack is only partly embedded, samples can be cut from the embedded and the remaining part. Thus, for example, for special characterizations, where cell parts for chemical analysis are necessary, these can be extracted in addition to embedded parts, which will allow the analysis of, for example, the contacting zone.

Figure 17.4 shows water-jet cutting.

(a)   (b)

Figure 17.3 Photographs of a partly embedded two-plane stack, here parallel to the gas flow direction for cutting in the perpendicular direction.

Figure 17.4 Water-jet cutting of embedded SOFC stack.

17.2.3
Photography and Distance Measurements

If a nonembedded stack is to be dissected plane-by-plane, starting with the upper side of the top plate and ending with the lower side of the bottom plate, which is the gas connection side, photographs are taken and directly fed into PC-based image-collecting software. Before dissection, especially when glass–ceramic sealants are used for stack sealing, the distances between the planes (interconnect–frame–interconnect) are measured on all four stack edges by a feeler gauge. The distance control and the total sealing heights of all sealing levels show whether there are significant differences within one plane or within the whole stack. In this way, uniform stack settling during the stack joining procedure at the SOFC start-up can also be controlled.

Figure 17.5 shows two photographs of a stationary design stack. After plane-by-plane dismantling every plane side – interconnect or frame with cell – is photographed irrespective of whether there are areas of special interest or which seem to have changed during stack operation to completely characterize the stack.

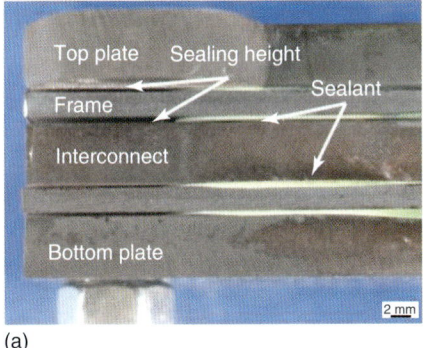
(a)

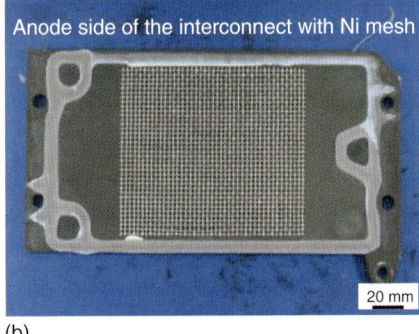
(b)

Figure 17.5 Photographs of stacks: (a) side view for distance measurements with feeler gauge and (b) top view of the interconnect anode side after plane dismantling.

These first photographs can answer various questions:

- Was the stack assembly correct?
- Does every plane side (anode, cathode) look alike?
- Are there special color changes, morphological changes, or irregularities?
- Is it necessary to post-analyze stack parts or surfaces?
- Did foreign phases form?
- Are the cells unchanged (reoxidation, cracking, coking) and undamaged?
- Are there leakages (visible by remnants of burning or gas streams)?
- At which interface does the sealant crack during dismantling (within the glass–ceramic or at a metal oxide–glass interface)?

If during stack dismantling areas of special interest are detected, the plane is moved directly for optical microscopy inspection, where further more highly magnified characterizations are carried out. If necessary, photographs are taken or parts are marked for further characterization, for example, SEM/EDX.

The "color intrusion method" has evolved as one of the most powerful tools for characterizing, in particular, stack leakages. Every stack is tested after operation with respect to leakage. The leak test result is split into internal leakage (fuel to air side) and external leakage (fuel and air side to surrounding environment). Primarily for the localization of external leakages, the color intrusion method functions excellently and with high accuracy so that even small leaks or cracks can be detected better than with a bubble test. For this type of characterization, after leakage testing, the stack is painted red, which is normally used for crack detection of welded seams. After painting and waiting for a few minutes for color intrusion, the outside of the stack is rinsed with water to remove the excess color. Subsequently, the stack is dried either under ambient conditions overnight or at slightly higher temperatures so that it can be dismantled earlier. During plane-by-plane stack dissection, the areas of leakage in the glass–ceramic sealant can be easily differentiated from the gas-tight regions by color. Additionally, it is also possible to differentiate between complete cracking, which means that the cracks run through the whole sealant, thus bridging the stack interior to the outside, and only partial cracking (gas tightness still maintained). Through this optical visualization procedure, it is also possible to verify whether leakages are related to special regions or whether they are distributed randomly. Therefore, this method and analysis can also be used as a tool to detect geometric problem areas or nonideal structures. Figure 17.6 presents some stack photographs showing various types of leakages after stack dismantling. In Figure 17.6a,b, two planes from the same F20 design stack after a thermocycling test are shown. One plane shows significant leakages at the north and south sides, whereas the second plane appears to be gas tight. The reason for this untypical result is a strong thermal gradient in the z-direction, which was consciously applied to characterize the sealing behavior. Figure 17.6c presents another plane from an F20 design stack which shows significant leakage at one of the air-inlet channels. Possible reasons for this kind of leakage are also thermal gradients and non-ideal sealant trails, like 90° angles. In Figure 17.6d, a plane from a cassette-type stack is presented,

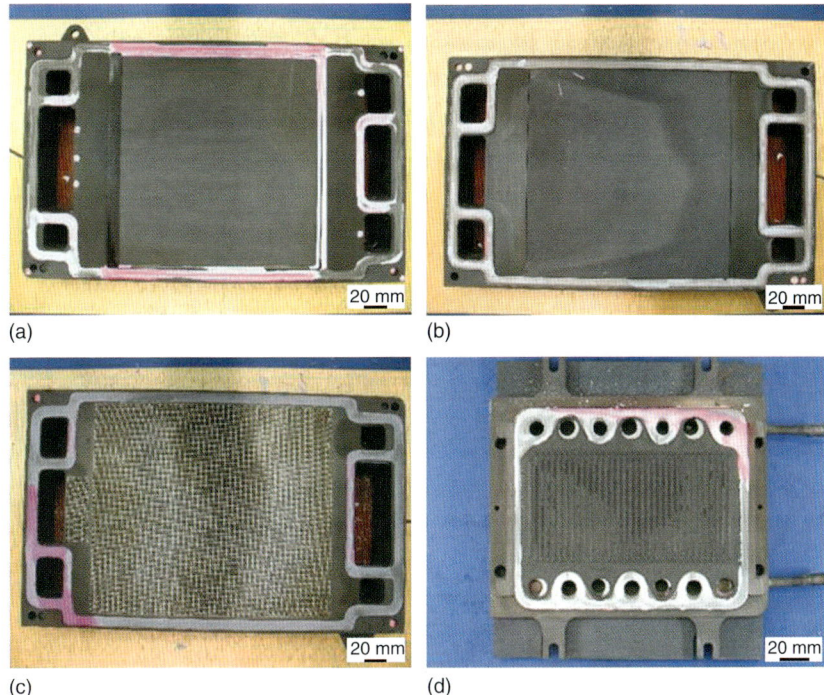

Figure 17.6 Photographs of stack planes after applying the color intrusion method and plane-by-plane dismantling: (a) plane of an F20 design stack with massive leakages on the north and south sides; (b) plane from the same stack as in (a) but without any leakage; (c) F20 design stack plane showing leakage in one air inlet channel at the southwest side; and (d) cassette-type stack plane with leakage near the air inlet channels.

which was harshly thermocycled with a heating rate of 25 K min^{-1} and then with a relatively cold air stream. The sealing cracks are at the region where the cold air enters the stack; this is the region of the major thermal differences in either the z- or x–y-direction. Summarizing these results, it can be concluded that severe temperature gradients within the stack, resulting in strong tensile stresses, lead to sealant cracking at the most stressed areas [17].

17.2.4
Optical Microscopy

Optical microscopy is used for stack analysis twice. First, it is used directly during stack dissection if surface-related changes or irregularities are observed. Second, it is used for characterizing the contact zones or to obtain an overview of the stack or stack parts with respect to geometric issues.

For the latter, stack parts need to be cut and prepared microscopically by embedding, grinding, and polishing. Figure 17.7 shows an optical photograph taken during stack dissection. Here, a color change at the anode side of the cell

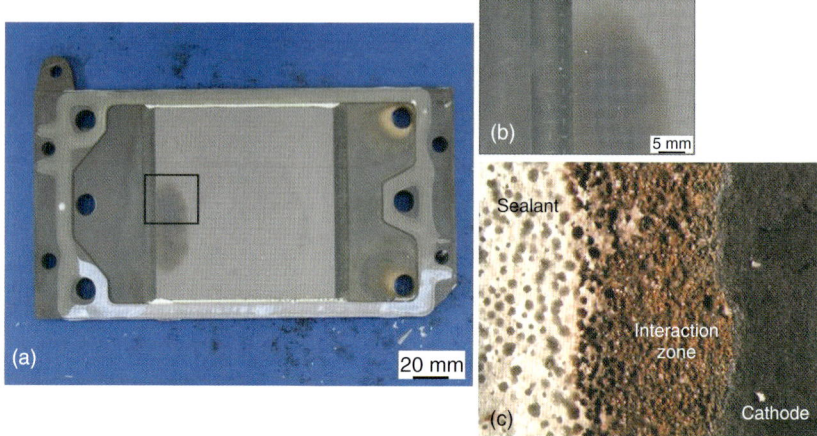

Figure 17.7 Characterization of areas of special interest during stack dissection by optical microscopy: (a) stack overview photograph; (b) higher magnification of marked rectangle in (a); (c) optical microscopy photograph from the rear side of the color-changed zone showing unwanted interaction between the cathode and the glass–ceramic sealant.

was detected in one plane. By optical microscopy, it was directly confirmed that the color change is due to an unwanted interaction between the cell cathode material and the glass–ceramic sealant. The reason for the interaction was an imprecise application of the glass paste. Both materials should have no contact during stack assembly or testing. An interaction causes stresses within the sealing, subsequently the material cracks, and a gas crossover between air and hydrogen leads to burning. Additionally, the interdiffusion phase thus formed is electrically conductive and may cause short-circuiting if it bridges the sealing gap.

Another example of obviously visible changes or interactions during stack testing is reoxidation phenomena. They are obvious because the anode color changes from gray (reduced Ni) to green (oxidized Ni), as shown in Figure 17.8a [15]. A typical reoxidation phenomenon is presented in Figure 17.8b. The electrolyte cracks due to tensile stresses caused by anodic Ni reoxidation and a subsequent volume change to NiO forming a porous NiO structure [5]. This fine-structured electrolyte crack network is typical visible proof of cracking due to anodic reoxidation.

In Figure 17.9, a cell defect is shown. This defect causes cell cracking and subsequently anode reoxidation, which finally leads to stack failure. Beneath the basic defect, various typical electrolyte cracks due to anode reoxidation can be detected. Therefore, stack dissection and characterization can also be used to discover the reason for stack failure.

Optical microscopy is also used for post-test analysis based on cut, ground, and polished samples. Fields of application are the analysis of layer thicknesses for layers thicker than ∼10 μm, contact areas, microstructures, geometric aspects such as tolerances and distances, and overview pictures for further SEM/EDX

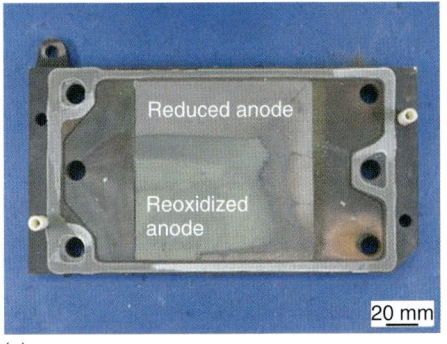 

(a) (b)

Figure 17.8 (a) Top view of the anode side from a reoxidized stack which is visible by the green color of the reoxidized Ni and the gas stream in the gas-out manifold (left side); (b) typical cracking structure of the electrolyte after anodic reoxidation (top view from the cathode side).

Figure 17.9 Optical microscopy top view of a cell defect (view from the cathode side).

characterization. In Figures 17.10 and 17.11, some examples of cross-sections are presented.

In Figure 17.10, an overview of three repeating units of one embedded stack is shown. With this photograph, it was possible to measure directly the anodic and cathodic distances and the sealant thickness. It is an easy, but destructive, technology to verify good stack assembly. In Figure 17.11, the cross-section of a special stack is shown. In this stack with five repeating units, the cathodic contact area was deliberately changed. Normally, cathodic contact is ensured by a channeled structure. On top of the channels, a perovskitic contact layer material ensures contact between the metallic channel and the cell cathode. The contact area was deliberately reduced by milling some of the channels. In this five-plane stack, the contact area of two planes was kept constant and the others were reduced by 25, 50, or 67%, respectively. After stack testing, the stack was embedded completely and then cut transversely to the channels to characterize the contact zones. The reason for reducing the channel size only by milling was so as not to influence the gas stream too much, that is, so as not to change two parameters at the same time

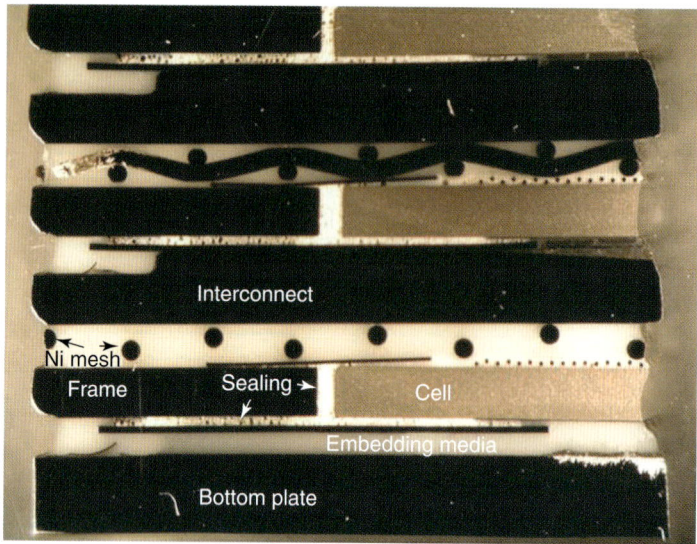

Figure 17.10 Optical microscopy photograph showing part of three repeating units in one embedded stack: cross-section parallel to gas channels.

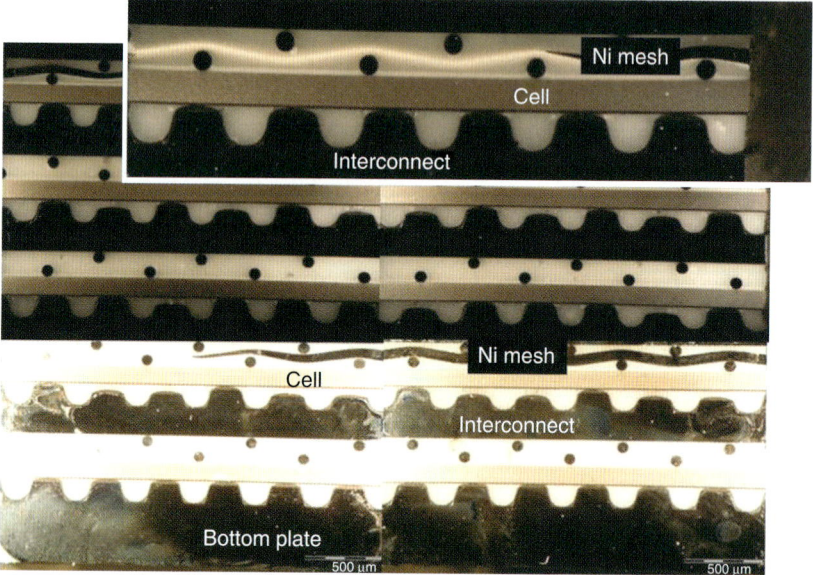

Figure 17.11 Optical microscopy photograph showing five repeating units of one embedded stack: cross-section perpendicular to gas channels; the inset at the top shows a higher magnification.

(contact area and air stream). In the inset photograph at the top of Figure 17.11, a higher magnification of one of the unchanged planes is shown. It can be seen that the gas channels do not contact ideally in-plane, but by a lens-like structure. This is due to the contact layer applied by wet powder spraying. The thicker the contact layer becomes, the more it tends to form a lens-like structure.

This embedding, cutting, polishing, and characterization by optical microscopy can also be used to describe the contacting, tolerances, and dimensions in the real stack in comparison with the blueprint. This is shown in Figure 17.12, which is an optical photograph of a cassette-type light-weight stack. In this picture, various on-line measurement points were devised in order to characterize the distances of the corrugated interconnect after stack assembly and testing and to verify the contact thickness and width of the contact layer applied by a dispenser system. The measurement system within the optical microscope can thus be used as a quality assurance system and, if necessary, it can be automated; however, this is also a destructive method.

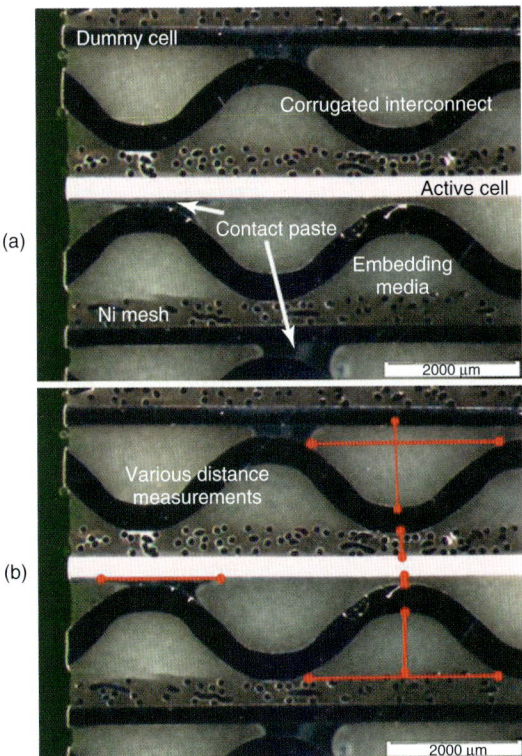

Figure 17.12 Cross-section of two repeating units of a cassette-type stack: (a) overview picture and (b) marking of various distance measurement points for quality assurance.

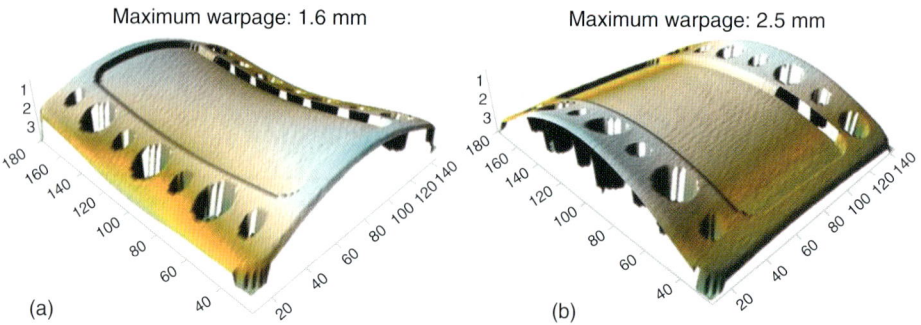

Figure 17.13 Laser topography of two cassette-type stack planes representing two kinds of component warpage (cell + metal frame).

17.2.5
Topography

The curvature of a component, its bending, layer thicknesses, and the surface roughnesses can be detected by laser topography. After stack dissection, it is used, for instance, to characterize the geometric changes of the corrugated interconnects which may occur due to pressing and creeping or to measure distances and heights. In Figure 17.13, two examples of a cassette-type stack plane with different sealed cells before stack assembly are presented. The topography is carried out by using an automatic x–y table mounted underneath a laser head, which works according to the laser triangulation method [18]. By using different sensors, either sample warpage (up to the millimeter range) or surface roughness (micrometer range) can be measured. One plane bends parallel to the gas flow direction and the other perpendicular to the former. There are numerous reasons for these different bending types; examples are cell thickness, type of substrate manufacturing process, application of oxide layers on the interconnects, and ratio between cell and interconnect thickness.

To minimize plane warpage caused by the dissection procedure, the stack can also be measured topographically as a whole stack or as stack parts, and it can be measured before and after stack testing.

17.2.6
Scanning Electron Microscopy (SEM) and Energy-Dispersive X-Ray (EDX) Analysis

SEM is used on various cell parts, such as the cell, sealant, and metal, after samples have been cut, embedded, ground, and polished. Mostly microstructures, layer thicknesses, materials interactions, and the formation of, for example, foreign phases are characterized by SEM/EDX. By using EDX analysis, chemical compositions and interdiffusions can also be studied.

In Figure 17.14, four examples of different post-test SEM analysis are presented. In Figure 17.14a, the interaction between the metallic interconnect material and

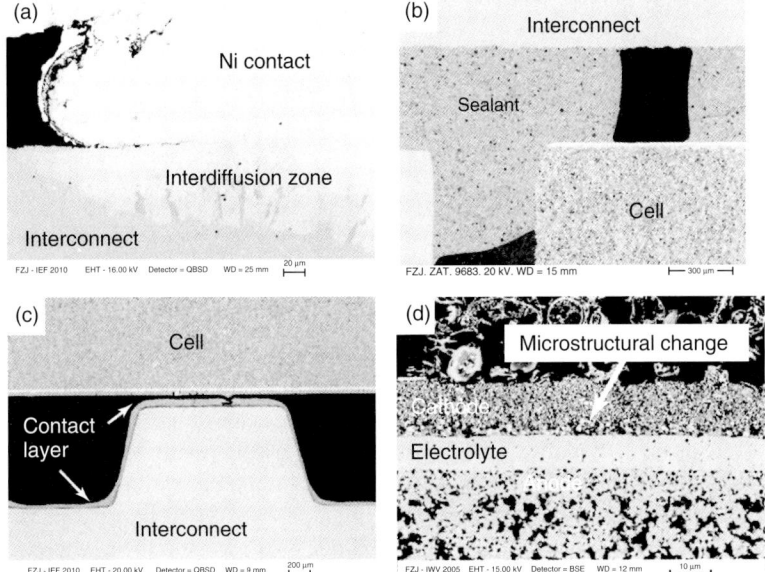

Figure 17.14 SEM cross-sectional micrographs of various stack zones after stack testing: (a) interdiffusion between Crofer22APU interconnect and metallic Ni mesh; (b) sealing situation between interconnect, cell, and sealant; (c) cathodic contact zone between the interconnect and the cell; (d) microstructural changes within the LSM cathode after interaction with volatile chromium species.

the anodic contacting Ni mesh after more than 17 000 h of stack operation is shown [19]. It is obvious that an interdiffusion has taken place. Iron and chromium from the interconnect diffuse into the Ni and, vice versa, nickel diffuses into the Crofer22APU. Due to Ni diffusion into the Crofer22APU, which is a ferritic material in its original state, it becomes austenitic. The austenitic phase is typically more brittle than the ferritic phase. This austenitization may lead to loss of contact during stack thermocycling.

In Figure 17.14b, the sealing situation on the cathode side is shown. There are two sealing zones: one is the cell sealing to the interconnect and the other is the sealing of the interconnect to the metallic frame, which bears the cell. It is obvious that the sealing microstructure is very homogeneous and that the sealant bonds well at all interfaces.

In Figure 17.14c, the contact situation on the cathode side between the interconnect channel and the cathode is highlighted. The contact layer applied by wet powder spraying covers all channel areas (bottom, top, and flanks) and on the top side it looks fairly homogeneous and flat, ensuring good cathodic contact (the gap is due to sample preparation).

In Figure 17.14d, a microstructural change within the cathode at the boundary to the electrolyte is presented. This microstructural change is caused by the interaction of the LSM (lanthanum strontium manganate) cathode with volatile chromium species. A foreign phase, $MnCrO_3$ or Cr_2O_3, is formed and subsequently the cathode structure coarsens [13].

If local materials analysis or concentration profiles are necessary, EDX analysis can be carried out as dot, line, or area analysis. In some cases, wavelength-dispersive X-ray (WDX) analysis can also be conducted, for example, to verify whether a phase consists of minimal amounts of Cr, or whether the Cr $K\alpha_1$ line overlaps with the La $L\beta_2$ line from the cathode material.

The interaction of volatile chromium species is regarded as a major cause of SOFC stack degradation [6–9]. The reaction between the volatile Cr species, mostly $CrO_2(OH)_2$, and the cathode depends, from a materials point of view, on the cathode type. If LSM is used, the Cr reacts at the interface between the cathode and the electrolyte forming chromia or a Cr–Mn spinel. In the case of an LSCF (La–Sr–Co–Fe)-type cathode, the chemical reaction is different. Whereas between Cr and LSM the main interaction is the reduction reaction of Cr(VI) to Cr(III) at the triple-phase boundaries and the subsequent reaction with Mn from the LSM, the chemical reaction between Cr and LSCF is not based on the reduction reaction. The volatile Cr species reacts with the Sr in the LSCF in a gas-phase interaction. During this reaction, the stable $SrCrO_4$ is formed and the new phase does not precipitate at the cathode/electrolyte interface, but at the top of the cathode surface (see also Section 17.2.7). The fact that the reaction between Cr and Sr is a gas-phase reaction was supported by the post-test findings of one particular stack, in which LSCF-based cathodes were used and which was thermocycled a number of times. In Figure 17.15, SEM surface micrographs of the zone between the cathode–contact layer area and the gas channel are presented. In this area where the air stream is hindered, a multitude of $SrCrO_4$ needles are formed. Higher magnifications of the needles show a layered microstructure which is typical of evaporation–condensation reactions. Obviously, with every thermocycle during the cooling regime the $SrCrO_4$ precipitates preferentially at the surface of the needles still being formed. The micrographs also show that the Sr chromate needles form primarily at the surface cracks of the cathode material. These cracks originate from the drying step of the screen-printing paste. The needles show some kind of outward growth from the cracks into the gas phase. EDX dot analysis confirms the formation of $SrCrO_4$.

This gas-phase chemical interaction between the LSCF cathode and volatile Cr species was observed and is presented here for the first time. This post-test analytical result was the starting point for ongoing R&D with respect to Cr and SOFC cathode interaction, which is now being conducted basically in single-cell test units.

17.2.7
X-Ray Diffraction (XRD)

XRD is used after stack dissection and sample cutting to characterize the crystal structure of the various stack materials. Additionally, their changes due to interactions, segregation, incorporation of foreign elements, or decompositions can

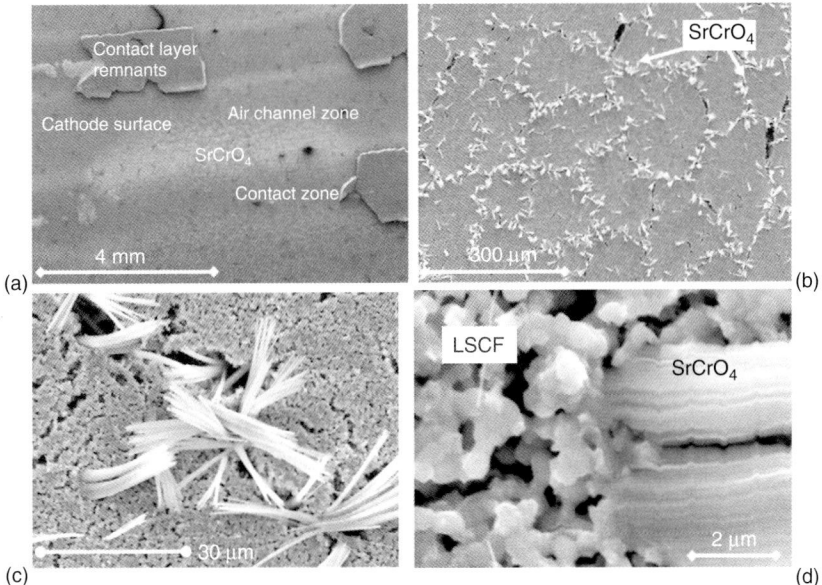

Figure 17.15 SEM images of the LSCF cathode surface of a thermocycled stack showing SrCrO$_4$ formation in the intermediate zone between the contact area and the air channel.

be investigated. The characterization of the cathode surface of a stack which has experienced long-term operation with an La–Sr–Fe–cobaltite cathode is presented in Figure 17.16 as an example of X-ray analysis. The X-ray analysis is taken from two different areas. Area 1 is the surface adjacent to the contact bar and area 2 is adjacent to the gas channels. The X-ray analysis reveals three crystallographic phases: copper, which is a signal from the copper strip used for covering the other areas which were not analyzed; an orthorhombic perovskite from the cathode material; and a third phase, monoclinic SrCrO$_4$. Comparison of the two diffractograms (at the bar and at the channel) shows that the Sr chromate reflexes are more pronounced in the area beneath the channel bar. Here, the secondary phase of Sr chromate is formed between the Sr from the cathode and the volatile Cr species. The amount of the foreign phase at the cathode surface beneath the gas channel is considerably lower, which leads to the conclusion that the reaction between the Sr and the Cr species is a gas-phase reaction and that both gas phases are transported outside of the stack by the air stream [8], while the species precipitates in the porous interface of the contact area, where no air stream exists.

Other interesting areas for carrying out XRD are, for example, deterioration of the 8YSZ electrolyte (8YSZ: with 8 mol% yttria-stabilized zirconia) from a cubic to monoclinic/tetrahedral structure after long-term operation or the formation of austenite in the metal by Ni diffusion.

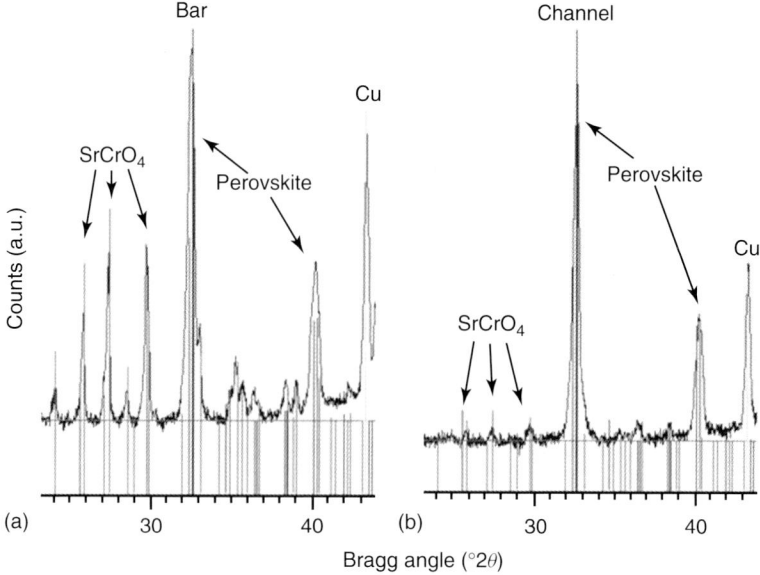

Figure 17.16 XRD of two cathode surface areas of a stack cell after long-term operation: (a) surface at the channel bar; (b) surface at the gas channel.

17.2.8
Wet Chemical Analysis

Wet chemical analysis is used, for example, to characterize the incorporation of foreign elements into the anode or cathode structure. Its advantage is that quantitative amounts can be analyzed, but the method is (i) destructive, (ii) cannot locally resolve the analysis, such as characterization of the cathode but not whether the foreign element is in the cathode, at a phase boundary, or in the current collector, and (iii) complex because it has to be discovered what sample preconditioning is necessary, for example, what acid leaching works, what acid or acid mixture dissolves all the material, and whether other components that should not be dissolved are also in the solution; for example, if the cathode material is dissolved, Ni may also be partly dissolved.

One major analysis that was carried out by wet chemical analysis in the past was the measurement of chromium, which is incorporated into the cathode during stack operation with metallic interconnects [10]. Here, cell parts with typical dimensions of 1 cm^2 are cut by laser cutting. Subsequently, the cathode is removed by leaching in perchloric acid [11, 13]. By doing so, the perovskite itself and the foreign phases formed such as chromia or Cr–Mn spinel are completely dissolved. Hardly any Y or Zr from the electrolyte and only small amounts of Ni from the anode dissolve. By taking the starting cathode composition as a reference and by taking a given sample area, the amount of foreign element incorporation can be calculated as the

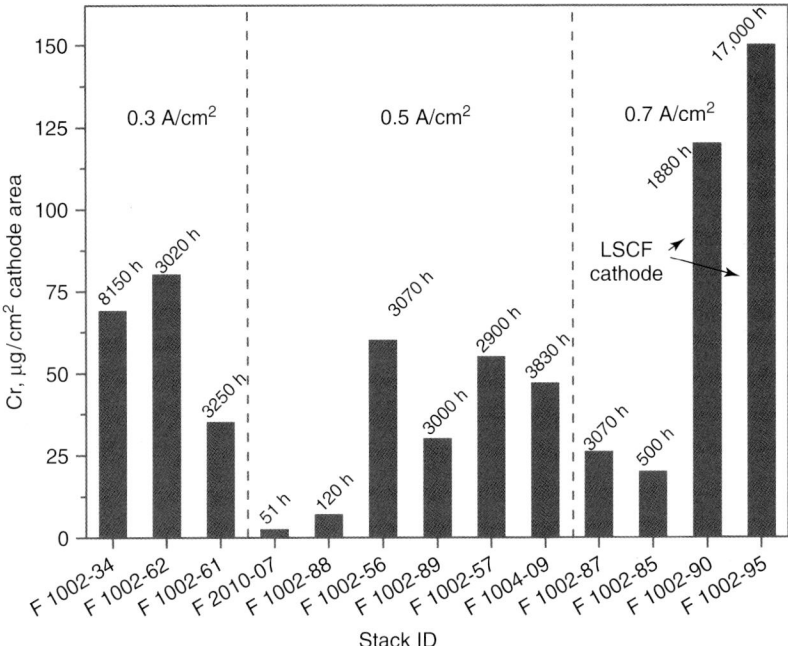

Figure 17.17 Wet chemical analysis of Cr incorporation in the cathode after stack operation; from each cell one or more 1 cm² cell samples were taken, the cathode was dissolved, and the solution was subsequently analyzed by inductively coupled plasma optical emission spectroscopy.

amount per square centimeter or by stoichiometry. An additional comparison with SEM images taken from the same cell sample can give twofold information: first, the amount from the wet chemical analysis, and second, the local resolution of the detection of Cr within the cathode, the current collector, or other cell or contact layers. Thus optical and elemental analysis can be combined.

Figure 17.17 presents an overview of analyzed Cr incorporation into cathodes operated for various times and operational conditions from stacks.

From the analyses in Figure 17.17, it was possible to draw some conclusions concerning the interaction of chromium species with SOFC cathodes. First, it is obvious that a time dependence exists, but additionally there seems to be some saturation effect: see, in particular, the results obtained at a current density of 0.5 A cm^{-2}. After ∼3000 h of operation there is no additional Cr incorporation into the cathode. Another result is that LSCF cathodes incorporate more Cr than LSM cathodes and that there is little difference in the amount of Cr after stack operation times of 1900 h and more than 17 000 h. The results obtained by these analyses were the basis for further R&D in the field of Cr-related degradation [11, 13] in order to understand better the basic interactions, chemical/electrochemical reactions, dependences on temperature, time, current density, and chromium partial pressure, and the impact on cell/stack performance.

17.2.9
Other Characterization Techniques

In addition to the widely used techniques of photography, optical microscopy, SEM, thermography, topography, measurement by tools, XRD, and wet chemical analysis, there are various other techniques that can be applied to post-test SOFC stack analysis. A few examples are given below, but this list does not claim to be exhaustive:

- TEM, including focused ion beam (FIB) for high-resolution investigations concerning particle size and structures, chemical analysis, crystallographic analysis, detection of foreign phases or inclusions, and so on; destructive method
- *tomography* to characterize sealing regions, porosities, or geometric situations (the technique is only limited by low resolution); nondestructive method
- *atomic force microscopy* (AFM) and *confocal microscopy* for the higher magnification of surface structures, roughness (including micro-roughness), and small defects; destructive method
- techniques for z-directional chemical analysis such as *LIBS* (laser-induced breakdown spectroscopy), *GDOS* (glow discharge optical emission spectroscopy), *SIMS* (secondary ion mass spectrometry), and *SNMS* (secondary neutron mass spectrometry); mostly destructive due to sample size restrictions.

17.2.10
Lessons Learned from Post-Test Stack Dissection and Analysis

SOFC stack tests are complex, time and material consuming, but irrecoverable. From single cell tests, a lot of information with respect to mimicking stack testing can be collected, for example, basic material interactions, interdiffusions, material coarsening, and poisoning. However, the stack environment has its own special features, for example: the real geometric relationships, the real anodic and cathodic contacting, the air and fuel gas streams, the degree of gas utilization, and the pressure differences between stack and environment or between anode and cathode. Therefore, some interactions, difficulties, events, and operational results only appear in stack testing. Hence careful quality assurance before stack assembly, careful recording of stack assembly, stack testing, and stack post-test analysis are indispensable for understanding SOFC operation and degradation. In this respect, post-test analysis is one of the keys to enhancing lifetime and minimizing stack degradation and thus to minimizing the gap between R&D to market entry of the SOFC.

Many events during stack testing cannot be investigated postmortem and in many cases no correlations could be found between the electrochemical events during operation and the post-test analysis. However, for some events, degradation phenomena, and interactions, a careful, reproducible, and scientific post-test analysis answers many questions.

Examples are the interaction of volatile chromium species with various cathode materials. The influences of design, air steam, current density, and so on, can only be adapted by using stack testing. Many of the results obtained so far were first observed within a stack and subsequently single-cell tests have been established to understand the basic reaction mechanisms and dependences.

Another example is the interaction of metallic interconnects with special glass–ceramics, which caused short-circuiting after short stack operation times [19]. This was a typical stack result that could not be found in single-cell tests because here the sealing is done with gold or silver wires and not with the glass–ceramic. Again, this example shows that a first result from the stack was followed by basic R&D and the reason for this severe interaction was then found and could be avoided; in this example the reasons for the severe interaction were small amounts of Al and Si in the metal and Pb in the glass, which caused major interactions and the formation of an outward-growing conductive phase, which led to short-circuiting after sealing gap bridging.

A third and final example of lessons learnt was the lens-like structure of the contact material on the channel bars. If only 50% of the channel bar contacts the cathode, the real current density is very different from the current density applied and calculated for the contact area. Optimizing the channel bar structure and the coating process reveals better flat-like contact zones (compares Figures 17.11 and 17.14c). Most of these findings would not have been revealed if the stacks had not been characterized after testing.

17.3
Conclusion and Outlook

Modern analytical methods are being increasingly used to characterize SOFC stacks after operation. In addition to nondestructive techniques such as tomography, topography measurement, and thermography, a wide variety of destructive analytical methods have been applied to SOFC stack components. Typically, intensive post-test characterization needs more than one analytical method because every method has its advantages and limitations – for example, only numerical information such as the amount of elements or the thickness, but no information about the (micro-)structure or location; structural information but no elemental information or microstructural and elemental but no crystallographic information; and so forth. Therefore, in most cases, it is necessary to combine different analytical methods and, after careful single result interpretation, conclusions can be drawn concerning degradation effects, materials incompatibilities, quality assurance efforts, or envisaged and real tolerances.

Before a decision is made about which analytical methods will be used to characterize the stack components individually, a basic decision must be made with respect to embedding or not embedding the stack. Figure 17.18 presents a flow chart of these decisions and, depending on the decision, the analytical possibilities.

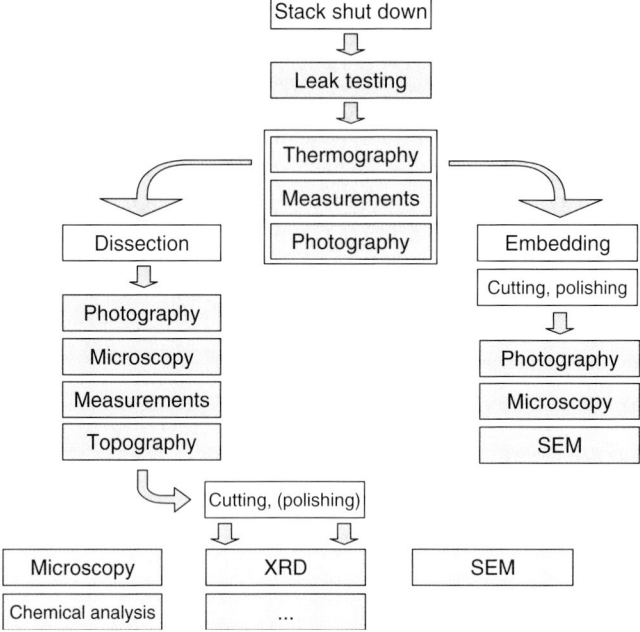

Figure 17.18 Flow chart of possible stack post-test analyses.

In the future, not only will post-test analysis become more and more important, but also *in situ* measurements during stack operation. In addition to classical measurement of temperature at various stack positions, also optical measurement techniques will be applied. With such techniques, the stack settling, warpages, creeping, and so on, can be observed during special operating conditions. Additionally, outlet gases can be analyzed, a high-temperature leak test can be applied, and acoustic tests to analyze cracking, depending on, for example, thermocycling, are of interest.

If enhanced quality assurance methods are applied to the starting materials, processes, and components, if the SOFC stack test stands are better equipped with *in situ* measuring methods, and if post-test analysis becomes more sophisticated and standardized, this will be a helpful set of instruments to ensure longer lifetime, lower degradation, and thus greater marketability of SOFCs.

Acknowledgments

The authors gratefully acknowledge support, cooperation, and discussions within the Forschungszentrum Jülich SOFC Stack Dissection Group (Prof. Blum, Dr. Ettler, Dr. Gross, Dr. Haanappel, Dr. Malzbender, Dr. Neumann, Dr. Shemet, Dr. Steinbrech, and Dr. Vinke), and also support from Dr. Fischer (X-ray analysis), Dr. Glückler (thermography), and Dr. Sebold (SEM).

References

1. Benhaddad, S. and Protkova, I. (2005) in *Proceedings of the 9th International Symposium on Solid Oxide Fuel Cells (SOFC IX)* (eds. S.C. Singhal and J. Mizusaki), Electrochemical Society, Pennington, NJ, pp. 923–930.
2. Batfalsky, P., Haanappel, V.A.C., Malzbender, J., Menzler, N.H., Shemet, V., Vinke, I.C., and Steinbrech, R.W. (2006) Chemical interaction between glass–ceramic sealants and interconnect steels in SOFC stacks. *J. Power Sources*, **155**, 128–137.
3. Menzler, N.H., Batfalsky, P., Gross, S.M., Shemet, V., and Tietz, F. (2011) Post-test characterization of an SOFC stack after 17,000 hours of steady operation. *ECS Trans.*, **35** (1), 195–206.
4. Menzler, N.H., Tietz, F., Bram, M., Vinke, I.C., and de Haart, L.G.J. (2008) Degradation phenomena in SOFCs with metallic interconnects. *Ceram. Eng. Sci. Proc.*, **29** (5), 91–102.
5. Ivers-Tiffée, E., Timmermann, H., Leonide, A., Menzler, N.H., and Malzbender, J. (2009) in *Handbook of Fuel Cells – Fundamentals, Technology and Applications*, vol. **6**, Chapter 64 (eds. W. Vielstich, H. Yokokawa, and H.A. Gasteiger), John Wiley & Sons, Ltd., Chichester, pp. 933–956.
6. Menzler, N.H., Mai, A., and Stöver, D. (2009) in *Handbook of Fuel Cells – Fundamentals, Technology and Applications*, vol. **5**, Chapter 38, (eds W. Vielstich, H. Yokokawa, and H.A. Gasteiger), John Wiley & Sons, Ltd., Chichester, pp. 566–578.
7. Hilpert, K., Das, D., Miller, M., Peck, D.H., and Weiss, R. (1996) Chromium vapor species over solid oxide fuel cell interconnect materials and their potential for degradation processes. *J. Electrochem. Soc.*, **143** (11), 3642–3647.
8. Yokokawa, H., Sakai, N., Horita, T., Yamaji, K., Brito, M.E., and Kishimoto, H. (2008) Thermodynamic and kinetic considerations on degradation in solid oxide fuel cell cathodes. *J. Alloys Compd.*, **452**, 41–47.
9. Yokokawa, H., Horita, T., Sakai, N., Yamaji, K., Brito, M.E., Yiong, Y.-P., and Kishimoto, H. (2006) Thermodynamic considerations on Cr poisoning in SOFC cathodes. *Solid State Ionics*, **177**, 3193–3198.
10. Menzler, N.H., Vinke, I., and Lippert, H. (2009) Chromium poisoning of LSM cathodes – results from stack testing. *ECS Trans.*, **25** (2), 2899–2908.
11. Neumann, A., Menzler, N.H., Vinke, I., and Lippert, H. (2009) Systematic study of chromium poisoning of LSM cathodes – single cell test. *ECS Trans.*, **25** (2), 2889–2898.
12. De Haart, L.G.J., Mougin, J., Posdziech, O., Kiviaho, J., and Menzler, N.H. (2009) Stack degradation in dependence of operation parameters; the REAL-SOFC sensitivity analysis. *Fuel Cells*, **6**, 794–804.
13. Neumann, A. (2010) Aufklärung der Chrom-bezogenen Degradation von Festoxid-Brennstoffzellen. PhD thesis, Ruhr-University Bochum, Faculty of Mechanical Engineering (available in German only).
14. Sarantaridis, D. and Atkinson, A. (2007) Redox cycling of Ni-based solid oxide fuel cell anodes: a review. *Fuel Cells*, **7** (3), 246–258.
15. Ettler, M., Timmermann, H., Malzbender, J., Weber, A., and Menzler, N.H. (2010) Durability of Ni anodes during reoxidation cycles. *J. Power Sources*, **195**, 5452–5467.
16. Menzler, N.H., Blaß, G., Giesen, S., and Buchkremer, H.P. (2002) Processing and quality control of planar SOFC components. *Matwiss. Werkstofftech.*, **33**, 367–371.
17. Blum, L., Gross, S.M., Malzbender, J., Pabst, U., Peksen, M., Peters, R., and Vinke, I.C. (2011) Investigation of SOFC sealing behavior under stack relevant conditions at Forschungszentrum Jülich. *J. Power Sources*, **196**, 7175–7181.
18. Mücke, R., Menzler, N.H., Kemnitzer, F., and Blöchl, K. (2008) The application of laser topography in SOFC development. In Proceedings 8th European SOFC Forum, A03101, Abstract 031, 30 June–4 July 2008, Lucerne, Switzerland, pp. 1–23.

19. Menzler, N.H., Batfalsky, P., Blum, L., Bram, M., Gross, S.M., Haanappel, V.A.C., Malzbender, J., Shemet, V., Steinbrech, R.W., and Vinke, I. (2007) Studies of material interaction after long-term stack operation. *Fuel Cells*, **5**, 356–363.

18
In Situ Imaging at Large-Scale Facilities
Christian Tötzke, Ingo Manke, and Werner Lehnert

18.1
Introduction

The importance of water management to successful cell operation of polymer electrolyte fuel cells (PEFCs) has directed the focus of extensive research activity on liquid water transport and its effect on cell performance, reliability, and durability of cell components. Water produced from the electrochemical reaction together with water from the humidified inlet gases maintains the hydration level of the membrane which is crucial for high proton conductance. However, excess liquid water can cause flooding of gas diffusion layer (GDL) regions and flow field channels, entailing oxidant starvation and significant power losses. Well-balanced water management is a major challenge to achieve the optimal efficiency and lifetime of these cells [1–7]. In addition to water management, the formation and transport of CO_2 bubbles on the anode side in direct methanol fuel cells (DMFCs) has to be understood in order to prevent blockage of anode-side gas channels [8, 9]. In recent years, phosphoric acid-based high-temperature polymer electrolyte fuel cells (HT-PEFCs) have attracted much attention from the fuel-cell community because of the high CO tolerance. Owing to the high operating temperature of 160 °C, no liquid water is present in the cells and there is no need to humidify the gases. In order to achieve optimum power densities, the distribution of the electrolyte in the electrodes and the membrane plays a crucial role [10, 11].

In the recent past, imaging methods have contributed a great deal to the advances in these research fields. Neutron and synchrotron X-ray imaging have become established as indispensable diagnostic tools for the optimization of fuel-cell components, for example, flow field geometry, GDLs or electrode structures, as they are able to reveal water transport processes in operating cells, CO_2 bubble formation in DMFCs, or the dynamics of the phosphoric acid distribution in HT-PEFCs in a non-destructive way. In the following, basic principles of these methods and a representative selection of practical applications in fuel-cell research are presented.

Fuel Cell Science and Engineering: Materials, Processes, Systems and Technology,
First Edition. Edited by Detlef Stolten and Bernd Emonts.
© 2012 Wiley-VCH Verlag GmbH & Co. KGaA. Published 2012 by Wiley-VCH Verlag GmbH & Co. KGaA.

18.2
X-Rays and Neutrons

18.2.1
Complementarity of X-Rays and Neutrons

X-ray and neutron imaging are complementary techniques for materials research. X-rays interact mainly with the electronic shell of atoms whereas neutrons as charge-neutral particles interact with the nuclei (Figure 18.1a,b). The different interaction mechanisms yield different beam attenuation properties. Figure 18.1c shows the values for the attenuation coefficients of X-rays and neutrons for different element numbers. In the case of X-rays, the attenuation increases with the number of electrons in the atom and, therefore, with the element number. In case of neutrons, no clear dependence on the amount of nuclei within the atomic core can be found. In contrast to X-rays, some light elements such as H and Li have a very

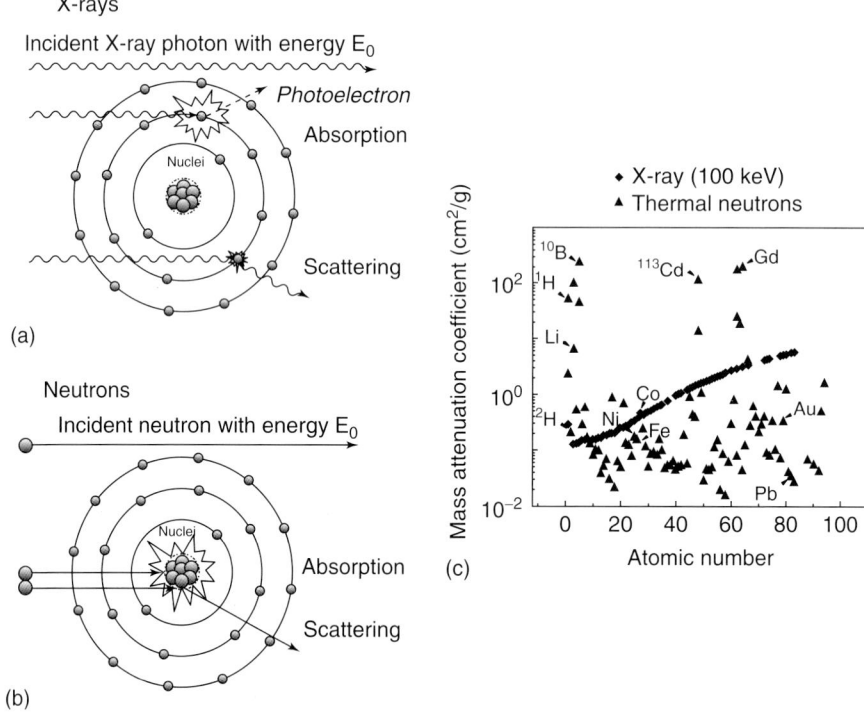

Figure 18.1 Interaction of matter with (a) X-rays and (b) neutrons, and (c) mass attenuation coefficient as a function of atomic number for all elements. Adapted from [12, 13] by permission of IOP Publishing and DGZfP – Deutsche Gesellschaft für Zerstörungsfreie Prüfung.

Table 18.1 Comparison of typical parameters between neutron and X-ray imaging in fuel cell and battery research.

	Neutron imaging	Synchrotron X-ray imaging	Laboratory X-ray imaging
Pros	Visualization of light elements such as H and Li in the presence of thick metallic components	High spatial resolution	Available at many locations
	Metal penetration	Fast measurements	Comparatively cheap, easily accessible
	Larger sample size up to several tens of centimeters	High image quality and easy quantification due to use of a monochromatic beam	–
Cons	Low beam intensities (low time and spatial resolution)	Visualization of light elements in the presence of metals	Low beam intensities
	Polychromatic beam	Electron storage ring necessary	Polychromatic beam
	Large-scale facility	–	Visualization of light elements in the presence of metals
Typical fluxes (intensity)	Typically 10^6–10^9 cm^{-2}	Typically 10^{11}–10^{15} cm^{-2}	10^6–10^8 cm^{-2}
Typical spatial resolution (μm)	20–200	0.6–5	1–30
Typical time resolution (single radiography)	0.1–100 s	0.001–5 s 1–50 ms (short periods)	0.2–100 s
Typical time resolution at maximum spatial resolution	3–20 min	1–5 s 10–200 ms (short periods)	5–20 s

high attenuation coefficient whereas many of the heavy elements such as Pb, are only weakly attenuating.

Furthermore, neutrons can distinguish between different isotopes. For example, deuterium (an isotope of hydrogen) has a much smaller attenuation coefficient than hydrogen, which can be exploited in imaging experiments. For example, H_2O can be exchanged by D_2O to study water exchange processes. Most neutron imaging facilities use cold or thermal neutrons which have energies around 4 and 25 meV, respectively. Cold neutrons give a better contrast because the attenuation is higher. However, thermal neutrons have some advantages for the investigation of large objects. In Table 18.1, advantages, disadvantages, and key properties of neutron and X-ray imaging are summarized.

18.2.2
Principles of Radiography and Tomography

18.2.2.1 Transmission and Attenuation

Imaging contrast is caused by the varying attenuation of a neutron or X-ray beam while penetrating a sample (Figure 18.2). The transmission signal intensity is detected with a 2D detector [14, 15].

The attenuation coefficient $\mu(x, y, z)$ is a product of various physical events which contribute to the overall beam attenuation such as coherent and incoherent scattering and absorption.

Because of the low beam fluxes (intensities) available at neutron sources, normally the whole available neutron spectrum (a "white beam") is used for real-time imaging applications, that is, a polychromatic beam is used especially for fast *in situ* radiography measurements on PEFCs.

The intensity $I_n(x, y)$ of a neutron or X-ray beam passing a sample with a thickness of $d = n\Delta z$ in the z-direction is given by:

$$I_n(x, y) = I_0(x, y) \exp\left[\sum_{i=1}^{n} -\mu_i(x, y, z_n) \Delta z\right] \quad (18.1)$$

or

$$I_n(x, y) = I_0(x, y) \exp\left[\int -\mu(x, y, z) dz\right] \quad (18.2)$$

where $\mu(x, y, z)$ denotes the distribution of the local attenuation coefficient in the sample, Δz the width of the corresponding voxel (volume pixel) within the sample, and $I_0(x, y)$ the initial beam intensity distribution.

18.2.2.2 Synchrotron X-Ray Sources and X-Ray Tubes

X-rays are mostly produced by accelerated electrons that hit a metallic target (e.g., tungsten) within an X-ray tube. The energy spectrum of the X-ray beam depends on the maximum acceleration potential, typically ranging from a few tens to some hundreds of kilovolts. The available beam intensities are limited, because of the local heating of the target, especially in micro focus tubes which are mostly used for imaging applications to achieve high spatial resolution.

Especially for fast *in situ* investigations of water transport mechanisms in PEFCs the beam intensity is often too low. Furthermore, the broad energy spectrum makes data quantification difficult because the attenuation depends strongly on the X-ray energy.

At synchrotron facilities, X-rays are produced in a different way: electrons moving at almost the speed of light are deflected by strong magnets. Thereby, so-called synchrotron radiation is emitted that comprises (for typical imaging applications) hard X-rays with an energy range of 10–200 keV. The high intensities (several orders

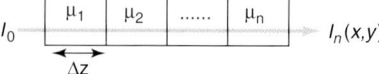

Figure 18.2 Schematic description of beam attenuation.

of magnitude higher compared with a laboratory X-ray tube) not only allow fast *in situ* measurements but also monochromatization of the beam and, therefore, much better data quantification. Furthermore, the beam has a high coherence, which can be used for phase contrast or holotomography measurements. It should be noted that at some facilities the beam intensities are extremely high. For this reason, it is necessary to attenuate the beam in some cases to protect sensitive samples from radiation damage.

18.2.2.3 Tomography and Tomographic Reconstruction

A tomographic measurement is based on the acquisition of several radiographic projection of the sample from different viewing angles. For this purpose, the sample is rotated around an axis perpendicular to the beam (Figure 18.3). A complete tomographic measurement typically contains between 300 and 3000 images.

It can be shown by mathematical means that it is possible to reconstruct a 3D image (tomogram) from this radiographic series that represents the interior structure of the investigated object, that is, the tomogram represents a 3D image of the local attenuation coefficients in the sample. The mathematical basis was first published by Radon in 1917 (Radon transform). The 3D image contrast obtained depends strongly on the probe used (neutrons, polychromatic/monochromatic X-rays). One of the most commonly used reconstruction algorithms is the so-called

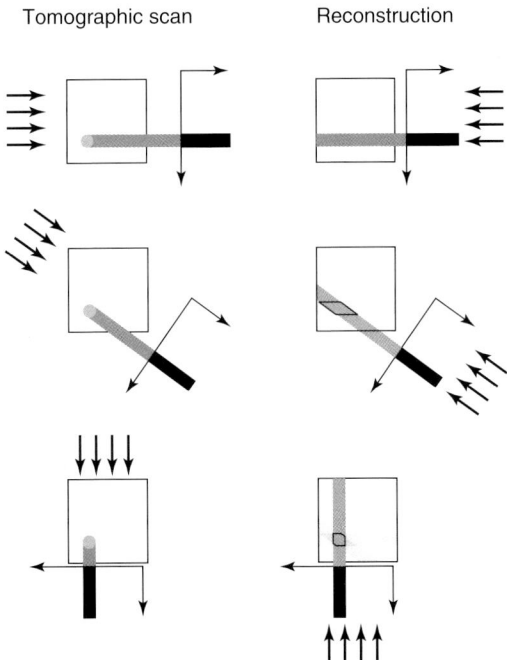

Figure 18.3 Principle of procedure for tomographic measurements and reconstructions.

back-projection or filtered back-projection. The principle of this method is shown in Figure 18.3. The main reason for using this algorithm is the short reconstruction times. With a currently state-of-the-art computer, a 3D data set with 2048 × 2048 × 2048 pixels can be reconstructed within a few minutes to a few hours. It should be mentioned that the size of a typical tomogram taken with a 2048 × 2048 or 4096 × 4096 charge-coupled device (CCD) is 16 or 128 GB, respectively. Data handling needs special hardware and software. Since the amount of detector pixels is increasing in the same way as the speed of computers, the problem will still be present in future.

18.2.2.4 Artifacts

In radiographic and especially in tomographic measurements, several different artifacts can occur, that is, the 2D or 3D image does not display the real distribution of the attenuation coefficients as it should, but shows some deviations. Typical artifacts that can be found in radiographic and tomographic imaging are as follows:

- Small-angle scattering and refraction effects in the sample produce artificial image structures/features.
- Incoherent scattering, for example, of neutrons at hydrogen atoms (problematic for water quantification).
- Fluctuations of the incident beam intensity during data acquisition (problematic for data quantification).
- Varying response of CCD detector pixels.
- Electronic noise of detectors.

Furthermore, tomographic reconstruction might yield to the following additional artifacts:

- Ring artifacts caused by defects or impurities of the scintillator crystals or imperfect detector elements.
- Motion artifacts caused by sample movements during data acquisition.
- Ring/crescent-like artifacts caused by a non-perfect rotation axis.
- Missing wedge/limited angles artifacts: if a fuel cell is too large for the X-ray or neutron beam, only part of the cell can be investigated and a fraction of the radiographic projections cannot be used for data reconstruction because of zero transmission. This effect causes strong blurring of all structures oriented perpendicular to the missing projections.
- Beam hardening artifacts (for polychromatic X-rays or neutrons) prevent exact quantification of the attenuation coefficients.

In order to correct such artifacts, several tools for data treatment are available. Appropriate experimental optimization, imaging filter procedures, and optimized reconstruction algorithms can reduce most of these effects.

It should be noted that refraction (phase contrast) artifacts can be exploited in phase contrast radiography/tomography to enhance the image contrast or to calculate "phase" maps or holo-tomograms.

A special problem is quantification of the amount of water in neutron imaging: Today, scattering correction tools are available [14, 16].

18.2.2.5 Image Normalization Procedure

In case of *in situ* radiography for the investigation of water distributions in fuel cells, a general normalization procedure is applied. First, a reference image of the water-free cell and a darkfield image (with the beam switched off) are taken. The darkfield image is subtracted from all other images to account for the contribution of the detector dark current signal:

$$I_{dry} = I_{in\,situ,dry} - I_{dark} = I_0 \times e^{-\mu_{cell} d_{cell}} \tag{18.3}$$

$$I_{water} = I_{in\,situ,water} - I_{dark} = I_0 \times e^{-\mu_{water} d_{water} - \mu_{cell} d_{cell}} \tag{18.4}$$

where I_0 is the intensity of the beam, $I_{in\,situ,dry}$ and $I_{in\,situ,water}$ denote the intensity of the detected transmission signal through the dry and wet (operating) cell, I_{dark} is the contribution of the darkfield signal (beam switched off), I_{dry} and I_{water} are the respective darkfield-corrected transmission intensities, d_{water} and d_{cell} are the material thickness of water and the cell attenuating the X-ray or neutron beam, respectively, and μ_{cell} and μ_{water} are the mean attenuation coefficients of the cell and water, respectively.

All radiographic images taken during the *in situ* investigation are then divided by the reference image of the water-free cell. Taking logarithms and rearranging yield the value of the detected water thickness:

$$\log\left(\frac{I_{water}}{I_{dry}}\right) = \log\left(\frac{e^{-\mu_{water} d_{water} - \mu_{cell} d_{cell}}}{e^{-\mu_{cell} d_{cell}}}\right) = -\mu_{water} d_{water} \tag{18.5}$$

The procedure is illustrated in Figure 18.4.

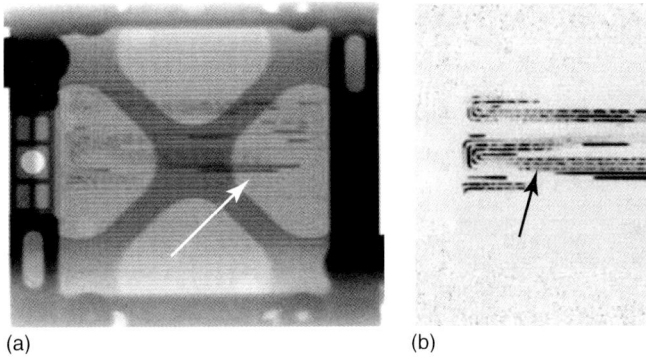

(a)   (b)

Figure 18.4 (a) Neutron radiographs (raw image) of an operating PEMFC. Water in flow field channels appears as dark horizontal lines due to the strong attenuation. (b) Two-dimensional liquid water distribution as obtained by the normalization procedure. Adapted from [17] by permission of Carl Hanser Verlag.

18.3
Application of *In Situ* 2D Methods

18.3.1
PEFCs

Low-temperature PEFCs are expected to play a crucial role for powering future mobile and stationary applications [18], [19]. Optimized water management is a crucial prerequisite for successful and reliable operation and for long-term stability of the cells. X-ray and neutron imaging techniques have become established as powerful tools to address different questions of water management.

18.3.1.1 X-Rays

Imaging instruments at synchrotron X-ray sources provide the opportunity to study water distributions within small areas (around a few millimeters) at high spatial resolution up to <1 μm. The high photon flux of synchrotron sources (typically between 10^{10} and 10^{12} photons mm^{-2} s^{-1}) facilitates monochromatic measurements with short exposure times (typically a few seconds or milliseconds) and offers excellent conditions for precise quantitative analyses of even the smallest water agglomerations.

Since X-rays are strongly attenuated by metals and other components used in fuel cells, slight design adaptations are necessary to allow for sufficient beam transmission. Small sealed holes are drilled into the metallic end plates while all other components remain unchanged. The cross-section of a fuel cell prepared for studies in through-plane perspective is sketched in Figure 18.5. Realistic conditions for heat and mass transfer can be retained because the cell design modification is kept to a minimum.

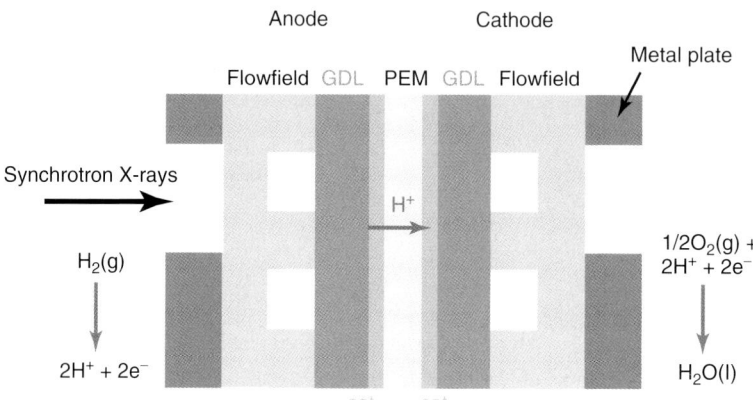

Figure 18.5 Typical test cell design modified for synchrotron X-ray tests in a through-plane perspective. Reproduced from [20] by permission of John Wiley and Sons, Inc.

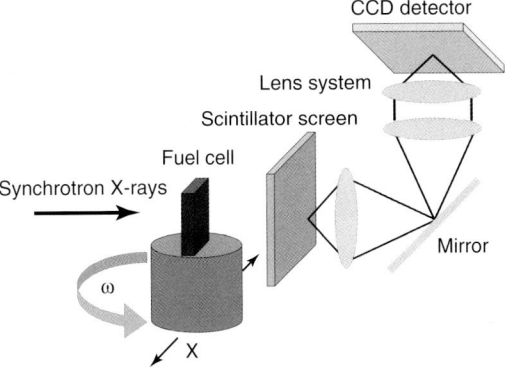

Figure 18.6 Principle of experimental setup for radiographic synchrotron X-ray measurements. Reproduced from [20] by permission of John Wiley and Sons, Inc.

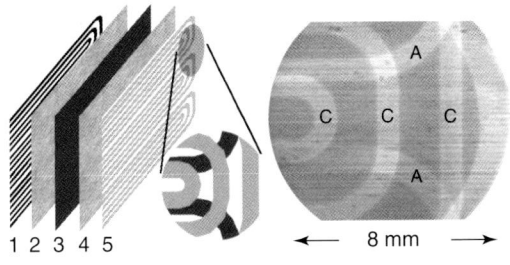

(a) Repeat unit/single cell   (b) Radiographic image

Figure 18.7 (a) Assembly of the most important fuel cell components (1, anodic flow field; 2, anodic GDL; 3, membrane electrode assembly (MEA); 4, cathodic GDL; and 5, cathodic flow field) and (b) radiographic through-plane view. Reproduced from [21] by permission of the American Institute of Physics.

The principle set-up for radiographic studies is illustrated in Figure 18.6. The fuel cell is mounted on translation and rotation stages to approach the desired measuring position. This includes cell movements towards the scintillator screen (z-direction), movements to the side (x-direction) for the acquisition of flat field images, x- and y-translations to bring desired cell details into the field of view, and rotation movements to align the cell in through-plane or in-plane perspective.

Figure 18.7a shows the scheme of a fuel-cell assembly in a perspective view and the corresponding 2D through-plane view of a selected cell section (Figure 18.7b).

The flow field channels on the anode and cathode sides can be identified as superposed bright structures. The amount of liquid water formed in the cell depends strongly on the operating conditions, such as electric current density, gas utilization rates (denoting the fraction of total gas input consumed in the cell), and the degree of humidification of the gas inlet streams. Figure 18.8 shows

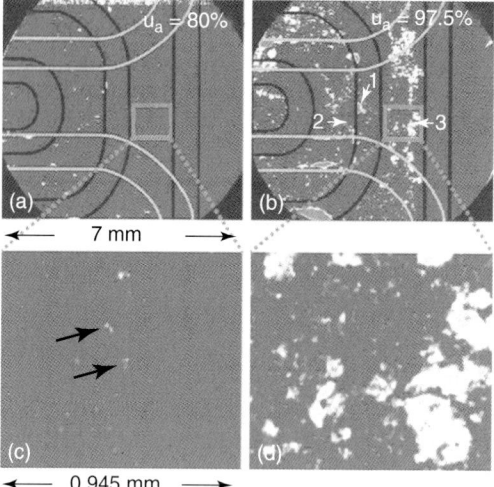

Figure 18.8 (a, b) Normalized radiographs showing the water distribution at different operating conditions. Flow field channels are marked by black lines. Parts (c) and (d) shows magnified details of (a) and (b). Reproduced from [21] by permission of the American Institute of Physics.

the water distributions in a cell operated at different anodic utilization rates and current densities. A selected area between cathode channels is shown at higher magnification.

The images, which are normalized with respect to an image of the dry cell, reveal the water distribution in the cell. The water thickness can be determined by means of the attenuation coefficient of water at the chosen X-ray energies. When operated with a utilization rate of $u_A = 80\%$ and a current density of $i_0 = 300$ mA cm^{-2} (Figure 18.8a), only a few small water clusters are present in the cell (indicated by arrows in the magnified image), preferably located in the GDL pores. These agglomerations have typical diameters of about 10–20 µm and contain only a few picoliters of water. The absence of larger water clusters suggests that most water is transported as vapor. Increasing the anodic utilization rate and current density to $u_A = 97.5\%$ and $i_0 = 500$ mA cm^{-2} (Figure 18.8b), respectively, leads to enhanced formation and agglomeration of liquid water. Many clusters with diameters of up to 300 µm and a volume of several hundred or thousand picoliters are detected. Most clusters are found between the cathodic flow field channels, indicating preferred water agglomeration in the GDL pores at the cathode side underneath the ribs. Subsequently, the liquid water migrates towards the channels, where it is removed by the gas stream. The different amount of water present in the cell can be easily understood in terms of the varied operating conditions. A higher electric current density results in enhanced electrochemical formation of water according to Faraday's law whereas increased utilization rates are realized by reducing the gas flow, which in turn reduces the water removal rates.

18.3.1.1.1 Dynamic Radiographic Studies of Water Transport

Synchrotron X-ray imaging is not only used to determine the water distribution in fuel cells under various conditions, but is also applied to study the dynamics of water transport. Of particular interest is the transition of liquid water from the GDL into the channels. In several studies, experimental evidence for the eruptive nature of the water transport was obtained [20, 21]. The mechanism is illustrated schematically in Figure 18.9. Water is ejected from the GDL pores to form a droplet at the edge of the flow field channel. In Figure 18.10, a time series of radiographs is presented that documents this transport. The images were normalized with respect to the image taken immediately before the ejection. The location of the channel wall is marked by the curved black line. Water which had been present in the GDL before but was then removed by the ejection and transferred into the channel appears as dark spots. Upon arrival in the channel, the water forms a droplet that appears bright in the image. The time series documents how the water is quickly ejected (between the acquisition of two consecutive images) from the GDL, forming a droplet that sticks at the channel wall where it is exposed to the (unsaturated) gas stream. Consequently, evaporation shrinks the droplet. Meanwhile the pores in the GDL are subjected to refilling until they finally eject a new droplet.

Figure 18.9 Schematic diagram explaining eruptive water transport. Water in the GDL is expelled within less than 5 s into the flow field, where it forms a droplet. Reprinted from [20] by permission of John Wiley and Sons, Inc.

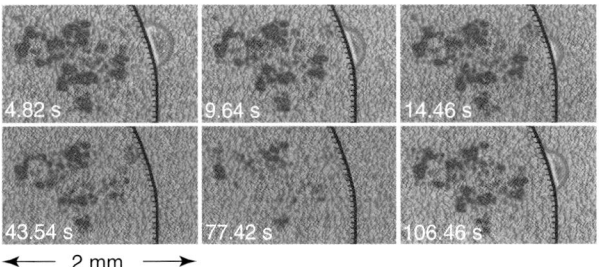

Figure 18.10 Normalized radiographic series documenting a cycle of droplet ejection and evaporation. Reproduced from [21] by permission of the American Institute of Physics.

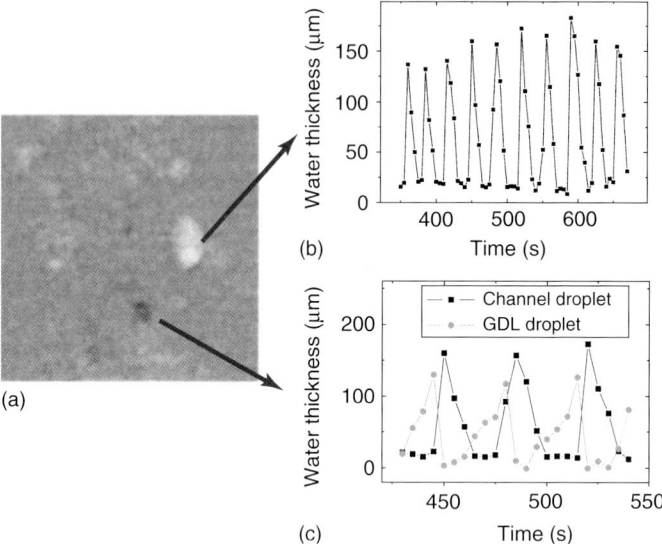

Figure 18.11 (a) Temporal correlation between the water content in GDL pores and in the corresponding flow field channel. Water amount (in terms of transmission thickness) (b) at the droplet location and (c) at the corresponding location within the GDL. For comparison, the graph for the droplet position is also included. Reproduced from [20] by permission of John Wiley and Sons, Inc.

Figure 18.11 illustrates the cyclic nature of the water ejection into the channel. By monitoring the water thickness at the droplet position in the channel and a corresponding location inside the GDL, the periodicity of the droplet life cycle and its temporal correlation with the water content in GDL pores is revealed. The GDL pore discharge and the droplet formation take place rapidly at the same time. Subsequently, the droplet shrinks and disappears while the GDL pore continues to recharge with water until it is saturated. Eventually, a new droplet is ejected, indicating the start of the next cycle.

18.3.1.2 Neutron Radiography

Similarly to X-ray radiography, neutron radiography has the potential to quantify the water distribution in all three dimensions. Because neutrons are strongly attenuated by water relative to other materials used in fuel-cell construction, for example, carbon or aluminum, studies of the liquid water content in PEFCs require less cell design modifications than in X-ray tests. However, spatial and temporal resolutions are inherently limited because the fluxes of neutron sources are smaller by several orders of magnitude than those of synchrotron X-ray sources. Recently, the development of novel high-resolution neutron detector systems [22] considerably improved the spatial resolution. Taking advantage of the improvement of typical resolutions to <20 µm, Hickner *et al.* were able to resolve the water distribution across the membrane electrode assembly (MEA), the GDL, and the flow field channels [23]. The cross-sectional water content was quantified as

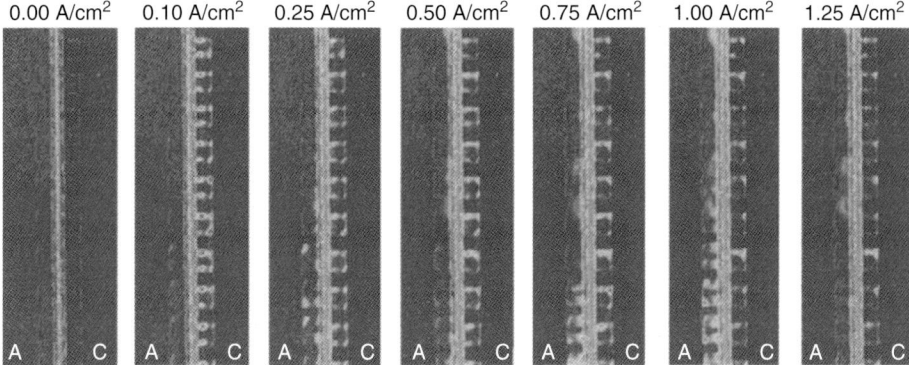

Figure 18.12 Radiographic series of water evolution for different current densities for the cell at 60 °C and 100% relative humidity inlet gas feed. Liquid water agglomerations appear bright. Reproduced from [23] by permission of ECS – The Electrochemical Society.

a function of different operating parameters at steady-state conditions. The test cell with an active area of 21×77 mm was equipped with a single meander flow field. The cell was imaged at half-height with the field of view rendering an area of about 14×14 mm. In Figure 18.12, the water distribution of the cell at 60 °C with 100% relative humidity of the inlet gas feeds is shown for a series of increasing current densities. At low current densities, heat and water formation are minimal. Liquid water is predominantly found in the cathode gas flow channel (GFC) (which is to the right of the membrane). indicating that the transport path of lowest resistance was through the cathode GDL into the channel. As the current density is increased to 0.75 and 1.0 A cm^{-2}, the water generation rate exceeds the transport capacity of the cathodic GDL. Water is transported to the anode side by back-diffusion and agglomerates in the anode channels. At a current density of 1.25 A cm^{-2}, a decrease in the water content is observed, indicating that the enhanced heat generation leads to significant evaporation.

Figure 18.13 shows the quantitative analysis of the water content for the anodic and cathodic channels and the MEA (including GDLs). The water content is minimal in the MEA for current densities lower than 0.25 A cm^{-2} due to the low water formation rates. In the range of 0.25–1 A cm^{-2} it remains relatively constant but is slightly depressed at 1.25 A cm^{-2}.

When combined with complementary measurements, such as locally resolved current measurements, the information value of neutron radiographic measurements can be further enriched. Taking advantage of the penetration power of neutrons, a dedicated cell design including a multi-layer printed circuit board with an embedded matrix of shunt resistors was used to measure the local current densities and the liquid water distributions concurrently [24]. With the combination of these methods, it is possible to correlate the local water content with the local electric performance of the cell. Figure 18.14 shows the superposition of the

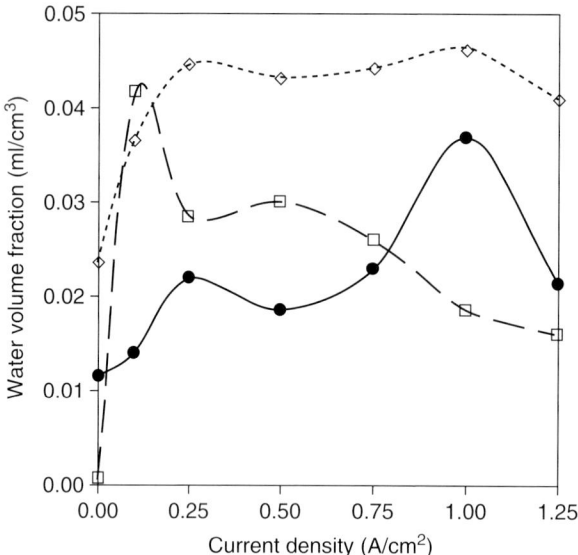

Figure 18.13 Water content volume fraction of MEA and GFCs as a function of current density at 60 °C and 100% relative humidity inlet gas feeds: (◇) MEA (including GDLs); (●) anode gas flow channel; and (□) cathode gas flow channel. Reproduced from [23] by permission of ECS – The Electrochemical Society.

areal liquid water distribution and the local current distribution derived by a 5 × 5 measure matrix for a PEFC with an active area of 10 × 10 cm. The cell was operated at a temperature of 60 °C, an overall current density of 700 A cm^{-2}, and anodic and cathodic utilization rates of 80 and 25%, respectively. Dark segments in the measure matrix refer to low current densities and bright segments refer to high current densities. As is obvious in Figure 18.14a,b, in the lower third of the cell the water channels are flooded, thereby forming an obstacle for the gas supply to the reactive spots. Consequently, the local performance in this area is low. The segments at the upper corners also show only lower local current densities, however, no excess liquid water can be accounted for the lower local performance. Apparently, at these spots the water content is too low to humidify the membrane sufficiently so that the performance is reduced as compared with the central cell area. The nonuniform current distribution underlines the unbalanced water management for the cell under these operating conditions.

In Figure 18.14c, it is demonstrated how a change in local water content influences the local current density. Within the highlighted measure (circle in Figure 18.14a), at least four GFCs are flooded. This implicates back-pressure of the gas stream in the affected channels until the water in one of these channels is pushed towards the outlet, giving way for an enhanced gas supply to the reactive area in the field, which results in an increase in current density.

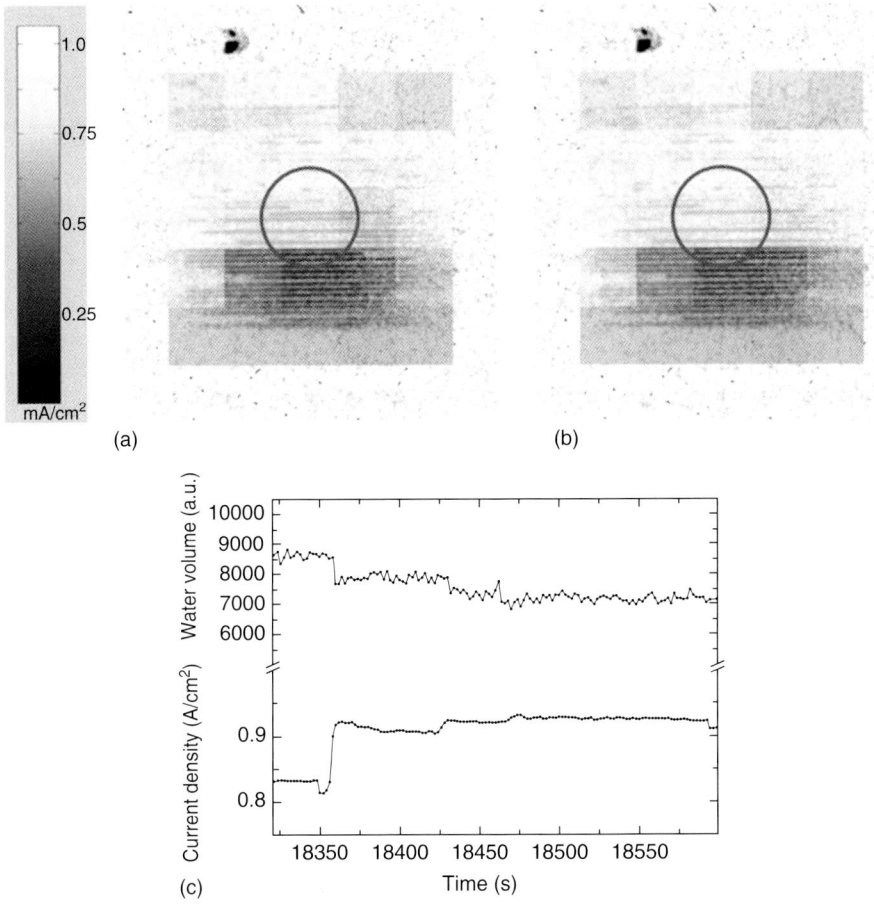

Figure 18.14 (a–c) Water distribution and corresponding current density obtained by combined radiographic and locally resolved current measurements. Reproduced from [24] by permission of Elsevier.

18.3.2
DMFCs

DMFCs are-low temperature cells with typical operating temperatures around 70 °C, fueled with an aqueous methanol solution at the anode side while air is fed to the cathode. The electrochemical reactions form carbon dioxide at the anode and liquid water at the cathode side. CO_2 formed at the catalytic layer migrates through the GDL and forms bubbles in the anodic flow field channels which are carried away by the methanol stream. At the cathode side, the water is transported through the GDL to form droplets in the cathodic flow field channels which must be removed by the air stream. As neutrons and X-rays offer undistorted insights into operating fuel cell setups, they can be used to study carbon dioxide evolution and bubble formation at the anode side and liquid water distribution at the anode side [25–27].

18.3.2.1 CO_2 Evolution Visualized by Means of Synchrotron X-Ray Radiography

Carbon dioxide evolution and bubble formation can be studied at high resolution by means of synchrotron X-ray radiography [25]. By combining perpendicular viewing perspectives a comprehensive picture of gas formation and transport can be derived. Measurements in a through-plane perspective reveal the areal CO_2 distribution in the cell, which represents the summed signal contributions of MEA, anodic, and cathodic GDLs. The electrochemical active area under the ribs can be clearly distinguished from that located under the flow field channels. Correspondingly, in-plane measurements provide the CO_2 distribution perpendicularly to the electrochemically active area. In this perspective, the media transport through the individual components can be followed. Figure 18.15a depicts the radiographic in-plane perspective on a test cell showing the MEA enclosed by the GDLs and flow field channels. CO_2 bubbles emerge from the interface of the GDL and are, subsequent to detachment, carried away by the methanol solution. Gas-filled spaces in the anodic GDL appear dark in the image and form a regular pattern with a characteristic distance of 500 µm. This pattern is again found in the through-plane perspective (Figure 18.15b) and can be assigned to the weaving pattern of the GDL, as demonstrated in the scanning electron microscopy (SEM) image (Figure 18.15c). The weaving pattern suggests that the gas resides under the loops of the fiber strands whereas the crossings facilitate migration paths for

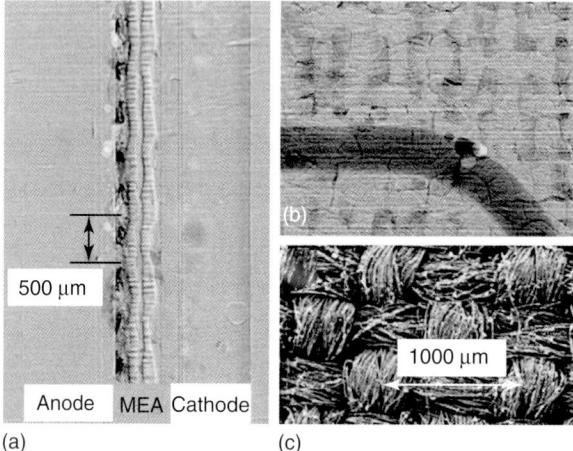

Figure 18.15 (a) Radiographic in-plane view on a DMFC revealing CO_2 evolution and bubble formation. (b) Radiographic through-plane view. Gas agglomerates in the bends of carbon cloth. (c) SEM image showing the weaving pattern of the GDL. Reproduced from [25] by permission of Elsevier.

CO_2 to exit into the flow field channels. It was found that gas bubbles ejected into the flow field show an eruptive cyclic behavior when a threshold current density of 20 mA cm^{-2} is exceeded.

18.3.2.2 Combined Approach of Neutron Radiography and Local Current Density Measurements

The surface properties of the individual cell components have a major influence on the water management and, hence, the performance of fuel cells. The influence of GDL wettability on the cell performance was studied *in situ* by Schröder *et al.* [26]. The combination of neutron radiography and local current measurements was shown to be a suitable approach to analyze differences in cell performance caused by an altered GDL hydrophobicity. The study took advantage of the large field of view provided by the neutron imaging instrument as the water distribution for the entire test cell with an active area of 4.2 × 4.2 cm^2 can be captured. This provides the opportunity to study the influence of a different GDL wettability inside a single cell by partitioning the GDL into a more and a less hydrophobic part and to compare the local cell performance under otherwise identical operating conditions. The influence of different GDL hydrophobicities was tested for the anode and cathode in separate experiments but using a similar design approach to that demonstrated in Figure 18.16.

Selected results are shown in Figure 18.17. Neutron radiographs of the cell and the corresponding local current densities are displayed for three different mean current densities under steady-state conditions. The images were normalized with respect to a reference state where the anode is flooded by methanol solution in bubble-free conditions whereas the cathode side is dry. Due to the applied normalization procedure, water agglomerations at the cathodic flow field appear dark whereas the emergence of bright spots in the images indicates the evolution of CO_2 at the anode side.

In Figure 18.17a,b, the test cell with a split anode GDL is shown at overall current densities of 50 and 300 mA cm^{-2}, respectively. Hardly any effect of the

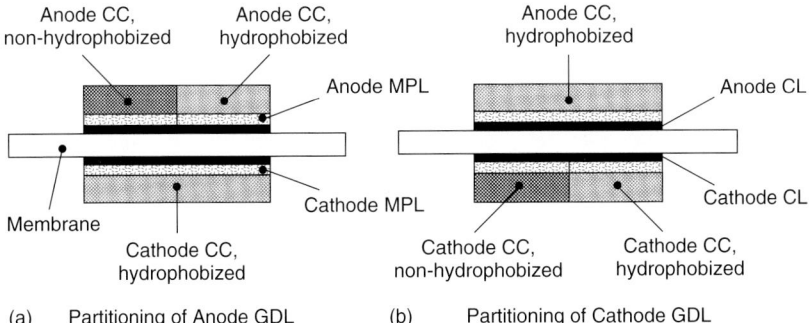

Figure 18.16 Scheme of the test cells equipped with either partitioned anodic (a) or cathodic (b) GDL. Reproduced from [26] by permission of Elsevier.

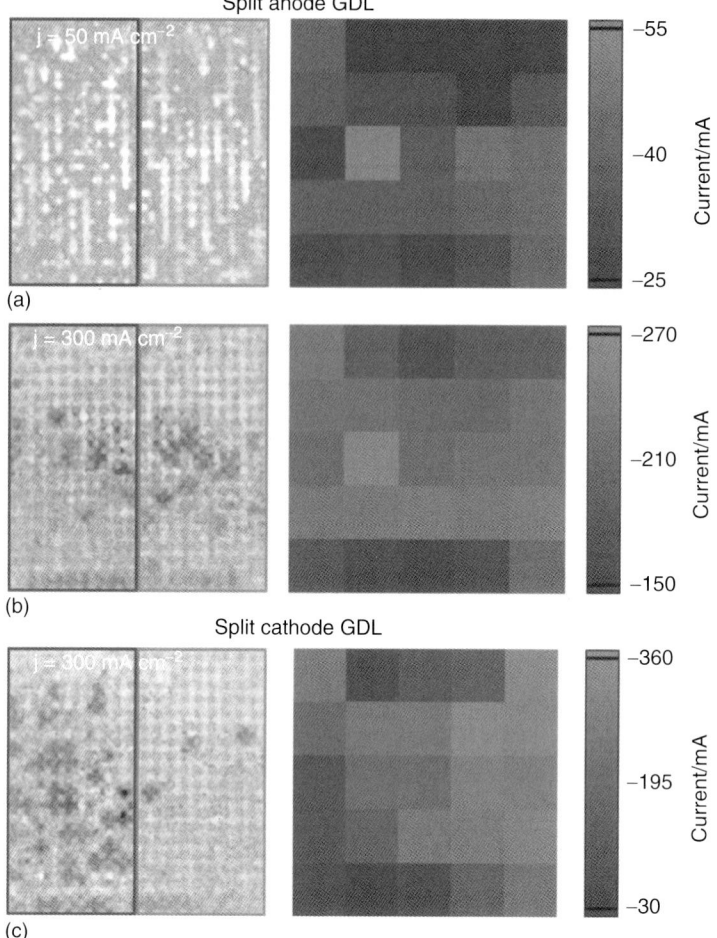

Figure 18.17 Water and CO_2 distribution of the partitioned MEA at different overall current densities and the respective current density distributions. (a, b) The left half of the anode is equipped with an untreated GDL and the right half with a hydrophobized GDL. (c) The left half of the cathode is equipped with an untreated GDL and the right half with a hydrophobized GDL. Reproduced from [26] by permission of Elsevier.

GDL hydrophobicity can be observed as CO_2, water, and current density are similar in both partitions of the cell. However, when the cathode GDL is split instead, the current density on the side with the hydrophilic GDL (Figure 18.17c) is strongly reduced. The reason is obvious in the radiographic image: flooding on the left side due to excessive liquid water agglomeration. On the hydrophobic side, the water is still balanced and the current density is about 60% higher than that on the hydrophilic side.

18.3.3
HT-PEFCs

Synchrotron X-ray radiography can be used as an efficient *in situ* analytical tool to elucidate changes in structure and composition of the MEA in operating HT-PEFCs. The membrane conductivity of this fuel cell type relies on the proton transport within phosphoric acid. Maier *et al.* investigated an HT-PEFC equipped with a polybenzimidazole-type membrane [poly(2,5-benzimidazole) (ABPBI)] doped with phosphoric acid [28]. During cell operation, the formation of product water leads to membrane swelling and hydration of the phosphoric acid. Figure 18.18 shows a series of in-plane radiographs taken at different current densities. When the current density was increased from 0 [i.e., open-circuit voltage (OCV)] to 140 mA cm^{-2}, a membrane swelling of about 10 μm was observed, which corresponds to an

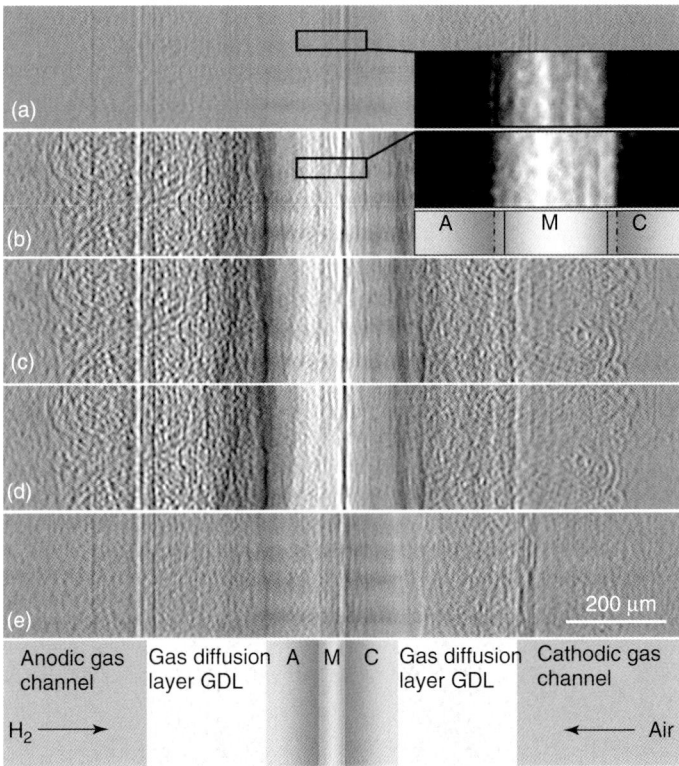

Figure 18.18 In-plane views of the HT-PEFC MEA at different current densities j: (a) 0 mA cm^{-2} (OCV before); (b) 140 mA cm^{-2}; (c) 300 mA cm^{-2}; (d) 550 mA cm^{-2}; and (e) 0 mA cm^{-2} (OCV after). Inset: non-normalized enlarged radiographs of the membrane showing the swelling after the current density jump to $j = 140$ mA cm^{-2}. A, anode; M, membrane; C, cathode. Reproduced from [28] by permission of Elsevier.

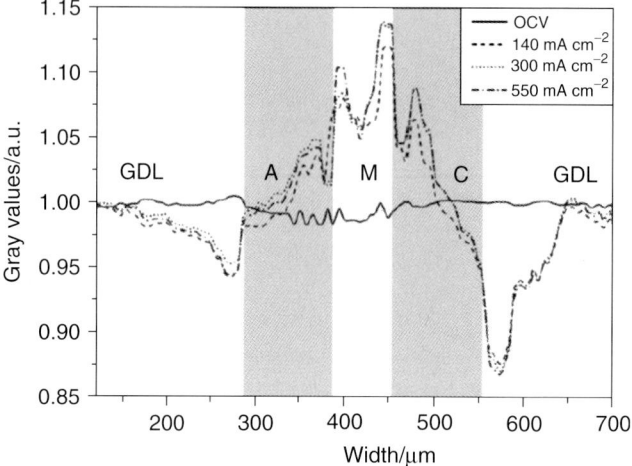

Figure 18.19 Relative transmission values of the GDL and MEA components (A, anode; M, membrane; C, cathode) at different current densities. Reproduced from [28] by permission of Elsevier.

expansion by 18%. A further increase in current density to 300 and 550 mA cm^{-2} did not result in additional membrane expansion.

Along with the membrane swelling, a clear change in transmission signals of the MEA and GDL was detected (Figure 18.19). When current is drawn from the cell, transmission values increase in the membrane and the adjacent zones of the electrodes. Further away from the membrane, the value of the relative transmission drops below 1. The minimum transmission is found at a position inside the GDL close to the electrode interfaces. To understand the evolution of the transmission signal, it is useful to recall the specific composition of the MEA. The required proton conductivity of the membrane is achieved by doping the ABPBI membrane with concentrated phosphoric acid. The distribution and local concentration of the phosphoric acid in the MEA depend strongly on the formation of product water in the cell. The attenuation coefficient of water is about one order of magnitude smaller than that of phosphoric acid (0.1572 compared with 1.0198 cm^{-1} at 30 keV). The evolution of the transmission signal within the membrane and its immediate vicinity can be explained by a change in chemical composition. The formation of water at the anode results in dilution of the phosphoric acid and, thus, in increased transmissibility. Further away from the membrane, the dropping transmission signal suggests that the hydrated acid spreads out and enters gas-filled voids of the electrodes and GDL. The filling of voids and pores with phosphoric acid leads to stronger attenuation, and hence relative transmission values smaller than 1.

18.4
Application of 3D Methods

18.4.1
Neutron Tomography

A basic requirement for meaningful reconstruction of a fuel cell is a steady water distribution during the acquisition of the tomographic scan. For large fuel cells, typical overall acquisition times range between 30 min and several hours. This complicates the tomographic measurements since during cell operation a shift of the water distribution occurs on this time scale and makes *in situ* measurements almost impossible. However, in the case of smaller fuel cells, tomographic scans can be performed much faster.

Figure 18.20 shows the result of a tomographic study of the water distribution in a small PEFC performed by Hussey *et al.* [29]. A set of selected slices was fanned out to reveal the water distribution in all parts of the cell, that is, the anode and cathode parts can be visualized separately. In this study, water was found to be concentrated in the anode GDL even in dry conditions, indicating that back-diffusion plays a significant role in the water distribution of the cell.

A suitable concept for tomographic studies of the water distribution in large PEFCs is the quasi-*in situ* approach. To get around the shifting of local water distribution during the scan, the gas supply and temperature regulation are shut off prior to the measurements. In this way, the local water distribution in the cell can be conserved for several hours.

Using the quasi-*in situ* approach, not only single cells but also several-fold fuel-cell stacks can be investigated [30]. Results for the study of a threefold fuel-cell stack are given in Figure 18.21. The three-dimensional information about the water distribution allows for quantification of the individual amount of water in all three cells of the stack. Furthermore, the water content of an individual cell

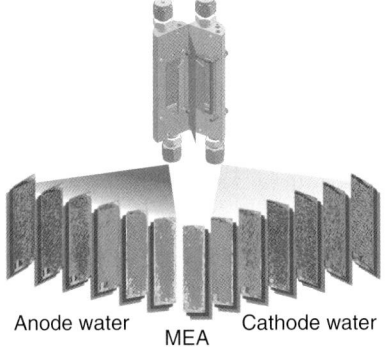

Anode water   MEA   Cathode water

Figure 18.20 Images of the water distribution obtained from tomographic measurement of a test cell. Width of each slice: 0.125 mm. Reproduced from [29] by permission of DEStech Publications.

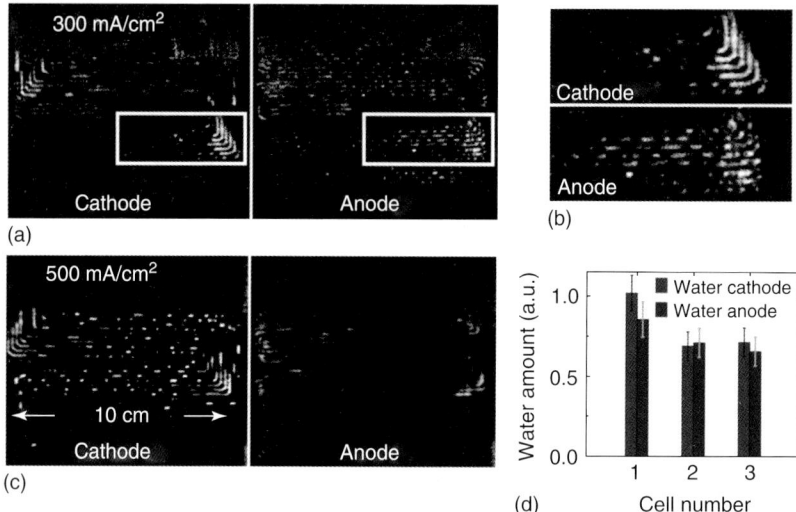

Figure 18.21 (a) Water distributions in the anodic (left) and cathodic (right) flow field channels of the first cell in a triple fuel-cell stack; (b) corresponding enlargements of the marked areas; (c) water distributions at higher current density; and (d) total water content in the anodic and cathodic flow fields of all three cells at $i_0 = 300$ mA cm^{-2}. Reproduced from [30] by permission of the American Institute of Physics.

can be split into anodic and cathodic contributions, so that intra-cell transport effects, such as back-diffusion and electroosmotic drag, can be analyzed. The water distribution at different current densities on the anode and cathode sides of the first cell in the stack is displayed in Figure 18.21a–c. At 300 mA cm^{-2}, water is found in almost equal proportions on both sides. Notably, the anode is sufficiently humidified although only the cathodic gas stream was externally humidified and the water production takes place at the cathode. Back-diffusion is responsible for the transport from the cathode to anode, ensuring uniform humidification of the membrane. However, as the current density is increased to 500 mA cm^{-2}, the anode is found to be in much dryer conditions. At higher current density, the effect of electroosmotic drag counteracting back-diffusion is more pronounced, as is obvious in Figure 18.21c for the case shown here. An appropriate balance of electroosmotic drag and back-diffusion can be achieved by proper adjustment of membrane transport properties, for example, by variation of the membrane thickness.

18.4.2
Synchrotron X-Ray Tomography

Synchrotron X-ray tomography can provide highly resolved, 3D insights into fuel cells. In contrast to neutron measurements, the brilliance of the synchrotron

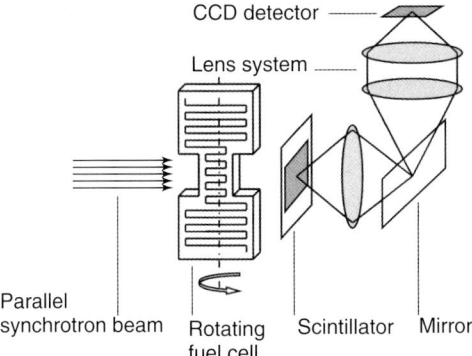

Figure 18.22 Schematic setup of a tomography of a fuel cell. Reproduced from [32] by permission of Elsevier.

radiation allows the study of the water distribution inside the GDL, providing access to even smallest liquid water agglomerations. A basic prerequisite for the meaningful reconstruction of a 3D volume is that no fuel cell parts drift out of the field of view during the scanning procedure. For this reason, often miniaturized fuel cells, with an active area much smaller than $1\,cm^2$, are used for tomographic studies [31]. However, results obtained with miniaturized setups may have limited significance as the operating conditions, for example, local gas utilization rates and temperature distributions, differ significantly from those of larger cells employed in practical applications. Larger fuel cells with more realistic utilization rates can be studied when the cell design is appropriately adapted [32, 33]. Figure 18.22 shows the experimental setup of a PEFC with an active cell area of about $12.5\,cm^2$. At half-height, the test cell width is reduced to fit into the field of view. Within this region, metallic end plates were replaced with acrylic glass to achieve sufficient beam transmission. The acquisition of a tomographic scan (typically comprising about 1800 projections) needs between 20 and 100 min for a fuel cell of this size. Therefore, quasi-*in situ* measurements, that is, with the gas supply and temperature regulation shut off during the scan, are necessary to conserve the water distribution.

A perspective view of the interior of the visualized fuel-cell section is depicted in Figure 18.23. The fiber structure of the GDL is clearly visible, and also liquid water agglomerations in the flow field channels at both the anode and cathode sides. At the back side of the channel, water agglomerates to form a film. Most droplets gather at the bottom of the channel, reflecting the gravitational influence.

The reconstructed 3D volume can be sliced to obtain cross-sections of the test cell for any desired position and perspective. In Figure 18.24, cross-sections of the flow field channel structure and the GDL were generated to analyze the through-plane water distribution in the channel and inside the GDL at positions close to the interfaces to the flow field and microporous layer (MPL), respectively, and also for the central region of the GDL. Most water present in the GDL resides close to the flow field interface underneath the land area from where it migrates into the flow

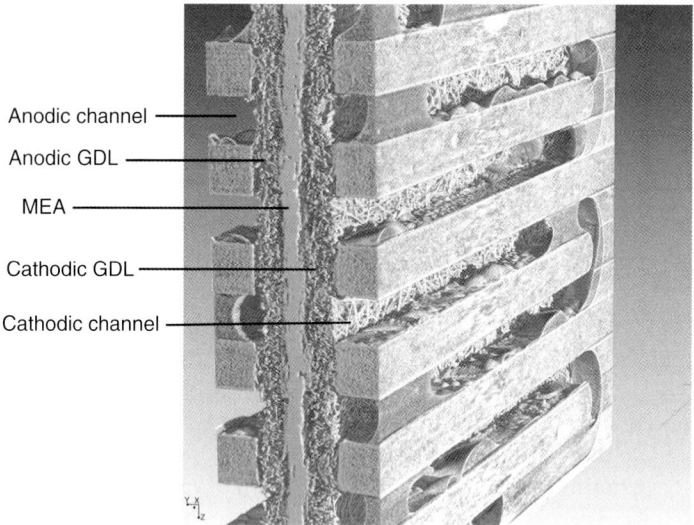

Figure 18.23 Visualization of the tomographed fuel cell part. The cathode is shown on the front side and the anode on the back side of the 3D image. Water films and separated droplets in the flow field channels could be visualized. The carbon fiber structure of the applied GDL is visible. Reproduced from [32] by permission of Elsevier.

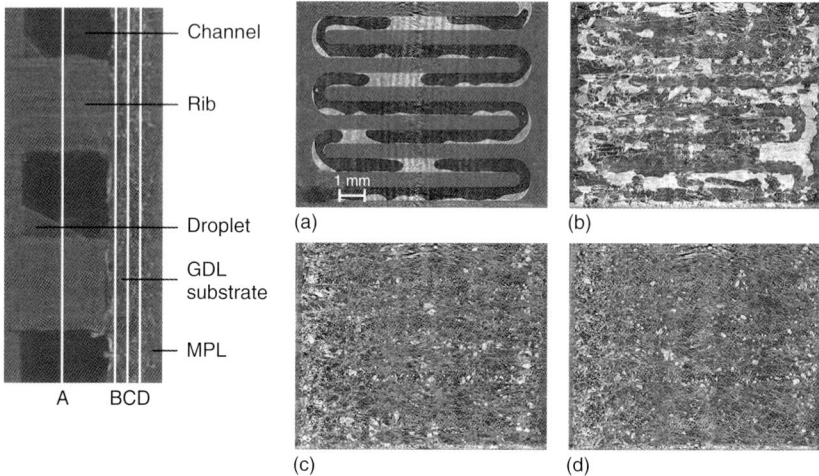

Figure 18.24 Cross-sections of the flow field structure (a) and the GDL (b–d) at different depths. The positions of the cross sections (a–d) are highlighted in the side view of the cell on the left. Reproduced from [32] by permission of Elsevier.

field channel. Smaller agglomerations are found at the center and close to the MPL interface. Here, no obvious influence of the alternating channel land pattern on the liquid water distribution is found. Instead, the distribution seems to be more related to the local pore geometry and the distribution of the Teflon coating. At both positions the amount of water does not differ significantly.

18.5
Conclusion

Neutron and synchrotron X-ray imaging are powerful diagnostic tools suitable for addressing various problems in fuel cell research. With respect to the water management issue of PEFCs and DMFCs, neutron radiography can provide insights into the liquid water distribution of large cell areas or even entire cells. Owing to the high penetration potential of neutrons, no significant cell design modifications are necessary to perform the imaging procedure. At the same time, the excellent sensitivity for water allows precise quantification of the cell water content. When applied in a through-plane perspective (beam propagation perpendicular to the active layer), the influence of different flow field geometries and/or GDLs on the water management can be studied. In combination with local current density measurements, the effects of unbalanced water management on the local cell performances are directly detectable. Owing to recent resolution improvements to <20 μm, visualization of the cross-sectional water distribution in the MEA has also become possible. This allows for in-plane studies of the through-plane water transport in MEAs and GDLs. When applied in the tomographic mode, neutron imaging provides 3D insights into the stationary water distribution of cells or even entire stacks.

Synchrotron X-ray imaging is a complementary method to neutron imaging, providing higher spatial and temporal resolution. Using this method, even the smallest water agglomerations, for example, present in the GDL pores, can be resolved. Furthermore, taking advantage of the high temporal resolution, it is possible to study the water transport dynamics, for example, the eruptive ejection of water droplets from the GDL into the flow field channels. Synchrotron X-ray tomography provides highly resolved, 3D information on the porous transport structure inside fuel cells and the stationary water distribution. In combination with radiographic studies, dynamic phenomena of the water transport can be related to the underlying transport structures.

As shown in the preceding sections, the application of the presented imaging methods is not restricted to studies of the water management of PEFCs and DMFCs. Neutron and synchrotron X-ray imaging were also successfully applied to study the media distribution and CO_2 bubble formation in DMFCs and the variation of membrane properties in an operating HT-PEFC.

Neutron and synchrotron X-ray imaging have proven to be indispensable tools to optimize water management with respect to high cell performance, reliable operation, and maximum durability.

References

1. Vielstich, W., Lamm, A., and Gasteiger, H.A. (eds.) (2003) *Handbook of Fuel Cells – Fundamentals, Technology and Applications*, John Wiley & Sons, Ltd., Chichester.
2. Garche, J., Dyer, C.K., Moseley, P.T., Ogumi, Z., Rand, D.A.J., and Scrosati, B. (2009) *Encyclopedia of Electrochemical Power Sources*, Elsevier, Amsterdam, p. 4538.
3. Wang, C.-Y. (2003) in *Handbook of Fuel Cells – Fundamentals, Technology and Applications* (eds. W. Vielstich, A. Lamm, and H.A. Gasteiger), John Wiley & Sons, Ltd., Chichester, pp. 337–347.
4. Carrette, L., Friedrich, K.A., and Stimming, U. (2001) Fuel cells – fundamentals and applications. *Fuel Cells*, **1**, 5–39.
5. Bazylak, B.A. (2009) Liquid water visualization in PEM fuel cells: a review. *Int. J. Hydrogen Energy*, **34** 3845–3857.
6. Schmittinger, W. and Vahidi, A. (2008) A review of the main parameters influencing long-term performance and durability of PEM fuel cells. *J. Power Sources*, **180**, 1–14.
7. Yousfi-Steiner, N., Moçotéguy, P., Candusso, D., Hissel, D., Hernandez, A., and Aslanides, A. (2008) A review on PEM voltage degradation associated with water management: impacts, influent factors and characterization. *J. Power Sources*, **183**, 260–274.
8. Hartnig, C., Jörissen, L., Lehnert, W., and Scholta, J. (2008) Direct methanol fuel cells (DMFC), in *Materials for Fuel Cells* (ed. M. Gasik), Woodhead Publishing., Cambridge, pp. 185–208.
9. Liao, Q., Zhu, X., Zheng, X., and Ding, Y. (2007) Visualization study on the dynamics of CO_2 bubbles in anode channels and performance of a DMFC. *J. Power Sources*, **171**, 644–651.
10. Lehnert, W., Wannek, C., and Zeis, R. (2010) in *Innovations in Fuel Cell Technology* (eds. R. Steinberger-Wilckens and W. Lehnert), RSC Publishing, Cambridge, pp. 45–75.
11. Wannek, C. (2010) High-temperature PEM fuel cells: electrolyte, cells and stack, in *Hydrogen and Fuel Cells* (ed. D. Stolten), Wiley-VCH Verlag GmbH, Weinheim, pp. 17–40.
12. Strobl, M., Manke, I., Kardjilov, N., Hilger, A., Dawson, M., and Banhart, J. (2009) Advances in neutron radiography and tomography. *J. Phys.D: Appl. Phys.*, **42**, 243001.
13. Banhart, J., Borbely, A., Dzieciol, K., Garcia-Moreno, F., Manke, I., Kardjilov, N., Kaysser-Pyzalla, A.R., Strobl, M., and Treimer, W. (2010) X-ray and neutron imaging – complementary techniques for materials science and engineering. *Int. J. Mater. Res.*, **101**, 1069–1079.
14. Banhart, J. (2008) *Advanced Tomographic Methods in Materials Research and Engineering*, Oxford University Press, Oxford.
15. Schillinger, B., Lehmann, E., and Vontobel, P. (2000) 3D neutron computed tomography: requirements and applications. *Physica B: Condens. Matter*, **276–278**, 59–62.
16. Heller, A.K., Shi, L., Brenizer, J.S., and Mench, M.M. (2009) Initial water quantification results using neutron computed tomography. *Nucl. Instrum. Methods Phys. Res. A*, **605**, 99–102.
17. Manke, I., Hartnig, C., Kardjilov, N., Hilger, A., Lehnert, W., and Banhart, J. (2008) Neutronen-Radiographie und -Tomographie in der Brennstoffzellenforschung. *ZfP-Ztg.*, **109**, 24–28.
18. Froeschle, P. and Wind, J. (2010) Fuel cell power trains, in *Hydrogen and Fuel Cells* (ed. D. Stolten.), Wiley-VCH Verlag GmbH, Weinheim, pp. 793–810.
19. Elter, J.F. (2010) Fuel cells for buildings, in *Hydrogen and Fuel Cells* (ed. D. Stolten), Wiley-VCH Verlag GmbH, Weinheim, pp. 755–792.
20. Manke, I., Hartnig, C., Kardjilov, N., Riesemeier, H., Goebbels, J., Kuhn, R., Krüger, P., and Banhart, J. (2010) In situ synchrotron X-ray radiography investigations of water transport in PEM fuel cells. *Fuel Cells*, **10**, 26–34.
21. Manke, I., Hartnig, C., Grunerbel, M., Lehnert, W., Kardjilov, N., Haibel, A., Hilger, A., Banhart, J., and Riesemeier, H. (2007) Investigation of water evolution and transport in fuel cells with high

resolution synchrotron X-ray radiography. *Appl. Phys. Lett.*, **90**, 174105.

22. Tötzke, C., Manke, I., Hilger, A., Choinka, G., Kardjilov, N., Arlt, T., Markötter, H., Schröder, A., Wippermann, K., Stolten, D., Hartnig, C., Krüger, P., Kuhn, R., and Banhart, J. (2011) Large area high resolution neutron imaging detector for fuel cell research. *J. Power Sources*, **196**, 4631–4637.

23. Hickner, M.A., Siegel, N.P., Chen, K.S., Hussey, D.S., Jacobson, D.L., and Arif, M. (2008) *In situ* high-resolution neutron radiography of cross-sectional liquid water profiles in proton exchange membrane fuel cells. *J. Electrochem. Soc.*, **155**, B427–B434.

24. Hartnig, C., Manke, I., Kardjilov, N., Hilger, A., Grünerbel, M., Kaczerowski, J., Banhart, J., and Lehnert, W. (2008) Combined neutron radiography and locally resolved current density measurements of operating PEM fuel cells. *J. Power Sources*, **176**, 452–459.

25. Hartnig, C., Manke, I., Schloesser, J., Krüger, P., Kuhn, R., Riesemeier, H., Wippermann, K., and Banhart, J. (2009) High resolution synchrotron X-ray investigation of carbon dioxide evolution in operating direct methanol fuel cells. *Electrochem. Commun.*, **11**, 1559–1562.

26. Schröder, A., Wippermann, K., Lehnert, W., Stolten, D., Sanders, T., Baumhöfer, T., Kardjilov, N., Hilger, A., Banhart, J., and Manke, I. (2010) The influence of gas diffusion layer wettability on direct methanol fuel cell performance: a combined local current distribution and high resolution neutron radiography study. *J. Power Sources*, **195**, 4765–4771.

27. Schröder, A., Wippermann, K., Mergel, J., Lehnert, W., Stolten, D., Sanders, T., Baumhöfer, T., Sauer, D.U., Manke, I., Kardjilov, N., Hilger, A., Schloesser, J., Banhart, J., and Hartnig, C. (2009) Combined local current distribution measurements and high resolution neutron radiography of operating direct methanol fuel cells. *Electrochem. Commun.*, **11**, 1606–1609.

28. Maier, W., Arlt, T., Wannek, C., Manke, I., Riesemeier, H., Krüger, P., Scholta, J., Lehnert, W., Banhart, J., and Stolten, D. (2010) *In-situ* synchrotron X-ray radiography on high temperature polymer electrolyte fuel cells. *Electrochem. Commun.*, **12**, 1436–1438.

29. Hussey, D.S., Owejan, J.P., Jacobson, D.L., Trabold, T.A., Gagliardo, J., Baker, D.R., Caulk, D.A., and Arif, M. (2008) in *Neutron Radiography: Proceedings of the Eighth World Conference* (eds. M. Arif and R.G. Downing), DEStech Publications, Lancaster, PA, pp. 470–479.

30. Manke, I., Hartnig, C., Grunerbel, M., Kaczerowski, J., Lehnert, W., Kardjilov, N., Hilger, A., Banhart, J., Treimer, W., and Strobl, M. (2007) Quasi-*in situ* neutron tomography on polymer electrolyte membrane fuel cell stacks. *Appl. Phys. Lett.*, **90**, 184101.

31. Schneider, A., Wieser, C., Roth, J., and Helfen, L. (2010) Impact of synchrotron radiation on fuel cell operation in imaging experiments. *J. Power Sources*, **195**, 6349–6355.

32. Krüger, P., Markötter, H., Haussmann, J., Klages, M., Arlt, T., Banhart, J., Hartnig, C., Manke, I., and Scholta, J. (2011) Synchrotron X-ray tomography for investigations of water distribution in polymer electrolyte membrane fuel cells. *J. Power Sources*, **196**, 5250–5255.

33. Markötter, H., Manke, I., Krüger, P., Arlt, T., Haussmann, J., Klages, M., Riesemeier, H., Hartnig, C., Scholta, J., and Banhart, J. (2011) Investigation of 3D water transport paths in gas diffusion layers by combined *in situ* synchrotron X-ray radiography and tomography. *Electrochem. Commun.*, **13**, 1001–1004.

19
Analytics of Physical Properties of Low-Temperature Fuel Cells
Jürgen Wackerl

19.1
Introduction

The determination of physical properties is one of the main tasks in the development of low-temperature fuel cells. It starts from the synthesis of the single components and ends in the postmortem analysis of the used and aged compounds. The data derived can be valuable in both the technical and scientific areas, since these data are the basis for comparison and steady improvement.

For the technical area, the quality of the product needs to be monitored and can therefore be maintained. Another important issue is the optimization of fabrication routes and handling to provide a less costly and thus more attractive product. Additionally, lifetime predictions can be made [1, 2], providing a basis for economic calculations. For the scientific area, these basic methods provide a starting point for more sophisticated – and mostly more complex and expensive – examinations. Here the knowledge about mechanisms and interactions is the key to finding more suitable materials and materials combinations for a specific application.

Many methods exist to accomplish the task of acquiring physical data for a material. In this chapter, the focus is set on following topics:

- gravimetric properties
- caloric properties
- mechanical properties
- structural properties in terms of porosimetry.

To be realistic, it is impossible to give a full outline of each of the topics listed above, as the complexity of each single topic can be enormous, as is obvious from the extensive literature available. Therefore, only the main aspects with respect to state-of-the-art results on fuel cells will be considered here. It should also be kept in mind that for a complete fuel cell very different material classes are examined – from polymers to metals and even to ceramics. Each of these classes is found in a modern polymer fuel cell, and having either a functional (active) or structural (passive) role: polymers as the main

Fuel Cell Science and Engineering: Materials, Processes, Systems and Technology, First Edition. Edited by Detlef Stolten and Bernd Emonts.
© 2012 Wiley-VCH Verlag GmbH & Co. KGaA. Published 2012 by Wiley-VCH Verlag GmbH & Co. KGaA.

component, especially for the membrane, but also the gas diffusion layer (GDL); metals as catalysts [3], but also as current collector or cell/stack housing [4]; and ceramics as fillers [5, 6] or sometimes acting as (co-)catalysts [7, 8]. This breakdown is, of course, not fixed. However, for each material to be examined, only certain examination methods or measurement conditions are reasonable. This makes the application of a specific method to all materials used nearly impossible.

Starting with the "core" of a fuel cell, the membrane consists mainly of a specialized polymer. Since not all properties of this polymer fulfill perfectly the requirements in terms of functionality, durability and workability – if they did then we would not need any further research! – blends, for example, mixtures with other materials, are produced. In the simplest case, just another polymer is used. In more complex cases, ceramic materials can be added to enhance specific properties. Among these are titania [9, 10] and zirconate phosphates [11], commonly used to enhance mechanical properties but also to enhance catalytic properties and/or alter ion transport properties. Attached to the membrane are the electrodes or the electrode layers (ELs). They consist mainly of the material carbon – often denoted a polymer, sometimes a ceramic. For the functionality, catalytically active metals are present. To stabilize the layer, mostly polymers are introduced. Various material combinations have already been tested [12]. On top of the electrodes, the GDL is attached. It has to provide good gas permeability but also sufficient electrical conductivity combined with reasonable structural stability. Carbon is currently the first choice for most cases. On the GDL, the flow fields are mounted, supplying the combined membrane–EL–GDL [membrane electrode assembly (MEA)] with fuel educts, and release channels for the products. Typical materials used here are again carbon or metals. However, when dealing with metals, one has also to consider ceramics which act as a corrosion protective layer. When the MEA is combined with the flow fields, the fuel cell is nearly complete. For a stack, several cells are combined and metals often act as electrical contacting, housing, and fixing materials.

There are, of course, far more analytical methods than those mentioned here, such as X-ray diffraction (XRD) and infrared spectroscopy. Some are specialized for specific materials or classes, others are suitable only in combination with other methods. The last point is important for all methods applied to materials. Precise and reliable evaluation and processing of the data obtained can only be achieved when the measurement conditions and steps are selected carefully and the state of the material before testing is already well known. If two or more methods are used simultaneously, the value of the data obtained can increase. Furthermore, true correlation of data is guaranteed, easing the interpretation of the experiment. A decrease in the time required for measurements is often a beneficial side effect, gravimetry being a good example. In most cases, thermogravimetry is coupled with calorimetry. For other purposes, it can be coupled with Fourier transform infrared (FTIR) spectroscopy, Raman spectroscopy [13], or mass spectrometry (MS) [14]. The possibilities are vast, depending on the capabilities of a laboratory and its level of experience.

Research in the field of low-temperature fuel cells in terms of materials is currently focused mainly on the following aspects:

- **Membrane materials**. High specific ionic conductivity combined with low gas and liquid permeability is required [15]. Aging induced by chemical degradation or contamination [16] or morphology changes should also be minimized. Additionally, the material must be sufficiently mechanically stable.
- **Catalyst materials**. Especially for polymer electrolyte membranes (PEMs), high selectivity for hydrogen and a high tolerance of carbon monoxide are necessary [17]. Since noble metals such as platinum are used, high activity but also a reasonable distribution are needed to reduce the amount of catalyst required [18, 19]. Degradation due to chemical stability and morphology is an issue.
- **GDL**. The main aim is to obtain mechanically stable structures which allow high permeation of fluids and gases. The pore size distribution and tortuosity of the channels play an important role. Additionally, hydrophobic or hydrophilic behavior is of interest.
- **Flow field**. The material should be of good workability since complex structures are needed. Similarly to the GDL, the surface properties are important to assure low pressure drops for operating a cell.

The material used for the components are described in more detail in the following. One of the most commonly used membrane materials for the PEM and direct methanol fuel cell (DMFC) is a persulfonated ionomer material, Nafion (DuPont). It is composed of a hydrophobic polytetrafluoroethylene backbone and regularly placed hydrophilic perfluorovinyl ether side chains. The termination consists of sulfonic acid groups. These end groups are highly susceptible to protons but also cations like Na^+, K^+, Ca^{2+} etc. Therefore, this material is used mainly for ion exchangers; however, its high proton conductivity makes it a reference membrane material [20]. One of the drawbacks of this material is its structural complexity. Although having a fairly simple chemical structure, the material forms channels, clusters, and domains as proposed by Gierke *et al.* [21]. This model is commonly accepted and extended based on the specific effects observed [22, 23]. The transport of water molecules through the polymer itself appears to be due to the persulfonic end groups [24]. On the one hand, this is not desirable since it reduces the efficiency of the proton conduction. By changing the synthesis of the Nafion membrane, other structures were obtained leading to improved water retention [25]. On the other hand, water inside the membrane is needed to achieve high proton conduction [26] via the Grotthuss mechanism additionally to the conduction mechanism via the sulfonic groups [27, 28]. Mechanical stress applied on the polymer can, however, change the electrical properties due to structural rearrangement [29]. The electrical conductivity of Nafion is predicted to be increased when the material is exposed to mechanical stress – but unfortunately only in the direction of the stress, which can hardly be used since it would lead to accelerated delamination. A list of currently used common PEMs with some characteristic data can be found elsewhere [30–32].

For the catalyst, platinum and platinum–ruthenium alloys are currently the materials of choice. At present, the high effectiveness of the catalytic activity still prevails over the materials costs. However, high and rising prices of the noble metals demand less costly materials – especially when PEMs become a mass product. Nickel-based catalysts seem to be an option, and also noble metal-coated particles [33]. The major problems concerning the catalysts are the long-term stability and tolerance issues regarding catalyst poisoning species [34]. The stability is governed by the catalytic activity and the particle size of the catalyst. The smaller a particle, the larger is the surface area, and therefore also its effectiveness related to its weight. However, the particle becomes more prone to material loss or reshaping. Therefore, a trade-off between the amount of catalyst, lifetime, and costs has to be made. Poisoning contributes to its sensitivity to other species. If they permanently cover a significant amount of the catalysts surface, more material is needed to achieve reasonable activity.

For the GDL, high permeability of the gas species is required to allow high mass transport from and to the catalyst layer. Therefore, a porous structure is needed, although structural stability combined with high electrical conductivity is essential. Therefore, carbon cloth, fleece, and paper are the most common materials. The cloths and fleeces are made of carbon fibers, which can induce degradation of the membrane material due to penetration. However, of greater interest is a balanced distribution of pore sizes and gas diffusion paths. Modeling of existing structures is performed to understand and improve gas diffusion issues [35]. Additionally, this layer has to provide an appropriate behavior for water evaporation, which can be influenced by the pore structure and surface properties [36].

Several examination methods exist for all these problems and the most common ones are discussed more specifically in the following.

19.2
Gravimetric Properties

The determination of the mass of an object can be carried out using two separate approaches, both based on the physical properties of a mass: inertia and gravitational attraction. The inertia can be deducted from the momentum of a body, defined as

$$\mathbf{p} = m\mathbf{v} \tag{19.1}$$

where $\mathbf{p}$ is the momentum (vector), m the mass, and $\mathbf{v}$ the velocity (vector). When integrated with time, the well-known relation

$$E = \frac{1}{2}m\mathbf{v}^2 \tag{19.2}$$

where E is the energy is derived. The mass of a body can therefore be deduced by measuring the energy needed to change the speed of the body or, for example, by transferring a certain momentum. The advantage of this method is the lack of a reference mass. However, the setups are complicated or difficult to construct with reasonable resolution. Nevertheless, for the very small or large masses to be dealt with in quantum physics or astrophysics, this approach is often the only one applicable.

The standard method of determining the mass of a body uses the gravimetric attraction force of two masses:

$$F(r) = G\frac{m_1 m_2}{r^2} \frac{r}{|r|} \tag{19.3}$$

where F is the force (vector) acting between the two masses m_1 and m_2 with the distance (vector) r and G is the gravity constant. It should be noted that two values for G exist, depending on the way in which it is derived. For experimental data, the value is $6.67428(67)$ Nm^2 kg^{-1}; when calculated from other natural constants, the value amounts to $6.70881(67)$ Nm^2 kg^{-1}, according to CODATA2006. However, this difference has little relevance in practice, since these kinds of measurements rely on a reference mass – which has to be known precisely enough or is being defined. In normal circumstances, one of the masses is the Earth itself. Although the latter is extremely and the distance between the Earth's center and the mass to be determined is large, this introduces some problems when high precision of the measurement is required. Commercially available balances offer precisions of up to factor of 10^7; for example, a total mass to $10\,g$ can be determined with a resolution of $1\,\mu g$ or even better. Balances of this kind are not symmetrical like traditional ones, where the mass to be measured is directly compensated with a proportional mass. Nowadays, the compensation is done by generating a mechanical force using a magnetic or electrical field or by bending a material with known stiffness. However, these force generators are decoupled from the Earth's mass. This allows temporal and spatial fluctuations of the Earth's local gravity, for example, due to seismic activity, and drifts – for example, a change in the local distance to the Earth's mass center due to the position of the Moon – to take effect on the measurement result. Therefore, and of course for some other reasons, calibrations with at least one reference mass have to be carried out from time to time. Another source of inaccurate measurement data arises from the atmosphere in which the objects of interest are examined and the volume of the object itself. This leads to buoyancy effects. Unless the mass measurement is carried out in a perfect vacuum, the volume of the sample replaces the same volume of atmosphere present there before the measurement – which itself has a certain mass. The mass determined without correction is therefore always too small; the buoyancy-corrected mass, or the mass determined in vacuum, is the so-called true mass. Although the density of the atmosphere is relatively low, for example, in the range $\rho = 1.2\,\text{mg cm}^{-3}$, this effect becomes more and more pronounced for samples with low density themselves, especially for polymer, and the more precise the measurements are. For the correction, various influences have to be considered. To elucidate the main influences, the law of ideal gases is considered:

$$pV = nRT \tag{19.4}$$

where p is the (atmospheric) pressure, V the exchanged volume, n the number of gas molecules in the volume V, T the absolute temperature, and R the ideal gas constant, $8.314472(15)$ J mol^{-1} K^{-1}. Expanding the relation with the molar mass M of the gas, one obtains to the expression

$$\rho = \frac{pM}{RT} \tag{19.5}$$

From this equation, it becomes evident that the temperature, pressure, and also the composition of the atmosphere – in the form of the molar mass – have some influence on the measurement. The last parameter is often difficult to determine. In most cases, humidity introduces the largest effect. Sometimes, even the CO_2 level can have a significant effect. Making precise mass measurements in an environment with controlled climate and pressure is therefore mandatory. It should be noted that this effect is important for all mass measurements where the density of the sample differs from that of the reference used for calibration. However, a mass correction for buoyancy is reasonable only for cases where the density of the sample is well enough known. For thermogravimetric analysis (TGA), for example, in measurements with changing temperature the thermal expansion behavior should also be considered. The correction of mass becomes complex at this point and is therefore mostly neglected. Fortunately, other sources of error have more influence for these measurements, superimposing the volume changing effect in most cases such as convection issues. These influences are normally covered sufficiently precisely by carrying out a measurement with the same parameters but without a sample – the so-called baseline. This is then subtracted or used to correct the measurement with the sample, leading to data solely from the sample in the ideal case.

Gravimetric measurements under isothermal conditions are often used to examine drying or decomposition processes. For this, first the sample is conditioned, for example, wetted, and then the mass versus time is recorded. From such simple experiments, it is possible to determine different stages of a drying process, the surface-controlled and the diffusion-controlled stages, for example, since they can differ in the rate of mass loss. For other tests, the atmosphere is changed to examine the oxidation or reduction stability of a material. The time that a material needs to reach equilibrium with a certain condition is another important property that can also be acquired.

For the experimental part, the water uptake of fuel-cell components is one of the most common gravimetric data provided, especially for the membrane materials. The definition of this is

$$\text{Water uptake} = WU = \frac{m_{\text{wet}}}{m_{\text{dry}}} - 1 \tag{19.6}$$

where m_{dry} is the mass of the totally dried sample and m_{wet} the mass of the humidified sample. At first sight, the determination of this parameter seems to be trivial. However, precise control of temperature and humidity during the equilibration of the material is crucial. It is important to check if an equilibrium state is reached when the measurements were carried out. In the case of Nafion, these times can be fairly long, ranging from days to weeks [37]. On the other hand, assuring a dry state is also not trivial. Since the bonding of water molecules to polar groups of polymers can be fairly strong, remaining water on drying can be released only upon decomposition of the polymer [38]. Comparing literature data therefore always depends on the drying conditions applied and defined. Alberti et al. [37] carried out thorough measurements of water uptake for Nafion under different temperature and humidity conditions. It was found that up to 71 water molecules per $-SO_3H$ group can be bound in Nafion. Similar data on the water uptake in other membrane materials such as polybenzimidazole (PBI) were provided by, for example, Gu et al. [39], especially for relevant high-temperature PEM operating conditions. For these kinds of membranes, the number of water molecules per acid group amounts to about 10 – far less than that for Nafion.

When studying the literature on water uptake, the special case of total humidification of a sample seems to cause some additional problems. Confusion exists as to whether the exposure of a sample to 100% relative humidity (RH) gives the same results as exposure to water directly. This is often termed Schröder's paradox. Nevertheless, Jeck et al. [40] showed that the history of the polymer and the activity of the solvent – here water – are important and that such a paradox does not exist. The difference in the data obtained for total humidification can therefore be simply explained by different water activity. This activity is different for the wetting condition, for example, direct contact with water, and the humidifying condition, where the contact is with the water vapor.

The thermal stability of the fuel-cell components is another very important point when gravimetric measurements are carried out. The higher the temperature that a material can withstand without significant alteration, the more likely is a long lifetime. Additionally, evolving decomposition products can introduce additional sources of degradation issues [41]. This is important since during fabrication high temperatures are sometimes applied to the materials. Assuring that this has no significant negative effect on the lifetime or operational behavior of the final product is an important issue.

A great enhancement of the interpretation of data can be achieved when TGA is coupled with other analysis techniques such as calorimetry, FTIR spectroscopy, and MS. Especially the coupling with calorimetry has a long history and is well established. This is because caloric data are strongly coupled with mass changes. The combination of the data obtained allows in most cases more sophisticated conclusions than from single measurements. Moreover, the calorimetric data are often more sensitive to decomposition effects than the mass data. When TGA is combined with MS, the products of decomposition or water release can be identified and distinguished, so that more accurate conclusions can be drawn. For Nafion, a three-step decomposition/degradation process could be confirmed [14] by these means.

19.3
Caloric Properties

For the calorimetric measurements, the heat flux of a sample to or from its environment is measured. For the ideal case, the sample is thermally coupled solely by a defined connection with a large thermal reservoir. One of the ways in which the heat flux is determined is by measuring the temperature difference between the two endings of the thermal connection – in most cases a pair of thermocouples. The heat flux can be induced by a change in temperature, mass, or simply time, since it is related to the inherent heat capacity of a sample. Two expressions exist for the heat capacity, depending whether a constant pressure (c_p) or volume (c_v) for the sample exists during measurement:

$$c_p(T) = \frac{\partial}{\partial T} H(T)$$
$$c_v(T) = \frac{\partial}{\partial T} U(T) \tag{19.7}$$

where H is the enthalpy and U the internal energy of the sample, both being a function of temperature T. For the sake of readability, this temperature dependence is implied when H and U are mentioned further. For most cases, c_p is determined, since it easier for most cases to keep the pressure constant than the volume. The main difference between H and U is that U includes the energy due to volume work. Therefore, different results are obtained when the ability of the sample to expand or compress is hindered.

For an inert sample without any phase changes, H and U are continuous, but not necessarily linear, with temperature. However, H and U are extensive properties; for example, they are dependent on the mass of the sample, in contrast to an intensive property such as temperature. This means that a mass change due to chemical reactions, decomposition, hydration, drying, or even degassing changes H and U. This explains why the calorimetric measurements are very sensitive to mass changes. However, every change of state of the sample also changes the heat capacity. Upon melting or vaporization, (heat) energy is required to transfer the material from one state to the other. Such behavior is called endothermic and the opposite exothermic. Typical exothermic processes are crystallization and condensation. When a process occurs upon heating and cooling at the same temperature with the same energy content, it is called reversible. The process can therefore be triggered repeatedly without alteration. However, for real processes the condition of taking place at the same temperature is often not fully fulfilled. Especially for the cooling condition, one often experiences the supercooling effect [42]. For water, for example, the freezing point can be shifted to below $-30\,°\mathrm{C}$ [43, 44]. Superheating, in contrast, is seldom observed. When superheating is observed, one first has to consider a decrease in mass for the measurement, to exclude a possible effect of thermal resistance or conductivity. Further, superheating may occur in polymers due to kinetic issues [45]. Other reversible processes include phase transitions such as changes in crystallographic structure.

Irreversible or apparently irreversible processes, in contrast, show different behavior in heating and cooling. Examples of irreversible processes are decomposition and (de-)alloying [46]. Drying, or more generally desorption, is an example of an apparently irreversible processes. The difference can be distinguished when the measurement conditions are changed and the reverse reaction (e.g., adsorption) is induced or not. If so, only an apparently irreversible process is present. This is of interest when the glass transition temperature T_g needs to be determined. When a molten material is cooled fast enough, it can solidify in an amorphous state. In that state, the material possesses interesting properties or features that are not available when long-range ordering due to crystallization is present. However, there are two stages for the amorphous state. When below T_g, the material behaves elastically. This means that no obvious differences in the mechanical behavior are found compared with the crystallized state. When above T_g, the material becomes rubbery. This stage is of interest for machining, since it allows mechanical deformation with a low risk of crack formation. However, when such an amorphous material is heated above its T_g, crystallization can also take place. This can eliminate the desired rubbery state and is therefore not desirable in most cases. Therefore, a glass transition is observed in many cases as an irreversible process – unless the material is molten and then frozen again. Additionally, this effect has two challenges: the first arises from the fact that it is mainly entropic. The energy changes and therefore also measurable effects are often very small and superimposed or masked by other effects. Second, it is often hindered by its kinetics. This means that depending on the time scale on which the effect is examined, the glass transition can be more or less pronounced and even shifted in temperature. Therefore, such transitions appear in many cases as a "smeared" effect and can hardly be determined precisely. As already mentioned, T_g is important for casting of the membrane. Data provided by Sumner et al. [47] indicate that sulfonated poly(arylene ether) copolymers are promising candidates since their T_g is in the range 170–260 °C and far below its decomposition temperature. Also, Nafion undergoes morphology changes due to temperature and annealing conditions. These can be assessed using differential scanning calorimetry [48]. Nores-Pondal et al. [49] showed that even small amounts of water can have a tremendous effect on the T_g of PBI in the temperature range −120 to −80 °C. The phosphoric acid bound in the PBI was found to be responsible for this behavior. For the GDL materials, the effect of temperature and humidity on the hydrophobicity of carbon materials was demonstrated by Kinumoto et al. [50].

Another topic of interest for PEMs is water freezing in the membrane materials. Water incorporated in polymers can be in a noncrystallized state upon freezing [51]. The supercooling of water is believed to be due to interactions of polar hydroxyl or other groups of the polymer with closely attached water molecules. The higher the number of polar groups in the polymer, the more water can be incorporated in the polymer without freezing thereof at the melting point. The stronger the interactions are, the lower is the melting temperature of the bound water. However, owing to the different local environments of the bound water molecules, the melting thereof becomes a broad range rather than a sharp peak. The example of cross-linked poly(vinyl alcohol) was presented by Hwang et al. [52]. For Nafion, Mendil-Jakani et al. [53] showed, using X-ray scattering, that the freezing becomes apparent only at high swelling degrees, which are reached only for high temperatures and humidity where the ratio between water and $-SO_3H$ groups is in the region of 22 or higher. Contaminants also influence the freezing behavior. Sodium, for example, preserves the supercooled state of water in Nafion [54]. However, one should always keep in mind that a lowered freezing point of a fluid might also be induced for only geometric reasons [55].

19.4
Structural Information: Porosity

Little information is currently available on the porosity or even pore size distribution of fuel-cell components such as the GDL. Only a few methods are capable of obtaining this information since a large range of pore size radii have to be measured. The range starts from pores a few nanometers in diameter, the so-called nanopores, which are essential especially close to the membrane, up to macropores with diameters of several micrometers, which are needed for a good supply of gas from the flow-field channels to the electrode layer. Among these methods are BET (Brunauer–Emmett–Teller) measurements, mercury intrusion, and some specialized methods.

When talking of porosity, one has first has to consider the accessibility of pores. Closed pores, incorporated in a material, are hardly accessible from the outside. However, they can act as a reservoir or even as transport paths inside the material [20]. Such pores can only be examined by destructive methods such as sectioning and imaging. Nondestructive methods such as X-ray or synchrotron imaging can be used only for certain conditions [56, 57]. In contrast, there are open pores which are summarized in the so-called open porosity and are accessible from the surface. The inside of the open pores is therefore in contact with the outer atmosphere of the sample. This gives the basis for several porosity measurement methods. One effect uses Washburn's equation, which relates the pressure of a gas to the pore diameter. Smaller diameters lead to higher pressures, meaning that more energy is needed to empty smaller pores than larger pores. This is mainly used for isothermal nitrogen adsorption measurements when using the BJH (Barrett–Joyner–Halenda) method [58]. The other effect used is based on the free surface energy acting between the liquid adsorbent and gas phase. Owing to this energy, a certain contact angle exists, so certain pores are blocked from intrusion and are not accessible without applying an additional force. Mercury intrusion porosimetry (MIP) measurements are based on that effect.

The determination of a pore size distribution, however, relies on several assumptions. For most distributions, spherical pores are assumed. The first problem here derives from the accessibility of the pores. The larger the entrance thereof, the faster the pores can be filled. Pores with small entrance holes therefore need longer times to be detected correctly. The next problem arises from the shape of a pore. In Nature, everything between a spherical and a cylindrical pore can appear. Cylindrical pores show a different filling behavior due to capillary pressure, resulting in an incorrect pore size distribution [59]. A very common problem arising from that is the so-called ink-bottle effect. Larger pores are connected here with smaller (cylindrical) pores and can form large-scale networks. Such networks show significant hysteresis behavior between intrusion (filling) and extrusion (draining). In such networks, the pore size distribution is shifted towards the smallest channel diameter [60]. Since this effect also has a strong influence on freezing and melting behavior [61], so-called thermoporosity measurements can be used to determine such networks. Getting even closer to reality is achieved with the introduction of tortuosity. This factor determines the channel width distribution of a pore network structure [62]. In addition to tomographic methods, nitrogen adsorption is the state-of-the-art method to access this factor. This factor has great importance when permeation values are calculated from pore size distributions since the effective diffusion is governed by the tortuosity [63].

However, the porosity and distribution derived are also dependent on the "solvent" used, as demonstrated by Harkness et al. [64] using a special water capillary pressure porosimeter. The reason was found in the contact angle dependence of the pressure and the pore radii. The deviations depending on the solvent increase for the case of nonideal, nonspherical pores. Especially for carbon, the contact angle of water on carbon can be shifted between 55° and 80° depending on the pretreatment [50]. To overcome some of the problems, the method of standard contact porosimetry (MSCP) can be used. The sample of interest is first flooded with the appropriate solvent and then depleted by vaporization while being in contact with known standards. Since the sample and the standards are in contact, a capillary equilibrium exists. First the larger pores will be depleted when evaporation takes place. The pore volume for a specific pore size range is determined by checking the mass of the standards from time to time in comparison with the sample. The main advantages of this method are as follows. (i) It is nearly nondestructive. Only a small pressure needs to be applied on the sample to keep it in contact with the standards. Therefore, a more realistic pore size distribution can be obtained than for the MIP technique – especially for small pore sizes. (ii) An appropriate solvent is applied: nearly every volatile solvent can be used, thus preventing the problem of different contact angles compared with the real conditions. (iii) Reusability: one sample can be measured repetitively and even with different solvents. The power of the MSCP was illustrated on different MEA components by Volfkovich et al. [65].

For GDL carbon materials, Caston et al. [66] showed using MIP on hand-made woven fabrics that a relationship exists between the amount of macropores and the through-plane gas permeability. They also showed that woven materials can exhibit high in-plane permeability, which means also a low pressure drop. For a non-woven carbon paper, Hiramitsu et al. [67] showed that even small amounts of hydrophobized carbon black introduced in the GDL reduce flooding significantly. The hindering of the formation of a continuous water film blocking the catalyst completely seems to be the reason for this positive effect.

19.5
Mechanical Properties

At first sight, the determination of mechanical properties seems to be trivial. The principle is fairly easy – stress is applied on a sample and the corresponding mechanical deformation in the form of strain is measured. However, there are several ways in which this stress-strain correlation can be measured, and what kind of data can be obtained. The situation here is similar to the determination of electrical properties. Nafion, for example, exhibits a strong dependence between time, hydration, temperature, and mechanical deformation. Changing the operating conditions of an MEA changes not only the electrical characteristics, but also mechanical stresses. Silberstein and Boyce [68] provided two models describing accurately the mechanical behavior of Nafion for a wide range of climatic and also mechanical loading conditions. Accessing reliable basic data for such models, however, is far from trivial – especially for polymers [69].

The most fundamental parameter used in mechanical testing is the strain, which is calculated from the elongation or compression related to the original length of the sample l_0 (commonly measured in millimeters):

$$\varepsilon = \frac{\Delta l}{l_0} = \frac{l - l_0}{l_0} = \frac{l}{l_0} - 1 \tag{19.8}$$

where l and l_0 are in millimeters and ε is hence a dimensionless value, often given in percentiles. The other fundamental quantity is the stress, defined as

$$\sigma(\varepsilon) = \frac{F(\varepsilon)}{A(\varepsilon)} \tag{19.9}$$

where σ is the (strain-dependent) stress in newtons per square millimeter, F the (strain-dependent) force in newtons, and A the (strain-dependent) cross-sectional area in square millimeters for the specific force applied. This is the common definition and called the *true stress*, which is not applicable, however, for most measurements. The reason is that the cross-sectional area changes during testing and can hardly be measured simultaneously. Therefore, the stress is related to the cross-sectional area at the beginning of the test (A_0) and is termed the *technical stress*:

$$\sigma = \frac{F}{A_0} \tag{19.10}$$

With these two data, one of the best-known mechanical data for materials, the so-called elastic modulus, E, is calculated. It describes the relation between stress and strain when the strain and stress are both axially aligned:

$$\sigma = \varepsilon E \tag{19.11}$$

and has the same units as the stress. It is common to provide this value in GPa or MPa (1 MPa = 1 N mm^{-2}), since the values are rather large. For metals and ceramics, values of several hundred GPa are common, whereas for polymers several hundred MPa are normal. The electrical counterpart is the specific electrical conductivity. As for the electrical conductivity, the mechanical values of ε and σ are dependent on the orientation to be measured and are therefore vectors:

$$\sigma = \varepsilon C \tag{19.12}$$

hence C is in fact a tensor. For the special case when stress and strain are measured in the same direction, for example, in parallel, C is reduced to a scalar value E called *Young's modulus*. Since most of the measurements are carried out on macroscopic-sized samples and polycrystals, the orientation dependence also vanishes. However, when foils are examined, the orientation can regain importance. When the stress and strain measured are orthogonal, the resulting reduction of C represents the shear modulus G, which amounts in general to about half to one-third of E. For the linear-elastic behavior of homogeneous materials, the relation between E and G is

$$\nu = \frac{E}{2G} - 1 \tag{19.13}$$

This is derived from the common definition of the Poisson's ratio as

$$\nu = -\frac{\frac{\Delta l}{l}}{\frac{\Delta d}{d}} \tag{19.14}$$

where l and d are the normal and orthogonal dimensions, respectively, and Δl and Δd are the changes thereof. The value of ν normally ranges from 0 to 0.5, and describes the behavior of the material upon applying stress. For the upper limit of $\nu = 0.5$, the material responds in an incompressible way to the stress. At this point it should be noted that even negative ratios are known to exist with a limit of -1, which is due to thermodynamics [70]. Materials with negative ratios behave as auxetics and normally have foam-like structures, for example, and should not be considered as homogeneous. When dealing with Poisson's ratio, there is contradictory information available for Nafion. Commonly a value of $\nu = 0.25$ is used, which is normally a good starting point for polymers. Solasi et al. [71] showed that a value of 0.4 is more appropriate. One of the main problems arises from the fact that the time for the polymer to reach thermodynamic equilibrium is fairly long and small deviations may have a significant influence on the mechanical properties [72].

The stress and strain are normally measured in a tensile testing machine. These are usually designed to measure uniaxially; for example, the stress is applied only in one special direction (axis), and the strain is measured in the same direction. One of the rarely conducted biaxial testings was reported by Silberstein et al. [73]. An additional axis where strain is applied was introduced. These biaxial tests reflect real conditions that are far better, especially for films. Nevertheless, Silberstein et al. also demonstrated that data obtained from uniaxial tests can be used for the prediction of the behavior of materials under biaxial strain conditions.

Another main influence on the mechanical properties of membrane materials such as Nafion apart from temperature is humidity [74]. As mentioned earlier, swelling due to water causes significant dimensional changes of the material. For example, on freezing, free water in a membrane can form ice, leading to high internal mechanical stresses for the polymer [75], and this is currently of great interest for membrane materials. For Nafion, the mechanical behavior due to water uptake or release has been thoroughly investigated in recent years [37, 76, 77]. Especially data for relaxation times of up to 10^6 s (about 11.5 days) were reported by Satterfield and Benziger [78].

However, the elastic properties of a material are only one side of the coin. The other side concerns plasticity, namely creep and time–dependent strain. Some modeling parameters for Nafion at 25 °C/50% RH, 25 °C/80% RH, and 65 °C/75% RH were provided by Yoon and Huang [75]. Owing to the complexity of the membrane material, nonlinear behavior of the mechanical properties is also observed for both the elastic and plastic responses. However, this is fairly common for polymers [69, 79, 80]. Owing to the plasticity of the materials, the mechanical properties measured depend on the thermal or mechanical treatment history, which makes the measurement of repetitive and comparable data challenging. Chen et al. showed for Nafion that even the direction of a treatment cycle may have an impact on the material parameter measured [81].

However, not only the history, humidity, and temperature may have a significant effect on membrane materials. For the case of the DMFCs, a softening of Nafion by a factor of nearly six is observed when the methanol concentration in water reaches 80% [82].

Corresponding to electrical tests, the mechanical tensile test can be compared with electrical measurements conducted with constant current or voltage (DC measurements). On introducing an excitation with variable frequency, impedance analysis can be carried out. For electrical testing, this is widely known as electrochemical impedance spectroscopy (EIS). A similar approach can be taken for mechanical testing, and in that case it is called dynamic mechanical analysis (DMA). For these tests, the relation between amplitude and phase shift for a mechanical excitation is recorded. When frequency spectra are recorded for constant conditions, an analysis similar to EIS can be made. The results thereof are the static properties such as E, but also the viscous properties. One of the major advantages of this method is the high sensitivity to processes in the material. Structural changes or changes in the bonding character have a direct influence on the mechanical properties, especially the dynamic properties. Majsztrik et al. [83] demonstrated this for Nafion, providing a kind of phase diagram for the state of structure depending on temperature and water activity. Phase transitions such as the T_g of a material can be detected accurately with DMA, even when calorimetry fails due to a lack of signal. Sgreccia et al. [84] used DMA on sPEEK [sulfonated poly(ether ether ketone)] membrane materials to investigate the influence of plasticizers and thermal treatment on T_g, which was found to shift between 105 and 205 °C with an accuracy of better than 5 °C. A correlation between the maximum stiffness and T_g was also shown. For PBI-based membrane materials, Sannigrahi et al. [85] investigated the influence of the chemical structure on the structural polymer backbone. Mixed *m*- and *p*-pyridine variants were studied. The *meta*-form polymers showed the highest T_g values of up to about 410 °C, whereas the *para*-form polymers showed the lowest T_g values of down to about 360 °C.

However, the tensile test and DMA are not the only methods for determining mechanical properties. Another approach uses a so-called nanoindenter. This method is especially suitable for the investigation of thin layers. The surface of the sample is penetrated by a small, stiff tip to a depth amounts of the order of several micrometers. Therefore, deformation of only a small area of the material is introduced. This allows measurements on the sample with a high spatial resolution or when material is scarce. However, only the (near-)surface properties are measured. Poornesh et al. [86] obtained data on a catalyst layer of 10 μm thickness with this method. One effect that can be studied with this method is the surface hardening of polymers [72], since the penetration resistance is mainly governed by surface effects. An interesting application of nanoindenter measurements using the micro-hardness of a material was described by Flores *et al.* [87]. Measuring the micro-hardness in relation to temperature can provide the phase-change temperature, especially the glass transition temperature T_g. A special case of nanoindenter measurements uses atomic force microscopy. Here the penetration depth is in the range of several tens of nanometers. Although the penetration depth is extremely small, this method is suitable for measurements on highly hydrated, soft membrane materials. Franceschini and Corti [88] demonstrated the feasibility of this method on doped PBI and Nafion. They found an increase in the stiffness of the membrane material due to doping for both membrane types. Additionally, dragging was observed for Nafion when the tip of the atomic force microscope was removed. This behavior could be relevant to damage to the membrane upon cyclic swelling.

Another method for measuring mechanical properties on the macroscopic scale uses the relation between mechanical properties and the propagation of acoustic waves [89]. The velocity of sound waves and also the damping thereof can be directly deduced from the elastic and viscous properties. For polymers, ultrasound can be used since the damping of the acoustic waves is decreased at high frequencies. However, this method seems not to have been applied to fuel cell-related membrane materials so far.

Related to the measurement of the sound velocity is Brillouin light scattering (BLS). This method measures the mechanical properties of polymers due to acoustic phonon activity. The interesting point of this method is the capability of measuring both the shear (transversal) and also the tensile (longitudinal) modulus simultaneously. This was demonstrated by Roberti *et al.* [90] for Nafion. The Young's modulus obtained by this method is about 10 times higher than that derived from classical tensile measurements. This difference was deduced from the high frequency used for BLS, which is in the range of several tens of gigahertz. The viscous properties have little effect on that time scale, and the results are in good agreement with the predictions from data available so far.

19.6
Conclusion

The determination of the physical properties of a material is essential for the development of materials and also their application. In this chapter, gravimetry, calorimetry, porosimetry, and mechanical testing were briefly explained, and common problems were also outlined. This background is necessary in order to be able to compare available data from different sources adequately. Based on the differences in testing conditions, not only could the behavior of materials, especially the membrane material Nafion, be explained, but also models describing the behavior with consistent parameters were developed in recent years. This allows reliable predictions and provides a better understanding of the underlying processes. By understanding these processes, such as hydration and swelling, materials can be improved. Here gravimetry plays an important role in determining primarily whether a material is thermally and chemically stable under the conditions of the application being considered. Therefore, and because it is easy to handle, it is applied mainly in the first stage of materials development. Calorimetry is also often used in this first stage. Phase transitions can be determined with this method. The focus here in recent years has been on the glass transition temperature, T_g. This parameter is important for the usability of especially membrane materials, since most polymers used in fuel cells are brittle at room temperature or in the dry state. Both handling and also applications are often affected directly by this parameter. Here also the mechanical properties have attracted attention. In recent years, the amount of available data has increased, not only for standard conditions but also for the less accessible operating conditions of fuel cells. Several consistent mechanical models for the elastic and viscous behavior have been developed, at least for Nafion, covering not only temperature but also the humidity/hydration level. These are prerequisites for application design. For porosity measurements, the current focus is set on the modeling and characterization of real structures. Visualization of gas diffusion paths and the modeling may help to improve the performance of fuel cells, especially at high power densities.

There are numerous physical properties of a specific material and hence numerous methods for their measurement. Since some materials are more complex than others, there is no golden rule as to which method should be applied. However, the sum of all properties combined with exactly specified conditions provides a good basis for application and further materials development.

References

1. Marrony, M., Barrera, R., Quenet, S., Ginocchio, S., Montelatici, L., and Aslanides, A. (2008) Durability study and lifetime prediction of baseline proton exchange membrane fuel cell under severe operating conditions. *J. Power Sources*, **182**, 469–475.
2. Bae, S.J., Kim, S.-J., Park, J.I., Lee, J.-H., Cho, H., and Park, J.-Y. (2010) Lifetime prediction through accelerated degradation testing of membrane electrode assemblies in direct methanol fuel cells. *Int. J. Hydrogen Energy*, **35**, 9166–9176.
3. Shao, M. (2011) Palladium-based electrocatalysts for hydrogen oxidation and oxygen reduction reactions. *J. Power Sources*, **196**, 2433–2444.
4. Dur, E., Cora, Ö.N., and Koç, M. (2011) Effect of manufacturing conditions on the corrosion resistance behavior of metallic bipolar plates in proton exchange membrane fuel cells. *J. Power Sources*, **196**, 1235–1241.
5. Namazi, H. and Ahmadi, H. (2011) Improving the proton conductivity and water uptake of polybenzimidazole-based proton exchange nanocomposite membranes with TiO_2 and SiO_2 nanoparticles chemically modified surfaces. *J. Power Sources*, **196**, 2573–2583.
6. Zheng, H. and Mathe, M. (2011) Enhanced conductivity and stability of composite membranes based on poly(2,5-benzimidazole) and zirconium oxide nanoparticles for fuel cells. *J. Power Sources*, **196**, 894–898.
7. He, X. and Hu, C. (2011) Building three-dimensional Pt catalysts on TiO_2 nanorod arrays for effective ethanol electrooxidation. *J. Power Sources*, **196**, 3119–3123.

8. Miecznikowski, K. and Kulesza, P.J. (2011) Activation of dispersed PtSn/C nanoparticles by tungsten oxide matrix towards more efficient oxidation of ethanol. *J. Power Sources*, **196**, 2595–2601.
9. Matos, B.R., Santiago, E.I., Rey, J.F.Q., Ferlauto, A.S., Traversa, E., Linardi, M., and Fonseca, F.C. (2011) Nafion-based composite electrolytes for proton exchange membrane fuel cells operating above 120 °C with titania nanoparticles and nanotubes as fillers. *J. Power Sources*, **196**, 1061–1068.
10. Thiam, H.S., Daud, W.R.W., Kamarudin, S.K., Mohammad, A.B., Kadhum, A.A.H., Loh, K.S., and Majlan, E.H. (2011) Overview on nanostructured membrane in fuel cell applications. *Int. J. Hydrogen Energy*, **36**, 3187–3205.
11. Hogarth, W.H.J., Diniz da Costa, J.C., and Lu, G.Q. (2005) Solid acid membranes for high temperature (>140 °C) proton exchange membrane fuel cells. *J. Power Sources*, **142**, 223–237.
12. Ermete, A. (2010) Composite materials: an emerging class of fuel cell catalyst supports. *Appl. Catal. B*, **100**, 413–426.
13. Gouadec, G. and Colomban, P. (2007) Raman spectroscopy of nanomaterials: how spectra relate to disorder, particle size and mechanical properties. *Prog. Cryst. Growth Charact. Mater.*, **53**, 1–56.
14. Bas, C., Flandin, L., Danerol, A.S., Claude, E., Rossinot, E., and Alberola, N.D. (2010) Changes in the chemical structure and properties of a perfluorosulfonated acid membrane induced by fuel-cell operation. *J. Appl. Polym. Sci.*, **117**, 2121–2132.
15. Seddiq, M., Khaleghi, H., and Mirzaei, M. (2006) Numerical analysis of gas cross-over through the membrane in a proton exchange membrane fuel cell. *J. Power Sources*, **161**, 371–379.
16. Kundu, S., Simon, L.C., Fowler, M., and Grot, S. (2005) Mechanical properties of Nafion™ electrolyte membranes under hydrated conditions. *Polymer*, **46**, 11707–11715.
17. Zamel, N. and Li, X. (2011) Effect of contaminants on polymer electrolyte membrane fuel cells. *Prog. Energy Combust. Sci.*, **37**, 292–329.
18. Cho, Y.-H., Yoo, S.J., Cho, Y.-H., Park, H.-S., Park, I.-S., Lee, J.K., and Sung, Y.-E. (2008) Enhanced performance and improved interfacial properties of polymer electrolyte membrane fuel cells fabricated using sputter-deposited Pt thin layers. *Electrochim. Acta*, **53**, 6111–6116.
19. Kulikovsky, A.A. (2009) Optimal shape of catalyst loading along the oxygen channel of a PEM fuel cell. *Electrochim. Acta*, **54**, 7001–7005.
20. Mauritz, K.A. and Moore, R.B. (2004) State of understanding of Nafion. *Chem. Rev.*, **104**, 4535–4586.
21. Gierke, T.D., Munn, G.E., and Wilson, F.C. (1981) The morphology in Nafion perfluorinated membrane products, as determined by wide- and small-angle X-ray studies. *J. Polym. Sci. Polym. Phys. Ed.*, **19**, 1687–1704.
22. Barbi, V., Funari, S.S., Gehrke, R., Scharnagl, N., and Stribeck, N. (2003) Nanostructure of Nafion membrane material as a function of mechanical load studied by SAXS. *Polymer*, **44**, 4853–4861.
23. Kong, X. and Schmidt-Rohr, K. (2011) Water–polymer interfacial area in Nafion: comparison with structural models. *Polymer*, **52**, 1971–1974.
24. Conti, F., Negro, E., and Di Noto, V. (2009) First time-resolved EPR observation of Nafion photochemistry. *Chem. Commun.*, 7006–7008.
25. Lu, J., Lu, S., and Jiang, S.P. (2011) Highly ordered mesoporous Nafion membranes for fuel cells. *Chem. Commun.*, 3216–3218.
26. Aleksandrova, E., Hiesgen, R., Andreas Friedrich, K., and Roduner, E. (2007) Electrochemical atomic force microscopy study of proton conductivity in a Nafion membrane. *Phys. Chem. Chem. Phys.*, **9**, 2735–2743.
27. Sagarik, K., Phonyiem, M., Lao-ngam, C., and Chaiwongwattana, S. (2008) Mechanisms of proton transfer in Nafion: elementary reactions at the sulfonic acid groups. *Phys. Chem. Chem. Phys.*, **10**, 2098–2112.
28. Choe, Y.K., Tsuchida, E., Ikeshoji, T., Yamakawa, S., and Hyodo, S.A. (2009)

Nature of proton dynamics in a polymer electrolyte membrane, Nafion: a first-principles molecular dynamics study. *Phys. Chem. Chem. Phys.*, **11**, 3892–3899.

29. Allahyarov, E. and Taylor, P.L. (2008) Simulation study of the correlation between structure and conductivity in stretched Nafion. *J. Phys. Chem. B*, **113**, 610–617.

30. Peighambardoust, S.J., Rowshanzamir, S., and Amjadi, M. (2010) Review of the proton exchange membranes for fuel cell applications. *Int. J. Hydrogen Energy*, **35**, 9349–9384.

31. Neburchilov, V., Martin, J., Wang, H., and Zhang, J. (2007) A review of polymer electrolyte membranes for direct methanol fuel cells. *J. Power Sources*, **169**, 221–238.

32. Bose, S., Kuila, T., Nguyen, T.X.H., Kim, N.H., Lau, K.T., and Lee, J.H. (2011) Polymer membranes for high temperature proton exchange membrane fuel cell: recent advances and challenges. *Prog. Polym. Sci.*, **36**, 813–843.

33. Kaplan, D., Burstein, L., Rosenberg, Y., and Peled, E. (2011) Comparison of methanol and ethylene glycol oxidation by alloy and core–shell platinum based catalysts. *J. Power Sources*, **196**, 8286–8292.

34. Andreasen, S.J., Vang, J.R., and Kær, S.K. (2011) High temperature PEM fuel cell performance characterisation with CO and CO_2 using electrochemical impedance spectroscopy. *Int. J. Hydrogen Energy*, **36**, 9815–9830.

35. Thiedmann, R., Hartnig, C., Manke, I., Schmidt, V., and Lehnert, W. (2009) Local structural characteristics of pore space in GDLs of PEM fuel cells based on geometric 3D graphs. *J. Electrochem. Soc.*, **156**, B1339–B1347.

36. Guo, Z., Liu, W., and Su, B.L. (2011) Superhydrophobic surfaces: from natural to biomimetic to functional. *J. Colloid Interface Sci.*, **353**, 335–355.

37. Alberti, G., Narducci, R., and Sganappa, M. (2008) Effects of hydrothermal/thermal treatments on the water-uptake of Nafion membranes and relations with changes of conformation, counter-elastic force and tensile modulus of the matrix. *J. Power Sources*, **178**, 575–583.

38. Laporta, M., Pegoraro, M., and Zanderighi, L. (1999) Perfluorosulfonated membrane (Nafion): FT-IR study of the state of water with increasing humidity. *Phys. Chem. Chem. Phys.*, **1**, 4619–4628.

39. Gu, T., Shimpalee, S., Van Zee, J.W., Chen, C.Y., and Lin, C.W. (2010) A study of water adsorption and desorption by a PBI–H_3PO_4 membrane electrode assembly. *J. Power Sources*, **195**, 8194–8197.

40. Jeck, S., Scharfer, P., and Kind, M. (2009) Water sorption in physically crosslinked poly(vinyl alcohol) membranes: an experimental investigation of Schroeder's paradox. *J. Membr. Sci.*, **337**, 291–296.

41. Collier, A., Wang, H., Zi Yuan, X., Zhang, J., and Wilkinson, D.P. (2006) Degradation of polymer electrolyte membranes. *Int. J. Hydrogen Energy*, **31**, 1838–1854.

42. Schülli, T.U., Daudin, R., Renaud, G., Vaysset, A., Geaymond, O., and Pasturel, A. (2010) Substrate-enhanced supercooling in AuSi eutectic droplets. *Nature*, **464**, 1174–1177.

43. Debenedetti, P.G. (2003) Supercooled and glassy water. *J. Phys. Condens. Matter*, **15**, R1669.

44. Guzman, J. and Braga, S. (2005) Supercooling water in cylindrical capsules. *Int. J. Thermophys.*, **26**, 1781–1802.

45. Toda, A., Hikosaka, M., and Yamada, K. (2002) Superheating of the melting kinetics in polymer crystals: a possible nucleation mechanism. *Polymer*, **43**, 1667–1679.

46. Swier, S., Ramani, V., Fenton, J.M., Kunz, H.R., Shaw, M.T., and Weiss, R.A. (2005) Polymer blends based on sulfonated poly(ether ketone ketone) and poly(ether sulfone) as proton exchange membranes for fuel cells. *J. Membr. Sci.*, **256**, 122–133.

47. Sumner, M.J., Harrison, W.L., Weyers, R.M., Kim, Y.S., McGrath, J.E., Riffle, J.S., Brink, A., and Brink, M.H. (2004) Novel proton conducting sulfonated

48. Lin, H.-L., Yu, T.L., Huang, C.-H., and Lin, T.-L. (2005) Morphology study of Nafion membranes prepared by solution casting. *J. Polym. Sci. Part B: Polym. Phys.*, **43**, 3044–3057.
49. Nores-Pondal, F.J., Buera, M.P., and Corti, H.R. (2010) Thermal properties of phosphoric acid-doped polybenzimidazole membranes in water and methanol–water mixtures. *J. Power Sources*, **195**, 6389–6397.
50. Kinumoto, T., Nagano, K., Tsumura, T., and Toyoda, M. (2010) Thermal and electrochemical durability of carbonaceous composites used as a bipolar plate of proton exchange membrane fuel cell. *J. Power Sources*, **195**, 6473–6477.
51. Nakamura, K., Hatakeyama, T., and Hatakeyama, H. (1983) Relationship between hydrogen bonding and bound water in polyhydroxystyrene derivatives. *Polymer*, **24**, 871–876.
52. Hwang, B.-J., Joseph, J., Zeng, Y.-Z., Lin, C.-W., and Cheng, M.-Y. (2011) Analysis of states of water in poly (vinyl alcohol) based DMFC membranes using FTIR and DSC. *J. Membr. Sci.*, **369**, 88–95.
53. Mendil-Jakani, H., Davies, R.J., Dubard, E., Guillermo, A., and Gebel, G. (2011) Water crystallization inside fuel cell membranes probed by X-ray scattering. *J. Membr. Sci.*, **369**, 148–154.
54. Corti, H.R., Nores-Pondal, F., and Pilar Buera, M. (2006) Low temperature thermal properties of Nafion 117 membranes in water and methanol–water mixtures. *J. Power Sources*, **161**, 799–805.
55. Fagerlund, G. (1973) Determination of pore-size distribution from freezing-point depression. *Mater. Struct.*, **6**, 215–225.
56. Fishman, Z., Hinebaugh, J., and Bazylak, A. (2010) Microscale tomography investigations of heterogeneous porosity distributions of PEMFC GDLs. *J. Electrochem. Soc.*, **157**, B1643–B1650.
57. Becker, J., Fluckiger, R., Reum, M., Buchi, F.N., Marone, F., and Stampanoni, M. (2009) Determination of material properties of gas diffusion layers: experiments and simulations using phase contrast tomographic microscopy. *J. Electrochem. Soc.*, **156**, B1175–B1181.
58. Barrett, E.P., Joyner, L.G., and Halenda, P.P. (1951) The determination of pore volume and area distributions in porous substances. I. Computations from nitrogen isotherms. *J. Am. Chem. Soc.*, **73**, 373–380.
59. Giesche, H. (2006) Mercury porosimetry: a general (practical) overview. *Part. Part. Syst. Charact.*, **23**, 9–19.
60. Moro, F. and Böhni, H. (2002) Ink-bottle effect in mercury intrusion porosimetry of cement-based materials. *J. Colloid Interface Sci.*, **246**, 135–149.
61. Khokhlov, A. et al. (2007) Freezing and melting transitions of liquids in mesopores with ink-bottle geometry. *New J. Phys.*, **9**, 272.
62. Salmas, C.E. and Androutsopoulos, G.P. (2000) A novel pore structure tortuosity concept based on nitrogen sorption hysteresis data. *Ind. Eng. Chem. Res.*, **40**, 721–730.
63. Zalc, J.M., Reyes, S.C., and Iglesia, E. (2004) The effects of diffusion mechanism and void structure on transport rates and tortuosity factors in complex porous structures. *Chem. Eng. Sci.*, **59**, 2947–2960.
64. Harkness, I.R., Hussain, N., Smith, L., and Sharman, J.D.B. (2009) The use of a novel water porosimeter to predict the water handling behaviour of gas diffusion media used in polymer electrolyte fuel cells. *J. Power Sources*, **193**, 122–129.
65. Volfkovich, Y.M., Sosenkin, V.E., and Bagotsky, V.S. (2010) Structural and wetting properties of fuel cell components. *J. Power Sources*, **195**, 5429–5441.
66. Caston, T.B., Murphy, A.R., and Harris, T.A.L. (2011) Effect of weave tightness and structure on the in-plane and through-plane air permeability of woven carbon fibers for gas diffusion layers. *J. Power Sources*, **196**, 709–716.
67. Hiramitsu, Y., Sato, H., and Hori, M. (2010) Prevention of the water flooding by micronizing the pore structure of

gas diffusion layer for polymer electrolyte fuel cell. *J. Power Sources*, **195**, 5543–5549.
68. Silberstein, M.N. and Boyce, M.C. (2010) Constitutive modeling of the rate, temperature, and hydration dependent deformation response of Nafion to monotonic and cyclic loading. *J. Power Sources*, **195**, 5692–5706.
69. Smith, T.L. (1973) Physical properties of polymers – an introductory discussion. *Polym. Eng. Sci.*, **13**, 161–175.
70. Lakes, R. (1987) Foam structures with a negative Poisson's ratio. *Science*, **235**, 1038–1040.
71. Solasi, R., Zou, Y., Huang, X., Reifsnider, K., and Condit, D. (2007) On mechanical behavior and in-plane modeling of constrained PEM fuel cell membranes subjected to hydration and temperature cycles. *J. Power Sources*, **167**, 366–377.
72. Tan, J., Chao, Y.J., Van Zee, J.W., Li, X., Wang, X., and Yang, M. (2008) Assessment of mechanical properties of fluoroelastomer and EPDM in a simulated PEM fuel cell environment by microindentation test. *Mater. Sci. Eng. A*, **496**, 464–470.
73. Silberstein, M.N., Pillai, P.V., and Boyce, M.C. (2011) Biaxial elastic–viscoplastic behavior of Nafion membranes. *Polymer*, **52**, 529–539.
74. Tang, Y., Karlsson, A.M., Santare, M.H., Gilbert, M., Cleghorn, S., and Johnson, W.B. (2006) An experimental investigation of humidity and temperature effects on the mechanical properties of perfluorosulfonic acid membrane. *Mater. Sci. Eng. A*, **425**, 297–304.
75. Yoon, W. and Huang, X. (2011) A nonlinear viscoelastic–viscoplastic constitutive model for ionomer membranes in polymer electrolyte membrane fuel cells. *J. Power Sources*, **196**, 3933–3941.
76. Silberstein, M.N. and Boyce, M.C. (2011) Hygro-thermal mechanical behavior of Nafion during constrained swelling. *J. Power Sources*, **196**, 3452–3460.
77. Bauer, F., Denneler, S., and Willert-Porada, M. (2005) Influence of temperature and humidity on the mechanical properties of Nafion 117 polymer electrolyte membrane. *J. Polym. Sci. Part B: Polym. Phys.*, **43**, 786–795.
78. Satterfield, M.B. and Benziger, J.B. (2009) Viscoelastic properties of Nafion at elevated temperature and humidity. *J. Polym. Sci. Part B: Polym. Phys.*, **47**, 11–24.
79. Rand, J.L., Henderson, J.K., and Grant, D.A. (1996) Nonlinear behavior of linear low-density polyethylene. *Polym. Eng. Sci.*, **36**, 1058–1064.
80. Matsuoka, S., Bair, H.E., Bearder, S.S., Kern, H.E., and Ryan, J.T. (1978) Analysis of non-linear stress relaxation in polymeric glasses. *Polym. Eng. Sci.*, **18**, 1073–1080.
81. Chen, X., Yan, L., Wang, Z., and Liu, D. (2011) Out-of-phase thermo-mechanical coupling behavior of proton exchange membranes. *J. Power Sources*, **196**, 2644–2649.
82. Jung, B., Moon, H.M., and Baroña, G.N.B. (2011) Effect of methanol on plasticization and transport properties of a perfluorosulfonic ion-exchange membrane. *J. Power Sources*, **196**, 1880–1885.
83. Majsztrik, P.W., Bocarsly, A.B., and Benziger, J.B. (2008) Viscoelastic response of Nafion. Effects of temperature and hydration on tensile creep. *Macromolecules*, **41**, 9849–9862.
84. Sgreccia, E., Chailan, J.F., Khadhraoui, M., Di Vona, M.L., and Knauth, P. (2010) Mechanical properties of proton-conducting sulfonated aromatic polymer membranes: stress–strain tests and dynamical analysis. *J. Power Sources*, **195**, 7770–7775.
85. Sannigrahi, A., Ghosh, S., Maity, S., and Jana, T. (2010) Structurally isomeric monomers directed copolymerization of polybenzimidazoles and their properties. *Polymer*, **51**, 5929–5941.
86. Poornesh, K.K., Cho, C., Kim, D.Y., and Tak, Y. (2010) Effect of gas-diffusion electrode material heterogeneity on the structural integrity of polymer electrolyte fuel cell. *Energy*, **35**, 5241–5249.
87. Flores, A., Ania, F., and Baltá-Calleja, F.J. (2009) From the glassy state to ordered polymer structures: a microhardness study. *Polymer*, **50**, 729–746.

88. Franceschini, E.A. and Corti, H.R. (2009) Elastic properties of Nafion, polybenzimidazole and poly[2,5-benzimidazole] membranes determined by AFM tip nano-indentation. *J. Power Sources*, **188**, 379–386.
89. Lévesque, D., Legros, N., and Ajji, A. (1997) Ultrasonic determination of mechanical moduli of oriented semicrystalline polymers. *Polym. Eng. Sci.*, **37**, 1833–1844.
90. Roberti, E., Carlotti, G., Cinelli, S., Onori, G., Donnadio, A., Narducci, R., Casciola, M., and Sganappa, M. (2010) Measurement of the Young's modulus of Nafion membranes by Brillouin light scattering. *J. Power Sources*, **195**, 7761–7764.

20
Degradation Caused by Dynamic Operation and Starvation Conditions

Jan Hendrik Ohs, Ulrich S. Sauter, and Sebastian Maass

20.1
Introduction

Fuel cells are a promising technology for clean and efficient energy conversion. The polymer electrolyte membrane fuel cell (PEMFC) is the preferred type for automotive application due to its low operating temperature, fast start-up, and dynamic behavior. Apart from the costs, long-term stability is the main challenge for developing a commercial fuel cell system competitive with the combustion engine. In automotive application, PEMFCs are exposed to highly dynamic operating conditions that affect the long-term stability: according to a US Department of Energy (DOE) report [1], an overall durability of 5000 operating hours is required until 2015. This includes 20 000–30 000 start-up/shut-down cycles [2, 3], with more than 1000 startups from $<-5\,°C$ and more than 75 start-ups from $<-25\,°C$ during the lifetime of an automobile. About 1 million rapid load changes and 300 000 large load cycles [4] occur during operation. The system is confronted with ambient temperatures from -40 to $+45\,°C$ and should be capable of running at low stoichiometry ($\lambda < 2$) in order to maximize the reactant gas efficiency. Regarding these general conditions, the power loss over the lifetime must not exceed 10% or $10\,\mu V\,h^{-1}$ [5, 6].

The reasons for the deterioration of cell performance can be distinguished in reversible and irreversible power loss. Inevitable irreversible performance loss is caused by carbon oxidation, platinum dissolution, and chemical attack of the membrane by radicals [7]. Reversible power loss can be caused by flooding of the cell, dehydration of the membrane electrode assembly (MEA), or change of the catalyst surface oxidation state [8]. If corrective actions are not started immediately, reversible effects lead to irreversible power loss that we define as degradation. In this chapter, we focus on the degradation of the catalyst layer due to undesired side reactions.

The catalyst layer consists of the platinum catalyst, carbon as porous support material and an ionomer coating that is usually a polysulfonic acid, forming a three-phase boundary between the electrically conductive material, the ionic conductive material, and the gas phase. The hydrogen oxidation reaction (HOR)

Fuel Cell Science and Engineering: Materials, Processes, Systems and Technology,
First Edition. Edited by Detlef Stolten and Bernd Emonts.
© 2012 Wiley-VCH Verlag GmbH & Co. KGaA. Published 2012 by Wiley-VCH Verlag GmbH & Co. KGaA.

and oxygen reduction reaction (ORR) occur within the catalyst layer of the anode and cathode side, respectively, as follows:

$$H_2 \rightleftharpoons 2H^+ + 2e^- \quad \varphi_{00} = 0 \text{ V} \tag{20.1}$$

$$O_2 + 4H^+ + 4e^- \rightleftharpoons 2H_2O \quad \varphi_{00} = 1.23 \text{ V} \tag{20.2}$$

In state-of-the-art catalyst layers, fine platinum particles of diameter $d_{Pt} < 4$ nm are dispersed on a porous substrate material in order to increase the platinum surface area per unit volume and reduce costs. In current MEAs, the platinum loading of the anode catalyst layer is about 0.05 mg cm^{-2} of geometric surface area while the cathode catalyst loading is as high as 0.4 mg cm^{-2} due to the sluggish reaction kinetics of oxygen reduction [9]. Platinum alloys such as Pt–Co, Pt–Cr–Ni, and Pt–Ru–Ir–Sn are expected to reduce the platinum loading and enhance activity but are still in the research stage [10].

Carbon is generally used as catalyst support material because of its high electric and thermal conductivity, chemical stability, and porous structure [11]. The catalytic activity of the catalyst layer increases with increasing carbon surface area due to better platinum dispersion. High surface area carbon blacks such as Ketjenblack and Vulcan are therefore preferred in PEMFC application. However, carbon is thermodynamically unstable at normal cathode potentials between 0.5 and 1 V. As shown in Figure 20.1a, carbon is oxidized to carbon dioxide (CO_2) or carbon monoxide (CO) at high electrode potentials whereas it is reduced to methane (CH_4) at low electrode potentials. The following reactions are relevant for fuel-cell operation:

$$C + 2H_2O \rightleftharpoons CO_2 + 4H^+ + 4e^- \quad \varphi_{00} = 0.207 \text{ V} \tag{20.3}$$

$$C + H_2O \rightleftharpoons CO + 2H^+ + 2e^- \quad \varphi_{00} = 0.518 \text{ V} \tag{20.4}$$

The carbon catalyst support reacts to give CO and CO_2, leaving the cell with the gas stream. This results in a reduced electric conductivity of the electrode, the isolation of platinum particles from the solid phase, or agglomeration of platinum, the last effect leading to particle growth [10]. Fortunately, carbon oxidation is strongly kinetically hindered within the normal operating range of PEMFCs ($T < 120\,°C$, $U_{cell} < 1$ V). At high electrode potentials above 0.9 V, carbon oxidation is reported to occur at slow but non-negligible reaction rates [13]. That is why stationary operation at open-cell voltage (OCV) should be avoided. The stability of carbon can be increased by graphitization, reducing carbon oxidation by one order of magnitude [6]. However, graphitization causes additional costs and reduces the surface area of the catalyst support material. Graphitized carbons are used in phosphoric acid fuel cells (PAFCs) because they allow for higher operating temperatures of $T \approx 200\,°C$ [4].

The platinum catalyst is also involved in undesired side reactions. Figure 20.1b shows the equilibrium states of platinum according to Pourbaix [12]. Elemental platinum is stable below the indicated potential limit. At higher potentials the platinum oxides PtO and PtO_2 and also Pt^{2+} ions become thermodynamically

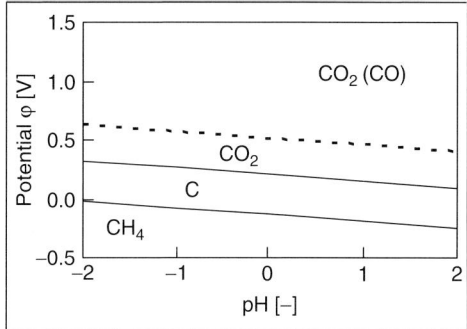

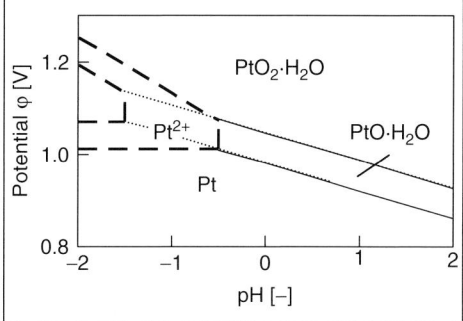

Figure 20.1 Pourbaix diagram of carbon and platinum [12]. In (b), the region for Pt^{2+} ions expands with decreasing Pt^{2+} concentration within the electrolyte, as illustrated by the dashed lines.

stable depending on the pH value. The following reactions are relevant in PEMFC operation [14]. Note that the first two are electrochemical reactions whereas the last two are chemical reactions.

$$Pt \rightleftharpoons Pt^{2+} + 2e^- \quad \varphi_{00} = 1.188 \text{ V} \tag{20.5}$$

$$Pt + H_2O \rightleftharpoons PtO + 2H^+ + 2e^- \quad \varphi_{00} = 0.98 \text{ V} \tag{20.6}$$

$$PtO + 2H^+ \rightleftharpoons Pt^{2+} + 2H_2O \tag{20.7}$$

$$Pt^{2+} + H_2 \rightleftharpoons Pt + 2H^+ \tag{20.8}$$

The formation of a platinum oxide surface layer as given in Eq. (20.6) prevents dissolution or precipitation of platinum according to Eq. (20.5), since dissolution of platinum oxide as shown by Eq. (20.7) is rather slow [14, 15]. Precipitation of Pt^{2+} ions by reaction with hydrogen may occur within the membrane, forming an electrically insulated Pt band [16]. Platinum losing electrical contact with the electrode is referred to as *catalyst islanding* [17]. The electrochemically active platinum surface area decreases because of these mechanisms, resulting in performance loss due to a higher platinum surface specific current density and therefore an increased overpotential.

In addition to carbon corrosion and platinum dissolution, water electrolysis is a further undesired side reaction in PEMFCs. It is the back reaction of the oxygen reduction that is given in Eq. (20.1). Water electrolysis does not cause cell degradation directly but results in dehydration of the MEA, temporarily reducing the electrolyte conductivity. According to Eq. (20.2), it occurs at electrode potentials $\varphi > 1.23$ V.

As described above, undesired side reactions resulting in cell degradation occur at high electrode potentials. In the next section, we introduce measurement techniques that allow the measurement of the electrode potentials, detection of inhomogeneous cell properties, and characterization of the cell. Cell degradation can be detected and evaluated by these techniques. Not only are high electrode potentials leading to undesired side reactions present during OCV operation, but undesired side reactions are also observed under dynamic operation and under starvation conditions. The relevant mechanisms are considered in Sections 20.3 and 20.4. Finally, we propose mitigation strategies in order to optimize the long-term stability of PEMFCs.

20.2
Measurement Techniques

In this section, measurement techniques for the detection of starvation conditions and for cell characterization during the degradation processes are presented.

20.2.1
Reference Electrode

Undesired side reactions occur at high electrode potentials in PEMFCs. In this section, reference electrodes are introduced as a measurement technique in order to sense the electrode potentials and detect critical cell states.

The cell voltage U_{cell} is defined as the potential difference between the cathode and the anode. It is usually measured during fuel-cell operation. The potential difference between the electrode and the electrolyte, which is called the anode or cathode potential in the following, is responsible for the electrochemical reaction occurring within the catalyst layers but cannot be measured directly. In the further text, we use electrolyte and membrane as equivalent expressions. While the electrode potential can be sensed from the bipolar plates, it is not feasible to sense the membrane potential directly, since each measurement equipment forms an interface between the membrane and the metal contact. Two methods for the installation of a reference electrode within the cell have been discussed in the literature, namely the reverse hydrogen electrode (RHE) [18] and the dynamic hydrogen electrode (DHE). In addition to cell internal methods, a conventional reference electrode can be connected to the cell from the outside via an electrolyte junction [19].

Several authors have used an electrically insulated part of the catalyst layer next to the anode gas inlet as an RHE [20–22]. When hydrogen is fed to the cell, it passes the reference electrode before entering the active area. At the same time, the reference electrode is in contact with the membrane, resulting in a stable reference signal. Alternatively, the reference electrode can be laminated on to a membrane strip outside the active area [23, 24]. In this case, the reference electrode needs its own hydrogen supply. This can cause difficulties concerning the gas tightness of the system.

To overcome these problems, Giner [25] first proposed the DHE as a reference electrode in 1964. In this setup, the membrane is contacted by two platinum wires at a distance of 1–4 mm [26, 27]. A small electric current between 6 and 100 µA [26–28] is applied to these wires in order to produce hydrogen and oxygen from water within the membrane via electrolysis. The platinum wire where hydrogen evolves can be used as reference electrode since it is in contact with H^+ ions from the membrane and hydrogen gas that is produced at the platinum surface. In contrast to the RHE, the DHE is not in equilibrium owing to the overpotential created by the enforced electrolysis current. However, the offset to equilibrium is small owing to the fast reaction kinetics of hydrogen evolution and it is constant if a constant electric current is applied.

With respect to the position of the reference electrode, two types of DHE can be distinguished, as illustrated in Figure 20.2, which are called sandwich configuration and edge configuration in the following. In the edge configuration, the reference is placed on a membrane strip outside the active area [26–32]. The strip has to be kept humidified in order to prevent local dehydration. The advantage of this position is the easy installation and only slight modifications from the standard buildup of the cell. However, several authors have reported inhomogeneities in the membrane potential distribution at the edge of the active area [33–37]. Electrode misalignment of more than three times the membrane thickness (~50–150 mm) intensifies this effect. Hence the reference signal from the edge of the membrane is not necessarily consistent with the potential within the active area. This is also

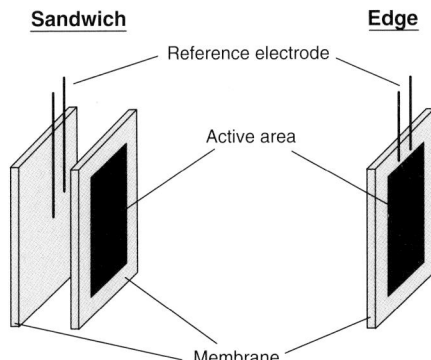

Figure 20.2 Dynamic hydrogen electrode (DHE) as reference electrode in sandwich configuration and edge configuration.

valid for the signal of an RHE and an external reference electrode, which are always located outside the active area of the cell.

In the sandwich configuration, the MEA consists of two membrane layers with the reference electrode sandwiched in between [28]. This requires a minimum membrane thickness and therefore may lead to an increased ohmic resistance compared with state-of-the-art ultrathin membranes. There is a uniform membrane potential distribution at the location of the reference electrode within the active area that decreases from the anode to the cathode. Therefore, the reference signal is not disturbed by edge effects. If a DHE in the sandwich configuration is applied, the electrolysis is compensated by product water from the cathode preventing local membrane dehydration. Alternatively, Dolle et al. [38] and Büchi and Scherer [39] used single copper and gold wires as reference electrodes in the sandwich configuration without applying an electric current. In this case, the reference signal might be unstable since the surface of the electrode comes in contact with hydrogen and oxygen permeating through the not perfectly gas-tight membrane.

20.2.2
Current Density Distribution

Undesired side reactions occur under starvation conditions, as discussed in detail in Section 20.4. These cell states are caused by an inhomogeneous distribution of reactant gases, temperature, and local water content across the active area during fuel-cell operation. Measurement of the current density distribution allows the detection of these critical conditions. However, gradients of the electrochemical reaction rates are mostly compensated by in-plane currents within the catalyst layer, gas diffusion layer, and bipolar plates. Therefore, for measurement of the current density distribution, the local current flow has to be separated into individual pathways.

Stumper et al. [40] fabricated an MEA with electrically insulated circular sections of the catalyst layer in order to measure the current density of these subcells individually. Cleghorn et al. [41] segmented the catalyst layer, gas diffusion layer, and collector plate of the anode by embedding them in a silicon gasket in a grid structure. Hakenjos et al. [42] used only a segmented gas diffusion layer in order to minimize in-plane currents since the gas diffusion layer has a high ratio of in-plane to through-plane conductivity.

Segmentation of the entire electrode including catalyst layer, gas diffusion layer, and bipolar plate is complex and the integration of insulating material for segmentation of the MEA involves significant modifications from the standard buildup of a PEMFC, influencing its performance. Therefore, most authors do not modify the catalyst and gas diffusion layer but segment the bipolar and current collector plate [43–53]. In this case, it has to be considered that the measured current density distribution is smoothed due to cross-currents within the MEA. Natarajan and Nguyen [43] and Sun et al. [44, 45] machined their flow field in an electrically insulated plate. The flow field area was segmented in stripes conducting the electric current sideways out of the cell. In order to increase spatial resolution,

Noponen and co-workers [46, 47] and Gülzow et al. [48] used a collector plate consisting of electrically insulated metal blocks measuring the current flow across each segment individually. Identical contact pressure for each block has to be assured in order to provide uniform contact resistance.

For measurement of the current density distribution within a stack, it is not feasible to conduct the electric current out of the cell to an external measurement system. To overcome this problem, Stumper et al. [40] segmented a collector plate into 121 graphite elements used as passive resistors. The current flow through these resistors causes a voltage drop that correlates with the electric current. An advancement of this method is the printed circuit board (PCB) approach, replacing the current collector and flow field plate [51–53]. The PCB surface consists of current collector segments flexible in shape and size whereas the flow field channels are machined directly into the plate. Passive resistors or coils [54] are integrated into the plate in order to sense the voltage drop across each segment. Advantages of the PCB technique are the high spatial resolution, real-time measurement, reduced thickness of the plate, and flexible position within a stack.

Alternatively, a noncontact magnetic loop array containing Hall sensors can be used to measure the current density distribution [55–57]. Each segment of this array consists of an electric conductor inducing a voltage within a ferromagnetic ring. From the voltage signal, the electric current in each segment can be determined. This technique is also suitable for application in a stack and it does not cause any additional resistance to the system. Disadvantages are the high cost of the Hall sensors and the limited spatial resolution, since Hall sensors influence each other if the segments are too small.

20.2.3
Cyclic Voltammetry

Cell degradation results in performance loss of the fuel cell. Appropriate measurement techniques are required in order to characterize the cell and determine the prevailing degradation process. Cyclic voltammetry (CV) is a common diagnostic tool for the characterization of electrochemical cells [18, 58]. With respect to PEMFCs, it provides information about the electrochemical active area, the double-layer characteristics, and the hydrogen fuel crossover through the membrane.

During CV, fully humidified hydrogen and nitrogen are fed to the anode and cathode, respectively. The cathode potential is swept reversely at a constant potential ramp rate between the potentials of hydrogen and oxygen evolution while the current response of the cell is measured. The cathode is the working electrode whereas the anode is used as counter and reference electrode based on the assumption that its polarization is small with respect to the electric current flow. Figure 20.3 shows a typical cyclic voltammogram of a PEMFC. The upper and lower halves of the plot are called the anodic and cathodic sweep, respectively. At potentials in the range $\varphi_C = 400$–600 mV, the anodic and cathodic parts of the cyclic voltammogram are close to each other and the current is almost constant. The platinum surface area within the catalyst layer is free of adsorbates in this

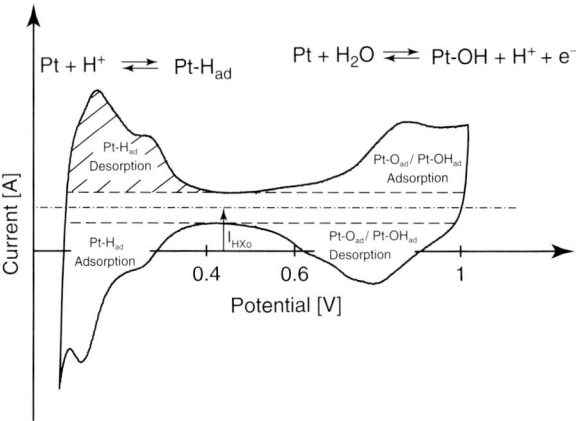

Figure 20.3 Cyclic voltammogram of a PEMFC.

range. The measured electric current can be attributed to charge or discharge the Helmholtz double layer of the electrode. For $\varphi_C > 600\,\text{mV}$ in the anodic part, platinum is oxidized, forming Pt–OH. At the beginning of the cathodic sweep, a negative current is measured due to the reduction of platinum oxides formed in the anodic part. For $\varphi_C < 400\,\text{mV}$, the adsorption of hydrogen starts, forming Pt–H_{ad}. It can be assumed that at the minimum potential, the platinum surface area is completely covered with hydrogen adsorbates. By integration of the electric current measured in the hydrogen desorption peak that is highlighted by the shaded area in the plot, the available electrochemically active platinum surface area of the electrode can be determined [59]. If compared with CV in a three-electrode setup, the cyclic voltammogram of a fuel cell shows a shift to positive currents over the whole potential range. This effect results from hydrogen permeation through the membrane. The permeated hydrogen is oxidized at the cathode due to the positive electrode potential $\varphi_C > 0\,\text{V}$. The amount of hydrogen crossover can be determined from the hydrogen permeation current I_{HXo}.

20.3
Dynamic Operation at Standard Conditions

Various degradation mechanisms based on undesired side reactions of the platinum catalyst and carbon support were introduced in Section 20.1. In this section, we focus on degradation under dynamic operation at standard conditions.

Platinum has a low but non-negligible solubility within the polysulfonic acid membranes that are usually used in PEMFCs. The equilibrium concentration increases with rising temperature and electrode potential, as shown in Figure 20.1. Carbon, which is usually used as catalyst support, is thermodynamically unstable under normal operating conditions of the cathode [$\varphi_C > 0.208\,\text{V}$, see Eqs. (20.3) and (20.4)]. Nonetheless, it is one of the most stable catalyst support materials for

PEMFC application owing to very slow kinetics of carbon oxidation. By means of the electrochemical quartz crystal microbalance (EQCM) on single carbon electrodes, Hung et al. [11] found that for electrode potentials $\varphi_C > 0.208$ V, first carbon surface oxides ($C_{surf}O$) are formed. With respect to carbon blacks (Vulcan XC-72), they observed CO_2 production at a potentials of $\varphi > 1.05$ V. In contrast, if graphitized carbon was used, they observed formation of CO_2 at $\varphi > 1.62$ V, underlining the stability of heat-treated carbons. During fuel-cell operation, CO_2 evolution from the carbon support (Ketjenblack) was detected for $U_{cell} > 0.8$ V while the reaction rate increased exponentially with rising potential [60]. Several authors have reported that carbon corrosion is catalyzed by platinum [61, 62]. Takeuchi et al. [63] proposed reaction pathways for this catalysis. In dynamic operation, carbon corrosion can be enhanced due to the reaction of carbon with oxygen adsorbates that are desorbing from the platinum surface [64].

Ferreira et al. [65] compared cell degradation under stationary operation at OCV with cell degradation under cathode potential cycles between 0.6 and 1 V. They found that loss of the electrochemically active platinum surface area was accelerated by one order of magnitude under potential cycling conditions. Young et al. [17] cycled a cell between 0.15 and 1.5 V for 30 h. This treatment led to a reduction of the cathode catalyst layer thickness to one-third of its original value and a reduction of the electrochemically active platinum surface area by 55% whereas no degradation was observed at the anode side. Chizawa et al. [66] cycled a cell between OCV and 0.2 A cm^{-2} while measuring the CO_2 content in the cathode exhaust gas. Each load step from 0.2 A cm^{-2} to OCV was followed by a CO_2 concentration peak decaying after a few seconds. Yasuda et al. [67] investigated the influence of the upper and lower potential limits during load cycling. They found that only the upper limit is significant with respect to cell degradation. At upper limits of 1 V or higher, platinum particles are detected within the membrane close to the cathode. It is assumed that platinum is dissolved in the membrane phase at such high potentials according to Eq. (20.5) and precipitate by reaction with permeating hydrogen according to Eq. (20.8). Further experiments showed that if the anode is fed with nitrogen instead of hydrogen, no platinum particles are observed within the membrane but instead the platinum loading in the anode catalyst layer increases. This means that Pt^{2+} ions are transported through the membrane to the anode if hydrogen for precipitation is not available. Makharia et al. [4] compared the stability of Pt/C and platinum alloys (PtCo/C) under voltage cycling between 0.7 and 0.9 V. Their results showed that platinum alloys are more stable than standard Pt/C under these conditions. The stability of Pt/C can be increased by increasing the particle size from 2.5 to 4.5 nm, which is a common value for Pt alloys. Further improvements resulting in a stability comparable to that of Pt alloys were achieved when Pt/C was additionally heat treated. The improved durability of Pt alloys is therefore not explained by intrinsically higher stability but by its large particle size and heat treatment during the fabrication process. However, heat-treated Pt/C does not meet the mass activity of state-of-the-art Pt alloys.

The effect of temperature on degradation during potential cycles between 0.87 and 1.2 V was investigated by Bi and Fuller [68]. Experiments were conducted at 40, 60, and 80 °C, each for 7000 cycles. The results showed that loss of the electrochemically active platinum surface area (ECA) and particle growth are more pronounced at higher temperatures. The ECA loss rate can be described by an Arrhenius approach.

By means of a transient 1D model, Darling and Meyers [15] studied the kinetics of platinum dissolution under constant load and load cycling conditions. They followed Eqs. (20.5–20.7) assuming that the oxidation of Pt to PtO was the dominant effect and assuming slow kinetics for the chemical dissolution of platinum. The formation of platinum oxide was allowed to exceed one monolayer on the platinum particles, leading to a surface coverage of $\theta > 1$. This assumption is based on experimental results showing that $\theta = 1$ is reached at an electrode potential of -1.15 V [69–71]. Once a PtO monolayer is formed, further dissolution or precipitation of platinum is inhibited.

The equilibrium potential of the electrochemical dissolution depends on the radius r of the monodisperse platinum particles:

$$\varphi_{00} = \varphi_{00}^{\text{bulk}} - \frac{\sigma_{Pt} M_{Pt}}{2 F r \rho_{Pt}} \tag{20.9}$$

where F is the Faraday constant, σ_{Pt} the surface tension, M_{Pt} the molar mass, and ρ_{Pt} the density of the platinum particles. According to Eq. (20.9), platinum dissolution is faster from smaller particles whereas precipitation is faster on larger particles, resulting in an overall particle growth. This process is called *Ostwald ripening* [10]. In order to account for this effect, Darling and Meyers [72] and also Bi and Fuller [14] expanded the model to a bimodal particle size distribution.

The results for constant potential holds over 400 h show that the loss of electrochemically active platinum surface area is negligible at 0.87 and 1.2 V, respectively, whereas it is significant at 1.05 V [72]. The reason is that at 0.87 V the reaction kinetics are slow whereas at 1.2 V a PtO monolayer is formed, blocking any dissolution or precipitation. When the electrode is held at intermediate potentials, significant catalyst degradation occurs.

An even greater loss of the electrochemically active platinum surface area within a few hours is observed if the potential is cycled between 0.87 and 1.2 V. This is explained by the delayed PtO formation during potential transients as depicted in Figure 20.4. The diagram shows a linear potential sweep between 0 and 1.2 V, the corresponding actual PtO coverage, and the equilibrium coverage. At rising potentials, the actual oxide coverage follows the equilibrium coverage with a significant time shift. This leads to $\theta < 1$ at high potentials and consequently to platinum dissolution during the complete positive potential sweep since there is no inhibition by the platinum oxide layer. Shortly after the beginning of the negative potential sweep at about 120 s, the PtO surface coverage reaches its maximum slightly below $\theta = 1$ and stays there for several tens of seconds. This hinders Pt^{2+} ions from being deposited on the PtO-covered catalyst particles. Since the equilibrium concentration of Pt^{2+} within the membrane decreases with decrease in potential (see Figure 20.1), platinum is deposited elsewhere in the MEA, for example, within the membrane, leading to isolation of catalyst particles from the

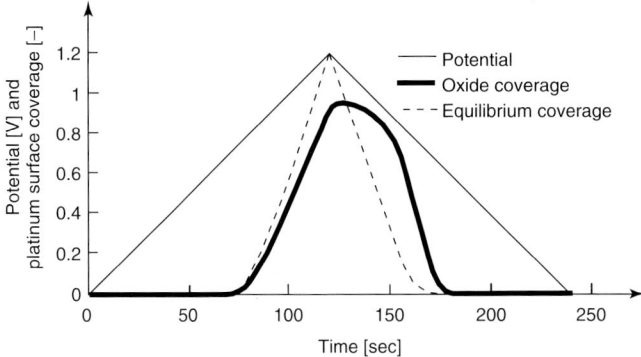

Figure 20.4 Comparison of the platinum oxide coverage and the equilibrium coverage during a linear potential sweep. Figure modified from Darling and Meyers [72], reproduced by permission of The Electrochemical Society.

electrode. Since the system never reaches equilibrium, the formation of a protective PtO monolayer is prevented as long as the potential is swept continuously.

In summary, it can be concluded that platinum dissolution and catalyst particle growth are particularly fast during potential transients because of the delayed PtO formation. Furthermore, accelerated oxidation of the carbon support is observed under cycling conditions. These mechanisms result in cell degradation lowering the durability of the fuel cell significantly.

20.4
Starvation Conditions

When the anode or cathode is insufficiently supplied with reactant gases, we speak of starvation conditions. These conditions can lead to severe cell degradation, as will be shown in the following. Starvation on the anode side is referred to as *hydrogen starvation* whereas starvation at the cathode side is referred to as *oxygen starvation*. With respect to hydrogen starvation, three cases can be distinguished: overall hydrogen starvation, hydrogen starvation during start-up/shut-down, and local hydrogen starvation.

20.4.1
Overall Hydrogen Starvation

Overall hydrogen starvation is defined as a state where the hydrogen supply is not sufficient to maintain the current demand. Provided that there are normally operating cells in a stack or there is an external power supply, the current is forced through the hydrogen-starved cell. As a consequence, alternative oxidation reactions have to take place in order to maintain the required flux of electrons and protons. Possible reactions are the oxidation of the carbon support and the electrolysis of water. These species do not react spontaneously. Therefore, in the case of overall hydrogen starvation with carbon or water present at the anode, the cell voltage breaks down to negative values. This state is called *cell reversal*.

For simulation of hydrogen starvation conditions, Baumgartner et al. [19] diluted the fuel supply with nitrogen in order to decrease the anode stoichiometry λ_A. For $\lambda_A < 1$, the cell voltage became negative and CO_2 was detected in the anode exhaust gas. Knights et al. [73] detected oxygen in the anode exhaust gas as soon as the cell voltages fell below $U_{cell} = -0.55$ V, underlining that water has been oxidized. Mitsuda and Murahashi [74] reporteed on cell reversal as soon as the hydrogen consumption exceeded 97% ($\lambda_A < 1.03$). In this state, they detected CO_2 and also small amounts of CO in the anode exhaust gas. Taniguchi et al. [75, 76] conducted experiments using an RHE as reference electrode that was contacted with the membrane by an electrolyte junction. Under overall fuel starvation, the cell voltage became negative and the potential difference between the anode and membrane exceeded 1.5 V whereas the cathode potential showed only minor variations. Under these conditions, tremendous power loss and particle growth of the platinum catalyst at the anode were observed after only 7 min of hydrogen starvation. Liang [22] integrated two DHEs as reference electrodes in a sandwich configuration next to the cell inlet and outlet. For measurement of the current density distribution, the cathode catalyst layer and collector plate were segmented. Under normal operating conditions ($\lambda > 2$), gradients in the electrode polarization or current distribution were not observed. However, if the anode stoichiometry was decreased to $\lambda_A = 0.91$, the electric current increased near the fuel inlet whereas it fell close to zero near the outlet. The anode potential rose to 1 and 1.8 V at the inlet and outlet, respectively, and the cell voltage became negative ($U_{cell} = -1.1$ V). In another experiment conducted by the same authors, the anode stoichiometry was reduced stepwise from $\lambda_A = 1.09$ to 0.55. The results showed that the lower the stoichiometry, the more negative the cell voltage became ($U_{cell} = -0.5$ to -2 V) and the higher the potential difference between the anode and membrane became ($\varphi_A = 0.7$–2.2 V). They found that during cell reversal, the CO_2 content in the anode exhaust gas remained at a constant level whereas the oxygen content increased with decreasing stoichiometry.

Concluding the observations from the experiments above, a cell state as depicted in Figure 20.5 can be deduced. The upper part shows a cell under insufficient hydrogen supply with respect to the given current demand I_{cell}. The rear part of the anode is starved of hydrogen. In order to meet the current demand, electrons and protons are provided from carbon oxidation or water electrolysis supporting the cell current. The lower part shows the potential distribution along the electrodes and membrane. The electrodes can be regarded as equipotential areas due to their high in-plane electric conductivity. In order to maintain the cell current, significant carbon corrosion reaction rates are required, leading to a potential difference between the anode and membrane of $\varphi_A - \varphi_M > 1$ V [11]. At the cathode, oxygen has to be reduced, which requires a potential difference between the cathode and membrane of $\varphi_C - \varphi_M < 1$ V owing to the high overpotential of the oxygen reduction reaction given in Eq. (20.2). Therefore, in order to balance the reaction

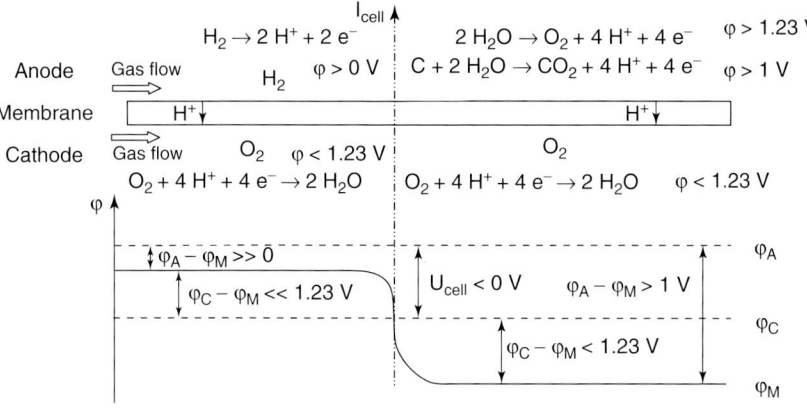

Figure 20.5 Gas and potential distribution within a PEMFC at overall hydrogen starvation.

rates on the anode and cathode, the cell voltage has to become negative. The lower the fuel stoichiometry, the higher are the reaction rates of water electrolysis and carbon oxidation and the more negative is the cell voltage. Carbon is oxidized at $\varphi_A - \varphi_M > 1\,\text{V}$ and water at $\varphi_A - \varphi_M > 1.23\,\text{V}$.

At the transition to the front part of the cell where hydrogen is available, a shift to higher membrane potentials can be observed. This can be reasoned by considering that hydrogen oxidation at the anode is coupled to oxygen reduction at the cathode by the flux of protons through the electrolyte. By lifting the membrane potential in the front part, hydrogen oxidation there becomes slower and oxygen reduction becomes faster until both reaction rates are balanced. Because of the relatively low in-plane conductivity of the membrane, a gradient in the membrane potential can be maintained along the cell.

As a summary, we conclude that overall hydrogen starvation results in negative cell voltage and carbon corrosion in addition to water oxidation within the anode catalyst layer. Degradation reaction rates increase with decreasing fuel stoichiometry.

20.4.2
Hydrogen Starvation During Start-up/Shut-down

During shut-down of an automotive fuel-cell system, both the anode and cathode fill with air due to the abreaction of residual hydrogen and diffusion of air into the cell from the environment. At restart, a transient H_2/air front is formed when fresh hydrogen is fed into the anode for restart. It takes from several hundred milliseconds to a few seconds [3, 6] until the entire anode is filled with hydrogen, depending on the gas velocity and channel volume.

In order to investigate this start-up/shut-down process, Yu et al. [3] purged the anode of a test cell with air and subsequently supplied fresh hydrogen for restart. A tremendous loss of cell voltage, particularly at high current density, was observed after 200 cycles. The mass transport resistance was increased in consequence of carbon corrosion. They stated that there is a linear relation between the residence time of the H_2/air front and the performance loss. Kim et al. [77] analyzed the influence of gas humidity on degradation during start-up/shut-down cycles. They found that performance loss and reduction of the electrochemically active platinum surface area increased with higher gas humidity. This effect of the water content was reasonable since carbon is oxidized by reaction with water according to Eqs. (20.3) and (20.4). Shen et al. [21] integrated DHEs as reference electrodes into a test cell at the inlet and outlet in a sandwich configuration. A temporary difference in the membrane potential between the inlet and outlet of up to 0.8 V was detected after switching the anode supply from oxygen to hydrogen. This potential difference disappeared as soon as the entire anode was filled with hydrogen. Kim et al. [78] conducted experiments using a cell with several DHEs in an edge configuration. A peak in the cathode potential of about 1.4 V was observed when the anode supply was switched from hydrogen to air or vice versa. At the same time, a peak in the CO_2 concentration of the cathode exhaust gas was detected, indicating carbon oxidation at the cathode.

In order to explain these observations, Reiser et al. [79] proposed the reverse current decay (RCD) mechanism as explained in the following and illustrated in Figure 20.6. The upper part shows a transient state where the first half of the anode is filled with hydrogen and the second half is filled with air. Such a state is observed when hydrogen enters an air-filled cell during start-up. The anode potential drops to zero with respect to the membrane potential owing to the fast reaction kinetics of the hydrogen oxidation. The lower left-hand side in Figure 20.6 shows the corresponding potential distribution resulting in an OCV of $U_{cell} \approx 1$ V. Electrons move freely within the highly conductive electrodes, compensating electric potential gradients. Hence the electrodes can be considered as equipotential areas. In the rear part of the anode that is still filled with air in this transient state, the electrochemical system reacts by a drop in the membrane potential. This results in a high cathode potential with respect to the membrane in the hydrogen-starved region, leading

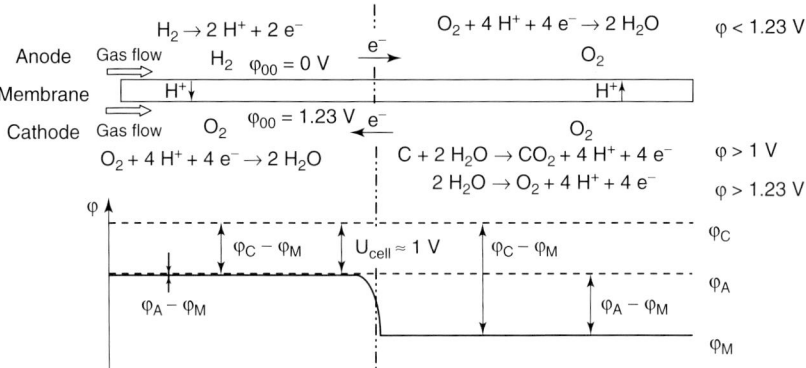

Figure 20.6 Gas and potential distribution during startup of a PEMFC resulting in the reverse current decay mechanism.

to carbon corrosion and water electrolysis, as shown on the right-hand side of Figure 20.6. Carbon corrosion and water electrolysis at the cathode side are coupled to oxygen reduction at the anode side by the flux of protons through the membrane. The membrane potential therefore decreases to a level balancing the reaction rates of the anode and cathode reactions. Note that there are anodic reactions at the cathode and cathodic reactions at the anode in the hydrogen-starved region of the cell, resulting in a locally reversed current. Electrons involved in the electrochemical reactions are transported in-plane within the electrodes while the overall cell current is zero. There is no external driving force necessary for this mechanism.

The mechanism also occurs during shut-down of the cell. The H_2/air front is created by permeation of air into the anode from the outside or by an air purge of the anode in order to remove residual hydrogen [21, 80, 81]. Cell degradation during shutdown may be accelerated, since the cell is still at the operating temperature where reaction kinetics are particularly fast [6].

In order to support their theory of an RCD mechanism, Reiser et al. [79] conducted experiments supplying half of the anode with hydrogen while the other half was fed with nitrogen. The entire cathode was supplied with air. A small amount of oxygen was present at the anode side due to permeation of oxygen through the membrane. Severe cell voltage loss was observed after 5 h of operation at OCV. The cathode catalyst layer was thinned by 50% in the region starved of hydrogen whereas degradation was not observed in the hydrogen-rich part of the cell. Farooque et al. [82] chose a simular experimental approach by supplying the anode of a PAFC partially with hydrogen and partially with air. They also detected severe degradation of the cathode catalyst layer in the hydrogen-starved region of the cell after 20 h of operation. Tang et al. [83] experimentally proved the RCD mechanism by electrically connecting the anodes and cathodes of two cells. The anode/cathode exhaust gas of the first cell was fed into the anode/cathode inlet of the second cell and the electric current between the anodes of the first and second cells was measured by a multimeter. When the anode feed gas was switched from hydrogen to air, an electron flow from the second to the first anode could be measured. After switching the anode supply back to hydrogen, an electron flow in the reverse direction was measured. As soon as the entire anodes of both cells were filled with hydrogen or air, the electronic current between the two cells stopped. The detected flow of electrons matched the in-plane electron transport expected from the RCD mechanism (see Figure 20.6). Siroma et al. [84] measured the effect of the RCD mechanism on the current density distribution by segmentation of the anode current collector plate and by integration of an RHE as reference electrode in the edge configuration. After switching the anode supply from air to hydrogen, a transient positive electric current was detected in the front part of the cell whereas a negative current was detected in the rear part of the cell, as expected from the RCD mechanism. At the same time, the cathode potential increased to 1.5 V.

Meyers and Darling [16] built a transient 1D model for the simulation of cell start-ups. For their cell design and start-up/shut-down conditions, they found that 0.002% of the catalyst support was lost in each cycle. Extrapolation indicated that all of the carbon was oxidized after 5500 cycles. Takeuchi and Fuller [85] simulated a sharp H_2/air front at the anode of a cell by means of a stationary 2D model. In the hydrogen-filled part of the cell, the electrolyte potential decreased linearly from the anode to the cathode. A more complex potential distribution was observed at the transition from hydrogen to air. The isopotential lines bent and become almost perpendicular to the electrodes at the H_2/O_2 front, resulting in a flux of protons from the H_2-rich to the H_2-starved part of the cell. At a certain distance to the gas transition, the membrane potential gradient was reversed and protons were transported from the cathode to the anode in the hydrogen-starved region of the cell. The authors stated that the amount of oxygen in the anode catalyst determines the amount of carbon corrosion. The higher the partial pressure of O_2 at the anode, the more CO_2 evolution at the cathode is predicted, since the reaction rates on anode an cathode are coupled.

The described experiments and simulations support Reiser et al.'s theory of the RCD mechanism. This mechanism occurs during start-up and shutd-own of the cell provided that the anode is not completely filled with hydrogen or air. Carbon and water are oxidized in the cathode catalyst layer, resulting in a reverse current and cell degradation.

20.4.3
Local Hydrogen Starvation

Local hydrogen starvation is defined as a state where the hydrogen concentration locally drops to zero while the global hydrogen supply is sufficient to maintain the global cell current. In contrast to the start-up/shut-down process, local hydrogen starvation is also possible in stationary states due to an inhomogeneous gas distribution across the active area. Local hydrogen starvation occurs either during operation at fuel stoichiometry close to $\lambda_A = 1$ or due to partial blockage of the gas pathways within the MEA.

Several authors analyze cell operation at an anode stoichiometry close to $\lambda_A = 1$ [19, 27, 86–88]. Mitsuda and Murahashi [86] conducted experiments with a PAFC containing four RHE reference electrodes in an edge configuration. They observed that anode and cathode potentials near the cell outlet increased as soon as the hydrogen utilization exceeded 90% ($\lambda_A = 1.11$). Overall hydrogen starvation resulting in a negative cell voltage as described in Section 20.4.1 was not observed until the hydrogen utilization exceedede 97% ($\lambda_A = 1.03$). Lauritzen et al. [27] integrated two DHEs as reference electrodes at the inlet and outlet regions by removing a portion of the cathode and anode catalyst layer from the active area. Whereas the anode stoichiometry was slowly reduced to $\lambda_A = 1.1$, the anode and cathode potentials at the cell outlet increased to about 0.5 and 1.3 V, respectively, and CO_2 was detected in the cathode exhaust gas. Baumgartner et al. [19] integrated four reference electrodes into a PEMFC contacted with the membrane by an electrolyte junction. While the anode outlet was kept closed, a constant current was drawn from the cell and a constant pressure was applied at the anode inlet in order to replace the consumed hydrogen. Air and water accumulated in the rear part of the anode due to permeation through the membrane. Starting at a fuel stoichiometry of $\lambda_A = 2.2$, the cell voltage collapsed and a significant amount of CO_2 was detected in the cathode exhaust gas after 300 s at 0.4 A cm^{-2}. At the same time, both electrode potentials rose and the cathode potential exceeded 1 V. Experiments were stopped when the cell voltage fell below 0.4 V in order to avoid cell reversal. Liu *et al.* [87] measured the current density distribution by means of a segmented current collector plate at the anode side. They reduced the cell voltage stepwise while the hydrogen feed gas flow was kept constant. The local current density in the outlet region dropped to zero with decreasing cell voltage whereas the cell performance in the inlet region was not affected. Note that the stoichiometry of a single cell was always $\lambda \geq 1$ in potentiostatic operation since the cell current was not fixed. In our own work [88], we simulated starvation conditions in potentiostatic operation by means of a steady-state 2D model. Hydrogen was diluted stepwise with nitrogen in order to simulate the conditions in an anode recycle loop. A decrease in the membrane potential was detected as soon as the hydrogen concentration fell to zero near the cell outlet. Carbon corrosion and water electrolysis occurred at the cathode side. It was shown that the corrosion rates were independent of the cell voltage but depended on the permeation flux of oxygen through the membrane.

From the results presented above, we can conclude that the RCD mechanism as explained in Section 20.4.2 is also present at local hydrogen starvation. In stoichiometric operation, the hydrogen concentration falls to zero close to the cell outlet. An H_2/air front is formed at the anode side due to permeation of oxygen and nitrogen through the membrane. The decrease in the membrane potential in the hydrogen-starved part of the cell results in carbon corrosion and water electrolysis at the cathode catalyst layer. In contrast to start-up/shut-down procedures, cell degradation is limited by the permeation flux of oxygen through the membrane since the reaction rates on the anode and cathode are coupled. Local hydrogen starvation can exist at stationary operation states resulting in severe cathode catalyst degradation over time. This state is unstable and it turns into cell reversal as soon as the stoichiometry falls below $\lambda_A = 1$ [19, 74].

Local hydrogen starvation not only occurs at operation close to $\lambda_A = 1$ as shown above, but it can also be caused by local blockage of the gas pathways at the anode side. Accumulation of liquid water within the diffusion media or a foreign object in the gas channel can prevent hydrogen from being evenly distributed over the entire catalyst layer during operation of a fuel cell [89]. In order to analyze experimentally local blocking of the anode under well-defined conditions, Liu *et al.* [90] used an anode flow field plate where there were no channels machined into one-third of the plate in the middle of the active area. The limiting current density in the hydrogen-blocked region of the cell was severely reduced after 16 h of operation whereas it was unaffected in the rest of the active area. Scanning electron micrographs showed that the cathode catalyst layer is less porous and thinned to 40% of its initial thickness in the H_2-starved region of the cell. Patterson and Darling [91] impregnated 12.5% of the anode gas diffusion layer with a circular Kynar film in order to block the active area from hydrogen. The catalyst layer in the H_2-starved region was thinned by 50% and appeared brighter due to a higher

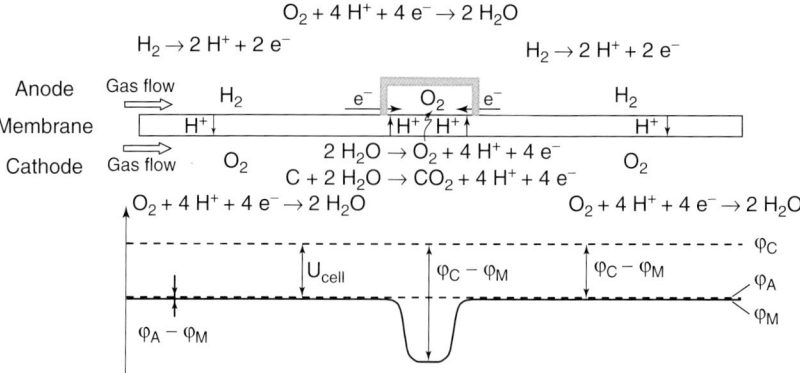

Figure 20.7 Gas and potential distribution at local hydrogen starvation resulting in the reverse current decay mechanism.

Pt/C ratio after 100 h of operation. They claimed that cathode potentials above $\varphi_C > 1.4\,\text{V}$ are required in order to result in the detected degradation. It is noticeable that degradation was less severe at the border area of the impregnated region than at its center.

From the experimental results described above, the potential distribution under local hydrogen starvation can be deduced qualitatively as depicted in Figure 20.7. The upper part shows a cell that is operated at a sufficient stoichiometry of hydrogen and air. A small part in the middle of the anode gas diffusion medium is blocked for gas transport. Consequently, this region is starved of hydrogen. Both oxygen and nitrogen accumulate there due to permeation through the membrane. The lower part shows the corresponding electrode and membrane potential profile. There is a significant decrease in the membrane potential and therefore a peak in the cathode potential in the hydrogen-starved part of the cell. The RCD mechanism as explained in Section 20.4.2 occurs in the hydrogen-starved region, resulting in local carbon corrosion and water electrolysis at the cathode side. The corrosion current is limited by the amount of oxygen at the anode side since the reaction rates on anode and cathode are coupled by the proton transport.

By means of a 1D model, Gu et al. [92] found that a minimum extension of several millimeters is necessary for a blocked area in order to exceed cathode potentials of 1 V leading to local degradation. Fuller and Gray [2] use a stationary 2D model along the channel and in the through-plane direction for simulation of a sharp H_2/air front within the anode catalyst layer. They found that close to the H_2/air transition, electrons and protons for oxygen reduction are provided by in-plane transport from the H_2-rich to the O_2-rich region of the anode. Further away from the transition, the in-plane conductance of the membrane is too low to meet the proton demand. Missing protons are provided by the RCD mechanism, leading to carbon corrosion and water electrolysis at the cathode. In accordance with Gu et al.'s findings [92], Fuller and Gray [2] stated that the hydrogen-starved region has to exceed a certain geometric extension in order to result in cell degradation, depending on the membrane thickness and conductivity. Degradation at the center of the starved region is more severe than degradation close to the H_2/air front.

In summary, we conclude that the RCD mechanism also occurs under local hydrogen starvation due to partial blockage of the gas pathways. The mechanism cannot be detected by measurement of the cell voltage or cell current since the electrical behavior of the cell is dominated by the cell area operating in normal fuel-cell mode providing the required cell current. Degradation rates are low compared with a start-up/shut-down process due to the limited availability of oxygen in the anode catalyst layer. However, this state is stable and persists as long as the gas pathways are blocked, leading to severe degradation over time. The statistical distribution of local blockage of the gas pathways results in degradation of the entire catalyst layer.

20.4.4
Oxygen Starvation

Oxygen starvation occurs when the amount of oxygen supplied to the cathode is insufficient. Other species have to be reduced at the cathode side in order to maintain the current demand of the cell. Kim *et al.* [93] analyzed the effect of fuel and oxygen stoichiometry on polarization curves. For oxygen excess ($\lambda_C = 3$), mass transport limitation was not observed even at high current densities. If the oxygen supply was reduced, the polarization curve dropped for $i > 1$ A cm^{-2} independently of the fuel stoichiometry. The cell voltage fell to zero as soon as the cell was operated under oxygen starvation [87, 94].

Taniguchi *et al.* [75] contacted a reference electrode with the membrane of a PEMFC by an electrolyte junction. The cathode potential decreased and fell slightly below the anode potential after increasing the air utilization to 120% ($\lambda_C = 0.83$). In this state, they observed a slightly negative cell voltage. Mitsuda and Murahashi [74] integrated an RHE as reference electrode in the edge configuration into a PAFC. The cathode potential decreased strongly when the oxygen consumption exceeded 80%, indicating an increase in the cathode polarization. The cathode potential became negative as soon as the air utilization exceeded 100%. At the same time, the anode potential remained constant, resulting in a slightly negative cell voltage. In this state, they detected a significant amount of hydrogen in the cathode exhaust gas. Its concentration increased with increasing overall cell current under oxygen starvation conditions. This indicates that hydrogen evolution occurs as an alternative reduction reaction according to the back-reaction of Eq. (20.1).

From the experiments described above, a potential distribution as shown in Figure 20.8 can be deduced. The upper part shows a cell that is insufficiently supplied with oxygen with respect to the current demand I_{cell}. As an alternative reduction reaction, protons from the membrane are reduced to hydrogen in the oxygen-starved region of the cathode. This reaction occurs at a slightly negative potential difference between cathode and membrane ($\varphi_C - \varphi_M < 0$ V) due to the fast reaction kinetics of hydrogen evolution and oxidation according to Eq. (20.1).

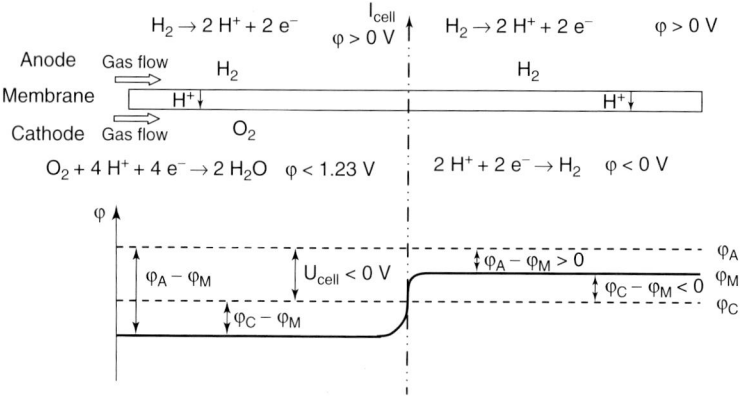

Figure 20.8 Gas and potential distribution within a PEMFC cell at oxygen starvation.

The protons are delivered by the standard hydrogen oxidation reaction at the anode side since the anode is supplied with sufficient hydrogen. The process of hydrogen oxidation at the anode coupled with hydrogen evolution at the cathode is called a *proton pump*. Note that the anode potential exceeds the cathode potential, resulting in a slightly negative cell voltage. High electrode potentials causing degrading reactions are not present due to the small overpotential of the hydrogen evolution and oxidation reaction keeping the anode and cathode potentials close to 0 V.

In the front part of the cell, oxygen is available. Hydrogen oxidation there becomes faster and oxygen reduction becomes slower by reducing the membrane potential until both reaction rates are balanced. Because of the relatively low in-plane conductivity of the membrane, a gradient in the membrane potential can be maintained along the cell. A gradient in the membrane potential from the starved to the well-supplied region is observed for the same reasons as under overall hydrogen starvation as shown in Figure 20.5.

In summary, we conclude that oxygen starvation results in a slightly negative cell voltage and hydrogen production at the cathode side. The simultaneous power loss is reversible and the mechanism does not lead to cell degradation since high electrode potentials required for carbon corrosion or platinum dissolution do not occur.

20.5
Mitigation

In the previous sections, degradation phenomena due to dynamic operation and starvation conditions were discussed. Here, we focus on mitigation strategies in order to reduce cell degradation and optimize long-term stability during fuel-cell operation. We distinguish between improvements concerning materials and design on the one hand and operation strategies on the other hand.

20.5.1
Materials and Design

A homogeneous gas distribution among all cells within a stack and also a uniform gas supply of the entire active area of each cell are essential in order to achieve low degradation rates. Design challenges exist particularly for end cells within a stack and the boundary area of the MEA. Perry *et al.* [6] detected large differences in cell voltage between cells in a stack using a conventional fuel manifold. At an inhomogeneous cell voltage distribution within a stack, single cells at low cell voltage tend to turn into cell reversal. The voltage differences are not observed when an improved fuel manifold is applied. Optimization of the reactant gas distribution is therefore necessary to prevent starvation conditions.

Bearing in mind that it is hardly possible to prevent every starvation condition during fuel-cell operation, a highly corrosion-resistant catalyst support is desired. Improvements are achieved by heat treatment of carbon blacks. Yu *et al.* [3] showed that graphitization of the MEA can reduce cell degradation during start-up/shut-down by a factor of five. However, stabilization of carbon against oxidation conflicts with the need for high catalytic activity [75]. Low surface area carbons are generally more stable than high surface area carbons but also require a higher platinum loading. Furthermore, heat treatment causes additional costs [6].

A selective catalyst preferring water oxidation to carbon corrosion could reduce harmful catalyst support degradation since these undesired side reactions are concurrent processes, as explained in detail in Section 20.4. The preference of water electrolysis over carbon oxidation also requires enhanced water retention within the catalyst layer and therefore modifications of the gas diffusion materials. Promising developments are based on Pt/Ru catalysts [73].

Takeuchi and Fuller [95] proposed a reduction of the platinum loading at the anode side in order to slow the abreaction of oxygen. This results in a reduced cathode degradation during start-up/shut-down and under local hydrogen starvation, since carbon corrosion at the cathode side is coupled to oxygen reduction at the anode side, as explained in Section 20.4.2. Reduction of the oxygen permeation rate through the membrane is another approach [96]. However, a reduced permeability usually goes along with a thicker or denser membrane layer, resulting in an increased ohmic loss of the cell.

20.5.2
Operation Strategies

Undesired side reactions occur at high electrode potentials as it is described in Sections 23.3 and 23.4. Long idle phases with cell voltages near OCV should therefore be avoided. This is particularly valid for the upper potential limit of potential cycles [67], since dynamic operation leads to accelerated cell degradation as explained in Section 20.3. A minimum cell current is required in order to keep the cell voltage significantly lower than OCV. If constant electric consumers are not present in the fuel-cell system, a dummy load can be integrated.

A large pressure drop along the gas channels can prevent local starvation conditions by dragging out liquid water from the gas pathways. This can be achieved by increasing the total volume flux, for example, by use of an anode recycle loop. Furthermore, sensitive control of the temperature profile can avoid condensation and therefore liquid water blockage within the cell [6].

Measurement of the single cell voltages in a stack allows the detection of cell reversals due to overall hydrogen or oxygen starvation. As soon as the minimum cell voltage falls below a given lower limit, it is possible to prevent cell degradation by reduction of the cell current.

Local hydrogen starvation is usually observed in the cell before cell reversal occurs. By measurement of the CO_2 concentration in the cathode exhaust gas, carbon corrosion can be detected. However, it is extremely difficult to distinguish between CO_2 evolution within the cell and its amount present in air due to low concentrations. Alternatively, critical conditions can be detected by integration of a reference electrode (see Section 20.2.1). Since the reference electrode has to be positioned within the starved region, several reference electrodes have to be used per cell in order to make detection of the starved region probable. This can only be realized in a laboratory cell and is not practical for automotive stacks.

Cell degradation during start-up/shut-down can be reduced by decreasing the residence time of the transient H_2/air front within the anode [3, 21]. Kim et al. [80] showed that by closing the inlets and outlets during shutdown, cell degradation can be reduced effectively. Hydrogen reacts with permeated oxygen from the cathode and the creation of new H_2/air fronts due to permeation of air into the anode from the environment is avoided. Perry et al. [6] controlled the voltages of all cells in a stack during start-up and shut-down. Their stack was short-circuited via a dummy load in order to reduce the cell voltages to a low but positive level. They proposed the following start-up procedure for an air-filled stack. (i) Keep the dummy load activated and feed hydrogen to the anode at high gas velocity. (ii) Keep the cathode closed in order to limit power dissipation to the amount of oxygen available in the cathode gas channels. (iii) Once the entire anode is filled with hydrogen, switch off the dummy load and supply air to the cathode. For shutdown: (i) stop the air supply while maintaining the fuel supply. (ii) Activate the dummy load in order to remove residual oxygen at the cathode and reduce the cell voltages. Subsequently there are three options depending on the fuel cell application: (iii) if hydrogen has to be removed for safety reasons, purge the anode with air. Degradation due to the created H_2/air front can be reduced by an activated dummy load resulting in a lower cathode potential. (iv) If residual hydrogen can remain within the anode, activate the dummy load until power dissipation drops to zero because all hydrogen is consumed by abreaction with air permeating through the membrane or entering the cell from the outside. This process can take several hours. (v) If fuel supply can continue after shutdown, feed hydrogen to the anode periodically in order to prevent air accumulation. In the latter case, an H_2/air front can be prevented during both start-up and shut-down. Perry et al. [6] reported a reduction of performance loss from 100 to 4 µV per cycle when a dummy load was applied. Kim et al. [97] observed a reduction of the CO_2 evolution rate by a factor of 1.8 due to application of a dummy load. By use of a single dummy load for a stack, it is challenging to control all cell voltages at a low but positive level in order to prevent cell reversals. A single dummy load for each cell in the stack overcomes that problem.

Shen et al. [21] showed that cell degradation is strongly reduced when the anode is purged with nitrogen before start-up and shut-down. However, the use of nitrogen is not practical in automotive application.

In summary, we conclude that a uniform reactant gas distribution across the entire stack is most effective in order to prevent starvation conditions. In addition, high electrode potentials $\varphi > 0.9$ V under cycling conditions and stationary operation have to be avoided. There are effective start-up and shut-down procedures that strongly reduce cell degradation induced by the RCD mechanism. By means of these mitigation strategies, long-term stability is optimized.

20.6
Conclusion

This chapter was focused on electrode degradation in PEMFCs for automotive application. It has been pointed out that undesired side reactions are responsible for the limited long-term stability of a fuel-cell system. These reactions occur at high electrode potentials which therefore have to be avoided.

We have shown that carbon corrosion and platinum dissolution reaction rates are particularly fast under potential transients. Hydrogen starvation conditions lead to accelerated cell degradation by the following mechanisms: if the fuel supply is insufficient, carbon and water are oxidized at the anode in order to maintain the cell current. The RCD mechanism occurs under local hydrogen starvation, for example, when an H_2/O_2 front is present at the anode. It results in corrosion of the cathode catalyst layer induced by high cathode potentials. The development of more stable catalysts and support materials has made huge progress but is still not sufficient for automotive requirements. Mitigation strategies therefore aim at a uniform reactant gas distribution across the entire fuel cell stack by a proper design of the gas distribution structures combined with the implementation of sophisticated start-up and shut-down procedures.

References

1. Borup, R., Davey, J., Xu, H., Ofstad, A., Garzon, F., and Pivovar, B. (2008) PEM Fuel Cell Durability, DOE Hydrogen Program, Annual Progress Report 2008, US Department of Energy, Washington, DC.
2. Fuller, T.F. and Gray, G. (2006) Carbon corrosion induced by partial hydrogen coverage. ECS Trans., **1** (8), 345–353.
3. Yu, P.T., Gu, W., Makharia, R., Wagner, F.T., and Gasteiger, H.A. (2006) The impact of carbon stability on PEM fuel cell startup and shutdown voltage degradation. ECS Trans., **3** (1), 797–809.
4. Makharia, R., Kocha, S.S., Yu, P.T., Sweikart, M.A., Gu, W., Wagner, F.T., and Gasteiger, H.A. (2006) Durable PEM fuel cell electrode materials:

requirements and benchmarking methodologies. *ECS Trans.*, **1** (8), 3–18.

5. Schmittinger, W. and Vahidi, A. (2008) A review of the main parameters influencing long-term performance and durability of PEM fuel cells. *J. Power Sources*, **180**, 1–14.

6. Perry, M.L., Patterson, T.W., and Reiser, C. (2006) Systems strategies to mitigate carbon corrosion in fuel cells. *ECS Trans.*, **3** (1), 783–795.

7. Xiao, S., Zhang, H., Bi, C., Zhang, Y., Zhang, Y., Dai, H., Mai, Z., and Li, X. (2010) Degradation location study of proton exchange membrane at open circuit operation. *J. Power Sources*, **195**, 5305–5311.

8. Qi, Z., Tang, H., Guo, Q., and Du, B. (2006) Investigation on "saw-tooth" behavior of PEM fuel cell performance during shutdown and restart cycles. *J. Power Sources*, **161**, 864–871.

9. Mathias, M.F., Makharia, R., Gasteiger, H.A., Conley, J.J., Fuller, T.J., Gittleman, C.J., Kocha, S.S., Miller, D.P., Mittelsteadt, C.K., Xie, T., Yan, S.G., and Yu, P.T. (2005) Two fuel cell cars in every garage? *Electrochem. Soc. Interface*, **14** (3), 24–35.

10. Wu, J., Yuan, X.Z., Martin, J.J., Wang, H., Zhang, J., Shen, J., Wu, S., and Merida, W. (2008) A review of PEM fuel cell durability: degradation mechanisms and mitigation strategies. *J. Power Sources*, **184**, 104–119.

11. Hung, C.C., Lim, P.Y., Chen, J.R., and Shih, H.C. (2011) Corrosion of carbon support for PEM fuel cells by electrochemical quartz crystal microbalance. *J. Power Sources*, **196**, 140–146.

12. Pourbaix, M. (1974) *Atlas of Electrochemical Equilibria in Aqueous Solutions*, National Association of Corrosion Engineers, Houston, TX, Chapter IV, Carbon: 17.1, Platinum: 13.6.

13. Yu, X. and Ye, S. (2007) Recent advances in activity and durability enhancement of Pt/C catalytic cathode in PEMFC. Part II: degradation mechanism and durability enhancement of carbon supported platinum catalyst. *J. Power Sources*, **172**, 145–154.

14. Bi, W. and Fuller, T.F. (2008) Modeling of PEM fuel cell Pt/C catalyst degradation. *J. Power Sources*, **178**, 188–196.

15. Darling, R.M. and Meyers, J.P. (2003) Kinetic model of platinum dissolution in PEMFCs. *J. Electrochem. Soc.*, **150** (11), A1523–A1527.

16. Meyers, J.P. and Darling, R.M. (2006) Model of carbon corrosion in PEM fuel cells. *J. Electochem. Soc.*, **153** (8), A1432–A1442.

17. Young, A.P., Stumper, J., and Gyenge, E. (2009) Characterizing the structural degradation in a PEMFC cathode catalyst layer: carbon corrosion. *J. Electrochem. Soc.*, **156** (8), B913–B922.

18. Hamann, C.H. and Vielstich, W. (2005) *Elektrochemie*, 4th edn., Wiley-VCH Verlag GmbH, Weinheim.

19. Baumgartner, W.R., Parz, P., Fraser, S.D., Wallnfer, E., and Hacker, V. (2008) Polarization study of o PEMFC with four reference electrodes at hydrogen starvation conditions. *J. Power Sources*, **182**, 413–421.

20. Gerteisen, D. (2007) Realising a reference electrode in a polymer electrolyte fuel cell by laser ablation. *J. Appl. Electrochem.*, **37**, 1447–1454.

21. Shen, Q., Hou, M., Liang, D., Zhou, Z., Li, X., Shao, Z., and Yi, B. (2009) Study on the processes of start-up and shutdown in proton exchange membrane fuel cells. *J. Power Sources*, **189**, 1114–1119.

22. Liang, D. (2009) Study of the cell reversal process of large area proton exchange membrane fuel cells under fuel starvation. *J. Power Sources*, **194**, 847–853.

23. Dross, R. and Maynard, B. (2007) In-situ electrode testing for cathode carbon corrosion. *ECS Trans.*, **11** (1), 1059–1068.

24. Yamauchi, S., Mitsuda, K., Maeda, H., and Takai, O. (2000) Application of a hydrogen reference electrode to a solid state water removal device. *J. Appl. Electrochem.*, **30**, 1235–1241.

25. Giner, J. (1964) A practical reference electrode. *J. Elechtrochem. Soc.*, **111** (3), 376–377.

26. Siroma, Z., Kakitsubo, R., Fujiwara, N., Ioroi, T., Yamazaki, S.-I., and Yasada, K. (2006) Compact dynamic

hydrogen electrode unit as a reference electrode for PEMFCs. *J. Power Sources*, **156**, 284–287.
27. Lauritzen, M., He, P., Young, A.P., Knights, S., Colbow, V., and Beattie, P. (2007) Study of fuel corrosion processes using dynamic hydrogen reference electrode. *J. New Mater. Electrochem. Syst.*, **10**, 143–145.
28. Li, G. and Pickup, P.G. (2004) Measurement of single electrode potentials and impedances in hydrogen and direct methanol PEM fuel cells. *Electrochim. Acta*, **49**, 4119–4126.
29. Kuver, A., Vogel, I., and Vielstich, W. (1994) Distinct performance evaluation of a direct methanol SPE fuel cell. A new method using a dynamic hydrogen reference electrode. *J. Power Sources*, **52**, 77–80.
30. Li, G. and Pickup, P.G. (2006) Analysis of performance losses of direct ethanol fuel cells with the aid of a reference electrode. *J. Power Sources*, **161**, 256–263.
31. Li, G. and Pickup, P.G. (2006) Dependence of electrode overpotentials in PEM fuel cells on the placement of the reference electrode. *Electrochem. Solid-State Lett.*, **9** (5), A249–A251.
32. Carmo, M., Roepke, T., Scheiba, F., Roth, C., Moeller, S., Fuess, H., Poco, J.G.R., and Linardi, M. (2009) The use of a dynamic hydrogen electrode as an electro-chemical tool to evaluate plasma activated carbon as electrocatalyst support for direct methanol fuel cell. *Mater. Res. Bull.*, **44**, 51–56.
33. He, W. and Nguyen, T.V. (2004) Edge effects on reference electrode measurements in PEM fuel cells. *J. Electrochem. Soc.*, **151** (2), A185–A195.
34. Adler, S.B., Hendersonm, B.T., Wilson, M.A., Taylor, D.M., and Richards, R.E. (2000) Reference electrode placement and seals in electrochemical oxygen generators. *Solid-State Ionics*, **134**, 35–42.
35. Winkler, J., Hendriksen, P.V., Bonanos, N., and Mogensen, M. (1998) Geometric requirements of solid electrolyte cells with a reference electrode. *J. Electrochem. Soc.*, **145** (4), 1184–1192.
36. Liu, Z., Wainright, J.S., Huang, W., and Savinell, R.F. (2004) Positioning the reference electrode in proton exchange membrane fuel cells: calculations of primary and secondary current distribution. *Electrochim. Acta*, **49**, 923–935.
37. Nagata, M., Itoh, Y., and Iwahara, H. (1994) Dependence of observed overvoltages on the positioning of the reference electrode on the solid electrolyte. *Solid-State Ionics*, **67**, 215–224.
38. Dolle, M., Orsini, F., Gozdz, A.S., and Tarascon, J.-M. (2001) Development of a reliable three-electrode impedance measurements in plastic Li-ion batteries. *J. Electrochem. Soc.*, **148** (8), A851–A857.
39. Büchi, F.N. and Scherer, G.G. (2001) Investigation of the tranversal water profile in Nafion membranes in polymer electrolyte fuel cells. *J. Electrochem. Soc.*, **148** (3), A183–A188.
40. Stumper, J., Campbell, S.A., Wilkinson, D.P., Johnson, M.C., and Davis, M. (1998) *In-situ* methods for the determination of current distribution in PEM fuel cells. *Electrochim. Acta*, **43** (24), 3773–3783.
41. Cleghorn, S.J.C., Derouin, C.R., Wilson, M.S., and Gottesfeld, S. (1998) A printed circuit board approach to measuring current distribution in a fuel cell. *J. Appl. Electrochem.*, **28**, 663–672.
42. Hakenjos, A., Muenter, H., Wittstadt, U., and Hebling, C. (2004) A PEM fuel cell for combined measurement of current and temperature distribution and flow field flooding. *J. Power Sources*, **131**, 213–216.
43. Natarajan, D. and Nguyen, T.V. (2004) Effect of electrode configuration and electronic conductivity on current density distribution measurements in PEM fuel cells. *J. Power Sources*, **135**, 95–109.
44. Sun, H., Zhang, G., Guo, L., and Liu, H. (2006) A novel technique for measuring current distributions in PEM fuel cells. *J. Power Sources*, **158**, 326–332.
45. Sun, H., Zhang, G., Guo, L., and Liu, H. (2009) A study of dynamic characteristics of PEM fuel cells by measuring local currents. *Int. J. Hydrogen Energy*, **34**, 5529–5536.
46. Noponen, M., Mennola, T., Mikkola, M., Hottinen, T., and Lund, P. (2002) Measurement of current distribution in a

free-breathing PEMFC. *J. Power Sources*, **106**, 304–312.

47. Noponen, M., Hottinen, T., Mennola, T., Mikkola, M., and Lund, P. (2002) Determination of mass diffusion overpotential distribution with flow pulse method from current distribution measurements in a PEMFC. *J. Appl. Electrochem.*, **32**, 1081–1089.

48. Gülzow, E., Kaz, T., Reissner, R., Sander, H., Schilling, L., and von Bradke, M. (2002) Study of membrane electrode assemblies for direct methanol fuel cells. *J. Power Sources*, **105**, 261–266.

49. Knöri, T. and Schulze, M. (2009) Spatially resolved current density measurements and real-time modelling as a tool for the determination of local operating conditions in polymer electrolyte fuel cells. *J. Power Sources*, **193** (1), 308–314.

50. Noponen, M., Ihonen, J., Lundblad, A., and Lindbergh, G. (2004) Current distribution measurements in a PEFC with net flow geometry. *J. Appl. Electrochem.*, **34**, 255–262.

51. Schönbauer, S., Kaz, T., Sander, H., and Gulzow, E. (2003) Segmented bipolar plate for the determination of current distribution in polymer electrolyte fuel cells, in 2nd European PEFC Forum 2003, Lucerne/Switzerland, Proceedings, vol. 1, pp. 231–237.

52. Schönbauer, S. (2009) Stromdichteverteilung im Polymer-Elektrolyt-Brennstoffzellenstapel, Dissertation, Shaker Verlag, Aachen.

53. Brett, D.J.L., Atkins, S., Brandon, N.P., Vesovic, V., Vasileiadis, N., and Kucernak, A.R. (2001) Measurement of the current distribution along a single flow channel of a solid polymer fuel cell. *Electrochem. Commun.*, **3**, 628–632.

54. Kraume, R. (2003) Verfahren und Vorrichtung zur Bestimmung der Stromdichteverteilung. Deutsche Patentanmeldung DE 102 13 479 A1.

55. Wieser, Ch., Helmbold, A., and Glzow, E. (2000) A new technique for two-dimensional current distribution measurements in electrochemical cells. *J. Appl. Electrochem.*, **30**, 803–807.

56. Schulze, M., Gülzow, E., Schönbauer, S., Knöri, T., and Reissner, R. (2007) Segmented cells as tool for development of fuel cells and error prevention/prediagnostics in fuel cell stacks. *J. Power Sources*, **173**, 19–27.

57. Hwnag, J.J., Chang, W.R., Peng, R.G., Chen, P.Y., and Su, A. (2008) Experimental and numerical studies of local current mapping on a PEM fuel cell. *Int. J. Hydrogen Energy*, **33**, 5718–5727.

58. Heinze, J. (1984) Cyclovoltammetrie – die "Spektroskopie" des Elektrochemikers. *Angew. Chem. Int. Ed. Engl.*, **11**, 823–916.

59. Bett, J., Kinoshita, K., Routsis, K., and Stonehart, P. (1973) A comparison of gas-phase and electrochemical measurements for chemisorbed carbon monoxide and hydrogen on platinum crystallites. *J. Catal.*, **29** (1), 160–168.

60. Takeuchi, N. and Fuller, T.F. (2010) Modeling and investigation of carbon loss on the cathode electrode during PEMFC operation. *J. Electrochem. Soc.*, **157** (1), B135–B140.

61. Maass, S., Finsterwalder, F., Frank, G., Hartmann, R., and Merten, C. (2008) Carbon supported oxidation in PEM fuel cell cathodes. *J. Power Sources*, **176**, 444–451.

62. Roen, L.M., Paik, C.H., and Jarvi, T.D. (2004) Electrocatalytic corrosion of carbon support in PEMFC cathodes. *Electrochem. Solid-State Lett.*, **7** (1), A19–A22.

63. Takeuchi, N., Jennings, E.N., and Fuller, T. (2009) Investigation and modelling of carbon oxidation of Pt/C under dynamic potential condition. *ECS Trans.*, **25** (1), 1045–1054.

64. Maass, S. (2007) Langzeitstabilität der Kathoden-Katalysatorschicht in Polymerelektrolyt-Brennstoffzellen, Dissertation, Cuvillier Verlag, Göttingen.

65. Ferreira, P.J., La O', G.J., Shao-Horn, Y., Morgan, D., Makharia, R., Kocha, S., and Gasteiger, H.A. (2005) Instability of Pt/C electrocatalysts in proton exchange membrane fuel cells. *J. Electrochem. Soc.*, **152** (11), A2256–A2271.

66. Chizawa, H., Ogami, Y., Naka, H., Matsunaga, A., Aoki, N., and Aoki, T.

(2006) Study of accelerated test protocol for PEFC focusing on carbon corrosion. *ECS Trans.*, **3** (1), 645–655.

67. Yasuda, K., Taniguchi, A., Akita, T., Ioroi, T., and Siroma, Z. (2006) Platinum dissolution and deposition in the polymer electrolyte membrane of a PEM fuel cell as studied by potential cycling. *Phys. Chem. Chem. Phys.*, **8**, 746–752.

68. Bi, W. and Fuller, T.F. (2007) Temperature effects on PEM fuel cell Pt/C catalyst degradation. *ECS Trans.*, **11** (1), 1235–1246.

69. Tilak, B.V., Conway, B.E., and Angerstein-Kozlowska, H. (1973) The real condition of oxidized Pt electrodes, Part III. Kinetic theory of formation and reduction of surface oxides. *J. Electroanal. Chem.*, **48**, 1–23.

70. Jerkiewicz, G., Vatankhah, G., Lessard, J., Soriaga, M.P., and Park, Y.-S. (2003) Surface-oxide growth at platinum electrodes in aqueous H_2SO_4. Reexamination of its mechanism through combined cyclic-voltammetry, electrochemical quartz-crystal nanobalance and Auger electron spectroscopy measurements. *Electrochim. Acta*, **49**, 1451–1459.

71. Nagy, Z. and You, H. (2002) Application of surface X-ray scattering to electrochemistry problems. *Electrochim. Acta*, **47**, 3037–3055.

72. Darling, R.M. and Meyers, J.P. (2005) Mathematical model of platinum movement in PEM fuel cells. *J. Electrochem. Soc.*, **152** (1), A242–A247.

73. Knights, S.D., Colbow, K.M., St-Pierre, J., and Wilkinson, D.P. (2004) Aging mechanisms and lifetime of PEFC and DMFC. *J. Power Sources*, **127**, 127–134.

74. Mitsuda, K. and Murahashi, T. (1991) Air and fuel starvation of phosphoric acid fuel cells: a study using a single cell with multi-reference electrodes. *J. Appl. Electrochem.*, **21**, 524–530.

75. Taniguchi, A., Akita, T., Yasuda, K., and Miyazaki, Y. (2004) Analysis of electrocatalyst degradation in PEMFC caused by cell reversal during fuel starvation. *J. Power Sources*, **130**, 42–49.

76. Taniguchi, A., Akita, T., Yasuda, K., and Miyazaki, Y. (2008) Analysis of degradation in PEMFC caused by cell reversal during air starvation. *Int. J. Hydrogen Energy*, **33**, 2323–2329.

77. Kim, J.H., Cho, E.A., Jang, J.H., Kim, H.J., Lim, T.H., Oh, I.H., Ko, J.J., and Oh, S.C. (2010) Effects of cathode inlet relative humidity on PEMFC durability during startup–shutdown cycling. *J. Electrochem. Soc.*, **157** (1), B104–B112.

78. Kim, J., Lee, J., and Tak, Y. (2009) Relationship between carbon corrosion and positive electrode potential in a proton-exchange membrane fuel cell during start/stop operation. *J. Power Sources*, **192**, 674–678.

79. Reiser, C.A., Bregoli, L., Patterson, T.W., Yi, J.S., Yang, J.D., Perry, M.L., and Jarvi, T.D. (2005) A reverse-current decay mechanism for fuel cells. *Electrochem. Solid State Lett.*, **8** (6), A273–A276.

80. Kim, H.-J., Lim, S.J., Lee, J.W., Min, I.-G., Lee, A.-Y., Cho, E.A., Oh, I.-H., Lee, J.H., Oh, S.-C., Lim, T.-W., and Lim, T.-H. (2008) Development of shut-down process for a proton exchange membrane fuel cell. *J. Power Sources*, **180**, 814–820.

81. Lee, S.-Y., Cho, E.A., Lee, J.-H., Kim, H.-J., Lim, T.-H., Oh, I.-H., and Won, J. (2007) Effects of purging on the degradation of PEMFCs operating with repetitive on/off cycles. *J. Electrochem. Soc.*, **154** (2), B194–B200.

82. Farooque, M., Kush, A., and Christner, L. (1990) Novel explanation of unusual localized corrosion in energy conversion devices. *J. Electrochem. Soc.*, **137** (7), 2025–2028.

83. Tang, H., Qi, Z., Ramani, M., and Elter, J.F. (2006) PEM fuel cell cathode carbon corrosion due to the formation of air/fuel boundary at the anode. *J. Power Sources*, **158**, 1306–1312.

84. Siroma, Z., Fujiwara, N., Ioroi, T., Yamazaki, S., Senoh, H., Yasuda, K., and Tanimoto, K. (2007) Transient phenomena in a PEMFC during the start-up of gas feeding observed with a 97-fold segmented call. *J. Power Sources*, **172**, 155–162.

85. Takeuchi, N. and Fuller, T.F. (2008) Modeling and investigation of design factors and their impact on carbon corrosion of PEMFC electrodes. *J. Electrochem. Soc.*, **155** (7), B770–B775.
86. Mitsuda, K. and Murahashi, T. (1990) Polarization study of a fuel cell with four reference electrodes. *J. Electrochem. Soc.*, **137** (10), 3079–3085.
87. Liu, Z., Yang, L., Mao, Z., Zhuge, W., Zhang, Y., and Wang, L. (2006) Behavior of PEMFC in starvation. *J. Power Sources*, **157**, 166–176.
88. Ohs, J.H., Sauter, U., Maass, S., and Stolten, D. (2011) Modeling hydrogen starvation conditions in proton-exchange membrane fuel cells. *J. Power Sources*, **196** (1), 255–263.
89. Bazylak, A. (2009) Liquid water visualization in PEM fuel cells: a review. *Int. J. Hydrogen Energy*, **34**, 3845–3857.
90. Liu, Z.Y., Brady, B.K., Carter, R.N., Litteer, B., Budinski, M., Hyun, J.K., and Muller, D.A. (2008) Characterization of carbon corrosion-induced structural damage of PEM fuel cell cathode electrodes caused by local fuel starvation. *J. Electrochem. Soc.*, **155** (10), B979–B984.
91. Patterson, T.W. and Darling, R.M. (2006) Damage to the cathode catalyst of a PEM fuel cell caused by localized fuel starvation. *Electrochem. Solid State Lett.*, **9** (4), A183–A185.
92. Gu, W., Makharia, R., Yu, P.T., and Gasteiger, H.A. (2006) Predicting local H_2 starvation in a PEM fuel cell: origin and materials impact. *Prepr. Pap. Am. Chem. Soc. Div. Fuel Chem.*, **51** (2), 692–695.
93. Kim, S., Shimpalee, S., and Van Zee, J.W. (2004) The effect of stoichiometry on dynamic behaviour of a proton exchange membrane fuel cell (PEMFC) during load change. *J. Power Sources*, **135**, 110–121.
94. Song, R.-H., Kim, C.-S., and Shin, D.R. (2000) Effects of flow rate and starvation of reactant gases on the performance of phosphoric acid fuel cells. *J. Power Sources*, **86**, 289–293.
95. Takeuchi, N. and Fuller, T.F. (2008) Investigation of carbon loss on the cathode during PEMFC operation. *ECS Trans.*, **16** (2), 1563–1571.
96. Takeuchi, N. and Fuller, T.F. (2007) Modeling of transient state carbon corrosion for PEMFC electrode. *ECS Trans.*, **11** (1), 1021–1029.
97. Kim, J.H., Cho, E.A., Jang, J.H., Kim, H.J., Lim, T.H., Oh, I.H., Ko, J.J., and Son, I.-J. (2010) Development of a durable PEMFC start-up process by applying a dummy load. *J. Electrochem. Soc.*, **157** (1), B118–B124.

Part IV
Quality Assurance

21
Quality Assurance for Characterizing Low-Temperature Fuel Cells

Viktor Hacker, Eva Wallnöfer-Ogris, Georgios Tsotridis, and Thomas Malkow

21.1
Introduction

Fuel cells represent a recent energy technology that can provide both highly efficient power and heat in a number of application areas for the end user. A multitude of different aspects of fuel cells need to be regulated and standardized in order to ensure the quality and safety of such power generation systems. Regulatory codes and standards are developing around a number of areas and applications, including product requirements, manufacturing, interconnection requirements, stationary specific, mobile specific, portable specific, and electric utility requirements, performance, reliability, recyclability, hydrogen storage, transport and installation, maintenance, and site requirements. Several publications are available that give an excellent overview of legal and standard specifications and for proposed test methodologies for fuel-cell systems for different applications at European and international levels [1–6].

Extensive work to harmonize testing procedures was performed in the Research and Training Network "Fuel Cell TEsting and STandardization thematic NETwork (FCTESTNET)" (*http://ie.jrc.ec.europa.eu/fctestnet*, contract ENK5-CT-2002-20657) under the Fifth Framework Programme of the European Community for research, technological development, and demonstration activities (FP5) and in the Specific Targeted Research project "Fuel Cell Systems Testing, Safety and Quality Assurance (FCTES^{QA})" (*http://fctesqa.jrc.ec.europa.eu/*, contract 020161), co-financed by the European Commission through FP6. The main aim of FCTES^{QA} was to address the aspects of pre-normative research, benchmarking, and validation through round-robin testing of harmonized, industry-wide test protocols, and testing methodologies for fuel cells. This activity provided support for the essential pre-normative research efforts towards standardization, thereby contributing to the early and market-oriented development of specifications and pre-standards [7].

In this chapter, extracts of results of standardized evaluation protocols of FCTES^{QA} for the performance of polymer electrolyte fuel cell (PEFC) single cells [polymer electrolyte membrane fuel cells (PEMFCs) or proton exchange membrane fuel cells] under various operating conditions, standardized single cells, lifetime investigations, and the design of experiments (DoEs) are discussed.

21.2
Test Procedures/Standardized Measurements

21.2.1
Preconditioning of the Fuel Cell

For PEFCs to function properly, a so-called conditioning step is required mainly to attain the desired level of membrane humidification and to obtain a quasi-steady performance that is reproducible under the applied test conditions. The start-up of the fuel cell and conditioning step can be performed following one of the following procedures:

- proposed by the manufacturer of the test object,
- proposed by the manufacturer of a fuel cell component,
- the one that is common practice at the testing organization, or
- as recommended below.

The most important factor, regardless of conditioning procedure, is that the cell voltage is considered stable before the actual measurement step starts. A stability criterion can be defined based on the deviation of the fuel-cell voltage measured over a fixed period of time. It is recommended that the variations in the cell voltage be lower than ± 5 mV from the average during the last hour before conditioning ends.

The test starts by bringing the operating conditions (inputs) to the values specified for the conditioning of the cell. The conditioning consists in keeping the previous conditions stable until the cell voltage reaches a stable value (normally corresponding to an optimized humidified state of the polymer membrane electrolyte). Cell conditioning can be part of the start-up procedure. If not, it is recommended to operate the cell in galvanostatic mode at the selected operating temperature and gas conditions, by increasing the current density in steps of 0.1 A cm^{-2} while maintaining the cell voltage at or above 0.5 V, until the demanded current density for conditioning is reached. The current density for the conditioning of the cell will correspond either to the maximum current attainable for a cell voltage of 0.5 V or to the current density specified by the objective of the test undertaken. The conditioning step should last at least 24 h [8].

21.2.2
Humidification Sensitivity Test

21.2.2.1 Setting the Test Conditions (Test Inputs)
Four levels of relative humidity (RH) are considered for hydrogen (or fuel) and three for air (or oxidant) in this humidification sensitivity test. Among the inputs,

Table 21.1 Variable test inputs.

Input	Value/range	Control accuracy	Sample rate (Hz)
i	0.8; 0.6; 0.4; 0.2 A cm^{-2}	±2% FS for $i < 0.1$ A cm^{-2} ±1% FS for $i \geq 0.1$ A cm^{-2}	≥1
$Q_{v,\lambda\text{fuel}}$ [a]	Corresponding to the fuel stoichiometry λ_{fuel}	±1% FS[b]	≥1
$Q_{v,\lambda\text{ox}}$ [a]	Corresponding to the oxidant stoichiometry λ_{ox}	±1% FS[b]	≥1
RH$_{\text{fuel}}$	0–75% (see Table 21.3)	–	≥1
RH$_{\text{ox}}$	25–75% (see Table 21.3)	–	≥1

[a] $Q_{v,\lambda\text{fuel}}$ and $Q_{v,\lambda\text{ox}}$ are the stoichiometry controlled volumetric flow rates of fuel and oxidant, respectively, unless they are smaller than their minimum flow rates, $Q_{v,\text{fuel,min}}$ and $Q_{v,\text{ox,min}}$.
[b] A digital mass flow meter usually used on the test bench typically provides for an accuracy of 1% of full scale (FS) or maximum flow and a minimum measurable flow of 2% of FS. It means that the measurement uncertainty decreases with increasing flow rate or increasing current density when operating at a given reactant stoichiometry.

Table 21.2 Static test inputs.

Input	Value/range	Control accuracy	Sample rate (Hz)
X_{fuel}	H$_2$ (or reformate, H$_2$/CO$_2$/N$_2$)	±0/ − 0.001% H$_2$	–
X_{ox}	Air or pure O$_2$	–	–
p_{ox} [a]	Ambient–120 kPa	±2%	≥1
p_{fuel} [a]	Ambient–120 kPa	±2%	≥1
λ_{fuel}	1.2	–	–
λ_{ox}	2	–	–
T_{cell}	T_{amb}–80 °C	±2 °C	≥1

[a] Absolute reactant gas pressure at either cell inlet or outlet.

the cell operating temperature T_{cell} should be variable only during the test steps of start-up and conditioning. The other inputs (see Table 21.1) are variable during the humidification sensitivity measurement step of the test. The inputs given in Table 21.2 should be static and fixed at a single value in the given ranges during all the steady measurement steps. The sequence of measurements includes several humidification steps (as described in Table 21.3) to cover the entire RH range given in Table 21.1. The fuel RH is set at a fixed value, starting at the maximum of the selected range, while the oxidant RH increases each time stepwise from minimum to maximum as provided for in Table 21.3. At each pair of reactant RH, four current densities, i (see Table 21.1), are applied following the order given in Table 21.4 and shown in Figure 21.1 [8].

The conditioning step should be applied only once just before setting the specific conditions of the test program: operating conditions given in Table 21.1; current

Table 21.3 Sequence of fuel RH and corresponding three different levels of oxidant RH applied with the operating conditions.

Humidification step number	Fuel RH (%)	Oxidant RH (%)
1	75	25
2	75	50
3	75	75
4	75	75
5	50	50
6	50	75
7	25	25
8	25	50
9	25	75
10	0	25
11	0	50
12	0	75

Table 21.4 Successive order of descending current densities applied for each humidification level (see Figure 21.1).

Current density stage	Current density (A cm^{-2})
1	0.8
2	0.6
3	0.4
4	0.2

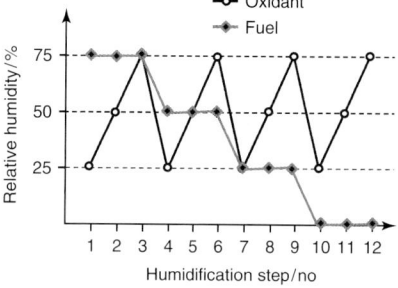

Figure 21.1 Relative humidity of the gases for each measurement step.

density (1); humidification step number 1. At each humidification level, the test lasts at least 2 h and for as long as the cell voltage is not complying with the stability criterion that the variation in the cell voltage does not exceed 5 mV from the average during the last 20 min of the measurement before starting the next step in the humidification. An overview of the test procedure is shown in Figure 21.2 [8].

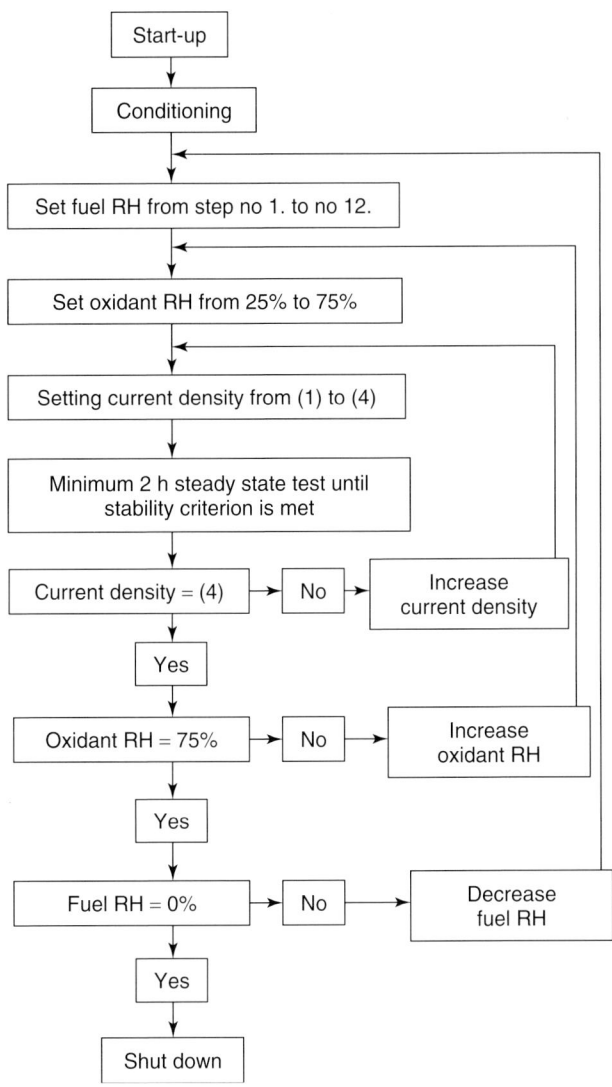

Figure 21.2 Schematic overview of the successive steps of progression of the RH sensitivity test including decision tree.

21.2.2.2 Measuring the Test Outputs

The main objective of the humidity sensitivity test is to determine the cell voltage under different RH levels for hydrogen (or fuel) and air (or oxygen) at four different current densities. During the test, the static test inputs (reactant temperatures, stoichiometries, composition, and pressures) should be kept at the values selected within the ranges and with the accuracy specified. All functional inputs and outputs are measured versus time (test duration). When the cell voltage drops below 0.3 V (minimum cell voltage), the current test step has to be interrupted, for example, by lowering the current density until the voltage recovers to a value above 0.3 V; then, one may proceed to the next test step as appropriate or end the test when test step number 12 is reached (shut-down). If the test forms part of a test program, the next module is proceeded rather than shut-down (load and reactant flow switch off) of the cell [8].

21.2.2.3 Data Post Processing

The evolution of the voltage V over time (test duration) is the main output of this type of test. The power density can be calculated as another test output and also the performance (voltage or power) loss between the best (highest cell voltage) level of reactant humidification (usually corresponding to humidification step number 12) and the other levels to compare the cell sensitivity with reactant humidification. The performance loss[1] is calculated as a fraction (or percentage) as follows:

$$\frac{V_{(\text{humidification no. 12 or best})} - V_{(\text{humidification no. 1 to 11 excl. no. 12 or best})}}{V_{(\text{humidification no. 12 or best})}} (\times 100\%)$$

Similarly, the loss in cell performance at one level of reactant RH can be compared with the RH levels of the other reactant. The cell has passed the humidification sensitivity test when the acceptance criterion is met; for example, the cell does not exceed a specified loss in performance (voltage or power) for the given range of reactant RH at the applied current densities and the other operating conditions (cell temperature, pressure, or reactant stoichiometry) [8].

21.2.3
On/Off Aging Test

21.2.3.1 Setting the Test Conditions (Test Inputs)

The test is under galvanostatic control with a load profile (on/off load cycling) at a steady operating temperature and reactant parameter. This step starts by setting the current density to i_{load} (corresponding to the "on" phase of the cycle) and then the operating conditions to the values specified for the voltage measurement versus time (selected within the ranges specified). A first value of the cell voltage at i_{load} is measured when the operating conditions have all met their stability criterion. The conclusion of the test referring to the qualification of the cell tested is partially based on this initial value [9].

21.2.3.2 Measuring the Test Outputs

During the test, the static test inputs (temperature, pressure, and RH) should be kept at the values selected within the ranges and for the accuracy specified

---

1) A performance gain would normally indicate insufficient cell conditioning, since the cell should exhibit maximum performance at the end of conditioning.

Table 21.5 Variable test inputs.

Input	Value/range	Control accuracy	Sample rate (Hz)
t	0–5000 h	–	–
i	0–2 A cm^{-2}	±2% FS for $i < 0.1$ A cm^{-2} ±1% FS for $i > 0.1$ A cm^{-2}	≥1
i_{load}	0 A cm^{-2} (during "off" phase) $0.5 < i_{load} < 0.8$ A cm^{-2} (during "on" phase)	See above	≥1
T_{cell}	$T_{amb.}$–80 °C	±2 °C	≥1
$Q_{v,\lambda fuel}$[a]	Corresponding to λ_{fuel}	±1% FS[a]	≥1
$Q_{v,\lambda ox}$[a]	Corresponding to λ_{ox}	±1% FS[a]	≥1

[a] See the remarks in Table 21.1.

Table 21.6 Static test inputs.

Input	Value/range	Control accuracy	Sample rate (Hz)
X_{fuel}	H$_2$ (or reformate, H$_2$/CO$_2$/N$_2$)	±0/ – 0.001% H$_2$	–
X_{ox}	Air or pure O$_2$	–	–
$p_{fuel/ox}$[a]	Ambient–300 kPa	±2%	≥1
RH$_{ox}$	0–100%	–	≥1
RH$_{fuel}$	0–100%	–	≥1
λ_{fuel}	1.2	–	–
λ_{ox}	2	–	–
T_{cell}	$T_{amb.}$–80 °C	±2 °C	≥1

[a] Absolute reactant gas pressure at either cell inlet or outlet.

(see Tables 21.5 and 21.6). All the functional inputs and outputs are measured versus time (test duration). An initial polarization curve test (see also Section 21.2.4) is performed prior to the on/off cycle measurement step. The polarization curve stops at the maximum current density or at the current density where the cell voltage remains above 0.3 V (to avoid irreversible cell damage).

The main objective of this on/off load cycling or accelerated aging test is to determine the evolution of both the open-circuit voltage (OCV) and the on-load voltage of the cell in terms of a performance progression or degradation rate calculated as voltage difference per unit time (test duration) when the cell is subjected to a specific load profile which includes "on" and "off" phases of 15 min each. The first 15 min "on" phase at i_{load} follows immediately the initial polarization curve measurement; then the on/off cycling proceeds as follows (load profile):

- "off" phase = 15 min at 0 A cm^{-2}
- "on" phase = 15 min at i_{load} A cm^{-2}.

During the "on" phase, it is recommended to increase the current density stepwise as follows:

- 0–25% of i_{load} in 10 s
- 25–50% of i_{load} in 10 s
- 50–75% of i_{load} in 10 s
- 75–100% of i_{load} in 10 s

which results in about 14 min of testing at i_{load}. During the two phases, the reactant flow rates Q should be controlled as follows:

- "on" phase: $Q_{fuel} = Q_{\lambda,fuel}$ and $Q_{ox} = Q_{\lambda,ox}$
- "off" phase: $Q_{fuel} = Q_{fuel,min}$ and $Q_{ox} = Q_{ox,min}$

where $Q_{fuel,min}$ and $Q_{ox,min}$ correspond to the flow of fuel and oxidant, respectively, necessary to maintain the reactant pressures, p_{fuel} and p_{ox}, during the "off" phase. Unless the cell voltage drops below the minimum cell voltage (e.g., 0.3 V), upon which the test is interrupted to avoid irreversible cell damage, the test duration depends on the objective of the test and on the test end criterion, for example, a specified period or upon exceeding a defined decrease in cell performance (power, voltage, or OCV). Cell performance permitting, a final polarization curve is performed at the end of the on/off cycle measurements which immediately follows the last "on" phase. The comparison of the two polarization curves is used to qualify the cell performance at each current density to assist in the analysis of the degradation of any cell.

The test schematic is sketched in Figure 21.3, which describes the evolution of some of the variable test inputs and of the cell voltage from the test steps of start-up and conditioning to the beginning of the on/off cycle measurements [9].

21.2.3.3 Data Post-Processing

The evolution of the cell voltage with time (test duration) is the main output of this accelerated aging test. The calculated power density P can serve as another test output.

The voltage deviations ΔV are calculated at OCV and on-load ("on" phase) over the total duration of the cycle measurements to evaluate a "performance progression rate" (normally a "performance degradation rate") $\Delta V_{OCV}/dt$ and $\Delta V_{iload}/dt$ as follows:

- $\Delta V_{OCV} = OCV_{\text{of the first "off" phase}} - OCV_{\text{of the last "off" phase}}$
- $\Delta V_{iload} = V_{\text{at iload of the first "on" phase}} - V_{\text{at iload of the last "on" phase}}$.

The on-load performance loss in cell voltage and power is calculated as a fraction (or percentage) at i_{load} as follows:

- $[V_{\text{at iload of the first "on" phase}} - V_{\text{at iload of the last "on" phase}}]/V_{\text{at iload of the first "on" phase}}$ ($\times 100\%$)
- $[P_{\text{at iload of the first "on" phase}} - P_{\text{at iload of the last "on" phase}}]/P_{\text{at iload of the first "on" phase}}$ ($\times 100\%$).

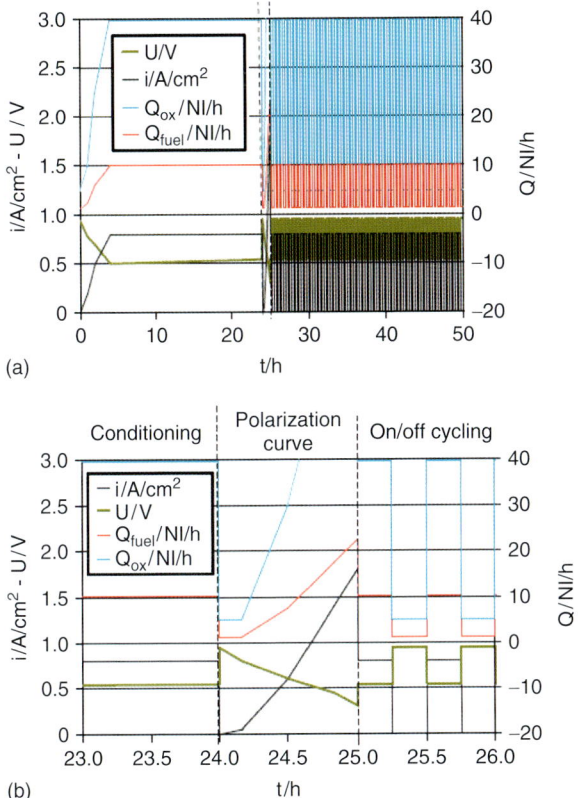

Figure 21.3 Schematic representation of the on/off cycle measurement steps of the accelerated aging test [(a) including start-up and (b) detail with end of conditioning and beginning of cycling] where $i_{load} = 0.8$ A cm^{-2}, $\lambda_{fuel}/\lambda_{ox} = 1.2/2$, $Q_{fuel,min} = Q_{\lambda fuel}$ and $Q_{ox,min} = Q_{\lambda ox}$ for $i = 0.1$ A cm^{-2}. Courtesy of FCTES^{QA}.

Similarly, the performance losses are calculated for each current density using the two polarization curves. The cell has passed the accelerated aging test when the acceptance criterion is met, for example, when the cell does not exceed a specified power loss for a number of specified on/off cycles or a specified test duration, or the loss rate in cell voltage remains below a specified value for the applied operating conditions [9].

21.2.4
Performance Test

21.2.4.1 Setting the Test Conditions (Test Inputs)

The polarization curve is performed under galvanostatic control at given operating temperatures, reactant pressures, stoichiometries, and RH. It starts by bringing

Table 21.7 Variable test inputs.

Input	Value/range	Control accuracy	Sample rate (Hz)
i	0–2 A cm^{-2}	±2% FS for $i < 0.1$ A cm^{-2}	≥1
		±1% FS for $i > 0.1$ A cm^{-2}	
$Q_{v,\lambda\text{fuel}}$[a]	Corresponding to λ_{fuel}	±1% FS[a]	≥1
$Q_{v,\lambda\text{ox}}$[a]	Corresponding to λ_{ox}	±1% FS[a]	≥1

[a] See the remarks in Table 21.1.

Table 21.8 Static test inputs.

Input	Value/range	Control accuracy	Sample rate (Hz)
X_{fuel}	H$_2$ (or reformate, H$_2$/CO$_2$/N$_2$)	±0/−0.001% H$_2$	–
X_{ox}	Air or pure O$_2$	–	–
T_{cell}	T_{amb}–80 °C	±2 °C	≥1
$p_{\text{fuel/ox}}$[a]	Ambient–300 kPa	±2 °C	≥1
λ_{fuel}	1.1–2	–	≥1
λ_{ox}	2–3	–	≥1
$Q_{v,\,\text{fuel/ox min}}$	According to the fuel-cell manufacturer's recommendations or as limited by test bench	±1% FS[b]	≥1
$RH_{\text{fuel/ox}}$	0–100%	–	–

[a] Absolute reactant gas pressure at either cell inlet or outlet.
[b] See the remark in Table 21.1.

the operating conditions to the values specified for the measurement when not already corresponding to cell conditioning or those of another test. Usually, it means that the cell voltage is brought to OCV, implying that the reactant flow rates are maintained at their minimum values, $Q_{\text{fuel,min}}$ and $Q_{\text{ox,min}}$, for at least 30 s but for less than 1 min [10].

21.2.4.2 Measuring the Test Outputs

The main objective of the polarization curve (performance) test is to determine the change in the cell voltage (and in the calculated power density) generated by variation of the current density (see Table 21.7). During this test step, the static test inputs (see Table 21.8) are maintained at their values within the specified ranges (see Tables 21.7 and 21.8). The test preferably starts at OCV and increases to the specified maximum current. The change in current density is either instantaneous or continuous at a given rate. The test inputs and outputs are measured versus time (test duration) at each current density for a minimum of 5 min. At the end of the test, either OCV is set or to avoid possible irreversible damage under OCV conditions, the current is set to a value specified for the next test of a test program.

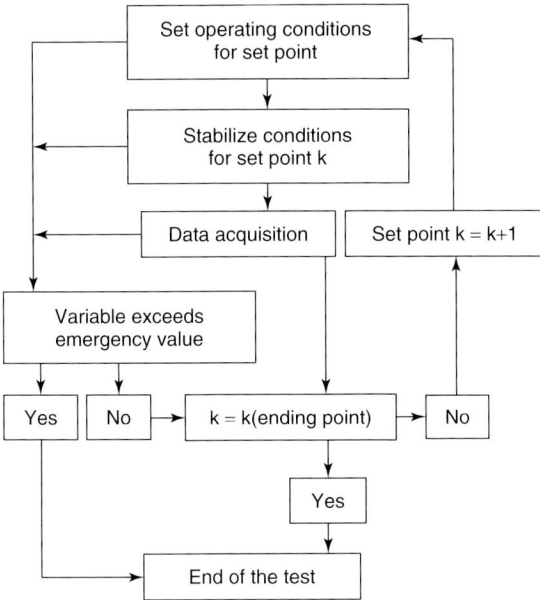

Figure 21.4 Schematic for measuring the test inputs and outputs; a set point k corresponds to a specified current density.

The schematic of the test procedure is shown in Figure 21.4. The test is aborted when the cell voltage drops below 0.3 V [10].

21.2.4.3 Data Post-Processing

The polarization curve (voltage and power density versus current density) is based on the cell voltage (and power density) averaged for the minimum 5 min period at each current density. In addition, the standard deviation of the voltage (and power density) is calculated. The cell has passed the performance test when the acceptance criterion is met, for example, the cell maintains a specified voltage at a given current density or exhibits a specified power density for the applied operating conditions [10].

21.2.5
Long-Term Durability Test

21.2.5.1 Setting the Test Conditions (Test Inputs)

The test is conducted under galvanostatic control at a given current density i_{load}, operating temperature, and reactant conditions for all the steady test phases and with varying reactant flow rates and hence varying current densities for the polarization curve measurements. This step starts by setting the operating conditions to the values specified. An initial measurement of the cell voltage is performed at i_{load} [$V_{iload}(t = 0)$] provided that the input values (operating conditions,

Table 21.9 Variable test inputs; the static inputs are given in Table 21.6.

Input	Value/range	Control accuracy	Sample rate (Hz)
t	0–10 000 h	–	–
i	0–2 A cm^{-2}	±2% FS for $i < 0.1$ A cm^{-2}	≥1
		±1% FS for $i > 0.1$ A cm^{-2}	
$Q_{v,\lambda\text{fuel}}$	Corresponding to λ_{fuel}	±1% FS[a]	≥1
$Q_{v,\lambda\text{ox}}$	Corresponding to λ_{ox}	±1% FS[a]	≥1

[a] A digital mass flow meter usually used on the test bench typically provides for an accuracy of 1% of full scale (FS) or maximum flow and a minimum measurable flow of 2% of FS. It means that the measurement uncertainty decreases with increasing flow rate or increasing current density when operating at a given reactant stoichiometry.

see Tables 21.9 and 21.6) are considered steady (stabilization). The conclusion of the test is based on this initial value [11].

21.2.5.2 Measuring the Test Outputs

The main objective of this long-term durability test is to determine the evolution of the voltage of the cell in terms of a performance degradation rate expressed as voltage difference per unit time when subjecting the cell to a constant load (current) for a long period. The test includes long-term steady steps and polarization curves. During the test, the static test inputs (reactant temperature, pressure, and RH) are kept at the values selected within the ranges and with the accuracy specified (see Tables 21.6 and 21.9). All the functional inputs and outputs are measured versus time (test duration). The polarization curves are performed at given intervals corresponding to $t_{\max}/10$, where $t_{\max}$ is the maximum duration of the test as defined by the objective of the test, but usually ranges between 500 and 10 000 h. The measurement steps are performed in the following sequence:

- initial polarization curve starting at $t = 0$ after stabilization at i_{load}
- long-term steady test phase number 1
- polarization curve number 2 at $t = t_{\max}/10$
- long-term steady test phase number 2
- polarization curve number 3 at $t = 2 \times t_{\max}/10$
- long-term steady test phase number n
- polarization curve number $n + 1$ at $t = n \times t_{\max}/10$ with $1 \leq n \leq 10$.

Unless the test is aborted due to a drop in cell voltage below 0.3 V, the test ends after the final polarization curve measurement. Another test end criterion can be the maximum permissible performance (cell voltage or power) loss. Comparison of the results of the final and initial polarization curves is used to analyze the performance loss (degradation) of the cell, if any, for the entire range of current densities [11].

21.2.5.3 Data Post-Processing

The evolution of the voltage over time (test duration) is the main output of this test. The power density is calculated as another output. The performance loss of the cell in terms of voltage difference and relative change in voltage and power is calculated as follows:

- $\Delta V_{iload} = V_{iload}(t=0) - V_{iload}(t=t_{max})$
- $\Delta V_{iload}/V_0 = [V_{iload}(t=0) - V_{iload}(t=t_{max})]/V_{iload}(t=0)$
- $\Delta P_{iload}/P_0 = [P_{iload}(t=0) - P_{iload}(t=t_{max})]/P_{iload}(t=0)$

The cell performance degradation in terms of voltage and power (or power density) can also be calculated as loss rates for selected periods ($dt = t_2 - t_1$) other than t_{max} as follows:

- $\Delta V_{iload}/dt = [V_{iload}(t=t_2) - V_{iload}(t=t_1)]/(t_2 - t_1)$
- $\Delta P_{iload}/dt = [P_{iload}(t=t_2) - P_{iload}(t=t_1)]/(t_2 - t_1)$.

Here, the voltage and power values may refer either to the long-term steady test phase or the polarization curve measurements. The cell has passed the long-term durability test when the acceptance criterion is met, for example, the cell degradation is below a specified value for the applied operating conditions [11].

21.2.6
Dynamic Load Cycling Aging Test

21.2.6.1 Setting the Test Conditions (Test Inputs)

The test is conducted under galvanostatic control with a load profile at constant operating temperature and reactant conditions. This step starts by setting the operating conditions including the current density to $i_{load\ max}$, which is the current density at P_{max}, where P_{max} is defined as the maximum power of the cell under normal operation ($i_{load\ max} = i_{@\ 100\%\ Pmax}$) to their specified values and ranges. An initial measurement of the cell voltage is performed at $i_{load\ max}(V_{initial})$ provided that the input values (operating conditions, see Tables 21.6 and 21.9) are considered steady (stabilization). The conclusion of the test is based on this initial value [12].

21.2.6.2 Measuring the Test Outputs

The main objective of this test is to determine the evolution of the cell voltage and power in terms of a performance loss expressed as normalized voltage (or power) difference when subjecting the cell to a specified load profile including low-power and high-power phases, corresponding to two current densities, $i_{load\ min}$ and $i_{load\ max}$, respectively. The test includes load cycling steps and polarization curves. An initial polarization curve is performed after stabilization at $i_{load\ max}$, which is followed by a 10 min high-power phase at constant current density $i_{load\ max}$ before a repetitive dynamic cycling phase is performed as follows (see also the schematic of the test in Figure 21.5):

- low-power steady phase for 40 s at $i_{@\ 20\%\ Pmax}$
- high-power dynamic phase for 20 s where i continuously increases to $i_{@\ 100\%\ Pmax}$.

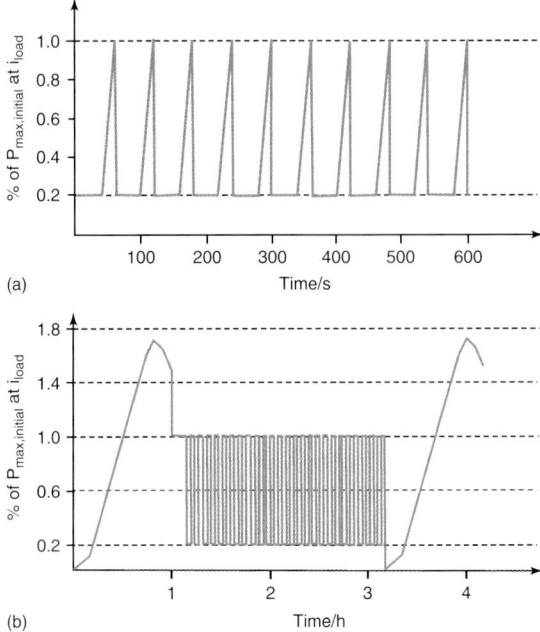

Figure 21.5 Example of a load cycling aging test profile (power versus test duration) illustrating the relative power during dynamic cycling phase for 10 min (a) and an initial 1 h polarization curve measurement followed by 2 h dynamic cycling phase and another polarization curve measurement (b). Courtesy of FCTES^{QA}.

The other polarization curves are performed at constant intervals corresponding to $t_{max}/10$ with t_{max}, the maximum duration of the test, usually ranging between 500 and 10,000 h. It assumes that the sequence of measurements is as follows:

- initial polarization curve starting at $t = 0$
- dynamic cycling phase number 1
- polarization curve number 2 at $t = t_{max}/10$
- dynamic cycling phase number 2
- polarization curve number 3 at $t = 2 \times t_{max}/10$
- dynamic cycling phase number n
- polarization curve number $n + 1$ at $t = n \times t_{max}/10$ with $1 \leq n \leq 10$.

Unless the cell voltage drops below 0.3 V, the test normally ends with the final polarization curve, although another test end criterion can also be defined [12].

21.2.6.3 Data Post-Processing

The evolution of the cell voltage is the main output of this test. The power density is calculated as another test output. The voltage deviation, ΔV, is calculated at

OCV from the polarization curve measurements and on load from the dynamic cycling measurements. The deviations as per phase or for the total test duration are evaluated as performance progression and degradation rates, $\Delta V_{OCV}/dt$ and $\Delta V_{iload}/dt$, respectively. The losses in performance (voltage and power) of the cell are calculated as fractions (or percentages) at i_{load} as follows:

- $[V_{\text{initial at } i @ 20\% \, Pmax} - V_{\text{final at } i @ 20\% \, Pmax}]/V_{\text{initial at } i @ 20\% \, Pmax} \; (\times 100\%)$
- $[V_{\text{initial at } i @ 100\% \, Pmax} - V_{\text{final at } i @ 100\% \, Pmax}]/V_{\text{initial at } i @ 100\% \, Pmax} \; (\times 100\%)$
- $[P_{\text{initial at } i @ 20\% \, Pmax} - P_{\text{final at } i @ 20\% \, Pmax}]/P_{\text{initial at } i @ 20\% \, Pmax} \; (\times 100\%)$
- $[P_{\text{initial at } i @ 100\% \, Pmax} - P_{\text{final at } i @ 100\% \, Pmax}]/P_{\text{initial at } i @ 100\% \, Pmax} \; (\times 100\%)$.

The cell has passed the dynamic load cycling aging test when the acceptance criterion is met, for example, the cell degradation (rate) is below a specified value for the applied operating conditions [12].

21.3
Standardized Test Cells

The performances of the components of the fuel cell, such as electrode catalysts and electrolytes, are influenced by the test cell used to evaluate the power generation performance itself. Different test cells lead to different results of material characterization. The exact description of the test cell used is therefore a requirement to reproduce measured material properties.

Examples of standardized test cells for PEFC are the Japan Automobile Research Institute (JARI) standard single cell [13] and the test cell according to the specifications of the International Electrotechnical Commission IEC/TS 62282-7-1 [14]. The Technical Specification TS 62282-7-1 describes standard single-cell test methods for PEFCs and standard test cells. It provides consistent and repeatable methods to test the performance of single cells. It is designed to be used by component or stack manufacturers and by fuel suppliers. The scope of the specification is the evaluation of the performance of membrane electrode assemblies (MEAs) in PEFCs, materials, or structures of other components of PEFCs, or the influence of impurities in fuel and/or in air on the fuel-cell performance. The IEC is the international standards developing organization comprising the different national electrotechnical standardization committees.

21.4
Degradation and Lifetime Investigations

21.4.1
Analysis of MEA Aging Phenomena

Degradation and lifetime investigations are intended to characterize the membrane, the catalyst, and the corrosion of the carbon support as a function of time and to observe the influence of the operating conditions on the lifetime of the MEA [15–17].

Concerning membrane aging, the change in the resistance can be observed with electrochemical impedance spectroscopy or with the change in slope of the polarization curve in the ohmic region. The appearance of pinholes in the membrane can be detected with hydrogen diffusion current measurement or by a decline in OCV. The amount of fluoride or acid released into the water exiting the cell indicates the progress of chemical degradation of the membrane, for example, of perfluorinated sulfonic acid or phosphoric acid-doped polybenzimidazole type.

The degradation and the loss of activity of the (platinum) catalyst can be observed with cyclic voltammetry. A certain loss of performance also indicates declining catalyst activity due to higher reaction overvoltage. Changes in platinum particle size or platinum dissolution and deposition in the membrane can be observed postmortem with scanning or transmission electron microscopy combined with energy- or wavelength-dispersive diffraction analysis.

Carbon corrosion through evolution especially of carbon dioxide and a corresponding reduction in catalyst support surface area can be detected with exhaust gas analysis. The increase in the cathode resistance can be observed with impedance spectroscopy.

21.4.2
Load Cycling

Load cycling is a relevant stressor for the membrane, the platinum catalyst, and carbon support material of PEFC compared with holds at constant potential. It is carried out by cycling a changing potential profile. Load cycling affects degradation of the MEA because of the changes in potential, humidity due to product water formation, and electrical charge quantity. Experimental data observed with the methods mentioned above are intended to characterize MEA degradation as a function of time [18].

Liu and Case [19] investigated MEAs with different load cycling conditions. One MEA (MEA1) was cycled for 10 periods, where one period includes 1,000 cycles (~100 h) with the profile given in Figure 21.6 at 80 °C and 100% RH. Another MEA (MEA2) was operated with constant maximum load at 1,060 A cm^{-2} for the same time. The MEAs were electrochemically characterized every 100 h. It was demonstrated that membrane thinning and/or pinhole formation of MEA1 was the most dominant degradation for cyclic current aging after 500 h. Constant current conditions affected mass transport limitations as the major degradation source of MEA2, whereas membrane degradation played a secondary role. A semiempirical phenomenological durability model was successfully established to incorporate the aging observations and describe the cell performance with time. The results illustrated the demand for a standard fuel-cell durability test protocol.

Borup et al. [20] observed with aging tests at different relative gas humidities at 80 °C an accelerated platinum particle growth. A decrease in hydrophobicity of the gas diffusion layer and a change in the water uptake of the membrane were also

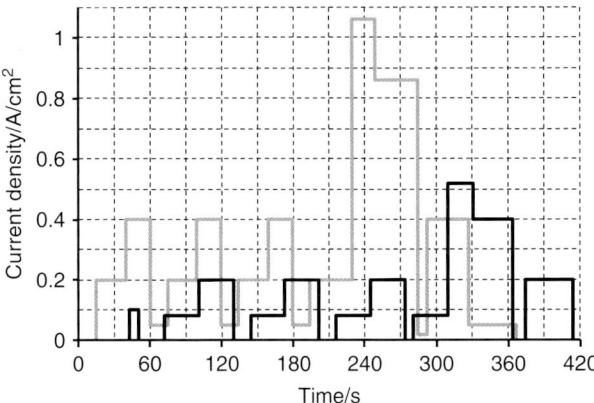

Figure 21.6 Load cycle profile of MEA1 based on data of Liu and Case [19] (gray line) and Borup et al. [20] (black line).

measured. An example of a load cycle as defined by the US Department of Energy is shown in Figure 21.6. It is notable that the profile is similar to that of Liu and Case [19]. In another study, Borup et al. [21] examined the catalyst surface decrease with four different voltage cycling tests between 0.1 and 1.2 V, 0.1 and 1.0 V, 0.1 and 0.96 V, and 0.1 and 0.75 V for 1500 cycles each and characterization after every 300 cycles. The results showed that the higher the voltage difference, the higher is the loss of initial platinum surface and the higher is the resulting particle size at the cathode. The platinum particle size at the anode was not influenced by the applied voltage. In this study, a simulated vehicle drive test with dynamic load cycling was described (see Figure 21.7).

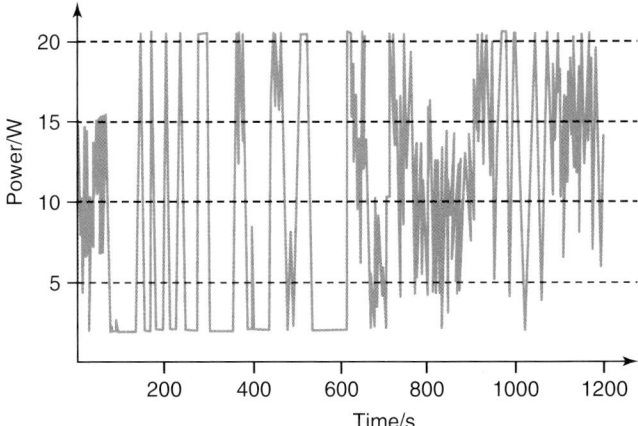

Figure 21.7 Fuel-cell test drive cycle based on a fuel-cell hybrid vehicle operating on the US06 drive cycle [21].

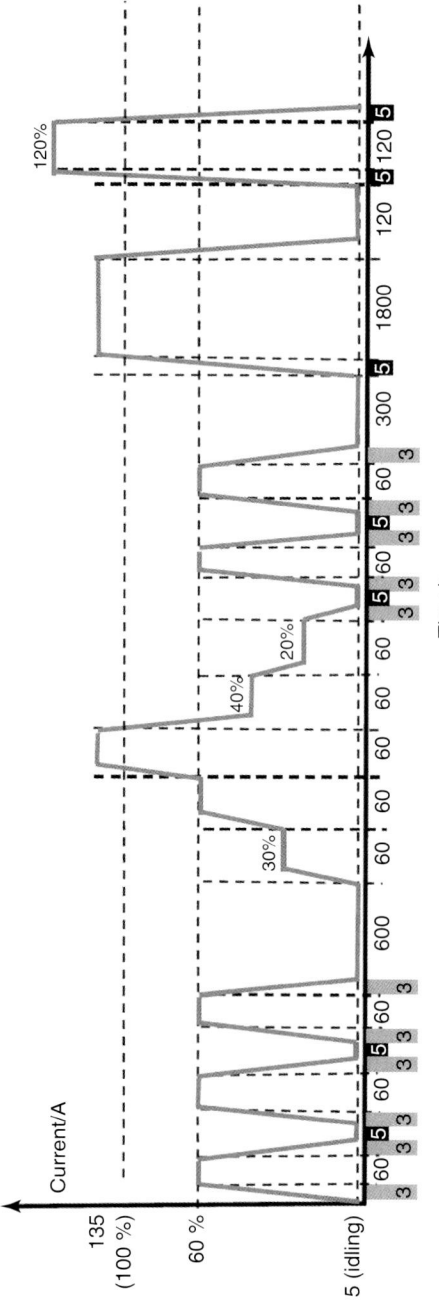

Figure 21.8 PEMFC stack test driving cycle used by Lu et al. [22].

Lu et al. [22] presented a semi-experimental voltage degradation model for PEFC stacks used in buses. The constants of the model were derived from a 5 kW stack, operated in the driving cycle given in Figure 21.8 for 640 h. Analyses showed that the activation overvoltage (kinetics loss) dominates the total losses with almost 80% followed by the ohmic loss. The concentration loss did not change with aging in the driving cycle.

Janssen et al. [23] cycled different types of MEAs at 80 °C and 90–100% RH with the load profile shown in Figure 21.9 for 600 h of operation. Before the cycling and after each set of 10 cycles, electrochemical diagnostics were carried out. After the load cycling test, the kinetic activity of the MEAs showed a decrease due to the gradual loss of active surface area. Membrane thinning was observed by a drop in OCV, although the membrane resistance was not affected by voltage cycling.

Load cycling under relevant automotive conditions was reported by Makharia et al. [24]. The MEAs were operated between peak power voltage at 0.7 V and idle load at 0.9 V for 300,000 cycles and 5,500 h of testing, respectively, at 80 °C with air and hydrogen at 100% RH (see Figure 21.10). Voltage cycling affected an accelerated loss of active catalyst surface. The surface loss of pure platinum was significantly higher in comparison with a Pt–Co alloy catalyst.

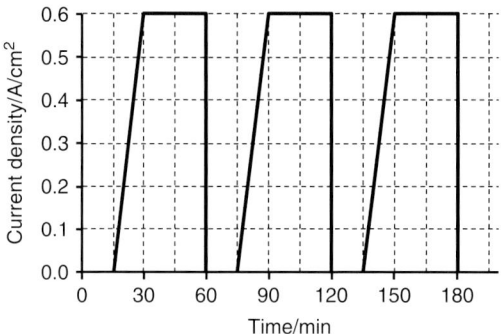

Figure 21.9 Three load cycles from the stress test based on data of Janssen et al. [23].

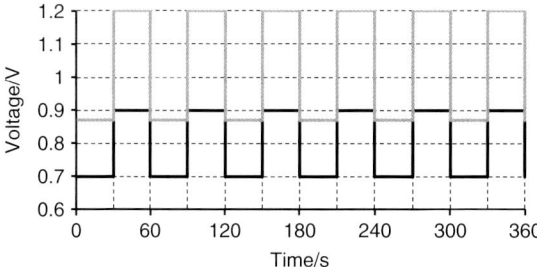

Figure 21.10 Six load cycles from the stress test based on data of Makharia et al. [24] (black line). In their study, another load cycling test profile by Patterson [25] is cited (gray line).

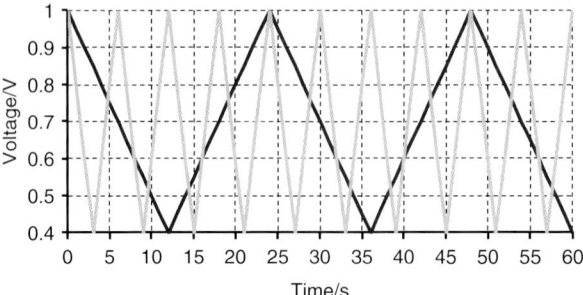

Figure 21.11 Different load cycling profiles based on data of Ohyagi et al. [26]. The profile with a scan rate of 50 mV s^{-1} was cycled 10,000 times (black line). The profile with a scan rate of 200 mV s^{-1} (gray line) was carried out for 50,000 cycles to simulate a vehicle operation mode.

Ohyagi et al. [26] carried out load cycling tests between 1.0 and 0.4 V under different humidification levels (20, 100, and 189% RH[2)]) of the oxygen–nitrogen or nitrogen gas supply at the cathode side (see Figure 21.11). Nitrogen was selected to examine the effect of humidity in the absence of product water. It was shown that the active catalyst surface area decreased under higher humidity conditions due to particle agglomeration. The presence of liquid water or oxygen accelerated the platinum degradation. Furthermore, it was shown that catalyst agglomeration was dependent on the number of oxidation and reduction cycles rather than on the voltage scan rate.

21.5
Design of Experiments in the Field of Fuel-Cell Research

PEFCs have a high system complexity due to their multivariate nature. Modeling is dependent on the estimation of unknown parameters which influence the measurement results. The standard modeling method is the analysis of experimental data from measurements. Experimental investigations are an expensive and time-consuming task, because most studies are carried out by varying one factor at a time and/or by using full-factorial designs. Therefore, it is important to minimize the number of experiments while achieving the best possible parameter estimation. The quality of the parameter estimation depends on the measurement points used and hence on the DoE. DoE is an important contribution to provide efficiently high-quality models and to speed up the commercialization of PEFCs [27–29].

Meiler et al. [27] adopted the general method of optimizing nonlinear experimental designs by the minimization of the covariance matrix of the least-squares

---

2) The cathode side was maintained under supersaturated conditions (humidified at 91 °C dew point against 75 °C cell temperature.

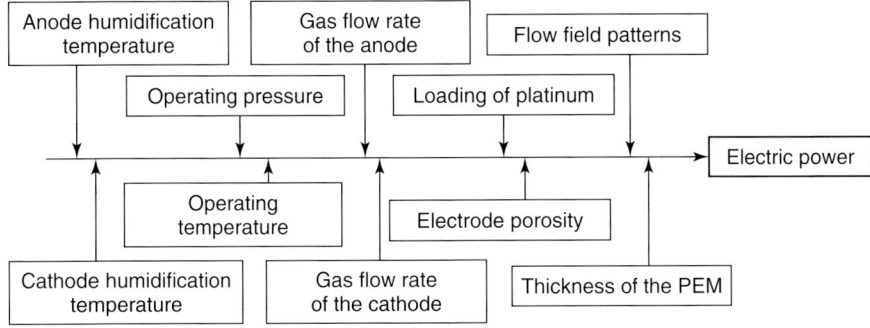

Figure 21.12 Cause–effect diagram of the electrical power for fuel cells [29].

estimate in order to investigate its ability for application in PEFC characterization. By using this method, already a number of experimental data points which is as high as four times the number of model parameters delivers an accuracy model simulation result. Further experiments would provide for minor improvements and are in most cases not recommendable in view of increasing experimental efforts. With the described method, the cost and efforts can be reduced by a factor of more than eight in comparison with a traditional characterization with performance curves. Wahdame et al. [28] concluded that DoE is able to highlight the impacts of the factors on the response of simple and exact models and detect possible interactions between parameters. Furthermore, DoE-based models can be incorporated as sub-models into global physical models to represent a local or a particular phenomenon which cannot be directly and easily described using a mechanistic approach. Another advantage of such models is that they can easily be used for optimization purposes. Yu et al. [29] presented an integrated approach that combines the fractional factorial DoE and the Taguchi method to optimize the operating conditions for the PEFC. Figure 21.12 presents control factors that may influence the electrical power. The number of experiments to identify the interactions between the factors can be reduced considerably by using the Taguchi method.

References

1. Chaudourne, S., Tombini, C., Perette, L., and Junker, M. Standardization and regulation on hydrogen systems in Europe and in the World, presented at the 1st European Hydrogen Energy Conference, EHEC 2003, Grenoble, September 2003, Paper CO5/147.
2. Davis, M.W., Ellis, M.W., and Doughtery, B.P. (2006) Proposed Test Methodology and Performance Rating Standard for Residential Fuel Cell Systems, NISTIR7131, National Institute of Standards and Technology, Washington, DC.
3. Tsotridis, G., Podias, A., Winkler, W., and Scagliotti, M. (eds.) (2006) Fuel Cells Glossary, EUR 22295 EN, European Commission, Joint Research Centre, Institute of Energy, Petten.
4. Honselaar, M. and Tsotridis, G. (2011) The dynamics of the stationary fuel cell standardisation framework. *Int. J. Hydrogen Energy*, **36** (16), 10255–10262.

5. Bove, R. and Malkow, T. (2008) PEM fuel cell stack testing in the framework of an EU-harmonized fuel cell testing protocol: results for an 11 kW stack. *J. Power Sources*, **180**, 452–460.
6. Malkow, T., Saturnio, A., Pilenga, A., De Marco, G., Honselaar, M., and Tsotridis, G. (2011) Assessment of PEFC performance by applying harmonized testing procedure. *Int. J. Energy Res.*, **35** (12), 1075–1089.
7. European Commission, Joint Research Centre (2011) Fuel Cell Systems Testing, Safety & Quality Assurance (FCTES^{QA}), http://fctesqa.jrc.ec.europa.eu/, European Commission, Joint Research Centre, Institute for Energy, Petten (last accessed 10 December 2011).
8. Malkow, T., De Marco, G., Pilenga, A., Honselaar, M., Tsotridis, G., Escribano, S., Antoni, L., Reissner, R., Thalau, O., Sitters, E., and Heinz, G. (2010) Testing the Humidification Sensitivity of a Single PEFC, Characterisation of the performances of a PEFC operating with fuel and oxidant at various relative humidity, Test Module PEFC SC 5-1, *http://fctesqa.jrc.ec.europa.eu/downloads/ PEMSCLV/Umidification_sensitivity_ TestProcedure.pdf*, European Commission, Joint Research Centre, Institute for Energy, Petten (last accessed 10 December 2011).
9. Malkow, T., De Marco, G., Pilenga, A., Honselaar, M., Tsotridis, G., Escribano, S., Antoni, L., Reissner, R., Thalau, O., Sitters, E., and Heinz, G. (2010) Testing the Voltage and the Power as a Function of the Current Density Following an On/Off Profile Versus Time, Accelerated ageing on/off cycling test for a PEFC single cell, Test Module PEFC SC 5-4, *http://fctesqa.jrc.ec.europa.eu/downloads/ PEMSCLV/OnOff_ageing_cycle_ TestProcedure.pdf*, European Commission, Joint Research Centre, Institute for Energy, Petten (last accessed 10 December 2011).
10. Malkow, T., De Marco, G., Pilenga, A., Honselaar, M., Tsotridis, G., Escribano, S., Antoni, L., Reissner, R., Thalau, O., Sitters, E., and Heinz, G. (2010) Testing the Voltage and Power as Function of Current Density, Polarisation curve for a PEFC single cell, Test Module PEFC SC 5-2, *http://fctesqa.jrc.ec.europa.eu/downloads/ PEMSCLV/Polarisation_curve_ TestProcedure.pdf*, European Commission, Joint Research Centre, Institute for Energy, Petten (last accessed 10 December 2011).
11. Malkow, T., De Marco, G., Pilenga, A., Honselaar, M., Tsotridis, G., Escribano, S., Antoni, L., Reissner, R., Thalau, O., Sitters, E., and Heinz, G. (2010) Testing the Voltage and the Power as a Function of Time at a Fixed Current Density, Long term durability steady test for a single PEFC, Test Module PEFC SC 5-6, *http://fctesqa.jrc.ec.europa.eu/downloads/ PEMSCLV/Durability_SteadyState_ TestProcedure.pdf*, European Commission, Joint Research Centre, Institute for Energy, Petten (last accessed 10 December 2011).
12. Malkow, T., De Marco, G., Pilenga, A., Honselaar, M., Tsotridis, G., Escribano, S., Antoni, L., Reissner, R., Thalau, O., Sitters, E., and Heinz, G. (2010) Testing the Voltage and the Power as a Function of the Current Density Following a Dynamic Profile Versus Time, Dynamic load cycling ageing test for a PEFC single cell, Test Module PEFC SC 5-7, *http://fctesqa.jrc.ec.europa.eu/downloads/ PEMSCLV/DynamicLoad_ageing_cycle_ TestProcedure.pdf*, European Commission, Joint Research Centre, Institute for Energy, Petten (last accessed 10 December 2011).
13. Hashimasa, Y., Numata, T., Moriya, K., and Watanabe, S. (2006) Study of fuel cell structure and heating method: development of JARI's standard single cell. *J. Power Sources*, **155**, 182–189.
14. IEC (2010) IEC/TS 62282-7-1. *Fuel Cell Technologies – Part 7-1: Single Cell Test Methods for Polymer Electrolyte Fuel Cell (PEFC)*, International Electrotechnical Commission, Geneva.
15. Wu, J., Yuan, X.Z., Martin, J.J., Wang, H., Zhang, J., Shen, J., Wu, J., and Merida, W. (2008) A review of PEM fuel cell durability: degradation mechanisms and mitigation strategies. *J. Power Sources*, **184**, 104–119.

16. Wu, J., Yuan, X.Z., Wang, H., Blanco, M., Martin, J., and Zhang, J. (2008) Diagnostic tools in PEM fuel cell research. Part I: electrochemical techniques. *Int. J. Hydrogen Energy*, **33**, 388–404.
17. Wu, J., Yuan, X.Z., Wang, H., Blanco, M., Martin, J., and Zhang, J. (2008) Diagnostic tools in PEM fuel cell research. Part II: physical/chemical methods. *Int. J. Hydrogen Energy*, **33**, 1747–1757.
18. Zhang, S., Yuan, X., Wang, H., Mérida, W., Zhu, H., Shen, J., Wurde, S., and Zhang, J. (2009) A review of accelerated stress tests of MEA durability in PEM fuel cells. *Int. J. Hydrogen Energy*, **34**, 388–404.
19. Liu, D. and Case, S. (2006) Durability study of proton exchange membrane fuel cells under dynamic testing conditions with cyclic current profile. *J. Power Sources*, **162**, 521–531.
20. Borup, R.L., Davey, J.R., Garzon, F.H., Wood, D.L., Welch, P.M., and Morec, K. (2006) PEM fuel cell durability with transportation transient operation. *ECS Trans.*, **3**, 879–886.
21. Borup, R., Davey, J., Garzon, F., Wood, D., and Inboy, M. (2006) PEM fuel cell electrocatalyst durability measurements. *J. Power Sources*, **163**, 76–81.
22. Lu, L., Ouyang, M., Huang, H., Pei, P., and Yang, F. (2007) A semi-empirical voltage degradation model for a low-pressure proton exchange membrane fuel cell stack under bus city driving cycles. *J. Power Sources*, **164**, 306–314.
23. Janssen, G.J.M., Sitters, E.F., and Pfrang, A. (2009) Proton-exchange-membrane fuel cells durability evaluated by load-on/off cycling. *J. Power Sources*, **191**, 501–509.
24. Makharia, R., Kocha, S.S., Yu, P.T., Sweikart, M.A., Gu, W., Wagner, F.T., and Gasteiger, H.A. (2006) Durable PEM fuel cell electrode materials: requirements and benchmarking methodologies. *ECS Trans.*, **1**, 3–18.
25. Patterson, T. (2002) in *Fuel Cell Technology, Topical Conference Proceedings. AIChE Spring National Meeting, New York* (eds. G.J. Igwe and D. Mah), American Institute of Chemical Engineers, New York, p. 313.
26. Ohyagi, S., Matsuda, T., Iseki, Y., Sasaki, T., and Kaito, C. (2011) Effects of operating conditions on durability of polymer electrolyte membrane fuel cell Pt cathode catalyst layer. *J. Power Sources*, **196**, 3743–3749.
27. Meiler, M., Andre, D., Pérez, Á., Schmid, O., and Hofer, E.P. (2009) Nonlinear D-optimal design of experiments for polymer-electrolyte-membrane fuel cells. *J. Power Sources*, **190**, 48–55.
28. Wahdame, B., Candusso, D., François, X., Harel, F., Kauffmann, J.M., and Coquery, G. (2009) Design of experiment techniques for fuel cell characterization and development. *Int. J. Hydrogen Energy*, **34**, 967–980.
29. Yu, W.L., Wu, S.J., and Shiah, S.W. (2008) Parametric analysis of the proton exchange membrane fuel cell performance using design of experiments. *Int. J. Hydrogen Energy*, **33**, 2311–2322.

22
Methodologies for Fuel Cell Process Engineering
Remzi Can Samsun and Ralf Peters

22.1
Introduction

The development of new technologies, such as fuel cells, requires a huge effort. Fuel cell process engineering aims to develop reliable energy conversion systems which fulfill the requirements defined by the application. In many applications, fuel cell-based energy conversion must compete with conventional energy conversion-technology, or even with a further modified future version of this technology. Therefore, new systems based on fuel cells must perform economically and technically as well as the conventional technology with which they compete. In niche applications, the fuel cell system can be more expensive than the technology it replaces, provided that it offers additional advantages. However, technical reliability must still be guaranteed. Intelligent engineering approaches must therefore accompany fuel cell systems development. Advanced methods are necessary for development and characterization. This chapter outlines methods that can be used in fuel cell process engineering. After the introduction of each method and a brief discussion of the corresponding theory, examples illustrate the application of these methods in fuel cell systems. The chapter is divided into two main sections: verification and analysis. In terms of verification methods, experimental techniques and methods for the evaluation of experiments are discussed. The analysis methods deal with methods for theoretical calculations in systems development.

22.2
Verification Methods in Fuel Cell Process Engineering

In this section, selected verification methods are highlighted that can be used effectively in the development of fuel cell systems. Fuel cells are characterized by their high efficiency and low emissions. However, different applications have different requirements on fuel cell systems. For example, the system may have to be designed for maximum lifetime, minimum weight and volume, and maximum efficiency, or for additional benefits, such as water production or tank inerting as

Fuel Cell Science and Engineering: Materials, Processes, Systems and Technology,
First Edition. Edited by Detlef Stolten and Bernd Emonts.
© 2012 Wiley-VCH Verlag GmbH & Co. KGaA. Published 2012 by Wiley-VCH Verlag GmbH & Co. KGaA.

required in aircraft application. Apart from these requirements, the interaction of the fuel cell stack with the other system components, such as those for fuel supply or reforming, air supply, heat recovery, or off-gas treatment, are also of major importance. An effective approach is therefore required for the characterization and optimization of fuel cell systems.

One option involves using the design of experiments (DOE) methodology. This approach is widely used in research, development, and industry to design experiments and analyze their results in a systematic way. A reliable analysis of the experimental results using measured quantities is only possible if the measurement uncertainty is also evaluated. A well-established and internationally recognized approach for this evaluation was proposed by the Joint Committee for Guides in Metrology (JCGM). A special application of this general approach, which is suitable for fuel cell systems, involves the use of Monte Carlo techniques. As a further verification method, we also introduce approaches for conversion determination in reforming processes.

22.2.1
Design of Experiments

The DOE methodology is based on varying all experimental parameters in an intelligent and balanced fashion so that maximum information can be gained from the analysis of the experimental results [1]. It is an important tool for quality engineering, which facilitates cost-effective experiments with a high level of information. Using DOE, an optimal ratio can be achieved between the number of trials and the information content of the results [1]. Such an approach also allows experimental error to be distinguished from real effects caused by changing parameters [1].

The basic principles of DOE were first published by Fisher in 1935 in *The Design of Experiments*, and many editions of this book have been published over the years [2]. Scheffler provided a good overview of these principles [3]:

- **Replication**: average values are more reliable than single values. Only replication gives information about variance.
- **Randomization**: parasitic effects and trends flow into variance and do not lead to errors in the analyzed effect.
- **Block generation**: via systematic restriction of randomization, differences in charges and trends affect neither the analyzed parameters nor the variance. They are considered as block effects.
- **Symmetry**: symmetrical design of experimental points will make the analysis easier and the result sharper.
- **Using the region of independent variables**: synchronous variation of multiple parameters reduces the number of experiments, and the optimal distribution of experimental points enhances the conclusion.
- **Confounding**: systematic overlapping of main effects and side effects reduces the number of experiments.

- **Sequential experimenting**: stepwise planning, experimenting, analyzing, and so on will make it possible to improve the conclusion in stages. The experiment can then be stopped, as the required information level has been reached.

In the following sections, different designs will be explained using fuel cell systems as examples.

22.2.1.1 2^2 Factorial Design

A 2^2 factorial design is a two-level analysis consisting of a lower level (−) and an upper level (+) using two factors. Four experiments are carried out combining the upper and lower levels of each factor. This design can only be used if the response function is linearly dependent on the observed factors. An additional experiment can be designed at the center point using the average value. The center point experiment can be used either for replication purposes or in order to check the validity of the assumption of using a linear model. Table 22.1 presents the settings for an experiment with a 2^2 factorial design including a center point. Since there are only two factors, there is only one interaction. The algebraic sign of the interaction is determined by multiplying the algebraic sign of each factor in each experiment. A visual presentation of the design points is given in Figure 22.1.

An example of an experiment with a 2^2 factorial design is the analysis of the CO concentration after a water gas shift (WGS) reactor in a fuel cell system operating with reformate. In the example, the factors were selected as the inlet temperature of the shift reactor and the flow rate of quench water fed between the high- and low-temperature shift stages. The response of these values to the CO concentration as the target parameter was analyzed. In order to simplify the discussion, the values at the lower and upper levels were standardized as −1 and +1, respectively. The response functions were derived using the standardized values for each factor.

Table 22.1 Settings for an experiment with a 2^2 factorial design including a center point.

Experiment No.	Factor 1	Factor 2	Interaction of 1 and 2
1	−	−	+
2	−	+	−
3	+	−	−
4	+	+	+
5 (center point)	0	0	0

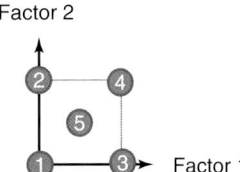

Figure 22.1 2^2 factorial design with center point according to factor combinations in Table 22.1.

This approach also simplified the determination of the coefficients of the response function. The response function is an equation that outputs the value of the target parameter, the CO concentration, for any inlet temperature, and water feed rate in a region defined by the lower and upper level of each parameter. This equation can later be used as a characteristic curve to predict reactor performance for parameter combinations which were not tested.

The standard linear model used for the prediction of response using the experimental results is given below; note that the interaction of factors 1 and 2 is also included:

$$y = a_1 x_1 + a_2 x_2 + a_3 x_1 x_2 + a_4 \tag{22.1}$$

In the actual example, the model parameters were predicted as $a_1 = -0.46$, $a_2 = 0.415$, $a_3 = -0.34$, and $a_4 = 1.455$. The parameters a_1 and a_2 represent the main effects, a_3 is the interaction between the factors, and a_4 is the constant term calculated using the average value of all four results.

Table 22.2 gives the CO concentrations measured during the experiment and the predicted results using the developed response function model. The model was predicted using the four parameter combinations with the upper (+) and lower (−) levels. The last experiment with the center point (0 0) was not used in the model. Figure 22.2 shows the response surface drawn using the model. Using such a response surface, the experimental results can be used to predict the system performance for the complete region for which boundaries are defined by the four points of the experimental design.

Since the model was fitted using the first four results, it calculates the reactor performance for the parameters used in these experiments with maximum accuracy. The fifth experiment with the center point was not taken into account in the model. The requirement from the model is to be able to use this model to predict the reactor performance in the complete region of parameter combinations which are bordered by the upper and lower levels. The fifth experiment can now be used to check the quality of the model, since the parameter combination at the center point

Table 22.2 Example of an experiment with a 2^2 factorial design: CO concentrations after a two-stage WGS reactor as a function of inlet temperature (factor 1) and water feed (factor 2).

Factor 1	Factor 2	Measured value	Predicted value
Inlet temperature	Water feed	CO concentration (vol.%)	CO concentration (vol.%)
−1	−1	1.16	1.16
−1	1	2.67	2.67
1	−1	0.92	0.92
1	1	1.07	1.07
0	0	1.00	1.46

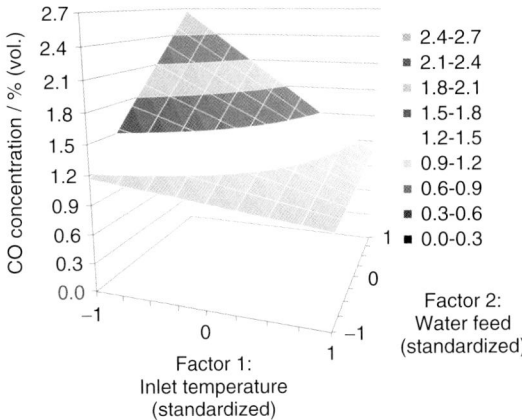

Figure 22.2 The response surface for the CO concentration predicted using a 2^2 factorial design.

is in a region covered by the model. Using the linear model with the parameters below Eq. (22.1), the CO concentration was calculated as 1.46 vol.% for the center point. The measured value, however, corresponded to 1 vol.%. This resulted in a deviation of 45.5%. The number of experiments and hence the experimental design with 2^2 factors is therefore insufficient to predict the performance of the WGS reactor in the tested region. This does not necessarily mean that a 2^2 design is not useful; it is just not the right choice for the example analyzed here.

At this point, the experiment must be extended. This does not mean that the five experiments which were carried out are useless. As mentioned above, the DOE approach makes it possible to improve the conclusion in stages.

22.2.1.2 3^2 Factorial Design

The high deviation between the measured and the predicted values using a 2^2 factorial design showed that the reactor performance cannot be predicted accurately using a linear interaction model. To improve the quality of the conclusions, a quadratic model was developed. The number of experiments was therefore increased in order to derive the necessary parameters for such a complicated model. The required information can be generated in two ways using the DOE approach.

In the first method, the 2^2 factorial design can be extended to a central composite design ([1], p. 22; [3], p. 230). In a central composite design, the factorial design is extended to include the "star points" and additional replicates at the center points in the design, which in turn allows the quadratic coefficients to be estimated ([1], p. 22). Apart from the already defined 2^n experiments using the factorial design, the central composite design consists of additional 2^n star point experiments and one center point experiment ([3], p. 231), where n is the number of factors (in the present example, $n = 2$). In ([1], p. 22), it is recommended that the number of replicates at the center point be selected so that they are roughly equal to the number of factors. A central composite design with star and center points using two factors is presented in Figure 22.3a.

In this type of design, the distance α between the star points and the center points must be defined. This value can be calculated using the following equation (adapted from [3], p. 236):

$$\alpha^2 = 0.5\left(\sqrt{n_{\text{total}}2^n} - 2^n\right) \qquad (22.2)$$

where n_{total} denotes the number of experiments and n the number of factors. The second method is based on the principle that at least three levels are required in order to include the non-linearity ([3], p. 217). This can be realized with a 3^n factorial design using three levels ($-$, 0, $+$). Here, we use a 3^2 factorial design with two factors, as shown in Figure 22.3b.

It must be noted that both methods result in the same experimental design when two factors are involved and α is defined as 1.

In this example, the previous experiment with a 2^2 factorial design has been extended to a 3^2 factorial design using the new settings given in Table 22.3. Two further experiments were carried out at the center point.

Using the measured CO concentrations as a function of the parameter combinations of the inlet temperatures and water feeds, two different models were

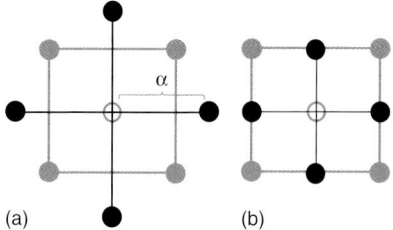

Figure 22.3 Central composite design with star and center points (a) and 3^2 factorial design (b).

Table 22.3 Settings for an experiment with a 3^2 factorial design[a].

x_1	x_2	$x_1 x_2$	x_1^2	x_2^2	$x_1^2 x_2$	$x_1 x_2^2$	$x_1^2 x_2^2$
−	−	+	+	+	−	−	+
−	+	−	+	+	+	−	+
+	−	−	+	+	−	+	+
+	+	+	+	+	+	+	+
−	0	0	+	0	0	0	0
+	0	0	+	0	0	0	0
0	−	0	0	+	0	0	0
0	+	0	0	+	0	0	0
0	0	0	0	0	0	0	0

[a]The shaded region represents the settings with a 2^2 factorial design.

Table 22.4 Example of an experiment with a 3^2 factorial design: CO concentrations after a two-stage WGS reactor as a function of inlet temperature (factor 1) and water feed (factor 2)[a].

Factor 1	Factor 2	Measured value	Predicted value (first model)	Predicted value (second model)
Inlet temperature	Water feed	CO concentration (vol.%)	CO concentration (vol.%)	CO concentration (vol.%)
−1	−1	1.16	0.89	1.16
−1	1	2.67	2.40	2.67
1	−1	0.92	1.19	0.92
1	1	1.07	0.80	1.07
−1	0	2.25	1.54	2.25
1	0	0.85	0.89	0.85
0	−1	0.86	0.86	0.86
0	1	1.42	1.42	1.42
0	0	1.00	1.04	1.03
0	0	1.03	1.04	1.03
0	0	1.07	1.04	1.03

[a]The shaded region represents the replication experiments at the center point.

developed. Table 22.4 gives an overview of the measured and calculated values. In the first model, the results were fitted using the most important information. In this case, the equation for the response surface can be expressed as

$$y = a_1 x_1 + a_2 x_2 + a_3 x_1 x_2 + a_4 x_1^2 + a_5 x_2^2 + a_6 \tag{22.3}$$

The results were fitted using six experimental results from the first eight experiments and one result from the average value of the last three center point experiments. Comparing the measured values with the predicted values using model 1, the maximum deviation was calculated as 32%. The deviation for the center point was calculated as 4% in the worst case. The new quadratic model represents the experimentally observed behavior better than the first model. However, an exact fit can also be achieved using all 11 experiments. In the second model, the equation for the response surface can be expressed as

$$y = a_1 x_1 + a_2 x_2 + a_3 x_1 x_2 + a_4 x_1^2 + a_5 x_2^2 + a_6 x_1^2 x_2 \\ + a_7 x_1 x_2^2 + a_8 x_1^2 x_2^2 + a_9 \tag{22.4}$$

The coefficient a_9 is calculated using the average value from the three center point experiments. The remaining eight experiments are used to calculate the coefficients $a_1 - a_8$. The fitted response surface is presented in Figure 22.4.

Comparison of the response surface using a 2^2 factorial design (Figure 22.2) with the surface using a 3^2 factorial design (Figure 22.4) clearly shows the advantage of

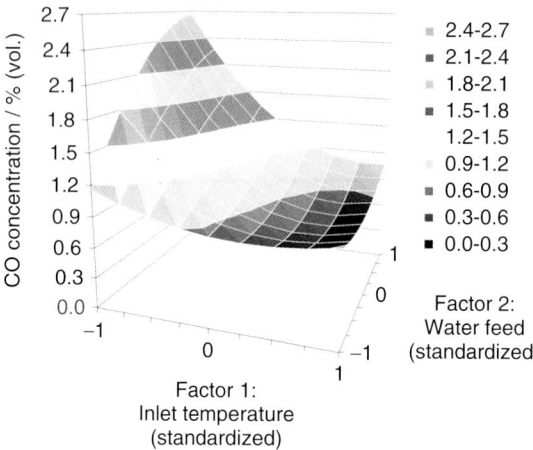

Figure 22.4 The response surface for the CO concentration predicted using a 3^2 factorial design.

extending the experiment. The quadratic response function helps us to interpret the experimental observations more accurately. In the concrete example here, the CO concentration in the hydrogen-rich reformate played an essential role in the performance of the high-temperature polymer electrolyte fuel cell (HT-PEFC). With the help of the quadratic response surface, a very wide range of parameter combinations could be defined in which a CO concentration of less than 1.2% was achieved. The linear response function in Figure 22.2, in contrast, defines a smaller region for parameter combinations, which achieves CO concentrations less than 1.2%. If the fuel cell system is designed according to the results from the 2^2 factorial design, the system will be less flexible in operation. If the more accurate results from the 3^2 factorial design are used, the system will be more flexible. Increased flexibility in system design leads to lower costs, more compact systems, and less complicated control. In the worst case, results could also be used from a model prediction, in which the predicted values are better than those observed in reality. However, this would lead to subsequent unsatisfactory system performance. In the worst case, the system would not work properly, as the optimum parameters have been defined using the model. Therefore, it is extremely important that the quality of response functions be correctly interpreted and that the experiment is extended until a high quality has been achieved.

22.2.1.3 2^3 Factorial Design

Until this point, we have only considered full factorial designs with two factors. Often, there are more parameters that influence the studied response variable. In many cases, it makes sense to design an experiment with three factors. This can be done with a 2^3 factorial design. In this case, it is also advantageous to carry out replication experiments at the center point. Typical settings for a 2^3 factorial

Table 22.5 Settings for an experiment with a 2^3 factorial design.

Experiment No.	x_1	x_2	x_3	x_1x_2	x_1x_3	x_2x_3	$x_1x_2x_3$
1	−	−	−	+	+	+	−
2	−	+	−	−	+	−	+
3	−	−	+	+	−	−	+
4	−	+	+	−	−	+	−
5	+	−	−	−	−	+	+
6	+	+	−	+	−	−	−
7	+	−	+	−	+	−	−
8	+	+	+	+	+	+	+

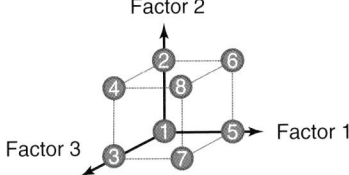

Figure 22.5 2^3 factorial design according to factor combinations in Table 22.5.

design are given in Table 22.5. The parameter combinations are shown visually in Figure 22.5.

As in the case with a 2^2 factorial design, the results of a 2^3 factorial design can be used to predict the value of response variables with acceptable accuracy only if the factors have a linear effect on the response. If this is not the case, the experiment must be extended to a 3^3 factorial design, which results in 27 experiments excluding the replication experiments. It is also possible to select a hybrid design with mixed factors if some factors have a linear effect and some have a nonlinear effect on the response parameter. For example, if one of the factors has a linear effect and others linear and nonlinear effects, a hybrid design with 2×3^2 factors can be used, resulting in 18 parameter combinations excluding the replication experiments.

In the following example, a complicated case was characterized using a 2^3 factorial design. A fuel cell system operating with reformate was designed. In such a system, the number of necessary components is typically very high. Therefore, it is necessary to integrate the heat exchangers into the reactors to recover heat for educt conditioning and to minimize system volume and weight. In the sub-system according to Figure 22.6, a certain amount of water had to be completely vaporized and superheated to a certain temperature level. To maximize the system efficiency, the necessary heat was recovered from the reaction heat. However, the necessary amount of heat could not be recovered from a single reactor. In the proposed strategy, some of the cold water was vaporized completely in the first reactor

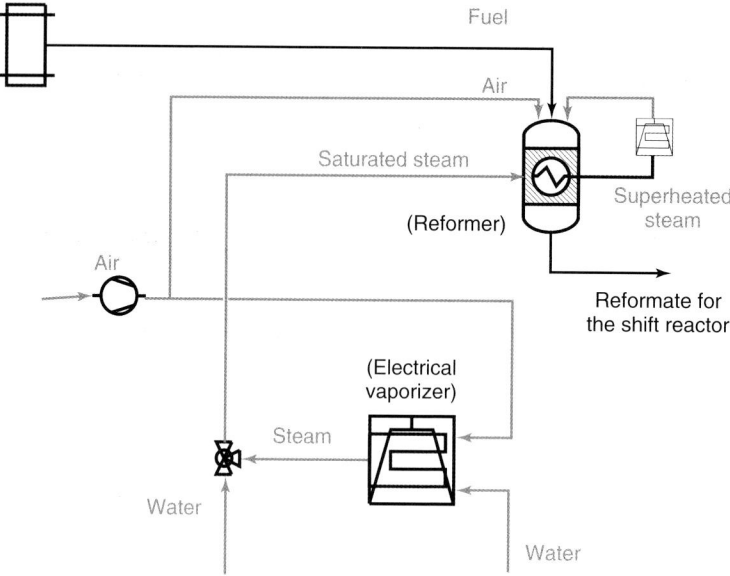

Figure 22.6 Experimental setup used in the example with 2^3 factorial design.

and later mixed with the remaining water before the second reactor, resulting in saturated steam. The first reactor was simulated by an electrical vaporizer in the experiment. The saturated steam was subsequently fed into the second reactor in which the reforming reaction took place. Saturated steam was vaporized completely and superheated here. The first factor (x_1) was defined as the ratio of water vaporized in the electrical vaporizer to the total amount of water in the system. The system design allowed some of the process air, which was also used for reforming, to be mixed with the cold water, which was fed into the electrical heater. The remaining air was then fed cold into the reactor. Alternatively, the total amount of air required for the reformer can be fed cold into the reformer. In this case, the second factor (x_2) was defined as the ratio of cold air to the total amount of air. As the last parameter, the reformer load was varied to allow changes between full-load and part-load operation to be observed. The amount of fuel feed was therefore defined as the third factor (x_3). All other parameters were kept constant during the experiment to minimize errors. The response variables here were the steam temperature and the temperature of the reformate, and both were measured at the exit of the integrated heat exchanger.

For simplicity, the factors x_1–x_3 are standardized and therefore the lower and upper values are defined as -1 and 1, respectively. The replication experiments are carried out using the mean value for each factor. The main effects, the two-factor and the three-factor interactions, can easily be calculated. The experimental results are presented for the superheated steam temperature at the inlet of the reformer as a function of different parameter combinations in Table 22.6.

Table 22.6 Example of an experiment with a 2^3 factorial design: steam temperature as a function of three factors[a].

Factor 1 (x_1)	Factor 2 (x_2)	Factor 3 (x_3)	Measured value
Ratio of water: vaporizer/total	Ratio of air: cold air/complete	Fuel flow rate	Steam temperature (°C)
−1	−1	−1	364.3
−1	+1	−1	345.2
−1	−1	+1	373.1
−1	+1	+1	343.0
+1	−1	−1	422.3
+1	+1	−1	398.2
+1	−1	+1	432.7
+1	+1	+1	403.9
0	0	0	388.7
0	0	0	386.6

[a] The shaded region represents the replication experiments at the center point.

Table 22.7 Main effects, two-factor interactions, and the three-factor interaction in an experiment with 2^3 factorial design.

Main effects		
Main effect of factor 1	x_1	57.9
Main effect of factor 2	x_2	−25.5
Main effect of factor 3	x_3	5.7
Two-factor interactions		
Interaction between 1 and 2	$x_1 x_2$	−0.9
Interaction between 1 and 3	$x_1 x_3$	2.4
Interaction between 2 and 3	$x_2 x_3$	−3.9
Three-factor interaction		
Three-factor interaction	$x_1 x_2 x_3$	1.6

Table 22.7 presents the calculated effects. As can clearly be seen, none of the main effects can be neglected. The first factor has the largest influence on the response variable. This is the ratio of water between the vaporizer and the total amount of water. Since vaporization is very energy intensive, vaporizing more water in the electrical vaporizer should result in a higher heat input to the reformer. Increasing the ratio of cold air via factor 2 has a relatively smaller effect than with factor 1. The effect is negative as the lower value of this factor leads to more heat input to the system, which increases heat recovery and the resulting steam temperature. The fuel flow rate, factor 3, has a smaller effect than the other two factors, but still

cannot be neglected. The interaction between factors 1 and 2 is relatively small. It is even smaller than the three-factor interaction, which is often neglected in 2^3 factorial designs. Since the calculation of single effects is not combined with a high effort in the case of a linear model, the regression model can be built using all observed effects.

The regression coefficients can be calculated in a similar way. In this case, the sum is divided by the number of experiments. The regression coefficient a_1 for the main effect of factor 1 can be calculated as

$$a_1 = \frac{-364.3 - 345.2 - 373.1 - 343.0 + 422.3 + 398.2 + 432.7 + 403.9}{8}$$
$$= 28.9 \qquad (22.5)$$

The regression coefficient a_8 is calculated from the average value of eight experiments:

$$a_8 = \frac{+364.3 + 345.2 + 373.1 + 343.0 + 422.3 + 398.2 + 432.7 + 403.9}{8}$$
$$= 385.4 \qquad (22.6)$$

The response function has the following form:

$$y = a_1 x_1 + a_2 x_2 + a_3 x_3 + a_4 x_1 x_2 + a_5 x_1 x_3 + a_6 x_2 x_3 + a_7 x_1 x_2 x_3 + a_8 \qquad (22.7)$$

The response surfaces in Figures 22.7 and 22.8 show the fitted data using the response function. The results are presented in two diagrams. Factor 3 was kept constant at its lower ($x_3 = -1$) or upper value ($x_3 = 1$). The temperature at the

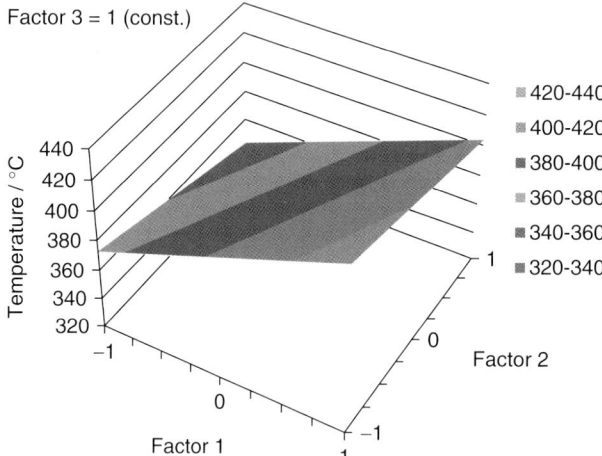

Figure 22.7 The response surface for the steam temperature predicted using a 2^3 factorial design. Factor 1 (x_1): ratio of water vaporized in the electrical vaporizer to the total amount of water. Factor 2 (x_2): ratio of cold air fed into the reformer to the total amount of air. Factor 3 (x_3): fuel flow rate kept constant at its upper value (1).

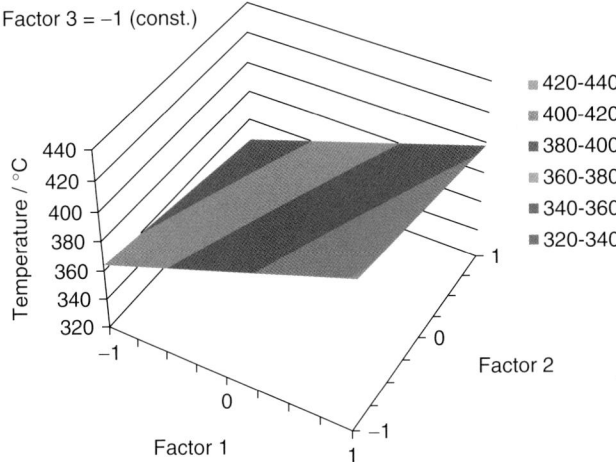

Figure 22.8 The response surface for the steam temperature predicted using a 2^3 factorial design. Factor 1 (x_1): ratio of water vaporized in the electrical vaporizer to the total amount of water. Factor 2 (x_2): ratio of cold air fed into the reformer to the total amount of air. Factor 3 (x_3): fuel flow rate kept constant at its lower value (-1).

center point predicted using the derived response function was 385.3 °C. The deviation between the predicted value and the measured values was less than 1%. This shows that a linear model is sufficient to describe the temperature behavior in the selected region.

22.2.1.4 2^{n-k} Fractional Factorial Designs

Up to now, we have only dealt with full factorial designs. In full factorial designs, the number of experiments increases exponentially with the number of factors to be analyzed. This leads to the difference between the number of factors and the number of experiments becoming unacceptably high. Four factors require 16 experiments for a full factorial design. In a fuel cell system, it is not unusual to have five factors, which would then require 32 experiments in a 2^5 factorial design. A high number of experiments results in a high statistical accuracy in determining the response of each parameter combination. However, this is not justified for the solution of practical problems [4]. It is still possible to gain enough information using a reduced number of experiments. In so doing, the number of experiments must be reduced in a systematic way, ensuring that any information lost is insignificant and that the most important information can still be generated. If there are more than four factors, a fractional factorial design is often useful. In a fractional factorial design, 2^{n-k} experiments are necessary, where n denotes the number of factors and k the number of times by which the number of settings has been halved compared with the corresponding complete 2^n design. Such a design reduces experimental effort, and can be adapted to the complexity of the system under investigation and the information required [1].

An analysis of the effects of an experiment with a fractional factorial design shows that they are confounded. Confounding is defined as a situation where an effect cannot unambiguously be attributed to a single main effect or interaction [1]. The calculated main effects can also be two-factor interactions of other factors. Therefore, care must be taken in interpreting the results. For example, if the effect of the three-factor interaction is negligible in a 2^3 factorial design, a fourth factor can be added in its place. Similarly, if the two-factor interaction of factors 1 and 2 is negligible, a fifth factor can be introduced [3]. The newly introduced factors 4 and 5 are, however, not independent. They are defined based on the other three factors according to the effect they replace.

In most cases, a statistical approach is required to analyze the results of experiments performed using the DOE methodology presented above. A common statistical technique is analysis of variance (ANOVA), which allows variations within sets of data to be isolated and estimated [5]. A brief discussion of this methodology, including manual calculations, can be found in [5] (p. 59). As discussed above, the response function with its coefficients can be derived using a linear or polynomial regression. Details and methods for regression are given in [5] (p. 92). Further details about the DOE can be found in [1–6]. In Section 22.3.1, the same method is applied for systems analysis illustrated with an example.

22.2.2
Evaluation of Measurement Uncertainty

Earlier, we stated that the characterization of fuel cell systems using experiments is a key issue in system development. In reporting the results of experiments, it is not sufficient to give a single number or value. Information on the quality of the result is also necessary, as is additional quantitative information, such as a measure of reliability. The *Guide to the Expression of Uncertainty in Measurement* (GUM) [7], produced by the JCGM, is the internationally accepted guide for the evaluation of uncertainty. The uncertainty of measurement is defined in GUM as a "parameter, associated with the result of a measurement, which characterizes the dispersion of the values that could reasonably be attributed to the measurand." This formal definition is considered to be an operational one focusing on the measurement result and its evaluated uncertainty. The guide also states that it is not inconsistent with other concepts of uncertainty of measurement, such as:

- measure of the possible error in the estimated value of the measurand as provided by the result of a measurement
- an estimate characterizing the range of values within which the true value of a measurand lies.

The guide considers these two traditional concepts valid as ideals, since they focus on unknowable quantities: the "error" of the result of a measurement and the "true value" of the measurand in contrast to its estimated value.

According to GUM, a measurement has imperfections which give rise to errors in the measurement result. A random error presumably arises from unpredictable or stochastic temporal and spatial variations of influence quantities. Although it is not possible to compensate for random error, it can usually be reduced by increasing the number of observations. Systematic error, like random error, cannot be eliminated but it too can often be reduced. Once the effect causing the systematic error has been recognized, the effect can be quantified and a correction can be applied to compensate for the effect. The uncertainty of the result of a measurement reflects the lack of exact knowledge of the value of the measurand. The result of a measurement after corrections is still only an estimate of the value of the measurand because of the uncertainty arising from random effects and from imperfect correction of the result for systematic effects [7].

GUM therefore believes that it is important to distinguish between error and uncertainty. In the following, a general approach will be presented for an uncertainty assessment. Subsequently, the use of Monte Carlo techniques will be introduced to evaluate measurement uncertainty.

22.2.2.1 Summary of Procedure to Evaluate and Express Uncertainty

The principles of GUM for the expression of uncertainty in a measurement are explained briefly in [5] (p. 162). To use and combine different uncertainties, a stepwise approach is introduced here.

In the first step, the measurand must be clearly specified. This consists of a clear definition of what is being measured, including all boundary conditions. Second, all relevant sources of uncertainty must be identified. This results in the measurement equation, which relates different sources to the final result. The equation is extended by further parameters required for uncertainty estimation, such as a precision term. The measurement equation can be written as

$$y = f(x_1, x_2, \ldots, x_n) \tag{22.8}$$

In the next step, standard uncertainties are defined for each source of uncertainty. GUM defines two different methods for estimating uncertainty. Type A is a method of evaluation by the statistical analysis of series of observations. Standard deviations can be calculated through repeated observations. Type B is a method of evaluation of uncertainty by means other than the statistical analysis of series of observations. Calibration results or tolerances given in manuals can be used here. They are usually expressed in the form of limits or confidence intervals. Typical rules for converting such information to an estimated standard uncertainty u are introduced in [5] (p. 164).

After this step, uncertainties in influence quantities must be converted to uncertainties in the analytical results. In simple cases, when measurement equations involve one algebraic operation, the following rules apply: if a quantity x_i is simply added to or subtracted from others to obtain the result y, the contribution to the uncertainty in y is simply the uncertainty $u(x_i)$ in x_i. In case of a division or multiplication, the contribution of the relative uncertainty in y, $u(y)/y$, is the relative uncertainty $u(x_i)/x_i$ in x_i. These rules do not apply in complicated cases with a mixture of different algebraic signs. In order to ascertain the change in the result when the input quantity changes by its uncertainty, a general answer in the form "the uncertainty in x_i times the rate of change of y with x_i" can be obtained using one of the following methods:

1) An experiment is performed to measure the effect of a variable on the result.
2) The input quantity is changed by its standard uncertainty and the result is recalculated.
3) Algebraic differentiation of the measurement equation.

Finally, different contributions to uncertainty, which are determined in the last step in the form of $u_i(y) = c_i u(x_i)$, combine as the root sum of their squares. This is sometimes called "the law of propagation of uncertainty":

$$u(y) = \sqrt{\sum_{i=1,n} c_i^2 u(x_i)^2} \qquad (22.9)$$

This equation is only valid where the uncertainties lead to independent effects on the result. If this is not the case, the equation is more complex:

$$u[y(x_{i,j,\ldots})] = \sqrt{\sum_{i=1,n} c_i^2 u(x_i)^2 + \sum_{\substack{i,k=1,n \\ i \neq k}} c_i c_k u(x_i, x_k)} \qquad (22.10)$$

where $u(x_i, x_k)$ is the covariance between x_i and x_k.

An expanded uncertainty can be used to report results. It is obtained by reporting the standard uncertainty by a coverage factor. The coverage factor is chosen based on the desired level of confidence.

22.2.2.2 The Use of the Monte Carlo Method to Evaluate Uncertainty

Supplement 1 to the GUM defines the Monte Carlo method as a "method for the propagation of distributions by performing random sampling from probability distributions" [8]. This method is particularly suitable for models that cannot be linearized or solved with classical methods. How the Monte Carlo method can be used to evaluate uncertainty is explained briefly in [9]. As discussed above, the first step involves defining the dependence of the output quantity as a function of all possible input quantities. This results in the measurement equation Eq. (22.8). After the measurement equation has been formulated, m random samples are generated for each input quantity with the help of a random number generator and probability distribution functions (PDFs) $\rho_{i,m}$:

$$\xi_{i,m} = \xi_{i,\min} + (\xi_{i,\max} - \xi_{i,\min}) \rho_{i,m} \qquad (22.11)$$

In doing this, the maximum and minimum values for each input quantity define the upper and lower limits for the PDF.

The mth random sample delivers a vector of values for the input quantities $x_1, \ldots, x_n$:

$$\xi_m = \{\xi_{1,m}, \ldots, \xi_{n,m}\} \qquad (22.12)$$

Using each random sample generated, a random sample value for the output is calculated with the measurement equation:

$$\eta_m = f(\xi_{1,m}, \xi_{2,m}, \ldots, \xi_{n,m}) \tag{22.13}$$

As a result, a set of m values are produced for the output value. Now, these results can be analyzed using statistical methods. For example, the average

$$\mu_y \equiv y \cong \bar{y} = M^{-1} \sum_{M=1}^{m} \eta_m \tag{22.14}$$

and standard deviation determined from

$$\sigma_y^2 \equiv u^2(y) \cong s_y^2 = (M-1)^{-1} \sum_{m=1}^{M} (\bar{y} - \eta_m)^2 \tag{22.15}$$

are taken as an estimate y of the output quantity Y and the standard uncertainty $u(y)$ associated with y, respectively [8].

22.2.2.3 Practical Example of the Use of the Monte Carlo Method to Evaluate Uncertainty

In the following, we explain the application of the Monte Carlo method using a simple example. The aim of the experiment in this example is to determine the CO conversion in a WGS reactor. Since the conversion is the result of an experiment, its uncertainty must also be expressed when the result is reported. We assume that only the following reaction takes place in the reactor:

$$CO + H_2O \rightleftharpoons CO_2 + H_2 \tag{22.16}$$

The CO conversion is calculated using the inlet and outlet molar flow rates of CO according to the following equation:

$$X_{CO} = \frac{\dot{n}_{CO,in} - \dot{n}_{CO,out}}{\dot{n}_{CO,in}} \tag{22.17}$$

The volumetric flow rate of the gas mixture and the CO concentration in the mixture at the inlet and outlet of the reactor were determined in the experiment. The above equation can be written as

$$X_{CO} = \frac{\dot{V}_{CO,in} - \dot{V}_{CO,out}}{\dot{V}_{CO,in}} = \frac{x_{CO,in} \dot{V}_{in} - x_{CO,out} \dot{V}_{out}}{x_{CO,in} \dot{V}_{in}} \tag{22.18}$$

This is the measurement equation that is used as the basis in the Monte Carlo simulation. In the next step, the upper and lower limits are defined for each input quantity. In our example, the upper and lower limits for the volumetric flow rates were calculated assuming a relative uncertainty of $\pm 1\%$. An absolute uncertainty of ± 0.13 was assumed for the measured CO concentrations. Using a random number generator, each input quantity was varied between its higher and lower values. We used the MATLAB/Simulink random number generator block for this purpose. A total of 1334 vectors of values were generated as the input quantities for the measurement calculation. The results are shown in the histogram in Figure 22.9.

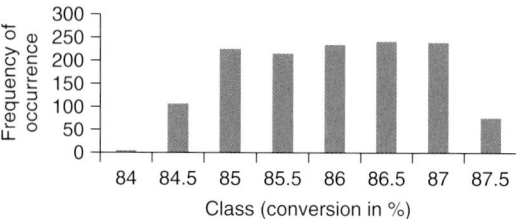

Figure 22.9 Distribution of calculated conversion values using the Monte Carlo method to evaluate measurement uncertainty.

A statistical analysis of the calculated conversion values resulted in a mean value of 85.74% for CO conversion. The average absolute deviation was 0.73, and the standard deviation was 0.85. With the help of these values, the measured conversion can be expressed with its uncertainty in a region instead of as a single value.

The selection of the upper and lower values for each input quantity is extremely important in order to express the uncertainty correctly. It is also important to select the correct measurement equation. In the following example, the same system as above is analyzed using a different approach. In Eq. (22.18), only two input factors are used: the concentration and the volumetric flow rate. Therefore, the resulting uncertainty expression is influenced by uncertainties in both input types. However, the analysis can be made more accurate if more information is incorporated in the measurement equation.

Coming back to the definition of CO conversion, it is possible to rewrite the equation of conversion in terms of the change in the molar flow rate of CO:

$$X_{CO} = \frac{\dot{n}_{CO,in} - \dot{n}_{CO,out}}{\dot{n}_{CO,in}} = \frac{\Delta \dot{n}_{CO}}{\dot{n}_{CO,in}} \tag{22.19}$$

Since the number of moles on each side of the shift reaction equation is the same, the absolute value of the change in the molar flow rate of each substance is also the same:

$$|\Delta \dot{n}_{CO}| = |\Delta \dot{n}_{H_2O}| = |\Delta \dot{n}_{H_2}| = |\Delta \dot{n}_{CO_2}| \tag{22.20}$$

Using this correlation, the CO conversion can be calculated using the rate of change of the molar flow rate of steam:

$$X_{CO} = \frac{\Delta \dot{n}_{H_2O}}{\dot{n}_{CO,in}} = \frac{\dot{n}_{H_2O,in} - \dot{n}_{H_2O,out}}{\dot{n}_{CO,in}} \tag{22.21}$$

The mass flow rate of water at the entrance and exit of the reactor are known from the experiment. Furthermore, the amount of water added for cooling was recorded. Using this information, Eq. (22.18) can be reformulated as

$$X_{CO} = \frac{\dfrac{(\dot{m}_{H_2O,in} + \dot{m}_{H_2O,quench}) - \dot{m}_{H_2O,out}}{M_{w,H_2O}}}{\dfrac{x_{CO,in} \dot{V}_{out}}{V_m}} \tag{22.22}$$

The molar weight of water M_w and the standard volume for ideal gases V_m were taken as constant. A relative measurement uncertainty of $\pm 1\%$ was assumed for the mass flow rate measurements of water. A value of $\pm 0.5\%$ was assumed for the mass flow rate of quench water delivered by a mass flow controller. The uncertainties for CO concentration and the volumetric flow rate of the gas mixture were kept on the same level as in the first example. Using the same procedure as described above, the distribution in Figure 22.10 was achieved with the measurement equation Eq. (22.22).

Comparing Figure 22.10 with Figure 22.9, it can be seen that the calculated conversion results using the alternative equation have a broader distribution than the conversion values calculated with the first equation. The average value was calculated as 88.05%, which is in agreement with the mean value calculated in the first case considering measurement uncertainties. The average absolute deviation of 1.85 was higher than in the first case. As can be seen in the histogram, the deviation from the average value is fairly high, with a calculated value of 5.04 for variation resulting in a standard deviation of 2.25.

A first interpretation of the actual results can lead to the argument that the increased uncertainty using the alternative measurement equation is caused by the single uncertainties coming from the new terms. However, a more detailed analysis with further simulations will show that the main effect for the increased uncertainty comes from the term in the denominator, which is the same in both equations. In the first case, however, the change in the denominator is compensated, as the same term is repeated in the nominator. In other words, even if the deviation between the measured value and the real value is large in the first equation, it will not be recognized by the simulation as a source of uncertainty because the same deviation is also present in the nominator. For example, if the uncertainty in the measured CO concentration at the entrance of the reactor is large, this will not have a strong effect on the conversion because the change in the amount of CO is also calculated using the same value. In the second equation, the terms in the nominator are independent of those in the denominator. The reference is calculated using the CO concentration at the inlet, whereas the change in the amount of CO is calculated using a completely different term, namely the mass flow rate of water. This leads to a large deviation in the calculated conversion.

The application of the Monte Carlo method to express measurement uncertainty has been demonstrated in the above examples. Despite the application of the method

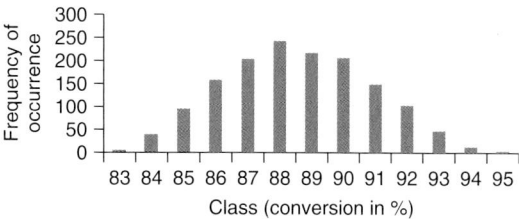

Figure 22.10 Distribution of calculated conversion values using the alternative measurement equation.

being fairly straightforward, special attention must be given to the formulation of the measurement equation. If possible, the measurement equation must be derived in such a way that all available experimental results are integrated together with their uncertainties in order to achieve a reliable expression of the measured quantity with its uncertainty.

22.2.3
Determination of Conversion in Reforming Processes

In this section, it will be shown that analytical and experimental tools should be selected carefully. Different methods lead to different qualities of accuracy. To elucidate this, we will look at the determination of fuel conversion in a reformer. Generally, conversion can be determined by the relation between the product obtained and the feedstock for the reformer. Conversion is defined as

$$\zeta_j = \frac{\dot{n}_j^{in} - \dot{n}_j^{out}}{\dot{n}_j^{in}} = \frac{\dot{m}_j^{in} - \dot{m}_j^{out}}{\dot{m}_j^{in}} \tag{22.23}$$

If liquid-petroleum gas (LPG) is used as a feedstock, the two gases propane (C_3H_8) and butane (C_4H_{10}) must be monitored. Analysis of the product gas can be performed by gas chromatography (GC). The experimental results lead to two conversion values, one for propane and the other for butane:

$$\zeta_{C_3H_8} = 1 - \frac{y_{C_3H_8}^{out} \dot{n}^{out}}{\dot{n}_{C_3H_8}^{in}} = 1 - \frac{y_{C_3H_8}^{out} \dot{V}^{out}}{\dot{V}_{C_3H_8}^{in}} \tag{22.24}$$

$$\zeta_{C_4H_{10}} = 1 - \frac{y_{C_4H_{10}}^{out} \dot{n}^{out}}{\dot{n}_{C_4H_{10}}^{in}} = 1 - \frac{y_{C_4H_{10}}^{in} \dot{V}^{out}}{\dot{V}_{C_4H_{10}}^{in}} \tag{22.25}$$

In Eqs. (22.24) and (22.25), ideal gas law is assumed for the determination of molar flow. The desired product gas of a steam reformer for hydrocarbons C_nH_m consists of hydrogen (H_2), carbon monoxide (CO), carbon dioxide (CO_2), and steam (H_2O) that is added to the mixture in excess. Partial oxidation uses air for fuel conversion, leading to nitrogen (N_2) as part of the product gas. It can be assumed that oxygen (O_2) reacts completely. Methane (CH_4) can always be found in reformates due to chemical equilibrium. Finally, the product gas of an autothermal reformer contains H_2, CO, CO_2, H_2O, N_2, and CH_4. The carbon balance (C) for an idealized reforming process of any C_nH_m without byproducts results in

$$n\, \dot{n}_{C_nH_m}^{in} = n\, \dot{n}_{C_nH_m}^{out} + \dot{n}_{CO}^{out} + \dot{n}_{CO_2}^{out} + \dot{n}_{CH_4}^{out} \tag{22.26}$$

By applying Eq. (22.26), the conversion can be determined for fuels with a complex composition using the amount of carbon in the desired product gas.

$$\zeta_{C_nH_m} = \frac{\dot{n}_{CO}^{out} + \dot{n}_{CO_2}^{out} + \dot{n}_{CH_4}^{out}}{n\, \dot{n}_{C_nH_m}^{in}} \tag{22.27}$$

If dodecane with the molecular formula $C_{12}H_{26}$ is considered as fuel and only byproducts such as alkanes C_nH_{2m+2}, alkenes C_nH_{2m}, and dienes C_nH_{2m-2} are allowed, the carbon balance results in

$$12\dot{n}^{in}_{C_{12}H_{26}} = \dot{n}^{out}_{CO} + \dot{n}^{out}_{CO_2} + \dot{n}^{out}_{CH_4} + 2\left(\dot{n}^{out}_{C_2H_6} + \dot{n}^{out}_{C_2H_4}\right)$$
$$+ 3\left(\dot{n}^{out}_{C_3H_8} + \dot{n}^{out}_{C_3H_6}\right) + 4\left(\dot{n}^{out}_{C_4H_{10}} + \dot{n}^{out}_{C_4H_8} + \dot{n}^{out}_{C_4H_6}\right)$$
$$+ \sum_{5}^{12} n\left(\dot{n}^{out}_{C_nH_{2n+2}} + \dot{n}^{out}_{C_nH_{2n}} + \dot{n}^{out}_{C_nH_{2n-2}}\right) \quad (22.28)$$

Together, Eqs. (22.27) and (22.28) only consider the origin molecule $C_{12}H_{26}$ and not the quality of reforming. It might be possible that $C_{12}H_{26}$ is completely converted into a hydrogen-rich gas, but fairly large amounts of ethane, ethene, propane, and propene are still present in the reformate. If a fossil fuel such as gasoline, kerosene (jet fuel), or diesel is analyzed, a huge number of different species, such as alkanes, naphthenes, and aromatic molecules, are found. Dodecane ($C_{12}H_{26}$) reacts almost completely in a reformer, that is, $\zeta_{C_{12}H_{26}} \approx 100\%$. Nevertheless, a number of different byproducts, such as alkenes, ketones, and aldehydes, can also occur. To evaluate the quality of a chemical reaction, selectivity is used, which gives the relation between the desired product and the feed material (educt). For reforming, hydrogen is the desired product, and CO, CO_2, and CH_4 are the essential byproducts. In this sense, Eq. (22.27) is more a kind of selectivity than the pure conversion of a pure species. In the scientific community, Eq. (22.27) is used as the definition of conversion ζ_C.

O'Connell et al. [10] accounted for all hydrocarbons with a chain length <9 in the product gas. They defined conversion as

$$\zeta_{(C_{9+})} = \frac{\dot{n}_{C_nH_m} - \dot{n}_{C_{9+}}}{\dot{n}_{C_nH_m}} \quad (22.29)$$

Using this equation, a conversion between 98.3 and 100% can be calculated for diesel steam reforming, despite the fact that hydrocarbons such as C_2, C_3, C_4, and C_{5-9} are present with maximum concentrations of 3100, 1770, 1040, and 2300 ppmv, respectively.

Several measurements are necessary before Eq. (22.27) can be used to determine conversion. The gas composition can be determined by GC or by mass spectrometry (MS). The type of fuel must also be known. Equation (22.30) demands the molar mass of a fuel and its number of carbon molecules. In order to calculate the molar flow using an ideal gas equation, temperature and pressure must first be measured.

$$\zeta_C = \frac{y_{CO} + y_{CO_2} + y_{CH_4}}{n \, \dot{m}_{C_nH_m}} \dot{V}_{gas,dry} \, M_{C_nH_m} \frac{p}{RT} \quad (22.30)$$

All of these quantities can only be measured with a certain accuracy, leading to an uncertainty for conversion.

Table 22.8 shows typical values and estimated errors. The type of fuel is incorporated in Eq. (22.30) by the ratio $M_{C_nH_m}/n$. This ratio varies only slightly for different alkanes, for example, from 14.23 for decane ($C_{10}H_{12}$) to 14.13 for

Table 22.8 Quantities used to determine conversion with Eqs. (22.30) and (22.45) [simplified as Eqs. (22.33), (22.46), and (22.47)].

Quantity	Expected value[a]	Accuracy	Equations	Remarks/measuring method
$\dot{m}_{C_nH_m}$	2.48 kg h^{-1}	±1%	(22.30) and (22.45)	Flow controller, weighing
y_{CO}	10.50%	±0.0015	(22.30)	MS, GC, FTIR
y_{CO_2}	11.98%	±0.0015	(22.30)	MS, GC
y_{CH_4}	0.10%	±1.5 × 10^{-5}	(22.30)	MS, GC
$\dot{V}_{gas,dry}$	18.92 m^3 h^{-1}	±1%	(22.30) and (22.45)	Flow controller
$\frac{M_{C_nH_m}}{n}$	14.195	±1.2%	(22.30) and (22.45)	Specification or GC–MS
p	1.013 bar	±2%	(22.30) and (22.45)	Ambient pressure
T	298.15 K	±1 K	(22.30) and (22.45)	Thermocouple
$\frac{M_{C_qH_p}}{q}$	14.195	±1.2%	(22.45)	GC–MS
$\dot{m}_{H_2O}^{aq}$	4.02 kg h^{-1}	±1%	(22.45)	Weighing

[a] Expected values were determined in a process analysis for a 10 kW$_e$ HT-PEFC combined with autothermal reforming of dodecane ($C_{12}H_{26}$) at a mixture formation with $H_2O:C = 1.9$ and $O_2:C = 0.47$.

eicosane ($C_{20}H_{42}$). A stronger, albeit still weak, effect is caused by the change from alkanes to alkenes. The ratios are 14.23 for decane and 14.03 for decene. How air is provided to the experimental equipment is also important. Synthetic air provided by gas bottles might contain a 20:80 oxygen–nitrogen mixture. Dry air contains 78.08% nitrogen, 20.95% oxygen, 0.93% argon, and 0.037% carbon dioxide. Calculations based on an oxygen-to-nitrogen ratio of 21:79 do not cause crucial errors. If ambient air is used, humidity must be taken into account. At 298 K and an assumed humidity of 65%, the water content in air amounts to 2%. The oxygen content decreases to 20.5%. An important effect occurs when the product gas is dried. Reformate leaves the autothermal reformer at about 973 K. The gas contains nearly 23% steam. GC and MS demand dry gases to perform proper measurements. Therefore, reformate is cooled to 280–288 K and water is withdrawn from the gas after condensation. According to thermodynamics, a residual amount of water remains in the gas, that is, 1.67% at 288.15 K and 1% at 280.85 K. Although measured dry concentrations for H_2, CO, CO_2, CH_4, and N_2 are determined, the "dry" volume flow must be corrected, otherwise a systematic error occurs depending on the condensing temperature.

Figure 22.11 shows the conversion calculated by a Monte Carlo analysis, taking the deviations for the measured values in Table 22.8 into account. Different conversion rates were calculated varying from 90 to 99%. The standard deviation is 0.79–0.89%. A systematic error is considered for the residual steam content between 1.02 and 1.67%. Cases (a)–(c) consider 1.02% and case (d) 1.67%. Finally, the calculated average values differ from the postulated conversion by 1–1.7%.

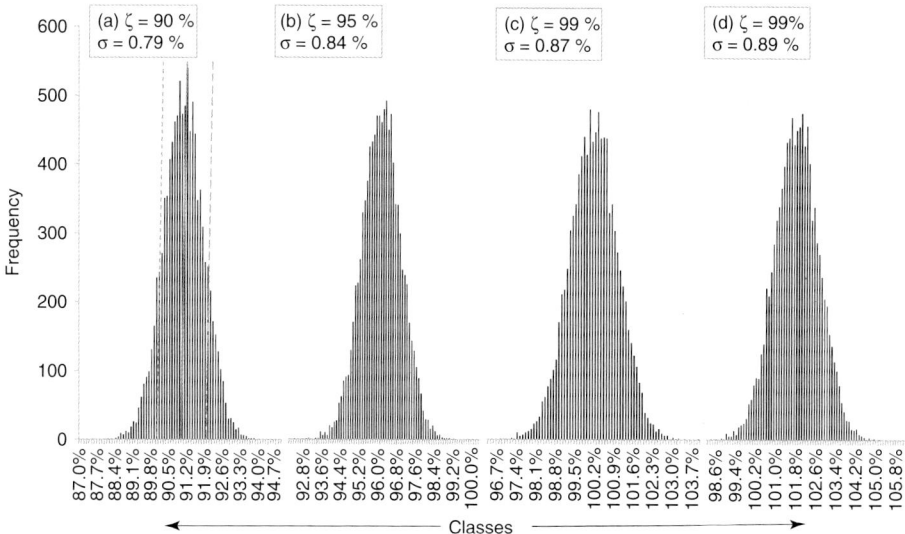

Figure 22.11 Distribution of conversion calculated by a Monte Carlo simulation with 10 000 variations of the experimental values as required by Eq. (22.30). Four different cases were considered: case (a), $\zeta = 90\%$; case (b), $\zeta = 95\%$; and cases (c) and (d), $\zeta = 99\%$. The statistical spreads for the required values are given in Table 22.8. Systematic deviation by incomplete water condensation: (a–c) 1% and (d) 2%.

As can be seen from Figure 22.11, conversion determined by Eq. (22.30) ranges from 90 to 95%. At high conversion, errors of 1% of the measured values lead to unacceptable uncertainties, especially if the residual steam contents are not corrected properly. For cases (c) and (d), conversion exceeds partially the theoretical limit of 100%.

At high conversion, it is not only the product gas but also the residuals that are of special interest for the analysis of reactor performance. Conversion can also be determined by the residual amount of fuel. The carbon (C) balance can be applied as follows:

$$n \, \dot{n}^{\text{in}}_{C_nH_m} = \dot{n}^{\text{out}}_{CO} + \dot{n}^{\text{out}}_{CO_2} + \dot{n}^{\text{out}}_{CH_4} + q \, \dot{n}^{\text{out}}_{C_qH_p} \tag{22.31}$$

This equation takes a change in the composition of the fuel from C_nH_m to C_qH_p into account. Therefore, the indices "in" and "out" are no longer necessary:

$$\zeta_C = \frac{n \, \dot{n}_{C_nH_m} - q \, \dot{n}_{C_qH_p}}{n \, \dot{n}_{C_nH_m}} = 1 - \frac{q \, \dot{n}_{C_qH_p}}{n \, \dot{n}_{C_nH_m}} \tag{22.32}$$

If the amount of residuals can be determined by weighing, a liquid conversion is given by

$$\zeta_C = 1 - \frac{q \, \dot{m}_{C_qH_p} \, M_{C_nH_m}}{n \, \dot{m}_{C_nH_m} \, M_{C_qH_p}} \tag{22.33}$$

Several papers have reported that aromatic, cyclic, or branched molecules are more challenging for reforming than n-alkanes [11, 12]. Therefore, an increasing amount of these species can be assumed in the condensate. Pasel *et al.* [13] found a concentration of about 50–60% for aromatics in the condensate compared with 15–20% in the inserted fuel. As an example, the exact value for a $15:85$ C_7H_8–$C_{12}H_{26}$ mixture as fuel and a $50:50$ C_7H_8–$C_{14}H_{30}$ mixture as a residue is 1.019. Owing to the minor changes in the H/C ratio between the original fuel and the residue, the determination of conversion by Eq. (22.33) can be simplified as a first approximation:

$$\frac{q\, M_{C_nH_m}}{n\, M_{C_qH_p}} \approx 1 \tag{22.34}$$

$$\zeta_C = 1 - \frac{\dot{m}_{\text{Residue}}}{\dot{m}_{\text{Fuel}}} \tag{22.35}$$

As can be found in several papers [10, 13–16], part of the converted fuel forms various hydrocarbon species other than CH_4, especially at lower conversion. Some species are found in the gaseous phase, whereas others are liquefied completely at 280–298 K.

$$q\, \dot{n}_{C_qH_p} = \left(q\, \dot{n}_{C_qH_p}\right)_{\text{gas}} + \left(q\, \dot{n}_{C_qH_p}\right)_{\text{liq}} \tag{22.36}$$

Most of the species are only partially liquefied. The ratio between the occurrence of a species in the liquid and gaseous phases can be approximated by simplified thermodynamic calculations. For a first approximation, an ideal gas state and an ideal mixing behavior are assumed. The condensed hydrocarbons are assumed to be immiscible in the aqueous phase and to form a second liquid organic phase. A gas–liquid equilibrium describes the ratio between the concentrations in the gaseous phase $y_{C_qH_p}$ and the liquid phase $x_{C_qH_p}$. Boiling pressures are calculated as a function of the temperature $p^S_{C_qH_p}$ [17, 18]:

$$y_{C_qH_p}\, p = x_{C_qH_p}\, p^S_{C_qH_p} \tag{22.37}$$

Equation (22.37) is a simplified formula describing vapor–liquid equilibrium (VLE). A more accurate but iterative calculation scheme will be discussed in the next section. For the organic phase, an ideal mixture can be assumed as

$$\left(y_{H_2} + y_{CO_2} + y_{CO} + y_{CH_4} + y_{N_2} + y_{H_2O}^{\text{gas}} + \sum_i y_i\right) p = \sum_i x_i\, p_i^S \tag{22.38}$$

To elucidate this approach, two cases were calculated as examples. In the first case, the gas composition of the product gas mixture is taken from a measurement of a gas phase of a steam reformer after 36 h of operation [10]. The following mixture of residual hydrocarbons is assumed: 2000 ppmv ethane (C_2H_6), 1100 ppmv propene (C_3H_6), 650 ppmv butadiene (C_4H_6), 700 ppmv pentane (C_5H_{12}), 100 ppmv octane (C_8H_{18}), 150 ppmv decane ($C_{10}H_{22}$), and 100 ppmv dodecane ($C_{12}H_{26}$). It is important to note that the steam content is 38.46%. The conversion according to

Eq. (22.27) is 87%, whereas the conversion ζ_{9+} defined in Eq. (22.29) is 98.4%. Decane and dodecane are chosen here as C_{9+} species to illustrate the thermodynamic calculations. The fraction of a species that is liquefied is calculated using

$$x_{\text{org,L}} = \frac{\dot{n}_j^{\text{org,L}}}{\dot{n}_j^{\text{org,L}} + \dot{n}_j^{\text{gas}}} \qquad (22.39)$$

Figure 22.12 shows that dodecane is present at lower temperatures in the liquid organic phase, whereas decane splits nearly 50:50 at 273 K. The organic liquid fraction of decane decreases progressively with increase in temperature. A second case considers a residual mixture with 25 ppmv propene (C_3H_6), 25 ppmv butadiene (C_4H_6), 100 ppmv pentane (C_5H_{12}), 100 ppmv hexane (C_6H_{14}), 200 ppmv benzene (C_6H_6), 200 ppmv toluene (C_7H_8), and 70 ppmv dodecane ($C_{12}H_{26}$) in the product gas of an autothermal reformer (see Table 22.9).

The conversion according to Eq. (22.27) is 97.3%. As can be seen from Figure 22.12, the liquid organic phase only contains dodecane. At 280 K, about 64% dodecane is in liquid form. Therefore, to determine the conversion accurately, a gas measurement up to C_{12} is necessary. The maximum concentration of a species that will not lead to start of condensation can be calculated by thermodynamics,

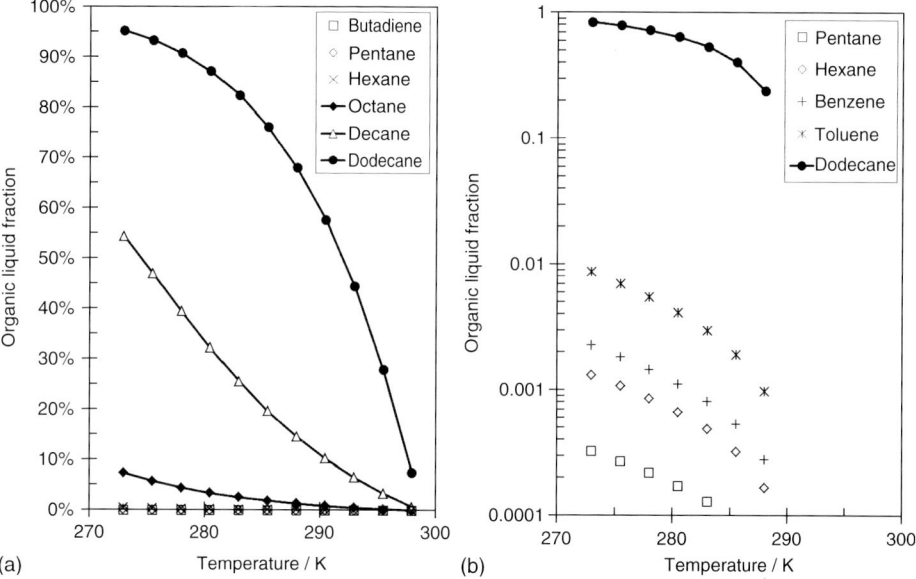

Figure 22.12 Distribution of selected species in the organic liquid and gaseous phases after cooling and condensing as approximated by Eqs. (22.37) and (22.38). (a) Reformate from steam reforming with 2000 ppmv C_2H_6, 1100 ppmv C_3H_6, 650 ppmv C_4H_6, 700 ppmv C_5H_{12}, 100 ppmv C_8H_{18}, 150 ppmv $C_{10}H_{22}$, and 100 ppmv $C_{12}H_{26}$. (b) Reformate from autothermal reforming with 25 ppmv C_3H_6, 25 ppmv C_4H_6, 100 ppmv C_5H_{12}, 100 ppmv C_6H_{14}, 200 ppmv C_6H_6, 200 ppmv C_7H_8, and 70 ppmv $C_{12}H_{26}$.

Table 22.9 Gas concentrations of different hydrocarbons that lead to the formation of an organic liquid phase at 280.65 and 298.15 K as calculated by Eq. (22.40).

n (C_nH_m)	Compound	C_nH_m	y_i^S (280.65 K)	y_i^S (298.15 K)	Solubility (ppmw at 25 °C)[a]
4	Butane	C_4H_{10}	Gaseous	Gaseous	61
5	Pentane	C_5H_{12}	33.9%	67.2%	40
6	Hexane	C_6H_{14}	8.90%	20.0%	12–50
7	Heptane	C_7H_{16}	2.35%	5.93%	2.2–2.4
8	Octane	C_8H_{18}	0.64%	1.84%	0.7
10	Decane	$C_{10}H_{24}$	487 ppmv	1780 ppmv	0.05
12	Dodecane	$C_{12}H_{26}$	34 ppmv	170 ppmv	n.d.v.[b]
14	Tetradecane	$C_{14}H_{30}$	3 ppmv	9 ppmv	n.d.v.
5	Pentene	C_5H_{10}	42.9%	83.4%	148–203
6	Hexene	C_6H_{12}	11.1%	24.3%	50–100
7	Heptene	C_7H_{14}	2.42%	6.01%	20
4	1,3 Butadiene	C_4H_6	Gaseous	Gaseous	735–1000
5	1,4-Pentadiene	C_5H_8	49.9%	96.3%	n.d.v.
6	Benzene	C_6H_6	5.32%	12.5%	1740–1800
6	Cyclohexane	C_6H_{12}	5.62%	12.9%	50
6	Cyclohexene	C_6H_{10}	4.27%	10.1%	50–160
7	Toluene	C_7H_8	1.45%	3.75%	470
8	Ethylbenzene	C_8H_{10}	0.43%	1.25%	140
8	m-Xylene	C_8H_{10}	0.38%	1.10%	180
9	1,3,5-Trimethylbenzene	C_9H_{12}	0.10%	0.33%	2900
10	Naphthalene	$C_{10}H_8$	0.009%	0.035%	32

[a] Solubilities in water at 25 °C were taken from [19, 20].
[b] n.d.v., not determined value.

assuming an ideal gas behavior. Boiling pressures can be determined using standard methods from the literature [17, 18].

$$y_{C_qH_p}^S = p_{C_qH_p}^S \frac{T_{dew}}{p} \tag{22.40}$$

Table 22.9 shows the concentration of selected species from such a dew point calculation. Most of the species offer high volatility. The comparison of the dew point concentration of steam, namely 1% at 280.65 K and 3% at 298.15 K, with octane, namely 0.64 and 1.84%, shows similar boiling behavior. It is also important to note that the dodecane concentration in the original steam–fuel–air mixture for the reported studies was 2 vol.%. Table 22.9 also gives the maximum solubility of these species in water at 25 °C, as taken from a compilation given by NIST and GESTIS [19, 20]. It is obvious that alkanes become less soluble in water with increasing chain length. Alkenes, naphthenes, and aromatic species can be dissolved in water fairly well. Therefore, it can be expected that the solubility of hydrocarbons in water at low fuel residues plays a role in determining conversion.

In the next section, a more detailed analysis offers new insights on solubility. In order to calculate conversion, the residual amounts of hydrocarbons must be analyzed in three phases:

$$q\, \dot{n}_{C_qH_p} = \left(q\, \dot{n}_{C_qH_p}\right)_{gas} + \left(q\, \dot{n}_{C_qH_p}\right)_{liq,org} + \left(q\, \dot{n}_{C_qH_p}\right)_{liq,aq} \quad (22.41)$$

In the gas phase, alkanes (C_qH_{2q+2}), alkenes and cycloalkanes (C_qH_{2q}), dienes (C_qH_{2q-2}), and the aromatic species ($C_{6+x}H_{6+2x}$) must be analyzed in detail:

$$\left(q\, \dot{n}_{C_qH_p}\right)_{gas} = \sum_{q=2}^{12} q\, \dot{n}_{C_qH_{2q+2}} + \sum_{q=2}^{12} q\, \dot{n}_{C_qH_{2q}}$$

$$+ \sum_{q=2}^{12} q\, \dot{n}_{C_qH_{2q-2}} + \sum_{x=1}^{n} (6+x)\, \dot{n}_{C_{6+x}H_{6+2x}} \quad (22.42)$$

Standard GC measures with an accuracy in the parts per million range. With the aid of GC–MS coupling, it is possible to identify single species from each other, such as cyclohexane and benzene. Often measurements only offer concentrations for C_2, C_3, C_4, and so on, which is sufficient for Eq. (22.42). GC–MS is also able to determine concentrations for chain lengths higher than C_4.

The organic liquid phase can be measured by a simple determination of masses using Eq. (22.43). It is important that the organic phase be properly separated from the aqueous phase. As an example, a conversion of 99.5% caused by autothermal reforming of dodecane and a complete liquefaction of the residual amount, that is, 72.6 ppm $C_{12}H_{26}$ in the gas phase, lead to $12.4\,g\,h^{-1}$ organic liquid and $4\,kg\,h^{-1}$ water. Assuming densities of $750\,g\,l^{-1}$ for $C_{12}H_{26}$ and $1000\,g\,l^{-1}$ for water, an organic phase of 3.5 ml must be separated from 1 l of water.

$$\left(q\, \dot{n}_{C_qH_p}\right)_{liq,org} = \frac{q\, \dot{m}_{C_qH_p}}{M_{C_qH_p}} \quad (22.43)$$

Figure 22.13 shows the distribution of conversion calculated using a Monte Carlo simulation with 10 000 variations of the experimental values as required by Eq. (22.33). The uncertainty for weighing of the organic phase was estimated as $\pm 1\%$. Two different cases with conversions of 99 and 99.5% were considered and are indicated by the bars in Figure 22.13 with a frequency of 10 000 units. Further variables by applying Eq. (22.33) are given in Table 22.8. Only dodecane occurs as a residual hydrocarbon in the gas phase and the liquid organic phase. Dodecane was neglected in the aqueous phase. Considering the data in Table 22.9, it is obvious that dodecane does not condense completely. The part that remains in the gas phase depends on the temperature (see also Figure 22.12). This causes a systematic deviation, leading to an increase in conversion of $\Delta\zeta = 0.1\%$ at 273.15 K and $\Delta\zeta = 0.2\%$ at 280.65 K. A conversion of 99.5% is calculated by this method as 99.7% when the condensate is drawn at 280.65 K. By neglecting to analyze the residual dodecane in the gas phase, the error increases with increase in temperature.

An organic phase above an aqueous phase can be detected easily by shaking the flask after sampling. A clouding offers two immiscible phases. According to the VLE

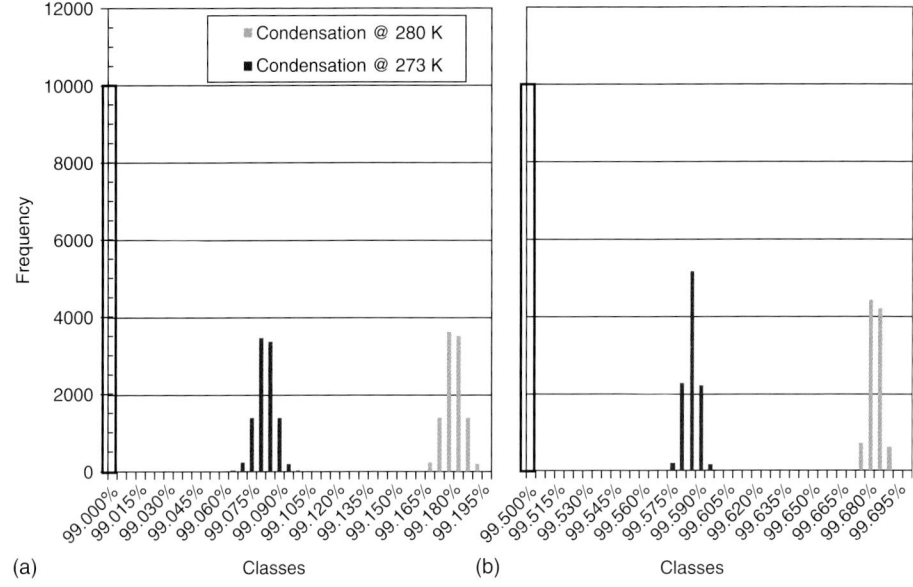

Figure 22.13 Distribution of conversion calculated by a Monte Carlo simulation with 10 000 variations of the experimental values as required by Eq. (22.33). Two different cases were considered: case (a), $\zeta = 99\%$, 144 ppm $C_{12}H_{26}$; and case (b), $\zeta = 99.5\%$, 72.6 ppm $C_{12}H_{26}$.

calculations at 280.65 K and 1.013 bar (see Table 22.9), dodecane forms an organic liquid phase for concentrations higher than 34 ppmv. This residual corresponds to a conversion of 99.8%. Analogous calculations can be performed for tetradecane and decane, resulting in conversions of 99.98 and 97.8%, respectively. Commercial fuels, such as gasoline, Jet A-1, and diesel, contain a mixture of hydrocarbons with different chain lengths. After reforming, organic residuals show a slight shift to higher boiling temperatures, indicating that long-chain hydrocarbons are still present. For the use of Jet A-1 or diesel, clouding effects in the aqueous phase could be expected for conversion lower than 99.8%.

To determine organic components in an aqueous phase, a water quality control method can be applied. This involves treating organic components in water by acidification and oxidation, before subsequently detecting and quantifying them as total organic carbon (TOC):

$$\left(q\, \dot{n}_{C_qH_p} \right)_{\text{liq,aq}} = \frac{\text{TOC}}{M_C} \dot{V}^{\text{aq}} = \frac{\text{TOC}}{M_C} \frac{\dot{m}_{H_2O}^{\text{aq}}}{\rho_{H_2O}} \tag{22.44}$$

The accuracy of this method is 0.5 mg TOC l^{-1}. Potable water has a TOC < 1 mg l^{-1}, whereas rivers and creeks have TOC in the range 2–5 mg l^{-1} due to pollution and bacteria. Typical measured values are 0.5–140 mg l^{-1} [13]. Assuming 9.6 ppmv toluene (C_7H_8) in the gas phase, a TOC of 200 mg l^{-1} after complete dissolution in water is expected. Such concentrations correspond to a conversion of 99.96%.

The amount of residual hydrocarbon is smaller by a factor of 13 compared with the example applying Eq. (22.43).

Finally, the following equation describes the complete determination method for conversion using residual amounts of hydrocarbons:

$$\zeta_C = 1 - \frac{\sum_{q=1}^{12} q\, y_{C_qH_p} \dfrac{p\,\dot{V}_{Dry}}{RT} + \dfrac{q\,\dot{m}_{C_qH_p}}{M_{C_qH_p}} + \dfrac{TOC}{M_C}\dfrac{\dot{m}_{H_2O}^{aq}}{\rho_{H_2O}}}{n\,\dfrac{\dot{m}_{C_nH_m}}{M_{C_nH_m}}} \qquad (22.45)$$

The different analysis methods have varying levels of accuracy. Table 22.10 shows different conversion values corresponding to residual amounts of hydrocarbons in either the gaseous or liquid phase. The TOC value gives no information about single species. GC–MS measurements, in particular, can be used to clarify the chemical structure of the hydrocarbon mixture. If the accuracy of all species with the same chain length from C_2 to C_{12} is assumed to be 1 ppmv, then the residual carbon amounts to 77 ppmv. The highest conversion is 99.966%.

As an alternative, Lindström et al. [16] used a Fourier transform infrared (FTIR) spectrometry with flame ionization detection to determine the sum of hydrocarbons in the gas phase as NMHCs (non-methane hydrocarbons). The best result for diesel slip indicates a background signal of 10 ppmv corresponding to 120–200 ppmv carbon. A higher accuracy of 100 ppbv for each C_y fraction leads to the best case conversion of 99.9966%. TOC measurements in the range 0.5–2 mg l^{-1} are definitely more accurate. Conversion higher than 99.99966% demands a TOC lower than 1.75 mg l^{-1} or a residual amount of carbon in the gas phase of 0.77 ppmv. In principle, accuracies in the ppb range are possible with specialized GC techniques [21, 22].

In order to show the accuracy of the measurements, the gas analysis and TOC measurements can be analyzed separately. Equation (22.46) shows the determination for a pure gas-phase analysis and Eq. (22.47) that for a TOC measurement with

Table 22.10 Demanded accuracies for determining residual hydrocarbons by GC–MS measurement of the gas phase and by TOC in the aqueous phase for three different conversions[a].

Conversion (%)	Demanded accuracy of GC–MS in ppbv for C_2–C_{12}	TOC (mg l^{-1})
99.966	1000/[Σ(2–12): 77 ppmv]	175
99.9966	100/[Σ(2–12): 7.7 ppmv]	17.5
99.99966	10/[Σ(2–12): 0.77 ppmv]	1.75

[a] Estimated TOC and GC–MS data are used as separate values and not as a combination of both.

no residuals in the gas phase:

$$\zeta_C = 1 - \frac{\sum_{q=1}^{12} q \, y_{C_qH_p}}{n \frac{\dot{m}_{C_nH_m}}{M_{C_nH_m}}} \frac{p \, \dot{V}_{Dry}}{RT} \tag{22.46}$$

$$\zeta_C = 1 - \frac{TOC}{n \, \rho_{H_2O}} \frac{\dot{m}_{H_2O}^{aq}}{\dot{m}_{C_nH_m}} \frac{M_{C_nH_m}}{M_C} \tag{22.47}$$

Monte Carlo simulations with 10 000 variations of the experimental values were performed for four exemplary cases. Cases (a) and (b) used Eq. (22.46), case (c) Eq. (22.47), and case (d) Eq. (22.45). Case (a) considers residual amounts of propene, butadiene, and benzene with 55, 25, and 13 ppmv, corresponding to values published by Porš et al. [15]. Some developers have reported on residuals of C_2 and C_3 components [10, 14]. Case (b) focused on ethane, propane, and butadiene with 400 ppmv C_2, 300 ppmv C_3, and 200 ppmv C_4. Case (c) was chosen as an example of a TOC measurement of 150 mg l^{-1}. Such a value corresponds to an increasing amount of unconverted hydrocarbons dissolved in the aqueous phase. Case (d) is the limiting case for TOC and GC–MS measurements. Residual hydrocarbons were set extremely low, that is, 100 ppbv C_3H_8 and 1 mg l^{-1} TOC. The statistical spread for these concentrations was ±3% for the gas-phase concentrations, except in case (d) at 50%, ±5% for TOC in case (c) and ±50% for TOC in case (d). Further values are given in Table 22.8. A systematic deviation of 1% caused by incomplete water condensation was considered for all cases (a)–(d). Figure 22.14 shows the distribution of conversion for these calculations. The uncertainty for weighing of the aqueous phase is estimated as ±1%. True values for conversion are given for all cases in Figure 22.14. For all Monte Carlo studies, 53 classes were defined with a chosen Δ of $(0.01–0.3) \times 10^{-5}$. The standard deviation σ is also given for each case. Owing to incomplete water condensation, the measured gas flows were slightly too high and the calculated conversion was slightly too low. This effect leads to a minor deviation, expressed by an average conversion of 99.862% compared with a true value of 99.863%. The width of the distribution curve for TOC measurements was smaller than that for the GC measurements. Although the relative deviations for extremely low residues were very high, that is, ±50%, the resulting conversion range was extremely small, 99.9996–99.9998%. These examples show the robustness of the method based on Eq. (22.45).

Equation (22.45) must not necessarily be used with all its terms for all possible experimental investigations. Conversion is a quantity that is used for different purposes:

- During catalyst screening, conversion amounts to 50–90% [23–25]. Here conversion is used to find the most active and selective catalyst. For this purpose, Eq. (22.30) can be applied.
- A number of papers have reported on reactor performance during reactor development for diesel reforming with a conversion of 90–99%. These results were caused by an improper functioning of reactors at a lower development level

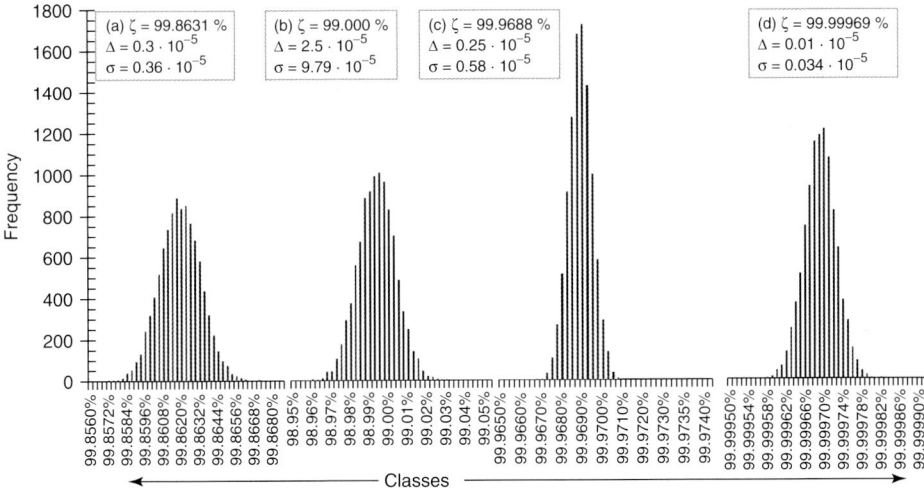

Figure 22.14 Distribution of conversion calculated by a Monte Carlo simulation with 10 000 variations of the experimental values as required by Eqs. (22.46) and (22.47). Four different cases were considered: case (a), 55 ppm C_3H_6, 25 ppm C_4H_6, and 13 ppm C_6H_6; case (b), 400 ppm C_2H_6, 300 ppm C_3H_6, and 200 ppm C_4H_6; case (c), TOC 150 mg l^{-1}; and case (d), 100 ppbv C_3H_6, TOC 1 mg l^{-1}. The statistical spreads for these concentrations were $\pm 3\%$ for the gas-phase concentrations, except for case (d) at 50 and $\pm 5\%$ for TOC in case (c) and $\pm 50\%$ in case (d). Further values are given in Table 22.8. The systematic deviation caused by incomplete water condensation for cases (a–d) was 1%.

or by unfavorable operating conditions. Residual amounts of fuel were reported as unconverted diesel slip, that is, 700–800 ppmv [16], or as broad spectra of C_1–C_8 species of 10–1000 ppmv [10].
- A large number of papers and conference contributions quote conversion rates of 100% instead of giving accurate HC residue amounts [26, 27]. Only a few developers [13, 14] have determined and published residual amounts of hydrocarbons and a conversion higher than 99.9%.

Various experimental results offer different combinations of residues leading to different methods for the determination of conversion:

- At a conversion lower than 95%, reliable results can be determined by Eq. (22.30).
- During reforming, an organic phase occurs after cooling. If hydrocarbons are not present in the gas phase, conversion can be calculated by Eq. (22.33). Otherwise, Eq. (22.45) should be applied. It could be feasible to neglect hydrocarbons in the aqueous phase.
- During reforming, a large amount of hydrocarbons can be found in the gas phase. The formation of an organic liquid phase can be avoided by heating. The amount of hydrocarbons in the aqueous phase can be neglected. Equation (22.46) should be chosen in this case.

- Only a small amount of hydrocarbons was analyzed in the gas phase. In parallel, a TOC measurement was performed. Equation (22.45) should be chosen.
- Hydrocarbons in the gas phase are below the detection limit. A TOC measurement was carried out, detecting small residuals in the range $1-10\,\mathrm{mg\,l^{-1}}$. A first approximation can be made by Eq. (22.47), considering the results given in Table 22.10.

22.3
Analysis Methods in Fuel Cell Process Engineering

In this section, selected analysis methods will be highlighted that can be used efficiently in the development of fuel cell systems. The first method deals with systems analysis via statistical methods. In the previous section, the DOE approach was discussed as a verification method in fuel cell process engineering. This methodology is useful not just for designing effective experiments, but also for designing detailed process and systems analyses, which are often used for the development of fuel cell systems.

The second method involves a predictive method that determines VLE and liquid–liquid equilibrium (LLE), which are important for designing evaporators and condensers. Another method is a powerful tool for parameter determination and optimization by Levenberg and Marquardt [28]. This method can be used to develop complex models for process engineering purposes which require an adaptation of parameters to experimental results. The final analysis method deals with the pinch-point method developed by Linnhoff *et al.* [29]. This technique aims to identify the maximum possible heat recovery and the minimum energy requirement of a thermal or chemical process [30–32]. The application of this method to fuel cell systems is explained in the final part of this section.

22.3.1
Systems Analysis via Statistical Methods

Fuel cell systems are composed of several components. Each component contains different parameters, which influence not only the performance of the particular component, but also the performance of the complete system. Fuel cell process engineering deals with the optimized design of fuel cell systems, including the optimized operation of each component in the system. This demands a characterization tool covering not only the influence of each individual parameter on the target values defined by the requirements of the special application, but also the influence caused by parameter interactions. One possible solution is to use the statistical approach discussed in the previous section.

In the following example, the use of a fractional factorial design will be explained. This time, the design will not be applied to an experiment. Instead, it will be used to analyze a fuel cell system. For this purpose, a polymer electrolyte fuel cell (PEFC) system based on the autothermal reforming of kerosene was designed. The effects of different parameters, such as system pressure, cathode air ratio, cell voltage, hydrogen utilization, and the air ratio in the catalytic burner were analyzed. The target parameters were system efficiency, required heat exchange area, amount of water produced by the system, critical temperature to close the water balance, off-gas amount, and the oxygen content in the off-gas. In this analysis, the first four parameters were defined independently, since it was expected that all of these parameters would have a significant effect on the observed target parameters. The last parameter, the air ratio of the catalytic burner, had a relatively smaller effect on the observed parameters. Therefore, it was defined in place of the four-factor interaction. The fractional factorial design then had the form shown in Table 22.11.

As can be seen in Table 22.11, the variation of four factors resulted in 16 different parameter combinations. The definition of the fifth factor as a function of the other four made it possible to include the effect of this parameter without increasing the number of parameter combinations.

The numerical results of the model are presented in Figure 22.15. A brief analysis of the results shows that the maximum efficiency was reached in case 13, when the upper levels of factor 3 (cell voltage) and factor 4 (hydrogen utilization) are combined with the lower level of factor 1 (system pressure). In case 15, a very high efficiency was also achieved, showing that the effect of factor 2 (cathode air ratio) does not have a significant influence on the efficiency. However, the aim was to analyze the system as a whole and not just to maximize its efficiency. According to the calculated amounts of water produced by the system, maximum water production was reached in case 4 with the lower levels of factors 3 and 4 and the higher level of factor 1. It is therefore impossible to maximize simultaneously the system efficiency and water production rate. Factor 2 directly influences the off-gas amount and the oxygen concentration in the off-gas.

The influence of factor 5, air ratio in the catalytic burner, has a relatively smaller influence than expected, thus allowing it to be selected freely without causing the results to change. In a further analysis, the results were standardized, allowing

Table 22.11 Example of a 2^{5-1} fractional factorial design.

		1	2	3	4	5	6	7	8	9	10	11	12	13	14	15	16
Factor 1(x_1)	System pressure	−1	1	−1	1	−1	1	−1	1	−1	1	−1	1	−1	1	−1	1
Factor 2(x_2)	Air ratio cathode	−1	−1	1	1	−1	−1	1	1	−1	−1	1	1	−1	−1	1	1
Factor 3(x_3)	Cell voltage	−1	−1	−1	−1	1	1	1	1	−1	−1	−1	−1	1	1	1	1
Factor 4(x_4)	Hydrogen utilization	−1	−1	−1	−1	−1	−1	−1	−1	1	1	1	1	1	1	1	1
Factor 5(x_5)	Air ratio catalytic burner	1	−1	−1	1	1	1	1	−1	−1	1	1	−1	1	−1	−1	1

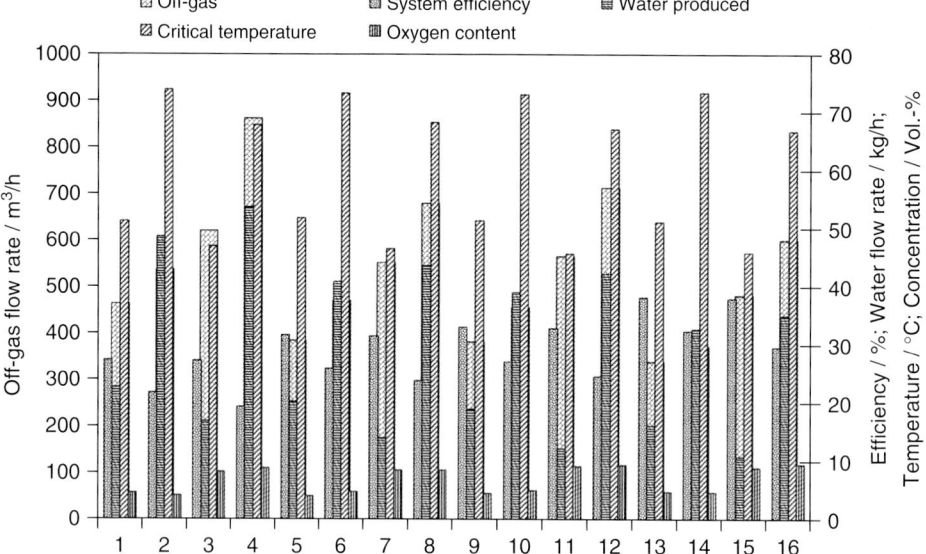

Figure 22.15 Results of the systems analysis as a function of parameter combinations using five parameters defined by the 2^{5-1} fractional factorial design.

the main effects and the two-factor interactions to be analyzed for each response variable. Figure 22.16 presents the general results of this analysis. This analysis method can be used to determine which parameters have a positive effect on the desired target value. The z-axis presents the effects in a standardized form. Value 1 on the z-axis represents the highest possible positive effect, whereas the value -1 represents the highest possible negative effect. For example, the figure shows that the main effect of system pressure x_1 has a large influence on the target values y_1–y_4, whereas this effect is negative for y_1 (system efficiency) and positive for y_2 (amount of water produced), y_3 (critical temperature) and y_4 (heat exchange area). The main effect of the same parameter on y_5 (off-gas amount) is still strong but smaller than on the first four parameters. It can also be clearly seen that x_1 has a negligible effect on the last target parameter y_6 (oxygen content). Similarly, the main effects of the other factors and also two-factor interactions can be observed and analyzed using Figure 22.16.

22.3.2
Predictive Method to Determine Vapor–Liquid and Liquid–Liquid Equilibria

The determination of VLE or LLE is important for designing evaporators and condensers. Bubble formation must be avoided to guarantee proper nozzle functionality. Equations (22.37) and (22.38) involve strong simplifications, such as the assumption of the ideal gas and ideal liquid phase. The complete description of the

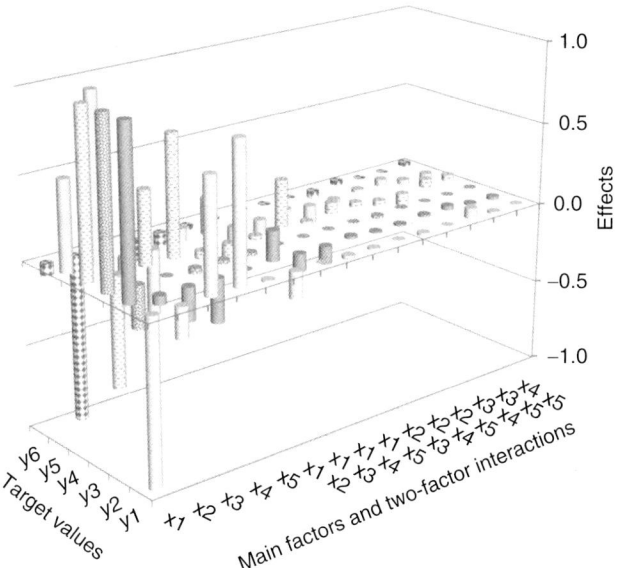

Figure 22.16 The influence of the main effects and two-factor interactions on selected system parameters calculated in a systems analysis using a 2^{5-1} fractional factorial design. The effects are presented on the z-axis in the standardized form. Factors x_1-x_5 are given in Table 22.11; y_1, system efficiency; y_2, amount of water produced; y_3, critical temperature to close the water balance; y_4, required heat exchange area; y_5, off-gas amount; y_6, oxygen content in the off-gas.

VLE of a species j is

$$y_j \, \varphi_j \, p = x_j \, \gamma_j \, \varphi_j^\circ \, p_j^S \, \exp\left(\int_{p_j^S}^{p} -\frac{V_j \, dp}{RT} \right) \tag{22.48}$$

where fugacity coefficients φ_j and φ_j° describe the deviation of real gas behavior in relation to ideal gas for species j in a mixture and as a pure substance at boiling conditions. At pressures lower than 1 atm, the fugacity coefficients are almost $\varphi_j = 1$. At moderate pressures, their influence tends to be cancelled out by the ratio $\varphi_j^\circ/\varphi_j$, leading to an ideal mixture of real gases. The Poynting factor, represented by the exponential function, describes the compressibility of the boiling liquid phase and differs significantly less from unity by only a few parts per thousand far away from the critical point and at moderate pressures. An important correction must be made using the activity coefficient of the liquid phase, γ_j. The fugacity coefficient is determined by an equation of state (EOS) and by its specific mixing rules for coefficients. A well-known EOS was proposed by Soave [33]. The Redlich–Kwong–Soave (RKS) EOS can be written as

$$p = \frac{RT}{v-b} - \frac{a(T)}{v(v+b)} \tag{22.49}$$

More details on determining the coefficients $a(T)$ and b for the mixture and the single species j and evaluating fugacity coefficients can be found elsewhere [33, 34].

Activity coefficients can be determined by specific methods for a defined thermodynamic system or for general purposes by a predictive method. The best known method is the UNIFAC method [35–37], which splits the activity coefficient into two parts, a combinatorial and a residual part:

$$\ln \gamma_j = \ln \gamma_j^C + \ln \gamma_j^R \tag{22.50}$$

Each molecule is separated into functional groups. The combinatorial part considers the properties of a species j in the mixture on the basis of its structure. The residual part considers the interactions of all functional groups in the mixture. Specific parameters of this method can be found in the literature [35–37].

The proposed methods for describing real fluid behavior are complex and a detailed discussion of the mathematical functions is beyond the scope of this chapter. Therefore, the original papers should be consulted for more in-depth studies [33, 35]. The performance of the methods is discussed using two examples in the next sections.

22.3.2.1 Residual Hydrocarbons in the Reformer Product Gas

As discussed in Section 22.2.3, residual hydrocarbons in the reformate can be found as a separated organic liquid phase in the gas phase and in soluble form in the liquid aqueous phase. The maximum solubilities of species such as benzene and alkanes are given in Table 22.9. These values, however, are misleading as the real behavior differs from that of ideal liquids. For an infinite dilution of hydrocarbons in water and for an ideal mixing in the gas phase, Eq. (22.48) can be written as

$$y_{C_qH_p} \, p = x_{C_qH_p} \, \gamma_{C_qH_p}^\infty \, p_{C_qH_p}^S \tag{22.51}$$

Activity coefficients at infinite dilution were determined by experimental methods and published by Atik et al. [38]. They measured activity coefficients for $x_j < 10^{-4}$, that is, <100 ppmv, of 2470 for benzene, 8570 for toluene, 5 for ethanol, and 8 for acetone. Extremely high values led to a strong stripping effect. Thermodynamic calculations describing such behavior are based on the UNIFAC method. Figure 22.17 shows the resulting phase diagrams for VLE and LLE of the systems water–benzene, water–1-butanol, and water–acetone at 1.013 bar. LLE were also determined by Eq. (22.52). The calculations resulted in a benzene concentration in the aqueous phase of about 1900 ppm at 298 K and in a 1-butanol concentration of 114 g kg^{-1}, respectively. This agrees well with the values of 1800 ppmw and 90 g l^{-1} in [20].

$$x'_{C_qH_p} \, \gamma'_{C_qH_p} = x''_{C_qH_p} \, \gamma''_{C_qH_p} \tag{22.52}$$

The water content in the organic phases was predicted with less accuracy, that is, 1 mol% water in benzene instead of 2.5% [19]. It is important to note that the maximum solubility is the result of LLE. The potential for dissolving hydrocarbons

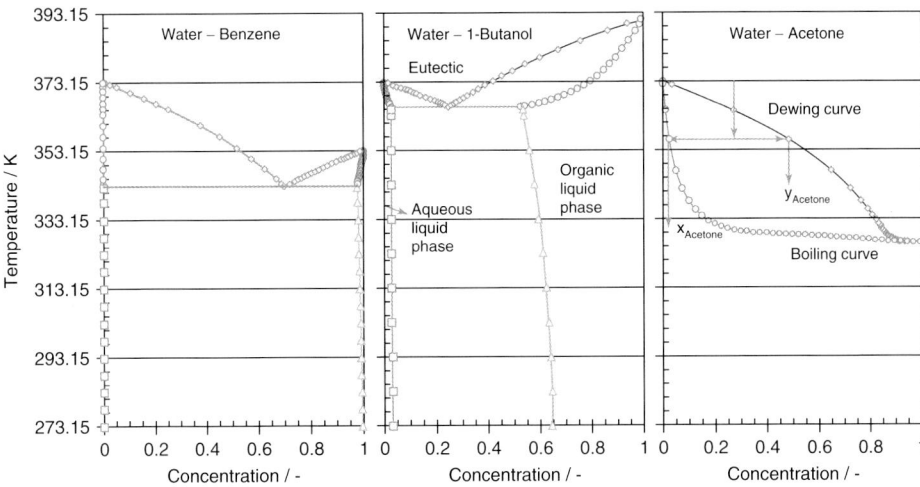

Figure 22.17 Phase diagrams for vapor–liquid and liquid–liquid equilibria of the systems water–benzene, water–1-butanol and water–acetone at 1.013 bar calculated with UNIFAC for an ideal gas phase.

in water can be estimated by the activity coefficient at infinite dilution. Fredenslund et al. [35] calculated these values for several alkanes, alkenes, and dienes using the UNIFAC method. For carbon numbers from 4 to 8, they calculated activity coefficients from 3000 to 10^7. It can be assumed that alkanes, alkenes, and aromatic components, such as benzene and toluene, are not present in the aqueous phase. Furthermore, a VLE calculation was performed to determine the concentrations of several hydrocarbons in the aqueous phase for a condenser operated with reformate in a fuel processor. Figure 22.18 shows the distribution of various types of hydrocarbons in the gaseous and aqueous phases as a function of temperature and carbon number. At a condenser temperature of 280 K, acids and alcohols are preferentially present in the aqueous phase, whereas formates and phenol were not dissolved in water and remained in the gaseous phase. Aldehydes and ketones prefer the gaseous phase, whereby the acetone fraction in the aqueous phase is three times higher than that for acetaldehyde at low temperatures. It can be estimated that the presence of soluble carbon dioxide in the aqueous phase has an influence on the solubility of hydrocarbons. In the UNIFAC calculations performed here, such effects were not considered.

Such VLE calculations are helpful in interpreting the results of fuel processors operating at nearly complete conversion. A TOC measurement of the condensate with $2\,\mathrm{mg\,l^{-1}}$ organic carbon in the aqueous phase leads to a volume fraction of 3 ppmv C. VLE calculations using the UNIFAC method allocate different oxygenates, such as alcohols, acids, and ketones, to the aqueous phase accompanied by residual amounts in the gaseous phase. Porš [39] gave an overview of oxygenates

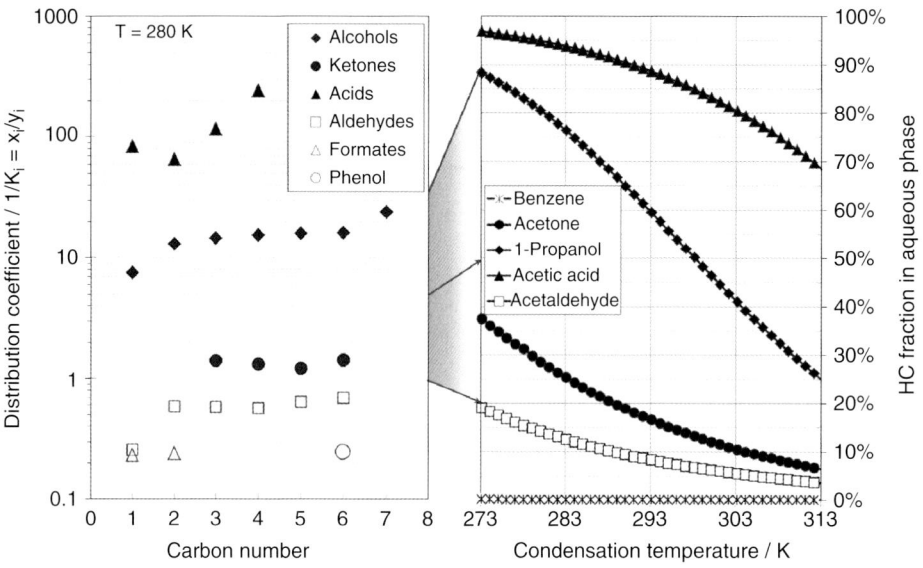

Figure 22.18 Distribution of various types of hydrocarbons in the gaseous and aqueous phases as a function of temperature and carbon number. Left part,: distribution coefficient of the concentration in the gaseous and liquid phases, that is, y_i and x_i. Right part: hydrocarbon fraction in relation to the molar flow in the gaseous and aqueous phases. Values determined at 10 ppmv.

as intermediate species during combustion and reforming processes. Especially formaldehyde (CH_2O), acetone (CH_3–CO–CH_3), and acrolein (CH_2=CH–CHO) occur during combustion [40]. According to Naidja et al., residuals in the product gas of reformers comprise 60% alkenes and 40% oxygenates [41]. At high conversion, that is, low amounts of residuals, it is challenging to identify a single component in the broad spectrum of these species using GC–MS measurements. If acetone is assumed to be a single source, a concentration of 1 ppmv C_3H_7O should be found in the aqueous phase in order to reach 3 ppmv carbon. However, 700 ppb would still be present in the gaseous phase predicted by the UNIFAC model. The TOC measurement results in a conversion of 99.9996%, whereas residues in both phases lead to 99.9987%. Other model substances, such as acids and alcohols, would improve the results as UNIFAC predicts higher distribution factors x_i/y_i. Such considerations should consider that the occurrence of alkenes as intermediates must be checked in the gas phase taking the necessary detection accuracy into account (see Table 22.10).

22.3.2.2 Evaporation of Model Fuels

The UNIFAC method can be applied to predict the evaporation behavior of fuels such as gasoline, kerosene, and diesel. In the following example, the evaporation behavior of gasoline is analyzed. Applying a boiling method according to EN ISO

3405, fuel will be heated at a rate of 1 K min^{-1}. The resulting boiling curve is very significant for practical purposes. It is not exact from a physical or thermochemical point of view. Components with a higher volatility are dragged by evaporated bubbles and components with a low volatility were restrained in the liquid phase due to the high distillation velocity. A more exact method is ASTM D-2892 or a simulated curve based on a GC method. Owing to the great effort needed, the last methods are not applicable for the quality control of fuels. More information can be found in [42].

Figure 22.19 shows the boiling curves for gasoline determined by the ASTM D-2892 and EN ISO 3405 methods. The EN ISO 3405 method provides a fairly flat curve between 310 and 473 K whereas ASTM D-2892 delivers experimental data from 253 K up to 523 K and higher. The boiling curve shown in Figure 22.19 is cut at 303 K. Considering fuel processing for fuel cell systems, fuel composition is important for the chemistry of reforming and also for fuel spraying and evaporation. A typical gasoline quality offers, for example, a hydrocarbon mixture with up to 42 vol.% aromatic components, up to 18 vol.% alkenes and 2.7 mass% oxygen bonded in oxygenates such as alcohols and ethers. Therefore, the simulated mixtures A and C with 17 different species are composed of 35–40% aromatics, 7% naphthenes, 13% alkenes, 11.5% oxygenates, that is, ethanol

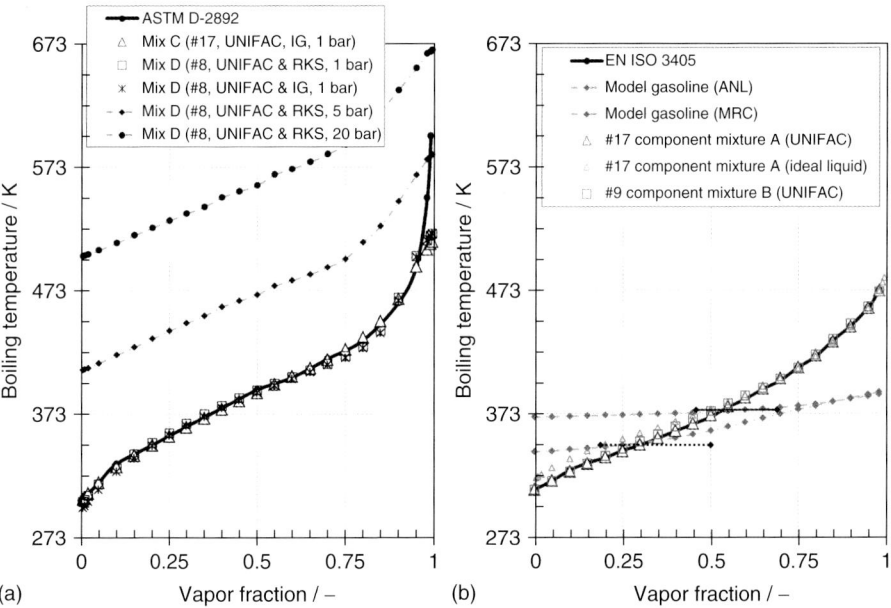

Figure 22.19 Boiling curves according to ASTM D-2892 (a) and EN ISO 3405 (b). Simulated boiling curves determined by UNIFAC for mixtures A–D (see Table 22.12) in relation to curves at elevated pressure (a) and model mixtures from literature [12, 43] (b). Model fuels were taken from Kopasz et al. (Argonne National Laboratory, ANL) [12] and from Ersoz et al. (Marmara Research Centre, MRC) [43].

Table 22.12 Composition of the simulated gasoline mixtures A–D.

No.	Compound	Formula	Class	A (%)	B (%)	C (%)	D (%)
1	Propane	C_3H_8	n-Alkanes	0.2	–	1.5	–
2	n-Tetradecane	$C_{14}H_{26}$	n-Alkanes	6	13.5	5	–
3	n-Eicosane	$C_{20}H_{42}$	n-Alkanes	1.5	1	5.5	7.6
4	2-Methylpropane	C_4H_{10}	Isoalkanes	8.7	11	4	4.2
5	2,3-Dimethylbutane	C_6H_{14}	Isoalkanes	12	–	6.3	–
6	2,2,4-Trimethylpentane	C_8H_{18}	Isoalkanes	5.2	5	6.2	22.6
7	Cyclopentane	C_5H_{10}	Naphthenes	4	10	3	8.1
8	Cyclopentene	C_5H_8	Naphthenes	3	–	4	–
9	Toluene	C_7H_8	Aromatics	9	–	7.7	–
10	Xylene	C_8H_{10}	Aromatics	7.2	35	11.3	40
11	1,3,5 Trimethylbenzene	C_9H_{12}	Aromatics	13.8	–	18	–
12	Biphenyl	$C_{12}H_{10}$	Aromatics	5	–	3	–
13	1-Propene	C_3H_6	Alkenes	0.8	–	3	–
14	1-Hexene	C_6H_{12}	Alkenes	7.9	13	4	6
15	1-Decene	$C_{10}H_{22}$	Alkenes	4.3	–	6	–
16	Ethanol (EtOH)	C_2H_5OH	Oxygenates	5	5	5	5
17	Methyl tert-butyl ether (MTBE)	$C_5H_{12}O$	Oxygenates	6.5	6.5	6.5	6.5

and methyl *tert*-butyl ether (MTBE), and 33.5–28.5% alkanes and isoalkanes. Table 22.12 shows the composition for mixtures A–D which were applied for the VLE calculations.

At elevated pressures, the Poynting correction [in Eq. (22.48)] can be integrated for incompressible fluids. This assumption cannot be applied near the critical point.

$$\exp\left(\int_{p_j^S}^{p} -\frac{V_j\, dp}{RT}\right) = \exp\left[-\frac{V_j\left(p - p_j^S\right)}{RT}\right] \quad (22.53)$$

At low pressures, such calculations can be performed for ideal gases or as ideal mixture for real gases without Poynting correction. Equation (22.54) was used for all VLE calculations with the UNIFAC method for 9–17 model species (see mixtures A–C in Table 22.12):

$$y_j\, p = x_j\, \gamma_j\, p_j^S \quad (22.54)$$

At moderate pressures, Eq. (22.53) gives a good approximation for incompressible fluids. At high pressures and especially in the region around the critical point, this assumption no longer holds. The integral in Eq. (22.48) and Eq. (22.53) can be solved for the EOS by RKS [33] [see Eq. (22.48)], resulting in Eq. (22.55). All basic thermodynamic data were taken from [18]. Equation (22.55) was applied for all calculations for mixture D.

$$\exp\left(\int_{p_j^S}^{p} -\frac{V_j\,dp}{RT}\right) = \exp\left(\frac{v_j p - v_j^S p_j^S}{RT}\right) \frac{v_j^S - b}{v - b} \left(\frac{v}{v_j^S} \frac{v_j^S + b}{v + b}\right)^{-\frac{b}{a}RT} \quad (22.55)$$

The compositions of the mixtures A–D were fitted to boiling curves according to the EN ISO 3405 and ASTM D-2892 methods based on the UNIFAC model for the liquid phase. As can be seen in Figure 22.19, there is good agreement between the fitted curves and the base curves. Slight deviations are visible for ideal liquid behavior. The boiling curve given by ASTM D-2892 cannot be simulated satisfactorily at high vapor fractions of 98% due to missing property data for hydrocarbons with chain lengths above 20 (higher than eicosane). The calculations clearly show that model fuels for gasoline, such as 74% isooctane (C_8H_{18}), 20% dimethylbenzene (C_8H_{10}), 5% methylcyclohexane (C_7H_{14}), and 1% pentene (C_3H_6) [12] or 33.6% hexane (C_6H_{14}), 28% hexane (C_6H_{12}), and 38.4% xylene (C_6H_{12}) [43] are not applicable for the design of an evaporator. Simulations with mixture D give good agreement with ASTM D-2892 data at 1.013 bar. If gasoline is to be used as a fuel for reformers, it must be sprayed by a nozzle into a mixture chamber. These nozzles were heated by the surroundings, leading to boiling of fuel in the nozzle and to irregular spray formation. The functionality of such nozzles demands a certain absolute pressure. Figure 22.19 shows that the predicted VLE data by UNIFAC and RKS lead to a maximum nozzle temperature of 410 K at 5 bar and of 500 K at 20 bar.

If the chemical composition needs to be analyzed in detail, more component mixtures such as A must be used. Mixture C is not applicable because it corresponds to a less exact boiling curve given by EN ISO 3405. Figure 22.20 shows the evaporation behavior as a rate of moles per kelvin as function of temperature. Corresponding to the selected species, aromatics evaporate mainly around the maximum of 423 K and oxygenates and naphthenes between 323 and 373 K. The series of aromatic molecules starts with benzene and its boiling point of 353 K at 1 bar. Benzene is limited to 1% in gasoline. Therefore, in this simulation toluene with 384 K is the aromatic species that should evaporate first. Finally, trimethylbenzene evaporates at 438 K and 1 bar. It is obvious that owing to the constant slope of the boiling curve, the evaporation rate is constant between 333 and 491 K. Due to the high content of aromatics, alkenes and alkanes must be selected carefully to guarantee such a boiling curve. Therefore, less alkanes were chosen with a chain length between C_7 and C_{13}.

22.3.3
Model Evaluation for Nonlinear Systems of Equations

The development of complex models for process engineering purposes often requires parameters to be adapted to experimental results. These models generally result in a nonlinear system of mathematical equations. A powerful tool for

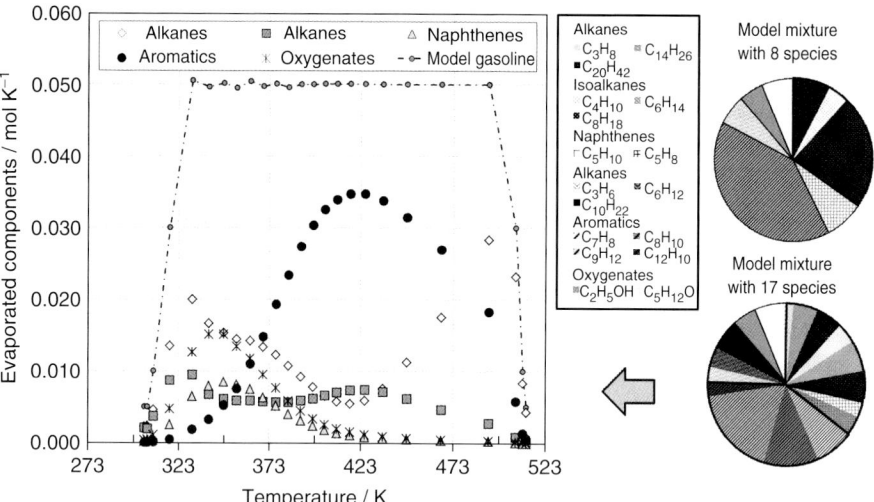

Figure 22.20 Evaporation of gasoline model fuel (mixture A, see Table 22.12) as a function of temperature.

parameter determination and optimization is the method developed by Levenberg and Marquardt [28]. A set of parameters $\vec{x}$ is determined by a vector calculation and the parameter changes are expressed by a vector $\Delta \vec{x}$:

$$\Delta \vec{x} = \frac{\vec{J}^T \times \vec{F}}{\mu \vec{E} + \vec{J}^T \times \vec{J}} \tag{22.56}$$

$$\vec{J} = \begin{bmatrix} \frac{\partial f_1}{\partial p_1} & \frac{\partial f_1}{\partial p_2} & \frac{\partial f_1}{\partial p_3} & \cdots & \frac{\partial f_1}{\partial p_n} \\ \frac{\partial f_2}{\partial p_1} & \frac{\partial f_2}{\partial p_2} & \frac{\partial f_2}{\partial p_3} & & \frac{\partial f_2}{\partial p_n} \\ \frac{\partial f_3}{\partial p_1} & \frac{\partial f_3}{\partial p_2} & \frac{\partial f_3}{\partial p_3} & & \frac{\partial f_3}{\partial p_n} \\ \cdots & \cdots & \cdots & & \cdots \\ \frac{\partial f_m}{\partial p_1} & \frac{\partial f_m}{\partial p_2} & \frac{\partial f_m}{\partial p_3} & \cdots & \frac{\partial f_m}{\partial p_n} \end{bmatrix} \tag{22.57}$$

The interaction between parameters $p_1 - p_n$ and the evaluation criteria $f_1 - f_m$ is expressed by a Jacobi matrix [see Eq. (22.57)]. The vector $\vec{E}$ is the unit vector and μ is a damping value.

The development of UNIFAC was also performed by derivations of this method [37, 44]. A parameter optimization requires a certain number of experimental data to produce a serious model. The minimum number of experimental data points is given by

$$m \geq \frac{1}{2} n (n + 1) \tag{22.58}$$

The modified UNIFAC method was applied for 45 different functional groups described by 530 parameters [37]. Following Eq. (22.58), a set of 140 000 data points was required. The quality of such a model is defined by minimization of the least-squares sum. Gmehling et al. [37] used thermodynamic data for VLE, that is, with different descriptive variables, LLE, excess heats of mixing (h^E), heat capacities c_p^E, and activity coefficients at infinite dilution, as shown in Eq. (22.59). For the parameter determination of the UNIFAC method by Gmehling et al. in 1993, about 90 000 isobars and isotherms were taken into account, guaranteeing the condition given by Eq. (22.58) [37].

$$\sum_{i=1,\ldots,N} F = \left(\frac{\gamma_i^\infty - \gamma_{i,\text{calc}}^\infty}{\gamma_i^\infty} \right)^2 \to \text{Min}. \tag{22.59}$$

The Dortmund Data Bank is still growing and currently contains about 4.78 million data points. Models with a high number of free variables fitted to a limited number of experimental data points, such as a kinetic model for methanol steam reforming with 13 parameters determined by 43 experiments [45], must be evaluated critically.

The accuracy of the predicted thermodynamic data depends on the type of system and also on the thermodynamic properties. Group predictive models often predicted activity coefficients at infinite dilution with less accuracy. A comparison between a UNIFAC prediction and values measured by Atik et al. [38] shows relative deviations of 2.4, 8, 15, 17, and 29% for activity coefficients at infinite dilution ($x = 10^{-4}$) of ethanol, acetone, benzene, 2-butanone, and 2-pentanone in water, respectively.

22.3.4
Pinch-Point Analysis

Pinch-point analysis is an important element in process analysis and a powerful engineering tool for designing heat exchanger networks [46]. A first approach considers only basic thermodynamic data, such as operating temperatures and pressures for different chemical and electrochemical reactions. Figure 22.21 shows a flow sheet diagram for a 10 kW$_e$ HT-PEFC system based on kerosene under simplified conditions. Water, kerosene, and air were mixed in the gas phase after each of these flows had been heated to 623 K. This mixture is described by characteristic ratios. They were chosen here as $H_2O : C = 2.5$ and $O_2 : C = 0.4$. These values lead to efficient fuel processing without accounting for undesired processes, such as incomplete conversion or carbon deposition. A WGS reaction was assumed as an isothermal reaction at 523 K. The heat exchangers were not connected.

Figure 22.22 shows a flow sheet diagram for a 10 kW$_e$ HT-PEFC system based on kerosene for a process analysis under realistic conditions verified by experiments. Kerosene was heated in the liquid phase only to 400 K to avoid undesired pre-reactions. A heat exchanger for air preheating was omitted. Experimental results showed that a certain heat exchange occurs within the metallic construction with no special heat exchanger areas. Water was heated, evaporated, and superheated to 798 K. The mixture of water, air, and fuel led to a mixing temperature of 623 K.

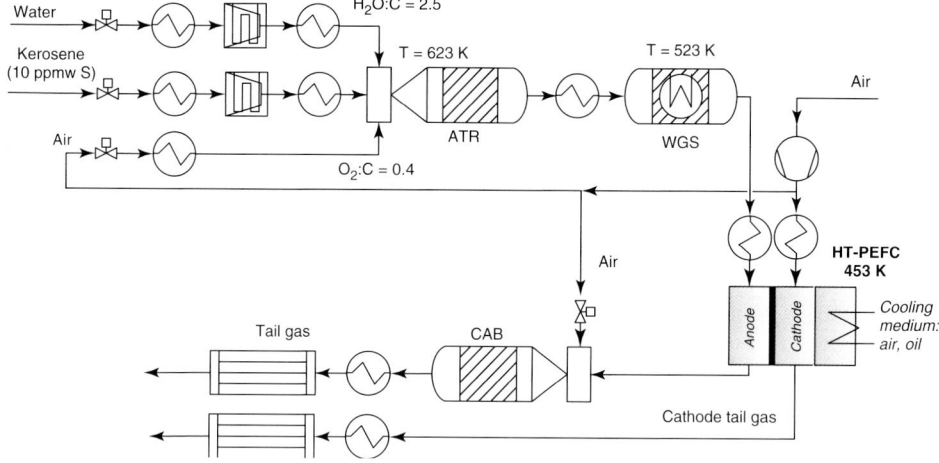

Figure 22.21 Flow sheet diagram for a 10 kW$_e$ HT-PEFC system based on kerosene for a process analysis under simplified conditions.

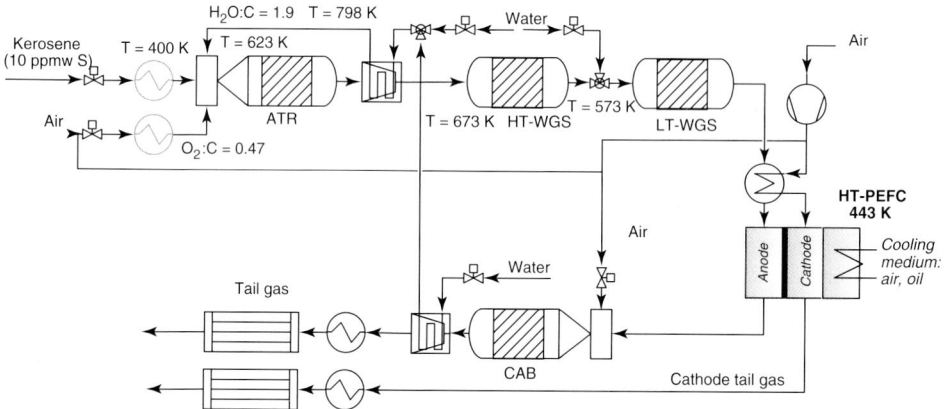

Figure 22.22 Flow sheet diagram for a 10 kW$_e$ HT-PEFC system based on kerosene for a process analysis under realistic conditions verified by experiments.

The mixture composition was chosen as $H_2O : C = 1.9$ and $O_2 : C = 0.47$ to achieve complete conversion and to avoid carbon deposition. The WGS reaction was performed in two adiabatic reactor stages with inlet temperatures of 673 and 573 K. An intermediate heat exchanger was not required as cold water was injected into the system. The catalytic burner and reformer were both required for water evaporation and superheating to cover the complete heat demand and to guarantee the inlet temperature for the first WGS reactor stage. When the operation temperature of

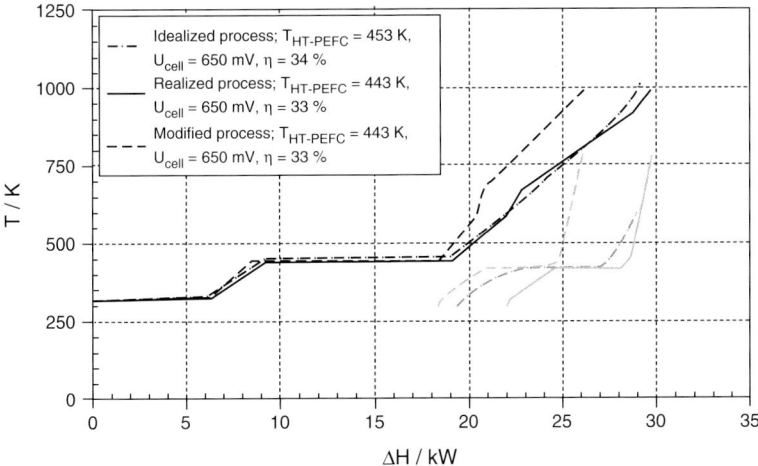

Figure 22.23 Pinch-point diagram for a 10 kW$_e$ HT-PEFC system based on the autothermal reforming of tetradecane (C$_{12}$H$_{26}$) as a model fuel for kerosene. Idealized process according to Figure 22.21; realized process according to Figure 22.22. The modified process considers a side stream of 62% from the catalytic burner off-gas for heating purposes (not depicted in the graph).

the fuel cell was decreased by 10 K, only one heat exchanger was necessary to preheat the cathode air and cool the reformate gas. The residual heat can be used for other purposes.

Figure 22.23 illustrates the different processes in a pinch-point diagram. An enthalpy of 29–30 kW$_{th}$ must be transferred in the heat exchanger network. Owing to differences in preheating air, fuel, and water in the idealized (see Figure 22.21) and realized processes (see Figure 22.22), the cold composite curve ends at a higher temperature for the realized system.

According to the flow sheet in Figure 22.22, two heat exchangers cool the burner off-gas. The first evaporates water and the second is cooled with ambient air. Instead of a serial setup, a parallel setup can be foreseen. When only some of the burner off-gas is used, the pinch-point diagram determines a splitting factor up to which the actual heat exchanger network can be used with minor changes. If more than 62% of the off-gas is used for heating purposes, water must be evaporated by the released heat of the fuel cell. This would lead to a fairly large change in the heat exchanger design.

22.4 Conclusion

Selected methodologies for fuel cell process engineering were introduced in this chapter. The use of the DOE theory was discussed as a powerful tool for system verification using examples from fuel cell subsystems. Measurement uncertainty was discussed according to the internationally recognized GUM procedure. The use of Monte Carlo methods for expressing measurement uncertainty was also explained using examples. Different approaches for determining conversion in reforming processes were outlined as a further verification methodology. A statistical approach was introduced to analyze different operation parameters for selected system characteristics. A predictive method for VLE and LLE was also introduced. This method was explained using calculations for residual hydrocarbons in reformer product gas and the evaporation of model fuels. Finally, pinch-point analysis was described using an example from fuel cell systems development. The examples showed that the methods introduced in this chapter can be used effectively in fuel cell process engineering.

Acknowledgments

The authors would like to thank the Department of Fuel Processing and Systems at Forschungszentrum Jülich GmbH, Institute of Energy and Climate Research–Fuel Cells (IEK-3) for their excellent cooperation in recent years. The authors would also like to thank Christiane Wiethege for her support during the experiments used as examples in this work to describe the DOE method.

References

1. Soravia, S. and Orth, A. (2009) *Design of Experiments Ullmann's Encyclopedia of Industrial Chemistry*, Wiley-VCH Verlag GmbH, Weinheim.
2. Fisher, R.A. (1960) *The Design of Experiments*, 7th edn., Oliver and Boyd, London.
3. Scheffler, E. (1997) *Statistische Versuchsplanung und – Auswertung: eine Einführung für Praktiker*, 3rd edn, Deutscher Verlag für Grundstoffindustrie, Stuttgart.
4. Bandemer, H. and Bellmann, A. (1994) *Statistische Versuchsplanung*, 4th edn., Teubner, Leipzig.
5. Ellison, L.R.S., Barwick, V.J., and Farrant, T.J.D. (2009) *Practical Statistics for the Analytical Scientist, a Bench Guide*, 2nd edn., Royal Society of Chemistry, Cambridge.
6. Toutenburg, H. (1994) *Versuchsplanung und Modellwahl*, Physica Verlag, Darmstadt.
7. JCGM (2008) Evaluation of Measurement Data–Guide to the Expression of Uncertainty in Measurement, 1st edn., JCGM 100:2008, Joint Committee for Guides in Metrology.
8. JCGM (2008) Evaluation of Measurement Data–Supplement 1 to the "Guide to the Expression of Uncertainty in Measurement"–Propagation of Distributions Using a Monte Carlo Method, 1st edn., JCGM 101:2008, Joint Committee for Guides in Metrology.
9. Siebert, B.R.L. and Sommer, K.-L. (2004) New developments of the GUM and Monte Carlo techniques. *Tech. Mess.*, **71** (2), 67–80.
10. O'Connell, M., Kolb, G., Schelhaas, K.P., Schuerer, J., Tiemann, D., Ziogas, A., and Hessel, V. (2009) Development and evaluation of a microreactor for the reforming of diesel fuel in the kW range. *Int. J. Hydrogen Energy*, **34** (15), 6290–6303.
11. Borup, R., Inbody, M., Hong, J., Morton B., and Tafoya, J. (2000) Fuels and fuel impurity effects and fuel processing catalysts, in Proceedings of the Fuel Cell Seminar, Portland, OR, p. 288.

12. Kopasz, J.P., Applegate, D., Ruscic, L., Ahmed, S., and Krumpelt, M. (2000) Effects of gasoline components on fuel processing and implications for fuel cell fuels, in Proceedings of the Fuel Cell Seminar, Portland, OR, p. 284.
13. Pasel, J., Meissner, J., Porš, Z., Samsun, R.C., Tschauder, A., and Peters, R. (2007) Autothermal reforming of commercial Jet A-1 on a 5 kW$_e$ scale. *Int. J. Hydrogen Energy*, **32**, 4847–4858.
14. Roychoudhury, S., Junaedi, C., Walsh, D., Mastanduno, R., Spence, D., DesJardins, J., and Morgan, C. (2008) *Proceedings of the Fuel Cell Seminar (CD), October 28–30, 2008, Palm Springs, CA*, Courtesy Associates, Washington, DC.
15. Porš, Z., Pasel, J., Tschauder, A., Dahl, R., Peters, R., and Stolten, D. (2008) Optimized mixture formation for diesel fuel processing. *Fuel Cells*, **8** (2), 129–137.
16. Lindström, B., Karlsson, J.A.J., Ekdunge, P., De Verdier, L., Häggendal, B., Dawody, J., Nilsson, M., and Petterson, L.J. (2009) Diesel fuel reformer for automotive fuel cell applications. *Int. J. Hydrogen Energy*, **34** (8), 3367–3381.
17. VDI-Gesellschaft (2006) *VDI-Wärmeatlas*, 10th edn., Springer, Berlin, p. DCA 1.
18. Perry, R.H. and Green, D.W. (1997) in *Perry's Chemical Engineers' Handbook*, 7th edn. (eds. R.H. Perry and D.W. Green), McGraw-Hill, New York, pp. 2–50.
19. IUPAC (2007) IUPAC–NIST Solubility Data Series: IUPAC–NIST Solubility Database, Version 1.0. NIST Standard Reference Database 106, http://srdata.nist.gov/solubility/ (last accessed 21 September 2011).
20. IFA: Institut für Arbeitsschutz der Deutschen Gesetzlichen Unfallversicherung (2011) GESTIS-Stoffdatenbank. Gefahrstoffinformationssystem der Deutschen Gesetzlichen Unfallversicherung, http://www.dguv.de/ifa/de/gestis/stoffdb/index.jsp (last accessed 21 September 2011).
21. Mühle, J. (2001) GC/MS-Messstem für Nicht-Methan-Kohlenwasserstoffe, PhD thesis, University of Mainz.
22. Perbelli, L., Pasini, F., Romani, S., Princivalle, A., and Brugnone, F. (2002) Analysis of benzene, toluene, ethylbenzene and *m*-xylene in biological samples from the general population. *J. Chromatogr. B*, **778**, 199–210.
23. Palm, C., Cremer, P., Peters, R., and Stolten, D. (2002) Small-scale testing of a precious metal catalyst in the autothermal reforming of various hydrocarbon feeds. *J. Power Sources*, **106**, 231–237.
24. Qi, A., Wang, S., Fu, G., and Wu, D. (2005) Autothermal reforming of *n*-octane on Ru-based catalysts. *Appl. Catal. A: Gen.*, **293**, 71–82.
25. Krumpelt, M., Krause, T.R., Carter, J.D., Kopasz, J.P., and Ahmed, S. (2002) Fuel processing for fuel cell systems in transportation and portable power applications. *Catal. Today*, **77**, 3–16.
26. Maeda, S., Kikunaga, S., Akoi, T., Nishikawa, S., Yamamoto, S., Akimoto, J., Anzai, I., and Ikeda, T. (2004) Development and operational study of kW-class PEFC cogeneration system using kerosene, in Proceedings of the Fuel Cell Seminar (CD), San Antonio, TX.
27. Saito, K., Matsumoto, H., Kisen, T., Takahashi, O., and Katsuno, H. (2004) Development of fuel processing technologies for kerosene fuel cell co-generation system. Proceedings of the Fuel Cell Seminar (CD), San Antonio, TX.
28. Brown, K.M. and Dennis, J.E. Jr. (1972) Derivative free analogues of the Levenberg–Marquardt and Gauss algorithms for non-linear least square approximation. *Numer. Math.*, **18**, 289–297.
29. Linnhoff, B., Townsend, D.W., Boland, D., Hewitt, G.F., Thomas, B.E.A., Guy, A.R., and Marsland, R.H. (1982) *User Guide on Process Integration for the Efficient Use of Energy*, IChemE, Rugby.
30. Linnhoff, B. and Lenz, W. (1987) Thermal integration and process optimization. *Chem. Ing. Tech.*, **59** (11), 851–857.

31. Godat, J. and Marechal, F. (2003) Optimization of a fuel cell system using process optimization techniques. *J. Power Sources* **118** (1–2), 411–423.
32. Göll, S., Samsun, R.C., and Peters, R. (2011) Analysis and optimization of SOFC-based auxiliary power units using a generic zero-dimensional fuel cell model. *J. Power Sources*, **196**, 9500–9509.
33. Soave, G. (1972) Equilibrium constants form a modified Redlich–Kwong equation of state. *Chem. Eng. Sci.*, **27**, 1197–1203.
34. Smith, J.M. and Van Ness, H.C. (1987) *Introduction to Chemical Engineering Thermodynamics*, McGraw-Hill, New York.
35. Fredenslund, A.A., Jones, R.L., and Prausnitz, J.M. (1975) Group-contribution estimation of activity coefficients in nonideal liquid mixtures. *AIChE J.*, **21** (6), 1086–1098.
36. Maurer, G. (1986) Vapor–liquid equilibrium of formaldehyde- and water-containing multicomponent mixtures. *AIChE J.*, **32** (6), 932–948.
37. Gmehling, J., Li, J., and Schiller, M. (1993) A modified UNIFAC model, 2. Present parameter matrix and results for different thermodynamic properties. *Ind. Eng. Chem. Res.*, **32**, 178–193.
38. Atik, Z., Gruber, D., Krummen, M., and Gmehling, J. (2004) Measurement of activity coefficients at infinite dilution of benzene, toluene, ethanol, esters, ketones, and ethers at various temperatures in water using the diluter technique. *J. Chem. Eng. Data*, **49**, 1429–1432.
39. Porš, Z. (2005) Eduktvorbereitung und Gemischbildung in Reaktionsapparaten zur autothermen Reformierung von dieselähnlichen Kraftstoffen, PhD thesis, RWTH Aachen.
40. Günter, R. (1974) Das entwerfen von flammen und flammenräumen. *Chem. Ing. Tech.* **46** (2), 56–62.
41. Naidja, A., Krishna, C.R., Butcher, T., and Mahajan, D. (2003) Cool flame partial oxidation and its role in combustion and reforming of fuels for fuel cell systems. *Prog. Energy Combust. Sci.*, **29**, 155–191.
42. van Basshuysen, R. and Schäfer, F. (2006) *Lexikon Motorentechnik*, ATZ-MTZ Series, Vieweg & Teubner, Wiesbaden.
43. Ersoz, A., Olgun, H., Ozdogan, S., Gungor, C., Akgun, F., and Tırıs, M. (2003) Autothermal reforming as a hydrocarbon fuel processing option for PEM fuel cell. *J. Power Sources*, **118**, 384–392.
44. Hansen, H.K., Rasmussen, P., Fredenslund, A., Schiller, M., and Gmehling, J. (1991) Vapor–liquid equilibria by UNIFAC group contribution. 5. Revision and extension. *Ind. Eng. Chem. Res.*, **30** (10), 2352–2355.
45. Peppley, A., Amphlett, J.C., Kearns, L.S., and Mann, R.F. (1999) Methanol–steam reforming on $Cu/ZnO/Al_2O_3$ catalysts. Part 2. A comprehensive kinetic model. *Appl. Catal. A*, **179** (1–2), 31–49.
46. Linnhoff, B. (1989) Pinch technology for the synthesis of optimal heat and power systems. *J. Energy Res. Technol.*, **111**, 137–147.

Edited by Detlef Stolten and Bernd Emonts

Fuel Cell Science and Engineering

Related Titles

Li, X.

Polymer Electrolyte Membrane Fuel Cells

ISBN: 978-0-470-87110-2

Stolten, D., Scherer, V. (eds.)

Efficient Carbon Capture for Coal Power Plants

2011
ISBN: 978-3-527-33002-7

Stolten, D. (ed.)

Hydrogen and Fuel Cells
Fundamentals, Technologies and Applications

2010
ISBN: 978-3-527-32711-9

Hirscher, M. (ed.)

Handbook of Hydrogen Storage
New Materials for Future Energy Storage

2010
ISBN: 978-3-527-32273-2

Barbaro, P., Bianchini, C. (eds.)

Catalysis for Sustainable Energy Production

2009
ISBN: 978-3-527-32095-0

Mitsos, A., Barton, P. I. (eds.)

Microfabricated Power Generation Devices
Design and Technology

2009
ISBN: 978-3-527-32081-3

Vielstich, W.

Handbook of Fuel Cells

6 Volume Set
2009
ISBN: 978-0-470-74151-1

Edited by Detlef Stolten and Bernd Emonts

Fuel Cell Science and Engineering

Materials, Processes, Systems and Technology

Volume 2

WILEY-VCH Verlag GmbH & Co. KGaA

The Editors

Prof. Detlef Stolten
Forschungszentrum Jülich GmbH
IEF-3: Fuel Cells
Leo-Brandt-Straße
52425 Jülich
Germany

Dr. Bernd Emonts
Forschungszentrum Jülich GmbH
IEF-3: Fuel Cells
Leo-Brandt-Straße
52425 Jülich
Germany

We would like to thank the following institutions for providing us with the photographic material used in the cover illustration: IdaTech Fuel Cells GmbH, EnergieAgentur.NRW, and Forschungszentrum Jülich GmbH.

All books published by **Wiley-VCH** are carefully produced. Nevertheless, authors, editors, and publisher do not warrant the information contained in these books, including this book, to be free of errors. Readers are advised to keep in mind that statements, data, illustrations, procedural details or other items may inadvertently be inaccurate.

Library of Congress Card No.: applied for

British Library Cataloguing-in-Publication Data
A catalogue record for this book is available from the British Library.

Bibliographic information published by the Deutsche Nationalbibliothek
The Deutsche Nationalbibliothek lists this publication in the Deutsche Nationalbibliografie; detailed bibliographic data are available on the Internet at <http://dnb.d-nb.de>.

© 2012 Wiley-VCH Verlag & Co. KGaA, Boschstr. 12, 69469 Weinheim, Germany

All rights reserved (including those of translation into other languages). No part of this book may be reproduced in any form – by photoprinting, microfilm, or any other means – nor transmitted or translated into a machine language without written permission from the publishers. Registered names, trademarks, etc. used in this book, even when not specifically marked as such, are not to be considered unprotected by law.

Print ISBN: 978-3-527-33012-6
ePDF ISBN: 978-3-527-65027-9
ePub ISBN: 978-3-527-65026-2
mobi ISBN: 978-3-527-65025-5
oBook ISBN: 978-3-527-65024-8

Cover Design Formgeber, Eppelheim
Typesetting Laserwords Private Limited, Chennai, India
Printing and Binding betz-druck GmbH, Darmstadt

Printed in the Federal Republic of Germany
Printed on acid-free paper

Contents to Volume 1

List of Contributors XIX

Part I Technology 1

1 Technical Advancement of Fuel-Cell Research and Development 3
Bernd Emonts, Ludger Blum, Thomas Grube, Werner Lehnert, Jürgen Mergel, Martin Müller, and Ralf Peters

2 Single-Chamber Fuel Cells 43
Têko W. Napporn and Melanie Kuhn

3 Technology and Applications of Molten Carbonate Fuel Cells 67
Barbara Bosio, Elisabetta Arato, and Paolo Greppi

4 Alkaline Fuel Cells 97
Erich Gülzow

5 Micro Fuel Cells 131
Ulf Groos and Dietmar Gerteisen

6 Principles and Technology of Microbial Fuel Cells 147
Jan B.A. Arends, Joachim Desloover, Sebastià Puig, and Willy Verstraete

7 Micro-Reactors for Fuel Processing 185
Gunther Kolb

8 Regenerative Fuel Cells 219
Martin Müller

	Part II	Materials and Production Processes 247
9		Advances in Solid Oxide Fuel Cell Development Between 1995 and 2010 at Forschungszentrum Jülich GmbH, Germany 249
		Vincent Haanappel
10		Solid Oxide Fuel Cell Electrode Fabrication by Infiltration 275
		Evren Gunen
11		Sealing Technology for Solid Oxide Fuel Cells 301
		K. Scott Weil
12		Phosphoric Acid, an Electrolyte for Fuel Cells – Temperature and Composition Dependence of Vapor Pressure and Proton Conductivity 335
		Carsten Korte
13		Materials and Coatings for Metallic Bipolar Plates in Polymer Electrolyte Membrane Fuel Cells 361
		Heli Wang and John A. Turner
14		Nanostructured Materials for Fuel Cells 379
		John F. Elter
15		Catalysis in Low-Temperature Fuel Cells – an Overview 407
		Sabine Schimpf and Michael Bron
	Part III	Analytics and Diagnostics 439
16		Impedance Spectroscopy for High-Temperature Fuel Cells 441
		Ellen Ivers-Tiffée, André Leonide, Helge Schichlein, Volker Sonn, and André Weber
17		Post-Test Characterization of Solid Oxide Fuel-Cell Stacks 469
		Norbert H. Menzler and Peter Batfalsky
18		*In Situ* Imaging at Large-Scale Facilities 493
		Christian Tötzke, Ingo Manke, and Werner Lehnert
19		Analytics of Physical Properties of Low-Temperature Fuel Cells 521
		Jürgen Wackerl
20		Degradation Caused by Dynamic Operation and Starvation Conditions 543
		Jan Hendrik Ohs, Ulrich S. Sauter, and Sebastian Maass

| Part IV | Quality Assurance 571 |

21 Quality Assurance for Characterizing Low-Temperature Fuel Cells *573*
Viktor Hacker, Eva Wallnöfer-Ogris, Georgios Tsotridis, and Thomas Malkow

22 Methodologies for Fuel Cell Process Engineering *597*
Remzi Can Samsun and Ralf Peters

Contents to Volume 2

List of Contributors *XIX*

| Part V | Modeling and Simulation 645 |

23 Messages from Analytical Modeling of Fuel Cells *647*
Andrei Kulikovsky
23.1 Introduction *647*
23.2 Modeling of Catalyst Layer Performance *648*
23.2.1 The Basic Equations *648*
23.2.2 Ideal Transport of Feed Molecules *650*
23.2.3 Polarization Curve *651*
23.2.4 The Critical Current Density *652*
23.2.5 The x-Shapes *653*
23.2.6 A Model for Cr Poisoning of the SOFC Cathode *654*
23.2.7 Optimum Catalyst Loading *657*
23.3 Polarization Curve of PEMFCs and HT-PEMFCs *658*
23.3.1 Oxygen Transport in the GDL and the Polarization Curve *659*
23.3.2 Low-Current Regime *660*
23.3.3 High-Current Regime *661*
23.3.4 One-Dimensional Cell Polarization Curve *662*
23.3.5 Oxygen Consumption in the Channel and the Quasi-Two-Dimensional Polarization Curve *663*
23.4 Conclusion *665*
List of Symbols *665*
References *667*

24 Stochastic Modeling of Fuel-Cell Components *669*
Ralf Thiedmann, Gerd Gaiselmann, Werner Lehnert, and Volker Schmidt
24.1 Multi-Layer Model for Paper-Type GDLs *670*
24.1.1 Modeling of Fibers *671*
24.1.2 Modeling of Binder *672*
24.1.3 Fitting of Model Parameters *674*
24.1.4 Further Results *675*

24.2	Time-Series Model for Non-Woven GDLs 676
24.3	Stochastic Network Model for the Pore Phase 677
24.3.1	Pore Phase Graph 678
24.3.1.1	Detection of Pores 678
24.3.1.2	Modification of Pore Phase Graph 679
24.3.2	Stochastic Modeling of Vertices 680
24.3.2.1	Multi-Layer Representation 680
24.3.2.2	Construction and Fitting of Point Process Model 680
24.3.3	Validation of Vertex Model 684
24.3.4	Marking of Vertices 685
24.3.4.1	Moving-Average Model for Dependent Marking 685
24.3.4.2	Degrees of Vertices 687
24.3.5	Stochastic Modeling of Edges 688
24.3.5.1	MCMC Simulation for Edge Rearrangement 689
24.4	Further Results 690
24.4.1	Classical Random Graph Models 690
24.4.2	Transport Simulations along Edges of Graphs 691
24.5	Structural Characterization of Porous GDL 692
24.5.1	Tortuosity 692
24.5.2	Pore Size Distributions 694
24.5.3	Connectivity 695
24.5.4	Validation of Multi-Layer Model 696
24.5.5	Validation of Graph Model 698
24.6	Conclusion 698
	References 699
25	**Computational Fluid Dynamic Simulation Using Supercomputer Calculation Capacity** *703*
	Ralf Peters and Florian Scharf
25.1	Introduction 703
25.2	High-Performance Computing for Fuel Cells 705
25.3	HPC-Based CFD Modeling for Fuel-Cell Systems 711
25.3.1	Principles of Computational Fluid Dynamics 712
25.3.2	Physical Model Principles 715
25.3.2.1	Turbulence 715
25.3.2.2	Heat Transfer 717
25.3.2.3	Mixtures and Reactions 717
25.3.2.4	Multiphase Flows 719
25.3.2.5	Porous Media 720
25.3.3	CFD Modeling of the Core Components of an HT-PEFC Auxiliary Power Unit 721
25.4	CFD-Based Design 728
25.5	Conclusion and Outlook 730
	Acknowledgments 731
	References 731

26	**Modeling Solid Oxide Fuel Cells from the Macroscale to the Nanoscale** *733*	
	Emily M. Ryan and Mohammad A. Khaleel	
26.1	Introduction *733*	
26.2	Governing Equations of Solid Oxide Fuel Cells *735*	
26.2.1	Mass Conservation *736*	
26.2.2	Momentum Conservation *738*	
26.2.3	Energy Conservation *739*	
26.2.4	Electrochemistry *740*	
26.2.4.1	Continuum-Level Electrochemistry Approach *741*	
26.2.4.2	Mesoscale Electrochemistry Approach *742*	
26.2.5	Chemical Reactions *745*	
26.3	Macroscale SOFC Modeling *747*	
26.3.1	System-Level Modeling *747*	
26.3.2	Stack-Level Modeling *750*	
26.3.3	Cell-Level Modeling *755*	
26.4	Mesoscale SOFC Modeling *758*	
26.5	Nanoscale SOFC Modeling *761*	
26.6	Conclusion *761*	
	References *762*	
27	**Numerical Modeling of the Thermomechanically Induced Stress in Solid Oxide Fuel Cells** *767*	
	Murat Peksen	
27.1	Introduction *767*	
27.2	Chronological Overview of Numerically Performed Thermomechanical Analyses in SOFCs *768*	
27.3	Mathematical Formulation of Strain and Stress Within SOFC Components *773*	
27.3.1	Cell, Sealant, and Wire Mesh Components *773*	
27.3.2	Metallic Components *776*	
27.4	Effect of Geometric Design on the Stress Distribution in SOFCs *778*	
27.4.1	Computational Fluid Dynamics (CFD) Analysis *779*	
27.4.2	Thermomechanically Induced Stress Analysis *782*	
27.4.2.1	Thermomechanically Induced Stress Within the Sealant Components *783*	
27.4.2.2	Thermomechanically Induced Stress Within the Metal Components *783*	
27.5	Conclusion *788*	
	References *789*	
28	**Modeling of Molten Carbonate Fuel Cells** *791*	
	Peter Heidebrecht, Silvia Piewek, and Kai Sundmacher	
28.1	Introduction *791*	
28.2	Spatially Distributed MCFC Model *794*	

28.2.1	General Assumptions *794*	
28.2.2	Anode Gas Channels *795*	
28.2.3	Cathode Gas Channels *798*	
28.2.4	Solid Phase *799*	
28.2.5	Potential Field Model *800*	
28.2.6	Interface to Electrode Models *804*	
28.3	Electrode Models *804*	
28.3.1	Spatially Lumped Models *806*	
28.3.2	Thin-Film Models *808*	
28.3.3	Agglomerate Models *809*	
28.3.4	Volume-Averaged Models *810*	
28.4	Conclusion *811*	
	List of Symbols *812*	
	References *814*	
29	**High-Temperature Polymer Electrolyte Fuel-Cell Modeling** *819*	
	Uwe Reimer	
29.1	Introduction *819*	
29.2	Cell-Level Modeling *821*	
29.3	Stack-Level Modeling *825*	
29.4	Phosphoric Acid as Electrolyte *827*	
29.5	Basic Modeling of the Polarization Curve *829*	
29.5.1	Activation Overpotential *830*	
29.5.2	Ohmic Resistance *831*	
29.5.3	Mass Transport *833*	
29.6	Conclusion and Future Perspectives *834*	
	References *835*	
30	**Modeling of Polymer Electrolyte Membrane Fuel-Cell Components** *839*	
	Yun Wang and Ken S. Chen	
30.1	Introduction *839*	
30.2	Polymer Electrolyte Membrane *842*	
30.3	Catalyst Layers *845*	
30.4	Gas Diffusion Layers and Microporous Layers *850*	
30.5	Gas Flow Channels *859*	
30.6	Gas Diffusion Layer-Gas Flow Channel Interface *864*	
30.7	Bipolar Plates *868*	
30.8	Coolant Flow *869*	
30.9	Model Validation *869*	
30.10	Conclusion *871*	
	List of Symbols *872*	
	References *874*	

31	**Modeling of Polymer Electrolyte Membrane Fuel Cells and Stacks**	*879*
	Yun Wang and Ken S. Chen	
31.1	Introduction *879*	
31.2	Cell-Level Modeling and Simulation *881*	
31.2.1	Dimensionality *882*	
31.2.2	Transient Operation *884*	
31.2.3	Nonisothermal Modeling *888*	
31.2.4	Two-Phase Flow *891*	
31.2.5	Cold Start Operation *893*	
31.2.6	Large-Scale Fuel-Cell Simulation *898*	
31.2.7	Flow Maldistribution *900*	
31.2.7.1	Single-Phase Flow *901*	
31.2.7.2	Two-Phase Flow *902*	
31.2.8	Model Validation *903*	
31.3	Stack-Level Modeling and Simulation *906*	
31.3.1	Why Is Stack-Level Modeling Needed? *906*	
31.3.2	Modeling and Simulation of Fuel-Cell Stacks *907*	
31.3.3	Model Validation *910*	
31.4	Conclusion *911*	
	List of Symbols *912*	
	References *913*	

Part VI   Balance of Plant Design and Components *917*

32	**Principles of Systems Engineering** *919*	
	Ludger Blum, Ralf Peters, and Remzi Can Samsun	
32.1	Introduction *919*	
32.2	Basic Engineering *920*	
32.2.1	General Considerations *920*	
32.2.2	Chemical Equilibrium *923*	
32.2.3	Analytical Methods for Heat Management *926*	
32.2.3.1	System Set-Up *926*	
32.2.3.2	Gibbs Energy Function *927*	
32.2.3.3	Pinch Point Diagram *928*	
32.2.3.4	Exergy Analysis *930*	
32.2.3.5	Process Optimization *932*	
32.2.4	Process Analysis and Design *940*	
32.3	Detailed Engineering *945*	
32.3.1	Piping and Instrumentation Diagram *948*	
32.3.2	FMEA *950*	
32.3.3	Selection of Peripheral Components *953*	
32.3.4	Drawings and Piping *954*	
32.4	Procurement *956*	
32.5	Construction *956*	
32.6	Conclusion *957*	

List of Symbols and Abbreviations *958*
Subscripts and Superscripts *958*
References *959*

33 System Technology for Solid Oxide Fuel Cells *963*
Nguyen Q. Minh
33.1 Solid Oxide Fuel Cells for Power Generation *963*
33.2 Overview of SOFC Power Systems *965*
33.2.1 General *965*
33.2.2 Type of SOFC Power System *968*
33.2.3 SOFC Power System Design *969*
33.3 Subsystem Design for SOFC Power Systems *970*
33.3.1 Power Generation Subsystem *970*
33.3.1.1 SOFC Stack *970*
33.3.1.2 Other Power Generating Equipment *977*
33.3.2 Fuel Processing Subsystem *979*
33.3.3 Fuel, Oxidant, and Water Delivery Subsystem *982*
33.3.4 Thermal Management Subsystem *983*
33.3.5 Power Conditioning Subsystem *987*
33.3.6 Control Subsystem *989*
33.4 SOFC Power Systems *991*
33.4.1 Portable Systems *991*
33.4.2 Transportation Systems *993*
33.4.2.1 SOFC-Based APUs for Automobiles and Trucks *993*
33.4.2.2 SOFC-Based APUs for Aircraft *994*
33.4.3 Stationary Systems *997*
33.4.3.1 Stationary Simple Cycle SOFC Systems *997*
33.4.3.2 SOFC/GT Hybrid Systems *998*
33.4.3.3 Integrated Gasification Fuel Cell (IGFC) Systems *1001*
Acknowledgments *1006*
References *1006*

34 Desulfurization for Fuel-Cell Systems *1011*
Joachim Pasel and Ralf Peters
34.1 Introduction and Motivation *1011*
34.2 Sulfur-Containing Molecules in Crude Oil *1011*
34.2.1 Crude Oil *1011*
34.2.2 Routes for Inserting Sulfur into the Molecules in Crude Oil *1012*
34.2.3 Different Chemical Classes of Sulfur-Containing Substances in Crud e Oil *1013*
34.2.4 Catalyst Poisoning by Sulfur-Containing Substances in Crude Oil Fractions *1015*
34.3 Desulfurization in the Gas Phase *1016*
34.3.1 Absorption *1017*
34.3.2 Adsorption *1017*

34.3.3	Chemisorption	*1018*
34.3.3.1	H_2S Removal	*1018*
34.3.3.2	S-Zorb Process	*1020*
34.3.3.3	SO_2 Removal	*1020*
34.3.4	Hydrofining	*1021*
34.3.5	Sulfur Recovery	*1022*
34.4	Desulfurization in the Liquid Phase	*1022*
34.4.1	Hydrodesulfurization with Presaturator	*1022*
34.4.2	Adsorption	*1024*
34.4.3	Ionic Liquids	*1026*
34.4.4	Selective Oxidation	*1028*
34.4.4.1	Plasma Desulfurization	*1028*
34.4.4.2	Photo-oxidation	*1028*
34.4.4.3	Oxidation with Peroxides	*1029*
34.4.4.4	Biological Processes	*1029*
34.4.5	Desulfurization with Overcritical Fluids	*1030*
34.4.6	Distillation	*1031*
34.4.7	Membrane Processes	*1032*
34.4.7.1	Processes with Porous Membranes	*1032*
34.4.7.2	Processes with Nonporous Membranes	*1032*
34.5	Application in Fuel-Cell Systems	*1034*
34.6	Conclusion	*1038*
	Acknowledgments	*1039*
	References	*1039*

35 Design Criteria and Components for Fuel Cell Powertrains *1045*
Lutz Eckstein and Bruno Gnörich

35.1	Introduction	*1045*
35.2	Vehicle Requirements	*1045*
35.2.1	Driving Resistance	*1045*
35.2.2	Energy Conversion and Driving Cycles	*1046*
35.3	Potentials and Challenges of Vehicle Powertrains	*1049*
35.3.1	Overview of Propulsion Systems	*1049*
35.3.2	Powertrain Comparison	*1055*
35.3.3	Fuel Cell Powertrains	*1058*
35.3.3.1	Non-Hybrid Fuel Cell Powertrains	*1058*
35.3.3.2	Hybrid Electric Fuel Cell Vehicles	*1058*
35.3.3.3	Triple-Hybrid Fuel Cell Vehicles	*1060*
35.4	Components of Fuel Cell Powertrains	*1061*
35.4.1	Hydrogen Storage	*1061*
35.4.2	Fuel Cell Systems for Automotive Applications	*1063*
35.4.3	Electrical Storage	*1065*
35.4.4	Electric Machines	*1067*
35.4.5	Cost Comparison of Vehicle Drivetrains	*1070*

35.5	Conclusion *1072*
	Acknowledgment *1073*
	References *1073*

36	**Hybridization for Fuel Cells** *1075*
	Jörg Wilhelm
36.1	Introduction *1075*
36.2	The Fuel-Cell Hybrid *1076*
36.2.1	Reasons for Hybridizing a Fuel Cell *1076*
36.2.2	Different Types of Fuel-Cell Hybrids *1077*
36.2.2.1	Series and Parallel Hybrids *1077*
36.2.2.2	Active and Passive Hybrids *1078*
36.2.3	Hybridization Degree *1081*
36.3	Components of a Fuel-Cell Hybrid *1081*
36.3.1	Fuel Cell *1082*
36.3.2	Energy Storage *1082*
36.3.3	Power Electronics *1083*
36.3.4	Control Unit *1084*
36.4	Hybridization Concepts *1085*
36.4.1	Overview *1085*
36.4.2	Basic Types *1085*
36.4.3	Possible Concepts *1087*
36.5	Technical Overview *1088*
36.5.1	Fuel-Cell Powertrains *1088*
36.5.1.1	Passenger Cars *1089*
36.5.1.2	Buses *1092*
36.5.2	Light Traction Applications *1092*
36.5.2.1	Scooters, Wheelchairs, and Electromobiles *1093*
36.5.2.2	Commercial Vehicles *1093*
36.5.2.3	Forklift Trucks *1094*
36.5.3	Other Applications *1096*
36.6	Systems Analysis *1096*
36.7	Conclusion *1098*
	References *1098*

	Part VII	**Systems Verification and Market Introduction** *1105*

37	**Off-Grid Power Supply and Premium Power Generation** *1107*
	Kerry-Ann Adamson
37.1	Introduction *1107*
37.2	Premium Power Market Overview *1107*
37.3	Off-Grid *1109*
37.3.1	Homes *1109*
37.3.2	Off-Grid Base Stations *1111*
37.4	Portable Applications *1113*

37.4.1	Remote Monitoring/Remote Sensing	*1113*
37.4.2	Military Applications for Prime Power	*1115*
37.4.2.1	Soldier Power	*1115*
37.4.3	Portable Generators –Military	*1116*
37.5	Discussion	*1117*
	References	*1117*

38 Demonstration Projects and Market Introduction *1119*
Kristin Deason

38.1	Introduction	*1119*
38.2	Why Demonstration?	*1119*
38.3	Transportation Demonstrations	*1120*
38.3.1	Germany	*1122*
38.3.1.1	Clean Energy Partnership	*1122*
38.3.1.2	Activities in North Rhine-Westphalia	*1124*
38.3.1.3	H_2 Mobility	*1125*
38.3.1.4	Additional Resources	*1125*
38.3.2	Japan	*1126*
38.3.2.1	Japan Hydrogen and Fuel Cell Demonstration Project (JHFC)	*1126*
38.3.2.2	Hydrogen Highway Project	*1129*
38.3.2.3	Activities in Fukuoka Prefecture	*1129*
38.3.2.4	Additional Resources	*1129*
38.3.3	United States	*1130*
38.3.3.1	The DOE Technology Validation Program	*1130*
38.3.3.2	State Activities	*1132*
38.3.3.3	Additional Resources	*1133*
38.3.4	European Union	*1133*
38.3.4.1	Fuel-Cell Bus Projects	*1134*
38.3.4.2	H2moves Scandinavia	*1135*
38.3.4.3	Regional and Member Country Activities	*1136*
38.3.4.4	Additional Resources	*1136*
38.3.5	Canada	*1136*
38.3.6	South Korea	*1137*
38.3.6.1	Domestic Fleet Program	*1137*
38.3.7	China	*1137*
38.3.7.1	GEF/UNDP-China FCB Demonstration	*1138*
38.3.7.2	Beijing Olympics and Shanghai World Expo	*1138*
38.3.8	Auto Maker Demonstration Programs	*1138*
38.4	Stationary Power and Early Market Applications	*1139*
38.4.1	Japan	*1140*
38.4.1.1	Regional Activities	*1141*
38.4.1.2	Additional Resource	*1141*
38.4.2	Denmark	*1142*
38.4.3	Germany	*1142*
38.4.3.1	Regional Activities	*1143*

38.4.3.2	Additional Information	*1143*
38.4.4	European Union	*1143*
38.4.5	United States	*1144*
38.4.6	South Korea	*1145*
	References	*1146*
	Further Reading	*1150*

Part VIII   Knowledge Distribution and Public Awareness *1151*

39   A Sustainable Framework for International Collaboration: the IEA HIA and Its Strategic Plan for 2009–2015 *1153*
Mary-Rose de Valladares

39.1	Introduction	*1153*
39.2	The IEA HIA Strategic Framework: Overview	*1154*
39.2.1	Theme 1: Collaborative RD&D	*1155*
39.2.1.1	Production Portfolio	*1157*
39.2.1.2	Storage Portfolio	*1158*
39.2.1.3	Integrated Systems Portfolio	*1159*
39.2.1.4	Integration in Existing Infrastructure Portfolio	*1160*
39.2.2	Theme 2: Analysis That Positions Hydrogen	*1161*
39.2.2.1	Technical Portfolio	*1161*
39.2.2.2	Market Portfolio	*1162*
39.2.2.3	Support for Political Decision-Making Portfolio	*1162*
39.2.3	Theme 3: Hydrogen Awareness, Understanding, and Acceptance	*1162*
39.2.3.1	Information Dissemination Portfolio	*1162*
39.2.3.2	Safety Portfolio	*1163*
39.2.3.3	Outreach Portfolio	*1164*
39.3	The Work Program: Issues and Approaches	*1166*
39.4	IEA HIA: the Past as Prolog	*1166*
39.5	The 2009–2015 IEA HIA Work Program Timeline	*1173*
39.6	Conclusion and Final Remarks	*1177*
	References	*1179*
	Further Reading	*1179*

40   Overview of Fuel Cell and Hydrogen Organizations and Initiatives Worldwide *1181*
Bernd Emonts

40.1	Introduction	*1181*
40.2	International Level	*1181*
40.2.1	International Partnership for Hydrogen and Fuel Cells in the Economy	*1182*
40.2.2	International Energy Agency	*1183*
40.2.2.1	Implementing Agreement on Advanced Fuel Cells	*1184*
40.2.2.2	Hydrogen Implementing Agreement	*1185*

40.3	European Level	*1187*
40.3.1	Fuel Cells and Hydrogen Joint Undertaking	*1187*
40.3.1.1	FCH JU Members	*1189*
40.3.1.2	Governance Structure	*1190*
40.3.2	European Hydrogen Association	*1193*
40.4	National Level	*1196*
40.4.1	US Fuel Cell and Hydrogen Energy Association	*1196*
40.4.1.1	Working Groups and Committees	*1197*
40.4.1.2	Resources	*1197*
40.4.2	Canadian Hydrogen and Fuel Cell Association	*1198*
40.4.3	German National Organization for Hydrogen and Fuel Cell Technology	*1200*
40.5	Regional Level	*1201*
40.5.1	European Regions and Municipalities Partnership for Hydrogen and Fuel Cells	*1201*
40.5.2	Hydrogen and Fuel-Cell Activities in Germany's Federal States	*1202*
40.6	Partnerships, Initiatives, and Networks with a Specific Agenda	*1204*
40.6.1	The California Fuel Cell Partnership	*1204*
40.6.2	UK Hydrogen Energy Network	*1206*
40.6.3	Initiative Brennstoffzelle	*1206*
40.7	Conclusion	*1208*
	References	*1209*
41	**Contributions for Education and Public Awareness**	*1211*
	Thorsteinn I. Sigfusson and Bernd Emonts	
41.1	Introduction	*1211*
41.2	Information for Interested Laypeople	*1212*
41.3	Education for School Students and University Students	*1213*
41.4	Electrolyzers and Fuel Cells in Education and Training	*1215*
41.5	Training and Qualification for Trade and Industry	*1216*
41.6	Education and Training in the Scientific Arena	*1218*
41.7	Clarification Assistance in the Political Arena	*1219*
41.8	Analysis of Public Awareness	*1220*
41.9	Conclusion	*1221*
	References	*1221*

Index *1223*

List of Contributors

Kerry-Ann Adamson
Pike Research – Cleantech
Market Intelligence
180–186 Kings Cross Road
London WC1X 9DE
UK

Elisabetta Arato
University of Genoa
PERT, Process Engineering
Research Team
Via Opera Pia 15
16145 Genoa
Italy

Jan B.A. Arends
Ghent University
Faculty of Bioscience Engineering
Laboratory of Microbial Ecology
and Technology (LabMET)
Coupure Links 653
9000 Ghent
Belgium

Peter Batfalsky
Forschungszentrum Jülich
GmbH
ZAT
Leo-Brandt-Straße
52425 Jülich
Germany

Ludger Blum
Forschungszentrum Jülich
GmbH, IEK-3
Leo-Brandt-Straße
52425 Jülich
Germany

Barbara Bosio
University of Genoa
PERT, Process Engineering
Research Team
Via Opera Pia 15
16145 Genoa
Italy

Michael Bron
Martin-Luther-Universität
Halle-Wittenberg
Naturwissenschaftliche Fakultät
II – Chemie, Physik, und
Mathematik
Institut für Chemie – Technische
Chemie I
von-Danckelmann-Platz 4
06120 Halle
Germany

Ken S. Chen
Sandia National Laboratories
7011 East Avenue
MS9154, Livermore
CA 94550
USA

Kristin Deason
NOW GmbH
Nationale Organisation
Wasserstoff- und
Brennstoffzellentechnologie
Fasanenstraße 5
10623 Berlin
Germany

Joachim Desloover
Ghent University
Faculty of Bioscience Engineering
Laboratory of Microbial Ecology
and Technology (LabMET)
Coupure Links 653
9000 Ghent
Belgium

Mary-Rose de Valladares
International Energy Agency
Hydrogen Implementing
Agreement (IEA HIA)
9650 Rockville Pike
Bethesda
MD 20814
USA

Lutz Eckstein
RWTH Aachen University
Institut für Kraftfahrzeuge (IKA)
Steinbachstraße 7
52074 Aachen
Germany

John F. Elter
Sustainable Systems LLC, 874
Old Albany Shaker Road, Latham
NY 12110
USA

and

University of Albany, State
University of New York
College of Nanoscale Science and
Engineering
NanoFab 300 East, 257 Fuller
Road, Albany
NY 12222
USA

Bernd Emonts
Forschungszentrum Jülich
GmbH, IEK-3
Leo-Brandt-Straße
52425 Jülich
Germany

Gerd Gaiselmann
Universität Ulm
Institut für Stochastik
HelmholtzStraße 18
89069 Ulm
Germany

Dietmar Gerteisen
Fraunhofer Institute for Solar
Energy Systems ISE
Department of Fuel Cell Systems
Heidenhofstraße 2
79110 Freiburg
Germany

Bruno Gnörich
RWTH Aachen
Institut für Kraftfahrzeuge (IKA)
SteinbachStraße 7
52074 Aachen
Germany

Paolo Greppi
University of Genoa
PERT, Process Engineering
Research Team
Via Opera Pia 15
16145 Genoa
Italy

Ulf Groos
Fraunhofer Institute for Solar
Energy Systems ISE
Department of Fuel Cell Systems
Heidenhofstrasse 2
79110 Freiburg
Germany

Thomas Grube
Forschungszentrum Jülich
GmbH, IEK-3
Leo-Brandt-Straße
52425 Jülich
Germany

Erich Gülzow
Deutsches Zentrum für Luft- und
Raumfahrt eV (DLR)
Institut für Technische
Thermodynamik
Pfaffenwaldring 38–40
70569 Stuttgart
Germany

Evren Gunen
TUBITAK Marmara Research
Center
Energy Institute
Dr. Zeki Acar Cad.
Baris mah. No: 1
Gebze
Kocaeli 41470
Turkey

Vincent Haanappel
Forschungszentrum Jülich
GmbH, IEK-3
Leo-Brandt-Straße
52425 Jülich
Germany

Viktor Hacker
Graz University of Technology
Institute of Chemical Engineering
and Environmental Technology
Inffeldgasse 25/C/II
8010 Graz
Austria

Peter Heidebrecht
Max Planck Institut
Dynamics of Complex Technical
Systems
Sandtorstraße 1
39106 Magdeburg
Germany

Ellen Ivers-Tiffée
Karlsruher Institut für
Technologie (KIT)
Institut für Werkstoffe der
Elektrotechnik (IWE)
Adenauerring 20b
Gebäude 50.40
76131 Karlsruhe
Germany

Mohammad A. Khaleel
Boston University
Department of Mechanical
Engineering
110 Cummington Street
Boston
MA 02215
USA

Gunther Kolb
Institut für Mikrotechnik Mainz
GmbH
Energietechnik und Katalyse
Carl-Zeiss-Straße 18–20
55129 Mainz
Germany

Carsten Korte
Forschungszentrum Jülich
GmbH, IEK-3
Leo-Brandt-Straße
52425 Jülich
Germany

Melanie Kuhn
Massachusetts Institute of
Technology
Department of Materials Science
and Engineering
77 Massachusetts Avenue
Cambridge
MA 02139
USA

Andrei Kulikovsky
Forschungszentrum Jülich
GmbH, IEK-3
Leo-Brandt-Straße
52425 Jülich
Germany

Werner Lehnert
Forschungszentrum Jülich
GmbH, IEK-3
Leo-Brandt-Straße
52425 Jülich
Germany

Werner Lehnert
Forschungszentrum Jülich
GmbH, IEK-3
Leo-Brandt-Straße
52425 Jülich
Germany

André Leonide
Karlsruher Institut für
Technologie (KIT)
Institut für Werkstoffe der
Elektrotechnik (IWE)
Adenauerring 20b
Gebäude 50.40
76131 Karlsruhe
Germany

Sebastian Maass
Robert Bosch GmbH
Corporate Sector Research and
Advance Engineering
CR/ARC1 – Energy Storage and
Conversion
Robert-Bosch-Platz 1
70839 Gerlingen-Schillerhöhe
Germany

Thomas Malkow
European Commission
Directorate-General Joint
Research Centre
Institute for Energy and
Transport
Westerduinweg 3
1755 LE Petten
The Netherlands

Ingo Manke
Helmholtz-Zentrum Berlin
Hahn-Meitner-Platz 1
D-14109 Berlin
Germany

Norbert H. Menzler
Forschungszentrum Jülich
GmbH, IEK-1
Leo-Brandt-Straße
52425 Jülich
Germany

Jürgen Mergel
Forschungszentrum Jülich
GmbH, IEK-3
Leo-Brandt-Straße
52425 Jülich
Germany

Nguyen Q. Minh
University of California,
San Diego
Center for Energy Research
9500 Gilman Drive
La Jolla
CA 92093-0417
USA

Martin Müller
Forschungszentrum Jülich
GmbH, IEK-3
Leo-Brandt-Straße
52425 Jülich
Germany

Téko W. Napporn
Université de Poitiers
Electrocatalysis Group (e-lyse),
IC2MP UMR 7285 CNRS
4 rue Michel Brunet
86022, Poitiers
France

Jan Hendrik Ohs
Robert Bosch GmbH
Corporate Sector Research and
Advance Engineering
CR/ARC1 – Energy Storage and
Conversion
Robert-Bosch-Platz 1
70839 Gerlingen-Schillerhöhe
Germany

Joachim Pasel
Forschungszentrum Jülich
GmbH, IEK-3
Leo-Brandt-Straße
52425 Jülich
Germany

Murat Peksen
Forschungszentrum Jülich
GmbH, IEK-3
Leo-Brandt-Straße
52425 Jülich
Germany

Ralf Peters
Forschungszentrum Jülich
GmbH, IEK-3
Leo-Brandt-Straße
52425 Jülich
Germany

Silvia Piewek
Max Planck Institute
Dynamics of Complex Technical
Systems
Sandtorstraße 1
39106 Magdeburg
Germany

Sebastiá Puig
University of Girona
Faculty of Sciences
Institute of the Environment
Laboratory of Chemical and
Environmental Engineering
(LEQUIA-UdG)
Campus Montilivi s/n
17071 Girona
Spain

Uwe Reimer
Forschungszentrum Jülich
GmbH, IEK-3
Leo-Brandt-Straße
52425 Jülich
Germany

Emily M. Ryan
Boston University
Department of Mechanical
Engineering
110 Cummington Street
Boston
MA 02215
USA

and

Pacific Northwest National
Laboratory
902 Battelle Boulevard
Richland
WA 99352
USA

Remzi Can Samsun
Forschungszentrum Jülich
GmbH, IEK-3
Leo-Brandt-Straße
52425 Jülich
Germany

Ulrich S. Sauter
Robert Bosch GmbH
Corporate Sector Research and
Advance Engineering
CR/ARC1 – Energy Storage and
Conversion
Robert-Bosch-Platz 1
70839 Gerlingen-Schillerhöhe
Germany

Florian Scharf
Forschungszentrum Jülich
GmbH, IEK-3
Leo-Brandt-Straße
52425 Jülich
Germany

Helge Schichlein
Karlsruher Institut für
Technologie (KIT)
Institut für Werkstoffe der
Elektrotechnik (IWE)
Adenauerring 20b
Gebäude 50.40
76131 Karlsruhe
Germany

Sabine Schimpf
Martin-Luther-Universität
Halle-Wittenberg
Naturwissenschaftliche Fakultät
II – Chemie, Physik, und
Mathematik
Institut für Chemie – Technische
Chemie I
von-Danckelmann-Platz 4
06120 Halle
Germany

Volker Schmidt
Universität Ulm
Institut für Stochastik
HelmholtzStraße 18
89069 Ulm
Germany

Thorsteinn I. Sigfusson
University of Iceland
Sæmundargötu 2
101 Reykjavík and Innovation
Centre Iceland
Keldnaholt
112 Reykjavik
Iceland

Volker Sonn
Karlsruher Institut für
Technologie (KIT)
Institut für Werkstoffe der
Elektrotechnik (IWE)
Adenauerring 20b
Gebäude 50.40
76131 Karlsruhe
Germany

Kai Sundmacher
Max Planck Institute
Dynamics of Complex Technical
Systems
Sandtorstraße 1
39106 Magdeburg
Germany

Ralf Thiedmann
Universität Ulm
Institut für Stochastik
HelmholtzStraße 18
89069 Ulm
Germany

Christian Tötzke
Technische Universität Berlin
StraBe des 17. Juni 135
D-10623 Berlin
Germany

and

Helmholtz-Zentrum Berlin
Hahn-Meitner-Platz 1
D-14109 Berlin
Germany

Georgios Tsotridis
European Commission
Directorate-General Joint
Research Centre
Institute for Energy and
Transport
Westerduinweg 3
1755 LE Petten
The Netherlands

John A. Turner
National Renewable Energy
Laboratory
1617 Cole Boulevard
Golden
CO 80401
USA

Willy Verstraete
Ghent University
Faculty of Bioscience Engineering
Laboratory of Microbial Ecology
and Technology (LabMET)
Coupure Links 653
9000 Ghent
Belgium

Jürgen Wackerl
Forschungszentrum Jülich
GmbH, IEK-3
Leo-Brandt-Straße
52425 Jülich
Germany

Eva Wallnöfer-Ogris
Graz University of Technology
Institute of Chemical Engineering
and Environmental Technology
Inffeldgasse 25/C/II
8010 Graz
Austria

Heli Wang
National Renewable Energy
Laboratory
1617 Cole Boulevard
Golden
CO 80401
USA

Yun Wang
University of California
Department of Mechanical and
Aerospace Engineering
4231 Engineering Gateway
Irvine
CA 92697
USA

André Weber
Karlsruher Institut für
Technologie (KIT)
Institut für Werkstoffe der
Elektrotechnik (IWE)
Adenauerring 20b
Gebäude 50.40
76131 Karlsruhe
Germany

K. Scott Weil
Pacific Northwest National
Laboratory
902 Battelle Boulevard
Richland
WA 99352
USA

Jörg Wilhelm
Forschungszentrum Jülich
GmbH, IEK-3
Leo-Brandt-Straße
52425 Jülich
Germany

Part V
Modeling and Simulation

23
Messages from Analytical Modeling of Fuel Cells
Andrei Kulikovsky

23.1
Introduction

The fuel-cell effect was discovered in 1838, and for more than 100 years after this event, low-temperature fuel cells have utilized liquid electrolyte as an ionic conductor between the anode and the cathode [1]. The situation changed with invention of Nafion, the first stable proton-conducting solid polymer. Nafion has radically modified the design of low-temperature cells and stacks. A liquid electrolyte makes cells bulky and unsafe, and it requires sophisticated sealing. In Nafion-based cells, the electrodes are separated by a thin, 20–100 μm thick polymer film. In addition, Nafion allowed cell electrodes that contain electrolyte to be much thinner.

Nafion was discovered in the mid-1960s; however, it took more than 20 years to understand the tremendous potential of this material for fuel cells and to introduce its mass production. Scientific databases reveal the start of the exponential growth of publications on low-temperature cell science and technology in the late 1980s.

Early modeling of fuel cells with a liquid electrolyte was focused mainly on the electrochemical problems of cell function [2]. The invention of Nafion generated high expectations and investments, which supported study of numerous physical and engineering aspects of cell operation. Today's fuel-cell science and technology include overlapping domains of electrochemistry, physics, fluid dynamics, and chemical and mechanical engineering.

This "new age" of low-temperature cells R&D concurs with Moore's exponential growth of computer power and software development. This growth has stimulated extensive numerical modeling of fuel cells. Since 1990, the number of papers on numerical simulations of fuel cells has been growing at an astounding rate.

Numerical modeling is an invaluable tool for understanding cell function and testing various cell designs. However, the information resulting from numerical calculations is in many respects limited. The greatest challenge for numerical fuel-cell models is the large number of parameters that determine cell operation.

The light bulb was invented by Thomas Edison without any simulations. The reason was simple: just three parameters determine bulb function: wire material,

Fuel Cell Science and Engineering: Materials, Processes, Systems and Technology, First Edition. Edited by Detlef Stolten and Bernd Emonts.
© 2012 Wiley-VCH Verlag GmbH & Co. KGaA. Published 2012 by Wiley-VCH Verlag GmbH & Co. KGaA.

wire resistance, and air pressure in the bulb. An experimental trial-and-error search in a reasonable subspace of the three-parameter space proved to be successful.

A fuel cell is a much more complicated system: its function depends on about 100 geometric, operational, physical, and electrochemical parameters. Numerical simulation gives a snapshot of fuel-cell operation at just one point in this multidimensional space. Obviously, trial-and-error wandering in a space of 100 coordinates is doomed.

This suggests analytical modeling of fuel cells. Analytical studies aim at understanding rather than a complete description. They ignore secondary details in order to capture the essential features of a process or device. The nearest analogy is an avant-garde portrait, which fixes only the characteristic features of a person's face. Being far from a detailed classical portrait, an avant-garde image often tells more about the person's character.

The simple analytical models discussed below demonstrate that in many cases, the cell operation depends on a composite of several parameters, rather than on the separate terms (factors) of this composite. This greatly lowers the dimensionality of the parameter space. Analytical models explicitly show the parametric dependences, indicating the direction to a cheap and robust fuel cell in this space. These models can be used to characterize the cells. Further, analytical models serve as building blocks for growing hybrid "analytical + computational fluid dynamics" modeling of cells and stacks.

In this chapter, we report on several analytical models for the performance of catalyst layers and fuel cells. The simple models discussed below are illustrated with examples demonstrating these models in use.

23.2
Modeling of Catalyst Layer Performance

23.2.1
The Basic Equations

At open-circuit, a fuel cell provides between the electrodes a thermodynamic voltage in the order of 1 V. Drawing current from the cell "costs" some potential and the cell voltage decreases.

Key components of any fuel cell are the anode and cathode catalyst layers (CLs). One of the layers converts the electron current into ionic current, and the other performs the reverse conversion. The dependence of voltage loss (overpotential) in the electrode on current is of primary interest for any application.

Regardless of the cell type, all catalyst layers do the same job of current-to-current conversion. Therefore, it is advisable to consider a model of a *generic catalyst layer*. A schematic diagram of a generic CL is shown in Figure 23.1. Ions come to/from the left (electrolyte) side, whereas feed molecules arrive from the right [gas diffusion layer (GDL)] side. The ionic current j decays and the electron current j_e grows with the distance from the electrolyte (Figure 23.1). The driving force for

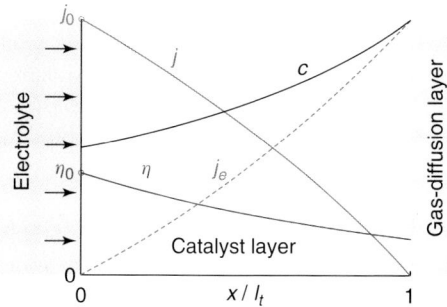

Figure 23.1 Schematic diagram of the generic catalyst layer. Shown are the shapes of ion (proton) current density j, electron current density j_e, feed molecules concentration c, and overpotential η. The CL polarization curve is $\eta_0(j_0)$.

the electrochemical conversion is the overpotential η, which decreases with x. The conversion reaction consumes neutral feed molecules, the concentration of which decreases towards the electrolyte (Figure 23.1).

The total cell current is equal to the ionic current in the electrolyte, j_0. The voltage required to perform complete conversion of j_0 from ionic into electronic form is η_0 (Figure 23.1) (see [3] for more details). The *polarization curve of the CL* is the equation $\eta_0(j_0)$ and the task of modeling is to rationalize this dependence.

The simplest model that describes the conversion includes three equations [4, 5]:

$$\frac{\partial j(x)}{\partial x} = -i_* \left(\frac{c}{c_h^0}\right)^\gamma \left[\exp\left(\frac{\eta}{b}\right) - \exp\left(-\frac{\eta}{b}\right)\right] \qquad (23.1)$$

$$j = -\sigma_t \frac{\partial \eta(x)}{\partial x} \qquad (23.2)$$

$$D\frac{\partial c(x)}{\partial x} = \frac{j_0 - j}{nF} \qquad (23.3)$$

where x is the coordinate across the CL (Figure 23.1), i_* is the volumetric exchange current density, that is, a total charge converted in unit volume of the CL per unit time, A m^{-3}, σ_t is the CL ionic conductivity, D is the effective diffusion coefficient of feed molecules in the CL, γ is the feed molecules reaction order, n is the number of electrons transferred in the reaction, b, given by

$$b = \frac{RT}{\alpha F} \qquad (23.4)$$

is the Tafel slope, and α is the transfer coefficient [6].

The first equation expresses the decay of ionic current with x at a Butler–Volmer rate (the right-hand side of this equation) [6]. The second is Ohm's law relating the ionic current to the overpotential gradient. The third indicates that the local diffusion flux of feed molecules is equal to the molar flux of electrons to be converted into ions.

After pioneering works [4, 5], the system of equations [Eqs. (23.1–23.3)] has been extensively studied analytically [7–11]. So far, however, no exact analytical solution to this system has been found. Nonetheless, in the limiting cases of ideal transport of feed molecules or ions, the system can be solved. For practical applications, most interesting is the case of ideal feed transport, which is a reasonable approximation for well-designed CLs in major types of cells.

23.2.2
Ideal Transport of Feed Molecules

In that case, Eq. (23.3) can be omitted. It is convenient to introduce the dimensionless variables

$$\tilde{x} = \frac{x}{l_t}, \quad \tilde{\eta} = \frac{\eta}{b}, \quad \tilde{c} = \frac{c}{c_h^0}, \quad \tilde{j} = \frac{jl_t}{\sigma_t b} \tag{23.5}$$

where l_t is the CL thickness and c_h^0 is the molar concentration of feed molecules at the channel inlet (see below).

With these variables, Eqs. (23.1) and (23.2) take the form

$$\varepsilon^2 \frac{\partial \tilde{j}}{\partial \tilde{x}} = -\sinh \tilde{\eta} \tag{23.6}$$

$$\tilde{j} = -\frac{\partial \tilde{\eta}}{\partial \tilde{x}} \tag{23.7}$$

where

$$\varepsilon = \sqrt{\frac{\sigma_t b}{2 i_* \tilde{c}_t^\gamma l_t^2}} \equiv \frac{l_N}{l_t} \tag{23.8}$$

is the dimensionless Newman's reaction penetration depth [12], which in the dimensional form is l_N, and c_t is the molar concentration of feed molecules in the CL. Physically, the parameter ε shows how deep the conversion reaction penetrates into the catalyst layer under ideal transport of all reactants, including protons.

To analyze Eqs. (23.6) and (23.7), it is convenient to eliminate the overpotential from this system [13]. Differentiating Eq. (23.6) over $\tilde{x}$, using Eq. (23.7) and the identity $\cosh^2(\tilde{\eta}) - \sinh^2(\tilde{\eta}) = 1$, we obtain

$$\varepsilon^2 \frac{\partial^2 \tilde{j}}{\partial \tilde{x}^2} = \tilde{j}\sqrt{1 + \varepsilon^4 \left(\frac{\partial \tilde{j}}{\partial \tilde{x}}\right)^2} \tag{23.9}$$

The boundary conditions for this equation are evident:

$$\tilde{j}(0) = \tilde{j}_0, \quad \tilde{j}(1) = 0 \tag{23.10}$$

In the limiting cases of small and large currents, Eq. (23.9) can be solved [11].

23.2.3
Polarization Curve

Solution of Eq. (23.9) leads to the following general equation for the CL polarization curve [11]:

$$\varepsilon \tilde{j}_0 = \sqrt{2 \sinh \tilde{\eta}_0} \tanh \left(\frac{\sqrt{2 \sinh \tilde{\eta}_0}}{2 \coth(\varepsilon^{-1})} \right) \quad (23.11)$$

This polarization curve is shown in Figure 23.2 for large and small ε (dashed curves), together with the solid curves for the low- and high-current limiting cases. The low- and high-current curves follow from Eq. (23.11) if we set $\tanh(x) \simeq x$ and $\tanh(x) \simeq 1$, respectively (see below).

Figure 23.2 shows that in general, there are three regimes of CL operation. In the low-current regime, the CL polarization curve is linear or Tafel like. If the cell current is high, the CL works in the high-current mode, featuring doubling of the Tafel slope [4, 5, 11]. The polarization voltage (the cost of current conversion) in this double-Tafel region is very high.

The low- and high-current regimes are separated by the transition region (Figure 23.2). In this region, the exact solution for the shapes of local parameters in the CL does not exist. However, Eq. (23.11) approximates well the exact numerical curve in this region.

The position of the transition region is determined by the following equation [11]:

$$\coth\left(\frac{1}{\varepsilon}\right) \leq \varepsilon \tilde{j}_0 \leq 4 \coth\left(\frac{1}{\varepsilon}\right) \quad (23.12)$$

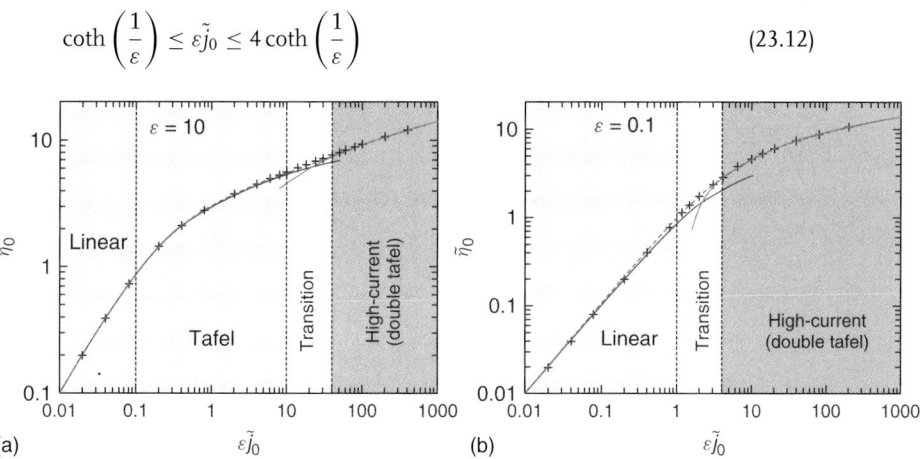

Figure 23.2 The general polarization curve of the catalyst layer with ideal feed transport. Solid lines, analytical solutions for the low- and high-current regimes; points, the exact numerical solution to the system [Eqs.(23.6) and (23.7)]; dashed line, the approximate analytical curve, Eq. (23.11), valid in the whole range of $\varepsilon \tilde{j}_0$. Parameter $\varepsilon = 10$ (a) and 0.1 (b). Linear domain is described by $\tilde{\eta}_0 = \varepsilon \tilde{j}_0$, the Tafel region is given by $\tilde{\eta}_0 = \text{arcsinh}\left[\varepsilon \tilde{j}_0 \coth(1/\varepsilon)\right]$, and the double-Tafel law is $\tilde{\eta}_0 = 2 \ln(\varepsilon \tilde{j}_0)$.

For small and large ε, these inequalities simplify to

$$1 \le \frac{j_0}{j_{crit}^0} \le 4, \quad \varepsilon \ll 1 \tag{23.13}$$

$$1 \le \frac{j_0}{j_{crit}^\infty} \le 4, \quad \varepsilon \gg 1 \tag{23.14}$$

where the critical current densities are given by

$$j_{crit}^0 = \sqrt{2i_* \sigma_t b \tilde{c}_t^\gamma}, \quad \varepsilon \ll 1 \tag{23.15}$$

$$j_{crit}^\infty = \frac{\sigma_t b}{l_t}, \quad \varepsilon \gg 1 \tag{23.16}$$

At small ε, the Tafel region disappears and the transition region directly links the linear and double-Tafel domains (Figure 23.2b). This situation is typical of a thick solid oxide fuel cell (SOFC) anode (anode-supported design). For such an anode, l_t is large and ε is small, which corresponds to the polarization curve in Figure 23.2b. The well-known linearity of a SOFC polarization curve means that the anode operates below the transition region.

23.2.4
The Critical Current Density

The critical current density [Eqs. (23.15) and (23.16)] indicates the left (low-current) boundary of the transition region ($j_0/j_{crit} = 1$). For the estimate, we may neglect the finite width of the transition region and assume that j_{crit} separates the low- and high-current regimes of CL operation.

Table 23.1 lists the typical parameters and critical current densities for the three electrodes. As can be seen, for the polymer electrolyte membrane fuel cell (PEMFC) cathode, j_{crit} is large and under a typical current density of about $1\,A\,cm^{-2}$ the cathode works in the Tafel regime. However, if the CL proton conductivity were to decrease by a factor of 3, j_{crit} would be three times lower and the CL would enter the transition region, in which the polarization voltage increases.

Much worse is the situation with a direct methanol fuel cell (DMFC) anode. For this electrode, j_{crit} is very low, about $20\,mA\,cm^{-2}$ (Table 23.1). This means that at a working current of $100\,mA\,cm^{-2}$, a typical DMFC anode operates in the double-Tafel regime, which dramatically increases the anode polarization voltage.

Table 23.1 Typical Parameters and Critical Current Densities for Electrodes in Various Fuel Cells.

	b (V)	σ_t ($\Omega^{-1}\,cm^{-1}$)	l_t (μm)	i_* (A cm^{-3})	ε	j_{crit} (A cm^{-2})
PEFC cathode	0.05	0.03	10	10^{-3}	10^3	1.5
DMFC anode	0.05	0.003	100	1	1	0.02
SOFC cathode	0.15	0.01	100	10	1	0.2

A SOFC cathode usually operates in the transition region. With $\varepsilon \simeq 1$, the critical current density for this electrode is about 200 mA cm^{-2} (Table 23.1), that is, at a working current of 500 mA cm^{-2}, the cathode regime falls into the transition region. Note that the SOFC working temperature may vary in a wide range. Owing to the exponential dependence of i_* and σ_t on T, with increase in T, ε decreases whereas j_{crit} increases rapidly, and the electrode enters the linear regime.

23.2.5
The x-Shapes

In the high-current regime ($j > 4j_{crit}$), the solution to the system [Eqs. (23.6) and (23.7)] is[1]

$$\tilde{j} = \beta \tan\left[\frac{\beta}{2}(1-\tilde{x})\right] \tag{23.17}$$

$$\tilde{S} = \frac{\beta^2 + \tilde{j}^2}{2} \tag{23.18}$$

$$\tilde{\eta} = \operatorname{arcsinh}\left(\varepsilon^2 \tilde{S}\right) \tag{23.19}$$

where $\tilde{S} \equiv -\partial \tilde{j}/\partial \tilde{x}$ is the rate of electrochemical conversion and the parameter $\beta < \pi$. For this parameter, the following asymptotic equation is valid:

$$\beta \simeq \frac{\pi \tilde{j}_0}{2 + \tilde{j}_0} \tag{23.20}$$

The curves for Eqs. (23.17–23.19) are shown in Figure 23.3. The reaction rate $\tilde{S}$ is maximal at the electrolyte interface; furthermore, most of the electrochemical conversion occurs in a small *conversion domain* at this interface (Figure 23.3). For the thickness l_* of this domain, calculation gives [11]

$$l_* = \frac{\sigma_t b}{j_0} \tag{23.21}$$

Importantly, l_* decreases with the increase in j_0.

The peak of $\tilde{S}$ at the electrolyte interface corresponds to the peak of local overpotential there (Figure 23.3). This suggests that the secondary electrochemical reactions, which may run in the CL, also peak at $\tilde{x} = 0$. This idea can be applied to model the dynamics of CL degradation, as discussed in the next section.

---

[1] Recently, it has been shown that the results of this section are valid also in the transition and normal Tafel regions. These results will be published elsewhere.

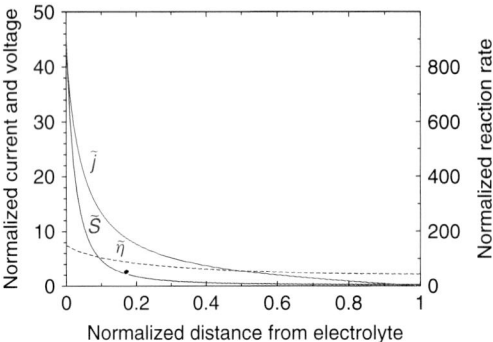

Figure 23.3 The shapes of the dimensionless ionic current density $\tilde{j}$, overpotential $\tilde{\eta}$, and the rate of the electrochemical conversion $\tilde{S}$ in the catalyst layer with ideal feed transport for high cell current. Parameters $\beta = 3$, $\varepsilon = 1$.

23.2.6
A Model for Cr Poisoning of the SOFC Cathode

In a planar SOFC stack, the bipolar plates are usually made of chromium-containing stainless steel. At high temperature, this steel is covered by a thin oxide film of Cr_2O_3, which reacts with traces of water to form volatile Cr-containing hydroxide (VCH). The VCH diffuses to the cathode catalyst layer, which in the SOFC community is called the *functional layer* (FL). In the FL, VCH is reduced electrochemically with the deposition of solid Cr_2O_3:

$$CrO_2(OH)_2(g) + 6e^- \rightarrow 3O^{2-} + Cr_2O_3(s) + 2H_2O(g) \qquad (23.22)$$

where the symbols (g) and (s) represent the gaseous and solid phase, respectively. Solid Cr_2O_3 blocks the triple-phase boundary in the FL, thereby lowering the electrode performance.

The x-shapes [Eqs. (23.17–23.19)] allow us to construct a model for this process [14]. Our main assumption will be that the Tafel slope of the reaction in Eq. (23.22) is close to the Tafel slope of the oxygen reduction reaction (ORR) running in the electrode. This means that the shape of the overpotential which drives the VCH reduction follows the shape of the ORR. The equilibrium potential of the VCH reduction differs from that of the ORR; however, this difference only shifts the overpotential as a whole along the potential axis. Hence the rate of Cr_2O_3 deposition also peaks at the electrolyte interface and the thickness of the domain of fast Cr deposition is equal to the thickness of the ORR conversion domain l_*.

Let the time of poisoning of the conversion domain be τ_d. In a time τ_d, the conversion domain is almost "dead" for the ORR and the ORR peak shifts from the electrolyte by a distance l_* (we assume that some small residual electrochemical activity exists in the poisoned domain). From this moment on, a fresh conversion domain is subject to Cr poisoning. Evidently, this leads to the traveling wave

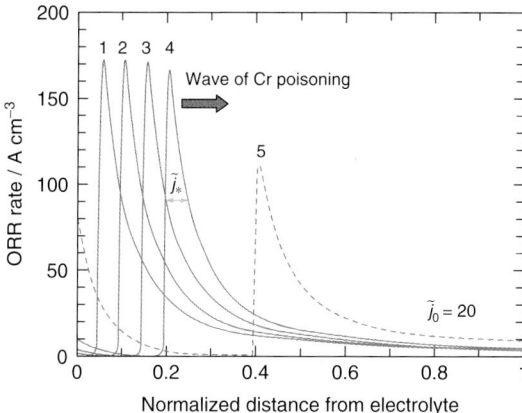

Figure 23.4 The wave of Cr poisoning in the functional layer. The numbers 1–4 represent successive frames of the wave motion. The initial distribution $\tilde{S}(\tilde{x})$ is shown in Figure 23.3. The dimensionless cell current density is $\tilde{j}_0 = 20$.

of Cr poisoning, which propagates from the electrolyte to the current collector (Figure 23.4).

Calculation based on the equations in Section 23.2.5 gives the time of complete FL poisoning, that is, the cell lifetime [14]:

$$t^{\lim} = \frac{12 F c_s^{\lim} l_t}{\gamma_s j_0} \quad (23.23)$$

where $c_s^{\lim}$ is the steady-state concentration of deposited Cr_2O_3 and γ_s is the small constant parameter (see [14] for details). Importantly, $t^{\lim}$ is inversely proportional to the cell current density j_0, which correlates with the recent experiments of Neumann [15].

Propagation of the poisoning wave increases the CL polarization voltage. Solution of auxiliary problem of performance of a partially degraded CL [16] allows us to plot the CL activation polarization (overpotential) as a function of time (Figure 23.5) [14]. As can be seen, for lower current densities, the FL polarization voltage increases almost linearly with time. However, for higher currents, η_0 exhibits rapid initial growth followed by a plateau. At high currents, already a 30–40% level of poisoning leads to a large polarization loss. Physically, owing to poor ionic conductivity, at large $\tilde{j}_0$ the conversion runs in the poisoned domain, where the "cost" of conversion in terms of potential is very high. Thus, at high $\tilde{j}_0$, a 30–40%-poisoned FL performs almost as badly as a fully poisoned FL.

Figure 23.6 shows the dependence of the half-cell voltage (η in our notation) on time in the poisoning experiments of Konysheva et al. [17]. Comparison of Figures 23.5 and 23.6 shows that the model qualitatively well reproduces the experimental data. At low cell currents, the experimental cell voltage increases linearly with time; cf. the curves FL13Cr-0.07, FL13Cr-0.18, and the lower curve

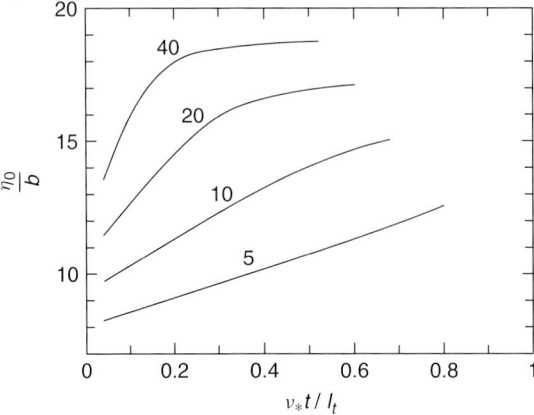

Figure 23.5 The time dependence of the half-cell polarization voltage $\tilde{\eta}_0$ for the indicated dimensionless current densities $\tilde{j}_0$.

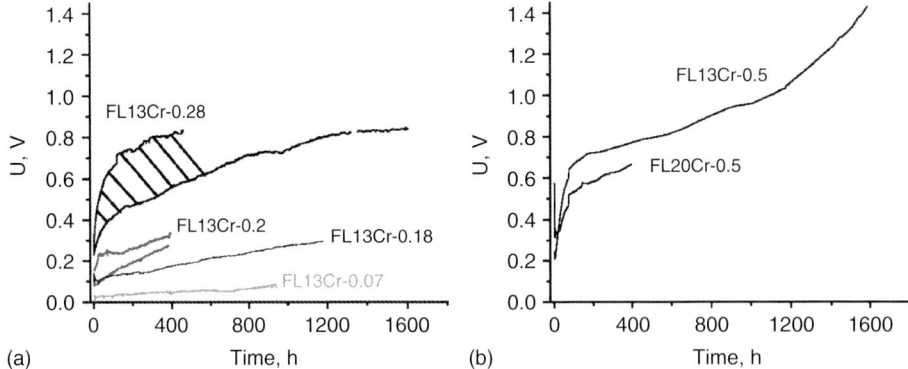

Figure 23.6 Experimental data of Konysheva et al. [17]. Cathode polarization voltage versus time for the functional layers of a thickness 13, 20, and 50 μm (FL13, FL20, and FL50, respectively). The cell current (A cm^{-2}) is indicated by the number at the end of the name of the curve. Note the initial phase of rapid voltage growth with time for currents above 200 mA cm^{-2}. Reprinted from [17] by permission of The Electrochemical Society.

FL13Cr-0.2 in Figure 23.6a. The corresponding model dependence of the cell voltage on time is represented in Figure 23.5 by the curve $\tilde{j}_0 = 5$.

At higher currents, the half-cell voltage increases with time much faster. A phase of rapid initial growth arises, followed by a phase of relatively slow, most probably resistive growth (Figure 23.6), curves FL13Cr-0.28, FL20Cr-0.5, and FL50Cr-0.5. The corresponding model dependence of the cell voltage on time is represented by the curves for $\tilde{j}_0 = 20$ and 40 (Figure 23.5).

The model suggests that in order to avoid triggering of the poisoning wave, the overpotential should be homogenized through the FL thickness. This can be done by increasing the FL ionic conductivity, or by shaping the distribution of the density of the triple-phase boundary (catalyst sites) across the CL, as discussed in the next section.

23.2.7
Optimal Catalyst Loading

Figure 23.3 suggests that in order to homogenize the overpotential distribution through the CL, it is beneficial to load more catalyst close to the electrolyte. Most catalyst-layer and fuel-cell models assume a uniform catalyst loading across the CL (see, e.g., [8, 18, 19]). Experimental and theoretical studies of the effects due to the catalyst gradient in the cathode catalyst layer of a PEMFC have been performed in [20, 21]. Both studies found an improvement in the performance of the CL with the catalyst loading increasing towards the membrane. However, the shapes of loading tested in both studies do not follow from theory and hence they may be not optimal.

In this section, we are interested in the high-current regime of CL operation. In this regime, the second exponent in Eq. (23.6) can be neglected and the equation takes the form

$$2\varepsilon^2 \frac{\partial \tilde{j}}{\partial \tilde{x}} = -g(\tilde{x}) \exp \tilde{\eta} \tag{23.24}$$

where the factor $g(\tilde{x})$ describes the normalized concentration (loading) of catalyst particles at $\tilde{x}$. Evidently, since for all shapes of loading the total amount of catalyst should be the same, the following constraint holds:

$$\int_0^1 g(\tilde{x}) d\tilde{x} = 1 \tag{23.25}$$

The function $g(\tilde{x})$ in the system of Eqs. (23.24) and (23.7) should be optimized to maximize the CL performance, that is, to minimize $\tilde{\eta}_0$ for given $\tilde{j}_0$. Figure 23.7 compares the x-shapes of the local proton current and overpotential for the uniform and nonuniform (optimal) loadings [22]. As can be seen, the optimal loading nearly doubles the cell current density.

The polarization curves for optimal and non-optimal (uniform) catalyst loading are compared in Figure 23.8. The benefit from optimal loading increases with the cell current. With the typical $b = 50$ mV, this benefit reaches 150 mV on the right side of this plot.

The region of current densities in Figure 23.8 includes the transition and the high-current (double-Tafel) domains (cf., Figure 23.2). Thanks to the high catalyst concentration at the membrane interface, the optimal loading effectively switches the CL function to the low-current mode with the "normal" Tafel kinetics. In other words, optimal loading shifts the transition region to higher currents.

Numerical tests show that the cell polarization curve is rather insensitive to the shape of $g(x)$: a small variation of $g(x)$ leads to a small variation of $\tilde{\eta}_0(\tilde{j}_0)$. This means that in practical applications, there is no need to reproduce the theoretical shape $g(x)$ with high accuracy.

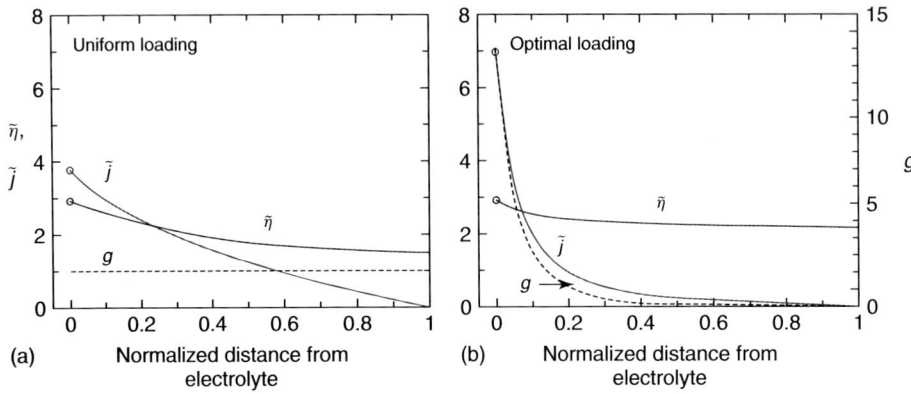

Figure 23.7 (a) Current $\tilde{j}$ and overpotential $\tilde{\eta}$ distribution for uniform loading [$g(\tilde{x}) = 1$]. (b) $\tilde{j}$ and $\tilde{\eta}$ for the optimal shape of catalyst loading $g(\tilde{x})$ (dashed line). Note that the total voltage loss $\tilde{\eta}(0)$ in (a) and (b) is the same. Note also the almost twice higher total current $\tilde{j}(0)$ in the case of optimal loading.

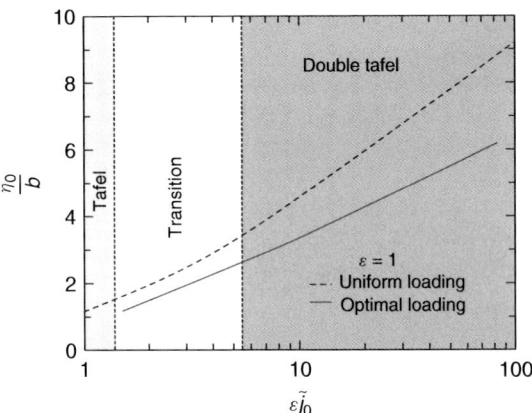

Figure 23.8 Dashed line, the exact numerical polarization curve for uniform loading; solid curve, polarization curve of the active layer with the optimal shape of catalyst loading.

23.3
Polarization Curve of PEMFCs and HT-PEMFCs

In this section, we construct several analytical polarization curves of PEMFCs and high-temperature polymer electrolyte membrane fuel cells (HT-PEMFCs). In these types of cell, owing to the excellent kinetics of the hydrogen oxidation reaction, the polarization voltage of the anode is negligible. The voltage loss in a PEMFC is determined by the oxygen transport, ORR kinetics, and the cell resistivity.

Several approximations can be made to take into account oxygen transport in the cell. In the simplest case, the oxygen concentration in the cathode channel, c_h, can be assumed constant. This corresponds to a high stoichiometry λ of the oxygen (air) flow. The transport term in the cell polarization curve then depends on the transport properties of the GDL only (Sections 23.3.1–23.3.4).

In real applications, the oxygen stoichiometry is usually not high, and the variation of c_h along the channel cannot be ignored. Calculation gives the dependence of the polarization curve on λ (Section 23.3.5). We complete this section with reference to the paper demonstrating how these simple equations work.

23.3.1
Oxygen Transport in the GDL and the Polarization Curve

To reach the catalyst layer, oxygen on the cathode side must be transported through the channel and the GDL (Figure 23.9). In this section, we assume that the oxygen concentration in the channel is constant. Transport of oxygen through the GDL obeys the following equation:

$$D_b \frac{\partial c}{\partial x} = \frac{j_0}{4F} \tag{23.26}$$

which means that the diffusion flux of oxygen in the GDL equals the stoichiometry flux required to convert the current density j_0.

Equation (23.26) is a linear first-order equation with a constant on the right-hand side. Hence $c(x)$ is a linear function of x and we may write

$$D_b \frac{c_h - c_t}{l_b} = \frac{j_0}{4F}$$

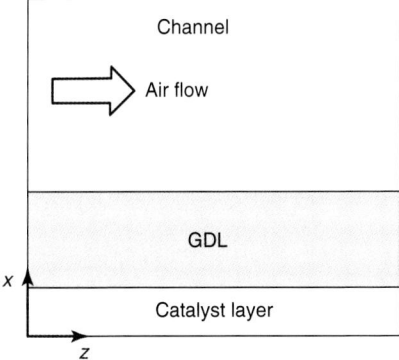

Figure 23.9 Cross-section of the cathode side. To reach the catalyst layer, oxygen in the channel must diffuse through the GDL.

where c_t is the oxygen concentration at the CL/GDL interface and l_b is the GDL thickness. Solving this equation for c_t, we find

$$c_t = c_h \left(1 - \frac{j_0}{j_{\lim}}\right) \tag{23.27}$$

where

$$j_{\lim} = \frac{4FD_b c_h}{l_b} \tag{23.28}$$

is the oxygen-limiting current density. If $j_0 = j_{\lim}$, oxygen concentration in the CL is zero and no more current can be produced. The dimensionless form of Eq. (23.27) is obtained is we place a tilde ($\sim$) over all variables:

$$\tilde{c}_t = \tilde{c}_h \left(1 - \frac{\tilde{j}_0}{\tilde{j}_{\lim}}\right) \tag{23.29}$$

Equation (23.29) should be substituted into the general equation for the CL polarization voltage, Eq. (23.11). In this equation, the concentration factor appears in the parameter ε given by Eq. (23.8). Substituting Eq. (23.29) into Eq. (23.8), we obtain

$$\varepsilon = \varepsilon_h \left(1 - \frac{\tilde{j}_0}{\tilde{j}_{\lim}}\right)^{-\gamma/2} \tag{23.30}$$

where

$$\varepsilon_h = \sqrt{\frac{\sigma_t b}{2 i_* \tilde{c}_h^\gamma l_t^2}} \tag{23.31}$$

Substituting Eq. (23.30) into Eq. (23.11), we obtain the general polarization curve of the cathode side, which takes into account oxygen transport in the GDL:

$$\varepsilon_h \tilde{j}_0 \left(1 - \frac{\tilde{j}_0}{\tilde{j}_{\lim}}\right)^{-\gamma/2} = \sqrt{2} \sinh \tilde{\eta}_0 \tanh \left\{ \frac{\sqrt{2} \sinh \tilde{\eta}_0}{2 \coth \left[\varepsilon_h^{-1} \left(1 - \frac{\tilde{j}_0}{\tilde{j}_{\lim}}\right)^{\gamma/2}\right]} \right\} \tag{23.32}$$

This equation implicitly relates $\tilde{\eta}_0$ and $\tilde{j}_0$. In the limiting cases Eq. (23.32) can be simplified.

23.3.2
Low-Current Regime

If $\varepsilon_h \gg 1$, the argument y of $\coth(y)$ in Eq. (23.32) is small and $\coth(y) \simeq 1/y$. If, in addition, $\sinh \tilde{\eta}_0$ is not large (low cell current), we may use the expansion $\tanh(x) \simeq x$. With this, the tanh function on the right-hand side of Eq. (23.32) transforms to

$$\tanh\left\{\frac{\sqrt{2}\sinh\tilde{\eta}_0}{2\coth\left[\varepsilon_h^{-1}\left(1-\frac{\tilde{j}_0}{\tilde{j}_{\lim}}\right)^{\gamma/2}\right]}\right\} \simeq \frac{\sqrt{2}\sinh\tilde{\eta}_0}{2\varepsilon_h}\left(1-\frac{\tilde{j}_0}{\tilde{j}_{\lim}}\right)^{\gamma/2}$$

Using this in Eq. (23.32) and solving the resulting equation for $\tilde{\eta}_0$, we come to

$$\tilde{\eta}_0 = \text{arcsinh}\left[\varepsilon_h^2 \tilde{j}_0 \left(1-\frac{\tilde{j}_0}{\tilde{j}_{\lim}}\right)^{-\gamma}\right] \qquad (23.33)$$

If, in addition, the argument of arcsinh exceeds 2, we may set $\text{arcsinh}(x) \simeq \ln(2x)$ and Eq. (23.33) simplifies to

$$\tilde{\eta}_0 = \ln\left(2\varepsilon_h^2 \tilde{j}_0\right) - \gamma \ln\left(1-\frac{\tilde{j}_0}{\tilde{j}_{\lim}}\right) \qquad (23.34)$$

In dimension form, this equation reads

$$\eta_0 = b\ln\left[\frac{j_0}{i_* l_t (c_h/c_h^0)^\gamma}\right] - b\gamma \ln\left(1-\frac{j_0}{j_{\lim}}\right) \qquad (23.35)$$

This equation and its analogs have been widely used in fuel-cell studies (see, e.g., [23–26]). Equation (23.35) shows that the CL polarization voltage includes the ORR activation term (the first logarithm) and the transport polarization (the second logarithm). Note that the reaction order of the feed molecules appears as a factor at the transport logarithm.

23.3.3
High-Current Regime

If $\tilde{\eta}_0$ is large, the argument of the tanh function in Eq. (23.32) is also large, we may set $\tanh(x) \simeq 1$ and Eq. (23.32) reduces to

$$\varepsilon_h \tilde{j}_0 \left(1-\frac{\tilde{j}_0}{\tilde{j}_{\lim}}\right)^{-\gamma/2} = \sqrt{2}\sinh\tilde{\eta}_0$$

Solving this equation for $\tilde{\eta}_0$ and replacing arcsinh by the logarithm of the double argument, we obtain

$$\tilde{\eta}_0 = 2\ln(\varepsilon_h \tilde{j}_0) - \gamma \ln\left(1-\frac{\tilde{j}_0}{\tilde{j}_{\lim}}\right) \qquad (23.36)$$

In dimension form, this equation reads

$$\eta_0 = 2b\ln\left[\frac{j_0}{\sqrt{2i_* \sigma_t b (c_h/c_h^0)^\gamma}}\right] - b\gamma \ln\left(1-\frac{j_0}{j_{\lim}}\right) \qquad (23.37)$$

The factor 2 on the right-hand side manifests the effect of Tafel slope doubling. As discussed above, at high currents, owing to poor ionic conductivity of the CL, ions do not penetrate deep into the catalyst layer and the electrochemical conversion runs at the electrolyte (membrane) interface. This regime of conversion requires a much higher polarization voltage. Note that the transport terms in Eqs. (23.35) and (23.37) coincide.

23.3.4
One-Dimensional Cell Polarization Curve

The cell polarization curve is given by

$$V_{cell} = V_{oc} - \eta_0 - R_{cell}\, j_0 \tag{23.38}$$

where η_0 is given by one of Eqs. (23.32), (23.35), or (23.37), and the last term accounts for the resistive losses in a cell. Here R_{cell} includes membrane and contact resistances.

The general and low-current polarization curves for the set of parameters in Table 23.1 are shown in Figure 23.10. As can be seen, for this set of parameters, the full Eq. (23.32) and the low-current approximation [Eq. (23.35)] almost coincide, and the difference between the two curves is marginal. In the range of current densities in Figure 23.10, the cell works at the boundary between the low-current and transition regions (Figure 23.2a) and the low-current limit describes the general curve well.

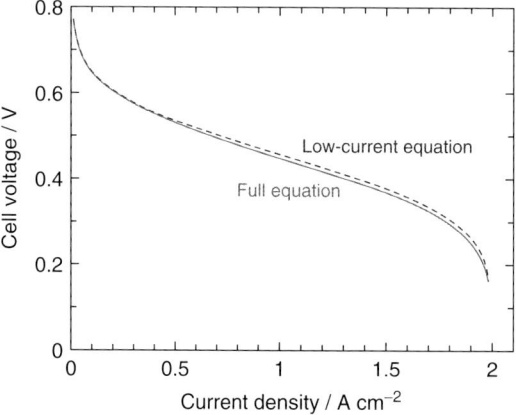

Figure 23.10 PEMFC polarization curve. Lower solid curve, overpotential η_0 is calculated with the full equation [Eq. (23.11)]; upper dashed curve, η_0 is from the low-current equation [Eq. (23.35)]. The limiting current density $j_{lim}^0 = 2\,\text{A cm}^{-2}$, the cell resistivity $R_{cell} = 0.05\,\Omega\,\text{cm}^2$, and the oxygen reaction order $\gamma = 1$. The other parameters are listed in Table 23.1.

23.3.5
Oxygen Consumption in the Channel and the Quasi-Two-Dimensional Polarization Curve

So far, we have assumed that the oxygen concentration in the cathode channel is constant. In this section, we will relax this assumption. Suppose that the cell is equipped with the single straight air channel; let the axis z be directed from the channel inlet to the outlet (Figure 23.9). We will assume that the flow in the channel is a plug flow, that is, a well-mixed flow with a constant velocity v. The validity of this approximation is discussed in [3]. The oxygen mass conservation equation in the channel then reads

$$v \frac{\partial c_h}{\partial z} = -\frac{j_0}{4Fh} \tag{23.39}$$

where h is the channel height above the surface of the membrane electrode assembly. This equation means that the oxygen concentration in the channel decreases at a rate given by the local current density (note that j_0 is now a function of z).

With the dimensionless variables [Eq. (23.5)], Eq. (23.39) takes the form

$$\lambda \tilde{J} \frac{\partial \tilde{c}_h}{\partial \tilde{z}} = -\tilde{j}_0 \tag{23.40}$$

where J is the mean current density in the cell, and the dimensionless distance is

$$\tilde{z} = \frac{z}{L}$$

where L is the channel length.

Suppose that the cell operates in the Tafel regime, that is, the cathode polarization voltage is given by Eq. (23.34). Further, we will assume, that η_0 is constant along z. To justify this assumption, we note that the resistive voltage loss is usually not large compared with the activation polarization η_0. A reasonable approximation of resistive term is $R_{\text{cell}} J$ and Eq. (23.38) takes the form

$$V_{\text{cell}} = V_{oc} - \eta_0 - R_{\text{cell}} J \tag{23.41}$$

The cell electrodes are equipotential, hence V_{cell} is independent of z. Since V_{oc} and $R_{\text{cell}} J$ also are constant along z, we see that $\eta_0 = V_{oc} - V_{\text{cell}} - R_{\text{cell}} J = \text{constant}$.

To include the effect of a finite λ, it is advisable to rewrite Eq. (23.34) in the following form:

$$\tilde{\eta}_0 = \ln \left[2(\varepsilon_h^0)^2 \frac{\tilde{j}_0}{\tilde{c}_h} \right] - \ln \left(1 - \frac{\tilde{j}_0}{\tilde{j}_{\lim}^0 \tilde{c}_h} \right) \tag{23.42}$$

where the oxygen reaction order γ is set to 1. Here ε_h^0 and $\tilde{j}_{\lim}^0$ are obtained by replacing c_h by the oxygen concentration at the channel inlet c_h^0 in Eqs. (23.31) and (23.28), respectively. Note that ε_h^0 and $\tilde{j}_{\lim}^0$ are independent of z.

The only $\tilde{z}$-dependent values in Eq. (23.42) are $\tilde{j}_0$ and $\tilde{c}_h$, which appear in this equation as a ratio, $\tilde{j}_0/\tilde{c}_h$. Since the left-hand side of Eq. (23.42) is independent of $\tilde{z}$, the ratio $\tilde{j}_0/\tilde{c}_h$ must be constant. This means that the right-hand side of Eq. (23.40) is $-a\tilde{c}$, where a is constant. In addition, $\tilde{j}_0(\tilde{z})$ must obey the constraint $\int_0^1 \tilde{j}_0 d\tilde{z} = \tilde{J}$. Elementary calculation gives

$$\tilde{c}_h = \left(1 - \frac{1}{\lambda}\right)^{z/L} \tag{23.43}$$

$$\tilde{j}_0 = \tilde{J} f_\lambda \cdot \left(1 - \frac{1}{\lambda}\right)^{z/L} \tag{23.44}$$

where

$$f_\lambda = -\lambda \ln\left(1 - \frac{1}{\lambda}\right) \tag{23.45}$$

From Eqs. (23.43) and (23.44), it follows that $\tilde{j}_0/\tilde{c}_h = \tilde{J} f_\lambda$. Using this in Eq. (23.42), we finally obtain

$$\tilde{\eta}_0 = \ln\left[2(\varepsilon_h^0)^2 f_\lambda \tilde{J}\right] - \ln\left(1 - \frac{f_\lambda \tilde{J}}{\tilde{j}_{\lim}^0}\right) \tag{23.46}$$

In dimension form this equation reads

$$\eta_0 = b\ln\left(\frac{f_\lambda J}{i_* l_t}\right) - b\ln\left(1 - \frac{f_\lambda J}{j_{\lim}^0}\right) \tag{23.47}$$

With Eq.(23.47), the cell polarization curve Eq. (23.41) takes the form

$$V_{\text{cell}} = V_{oc} - b\ln\left(\frac{f_\lambda J}{j_*}\right) + b\ln\left(1 - \frac{f_\lambda J}{j_{\lim}^0}\right) - R_{\text{cell}} J \tag{23.48}$$

where $j_* = i_* l_t$ is the superficial exchange current density.

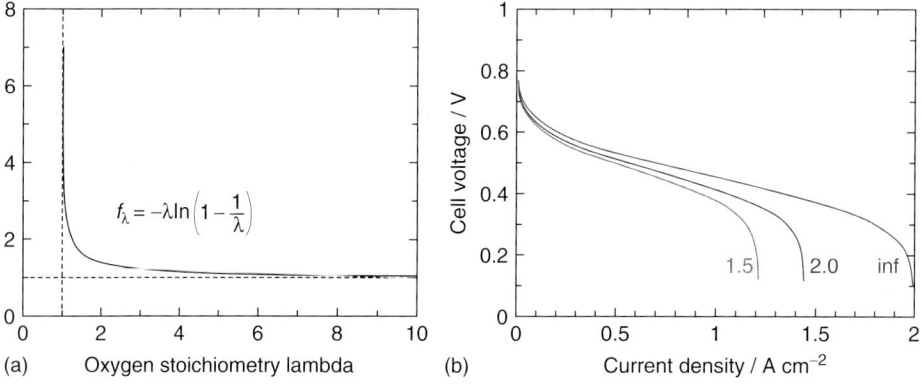

Figure 23.11 (a) The function f_λ. Note that this function tends to infinity as $\lambda \to 1$ and it tends to 1 as $\lambda \to \infty$. (b) The cell polarization curves for indicated values of λ. The limiting current density at infinite λ is $j_{\lim}^0 = 2$ A cm^{-2}, the cell resistivity $R_{\text{cell}} = 0.05$ Ω cm^2, and the oxygen reaction order $\gamma = 1$. The other parameters are listed in Table 23.1.

We see that in Eq. (23.47), finite oxygen stoichiometry rescales the mean current density J by a factor f_λ. In particular, from Eq. (23.47) it follows that the limiting current density is lowered by a factor f_λ. The function f_λ and the effect of λ on the cell polarization curve [Eq. (23.48)] are depicted in Figure 23.11. As can be seen, a small λ dramatically lowers the limiting current density.

Equation (23.48) was used to characterize PEMFCs by engineers at Ford Motor company [27]. Huang *et al.* [27] stated that "Insight understandings of fuel cell performance are gained by using this model-based methodology. This methodology can also help to characterize cell performance by limited or less testing data. This work will help to reduce testing time and cost. It will provide a fast evaluation of fuel cells characterization and their capability."

23.4 Conclusion

A fuel cell is easy to build at home: a children's kit for assembling a fuel-cell-based mini-car costs about US$100. However, the real fuel-cell car is expected on the market by 2015, after several decades of very expensive R&D.

The reason for the delay is a number of yet unresolved problems in cell and stack design. The greatest problem is relatively fast stack aging under real operating conditions. On pressing the accelerator and then braking, the driver moves the working point in Figure 23.10 to the limiting current density and then back to the open-circuit voltage. These jumps are very detrimental for the stack health: they destroy carbon support, spoil membrane conductivity, worsen transport properties of the GDL, and accelerate parasitic electrochemical processes, which reduce the CL conversion activity. Similar negative effects induce start–stop cycles. These problems could partly be mitigated by using a hybrid fuel cell–battery system, which is expected to appear on the market first.

Nonetheless, the development of reliable fuel-cell stacks is hardly possible without understanding the details of aging processes, and simple analytical models could contribute substantially in this process. Overall, the message from analytical modeling remains the same: "Fuel cells can be made much better" [28].

List of Symbols

~	indicates dimensionless variables
b	Tafel slope (V)
c	molar concentration (mol m^{-3})
c_s	molar concentration of Cr_2O_3 (mol m^{-3})
c_t	feed molar concentration in the catalyst layer (mol m^{-3})
c_h	feed molar concentration in the channel (mol m^{-3})
D	diffusion coefficient of feed molecules in the CL (m^2 s^{-1})

D_b	diffusion coefficient of feed molecules in the GDL (m² s⁻¹)
F	Faraday constant
f_λ	dimensionless function, Eq. (23.45)
g	normalized catalyst loading ($0 \le g \le 1$)
J	mean current density (A m⁻²)
j	proton current density in the CL (A m⁻²)
j_0	cell current density (A m⁻²)
j_{crit}	critical current density, Eqs. (23.15) and (23.16) (A m⁻²)
j_{lim}	oxygen-limiting current density, Eq. (23.28) (A m⁻²)
h	channel height (m)
i_*	volumetric exchange current density (A m⁻³)
L	channel length (m)
l_b	GDL thickness (m)
l_m	membrane thickness (m)
l_N	Newman's reaction penetration depth, Eq. (23.8)
l_t	CL thickness (m)
n	number of electrons transferred in the reaction
R	gas constant
R_{cell}	cell resistivity (Ω m²)
S	reaction rate (A m⁻³)
T	cell temperature (K)
t^{lim}	cell lifetime (s)
V_{cell}	cell voltage (V)
V_{oc}	cell open-circuit voltage (V)
v	flow velocity (m s⁻¹)
x	through-plane coordinate (m)
z	coordinate along the channel (m)

Subscripts

0	membrane/CL interface
crit	critical
h	channel
lim	limiting
m	membrane
t	catalyst layer
b	gas–duffusion layer

Superscripts

0	channel inlet

Greek Letters

α	transfer coefficient
β	dimensionless parameter, Eq. (23.20)
γ	oxygen reaction order
γ_s	dimensionless parameter, Eq. (23.23)
η	local overpotential (V)
η_0	half-cell overpotential (V)
λ	feed molecules (oxygen) stoichiometry
ε	dimensionless Newman's reaction penetration depth, Eq. (23.8)
ε_h	ε with $c = c_h$, Eq. (23.31)
σ_t	membrane phase proton conductivity in the CL (Ω^{-1} m^{-1})

References

1. Perry, M.I. and Fuller, T.F. (2002) A historical perspective of fuel cell technology in the 20th century. *J. Electrochem. Soc.*, **149**, S59–S67.
2. Newman, J. and Tiedemann, W. (1975) Porous-electrode theory with battery applications. *AIChE J.*, **21**, 25–41, and references therein.
3. Kulikovsky, A.A. (2010) *Analytical Modelling of Fuel Cells*, Elsevier, Amsterdam.
4. Perry, M.L., Newman, J., and Cairns, E.J. (1998) Mass transport in gas-diffusion electrodes: a diagnostic tool for fuel-cell cathodes. *J. Electrochem. Soc.*, **145**, 5–15.
5. Eikerling, M. and Kornyshev, A.A. (1998) Modelling the performance of the cathode catalyst layer of polymer electrolyte fuel cells. *J. Electroanal. Chem.*, **453**, 89–106.
6. Bard, A.J. and Faulkner, L.R. (2001) *Electrochemical Methods: Fundamentals and Applications*, John Wiley & Sons, Inc., New York.
7. Kulikovsky, A.A. (2002) Performance of catalyst layers of polymer electrolyte fuel cells: exact solutions. *Electrochem. Commun.*, **4**, 318–323.
8. Eikerling, M. (2006) Water management in cathode catalyst layers of PEM fuel cells. *J. Electrochem. Soc.*, **153**, E58–E70.
9. Wang, Y. and Feng, X. (2008) Analysis of reaction rates in the cathode electrode of polymer electrolyte fuel cell. I. Single-layer electrodes. *J. Electrochem. Soc.*, **155**, B1289–B1295.
10. Wang, Y. and Feng, X. (2009) Analysis of reaction rates in the cathode electrode of polymer electrolyte fuel cell. II. Dual-layer electrodes. *J. Electrochem. Soc.*, **156**, B403–B409.
11. Kulikovsky, A.A. (2010) The regimes of catalyst layer operation in a fuel cell. *Electrochim. Acta*, **55**, 6391–6401.
12. Newman, J. (1991) *Electrochemical Systems*, Prentice Hall, Englewood Cliffs, NJ.
13. Kulikovsky, A.A. (2009) A model of SOFC anode performance. *Electrochim. Acta*, **54**, 6686–6695.
14. Kulikovsky, A.A. (2011) A model for Cr poisoning of SOFC cathode. *J. Electrochem. Soc.*, **158**, B253–B258.
15. Neumann, A. (2011) Chrom-bezogene Degradation von Festoxid-Brennstoffzellen. PhD thesis, Forschungszentrum Jülich.
16. Kulikovsky, A.A. (2010) Polarization curve of partially poisoned catalyst layer. *Electrochem. Commun.*, **12**, 1780–1783.
17. Konysheva, E., Mertens, J., Penkalla, H., Singheiser, L., and Hilpert, K. (2007) Chromium poisoning of the porous composite cathode. Effect of cathode thickness and current density. *J. Electrochem. Soc.*, **154**, B1252–B1264.
18. Wang, C.-Y. (2004) Fundamental models for fuel cell engineering. *Chem. Rev.*, **104**, 4727–4766.

19. Weber, A. and Newman, J. (2004) Modeling transport in polymer-electrolyte fuel cells. *Chem. Rev.*, **104**, 4679–4726.
20. Antoine, O., Bultel, Y., Ozil, P., and Durand, R. (2000) Catalyst gradient for cathode active layer of proton exchange membrane fuel cell. *Electrochim. Acta*, **45**, 4493–4500.
21. Taylor, A.D., Kim, E.Y., Humes, V.P., Kizuka, J., and Thompson, L.T. (2007) Inkjet printing of carbon supported platinum 3-D catalyst layers for use in fuel cells. *J. Power Sources*, **171**, 101–106.
22. Kulikovsky, A.A. (2009) Optimal shape of catalyst loading across the active layer of a fuel cell. *Electrochem. Commun.*, **11**, 1951–1955.
23. Squadrito, G., Maggio, G., Passalacqua, E., Lufrano, F., and Patti, A. (1999) An empirical equation for polymer electrolyte fuel cell (PEFC) behaviour. *J. Appl. Electrochem.*, **29**, 1449–1455.
24. Mann, R.F., Amphlett, J.C., Hooper, M.A.I., Jensen, H.M., Peppley, B.A., and Roberge, P.R. (2000) Development and application of a generalized steady-state electrochemical model for a PEM fuel cell. *J. Power Sources*, **86**, 173–180.
25. Pisani, L., Murgia, G., Valentini, M., and D'Aguanno, B. (2002) A new semi-empirical approach to performance curves of polymer electrolyte fuel cells. *J. Power Sources*, **108**, 192–203.
26. Kim, G.S., St-Pierre, J., Promislow, K., and Wetton, B. (2005) Electrical coupling in proton exchange membrane fuel cell stacks. *J. Power Sources*, **152**, 210–217.
27. Huang, Ch., Wang, T., and Hirano, Sh. (2008) A model-based methodology for fuel cell peak power prediction. Meeting Abstracts– MA 2008-01, 213th ECS Meeting, Phoenix, AZ, 18–22 May, Abstract No. 488.
28. Eikerling, M., Kornyshev, A., and Kulikovsky, A.A. (2005) Can theory help to improve fuel cells? *Fuel Cell Rev.*, 15–24.

24
Stochastic Modeling of Fuel-Cell Components

Ralf Thiedmann, Gerd Gaiselmann, Werner Lehnert, and Volker Schmidt

The microstructure of porous media is closely related to their physical properties. In particular, transport of fluids through the pore phase depends on the morphology of the material. Hence, the morphology of the considered material influences its functionality to a large extent. Therefore, the systematic development of "designed" morphologies with improved physical properties is an important issue which has various applications in the development of advanced materials used for fuel cells, and, for instance, also for batteries and solar cells.

Mathematical models from stochastic geometry are useful tools to achieve this goal since they provide methods allowing for a quantitative description of the correlation between microstructure and functionality. Moreover, systematic modifications of model parameters, in combination with numerical transportation models, offer the opportunity to identify morphologies with improved physical properties by model-based computer simulations, that is, to perform a *virtual material design*.

The focus of this chapter is on analyzing and modeling the microstructure of gas diffusion layers (GDLs) in polymer electrolyte membrane fuel cells (PEMFCs). GDLs are fiber-based materials and one of their main tasks is the transportation of hydrogen and oxygen towards the electrodes where the electrochemical reaction takes place, that is, electricity is generated and the transport of the byproduct water from the electrode towards the channel is accomplished [1].

Two different approaches are presented in order to describe the microstructure of GDLs. First, the solid phase of paper- and non-woven GDLs is modeled. This is described in Sections 24.1 and 24.2. Then, in Section 24.3, the pore phase is represented by a stochastic network model which is based on the notion of a random geometric graph. Note that the parameters of the models are fitted to real image data using structural information from 2D and 3D images gained by scanning electron microscopy (SEM) and synchrotron tomography, respectively.

In Section 24.5, image characteristics are introduced which allow for an investigation of GDL materials with respect to transport-relevant structural properties that are necessary to describe transport properties in the frame of continuum models. In addition, model validation is performed, whereby image characteristics computed from the simulated and the corresponding real data are compared with each other. Note that by the validation of a stochastic simulation model it is checked whether the virtual reconstruction of realistic structures on the computer is possible or not. Hence, the fitting of model parameters can be interpreted as a model adjustment, which puts us in a position to simulate virtual morphologies with structural properties similar to those of original data.

The fields of application of the fitted model are widespread. They can be used, for example, to simulate realistic morphologies in sampling windows which are larger than those obtained by microscopic imaging techniques. This can be seen as data extrapolation. Examples regarding the use of a stochastic model for extrapolation of image data can be found elsewhere, for example [2, 3].

24.1
Multi-Layer Model for Paper-Type GDLs

In this section, we describe a stochastic 3D model for paper-type GDLs which is based on a multi-layer approach and uses the idea of randomly thrown line systems. This idea is inspired by visual impressions of 2D SEM images such as Figure 24.1a, leading to the conclusion that single fibers can be approximated by dilated straight lines. Furthermore, it is assumed that the straight lines are horizontally oriented. This is a valid assumption for paper- and non-woven GDLs; see the cross-sectional view in Figure 24.1b. Another main assumption of the model is that we consider the material to be a stack of several disjoint and independent thin sections with a thickness equal to the diameter of fibers, that is, we assume the fibers to be mutually penetrating cylinders within each thin section.

However, paper-type GDLs also contain binder for fixing the fibers together. This is included in the model by filling randomly chosen cells formed by the intersecting lines; see also [4–6] for further details of the model described in the following.

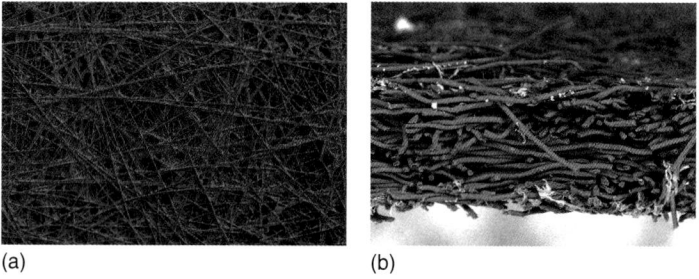

Figure 24.1 (a) 2D SEM images of a paper-type GDL (manufacturer: Toray); (b) cross-sectional view of a non-woven GDL (manufacturer: Freudenberg).

24.1.1
Modeling of Fibers

As already mentioned, the fiber system of GDLs is modeled by a multi-layer approach, that is, a stack of thin sections, where the fibers are modeled by planar Poisson line systems within each section.

A planar line tessellation is built by the cells $\xi_1, \xi_2, \ldots$ which arise from intersecting lines $l_1, l_2, \ldots$ scattered randomly in the plane. Note that a line l_i can be described by its normal form, that is, its (orthogonal) signed distance s_i from the origin and a direction m_i (given as the angle to a predetermined direction).

Then, a *Poisson line tessellation* (PLT) with parameter γ is a random tessellation which can be identified with a marked point process $\{(S_i, M_i)\}$ on $\mathbb{R}$ with marks in $(0, \pi]$. The points $\{S_i\}$ form a stationary Poisson point process on $\mathbb{R}$ with intensity $\gamma > 0$ and the marks $\{M_i\}$ are a sequence of independent and $U(0, \pi)$-distributed random variables. Moreover, the sequences $\{S_i\}$ and $\{M_i\}$ are independent. Hence, every marked point (s_i, m_i) represents a line with perpendicular distance $|s_i|$ and angle m_i to the x-axis measured counterclockwise. The family of all cells $\Xi_1, \Xi_2, \ldots$ formed by all the intersecting random lines $l_{(S_1, M_1)}, l_{(S_2, M_2)}, \ldots$ is then called a (planar) Poisson line tessellation with parameter γ. An illustration of the Poisson line tessellation is shown in Figure 24.2.

The 2D PLT is fully described by a single parameter, the parameter γ of the underlying Poisson process $\{S_i\}$, which is equal to the mean total length of the (random) set $\bigcup_{i=1}^{\infty} \partial \Xi_i$ of edges per unit area.

Since the fibers in GDLs are 3D objects, we dilate each line of the PLT using a 3D structuring element to obtain real 3D objects. Metaphorically speaking, dilation means an expansion of the structure as a function of the structuring element. If the structuring element is indicated by a ball, the dilation means a blowing up of the structure. Hence, in order to model a single thin section of fibers, the edge set $\bigcup_{i=1}^{\infty} \partial \Xi_i$ of the underlying (planar) PLT $\{\Xi_i\}$ is interpreted as a set of lines in $\mathbb{R}^3$ parallel to the xy-plane. Subsequently, it is dilated with respect to 3D, that is, the Minkowski sum:

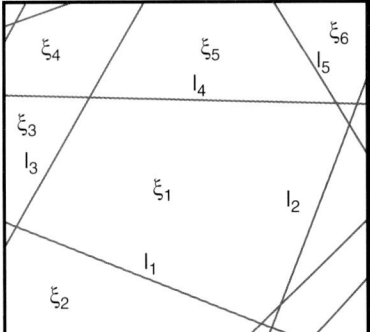

Figure 24.2 Realization of a Poisson line tessellation.

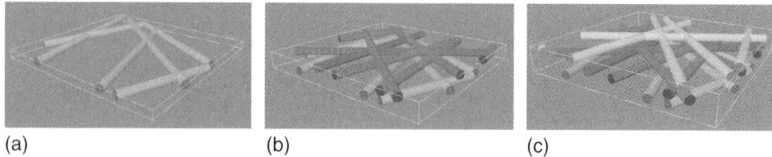

(a) (b) (c)

Figure 24.3 Schematic display of the multi-layer model: (a) one layer; (b) two layers; (c) several layers. Reprinted with permission from [4]. Copyright (2008) *Journal of the Electrochemical Society.*

$$\left(\bigcup_{i=1}^{\infty} \partial \Xi_i\right) \oplus B = \bigcup_{i=1}^{\infty} \partial \Xi_i \oplus B \tag{24.1}$$

is considered for some set $B \in \mathcal{B}(\mathbb{R}^3)$, leading to an object in 3D representing one thin section. The stack of n of these 3D dilated PLT forms the 3D multi-layer model for the fiber system; see Figure 24.3 for a schematic illustration.

For dilating the edges of the PLTs, the structuring element B modeling the profile of the fibers can, for example, be chosen as a 3D sphere $b(o, r_F)$ or a cube $C_{r_F} \subset \mathbb{R}^3$ with diameter or side length $2r_F > 0$, respectively, and centered at the origin (see Figure 24.4).

In addition to the space continuous representation of the model described above, a discretization of the dilated lines of the PLT is necessary, for example, to simulate transport processes in realizations of the model or to apply the characterization methods of Section 24.5.

Figure 24.4 shows three possible versions of discretized cross-sections of possible structuring elements. Note that all of these profiles can be seen as discretizations of fibers dilated by a sphere when the fiber diameter is represented by 5 pixels. In the following, we use the discretization shown in Figure 24.4d.

24.1.2
Modeling of Binder

The binder has an essential influence on transport processes through a material, since it blocks many paths through the pore phase.

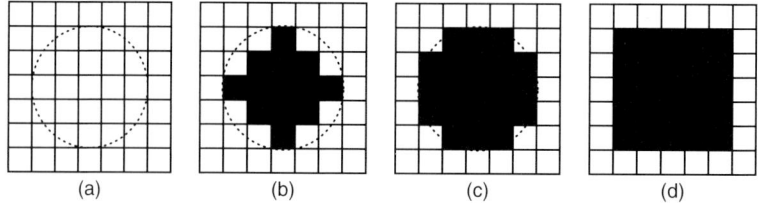

(a) (b) (c) (d)

Figure 24.4 Different possibilities for discretizing a spherical profile of fibers: (a) sphere; (b) lozenge; (c) square without corners; (d) square.

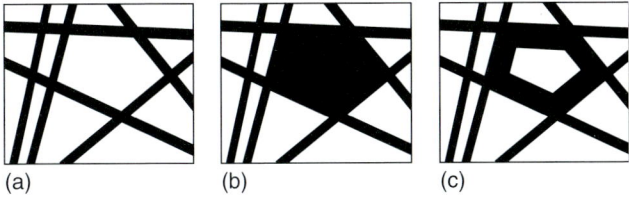

Figure 24.5 Schematic illustration of binder modeling: (a) fiber system; (b) binder modeled by complete filling; (c) binder modeled by partial filling.

We model the binder by a so-called *Bernoulli filling*, where in each layer Bernoulli experiments with probability $p > 0$ are performed for each cell of the PLT independently. If the Bernoulli experiment is successful, the cell is filled with binder, either completely or partially. An example can be seen in Figure 24.5, where in (a) the fiber system without binder is shown, (b) depicts the fiber system with a cell completely filled with binder, and in (c) the fiber system with a partially filled cell can be seen. A 3D realization of the stochastic multi-layer model including binder is shown in Figure 24.6.

The partial filling of a cell chosen at random can be realized as follows. Informally said, the segments forming the boundary of the cell, that is, the lines of the basic PLT, are dilated towards the interior of the cell to be filled (at least partially). More precisely, the lines representing the fiber system are additionally dilated by cubes C_{r_B}, centered at the origin with side length $2r_B > 2r_F > 0$, where only the part within the chosen cell is taken into account. A more formal definition can be found for example in [5, 6].

Note that the binder model, where chosen cells are filled completely with binder, is a special case of the partial filling of cells with $r_B = \infty$.

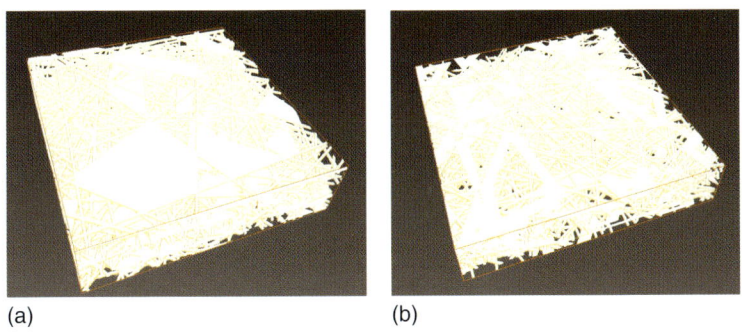

Figure 24.6 Realizations of the 3D multi-layer model: (a) binder modeled by complete filling of chosen cells; (b) binder modeled by partial filling of chosen cells.

24.1.3
Fitting of Model Parameters

The multi-layer model depends on five parameters, γ, r_F, r_B, p, and n, where $\gamma > 0$ is the intensity of the underlying PLT modeling the fibers of the sub-layers, $2r_F > 0$ is the diameter of the fibers, $r_B > 0$ is the dilation parameter for the binder model, $p \in [0, 1]$ is the probability that a cell is chosen to contain binder, and n is the number of sub-layers in the 3D model.

The parameter $2r_F$, that is, the fiber diameter, is known from the manufacturer and in the case of paper-type GDLs is about 7.5 µm. Note that this is equal to the thickness of one thin section or one layer in the model. The thickness of the material, which is given in the model by $2r_F \times n$, is also a well-known characteristic or is easy to measure. Then, the number n of sub-layers is the ratio of the thickness of the material to the diameter $2r_F$ of the fibers. For further computations, we use $n = 20$, that is, we assume the GDL to have a thickness of 150 µm. Furthermore, a natural estimator for γ is based on counting the lines from the top thin section, which can, for example, be non-interactively detected from 2D SEM images such as Figure 24.1 according to a procedure described in [4, 6]. Hence, we consider the following estimator for γ:

$$\widehat{\gamma} = \frac{\pi}{|\partial W|} \#\{n : l_{(S_n, M_n)} \cap W \neq \emptyset\} \tag{24.2}$$

where $\#\{n : l_{(S_n, M_n)} \cap W \neq \emptyset\}$ is the number of lines detected from the first thin section of the GDL and $|\partial W|$ denotes the length of the boundary of the sampling window W; see, for example [7, 8].

In order to determine the parameters p and r_B for modeling the binder, the following expressions for volume fractions of fibers and binder, respectively, are useful.

The volume fraction of binder $V_{\text{binder}}(\gamma, p, r_F, r_B)$ depends on four parameters. It can be shown that the volume fraction of binder is equal to the probability that the origin is covered by binder. Thus, if the profile of the discretization of the dilated fibers is a cube C_{r_F} with side length $2r_F$, we have

$$V_{\text{binder}}(\gamma, p, r_F, r_B) = p\left[\exp(-2\gamma r_F) - \exp(-2\gamma r_B)\right] \tag{24.3}$$

where γ denotes the intensity of the PLT, $p > 0$ is the probability that a cell contains binder, r_F is the dilation parameter of the fibers, and r_B denotes the parameter describing the amount of binder. The volume fraction of fibers is given by $V_{\text{fiber}}(\gamma, r_F) = 1 - \exp(-2\gamma r_F)$.

In addition, the following relationship holds:

$$\varepsilon = 1 - \left[V_{\text{fiber}}(\gamma, r_F) + V_{\text{binder}}(\gamma, p, r_F, r_B)\right] \tag{24.4}$$

following directly from the definition of porosity, which is just the complete volume fraction minus the volume fraction of material (fiber and binder).

For determining the probability p, that is, the fraction of cells containing binder, we plug in a predefined value $r_B > 0$ and the estimated values $\widehat{\gamma}$, $\widehat{\varepsilon}$, and $\widehat{r}_F$ into Eq. (24.4) and solve it for p. In this way, we obtain an estimator $\widehat{p}$ for p.

The structural fitting of the multi-layer model to paper-type GDLs leads to the parameter γ of the PLT averaged over 10 non-interactively segmented 2D SEM images of GDLs as described by Thiedmann *et al.* [4], namely $\widehat{\gamma} = 0.025$. To obtain the missing model parameters, we compare the porosity ε of the model with the porosity $\widehat{\varepsilon}$ of real 3D data, which is estimated/measured as 0.78 (see, e.g., [4, 9]).

It is found that 17% of the considered GDL material consists of fibers and 5% of binder, which is in line with experimental findings. Note that the value for r_B is chosen arbitrarily, that is, other values are possible, hence, additional combinations of r_B and p are investigated in Section 24.5.4 using structural characteristics.

For validating the multi-layer model fitted to paper-type GDLs, we compare it with real 3D data gained by synchrotron tomography. These images have a resolution of 1.5 µm and we use the same resolution to discretize the realizations of the model. Thus, the profile of the structuring element has to be represented by 5 pixels (or 7.5 µm), as can be seen in Figure 24.4. We choose profile (d), that is, the square, for further analysis and validation. The results of the validation are shown in Section 24.5.4, leading to the conclusion that the chosen model fits structurally well to real data, at least with respect to the considered structural characteristics describing the pore phase.

24.1.4
Further Results

Recall the consideration of Figure 24.1 and the knowledge of the production process of GDL leading to the assumptions of the multi-layer model for paper-type GDLs. By means of the validation of the multi-layer model in Section 24.5.4, single fibers can be approximated by dilated straight lines oriented horizontally and the GDL can be considered to be a stack of several disjoint and independent thin sections with similar properties.

We are aware that these assumptions are simplifications of the high complexity of real GDL structures. However, as will be shown in Section 24.5, the fitted model coincides fairly well with real data. Note that the characteristics for the model validation are mainly focused on properties of the pore phase, that is, the multi-layer model seems adequate to investigate processes in the pore phase such as gas and water transport.

Wang *et al.* [10] combined the stochastic model of GDL structures described in Section 24.1 and direct numerical simulations to investigate pore-level single-phase transport within GDLs. The results of such computations can be used to describe the impact of the microstructure of a GDL on pore-level transport. The drawback of the approach of direct numerical simulation of transport processes is that only relatively small parts of GDL microstructures can be considered owing to the high computational complexity. Hence, only parts consisting of $140 \times 140 \times 100$ voxels and $210 \times 210 \times 150\,\mu m^3$, respectively, are considered for the computations of Wang *et al.* [10]. This yields a high variability in the results for, for example, the simulated permeability, for different realizations of the GDL structure due to the influence of the binder modeling with complete cell filling.

A possible solution for simulating transport processes in larger domains can be the lattice Boltzmann approach; see, for example, [11, 12]. The great advantage of this method is that it allows for a massive parallelization of the implementation, which permits computation on large domains.

In the literature, various other modeling approaches for fiber systems in GDL have also been proposed; see, for example, [13–17]. For instance, Schulz et al. [15] based the proposed modeling approach for GDL structures on a 3D Poisson line process; see, for example, [18]. Since the fibers in GDLs have a preferred direction, that is, they are mainly horizontally oriented, the directional distribution of the lines has to be analyzed in detail. The directional distribution of the lines as proposed in Schladitz et al. [19] is used. Considering polar coordinates, the density of the directional distribution is a function $p(\eta, \varphi)$, where $\eta \in [0, \pi)$ and $\varphi \in [0, 2\pi)$ denote altitude and longitude, respectively. Since in this model, as already in the multi-layer model described above, the orientation of the fibers is assumed to be isotropic in the xy-plane, the density of the directional distribution is given by

$$p(\eta, \varphi) = \frac{1}{4\pi} \frac{\beta \sin(\eta)}{\left[1 + (\beta^2 - 1) \cos^2(\eta)\right]^{\frac{3}{2}}} \qquad \eta \in [0, \pi), \varphi \in [0, 2\pi) \qquad (24.5)$$

which is independent of φ. The parameter $\beta > 0$ is called the *anisotropy parameter*. If $\beta = 1$, the fiber process is isotropic. With increasing β, the fibers tend to become more and more parallel to the xy-plane. Schulz et al. [15] chose the value $\beta = 10\,000$. Then, the lines are dilated in 3D to generate 3D objects, in this case cylinders, modeling the fibers. Spiess and Spodarev [16] considered 3D Poisson cylinder processes to describe the fiber system in paper-type GDLs, where the structuring element is random. In contrast to the other modeling approaches mentioned above, our multi-layer model is fitted to real data and moreover the model is structurally validated by means of a 3D synchrotron tomographic image.

24.2
Time-Series Model for Non-Woven GDLs

The multi-layer approach described above can easily be extended to curved fibers as occurring in non-woven GDLs. Again, it is assumed that the GDL consists of disjoint and independent thin sections. Moreover, the algorithm for the non-interactive extraction of fibers can be adapted to gain information about fiber properties such as curvature [2, 6]. This information can be applied to fit a multi-layer model for curved fibers to real data [2]. The latter model is based on the idea of representing each single fiber by a 2D polygonal track which is interpreted as a two-dimensional time series. The components of this time series are the length of the current line segment and the change of direction to the next line segment.

As a multivariate time series model, the vectorial autoregressive model of order p (VAR(p)) is used (see, e.g., [20, 21]).

The VAR(p) model is a stochastic process of random vectors and is defined as

$$Y_t = \tau + A_1 \cdot Y_{t-1} + \ldots + A_p \cdot Y_{t-p} + \delta_t \qquad t = 0, 1, 2, \ldots \qquad (24.6)$$

where $\tau \in \mathbb{R}^2$ is an intercept vector allowing for the possibility of non-zero mean $\mathbb{E} Y_t = \tau$; $A_i \in \mathbb{R}^{2\times 2}, i \in \{1, \ldots, p\}$, denote the coefficient matrices. Moreover, the residuals $\{\delta_t\}_{t=0,1,2,\ldots}$ form a two-dimensional white noise process. This leads to a model for single fibers.

One layer of the multi-layer model is modeled by randomly placed fibers which are dilated with respect to 3D as described in Eq. (24.1), where the starting points of the polygonal tracks follow the principle of complete spatial randomness. Note that due to the multi-layer approach, only a 2D polygonal track has to be modeled, which reduces the complexity dramatically compared with a (real) 3D model of curved fiber trajectories.

Finally, the model for the 2D polygonal tracks is established by means of geometric comparisons of the curvature of extracted and simulated polygonal tracks representing the curved fibers.

In summary, the multi-layer approach with curved fibers (Figure 24.7) provides an easy to handle model for materials such as non-woven GDLs, including the possibility of a (structural) fitting to only 2D SEM images. Further details can be found in [2].

24.3
Stochastic Network Model for the Pore Phase

In this section, a stochastic network model for the pore phase of non-woven GDLs is presented, where we closely follow Thiedmann *et al.* [6, 22]. It is based on methods from stochastic geometry and spatial statistics combined with tools from graph theory and Markov chain Monte Carlo (MCMC)-simulation. The model type for random geometric graphs described in this section is rather different from random graphs considered in the literature (see Section 24.4.1). Instead of modeling the fiber phase of the GDL and regarding the pore phase as its complement, the pore phase is modeled directly in this section. Since important transport processes take place in the pore phase, it could be assumed that structural properties which are

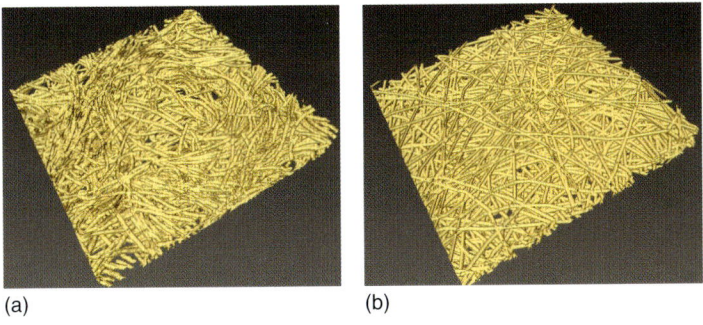

(a)　　　　　　　　　　(b)

Figure 24.7 (a) Binarized synchrotron data; (b) simulated non-woven GDL drawn from the 3D multi-layer model.

relevant to transport processes are fitted better with this procedure. This approach allows for the simulation of virtual pore phases with similar structural properties to pore phases observed in real GDLs.

The construction of the model for random geometric graphs can be divided into three main components. First, the vertices of the graph are modeled using tools from point process theory. In the second step, a given realization of the model for the vertices is (dependently) marked, whereby a moving-average procedure is applied. The edges are constructed combining tools from graph theory and MCMC simulation. Note that for each modeling component, a method is proposed to fit the model to real data with respect to the structure.

The model parameters are fitted to real 3D image data gained by means of X-ray synchrotron tomography [23, 24], where a pore phase graph is considered as described by Thiedmann et al. [5]. In particular, the parameters are specified in such a way that the distributions of vertex degrees and edge lengths coincide to a large extent for real and simulated data.

Note that the abstract representation of a pore phase by a graph has several advantages compared with modeling the pore phase with all its details as described, for example in Section 24.1. Since a complete representation of the pore phase mostly results in very complex geometric structures, that is, it is described by a huge set of voxels, numerical simulations of transport processes are quite complicated and computer time consuming. In particular, for the investigation of processes in GDLs on very large domains, a graph representation of the pore phase provides a suitable alternative approach. It can be applied, for example, to investigate transport processes in GDLs on a large scale since solving the required equations along the edges can be done very efficiently because mainly one-dimensional equations have to be solved; see also Section 24.4.2. This approach is only meaningful if we consider huge domains: the transport processes in GDLs calculated on the graph are approximate solutions which describe the behavior of the whole system correctly on average. Hence, it is essential to consider huge domains on which the calculation based on voxels reaches its limit owing to the enormous computational time and the calculation based on edges increases its accuracy.

Furthermore, the graph representation of the pore phase can be used to introduce a morphology-based definition of pores and their sizes; see, for example, [5, 22].

24.3.1
Pore Phase Graph

The underlying data basis of the subsequent modeling is a graph (V_r, E_r) extracted from 3D image data by a skeletonization as described, for example, by Fourard et al. [25]. Hence (V_r, E_r) is the basis for fitting the parameters of the model to real data.

24.3.1.1 Detection of Pores
The proposed model for random geometric graphs representing pore phases can be used, for example, for morphological characterizations and computations of

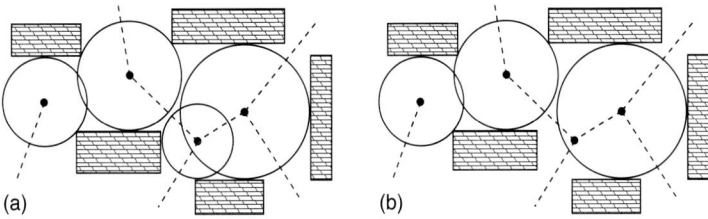

Figure 24.8 Definition of pores: (a) all potential pores; (b) final set of pores.

transport processes. Hence, an adequate marking with volume-describing properties is required. In addition, the vertices of the final graph model will be interpreted as pore centers, which requires a definition of the notion "pore." Therefore, we apply the definition of Thiedmann et al. [5], where each vertex of the graph is considered as a potential pore center. Its corresponding pore size is the spherical distance to the solid phase, that is, the sphere centered at the vertex with radius equal to the spherical distance to the solid phase is considered as a potential pore. However, if all vertices were taken as pore centers, some pores would be contained partially or completely in other pores. Hence, only those vertices which are not contained in larger pores are considered as (final) pore centers (see Figure 24.8).

Note that this definition can also be used to estimate a pore size distribution from 3D image data, which is performed in Section 24.5.2.

24.3.1.2 Modification of Pore Phase Graph

According to the definition of pores above, we modify the extracted graph (V_r, E_r), where we delete all those vertices which have not been classified as pore centers. This implies that the edges from which at least one endpoint is deleted also have to be changed. Their endpoints are then shifted toward the vertices classified as pore centers in whose pores they are located. This is done in such a way that all pores which were connected before are still connected in the modified graph (see Figure 24.9a). Furthermore, if there are overlapping pores which have no common edge, we add such an edge to the graph (see Figure 24.9b). The modified graph, where each vertex can be interpreted as a pore and where the random geometric graph model is fitted, is denoted by (V'_r, E'_r) in the following.

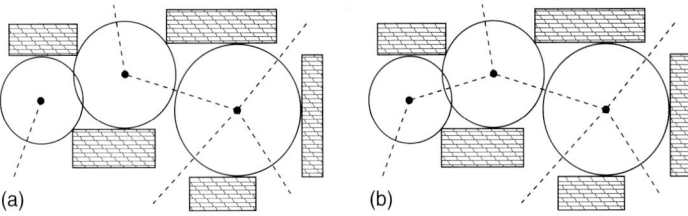

Figure 24.9 Modification of graph: (a) elimination of vertices that are not pore centers; (b) adding of edges if pores overlap.

24.3.2
Stochastic Modeling of Vertices

In this section, a model for the vertices of the random geometric graph for the pore phase of non-woven GDLs is proposed. The construction of the vertices is performed in two main steps. First, we fit a point process model to the vertex set V_r of the extracted graph (V_r, E_r) which is described in this section. Subsequently, we mark these points with spherical distances to the solid phase (see Section 24.3.4). Finally, we apply the definition of a pore from above, leading to a thinned set of vertices which is considered as a final set of vertices for the random geometric graph model. Then, for a given set of vertices, the edge set is constructed in a separate step which is explained in Section 24.3.5.

24.3.2.1 Multi-Layer Representation

The basic idea for modeling the vertices V'_r of the 3D graph (V'_r, E'_r) described above is to use a multi-layer approach similarly as in Section 24.1.

It is known from the manufacturer that the fibers of velt-type GDLs have a thickness of about 9 μm. Hence, we assume that the fiber system forms a stack of thin sections (parallel to the xy-plane), each with a thickness of 9 μm. Furthermore, we decompose the 3D point pattern of vertices V_r of the unmodified graph into the same type of thin sections, that is, with the same thickness of 9 μm. In order to model these thin sections of vertices, we project all points of a given thin section on to its base, being parallel to the xy-plane, say. These 2D point patterns are the data basis for the vertex model described in the next section. Note that for modeling the set of vertices of the modified graph V'_r, that is, the set of vertices that is later used for the graph construction, we start with modeling the original set V_r. Subsequently, this point process is marked with spherical distances and afterwards thinned using the same techniques as for modifying the graph in Section 24.3.1.

24.3.2.2 Construction and Fitting of Point Process Model

For modeling the 3D point pattern, that is, the set of vertices of the extracted graph (V_r, E_r), we first consider the pair-correlation function; see Illian et al. [26] for a formal definition and techniques for how to estimate it from a given point pattern.

The pair-correlation function $g(r)$ is proportional to the frequency of point pairs with distance $r > 0$ from each other. In addition, $g(r)$ provides information about the frequency of possible configurations of point pairs with respect to the Poisson point process. More precisely, in the Poisson case, it holds that $g(r) = 1$. Hence, $g(r) < 1$ clearly indicates repulsion of point pairs with distance r to each other and $g(r) > 1$ stands for a clustering of point pairs with distance r. If $g(r)$ vanishes, no point pair with a distance r to each other occurs.

Figure 24.10, where the dashed lines show the estimated pair-correlation function of the set of vertices V_r from the extracted graph, indicates a strong clustering of vertices with an unusually high peak of the pair-correlation function at small distances of about 4–5 μm. This suggests the idea of fitting a clustered point-process model with narrow and, simultaneously, elongated clusters.

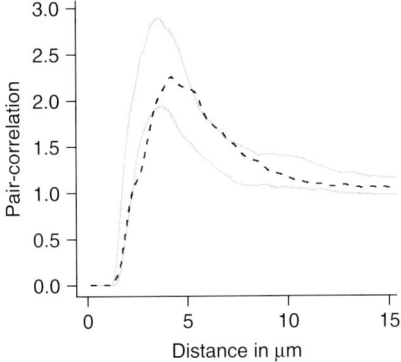

Figure 24.10 Pair-correlation functions for real (dashed line) and simulated data (gray solid lines; pointwise 96% confidence bands). Copyright (2011) J. Mater. Sci.

We model the 3D set of vertices by a multi-layer approach, whereby all points of one thin section are projected on to their bases, that is, we have to find a 2D model for the projected points. Therefore, we use a generalized Thomas process with elliptically shaped clusters, which has the following structure; see also, for example, [27]. The parent points form a stationary Poisson point process with intensity $\lambda_p > 0$, that is, they follow the principle of complete spatial randomness. The random number of child points per cluster is Poisson distributed with expectation c, and the random deviations of child points from their parent points are given via a 2D normal distribution $N(0, C)$, with expectation vector 0 and covariance matrix

$$C = \begin{pmatrix} \sigma_1^2 & 0 \\ 0 & \sigma_2^2 \end{pmatrix} \tag{24.7}$$

In addition, according to the uniform distribution on the interval $[0, 2\pi)$, the child points of each cluster are jointly rotated around their parent point. Hence, the considered generalized Thomas process is isotropic, although its clusters have elliptical shapes. Note that only the child points are considered as points of the generalized Thomas process. For an illustration, see Figure 24.11, where the red points represent the parent points and the black points build a realization of the generalized Thomas process.

The fitting of the generalized Thomas process is done by the method of minimum contrast (see below), whereby its pair-correlation function $g_\theta : (0, \infty) \to [0, \infty)$ is considered, which depends on the parameter vector $\theta = (\lambda_p, \sigma_1^2, \sigma_2^2)$.

The theoretical equation for the pair-correlation function of the generalized Thomas process is

$$g_\theta(r) = 1 + \frac{1}{4\pi \lambda_p \sigma_1 \sigma_2} \exp\left(-r^2 \frac{\sigma_1^2 + \sigma_2^2}{8\sigma_1^2 \sigma_2^2}\right) I_0\left(r^2 \frac{\sigma_1^2 - \sigma_2^2}{8\sigma_1^2 \sigma_2^2}\right) \quad r \geq 0 \tag{24.8}$$

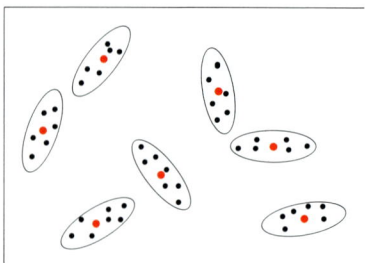

Figure 24.11 Realization of a generalized Thomas process.

where I_0 denotes the modified Bessel function, which can be evaluated by

$$I_0(z) = \sum_{k=0}^{\infty} \frac{(1/4z^2)^k}{(k!)^2} \quad z \in \mathbb{R} \tag{24.9}$$

see also [27, 28].

The idea of minimum-contrast estimation is the minimization of a distance measure between a theoretical characteristic depending on the model parameters and the corresponding estimated characteristic from the data. For this minimization, we use the pair-correlation function, since this second-order characteristic contains comprehensive information; see, for example, [29].

Note that for the fitting, the vertex set V_r from the originally extracted graph (V_r, E_r) is used. As already mentioned, the real data are divided into thin sections with a thickness of 9 µm and the vertices are projected on to their bases. The pair-correlation function is then estimated for all these 2D data sets separately and the pointwise average of the estimated pair-correlation functions is computed, which is denoted by $\widehat{g}$ in the following.

For fitting the generalized Thomas process to the data, its four parameters have to be determined: λ_p, c, σ_1^2, and σ_2^2. The fitting of these parameters by the method of minimum contrast with respect to the pair-correlation function leads to the following minimization problem:

$$f(\theta) = \int_{r_1}^{r_2} [\widehat{g}(r) - g_\theta(r)]^2 \, dr \to \min \tag{24.10}$$

for appropriately chosen $r_2 > r_1 > 0$. The minimum-contrast estimator $\hat{\theta} = (\hat{\lambda}_p, \hat{\sigma}_1^2, \hat{\sigma}_2^2)$ for θ is then given by $\hat{\theta} = \operatorname{argmin}_\theta f(\theta)$. Since the pair-correlation function g_θ given in Eq. (24.8) does not depend on the mean number c of child points per cluster, we use the relationship

$$c = \frac{\lambda}{\lambda_p} \tag{24.11}$$

which means that c can be estimated using $\hat{c} = \frac{\hat{\lambda}}{\hat{\lambda}_p}$, where $\hat{\lambda}$ denotes the natural estimator of the over all intensity λ, which can be estimated fairly easily, just by counting the number of all points in the sampling window divided by its volume.

On a scale where one distance unit corresponds to 1.0 μm, the result of this fitting is $\widehat{\lambda}_p = 0.000533$, $\hat{\sigma}_1^2 = 4.50$, and $\hat{\sigma}_2^2 = 78.75$. Based on this result, it follows immediately that $\widehat{c} = 2.28$. Hence, the estimated variances $\hat{\sigma}_1^2$ and $\hat{\sigma}_2^2$ are rather different, which means that the fitted Thomas process has clusters with clearly elongated shapes.

Finally, the projection of vertices on to the bases of thin sections has to be reversed. To incorporate this reversal step into the vertex model, we propose the following procedure.

In addition to clustering, a hard-core effect of 2–3 μm is observed in the point pattern of vertices V_r of the 3D graph (V_r, E_r) (see Figure 24.10), resulting from the skeletonization and subsequent transformation into vector data. Furthermore, an analysis of the z-coordinates observed in the point pattern of vertices of the 3D graph (V_r, E_r) shows that they are almost uniformly distributed (see Figure 24.12). Further, looking at the pair-correlation function given in Figure 24.10 indicates that there are many point pairs with a distance of about 4–5 μm. Since an independent shift of the points from the 2D Thomas process would cause a loss of the cluster structure, we apply a dependent shifting along the z-axis. This is based on the following property of the exponential distribution: let $Z_1, \ldots, Z_4 \sim \exp(1/4)$ be independent and exponentially distributed random variables. Then, it holds that $\exp(-\min\{Z_1, \ldots, Z_4\}) \sim U(0, 1)$.

Thus, considering a sample of a Thomas process which has $n \geq 4$ points in the sampling window, we first associate these points with independent random variables $Z_1, \ldots, Z_n \sim \exp(1/4)$. Then, for the ith point of these n points, $i = 1, \ldots, n$, we consider its three nearest neighbors with indices $i_1, i_2, i_3 \in \{1, \ldots, n\} \setminus \{i\}$, say. Then, we shift the ith point within the corresponding layer along the z-axis, according to $\exp\left(-\min\left\{Z_i, Z_{i_1}, Z_{i_2}, Z_{i_3}\right\}\right) \sim U(0, 1)$ (suitably scaled to the thickness of the layer). This dependent shifting along the z-axis ensures that the principle structure of the generalized Thomas processes is not changed dramatically.

Finally, to incorporate the hard-core distance into the model, we apply a subsequent shift of the points along the z-axis if two points are too close to each other. Hence, we look at that pair of points of the complete 3D point pattern which has the

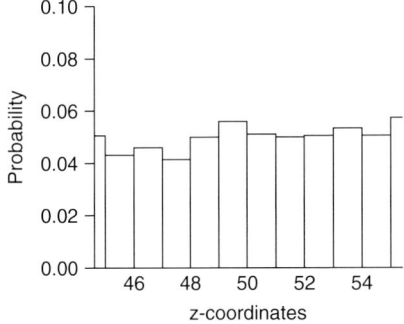

Figure 24.12 Estimated distribution of z-coordinates of vertices for some layers. Copyright (2011) J. Mater. Sci.

smallest distance from each other and choose one of these two points at random. Then, we shift this point again along the z-axis within the corresponding thin layer, according to a uniformly distributed random variable. We repeat this procedure until the required hard-core distance of 3 µm is achieved for almost all point pairs or if no further improvement is possible.

24.3.3
Validation of Vertex Model

In order to validate the point-process model introduced in Section 24.3.2, we consider three different characteristics of stationary point processes: the distribution function of (spherical) contact distances $H_S : [0, \infty) \to [0, 1]$, the nearest-neighbor distance distribution function $D : [0, \infty) \to [0, 1]$, and the pair-correlation function $g : [0, \infty) \to [0, \infty)$, which can be found, for example, in Illian et al. [26].

Note that these characteristics are all considered in 3D, whereas for the fitting of the generalized Thomas process the pair-correlation function of the projected point patterns has been used, that is, it has been computed in 2D. Note that $H_S(r)$ is the probability that the distance from an arbitrary location in $\mathbb{R}^3$, chosen at random, to the closest point of the point process is not larger than $r > 0$. Similarly, $D(r)$ is the probability that the distance from an arbitrary point of the point process, chosen at random, to its nearest neighbor within the point process is not larger than $r > 0$.

To verify whether the 3D point process model fits real data sufficiently well, we estimate H_S, D, and g for 50 cutouts of vertex sets extracted from synchrotron data, whereby standard (boundary-corrected) estimators are used; see, for example, [18, 26, 30]. The pointwise averages of these estimates are denoted by $\widehat{H}_S$, $\widehat{D}$, and $\widehat{g}$. Then, we compute pointwise 96% confidence bands for the three considered characteristics of point processes mentioned above, whereby we generate 50 samples of the 3D point process model with the estimated parameters as given in Section 24.3.2 in a sampling window of $768 \times 768 \times 195$ µm. The bands estimated from these realizations are plotted as gray solid lines.

The results for the functions H_S and D illustrated in Figure 24.13 show that the empirical distribution functions $\widehat{H}_S$ and $\widehat{D}$ computed from real data (plotted as a black dashed line) are more or less within the confidence bands obtained from simulated data (gray solid lines). However, the estimated pair-correlation function $\widehat{g}$ from real data does not match the confidence band of simulated data perfectly (see Figure 24.10). However, the main structural properties of $\widehat{g}$ such as the hard-core distance, the large peak at about 4 µm, and the declining rate of the tail towards the level of 1 are reflected fairly well by the model.

Although not all considered characteristics of the 3D point process model fit perfectly to real data, that is, to the set of vertices V_r of the extracted graph (V_r, E_r), we can conclude that the 3D vertex model introduced in Section 24.3.2 fits since the main structural properties are reflected fairly well.

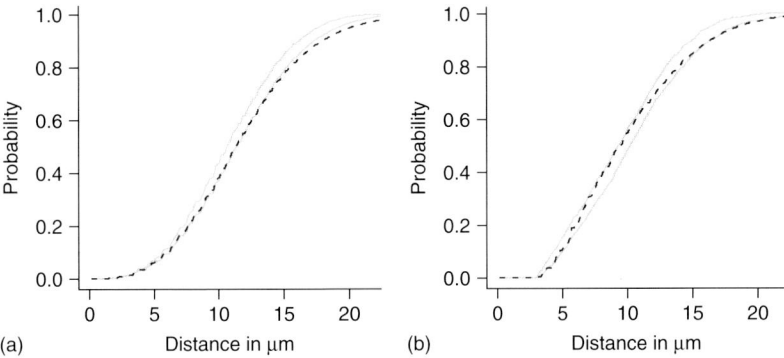

Figure 24.13 Spherical-contact (a) and nearest-neighbor (b) distance distribution functions for real (dashed line) and simulated data (gray solid lines; pointwise 96% confidence bands). Copyright (2011) J. Mater. Sci.

24.3.4
Marking of Vertices

Since the final graph will be marked with volume-describing properties, we extend the stochastic point process model introduced in Section 24.3.2 to a marked point process. Therefore, we consider two different types of marks for each vertex: its spherical distance from the solid phase and the number of emanating edges. Note that the spherical distance is closely related to the notion "pore" in a porous media (see Section 24.3.1). The second kind of marks is the degree of vertices, also called the *coordination number* in physics and geology, which is used later in Section 24.3.5 for modeling the edges of the random geometric graph.

24.3.4.1 Moving-Average Model for Dependent Marking

For the considered pore phase of non-woven GDLs, it turns out that the spherical contact distances of vertices, that is, the nearest distance of a vertex to the solid phase, can be modeled by a Γ-distribution. This is visualized in the histogram in Figure 24.14a, showing the spherical contact distances to the solid phase for the vertices extracted from synchrotron data. It can be nicely fitted by a Γ-distribution $\Gamma(\rho, \zeta)$ with parameters $\rho > 0$ (rate) and $\zeta > 0$ (shape), using maximum-likelihood estimation or the method of moments; see, for example, [31]. For the parameters of this Γ-distribution (black curve in Figure 24.14), the averaged values of $\rho = 1.077$ and $\zeta = 7.331$ have been obtained, where the average is taken from all 50 cutouts from the synchrotron data. However, a closer analysis of these marks from the network extracted from synchrotron data shows that the contact distances of neighboring vertices are strongly (positively) correlated. This observation is indicated by the mark-correlation function $\kappa : (0, \infty) \to [-1, 1]$ of stationary marked point processes, where $\kappa(r)$ is the correlation of the marks of an arbitrary pair of points, chosen at random, with distance $r > 0$ from each other; see [26] for a formal definition. It is plotted in Figure 24.14b. Hence, vertices which are

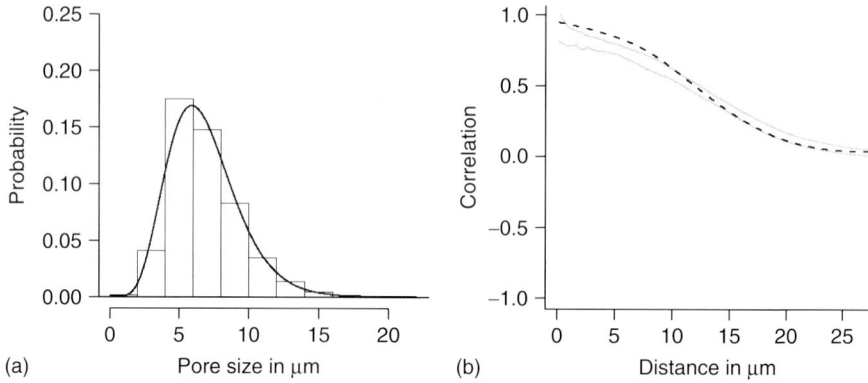

Figure 24.14 (a) Histogram of spherical distances of vertices to the solid phase and fitted Γ-distribution (black solid line). (b) Mark-correlation functions for spherical distances of real (black dashed line) and simulated (gray solid lines; pointwise 96% confidence bands) data. Copyright (2011) J. Mater. Sci.

located close to each other show strongly (positively) correlated contact distances and, vice versa, the spherical contact distances of vertices which are far away from each other are more or less uncorrelated. Hence, we propose a moving-average procedure to mimic this correlation structure in the model, whereby we proceed similarly as in Section 24.3.2.2 with the dependent shifting of vertices in the z-direction.

Here, we use a well-known stability property with respect to convolution for the family of Γ-distributions. More precisely, if there are $n \geq 3$ points in the sampling window, we first associate these points with independent random variables $Z_1, \ldots, Z_n \sim \Gamma(\rho, \zeta/3)$, distributed according to the Γ-distribution shown in Figure 24.14, where $\rho = 1.077$ and $\zeta = 7.331$. Then, for the ith of these n points, $i = 1, \ldots, n$, we consider its two closest neighbors with indices $i_1, i_2 \in \{1, \ldots, n\} \setminus \{i\}$, say. Then, as the mark of the ith point, we finally choose the sum

$$Z_i + Z_{i_1} + Z_{i_2} \sim \Gamma(\rho, \zeta) \tag{24.12}$$

This dependent marking of points ensures that the principle structure of the empirical mark-correlation function $\widehat{\kappa}$ is reconstructed fairly well. The result is displayed in Figure 24.14, showing pointwise 96% confidence bands (gray solid lines), which were computed from 50 samples of the 3D point process model with the moving-average marking as described above.

After marking the vertices with spherical distances, a thinning of this set of marked points has to be performed. So far, the developed and fitted model for the vertices is for the original pore phase graph (V_r, E_r). To obtain a model of the modified pore phase graph (V'_r, E'_r) in Section 24.3.1.2, where each vertex can be interpreted as a pore center, the same thinning algorithm as explained in Section 24.3.1.1 is applied. After this thinning, the remaining set of marked points can be interpreted as a set of pores according to the applied definition used in the following for the construction of the random geometric graph model.

24.3.4.2 Degrees of Vertices

Regarding the description of connectivity of a graph, the degrees or coordination numbers of its vertices are of great importance, that is, the number of edges emanating from each vertex. Hence, we consider the vertex degrees as additional marks of the point process model for the vertices introduced above.

Note that the analysis and fitting of this mark are not based directly on the graph (V_r, E_r) extracted from synchrotron data, but on the modified graph (V'_r, E'_r) in Section 24.3.1.2.

For marking the vertices with their degrees, we proceed in the same way as in the case of the spherical marks. First, the empirical distribution of vertex degrees for the modified graph (V'_r, E'_r) is determined, which is shown in Figure 24.15a. Subsequently, we analyze their correlation structure, whereby we compute the mark-correlation function of vertex degrees (see Figure 24.15b). Observing no correlation between vertex degrees, it seems that the degrees of vertices can be modeled in an independent and identically distributed (iid) way, according to the distribution shown in Figure 24.15a. Note that this could come into conflict with the fact that not for every configuration of vertex degrees can a graph be constructed. Hence, we propose a combination of iid sampling from the distribution shown in Figure 24.15a with a certain acceptance–rejection procedure which works as follows.

Supposing that the sample of the random graph to be constructed has n vertices in the observation window. Then, we generate an iid sample $d_1, \ldots, d_n > 0$ of candidates for vertex degrees according to the distribution shown in Figure 24.15a. The Erdös–Gallai theorem of graph theory (see, for example, [32, 33]) allows one to test whether $d_1, \ldots, d_n$ is an admissible configuration of vertex degrees or not.

Therefore, we rearrange the numbers $d_1, \ldots, d_n > 0$ in descending order, obtaining the sequence $d'_1 \geq d'_2 \geq \ldots \geq d'_n > 0$, say. Then, a simple graph, that is, each pair of vertices has at most one direct connecting edge, can be constructed

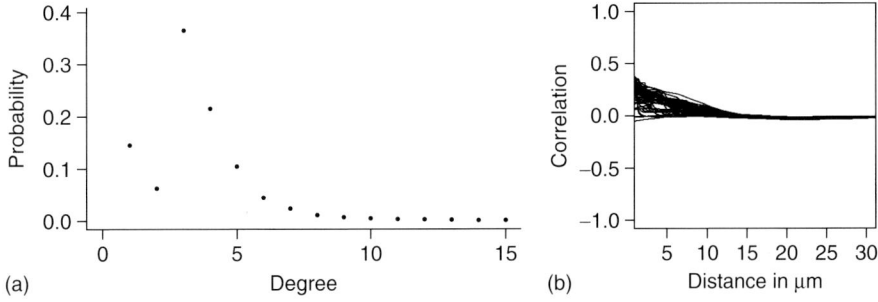

Figure 24.15 Coordination number analysis: (a) estimated distribution of the coordination number; (b) estimated mark-correlation functions. Copyright (2011) J. Mater. Sci.

possessing the configuration $d_1, \ldots, d_n > 0$ of vertex degrees if and only if $d_1 + d_2 + \ldots + d_n$ is even, and for all $k = 1, \ldots, s$, where s is determined by $d'_s \geq s$ and $d'_{s+1} < s+1$, it holds that

$$\sum_{i=1}^{k} d'_i \leq k(k-1) + \sum_{i=k+1}^{n} \min\{k, d'_i\} \qquad (24.13)$$

If the sequence $d_1, \ldots, d_n > 0$ of potential vertex degrees fulfills the conditions of the Erdös–Gallai theorem, a graph with n vertices and vertex degrees $d_1, \ldots, d_n > 0$ can be constructed. Otherwise, we reject the sample $d_1, \ldots, d_n > 0$ and generate a new one according to the distribution shown in Figure 24.15a. This procedure is repeated until a sequence of vertex degrees is generated which fulfills the conditions of the Erdös–Gallai theorem, that is, we have an admissible configuration of vertex degrees. Then, according to the given degrees, a graph is constructed using the proposed algorithms in Section 24.3.5.

24.3.5
Stochastic Modeling of Edges

In the previous sections, we have presented a model for the vertices of the graph and marked them among others by their degrees. If a realization of the random marked point process from above is given, or more precisely a finite number of marked points is given in some observation window W, we describe a stochastic model for generating the edges of the random geometric graph. Hence, we combine tools from graph theory and MCMC simulation. In particular, the model is constructed in such a way that the distributions of vertex degrees and edge lengths coincide to a large extent for real and simulated data.

As described in Section 24.3.4.2, candidates for vertex degrees are sampled in an iid manner, according to the distribution shown in Figure 24.15a. Recall that this is followed by an acceptance–rejection procedure which ensures that the conditions of the Erdös–Gallai theorem are fulfilled.

Then, under the condition that an admissible configuration of vertex degrees is given, edges are included using the well-known Hakimi–Havel algorithm of graph theory. The Hakimi–Havel algorithm provides a tool to construct a random 3D graph from a given configuration of vertices and a degree sequence under certain conditions. This construction leads to a random graph whose distribution of vertex degrees fits the predefined distribution of vertex degrees. For further information about the Hakimi–Havel algorithm, see, for example, [32]. However, this algorithm does not take into account the locations of vertices, which means that in general the distribution of edge lengths computed, for example, from real data is not fitted well.

In order to minimize this discrepancy, the Hakimi–Havel algorithm is supplemented by an MCMC procedure to rearrange edges in such a way that the distribution of vertex degrees is kept fixed and, simultaneously, the fit to the predefined distribution of edge lengths computed from real data is improved.

24.3.5.1 MCMC Simulation for Edge Rearrangement

Since the Hakimi–Havel algorithm does not take into account the locations of the vertices, it is obvious that a predefined distribution of edge lengths, which can be gained, for example, by estimation from real data, is not met in general. As can be seen in Figure 24.16b, the distribution of edge lengths of the graph constructed with the Hakimi–Havel algorithm differs, although the model for the vertices fits structurally fairly well to the real data (see Section 24.3.3). Therefore, in order to obtain a better fit, the Hakimi–Havel algorithm is supplemented by an MCMC procedure which rearranges the edges in such a way that the distribution of vertex degrees is kept fixed and, simultaneously, the fit of the predefined distribution of edge lengths is improved. For a detailed description of this MCMC procedure see [6], and for further information on Markov chains and MCMC simulation see, for example, [34–36].

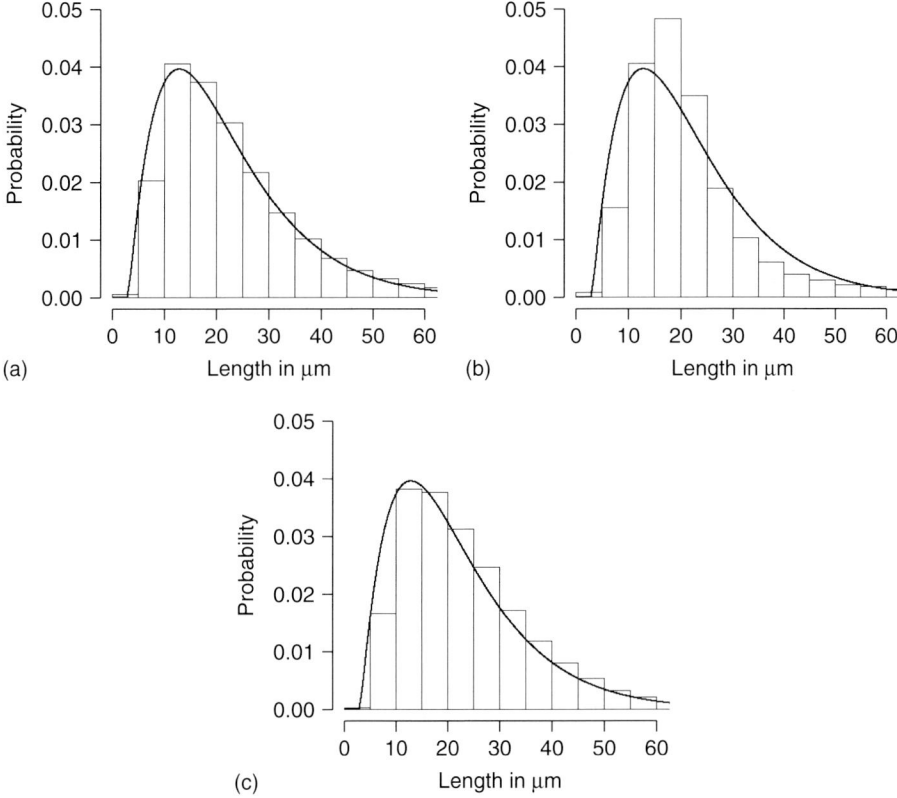

Figure 24.16 Edge length distributions: (a) from the modified graph and a fitted (shifted) Γ-distribution (black solid line); (b) from the graph generated by the Hakimi–Havel algorithm; (c) from the graph after MCMC-simulation. Copyright (2011) J. Mater. Sci.

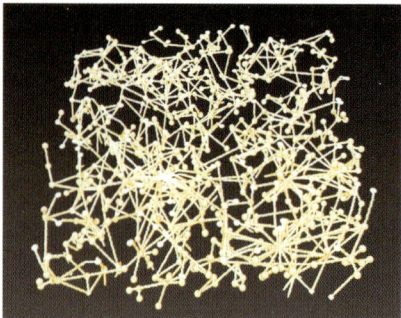

Figure 24.17 Small cutout of a realization of a 3D random geometric graph.

The resulting distribution of edge lengths after the MCMC simulation as described in the previous section can be seen in Figure 24.16c, where the fitted gamma distribution is also plotted.

A realization of the random geometric graph model, where the edge lengths have been fitted to real data by the above-described MCMC simulation, can be seen in Figure 24.17. Note that Figure 24.17 just shows a small cutout of a realization of the network model describing the pore phase of non-woven GDL.

So far, a validation of the final, that is, complete, graph model is lacking, although the different parts of the construction have already been compared with real data. Since the main application of such a pore phase graph lies in the field of efficient transport simulations along the edges of the graph (see also Section 24.4.2), we use structural characteristics that are related to transport properties. A detailed description of these characteristics is given in Section 24.5 and their application for validating the above-constructed and fitted graph model is discussed in Section 24.5.5.

24.4
Further Results

24.4.1
Classical Random Graph Models

Random graph models have been considered in the literature for a long time, starting in the middle of the last century; see, for example, [37–39].

The model type for random geometric graphs described in the previous sections is rather different from random graphs considered in the literature; see, for example, [40–45] for an overview.

Often, there is no geometric model for the vertices taken into account, that is, the underlying geometric structure of the graph is taken into account at most partially. Such models are widely used, for example, for telecommunication networks describing properties of the Internet. Many models for random geometric graphs containing an explicit model for the vertices suppose them to be a Poisson point process. Although this assumption allows for theoretical results, it only meets the complex structures occurring in reality in some special cases.

The degrees of the vertices are also of great interest for applications since they describe the connectivity of a graph to a large extent. However, some graph models do not include the degrees of vertices explicitly, that is, the required distribution of degrees cannot be predefined in the model. Most models for random graphs lead to a binomial or Poisson distribution of the degrees. For example, the construction of the edges follows some rules such as: all vertices are connected with their k-nearest neighbors or the connecting probability depends only on the distance between the vertices; see, for example, [46] for an explicit description of such models.

However, some models for random geometric graphs exist in the literature which produce a predefined power-law distribution of the degree sequence of the vertices. Such graphs are also called *scale-free networks*, see, for example, [47].

24.4.2
Transport Simulations along Edges of Graphs

The investigation of pore phases using a graph representation has already been performed in the literature; see, for example, [5, 48]. In particular, some authors have considered grid-based graph approaches for the analysis of GDLs used in PEMFCs; see, for example, [49–51]. More precisely, the pores/vertices in these models are located on a grid and the models are calibrated with respect to global physical characteristics such as permeability. Despite their (structural) simplicity, these pore network models for pore phases in GDLs are useful tools for simulating transport processes through GDLs and analyzing, for example, the influence of different PTFE treatments on the GDL functionality, that is, to investigate different hydrophilicities of (internal) GDL surfaces.

In contrast to this global approach for model fitting, the model for random geometric graphs described above is fitted with respect to local structural characteristics of the pore phase. Hence it provides a much more detailed mapping of the structural properties of the pore phase that will be investigated. Subsequently, the computation of transport processes along the edges of such a structurally fitted graph allows for a more detailed investigation of the influence of the microstructure on the transport properties than grid-based graphs.

Note that the simulation of transport processes through GDLs is often based on solving differential equations, which is computationally very time consuming in 3D. Using a graph representation of the pore phase, the differential equations have (mainly) to be solved for the edges of the graph, that is, it is reduced to 1D problems that can be solved fairly quickly or where even analytical solutions are known; see, for example, [52–54]. In summary, transport simulations on graphs representing pore phases can be performed very efficiently and, therefore, in larger domains than simulations of transport processes using 3D voxel-given descriptions of the pore phase.

24.5
Structural Characterization of Porous GDL

As already mentioned, the nano- or micromorphology of a porous material has an essential influence on its physical behavior. One possibility to describe quantitatively the correlation between morphology and functionality is the investigation of 3D (image) data. Therefore, several characteristics are presented in this section with special regard to transport properties. Note that also in physics various experimental methods exist which can be applied for a characterization; however, in most cases these approaches provide only so-called integrated characteristics, that is, it is a global point of view gained by an averaging over the whole sample. These characteristics are essential, for example, in order to describe fluid flow in porous media in the frame of continuum models. In contrast, techniques using information from 3D images allow for a local characterization of morphologies providing much more information.

24.5.1
Tortuosity

We present a local structural characteristic which describes the detour of, for example, a gas flowing through a porous material, the so-called *tortuosity*, usually defined as the ratio of the mean effective path length of the fluid through a porous material and the material thickness. Note that a unique definition in physics is missing; see also the discussion in Wang et al. [10].

An alternative approach to describe porous media with respect to tortuosity properties is the consideration of the so-called *geometric tortuosity*, where effective pathlengths are replaced by shortest geometric pathlengths. Such characteristics can be determined for porous media from 3D images; see, for example, [5, 6, 55–58]. The main difference between these approaches is the representation of the pore phase on which the shortest paths is determined. Note that the effective pathlength which is used in the standard (physical) definition of tortuosity is just a mean value. In addition, it depends not only on the microstructure but also on the side conditions of the performed experiment. In contrast to the calculation of tortuosity, the lengths of the shortest paths can be determined uniquely and depend only on the given microstructure.

In the first method presented, the shortest paths are computed along edges of a pore phase graph which can be obtained, for example, by a skeletonization as described by Fourard et al. [25], or a direct modeling of the graph as described in Section 24.3.

The second approach is directly based on the voxel representation of the pore phase, that is, it can be applied directly to 3D binary images or realizations of stochastic models as proposed, for example, in Section 24.1.

In order to compute the shortest paths through a porous medium along edges of a 3D graph representing the pore phase, we first have to determine the starting points of these paths. Therefore, a two-dimensional stationary Poisson point process $\{S_i\}$

with some intensity $\lambda > 0$ is generated on top (by definition) of the porous medium and connected to the graph by additional edges.

This model for starting points of shortest paths is chosen since gas molecules can start their diffusion or migration at any point on the surface of a porous medium with the same probability (in *ex situ* experiments).

For the computation of shortest paths, we use Dijkstra's algorithm; for details see, for example, [59, 60]. The length of a shortest path normalized with the thickness of the considered material can then be interpreted as mark L_i of the corresponding starting point. Thus, we consider a 2D marked point process $\{(S_i, L_i)\}$ with marks in $\mathbb{R}_+$. The estimated distribution of the marks can be seen as a distribution of the geometric tortuosity, containing much more information than just a mean value. Note that for minimizing the influence of edge effects in the estimation, a (2D) minus sampling is applied to $\{(S_i, L_i)\}$, that is, only marks L_i of points S_i in a (smaller) sub-window on top of the porous medium are considered. In other words, only marks L_i of points S_i which are not too close to the boundary of the observation window with respect to the xy-plane are taken into account.

Since 3D images are given by voxels, that is, a 3D image can be seen as a 3D array, another possibility to define geometric tortuosity is based on the voxel representation of the pore phase; see, for example, [55, 57, 61].

Decker *et al.* [55] and Demarty *et al.* [61] considered the voxel representation as a 3D (cubic) graph which is constructed using the six-neighborhood of 3D images. More precisely, each voxel of the 3D (binary) image is considered as one vertex of the graph which is connected to all six neighbors with common two-dimensional facets via edges. To take into account the voxel-given microstructure, only edges in the pore phase are considered.

In contrast to the graph representation gained by, for example, a skeletonization as described by Thiedmann *et al.* [5], the voxel-based graph contains the real microstructure in more detail. However, the runtime for the computations of the shortest paths rise since the numbers of edges and vertices increase. In addition, graphs which are extracted from 3D images by, for example, skeletonization [25] can be seen as an averaging over space since it is located more or less in the center of the pore phase. Hence, it can be expected that the real flow, for example the results of numerical transport simulations, are approximated better.

A slightly different approach to determine geometric tortuosities is described by Jørgensen [57], which is also based on a voxel representation of the microstructure. The so-called *fast marching method* (see, for example, [62–64]) is used to determine the shortest paths through voxel-given structures. In addition, Jørgensen [57] discussed a modified version which takes the volume of the considered paths into account. Although it is also a purely structural characteristic, its results seem to be closer to physics than just a shortest path approach. Note that a short discussion of this method with respect to its physical interpretation was given by Shearing *et al.* [65].

24.5.2
Pore Size Distributions

In this section, we introduce characteristics of porous materials describing *pore sizes* and *pore size distributions*, which is of particular interest in the physical and engineering literature; see, for example, [66, 67]. Note that no unique definition for their determination from 3D images is given in the literature; see, for example, [5, 68–70], where a selection is presented in the following. In addition, Münch and Holzer [69] presented a method which can be applied to images and which mimics porosimetry measurements. Note that several measurement techniques are also available for determining pore size distributions experimentally; see, for example, [9, 66, 67, 71], and references therein. However, no systematic investigation of the correlation between pore size distributions resulting from physical experiments and from image analysis has been performed (at least so far). In general, the results of physical porosimetry measurements do not coincide with pore size distributions extracted from 3D images. One reason is that the analysis of results from, for example, mercury porosimetry experiments to determine a pore size distribution uses many assumptions about the structure of pores which are in most cases not fulfilled for real nano- or microstructures such as GDL materials; see, for example, [72].

The representation of the (voxel-given) pore phase by a geometric 3D graph permits a definition of pores. Hence, all vertices of the graph are considered as potential pore centers, where the pore sizes or pore radii are defined as the spherical distances of the pore centers to the solid phase, which can be determined very efficiently by a Euclidean distance transformation; see, for example, [73]. However, the definition of pore proposed by Thiedmann *et al.* [5] does not consider all vertices as pore centers, because otherwise some pores could be contained predominantly or even completely in other pores. Hence, only those vertices which are not contained in larger pores are interpreted as pore centers. The resulting set of pores can mutually overlap, but no pore contains a center of another pore. An example in 2D can be seen in Figure 24.8, where all potential pores (a) and the remaining, that is, final, set of pores (b) are displayed. This avoids the overestimation of small pores; see also Section 24.3.1.1.

A further possible definition to describe the notion "pore" from 3D images is based on a distance transformation and a subsequent watershed transformation; see, for example, [70].

From a binary image, where the pore phase is assumed to be the foreground, the detection of pores proceeds as follows. In the first step, a distance transformation is applied to the foreground. The local maxima of the resulting distance map are assumed to be the centers of the pores and the subsequent watershed transformation uses these local maxima as initial points. The result is a partitioning of the pore phase in disjoint basins, where each basin is interpreted as a pore; see Figure 24.18 for an example in 2D.

Note that in some cases an additional smoothing of the distance map may be appropriate to avoid over-segmentation, that is, to reduce the number of very

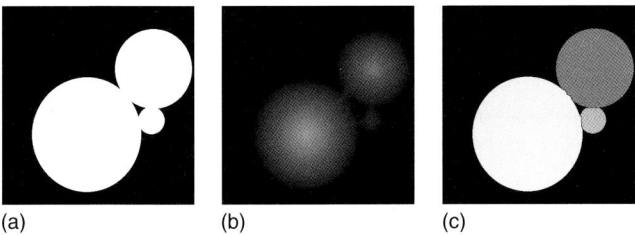

Figure 24.18 2D example of pore definition based on a watershed segmentation: (a) binary image of porous structure; (b) distance transformation; (c) three detected pores (different gray values).

small pores which have no physical interpretation but are only artifacts; see, for example, [74].

A completely different approach is the so-called *continuous pore size distribution* proposed by Münch and Holzer [69], which is (from a statistical point of view) not a distribution function but a function of volume fractions. It is driven by the idea of a geometric implementation of porosimetry measurements with respect to image data. It is based on the determination of volume fractions of the pore phase which can be covered by a sphere with fixed radius $r_S \geq 0$.

It is obvious that the volume fraction which can be covered by spheres with radius r_S increases with decreasing radius r_S. This can physically be related, for example, to mercury porosimetry measurements. Using high pressures, mercury occupies more or less the complete pore phase, which corresponds to small radii r_S in the definition of continuous pore size distribution. On the other side, low pressures correspond to large radii r_S. The reason is that for low pressures, the hydrophobicity dominates the pressure to fill the pore phase. Hence for low pressures, only larger pores are filled. More detailed information on this topic is available elsewhere ([9, 66, 67, 69, 71] and references therein).

24.5.3
Connectivity

For transport processes in porous media, the connectivity of the pore phase plays an essential role; see, for example, [66, 75] and references therein.

One possibility for its quantification is the consideration of the minimum spanning tree (MST) of a graph representing the pore phase of a porous medium. The basic idea of the MST is to remove as many edges as possible from a connected graph in such a way that the set of vertices is still connected but the remaining set of edges have minimum total length. For the computation of the MST, Prim's algorithm can be used; see, for example, [59, 60] for further details.

For describing the connectivity of the pore phase, we consider the relative length of the MST, which is the ratio of the total edge length of the MST to the total edge length of the original graph. The interpretation is as follows: the closer the characteristics is to 1, the less connected is the graph; on the other hand, if the relative length is close to zero, the graph is well connected.

Another characteristic that can be applied is the coordination number, which is related to the definition of pores based on a graph representation of the pore phase.

The *coordination number of a porous media* is defined as the number of emanating *necks*, *throats*, or *edges* from a pore and is applied, for example, in geology [48].

24.5.4
Validation of Multi-Layer Model

In this section, the structural model for paper-type GDLs is compared with real 3D data gained by means of synchrotron tomography [23, 24]. This comparison is based on the above-described structural characteristics for porous media and can be seen as a model validation. In particular, characteristics are applied which are related to transport properties of the pore phase, since the investigation of transport processes is the main area of application of the multi-layer model; see also the results and discussions in Thiedmann et al. [5].

Note that the model fitting in Section 24.1.3 fix the porosity ε, the parameter γ of the PLT, and the radius of the fibers r_F. Hence, only the parameters for binder modeling can be varied, that is, the probability p that a cell contains binder and the amount of binder per chosen cell, which is controlled by r_B. Note that these parameters are not independent and the considered combinations are given in Table 24.1 (first column).

As can be seen in Table 24.1, the structural comparison between real 3D data and realizations of the multi-layer model shows that the model fits fairly well to reality, although the pore size distribution of the model (for all approaches of binder modeling) does not match real data perfectly, and the geometric tortuosity (see also Figure 24.19) is slightly different. However, the connectivity describing properties of the model fit almost perfectly. Summarizing, we conclude that apart from some discrepancies, the model fits the real structure of paper-type GDLs fairly well and,

Table 24.1 Mean values and standard deviations (SDs) of geometric tortuosity, pore size, and coordination number, and mean relative length of MST.

Data	Porosity	Geometric tortuosity		Pore size		Coordination number		Relative length of MST
		Mean	SD	Mean	SD	Mean	SD	Mean
$r_B = \infty, p = 0.059$	78%	1.73	0.25	10.16	4.38	4.37	2.01	0.46
$r_B = 30, p = 0.081$	78%	1.66	0.16	10.11	4.36	4.36	2.00	0.47
$r_B = 18, p = 0.116$	78%	1.64	0.15	10.13	4.36	4.36	1.99	0.47
$r_B = 6, p = 0.555$	78%	1.62	0.14	10.12	4.37	4.34	1.97	0.47
3D X-ray image	78%	1.51	0.16	8.05	3.74	4.38	5.31	0.44

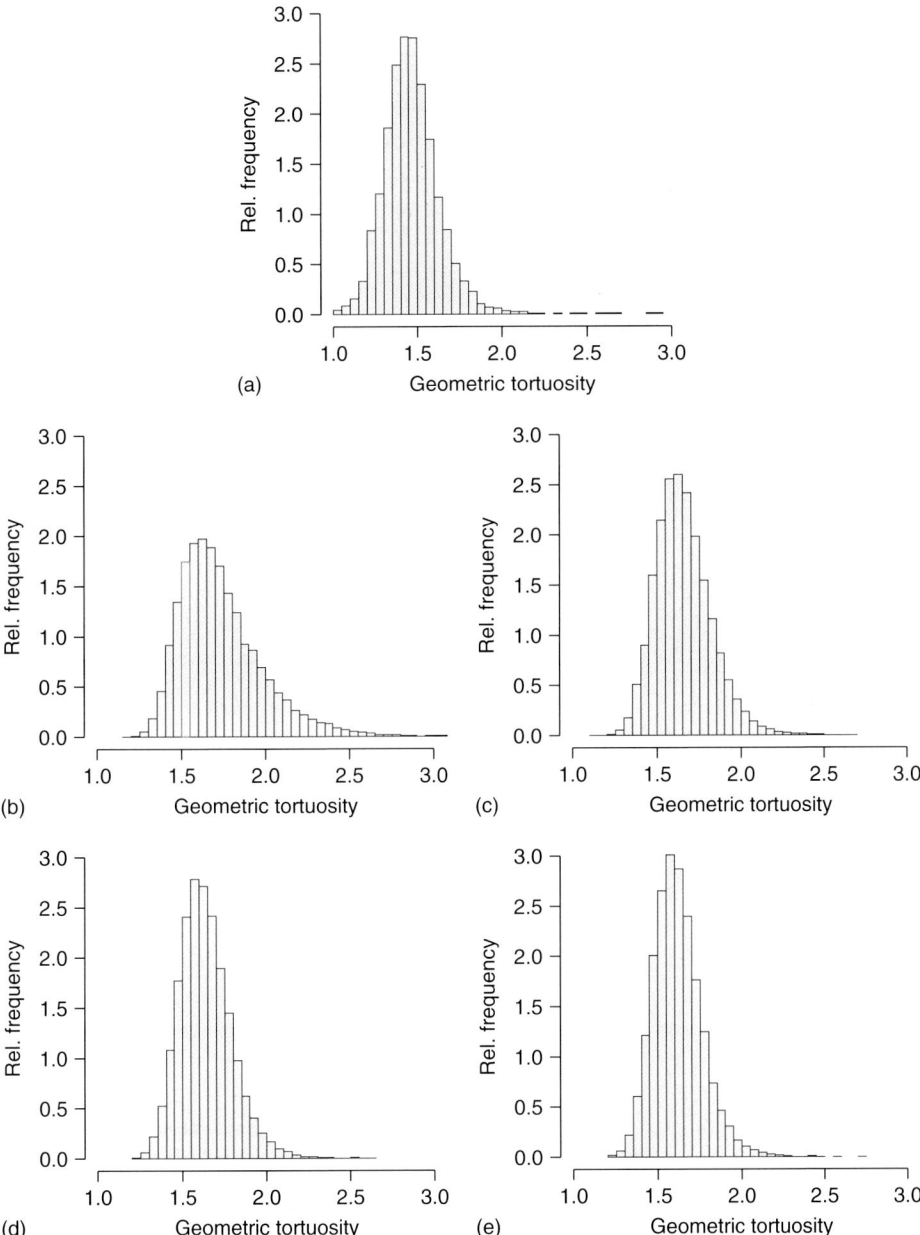

Figure 24.19 Histograms of geometric tortuosities with varying binder modelings: (a) original data; (b) complete cell filling; (c) partial cell filling with $r_B = 30$; (d) partial cell filling with $r_B = 18$; (e) partial cell filling with $r_B = 6$. Reprinted with permission from [5]. Copyright (2008) *Journal of the Electrochemical Society*.

Table 24.2 Tortuosity and MST results of (modified) real and simulated graphs.

	Tortuosity		Relative length of MST
	Mean	SD	
Graph extracted from real data	1.35	0.11	0.38
Random geometric graph model	1.41	0.15	0.40

hence, it is a useful tool for investigating GDLs with respect to transport processes; see, for example, [10].

24.5.5
Validation of Graph Model

In this section, we consider the 3D model for the random geometric graph representing the pore phase of non-woven GDLs presented in Section 24.3. Note that it is not fitted to the extracted pore phase graph directly but to a modified version of it, where the set of vertices is thinned in such a way that each vertex can be interpreted as a pore center according to the definition in Section 24.5.2 based on graphs. The modification of the set of vertices implies that the set of edges also has to be changed. This version of the pore phase graph, described in Section 24.3.1.2, is subsequently used for validating the fitted random geometric graph model.

In Section 24.3, the different modeling steps have already been validated, that is, the model for the vertices, the vertex degrees, the spherical marking, and the edge lengths. So far, only the consideration of the complete, that is, final graph model is lacking. As already mentioned above and in Section 24.4.2, the main application of the model for random graphs representing the pore phase of GDLs is the efficient computation of transport process through the GDL. Hence, for the validation we consider the geometric tortuosity and the MST, which are relevant for transportation properties of porous media. Note that the considered characteristics have not been used for model fitting. Even though there is no perfect match between these characteristics computed from real and simulated data, as can be seen in Table 24.2, they are in both cases at least fairly similar. Hence, we conclude that the fitted model for random geometric graphs fits reality fairly well.

24.6
Conclusion

In this chapter, we have presented models for the solid and pore phase of GDL materials. The main application of these models is the investigation of correlations between the morphology and transport properties of GDLs. Hence, the stochastic simulation models serve as input for numerical simulations of transport processes. In addition, the models can be used for generating virtual materials, that is, they can generate microstructures of GDLs which have not been recorded so far. The combination of stochastic simulation of microstructures and numerical simulation of functionality leads to the concept of the so-called *virtual material design*, that is, the investigation and/or optimization of GDL morphologies based on computer experiments. This approach can be a valuable extension to physical experiments since computer experiments can be executed fairly fast and cheaply. Hence, many more scenarios than with physical experiments can be generated and analyzed in detail.

In addition, structural characteristics which are related to the transport properties of GDLs are introduced. In addition to their application for validating structural models as shown above, they can be used for comparing different GDL materials. For instance, Thiedmann *et al.* [76] made a structural comparison between pore phases of paper- and non-woven GDLs. Despite the obvious differences of the fiber systems, the pore phases possess some similarities. Note that this is in accordance with physical experiments showing that the performances of fuel cells under the same operating conditions, but with different GDL types, vary [77].

References

1. Mathias, M.F., Roth, J., Fleming, J., and Lehnert, W. (2003) Diffusion media materials and characterisation, in *Handbook of Fuel Cells*, vol. III (eds. W. Vielstich, A. Lamm, and H. Gasteiger), John Wiley & Sons, Ltd., Chichester, Chapter 42, pp. 517–537.
2. Gaiselmann, G., Thiedmann, R., Manke, I., Lehnert, W., and Schmidt, V. (2012) Stochastic 3D modeling of fiber-based materials.. *Comp. Mat. Sci.*, (to appear).
3. Stenzel, O., Koster, L.J.A., Thiedmann, R., Oosterhout, S.D., Wienk, M.M., Janssen, R.A.J., and Schmidt, V. (2011) A new approach to model-based simulation of disordered-polymer-blend solar cells. *Adv. Funct. Mater.*, to appear.
4. Thiedmann, R., Fleischer, F., Hartnig, C., Lehnert, W., and Schmidt, V. (2008) Stochastic 3D modeling of the GDL structure in PEM fuel cells based on thin section detection. *J. Electrochem. Soc.*, **155**, B391–B399.
5. Thiedmann, R., Hartnig, C., Manke, I., Schmidt, V., and Lehnert, W. (2009) Local structural characteristics of pore space in GDL's of PEM fuel cells based on geometric 3D graphs. *J. Electrochem. Soc.*, **156**, B1339–B1347.
6. Thiedmann, R. (2011) New approaches to stochastic image segmentation and modeling of complex microstructures – application to the analysis of advanced materials, PhD thesis, Ulm University.
7. Baddeley, A.J. and Cruz-Orive, L.M. (1995) The Rao–Blackwell theorem in stereology and some counterexamples. *Adv. Appl. Probab.*, **27**, 2–19.
8. Schladitz, K. (2000) Estimation of the intensity of stationary flat processes. *Adv. Appl. Probab.*, **32**, 114–139.
9. Gostick, J.T., Fowler, M.W., Ioannidis, M.A., Pritzker, M.D., Volfkovich, Y.M., and Sakars, A. (2006) Capillary pressure and hydrophilic porosity in gas diffusion layers for polymer electrolyte fuel cells. *J. Power Sources*, **156**, 375–387.
10. Wang, Y., Cho, S., Thiedmann, R., Schmidt, V., Lehnert, W., and Feng, X. (2010) Stochastic modeling and direct numerical simulation of the diffusion media for polymer electrolyte fuel cells. *Int. J. Heat Mass Transfer*, **53**, 1128–1138.

11. Froning, D., Lehnert, W., Thiedmann, R., Schmidt, V., and Manke, I. (2010) 3D analysis, modeling and simulation of transport processes in fibrous microstructures using the lattice Boltzmann method, in Proceedings of the International Symposium on Transport in Porous Materials and in Networked Microstructures, with Special Focus on the Link Between Microscopy and Modelling, Villigen.
12. Hao, L. and Cheng, P. (2009) Lattice Boltzmann simulations of anisotropic permeabilities in carbon paper gas diffusion layers. *J. Power Sources*, **186**, 104–114.
13. Inoue, G., Matsukuma, Y., and Minemoto, M. (2007) Numerical analysis of two-phase transport in GDL of polymer electrolyte fuel cells, in Proceedings of the 2nd European Fuel Cell Technology and Applications Conference, EFC2007-39024.
14. Inoue, G., Yoshimoto, T., Matsukuma, Y., and Minemoto, M. (2008) Development of simulated gas diffusion layer of polymer electrolyte fuel cells and evaluation of its structure. *J. Power Sources*, **175**, 145–158.
15. Schulz, V.P., Becker, J., Wiegmann, A., Mukherjee, P.P., and Wang, C.Y. (2007) Modeling of two-phase behavior in the gas diffusion medium of PEFCs via full morphology approach. *J. Electrochem. Soc.*, **154**, B419–B426.
16. Spiess, M. and Spodarev, E. (2010) Anisotropic Poisson processes of cylinders. *Methodol. Comput. Appl. Probab.*, **13** (4), 801–819.
17. Yoneda, M., Takimoto, M., and Koshizuka, S. (2007) Effects of microstructure of gas diffusion layer on two-phase flow transport properties. *ECS Trans.*, **11**, 629–635.
18. Stoyan, D., Kendall, W.S., and Mecke, J. (1995) *Stochastic Geometry and its Applications*, 2nd edn., John Wiley & Sons, Ltd., Chichester.
19. Schladitz, K., Peters, S., Reinel-Bitzer, D., Wiegmann, A., and Ohser, J. (2006) Design of acoustic trim based on geometric modeling and flow simulation for non-woven. *Comput. Mater. Sci.*, **38**, 56–66.
20. Lütkepohl, H. (2006) *New Introduction to Multiple Time Series Analysis*, Springer, Berlin.
21. Fuller, W.A. (1996) *Introduction to Statistical Time Series*, 2nd edn, John Wiley & Sons, Inc., New York.
22. Thiedmann, R., Manke, I., Lehnert, W., and Schmidt, V. (2011) Random geometric graphs for modelling the pore space of fibre-based materials. *J. Mater. Sci.*, **46**, 7745–7759.
23. Hartnig, C., Kuhn, R., Krüger, P., Manke, I., Kardjilov, N., Goebbels, J., Müller, B.R., and Riesemeier, H. (2008) Water management in fuel cells – a challenge for non-destructive high resolution methods. *MP Mater. Test.*, **50**, 609–614.
24. Manke, I., Hartnig, C., Grünerbel, M., Lehnert, W., Kardjilov, N., Haibel, A., Hilger, A., and Banhart, J. (2007) Investigation of water evolution and transport in fuel cells with high resolution synchrotron X-ray radiography. *Appl. Phys. Lett.*, **90**, 174105-1–174105-3.
25. Fourard, C., Malandain, G., Prohaska, S., and Westerhoff, M. (2006) Blockwise processing applied to brain microvascular network study. *IEEE Trans. Med. Imaging*, **156**, B1339–B1347.
26. Illian, J., Penttinen, A., Stoyan, H., and Stoyan, D. (2008) *Statistical Analysis and Modelling of Spatial Point Patterns*, John Wiley & Sons, Ltd., Chichester.
27. Daley, D.J. and Vere-Jones, D. (2003) *An Introduction to the Theory of Point Processes. Volume I: Elementary Theory and Methods*, Springer, New York.
28. Weil, H. (1954) The distribution of radial error. *Ann. Math. Stat.*, **25**, 168–170.
29. Stoyan, D. and Stoyan, H. (1996) Estimating pair-correlation functions of planar cluster processes. *Biometr. J.*, **38**, 259–271.
30. Daley, D.J. and Vere-Jones, D. (2008) *An Introduction to the Theory of Point Processes. Volume II: General Theory and Structure*, Springer, New York.
31. Casella, G. and Berger, R.L. (2002) *Statistical Inference*, 2nd edn, Duxbury Press, Pacific Grove, CA.

32. Thulasiraman, K. and Swamy, M.N.S. (1992) *Graphs: Theory and Algorithms*, John Wiley & Sons, Ltd., Chichester.
33. Tripathi, A. and Vijay, S. (2003) A note on a theorem of Erdös and Gallai. *Discrete Math.*, **265**, 417–420.
34. Asmussen, S. and Glynn, P. (2007) *Stochastic Simulation: Algorithms and Analysis*, Springer, Berlin.
35. Kroese, D.P., Taimre, T., and Botev, Z.I. (2011) *Handbook of Monte Carlo Methods*, John Wiley & Sons, Inc., Hoboken, NJ.
36. Rubinstein, R.Y. and Kroese, D.P. (2008) *Simulation and Monte Carlo Method*, 2nd edn, John Wiley & Sons, Inc., Hoboken, NJ.
37. Erdös, P. and Rényi, A. (1959) On random graphs. *Publ. Math.*, **6**, 290–297.
38. Erdös, P. and Rényi, A. (1960) On the strength of connectedness of a random graph. *Acta Math. Hung.*, **12**, 261–267.
39. Gilbert, E.N. (1959) Random graphs. *Ann. Math. Stat.*, **30**, 1141–1144.
40. Alon, N. and Spencer, J. (2000) *The Probabilistic Method*, John Wiley & Sons, Inc., New York.
41. Bollobas, B. (2001) *Random Graphs*, 2nd edn, Cambridge University Press, Cambridge.
42. Franceschetti, M. and Meester, R. (2008) *Random Networks for Communication*, Cambridge University Press, Cambridge.
43. Janson, S., Luczak, T., and Rucinski, A. (2000) *Random Graphs*, John Wiley & Sons, Inc., New York.
44. van der Hofstad, R. (2010) Percolation and random graphs, in *New Perspectives in Stochastic Geometry* (eds. W.S. Kendall and I. Molchanov), Oxford University Press, Oxford, pp. 173–247.
45. Penrose, M.D. (2003) *Random Geometric Graphs*, Oxford University Press, Oxford.
46. Meester, R. and Roy, R. (1996) *Continuum Percolation*, Cambridge University Press, Cambridge.
47. Aiello, W., Chung, F., and Lu, L. (2000) In *Proceedings of the 32nd Annual ACM Symposium on Theory of Computing*, Association of Computing Machinery, New York, pp. 171–180.
48. Blunt, M.J., Jackson, M.D., Piri, M., and Valvatne, P.H. (2002) Detailed physics, predictive capabilities and macroscopic consequences for pore-network models of multiphase flow. *Adv. Water Res.*, **25**, 1069–1089.
49. Gostick, J.T., Ioannidis, M.A., Fowler, M.W., and Pritzker, M.D. (2007) Pore network modeling of fibrous gas diffusion layers for polymer electrolyte membrane fuel cells. *J. Power Sources*, **173**, 277–290.
50. Sinha, P.K., Mukherjee, P.P., and Wang, C.-Y. (2007) Impact of GDL structure and wettability on water management in polymer electrolyte fuel cells. *J. Mater. Chem.*, **17**, 3089–3103.
51. Sinha, P.K. and Wang, C.-Y. (2007) Pore-network modeling of liquid water transport in gas diffusion layer of a polymer electrolyte fuel cell. *Electrochim. Acta*, **52**, 7936–7945.
52. Dorn, B., Kramar Fijavz, M., Nagel, R., and Radl, A. (2010) The semigroup approach to transport processes in networks. *Physica D: Nonlin. Phenom.*, **239**, 1416–1421.
53. Hante, F.M., Leugering, G., and Seidman, T.I. (2009) Modeling and analysis of modal switching in networked transport systems. *Appl. Math. Optim.*, **59**, 275–292.
54. Kramar, M. and Sikolya, E. (2005) Spectral properties and asymptotic periodicity of flows in networks. *Math. Z.*, **249**, 139–162.
55. Decker, L., Jeulin, D., and Tovena, I. (1998) 3D morphological analysis of the connectivity of a porous medium. *Acta Stereol.*, **17**, 107–112.
56. Decker, L., Jeulin, D., and Tovena, I. (1999) Transport properties of compacted waste, in Proceedings of the 7th International Conference on Radioactive Waste Management and Environmental Remediation – ASME 1999 (ICEM'99), Nagoya.
57. Jørgensen, P. (2010) Quantitative data analysis methods for 3D microstructure characterization of solid oxide cells, PhD thesis, Technical University of Denmark, Lyngby.
58. Tovena, I., Decker, L., Jeulin, D., Pocachard, J., and Ragot, C. (1998) Simulation of water diffusion in packed

metallic waste. Presented at the Waste Management Symposium, Tucson, AZ.
59. Jungnickel, D. (2007) *Graphs, Networks and Algorithms*, 3rd edn, Springer, Berlin.
60. Smith, D.K. (2003) *Networks and Graphs – Techniques and Computational Methods*, Horwood Publishing, Chichester.
61. Demarty, C., Grillon, F., and Jeulin, D. (1996) Study of the contact permeability between rough surfaces from confocal microscopy. *Microsc. Microanal. Microstruct.*, **7**, 505–511.
62. Popovici, A.M. and Sethian, J.A. (1998) Three-dimensional travel-time computation using the fast marching method. *Proc. SPIE*, **3453**, 82–93.
63. Sethian, J.A. (1996) A fast marching level set method for monotonically advancing fronts. *Proc. Natl. Acad. Sci. U. S. A.*, **93**, 1591–1595.
64. Sethian, J.A. (1999) Fast marching methods. *SIAM Rev.*, **41**, 199–235.
65. Shearing, P.R., Howard, L.E., Jørgensen, P.S., Brandon, N.P., and Harris, S.J. (2010) Characterization of the 3-dimensional microstructure of a graphite negative electrode from a Li-ion battery. *Electrochem. Commun.*, **12**, 374–377.
66. Armatas, G.S. (2006) Determination of the effects of the pore size distribution and pore connectivity distribution on the pore tortuosity and diffusive transport in model porous networks. *Chem. Eng. Sci.*, **61**, 4662–4675.
67. Maheshwaria, P.H., Mathur, R.B., and Dhamia, T.L. (2008) The influence of the pore size and its distribution in a carbon paper electrode on the performance of a PEM fuel cell. *Electrochim. Acta*, **54**, 655–659.
68. Dong, H. and Blunt, M.J. (2009) Pore-network extraction from micro-computerized-tomography images. *Phys. Rev. E*, **80**, 36307-1–36307-11.
69. Münch, B. and Holzer, L. (2008) Contradicting geometrical concepts in pore size analysis attained with electron microscopy and mercury intrusion. *J. Am. Ceram. Soc.*, **91**, 4059–4067.
70. Schladitz, K., Redenbach, C., Sych, T., and Godehardt, M. (2008) Microstructural characterisation of open foams using 3D images. Berichte des Fraunhofer ITWM, Technical Report 148, Fraunhofer-Institut fur Techno- und Wirtschaftsmathematik ITWM, Kaiserlautern.
71. Giesche, H. (2006) Mercury porosimetry: a general (practical) overview. *Part. Part. Syst. Charact.*, **23**, 9–19.
72. León y León, C.A. (1998) New perspectives in mercury porosimetry. *Adv. Colloid Interface Sci.*, **76-77**, 341–372.
73. Saito, T. and Toriwaki, J.I. (1994) New algorithms for Euclidean distance transformation of an n-dimensional digitized picture with applications. *Pattern Recognit.*, **27**, 1551–1565.
74. Ohser, J. and Schladitz, K. (2009) *3D Images of Materials Structures – Processing and Analysis*, Wiley-VCH Verlag GmbH, Weinheim.
75. Laudone, G.M., Matthews, G.P., and Gane, P.A.C. (2008) Modelling diffusion from simulated porous structures. *Chem. Eng. Sci.*, **63**, 1987–1996.
76. Thiedmann, R., Manke, I., Lehnert, W., and Schmidt, V. (2010) Structural analysis of pore space of gas diffusion layers in fuel cells by means of geometrical 3D graphs. *MP Mater. Test.*, **52**, 736–743.
77. Hartnig, C., Jörissen, I., Kerres, J., Lehnert, W., and Scholta, J. (2008) in *Materials for Fuel Cells* (ed. M. Gasik), Woodhead Publishing, Cambridge, pp. 101–184.

25
Computational Fluid Dynamic Simulation Using Supercomputer Calculation Capacity

Ralf Peters and Florian Scharf

25.1
Introduction

Supercomputers can be defined as the front line of current processing capacity. They were introduced in the 1960s, having been primarily designed by Seymour Cray for Control Data Corporation. These early machines were simply very fast scalar processors, some 10 times the speed of the fastest machines offered by other companies at the time. In the 1970s, most supercomputers were dedicated to running a vector processor. During the 1980s, machines with a modest number of vector processors working in parallel became the standard. Typical numbers of processors were in the range 4–16. In the later 1980s and the 1990s, attention turned from vector processors to massive parallel processing systems with thousands of "ordinary" central processing units (CPUs), some being off-the-shelf units and others being custom designs. Today, parallel designs are based on off-the-shelf server-class microprocessors. Most modern supercomputers are highly tuned computer clusters that use commodity processors combined with custom interconnects.

Since June 2005, a ranking list has been published by the Universities of Mannheim (Germany) and Tennessee (USA) and the National Energy Research Scientific Computing Center (USA). Within this time period of 5 years – between June 2005 and June 2010 – the maximum processor speed has been enhanced by a factor of 13 and peak performance has increased by a factor of 16. The processor speed is measured in special test procedures such as the Linpack Benchmark [1]. It is given as teraflops per second and not in megahertz as for single CPUs; flops s^{-1} is the abbreviation for floating point operations per second. For example, a performance of 1000 teraflops s^{-1} leads to 10^{15} additions per seconds. An Intel Core i7 processor has a peak performance of about 33 Gflops s^{-1}.

Table 25.1 shows the current TOP500 list of supercomputers. At present, most of the supercomputers in the TOP20 are located in the USA, China, and Germany. All of the large national laboratories in the USA are within the TOP10. It should be noted that current progress leads to great variance in the rankings. The two supercomputers at Forschungszentrum Jülich dropped from 3rd to 5th and from 10th to 14th place between June 2009 and June 2010. Nevertheless, this chapter

Fuel Cell Science and Engineering: Materials, Processes, Systems and Technology, First Edition. Edited by Detlef Stolten and Bernd Emonts.
© 2012 Wiley-VCH Verlag GmbH & Co. KGaA. Published 2012 by Wiley-VCH Verlag GmbH & Co. KGaA.

Table 25.1 TOP500 list (June 2010) [2].

Rank	Site	Manufacturer	Cores	R_{max} (teraflops s^{-1})	R_{peak} (teraflops s^{-1})	Power (kW)
1	Oak Ridge National Laboratory (USA)	Cray	224 162	1759.0	2331.0	6950.6
2	National Supercomputing Center in Shenzhen (NSCS) (China)	Dawning	120 640	1271.0	2984.3	–
3	DOE/NNSA/Los Alamos National Laboratory (LANL) (USA)	IBM	122 400	1042.0	1375.8	2345.5
4	National Institute for Computational Sciences/University of Tennessee (USA)	Cray	98 928	831.7	1028.9	–
5	Forschungszentrum Jülich (FZJ) (Germany)	IBM	294 912	825.0	1002.7	2268
14	Forschungszentrum Jülich (FZJ) (Germany)	Bull	26 304	274.8	308.3	1549

will show that very efficient design aspects of engineering can be calculated on two compute nodes out of the 2208 which were implemented in total on JuRoPA (see Figure 25.1).

An overview of the beginning of supercomputing and a broad range of modeling examples from the early 1990s can be found in a book by Kaufmann and Smarr [3]. Today's supercomputers have been applied to the simulation of limited emissions in the Earth's atmosphere, the structure of fibrous protein molecules, the formation of proteins on Earth under high pressure in hot water [4], computational fluid dynamic (CFD) simulations to enhance the performance for Formula 1 racing [5], the food web complexity in ecological systems and the resulting chaotic population dynamics [6], shape effects on the electronic states of nanocrystals [7], the migration behavior of birds in the USA [8], and astronomical problems [9]. This chapter describes a method aimed at improving chemical reactors for fuel-cell systems using CFD simulations on a supercomputer.

CFD is by far the largest user of high-performance computing (HPC). The main scientific challenge is to increase our understanding of turbulence and its interactions with engineering design aspects. The transfer of momentum, heat, and mass are particularly interesting for aerodynamics, industrial flows, and for combustion and chemical reactor systems. HPC refers to the use of high-speed

Figure 25.1 Supercomputer JuRoPA. Hardware characteristics: 2208 compute nodes with two Intel Xeon X5570 (Nehalem.EP) quad-core processors, 2.93 GHz, and 24 GB memory, in total 17 764 cores total resulting in 207 teraflops peak performance and 183.5 teraflops Linpack performance. JuRoPA (Sun) and HPCFF (Bull) together form the supercomputer ranked 14th in Table 25.1.

processors and related technologies to solve computationally intensive problems. The term *HPC* has often been used in conjunction with its implementation on supercomputers. Since fast computer clusters with a certain number of cores can also be used for HPC, supercomputers are not a prerequisite for HPC.

25.2
High-Performance Computing for Fuel Cells

Fuel cells and fuel-cell systems involve a number of tasks for modeling on HPC machines. The different tasks cover different scales, ranging from nano- to decimeters. Large-scale single-phase flow simulations are state-of-the-art in order to describe the cell or stack behavior [10–12]. One major concern of polymer electrolyte fuel-cell developers is water management in polymer electrolyte fuel cells (PEFCs). Continuum models which describe water saturation in the porous structures are often used in combination with commercial CFD tools [13]. More detailed models describing the two-phase flow behavior inside the porous structures of PEFCs require different methods such as lattice Boltzmann, Monte Carlo, and molecular dynamic methods [14, 15]. These detailed models demand a high capacity from scientific computing.

The chemical reactions on the surface of catalytically active materials in fuel-cell processors and fuel cells are complex [16, 17]. Theoretical models can be developed on a nano- to micrometer scale – on the one hand covering surface reactions on active centers and the interaction of intermediate species with the support material, and on the other mass transport phenomena.

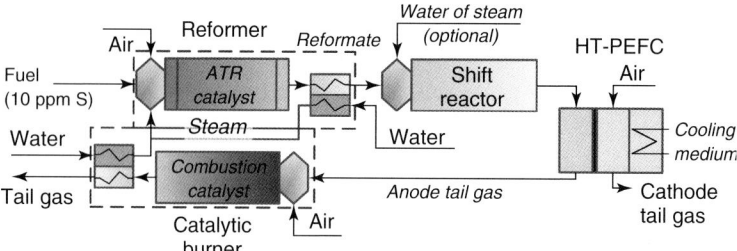

Figure 25.2 Simplified flow sheet for an HT-PEFC system with autothermal reforming, water gas shift reaction, HT-PEFC, and catalytic combustion.

This chapter focuses on the engineering aspects of fuel processor and stack design. Modeling activities at Forschungszentrum Jülich began with CFD calculations for fuel processors using the commercial ANSYS/FLUENT software. The first models were developed in 2003 for the mixture formation in an autothermal reformer (ATR) [18]. More complex modeling items, such as the consideration of an evaporating fuel spray and/or large eddies, require HPC. It is also very important to note that a validity proof of the calculation results must be performed, for example, by visualization methods for CFD modeling. Finally, hardware is designed on the basis of the modeling results and an experimental proof of the reformer shows the high precision of the proposed method [19, 20]. Therefore, the development of fuel processor components by HPC forms the central example in this chapter.

A high-temperature polymer electrolyte fuel cell (HT-PEFC) is one possible type of fuel cell that is considered. Figure 25.2 shows a simplified flow sheet of an HT-PEFC system in combination with a fuel processor. Fuel is mixed with steam and air in a mixing chamber in front of an ATR. Steam is generated by both the ATR cooling the product gas to 673 K and by the tail gas of the catalytic burner. Subsequently, the reformer product gas (reformate) is reduced to 1% carbon monoxide concentration within a water gas shift (WGS) reactor. The WGS reaction is limited by chemical equilibrium in the high-temperature regime and by slow kinetics at low temperatures. In order to optimize the reactor performance and to overcome the need for a further heat exchanger, an intermediate injection of water is applied. The catalytic burner must convert the residual hydrogen and carbon monoxide from the fuel cell residual gases into steam and carbon dioxide at low emissions. Several two-phase liquid–gas flows have been identified in the system, leading to spray modeling of the evaporating species, that is, fuel in the ATR mixing chamber, water in the mixing zone of the WGS reactor, and water in the steam generator of the catalytic burner.

There are substantial differences between CFD modeling of the chemical reactors for fuel processing and that of the fuel cells. Low velocities over the bipolar plates of a fuel cell lead to laminar flow in the channels. As shown in Figure 25.3a, cross flow can be neglected for laminar flow. If fluid mixing between neighboring volume elements is demanded for laminar flow, it must be realized by mixing devices or in

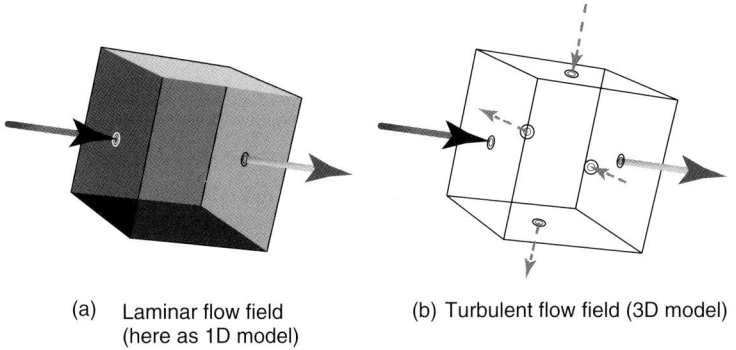

(a) Laminar flow field (here as 1D model)

(b) Turbulent flow field (3D model)

Figure 25.3 Finite fluid volume (FV) – sketched here as three-dimensional cubes – showing the main flow directions: (a) for laminar flow and (b) for turbulent flow field.

the redirection zones of the bipolar plate. Compact chemical reactors are indicated by short residence times in a fixed fluid volume (FV) leading to a turbulent flow regime with a strong effect of eddies. For each of these finite FVs, a mass and momentum balance must be programmed and compiled with all other elements, as described in the next section.

Within a fuel processor, further physical and chemical processes can occur. Figure 25.4 shows an inner and an outer infinite fluid element. The chemical reaction is modeled as a volumetrically considered source/sink term for participating species. It must be noted that the strength of the volumetric source/sink term is derived by converting the area-specific reaction rate. For fuel-cell modeling, the chemical reaction is directly modeled as a surface reaction with a 2D source/sink term at one border area. In addition to the main convective mechanism, the heat transfer in a fuel processor accounts for radiation between different elements which might not necessarily be the nearest neighboring elements. The radiation effect can be neglected at operating temperatures below 473 K, such as those in HT-PEFC

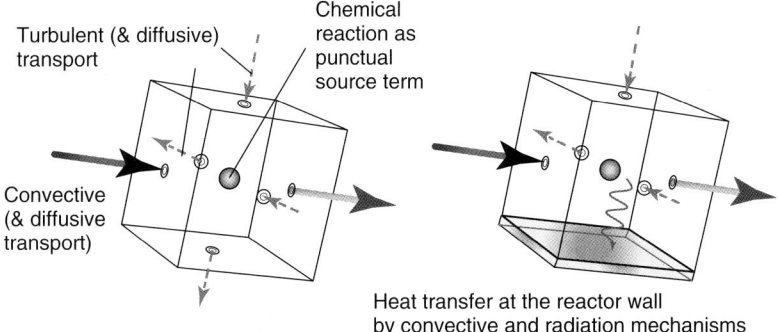

Figure 25.4 Finite fluid volume (FV) for an inner element of a fuel processor including a chemical reaction and an outer element with heat transfer.

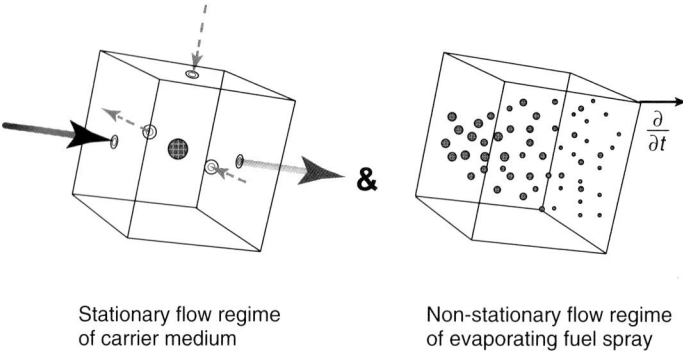

Figure 25.5 Finite fluid volume (FV) for an inner element of spray modeling, with an overlap between a continuous steady-state flow regime of the carrier medium and a non-stationary modeling of evaporating fluid elements by moving a second finite fluid volume.

and PEFC stacks. In such models, an energy balance must be considered as a third balance equation for each element.

Figure 25.5 shows a further complex item of fuel processor modeling. Liquids are injected as spray into a gas flow while the small droplets are evaporated by the energy content of the surrounding gases, that is, steam and air. The droplets are modeled in a moving finite FV which is superimposed on the steady-state gaseous flow field. Modeling of chemical reactions with hydrocarbons is currently limited to some prereactions. Diesel fuel consists of thousands of components and, in combination with oxygen and steam, it offers a huge number of imaginable intermediate species, such as aldehydes, cycloalkanes, branched alkanes, alkylated aromatics, and so on.

A fuel cell stack must be divided into a large number of finite volumes. Figure 25.6 shows a grid point mesh with 830 000 finite fluid elements for a two-cell PEFC short stack used for preliminary studies (Spiller and de Haart, personal communication). One decisive difference between modeling of a chemical reactor and modeling of

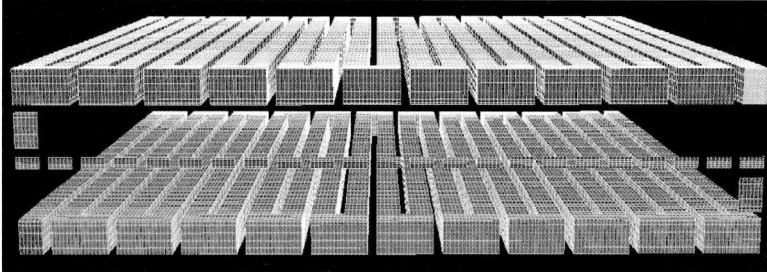

Figure 25.6 Grid point mesh with 830 000 finite fluid elements for a two-cell PEFC short stack for preliminary studies (Spiller and de Haart, personal communication).

a stack is the connection between the grid points. Chemical reactors used in fuel processing have a tubular design in most cases. Fuel cells have a plate-type design with a meander-type configuration of flow channels on the anode and cathode sides. For electrochemical models, source terms and physical properties are mostly implemented in user-defined functions (UDFs). The single fuel-cell assemblies are coupled with each other by these UDFs, leading to a complex network of balance equations, whereby the electrochemical effect is directed orthogonally to the flow field.

A number of papers have reported on CFD modeling with the commercial ANSYS/FLUENT software [13]. Several terms must be defined in conjunction with massive parallel processing prior to discussing different HPC applications. Cortex is a process that provides the user interface and graphics for FLUENT, that is, the graphic user interface (GUI) of ANSYS/FLUENT. ANSYS/FLUENT interacts with the host processor, which organizes and delegates the numerical tasks to different nodes. Each node possesses its own random-access memory (RAM), that is, 24 GB per node and 3 GB per core for the JuRoPA architecture. The host sends a request to the nodes, which store their results directly in parallel on the hard disk.

The first attempts at CFD modeling for fuel cells by Wang and Wang [10] and Liu et al. [12] showed that HPC must be applied to reduce CPU time. Wang and Wang [13] applied a 2D model with 120 000 FVs for a single cell to elucidate the interactions between two-phase transport and phase-change heat transfer. They also published results for a single PEFC cell with 23.5 million FVs [10]. The model was implemented on a LINUX PC cluster with 32 nodes of Intel Pentium IV CPUs (Pentium IV, 2.8 GHz, 1 GB RAM). The solution demanded 600 iterations and 20 h of CPU time. Liu et al. [12] modeled a PEFC stack with six cells with less than 200 000 FVs. The solution method would take 30 h and 2500 iterations on a PC (Pentium IV, 2.4 GHz, 1 GB RAM).

Spiller and de Haart (personal communication) checked the capability of the forerunner of JuRoPA, known as *JUMP*, to model PEFC stacks with between two and 32 cells. A five-cell stack demands about 9.3 million FVs and 0.61 GB RAM. A 55-cell PEFC stack, such as that developed for 5 kW$_e$ applications, would require a grid with 99 million FVs. In order to check the influence of the electrochemical modeling on CPU time, pure CFD simulations without UDFs were first computed. By increasing the number of cores per FV from one to eight CPUs per million FVs, the performance was increased by a factor of 7.5. It is remarkable that for the case of four and 32 cells, this factor drops to a value of about 6. Implementing UDFs leads to a lower performance than pure CFD modeling in terms of time demand per iteration for between two and 16 cells. A 16-cell design with UDFs requires about 200 s per iteration at 1 million FVs per core, that is, 15.8 million FVs on 16 cores, and 50 s for 128 cores. The performance gain only amounts to between 4 and 5, while the number of cores increases by a factor of 8. Unfortunately, for a 32-cell stack with about 31.7 million FVs, it makes no difference if the calculations are run on 64 or 128 cores. The time demand per iteration remains at nearly 250 s. It must be noted that the application of UDFs for massive parallel processing demands intensive adaptation to the computer architecture. Instead of using only one UDF, different UDFs for each element, namely for cortex, host, and nodes, must be implemented to achieve effective parallelization. Despite this measure, a rather high rate of communication between cortex, host, and nodes is demanded, which leads to an increase in calculation time per iteration. It must also be considered that FVs, which exchange UDF variables, should be calculated on the same core. This recommendation is easier to realize for FVs that are direct neighbors.

In general, the partitioning of FVs to the cores should be performed in relation to iteration times and not to grid size. Otherwise, processes on the host have to wait on the slowest participant of the parallelized network. It could be shown that a linear effect on computational performance is possible, even for CFD models with a grid size of 111 million FVs, that is, by increasing the number of cores from 64 to 256, the rating is also increased by a factor of 8 [21]. Nevertheless, an increasing number of cores leads to a certain performance loss, indicated by a decreasing slope of the performance chart. When 256 cores were applied, a rating of 92% could be gained for the biggest model. If the number of cores is increased further, the slope of the performance curve becomes degressive.

Table 25.2 shows the typical performance data of different ANSYS/FLUENT models considered at Forschungszentrum Jülich. Porš [18] optimized the mixing chamber of an ATR of the eighth generation by a model with nearly 2 million FVs in 2006. As shown in Figure 25.5, spray modeling comprises a stationary gas flow and a non-stationary flow of liquid droplets. During the numerical processing of

Table 25.2 Performance data of different ANSYS/FLUENT models, namely for a 10-cell PEFC stack in relation to different reformer tasks (Source: http://www.ansys.com/Support/Platform+Support/Benchmarks+Overview/Archives/ANSYS+Fluent/12.0/).

	Mixing chamber ATR 8, 5 kW$_e$	Mixing chamber ATR 10, 50 kW$_e$	ATR with heat exchange and water evaporation in the discrete phase	Ten-cell PEFC stack
Number of FVs	1 805 915	2 663 202	10 000 000	~18 000 000
Number of CPUs	16	16	16	128
Iteration step for a single solution	250–500	500–1000	7500–10 000	300–500
Size of case file	36 098 kB	55 350 kB	n.a.	1 GB
Memory demand for data file	258 822 kB	424 882 kB	n.a.	1.3 TB (130 GB)
CPU processing time (h)	12.5	40	400	850
RAM demand during processing	–	4 GB	–	–

the injection models for small liquid droplets, the discrete phase model (DPM) is superimposed every tenth iteration step in relation to the continuous gas phase. A single solution for this example demands 250–500 iteration steps and currently needs about 12.5 h of CPU time. Larger grids for a scale-up from 5 to 50 kW$_e$ demand 2.7 million FVs and ~40 h of CPU time. A 10-cell PEFC stack differs from ATR modeling by a large number of FVs and a lower number of iterations steps. Finally, a CPU time of 850 h is expected (Spiller and de Haart, personal communication). The modeling of large 64-cell PEFC stacks is limited by the RAM size (Spiller and de Haart, personal communication). As shown in Table 25.2, the CFD modeling of an ATR 10 mixing chamber (column 2) demands about 4 GB of RAM. In general, ANSYS CFX solvers provide high memory efficiency. A mesh consisting of 1 million tetrahedral elements can be run on 400 MB of RAM [22]. The implementation of UDFs and the analysis of dynamic processes (see Figure 25.5) greatly increase the required RAM size.

Based on today's supercomputer performance, the analysis of an ATR-10 design demands only 2.5 h using 16 CPUs in parallel. However, even more time is demanded by pre- and post-processing, that is, design, grid implementation, model set-up, and evaluation of results. In 2003, when the CFD modeling activities began, an Intel Pentium 4 (3.2 GHz) offered a peak performance of 3.1 Gflops and workstation gains of about 6.4 Gflops [23]. A CFD simulation back then on a normal workstation would have required ~128 Gflops (peak performance) on 16 CPUs and would have lasted 2 days. Therefore, the use of a supercomputer was essential. In the future, CFD calculations could be performed on PCs with a certain RAM size. However, more detailed reactor systems and larger stacks should be designed with the aid of CFD modeling on a supercomputer.

In addition to the aspect of time demand, cost considerations for the design, engineering, and construction of hardware components speak in favor of HPC. A CFD design with 10 different runs for a new reformer type requires 25 h of CPU time and leads to costs lower than €50. This value is low in relation to the manufacturing costs for the corresponding hardware component, which costs €30 000 per unit [24]. The high amount of energy required at a set site to operate a supercomputer must also be taken into account (see Table 25.1). Therefore, new supercomputers with much lower energy consumption are currently being developed [25].

25.3
HPC-Based CFD Modeling for Fuel-Cell Systems

In the previous sections, the boundary conditions of HPC-based CFD were presented. The modeling requirements and HPC specifications depend on the flow system being analyzed. In this section, the numerical basis for flow modeling is explained along with physical models for the representation of complex flow phenomena. In conclusion, an ATR will be used as an example to illustrate the application of HPC calculation capacity in CFD. This flow system is characterized by strong flow gradients, complex physical processes, and interactions, and also by its geometric parameters.

25.3.1
Principles of Computational Fluid Dynamics

Mathematical modeling for real technical flow systems is based on the balance of a globally conserved quantity $\vec{\phi}$ [26]. In order to visualize fully the spatial and temporal processes of non-stationary inhomogeneous flow phenomena, the conserved balances are written in differential form as shown in Figure 25.7:

$$\underbrace{\frac{\partial\left(\rho\vec{\phi}\right)}{\partial t}}_{\text{Accumulation}} + \underbrace{\left(\vec{\nabla}\cdot\rho\vec{\phi}\vec{v}\right)}_{\substack{\text{Convective}\\\text{transport}}} - \underbrace{\vec{\nabla}\left(\Gamma_\phi\vec{\nabla}\vec{\phi}\right)}_{\substack{\text{Diffusive}\\\text{transport}}} = \underbrace{S_\phi}_{\text{Sources}} \qquad (25.1)$$

where the accumulation describes the time derivative of the total amount of the conserved quantity stored within the balance element $\vec{\phi}$. The transport of the conserved quantity is composed of a convective term and a diffusive term together with the generalized diffusion coefficient Γ_ϕ. The general source term S_ϕ accounts for both the production density inside the balance element σ_ϕ^p and the applied density σ_ϕ^f, which describes the influence of external fields (e.g., gravitation). The basic system of equations in CFD comprises conservation equations for mass and momentum. The mass balance according to Eq. (25.1) provides us with the continuity equation for a continuous phase:

$$\underbrace{\frac{\partial\left(\rho\right)}{\partial t}}_{\text{Accumulation}} + \underbrace{\left(\vec{\nabla}\cdot\rho\vec{v}\right)}_{\substack{\text{Convective}\\\text{transport}}} = \underbrace{S_m}_{\text{Sources}} \qquad (25.2)$$

The source term S_m here represents a general mass source or sink and therefore also allows for phase transition phenomena. The momentum conservation equation is derived from Newton's second law as a fixed control volume, whereby the

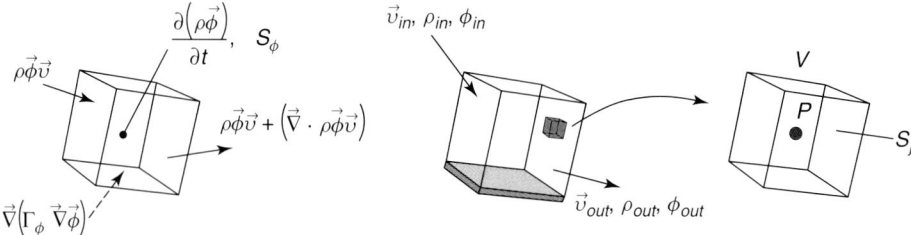

Figure 25.7 General balance and discretization of a flow system and the flow boundary conditions using a finite volume method.

change in the momentum of a mass is equal to the force working on this mass [27]. The transport term is composed of convective momentum transport and dissipative momentum transport together with the viscous stress tensor $\bar{\bar{\tau}}$ and the thermodynamic pressure p. The terms of dynamic and shear viscosity contained in the stress tensor are treated in accordance with Newton's law of viscosity [28]. If we take gravitation and external field forces $\vec{F}$ into account, we obtain the Navier–Stokes equations, the fundamental equations in fluid dynamics:

$$\underbrace{\frac{\partial (\rho \vec{v})}{\partial t}}_{\text{Accumulation}} + \underbrace{\left(\vec{\nabla} \cdot \rho \vec{v} \vec{v}\right)}_{\substack{\text{Convective} \\ \text{transport}}} = \underbrace{-\vec{\nabla}p + \vec{\nabla} \cdot (\bar{\bar{\tau}})}_{\substack{\text{Dissipative momentum} \\ \text{transport}}} + \underbrace{\rho \vec{g}}_{\text{Gravitation}} + \underbrace{\vec{F}}_{\text{Sources}} \qquad (25.3)$$

The Navier–Stokes equations form a system of nonlinear coupled differential equations. To date, the existence and regularity of global solutions for the three-dimensional incompressible Navier–Stokes equations have not been proven [29].

In CFD, the Navier–Stokes equations are solved by numerical methods. The mathematical problem is transformed into a system of algebraic equations for discrete points in space and time [27, 30, 31]. The CFD package FLUENT is based on the discretization approach of the finite volume method [32]. In addition to the general balance element, Figure 25.7 also shows the global representation of a cubic flow system including the flow boundary conditions (cf., Figures 25.3–25.5). In the finite volume method, the flow system is broken down into a finite amount of continuous control volume with the volumes V and the sides S (see Figure 25.6). The physical variables are located at the midpoint P of the control volumes. The description of a flow system using dispersed spatially fixed control volumes corresponds to the field-based Eulerian flow description. The differential conservation equations can be formulated for each individual control volume. For the numerical solution of the system of equations, the control volume conservation equations are integrated, whereby the convective and diffusive transport terms are converted into surface integrals in accordance with Gauss's divergence theorem:

$$\underbrace{\frac{\partial}{\partial t} \int_{V(t)} \rho \phi \, dV}_{\text{Accumulation}} + \underbrace{\int_{S_c} \left(\rho \phi \vec{v}\right) d\vec{A}}_{\substack{\text{Convective} \\ \text{transport}}} - \underbrace{\int_S \left(\Gamma_\phi \vec{\nabla} \phi\right) d\vec{A}}_{\substack{\text{Diffusive} \\ \text{transport}}} = \underbrace{\int_V \left(S_\phi\right) dV}_{\text{Source}} \qquad (25.4)$$

The surfaces S_c stand for the sides of the balance element through which a convective flow can pass. The integral terms of mass transfer can be approximated by means of Cauchy's mean value theorem in a first approximation using calculated mean values of the conserved quantities on the control volume surfaces. The mean surface values are usually determined by the central difference method or the upwind method from the function values in the control volume midpoint [32–34]:

$$\int_S \left(\rho v_i \vec{\phi} - \Gamma_\phi \vec{\nabla}\phi\right) d\vec{A} = \sum_j \left[\rho \vec{v}\vec{\phi}_{jm} - \Gamma_\phi \left(\vec{\nabla}\phi\right)_{jm}\right]\delta \vec{A} \approx \sum_j \rho \vec{v}\vec{\phi}_j\, \delta \vec{A}$$
$$- \sum_j \Gamma_\phi \left(\vec{\nabla}\phi\right)_j \delta \vec{A} \qquad (25.5)$$

Analogously, the volume integrals are approximated using the midpoint rule, the trapezoidal rule, Simpson's rule, or other integration formulae. Time discretization is then performed using implicit or explicit time stepping [26]. A discretization of the flow boundary conditions is performed analogously on the boundary elements of the flow system. In systems of equations describing flow, momentum equations serve as conditional equations for the velocity components. The continuity equation, in contrast, is not assigned to any flow variables. The system of equations therefore does not include an independent conditional equation for pressure. In order to complete the system of equations, pressure is determined using an artificial compressibility or a pressure-correction method. FLUENT provides the SIMPLE, SIMPLEC, and PISO techniques for pressure correction [32]. The algebraic system of equations thus formed can be solved iteratively using numerical methods. More detailed descriptions of the different discretization and pressure-correction methods can be found elsewhere [27, 31, 33, 34].

The discretization quality (time step length, cross-linking) has a decisive influence on the stability and accuracy of the numerical method. Non-stationary systems, in particular, require an adaptive discretization using the gradients in the flow system. Methods and guidelines on grid generation and evaluation can be found elsewhere [31, 35, 36]. The stability of the numerical computational method can be additionally increased via what is known as the *under-relaxation factor* α, which limits the changes in a variable between the individual iteration steps [32]:

$$\phi^{k+1} = \phi^k + \alpha \Delta \phi \qquad (25.6)$$

As shown in Figure 25.8, reducing the under-relaxation factor leads to increased computational cost because a larger number of iteration steps are required to account fully for the relevant limiting factor in the system.

The application of HPC calculation capacity allows the calculation of complex, finely discretized, and geometrically large flow systems. The system of equations is parallelized for this purpose. Parallelization is performed by means of the geometric partitioning of the flow system. Mass, momentum, and turbulence persist across the interfaces as the inner boundary conditions of the interfaces thus created. The partitioning of the flow system has a significant influence on how much computing time is needed for the simulation, particularly in inhomogeneous flow systems, such as the core components of fuel processing. The flow system should be partitioned in such a way that the individual partitions require almost the same amount of computing time and storage space, while simultaneously keeping the need for communication between the partitions to a minimum. The finite volume method is therefore particularly suitable for HPC with a massive multiprocessor architecture, such as the JuRoPA supercomputer installed at Forschungszentrum Jülich [37].

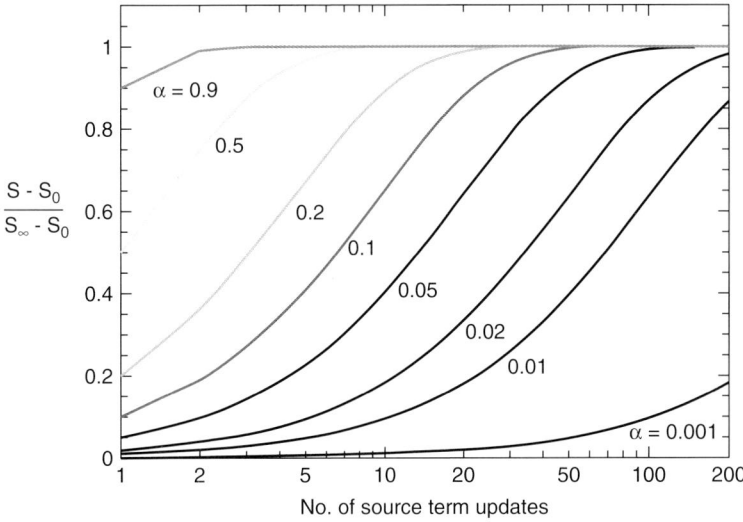

Figure 25.8 The influence of under-relaxation factors on the integration of flow variables and of the resulting iteration steps required for full integration [32].

25.3.2
Physical Model Principles

In order to describe fully a real technical flow problem, the system of equations of the underlying Navier–Stokes equations is expanded by the corresponding physical submodels. In general, an increase in the degree of accuracy of the model leads to an increased need for computing time. The application of HPC calculation capacity here allows a higher degree of accuracy of the submodels. Unless stated otherwise, the physical model equations presented in the following are taken from or are based on the FLUENT model library [32, 36].

25.3.2.1 Turbulence

Turbulent flows have a strong non-stationary, chaotic character, and are therefore one of the greatest challenges in flow simulation. The direct numerical simulation (DNS) of turbulent flow effects represents a complete solution of the non-stationary Navier–Stokes equations [38]. DNS requires a very fine discretization in space and time in order also to resolve the small-scale turbulent flow structures. This leads to a very high computational cost and is possible only for simple flows with low Reynolds numbers. From an engineering point of view, macroscopic flow phenomena are particularly interesting, which means that statistical turbulence models are usually used to describe turbulent flow phenomena. A general modeling approach for this is Reynolds decomposition [26]. This involves separating the flow variables $\psi(t)$ into an averaged term $\overline{\psi}$ and a superimposed statistical perturbation (variance) $\psi'(t)$. The use of Reynolds decomposition for velocity and pressure in

Eq. (25.3) and the subsequent time averaging yield the Navier–Stokes equations for averaged flow variables [Reynolds-averaged Navier–Stokes (RANS) equations]:

$$\frac{\partial\left(\rho\vec{\bar{v}}\right)}{\partial t} + \left(\vec{\nabla}\cdot\rho\vec{\bar{v}}\,\vec{\bar{v}}\right) - \bar{\bar{R}} = -\vec{\nabla}\bar{p} + \vec{\nabla}\cdot(\bar{\bar{\tau}}) + \rho\vec{g} + \vec{\bar{F}} \quad (25.7)$$

These contain what is known as Reynolds stress $\bar{\bar{R}}$ as an additional term describing turbulence. To complete the system of equations, Reynolds stress is usually determined by an heuristic turbulence model. To do so, FLUENT provides one-equation models (Sparlat–Allmaras), two-equation models (k–ε model, k–ω model), and the closure approaches k–kl–ω transition model, SST transition model, v^2–f model, and the Reynolds stress model (RSM) [32, 36]. A two-equation model widely used in practice is the standard k–ε model, which is a turbulence viscosity model and represents a good compromise between accuracy and computational cost. The flow turbulence is described here by the turbulent kinetic energy k and its degree of dissipation ε. In accordance with Eq. (25.1), these two variables are determined using two additional conservation equations.

Due to the no-slip conditions near the walls, the viscous flows form interfaces, which significantly influence the main flow. In a flow with high Reynolds numbers and finite viscosity, the interface as a function of the dimensionless wall distance y^+ can be separated into the purely viscous sublayer, the transition layer, and the overlap layer [26, 38]. The k–ε model does not contain a geometry-dependent term, which means that the flow effects in the interfaces must be taken into account in more advanced models. Low Reynolds number models expand the validity of the turbulence model by introducing an additional geometry-dependent damping function up to the wall. The resolution of the wall effect requires very fine meshing in the vicinity of the interface. Standard wall functions calculate the relationship between wall distance and velocity using a logarithmic wall law. The fineness of the mesh is selected so that the first near-wall cell includes the interface. The use of the standard wall function means that fine meshing in the interface required for the low Reynolds number model is no longer necessary and this in turn significantly reduces the computing time.

A compromise between the exact resolution of turbulent flow phenomena using DNS and the easy-to-calculate turbulence viscosity models is embodied by large eddy simulations (LESs). By means of a time and space low-pass filter, a non-stationary and spatially resolved calculation of the large-scale turbulent flow phenomena is performed, while small-scale turbulent flow phenomena are modeled using turbulence models. This allows LES to provide a more exact description of the turbulent flow phenomena than the purely statistical methods, albeit with a higher computational cost. Detached eddy simulations (DESs) represent a compromise between LES and the statistical RANS models. The turbulence is represented as a function of the grid granularity and the wall distance in large eddy models (free flow) or in RANS models (near-wall). A comprehensive overview of the different turbulence models can be found elsewhere [26, 28].

25.3.2.2 Heat Transfer

In order to simulate temperature-dependent flow systems, in addition to the continuity equation and the Navier–Stokes equations, the conservation of energy is introduced as an additional descriptive fundamental equation of the flow problem. The balance of total energy according to Eq. (25.1) comprises the three heat transfer mechanisms of conduction, convection, and radiation:

$$\underbrace{\frac{\partial (\rho e)}{\partial t} + \vec{\nabla} \cdot [\vec{v}(\rho e + p)]}_{\text{Convection}} = \vec{\nabla} \cdot \left[\underbrace{k_{\text{eff}} \vec{\nabla} T}_{\text{Conduction}} - \underbrace{\sum_i \vec{j}_i h_i}_{\text{Diffusion}} + \underbrace{(\overline{\overline{\tau}}_{\text{eff}} \cdot \vec{v})}_{\text{Dissipation}} \right]$$
$$+ \underbrace{\sum_i \rho_i \sigma_i + S_h}_{\text{Field effects}} \qquad (25.8)$$

The laws of convection and conduction are contained directly in Eq. (25.8), whereas radiation requires an additional radiation model and is taken into account using the energy source density σ_i or an additional source term. The terms k_{eff} and $\overline{\overline{\tau}}_{\text{eff}}$ describe the effective conductivity and effective stress tensor, respectively, and comprise a laminar and a turbulent part. The other production and applied densities are summarized in a simplified form in the energy source term S_h in Eq. (25.8).

To describe the heat transfer by radiation, FLUENT provides the following radiation models: Rosseland model, P1 model, discrete transfer radiation model (DTRM), surface to surface model, and discrete ordinates (DOs) model [32, 36]. These are explained in more detail in [38]. Due to its suitability for the entire range of optical density and the justifiable cost of computation, the DOs model is used to model the reactors used to process fuel. In contrast to other radiation models, the DOs model does not track individual heat rays but solves the radiative transfer equation (RTE) in the discrete directions.

25.3.2.3 Mixtures and Reactions

For the description of mixtures of substances, FLUENT provides the species model. This model calculates the convection, diffusion, and reaction equations for each component in a mixture. This allows the volumetric reactions, surface reactions, and reactions at phase boundaries to be modeled. For the analysis of one-phase mixtures, the conservation of mass equation for a component i can be formulated by accounting for the local mass fraction Y_i:

$$\underbrace{\frac{\partial (Y_i \rho)}{\partial t}}_{\text{Accumulation}} + \underbrace{(\vec{\nabla} \cdot Y_i \rho \vec{v})}_{\text{Convection}} + \underbrace{(\vec{\nabla} \cdot \vec{j}_i)}_{\text{Diffusion}} = \underbrace{r_i^V}_{\text{Reactions}} + \underbrace{S_i^m}_{\text{General source term}} \qquad (25.9)$$

where $\vec{j}_i$ is the diffusion mass flow and r_i^V is the reaction rate per unit mass of component i. S_i^m describes other mass sources of the component. The diffusion term is generally based on Fick's laws of diffusion.

The net source term per unit mass of component i resulting from chemical reactions r_i can be calculated from the molar mass of the component and the sum of the rate of change of the amount of substances in the single reactions $R_{i,r}$:

$$R_i = M_{w,i} \sum_{r=1}^{N_R} \hat{R}_{i,r} \tag{25.10}$$

The general form of the reactions r is

$$\sum_i^N v'_{i,r} M_i \underset{k_{b,r}}{\overset{k_{f,r}}{\rightleftarrows}} \sum_i^N v''_{i,r} M_i \tag{25.11}$$

where $v'_{i,r}$ and $v''_{i,r}$ are the stoichiometric coefficients of the educts and products and $k_{f,r}$ and $k_{b,r}$ are the reaction rate constants of the direct reaction and the back reaction. M_i stands for component i. The rate constant of the direct reaction $k_{f,r}$ is calculated using the Arrhenius equation. For an equilibrium reaction according to Eq. (25.11), the rate of change of the amount of substances can be calculated using

$$\hat{R}_{i,r} = \Gamma \left(v''_{i,r} - v'_{i,r}\right) \left(k_{f,r} \prod_{j=1}^N [C_{j,r}]^{\left(\eta'_{j,r}\right)} - k_{b,r} \prod_{j=1}^N [C_{j,r}]^{\left(v''_{j,r}\right)} \right) \tag{25.12}$$

where $C_{i,j}$ is the molar concentration of the components j in the reaction and $\eta_{j,r}$ is the reaction order exponent of the direct reaction and the back reaction. Γ stands for the influence of the third bodies. It should be ensured that the reaction order exponents of the back reaction here agree with the stoichiometric coefficients of components j in the product. For a pure direct reaction, the reaction rate of the back reaction is $k_{b,r} = 0$. The reaction order exponent of the direct reaction is given by the sum of the exponents of the educts and products. The rate constant of the back reaction is given by the quotient of the rate constant of the direct reaction and the equilibrium constant K_r. In an ideal system (fugacity coefficients $F = 1$), K_r can be calculated from the thermodynamic equilibrium constant K. This is calculated from the change in the Gibbs free reaction enthalpy ΔG_r^0, which is determined from the known standard enthalpy of formation and the standard entropy of formation of the reaction educts and products. The general approach for determining the reaction kinetics discussed here is used to model volumetric reactions. In fuel processing reactors, however, heterogeneously catalyzed surface reactions occur. To model them, the volumetric reaction source term in the conservation equations is replaced by a surface-related source term. The general form of a surface reaction r accounts for the gas-phase components, the components that are adsorbed at the surface (site) and the surface material components (bulk):

$$\sum_{i=1}^{N_g} g'_{i,r} G_i + \sum_{i=1}^{N_b} b'_{i,r} B_i + \sum_{i=1}^{N_s} s'_{i,r} S_i \overset{K_r}{\rightleftarrows} \sum_{i=1}^{N_g} g''_{i,r} G_i + \sum_{i=1}^{N_b} b''_{i,r} B_i + \sum_{i=1}^{N_s} s''_{i,r} S_i \tag{25.13}$$

25.3.2.4 Multiphase Flows

In a multiphase flow, the conservation equations can be formulated separately for each individual phase. In doing so, the volume fraction of the phase α_l is taken into account in the balance for the continuity equation. Correspondingly, the modified momentum balance is

$$\frac{\partial (\alpha_i \rho_i \vec{v}_i)}{\partial t} + \left(\vec{\nabla} \cdot \alpha_i \rho_i \vec{v}_i \vec{v}_i\right) = -\vec{\nabla} p + \vec{\nabla} \cdot (\overline{\overline{\tau_i}}) + \rho_i \vec{g} + \vec{F} \tag{25.14}$$

where ρ_l is the density and $\vec{v}_l$ is the velocity of the phase i. In modeling, a distinction is made between dispersed and separated multiphase systems.

FLUENT provides three different models for the description of dispersed multiphase systems. The Euler–Euler approach models each phase as a continuum. For each of the continuous interpenetrating phases, the Navier–Stokes equations modified by the phase fraction α_l are solved in the Eulerian flow description. Phase interactions such as buoyancy, resistance, and phase transition are incorporated into the Navier–Stokes equations as additional source terms S_l^k. The size and shape of the dispersed particles (particles, droplets, and bubbles) are input parameters for the model equations and must be determined by means of assumptions, measurements, or calculations. The mixture model represents a simplification of the Euler model, in which the conservation equations are calculated as an average only once for the entire mixture. The flow behavior of the individual phases is determined by introducing differential velocities. The mixture model is therefore an algebraic slip model. The mixture model is based on the assumption of a low phase fraction of the dispersed phase and on a low inertia of the particles. Both the Euler–Euler approach and the algebraic slip models fail to resolve the phase interface of the dispersed phase. Dispersed particle flows can be described using the Euler–Lagrange model by means of resolved phase interfaces. FLUENT provides the DPM for this purpose. The flow behavior of the continuous phase is still determined by means of the Eulerian flow description, which is based on the field concept. The second discrete phase, consisting of particles, droplets, or bubbles, is described by means of the Lagrangian flow description in an associated coordinate system. The differential equations for space, force, and momentum are solved here for representative particles along their trajectories. For every time step Δt, the interactions and equilibrium between the discrete and continuous phases, mass and heat transfer, and the trajectory of the particles resulting from the acting forces are calculated in the corresponding finite elements. The momentum and heat exchange interaction terms are incorporated as additional source terms in the corresponding conservation equations for the continuous phase.

FLUENT provides the volume of fluid model (VOF) for the description of separated multiphase flows. The VOF model is based on the resolution of the phase interface in a fixed Eulerian mesh. The conservation equations here are not solved separately for the individual phases but rather for the entire calculation domain with material properties averaged across the phases. For this purpose, an additional conservation equation is introduced for the volume fraction f in the continuous phase. A cell contains either the dispersed phase only ($f = 0$), the continuous phase only ($f = 1$), or the phase interface ($0 < f < 1$). In order to avoid blurring the interface, the mesh in the area near the phase boundary must be as fine as possible. The resulting high cost of computation can be reduced by non-stationary modeling using automatic adaptive mesh refinement. An overview of the various multiphase models can be found elsewhere [38, 39].

The steam generators of the fuel processing system contain wet steam with a minimum steam fraction of 50%. Owing to the large average interparticle distance ($L/d = 9.5$) and the low loading ($\gamma = 1$), the particle interactions can be neglected in such a two-phase flow. The carrier medium influences the particles through the flow resistance and the turbulence; the particles themselves, however, have no influence on the carrier medium. Three CFD models can be used to describe such a system: DPM, Euler, and mixture. The VOF model can be used within the scope of the CFD modeling of a fuel processing system for the highly resolved modeling of an injector nozzle, including the upstream fluid line and the downstream droplet decay.

25.3.2.5 Porous Media

The fuel processing reactors contain different types of flow internals in order to maximize contact surfaces or increase flow turbulence. These internals, made of metal or ceramic, have either an ordered channel structure or a chaotic character. Fully resolving the small-scale porous structures is extremely time consuming and computationally intensive. These structures are therefore simplified and represented as porous bodies. A porous structure is not modeled in FLUENT as a solid body penetrated by a fluid, but is only taken into account in terms of its flow interactions and a heat balance averaged over the control volume. Taking the porosity γ into account, the Navier–Stokes equations can be written as

$$\frac{\partial (\gamma \rho \vec{v})}{\partial t} + \left(\vec{\nabla} \cdot \gamma \rho \vec{v} \vec{v} \right) = -\gamma \vec{\nabla} p + \vec{\nabla} \cdot (\gamma \overline{\overline{\tau}}) + \gamma \vec{B}_f + S_i \tag{25.15}$$

The additional vector term S_i in the momentum balance describes the influence of the porous body on the flow. In a multiphase flow, analogous to Eq. (25.14), the volume fraction α is additionally introduced. The energy balance in a porous body is then given by

$$\frac{\partial \left[\gamma \rho_f e_f + (1-\gamma) \rho_s e_s \right]}{\partial t} + \nabla \cdot \left[\vec{v} \left(\rho_f e_f + p \right) \right]$$
$$= \vec{\nabla} \cdot \left[k_{\text{eff}} \nabla T - \sum_i \vec{j}_i h_i + \left(\overline{\overline{\tau}} \cdot \vec{v} \right) \right] + S_f^h \tag{25.16}$$

Deviating from Eq. (25.8), the accumulation term does not just contain the total energy of the free flow but also the total energy of the solid porous matrix. The conduction term k_{eff} is also composed of a free-flow term and a porous-matrix term. As shown by Eq. (25.16), the finite volume is not resolved into a flow and a solid body. The flow and conduction effects between two neighboring control volumes are calculated based on averaged values, which leads to a blurring of the temperature gradient. Conduction as transverse conduction through the solid porous matrix is depicted as weaker. The influence of the porous body on the continuous flow is taken into account via the flow resistance. The flow resistance is broken down into the two factors viscous resistance coefficient ($1/\alpha$) and inertial resistance coefficient (C_2). The real flow rate $\vec{v}_{physical}$ in the porous body is derived from the product of the porosity and the velocity in the free cross-section $\vec{v}_{superficial}$. The pressure loss in the porous medium is usually calculated on the basis of $\vec{v}_{superficial}$, as experimentally determined pressure loss coefficients are generally related to the flow in the free cross-section. The pressure loss in the porous medium is calculated in accordance with the Darcy–Forchheimer law:

$$S_i = -\left(\frac{\eta}{\alpha} + \frac{C_2 \rho}{2} |\vec{v}|\right) \vec{v} \qquad (25.17)$$

where the quotient of dynamic viscosity η and permeability α describes the permeability effects and complies with the Darcy approach. This is expanded by the momentum loss term C_2. When modeling fine channel structures as a porous medium, it should be ensured that in a laminar flow there is a corresponding linear correlation between pressure loss and flow rate. The C_2 momentum loss term can be neglected.

25.3.3
CFD Modeling of the Core Components of an HT-PEFC Auxiliary Power Unit

The numerical and physical models presented here provide the basis for the CFD modeling of real flow systems. The selection and parameter setting of the models varies considerably depending on the specific application in question. This results in varying demands on the applied HPC. In the following, the applications of HPC and the corresponding specific model parameters are presented with the example of the design of an ATR in the 50 kW_{el} power class (ATR 10). Also, an overview is given of the possibilities and the limitations and requirements of CFD-based component development. The efficient use of HPC calculation capacity for CFD modeling is also outlined.

The ATR is composed of three functional areas that interact with each other. In the first functional area, the educts are prepared and then converted into a mixture that is homogeneous over the flow cross-section. This mixture enters the monolithic reaction zone, in which the chemical reactions occur. After the reaction zone, the hot product gas is cooled in a heat exchanger. In the heat exchanger, steam is simultaneously made available as the reforming educt. The three functional areas do not form a linear system connected in series; instead, they interact thermally with each other because of the integrated construction. The use of the ATR 10 autothermal reformer as the core component of an HT-PEFC APU in the mobile sector necessitates a compact and self-sufficient design. The development of the ATR 10 is based on successfully tested prototypes of ATRs in the 5 kW_{el} power class [40]. The basic design of the educt supply line and the shape of the mixing chamber were optimized and validated in the 5 kW_{el} class. However, the high power class of 50 kW_{el} requires the incorporation of flow bodies and mixture recirculation in the mixing chamber of the ATR 10 in order to ensure a homogeneous mixture over

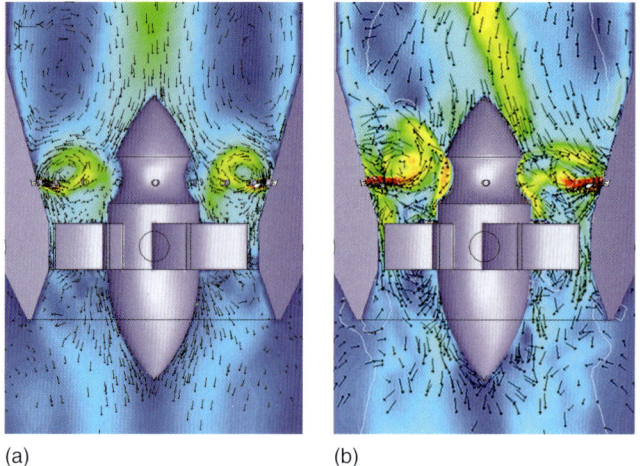

Figure 25.9 Flow pattern and velocity profile in the ATR 10 mixing chamber calculated with the $k-\varepsilon$ turbulence model (a) and the LES turbulence model (b) [40].

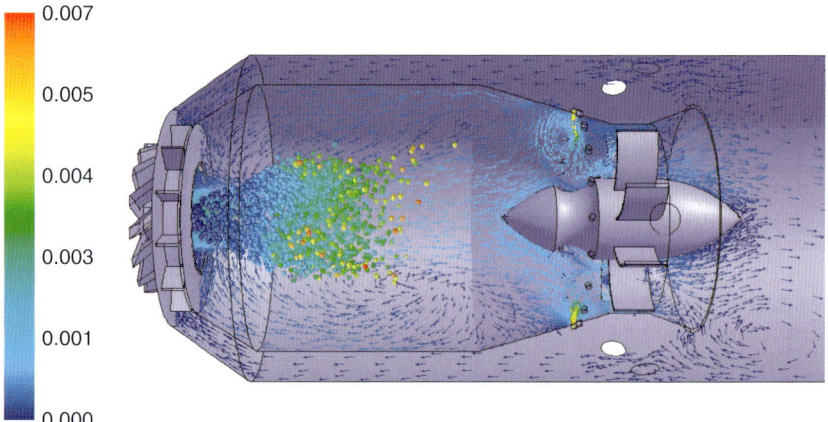

Figure 25.10 Flow pattern and spray pattern in the ATR 10 mixing chamber.

the flow cross-section at the entry into the reaction zone (cf., Figures 25.9–25.13). The geometry of the mixing chamber and the flow internals were optimized for this purpose using CFD simulations. Within the framework of the scale-up of the reformer power class, a parameter variation and geometry optimization were undertaken. The CFD model of the ATR 10 mixing chamber comprises 2.6 million finite volume elements, as shown in Table 25.2.

For a corresponding CFD model, Figure 25.14 shows, as a function of the number of CPU cores used for the calculation, the number of iteration steps that can be performed per day on a dual-core (Intel Core 2 DUO CPU E8400) and quad-core

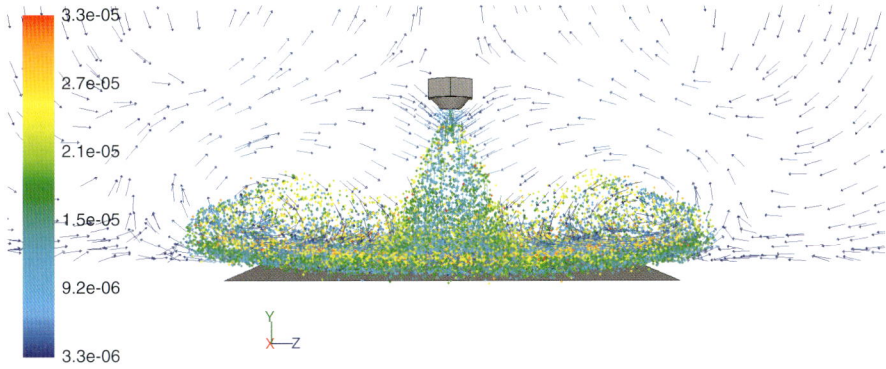

Figure 25.11 Flow pattern and spray pattern in a test set-up for the determination of droplet–wall interactions.

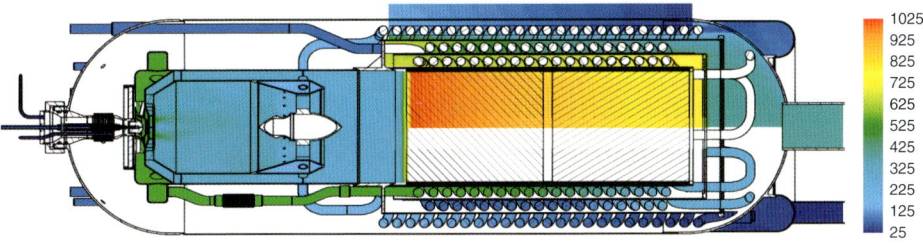

Figure 25.12 Assembled complete model of the ATR 10.

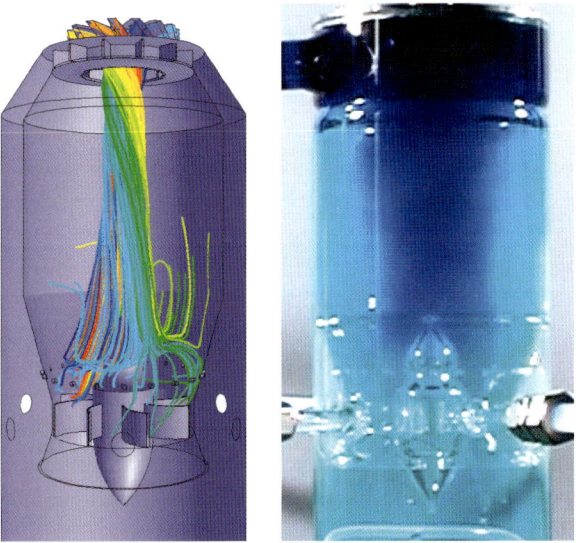

Figure 25.13 Flow lines in the ATR 10 mixing chamber and the visualization experiment conducted with a glass reactor for validation [40].

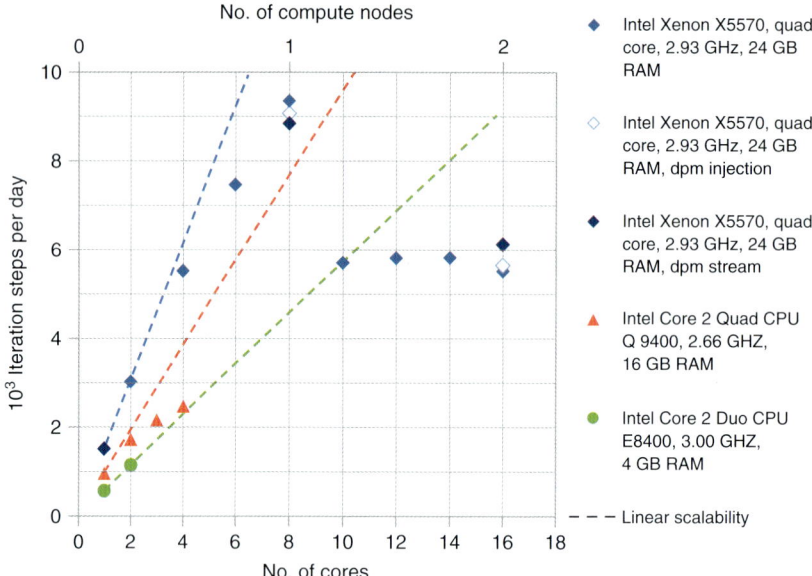

Figure 25.14 Number of iteration steps that can be performed per day on desktop PCs with Intel Core microarchitecture and on HPC with Intel Nehalem microarchitecture as a function of the number of CPU cores used for the calculation of a one-phase flow, a flow with injection (DPM injection), and a dispersed two-phase flow (DPM stream) taking the ATR 10 mixing chamber as an example (cf. Table 25.2).

(Intel Core 2 Quad CPU Q 9400) desktop PC with Intel Core microarchitecture and on the JuRoPA supercomputer (Intel Xeon X5570 [37]) with Intel Nehalem microarchitecture. The up to 1000 iteration steps needed to achieve convergence can be performed on two cores in the dual-core desktop PC in 1 day. Within the same period of time, JuRoPA can perform up to 10 complete simulations on eight cores. The large difference in the calculation speed of the two computer systems when one and/or two CPU cores are used is mainly due to the more efficient chip and computer architecture of the JuRoPA supercomputer. Another important factor is the RAM available for the calculation. Despite the lower clock speed, when the same number of computing cores are used, the quad-core PC has a calculation speed that is approximately 1.5 times faster than the dual-core PC based on the same chip microarchitecture. The 4 GB of RAM available in the dual-core PC already proves insufficient for a simulation on one core, meaning that the data must be swapped to the virtual RAM on the hard drive. In the quad-core PC, 16 GB of RAM are available for use. On this system, the partitioning of the simulation calculation increased the need for RAM from 4.2 GB on one core to 4.8 GB for four cores. The JuRoPA supercomputer, in contrast, has access to 24 GB of RAM per compute node.

For a calculation on more than eight cores, there is a sharp decrease in the number of iteration steps that can be performed per day on the JuRoPA supercomputer down to the level of four cores. Even when the calculation capacity is increased further to up to 16 cores, the number of iteration steps per day does not increase. This behavior can be explained by an inadequate adaptation of the CFD software to the JuRoPA supercomputer. This results in problems in the area of domain decomposition, for example, and in the communication between the individual compute nodes. In the case of the simulation calculations outlined here, parallel processing on two compute nodes proved unsuccessful. Although the model is divided into the required 10–16 partitions, they are calculated on a single node with eight cores. The calculation capacity available for the calculation is therefore not fully exploited. The maximum number of iteration steps that can be performed per day is therefore reached using eight cores.

A reduction in the scalability is generally expected on changing from one to two compute nodes. Part of the communication between the partitions no longer occurs internally on the compute node but rather over the InfiniBand connection. This slow data transmission limits the calculation speed, which in turn reduces the number of possible iteration steps per day. Furthermore, hard partitioning leads to small partitions with an increased need for communication, which in turn means that the time required for data transmission dominates the actual computing time.

From the benchmark tests shown in Figure 25.14, we can see that in contrast to the ideal linear behavior, the JuRoPA supercomputer has a scalability of 99% for two cores and a maximum of 76% for eight cores. The simulation calculations shown here were performed using the CFD software FLUENT 12 with automatic non-optimized partitioning. For a similar simulation calculation with the program package ANSYS 13, the use of eight cores led to a scalability of 78% of the ideal behavior. For the software version ANSYS 13 on a dual-socket quad-core system with Intel Nehalem microarchitecture (Xeon 5500), ANSYS specifies a scalability of 80–95% of the ideal linear behavior [41].

The accuracy of the simulation results for the ATR mixing chamber depends decisively on the representation of the turbulent flow phenomena.

Figure 25.9 shows the simulation results for a section of the ATR 10 mixing chamber for different turbulence models. Figure 25.9a shows the results of the stationary calculation of the flow velocity using the $k-\varepsilon$ turbulence model and the flow profile around the area of the educt supply line and the flow body incorporated for homogenization. Figure 25.9b shows the corresponding results of a non-stationary simulation performed using the LES Smagorinsky–Lilly turbulence model implemented in FLUENT. The numerical time step of the simulation calculation was 2×10^{-4} s. In general, the flows calculated with the different turbulence models show the same character. The flow profiles calculated with the LES turbulence model, however, exhibit stronger non-stationary turbulence. The results of the $k-\varepsilon$ model, on the other hand, lead to stronger symmetries than is actually the case in reality. Non-stationary turbulence structures are not sufficiently taken into account. The solution of an LES model corresponds better to reality than simulation with the $k-\varepsilon$ model. The issue associated with the LES model, however, is that if a model grid is not meshed finely enough, the intensity of the non-stationary turbulence structures can be overestimated, which in turn leads to unrealistic results. The LES model therefore requires a fine discretization in space and time, which significantly increases the cost of computation compared with the non-stationary calculation of the $k-\varepsilon$ model. The use of HPC calculation capacity here allows the more accurate turbulence model to be chosen and thus increases the degree of accuracy of the simulation.

In addition to optimizing the mixture formation, the vaporization of the injected fuel is also important in terms of the design of the mixing chamber. The fuel is modeled as a discrete particle with the DPM multiphase model, as shown in Figure 25.10. The droplet particle type is used to model the droplets. The heating and vaporization of a droplet as a function of the droplet temperature are described by the three laws of heating ($T_p < T_{\text{Vap}}$), vaporization ($T_{\text{Vap}} < T_p < T_b$), and boiling ($T_p > T_b$). Based on the assumed large average interparticle distance and the low loading, the DPM model disregards particle interactions and an influence of the particle flow on the continuous phase. Because local high loading and low average interparticle distances occur in discrete multiphase flows, particle interaction phenomena can be taken into account using additional models [32].

Figure 25.10 shows the results of the stationary modeling of the flow pattern and the spray pattern in the ATR 10 mixing chamber calculated using the $k-\varepsilon$ turbulence model and the DPM multiphase model. The injected kerosene is fully vaporized directly behind the nozzle, which means that no droplet-wall contact occurs. The optimized flow body leads to a stronger turbulence and mixing of the flows in the area around the middle educt supply line. At the same time, the mixture recirculation ensures the formation of a homogeneous mixture. The simulation of kerosene injection leads locally to an increased cost of calculation in the system, which in turn increases the computing time in the relevant partitions. In the modeled example, approximately 2000 discrete particles were continuously present in the system. This leads to a discernable decrease in the number of iteration steps that can be performed per day (DPM injection), as shown in Figure 25.14. This effect is significantly amplified by a more accurate image of the spray pattern using a larger number of modeled discrete particles and by reducing the DPM time step length.

Within the framework of design and development, the flow and the vaporization in the optimized ATR 10 mixing chamber were simulated for different load ranges, nozzle positions, and educt flows using HPC calculation capacity. A total of more than 20 design variations were modeled using HPC within the scope of the parameter study and the geometry optimization of the ATR 10 mixing chamber. The next stage of development involved the simulation of different modes of operation and load ranges with the optimized geometry. Both stationary and non-stationary analyses were conducted.

In addition to the vaporization of the kerosene spray, the DPM model was used to simulate the wet steam flows in the system vaporizers. Here, the discrete particles are dispersed in the system, in contrast to injection. For the individual partitions, this results in a similar computing time, but the flow of the discrete phase causes a significant increase in the need for communication between the partitions. For Figure 25.14, this effect was calculated using a dispersed flow of

2000 DPM particles through the ATR 10 geometry (DPM stream). For an accurate modeling of a vaporizer component, far more than 10 000 DPM particles may be present in the system, which again causes a significant decrease in the number of possible computing operations per day. At the same time, the simulation of water vaporization requires a decrease in the under-relaxation factor, which leads to an increase in the number of necessary iteration steps to up to 10 000. According to Figure 25.14, a vaporizer model in the order of magnitude of 2–3 million finite volume elements could be calculated in around 1 day with eight partitions on JuRoPA. The same calculation would take more than 1 week on two cores on the tested desktop PC. However, when modeling real vaporizers, in contrast to fuel vaporization, droplet–wall interactions play a decisive role in designing the heat exchanger surfaces. In the vaporizers, wet steam in a pipe coil flows over a hot heat exchanger surface. To describe the droplet–wall interactions, FLUENT provides five interaction models for the DPM model [32, 36]:

- In the escape model, the droplets are removed from the system without any further interactions.
- In the trap model, the droplets vaporize abruptly independently of the thermodynamic boundary conditions. The total mass is transferred to the neighboring cell and the energy required for vaporization is simultaneously withdrawn.
- In the reflect model, the droplets are reflected from the surface as a function of the angle of impact and the wall properties.
- The wall-jet model corresponds to a tangential reflection of the droplets and leads to a droplet flow over the surface. No direct droplet–wall contact occurs, meaning that the wall-jet model conforms to a droplet flow during film boiling.
- In the wall-film model, a water film is formed on the surface and includes vaporization on the surface, droplet reflection, and the ejection of secondary droplets. Direct interactions occur here between droplets and the wall. The wall-film model therefore corresponds to a flow for purely convective boiling.

Selecting the correct interaction model has a significant impact on the accuracy of the simulation. The droplet–wall interaction must therefore be determined experimentally. Figure 25.11 shows the CFD simulation of a test stand for the determination of the flow behavior and heat transfer behavior of water droplets on a hot surface. Water droplets are sprayed on to a heating plate with a constant surface temperature. The heating capacity required to keep the constant temperature is measured. The flow behavior of the water droplets is simultaneously examined with a microscope–high-speed camera system. This allows the flow behavior of the droplets and the heat transfer coefficient between droplets and heating surface to be determined as a function of the surface temperature. The modeling of this test stand shown in Figure 25.11 is used to validate and modify the droplet–wall interaction model. The imaged droplet behavior was calculated with the DPM multiphase model and the wall-jet wall function. The experimentally determined models that were validated with such detailed simulations will be used in the course of development to optimize the complete model of the ATR. The spray pattern and dynamic behavior of the injection nozzles can also be investigated in similar laboratory experiments. The size of the droplets and the geometry of the spray pattern influence the performance of the vaporization process and constitute the input parameters of the droplets in the DPM multiphase description. For the analysis of the nozzle behavior, the VOF multiphase model is used to calculate the nozzle in a non-stationary, finely discretized detailed model. This detailed approach and the experiments performed in parallel allow the droplet input parameters to be calculated in the DPM model along with the precise design of the hydraulic pipe in front of the nozzle.

The results of the detailed studies are then fed into the CFD submodels of the ATR 10. The submodels can be assembled to form a complete model, as shown in Figure 25.12. The complete model shown was assembled from the results of the 3D mixing chamber simulations, simple 2D simulations of the heat exchangers, and 3D simulations of the educt supply. In addition to the turbulence and multiphase models discussed, heat transfer, the chemical reaction heat input, and the porous catalyst support also have to be taken into account. The complete model shown here does not take into account any thermal interactions between the individual functional zones. The heat exchange between mixing chamber and the steam supply pipe, for example, is therefore not imaged. In order to investigate the thermal interactions in the fuel processing core components, more detailed simulations will be performed at IEK-3 in order to model the complete geometry of the individual reactors in three dimensions. The interaction effects are therefore particularly interesting for start-up procedures whereby a non-stationary calculation is necessary. Such a geometrically large, complex, non-stationary, and adaptively finely discretized CFD model can only be calculated using HPC calculation capacity.

25.4
CFD-Based Design

In the previous section, the ATR 10 ATR was used as an example to show parts of the CFD-based development process of a technical flow and reaction system. Within the framework of an integrated R&D approach, this process is characterized by the close combination of CFD, experiments and design. By incorporating the strong interactions between these classically separate disciplines, efficient component development can be facilitated. Figure 25.15 shows schematically the iterative process of CFD-based component development at IEK-3. Experiments are not just used to validate the simulations; within the framework of modeling, they also provide an important basis for numerical descriptions. CFD simulations also allow experiments to be developed and optimized in a targeted manner. In general, CFD simulation results must be validated in visualization experiments and prototype tests. The selected models and assumptions must also be carefully evaluated in order to be able to assess the significance of the simulation results. Parameter studies and flow optimization, on the other hand, require close coordination with design in order to achieve a good balance between enhanced efficiency and the cost of design. The close interplay between CFD simulations and design leads to an

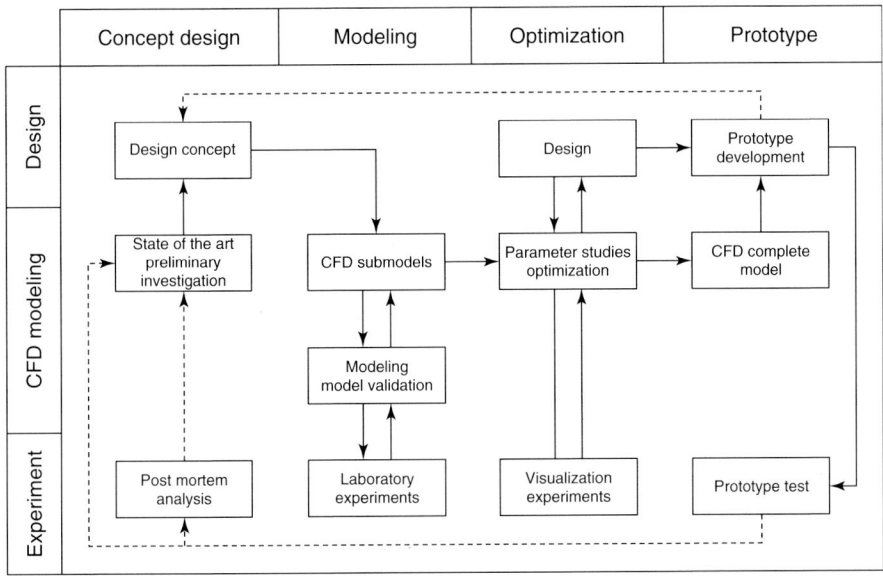

Figure 25.15 Diagram showing CFD-based iterative development process.

application-oriented design and speeds up the designing of the flow components. The results of the prototype test, the empirical values of the design, and the postmortem analyses form the basis for the next stage of development. The same CFD models can be used as a basis for the next iterative stage of development, which significantly reduces the cost of modeling and validation. If further system optimization requires more accurate modeling, however, then the appropriate experiments and detailed simulations need to be conducted.

Within the framework of component development, CFD is used for scientific modeling and model validation in addition to the classical engineering parameter studies and optimization processes. Both approaches are based on the use of HPC calculation capacity. Within the framework of modeling and validation, HPC facilitates a complex representation of the physical phenomena with fine space and time discretization. With the aid of such submodels and appropriate laboratory experiments, models for nozzles, heat transfer phenomena, two-phase flow, and so on can be derived and validated. CFD models thus selected and validated form the basis for the CFD-based design and optimization of flow systems. The classical engineering problem of parameter variation and optimization requires a large number of simulation calculations and therefore leads to an extremely high cost of computation. HPC allows the parallelization of individual simulations, which in turn makes it possible to calculate several simulations simultaneously and thus enables comprehensive parameter studies and flow optimizations to be completed in an acceptable time frame. In the ATR 10 development process, CFD simulations were conducted on up to 16 cores of the JuRoPA supercomputer simultaneously. This meant that when two simulation calculations were simultaneously performed on eight cores each, 20 times as much time could be saved than on a single dual-core desktop PC. HPC is therefore a valuable tool in terms of CFD-based, application-oriented design. Commercial CFD software, however, tends to be optimized for the use of simple cluster computers. HPC with massive multiprocessor architecture, such as the Jülich supercomputer JuRoPA, therefore requires software customization for supercomputer architecture in order to avoid the scaling problems outlined in Figure 25.14. The basis for efficient parallelization on the hardware side is above all fast communication and sufficiently large RAM. Against the background of hardware development, it can be assumed that in the near future, simple desktop PCs will also permit efficient parallelization corresponding to the characteristics shown in Figure 25.14.

With every new stage of development, the iterative development process requires an assessment of the selected models and where necessary an adjustment of the accuracy. A continuous development and optimization process therefore necessarily leads to increases in model accuracy and the cost of computation. In parallel with improving the hardware and the HPC needed for CFD, the requirements of the corresponding systems also increase.

25.5
Conclusion and Outlook

High-performance computers allow numerical tasks to be processed that have a greater need for computing power and a high memory consumption. However, HPC is becoming increasingly important in the scientific sector as a tool for the calculation, modeling, and simulation of complex systems. One of the main fields where HPC is used is CFD modeling. In addition to the scientific sector, HPC is also used as an engineering tool to solve complex design and optimization problems. Today, cluster computers are standard for the calculation of massively parallel CFD simulations. As a result, commercial CFD software packages have been optimized for the use of such systems. Owing to the different system architecture, the use of supercomputers can mean that the theoretically available computing power is not fully utilized. This results in issues in the area of domain decomposition, for example, and in the communication between the individual compute nodes. Therefore, if CFD software has not been optimized for application on supercomputers, it can in fact be more efficient to use simple cluster computers than HPC.

The numerical description of real flow systems is based on the mathematical approaches and physical models discussed here. Taking these as a basis, the use of HPC in CFD was outlined using the example of an ATR for a fuel cell system. The CFD-based development process is centered on close interlinkage of scientific and engineering CFD analyses and on experiments performed for design, modeling, and validation. From this perspective, HPC facilitates the highly discretized analysis of physical phenomena and the processing of complex optimization problems. The use of HPC therefore provides the foundation for a scientifically based and application-oriented development process. Against the background of the further development and proliferation of HPC technology, an increase in the significance of parallelized CFD is expected in both science and technology. CFD software must therefore be customized for HPC. At the same time, desktop PCs will play an ever greater role in the calculation of weakly parallelized problems. It has been shown that in general, the numerical task, CFD software, and the hardware used must be coordinated with each other in order to ensure an efficient calculation. As a separate tool in the development process, HPC-based CFD modeling demands that experiments be closely combined with design. The CFD-based iterative development process presented here could provide the basis for an efficient approach.

Acknowledgments

The CFD modeling of droplet vaporization presented in this paper was conducted within the framework of the ADELHEID project, which was funded by the Federal State of North Rhine-Westphalia (funding program: Efficient energy use, renewable energies, and energy savings; progres.nrw). The computing time on the JuRoPA supercomputer was provided by the Jülich Supercomputer Centre (JSC) within the framework of the VSR project JIEF32. The authors would like to thank W. Lehnert, L.G.J. de Haart, and B. Körfgen for valuable discussions.

References

1. Petitet, A., Whaley, R.C., Dongarra, J., and Cleary A. (2008) HPL – a Portable Implementation of the High-Performance Linpack Benchmark for Distributed-Memory Computers, http://www.netlib.org/benchmark/hpl (last accessed 3 November 2010).
2. TOP500. TOP500 List June 2010, http://www.top500.org/list/2010/06/100 (last accessed 3 November 2010).
3. Kaufmann, W.J. and Smarr, L.J. (1992) *Supercomputing and the Transformation of Science*, Freeman, New York.
4. Forschungszentrum Jülich. Internal Report, Research in Jülich, 2/2008, http://www.fz-juelich.de/portal/EN/Press/Publications/research-in-juelich/_node.html (last accessed 19 January 2012).
5. Hanna, K. (2007) *ANSYS Adv.*, **1** (1), s2–s3.
6. Fussmann, G.F. and Heber, G. (2002) *Ecol. Lett.*, **5**, 394–401.
7. Li, J. and Wang, L.W. (2003) *Nano Lett.*, **3** (10), 1357–1363.
8. Marris, E. (2010) *Nature*, **466** (8), 807.
9. Ocvirk, P., Pichon, C., and Teyssier, R. (2008) *Mon. Not. R. Astron. Soc.*, **390**, 1326–1338.
10. Wang, Y. and Wang, C.-Y. (2006) *J. Power Sources*, **153**, 130–135.
11. Lehnert, W. and Froning, D. (2009) in *Encyclopedia of Electrochemical Power Sources*, vol. 3 (eds. J. Garche, C. Dyer, P. Moseley, Z. Ogumi, D. Rand, and B. Scrosati), Elsevier, Amsterdam, pp. 135–147.
12. Liu, X., Mao, Z., Wang, C., Zhuge, W., and Zhang, Y. (2006) *J. Power Sources*, **160**, 1111–1121.
13. Wang, Y. and Wang, C.-Y. (2006) *J. Electrochem. Soc.*, **153** (6), 1193–1200.
14. Shah, A.A., Luo, K.H., Ralph, T.R., and Walsh, F.C. (2011) *Electrochim. Acta*, **56**, 3731–-3757.
15. Jiao, K. and Li, X. (2011) *Prog. Energy Combust. Sci.*, **37**, 221–291.
16. Peters, R. (2008) in *Handbook of Heterogeneous Catalysis* (eds. G. Ertl, H. Knözinger, F. Schüth, and J. Weitkamp), Wiley-VCH Verlag GmbH, Weinheim, pp. 3045–3080.

17. Gasteiger, H.A. and Garche, J. (2008) in *Handbook of Heterogeneous Catalysis* (eds. G. Ertl, H. Knözinger, F. Schüth, and J. Weitkamp), Wiley-VCH Verlag GmbH, Weinheim, pp. 3081–3121.
18. Porš, Z. (2006) *Eduktvorbereitung und Gemischbildung in Reaktionsapparaten zur Autothermen Reformierung von Dieselähnlichen Kraftstoffen*, Forschungszentrum Jülich, Zentralbilbliothek, Jülich.
19. Pasel, J., Meissner, J., Porš, Z., Samsun, R.C., Tschauder, A., and Peters, R. (2007) *J. Hydrogen Energy*, **32** (18), 4847–4858.
20. Porš, Z., Pasel, J., Tschauder, A., Dahl, R., Peters, R., and Stolten, D. (2008) *Fuel Cells*, **8** (2), 129–137.
21. ANSYS, Inc. (2010) *ANSYS FLUENT UDF Manual, Release 13.0*, ANSYS, Canonsburg, PA.
22. ANSYS, Inc. http://www.ansys.com/Resource+Library?type=Technical%20Brief (last accessed 25 January 2012).
23. Creel, M. and Goffe, W.L. (2008) *Comput. Econ.*, **32**, 353–382.
24. Samsun, R.C., Werhahn, J., Reichardt, A., Grube, T., Tschauder, A., and Peters, R. (2008) Production cost analysis for an ATR, in Proceedings of the Lucerne SOFC Forum Lucerne, 30 June– 4 July 2008, European Fuel Cell Forum, Oberrohrdorf (ed. R. Steinberger-Wilckens).
25. The Green500. The Green Lists, http://www.green500.org/lists.php (last accessed 12 December 2010).
26. Ferziger, J.H. and Peric, M. (2008) *Numerische Strömungsmechanik*, Springer, Berlin.
27. Ansorge, R. and Sonar, T. (2009) *Mathematical Models of Fluid Dynamics, Modelling, Theory, Basic Numerical Facts*, Wiley-VCH Verlag GmbH, Weiheim.
28. Herwig, H. (2004) *Strömungsmechanik, eine Einführung in die Physik und die Mathematische Modellierung von Strömungen*, Springer, Berlin.
29. Feffermann, C.L. Existence and Smoothness of the Navier–Stokes Equation, http://www.claymath.org/millennium/Navier–Stokes_Equations/navierstokes.pdf (last accessed 12 December 2010).
30. Brebbia, C.A. and Ferrante, A.J. (1986) *Computational Methods for the Solution of Engineering Problems*, 3rd revised edn., Pentech Press, London.
31. Schönung, B.E. (1990) *Numerische Strömungsmechanik, Inkompressible Strömungen mit komplexen Berandungen*, Springer., Berlin.
32. ANSYS, Inc. (2009) *ANSYS FLUENT Theory Guide, Release 12.0*, ANSYS, Canonsburg, PA.
33. Fletcher, C.A.J. (1991) *Computational Techniques for Fluid Dynamics, Fundamental and General Technique*, vol. **1**, Springer, Berlin.
34. Kistner, B. (2000) Modellierung und numerische Simulation der Nachlaufstruktur von Turbomaschinen am Beispiel einer Axialturbinenstufe, Dissertation, Technische Universität Darmstadt.
35. Fletcher, C.A.J. (1991) *Computational Techniques for Fluid Dynamics, Specific Techniques for Different Flow Categories*, vol. **2** Springer, Berlin.
36. ANSYS, Inc. (2009) *ANSYS FLUENT User's Guide, Release 12.0*, ANSYS, Canonsburg, PA.
37. Detert, U. (2010) JuRoPA – JuRoPA-JSC/HPC-FF an Overview, http://www.fz-juelich.de/SharedDocs/Downloads/IAS/JSC/docs/presentations/supercomputer-ressources/sc-juropa-introduction.pdf?__blob=publicationFile (last accessed 25 January 2012).
38. Herwig, H. (2002) *Strömungsmechanik A–Z, eine systematische Einordnung von Begriffen und Konzepten der Strömungsmechanik*, Vieweg, Wiesbaden.
39. Paschedag, A.R. (2004) *CFD in der Verfahrenstechnik: Allgemeine Grundlagen und Mehrphasige Anwendungen*, Wiley-VCH Verlag GmbH, Weinheim.
40. Pors, Z. (2006) Strömungstechnische Analysen der Mischkammer der geplanten Kerosinreformer ATR-10 in der 50 kWel Leistungsklasse. Internal Report, IEK-3, Forschungszentrum Jülich, Jülich.
41. Löffler, R. (2010) High productivity computing, in Conference Proceedings, ANSYS Conference and 28th CADFEM Users' Meeting 2010.

26
Modeling Solid Oxide Fuel Cells from the Macroscale to the Nanoscale

Emily M. Ryan and Mohammad A. Khaleel

26.1
Introduction

Mathematical modeling of solid oxide fuel cells (SOFCs) provides insight into the multi-physics occurring within the fuel cell and how the design and operation of the fuel cell affect its performance. Although the design and operation of an SOFC appears simple, many of the phenomena dominating the performance of the SOFC are complex, competing, and poorly understood. Experimental insights into the operation and performance of SOFCs are limited because of the expense of producing test cells, the time-consuming nature of long-term testing, and the number of cases needed to consider a range of designs and operating conditions. To accelerate the development and understanding of SOFCs, it is highly desirable to establish mathematical models that incorporate the known physics and behavior of SOFC materials to predict and improve performance.

SOFC operations involve multiple chemistry and physics processes and require versatile multi-physics tools for realistic descriptions of SOFC performance. A mathematical simulation of an SOFC can aid in examining the operating conditions of the SOFC, such as temperatures, potential, gas species distributions, and stresses, and in considering the effects of materials, geometries, dimensions, and fuels on SOFC performance. Such a simulation can be used to determine the effects of varying the design and operating parameters on the power generated, maximum cell temperature reached, fuel conversion efficiency, and stresses caused by temperature gradients, and the effects of different material properties, such as coefficients of thermal expansion, on the electrolyte, electrodes, and interconnect. Mathematical simulations can also help to answer questions about which properties of the fuel cell have the greatest effect on performance or what air and fuel flow rates are needed to avoid excessive temperature and/or pressure drops across the fuel cell. Thus, mathematical simulations offer the potential to direct technology development, test the significance of various design features, investigate the effectiveness of developments in materials or fabrication processes, and select operating conditions to optimize the SOFC's performance over its operating lifetime. Modeling of SOFCs is a critical aspect of the SOFC technology development process.

Mathematical modeling of SOFCs can occur at various levels of spatial and temporal resolution (Figure 26.1). From the electrochemistry of an SOFC to the operation of an SOFC power system, length scales can range from nanometers to tens of meters. SOFC power systems consist of hundreds or thousands of SOFC cells placed in series to form large stacks, and modules of stacks, wired and manifolded in series, and/or in parallel to produce a specific voltage or power [1, 2], while the electrochemistry of the SOFC takes place at the surfaces and interfaces within the electrodes of the fuel cell [3, 4]. The time scales of interest in SOFCs also span various scales. To compete with traditional fossil fuel power plants, SOFC systems are being designed for a 40 000 h plus lifetime [5]; in contrast, the electrochemistry of the fuel cell occurs on the order of milli- to nanoseconds. Because of the complex, multi-scale nature of SOFCs, it is important to study these systems at a variety of scales with a variety of modeling tools.

Mathematical models of SOFCs have been developed at scales from the molecular/atomic level through the system level. At the nanoscale, molecular dynamics and density functional theory (DFT) models have been developed to investigate the transport of species on the surfaces of the electrodes [6]; while at the macroscale, system-level models are used to consider the controls of an SOFC system [7]. The

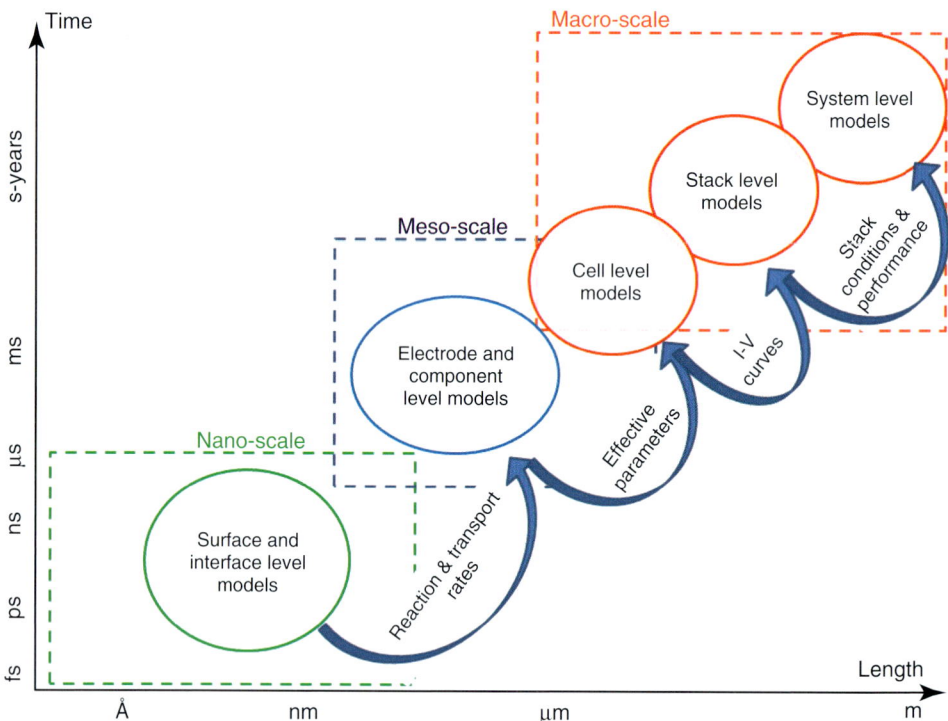

Figure 26.1 Schematic of the time and length scales involved in modeling SOFCs at the nano-, meso-, and macroscales.

level of detail included in the models depends on the aims of the model and the specific problem they are considering. Many SOFC models focus on individual thermal–mechanical, flow, chemical, and electrochemical subsystems and as such only model the system of interest. Other models investigate the effects of operating conditions on the overall performance of the SOFC systems. The various spatial and temporal scales of the SOFC can be investigated using models at multiple scales coupled together through the model parameters, such as reaction rates, effective parameters, or electrochemical performance (Figure 26.1). In this chapter, we discuss mathematical models that span various spatial and temporal scales to study the operation, design, performance, and degradation of SOFCs.

This chapter discusses SOFC modeling at the macro-, meso-, and nanoscales; discussing models ranging from the system level through the stack, cell, and surface level. Most mathematical modeling of SOFCs has focused on the cell and stack level with a variety of macro- and mesoscale models that have been used to investigate SOFC design and operation. Hence the majority of the chapter is dedicated to cell- and stack-level models; with a brief discussion of system-level models and nanoscale models. An introduction to the governing equations of SOFCs is presented in Section 26.2, which includes a discussion of the different methods of incorporating the electrochemistry of the SOFC in mathematical models. Next, the macroscale models are examined from the system level through the cell level, including a discussion of distributed electrochemistry modeling, which attempts to model the electrochemistry through the thickness of the SOFC electrodes using a continuum-scale description of the electrodes and electrolyte. This is followed in Section 26.4 by a discussion of mesoscale modeling of the SOFC electrodes in which the SOFC electrodes are explicitly resolved and the detailed reactive transport and electrochemistry is modeled. Section 26.5 briefly describes nanoscale approaches for modeling the transport and reactions of species in the SOFCs, which are suitable for elucidating kinetic and mechanistic issues relevant to SOFC performance.

26.2
Governing Equations of Solid Oxide Fuel Cells

The multi-physics of SOFCs are governed by the mass, momentum, and energy conservation equations, and the chemistry and electrochemistry. The governing equations of SOFCs are tightly coupled and changes to one aspect of the fuel cell can drastically affect another. For example, the rate and composition of the fuel flow in the anode will affect the temperature distributions in the cell, which can induce stresses due to mismatches between the coefficients of thermal expansion of the various layers in the SOFC. The fuel flow will also affect the overall performance of the fuel cell based on the distribution of species in the anode and the electrochemical reactions.

Three distinct domains can be considered when discussing the governing equations of SOFCs: channel, pore space, and solid (Figure 26.2). The fuel and air

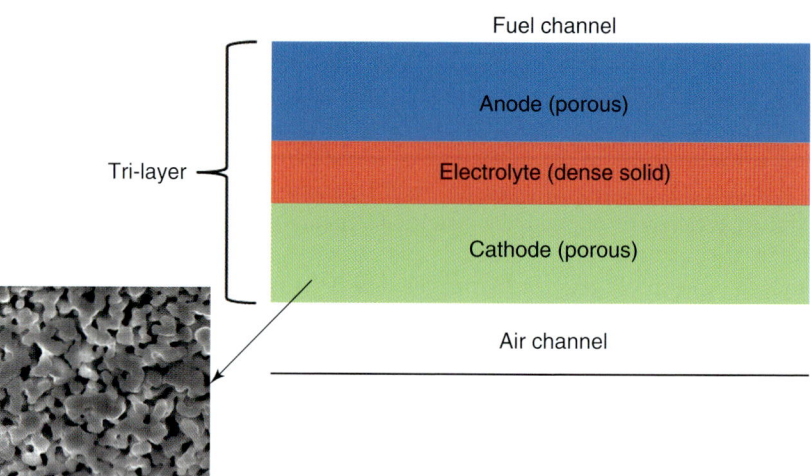

Figure 26.2 Schematic of a SOFC cell showing the channel, porous media, and solid domains. An SEM image of a SOFC cathode is shown in the detail. Note that the schematic is not to scale.

channels are the open space above the porous electrodes where the fuel and air enter the SOFC. The channels are typically rectangular openings formed by the ribs of the interconnect plates of the SOFC. The pore space is the open regions of the porous electrodes (see detail in Figure 26.2) which are filled with the fuel (anode) and air (cathode). The solid domain is the interconnected solid of the tri-layer which includes the electrodes and electrolyte. In state-of-the-art SOFCs, composite electrodes are typically used. In a composite electrode, the solid consists of the ion-conducting material of the electrolyte mixed with the electron-conducting material of the electrode; this allows the electrochemical reactions to occur further away from the electrode–electrolyte interface.

26.2.1
Mass Conservation

In a reacting system, the mass of individual species is solved via the species conservation equation:

$$\frac{\partial c_i}{\partial t} = \nabla J + \vec{u} \cdot \nabla c_i + S_i \tag{26.1}$$

where c_i is the concentration of species i, J is the diffusive flux, $\vec{u}$ is the velocity vector, and S_i is the species source term. S_i is calculated from the rate of production or consumption of species i due to the chemical and electrochemical reactions in the SOFC and will be discussed further in Sections 26.2.4 and 26.2.5.

In the air and fuel channels of the SOFC, the gas is well mixed and species transport is dominated by the advection of species due to the velocity. In the pore space of the electrodes, diffusive transport of the species is the dominant mechanism. The diffusive flux (J) can take several forms depending on the number of species in the system and the level of complexity desired in the model. The simplest and most commonly used form is Fick's law:

$$J_i = -D_{ij} \nabla c_i \qquad (26.2)$$

where the diffusion coefficient, D_{ij}, is the binary molecular diffusion coefficient, which can be found from experimental data or calculated from theoretical and empirical equations [8]. Fick's law is formulated for a binary system; in systems with more than two gas species, multi-component molecular diffusion occurs in which the effects of all gas species' interactions on diffusion must be accounted for. This is often the case in the pore space of the SOFC electrodes, where air in the cathode and the fuel mixtures in the anode contain numerous species. Under certain conditions, the effects of multicomponent molecular diffusion can be approximated as binary diffusion, such as in dilute solutions [8]. In this case, an effective diffusion coefficient is used in Eq. (26.2) to account for the effects of all species on the transport of species i.

To rigorously model the effects of multicomponent molecular diffusion, an alternative diffusion equation should be used, such as the Stefan–Maxwell equation [9, 10]:

$$-\nabla c_i = \sum_{\substack{j=1 \\ j \neq i}}^{n} \frac{N_i c_j - N_j c_i}{c_T D_{ij}} \qquad (26.3)$$

where c_T is the total molar concentration of the mixture and N_i is the molar flux of species i relative to a stationary coordinate system. The diffusion coefficients in the Stefan–Maxwell equation are the binary molecular diffusion coefficients as in Fick's law. In SOFC models, both the Fick and Stephan–Maxwell equations are commonly used to simulate the molecular diffusion of species in the pore space of the anode and cathode.

In addition to molecular diffusion, Knudsen diffusion can also be a significant transport mechanism in the pore space of SOFC electrodes. In Knudsen diffusion, the interactions of gas molecules with the pore walls are of the same frequency as the interactions between gas molecules. Knudsen diffusion is typically formulated as Fickian diffusion [Eq. (26.2)], with the Knudsen diffusion coefficient being used in place of the binary diffusion coefficient. The Knudsen diffusion coefficient of a species is independent of the other species in the system and is derived from the molecular motion of the gas molecules and the geometry of the pores [8, 11, 12]. Owing to the small average pore radii of SOFCs ($\sim 10^{-6}$ m [13–15]), diffusion in the pore space of the electrodes usually falls within a transition region where both molecular and Knudsen diffusion are important [16]. To model the transition region, Fick's law can be used with an effective diffusion coefficient to account for both molecular and Knudsen diffusion. The Bosanquet equation can be used to calculate the effective total diffusivity [15]:

$$D_i^{\text{eff}} = \left(\frac{1}{D_{ij}} + \frac{1}{D_i^{\text{Kn}}}\right)^{-1} \quad (26.4)$$

The inclusion of Knudsen diffusion in simulations of SOFCs can have a significant effect on the transport in the pore space of the electrodes [17] and should be included in macroscale models to ensure an accurate simulation of the SOFC performance.

Depending on the resolution of the mathematical model, different forms of the species conservation equations may be considered in the porous electrodes. For instance, in the multi-scale modeling of Khaleel *et al.* [18], a mesoscale lattice Boltzmann model of the electrodes resolves the species transport in the gas, on the surface of the electrode, and through the bulk solid of the electrode. In this model, Eq. (26.1) is solved in three separate domains with corresponding transport properties and source terms. In contrast, in the macroscale distributed electrochemistry model of Ryan *et al.* [19], the porous medium of the SOFC electrodes is not explicitly resolved but is included in the model via effective properties. In the effective properties model, the diffusion coefficient of Eq. (26.1) is replaced with an effective diffusion coefficient, which is discussed in Section 26.3.3.

26.2.2
Momentum Conservation

The gas flow in the SOFC is governed by the momentum conservation equation (Navier–Stokes equation):

$$\rho \frac{\partial \vec{u}}{\partial t} + \rho \vec{u} \cdot \nabla \vec{u} = -\nabla P + \nabla \cdot \left(\mu \nabla \vec{u}\right) + \rho \vec{g} \quad (26.5)$$

where ρ is density, μ is viscosity, and $\vec{g}$ is the gravity vector. The momentum conservation equation calculates the flow fields in the SOFC which affect the distribution of species and the heat transfer in the SOFC. In the channels, the transport of species is dominated by advection and so the solution of Eq. (26.5) is vital to the channel transport. In the pore space, transport is dominated by diffusion and Eq. (26.5) can be excluded from the pore space domain without affecting the accuracy of the solution. This reduces the number of equations solved in the electrodes and has been shown to be a valid assumption in the electrodes under most conditions [20].

Solving the full Navier–Stokes equations in the channels requires a rigorous computational fluid dynamics (CFD) simulation. During transient operation, such as start-up and shut-down, the flow fields can have a significant effect on the concentration and temperature profiles in the system. Under normal operation, it may be desirable to assume fully developed laminar flow to reduce the computational time and quickly estimate flow parameters based on fluid dynamics correlations.

In the fuel and air channel, the pressure drop through the channel due to viscosity (assuming laminar flow) can be estimated by

$$\Delta P = \frac{1}{2} \frac{\rho U^2 l f}{Re D_h} \tag{26.6}$$

where Re is the hydraulic diameter-based Reynolds number, D_h is the hydraulic diameter, l is the length of the channel, U is the magnitude of the velocity in the channel, and f depends on the shape of the cross-section of the channel; for example, $f = 56.8$ and 64 for a square and a round channel, respectively [21]. This type of simplification has been used to reduce significantly the computational cost in many stack-level SOFC models [22].

26.2.3
Energy Conservation

The temperature distributions and heat fluxes in the SOFC are governed by the energy conservation equation:

$$\frac{\partial (\rho c_p T)}{\partial t} + \nabla \cdot \left(\rho c_p T \vec{u} \right) = \nabla \cdot (k \nabla T) + Q + \frac{\partial P}{\partial t} \tag{26.7}$$

where k is thermal conductivity, c_p is specific heat, and Q is the volumetric heat source term. The volumetric heat source term, Q, represents heat generation by the electrochemical reactions, chemical reactions (e.g., hydrocarbon reforming and CO water gas shift reaction), and Joule heating (due to the ohmic resistance of the electrolyte and electrodes). The full form of Eq. (26.7) applies to the gas phase in the channels and pore space of the SOFC; by removing the final term on the right-hand side of Eq. (26.7), the equation also applies to the solid of the SOFC. In SOFC modeling of energy conservation, appropriate boundary conditions must be applied within the SOFC. This includes the boundaries between the gas and solid phases and the boundaries between solid phases of different materials. Typically, a convective heat transfer boundary condition is appropriate for gas–solid boundaries, such as the interface between the anode and the fuel channel above the anode. Solid–solid boundaries occur between the different layers of the tri-layer such as the interface of the cathode with the electrolyte and between the tri-layer and the interconnect. With solid–solid boundaries, lumped effective thermal conductivities are typically applied at the interfaces between materials. Additionally, in stack-level models, the tri-layer is often considered as a single lump material with effective thermal properties based on the individual layers of the tri-layer.

Many of the structural stresses in the SOFC are created by the thermal distributions in the cell. Mismatches in material properties between the different layers of the tri-layer and other components of the SOFC, such as the interconnects and seals, can lead to large stresses in the SOFC during thermal cycling. In designing SOFCs, the thermal stresses must be considered to ensure that the maximum stresses in the system are low enough that they do not exceed the material strengths as determined by appropriate failure criteria. The SOFC materials must not debond, delaminate, crack, or deform significantly in order to maintain the structural integrity of the stack. Since thermal mismatch between different cell components can be a major engineering problem in SOFCs, accurate numerical models should include the thermal solution through the thickness of the fuel cell.

For steady-state cases, the energy conservation equation can be simplified by removing the time-dependent terms, which decreases the computational cost of the thermal solution. However, the general form of Eq. (26.7) is necessary for the simulation of transient operating conditions, such as start-up and load change, when the thermal stresses in the system will be greatest.

26.2.4
Electrochemistry

The electrochemistry of the SOFC generates the electrical power of the SOFC system. In the SOFC, the oxidation reaction of H_2 with oxygen ions (O^{2-}) in the anode:

$$H_2(g) + O^{2-}(el) \rightleftharpoons H_2O(g) + 2e^-(an) \tag{26.8}$$

and the reduction reaction of O_2 with electrons (e^-) in the cathode:

$$\frac{1}{2}O_2(g) + 2e^-(ca) \rightleftharpoons O^{2-}(el) \tag{26.9}$$

govern the main electrochemistry of the SOFC. SOFCs are also able to oxidize CO to CO_2 and CH_4 to H_2O and CO_2 as alternative electrochemistry pathways. In this discussion, we will focus on the H_2 to H_2O pathway; however, the same principles apply to the other electrochemical pathways.

In SOFC modeling, the electrochemistry of the fuel cell can be included in the model at various levels of detail. In a continuum-scale approach, empirical current–potential (I–V) relations are typically used to model the electrochemistry of the SOFC. In a mesoscale approach, the electrochemical reactions and the transport of electrons and ions in the SOFC can be modeled explicitly. The continuum-scale approach allows for a quick evaluation of the I–V performance of the SOFC by assuming that the electrochemistry occurs only at the interface of the electrode and electrolyte. In the mesoscale approach, the electrochemistry of the SOFC is resolved through the thickness of the electrodes based on the local conditions in the cell. In this section, we discuss the details of both approaches.

For a given electrochemical reaction, the theoretical open-circuit voltage, the Nernst potential, is an important, if not the most important, parameter that affects and measures a fuel cell's performance. The Nernst potential is affected by the operating conditions of the SOFC, such as the temperature, pressure, and fuel composition, and is calculated from the maximum electrical work obtainable, the Gibbs free energy of the reaction (ΔG):

$$E_{eq} = -\frac{\Delta G}{nF} \tag{26.10}$$

where F is Faraday's constant (95 484.56 C mol^{-1}) and n is the number of electrons exchanged in the reaction. Using the ideal gas assumption, which is a good approximation for the gases in the SOFC, Eq. (26.10) can be written as

$$E_{eq} = E° + \frac{RT}{2F} \ln\left(\frac{p_{H_2O}}{p_{H_2} p_{O_2}^{\frac{1}{2}}}\right) \tag{26.11}$$

where $E° = \frac{-\Delta G°}{2F}$ and $\Delta G°$ is the Gibbs free energy of the reaction when all the species are at the standard pressure of 1 atm. p_{H_2} and p_{H_2O} are the partial pressures of H_2 and H_2O in the anode and p_{O_2} is the partial pressure of O_2 in the cathode. In many macroscale system- and stack-level models, the partial pressures are based on the values in the channels. E_{eq} is the ideal or equilibrium (open-circuit) voltage of the fuel cell, and $E°$ is often referred to as the *standard potential of the cell* and is only a function of temperature.

26.2.4.1 Continuum-Level Electrochemistry Approach

Continuum-level electrochemistry calculates the cell voltage (V) based on the Nernst equation and the losses (polarizations) in the cell:

$$V(i) = E_{eq} - iR_i - \eta_C - \eta_A \tag{26.12}$$

where the polarizations include the ohmic (iR_i), concentration (η_C), and activation (η_A) polarizations. The concentration and activation polarizations can be further divided into anodic and cathodic contributions.

The ohmic polarization in Eq. (26.12) represents the total area specific ohmic resistance of the cell. R_i is the sum of the anode, cathode, electrolyte, interconnect, and contact ohmic resistances. Typically, the ohmic resistance is dominated by the electrolyte resistance and decreases with increasing operating temperature. The reduction in ohmic polarization is part of the reason why anode-supported cells have become the standard design in current high-performance SOFCs.

The concentration polarization is the voltage loss associated with the transport resistance of the porous electrodes to the transport of reactant species to and product species from the reaction sites. Diffusion is the dominant transport mechanism in the electrodes and is limited by the diffusion rates of the gas species and the concentration gradients in the electrode. The concentration difference between the bulk gas in the electrode and the gas contacting the reaction sites forms a concentration cell whose cell voltage opposes the overall SOFC voltage and is seen as a voltage loss or polarization. In the cathode, this is seen as a lower partial pressure of oxygen in the cathode pores near the cathode–electrolyte interface when compared with the partial pressure in the air channel. Typically, the concentration polarization of the cathode is much greater than that of the anode. This is due to the binary diffusion coefficients of H_2 in the anode and O_2 in the cathode. The diffusion coefficient of O_2 is almost four times smaller than that of H_2, which leads to a larger concentration polarization. This is another reason why an anode-supported cell design is used in state-of-the-art SOFCs. To decrease the concentration polarization of the cathode, a thin cathode should be used with the porosity and pore sizes as large as possible.

The activation polarization represents the voltage loss due to the activation necessary for the charge-transfer reactions. The phenomenological Butler–Volmer relation is typically used to relate the activation polarization to the current density of the cell [23]:

$$i = i_0 \left\{ \exp\left(\frac{-\alpha_e n F \eta_A}{RT}\right) - \exp\left[\frac{(1-\alpha_e) n F \eta_A}{RT}\right] \right\} \quad (26.13)$$

where i_0 is the exchange current density and α_e is the electrode transfer coefficient. Note that the relation between the current density (i) and the activation polarization is nonlinear and implicit, which does not allow for an explicit calculation of the activation polarization from the current density. Simplified expressions can be written for cases with high or low current density, such as the Tafel equation for the high-current regime [24].

The exchange current density is the dynamic rate of the forward and reverse reaction of the fuel cell at equilibrium. i_0 can be expressed as $i_0 = P_x \exp(-E_{act}/RT)$, where P_x is a pre-exponential factor, which accounts for the dependence of the current on reactant and product surface concentrations, and the surface area available for the electrochemical reactions, and E_{act} is the energy barrier between the reactants and the activated sites of the electrode [25].

In continuum-scale electrochemistry, the materials and microstructure of the SOFC tri-layer are embedded in the parameters of the polarization losses. The electrochemistry of the SOFC is inherently dependent on the SOFC microstructure, including the surface area available for the electrochemical reactions, the porosity, tortuosity, and permeability of the porous media, and the material properties of the tri-layer. All these properties affect the rate of reactions in the electrodes and thus the overall voltage produced by the SOFC. Although continuum-scale electrochemistry does not resolve the explicit electrochemical reactions, they are able to accurately model the performance of the SOFC when using experimental data to estimate the parameters of the electrochemistry model. Often the parameters of the continuum-scale electrochemistry model, such as the pre-exponential factors, and activation energies and polarizations are used to fit the continuum-scale electrochemistry to experimental I–V curves.

26.2.4.2 Mesoscale Electrochemistry Approach

The electrochemistry of the SOFC can be modeled at the mesoscale in cell- and electrode-level models. In these models, the electrochemistry is calculated based on the local conditions within the SOFC tri-layer and the local Faradaic current density is resolved through the thickness of the cell. In mesoscale electrochemistry, both the local electrochemistry and the global electrochemistry are considered. The local electrochemistry is the local reactions within the electrodes which produce a local Faradaic current density; the global electrochemistry is the integral of the local electrochemistry over the tri-layer and produces the common I–V curves that are used to describe the performance of the SOFC.

There are two approaches to modeling the SOFC electrochemistry at the mesoscale: an elementary kinetics-based model and a modified Butler–Volmer model. In the elementary kinetics-based model, the electrochemical reactions of the SOFC are modeled exactly, whereas in the modified Butler–Volmer model, the phenomenological Butler–Volmer equation is solved based on the local Faradaic current density.

In the elementary kinetics-based electrochemistry models, the electrochemical reactions of the gas species are described by an electrochemical reaction mechanism. For example, in the anode, the electrochemical reactions can be described as [26]

$$H_2(g) \rightleftharpoons 2H(an) \tag{26.14}$$

$$H(an) + O^{2-}(el) \rightleftharpoons OH^-(el) + e^-(an) \tag{26.15}$$

$$H(an) + OH^-(el) \rightleftharpoons H_2O(el) + e^-(an) \tag{26.16}$$

$$H_2O(el) \longleftrightarrow H_2O(g) \tag{26.17}$$

The local Faradaic current density of the SOFC is formulated from the electrochemical reactions, Eqs. (26.14–26.17), as [27]

$$i_{e,F}(\vec{x}) = nFl_{TPB} \left[k^f_{CT} \prod_j^{R,ct} \theta_j^{v^f_j}(\vec{x}) - k^r_{CT} \prod_j^{P,ct} \theta_j^{v^r_j}(\vec{x}) \right] \tag{26.18}$$

where l_{TPB} is the volumetric triple-phase boundary length, which describes the reactive surface area of the electrodes, θ_i is the surface coverage of species i, $v_j^{f/r}$ is the stoichiometric coefficients of the reactions, and the charge-transfer reaction rates are defined as [27]

$$k^f_{CT} = k^\circ_{f,CT} \exp\left(\frac{-E^{act}_{f,CT}}{RT}\right) \exp\left[-(1-\alpha_e)\frac{nF}{RT}E_e\right] \tag{26.19}$$

$$k^r_{CT} = k^\circ_{r,CT} \exp\left(\frac{-E^{act}_{r,CT}}{RT}\right) \exp\left(\alpha_e \frac{nF}{RT}E_e\right) \tag{26.20}$$

where k° are the pre-exponential factors, E^{act}_{CT} is the activation energy, and E_e is the potential difference. The potential difference is defined as

$$E_e(\vec{x}) = \phi_e(\vec{x}) - \phi_{e,elyt}(\vec{x}) \tag{26.21}$$

where ϕ_e is the electric potential in the electrode and $\phi_{e,elyt}$ is the potential of the electrolyte phase of a composite electrode.

Additionally, the species source terms of Eq. (26.1) can be calculated from the reaction mechanism as

$$S_i(\vec{x}) = \sum_m^{reactions} k^f_k \prod_j^R c_j^{v^f_j}(\vec{x}) - k^r_k \prod_j^P c_j^{v^r_j}(\vec{x}) \tag{26.22}$$

which is based on the reaction mechanisms of each reaction step m via the forward reaction rate (k^f_i), and the reverse reaction rate (k^r_i). The forward reaction rates are calculated as an Arrhenius expression and the reverse reaction rates are calculated from the forward reaction rate and the thermodynamic properties of the reactions [27].

Determining the reaction rate coefficients is one of the main challenges in elementary kinetics-based models. In SOFCs, the electrochemical reactions occur on the surface of exotic materials, such as strontium-doped lanthanum manganese oxide, at high temperatures where it is difficult to obtain experimental data. There has been extensive research on defining the main reactions of the electrochemical reactions of SOFCs and quantifying and validating the reaction rate coefficients of these reactions [26, 28, 29]. However, questions still remain on the exact steps and pathways of the electrochemistry in SOFCs [30].

Many SOFC models chose not to model explicitly the electrochemical reactions. Instead, a modified Butler–Volmer relation based on the local conditions within the tri-layer can be used to solve for the current density of the fuel cell in a mesoscale electrochemistry approach [25]. The local Faradaic current density can be calculated from the modified Butler–Volmer relation as [31]

$$i_{e,F}(\vec{x}) = i_{0,e}(\vec{x}) l_{TPB} \left\{ \exp\left[\frac{\alpha_e F \eta_A(\vec{x})}{RT}\right] - \exp\left[-\frac{(1-\alpha_e) F \eta_A(\vec{x})}{RT}\right] \right\}$$

(26.23)

where $i_{0,e}$ is the local exchange current density and η_A is the local activation polarization. The activation polarization is calculated as

$$\eta_A(\vec{x}) = E_e(\vec{x}) - E_e^{Eq}(\vec{x})$$

(26.24)

where E_e is the local potential of the electrode, Eq. (26.21), and E_e^{Eq} is the local equilibrium potential, which is calculated by Eq. (26.11) with the local partial pressures of the gas.

Both mesoscale electrochemistry modeling approaches calculate the potential distribution in the tri-layer based on the local Faradaic current density via Poisson's equation:

$$\sigma^{eff} \frac{\partial^2 E_e}{\partial \vec{x}^2} = -i_{e,F}(\vec{x})$$

(26.25)

where σ^{eff} is the effective ionic conductivity of the composite electrode. In the electrolyte, Poisson's equation reduces to the Laplacian since there is no Faradaic current density source:

$$\sigma \frac{\partial^2 E_e}{\partial \vec{x}^2} = 0$$

(26.26)

and σ is the ionic conductivity of the electrolyte material.

26.2.5
Chemical Reactions

In addition to the electrochemical reactions of the SOFC, chemical reactions also occur in the electrodes. These reactions are typically related to on-cell reforming (anode) or degradation (anode or cathode). On-cell reforming is the process by which methane in the fuel stream is reformed with steam in the anode to generate hydrogen. Degradation reactions occur when a contaminant in the fuel or air stream reacts with the electrodes [32–34]. There are a variety of degradation mechanisms which occur in SOFCs and to cover them properly would require a separate chapter. Here we focus our discussion on the on-cell reforming reactions in the anode.

On-cell reforming allows the SOFC to run on alternative fuels and helps to reduce the temperatures of the SOFC. In the case of steam–methane reforming, reformation occurs on the surface of the Ni in a Ni–YSZ anode. The methane reacts with steam to form H_2 and CO by an endothermic reaction:

$$CH_4 + H_2O \rightleftharpoons 3H_2 + CO \quad \Delta H = 206 \text{ kJ mol}^{-1} \quad (26.27)$$

which occurs in conjunction with the water gas shift reaction:

$$CO + H_2O \rightleftharpoons H_2 + CO_2 \quad \Delta H = -41.1 \text{ kJ mol}^{-1} \quad (26.28)$$

Chemical reactions affect the species concentrations in the system and the temperatures of the SOFC. The endothermic reactions of on-cell reforming help to remove heat from the system and reduce the temperatures inside the cell. The heat source of on-cell reforming is calculated from the enthalpy change of the reactions and is coupled with the energy [Eq. (26.7)] through the volumetric heat source, Q. If chemical equilibrium is achieved in the SOFC, then the fuel composition and heat generation can be determined from thermodynamics. However, this is not typically the case, and therefore the fuel composition and heat generation must be determined from approximations. To model the chemical reactions, apparent kinetic parameters are used, which are based on experimental data. Many of the chemical and electrochemical reactions in SOFCs involve multiple reaction steps and intermediate species, making it difficult to know the exact reaction mechanism. Instead, experimental data are used to determine effective rate constants and other reaction kinetic parameters for the reaction.

For the on-cell reforming of methane, the reaction rate expression for CH_4 can be written as [35]

$$S_{CH_4} = -2.188 \times 10^8 \exp\left(-\frac{E_a}{RT}\right) c_{CH_4} c_{CO_2}^{-0.0134} \quad (26.29)$$

where E_a is the activation energy and c_i are the concentrations of CH_4 and CO_2 in the gas. Equation (26.29) is the result of detailed experimental tests performed on a plug flow reactor. Chemical rate equations such as Eq. (26.29) are coupled with the species and energy conservation equations in cell- and stack-level models to investigate the effects of on-cell reforming on the operation and performance of SOFCs. Recknagle et al. [35] modeled on-cell reforming in a single cell stack operating under pressurized conditions. To model the system at pressures other than atmospheric, Eq. (26.29) was modified to include the effects of the reverse reactions which can be favored at high pressures [35]. They showed that increasing the operating pressure of the SOFC helped to further reduce the temperatures in the stack and increase the electrochemical performance (Figure 26.3).

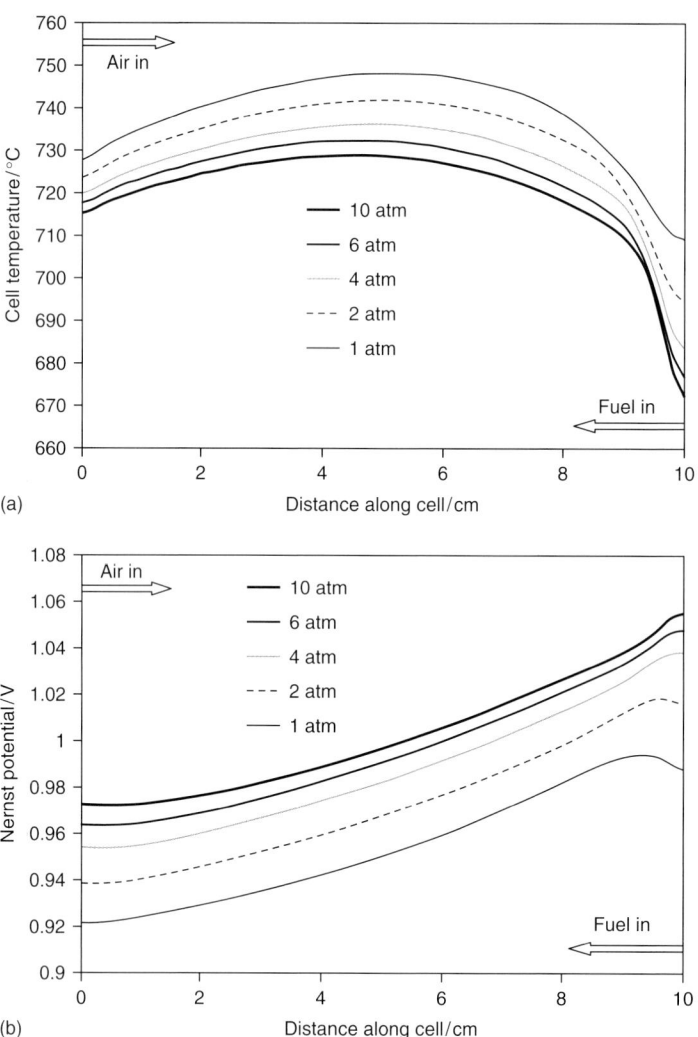

Figure 26.3 Temperature (a) and potential (b) along the channel flow path of a single cell stack operated under pressurization [35].

26.3
Macroscale SOFC Modeling

Traditionally, the majority of SOFC modeling efforts have taken place at the macroscale and have provided valuable insight into the operation of SOFC cells and stacks [15, 36–38]. In macroscale modeling, simplified models of the SOFC multi-physics are used to simulate the operating conditions and performance of the SOFC. Macroscale models are used to investigate the performance of experimental cells and stacks [35, 39], SOFC system-level operation and controls [40, 41], and thermal stresses and strains in the cells and stacks [42–44].

Macroscale modeling of SOFCs occurs at the system, stack, and cell levels. These models can help to improve our understanding of the complex interactions between the fluid dynamics and thermal, chemical, and electrochemical phenomena of SOFCs. Macroscale models can also help to maximize the efficiency or power density of the SOFC by optimizing the tri-layer design, cell configuration, and stack architecture for a given set of operating conditions. They can also add insight into the long-term operation of SOFCs and degradation issues which can occur at the cell and stack levels.

26.3.1
System-Level Modeling

System-level models consider the entire SOFC power system. This includes the SOFC cells and stacks of the system, the balance of plant (BoP) equipment, such as heat exchangers, reformers, blowers and compressors, and topping and bottoming turbines. Several power system applications are being considered for SOFCs, including hybrid power plants, gasified coal power plants [45, 46], and auxiliary power units (APUs) [7, 47]. In these systems, there are hundreds to thousands of individual SOFC cells, which are placed in stacks connected in series or parallel to produce the desired output voltage of the system. System-level models simulate the SOFC system by combining a one- or two-dimensional stack-level model with thermodynamics- and heat transfer-based process models of the entire system [7, 47]. A schematic of a system-level model is shown in Figure 26.4, where the stack model (light blue block) is coupled with the system-level components such as an external fuel reformer (orange block), heat exchangers (red blocks), and gas conditioners (gray blocks). The principal objectives of system-level models are to aid in the design and sizing of the system and to determine the operating requirements, energy efficiency, and heat/power ratio of the system. System-level models also allow researchers to test out different control schemes and to address the technical challenges of controller design in SOFCs, including thermal and fuel management, and transient and load following characteristics [47].

SOFC systems have numerous operating requirements which are necessary to ensure reliable operation of the SOFC system over its lifetime. Some of these requirements include the following [46, 47]:

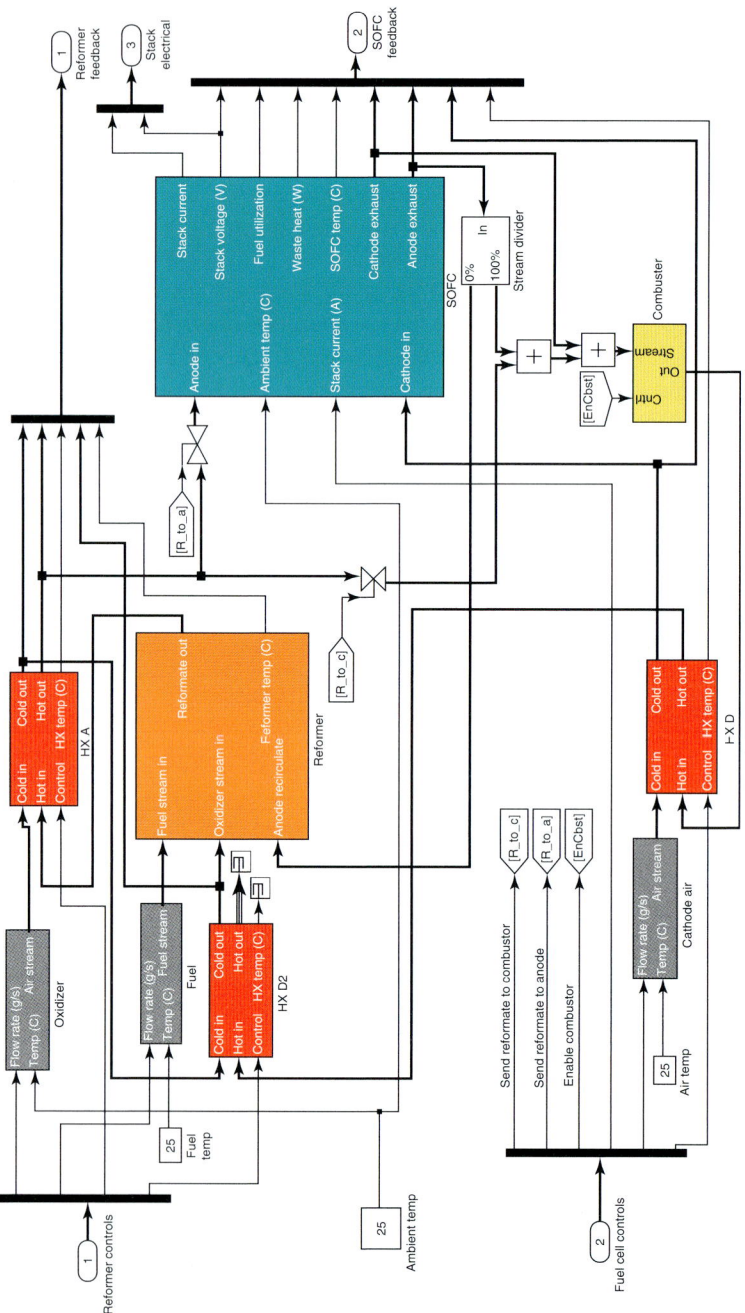

Figure 26.4 Schematic of system-level modeling of an APU SOFC system.

- Small temperature variations throughout the cells and stack. Changes in temperature of more than 100 K across the cells can lead to thermal stresses, which can cause cracking and delamination of the cell.
- Sufficient fuel and oxidant flows throughout the SOFC stacks. Depletion of fuel or oxidant can cause oxidation and reduction of the SOFC electrodes.
- Relatively uniform current through the stacks. Degradation can occur due to increases in local heating within the cells.
- Clean fuel flow. Contaminants in the fuel flow of coal gas are known to cause degradation in the SOFC anode that can lead to SOFC failure [33].

System-level modeling of SOFCs aims to ensure that nearly uniform temperatures, currents, and fuel and air flows are seen by each individual SOFC cell in the system. This can be a daunting task in large power systems; however, modeling can help in the design of the systems to ensure that there are a sufficient number of sensors, valves, and so on, to control the system and its operation correctly.

A good example of system-level modeling for the design and control of an SOFC power system was presented by Mueller et al. [46]. They used a MATLAB-SIMULINK-based system-level model to provide insight into effective system design and integrated control of a 100 MW hybrid power plant running on gasified coal [46]. Their paper presents a modular plant design that allows the number of cells controlled by a single actuator to be varied; the design also allows sections of the system to be taken down for maintenance without affecting the overall operation of the system. The system-level model of the modular design is used to optimize the system efficiency by varying the number of cells and the current density of the system, and taking into account the changes in power necessary for recirculation of the gas. When considering the system efficiency of an SOFC power system, it is important to consider the power needs of the BoP equipment, which will take away from the total system efficiency of the SOFC system.

Mueller et al.'s [46] system-level model simulates the primary components of the SOFC system (SOFC, turbine, compressor, heat exchangers, etc.) individually based on the conservation equations of the component. Each individual model is then integrated into the system-level model to consider the design and control of the power plant. A similar approach was used by Mazumder et al. [7] in the simulation of an SOFC vehicular APU. Although simpler then detailed cell-level models of SOFCs, modeling the multi-physics in the SOFC and system components in system-level modeling is still computationally expensive. The use of simplifying assumptions and reduced order models can decrease the computational cost of system-level models and still provide accurate insight into the operation and control of SOFC systems. Mazumder et al. [7] compared their more detailed system-level modeling with reduced order models of an APU system. By reducing the SOFC model from two to one dimensional, simplifying the power electronics model, and reducing the BoP models to polynomial approximations, they were able reduce their computational time by a significant amount while increasing their computational error by less than 10% [7].

SOFC system-level models provide valuable insight into the overall operation and design of SOFC power systems. They provide a means to investigate an SOFC system quickly without resolving the detail that is found in cell-level models. This approach is helpful in the initial design of a system and should be included as part of a larger modeling procedure for the development of SOFC technologies and power systems.

26.3.2
Stack-Level Modeling

Stack-level modeling considers the distributions of temperature, species, current, and pressure through multiple cells connected in series or parallel to form an SOFC stack. In addition to the SOFC tri-layer, stack-level models may also include the interconnects (separator plates), seals, cell frames, and manifolds, which are integral parts of the stack design (Figure 26.5). Stack-level modeling typically uses a finite element or finite volume modeling approach to solve the SOFC multi-physics [1, 48] and considers the cell in one, two, or three dimensions (1D, 2D, 3D). For co- and counter-flow planar SOFC designs, the symmetry of the cell allows for axisymmetric 2D models, which are able to provide quick simulations of the basic physics and can provide accurate results [1]. For more complex cell designs and flow configurations, such as the cross-flow planar cell, a 3D model is more appropriate.

Stack-level models solve the basic species [Eq. (26.1)], momentum [Eq. (26.5)], and energy [Eq. (26.7)] equations in the SOFC cells based on simplifying assumptions such as a lumped tri-layer which does not resolve the individual electrodes and electrolyte, or the use of simplified flow equations based on the assumptions of fully developed laminar flow in the air and fuel channels [Eq. (26.6)]. The electrochemistry is solved with the continuum electrochemistry equations (Section 26.2.4.1) and is based on experimental performance data. Often the thermal fluids solution of the stack model is coupled with a mechanical analysis to consider the stresses in the stack [48]. Owing to the mismatched properties of the materials in the SOFC stack, stresses can arise during operation and thermal cycling which reduce the stability and performance of the SOFC (Figure 26.5). Hence an accurate solution of the energy equation is vital to understanding the operation and durability of SOFC stacks.

To study the stresses in the system, it is first necessary to calculate the temperature distributions of the SOFC stack. Owing to the coupled nature of the SOFC multi-physics, the temperatures in the stack will affect both the electrochemical performance and the mechanical stresses of the stack [49]. The electrochemical performance of the SOFC is coupled to the temperature through the Nernst equation [Eq. (26.11)]. Stack-level models are often used to consider the temperature distributions and how the operating conditions and design of the stack affect the temperatures [1, 48, 49]. In these models, the energy conservation equation [Eq. (26.7)] is solved in the gas and solid phases, and includes the effects of convection in the fuel and air channels, radiation between the solid tri-layer and the gas, radiation between the stack and its surroundings, conduction through the tri-layer, and heat sources due to chemical and electrochemical reactions [1, 50]. The balance

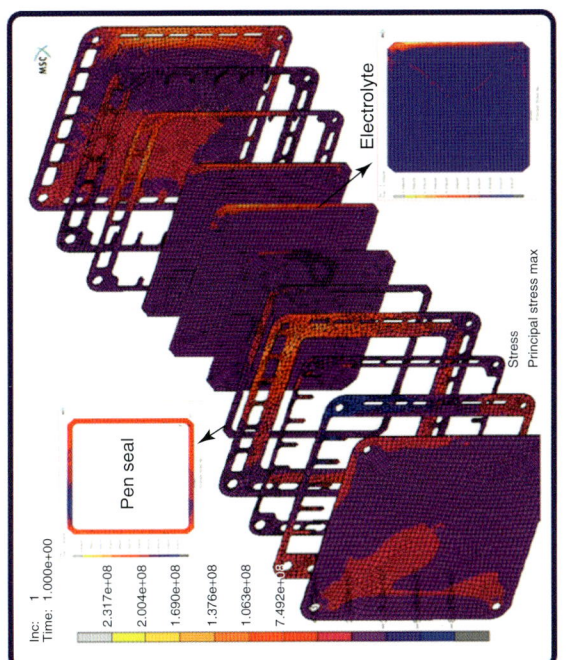
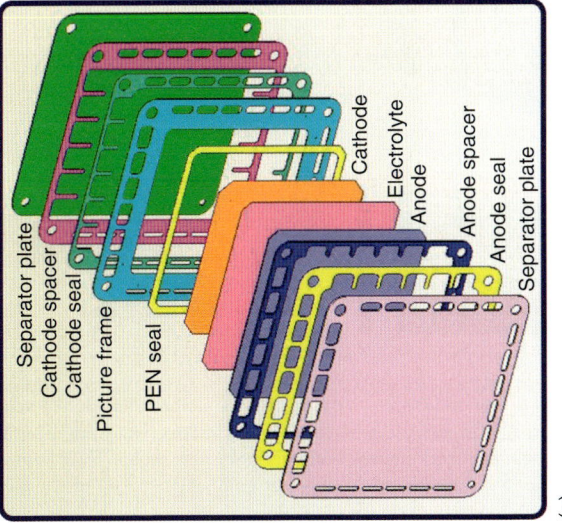

Figure 26.5 Components of an SOFC stack (a) and the stresses in the components (b) at operating temperature modeled with a coupled thermomechanical model using SOFC-MP and MSC Marc.

between the heat transfer through the cells and the thermal sources in the stack can lead to large temperature gradients in the SOFC that lead to thermal stresses and deformation [48, 49] and high local temperatures, which can decrease the stability and durability of the SOFC stack and its materials.

The structural solution computes the full 3D elastic–plastic deformation and stress fields for the solid components of the stack. The primary stress-generation mechanism in the SOFC is thermal strain, which is calculated using the coefficient of thermal expansion (CTE) and the local temperature difference from the material's stress-free temperature. These thermal strains and mismatches in thermal strains between different joined materials cause the components to deform and generate stresses. In addition to the thermal load, the stack will have boundary conditions simulating the mechanical constraints from the rest of the system and may also have external mechanical preloading. The stress solution is obtained based on the imposed mechanical constraints and the predicted thermal field. Figure 26.6 shows

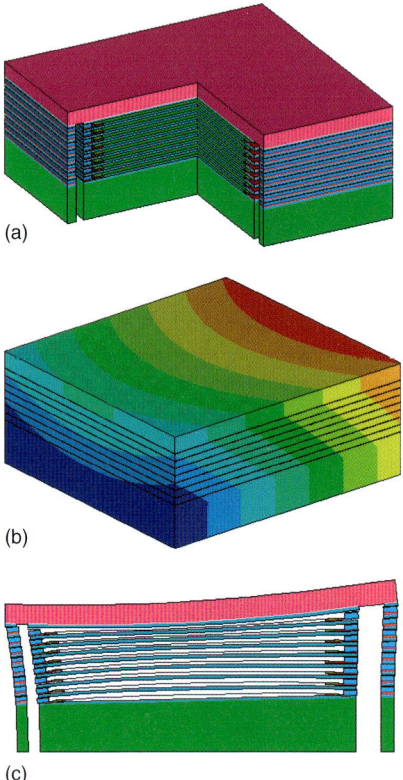

Figure 26.6 Example of a four-cell stack simulation which was used to investigate the thermomechanical stresses in a stack under a load. (a) A schematic of the baseline stack; (b) the temperature distributions through the stack; (c) the deformation of the stack due to thermal stresses. Note the deformations of the stack cross-section in (c) are exaggerated for emphasis.

a sample four-cell stack simulation which considers the effects of temperature distribution on the deformation of an SOFC stack which is under preload. A uniform pressure was applied to the stack to simulate clamping or the load of additional cells on top of the four cells considered. The temperature distributions through the stack were calculated based on the electrochemistry and gas transport and are shown by the contour plot in Figure 26.6b. The temperature distributions were then used as input to a stress simulation to consider the effects of temperature on the mechanical stability. Figure 26.6c shows the resulting deformation of the stack due to a lack of support. Note that the bottom image shows exaggerated deformations of the stack for emphasis. In this case, the performance of the stack could be improved by increasing the stiffness of the top plate or moving the stack load to the edges of the top plate only.

At Pacific Northwest National Laboratory, a stack-level model called solid oxide fuel cell-multiphysics (SOFC-MP) [1] has been developed to model stacks in 2D and 3D. SOFC-MP is used to investigate the effects of stack design on the operation and stability of the SOFC stack. SOFC-MP considers the thermo-fluids and electrochemistry of the stack and can be coupled with mechanical models to investigate stresses. Koeppel et al. [49] used SOFC-MP to investigate the effects of stack geometry and operation on the maximum temperature, the temperature difference through the stack, and the maximum temperature difference in a cell. Using the 2D SOFC-MP stack model, the authors considered fuel reformation in the SOFC anode. Fuel reforming presents a trade-off for SOFC designers: it can help to reduce the temperatures of the stack (as discussed in Section 26.2.5), but also reduces the electrochemical performance of the SOFC (Figure 26.7). By controlling the rate and amount of CH_4 reformation in the anode, Koeppel et al. [49] showed that one can control the temperatures in the stack. This type of modeling can be used by designers to weigh the trade-offs between mechanical stability and the electrochemical performance of the SOFC (Figure 26.8).

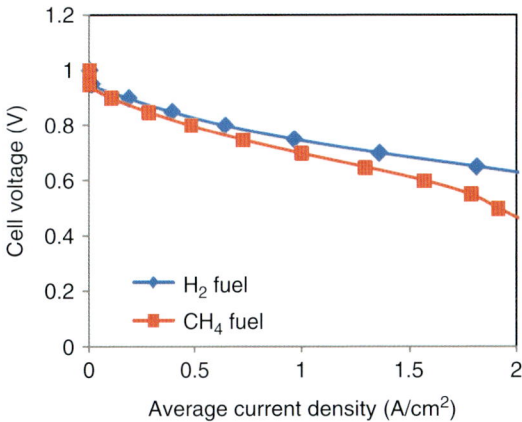

Figure 26.7 Current–voltage response of a single cell operating on CH_4 and H_2 [49].

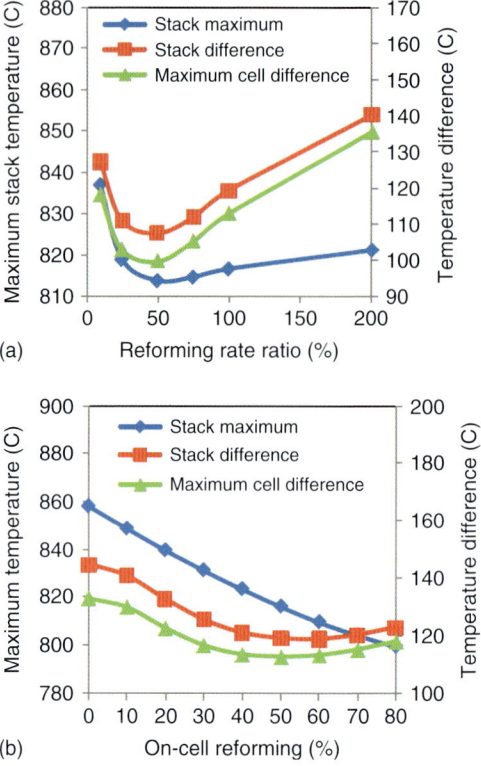

Figure 26.8 The effects of reforming rate (a) and percentage of reforming (b) on the temperatures in a 96-cell co-flow SOFC stack [49].

The temperature distributions in the SOFC stack are also affected by the design of the cells and stack components. The interconnect plates, which act as current collectors and separator plates between cells, are typically made of stainless steel and can be used to distribute the heat in the stack. Increasing the thickness of the interconnect plates decreases the temperatures and temperature differences in the stack [49]. However, increasing the thickness of all interconnect plates increases the size, weight, and cost of an SOFC stack. Instead, selected interconnect plates throughout the stack can be increased to reduce the temperature strategically in certain areas of the stack. Koeppel et al. [49] studied the effects of increasing the mass of interconnect plates in a 96-cell stack. They found that with only 10–20 interconnect plates (with increased mass) they were able to achieve the greatest reduction in temperature (Figure 26.9).

Stack-level modeling is a valuable tool for the design of SOFC systems. The stack-level models provide a quick analysis of design and operating changes without resolving the detailed multi-physics within the individual cells of the stack. They are able to provide design guidance such as the trade-offs between

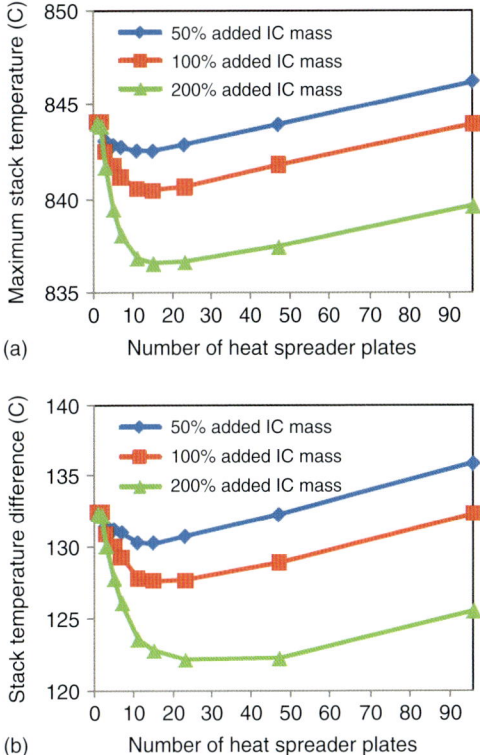

Figure 26.9 Effects of increasing the mass of interconnector plates in a 96-cell stack on the maximum stack temperature (a) and on the temperature difference in the stack (b) [49].

temperature and performance in on-cell reforming, or design options for distributing the heat in the stack. However, the lack of cell-level detail limits stack-level models to considering changes to known systems and materials. To consider the effects of changing, the materials and microstructure of SOFC cells requires a more detailed cell-level model.

26.3.3
Cell-Level Modeling

Cell-level macroscale models consider the heat transfer, species transport, chemical reactions, and electrochemistry within the SOFC cell [27, 31, 51, 52]. In cell-level models, the detailed transport of gas in the fuel and air channels and in the porous electrodes are simulated on a macroscale. This requires a rigorous CFD simulation and commercial codes, such as FLUENT, COMSOL, and Star-CD, are used for cell-level models. Cell-level models consider the electrodes and electrolyte on a continuum scale, which means that the models do not explicitly resolve the microstructures of the tri-layer but instead include it via effective parameters based on the porosity, tortuosity, volume fractions, and so on, of the porous electrodes. Although cell-level models do not resolve the tri-layer microstructure, they do model the individual layers of the tri-layer explicitly. This allows for a more computationally efficient code than mesoscale models, which explicitly resolve the microstructure, but still provides sufficient detail to allow investigation of the SOFC multi-physics at the cell level.

In an effective properties model, the porous microstructures of the SOFC electrodes are treated as continua and microstructural properties such as porosity, tortuosity, grain size, and composition are used to calculate the effective transport and reaction parameters for the model. The microstructural properties are determined by a number of methods, including fabrication data such as composition and mass fractions of the solid species, characteristic features extracted from micrographs such as particle sizes, pore size, and porosity, experimental measurements, and smaller meso- and nanoscale modeling. Effective transport and reaction parameters are calculated from the measured properties of the porous electrodes and used in the governing equations of the cell-level model. For example, the effective diffusion coefficients of the porous electrodes are typically calculated from the diffusion coefficient of Eq. (26.4), and the porosity (ϕ_{gas}) and tortuosity (τ_{gas}) of the electrode:

$$D_i^{eff} = \frac{\phi_{gas}}{\tau_{gas}} \left(\frac{1}{D_{ij}} + \frac{1}{D_i^{Kn}} \right)^{-1} \quad (26.30)$$

Percolation theory [53] is also used to calculate the effective properties such as the ionic conductivity in the SOFC electrodes. The effective conductivity of a composite electrode is less than that of the pure material due to the composite structure and porosity of the electrode. Percolation theory calculates an effective ionic conductivity that accounts for the tortuous path of the electrolyte phase in the electrodes and is based on the probability of finding a percolated chain of the electrolyte phase through the electrode [53].

Cell-level models solve the species [Eq. (26.1)], momentum [Eq. (26.5)], and energy [Eq. (26.7)] conservation equations using the effective properties of the electrodes and can include the electrochemistry using a continuum-scale (Section 26.2.4.1) or a mesoscale (Section 26.2.4.2) approach. Traditionally, cell-level models use a continuum-scale electrochemistry approach, which includes the electrochemistry as a boundary condition at the electrode–electrolyte interface [17, 51, 54] or over a specified reaction zone near the interface. The electrochemistry is modeled via the Nernst equation [Eq. (26.12)] using a prescribed current density and assumptions for the polarizations in the cell. The continuum-scale electrochemistry is then coupled to the species conservation equation [Eq. (26.1)] using Faraday's law:

$$S_i = \frac{i}{nF} \quad (26.31)$$

as the species source term for H_2, H_2O, O_2, and any other electrochemically active species in the system.

The continuum-scale electrochemistry approach has been used to study a variety of problems in SOFC cells, such as cell design and performance, experimental setup, and the use of coal gas in the anode. Barzi *et al.* [54] used a macroscale cell-level model with the continuum electrochemistry approach to study the nonuniformity of gas species, current density, and temperature around the fuel and air tubes of an experimental SOFC button cell apparatus. The model considers the anode, cathode, and electrolyte as a continuum and focuses on how the experimental setup of a button cell will affect the operation of an SOFC [54]. Gemmen and Trembly [51] used cell-level models to investigate the effects of coal gas on the performance of the SOFC anode using a reactive transport model of the anode. The model simulates the transport of coal gas species using the dusty gas model (DGM), which is based on the Stefan–Maxwell equation [Eq. (26.3)], and considers the reactions of the gas species in the anode. The use of the DGM and other multicomponent effective transport models has been the focus of several modeling efforts [15, 52] which focused on understanding the transport limitations in the electrodes and the effects of operating conditions on the species distributions in the electrodes; while using simplified models of the electrochemistry.

Recently, several groups have taken cell-level macroscale models a step further to investigate the electrochemistry through the thickness of the electrodes using the mesoscale electrochemistry approach [19, 27, 31]. In these models, no assumptions are made about a reactive zone for the electrochemical reactions; instead, the electrochemistry is modeled through the thickness of the electrodes based on a mesoscale electrochemistry approach (Section 26.2.4.2) in which the explicit charge-transfer reactions [27] or a modified Butler–Volmer approach [19, 31] are modeled. This extends the effects of the electrochemical reactions away from the electrolyte interface into the electrodes. In these cell-level models, the electrochemistry is coupled to the local species concentrations, pressures, and temperatures, and provides a more detailed view into the local conditions within the fuel cell and how these local conditions affect the overall SOFC performance.

These mesoscale electrochemistry cell-level models can be used to consider the effects of heterogeneity on the overall SOFC performance [19]. Stack and cell models, which apply the electrochemistry as a boundary condition, cannot reflect the effects of heterogeneity through the electrodes on the performance of the cell. The mesoscale electrochemistry models are able to include this and can use this ability to consider the effects of graded electrodes, manufacturing flaws, or degradation on the performance of the SOFC. Ryan *et al.* [19] investigated the effects of improved transport and electrochemical properties, such as porosity, TPB length, and ionic conductivity, near the cathode–electrolyte interface and showed that changes to the local properties in a small area of the cathode can have a significant effect on the performance of an SOFC cell. The same group [55] used a mesoscale electrochemistry model to optimize the microstructural properties and composition of a graded, composite anode. They considered the cumulative effects of changing the properties of the anode to obtain the highest performance possible. Changes in the local Faradaic current density resulted in increases in the overall current density of the cell (Figure 26.10) [55].

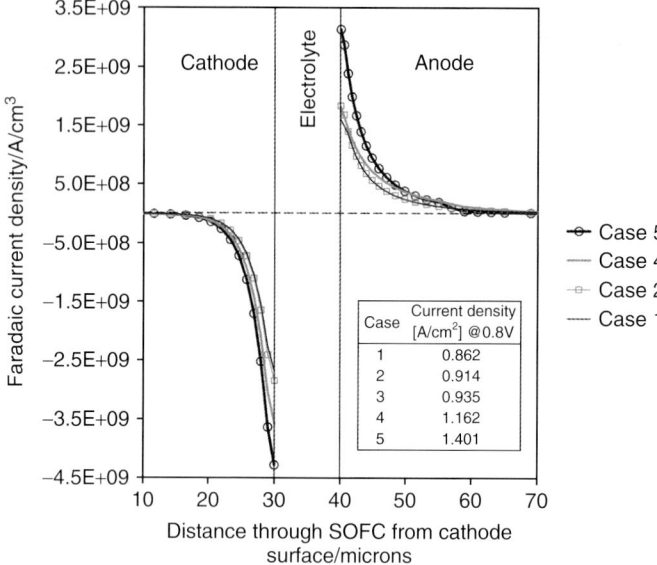

Figure 26.10 Local Faradaic current density through the tri-layer for several optimization cases. The table inset gives the overall cell current density for each case. Note that Case 3 is not shown in the plot as it overlaps Case 2 [55].

Macroscale cell-level models are able to provide a great amount of insight into the operation and performance of SOFCs. With the newer mesoscale electrochemistry models, information about the conditions within the SOFC electrodes and electrolytes can even be resolved. However, due to the continuum-scale treatment of the SOFC, these models still rely on effective parameters, which need to be determined through smaller scale modeling or by fitting the models to experimental data.

26.4
Mesoscale SOFC Modeling

To resolve greater details in SOFCs, mesoscale models are used to study the electrode microstructures and the transport and reactions of species in the SOFC electrodes [3, 34, 56]. Mesoscale models provide more insight into the physics of the electrodes than macroscale models but are limited in the spatial and temporal scales that they can simulate owing to their computational intensity. Owing to the complex geometry of the SOFC electrodes and the significance of surface and interfacial phenomena, mesh-free modeling methods, such as smoothed particle hydrodynamics [34] and the lattice Boltzmann method [3, 57–59], have been used to investigate SOFCs at the mesoscale.

Mesoscale modeling of SOFCs focuses on modeling the transport and reactions of gas species in the porous microstructures of the electrodes [3, 34, 56–59]. In these models, the porous microstructure is explicitly resolved, which negates the need for the effective parameters of macroscale models. The transport and reactions of species in mesoscale models are described by the species [Eq. (26.1)], momentum [Eq. (26.5)], and energy [Eq. (26.7)] conservation equations, which are solved at the pore scale. At the pore scale, the conservation equations are solved in two separate domains: the solid domain of the tri-layer and the gas domain of the pore space within the tri-layer. Mesoscale models aim to understand the effects of microstructure and local conditions near the electrode–electrolyte interface on the SOFC physics and performance. These models have been used to investigate a number of design and degradation issues in the electrodes such as the effects of microstructure on the transport of species in the anode [19, 56] and the reactions of chromium contaminants in the cathode [34].

To model the porous microstructures of the SOFC electrodes, mesoscale models require the imaging of actual electrode microstructures or the use of idealized microstructures. Shikazono et al. [60] investigated reactive transport in an actual SOFC anode by reconstructing the 3D anode microstructure from 2D focused ion beam–scanning electron microscopy (FIB–SEM) imaging data. The use of actual SOFC anode microstructures in mesoscale modeling has been investigated by a number of groups [60–63]. Reconstructing the electrodes from experimental data allows for the exact details of the microstructure to be incorporated into mesoscale models; however, reconstruction also requires a large amount of experimental data and is specific to the electrode sample that is imaged [63]. Using statistical correlations based on experimental imaging data allows for a statistically similar microstructure to be used with less intensive experimental work [63]. Using statistical representations also allows for the investigation of the effects of microstructural design on the performance and multi-physics within the electrodes by simply modifying the properties (statistics) of the statistical representation.

Mesoscale modeling has been used to investigate the reactive transport of species in the anode and the cathode. Ryan et al. [34] modeled a section of the SOFC cathode to investigate the chromium poisoning degradation mechanism. They considered the reactive transport of oxygen and chromium through the cathode thickness and investigated how chromium reacts with the cathode and what the driving forces for chromium poisoning are. Using a reaction rate which is an empirical function of the local current density, the mesoscale model was able to predict distributions of chromium in the cathode which qualitatively match those seen experimentally (Figure 26.11) [34]. Grew et al. [3] preformed similar modeling in the anode to investigate the effects of microstructure on the electrochemical reactions in the anode. Both models consider a 2D section of an electrode. The average pore radii in the SOFC electrodes is ~1 μm; a high-resolution model is needed to resolve the microstructure sufficiently in mesoscale models, which in turn limits the allowable domain size of the models.

Mesoscale modeling can also be used to investigate the stresses in the electrodes and their effect on the structural stability of the electrode [64]. Phosphorus poisoning

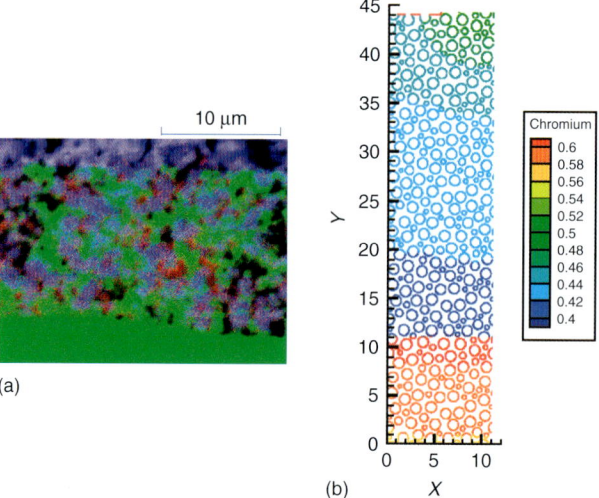

Figure 26.11 (a) An energy-dispersive spectroscopy (EDS) image of a chromium-poisoned SOFC cathode and (b) a contour plot of the chromium distribution in the cathode predicted by mesoscale modeling [34]. The EDS image shows the interface between the YSZ electrode (green) and the LSM cathode (blue), with chromium deposits shown in red (within ~10 μm of the interface). The contour plot shows concentrations of chromium adsorbed on the surface of the cathode, which is represented as circular solid grains. The units of the X and Y axes of the contour plot are micrometers and the concentration of adsorbed chromium is normalized by the maximum chromium concentration in the cathode [34].

in the SOFC anode occurs when phosphorus in gasified coal reacts with nickel in the anode to form Ni_xP_y, which has been shown to cause microcracks in the anode [64]. A joint experimental and computational study investigated the different forms of Ni_xP_y formed during phosphorus poisoning, and the local stresses in the anode caused by volume changes due to Ni_xP_y formation [64]. In the experimental results, cracks oriented in the planar direction were seen in the anode after exposure to phosphorus. Computational simulations of the anode microstructure with increased volumes of the Ni particles due to reactions with the phosphorus showed large horizontal stresses in the anode which corresponded to the horizontal cracking seen in the experimental data [64]. The mesoscale modeling was able to recreate the experimental behavior of the anode and provide a deeper understanding of the cause of cracking.

Mesoscale models provide valuable insight into the operation of SOFCs and how the micrometer-scale phenomena translate into the macroscale behavior of the SOFC. By discretely modeling the gas phase and solid phase of the SOFC electrodes, they can investigate the surface reactions and transport in SOFCs, which could lead to advances in the design of the electrodes to improve the electrochemical performance of the SOFC. They are also able to provide macroscale models with effective properties for the transport and reaction parameters based on the local microstructure and physics of the SOFC.

26.5
Nanoscale SOFC Modeling

Nanoscale modeling methods, such as molecular dynamics (MD) and DFT, aim to understand the kinetics of the electrochemical and chemical reactions on the electrode surfaces and at the interfaces between the electrodes and electrolyte, and the conduction processes in the electrolyte and the electrodes.

Modeling of SOFCs at the nanoscale has not attracted as much research interest as macro- and mesoscale modeling. Nanoscale modeling of SOFCs is challenging due to the complicated physics and complex materials of SOFCs; however, it can provide valuable insight into the reaction mechanisms and reaction rates of species with the SOFC electrode surfaces. MD modeling of SOFC electrodes and electrolytes has been performed by a number of groups, with the work concentrating on understanding the transport of species in the bulk SOFC materials [65, 66], exploring alternative materials [67], and investigating the effects of doping on transport properties [68]. Shishkin and Zeigler [67] investigated the use of alternative anode and electrolyte materials to reduce the risk of coking in an SOFC running under reforming conditions. Using a DFT model, they showed that using $Ni-CeO_2$ as an alternative anode–electrolyte combination reduces the likelihood of coking in the anode.

Nanoscale modeling of degradation mechanisms, such as coking, in SOFCs can aid in the understanding of the degradation rates in the electrodes, mechanisms of degradation, and the surface conditions which facilitate degradation. Very little nanoscale research on SOFC degradation has been done. Apart from the work on coking by Shishkin and Zeigler [67], only one other SOFC degradation mechanism has been studied on the nanoscale, sulfur poisoning. Sulfur poisoning occurs in the SOFC anode when trace sulfur species in gasified coal adsorb on the nickel in the anode. Marquez *et al.* [69] studied the mechanisms of sulfur poisoning in the anode using MD and DFT modeling. The study was able to determine the effects of H_2S on the transport and reactions of hydrogen with the anode by comparing the binding energies and radial distribution functions of hydrogen with and without H_2S. Galea *et al.* [70] used DFT simulations to investigate the removal of sulfur from a poisoned anode by the introduction of oxygen to the anode.

Further applications of nanoscale modeling to sulfur poisoning and other contaminants in the electrodes could lead to a better understanding of what causes degradation and the properties and conditions that could minimize degradation. With continuing improvements in hardware and advanced software tools, nanoscale modeling will gain importance because of its ability to deal with increasingly complex systems.

26.6
Conclusion

Rapid advances in computational power over the last decade and the development of powerful modeling tools have facilitated SOFC modeling for quick predictions of SOFC performance at a number of scales, and has aided in the design of SOFC systems. Owing to the variety of spatial and temporal scales in SOFCs, a multi-scale modeling approach is needed to understand fully the multi-physics of the fuel cell (Figure 26.1). Combining the details of the mesoscale and nanoscale models with the computational efficiency and SOFC system-level information of macroscale models will help to advance SOFC technologies further. As discussed in the previous sections, each of the modeling scales have their own advantages and disadvantages and none of the modeling scales are able to incorporate all the details necessary to understand SOFCs fully. Macroscale modeling provides efficient, powerful tools for investigating the overall operation and performance of SOFC cells, stacks, and systems but is unable to resolve the detailed physics within the cells and relies on effective properties to model the SOFC. Mesoscale models are able to provide more insight into the local conditions within the SOFC electrodes and electrolytes but owing to their computational intensity are unable to model the whole cell. Nanoscale modeling allows the detailed reactions and transport of species at the surfaces and interfaces of the SOFC to be investigated and can provide the parameters needed for larger scale models; however, nanoscale modeling has not been widely applied to SOFC problems and is also computationally expensive, limiting the domains which can be investigated. By combining these various modeling scales into an integrated multi-scale modeling approach, a more detailed understanding of SOFC operation and the challenges facing SOFCs can be investigated. As SOFC modeling continues to advance, attention should be focused on coupling the different scales of SOFC modeling to gain a fuller understanding of the SOFC physics and overall performance.

References

1. Lai, K., Koeppel, B.J., Kyoo Sil, C., Recknagle, K.P., Xin, S., Chick, L.A., Korolev, V., and Khaleel, M. (2011) A quasi-two-dimensional electrochemistry modeling tool for planar solid oxide fuel cell stacks. *J. Power Sources*, **196**, 3204–3222.
2. Agnew, G.D., Collins, R.D., Jorger, M., Pyke, S.H., and Travis, R.P. (2007) The components of a Rolls-Royce 1 MW SOFC system, in *Solid Oxide Fuel Cells 10 (SOFC-X) – 10th International Symposium on Solid Oxide Fuel Cells (SOFC-X), June 3–8, 2007, Nara, Japan*, ECS Transactions, vol. 7 (1) (eds. K. Eguchi, J. Mizusaki, S. Singhal, and H. Yokokawa), Electrochemical Society, Pennington, NJ, p. 105.
3. Grew, K.N., Joshi, A.S., Peracchio, A.A., and Chiu, W.K.S. (2010) Pore-scale investigation of mass transport and electrochemistry in a solid oxide fuel cell anode. *J. Power Sources*, **195** (8), 2331–2345.
4. Ryan, E.M., Recknagle, K.P., Liu, W., and Khaleel, M.A. (2011) The need for nano-scale modeling in solid oxide fuel cells. *J. Nanosci. Nanotechnol.* In Press.
5. Williams, M.C. (2007) Solid oxide fuel cells: fundamentals to systems. *Fuel Cells*, **7** (1), 78–85.
6. Van Duin, A.C.T., Merinov, B.V., Jang, S.S., Goddard Iii, W.A. (2008) ReaxFF reactive force field for solid oxide fuel cell systems with application to oxygen ion transport in yttria-stabilized zirconia. *J. Phys. Chem. A*, **112** (14), 3133–3140.
7. Mazumder, S.K., Pradhan, S.K., Hartvigsen, J., Rancruel, D., von Spakovsky, M.R., and Khaleel, M. (2008) A multidiscipline and

multi-rate modeling framework for planar solid-oxide-fuel-cell based power-conditioning system for vehicular APU. *Simulation*, **84** (8–9), 413–426.
8. Cussler, E.L. (1997) *Diffusion: Mass Transfer in Fluid Systems*, 2nd edn., Cambridge University Press, Cambridge.
9. Arnost, D. and Schneider, P. (1995) Dynamic transport of multicomponent mixtures of gases in porous solids. *Chem. Eng. J. Biochem. Eng. J.*, **57** (2), 91–99.
10. Bird, B., Stewart, W., and Lightfoot, E. (2002) *Transport Phenomena*, 2nd edn., John Wiley & Sons, Inc., New York.
11. Mason, E.A. and Malinauskas, A.P. (1983) *Gas Transport in Porous Media: the Dusty-Gas Model*, Chemical Engineering Monographs, vol. 17, Elsevier, Amsterdam.
12. Coppens, M.O. and Froment, G.F. (1995) Knudsen diffusion in porous catalysts with a fractal internal surface, *Fractals*, Dutch Antilles, Curacao. **3** (4), 807–820.
13. Hwang, J.J., Chen, C.K., and Lai, D.Y. (2005) Computational analysis of species transport and electrochemical characteristics of a MOLB-type SOFC. *J. Power Sources*, **140** (2), 235–242.
14. Ma, L., Ingham, D.B., and Pourkashanian, M.C. (2005) in *Transport Phenomena in Porous Media III* (ed. D.B.P. Ingham), Elsevier, Amsterdam, pp. 418–439.
15. Suwanwarangkul, R., Croiset, E., Fowler, M.W., Douglas, P.L., Entchev, E., and Douglas, M.A. (2003) Performance comparison of Fick's, dusty-gas and Stefan–Maxwell models to predict the concentration overpotential of a SOFC anode. *J. Power Sources*, **122** (1), 9–18.
16. Bravo, M.C. (2007) Effect of transition from slip to free molecular flow on gas transport in porous media. *J. Appl. Phys.*, **102** (7), 074905.
17. Ackmann, T., De Haart, L.G.J., and Stolten, D. (2002) Modelling of mass transport in planar substrate type SOFCs, presented at the 5th European Solid Oxide Fuel Cell Forum, Lucerne, Switzerland.
18. Khaleel, M.A., Rector, D.R., Lin, Z., Johnson, K., and Recknagle, K.P. (2005) Multiscale electrochemistry modeling of solid oxide fuel cells. *Int. J. Multiscale Comput. Eng.*, **3** (1), 33–48.
19. Ryan, E.M., Recknagle, K.P., and Khaleel, M.A. (2011) Modeling the electrochemistry of an SOFC through the electrodes and electrolyte. *Solid Oxide Fuel Cells 12 (SOFC-XII)*, Montreal, CA.
20. Resch, E. (2008) Numerical and Experimental Characterisation of Convective Transport in Solid Oxide Fuel Cells, in *Mechanical Engineering*, Queen's University, Kingston, ON, p. 161.
21. Incropera, F.P. and DeWitt, D.P. (2002) *Fundamentals of Heat and Mass Transfer*, 5th edn., John Wiley & Sons, Inc., New York.
22. Khaleel, M.A., Lin, Z., Singh, P., Surdoval, W., and Collin, D. (2004) A finite element analysis modeling tool for solid oxide fuel cell development: coupled electrochemistry, thermal and flow analysis in MARC®. *J. Power Sources*, **130** (1–2), 136–148.
23. Oldham, K.B. and Myland, J.C. (1994) *Fundamentals of Electrochemical Science*, Academic Press, San Diego, CA.
24. Singhal, S. and Kendall, K. (2006) *High Temperature Solid Oxide Fuel Cells*, Elsevier, Oxford.
25. O'Hayre, R., Cha, S.-W., Colella, W., and Prinz, F. (2006) *Fuel Cell Fundamentals*, John Wily & Sons, Inc., New York.
26. Zhu, H., Kee, R.J., Janardhanan, V.M., Deutschmann, O., and Goodwin, D.G. (2005) Modeling elementary heterogeneous chemistry and electrochemistry in solid-oxide fuel cells. *J. Electrochem. Soc.*, **152** (12), A2427–A2440.
27. Bessler, W.G., Gewies, S., and Vogler, M. (2007) A new framework for physically based modeling of solid oxide fuel cells. *Electrochim. Acta*, **53**, 1782–1800.
28. Zhu, H. and Kee, R.J. (2003) A general mathematical model for analyzing the performance of fuel-cell membrane-electrode assemblies. *J. Power Sources*, **117**, 61–74.
29. Hecht, E.S., Gupta, G.K., Zhu, H., Dean, A.M., Kee, R.J., Maier, L., and Deutschmann, O. (2005) Methane reforming kinetics within a Ni-YSZ SOFC

anode support. *Appl. Catal. A: Gen.*, **295** (1), 40–51.

30. Co, A.C. and Birss, V.I. (2006) Mechanistic analysis of the oxygen reduction reaction at (La, Sr)MnO$_3$ cathodes in solid oxide fuel cells. *J. Phys. Chem. B*, **110** (23), 11299–11309.

31. Zhu, H. and Kee, R.J. (2008) Modeling distributed charge-transfer processes in SOFC membrane electrode assemblies. *J. Electrochem. Soc.*, **155** (7), B715–B729.

32. Marina, O.A., Coyle, C.A., Thomsen, E.C., Edwards, D.J., Coffey, G.W., Pederson, L.R. (2010) Degradation mechanisms of SOFC anodes in coal gas containing phosphorus. *Solid State Ionics*, **181** (8–10), 430–440.

33. Marina, O.A., Pederson, L.R., Edwards, D.J., Coyle, C.A., Templeton, J.W., Engelhard, M.H., and Zhu, Z. (2008) Effect of coal gas contaminants on solid oxide fuel cell operation. *Solid-State Ionic Devices 5 – 212th Electrochemical Society Meeting, October 7–12, 2007, Washington, DC*, Electrochemical Society Inc, United states.

34. Ryan, E.M., Tartakovsky, A.M., Recknagle, K.P., Amon, C.H., and Khaleel, M.A. (2011) Pore scale modeling of the reactive transport of chromium in the cathode of a solid oxide fuel cell. *J. Power Sources*, **196**, 287–300.

35. Recknagle, K.P., Ryan, E.M., Koeppel, B.J., Mahoney, L.A., and Khaleel, M.A. (2010) Modeling of electrochemistry and steam–methane reforming performance for simulating pressurized solid oxide fuel cell stacks. *J. Power Sources*, **195**, 6637–6644.

36. Recknagle, K.P., Koeppel, B.J., Sun, X., Khaleel, M.A., Yokuda, S.T., and Singh, P. (2007) Analysis of percent on-cell reformation of methane in SOFC stacks and the effects on thermal, electrical, and mechanical performance, in *30th Fuel Cell Seminar 30th Fuel Cell Seminar, Honolulu, HI*, Electrochemical Society Inc, United states.

37. Bove, R. and Ubertini, S. (2006) Modeling solid oxide fuel cell operation: approaches, techniques and results. *J. Power Sources*, **159** (1), 543–559.

38. Recknagle, K.P., Williford, R.E., Chick, L.A., Rector, D.R., and Khaleel, M.A. (2003) Three-dimensional electrochemical modeling of planar SOFC stacks. *J. Power Sources*, **113**, 109–114.

39. Barzi, Y.M., Raoufi, A., and Lari, H. (2010) Performance analysis of a SOFC button cell using a CFD model, *Int. J. Hydrogen Energy*, **35** (17), 9468–9478.

40. Mu, L., Powers, J.D., and Brouwer, J. (2010) A finite volume SOFC model for coal-based integrated gasification fuel cell systems analysis. *J. Fuel Cell Sci. Technol.*, **7**, 041017.

41. Vincent, T.L., Sanandaji, B., Colclasure, A.M., Zhu, H., and Kee, R.J. (2009) Physically based model-predictive control for SOFC stacks and systems. *Solid Oxide Fuel Cells 11 (SOFC-XI) – 216th ECS Meeting*, Vienna, Austria.

42. Joulaee, N., Makradi, A., Ahzi, S., Khaleel, M.A., and Koeppel, B.K. (2009) Prediction of crack propagation paths in the unit cell of sofc stacks. *Int. J. Mech. Mater. Des.*, **5** (3), 217–230.

43. Malzbender, J., Wakui, T., and Steinbrech, R.W. (2006) Curvature of planar solid oxide fuel cells during sealing and cooling of stacks. *Fuel Cells*, **6**, 123–129.

44. Nakajo, A., Wuillemin, Z., Van herle, J., and Favrat, D. (2009) Simulation of thermal stresses in anode-supported solid oxide fuel cell stacks. Part II: loss of gas-tightness, electrical contact and thermal buckling. *J. Power Sources*, **193** (1), 216–226.

45. Li, M., Brouwer, J., Rao, A.D., and Samuelsen, G.S. (2011) Application of a detailed dimensional solid oxide fuel cell model in integrated gasification fuel cell system design and analysis. *J. Power Sources*, **196** (14), 5903–5912.

46. Mueller, F., Tarroja, B., Maclay, J., Jabbari, F., Brouwer, J., and S., Samuelsen (2010) Design, simulation and control of a 100 MW-class solid oxide fuel cell gas turbine hybrid system. *J. Fuel Cell Sci. Technol.*, **7** (3), 031007.

47. Lu, N., Li, Q., Sun, X., and Khaleel, M.A. (2006) The modeling of a standalone solid-oxide fuel cell auxiliary power unit. *J. Power Sources*, **161** (2), 938–948.

48. Peksen, M. (2011) A coupled 3D thermofluid–thermomechanical analysis of a planar type production scale SOFC stack. *Int. J. Hydrogen Energy*, **36** (18), 11914–11928.
49. Koeppel, B.J., Lai, K., and Khaleel, M.A. (2011) Effects of geometry and operating parameters on simulated SOFC stack temperature uniformity, presented at the ASME 2011 9th Fuel Cell Science, Engineering and Technology Conference, Washington, DC, 7–10 August 2011.
50. Kulikovsky, A.A. (2010) Temperature and current distribution along the air channel in planar SOFC stack: model and asymptotic solution. *J. Fuel Cell Sci. Technol.*, **7** (1), 011015.
51. Gemmen, R.S. and Trembly, J. (2006) On the mechanisms and behavior of coal syngas transport and reaction within the anode of a solid oxide fuel cell. *J. Power Sources*, **161** (2), 1084–1095.
52. Tseronis, K., Kookos, I.K., and Theodoropoulos, C. (2008) Modelling mass transport in solid oxide fuel cell anodes: a case for a multidimensional dusty gas-based model. *Chem. Eng. Sci.*, **63**, 5626–5638.
53. Chen, D., Lin, Z., Zhu, H., and Kee, R.J. (2009) Percolation theory to predict effective properties of solid oxide fuel-cell composite electrodes. *J. Power Sources*, **191** (2), 240–252.
54. Barzi, Y.M., Raoufi, A., and Lari, H. (2010) Performance analysis of a SOFC button cell using a CFD model. *Int. J. Hydrogen Energy*, **35** (17), 9468–9478.
55. Recknagle, K.P., Ryan, E.M., and Khaleel, M.A. (2011) Numerical modeling of the distributed electrochemistry and performance of solid oxide fuel cells, presented at the ASME 2011 International Mechanical Engineering Congress and Exposition, Denver, CO, 11–17 November 2011.
56. Asinari, P., Quaglia, M.C., von Spakovsky, M.R., and Kasula, B.V. (2007) Direct numerical calculation of the kinematic tortuosity of reactive mixture flow in the anode layer of solid oxide fuel cells by the lattice Boltzmann method. *J. Power Sources*, **170** (2), 359–375.
57. Joshi, A.S., Grew, K.N., Izzo, J.J.R., Peracchio, A.A., and Chiu, W.K.S. (2010) Lattice Boltzmann modeling of three-dimensional, multicomponent mass diffusion in a solid oxide fuel cell anode. *J. Fuel Cell Sci. Technol.*, **7** (1), 011006–011008.
58. Joshi, A.S., Grew, K.N., Peracchio, A.A., and Chiu, W.K.S. (2007) Lattice Boltzmann modeling of 2D gas transport in a solid oxide fuel cell anode. *J. Power Sources*, **164** (2), 631–638.
59. Joshi, A.S., Peracchio, A.A., Grew, K.N., and Chiu, W.K.S. (2007) Lattice Boltzmann method for continuum, multi-component mass diffusion in complex 2D geometries. *J. Phys. D: Appl. Phys.*, **40** (9), 2961–2971.
60. Shikazono, N., Kanno, D., Matsuzaki, K., Teshima, H., Sumino, S., and Kasagi, N. (2010) Numerical assessment of SOFC anode polarization based on three-dimensional model microstructure reconstructed from FIB–SEM images. *J. Electrochem. Soc.*, **157** (5), B665–B672.
61. Izzo, J.R. Jr., Joshi, A.S., Grew, K.N., Chiu, W.K.S., Tkachuk, A., Wang, S.H., and Yun, W. (2008) Nondestructive reconstruction and analysis of SOFC anodes using X-ray computed tomography at sub-50 nm resolution. *J. Electrochem. Soc.*, **155** (5), B504–B508.
62. Wilson, J.R., Kobsiriphat, W., Mendoza, R., Chen, H.Y., Hiller, J.M., Miller, D.J., Thornton, K., Voorhees, P.W., Adler, S.B., and Barnett, S.A. (2006) Three-dimensional reconstruction of a solid-oxide-fuel-cell anode. *Nat. Mater.*, **5**, 541–544.
63. Groeber, M.A., Uchic, M.D., Dimiduk, D.M., Bhandari, Y., and Ghosh, S. (2008) A framework for automated 3D microstructure analysis and representation. *J. Comput.-Aided Mater. Des.*, **14** (Suppl. 1), 63–74.
64. Liu, W., Sun, X., Pederson, L.R., Marina, O.A., and Khaleel, M.A. (2010) Effect of nickel–phosphorus interactions on structural integrity of anode-supported solid oxide fuel cells. *J. Power Sources*, **195** (21), 7140–7145.

65. Cheng, C.H., Chang, Y.W., and Hong, C.W. (2005) Multiscale parametric studies on the transport phenomenon of a solid oxide fuel cell. *J. Fuel Cell Sci. Technol.*, **2**, 219–225.
66. Van Duin, A.C.T., Merinov, B.V., Jang, S.S., and Goddard Iii, W.A. (2008) ReaxFF reactive force field for solid oxide fuel cell systems with application to oxygen ion transport in yttria-stabilized zirconia. *J. Phys. Chem. A*, **112** (14), 3133–3140.
67. Shishkin, M. and Zeigler, T. (2010) The electronic structure and chemical properties of a Ni/CeO$_2$ anode in a solid oxide fuel cell: a DFT + U study. *J. Phys. Chem. C*, **114**, 21411–21416.
68. Lee, S.F. and Hong, C.W. (2010) Multi-scale design simulation of a novel intermediate-temperature micro solid oxide fuel cell stack system. *Int. J. Hydrogen Energy*, **35**, 1330–1338.
69. Marquez, A.I., De Abreu, Y., and Botte, G.G. (2006) Theoretical investigations of NiYSZ in the presence of H$_2$S. *Electrochem. Solid-State Lett.*, **9** (3), A163–A166.
70. Galea, N.M., Lo, J.M.H., and Ziegler, T. (2009) A DFT study on the removal of adsorbed sulfur from a nickel(111) surface: reducing anode poisoning. *J. Catal.*, **263** (2), 380–389.

27
Numerical Modeling of the Thermomechanically Induced Stress in Solid Oxide Fuel Cells

Murat Peksen

27.1
Introduction

Solid oxide fuel cells (SOFCs) represent an increasingly promising technology. However, despite the wide range of advantages, including high efficiency, low pollutant emission, and high volumetric and gravimetric power densities, there are still various problems to be solved. For a reliable and robust SOFC design, hermetic sealing is required that prevents gas leakage, separates the fuel and oxidant within the fuel cell, and bonds the fuel cell components. Currently, three types of sealants are utilized: compressive, compliant, and rigid. Compressive sealants require an externally applied load, whereas compliant sealants are susceptible to chemical reactions and are electrically conductive. Rigid sealants such as glass and glass–ceramics are rigidly bonded to the cell components, can prevent leakage and mixing of gases, and are electrically insulating. Moreover, they are flexible in design, easy to manufacture, and cost competitive.

One of the major challenges in sealing is hermetic and structural reliability. The sealants must be chemically and mechanically compatible with different oxide and metallic cell components and must be electrically insulating. Further, they must withstand thermal cycling. These prerequisites have to be fulfilled because the sealant material can exhibit static, dynamic, cyclic, and thermal fatigue [1]. Extended overviews of sealant requirements and current technologies of sealants in SOFC technology have been given recently by Fergus [2] and Mahapatra and Lu [3]. The efforts to increase the reliability of SOFC components cover overlapping research fields in which numerical modeling has particular importance. Among various models introduced in the literature, particular focus has been given to the thermofluid flow behavior [4–7], whereas less attention has been given to the thermomechanically induced stress within SOFCs.

However, to compete with and ensure the same reliability and long-term stability that traditional power systems provide, better understanding and management of the component regions susceptible to failure are required. Accordingly, this chapter is intended to elucidate the status of the current numerical modeling activities considering the thermomechanically induced stress within SOFCs. In

Fuel Cell Science and Engineering: Materials, Processes, Systems and Technology,
First Edition. Edited by Detlef Stolten and Bernd Emonts.
© 2012 Wiley-VCH Verlag GmbH & Co. KGaA. Published 2012 by Wiley-VCH Verlag GmbH & Co. KGaA.

addition, a complementary three-dimensional (3D) study is described in order to shed light on some modeling issues and the effect of geometric design on the thermomechanical behavior of SOFCs.

27.2
Chronological Overview of Numerically Performed Thermomechanical Analyses in SOFCs

In 2001, Yakabe et al. [8] presented a 3D computational study employing a single unit model with bipolar channels that was considered to be taken from the center of a whole SOFC stack. The thermal boundaries to adjacent units were set as adiabatic. The calculated temperature profile was obtained for co-flow and countercurrent flow. The resulting temperature profile obtained using the commercial tool STAR CD was transferred on to a finite element mesh, in order to perform a structural analysis using the commercial code ABAQUS. The analysis neglected the mechanical constraints among the channels and also no external force was applied. Thus, only thermally induced free elongation was considered. Both the electrolyte and the interconnector results were evaluated according the principal stresses. According to their approach, the internal reforming of the fuel cell would induce a steep drop in fuel temperature near the inlet, resulting in large stresses in the electrolytes. The material models and properties used were not mentioned.

Two further studies by Yakabe et al. [9, 10] were presented in 2004. The first was a numerical study to predict the residual stresses within the electrolyte. Constant thermal boundary conditions were assumed between connecting neighboring components, so as to obtain the temperature distribution. Experimental measurements were presented to demonstrate the agreement of the results. The second study focused on the effects of the cell geometry, fabrication method, and the flattening treatment of the electrolyte.

Selimovic et al. [11] in 2005 introduced a coupled thermal structural analysis, with emphasis on the thermal stresses caused by temperature gradients and the effect of the thermal expansion coefficient on the cell components. They used a FORTRAN code for the solution of current density, species transport, and the flow within the air channels. The temperature distribution obtained was mechanically analyzed using the commercial code FEMLAB. The mechanical analysis was performed solely on the cell components, having neglected the interconnector plates. An elastic approach was used and the cell components were assumed to be free of constraint. Material properties based on the literature were used. Stress during operation was elucidated. Both steady-state and transient analyses were conducted.

Nakajo et al. [12] in 2006 carried out a numerical analysis, studying the effect of the temperature profile characteristics on the stress field of the cell components in a tubular SOFC design. Radial thermal gradients were of particular interest. A simplified Weibull analysis was performed to evaluate the evolution of the probability of survival of the cell components in steady and transient state. Sensitivity analysis was performed for the thermal expansion coefficient effect on the stress using gPROMS and FEMLAB to verify the code results. Their results revealed that high stress values are achieved within the whole anode and the edges of the electrolyte. This was attributed to the differences in thermal expansion coefficients. Critical values were obtained for the probability of survival.

In the same year, Weil *et al.* [13] introduced a sealing concept for planar SOFCs. The finite element method (FEM) was used to aid in scaling up a bonded compliant sealant design to a 120 × 120 mm component. The stresses of the cell, foil, brazes, and frame were calculated and compared with experimental fracture and yield stress results. A quarter symmetrical model was used. The commercial software ANSYS was utilized. The tensile stress of the component was predicted, considering thermal cycling from elevated temperature to room temperature. The materials used were mentioned, but no properties were given. Regarding the structural analysis boundary conditions and the failure criteria employed, material models were not depicted.

The following year, Lin *et al.* [14] presented thermal stress results obtained utilizing a 3D multiple-cell SOFC stack. The thermal stress distribution of the planar SOFC stack was investigated for various process stages, including operational and shut-down stages. The effects of bottom support conditions, thermal expansion coefficient, temperature profiles, and viscous behavior of the glass–ceramic sealant considered were investigated. The commercial FEA code ABAQUS was used for the study. Half of the stack geometry was utilized, including the cell assembly, two interconnector plates with air channels, the glass–ceramic sealants, and the nickel mesh. Continuum shell elements were used. A countercurrent flow thermofluid analysis, including the thermoelectrochemical behavior of the stack, was performed to obtain a realistic temperature profile during the start-up and steady operating stages. The predicted results were then read into the FEA code to execute the stress analysis. Plastic deformation of the metal parts was implemented using in-house experimental data. Different boundary conditions were tested to prevent rigid body motion. They assumed that the elastic modulus of the glass–ceramic used would decrease by one-third owing to its viscous behavior. Failure was judged according to the maximum principal stress, exceeding the ultimate glass–ceramic stress. For the interconnector and frame components, the Tresca equivalent stress exceeding the yield strength of the steel was considered to result in failure. Their predictions revealed that the glass–ceramic is the most critical part of the investigated stack. The commercialization of the stack was mentioned as not being feasible. This was the first example in which the glass–ceramics were considered within the analysis. The results implied that the maximum stress occurs around the borders of the glass–ceramic and cell components.

Weil and Koeppel [15] reported a study in 2008 involving bonded compliant sealants. The thermally induced stress–strain behavior of the cell, sealant, and frame components was investigated using the 3D FEM. ANSYS was used for the calculations. The assembly was heated and cooled uniformly. A one-quarter section was modeled due to the fourfold symmetry of the component. The cell was treated computationally as a single material, considering the anode properties.

A bilinear elastic–plastic constitutive model was employed for the interconnector plate. Material properties obtained from the literature were used. The sealant was treated as elastic. Isothermal conditions were set. They found that some bowing occurred during cooling, and that the stress is transferred to the sealing foil and soft silver braze that were used in the study. The component stresses were found to be lower in a second thermal cycle compared with the first cycle.

In 2008, Weil and Koeppel [16] investigated three different bonded sealant designs, namely bonded compliant, glass–ceramic, and air-brazed silver–copper oxide. An FEM including an anode structure, frame, and the sealant was used to calculate the mechanical behavior of the assembly. A uniform thermal load was applied for the operational and shut-down stage simulations. A slow heating rate furnace was imitated. The same boundary conditions as employed in their previous studies were adopted. The interconnector plates were assumed to offer little resistance to the out-of-plane deformation, thus interpreted as negligible. The metallic components were treated using a bilinear kinematic hardening model, whereas the cell was treated as elastic. The material properties were given in graphical representations. The results showed highly overpredicted stress values, in particular when using glass–ceramics.

In 2009, Lin et al. [17] investigated the thermal stress distribution of a multi-cell SOFC stack, consisting of three cells joined together. Interconnector plates, cell, frame, nickel mesh (as continuum), and glass sealants were considered in the 3D FEM developed using ABAQUS. Mica-based compressive gaskets were used to seal the cell, supporting frame, and the interconnectors. Continuum shell elements were used. A countercurrent flow configuration was used to obtain the thermal field of the model used. A mechanical load ranging from 0.06 to 6 MPa was applied. Various stages during an operating cycle were investigated. For steady operation, results from their earlier work [14] were imported. Information on how the model was cooled to room temperature or heated was not given. The stress results for the cooling and heating stages were given for the end time of each stage, thus considering a uniform temperature distribution. Tight bonding conditions were applied between the glass sealant and the metallic components. The other connecting pairs of components were constrained using a contact constraint.

Failure of bulk material was considered. Plastic deformation of the metallic parts was evaluated using the Tresca criterion. Glass sealants were considered to be linear at room temperature and nonlinear at operating temperature. The results implied that a mechanical load of 0.6 MPa results in optimum bending behavior and maintains critical stress values. Room temperature stresses showed lower values than the fracture strength, whereas steady operation stresses were locally higher than the failure values. The stresses of cell, frame, interconnector plates, and nickel mesh were claimed to result mainly from thermal expansion differences and thermal gradients. The use of a mica sealant gasket joint resulted in lower stress values, but considerably higher values compared with glass–ceramic sealants were achieved at the operating stage.

A comprehensive numerical study investigating thermal stress using a bonded compliant sealant design was introduced by Jiang and Chen [18] in the same year. A unit cell composed of single air and fuel channels, cell, and interconnector plates was numerically analyzed. The temperature distribution of the unit cell was obtained. It was assumed that the whole plate had the same distribution in width, hence the results were extended to the remaining channels of a single plate. A co-flow design was selected, employing half of the plate and considering 20 unit cells for the solution of the structural analysis. The cell, considered as one layer with anode properties, was assumed to behave elastically; whereas the metallic parts were considered to behave elastic–plastically, following a bilinear stress–strain relationship. Solid elements were used, where ANSYS was employed. Current–voltage curves were used to justify the thermal results. The authors mentioned that the nonuniform temperature distribution has a significant effect on the stress distribution. The temperature gradient effect on the thermal stress of the compliant sealant would be comparable to that in the glass–ceramic sealant design.

Fischer and Seume [19] also in 2009 investigated the impact of the temperature profile on the thermal stress arising in tubular SOFCs. A 2D FEM model was employed. The commercial tool COMSOL was used for the analyses. An axisymmetric model considering the membrane electrode assembly (MEA) was used. Constant material properties were used for specified temperatures. Free expansion was allowed. The effect of absolute temperature and thermal gradients together with the internal reforming (direct and indirect) was elucidated. The results suggested a strong coupling between the reforming method and the stress distribution. The authors mentioned that thermal gradients in tubular SOFCs are less relevant to thermal stress.

Wening and co-workers [20, 21] in 2010–2011 investigated the creep behavior of glass–ceramic sealants and interconnector plates by employing a one-cell [20] and a three-cell SOFC stack model [21], operating with cross-flow. A continuum approach was attempted, as the presented geometry showed no physical resolution details of the components. The results implied that the creep effect of the glass and interconnector would change the channel dimensions in long-term operating conditions. MSC Marc Mentat was used for the studies.

In 2010, Chiang et al. [22] reported a 3D study investigating the thermofluid and thermal stress behavior of a single SOFC cell. The model used included an electrolyte, sandwiched between two aluminum gas distributors. The commercial software STAR-CD and MSC Marc were used for the thermal field and the thermal stress analyses, respectively. The cell was supported from its bottom surface; friction was defined among the cell and the gas distributors. A uniform load was applied on the top surface. The edge points were constrained. Experimental current–voltage curves were used to show that the results were realistic. The predicted thermal stress within the cell was lower at lower temperature gradients. The results implied that the stress within the cell is dominated by the thermal gradients and the thermal expansion coefficient.

Again in 2010, a sequentially solved thermofluid, thermomechanical analysis using the commercial software COMSOL and utilizing triangular elements was introduced for a micro tubular SOFC by Serincan et al. [23]. Thermal stresses arising from the fabrication of the cell, exterior constraints, and the fuel cell operation were considered. An axisymmetric 2D approach was presented. The model considered aluminum tubes and also the anode, cathode, and electrolyte. The predicted thermal field was used in a mechanical model that considered the ceramic sealant. Materials were considered to behave elastically. The properties were calculated using the composite sphere method. The authors mentioned the importance of accounting for the sealants within the modeling procedure. The results suggested that both compressive and tensile stresses may arise simultaneously in the same layer, due to interactions among the cell and the supporting structures.

Peksen [24] filed a patent in 2011 describing the geometric design relations of fuel cell components affecting the thermomechanically induced stress behavior. All dimensions of an interconnector plate were considered and described as a function of the width of the plate. The relations clarify the requirements for a robust and reliable geometric design.

Liu et al. [25] carried out a numerical study in 2010 that considered thermal stress and crack nucleation of a three-layer SOFC model. The model comprised the anode, electrolyte, and a thin interface layer between them. The shear stress and peeling stress were investigated using the Paris law. It was attempted to evaluate the results using literature data. The model considered solely the cell components, neglecting the remaining metallic and sealant parts. It was aimed to estimate the lifetime of the cell under thermal cycling for different cell configurations.

Also in 2010, a systematic approach combining experimental measurements and postmortem stack analyses coupled with numerical analyses was described by Blum et al. [26]. The results revealed that even small thermal gradients lead to stress zones within the stack, in particular in the vicinity of the manifolds. Furthermore, the stack operating in a furnace would not explicitly reflect the stack behavior operating during a real process.

In 2011, an advanced coupled 3D computational fluid dynamics (CFD)–FEM analysis of a SOFC stack was presented by Peksen [27]. The study gave a detailed insight into the limitations of the usual assumptions employed in the literature, including a continuum approach versus a discrete approach, and also a production-scale thermofluid, thermomechanical analysis. This study represents the first published investigation of a full production-scale 36-layer stack considering all of its components including the cell, sealant, and metal components. The manifold regions and the channel regions were presented in physical resolution. Moreover, a broad comparison of the material nonlinearity effect was presented, including the thermoelastic and plastic strain effect on the thermomechanical behavior of the SOFC stack.

The above literature review reveals that most of the studies performed were concentrated on the thermomechanical behavior of the cell components, and little effort has been made to include typical SOFC components within the models, even though those components contribute to the overall thermomechanical behavior.

As most experiments show that hermetic problems are present when operating in real conditions, it is essential to account in the modeling activities for the metallic parts in addition to sealants, if any are present within the assembly.

Moreover, care should be taken with the geometric simplifications performed, if they are targeted at improving the overall understanding of the thermomechanical behavior of SOFCs. Choosing the right material properties and the material models also has a great impact on the overall results, which may lead to inadequate evaluations. Apart from the commercially available material properties, an overview including the thermomechanical material properties used in the literature cited above is given in Table 27.1. Further, to give an insight into the numerical models utilized, Figure 27.1 summarizes the currently employed thermomechanical model status with regard to SOFCs.

27.3
Mathematical Formulation of Strain and Stress Within SOFC Components

The way in which each individual SOFC component responds to stress varies widely depending on the physical conditions under which the deformation takes place and the mechanical properties of the materials. As the stiffness and strength of the material are closely interrelated, it is exceedingly important to describe the material behavior of the components to be modeled. The components usually employed in SOFC technology comprise the cell components that are usually investigated as a continuum layer, sealants, wire mesh, and the metallic components, that is, the interconnector plate and frame. This section outlines the mathematical formulation of the thermomechanically induced stress within each individual component.

27.3.1
Cell, Sealant, and Wire Mesh Components

The ceramic cell components within SOFC technology feature a high elastic stiffness and high strength, whereas the sealants utilized usually exhibit high crystalline behavior, that is, having a particular melting point. Thus, unlike amorphous glass–ceramics, they do not behave like a viscous fluid at elevated temperatures. Owing to the nature of the chemical bonds, they show brittle behavior and are susceptible to failure. The wire mesh, which should enhance the contact within the fuel cell components, is usually made of nickel. Furthermore, it is flexible. Hence all three components can be regarded as isotropic elastic materials. The mathematical formulation of the three macroscopic components is devoted to the elastic behavior, in which the total strain $\{\varepsilon_{tot}\}$ can be described as

$$\{\varepsilon_{tot}\} = \{\varepsilon_m\} + \{\varepsilon_{th}\} \tag{27.1}$$

Table 27.1 Overview of the thermomechanical material properties used in some of the studies cited.

Ref.	Item	Modulus of elasticity (GPa)				Poisson's ratio	Thermal expansion coefficient (K^{-1})			
		RT[a]	400°C	600°C	800°C		RT[a]	400°C	600°C	800°C
[8]	Interconnector									
	Electrolyte			206		0.3			10.56 × 10^{-6}	
	Anode			96		0.3			12.22 × 10^{-6}	
	Cathode									
	Sealant									
[11]	Interconnector									
	Electrolyte			215		0.32			10.0 × 10^{-6}	
	Anode			55		0.17			13.0 × 10^{-6}	
	Cathode			35		0.25			11.0 × 10^{-6}	
	Sealant									
[12]	Interconnector									
	Electrolyte	212			183	0.32			10.8 × 10^{-6}	
	Anode		57			0.28			12.12 × 10^{-6}	
	Cathode		35			0.25			11.7 × 10^{-6}	
	Sealant									
[19]	Interconnector									
	Electrolyte	216			155 (900°C)	0.316			10.3 × 10^{-6}	
	Anode	81			64 (900°C)	0.25			12.0 × 10^{-6}	
	Cathode	65			58 (900°C)	0.28			10.7 × 10^{-7}	
	Sealant									
[23]	Interconnector									
	Electrolyte								11 × 10^{-6}	
	Anode								11.8 × 10^{-6}	
	Cathode								13 × 10^{-6}	
	Sealant			90		0.31			7.74 × 10^{-6}	
[25]	Interconnector									
	Electrolyte			205		0.31			10.3 × 10^{-6}	
	Anode			220		0.3			12.5 × 10^{-6}	
	Cathode			114		0.28			12.4 × 10^{-6}	
	Sealant									

[a] RT, room temperature.

where $\{\varepsilon_{th}\} = \{\alpha\,\alpha\,\alpha\,0\,0\,0\}\,\Delta T$ is the thermal strain vector and α denotes the coefficient of thermal expansion. ΔT refers to the increase in temperature in the solid, that is, the difference between the current temperature and the reference temperature at zero initial strain. $\{\varepsilon_m\}$ is the mechanical strain, considering the elastic strain vector. The thermomechanically induced strain is related to stress through

27.3 Mathematical Formulation of Strain and Stress Within SOFC Components

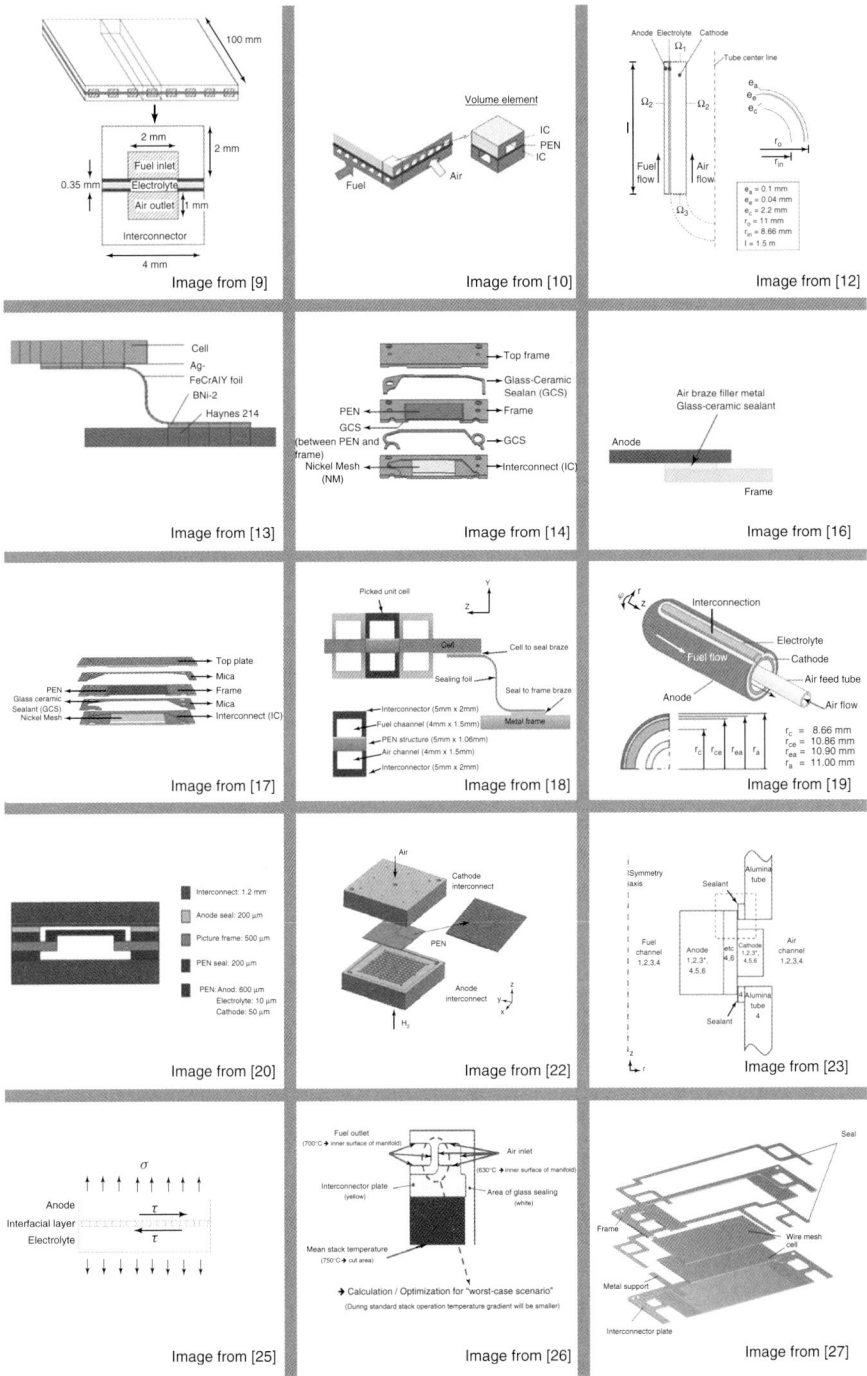

Figure 27.1 Overview of some computational models that have been employed.

$$\begin{Bmatrix} \varepsilon_{xx} \\ \varepsilon_{yy} \\ \varepsilon_{zz} \\ 2\varepsilon_{yz} \\ 2\varepsilon_{xz} \\ 2\varepsilon_{xy} \end{Bmatrix} = \frac{1}{E} \begin{bmatrix} 1 & -v & -v & 0 & 0 & 0 \\ -v & 1 & -v & 0 & 0 & 0 \\ -v & -v & 1 & 0 & 0 & 0 \\ 0 & 0 & 0 & 2(1+v) & 0 & 0 \\ 0 & 0 & 0 & 0 & 2(1+v) & 0 \\ 0 & 0 & 0 & 0 & 0 & 2(1+v) \end{bmatrix} \begin{Bmatrix} \sigma_{xx} \\ \sigma_{yy} \\ \sigma_{zz} \\ \sigma_{yz} \\ \sigma_{xz} \\ \sigma_{xy} \end{Bmatrix}$$

$$+ \alpha \Delta T \begin{Bmatrix} 1 \\ 1 \\ 1 \\ 0 \\ 0 \\ 0 \end{Bmatrix} \tag{27.2}$$

where $\varepsilon_{xx}, \varepsilon_{yy}$, and ε_{zz} are the normal strain values and $\varepsilon_{yz}, \varepsilon_{xz}$, and ε_{xy} denote the shear strains in each plane, v is Poisson's ratio, and E is the modulus of elasticity. The resulting stress can be calculated using the inverse relationship between the stress and strain:

$$\begin{Bmatrix} \sigma_{xx} \\ \sigma_{yy} \\ \sigma_{zz} \\ \sigma_{yz} \\ \sigma_{xz} \\ \sigma_{xy} \end{Bmatrix} = \frac{E}{(1+v)(1-2v)} \begin{bmatrix} 1-v & v & v & 0 & 0 & 0 \\ v & 1-v & v & 0 & 0 & 0 \\ v & v & 1-v & 0 & 0 & 0 \\ 0 & 0 & 0 & \frac{(1-2v)}{2} & 0 & 0 \\ 0 & 0 & 0 & 0 & \frac{(1-2v)}{2} & 0 \\ 0 & 0 & 0 & 0 & 0 & \frac{(1-2v)}{2} \end{bmatrix}$$

$$\begin{Bmatrix} \varepsilon_{xx} \\ \varepsilon_{yy} \\ \varepsilon_{zz} \\ 2\varepsilon_{yz} \\ 2\varepsilon_{xz} \\ 2\varepsilon_{xy} \end{Bmatrix} - \frac{E\alpha \Delta T}{1-2v} \begin{Bmatrix} 1 \\ 1 \\ 1 \\ 0 \\ 0 \\ 0 \end{Bmatrix} \tag{27.3}$$

27.3.2
Metallic Components

The most frequent SOFC materials for the design of metallic frame and interconnector plates are made of high-Cr ferritic steels. Owing to their high chromium contents (22 wt% or more), the material possesses the required long-term oxidation resistance and a thermal expansion coefficient similar to that of the ceramic components [28]. The ability of the metal to deform under stress and elevated temperature represents its ductility. At high temperatures the ductility increases, causing the material to become weaker. Moreover, the deviation from the ideal elastic state increases, rendering the elastic appearance inadequate. The real material behavior can be described using the theory of plasticity. This requires an understanding of the relationship between the stress and strain within the metals used. To explore this issue, it is essential to know how the stress induced from particular loads characterizes the deformation of the material. Experimental data need to be acquired to imitate the behavior and implement in numerical models. Ferritic steels employed in SOFCs are typically subjected to temperatures of 20–800 °C, showing a smooth elastic–plastic transition that can be described by the Ramberg–Osgood equation [29]. The numerical predictions demonstrated in this study are based on the nonlinear finite element analysis, employing a multilinear form of this equation. The total strain vector $\{\varepsilon_{tot}\}$ within the metal component is given by

$$\{\varepsilon_{tot}\} = \{\varepsilon_{el}\} + \{\varepsilon_{pl}\} + \{\varepsilon_{th}\} \tag{27.4}$$

where $\{\varepsilon_{el}\}$ and $\{\varepsilon_{pl}\}$ are the elastic and plastic strain vectors, respectively, which form the mechanical strain $\{\varepsilon_m\}$. The thermal strain vector $\{\varepsilon_{th}\}$ is expressed as given in Eq. (27.1).

Based on the mathematical theory of plasticity, the plastic deformation behavior of the material can be described by the three components of the rate-independent plasticity model, namely yield criterion, flow rule, and hardening rule. The yield criterion determines the stress level at which yielding is initiated. This is represented by the equivalent stress σ_{eq}, which is a function of the individual stress vector components $\{\sigma\}$. Plastic strain is developed in the metal parts when the equivalent stress is equal to a material yield parameter σ_y; finally, the flow rule determines the direction of plastic straining:

$$\{d\varepsilon_{pl}\} = \lambda \left\{\frac{\partial Q}{\partial \sigma}\right\} \tag{27.5}$$

where the plastic multiplier λ determines the amount of plastic straining and Q denotes the plastic potential that establishes the direction of plastic straining. The hardening rule describes the stress–strain relationship to increase further the amount of stress to produce additional strain after having reached the elastic limit. The isotropic (or work) hardening rule is used where the yield surface remains centered about its initial centerline and expands in size as the plastic strains develop. Hence in practice this means that if the equivalent stress calculated using elastic properties exceeds the material yield, then plastic straining must occur. Plastic strains reduce the stress state so that it satisfies the yield criterion. The hardening rule states that the yield criterion changes with isotropic hardening. Considering these points, the stress increment can be calculated using the elastic stress–strain relations

$$\{d\sigma\} = [D]\{d\varepsilon_{el}\} \tag{27.6}$$

where the inverse of the elastic stiffness matrix is expressed as

$$[D]^{-1} = \begin{bmatrix} 1/E & -v/E & -v/E & 0 & 0 & 0 \\ -v/E & 1/E & -v/E & 0 & 0 & 0 \\ -v/E & -v/E & 1/E & 0 & 0 & 0 \\ 0 & 0 & 0 & 1/G & 0 & 0 \\ 0 & 0 & 0 & 0 & 1/G & 0 \\ 0 & 0 & 0 & 0 & 0 & 1/G \end{bmatrix} \tag{27.7}$$

and

$$\{d\varepsilon_{el}\} = \{d\varepsilon_{tot}\} - \{d\varepsilon_{pl}\} \tag{27.8}$$

The equivalent stress σ_{eq} of the metal components is obtained based on the von Mises yield criterion:

$$\sigma_{eq} = \left(\frac{3}{2}\{S\}^T[M]\{S\}\right)^{\frac{1}{2}} \tag{27.9}$$

where [M], the factor matrix, is given by

$$[M] = \begin{bmatrix} 1 & 0 & 0 & 0 & 0 & 0 \\ 0 & 1 & 0 & 0 & 0 & 0 \\ 0 & 0 & 1 & 0 & 0 & 0 \\ 0 & 0 & 0 & 2 & 0 & 0 \\ 0 & 0 & 0 & 0 & 2 & 0 \\ 0 & 0 & 0 & 0 & 0 & 2 \end{bmatrix} \tag{27.10}$$

and the deviatoric stress $\{S\}$ is expressed as

$$\{S\} = \{\sigma\} - \frac{1}{3}\left(\sigma_x + \sigma_y + \sigma_z\right)\begin{Bmatrix} 1 & 1 & 1 & 0 & 0 & 0 \end{Bmatrix}^T \tag{27.11}$$

27.4
Effect of Geometric Design on the Stress Distribution in SOFCs

Most of the experimental observations show that thermomechanically induced hermetic problems, or sealant failures within the SOFC assembly, occur mostly in the vicinity of the manifold ports [26]. In order to understand and improve the knowledge about the reliability of the component, it is exceedingly important to determine the stress distribution and the associated locations susceptible to stress. As the component is subjected to thermomechanical stress during heating, cooling, and operation, stress concentrations arise at the corners of the manifold ports. These can lead to fracture and failure of the sealant. From a computational modeling point of view, it is very important to investigate and understand the limits of geometric simplifications, as these may affect the analysis results drastically. The thermofluid flow models employed may consider simplifications that may lead to inaccurate thermomechanical analysis results and vice versa [30]. To visualize the critical locations and the stress levels that arise, it is essential to employ 3D models for both thermofluid and thermomechanical analyses within the SOFC technology.

Most of the numerical studies published on thermomechanical SOFC analyses, have been focused on thermofluid flow-based effects, such as process conditions and thermal gradients, or on typical known effects such as the thermal expansion of materials, and so on. The demonstrated study aims to give a flavor about the limits of geometrical modeling simplifications. Thereby, it will shed light on the geometrical effect of SOFC designs on the thermomechanically induced stress. Hence a fundamental coupled CFD–FEM approach is presented below; in particular, the effect of fillet radius on the thermomechanically induced stress within SOFCs is addressed.

27.4.1
Computational Fluid Dynamics (CFD) Analysis

The computational study considers the quarter manifold region of an interconnector plate, metal frame, and two sealants located between the metal parts that have been virtually modeled in 3D. The approach demonstrates the geometric effect of the manifold regions subjected to the same thermofluid flow boundary conditions. The model accurately presents the configuration of the physical fuel cell manifold region. For this study, three geometric designs with different fillet radii are used. Figure 27.2 illustrates the approach employed together with the different geometric designs.

The study utilized the thermofluid boundary conditions given in Table 27.2. A constant temperature was applied to the channel walls at the cut region, imitating the hotter thermal zone due to the cell component, which at this location approaches approximately a constant value. Symmetry boundary conditions were applied for the symmetry axis. More accurate results can be obtained using advanced finite element sub-modeling techniques, implementing the results from a whole component solution and concentrating on the particular zone, using a finer numerical grid resolution [30–32].

The numerical analysis results predicted for the three cases are illustrated in Figure 27.3. The contour lines of the results show the constant-temperature value of the solid along each curve. The tightness of the lines indicates a high gradient of the temperature, that is, the variation is steep. The predictions reveal that case 1 with no fillet radius has the highest thermal gradients compared with case 2, whereas case 3 shows the lowest thermal gradients visible from the smooth transitions and horizontal behavior of the lines. This suggests that the use of a higher fillet radius reduces the thermal gradients favorably, whereas the use of small radii or no radius increases the thermal gradients.

Table 27.2 Applied boundary conditions for thermofluid analysis.

Boundary name	Mass flow rate (kg s^{-1})	Temperature (°C)
Air inlet	1.8×10^{-4}	630
Fuel inlet	1.096×10^{-5}	500
Wall at cut region	–	635

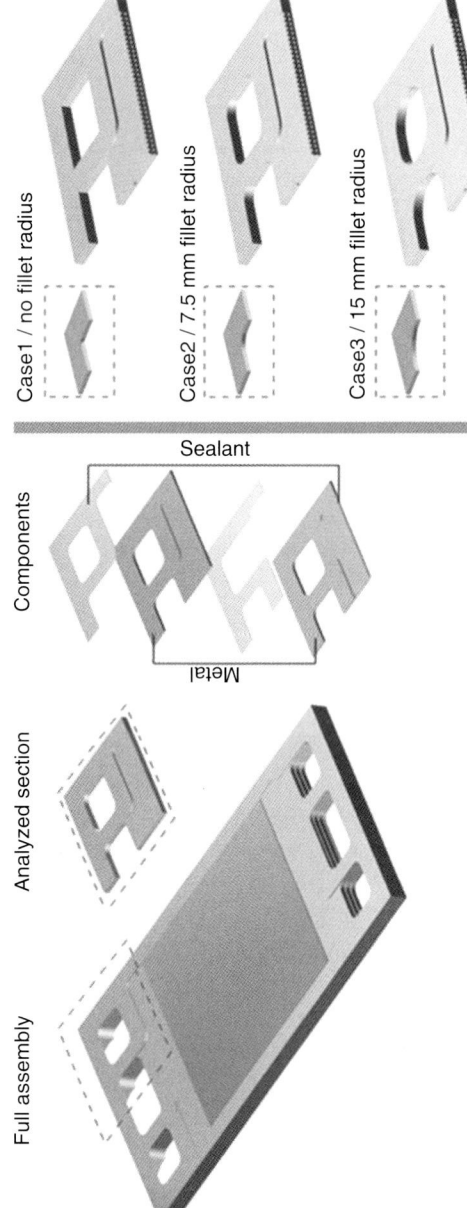

Figure 27.2 Models employed for the coupled CFD–FEM analyses.

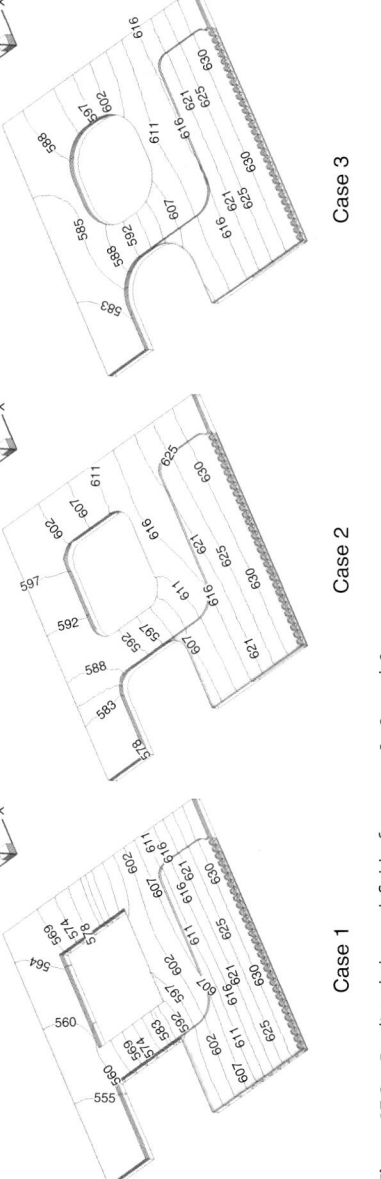

Figure 27.3 Predicted thermal fields for cases 1, 2, and 3.

27.4.2
Thermomechanically Induced Stress Analysis

To investigate the thermomechanically induced stress within the three components investigated, direct coupling of the predicted CFD results was applied. Each predicted thermal field was used in the associated finite element analyses as thermal load boundary conditions. The computational solid mechanics analysis performed in this section utilized a standard glass–ceramic sealant [26], whereas the metal components including the frame and interconnector plate were assumed to be made of Crofer 22APU. As described, the sealant was regarded as elastic, whereas the material behavior of the metal components was described using the elastic–plastic stress–strain behavior shown in Figure 27.4.

Owing to the symmetry plane, the components were constrained with frictionless support so as to avoid perpendicular movement in the direction of the symmetry plane. The pressure force on the component walls induced by the fluid inside the fluid path was not considered, as it is negligible compared with the thermal and mechanical loads applied. The analyses assume that the mechanical stress caused by the assembly weight is negligible compared with the coupled mechanical–thermal stress due to the mechanical constraints between the plates and the thermal field, and hence is neglected. A bottom support was applied to each model, in order to avoid a translation downwards and causing rigid body motion. The analyses considered a mechanical load of 0.064 MPa applied on top of the assembly, which normally is used to provide full contact of the gaskets present among integrated SOFC module components. The numerical solution for the thermomechanically induced stress within each individual component and design was investigated.

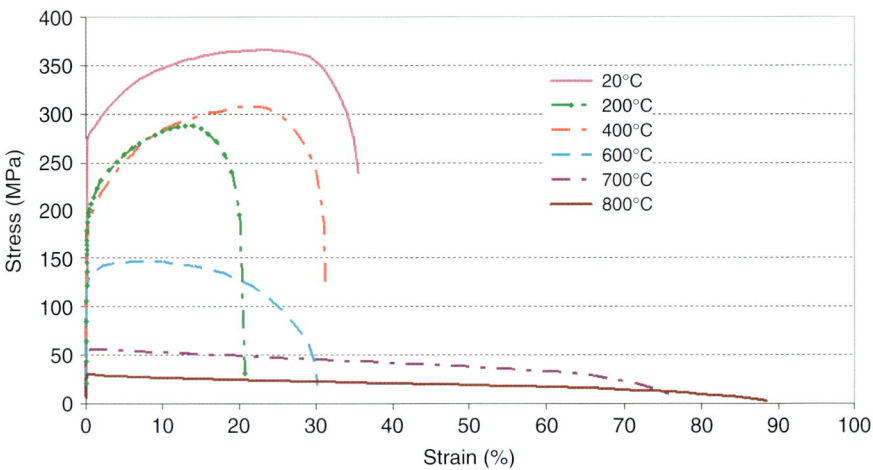

Figure 27.4 Utilized nonlinear material behavior curves of the metal components [32].

27.4.2.1 Thermomechanically Induced Stress Within the Sealant Components

The results for the sealant component were evaluated using the maximum principal stress theory valid for brittle materials. The stress analyses were performed for all three cases used in the CFD analyses. The stress results are illustrated at the fillet regions, in order to evaluate the geometric effect on the stress behavior. Path lines are defined at specified locations, and all three cases are superimposed geometrically, so as to ensure that results are obtained along the same path within each component. Figure 27.5 illustrates the maximum principal stress results and the specified locations utilized for the comparison of cases 1, 2, and 3.

The results show that the highest stress values arise at Line_A and Line_C with maximum values achieved in case 1, having no fillet radius. Locations B and D are subjected to lower stress, showing clearly favorable values for cases 2 and 3. It is noticeable that the maximum values for cases 2 and 3 reduce to approximately one-quarter of the value predicted for case 1. The predictions for cases 2 and 3 show similar values at location A and C compared at the same path length, whereas substantial differences are observed at location B, where the use of a longer radius results in low tensile stress (<1 MPa) compared with case 2, which shows a tensile stress of nearly 6 MPa. Figure 27.6 shows the results simulated for the top sealant.

The results for this layer show that the maximum principal stress along Line_A and Line_D are the critical locations for the component with lower maximum values compared with the bottom sealant, which is attributed to the single side lap compared with the double side lap of the bottom sealant. Furthermore, the results for case 3 show slightly lower stress values than for case 2. As the presence of sealant material is lacking at location B, the effect of the fillet radius difference is less compared with the sealant used at the bottom of the assembly.

27.4.2.2 Thermomechanically Induced Stress Within the Metal Components

The computational predictions for the metal components were evaluated according to the von Mises theory, subject to ductile materials. Figure 27.7 depicts the results for the interconnector plate located at the bottom of the assembly (Figure 27.2). The results for the metal components comprise all six path lines defined, hence they are the same locations as used in the sealant analyses and an additional representation is omitted.

The results imply that fillets A and C are subject to the highest stress loads as for the sealant. The higher values for case 2 compared with case 1 at location B are due to the distributed stress at this location. The remaining fillets show clearly the higher stress resulting when no fillet radius is present. Location B reflects approximately that the stress increases threefold when utilizing a fillet radius of 7.5 mm compared with 15 mm. The numerical analysis results for the top metal component are depicted in Figure 27.8.

The predictions show that the same tendencies at specified locations are visible at location B, showing the expected order, that is, case 1 giving higher stress values than case 2. The results imply that the use of a longer fillet radius improves the thermomechanical response of the components as lower tensile values are obtained. In general, it is clearly demonstrated that the thermomechanically induced stress

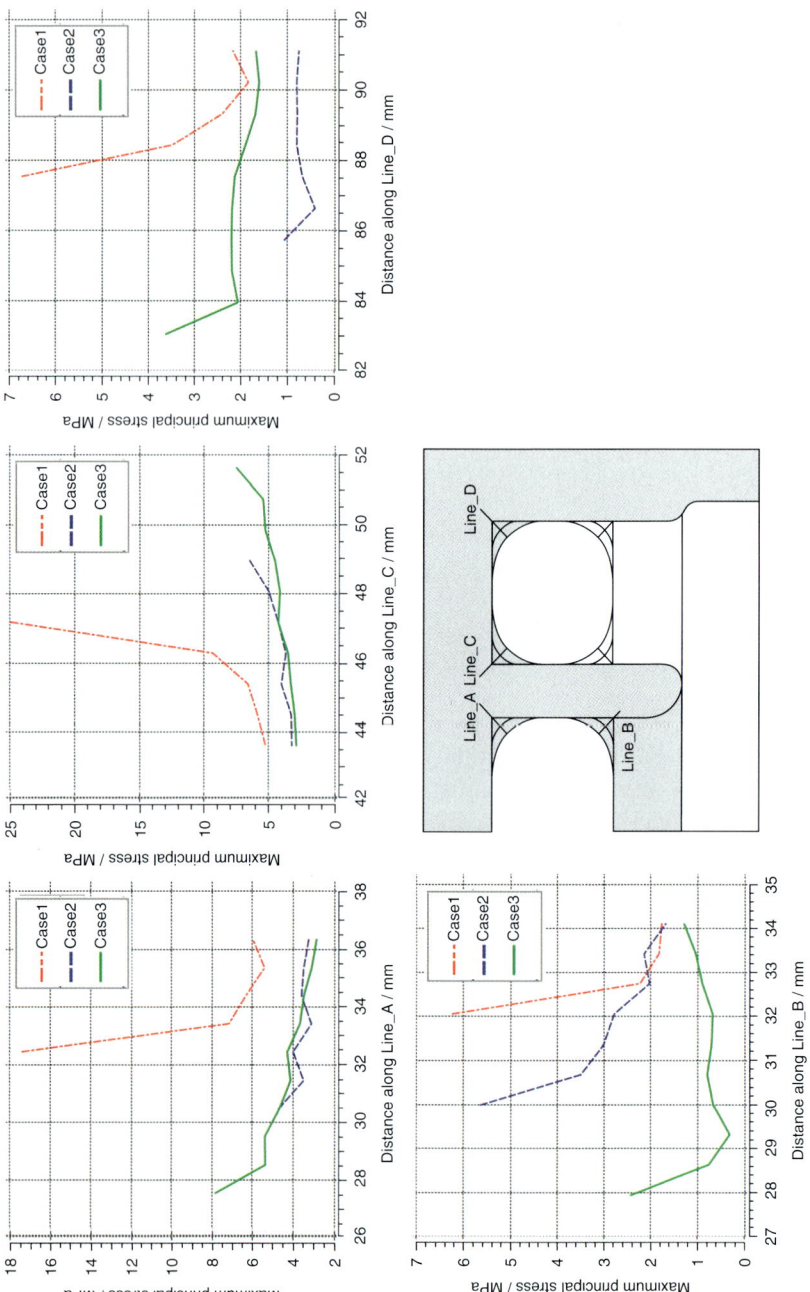

Figure 27.5 Maximum principal stress results for the bottom sealant.

27.4 Effect of Geometric Design on the Stress Distribution in SOFCs

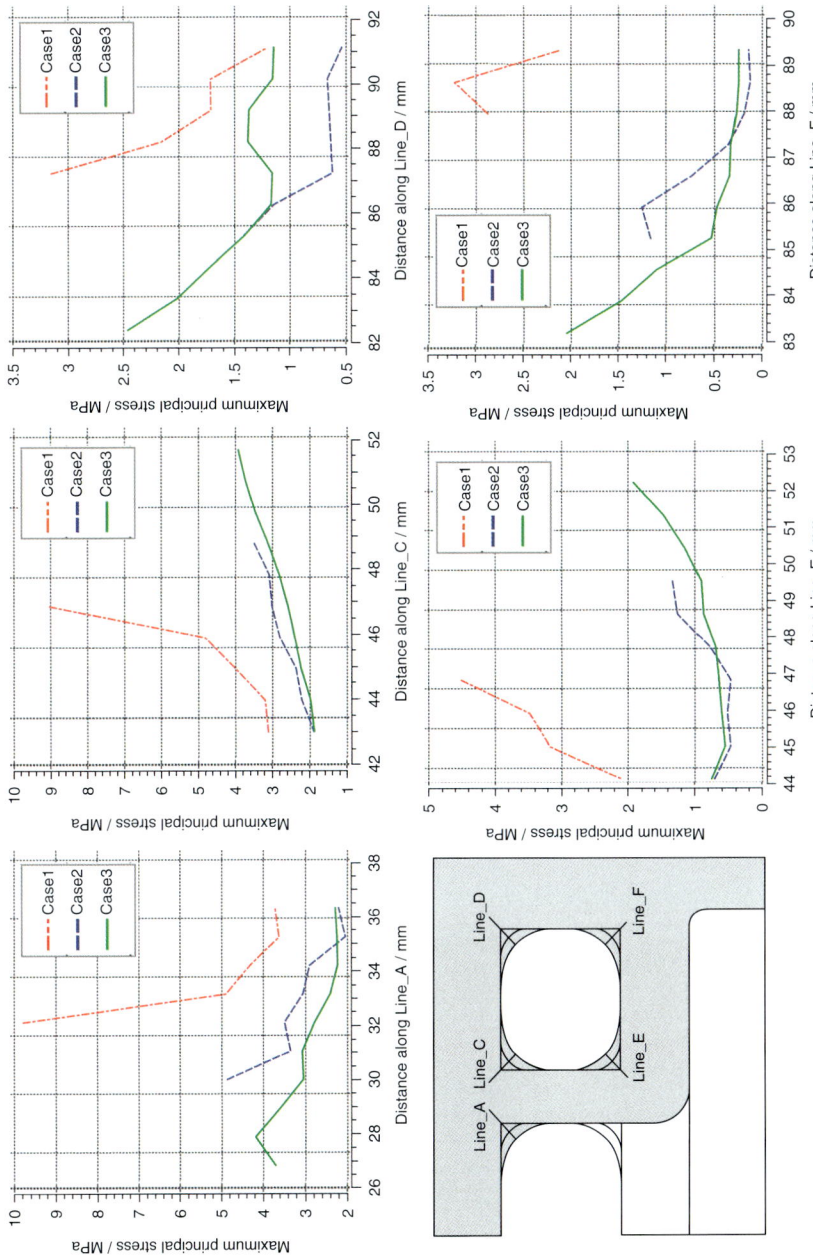

Figure 27.6 Maximum principal stress results for the top sealant.

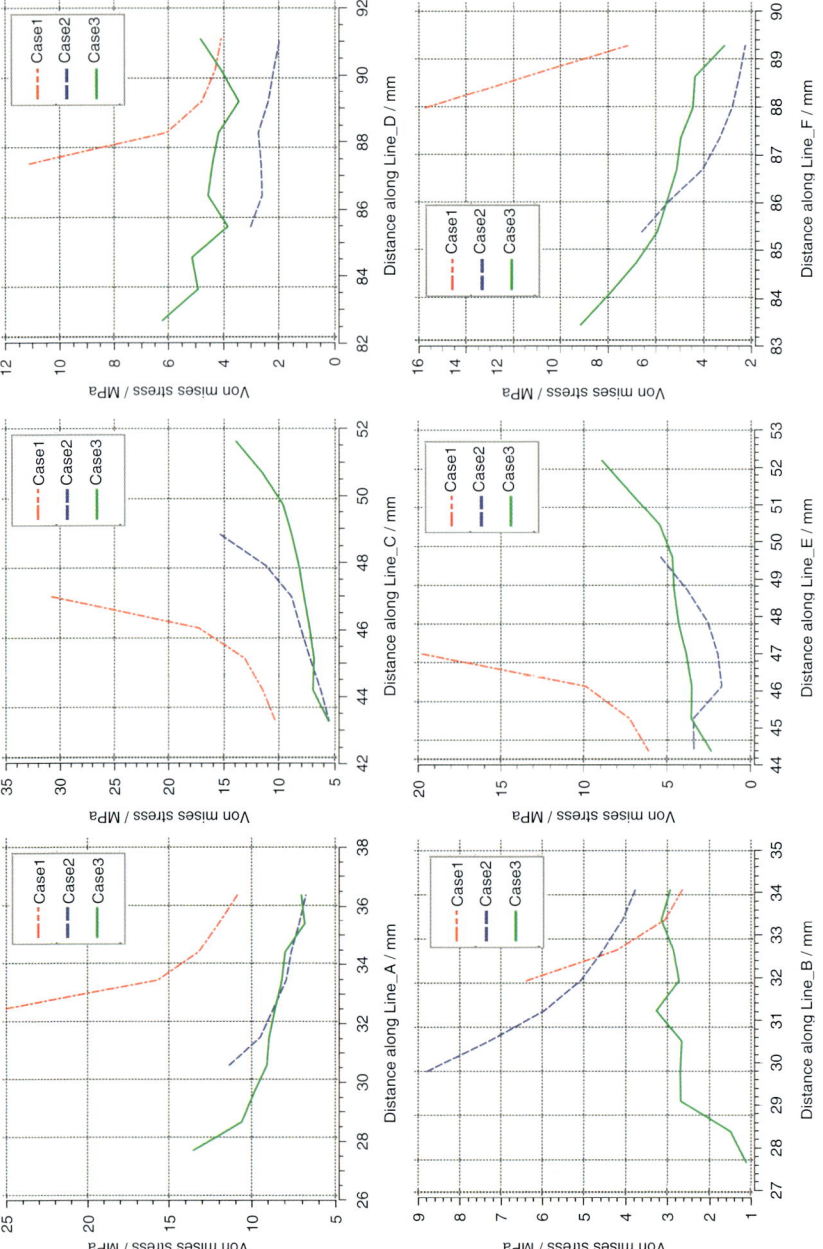

Figure 27.7 Von Mises stress results for the bottom interconnector plate.

27.4 Effect of Geometric Design on the Stress Distribution in SOFCs | 787

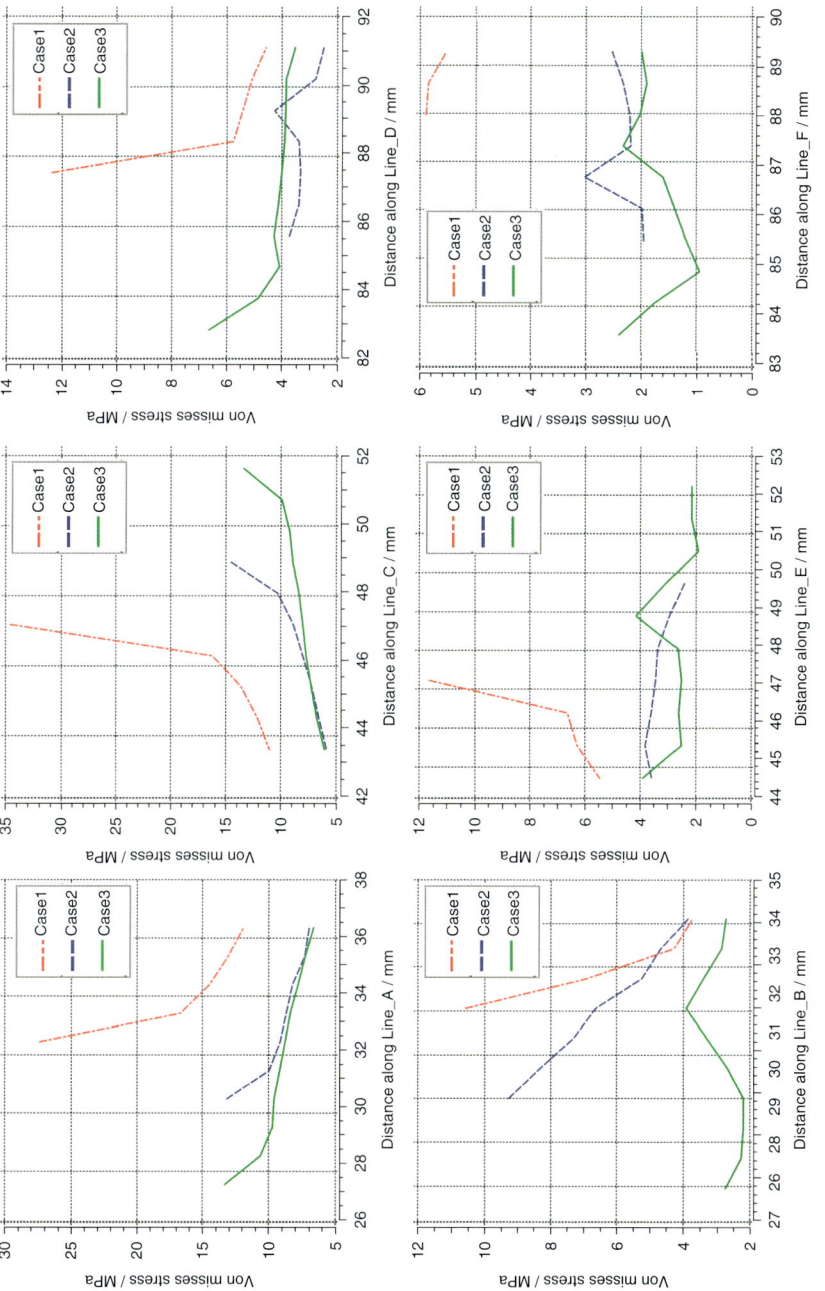

Figure 27.8 Von Mises stress results for the top interconnector plate.

differs considerably when using different manifold port geometries. This also indicates that several modeling simplifications performed to mitigate computational difficulties have to be set carefully, as the resulting thermomechanical behavior is influenced drastically.

27.5
Conclusion

The hermetic and structural reliability of SOFCs is still an impediment for commercialization, which needs to be improved. This is required so as to ensure satisfactory long-term performance and to compete with traditional power systems. This requires a thorough understanding and management of the component regions susceptible to failure. As high-temperature experiments are prohibitive, computational techniques have been attractive with the aim of improving knowledge in complex SOFC multi-physics.

However, it is essential to employ both thermofluid and thermomechanical analyses on a reasonable geometric level, preferably coupled. Ignoring the coupling of results will lead to inaccurate interpretation of the overall results. A robust and reliable SOFC can only be designed if both approaches are coupled so as to set an optimum, fulfilling both thermofluid and thermomechanical requirements. Moreover, for a comprehensive understanding of component failure, 3D analyses are essential, as the induced stress is in three dimensions. Another important issue has been the use of nonlinearities and temperature-dependent material properties. The material models employed and temperature-dependent material properties influence the computational results considerably, which needs particular care. The literature review of the field of numerical modeling in SOFCs reveals a large amount of thermofluid flow model studies, whereas little attention has been paid to the thermomechanically induced stress behavior of SOFCs.

To contribute to a better understanding of the thermomechanically induced stress within SOFCs, it was aimed to present a study shedding light on the geometric effects on the thermomechanical stress. Moreover, this would give readers an overview of the geometric simplification limit of their models, and how it would affect the evaluation of their results. Hence a coupled 3D CFD–FEM analysis was performed, using three discretely modeled geometric sections with different manifold port radii. All three were subjected to the same thermofluid flow boundary conditions that were simulated using CFD. Subsequently, the thermomechanically induced stress within the geometries was analyzed using FEM and the thermal load predicted by CFD.

In general, the results suggest the use of a larger fillet radius at the manifold regions, as it had a considerable impact on the thermal stress results. Furthermore, the use of larger fillet radii reduces the thermal gradients. In addition to allowing visualization of locations susceptible to stress, the study contributed to a better understanding of the geometric simplification limits, a topic that usually has been neglected in the literature. Accordingly, it is obvious that for the same thermofluid flow conditions, the CFD results yield different thermal fields by applying different levels of simplifications, which affects the thermomechanically induced stress values and distributions.

References

1. Singh, J.P., Niihara, K., and Hasselman, D.P.H. (1981) Analysis of thermal fatigue behaviour of brittle structural materials. *J. Mater. Sci.*, **16**, 2789–2797.
2. Fergus, J.W. (2010) Sealants for solid oxide fuel cells. Review. *J. PowerSources*, **147**, 46–57.
3. Mahapatra, M.K. and Lu, K. (2010) Seal glass for solid oxide fuel cells. Review. *J. Power Sources*, **195**, 7129–7139.
4. Bove, R. and Ubertini, S. (2006) Modeling solid oxide fuel cell operation: approaches, techniques and results. *J. Power Sources*, **159** (1), 543–559.
5. Kakac, S., Pramuanjaroenkij, A., and Zhou, X. (2007) A review of numerical modeling of solid oxide fuel cells. *Int. J. Hydrogen Energy*, **32**, (7), 761–786.
6. Mench, M.M. (2010) *Hydrogen and Fuel Cells*, Wiley-VCH Verlag GmbH, Weinheim, pp. 89–113.
7. Secanell, M., Wishart, J., and Dobson, P. (2011) Computational design and optimization of fuel cells and fuel cell systems: a review. *J. Power Sources*, **196** (8), 3690–3704.
8. Yakabe, H., Ogiwara, T., Hishinuma, M., and Yasuda, I. (2001) 3-D model calculation for planar SOFC. *J. Power Sources*, **102** (1–2), 144–154.
9. Yakabe, H., Baba, Y., Sakurai, T., Satoh, M., Hirosawa, I., and Yoda, Y. (2004) Evaluation of residual stresses in a SOFC stack. *J. Power Sources*, **131** (1–2), 278–284.
10. Yakabe, H., Baba, Y., Sakurai, T., Satoh, M., Hirosawa, I., and Yoda, Y. (2004) Evaluation of residual stresses for anode-supported SOFCs. *J. Power Sources*, **135** (1–2), 9–16.
11. Selimovic, A., Kemm, M., Torisson, T., and Assadi, M. (2005) Steady state and transient thermal stress analysis in planar solid oxide fuel cells. *J. Power Sources*, **145** (2), 463–469.
12. Nakajo, A., Stiller, C., Härkegård, G., and Bolland, O. (2006) Modeling of thermal stresses and probability of survival of tubular SOFC. *J. Power Sources*, **158** (1), 287–294.
13. Weil, K.S., Hardy, J.S., and Koeppel, J.B. (2006) New sealing concept for planar solid oxide fuel cells. *J. Mater. Eng. Perform.*, **15** (4), 427–432.
14. Lin, C.K., Chena, T.T., Chyou, Y.P., and Chiang, L.K. (2007) Thermal stress analysis of a planar SOFC stack. *J. Power Sources*, **164**, 238–251.
15. Weil, K.S. and Koeppel, J.B. (2008) Thermal stress analysis of the planar SOFC bonded compliant seal design. *Int. J. Hydrogen Energy*, **33**, 3976–3990.
16. Weil, K.S. and Koeppel, J.B. (2008) Comparative finite element analysis of the stress–strain states in three different bonded solid oxide fuel cell seal designs. *J. Power Sources*, **180**, 343–353.
17. Lin, C.K., Huang, L.H., Chiang, L.K., and Chyou, Y.P. (2009) Thermal stress analysis of planar solid oxide fuel cell stacks: effects of sealing design. *J. Power Sources*, **192**, 515–524.
18. Jiang, T.L. and Chen, M.H. (2009) Thermal-stress analyses of an operating planar solid oxide fuel cell with the bonded compliant seal design. *Int. J. Hydrogen Energy*, **34**, 8223–8234.
19. Fisher, K. and Seume, J. (2009) Impact of the temperature profile on thermal stress in a tubular solid oxide fuel cell. *J. Fuel Cell Sci. Technol.*, **6**, 011017.
20. Wenning N. L., Sun, X., Koeppel, B., and Stephens, E. (2011) Creep behavior of glass/ceramic sealant and its effect on long-term performance of solid oxide fuel cells. *Int. J. Appl. Ceram. Technol.*, **8** (1), 49–59.
21. Wenning, N.L., Sun, X., and Khaleel, M.A. (2010) Effect of creep of ferritic interconnect on long-term performance

of solid oxide fuel cell stacks. *Fuel Cells*, **10** (4), 703–717.
22. Chiang, L.K., Liu, H.C., Shiu, Y.H., Lee, C.H., and Lee, R. (2010) Thermal stress and thermo-electrochemical analysis of a planar anode-supported solid oxide fuel cell: effects of anode porosity. *J. Power Sources*, **195**, 1895–1904.
23. Serincan, M.F., Pasaogullari, U., and Sammes, N.M. (2010) Thermal stresses in an operating micro-tubular solid oxide fuel cell. *J. Power Sources*, **195**, 4905–4914.
24. Peksen, M. (2011) Fuel cell module, European patent application EP 2 372 825 A1.
25. Liu, L., Kim, G.K., and Chandra, A. (2010) Modeling of thermal stresses and lifetime prediction of planar solid oxide fuel cell under thermal cycling conditions. *J. Power Sources*, **195**, 2310–2318.
26. Blum, L., Gross, S.M., Malzbender, J., Pabst, U., Peksen, M., Peters, R., and Vinke, I.C. (2010) Investigation of solid oxide fuel cell sealing behavior under stack relevant conditions at Forschungszentrum Jülich. *J. Power Sources*, **196** (17), 7175–7181.
27. Peksen, M. (2011) A coupled 3D thermofluid–thermomechanical analysis of a planar type production scale SOFC stack. *Int. J. Hydrogen Energy*, **36** (18), 11914–11928.
28. Niewolak, L., Wessel, E., Singheiser, L., and Quadakkers, W.J. (2010) Potential suitability of ferritic and austenitic steels as interconnect materials for solid oxide fuel cells operating at 600 °C. *J. Power Sources*, **195** (22), 7600–7608.
29. Ramberg, W. and Osgood, W.R. (1943) Description of Stress–Strain Curves by Three Parameters. Technical Note No. 902, National Advisory Committee for Aeronautics, Washington, DC.
30. Peksen, M., Peters, R., Blum, L., and Stolten, D. (2011) Hierarchical 3D multiphysics modelling in the design and optimisation of SOFC system components. *Int. J. Hydrogen Energy*, **36**, 4400–4408.
31. Peksen, M., Peters, R., Blum, L., and Stolten, D. (2010) Component design in SOFC technology: 3D computational continuum mechanics modelling and experimental validation of a planar type Pre-heater, presented at the 7th Symposium on Fuel Cell Modelling and Experimental Validation, Morges (Lausanne), Switzerland.
32. Peksen, M., Peters, R., Blum, L., and Stolten, D. (2010) 3D coupled CFD/FEM modelling and experimental validation of a planar type air pre-heater used in SOFC technology. *Int. J. Hydrogen Energy*, **36** (11), 6851–6861.

28
Modeling of Molten Carbonate Fuel Cells

Peter Heidebrecht, Silvia Piewek, and Kai Sundmacher

28.1
Introduction

In the development of molten carbonate fuel cells (MCFCs), many issues require mathematical models. Some of them, for example, the design of controllers and the integration of an MCFC stack in a larger plant system, can be solved with spatially lumped models. Other questions such as the analysis of an inhomogeneous current density profile or the optimal design and operation of a fuel cell with respect to temperature limitations, need spatially distributed models. Because the latter are usually the more complex models, this chapter is focused on these models.

A selection of spatially distributed MCFC models is given in Table 28.1. The first such models were proposed by Sampath *et al.* [1] and Wolf and Wilemski [2, 3]. Despite the limited computational capacity at that time, they already solved relatively complex sets of partial differential equations in two spatial dimensions. Modeling activities increased around the end of the last century and are still continuing. In principle, all of these models are based on the same set of balances of mass, momentum, enthalpy, and charges. In many models, the momentum balance has been replaced by the assumption of isobaric conditions (fourth column), and most consider only steady-state conditions (third column). Some of the models have also been used to simulate conditions in an arrangement of several cells (fifth column). Obviously, most of the models consider the total cell current or the cell current density as an input variable (galvanostatic operation); only a few are designed to reflect potentiostatic conditions (eighth column).

The exact formulation of the balance equations varies. For example, the model of Wolf and Wilemski [2, 3] describes the gas composition via conversions of chemical reactions rather than in terms of concentrations, partial pressure, or mole fractions. Although this transformation elegantly reduces the number of state variables of the model, it does not change the character of the model. The most important differences lie in the approaches to model the internal reforming process (sixth column) and the porous electrodes (seventh column).

The reforming reactions can be reflected in a rate-based approach, under an equilibrium assumption, or they are disregarded in the model. With regard to

---

Fuel Cell Science and Engineering: Materials, Processes, Systems and Technology, First Edition. Edited by Detlef Stolten and Bernd Emonts.
© 2012 Wiley-VCH Verlag GmbH & Co. KGaA. Published 2012 by Wiley-VCH Verlag GmbH & Co. KGaA.

Table 28.1 Selection of spatially distributed MCFC models.[a]

Model	Dimensions	Dynamic	Pressure	Cell/stack	Reforming	Electrode model	Operating mode	Validation	Remarks
Sampath et al. [1]	2D cross flow	–	–	Cell	WGS (equil.)	Resistance model, constant resistances	V given	3 cm² cell	–
Wolf and Wilemski [2, 3]	2D cross/co-/counter flow	–	–	Cell	SMR (equil.) WGS (equil.)	1D thin-film electrode model	I given	94 cm² cell	Most cited Detailed electrode model
Watanabe et al. [4–6]	2D cross/co-/counter flow	–	–	Cell/stack	WGS (equil.)	Resistance function, from thin-film model [4]	I given	1 m² cell, 100 kW stack	–
Bosio et al. [7, 8]	2D cross flow	–	Darcy law	Cell/stack	WGS (equil.)	Resistance function, derived from volume-averaged model [9]	I given	100 cm² cell; stack: 20 cells @ 0.75 m², 13 kW	–
Koh et al. [10–12]	3D/2D (length, height) co-flow	–	–	Stack	–	Resistance function according to Yuh and Selman [13]	I given	100 cm² cell, stack: 20 cells @ 0.3 m²	Constant current density
Park et al. [14]	2D cross flow with IIR	–	–	Cell	IIR: SMR (irreversible), WGS (equil.)	Resistance function according to Yuh and Selman [13]	I given	–	Includes IIR

Fermeglia et al. [15]	2D cross flow	Dynamic	Reduced momentum balance	Cell	WGS (equil.)	Overall Butler–Volmer kinetics	Unclear	–	Overdetermined electrode model
Heidebrecht et al. [16–18]	2D cross flow with IIR	Dynamic	–	Cell/stack	SMR (rate) WGS (rate)	Lumped model with Butler–Volmer kinetics and mass transport	I given	250 kW stack, 342 cells at 0.79 m^2	Includes IIR Coupling A/C via burner Cathode gas recycle Transport limitation in IIR
Bittanti et al. [19, 20]	2D cross flow	Dynamic	Momentum balance	Cell	WGS (rate)	Gouy–Chapman double layer, transient balance of adsorbed species	V/I (resistance) given	–	–
Iora and Campanari [21]	2D cross flow	–	Darcy law	Stack	Unclear	Resistance function as used in [7]	Unclear	Stack: 14 cells @0.1 m^2	Incomplete information about model equations
Lee et al. [22]	2D co-flow	–	–	Stack	WGS (equil.)	Resistance function according to Yuh et al., but unclear reference	I given	5 kW stack (9 cells)	–
Liu and Weng [23]	1D co-flow	Dynamic	Momentum balance	Cell	WGS (rate)	Resistance function, from thin-film model [4]	V given	3 cm^2 cells	–

^{a}WGS, water gas shift reaction; SMR, steam methane reforming reaction; IIR, indirect internal reforming.

the electrode models, a wide variety of modeling approaches have been applied, ranging from constant resistance coefficients [1] to spatially distributed electrode pore models (e.g., [2, 19]) which are simultaneously solved with the other model equations.

In Section 28.2, a representative formulation of the balance equations of an MCFC model is given and briefly discussed. The difference between galvanostatic and potentiostatic operating modes is also illustrated (Section 28.2.5). Owing to the wide variety of electrode models, the description of the electrode behavior is intentionally omitted in that section. Instead, the electrode model is separated from the other balance equations of the fuel-cell model. For this purpose, an interface between these models is defined (Section 28.2.6) which shows how they are interconnected. In Section 28.3, an overview of the most important electrode models is given, and it is indicated how they fit to the interface.

28.2
Spatially Distributed MCFC Model

In the following, a mathematical model of a single MCFC is presented and discussed. In order to keep the complexity of the model at a reasonable level, the focus is only on the fuel cell itself here. Furthermore, no model of the electrode processes is given here; instead, an interface between the fuel-cell model and any arbitrary electrode model is defined, and different electrode models are discussed afterwards (Section 28.3).

28.2.1
General Assumptions

The model reflects a cross-flow cell, where the main flow directions of the anode and the cathode gases are perpendicular to each other. Figure 28.1 indicates this configuration, together with the applied coordinate system and the inlet and outlet fluxes.

The following general assumptions are applied to derive the model:

- The fuel-cell model considers three compartments: the anode, the cathode gas channels, and the solid phase. The solid phase includes all immobile parts of the real fuel cell: the solid part of the electrodes, the electrolyte, the channel walls, and the bipolar plates.
- In each of the compartments, all gradients in the stack direction (perpendicular to the cell plane, along z_3) are neglected. This assumption is justified because the differences in temperature and gas composition along z_3 are much smaller than those along the other two coordinates, and they do not play an important role in this fuel cell. This reduces the model to two spatial dimensions.
- Pressure gradients are neglected. Typically, pressure differences in MCFCs are in the order of a few millibars, so the pressure distribution does not have a significant impact on the cell performance. If necessary, the pressure drop can be

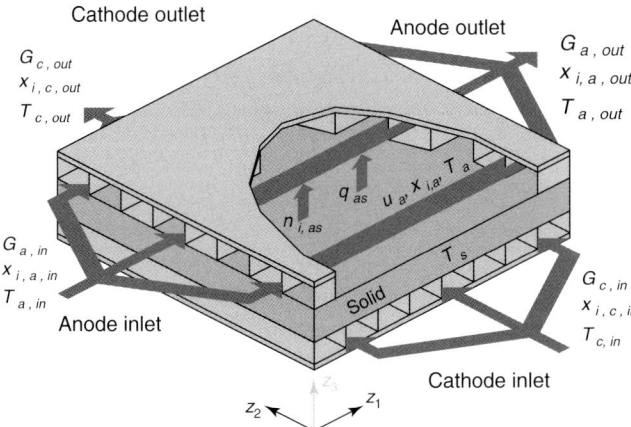

Figure 28.1 Schematic diagram of a cross-flow MCFC.

estimated with little effort from the simulated gas velocity and an experimentally determined pressure drop coefficient.
- The ideal gas law is applied to all gases.
- The heat capacity of each gas species is assumed to be constant.

28.2.2
Anode Gas Channels

In addition to the general assumptions, several additional assumptions are made regarding the modeling of the anode gas channel:

- The gas flows only along the main gas flow direction (z_1 as indicated in Figure 28.1). No mass exchange perpendicular to this direction (i.e., along z_2) is considered.
- Owing to the high gas velocity that is usually applied in MCFC systems and to the low tortuosity in typical channel structures, dispersion can be neglected. The same applies for heat conduction in the gas phase.
- Two reforming reactions are modeled as quasi-homogeneous gas-phase reactions: the steam methane reforming reaction (SMR, subscript "ref1") and the water gas shift reaction (WGSR, subscript "ref2"). Reversible kinetics are applied to describe the rates of these reactions. Their heat of reaction is completely attributed to the gas phase.
- Mass exchange is considered between the anode gas channel and the anode, which is part of the solid compartment. This mass flux also causes an enthalpy exchange between these two compartments.
- Heat exchange between the gas channels and the solid parts of the fuel cell is modeled by linear kinetics. The heat exchange coefficient also includes the contribution due to radiation.

28 Modeling of Molten Carbonate Fuel Cells

The following differential equations describe the gas composition in terms of molar fractions, $x_{i,a}$, the gas velocity, u_a, and the gas temperature, T_a. They are obtained from the partial mass balances of the gas species (subscript i), the total mass balance, and the enthalpy balance in temperature form:

$$c_{t,a} \frac{\partial x_{i,a}}{\partial t} = -c_{t,a} u_a \frac{\partial x_{i,a}}{\partial z_1} + \frac{n_{i,as}}{h_a} - x_{i,a} \frac{n_{t,as}}{h_a} + \sum_{j=\text{ref}} \left(v_{i,j} - x_{i,a} \bar{v}_j \right) r_j \quad (28.1)$$

$$c_{t,a} \bar{c}_{p,a} \frac{\partial T_a}{\partial t} = -c_{t,a} \bar{c}_{p,a} u_a \frac{\partial T_a}{\partial z_1} + q_{\text{ref}} + \frac{q_{as}}{h_a} + \frac{q^+_{m,as}}{h_a} \quad (28.2)$$

$$0 = -\frac{\partial u_a}{\partial z_1} + \frac{1}{c_{t,a} \bar{c}_{p,a} T_a} \left(q_{\text{ref}} + \frac{q_{as}}{h_a} + \frac{q^+_{m,as}}{h_a} \right) + \frac{1}{c_{t,a}} \frac{n_{t,as}}{h_a}$$

$$+ \frac{1}{c_{t,a}} \sum_{j=\text{ref}} \bar{v}_j r_j \quad (28.3)$$

The total molar concentration is calculated according to the ideal gas law:

$$c_{t,a} = \frac{p}{RT_a} \quad (28.4)$$

The specific molar exchange flux density between the anode gas channel and the electrode is obtained from the electrode model:

$$n_{ias} \longrightarrow \text{see electrode model (Sections 28.2.6 and 28.3)} \quad (28.5)$$

The total molar exchange flux density is the sum of the specific exchange fluxes:

$$n_{t,as} = \sum_i n_{i,as} \quad (28.6)$$

The rate expressions of the reforming reactions follow the power law kinetics with an Arrhenius term. They may be replaced by other expressions [24].

$$r_{\text{ref1}} = k^0_{\text{ref1}} \exp\left(-\frac{E_{\text{ref1}}}{RT_a}\right) \left[x_{CH_4,a} x_{H_2O,a} - \frac{x^3_{H_2,a} x_{CO,a}}{K_{eq,\text{ref1}}(T_a)} \right] \quad (28.7)$$

$$r_{\text{ref2}} = k^0_{\text{ref2}} \exp\left(-\frac{E_{\text{ref2}}}{RT_a}\right) \left[x_{CO,a} x_{H_2O,a} - \frac{x_{H_2,a} x_{CO_2,a}}{K_{eq,\text{ref2}}(T_a)} \right] \quad (28.8)$$

The average heat capacity of the gas phase is calculated as

$$\bar{c}_{p,a} = \sum_i x_{i,a} c_{p,i} \quad (28.9)$$

The heat of the reforming reactions is the product of the reaction enthalpies and the corresponding rates:

$$q_{\text{ref}} = \sum_{j=\text{ref}} \left[-\Delta_R h^\theta_j (T_a) \right] r_j \quad (28.10)$$

The heat transfer between the electrode and the gas is modeled by a linear rate expression:

$$q_{as} = k^h_{as} (T_s - T_a) \quad (28.11)$$

The enthalpy transfer due to mass exchange with the electrode is obtained from

$$q_{m,as}^+ = \sum_i n_{i,as}^+ c_{p,i} (T_s - T_a) \qquad (28.12)$$

In this expression, only the mass fluxes from the electrode into the gas channel are considered:

$$n_{i,as}^+ = \begin{cases} n_{i,as}, & \text{if } n_{i,as} > 0 \\ 0, & \text{else} \end{cases} \qquad (28.13)$$

The enthalpy of reaction and the equilibrium constants can be obtained from thermodynamic correlations (e.g., [25]).

Equation (28.1) contains a convective term on the right-hand side, followed by two terms considering the mass exchange with the electrode and one term considering the effect of the reforming reactions. The temperature equation [Eq. (28.2)] also has a convective term plus source terms regarding the heat of the reforming reactions, the heat exchange with the electrode, and the enthalpy exchange due to mass transport.

Equation (28.3)) is obtained from the total mass balance in combination with the ideal gas law and the isobaric assumption. The result shown here is obtained after several manipulation steps that also require the enthalpy balance. This equation is an ordinary differential equation with respect to the spatial coordinate. It describes the change of the gas velocity due to a local change in temperature (second term on the right-hand side) and due to changes in the total mole number due to mass exchange with the electrode (third term) or the reforming process (last term).

The appropriate initial conditions for Eqs. (28.1–28.3) include spatial profiles for the molar fractions and the temperature at initial time (not explicitly shown here). The boundary conditions correspond to the feed conditions at the inlet position ($z_1 = 0$):

$$x_{i,a}(z_1 = 0, z_2, t) = x_{i,a,\text{in}}(t) \qquad (28.14)$$

$$T_a(z_1 = 0, z_2, t) = T_{a,\text{in}}(t) \qquad (28.15)$$

$$u_a(z_1 = 0, z_2, t) = u_{a,\text{in}}(t) = \frac{G_{a,\text{in}}(t)}{c_t(T_{a,\text{in}}) L_2 h_a} \qquad (28.16)$$

The gas at the anode exit has spatially distributed profiles of velocity, composition, and temperature. Usually, the exhaust gas is mixed and fed to a subsequent unit, for example, a catalytic burner. The average outlet conditions in terms of total molar flow, $G_{a,\text{out}}$, average gas composition, $x_{i,a,\text{out}}$, and average gas temperature, $T_{a,\text{out}}$, are obtained from the following equations:

$$G_{a,\text{out}} = \int_0^{L_2} u_a(z_1 = L_1, z_2) c_{t,a}(z_1 = L_1, z_2) h_a \, dz_2 \qquad (28.17)$$

$$G_{a,\text{out}} x_{i,a,\text{out}} = \int_0^{L_2} u_a(z_1 = L_1, z_2) c_{t,a}(z_1 = L_1, z_2) x_{i,a}(z_1 = L_1, z_2) h_a \, dz_2 \qquad (28.18)$$

28.2.3
Cathode Gas Channels

The assumptions for the cathode gas channels are similar to those applied to model the anode gas channels, with two exceptions. The first difference is that no reforming reactions occur in the cathode gas channels, and the second is that the main gas flow direction is along z_2 instead of z_1. Consequently, the equations are similar, so they are given here without further comments:

$$c_{t,c} \frac{\partial x_{i,c}}{\partial t} = -c_{t,c} u_c \frac{\partial x_{i,c}}{\partial z_2} + \frac{n_{i,cs}}{h_c} - x_{i,c} \frac{n_{t,cs}}{h_c} \tag{28.20}$$

$$c_{t,c} \bar{c}_{p,c} \frac{\partial T_c}{\partial t} = -c_{t,c} \bar{c}_{p,c} u_c \frac{\partial T_c}{\partial z_2} + \frac{q_{cs}}{h_c} + \frac{q_{m,cs}^+}{h_c} \tag{28.21}$$

$$0 = -\frac{\partial u_c}{\partial z_2} + \frac{1}{c_{t,c} \bar{c}_{p,c} T_c} \left(\frac{q_{cs}}{h_c} + \frac{q_{m,cs}^+}{h_c} \right) + \frac{1}{c_{t,c}} \frac{n_{t,cs}}{h_c} \tag{28.22}$$

where

$$c_{t,c} = \frac{p}{R T_c} \tag{28.23}$$

$$n_{ics} \longrightarrow \text{see electrode eodel (Sections 28.2.6 and 28.3)} \tag{28.24}$$

$$n_{t,cs} = \sum_i n_{i,cs} \tag{28.25}$$

$$\bar{c}_{p,c} = \sum_i x_{i,c} c_{p,i} \tag{28.26}$$

$$q_{cs} = k_{cs}^h (T_s - T_c) \tag{28.27}$$

$$q_{m,cs}^+ = \sum_i n_{i,cs}^+ c_{p,i} (T_s - T_c) \tag{28.28}$$

$$n_{i,cs}^+ = \begin{cases} n_{i,cs}, & \text{if } n_{i,cs} > 0 \\ 0, & \text{else} \end{cases} \tag{28.29}$$

The appropriate boundary conditions correspond to the inlet conditions:

$$x_{i,c}(z_1, z_2 = 0, t) = x_{i,c,\text{in}}(t) \tag{28.30}$$

$$T_c(z_1, z_2 = 0, t) = T_{c,\text{in}}(t) \tag{28.31}$$

$$u_c(z_1, z_2 = 0, t) = u_{c,\text{in}}(t) = \frac{G_{c,\text{in}}(t)}{c_t(T_{c,\text{in}}) L_1 h_c} \tag{28.32}$$

The average properties at the outlet are

$$G_{c,out} = \int_0^{L_1} u_c(z_1, z_2 = L_2) c_{t,c}(z_1, z_2 = L_2) h_c dz_1 \qquad (28.33)$$

$$G_{c,out} x_{i,c,out} = \int_0^{L_1} u_c(z_1, z_2 = L_2) c_{t,c}(z_1, z_2 = L_2) x_{i,c}(z_1, z_2 = L_2) h_c dz_1 \quad (28.34)$$

$$G_{c,out} \sum_i x_{i,c,out} c_{p,i} T_{c,out}$$
$$= \int_0^{L_1} u_c(z_1, z_2 = L_2) c_{t,c}(z_1, z_2 = L_2) \sum_i x_{i,c}(z_1, z_2 = L_2) c_{p,i} T_c(z_1, z_2 = L_2) h_c dz_1$$

$$(28.35)$$

28.2.4
Solid Phase

Only the enthalpy balance is considered in the solid phase. It is derived according to the following assumptions:

- Heat conduction is considered in both spatial directions, z_1 and z_2. The heat conductivity and the heat capacity are averaged parameters which combine the properties of all fuel cell parts that are lumped in the solid phase (see Section 28.2.1). The heat capacity is assumed to be constant.
- The reaction heats from the electrochemical reactions at the anode and cathode are completely attributed to the solid phase. Also, the losses due to ion transport through the electrolyte are considered.
- Heat losses across the boundary of the solid phase are neglected. This corresponds to the assumption of a thermally isolated fuel-cell stack. This assumption may be replaced by a more realistic heat-loss model if data are available regarding the surrounding temperature.

Based on these assumptions, the enthalpy balance in temperature form yields the following partial differential equation:

$$c_{p,s} \frac{\partial T_s}{\partial t} = \lambda_s \frac{\partial^2 T_s}{\partial z_1^2} + \lambda_s \frac{\partial^2 T_s}{\partial z_2^2} + \frac{1}{h_s} \left(\bar{q}_{m,as} + \bar{q}_{m,cs} - q_{as} - q_{cs} + q_s + q_{ext} \right) \quad (28.36)$$

The first two terms on the right-hand side consider the heat conduction in the solid phase in both directions. The first two heat sources in the bracketed expression consider the enthalpy flux due to gases that enter the electrodes:

$$\bar{q}_{m,as/cs} = \sum_i \bar{n}_{i,as/cs} c_{p,i} (T_s - T_{a/c}) \qquad (28.37)$$

where

$$n^-_{i,as/cs} = \begin{cases} n_{i,as/cs}, & \text{if } n_{i,as/cs} < 0 \\ 0, & \text{else} \end{cases} \quad (28.38)$$

The next two source terms describe the heat exchange between the solid and the gas channels [see Eqs. (28.11) and (28.27)].

The fifth term considers the heat sources due to the electrochemical processes at both electrodes, q_a and q_c, which are obtained from the electrode model. It also includes the heat released due to the ion conduction in the electrolyte layer:

$$q_s = q_a + q_c + i_e(\phi^L_c - \phi^L_a) \quad (28.39)$$

where

$$q_a, q_c \longrightarrow \text{see electrode model (Sections 28.2.6 and 28.3)} \quad (28.40)$$

$$i_e, \phi^L_{a/c} \longrightarrow \text{see potential field model (Section 28.2.5)} \quad (28.41)$$

The last term in Eq. (28.36), q_{ext}, considers all heat fluxes in the stack direction, for example, heat exchange with a neighboring fuel cell. According to the general assumptions (see Section 28.2.1), this term is zero. However, this assumption may be lifted if this cell model is extended to a stack model (e.g., [18]).

As initial conditions, Eq. (28.36) requires any arbitrary temperature profile at the initial time. The boundary conditions correspond to isolation conditions at all boundaries:

$$\frac{\partial T_s}{\partial z_1}(z_1 = 0, z_2, t) = \frac{\partial T_s}{\partial z_1}(z_1 = L_1, z_2, t) = 0 \quad (28.42)$$

$$\frac{\partial T_s}{\partial z_2}(z_1, z_2 = 0, t) = \frac{\partial T_s}{\partial z_2}(z_1, z_2 = L_2, t) = 0 \quad (28.43)$$

28.2.5
Potential Field Model

The electrochemical reaction rates at both electrodes depend strongly on the local gas compositions, the temperature, and the electrochemical potential difference between the electron-conducting phase and the ion-conducting phase. In an MCFC, the gas composition and the temperature are clearly spatially distributed, and this also induces a spatial distribution of the potential field. The electrical potential is described under the following assumptions:

- The potential differences in each bipolar plate are significantly smaller than the typical cell voltage, and are therefore neglected. This assumption is valid in either of the following two cases. The first is if the bipolar plate is a massive end plate, in which even high electric currents perpendicular to the stack direction (z_1 and z_2 directions) cause negligible voltage losses. The second is if the fuel cell is placed inside a stack where the neighboring fuel cells have a similar current density distribution, so that the cross-currents in the bipolar plates (again, in the z_1 and z_2 directions) are small.

28.2 Spatially Distributed MCFC Model

Table 28.2 Potential field models for different modes of operation.

Comments	Galvanostatic operation	Potentiostatic operation
Given quantities	Total cell current, $I(t)$	Cell voltage, gradient, $\phi_c^S, \dot{\phi}_c^S$
AE current density[a]	$i(z) = \left(\dfrac{1}{c_a}+\dfrac{1}{c_e}+\dfrac{1}{c_c}\right)^{-1} \left(\dfrac{\dfrac{i_a}{c_a}+\dfrac{i_e}{c_e}+\dfrac{i_c}{c_c}}{-\dfrac{1}{L_1 L_2}\left(\dfrac{I_a}{c_a}+\dfrac{I_e}{c_e}+\dfrac{I_c}{c_c}\right)}+\dfrac{I}{L_1 L_2}\right)$ (28.44)	$i(z) = \left(\dfrac{1}{c_a}+\dfrac{1}{c_e}+\dfrac{1}{c_c}\right)^{-1}\left(\dfrac{i_a}{c_a}+\dfrac{i_e}{c_e}+\dfrac{i_c}{c_c}-\dot{\phi}_c^S\right)$ (28.45)
		$\phi_a^S = 0$ (28.46)
		$\dot{\phi}_a^L = -\dfrac{i-i_a}{c_a}$ (28.47)
		$\dot{\phi}_c^L = -\dfrac{i-i_a}{c_a}-\dfrac{i-i_e}{c_e}$ (28.48)
ODE voltage / AE current density	$\dot{\phi}_c^S = -\dfrac{I-I_a}{c_a}-\dfrac{I-I_e}{c_e}-\dfrac{I-I_c}{c_c}$ (28.49)	$I = \int_A i(z)\, dz$ (28.50)
Abbreviations	$I_a = \int_A i_a(z)\, dz;\ I_e = \int_A i_e(z)\, dz;\ I_c = \int_A i_c(z)\, dz$	

Abbreviation: AE, algebraic equation; ODE, ordinary differential equation.
[a]The current densities at both electrodes, i_a and i_c, are obtained from electrode models; the ionic current density through the electrolyte, i_e, can be calculated from Eq. (28.57).

- The electric potential in the electron-conducting phase at the anode is arbitrarily set to zero.
- The electric potentials in the ion-conducting phase and in the electron-conducting phase of the electrode are reflected by representative potentials, ϕ^L and ϕ^S, respectively.
- Owing to the small thickness of the electrolyte layer, ion transport rates in the cell plane are negligible compared with transport rates through the electrolyte layer. Therefore, ion transport along the z_1 and z_2 directions is neglected.

These potentials are illustrated in Figure 28.2. The electric potentials in the electron-conducting (solid) phases, ϕ_a^S and ϕ_c^S, are transient, but spatially lumped. Note that the difference between these two potentials corresponds to the cell voltage; with the second assumption, ϕ_c^S is actually identical with the cell voltage. The representative electric potentials in the ion-conducting (liquid) phase at the anode and cathode, ϕ_a^L and ϕ_c^L, are transient and spatially distributed.

The potentials can be described by a discretized form of the Poisson equation combined with transient charge balances at the charged double layers. In order to clarify the variables' spatial and temporal dependences, they are explicitly given in the following formulations:

$$\phi_a^S = 0 \tag{28.51}$$

$$\frac{\partial \phi_a^L(z,t)}{\partial t} = -\frac{i(z,t) - i_a\left[\phi_a^L(z,t)\right]}{c_a} \tag{28.52}$$

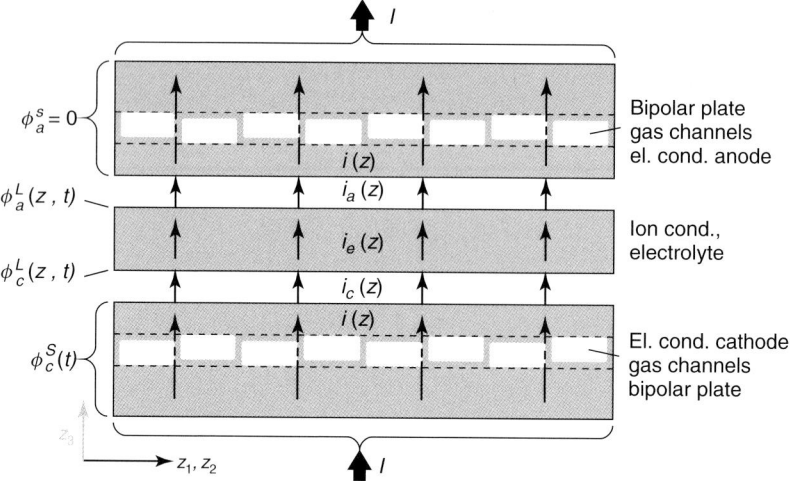

Figure 28.2 Schematic diagram of the electric potential model, including electric and ionic current densities.

$$\frac{\partial \phi_c^L(z,t)}{\partial t} = -\frac{i(z,t) - i_a[\phi_a^L(z,t)]}{c_a} - \frac{i(z,t) - i_e[\phi_a^L(z,t), \phi_c^L(z,t)]}{c_e} \quad (28.53)$$

$$\frac{d\phi_c^S(t)}{dt} = -\frac{i(z,t) - i_a[\phi_a^L(z,t)]}{c_a} - \frac{i(z,t) - i_e[\phi_a^L(z,t), \phi_c^L(z,t)]}{c_e}$$
$$- \frac{i(z,t) - i_c[\phi_c^L(z,t), \phi_c^S(t)]}{c_\iota} \quad (28.54)$$

$$I(t) = \int_A i(z,t)\,dz \quad (28.55)$$

where the local current densities are obtained from the electrode model (Sections 28.2.6 and 28.3).

In these equations, i represents the current density distribution through the current collectors at the anode and cathode and i_a and i_c are the current densities obtained from the electrode. They depend on the local potential differences at the respective electrode and are obtained from the electrode model:

$$i_a, i_c \longrightarrow \text{see electrode model (Sections 28.2.6 and 28.3)} \quad (28.56)$$

The current density through the electrolyte, i_e, is calculated from a linear kinetic relationship:

$$i_e(z,t) = \frac{\kappa_e}{h_e}[\phi_a^L(z,t) - \phi_c^L(z,t)] \quad (28.57)$$

This equation system only needs a suitable set of initial conditions. It can be applied for both galvanostatic and potentiostatic modes of operation. Under galvanostatic operating conditions, the total cell current, I, is given. The electric potentials can be obtained from Eqs. (28.52–28.55). However, the unknown current density distribution, i, cannot be obtained explicitly from these equations. This can be amended by integration of Eq. 28.54 over the whole cell area and subsequent insertion of Eq. 28.55. The resulting set of equations is given in Table 28.2 on the left-hand side. With potentiostatic operation, the cell voltage is given as a function of time, hence its gradient is also known. With that, a slightly different set of equations can be obtained from Eqs. (28.51–28.55), which are given in Table 28.2 on the right-hand side.

Both model formulations in Table 28.2 are explicit in terms of all unknown variables, which is numerically favorable. However, they describe transient potentials, which have a very small time constant. In combination with the mass and energy balances of the fuel-cell model, this leads to a stiff system, which requires appropriate numerical solvers. In case if this is not desired, the potential field equations can be set to their quasi-steady state. This leads to the following set of algebraic equations:

$$i_a\left[\phi_a^L(z,t)\right] = i_e\left(\phi_a^L(z,t), \phi_c^L(z,t)\right) = i_c\left[\phi_c^L(z,t), \phi_c^S(t)\right] \tag{28.58}$$

$$I = \int_A i_a\left[\phi_a^L(z,t)\right]dz \tag{28.59}$$

$$V = \phi_c^S - \phi_a^S \tag{28.60}$$

In the case of only one electrochemical reaction at each electrode, the potential differences can be replaced by their deviation from equilibrium potential differences, the overpotentials η:

$$\begin{aligned} V &= \phi_c^S - \phi_a^S \\ &= \Delta\phi_{\mathrm{red}}^\theta(T) - \Delta\phi_{\mathrm{ox},H_2}^\theta(T) + \left[\phi_c^S - \phi_c^L - \Delta\phi_c^\theta(T)\right] - \left(\phi_a^L - \phi_c^L\right) \\ &\quad - \left[\phi_a^S - \phi_a^L - \Delta\phi_a^\theta(T)\right] \\ &= V_{\mathrm{Nernst}}(T) + \left[\Delta\phi_c - \Delta\phi_c^\theta(T)\right] - \Delta\phi_e - \left[\Delta\phi_a - \Delta\phi_a^\theta(T)\right] \\ &= V_{\mathrm{Nernst}}(T) + \eta_c - \eta_e - \eta_a \end{aligned} \tag{28.61}$$

This is a widely used expression in many fuel-cell models. As shown here, it is the steady-state case of the more general dynamic potential field model proposed above. Introducing the overpotentials into Eqs. (28.58) and (28.59) completes this model:

$$i_a\left[\eta_a(z)\right] = i_e\left[\eta_e(z)\right] = i_c\left[\eta_c(z)\right] \tag{28.62}$$

$$I = \int_A i_a\left[\eta_a(z)\right]dz \tag{28.63}$$

This is a highly nonlinear, implicit system of algebraic equations. The integral in Eq. (28.63) contains a function of an unknown variable; usually, Eq. (28.62) cannot be rearranged to explicit equations in terms of the unknown overpotentials; and Eq. (28.61) has a spatially lumped quantity on the left-hand side, but spatially distributed quantities on the right-hand side. This makes the numerical solution of this equation system difficult. This can also be seen in many publications

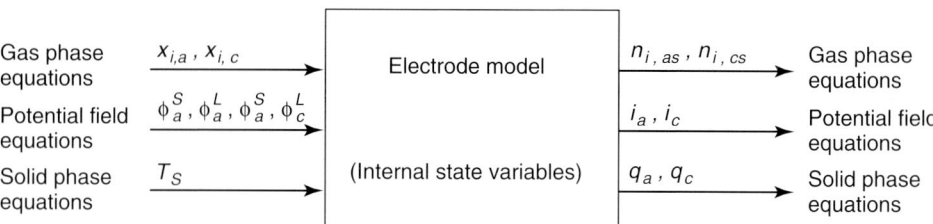

Figure 28.3 Input/output scheme of electrode models.

(e.g., [3, 21, 22]), where the authors propose various strategies to solve this equation system. The transient model equations (Table 28.2), however, can be solved straightforwardly by a numerical integrator.

28.2.6
Interface to Electrode Models

Owing to the variety of existing electrode models, they are discussed separately in Section 28.3. In order to establish a clear connection between the balance equations discussed in this section and the electrode models, an interface between both models is defined. This interface is illustrated in Figure 28.3 in the sense of an input/output scheme of the electrode model.

Any arbitrary electrode model is required to deliver values for several quantities, which are listed on the right of the box in Figure 28.3. They represent the fluxes of mass, current, and thermal energy between the electrodes and the surrounding compartments (solid phase, gas channels). They are the output variables of the electrode model, and are used at various places in the fuel-cell model [Eqs. (28.5), (28.24), (28.40), and (28.56)]. The values of these variables depend on state variables which are listed on the left of the box in Figure 28.3. These input variables of the electrode model include the local gas composition in the channels, the electric potentials in the electron- and ion-conducting phases at each electrode, and the local fuel cell temperature. Any model that describes the depicted output variables as a function of these input variables can be combined with the balance equations in this section.

28.3
Electrode Models

Table 28.3 gives a chronological overview of selected publications on electrode models. Most of them have been developed especially for cathodes, some have been applied to cathodes and anodes, and only two of them have been applied solely to simulate anodes. Most electrode models are validated with data obtained from laboratory-scale half-cells.

Table 28.3 Selection of MCFC electrode models.

Model	Classification	Electrode	Validation	Remarks
Wilemski et al. [2, 3, 26]	Thin-film model	A, C	94 cm^2 cross-flow cell [27]	–
Jewulski and Suski [28]	Thin-film model	A	Literature data [27]	Comparison with model results [2]
Kunz et al. [29]	Agglomerate model	C	Half-cell [30]	–
Yuh and Selman [31, 32]	Agglomerate model	A, C	3 cm^2 cell [13, 33]	Anode, dry agglomerate model; cathode, film agglomerate model
Jewulski [34]	Thin-film model	C	Literature data [27]	Dusty gas model equations for gas-phase transport
Christensen and Livbjerg [35]	Agglomerate model	A, C	Literature data [30, 31]	Based on capillary theory [36]
Fontes et al. [37, 38]	Agglomerate model	C	3 cm^2 cell	Based on [29]
Prins-Jansen et al. [39]	Volume-averaged model	C	AC impedance study [40]	Comparison with own agglomerate model [41, 42]
Prins-Jansen et al. [41, 42]	Agglomerate model	(A), C	Literature data [43]	Different reaction mechanisms; parameter studies
Morita et al. [4]	Thin-film model	C	258/108/100 cm^2 cells	–
Lim et al. [44]	Agglomerate model	A	Laboratory-scale half-cell	Filmed agglomerates; based on [26]
Bosio et al. [45]	Spatially lumped model	A, C	Literature data [7]	–
Subramanian et al. [46]	Volume-averaged model	C	3 cm^2 half-cell	Commented on by Berg and Findlay [47]
Hong and Selman [48, 49]	Stochastic structure model	A, C	No	Used to identify structure parameters for agglomerate models

(continued overleaf)

Table 28.3 (continued).

Model	Classification	Electrode	Validation	Remarks
Heidebrecht and Sundmacher [16]	Spatially lumped model	A, C	No	–
Brouwer et al. [50]	Agglomerate model	A, C	100 cm² cell	Based on [13, 33]

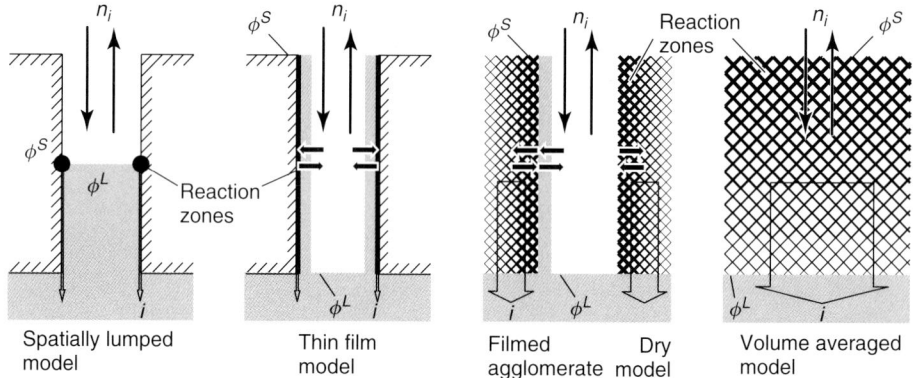

Figure 28.4 Schematic diagrams of the most common electrode models.

Four categories of electrode models can be identified from Table 28.3: the spatially lumped model, the thin-film model, the agglomerate model, and the volume-averaged model. Schemes of the basic concepts of these four model categories are depicted in Figure 28.4. Each of the schemes shows an electrode pore, with the gas channels located at the top and the liquid electrolyte, depicted in gray, at the bottom. In some models, electrolyte is also present in the pore. The reaction zones are indicated by a black face (spot, line, or grid structure). Fluxes of mass and ions are indicated by arrows.

In the following, a short overview of the most important assumptions and equations of the four electrode model categories is given. In addition, it is pointed out how the electrode models comply with the input/output scheme in Figure 28.3. The equations presented in this chapter are formulated in a general notation, which does not discriminate between anode and cathode models.

28.3.1
Spatially Lumped Models

These models combine Butler–Volmer reaction kinetics with mass transport kinetics in a simple approach [16, 45]. The mass transport between the gas phase in the channels and the reaction zone is described by a linear approach, which lumps all contributions of the complex mass transport in a single parameter:

$$n_i = D_i(\varphi_i - x_i) \tag{28.64}$$

where D_i are the lumped mass transport coefficients of component i including all transport resistances between the gas channel and the reaction site. φ_i represents the relative partial pressure of this component at the reaction site, and x_i is the molar fraction of the species in the channel gas phase and is obtained from the component mass balances [Eqs. (28.1) and (28.20)]. The reaction rates at the electrodes are described by generalized Butler–Volmer equations combined with an Arrhenius term to include the effect of the solid temperature. This model also provides the opportunity to consider parallel oxidation reactions, as they occur at the anode:

$$r_{ox,H_2} = k_{ox,H_2}(T_s) \left\{ \varphi_{H_2,ac} \exp\left(\alpha_{ox,H_2} \frac{nF}{RT_s} \left[\left(\phi_a^S - \phi_a^L \right) - \Delta\phi_{ox,H_2}^\theta(T_s) \right] \right) \right.$$
$$\left. - \varphi_{H_2O,ac} \varphi_{CO_2,ac} \exp\left(-\left(1 - \alpha_{ox,H_2}\right) \frac{nF}{RT_s} \left[\left(\phi_a^S - \phi_a^L \right) - \Delta\phi_{ox,H_2}^\theta(T_s) \right] \right) \right\} \tag{28.65}$$

$$r_{ox,CO} = k_{ox,CO}(T_s) \left\{ \varphi_{CO,ac} \exp\left(\alpha_{ox,CO} \frac{nF}{RT_s} \left[\left(\phi_a^S - \phi_a^L \right) - \Delta\phi_{ox,CO}^\theta(T_s) \right] \right) \right.$$
$$\left. - \left(\varphi_{CO_2,ac} \right)^2 \exp\left(-\left(1 - \alpha_{ox,CO}\right) \frac{nF}{RT_s} \left[\left(\phi_a^S - \phi_a^L \right) - \Delta\phi_{ox,CO}^\theta(T_s) \right] \right) \right\} \tag{28.66}$$

The kinetics of the reduction reaction at the cathode reads:

$$r_{red} = k_{red}(T_s) \left\{ \left(\varphi_{CO_2,cc} \right)^{\beta_1} \exp\left(\alpha_{red}^+ \frac{nF}{RT_s} \left[\left(\phi_c^S - \phi_c^L \right) - \Delta\phi_{red}^\theta(T_s) \right] \right) \right.$$
$$\left. - \left(\varphi_{O_2,cc} \right)^{\beta_2} \left(\varphi_{CO_2,cc} \right)^{\beta_3} \exp\left(-\alpha_{red}^- \frac{nF}{RT_s} \left[\left(\phi_c^S - \phi_c^L \right) - \Delta\phi_{red}^\theta(T_s) \right] \right) \right\} \tag{28.67}$$

In this notation, the reaction orders of the cathodic reaction can be adapted to different reaction mechanisms (see [41, 42]). The electric potentials, ϕ_a^S, ϕ_a^L, ϕ_c^S, and ϕ_c^L, are obtained from the potential field equations (Table 28.2). The local solid temperature, T_S, is obtained from the enthalpy balance in the solid phase [Eq. (28.36)].

The mass balance at the reaction zone connects the component mass flux density with the reaction rates:

$$n_i = \sum_j v_{i,j} r_j \tag{28.68}$$

where at the anode, both oxidation reactions [Eqs. (28.65) and (28.66)] are considered in the summation, and only one reduction reaction [Eq. (28.67)] is accounted for at the cathode.

The gas composition at the reaction zone, φ_i, the reaction rates, $r_{ox,H2}$, $r_{ox,CO}$, and r_{red}, and the mass flux densities, n_i, can be uniquely determined from the previous set of equations [Eqs. (28.64–28.68)]. The mass flux densities, n_i, which are output quantities of the electrode model (see Figure 28.3), are thus defined. Another output quantity, the current density, is calculated by Faraday's law:

$$i = \sum_j nFr_j \tag{28.69}$$

where the corresponding reaction rates are used, depending on the electrode.

The last output quantities, heat source density q_a and q_c, are obtained from enthalpy balances at the reaction sites:

$$q_s = \sum_j \left[-\Delta_R h_j^\theta(T_s) + (\phi^S - \phi^L)\right] r_j \tag{28.70}$$

This completes the spatially lumped electrode model, which obviously fits to the input/output interface described earlier (Figure 28.3). The electrode model can thus be combined with the balance equations of the MCFC model described in Section 28.2.

28.3.2
Thin-Film Models

The model of Wilemski and co-workers [2, 3, 26] is one of the first molten carbonate fuel-cell electrode models. Together with the models of Jewulski and Suski [28, 34], they form the basis for the class of thin-film electrode models (Figure 28.4). The following general assumptions are made in thin-film models:

- The electrode pores are represented by a single pore with a constant width.
- The pore walls are covered with an electrolyte film, usually of uniform thickness.
- The electrochemical reactions at the pore wall are described by generalized Butler–Volmer kinetics.
- The solid phase in the electrode has an infinitely high electrical conductivity.

Apart from these common assumptions, specific formulations of the thin-film electrode models differ in certain aspects. For instance, some thin-film models consider limited mass transport in the pore gas phase, whereas others assume a uniform gas composition along the pore depth. The following formulations follow the model presented by Wilemski [2]. The charge balance inside the electrolyte combined with Ohm's law delivers

$$\frac{d^2\eta(z)}{dz^2} = \frac{j[\eta(z)]}{\kappa_e \delta} \tag{28.71}$$

with the boundary conditions

$$\eta(0) = \eta_0 \tag{28.72}$$

$$\frac{d\eta(z=0)}{dz} = 0 \tag{28.73}$$

The current density at the pore wall, j, depends of the local overvoltage, η, according to some Butler–Volmer kinetics, which are not given here explicitly. The first boundary condition [Eq. (28.72)] is equivalent to the definition of $(\phi^S - \phi^L) - \Delta\phi_{eq}$ at the top of the electrode pore, so it defines the electric potential there. The second boundary condition [Eq. (28.73)] demands that no charge flux exits the electrolyte at the top of the pore. This differential equation can be solved in combination with a reaction rate expression, for example, Butler–Volmer kinetics.

The overall current density from the electrode, i, is obtained from the gradient of the overvoltage at the bottom of the electrode pore, multiplied by the ionic conductivity and the ratio of the cross-sectional area of the film and the electrode area:

$$i = \kappa_e \frac{A_{film}}{A_{electrode}} \frac{d\eta}{dz}\bigg|_{z=l} \tag{28.74}$$

This is the first output quantity according to Figure 28.3. The mass flux densities, n_i, can be obtained from Faraday's law combined with the component mass balance [Eqs. (28.68) and (28.69)]. The heat source density, q, is obtained from an enthalpy balance around the electrode pore, similar to Eq. (28.70). Corresponding equations were given by Wolf and Wilemski [2, 3], although not all of them are formulated as part of the electrode model.

28.3.3
Agglomerate Models

The agglomerate model is the most widely used electrode model for MCFCs. The fundamental assumption of these models is a bimodal pore size distribution. The wide pores, called macropores, are reflected by a single pore with a representative width. Similarly to the film model, the walls of the macropores can be wetted by electrolyte (wet or filmed agglomerate model), or they are nonwetted (dry agglomerate model). Between these large macropores, an irregular structure consisting of small micropores and solid electron conductor is located. The micropores are partially filled with electrolyte. In them, transport of mass and charge and also electrochemical reactions occur. This structure is referred to as agglomerate and indicated by the cross-hatched area in Figure 28.4. It is usually assumed to have a cylindrical or spherical shape.

The agglomerate is modeled based on a volume-averaging approach, which neglects the exact structure of the micropores [51–55]. It describes the average component concentrations and the electric potential in these pores. In order to reduce the complexity of this spatially two-dimensional model, mass transport in the agglomerate along the axial direction of the macropores is neglected. Furthermore, the electric potential is assumed to depend only on this axial coordinate. The volume-averaged component mass balance thus reads

$$D_i^{eff} \frac{\partial^2 c_i}{\partial z^2} = \frac{v_i a}{nF} j \tag{28.75}$$

where D_i^{eff} is the effective diffusivity of component i in the agglomerate taking into account the mass transport in the gas phase and the electrolyte phase in the micropores, c_i is the dissolved species concentration in the electrolyte, ν_i is the stoichiometric coefficient of component i, and a is the specific surface area of the reaction zone. The current density, j, at the electrode–electrolyte interface can be expressed by Butler–Volmer-type rate expressions.

The electric potential is also described by a second-order ordinary differential equation:

$$\frac{d^2 \phi^L}{dz^2} = -\frac{2 n D_i^{\text{eff}} F}{\nu_i \kappa_e^{\text{eff}} R_{\text{agg}}} \left(\frac{dc_i}{dr_{\text{agg}}} \right)\bigg|_{r_{\text{agg}} = R_{\text{agg}}} \tag{28.76}$$

This equation is derived from the charge balance in the agglomerate, combined with Ohm's law and with neglected potential gradients along the radius of the agglomerate, r_{agg}. The term on the right-hand side describes the flux of reactants into the agglomerate at a certain position z, which is proportional to the integral electrical current produced at that position over the whole agglomerate radius.

The output quantities required from this model are obtained in a similar approach as in the film model; the current density from the electrode is proportional to the potential gradient at the bottom of the agglomerate, the mass fluxes are obtained from mass balances in the macropores, and the heat flux is obtained from an enthalpy balance over the complete pore. However, in the publications listed in Table 28.3, these equations are not mentioned explicitly.

28.3.4
Volume-Averaged Models

The fourth type of electrode models [39, 46] is based on the volume-averaging approach for porous media [51–55]. As indicated in Figure 28.4, it is similar to the agglomerate model, but no macropores are considered. Hence the main direction of mass transport in these micropores is along the depth of the electrode.

As in the agglomerate model, the governing equations of the volume-averaged model describe the representative concentration, c_i, and the electric potential, here in the form of the overpotential, η, via spatially distributed partial differential equations (cf., [39]). The component mass balance is identical with Eq. (28.75):

$$D_i^{\text{eff}} \frac{\partial^2 c_i}{\partial z^2} = \frac{\nu_i a}{nF} j \tag{28.77}$$

The two appropriate boundary conditions of this partial differential equation determine the concentrations at the interface between the electrode and the gas channel and they demand that the concentrations at the bottom of the electrode become zero.

The charge balance leads to a partial differential equation which is very similar to Eq. (28.71) from the film model:

$$\frac{\partial}{\partial z}\left(\kappa_e^{\text{eff}} \frac{\partial \eta}{\partial z} \right) = aj \tag{28.78}$$

As in the film model, the first boundary condition defines the overpotential at the upper or lower boundary of the electrode [cf., Eq. (28.72)]. The second boundary condition reflects the fact that no ionic current may cross the interface toward the gas channel, so the potential gradient in the electrolyte is set to zero at the upper end of the electrode [cf., Eq. (28.73)].

The component mass flux density according to the volume-averaged model can be obtained from the concentration gradient at the boundary towards the gas channel:

$$n_i = -D_i^{\text{eff}} \left.\frac{\partial c_i}{\partial z}\right|_{z=0} \tag{28.79}$$

The electric current density is calculated from the gradient of the overpotential at the bottom of the electrode, similar to Eq. (28.74). An additional enthalpy balance, which is not given here in detail, is necessary to quantify the heat flux from the electrode.

28.4 Conclusion

The equations presented in this chapter are a common basis for many spatially distributed MCFC models. The full diversity of existing models cannot be reflected here; Table 28.1 and the references therein give a good overview of the spectrum of these models. A focus in MCFC modeling is on the description of the processes in the electrodes, which are illustrated in a separate section of this chapter. Also, the combination of electrode models and MCFC models has been shown in detail.

Many of the models can be used to describe not only the behavior of single cells, but also that of a whole stack. This extension of the model equations has not been discussed here but, as indicated in Table 28.1, this has been realized with many models. Other extensions and modifications, such as the application of equilibrium assumptions with regard to the reforming reactions, the modeling of a catalytic burner between the anode exhaust and the cathode inlet, or the addition of model equations describing an indirect internal reforming reactor, are frequently applied in MCFC models. For the sake of brevity, they have not been discussed in detail in this chapter.

No exemplary simulation results are presented here. Anyway, these would only be applicable for a certain MCFC system and under certain conditions, and they would not be representative for the broad range of available models. Nevertheless, MCFC models have been applied for various purposes; Yoshiba *et al.* [5] compared different flow configurations, Koh and Kang [10] predicted the impact of pressurized operation on fuel-cell performance, Park *et al.* [14] and Heidebrecht and Sundmacher [56] applied MCFC models to evaluate the effect of the reforming process on the fuel cell and to optimize it, and Bosio *et al.* [8] studied the application of nonuniform gas distributions with regard to the temperature distribution in MCFCs.

As these examples show, mathematical models have already contributed significantly to the development of these highly integrated, complex, and efficient systems and will certainly continue to do so in the future.

List of Symbols

Symbol	Meaning	Units
Roman symbols		
a	Specific surface area	m^{-1}
A_{film}, $A_{electrode}$	Cross-sectional area of electrolyte film, size of electrode	m^2
c_a, c_c, c_e	Capacitance of anodic/cathodic double layers, electrolyte layer	$C\,m^{-2}\,V^{-1}$
c_i	Effective concentration of species i in agglomerate	$mol\,m^{-3}$
$\bar{c}_{p,a}$, $\bar{c}_{p,c}$	Average molar heat capacity of anode/cathode gas channel	$J\,mol^{-1}\,K^{-1}$
$c_{p,i}$	Molar heat capacity species i	$J\,mol^{-1}\,K^{-1}$
$c_{p,s}$	Heat capacity of solid phase	$J\,m^{-3}\,K^{-1}$
$c_{t,a}$, $c_{t,c}$	Total molar concentration of anode/cathode gas channel	$mol\,m^{-3}$
D_i	Mass exchange coefficient of gas channel/reaction zone	$mol\,m^{-2}\,s^{-1}$
D_i^{eff}	Effective diffusion coefficient of species i in agglomerate	$m^2\,s^{-1}$
E_j	Activation energy of reaction j	$J\,mol^{-1}$
F	Faraday constant	$C\,mol^{-1}$
G	Total molar flow rate	$mol\,s^{-1}$
h_a, h_c, h_e	Height of anode/cathode gas channels, electrolyte layer	m
i	Electrical current density in current collectors	$A\,m^{-2}$
I	Total electrical cell current	A
I_a, I_c, I_e	Total current in anode/cathode double layer, electrolyte	A
i_a, i_c, i_e	Current density in anode/cathode double layer, electrolyte	$A\,m^{-2}$
j	Current density at charged double layer in agglomerate	$A\,m^{-2}$
$k_j(T)$	Temperature-dependent rate coefficient of reaction j	$mol\,m^{-3}\,s^{-1}$ or $mol\,m^{-2}\,s^{-1}$
k_j^0	Rate constant of reaction j	$mol\,m^{-3}\,s^{-1}$ or $mol\,m^{-2}\,s^{-1}$
k_{as}^h, k_{cs}^h	Heat exchange parameter of anode/cathode gas channel to solid	$W\,m^{-2}\,K^{-1}$
$K_{eq,j}$	Equilibrium constant of reaction j	–
L_1, L_2	Cell size (z_1, z_2 directions)	m

Symbol	Description	Units
n	Number of electrons transferred	–
$n_{i,as}, n_{i,cs}$	Molar flux density of species i between gas channels and electrode	mol m^{-2} s^{-1}
$n_{t,as}, n_{t,cs}$	Total molar flux density between gas channels and electrode	mol m^{-2} s^{-1}
p	Pressure	Pa
q_a, q_c	Heat source density of anode/cathode	W m^{-2}
q_{as}, q_{cs}	Heat flux density between gas channels and solid	W m^{-2}
$q_{m,as}^+, q_{m,cs}^+$	Mass-related heat flux density from solid to gas channels	W m^{-2}
$q_{m,as}^-, q_{m,cs}^-$	Mass-related heat flux density from gas channels to solid	W m^{-2}
q_{ext}	Heat source density from external sources	W m^{-2}
q_{ref}	Heat source density due to reforming process	W m^{-3}
q_s	Heat source density of solid phase	W m^{-2}
r_{agg}	Radial coordinate in the agglomerate	m
r_j	Rate of reaction j	mol m^{-3} s^{-1}
R	Universal gas constant	J mol^{-1} K^{-1}
R_{agg}	Agglomerate radius	m
t	Time	s
T_a, T_c, T_s	Temperature of anode/cathode gas channels, solid phase	K
u_a, u_c	Gas velocity of anode/cathode gas channels	m s^{-1}
V	Cell voltage	V
$x_{i,a}, x_{i,c}$	Molar fraction of species i in the anode/cathode gas channels	–
z	Spatial coordinate	m

Greek symbols

Symbol	Description	Units
$\alpha_{ox,H_2}, \alpha_{ox,CO}$	Charge-transfer coefficients of anodic reactions	–
$\alpha_{red}^+, \alpha_{red}^-$	Charge-transfer coefficients of forward/backward cathode reaction	–
$\beta_1, \beta_2, \beta_3$	Reaction orders of cathodic reaction kinetics	–
δ	Electrolyte film thickness	m
$\Delta_R h_j^\theta$	Temperature-dependent standard enthalpy of reaction j	J mol^{-1}
$\Delta\phi_e$	Potential difference at the electrolyte layer	V
$\Delta\phi_j^\theta(T)$	Temperature-dependent Nernst potential difference of reaction j	V
η_a, η_c, η_e	Overpotential at anode/cathode double layer, electrolyte layer	V
κ_e	Ionic conductivity of electrolyte layer	A V^{-1} m^{-2}
κ_e^{eff}	Effective ionic conductivity in agglomerate	A V^{-1} m^{-2}
λ_s	Heat conductivity of solid phase	W m^{-1} K^{-1}
$\nu_{i,j}$	Stoichiometric coefficient of species i in reaction j	–
$\bar\nu_j$	Cumulative stoichiometric coefficients of reaction j	–

φ_i	Relative partial pressure of gas component i at reaction zone	–
ϕ_a^L, ϕ_c^L	Electric potential of ion conductor at anode/cathode	V
ϕ_a^S, ϕ_c^S	Electric potential of electron conductor at anode/cathode	V

Subscripts

a	Anode/anode gas
as	Anode/anode gas interface
c	Cathode/cathode gas
cs	Cathode/cathode gas interface
e	Electrolyte
ext	External
i	Chemical species (CH_4, H_2O, H_2, CO, CO_2, O_2, N_2)
j	Reaction
in	Inlet
Nernst	Nernst/equilibrium voltage
out	Outlet
ref	Reforming
ref1, ref2	Steam methane reforming, water gas shift reaction
ox,H_2/ox,CO	Electrochemical oxidation of H_2, CO at anode
red	Electrochemical reduction at cathode
s	Solid phase

Superscripts

L	Liquid/ion-conducting part of electrode
ref	Reference value
S	Solid/electron conducting part of electrode
θ	Standard conditions

References

1. Sampath, V., Sammells, A.F., and Selman, J.R. (1980) Performance and current distribution model for scaled-up molten carbonate fuel-cells. *J. Electrochem. Soc.*, **127** (1), 79–85.
2. Wilemski, G. (1983) Simple porous-electrode models for molten carbonate fuel-cells. *J. Electrochem. Soc.*, **130** (1), 117–121.
3. Wolf, T.L. and Wilemski, G. (1983) Molten carbonate fuel-cell performance-model. *J. Electrochem. Soc.*, **130** (1), 48–55.
4. Morita, H., Mugikura, Y., Izaki, Y., Watanabe, T., and Abe, T. (1998) Model of cathode reaction resistance in molten carbonate fuel cells. *J. Electrochem. Soc.*, **145** (5), 1511–1517.
5. Yoshiba, F., Ono, N., Izaki, Y., Watanabe, T., and Abe, T. (1998) Numerical analyses of the internal conditions of a molten carbonate fuel cell stack: comparison of stack performances for various gas flow types. *J. Power Sources*, **71** (1–2), 328–336.
6. Yoshiba, F., Abe, T., and Watanabe, T. (2000) Numerical analysis of molten carbonate fuel cell stack performance: diagnosis of internal conditions using cell voltage profiles. *J. Power Sources*, **87** (1–2), 21–27.
7. Bosio, B., Costamagna, P., and Parodi, F. (1999) Modeling and experimentation of molten carbonate fuel cell reactors in a scale-up process. *Chem. Eng. Sci.*, **54** (13–14), 2907–2916.

8. Bosio, B., Marra, D., and Arato, E. (2010) Thermal management of the molten carbonate fuel cell plane. *J. Power Sources*, **195** (15), 4826–4834.
9. Costamagna, P., Costa, P., and Antonucci, V. (1998) Micro-modelling of solid oxide fuel cell electrodes. *Electrochim. Acta*, **43** (3-4), 375–394.
10. Koh, J.H. and Kang, B.S. (2001) Theoretical study of a molten carbonate fuel cell stack for pressurized operation. *Int. J. Energy Res.*, **25** (7), 621–641.
11. Koh, J.H., Kang, B.S., and Lim, H.C. (2001) Analysis of temperature and pressure fields in molten carbonate fuel cell stacks. *AIChE J.*, **47** (9), 1941–1956.
12. Koh, J.H., Seo, H.K., Yoo, Y.S., and Lim, H.C. (2002) Consideration of numerical simulation parameters and heat transfer models for a molten carbonate fuel cell stack. *Chem. Eng. J.*, **87** (3), 367–379.
13. Yuh, C.Y. and Selman, J.R. (1991) Polarization of molten carbonate fuel cell electrodes. I. Analysis of steady-state polarization data. *J. Electrochem. Soc.*, **138** (12), 3642–3648.
14. Park, N.K., Lee, Y.R., Kim, M.H., Chung, G.Y., Nam, S.W., Hong, S.A., Lim, T.H., and Lim, H.C. (2002) Studies of the effects of the reformer in an internal-reforming molten carbonate fuel cell by mathematical modeling. *J. Power Sources*, **104** (1), 140–147.
15. Fermeglia, M., Cudicio, A., DeSimon, G., Longo, G., and Pricl, S. (2005) Process simulation for molten carbonate fuel cells. *Fuel Cells*, **5** (1), 66–79.
16. Heidebrecht, P. and Sundmacher, K. (2005) Dynamic model of a cross-flow molten carbonate fuel cell with direct internal reforming. *J. Electrochem. Soc.*, **152** (11), A2217–A2228.
17. Gundermann, M., Heidebrecht, P., and Sundmacher, K. (2008) Parameter identification of a dynamic MCFC model using a full-scale fuel cell plant. *Ind. Eng. Chem. Res.*, **47** (8), 2728–2741.
18. Pfafferodt, M., Heidebrecht, P., and Sundmacher, K. (2010) Stack modelling of a molten carbonate fuel cell (MCFC). *Fuel Cells*, **10** (4), 619–635.
19. Bittanti, S., Canevese, S., De Marco, A., Errigo, A., and Prandoni, V. (2006) Molten carbonate fuel cell electrochemistry modelling. *J. Power Sources*, **160** (2), 846–851.
20. Bittanti, S., Canevese, S., De Marco, A., Giuffrida, G., Errigo, A., and Prandoni, V. (2007) Molten carbonate fuel cell dynamical modeling. *J. Fuel Cell Sci. Technol.*, **4** (3), 283–293.
21. Iora, P. and Campanari, S. (2007) Development of a three-dimensional molten carbonate fuel-cell model and application to hybrid cycle simulations. *J. Fuel Cell Sci. Technol.*, **4** (4), 501–510.
22. Lee, S.Y., Kim, D.H., Lim, H.C., and Chung, G.Y. (2010) Mathematical modeling of a molten carbonate fuel cell (MCFC) stack. *Int. J. Hydrogen Energy*, **35** (23), 13096–13103.
23. Liu, A.G. and Weng, Y.W. (2010) Modeling of molten carbonate fuel cell based on the volume-resistance characteristics and experimental analysis. *J. Power Sources*, **195** (7), 1872–1879.
24. Xu, J.G. and Froment, G.F. (1989) Methane steam reforming, methanation and water-gas shift. 1. Intrinsic kinetics. *AIChE J.*, **35** (1), 88–96.
25. Binnewies, M. and Milke, E. (2002) *Thermochemical Data of Elements and Compounds*, 2nd edn., Wiley-VCH Verlag GmbH, Weinheim.
26. Mitteldorf, J. and Wilemski, G. (1984) Film thickness and distribution of electrolyte in porous fuel-cell components. *J. Electrochem. Soc.*, **131** (8), 1784–1788.
27. Institute of Gas Technology (1977) *Fuel Cell Research on Second Generation Molten Carbonate Systems*, vol. **II**, Project 8984, Final Status Report, Institute of Gas Technology, US Department of Energy, Chicago.
28. Jewulski, J. and Suski, L. (1984) Model of the isotropic anode in the molten carbonate fuel cell. *J. Appl. Electrochem.*, **14** (2), 135–143.
29. Kunz, H.R., Bregoli, L.J., and Szymanski, S.T. (1984) A homogeneous/agglomerate model for molten carbonate fuel cell cathodes. *J. Electrochem. Soc.*, **131** (12), 2815–2821.
30. Bregoli, L.J. and Kunz, H.R. (1982) The effect of thickness on the performance of molten carbonate fuel-cell cathodes. *J. Electrochem. Soc.*, **129** (12), 2711–2715.

31. Yuh, C.Y. and Selman, J.R. (1984) Polarization of the molten carbonate fuel-cell anode and cathode. *J. Electrochem. Soc.*, **131** (9), 2062–2069.
32. Yuh, C.Y. and Selman, J.R. (1992) Porous-electrode modeling of the molten-carbonate fuel-cell electrodes. *J. Electrochem. Soc.*, **139** (5), 1373–1379.
33. Yuh, C.Y. and Selman, J.R. (1991) Polarization of molten carbonate fuel cell electrodes. II. Characterization by AC impedance and response to current interruption. *J. Electrochem. Soc.*, **138** (12), 3649––3365.
34. Jewulski, J. (1986) Process modelling in the porous molten carbonate fuel cell (MCFC) cathode. *J. Appl. Electrochem.*, **16** (5), 643–653.
35. Christensen, P.S. and Livbjerg, H. (1992) A new model for gas diffusion electrodes. Application to molten carbonate fuel cells. *Chem. Eng. Sci.*, **47** (9–11), 2933–2938.
36. Livbjerg, H., Christensen, T.S., Hansen, T.T., and Villadsen, J. (1987) Theoretical foundation of cluster formation in supported liquid-phase catalysts. *Sadhana*, **10** (1–2), 185–216.
37. Fontes, E., Fontes, M., and Simonsson, D. (1995) A heterogeneous model for the MCFC cathode. *Electrochim. Acta*, **40** (11), 1641–1651.
38. Fontes, E., Lagergren, C., and Simonsson, D. (1993) Mathematical modelling of the MCFC cathode. *Electrochim. Acta*, **38** (18), 2669–2682.
39. Prins-Jansen, J.A., Fehribach, J.D., Hemmes, K., and De Wit, J.H.W. (1996) A three-phase homogeneous model for porous electrodes in molten-carbonate fuel cells. *J. Electrochem. Soc.*, **143** (5), 1617–1628.
40. Prins-Jansen, J.A., Plevier, G.A.J.M., Hemmes, K., and De Wit, J.H.W. (1996) An ac-impedance study of dense and porous electrodes in molten-carbonate fuel cells. *Electrochim. Acta*, **41** (7–8), 1323–1329.
41. Prins-Jansen, J.A., Hemmes, K., and De Wit, J.H.W. (1997) An extensive treatment of the agglomerate model for porous electrodes in molten carbonate fuel cells – I. Qualitative analysis of the steady-state model. *Electrochim. Acta*, **42** (23–24), 3585–3600.
42. Prins-Jansen, J.A., Hemmes, K., and De Wit, J.H.W. (1997) An extensive treatment of the agglomerate model for porous electrodes in molten carbonate fuel cells – II. Quantitative analysis of time-dependent and steady-state model. *Electrochim. Acta*, **42** (23–24), 3601–3618.
43. Makkus, R.C. (1991) Electrochemical Studies on the Oxygen Reduction and NiO(Li) Dissolution in Molten Carbonate Fuel Cells. Dissertation, Delft University of Technology.
44. Lim, J.H., Yi, G.B., Suh, K.H., Lee, J.K., Kim, Y.S., and Chun, H.S. (1999) A simulation of electrochemical kinetics for gas–liquid–solid phase of MCFC anode. *Korean J. Chem. Eng.*, **16** (6), 856–860.
45. Bosio, B., Arato, E., and Costa, P. (2003) Concentration polarisation in heterogeneous electrochemical reactions: a consistent kinetic evaluation and its application to molten carbonate fuel cells. *J. Power Sources*, **115** (2), 189–193.
46. Subramanian, N., Haran, B.S., Ganesan, P., White, R.E., and Popov, B.N. (2003) Analysis of molten carbonate fuel cell performance using a three-phase homogeneous model. *J. Electrochem. Soc.*, **150** (1), A46–A56.
47. Berg, P. and Findlay, J. (2010) Comment on "Analysis of molten carbonate fuel cell performance using a three-phase homogeneous model" [*J. Electrochem. Soc.*, **150**, A46 (2003)]. *J. Electrochem. Soc.*, **157** (8), S13.
48. Hong, S.G. and Selman, J.R. (2004) A stochastic structure model for liquid-electrolyte fuel cell electrodes, with special application to MCFCs. I. Electrode structure generation and characterization. *J. Electrochem. Soc.*, **151** (5), A739–A747.
49. Hong, S.G. and Selman, J.R. (2004) A stochastic structure model for liquid-electrolyte fuel cell electrodes, with special application to MCFCs II. Effect of structure on performance. *J. Electrochem. Soc.*, **151**, (5), A748–A755.
50. Brouwer, J., Jabbari, F., Leal, E.M., and Orr, T. (2006) Analysis of a molten carbonate fuel cell: numerical modeling

and experimental validation. *J. Power Sources*, **158** (1), 213–224.

51. Slattery, J.C. (1969) Single-phase flow through porous media. *AIChE J.*, **15** (6), 866–872.
52. Slattery, J.C. (1971) *Momentum, Energy, and Mass Transfer in Continua*, McGraw-Hill Chemical Engineering Series, McGraw-Hill, New York.
53. Whitaker, S. (1973) The transport equations for multi-phase systems. *Chem. Eng. Sci.*, **28** (1), 139–147.
54. Bachmat, Y. and Bear, J. (1986) Macroscopic modeling of transport phenomena in porous-media. 1. The continuum approach. *Transp. Porous Media*, **1** (3), 213–240.
55. Kaviany, M. (1991) *Principles of Heat Transfer in Porous Media*, Mechanical Engineering Series, Springer, New York.
56. Heidebrecht, P. and Sundmacher, K. (2005) Optimization of reforming catalyst distribution in a cross-flow molten carbonate fuel cell with direct internal reforming. *Ind. Eng. Chem. Res.*, **44** (10), 3522–3528.

29
High-Temperature Polymer Electrolyte Fuel-Cell Modeling
Uwe Reimer

29.1
Introduction

The high-temperature polymer electrolyte fuel cell (HT-PEFC) is a polymer electrolyte fuel cell (PEFC) operated above 100 °C, that is, the product water occurs in the gas phase. The typical operating temperature of a phosphoric acid-based HT-PEFC is about 160 °C. A good overview of HT-PEFC technology can be found in a review article by Zhang *et al.* [1] and will not be discussed further here. The literature on HT-PEFC modeling is not very abundant at present. On the other hand, a number of example applications have been presented in the last few years, one of them being an aircraft powered exclusively by a commercial HT-PEFC [2]. Obviously, engineering knowledge is available at a high level. From the modeling point of view, many aspects can be taken from solid oxide fuel cells (SOFCs) and "classical" PEFCs [3–8]. The major challenge is to adopt the existing modeling approaches to the needs of the HT-PEFC.

In comparison with the PEFC, the HT-PEFC requires a description of the electrochemistry with modification to higher tolerance against carbon monoxide (CO) and a simpler approach to fluid flow because of the absence of liquid water. The CO tolerance requires special submodels that account for the reversible decrease in catalyst activity if the fuel is reformate gas. Compared with the SOFC, the HT-PEFC requires different electrochemical parameters because of the very different catalysts and operating temperatures and the use of H^+ instead of O^{2-} as charge carrier. Thermomechanical stress is less important because of the much more moderate operating temperature.

Currently, there are two major types of PEFC operated at elevated temperatures: one with polysulfonic acid (Nafion)-based membranes and the other with polybenzimidazole (PBI)–phosphoric acid-based membranes. For Nafion-type membrane PEFCs with a typical operating temperature of about 110–120 °C, already existing models can be readily used. A detailed discussion of these models tcan be found in review articles in the literature [9–12]. This chapter focuses on the HT-PEFC type based on PBI–phosphoric acid membranes, which type can be referred to as a phosphoric acid fuel cell (PAFC) with a polymer membrane. The "classical" PAFC

Fuel Cell Science and Engineering: Materials, Processes, Systems and Technology,
First Edition. Edited by Detlef Stolten and Bernd Emonts.
© 2012 Wiley-VCH Verlag GmbH & Co. KGaA. Published 2012 by Wiley-VCH Verlag GmbH & Co. KGaA.

was introduced commercially in 1991 [13] and since then has shown good long-term stability in the several hundred kilowatt range. The phosphoric acid is incorporated in a porous matrix of silicon carbide. The drawbacks are acid leakage and a limitation on the minimum size of the silicon carbide matrix, since the thickness has a major impact on the overall cell resistance. In HT-PEFCs, the phosphoric acid is incorporated in a thin polymer membrane, where acid leakage seems not to be an issue. Thus, the advantages of the PAFC such as higher tolerance towards CO, higher operating temperature, and easier water management are combined with the advantages of the PEFC such as robustness and lightweight design for mobile applications. Since this combination is fairly new, it still has some weak points that have to be improved. Major issues are inexpensive materials that can withstand the corrosive environment of the phosphoric acid and the hindrance of the oxygen reduction reaction on the cathode due to acid adsorption on the catalyst surface.

In contrast to PEFCs, the elevated operating temperature leads to different cooling strategies. The use of phosphoric acid as electrolyte permits operating temperatures up to 220 °C, whereas the typical operating temperature of an HT-PEFC is about 160 °C. This excludes liquid water as a cooling medium at ambient pressure, hence stacks are cooled with either air or oil.

Usually, the starting point of model derivation is either a physical description along the channel or across the membrane electrode assembly (MEA). For HT-PEFCs, the interaction of product water and electrolyte deserves special attention. Water is produced on the cathode side of the fuel cell and will either be released to the gas phase or become adsorbed in the electrolyte. As can be derived from electrochemical impedance spectroscopy (EIS) measurements [14], water production and removal are not equally fast. Water uptake of the membrane is very fast because the water production takes place inside the electrolyte, whereas the transport of water vapor to the gas channels is diffusion limited. It takes several minutes before a stationary state is reached for a single cell. The electrolyte, which consists of phosphoric acid, water, and the membrane polymer, changes composition as a function of temperature and water content [15–18]. As a consequence, the proton conductivity changes as a function of current density [14, 19, 20].

There is currently no information in the literature about a concentration difference of phosphoric acid between the anode and cathode. A rough estimate of water distribution as a function of stoichiometric factor λ is given in Table 29.1. The estimate is based on the assumptions that the cell is operated with hydrogen and air and that no significant amount of water is accumulated inside the cell. The expected distribution from Table 29.1 agrees well with in-house experiments and literature data, where for a value of $\lambda(H_2/air) = 2/2$, 15–20% of the product water was found in the anode off-gas [21]. Therefore, it can be reasoned that the concentration gradient of phosphoric acid between the anode and cathode is negligible, as the MEA is very thin (100–300 μm) and water distribution should be fast at 160 °C.

The water distribution along the channel in a cell is not homogeneous because the water vapor partial pressure increases from the inlet to outlet due to the water production in the electrochemical reaction. Furthermore, the current density is not homogeneous along the channel, as was shown by models independent of the

Table 29.1 Estimated distribution of water vapor in the gas stream for operation on pure hydrogen and air.

λ_{H_2}	λ_{air}	Water fraction anode (%)	Water fraction cathode (%)
1	1	0	100
1.2	2	5	95
2	2	19	81
3	3	23	77

fuel-cell type [8]. These two effects lead to the following consequences. Along the channel, the phosphoric acid concentration inside the MEA changes. Furthermore, as a result of product water adsorption the amount of electrolyte increases, which leads to different levels of flooding in the catalyst layer. As a consequence, the available catalyst surface area and local oxygen diffusion will change. Therefore, a gradient in local proton conductivity and local cell resistance should be considered. These effects are usually not described by existing models. It seems that experimental data on the behavior of phosphoric acid at high temperatures, high concentrations, and incorporated in a polymer network are still rare.

In the following sections, recent model approaches for HT-PEFCs are reviewed. A more detailed discussion about the role of the electrolyte is presented in Section 29.4 without claiming to be a perfect explanation. In Section 29.5, a very basic approach to modeling the polarization curve of an HT-PEFC is discussed using the example of a Celtec MEA from BASF. The intention is to demonstrate the consequences of the unique behavior of the electrolyte, which is discussed in Section 29.4.

29.2
Cell-Level Modeling

Most cell models use the description of physical phenomena across the MEA as the starting point. The catalyst layer is a thin (10–100 μm) three-dimensional region. The electrochemical reaction takes place at the catalyst surface, which is covered by a thin film of electrolyte. Electrochemical reaction rates vary across the thickness of the catalyst layer according to the interplay between the ionic conductivity of the electrolyte, the electronic conductivity of the electrode, and the species concentration [8], where the flooding of the electrolyte with phosphoric acid as a function of operating conditions changes this interplay.

One major goal of fuel-cell models is to match the experimental polarization curve for the operating conditions considered. The simplest approach is to treat the MEA as a reactive boundary between the anode and cathode sides. With increasing complexity, the two catalyst layers can be described separately as an effective boundary, a quasi-three-dimensional layer which is partially flooded with electrolyte, or as a three-dimensional volume that contains agglomerates of catalyst and carbon support. Additionally, the influence of varying concentrations along the gas channels can be considered implicitly or explicitly – ranging from pure boundary conditions to the full description of fluid flow. An overview of recent approaches in the literature is given in the following.

Cheddie and Munroe presented a one-dimensional model (across the MEA), where the effects of gas solubility are taken into account and a simple submodel for the blocking of catalyst surface sites by adsorbed phosphate species is incorporated [22]. The conductivity of the electrolyte and the solubility of hydrogen and oxygen are taken into account as functions of temperature. The numerical solution shows very good agreement with experimental results. The influence of the phosphoric acid doping level is discussed. One major conclusion of the model is that only 1% of the catalyst surface is utilized in fuel-cell operation.

The same authors developed this model further by transforming the volumetric catalyst layer source terms into interfacial boundary conditions for a full three-dimensional fuel-cell model [23]. The catalyst surface is represented as a two-dimensional plane, which is coupled to computational fluid dynamics (CFD) code. The modeling domain includes a channel pair with ribs and MEA.

A two-dimensional isothermal model, which describes the catalyst as spherical agglomerates with porous inter-agglomerate spaces, was introduced by Sousa *et al.* [24]. The model considers the MEA and gas flow channels. The inter-agglomerate species are filled with a mixture of phosphoric acid and polytetrafluoroethylene (PTFE). The model was validated against experimental data. The resulting utilization of catalyst particles at high current densities is very low. The authors pointed out that the amount of phosphoric acid in the catalyst layers is not constant in order to obtain good agreement between model and experiment. This is a remarkable modeling result. For optimum fuel-cell performance, a volume fraction of phosphoric acid between 30 and 55% in the catalyst layer is desirable. The model also demonstrates the poisoning of the anode catalyst by CO.

The same authors incorporated this model in finite element software [25]. The two-dimensional model includes endplates of the single cells and external heaters. Two geometries were considered: along the channel and across the channel direction. The geometry along the channel did not yield satisfactory results. The authors reported large temperature differences across the channel direction, which indicate that the catalyst layer was not used efficiently.

In a further article, dynamic simulations were presented Sousa *et al.* that focused on the time-dependent response of temperature and current with respect to load changes [26]. The results showed that a current overshoot could be detected at step changes in cell voltage. The overshoot is caused by the delayed change of local oxygen concentration, and can be counterbalanced by an increase in the double-layer capacitance of the model. The dynamic model was modified to account for degradation mechanisms of phosphoric acid loss and platinum sintering in the cathode catalyst layer. It was found that during the first period of $\sim$300 h, catalyst activity loss due to the change in mean particle size was the dominant effect. This finding is consistent with an earlier study by Hu *et al.*, in which a

500 h aging test was performed [27]. The experimental results by EIS showed clear changes in electrochemical surface area and high-frequency resistance with time. The particle size was analyzed by transmission and scanning electron microscopy. From the results, a one-dimensional model was constructed that fitted the observed polarization curves well.

The influence of CO poisoning at the anode of an HT-PEFC was investigated by Bergmann et al. [28]. The dynamic, nonisothermal model takes the catalyst layer as a two-dimensional plane between the membrane and gas diffusion layer into account. The effects of CO and hydrogen adsorption with respect to temperature and time are discussed in detail. The CO poisoning is analyzed with polarization curves for different CO concentrations and dynamic CO pulses. The analysis of fuel-cell performance under the influence of CO shows a nonlinear behavior. The presence of water at the anode is explicitly considered to take part in the electrooxidation of CO. The investigation of the current response to a CO pulse of 1.31% at the anode inlet showed a reversible recovery time of 20 min.

A three-dimensional model of a channel pair was also presented by Peng and Lee and implemented in CFD software [29]. The numerical study showed that the ratio between the width of the gas channel and the width of the land area is a key optimization parameter for fuel-ell operation. This model was extended by Peng et al. to describe the transient behavior of an HT-PEFC [30]. The predictions showed that the current density overshoots the value for the steady state when the cell voltage is abruptly decreased. The peak value of this overshoot seems to be related to the cathode stoichiometry. In subsequent work by Jiao and Li, a CFD model of a pair of channels was applied to investigate the influence of the phosphoric acid doping level and humidification of reaction gases [31]. The results of the nonisothermal model indicated that increasing both the operating temperature and the phosphoric acid doping level is favorable for the overall cell performance. Humidification of the feed gases had only minor effects and was not recommended by the authors.

Ubong et al. also presented a three-dimensional model of a channel pair [32]. The isothermal model incorporated a Butler–Volmer-type equation for electrochemistry and was solved with the finite element method. Simulations were validated against a single cell with triple serpentine flow field, which was operated in the temperature range 120–180 °C. The results showed that there is no drastic decrease in cell voltage at high current density due to mass transport limitation. This is explained by the absence of accumulation of liquid water. It was also concluded that reaction gases need not be humidified.

A CFD model that describes a complete single cell was introduced by Lobato et al. [33]. The MEA was implemented as a single plane separating the anode and cathode. The model considers only the cathodic part of the overpotential from the Butler–Volmer equation and the simulation results are presented for operation with pure hydrogen and oxygen. The impact of three different flow field designs, serpentine-like, parallel, and pin-type, on the overall fuel-cell performance were investigated. The best performance was observed for serpentine-like and pin-type flow fields. It should be noted that the bad performance of the selected parallel flow field may be a consequence of the small cell manifold used, which has an important influence [7].

Siegel et al. recently presented a CFD cell model with a six-channel serpentine flow field [34]. The model is isothermal and steady state. The description of the catalyst layer follows an agglomerate approach, which takes diffusivity and solubility of gases in phosphoric acid into account. The submodel for the temperature dependence of the conductivity of the phosphoric acid is critically discussed. In the range 150–160 °C, good agreement with experimental results was obtained.

Lobato et al. used an artificial neural network approach to describe the polarization curve of an HT-PEFC [35]. Four different neural network types were applied. Special attention was paid to describe the influence of the PTFE content in the gas diffusion layer. For this purpose, the tortuosity was used as a model parameter and the results showed good agreement with experimental polarization curves.

Based on an analytical model for PEFCs, Kulikovsky et al. presented a two-step procedure to evaluate the parameters Tafel slope, exchange current density, and cell resistance from two sets of polarization curves for an HT-PEFC [36]. The method was validated with experimental data. Shamardina et al. described an analytical model taht accounts for the crossover of gases through the membrane [37]. The model is pseudo-two dimensional and describes mainly the effects across the MEA. Temperature and pressure variations in the cell were not considered. From these analytical studies, it follows that the crossover effect has a considerable influence only at a low stoichiometry of oxygen.

The majority of the models discussed have in common that the electrochemical reaction at the electrodes is described by a Butler–Volmer- or Tafel-type equation. Therefore, all of these models share a very similar set of fundamental parameters, which are summarized in Table 29.2. These can be compared in order to check for model consistency. Nevertheless, all of these models are based on slightly different assumptions. The parameters that are obtained by curve-fitting procedures are

Table 29.2 Commonly used parameters for Butler–Volmer- and Tafel-type equations.

Parameter	Description	Units
η	Overpotential or potential loss	V
α	Charge-transfer coefficient	–
j	Current density	$A\,cm^{-2}$
j_0^{ref}	Exchange current density	$A\,cm^{-2}$
c_i	Mean concentration at catalyst surface of species i	–
c_{ref}	Reference concentration, that is, the concentration at which the exchange current density j_0^{ref} is reckoned	–
γ	Exponent related to the kinetic reaction order, mostly considered to be 1	–
E_A	Activation energy	$J\,mol^{-1}$

Table 29.3 Selection of Butler–Volmer-type equations from the models discussed.

Equation	Ref.
$j = a j_0^{\text{ref}} \left(\dfrac{c_i}{c_{\text{ref}}} \right)^{\gamma_i} \left\{ \exp\left(\dfrac{\alpha F}{RT} \eta \right) - \exp\left[\dfrac{(1-\alpha) F}{RT} \eta \right] \right\}$	[22]
$j = a j_0^{\text{ref}} m_i^{\gamma_i} \left[\exp\left(\dfrac{\alpha_a F}{RT} \eta \right) - \exp\left(\dfrac{\alpha_c F}{RT} \eta \right) \right]$	[23]
$j = j_0^{\text{ref}} \left[\exp\left(\dfrac{-\alpha_{\text{red}} F}{RT} \eta \right) - \exp\left(\dfrac{\alpha_{\text{ox}} F}{RT} \eta \right) \right]$	[24]
$j = a_{\text{Pt}} j_0^{\text{ref}} \left[\exp\left(\dfrac{-\alpha_{\text{red}} F}{RT} \eta \right) - \exp\left(\dfrac{\alpha_{\text{ox}} F}{RT} \eta \right) \right]$	[26]
$j = a j_0^{\text{ref}} \left(\dfrac{c_i}{c_{\text{ref}}} \right)^{\gamma_i} \left[\exp\left(\dfrac{\alpha_c F}{RT} \eta \right) - \exp\left(\dfrac{-\alpha_a F}{RT} \eta \right) \right]$	[27]
$j = j_0^{\text{ref}} \left[\dfrac{c_i(t)}{c_{\text{ref}}} \right]^{\gamma_i} \left\{ \exp\left[\dfrac{\alpha 2F}{RT} \eta(t) \right] - \exp\left[\dfrac{(1-\alpha) 2F}{RT} \eta(t) \right] \right\}$	[28]

effective parameters; in most cases several effects are described as a total by the particular model. This demonstrates that collecting parameters from different sources without careful examination of the underlying models or experiments should be avoided because this may lead to an inconsistent set of parameters. A detailed discussion of the procedure for extracting and comparing literature values was given by Sousa *et al.* [24]. To illustrate this proposition, a selection of equations from the models discussed above are presented in Table 29.3, and their respective parameters are compared in Table 29.4. It is obvious that the parameters in Table 29.4 cannot be exchanged directly between the different models.

29.3
Stack-Level Modeling

As became obvious in the previous section, a detailed description of a complete fuel cell is computationally very demanding. The stack models thatt are discussed in this section are on a higher abstraction level. They serve mainly as one component of a complete fuel-cell system. The discussion of system simulation is beyond the scope of this chapter, and at this point only the characteristics of the stack models are mentioned.

Among the early work that considered HT-PEFC stacks is a study by Ahluwalia *et al.*, who compared the impact of a "hypothetical" high-temperature stack on

Table 29.4 Parameters of Butler–Volmer-type equations from the models discussed.

Parameter	Anode	Cathode	Units	Ref.
α	0.50	2.00	–	[22], taken from PEFC [38]
α	0.50	0.73	–	See discussion in [24]
α	0.50	10.00	–	[27]
α	0.50	0.50	–	[28]
α	–	0.78	–	[36]
γ	0.50	1.00	–	[22], taken from PEFC [38]
γ	1.00	1.00	–	[24, 27, 28], see discussion in [24]
j_0^{ref}	1.00×10^5	1.00×10^{-4}	A m^{-2}	[22]
j_0^{ref}	14.40×10^{-2}	2.63×10^{-8}	A m^{-2}	Taken from rotating disc experiments [39], see discussion in [24]
j_0^{ref}	–	3.50×10^{-7}	A m^{-2}	[27]
j_0^{ref}	–	32.80×10^{-3}	A m^{-2}	[28]
j_0^{ref}	–	1.50×10^{-5}	A m^{-2}	[36]
E_A	16.90×10^3	72.40×10^3	J mol^{-2}	See discussion in [24]
E_A	–	102.86×10^3	J mol^{-2}	[28]

system performance with a conventional PEFC [40]. The PEFC was operated at 80 °C and a parametric study was conducted to compare the performance of a high-temperature stack operating at 150–200 °C. The results indicated that a high-temperature system has a higher efficiency compared with a PEFC system.

A semiempirical model for an HT-PEFC polarization curve intended to be used in system modeling was presented by Korsgaard et al. [41]. The cell voltage was calculated as a function of current density, temperature, and cathode stoichiometry. The influence of cathode stoichiometry was modeled by an effective resistance term. As fuel gases, pure hydrogen and reformate gas were considered. The results showed excellent agreement with experimental data, which were also presented. It was found that it is possible to operate the fuel cell at 160 °C with a CO content as high as 2%. The prediction window ranges from 120 to 180 °C and with athode stoichiometric ratios from 2 to 5. In subsequent work, the same group successfully integrated this model into a model for a complete HT-PEFC-based fuel-cell system [42, 43]. The same approach was used by Chrenko et al. for static and dynamic modeling of a diesel-fed fuel-cell power supply [44].

Andreasen et al. introduced a stack model that is suitable for prediction and analysis using EIS [45]. The typical output of such a measurement is a Nyquist plot, which shows the imaginary and real parts of the impedance of the measured system. The full stack impedance depends on the impedance of each of the single cells of the stack. Equivalent circuit models for each single cell can be used to predict the stack impedance at different temperature profiles of the stack. The results showed that a simple equivalent circuit model can be used to simulate the stack behavior. It was concluded that a more thorough characterization is required to predict the voltage dynamics under all operating conditions.

A CFD model of a full stack was reported by Kvesić et al. [46], in which the model comprises an averaged volume approach. The cell flow fields are described by an effective porous medium, where the volume of the channels and lands are combined. Hence fluid flow inside the cells is described by Darcy's law. Electrochemical conversion is described by a standard approach s based on the Tafel equation. The model encompasses fuel cells, cooling cells, stack manifold, and endplates. The results were validated against measurement of local temperature and local current density inside an HT-PEFC short stack consisting of five cells with an active area of 200 cm^2. For the temperature distribution, almost perfect agreement was observed between model and experiment. The authors noted that for the given operating conditions the temperature inside the stack is nearly homogeneous. The measured temperature variation is within the experimental error. For operation with pure hydrogen and air, the model predicts that near the cell inlet the current density is almost twice as high as in the area near the cell outlet. The simulated values for local current density matched the experimental values well within the experimental error.

29.4
Phosphoric Acid as Electrolyte

HT-PEFCs and PEFCs mainly differ in the nature of the polymer membrane and electrolyte. In an HT-PEFC, a combination of PBI, phosphoric acid, and water is used, whereas in a PEFC, Nafion and water are employed. Although phosphoric acid has been used for a longer time in PAFCs, many aspects of this substance remain unknown. The most important facts are summarized below to give an overview of this fairly complex matter; which is discussed in greater detail in a separate chapter of this book [47].

Phosphoric acid is an almost ideal electrolyte. It is available on an industrial scale while also being reasonably cheap. It is stable up to 220 °C, thus allowing operating temperatures that are suitable to avoid the formation of liquid water and to take advantage of the higher CO tolerance of the platinum catalyst at this temperature. The vapor pressure of phosphoric acid is almost zero – phosphorus pentoxide is formed at higher temperatures, which is a solid – hence evaporation losses can be neglected and the amount of phosphoric acid inside the fuel cell remains unchanged over a very long period. Further, it is an almost ideal proton conductor. Pure phosphoric acid is a liquid with a low diffusion coefficient of phosphate species [48] but with an extremely high proton mobility, which involves proton transfer between phosphate species and some structural rearrangements. The contribution of protons to the total conductivity is more than 90%, as shown both experimentally [49] and theoretically [50, 51]. The autoprotolysis of phosphoric acid seems to be of very little importance since the diffusion of phosphate species is very low [48]. The overall conductivity is clearly dominated by the proton transfer via the shift of hydrogen bonds (Grotthuss mechanism) [49, 50, 52], which is shown in Figure 29.1.

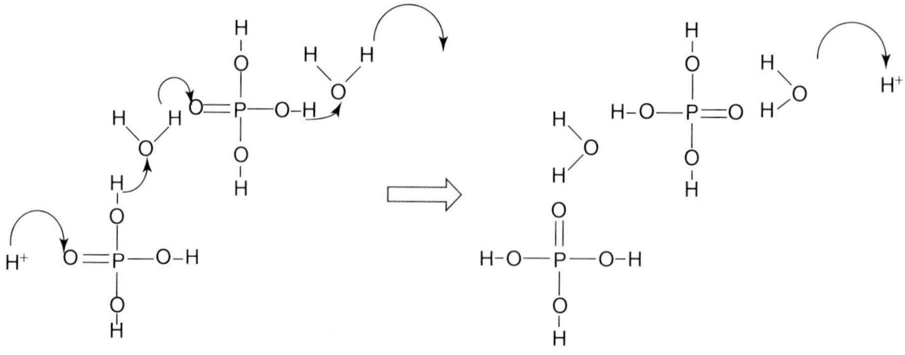

Figure 29.1 Scheme of the proton transfer mechanism in phosphoric acid solutions (Grotthuss mechanism).

Another striking feature is the ability to bind and release water very fast over a wide concentration range. At room temperature, concentrated phosphoric acid (85 wt%) has a molar ratio of H_3PO_4 to H_2O of nearly 1:1. At higher temperatures, the remaining water evaporates. As the concentration increases, additional water can be released by condensation reactions, which lead from phosphoric acid through pyrophosphate species to phosphorus pentoxide in the final stage. Especially the first condensation step [Eq. (29.1)] seems to be very fast in both directions [18].

$$H_3PO_4 + H_3PO_4 \rightleftharpoons H_4P_2O_7 + H_2O \tag{29.1}$$

In order to obtain a detailed model of the conductivity for an operating HT-PEFC, the above-described effects of phosphoric acid and water should be combined with the interaction of the membrane material, the dynamic water production at the catalyst sites, and the dynamic water removal due to the gas flow in the channels. A fully featured model with atomistic resolution seems to be impossible at present. Recent modeling approaches can describe the proton conductivity mechanism at a single concentration with atomistic resolution [50, 51].

For fuel-cell catalysts, there are two models available that incorporate the above-described effects implicitly [22, 24]. Both models are based on the reasonable assumption that the surface of the catalyst is covered by a thin film of electrolyte, which includes another effect, namely the wetting behavior of the electrolyte on the catalyst. The catalyst itself can be described to a first approximation like a single particle [22]. A further advance is the description of the catalyst as spherical agglomerates with porous inter-agglomerate spaces [24].

At the cell level, the above-described mechanisms lead to the following consequences. The acid composition is a function of current density and cell temperature due to the delicate balance between internal water production and removal. Hence the conductivity changes as a function of current density, which can be observed experimentally during load changes in HT-PEFC operation [14]. As the phosphoric acid takes up water, the volume of the membrane increases, that is, it swells as shown by synchrotron radiography experiments [53]. The resulting situation is depicted in Figure 29.2. At present, there is no model for an HT-PEFC which also

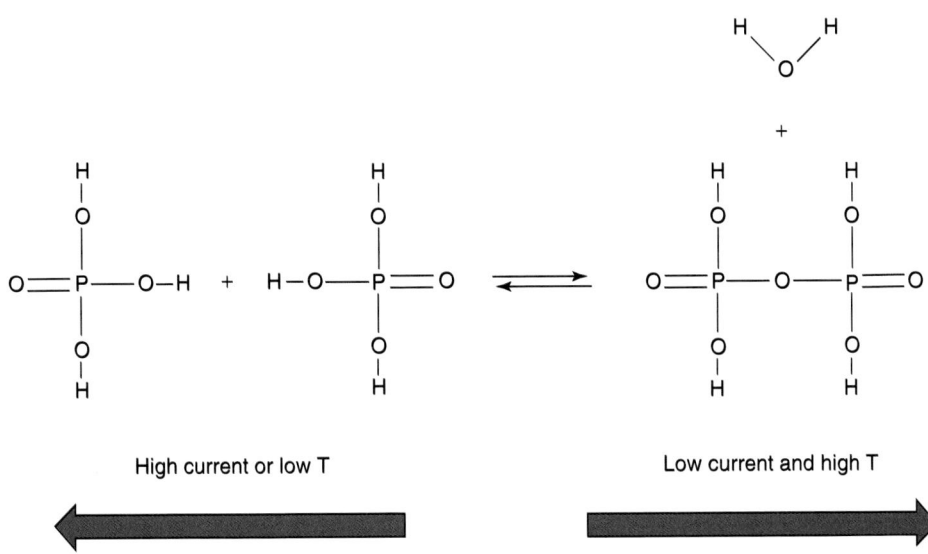

Figure 29.2 Assumed change of composition of phosphoric acid species during fuel-cell operation.

accounts for the resulting flooding effect inside the catalyst layer as a function of current density. Another model exists for a similar scenario for a PEFC, where water floods the pores as a function of current density [54]. These effects are summarized in a recent review on PEFCs [55].

Of course, there are also some disadvantages. The use of phosphoric acid imposes severe constraints on the fuel-cell material owing to the corrosive nature of the acid. Additionally, the oxygen reduction reaction is hindered at the catalyst surface. This seems to be a combination of adsorbed species partially blocking catalytic sites [56, 57] and the low oxygen solubility [58]. Therefore, other electrolytes have been investigated that could replace phosphoric acid. Among promising candidates that are currently under consideration are phosphates such as cesium dihydrogenphosphate (CsH_2PO_4) [59, 60] and sulfonates such as diethylmethylammonium trifluoromethanesulfonate ([dema][TfO]) [61]. At present, phosphoric acid is still superior for the use in HT-PEFCs.

29.5
Basic Modeling of the Polarization Curve

In the previous section, the underlying principles of the behavior of the electrolyte were laid out. The main intention of this section is to discuss the consequences following from these considerations on the parameters of fuel-cell models and mediate a critical discussion. For this purpose, the widely used standard approach is applied, where the polarization curve of a HT-PEFC is described by the following equation [3, 6, 7]:

$$E_{cell} = E_{Nernst} - R_\Omega j - \eta_{act} - \eta_{trans} \tag{29.2}$$

The cell voltage E_{cell} is obtained by calculating the Nernst voltage E_{Nernst} and subtracting the losses caused by ohmic resistance R_Ω, activation overpotential η_{act}, and mass transport limitations η_{trans}. The nomenclature follows that used in Section 29.2 (see Table 29.2). The following assumptions were made in order to discuss a very simple model:

- Operation with pure hydrogen and air is considered.
- Anodic losses are neglected, since the cathodic overpotential is the dominant effect.
- Current density and voltage are assumed to be constant throughout the cell (zero-dimensional model).

The fitted values for the parameters were obtained from experiments with a single testing cell, operated at 160 °C. The MEA (Celtec) was obtained from BASF, which is one of the most widely used at present. The active area was 16.65 cm^2 and the mean resistance was $R_\Omega = 0.1 \, \Omega \, cm^2$.

The values obtained in this way are by no means "absolute" or "correct." They inherit from the selected model the advantage of being some kind of minimum set. Therefore, they are useful as an aid in order to assess the broad range of values that can be found in the literature and to judge which one is more likely to suit the modeling task at hand.

29.5.1
Activation Overpotential

The activation overpotential describes the voltage losses that are connected to the kinetics of the electrochemical reaction at the catalyst surface [3, 6, 7]. It is described by the Butler–Volmer equation. If very small current densities can be neglected, the Tafel equation is an exact approximation:

$$j = j_0 \exp\left(\frac{\alpha F \eta}{RT}\right) \tag{29.3}$$

The most interesting parameter here is the exchange current density j_0, which describes a hypothetical limit of how much current could be produced by the catalyst in the absence of all mass transport effects (i.e., very close to the open cell voltage). The value of j_0 depends on the oxygen concentration at the catalyst surface and the temperature. The concentration dependence can be expressed as follows [62–64]:

$$j_0 = j_0^{ref} \left(\frac{c_i}{c_{ref}}\right)^{\gamma_i} \tag{29.4}$$

The temperature dependence can be given as an Arrhenius-type equation:

$$j_0^{ref} = j_* \exp\left(-\frac{E_A}{RT}\right) \tag{29.5}$$

From the literature, is is known that the value of the activation energy E_A depends also on the electrode potential (for hydrochloric acid and trifluoromethane sulfonic acid (TFMSA) in [65] and for phosphoric acid in [66]). From the data in [66], the following expression can be obtained:

$$E_A = \phi \times 248.68 \text{ kJ mol}^{-1} \text{ V}^{-1} - 172.61 \text{ kJ mol}^{-1} \qquad (29.6)$$

where ϕ is the electric potential of the cathode in volts versus RHE (reversible hydrogen electrode). For a fuel cell in operation, the value of ϕ is equal to the Nernst voltage, that is, $\phi = E_{\text{Nernst}}$. It has to be kept in mind that the data in [66] relate to a half-cell experiment, where the supported catalyst floods freely in phosphoric acid. The following assumptions have to be made:

- The concentration of phosphoric acid is constant (constant water content).
- The setup of the experiment in [66] is comparable to that of the fuel cell.
- The catalyst used in [66] is comparable to that used in the fuel cell.

If all of the above are considered carefully, the activation energy can be estimated as $E_A = 100$ kJ mol^{-1} using Eq. (29.6) with $T = 160\,°C$ (433.15 K) and $E_{\text{Nernst}} = 1.0940$ V. Especially the first assumption of constant electrolyte concentration is only justified for higher current densities, where the balance between internal water production and removal seems to lead to very similar concentration values. This can be deduced from EIS measurements [14].

From curve fitting to the data obtained by single-cell operation of a Celtec MEA, we obtained $\alpha = 0.7$, $j_0 \approx j_0^{\text{ref}} \approx 5 \times 10^{-5}$ A cm^{-2}, and subsequently $j_* = 57.3$ A cm^{-2}. Again, it must be emphasized that these are not absolute values. They are strongly dependent on the modeling approach chosen, that is, on the number of parameters which are considered in Eq. (29.2). Compared with the literature data summarized in Table 29.4, the value for j_0^{ref} is a good estimate. The values in Table 29.4 range from 10^{-4} to 10^{-8} A m^{-2} which is a strong indication that j_0^{ref} is more of a fitting parameter of the given model rather than a physical material property. Therefore, care must be taken in interpreting its value.

29.5.2
Ohmic Resistance

The ohmic resistance R_Ω depends mainly on the proton conductivity of the electrolyte. The conductivity of phosphoric acid as function of temperature can be obtained from the literature [52]. Unfortunately, the exact composition of the electrolyte during current production is unknown. EIS data for HT-PEFC show that the overall resistance changes significantly between the open-cell voltage and current production regime [14]. It can be assumed that under low current conditions, the composition of the electrolyte is shifted towards the pyrophosphate species. Under high current conditions, the composition shifts towards the orthophosphate species, which exhibits a higher conductivity owing to the increased water content. This would be a dynamic effect. The temperature dependence of the conductivity

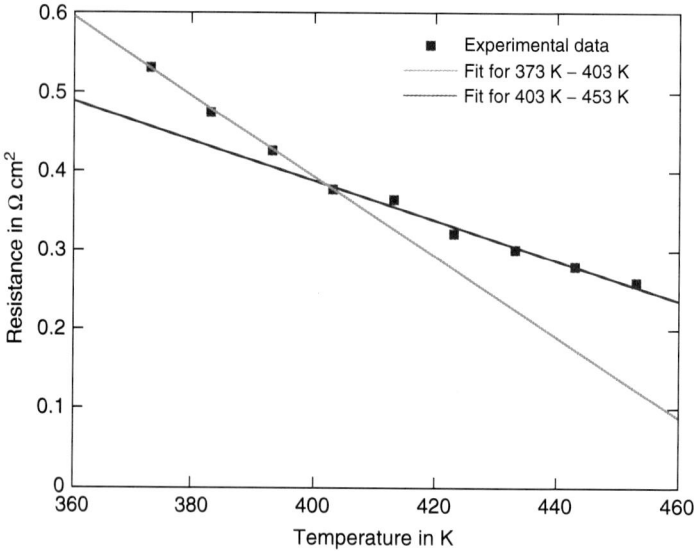

Figure 29.3 Ohmic resistance calculated from conductivity data in [52] and fitted equations.

also shows this conversion as a static effect. At about 120 °C (393 K), orthophosphoric acid starts to release water according to Eq. (29.1) and the composition is shifted towards pyrophosphoric acid [67].

Since there is a lack of experimental data for the composition change of phosphoric acid as a function of current density, a good starting point is to use data that are available for concentrated phosphoric acid. In order to obtain a temperature-dependent function for R_Ω, the conductivity data for orthophosphoric acid (100 wt%) can be used [52]. These data also show a significant change in conductivity at about $T = 400$ K (see Figure 29.3). Two linear functions can be fitted to these data based on the idea that the lower temperature part belongs mainly to orthophosphate species and the higher temperature part belongs to pyrophosphate species.

Two linear equations are obtained. Based on these assumptions, Eq. (29.7) describes the temperature dependence of the orthophosphate species and Eq. (29.8) the temperature dependence of the pyrophosphate species. The fitting equations are further corrected to yield the experimental value of $R_\Omega = 0.1$ Ω cm² for $T = 160$ °C (433 K).

$$R_\Omega (T) = (2.42942 \text{ Ω cm}^2 - 0.00509575 \text{ Ω cm}^2 \text{ K}^{-1} \times T) - 0.12296025 \text{ Ω cm}^2$$
$$(\text{for } 373 \text{ K} \leq T \leq 403 \text{ K}) \tag{29.7}$$

$$R_\Omega (T) = (1.39386 \text{ Ω cm}^2 - 0.002516 \text{ Ω cm}^2 \text{ K}^{-1} \times T) - 0.204432 \text{ Ω cm}^2$$
$$(\text{for } 403 \text{ K} < T \leq 453 \text{ K}) \tag{29.8}$$

Now the dilemma of which function to choose arises. At the operating temperature of $T = 160\,°C$ we assume that for medium and high current densities we have orthophosphoric acid inside the cell. For this temperature region, there is only a valid fit for the proposed pyrophosphate species available. Additionally, it should be kept in mind that the ohmic resistance is influenced by the interplay of at least three components: phosphoric acid species, water, and the polymer network.

From practical considerations, it follows that the HT-PEFC fortunately has a fairly narrow operating window, if the startup procedure is neglected. Therefore, it seems to be convenient to use Eq. (29.8), which works well for cell and stack models. It has to be kept in mind that usually the temperature influence is less important, since large temperature gradients inside the fuel cell are normally avoided, for several reasons. First, a large temperature gradient imposes thermal stress on the respective materials and is a source of accelerated degradation or failure. Second, the electrochemical reaction is sensitive to temperature. An increase in temperature leads to an increase in current density. This in turn would amplify a nonideal current distribution, leading to an increase in losses connected with cross-currents.

29.5.3
Mass Transport

Liu and Eikerling presented a model for PEFCs that describes two valid scenarios for fuel cells in operation [54]. At low current, the catalyst is connected via a pore network to the gas phase. Therefore, oxygen diffusion is fast. At high current, the catalyst can be flooded by product water. In this case, oxygen has to diffuse through the liquid phase, which reduces the effective diffusion coefficient drastically.

A similar scenario can be assumed for HT-PEFCs. At high current density, a larger amount of water is produced. This leads to swelling of the total volume of phosphoric acid, since water is readily adsorbed. Depending on how fast the water can be removed by the streaming gases, an increase in volume of the liquid electrolyte will eventually lead to flooding of the catalyst pore space. Again, oxygen has to diffuse through the liquid phase, which reduces the effective diffusion coefficient. Experimental proof is still required in order to justify this assumption.

The standard equation for the overpotential of mass transport limitation η_{trans} is given as follows as a function of limiting current density j_{lim}:

$$\eta_{trans} = \frac{RT}{\alpha F} \ln\left(\frac{j_{lim}}{j_{lim} - j}\right) \qquad (29.9)$$

In order to fit the required polarization curve, a value of $j_{lim} = 1.6\,\text{A cm}^{-2}$ was obtained from our own experiments. The limiting current density is often the last parameter of a series to be fitted. In that case, it is the least reliable value and a sound physical meaning should not be expected.

The polarization curve of the resulting model is shown in Figure 29.4 together with two experimental curves. The experimental curves differ only in the charge number of the commercial MEA and the second one was obtained 2 months after

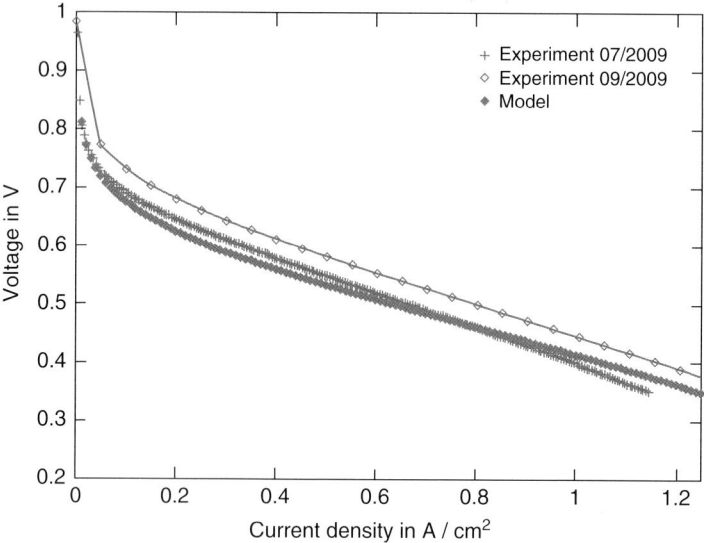

Figure 29.4 Polarization curves for a single-cell HT-PEFC with Celtec MEA, operated at 160 °C with pure hydrogen–air and a stoichiometric ratio of 2:2.

the first. The fit was based on the first curve and the quality of the fitting is far from perfect. Figure 29.4 clearly illustrates that despite all modeling efforts, the experimental error should not be neglected.

29.6
Conclusion and Future Perspectives

At present, a typical HT-PEFC requires a higher catalyst loading and yields a lower power density compared with a classical PEFC. The key issue to overcome these limitations lies in a deeper understanding of the properties of the electrolyte and its interaction with the catalyst. The first steps have already been taken by introducing methods to include the shape of the catalyst layer in fuel-cell modeling. The dynamic interaction of phosphoric acid with water and the network of the polymer membrane is far from being understood. This field requires modeling and experiments to go hand in hand in order to achieve a deeper understanding. The future will show whether phosphoric acid with its remarkable properties has to be replaced by another electrolyte or whether major improvements can be achieved by utilizing meticulous structured catalyst layers. One major modeling task will be to relate the complex, stochastic, and sometimes dynamic structure of the membrane and also the catalyst and gas diffusion layer to physical properties. These features will improve the understanding of all polymer membrane-based fuel cells and hopefully lead to the design of low-cost and highly efficient energy converters.

References

1. Zhang, J., Xie, Z., Zhang, J., Tang, Y., Songa, C., Navessin, T., Shi, Z., Songa, D., Wang, H., Wilkinson, D.P., Liu, Z.S., and Holdcroft, S. (2006) High temperature PEM fuel cells. *J. Power Sources*, **160**, 872.
2. Grotelüschen, F. (2010) Mit Strom über den Atlantik. *Bild Wiss.*, **9**, 80.
3. O'Hayre, R.P., Cha, S.W., Colella, W.G., and Prinz, F.B. (2009) *Fuel Cell Fundamentals*, 2nd edn., John Wiley & Sons, Inc., Hoboken, NJ.
4. Mench, M.M. (2008) *Fuel Cell Engines*, 1st edn., John Wiley & Sons, Inc., Hoboken, NJ.
5. Larminie, J. and Dicks, A. (2003) *Fuel Cell Systems Explained*, 2nd edn., John Wiley & Sons, Ltd., Chichester.
6. EG&G Technical Services (2004) *Fuel Cell Handbook*, 7th edn., National Technical Information Service, US Department of Commerce, Springfield, VA.
7. Barbir, F. (2005) *PEM Fuel Cells*, 1st edn., Elsevier, Amsterdam.
8. Kulikovsky, A.A. (2010) *Analytical Modelling of Fuel Cells*, 1st edn., Elsevier, Amsterdam.
9. Djilali, N. (2007) Computational modelling of polymer electrolyte membrane (PEM) fuel cells: challenges and opportunities. *Energy*, **32**, 269.
10. Siegel, C. (2008) Review of computational heat and mass transfer modeling in polymer-electrolyte-membrane (PEM) fuel cells. *Energy*, **33**, 1331.
11. Mench, M.M. (2010) in *Hydrogen and Fuel Cells*, 1st edn. (ed. D. Stolten), Wiley-VCH Verlag GmbH, Weinheim, p. 89.
12. Wang, C.Y. (2004) Fundamental models for fuel cell engineering. *Chem. Rev.*, **104**, 4727.
13. UTC Power (2011) What are Phosphoric Acid Fuel Cells? http://www.utcpower.com/knowledge-library/what-are-phosphoric-acid-fuel-cells (last accessed 28 January 2012).
14. Wippermann, K., Wannek, C., Oetjen, H.F., Mergel, J., and Lehnert, W. (2010) Cell resistances of poly(2,5-benzimidazole)-based high temperature polymer membrane fuel cell membrane electrode assemblies: time dependence and influence of operating parameters. *J. Power Sources*, **195**, 2806.
15. Brown, E.H. and Whitt, C.D. (1952) Vapor pressure of phosphoric acids. *Ind. Eng. Chem.*, **44**, 615.
16. Jameson, R.F. (1959) The composition of the 'strong' phosphoric acids. *J. Chem. Soc.*, 752.
17. MacDonald, D.I. and Boyack, J.R. (1969) Density, electrical conductivity, and vapor pressure of concentrated phosphoric acid. *J. Chem. Eng. Data*, **14**, 380.
18. Munson, R.A. (1965) Kinetics of the ortho–pyro interconversion in 100% phosphoric acid. *J. Phys. Chem.*, **69**, 1761.
19. Lehnert, W., Wipperman, K., and Wannek, C. (2010) in *18th World Hydrogen Energy Conference 2010 – WHEC 2010, Parallel Sessions Book 1: Fuel Cell Basics/Fuel Infrastructures*, 1st edn. (eds. D. Stolten and T. Grube), Forschungszentrum Jülich GmbH, Zentralbibliothek, Verlag, Jülich, p. 109.
20. Chen, C.Y. and Lai, W.H. (2010) Effects of temperature and humidity on the cell performance and resistance of a phosphoric acid doped polybenzimidazole fuel cell. *J. Power Sources*, **195**, 7152.
21. Wannek, C., Kohnen, B., Oetjen, H.F., Lippert, H., and Mergel, J. (2008) Durability of ABPBI-based MEAs for high temperature PEMFCs at different operating conditions. *Fuel Cells*, **8**, 87.
22. Cheddie, D.F. and Munroe, N.D.H. (2007) A two-phase model of an intermediate temperature PEM fuel cell. *Int. J. Hydrogen Energy*, **32**, 832.
23. Cheddie, D.F. and Munroe, N.D.H. (2008) Semi-analytical proton exchange membrane fuel cell modeling. *J. Power Sources*, **183**, 164.
24. Sousa, T., Mamlouk, M., and Scott, K. (2010) An isothermal model of a laboratory intermediate temperature fuel cell using PBI doped phosphoric acid membranes. *Chem. Eng. Sci.*, **65**, 2513.
25. Sousa, T., Mamlouk, M., and Scott, K. (2010) A non-isothermal model of a

26. Sousa, T., Mamlouk, M., and Scott, K. (2010) A dynamic non-isothermal model of a laboratory intermediate temperature fuel cell using PBI doped phosphoric acid membranes. *Int. J. Hydrogen Energy*, **35**, 12065.

27. Hu, J., Zhang, H., Zhai, Y., Liu, G., Hu, J., and Yi, B. (2006) Performance degradation studies on PBI/H_3PO_4 high temperature PEMFC and one-dimensional numerical analysis. *Electrochim. Acta*, **52**, 394.

28. Bergmann, A., Gerteisen, D., and Kurz, T. (2009) Modelling of CO poisoning and its dynamics in HTPEM fuel cells. *Fuel Cells*, **10**, 278.

29. Peng, J. and Lee, S.J. (2006) Numerical simulation of proton exchange membrane fuel cells at high operating temperature. *J. Power Sources*, **162**, 1182.

30. Peng, J., Shin, J.Y., and Song, T.W. (2008) Transient response of high temperature PEM fuel cell. *J. Power Sources*, **179**, 220.

31. Jiao, K. and Li, X. (2010) A three-dimensional non-isothermal model of high temperature proton exchange membrane fuel cells with phosphoric acid doped polybenzimidazole membranes. *Fuel Cells*, **10**, 351.

32. Ubong, E.U., Shi, Z., and Wang, X. (2009) Three-dimensional modeling and experimental study of a high temperature PBI-based PEM fuel cell. *J. Electrochem. Soc.*, **156**, B1276.

33. Lobato, J., Cañizares, P., Rodrigo, M.A., Pinar, F.J., Mena, E., and Úbeda, D. (2010) Three-dimensional model of a 50 cm^2 high temperature PEM fuel cell. Study of the flow channel geometry influence. *Int. J. Hydrogen Energy*, **35**, 5510.

34. Siegel, C., Bandlamudi, G., and Heinzel, A. (2011) Systematic characterization of a PBI/H_3PO_4 sol–gel membrane – modeling and simulation. *J. Power Sources*, **196**, 2735.

35. Lobato, J., Cañizares, P., Rodrigo, M.A., Piuleac, C.G., Curteanu, S., and Linares, J.J. (2010) Direct and inverse neural networks modelling applied to study the influence of the gas diffusion layer properties on PBI-based PEM fuel cells. *Int. J. Hydrogen Energy*, **35**, 7889.

36. Kulikovsky, A.A., Oetjen, H.F., and Wannek, C. (2010) A simple and accurate method for high-temperature PEM fuel cell characterisation. *Fuel Cells*, **10**, 363.

37. Shamardina, O., Chertovich, A., Kulikovsky, A.A., and Khokhlov, A.R. (2010) A simple model of a high temperature PEM fuel cell. *Int. J. Hydrogen Energy*, **35**, 9954.

38. Bernardi, D.M. and Verbrugge, M.W. (1992) A mathematical model of the solid-polymer-electrolyte fuel cell. *J. Electrochem. Soc.*, **139**, 2477.

39. McBreen, J., O'Grady, W.E., and Richter, R. (1984) A rotating disk electrode apparatus for the study of fuel cell reactions at elevated temperatures and pressures. *J. Electrochem. Soc.*, **131**, 1215.

40. Ahluwalia, R.K., Doss, E.D., and Kumar, R. (2003) Performance of high-temperature polymer electrolyte fuel cell systems. *J. Power Sources*, **117**, 45.

41. Korsgaard, A.R., Refshauge, R., Nielsen, M.P., Bang, M., and Kær, S.K. (2006) Experimental characterization and modeling of commercial polybenzimidazole-based MEA performance. *J. Power Sources*, **162**, 239.

42. Korsgaard, A.R., Nielsen, M.P., and Kær, S.K. (2008) Part one: a novel model of HTPEM-based micro-combined heat and power fuel cell system. *Int. J. Hydrogen Energy*, **33**, 1909.

43. Korsgaard, A.R., Nielsen, M.P., and Kær, S.K. (2008) Part two: control of a novel HTPEM-based micro combined heat and power fuel cell system. *Int. J. Hydrogen Energy*, **33**, 1921.

44. Chrenko, D., Lecoq, S., Herail, E., Hissel, D., and Péra, M.C. (2010) Static and dynamic modeling of a diesel fed fuel cell power supply. *Int. J. Hydrogen Energy*, **35**, 1377.

45. Andreasen, S.J., Jespersen, J.L., Schaltz, E., and Kær, S.K. (2009) Characterisation and modelling of a high temperature

PEM fuel cell stack using electrochemical impedance spectroscopy. *Fuel Cells*, **4**, 463.
46. Kvesić, M., Reimer, U., Lehnert, W., Froning, D., Lüke, L., and Stolten, D. (2012) 3D modeling of a 200 cm^2 HT-PEFC short stack. *Int. J. Hydrogen Energy*, **37**, 2430.
47. Korte, C. (2012) Phosphoric acid, an electrolyte for intermediate-temperature fuel cells – temperature and composition dependence of vapor pressure and proton conductivity, in *Fuel Cell Science and Engineering*, 1st edn. (eds. D. Stolten and B. Emonts), Wiley-VCH Verlag GmbH, Weinheim, Chapter 11.
48. Das, A. and Changdar, S.N. (1995) Tracer diffusion in the system phosphoric acid–disodium hydrogen phosphate–water by a radioactive method. *Radiat. Phys. Chem.*, **45**, 773.
49. Chung, S.H., Bajue, S., and Greenbaum, S.G. (2000) Mass transport of phosphoric acid in water: a ^{1}H and ^{31}P pulsed gradient spin-echo nuclear magnetic resonance study. *J. Chem. Phys.*, **112**, 8515.
50. Kreuer, K.D., Paddison, S.J., Spohr, E., and Schuster, M. (2004) Transport in proton conductors for fuel-cell applications: simulations, elementary reactions, and phenomenology. *Chem. Rev.*, **104**, 4637.
51. Vilciauskas, L., Paddison, S.J., and Kreuer, K.D. (2009) *Ab initio* modeling of proton transfer in phosphoric acid clusters. *J. Phys. Chem. A*, **113**, 9193.
52. Chin, D.T. and Chang, H.H. (1989) On the conductivity of phosphoric acid electrolyte. *J. Appl. Electrochem.*, **19**, 95.
53. Maier, W., Arlt, T., Wannek, C., Manke, I., Riesemeier, H., Krüger, P., Scholta, J., Lehnert, W., Banhart, J., and Stolten, D. (2010) *In-situ* synchrotron X-ray radiography on high temperature polymer electrolyte fuel cells. *Electrochem. Commun.*, **12**, 1436.
54. Liu, J. and Eikerling, M. (2008) Model of cathode catalyst layers for polymer electrolyte fuel cells: the role of porous structure and water accumulation. *Electrochim. Acta*, **8**, 87.
55. Jiao, K. and Li, X. (2011) Water transport in polymer electrolyte membrane fuel cells. *Prog. Energy Combust. Sci.*, **37**, 221.
56. Scharifker, B.R., Zelenay, P., and Bockris, J.O'M. (1987) The kinetics of oxygen reduction in molten phosphoric acid at high temperatures. *J. Electrochem. Soc.*, **134**, 2714.
57. Zelenay, P. (1992) *In situ* SNIFTIRS and radiotracer study of adsorption on platinum: trifluoromethane sulfonic acid and phosphoric acid. *Electrochem. Transit.*, **81**, 9.
58. Klinedinst, K., Bett, J.A.S., Macdonald, J., and Stonehart, P. (1974) Oxygen solubility and diffusivity in hot concentrated H_3PO_4. *J. Electroanal. Chem. Interfacial Electrochem.*, **57**, 281.
59. Chisholm, C.R.I., Boysen, D.A., Papandrew, A.B., Zecevic, S., Cha, S., Sasaki, K.A., Varga, Á., Giapis, K.P., and Haile, S.M. (2009) From laboratory breakthrough to technological realization: the development path for solid acid fuel cells. *Electrochem. Soc. Interface*, **18**, 53.
60. Varga, Á., Brunelli, N.A., Louie, M.W., Giapis, K.P., and Haile, S.M. (2010) Composite nanostructured solid-acid fuel-cell electrodes via electrospray deposition. *J. Mater. Chem.*, **20**, 6309.
61. Lee, S.Y., Ogawa, A., Kanno, M., Nakamoto, H., Yasuda, T., and Watanabe, M. (2010) Nonhumidified intermediate temperature fuel cells using protic ionic liquids. *J. Am. Chem. Soc.*, **132**, 9764.
62. Wedler, G. (1987) *Lehrbuch der Physikalischen Chemie*, 3rd edn., Wiley-VCH Verlag GmbH, Weinheim.
63. Kulikovsky, A.A. (2002) The voltage–current curve of a polymer electrolyte fuel cell: 'exact' and fitting equations. *Electrochem. Commun.*, **4**, 845.
64. Kulikovsky, A.A. (2002) Performance of a polymer electrolyte fuel cell with long oxygen channel. *Electrochem. Commun.*, **4**, 527.
65. Anderson, A.B., Roques, J., Mukerjee, S., Murthi, V.S., Markovic, N.M., and Stamenkovic, V. (2005) Activation energies for oxygen reduction on platinum

alloys: theory and experiment. *J. Phys. Chem. B*, **109**, 1198.
66. Maoka, T. (1988) Electrochemical reduction of oxygen on small platinum particles supported on carbon in concentrated phosphoric acid – I. Effects of platinum content in the catalyst layer and operating temperature of the electrode. *Electrochim. Acta*, **33**, 371.
67. Ma, Y.L., Wainright, J.S., Litt, M.H., and Savinell, R.F. (2004) Conductivity of PBI membranes for high-temperature polymer electrolyte fuel cells. *J. Electrochem. Soc.*, **151**, A8.

30
Modeling of Polymer Electrolyte Membrane Fuel-Cell Components

Yun Wang and Ken S. Chen

30.1
Introduction

Polymer electrolyte membrane fuel cells (PEMFCs), also called polymer electrolyte fuel cells (PEFCs), convert the chemical energy stored in hydrogen fuel directly and efficiently to electrical energy with water as the only byproduct, and have the potential to reduce our energy use, pollutant emissions, and dependence on fossil fuels. Computational or numerical modeling of PEFCs has been a rapidly growing field of research and plays a critically important role in the fundamental study and advancement of PEFC technology and also the engineering design and optimization of PEFCs. Computational modeling of PEFCs refers to the mathematical description of the relevant physical phenomena and the consequent numerical solution of the equations that govern a fuel-cell system.

The two greatest barriers to the widespread deployment of PEFCs at present are durability and cost [1, 2]. Developing high-fidelity PEFC models is important for fuel-cell cost reduction and durability improvement. For example, detailed operating information within a fuel cell, such as temperature/species concentration distributions and local electrochemical reaction rate, can be revealed by model prediction. Hot-spot formation or low-humidity operation, which can degrade fuel-cell components, can also be predicted by appropriate models. In addition, dimensionless groups, developed through analysis and modeling, can be used to guide PEFC design and material selection. For example, the dimensionless group $\hbar$ characterizes the spatial variation of local reaction, and therefore aids in optimizing the ionomer/Pt loading for electrode fabrication. Transient phenomena and dynamic responses can be predicted by time-dependent models and the simulation results can improve our understanding of material failure arising from transient operation. Fuel cells may suffer performance decay during cold start (or freeze–thaw) cycles. Solid water formation and distribution, which can be revealed through detailed model prediction, are important for probing the degradation mechanisms during cold start. Furthermore, computational modeling is important for improving PEFC performance, and thus reducing costs (e.g., by use of a catalyst

Fuel Cell Science and Engineering: Materials, Processes, Systems and Technology,
First Edition. Edited by Detlef Stolten and Bernd Emonts.
© 2012 Wiley-VCH Verlag GmbH & Co. KGaA. Published 2012 by Wiley-VCH Verlag GmbH & Co. KGaA.

with less Pt loading or simply a less expensive catalyst): the improved performance yields higher power density, thereby lowering the fuel-cell cost per kilowatt.

Phenomena in PEMFC operation involve heat transfer, multi-species and charge transport, multi-phase flows, and electrochemical reactions. These phenomena occur in various components, namely the membrane electrode assembly (MEA) consisting of the catalyst layers (CLs) and membrane, gas diffusion layers (GDLs) and microporous layers (MPLs) [together referred to as diffusion media (DM)], gas flow channels (GFCs), and bipolar plates (BPs). Specifically, as shown schematically in Figure 30.1, the following multi-physics, highly coupled and nonlinear transport and electrochemical phenomena take place during fuel cell operation (Table 30.1) [2]: (i) hydrogen gas and air are forced (by pumping) to flow down the anode and cathode GFCs, respectively; (ii) H_2 and O_2 flow through the respective porous GDLs/MPLs and then flow/diffuse into the respective CLs; (iii) H_2 is oxidized in the anode CL, forming protons and electrons; (iv) protons migrate (driven by the

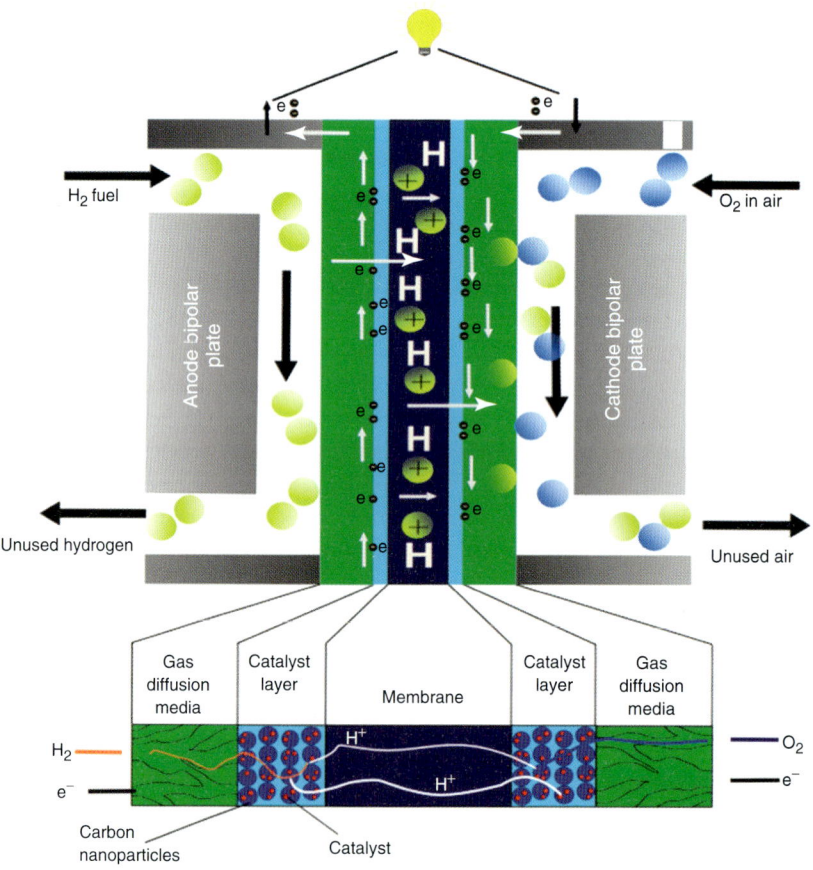

Figure 30.1 Schematic of a PEFC.

Table 30.1 Frequently used electrochemical and transport properties [2].

Description	Units	Value
Electrochemical kinetics:		
Exchange current density (anode/cathode)	$A\,m^{-3}$	$10^9/10^3-10^4$
Faraday constant	$C\,mol^{-1}$	96 487
Electrical conductivity of DMs/BPs	$S\,m^{-1}$	300/20 000
Species transport properties:		
H_2/H_2O diffusivity (H_2–H_2O) at standard conditions	$m^2\,s^{-1}$	$8.67/8.67 \times 10^{-5}$
O_2/H_2O (v) diffusivity in air at standard conditions	$m^2\,s^{-1}$	$1.53/1.79 \times 10^{-5}$
Viscosity at 80 °C (H_2/air)	$m^2\,s^{-1}$	$9.88 \times 10^{-6}/1.36 \times 10^{-5}$
Thermal properties:		
$H_2/N_2/O_2/H_2O$ (v) thermal conductivity	$W\,m^{-1}\,K^{-1}$	0.170/0.024/0.024/0.024
Anode/cathode GDL conductivity	$W\,m^{-1}\,K^{-1}$	0.3–3
Anode/cathode CL conductivity	$W\,m^{-1}\,K^{-1}$	0.3–1.5
Membrane thermal conductivity	$W\,m^{-1}\,K^{-1}$	0.95
Anode/cathode bipolar plate thermal conductivity	$W\,m^{-1}\,K^{-1}$	>10.0
$H_2/N_2/O_2/H_2O$ (v) specific heat at 80 °C	$J\,kg^{-1}\,K^{-1}$	14 400/1041/917/2000
Anode/cathode GDL heat capacity	$J\,K^{-1}\,m^{-3}$	5.68×10^5
Anode/cathode CL heat capacity	$J\,K^{-1}\,m^{-3}$	1.69×10^6
Membrane heat capacity	$J\,K^{-1}\,m^{-3}$	1.65×10^6
Anode/cathode bipolar plate heat capacity	$J\,K^{-1}\,m^{-3}$	1.57×10^6
Latent heat of sublimation	$J\,mol^{-1}$	5.1×10^{-4}
Material properties:		
Permeability of anode/cathode GDL	m^2	1.0×10^{-12}
Permeability of anode/cathode CL	m^2	1.0×10^{-13}
Anode/cathode GDL porosity	–	0.4–0.8
Anode/cathode CL porosity	–	0.3–0.5
Ionomer volume fraction in CL	–	0.13–0.4
Equivalent weight of ionomers	$kg\,mol^{-1}$	0.9, 1.1, or 1.2[a]
Dry density of membrane	$kg\,m^{-3}$	1.98×10^{3}[a]

[a] Several typical Nafion membranes.

gradient of electrical potential in the electrolyte) and water is transported through the membrane; (v) electrons are conducted via carbon support to the anode current collector, and then to the cathode current collector via an external circuit; (vi) O_2 is reduced with protons and electrons in the cathode CL to form water; (vii) product water is transported out of the cathode CL, through the cathode GDL/MPL and the GDL/channel interface, and eventually out of the cathode GFC; and (viii) waste heat is generated due to chemical energy-to-electricity conversion, mainly in the cathode CL due to the sluggish oxygen reduction reaction (ORR), and it is conducted out of the cell via carbon support and BPs. The transport phenomena are three-dimensional because the flows of fuel (H_2) in the anode GFC and oxidant (O_2) in the cathode GFC are usually normal to proton transport through the membrane and gas transport through the respective GDLs/MPLs and CLs. The frequently used electrochemical and transport properties are listed in Table 30.1.

This chapter is intended to provide a comprehensive review of PEFC modeling at the component level. The focus is placed on describing and discussing the macroscopic or continuum-level models, which can be directly applied to predict cell performance and optimize fuel-cell design/control and also explore fundamentals. The recent development of mesoscopic models based on the lattice Boltzmann methods (LBMs) and direct numerical simulations (DNSs), and approaches based on volume of fluids (VOFs) will also be discussed. Although we attempt to cover the majority of the literature on the topic of PEFC modeling, there are undoubtedly some that may be left out. Prior to the present work, two comprehensive reviews on fuel-cell modeling through early 2004, including PEFCs and solid oxide fuel cells (SOFCs), and transport in PEFCs, were provided by Wang [3] and Weber and Newman [4], respectively. More recently, Gurau and Mann [5], Siegel [6], and Jiao and Li [7] reviewed the modeling aspects related to heat/mass transport in PEFCs. Djilali [8] reviewed water transport and its modeling in PEFCs. This chapter differs from the previous reviews by focusing on a review of PEFC modeling at the component level; and it also reviews newly developed approaches such as mesoscopic studies, two-phase channel flow, and detailed model validation. In terms of time frame, this review covers mainly the research findings that have been published through early 2011. Lastly, owing to space limitations, numerical algorithms and boundary conditions (which often are geometry/operation specific) are not covered in this chapter.

30.2
Polymer Electrolyte Membrane

Membrane refers to the thin layer of electrolyte (usually ~ 10–100 μm, e.g., 18 μm for Gore 18 and 50 μm for Nafion 112) that allows the protons [which are produced in the anode's hydrogen oxidation reaction (HOR)] to transport from the anode to the cathode. Desirable membrane materials are those that exhibit high ionic conductivity, while preventing electron transport and the crossover of hydrogen fuel from the anode and oxygen reactant from the cathode. Two major transport processes take place in membranes, namely proton and water transport. Gierke and Hsu [9] employed a cluster model to describe the polymeric membrane in terms of an inverted micellar structure in which the ion-exchange sites are separated from the fluorocarbon backbone, forming spherical clusters (pores), connected by short, narrow channels. The cluster sizes depend on the local water content. The main driving force for proton transport is the gradient of electrolyte phase potential. That is, protons transport across the membrane mainly due to the existence of an electrolyte potential gradient, and this transport mechanism is generally referred to as electromigration. Water in the membrane is essential for proton transport. One mechanism is called the "vehicular" diffusion. By forming hydronium ions (H_3O^+), protons can be transported from high to low water concentration regions due to water diffusion, that is, water serves as a vehicle for proton transport [10]. Another important process for protons to transport within the membrane is through the mechanism of "hopping," which takes place when sufficient water content is present so that the side chains of sulfonic groups are connected, in which protons can move directly from one site to another by hopping along the side chains of the sulfonic groups [11, 12]. In addition, protons can be transported via diffusion (driven by the concentration gradient of protons), but this is usually small in comparison with migration, and therefore can be neglected in most situations. Proton transport via electromigration is described by the following electrolyte phase potential equation, which is based on the Ohmic law:

$$0 = \nabla \cdot \left(\sigma_m^{\text{eff}} \nabla \Phi^{(m)} \right) \quad (30.1)$$

where σ_m^{eff} is the effective ionic conductivity. Several studies have been carried out to determine σ_m^{eff} experimentally, including that measured at 30 °C for Nafion 117 [13] and at subfreezing temperatures (−30 to 0 °C) [14] as shown in Figure 30.2. The following correlation is due to Springer et al. [13]:

$$\sigma_m^{\text{eff}} = (0.5139\lambda - 0.326) \exp\left[1268 \left(\frac{1}{303} - \frac{1}{T} \right) \right] \quad (30.2)$$

where the water content λ is defined as the number of water molecules per sulfonic group and is given by:

$$\lambda = 0.043 + 17.81a - 39.85a^2 + 36.0a^3 \quad \text{for } 0 < a \leq 1$$
$$\lambda = 14 + 1.4(a - 1) \quad \text{for } 1 \leq a \leq 3 \quad (30.3)$$

In the above, the water activity is determined by

$$a = \frac{C_w}{C_{\text{sat}}} \text{ and } C_{\text{sat}}(T) = \frac{P_{\text{sat}}(T)}{RT}$$

where the saturated vapor pressure, P_{sat}, is a strong function of temperature and can be expressed by the following empirical correlation [13]:

$$\log_{10} P_{\text{sat}} = -2.1794 + 0.02953(T - 273.15) - 9.1837 \times 10^{-5}(T - 273.15)^2$$
$$+ 1.4454 \times 10^{-7}(T - 273.15)^3 \quad (30.4)$$

The empirical correlation of Springer et al. has been widely adopted in modeling transport in membranes for temperatures above freezing. Recently, Chen and co-workers [15, 16] formulated a new constitutive model for predicting proton conductivity in polymer electrolytes. Their conductivity model depends on the molar volumes of dry membrane and water but otherwise requires no adjustable parameter. Predictions computed from Chen and co-workers' conductivity model yield good agreement with experimental data from the literature and those from their own measurements for a wide range of water contents. Weber and Newman [17, 18] developed a comprehensive membrane model that treats membrane swelling, and seamlessly and rigorously accounts for both vapor- and liquid-equilibrated transport modes using a single driving force of chemical potential. The transition between the two modes is determined based on the energy needed to swell and connect the water filled the nano-domains. However, there are still some discrepancies that need to be resolved, such as an underestimation of the interfacial water

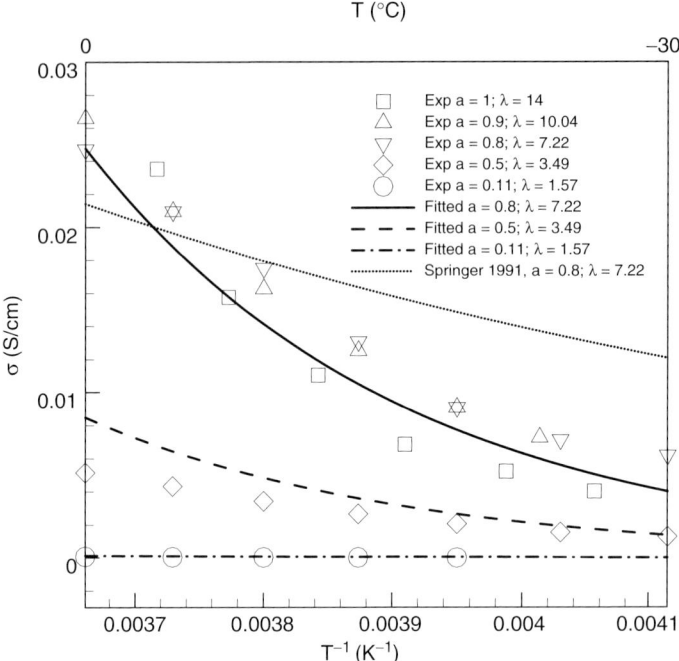

Figure 30.2 Ionic conductivity of the membrane at subfreezing conditions [14].

mass-transport resistance and a lack of consideration of membrane state or history. Weber and Newman [19] also performed model validation of their membrane sub-model by including it in a simple fuel-cell model.

At subfreezing temperatures, a portion of the water in membranes will freeze and has a negligible contribution to the ionic conductivity. Wang *et al.* [14] experimentally measured the conductivity and developed the following correlation of the ionic conductivity as a function of temperature and water content (also see Figure 30.2):

$$\sigma_m^{\text{eff}} = (0.01862\lambda - 0.02854) \exp\left[4029\left(\frac{1}{303} - \frac{1}{T}\right)\right] \text{ or}$$
$$= (0.004320\lambda - 0.006620) \exp\left[4029\left(\frac{1}{273} - \frac{1}{T}\right)\right]$$
$$= \sigma_{m,0}(\lambda) \exp\left[4029\left(\frac{1}{273} - \frac{1}{T}\right)\right]$$
$$\text{for } \lambda \leq 7.22 \quad \sigma_m = \sigma_m(\lambda = 7.22) \quad \text{for} \quad \lambda > 7.22 \tag{30.5}$$

The general equation that governs water transport through membrane can be written as

$$\varepsilon_m \frac{\partial C_w^m}{\partial t} = \nabla \cdot \left(D_w^{m,\text{eff}} \nabla C_w^m + \vec{G}_{w,\text{perm}}\right) - \frac{1}{F} \nabla \cdot \left(n_d \vec{i}_e\right) \tag{30.6}$$

where C_w^m is the equivalent water concentration in the membrane and $\vec{G}_{w,\text{perm}}$ is the permeation flux given by Eq. (30.12) below. The water diffusion coefficient can be determined experimentally and a correlation was given by Motupally et al. [20]:

$$D_w^m = \begin{cases} 3.1 \times 10^{-3} \lambda \left(e^{0.28\lambda} - 1\right) e^{(-2436/T)} & \text{for } 0 < \lambda \leq 3 \\ 4.17 \times 10^{-4} \lambda \left(1 + 161 e^{-\lambda}\right) e^{(-2436/T)} & \text{otherwise} \end{cases} \quad (30.7)$$

Another diffusion coefficient correlation was given by Springer et al. as follows [13]:

$$D_w^m = \begin{cases} 2.693 \times 10^{-10} \\ (2.08\lambda - 3.29) \times 10^{-10} e^{(2416/303.15 - 2426/T)} & \text{for } 2 < \lambda \leq 3 \\ (6.84 - 1.3\lambda) \times 10^{-10} e^{(2416/303.15 - 2426/T)} & \text{for } 3 < \lambda \leq 4 \\ \left(2.563 - 0.33\lambda + 0.0264\lambda^2 - 0.000671\lambda^3\right) \\ \times 10^{-10} e^{(2416/303.15 - 2426/T)} & \text{otherwise} \end{cases} \quad (30.8)$$

For Gore membranes, the diffusivity expression can be modified by accounting for the portion of the ionomer ε_m:

$$D_w^{m,\text{eff}} = \varepsilon_m^{\tau_m} D_w^m \quad (30.9)$$

In addition to diffusion, protons carry water molecules with them when moving from the anode to cathode – this process is generally referred to as water transport via electroosmosis. The electroosmotic drag coefficient has been determined experimentally by several groups and the following value has been widely used in fuel cell models [21]:

$$n_d = \begin{cases} 1.0 & \text{for } \lambda \leq 14 \\ \dfrac{1.5}{8}(\lambda - 14) + 1.0 & \text{otherwise} \end{cases} \quad (30.10)$$

The protonic current flux in the membrane can be calculated by

$$\vec{i}^{(m)} = -\sigma_m^{\text{eff}} \nabla \Phi^{(m)} \quad (30.11)$$

Yet another mechanism for water transport in membranes is through hydraulic permeation, which is driven by the liquid pressure difference between the anode and cathode. The permeation flux is controlled by the membrane permeability K_m, and the liquid pressure gradient across the membrane [22]:

$$\vec{G}_{w,\text{perm}} = -\frac{K_m}{M_w v_l} \nabla P^l \quad (30.12)$$

30.3 Catalyst Layers

The CLs are the most important components in a PEFC in which the hydrogen oxidation reaction (HOR) and Oxygen reduction reaction (ORR) take place in the anode and cathode, respectively. CLs are usually thin with a thickness of about 10 μm. Several materials contained in a CL are critical to the electrochemical activities: (i) carbon support with Pt particles dispersed on the carbon surface, (ii) ionomer, and (iii) void space (see Figure 30.3). The catalyst employed in a PEFC (usually platinum or a platinum alloy) plays the critical role of reducing the reaction activation barrier, and the following electrochemical reactions take place in a PEFC:

$$\text{HOR in the anode}: \quad H_2 \rightarrow 2H^+ + 2e^-$$
$$\text{ORR in the cathode}: O_2 + 4e^- + 4H^+ \rightarrow 2H_2O \tag{30.13}$$

Because the catalyst used for both the ORR and HOR is usually platinum or a platinum alloy, the CLs contribute a significant portion of the cost for a PEMFC [2]. Consequently, improving Pt utilization is important for achieving cost reductions. The HOR and ORR take place at the triple-phase boundaries and in the areas within which active catalyst surface is usually large, providing an effective way to improve the CL reaction kinetics. This can be directly seen from the well-known Butler–Volmer equation:

$$j = ai_0 \left[\exp\left(\frac{\alpha_a}{RT}F\eta\right) - \exp\left(\frac{\alpha_c}{RT}F\eta\right) \right] \tag{30.14}$$

where j is the reaction current or transfer current per unit volume (which represents the rates of HOR and ORR), i_0 the exchange current density, a the specific active area per unit volume, F the Faraday constant, and R the universal gas constant. The value of ai_0 is usually on the order of 10^9 A m^{-3} for the anode and 10^4 A m^{-3} for the cathode. The HOR is usually fast, thus yielding a relatively small anode overpotential, and the anode reaction can be adequately approximated by the following linearized Butler–Volmer kinetic equation:

$$\text{In the anode}: \quad j_c = ai_{0,a}^{\text{ref}} \left(\frac{C_{H_2}}{C_{H_2}^{\text{ref}}}\right)^{\frac{1}{2}} \left(\frac{\alpha_a + \alpha_c}{RT}F\eta\right) \tag{30.15}$$

For ORR, sluggish kinetics result in a large cathode overpotential. Hence the Butler–Volmer equation can be simplified to yield the following Tafel kinetic equation:

$$\text{In the cathode}: \quad j_c = -ai_{0,c}^{\text{ref}} \left(\frac{C_{O_2}}{C_{O_2}^{\text{ref}}}\right) \exp\left(-\frac{\alpha_c F}{RT}\eta\right) \tag{30.16}$$

The surface overpotential is defined as

$$\eta = \Phi^{(s)} - \Phi^{(m)} - U_0 \tag{30.17}$$

where the equilibrium potential, U_0, is zero by definition when hydrogen is used in the anode, whereas in the cathode it can be calculated by the following correlation (where T is in kelvin and U_0 is in volts):

$$U_0 = 1.23 - 0.0009\,(T - 298) \tag{30.18}$$

Although thin, the microstructure in the anode or cathode CL is complex, generally consisting of several interconnected networks or phases for proton, electron, and reactant transport. Mukherjee and Wang [27] and Kim and Pitsch [25] proposed

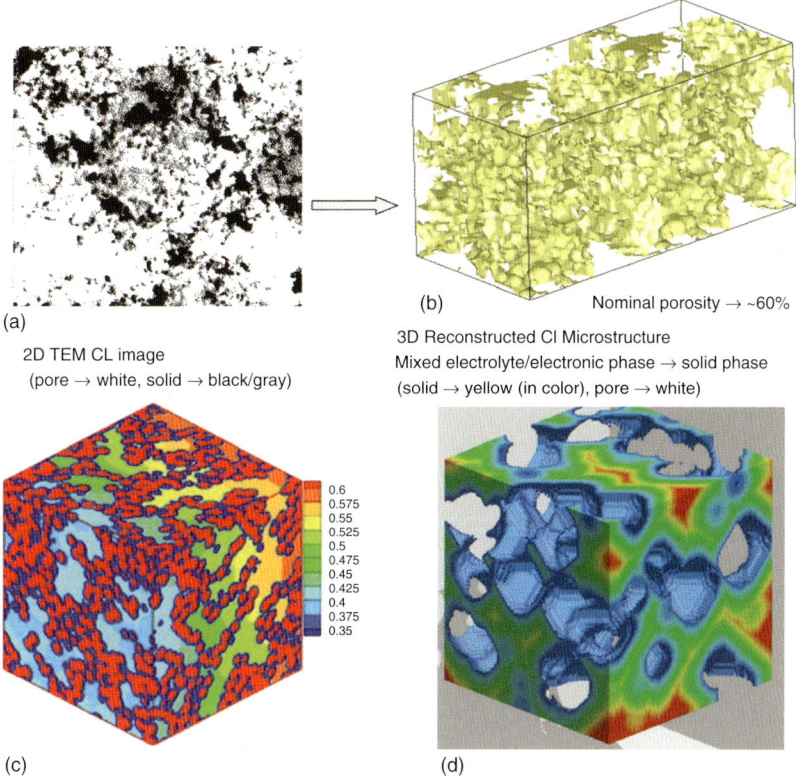

Figure 30.3 (a) 2D TEM CL image and (b) 3D reconstructed CL microstructure using a stochastic reconstruction method [23, 24]; (c) distribution of the mass fraction Y in the reconstructed CL (the red and dark-blue colors represent the solid phase and the electrolyte phase, respectively) [25]; and (b) gas phase structure of the CL [26].

numerical techniques to reconstruct this microstructure digitally (see Figure 30.3). Based on the reconstructed CL, DNS can be further carried out [27]. In the DNS, the transport equations, in conjunction with the electrochemical reaction kinetics, are numerically solved in a computational domain with fine mesh. Because different phases in CLs are denoted by different computational grids, the resolved equations are free of any macroscopic parameters such as porosity or effective coefficients. For example, the equation that describes oxygen transport in the void space and proton transport in the Nafion ionomer becomes

$$\nabla \cdot \left(\vec{u} C_{O_2} \right) = \nabla \cdot \left(D^g_{O_2} \nabla C^g_{O_2} \right) \quad \text{and} \quad 0 = \nabla \cdot \left(\sigma_m \nabla \Phi^{(m)} \right) \tag{30.19}$$

The reaction rate can be treated as surface flux at the catalyst surface if the computational domain is fine enough to describe individual catalyst particles (otherwise at the carbon surface), or volumetric source/sink in the computational

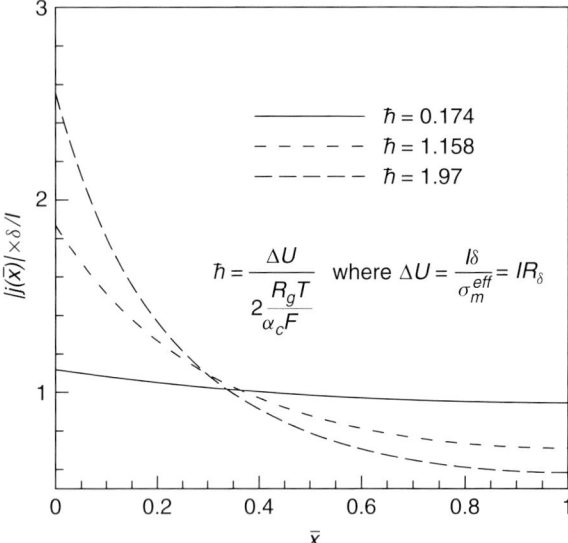

Figure 30.4 Spatial variation of the reaction rate across the catalyst layer [28].

grids near the surface. The DNS prediction indicated that the reaction rate can vary greatly across CLs in spite of their thinness. A dimensionless parameter $\hbar$ was identified by Wang and Feng to quantify the degree of the reaction spatial variation (see Figure 30.4) [28].

In full fuel-cell modeling and simulation, it is difficult to incorporate the DNS-based CL sub-model owing to the computational burden, that is, the large number of unknowns involved. A macroscopic approach is more suitable in this regard. The CL is usually treated as a homogeneous medium with macroscopic properties such as effective permeability and diffusion coefficients. Darcy's law usually applies to describe the flow. In contrast to the common flow equation, mass exchange exists at the solid–fluid interface in CLs, and a mass term must therefore be added to the conservation equation as follows [29]:

$$S_m = \begin{cases} M^w \nabla \cdot \left(D_w^{m,\text{eff}} \nabla C^w + G_{\text{perm}}^w \right) - M^{H_2} \frac{j}{2F} - M^w \nabla \cdot \left(\frac{n_d}{F} \vec{i}^{(m)} \right) & \text{in the anode} \\ M^w \nabla \cdot \left(D_w^{m,\text{eff}} \nabla C^w + G_{\text{perm}}^w \right) + M^{O_2} \frac{j}{4F} - M^w \frac{j}{2F} - M^w \nabla \cdot \left(\frac{n_d}{F} \vec{i}^{(m)} \right) & \text{in the cathode} \end{cases}$$

(30.20)

The species transport equations in CLs can be unified as

$$\varepsilon^{\text{eff}} \frac{\partial C_k}{\partial t} + \nabla \cdot \left(\gamma_c \vec{u} C_k \right) = \nabla \cdot \left(D_k^{g,\text{eff}} \nabla C_k^g + \vec{G}_k^l \right) + S_k \qquad (30.21)$$

where the convection corrector γ_c and flux $\vec{G}_k^l$ are due to liquid-phase transport, and will be discussed in detail in the next section. In absence of liquid water, the effective diffusion coefficients are modified to account for the porosity and tortuosity:

$$D_k^{g,\text{eff}} = \frac{\varepsilon}{\tau} D_k^g \sim \varepsilon^{\tau d} D_k^g \quad \text{where} \quad D^g = D_0 \left(\frac{T}{353}\right)^{\frac{3}{2}} \left(\frac{1}{P}\right) \quad (30.22)$$

In addition, water is transported through both void space and ionomer phase; consequently, the effective diffusion coefficient can be defined as [30]

$$D_w^{\text{eff}} = \varepsilon^{\tau d} D_w^g + \varepsilon_m^{\tau m} \frac{\rho^m}{EW} \frac{RT}{P_{\text{sat}}} \frac{d\lambda}{da} D_w^m \quad (30.23)$$

Furthermore, avoiding CL flooding is of critical importance for optimal PEFC performance and durability; however, it is not well understood yet. Most two-phase flow models in CLs follow the approach in the GDLs, adopting the multi-phase mixture (M^2) model to describe the transport of liquid water and gaseous air. This approach will be discussed in detail in the next section (GDLs and MPLs). Again, the ability to model the transport and electrochemical reactions in CLs is critical, particularly for the cathode, in which the ORR is sluggish and inefficient and water is generated. There is also a great need to elucidate mechanisms of liquid water transport/evaporation in CLs and the interactions with the microstructure and wettability. To elucidate effects of the CL, Harvey *et al.* [31] compared three different approaches for describing the cathode CL, namely a thin-film model, a discrete-catalyst volume model, and an agglomerate model (see Figure 30.5). They indicated that for a given electrode overpotential, the thin-film model significantly over-predicts the current density and exaggerates the variation in current density both along and across the channel, whereas the agglomerate model predicts noticeable mass transport losses. In addition, the CL is usually thin, but may be

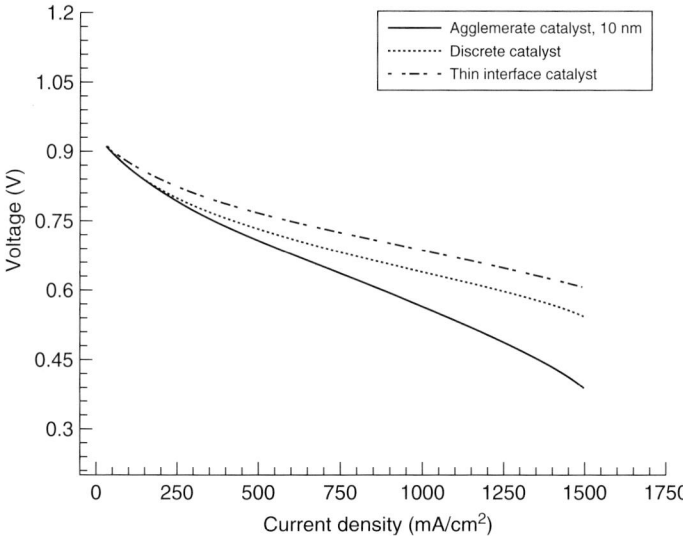

Figure 30.5 Cathode polarization curve for the three catalyst models [31].

subjected to mass transport limitation or a considerable ohmic loss. In this regard, further reducing the CL thickness is necessary to improve its performance. A CL model that properly captures the key transport phenomena and the HOR or ORR reaction at the three-phase interface can be employed to optimize the CL thickness. Specifically, such a model can help elucidate the effect of CL thinning on PEFC performance. Furthermore, coupling the continuum models with the microscopic models that describe the phenomena in the CL in much more detail is greatly needed. Such a coupling will enable the effect of CL microstructures on PEFC performance to be predicted.

30.4
Gas Diffusion Layers and Microporous Layers

The most important role of GDLs is to supply and distribute the hydrogen fuel and oxygen reactant to the reaction sites. GDLs also play the following roles: (i) electronic connection between the BP with channel–land structure and the electrode, (ii) passage for heat and water removal, (iii) mechanical support to the MEA, and (iv) protection of the CL from corrosion or erosion caused by flows or other factors. The physical processes in GDLs, in addition to diffusive transport, include bypass flow induced by in-plane pressure difference between neighboring channels, through-plane flow induced by mass source/sink due to electrochemical reactions, heat transfer such as the heat pipe effect, two-phase flow, and electron transport.

In most studies, GDLs are taken to be homogeneous media; therefore, the numerical treatment can follow that for transport in porous media, similarly to CLs. The species transport in GDLs can be unified as Eq. (30.21), with the effective coefficients defined by Eq. (30.22). GDLs are usually 100–300 μm thick. A popular GDL material is carbon-fiber based porous media: the fibers are either woven together to form a cloth, or bonded together by resins to form a paper (see Figure 30.6). Ralph *et al.* [32] showed that carbon cloth exhibits a better performance than paper at high current (>0.5 A cm^{-2}) with internal humidification. Wang *et al.* [33] investigated the structural features of carbon cloth and paper and explained the distinct performance observed experimentally (see Figure 30.7).

Several stochastic models have been developed to reconstruct the GDL microstructure. With a reconstructed GDL, Wang *et al.* [34] further presented a detailed DNS study to investigate the transport phenomena of mass, reactant, electron, and heat inside a GDL. In this approach, the Navier–Stokes equation was applied directly to the void space with no slip boundary condition at the solid matrix surface. The oxygen and water transport can be expressed in a similar way to that in the DNS of CLs. Figure 30.8 shows the mass flow through the microstructure of the GDL. The results were used to evaluate the GDL properties, such as the permeability and tortuosity. They found that the tortuosity of the considered pore structure is 1.2 whereas the permeability K is 3.1×10^{-12} m^2 [34].

Multi-phase flow, originating from the water production by the ORR, is critical to fuel cell water management. Figure 30.9 shows the scanning electron microscopy

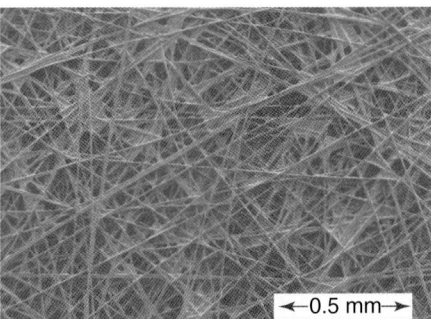
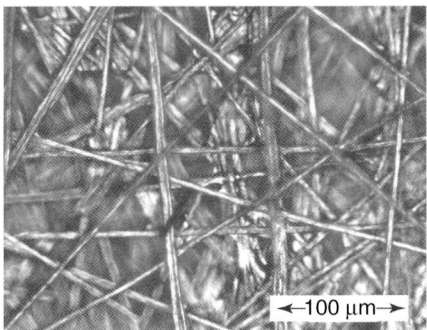

Figure 30.6 The carbon paper GDL [34].

(SEM) images of liquid water in GDLs in an *ex situ* study [35]. The presence of liquid water hinders the reactant delivery to the reaction sites, increasing the concentration polarization. "Flooding" is referred to the situation in which the presence of liquid water causes severe consequences, significantly reducing durability and performance due to reactant starvation. To avoid flooding, the GDL materials are usually rendered hydrophobic to facilitate liquid water drainage by adding polytetrafluoroethylene (PTFE, e.g., DuPont's Teflon) to the carbon paper or carbon cloth substrate.

Wang and co-workers [36, 37] used the LBM to study the meso-scale transport of liquid, based on the reconstructed GDL structure from stochastic modeling. The Lattice Boltzmann methods (LBM) is a powerful technique for simulating transport and fluid flows involving interfacial dynamics and complex geometries. It is based on first principles and considers flows to be composed of a collection of pseudo-particles residing on the nodes of an underlying lattice structure. The governing equations used in the LBM are different from the conventional Navier–Stokes equations, which are based on the macroscopic continuum description of flow phenomena. In their work, the model (interaction potential based model), originally proposed by Shan and Chen [38, 39], introduces k distribution functions for a fluid mixture comprising k components. Each distribution function represents a fluid component and satisfies the evolution equation. The non-local interaction between particles at neighboring lattice sites is included in the kinetics

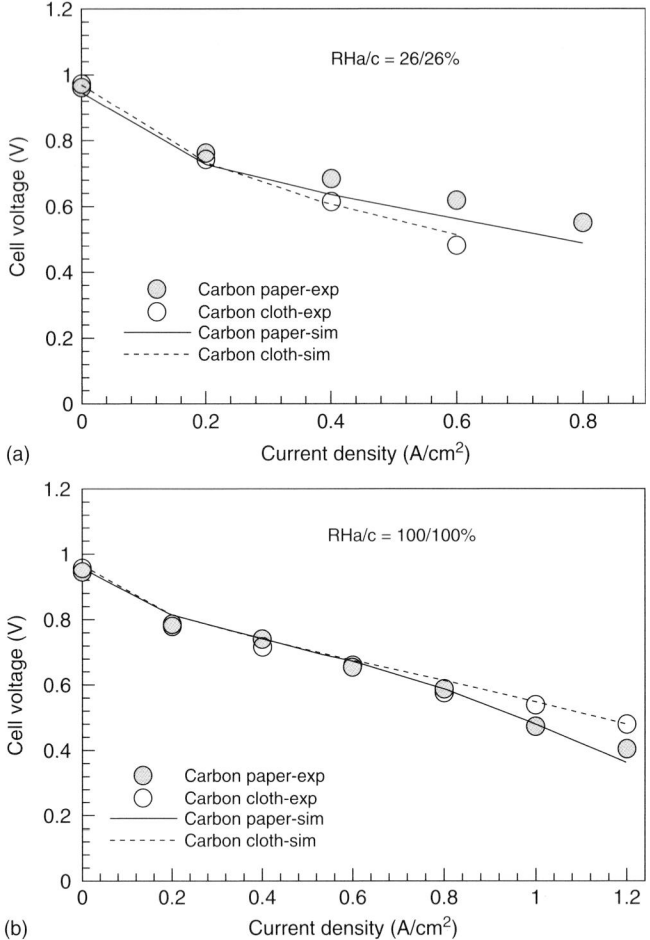

Figure 30.7 Cell performance using different DMs at 80°C/75% RH cathode inlet conditions: (a) 25/25% RH for anode/cathode and 80°C and (b) 100/100% RH for anode/cathode and 80°C [33].

through a set of potentials. The evolution equation for the kth component can be written as

$$f_i^k(\mathbf{x} + \mathbf{e}_i \delta_t, t + \delta_t) - f_i^k(\mathbf{x}, t) = -\frac{f_i^k(\mathbf{x}, t) - f_i^{k(\text{eq})}(\mathbf{x}, t)}{\tau_k} \quad (30.24)$$

where $f_i^k(\mathbf{x}, t)$ is the density distribution function for the kth component in the ith velocity direction at position $\mathbf{x}$ and time t, and δ_t is the time increment. The right-hand side of the above equation represents the collision term based on the Bhatnagar–Gross–Krook (BGK) or the single-time relaxation approximation [40]. τ_k is the relaxation time of the kth component in lattice unit and $f_i^{k(\text{eq})}(\mathbf{x}, t)$ is the

30.4 Gas Diffusion Layers and Microporous Layers | 853

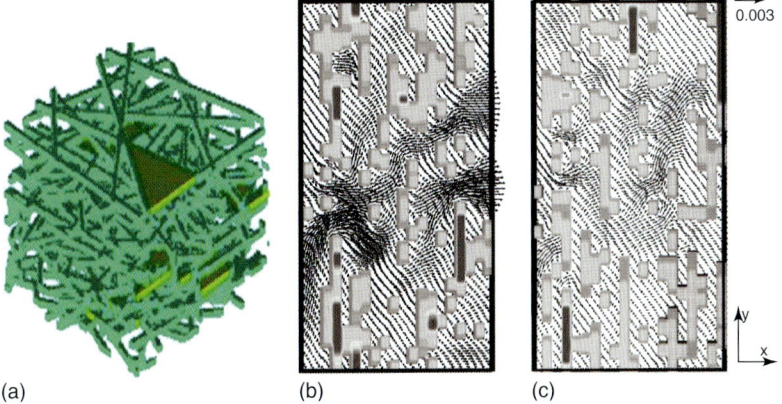

(a) (b) (c)

Figure 30.8 Mass flow at different portions of the GDL: $\bar{z} = 0.5$ (a) and 0.75 (b). $\bar{z}$ is the dimensionless distance in the z-direction ranging from 0 to 1. The gray region denotes the solid with the light gray being the carbon fibers and the dark the binders [34].

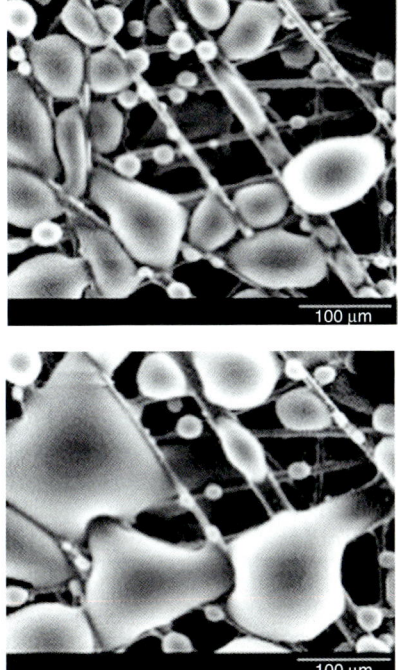

Figure 30.9 Two consecutive environmental SEM images of a diffusion medium exposed to water vapor saturated atmosphere [35].

corresponding equilibrium distribution function. The phase separation between different fluid phases, the wettability of a particular fluid phase to the solid, and the body force can be taken into account by modifying the velocity used to calculate the equilibrium distribution function. An extra component-specific velocity due to the inter-particle interaction is added on top of a common velocity for each component. Inter-particle interaction is realized through the total force $\mathbf{F}_k$ acting on the kth component, including fluid–fluid interaction, fluid–solid interaction, and external force. More details can be found elsewhere [37]. Compared with the VOF method, the LBM is advantageous in simulating multi-phase flows because of its inherent ability to incorporate particle interactions to yield phase segregation and thus eliminate explicit interface tracking. As most VOF studies are focused on channel droplet dynamics, it will be presented in a later part of this chapter. An example of model prediction using the LBM is shown in Figure 30.10 [37].

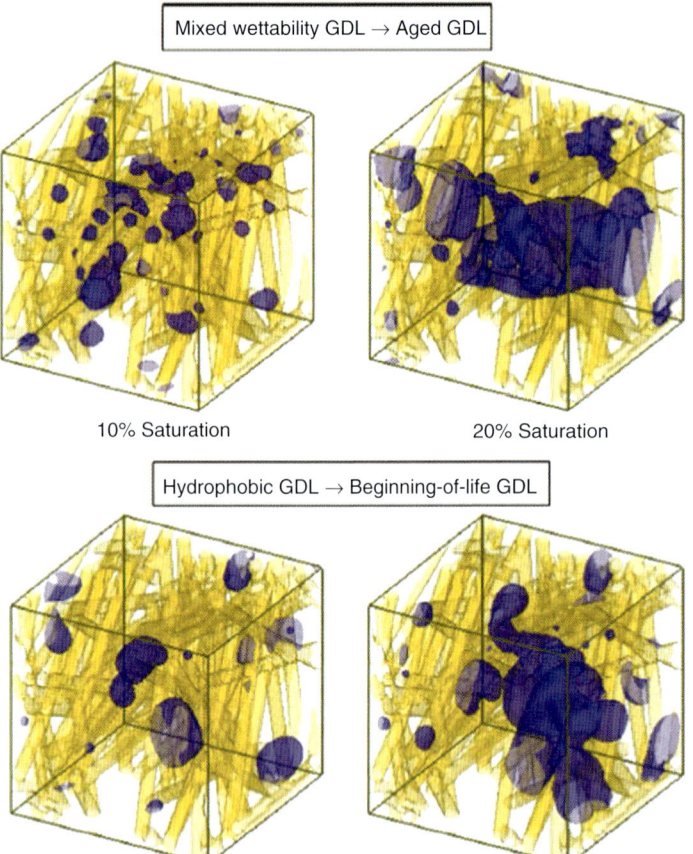

Figure 30.10 3D liquid water distributions in a hydrophobic and a mixed wettability GDL from the two-phase LBM simulations [37].

It is difficult to incorporate the LBM-based GDL model with other PEFC component models to predict full-cell performance. Macroscopic modeling is a more popular approach adopted when the entire fuel cell is the focus. A number of macroscopic models have been developed to capture the two-phase characteristics in GDLs [2–5]. They mostly treat the GDL as a uniform hydrophilic or hydrophobic medium. The gradient of capillary pressure is found to be the major driving force for liquid flow in GDLs. The capillary pressure can be expressed as a function of saturation via the Leverett function (see Figure 30.11) [3]:

$$P_c = P^g - P^l = \tau \cos(\theta_c) \left(\frac{\varepsilon}{K}\right)^{\frac{1}{2}} J(s) \qquad (30.25)$$

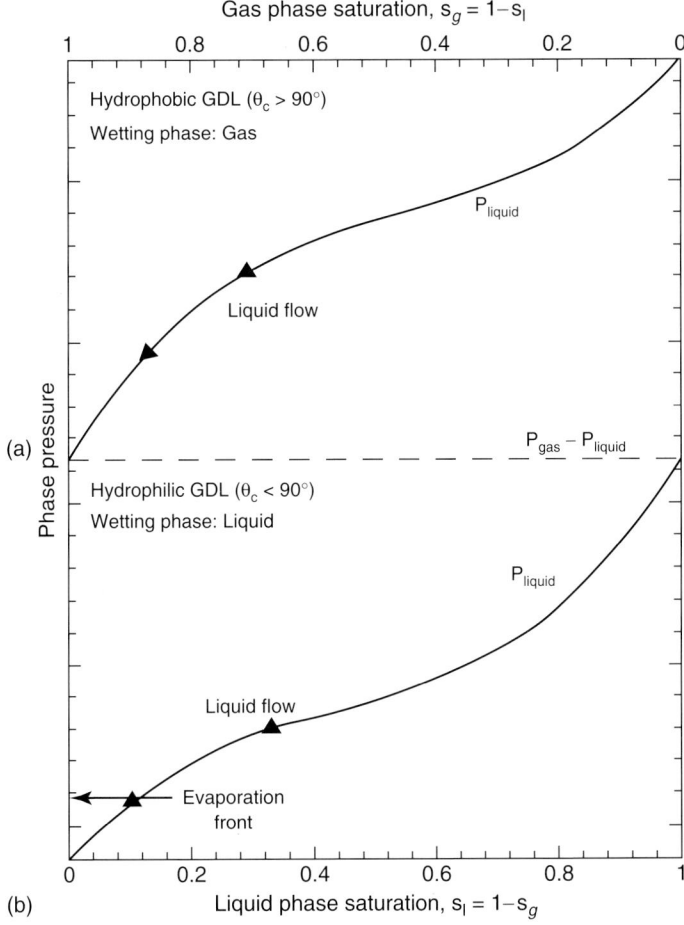

Figure 30.11 Schematic illustration of liquid- and gas-phase pressure profiles in (a) hydrophobic and (b) hydrophilic porous media [3].

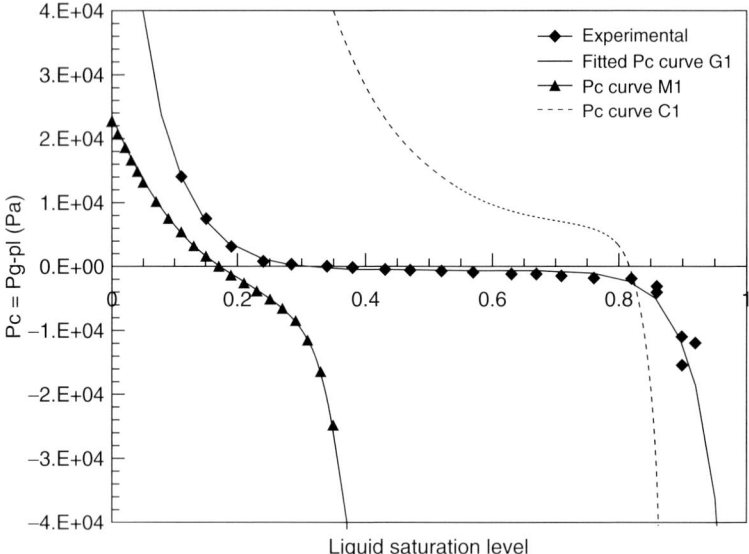

Figure 30.12 Capillary curves used in the case with and without the MPL. The fitting parameters of a_1, a_2, b, c, and d are -22.7, -16.2, -644.9, 0.58, and -7.6 for G1, -30.0, -7.0, -9878.4, 0.3, and -4000.0 for M1, and -45.0, -8.0, 5000.0, 0.78, and -1106.56 for C1, respectively [41].

where σ is the surface tension and $J(s)$ is given by

$$J(s) = \begin{cases} 1.417(1-s) - 2.120(1-s)^2 + 1.263(1-s)^3 & \text{for } \theta_c < 90° \\ 1.417s - 2.120s^2 + 1.263s^3 & \text{for } \theta_c > 90° \end{cases} \quad (30.26)$$

It should be pointed out that the Leverett function was originally developed for liquid water transport in soils and, as such, it may not be directly applicable to liquid-water transport in the GDLs owing to their unique pore characteristics. To take this into account, Wang and Nguyen [41] proposed an empirical correlation based on experimental data (see Figure 30.12).

In reality, GDLs are not homogeneous media; rather, they are highly heterogeneous in the thickness dimension. Recently, Hinebaugh *et al.* [42] measured experimentally the local porosity of a carbon paper GDL, indicating that this is a spatially varying property (see Figure 30.13) [43]. As a result, the permeability, a factor that has a great impact on two-phase flow, varies spatially; see the Blake–Kozeny equation. To explain the effect of the spatial varying DM properties, the water flux driven by the capillary pressure accounting for spatially varying GDL properties can be written as follows [43]:

$$\frac{\lambda^l \lambda^g}{\nu} K \nabla P_c(\tau, \theta_c, \varepsilon, K, s) = \frac{\lambda^l \lambda^g}{\nu} K \left(\frac{\partial P_c}{\partial \tau} \nabla \tau + \frac{\partial P_c}{\partial \theta_c} \nabla \theta_c + \frac{\partial P_c}{\partial \varepsilon} \nabla \varepsilon + \frac{\partial P_c}{\partial K} \nabla K + \frac{\partial P_c}{\partial s} \nabla s \right) \quad (30.27)$$

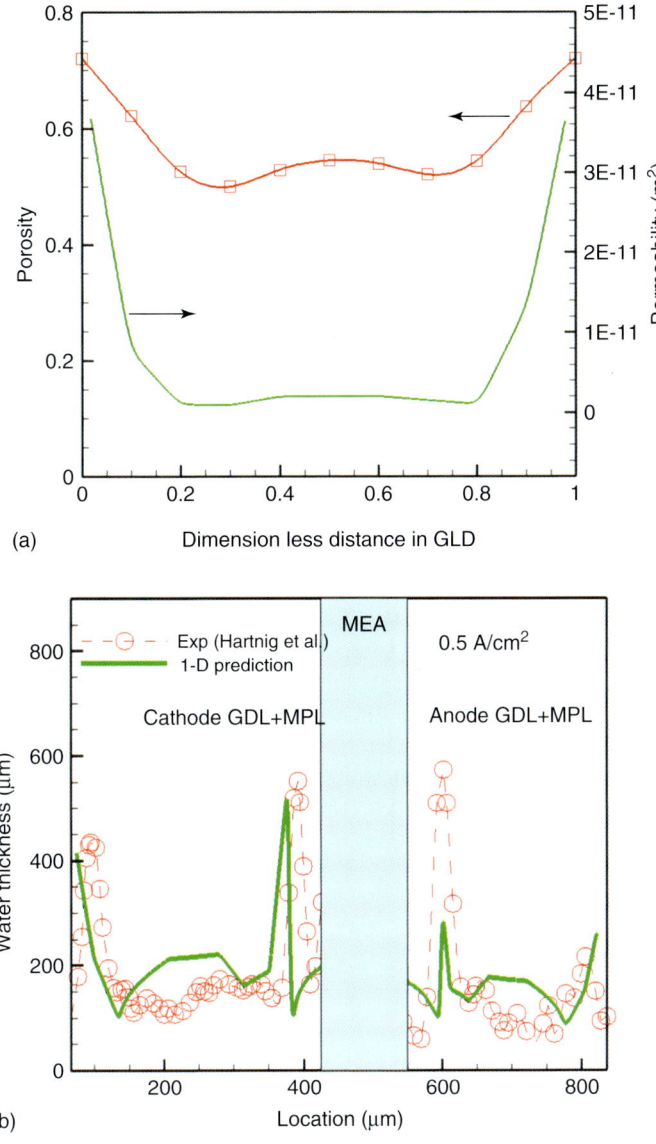

Figure 30.13 (a) The spatial variation of the GDL porosity [42] and permeability and (b) the profile of liquid water across the anode/cathode GDL and MPL [43].

Owing to the extra terms arising from the spatially varying property on the right, local maxima/minima in water content may occur (see Figure 30.13).

To improve the multi-phase, particularly liquid, flow characteristics, an MPL can be added and placed between the GDL and CL. This layer is composed of carbon black powder with a fine pore structure. Studies have shown that adding

MPLs leads to better water drainage characteristics and fuel cell performance. Weber and Newman [44], Pasaogullari et al. [45], and Wang and Chen [46] asserted that the MPL acts as a valve that drives water away from electrodes to reduce the electrode flooding. Modeling MPLs can follow the framework for GDLs by considering MPL transport properties. In addition, at the MPL–GDL interface under certain conditions, Pasaogullari et al. [45] and Wang and Chen [46] consider that the following relation (capillary pressure being continuous at the GDL/MPL interface holds:

$$\tau \cos\left(\theta_c^{GDL}\right) \left(\frac{\varepsilon^{GDL}}{K^{GDL}}\right)^{\frac{1}{2}} J(s) = P_c^{GDL} = P_c^{MPL} = \tau \cos\left(\theta_c^{MPL}\right) \left(\frac{\varepsilon^{MPL}}{K^{MPL}}\right)^{\frac{1}{2}} J(s) \tag{30.28}$$

This equation adopts the Leverett relation. Generally, the MPL porosity and mean pore size are much smaller than that of the GDL, therefore leading to a sharp discontinuity at the MPL–GDL interface. In experiment, Mukundan et al. [47] employed neutron radiography to investigate the effects of PTFE loadings on the water content within both GDLs and MPLs, and indicated that lower PTFE loadings in MPLs may result in better performance and lower transport resistance. Hickner et al. [48] also applied neutron imaging to quantify the liquid water content within MPLs and GDLs. It should be pointed out that the above equation is valid only when the properties of the GDL and MPL are similar and it may need to be modified when the properties are significantly different [49]. Indeed, what conditions should be specified at the GDL–MPL interface is still under active research.

The macroscopic two-phase flow approach has been widely employed to model liquid water transport through the GDL and MPL, mostly using the Leverett function, which likely results in inaccuracy because it was formulated originally for water transport in soils. Therefore, a new correlation is required to provide the relationship between the capillary pressure and saturation for fuel-cell GDLs. Two types of information are strongly needed. One is the experimentally measured capillary pressure as a function of saturation as demonstrated by Ohn et al. [50], Nguyen et al. [51], Fairweather et al. [52], and Sole and Ellis [53]. This will address the concern that the Leverett function was originally developed for describing gas and liquid transport in soils and rocks, which have vastly different pore-size distributions and shapes to those in carbon paper or cloth GDLs. Recently, Wang and Nguyen [41] presented a capillary-pressure correlation determined by experiment for CLs, GDLs, and MPLs. They used the following form of capillary pressure correlation, which involves five parameters:

$$P_c = d\left[e^{-a_1(s-c)} - e^{a_2(s-c)}\right] + b \tag{30.29}$$

where a_1, a_2, b, c, and d are the adjustable parameters (see Figure 30.12). Another approach to calculating the capillary pressure is to relate the local capillary pressure to the local radius of curvature and contact angle by employing the Young–Laplace equation:

$$P_c = \frac{2\tau \cos\theta_c}{r} \tag{30.30}$$

where P_c is the capillary pressure in a circular pore and r is the radius of curvature, which can differ from the pore radius. Obviously, the capillary pressure gradient requires an existing gradient of the radius of curvature and/or a gradient of contact angle. The challenge here is how to measure the spatial distributions of radius of curvature and contact angle in the direction of water transport flowing mainly in the through-plane direction, so that the gradients of radius of curvature and/or contact angle can be determined. This equation can be used to account for non-uniformity in terms of its hydrophilic and hydrophobic properties. Further studies are needed to characterize the pore-size distribution and also hydrophilicity and hydrophobicity distributions and to use this information to develop pore-level models. This type of work can aid in allowing the realistic and accurate simulation of liquid water and gas transport through the GDLs with highly non-uniform pore sizes and wettability and a complete understanding of how GDL properties influence fuel-cell performance. In addition, it may be numerically challenging when treating the MPLs as a distinct physical region in comparison with the GDL – it is difficult to achieve convergence in some cases. This numerical issue mostly arises from the sharp change in liquid saturation at the GDL–MPL interface and also in the along-channel direction. Therefore, how to account for the effect of the MPL on PEFC performance without having the numerical non-convergence or divergence problem is an important research issue.

30.5
Gas Flow Channels

GFCs are an important component, supplying and distributing hydrogen fuel to the anode and oxygen reactant to the cathode for the HOR and the ORR reactions, respectively. The channels also provide a means to remove by-product water. The PEFC channels are located within the BPs with a typical cross-sectional dimension of around 1 mm. Insufficient supply of reactants will lead to hydrogen/oxygen starvation, reducing cell performance and durability. Several types of flow fields have been developed, namely parallel, serpentine, pin-type, interdigitated, porous media designs, and zigzag flow fields [2] (see Figure 30.14). Studies have also been carried out to investigate the cross-sectional dimension of GFCs. Inoue *et al.* [54] examined the channel height and found that shallow channels may enhance oxygen transport to the electrodes. Wang [55] analyzed the channel in-plane dimension by examining heat and electron transport characteristics. Wang *et al.* [56] investigated the channel aspect ratio for a serpentine flow field.

Convection is the dominant process for species transport in GFCs, and the flow has customarily been treated using the single-phase approach: either considering the vapor phase as supersaturated or treating it as mist flow (in which tiny water droplets remain so that they travel along with the main gas flow) – neither of these two approaches describes the reality of flow in GFCs. The Navier–Stokes equations

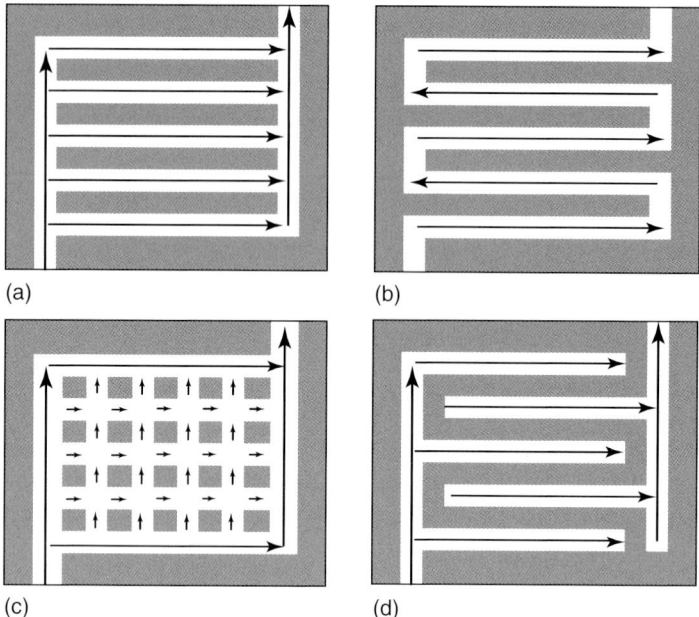

Figure 30.14 Typical flow fields of PEFCs. (a) Parallel flow field; (b) serpentine flow field; (c) pin flow field; and (d) interdigitated flow field.

are usually employed to describe flow phenomena in PEFC channels:

Continuity equation: $\nabla \cdot (\rho \vec{u}) = 0$

Momentum conservation: $\dfrac{1}{\varepsilon}\left[\dfrac{\partial \vec{u}}{\partial t} + \dfrac{1}{\varepsilon}\nabla \cdot (\vec{u}\vec{u})\right] = -\nabla\left(\dfrac{p}{\rho}\right) + \nabla \cdot \tau$ (30.31)

In the channel, the velocity usually ranges from 0.1 to 10 m s^{-1} such that the Reynolds number (Re) ranges from o(1) to about 1000, which is below 2300, the critical value of Re for turbulent flow in a duct or pipe; consequently, the flow falls in the laminar regime. Wang and Wang [29] further indicated that the flow rate will change due to the mass exchange between the anode and cathode, resulting in a large variation in the anode stream but little in the cathode. Since there is no chemical reaction occurring in the channel, the standard diffusion–convection governing equations can be applied. When operating at a low stoichiometric flow ratio or simply stoichiometry, high relative humidity (RH), or high current density, channel flow falls in the two-phase regime owing to water addition from the ORR and the channel's inability to remove water efficiently. Liquid may block channels, hampering reactant supply and causing unstable fuel cell operation. Figure 30.15 shows the cell voltage variation over time, indicating that the voltage becomes oscillatory at a stoichiometry of 2. Cathode flooding can result in a

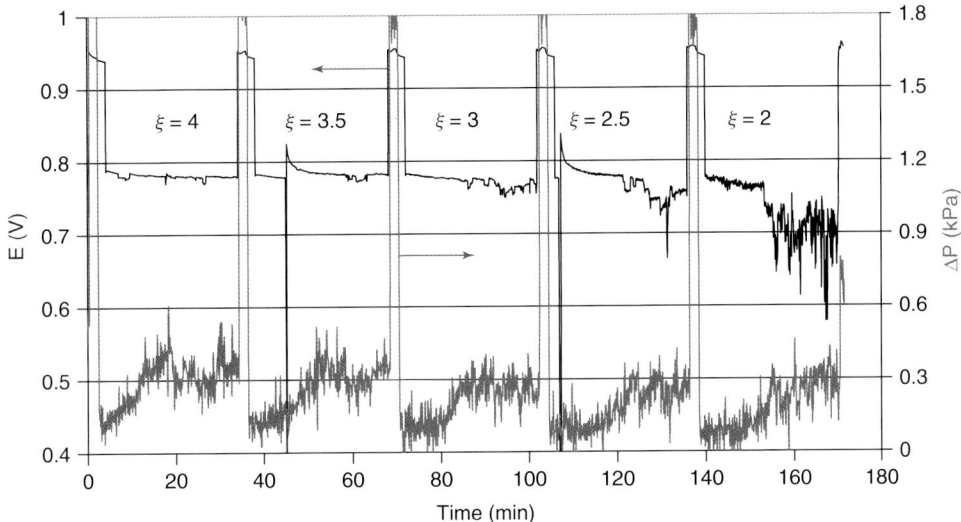

Figure 30.15 Cell voltage and cathode-side pressure drop measured under various air stoichiometric ratios (ξ) in a 14 cm² fuel cell for 0.2 A m⁻², 150 kPa and 80 °C with the dew point of both H$_2$ and air at 70 °C. As the air stoichiometry decreases, the pressure drop does not decrease proportionally, indicating the onset of severe water flooding in cathode channels. Simultaneously, the cell voltage drops by as much as 120 mV and fluctuates more [57].

performance loss (~120 mV) that completely negates any potential improvement from catalyst development: for instance, a fourfold increase in catalytic activity yields only ~45 mV gain in cell voltage [57]. Moreover, the voltage fluctuation induced by channel flooding may set up voltage cycling at high potentials, raising durability concerns. The wettability of the GFC wall, that is, the hydrophilicity or hydrophobicity, may have significant effects on the channel two-phase flow: hydrophilic GFC walls seem to be favored by practitioners since they facilitate the formation of a thin liquid-water film and provide a steady flow of air and thus O$_2$ to the reaction sites, whereas the hydrophobic GFC walls can result in unsteady cell operation due to large water droplets that can partially and completely block flow in the channel.

Modeling two-phase flow in the PEFC channel is numerically very challenging. Wang et al. [57] envisaged the mini-channels as structured and ordered porous media. A two-phase channel flooding model was developed based on the two-phase mixture description. They also developed a one-dimensional analysis and obtained an analytical solution for the liquid water profile (in terms of liquid saturation) along flow channels:

$$\frac{s - s_{ir}}{1 - s_{ir}} = \frac{1}{\left(\frac{1-\lambda^l}{\lambda^l}\frac{v_g}{v_l}\right)^{\frac{1}{n_k}} + 1} \tag{30.32}$$

where s_{ir} denotes the irreducible saturation and λ^l is the liquid mobility. λ^l is given by

$$\lambda^l = \frac{\frac{\xi_c}{2}\left(\frac{c^{H_2O}}{c^{O_2}}\right)\bigg|_{in} + \frac{\int_0^Y I(\overline{Y})d\overline{Y}}{l_{av}} - \frac{1}{\rho_g}c_{sat}^{H_2O}\left[\frac{\xi_c}{2}\left(\frac{\rho}{c^{O_2}}\right)\bigg|_{in} + \frac{\int_0^Y I(\overline{Y})d\overline{Y}}{l_{av}}M^{H_2}\right]}{\left(\frac{1}{M^{H_2O}} - \frac{1}{\rho_g}c_{sat}^{H_2O}\right)\left[\frac{\xi_c}{2}\left(\frac{\rho}{c^{O_2}}\right)\bigg|_{in} + \frac{\int_0^Y I(\overline{Y})d\overline{Y}}{l_{av}}M^{H_2}\right]}$$

(30.33)

Wang and co-workers [58, 59] followed their earlier work [57] and studied the flow maldistribution among fuel cell channels. Wang [55] further proposed a concept of porous media channels and examined the characteristics of reactant flows, heat transfer, species transport, and two-phase transport. Various correlations of the relative permeability were discussed and found to affect liquid profiles significantly (see Figure 30.16).

Studies have been proposed to investigate liquid transport using the VOF method and LBM. Figure 30.17 shows the slug formation and droplet removal calculated using the VOF method [61, 62]. Most VOF studies have been focused on the dynamics of liquid droplets, which will be detailed in the next subsection. However, models that simulate two-phase behavior in channels and can be incorporated in a full PEFC fuel-cell model still remains as a key challenge owing to the lack of efficient

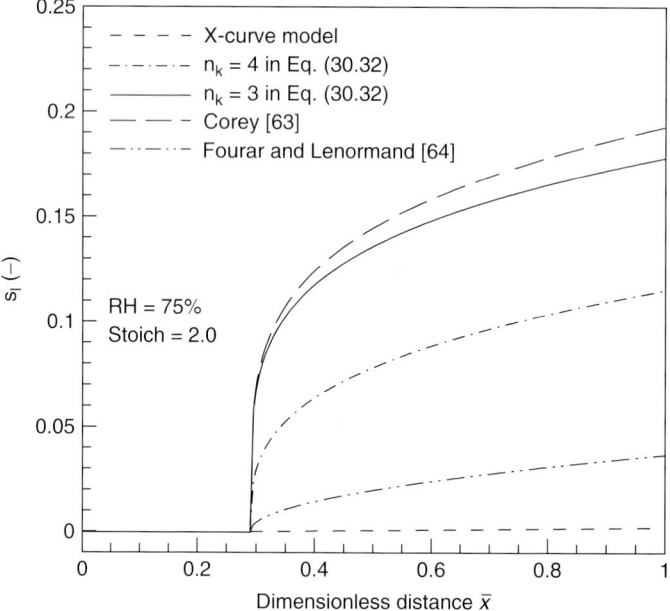

Figure 30.16 Liquid water profiles using different models for the relative permeability [60].

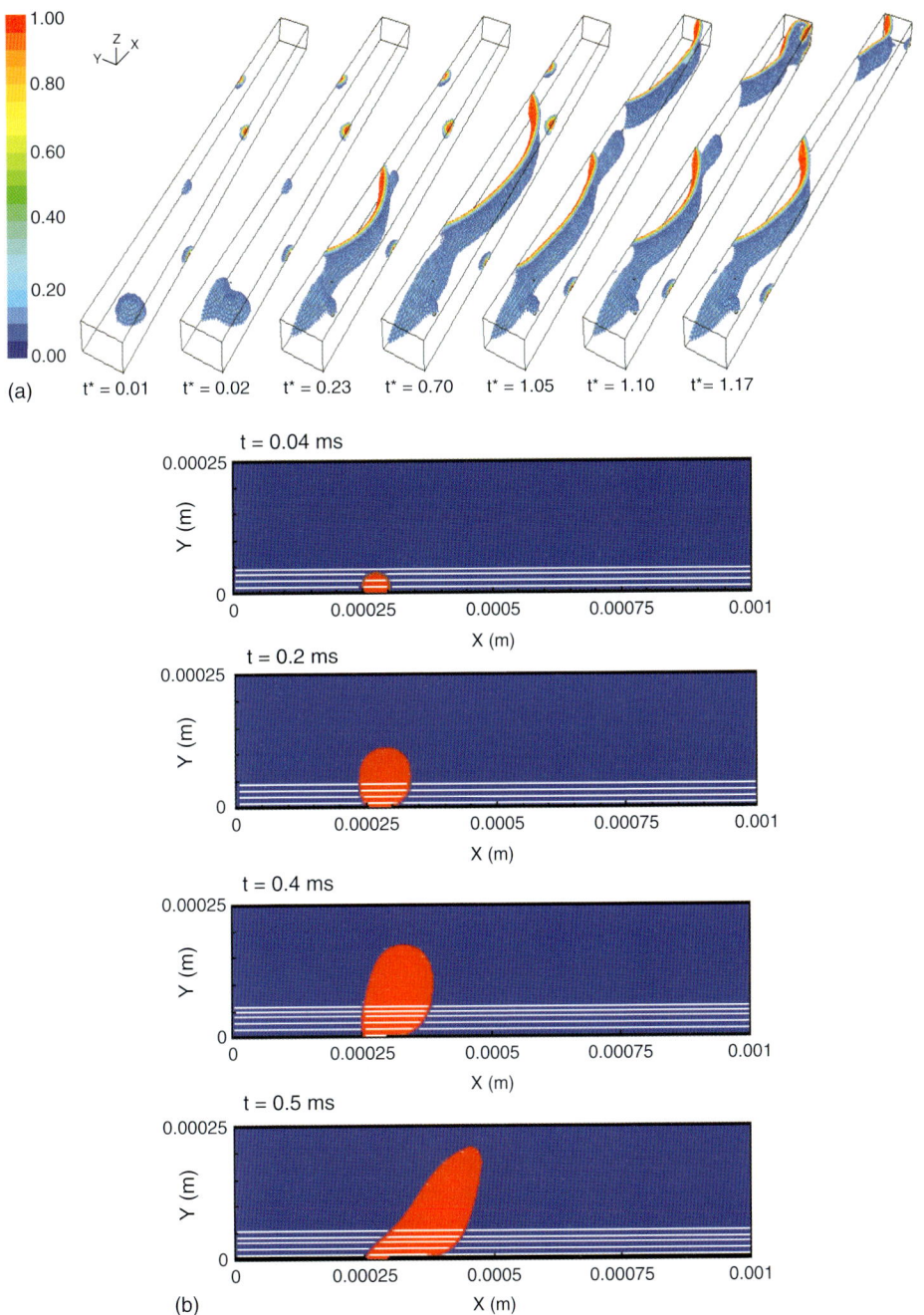

Figure 30.17 Water droplet evolution in a microchannel with smooth hydrophilic surface predicted by VOF: (a) [61] and (b) [62].

numerical methods to track the two-phase interface and capture multicomponent transport. Further studies are required to characterize the two-phase flow regime of fuel-cell operation with slug flow and slug–annulus transition. Anderson et al. [65] recently provided a comprehensive review on the two-phase flow in fuel-cell channels and indicated several possible flow patterns, such as droplets, films, and slug flows. Each pattern may be dominated by different mechanisms; therefore, modeling the channel two-phase transport can take different modeling approaches, depending on the flow regime. In the case of high gas velocity, liquid water likely exists as small droplets traveling with the gas flow; therefore, the well-dispersed mixture treatment can be applied. In the case of a lower gas flow, the gas is unable to remove and carry liquid droplets; consequently, liquid tends to attach to and spread on the channel wall, resulting in film flow. This flow regime may need a model with interface tracking capability to describe properly the transition from individual droplets merging and spreading on the channel wall surface to form a liquid film. Modeling challenges include the numerical treatments of various channel two-phase flow scenarios, in particular the transition between flow patterns, and the coupling of channel flow with phenomena that take place in other fuel-cell components.

30.6
Gas Diffusion Layer-Gas Flow Channel Interface

At the cathode GDL–GFC interface, oxygen transports towards the electrode or CL, where it reacts with protons and electrons to produce water. The product water also enters the channel through the GDL–GFC interface. The presence of liquid water can considerably increase the interfacial resistance of reactant transport. Optical visualizations (see Figure 30.18) [60, 66] show that liquid water exists as droplets on the GDL surface, being taken away by the gas flow or attached to the channel wall. The behavior of liquid water droplets at the GDL–GFC interface consists of three sub-processes: (i) liquid water transport from the CL through the GDL to the GDL–GFC interface via capillary action; (ii) removal at the GDL–GFC interface via detachment or evaporation; and (iii) transport through the GFC in the form of films, droplets, and/or vapor. The growth and detachment of water droplets are mainly influenced by two factors: the operating conditions of the fuel cell and the physical (e.g., surface roughness) and chemical (e.g., wettability) material characteristics of the GDL surface (e.g., in terms of the hydrophilic/hydrophobic properties). Chen et al. [67] pioneered the analysis of water-droplet instability at and detachment from the GDL/channel interface. They indicated that the static contact angle (θ_s) and contact angle hysteresis – the difference between advancing and receding contact angles, that is, $\theta_A - \theta_R$, – are both important parameters in describing the surface tension force that tends to hold the droplet onto the GDL surface. Instability diagrams were developed to explore the operating conditions under which droplets become unstable (an example of such instability diagram is shown in Figure 30.19). Unstable conditions are desirable to operate the fuel cell under such conditions that droplets can be removed instantaneously from the

Figure 30.18 Visualization of the two-phase flow inside a hollow channel in the PEFC cathode. (a) Droplets in a channel. (b) Droplets in the lower channel and droplet attaching the channel wall in the upper channel. The channel cross-sectional dimension is 1 × 1 mm [60].

GDL/GFC interface so as to prevent blockage of pathways for oxygen transport to the three-phase reaction sites.

Wang and co-workers [68, 69] adopted the liquid coverage model to account for the droplet influence. They treated the droplet covering the GDL surface as a boundary condition for the liquid transport equation in GDLs, that is, the ratio of the liquid-covered surface area to the total area. As the liquid droplet detachment is related to several major factors, such as gas flow rates, wettability, and liquid production rate, the saturation at the surface can be written as a function of them:

$$s_{int} = S\left(\vec{u}, \theta_c, I\right) \tag{30.34}$$

Based on experimental data, the following formulation was proposed:

$$s_{int} = C \frac{I^{0.656}}{|\vec{u}|^{3.32}} \tag{30.35}$$

where the coefficient C can be determined by a single-point calibration, and the parameters $\vec{u}$ and θ_c are the contact angle and channel velocity, respectively.

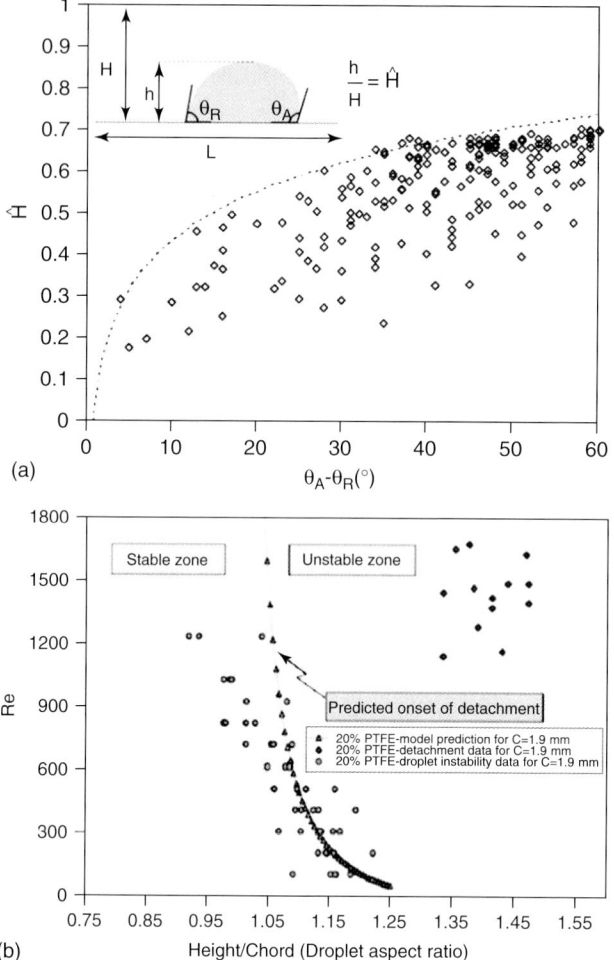

Figure 30.19 (a) Droplet instability diagram for an air flow velocity of 800 cm s^{-1}, $L = 7$ cm, and $\theta_s = 140°$. (◊) Experimental data points; (---) upper bound of experimental data [67]. (b) Critical Reynolds number versus droplet aspect ratio for a droplet with a specified chord length of 1.9 mm on 20% PTFE DM [66].

Further work of interest includes development of 3D fundamental models to predict accurately droplet behaviors at the interface, particularly the effect of GDL surface properties on droplet dynamics. Since the droplets appear randomly at the GDL surface, experiments are needed to collect statistical data, for given GDL characteristics and operating conditions, on the distribution and coverage of droplets on the GDL surface, and statistical methods may be adopted to evaluate the proportion of area covered by liquid. Also, since the GFC–GDL interface bridges the transport in channels and GDL, a fundamental understanding of and a mathematical model that can describe this coupling will be required. A simplified explicit model was further developed by Chen [70] for analyzing water droplet detachment in the inertia-dominant regime. Chen also carried out 3D numerical simulations of droplet detachment in the inertia-dominant regime using the VOF method [70]. The VOF method, which tracks the free surface area of two different fluids, has been widely applied to study liquid flows, including that in fuel-cell gas channels (e.g., [59, 62, 70]). The 3D governing equations employed in the VOF numerical treatment can be expressed as follows [71]:

$$\frac{\partial}{\partial t}\rho + \nabla \cdot (\rho \vec{v}) = S_m$$

$$\frac{\partial}{\partial t}(\rho \vec{v}) + \nabla \cdot (\rho \vec{v} \vec{v}) = -\nabla p + \nabla \cdot \left[\mu \left(\nabla \vec{v} + \nabla \vec{v}^T\right)\right] + \rho \vec{g} + \vec{F}_{\text{vol}} \quad (30.36)$$

where ρ is the density, μ the viscosity, $\vec{v}$ the velocity vector, S_m the mass source, p pressure, $\vec{g}$ the gravitational force, and $\vec{F}$ the momentum source. The momentum source term $\vec{F}_{\text{Vol}}$ takes into account the surface tension force that is calculated through the surface tension τ, surface curvature and gradient of volume fraction $\nabla \alpha_k$. The volume fraction α_k is defined as the portion of a computational cell filled with kth phase of fluid. In this study, the primary fluid is set as air and the secondary fluid is water. To satisfy the continuum theory of fluid, the summation of α in a cell should be unity:

$$\alpha_{\text{water}} + \alpha_{\text{air}} = 1 \quad (30.37)$$

and satisfies the following equation:

$$\frac{\partial}{\partial t}(\alpha_k \rho_k) + \nabla \cdot (\alpha_k \rho_k \vec{v}_k) = 0 \quad (30.38)$$

The properties of fluids at the interface between gas and liquid can be evaluated based on the volume fraction of the individual phase:

$$\rho = \rho_{\text{air}} + \alpha_{\text{water}}(\rho_{\text{water}} - \rho_{\text{air}}) \quad (30.39)$$

The pressure drop across the interface can be evaluated through surface tension and curvatures:

$$\Delta p = \tau \left(\frac{1}{R_1} + \frac{1}{R_2}\right) \quad (30.40)$$

where R_1 and R_2 are the measured curvatures of two radii in orthogonal directions. Figure 30.17 shows the liquid water formation and removal as computed by the VOF method.

Lastly, there is an urgent need to develop a GDL–channel interface sub-model for implementation in the full fuel-cell model. Although the coverage model has been developed by Meng and Wang [68], it only specifies the interfacial saturation. The ability to specify the water flux at the GDL–channel interface will permit a more realistic and accurate coupling between transport in the through-plane direction (which takes place in the MEA and the GDL–MPL) and transport along the channel. This water flux at the GDL–channel interface can be formulated as a function of localized GDL surface wettability, local water content on both the GDL side and the channel side at the GDL–channel interface, and so on.

30.7
Bipolar Plates

BPs provide mechanical support over GDLs and conductive passages for both heat and electron transport. Fabrication of BPs, together with GFCs, may contribute a considerable portion of a fuel cell's cost [2]. Cooling channels can be machined within the BPs, and are critical for waste heat removal, in particular for large-scale fuel cells. Comprehensive reviews on flow fields and BPs were provided by Wilkinson and Vanderleeden [72], EG&G Technical Service [73], and Li and Sabir [74]. Two important transport processes occur in the BPs during PEFC operation: electrical current flow and heat transfer. The electrical current equation can be formulated based on Ohm's law:

$$0 = \nabla \cdot \left(\sigma_s^{\text{eff}} \nabla \Phi^{(s)} \right) \tag{30.41}$$

Figure 30.20 shows the contours of electrical potential in the BPs (current collector) and GDL. It is seen that the variation is small throughout the BPs given their high conductivity. In addition, incorporating the electron transport equation enables the current density to be readily preset as an input parameter in numerical simulation by setting the electrical flux boundary condition at the outer BP surface; an

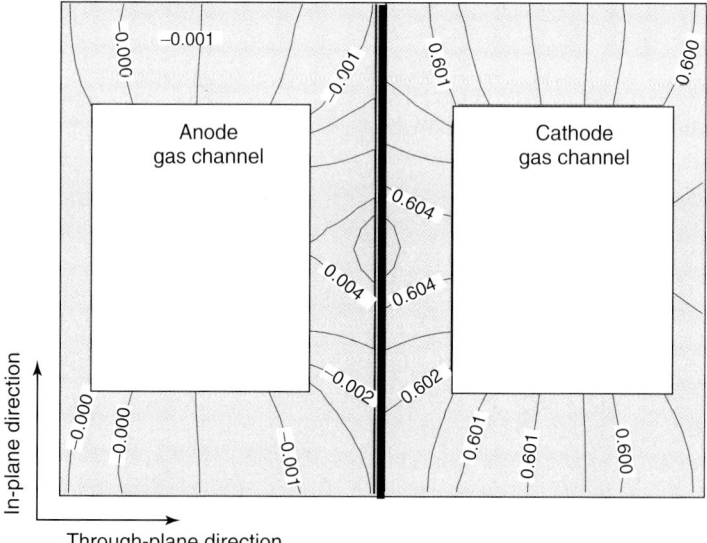

Figure 30.20 The contours of electronic phase potential in the current collectors and GDLs at the middle section of the PEFC (50% fraction distance from the inlet). Units: volts.

example is

$$\frac{\partial \Phi^{(s)}}{\partial n}\bigg|_c = -\frac{IA_m}{\sigma^{\text{eff}} A_{c,\text{wall}}} \tag{30.42}$$

where A_c, wall is the area of the cathode outer surface.

30.8
Coolant Flow

Cooling channels must be added to keep fuel cells at their optimal temperature when a large amount of waste heat is generated. Cooling channel modeling has received a relatively low level of attention in the past, compared with other cell components. Wang and Wang [75] indicated that the cooling channel design and control can be optimized for better water/thermal management (see Figure 30.21). Yu *et al.* [76] and Inoue *et al.* [77] also presented a study on cooling channels or units for PEFCs. The main focus of coolant flow models is placed on fluid flow and energy transfer, which is important for evaluating the pressure drop and heat removal.

30.9
Model Validation

Although detailed models have been developed for all of the major components, few validations have been presented to compare the modeling predictions with

Figure 30.21 Surface temperature at the cathode flow plate for a coolant flow rate of 2.9×10^{-6} m s^{-1} and a coolant inlet temperature of 345.15 K (72 °C) under 1.0 A cm^{-2} and 0.616 V [75].

experimental data at the component level [19, 78–80], partly owing to the lack of experimental data for individual cell components.

In the last few years, neutron imaging has emerged as a powerful tool for detecting liquid water in fuel cells, and key findings have been reported, including the through-plane liquid-water profiles recently obtained through the high-resolution neutron radiography. In addition to neutron radiography, other techniques such as magnetic resonance imaging and X-ray imaging can also be employed to probe liquid water in PEFCs.

A direct comparison between the predicted water through-plane profiles and neutron imaging data was attempted by Weber and Hickner [78]. Their model prediction agrees reasonably well for the 80 °C case, but a significant discrepancy exists for the 60 °C case [81]. Weber and Hickner attributed the large discrepancy to their pseudo-2D (or 1D + 1D) approach in which the effect of the ribs was neglected; and they pointed out that improvement to the model is needed for an accurate prediction. Wang and Chen [82] recently conducted a comparison using their full 2D model, and achieved reasonably good agreements with the neutron imaging data from both Sandia National Laboratories and Las Alamos National Laboratory (see Figure 30.22) [82]. They pointed out that in the membrane a correction factor, α, may be used to account for factors such as membrane swelling and neutron scattering, and Gaussian smoothing may be employed to post-process the model prediction in order to account for the geometric blur in neutron imaging. In GDLs, the general trend of water profiles and also the level of water content agree well with the experimental data, but the local maximum and minimum inside GDLs as seen in the neutron data were not captured. In their follow-on work, Wang and Chen [79] performed model validation using the experimental X-ray data reported by Hartnig *et al.* [83], and they explained that the spatially varying GDL property may lead to the local maximum and minimum water content observed experimentally. With the varying GDL property accounted for, the model prediction obtained by Wang and Chen shows much better agreement in capturing the local maximum and minimum in water content; see Figure 30.13 [79]. Another model validation was recently conducted by Tabuchi *et al.* [80], who compared their 2D prediction with neutron imaging data for liquid water profiles in the in-plane direction, in particular under the channel versus under the land. They also compared their predicted cell performance with experimental data.

It should be pointed out that only limited model validation has been reported in the literature. Consequently, more comprehensive model validation is needed. First, experimental data for the liquid water profile across the CL and MPL are lacking. Second, temperature and species concentration distributions in all fuel cell components have not been reported experimentally. Third, models should also be validated in terms of several sets of experimental data at the same time, for example, water profiles in three dimensions.

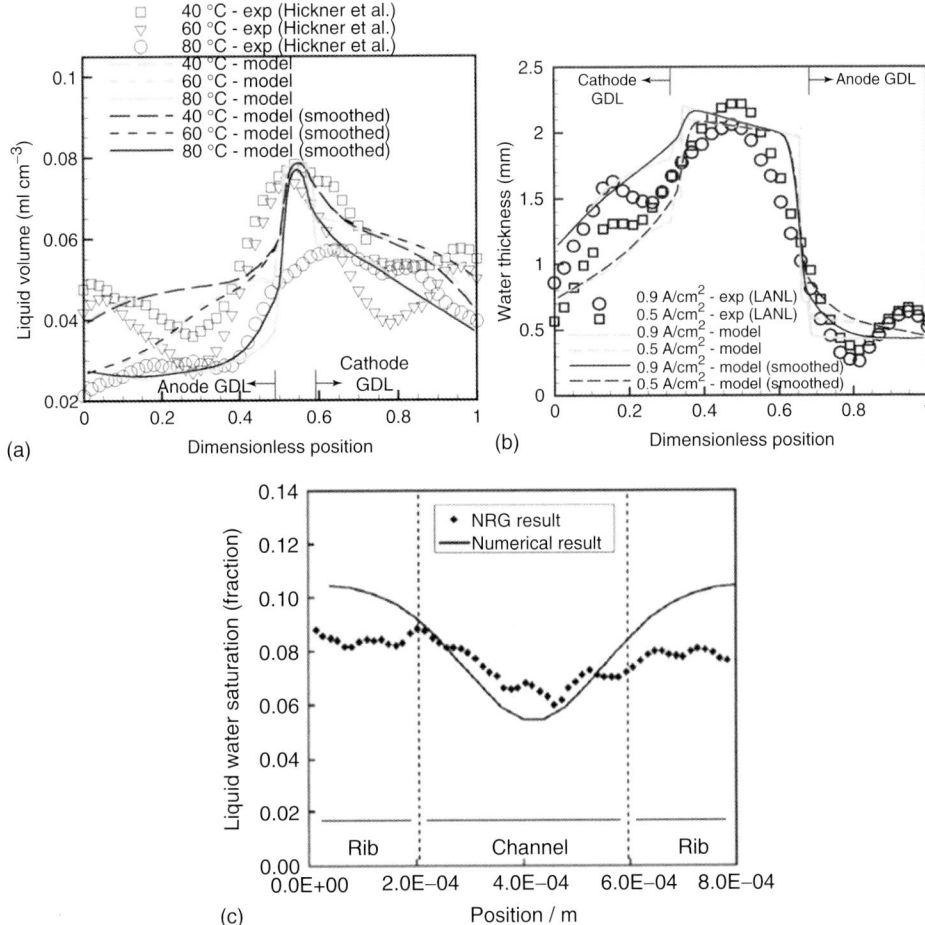

Figure 30.22 Comparison of the predicted water profiles and neutron imaging experimental data for (a) RH(anode/cathode) = 100/100% and (b) RH(anode/cathode) = 100/50% and 40 °C at 0.75 A cm^{-2} [82]. (c) Numerical validation results of average liquid water saturation distribution in MEA between rib and channel area at 2.5 A cm^{-2} [80].

30.10
Conclusion

In this chapter, the current status and several major aspects of PEFC component modeling are described and discussed. Mathematical and computational modeling (both analytical and numerical) play important roles in the technology development and optimal operation of PEFCs: from a fundamental understanding of underlying phenomena to engineering design and optimization that can lead to cost reductions and durability improvements. At present, the macroscopic descriptions of phenomena in the individual fuel-cell components have been formulated and corresponding models developed with validation against experimental data being carried out to some extent. The major transport and electrochemical phenomena have been modeled up to three dimensions within the individual components: membrane, CL, MPL/GDL, GFCs, and BPs/cooling flow. The essential phenomena include two-phase flow, heat transfer, multi-species mass and charge transport, and hydrogen oxidation and oxygen reduction electrochemical reactions. In addition, mesoscopic approaches such as LBM and DNS have been proposed and employed to investigate pore-level phenomena, which can help the macroscopic modeling and enhance our understanding of the physical processes that take place in the porous media of fuel cells. VOF methods have also been applied to investigate the liquid water dynamics in fuel-cell channels and at the GDL–channel interface, which is important for the macroscopic modeling of two-phase flow in GFCs. Moreover, agglomerate CL models have also been developed and used to examine the effects of the microstructure of the CL on fuel-cell performance. Model validation, although limited in scope, has been conducted in terms of local species concentration and liquid water profiles.

In spite of the significant advances made in PEFC component modeling, several challenges remain: first, more accurate physical/chemical/transport properties of each cell component are needed to improve model prediction; for example, more general correlations are needed for the membrane ionic conductivity as a function of water content and the capillary pressure as a function of saturation in the GDL with a mixed wettability. Second, a sub-model of channel two-phase flows that can be incorporated in a 3D fuel-cell model is still lacking. Although great efforts in modeling the dynamic behavior of water droplets using the VOF method and the like and experimental visualization have been made, it remains a major challenge to develop a computationally efficient sub-model for channel two-phase flow, partly owing to the numerical difficulty in tracking the two-phase interfaces and describing the occurrence of several flow regimes (e.g., the merging of droplets to form larger droplets or spread to form a thin film) in the GFC. Lastly, there is a critical need to couple, in some computationally efficient way, the pore-level or particle-level sub-models with the macroscopic models in order to take into account the effects of the microstructures of GDL/MPL and CL.

List of Symbols

A	electrode area, m^2
a	water activity; effective catalyst area per unit volume, m^2 m^{-3}
a_0	catalyst surface area per unit volume, m^2 m^{-3}
C	molar concentration of species k, mol m^{-3}
D	species diffusivity, m^2 s^{-1}
EW	equivalent weight of dry membrane, kg mol^{-1}
F	Faraday's constant, 96 487 C mol^{-1}
$\vec{G}$	species diffusion/permeation flux, mol m^{-2}
I	current density, A cm^{-2}
$\vec{i}_e$	superficial current density, A cm^{-2}
j	transfer current density, A cm^{-3}

$\vec{j}^{(l)}$	mass flux of liquid phase, $\text{kg m}^{-2}\,\text{s}^{-1}$
K	permeability, m^2
k_r	relative permeability
L	length, m
M	molecular weight, kg mol^{-1}
$mf_k^{(l)}$	mass fraction of species k in liquid phase
n	the direction normal to the surface
n_d	electroosmotic coefficient, H_2O/H^+
P	pressure, Pa
R	universal gas constant, $8.134\,\text{J mol}^{-1}\,\text{K}^{-1}$
S	source term
s	liquid saturation
t	time, s
T	temperature, K
U_0	equilibrium potential, V
$\vec{u}$	velocity vector, m s^{-1}

Greeks

α	transfer coefficient; net water flux per proton flux
ρ	density, kg m^{-3}
v	kinematic viscosity, $\text{m}^2\,\text{s}^{-1}$
θ_c	contact angle, °
Φ	phase potential, V
ξ	stoichiometric flow ratio
λ	membrane water content
$\lambda^{(k)}$	mobility of phase k
ε	porosity
η	surface overpotential, V
$\vec{\vec{\tau}}$	shear stress, N m^{-2}
τ	surface tension, N m^{-1}
γ_c	correction factor for species convection
δ	thickness, m
σ	electronic or ionic conductivity, S m^{-1}

Superscripts and Subscripts

a	anode
c	cathode; capillary
CL	catalyst layer
D	diffusion
DM	diffusion media
eff	effective value
g	gas phase
GDL	gas diffusion layer

in	inlet
k	species; liquid or gas phase
l	liquid
m	membrane phase
0	gas channel inlet value; reference value
ref	reference value
s	solid
sat	saturated value
w	water

References

1. Papageorgopoulos, D. (2010) DOE fuel cell technology program overview and introduction to the 2010 Fuel Cell Pre-Solicitation Workshop. Presented at DOE Fuel Cell Pre-Solicitation Workshop, Department of Energy, Lakewood, CO.
2. Wang, Y., Chen, K.S., Mishler, J., Cho, S.C., and Adroher, X.C. (2011) A review of polymer electrolyte membrane fuel cells: technology, applications, and needs on fundamental research. *Appl. Energy*, **88**, 981–1007.
3. Wang, C.Y. (2004) Fundamental models for fuel cell engineering. *Chem. Rev.*, **104**, 4727–4766.
4. Weber, A.Z. and Newman, J. (2004) *Chem. Rev.*, **104**, 4679.
5. Gurau, V. and Mann, J.A. (2009) A critical overview of computational fluid dynamics multi-phase models for proton exchange membrane fuel cells. *SIAM J. Appl. Math.*, **70**, 410–454.
6. Siegel, C. (2008) Review of computational heat and mass transfer modeling in polymer-electrolyte-membrane (PEM) fuel cells. *Energy*, **33** (9), 1331–1352.
7. Jiao, K. and Li, X. (2011) Water transport in polymer electrolyte membrane fuel cells. *Prog. Energy Combust. Sci.*, **37**, 221–291.
8. Djilali, N. (2007) Computational modelling of polymer electrolyte membrane (PEM) fuel cells: challenges and opportunities. *Energy*, **32** (4), 269–280.
9. Gierke, T.D. and Hsu, W.Y. (1982) in *Perfluorinated Ionomer Membranes*, ACS Symposium Series, Vol. 180 (eds. A. Eisenberg and H.L. Yeager), American Chemical Society, Washington, DC.
10. Kreuer, K.D. and Weppner, R.A.W. (1982) Vehicle mechanism, a new model for the interpretation of the conductivity of fast proton conductors. *Angew. Chem. Int. Ed. Engl.*, **21**, 208–209.
11. Kornyshev, A.A., Spohr, K.A.E., and Ulstrup, J. (2003) Kinetics of proton transport in water. *J. Phys. Chem. B*, **107**, 3351–3366.
12. Marx, D. Tuckerman, M.E., Hutter, J., and Parrinello, M., (1999) The nature of the hydrated excess proton in water. *Nature*, **397** (6720), 601–604.
13. Springer, T.E., Zawodinski, T.A., and Gottesfeld, S. (1991) *J. Electrochem. Soc.*, **138**, 2334–2341.
14. Wang, Y., Mukherjee, P.P., Mishler, J., Mukundan, R., and Borup, R.L. (2010) Cold start of polymer electrolyte fuel cells: three-stage startup characterization. *Electrochim. Acta*, **55**, 2636–2644.
15. Chen, K.S., Hickner, M.A., Siegel, N.P., Noble, D.R., and Pasagullari, U. (2006) Final Report on LDRD Project: Elucidating Performance of Proton-Exchange-Membrane Fuel Cells via Computational Modeling with Experimental Discovery and Validation. Sandia Technical Report No. SAND2006-6964, Sandia National Laboratories, Albuquerque, NM.
16. Chen, K.S. and Hickner, M.A. (2004) A new constitutive model for predicting proton conductivity in polymer electrolytes. Proceedings of 2004 International Mechanical Engineering Congress and Exhibits.

17. Weber, A.Z. and Newman, J. (2003) Transport in polymer-electrolyte membranes I. Physical model. *J. Electrochem. Soc.*, **150** (7), A108–A115.
18. Weber, A.Z. and Newman, J. (2004) Transport in polymer-electrolyte membranes II. Mathematical model. *J. Electrochem. Soc.*, **151** (2), A311–A325.
19. Weber, A.Z. and Newman, J. (2004) Transport in polymer-electrolyte membranes. III. Model validation in a simple fuel-cell model. *J. Electrochem. Soc.*, **151**, A326.
20. Motupally, S., Becker, A.J., and Weidner, J.W. (2000) *J. Electrochem. Soc.*, **147**, 3171.
21. Zawodzinski, T.A., Davey, J., Valerio, J., and Gottesfeld, S. (1995) *Electrochim. Acta*, **40**, 297–302.
22. Wang, Y. (2008) Modeling of two-phase transport in the diffusion media of polymer electrolyte fuel cells. *J. Power Sources*, **185**, 261–271.
23. Mukherjee, P.P. and Wang, C.Y. (2006) Stochastic microstructure reconstruction and direct numerical simulation of the PEFC catalyst layer. *J. Electrochem. Soc.*, **153**, A840.
24. Chen, J., Wang, Y., and Mukherjee, P.P. (2008) One dimensional analysis of sub-zero start-up for polymer electrolyte fuel cells. *ECS Trans.*, **16** (2), 273–284.
25. Kim, S.H. and Pitsch, H. (2009) Reconstruction and effective transport properties of the catalyst layer in PEM fuel cells. *J. Electrochem. Soc.*, **156** (6), B673–B681.
26. Djilali, N. and Sui, P.C. (2008) Transport phenomena in fuel cells: from microscale to macroscale. *Int. J. Comput. Fluid Dyn.*, **22**, 115–133.
27. Mukherjee, P.P. and Wang, C.Y. (2007) Direct numerical simulation modeling of bilayer cathode catalyst layers in polymer electrolyte fuel cells. *J. Electrochem. Soc.*, **154** (11), B1121–B1131.
28. Wang, Y. and Feng, X. (2008) Analysis of reaction rates in the cathode electrode of polymer electrolyte fuel cells. Part I: single-layer electrodes. *J. Electrochem. Soc.*, **155** (12), B1289–B1295.
29. Wang, Y. and Wang, C.Y. (2005) Modeling polymer electrolyte fuel cells with large density and velocity changes. *J. Electrochem. Soc.*, **152**, A445–A453.
30. Wang, Y. and Wang, C.Y. (2005) Transient analysis of polymer electrolyte fuel cells. *Electrochim. Acta*, **50**, 1307–1315.
31. Harvey, D., Pharoah, J.G., and Karan, K. (2008) A comparison of different approaches to modelling the PEMFC catalyst layer. *J. Power Sources*, **179** (1), 209–219.
32. Ralph, T.R. Hards, G.A., Keating, J.E., Campbell, S.A., Wilkinson, D.P., Davis, M., St-Pierre, J., and Johnson, M.C., (1997) Low cost electrodes for proton exchange membrane fuel cells. *J. Electrochem. Soc.*, **144** (11), 3845–3857.
33. Wang, Y., Wang, C.Y., and Chen, K.S. (2007) Elucidating differences between carbon paper and carbon cloth in polymer electrolyte fuel cells. *Electrochim. Acta*, **52**, 3965–3975.
34. Wang, Y., Cho, S., Thiedmann, R., Schmidt, V., Lehnert, W., and Feng, X. (2010) Stochastic modeling and direct simulation of the diffusion media for polymer electrolyte fuel cells. *Int. J. Heat Mass Transfer*, **53**, 1128–1138.
35. Nam, J.H. and Kaviany, M. (2003) Effective diffusivity and water-saturation distribution in single- and two-layer PEMFC diffusion medium. *Int. J. Heat Mass Transfer*, **46** (24), 4595–4611.
36. Sinha, P.K., Mukherjee, P.P., and Wang, C.Y. (2007) Impact of GDL structure and wettability on water management in polymer electrolyte fuel cells. *J. Mater. Chem.*, **17** (30), 3089–3103.
37. Mukherjee, P.P., Wang, C.Y., and Kang, Q. (2009) Mesoscopic modeling of two-phase behavior and flooding phenomena in polymer electrolyte fuel cells. *Electrochim. Acta*, **54** (27), 6861–6875.
38. Shan, X. and Chen, H. (1993) *Phys. Rev. E*, **47**, 1815.
39. Shan, X. and Chen, H. (1994) *Phys. Rev. E*, **49**, 2941.
40. Bhatnagar, P., Gross, E., and Krook, M. (1954) *Phys. Rev.*, **94**, 511.
41. Wang, X. and Nguyen, T.V. (2010) Modeling the effects of the microporous layer on the net water transport rate across the membrane in a PEM fuel cell. *J. Electrochem. Soc.*, **157**, B496–B505.

42. Hinebaugh, J., Fishman, Z., and Bazylak, A. (2010) Proceedings of Fuel Cell 2010, the ASME 2010 Eighth International Fuel Cell Science, Engineering and Technology Conference, Brooklyn, NY, 14–16 June 2010, paper 33097.
43. Wang, Y. and Chen, K.S. (2010) Elucidating through-plane liquid water profile in a polymer electrolyte membrane fuel cell. ECS Trans., 33 (1), 1605–1614.
44. Weber, A.Z. and Newman, J. (2005) Effects of microporous layers in polymer electrolyte fuel cells. J. Electrochem. Soc., 152 (4), A677–A688.
45. Pasaogullari, U., Wang, C.Y., and Chen, K.S. (2005) Two-phase transport in polymer electrolyte fuel cells with bilayer cathode gas diffusion media. J. Electrochem. Soc., 152 (8), A1574–A1582.
46. Wang, Y. and Chen, K.S. (2010) Predicting through-plane water distribution in the MEA–GDL components of a PEM fuel cell. Proceedings of Fuel Cell 2010, the ASME 8th International Fuel Cell Science, Engineering and Technology Conference, Brooklyn, NY, 14–16 June 2010, paper 33029.
47. Davey, J. R., Rockward, T., Spendelow, J.S., Pivovar, B., and Hussey, D.S., Jacobson, D.L., and Arif, M., (2007) Imaging of water profiles in PEM fuel cells using neutron radiography: effect of operating conditions and GDL composition. ECS Trans., 411 (1), 11–422.
48. Hickner, M.A., Siegel, N.P., Chen, K.S., Hussey, D.S., Jacobson, D.L., and Arif, M., (2008) In situ high-resolution neutron radiography of cross-sectional liquid water profiles in proton exchange membrane fuel cells. J. Electrochem. Soc., 155 (4), B427–B434.
49. Ji, Y., Luo, G., and Wang, C.Y. (2010) Computer simulation of liquid water transport at pore level in MPL and GDL and their interface. Presented at the ASME 8th International Fuel Cell Science, Engineering and Technology Conference, Brooklyn, 14–16 June 2010.
50. Ohn, H., Nguyen, T., Hussey, D., Jacobson, D., and Arif, M., (2006) Capillary pressure properties of gas diffusion materials used in PEM fuel cells. ECS Trans., 1 (6), 481–489.
51. Nguyen, T.V., Lin, G., Ohn, H., Wang, X., Hussey, D.S., Jacobson, D.L., and Arif, M. (2006) Measurements of two-phase flow properties of the porous media used in PEM fuel cells. ECS Trans., 3 (1), 415–423.
52. Fairweather, J.D., Cheung, P., St-Pierre, J., and Schwartz, D.T. (2007) A microfluidic approach for measuring capillary pressure in PEMFC gas diffusion layers. Electrochem. Commun., 9 (9), 2340–2345.
53. Sole, J.D. and Ellis, M.W. (2008) Determination of the relationship between capillary pressure and saturation in PEMFC gas diffusion media. Proceedings of Fuel Cell 2008, the Sixth International Conference on Fuel Cell Science, Engineering and Technology, Denver, CO.
54. Inoue, G., Matsukuma, Y., and Minemoto, M. (2006) Effect of gas channel depth on current density distribution of polymer electrolyte fuel cell by numerical analysis including gas flow through gas diffusion layer. J. Power Sources, 157 (1), 136–152.
55. Wang, Y. (2009) Porous-media flow fields for polymer electrolyte fuel cells, I. Low humidity operation. J. Electrochem. Soc., 156 (10), B1124–B1133.
56. Wang, X.D., Duan, Y., Yan, W., Lee, D., Su, A., and Chi, P. (2009) Channel aspect ratio effect for serpentine proton exchange membrane fuel cell: role of sub-rib convection. J. Power Sources, 193 (2), 684–690.
57. Wang, Y., Basu, S., and Wang, C.Y. (2008) Modeling two-phase flow in PEM fuel cell channels. J. Power Sources, 179, 603–617.
58. Basu, S., Li, J., and Wang, C.Y. (2009) Two-phase flow and maldistribution in gas channels of a polymer electrolyte fuel cell. J. Power Sources, 187 (2), 431–443.
59. Basu, S., Wang, C.Y., and Chen, K.S. (2009) Two-phase flow maldistribution and mitigation in polymer electrolyte fuel cells. J. Fuel Cell Sci. Technol., 6 (3), 031007.
60. Wang, Y. (2009) Porous-media flow fields for polymer electrolyte fuel cells, II. Analysis of channel two-phase

flow. *J. Electrochem. Soc.*, **156** (10), B1134–B1141.
61. Cho, S. and Wang, Y. Two-phase flow characteristics in a gas flow channel of polymer electrolyte fuel cells I: carbon paper diffusion media, submitted.
62. Zhu, X., Sui, P.C., and Djilali, N. (2007) Dynamic behavior of liquid water emerging from a GDL pore into a PEMFC gas flow channel. *J. Power Sources*, **172**, 287.
63. Corey, A.T., (1954) Producers Monthly, **19**, p. 38.
64. Fourar, M. and Lenormand, R. (1998) Paper SPE 49006 presented at the SPE Annual Technical Conference and Exhibition, New Orleans, LA.
65. Anderson, R., Zhang, L., Ding, Y., Blanco, M., Bi, X., and Wilkinson, D.P. (2010) A critical review of two-phase flow in gas flow channels of proton exchange membrane fuel cells. *J. Power Sources*, **195**, 4531–4553.
66. Kumbur, E.C., Sharp, K.V., and Mench, M.M. (2006) Liquid droplet behavior and instability in a polymer electrolyte fuel cell flow channel. *J. Power Sources*, **161**, 333.
67. Chen, K.S., Hickner, M.A., and Noble, D.R. (2005) Simplified models for predicting the onset of liquid water droplet instability at the gas diffusion layer/gas flow channel interface. *Int. J. Energy Res.*, **29** (12), 1113–1132.
68. Meng, H. and Wang, C.Y. (2005) Model of two-phase flow and flooding dynamics in polymer electrolyte fuel cells. *J. Electrochem. Soc.*, **152** (9), A1733–A1741.
69. Wang, Y., Wang, C.Y., and Chen, K.S. (2007) Elucidating differences between carbon paper and carbon cloth in polymer electrolyte fuel cells. *Electrochim. Acta*, **52** (12), 3965–3975.
70. Chen, K.S. (2008) Modeling water-droplet detachment from GDL/channel interface in PEM fuel cells. Proceedings of Fuel Cell 2008, the Sixth International Conference on Fuel Cell Science, Engineering and Technology, Denver, CO.
71. Hirt, C.W. and Nichols, B.D. (1981) Volume of fluid (VOF) method for the dynamics of free boundaries. *J. Comput. Phys.*, **39** (1), 201–225.
72. Wilkinson, D.P. and Vanderleeden, O. (2003) in *Handbook of Fuel Cells: Fundamentals, Technology and Applications* (eds. W. Vielstich, H. Gasteiger, and A. Lamm), John Wiley & Sons, Ltd., Chichester, Chapter 27, 315.
73. EG & G Technical Services (2004) *Fuel Cell Handbook*, vol. 7, US Department of Energy, Office of Fossil Energy, National Energy Technology Laboratory.
74. Li, X. and Sabir, I. (2005) Review of bipolar plates in PEM fuel cells: flow-field designs. *Int. J. Hydrogen Energy*, **30** (4), 359–371.
75. Wang, Y. and Wang, C.Y. (2006) Ultra large-scale simulation of polymer electrolyte fuel cells. *J. Power Sources*, **153** (1), 130–135.
76. Yu, S.H., Sohn, S., Nam, J.H., and Kim, C.J. (2009) Numerical study to examine the performance of multi-pass serpentine flow-fields for cooling plates in polymer electrolyte membrane fuel cells. *J. Power Sources*, **194** (2), 697–703.
77. Inoue, G., Yoshimoto, T., Matsukuma, Y., Minemoto, M., Itoh, H., and Tsurumaki, S. (2006) Numerical analysis of relative humidity distribution in polymer electrolyte fuel cell stack including cooling water. *J. Power Sources*, **162** (1), 81–93.
78. Weber, A.Z. and Hickner, M.A. (2008) Modeling and high-resolution-imaging studies of water-content profiles in a polymer-electrolyte-fuel-cell membrane-electrode-assembly. *Electrochim. Acta*, **53**, 7668–7674.
79. Wang, Y. and Chen, K.S. (2011) Effect of spatially-varying GDL properties and land compression on water distribution in PEM fuel cells.. *J. Electrochem. Soc.*, **158** (11), B1292–B1299.
80. Tabuchi, Y., Shiomi, T., Aoki, O., Kubo, N., and Shinohara, K. (2010) Effects of heat and water transport on the performance of polymer electrolyte membrane fuel cell under high current density operation. *Electrochim. Acta*, **56**, 352–360.
81. Borup, R. (2009) Water transport exploratory studies. Presented at 2009

DOE Hydrogen Program Review, 18–22 May 2009.

82. Wang, Y. and Chen, K.S. (2010) Through-plane water distribution in a polymer electrolyte fuel cell: comparison of numerical prediction with neutron radiography data. *J. Electrochem. Soc.*, **157** (12), B1878–B1886.

83. Hartnig, C., Manke, I., Kuhn, R., Kardjilov, N., Banhart, J., and Lehnert, W. (2008) *Appl. Phys. Lett.*, **92**, 134106.

31
Modeling of Polymer Electrolyte Membrane Fuel Cells and Stacks

Yun Wang and Ken S. Chen

31.1
Introduction

The individual components must be assembled together in order to convert energy continuously and efficiently. As a result, the multi-physics phenomena that occur in each component are coupled with those in the adjacent components via their interfaces. Figure 31.1a displays schematically the connections of the various polymer electrolyte fuel cell (PEFC) components and indicates several major phenomena. To describe the multi-physics transport mechanisms and electrochemical reactions so as to predict fuel-cell performance, a set of highly coupled and nonlinear governing equations are usually required. In practice, a number of fuel cells are normally placed in a stack (Figure 31.1b) to generate the desired power output. The cell–cell interactions can greatly affect the stack performance [1].

Using the schematic of phenomena depicted in Figure 31.1a as an example, air is injected into the cathode gas flow channel (GFC). Oxygen in air diffuses toward the catalyst layer (CL) via the gas diffusion layer (GDL), where it combines with electrons and protons to produce water. The electrons and protons, produced by the hydrogen oxidation reaction, are transported from the anode CL via the GDL-bipolar plate (BP) and membrane, respectively, to the cathode CL. The product water can either be transported to rehydrate the anode through the membrane or enter the cathode GFC through the GDL. Waste heat produced from the electrochemical reactions in the CLs during fuel-cell operation is transferred towards the BP via the GDL and is further removed by the cooling flow. At the stack level, reactant flow fields share the same inlet and outlet manifolds; consequently, local blockage in one fuel cell will cause bypass reactant flows in other cells. In addition, electrons produced in individual cells are transported through all the cells in the stack. One aspect that may be of practical interest is what happens when a particular cell malfunctions.

To distinguish from the previous chapter, which focuses on modeling at the component level, that is, emphasizing the differences between components and describing phenomena in a particular component or components, this chapter

Fuel Cell Science and Engineering: Materials, Processes, Systems and Technology,
First Edition. Edited by Detlef Stolten and Bernd Emonts.
© 2012 Wiley-VCH Verlag GmbH & Co. KGaA. Published 2012 by Wiley-VCH Verlag GmbH & Co. KGaA.

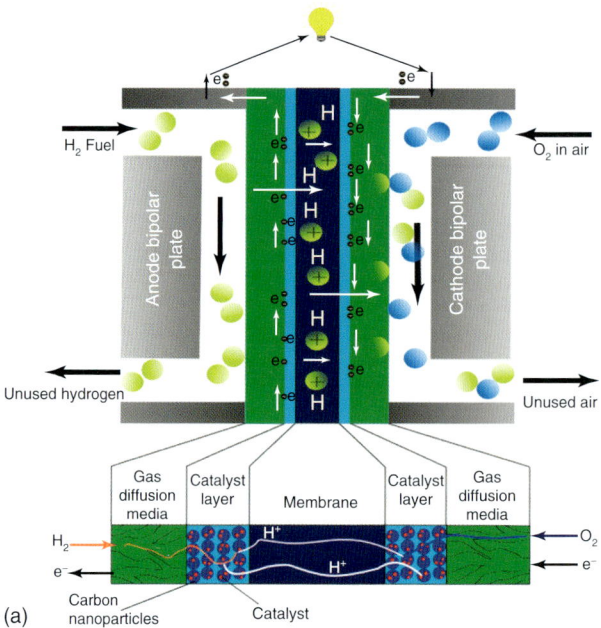
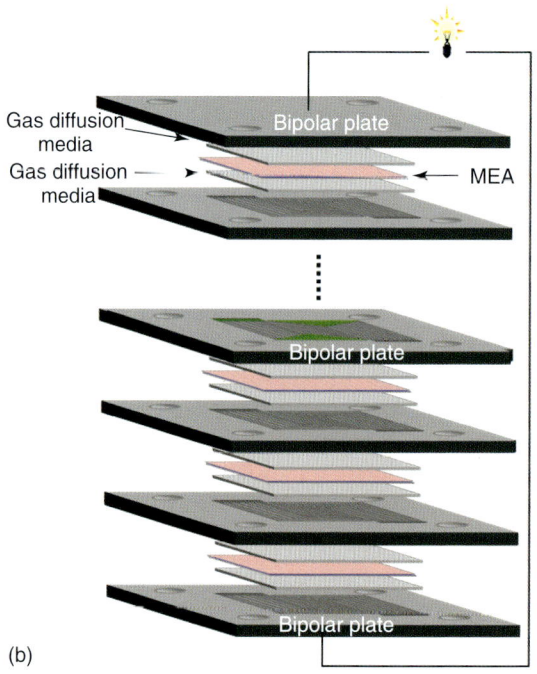

Figure 31.1 (a) Schematic diagram of a PEFC and coupled phenomena among components and (b) schematic diagram of a fuel-cell stack and each cell.

aims at providing a review at the two levels with larger length scales, namely the cell level, which takes into account contributions from all essential components and their interactions or coupling, and the stack level, which accounts for interactions or coupling among all cells. The focus is placed on the macroscopic or continuum models, which can be directly applied to predict cell performance and optimize fuel-cell design and operation and also explore fundamentals. The cell- and stack-level models are the natural extension from that at the component level presented in the previous chapter. Therefore, there will unavoidably be some overlaps in the model description. Those overlaps are minimized as much as possible by citing appropriate texts presented in the previous chapter. It is also worth mentioning that previously cited reviews by Wang [2], Weber and Newman [3], Gauru and Mann [4], Siegel [5], Jiao and Li [6], and Djilali [7] cover the topics of cell- and stack-level studies to some extent. To distinguish from these previous studies, this chapter includes newly developed approaches such as cold start analysis, two-phase channel flow, phase change, two-phase transient, and detailed model validation, providing a focused and unique review on PEFC modeling at the cell and stack levels. In terms of time frame, the present review focuses its scope on recent work published from 2004 to early 2011.

31.2
Cell-Level Modeling and Simulation

A polymer electrolyte membrane fuel cell (PEMFC) consists of various components as discussed in the previous chapter, therefore individual component submodels can be combined or coupled to form a cell-level model for simulating full fuel-cell operation. In fact, even in the very early stages, most modeling studies were targeted at the cell level, in view of the importance of predicting fuel-cell performance. The component-level models can be unified by employing the following governing equations at the cell level [1–3, 8]:

Continuity equation: $\quad \varepsilon \dfrac{\partial \rho}{\partial t} + \nabla \cdot (\rho \vec{u}) = S_m \quad$ (31.1a)

Momentum conservation: $\quad \dfrac{1}{\varepsilon}\left[\dfrac{\partial \vec{u}}{\partial t} + \dfrac{1}{\varepsilon}\nabla \cdot (\vec{u}\vec{u})\right] = -\nabla\left(\dfrac{p}{\rho}\right) + \nabla \cdot \tau + S_u \quad$ (31.1b)

Energy conservation: $\quad \dfrac{\partial \overline{\rho c_p} T}{\partial t} + \nabla \cdot (\gamma_T \rho c_p \vec{u} T) = \nabla \cdot \left(k^{\text{eff}} \nabla T\right) + S_T \quad$ (31.1c)

Species conservation $(H_2O/H_2/O_2)$: $\quad \varepsilon^{\text{eff}} \dfrac{\partial C_k}{\partial t} + \nabla \cdot (\gamma_c \vec{u} C_k)$

$$= \nabla \cdot \left(D_k^{g,\text{eff}} \nabla C_k^g\right) - \nabla \cdot \left[\left(\dfrac{mf_l^k}{M^k} - \dfrac{C_k^g}{\rho_g}\right)\vec{j}_l\right] + S_k \quad (31.1d)$$

Charge conservation (electrons): $\quad 0 = \nabla \cdot \left(\sigma_s^{\text{eff}} \nabla \Phi^{(s)}\right) + S_{\phi_s} \quad$ (31.1e)

Charge conservation (protons): $\quad 0 = \nabla \cdot \left(\sigma_m^{\text{eff}} \nabla \Phi^{(m)}\right) + S_{\phi_e} \quad$ (31.1f)

Table 31.1 Source Terms for the Conservation Equations in Each Region [1].

	S_m	S_u	$S_k{}^a$	S_ϕ	S_T
GFC	0	0	0	–	S_{fg}
DM	0	$-\dfrac{\mu}{K_{GDL}}\vec{u}$	0	0	$\dfrac{i^{(s)2}}{\sigma_s^{eff}} + S_{fg}$
Catalyst layer	$M^w \nabla(D_w^m \nabla C_w) + \sum_k S_k M^k$	$-\dfrac{\mu}{K_{CL}}\vec{u}$	$-\nabla\left(\dfrac{n_d}{F} i^{(m)}\right) - \dfrac{s_k j}{n_k F}$	j	$j\left(\eta + T\dfrac{dU_0}{dT}\right) + \dfrac{i^{(m)2}}{\sigma_m^{eff}} + \dfrac{i^{(s)2}}{\sigma_s^{eff}} + S_{fg}$
Membrane	0	–	0	0	$\dfrac{i^{(m)2}}{\sigma_m^{eff}}$

Electrochemical reaction:

$\sum_k s_k M_k^z = ne^-$

In PEM fuel cells, there are:

(anode) $H_2 - 2H^+ = 2e^-$

(cathode) $2H_2O - O_2 - 4H^+ = 4e^-$

where

$\begin{cases} M_k \equiv \text{chemical formula of species } k \\ s_k \equiv \text{stoichiometry coefficient} \\ n \equiv \text{number of electrons transferred} \end{cases}$

$^a n_d$ is the electroosmotic drag coefficient for water. For H_2 and O_2, $n_d = 0$.

where ρ is the multiphase mixture density, $\vec{u}$ the superficial fluid velocity vector, p the pressure, C_k/C_w the molar concentration of reactant/water, T the temperature, and $\Phi^{(m)}$ and $\Phi^{(s)}$ the electrolyte and electronic phase potentials, respectively. Table 31.1 lists the expressions for the major source terms. The model parameters and physical properties can be found in Table 30.1 in the previous chapter. The above general form of governing equations with standard diffusion and convection terms fits most commercial CFD (computational fluid dynamics) software, for example, Fluent, COMSOL, CFX, and Star CD.

31.2.1
Dimensionality

Grossly, models reported in the open literature so far can be categorized into three groups based on their dimensionality. The first group is based on the system level without distinguishing individual fuel-cell components. The governing equations are developed based on the conservations of mass and energy over the entire fuel cell in conjunction with the electrochemical reaction kinetics, and only the time dimension is considered. An example of such a lumped approach with zero spatial dimension is given in the following equation for fuel-cell stack energy balance [9]:

$$MC\frac{dT_{stack}}{dt} = \sum_i \dot{q}_i \qquad (31.2)$$

where several mechanisms affecting stack thermal energy, for example, heat production and sensible heating, were taken into account on the right-hand side. The energy balance of a fuel cell during cold start can be expressed as follows [10]:

$$\int_0^{\tau_{T,1}} (E_0 - V_{cell}) dt = \frac{(m_m Cp_m + m_{CL} Cp_{CL} + m_{GDL} Cp_{GDL} + m_{BP} Cp_{BP})(273.15 - sT_0)}{A_m I} \qquad (31.3)$$

where T_0 is the initial cell temperature. Simplified models in this category can be used to obtain analytical solutions, to study fuel-cell dynamics and control, and to combine with other subsystems in a PEFC system.

Models in the second group focus on electrochemical and transport modeling in one or pseudo-two spatial dimensions. Gas flow and density along anode and cathode channels are either ignored or assumed to remain uniform throughout a PEFC. Models in this category are able to distinguish various fuel-cell components and account for their unique properties. Due to the reduced dimension of 1D or pseudo-2D, a detailed or fine computational grid can be implemented, yielding relatively high numerical accuracy in the dimension(s) modeled. In addition, the simplified 1D approach also allows one to elucidate new mechanisms and examine their effects. For example, an analytical solution for the following governing equation that describes liquid transport across the GDL was obtained by Pasaogullari and Wang [11]:

$$(1.417 - 4.24s + 3.789s^2) s^3 \frac{ds}{dx} = \frac{I}{2F} M^w \frac{\nu}{\sigma \cos \theta_c (\varepsilon K)^{\frac{1}{2}}} \qquad (31.4)$$

Wang and coworkers [12–14] obtained solutions to single-, dual-, and multi-layer electrodes. For example, they provided the following explicit solution for single-layer electrodes:

$$\Phi^{(m)} = \frac{RT}{\alpha_c F} \ln\left[\Pi\left(\Delta U^{j\delta}, \bar{x}\right) + 1\right] + \Phi_\delta^{(m)} \text{ and } \frac{j(\bar{x}) - j_\delta}{j_\delta} = \Pi\left(\Delta U^{j\delta}, \bar{x}\right) \qquad (31.5a)$$

where

$$\Pi\left(\Delta U^{j\delta}, \bar{x}\right) = \left\{\tan\left[\pm\left(\frac{\alpha_c F}{2RT}\Delta U^{j\delta}\right)^{\frac{1}{2}}(1-\bar{x})\right]\right\}^2 \text{ and } \bar{x} = \frac{x}{\delta} \qquad (31.5b)$$

Models in the third group are based primarily on the CFD approach, in which two- or three-dimensional solutions can be obtained by solving governing equations that are derived from the laws of conservation of mass, momentum, species, energy, and charge. Models that fully incorporate the multiple transport processes may encounter numerical convergence issues in simulation and therefore require considerable computational memory in large-scale simulations. It has been observed

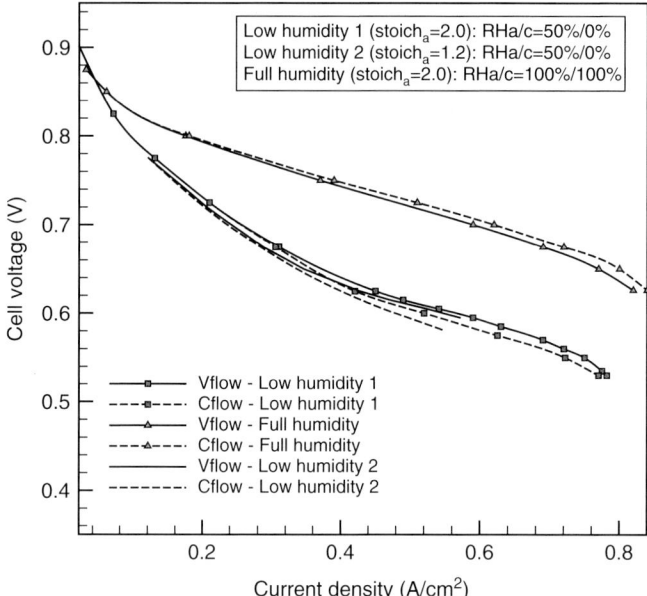

Figure 31.2 Comparison of polarization curves for the two models: the varying flow (Vflow) and constant flow (Cflow) models [15].

that, under normal operating conditions, flow and density fields in gas channels remain approximately invariable. Hence simplifications can be made to neglect the mass source or sink term in the continuity equation and assume a constant gas density in the momentum equation. This approximation permits the decoupling of the flow field from the species, electric potential, and temperature fields, thereby significantly accelerating the computations. A possible inaccuracy introduced by this splitting of the problem may occur on the anode side, however, where the hydrogen concentration profile is relatively unimportant as the anode overpotential is typically negligible. A detailed derivation of the full coupling and evaluation of the inaccuracy due to the simplification were given by Wang and Wang [15]. As an example, Figure 31.2 compares the predictions by the decoupled (Cflow) and fully coupled models (Vflow).

31.2.2
Transient Operation

While most modeling efforts have focused on steady-state operation, the dynamic behavior is of paramount importance for fuel-cell transportation applications due to the inherent load variation involved. Transient phenomena in automotive fuel cells are not yet fully understood. In addition to the complex dynamic response involving various time scales, severe degradation of membrane electrode assemblies has been observed and attributed to transient operation; these include fuel/oxidant starvation, membrane dryout, electrode flooding, and voltage reversal.

In low-humidity operation, water produced from the oxygen reduction reaction (ORR) can help hydrate membrane for better ionic conductance. Therefore, the operation involves profound interactions between water transport and dynamic response [16–18]. First, a time scale for membrane hydration by the reaction water can be estimated by:

$$\tau_{m,H} = \frac{\frac{\rho \delta_m \Delta \lambda}{EW}}{\frac{I}{2F}} \qquad (31.6)$$

Second, the two major mechanisms of water transport through the membrane, the electroosmotic drag and back-diffusion, can create complex transient behavior involving different time scales. For example, during a step change in current density, the electroosmotic drag will immediately remove water from the anode side of the membrane before water back-diffusion from the cathode to anode takes effect. The time constant of water diffusion across a membrane can be evaluated by

$$\tau_{m,D} = \frac{\delta_m^2}{D_m^{\text{eff}}} \approx 0.2 \text{ s at } \lambda = 3 \qquad (31.7)$$

This can cause a temporary dryout on the anode side of the membrane and hence a jump in membrane resistance or a sharp decrease in cell voltage. This voltage drop is, however, recoverable within a period of the time constant that is characteristic of water back-diffusion through the membrane.

In addition, voltage variation involves the electrochemical double-layer dynamics, for example, releasing or accumulating charges. The time constant for this process can be estimated by

$$\tau_{dl} = \delta_{CL}^2 a C \left(\frac{1}{\kappa} + \frac{1}{\sigma} \right) \qquad (31.8)$$

Current density changes can lead to variations in reactant consumption. The diffusion time scale across the GDL ranges from 0.01 s to 0.1 s as estimated by

$$\tau_k = \frac{\delta_{GDL}^2}{D_g^{\text{eff}}} \qquad (31.9)$$

To describe the time-dependent behavior, a transient term should be added to the governing equations, as shown in Eq. (31.1), to account for the storage rates of mass, momentum, species, energy, and charges. In the catalyst layer, both the ionomer phase and void space can hold water, hence an effective factor, ε^{eff}, in Eq. (31.1) can be introduced to simplify the model expression [16]:

$$\varepsilon^{\text{eff}} = \varepsilon_g + \varepsilon_m \frac{dC_w^m}{dC_w} = \varepsilon_g + \varepsilon_m \frac{\rho_m}{EW} \frac{RT}{p^{\text{sat}}} \frac{d\lambda}{da} \qquad (31.10)$$

where ρ_m is the density of a dry membrane. Figure 31.3 plots the effective factor in an operating fuel cell, indicating that the factor can be as high as ~1000.

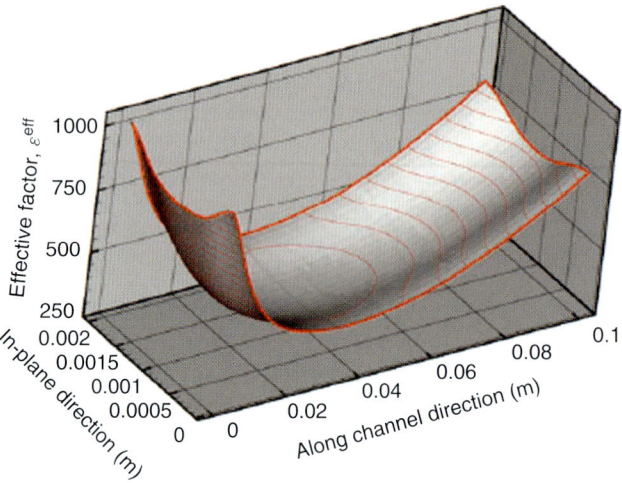

Figure 31.3 The effective factor, ε^{eff}, in Eq. (31.10) in the middle of the membrane, under 0.65 V and relative humidity RHa/c = 100/0% [16].

Figure 31.4 shows a typical variation of current density when altering cell voltage, indicative of complex dynamic responses occurring at different time scales. Another important transient phenomenon is liquid water dynamics in GDLs. The liquid water content in a GDL will change when the operating conditions are altered, either external humidification or current density. The liquid content is important to determine the transport polarization. In addition, residual liquid can degrade fuel-cell components; consequently, understanding dewetting and two-phase transient becomes important particularly for areas that are subject to freezing winter. Figure 31.5 schematically illustrates the drying processes in the typical land-channel structure of a fuel cell. As shown, the GDL dewetting process can be decoupled by two sub-problems, one the in-plane dewetting and the other the through-plane dewetting [18]. Each sub-problem can be described using an ordinary differential equation, and hence an analytical solution is possible. In the through-plane direction, assuming there exist a water supply from the catalyst layer and water vapor removal from the evaporation front, the water balance gives rise to

$$(1+2\alpha)\frac{I}{2F} - D_w^{g,\text{eff}}\frac{\Delta C}{\delta} = \frac{d}{dt}\left[(\delta_{GDL} - \delta)\frac{\varepsilon s_0 \rho_l}{M^w}\right] \tag{31.11}$$

where s_0 is the average liquid water saturation in the two-phase zone prior to drying (or the residual liquid saturation). We define $Y = \delta/\delta_{GDL}$ and rearrange the above equation as

$$(1+2\alpha)\frac{I}{2F} - D_w^{g,\text{eff}}\frac{\Delta C}{\delta_{GDL}}\frac{1}{Y} = \frac{\varepsilon s_0 \rho_l \delta_{GDL}}{M^w}\left(-\frac{dY}{dt}\right) \tag{31.12}$$

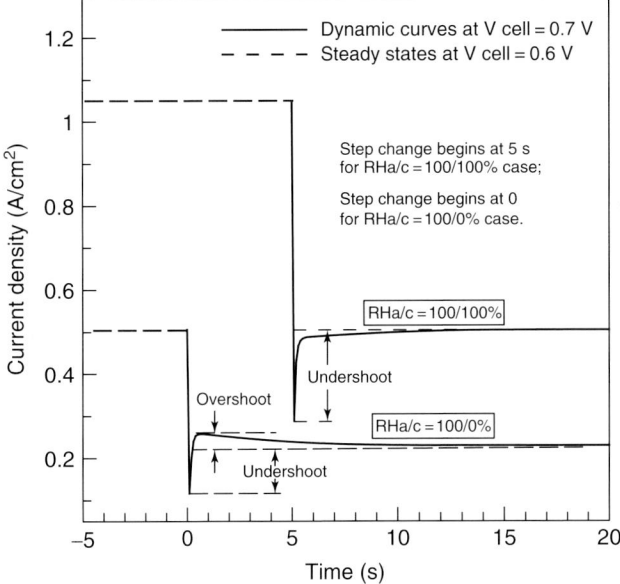

Figure 31.4 Dynamic responses of average current density to the step change of cell voltages from 0.6 to 0.7 V, under RHa/c = 100/100% and 100/0% [16].

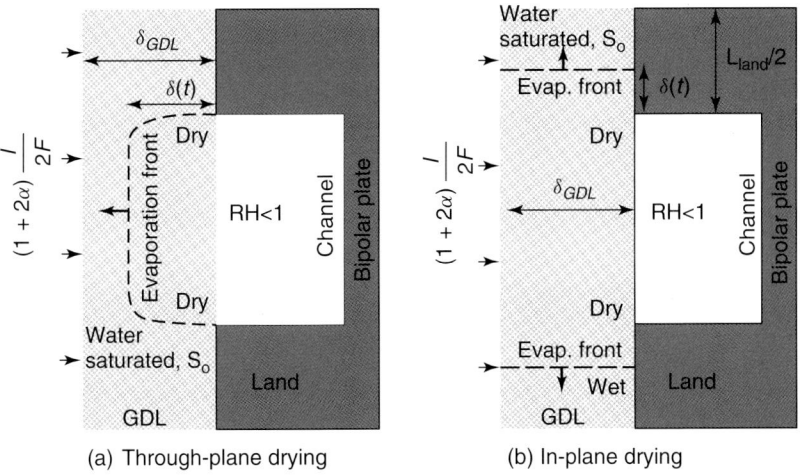

Figure 31.5 Schematics of GDL (a) though-plane and (b) in-plane drying [18].

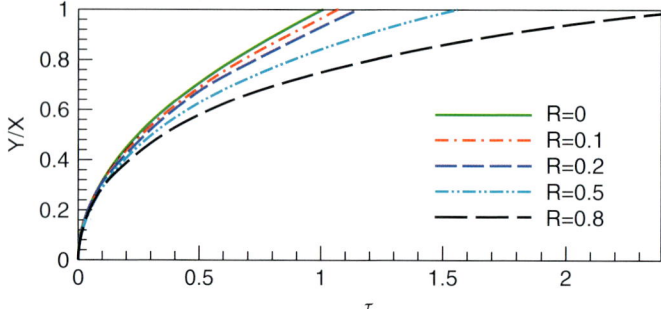

Figure 31.6 Y/X as a function of dimensionless time, τ [18].

By introducing the following three parameters:

$$\tau = \frac{t}{t_1}, t_1 = \frac{\varepsilon s_0 \rho_l (\delta_{GDL})^2}{2 D_w^{g,\text{eff}} M^w \Delta C}, \text{ and } R = \frac{(1+2\alpha) I \delta_{GDL}}{2 F D_w^{g,\text{eff}} \Delta C} \quad (31.13)$$

we reach a simplified equation:

$$R - \frac{1}{Y} = -2 \frac{dY}{d\tau} \quad (31.14)$$

The in-plane transient follows the same equation as above after scaling by different reference time and length scales. Figure 31.6 plots the X/Y as a function of dimensionless time, where X is for in-plane dewetting.

In PEFC operation, the 2D two-phase transient is coupled with the electrochemical reactions and other transport phenomena. Figure 31.7 displays the four liquid dewetting stages and their impacts on the cell voltage, predicted by a 2D transient two-phase model [18].

31.2.3
Nonisothermal Modeling

During PEFC operation, only a fraction of the chemical energy can be converted to electrical energy, with the rest being released in form of waste heat. The waste heat production rate is not small, usually comparable to the electric power generated. The heating comes from various sources: the major mechanisms include the entropic heat of the reactions, the irreversibility of the electrochemical reactions, Joule heating due to ohmic resistance, and latent heat release/absorption during phase changes. Heat transfer occurs in all of the fuel-cell components and is critical to fuel-cell thermal and water management.

In terms of location, most of the heating takes place in the catalyst layer, in particular in the cathode side arising from the relatively large overpotential due to the sluggish ORR. Joule heating in a PEFC is primarily due to the ionic resistance of the electrolyte membrane and catalyst layers. In some cases, the electric resistance in GDLs can be significant, particularly for thin GDLs, which may raise the Joule

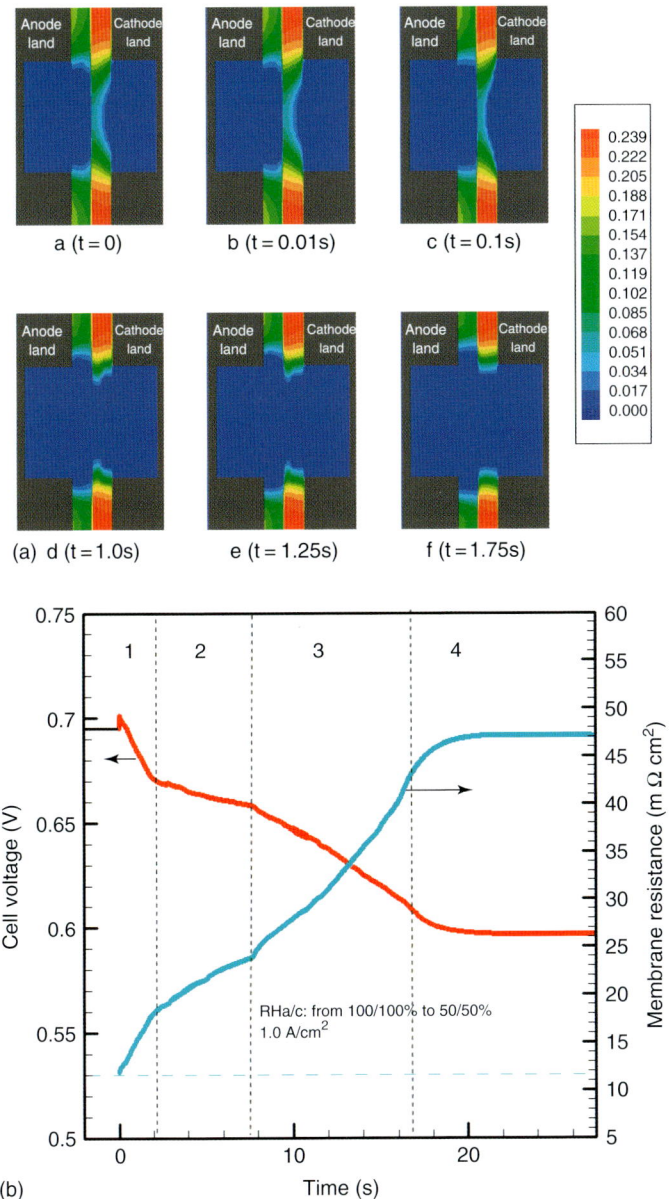

Figure 31.7 Liquid water saturation distributions at time instants of 0, 0.01, 0.1, 1.0, 1.25, and 1.75 s, in transient of changing the RH from 100 to 50% for both sides (a) and time evolutions of cell voltage and membrane resistance (b) [18].

heating due to the electronic resistance. Phase change takes place in most fuel-cell components, including the membrane, catalyst layers, GDLs, and channels. In the membrane, phase change occurs during membrane water absorption/release. In the void space, the vapor–liquid water or liquid–solid water phase change may occur.

Mathematically, the following expression for the entropic heat of reactions, also called *reversible heat release*, can be derived based on the entropic change of the electrochemical reactions [19, 20]:

$$\dot{q}_{\text{rev}} = -T\mathrm{d}s\frac{j}{2F} = jT\frac{\mathrm{d}U_0}{\mathrm{d}T} \tag{31.15}$$

The irreversibility of the electrochemical reactions arises from the reaction overpotential at the interface and can be given by $j\eta$. The Joule heating can be derived by considering the local protonic or electronic current flux, $i^{(m)2}/\sigma_m^{\text{eff}}$ or $i^{(s)2}/\sigma_s^{\text{eff}}$.

The latent heat release/absorption during phase change is determined by the local phase change rate:

$$S_T^{fg} = h_{fg}\dot{m}_{fg} \tag{31.16}$$

where h_{fg} and $\dot{m}_{fg}$ are the latent heat of vapor–liquid phase change and the phase change rate, respectively. The latter is readily calculated from the liquid continuity equation:

$$\dot{m}_{fg} = \nabla \cdot \left(\rho_l \vec{u}_l\right) \tag{31.17}$$

where the liquid-phase velocity in the two-phase mixture model is calculated from

$$\rho_l \vec{u}_l = \vec{j}_l + \lambda^l \rho \vec{u} \tag{31.18}$$

Note that $\dot{m}_{fg}$ is the interfacial mass transfer rate between the liquid and gas phases. In the absence of any external mass source or sink, which holds true in GDLs, the sum of these interfacial mass transfer rates over all the phases is zero. This phase change rate can be affected by several parameters, such as the relative humidity and GDL thermal conductivity. Figure 31.8 compares the phase change rates for different GDL thermal conductivities [21]. Although production and transfer of heats are the same for the two cases, the phase change rate contours show a much higher level of condensation on the cathode land surface for the case with a lower GDL conductivity, indicating a stronger heat pipe effect.

Numerically, the energy conservation equation can be used with the source terms describing the waste heat production. For porous components, the model parameters need to account for the constituent phases; for example, the specific heat capacity is given by

$$\overline{\rho c_p} = (1-\varepsilon)\left(\rho c_p\right)_s + \varepsilon s \left(\rho c_p\right)_l + \varepsilon (1-s) \left(\rho c_p\right)_g \tag{31.19}$$

Figure 31.9 shows an example of the temperature profiles in a fuel cell predicted by a 3D fuel cell model [22].

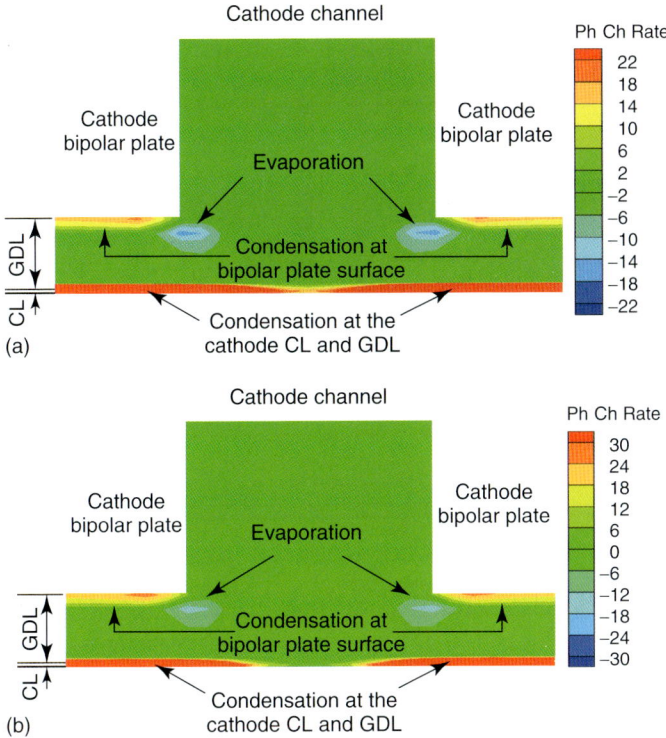

Figure 31.8 Phase change rate (kg m^{-3} s^{-1}) contours in the cathode side, where positive and negative values denote condensation and evaporation, respectively, under conditions of $I = 1.0$ A cm^{-2}, stoichiometry (A/C) = 2.0, and RH = 100% for K_{gdl} = (a) 3.0 and (b) 2.0 W mK^{-1} [21].

31.2.4
Two-Phase Flow

One of the most complex phenomena in PEFCs is two-phase transport, originating from water production by the ORR in the cathode. Liquid may block pore paths of mass transport through porous diffusion media and catalyst layers, thereby reducing the PEFC performance and further leading to cell degradation. Since liquid water formation depends on the vapor saturation pressure, which is a strong function of temperature, heat transfer and its coupling with water condensation and/or evaporation are critical to the study of two-phase transport and the ensuing flooding. Numerically, modeling multi-phase transport in fuel cells is very challenging due to steep gradients in liquid saturation, velocity, and so on. Consequently, two-phase transport has been excluded in most early numerical studies. Pioneering work was conducted by Wang et al. [23], He et al. [24], You and Liu [25], Mazumder and Cole [26], and Berning and Djilali [27]. Most of these modeling

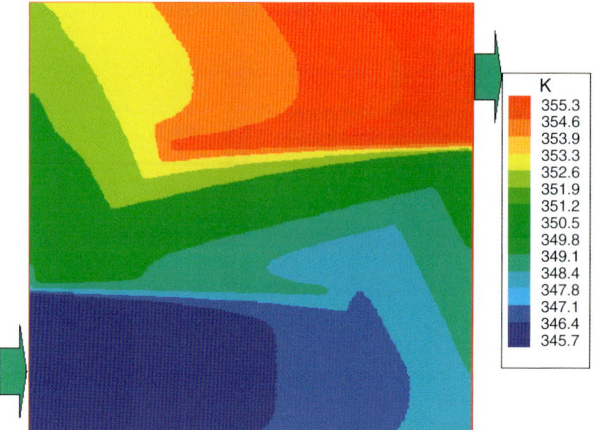

Figure 31.9 Surface temperature at the cathode flow plate for a coolant flow rate of 2.9×10^{-6} m s^{-1} and coolant inlet temperature of 345.15 K (72 °C) under 1.0 A cm^{-2} and 0.616 V [22].

efforts employed the multiphase mixture or M^2 formulation [28], which significantly simplifies the governing equation for two-phase flow, particularly making it possible to incorporate it readily into commercial software through its user-defined function capability. The M^2 formulation allows the establishment of the framework of two-phase flow for many subsequent two-phase fuel-cell models.

In the M^2 formulation, the two-phase mixture density is defined as

$$\rho = s\rho_l + (1-s)\rho_g \tag{31.20}$$

where s is the liquid water saturation. In addition, ρ_g is determined by the constituent species and their contents in the gas, and its values on the anode and cathode sides are quite different. The liquid saturation s is obtained from its relation with the mixture water concentration:

$$s = \begin{cases} 0 & C_w \leq C_{\text{sat}} \\ \dfrac{C_w - C_{\text{sat}}}{\rho_l/M^w - C_{\text{sat}}} & C_w > C_{\text{sat}} \end{cases} \tag{31.21}$$

The interaction of the two phase flows in diffusion media is described through the relative permeabilities, k_r^l and k_r^g, defined as the ratio of the intrinsic permeability of liquid and gas phases, respectively, to the total intrinsic permeability of a porous medium. Physically, these parameters describe the extent to which one fluid is hindered by others in pore spaces, and hence can be formulated as a function of liquid saturation. Most of previous work adopted the following correlations [11, 29]:

$$k_r^l = s^{\tau_k} \text{ and } k_r^g = (1-s)^{\tau_k} \tag{31.22}$$

where the coefficient τ_k is constant and usually set as 3 or 4. In the multiphase region, assuming liquid spreads on the pore wall and follows the solid matrix morphology, the effective gas diffusion coefficient is then as modified by Wang et al. [30]:

$$D^{g,\text{eff}} = [\varepsilon(1-s)]^{\tau_d} D^g \qquad (31.23)$$

Liquid flow also affects species transport and the impact can be included through defining the convection corrector factor γ [28]:

$$\begin{cases} \gamma_k = \frac{\rho \lambda^g}{\rho_g(1-s)} \\ \gamma_w = \frac{\rho}{C_w}\left(\frac{\lambda^l}{M^w} + \frac{\lambda^g}{\rho_g} C_{\text{sat}}\right) \end{cases} \qquad (31.24)$$

where the relative mobilities of individual phases $\lambda^{l/g}$ are defined as

$$\lambda^l = \frac{k_r^l/v_l}{k_r^l/v_l + k_r^g/v_g} \text{ and } \lambda^g = \frac{k_r^g/v_g}{k_r^l/v_l + k_r^g/v_g} \qquad (31.25)$$

The flux $\vec{j}_l$ in the water equation of Eq. (31.1) can be obtained through

$$\vec{j}_l = \frac{\lambda^l \lambda^g}{v} K\left[\nabla P_c + (\rho_l - \rho_g)\vec{g}\right] \qquad (31.26)$$

where the capillary pressure P_c can be evaluated by the Leverett function. The above two-phase equation can be applied to the general porous media, and has been extended to describe the two-phase phenomena in the catalyst layers, microporous layers (MPLs) and GDLs on both the anode and cathode sides, and even gas flow channels [31, 32]. Some also adopted two fluids models. Figure 31.10 shows the liquid water effect on fuel cell performance [33]. Liquid water increases the reactant transport resistance, therefore reducing the fuel-cell performance in the high current density regime where the mass transport polarization becomes dominant.

31.2.5
Cold Start Operation

PEMFCs must have the ability to survive and start up from sub-freezing temperatures, also called *cold start*, to be deployed successfully in automobiles. Under freezing environmental conditions, water produced has a tendency to freeze in open pores in the catalyst layer and GDL, rather than being removed from the fuel cell, thus creating mass transport limitations that eventually result in the shutdown of PEFC operation. Figure 31.11 shows the accumulation of solid water in a fuel cell in sub-freezing operation detected by neutron imaging [34].

Cold start is essentially a transient operation. Research on cold start is relatively new. Hishinuma et al. [35], Wilson et al. [36], McDonald et al. [37], and Cho et al. [38] conducted early studies on cold start. Wang [10] presented theoretical analyses, identifying three stages of cold start. Considering a fuel-cell lumped system, the cell temperature is then proportional to the waste heat production rate. The time scale that it takes to heat the cell to $0\,^\circ\text{C}$ can be evaluated by

$$\tau_{T,1} = \frac{m\bar{c}_p}{I(E_0 - V_{\text{cell}})A_m}(273.15 - T_0) \qquad (31.27)$$

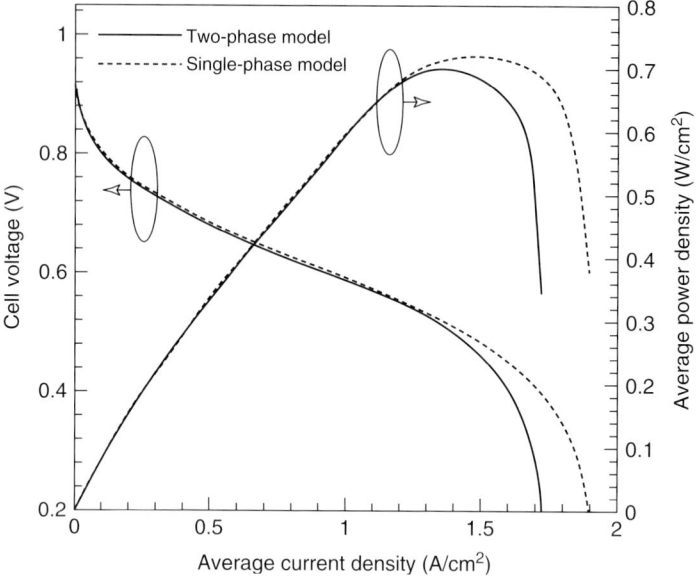

Figure 31.10 Calculated polarization and power curves of two-channel serpentine PEFC at a cell temperature of 80 °C, a pressure of 1.5 atm, fully humidified inlets, and anode:cathode stoichiometry 2 at $1\,A\,cm^{-2}$ [33].

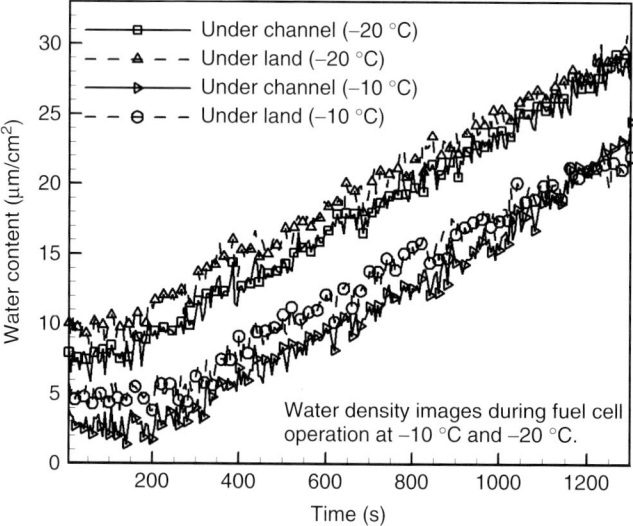

Figure 31.11 Water content evolution from the neutron images during fuel-cell operation at two sub-freezing temperatures, -10 and $-20\,°C$ [34].

where m is the mass of the fuel cell, $\bar{c}_p$ the average specific heat, and A_m the reaction area. The thermal mass of the BPs usually dominates; and there are several BP material choices; for example, graphite has $\rho_{BP}\,c_{p,BP} \approx 1600\,\text{kJ K}^{-1}\,\text{m}^{-3}$, then a δ_{BP} of $\sim 2\,\text{mm}$ yields $\tau_{T,1} \approx 100\,\text{s}$ for cold start at $-30\,°\text{C}$ at $0.1\,\text{A cm}^{-2}$, whereas for stainless steel (e.g., 316 grade), $\rho_{BP}\,c_{p,BP} \approx 4000\,\text{kJ K}^{-1}\,\text{m}^{-3}$, leading to $\tau_{T,1} \approx 300\,\text{s}$. When the cold start current is lowered to $0.04\,\text{A cm}^{-2}$, $\tau_{T,1}$ will increase to over 10 min for the BPs made of stainless steel.

In the regime of small variations in cell voltage, a constant heat generation rate can be assumed, and the cell temperature can be estimated by

$$T = \frac{t}{\tau_{T,1}}(273.15 - T_0) + T_0 \tag{31.28}$$

Water in the cathode catalyst layer may exist in multiple states such as absorbed in the ionomer, as vapor, and as solid phases at sub-freezing conditions. Assuming phase equilibrium, solid water can emerge when the vapor pressure reaches the saturated level. Before that, most water produced from the ORR can be absorbed in the ionomer, which hereby can be defined as the first stage. The second stage is characterized by solid production and ice volume growth within the catalyst layer. The third stage starts when the fuel cell reaches $0\,°\text{C}$, melting the ice produced in the catalyst layer by the waste heat of fuel cells.

In the first stage, the ionomer is able to hold a large amount of water compared with that in the gaseous phase; therefore, the time constant $\tau_{\text{ice},1}$ for the first stage can be estimated by [10]

$$\tau_{\text{ice},1} = \frac{2F\rho_m \varepsilon_m \delta_{CL}(14-\lambda_0)}{EW(1+2\alpha)I} \tag{31.29}$$

The ionomer volumetric content ranges from 13 to 40%, leading to $\tau_{\text{ice},1} \approx 0-30\,\text{s}$ with $\delta_{CL} = 10\,\mu\text{m}$, $\alpha = 0.1$, and $I = 0.1\,\text{A cm}^{-2}$. Note that the above treatment considers an overall water loss of the anode and membrane to the cathode due to the electroosmotic drag, that is, a positive net water transfer coefficient α. In transient, α may be a function of time, as discussed by Wang [10]. At sub-freezing temperatures, the water diffusivity in the membrane becomes very low, which may result in a net water flow to the cathode. This can be qualitatively verified through the isothermal cold start experiment; see Figure 31.12 [34], which shows the high-frequency resistance (HFR) in cold start from -10 to $-20\,°\text{C}$. The initial HFR decrease may be attributed to the water addition on the cathode side. In the second stage, added water is in the solid phase, while the anode side continues to lose water to the cathode, increasing the HFR.

In the second stage, $t > \tau_{\text{ice},1}$, solid water will be produced and the ice volume fraction in the void space, s_{ice}, can be evaluated by:

$$s_{\text{ice}}(t) = \frac{t}{\tau_{\text{ice},2}} - k_\tau \quad \tau_{\text{ice},1} < t \leq \tau_{\text{ice},2} + \tau_{\text{ice},1} \tag{31.30}$$

where $\tau_{\text{ice},2}$ and k_τ are defined as

$$\tau_{\text{ice},2} = \frac{2F\varepsilon_{CL}\rho_{\text{ice}}\delta_{CL}}{(1+2\alpha)M^w I} \quad \text{and} \quad k_\tau = \frac{\tau_{\text{ice},1}}{\tau_{\text{ice},2}} = \frac{\rho_m \varepsilon_m (14-\lambda_0) M^w}{\rho_{\text{ice}}\varepsilon_{CL} EW} \tag{31.31}$$

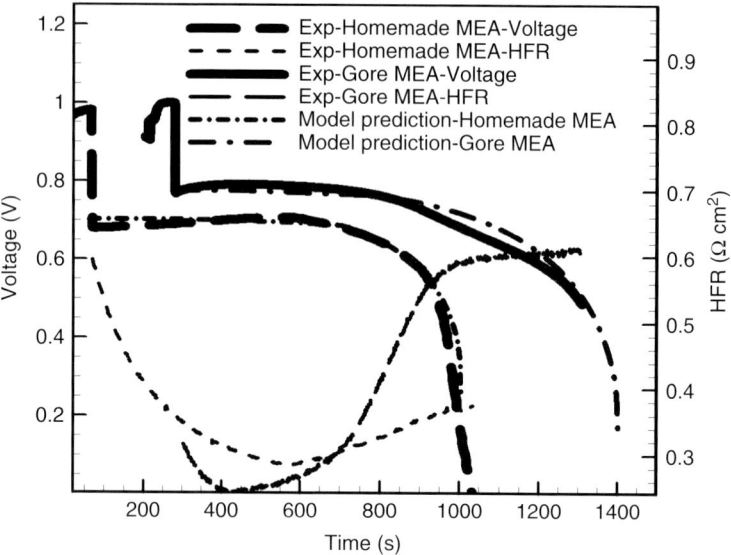

Figure 31.12 Transient response of cell voltages and HFR during sub-freezing operations. All cells are operated at 0.02 A cm^{-2} and $-10\,°$C [39].

Applying the typical ranges of the parameters will lead to $\tau_{ice,2} \approx 50$–150 s at 0.1 A cm^{-2} and k_τ ranging from 0 to 0.35. Note that the physical meaning of $\tau_{s_{ice}}$ is the time of ice fully occupying the void in the cathode catalyst layer (i.e., $s_{ice} = 1$). A dimensionless parameter β_2 can then be defined [10] to compare the time constants for ice formation and temperature increase:

$$\beta_2 = \frac{\tau_{T,1}}{\tau_{s_{ice}}} = \frac{(273.15 - T_0)(1+2\alpha)\bar{c}_p}{2F(E_0 - V_{cell})} \frac{1}{\frac{\rho_m \varepsilon_m (14-\lambda_0)}{\rho_{BP} EW} + \frac{\varepsilon_{CL}\rho_{ice}}{M^W \rho_{BP}}} \frac{\delta_{BP}}{\delta_{CL}} \quad (31.32)$$

In the above, we approximate the thermal mass of a fuel cell to that of the BP, that is, $m\bar{c}_p/A = \rho_{BP}\bar{c}_p\delta_{BP}$. At $\beta_2 < 1$, the fuel cell may start up successfully, otherwise it will fail. For $\beta_2 < 1$, the maximum ice volume fraction s_{ice}^{max} can be estimated by

$$s_{ice}^{max} = \frac{\tau_{T,1}}{\tau_{ice,2}} - k_\tau \quad (31.33)$$

Note that even though physically s_{ice}^{max} must not be >1, the above expression can mathematically extend to $\beta_2 > 1$. In that case, similarly to β_2, the above equation can be used as another criterion to evaluate the cold start operation; the cold start will fail when $s_{ice}^{max} > 1$.

For $\beta_2 < 1$, the fuel cell will experience the third stage, that is, ice melting. This stage is characterized by a constant cell temperature of $0\,°$C. The characteristic time scale $\tau_{T,2}$ can be estimated by comparing the heat generation rate with the melting latent heat:

$$\tau_{T,2} = \frac{\rho_{ice} h_{fusion} \delta_{CL} \varepsilon_{CL}}{I(E_0' - V_{cell})} s_{ice}^{max} \quad (31.34)$$

The ratio of $\tau_{T,2}$ to $\tau_{ice,2}$ is then expressed as

$$\frac{\tau_{T,2}}{\tau_{ice,2}} = \frac{h_{fusion}(1+2\alpha)M^w}{2F(E'_0 - V_{cell})} s_{ice}^{max} \tag{31.35}$$

The third region is usually short, with the above ratio varying from 0 to 0.04. The ice volume fraction in this stage can be expressed as

$$s_{ice} = s\left(\frac{\tau_{T,1}}{\tau_{ice,2}} - k_\tau\right)\frac{\tau_{T,2} + \tau_{T,1} - t}{\tau_{T,2}} \quad \tau_{T,1} < t \le \tau_{T,1} + \tau_{T,2} \tag{31.36}$$

For single-layer electrodes, the reaction rate across the catalyst layer is almost uniform at small $\hbar$s (e.g., low current density, see Figure 30.4), which is usually satisfied for most cold start operations. When oxygen transport becomes a limiting factor due to solid water buildup, the local reaction will differ spatially. In cold start, the reaction rate in the cathode can be expressed as

$$j = -a_0 i_{0,c}^{ref}(1 - s_{ice})^{\tau_a} \exp\left[-\frac{E_a}{R}\left(\frac{1}{T} - \frac{1}{353.15}\right)\right] \frac{C^{O_2}}{C^{O_2,ref}} \exp\left(-\frac{\alpha_c F}{RT}\eta\right) \tag{31.37}$$

The oxygen profile can be approximated by only considering oxygen diffusive transport:

$$\frac{C^{O_2}}{C_{cCL}^{O_2}} = 1 - Da\frac{1-\bar{x}^2}{\varepsilon_{CL}^{\tau_d - \tau_{d,0}}[(1-s_{ice})]^{\tau_d}} \quad \text{where } \bar{x} = 1 - \frac{x_{cCL} - x}{\delta_{CL}} \tag{31.38}$$

where $\bar{x}$ is the dimensionless distance from the interface between the membrane and catalyst layer, and the dimensionless group Da is called the Damköhler number [10]. Severe local oxygen starvation may occur at the side of $\bar{x} = 0$ when C^{O_2} first reaches around zero, that is,

$$1 = Da\frac{1}{\varepsilon_{CL}^{\tau_d - \tau_{d,0}}\left[(1-s_{ice}^{starvation})\right]^{\tau_d}} \tag{31.39}$$

Assuming $\tau_d = \tau_{d,0}$, the above equation can be simplified to

$$s_{ice}^{starvation} = 1 - \sqrt[\tau_d]{Da} \tag{31.40}$$

For $\tau_d = 2$ and $Da \approx 1.0 \times 10^{-4}$, $s_{ice}^{starvation} \approx 99\%$. To evaluate the overpotential, the oxygen profile can be substituted in Eq. (31.37), yielding

$$\eta(s_{ice}, \bar{x}) = -\frac{RT}{\alpha_c F}\ln\left\{\frac{IC_{O_2,ref}}{a_0 i_{0,c}^{ref}\exp\left[-\frac{E_a}{R}\left(\frac{1}{T_0} - \frac{1}{353.15}\right)\right]C_{cCL}^{O_2}\delta_{CL}}\exp\left[\frac{E_a}{R}\left(\frac{1}{T} - \frac{1}{T_0}\right)\right]\right\}$$
$$+ \frac{RT}{\alpha_c F}\ln\left\{(1-s_{ice})^{\tau_a}\left[1 - Da\frac{1-\bar{x}^2}{\varepsilon_{CL}^{\tau_d - \tau_{d,0}}(1-s_{ice})^{\tau_d}}\right]\right\} \tag{31.41}$$

Note that the impacts of ice are all included in the second term on the right while the first accounts primarily for the temperature dependence of the exchange current density. The overpotential can further be written as

$$\eta(s_{ice}, T, \bar{x}) = \eta_0 + \Delta\eta_T + \Delta\eta_{c,1} + \Delta\eta_{c,2} \tag{31.42}$$

where

$$\eta_0 = -\frac{RT}{\alpha_c F} \ln \left\{ \frac{IC^{O_2,\text{ref}}}{a_0 i_{0,c}^{\text{ref}} \exp\left[-\frac{E_a}{R}\left(\frac{1}{T_0} - \frac{1}{353.15}\right)\right] C_{cCL}^{O_2} \delta_{CL}} \right\} \quad (34.43\text{a})$$

$$\Delta \eta_T = -\frac{TE_a}{\alpha_c F}\left(\frac{1}{T} - \frac{1}{T_0}\right) \quad (34.43\text{b})$$

$$\Delta \eta_{c,1} = \frac{RT\tau_a}{\alpha_c F} \ln(1 - s_{\text{ice}}) \quad (34.43\text{c})$$

and

$$\Delta \eta_{c,2} = \frac{RT}{\alpha_c F} \ln\left[1 - Da \frac{1 - \bar{x}^2}{\varepsilon_{CL}^{\tau_d - \tau_{d,0}}(1 - s_{\text{ice}})^{\tau_d}}\right] \quad (34.43\text{d})$$

The above overpotential can be incorporated in a fuel cell model to predict the cell voltage evolution. Figure 31.12 plots the model validation with experimental curve [39].

Three-dimensional cold start models have also been developed based on the framework for modeling fuel-cell operation over 0 °C. In addition to the electrochemical and transport properties, which may deviate from usual correlations such as the membrane ionic conductivity [34], the mass conservation equation must be derived to account for the solid water formation; an example is as follows [40]:

$$\frac{\partial(\varepsilon_{\text{ice}} \rho_{\text{ice}})}{\partial t} + \frac{\partial(\varepsilon \rho)}{\partial t} + \nabla \cdot (\rho \vec{u}) = S_m \quad (31.44)$$

Figure 31.13 shows the solid water saturation distribution [41]. In addition, Jiang et al. [42] employed a model to examine current ramping for rapid start-up from a sub-freezing environment, see Figure 31.14.

31.2.6
Large-Scale Fuel-Cell Simulation

Owing to the structural features, most computational or numerical studies are focused on a single straight flow channel design in the hope of extending the conclusion to a real fuel cell consisting of a number of gas flow channels; see Figure 31.15 for an example of an industry-sized fuel cell with a serpentine flow field. Note that the polarization curve is usually defined based on current density instead of total current. The single-channel approach is preferred in most numerical studies because it is geometrically simple and therefore the computational grid can be easily constructed, and the numerical simulation is more stable. Also, most important phenomena take place in the through-plane direction; therefore, a fine mesh can be set in this dimension for a single-channel approach. Figure 31.16 shows the velocity distribution inside the cross-section of a single-channel fuel cell [15]. However, the single straight-channel approach is unable to capture several

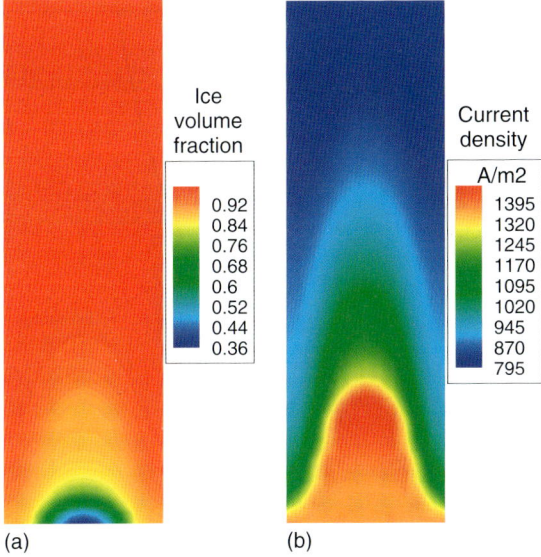

Figure 31.13 (a) Ice volume fraction in the middle plane of the cathode catalyst layer. (b) Current density distribution in the middle plane of the membrane at 30 s or near the end of the cold start. The cell voltage is 0.51 V [41].

phenomena. One is the channel communication due to the pressure difference. Figure 31.17 shows the flow from one channel to another via the GDL under the land [43]. Owing to this bypass flow, it is difficult to remove liquid water in the middle region of the fuel cell. Another is the side effect. Even though gas channels are parallel, it is not totally symmetrical due to the side effect. Figure 31.9 shows the temperature distribution in a fuel cell, predicted by a large-scale simulation involving ~23 million grid points. It is evident that differences exist between the channels in the middle and those near the cell edge. The third phenomenon is the flow maldistribution: Figure 31.18 shows the visualization of liquid water in channels, indicating different two-phase flows among channels due to flow maldistribution [31, 44].

Large-scale simulation can employ the same numerical treatment as that used in the single-channel approach, except that the former usually requires parallel computation. In this regard, the communication among parallel processing nodes will affect the computing efficiency. Meng and Wang [45] were among the first researchers who proposed parallel computing methodology to handle large-scale PEFC simulations involving millions of gridpoints. They benchmarked the parallel computing performance to be more than a sevenfold speed-up with 10 processors. Currently, the capability of large-scale PEFC simulation is available and routine numerical runs can be carried out to optimize fuel-cell design in most major fuel-cell companies and research/national laboratories.

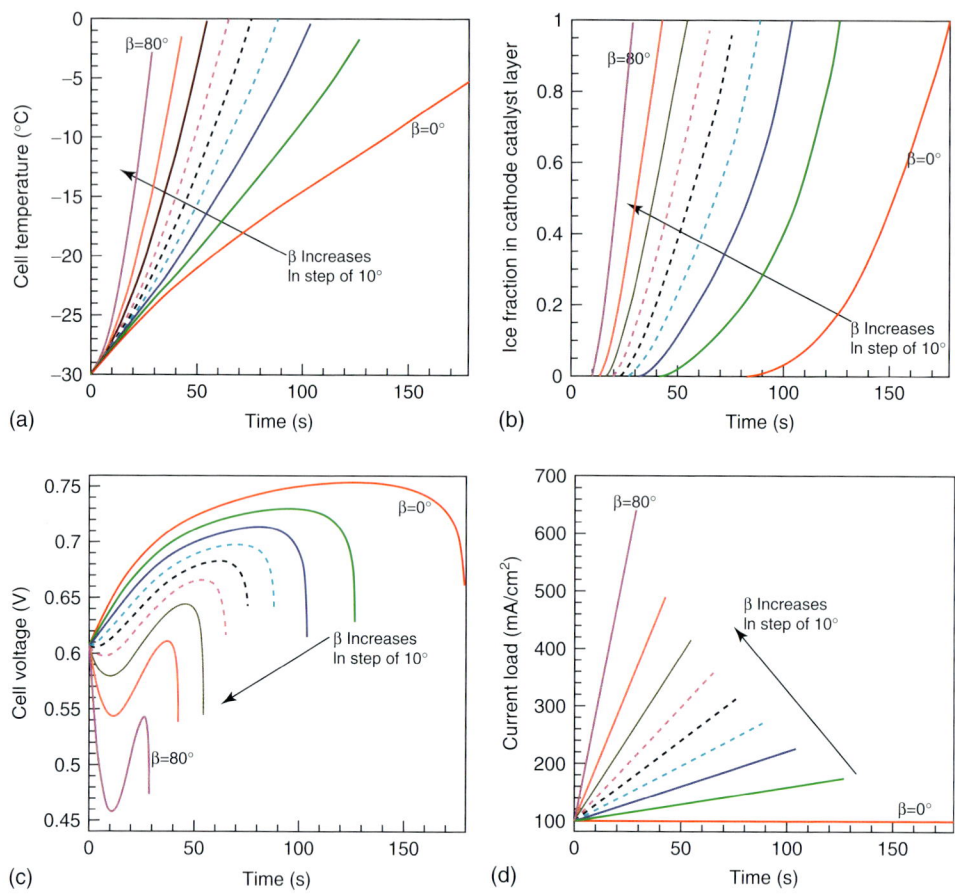

Figure 31.14 Current-ramping start-ups with an initial current density of 100 mA cm^{-2}. (a) Cell temperature versus time, (b) ice fraction in cathode catalyst layer versus time, (c) cell voltage versus time, and (d) current density versus time. Dashed lines indicate successful and solid lines indicate failed start-up cases [42].

31.2.7
Flow Maldistribution

The flow field with parallel channels is a popular design for PEFCs, and it is meant to supply each channel with same amount of reactant. However, owing to asymmetry among channels or two-phase flow, the flow rates may differ among channels. The standard deviation of the normalized flow through the channels can be a good indicator of the performance of the system in distributing reactants uniformly. The lower the standard deviation, the better the system performance is in terms of flow uniformity.

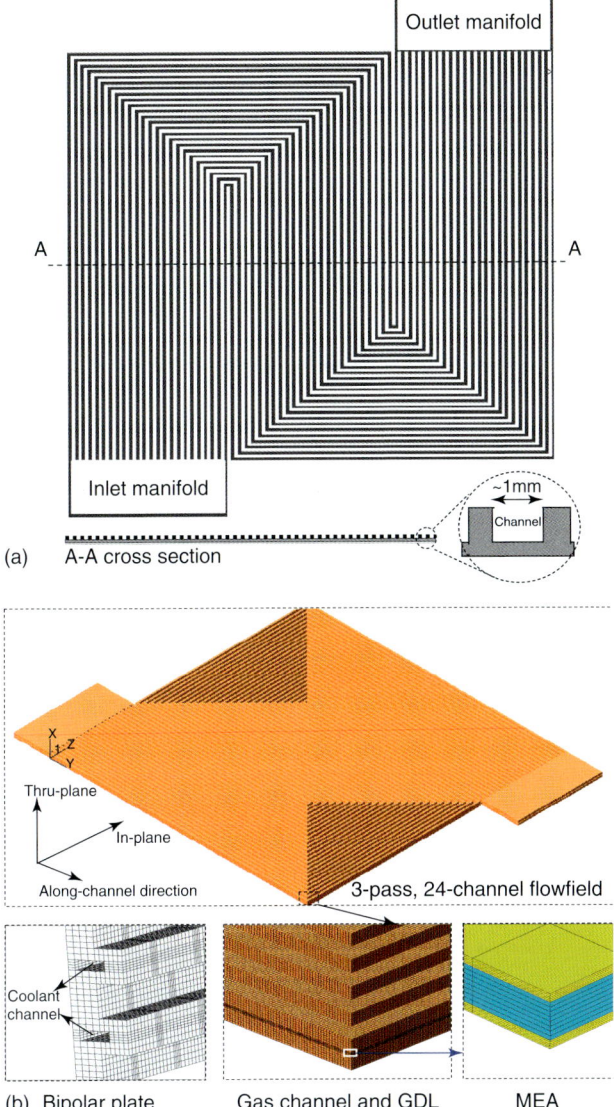

Figure 31.15 (a) Flow field of and (b) computational domain and mesh of a 200 cm^2 PEFC [22].

31.2.7.1 Single-Phase Flow

An important factor leading to asymmetry is the variations in the cross-sectional area among channels. A main cause for this is the GDL intrusion into the channel space. Figure 31.19a presents the velocity contours in flow channels with 20% area maldistribution, which shows the flow maldistribution clearly. The standard deviation of the normalized flow through the channels is 0.315 [46].

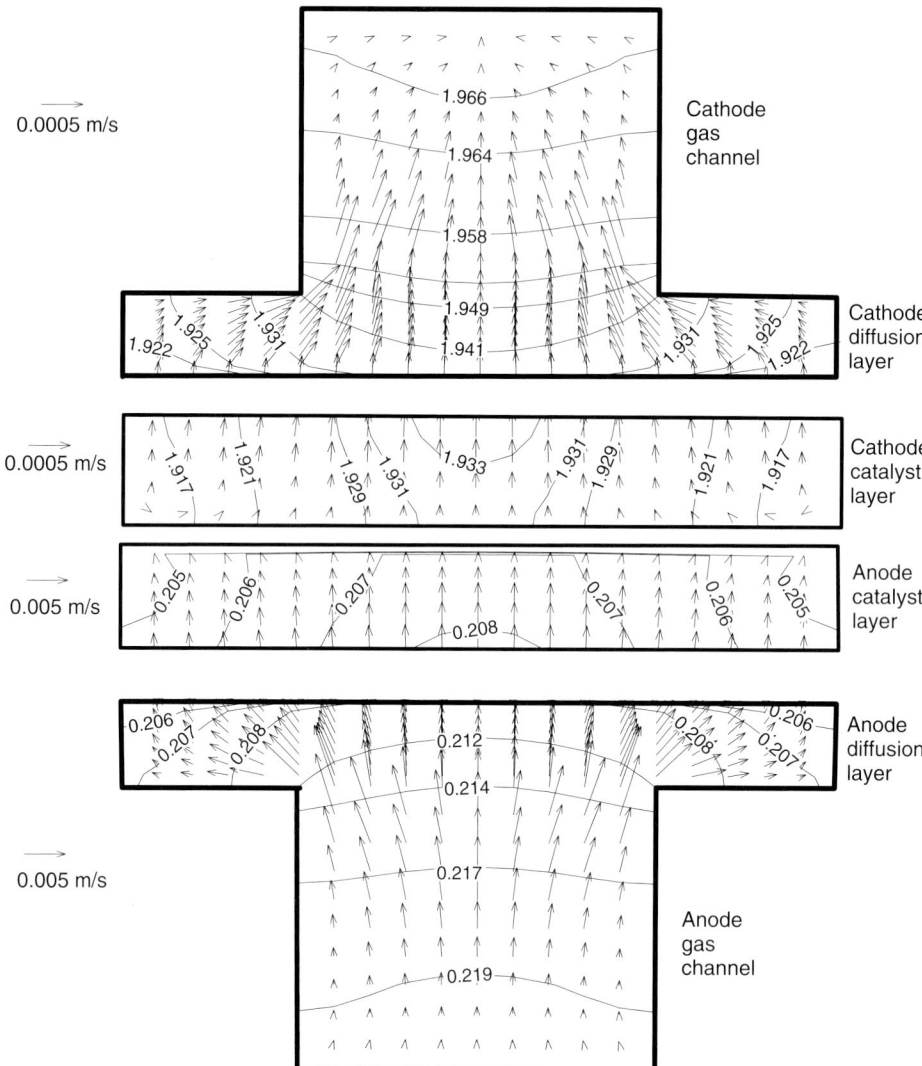

Figure 31.16 Distributions of velocity and density in the cross-section near the inlet, predicted by the variable-flow model for stoich$_a$ = 2.0 and 0.625 V. The catalyst layers are expanded for clarity [15].

31.2.7.2 Two-Phase Flow

When liquid emerges in channels, the liquid may cause flow maldistribution. A major reason for this is that liquid transport is not symmetrical from inlet to outlet; in other words, more liquid appears near the outlet. Therefore, for the parallel channels that share the same headers, the liquid distributions may differ among channels because liquid flows through individual channels plus corresponding

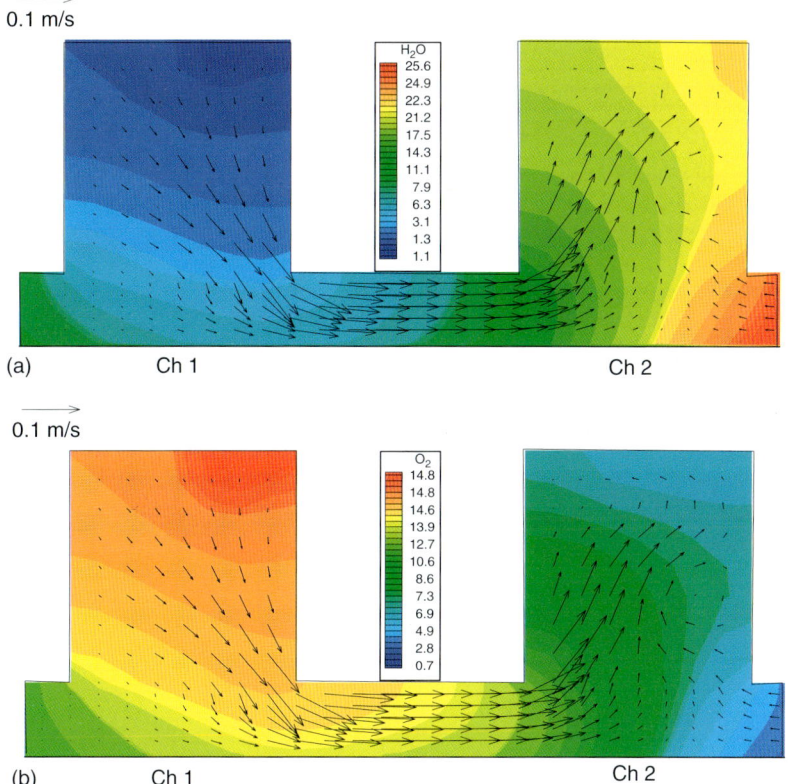

Figure 31.17 (a) H_2O and (b) O_2 distributions and velocity in the cathode side at the mid-length cross-section between inlet and outlet channels (the two parallel serpentine channels are operated in counter-flow pattern) with convection in the GDL at $V_{cell} = 0.65$ V and $I = 0.88$ A cm^{-2} [43].

header portions. Figure 31.19b shows the liquid saturation contours for the case of 0.2 A cm^{-2} and a stoichiometry flow ratio of 4.0 with perfect channels, indicating different liquid contents among channels. It can be seen that due to the liquid in the outlet header, the last channel on the left experiences less gas transport resistance, and hence more gas flow, which in turn reduces the liquid content in the channel and thus the transport resistance.

31.2.8
Model Validation

The cell-level numerical models can directly predict the fuel-cell performance, that is, the power output. The cell performance can be evaluated in two respects: one is the overall polarization curve that plots the cell voltage as a function of the current density, and the other is the local fuel-cell performance as measured

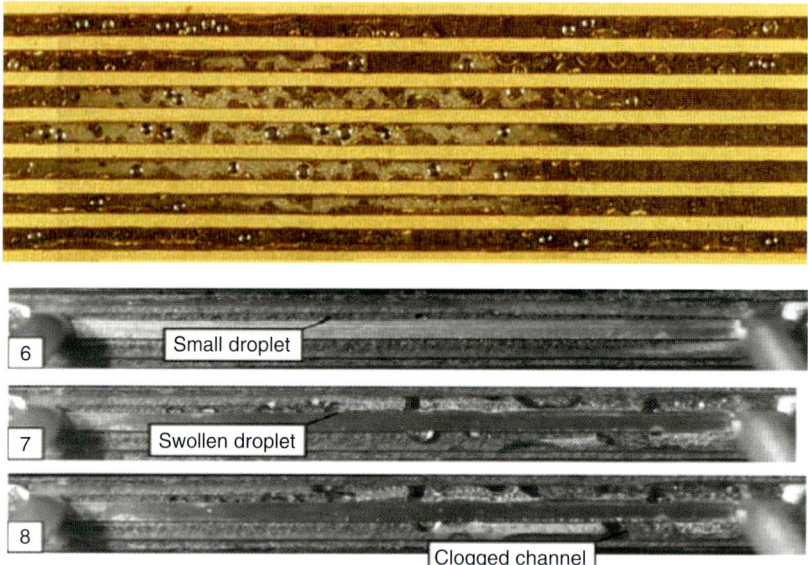

Figure 31.18 Visualization of two-phase flow and flow maldistribution in PEM fuel cell channels: (a) [31] and (b) [44].

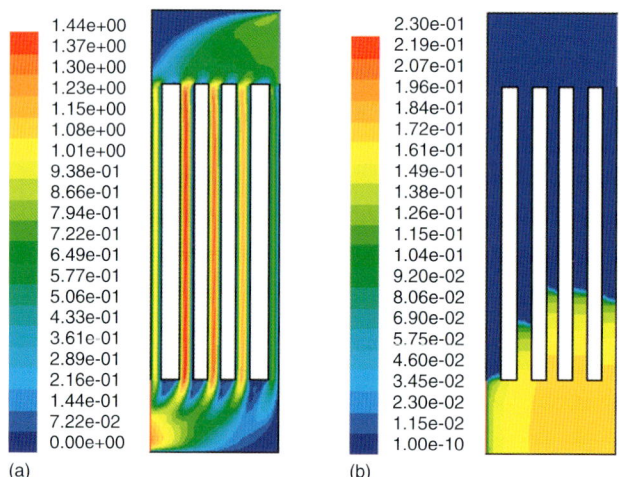

Figure 31.19 (a) Velocity contours (m s^{-1}) in flow channels with 20% area maldistribution predicted by a single-phase model and (b) the liquid saturation distribution without varying cross-sectional area among channels predicted by a two-phase model [46].

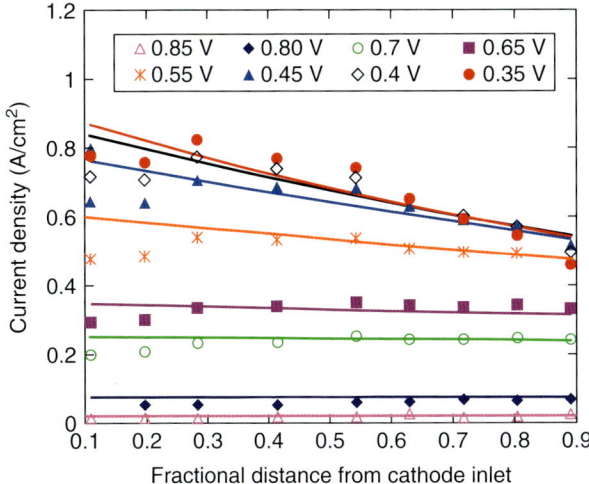

Figure 31.20 Comparison of model prediction (lines) and experimental (symbols) results for cathode stoichiometry of 2.0 at 0.75 A cm^{-2} and full humidification [48].

by the local current density distribution. The first type of performance metrics, namely the overall polarization (or $I-V$) curves, can be obtained by measuring the current density and cell voltage of a fuel cell. The overall polarization curve is important in evaluating the fuel-cell power density and efficiency, and thus overall cell performance. The second type of data are usually obtained by designing segmented fuel cells and measuring the current from each segment.

Even in the early stages, validation was provided to some extent when presenting models to verify their fuel-cell performance prediction. An important function of models is for fuel-cell design and engineering, such as material selection. Therefore, it is important to predict the fuel-cell performance using different materials. Wang et al. [30] validated the model prediction for both the carbon paper and cloth GDLs under various conditions. Some transient validations were also provided [39]; see Figure 31.12. Tabuchi et al. [47] also performed model validation using polarization data.

However, it has been pointed out that the overall polarization curve is insufficient to reflect real fuel-cell operation. Wang [2] explained that the fuel cell may operate with totally different reaction distributions even under the same polarization curve, indicating that a detailed validation based on local current density is more important. Ju and Wang [48] presented pioneering work in this area; see Figure 31.20.

In addition to cell performance, Weber and Newman [49] validated their membrane model using a simplified fuel-cell model in terms of net water flux per proton as a function of current densities, stoichiometries, and cell temperature. This net water flux is determined by a number of factors and its validation must employ cell-level models. Wang and Chen [50] provided validation of their cell-level model in terms of through-plane liquid water profiles across multiple layers, that is, GDL–CL–membrane. Challenges in cell-level model validation include the following. (i) More validations are needed for quantities other than cell voltage and current density, such as species and temperature distributions; in particular, distributions in all three dimensions and all the components are highly desirable; currently, most validation focuses only on one dimension or even zero-dimensional (i.e., polarization data). (ii) Cell-model validation in terms of several quantities at the same time, for example, the local current density and water distributions. Indeed, challenges also exist on the experimental side, namely accurate high-resolution experimental data need to be obtained for detailed model validation. (iii) Cell-model validation in a wide range of operating conditions, such as steady state versus transient, and dry operation versus full humidification. At present, a model can only be validated under very limited operating conditions.

31.3
Stack-Level Modeling and Simulation

31.3.1
Why Is Stack-Level Modeling Needed?

A single fuel cell is only able to produce a voltage up to $\sim$1.2 V (usually 0.6–0.9 V). In practice, for example, regarding automobile fuel cells, hundreds of fuel cells are assembled in a stack, sharing with one or several inlets/outlets through manifolds; see Figure 31.21 for an example of fuel cells and stacks at different scales

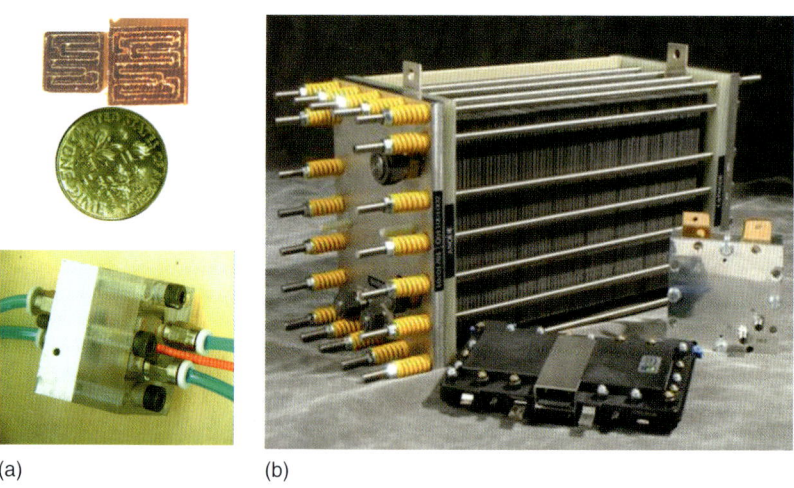

(a)   (b)

Figure 31.21 (a) A 0.1 W fuel cell [51]. (b) A 5 kW fuel cell manufactured by PlugPower (large cell), a 25 W fuel cell (three-cell stack) manufactured by H2 Economy (smaller silver cell), and a 30 W fuel cell manufactured by Avista Laboratories [52].

[51, 52]. At the stack level, water and heat management becomes more complex due to the interactions of constituent cells. The cells can communicate in many ways within a stack. One is the electrical connection, that is, the electrical current flows through all the individual cells connected in series; therefore, the high electronic resistance in one cell will significantly affect the entire stack performance. Another is through-flow field. In practice, several fuel cells share one inlet/outlet manifold in a stack. Therefore, a fuel cell with high flow resistance receives smaller amounts of reactants, causing local reactant starvation and consequent cell performance decay and material degradation. Yet another is heat transfer. A fuel cell exhibiting a larger thermal resistance or exposure to insufficient cooling will be subject to higher temperature operation, disposing of its extra waste heat to neighboring fuel cells. Lastly, the failure of a particular cell in a stack can have undesirable consequences. Extending numerical study from a single fuel cell to a stack is straightforward since it does not involve any new physics, but it is numerically challenging to simulate stacks with a detailed 3D cell-level model owing to the computational burden due to the large number of unknowns involved. Another challenge is to simulate the flow field communication between cells, in particular accounting for two-phase channel flow.

31.3.2
Modeling and Simulation of Fuel-Cell Stacks

Stack models have been attempted by researchers in several studies and most of them were simplified to a large extent. For example, Promislow and Wetton [52] developed a model for describing steady-state thermal transfer in stacks. The model is appropriate for fuel cells with straight coolant channels. It considered averaged quantities in the cross-channel direction, ignoring the effect of the gas and coolant channel geometries. Kim et al. [53] developed an electrical interaction model for stacks and validated it using two types of anomalies. Their unit cells were described by a simple, steady-state, 1 + 1-dimensional model appropriate for fuel cells with straight reactant gas channels. The voltage for each cell, denoted by $V_j(x)$, is derived by local current (through-plane current density i and in-plane length-specific current I):

$$\frac{dI_{j/j+1}(x)}{dx} = i_j(x) - i_{j+1}(x)$$
$$\frac{dV_j(x)}{dx} = \lambda \left[I_{j-1/j}(x) - I_{j/j+1}(x) \right] \tag{31.45}$$

where $\lambda = \frac{\rho_s}{L_t}$ represents the length-specific resistivity of the bipolar plate and j the cell number. Differentiation of the voltage equation yields

$$\frac{d^2 V_j(x)}{dx^2} - \lambda \left[i_{j-1}(x) - 2i_j(x) + i_{j+1}(x) \right] = 0 \tag{31.46}$$

This equation was solved by considering the modified Butler–Volmer relationship and boundary conditions, with modifications to voltage and current balances at the end plates. The channel oxygen concentrations change due to consumption,

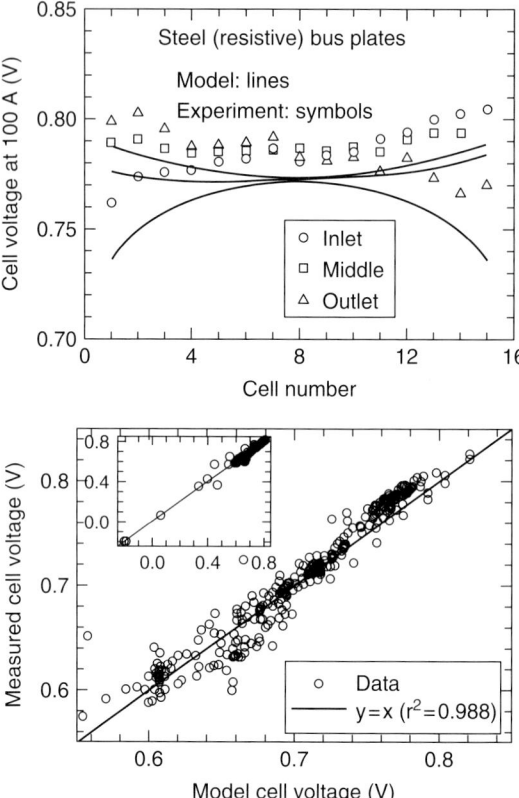

Figure 31.22 Comparison of stack performance between experimental data and model prediction [53].

determined by the local current density $i_j(x)$. In Figure 31.22, the experimental stack voltage distribution data are compared with the model prediction.

Berg et al. [54] presented a similar stack approach with the unit cells described by 1D models appropriate for straight gas channel design. Karimi et al. [55] used flow networks to determine the pressure and flow distributions. The results were then incorporated into the individual cell model developed by Baschuk and Li [56]. Chang et al. [57] used a flow distribution model to examine the sensitivity of stack performance to operating conditions (inlet velocity and pressure) and design parameters (manifold, flow configuration and friction factor). Park and Li [58] presented a flow model and concluded that flow uniformity can be enhanced by a large manifold. Chang et al. [57] developed a stack model incorporating flow distribution effects and a reduced-dimension individual cell model (Figure 31.23). The mass and momentum conservations are applied throughout the stack. Flow splitting and recombination are considered at each tee junction, while along cell channels, reactant consumption, and by-product production are accounted for. The

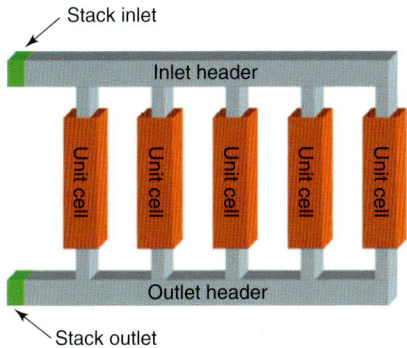

Figure 31.23 U-type manifold. The reactants are fed through the stack inlet and distributed in parallel among the unit cells. The outflows from the unit cells are combined in the outlet header and exit through the stack outlet [57].

mass and momentum equations take the following forms:

$$\frac{\partial (\rho u)}{\partial x} + \frac{1}{A}\oint_{\partial A} \dot{m}dl = 0$$

$$\frac{\partial \left(\frac{\rho u^2}{2}\right)}{\partial x} + \frac{1}{A}\oint_{\partial A} \dot{m}udl + \frac{\partial P}{\partial x} = \frac{f\rho u^2}{2D_h} \tag{31.47}$$

where $x = y$ represents along the unit cell channels and $x = z$ along the manifolds. The velocity u is related to the molar flow rate Q and the pressure P by

$$u = \frac{QRT}{PA}\oint_{\partial A} \dot{m}udl \tag{31.48}$$

where $\oint_{\partial A} \dot{m}dl$ represents the source or sink of mass flux due to flow splitting or combination at the header tee junction and mass exchange with GDLs along the channels. The term $\frac{f\rho u^2}{2D_h}$ accounts for the friction loss, where the friction factor f may differ among both channels and headers. The unit cell conditions, including current density, voltage, channel pressure, and channel molar flow rates, are computed using a unit cell model. For the voltage, the following is used:

$$V = E_0 - iR - \frac{RT}{\alpha_c F}\ln\left[\frac{iC_{\text{ref}}}{i_0(C_0 - \delta i)}\right] \tag{31.49}$$

In Figure 31.24, the experimental stack voltage distribution data are compared with the model prediction for the case of two stainless-steel bus plates.

In addition, Yu et al. [59] proposed a water and thermal management model of a Ballard fuel cell stack which takes a set of gas input conditions and stack parameters such as channel geometry, heat transfer coefficients, and operating current. The model can be used to optimize the stack thermal and water management. Chen et al. [60] investigated numerically the flow distribution in a stack, and concluded

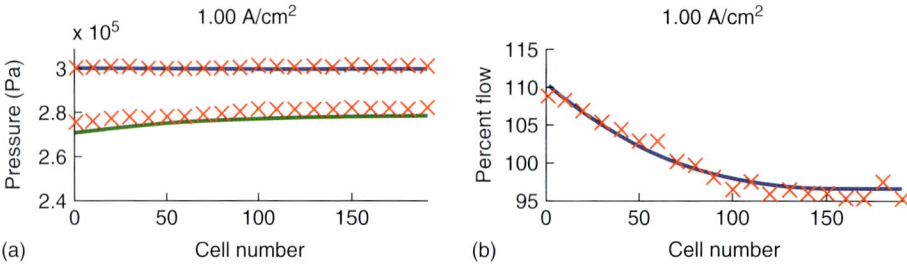

Figure 31.24 (a) Comparison of the computed (solid line) and measured (crosses) static pressures along the cathode inlet and outlet headers of a Mk 7 fuel cell stack for current densities of 1, 0.88, and 0.77 A cm^{-2}. (b) Comparison of the air flow distribution computed with the model (solid line). Other operating conditions: $T = 343$ K, air:H$_2$ stoichiometry $= 2.0:1.5$, dew point $= 343/347$ K (cathode/anode) [57].

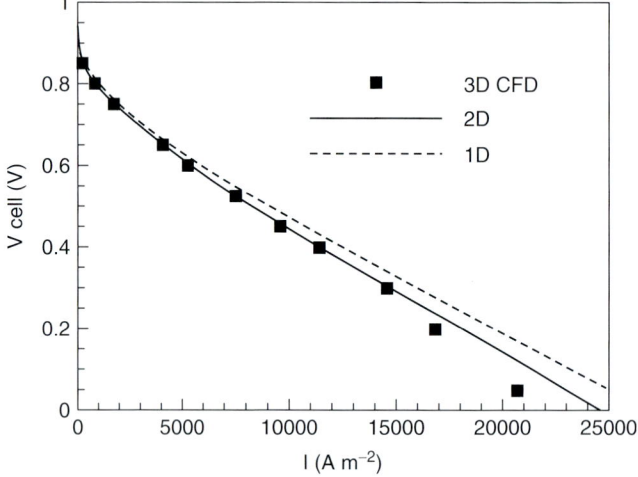

Figure 31.25 A comparison between the 1D, 2D, and 3D results for the baseline conditions [62].

that the channel resistance, manifold dimensions, and gas feed rate may affect the flow distribution. Chang et al. [61] proposed the separation of the complex model into computationally manageable parts. Their computational method was supported by simplified analysis.

31.3.3
Model Validation

The stack-level models are not used as frequently as the other two levels (component and cell). However, they are important when the stack design and control are the focus. Validation has been conducted to some extent. In Figures 31.22 and 31.24, experimental data are compared with model predictions, showing good agreement.

Comprehensive models fully coupling the fuel and reactant flow in GFCs and manifolds and the transport within fuel cells in conjunction with electrochemical reaction are strongly needed. One critical part is the two-phase flow in the complex flow field of stacks, which is essential to capture the flow maldistribution phenomena. In addition, computational studies based on a comprehensive model are still computationally too expensive at present, so efficient numerical schemes are required. Kim *et al.* [62] recently proposed a reduced-dimensional model for straight-channel PEFCs, and reported good agreement with 3D calculations (Figure 31.25). However, its applicability to serpentine channel fuel cells and stacks needs further evaluation. In addition, numerical computation has advanced significantly with respect to both hardware and software in recent years, such that a problem involving a couple of million unknowns can be solved in a standalone workstation. Consequently, the stack modeling challenge can be addressed in two major ways: (i) by developing reduced models and (ii) by developing efficient numerical algorithms to take advantage of the rapidly advancing computer technology.

31.4 Conclusion

In this chapter, the current status and several major aspects of PEFC modeling at the cell and stack levels have been described and discussed. At present, the macroscopic descriptions of phenomena in the entire fuel cell and the stack have been formulated and corresponding models developed with validation against experimental data being carried out to some extent. The essential phenomena include two-phase flow, heat transfer, multi-species mass and charge transport, hydrogen oxidation and oxygen reduction electrochemical reactions, solid water formation, and transient operation. Large-scale numerical simulations, employing millions of computational cells or grid points, have also been carried out to elucidate detailed phenomena in industrial-sized fuel cells and to perform their design/optimization. Model validation, although limited in scope, has been conducted in terms of polarization curves, local species concentration, and liquid water profiles. At the stack level, most modeling efforts have been focused on simplified approaches to predict the stack performance or individual cell temperature and performance.

In spite of the advances made in PEFC modeling, many challenges remain. First, there is a critical need to couple, in some computationally efficient way, the pore-level or particle-level submodels with the macroscopic cell-level models in order to take into account the effects of the microstructures of GDL/MPL and CL. Second, further efforts are also needed to model the cold start, transient, and two-phase transport at the cell level. At present, a framework of single fuel-cell modeling has been developed, but the physics is not yet completely understood. For example, ice formation within the catalyst layer and its impact on the electrochemical reaction need further study. The two-phase transport at the CL–MPL and MPL–GDL interfaces is not clearly understood at present. Third, current stack models are still overly simplified: most are either zero- or one-dimensional. Therefore, there is a great need for a more detailed, multi-dimensional stack model that can be employed for the design, optimization, and control of fuel-cell stacks. This requires efficient numerical algorithms for handling ultra-large computational domains. The challenge here is how to balance the model fidelity required with the computational costs. One option is to develop reduced-dimension models with sufficient fidelity for stacks. Last, more comprehensive and systematic comparisons between model predictions and experimental data are required to validate the models. There are three aspects of desired validation: one is to validate a model in all three dimensions at the same time; another is to compare with experimental data in terms of several key quantities together, such as local current densities, liquid water distribution, and oxygen concentration profiles; and the third is to validate a model in a wide range of operating conditions, such as steady state versus transient, and dry operation versus full humidification.

List of Symbols

A	electrode area, m^2
a	water activity; effective catalyst area per unit volume, $m^2\,m^{-3}$
a_0	catalyst surface area per unit volume, $m^2\,m^{-3}$
C	molar concentration of species k, $mol\,m^{-3}$
D	species diffusivity, $m^2\,s^{-1}$
EW	equivalent weight of dry membrane, $kg\,mol^{-1}$
F	Faraday's constant, $96\,487\,C\,mol^{-1}$
$\vec{G}$	species diffusion/permeation flux, $mol\,m^{-2}\,s^{-1}$
I	current density, $A\,cm^{-2}$
$\vec{i}_e$	superficial current density, $A\,cm^{-2}$
j	transfer current density, $A\,cm^{-3}$
$\vec{j}^{(l)}$	mass flux of liquid phase, $kg\,m^{-2}\,s^{-1}$
K	permeability, m^2
k_r	relative permeability
L	length, m
M	molecular weight, $kg\,mol^{-1}$
$mf_k^{(l)}$	mass fraction of species k in the liquid phase
n	direction normal to the surface
n_d	electroosmotic coefficient, H_2O/H^+
P	pressure, Pa
R	universal gas constant, $8.134\,J\,mol^{-1}\,K^{-1}$
S	source term
s	liquid saturation
t	time, s
T	temperature, K
U_0	equilibrium potential, V
$\vec{u}$	velocity vector, $m\,s^{-1}$

Greeks

α	transfer coefficient; net water flux per proton flux
ρ	density, kg m^{-3}
υ	kinematic viscosity, m^2 s^{-1}
θ_c	contact angle, °
Φ	phase potential, V
ξ	stoichiometric flow ratio
λ	membrane water content
$\lambda^{(k)}$	mobility of phase k
ε	porosity
η	surface overpotential, V
$\vec{\vec{\tau}}$	shear stress, N m^{-2}
τ	surface tension, N m^{-1}
γ_c	correction factor for species convection
δ	thickness, m
σ	electronic or ionic conductivity, S m^{-1}

Superscripts and Subscripts

a	anode
c	cathode; capillary
CL	catalyst layer
D	diffusion
DM	diffusion media
eff	effective value
g	gas phase
GDL	gas diffusion layer
in	inlet
k	species; liquid or gas phase
l	liquid
m	membrane phase
0	gas channel inlet value; reference value
ref	reference value
rev	reversible
s	solid
sat	saturated value
w	water

References

1. Wang, Y., Chen, K.S., Mishler, J., Cho, S.C., and Adroher, X.C. (2011) A review of polymer electrolyte membrane fuel cells: technology, applications, and needs on fundamental research. *Appl. Energy*, **88**, 981–1007.
2. Wang, C.Y. (2004) Fundamental models for fuel cell engineering. *Chem. Rev.*, **104**, 4727–4766.
3. Weber, A.Z. and Newman, J. (2004) *Chem. Rev.*, **104**, 4679.
4. Gurau, V. and Mann, J.A. (2009) A critical overview of computational fluid dynamics multiphase models for proton exchange membrane fuel cells. *SIAM J. Appl. Math.*, **70**, 410–454.
5. Siegel, C. (2008) Review of computational heat and mass transfer modeling

in polymer-electrolyte-membrane (PEM) fuel cells. *Energy*, **33** (9), 1331–1352.

6. Jiao, K. and Li, X. (2011) Water transport in polymer electrolyte membrane fuel cells. *Recent Prog. Energy Combust. Sci.*, **37** (3), 221–291.

7. Djilali, N. (2007) Computational modelling of polymer electrolyte membrane (PEM) fuel cells: challenges and opportunities. *Energy*, **32** (4), 269–280.

8. Wang, Y. (2008) Modeling of two-phase transport in the diffusion media of polymer electrolyte fuel cells. *J. Power Sources*, **185**, 261–271.

9. Amphlett, J.C., Mann, R.F., Peppley, B.A., Roberge, P.R., and Rodrigues, A. (1996) *J. Power Sources*, **61**, 183.

10. Wang, Y. (2007) Analysis of the key parameters in the cold start of polymer electrolyte fuel cells. *J. Electrochem. Soc.*, **154**, B1041–B1048.

11. Pasaogullari, U. and Wang, C.Y. (2004) Liquid water transport in gas diffusion layer of polymer electrolyte fuel cells. *J. Electrochem. Soc.*, **151** (3), A399–A406.

12. Wang, Y. and Feng, X. (2008) Analysis of reaction rates in the cathode electrode of polymer electrolyte fuel cells. Part I: single-layer electrodes. *J. Electrochem. Soc.*, **155** (12), B1289–B1295.

13. Wang, Y. and Feng, X. (2009) Analysis of reaction rates in the cathode electrode of polymer electrolyte fuel cells. Part II: dual-layer electrodes. *J. Electrochem. Soc.*, **156** (3), B403–B409.

14. Feng, X. and Wang, Y. (2010) Multi-layer configuration for the cathode electrode of polymer electrolyte fuel cell. *Electrochim. Acta*, **55**, 4579–4586.

15. Wang, Y. and Wang, C.Y. (2005) Modeling polymer electrolyte fuel cells with large density and velocity changes. *J. Electrochem. Soc.*, **152**, A445–A453.

16. Wang, Y. and Wang, C.Y. (2005) Transient analysis of polymer electrolyte fuel cells. *Electrochim. Acta*, **50**, 1307–1315.

17. Wang, Y. and Wang, C.Y. (2006) Dynamics of polymer electrolyte fuel cells undergoing load changes. *Electrochim. Acta*, **51**, 3924–3933.

18. Wang, Y. and Wang, C.Y. (2007) Two-phase transients of polymer electrolyte fuel cells. *J. Electrochem. Soc.*, **154**, B636–B643.

19. Ju, H., Meng, H., and Wang, C.Y. (2005) *Int. J. Heat Mass Transfer*, **48**, 1303.

20. Wang, Y. and Wang, C.Y. (2006) A non-isothermal, two-phase model for polymer electrolyte fuel cells. *J. Electrochem. Soc.*, **153**, A1193–A1200.

21. Basu, S., Wang, C.Y., and Chen, K.S. (2009) Phase change in a polymer electrolyte fuel cell. *J. Electrochem. Soc.*, **156** (6), B748.

22. Wang, Y. and Wang, C.Y. (2006) Ultra large-scale simulation of polymer electrolyte fuel cells. *J. Power Source*, **153** (1), 130–135.

23. Wang, Z.H., Wang, C.Y., and Chen, K.S. (2001) *J. Power Sources*, **94**, 40.

24. He, W., Yi, J.S., and Nguyen, T.V. (2000) *AIChE J.*, **46**, 2053.

25. You, L. and Liu, H. (2002) *Int. J. Heat Mass Transfer*, **45**, 2277.

26. Mazumder, S. and Cole, J.V. (2003) *J. Electrochem. Soc.*, **150**, A1510.

27. Berning, T. and Djilali, N.J. (2003) *J. Electrochem. Soc.*, **150**, A1598.

28. Wang, C.Y. and Cheng, P. (1997) *Adv. Heat Transfer*, **30**, 93.

29. Wang, Y. (2009) Porous-media flow fields for polymer electrolyte fuel cells, II. Analysis of channel two-phase flow. *J. Electrochem. Soc.*, **156** (10), B1134–B1141.

30. Wang, Y., Wang, C.Y., and Chen, K.S. (2007) Elucidating differences between carbon paper and carbon cloth in polymer electrolyte fuel cells. *Electrochim. Acta*, **52**, 3965–3975.

31. Wang, Y., Basu, S., and Wang, C.Y. (2008) Modeling two-phase flow in PEM fuel cell channels. *J. Power Sources*, **179**, 603–617.

32. Basu, S., Li, J., and Wang, C.Y. (2009) Two-phase flow and maldistribution in gas channels of a polymer electrolyte fuel cell. *J. Power Sources*, **187** (2), 431–443.

33. Pasaogullari, U. and Wang, C.Y. (2005) Two-phase modeling and flooding prediction of polymer electrolyte fuel cells. *J. Electrochem. Soc.*, **152**, A380–A390.

34. Wang, Y., Mukherjee, P.P., Mishler, J., Mukundan, R., and Borup, R.L. (2010) Cold start of polymer electrolyte fuel cells: three-stage startup characterization. *Electrochim. Acta*, **55**, 2636–2644.

35. Hishinuma, Y., Chikahisa, T., Kagami, F., and Ogawa, T. (2004) *JSME Int. J., Ser. B*, **47**, 235.
36. Wilson, M.S., Valerio, J.A., and Gottesfeld, S. (1995) *Electrochim. Acta*, **40**, 355.
37. McDonald, R.C., Mittelsteadt, C.K., and Thompson, E.L. (2004) *Fuel Cells*, **4**, 208.
38. Cho, E., Ko, J.J., Ha, H.Y., Hong, S.A., Lee, K.Y., Lim, T.W., and Oh, I.H. (2004) *J. Electrochem. Soc.*, **151**, A661.
39. Mishler, J., Wang, Y., Mukherjee, P.P., Mukundan, R., and Borup, R.L. (2011) Subfreezing operation of polymer electrolyte fuel cells: Ice formation and cell performance loss, Electrochimica Acta, http://dx.doi.org/10.1016/j.electacta.2012.01.020.
40. Mao, L. and Wang, C.Y. (2007) *J. Electrochem. Soc.*, **154**, B341.
41. Wang, Y. and Mishler, J. (2010) Modeling and analysis of polymer electrolyte fuel cell cold-start, in Proceedings of ES2010-90139, ASME 2010 4th International Conference on Energy Sustainability, Phoenix, AZ, May 17–22, 2010.
42. Jiang, F.M., Wang, C.Y., and Chen, K.S. (2010) Current ramping: a strategy for rapid start-up of PEMFCs from subfreezing environment. *J. Electrochem. Soc.*, **157**, B342.
43. Wang, Y. and Wang, C.Y. (2005) Simulations of flow and transport phenomena in a polymer electrolyte fuel cell under low-humidity operations. *J. Power Source*, **147**, 148.
44. Tüber, K., Pocza, D., and Hebling, C. (2003) *J. Power Sources*, **124**, 403–414.
45. Meng, H. and Wang, C.Y. (2004) *Chem. Eng. Sci.*, **104**, 4727–4766.
46. Basu, S., Wang, C.Y., and Chen, K.S. (2009) Two-phase flow maldistribution and mitigation in the cathode channels of a polymer electrolyte fuel cell. *Trans. ASME J. Fuel Cell Sci. Technol.*, **6**, 031007-1–031007-11.
47. Tabuchi, Y., Shiomi, T., Aoki, O., Kubo, N., and Shinohara, K. (2010) Effects of heat and water transport on the performance of polymer electrolyte membrane fuel cell under high current density operation. *Electrochim. Acta*, **56**, 352–360.
48. Ju, H. and Wang, C.Y. (2004) Experimental validation of a PEM fuel cell model by current distribution data. *J. Electrochem. Soc.*, **151**, A1954–A1960.
49. Weber, A.Z. and Newman, J. (2004) Transport in polymer-electrolyte membranes. III. Model validation in a simple fuel-cell model. *J. Electrochem. Soc.*, **151**, A326.
50. Wang, Y. and Chen, K.S. (2010) Through-plane water distribution in a polymer electrolyte fuel cell: comparison of numerical prediction with neutron radiography data. *J. Electrochem. Soc.*, **157** (12), B1878–B1886.
51. Wang, Y., Pham, L., Salerno de Vasconcellos, G.P., and Madou, M. (2010) Fabrication and characterization of micro PEM fuel cells using pyrolyzed carbon current collector plates. *J. Power Sources*, **195**, 4796–4803.
52. Promislow, K. and Wetton, B. (2005) A simple, mathematical model of thermal coupling in fuel cell stacks. *J. Power Sources*, **150**, 129–135.
53. Kim, G.-S., St-Pierre, J., Promislow, K., and Wetton, B. (2005) Electrical coupling in proton exchange membrane fuel cell stacks. *J. Power Sources*, **152**, 210–217.
54. Berg, P., Caglar, A., Promislow, K., St-Pierre, J., and Wetton, B. (2006) Electrical coupling in proton exchange membrane fuel cell stacks: mathematical and computational modeling. *IMA J. Appl. Math.*, **71** (2), 241–261.
55. Karimi, G., Baschuk, J.J., and Li, X. (2005) Performance analysis and optimization of PEM fuel cell stacks using flow network approach. *J. Power Sources*, **147**, 162–177.
56. Baschuk, J.J. and Li, X. (2003) Mathematical model of a PEM fuel cell incorporating CO poisoning and O_2 (air) bleeding. *Int. J. Global Energy Issues*, **20** (3), 245–276.
57. Chang, P.A.C., St-Pierre, J., Stumper, J., and Wetton, B. (2006) Flow distribution in proton exchange membrane fuel cell stacks. *J. Power Sources*, **162** (1), 340–355.

58. Park, J. and Li, X. (2006) Effect of flow and temperature distribution on the performance of a PEM fuel cell stack. *J. Power Sources*, **162** (1), 444–459.
59. Yu, X., Zhou, B., and Sobiesiak, A. (2005) Water and thermal management for Ballard PEM fuel cell stack. *J. Power Sources*, **147** (1–2), 184–195.
60. Chen, C.H., Jung, S.P., and Yen, S.C. (2007) Flow distribution in the manifold of PEM fuel cell stack. *J. Power Sources*, **173** (1), 249–263.
61. Chang, P., Kim, G.S., Promislow, K., and Wetton, B. (2007) Reduced dimensional computational models of polymer electrolyte membrane fuel cell stacks. *J. Comput. Phys.*, **223** (2), 797–821.
62. Kim, G.S., Sui, P.C., Shah, A.A., and Djilali, N. (2010) Reduced-dimensional models for straight-channel proton exchange membrane fuel cells. *J. Power Sources*, **195** (10), 3240–3249.

Part VI
Balance of Plant Design and Components

32
Principles of Systems Engineering

Ludger Blum, Ralf Peters, and Remzi Can Samsun

32.1
Introduction

Systems engineering covers a wide range of engineering tasks and involves a plurality of different methods for designing systems. Systems engineering comprises both systems analysis and systems technology. Systems analysis considers social and technological aspects of systems from a community perspective. Life-cycle analysis and risk management are important elements of evaluation. Energy systems must be assessed using criteria such as efficiency, environmental impacts, and costs. Systems analysis provides recommendations for politics and industry. It can be performed using top-down or bottom-up approaches. Bottom-up models use information from the technology development of existing systems and related technologies. Top-down models give an overview of a system. This is segmentally specified to gain insight into its compositional subsystems.

Systems technology requires different intermediate steps for successful system development. In the 1980s, the air and space industry developed a scale for evaluating the distance to commercialization [1, 2]. This scale, known as the technology readiness level (TRL), contains nine different levels.

- In TRL 1, the basic principles of a system are analyzed. Often, only an idea has been reported or patented.
- The next step (TRL 2) involves checking the practicability of the technology concept. Systems analysis supports the definition of crucial items to be checked by basic examinations in the next level.
- TRL 3 initiates research and development with laboratory-based methods. The analytical predictions are physically validated. A proof of concept of the system is performed.
- After TRL 3, the basic components are integrated into a system in a laboratory environment (TRL 4). Peripheral components are not developed for the specific application. Such a system is often referred to as a breadboard or brassboard system. The system applicability of the different devices is checked. Components requiring further development are identified.

Fuel Cell Science and Engineering: Materials, Processes, Systems and Technology,
First Edition. Edited by Detlef Stolten and Bernd Emonts.
© 2012 Wiley-VCH Verlag GmbH & Co. KGaA. Published 2012 by Wiley-VCH Verlag GmbH & Co. KGaA.

- During the next level of development (TRL 5), systems are tested in a real or simulated environment under test conditions. The maturity of system components is increased significantly. Peripheral components, which have been integrated into the system, should be at an advanced stage of development in terms of the targeted application.
- TRL 6 requires system verification in a relevant environment under realistic conditions. The main components and the peripheral devices are developed to high maturity.
- TRL 7 demands a prototyping demonstration of the system in its regular environment. The scale of this system should be close to the scale of the planned application.
- In TRL 8, the actual system or technology is completed and qualified using reliable tests and demonstrations.
- If a system has been proven to work through successful operation, it has reached TRL 9.

Each of these levels demands different development steps depending on the application and its specific conditions. System development covers the steps of basic engineering, detailed engineering, procurement, construction, and testing.

In the context of numerous application areas, different energy conversion technologies and the nine maturity levels of development, this chapter focuses on fuel-cell systems based on carbonaceous energy carriers. Such systems are ideal for explaining most of the aspects of systems engineering. The developments described here belong to TRL 1–6.

Section 32.2 discusses process analysis in detail as a powerful basic engineering tool. At this stage, the system is categorized as TRL 1–2, and the analysis forms the basis for the specification list of systems at a higher level of development. Numerous papers have reported on the process performance of different types of fuel cells and various fuels over the last 10 years (e.g., [3–7]), In Sections 32.3–32.5, further steps and examples are discussed to illustrate important stages of system development for TRL 4 and upwards. With a higher level of maturity, the number of tasks which must be handled increases. Depending on the application, further tasks may have to be taken into account. Therefore, this chapter can only give a brief, general overview of the principles of systems engineering.

32.2
Basic Engineering

32.2.1
General Considerations

The tasks of basic engineering can be divided into several sequential steps. The first step involves considering some general calculations. The core of a fuel cell system is the fuel cell itself. A fuel cell is characterized by its voltage-current characteristics. For stationary applications, a design point must be defined. For

32.2 Basic Engineering

mobile applications, in contrast, a basic layout can be used with a possible reserve in terms of maximum power. The most important value at this stage is the cell voltage at which the stack will be operated. An often preferred cell voltage for fuel-cell stacks is 750 mV. In order to calculate a mass flow of hydrogen, a relation between cell voltage and efficiency is required. Equation (32.1) determines the efficiency of the electrochemical cell using the relation between cell voltage and heating voltage E_H°.

$$\eta_{Cell} = \frac{E_{Cell}}{E_H^\circ} \qquad (32.1)$$

Efficiency can also be related to the lower or the higher heating value of the fuel. The heating voltage is 1253 mV for steam as product and 1482 mV for water.

$$H_2 + \frac{1}{2}O_2 \rightarrow H_2O \text{ (gas)} \qquad (E_H^\circ = 1253 \text{ mV}) \qquad (32.2)$$

$$H_2 + \frac{1}{2}O_2 \rightarrow H_2O \text{ (liquid)} \qquad (E_H^\circ = 1482 \text{ mV}) \qquad (32.3)$$

For the gaseous state of water, an efficiency of 59.86% is calculated at a cell voltage of 750 mV. This is somewhat lower than the theoretical value η_{Th}, which is defined as

$$\eta_{Th} = \frac{\Delta G^\circ}{\Delta H^\circ} = \frac{E^\circ}{E_H^\circ} \qquad (32.4)$$

that is, the relation between the standard Gibbs energy and standard reaction enthalpy. With regard to stack design and stack operation, several effects due to mass transfer must be taken into account. Hydrogen is converted by the electrochemical reaction and water is formed on the anode or cathode side of the fuel cell, depending on the type of electrolyte. When a hydrogen-rich gas mixture is used as the fuel, the partial pressure of hydrogen decreases in the flow direction of a channel-type flow field structure. In addition, the composition of a gas mixture affects the Nernst voltage at which the characteristic voltage-current curve starts. Lower partial pressures lead to a lower Nernst voltage. At the stack outlet, the mass transfer of hydrogen is severely limited when the hydrogen concentration in the mixture is too low. Without purging, inert components will also be located at the outlet region of the stack. The amount of hydrogen that will be purged can be calculated using a purge rate or the hydrogen utilization can be defined for the stack. Utilization is defined as

$$u_{H_2} = \frac{\Delta \dot{m}_{H_2}}{\dot{m}_{H_2}^{in}} \qquad (32.5)$$

With regard to hydrogen utilization, the stack efficiency can be calculated as

$$\eta_{Stack} = u_{H_2} \eta_{Cell} \qquad (32.6)$$

At 85% utilization, then stack efficiency is 50.9%. A 10 kW$_e$ fuel-cell system therefore requires 19.65 kW$_{th}$ [lower heating value (LHV) H$_2$] hydrogen. Using the LHV of hydrogen, that is, 241.8 kJ mol^{-1}, a molar flow of 0.0813 mol s^{-1} is determined.

Several conversion technologies can be used to generate hydrogen from natural gas or diesel fuel. If we take methane as the main component of natural gas, the following chemical reactions are possible in an external reactor or internally in a solid oxide fuel cell (SOFC) or an molten carbonate fuel cell (MCFC):

$$CH_4 \rightarrow C + 2H_2 \text{(pyrolysis)} \tag{32.7}$$

$$CH_4 + H_2O \rightarrow CO + 3H_2 \text{ (steam reforming)} \tag{32.8}$$

$$CH_4 + \frac{1}{2}O_2 \rightarrow CO + 2H_2 \text{ (partial oxidation)} \tag{32.9}$$

Additional hydrogen can be gained by the water gas shift (WGS) reaction:

$$CO + H_2O \rightarrow CO_2 + H_2 \text{ (water gas shift reaction)} \tag{32.10}$$

Chemical reactions (32.7), (32.9), and (32.10) are exothermic, whereas steam reforming [Eq. (32.8)] is endothermic, which means that heat must be supplied. Equation (32.7) requires 1 mol of methane to form 2 mol of hydrogen. For the considered 10 kW$_{el}$ system, about 0.0406 mol s^{-1} of methane is necessary. Taking the LHV of methane into account, that is, 802.15 kJ mol^{-1}, an enthalpy flow of 32.6 kW$_{th}$ (LHV CH$_4$) can be determined, leading to a gross efficiency of 30.7% based on Eq. (32.11):

$$\eta_{\text{System, gross}} = \frac{P_{el}}{\dot{H}_{CH_4, \text{LHV}}} \tag{32.11}$$

A strong improvement in efficiency can be made by combining Eqs. (32.8) and (32.9) with the WGS reaction [Eq. (32.10)], giving an additional amount of hydrogen produced. Partial oxidation [Eq. (32.9)] combined with the completed WGS reaction [Eq. (32.10)] leads to a flow of 0.0271 mol s^{-1} of methane and an efficiency of 46% [21.7 kW$_{th}$ (LHV CH$_4$)]. The corresponding calculations for combining steam reforming [Eq. (32.8)] and the WGS reaction [Eq. (32.10)] lead to 0.0203 mol s^{-1} methane, 16.3 kW$_{th}$ (LHV CH$_4$) and an efficiency of 61.3%. To generate heat for steam reforming, hydrogen can be burned by total oxidation according to Eq. (32.2) or (32.3). Assuming a hydrogen utilization of 85% in the stack, the heat released by the combustion process is nearly 3 kW without using the enthalpy of the condensation of water. Taking the higher heating value into account, a maximum heat of 3.5 kW can be determined. These examples show that the processes must be understood in more detail and that efficient heat recuperation is essential if high electrical system efficiencies are to be achieved. Further thermodynamic calculations will lead to limitations and a decrease in the supposed efficiency numbers. Therefore, steam reforming does not have the potential for efficiencies higher than 60%. The potential must be checked by additional thermodynamic methods.

The type of fuel used strongly affects predicted efficiencies. If n-tetradecane (C$_{14}$H$_{30}$) is used as a model fuel for diesel, a molar flow of 0.0054 mol s^{-1} C$_{14}$H$_{30}$ is necessary for partial oxidation. A flow of 0.0019 mol s^{-1} C$_{14}$H$_{30}$ is required for steam reforming combined with a WGS reaction. The corresponding enthalpy flows, related to LHVs, are 47.4 and 16.5 kW$_{th}$, respectively. The heat required for steam generation is 4.9 kW$_{th}$. In addition, heat is required for endothermic steam reforming. The enthalpy flow of the combusted residual gases from the afterburner are too low to provide the complete heat required by the system. Therefore, an intensive heat recovery system is necessary. Under severe conditions, an additional amount of fuel must be burned to cover the conditions governed by the first and second laws of thermodynamics. In the fuel cell, the electrochemical potential in a gas mixture changes in relation to pure hydrogen. This in turn will lead to changes in the cell voltage, as it is not the same for a hydrogen-rich fuel gas as for pure hydrogen.

In the sections that follow, an increasingly more detailed analysis of fuel-cell systems is presented, and established and new methods for systems design are discussed.

32.2.2
Chemical Equilibrium

The chemical equilibrium composition is often calculated for gas mixtures to determine product gas quality depending on the operating conditions of a chemical apparatus. Several examples can be found in the literature [3, 4, 6]. Figure 32.1 shows the gas composition for natural gas steam reforming in the pressure range from 1 to 10 bar and at temperatures between 473 and 1473 K. The molar ratio between methane and steam was kept constant at 2.5. Natural gas was assumed to consist of 91.24% methane, 8.44% nitrogen, and 0.32% carbon dioxide. Higher hydrocarbons were neglected in these calculations. The steam reforming of natural gas was considered in parallel to the WGS reaction.

As can be seen in Figure 32.1, steam reforming starts to produce hydrogen at temperatures between 873 and 1073 K. The pressure dependence of the chemical equilibrium is rather low. A strong increase in H_2 concentration was observed with decreasing pressure at the shoulder of the H_2 surface only (see Figure 32.1). However, this area is of special interest because lower reaction temperatures between 973 and 1073 K ultimately lead to less demanding requirements on the reactor materials.

Another challenging task is to avoid carbon deposition during the reforming of higher hydrocarbons, while achieving a high hydrogen yield as moles H_2 per mole hydrocarbon. Figure 32.2 shows the hydrogen yield as a function of the mixture conditions, that is, the oxygen-to-carbon and the steam-to-carbon ratios for the reforming of $C_{14}H_{30}$. These results were obtained for a chemical equilibrium at 1023 K assuming methanation and a WGS reaction. Carbon deposition was neglected for the calculation of the product gas composition.

In order to increase the hydrogen yield, an additional WGS reactor was considered at 573 K. Another option not considered here would be to add water. Methane will not react at such low temperatures. In principle, three different reactions are possible:

1) pyrolysis with 15 mol H_2 mol^{-1} $C_{14}H_{30}$:

$$C_nH_m \rightarrow nC + \frac{m}{2}H_2 \quad (O_2{:}C = H_2O{:}C = 0) \tag{32.12}$$

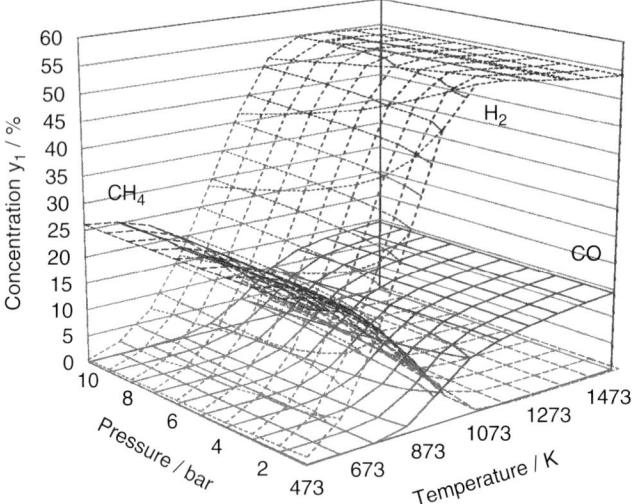

Figure 32.1 Hydrogen, carbon monoxide, and methane concentrations in chemical equilibrium based on the steam reforming of natural gas and the WGS reaction for a steam-to-methane ratio of 2.5. Carbon deposition was not considered in these calculations. Natural gas content (here without higher alkanes): 91.24% methane (CH_4), 8.44% nitrogen (N_2), and 0.32% carbon dioxide (CO_2).

2) partial oxidation with 15 mol H_2 mol^{-1} $C_{14}H_{30}$ as a maximum:

$$C_nH_m + \frac{n}{2}O_2 \rightarrow nCO + \frac{m}{2}H_2 \quad (O_2{:}C = 0.5; H_2O{:}C = 0) \quad (32.13)$$

3) steam reforming with 29 mol H_2 mol^{-1} $C_{14}H_{30}$ as a maximum without WGS, see Eq. (32.10):

$$C_nH_m + nH_2O \rightarrow nCO + \left(\frac{m}{2} + n\right)H_2 \quad (O_2{:}C = 0; H_2O{:}C = 1) \quad (32.14)$$

and 43 mol H_2 mol^{-1} $C_{14}H_{30}$ as a maximum with completed WGS:

$$CO + H_2O \rightarrow CO_2 + H_2 \quad (O_2{:}C = 0; H_2O{:}C > 1) \quad (32.15)$$

Autothermal reforming (ATR) results in values between these limitations, for example, 31 mol H_2 mol^{-1} $C_{14}H_{30}$ for $O_2/C = 0.35$ and $H_2O/C = 1.5$. Figure 32.2 indicates that steam reforming is favorable in terms of high yields.

Unfortunately, steam reforming today is often faced with the problem of carbon deposition [10–12]. Few data exist for operating times >300 h for JP-8 or diesel fuel [13]. Often, experiments are limited to operating times of <50 h [14, 15]. The thermodynamic conditions for carbon deposition can also be calculated (gray area circumscribed by the bold dotted line in Figure 32.2). For pure steam reforming, a minimum H_2O/C ratio of 1.1 is theoretically necessary, and a value of 4.5 can also lead to serious problems [11]. Even at a high O_2/C ratio of 0.55, the partial oxidation process still shows a small potential for carbon deposition at 1023 K.

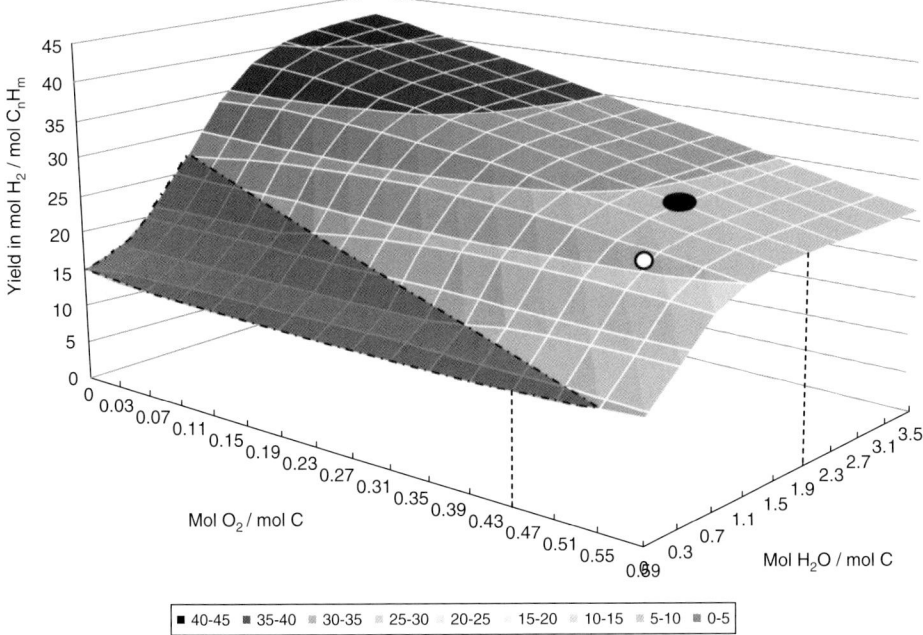

Figure 32.2 Hydrogen yield from 1 mol C_nH_m (here $C_{14}H_{30}$) in relation to the steam-to-carbon ratio and oxygen-to-carbon ratio for reforming. Methane concentrations were determined using an equilibrium calculation assuming a WGS reaction and methanation at 1023 K. In order to extend the hydrogen yield, a WGS reactor operated in equilibrium at 573 K was considered. An additional water injection between the reformer and WGS was not considered. The black circle represents the conditions favored by Pasel et al. [8], and the white circle those favored by Roychoudhury et al. [9]. The gray area circumscribed by the bold dotted line represents favored conditions for carbon deposition.

Steam reforming offers the highest hydrogen yield of about 40–42 mol H_2 mol^{-1} $C_{14}H_{30}$ at H_2O/C ratios >2.3. The hydrogen yield is somewhat lower than the theoretical value owing to the incomplete WGS equilibrium. Although carbon deposition can be neglected with thermodynamic conditions, several experimental setups suffer from it [10–12]. Few investigators have reported successful long-term experiments with kerosene steam reforming [16, 17] for domestic heating systems. The two cited claimed 100% conversion for 10 000 and 30 000 h, respectively. Unfortunately, attention was not paid to residual substances and no details were published.

Two operation points of an ATR unit with a durability of more than 1000 h are also included in Figure 32.2. Pasel et al. [8] tested their reformers successfully with the parameters $O_2/C = 0.47$ and $H_2O/C = 1.9$ for premium diesel and Jet A-1. Roychoudhury et al. [9] chose $O_2/C = 0.51$ and $H_2O/C = 0.9$–1.5 for their ATR unit operated with JP-8. As can be seen in Figure 32.2, the first operation point leads to a higher hydrogen yield of 29.5 mol H_2 mol^{-1} $C_{14}H_{30}$ compared with 26.5 mol H_2 mol^{-1} $C_{14}H_{30}$ for lower H_2O/C ratios. This effect is caused by the WGS equilibrium at 573 K, which results in different CO concentrations of 0.9 and 3.6 vol.%, respectively. The lower CO concentration corresponds to a design for low-temperature fuel cells, such as the polymer electrolyte fuel cell (PEFC) and high-temperature polymer electrolyte fuel cell (HT-PEFC), and the higher concentration to an SOFC design. In order to avoid water recovery, anode gas recycling could be considered for SOFC systems, which would result in low H_2O/C ratios of 0.3–1. Considering the general remarks in Section 32.2.1, ATR units operated under different conditions lead to enthalpy flows based on the LHV of diesel between 24.5 and 26.8 kW_{th} for a 10 kW_{el} fuel-cell system. The efficiencies for fuel gas production vary between 73 and 80%.

32.2.3
Analytical Methods for Heat Management

As outlined in the last two sections, a fuel-cell system must be designed in a manner that also considers the stack and fuel-cell system performance and the conditions for fuel processing. In this section, all processes with heat transfer or those for which heat of chemical reaction occurs are analyzed to find an optimum system design.

32.2.3.1 System Set-Up
In order to design a basic flow sheet, all operating conditions must be defined for chemical reactors and for the fuel cell. Different operating temperatures require a set of different heat exchangers. During the first phase of process analysis, the exchangers are not connected to each other. An HT-PEFC system based on jet fuels as the energy carrier is analyzed here as an example. The thermodynamic conditions must be defined for each flow line:

- Water is fed into the system by a pump at a pressure of 4.5 bar. It must be completely evaporated at 420 K and superheated to 623 K.
- ATR air is compressed to 1.5 bar and heated to 623 K.
- Dodecane, as a model jet fuel ($C_{12}H_{26}$), is heated, evaporated at 604 K and superheated to 623 K.
- Cathode air is compressed to 1.2 bar and heated to 453 K.
- Catalytic burner air is compressed to 1.2 bar and heated to 453 K.
- The reactors and the fuel cell operate under the following conditions:
  - The ATR unit is operated adiabatically with a mixture corresponding to $O_2/C = 0.40$ and $H_2O/C = 2.5$. The adiabatic peak temperature in the ATR unit is estimated as 1150 K and the adiabatic temperature at the reactor exit is determined as 922 K. The ATR product gas corresponds to chemical equilibrium with regard to methane formation (methanation) and the WGS reaction [see Eqs. (32.8) and (32.10)].
  - The WGS reactor is operated isothermally at 523 K without an additional water supply. WGS equilibrium is assumed in the product gas.

- The fuel cell (HT-PEFC) is operated at 453 K with an air ratio of 2 and a hydrogen utilization of 90%, which is rather challenging for reformate operation. The cell voltage is 650 mV.
- The catalytic burner is operated adiabatically with an air ratio of 1.3, leading to a peak temperature of 931 K.

These conditions lead to a data set of 44 thermodynamic states (i: 1–44) which can be described by enthalpy $H^i(p, T)$, entropy $S^i(p, T)$, and Gibbs energy $G^i(p, T)$. Nine heat exchangers for heating purposes and seven for cooling purposes are required to analyze the different operating temperatures. A steam generator consists, in principle, of three basic units: heater, evaporator, and superheater. In the following section, different methods are introduced to analyze and optimize these systems.

32.2.3.2 Gibbs Energy Function

Based on thermodynamic functions for the specific heat capacity $c_p(T)$ [18] and under the assumption that all gases, that is, O_2, N_2, CO, CO_2, H_2, and so on, can be described as ideal gases, the thermodynamic state variables $\Delta H(T)$ and $\Delta S(T)$ can be determined for each component. Here, the absolute enthalpy values of O_2, N_2, H_2, and C were set to zero at 1.013 bar and 298 K. In combination with the tabulated heat of formation and the given flows $\dot{n}_i$ all absolute enthalpy data $\dot{H}(T)$ can be described. Considering entropy data from the literature and the standard state [18], $\dot{S}(T)$ and $\dot{G}(T)$ can also be determined:

$$\dot{G}(T) = \dot{n} \times g = \dot{n} \times [h(T) - T \times s(T)] \quad (32.16)$$

In order to analyze the processes in a fuel-cell system, all units with a positive change in Gibbs energy and those with a negative change were summarized in two composite curves. Figure 32.3 shows the composite curves for a 10 kW$_{el}$ HT-PEFC system based on $C_{12}H_{26}$ as a model fuel for kerosene. At 298 K, the black curve

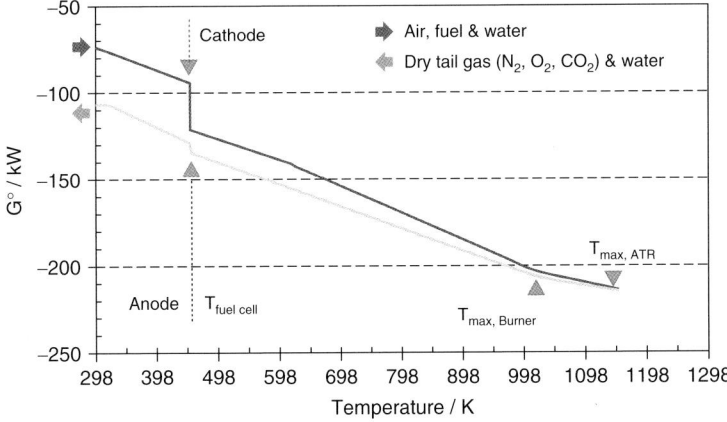

Figure 32.3 Gibbs energy as a function of temperature for a 10 kW$_{el}$ HT-PEFC system based on $C_{12}H_{26}$ as a model fuel for kerosene.

starts with the sum of all Gibbs energy data for the flows of air, steam, and water. The warming, evaporation, and superheating processes led to a decrease in Gibbs energy with increasing temperature. Chemical processes were determined by the Gibbs reaction energy $\Delta \dot{G}_R(T)$, which was negative, for example, in the case of combustion reactions caused by the formation of carbon dioxide and water. A large drop can be identified at 453 K, the operating temperature of the HT-PEFC, which is caused by the formation of water from oxygen and hydrogen on the cathode side. The hot composite curve and the cold composite curve converge at the maximum temperature of the reforming process. At this point, exothermic partial oxidation ends and endothermic steam reforming starts.

As can be seen in Figure 32.3, the difference in Gibbs energy between incoming air, fuel, and water and the exhausted tail gases is $\Delta \dot{G}(298\text{ K}) = 30\text{ kW}$. The exergetic efficiency can be determined directly at ambient conditions. Exergetic efficiency is defined by the net electricity of the system divided by the chemical exergy of the fuel at ambient conditions:

$$\eta_{\text{ex}} = \frac{P_{\text{el, net}}}{\Delta \dot{E}_{\text{Chem}}(298\text{ K})} = \frac{P_{\text{el, net}}}{\Delta \dot{G}(298\text{ K})} \tag{32.17}$$

In our example, the exergetic efficiency was 33%.

32.2.3.3 Pinch Point Diagram

Pinch point analysis was introduced by Linnhoff [19] to optimize heat exchanger networks for industrial energy conversion processes, especially for power plants. This method focuses on the change in enthalpy in a process caused by heat exchange. Processes with a positive change in enthalpy and those with a negative change are superimposed on each other, resulting in two composite curves, that is, a hot and a cold composite curve. These hot and the cold composite curve are then sketched in a $T(\Delta \dot{H})$ diagram. The first and second laws of thermodynamics can be checked by visualization. The position of the curves in relation to each other can be set to a certain extent. If the hot composite curve is larger than the cold one, the cold curve should be moved to the right, that is, towards the maximum enthalpy value of the hot composite curve. The hypothetical heat exchanger networks are closed by an additional heat exchanger, which cools the rest of the hot composite curve by ambient air. The first law of thermodynamics is therefore fulfilled. If the hot composite curve has temperatures that are lower in the diagram than those of the cold composite curve, the second law of thermodynamics is deemed to be violated, that is, heat cannot be transferred from the cold to the hot side of the networks. In such a case, the cold composite curve must be moved to the right until both curves converge at a single point – the pinch point. This pinch point is only of theoretical interest. The area for a heat exchanger is determined by

$$A_{\text{Hex}} = \int_{T_{\text{min}}}^{T_{\text{max}}} \frac{\text{d}\dot{H}}{k\Delta T} = \int_{T_{\text{min}}}^{T_{\text{max}}} \frac{\text{d}\dot{H}}{k(T_{\text{Hot}} - T_{\text{Cold}})} \tag{32.18}$$

The appearance of a pinch point leads to an infinitely large heat exchanger area. Therefore, a minimum temperature difference between the hot and cold composite curves must be fixed. An intercept on the x-axis of the cold curve (ΔH_{gap}), which is not covered by the hot curve, must be closed by an electrical heating cartridge in order to fulfill the first law of thermodynamics. In a fuel-cell system, the hydrogen fuel utilization in the fuel cell can be decreased or fresh fuel can be burned to fulfill this demand, ΔH_{gap}. This would also change the characteristics of the hot composite curve.

Chemical processes are easier to identify in a pinch point diagram than in a Gibbs energy analysis. When the reactor is operated adiabatically, the heat exchange is zero, that is, $\Delta \dot{H} = 0$. When the reactor is operated isothermally, $\Delta T = 0$. Furthermore, the effort of evaporation is much clearer. At constant pressure and constant temperature, ΔG^{LV} for a single component is zero, whereas ΔH^{LV} for water is fairly large and is related to the enthalpy change of the heating water or steam.

The main objective of the pinch point method is to clarify the effort of the heat exchanger networks without knowing their exact design. If the minimum temperature difference is increased, the heat exchanger area decreases according to Eq. (32.18). In contrast, more heat may be required on the hot side, which would also lead to a larger heat exchanger area for cooling purposes and thus to a lower system efficiency. This shear can be solved by a numerical integration solving Eq. (32.18). With regard to different fuel-cell system designs, most of such examples will occur for systems with endothermic steam reforming due to the high heat demand at elevated temperatures. Figure 32.4 shows the pinch diagram for a 10 kW$_{el}$ HT-PEFC system based on $C_{12}H_{26}$ as a model fuel for kerosene. The fuel-cell system was calculated using the parameters given in Section 32.2.3.1.

A pinch diagram does not show any single component but it does allow some processes to be identified by the composite curves based on their special

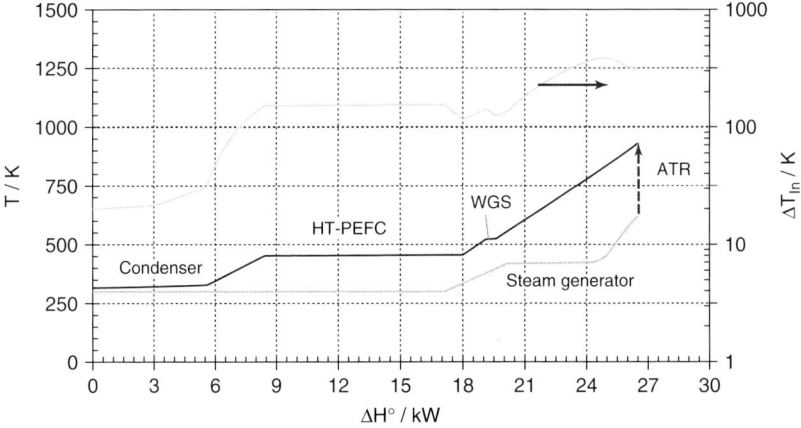

Figure 32.4 Pinch diagram for a 10 kW$_{el}$ HT-PEFC system based on the ATR of tetradecane ($C_{12}H_{26}$) as a model fuel for kerosene.

characteristics. In order to generate 10 kW of electricity, about 29 kW must be transferred by heat exchange from a heat source to a heat sink. Nearly 10 kW$_{th}$ comes from the heat exchange between different flows in a fuel processor, that is, the hot reformate and hot tail gas of the catalytic burner on the hot side, and the steam generation, air heating, and diesel evaporation on the cold side of a giant heat exchanger. A fuel-cell stack must also be cooled by a heating medium, which itself will be cooled by the surrounding air. The cooling effort contributes 10 kW$_{th}$ to the hot composite curve. Finally, 9 kW$_{th}$ is required for the cooling and condensing tasks of the cathode air off-gas and the tail gas of the catalytic burner. In addition, the heat release of the WGS reaction and the heat demand of the steam generation can be identified by their isothermal characteristics. Autothermal reforming connects the right-hand ends of the curves with each other.

32.2.3.4 Exergy Analysis

A number of papers propose using the thermodynamic property of exergy to evaluate processes of fuel-cell systems [20–23]. Exergy is defined by

$$\dot{E} = \left(\dot{H} - \dot{H}_0\right) - T_0 \left(\dot{S} - \dot{S}_0\right) = \left(\dot{G} - \dot{G}_0\right) + \dot{S}\left(T - T_0\right) \tag{32.19}$$

Furthermore, Cownden *et al.* [20] described chemical exergy as a compositional imbalance between a substance and its environment. They compared the chemical potential of a species j in a flow, that is, μ_{j0}, with its potential after the chemical reaction with its environment, μ_{j00}. A summation was then carried out for all reacting species.

$$\dot{E}_{chem} = \dot{n} \left[\sum_j x_j \left(\mu_{j0} - \mu_{j00}\right) \right] \tag{32.20}$$

This method allows all process steps to be evaluated. Figure 32.5 shows the exergy flow in the 10 kW$_{el}$ HT-PEFC system, which is also the basis for the pinch and G,T diagrams (see Figures 32.3 and 32.4). In contrast to a pinch analysis, single components must be defined in the exergy flow. These are sketched as boxes in Figure 32.5. In each box, the enthalpy change, the exergy loss, and the exergetic efficiency are given. As can be seen in Figure 32.5, an exergy flow of 30.5 kW for diesel is pumped into the fuel evaporator. It is mixed with superheated steam and compressed air at 623 K. The exergetic value of the mixture is determined by the chemical exergy of diesel. Autothermal diesel reforming leads to a high exergetic efficiency of 98.7%. In addition, the subsequent steps of cooling and gas cleaning offer high exergetic efficiencies. The high chemical exergy of the reformate gas and the low exergetic losses by gas cleaning determine the efficiency calculation. The fuel cell converts 85% of the hydrogen into electricity with a cell efficiency of about 52%, assuming a cell voltage of 650 mV. Without using the released heat of the electrochemical reaction, an exergetic efficiency of 62.8% can be calculated. Finally, an exergy flow of 6.2 kW leaves the fuel cell on the anode side. The combustion process, which occurs in the catalytic burner at an adiabatic temperature of 1022 K, leads to a high exergetic loss of nearly 3 kW, that is, $\eta_{Ex} = 53.6\%$. The burner off-gas

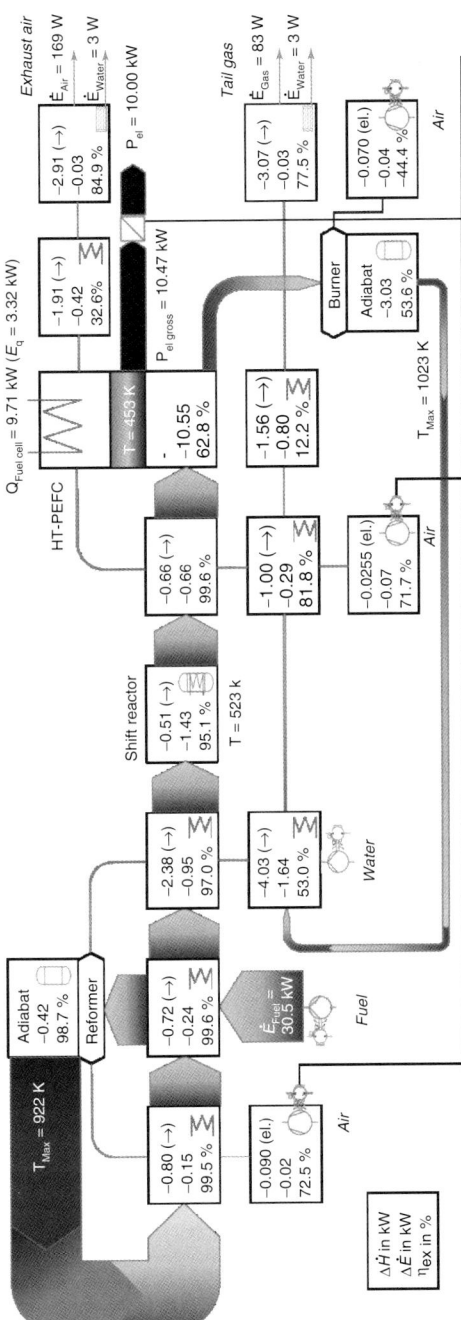

Figure 32.5 Exergy flow diagram for a 10 kW$_{el}$ HT-PEFC system based on $C_{12}H_{26}$ as a model fuel for kerosene.

is used in this scheme to warm the water and to achieve partial evaporation. The temperature difference between the hot burner off-gas and the boiling water is fairly high and leads to a high exergy loss, that is, $\eta_{Ex} = 53.0\%$. The evaporator outlet temperature of the tail gas is 592 K. A subsequent heat exchanger recovers heat from the tail gas to warm the cathode air from 319 to 400 K. The inlet temperature demanded for the cathode air is realized with a countercurrent heat exchanger, which cools the shift reactor off-gas. The residual heat of the cathode off-gas and tail gas is transferred to the surroundings, resulting in high exergy losses, that is, $\eta_{Ex} = 12$ and 32%. Finally, water is condensed to realize a closed water circuit.

The focal point of this method is the optimization of the electricity production rate. Important exergy losses were found for steam generation, catalytic combustion, and the fuel cell.

32.2.3.5 Process Optimization

The usefulness of various evaluation methods depends strongly on the system configuration. Different fuels and fuel-cell types and varying targets for diversified applications lead to a huge number of possible examples for combined systems. In this section, systems of particular interest are discussed to elucidate some of their typical principles. The focal point for a pinch point analysis is the optimization of the effort for heat exchanger networks. Figure 32.6 shows three different system setups: a PEFC operated with diesel, an HT-PEFC for methanol, and an SOFC combined with partial oxidation of gasoline. Tetradecane ($C_{14}H_{30}$) was used as the model fuel for diesel. All systems had an electric power of 10 kW. In the PEFC system, about 32 kW was transferred by heat exchange from the heat source to the heat sink. The parameters for ATR were $O_2/C = 0.47$, $H_2O/C = 1.9$, $T_{Inlet} = 683$ K, and $p = 2$ bar [8]. The elevated pressure was necessary to close the water balance at a temperature for condensing of about 60 °C. The fuel cell was operated at a

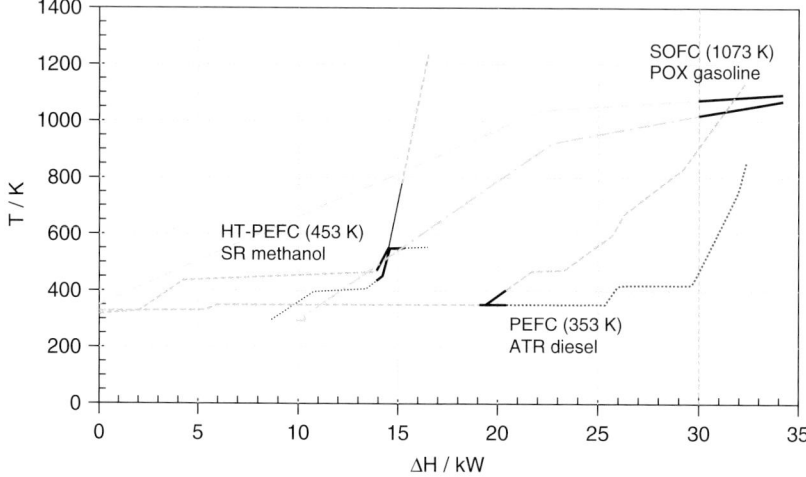

Figure 32.6 Pinch diagram for different 10 kW$_{el}$ fuel-cell systems based on selected fuels.

Table 32.1 Data collection from a basic pinch analysis for four different fuel-cell systems and four different fuels.

Evaluation criteria	SOFC	HT-PEFC	HT-PEFC	PEFC
Stack data: U_{Cell} (mV); u_{H_2} (%); p (bar)	750; 80; 1.1 anode and cathode recycling	650; 90.5; 1.1	650; 85; 1.2	750; 85; 2
Fuel; kind of reforming; mixture; T_{Inlet} (K)	C_8H_{18}; POX; $H_2O/C_8H_{18} = 8$, $O_2/C = 0.475$; 900	CH_3OH; SR; $H_2O/C = 1.3$; 553	$C_{12}H_{26}$; ATR; $H_2O/C = 2.5$, $O_2/C = 0.4$; 623	$C_{14}H_{30}$; ATR; $H_2O/C = 1.9$, $O_2/C = 0.47$; 683
System efficiency, η_{sys} (%)	37.2	52.1	36.0	32.7
Heat exchanger surface area, A (m^2) [Eq. (32.18)]	3.0	2.5	4.1	5.2
Heat exchange (kW)	34.3	16.5	29.2	32.2
ΔT_{max} (K)	375	621	400	400
ΔT_{min} (K)	27	16	20	16

POX = partial oxidation; SR = steam reforming.

cell voltage of 750 mV, an air ratio of 2, and a hydrogen utilization rate of 85%. Further results are given in Table 32.1. As indicated in Figure 32.6, the pinch temperature was about 353 K, which is the operating temperature of the PEFC. The analysis of the corresponding data resulted in a minimum temperature difference of $\Delta T_{min} = 16$ K and a maximum of $\Delta T_{max} = 400$ K. The system design can be optimized by detailed engineering. As an example of a possible improvement, the moistening of the cathode air could be better realized by a humidification membrane instead of spraying condensate into air, which requires a heat of evaporation of about 5.7 kW$_{th}$.

Our second example is an HT-PEFC combined with methanol steam reforming. The conditions for steam reforming were $H_2O/CH_3OH = 1.3$, $T = 553$ K, and $p = 1.1$ bar. The fuel cell was operated at a cell voltage of 650 mV, an air ratio of 2.5, and a hydrogen utilization rate of 95%. The high utilization rate was selected to test a system in a critical region with respect to the pinch point. The temperature difference between the hot and cold composite curves was between a minimum of 16 K and 621 K at the hot end of the catalytic burner. Such large temperature differences are difficult to realize with economic materials in a real system. The adiabatic temperature of the catalytic burner was reduced from 1253 to 950 K by increasing the air ratio from 1.1 to 3. This measure led to a conflict with the second law of thermodynamics and also to cutting of both composite curves. As a counter measure, the hydrogen utilization rate was reduced to 85%. This led to a lower efficiency, that is, $\Delta\eta = -3\%$, and a lower adiabatic temperature of 1030 K at $\lambda = 3$. Finally, constructive measures, such as the implementation of a steam circuit as a heat transfer medium [24] or combined burner/reformer devices [25, 26], were found to be the most effective solution. Methods such as pinch analysis make these problems more obvious.

The third example in Figure 32.6 shows the composite curves for a SOFC system combined with the partial oxidation of octane (C_8H_{18}). The conditions were $O_2/C = 0.475$, $H_2O/C_8H_{18} = 8$ [steam to carbon ratio (S/C) = 1], $T_{Inlet} = 683$ K, and $p = 2$ bar. In practice, developers tend to realize anode gas recycling with S/C ratios of about 0.3, which lead to a lower effort for recycling [27]. As can be seen from Figure 32.6 and Table 32.1, the exchanged heat was fairly high, that is, 34.3 kW_{th}, and the pinch temperature corresponded to the operation temperature of the fuel cell. SOFCs are cooled by the inflowing gases. The heat of the electrochemical reaction is therefore used to heat the anode and cathode gases. The temperature profile of the anode and cathode gas flows can be determined by extended computational fluid dynamics (CFD) models. The difference in temperature for the incoming gases is critical due to the maximum stresses in the composite of ceramic and metallic materials. For a first approximation, this difference was fixed to 150 K. The results for the SOFC system in Figure 32.6 and Table 32.1 correspond to an improved design including an anode and a cathode gas recycle loop. An anode gas recycling loop allows steam from the SOFC to be used for ATR. A recycling rate of 51.5% for the outlet anode gas flow at operation conditions of $H_2O/C_8H_{18} = 8$ and $O_2/C = 0.475$ therefore avoids the use of a steam generator. A cathode gas recycling loop leads to a significantly smaller air preheater. A recycling rate of 70% for the outlet cathode gas flow leads to an air ratio of 3.2 for fresh air at an overall air ratio of 8.6 and a tolerable temperature difference at the entrance of 150 K. Without both cycles, an air ratio of 8.1 at $\Delta T = 150$ K and an exchanged heat of 66 kW_{th} for all of the heat exchanger networks was calculated. It must be stressed that a pinch analysis is not compulsory for these findings.

Table 32.1 shows the basic results for four different fuel-cell systems. Although methanol offers the highest efficiency as an energy carrier for a fuel-cell system, the efficiency of methanol production from natural gas is only 62%, whereas diesel can be produced with an efficiency of 86% [28]. However, small methanol-based systems in combination with a feasible infrastructure require less effort for reforming and heat exchange than diesel-based systems. The choice between different types of fuel cells depends strongly on the type of application. As an example, airborne auxiliary power unit (APU) applications are discussed today as multifunctional systems involving water production, the use of the cathode tail gas for tank inerting (O_2 contents less than 12%) and electricity production [29]. SOFCs, with their high air ratios, cannot be used in systems for tank inerting. Additionally, under severe ambient conditions, they lead to a non-closed water balance. A PEFC system operated at 2 bar suffers from the peripheral losses of the air compressor. An elevated pressure is therefore necessary to close the water balance [30]. Table 32.1 shows the results for an HT-PEFC system based on kerosene as the fuel. By

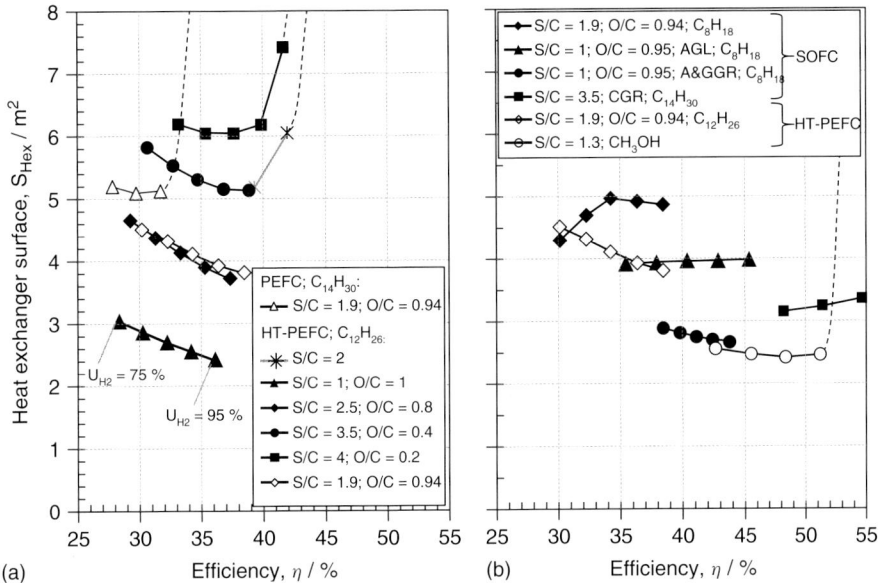

Figure 32.7 Estimated heat exchanger areas for different systems as a function of efficiency; S_{Hex} in both diagrams was determined for hydrogen utilization from 75 to 95% in 5% steps.

comparing the pinch diagrams in Figures 32.4 and 32.6, it seems that it is not necessary in all cases to perform a complete pinch analysis.

In a pinch diagram, hot and cold composite curves can be shifted towards each other using additional heating and cooling devices. In a fuel-cell system, the amount of heat provided varies with hydrogen utilization. A higher utilization in the fuel cell leads to lower adiabatic temperatures in the catalytic burner and a lower enthalpy flow for heating purposes. It is expected that optimum utilization exists in terms of heat exchanger effort.

Figure 32.7 shows different examples of how the pinch point method is applied. As can be seen in (a), most ATR/HT-PEFC systems for kerosene have a heat exchanger area that decreases with increasing utilization. Increased efficiency overrules a higher effort due to a smaller temperature difference. In the literature, operating parameters can be found close to those of steam reforming [31], that is, S/C = 3.5–4 and O/C = 0.2–0.4. By assuming such values for an ATR process, a minimized heat exchanger surface appears at utilization rates of 80–85%. Higher rates at 94% lead to the existence of a pinch point.

Diesel, air, and steam were heated to a uniform temperature for these calculations. The inlet temperature was 623 K for the mixtures with S/C = O/C = 1 and S/C = 2.5, O/C = 0.8. For mixtures with S/C = 3.5, O/C = 0.4 and S/C = 4.0, O/C = 0.2, an inlet temperature of 1123 K was required to gain an adiabatic outlet temperature of 860 K. Experimental results have shown that diesel and kerosene should not be heated to temperatures much higher than 473 K as carbon deposition occurs. Therefore, the steam was superheated to elevated temperatures compared to the final mixture to gain a defined mixing temperature, while diesel was heated in the liquid state to 400 K only. For such a process route, diesel at 400 K must be sprayed and evaporated in steam at 798 K to gain a mixture with S/C = 1.9 at 623 K. The chemical equilibrium of the reformer product gas leads under adiabatic conditions to a temperature of 993 K. Temperatures higher than 860 K are required to achieve a methane concentration of around 2 vol.% and 930 K for below 0.5 vol.%. In low-temperature fuel cells, such as the PEFC and HT-PEFC, methane formation is considered a loss in system efficiency, whereas in the MCFC and SOFC, methane is converted by internal reforming processes.

It is important to note that experimental findings determine the feasible operating range of a fuel processor. To avoid carbon deposition and to achieve complete fuel conversion, a certain amount of oxygen must be added to the fuel–steam mixture. An operating temperature between 1123 and 1223 K is most favorable for ATR. Pasel *et al.* [8] reported on long-term experiments with the operating parameters S/C = 1.9 and O/C = 0.94. Figure 32.7 shows the relation between efficiency and heat exchanger surface for an ATR process operated at S/C = 1.9 and O/C = 0.94, where the hydrogen utilization was varied between 75 and 95%. The characteristic curve here corresponds well with that for the parameters S/C = 2.5 and O/C = 0.8. A parameter set with less oxygen and more steam, that is, S/C = 3.5 and O/C = 0.4, leads to a small advantage in efficiency at constant hydrogen utilization, while the required heat exchanger surface must be extended by 25%.

Figure 32.7 shows further curves for various system configurations. An HT-PEFC system based on kerosene ATR ($C_{12}H_{26}$; S/C = 1.9 and O/C = 0.94) is used as a reference curve. PEFC systems based on the ATR of diesel (here $C_{14}H_{30}$; S/C = 1.9 and O/C = 0.94) are limited by a pinch point for hydrogen utilization rates higher than 85%. To close the water balance at a condensing temperature of 55 °C in the tail gas, a pressure of 2 bar is necessary. This results in a remarkable peripheral loss by the air compressor. Although the PEFC cell voltage chosen here was somewhat above that of an HT-PEFC (750 mV compared with 650 mV), the system efficiency was about 1–2% lower, and the demanded heat exchanger area was about 25% higher.

SOFC systems have a sophisticated behavior depending on the system design. The basic design results in an increasing demand on the heat exchanger surface with an increasing hydrogen utilization and increasing system efficiency. By introducing an anode gas recycling loop, a certain amount of steam will be recycled. The reformer operates in ATR mode with S/C = 1 and O/C = 0.95, but without an external steam supply it adapts to the fuel supply of partial oxidation. An analysis of the gas composition in the process showed a clear difference in the chemical reactions on the anode side of an SOFC. With anode gas recycling, about 12.6% CO, 12.1% H_2O, and 0.1% CH_4 flow to the anode, whereas without a recycle loop, the reformer off-gas consists of 8.7% CO, 20.7% H_2O, and 0.6% CH_4. In principle, an SOFC cell can convert H_2, CO, and CH_4 directly into electricity by electrochemical reactions. In practice, methane steam reforming and a WGS reaction lead to a chemical conversion of CH_4 and CO into H_2 and CO_2, respectively. Owing to the limited steam content in the case of anode gas recycling, a certain amount of CO cannot be converted. Steam, which suppresses carbon deposition, will be depleted by the WGS reaction. At 75–85% hydrogen utilization, the heat exchanger surface area demanded increases from 3.5 to 4.3 m² for a 10 kW_{el} system. At higher utilization rates, it remains at 4.3 m². Hydrogen utilization is limited by the recycle loop. A rate of 80% in the stack leads to a rate of 89% for the system. Higher utilization rates in the system lead to further decreases in steam concentration in the stack. One way to reduce the air preheater size is to introduce a cathode recycling loop. The S_{Hex}, η curve in Figure 32.7 offers high efficiencies at low effort for heat exchange. Here, the heat exchanger surface area demanded decreases slightly with increasing efficiency. A hydrogen utilization of 80% leads to a 48% system efficiency and a heat exchanger surface area of 2.4 m². The main challenge is to develop an air blower or an air ejector for operation at temperatures between 923 and 1100 K.

Reflecting the results in Figure 32.7, steam reforming has a maximum yield for diesel, that is, 43 mol H_2 mol^{-1} tetradecane. A pinch point analysis for steam reforming combined with an HT-PEFC or an SOFC results in opposite findings. For an HT-PEFC system, the operating conditions are fairly challenging. At a hydrogen utilization rate of 75%, an S/C ratio of 2 is required for system configuration. The mixing chamber conditions are critical for carbon deposition owing to a low steam temperature of 603 K and a high fuel temperature of 573 K. The reforming process can only be realized for such an HT-PEFC system if the reforming catalyst ignites the gas mixture at 583 K and a catalyst temperature of about 1173 K is reached at the exit. Furthermore, the low-temperature shift stage must be operated at a low temperature (473 K) to gain an equilibrium concentration of 1.7%. At such a low temperature, poor WGS kinetics give rise to a huge reactor size. At a hydrogen utilization rate of 70%, these conditions could be slightly improved, but the advantage of 39% in efficiency is rather small compared with 36% for diesel ATR.

An impressive step forward could be made if the steam reforming of diesel was feasible for SOFC systems. Owing to the high operating temperatures of an SOFC, that is, higher than 873 K, good heat integration of steam reformers is possible. Diesel can be injected at 400 K and evaporated in superheated steam with a temperature of 823 K. Even at a high S/C ratio of 3.5, no pinch point occurs. Reforming should be performed at 923–1073 K. The reformer must be integrated directly into the SOFC to guarantee that the heat for the steam reformer can be provided by the cathode gas recycling loop. System efficiencies of 48–55% are favorable at hydrogen utilization rates of 75–85%. In principle, higher rates could be assumed but they are not realistic and would impair the visual comparison of the different systems in Figure 32.7. If the reforming catalyst is not active enough to achieve complete conversion at 1073 K, more thermal energy must be delivered by combustible residual gases. At reforming temperatures between 1023 and 1173 K, the burner device must be coupled to the steam reformer. Heat exchange is only realized at utilization rates lower than 70%, leading to a system efficiency of 45%.

Following the results of an exergy analysis, processes such as catalytic combustion, steam generation, and the fuel cell should be improved with regard to their exergetic efficiency. Several papers have proposed hybrid systems combining fuel cells – mostly high-temperature fuel cells (SOFC) – with small turbo machines [32–37]. Such systems have been designed in most cases for fairly large systems ($P_{el} > 400$ kW) operated as APUs in airplanes or ships. Multifunctional systems for aircraft using SOFCs have decisive disadvantages [29]. Applying the cathode off-gas for tank inerting purposes requires an oxygen content below 12 vol.%. SOFCs are cooled by inflowing air on the cathode side at higher air ratios, that is, $\lambda > 3-6$, which lead to low oxygen depletion. In addition, SOFC systems combined with ATR or catalytic partial oxidation only offer a low potential for water condensation [30].

Therefore, an HT-PEFC is analyzed as a fuel cell in a hybrid system in the following section. In order to compare the results with those in Figure 32.5, the fuel mass flow is held constant. Figure 32.8 shows an exergy flow chart illustrating the substantial changes brought about by such a system design.

- During flight – in avionics denoted mission – the ambient temperature is 216.5 K and ambient pressure is 256 mbar. The exergy calculations were therefore adapted to these conditions.
- During mission the cabin pressure drops to 750–800 mbar and the humidity of the cabin air is 15% for a long-haul flight [38]. It is recommended that cabin air be used for a fuel-cell system in aircraft [39] instead of pressurizing ambient air using an additional air compressor. For our calculations, an air compression of 0.8–1.013 bar was assumed. The air compressor operated at an isentropic efficiency of 70%. The energy loss for pressurizing the cabin air to 0.8 bar was allocated to the bleed air of the turbine.
- The burner off-gas cannot be used to generate steam. Water must be evaporated by the heat produced by the electrochemical reaction in the HT-PEFC. In Figure 32.8, the evaporation is performed directly in the fuel cell. In practice, the use of a heating oil cycle would be more realistic.
- For a closed water circuit, the condensation of water at 313 K from both fuel-cell off-gases is necessary. The catalytic burner is therefore fed with a fuel–air mixture at 313 K. This temperature is fairly low for ignition of the combustible species in the fuel–gas mixture.
- In order to simplify the system by reducing the number of turbo machines, the cathode off-gas is used for the combustion process. This measure leads to a high air ratio of 4.4 and a low adiabatic temperature of about 640 K. With regard to CH_4 emissions, complete methane combustion requires a temperature of 723–773 K.
- The expander efficiency was calculated with aid of a polytropic expansion coefficient of 1.67, that is, an isotropic efficiency of 70%. A reversible process [22] is totally misleading for determining the potential of such a hybrid system. In addition, the mechanical and electrical transmission losses were evaluated with a transmission efficiency of 95%.

As a result, 1.54 kW of electricity was supplied to the turbo machine and 10.47 kW to the fuel cell. Peripheral components, such as pumps and compressors, demanded

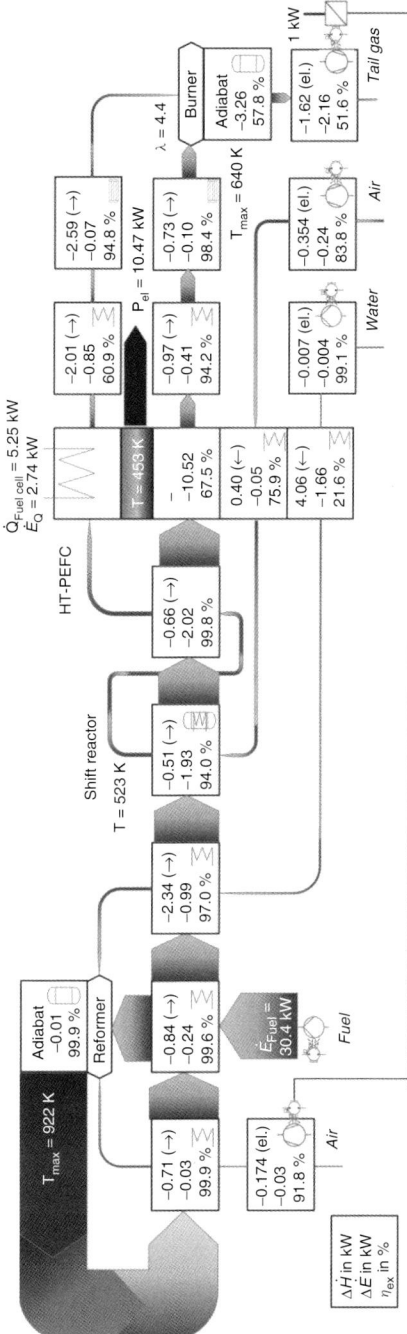

Figure 32.8 Exergy flow diagram for a power-optimized 10 kW$_{el}$ HT-PEFC system based on C$_{12}$H$_{26}$ as a model fuel for kerosene designed as a fuel cell/turbine hybrid.

0.58 kW of electricity, leading to an efficiency of 38.9%. This is an improvement of 4.5% compared with 34.4% for the calculations shown in Figure 32.5. Whereas the exergetic efficiency without an expander was 32.8%, the improved system offers 34.5%. This small change includes the transfer of an exergy flow of 2.7 kW from the jet engines to the APU system by cabin air. Finally, several critical aspects must be taken into account regarding the realization of such a system. In addition to the negative effects on methane combustion, the system size must be scaled up to several hundred kilowatts to find components for an applicable design. For example, the HT-PEFC should have a gross electric power of 400 kW_{el} combined with a 60 kW_{el} turbo expander.

32.2.4
Process Analysis and Design

Before starting work on the detailed design of a system, the basic concept and arrangement of the main components must first be decided with the help of various flow diagrams. According to international standards, for example, ANSI/ISA [40], a flow diagram is the simplified description of the structure and function of a process engineering plant using design marks. It helps the people involved in development to understand the planning, assembly, and operation of such plants.

Depending on the necessary information and its presentation, different types of flow diagrams are possible.

The simplest one is the *block [flow] diagram*, which describes the system in a simple form using rectangles connected by lines. This type of diagram contains the following basic information:

- designation of the rectangles
- media going in and out
- flow direction of the main media.

Figure 32.9 shows a possible SOFC power plant structure as an example. It shows the fuel side with gas supply, for example, flow control units, valves and measurement parts, and the gas preparation and preheating, which may involve desulfurization, pre-reforming, and recuperative heating using the hot anode off-gas. The air side is composed of a supply system (mainly consisting of a filter and blower) and preheating using the hot cathode off-gas. The cooled off-gases are fed into an off-gas burner (also called afterburner), where the remaining fuel which was not electrochemically converted in the SOFC is (catalytically) burned, increasing the off-gas temperature. This energy can be used for heating purposes (district heating) or to drive a gas turbine to increase the electrical output of the system [41, 42]. Finally, humid waste gas leaves the system at a different temperature than ambient temperature. This difference determines the exergy loss of the system. As the power output is typically connected to the AC grid, an inverter is needed to transform the DC power of the fuel cell into AC. This component can easily consume up to 5% of the produced electric power. For measurement and control, special equipment is necessary to operate the system in a flexible and

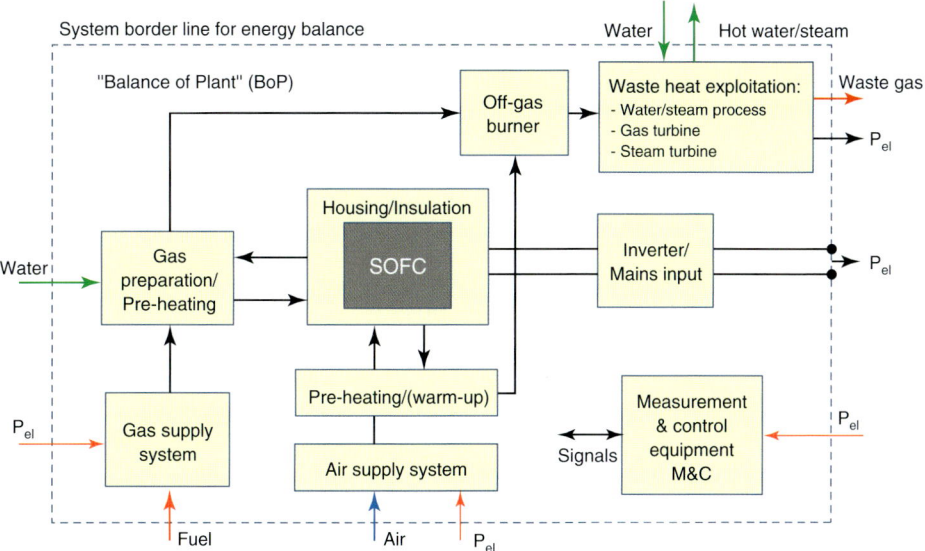

Figure 32.9 Block diagram of an SOFC power plant.

safe way. The dotted line surrounding the rectangles in Figure 32.9 represents the borderline for performing the energy balance of the system. To determine the efficiency of the system, all energy fluxes entering or leaving the system are considered.

Having decided on the basic arrangement of the system components, the next step is the detailed layout of the system. For this purpose, the second type of flow diagram is used, the *process flow sheet*, which describes the process using defined design marks for apparatus, armatures, and flow lines. This type of diagram contains the following basic information:

- all apparatus and machines necessary for the process and the main media lines
- designation of inlet and outlet media with their flow rates
- designation of energy and respective types of energy
- characteristic operating parameters.

Before creating the process flow sheet, some parameters must first be fixed. These parameters are necessary to determine the system configuration and to perform the systems energy balance. This energy balance will provide the characteristic operating parameters of the various components. An example of such a parameter list is given in Table 32.2. Some of these values are fixed based on already available measurements or on experience, for example, the pressure drop of the system. The exact value is available at the earliest after the detailed design of all components and piping has been completed.

A possible process flow sheet for an SOFC power plant [43] is shown in Figure 32.10. On the fuel side, water is injected into the methane supply line in front of a boiler which is heated by the hot off-gas. The amount of water is derived

Table 32.2 Parameter list as a basis for system calculations.

Input parameter	Basic system data
Fuel	Methane
Air composition	N_2 79 vol.%, O_2 21 vol.%
Air blower efficiency	Isentropic efficiency 60%, mechanical efficiency 50%
S/C value	2
Reformer	Steam reformer in equilibrium
Reformer equilibrium temperature	450 °C
Mean cell voltage	750 mV
Air temperature stack inlet	625 °C
Air temperature stack outlet	700 °C
Δp_{system} (cathode side)	200 mbar
Heat loss stack and components	0 (means neglected in the first run)
Condensation temperature	50 °C (off-gas temperature)
TTD heat exchangers	Gas/gas 50 K, gas/water 10 K
System gross AC power	20 kW

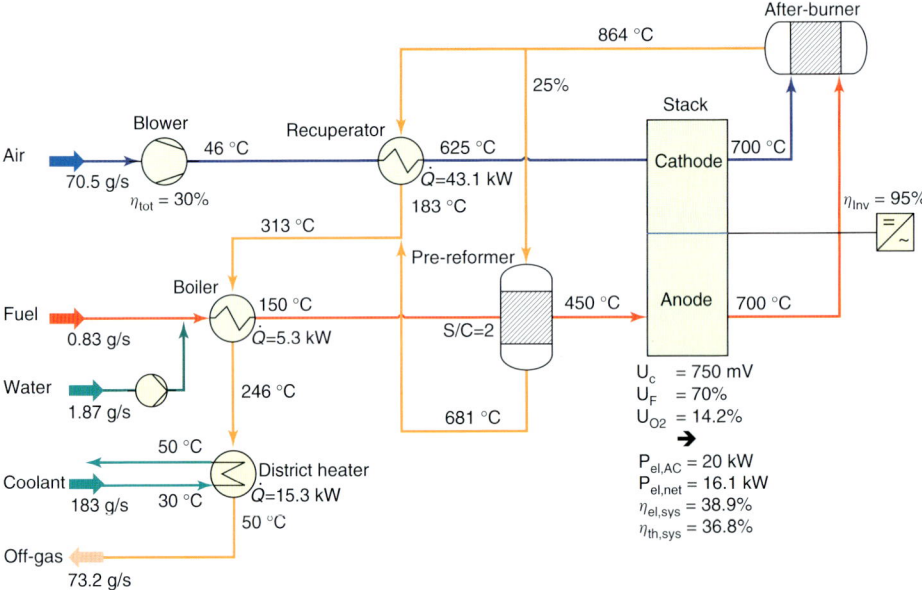

Figure 32.10 Process flow sheet of an SOFC power plant.

from the required S/C ratio of 2. This humidified fuel is overheated to 150 °C to avoid condensation in the piping. A certain amount of methane is converted in a pre-reformer into hydrogen, which is necessary to operate the SOFC without running the risk of damaging the anode [44]. The composition in equilibrium is determined by the outlet temperature, which is set at 450 °C. This is also the inlet temperature to the anode of the SOFC. The main consumer of electrical energy in the system is the air blower. Therefore, a pressure drop in the system on the cathode side and the efficiency of the blower are important parameters for the system calculation. The air is preheated to the required stack inlet temperature recuperatively. In contrast to the arrangement shown in Figure 32.9, air is not heated directly by the cathode off-gas but with the gas leaving the afterburner. Burning the anode and cathode off-gases directly after they have left the stack increases the off-gas temperature. This leads to a larger temperature difference in the air preheater, which in turn cause the terminal temperature difference (TTD) to increase, resulting in a smaller heat exchanger. A certain portion of the off-gas must be fed into the pre-reformer to provide sufficient heat for the endothermic reforming process and to heat the fuel to 450 °C. Minimizing the flow via the pre-reformer is also an issue of optimization, which in turn minimizes the size of the air preheater. The two off-gas streams are remixed to provide the heat for the steam generator (including overheating). Finally, heat can be transferred into a district heating loop. The amount of heat used depends on the inlet temperature of the coolant. In this case, the inlet temperature has been set to 30 °C, which represents an optimal case (low-temperature heating system). The effect of this temperature on the thermal efficiency has been investigated, for example, by Blum et al. [7]. Based on all of these input data, a thermodynamic process analysis can be performed, using commercial process modeling tools, such as Cycle-Tempo (already includes a fuel-cell stack model) or ASPEN Plus (a fuel-cell stack model has to be added). Some of the data resulting from these calculations are depicted in the process flow sheet in Figure 32.10. As can be seen, the total electrical net efficiency here was 38.9%.

In the next step, different parameters can be varied to determine which provide the highest potential for efficiency improvement. Parameters that can be varied are cell voltage (influences the cell efficiency), reformer temperature (influences the amount of internal reforming and thus the amount of air required for cooling), air inlet temperature to the stack (determines the temperature difference along the stack and thus also the amount of air required for cooling), the fuel utilization, and the compressor efficiency (influences the internal consumption of electric power). Three different values are taken for each parameter (see Table 32.3), which would result in $5^3 = 125$ combinations. Twenty-seven combinations are shown in Table 32.3, where the optimum value was always taken for one variant and combined with three parameters from the other three variants. These combinations were then calculated for three cell voltages. The electrical and thermal net efficiency are given as results. The electrical net efficiency is given as a function of cell voltage for all of these combinations in Figure 32.11. Depending on the parameters chosen, the electrical net efficiency of the system varies from 34.1 to 58.7%.

Based on these calculations, the largest effect on system efficiency is caused by increasing the cell voltage from 750 to 850 mV, which increases the efficiency by up to 10.5%. The increase in fuel utilization improves the efficiency by up to 5.8%. Reducing the reformer temperature from 550 to 450 °C improves the efficiency by 4.8%. Increasing the temperature difference along the stack from 75 to 125 °C

Table 32.3 Efficiencies as the result of system calculations for five variants, each having three values.

U_C (mV)	T_{ref} (°C)	ΔT_{st} (°C)	u_F (%)	η_c (%)	$\eta_{el,net}$ (%)	$\eta_{th,net}$ (%)	η_{tot} (%)
750	550	75	70	30	34.1	36.4	70.5
800					39.4	33.2	72.6
850					44.6	30.0	74.6
750	500	75	70	30	36.8	36.6	73.4
800					42.0	33.4	75.4
850					47.3	30.2	77.5
750	450	75	70	30	38.9	36.8	75.7
800					44.1	33.6	77.7
850					49.4	35.1	84.5
750	450	100	70	30	41.0	37.0	78.0
800					45.8	36.9	82.7
850					50.5	39.5	90.0
750	450	125	70	30	42.4	37.1	79.5
800					46.8	40.8	87.6
850					51.2	42.2	93.4
750	450	125	75	30	44.7	33.4	78.1
800					49.5	33.5	83.0
850					54.2	35.0	89.2
750	450	125	80	30	46.9	30.1	77.0
800					51.9	26.4	78.3
850					57.0	28.4	85.4
750	450	125	80	40	48.7	29.2	77.9
800					53.4	25.6	79.0
850					58.1	27.8	85.9
750	450	125	80	50	49.8	29.2	79.0
800					54.3	25.6	79.9
850					58.7	27.8	86.5

improves the efficiency by up to 3.5%, with a lower effect at higher cell voltages. The same applies for the compressor efficiency, where an improvement from 30 to 50% results in an efficiency that is up to 2.9% higher. This effect would be higher at lower temperature differences along the stack. Based on the results of these system calculations together with the available stack characteristics (e.g., allowed temperature difference along the stack, power density at a certain voltage), the optimum parameter values can be set. Furthermore, all necessary layout data are available for a detailed design of the various plant components (e.g., heat exchanger, afterburner).

The process flow sheet presented in Figure 32.10 provides the minimum information that should be given. Very often, additional information is implemented in the form of a rectangle with four different values, for example, pressure, enthalpy, mass flow, and temperature. The orientation, together with the units used, should also be shown in the diagram. As an example, the configuration with the highest

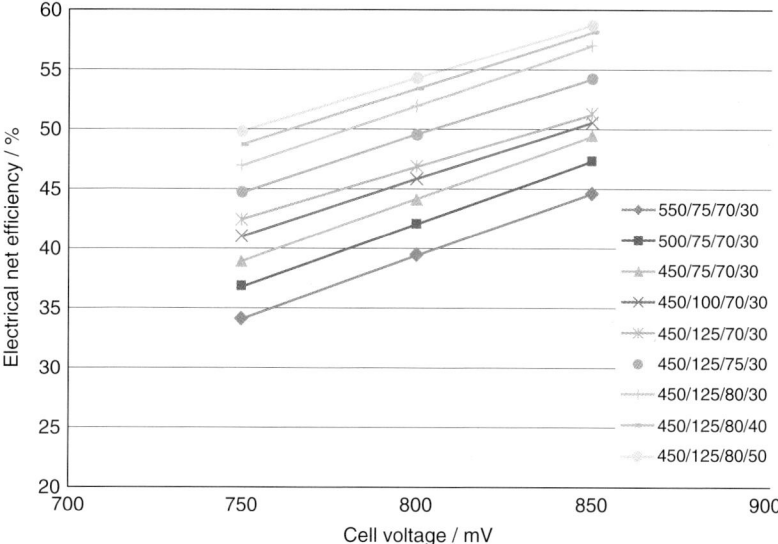

Figure 32.11 Process flow sheet of an SOFC power plant (parameters are $T_{ref}/\Delta T_{stack}/u_F/\eta_c$).

electrical net efficiency of 58.7% from the examples from Table 32.3 is shown in Figure 32.12.

32.3
Detailed Engineering

Detailed engineering builds on the work completed in basic engineering and all of the planning details are finalized.

Detailed engineering is composed of the following work packages:

- diagrams and plans
- failure mode and effect analysis (FMEA)
- selection of peripheral components
- drawings and piping.

Basic engineering uses block diagrams and process flow sheets to provide general information about the fuel-cell system to be developed. It covers the main components, temperatures, and flow rates, which are used later as input for detailed engineering. The properties of single components such as reformer, stack, and so on, and also experimental results for single components, are input here. Furthermore, experimental results from other systems, including experience with the predecessor system, can be used in the design. The results from modeling efforts, such as data for heat exchangers and compressors, are also useful information sources and are therefore incorporated into detailed engineering.

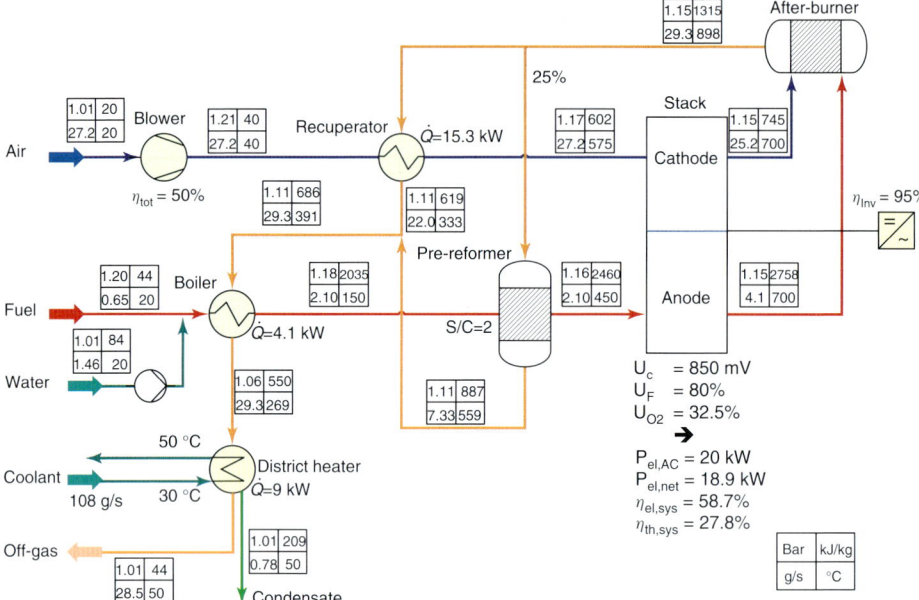

Figure 32.12 Process flow sheet of an SOFC power plant with additional information.

The first task in detailed engineering is to finalize the piping and instrumentation (P&I) diagram, starting with the information given in the block diagram, process flow sheet, and the preliminary P&I diagram. The P&I diagram contains all of the necessary information on the technical design of the system. The P&I diagram should include the following items [45]:

- All apparatus, machines, piping, controls, and instruments should be presented using standardized symbols.
- Nominal diameters, pressure levels, materials, and the design of piping must be given in their entirety. Piping, controls, and instruments should be drawn with respect to their proper functioning and location.
- Insulations must be indicated and marked.
- Measurement and control devices must be drawn schematically.
- All apparatus, controls, instruments, piping, and so on, must be labeled with distinguishing letters and consecutive numbers.
- The diagrams must be accompanied by specification sheets for the main apparatus.

The development of the P&I diagram is a continuous effort and the diagram is often updated during the subsequent steps. After the first outline of the P&I diagram is ready, an FMEA is carried out. FMEA is part of quality engineering. During the FMEA, the function of every component in the P&I diagram is checked in different operating modes. The failure of the components and the consequences of this on the system are also investigated.

At this stage, the P&I diagram is frozen. The next step involves preparing the parts list with design calculations. The parts list comprises detailed information about every single component in the P&I diagram. It is important to use a systematic marking procedure during the development of the P&I diagram so that the parts list can also be developed in an organized manner. Otherwise, both the P&I diagram and the parts list could become so complicated that further system development steps would be very difficult. In Europe, the standard ISO 10628 [46] is used to prepare flow diagrams. Code letters for apparatus, machines, instrumentation, and piping, and also examples of flow sheet diagrams, are standardized by this document. Another commonly used standard is ISA S5.1 from the International Society of Automation [40].

The parts list is composed of single sheets for valves and fittings, heating systems and heat exchangers, pumps, pressure vessels, piping, mechanically operated controllers, flow controllers (CFs), pressure measurement units, temperature sensors, and special units, such as fuel-cell stacks or cell voltage monitoring. Each sheet is built up with the necessary information for each of these units. For instance, the necessary information for a fitting will be its identification mark, supplier and type, medium, density, maximum flow rate, connections at inlet and outlet, maximum pressure and operating pressure and pressure difference, flow coefficient, and material. It is common practice to extend the parts list with information about procurement. If this is the case, the parts list also contains additional information about the actual status of inquiry, bidding, ordering, and delivery. A pressure drop calculation for the complete system is also included in the parts list. Peripheral components are selected at this stage.

After the parts list has been completed, several work packages are started and run in parallel. These include the list of metering points and procurement. The list of metering points consists of information for the electrical diagram, and also for the development of control and monitoring systems. Procurement can start once the parts list is ready. When the list of metering points is complete, procurement will be extended to include the electrical parts.

The next work package is the preparation of drawings. In this work package, the fuel-cell system is designed using computer-aided design (CAD). The parts list and the list of metering points together with the P&I diagram deliver the most important information for the development of drawings. Other important information comes from 3D CAD models of single components, external and internal requirements such as dimensions and arrangements, and the 3D CAD models or the dimensions of purchased parts. The first step involves developing the optimal design concept using the available information. Since this step is decisive for the final system design, it is advantageous to carry out a feasibility study or a preliminary study with several concept ideas and to select the best concept before starting detailed design. After the design concept has been finalized, the base frame is designed, incorporating the position of the main components. This is followed by the spatial arrangement of the prefabricated parts. In the last stage, the components are combined with each other to form the complete system.

Detailed engineering delivers all of the necessary documentation for procurement and construction and also for commissioning and normal operation. These documents are used as a basis for the following [47]:

- procurement of all necessary components
- construction of the system
- functional testing and safety-related testing
- confirmation of proper construction and operation
- use of the system in accordance to regulations
- maintenance and if necessary extension of the system
- warranty and confirmation of due diligence of responsible persons and institutions.

During the detailed engineering phase, the final planning documents are prepared. These will be summarized later in the technical documentation of the system. Detailed engineering is a continuous effort, which finishes only with the commissioning of the system.

32.3.1
Piping and Instrumentation Diagram

The main content of a P&I diagram was outlined in the previous section, namely all of the details about the system to be built. However, it is impossible to record all process details on the diagram. Therefore, the P&I diagram and the parts list together provide the information required for the remaining steps. Different standards exist that define the graphical symbols to be used. However, it should also be noted that many institutions have their own system for the selection of graphic symbols and code letters. The first letter always indicates the property measured. The letters given in Table 32.4 are typically used in P&I diagrams.

If the instrument only has an indicating function, it should be denoted with I, such as TI for temperature indication. If it only has a recording function, the second letter is R. A controlling function is denoted by C, whereas S denotes steering. If an instrument has mixed functions, such as recording (R) and controlling (C), it must be denoted with both letters. FRC denotes flow rate recording and controlling. Similarly, indicating and controlling is denoted with IC such as FIC for flow rate indication and controlling. The letter A is typically used to indicate an alarm. An upper alarm level is indicated with (+) and (−) denotes a lower alarm level. In complicated diagrams, different flow media are indicated with different colors and are assigned a letter code. Using this system, it becomes easier to find the position of the particular component, since it is obvious in which string it is located. In addition, standard symbols are used to define measurement and control units for volumetric and mass flow rates, temperature, and pressure. The piping must also be identified with a clear code, as should the pipe diameter and thickness. Further details on codes and symbols can be found in standards [40, 46, 48, 49]. Some commonly used symbols and identification methods are given below as an introduction to the topic.

Table 32.4 Letters typically used to indicate measured properties in P&I diagrams.

Letter used	Measured property
F	Flow rate
L	Level
P	Pressure
Q	Quality, analysis, quantity
R	Radiation
T	Temperature
W	Weight
A	Analysis
B	Burner, combustion
D	Density
E	Electrical variables, voltage
G	Position, length
H	Hand
I	Current
J	Power
K	Time
M	Humidity
S	Speed, frequency
V	Viscosity, vibration
C, X, O, Z	Any other property

Figure 32.13 Examples for measuring and display in P&I diagrams.

In Figure 32.13, different examples are given for measuring and display. The first example (FIRCA +/++) is for flow indication (FI), registration (R), control (C), and alarm (A). It includes two alarm levels, both at the upper level: a first alarm (+) and a second alarm (++). The letter combination CF is used for flow controller. The first number after letter codes (4) shows the number code for the actual system. It is advantageous to use such a code if more than one system are operated in parallel or series. The letter A denotes the medium that flows through the particular string. The last two numbers (01) denote that this is the first CF on the actual string (A). The number in the parallelogram shows the flow rate of gas. According to the type of diagram, this number can be the maximum, minimum, or design value. The unit is additionally drawn at a proper position in the diagram. The second example (PIAS -/-) is of a pressure switch, and deals with pressure indication, alarm, and

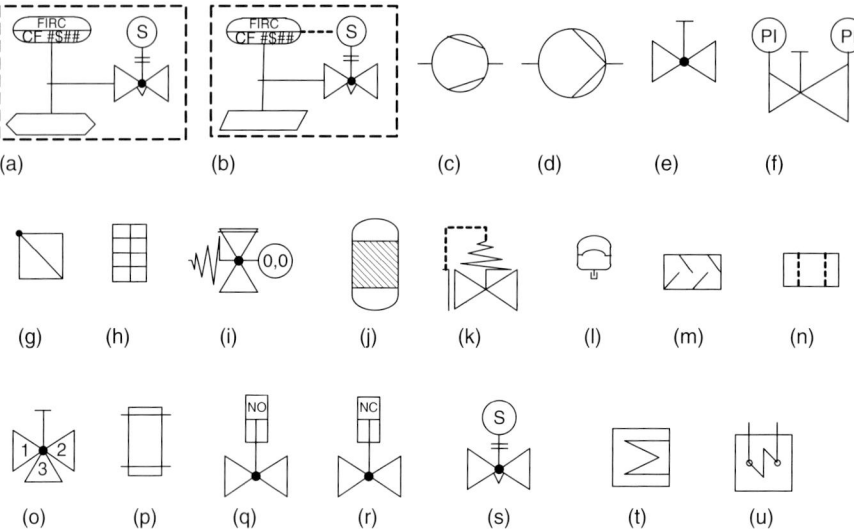

Figure 32.14 Examples of commonly used symbols in P&I diagrams: (a) mass flow controller; (b) volumetric flow controller; (c) compressor (general); (d) pump (general); (e) isolation valve; (f) pressure regulator; (g) back-pressure valve; (h) flame trap; (i) safety valve; (j) reaction vessel; (k) quick coupler; (l) membrane reservoir; (m) mixer; (n) liquid filter; (o) 3/2-way shut-off valve; (p) barrel; (q) pneumatic diaphragm valve (normally open); (r) (normally closed); (s) regulating valve; (t) heat exchanger (general); and (u) evaporator.

steering. The alarm levels are both for the lower level. The third example (TIRAS +/++) represents temperature indication, registration, alarm, and steering, as in a thermocouple, with both alarms at the upper level.

Examples of commonly used symbols for P&I diagrams are given in Figure 32.14. It should be taken into account that this is not a complete list and only gives an overview. Figure 32.15 shows the P&I diagram of a test module for fuel-cell stacks. Owing to its complexity, only a sectional drawing is included here.

32.3.2
FMEA

FMEA is a quality engineering tool that can be used to distinguish possible errors and their effects during the design phase of a fuel-cell system. It can be very advantageous to carry out an FMEA during the detailed engineering phase, before the P&I diagram is frozen and procurement begins. An FMEA will make it possible to acquire an overview of the function of the designed system as a result of the planned concept and of selected apparatus and also controls and instruments. An FMEA typically results in some changes in the P&I diagram, with the aim of minimizing errors during later operation. In particular, it identifies errors affecting the optimal operation of the system and also errors that could lead to deficits in the safety concept, thus allowing them to be eliminated. It is advantageous to carry

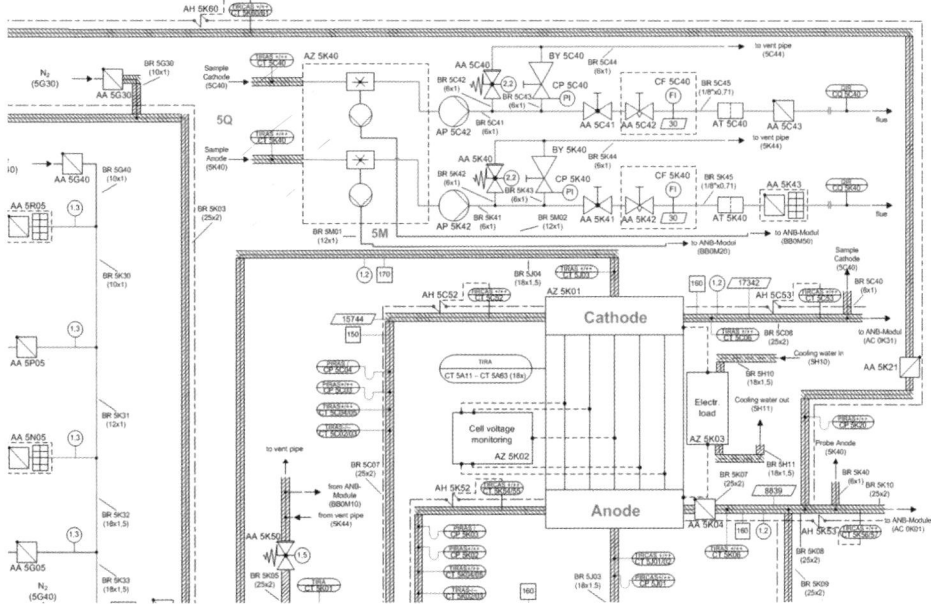

Figure 32.15 Sectional drawing of the P&I diagram of a test module for fuel-cell stacks. Source: Forschungszentrum Jülich, IEK-3.

out the FMEA with the participation of additional personnel to those who were involved in the actual planning.

The benefits of an FMEA can be summarized as follows (adapted from [50]):

- identification of weak points
- early identification and elimination of possible errors during different phases
- estimation and quantification of risk of errors
- target-oriented utilization of available expertise
- reduction in development time with the help of better preparation
- reduction in the development of errors with the help of termination at the right time
- ease of handling, broad range of application areas
- supplement for well-proven and new working methods
- proof of robustness of systems and processes against errors
- improved reliability.

Different FMEA types are discussed in the literature [51], and can be classified as *system FMEA*, *construction FMEA*, and *process FMEA*. In the context of fuel-cell systems development, the system FMEA will be considered here. Such an analysis deals with the proper interaction between system components and also their interfaces. However, it should be noted that all FMEA types follow almost the same procedure. It is possible to define the following phases for a system FMEA [51]:

- structure analysis
- functional analysis
- error analysis
- risk assessment
- optimization.

The structure analysis deals with the design of a structure based on the requirements of a system. In terms of functional analysis, the functions of system components are defined and fixed. Subsequently, functional dependences are identified in preparation for the error analysis. Potential failures are defined in the error analysis. First, the failures are derived from functional analysis. Second, the effects of failures and their causes are defined. The information collected is then analyzed in terms of risk in the risk assessment. The risk is estimated using three criteria: severity, probability, and detection. Severity defines the importance of the effects of this failure. Probability defines how often this failure can happen. Detection is a measure that defines the probability of detection of the failure. Typically, these three risk criteria are multiplied by each other, resulting in a risk priority number (RPN). These numerical ratings vary widely in practice. Therefore, the single rating for each risk criteria must be defined at the beginning of the FMEA.

In the following, the FMEA procedure outlined above will be discussed in more detail with the help of a practical example. For this purpose, one component will be analyzed using the chart in Figure 32.16. For the sake of simplicity, the component chosen was a fuel-cell stack, its function being electrochemical conversion. In detailed systems, it is advantageous to follow a systematic approach. For example, the FMEA can be carried out analyzing each string in the P&I diagram. In our case, the fuel cell is a component of the reformate string. A fuel cell can undergo a number of failures. One possible failure type is the leakage of air to the outside. A reason for the leakage could be a breakdown in the sealing. This would result in an air deficit at the cathode, a decrease in cell voltage, or a decrease in pressure. Such a failure can be detected via pressure difference measurements or cell voltage monitoring. If we assume that the stack was leak-proof at the beginning of its operation, it can be concluded that such a failure cannot be prevented as it occurs during operation. In the next step, the numerical ratings are defined for the considered failure. A possible rating scale is given in Table 32.5.

Following the procedure in Figure 32.16, two ratings must be defined for each criterion. This is usually the case if the failure has more than one possible reason, or if there are alternative methods of detection. Coming back to our example of air leakage, we can define the probability as between 2 and 3. The possibility of detection is defined as between 4 and 1: if the leakage is small, it will never be detected; larger leakages are usually detected. The rating for severity can be defined as 1 for very small leakages to 3 for larger leakages. Multiplying the single ratings by each other, we get a RPN between 8 and 9. These numbers should be considered only as an indication and not as a decisive factor [51]. If high RPNs are calculated, the system must be modified to reduce the risk of failure during later operation.

Project:

FMEA Meeting nr.:	
Date / Place	
FMEA participants:	
System	
Operation mode:	

Nr.	Component	Function	Type of failure	Reason for failure	Effect of failure	Failure detection	Failure prevention	Probalitiy		Detection		Severity		Risk priority number		Remarks / Measures
								from	to	from	to	from	to	from	to	

Figure 32.16 Worksheet for the FMEA analysis. Forschungszentrum Jülich, IEK-3.

Table 32.5 Rating scale for risk assessment.

Rating	Probability	Detection	Severity
0	Never	Always	None
1	Very rarely	Usually	Very small
2	Rarely	Unlikely	Small
3	Sometimes	Very unlikely	Medium
4	Often	Never	Hard
5	–	–	Disastrous

32.3.3
Selection of Peripheral Components

Peripheral components, which are classified under the term "balance of plant" (BoP), must be selected depending on requirements such as compactness, reliability, high efficiency, power class, load profile and operating characteristics, low noise emissions, and non-reactivity. The field test results of fuel-cell systems in use show that the rate of system breakdowns are often caused by the BoP components and not by the fuel cell or major components such as the reformer. Typical BoP components in a fuel cell system are as follows:

- heat exchangers for heat recovery and educt conditioning, enabling operation with low pressure drops, high temperature differences, and high operating temperatures
- humidifiers especially for PEFCs
- pumps for water, fuel, or coolant supply
- blowers or compressors for the cathode, burner, and reformer; high-temperature blowers for systems with recycling such as SOFC systems
- recuperators for water recovery

- power conditioning units, such as DC/DC converters and DC/AC inverters based on the application and its requirements
- light batteries for hybridization
- sensors for various gases in the system, especially hydrogen sensors
- fuel storage, especially hydrogen storage
- valves and actuators.

In terms of reliability, the main challenges to be coped with are long operation times, a high temperature tolerance, and robustness during thermal cycles. High efficiency is the key issue for power conditioning units and cathode blowers. Electrical turbochargers offer very good efficiencies. Another major problem in the selection of BoP components for fuel-cell systems is their scalability. Most components are over-dimensioned and therefore do not run at their design level, since they have been developed for other applications.

A very helpful tool for the selection of peripheral components is the use of dynamic system models, including start-up, shut-down, and load change characteristics. With the help of such models, the requirements on the BoP components can be defined accurately in relation to real operating behavior.

32.3.4
Drawings and Piping

A piping model is used for the appropriate configuration and critical control of connections. Points of intersections and collisions can be identified and control elements can be arranged easily. Such a model contains not only the piping and instruments, but also fittings and components for measuring and control. The complete model delivers information about the manufacturing of the pipes and their assembly [45].

Instead of piping models, 3D CAD drawings are increasingly preferred. CAD drawings visualize the complete system to be built and include almost all details in a realistic way. With the help of such drawings, the accessibility and ease of handling of single components can be verified before the complete system is built [45]. Working with 3D CAD drawings also makes an iterative design approach possible.

As discussed above, a three-level approach can be used in design. Figure 32.17 can be used to explain this approach. First, the main frames are designed on the basis of the main components and larger components. After this, subassemblies (Figure 32.17b) are designed in 3D. The subassemblies are then combined with each other using appropriate piping and fittings, resulting in a complete system. A fuel-processing system (upper right corner in Figure 32.17c) is shown together with the complete testing facility designed in modular plug-and-play fashion as an example of a complete system. However, to be able to build this system, more information is required. From the complete model, single part drawings must be prepared in 2D and they must include all details required for manufacturing. Such a drawing is shown in Figure 32.17a.

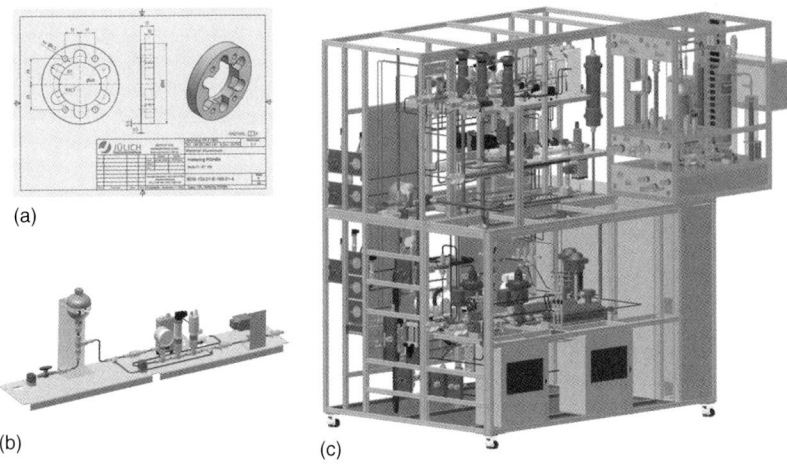

Figure 32.17 Three levels of drawings used in 3D CAD design: (a) example of a single part drawing for a part to be produced; (b) subassembly; and (c) complete system presenting the arrangement and combination of all parts and subassemblies.

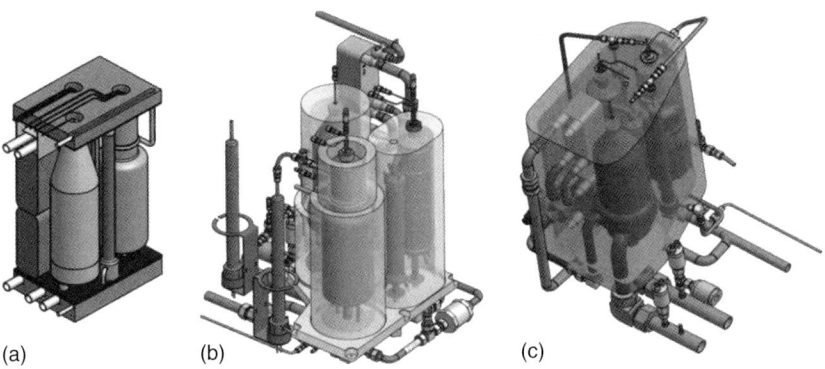

Figure 32.18 Concept based on the design study: (a) basic designs; (b, c) two versions using 3D CAD for a fuel processor system.

Another common usage of 3D CAD drawings is during the predesign phase. The same tool can be used in less detail to develop system concepts parallel to other planning activities, for example, after an initial P&I diagram is ready. This will help a compact system design to be achieved. An example of such a use is shown in Figure 32.18. A rough concept (Figure 32.18a) was developed in the context of the design study. The concept was later implemented during the design of the fuel processor, taking additional details into consideration. In two stages (Figure 32.18b,c), a similar concept was realized with the help of 3D CAD.

32.4
Procurement

After the P&I diagram and accompanying lists have been finalized in detailed engineering, these are not usually changed unless there is an exceptional reason, for example, something that might result in malfunction of the system or cause a safety risk. This permits an ideal procurement process. The main steps of procurement are summarized (adapted from [52]) as follows:

- preparation of bid invitations using the information from the parts list
- sending inquiries to possible suppliers
- handling queries from bidders
- checking bids for completeness and comparability
- bid comparison and order recommendation by purchasing department
- negotiations
- preparation of order documents
- checking order confirmation
- approval of bill payments after checking the delivery.

In terms of the selection of suppliers, not only the technical ability of the supplier should be considered, but also the economic capacity, customer service, internal or external experience and references, and reliability concerning confidentiality [50]. The most important part of procurement is the preparation of bid invitations. The information on the bid invitation must cover the specifications and requirements in detail. This allows the bids to be checked for completeness and comparability. The purchasing department adds relevant business conditions to the invitations, makes the commercial bid comparison, and is responsible for order recommendation [52]. Diligent preparation of the information in the bid invitation will make it possible to use the same information for the order [52]. Later, the delivery must be checked for completeness. This includes examining the documentation and certificates, which will subsequently play an important role in the documentation of the project as a whole.

32.5
Construction

The construction phase begins when the detailed engineering and procurement phases have been completed. Construction can be classified into following steps (adapted from [50]):

- organization and preparation
- rough assembly of machines and apparatus
- subassembly of parts
- connection of subassemblies using piping
- assembly of electronic devices, controls, and instruments
- pressure test, labeling of components and piping

- insulation of the system
- preparation for commissioning.

In the detailed engineering phase, the system design was carried out using 3D CAD. The next step involves preparation for construction. Once the complete system has been designed, single part drawings are generated for the parts to be produced. In the next step, subassembly drawings with parts lists are prepared in the form of assembly schedules. Finally, a complete assembly schedule is generated, presenting the arrangement and combination of all parts and subassemblies. Using this approach, preparation for construction is integrated into the design phase. Since construction has already been dealt with and scheduled in the design phase, it can be carried out in a systematic and time-effective approach. First, the main frames are built. Then, the single parts are manufactured and the subassemblies are mounted. In the final step, all components and subassemblies are combined with each other via piping. This concludes mechanical construction. Parallel to the mechanical phase, two work packages are carried out: electrical construction and preparation for commissioning. In terms of electrical construction, the switch cabinet is prepared and the wiring inside the system is performed. In preparation for commissioning, the electronic wiring is checked and a pressure test is performed. The construction phase finishes with the insulation of the system.

32.6 Conclusion

Systems engineering covers a wide range of engineering tasks and involves a plurality of different methods for designing systems. It requires different intermediate steps for successful system development depending on the application and its specific conditions. System development covers the steps of basic engineering, detailed engineering, procurement, construction, and commissioning.

This chapter focused on the principles of systems engineering for fuel-cell systems based on carbonaceous energy carriers. Such systems are ideal for explaining most of the aspects of systems engineering. Systems technology requires different intermediate steps for successful system development. In the 1980s, the air and space industries developed a scale to evaluate the distance to commercialization. This scale was named the technology readiness level, and contains nine different steps. The developments described in this chapter belong to technology readiness levels 1–6. General basic engineering considerations were presented, dealing with the definition of efficiencies and reaction mechanisms, particularly those concerning the transformation of hydrocarbon fuels into hydrogen-rich gases. Analytical methods for heat management, such as pinch point diagrams and exergetic analyses, were also described and used for process optimization. Process analysis was then explained by means of block diagrams and process flow sheets using an SOFC plant concept as an example. It was shown that varying the operating parameters can improve the electrical net efficiency of such a system from 34.1 to 58.7%.

We then moved on to detailed engineering, which involved finalizing all details of the planning. This was explained using a P&I diagram and an FMEA. The latter is a quality engineering tool that can be used to distinguish possible errors and their effects during the design phase of a fuel-cell system. Detailed engineering was accomplished by elaborating the parts list and drawings, by selecting components, and by procurement and construction.

List of Symbols and Abbreviations

A	area
ATR	autothermal reforming
E	exergy
FMEA	failure mode and effect analysis
H	enthalpy
HT-PEFC	high-temperature polymer electrolyte fuel cell
LHV	lower heating value
MCFC	molten carbonate fuel cell
$\dot{m}$	mass flow
g	molar specific Gibbs energy
h	molar specific enthalpy
s	molar specific entropy
P&I	piping and instrumentation
PEFC	polymer electrolyte fuel cell
POX	partial oxidation
S	entropy
SOFC	solid oxide fuel cell
T	temperature
TRL	technology readiness level
TTD	terminal temperature difference
u	utilization
WGS	water gas shift
ΔG	Gibbs energy
ΔH	reaction enthalpy
η	efficiency
μ	chemical potential

Subscripts and Superscripts

chem	chemical
Hex	heat exchanger
HV	higher heating value
LV	lower heating value
Th	theoretical
th	thermal
el	electric

References

1. Sadin, S.R., Povinelli, F.P., and Rosen, R. (1989) The NASA technology push towards future space mission systems. *Acta Astronaut.*, **20**, 73–77.
2. Mankins, J.C. (1995) Technology Readiness Level – A White Paper, http://ehbs.org/trl/Mankins1995.pdf (last accessed 1 August 2011).
3. Amphlett, J.C., Mann, R.F., Peppley, B.A., Roberge, P.R., Rodrigues, A., and Salvador, J.P. (1998) *J. Power Sources*, **71**, 179–184.
4. Höhlein, B., von Adrian, S., Grube, T., and Menzer, R. (2000) *J. Power Sources*, **86**, 243–249.
5. Ahmed, S. and Krumpelt, M. (2001) *Int. J. Hydrogen Energy*, **26**, 291–301.
6. Ersoz, A., Olgun, H., and Ozdogan, S. (2006) *J. Power Sources*, **154**, 67–73.
7. Blum, L., Deja, R., Peters, R., Stolten, D. (2011) Comparison of efficiencies of low, mean and high temperature fuel cell systems. *Int. J. Hydrogen Energy*, **36**, 11056–11067.
8. Pasel, J., Meissner, J., Porš, Z., Samsun, R.C., Tschauder, A., and Peters, R. (2007) Autothermal reforming of commercial Jet A-1 on a 5 kW$_e$ scale. *Int. J. Hydrogen Energy*, **32**, 4847–4858.
9. Roychoudhury, S., Junaedi, C., Walsh, D., Mastanduno, R., Spence, D., DesJardins, J., and Morgan, C. (2008) in *Proceedings of the Fuel Cell Seminar, Palm Springs, CA, October 28–30, 2008*, Courtesy Associates, Washington, DC.
10. Krummrich, S., Tunistra, B., Kraaij, G., Roes, J., and Olgun, H. (2006) *J. Power Sources*, **160** (1), 500–504.
11. Mengel, C., Konrad, M., Wruck, R., Lucka, K., and Köhne, H. (2008) Diesel steam reforming for PEM fuel cells. *J. Fuel Cell Sci. Technol.*, **5** (2), 021005. d
12. Thormann, J., Pfeifer, P., Schubert, K., and Kunz, U. (2008) Reforming of diesel fuel in a microreactor for APU systems. *Chem. Eng. J.*, **135S**, S74–S81.
13. Namazian, M., Sethuraman, S., Venkataraman, G., Bhalerao, A., Lux, K., Elder, W., Centeck, S., and Maslach, D. (2008) Complete 10 kWe JP-8-to-Power solution for APUs, in *Proceedings of the Fuel Cell Seminar 2008, Phoenix, AZ*, Courtesy Associates, Washington, DC.
14. Roychoudhury, S. (2009) Proceedings of the 10th Annual SECA Workshop, Pittsburgh, July 16 2009, http://www.netl.doe.gov/publications/proceedings/09/seca/ (last accessed 16 November 2009).
15. O'Connell, M., Kolb, G., Schelhaas, K.P., Schuerer, J., Tiemann, D., Ziogas, A., and Hessel, V. (2009) Development and evaluation of a micro reactor for the reforming of diesel fuel in the kW range. *Int. J. Hydrogen Energy*, **34** (15), 6290–6303.
16. Maeda, S., Kikunaga, S., Akoi, T., Nishikawa, S., Yamamoto, S., Akimoto, J., Anzai, I., and Ikeda, T. (2004) Development and operational study of kW-class PEFC cogeneration system using kerosene, in *Proceeding of the Fuel Cell Seminar, San Antonio, TX, November 1–5, 2004*, Courtesy Associates, Washington, DC.
17. Saito, K., Matsumoto, H., Kisen, T., Takahashi, O., and Katsuno, H. (2004) Development of fuel processing technologies for kerosene fuel cell co-generation system, im *Proceeding of the Fuel Cell Seminar, San Antonio, TX, November 1–5, 2004*, Courtesy Associates, Washington, DC.
18. Perry, R.H., Green, D.W., and Maloney, J.O. (1997) *Perry's Chemical Engineerings' Handbook*, 7th edn., McGraw-Hill, New York.
19. Linnhoff, B. (1989) Pinch technology for the synthesis of optimal heat and power systems. *J. Energy Resour. Technol.*, **111**, 137.
20. Cownden, R., Nahon, M., and Rosen, M.A. (2001) Exergy analysis of a fuel cell power system for transport applications. *Exergy Int. J.*, **1** (2), 112–121.
21. Chan, S.H., Low, C.F., and Ding, O.L. (2002) Energy and exergy analysis of simple solid-oxide fuel-cell power systems. *J. Power Sources*, **103**, 188–200.
22. Winkler, W., Nehter, P., and Williams, M.C. (2010) Basic thermodynamic studies and exergetic analysis of fuel cell steam reforming for natural gas in

waste landfill and anaerobic digestor gases. *Resour. Process.*, **57**, 3–7.
23. Nehter, P., Bøgild-Hansen, J., and Larsen, P.K. (2010) A techno-economic comparison of fuel processors utilizing diesel for SOFC APUs, in *Proceedings of the 9th European Solid Oxide Fuel Cell Forum, Lucerne, Switzerland, 29 June–2 October, 2010* (ed. P. Connor), European Fuel Cell Forum, Oberrohrdorf, 4.1–4.17.
24. Emonts, B., Bøgild-Hansen, J., Schmidt, H., Grube, T., Höhlein, B., Peters, R., and Tschauder, A. (2000) Fuel cell drive system with hydrogen generation in test. *J. Power Sources*, **86**, 228–236.
25. Men, Y., Gnaser, H., Zapf, R., Hessel, V., Ziegler, C., and Kolb, G. (2004) Steam reforming of methanol over $Cu/CeO_2/\gamma\text{-}Al_2O_3$ catalysts in a microchannel reactor. *Appl. Catal. A: Gen.*, **277**, 83–90.
26. Park, G.-G., Yim, S.-D., Yoon, Y.-G., Kim, C.-H., Seo, D.-J., and Eguchi, K. (2005) Hydrogen production with integrated microchannel fuel processor using methanol for portable fuel cell systems. *Catal. Today*, **110**, 108–113.
27. Lindermeir, A., Kah, S., Kavurucu, S., and Mühlner, M. (2007) On-board diesel fuel processing for an SOFC-APU – technical challenges for catalysis and reactor design. *Appl. Catal. B: Environ.*, **70**, 488–497.
28. Concawe (2008) Tank to Wheels Report Version 3.0, http://ies.jrc.ec.europa.eu/uploads/media/V3.1%20TTW%20Report%2007102008.pdf (last accessed 3 February 2012).
29. Peters, R. and Westernberger, A. (2010) in *Innovations in Fuel Cell Technologies* (eds. R. Steinberger-Wilckens and W. Lehnert), Royal Society of Chemistry, Cambridge, pp. 76–152.
30. Peters, R., Latz, J., Pasel, J., Samsun, R.C., and Stolten, D. (2009) Abschlussbericht Verbundvorhaben, ELBASYS, elektrische basissysteme in einem CFK-Rumpf, *Teilprojekt: Brennstoffzellenabgase zur Tankinertisierung, Schriften des Forschungszentrums Jülich*, Reihe Energie & Umwelt, Band 46, Forschungszentrum Jülich, Jülich.
31. Venkataraman, G., Sethuraman, S., Lux, K., Elder, W., Bhalerao, A., and Namazian, M. (2008) Distillate-fuel to power systems using high-temperature PEMFCs, in *Proceedings of the Fuel Cell Seminar, Palm Springs, CA, October 28–30, 2008*, Courtesy Associates, Washington, DC.
32. Srinivasan, H., Yamanis, J., Welch, R., Tilyani, S., and Hardin, L. (2006) *Solid Oxide Fuel Cell APU Feasibility Study for a Long Range Commercial Aircraft Using UTC ITAPS Approach*, Aircraft Propulsion and Subsystems Integration Evaluation, vol. I, NASA/CR-2006-214458/VOL1, National Technical Information Service, Springfield, VA.
33. Gummalla, M., Pandy, A., Braun, R., Carriere, T., Yamanis, J., Vanderspurt, T., Hardin, L., and Welch, R. (2006) *Fuel Cell Airframe Integration Study for Short-Range Aircraft*, Aircraft Propulsion and Subsystems Integration Evaluation, vol. I, NASA/CR-2006-214457/VOL1, National Technical Information Service, Springfield, VA.
34. Mak, A. and Meier, J. (2007) *Fuel Cell Auxiliary Power Study*, Raser Task Order 5, vol. I, NASA/CR-2007-214461/VOL1, National Technical Information Service, Springfield, VA.
35. Dollmeyer, J., Bundschuh, N., and Carl, U.B. (2006) *Aerosp. Sci. Technol.*, **10**, 686–694.
36. Bensaid, S., Specchia, S., Federici, F., Saracco, G., and Specchia, V. (2009) *Int. J. Hydrogen Energy*, **34** (4), 2026–2042.
37. Specchia, S., Saracco, G., and Specchia, V. (2008) *Int. J. Hydrogen Energy*, **33** (13), 3393–3401.
38. Lufthansa, Company information, Klima an Bord von Verkehrsflugzeugen, http://presse.lufthansa.com/index.php?id=104&tx_ttnews%5Btt_news%5D=840&no_cache=1 (last accessed 04 February 2012).
39. Campanari, S., Manzolini, G., Beretti, A., and Wollrab, U. (2008) *J. Eng. Gas Turbines Power*, **130**, 021701-1–021701-8.
40. ANSI/ISA (2009) ANSI/ISA-5.1-2009. *Instrumentation Symbols and Identification*,

The International Society of Automation, Research Triangle Park, NC.
41. Kazempoor, P., Dorer, V., and Ommi, F. (2009) Evaluation of hydrogen and methane-fuelled solid oxide fuel cell systems for residential applications: system design alternative and parameter study. *Int. J. Hydrogen Energy*, **34**, 8630–8644.
42. Veyo, S.E., Lundberg, W.L., Vora, S.D., and Litzinger, K.P. (2003) Tubular SOFC hybrid power system status. Proceedings of ASME Turbo Expo 2003: Power for Land, Sea, and Air, Atlanta, Georgia, ASME Paper 2003-GT-38943.
43. Blum, L., Peters, R., David, P., Au, S.F., and Deja, R. (2004) Integrated stack module development for a 20 kW system, in Proceedings of Sixth European Solid Oxide Fuel Cell Forum, Lucerne (ed. M. Mogensen), pp. 173–182.
44. Gubner, A., Landes, H., Metzger, J., Seeg, H., and Stübner, R. (1997) Investigations into the degradation of the cermet anode of a solid oxide fuel cell, SOFC-V, Aachen, pp. 844–850.
45. Christen, D.S. (2005) *Praxiswesen der Chemischen Verfahrenstechnik*, Springer, Berlin.
46. ISO (1997) 10628:1997. *Flow Diagrams for Process Plants – General Rules*, International Organization for Standardization, Geneva
47. Weber, K.H. (2006) *Inbetriebnahme Verfahrenstechnischer Anlagen*, 3rd edn., Springer, Berlin.
48. ISA (1985) ISA – 5.5. *Graphic Symbols for Process Displays*, The International Society of Automation, Research Triangle Park, NC.
49. ISO (1977) 3511-1:1977. *Process Measurement Control Functions and Instrumentation – Symbolic Representation – Part 1: Basic Requirements*, International Organization for Standardization, Geneva
50. Sattler, K. and Kasper, W. (2000) *Verfahrenstechnische Anlagen – Planung, Bau und Betrieb*, Wiley-VCH Verlag GmbH, Weinheim.
51. DGQ-Band 13-11 (2004) *FMEA – Fehlermöglichkeits- und Einflussanalyse* 3 Auflage (Hrsg. Deutsche Gesellschaft für Qualität e.V.), Beuth Verlag, Frankfurt.
52. Mosberger, E. (2005) Chemical plant design and construction, in *Ullmann's Encyclopedia of Industrial Chemistry* Vol. 8, Wiley-VCH Verlag GmbH, Weinheim, pp. 68–70.

33
System Technology for Solid Oxide Fuel Cells
Nguyen Q. Minh

33.1
Solid Oxide Fuel Cells for Power Generation

A solid oxide fuel cell (SOFC) is an all-solid-state energy conversion device that produces electricity by electrochemically combining a fuel with an oxidant across an oxide electrolyte [1, 2]. The fundamental building block of a SOFC is a cell or single cell. A SOFC cell or single cell is made of three layers: a fully dense ionic conducting oxide electrolyte sandwiched between a porous conducting (electronic or mixed ionic and electronic conducting) oxide cathode, and a porous conducting cermet (ceramic + metal) or oxide anode. In a stack of cells, a component commonly referred to as an interconnect (a fully dense electronic conducting oxide or metal) connects the anode of one cell to the cathode of the next cell in electrical series.

The attributes distinguishing the SOFC from other types of fuel cell are its solid-state construction (mainly ceramic or ceramic and metal) and high operating temperature (600–1000 °C). The combination of these features leads to a number of advantages for the SOFC, including flexibility in cell and stack designs, multiple manufacturing process options, and multi-fuel capability. Thus, the SOFC can be designed into unique configurations to achieve additional performance improvements, can be fabricated into lightweight and compact structures, and can be operated efficiently on a variety of fuels. Suitable fuels for the SOFC include a broad spectrum of practical fuels such as natural gas, biogas, alcohols, coal gas, gasoline, and diesel. Because of its capabilities for internal reformation [3] and direct oxidation [4] of hydrocarbons, the SOFC can be configured to operate with or without an external reformer, depending on the specific system design. No carbon monoxide cleanup is required for the SOFC since CO is also a fuel for this fuel cell technology [1, 2].

Current SOFCs, depending on the selected materials and designs, operate in the temperature range 600–1000 °C. The common materials used in the SOFC [5] are summarized in Table 33.1.

SOFC cell configurations can be classified in two broad categories: self-supporting and external supporting [6]. In the self-supporting configuration, one of the cell components (often the thickest layer) acts as the cell structural support.

Fuel Cell Science and Engineering: Materials, Processes, Systems and Technology,
First Edition. Edited by Detlef Stolten and Bernd Emonts.
© 2012 Wiley-VCH Verlag GmbH & Co. KGaA. Published 2012 by Wiley-VCH Verlag GmbH & Co. KGaA.

Table 33.1 Materials for SOFCs.

Component	Most common materials	Other materials
Electrolyte	Yttria-stabilized zirconia (YSZ)	Doped $LaGaO_3$, doped CeO_2, high-temperature proton conductors (e.g., doped $SrCeO_3$)
Cathode	Sr-doped $LaMnO_3$ (LSM) Sr-doped $LaCo_{0.2}Fe_{0.8}O_3$ (LSCF)	Several perovskite oxides (e.g., Sr-doped $LaFeO_3$ or LSF)
Anode	Ni/YSZ	Cu/doped CeO_2, several conducting oxides (e.g., doped $SrTiO_3$)
Interconnect	Stainless steels, doped $LaCrO_3$	Other high-temperature alloys and doped oxides

Figure 33.1 Photographs of anode-supported single cell and its microstructure.

Thus, single cells can be designed as electrolyte-supported, anode-supported, or cathode-supported. In the external supporting configuration, the single cell is configured as thin layers on the interconnect or a substrate. Figure 33.1 shows an example of an anode-supported SOFC single cell along with a micrograph of its fracture surface.

There are four common SOFC stack designs: the tubular design, the segmented-cells-in-series design, the monolithic design, and the planar design [2]. The designs differ in the pattern of current flow, thus the extent of dissipative losses, within the cells, in the manner of sealing between fuel and oxidant channels and gas manifolding, and in making cell-to-cell electrical connections in a stack of cells. At present, the planar design is the most common. In a SOFC stack, the gas flow configuration can be arranged to be crossflow, coflow, or counterflow and gas manifolds can be external or integral [6]. Figure 33.2 is a photograph of a planar 40-cell SOFC stack (with integral gas manifolds).

The SOFC has been considered for a wide range of power generation applications and markets. Potential markets for the SOFC cover portable, transportation, and

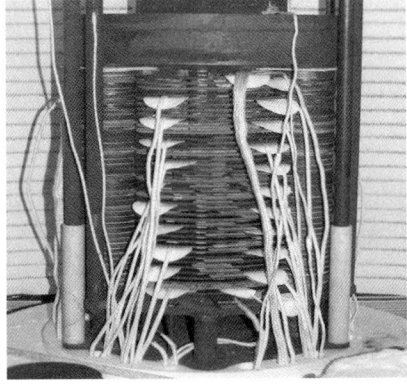

Figure 33.2 Photograph of a 40-cell planar SOFC stack (16 cm diameter cells).

Table 33.2 SOFC markets and applications.

Market	Example of application	Power size	Status
Portable	Soldier power	20–100 W	Demonstration
	Battery charger	500 W	Demonstration
Transportation	Automobile and truck APU	5–50 kW	Demonstration
	Aircraft APU	Up to 500 kW	Concept
Stationary	Residential	1–10 kW	Prototype
	CHP and DG	100 kW to 1 MW	Prototype and concept
	Base load	100–500 MW	Concept

stationary sectors. Examples of portable applications are 10–100 W soldier power and 500 W battery chargers. Examples of transportation applications are 5–50 kW automobile and truck auxiliary power units (APUs) and 500 kW APUs for aircraft. Examples of stationary applications include 1–10 kW residential, 100 kW–1 MW combined heat and power (CHP) and distributed generation (DG), and multi-MW base load power plants. Many of these applications have progressed to hardware demonstration and prototype stages while several applications, especially those with large power outputs, are at the conceptual stage (Table 33.2) [7]. Examples of several SOFC system hardware and prototypes are given in Figure 33.3.

33.2 Overview of SOFC Power Systems

33.2.1 General

A SOFC power system consists of fuel cell stack(s) and all other required components (referred to as balance of plant or BoP) for a fully functional unit. The

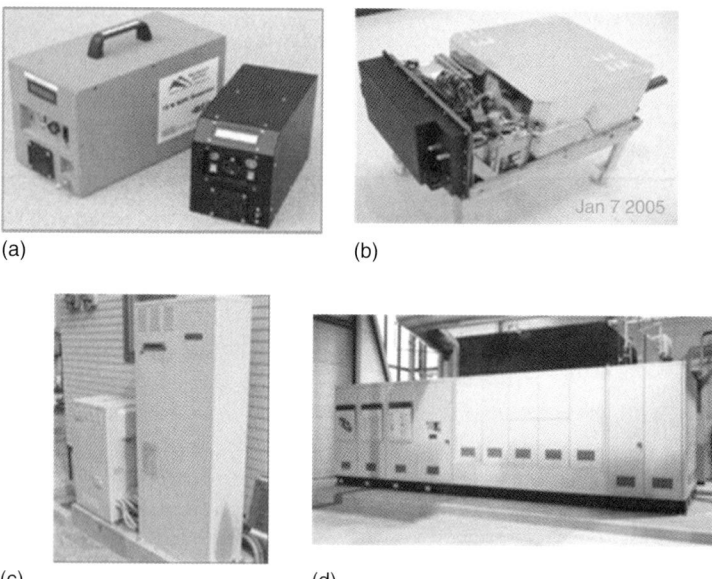

Figure 33.3 Photographs of SOFC systems: (a) portable; (b) automobile APU; (c) residential; (d) DG.

fuel-cell stack is the main constituent of the power generation subsystem and the BOP components can be grouped into different subsystems depending on their functions within the system: fuel processing, fuel, oxidant, and water delivery, thermal management, power conditioning, and control.

- **Power generation subsystem**: For the majority of SOFC power systems, the fuel-cell stack is the only component that produces electricity, thus constituting the power generation subsystem. In certain designs (hybrid designs), the system contains additional power-generating equipment such as gas turbine (GT) or steam turbine (ST) generators.
- **Fuel processing subsystem**: This subsystem is composed of a reformer (if the system design includes external reformation for fuels other than hydrogen) and other equipment for fuel cleanup, if any.
- **Fuel, oxidant, and water delivery subsystem**: This subsystem consists mainly of blowers/compressors and valves/orifices to deliver required reactants to the fuel processor and fuel and oxidant to the SOFC stack.
- **Thermal management subsystem**: The subsystem consists of a number of heat exchangers/recuperators (including steam generators) and combustors/burners to maintain the SOFC stack temperature at the required level and to control heat supply/removal for efficient operation of the fuel processor. Insulation is also an important element in thermal management of the system to contain heat losses.
- **Power conditioning subsystem**: Depending on the particular application, this subsystem may consist of power electronics (PE) (DC–DC converters, DC–AC

inverters) and transformers. The power conditioning subsystem converts variable DC from the fuel cell to regulated DC or AC power appropriate for the application.
- **Control subsystem**: This subsystem is a controller including control software that provides for system startup, shutdown, and normal operation while maintaining the system within its operating constraints when subjected to load changes or disturbances.

Figure 33.4 shows, as an example, a simplified schematic of a configuration for a 5 kW SOFC system operating on methane fuel with an autothermal reformer (ATR) (PE and controller not shown) for stationary applications and Table 33.3 summarizes the main components of this system [8]. A conceptual representation of such a system and a photograph of a laboratory prototype are shown in Figure 33.5.

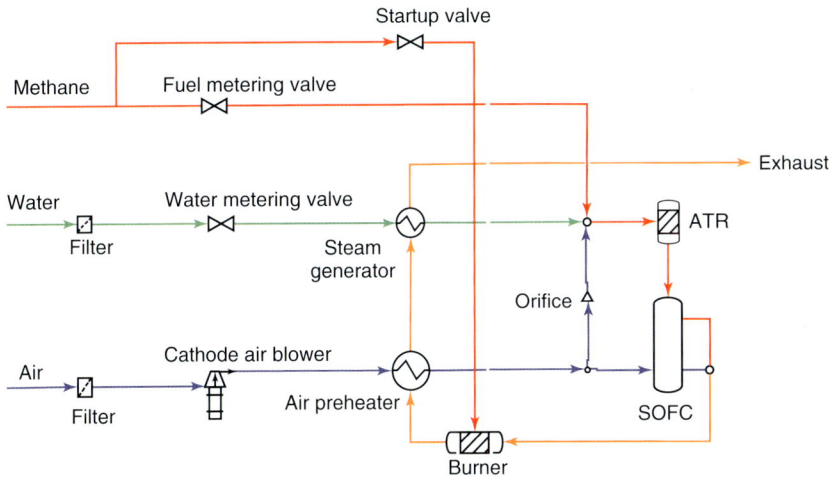

Figure 33.4 An example of a 5 kW SOFC system schematic.

Table 33.3 List of system components for 5 kW SOFC system (shown in Figure 33.4).

Subsystem	Component
Power generation	SOFC stacks
Fuel processing	ATR reformer
Fuel, oxidant, and water delivery	Fuel metering valve, fuel startup valve
	Water metering valve, water filter
	Air blower, air orifice for fuel reformer, air filter
Thermal management	Air preheater, steam generators, tailgas burner

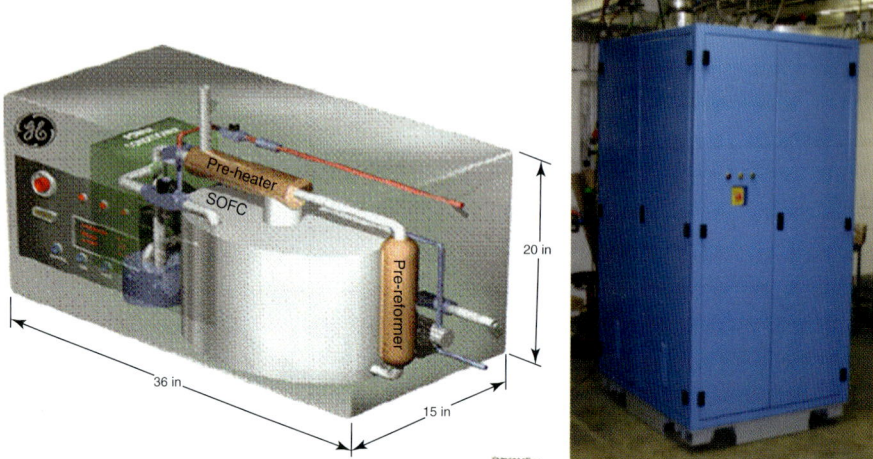

Figure 33.5 A 5 kW system concept and photograph of a laboratory prototype system.

33.2.2
Type of SOFC Power System

Current SOFC power systems can be divided into two general classes depending on their power generation cycles: simple cycle systems and hybrid cycle systems.

- **Simple cycle SOFC systems**: The SOFC is the only power-generating component of a simple cycle system. The simple cycle configuration is typically used for power systems with outputs ranging from tens of Ws to hundreds of KWs, although it is also considered for MW-size power plants [5]. Simple cycle SOFC systems have been shown to achieve high system efficiencies, for example, 45–60% for 1–2 kW systems [9, 10]. In large systems, the efficiency of simple cycles may be lower than that of hybrid cycles of similar power levels (hybrid cycles discussed below); however, the critical issues concerning integration of the fuel cell with a heat engine and pressurized operation, if required, can be avoided for simple cycle designs.
- **Hybrid cycle SOFC systems**: The SOFC can be combined with a heat engine to form a hybrid cycle power system [11, 12]. In a typical hybrid combination, the heat energy of the fuel cell exhaust is used to generate additional electricity in the heat engine. The heat engine can be GT, ST, or a combination of heat engines such as ST and GT combined cycle (CC) and integrated gasification combined cycle (IGCC) [13, 14]. Hybridization of the SOFC with a GT is the most common setup for hybrid cycle systems [15–24].

Figure 33.6 shows a simplified schematic of a SOFC/GT hybrid [23]. It can be seen that the tailgas (TG) from the SOFC is combusted and then fed to the turbine section of the GT to generate additional electricity. In a generic SOFC/GT hybrid design, the SOFC produces about 65–80% of the power and about 20–35% is from

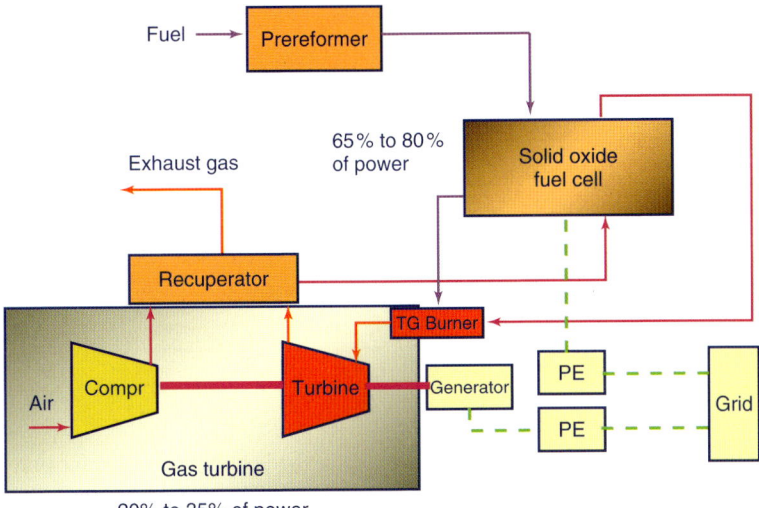

Figure 33.6 Simplified SOFC/GT hybrid system schematic (direct-fired design).

the GT. The SOFC/GT system schematic shown in Figure 33.6 is based on the direct-fired design. In this design, the SOFC operates under pressure. In addition to generating power, the GT also provides some BoP functions for the fuel cell such as supplying air under pressure and preheating fuel and air in a recuperator. In the indirect-fired design, the recuperator transfers the heat energy of the SOFC exhaust to compressed air supply for the turbine and the expanded air is supplied to the fuel cell. In this design, the SOFC operates under ambient pressure. In general, direct-fired hybrid systems exhibit higher system efficiencies than indirect-fired systems under equivalent design conditions [25].

Hybridization of the SOFC with a GT significantly improves the system efficiency and can be beneficial for use in large (hundreds of KWs and higher) power plants. Table 33.4 compares the estimated performances of three SOFC power systems: simple cycle, SOFC/GT hybrid, and SOFC/GT hybrid with anode and cathode gas recycle (AGR and CGR) [7]. The simple cycle system and hybrid systems (based on 800 °C planar SOFCs) operate at 1.3 and 4.6 atm, respectively and the fuel is methane. It can be seen that the hybrid system with gas recycle has an estimated efficiency of about 71%, about 26 points higher than the simple cycle system.

33.2.3
SOFC Power System Design

The focus in designing a fuel-cell power system is to develop and optimize the system configuration to meet the specifications of its intended application. These specifications could include the following: fuel specification, duty cycle, cost (purchase and installation), cycle efficiency, reliability, maintenance, size and weight, environmental interfaces, cogeneration, acoustic noise, power quality, and safety.

Table 33.4 Comparison of performance of simple cycle and SOFC/GT hybrid systems.

	Simple cycle	SOFC/GT hybrid	SOFC/GT hybrid with gas recycle
Efficiency (%)	44.8	61.1	71.0
SOFC power[a] (kW)	3709	3709	4099
GT power[a] (kW)	0	976	1447
Parasitic power (kW)	−389	−100	−288
Net plant power (kW)	3320	4585	5258
SOFC pressure (atm)	1.3	4.6	4.6

[a] After power conversion.

In general, SOFC systems are designed to fulfill the key requirements for practical/commercial products, namely performance, reliability, and cost. The design aims at establishing system configurations and defining system components, including component specifications, component performance characteristics, and effects of components and process variables on the system performance and reliability. The detailed system design can be used to estimate/determine system costs. A common system design process is illustrated schematically in Figure 33.7. Given the system requirements and knowledge of the primary technologies and components, a conceptual system design is developed, modeled, and analyzed using assumed or known performance. From the system modeling and analysis, specifications are created for each component given the existing technology base. Adjustments to assumed performance may be made to refine the system model. Other component characteristics are also compared with requirements. At this point, the overall system can be compared with the application requirements to identify technology gaps, if any, that need to be resolved in order for the system design to be realized. As shown in Figure 33.7, the inputs for this process include the system requirements and technology base and the major outputs include a system design, identified technology gaps to meet the requirements, and the definition of various system components.

33.3
Subsystem Design for SOFC Power Systems

33.3.1
Power Generation Subsystem

33.3.1.1 SOFC Stack
A stack or set of stacks is the core of the power generation subsystem. The stacks must be capable of meeting the (stack) requirements which are defined by the

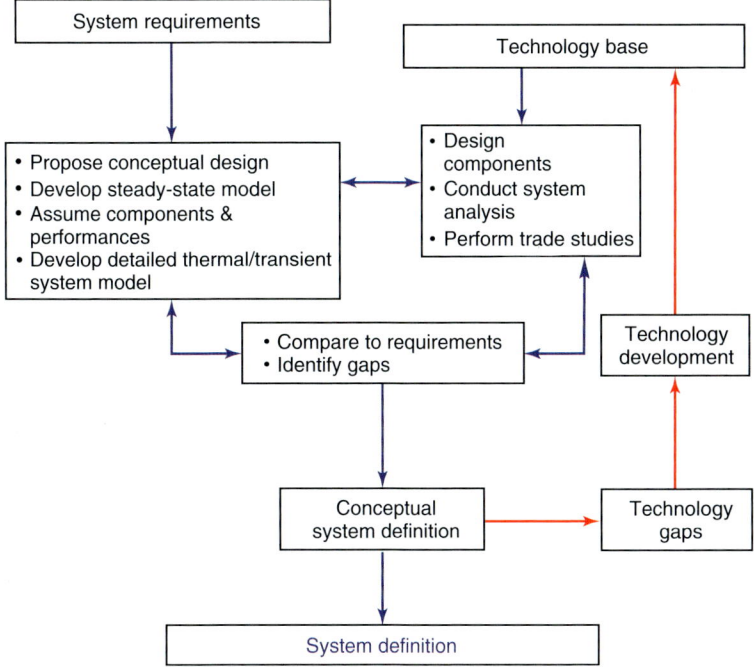

Figure 33.7 System design process.

system requirements and can interface and function effectively with the BoP. Several stack design parameters and operating variables are critical in the design of the system (to meet the performance, reliability, and cost targets of the intended application) and are discussed below.

33.3.1.1.1 Number of Stacks and Stack Arrangement

The number of stacks selected for a SOFC power system is dependent on several factors such as power level and other requirements (e.g., cost, reliability), stack design, and available technology. In general, the number of stacks in a system is preferably as small as feasible since a smaller number of stacks decreases part numbers (e.g., fewer cells) and reduces packaging (e.g., less insulation), resulting in lower system costs. Multiple stacks deployed in the system can be arranged either in parallel or in series (stages) or in a combination of parallel and series with respect to electrical connection and gas flow. These arrangements are a design option to optimize the system configuration. For example, in the series or staged airflow arrangement, the exhaust of one stack is the inlet to a subsequent stack. This staged arrangement potentially results in higher system efficiencies with a small cost penalty, as the byproduct heat of the upstream stage is used to preheat the subsequent stack air inlet. Similarly, a staged arrangement in fuel flow may result in a high overall fuel utilization (FU) in the system even when each stack's FU is low, thus resulting in a high system efficiency (*fuel utilization* is defined as the fraction of the inlet fuel flow that is spent).

For large SOFC power systems, defining the number of stacks and stack arrangement becomes more important as it is impractical from the cost, reliability, and technology standpoint to build, for example, one stack for MW-sized power plants. Therefore, multiple stacks have to be integrated in such systems and determination of the optimal number of stacks and the stack size is needed. One approach to guide this determination is to examine the effects of stack size (cell size and number of cells in the stack) and stack arrangement on the performance, cost, and reliability of the system. An example is given here on the design of a conceptual 25 MW planar SOFC/GT hybrid system (with 20 MW from the SOFC) [26]. Cost analysis of this system shows that the system cost is a weak function of the number of stacks and the system cost is only a function of cell size and not the number of cells in the stack [26]. Hence the optimal cell size can be determined solely by minimizing the SOFC stack cost. Figure 33.8 shows stack cost as a function of cell size derived from the stack cost analysis [26]. This figure shows an optimal cell size for a minimal stack cost. Other constraints, however, may limit the cell size. Technology capability is a constraint. Reliability considerations favor large cell diameters and a smaller cell count, but the cell reliability may decline rapidly with the cell diameter, thus placing a constraint on the cell size. Therefore, the cell size is generally selected through a stack cost and reliability optimization.

The number of cells in a stack is determined by consideration and optimization of the power conditioning subsystem (to minimize power conditioning losses). For the 25 MW SOFC/GT system example (with 20 MW from the SOFC), the optimal maximum stack voltage is 400 V, which translates into the optimal number of cells per stack of 400–500. Given the cell size and the number of cells per stack, the optimal stack building block for the 25 MW plant is estimated to have a nominal power rating of about 320 kW. This system thus needs 64 stack building blocks; these stacks, however, can be divided into modules to lower capital costs (multiple

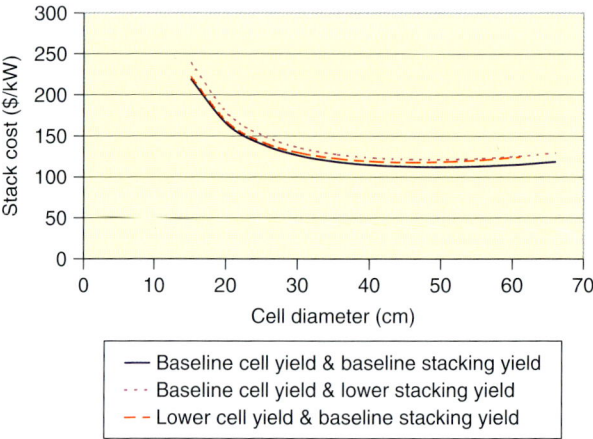

Figure 33.8 Stack cost as a function of cell diameter for a 25 MW SOFC/GT hybrid system.

stacks inside a single vessel is called a *module*). Based on reliability considerations, the optimum number of stacks per module, and hence modules per plant, can be determined. In this 25 MW SOFC/GT system, the selected stack arrangement scheme includes eight active stack modules, each module containing eight stack building blocks. In addition, the system may contain redundant module(s) to improve system reliability. Figure 33.9 is a schematic showing the stack/module arrangement for the system with a redundant module in a parallel redundancy scheme [26].

33.3.1.1.2 Stack Cell Voltage and Current Density

The performance of a stack, reflected in its cell voltage and current density, is an important operating parameter in the design of SOFC power systems. System efficiency is obviously a strong function of cell voltage/current. Figure 33.10 shows the effect of voltage and current density on the net system efficiency (defined as

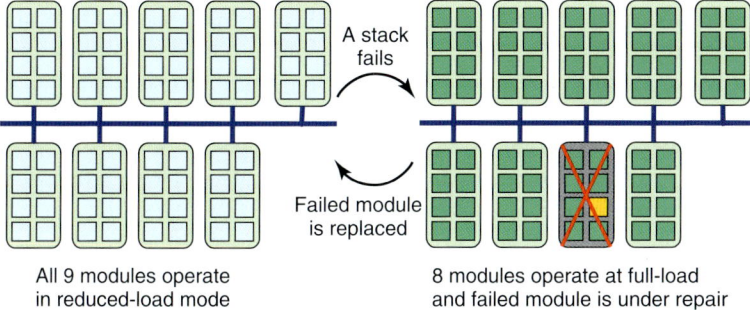

Figure 33.9 Stack module arrangement with parallel redundancy scheme.

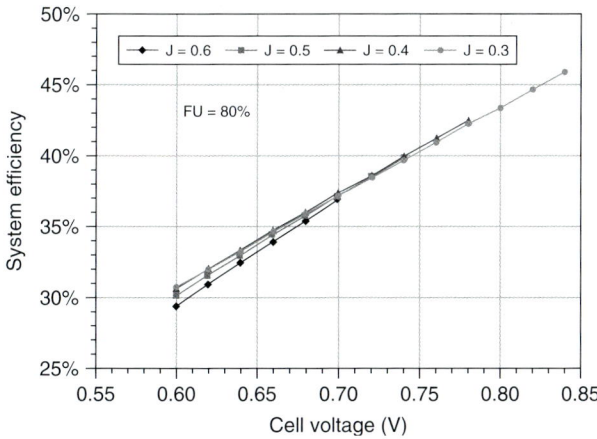

Figure 33.10 Effect of cell voltage on net system efficiency for different current densities J (in $A\,cm^{-2}$) at a FU of 80% for a 5 kW system.

the net AC power produced by the system divided by the low heating value or LHV of the fuel delivered to the system) of a 5 kW system operating on methane fuel. The cell voltage is traded with the SOFC power density (or current density) to achieve the most favorable system efficiency–system cost balance. This tradeoff requires knowledge of the SOFC stack polarization or voltage–current density curve, showing the average cell voltage as a function of average current density. This polarization curve is a reflection of the cell technology and stack design.

33.3.1.1.3 Fuel Utilization

FU is a stack operating parameter that requires definition and optimization in designing SOFC power systems. An example is given in Figure 33.11, showing the efficiency of a 5 kW SOFC system as a function of FU and operating voltage for current densities of 0.3 and 0.6 A cm^{-2} [8]. The uppermost line on the plot is for a current density of 0.3 A cm^{-2} and an operating voltage of 0.84 V. The lowest set of lines on the plot is for current densities of 0.3 and 0.6 A cm^{-2} with operating voltage of 0.6 V. In this case, at an operating voltage of 0.6 V, a 10% decrease in FU results in a 3.7% decrease in system efficiency. At an operating voltage of 0.84 V, a 10% decrease in FU results in a 6% decrease in system efficiency. It should be noted that cell operating voltage and FU are not independent parameters, and decreasing the FU will generally tend to increase the operating voltage of a cell. The increase in cell voltage will tend to offset the drop in system efficiency; however, the effect is considered to be minor from a system standpoint.

It should be noted that the level of FU has significant impact on other components of the system. Key system components most affected by operation at low FU include the tailgas combustor, the cathode air blower, and the steam generator. Operation at low FU impacts the tailgas combustor directly by increasing the temperature of the combustor and associated downstream components beyond the normal operating temperature limits of the materials used to manufacture these components. The

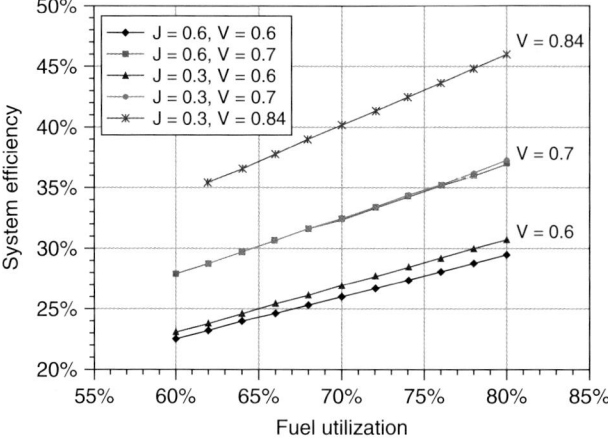

Figure 33.11 Effect of fuel utilization on system efficiency for a 5 kW SOFC system.

cathode air blower is indirectly affected by operation at low FU since it will be required to provide increased flow to the system in order to maintain combustor temperatures below a limit. If increased cooling air flow is provided to the system, the steam generator performance will also be affected. Increased cooling air flow to the system has the net effect of reducing the pinch temperature difference between the hot and cold streams supplied to the steam generator, thereby making steam generation more difficult.

33.3.1.1.4 Stack Pressure Drop

The SOFC system performance at the design point depends on component efficiencies, pressure drops, and heat losses. In terms of pressure drop, all major components of the system must be evaluated with regard to their contribution to the overall system pressure drop and, where feasible, design efforts must be made to lower their design pressure drop since high pressure drops result in high parasitic power consumption. In typical SOFC systems, the major contributor to the overall system pressure drop can be the pressure drop through the stack and the major parasitic power consumption is the cathode air delivery blower/compressor. Therefore, from a system standpoint, minimization of the parasitic power consumption in the system requires minimizing the pressure drop through the stack, which in turn lowers the power requirements for the cathode air blower. Figure 33.12 shows estimates for system pressure drop (in psid or pound per square inch differential) as a function of cell operating voltage (V) and current density (A cm^{-2}) at a constant FU of 80% for a 5 kW system [8]. As expected, the pressure drop is reduced as the cell operating voltage increases due to decreasing air flow requirements.

33.3.1.1.5 Stack Temperature Gradient

One important design parameter is the stack temperature gradient or stack air temperature rise (the difference between stack air outlet and inlet temperatures).

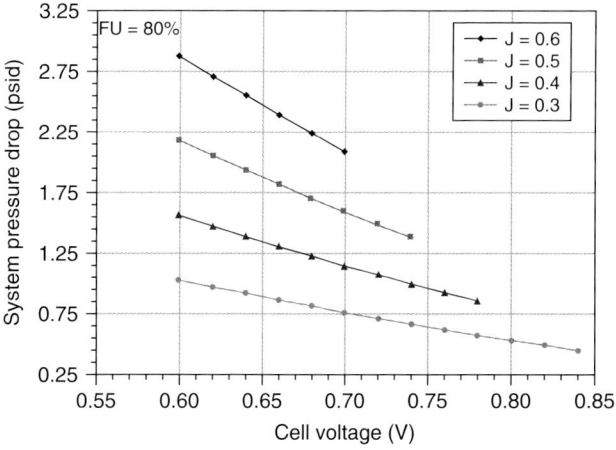

Figure 33.12 System pressure drop as a function of cell voltage in a 5 kW SOFC system.

A stack air temperature rise as high as possible is desired as it lowers air cooling requirements (hence less power is required to deliver cooling air to the stack), resulting in improved system efficiency. The main limiting factor in setting the stack air temperature rise for a system is the thermal stress (as a result of the temperature gradient within the cell caused by the air temperature rise) that can lead to stack structural failure. Therefore, it is desirable to develop stack designs that minimize cell temperature gradients while permitting increased air temperature rises. A typical stack air temperature rise of current SOFC systems is on the order of 100–200 °C.

Figure 33.13 shows an example of the estimated cooling airflow requirements necessary to limit the stack air temperature rise to 175 °C as a function of cell voltage and FU at a current density of 0.7 A cm^{-2} for a 5 kW SOFC system [8]. It can be seen that in this case, suitable operating points are those below the air blower upper limit of 60 scfm (standard cubic foot per minute).

33.3.1.1.6 Stack Operating Pressure

In simple cycle SOFC systems, the fuel cell typically operates under atmospheric pressure. In principle, the SOFC can be pressurized and pressurized operation can provide additional benefits. The benefits of pressurization include improved cell/stack performance, reduced heat losses and pressure drops, and smaller equipment sizes. Although a pressurized SOFC power system has several advantages relative to an atmospheric pressure system, pressurization is not commonly used in simple cycle systems because of the inherent cost and technical issues relating to pressurized operation, especially for systems with small power outputs.

In direct-fired SOFC/GT hybrid systems, the fuel cell operates under pressure; the stacks are generally placed inside pressure vessels. In this case, pressurized air from the compressor of the GT is supplied to the SOFC and the high-pressure

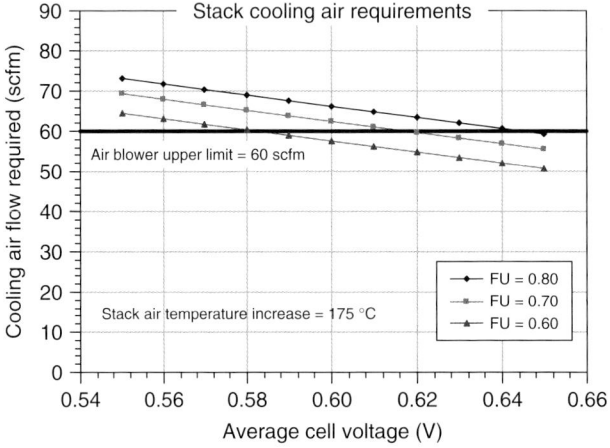

Figure 33.13 Stack air cooling requirements to limit air temperature rise to 175 °C for a 5 kW system.

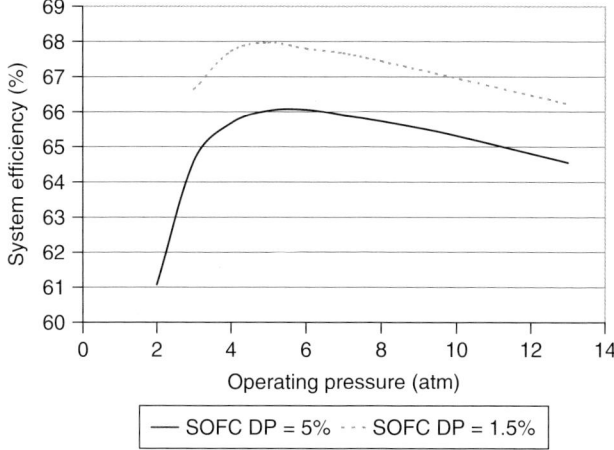

Figure 33.14 System efficiency as a function of operating pressure for different stack pressure drops (DPs) in a 25 MW SOFC/GT system.

exhaust from the fuel cell drives the turbine. The optimal operating pressure of the SOFC in the system can be determined via performance and cost optimizations. Figure 33.14 shows system efficiency as a function of operating pressure for a 25 MW SOFC/GT hybrid system design [26]. The figure also shows the effect of SOFC pressure drop (Delta P or DP) on system efficiency. It can be seen that the optimal pressure in this case is around 5 atm. As indicated earlier, although increasing SOFC pressure may increase the system efficiency, the possible negative effects of pressure on the SOFC degradation and reliability may offset the efficiency gains, from a total cost of electricity point of view.

Operation of the SOFC under pressure has been demonstrated and pressurized operation has been shown to improve cell performance [24, 27–33]. For example, when the pressure is increased from 1 to 6 atm, the maximum power density increases from 266.7 to 306 mW cm^{-2} at 800 °C [33]. The power enhancement by pressurization is due to increased open-circuit voltage and reduced polarization resistance [31]. On the other hand, pressurization may accelerate the degradation of cell performance [24]. Figure 33.15 is an example of performance curves of a planar SOFC tested under different pressures [24].

33.3.1.2 Other Power Generating Equipment

As indicated earlier, the most common hybrid cycle SOFC system is the SOFC/GT hybrid. In general, the GT provides 20–35% power of the system, as shown in Table 33.5. For smaller size systems (e.g., 1–10 MW), the GT pressure ratio (P/P) is 3 : 1 to 6 : 1 whereas for larger systems (e.g., 200 MW), the P/P is 8 : 1 to 12 : 1.

A critical factor in the design of a SOFC/GT hybrid system is the matching of design parameters between the two components, especially the turbine inlet temperature, the P/P, and the air flow rate. In general, the desired features of

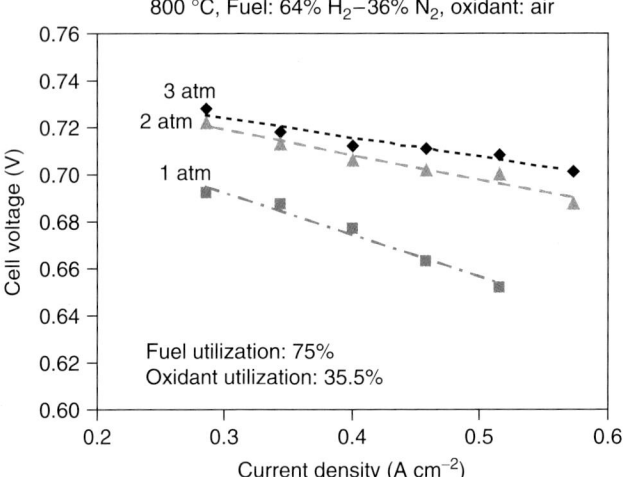

Figure 33.15 Performance curves of a SOFC under different pressures.

Table 33.5 Typical GT power of SOFC/GT hybrid systems.

System size	GT size
1 MW	200–350 kW
10 MW	2–3.5 MW
200 MW	40–75 MW

the GT for hybrid systems are (i) adequate performance on low turbine inlet temperatures and low P/Ps, (ii) ability to sustain long-duration thermal cycling and slow time-response output of the fuel cell, and (iii) large window of operation to allow for system turndown and avoid system shutdown [34]. It is well known that the design principles and operating characteristics of these two technologies are completely different. For example, high turbine inlet temperature and high P/P are desirable to improve the performance of stand-alone GTs but are not necessary for hybrid systems. Therefore, SOFC and GT technologies with matching designs may not be available in practice. To date, many conceptual hybrid designs have been developed with the underlying assumption that the GT has the corresponding parameters suitable for hybridization. Microturbine generators are the only GT that has been tested in a hybrid system. This type of GT has a number of beneficial operating characteristics such as relatively low turbine inlet temperatures and relatively low P/Ps, making it amenable to integration with the SOFC.

It is preferable to use existing GTs for SOFC/GT hybrid systems because it can be prohibitively costly to develop a new technology. When such an existing GT is deployed, constraints on system design and impacts on system performance

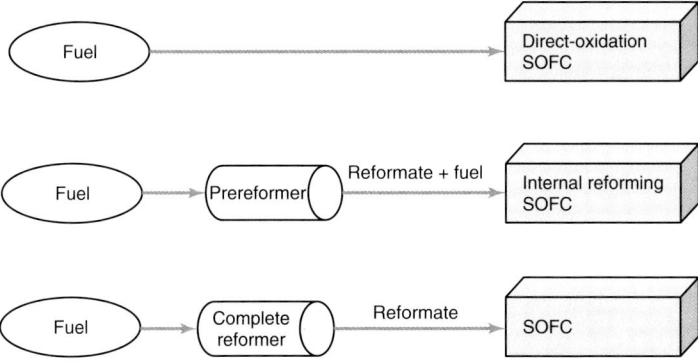

Figure 33.16 Options for SOFC operation on fuels other than hydrogen.

need to be fully analyzed and understood [35]. For example, in a hybrid system with a fixed GT design, the turbine operating parameters, such as turbine inlet temperature, shift from the design parameters. Thus, the power produced by the GT in the hybrid system is lower than the design power [35].

33.3.2
Fuel Processing Subsystem

There are three options for using fuels other than hydrogen in a SOFC system (Figure 33.16):

1) **Direct oxidation (or direct utilization)**: The fuel is oxidized directly in the SOFC without external reformation. The SOFC has been shown to have the capability for direct oxidation of different types of fuels [4, 36–38]. To address the carbon deposition issue associated with nickel commonly used in the anode composition, other metals such as copper have been tested. The ability of copper to resist carbon formation leads to the development of a composite anode composed of a ceria support and a copper phase [38]. The key technical challenges in the development of direct-oxidation SOFCs relate to the anode, especially the electrode's performance, stability, and direct-oxidation capability.

2) **Internal reforming**: A portion of the fuel is reformed in an external reformer (prereformer) and the resulting reformate plus the remaining fuel are fed to the SOFC where the fuel is internally reformed (via steam reforming) within the stack. This option is commonly employed to use the endothermic reforming reactions to reduce cooling requirements in thermal management of the SOFC. It is possible to have complete (100%) internal reforming; in this case, a prereformer is not needed. There are a number of different methods for carrying out internal reformation. One of the most straightforward is to carry out the reformation reaction directly upon the anode of the SOFC. This is possible since the primary component of the anode is nickel and nickel is a highly efficient catalyst for steam reformation. However, on-anode reformation requires appropriate control of the reforming reaction steps (Figure 33.17)

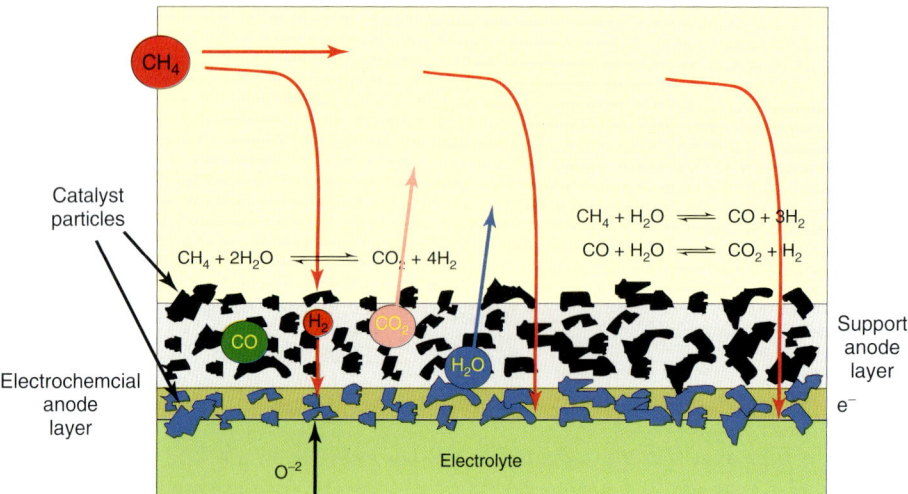

Figure 33.17 Main reaction steps in on-anode reformation of methane (with a bilayer anode).

along with careful consideration of anode thermal management. On-anode reformation has been demonstrated for SOFCs. For example, a 5 kW system was operated with an ATR reformate containing about 7 vol.% methane slip [8].

3) **Complete reformation**: In this case, fuel is completely reformed in an external reformer and the reformate is fed to the SOFC for power generation.

Hence one important consideration in the design of a SOFC power system with external reformation is the choice of the fuel reformer (FR) and its thermal integration, in addition to water management. Different kinds of FRs can be considered, namely steam reformers (SRs), partial oxidation reformers (POXs), and ATR. The overall reactions for the different reformation processes are as follows (written for methane):

$$CH_4 + H_2O \longrightarrow CO + 3H_2 \text{ (SR)}$$

$$CH_4 + \frac{1}{2}O_2 \longrightarrow CO + 2H_2 \text{ (catalytic POX or CPOX)}$$

$$2CH_4 + \frac{1}{2}O_2 + H_2O \longrightarrow 2CO + 5H_2 \text{ (ATR)}$$

The selection of a particular external reformer for a SOFC system is very much dependent on system size and design and other considerations/constraints. In general, SRs are used for large power systems; the capital cost of SRs is prohibitive for small and medium-sized applications because this technology does not scale down well. POXs have several advantages over SRs, notably their exothermicity and functionality without water. The main difference between SRs and ATRs is that SRs do not require oxygen. The main advantage of ATRs, in contrast to SRs and POXs, is that there is no need, in principle, to supply or dissipate heat to or from the reformer.

Steam is required for fuel processing based on SRs and ATRs and oxygen is required for ATRs and POXs. For steam supply, there are three options: (i) outside water supply with steam generation, (ii) a water pump with a steam generator providing steam from a water tank with a condenser at the system exhaust, and (iii) recycling of the SOFC anode outlet containing product water to the reformer. Options (i) and (iii) maintain water neutrality and are preferred if the system has fresh water supply limitations.

It is critical to determine and control the steam-to-carbon (S/C) and/or oxygen-to-carbon (O/C) ratio of the feed to the reformer and S/C to the stack (with internal reforming) to avoid carbon deposition. Thermodynamic analysis is commonly used to estimate the minimum ratios. For example, Figure 33.18 shows the equilibrium number of moles of carbon per mole of methane introduced into an ATR as a function of S/C and O/C at two reformer inlet temperatures of 150 and 400 °C [8]. It can be seen that for all values of O/C between 0 and 1.5, carbon deposition should not be a concern if an S/C > 1.2 is maintained in the fuel gas mixture entering the ATR (fully mixed inlet stream). It should be noted that many thermodynamic calculations (as in this example) assume adiabatic equilibrium reactions and do not take into account reaction kinetic effects. The inclusion of reaction kinetics in the analysis may lead to different results.

In addition to the S/C (and O/C), other key parameters to be defined in the design and operation of an external reformer in the system include gas space velocity, pressure drop, gas inlet temperature, and gas exit temperature.

In general, a reformer is designed to minimize gas-phase reactions, avoid heat loss, and reduce thermal mass to facilitate rapid startup. The integration of an external reformer and a stack requires definition of appropriate procedures and conditions to achieve stable and efficient operation of the system (to avoid carbon deposition and to ensure the ability of the system to accommodate transients in

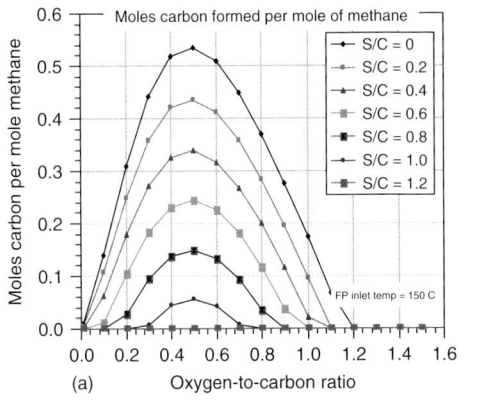 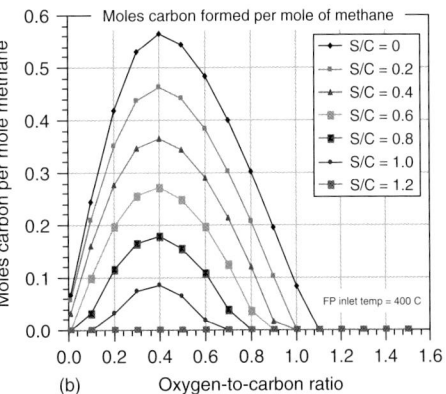

Figure 33.18 Carbon formation as a function of S/C and O/C in an ATR at reformer/fuel processor (FP) inlet temperatures of (a) 150 and (b) 400 °C from thermodynamic analysis.

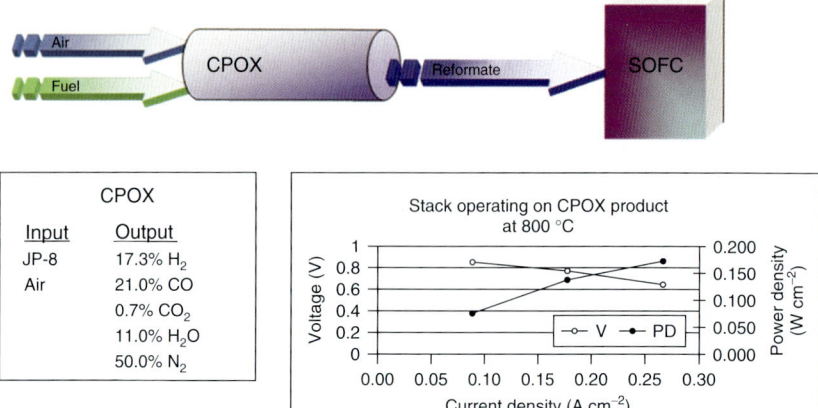

Figure 33.19 Integrated operation of a SOFC stack with a CPOX reformer.

fuel and/or steam feeds). Testing is often used to verify the defined parameters and obtain a clear understanding of how the reformer and the SOFC stack respond to system perturbations. Figure 33.19 shows an example of testing of integrated operation of a SOFC stack with an external CPOX (catalytic partial oxidation) reformer with JP-8 fuel [39].

In addition to the reformer, system designs may call for fuel cleanup. For hydrocarbon fuels, a desulfurizer (DS) is most common. For systems using syngas from a gasifier, fuel cleanup may include components to remove other elements (e.g., P, As) to certain levels to minimize their deleterious effects on the SOFC stack.

33.3.3
Fuel, Oxidant, and Water Delivery Subsystem

- **Fuel delivery**: Accurate and predictable delivery of fuel quantity is essential for achieving high overall system efficiency while maintaining an adequate safety margin to prevent an excessive combustor temperature. Also, the SOFC is operated at very specific levels of FU dictated by both stack reliability and efficiency requirements. Hence it is very important to have fast, accurate, and precise control of fuel flow to match the power demanded of the SOFC system. A fuel metering valve is commonly used for the system and selected based on the required fuel flows and other characteristics. Many system designs also employ a blower for anode gas recycle (AGR).
- **Oxidant delivery**: The main component for cathode air supply to the SOFC stack is a blower or compressor. The blower/compressor provides the required airflow and overcomes the system component pressure drops throughout the entire range of system operation. This is the largest electrical parasite on the system; an example is given in Figure 33.20 showing power breakdown for a

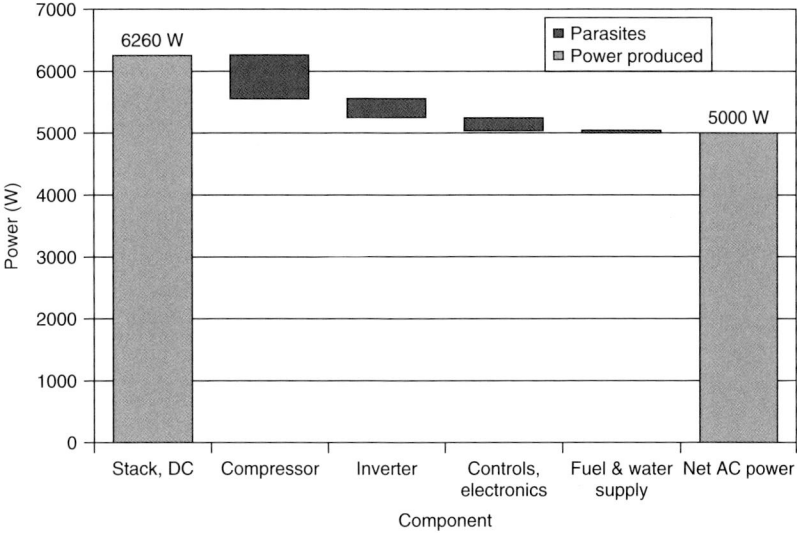

Figure 33.20 Power breakdown for a 5 kW system.

5 kW system [8]. Thus, the blower/compressor needs to operate with minimum power consumption in order for the overall system to meet efficiency targets. Additional high-level requirements for the blower/compressor are performance, high-volume production cost, safety, and reliability. Many system designs also use a blower for CGR.

In a system with an ATR or POX reformer, system designs may call for part of the cathode air stream to be diverted to the fuel processor via valving or a separate blower may be added so that a more controlled and decoupled air stream to the fuel processor can be assured.

- **Water delivery**: For a system that incorporates an SR or ATR, a pump capable of covering required flow rates is commonly used to deliver water to the reformer. A key consideration is the need for stable water delivery. Unsteady water delivery has the potential to introduce steam flow oscillations into the reformer that can lead to fluctuations in hydrogen production or fuel flow and thus potentially damaging voltage swings in the stack. A loss of steam also can lead to carbon deposition.

33.3.4
Thermal Management Subsystem

The thermal management subsystem of a SOFC system consists of (i) those components downstream of the stack whose primary functions are to react any remaining combustibles in the anode exhaust and to preheat the various streams that eventually find their way to the stack inlets [8, 40–42] and (ii) insulation. The primary components of this subsystem may include, but are not limited to, combustor, steam generator, and heat exchangers (e.g., cathode air preheater, fuel processor steam superheater, and reformer air preheater).

- **Combustor**: The combustor/burner can be either a catalytic type or a more conventional, diffusion-type combustor. In an operating SOFC system, the combustor is expected to perform over a wide range of fuel and oxidant flows; from moderately lean to extremely lean mixtures. Flow rates for each stream may vary by as much as 10-fold. Operating temperatures of the combustor are expected to be very high (especially at low stack FU), with little or no available supplementary cooling air. Furthermore, the pressure drop must be kept low, so as to not affect adversely the air delivery power requirements and, subsequently, the net system efficiency.
- **Steam generator**: An example of the steam generator design for a SOFC system is that composed of a compact cylindrical heat exchanger with a helically coiled, finned tube placed in the annulus region of concentric pipes. Such a design is commonly used for recovering waste heat from diesel engine exhaust gases. The design utilizes full counterflow heat transfer for maximum effectiveness. Water is introduced into the top of the unit and allowed to flow downhill across the hot heat exchanger tubing. The fins are included on the tubing to maximize the gas-side heat transfer surface area within the available space. A sketch of such a steam generator is shown in Figure 33.21.

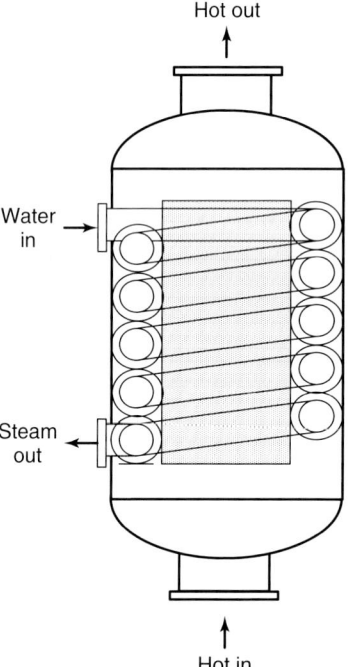

Figure 33.21 Helically coiled steam generator.

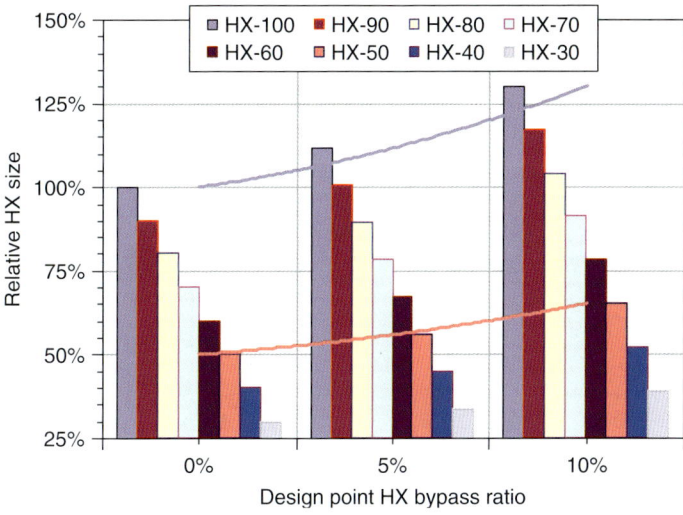

Figure 33.22 Relative heat exchanger size as a function of design point bypass ratios.

- **Heat exchanger**: In addition to the combustor and steam generator, the thermal management subsystem includes several heat exchangers (HXs). The key HX is the cathode air preheater that provides air at a specified temperature to the stack. It is important to control the air preheat temperature (i.e., cathode inlet temperature) and to size the HX properly. One approach for temperature control is to use a cold air bypass when the cathode air preheat temperatures have the potential to exceed the specified temperature. The HX must be sized for a number of different conditions corresponding to different power levels and different HX bypass ratios. An example is shown in Figure 33.22. The designations HX-100, HX-90, HX-80, and so on correspond to system design point power levels of 100, 90, 80%, and so on. The reference is an HX sized at conditions corresponding to 100% net system power, 80% FU, sufficient cathode air to provide a 100 °C temperature rise across the stack, and without the use of HX bypass [8]. For all cases, the HXs provide cathode air to the stack at a specified temperature of 750 °C. It can be seen from Figure 33.22 that if the HX is sized at conditions corresponding to lower power levels, the relative size of the HX decreases due to decreased heat duty requirements. If the HX is sized at conditions corresponding to higher levels of bypass flow, the relative HX size increases due to the reduction in mean temperature difference across the HX.

The selection of the proper design point for the HX is essential if temperature control of the cathode inlet gas stream is expected for a wide range of operating conditions. For example, Figure 33.23 shows the cathode inlet temperature as a function of system power for four separately sized HXs [8]. The design point for each HX is highlighted with larger symbols. The sizing conditions for each HX correspond to 80% FU and net system power levels of 100, 80, 60, and 40%. For an HX sized at 100% power (without the use of bypass air, i.e., HX-100-0), the cathode

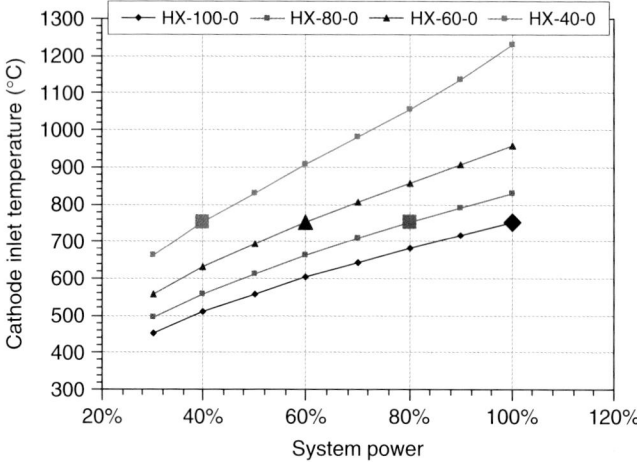

Figure 33.23 Performance of cathode air preheater at off-design conditions for separately sized HXs.

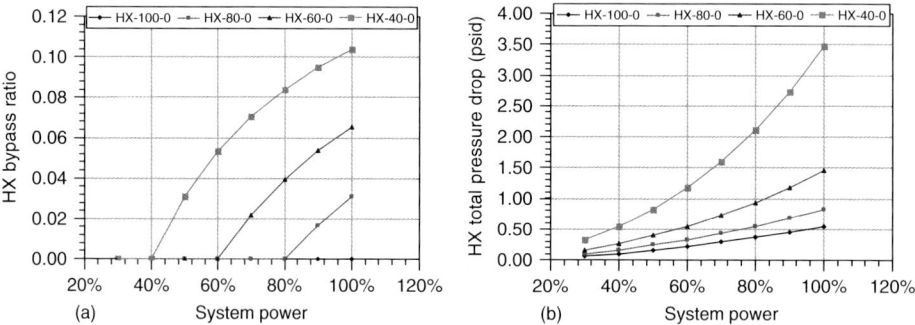

Figure 33.24 (a) Bypass air requirements and (b) HX total pressure drops to maintain 750 °C cathode inlet temperature.

air preheater will not be able to maintain 750 °C cathode inlet temperatures as system power is decreased. Alternately, if the HX is sized at 40% power (without the use of bypass air, i.e., HX-40-0), the cathode inlet temperatures will exceed 750 °C as the power level is increased. Although such temperatures can be moderated by initiating cold air bypass, the increase in pressure drop as one moves from low to high power is significant. This is shown in Figure 33.24, which plots the percentage of bypass air needed to lower the cathode inlet air supply temperatures to 750 °C and also the pressure drop across the HX.

As indicated earlier, insulation is an essential element of system thermal management to control heat losses. It is preferable to design the system with a small, thermally integrated package to minimize insulation. In designing system packaging, insulation materials must be selected and required thickness determined to ensure that the heat loss is less than the specification derived from the system design.

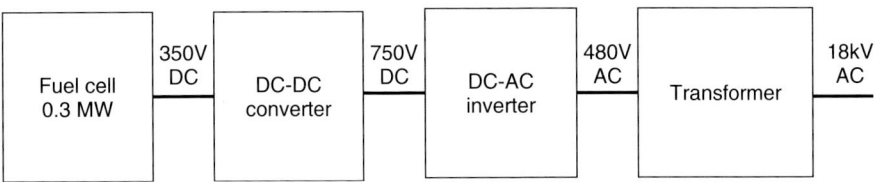

Figure 33.25 Power conditioning with stepup transformer for megawatt power systems.

33.3.5
Power Conditioning Subsystem

The power conditioning subsystem consists of several components (commonly referred to as power electronics) that convert variable DC from the fuel cell to regulated DC or AC power appropriate for the application [43–48]. The main PE components are DC–DC converters and DC–AC inverters. For certain applications, such as portable devices, the power conditioning subsystem includes only DC–DC converters. For large systems, such as MW class utility power plants, the power conditioning subsystem may include separate transformers, for example, a 60 Hz transformer to step up from 480 V AC (from inverter) to 18 kV AC (Figure 33.25) and a transformer to step up from 18 to 300 kV AC for delivery to the grid [43]. Like the fuel-cell stack, the key factors in selecting/developing a power conditioning subsystem mainly relate to efficiency, reliability, and cost. Other specific factors are DC isolation, fuel-cell ripple current, and electromagnetic interference (EMI) emission. The efficiency of a power conditioning subsystem depends on the conduction and switching losses. The conduction losses can be minimized by reducing the usage of components and their operating ranges. The switching losses can be reduced by soft switching techniques either by zero-voltage crossing or zero-current crossing. The power conditioning subsystem is designed to minimize fuel-cell ripple current for optimized system efficiency since the subsystem usually draws a current which has a low-frequency (LF) ripple and a high-frequency (HF) switching ripple. This can force the operating point of the fuel cell into the mass transport regime where the stack suffers from excessive performance losses [49].

The power conditioning system for DC power applications includes only a DC–DC converter. For AC power applications, the system can consist of a single DC–AC inverter, however, if isolation or a high ratio of the voltage conversion is required, a LF transformer placed at the output of the inverter is usually incorporated into the system. The drawback of this arrangement is that the transformer makes the system bulky and expensive. Thus, a DC–DC converter is usually put between the fuel cell and the inverter. The converter acts as the DC isolation for the inverter and provides sufficient voltage for the inverter input so the required AC voltage can be produced. Another possible power conditioning configuration includes an HF DC–AC inverter (fuel cell DC voltage to HF AC voltage) and a cycloconverter (HF AC voltage to LF AC voltage). In this case, the power conversion is more direct than with a conventional DC bus structure with an isolated DC–DC converter. Figures 33.26 shows an example of a power conditioning system configuration for

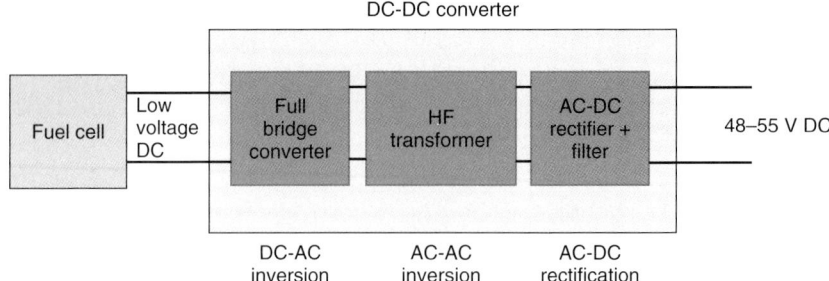

Figure 33.26 Power conditioning configuration of fuel-cell systems for telecom applications.

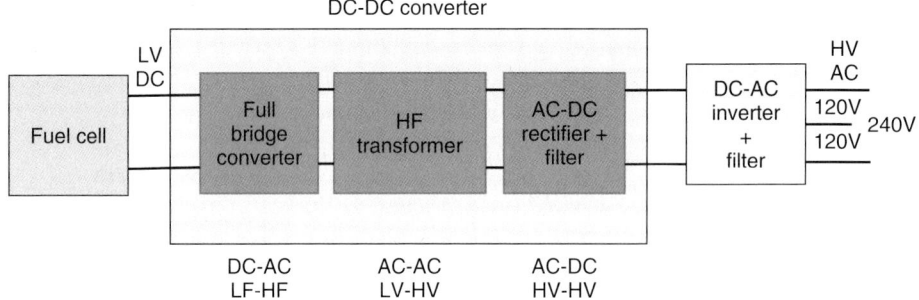

Figure 33.27 Power conditioning configuration of fuel-cell systems for residential applications.

telecom applications (DC–DC converter only) and Figure 33.27 a configuration for residential applications (DC–DC converter and DC–AC inverter) [44].

DC–DC converters convert DC voltage from the fuel cell to another DC voltage level. DC–DC converters can be non-isolated (no transformer) or isolated (with transformer) and have many topology options. DC isolation is needed for most fuel-cell power systems since isolation allows high voltage conversion ratios and avoids noise coupling issues while meeting required safety standards. Examples of DC–DC converters proposed and evaluated for fuel-cell systems include full bridge (isolated, relatively high switching loss), push–pull (isolated, lower switching loss but lower efficiency), and boost (non-isolated, lower switching loss but lower voltage boosting ratio).

DC–AC inverters of the power conditioning subsystem use solid-state semiconductor switches to convert DC to AC. Inverters can be classified as hard switching and soft switching and can be single-phase or three-phase, depending on the connection. Two popular inverter topologies for fuel-cell applications are the hard-switching three-phase voltage-source inverter (VSI) and the resonant DC link inverter. The VSI design is well proven and widely used in industrial applications, but it suffers from significant switching loss. The resonant phase leg inverter (RPLI) design is an improvement to the hard-switching inverter with zero-voltage switching. The auxiliary resonant commutated pole inverter (ARCPI) uses auxiliary transistors to assist the zero-voltage switching; this topology has the advantage of a wide range of load, but the component count is high. The active clamp resonant DC link inverter (ACRDI) design is an improvement to the classical resonant DC link inverter where the DC link voltage is twice as high as the original DC link voltage. With additional clamping devices, the DC link voltage can be controlled to 1.3 times the input voltage; however, only the delta modulation method can be used, causing some reliability issues [45].

33.3.6
Control Subsystem

A typical control subsystem includes a system controller, a signal conditioning module, and sensors measuring temperature, pressure, and flow rate, in addition to actuators such as valves. Designing the control subsystem for a SOFC system involves control algorithm development, software design, and hardware identification. System control design must consider the diverse time scales for physical phenomena throughout the system (Figure 33.28) [8]. The control design must be able to account for fast dynamic behavior in the PE and SOFC electrochemistry, slower thermo-fluid response, and long-term performance degradation effects. Before being deployed in the system, the whole control system, including all the software and hardware, needs to pass through a series of tests to ensure its reliability and performance.

In general, the control subsystem with an appropriate architecture is designed to manage various control tasks (e.g., startup, shutdown and emergency shutdown, normal operation, and other modes such as maintenance, idle, and power off)

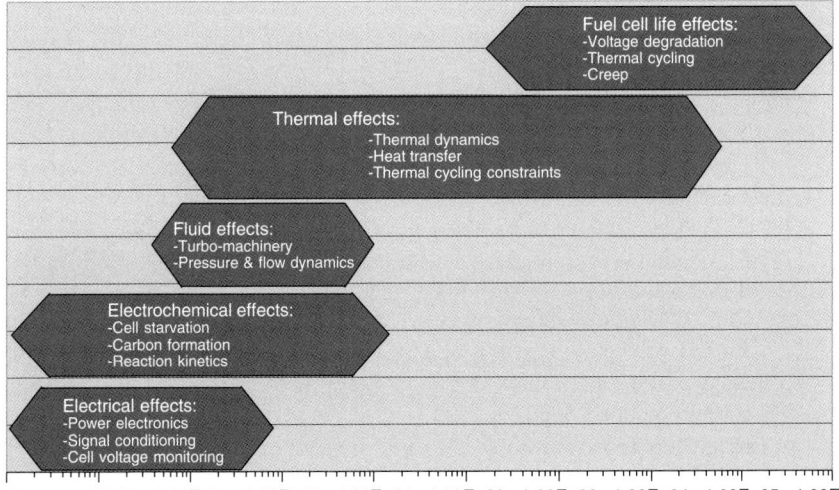

Figure 33.28 SOFC system time scales.

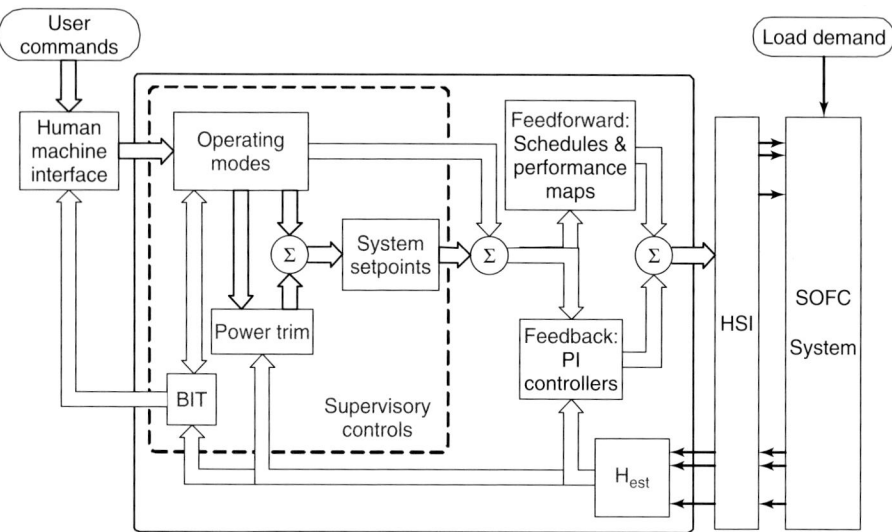

Figure 33.29 General control subsystem architecture.

while maintaining the system within its operating constraints when subjected to load changes or disturbances. A typical control architecture consists of top-level supervisory algorithms that determine setpoints based on user settings and system conditions. These setpoints are provided to a set of active controls that handle setpoint tracking and disturbance rejection. This architecture is shown schematically in Figure 33.29 [8]. The supervisory controls serve the function of coordinating system operation, providing the structure for the various operating modes of the system, handling the sequencing and transition between operating modes, monitoring the health and safe operation of the system, and optimizing system efficiency. The supervisory control may include a built-in-test (BIT) to monitor system health. Depending on the system configuration (i.e., with or without an external reformer), there are up to five key independent variables that govern the operation of the system: output power, system FU, system air utilization, O/C ratio, and S/C ratio. These key independent variables are set by the supervisory controls to maximize system efficiency and stability while meeting the required power command. These key variables are then interpreted and driven down to the lower level control loops as individual actuator setpoints. The active controls translate setpoint commands from the supervisory controls into signals that ultimately drive individual actuators throughout the system. The control design can employ or combine the output of feedforward and feedback algorithms.

- The feedforward algorithms utilize the setpoint information, either by itself or coupled with sensor data from the system, along with a map generated from models or empirical data. The map transforms the system-level setpoint targets and any feedback signals into a setpoint that is recognizable by the individual actuators such as speed, valve position, and so on.

- The control feedback employs single-loop proportional–integral (PI)-type compensation for improved tracking and disturbance rejection. The PI controllers also incorporate an anti-windup feature to prevent saturation problems. The gain values provided to the controllers may be scheduled to compensate for variations in the dynamic response of the system in different operating regimes. Feedback control is enhanced through state estimation using multiple measurements and sensor types where practical.

33.4
SOFC Power Systems

33.4.1
Portable Systems

Small SOFC systems in the 10–500 W power range operating on hydrocarbon fuels have been considered and developed for portable applications. Examples of portable applications are battery charging, remote power, and low-level auxiliary power [50, 51]. Portable SOFC systems are particularly suitable for tactical military applications (e.g., soldier power and unmanned vehicle power) due to its potential for operation on logistics fuels such as JP-8 [52].

Three factors critical in the design of SOFC power systems for portable applications are (i) stack design and stack power density, (ii) system thermal management, and (iii) reforming process selection and reformer design. The stack must be designed and configured for minimum size and weight, optimum power, and desired operating characteristics required for the portable system. For example, the microtubular stack design has been selected and developed for several portable SOFC systems because of its suitability for low power and rapid startup [53–55].

Thermal management is one of the key design aspects of portable SOFC systems. For large-scale applications, thermal management can be addressed by ancillary systems such as heat exchangers, heaters, and combustors. For weight- and size-critical portable applications, such peripherals must be reduced in size or eliminated and optimized for power consumption without degrading performance. In addition, the system design must be optimized to have a small, thermally integrated insulation package.

Another critical aspect in designing portable SOFC systems is selecting the appropriate reforming process and developing a compact design for the reformer [56]. The choice of the reaction process and the design of the reformer depend on the system requirements, thus the operating characteristics of the application (e.g., the type of fuel, startup/shutdown frequency, and varying power demand). The reformer most commonly used in portable SOFC systems is the CPOX. The advantages of this type of reformer include operation without water, rapid startup, and compact design. The main disadvantage is a high concentration of inert nitrogen in the reformate. Figure 33.30 shows an example of a CPOX reactor (1 in

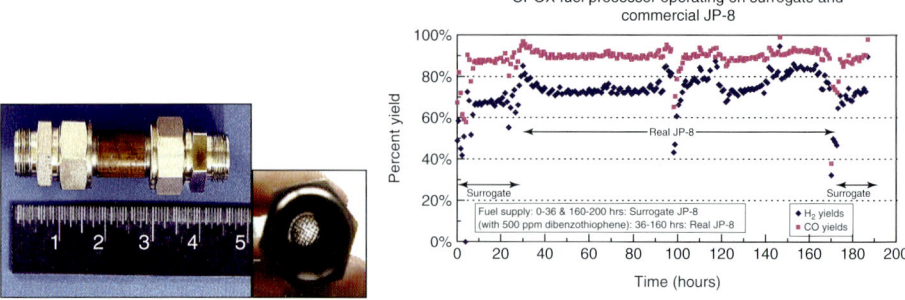

Figure 33.30 CPOX reactor and its operation on JP-8 fuel.

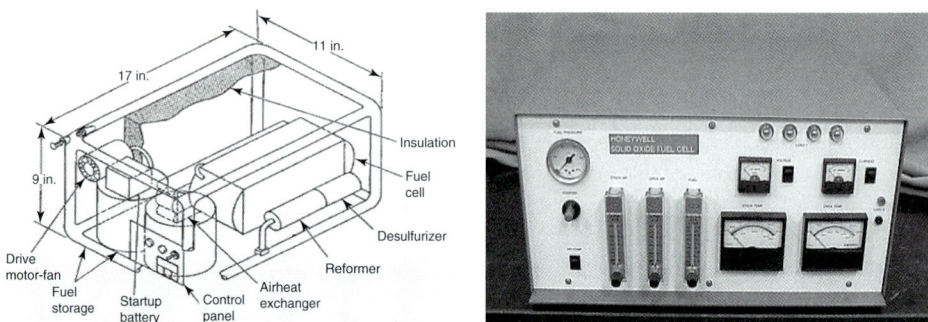

Figure 33.31 500 W battery charger concept and laboratory prototype.

or 2.5 cm diameter, 2 in or 5 cm length) capable of producing sufficient reformate to support the operation of a portable 500 W SOFC system on JP-8 fuel [57].

Two examples of portable SOFC system concepts are given here: a 500 W battery charger and a 20 W soldier power concept [57, 58]. A conceptual presentation of a 500 W SOFC battery charger showing the main components and dimensions (in inches) and a photograph of a laboratory system are presented in Figure 33.31. In this system design, the reformer is a CPOX. This 500 W system produces 28 V DC and weighs about 7 kg in a volume of $43 \times 28 \times 23$ cm. At a cell voltage of 0.65 V and FU of 0.84, the gross power of this portable system is 585 W and the estimated efficiency is 25.1%. Figure 33.32 shows a 20 W soldier power concept along with its projected properties and operating characteristics.

Two examples of portable SOFC system prototypes are a 75 W generator [59] and a 300 W portable APU [55]. The 75 W portable generator is $\sim 13 \times 18 \times 25$ cm with a dry mass of 3 kg and a fuel consumption of about 0.55 kg per day operating on propane or low-sulfur kerosene. This generator includes an internal hybrid battery and can provide peak power of up to 150 W at 12 V DC or 24 V DC output. The 300 W portable APU is $15.75 \times 8 \times 14$ in ($40.0 \times 20.3 \times 35.5$ cm) with a mass of 11.6 kg and a fuel consumption of 0.12 kg h^{-1} on propane. The performance of this APU is 300 W at 28.8 V and the system is capable of 200–300 on–off cycles.

- Power: 20 W, 12 VDC
- Estimated dimension: 8.5 × 11.5 × 20 cm (10-day mission)
- Estimated weight (excluding fuel): 0.5 kg
- Estimated energy density (JP-8 fuel):
  −2000 Wh kg^{-1} (3-day mission)
  −3700 Wh kg^{-1} (10-day mission)
- Estimated efficiency (LHV): 57%
- Fuel: JP-8

Figure 33.32 20 W SOFC soldier power concept.

33.4.2 Transportation Systems

33.4.2.1 SOFC-Based APUs for Automobiles and Trucks

SOFC power systems have been developed for automobile and truck APU applications [60–63]. Examples of automobile APU applications are automobile 5 kW APUs for engine-off power and 5 kW APUs in combination with lithium ion batteries for electric vehicle range extension. Examples of truck APU applications are Class 6–8 truck 3–10 kW APUs, recreational vehicle 3–5 kW APUs and long-haul Class 8 truck 10–30 kW APUs for refrigeration. SOFC based APUs for automobiles and trucks are designed to complement the internal combustion engine, serving as an efficient generator to provide power with the engine on or off. The expected benefits of SOFC APUs include (i) power supply with high efficiency and essentially zero or very low emissions with the engine on or off, (ii) operation of any electrical accessory, and (iii) possible enabler for high power consuming advancements. In addition, byproducts from SOFC APUs, that is, syngas and heat, may be very valuable. For instance, the syngas can be used for enhanced combustion and aftertreatment in an internal combustion engine. The heat can be used for vehicle heating and accelerated engine and catalyst warmup and may drive a bottoming cycle such as an expander to recover additional power.

The main components of an automobile/truck SOFC-based APU are SOFC stack, FR, and other BoP parts for process air supply, thermal management, waste energy recovery, power electronics/controls, and heating, ventilation, and air conditioning (HVAC). Since common fuels for automobile/truck SOFC APUs are gasoline and diesel, the selection and development of a suitable reformer or prereformer is an important element in the design of SOFC APUs [64, 65]. A schematic of a system design for a SOFC APU with a CPOX reformer and anode tailgas recycle is given in Figure 33.33 [66] and Figure 33.34 shows a simplified block diagram of PE and controls for an automobile SOFC APU [60].

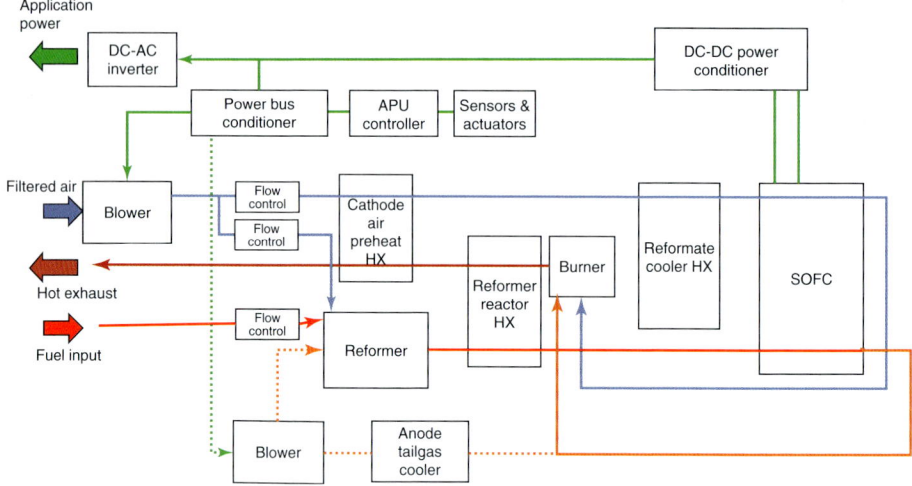

Figure 33.33 Schematic of an SOFC APU configuration with anode gas recycle.

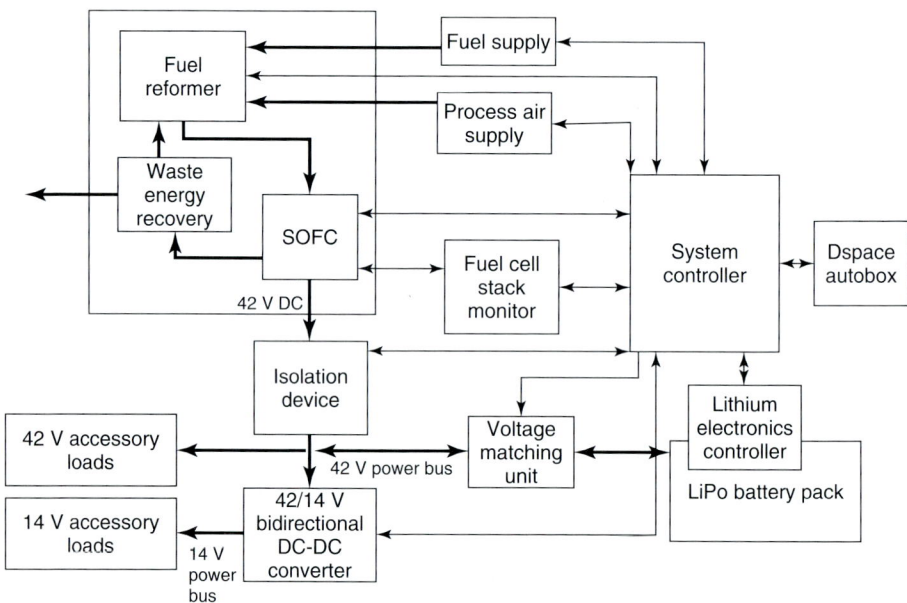

Figure 33.34 Block diagram of power electronics and controls of an automobile SOFC APU.

33.4.2.2 SOFC-Based APUs for Aircraft

SOFC power systems, especially SOFC/GT hybrid systems, have been considered as a possible replacement for conventional aircraft turbine-powered APUs in future more-electric aircraft. Figure 33.35 is a simplified diagram of an aircraft architecture with a fuel-cell APU [67]. The use of SOFC-based APUs has the

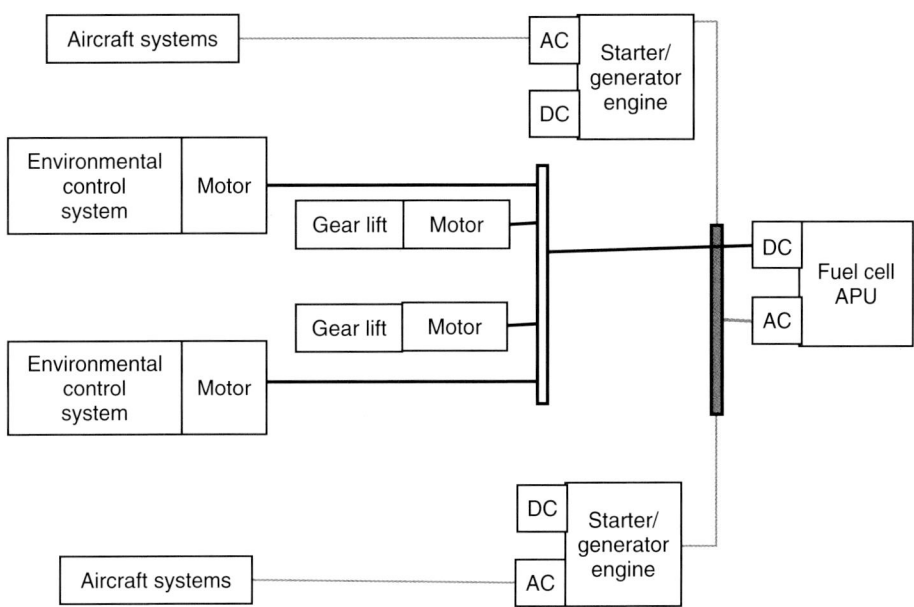

Figure 33.35 More-electric aircraft architecture with fuel-cell APU.

potential for reducing fuel consumption, noise, and exhaust emissions in aircraft in-flight and on-ground power generation [19, 67–71]. Projected fuel savings with SOFC APUs when compared with GT APUs are 40% less fuel used in flight and 75% less fuel used on the ground [67]. Noise and NO_x emissions with SOFC APUs are also significantly reduced [70]. The key issues regarding SOFC-based APUs relate to weight, startup time, and cost.

One aspect in the design of aircraft SOFC based APUs is the CHP configuration to use the high-temperature exhaust gas to supply aircraft thermal loads, thus reducing the electric power request and, therefore, the size and weight of the APU. Another aspect is the configuration of the feed air, especially since the external air is highly affected by altitude changes. Air can be supplied by either using cabin exhaust flow or taking the air directly from outside the aircraft. The approach of feeding the SOFC with cabin air yields high system efficiency at high-altitude conditions. The lower outside air pressure at high altitudes results in higher power turbine expansion ratios, therefore increased turbomachinery performance. The fuel cell can run at reduced power and at a more efficient operating point. Hence the overall system efficiency increases at high altitudes, particularly when using cabin air.

Figure 33.36 is an example of a 300 kW aircraft SOFC APU design [70]. In this system design, the power generation subsystem consists of twin 150 kW hybrid SOFCs and the fuel processing subsystem consists of an ATR FR and a DS. A fuel heat exchanger (HEX) is used to cool the reformate gas stream. A portion of exhaust gases from the SOFC are recycled via the CGR and AGR blowers.

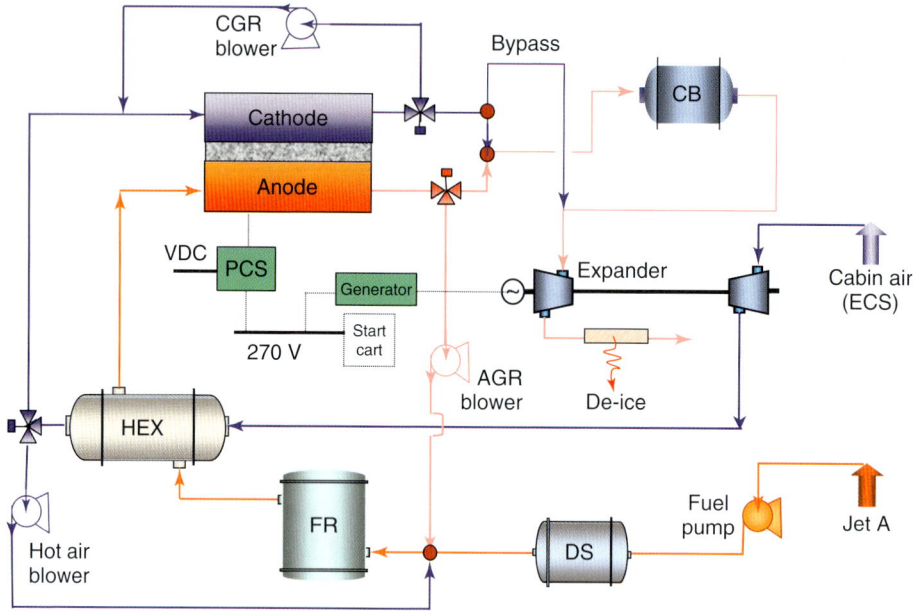

Figure 33.36 Schematic of 300 kW aircraft SOFC APU.

The remainder of the gases are mixed and sent to the catalytic burner (CB). A power conditioning system (PCS) converts DC voltage from the fuel cell to 270 V. In the operation of this APU, the air requirements are met entirely with the available cabin air. The SOFC stack is pressurized to 3 atm and operates at about 615 °C with an FU of about 80%. The projected efficiency of the system is 45% LHV under ground conditions and 64% LHV at cruise conditions. The efficiency advantage of this system over conventional APUs is 28% points, equivalent to a fuel saving of 70% over 1 day of ground operation that translates to about 3.3% of the total fuel burn during daily aircraft operations. The SOFC APU can provide some fuel burn savings during in-flight segments when the conventional APU is not typically operated due to its poor efficiency. Thus, operating the SOFC-based APU during the climb, cruise, and descent portions of the flight can provide a 1.3% saving in total aircraft fuel burn. The total fuel consumption savings of the SOFC APU from both ground and flight operations is therefore about 4.7% [70].

SOFC-based APUs can be located in the tail end of the airplane, the current location of turbine-powered APUs in commercial aircraft [67]. Another integration concept for SOFC APUs is to locate the fuel cell and the reformer in a fire compartment in the wing root since most of the electric consumers are in the center of the aircraft. An advantage of this arrangement is the option to use the hot exhaust for wing anti-ice. Air for the SOFC is taken from the recirculation and mixing plenum of the aircraft environmental control system (ECS).

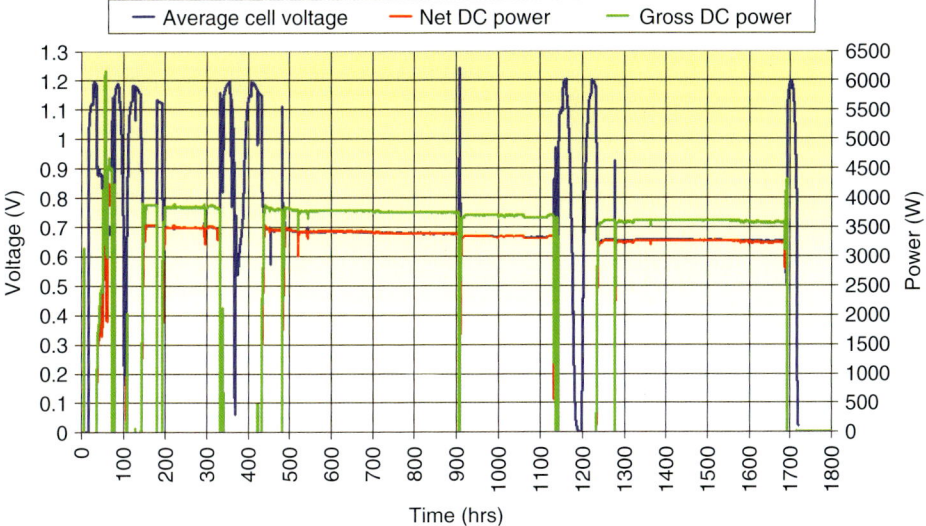

Figure 33.37 Average cell voltage and DC power of 5 kW SOFC prototype.

33.4.3
Stationary Systems

33.4.3.1 Stationary Simple Cycle SOFC Systems

Simple cycle SOFC systems at a power level of 1–200 kW have been developed for residential, CHP, and on-site power applications. An example of the design of a 5 kW SOFC system suitable for grid power connection and local load applications is given in Figure 33.4 (see Section 33.2.1). Table 33.3 summarizes the main components of this system and a conceptual representation of such a system and a photograph of a laboratory prototype are shown in Figure 33.5 (see Section 33.2.1). The performance of the laboratory prototype of this system during various operation events is shown in Figure 33.37 [8].

The initial peak DC efficiency of this prototype is 40.9% with a net DC power of 3.26 kW (projected AC efficiency of 38.0%, net AC power of 3.03 kW) and the initial peak DC power is 5.43 kW with a DC efficiency of 29.0% (net AC power of 5.10 kW, AC efficiency of 27.2 %). The operating conditions at the peak efficiency and peak power points are given in Table 33.6.

SOFC systems can be designed to include a heat recovery component such as an adsorption chiller heater for CHP applications [72–76]. An example is a small (1–10 kW) methane-fueled residential CHP SOFC system that integrates CGR, AGR, and internal reforming [73]. The system consists of a fuel-cell stack, steam preformer, various fluid delivery devices (blowers, ejectors, compressor, and water pump), heat exchangers, and catalytic combustor and power conditioning device along with a heat recovery component. Based on certain system parameters (50-cell stack of 81 cm^2 active area, nominal cell temperature of 800 °C, current density of 0.57 A cm^{-2}, power density of 0.40–0.43 W cm^{-2}, S/C ratio of 2.0, SOFC

Table 33.6 Operating conditions at peak efficiency and peak power of 5 kW system.

	Peak efficiency point Peak efficiency: 40.9% Gross DC power: 3.59 kW Net DC power: 3.26 kW	Peak power point DC efficiency: 29.0% Gross DC power: 6.13 kW Peak DC power: 5.43 kW
Fuel utilization (%)	78	67
Air utilization (%)	26	18
S/C ratio	1.2	1.3
O/C ratio	0.68	0.66
Current density (mA cm^{-2})	218	444
Average cell voltage (V)	0.73	0.63

FU of 85%, cathode air temperature rise of 100 °C, AGR of 62%, CGR of 77%, and 100% internal reforming), the projected performance of the system is an electric efficiency of 40.2% HHV (high heating value), a CHP efficiency of 79.3% HHV, and a thermal-to-electric ratio of 0.97. With internal reforming of 80%, the system electric efficiency is reduced to about 39.3% HHV.

33.4.3.2 SOFC/GT Hybrid Systems

SOFC/GT hybrid systems have been considered for DG and central power generation applications [24, 26, 77]. As indicated earlier, hybridization of the SOFC with a GT significantly improves the system efficiency and can be beneficial for use in large (hundreds of KWs and higher) power plants. Figure 33.38 is a schematic of a small (500 kW) SOFC/GT system for DG applications. This design is based on a planar SOFC, a microturbine, and a steam reformer for operation with natural gas [24]. The key components are given in Table 33.7.

The projected efficiency of this system (based on certain assumptions and constraints) is 67% LHV (at 800 °C operating temperature, 4 atm pressure, 75% FU, 0.75 V cell voltage and 100 °C temperature rise for the SOFC with CGR and AGR). The net power is estimated to be 517 kW with a SOFC power of 495 kW, turbine power of 78 kW, and parasitic power of 56 kW. System efficiencies at part loads are shown in Figure 33.39. It can be seen that the system efficiency remains relatively flat for a net power higher than about 250 kW. The system power in this region is a strong function of two parameters: (i) the cell voltage (the higher the voltage, the higher is the system efficiency) and (ii) the SOFC specific power, that is, the ratio of the SOFC power to the cooling air flow rate (the higher the specific power, the higher is the efficiency). The voltage rises at lower power loads because the SOFC stack operates at a lower current density. However, this rise is tempered by the decreasing fuel-cell temperature and pressure at part load. On the other hand, the specific power tends to decrease with decreasing speed owing to the recuperator temperature constraint. The combination of these effects results in a relatively flat efficiency line at higher

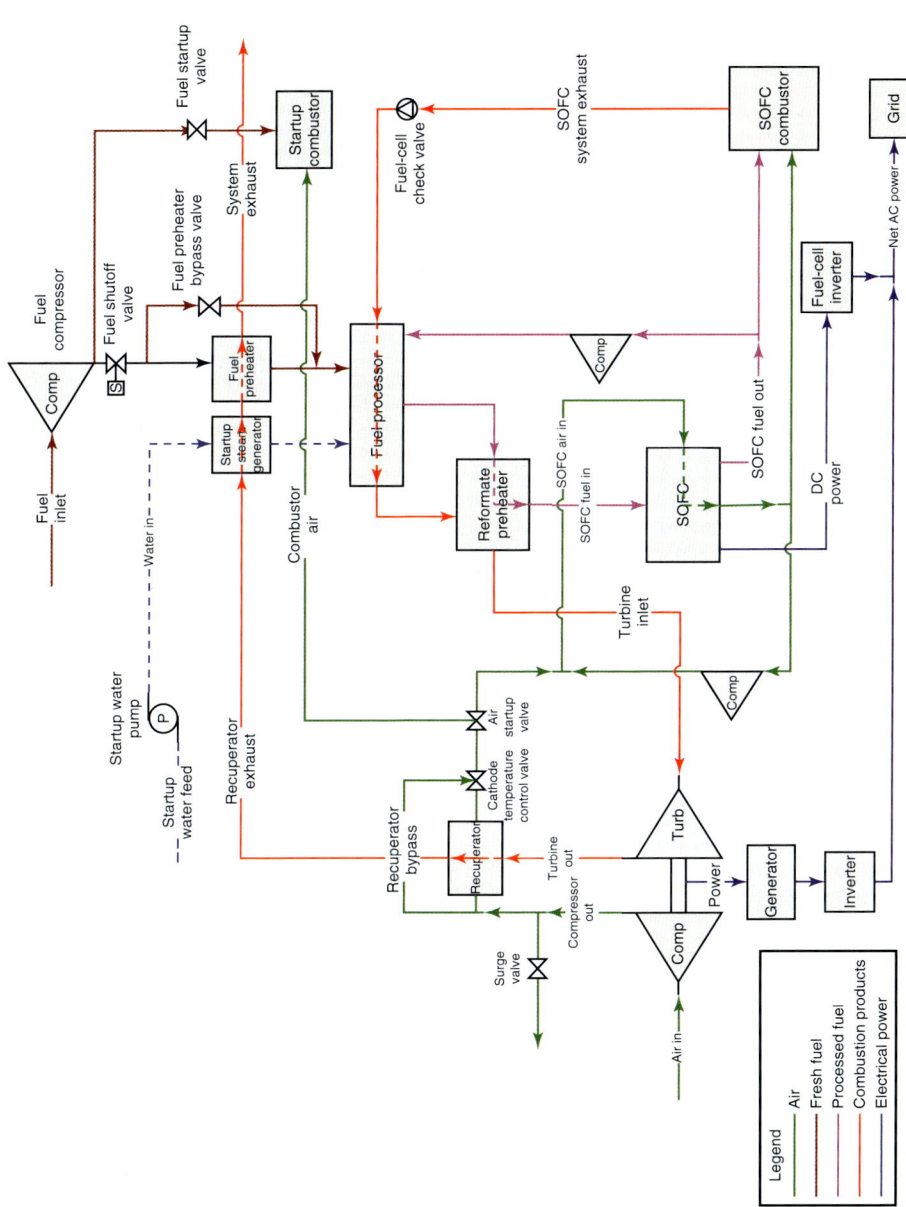

Figure 33.38 Schematic of SOFC/GT hybrid system.

Table 33.7 Key components of 500 kW SOFC/GT system.

Subsystem	Component
Power generation	SOFC stack, microturbine
Fuel processing	SR
Fuel, oxidant, and water delivery	Natural gas compressor, anode recycle blower/compressor, cathode recycle blower/compressor, startup water pump, valves
Thermal management	Fuel preheater, reformate preheater, recuperator, startup combustor, SOFC tailgas combustor, startup steam generator
Power conditioning	Fuel-cell inverter, turbine inverter
Control	Controller

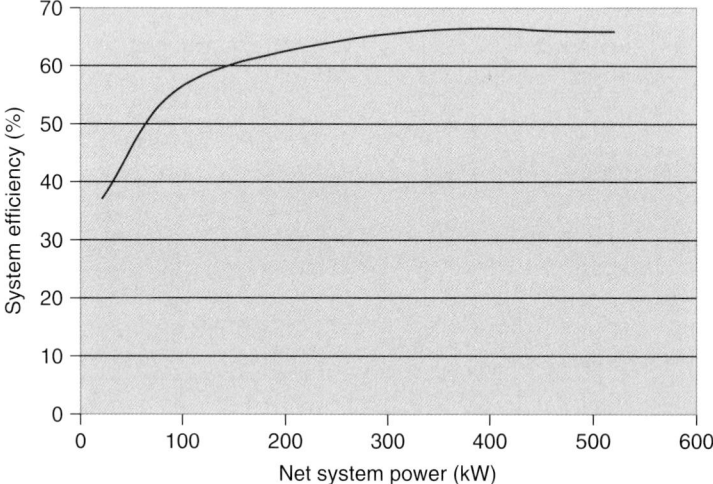

Figure 33.39 SOFC/GT system efficiency as a function of net system power.

part loads. However, the SOFC specific power decreases at a faster pace with decreasing system power, which eventually results in lower efficiencies at low part loads.

Figure 33.40 is a drawing of a 5 MW SOFC/GT hybrid power plant and Figure 33.41 shows the design of a 25 MW SOFC/GT hybrid system for central power generation applications [26]. The design of this 25 MW power plant uses a parallel arrangement of planar stacks with respect to both air and fuel. The air preheat is accomplished with the air recycle and the water management with the fuel recycle to the preformer. In this design, the SOFC operates at 800 °C on natural gas with internal reforming and the preformer is an SR. The projected system efficiency is about 65.5% LHV with a net power of 25 MW (SOFC power of

Figure 33.40 5 MW SOFC/GT hybrid system.

20.3 MW, GT power of 6.1 MW, and parasitic power of 1.4 MW). The key operating parameter assumptions for the SOFC in this case are 0.7 V cell voltage, 70% one-pass FU, 100 °C temperature rise, and 5 atm pressure. The optimal number of stacks and the stack size are determined for this system using an approach to examine stack size effects on the performance, cost, and reliability of the system. In this case, a cell diameter of 45.7 cm is chosen based on system cost and reliability projections [26]. From stack cost optimization and PE subsystem loss studies, the optimal SOFC stack configuration for this 25 MW hybrid has 400 cells 45.7 cm in diameter. Assuming a cell power density of 0.5 W cm^{-2}, this SOFC stack size translates into a 320 kW stack power. Given the 20 MW total SOFC stack subsystem power requirement, a minimum of 62 stacks is required to construct the 25 MW hybrid plant.

33.4.3.3 Integrated Gasification Fuel Cell (IGFC) Systems

A SOFC can be integrated with a coal gasifier to form the so-called integrated gasification fuel cell (IGFC) system for multi-MW base load power generation applications. The main advantages of the IGFC are its projected increased system efficiencies and comparable estimated costs compared with those of the state-of-the-art IGCC systems [78]. IGFC systems can operate under atmospheric pressure (with atmospheric SOFCs) or under pressure (with pressurized SOFCs or SOFC/GT hybrids) [79, 80]. A typical IGFC consists of the following sections: gasifier, gas cleanup, air separation unit (ASU), GT (if an SOFC/GT hybrid is included in the design), ST and heat recovery steam generation (HRSG) (if a steam bottoming cycle is included in the design), CO_2 separation (if included in the design), and SOFC. Figure 33.42 is a simplified schematic of a pressurized IGFC configuration and a conceptual presentation of an IGFC power plant is shown in Figure 33.43.

A key feature in the design of an IGFC system mainly relates to air and fuel flow arrangements in the SOFC (and GT) section(s). Figure 33.44 is an example of a SOFC and GT design for a 300 MW pressurized IGFC system [78]. In this design concept, the SOFC subsystem consists of several pressure modules; each module contains several fuel cell stages (denoted FC1, FC2, ..., FCN; each stage is a SOFC stack or several stacks placed side-by-side). The depleted fuel along with air from the last fuel-cell stage is combusted in a low-Btu combustor and fed into a GT for

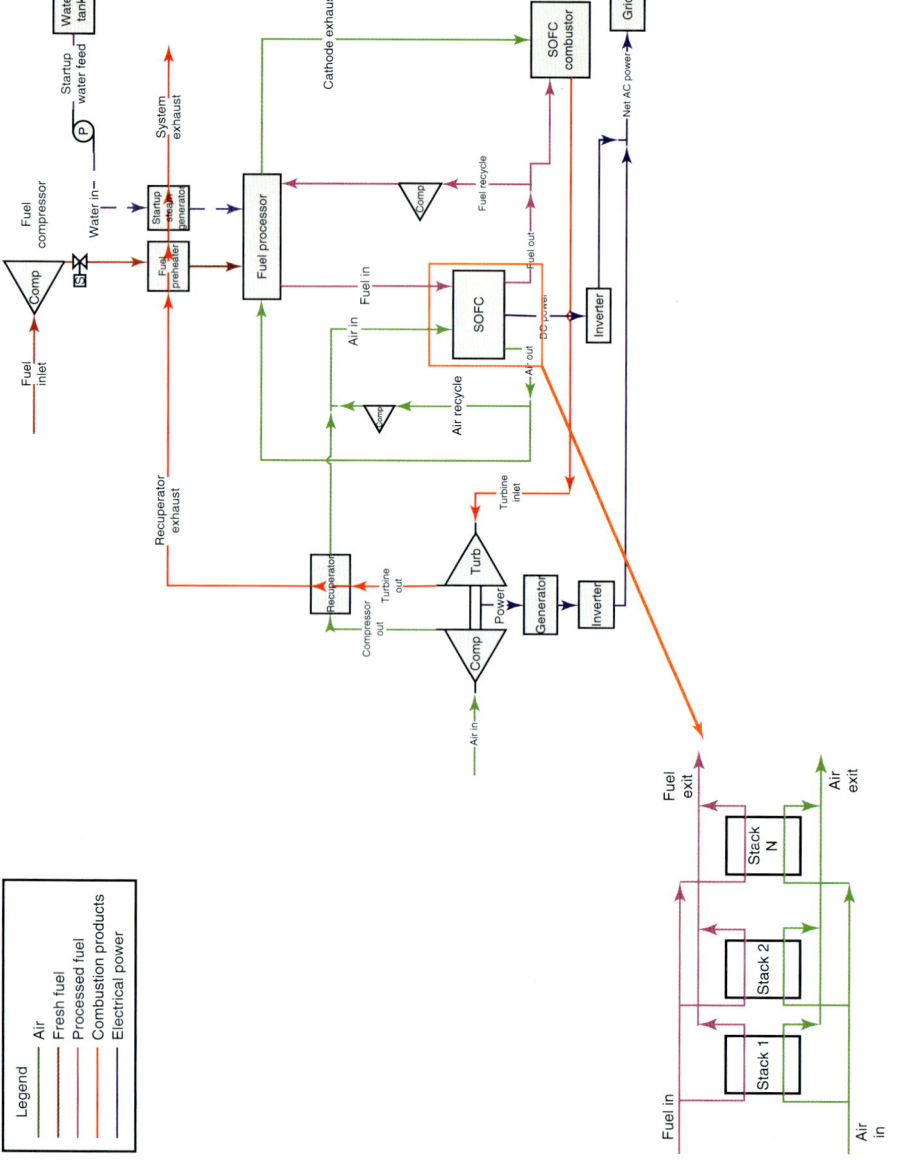

Figure 33.41 Schematic of 25 MW SOFC/GT hybrid system.

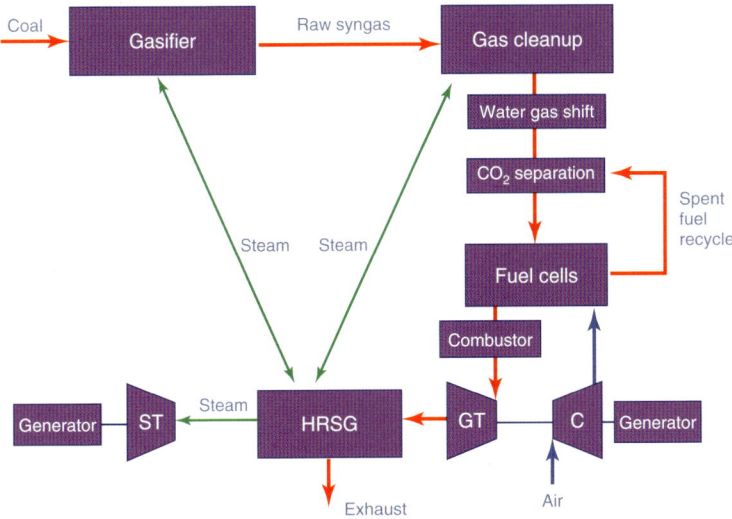

Figure 33.42 Simplified schematic of pressurized IGFC.

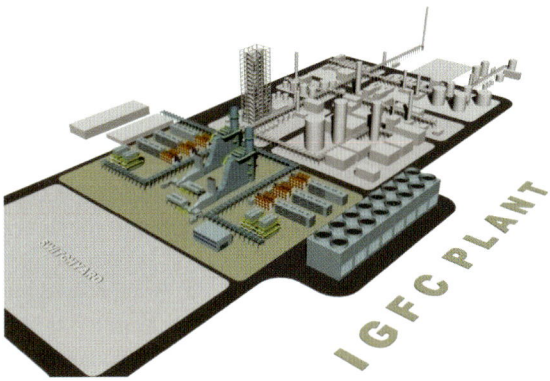

Figure 33.43 IGFC power plant.

expansion. Gases are supplied to the modules by one or two large GTs (C1, T1, B1) and a syngas expander (T2). Several modules are served by the same GT system. Compressed air from C1 is fed to a plenum and distributed equally to the modules. Similarly, vitiated air is collected at the hot plenum and fed to the burner (B1). On the fuel side, high-pressure syngas is expanded through T2 down to the fuel cell operating pressure and distributed to the various modules and cells inside the module in a similar way. The basic idea in this design concept is that air is used for inter-cooling between stages. The bulk of the compressed air stream 3 is used for inter-cooling; only a bleed (stream 4) is used to provide air for the first stage. The temperature of the air stream at station 4 is not high enough to be admitted directly into the stack, so a portion of the cathode exhaust is recycled, stream 55, and mixed to produce stream 5. Streams 55 and 4 mix in the mixer MR. (Note that MA

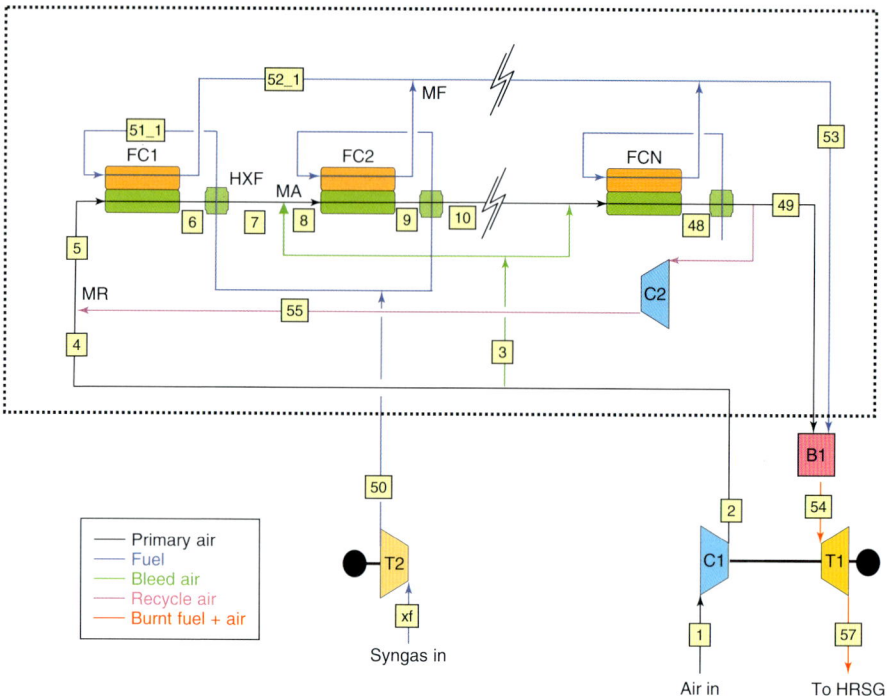

Figure 33.44 SOFC and GT design concept for a 300 MW IGFC.

and MF denote air mixer and fuel mixer, respectively.) The recycle stream is bled from the last-stage cathode exhaust and a recycle blower C2 is used to overcome the pressure differential. The fuel-cell stages have built-in fuel preheaters, shown as HXF. These units are integral to the stack design, and act as heat exchangers to raise the fuel stream temperature to the point where it can be admitted to the anode inlet. The heat source is the waste heat of the stack, which is adequate to raise the fuel temperature to the desired value. A maximum fuel-cell exit temperature of 775 °C is assumed in this case. The spent fuel from all the stages is collected at station 53 and sent to burner B1 to burn with the cathode exhaust 49.

Figure 33.45 shows the overall plant performance for the IGFC design concept (shown in Figure 33.44) versus fuel-cell operating pressure. The HRSG inlet temperatures are also shown with maximum temperature (Tmax) and minimum temperature (Tmin) limit lines. At operating pressures above 10 bar (about 10 atm), the HRSG inlet temperature drops below 1000 °F (537.8 °C). Below this temperature, the bottoming cycle performance begins to suffer as it becomes difficult to drive a two-pressure reheat ST. At low pressures (below 6 bar), the HRSG inlet temperature exceeds the assumed maximum (1280 °F or 693.3 °C). The design point is shown as a star. At the design point pressure of 10 bar, the net plant efficiency is 53.4% (coal HHV basis). Figures 33.46 shows the effect of SOFC FU on plant efficiency and HRSG inlet temperature for this IGFC configuration.

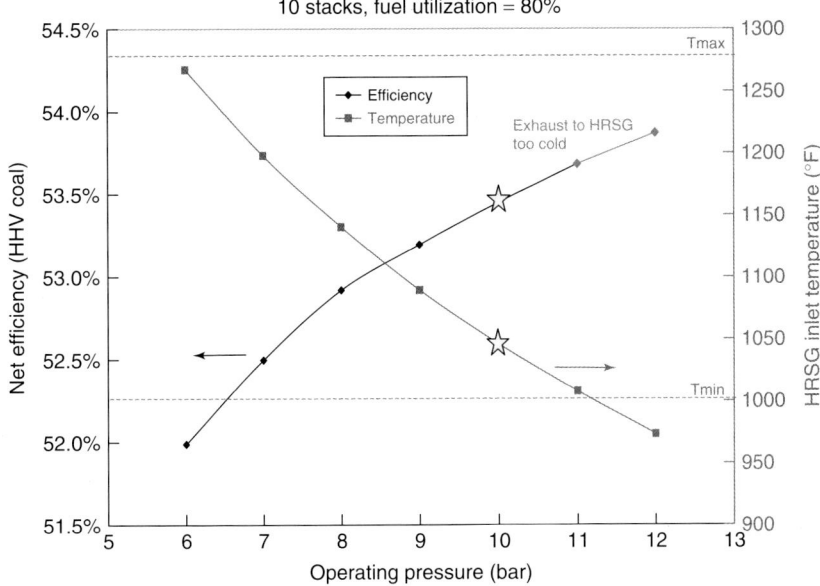

Figure 33.45 Overall performance of IGFC system (star: selected design point).

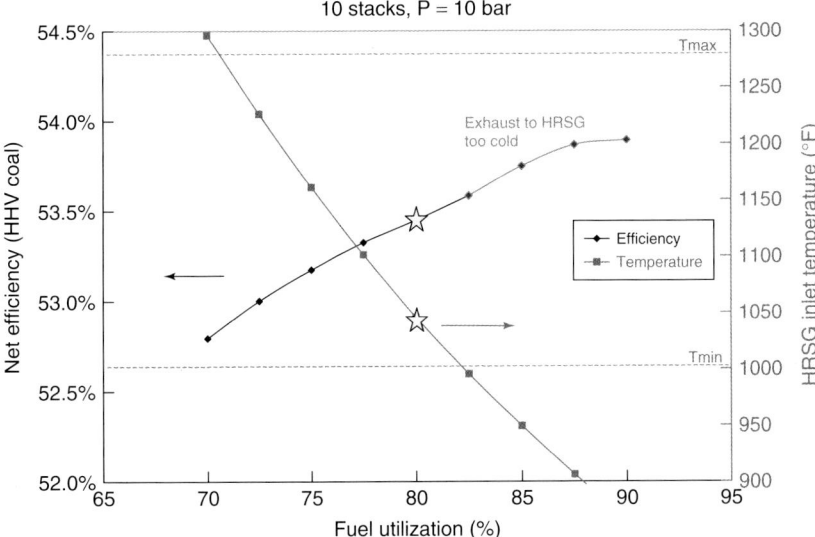

Figure 33.46 Effect of SOFC fuel utilization on IGFC system efficiency and HRSG inlet temperature (star: selected design point).

Acknowledgments

This chapter includes selected results from the SOFC programs supported by the US Department of Energy/National Energy Technology Laboratory (NETL) and US Department of Defense/Defense Advanced Research Projects Agency (DARPA) when the author was the Principal Investigator of the programs at AlliedSignal/Honeywell/GE. The work under these programs was performed by the AlliedSignal/Honeywell/GE fuel-cell teams. The author would like to thank Dr. Eric Armstrong of the Center for Energy Research, University of California, San Diego for reading the original manuscript and providing comments and suggestions.

References

1. Minh, N.Q. (1993) Ceramic fuel cells. *J. Am. Ceram. Soc.*, **76**, 563–588.
2. Minh, N.Q. and Takahashi, T. (1995) *Science and Technology of Ceramic Fuel Cells*, Elsevier, Amsterdam.
3. Lee, A.L. (1987) Internal reforming development for solid oxide fuel cells, final report, Technical Report DOE/MC/22045-2364, US Department of Energy, Office of Scientific and Technical Information, Oak Ridge, TN.
4. Park, S., Vohs, J.M., and Gorte, R.J. (2000) Direct oxidation of hydrocarbons in a solid-oxide fuel cell. *Nature*, **404**, 265–267.
5. Singhal, S.C. and Kendall, K. (eds.) (2003) *High Temperature Solid Oxide Fuel Cells: Fundamentals, Design and Applications*, Elsevier, Oxford.
6. Kendall, K., Minh, N.Q., and Singhal, S.C. (2003) in *High Temperature Solid Oxide Fuel Cells: Fundamentals, Design and Applications* ,(eds. S.C. Singhal and K. Kendall), Elsevier, Oxford, pp. 197–228.
7. Minh, N.Q., Singhal, S.C., and Williams, M.C. (2009) in *Fuel Cell Seminar 2008*, ECS Transactions, vol. **17** (1) (eds. M. Williams, K. Krist, and N. Garland), Electrochemical Society, Pennington, NJ, pp. 211–218.
8. GE Hybrid Power Generation Systems (2006) Solid State Energy Conversion Alliance (SECA) Solid Oxide Fuel Cell Program, DOE/NETL Cooperative Agreement DE-FE-26-01NT41245, Final Report.
9. Suzuki, M., Iwata, S., Higaki, K., Inoue, S., Shigehisa, T., Miyachi, I., Nakabayashi, H., and Shimazu, K. (2009) in *SOFC-XI, Part 1*, ECS Transactions, Vol. **25** (2) (eds. S.C. Singhal and H. Yokokawa), Electrochemical Society, Pennington, NJ, pp. 143–147.
10. Payne, R., Love, J., and Kah, M. (2009) in *SOFC-XI, Part 1*, ECS Transactions, Vol. **25** (2) (eds. S.C. Singhal and H. Yokokawa), Electrochemical Society, Pennington, NJ, pp. 231–239.
11. Winkler, W. (1994) in *Proceedings of First European Solid Oxide Fuel Cell Forum*, Vol. **2** (ed. U. Bossel), Lucerne, pp. 821–848.
12. Samuelsen, S. (2004) Fuel Cell/Gas Turbine Hybrid Systems, ASME International Gas Turbine Institute, Norcross, GA.
13. Zabihian, F. and Fung, A. (2009) A review on modeling of hybrid solid oxide fuel cell systems. *Int. J. Eng.*, **3**, 85–119.
14. Winkler, W. and Lorenz, H. (2000) in *Proceedings of Fourth European Solid Oxide Fuel Cell Forum*, vol. **1** (ed. U. Bossel), Lucerne, pp. 413–420.
15. Kurz, R. (2005) Parameter optimization on combined gas turbine-fuel cell power plants. *J. Fuel Cell Sci. Technol.*, **2**, 268–273.
16. Steffen, C.J. Jr., Freeh, J.E., and Larosiliere, L.M. (2005) Solid oxide fuel cell/gas turbine hybrid cycle technology for auxiliary aerospace power, NASA/TM-2005-213586, NASA Center

for Aerospace Information, Hanover, MD.
17. Himansu, A., Freeh, J.E., Steffen, C.J. Jr., Tornabene, R.T., and Wang, X.Y.J. (2005) hybrid solid oxide fuel cell/gas turbine system design for high altitude long endurance aerospace missions, NASA/TM-2006-214328, NASA Center for Aerospace Information, Hanover, MD.
18. Van Osdol, J., Liese, E., Tucker, D., Gemmen, R., and James, R. (2010) Scaling of a solid oxide fuel cell gas turbine hybrid system to meet a range of power demand. *J. Fuel Cell Sci. Technol.*, **7**, 015001.
19. Rajashekara, K., Grieve, J., and Dagget, D. (2006) Solid oxide fuel cell/gas turbine hybrid APU system for aerospace applications, in Industry Applications Conference, 2006. 41st IAS Annual Meeting, Conference Record of the 2006 IEEE, pp. 2185–2192.
20. Fathian, F. (2010) Thermodynamic analysis of a hybrid solid oxide fuel cell gas turbine power plant. *World Appl. Sci. J.*, **10**, 388–396.
21. Yang, J.S., Sohn, J.L., and Ro, S.T. (2007) Performance characteristics of a solid oxide fuel cell/gas turbine hybrid system with various part-load control modes. *J. Power Sources*, **166**, 155–164.
22. Yang, J.S., Sohn, J.L., and Ro, S.T. (2008) Performance characteristics of part-load operations of a solid oxide fuel cell/gas turbine hybrid system using air-bypass valves. *J. Power Sources*, **175**, 296–302.
23. Minh, N.Q. (2005) in *SOFC-IX*, vol. 1 (eds. S.C. Singhal and J. Mizusaki), Electrochemical Society, Pennington, NJ, pp. 76–81.
24. GE Hybrid Power Generation Systems (2004) Solid oxide fuel cell hybrid system for distributed power generation, DOE/NETL Cooperative Agreement DE-FC26-01NT40779, Phase 1 Topical Report.
25. Park, S. and Kim, T.S. (2006) Comparison between pressurized design and ambient pressure design of hybrid solid oxide fuel cell-gas turbine design. *J. Power Sources*, **163**, 490–499.
26. GE Hybrid Power Generation Systems (2004) Solid oxide fuel cell hybrid system for distributed power generation, SOFC scaleup for hybrid and fuel cell systems, DOE/NETL Cooperative Agreement DE-FC26-01NT40779, Final Topical Report.
27. Ray, E.R., Basel, R.A., and Pierre, J.F. (1995) Pressurized solid oxide fuel cell testing, DOE/MC/28055-96/C0600, US Department of Energy, Office of Scientific and Technical Information, Oak Ridge, TN.
28. Hashimoto, S., Nishino, H., Liu, Y., Asano, K., Mori, M., Furahashi, Y., and Fujishiro, Y. (2008) Effects of pressurization on cell performance of a microtubular SOFC with Sc-doped zirconia electrolyte. *J. Electrochem. Soc.*, **155**, B587–B591.
29. Seidler, S., Henke, M., Kallo, J., Bessler, W.G., Maier, U., and Friedrich, K.A. (2011) Pressurized solid oxide fuel cells: experimental studies and modeling. *J. Power Sources*, **196**, 7195–7202.
30. Bevc, F. (1997) Advances in solid oxide fuel cells and integrated power plants. *Proc. Inst. Mech. Eng. A: J. Power Energy*, **211**, 359–366.
31. Hashimoto, S., Liu, Y., Asano, K., Yoshiba, F., and Mori, M. (2011) Power generation properties of microtubular solid oxide fuel cell bundle under pressurized conditions. *J. Fuel Cell Sci. Technol.*, **8**, 021010.
32. Hashimoto, A., Kosaka, K., Matake, N., Yamashita, A., Kobayashi, Y., Kabata, T., and Tomida, K. (2010) Anode reaction in pressurized solid oxide fuel cells. *J. Power Energy Syst.*, **4**, 348–360.
33. Zhou, L., Cheng, M., Yi, B., Dong, Y., Cong, Y., and Yang, W. (2008) Performance of an anode-supported tubular solid oxide fuel cell (SOFC) under pressurized conditions. *Electrochim. Acta*, **53**, 5195–5198.
34. Brouwer, J. (2006) in *The Gas Turbine Handbook* (ed. R. Dennis), US Department of Energy, National Energy Technology Laboratory, Morgantown, WV, pp. 127–163.
35. Park, S.K., Oh, K.S., and Kim, T.S. (2007) Analysis of the design of a pressurized SOFC hybrid system using

a fixed gas turbine design. *J. Power Sources*, **170**, 130–139.
36. Murray, E.P., Tsai, T., and Barnett, S.A. (1999) Direct methane solid oxide fuel cell with ceria-based anode. *Nature*, **400**, 649–651.
37. Kim, H., Park, S., Vohs, J.M., and Gorte, R.J. (2001) Direct oxidation of liquid fuels in a solid oxide fuel cell. *J. Electrochem. Soc.*, **148**, A693–A695.
38. Gorte, R.J. and Vohs, J.M. (2003) Novel SOFC anodes for the direct electrochemical oxidation of hydrocarbons. *J. Catal.*, **216**, 477–486.
39. Minh, N., Anumakonda, A., Chung, B., Doshi, R., Ferrall, J., Guan, J., Lear, G., Montgomery, K., Ong, E., Schipper, L., and Yamanis, J. (1999) High-performance, reduced-temperature SOFC technology. *Fuel Cells Bull.*, 9–11.
40. Shah, R.K. (2003) Compact Heat Exchangers and Enhancement Technology for the Process Industries, Begell House, New York, pp. 205–215.
41. Magistri, L., Traverso, A., Massardo, A.F., and Shah, R.K. (2005) Heat exchangers for fuel cell and hybrid system applications, in *Proceedings of 3rd International Conference on Fuel Cell Science, Engineering and Technology*, FUEL CELL 2005-74176, ASME, Fairfield, NJ.
42. Yoshida, H. and Iwai, H. (2005) Thermal management in solid oxide fuel cell systems, in *Proceedings of Fifth International Conference on Enhanced, Compact and Ultra-Compact Heat Exchangers: Science, Engineering and Technology* FUEL CELL 2005-74176, (eds. R.K. Shah, M. Ishizuka, T.M. Rudy, and V.V. Wadekar), CHE2005-01, Engineering Conferences International, Hoboken, NJ.
43. Hefner, A. (2007) Power electronics for fuel cell based power generation systems, in *Proceedings of 8th SECA Annual Workshop*, FUEL CELL 2005-74176, US Department of Energy, National Energy Technology Laboratory, Morgantown, WV.
44. Lai, J.S. (2005) Power electronic technologies for fuel cell power systems, in *Proceedings of 6th SECA Annual Workshop*, FUEL CELL 2005-74176, US Department of Energy, National Energy Technology Laboratory, Morgantown, WV.
45. Cheng, K.W.E., Sutanto, D., Ho, Y.L., and Law, K.K. (2001) Exploring the power conditioning system for fuel cell. *IEEE Xplore*, 2197–2202.
46. Breg, J.J., Bordolla, C.R., Carrasco, J.M., Galvan, E., Jimenez, A., and Moreno, E. (2007) Power conditioning of fuel cell systems in portable applications. *Int. J. Hydrogen Energy*, **32**, 1559–1566.
47. Jeong, J.K., Lee, J.H., Han, B.M., and Cha, H.J. (2011) Grid-tied power conditioning system for fuel cell composed of three-phase current-fed DC–DC converter and PWM inverter. *J. Electr. Eng. Technol.*, **6**, 255–262.
48. Mazumder, S.K., Charya, K., Haynes, C.L., Williams, R., von Spakovsky, M.R., Nelson, D.J., Rancruel, D.F., Hartvigsen, J., and Gemmen, R.S. Jr. (2004) Solid-oxide-fuel-cell performance and durability: resolution of the effects of power-conditioning systems and application loads. *IEEE Trans. Power Electron.*, **19**, 1263–1278.
49. Mazumder, S.K., Burra, R.K., and Acharya, K. (2007) A ripple-mitigating and energy-efficient fuel cell power-conditioning system. *IEEE Trans. Power Electron.*, **22**, 1437–1452.
50. Weston, M. and Matcham, J. (2002) Portable power applications of fuel cells, DTI Pub Urn No. 02/1493. London, UK.
51. Traversa, E. (2009) Toward the miniaturization of solid oxide fuel cells. *Electrochem. Soc. Interface*, **18** (3), 49–52.
52. Nowak, R.J. (2001) A DARPA perspective on small fuel cells for the military, in *Proceedings of the 2nd SECA Annual Workshop*, FUEL CELL 2005-74176, US Department of Energy, National Energy Technology Laboratory, Morgantown, WV.
53. Hayashi, K., Yamamoto, O., and Minoura, H. (2000) Portable solid oxide fuel cells using butane gas as fuel. *Solid State Ionics*, **135**, 343–345.
54. Cheekatamarla, P.K., Finnerty, C.M., Robinson, C.R., Andrews, S.M., Brodie, J.A., Lu, Y., and DeWald, P.G. (2009) Design, integration and demonstration

of a 50 W JP8/kerosene fueled portable SOFC power generator. *J. Power Sources*, **193**, 797–803.
55. Rice, J. (2011) Solid oxide fuel cells to reduce the logistics burden, http://www.dtic.mil/ndia/2011power/Session14_12801Rice.pdf (last accessed 18 December 2011).
56. Krumpelt, M., Krause, T.R., Carter, J.D., Kopasz, J.P., and Ahmed, S. (2002) Fuel processing for fuel cell systems in transportation and portable power applications. *Catal. Today*, **77**, 3–16.
57. Minh, N., Anumakonda, A., Chung, B., Doshi, R., Ferrall, J., Guan, J., Lear, G., Montgomery, K., Ong, E., and Yamanis, J. (1999) in *SOFC-VI* (eds. S.C. Singhal and M. Dokiya), Electrochemical Society, Pennington, NJ, pp. 68–74.
58. Minh, N., Anumakonda, A., Doshi, R., Guan, J., Huss, S., Lear, G., Montgomery, K., Ong, E., and Yamanis, J. (2001) in *SOFC-VII* (eds. H. Yokokawa and S.C. Singhal), Electrochemical Society, Pennington, NJ, pp. 190–195.
59. Poshusta, J.C., Kulprathipanja, A., Martin, J.L., and Martin, C.M. (2006) Design and Integration of Portable SOFC Generators, http://aiche.confex.com/aiche/2006/techprogram/P69269.HTM (last accessed 18 December 2011).
60. De Minco, C. and Mukerjee, S. (2001) SOFC status and challenges, in *Proceedings of the 2nd SECA Annual Workshop*, FUEL CELL 2005-74176, US Department of Energy, National Energy Technology Laboratory, Morgantown, WV.
61. Botti, J.J., Grieve, M.J., and MacBain, J.A. (2005) Electric vehicle range extension using an SOFC APU, in *2005 SAE World Congress*, FUEL CELL 2005-74176, 2005-01-1172, SAE International, Warrendale, PA (SAE International).
62. Salameh, T. (2008) SOFC auxiliary power units (APUs) for vehicles, 2008 TRRF05 Fuel Cell Technology, Project Report, Lund University, Lund, Sweden.
63. Shaffer, S. (2004) Development update on Delphi's solid oxide fuel cell system, in *Proceedings of the 5th SECA Annual Workshop*, FUEL CELL 2005-74176, US Department of Energy, National Energy Technology Laboratory, Morgantown, WV.
64. Aicher, T., Lenz, B., Gschnell, F., Groos, U., Federici, F., Caprile, L., and Parodi, L. (2006) Fuel processors for fuel cell APU applications. *J. Power Sources*, **154**, 503–508.
65. Lindermeir, A., Kah, S., Kavurucu, S., and Muhlner, M. (2007) On-board diesel fuel processing for an SOFC–APU-Technical challenges for catalysis and reactor design. *Appl. Catal. B: Environ.*, **70**, 488–497.
66. Shaffer, S. (2007) Development update on Delphi's solid oxide fuel cell system, in *Proceedings of the 8th SECA Annual Workshop*, FUEL CELL 2005-74176, US Department of Energy, National Energy Technology Laboratory, Morgantown, WV.
67. Daggett, D. (2003) Commercial airplanes fuel cell APU overview, *Proceedings of the 4th SECA Annual Meeting*, FUEL CELL 2005-74176, US Department of Energy, National Energy Technology Laboratory, Morgantown, WV.
68. Eelman, S., Pozo del Poza, E., and Krieg, T. (2004) Fuel cell APU's in commercial aircraft an assessment of SOFC and PEM concepts, in 24th International Congress of the Aeronautical Sciences, pp. 1–10.
69. Gummalla, M., Pandy, A., Braun, R., Carriere, T., Yamanis, J., and Welch, R. (2006) Fuel cell airframe integration study for short-range aircraft: aircraft propulsion and subsystem integration evaluation, vol.1, NASA/CR-2006-214457/VOL1, NASA Center for Aerospace Information, Hanover, MD.
70. Braun, R.J., Gummalla, M., and Yamanis, J. (2009) System architectures for solid oxide fuel cell-based auxiliary power units in future commercial aircraft applications. *J. Fuel Cell Sci. Technol.*, **6**, 031015.
71. Santarelli, M., Cabrera, M., and Cali, M. (2010) Solid oxide fuel based auxiliary power unit for regional jets: design and mission simulation with different cell geometries. *J. Fuel Cell Sci. Technol.*, **7**, 021066.

72. Braun, R.J., Klein, S.A., and Reindl, D.T. (2006) Evaluation of system configurations for solid oxide fuel cell-based micro-combined heat and power generators in residential applications. *J. Power Sources*, **158**, 1290–1305.
73. Braun, R.J. (2010) Techno-economic optimal design of solid oxide fuel cell systems for micro-combined heat and power applications in the US. *J. Fuel Cell Sci. Technol.*, **7**, 031018.
74. Haga, T., Komiyama, N., Nakatomi, H., Konishi, K., Suton, T., and Kikuchi T. (2009) in *SOFC-XI, Part 1*, ECS Transactions, Vol. **25** (2) (eds. S.C. Singhal and H. Yokokawa), Electrochemical Society, Pennington, NJ, pp. 71–76.
75. Love, J., Amarasinghe, S., Selvey, D., Zheng, X., and Christiansen, L. (2009) in *SOFC-XI, Part 1*, ECS Transactions, Vol. **25** (2) (eds. S.C. Singhal and H. Yokokawa), Electrochemical Society, Pennington, NJ, pp. 115–124.
76. Suzuki, M., Iwata, S., Higaki, K., Inoue, S., Shigehisa, T., Miyachi, I., Nakabayashi, H., and Shimazu, K. (2009) in *SOFC-XI, Part 1*, ECS Transactions, Vol. **25** (2) (eds. S.C. Singhal and H. Yokokawa), Electrochemical Society, Pennington, NJ, pp. 143–147.
77. Siemens Westinghouse Power Corporation (2000) Pressurized solid oxide fuel cell/gas turbine power system, Contract DE-AC26-98FT40355, Final Report, US Department of Energy, Office of Scientific and Technical Information, Oak Ridge, TN.
78. GE Hybrid Power Generation Systems (2004) Coal integrated gasification fuel cell system study, DOE/NETL Cooperative Agreement DE-FC26-01NT40779, Final Report.
79. Grol, E. and Wimer, J. (2009) System analysis of an integrated gasification fuel cell combined cycle: technical assessment, vol.1, DOE/NETL-40/080609, US Department of Energy, National Energy Technology Laboratory, Morgantown, WV.
80. Gerdes, K., Grol, E., Keairns, D., and Newby, R. (2009) Integrated gasification fuel cell performance and cost assessment, DOE/NETL-2009/1361, US Department of Energy, National Energy Technology Laboratory, Morgantown, WV.

34
Desulfurization for Fuel-Cell Systems
Joachim Pasel and Ralf Peters

34.1
Introduction and Motivation

The option of using fuel-cell systems as so-called auxiliary power units (APUs) for on-board power supply in aircraft, fork lifts, trucks, and earth-moving equipment is of great interest to the respective industries. The application of fuel-cell systems in these mobile devices offers the potential to achieve individual tasks such as energy conversion or the production of water and inert gas by means of a single system, that is, the fuel-cell system. In aircraft applications, for example, water tanks, the conventional gas turbine-powered APU, and the tank inerting system can be dispensed with. The dimensions of generators and batteries can therefore be decreased. Such measures reduce fuel consumption, increase the overall efficiency of the mobile equipment, and permit low-emission operation. The fuels have to be desulfurized in order to operate these fuel-cell systems using middle distillates such as diesel, kerosene, and heating oil. If this is not done, the precious metal catalysts of the fuel processing unit will be irreversibly damaged. The upper value for the sulfur mass fraction of the fuel entering the fuel-cell system is ~10 ppm [1–3]. In the fuel processing unit, fuel is converted into a hydrogen-rich gas by means of autothermal reforming (ATR) and subsequent gas cleaning steps such as the water gas shift (WGS) reaction and catalytic combustion. It is well known that most of the sulfur contained in hydrocarbon fuels is converted to H_2S at low temperatures (300–400 °C) in a nonoxidizing atmosphere and to SO_2 at higher temperatures (>600 °C) in an oxidizing atmosphere [4].

34.2
Sulfur-Containing Molecules in Crude Oil

34.2.1
Crude Oil

Crude oil is formed over many millions of years from organic residues of plant and animal organisms. The residues are deposited in sea sediments under anaerobic

Fuel Cell Science and Engineering: Materials, Processes, Systems and Technology,
First Edition. Edited by Detlef Stolten and Bernd Emonts.
© 2012 Wiley-VCH Verlag GmbH & Co. KGaA. Published 2012 by Wiley-VCH Verlag GmbH & Co. KGaA.

Table 34.1 Mass fractions of sulfur in crude oil itself and in different crude oil fractions [7].

Product	North Africa	North Sea	Middle East	North America	South America
Crude oil	0.1	0.3	2.5	1	5.5
Gasoline (0–70 °C)	0.001	0.001	0.2	0.002	1
Naphtha (70–140 °C)	0.002	0.001	0.2	0.005	0.45
Petroleum (140–250 °C)	0.01	0.02	0.2	0.206	2.5
Gasoil (250–350 °C)	0.1	0.18	1.4	0.49	4.4
Residue (>350 °C)	0.3	0.6	4.1	1.5	6

conditions over many millions of years. During this time, they are chemically modified at increased temperatures so that today's crude oils consist of a broad variety of predominantly reduced substances. Aliphatic hydrocarbons dominate in most crude oils. However, aromatic hydrocarbons also occur in considerable quantities [5].

In refineries, crude oil is desalinated and fractionated. The crude oil stream is separated into product groups which differ in their boiling ranges. The components of gasoline boil below 70 °C, whereas naphtha has a boiling range between 70 and 140 °C. The fractions boiling between 140 and 350 °C are denoted middle distillates. For their part, middle distillates consist of the fractions of petroleum (140–250 °C) and gasoil (250–350 °C). Diesel fuel and heating oil are obtained by further upgrading [6, 7].

Crude oils with a sulfur mass fraction of 14 mass% are of no commercial interest. Commercial crude oils contain between 0.1 and 3 mass% sulfur depending on their geographic origin. The average sulfur mass fraction of the crude oil reserves worldwide amounts to 1.8 mass% [5]. Crude oils from North Africa and the North Sea often contain less than 0.1–0.3 mass% sulfur. These crude oils are therefore more valuable for applications in refineries than those from the Middle East or South America, which have sulfur mass fractions between 2.5 and 5 mass%. Table 34.1 gives an overview of crude oils from different geographic regions and their different sulfur mass fractions.

34.2.2
Routes for Inserting Sulfur into the Molecules in Crude Oil

Sulfur is a natural component of crude oil. After carbon and hydrogen, sulfur is the third most frequently occurring component. In fact, sulfur occurs in all living organisms. However, the quantities of sulfur in these organisms are too low to explain the considerable amounts of sulfur sometimes found in crude oil and its different fractions [5]. It is believed today that the sulfur in crude oils originates from sulfate ions in sea water. Microorganisms such as *Desulfovibrio* are able to transfer electrons to the sulfate ions. The sulfur is thereby reduced, forming molecules such as H_2S, hydrogen sulfide [8]. These reduced substances

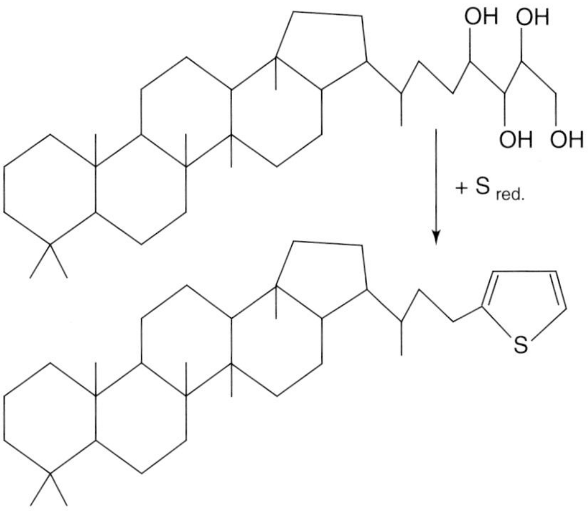

Figure 34.1 Possible formation route to a thiophene from bacteriohopantetrol.

are again oxidized by other bacteria such as *Thiobacillus*, forming elemental sulfur and polysulfides. Under anaerobic conditions, these species react fairly easily with the organic molecules embedded in the sediments. Sulfur is thus inserted into the structure [9].

First, aliphatic sulfur-containing compounds are formed. These then undergo further reactions, such as condensations, forming ring structures, or reactions leading to aromatic molecules or additional alkylations. Such reactions result in the formation of very complex mixtures of different sulfur-containing substances. Most of these reactions are as yet poorly understood. Nevertheless, compounds derived from vegetable substances can often be identified in crude oils. Figure 34.1 shows the possible formation route to a thiophene from bacteriohopantetrol [9].

34.2.3
Different Chemical Classes of Sulfur-Containing Substances in Crude Oil

These processes result in a large number of different sulfur-containing compounds in crude oil. The substances cover the whole range of molar masses present in crude oil. Therefore, almost any fraction of crude oil includes sulfur-containing substances and the sulfur content of the fractions rises with increasing boiling range [10].

The sulfur compounds in crude oil can be separated into two classes: aliphatic and aromatic hydrocarbons. The aliphatic hydrocarbons are characterized by open carbon chains, which can be branched. One or several sulfur atoms are inserted into these carbon chains in different ways, as shown in Table 34.2. These compounds are denoted sulfides, disulfides, and thiols [11].

The basis of aromatic sulfur-containing compounds in crude oil is the thiophene molecule, which consists of four carbon atoms and one sulfur atom (cf., Table 34.3).

34 Desulfurization for Fuel-Cell Systems

Table 34.2 Different classes of aliphatic sulfur compounds in crude oil.

Class	Boiling point (°C)	Structural formula
Sulfide	>37	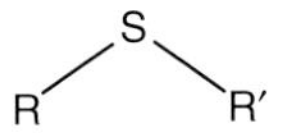 R–S–R'
Disulfide	>110	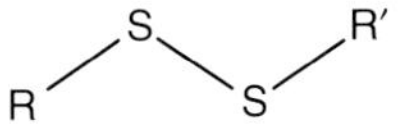 R–S–S–R'
Thiol	>6	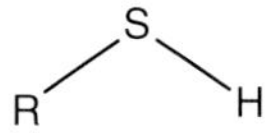 R–S–H

Table 34.3 Different classes of aromatic sulfur compounds in crude oil.

Class	Boiling point (°C)	Structural formula
Thiophene	84	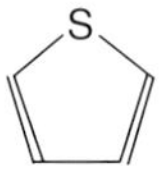 (thiophene)
Methylthiophene	113–115	
Dimethylthiophene	137–145	
Trimethylthiophene	–	
Benzothiophene	220	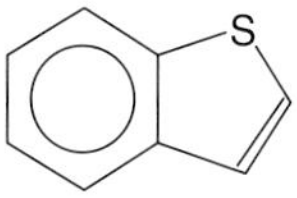 (benzothiophene)
Methylbenzothiophene	243–246	
Dimethylbenzothiophene	–	
Trimethylbenzothiophene	>240	
Dibenzothiophene	310	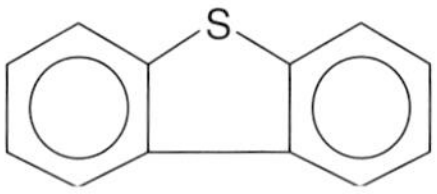 (dibenzothiophene)
Methyldibenzothiophene	316–327	
Dimethyldibenzothiophene	332–343	

The sulfur atom is able to provide one pair of free nonbinding electrons. Together with the four electrons from the two double bonds in thiophene, an aromatic system is formed according to Hückel's law. Five mesomeric structures of the thiophene molecule can be formulated (cf., Figure 34.2) [12]. The electrons are therefore evenly distributed over the whole planar ring system of the thiophene molecule. The resulting mesomeric energy amounts to 122 kJ mol^{-1}. This means that the energetic level of the five mesomeric structures of thiophene is lowered by 122 kJ mol^{-1} in comparison with the simple structure of thiophene in Table 34.3 [12].

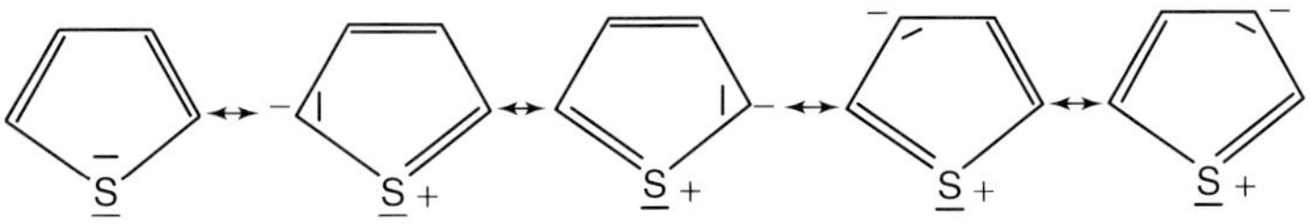

Figure 34.2 Mesomeric structures of thiophene.

Owing to this mesomeric effect, aromatic sulfur compounds such as thiophene are more stable, and therefore more difficult to remove from the fuel, than the aliphatic sulfur-containing hydrocarbons in Table 34.2, which are unable to form mesomeric structures owing to a lack of double-bond electrons. In any chemical process, including desulfurization, more severe reaction conditions, that is, higher temperatures or higher pressures, are necessary to activate the chemical bonds in aromatic substances in comparison with those in aliphatic hydrocarbons. Additional benzene rings or carbon chains can be bound to the thiophene molecule, leading to the highly condensed and alkylated compounds in Table 34.3 [11, 13, 14]. Benzothiophene and dibenzothiophene are even more stable than thiophene owing to the additional aromatic benzene rings, strongly increasing the mesomeric energy of the molecules.

34.2.4
Catalyst Poisoning by Sulfur-Containing Substances in Crude Oil Fractions

A number of studies have been published on this subject and some of the major findings are summarized here. Cheekatamarla and Lane [15] conducted experiments with respect to the ATR of synthetic diesel fuel and JP-8 in an adiabatic reactor using a Pt/ceria catalyst. The stability of the catalyst and its behavior in the presence of sulfur were investigated. The experiments showed that the catalytic reforming activity using synthetic diesel fuel with 10 ppm sulfur was stable over the time period tested. The observed loss in activity for JP-8 fuel with 1000 ppm sulfur was attributed in part to the sulfur-containing species in JP-8. When different mass fractions of SO_2 (0–400 ppm) were added to the feed, the hydrogen yield decreased dramatically from 75 to 40% while the mass fraction of sulfur increased to 200 ppm. The hydrogen yield stabilized at higher values (up to 400 ppm sulfur). In a series of experiments, the reforming reaction was first performed with a sulfur-free feed. After the reaction was stabilized, sulfur in form of H_2S or SO_2 was added. The feed was then switched back to the original pure feed. This process was repeated twice. When the sulfur-containing feeds were supplied to the reactor, the hydrogen yield decreased sharply. After a certain period of time, the decrease in concentration slowed and a stable H_2 yield was achieved. The poisoning appeared to be partly reversible, since the H_2 yield increased after switching back to the pure feed. The hydrogen yield increased to a certain extent, but the original level was not achieved. This finding may be due to irreversible adsorption of sulfur on the catalytic surface. In a subsequent paper, Cheekatamarla and Lane reported temperature-programmed reduction (TPR) studies in which they ascertained that the oxidation–reduction properties of ceria are affected by sulfur poisoning. Temperature-programmed desorption (TPD) and X-ray photoelectron spectroscopic (XPS) analysis confirmed the formation of chemisorbed sulfur species. This explains the observed irreversible poisoning [16].

Ferrandon *et al.* [17] investigated a 2 mass% Rh–La/Al_2O_3 catalyst for the ATR of sulfur-free gasoline and gasoline with 34 ppm sulfur. They found a considerable decrease in the hydrogen yield when the sulfur-containing gasoline was applied.

The decrease was much stronger during operation at lower temperatures. Complete recovery of the initial activity was achieved by switching from sulfur-containing gasoline to sulfur-free gasoline at 800 °C, while only 50% of the activity could be restored at 700 °C. The catalyst was characterized by scanning electron microscopy, elemental analyses, CO chemisorption, diffuse reflection infrared Fourier transform spectroscopy (DRIFTS), and X-ray absorption spectroscopy. It was found that sulfur increased the sintering of the Rh particles on the support due to an increase in the catalyst temperature. This temperature increased in the catalyst because the sulfur species in gasoline inhibited the endothermic steam reforming to a greater extent than the exothermic partial oxidation of the hydrocarbons in gasoline. An increase in the molar H_2O/C ratio from 2.0 to 3.0 or the addition of K to Rh significantly enhanced the sulfur tolerance of the catalysts.

Kaila et al. [18] tested both simulated and commercial low-sulfur diesel fuels (sulfur <10 ppm) for ATR on Rh and Pt and bimetallic RhPt catalysts supported on ZrO_2. Using simulated fuels, the catalysts were deactivated markedly in the presence of H_2S, whereas 4,6-dimethyldibenzothiophene (4,6-DMDBT) in the feed led to only a slight deactivation of the catalyst. Also, during ATR of commercial diesel, the catalyst was deactivated only slightly over several hours. Sulfur deposition on the catalyst was found to be strongest on the monometallic Rh/ZrO_2.

Mayne et al. [19] investigated the effect of sulfur on the deactivation of $Ni/Ce_{0.75}Zr_{0.25}O_2$ catalysts. They studied isooctane conversion to syngas in the presence of small amounts of thiophene. It was found that, depending on the reaction conditions, thiophene underwent different degrees of desulfurization. Under reaction conditions leading to nearly complete conversion of thiophene to H_2S, the nickel catalyst lost a small amount of its initial activity, and then remained stable over time on-stream. In contrast, reaction conditions under which thiophene emerged unconverted from the reactor led to severe deactivation of the catalysts.

Palm et al. [20] tested the ATR of a $C_{13}-C_{15}$-alkane mixture with benzothiophene on a precious metal catalyst. The sulfur mass fractions in the feed amounted to 11 and 30 ppm. The addition caused a sharp decrease in the hydrocarbon conversion and deactivation of the catalyst was found to be irreversible.

It is therefore obvious that a desulfurization unit upstream of the autothermal reformer is essential if sulfur-containing fuels are used. As mentioned previously, the upper value for the sulfur mass fraction of the fuel entering the fuel-cell system is ~10 ppm [1–3]. In the following sections, different approaches will be presented for the desulfurization of middle distillates in the gas and liquid phases.

34.3
Desulfurization in the Gas Phase

Desulfurization in the gas phase can be classified by different process routes or by the kind of species that is to be removed from the fuel. Typical gaseous sulfur species are sulfur dioxide, hydrogen sulfide, carbonyl sulfide, dimethyl and diethyl sulfide, and thiols (mercaptans). They can be removed by adsorption, absorption, and different types of chemical reactions, often in combination with the addition of hydrogen.

34.3.1
Absorption

Absorption is one of the most widely applied processes in industry for cleaning product gases. *Absorption* is defined as the transfer process of a component in a gas phase to a liquid phase in which it is soluble. It can be classified according to the interaction between absorbate and absorbent, that is, physical solution, reversible, and irreversible reaction. For H_2S removal, monoethanolamine (MEA), diethanolamine (DEA), and triethanolamine (TEA) are often favored as absorbents. MEA has the chemical formula $H_2NCH_2CH_2OH$. The choice of process solution is determined by the pressure and temperature conditions of the gas to be treated. Furthermore, the educt composition and the product quality required in terms of the maximum permissible H_2S mass fraction play a major role in process selection. Previously, aqueous MEA solutions were exclusively used to remove H_2S and CO_2 from natural and synthetic gases, particularly for maximum removal at low pressures. The disadvantages of MEA are the high corrosion potential in conjunction with a high loading of acid gases and the formation of irreversible reaction by-products such as COS and CS_2.

A typical feed gas composition varies between 2 and 7 mol% H_2S [21]. About 1.8–2.5 mol of MEA per mole of gas are required. The solvent-to-gas ratio is about 15–25 l of MEA solution to 1 l of gas. MEA is diluted in water, resulting in a concentration of 15–18 mass%. The reported product gas quality is 0.0057 mg $H_2S\,l^{-1}$ gas, that is, a minimum of about 3.8 ppm H_2S [21]. Additional regeneration by stripping with steam also has to be taken into account. MEA can be used in the temperature range 40–100 °C. The final H_2S concentration is lowest at lower temperatures and for low H_2S loadings of the MEA solution. Additional CO_2 loading of MEA plays a major role in the absorption capacity. By increasing the CO_2 loading from 0 to 0.5 mol CO_2 per mole of MEA, the partial pressure of H_2S above the MEA solution increases by a factor of 100 at low H_2S loading, that is, 0.03 mol H_2S per mole of MEA and 40 °C. Therefore, absorption columns offer an outlet composition of 600–1000 ppm H_2S.

34.3.2
Adsorption

Another well-known gas cleaning process is adsorption. Kohl and Nielsen reported on H_2S adsorption on molecular sieves at temperatures between 25 and 150 °C [21]. Low H_2S pressures, corresponding to 1300 ppm H_2S at 1 bar, are only possible for lower temperatures. Under these conditions, the maximum H_2S loading amounts to 0.05 g H_2S per gram of adsorbent.

Activated carbons are used as an adsorbent to remove methanethiol (methylmercaptan), CH_3SH, from air [22]. CH_3SH is used at low mass fractions in natural gas as an odorant for leak detection. Bashkova *et al.* [22] measured CH_3SH capacities between 0.03 and 0.35 g per gram of adsorbent. Functional groups, such as iron, catalyze the oxidation of adsorbed methanethiol to disulfides. In wet gases, there is competitive adsorption between water and disulfides.

34.3.3
Chemisorption

Chemisorption is a special type of adsorption, which is driven by a chemical reaction occurring at the exposed surface. A chemical species is generated at the adsorbent surface, for example, a metal sulfide. The strong interaction between the adsorbate and the substrate surface creates ionic or covalent electronic bonds [23].

34.3.3.1 H_2S Removal
An adsorbent must have an acceptable sulfur capacity and should preferably be regenerable. Generally, it should retain its activity and capacity through a large number of sulfidation/regeneration cycles. For fuel-cell applications, nonpyrophoric materials are recommended [24]. Target H_2S concentrations are below 0.1 ppmv for polymer electrolyte fuel cells (PEFCs) and below 20 ppmv for integrated gasification combined cycle (IGCC) power plants. Different adsorption materials have been examined for H_2S removal between 300 and 800 °C. Temperatures higher than 650 °C relate to Ba, Ca, Sr, Cu, Mn, Mo, and W, and temperatures between 300 and 350 °C to V, Zn, Co, and Fe.

Gangwal *et al.* [25] investigated zinc ferrite and zinc ferrite/copper oxide adsorbents for the removal of H_2S from coal gas for molten carbonate fuel cell (MCFC) applications at 540–800 °C. H_2S concentrations are much higher, that is, 13 400 ppmv, and must be below 10 ppmv in the product gas. A typical breakthrough performance was achieved at 600 °C for 3 ppmv, leading to a capacity of 11.8 g of S per 100 g of sorbent. Further investigations on improved adsorbents have been reported [26–28].

Regeneration of sorbents is an important issue. Lee *et al.* [29] reported on the regeneration of zinc titanate sorbent, which was used for H_2S removal from coal gas. Ideal conditions were achieved at 650 °C, oxygen concentrations in the regeneration gas higher than 5%, and steam between 10 and 20%.

The formation of COS plays an important role for H_2S removal from coal-based gases. Sasaoka *et al.* [30] reported the by-reaction

$$2H_2S + CO + CO_2 \rightleftharpoons 2COS + H_2 + H_2O \tag{34.1}$$

At 500 °C, a COS concentration of 6–10 ppmv is found, starting from an inlet mixture with 500 ppmv H_2S, 22% H_2, 33% CO, 10% CO_2, 18% H_2O, and N_2.

Novochinskii *et al.* [24, 31] reported on investigations with ZnO particles in fixed-bed reactors and as coated monoliths. The adsorption of H_2S on ZnO follows the reaction

$$H_2S + ZnO \longrightarrow H_2O + ZnS \qquad (34.2)$$

ZnS is generated at the adsorbent surface as a new chemical species. Therefore, Eq. (34.2) must be attributed to chemisorption, which is a sub-class of adsorption. The maximum possible sulfur loading is 33%, which corresponds to the complete conversion of ZnO to ZnS. If the amount of sulfur in the feed gas concentration is small enough, on-site regeneration can be dispensed with. After breakthrough, the absorbent bed must be replaced by a fresh one. In such a concept, ZnO is used as a polishing bed.

The H_2S reaction is a typical gas–solid reaction. External mass transfer or diffusion of H_2S through the ZnO bed could limit the reaction rate. Novichinskii et al. [24] reported that flake- or plate-type adsorbents offer lower mass transfer limitations compared with cube- or prism-type materials. Furthermore, an optimum ZnO particle size should be chosen with regard to capacity and pressure difference.

The experimental results of Novichinskii et al. [24] covered an operating range of 300–400 °C, 2660–8775 h^{-1}, 0–12% CO_2, and up to 20% H_2O. Higher H_2S capture was achieved at lower operating temperatures, lower space velocities, higher H_2S, low H_2O, and low CO_2 concentrations. Additionally, high steam concentrations can lead to a release of already captured sulfur from ZnS. The H_2S capacity of the modified ZnO is 0.2–0.4 g of S per 100 g of ZnO, depending on the operating conditions. Compared with these figures, commercial ZnO reaches up to 2.8 g of S per 100 g of ZnO, but the final H_2S concentration crosses the threshold of 0.1 ppmv very quickly.

For reactor design in mobile applications, coated structures are preferred to fixed beds. Novichinskii et al. [31] used wash-coated monoliths with ZnO particles. They covered an operating range of 300–400 °C, H_2S inlet concentrations of 0.5–8 ppmv, and space velocities from 3468 to 46 800 h^{-1}. The best performance was achieved for a monolith with 400 cpsi (cells per square inch) compared with 200, 600, and 900 cpsi. The use of a monolith led to a major performance improvement, resulting in an extension of breakthrough at a threshold of 0.1 ppmv from 4 to 14 h. At low space velocities, the H_2S capacity is 0.9–3 g of S per 100 g of ZnO for the range 1–8 ppm H_2S in the educt.

Li and King [32] also reported on COS formation above ZnO. Experiments were performed on a dry base at concentrations of 25 ppmv H_2S, at temperatures of 150–250 °C and at high space velocities of 75 000 h^{-1}. The sum of CO and CO_2 always amounted to 21%, whereas CO was varied between 0 and 12%. Their experiments showed that CO has a negative effect on the sulfur removal capacity, which was attributed to COS formation over ZnS. COS measurements were not performed. Equilibrium calculations for a mixture of 3% CO, 13% CO_2, 32% H_2, and 23% steam (balanced He as inert gas) led to a concentration of 234 ppmv COS via formation from CO and 7 ppmv COS via formation from CO_2. A coupling effect by the WGS reaction was neglected as ZnO does not catalyze this reaction.

As an alternative to ZnO, Lew et al. investigated the adsorbent ZnO–TiO$_2$ [33]. The sulfidation reactions that may occur with the various zinc titanate phases are

$$Zn_2TiO_4 + 2H_2S \longrightarrow 2ZnS + TiO_2 + 2H_2O \quad (34.3)$$

$$ZnTiO_3 + H_2S \longrightarrow ZnS + TiO_2 + H_2O \quad (34.4)$$

The adsorption processes were performed at 650 °C and led to an H$_2$S concentration below 5 ppmv in the product gas. Regeneration was performed at 700 °C with a mixture of 10 mol% air in nitrogen. Lew et al. reported losses of 25 mass% Zn per 1000 h during their experiments with ZnO. ZnO–TiO$_2$ sorbents were better with losses of only 5–8 mass% Zn per 1000 h. A structural change was observed from hexagonal to needle-like crystals due to the regeneration process.

34.3.3.2 S-Zorb Process

Aromatic sulfur components in liquid fuels such as gasoline, kerosene, and diesel can be removed in the gas phase by a reactive adsorption known as the *S-Zorb process*.

Since 2001, ConocoPhillips has applied the S-Zorb process for the commercial desulfurization of gasoline [34, 35], and investigations with a pilot plant have shown its applicability to diesel and kerosene [36]. The process is based on a solid sorbent in a moving-bed reactor with integrated sorbent regeneration. The operating range for middle distillates is temperatures from 393 to 404 °C and pressures from 33.7 to 37.2 bar at liquid hourly space velocities of 1.8–3 h^{-1}. The fuel is partially evaporated at these conditions and mixed with gaseous hydrogen. The reactive adsorption can be described by possible chemical reactions for an alkylated benzothiophene:

$$(CH_3)_n - C_6H_{4-n} - C_2H_2S + 2H_2 \longrightarrow (CH_3)_n - C_6H_{5-n} - C_2H_5 + S\# \quad (34.5)$$

$$(CH_3)_n - C_6H_{4-n} - C_2H_2S + H_2 \longrightarrow (CH_3)_n - C_6H_{5-n} - C_2H_3 + S\# \quad (34.6)$$

The sulfur atom is separated from the benzothiophene molecule and an alkylated ethyl benzene [Eq. (34.4)] or an alkylated styrene [Eq. (34.6)] is formed. The remaining sulfur adheres to the sorbent (S#). The sorbent must be regenerated once all active centers are occupied by sulfur. Similarly to the hydrofining process, hydrogen must be added in excess, and is therefore recycled. The amount of converted hydrogen is 35 m^3_Nt$^{-1}_{Fuel}$. As shown by ConocoPhillips [37], diesel fuel can be desulfurized from 523 to 6 ppmw and kerosene from 2000 to 1 ppmw. These figures indicate that the reactivity of benzothiophenes is higher than that of dibenzothiophenes.

The regeneration is performed at temperatures of 427–649 °C under oxidative conditions and sulfur dioxide is formed as a product. The sorbent must be activated in a subsequent step with hydrogen.

34.3.3.3 SO$_2$ Removal

SO$_2$ removal processes have become important in the last few decades for reducing SO$_2$ emissions in the flue gases of coal-fired power plants and other industrial processes. In 1997, Kohl and Nielsen described a huge number of different processes developed in the USA, Japan, and Europe [21]. Selection criteria were parameters such as gas flow rate, inlet and outlet SO_2 concentration, installation costs, type of sorbents, and the availability of water, steam, and power supply. The processes for SO_2 removal can be divided into regenerative and nonregenerative processes. The most frequently applied nonregenerative process uses limestone ($CaCO_3$) as a sorbent. Wet scrubbing uses limestone slurry following the reaction

$$CaCO_3(s) + SO_2(g) \longrightarrow CaSO_3(s) + CO_2(g) \tag{34.7}$$

As an alternative, wet scrubbing with $Ca(OH)_2$ (lime) slurry is also possible:

$$Ca(OH)_2(s) + SO_2(g) \longrightarrow CaSO_3(s) + H_2O(l) \tag{34.8}$$

$CaSO_3$ (calcium sulfite) is further oxidized to produce marketable $CaSO_4 \cdot 2H_2O$ (gypsum):

$$CaSO_3(s) + H_2O(l) + \frac{1}{2}O_2(g) \longrightarrow CaSO_4(s) + H_2O(l) \tag{34.9}$$

Other processes result in ammonium sulfate, $(NH_4)_2SO_4$ [21]. A regenerative process often applied in refineries is the Wellman–Lord process. Using an aqueous solution of sodium sulfate, the following reaction takes place:

$$NaSO_3(aq.) + H_2O(l) + SO_2(g) \rightleftharpoons 2NaHSO_3(aq.) \tag{34.10}$$

SO_2 can be recovered from flue gas and stored as sodium bisulfite in an aqueous solution. The process is reversible, leading finally to pure SO_2, which can be used in a Claus process to recover sulfur [21]. More details on different SO_2 removal processes and their technical applications can be found in the book by Kohl and Nielsen [21].

34.3.4
Hydrofining

Hydrofining is the most widely applied process in refineries for the removal of sulfur in gasoline, kerosene, and diesel. In a first step, the organic sulfur components react above a catalyst to give hydrogen sulfide. Commercial catalysts consist of Co–Mo/Al_2O_3 or Ni–Mo/Al_2O_3. The liquid fuel is compressed to 7–70 bar, mixed with hydrogen and heated to 205–425 °C [38]. Components such as diethyl sulfide, thiophene, and benzothiophene react to give ethane, butadiene, and styrene. Commercial jet fuels contain only minor amounts of pure benzothiophene and a mixture of multiple alkylated thiophenes and benzothiophenes. Figure 34.3 shows possible hydrofining reactions for these species.

The reaction conditions depend mainly on the kind of crude oil fraction and on the degree of desulfurization. High inlet sulfur concentration, high-boiling fuels, and low product sulfur concentrations lead to higher pressures and higher temperatures. Temperatures above 415 °C lead to carbon deposition and to degradation. Mixtures of air and steam or air and tail gases can be used for regeneration. Hydrogen is added in excess to the gaseous fuel and is therefore recycled after a subsequent cooling and condensing process [39]. The product gas still contains sulfur as hydrogen sulfide. For H_2S removal, see Section 34.3.3.1.

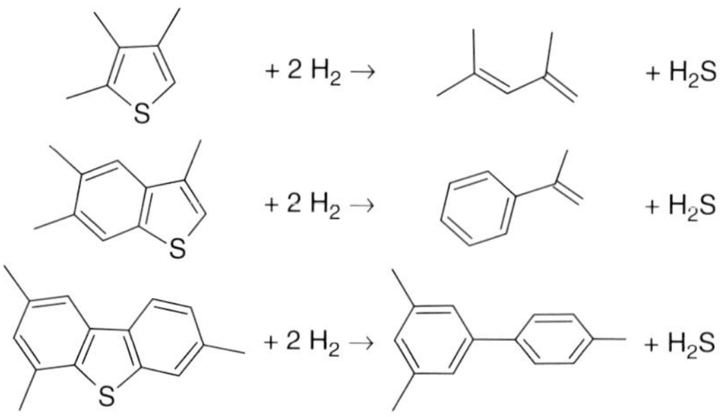

Figure 34.3 Hydrofining of alkylated forms of thiophene, benzothiophene, and dibenzothiophene.

34.3.5
Sulfur Recovery

The Claus process was discovered in 1883. It can be divided into two process steps: a thermal and a catalytic step. During the thermal step, hydrogen sulfide reacts in a sub-stoichiometric combustion ($\lambda < 1$) at temperatures above 850 °C and causes elemental sulfur to precipitate in the downstream process gas cooler. Claus gases with no further combustible contents apart from H_2S are burned in lances surrounding a central muffle according to the following reaction:

$$2H_2S + 3O_2 \longrightarrow 2SO_2 + 2H_2O \tag{34.11}$$

This is a strongly exothermic total oxidation of hydrogen sulfide, producing sulfur dioxide that reacts away in subsequent reactions. The most important of these is the Claus reaction:

$$2H_2S + SO_2 \longrightarrow 3S + 2H_2O \tag{34.12}$$

The Claus reaction continues in the catalytic step with activated aluminum(III) or titanium(IV) oxide, and serves to boost the sulfur yield. More hydrogen sulfide (H_2S) reacts with the SO_2 formed during combustion in the reaction furnace in the Claus reaction, and results in gaseous elemental sulfur. About one-third reacts via Eq. (34.11) and two-thirds via Eq. (34.12). Further process modification such as the COPE, Lurgi OxyClaus, BASF Catasulf, and Superclaus were described by Kohl and Nielsen [21].

34.4
Desulfurization in the Liquid Phase

34.4.1
Hydrodesulfurization with Presaturator

Hydrodesulfurization with a presaturator represents a further development of conventional hydrodesulfurization. As the fuel is desulfurized in the liquid state in

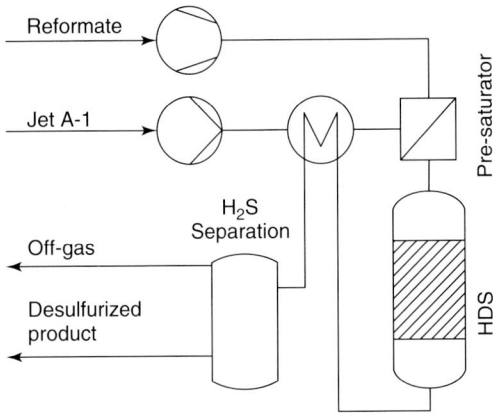

Figure 34.4 Process diagram for hydrodesulfurization with presaturator.

this process and hydrogen recirculation is no longer necessary, this method is an interesting option.

The basic difference from conventional hydrogenation is that the hydrogen required for the hydrogenation is dissolved in the liquid fuel phase before it enters the reactor. The diffusion of hydrogen into the liquid phase occurs externally, meaning that no gas phase occurs in the reactor, in contrast to conventional three-phase trickle-bed reactors. The system is therefore a pure two-phase system comprising saturated fuel and catalyst [40]. The hydrogen surplus in the fuel after hydrogenation is so small that no hydrogen recirculation is required [39]. The process is shown in Figure 34.4.

First, the fuel and the hydrogen are compressed to the reactor pressure between 15 and 30 bar. The gas is then fed into the presaturator and dissolved in the liquid fuel. Current applications use a saturator in the form of a vessel through which gas and liquid flow continuously. In order to achieve as high a distribution of the gas as possible in the fuel, the hydrogen is nebulized with a frit [39]. Special measures to intensify the mass transfer between the phases, however, mean that the overall dimensions of the saturator can be reduced. One option involves a jet loop reactor [41]. In order to minimize the required residence time for introducing the gas into the liquid in the saturator, the momentum of a liquid jet is used to generate a circulation flow, which maximizes the contact area between the liquid and the gas.

The liquid fuel containing the dissolved gas is then fed into the reactor. To prevent the appearance of a gas phase in the reactor, the temperature in the presaturator must be set so that the amount of dissolved gas is below the saturation limit in the reactor. In the catalyst bed, the sulfur compounds are converted to hydrogen sulfide, in analogy with conventional hydrogenation. After product relaxation, the excess gas stream can be removed. However, the residual H_2S content remains dissolved in the fuel and must be separated subsequently from the product stream, for example, in a stripper column or by adsorption [42].

To convert the sulfur fully in the injected fuel, sufficient hydrogen must be dissolved in the liquid for the hydrogenation. In a state of equilibrium, the dissolved gas content depends on the fuel used, the injected gas stream, the pressure, and the temperature. Previous experiments on hydrodesulfurization with a presaturator merely considered the addition of pure hydrogen. In a fuel-cell APU operated with middle distillates, however, no pure hydrogen is available. When an autothermal reformer is used, the reformate contains 36.5 vol.% H_2 at the reactor outlet in the dry state [3]. At the shift reactor outlet, the H_2 content in the dry state is 41 vol.% [43].

The principle of the hydrodesulfurization of middle distillates with a presaturator was patented by Esso in 1963 [44]. The patent involves the external saturation of the fuel with a gas containing at least 70 vol.% hydrogen before hydrogenation. Other patents from 1959 and 1969 describe similar processes, although they refer to the desulfurization of crude oil and desulfurization in radial flow reactors, respectively. There is no known technical application of the patented processes, however, and no experimental results have been published to date. Desulfurization with a presaturator is described in more detail in a patent by Datsevich and Muhkortov [45]. Detailed laboratory studies on the hydrodesulfurization of middle distillates were conducted at the Department of Chemical Engineering, University of Bayreuth. These studies led to another patent being registered by BP Oil International in 2003 [42]. This patent describes the deep desulfurization of diesel-like fuels and summarizes the process with subsequent ultrapurification and the separation of hydrogen sulfide. The adsorption and flushing of the product with a gas stream are described as examples of downstream purification processes.

The application of presaturator technology has clear advantages over conventional hydrodesulfurization and thus has the potential to be used in mobile fuel-cell APUs, whereas conventional hydrogenation does not. The energy-intensive recirculation of the hydrogen is no longer necessary owing to the very small hydrogen surplus. The reactor can be deployed regardless of position because it is fed only with liquid. Smaller catalyst particles can be used because the pressure loss during liquid reactions is insignificant for energy consumption [39]. Improved catalyst wetting and the smaller particle sizes lead to improved kinetics. This in turn gives rise to shorter residence times and thus to smaller reactor dimensions.

34.4.2
Adsorption

Adsorption is based on the binding of molecules from a liquid phase to the surface of a porous solid referred to as the adsorbing agent (see Figure 34.5).

A distinction is made between the bonding principles of chemisorption and physisorption, which are characterized by bonds of varying strengths. In the case of chemisorption, the adsorbed adsorpt chemically bonds with the adsorbing agent. This form of adsorption is usually irreversible or at least not fully desorbable. A relatively weak and reversible bond is the basis of physisorption. This bond is based on intermolecular forces with no electron transfer. Examples of this type of bond include van der Waals forces, dipole forces, dispersion forces, and induction forces [46].

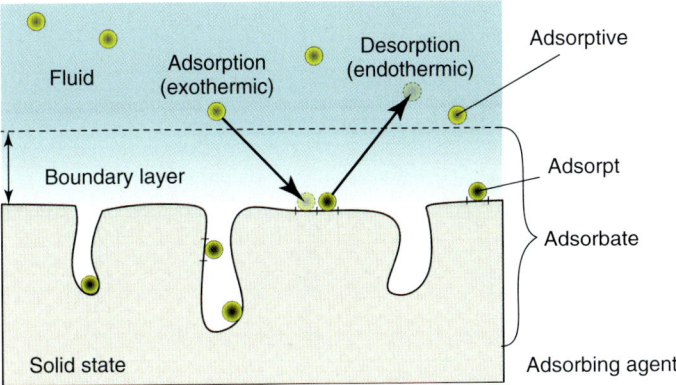

Figure 34.5 Scheme of adsorption–desorption processes.

One of the first attempts to commercialize adsorptive desulfurization was the Irvad process, which was patented in 1998 [47]. Here, activated aluminum doped with metal ions is used as the adsorbing agent. The adsorbing agent is distributed commercially by Alcoa under the name Selexsorb CD. The process is based on the fluidized-bed concept and comprises an adsorption and a regeneration reactor. The four-stage adsorption occurs at temperatures below 40 °C without the addition of hydrogen. Desorption also involves multiple stages and occurs under a hydrogen atmosphere at temperatures ranging from 70 to 270 °C. The target application for commercialization was the desulfurization of a gasoline fraction of 1276 ppm to below 100 ppm [48]. The stipulated targets were not achieved in the prescribed development period and activities were discontinued [49].

Intensive research work on the development of adsorbing agents for the desulfurization of liquid fuels was also pursued by Yang's group at Michigan University and Song's group at Pennsylvania State University. This work concentrated in particular on the desulfurization of gasoline, kerosene, and diesel fuels for fuel-cell APU systems [50]. The adsorbing agents developed at Michigan University bind the sulfur molecules at room temperature via π-complex bonds. As the bonding forces are stronger than the van der Waals bonds, higher selectivities can be achieved with π-complex bonds [51, 52]. Nevertheless, the bonding forces are weaker than is the case for chemisorption, and they can be broken up again by conventional desorption processes, such as increasing the temperature or decreasing the pressure [53]. A description of how JP-5 kerosene was desulfurized by means of these π-complex bonds to CuCl and $PdCl_2$ on MCM-41 and SBA-15 silicate substrates was given by Wang et al. [51].

At Pennsylvania State University, a variety of potential adsorbing agents were also systematically investigated [14, 54, 55]. Nickel-based materials also showed very promising results (see Table 34.4).

Only a low capacity of less than 0.8 mg g^{-1} was found for an Ni/SiO_2 adsorbing agent with commercial gasoline at room temperature. When the adsorption was

Table 34.4 Breakthrough capacities of adsorbing agents based on Ni for a product with 10 ppm S.

Adsorbing agent	Fuel	Sulfur mass fraction (ppm)	Temperature (°C)	Breakthrough capacity, 10 ppm sulfur (mg S g^{-1})
Ni/SiO$_2$–Al$_2$O$_3$	Gasoline	305	20	<0.8
Ni/SiO$_2$–Al$_2$O$_3$	Gasoline	305	200	7.3
Ni/SiO$_2$–Al$_2$O$_3$	JP-8	736	200	9.9
Ni/SiO$_2$–Al$_2$O$_3$	Light JP-8	380	200	14.2
KYNiE-3	Light JP-8	380	80	4.44
KYNiE-3	Light JP-8	380	80	1.48

performed with gaseous fuel at 200 °C, the capacity increased by a factor of more than nine [54]. Experiments with turbine jet fuels revealed the positive effects of coupling adsorption with distillative separation. A breakthrough capacity of 9.9 mg g^{-1} was achieved for JP-8 with 736 ppm sulfur. For a light 70 vol.% fraction of the same fuel with 380 ppm sulfur, the capacity can be increased by over 40% [14]. An obstacle to the application of the adsorbing agent in fuel-cell APUs is the fact that Ni/SiO$_2$–Al$_2$O$_3$ requires an adsorption temperature of 200 °C, which means that the fuel must be at least partially vaporized. Furthermore, the adsorbing agent must be activated after oxidative regeneration under a hydrogen atmosphere. Further investigations were conducted with N--Y zeolites (see Table 34.4). With a reduced Ni(I)-Y zeolite (KYNiE-3), a capacity of 4.44 mg g^{-1} was achieved at 80 °C. In the unreduced Ni(II) state, 1.48 mg g^{-1} was still achieved. It is advantageous that a complete regeneration is possible in the unreduced state with air at 300 °C in 1–2 h [55].

As a result of very good adsorption capacity, active carbon has great desulfurization potential. The adsorptive removal of thiophene and dibenzothiophene from liquid fuels was tested on modified active carbon. Activation with H$_2$SO$_4$ and HNO$_3$ resulted in an increased desulfurization capacity [56–58]. Wang and Yang [58] investigated CuCl, PdCl$_2$, and Pd on an active carbon for the removal of thiophene from a model kerosene. The saturated adsorbing agents were regenerated using ultrasound and a solvent of benzene and n-octane.

The US company Mesoscopic Devices patented a method for an adsorptive desulfurization process with integrated regeneration in mobile fuel-cell systems [59]. A multiport rotary valve is used to connect 12 fixed-bed reactors in a rotary adsorber for high utilization of the amount of adsorbing agent (see Figure 34.6).

34.4.3
Ionic Liquids

A new concept for the desulfurization of mineral oil fractions is liquid–liquid extraction with ionic liquids [60]. For applications in fuel-cell APUs, it is significant that the process is conducted at room temperature and ambient pressure and that

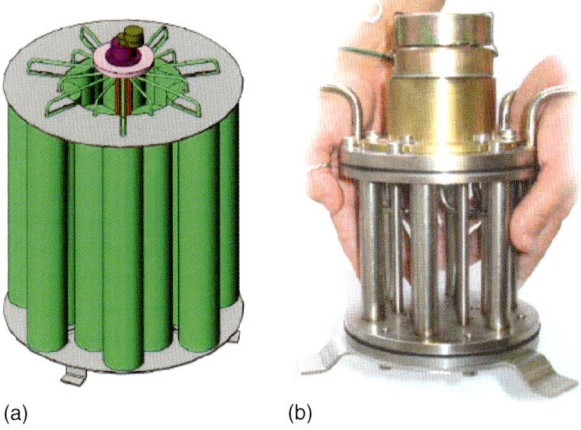

(a)   (b)

Figure 34.6 Desulfurization system of Mesoscopic Devices: (a) 50 kW unit and (b) 5 kW prototype.

Table 34.5 Distribution coefficient K_N as a measure of the extraction efficiency of different ionic liquids.

Ionic liquid	K_N [(mg$_S$ kg$_{IL}^{-1}$) (mg$_S$ kg$_{Oil}^{-1}$)$^{-1}$]	Solubility (mass%)	
		n-C$_{12}$	Cyclo-C$_6$
[BMIM]Cl/AlCl$_3$	4.0	–	–
[BEIM][EtSO$_4$]	2.0	<1	5
[BMIM][OcSO$_4$]	1.9	6.0	40.2
[EMIM][EtSO$_4$]	0.8	<1	2.5

neither hydrogen nor a catalyst is required [61]. Ionic liquids consist solely of ions and can be distinguished from conventional molten salts because they form relatively low-viscosity liquids at temperatures below 100 °C. Important properties such as melting point, thermal stability, and solubility can be varied over a wide range by the choice of cations and anions. Moreover, the distillative product separation is also significantly simplified because ionic liquids have practically no vapor pressure. Ionic liquids are therefore particularly suitable for use as solvents in extraction processes [61].

Esser investigated the extraction of sulfur compounds from a model fuel using 50 ionic liquids in order to determine which are suitable. The criteria here were a low solubility of hydrocarbons in the ionic liquid, industrial availability, and a high distribution coefficient K_N as a measure of the extraction efficiency [62]. The *distribution coefficient* is defined as the ratio of the mass percentage of sulfur in the ionic liquid to the mass percentage of sulfur in the fuel in the state of equilibrium. Data for four powerful ionic liquids are given in Table 34.5.

1-n-Butyl-3-methylimidazolium tetrachloroaluminate ([BMIM]Cl/AlCl$_3$) has the highest distribution coefficient. A high corrosiveness and low hydrolysis stability, in addition to environmental considerations, however, hinder the technical application of halogen-containing ionic liquids.

Huang et al. investigated a different desulfurization method using ionic liquids [63]. Here, an ionic liquid was fabricated with a Cu(I) group, which was used at the University of Michigan for solid–liquid adsorption. This ionic liquid, BMIMCu$_2$Cl$_3$, binds the sulfur compounds via a π-complex and has better desulfurization efficiency than the liquids investigated by Esser [62].

34.4.4
Selective Oxidation

Another desulfurization method involves selective oxidation. In the first processing step, the sulfur compounds are selectively oxidized to the corresponding sulfoxides and sulfones. The subsequent separation of the sulfur compounds from the treated fuel is made possible by the higher polarity of the oxidized compounds. This usually occurs by adsorption on a solid adsorbing agent or by means of extraction, which often necessitates the regeneration of the adsorbing agent or the extraction agent as an additional processing step. There are different approaches for the selective oxidation of sulfur compounds, yet none of these have been developed beyond the laboratory scale. They include oxidation with oxygen plasmas, photo-oxidation, oxidation with peroxides, and biological processes.

34.4.4.1 Plasma Desulfurization

Plasma is defined as ionized gases containing neutral particles in addition to free ions and electrons. The selective oxidation of sulfur compounds with oxygen plasmas produces a low-pressure oxygen plasma on a fuel surface through a high electrical charge [64]. At an operating pressure below 100 mbar, the fuel temperature must be decreased to −50 °C in order to lower the vapor pressure of the fuel because reactions in the gas phase inhibit the reaction. For the reaction in the plasma, only the atomic and singlet oxygen are relevant. While atomic oxygen does not exhibit selectivity for sulfur compounds and predominantly reacts with hydrocarbons to alcohols, singlet oxygen is inert to hydrocarbons and exclusively oxidizes sulfur compounds. The oxidation products are almost insoluble in the starting material and are removed as a separate phase [65].

34.4.4.2 Photo-oxidation

Photo-oxidation requires the use of photocatalysts, which generate singlet oxygen under UV radiation or exposure to sunlight. Examples of photocatalysts include the dyes Rose Bengal, Methylene Blue, and fullerenes in solvents such as water, methanol, and acetonitrile [66–68]. The oxidation products in turn are insoluble in fuel and can be separated with the catalyst phase.

Studies with Rose Bengal and polymerized fullerenes with model fluids resulted in conversion rates of up to 98% for oxidation with thioethers and sulfides. Either

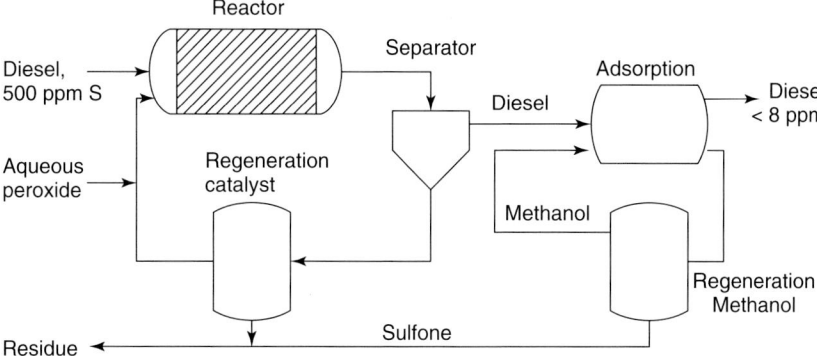

Figure 34.7 Schematic diagram of the Unipure process.

no or a very small conversion was observed for aromatic sulfur compounds [67, 68]. With *N*-methylquinolinium tetrafluoroborate, however, dibenzothiophene was almost fully oxidized, with reaction times between 2 and 4 h [66].

34.4.4.3 Oxidation with Peroxides

The oxidation of sulfur compounds with peroxides usually involves the use of two-phase systems in which aqueous hydrogen peroxide is mixed with a fuel. To accelerate the reaction, different catalysts, adsorbing agents, and solvents can be used, and the energy can be transferred with ultrasound or microwaves [69–72]. After the reaction, the aqueous phase must be separated and the oxidized sulfur compounds must be removed from the fuel. The most advanced process is the Unipure ASR-2 process [73], shown schematically in Figure 34.7.

The process has already been realized in a pilot plant. It is based on oxidation with hydrogen peroxide, which is added in the aqueous phase. The formic acid added as a catalyst must be stripped after oxidation and treated. The oxidation products are then extracted from the fuel with an aluminum oxide adsorbing agent. The adsorbing agent is purified with methanol, which in turn must be regenerated for reuse.

34.4.4.4 Biological Processes

Another alternative for the desulfurization of liquid fuels involves biotechnological and biocatalytic processes. These processes are based on bacterial strains, the metabolism of which can convert carbon-rich sulfur compounds. Through a series of enzyme catalytic reactions, the strains remove the sulfur from the hydrocarbon compounds without altering the carbon skeleton [74] (see Figure 34.8).

The sulfur compounds are first selectively oxidized before the carbon–sulfur bond is broken and the hydrocarbon is converted to the corresponding hydroxyphenylbenzenesulfonate. The inorganic sulfur compounds formed as a result are soluble in water and can be filtered out of the product. This facilitates a simple process for the separation of sulfur.

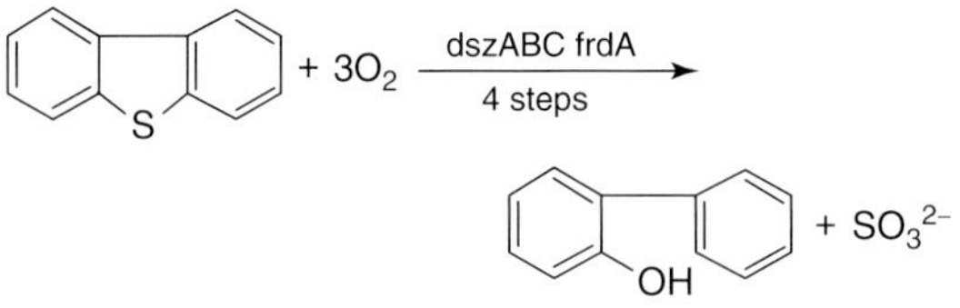

Figure 34.8 Four-step conversion for the removal of sulfur from dibenzothiophene using bacteria.

The operating conditions, which are easy to manage, make the application of biological processes for the desulfurization of middle distillates advantageous. The process involving the recovery of the bacterial solution does not require the addition of hydrogen or any other consumable. One factor impeding application in fuel cell APUs, however, is the insufficient activity of bacterial strains known to us today. The residence times required would result in unrealistic reactor dimensions.

34.4.5
Desulfurization with Overcritical Fluids

A fluid is in an overcritical state when the properties of the liquid phase and the gas phase have become so similar above a certain pressure and temperature state that there is only one phase. For water, this critical point is $T_C = 374\,°C$ and $p_C = 221$ bar [75].

Close to the critical point, the density changes, as does the viscosity, and, in the case of water, the dielectric constant. Whereas subcritical water is insoluble for nonpolar organic substances, overcritical water can be used like a nonaqueous solvent [76]. The pressure and temperature can be adjusted to set optimal properties of water for the reaction. This makes it possible to use overcritical water, which can be mixed with liquid fuels, for desulfurization. The free radicals, which appear more frequently at high temperatures, lead to the splitting of the sulfur compounds and the formation of hydrogen sulfide [77].

Experiments have shown that good reactivity can only be achieved with aliphatic compounds. The more stable aromatic compounds, in contrast, were only minimally converted [78]. Better results were achieved with the addition of a CoMo catalyst, usually used for hydrodesulfurization, and with the addition of hydrogen at 250 bar. In a gasoil with 8000 ppm sulfur, in which dibenzothiophene represented the main sulfur compounds, this allowed the sulfur mass fraction to be reduced by 85% with a residence time of 60 min [77].

As middle distillates predominantly contain heterocyclic aromatic sulfur compounds, which do not react with overcritical water, the simple process is unsuitable for the desulfurization of middle distillates. If a CoMo catalyst and hydrogen are also added, the process can be classified as hydrodesulfurization with overcritical water, where the even higher operating pressure compared with conventional hydrogenation causes further disadvantages.

Desulfurization with overcritical water only appears to be advantageous in conjunction with the simultaneous reforming of the liquid fuel. The Fraunhofer

Institute of Chemical Technology (ICT) conducts research in this area of fuel reforming with overcritical water. The operating conditions for reforming are ~450 °C and 250–550 bar [75]. The size and weight of the necessary high-pressure autoclave make it unsuitable for mobile applications.

34.4.6
Distillation

Distillative separation is based on fractionation whereby the light fuel fraction is removed by distillation. As the sulfur compounds contained in the middle distillate usually belong to the higher boiling range, the sulfur content in the light fractions is reduced, while the sulfur compounds accumulate in the distillation residue. After subsequent deep desulfurization, the light fraction with reduced sulfur content can then be used in the fuel-cell system [79]. The residue with a high sulfur mass fraction can be used for other purposes. When used in aircraft or ships, for example, the residue can be combusted by the propulsion unit, making the process suitable for fuel-cell APUs.

The process patented by Daimler was investigated experimentally by Altex Technologies and at Pacific Northwest National Laboratory (PNNL) [79–81].

The attainable sulfur mass fraction in the light fraction depends on the origin of the mineral oil and the sulfur compounds contained in it. For fuels with a high proportion of high-boiling sulfur compounds, a considerable decrease in the sulfur mass fraction is expected. Table 34.6 shows the experimentally determined reduction in the sulfur mass fraction for different fuels and the mass percentages of the light fraction.

The experimental findings reveal greater reductions in the sulfur mass fraction of the light fraction for lower mass percentages of the light fraction at the educt. Furthermore, the experimental values for diesel fuel were better than for JP-8. For further desulfurization, not only is the absolute reduction in the sulfur mass fraction of the light fraction important, but also the fact that the high-boiling,

Table 34.6 Reduction of the sulfur mass fraction by distillation applying different middle distillates [80, 81].

Fuel	Light fraction (mass%)	S mass fraction educt (ppm)	S mass fraction product (ppm)	Reduction S mass fraction (mass%)
Altex				
JP-8	70	736	380	48
Diesel	70	325	142	56
Diesel	20	325	34	90
PNNL				
JP-8	68	1341	850	37
JP-8	50	1341	621	54
JP-8	27	1341	415	69

low-reactive sulfur compounds, in particular, remain in the residue. Depending on the process used, this can make deep desulfurization of the light fraction easier.

Assuming that the residue stream can be used for another purpose in the overall system, distillative separation becomes a relevant process in mobile fuel-cell systems for rough desulfurization, provided that it considerably simplifies subsequent desulfurization with further processes.

34.4.7
Membrane Processes

Membrane processes make use of partially porous layers to separate a feed stream into a permeate and a retentate. The characteristic variables here are the transmembrane flow and the selectivity, which can be described by Eqs. (34.13) and (34.14) [82].

$$\dot{n}_k'' = \left(\frac{\text{permeability of component } k}{\text{effective membrane thickness}}\right)(\Delta p_i, \Delta \pi_i \text{ or } \Delta c_i) \quad (34.13)$$

$$\alpha_{ij} = \frac{c_i^{\text{permeate}}/c_j^{\text{permeate}}}{c_i^{\text{feed}}/c_j^{\text{feed}}} \quad (34.14)$$

34.4.7.1 Processes with Porous Membranes

Mass transport through porous membranes can be described with the pore model. In accordance with particle filtration, selectivity is determined solely by the pore size of the membrane and the particle or the molecular size of the mixture to be separated. This process is driven by the pressure difference between the feed and permeate sides [83]. The processes described by the pore model include microfiltration and ultrafiltration. Whereas membranes for microfiltration are characterized by their real pore size, membranes for ultrafiltration are defined according to the molar mass of the smallest components retained.

Sulfur compounds in liquid fuels cannot be separated by porous membranes. The molar masses of the sulfur compounds and also those of the hydrocarbons are spread over a wide range in liquid fuels, and there is no significant difference between them.

34.4.7.2 Processes with Nonporous Membranes

The remaining membrane processes are described by the solution–diffusion model. The model describes the solution and diffusion of permeands in a nonporous selective layer. The separating layer has a thickness between 0.3 and 2.5 µm, and is described as a real liquid in which permeating components are dissolved and then transported through the membrane in accordance with the gradient of the driving force [83]. The active layer is applied to a porous substrate to ensure sufficient mechanical strength. The substrate does not affect the selectivity. The layer has a thickness of ~50–250 µm. The solution–diffusion model is used to describe electrodialysis, reverse osmosis, and nanofiltration, and also gas permeation and pervaporation.

Pervaporation is used to separate the liquid mixture. A phase transition occurs at the phase boundary on the permeate side, allowing desorption by vaporization. According to the solution–diffusion model, selectivity is primarily achieved because not all components in the mixture of substances can be dissolved equally well in the membrane material. Pervaporation involves a second selectivity step as a result of the required vaporization of the permeating components. For this, the partial pressure on the permeate side of the components must be lower than the saturated steam pressure. If only some of the components dissolved by the membrane boil at the operating point, the remainder of the components are not desorbed.

In addition to mass transfer, heat transfer must also be considered during pervaporation. The necessary vaporization enthalpy is withdrawn from the feed stream in the form of heat. In order to decrease the partial pressure of the permeating components on the permeate side to below the saturated steam pressure, a vacuum is usually created.

The desulfurization of liquid fuels using pervaporation has been increasingly investigated over the last few years [84]. As middle distillates contain mainly aromatic sulfur compounds, desulfurization membranes tend to make use of developments in aromatic–aliphatic separation. The most frequently used membrane materials investigated for the desulfurization of liquid hydrocarbon mixtures are polyurea–polyurethane, polysiloxane, Nafion, cellulose triacetate, and polyimide [84]. In addition to a range of processes for the desulfurization of naphtha fractions patented by ExxonMobil, Transionics, and Marathon Oil, only the S-Brane process developed by W. R. Grace and Sulzer has been tested beyond the laboratory scale [84].

The S-Brane process for the desulfurization of naphtha fractions was developed for application in refineries. A polymer membrane separates a permeate enriched with sulfur compounds, which is added to the conventional hydrodesulfurization process. The low-sulfur retentate can be further processed with no additional desulfurization [85].

Intelligent Energy developed a process for the desulfurization of kerosene for fuel-cell APU systems. The project was conducted in cooperation with Boeing and W. R. Grace and involved the application of different membranes including the S-Brane membrane [86]. The approach involves desulfurizing Jet A with 3000 ppm sulfur to less than 100 ppm in a three-stage pervaporation process. The sulfur mass fraction is then reduced even further with an additional adsorption step [87] (see Figure 34.9).

The multistage process means that only 1% of the educt stream is required for the APU and 99% of the educt can be redirected back into the tank. The first two steps make use of membranes that lower the sulfur mass fraction in the permeate, which is a low-boiling fraction containing sulfur compounds mainly in the form of thiophene. The patent specification includes experimental results on the membrane steps. For Jet A with 1473 ppm sulfur, the first two steps together achieved an enrichment factor of 0.12. In the third step, an S-Brane membrane was used and led to a retentate with a low sulfur mass fraction ranging between 10 and 30 ppm [87].

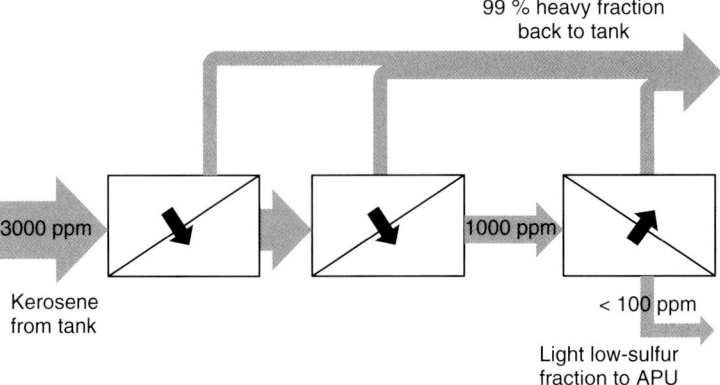

Figure 34.9 Schematic diagram for the desulfurization process from Intelligent Energy.

Latz [88] and Wang *et al.* [89] investigated the pervaporation process for the removal of sulfur-containing substances from jet fuel using different types of membranes. They found that pervaporation is suitable for a first cleaning step, for so-called rough desulfurization, which reduces the initial sulfur mass fraction of conventional jet fuel by half, in the range of several hundred parts per million sulfur. However, to reach the above-mentioned target value for a fuel-cell system of 10 ppm sulfur, rough desulfurization has to be combined with a fine cleaning step, for example, adsorption. Bettermann and Staudt [90] investigated the separation of benzothiophene–*n*-dodecane mixtures using different fluorinated copolyimide membranes.

The use of pervaporation has several advantages over distillative separation. The operating temperature for pervaporation is always below the boiling temperature of the fuel added and, depending on the thermal stability of the polymer membranes, is never more than 80–150 °C. Highly compact membrane modules can be industrially fabricated. Winding modules comprise ∼0.4–0.8 m^2 active membrane area per liter, and hollow-fiber modules, in the same form as already used for gas permeation, between 2 and 5 m^2 l^{-1} [84]. The compact and light construction method leads to a higher power density of the fuel-cell system. In contrast to a vaporizer, as used in distillative separation, pervaporation can be operated irrespective of position without any problems.

34.5
Application in Fuel-Cell Systems

The application of the different desulfurization processes in the various fuel-cell systems must be discussed in terms of process conditions, the required sulfur removal capacity, and the required sulfur mass fraction in the fuel or fuel gas. In general, all processes in fuel-cell systems must be highly efficient, reliable, and cost-effective, and they should offer high long-term stability. In addition, desulfurization must cover a number of broader conditions:

1) It must be adaptable to other processes in the complete fuel-cell system.
2) It should fulfill the requirements of the application; for example, it must withstand vibrations in mobile systems and should not be position dependent in aeronautic applications.
3) The sorbent should be regenerable. Regeneration must be performed with a gas stream generated by the system or with air. Possible regeneration fluids include air, steam, reformate, and tail gases from a burner.
4) The process should be performed without additives requiring an additional storage system and a metering pump.
5) The chosen desulfurization process should be as simple as possible.

As an example, Figure 34.10 shows the operating range in terms of temperature and mass fractions of eight selected desulfurization processes for liquid fuels. The diagram refers to the ATR of a liquid fuel that will be injected into a mixing zone. The maximum fuel temperature is limited owing to undesired pre-reactions such as pyrolysis and premature evaporation in a nozzle. In order to fulfill condition (1) above, the fuel must be cooled or heated to ~150 °C after desulfurization. The required sulfur mass fraction is 10 ppmw. As can be seen, processes (c), (d), (f), and in part (e) do not achieve the target value. They must therefore be combined with other processes in a multi-stage unit.

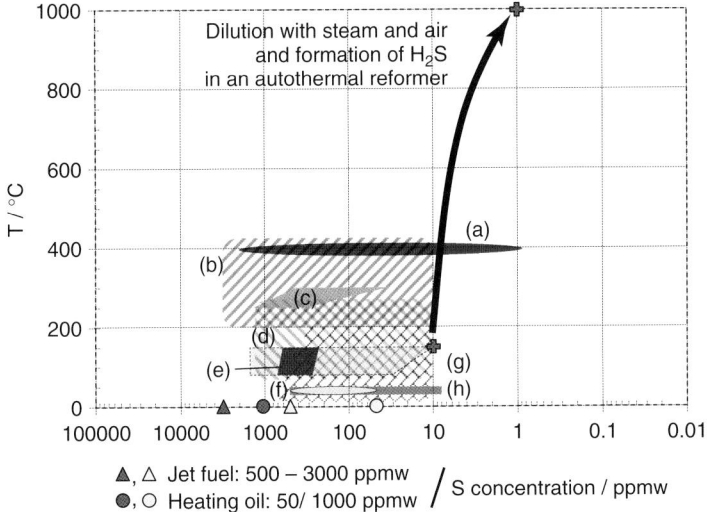

Figure 34.10 Operating range of various desulfurization processes for fuels. (a) S-Zorb at 33.7–37.2 bar for gasoline desulfurization; (b) hydrofining at 7–70 bar for desulfurization of gasoline, kerosene, and diesel in gaseous or liquid states; (c) fractional distillation; (d) adsorption for sulfur-containing components in gasoline; (e) pervaporation for kerosene desulfurization; (f) biological systems; (g) adsorption for kerosene; (h) selective oxidation for diesel desulfurization.

Process (a) demands a hydrogen recycling loop, an efficient heat-recovery system and partial evaporation of fuels. It has been commercialized only for gasoline. Process (b) in the gas phase has the same disadvantages as process (a). Furthermore, the use of small particles conflicts with the second condition listed above. Therefore, only hydrofining with presaturated hydrogen is an option [39, 40, 88]. Process (c) can be applied as rough desulfurization, leading to higher efficiency losses due to a more complex heat recovery [88]. Process (e) separates in a similar manner to (c) but the losses are much lower for operating temperatures of 80–120 °C [88]. Furthermore, it is possible to clean jet fuels down to 100 ppmw using multistage membrane processes, but only 1% of the original fuel can be used [87]. Adsorption processes [processes (d)) and (g)] could be an option for gasoline with an inlet mass fraction of 300 ppmw S and for kerosene with 200–300 ppmw S. Sulfur adsorption of diesel is more difficult because the sulfur is held within dibenzothiophene. With regard to the specified mass fraction of 3000 ppmw S for jet fuels, adsorption could only be applied after a rough desulfurization such as distillation (c) or pervaporation (e). Biological systems (f) offer a rather slow alternative, leading to huge reactor volumes. Oxidation with peroxides achieves the 10 ppm target, but it fails to fulfill condition (4).

Gasoline can be desulfurized by adsorption [14, 54, 55], whereby gasoline fuels in Europe are purified in refineries to mass fractions lower than 10 ppmw S. In North America and Turkey, gasoline contains about 30–80 ppmw S, in South-East Asia and Australia 150 ppmw S, and in Russia, Mexico, Argentina, and some parts of Africa 200–600 ppmw [91]. Only African, Arabian, and South American countries use gasoline with 1000–2500 ppmw S in their infrastructure. For US gasoline with 50 ppmw S and a 15 kW$_{th}$ fuel processor, an adsorbent mass of 7 g h^{-1} operating time is necessary. Considering these data, a design with two regenerable adsorbent reactors is preferable.

Jet fuel with a specified sulfur mass fraction of 3000 ppmw can be desulfurized by a hydrofining process in the liquid phase with subsequent separation of gaseous H$_2$S by decompression. As shown in Figure 34.10, this process requires the heating of fuel from ambient temperature to 350–400 °C and subsequent cooling to 80–150 °C. As an alternative, a two-stage concept with pervaporation at 80–110 °C and adsorption at 200 °C can be considered for jet fuels with a sulfur mass fraction lower than 1600 ppmw S [88]. Almost half of the sulfur-containing hydrocarbons are alkylated thiophenes. The remainder consists mainly of benzothiophenes with one to three methyl groups. This route requires more research work in the future.

Diesel fuel and heating oil extra light (EL) contain mainly dibenzothiophenes with one to three methyl groups. Dibenzothiophenes with methyl groups at positions 4 and 6 are particularly difficult to remove and first occur during sorbent degradation [92]. The sulfur mass fraction in diesel fuels is 10–15 ppmw in North America, Europe, Japan, and Australia, 50 ppmw in Turkey, Belarus, and Chile, and 300–500 ppmw in Mexico, Russia, Asia, India, and China [93]. Only African, Arabian, and South American countries use diesel with 1000–12 000 ppmw S in their infrastructure. Taking considerations (1)–(5) above into account, only a few options remain, most of which demand further research and development work [39, 73, 92]. Hydrofining in the liquid phase is the most favorable option for the broad range of existing diesel qualities.

Sulfur is present in the product gas of a reformer as H_2S. After passing through the reformer, the sulfur concentration is lower than 1 ppmv H_2S due to chemical conversion and dilution with air and steam. A further removal process for H_2S is required depending on the applied fuel cell type.

Several authors appear to disagree on the sulfur tolerance of solid oxide fuel cells (SOFCs) [94–98]. Even at low levels, H_2S has a negative effect on cell performance. This can be reversed after switching to clean fuels. At concentrations higher than 100 ppmv H_2S, the voltage drop is only partly recoverable [94]. Norheim et al. [95] observed a linear voltage drop at 800 °C from 0 to 60 ppmv H_2S and no further degradation in the range 80–100 ppmv. Rasmussen and Hagen [96] narrowed the reversible sulfur tolerance to 0–40 ppmv H_2S. They found a voltage drop of $0.2\,mV\,h^{-1}$ at 10 ppmv H_2S and $0.7\,mV\,h^{-1}$ at 100 ppmv H_2S at 850 °C. McPhail et al. [97] lowered the sulfur limit to 5–10 ppmv H_2S. The electrode activity appeared to be irreversibly affected by these levels during long-term exposure. The effects of sulfides depend on factors such as bulk concentrations, hydrogen concentration, humidity, electrical load, and temperature [97]. The degradation can be described by the adsorption of sulfur on nickel. Bøgild Hansen [98] fitted a series of experimental data by a Temkin isotherm in the sulfur range 0.05–50 ppmv H_2S. Finally, the cell performance loss amounted to 1–18%. In general, the tolerance limit for SOFCs cannot be firmly defined [94].

For MCFCs, sulfur should be limited to 0.5–1 ppmv H_2S [97]. This threshold is challenging for modified ZnO sorbents, which were developed especially for MCFCs operated with coal gas. Coal gas cleaned at 600–650 °C still contained 1–5 ppmv H_2S [26, 27]. The high-temperature polymer electrolyte fuel cell (HT-PEFC) can tolerate a feed gas with 30% H_2, 1% CO, and 10 ppm H_2S, and should be operated with dry cathode air below the dew point [99]. The voltage drop is 4% (from 630 mV) at 180 °C and a current density of 200 mA cm^{-2}. Similar experiments by Norheim et al. [95] for an SOFC at 800 °C, 200 mA cm^{-2}, and 80–100 ppmv H_2S resulted in a voltage drop of about 2.5% (from 810 mV). PEFCs require a fuel gas with low concentrations of carbon monoxide and hydrogen sulfide, that is, lower than 100 ppbv H_2S [100].

Figure 34.11 shows the proposed operating range of SOFCs, MCFCs, HT-PEFCs, and PEFCs in terms of temperature and H_2S concentration. As can be seen, sulfur levels of 50–1000 ppmv H_2S, which can be found in coal or biogases, are challenging for MCFCs and PEFCs using modified ZnO sorbents. As described by Gangwal et al. [101], a mixed-metal zinc adsorbent including Zn, Fe, and Cu could reduce the sulfur concentration to less than 1 ppmv H_2S at 600 °C and for 20% steam in the fuel gas. An increase in pressure from ambient to 10 bar raises the sulfidation rate. The operating temperature suits MCFC conditions very well and to some extent those of SOFCs.

Biological processes (c) are an alternative for tail gas cleaning processes, such as amine washing, due to their low operating temperature. Microbiological processes

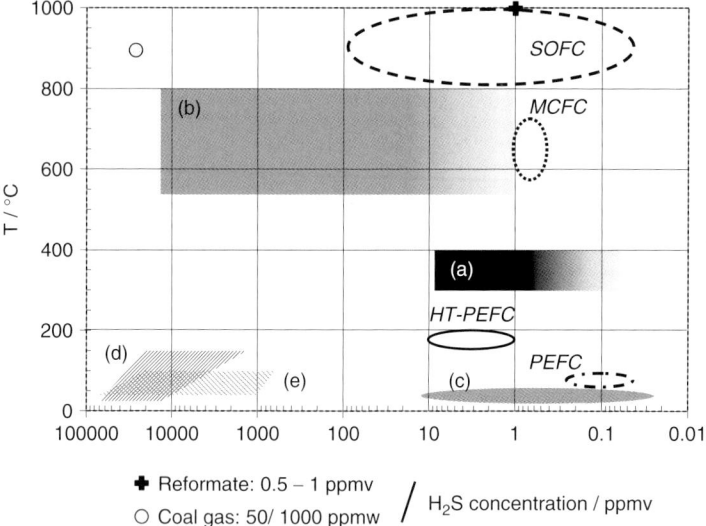

Figure 34.11 Operating range of various desulfurization processes for H_2S. (a) Chemisorption on low-temperature ZnO; (b) chemisorption on high-temperature ZnO; (c) biological systems; (d) adsorption on molecular sieves and activated carbon; (e) absorption with amines.

are rather slow compared with chemical reactions in industry. A review has been published [102].

The combination of feed gases from coal or biomass gasification with low-temperature fuel cells especially for PEFCs does not make sense owing to the great effort required for sulfur reduction and the limited usage of residual heat at an operating temperature of 80 °C.

Reformate from fuel processing can be directly used in all fuel cells except PEFCs. In PEFC systems, an additional ZnO adsorbent is required at operating temperatures of 350–400 °C. The sorbent capacity for a threshold of 0.1 ppmv H_2S is 9 mg of S per gram of ZnO at 1 ppmv H_2S in the reformate [31]. For European gasoline with 10 ppmv S and a 15 kW$_{th}$ fuel processor, an adsorbent mass of $1.3\,g\,h^{-1}$ operating time is necessary to decrease the sulfur concentration to 0.1 ppmv H_2S.

With regard to the application of desulfurization in fuel-cell systems, there are still challenges. It seems that SOFCs and HT-PEFCs offer some advantages over MCFCs and PEFCs. Further research is required on the long-term effects and the reasons for the sudden death of materials.

34.6
Conclusion

Desulfurization is a well-known process in the chemical industry. However, most desulfurization processes are not applicable or must be adapted for use in fuel-cell systems. If fossil fuels are used for fuel-cell systems, different types of reforming technologies can be applied for conversion into hydrogen-rich synthesis gases. Such fuel processors use catalysts that demand a sulfur mass fraction no higher than 10 ppmw. Liquid-phase desulfurization is preferred but it is not yet mature. Depending on the fuels used and their specific sulfur compounds and mass fractions, different kinds of processes could be applied in the future. Dibenzothiophenes – the main compound group in diesel fuels – with methyl groups at positions 4 and 6 are particularly difficult to remove.

Having passed through the reformer, the sulfur concentration is lower than 1 ppmv H_2S due to chemical conversion and dilution with air and steam. A further removal process for H_2S is required depending on the fuel cell type. The sulfur tolerance of different fuel cell types cannot be clearly defined and further studies are necessary with regard to long-term stability. A threshold of 0.1 ppmw H_2S for PEFC application is the most challenging task.

Acknowledgments

We are grateful to the German Federal Ministry of Economics and Technology (BMWi) within the National Aerospace Research Program (ELBASYS project) for financial support of desulfurization aspects. The authors would also like to acknowledge R. Dahl, J. Latz, B. Sobotta, and Y. Wang for valuable technical assistance.

References

1. Benz, W. (2005) Einfluss von Schwefelverbindungen in flüssigen Kohlenwasserstoffen auf ein Brennstoffzellen-Gesamtsystem am Beispiel eines katalytischen Crackers mit nach geschalteter PEMFC, PhD thesis, University of Duisburg.
2. Hagen, J. (1996) *Katalyse – Eine Einführung*, Wiley-VCH Verlag GmbH, Weinheim, p. 199.
3. Pasel, J., Meissner, J., Porš, Z., Samsun, R.C., Tschauder, A., and Peters, R. (2007) Autothermal reforming of commercial Jet A-1 on a 5 kWe scale. *Int. J. Hydrogen Energy*, **32**, 4847–4858.
4. Ahmed, S., Kumar, R., and Krumpelt, M. (1994) *Fuel Cells Bull.*, **2**, 4.
5. Andersson, T.A. (2005) *Chem, Unserer Zeit.*, **39**, 116–120.
6. Onken, U. and Behr, A. (1996) *Chemische Prozesskunde*, Lehrbuch der Technischen Chemie, Band **3**, Wiley-VCH Verlag GmbH, Weinheim, p. 226.
7. Förster, F. (1978) *Das Buch vom Erdöl*, Deutsche BP, Hamburg, p. 163.
8. Kleinjan, W.E., de Keizer, A., Janssen, and A.J.H. (2003) *Top. Curr. Chem.*, **230**, 167.
9. Orr, W.L. and Damsté, J.S.S. (1990) *ACS Symp. Ser.*, **429**, 2.
10. Roberz, B. (2005) Bestimmung polycyclischer aromatischer Schwefelheterozyklen aus der Verbrennung von Kraftstoffen, PhD thesis, University of Münster, p. 14.
11. Coleman, H.J., Hopkins, R.L., and Thompson, C.J. (1971) Highlights of some 50 man-years of petroleum sulfur studies by bureau of mines. *Int. J. Sulf. Chem. B*, **6**, 41–61.

12. Latscha, H.P., Kazmaier, U., and Klein, H.A. (2005) *Chemie für Biologen*, Springer-Lehrbuch, Part III, Springer, Berlin, p. 570.
13. Novak, W.J. and Stanley, J.F. (2003) ExxonMobil technologies for ultra low sulfur gasoline and diesel-an overview with recent commercial examples. Presented at Instituto Argentino del Petroleo y del Gas Meeting, ExxonMobil Research and Engineering, p. 15.
14. Velu, S., Ma, X., Song, C., Namazian, M., Sethuraman, S., and Venkataraman, G. (2005) Desulfurization of JP-8 jet fuel by selective adsorption over Ni-based adsorbent for micro solid oxide fuel cells. *Energy Fuels*, **19**, 1116.
15. Cheekatamarla, P.K. and Lane, A.M. (2005) Catalytic autothermal reforming of diesel fuel for hydrogen generation in fuel cells. I. Activity tests and sulfur poisoning. *J. Power Sources*, **152**, 256.
16. Cheekatamarla, P.K. and Lane, A.M. (2006) Catalytic autothermal reforming of diesel fuel for hydrogen generation in fuel cells. II. Catalyst poisoning and characterization studies. *J. Power Sources*, **154**, 223.
17. Ferrandon, M., Mawdsley, J., and Krause, T. (2008) Effect of temperature, steam-to-carbon ratio, and alkali metal additives on improving the sulfur tolerance of a Rh/La–Al$_2$O$_3$ catalyst reforming gasoline for fuel cell applications. *Appl. Catal. A: Gen.*, **342**, 69.
18. Kaila, R.K., Gutiérrez, A., and Krause, A.O.I. (2008) Autothermal reforming of simulated and commercial diesel: the performance of zirconia-supported RhPt catalyst in the presence of sulfur. *Appl. Catal. B: Environ.*, **84**, 324.
19. Mayne, J.M., Tadd, A.R., Dahlberg, K.A., and Schwank, J.W. (2010) Influence of thiophene on the isooctane reforming activity of Ni-based catalysts. *J. Catal.*, **271**, 140.
20. Palm, C., Cremer, P., Peters, R., and Stolten, D. (2002) Small-scale testing of a precious metal catalyst in the autothermal reforming of various hydrocarbon feeds. *J. Power Sources*, **106**, 231.
21. Kohl, A. and Nielsen, R. (1997) *Gas Purification*, 5th edn., Elsevier, Amsterdam.
22. Bashkova, S., Bagreev, A., and Bandosz, T.J. (2002) *Environ. Sci. Technol.*, **36**, 2777–2782.
23. Oura, K., Lifshits, V.G., Saranin, A.A., Zotov, A.V., and Katayama, M. (2003) *Surface Science: an Introduction*, Springer, New York.
24. Novochinskii, I.I., Song, C., Ma, X., Liu, X., Shore, L., Lampert, J., and Farrauto, R.J. (2004) *Energy Fuels*, **18**, 576–583.
25. Gangwal, S.K., Stogner, J.M., Harkins, S.M., and Bossart, S.J. (1989) *Environ. Prog.*, **8**, (1), 26–34.
26. Tamhankar, S.S., Bagajewicz, M., Gavalas, G.R., Sharma, P.K., and Flytzani-Stephanopoulus, M. (1986) Mixed-oxide sorbents for high-temperature removal of hydrogen sulfide. *Ind. Eng. Chem. Process Des. Dev.*, **25**, 429–437.
27. Abbasian, J. and Slimane, R.B. (1998) A regenerable copper-based sorbent for H$_2$S removal from coal gases. *Environ. Prog.*, **8**, (1), 26–34.
28. Jun, H.K., Lee, T.J., Ryu, S.O., and Kim, J.C. (2001) A study of Zn–Ti based H$_2$S removal sorbents promoted with cobalt oxides. *Ind. Eng. Chem. Res.*, **40**, 3547–3556.
29. Lee, T.J., Kwon, W.T., Chang, W.C., and Kim, J.C. (1997) *Korean J. Chem. Eng.*, **14**, (6), 513–518.
30. Sasaoka, E., Taniguchi, K., Hirano, S., Uddin, M. A., Kasaoka, S., and Sakata, Y. (1995) *Ind. Eng. Chem. Res.*, **34**, 1102–1106.
31. Novochinskii, I.I., Song, C., Ma, X., Liu, X., Shore, L., Lampert, J., and Farrauto, R.J. (2004) *Energy Fuels*, **18**, 584–589.
32. Li, L. and King, D.L. (2006) *Catal. Today*, **116**, 537–541.
33. Lew, S., Jothimurugesan, K., and Flytzani-Stephanopoulos, M. (1989) *Ind. Eng. Chem. Res.*, **28**, 535.
34. Gislason, J. (2001) Refiners' sulfur dilemma – Phillips sulfur-removal process nears commercialization. *Oil & Gas J.*, **99** (47), 72–76.

35. ConocoPillips S-Zorb SRT–S-Zorb in Action, *http://www.coptechnology solutions.com/sulfur_removal/szorb/szorb_ action/index.htm*, (last accessed 27 February 2008).
36. Johnson, M.M., Sughrue, E.L., and Owen, S.A. (2003) Desulfurization of middle distillates. World patent application WO 03/054117 A1.
37. ConocoPhillips (2001) S-Zorb sulfur removal technology. Presented at the World Fuels Conference, 16 May 2001, Brussels.
38. Speight, J.G. (2007) *The Chemistry and Technology of Petroleum*, 4th edn, CRC Press/Taylor & Francis, Boca Raton, FL.
39. Schmitz, C. (2003) Zur Kinetik und zur verbesserten Reaktionsführung der hydrierenden Tiefentschwefelung von Dieselöl, PhD thesis, University of Bayreuth.
40. Datsevich, L.B. and Muhkortov, D.A. (2004) Multiphase fixed-bed technologies – comparative analysis of industrial processes. *Appl. Catal. A: Gen.*, **261**, 143.
41. Benfer, R., Haarde, W., and Zehner, P. (1993) Mehrphasenströmungen in verfahrenstechnischen Apparaten. *Chem. Ing. Tech*, **65**, 1067.
42. Jess, A., Datsevich, and L.B., Gudde, N.J. (2003) Purification process. World patent WO 03/091363.
43. Pasel, J., Samsun, R.C., Tschauder, A., Schmitt, D., Peters, R., and Stolten, D. (2005) Design and test of a two-stage water-gas-shift reactor at a 5 kWe-scale. Presented at the 3rd European PECF Forum, Lucerne, Switzerland.
44. Berlin, N.H. and McCall, P.P. (1963) Improvements in the hydrofining of hydrocarbon liquids. British patent 934907.
45. Datsevich, L.B. and Muhkortov, D.A. (1997) Process for hydrogenation of organic compounds. Russian patent 2083540.
46. Wedler, G. (1970) Adsorption – Eine Einführung in die Physisorption und Chemisorption, Verlag Chemie, Weinheim/Bergstrasse.
47. Irvine, R.L. (1998) Process for desulfurizing gasoline and hydrocarbon feedstocks. US patent 5,730,860.
48. Irvine, R.L. and Varraveto, D.M. (1999) Adsorption process for removal of nitrogen and sulphur, *PTQ*, Summer, Q3.
49. Babich, I.V. and Moulijn, J.A. (2003) Science and technology of novel processes for deep desulfurization of oil refinery streams: a review. *Fuel*, **82**, 620.
50. Heinzel, J.M., Cervi, M., Hoffman, D., Nickens, and A. (2006) U.S. Navy sorbent development for liquid-phase removal of sulfur from logistics fuels for fuel cell applications. Presented at Fuel Cell Seminar, San Antonio, TX.
51. Wang, Y., Yang, R.T., and Heinzel, J.M. (2008) Desulfurization of jet fuel by π-complexation adsorption with metal hailides supported on MCM-42 and SBA-15 mesoporous materials. *Chem. Eng. Sci.*, **63**, 356.
52. King, C.J. (1987) Separation processes based on reversible chemical complexation, in *Handbook of Separation Process Technology* (ed. R.W. Rousseau), John Wiley & Sons, Inc., New York, p. 760.
53. Yang, R.T. (2003) *Adsorbents: Fundamentals and Application*, John Wiley & Sons, Inc., New York, p. 191.
54. Ma, X., Velu, S., Kim, J.H., and Song, C. (2005) Deep desulfurization of gasoline by selective adsorption over solid adsorbents and impact of analytical methods on ppm-level sulphur quantification for fuel cell applications. *Appl. Catal. B: Environ.*, **56**, 137.
55. Velu, S., Song, C., Engelhard, M.H., and Chin, Y.H. (2005) Adsorptive removal of organic sulfur compounds from jet fuel over K-exchanged NiY zeolites prepared by impregnation and ion exchange. *Ind. Eng. Chem.*, **44**, 5740.
56. Yang, Y., Lu, H., Ying, P., Jiang, Z., and Li, C. (2007) Selective dibenzothiophene adsorption on modified active carbons. *Carbon*, **45**, 3042.
57. Yu, C., Qiu, J.S., Sun, Y.F., Li, X.H., and Zhao, Z.B. (2008) Adsorption removal of thiophene and dibenzothiophene from oils with activated carbon

as sorbent: effect of surface chemistry. *J. Porous Mater.*, **15**, 151.
58. Wang, Y. and Yang, R.T. (2007) Desulfurization of liquid fuels by adsorption on carbon-based sorbents and ultrasound assisted sorbent regeneration. *Langmuir*, **23**, 3825.
59. Poshusta, J. and Martin, J.L. (2006) Desulfurization method, apparatus, and materials. US patent application 2006/0076270.
60. Jess, A., Wasserscheidt, P., and Esser, J. (2004) Einsatz ionischer Flüssigkeiten zur Entschwefelung von Produktströmen bei der Erdölverarbeitung. *Chem. Ing. Tech.*, **76**, 1407.
61. Jess, A. and Wasserscheidt, P. (2002) Tiefentschwefelte Kraftstoffe – konventionelle Verfahren und neue Methoden. *Z. Umweltchem. Ökotoxikol.*, **14** (3), 145.
62. Esser, J. (2006) Tiefentschwefelung von Mineralölfraktionen durch Extraktion mit ionischen Flüssigkeiten, PhD thesis, University of Bayreuth.
63. Huang, C., Chen, B., Zhang, J., Liu, Z., and Li, Y. (2004) Desulfurization of gasoline by extraction with new ionic liquids. *Energy Fuels*, **18**, 1862.
64. Walter, H.G. (1984) Untersuchung zur Entschwefelung von Erdöl mit Hilfe von Plasmabehandlung von Modellverbindungen, PhD thesis, University of Tübingen, p. 19.
65. Suhr, H. (1988) *Entschwefelung von flüssigen Erdöl und Erdölfraktionen durch Oxidation mit angeregtem Sauerstoff*, Forschungsbericht Kernforschungszentrum Karlsruhe, KfK-PEF 43, Kernforschungszentrum Karlsruhe, Karlsruhe, p. 5.
66. Che, Y., Ma, W., Ren, Y., Chen, C., Zhang, X., Zhao, J., and Zang, L. (2005) Photooxidation of dibenzothiophene and 4,6-dimethyldibenzothiophene sensitized by *N*-methylquinolinium tetrafluoroborate: mechanism and intermediates investigation. *J. Phys. Chem. B*, **109**, 8270.
67. Schenck, G.O. and Krauch, C.H. (1962) Zur photosensibilisierten O_2-Übertragung auf Schwefel-Verbindungen. Neuer Weg zu Sulfoxyden. *Angew. Chem.*, **74**, 510.
68. Latassa, D., Enger, O., Thilgen, C., Habicher, T., Offermanns, H., and Diederichs, F. (1993) *J. Mater. Chem.*, **12**, 1993.
69. Yen, T.F. (2002) Oxidative Entschwefelung von fossilen Kraftstoffen mit Ultraschall. World patent WO 0226916.
70. Murata, S., Murata, K., Kidena, K., and Nomura, M. (2004) A novel oxidative desulfurization system for diesel fuels with molecular oxygen in the presence of cobalt catalysts and aldehydes. *Energy Fuels*, **18**, 116.
71. Yu, G., Lu, S., Chen, H., and Zhu, Z. (2005) Oxidative desulfurization of diesel fuels with hydrogen peroxide in the presence of activated carbon and formic acid. *Energy Fuels*, **19**, 447.
72. Ali, M.F., Al-Malki, A., El-Ali, B., Martinie, G., and Siddiqui, M.N. (2006) Deep desulphurization of gasoline and diesel fuels using non-hydrogen consuming techniques. *Fuel*, **85**, 1354.
73. Levy, R.E., Rappas, A.S., Sudhakar, C., Nero, V.P., and Decanio, S.J. (1963) Hydrodesulfurization of oxidized sulfur compounds in liquid hydrocarbons. World patent WO 2003014266.
74. Schilling, B.M., Alvarez, L.M., Wang, D.I.C., and Cooney, C.L. (2002) Continuous desulfurization of dibenzothiophene with *Rhodococcus rhodochrous* IGTS8 (ATCC 53968). *Biotechnol. Prog.*, **18**, 1207.
75. Kruse, A. (2001) Reaktionen in nah- und überkritischem Wasser. *Nachr. Forschungszentrum Karlsruhe*, **33** (1), 59.
76. Bröll, D. (2001) Partialoxidationen in überkritischem Wasser mit molekularem Sauerstoff – die Reaktionen von Methanol, Methan und Propylen mit und ohne Silberkatalysatoren, PhD thesis, University of Darmstadt, p. 13.
77. Vogelaar, B.M., Makkee, M., and Moulijn, J.A. (1999) Applicability of supercritical water as a reaction medium for desulfurization and demetallisation of gasoil. *Fuel Process. Technol.*, **61**, 265.

78. Katritzky, A.R., Barcock, R.A., Balasubramanian, M., and Greenhill, J.V. (1995) Aqueous high-temperature chemistry of carbo- and heterocycles, reactions of sulfur-containing compounds in supercritical water at 460 °C. *Energy Fuels*, **8**, 498.
79. Beckmann, T. and Konrad, G. (2004) Verfahren zur Abtrennung einer schwefelarmen Kraftstofffraktion. German patent DE 10239361.
80. Sethuraman, S., Namazian, M., Venkataraman, G., Elder, E., Song, C., Ma, X., and Wantanabe, S. (2005) Fuel preprocessor and desulfurizer for fuel cell operation on distillate fuels. Prtesented at the Fuel Cell Seminar, Palm Springs, USA.
81. Huang, X. and King, D.L. (2006) Gas-phase hydrodesulfurization of JP-8 light fraction using steam reformate. *Ind. Eng. Chem. Res.*, **45**, 7050.
82. Ho, W.S.W. and Sirkar, K.K. (eds.) (1992) *Membrane Handbook*, Springer, New York, p. 4.
83. Melin, T. and Rautenbach, R. (2004) *Membranverfahren: Grundlagen der Modul- und Anlagenauslegung*, Springer, Berlin.
84. Baker, R.W. (2004) *Membrane Technology and Applications*, John Wiley & Sons, Ltd., Chichester.
85. Balko, J., Bourdillon, G., and Wynn, N. (2003) Membrane separation for producing ULS gasoline, *PTQ*, Spring, Q1, 17–25.
86. Durai-Swamy, K. (2006) Membrane desulfurization of logistic fuels for fuel cell auxiliary power units. Presented at the Aviation Technical Committee Meetings of the Coordinating Research Council, Alexandria, USA.
87. Durai-Swamy, K. and Woods, R.R. (2006) Multi-stage sulfur removal system and processor for an auxiliary fuel system. World patent WO 2006/084002.
88. Latz, J. (2008) Entschwefelung von Mitteldestillaten für die Anwendung in mobilen Brenstoffzellen-Systemen, PhD thesis, Forschungszentrum Jülich, Jülich.
89. Wang, Y., Latz, J., Dahl, R., Pasel, J., and Peters, R. (2009) Liquid phase desulfurization of jet fuel by a combined pervaporation and adsorption process. *Fuel Process. Technol.*, **90**, 458.
90. Bettermann, I. and Staudt, C. (2009) Desulphurisation of kerosene: pervaporation of benzothiophene/n-dodecane mixtures. *J. Membr. Sci.*, **343**, 119.
91. (2010) International Gasoline Rankings – Top 100 Sulfur., International Fuel Quality Center, http://208.72.1.66/UserFiles/file/Misc/GasolineRanking-May2010.pdf (last accessed 24 December 2010).
92. van Rheinberg, O., Lucka, K., Köhne, H., Schade, T., and Anderson, J.T. (2008) Selective removal of sulphur in liquid fuels for fuel cell application. *Fuels*, **87**, 2988–2996.
93. IFQC (2010) International Diesel Rankings – Top 100 Sulfur, http://www.ifqc.org/NM_Top5.aspx (last accessed 24 December 2010).
94. Minh, N.Q. and (1993) Ceramic fuel cells. *J. Am. Ceram. Soc.*, **76**, 563-588.
95. Norheim, A., Wærnhus, I., Broström, M., Hustad, J.E., and Vik, A. (2007) Experimental studies on the influence of H_2S on solid oxide fuel cell performance at 800 °C. *Energy Fuels*, **21**, 1098–1101.
96. Rasmussen, J. F.B. and Hagen, A. (2009), The effect of H_2S on the performance of Ni-YSZ anodes in solid oxide fuel cells. *J. Power Sources*, **191**, 534–541.
97. McPhail, S.J., Aarva, A, Devianto, H., Bove, R., and Moreno, A. (2011) SOFC and MCFC: commonalities and opportunities for integrated research. *Int. J. Hydrogen Energy*, **36** (16), 10337–10345.
98. Bøgild Hansen, J. (2008) Correlating sulfur poisoning of SOFC nickel anodes by a Temkin isotherm. *Electrochem. Solid-State Lett.*, **11** (10), B178–B180.
99. Schmidt, T.J. and Baurmeister, J. (2006) *ECS Trans.*, **3** (1), 861–869.
100. Song, C.S. (2002) Fuel processing for low-temperature and high-temperature fuel cells, challenges, and opportunities for sustainable development in the 21st century. *Catal. Today*, **77**, 17.

101. Gangwal, S.K., Harkins, S.M., Woods, M.C., Jain, S.C., and Bossart, S.J. (1989) *Environ. Prog.*, **8** (4), 265–269.

102. Jensen, A.B. and Webb, C. (1995), Treatment of H_2S-containing gases: a review of microbiological alternatives. *Enzyme Microb. Technol.*, **17**, 2–10.

35
Design Criteria and Components for Fuel Cell Powertrains
Lutz Eckstein and Bruno Gnörich

35.1
Introduction

The design of road vehicles is subject to a multitude of criteria concerning performance, dimensions, safety, operability, durability, and manufacturing and servicing constraints. In addition to these engineering tasks, economic and marketing criteria play an important role for successful vehicles. While customers and drivers tend to be hesitant about revolutionary vehicle concepts that offer limited mobility but enhance the eco-footprint of their driving, manufacturers face ever more stringent legislative boundary conditions for new vehicle generations. An early conclusion for advanced, eco-friendly vehicle development could therefore be to make new cars to be operable just like present cars. Driving experience should not be limited by the vehicle's technology. In this chapter, the focus is set on fuel cell powertrains. As a start for more detailed analyses of design criteria, the following formal requirements can be formulated in order to make fuel cell vehicles successful: fuel cell vehicles must be more efficient than conventional counterparts, yield the same or better performance and envisage similar costs. This non-exhaustive list forms the basis for the subsequent sections that give an insight into design criteria and relevant components for fuel cell powertrains, starting with basic vehicle requirements and an overview of different drivetrain concepts.

35.2
Vehicle Requirements

35.2.1
Driving Resistance

Different driving resistance forces need to be overcome by any vehicle to accelerate and to maintain driving velocity. They can be subdivided into two categories: first, resistance forces that include wheel, air, and climbing resistance and occur during constant driving or even at standstill; and second, acceleration resistance caused by the vehicle's inertia during acceleration. Table 35.1 gives an overview of these forces.

The individual resistances presented in Table 35.1 can be combined into a total resistance, which must be provided by the propulsion system at the propelled wheels. These forces can also be illustrated by the quotient of the driving torques at the front and rear axles and the dynamic radius of the wheels:

$$F_{req} = \frac{M_f + M_r}{r_{dyn}} = f_R F_Z + c_W A \frac{\rho_{air}}{2} v^2 + p F_Z + (e_i m_F + m_{add}) a_x \quad (35.1)$$

An illustration of the individual forces to the gross resistance is shown in Figure 35.1 in a single diagram, the resistance force diagram. The single driving resistances are summarized one by one, resulting in the gross resistance represented by the uppermost line. The resistance force curves correspond to the required torque which the propulsion system needs to provide at a certain operating point. The power output chart can be derived from the resistance force diagram.

The impact that the single resistances have on the gross resistance depends on the vehicle and the driving condition. Other criteria with regard to transmissible power include the vehicle proportions, drivetrain layout, and suspension. Figure 35.2 illustrates the horizontal and vertical forces acting on a vehicle during an accelerated uphill drive.

The top speed, the maximum acceleration, and climbing capability are characteristic for the driving performance of a vehicle. They can be calculated using the vehicle's maximum performance and the power actually required. Figures 35.3 and 35.4 show such an approach using the traction and the driving performance curves for a conventional and an electric vehicle.

In both cases shown here, a (manual) stepped transmission is also used to convert the propulsion system's output torque accordingly, although present electric vehicles often do not feature any shiftable torque converter.

In contrast to its conventional counterparts, electric vehicle drivetrains cover most of the traction and driving performance charts due to the electric machine's performance across much of their speed range. On the other hand, combustion engines cannot deliver their maximum power at the highest speeds, leading to performance gaps that need to be overcome by appropriate torque conversion.

35.2.2
Energy Conversion and Driving Cycles

To provide driving power, the drive needs to be supplied with energy (e.g., electricity or fuel), which is converted into mechanical energy by the propulsion system. In today's vehicles, liquid or gaseous fuel is mostly used as a chemical energy carrier. The fuel consumption is indicated as route fuel consumption (liters per 100 km) or as temporal fuel consumption (liters per hour).

An engine map can be used to predict the fuel consumption. It represents characteristics of the constant-specific fuel consumption (the so-called shell curves,

Table 35.1 Overview of vehicle driving resistance forces [1].

Force	Description		
Wheel resistance	$F_R = F_x \approx F_{R,\text{Roll}} = f_R F_Z$		
	Tires and road surface	f_R	
	Truck tires on tarmac	0.005 – 0.01	
	Car tires on tarmac	0.007 – 0.015	
	Car tires on concrete	0.007 – 0.02	
	Car tires on gravel	0.02	
	Car tires on cobblestones	0.015 – 0.03	
	Car tires on firm sand	0.04 – 0.08	
	Car tires on loose sand	0.15 – 0.4	
Air drag resistance	$F_{\text{drag}} = c_W A \dfrac{\rho_{\text{air}}}{2} v_x^2$		
	Vehicle	c_W	Area, A (m^2) / $c_W A$ (m^2)
	Citroën 2CV	0.51	1.65 / 0.85
	BMW 5 series (E60)	0.27	2.26 / 0.61
	Mercedes-Benz S class (W221)	0.26	2.39 / 0.62
Climbing resistance	$F_{\text{gr}} = mg \sin \alpha_{\text{gr}} \approx mg \tan \alpha_{\text{gr}} \approx mgp$		
	Design Speed (km h^{-1})	Permitted grade for roads $p_{\max}$ (%)	
		Outside of building areas (category A)	Within building areas (category BI/BII)
	50	9.0	12.0
	70	7.0	8.0
	90	5.0	6.0
	120	4.0	–
Acceleration resistance	$F_a = F_{a,\text{Trans}} + F_{a,\text{Rot}}$ $= \left(m_{\text{veh}} + m_{\text{add}} + \dfrac{\Theta_{\text{red},i}}{r_{\text{dyn}}^2} \right) a_x$ $= (e_i m_{\text{veh}} + m_{\text{add}}) a_x$ with $e_i = 1 + \dfrac{\Theta_{\text{red},i}}{m_{\text{veh}} r_{\text{dyn}}^2}$		
	Mass factor, e_{Gear}	e_1 / e_2 / e_3 / e_4	
	BMW 730iA	1.21 / 1.10 / 1.05 / 1.03	

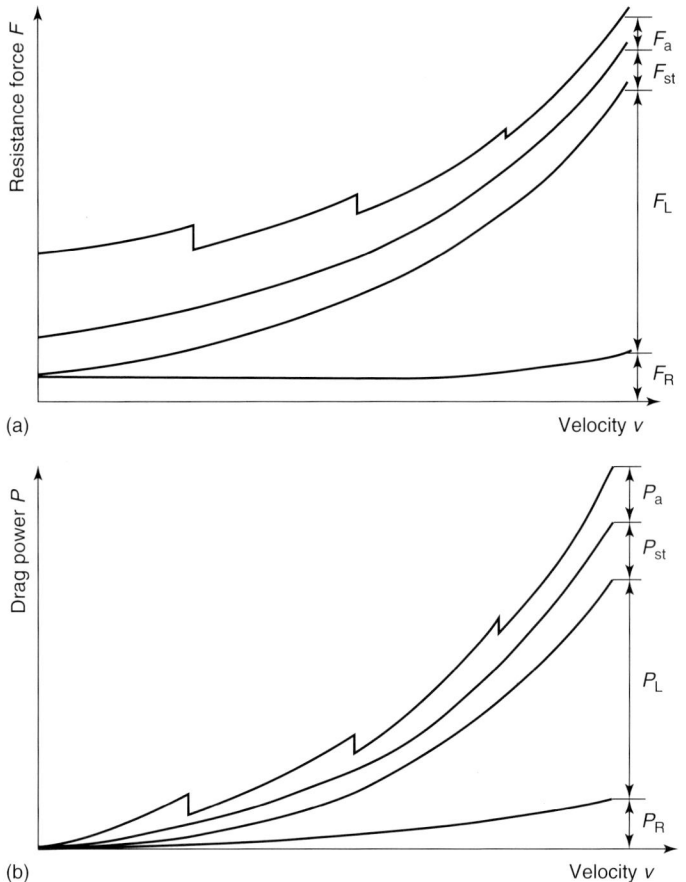

Figure 35.1 Resistance force and power chart with individual resistances (on an incline) [1].

Figure 35.5), usually given in grams of fuel per kilowatt-hour. In internal combustion engines, the lowest fuel consumption occurs at high torque and moderate engine speeds. This is a reason for the current trend towards smaller, supercharged engines. Maximum efficiency for passenger car diesel engines is ~40%, with part-load efficiency down to 20%.

Electric machines yield a much higher efficiency spreading over a wider area (Figure 35.6). It can exceed 90%, with a less pronounced drop towards part-load operation compared with internal combustion engines.

The standard consumption can also be measured under standardized test conditions. The "new European driving cycle" (NEDC) adds up five driving cycles. The first four cycles consist of ECE15-basic city drive cycles, to which the EUDC (extra urban driving cycle) is appended (Figure 35.7). With respect to the gross cycle, this amounts to a length of 11 km with a duration of 1180 s and an average speed of 33.5 km h^{-1} [1].

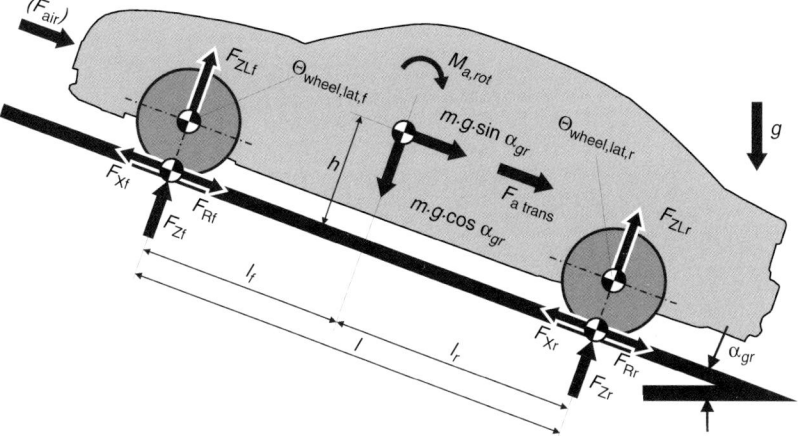

Figure 35.2 Forces on the vehicle (accelerated uphill drive) [1].

In the USA, the Supplemental Federal Test Procedure (SFTP) is used to evaluate and compare fuel economy. It consists of the Federal Test Procedure (FTP75, Figure 35.8) and of two further test cycles. The FTP75 consists of the city driving cycle and the highway driving cycle [1].

The driving cycle can be separated into three phases: the cold transitional phase (0–505 s), the steady phase (506–1372 s), and the warm transitional phase (1373–1877 s). First, it will be consecutively driven through the first two phases. After a 10 min break, at which the engine will be stopped, the warm transitional phase follows. During the course of the test, a route of 17.77 km is to be driven in a period of 1874 s, which corresponds to an average speed of 34.1 km h^{-1}. Losses in the components of the drivetrain occur during the conversion of chemical energy contained in the fuel into mechanical energy to overcome driving resistances. A schematic distribution of these losses is displayed in Figure 35.9.

About two-thirds of the energy is lost in the form of heat via the exhaust and cooling system. Only one-third of the energy contained in fuel is converted into mechanical energy. Part of this mechanical energy is lost due to energy requirements of auxiliaries and due to losses in the torque converter. In the end, only one-sixth of the fuel energy is actually used to drive the vehicle.

35.3
Potentials and Challenges of Vehicle Powertrains

35.3.1
Overview of Propulsion Systems

A general overview of the different (conventional and alternative) systems with a presentation of the total energy conversion process and the resulting classification

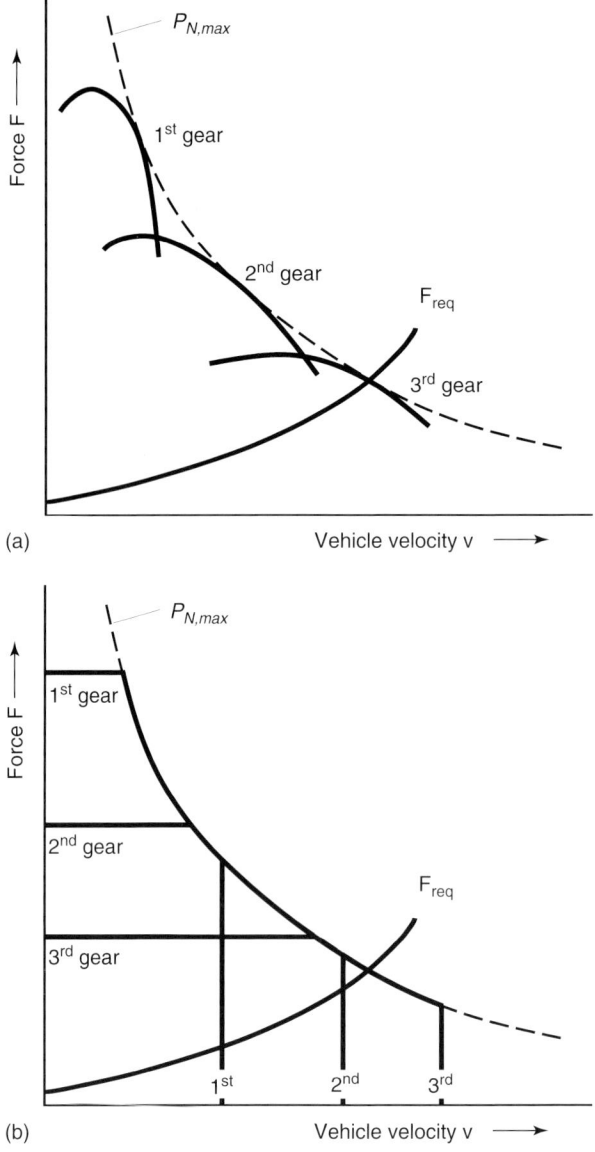

Figure 35.3 Traction diagrams for a conventional (a) and an electric vehicle (b).

of transport systems is shown in Figure 35.10. The conventional conversion chain is presented by the uppermost path in this figure. Besides conventional fuels such as gasoline and diesel, unconventional fuels include liquefied gases and alcohol fuels, which are refined from natural gas, coal, or biomass. Electricity is a secondary energy carrier and can be produced from all of the mentioned primary energy sources [2].

35.3 Potentials and Challenges of Vehicle Powertrains

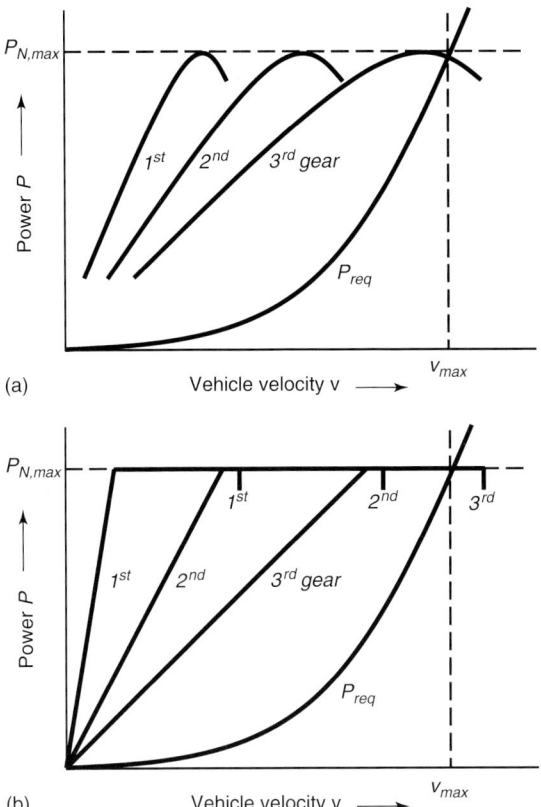

Figure 35.4 Driving performance diagrams for a conventional (a) and an electric vehicle (b).

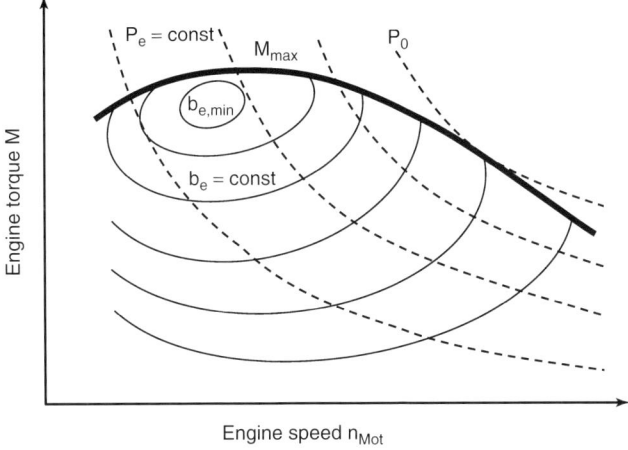

Figure 35.5 Schematic engine map.

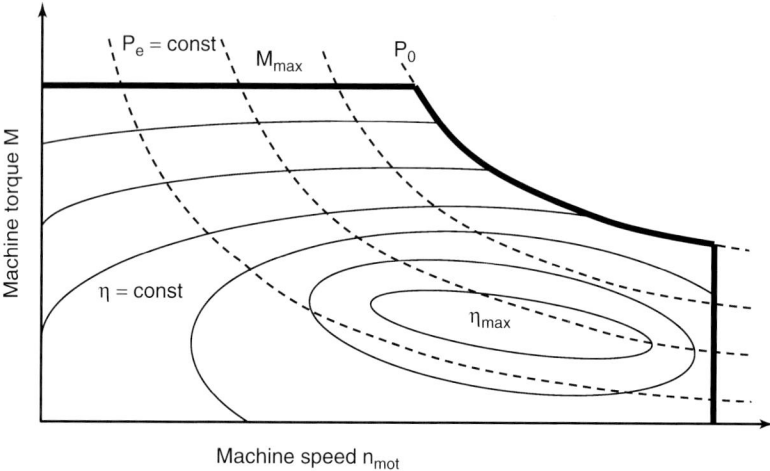

Figure 35.6 Schematic electric motor map.

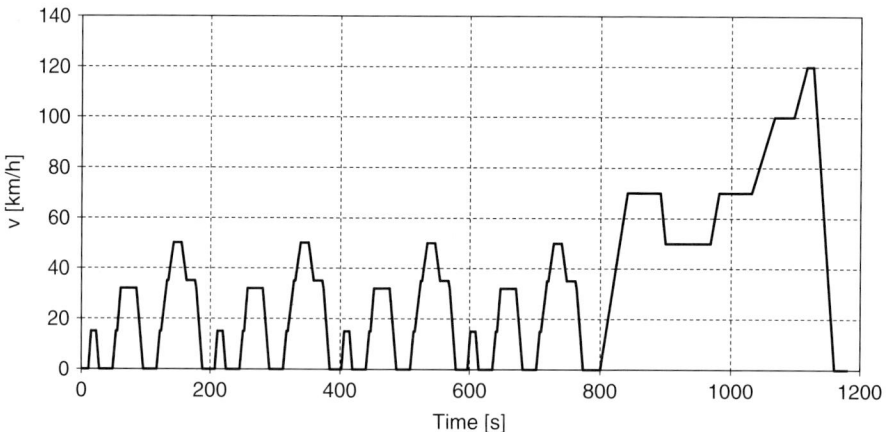

Figure 35.7 NEDC: New European Driving Cycle.

The distribution of primary energy sources for electricity generation (energy mix) varies substantially in different regions. Different economic and ecological boundary conditions exist for this secondary energy carrier depending on the energy mix used. Hydrogen generation is based on hydrocarbons or electrolysis. In short, reformers are used to split hydrocarbons into hydrogen and carbon dioxide, whereas electrolyzers split water into hydrogen and oxygen using electricity. As for electricity generation, similar economic and ecological boundary conditions apply for hydrogen as an energy carrier.

Both hybrid vehicles and electric vehicles allow feeding back of kinetic energy into their storage system, for example, while braking. In order to analyze the energy

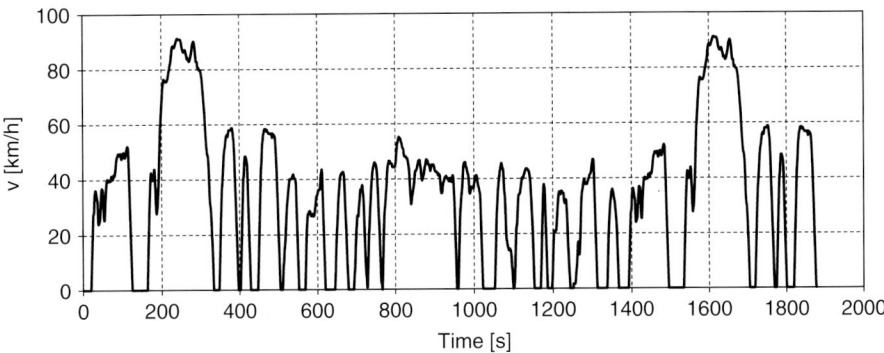

Figure 35.8 FTP75 city driving cycle [UDDS (Urban Dynamometer Driving Schedule)].

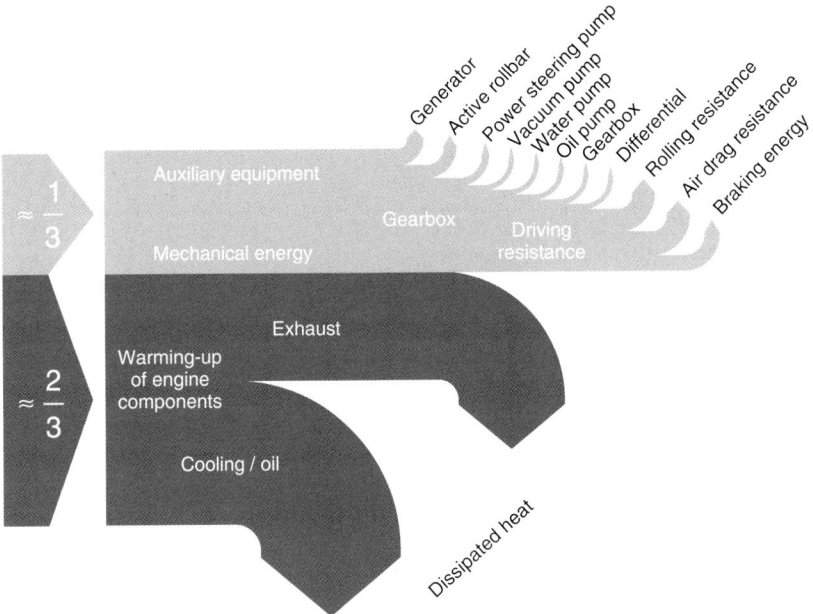

Figure 35.9 Loss distribution in the drivetrain [2].

conversion chain, there are three different approaches: *well to tank, tank to wheel,* and *well to wheel.*

In the first step, the well to tank analysis evaluates the energy conversion from its primary source to its storage in the reservoir. The tank to wheel analysis describes the conversion from the reservoir via energy converter and torque converter to the wheel of the transport system. The well to wheel analysis combines both approaches and displays the full picture with regard to the relevant parameters, for example, greenhouse gas emissions, efficiency, and others.

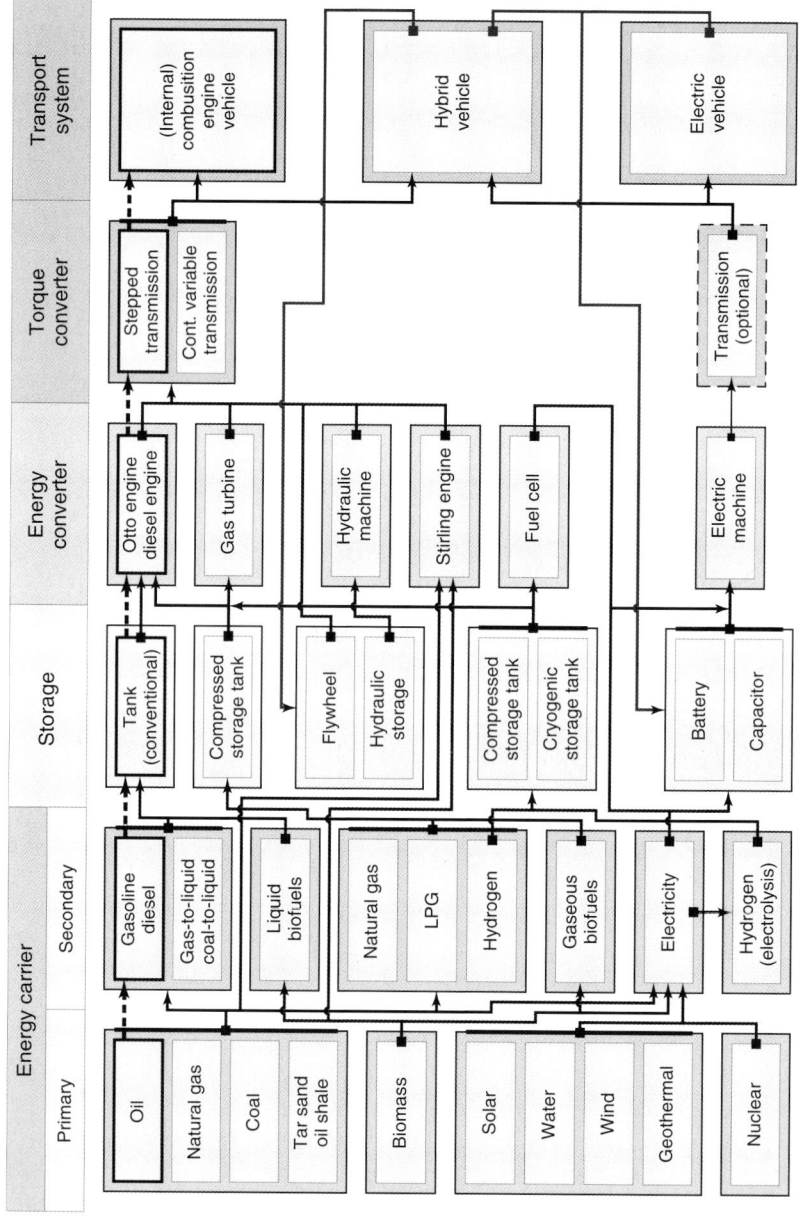

Figure 35.10 Well to wheel energy chain of road transportation [2].

35.3.2
Powertrain Comparison

Internal combustion engines have been the dominant propulsion system for road vehicles since the early years of the twentieth century. Their increasing efficiency combined with the ease of refueling made them the prime choice over electric drivetrains that were also available at the time. Today, a multitude of propulsion variants with internal combustion engines exists, with the most common ones shown in Figure 35.11.

Apart from the fact that such conventional drivetrains are technologically very mature, their costs are difficult to meet by any new system. Today's conventional drivetrains for passenger vehicles cost roughly €10–20 kW_{mech}^{-1}, with diesel engines towards the upper end owing to the complex exhaust gas after-treatment that contributes about one-third of the cost [3]. Four-wheel drive systems are slightly more expensive than two-wheel drive counterparts. Engines for commercial and heavy-duty vehicles are slightly more expensive because of their durability requirements and relatively low production volumes. Nonetheless, the benefit of this technology is the extensive field of players for production and service, and the excellent fuel infrastructure coverage. Finally, consumer habits have evolved around this technology. The robustness, the almost instantaneous start-up, and the driving range epitomize individual mobility. The drawback of conventional vehicles include their (local) emissions, both exhaust gases and noise, efficiency limits of the combustion engine, and the prospective increase in fuel costs.

To add a battery and an electric propulsion system to these drivetrains can help to overcome these problems. In such hybrid drivetrains, a variety of energy converter and energy storage technology layouts can be realized. Figure 35.12 shows

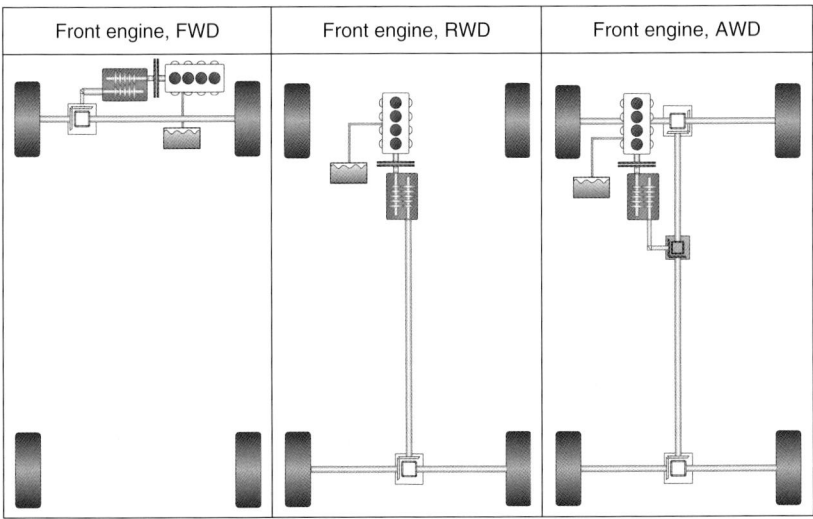

Figure 35.11 Drivetrain layout of conventional (internal combustion engine) vehicles [2].

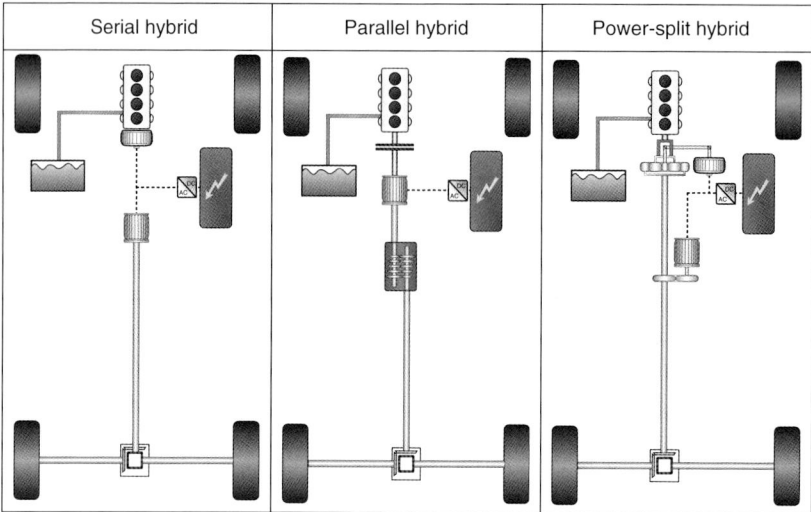

Figure 35.12 Drivetrain layout of ICE hybrid electric vehicles [2].

possible structures of internal combustion engine (ICE) hybrid electric drivetrains for passenger vehicles. The layouts vary in how the different types of energy (fuel or electricity) contribute to the propulsion of the vehicle.

In **serial hybrid vehicles**, the internal combustion engine is directly connected to a generator. The generator feeds both the battery and the electric motor(s) which propel the vehicle purely electric. Theoretically, the internal combustion engine can be run near its best operating point, while the battery acts as a buffer for power peaks. This allows using a smaller engine but requires a sufficiently large battery. Serial hybrids featuring significant electric driving range and plug-in capability are also called *range extender hybrids* or *electric vehicles with range extender*.

The internal combustion engine of **parallel hybrid vehicles** is mechanically connected to the driveline and also to one or more electric machines. In a common parallel hybrid drivetrain architecture, the electric machine is connected t either o the input or the output side of the vehicle's gearbox. The rotational speed of the internal combustion engine and the electric machine is defined by the vehicle's velocity, but still allows shifting of the ICE's torque towards areas of higher efficiency. The use of automated (e.g., double clutch) transmissions allows shifting of the operating point even closer to optimum efficiency. Electric propulsion is also possible if the electric motor is adequately powerful. In a road coupled parallel hybrid, the ICE drives one vehicle axle while the electric machine(s) drive the wheels of the other axle. Even this drivetrain architecture allows a torque shift from the combustion engine to the electric machines using the road as coupling.

The **power-split hybrid** configuration combines both serial and parallel hybrid structures, for example, by means of a planetary gear set. The power provided by an ICE is split into a mechanical path driving the vehicle via the gearbox and an electrical path via an electric generator, a battery, and one or more electric machines

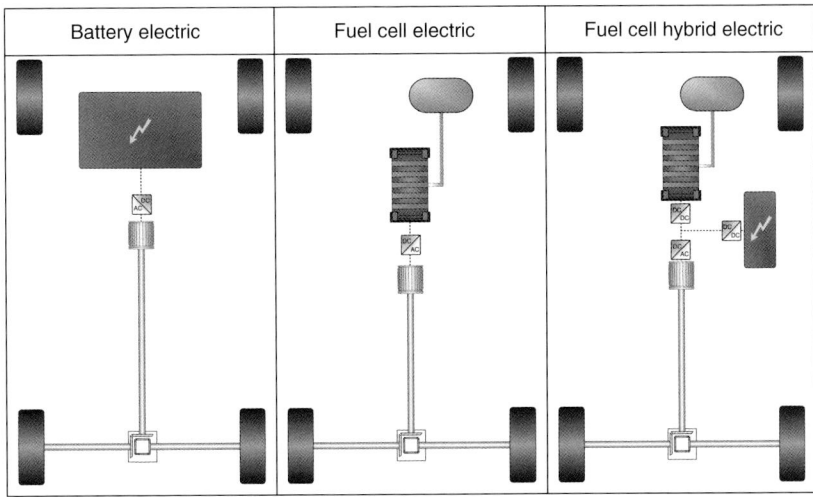

Figure 35.13 Drivetrain layout of battery and fuel cell electric vehicles [2].

propelling the wheels. Thus, propulsion can be provided directly by the ICE, or electrically via the two electric machines and the battery, or by a combination of both propulsion systems. In addition to shifting the ICE's torque, its operating point can be shifted without altering the power output. This allows the ICE to operate at optimum efficiency, if the energy management system and the battery capacity permit. An example vehicle featuring this drivetrain is the Toyota Prius, first presented in 1997.

Battery electric vehicles (BEVs) feature a battery storing energy from the grid, suitable power electronics and one or more electric machines to propel the vehicle (Figure 35.13).

Apart from balance-of-plant components, the only moving parts are the motor's rotor and the driveshafts. Battery electric vehicles produce hardly any noise or vibrations and are very efficient in converting chemical energy via electricity into mechanical energy. Consequently, they only offer limited waste heat to be used for the vehicle's heating. In order to avoid electrical heating reducing the range, small diesel-or ethanol-based heaters have been developed to counteract this energetic bottleneck. On the downside, the costs of common lithium-ion batteries are high (€1200 kWh^{-1} [4], €300 kWh^{-1} expected by 2015), making an electric vehicle with a usual range of 500 km extremely expensive. Moreover, the battery pack would use much space and cause substantial weight, since an energy density of about 100 Wh kg^{-1} at the system level and an energy consumption of at least 10–15 kWh per 100 km would result in a minimum battery mass of 500–750 kg. This makes BEVs unsuitable for longer journeys unless battery and/or recharging technology make substantial progress. A battery-changing infrastructure has been proposed several times over the past few years [5], mostly for island-like surroundings where vehicles do not travel long distances. Another option to provide electric energy for an electrically driven vehicle is a fuel cell system. Again, hybrid configurations are possible.

Adding a battery to a fuel cell electric vehicle has two advantages: (i) the longitudinal dynamics of the vehicle do not depend directly on the performance of the fuel cell system and (ii) kinetic energy can be recuperated and stored in the battery.

Fuel cells [i.e., polymer electrolyte membrane fuel cells (PEMFCs)] in vehicle drivetrains returned to the agenda in the early 1990s with the Mercedes-Benz NECAR van in 1994. A substantial number of concepts have been developed since then, and several manufacturers envisage market introduction within a few years. The following section describes in detail concepts and components of fuel cell (hybrid) electric vehicles.

35.3.3
Fuel Cell Powertrains

35.3.3.1 Non-Hybrid Fuel Cell Powertrains

The simplest concept of a fuel cell vehicle contains no battery or other energy storage device apart from a tank for compressed or liquefied hydrogen. Depending on the required power (top speed, acceleration), the vehicle components need to be dimensioned adequately. As an example of this concept, Figure 35.14 shows the GM HydroGen 3, which is based on the Opel Zafira and was tested by fleet operators in the early 2000s [6].

The drivetrain consists of a PEMFC with 94 kW which is fed with hydrogen by a 700 bar pressure tank or a liquid hydrogen (LH_2) tank. The first general disadvantage of this concept is that regenerative braking is not possible, since this energy cannot be stored. Moreover, the efficiency of this drivetrain concept is limited, since the operating point of the fuel cell systems is defined by the power required for propulsion. Especially at higher loads, the efficiency of the fuel cell system decreases.

35.3.3.2 Hybrid Electric Fuel Cell Vehicles

Fuel cell vehicles with additional energy storage are often equipped with a traction battery or a double-layer capacitor. The implementation has the following advantages:

Technical data	
FC stack	94 kW
el. motor	60 kW
H_2 tank	CGH 700 bar (3.1 kg) or liquefied (4.6 kg)
Fuel economy	1.15 kg/100 km
0-100 km/h	approx. 16 s
v_{max}	160 km/h
el. storage	-

Figure 35.14 GM HydroGen 3.

- recuperation of brake energy
- increased power by means of boosting
- lower energy consumption due to more effective operation of the components
- increased driving range.

The optimum energy management of the vehicle and its components is one of the primary challenges of fuel cell hybrid electric vehicles. The difficulties result from the high nonlinearity of the control path (fuel cell, power electronics, battery, and electric machine) and subsequently the complexity of the control architecture. In addition, the increasing vehicle mass and potentially higher costs need to be addressed. A few of the existing vehicles are presented below.

The Ford Focus FCV Hybrid (Figure 35.15) was presented in 2002 and since then has been produced in different variants [7].

The vehicle has a stack with a nominal power of 68 kW and a 18 kWh nickel metal hydride battery. The hydrogen is stored pressurized at 350 bar. Its 3.5 kg of hydrogen allows a range of 320 km. Because the vehicle has no gears, the top speed is limited to 128 km h^{-1} by the top speed of the electric motor. This vehicle also features a number of weight-reducing measures which result in total to a weight reduction of 300 kg. Lightweight body panels made of aluminum, a tailgate made of carbon fiber, and side and front windows made of polycarbonate are part of these measures.

In 2007, Mercedes-Benz presented the F-Cell based on the B-class (Figure 35.16). Approximately 200 vehicles of this type are currently in operation, a few having completed a world tour in 2011. Daimler announced market introduction for 2014. The electric machine of the F-Cell provides 100 kW and a maximum torque of 290 Nm and is supplied by an 80 kW stack and also by a 35 kW lithium-ion battery with a capacity of 1.4 kWh. The tank stores 4 kg of compressed hydrogen at 700 bar. With this technology, the F-Cell has a range of 400 km, which correlates to a hydrogen consumption of 0.97 l per 100 km.

The Volkswagen Bora Hy.Power (Figure 35.17) from 2002 features a hybrid concept with fuel cell and high-performance capacitor, which was developed in cooperation with the Swiss Paul Scherer Institute. The vehicle has a weight of 1900 kg and is based on the Volkswagen Bora Hy.Motion with a NiMH battery [8].

Technical data	
FC stack	68 kW
el. motor	230 Nm, 12,500 min^{-1}
H$_2$ tank	compr. (350 bar, 3.5 kg)
Fuel economy	approx. 1.1 kg/100 km
0-100 km/h	approx. 15 s
v$_{max}$	128 km/h
el. storage	Nickel metal hydride

Figure 35.15 Ford Focus FCV hybrid.

Technical data	
FC stack	80 kW
el. motor	100 kW
H$_2$ tank	compr. (700 bar)
Fuel economy	0.97 kg/100 km
0-100 km/h	n.a.
v$_{max}$	170 km/h
el. storage	Lithium ion

Figure 35.16 Mercedes-Benz F-cell.

Technical data	
FC stack	28 kW
el. motor	75 kW
H$_2$ tank	cryogen (3.5 kg)
Fuel economy	approx. 1.8 kg/100 km
0-100 km/h	approx. 15 s
v$_{max}$	115 km/h
el. storage	Supercap

Figure 35.17 Volkswagen Bora Hy.Power (Photograph: PSI).

The system features a 28 kW PEMFC stack and a 75 kW electric motor. The LH$_2$ tank has a capacity of 3.5 kg and allows a range of ∼200 km. The fuel consumption here is 1.8 kg per 100 km. This is a 40% reduction in energy consumption compared with conventional counterparts, but about 60% more than other fuel cell hybrid vehicle concepts, caused by the high weight of the vehicle.

35.3.3.3 Triple-Hybrid Fuel Cell Vehicles

The different characteristics of batteries and high-performance capacitors with regard to power and energy density lead to different operating characteristics of each particular hybrid vehicle. Vehicles with batteries are especially suitable for increasing the range, whereas vehicles equipped with capacitors are suitable for high power demands (boost operation). These different characteristics can be combined in fuel cell vehicles that are equipped with both types of energy storage. Such vehicles have not yet been produced as passenger cars, but have been presented as commercial and material-handling vehicles (Figure 35.18).

In October 2008, Proton Motor and Karmann presented a compact commercial car which contains the *triple-hybrid fuel cell system* designed by Proton Motor [9]. Target groups for this vehicle are municipalities and other fleet operators.

Figure 35.18 EcoCarrier HY3 with Proton Motor's triple-hybrid fuel cell system.

35.4
Components of Fuel Cell Powertrains

35.4.1
Hydrogen Storage

The technology for on-board storage of hydrogen plays a decisive role for the market introduction of fuel cell vehicles. The most important criteria for the selection of a compatible storage technology are shown in Table 35.2. They were formulated within with the European project *StorHy* in 2008 [10].

For the use in road vehicles, usually two technologies are used: pressure vessels with a maximum pressure of 700 bar and LH_2 tanks at 20 K. A third type, hybrid storage, is based on adsorption of hydrogen. These systems are being assessed for

Table 35.2 H_2 Storage design criteria for mobile applications [10].

Parameter	Units	Design criteria
Range	km	600
Hydrogen storage mass	kg	6–10
Gravimetric system storage density	$kWh\,kg^{-1}$	2
	mass%	6
Volumetric system storage density	$kWh\,l^{-1}$	1.5
	vol.%	4.5
Operation temperature	°C	−40 to +85
Storage fill rate	$kg\,H_2\,min^{-1}$	1.2
Maximum delivery rate	$g\,H_2\,s^{-1}$	Fuel cell: 2.0
		H_2 ICE: 5.5
Minimum pressure	bar	6
Maximum leakage	$Ncm^3\,h^{-1}\,l^{-1}$	1
Maximum boil-off rate	$G\,h^{-1}\,kg^{-1}$	1
Lifetime	Cycles	15 000
Cost	$€\,kWh^{-1}$	20

special types of fleet vehicles, which can or need to have a higher kerb weight, for example, as balance weight for fork lifts.

Tanks for compressed hydrogen are categorized into types I–IV. Type I–III tanks feature a metal liner (e.g., steel or aluminum). Types II–IV have carbon fiber filament windings and a maximum filling pressure of 300–700 bar. The liner in type IV tanks is made from polyamide to reduce weight and costs.

Type III tanks with a filling pressure of 700 bar have a theoretical volumetric energy density of 1.4 kWh l^{-1} H_2, which means that 5 kg of hydrogen can be stored in a 125 l tank. Marketed tanks have an energy density from $\sim 0.8 \text{ kWh l}^{-1}$ or 0.83 kWh kg^{-1} up to 1.02 kWh l^{-1} or 2.06 kWh kg^{-1}.

To build tanks with the characteristics required in Table 35.2, a special focus on materials and production processes is needed, especially regarding the inner liner and the carbon fiber shell, which is responsible for $\sim 50\%$ of the overall cost.

Current type III tanks (liner made from CrMo steel, up to 700 bar) do not feature any leakage, but are costlier to build because their manufacturing technology necessitates a high-precision welding process. Newer alternatives such as internal high-pressure forming (hydroforming) represent a very interesting alternative and are currently also being considered.

Type IV storage tanks (liner made of polyamide PA6) can be produced more easily and more cost-effectively ($\sim$€13 kWh^{-1} [10]), but suffer from a small leakage of about $2 \text{ mg h}^{-1} \text{ kg}^{-1}$). The production process consists of rotational molding with simultaneous polymerization. The higher elasticity compared with steel tanks potentially offers even better crash worthiness (bursting and impact strength).

The use of exchangeable systems is also possible, provided that safe coupling systems are available. Due to a much faster refilling process compared with battery charging duration, a hydrogen tank swapping technology seems unlikely for most applications.

Storage units for LH_2 are being built in the form of double-walled tanks with high vacuum and isolation between the inner and outer walls, to minimize heat transfer to the interior and to keep the inner temperature to a maximum of 20 K. Systems currently used in road vehicles have a gravimetric energy density of up to 5.6 wt% and a boil-off-rate of $1.25 \text{ g h}^{-1} \text{ kg}^{-1}$. A cylindrical tank geometry is the most advantageous regarding manufacturing, owing to a much easier realization of pressure resistance than for other geometries. In contrast, free-form surfaces offer the potential to increase the gravimetric energy density significantly. In 2008, a free-form tank with an energy density of 18% was presented in the StorHy project (Table 35.3).

Due to free-form surfaces, this type of storage is only 30% heavier than a gasoline tank with identical energy capacity (10 kg hydrogen and 30 kg gasoline). Although the leakage is 30% lower than for cylindrical tanks, it is still higher than for pressure tanks.

Reliable cost data are not available but it is likely that LH_2 tanks will not be cheaper than compressed hydrogen tanks. The energy density and the refueling fill rate are the deciding factors for the selection for road vehicles. A comparison

Table 35.3 Characteristics of cylindrical (BMW Hydrogen 7) and free-form LH$_2$ tanks [10].

Parameter	Units	Cylindrical	Free-form
Hydrogen storage mass	kg	10	10
System mass including hydrogen	kg	170	56
System volume	l	295	280
Gravimetric energy density	mass%	5.6	18
Volumetric energy density	kg H$_2$ per 100 l	3.3	3.6
Fill rate	kg H$_2$ min^{-1}	3	3
Leakage	g h^{-1} kg^{-1} H$_2$	1.25	0.8

Table 35.4 Comparison of different storage systems [10].

Legend:
- ■ High need for development
- ▨ Medium need for development
- □ Low/no need for development

	Volumetric energy density (kWh l^{-1})	Gravimetric energy density (kWh kg^{-1})	Fill-rate (kWh min^{-1})	Leakage (g h^{-1} kg^{-1} H$_2$)	Geometry (1 = cylindrical, 5 = complex)	Cost (1 = low, 5 = high)
Pressure tank type II (350 bar)	0.5	1.3	50	0	2	5
Pressure tank type III (700 bar)	0.8	1.3	50	0	2	4
Pressure tank type IV (700 bar)	0.8	1.5	50	0.002	2	3
Cylindrical cryo-tank (BMW)	1.2	2	100	1.3	1	5
Cylindrical cryo-tank (StorHy)	1.3	5	100	1	1	5
Free-form cryo-tank (StorHy)	1.2	5.9	100	0.8	4	4
Lithium ion battery (Mitsubishi)	0.2	0.1	0.5	0	4	4
Gasoline tank	7	8	>200	0	5	1

of the different storage systems is shown in Table 35.4. There is still a need for development, with regard to both technical issues and production costs.

35.4.2
Fuel Cell Systems for Automotive Applications

Fuel cells are galvanic cells which, when supplied with of fuel and oxidants, release their reaction enthalpy in form of heat and electric power. Different types of fuel

Table 35.5 Overview of types of fuel cells [12].

	Polymer electrolyte membrane fuel cell (PEMFC)	Phosphoric acid fuel cell (PAFC)	Molten carbonate fuel cell (MCFC)	Solid oxide fuel cell (SOFC)
Operating temperature	<100 °C	2000–220 °C	~650 °C	~1000 °C
Fuel	H_2	H_2	H_2, CO	H_2, CO
Ionic conductivity	H^+	H^+	O^{2-} via CO_3^{2-}	O^{2-}
Anode	Pt–C	Pt–C, 0.1 mg Pt cm^{-2}	Ni–10 wt% Cr	Ni–ZrO_2
Electrolyte	Polymer, H_2O	H_3PO_4	Li_2CO_3, K_2CO_3	ZrO_2, 8 mol%
Matrix	–	PTFE, SiC	γ-$LiAlO_2$	–
Cathode	Pt–C	Pt–C, 0.5 mg Pt cm^{-2}	NiO	$La_{1-x}Sr_xMnO_3$

cells exist, but all of them feature a gas-tight separation of fuel and oxidants via a membrane (electrolyte), which is coated with electrode material on both sides. The electrodes are linked via an electric load. Table 35.5 lists the most common types of fuel cells with their basic characteristics. Direct methanol fuel cells are a subgroup of PEMFCs and therefore are not listed separately. Furthermore, alkaline fuel cells are not considered due to their decreasing importance and their requirement for pure oxygen.

Because of their operating temperature below 100 °C, PEMFCs are best suitable for mobile applications which need a quick starting time. On the other hand, the liquid product water has to be removed without interfering with the humidification of the electrolyte. The functionality of a PEMFC is shown in Figure 35.19 [11].

Hydrogen is supplied at the anode, where it is adsorbed and ionized via a platinum catalyst. The protons reach the cathode through a membrane, where it reacts with oxygen or air, forming water. Because the membrane is impermeable to electrons, they pass through the electric load, for example, an electric machine. The power density of current PEMFCs is ~500 W kg^{-1}.

The costs for fuel cell systems are currently extremely high due to individual manufacturing, but a significant reduction is feasible today by mass production. A study by NREL and TIAX [13] calculated the costs for the components of fuel cell systems and the overall cost for mass production.

The costs for a 80 kW fuel cell system are shown in Figure 35.20. The current overall production costs amount to ~€70 kW^{-1} (500 000 units per year; conversion rate €1.00 = US$1.50).

The biggest share of 63% or €42 kW^{-1} is caused by stack components, followed by the cost of auxiliaries (34%). The contributors to the stack costs are shown on the pie-chart in Figure 35.20b. The membrane electrode assembly (MEA) has the

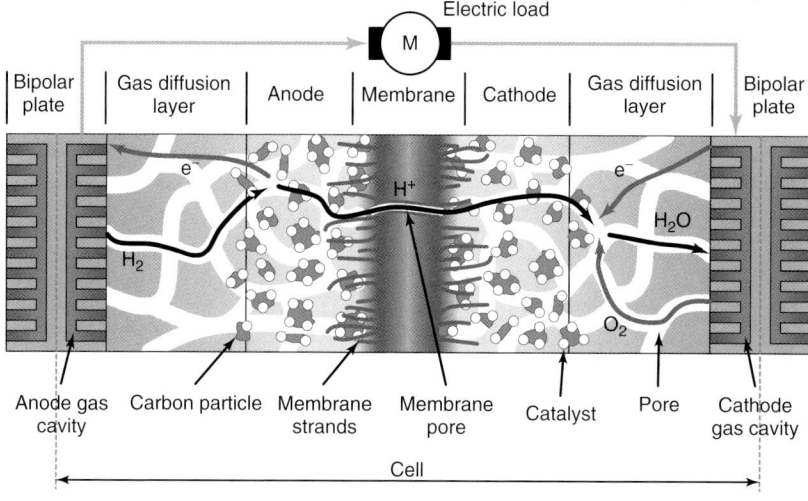

Figure 35.19 Schematic diagram of a PEMFC [11].

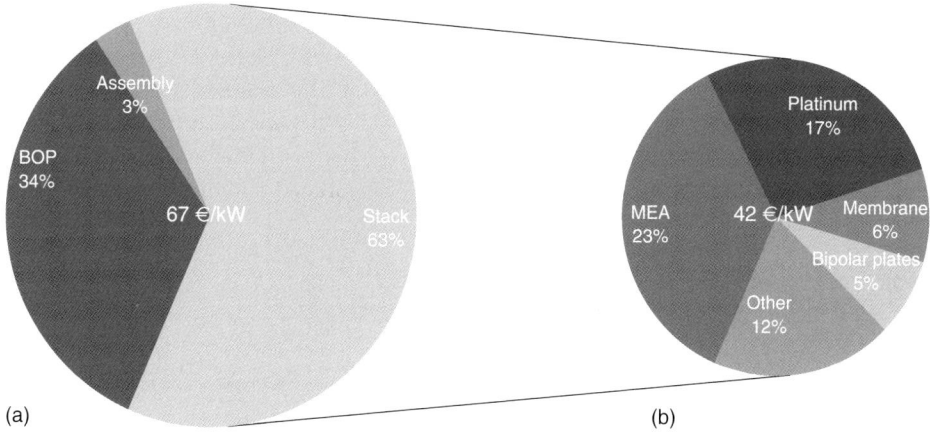

Figure 35.20 Production costs of an 80 kW PEMFC system (a) and its stack (b) [13].

highest share of 23%. Platinum accounts for about 17% of the total costs. Its share cannot be reduced by learning curves of mass production. It is therefore necessary to improve platinum efficiency and to reduce its loading.

35.4.3
Electrical Storage

Different technologies can be used to store electrical energy in road vehicles. Notable technologies today are lead acid, nickel metal hydride, and lithium ion batteries, and also double-layer capacitors (supercapacitors). Electrochemical high-temperature cells (e.g., ZEBRA) are no longer considered for most applications in passenger

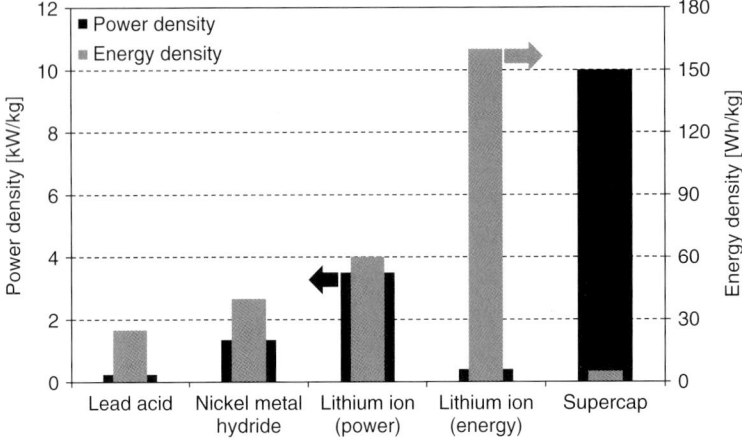

Figure 35.21 Power and energy density of different electric energy storage systems [4].

cars because of their critical thermal management. Apart from capacitors, all automotive battery technologies are secondary cells with reversible electrochemical energy conversion, an essential prerequisite for electric cars. Lacking a wide infrastructure for replacement, primary cells are unsuitable for traction purposes. They may only make sense in small areas or islands, where vehicles travel only moderate distances.

Choosing the right type of storage requires a thorough comparison of the properties of each type of storage. Figure 35.21 shows the power and energy densities of current technologies [4].

The double-layer capacitor presents the lowest average energy density; lithium-ion batteries can store up to 160 Wh kg^{-1} on the cell level. Lead acid batteries have an energy capacity of 25 Wh kg^{-1} and nickel metal hydride batteries 40 Wh kg^{-1}. The comparison of power density reveals a different picture. In this case, double-layer capacitors show the highest average by far (10 kW kg^{-1}). Lead acid batteries have the lowest power density of all technologies considered here.

The current costs for the different storage systems are shown in Figure 35.22, where the actual degree of discharge (DOD) is used as a basis for cost calculations. The diagram underlines the difficulties in designing appropriate lithium ion battery systems as a compromise between power and energy density. Lead acid batteries are comparatively expensive in this case, but double-layer capacitors are much more expensive than any other technology.

The comparison of costs can also be made by considering the lifetime (number of cycles or calendaric). In this case, double-layer capacitors are the cheapest solution concerning power and energy density. However, this conclusion must be put into the perspective of the different operating conditions. For example, the charging or discharging of a battery takes more time and takes place less often that full charging cycles of double-layer capacitors.

In principle, all of the energy storage technologies presented here are feasible for traction application in road vehicles. Lead acid batteries play only a role in

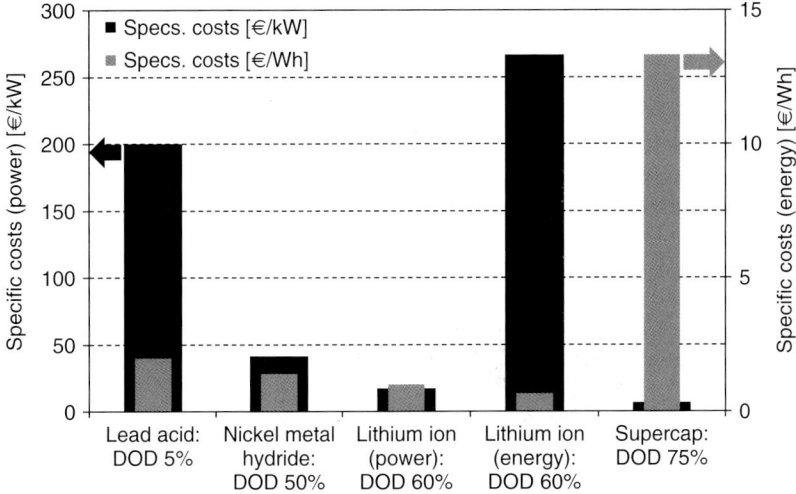

Figure 35.22 Specific power and energy costs for electric energy storage (realistic degree of discharge).

niche applications and two-wheelers. The most commonly used technology today is lithium ion batteries. The double-layer capacitor will be reserved for special applications in short-term or high-performance storage.

35.4.4
Electric Machines

Electric machines are electromechanical converters, where energy conversion takes place by means of a force on the mechanical and induced voltage on the electrical side. In principle, every electric machine can be operated as a motor or as a generator. For every electric machine, operational limits for speed, torque, and power exist (Figure 35.23). A distinction must be made between nominal values and maximum values. Nominal values (M_{Nom}, P_{Nom}) can be applied permanently; maximum values (M_{Max}, P_{Max}) only for a short period, otherwise mechanical and thermal failure may occur and affect the durability of the machine [2].

Operation of electric machines can be divided into two areas: base load range and field weakening range. In the base load range, the machine is capable of delivering maximum torque (M_{Nom} or M_{Max}) at all speeds from standstill. Nominal power is then available at nominal speed n_{Nom}:

$$n_{Nom} = \frac{P_{Nom}}{2\pi M_{Nom}} \qquad (35.2)$$

For permanent operation, the nominal power must not be exceeded. As a result, the output torque decreases with increase in speed:

$$M = \frac{P_{Nom}}{2\pi n} \qquad (35.3)$$

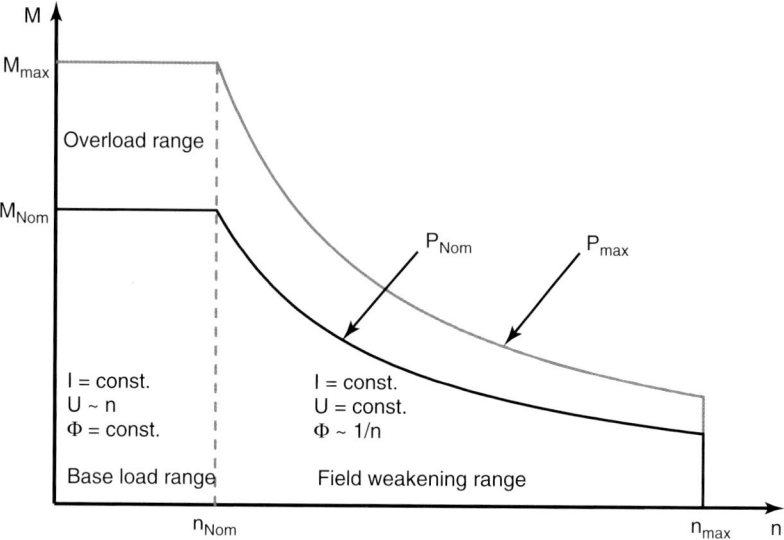

Figure 35.23 Operating range of electric machines [2].

with $P =$ constant. Equation (35.3) describes the area of constant power achieved by weakening of the magnetic field. Today, almost all machines used in electric vehicles are synchronous or asynchronous three-phase machines owing to their high efficiency. High efficiency can also be achieved with permanently excited machines, but their permanent magnets are expensive. Transversal flux or reluctance machines combine the characteristics of the other machine types.

Depending on the concept, electric machines can be placed at different positions of the drivetrain. There are many possibilities, from a simple connection to the gearbox to direct integration into the gearbox. A special configuration is the wheel hub drive, which represents integration of the electric machine directly into the wheel hub. This is beneficial with regard to the vehicle package, as differentials and drive shafts become superfluous, and in view of functionalities that can be implemented into the propulsion algorithm, such as ABS (automatic braking system), traction control, and torque vectoring. However, the higher unsprung mass needs to be addressed in suspension design. Wheel hub motors are not yet deployed in any series (passenger) vehicle. Integrative solutions featuring lower mass are currently being developed, for example, the *Active Wheel* system by Michelin, with a nominal power of 30 kW per wheel and a mass of 7 kg [14]. Table 35.6 summarizes the characteristics of the different machine types.

Electric machines for traction applications in road vehicles are currently substantially more expensive than combustion engines of identical power. Synchronous machines cost approximately €50 kW_{mech}^{-1} including voltage converter, which means a fourfold higher price than for combustion engines [15]. The mass of currently available machines is ~1 $kg\,kW_{mech}^{-1}$.

Table 35.6 Characteristics of different machine types [2].

Legend		DC motor		AC motor				
				Synchronous motor				Asynchronous
		Electrically excited	Permanently excited	Electrically excited	Permanently excited	Transversal flux	Switched reluctance	
++	Very good							
+	Good							
0	Average							
.	Poor							
−	Very poor							
Power density		0	+	+	++	++	++	+
Reliability		0	+	+	+	+	++	++
Efficiency		−	.	+	++	++	++	0
Controllability		++	++	+	+	+	++	0
Overload capacity		+	+	+	+	+	+	+
Noise level		.	.	+	+	+	+	+
Thermal overload protection		.	.	+	++	+	+	+
Costs		0	0	0	.	−	+	+
Costs for the machine		−		.	.	−	++	+
Costs of the control		++		.	−	0	0	.
Development status		++		0	0	.	0	+

35.4.5
Cost Comparison of Vehicle Drivetrains

Market success is an important criterion of any new drivetrain technology. Its benchmark is usually the cost of a conventional powertrain. In view of the different possible layouts, analyses should be based on production costs, not end-user price, and refer to mechanical power output to retain comparability in terms of driving performance. In a second step, life-cycle costs also need to be assessed in order to draft reliable conclusions. Figure 35.24 shows the specific costs of conventional ICE, hybrid, and BEV drivetrains for a vehicle with $90\,kW_{mech}$. Component costs are those stated in the previous sections.

Unsurprisingly, ICE drivetrains are the cheapest by far, and BEV drivetrains are the most expensive, even with a reduced battery cost of €800 kWh^{-1} compared with €1200 kWh^{-1} for the hybrid electric vehicle. On the other hand, BEVs with a significant driving range (500 km or more; e.g., 10 kWh per 100 km and 50 kWh) are unlikely to be of market relevance in the near future. Smaller vehicles will play a role in commuter and other short-distance transport. Full-range BEVs will not be competitive at these costs even if electricity is available free, because the ICE's cost benefit of €42 700 here translates into more than 25 000 l of petrol at current prices, which is easily enough for a travel distance of 300 000 km or more. Battery costs of €200 kWh^{-1} lead to drivetrain costs of about €160 kW_{mech}^{-1} [3].

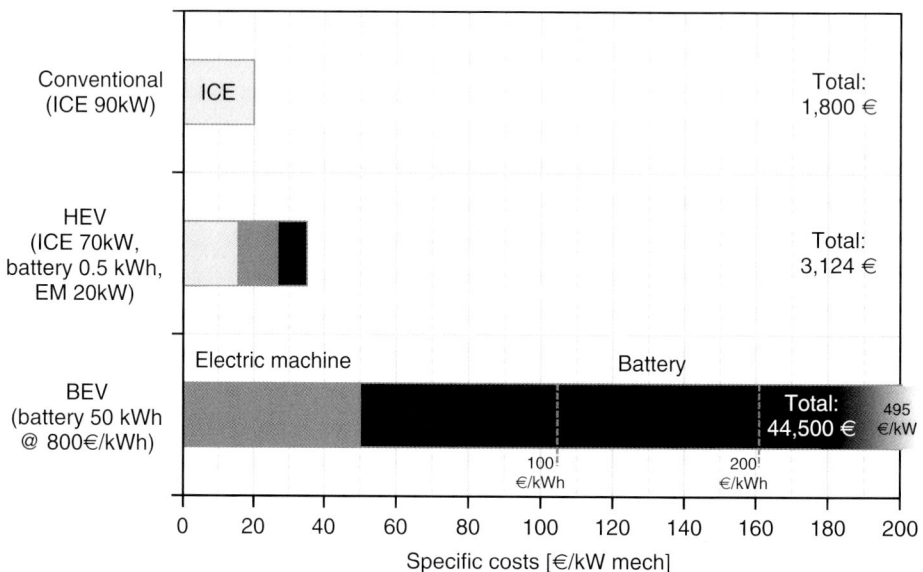

Figure 35.24 Specific costs of ICE, hybrid, and battery electric drivetrains.

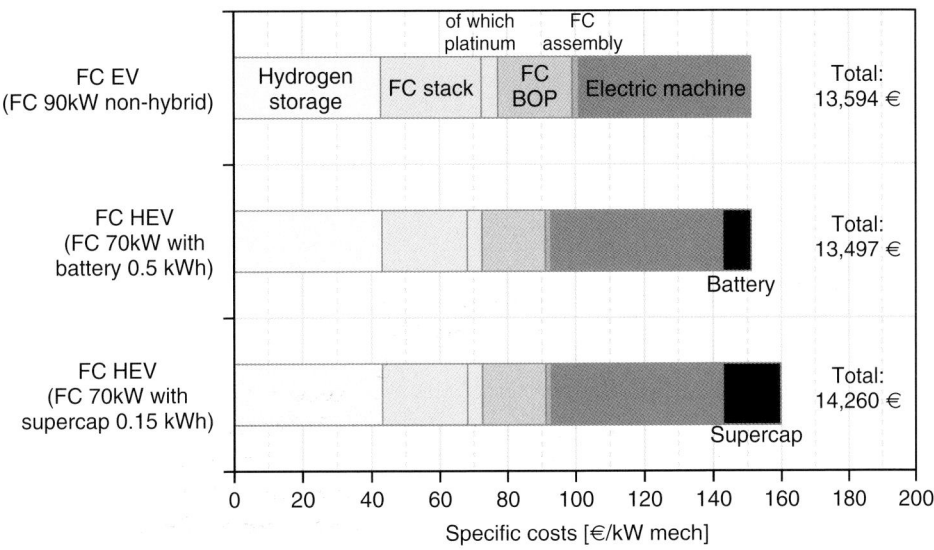

Figure 35.25 Specific costs of fuel cell drivetrains [3].

Hybrid electric drivetrains have good potential to compensate their higher drivetrain costs with lower fuel consumption and thus possibly lower total cost of ownership, provided that the fuel economy is superior to that of the conventional counterpart and the battery system has a competitive durability.

Fuel cell vehicle drivetrains are also substantially more expensive than conventional and hybrid drivetrains, as seen in Figure 35.25. All of the concepts shown feature a 90 kW electric machine to propel the vehicle and a 700 bar tank holding about 5.5 kg of hydrogen. Hybrid versions include either a lithium ion battery or a double-layer capacitor unit (supercap).

The cost of the non-hybrid drivetrain consists of one-third each of the cost of the tank, the electric machine, and the FC system, adding up to ~€150 kW_{mech}^{-1} or ~€13 600 in total for the drivetrain. By using a smaller stack with 70 kW nominal power and integrating a battery while using the same electric motor as before, the total cost remains almost the same, but with the added benefit of hybrid functionality (e.g., braking energy regeneration) with better energy efficiency. Therefore, a fuel cell hybrid electric vehicle will be cheaper to run than a non-hybrid fuel cell vehicle, provided that it has better fuel economy and the battery has acceptable durability. Whether it can compete with an ICE drivetrain in view of total costs depends on its efficiency, the cost of hydrogen fuel, the amortization driving distance, and other factors [3]. The same applies to the fuel cell hybrid electric drivetrain with an added supercap, whereas it must be significantly more efficient than the battery version because it is more expensive. In passenger vehicles, this aspect seems difficult to achieve due to the higher mass even when using small supercap systems.

As a conclusion to the cost analysis, it can be said that hybridization is essential for fuel cell drivetrains. It allows the use of smaller FC systems without compromising performance, and it can be beneficial to the competitiveness compared with conventional and other new drivetrain technologies.

35.5
Conclusion

Individual mobility is an indisputable asset of free and developed society. Road vehicles have played a vital role ever since they became affordable in the early twentieth century. Drivetrains based on combustion engines permit low costs and use fairly inexpensive fuel. For many reasons, the development of new propulsion technologies has become necessary and electric drivetrains – either single or hybrid configurations – are deemed promising concepts for future mobility.

Battery-electric drivetrains have been widely investigated and enjoy strong support among policy makers and electricity providers. However, it is mostly accepted that they will not be suitable for full-range vehicles that can drive long distances with only short refueling breaks. Other solutions are needed to overcome this bottleneck. Hydrogen fuel cells can act as an on-board electricity generator for electric vehicles and solve the range issue. Despite this, several other problems persist that need to be addressed prior to large-scale market introduction of hydrogen fuel cell vehicles.

First, the hydrogen infrastructure is not in place. There are fleet operators in some cities having access to hydrogen fueling stations, but public stations are rare. Naturally, hydrogen has to be affordable and ideally should come from a renewable source, although this is not required at the beginning.

From the technological point of view, the most important aspect is to bring drivetrain costs down, as fuel economy benefits take very long to compensate surplus costs compared with ICE hybrid drivetrains. The amortization can be accelerated by using hybrid drivetrain structures with an additional battery (or supercap) to increase drivetrain efficiency.

Components that are critical for success include the hydrogen tank (pressure vessels) and the electric machines, which can be responsible for two-thirds of fuel cell drivetrain costs. Production and assembly processes can lead to learning effects and possible cost reductions. Fuel cell systems need to be optimized for their automotive application, for example, in terms of durability, packaging and start-up behavior.

Finally, market success is not only a technological issue, but also depends on customer acceptance. This can only be achieved if customers feel comfortable and safe driving fuel cell electric vehicles and retain to their individual mobility without spending a fortune. If this can be achieved, hydrogen fuel cells will become an important item in future propulsion systems.

Acknowledgment

All photographs by B. Gnörich, if not stated otherwise.

References

1. Eckstein, L. (2011) Longitudinal Dynamics of Vehicles, Schriftenreihe Automobiltechnik, Aachen.
2. Eckstein, L. (2010) Alternative Vehicle Propulsion Systems, Schriftenreihe Automobiltechnik, Aachen.
3. Gnörich, B. (2010) Vergleichende Gesamtkostenanalyse von Brennstoffzellenfahrzeugen (Comparative life cycle cost analysis of fuel cell vehicles). Dissertation, RWTH Aachen, Schriftenreihe Automobiltechnik, Aachen.
4. Sauer, D. (2007) Electrochemical storage systems: tutorial "Propulsion systems for hybrid and fuel cell electric vehicles". Presented at the 12th European Conference on Power Electronics and Applications, Aalborg, September 2, 2007.
5. Better Place (2011) Better Place Unveils Network Deployment Roadmap for Israel, Offering Electric Car Drivers Complete Nationwide Coverage by End of Year. Press release.
6. General Motors (2001) Brennstoffzellen-Studie "HydroGen3" auf dem Weg zur Serienreife. Press release.
7. Ford Motor Company (2002) Ford Focus FCV Hybrid: Bringing Together Propulsion Systems. Data Sheet.
8. Paul Scherer Institute (2002) Unterwegs mit 40 Prozent weniger Energie: PSI-Wasserstoff-Auto am Erdgipfel in Johannesburg. Press release.
9. Proton Motor Fuel Cell (2009) 12m City Bus with PM Fuel Cell Triple Hybrid Drive. Product data sheet.
10. Strubel, V. (2008) StorHy – Hydrogen Storage Systems for Automotive Application. Final Report of EC Project No. 502667 "StorHy", October 2008.
11. Wehrhahn, J. (2009) Kosten von Brennstoffzellensystemen auf Massenbasis in Abhängigkeit von der Absatzmenge. Dissertation, RWTH Aachen.
12. Winkler, W. (2002) Brennstoffzellenanlagen. Springer, Hamburg, I.
13. Carlson, E., Kopf, P., Sinha, J., Sriramulu, S., and Yang, Y. (2005) Cost Analysis of PEM Fuel Cell Systems for Transportation. NREL/SR-560-39104, TIAX LLC, Cambridge, MA.
14. Oliva, P. (2009) Michelin arbeitet seit zwölf Jahren am neuen Rad. Interview, Automobiltechnische Zeitschrift (ATZ), March 2009.
15. Gnörich, B. (2009) Result of expert discussion. HySYS Technical Workshop, Aachen, May 2009.

36
Hybridization for Fuel Cells
Jörg Wilhelm

36.1
Introduction

In daily life, the demand for energy is rising continuously, since ongoing availability by cell phone and mobility by car is important for many users. As a consequence, the concentrations of greenhouse gases are rising. Therefore, reducing energy consumption or producing energy more efficiently is an important objective. One promising approach involves fuel cells, as they produce energy with high efficiency and low emissions. As will be shown below, fuel cells could be a substitute for internal combustion engines (ICEs) in passenger cars or for batteries in light traction applications. In the latter application, fuel cells can prove their superiority, as they have a higher energy density than batteries. However, they also have some limitations, which will be shown later in this chapter. As a consequence, they have to be hybridized in some cases. The reasons for hybridization and different possible fuel-cell hybrids will be described here. The chapter concludes with a technical overview in which technical details of different fuel-cell hybrids for different applications are presented.

First, it is important to know the origin of the word *hybrid*: it has a number of different meanings and is used in a wide range of areas of daily life. In English, the word hybrid can be used as both a noun and an adjective with the following meanings [1]:

- thing made by combining two different elements (noun)
- of mixed character (adjective).

The origin of the word hybrid is the Latin word *hibrida*, which can be translated as "a crossbred animal" [2]. The meaning of the English word is derived from this Latin meaning. The word hybrid is used in several fields of daily life with different meanings. Table 36.1 gives some examples.

Hybridization for fuel cells refers to the meaning of hybrid as a hybrid energy system, which is a system that combines two or more forms of energy or power to provide a particular energy service [4]. A fuel-cell hybrid is a special form of a hybrid energy system. Relating to vehicles, a fuel-cell hybrid is an electric vehicle

Table 36.1 Use of the word hybrid [3].

Field	Meaning
Biology	An offspring resulting from cross-breeding
Electronics	A computer combining analog and digital features
Linguistics	A word derived from more than one language

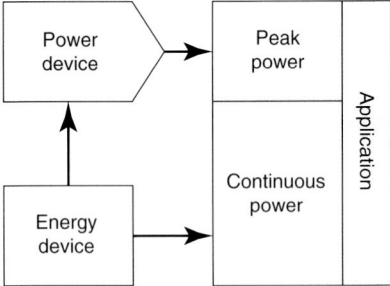

Figure 36.1 Energy and power source in a fuel-cell hybrid. Reproduced from [6].

in which a fuel cell and at least one energy storage device are used for traction [5]. The definition of a fuel-cell hybrid for electric vehicles can also be transferred to other applications. Each fuel-cell hybrid system consists of an energy source and a power source. The devices shown in Figure 36.1 have different functions [6]:

- energy source → continuous power
- power source → peak power.

36.2
The Fuel-Cell Hybrid

36.2.1
Reasons for Hybridizing a Fuel Cell

In a fuel-cell hybrid, the fuel cell is combined with an energy storage system. The reasons for hybridizing a fuel cell and the benefits derived from this can be explained with the help of a characteristic driving cycle. An example is shown in Figure 36.2. The application is a horizontal order picker [7]. The peak power for acceleration ($P_{motor} > 0$ W) is about three times higher than the constant power. The negative power is generated during braking. This driving cycle is characterized by a repetition of power peaks for acceleration, lower continuous power consumption for driving, and negative power peaks during braking.

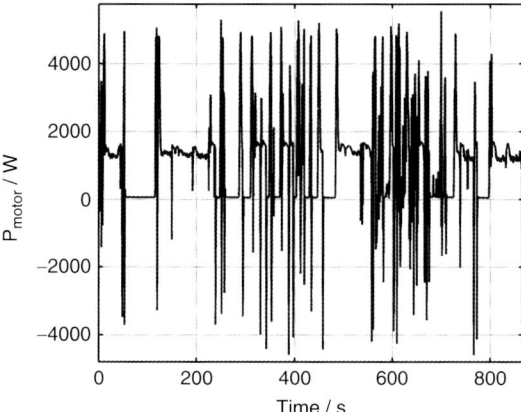

Figure 36.2 Driving cycle of a horizontal order picker. Reproduced from [7].

For the energy supply in an electric vehicle corresponding to the driving cycle in Figure 36.2, a pure fuel-cell system is not sufficient, as the power response time of a fuel cell is not quite enough and regenerative energy recovery during braking is not possible [8]. These two disadvantages of fuel cells constitute the main reasons for hybridizing them. Especially braking energy recovery improves the fuel economy [9].

When the power demand of a fuel-cell hybrid system is very high, the peak power will be supplied by the energy storage system. The fuel cell will be operated on average power, which is lower than the peak power. In comparison with a pure fuel-cell system, the size of the fuel cell could be reduced, thus reducing costs [10].

Normally, a fuel cell needs some time to reach the operating temperature after start-up. During this heating procedure, the power output of the fuel cell is limited. To cover the power consumption of the auxiliary components, and also the power demand of the load, the fuel cell has to be hybridized with an energy storage system [11].

Apart from these reasons for hybridizing, which also constitute the main advantages of a fuel-cell hybrid system, there are also some disadvantages [10]:

- increase in system complexity
- higher system weight
- more complex control system
- extra costs for the energy storage system.

36.2.2
Different Types of Fuel-Cell Hybrids

36.2.2.1 Series and Parallel Hybrids
Today, the word "hybrid" generally refers to hybrid vehicles with an ICE. These vehicles are already on the market. An overview of different hybrid vehicles can be found elsewhere [12]. The following two basic types exist:

- series hybrid
- parallel hybrid.

The architecture of a series hybrid and a parallel hybrid is depicted in Figure 36.3. The main difference between these two hybrids is the coupling of the ICE and the energy storage system (ESS). In a series hybrid, the coupling is done electrically via a DC bus [13]. Thus, the mechanical power output of the ICE has to be converted to electrical power via a generator (G). In contrast, the coupling in a parallel hybrid is carried out mechanically via a gear box [13]. In this case, the electrical power output of the ESS is converted to mechanical power via an electric motor (EM). In both cases, the fuel for the ICE is stored in a tank (T).

To apply this model to a fuel-cell hybrid vehicle, some components in Figure 36.3 are substituted as follows:

- series hybrid: tank (T) + internal combustion engine (ICE) + generator(G) → tank (T) + fuel cell (FC) + electric motor (EM)
- parallel hybrid: tank (T) + internal combustion engine (ICE) → tank (T) + fuel cell (FC) + electric motor (EM).

The resulting series hybrid and parallel hybrid for fuel-cell hybrid vehicles are shown in Figure 36.4. The setup of a series fuel-cell hybrid is very simple. The fuel cell and the energy storage system are connected here via a DC bus. In the case of a parallel fuel-cell hybrid, two electric motors and a gear box are needed for the coupling of fuel cell and energy storage system. This is why a fuel-cell hybrid is referred to as a series hybrid [15]. This definition is also used here in this chapter. In a series fuel-cell hybrid, the energy storage system assists the fuel cell with additional power output. The energy storage system is recharged via the fuel cell or by recuperation of braking energy. If no recuperation of braking energy is possible, this is called a semi-hybrid [16]. In this case, the energy storage system will be recharged only by the fuel cell.

36.2.2.2 Active and Passive Hybrids

It was shown that in a series fuel-cell hybrid, the fuel cell and the energy storage system are connected via a DC bus. This connection can be done actively or passively

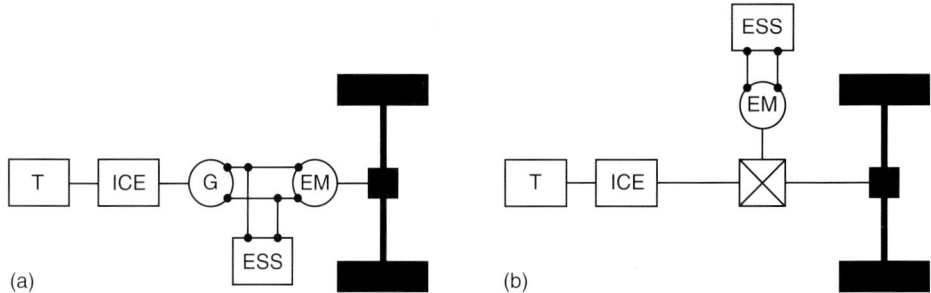

Figure 36.3 Series hybrid (a) and parallel hybrid (b) for an ICE hybrid vehicle. Reproduced from [14].

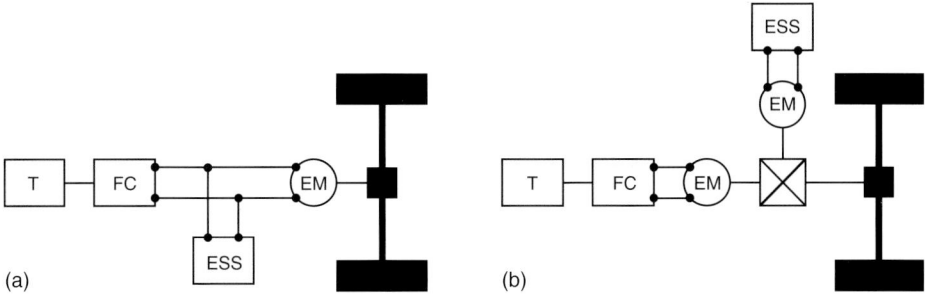

Figure 36.4 Series hybrid (a) and parallel hybrid (b) for fuel-cell hybrid vehicle. Reproduced from [14].

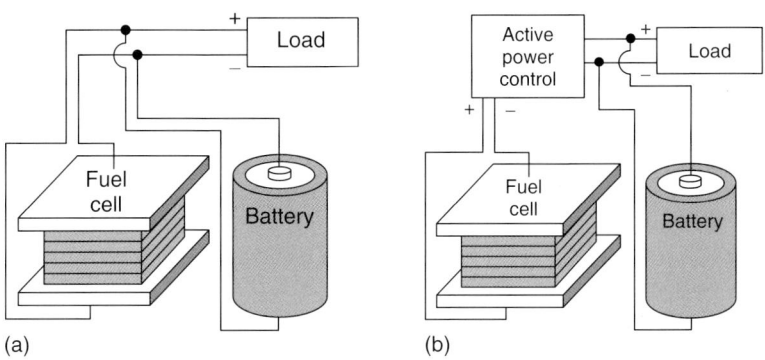

Figure 36.5 Passive (a) and active fuel-cell hybrid (b). Reprinted from [18] with permission from Elsevier.

[17]. This classification depends on whether or not power electronic components are used to control the power flow between the fuel cell and the energy storage system [18]. Figure 36.5 shows both configurations with a battery used for the energy storage. In a passive fuel-cell hybrid, the fuel cell and the energy storage system are connected directly without any active power control. The active fuel-cell hybrid has this type of active power control between the fuel cell and the energy storage system. The active power control may be a DC–DC converter, for example, which will be explained later.

The main characteristics of a passive fuel-cell hybrid and an active fuel-cell hybrid are shown in Table 36.2. Especially the voltage level of the fuel cell is very important, since a too low fuel cell voltage causes catalyst corrosion [19]. If the fuel cell and the energy storage system are coupled directly in the case of the passive hybrid, voltage limitation for degradation protection is not possible.

Each of the described topologies has several advantages and disadvantages. To compare them, a rating matrix is used, which is shown in Table 36.3. The matrix

Table 36.2 Characteristics of a passive fuel-cell hybrid and an active fuel-cell hybrid [17, 18].

Criterion	Passive hybrid	Active hybrid
Voltage level	Voltage level of the fuel cell and the energy storage system are equal	Different voltage level of the fuel cell and the energy storage system
Power flow control	Control of the power flow between the fuel cell and the energy storage system is not possible	Active control of the power flow between the fuel cell and the energy storage system
Pulse load protection	No pulse load protection of the fuel cell	Fuel cell is protected from the pulse load through the power converter

Table 36.3 Rating matrix for a pure fuel-cell system and a passive/active fuel-cell hybrid[a].

Criterion	Fuel-cell system	Passive hybrid	Active hybrid
Number of components	+	0	−
Overall system losses	+	0	−
Simplicity of the control system	+	+	0
Flexibility in system operation	0	0	+
Dynamics of the fuel cell	−	0	+
Lifetime of the fuel cell	−	0	+

[a] +, Best; 0, second best; −, worst.

compares a passive hybrid and an active hybrid with a pure fuel-cell system based on [20]. In contrast to a pure fuel-cell system, an active fuel-cell hybrid requires a DC–DC converter and an energy storage system as additional components, both of which have additional system losses. Although increasing the number of components increases the flexibility in the system operation, it also makes the entire control system more complicated. In an active fuel-cell hybrid, the operation of the fuel cell can be controlled independently of the operation of the energy storage system or the load. In the case of a pure fuel-cell system, the fuel cell is directly connected to the load. As a consequence, the fuel cell is operated very high dynamically. When coupling the fuel cell with an energy storage system with or without a DC–DC converter, the dynamics decrease. It has been reported that the fuel-cell current slope must be limited to $4\,\text{A}\,\text{s}^{-1}$ to prevent fuel starvation [21]. This may influence the lifetime. An active hybrid could overcome this.

36.2.3
Hybridization Degree

To compare different hybridization concepts for fuel cells, a characteristic number is needed. To establish this number, the hybridization degree (*HD*) is defined [22]:

$$HD = \frac{P_{ESS,max}}{P_{FC,max} + P_{ESS,max}} \qquad (36.1)$$

where $P_{ESS,max}$ is the maximum power of the energy storage system and $P_{FC,max}$ is the maximum power of the fuel cell. The hybridization degree is defined between 0 for a pure fuel-cell system and 1 for a pure energy storage system.

The application of Eq. (36.1) is shown with the help of an example described in [23], where a fuel-cell hybrid system for electric aircraft was designed. The distribution of fuel-cell power and battery power as a function of the hybridization degree is shown in Figure 36.6. Two horizontal lines indicate the power required for takeoff and the power required for a steady level flight. The sum of the fuel cell and the battery power equal the power required for takeoff for each hybridization degree. The hybridization degree in this case is about 45%.

36.3
Components of a Fuel-Cell Hybrid

As has been defined, a fuel-cell hybrid consists of a fuel cell with an energy storage system and optionally a DC–DC converter for power control. The fuel-cell system with all peripheral components such as pumps and blowers is controlled, in addition to the power distribution via the DC–DC converter. In the following, the main components of a fuel-cell hybrid system are described.

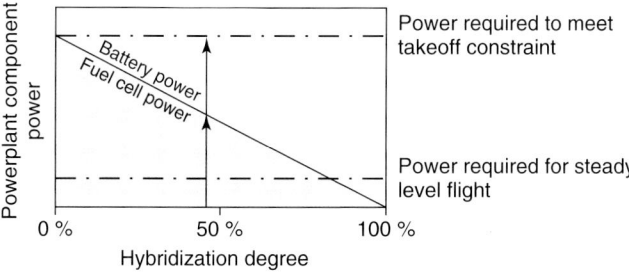

Figure 36.6 Trade-off between hybridization degree and power requirements of a fuel-cell hybrid aircraft. Reproduced from [23].

36.3.1
Fuel Cell

Fuel cells are used because they are efficient, simple, emission-free, and quiet [24]. Depending on the operating temperature, the following fuel-cell types exist [24]:

- direct methanol fuel cell (DMFC): 20–90 °C
- proton exchange membrane fuel cell (PEMFC): 30–100 °C
- alkaline fuel cell (AFC): 50–200 °C
- phosphoric acid fuel cell (PAFC): ∼220 °C
- molten carbonate fuel cell (MCFC): ∼650 °C
- solid oxide fuel cell (SOFC): 500–1000 °C.

Fuel cells are mainly used for combined heat and power (CHP), mobile power systems, vehicles, portable computers, mobile phones, and military communications equipment [24]. An overview is given in Figure 36.7.

36.3.2
Energy Storage

For energy storages systems, two main types exist:

- batteries: lead acid batteries, NiMH batteries, and lithium ion batteries
- supercapacitors.

The main difference between batteries and supercapacitors becomes clear in the Ragone plot in Figure 36.8. They mainly differ in their energy density and power density. Whereas batteries could have high energy density, and also high power density in the case of lithium ion batteries, supercapacitors only have high power density.

Figure 36.8 also shows that fuel cells are characterized by higher energy density than energy storage systems. Their main disadvantage is their low power density.

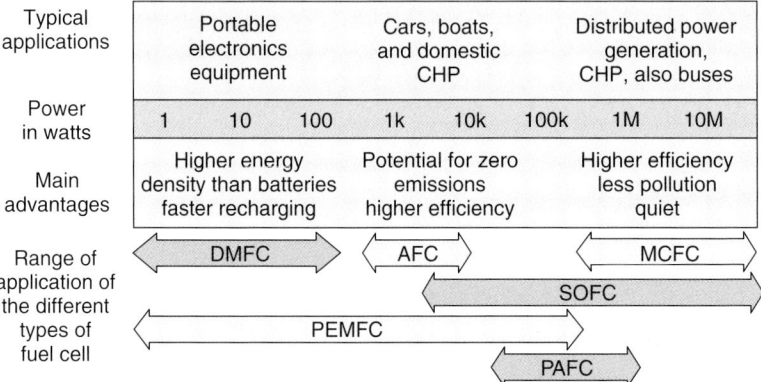

Figure 36.7 Different fuel-cell types for different applications. Reprinted from [24] with permission from John Wiley & Sons, Ltd.

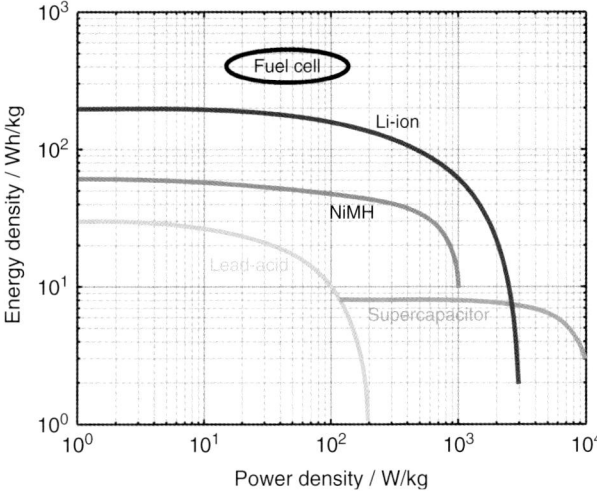

Figure 36.8 Ragone plot. Reproduced from [25].

Combining a fuel cell with an energy storage system results in longer operating times and better peak power performance compared with a pure fuel cell or energy storage system.

36.3.3
Power Electronics

In active fuel-cell hybrid systems, the coupling between fuel cell and energy storage system is done via power electronic devices. The main elements are DC–DC converters. The following basic types exist [26]:

- step-down converter (= buck converter)
- step-up converter (= boost converter)
- step-down/step-up converter (= buck–boost converter)
- full-bridge converter.

All of these types differ in the ratio of the DC–DC converter input voltage $U_{DCDC,in}$ to the output voltage $U_{DCDC,out}$, and also the direction of the power flow. For $U_{DCDC,in} > U_{DCDC,out}$ a step-down converter is used. The input voltage is stepped down to a lower voltage, whereas the output current $I_{DCDC,out}$ increases. If $U_{DCDC,in} < U_{DCDC,out}$, a step-up converter is used. To cover both cases, a combination is used. The buck–boost converter is a cascade connection of a step-down converter and a step-up converter [26]. For these three types, the power flow is only in the direction from input to output. If a bidirectional power flow is needed, a full-bridge converter is used. For this type, the magnitude and polarity of the output voltage $U_{DCDC,out}$, and also the magnitude and direction of the output current

$I_{DCDC,out}$, can be controlled [26]. The simplified equivalent circuits of the types described above are shown in Figure 36.9. The main elements of all converters are capacitors, inductors, diodes, and switches. In the simplified equivalent circuits, a capacitor is represented by an equivalent voltage source and an inductor is represented by an equivalent current source [26].

36.3.4
Control Unit

For controlling fuel-cell hybrid systems, two different types of controllers are used: a fuel-cell controller and a DC–DC controller. An example of a control structure is shown in Figure 36.10. The first controller is used to control the air and fuel supply

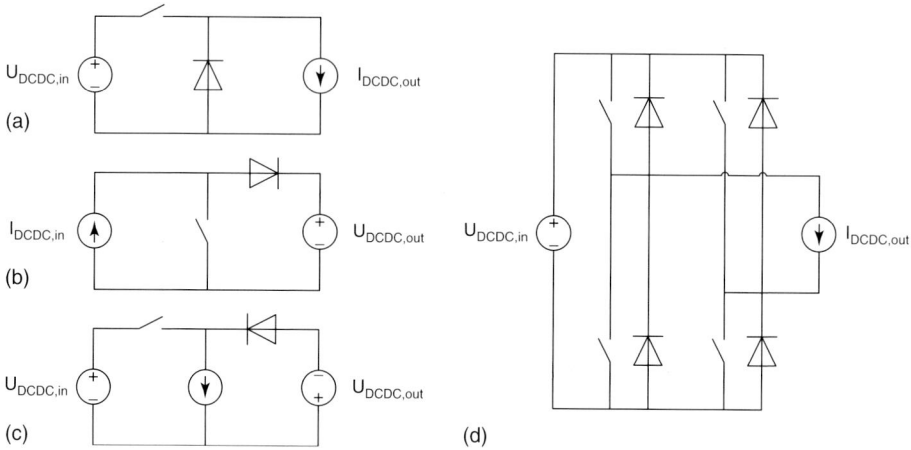

Figure 36.9 Equivalent circuits of different DC–DC converters. (a) Step-down; (b) step-up; (c) step-down/step-up; (d) full bridge. Reproduced from [26].

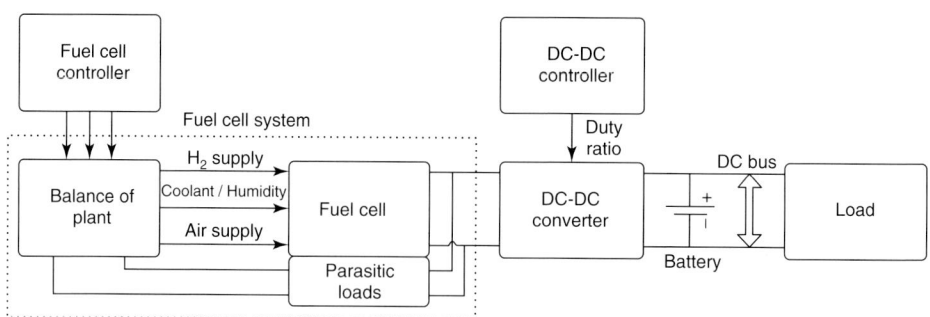

Figure 36.10 Block diagram of a fuel-cell hybrid system. Reprinted from [27] with permission from IEEE.

of a fuel cell. The second controller is used in case of an active fuel-cell hybrid to control the power flow between the fuel cell and the energy storage system.

An example of a DC–DC controller is shown in Figure 36.11. The pulse-width modulation (PWM) switching signals for the two DC–DC converters are calculated here with two PI controllers and a fuzzy logic controller. Controller input values are fuel-cell voltage and current, the state of charge (SOC) of the supercapacitors, the DC bus voltage, and the power demand of the drive train. The control variable is the DC bus voltage.

36.4 Hybridization Concepts

36.4.1 Overview

The classification of possible hybridization concepts will be made according to [7]. First, basic types for fuel-cell hybrids are defined. Second, these basic types are combined with additional elements to obtain further hybridization concepts. The following assumptions are taken into consideration:

- only one fuel cell
- minimum of one energy storage system and maximum of two different energy storage systems
- maximum of one DC–DC converter per fuel cell or energy storage system.

Figure 36.12 gives an overview of series fuel-cell hybrids that fulfill these assumptions. In the first step, they are divided into two groups: passive fuel-cell hybrids and active fuel-cell hybrids. Active series hybrids may have between one and three DC–DC converters. In the case of two or three DC–DC converters, they can be arranged in a parallel fashion or in a cascade connection.

36.4.2 Basic Types

Fulfilling the assumptions described above, the maximum stage of expansion for a fuel-cell hybrid has two energy storages and three DC–DC converters. In a first step, a fuel-cell system is combined with one energy storage system and up to two DC–DC converters. This results in four basic types of fuel-cell hybrids [7]. Block diagrams of these basic types are shown in Figure 36.13.

In Figure 36.13, four elements A, B, C, and D are also marked. These are the basic combination elements used to make a fuel-cell hybrid system out of a pure fuel-cell system. In the next step, these elements are combined with the basic types in Figure 36.13 to obtain a second stage of fuel-cell hybrid systems.

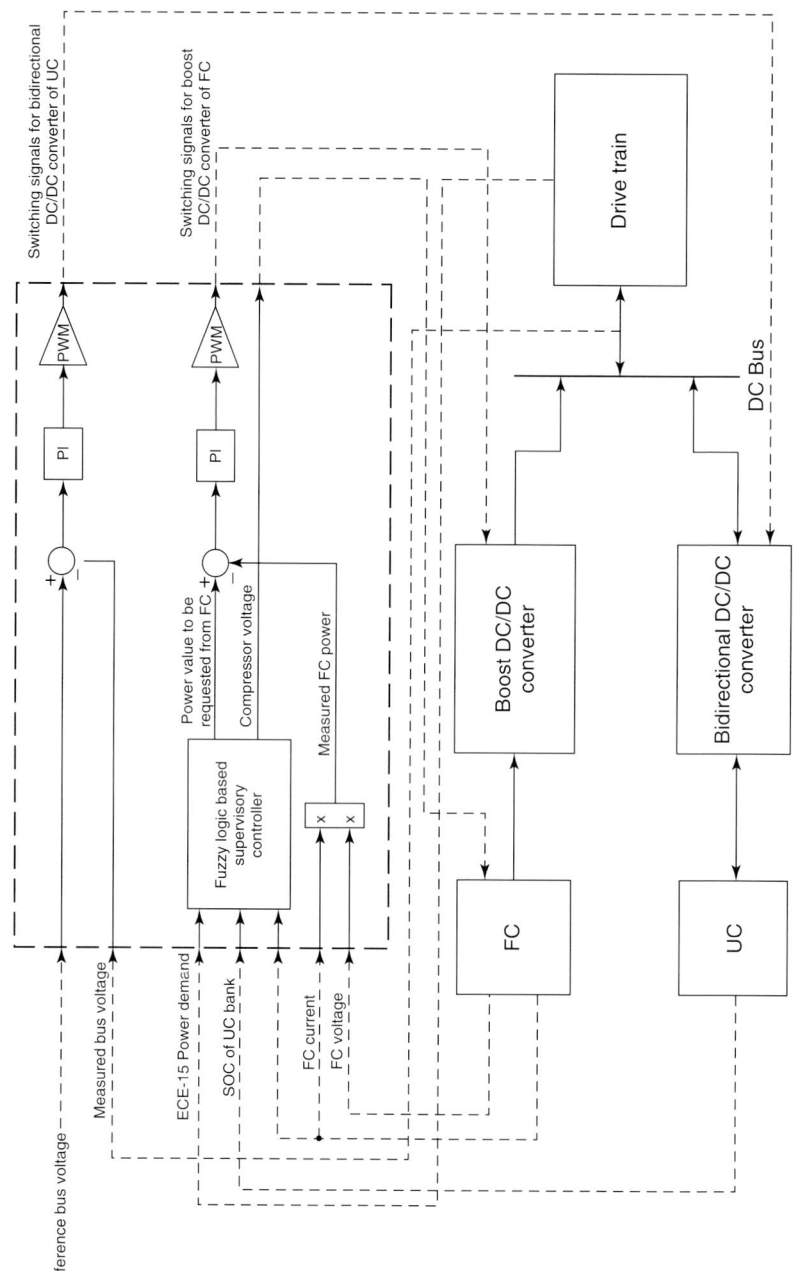

Figure 36.11 Controller for a fuel-cell hybrid system. Reprinted from [28] with permission from Elsevier.

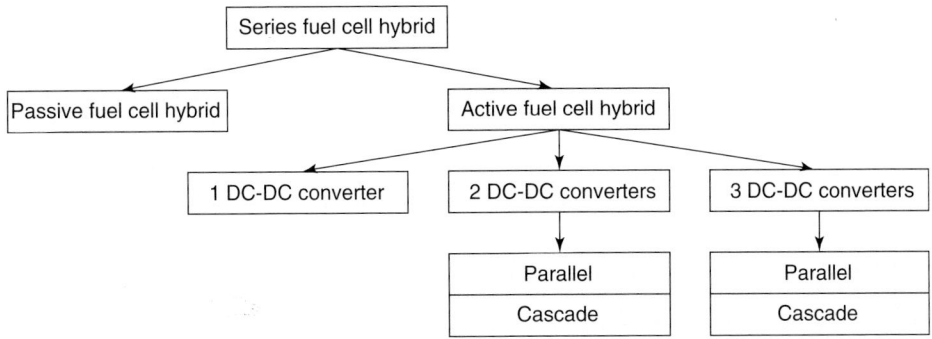

Figure 36.12 Concept overview of series fuel-cell hybrids. Reproduced from [7].

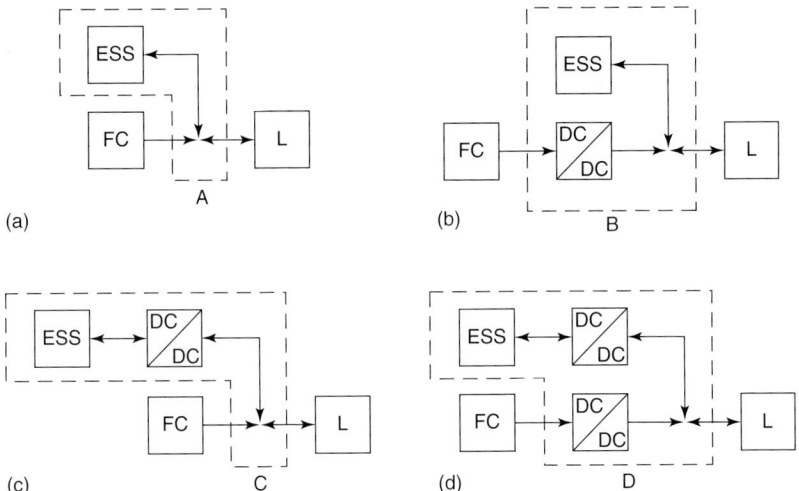

Figure 36.13 (a–d) Basic types for fuel-cell hybrids. FC, fuel cell; ESS, energy storage system; DC/DC, DC–DC converter; L, load. Reproduced from [7].

36.4.3
Possible Concepts

All possible hybridization concepts are derived from a pure fuel-cell system in two steps. This procedure is summarized in Figure 36.14. The rows of the matrix define the number of energy storage systems. The columns indicate the number of DC–DC converters used. In the first row in the first column, the pure fuel-cell system is indicated with the number 0.0 (first number, row; second number, column). The combination elements A, B, C, and D in Figure 36.13 are indicated in this matrix with arrows. Combining the pure fuel-cell system with one or two of these combination elements leads to 18 possible hybridization concepts for fuel cells.

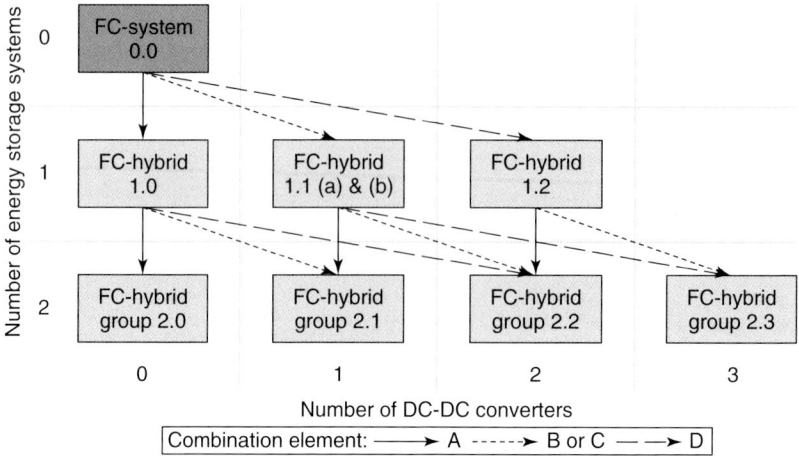

Figure 36.14 Matrix with possible fuel-cell hybrids. Reproduced from [7].

An overview of possible fuel-cell hybrids is presented in Table 36.4, indicating which initial concept is combined with which combination element. The resulting final concepts are enumerated according to the matrix in Figure 36.14. The basic types in Figure 36.13 result from the first combination step. For concept 1.2, only combination elements A and C are possible. Using combination element B will lead to a concept with two DC–DC converters in a row. They could be combined to make a single DC–DC converter. Combination element D is also not possible, as this would result in four DC–DC converters. Table 36.4 also indicates whether the final concept is a passive or an active hybrid or a mixture of both.

36.5
Technical Overview

This technical overview provides a short summary of hybridization concepts for different applications. They are characterized according to the fuel cell type used, the energy storage system used, and the hybridization concept. The concepts are numbered according to Table 36.4. To compare the different concepts, the defined hybridization degree [see Eq. (36.1)] is calculated, if possible.

36.5.1
Fuel-Cell Powertrains

The limited supply of fossil energy resources means that new automotive concepts are essential. Electric vehicles are a promising alternative, especially electric vehicles with a fuel cell, which have a number of advantages [34]. This technical overview presents some examples of fuel cell-driven passenger cars and buses.

Table 36.4 Overview of possible fuel-cell hybrids [7].

Initial concept	Combination element				Final concept	Passive hybrid	Active hybrid	Figure	Ref.
	A	B	C	D					
0.0	X				1.0	X		36.13a	[29, 30]
		X			1.1 (a)		X	36.13b	[27, 30]
			X		1.1 (b)		X	36.13c	[9, 30]
				X	1.2		X	36.13d	[21, 31]
1.0	X				2.0	X		36.15	[29]
		X			2.1 (a)	X	X	36.16a	
			X		2.1 (b)	X	X	36.16c	
				X	2.2 (a)	X	X	36.17a	
1.1 (a)	X				2.1 (c)		X	36.16b	
		X			2.2 (b)		X	36.17c	
			X		2.2 (c)		X	36.17b	[8]
				X	2.3 (a)		X	36.18a	
1.1 (b)	X				2.1 (d)		X	36.16d	
		X			2.2 (d)		X	36.17d	
			X		2.2 (e)		X	36.17e	[22]
				X	2.3 (b)		X	36.18b	[32]
1.2	X				2.2 (f)		X	36.17f	
			X		2.3 (c)		X	36.18c	[33]

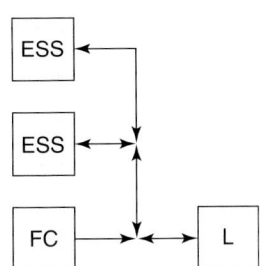

Figure 36.15 Fuel-cell hybrid group 2.0. Reproduced from [7].

36.5.1.1 Passenger Cars

The technical overview of passenger cars with fuel cells is presented in Table 36.5. All of the fuel cell passenger cars presented here use a PEMFC, as high power density is needed for this application. The fuel cell power is between 63 and 100 kW. The energy storage system is usually a lithium ion battery or an NiMH battery. A supercapacitor is only used once. Information about the hybridization concept is not provided in the literature. The hybridization degree, if it can be calculated, is between 0.17 and 0.30. For values lower than 0.50, the fuel cell delivers the majority of the traction power, for instance, during acceleration.

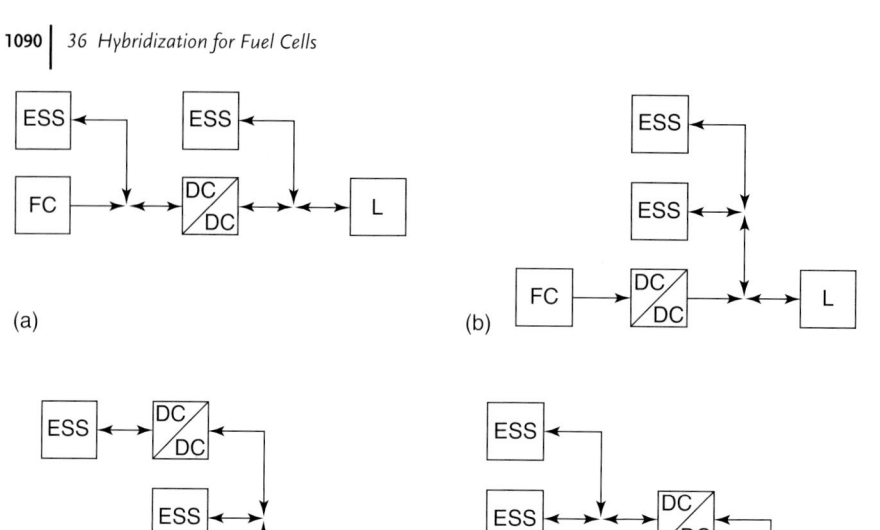

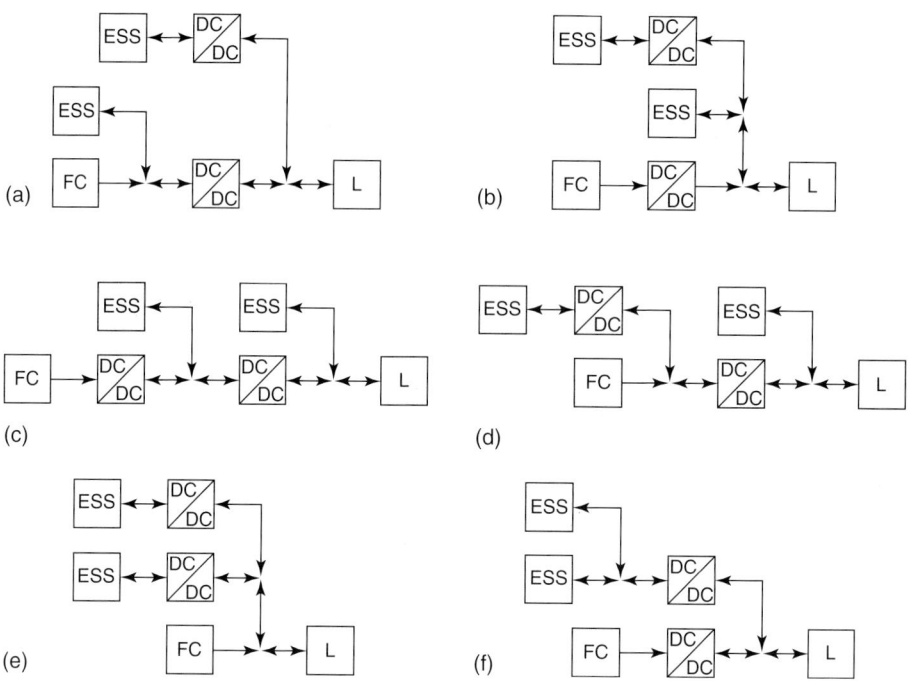

Figure 36.16 (a–d) Fuel-cell hybrid group 2.1. Reproduced from [7].

Figure 36.17 (a–f) Fuel-cell hybrid group 2.2. Reproduced from [7].

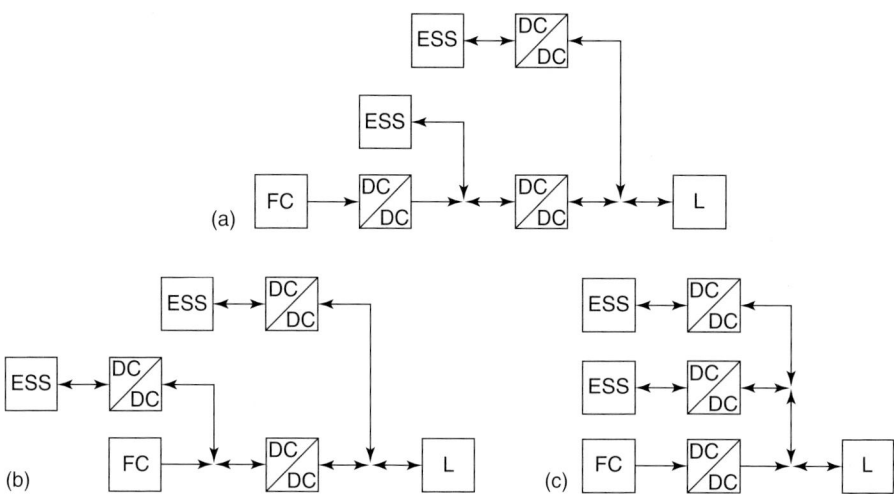

Figure 36.18 (a–c) Fuel-cell hybrid group 2.3. Reproduced from [7].

Table 36.5 Technical overview of passenger cars.

Model	Fuel cell	Energy storage system	Hybridization concept[a]	Hybridization degree[a]	Ref.
Daimler F-Cell B-Class	PEMFC 80 kW	Li ion 1.4 kWh, 35 kW	n/a	0.30	[35, 36]
Ford Focus FCV	PEMFC 72 kW	NiMH 20 kW	n/a	0.22	[37]
GM/Opel HydroGen4	PEMFC 93 kW	NiMH 1.8 kWh, 35 kW	n/a	0.27	[38]
Honda FCX Clarity	PEMFC 100 kW	Li ion	n/a	n/a	[39]
Hyundai Tucson ix FCEV	PEMFC 100 kW	Battery 21 kW	n/a	0.17	[40]
Kia Borrego FCEV	PEMFC 100 kW	Supercapacitor	n/a	n/a	[41]
Nissan X-Trail FCV	PEMFC 63 kW	Li ion	n/a	n/a	[42]
Toyota FCHV-adv	PEMFC 90 kW	NiMH 21 kW	n/a	0.19	[43, 44]
Volkswagen Tiguan HyMotion	PEMFC 80 kW	Li ion	n/a	n/a	[45]

[a] n/a, not applicable.

36.5.1.2 Buses

Some buses with fuel-cell hybrid systems are shown in Table 36.6. Similarly to passenger cars, a PEMFC is used. The fuel-cell power range is between 65 and 200 kW. Three concepts use a supercapacitor, in addition to a battery for the energy storage system. This is called a triple hybrid and corresponds to concepts 2.2 (c) and 2.3 (c). The coupling between fuel cell and energy storage systems is done here via either two or three DC–DC converters. The other concepts generally use a battery as the energy storage system. All of the hybridization concepts presented are active fuel-cell hybrids. The hybridization degree is higher than for passenger cars. A value of 0.6 for a concept means that the energy storage system delivers most of the traction power. This has to do with the fact that driving cycles of buses are characterized by a large number of acceleration and braking peaks. For passenger cars, the percentage of continuous driving is much higher.

36.5.2 Light Traction Applications

Mobile applications with a power range between 100 W and 5 kW are called light traction applications [51]. Similarly to automotive applications, fuel cells are also very attractive for these special vehicles as they are emission free and offer a very fast recharge [51]. The group of light traction applications includes scooters, forklift trucks, and commercial vehicles. The following technical overview gives some examples.

Table 36.6 Technical overview of buses.

Model	Fuel cell	Energy storage system	Hybridization concept[a]	Hybridization degree[a]	Ref.
Brazilian Hybrid Electric Fuel Cell Bus	PEMFC 77 kW	Supercapacitor + Li ion	2.2 (c)	n/a	[46]
Daimler Citaro FuelCELL-Hybrid	PEMFC 120 kW	Li ion 26.9 kWh, 180 kW	n/a	0.60	[35]
Fuel-cell hybrid bus	PEMFC 65 kW	Supercapacitor + lead acid	2.2 (c)	n/a	[8]
H_2 Bus NRW	PEMFC 150 kW	Supercapacitor + NiMH	2.3 (c)	n/a	[47]
Hyundai Fuel cell bus	PEMFC 200 kW	Supercapacitor 100 kW	n/a	0.33	[48]
New Flyer H40 LFR	PEMFC 150 kW	Li ion	n/a	n/a	[49]
Van Hool A330 fuel cell	PEMFC 120 kW	Battery 17.8 kWh	1.1 (a)	n/a	[50]

[a] n/a, not applicable.

36.5.2.1 Scooters, Wheelchairs, and Electromobiles

Some examples of scooters, wheelchairs, and electromobiles with fuel-cell hybrid systems are shown in Table 36.7. In contrast to fuel-cell powertrains, the fuel-cell power capacity is lower (130 W to 2.3 kW). The fuel cell is either a PEMFC or a DMFC. Because of the high energy density of methanol, the DMFC is especially well suited for light traction applications, as it allows longer operating times [52]. Hybridization is done here only with batteries, as the energy stored in the batteries is needed for start-up of the fuel-cell system. High power density of the energy storage system is not as essential as it is for fuel-cell powertrains; high energy density is more important here. The coupling of the fuel cell and the battery is carried out here either directly, which is concept 1.0, or with a DC–DC converter, which is concept 1.1 (a). Because of their simplicity, these two concepts can be realized cheaply and easily in the limited space of these light traction applications. The hybridization degree could be calculated only once. The calculated value shows that the majority of the power required for acceleration comes from the battery.

36.5.2.2 Commercial Vehicles

A technical overview of commercial vehicles is presented in Table 36.8. The fuel cell used for these vehicles is the PEMFC with power capacities between 2.5 and

Table 36.7 Technical overview of scooters, wheelchairs, and electromobiles.

Model	Fuel cell	Energy storage system	Hybridization concept[a]	Hybridization degree[a]	Ref.
Electric vehicle	PEMFC 2 kW	Li ion	1.1 (a)	n/a	[53]
HYCHAIN Cargobike	PEMFC 250 W	Battery	n/a	n/a	[54]
HYCHAIN Scooter	PEMFC 2 kW	Battery	n/a	n/a	[55]
HYCHAIN Wheelchair	PEMFC 350 W	Battery	n/a	n/a	[56]
HYSYRIDER	PEMFC 300 W	Battery 7.2 Ah	1.1 (a)	n/a	[57]
JuMOVe	DMFC 1.3 kW	Li ion 972 Wh	1.0	n/a	[58]
JuMOVe 2nd	DMFC 1.6 kW	Lead acid	1.1 (a)	n/a	[59]
Pro-Power Gen To I	DMFC 600 W	Li ion 1 kW	n/a	0.63	[60]
Scoophin	DMFC 1 kW	Li ion 1 kWh	n/a	n/a	[61]
Start Lab EV	DMFC 130 W	Lead acid	1.0	n/a	[62, 63]
Suzuki Fuel-cell Scooter	PEMFC 2.5 kW	Li ion	n/a	n/a	[64]
Wheelchair	PEMFC 300 W	Lead acid	1.1 (a)	n/a	[65]
Yamaha FC-AQEL	PEMFC	Li ion	1.1 (a)	n/a	[66, 67]
Yamaha FC-Dii	DMFC 580 W	Li ion	1.1 (a)	n/a	[67, 68]

[a] n/a, not applicable.

Table 36.8 Technical overview of commercial vehicles.

Model	Fuel cell	Energy storage system	Hybridization concept[a]	Hybridization degree[a]	Ref.
Aircraft baggage tractor	PEMFC 8 kW	Lead acid 26 kWh	1.1 (a)	n/a	[69]
HYCHAIN utility vehicle	PEMFC 2.5 kW	Battery	n/a	n/a	[70]
HY.MUVE municipal vehicle	PEMFC 20 kW	Li ion 12 kWh	1.1 (a)	n/a	[71, 72]
Proton Motor HyRange	PEMFC 5 kW	Li ion 40 kWh	n/a	n/a	[73]

[a]n/a, not applicable.

20 kW. For hybridization, concept 1.1 (a) is used with batteries and a DC–DC converter between the fuel cell and battery. The batteries used have a high energy content, which is between 12 and 40 kWh. This is important for continuous driving of these vehicles. The hybridization degree could not be calculated with the data indicated in the literature.

36.5.2.3 Forklift Trucks

Fuel-cell hybrid systems are also very attractive for material handling applications as a replacement for batteries in electric-powered forklift trucks. The classification of forklift trucks is shown in Figure 36.19 and can be divided into three classes [74]:

- **class I**: three- or four-wheeled counterbalanced rider trucks
- **class II**: rider trucks for narrow aisle operation
- **class III**: end rider and walk-behind pallet jacks.

Some examples of forklift trucks with fuel-cell hybrid systems are shown in Table 36.9. Further information on end-users and the number of units sold can be found in [74]. The technical overview shows that there are a number of fuel-cell hybrid systems for forklift trucks. For class I and II forklift trucks, only PEMFCs are used. The fuel-cell power is between 8 and 30 kW. For class III, the fuel cell power is lower (830 W to 3.2 kW). For this class, DMFC systems also exist. DMFCs are especially well suited for class III because of the high energy density. Hybridization is performed with either supercapacitors or batteries.

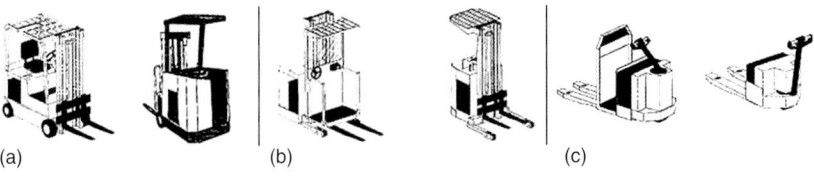

Figure 36.19 Forklift truck classification. (a): class I; (b) class II; (c) class III. Reprinted from [75] with permission.

Table 36.9 Technical overview of forklift trucks.

Model	Fuel cell	Energy storage system	Hybridization concept[a]	Hybridization degree[a]	Ref.
Class I					
H2 logic H2Drive	PEMFC 10 kW	Battery 8 kWh, 35 kW	n/a	0.78	[76, 77]
Hydrogenics HyPX 1-27 and HyPX 1-33	PEMFC 12.5 kW	Supercapacitor 30 kW	n/a	0.71	[78–80]
Nuvera C Series	PEMFC 10 kW	Li ion 40 kW (CM40) 48 kW (CM48)	n/a	0.80 (CM40) 0.83 (CM48)	[81]
Proton Motor	PEMFC 18 kW	Supercapacitor 38 kW	n/a	0.68	[82]
Plug Power GenDrive Series 1000	PEMFC 8 kW (1500, 1600, 1700) 10 kW (1600, 1700)	n/a	n/a	n/a	[83]
Toyota FCHV-F	PEMFC 30 kW	Supercapacitor	n/a	n/a	[84]
VTT	PEMFC 16 kW	Lead acid + supercapacitor	2.0	n/a	[85]
Class II					
Hydrogenics HyPX 2-21	PEMFC	Supercapacitor 22 kW	n/a	n/a	[78, 80]
Nuvera R Series	PEMFC 10 kW	Li ion 40 kW	n/a	0.80	[86]
Plug Power GenDrive Series 2000	PEMFC 8 kW (2300, 2400) 10 kW (2300, 2400)	n/a	n/a	n/a	[87]
Class III					
Forschungszentrum Jülich	DMFC 1.3 kW	Li ion 1.1 kWh, 6.8 kW	1.1 (a)	0.84	[7, 88]
Nuvera W Series	PEMFC	Battery 10 kW	n/a	n/a	[89]
OorjaPac Model III	DMFC 830 W	Lead acid	n/a	n/a	[90]
Plug Power GenDrive Series 3000	PEMFC 1.8 kW (3300) 3.2 kW (3300-D)	n/a	n/a	n/a	[91]
Pro-Power SS Drive	DMFC 1 kW	Li ion 5.9 kWh	n/a	n/a	[92]

[a] n/a, not applicable.

Supercapacitors are advantageous for class I and II, as they have high peak power performance. For class I, a triple hybrid corresponding to concept 2.0 is described. The other hybrid systems are realized with a DC–DC converter between fuel cell and energy storage system, which is concept 1.1 (a). For all systems, the hybridization degree is between 0.68 and 0.84, as the acceleration peak power delivered by the energy storage system is higher than the mean driving power delivered by the fuel cell.

36.5.3
Other Applications

In addition to the fuel-cell hybrid systems for powertrains and light traction applications described above, there are several other applications that use fuel-cell hybrid systems, some of which are described briefly below. These examples are only niche market applications. At present, the main application area for fuel-cell hybrid systems is mobile applications, as has been described.

An example is a fuel-cell hybrid system for maritime application. In [93] the first commercially used fuel-cell passenger ship (ZEMship) is described. The propulsion system for this ship consists of two 48 kW PEMFC stacks hybridized with 201.6 kWh lead acid batteries. Especially for maritime applications, the quiet and emission-free operation of fuel cells is an important factor.

A fuel-cell-powered hybrid electric aircraft was described in [23]. For this small aircraft, a 700 W PEMFC was hybridized with a battery and a DC–DC converter according to concept 1.1 (b).

As has been described, DMFCs are attractive for light traction applications with power capacities lower than 5 kW. Another low-power application is a humanoid robot, and an example was described in [94]. A DMFC stack of 405 W was coupled with a 200 Wh battery via a DC–DC converter.

Renewable energy, such as solar power and wind, is characterized by fluctuating power production depending on the weather conditions. To store excess energy or to deliver additional energy, renewable energy systems could be combined with fuel-cell hybrid systems. In [95] a photovoltaic panel is combined with a PEMFC supercapacitor hybrid system. Concept 1.2 was used here. Additionally, a water electrolyzer was connected to the DC bus to produce hydrogen from surplus energy.

36.6
Systems Analysis

The technical overview above presented several fuel-cell hybrid systems with respect to the fuel cells and energy storage systems used, and also the hybridization concept. In this section, some aspects regarding the systems analysis of these concepts are presented as examples.

In [96] two fuel-cell hybrid systems are analyzed with experiments on a test bench. The first hybrid system was concept 1.1 (a) with a battery, whereas the second hybrid system was concept 1.2 with a supercapacitor. It was shown that a fuel-cell hybrid with supercapacitors delivers better performance in terms of power assist for the fuel cell during transient power changes. Because of the higher energy density, the battery system has advantages during start-up of the fuel-cell system, which can be up to 10 min for a PEMFC.

A system analysis of different concepts with simulations was presented in [30]. Three hybridization concepts with supercapacitors or batteries were analyzed: concepts 1.0, 1.1 (a), and 1.1 (b). They were all compared with a pure fuel-cell system. The simulations showed that hybridization with supercapacitors significantly load-leveled the fuel cell operation. Fuel economy for the FUDS (Federal Urban Driving Schedule) cycle was improved up to 28% in case of supercapacitors. Because of the higher losses in the batteries, the improvement in fuel economy was only 6% for batteries. The highest fuel economy improvement was achieved for concept 1.0 with supercapacitors directly coupled to the fuel cell, as there are no additional losses for the DC–DC converter.

The fuel economy of a fuel-cell hybrid vehicle was compared with a fuel cell vehicle and ICE vehicle in [97]. The driving distance in miles per gallon was 1.5 times higher for a fuel-cell vehicle than an ICE vehicle. By hybridizing the fuel cell, the driving distance was 1.8 times higher. One reason for this is the recuperation of regenerative braking energy in the case of the fuel-cell hybrid.

The influence of energy recovery during braking was investigated in [9]. The tank to wheel (TTW) efficiency for an electric vehicle with a 120 kW fuel cell was calculated as 42% for the FUDS cycle. The same electric vehicle with a fuel-cell hybrid system consisting of a 65 kW fuel cell and a 55 kW battery had a TTW efficiency of 54%. This increase was achieved by using the braking energy.

An important criterion for the comparison of different fuel-cell hybrid systems is the maximum fuel-cell current ramp rate. Current changes that are too rapid could lead to fuel starvation, which would result in increased degradation of the fuel cell. In [85] a fuel-cell triple hybrid corresponding to concept 2.0 for a forklift truck is analyzed. For a characteristic driving cycle, the maximum fuel cell current ramp was 600 A s^{-1} in the case of a pure fuel-cell system. By hybridizing the fuel cell with only a battery according to concept 1.0, this could be reduced to about 200 A s^{-1}, which was also very fast. Only by combining the battery with supercapacitors was a further reduction achieved. For a 20 F supercapacitor, the calculated current ramp rate was 100 A s^{-1}. To obtain a current ramp rate of about 20 A s^{-1}, a 165 F supercapacitor was needed.

The difference between active and passive hybrids is the number of DC–DC converters. As passive hybrids realize the coupling of fuel cells and energy storage systems without any DC–DC converters, they constitute a cheap and simple solution. Simulations in [98] compared an active fuel-cell hybrid according to concept 1.1 (a) with a passive fuel-cell hybrid according to concept 1.0. In a passive hybrid, the voltages of fuel cell and battery are the same. Therefore, active power sharing is not possible. This can only be achieved by adjusting the operating pressure of the fuel cell. For a constant voltage, the fuel cell current decreases on decreasing the operating pressure in a PEMFC.

36.7 Conclusion

It was shown that for some applications, a fuel-cell system has to be hybridized with an energy storage system. The primary reasons for hybridization are the limited dynamic behavior of the fuel cell and the need for energy recovery during braking. The coupling of a fuel cell and energy storage system in this type of series hybrid can be undertaken with DC–DC converters in case of an active hybrid or without DC–DC converters in case of a passive hybrid. The passive fuel-cell hybrid is less complex and has lower overall losses, but the flexibility in system operation is higher for the active hybrid. Furthermore, the dynamics of the fuel cell are smaller, resulting in a longer lifetime of the fuel cell. For a fuel-cell hybrid system consisting of one fuel cell, up to two energy storage systems, and a maximum of one DC–DC converter per energy converter, there are 18 possible hybridization concepts. All of these concepts are based on four basic types. The derivation from these basic types was presented in the form of a matrix.

The technical overview presented characterized different fuel-cell hybrid systems for different applications according to the fuel cell type used, the energy storage system used, and the hybridization concept. Additionally, a classification number, referred to as the hybridization degree, was defined. By calculating the hybridization degree, it is possible to compare different fuel-cell hybrid systems with each other. At present, the main applications in which fuel-cell hybridization is used are mobile applications. The technical overview presented different concepts for fuel cell powertrains and light traction applications. For high-power applications ($>10\,kW$), the PEMFC is the fuel cell of the choice. For smaller power capacities, the DMFC has advantages because of the higher energy density. This is especially true for light traction applications. Hybridization is generally performed with batteries, as for some applications the energy stored in the batteries is required for fuel-cell start-up. For high acceleration power in buses and forklift trucks, supercapacitors may have advantages. As the control of the power flow between the fuel cell and the energy storage system is easier, most of the described concepts are realized with DC–DC converters.

In addition to the mobile applications described above, other applications such as planes, ships, and robots will become important in the future. Especially when used as a replacement for batteries, longer operating times can be achieved with a fuel-cell system. Fuel-cell hybrid systems are even an option for peak power shaving in the case of renewable energy such as solar or wind power.

References

1. OUP (2011) Oxford Dictionaries Online, http://oxforddictionaries.com/view/entry/m_en_gb0393270, Oxford University Press, Oxford (last accessed 28 March 2011).
2. Dictionary.com (2011) Hybrid http://dictionary.reference.com/browse/hybrid (last accessed 29 March 2011).
3. Wikipedia (2011) Wikipedia – The Free Encyclopedia http://en.wikipedai.org/

wiki/Hybrid (last accessed 29 March 2011).
4. Cleveland, C.J. and Morris, C. (2006) *Dictionary of Energy*, 1st edn., Elsevier, Oxford.
5. Treffinger, P., Gräf, M., and Goedecke, M. (2002) Hybridization of fuel cell powered drive trains. Presented at the VDI-Tagung "Innovative Fahrzeugantriebe", Dresden, October 2002.
6. Kötz, R., Dietrich, P., Hahn, M., and Büchi, F. (2005) Supercaps – properties and vehicle application. *VDI-Ber.*, **1874**, 175–188.
7. Wilhelm, J. (2010) *Hybridisierung und Regelung eines mobilen Direktmethanol-Brennstoffzellen-Systems*, Energy and Environment, Vol. 73, Schriften des Forschungszentrums Jülich, Jülich.
8. Gao, D., Jin, Z., and Lu, Q. (2008) Energy management strategy based on fuzzy logic for a fuel cell hybrid bus. *J. Power Sources*, **185**, 311–317.
9. Ahluwalia, R.K., Wang, X., and Rousseau, A. (2005) Fuel economy of hybrid fuel-cell vehicles. *J. Power Sources*, **152**, 233–244.
10. Jeong, K.S. and Oh, B.S. (2002) Fuel economy and life-cycle cost analysis of a fuel cell hybrid vehicle. *J. Power Sources*, **105**, 58–65.
11. Pesaran, A., Zolot, M., Markel, T., and Wipke, K. (2004) Fuel cell/battery hybrids: an overview of energy storage hybridization in fuel cell vehicles. Presented at the 9th Ulm Electrochemical Talks, Ulm, May 2004.
12. Wishart, J., Zhou, Y.L., and Dong, Z. (2008) Review of multi-regime hybrid vehicle powertrain architecture. *Int. J. Electr. Hybrid Veh.*, **1** (3), 248–275.
13. Gay, S.E., Gao, H., and Ehsani, M. (2002) Fuel cell hybrid drive train configurations and motor drive selection. *IEEE Veh. Technol. Conf.*, **56** (2), 1007–1010.
14. Maume, C. (2002) Systemanalyse und Simulation eines Brennstoffzellen-Hybrid-Fahrzeugs mit autothermer Methanolreformierung, PhD thesis, Technische Universität München.
15. Paganelli, G., Guezennec, Y., and Rizzoni, G. (2002) Optimizing control strategy for hybrid fuel cell vehicle. Presented at the SAE 2002 World Congress, Detroit, March 2002.
16. Cho, H.Y., Gao, W., and Ginn, H.L. (2004) A new power control strategy for hybrid fuel cell vehicles. Presented at the 8th IEEE Workshop on Power Electronics in Transportation, Detroit, October 2004.
17. Gao, L., Jiang, Z., and Dougal, R.A. (2004) An actively controlled fuel cell/battery hybrid to meet pulsed power demands. *J. Power Sources*, **130**, 202–207.
18. Blackwelder, M.J. and Dougal, R.A. (2004) Power coordination in a fuel cell–battery hybrid power source using commercial power controller circuits. *J. Power Sources*, **134**, 139–147.
19. Steinberger-Wilckens, R., Mergel, J., Glüsen, A., Wippermann, K., Vinke, I., Batfalsky, P., and Smith, M.J. (2008) in *Materials for Fuel Cells* (ed. M. Gasik), Woodhead Publishing, Cambridge, pp. 425–465.
20. Herb, F. (2010) Alterungsmechanismen in Lithium-Ionen-Batterien und PEM-Brennstoffzellen und deren Einfluss auf die Eigenschaften von daraus bestehenden Hybrid-Systemen, PhD thesis, University of Ulm.
21. Thounthong, P., Chunkag, V., Sethakul, P., Davat, B., and Hinaje, M. (2009) Comparative study of fuel cell vehicle hybridization with battery or supercapacitor storage device. *IEEE Trans. Veh. Technol.*, **58** (8), 3892–3904.
22. Gao, W. (2005) Performance comparison of a fuel cell-battery hybrid powertrain and a fuel cell–ultracapacitor hybrid powertrain. *IEEE Trans. Veh. Technol.*, **54** (3), 846–855.
23. Bradley, T.H., Moffitt, B.A., Parekh, D.E., Fuller, T.F., and Mavris, D.N. (2009) Energy management for fuel cell powered hybrid-electric aircraft. Presented at the 7th International Energy Conversion Engineering Conference, Denver, CO, August 2009.
24. Larminie, J. and Dicks, A. (2003) *Fuel Cell Systems Explained*, 2nd edn., John Wiley & Sons, Ltd., Chichester.

25. Conte, F.V., Gollob, P., and Lacher, H. (2009) Safety in the battery design: the short circuit. *World Electr. Veh. J.*, **3** (EVS24), 1–8.
26. Mohan, N., Undeland, T.M., and Robbins, W.P. (1995) *Power Electronics: Converters, Applications and Design*, 2nd edn., John Wiley & Sons, Inc., New York.
27. Suh, K.W. and Stefanopoulou, A. (2005) Coordination of converter and fuel cell controllers. Presented at the 13th Mediterranean Conference on Control and Automation, Limassol, Cyprus, June 2005.
28. Eren, Y., Erdinc, O., Gorgun, H., Uzunoglu, M., and Vural, B. (2009) A fuzzy logic based supervisory controller for an FC/UC hybrid vehicular power system. *Int. J. Hydrogen Energy*, **34**, 8681–8694.
29. Muller, J., Rotkopf, K., Sonntag, C., Bohm, C., and Rabenseifer, P. (2005) Hybrid power source. US patent application US 2005/0249985 A1.
30. Zhao, H. and Burke, A.F. (2010) Fuel cell powered vehicles using supercapacitors – device characteristics, control strategies and simulation results. *Fuel Cells*, **10** (5), 879–896.
31. Miller, A.R., Peters, J., Smith, B.E., and Velev, O.A. (2006) Analysis of fuel cell hybrid locomotives. *J. Power Sources*, **157**, 855–861.
32. Zandi, M., Payman, A., Martin, J.P., Pierfederici, S., Davat, B., and Meibody-Tabar, F. (2010) Flatness based control of a hybrid power source with fuel cell/supercapacitor/battery. Presented at the IEEE Energy Conversion Congress and Exposition, Atlanta, GA, September 2010.
33. Di Napoli, A., Crescimbini, F., Solero, L., Pede, G., Lo Bianco, G., and Pasquali, M. (2001) Ultracapacitor and battery storage system supporting fuel-cell powered vehicles. Presented at the 18th International Electric Vehicle Symposium, Berlin, October 2001.
34. Froeschle, P. and Wind, J. (2010) in *Hydrogen and Fuel Cells* (ed. D. Stolten), Wiley-VCH Verlag GmbH, Weinheim, pp. 793–810.
35. Niestroj, A. and Mohrdieck, C. (2010) The electrification of the automobile – technical and economical challenges. Presented at the 18th World Hydrogen Energy Conference, Essen, May 2010.
36. Daimler AG (2009) IAA 2009: The new Mercedes-Benz B-Class F-CELL, http://media.daimler.com/dcmedia/0-921-614307-1-1237106-1-0-0-1237175-0-1-11702-854934-0-1-0-0-0-0-0.html (last accessed 5 February 2012).
37. Flanz, S. (2010) 5 Years of experience with Ford fuel cell vehicle fleet operations. Presented at the 18th World Hydrogen Energy Conference, Essen, May 2010.
38. Thiesen, L.P., von Helmolt, R., and Berger, S. (2010) Hydrogen4 – the first year of operation in Europe. Presented at the 18th World Hydrogen Energy Conference, Essen, May 2010.
39. Brachmann, T. (2010) The Honda state of the art aromatic PEM V flow stack and its application to FCX clarity – a viable solution for FCEVs today and beyond 2015. Presented at the 12th Ulm Electrochemical Talk, Ulm, June 2010.
40. Hyundai (2011) Hyundai Unveils Tucson IX Hydrogen Fuel Cell Electric Vehicle, http://www.hyundaiusa.com/about-hyundai/news/Corporate_Tucson_ix_FCEV_Release-20110214.aspx (last accessed 29 April 2011).
41. Kia (2011) Brenstoffzellen, http://www.kia.de/Zukunft/Umwelt/Brennstoffzellen/ (last accessed 29 April 2011).
42. Nissan (2011) X-Trail FCV '03 Model, http://www.nissan-global.com/EN/TECHNOLOGY/INTRODUCTION/DETAILS/XTRAIL-FCV/index.html (last accessed April 29 2011).
43. Toyota (2011) Technology File: Fuel Cell Vehicle, http://www.toyota-global.com/innovation/environmental_technology/technology_file/fuel_cell_hybrid.html (last accessed 29 April 2011).
44. Toyota (2010) Fuel Cell Hybrid Vehicle – Advanced, http://www.toyota.com/esq/articles/2010/FCHV_ADV.html (last accessed 29 April 2011).
45. Volkswagen (2011) HyMotion – Touran/Tiguan, http://www.volkswagenag.com/vwag/vwcorp/content/de/innovation/

research_vehicles/HyMotion.html (last accessed 29 April 2011).

46. de Miranda, P.E.V. and Carreira, E.S. (2010) Brazilian hybrid electric fuel cell bus. Presented at the 18th World Hydrogen Energy Conference, Essen, May 2010.

47. Kaup, D., Bouwman, R., Schädlich, G., Sauer, D.U., and Lohner, A. (2010) H_2 bus NRW – the hybrid electric fuel-cell bus. Presented at the 18th World Hydrogen Energy Conference, Essen, May 2010.

48. Hyundai (2011) News Room: Hyundai's 2nd Generation Hydrogen Fuel Cell Electric Bus Unveiled, *http://worldwide. hyundai.com/web/News/View.aspx?idx= 181&nCurPage=1&strSearchColunm= &strSearchWord=&ListNum=119* (last accessed 29 April 2011).

49. New Flyer (2010) Hydrogen fuel cell technology, *http://www.newflyer.com/ index/cms-filesystem-action?file=news center/brochures/hydrogen fuel cell.pdf* (last accessed 29 April 2011).

50. Van Hool (2011) Hybrid fuel cell bus, *http://www.vanhool.be/ home fr/transport public/hybride pile a combustible/Resources/folderFuel Cell.pdf* (last accessed 29 April 2011).

51. Garche, J. (2010) in *Hydrogen and Fuel Cells* (ed. D. Stolten), Wiley-VCH Verlag GmbH, Weinheim, pp. 715–732.

52. Mergel, J., Glüsen, A., and Wannek, C. (2010) in *Hydrogen and Fuel Cells* (ed. D. Stolten), Wiley-VCH Verlag GmbH, Weinheim, pp. 41–60.

53. Corbo, P., Migliardini, F., and Veneri, O. (2010) Lithium polymer batteries and proton exchange membrane fuel cells as energy sources in hydrogen electric vehicles. *J. Power Sources*, **195**, 7849–7854.

54. HYCHAIN (2011) Fuel Cell Cargobike, *http://www.hychain.org/newhychain/ showroom/cargo.htm* (last accessed 9 May 2011).

55. HYCHAIN (2011) Fuel Cell Scooter, *http://www.hychain.org/newhychain/ showroom/scooter.htm* (last accessed 9 May 2011).

56. HYCHAIN (2011) Fuel Cell Wheelchair, *http://www.hychain.org/newhychain/ showroom/wheel.htm* (last accessed 11 May 2011).

57. HYSY LAB (2011) HYSYRIDER Hybrid Hydrogen Scooter, *http://www. hysylab.com/interne/pdf/HYSY_LEAF.pdf* (last accessed 6 May 2011).

58. Janßen, H., Blum, L., Kimiaie, N., Maintz, A., Mergel, J., Müller, M., and Stolten, D. (2005) Performance characterization of a 4-wheel DMFC scooter. Presented at the 3rd European PEFC Forum, Lucerne, July 2005.

59. Forschungszentrum Jülich (2007) *IEF-3 Report 2007. From Basic Principles to Complete Systems*, Energy Technology, Vol. **70**, Schriften des Forschungszentrums Jülich, Jülich.

60. Pro-Power (2011) Model GenTo I *http://www.propower/co.kr/eng/ download/Case_Study(Whellchair).pdf* (last accessed 11 May 2011).

61. Pro-Power (2011) Model Scoophin *http://www.propower/co.kr/eng/ download/Case Study(Scooter).pdf* (last accessed 6 May 2011).

62. Steckmann, K. (2009) Extending EV range with direct methanol fuel cells. *World Electr. Veh. J.*, 3 (EVS24), 1–4.

63. Steckmann, K. (2010) Die EFOY Brennstoffzelle – Nachrüstung für Elektromobile, *http://www.solarstrombau. de/energieladen/img/Solarmobil10-EFOY- Nachruestung.pdf* (last accessed 6 May 2011).

64. Suzuki (2011) Suzuki's Fuel-Cell Vehicles, *http://www. globalsuzuki.com/Burgman_Fuel- Cell_Scooter/index.html* (last accessed 9 May 2011).

65. Bouquain, D., Blunier, B., and Miraoui, A. (2008) A hybrid fuel cell/battery wheelchair modeling, simulation and experimentation. Presented at the IEEE Vehicle Power and Propulsion Conference, Harbin, China, September 2008.

66. Yamaha (2006) Detail – News Releases, *http://www.yamaha-motor.co.jp/ global/news/2006/10/19/fc-aqel.html* (last accessed 10 May 2011).

67. Yamaha (2007) Challenges to commercialization of direct methanol fuel cell-powered motorbikes. Presented at the 3rd International Hydrogen and

Fuel Cell Expo and Seminar, Tokyo, February 2007.
68. Yamaha (2007) Detail – News Releases, http://www.yamaha-motor.co.jp/global/news/2007/10/24/tms-05.html (last accessed 10 May 2011).
69. van Sterkenburg, S., van Rijs, A., and Hupkens, H. (2010) A fuel cell driven aircraft baggage tractor. Presented at the 18th World Hydrogen Energy Conference, Essen, May 2010.
70. HYCHAIN (2011) Fuel Cell Utility Vehicle, http://www.hychain.org/newhychain/showroom/vehicle.htm (last accessed 11 May 2011).
71. Bach, C. and Schlienger, P. (2009) HY.MUVE – die Schweiz bewegt. *HZwei*, **1**, 18–19.
72. Schlienger, P., Bach, C., and Büchi, F. (2010) Hydrogen driven municipal vehicle (Hy.Muve) – vehicle concept demonstration and field testing in Switzerland. Presented at the18th World Hydrogen Energy Conference, Essen, May 2010.
73. Proton Motor (2011) Smith Edison HyRange, http://www.proton-motor.de/fileadmin/downloads/FlyerHyRange-e-small.pdf (last accessed 11 May 2011).
74. McConnell, V.P. (2010) Fuel cells in forklifts extend commercial reach. *Fuel Cells Bull.*, **2010** (9), 12–19.
75. Industrial Truck Association (2011) Industrial Truck Association, https://www.indtrk.org/products.asp?id=rmp (last accessed 10 May 2011).
76. H2 Logic (2011) H2Drive Unlimited and zero emission power for forklifts, http://www.h2logic.com/com/h2drive/H2Logic_H2Drive_specification_5_ENG_web.pdf (last accessed 10 May 2011)
77. H2 Logic (2011) H2Drive Unlimited and zero emission for forklifts, http://www.h2logic.com/com/h2drive/H2Logic_H2Drive_brochure_5_ENG_web.pdf (last accessed 10 May 2011).
78. Hydrogenics (2011) HyPX Fuel Cell Power Packs, http://www.hydrogenics.com/assets/pdfs/NA-HyPx Fact Sheet.pdf (last accessed 11 May 2011).
79. Hydrogenics (2010) HyPM Fuel Cell Power Modules, http://www.hydrogenics.com/assets/pdfs/HyPM-Fuel Cell Power Module Brochure 2010.pdf (last accessed 11 May 2011).
80. Hydrogenics (2011) HyPM Fuel Cell Power Products: Success in Mobility Applications, http://www.hydrogenics.com/assets/pdfs/Successinmobilityapplications.pdf (last accessed 11 May 2011).
81. Nuvera (2011) PowerEdge C Series for Counterbalance Trucks, http://www.nuvera.com/pdf/Nuvera_PowerEdge_C.pdf (last accessed 11 May 2011).
82. Proton Motor (2011) Munich Airport: Forklift Truck With Fuel Cell System, http://www.proton-motor.de/forkliftruck.html (last accessed 11 May 2011).
83. Plug Power (2012) GenDrive, http://www.plugpower.com/Libraries/Documentation_and_Literature/Series_1000_Product_Spec_Sheet.sflb.ashx (last accessed 5 February 2012).
84. Toyota (2011) Toyota Material Handling Deutschland,http://www.toyotagabelstapler.de/tgd/mm_400/sm_1/art_135 (last accessed 11 May 2011).
85. Keränen, T.M., Karimäki, H., Viitakangas, J., Vallet, J., Ihonen, J., Hyötylä, P., Uusalo, H., and Tingelöf, T. (2011) Development of integrated fuel cell hybrid power source for electric forklift. *J. Power Sources*, **196** (21), 9058–9068.
86. Nuvera (2011) PowerEdge R Series for Reach Trucks, http://www.nuvera.com/pdf/Nuvera_PowerEdge_R.pdf (last accessed 11 May 2011).
87. Plug Power (2012) GenDrive, http://www.plugpower.com/Libraries/Documentation_and_Literature/Series_2000_Product_Spec_Sheet.sflb.ashx (last accessed 5 February 2012).
88. Janssen, H., Blum, L., Hehemann, M., Mergel, J., and Stolten, D. (2010) System technology aspects for light traction applications of direct methanol fuel cells. Presented at the 18th World Hydrogen Energy Conference, Essen, May 2010.
89. Nuvera (2011) PowerEdge W Series for Pallet Jacks, http://www.nuvera.com/pdf/Nuvera_PowerEdge_W.pdf (last accessed 11 May 2011).

90. Oorja Protonics (2010) OorjaPac Model III, http://www.oorjaprotonics.com/PDF/OorjaPac_Model_III_Product_Sheet.pdf (last accessed 11 May 2011).
91. Plug Power (2012) GenDrive, http://www.plugpower.com/Libraries/Documentation_and_Literature/Series_3000_Product_Spec_Sheet.sflb.ashx (last accessed 5 February 2012).
92. Pro-Power (2011) Model SS Drive, http://www.propower.co.kr/eng/download/CaseStudy(Forklift).pdf (last accessed 11 May 2011).
93. Proton Motor (2010) ZEMships Zero Emission Ships, http://www.proton-motor.de/fileadmin/downloads/Zemships_brochure_english.pdf (last accessed 12 May 2011).
94. Joh, H.I., Ha, T.J., Hwang, S.Y., Kim, J.H., Chae, S.H., Cho, J.H., Prabhuram, J., Kim, S.K., Lim, T.H., Cho, B.K., Oh, J.H., Moon, S.H., and Ha, H.Y. (2010) A direct methanol fuel cell system to power a humanoid robot. *J. Power Sources*, **195**, 293–298.
95. Uzunoglu, M., Onar, O.C., and Alam, M.S. (2009) Modeling, control and simulation of a PV/FC/UC based hybrid power generation system for stand-alone applications. *Renewable Energy*, **34**, 509–520.
96. Thounthong, P. and Rael, S. (2009) The benefits of hybridization. An investigation of fuel cell/battery and fuel cell/supercapacitor hybrid sources for vehicle applications. *IEEE Ind. Electron. Mag.*, **3** (3), 25–37.
97. Sisiopiku, V.P., Rousseau, A., Fouad, F.H., and Peters, R.W. (2006) Technology evaluation of hydrogen light-duty vehicles. *J. Environ. Eng.*, **132** (6), 568–574.
98. Bernard, J., Hofer, M., Hannesen, U., Toth, A., Tsukada, A., Büchi, F.N., and Dietrich, P. (2011) Fuel cell/battery passive hybrid power source for electric powertrains. *J. Power Sources*, **196**, 5867–5872.

Part VII
Systems Verification and Market Introduction

37
Off-Grid Power Supply and Premium Power Generation
Kerry-Ann Adamson

37.1
Introduction

This chapter looks at the provision of premium power to two key future market sectors: the off-grid market and the portable market.

We define *premium power* as power that is modular, distributed, and high quality. In other words, these are systems for producing, storing, and delivering power for applications without a central generating station.

Within this definition very clearly sits off-grid power and power for applications that are portable in nature. *Off-grid power* can be defined as the provision of power that is not connected to any utility through transmission cables or pipelines. Hence this definition includes the natural gas network. *Portable power*, on the other hand, is any unit that is designed to be independently moved. This includes personal electronics, external battery rechargers, remote monitoring, and generators. As the personal electronics market is so large in itself, has special operating characteristics, and will require different stack architecture from most other applications, in this chapter we only look at markets such as generators and remote monitoring.

37.2
Premium Power Market Overview

The current fuel-cell industry is still very small, with the most recent Pike Research Annual Report showing unit adoption in 2010 of under 16 000 units. These figures do not include the educational or toy market. As can be seen from Figure 37.1, taken from this report [1], stationary fuel cells are leading the way in building the industry.

If we break this dataset down into off-grid and portable premium power applications, however, we would see that, at present, this makes up less than 10% of all shipments. The reason for this is twofold:

Fuel Cell Science and Engineering: Materials, Processes, Systems and Technology, First Edition. Edited by Detlef Stolten and Bernd Emonts.
© 2012 Wiley-VCH Verlag GmbH & Co. KGaA. Published 2012 by Wiley-VCH Verlag GmbH & Co. KGaA.

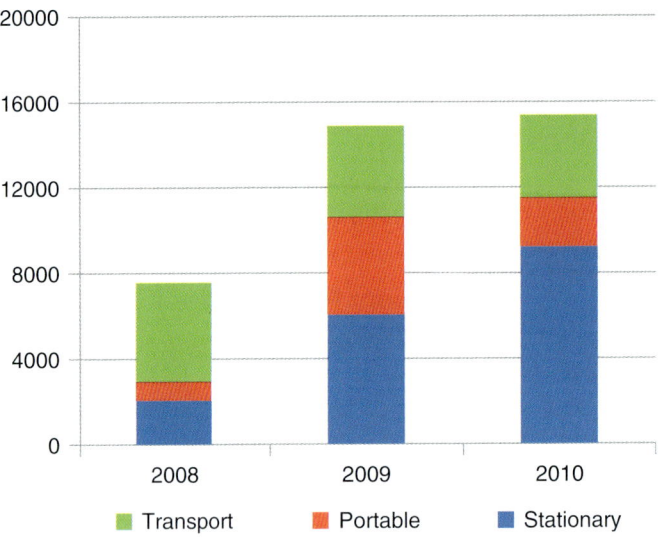

Figure 37.1 Fuel cell adoption, 2008–2010. Source: Pike Research.

1) It is technically easier to start off with applications that are grid tied. Here the grid, in the majority of cases, operates as the prime power source, with the fuel cell only operating as a back-up power source;
2) The market pull for premium power to date has come from countries with less access to a widespread grid. These include the African nations, India, and many Latin American countries. As these countries cannot afford to be first adopters of very high-cost technology, there is a market-induced lag between the pull and push of the technology.

Owing to the increasing market for premium power in key adopter groups in the Global North, although still very limited when compared with developing countries, we are likely to see initial roll-out in the next 5 year time frame there and then a trickle-down effect into Africa, India, and Latin America when costs come down with mass manufacturing and component standardization.

If we look just at markets with some pull in the Global North, we see that they can be both prime power and backup power products such as those for base stations, fuel-cell "generators" for use in military mobile hospitals, and prime power providers to off-grid villages. If we focus on just one example – the military – we can illustrate the levels of power we are discussing here.

The US Department of Defense (DOD) purchases more light refined petroleum products than any other single organization or company in the world. With an annual budget of US$3.5 billion, the DOD's Defense Logistics Agency (DLA) procures nearly 110 million barrels of petroleum products each year [2]. This demand includes jet fuels, aviation gasoline, automotive gasoline, heating oils, power generation, naval propulsion fuels, lubricants, natural gas, and coal. The specific military fuels procured include JP-5, a kerosene-based jet fuel primarily used for Navy carrier-based aircraft, JP-8, a kerosene-based fuel similar to Jet A-1 commercial jet fuel, and F-76, a US naval diesel similar to marine gas oil.

As such a large user of petroleum, the DOD estimates that every $1 increase in the cost of a barrel of oil costs the DOD around $130 million in increased annual energy costs [3]. This reliance on oil and an increasingly fragile commercial grid for delivering electricity to its 500-plus major installations have subsequently been recognized as a significant tactical and strategic vulnerability that poses a serious threat to the continuity of critical missions. From this, it is clear that the market pull for premium off-grid power here is strong and the DOD especially is investing heavily in new technologies that could provide this.

This is just one example of the increasingly commercial, and strategic, focus on energy and specifically premium energy around the world.

The remainder of this chapter looks at the two specific markets of off-grid and premium power for portable applications.

37.3 Off-Grid

The off-grid market covers a broad subset of applications; the two with the greatest potential for early adoption and using fuel-cell technology as a change agent are homes and base stations. Although in terms of deployment currently off-grid units are in the dozens of installations they represent, for many companies, markets included in long-term development plans.

37.3.1 Homes

Today it is estimated that 1.7 billion people worldwide live off-grid. By this we mean that 1.7 billion people have no access to power provided by a central utility. The vast majority of these 1.7 billion live in regions of the world that are less developed economically, but there is a growing trend in developed countries to choose to go off-grid. It is estimated that in the USA alone there are about one-quarter of a million off-grid homes, with one source estimating that this number is growing at 33% per annum. These off-gridders include many villages and towns where the extension of the electricity grid is either prohibitively expensive or geographically daunting.

Hence although in many countries government policy is provide all homes with some form of centralized power, in others people are actively turning their back on the utilities.

Both of these trends, however, have had the critical impact of sharply increasing the interest in distributed generation technologies. Distributed generation technologies are energy generation technologies which provide power at the point of use, rather than from a centralized point.

At present, most off-grid homes have a mix of a generator, micro-renewables, and potentially deep discharge batteries for energy storage. This mix of technologies means that the homes are not fully reliant on one energy source or fuel. This current mix of technologies, however, has proven to be a sub-optimal solution in many developing countries where, for example, there is a thriving black market in diesel and solar panels.

Fuel-cell technology for off-grid homes is not being developed as a silver bullet, however. At present most developers are focusing very clearly on residential products that are grid tied and do not work when the grid goes down. During the recent power crisis in Japan, for example, users of the Ene-Farm mCHP fuel cells were advised to switch off the fuel cells during all the planned blackouts. Developing an off-grid fuel cell for a single home is a very different technological challenge from this.

A fuel cell for residential use in an off-grid home would have to be able to work on opportunity fuels, whatever local fuels are to hand, such as propane or ammonia. Or the dream solution for many would be to use excess renewables, stored in hydrogen and then used in the fuel cell when there was no sun or wind, and so on. In both of these situations, the energy storage system and the fuel cell would need to be able to run for thousands of hours and be robust. If the fuel cell was being used as part of a storage system, where any excess renewable electricity was converted through an electrolyzer into hydrogen to be used later in a fuel cell, then the unit would also need to be able to cycle up and down many hundreds of times and not be susceptible to degradation when not in use.

As the cost of a grid tied mCHP unit is still in the tens of thousands of dollars range and the market for these products is in the millions, the developmental focus is much more likely to continue to be on this sector rather than the much more price-sensitive, and technologically more challenging, off-grid home market. That is not to say that the occasional luxurious off-grid home with a fuel cell will not be developed over the next decade, but that this will be a rarity and not the norm.

What we will see, and are already seeing, is the adoption and deployment of fuel-cell and hydrogen technology for off-grid villages. These collections of houses can be much more efficiently targeted by a renewable/hydrogen/fuel-cell solution and companies such as Hydrogenics are already active in this area. These "solutions in a box" are being sold as clean off-grid power plants providing tens to hundreds of kilowatts of power.

One highly successful example of this has been the installation in Bella Colla, Canada. The Hydrogen-Assisted Renewable Power System (HARP) project, as it is officially known, was the installation of electrolyzer and fuel cell to take the excess run of the river power and store it for use when needed. The system included a 200 kWe fuel cell unit. It should be made clear that the fuel cell – hydrogen electrolyzer unit did not fully replace the diesel generators but significantly reduced the power required by them. This first project was something of a demonstration, which in Canada has a real significance as there are an estimated 300 such remote communities [4] which are all looking to increase their energy resilience and reduce their energy costs.

Globally, it is safe to assume that there are over 1000 such communities in the Global North with many times this in the Developing South. Although some spreadsheets have been published that show a potential payback of a renewable – hydrogen – fuel-cell system over a diesel generator in less than 7 years, it is likely that costs will still need to come down significantly before the off-grid community market in the Developing South can be fully targeted.

37.3.2
Off-Grid Base Stations

Base stations (Figure 37.2) are wireless communications stations installed at a fixed location and used to communicate as part of a telecoms network. Although they are critical for the continued roll-out of mobile and increasingly smart phone use, they create part of the hidden cost of the move to an electron economy.

Base stations are heavy users of energy, with the GSMA, the trade association for the mobile communication industry, releasing figures which show that 85% of a network operator's costs come from the energy requirements at base stations. In 2011, Pike Research calculated that in the USA alone this demand topped 6613 GWh annually, with an associated 3 million tons of CO_2 annually.

Currently, most base stations are grid tied with the electricity grid providing prime power. Backup comes in the form of diesel generators, and to a much lesser extent some form of renewable – battery hybrid system. Off-grid base stations are much less common but when they are installed the norm is a two-diesel generator system, where one acts as the primary with the backup power provided by the other.

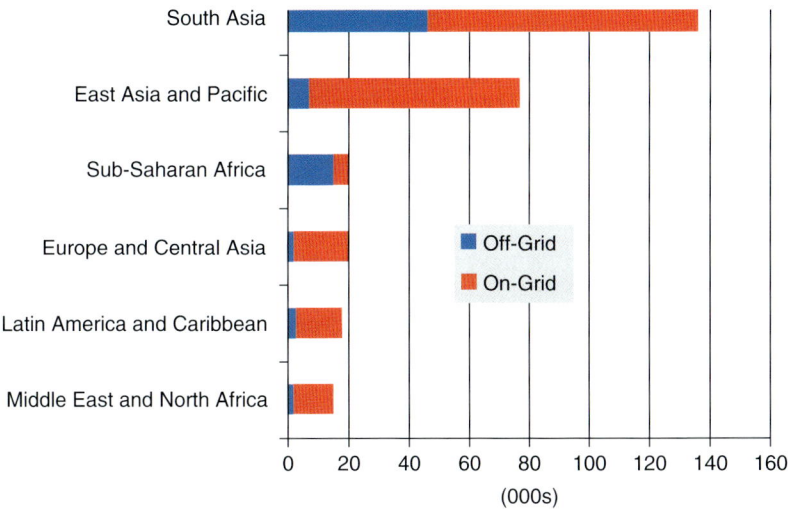

Figure 37.2 Annual growth in base stations, developing regions, 2007–2012. Source: GSMA.

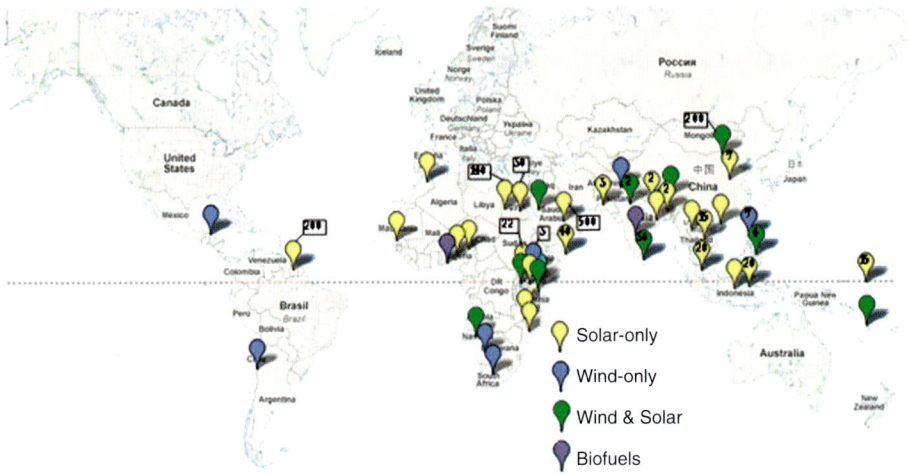

Figure 37.3 2009 trail sites of green off-grid base stations[5]. Source: GSMA.

For off-grid base stations to reduce energy demand, risk, and emissions, the option that has garnered the most interest is the roll-out of the prime power "green" sites which use a combination of renewable energy and deep discharge batteries, or fuel cells, or use some other lower carbon fuel source such as ammonia (Figure 37.3). The GSMA is piloting a number of these sites, collecting and analyzing data on performance. It should be noted that this scheme does not include fuel cells, as to date the GSMA considers fuel-cell technology not to be market ready for off-grid base stations.

As Figure 37.2 shows, the three regions with the projected highest growth of off-grid stations are South Asia, East Asia, and Pacific and Sub-Saharan Africa. In these regions, removing as much diesel from the equation as possible is seen as one of the key drivers to switching to fuel cells or other technology options. Diesel, especially in Africa, has a ready black market with an estimated 15–22% of diesel being lost to theft.

It is interesting that company literature regarding the deployment of fuel cells and other new technologies is often tagged to the reduction in emissions, whereas it is more likely that OPEX (operational expenditure) reduction and extended run time (XRT) capabilities are the likely economic drivers, with reductions in emissions being the proverbial icing on the cake. This is unlikely to change without some form of policy or legislation prompting the telecom carriers to target and reduce its carbon footprint.

Although the speed of change is slow, we are seeing deployments in southern Africa and Germany. In 2011, Germany saw the deployment of its first off-grid fuel-cell powered base station. E Plus and Nokia Siemens installed a FutureE "Jupiter" fuel-cell system, alongside a wind turbine, solar panels, and deep discharge batteries. South Africa has taken delivery of the first "PowerCube" unit from Diverse Energy. Its fully packaged system, which uses ammonia as a fuel,

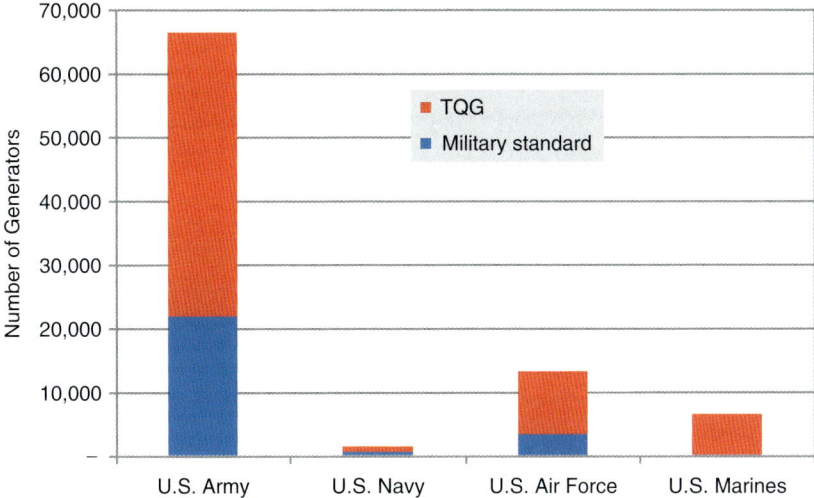

Figure 37.4 Penetration of portable generators by US Military Branch, 2010. Source: US DOD.)

claims to provides an 80% reduction in greenhouse gas emissions, a 74% reduction in energy use, and in local conditions in Africa a 25% reduction in total cost of ownership with a 2 year payback. As with FutureE, Diverse has shipped and installed one unit, the poetically named Ainra, which translates as "Eternal Power" in Swahili.

37.4
Portable Applications

The premium power portable market also covers both civilian and military applications. Here, however, the military is very clearly a driver due to the increased push for new technologies. The markets that are being developed for fuel-cell technology again range from watts to megawatts. In the very low wattage range there is personal electronics, which is being driven exclusively by consumer demand, not policy, right up the megawatt replacement generators for military installations such as hospitals for forward bases.

37.4.1
Remote Monitoring/Remote Sensing

The demand for data collection and real-time monitoring in public and private sectors is continually increasing. Government agencies, corporations, and businesses rely on these data for a wide variety of tasks beyond early warning alerts, such as:

- level gauges used in flood and tsunami early warning systems
- water quality meters

- biological and geological instruments
- remote servicing stations
- vehicle-based measurement instruments.

The size of the market globally is somewhat of a black box – we know that it exists but there appears to have been very little systematic research to quantify it. What we do know is that in the USA, for example, there are:

- 1500 weather stations (RAWs)
- 82 600 dam sites
- a minimum of 3000 wildlife monitoring sites
- 7000 seismic monitoring sites.

This does not take into account any military sites. Each of these sites requires a very low wattage of power to provide power to the monitoring equipment. At present this is produced by some form of off-grid prime power generation technology.

Until recently, operators of off-grid monitoring systems had to rely on conventional power supplies such as batteries, solar cells, or generators. Each of these power supplies, however, presents significant drawbacks, including the following:

- Batteries are heavy and contain relatively little power in relation to their weight. Additionally, battery exchange requires frequent and labor-intensive visits by maintenance personnel. These visits often present a logistical challenge, as battery exchanges may be required at inconvenient times and the cost required to transfer the battery to the equipment site may be considerable.
- Solar cells offer the convenience of no recharging, but their reliance on direct sunlight means they may be inoperable in bad weather or have limited power during the short days of winter and at critical times. Solar modules are also highly visible and could therefore betray the location of the monitoring equipment. This creates additional issues because solar modules are often subject to theft or vandalism.
- Generators are rarely used in off-grid applications, as they are maintenance intensive and require regular refills. At the same time, generators are also noisy and produce harmful exhausts, which make them unsuitable for use in environmentally protected wilderness areas.

For the remote monitoring market, it is the potential to have a significant increase in run time between visits and reduced overall maintenance costs, which have an increasing number of companies looking at this as an option.

As has already been mentioned, as of 2011, the commercial deployment into this space has been very limited. Here the market leader, or to be more accurate the market pioneer, is SFC Energy. Along with its distributor partner in the UK, UPS Systems, they have installed 20 EFOY units at the Elan Valley Reservoir to power telemetry equipment. The direct methanol fuel-cell systems use methanol, and with their 28 l canisters there is sufficient energy to power the equipment 24 h per day for a minimum of 300 days. This is the largest project of its kind to date.

37.4.2
Military Applications for Prime Power

37.4.2.1 Soldier Power

In the light of the current nature of military operations throughout the world, portable power and battery technologies have come under the increased focus of military systems, logistics, and operations planning. In a commercial sense, portable power (generators) and battery technologies have continued to evolve as new battery chemistries have been introduced into the marketplace to meet the demands of longer life portable electronics devices. Similarly, portable power generation for military forward operating bases and battery technologies for mobile soldier communications radios and tactical devices remain critical for the effectiveness of soldiers on the battlefield.

The modern soldier relies on an increasing array of sophisticated electronics equipment. Soldiers and commanders alike demand an improved capability to gather intelligence, communicate, and coordinate their actions due to advances in the equipment they carry. Military missions are generally limited by the cost and weight of batteries, in addition to the severe burden they place on individual soldiers. The popular military trend towards the use of wireless technologies, such as Wi-Fi, compounds these power problems because Wi-Fi is extremely power intensive. Fuel cell systems – which have much higher energy density than batteries and lower weight than electronic generators – will either replace or supplement batteries as the primary soldier power source and will result in more mobile soldiers, longer mission lengths, and lighter pack loads.

Today, batteries remain the primary sources of power for small hand-held and wireless devices carried by soldiers, who – for extended missions beyond the normal lives of these batteries – must either carry extra batteries as replacements or employ some method of recharging their original batteries.

Although battery energy density has generally increased in a fairly linear fashion, the energy requirements of the devices being carried have gone up more exponentially. The reason behind this gap is that whereas the technology of battery energy is developing relatively slowly, the devices being carried are quickly becoming smaller and more sophisticated, incorporating more technology, and requiring greater energy density.

As it currently stands, a single soldier is required to carry up to 10 BA 5590 lithium batteries for a 24 h mission and may be required to carry up to 30 primary batteries for a 90 h mission. At the same time, a single eight-bay 250 W battery charger will provide a squad with sufficient mission power for up to 90 h.

To address the significantly vital role that battery packs and portable power technologies fulfill on the modern battlefield, the DOD has begun to invest in alternative energy sources, fuels, and technologies to replace or complement existing portable power technologies. Fuel-cell technologies have subsequently been acknowledged as a potential solution for reducing weight and extending the mission time for soldiers in the field. For a typical 72 h mission at 20 W average power usage, a fuel cell with three cartridges today weighs 2.9 kg – one-third of the

Table 37.1 ATO fuel cell technology areas.

Technology areas	Technical objectives		
Soldier hybrid direct methanol fuel cell power source	25 W	0.68 kg	TRL 4/6
Soldier hybrid fuel cell power source	50–100 W	1.59 kg	TRL 4/5
Portable hybrid power sources and chargers (JP-8 fueled)	150–250 W	11.34 kg	TRL 4/6

Abbreviation: TRL, technology readiness level development - demonstration.
Source: CERDEC.

weight of lithium ion batteries. Fuel-cell systems are able to power the full range of devices that soldiers carry – from sensors to ruggedized computers to satellite phones.

Within the US Army's Communications – Electronics Research, Development, and Engineering Center (CERDEC), the Army Power Division (ATO) is devoted to testing and demonstrating potential portable soldier power solutions. The ATO's most recent areas of interest in fuel cells are shown in Table 37.1.

Moreover, a specialized Fuel Cell Team has been created to develop and transition suitable fuel-cell technologies rapidly to applications where they are most needed. These specific mission areas include:

- sensory and soldier power (1–100 W)
- man-portable power (100–500 W)
- auxiliary power units (500–10 kW).

Since 2008, CERDEC's fuel-cell programs have involved a number of manufacturers and institutions across a range of technologies. As CERDEC's activities suggest, those portable military applications with the greatest near-term potential for the integration and adoption of fuel-cell technology center predominantly upon soldier wearable and portable power and remote sensing and surveillance.

37.4.3
Portable Generators – Military

Outside of civilian use, the military represents a large, and lucrative, market for generators. With nearly 88 000 standard and tactical quiet generators (TQGs) in the US armed forces, this represents almost $1 billion in hardware. If we look at just the US Army, the heaviest user of portable electrical power, we can see that 83% of their generators are under 10 kW. The US military, understandably, has a practice of developing standardized products and supply lines and at present is using a so-called standard generator and a TQG.

TQGs represent the latest generation of generators to be integrated into the US armed forces. Mounted on skids or a trailer, these systems are known to offer the military substantial improvements over their predecessors, including improved reliability, maintainability, fuel efficiency, noise abatement, and survivability.

Since the introduction of TQGs in 1993, the DOD has introduced new 100 and 200 kW versions to power weapons systems, command posts (CPs), command, control, communications, computers, intelligence, surveillance, and reconnaissance (C4ISR) systems, and numerous other battlefield systems throughout the DOD. In terms of development, the TQGs represent the minimum benchmark that fuel-cell development needs to reach.

At present, most military generators use a petroleum-based liquid fuel or an AC electric source.

In reality, fuel-cell manufacturers face a difficult challenge in seeking to penetrate this market for military portable generators, given the performance and reliability associated with established diesel generator technologies. The unit has to be highly rugged, and also capable of being maintained in the field without specialist parts or highly specialized training.

37.5
Discussion

The market for premium power applications for fuel-cell technology is massive. It covers a broad range of applications, globally, and is being created by consumer demand in addition to policy pull.

To date, however, none of the applications where fuel cells could show a clear advantage over incumbent technologies have been systematically developed. This is starting to change, with a number of companies specifically developing off-grid technologies. The challenges, as with many other areas of fuel-cell interest, are the cost, durability, and being fit-for-purpose. As more companies focus on these markets, units that have a more robust cycling profile and longer durability and are cheaper will be produced. At present, however, this chapter has had to be focused on the future potential rather than the current market.

References

1. Adamson, K.A. (2011) Pike Research Fuel Cell Annual Report, 2011, Pike Research, London.
2. GlobalSecurity.org DLA Energy, *http://tinyurl.com/65n6ljq* (last accessed 18 December 2011).
3. Miles, D. (2008) Military Looks to Synthetics, Conservation to Cut Fuel Bills, American Forces Press Service, US Department of Defense, *http://www.defense.gov/news/newsarticle.aspx?id=50131* (last accessed 18 December 2011).
4. Ah-You, K. and Leng, G. (2009) Renewable Energy in Canada's Remote Communities, Natural Resources Canada.
5. GSMA (2009) Green Power for Mobile: Top 10 Findings, *www.gsmworld.com/greenpower* (last accessed 18 December 2011).

38
Demonstration Projects and Market Introduction
Kristin Deason

38.1
Introduction

In order to advance the technological and market readiness of hydrogen and fuel cell technologies, many countries have launched demonstration projects designed to display and evaluate the operation of these technologies in real-world environments. In contrast to research and development (R&D) projects, which seek to discover new concepts, methods, or materials and create new technologies that may or may not be targeted towards a specific application, demonstration projects are intended to prove that a newly developed technology is viable in the actual environment where it is intended for use. Other chapters in this book describe the state-of-the-art of R&D in various technology areas related to hydrogen and fuel cells. This chapter seeks to provide an overview of programs around the world aimed at demonstrating, deploying, and introducing these technologies, and to give an impression of the reasons why governments and other organizations undertake these projects and the benefits they receive from them. It should be noted that this chapter is not exhaustive – there are thousands of fuel cells currently in operation throughout the world, and this chapter merely provides an overview of the major demonstration and deployment programs.

38.2
Why Demonstration?

One may ask why, if enough R&D has been done so that the technology performs at an acceptable level, there is a need for demonstration projects. If the technology is good enough to compete with incumbent technologies, why can we not expect it to survive on its own in the market? The answer is complicated and has to do with the many barriers that a new technology faces when entering the market. Because most of these barriers occur at the junction between R&D and commercialization, it makes sense for governments and industry to jointly provide funding for these types of projects.

Fuel Cell Science and Engineering: Materials, Processes, Systems and Technology,
First Edition. Edited by Detlef Stolten and Bernd Emonts.
© 2012 Wiley-VCH Verlag GmbH & Co. KGaA. Published 2012 by Wiley-VCH Verlag GmbH & Co. KGaA.

Demonstration projects can take many different forms, and each project has its own unique set of goals that it aims to accomplish. This section describes some of the barriers to market entry that are commonly addressed by these projects.

One barrier that demonstration projects address is a lack of operating experience in real-world environments. In transitioning from the laboratory to the market, technologies must deal with variations in a myriad of different physical variables ranging from temperature extremes to changing usage patterns to interfacing with numerous external components. Often, an iterative process is undertaken to discover shortcomings and identify areas that need work through demonstration projects, followed by additional R&D to correct these shortcomings and further improve the technology. Similarly, demonstration projects can serve to measure the progress of a technology, validate models with real-world data, and test integration of components.

In many cases, the technologies being demonstrated are unfamiliar to the general public and other stakeholders, for example, permitting officials. Many projects address this barrier by educating and exposing potential future users to the technology, helping them become more familiar with it and thus more likely to use it in the future. Like many emerging technologies, hydrogen and fuel cells sometimes suffer from a lack of appropriate codes and standards, causing various issues, including longer permitting times for installations and high project costs. Demonstration projects can help alleviate this issue by educating officials, helping to develop and refine appropriate codes and standards, and trialing approval procedures to make them more efficient and effective, paving the way for faster adoption in the future.

Demonstration projects can also help develop markets for hydrogen and fuel cell technologies by establishing the needed infrastructure, such as hydrogen refueling stations, that will be needed to support the widespread roll-out of the products in the future.

Finally, demonstration projects can help to reduce the cost of hydrogen and fuel cell technologies by driving increased demand for products, increasing manufacturing volumes, and thereby reducing costs.

38.3
Transportation Demonstrations

In the transportation application area, several countries have run major demonstration programs in recent years, focused on demonstrating technology in real-world situations, collecting data on performance and range, identifying R&D needs, and building infrastructure to support future commercialization. These projects generally incorporate passenger vehicles and/or buses, along with the associated refueling infrastructure. Major programs include joint government- and industry-funded initiatives in Germany, the European Union (EU), Japan, the USA, and Korea. Some car manufacturers have also run their own early deployment programs in addition to participating in the government programs.

From the experience gained through demonstration projects, along with simultaneous R&D, fuel cell vehicles (FCVs) have shown marked improvements that are allowing them to approach competitiveness with conventional vehicles in terms of price and performance. In 2010, Toyota announced that it expects to be able to sell hydrogen vehicles for around USD 50 000 as it has reduced FCV production costs by 90% since the mid-2000s [1]. Daimler made a similar announcement in early 2011, stating that a fuel cell car would not cost more than a comparable diesel hybrid by 2015 [2], and General Motors (GM) has stated that it expects the costs of their fuel cell system to decline by 75% by 2015; their current vehicle production cost is around EUR 500 000 [3].

The performance of the vehicles has also improved. In 2010, Daimler stated that its current FCV had a stack lifetime of 110 000 km, and projected that this would improve to 250 000 km by 2020 [4]. GM similarly predicts a lifetime of 200 000 km and 5500 h [3]. Further, a 2010 study by a consortium of over 30 stakeholders from various industries concluded that current FCVs display performance, ranges and refueling times comparable to those of conventional vehicles on the market today, and that the total cost of ownership for all powertrains studied [internal combustion engine (ICE) vehicles, fuel cell vehicles (FCVs), plug-in hybrid electric vehicles (PHEVs), and battery electric vehicles (BEVs)] are expected to converge by 2025 [5].

Building on the success of recent demonstration programs and on the technical improvements made to the vehicles over recent years, car manufacturers are now targeting 2015 for commercialization of fuel cell passenger vehicles in specific geographic markets. In late 2009, a joint letter of understanding from the main FCV manufacturers (Daimler, Ford, GM/Opel, Honda, Hyundai/Kia, Renault, Nissan, and Toyota) stated that they "strongly anticipate" commercialization beginning in 2015 [6]. This was closely followed by the announcement of the H_2 Mobility Initiative, in which Daimler, along with several energy companies, agreed to begin joint evaluation and planning for a build-up of hydrogen infrastructure in Germany as a lead market [7]. In early 2011, a group of 13 major Japanese auto manufacturers and energy companies jointly announced their intention to launch mass produced FCVs in 2015, and to develop concurrently supporting infrastructure in four metropolitan areas in Japan [8].

On the bus side, costs are an obstacle owing to low production volumes, and an additional challenge is that the global market for buses is significantly smaller than that of passenger cars, meaning that achieving cost reductions through economies of scale will be more difficult [9]. Nevertheless, governments worldwide see zero-emission buses as key to reducing emissions in public transit, and are funding demonstration projects in hopes of bringing the technology to commercial readiness. During the last decade, manufacturers have made significant improvements to bus performance as a result of experience gained through these projects.

Most countries involved in developing hydrogen and fuel cell technology have launched demonstration projects in addition to R&D activities. These projects vary considerably in their size, objectives, funding source, and form. The following sections provide an overview of the major hydrogen and fuel cell demonstration projects currently in place around the world.

38.3.1
Germany

In Germany, hydrogen and fuel cell demonstration projects are mainly funded by the National Innovation Program (NIP), and administered by the National Organization for Hydrogen and Fuel Cell Technology (NOW). The NIP is a public–private partnership funded 50% by the federal government and 50% by industry, and aims to prepare the market for hydrogen and fuel technologies by funding both R&D and demonstration projects. Funding for demonstration projects makes up ~70% of the total funding [10].

NOW's demonstration projects focus on a specific application area, with the clean energy partnership (CEP) being the main project for transportation. In addition to funding demonstration projects, some of NOW's R&D funding is also allocated to "demonstration preparation," which supports companies in developing key components for next-generation vehicles.

Germany is also the focus region for the H_2 mobility initiative, a partnership of key industry stakeholders that is planning to study and jointly build an initial hydrogen infrastructure to support the market for FCVs.

38.3.1.1 Clean Energy Partnership

The CEP is one of the largest hydrogen demonstration projects in the world, and includes hydrogen vehicles and buses, refueling infrastructure, and sustainable hydrogen production and delivery. The project currently involves 13 partners, including automotive companies, energy companies, and public transportation providers, and is focused on validating the technology under real-world conditions (Figure 38.1). Vehicles from five different manufacturers are being used in everyday operation by real customers and fueled at stations that are integrated with existing refueling stations and open to the public. The project also incorporates hydrogen buses serving actual customers within public transit networks.

Started in 2003 with seven partners, the CEP has grown over the years to include additional partners and regions. The project expanded to include the Hamburg region in 2009, and the German federal states of North Rhine-Westphalia (NRW) and Baden-Württemberg in 2010. CEP has also been extended time-wise since its inception – it was originally designed to be a 5 year project, but was eventually extended and now is divided into three multi-year phases [4]. Phase I ran from 2003 to 2008 and focused solely on Berlin, demonstrating two public refueling stations and a fleet of vehicles and buses. In phase II (2008–2010), the number of vehicles, locations, and fueling stations was expanded, and upgraded vehicles were introduced. Phase III of the project will run from 2011 to 2016, and will aim to finalize preparation of the market to allow for commercial availability of FCVs by 2016.

Figure 38.1 The current CEP partners [11].

Like all NIP-funded projects, around half of the funding from CEP comes from the German federal government, with the remainder being provided by the participating companies. Hydrogen at the CEP fueling stations is offered at €8 kg^{-1} (for both gaseous and liquid), which reflects government subsidization of the supply and production costs [12].

CEP phase I achieved many steps toward demonstrating a real-world hydrogen transportation system. A fleet of 17 cars from five different manufacturers, along with four buses, was demonstrated in Berlin, driving a total of 374 000 km. Overall, the vehicles achieved a high level of reliability with a low breakdown rate, along with positive reviews from drivers. The main criticism was of the relatively limited range of the vehicles (ranging from 150 to 300 km) [12]. On the infrastructure side, two public fueling stations were put in place, using three different hydrogen production methods and two different types of hydrogen fuel (gaseous and liquid). The electricity used for electrolysis and liquefaction was certified as renewable using the renewable electricity certificates system (RECS). During the project, the stations demonstrated availability of between 90 and 95%. A joint service station was also established for vehicle repair and maintenance [12].

The successful operation of this fleet showed that production and supply of hydrogen vehicles was possible, and that the vehicles could be operated safely and reliably. Experience gained during the project also served to identify and remove technical, administrative, and economic barriers to using hydrogen for road transportation and also to further standardize and certify components that are needed for both vehicles and infrastructure. The project also introduced members of the public to hydrogen and fuel cell technologies, with over 3000 people having visited the project during this phase [12].

Phase II ran from 2008 to 2010 and expanded the project in terms of number of both vehicles and locations. The Berlin fleet of cars grew to 47, and vehicles and buses were introduced in Hamburg. Two new stations opened, one in Berlin and one in Hamburg. NRW also joined as a new CEP region, and Shell, Toyota, and Hamburger Hochbahn joined as partners. In addition, next-generation FCVs with 700 bar storage were introduced. Public outreach activities were expanded, and included a rally from Berlin to Hamburg showcasing the extended ranges of the new vehicles. The new fueling station at Berlin Holzmarktstrasse, opened in 2010, incorporates renewable energy in the form of hydrogen produced from biomass, photovoltaic arrays at the station, and micro CHP (combined heat and power) powered by residual hydrogen [4].

In phase III, focused on market preparation, the partnership will again expand geographically to include Baden-Württemberg (the Stuttgart region), Hessen, and possibly additional areas. At least 10 new fuel cell buses are planned to be deployed in Hamburg. The passenger vehicle fleet will be expanded to over 130 vehicles by 2013, and additional stations will be built to bring the total number to at least 15 in five regions. Additional renewable production of hydrogen will be incorporated into the new stations, with a target of producing 50% of the hydrogen from renewable sources by 2015. Among the new stations to be built will be a completely CO_2-neutral filling station at Berlin Brandenburg International Airport, which will produce electricity and hydrogen via electrolysis from a nearby purpose-built wind farm [13, 14].

These strong demonstration activities in Germany have not only served to improve the technologies and increase public awareness, but have also established a good base of infrastructure that has catalyzed the identification of Germany as one of the first target markets for commercialization by several car makers.

38.3.1.2 Activities in North Rhine-Westphalia

At the state level, there are also strong demonstration activities in North Rhine-Westphalia (NRW), in the western part of Germany. NRW produces 50% of the by-product hydrogen in Germany (about 31 000 tons per year or 1000 GWh, enough to fuel 260 000 cars or 6000 buses), and is also home to an existing 230 km hydrogen pipeline [15, 16]. NRW aims to take advantage of these resources by concurrently building a hydrogen infrastructure and demonstrating various types of vehicles. In 2008, the NRW state government began the NRW Hydrogen HyWay Program, which will provide over €60 million in funding (2008–2011) to support these projects [17]. The NRW Fuel Cell and Hydrogen Network, a nonprofit organization funded by the state and composed of 360 members from industry and science, is working to implement this program. It will include demonstrations of light utility service vehicles, midi buses, and articulated transit buses and passenger cars, together with stationary and special applications, in addition to projects in hydrogen production and infrastructure build-up strategies. Two hydrogen-powered micro buses have been in service since 2009, and there are six hydrogen stations already in place in the state [16, 18]. NRW recently joined the CEP, and plans to demonstrate

10 passenger vehicles and build one additional public fueling station as part of this program [13]. There are also plans to build additional stations as part of the H$_2$ Mobility program.

38.3.1.3 H$_2$ Mobility

There is an issue often called the "chicken and egg dilemma," which refers to the fact that it is difficult to deploy vehicles and infrastructure simultaneously because vehicle manufacturers are reluctant to offer FCVs for sale when a robust fueling infrastructure is not in place, but infrastructure providers are also reluctant to build fueling stations unless the cars are available to guarantee a market for their product. To help circumvent this problem, key stakeholders from the automotive, oil, utility, and industrial gas sectors, as well as NOW as a public–private partnership, put in place an agreement to investigate joint development of a hydrogen infrastructure in Germany to support commercialization of FCVs in Europe. The initiative builds on the accomplishments made in CEP and offers a way forward from demonstration to full deployment and commercialization. This approach is unique in that players from the various industries involved have agreed to work together to jointly develop a commercialization plan, which reduces the new technology risk for each individual company.

The original Memorandum of Understanding which created the H$_2$ Mobility initiative was signed in September 2009 by Daimler, EnBW, Linde, OMV, Shell, Total, Vattenfall, and NOW, and goes hand in hand with the Letter of Understanding signed previously by all seven auto manufacturers involved in FCV development targeting commercialization of FCVs by 2015. As a first step, the H$_2$ Mobility partners, along with numerous other stakeholders, commissioned a 2010 study using aggregated proprietary data to compare the performance, economics, and sustainability of several powertrains, including FCVs [5]. Following that, the H$_2$ Mobility initiative has defined two successive phases. Phase 1 (2009–2011) is focused on analyzing the business case for infrastructure development, developing a business plan, and setting up a joint venture to carry out this plan. This includes a techno-economic feasibility analysis, scenario development, and analysis of what government policies are needed to support a successful roll-out from a business perspective. Phase II (2011+) will involve implementation of the business plan developed in phase I. Following these two phases, the partners plan to formulate a roll-out scenario for the other regions in the EU in conjunction with the EU's Fuel Cell and Hydrogen Joint Technology Initiative, contingent on the success of the first two phases [19].

38.3.1.4 Additional Resources

- NOW: *http://www.now-gmbh.de/en/home.html*
- CEP: *http://www.cleanenergypartnership.de/en/news/*
- NRW Fuel Cell and Hydrogen Network: *http://www.fuelcell-nrw.de/*
- Report – A portfolio of powertrains for Europe: a fact-based analysis: *http://www.zeroemissionvehicles.eu/*.

38.3.2
Japan

Japan has established itself as one of the world's leaders in hydrogen and fuel cell technology. Home to three large automotive original equipment manufacturers (OEMs) that are all pursuing the technology, the government has made both R&D and demonstration projects a priority. In 2007, as part of the "Cool Earth" initiative, FCVs, residential fuel cells, and hydrogen production, storage, and transport were identified as among 21 innovative technologies to be given increased focus by the government. In 2009, Japan committed to reducing its greenhouse gas emissions by 25% by 2020 (relative to a 1990 level), and the Prime Minister has indicated that hydrogen and fuel cells will play a role in achieving this goal [20].

In terms of government support for hydrogen and fuel cell activities, the Ministry of Economy, Trade, and Industry (METI) funds the New Energy and Industrial Technology Development Organization (NEDO), which then funds and manages R&D and demonstration projects carried out by universities, industries, and national laboratories.

For a number of years, the Fuel Cell Commercialization Conference of Japan (FCCJ), an industry association, has targeted commercialization of FCVs in 2015. The Japan Hydrogen and Fuel Cell (JHFC) Demonstration Project, which ran from 2002 to 2010, was the country's main passenger vehicle demonstration program, and successfully demonstrated over 60 vehicles and 15 stations over this period (Figure 38.2). In early 2010, an industry announcement from 13 major Japanese auto manufacturers and energy companies stated that they are on track to meet the 2015 commercialization goal, and plan to construct 100 hydrogen stations by 2015 in four major metropolitan areas [8].

38.3.2.1 Japan Hydrogen and Fuel Cell Demonstration Project (JHFC)

Japan's major vehicle demonstration program, the JHFC Demonstration Project, took place in two phases: JHFC-1 from 2002 to 2005, and JHFC-2, from 2006 to 2010 [22]. The project featured demonstration of five models of FCVs, one hydrogen internal combustion vehicle, and one fuel cell bus (Figure 38.3). A follow-on project is expected to be announced sometime in 2011.

The JHFC was funded through METI, and became a NEDO project after 2009. It built on fundamental, basic research happening in NEDO, including hydrogen production, delivery, and storage systems, and consisted of a vehicle demonstration run by the Japanese Automobile Research Institute (JARI), and also an infrastructure demonstration run by the Engineering Advancement Association (ENAA) of Japan. The Japan Petroleum Energy Center (JPEC) and the Japan Gas Association (JGA) were also involved. The project involved eight automakers and 17 energy and infrastructure companies, and aimed to demonstrate the feasibility of FCVs and their infrastructure, in addition to collecting real-world data on hydrogen production, vehicle, and infrastructure performance, safety, efficiency, and environmental impacts. The project also supported the target of commercialization by 2015 by developing roadmaps for vehicle and infrastructure deployment [23].

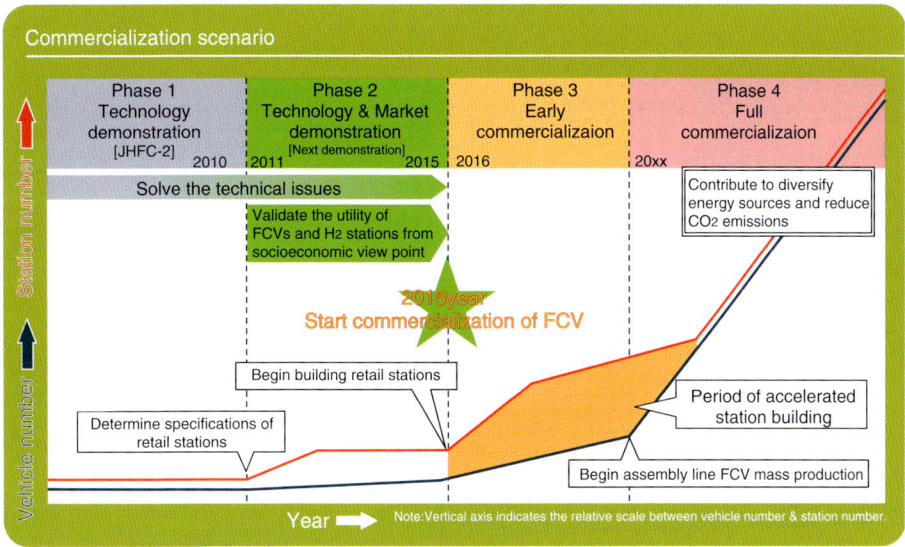

Figure 38.2 FCCJ's FCV commercialization scenario [21].

Figure 38.3 Hydrogen-powered vehicles demonstrated in the JHFC program [21].

In the first phase, over 50 FCVs were demonstrated, and 12 stations were built (Figure 38.4) [24]. Data collection verified the high efficiency of FCVs under actual conditions, and also "well to wheel" efficiency of the entire process including production and distribution of the fuel. In the second phase, the project was expanded and 700 bar fueling was introduced with upgrades to stations and new vehicles. Some of the vehicles were tested by fleet users, and new types of small hydrogen-powered vehicles were introduced, such as fuel cell wheelchairs and

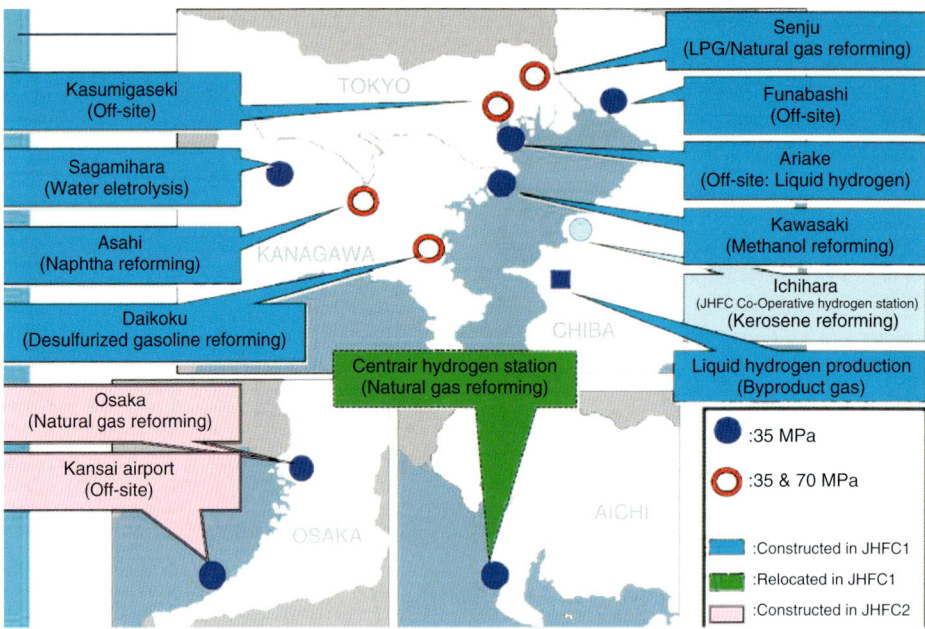

Figure 38.4 Hydrogen stations in the JHFC program [21].

bicycles [21]. Data continued to be collected, supporting continued efficiency and cost calculations, verification of fuel economy and environmental impacts, and also codes and standards development, which is very important in Japan where very strict regulations cause infrastructure costs to be significantly higher than in other countries. The vehicles in this phase showed significant improvements in fuel economy, and the 700 bar stations demonstrated an efficiency drop of only 1–2% [22].

An emphasis on raising public awareness was also included in the project. An example is the building of the JHFC park, a facility which served as the base of JHFC operations and also offered facility tours, events for children, and test drives. In late 2009, the project conducted a "Long Distance Demonstration Drive" event, in which three FCVs (Nissan FCHV-adv, Nissan X-Trail FCV, and Honda Clarity) traveled 1100 km from Tokyo to Fukuoka, stopping only twice to refuel. Media events at government offices in several cities along the way served to introduce the technology to the public, emphasizing the increased range of the new generation of vehicles [20].

The JHFC was significant in moving the state of the FCVs and infrastructure closer to commercialization. The stations and vehicles operated for over 8 years without any serious accidents, demonstrating the safety of the technology. Technical advances were demonstrated and measured on both the vehicle and infrastructure sides. Eventually, the project operated 15 stations clustered in five metropolitan areas, including four stations offering 700 bar refueling, and demonstrated a wide variety of production methods, including reforming of gasoline, naphtha, methanol, and natural gas, water electrolysis, and purification of industrial by-products. In total, the project developed specifications for 13 different types of fueling stations based on actual experience. Based on the work done on estimating infrastructure costs, targets were developed for cost reductions of various components that must be met for commercialization in 2015. Further, regulatory actions needed to support commercialization by 2015 were identified and prioritized.

This body of experience and data will be critical as the industry begins mass production and infrastructure expansion over the next several years, and the next phase of JHFC is expected to continue to support this process. Priorities for future work include developing an infrastructure model for commercialization, development of a process to validate infrastructure from a safety perspective, and continuing to lower costs for both vehicles and stations [22].

38.3.2.2 Hydrogen Highway Project
The Japanese Hydrogen Highway Project, started by METI in 2009, falls under the umbrella of the "Hydrogen Energy Social Infrastructure Development Demonstration Project," which "aims at creation of a hydrogen society in the future." The project is run by the Research Association of Hydrogen Supply/Utilization Technology (HySUT), an association of 13 private companies founded to run demonstration projects, and focuses on demonstrating FCVs – both buses and taxis – operating between downtown Tokyo and Haneda and Narita airports. The project also includes hydrogen production, transportation via high-pressure trailer, and construction and operation of three hydrogen stations [25, 26]. The first vehicles went into service in late 2010 and early 2011 [27, 28].

38.3.2.3 Activities in Fukuoka Prefecture
Hydrogen demonstration activities have also been undertaken by the regional government in Fukuoka Prefecture, an industrial area located on Kyushu Island. Similarly to the state of NRW in Germany, industrial plants in Fukuoka generate large amounts of by-product hydrogen (500 million m^3 annually), and there is an existing 10 km hydrogen pipeline in place. The Hy-Life project, started in 2004, encompasses R&D, demonstrations, and development of work forces and new industries based on hydrogen. In addition to stationary demonstrations (discussed below), this project includes the construction of a hydrogen highway between the cities of Kitakyushu and Fukuoka consisting of two 35 MPa stations. Both stations opened in 2009, with one producing hydrogen from renewable energy and the other using hydrogen from the existing pipeline. Beginning in 2009, three vehicles (two Toyota FCVs and one Mazda hydrogen rotary engine vehicle) began operation and are being used by municipal government officials and the electricity company [29].

38.3.2.4 Additional Resources

- JHFC: *http://www.jari.or.jp/jhfc/e/index.html*
- FCCJ: *http://fccj.jp/index_e.html*

- NEDO: http://www.nedo.go.jp/english/index.html
- HySut: http://hysut.or.jp/en/index.html.

38.3.3
United States

The USA has been a strong supporter of hydrogen and fuel cell technologies over the last several decades, despite a weakening of federal government support over the past few years. The USA was one of the first countries to implement a transportation demonstration program for FCVs, and was instrumental in helping several other countries to develop roadmaps for hydrogen and fuel cell technology and in facilitating international collaboration in this area.

The main funding body in the U.S. federal government is the Department of Energy (DOE)'s Fuel Cell Technologies Program, although some activities, specifically bus demonstrations, are supported by the Department of Transportation. The DOE's FCV and infrastructure demonstration program, called the Technology Validation Program, began in 2004 and is slated to conclude by 2012. Several U.S. states are also running demonstration programs, notably California, which is targeted by automakers as one of the lead markets for vehicles, and Hawaii, which in partnership with GM is in the first stages of implementing the Hawaii Hydrogen Initiative.

38.3.3.1 The DOE Technology Validation Program

The DOE's Fuel Cell Technologies program is largely focused on R&D, but since 2004 has included a demonstration program known as the National Learning Demonstration, or the Technology Validation Program. This program builds on R&D work to validate the technologies and gather both performance and economic data under real-world operating conditions in order to assess the status of the technology and its potential in the market. The program includes demonstrations in several application areas, including stationary power and early markets, but only the vehicle and infrastructure portions will be discussed here. Government and industry shared equally in the cost of the program.

The passenger vehicle and infrastructure portion of the program was structured into four teams led by either an auto maker or an energy provider, each focused on a specific geographic area representing three sets of climatic conditions. Since the project began in 2004, the demonstration vehicles have driven over 2.8 million miles, and over 130 000 kg of hydrogen has been produced or dispensed at the stations. Several models of vehicles were tested, as well as various station configurations including delivered hydrogen and on-site generation. In addition to testing the performance of the vehicles, the project also incorporated the development of safety plans, codes and standards development, and education and outreach activities.

As of 2010, the project had incorporated data from 144 vehicles and 23 fueling stations [30]. An enormous amount of data was collected and analyzed by the National Renewable Energy Laboratory (over 95 GB of on-road data representing over

Table 38.1 Status and targets of the US DOE Technology Validation program [33].

	Current status (demonstrated by technology validation)	Program targets: phase 1 – in progress	Ultimate targets
System efficiency (%)	53–59	60	60
Fuel cell system durability	2500 h (~75 000 miles)	2000 h (~60 000 miles)	5000 h (~150 000 miles)
Vehicle range (miles)	Up to 254	250	300
Fuel cost	$7.70–10.30 (projected, from on-site natural gas reforming)^a $10.00–12.90 (projected, from on-site electrolysis)^a	$3 per gge^b	$2–4 per gge^b
Refueling rate (kg min^{-1})	0.77	1.0	1.67

Abbreviation: gge, gasoline gallon equivalent.
^{a}Projections are based on estimates of costs provided by project partners, projected to a production rate of 1500 kg per day in an early market roll-out – they do not include significant cost-reductions that would result from deployments of fueling stations in a more mature market. For widespread market penetration (assuming a build-out rate of 500 fueling stations per year), DOE independent panels estimated the cost of producing hydrogen to be $2.75–3.00/kg using distributed electrolysis (2000).
^{b}Hydrogen cost targets assume a production rate of 1500 kg per day at each station and a build-out rate of 500 stations per year.

427 000 individual trips), generating many useful outputs for both industry and the DOE program [31]. Among other objectives, the National Learning Demonstration aims to validate that certain DOE targets have been met under realistic operating conditions. These targets and their current status can be seen in Table 38.1. The program has shown that fuel cell durability exceeds 2500 h, and that range exceeds 400 km. Additionally, in 2009 the program conducted a long-distance range test of the Toyota FCHV-adv FCV and demonstrated a range of 692 km without refueling [32]. Additional accomplishments of the program were verifying that efficiency was not sacrificed for improvements in durability and freeze capabilities in second-generation vehicles, projection of early market hydrogen production costs, and analysis of fuel cell power degradation [31].

In 2010, the projects led by Ford and Chevron were completed, and the GM and Mercedes-Benz North America projects are scheduled to be completed during 2011. Although some stations have been decommissioned, as of 2010 over half of the stations put into place as part of the National Learning Demonstration were still in operation [31].

The National Learning Demonstration program also included demonstrations of fuel cell buses. In collaboration with the U.S. Department of Transportation, the program has operated nine fuel cell buses at five sites, demonstrating fuel economies with improvements of 39–141% over traditional technologies such as diesel and natural gas [32].

Originally, a second phase of the program was planned for 2010–2015, but owing to recent funding cuts in the DOE program, it is unclear whether this will happen.

38.3.3.2 State Activities

Although the future of government-funded hydrogen vehicle and infrastructure demonstrations in the USA is somewhat in doubt currently at the federal level, several states continue to move forward boldly. California has had a strong program for several years, as described below. Other states also have industry-led plans – in Hawaii, GM and the Gas Company plan to build 20–25 hydrogen fueling stations by 2015 in cooperation with several other companies as part of the Hawaii Hydrogen Initiative [34]. SunHydro, a Connecticut-based company, plans to build at least nine stations along interstate highways on the east coast and opened their first solar-powered station in Hartford, CT in late 2010, with Toyota donating 10 vehicles for demonstration purposes [34].

38.3.3.2.1 California

Building on the State of California's past leadership in the promotion of clean vehicles, the California Fuel Cell Partnership (CaFCP) was formed in 1999 to demonstrate the potential and feasibility of FCVs. Initially, the partnership consisted of the California Air Resources Board and the California Energy Commission (both government agencies) and six companies: Ballard Power Systems, Daimler Chrysler, Ford, BP, Shell Hydrogen, and ChevronTexaco, and has since expanded to include 33 organizations from industry, government, and academia [35]. Since its inception, the partnership has played a key role in coordinating stakeholder actions in California to demonstrate and develop FCVs (both cars and buses) and their accompanying infrastructure. At its headquarters in Sacramento, CaFCP maintains an educational center, a fueling station and shared vehicle maintenance facilities [35].

During the initial phase of their demonstration activities between 1999 and 2003, the CaFCP partners concluded that FCVs were indeed technically viable and undertook a second phase (2004–2007), aiming to determine whether the technology had the potential to be successful in the marketplace as an alternative to conventional vehicles. Based on demonstration activities and analysis of the potential customer base and public policy, the partners determined that FCVs do have potential to be a commercially viable product. The third, and current, phase of the project is focused on developing a market for the commercialization of these technologies [35].

The CaFCP partners collaborate on a number of activities, including vehicle demonstrations, infrastructure development planning and analysis, codes and standards development, and educational activities. Since it began operations, the CaFCP has demonstrated 359 FCVs and buses, and over 155 continue to operate as of February 2011 [36]. As of February 2011, there are currently 24 hydrogen fueling stations in the state, although only four are publicly accessible. Eight stations are in development, with 11 additionally in the planning or funding phase [36]. These intense infrastructure development plans have attracted the attention of car manufacturers, who now see California as one of the leading early market areas for FCVs.

With the vehicles currently at a stage ready for commercialization, CaFCP has turned its recent focus more towards infrastructure development. Among their current and future activities are the development of next-generation hydrogen stations incorporating additional renewable technologies, such as on-site solar electrolysis (at the planned Los Angeles station) and hydrogen from biogas (at the planned Fountain Valley station). CaFCP is also playing a strong role in helping the industry move towards commercially viable retail fuel sales by assuring that necessary codes, standards, and regulations are in place, and helping to develop business models and customer experiences that are viable in the market. In this vein, they are now performing work on hydrogen quality, a vehicle authorization system, and a mobile phone application that will alert customers of station status [36].

38.3.3.3 Additional Resources

- DOE Fuel Cell Technology Validation Program: *http://www1.eere.energy.gov/ hydrogenandfuelcells/tech_validation/*
- California Fuel Cell Partnership: *http://www.fuelcellpartnership.org/*.

38.3.4
European Union

The EU has also emerged as a global leader in developing and demonstrating hydrogen and fuel cell technologies. The European Strategic Energy Technology (SET) Plan identifies fuel cells as one of several key technologies needed to meet the EU's 20–20–20 goal (20% increase in renewables, 20% reduction in emissions, and 20% increase in efficiency by 2020), and the EU has funded R&D and demonstrations in this area since 1986 through its Framework Program structure. Today, the main government funding body in the EU for hydrogen and fuel cell technologies is the Joint Undertaking on Fuel Cells and Hydrogen (FCH JU), which was created in 2008 and is a public–private partnership that aims to accelerate technology development by leveraging the combined resources of government, industry, and the research community [37]. The FCH JU has a total budget of €940 million over 2008–2013, half of which is government funding and half is industry in-kind contribution. About 41–46% of the budget is allocated for demonstration activities, most of which are large projects involving many partners and locations through the EU [38]. In the area of transportation and infrastructure, important projects are the Clean Hydrogen in European Cities (CHIC) project, which focuses on buses along with its predecessor projects, and the current H2moves Scandinavia project on hydrogen passenger cars. These demonstration projects are assisted by NextHyLights, a supporting project that performs analysis of technologies, demonstration projects, and roll-out strategies in order to guide planning and development of future large-scale demonstration activities in the EU [39].

38.3.4.1 Fuel Cell Bus Projects

The current CHIC project focuses on the commercialization of fuel cell buses, and builds on work done by two previous European fuel cell bus demonstration projects: CUTE/ECTOS and HyFLEET:CUTE. Together, the Clean Urban Transport for Europe (CUTE) and Ecological City Transport System (ECTOS) projects ran from 2001 to 2005 and demonstrated 30 fuel buses in 10 European cities, in addition to providing buses to partner programs in Perth, Australia, and Beijing, China. The project showed that fuel cell buses could be delivered using series production, and used safely and reliably in public transit routes. Fueling stations were constructed and operated in each of the project cities, incorporating on-site renewable production of hydrogen and achieving station availability of 80%. Further, bus drivers and customers accepted and welcomed the buses with a high level of satisfaction. Throughout the project, areas for future improvement were noted for follow-on work, including: required improvements in bus efficiency, reductions in vehicle weight and noise, increasing the reliability and efficiency of hydrogen production equipment, improving hydrogen purity, station availability, and time required for the approval process [40].

HyFLEET:CUTE was the next bus project run in the EU, from 2006 to 2009. It built upon the work done in CUTE/ECTOS, using the existing station infrastructure (plus one additional station built in Berlin) to demonstrate 47 hydrogen-powered buses (33 fuel cell and 14 ICE) in public transit in 10 cities on three continents. The project also covered the design, construction, and testing of next-generation fuel cell and ICE buses, in addition to improvements and optimization of the existing stations. The project was very wide ranging, consisting of 31 partners from the government, industry, research, and consulting sectors. In addition to demonstrating and evaluating the vehicle and infrastructure technology, the project also sought to develop recommendations for key European Commission and industry decision-makers on the impacts of using hydrogen for transportation, and also educating new EU Member States on the technology. The project eventually demonstrated more than 2.6 million km driven, over 555 000 kg of hydrogen dispensed, vehicle availability of 92%, and station availability of 89.5%, substantial improvements over previous projects. In addition, 79% of the hydrogen produced on-site at the stations was renewably produced, up from 56% in the previous project [41]. Although the project did show improvements in the performance of the technologies, the partners concluded that owing to the remaining technical and cost challenges, government-funded projects were still necessary to carry the technology forward towards commercialization [42, 43].

The CHIC project, launched in 2010, aims to address these challenges and pave the way for full commercialization starting in 2015. This project involves 25 partners and plans to deploy 26 fuel cell buses from three different manufacturers in five cities: London, England; Aargau, Switzerland; Bolzano/Bozen and Milan,

Italy; and Oslo, Norway. Two 200 kg per day fueling stations will be built in each city, with a goal of 98% station availability. On the bus side, the project is striving for a lifetime of over 6000 h, availability of at least 85% and mileage below 13 kg per 100 km [41].

The project is also undertaking additional activities designed to facilitate commercialization. Through a city partnership scheme, information and experiences will be shared between the cities in the program and several that are considering using fuel cell buses in the future. Assessment of the technology performance, environmental affects, economics, and social acceptance will also be performed [41].

38.3.4.2 H2moves Scandinavia

The H2moves Scandinavia project builds on previous work done in Norway, Sweden, and Denmark by the Scandinavian Hydrogen Highway Partnership (SHHP) and HyNor by linking and expanding existing hydrogen infrastructure networks and introducing new vehicles to utilize this infrastructure (Figure 38.5). The project is supported by Danish and Norwegian government funds and also the FCH JU. As of March 2011, there are currently seven hydrogen fueling stations in operation as part of previous projects, with three under construction and seven more in planning. H2moves will build one additional 700 bar fueling station in Oslo and also provide one mobile refueler to support five planned road tours. With the additional stations in place, the Scandinavian hydrogen infrastructure network will be linked with the CEP network in Germany. The 17 vehicles to be demonstrated (Daimler B-class FCEVs, Alfa Romeo Mito FCEVs and Th!nk plug-in

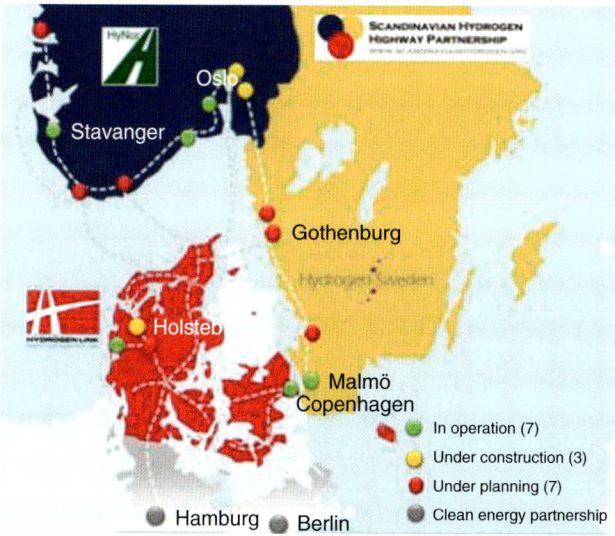

Figure 38.5 The hydrogen fueling station network in Scandinavia [44].

city cars with fuel cell range extenders) will be delivered in early 2011, driven by private customers and monitored for performance and driving experience. The project will also train maintenance workers, and perform an assessment of codes and standards affecting vehicles and station certification in order to make the process faster and more efficient [44].

38.3.4.3 Regional and Member Country Activities

On a regional level, the European Regions and Municipalities Partnership for Hydrogen and Fuel Cells (HyRaMP), founded in 2008, facilitates communication among projects in the EU member states and assists in coordinating regional infrastructure build-up in Europe.

Several member states fund hydrogen and fuel cell demonstration activities on their own in addition to the FCH JU funding scheme, and many of these programs contribute funding to the FCH JU projects described above in their country or region. Germany has extensive demonstration programs of its own, as described above. Other European countries and cities have smaller demonstration programs of various types, but are too numerous to list fully here. Examples are London's plans to deploy 20 fuel cell taxis by 2012, Italy's several projects demonstrating FCVs and buses, and the WaterstofNet project in The Netherlands, which will demonstrate several types of technology including hydrogen-powered ships, mobile fueling stations, and electricity generation using by-product hydrogen.

38.3.4.4 Additional Resources

- FCH JU: *http://www.fch-ju.eu/*
- H2moves.eu (information on past demonstration projects in Europe): *http://www.h2moves.eu/index.html*
- SHHP: *http://www.vatgas.se/shhp/*
- CHIC: *http://chic-project.eu/*
- NextHyLights: *http://www.nexthylights.eu/*
- HyRaMP: *http://www.hy-ramp.eu/*.

38.3.5
Canada

Canada has also implemented hydrogen and fuel cell demonstration programs for transportation applications, mainly through Natural Resources Canada's ecoENERGY Technology Initiative. At the 2010 Vancouver Olympics, 20 fuel cell buses were deployed in transit service, in addition to 15 fuel cell passenger vehicles used to transport VIPs and the media. Eight fueling stations were built to support this fleet.

In addition, Canada is also running several other projects, including five Ford Focus vehicles in Vancouver, three ICE shuttle buses used to ferry lawmakers in Ottawa, two ICE shuttle buses on Prince Edward Island, and an integrated waste hydrogen utilization project fueling 11 ICE shuttle buses and pick-up trucks [45].

38.3.6
South Korea

South Korea is another important country for hydrogen and fuel cell technology development and demonstration, with strong government programs and policy and industry expertise. As part of its "Low Carbon, Green Growth" program, the government is aiming to foster market development in the new and renewable energy industries, including hydrogen and fuel cells. Government hydrogen and fuel cell demonstration programs are funded through the Ministry of Knowledge Economy (MKE) and consist of an FCV demonstration program as well as residential power.

38.3.6.1 Domestic Fleet Program

South Korea's vehicle demonstration program focuses on demonstrating vehicles from its domestic automaker, Hyundai/Kia, and aims at preparing the market and technologies to support Hyundai/Kia's commercialization strategy targeting precommercial production (thousands of vehicles) beginning in 2012 and commercial production (tens of thousands of vehicles) beginning in 2015. The domestic fleet program has been rolled out in two phases, with the government providing 50% of the funding in the first phase and 30% of the funding in the second phase. The first phase, running from 2006 to 2010, demonstrated 30 passenger cars and four buses driving a total of 743 500 km. Vehicles fueled at five new stations that were built as part of the project, in addition to five existing stations that were also utilized. In the second phase of the project, running from late 2009 to late 2011, 80 passenger cars will be deployed in two cities (Seoul and Ulsan). Several additional stations will be built, giving a total of 13 nationwide by 2012. Three of these will be 700 bar stations. The focus of this second demonstration phase is on validating the 700 bar fueling technology in addition to resolving the remaining technical issues on the vehicle side [46, 47] (Figure 38.6).

38.3.7
China

With China's rapidly growing, technology-hungry population (it is now the world's largest passenger car market) and, air pollution and emissions challenges, it represents a possible large market for FCVs. Recognizing this, the Chinese government is funding R&D and demonstrations in this area and developing domestic production capability in the hope of becoming a global player in the automotive industry [48].

Hydrogen and fuel cells were named as one of seven key areas in the National Mid-to-Long Term Science and Technology Plan, and federal demonstration projects are funded by the Ministry of Science and Technology (MOST) under the "863 Program" [49].

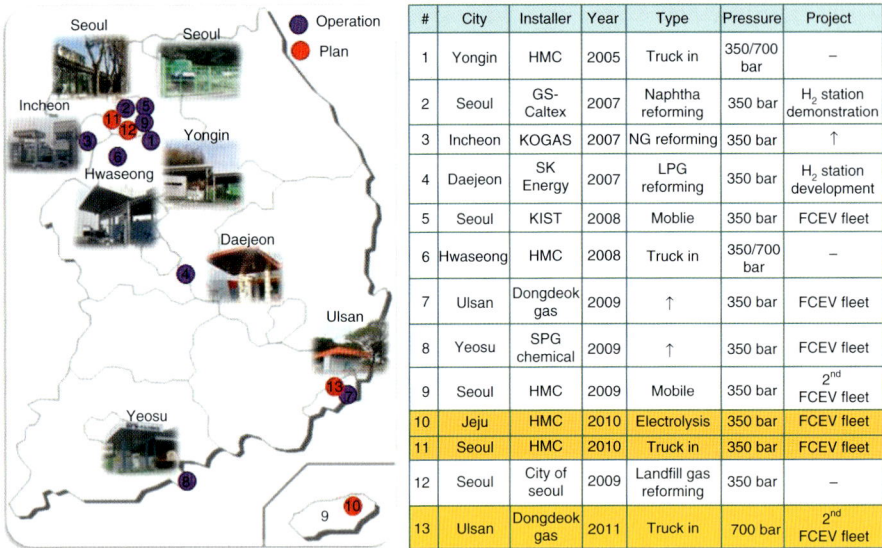

Figure 38.6 Current stations and plans in South Korea [47].

38.3.7.1 GEF/UNDP-China FCB Demonstration

Funded by MOST, in addition to municipal governments and GEF/UNDP, three fuel cell buses from Daimler were demonstrated in Beijing from 2006 to 2007. The buses were fueled at the Beijing Hydrogen Park, which was built in 2006 and consists of a fueling station along with an R&D center and vehicle maintenance facilities. This project was followed by a second phase demonstrating six buses in Shanghai produced by China's SAIC (Shanghai Automotive Industry Corporation) from 2007 to 2009 [48, 49].

38.3.7.2 Beijing Olympics and Shanghai World Expo

In recent years, China's FCV demonstrations have focused around major national events. In 2008, 20 fuel cell passenger vehicles and three buses (all manufactured in China) were demonstrated at the Beijing Olympics, where they transported VIPs and officials. The vehicles were fueled at the existing Beijing Hydrogen Park.

In 2010, China demonstrated a large fleet of various clean vehicles, including six fuel cell buses, 90 fuel cell cars, and 100 fuel cell "tourist cars" (small, lightweight vehicles) at the World Expo in Shanghai. As in their previous demonstrations, all of the vehicles were manufactured in China. Two mobile stations and two fixed stations (one using purified by-product hydrogen), were used to fuel the vehicles [48, 49].

38.3.8
Automaker Demonstration Programs

In addition to participating in government-led demonstration programs and conducting their own internal testing, several automakers have run their own demonstration/limited deployment programs in recent years. Focused on collecting vehicle data and customer feedback, these programs have put vehicles directly in the hands of celebrities, everyday families, and public agencies.

Toyota was the first automaker to release an FCV when it began leasing the FCHV in 2002. These vehicles were leased in very limited (unreleased) numbers in Japan and the USA. In September 2008, Toyota began leasing the next generation FCHV-adv, but numbers of vehicles and costs have not been published [50].

GM's Project Driveway demonstrated 119 Chevrolet Equinox vehicles from 2007 to 2010 with customers in New York, Los Angeles, and Washington, DC. Customers were able to use the vehicles for their daily driving needs for periods of a few days to several months. Vehicles, fuel, and insurance were provided free of charge in exchange for regular feedback. GM has stated that their learnings from the program have been significant, and they are currently continuing the program with a new set of customers driving previously used Equinoxes that have been refitted with upgraded powertrains and components. Comparisons between the performances of the new and old vehicles should help GM evaluate the effectiveness of the updates in preparation for roll-out of its next generation of vehicles [51, 52].

Honda has also placed a number of its Clarity FCVs with customers via a lease program that charges $600 per month. The three-year program started in 2008, and Honda hopes to trial about 200 vehicles over that period in the USA, Europe, and Japan, with the majority of the customers in southern California. However, the vehicle roll-out has been slower than planned owing to slower than expected station build-up [53, 54].

Daimler is also starting limited deployment of its new B-class FCVs, planning to lease 200 to customers in Europe and the USA through a program that began in 2010 [55]. In 2011, Daimler successfully completed the F-Cell World Drive, in which three B-class FCVs drove over 30 000 km each through 14 countries on four continents, in an effort to demonstrate the vehicles' capabilities and the need for infrastructure [56].

38.4
Stationary Power and Early Market Applications

Fuel cells are also being demonstrated and deployed in several different stationary and early niche market applications. They can be used in combined heat and power (CHP) systems to provide electricity and hot water to homes (generally around 1 kW in size) and to businesses and other installations such as grocery stores and hospitals ($\sim$200–400 kW). Larger fuel cells (MW scale) can be used for industrial power generation. Fuel cells are also being used as uninterruptible power supplies (UPS) or back-up power (in the under 10 kW range) for various applications including cell phone towers and data centers; and to power forklifts and other material handling equipment.

Thousands of fuel cells have already been deployed successfully in these applications throughout the world, with many products already commercially available [57]. Despite this, the technologies still need support to become fully established in the market, and governments are continuing to fund demonstration projects to further improve performance, lower costs, and develop markets – in addition to providing subsidies and incentives that support early commercial deployments. This section discusses current major demonstration projects in stationary power and early market applications, and government programs that support early market deployments. In the residential fuel cell sector, Germany, Japan, and Denmark have major programs, with the technology nearing commercial status in Japan. Strong policies in the USA are supporting commercial roll-outs in the areas of large stationary fuel cells, UPS, and forklifts, with only a few small demonstration programs in other countries.

38.4.1
Japan

Japan is currently the leader when it comes to demonstration and commercialization of residential fuel cell systems. In recent years, the government has led a coordinated approach involving collaboration between several companies to reduce the cost of residential fuel cell systems and market them to consumers in an attractive way.

From 2004 to 2008, the government funded a demonstration program that deployed over 3300 residential CHP polymer electrolyte membrane (PEM) units in homes, evaluating their performance, reliability, and durability and showing that the products were ready for a government-supported commercial launch. During the years of the project, the units also showed significant improvements in efficiency, energy conservation, and carbon dioxide emission reductions over conventional technologies [58].

Following this project, the current ENE-FARM program was launched in 2009, representing early commercialization of the technology, and supporting Japan's goal of full commercialization of residential fuel cell systems by 2015 (Figure 38.7). Coordinated by the government, three manufacturers, each in partnership with one of three energy companies, are offering consumers residential systems that are co-branded under the ENE-FARM logo. The 0.7–1 kW systems are fueled by either natural gas, propane, or kerosene and provide electricity and hot water to the homes. The government subsidizes half of the cost of the systems, with the other half paid for by the customer [59]. The program also encompasses codes and standards activities geared towards lowering the costs of future systems, and places a focus on the collection and analysis of data to fuel continuous improvement of the systems in order to bring costs down in future product generations.

Another objective of the program is to increase manufacturing volumes to decrease costs. Japan's target cost for residential fuel cell systems is 0.5–0.7 million yen (~€4100–5900) – current costs of the system are around 2–2.5 million yen (~€16 700–21 000) [58–60]. Despite the fact that current prices offer customers a payback period of over 10 years, even including the subsidy, the units are selling fairly well – as of January 2011, over 5200 units had been installed. The project partners credit this success to the environmentally beneficial attributes of the systems [60]. In the future, manufacturers plan to expand their market by reducing

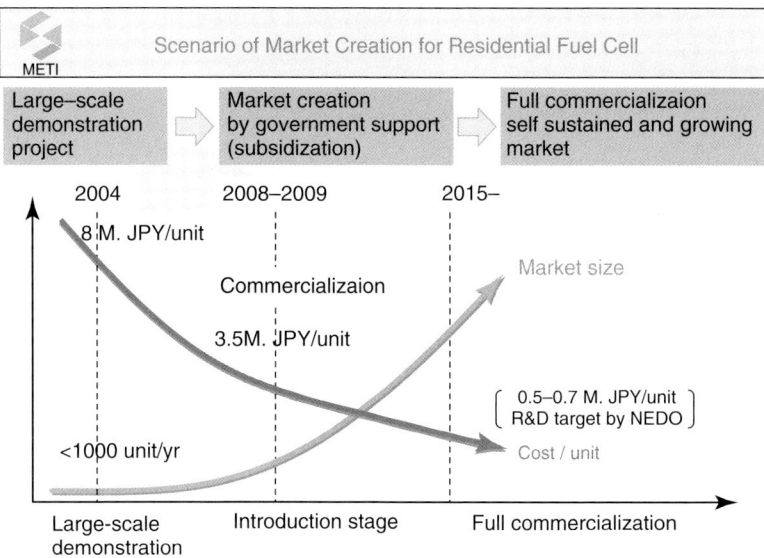

Figure 38.7 Japan's strategy for commercialization of residential fuel cells [59].

the size of the systems so that they can be used in apartments, adding solar power, and marketing them for back-up power, in addition to further decreasing costs [60]. Currently, manufacturers have a capacity to produce 10 000 units per year, with a target of 40 000 units per year by 2015 [58].

Japan is also running a residential demonstration program on solid oxide fuel cell (SOFC) units, with 36 units installed as of March 2011 [58, 59].

38.4.1.1 Regional Activities

As is the case with transportation applications, Fukuoka Prefecture has strong regional government demonstration projects for stationary fuel cells. In 2008, the Prefecture launched its "Fukuoka Hydrogen Town" project, which promoted the installation of 150 ENE-FARM units in the community [61].

In January 2011, Japan announced an additional project under the "Hydrogen Energy Social Infrastructure Development Demonstration Project" umbrella. The "Hydrogen Town Project," run by METI, will be located in Fukuoka Prefecture and will install pipelines to bring hydrogen from Nippon Steel Corporation to homes and businesses to be used in fuel cells. The initiative will focus on testing the feasibility of the system and also the business model, and on issues related to pipeline transport of hydrogen, hydrogen metering, use of odorants, and fuel cell performance [62].

38.4.1.2 Additional Resource

- ENE-FARM web site (in Japanese): *http://www.fca-enefarm.org/*.

Figure 38.8 Building the hydrogen pipe distribution network in Lolland, Denmark [63].

38.4.2
Denmark

In Denmark, the government is funding what was the first full-scale European demonstration project on residential CHP. The project, begun in 2007, will eventually install 35 2 kW PEM units in actual customers' homes, and takes advantage of the excess wind power that is produced on the island to generate hydrogen in an electrolyzer which is then piped to the homes (Figure 38.8). The project also built an educational center associated with the installations to provide information to the public. Although the systems are performing well, project leaders have concluded that cost reductions must be made to make them a competitive option in the market [63].

38.4.3
Germany

The German NIP, described above, also funds a relatively large stationary fuel cell program. The main demonstration program for residential applications is called Callux, an umbrella program overseeing demonstration projects by five utilities and three appliance suppliers to install 800 units (both SOFC and PEM) in single family homes by 2015, and also associated R&D work. As of January 2011, 111 systems have been installed (Figure 38.9). The natural gas-fueled installations are clustered in four metropolitan regions in Germany and are aimed at preparing the market for the commercial launch of residential CHP systems by developing supply chains, verifying the technical readiness and marketability of the systems, and training market partners. The projects also facilitate communications activities among stakeholders, including customers, project partners, scientists, installers, and policy makers, in an effort to discover synergies that will lead to accelerated innovation. Thus far the project has shown that the systems are not quite market

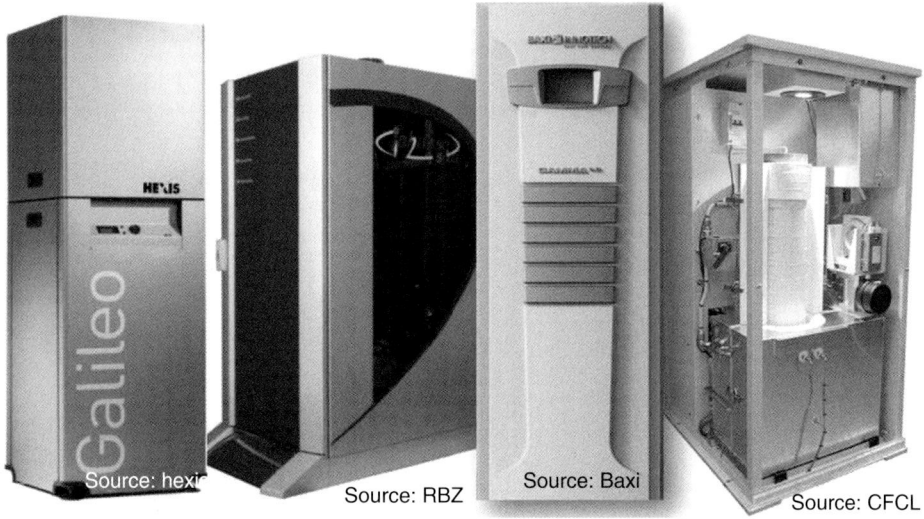

Figure 38.9 The residential units being demonstrated in Germany [64].

ready, but the practical experience gained during the project is expected to help identify issues and further improve the products [13, 64].

Germany's industrial fuel cell demonstration project, NEEDS, has demonstrated two molten carbonate fuel cell (MCFC) plants, but is now on hold following the major partner (MTU)'s decision to withdraw from the fuel cell business. Germany is continuing to demonstrate fuel cells for marine applications through its e4ships program, and for UPS systems through its Clean Power Net program [13, 64].

38.4.3.1 Regional Activities
As described above, the German state of NRW also funds demonstration projects as part of its Hydrogen HyWay program. A number of these projects have focused on stationary applications, for example, installation of six PEM units for UPS, two of which were used at stadium telecommunications towers during the 2006 World Cup [65].

38.4.3.2 Additional Information

- Callux: *www.callux.net*
- Clean Power Net: *www.cleanpowernet.de*.

38.4.4
European Union

Currently, the FCH JU is providing a relatively small amount of funding for demonstration projects in stationary applications, with the majority of funding in this area going to R&D. The program is currently participating in funding a 3 year project led by Diverse Energy of the UK to install 40 ammonia-based fuel cells to power cell phone towers in areas without grid connections. Future plans include the demonstration of a 1 MW PEM fuel cell powered with by-product hydrogen from a chlor-alkali plant and a project demonstrating 40 1–2 kW SOFC CHP units [66].

Individual countries within the EU are also funding their own projects. Germany and Denmark both have major programs demonstrating residential units, as described above. Italy and Finland, among others, also have smaller programs demonstrating fuel cells for power generation and auxiliary power on ships.

38.4.5
United States

In the USA, favorable policies including grants and tax incentives at the federal and state levels are fueling growing numbers of industry-funded deployments of stationary fuels for heat and power at installations such as supermarkets and food producers, and for telecom back-up and forklifts [67]. At the federal level, the USA offers an investment tax credit for the purchase of fuel cells for stationary power and early markets such as forklifts; and California, New York, Massachusetts, and Connecticut all have strong state-level incentives or grants that support deployment of fuel cell systems [57]. Further, the American Recovery and Reinvestment Act of 2009 provided funding for the deployment of over 1000 fuel cells for back-up power, lift trucks, and several other early market applications (Figure 38.10).

The DOE funds a small number of large CHP demonstrations through its Technology Validation program, and has also funded the deployment of over 200 smaller fuel cells between 2009 and 2010 for applications including forklifts

Figure 38.10 US early market fuel cell deployments [68].

and back-up power. Plans are in progress for a project deploying 50 5 kW CHP systems at industry locations, and to deploy fuel cells where feasible at national laboratories [68]. These projects are designed to continue the evaluation of the fuel cells' performance in real-world environments, and also to increase production volumes ordered from suppliers to support cost reductions through economies of scale.

The U.S. Department of Defense has also run several stationary fuel cell activities, having demonstrated phosphoric acid fuel cells (PAFCs) on military bases in the late 1990s and early 2000s, 91 small (1–5 kW) residential fuel cell systems through its PEM Residential Program between 2001 and 2004, and, most recently, is rolling out a program to demonstrate 44 PEM fuel cells for backup power at federal facilities [51].

38.4.6
South Korea

South Korea's high natural gas prices and low electricity prices present a barrier to economic use of fuel cells for stationary power, but the government has implemented strong programs and policies to help make them commercially viable in the market. The government's approach to commercializing technology in this area is similar to that of Japan, in that they have started with government-funded demonstration projects, and then moved on to government facilitated deployment of systems funded by customers, but supported by strong subsidies.

Taking advantage of its extensive natural gas infrastructure, South Korea installed over 200 1 kW natural gas-fueled PEM residential units between 2006 and 2010 as part of its residential fuel cell demonstration program. Units were also demonstrated at several public sites, including the President's residence, Seoul City Hall, and Gyungbook National University [69]. In 2010, the second phase of this program began, consisting of subsides up to 80% for residential installations. The government aims to install 1000 units by 2012 under this program. Starting in 2013, the subsidy will decrease to 50% [70]. The country also aims to install 100 000 residential units by 2020 as part of its "1 Million Green Homes Project." The target for this program is a cost of US$10–15 kWh^{-1}. Taking advantage of the subsidies listed above and also local incentives which can cover up to 10% of the cost, residential systems in South Korea currently cost customers around US$1500, down from the total current system cost of around US$20 000. South Korea is also looking at SOFC systems for residential use, but the program is still in the R&D phase [70].

In the area of larger stationary fuel cells for distributed power, between 2004 and 2007 the government demonstrated two 250 kW units, with a third unit operating under POSCO power funding [71]. Since then, around 36 MW of power has been installed at 15 sites (14 sites with MCFC and one with PAFC technology). These installations are supported by a feed-in tariff, a renewable portfolio standard that begins in 2012, and a requirement for public buildings to include new and renewable energy sources [47].

References

1. Ohnsman, A. (2010) Toyota Targets $50,000 Price for First Hydrogen Car, http://www.businessweek.com/news/2010-05-06/toyota-targets-50-000-price-for-first-hydrogen-car-update2-.html (last accessed 3 February 2011).
2. Hamprecht, H. (2011) Daimler Predicts Price Declines for EVs, Fuel Cells, http://www.autonews.com/apps/pbcs.dll/article?AID=/20110122/ANE/110129948/1193 (last accessed 3 February 2011).
3. Harris, S. (2010) GM to Streamline Production of fuel cell Systems, http://www.theengineer.co.uk/news/gm-to-streamline-production-of-fuel-cell-systems/1006443.article (last accessed 3 February 2011).
4. Ehret, O. and Dignum, M. (2010) Introducing hydrogen and fuel cell vehicles in Germany, in *Automobility in Transition? A Sociotechnical Analysis of Sustainable Transport* (eds. F. Geels, R. Kemp, G. Dudley, and G. Lyons), Routledge, London.
5. McKinsey & Company (2010) A Portfolio of Powertrains for Europe: a Fact-Based Analysis, http://www.zeroemissionvehicles.eu/ (last accessed 3 February 2011).
6. Daimler (2009) Automobile Manufacturers Stick up for Electric Vehicles with Fuel Cell, http://media.daimler.com/dcmedia/0-921-941776-1-1235424-1-0-0-0-0-0-11701-854934-0-1-0-0-0-0-0.html (last accessed 6 February 2012).
7. Daimler (2009) Initiative "H2 Mobility" – Major Companies Sign Up to Hydrogen Infrastructure Built-up Plan in Germany, Daimler, http://media.daimler.com/dcmedia/0-921-656186-1-1236407-1-0-0-0-0-0-11701-614316-0-1-0-0-0-0-0.html (last accessed 6 February 2012).
8. Toyota (2011) 13 Japanese Companies Eye Smooth Domestic Launch of FCVs, Toyota.
9. Jerram, L.C. (2008) 2008 Bus Survey, Fuel Cell Today.
10. Bonhoff, K. (2010) IPHE Workshop: Governmental Programmes on E-Mobility – Germany, http://www.iphe.net/docs/Events/uect/final_docs/Germany%20Bonhoff%20WS%20Ulm%20IPHE_150610.pdf (last accessed 5 January 2011).
11. Clean Energy Partnership (2011) http://www.cleanenergypartnership.de/ (last accessed 30 August 2011).
12. Clean Energy Partnership (2007) CEP Report 2002–2007. Clean Energy Partnership, Hamburg.
13. Bonhoff, K. (2011) Current progress of fuel cell and hydrogen demonstrations in Germany. Presented at the JHFC Seminar, Tokyo.
14. Berliner Flughafen (2010) World's First CO_2-Neutral Filling Station at Berlin Brandenburg International Airport BBI, http://www.berlin-airport.de/EN/Presse/Pressemitteilungen/2010/2010_03_04-CO2_freie_Tankstelle.html (last accessed 3 February 2011).
15. Ziolek, A. (2008) Activities of the fuel cell hydrogen network North Rhine-Westphalia. Presented at IPHE, http://www.iphe.net/docs/Meetings/Germany_2-08/Activities_Rhine-Westphalia.pdf (last accessed 8 February 2011).
16. Ziolek, A. (2010) The future of the electric mobility in the energy supply. Presented at the 10th Annual Meeting of the Fuel Cell and Hydrogen Network NRW, http://www.brennstoffzelle-nrw.de/fileadmin/daten/jahrestreffen/2010/Vortraege/Ziolek-EANRW_NBW_091210.pdf (last accessed 30 August 2011).
17. Kattenstein, T. (2011) Fuel cells and hydrogen in North Rhine-Westphalia. Presented at Fuel Cell and Hydrogen Energy 2011, Washington, DC.
18. EnergyRegion.NRW The EnergyEconomy Cluster and its Networks, https://services.nordrheinwestfalendirekt.de/broschuerenservice/download/70507/energyregiona4_final.pdf (last accessed 3 February 2011).
19. Williamson, I. (2010) H_2 Mobility – Towards Commercialization of Fuel Cell Vehicles. Presentation to US HTAC, http://www.hydrogen.energy.gov/pdfs/

4_williamson_0610.pdf (last accessed 8 February 2011).
20. Koguchi, H. (2010) Research and development of fuel cell vehicles and hydrogen in Japan. Presented at the 14th IPHE Steering Committee Meeting, September 2010, http://www.iphe.net/docs/Events/China_9-10/1-3_IPHE_ShangHai_10Sep21_Koguchi.pdf (last accessed 30 August 2011).
21. Tomuro, J. (2010) Current status of JHFC project. Presented at the IPHE Infrastructure Workshop, http://iphe.net/docs/2010_Infrastructure_Meeting/3.Japan.pdf (last accessed 30 August 2011).
22. Tomuro, J. (2010) JHFC project activities in FY 2009. Presented at the IPHE International Hydrogen Fuel Cell Technology and Vehicle Development Forum, http://iphe.net/docs/Events/China_9-10/2-1_Introduction%20of%20JHFC%20project%20activities%20at%20IPHE%20workshop%20in%20Shanghai.pdf (last accessed 30 August 2011).
23. (2010) Japan Hydrogen and Fuel Cell Demonstration Project, http://www.jari.or.jp/jhfc/e/index.html (last accessed 9 February 2011).
24. Takahara, I. (2005) Japan's approach to commercialization of fuel cell/hydrogen technology. Presented at the 4th IPHE Steering Committee Meeting, September 2005, http://iphe.net/docs/Meetings/Japan_9-05/Japan_Statement.pdf (last accessed 30 August 2011).
25. HySUT (2010) The Research Association of Hydrogen Supply/Utilization Technology, http://hysut.or.jp/en/index.html (last accessed 22 February 2011).
26. Kitanaka, M. (2010) Activities of the Research Association of Hydrogen Supply/Utilization Technology. Presented at the HTAC Open Meeting, June 3–4, 2010, http://www.hydrogen.energy.gov/pdfs/10_kitanaka_0610.pdf (last accessed 30 August 2011).
27. ANA (2011) ANA Introduces Fuel Cell Electric Passenger Courtesy Vehicles, http://www.ana.co.jp/eng/aboutana/press/2010/110121.html (last accessed 22 February 2011).
28. Toyota (2010) TMC, Hino to Provide fuel cell Bus for Tokyo Airport Routes, http://www2.toyota.co.jp/en/news/10/12/1207_1.html (last accessed 22 February 2011).
29. (2009) Fukuoka hydrogen strategy Hy-Life project. Presented at the Fukuoka Strategy Conference for Hydrogen Energy.
30. Marcinkoski, J. (2010) U.S. Department of Energy efforts in electrified vehicle power. Presented at the IPHE Workshop: Governmental Programs on E-Mobility, http://www.iphe.net/docs/Events/uect/final_docs/US%20Marcinkoski%20WS%20Ulm%20IPHE_150610.pdf (last accessed 30 August 2011).
31. Wipke, K. (2010) Controlled hydrogen fleet and infrastructure analysis. Presented at the 2010 DOE Hydrogen Program Annual Merit Review and Peer Evaluation Meeting, http://www.hydrogen.energy.gov/pdfs/review10/tv001_wipke_2010_o_web.pdf (last accessed 30 August 2011).
32. US Department of Energy (2010) DOE Hydrogen Program 2010 Progress Report, http://www.hydrogen.energy.gov/annual_progress10.html (last accessed 23 February 2011).
33. US Department of Energy (2010) The Department of Energy Hydrogen and Fuel Cells Program Plan (draft), http://www1.eere.energy.gov/hydrogenandfuelcells/pdfs/program_plan2010.pdf (last accessed 23 February 2011).
34. General Motors (2010) Hawaii Hydrogen Infrastructure Gets Boost, http://media.gm.com/content/media/us/en/news/news_detail.brand_gm.html/content/Pages/news/us/en/2010/Dec/1208_fuelcell (last accessed 30 August 2011).
35. California Fuel Cell Partnership (2011) California Fuel Cell Partnership, http://www.fuelcellpartnership.org (last accessed 14 March 2011).
36. Dunwoody, C. (2011) Hydrogen fuel cell vehicles: building market foundations

in California. Presented at the 7th International Hydrogen and Fuel Cell Expo 2011, Tokyo.

37. De Colvenaer, B. (2010) Fuel cells and hydrogen joint undertaking: deliverables and future perspectives. Presented at the FCH JU Stakeholders General Assembly 2010, http://ec.europa.eu/research/fch/pdf/bert_de_colvenaer.pdf#view=fit&pagemode=none (last accessed 30 August 2011).

38. De Colvenaer, B. (2011) Fuel cells and hydrogen joint undertaking. Presented at the JHFC Seminar.

39. Landinger, H. (2010) Overview and status quo of the NextHyLights project. Presented at the FCH JU Stakeholders General Assembly 2010, http://www.fch-ju.eu/sites/default/files/documents/ga2010/hubert_landinger.pdf (last accessed 30 August 2011).

40. CUTE (2006) CUTE – Clean Urban Transport for Europe: Detailed Summary of Achievements, http://ec.europa.eu/energy/res/fp6_projects/doc/hydrogen/deliverables/summary.pdf (last accessed 22 March 2011).

41. Kentzler, M. (2010) The CHIC project. Presented at the FCH JU Stakeholders' General Assembly, http://www.fch-ju.eu/sites/default/files/documents/ga2010/monika_kentzler.pdf (last accessed 30 August 2011).

42. Kentzler, M. (2009) The Achievements of the World's Largest Hydrogen Powered Bus Fleet, http://hyfleetcute.com/data/Kentzler_AchievementsWorldLargestH2Fleet.pdf (last accessed 23 March 2011).

43. HyFLEET:CUTE (2011) HyFLEET:CUTE, http://www.global-hydrogen-bus-platform.com/ (last accessed 23 March 2011).

44. Bünger, U. (2010) Status of the first European Lighthouse Project to demonstrate hydrogen fuel cell cars in Scandinavia. Presented at the FCH JU Stakeholders General Assembly 2010, http://ec.europa.eu/research/fch/pdf/ulrich_buenger.pdf#view=fit&pagemode=none (last accessed 15 August 2011).

45. Burrelle, C. (2010) Country update: Canada. Presented at the 14th IPHE Steering Committee Meeting, http://www.iphe.net/docs/Meetings/China_9-10/Canada_Statement.pdf (last accessed 15 August 2011).

46. Ahn, B.K. (2010) Status of FCV development and H_2 infrastructure in Korea. Presented at the IPHE Infrastructure Workshop, http://www.iphe.net/docs/2010_Infrastructure_Meeting/4.Korea.pdf (last accessed 15 August 2011).

47. Kim, C. (2010) R&D status and prospects of fuel cells in Korea. Presented at the Fuel Cell Seminar, http://www.fuelcellseminar.com/media/5505/kim_10_19_2010.pdf (last accessed 30 August 2011).

48. Zhang, F. and Cooke, P. (2010) Hydrogen and fuel cell development in China: a review. *Eur. Plann. Stud.*, **18** (7), 1153–1168, http://www.e-to-china.com/2010/1125/88737.html (last accessed 30 August 2011).

49. Jianing, C. and Pan, X. (2009) Chinese hydrogen update. Presented at the IPHE Joint ILC/SC Meeting December 2009, http://www.iphe.net/docs/Meetings/USA_12-09/Country_Presentations/China.pdf (last accessed 30 August 2011).

50. Green Car Congress (2008) Toyota to Begin Leasing New Fuel Cell Hybrid Vehicle, http://www.greencarcongress.com/2008/08/toyota-to-begin.html (last accessed 6 February 2012).

51. Abuelsamid, S. (2008) DC Auto Show: General Motors Introduces the First DC-Area Project Driveway Participants, Autobloggreen, http://green.autoblog.com/2008/01/22/dc-auto-show-general-motors-introduces-the-first-dc-area-projec/ (last accessed 27 March 2011).

52. Siler, S. (2010) GM Continues Project Driveway fuel cell Development Program, Phases in Second-Gen Technology, *Car and Driver*, http://blog.caranddriver.com/gm-continues-project-driveway-fuel-cell-development-program-phases-in-second-gen-technology/ (last accessed 27 March 2011).

53. Garrett, J. (2010) As Honda Ramps Up E.V.'s and Hybrids, Fuel Cell Program Lags, http://wheels.blogs.nytimes.com/2010/07/

54. Honda (2011) Drive FCX Clarity FCEV, http://automobiles.honda.com/fcx-clarity/drive-fcx-clarity.aspx (last accessed 30 August 2011).
55. Korzeniewski, J. (2010) Mercedes-Benz Launches B-Class F-Cell, Leases to Begin in Early 2010, *AutoBlog*, http://www.autoblog.com/2009/08/28/report-mercedes-benz-launches-b-class-f-cell-leases-to-begin-i/#comments (last accessed 23 March 2011).
56. Daimler, AG. (2011) Succesful Finish: F-CELL World Drive Reaches Stuttgart After Circling the Globe, http://www.daimler.com/dccom/0-5-1367004-1-1367069-1-0-0-0-0-0-0-7145-0-0-0-0-0-0.html (last accessed 30 August2011).
57. Curtin, S. and Gangi, J. (2010) The Business Case for Fuel Cells: Why Top Companies are Purchasing Fuel Cells Today, *Fuel Cells 2000*, http://www.fuelcells.org/BusinessCaseforFuelCells.pdf (last accessed 31 March 2011).
58. Okawara, A. (2011) Activities of NEDO for practical use of stationary fuel cell systems. Presented at the IPHE Workshop on Stationary Applications, http://www.iphe.net/docs/Events/Japan_3-11/1%20OKAWARA_Nedo.pdf (last accessed 30 August 2011).
59. Koguchi, H. (2010) Research and development of hydrogen and fuel cells in Japan. Presented at the FCH JU Stakeholders' General Assembly, http://www.fch-ju.eu/sites/default/files/documents/ga2010/haruhisa_koguchi.pdf (last accessed 30 August 2011).
60. Okamura, K. (2010) Tokyo Gas's fuel cell CHP business for residential use. Presented at the FCH JU Stakeholders' General Assembly, http://www.fch-ju.eu/sites/default/files/documents/ga2010/kiyoshi_okamura.pdf (last accessed 30 August 2011).
61. Japan for Sustainability (2009) World's Largest 'Hydrogen Town Project' Starts in Japan, http://www.japanfs.org/en/pages/028694.html (last accessed 1 April 2011).
62. METI (2011) Launch of "Hydrogen Town Project" under the "Hydrogen Energy Social Infrastructure Development Demonstration Project", http://www.meti.go.jp/english/press/2011/0113_02.html (last accessed 30 August 2011).
63. Grahl-Madsen, L. (2010) The Hydrogen Demonstration Society @ Lolland Island, Denmark. Presented at the FCH JU 2010 Stakeholders General Assembly, http://www.fch-ju.eu/sites/default/files/documents/ga2010/laila_grahl-madsen.pdf (last accessed 30 August 2011).
64. Klinder, K. (2011) Cooperation as success factor for the German National Innovation Program. Presented at the IPHE Workshop on Stationary Fuel Cells, http://www.iphe.net/docs/Events/Japan_3-11/1%20Klinder_NOW.pdf (last accessed 30 August 2011).
65. EnergieAgentur.NRW Fuel Cell and Hydrogen Network NRW, http://www.fuelcell-nrw.de/index.php?id=3 (last accessed 27 March 2011).
66. Atanasiu, M. (2011) Fuel cells and hydrogen joint undertaking. European public/private joint support for fuel cells and hydrogen activities. Presented at the IPHE Workshop on Stationary Applications, http://www.iphe.net/docs/Events/Japan_3-11/1%20Atanasiu_FCH%20JU.pdf (last accessed 15 August 2011).
67. Farmer, R. (2011) Overview of hydrogen and fuel cell activities. Presented at the IPHE Workshop on Stationary Applications, http://www.iphe.net/docs/Events/Japan_3-11/1%20Farmer_DOE.pdf (last accessed 30 August 3011).
68. Devlin, P. (2011) DOE Hydrogen and Fuel Cell Overview. U.S. Department of Energy.
69. Park, D. (2011) Current status of commercialization of residential fuel cells in Korea. Presented at the 7th International Hydrogen and Fuel Cell Expo.
70. Butler, J. (2010) 2010 Survey of Korea, *Fuel Cell Today*, http://www.fuelcelltoday.com/analysis/surveys/2010/

2010-survey-of-korea (last accessed 6 February 2012).

71. Kim, T.-H. (2007) Status of molten carbonate fuel cell development in Korea. Presented at the 8th IPHE Implementation-Liaison Committee Meeting, *http://www.iphe.net/ docs/Meetings/Korea_6-07/Status_of_ Molten_Carbonate_FC_Dev_in_Korea.pdf*.

residential fuel cells. Presented at the IPHE Workshop on Stationary Applications, *http://www.iphe.net/docs/Events/Japan_3-11/2%20Ramesohl_E.ON_IPHE_1Mar2011.pdf* (last accessed 30 August 2011).

Further Reading

Ramesohl, S. (2011) The German NIP Lighthouse Project Callux: field test of

Part VIII
Knowledge Distribution and Public Awareness

39
A Sustainable Framework for International Collaboration: the IEA HIA and Its Strategic Plan for 2009–2015

Mary-Rose de Valladares

39.1
Introduction

The Hydrogen Implementing Agreement (HIA) of the International Energy Agency (IEA), known as the *IEA HIA*, is an autonomous organization within the framework of the Organisation for Economic Co-operation and Development (OECD). Its purpose is to develop and promote hydrogen as a clean, renewable energy source by facilitating international cooperation and information exchange for hydrogen research, development, and demonstration (RD&D). The IEA HIA is the world's largest and longest lived collaboration in hydrogen energy research. This chapter outlines the Agreement's mission, strategic framework, and plans for the period 2009–2015, and also its accomplishments for the period 2004–2009. More information can be found at the IEA HIA web site at *www.ieahia.org*.

IEA HIA's tasks and activities encompass the full spectrum of research issues in hydrogen production, storage, conversion, safety, integrated systems, and infrastructure, and also analysis and outreach in support of its RD&D activities. Significant technical progress has been made in these areas as a result of IEA-HIA coordinated research. Over the course of its 30-plus-year history, the IEA HIA has created a broad portfolio of 31 tasks. After completing its most recent term of operation on 30 June 2009, the Agreement began a new term with a *Strategic Plan for 2009–2015*.[1] The Strategic Plans [1], and also all IEA HIA tasks and activities, are self-determined on a "bottom-up" basis by members of the collaboration. The IEA HIA has 23 members: 21 countries, the Commission of the European Union and the United Nations Industrial Development Organization (UNIDO). The Agreement is managed by a professional Secretariat. This organizational capacity,

---

1) On the advice of the IEA, the time period of the official term was legally changed from 2014 to 2015. Hence, although the Strategic Plan document title reads 2014, all future references in this document refer to 2015 as the official end time for the current planning cycle.

Fuel Cell Science and Engineering: Materials, Processes, Systems and Technology,
First Edition. Edited by Detlef Stolten and Bernd Emonts.
© 2012 Wiley-VCH Verlag GmbH & Co. KGaA. Published 2012 by Wiley-VCH Verlag GmbH & Co. KGaA.

coupled with the Agreement's long-standing tradition of international cooperation, position the IEA HIA for continued success.

The IEA HIA's capabilities have grown with each new member. Membership increased by ~60% during the 2004–2009 term. UNIDO's accession is of particular significance, as it extends the IEA HIA's outreach to the developing world that offers vast opportunity for sustainable energy.

Although the IEA HIA has not yet invited industry formally to join the Agreement, industry participation in tasks has occurred and is increasing. The role of industry will continue to grow in the next 4 years as the time frame for commercialization approaches. In principle, industry is eligible to join the Agreement as a "sponsor." The IEA HIA Executive Committee expects to formulate a pathway for official sponsor accession in the near future.

39.2
The IEA HIA Strategic Framework: Overview

The IEA-HIA vision is for a hydrogen future in which a clean, sustainable energy supply of global proportions plays a key role in all sectors of the economy.

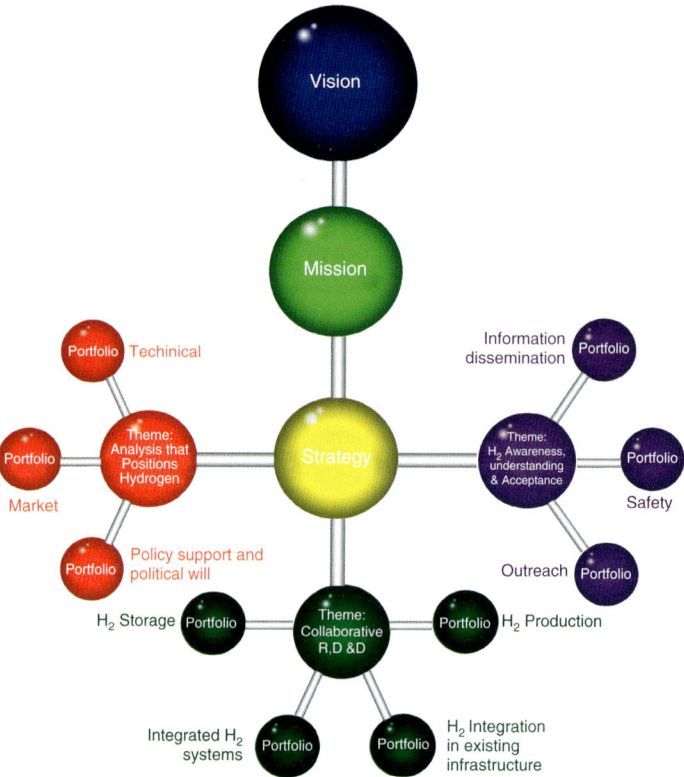

Figure 39.1 IEA HIA's mission and vision.

The Agreement operates in 5 year cycles. In each cycle, the IEA HIA's overall portfolio reflects members' direction, priorities, and RD&D requirements for that period.

Contemplating a new phase of expansion and progress, the IEA HIA Executive Committee adopted the following Mission Statement for 2009–2015:

> Accelerate hydrogen implementation and widespread utilization to optimize environmental protection, improve energy security, and promote economic development internationally while establishing the HIA as a premier global resource for expertise in hydrogen.

For the planning period 2009–2015, the IEA HIA has identified three major themes that stem from its mission and vision (Figure 39.1). These themes are at once goals and priorities. Each theme is associated with a set of portfolios that contain tasks and activities. The themes and portfolios are listed below and also depicted graphically in Figure 39.2.

Table 39.1 lists the current IEA HIA tasks by theme and portfolio.

39.2.1
Theme 1: Collaborative RD&D

Collaborative RD&D is the IEA HIA's core business. The RD&D is typically medium and long term in scope and pre-competitive in nature. The IEA HIA's

- Collaborative RD&D that advances hydrogen science and technology, including four portfolios:

  - Hydrogen production
  - Hydrogen storage
  - Integrated hydrogen systems
  - Hydrogen Integration in existing infrastructure

- Analysis that positions hydrogen for technical progress and optimization, for market preparation and deployment, and for support in political decision-making, including three portfolios:

  - Technical
  - Market
  - Support for political decision-making

- Hydrogen Awareness, Understanding and Acceptance that fosters technology diffusion and commercialization, including three portfolios:

  - Information dissemination
  - Safety
  - Outreach – Inform and engage critical subsets of HIA stakeholders

Figure 39.2 IEA HIA Strategic Framework 2009–2015.

Table 39.1 2009–2015: Themes, portfolios, and tasks.

Task	Task name / Operating agent (OA) name / Public task web site (if applicable)	Theme: Collaborative R,D&D				Theme: Analysis that positions hydrogen			Theme: Analysis that positions hydrogen		
		Production	Storage	Integrated H_2 systems	Integration in existing infrastructure	Technical	Market	Support for political decision making	Information dissemination	Safety	Outreach
21	BioHydrogen Dr. Michael Seibert	√									
22	Fundamental and applied H_2 storage materials development Dr. Bjorn Hauback http://www.hydrogenstorage.org/		√								
23	Small-Scale reformers for On-Site H_2 supply Dr. Ingrid Schjølberg	√		√	√		√				
24	Wind Energy and H_2 Integration Dr. Luis Correas & Dr. Ismael Aso http://task24.hidrogenoaragon.org/	√		√		√	√				
25	High temperature H_2 production Dr. Francois Le Naour http://www-prodh2-task25.cea.fr/	√									
26	Advanced Materials for waterphotolysis Dr. Eric Miller	√									
27	Near-Term market routes to H_2 Co-Utilization of biomass as a renewable Energy source with fossil fuels Dr. J. Hanssen & Ms. Berrin Bay	√				√	√				
28	Large-scale hydrogen delivery infrastructure Dr. Marcel Weeda				√						
29	Distributed and community hydrogen (DISCO H_2) Dr. Federico Villatico			√							
30	Global hydrogen systems analysis Mr. Jochen Linssen and Dr. Susan Schoenung					√	√	√			
31	Hydrogen safety Mr. William Hoagland									√	

study *Hydrogen Production and Storage: R&D Priorities and Gaps* [2] examined near-, mid-, and long-term research needs in hydrogen production and storage. The 2009–2015 Strategic Plan will address many of the research needs identified by the study. It will also address the needs and priorities separately identified by members during the 2008 strategic planning process that led to development of the *Strategic Plan for 2009–2015*. Salient aspects of hydrogen research needs and priorities are woven into the following discussion on themes, portfolios, and tasks.

39.2.1.1 Production Portfolio

Progress has been made over the last decade in new technologies for hydrogen *production*. Research in this area continues with full force. Throughout the Agreement's history, there have always been more IEA HIA tasks in the production area than in any other R&D area. *Hydrogen Production and Storage: R&D Priorities and Gaps* concluded that overall, there are significant needs for improvement in increased plant efficiency, reduction of capital costs, and reliability and operating flexibility for all production processes [3].

39.2.1.1.1 Near-Term Production Options

With respect to near-term options, *electrolysis and natural gas reforming* are proven technologies that can be used in the early phases of building a hydrogen infrastructure. *Task 24 – Wind Energy and Hydrogen Integration* is working to combine efficiently electrolyzers (a constant-input device) with variable-output wind turbines for production of hydrogen as a transportation fuel and for on-site conversion of hydrogen electricity for load balancing. This task is exploring the multifaceted (technical, economic, social, environmental, market, and legal) issues associated with hydrogen production via electrolysis of wind energy. *Small-scale natural gas reformers* remain a subject of research; there are several demonstration cases but limited commercial availability. *Task 23 – Small-Scale Reformers for On-Site Supply of Hydrogen* is investigating both technical and marketing issues related to small scale reformers, including carbon capture and storage. Task 23 seeks to establish a basis for harmonization of the technology for on-site hydrogen production. This effort has benefited from extensive industry involvement as businesses prepare for near-term commercialization. Both Task 23 and Task 24 are expected to influence near-term adoption and use of hydrogen technology. Both tasks are slated to conclude in 2011; their findings and conclusions will then be published for public dissemination.

39.2.1.1.2 Near-Mid-Term Production Options

Mid-term, *central fossil-based production with CO_2 capture and storage* could play a significant role in hydrogen production. It is generally expected that hydrogen production in the near and medium term will be based to a large extent on carbon-containing materials. Research is needed on absorption and other types of separation processes in addition to overall process layout and configuration. *Biomass to hydrogen processes* are also a mid-term option. More focus on feedstock preparation is needed. Logistics pose a challenge for this method and production appears economical only at large scale.

The potential for carbon capture and sequestration (CCS) and precombustion decarbonization expand possibilities for the sustainable use of conventional fossil energy carbon-containing resources. New efforts in this area during the 2009–2015 term may be gasification related. Currently, *Task 27 – Near-Term Market Routes to Hydrogen by Co-utilization of Biomass as a Renewable Energy Source with Fossil Fuels* addresses many of these issues. This task seeks to advance the development of hydrogen production based on renewable sources in the market place, focusing on biomass and its industrial market potential. It explores the "greening" of fossil fuel stocks via biomass, and also the potential for a renewable-based hydrogen supply chain utilizing upgraded biomass waste to produce a tradable intermediate (a biomass carrier). Task 27 is also investigating stand-alone biomass gasification technology for near-to-medium-term hydrogen markets.

39.2.1.1.3 Longer Term Production Options

Farther out on the time scale, basic and applied research is needed for both photoelectrolytic and biohydrogen production methods. Hydrogen production in *biological processes* that entail the use of hydrogenases (enzymes), and also genetically engineered organisms, are characterized by low conversion efficiencies. Fundamental research is needed to understand the natural processes and genetic regulation involved in these reactions. It is anticipated that this work will continue through much or all of this term. *Task 21 – BioHydrogen*, recently extended from 2010 to 2013, deals with a full range of biohydrogen topics: dark fermentation systems, light-driven biohydrogen production, biological electrochemical systems, and bio-inspired systems. High-temperature ($>500\,°C$) production of hydrogen is also an important area of investigation for the mid and long term, focusing on materials development, high-temperature membranes and heat exchangers, and high-temperature electrolysis. *Task 25 – High-Temperature Hydrogen Production Processes* examines solar and nuclear processes capable of producing massive quantities of hydrogen. This task, scheduled to conclude in 2011, will identify, classify, and assess each family of high-temperature production processes.

The need for continuing or additional *advanced materials research* was a recurring topic in the HIA's strategic assessment for 2009–2015. The need for research on catalysis was another recurring motif. Consequently, new basic and applied research activities in these areas are anticipated in this term.

Task 26 – Advanced Materials for Water Photolysis is concentrating on advanced materials research related to photoelectrolytic water splitting. Actual application of advanced materials for water photolysis will involve use of photoelectrochemical (PEC) devices. However, the magnitude of the challenge involved in materials research for water splitting is currently taking precedence over further research on PEC devices. In the future, the IEA HIA will resume work on the design and development of PEC devices for water photolysis.

As systems evolve for all types of hydrogen production methods, the IEA HIA expects heightened interest in *applied technology research and development* on the components necessary for various hydrogen production methods.

39.2.1.2 Storage Portfolio

Both *compressed gas and liquid hydrogen* are commercially available today, but research continues to improve performance and reduce costs. The mass market in automobiles is a topic of ongoing global discussions by industry, policy makers, government, and the public. Storage figures largely in these discussions. The really good news – confirmed in an independent study by McKinsey & Company – is that there are no major breakthroughs needed to commercialize hydrogen cars. Hence, while R&D in hydrogen storage needs to and will continue in earnest, the status of hydrogen storage is not a "showstopper" for the automobile [4].

R&D issues related to compressed gas include fracture mechanics, safety, compression energy, and volume. Important R&D issues for liquid hydrogen include more efficient liquefaction, lower cost/better insulated containers, automated boil-off capture (e.g., via hydrides), re-liquefaction, and volume.

For on-board hydrogen storage applications, research has focused on materials-based solid-state storage, which is currently in the development phase. The potential advantages of materials-based storage are lower volume, lower pressure (greater energy efficiency), and higher purity of hydrogen output. Important materials storage R&D issues include volume, weight, lower desorption temperatures, improved desorption kinetics, recharge time and pressure, heat management, cost, chemical and environmental reactivity, durability, container compatibility, and optimization [5].

Task 22 – Fundamental and Applied Hydrogen Storage Materials is the largest hydrogen storage collaboration in the world. With over 50 experimental, engineering, theoretical, and modeling projects to its credit, this task has recently been extended for a further 3 years. The research examines the following classes of materials: reversible metal hydrides, regenerative hydrogen storage materials, nanoporous materials, and rechargeable organic liquids and solids.

Later in this term, as storage systems meet technical targets, emphasis will shift to include R&D on components for all applications.

39.2.1.3 Integrated Systems Portfolio

Systems integration is an essential next step in collaborative hydrogen R&D. Systems integration brings component subsystems together, ensuring their efficient functioning.

Task 18 – Integrated Systems Evaluation ended in 2009. This task undertook modeling and evaluation of a broad collection of hydrogen demonstration projects for the express purpose of analyzing and modeling their overall design and performance. It merits inclusion among the current tasks because the Task 18 information bases remain the best resource for global information on all things hydrogen, including worldwide trends.[2] As such, they provide a solid foundation for other IEA HIA tasks in analysis (Task 30) and infrastructure (Task 28).

2) http://iea-hia-annex18.sharepointsite.net/Public/default.aspx: *Task 18 – Integrated Systems Evaluation* web site with information bases. Task 18 information may also be found by visiting www.ieahia.org and clicking on past tasks.

Task 23 – Small-Scale Reformers for On-Site Supply of Hydrogen was previously discussed as part of the production portfolio. It also falls into the Integrated Systems portfolio by virtue of its focus on integration of components for reformer systems, as well as its examination of technology performance and cost, emission handling, and CCS. Likewise, *Task 24 – Wind Energy and Hydrogen Integration*, also falls under this portfolio because it deals with system integration.

In late 2010, *Task 27 – Distributed and Community Hydrogen (DISCO-H$_2$)* was approved. This task was created expressly to further the optimization and replication of green hydrogen with distributed and community hydrogen systems. It encompasses three application categories: islands and remote sites, off-grid, and industrial (meaning literally industrial, or subdivisions of new or existing communities).

Market introduction and penetration of hydrogen technology require optimized, well-integrated systems. Therefore, as time goes on, the Agreement will focus increasing effort and attention on (i) the *components, devices, and sensors* that comprise the systems and (ii) their *engineering as integrated systems*. This will likely translate into new activities and new tasks during the 2009–2015 term.

This discussion would not be complete without mention of the *fuel cell*, arguably the most important hydrogen energy conversion technology. The IEA HIA understandably has an abiding interest in fuel cells. As systems integration efforts advance, the HIA will certainly investigate fuel-cell issues more closely, likely in cooperation with the IEA Advanced Fuel Cell Implementing Agreement.

39.2.1.4 Integration in Existing Infrastructure Portfolio

Future delivery of hydrogen requires the integration of hydrogen systems into the existing energy infrastructure. While that infrastructure services all sectors of the economy, the world's attention has focused first on the transport sector with its mass markets, followed by the stationary power sector. Expansion of the infrastructure will require coordination on many fronts: technology; finance, and insurance; market mechanisms and policy instruments; construction and engineering; operation and maintenance; and codes and standards. Although all these factors warrant attention, the HIA's focus is on RD&D issues and technical barriers – which require research activities that interface with conventional resource chains (the grid, pipelines, trucking, and other delivery systems). Likewise at issue are centralized and distributed hydrogen production, and also mass storage.

Task 23 fits into this portfolio because it deals specifically with integrating small-scale reformers directly into the existing infrastructure – literally the gas station near you!

A task on *Large-Scale Hydrogen Delivery Infrastructure – Task 28*, was approved in late 2010 to consider the transport and distribution aspects of hydrogen infrastructure for mass market applications. This task will use a modeling approach to improve the understanding of infrastructure needed to deliver the hydrogen needed based on different roll-out strategies. The transport part of the infrastructure network will consider liquefaction, storage, trucks, and pipelines. The distribution part of the network encompasses fueling stations and local grids.

39.2.2
Theme 2: Analysis That Positions Hydrogen

In less than a decade, the avalanche of interest in hydrogen has been met by roadmaps and a wide array of analytic efforts on RD&D, infrastructure, and market issues. Some fine analytic work on hydrogen is available today. However, a comprehensive analysis of global energy conditions that incorporates hydrogen in the world's energy future is a complex and challenging proposition, complicated by high levels of political, technical, environmental, and economic uncertainty. There is, in effect, *a hydrogen information gap* that needs to be filled with coherent and *balanced information*, providing a clearer picture of hydrogen R&D needs and the future of hydrogen in the economy. Furthermore, the information needs to be appropriately packaged for specific audiences.

Therefore, the Agreement has reaffirmed its commitment to rigorous, independent analysis that supports collaborative R&D efforts and addresses the larger issue of the transition to hydrogen in the economy. This commitment amounts, in no uncertain terms, to an *"Analytic Imperative."* As the premier global resource for technical expertise in hydrogen energy, the IEA HIA is better positioned to offer balanced analysis on these questions than any other organization. The Executive Committee expects that development of analysis products will not only contribute to filling the information gap but also more firmly establish the HIA as the leading technical resource in hydrogen. The Agreement expects to cooperate closely on IEA analytic efforts such as the well-regarded Energy Technology Perspectives (ETPs) and the World Energy Outlook (WEO) publications.

The Analysis theme contains three portfolios: technical, market, and support for political decision-making. Collectively, these three portfolios will cross-cut all IEA HIA tasks to provide the analyses needed to provide relevant stakeholders and policy-makers with balanced information that stimulates RD&D, market adoption, and widespread application of hydrogen.

During the last term, the IEA HIA began the process of organizing for analysis that will address important questions about hydrogen demand, supply, and infrastructure. This process, spearheaded by the Executive Committee Analysis Group, has evolved *Task 31 – Global Hydrogen Systems Analysis*. This task comprises global hydrogen resource analysis, updating and harmonizing the level of hydrogen knowledge, and collaboration with IEA Analytics. Its efforts encompass all three analysis portfolios: technical, market, and support for political decision-making.

39.2.2.1 Technical Portfolio
The technical portfolio consists of analysis intended, first and foremost, to promote advancement and optimization of the technology. This effort is also intended to ensure that the HIA provides a clear picture of evolving hydrogen RD&D needs.

There is a larger need to make a cross-cutting technical case for all hydrogen technologies, including production, storage, conversion (e.g., fuel cell), delivery, and infrastructure. This effort will be based in the Analysis Task. The foundation for the effort is a literature review and gap analysis developed by *Task 18 – Integrated Hydrogen Systems*. Technical analysis also takes place at the task level. Indeed, it is a standard feature of many IEA HIA tasks.

39.2.2.2 Market Portfolio

The market portfolio of analysis activities contemplates issues of market preparation and deployment. These issues include the topic of market transformation that supports the deployment of innovative technology, bridging the early, and often fatal, stage of market introduction and the later stage of market penetration and growth. The market portfolio effort will make the business case for hydrogen, positioning hydrogen for competitive advantage in the marketplace. This analysis will entail both supply and demand side assessments, including the nonenergy sector. The supply side analysis will incorporate a market perspective. At the task level, techno-economic analysis with a market perspective will also be performed on individual technologies.

39.2.2.3 Support for Political Decision-Making Portfolio

Recognizing that public policy will play a crucial role in the development of hydrogen technology and its deployment in the marketplace, and that support for political decision-making is indispensable to a future with hydrogen energy, the Analysis theme includes the Support for Political Decision-Making Portfolio. Activities in this portfolio feature analysis that aligns investment in hydrogen technology with global public policy concerns (notably climate change and emissions reduction). Wherever appropriate, this analysis will utilize findings and conclusions from the Agreement's technical and market analyses. The results will be presented in position papers and briefs.

39.2.3
Theme 3: Hydrogen Awareness, Understanding, and Acceptance

This theme complements the HIA's principal theme – Collaborative R&D, and its supporting theme – Analysis. It acknowledges that awareness, understanding, and acceptance are requisite to technology diffusion and commercialization. It recognizes that the benefits of hydrogen must be articulated to stakeholders and decision-makers. Also, it accepts a major role for the HIA in the communications process. Through this three-portfolio effort to foster technology diffusion and commercialization, the IEA HIA expects to increase its visibility as the reference institute for hydrogen.

39.2.3.1 Information Dissemination Portfolio

The ultimate success of the "Analysis Imperative" depends upon effective information dissemination. This function targets key stakeholders in the science, energy, and environmental communities, and also the media, government and industry. The IEA is itself an important target audience, and the Agreement will also disseminate information beyond the borders of IEA member countries. UNIDO, a new IEA HIA member, is expected both to contribute to and to benefit from information dissemination activities.

At the task level, Agreement Experts have produced over 1000 HIA-related publications/reports and made around 1000 HIA-related presentations during the 2004–2009 term. The trend towards increased production of information in publications/reports and presentations is expected to continue. As hydrogen progresses toward commercialization during the 2009–2014 term, the IEA HIA newsletter (*IEA HIA News*) will evolve to focus more on hydrogen demonstrations and the hydrogen marketplace. The IEA HIA Secretariat now develops and disseminates information via its web site (*www.ieahia.org*), Annual Report [6], special reports, newsletter, and brochures, in addition to the extensive technical reports and publications developed at the task level.

The Agreement's conference program facilitates preparation and delivery of abstracts, papers, exhibits, and related presentations. During the 2009–2015 term, the IEA HIA expects to create new technology platforms and channels for information dissemination (e.g., Webinars and podcasts). It is further its intention to institute end-of-task workshops in order to communicate the results of tasks directly to targeted audiences, policymakers, and the interested public.

39.2.3.2 Safety Portfolio

Safety and consumer comfort with hydrogen are vital ingredients for the acceptance of hydrogen. Hydrogen safety concerns cut across all HIA RD&D portfolios. *Task 19 – Hydrogen Safety*, and its successor, *Task 31, also called Hydrogen Safety*, deal explicitly with safety through analysis, testing, and the development of target information products. During this term, as safety information products become available, Task 31 plans to distribute them as broadly as possible. Plans are now under way to hold two workshops in the 2011–2012 time frame to disseminate the results of Task 19 and the progress in Task 31. One workshop will be held in Paris and the other in the USA at the campus of the IEA HIA office in Bethesda, MD. The Paris workshop will target audiences in three distinct yet related categories:

- Chairs, conveners, and task group leaders of relevant subcommittees and Working Groups of ISO TC 22 Road vehicles, ISO TC 58 Gas cylinders, ISO TC 197 Hydrogen technologies, ISO TC 220 Cryogenic vessels, IEC TC 31 Equipment for explosive atmospheres, and IEC TC 105 Fuel cells, and experts involved in developed of relevant EN standards.
- Chairs, conveners, and technical experts of the European Industrial Gas Association (EIGA) developing requirements for hydrogen use.
- Technical decision-makers/advisors within JTI/JTU of the European Commission who have input to or have funding authority over the funding of pre-normative research in the fields of hydrogen and fuel cells.

Invited technical experts to the Bethesda workshop will include similar three groups:

- Chairs, conveners, and task group leaders of all Standards Development Organizations in the USA and Canada developing codes and standards for hydrogen technologies such as NFPA, ICC, API, CSA America, ASTM, ASME, NIST, CGA, SAE, and BNQ (Canada).
- Chairs and members of US DOE Technical Advisory Groups and CACs (Canada) on relevant ISO and IEC mirror technical committees.
- Technical decision-makers/advisors within US DOE, DOT, and so on, and Natural Resources Canada who have input to or have funding authority over pre-normative research in the fields of hydrogen and fuel cells.

39.2.3.3 Outreach Portfolio

This portfolio goes beyond information dissemination to both inform and *engage* a critical subset of IEA-HIA *stakeholders and decision-makers*. Engagement may take several forms, including participation as an IEA-HIA Expert, a member, or possibly a sponsor. Engagement may also imply cooperation in a more limited time frame or for a particular purpose. As with Information Dissemination, the Outreach Effort operates out of the Secretariat at the direction of the Executive Committee.

To engage important target audiences, such as industry, government, and members of the renewable energy community, the Outreach Portfolio employs the full array of IEA HIA information products and all available channels, including *networking opportunities*. Active participation of the Executive Committee Members, Operating Agents, and the Secretariat, who are well positioned to carry out these activities in strategic situations around the world, is considered essential.

Many of the world's leading experts in hydrogen energy RD&D have partnered with the HIA since its inception. We are pleased to recognize their dedication and innovation with two new prizes: the *IEA HIA Individual Prize* and the *IEA HIA Project Prize*.

The IEA HIA Individual Prize was created to celebrate hydrogen research and development distinguished by technical excellence and harmony in international cooperation that contributes to the understanding and advancement of basic and applied science. The Agreement awarded its inaugural IEA HIA Individual Prize in June 2008 to Dr. Gary Sandrock (Figure 39.3). Although the Individual Prize was conceived as a single award, the Executive Committee decided that special circumstances warrant special measures: the late Dr. Tapan Kumar Bose (Figure 39.4), who passed away in 2008, was honored as the recipient of a IEA HIA Memorial Prize for lifetime achievement in hydrogen R&D.

The first IEA HIA Project Prize was awarded at the 2010 World Hydrogen Energy Conference in Essen, Germany (Figure 39.5). There were recipients in two categories: Fundamental Research and Technology Demonstration. In both cases, the awards were made:

Figure 39.3 Dr. Gary Sandrock, IEA HIA Individual Prize recipient.

Figure 39.4 Dr. Tapan Bose, IEA HIA Memorial Individual Prize recipient.

Figure 39.5 Winners of IEA HIA Project Prizes for fundamental and applied research during the award ceremony at the 2010 WHEC in Essen, Germany.

For research characterized by technical excellence and harmony in international cooperation that contributes to the understanding and advancement of basic and applied hydrogen science.

For Fundamental Research, the IEA HIA Project Prize winner was Fundamental Safety Testing and Analysis of Hydrogen Storage Materials and Systems (Task 22 H-25), a fundamental research project of IEA HIA Task 22, Fundamental and

Figure 39.6 Dr. James M. Ohi, IEA HIA Memorial Individual Prize recipient.

Applied Hydrogen Storage Materials Development. For Technology Demonstration, the IEA HIA Project Prize winner was the ITHER (Infraestructura Tecnológica del Hidrógeno y Energías Renovables) Project "Green Hydrogen from Wind and Solar Mobile Applications," a project of IEA HIA Task 24, Wind Energy and Hydrogen Integration. At the Canadian Hydrogen and Fuel Cell Conference in Vancouver, the IEA HIA awarded its 2011 Individual Prize to Dr. James M. Ohi for coordinating global development and harmonization of hydrogen and fuel codes and standards that advance the safe use and market deployment of hydrogen and fuel cell technologies (Figure 39.6).

39.3
The Work Program: Issues and Approaches

The Plan of Work 2009–2015 is given in Table 39.2. Taken directly from the Strategic Plan 2009–2015, the table is organized by theme and portfolio. It outlines the tasks' key issues and the broad "approach" to each task, differentiating what was in place in 2009 before adoption of the new strategic plan, and what is expected during the 2009–2015 term. The RD&D theme is divided into fundamental and technology categories. The other two themes – Analysis that Positions Hydrogen and Hydrogen Awareness, Understanding, and Acceptance – are cross-cutting and therefore constitute single categories.

39.4
IEA HIA: the Past as Prolog

To understand better the current status of IEA HIA activities and its plans for 2009–2015, it is helpful to review the outcome of the immediate past 5 year term from 2004 to 2009. This section provides descriptions and metrics on accomplishments, benefits, and success stories for the IEA HIA's immediate past term.

Table 39.2 Plan of work 2009–2015.

	Theme and portfolio	Key issues	Approach in place 2009	Approach proposed/potential
Fundamental	Theme: Collaborative R,D &D RD&D Production Portfolio H$_2$ Production	Anaerobic use of bacterial dark fermentations and photosynthetic microbes; increased yields; biomimetics; biohydrogen acceptance	Task 21, BioHydrogen	Extend Task 21 past 2010
		Advanced materials for photoelectrochemical (PEC) watersplitting	Task 26, Advanced Materials for Waterphotolysis	Possible extension of Task 26 past 2011
		Advanced materials (catalysts) for other production methods		Task on catalyst research for other production methods
		High-temperature production from nuclear and solar electrolysis	Task 25, High-Temperature Production	Possible extension of Task 25 High-temperature electrolysis activity
Technology		Biofuels for reformers; CCS and emission handling	Task 23, Small-Scale Reformers for On-Site Supply of Hydrogen	Possible 1 year Task 23 extension past 2009
		Gasification and gas clean-up		Task on Purification/separation, ICE for on-site reformers
		Design/development of photo-electrochemical (PEC) devices	Task 24, Wind Energy and Hydrogen Integration	Possible successor to Task 26 on development of PEC devices

(continued overleaf)

Table 39.2 (continued).

	Theme and portfolio	Key issues	Approach in place 2009	Approach proposed/potential
Fundamental	RD&D Storage H₂ Storage Portfolio	Fully integrated wind and H₂ application	Task 27, Near-Market Routes to Hydrogen by Co-utilisation of Biomass as a Renewable Energy Source with Fossil Fuels	Possible Task 24 extension
		Co-gasification of biomass with fossil fuels; tradable intermediates		Follow-up activities/successor task on low temperature electrolysis Follow-on efforts TBD
		Reversible/ regenerative H₂ storage media fulfilling int. targets	Task 22, Fundamental and Applied H₂ Storage Materials Development	Task 22 seeking 2–3 years
		Fundamental and engineering understanding		Extension past 2009; thereafter, disposition of activity TBD
		Materials for stationary applications Compression Metal embrittlement		Possible task/activities on H₂ interactions with materials
Technology		Applied aspects of H₂ storage systems in vehicles: compressed gas, liquid, and materials-based		New task examining

Table 39.2 (continued).

Theme and portfolio	Key issues	Approach in place 2009	Approach proposed/potential
RD&D Integrated Systems Portfolio: Integrated H$_2$ systems	All-purpose information on H$_2$ integration Harmonization of components for reformer systems; technology performance and cost; CCS and emission handling	Task 18, Integrated Systems Evaluation Task 23, Small-Scale Reformers for On-Site Supply of Hydrogen	Technologies for H$_2$ storage: Compressed gas, liquid and materials-based; techno-economic analysis of alternatives Component task Possible continuation of current modeling and analysis of demonstration systems Possible 1 year extension of Task 23
RD&D H$_2$ Integration in Existing Infrastructure	Specifications for Integrated Systems Geologic storage, pipelines, and "mass" storage	Task 24, Subtask B In Definition: Infrastructure and Mass Storage Task	New Task H$_2$ Communities Modeling and design: islands, remote, and rural communities Possible 1 year Task 23 extension past 2009 TBD

(continued overleaf)

Table 39.2 (continued).

	Theme and portfolio	Key issues	Approach in place 2009	Approach proposed/potential
Cross-cutting	Theme: Analysis that positions hydrogen Analyses that position H_2 Technical Portfolio Technical	"Where will the H_2 come from?" What role will H_2 play? Develop a balanced view Competition in transport sector	Analysis group Task 18 – H_2 Literature Review	Analysis task: Supply/Demand Comprehensive report that includes CO_2 reduction based on introduction of H_2 technology Proactive cooperation with IEA analytics Part of proposed storage task: techno-economic analysis of alternatives in automotive industry
	Analyses that position H_2 Market Portfolio Market	"Where will the H_2 come from?" What role will H_2 play? Include nonenergy sector	Analysis Group Task 23, Subtask 3 Market Studies Task 24, Subtask C and D, Business Concept Development Task 27, Subtask D, Roadmap Development and Verification	Analysis task: supply/demand from a market perspective – a comprehensive report that includes CO_2 reduction market, cap and trade Proactive cooperation with IEA analytics Successor task on market studies for on-site reformer technology

Table 39.2 (continued).

	Theme and portfolio	Key issues	Approach in place 2009	Approach proposed/potential
Cross-cutting	Analyses that position H_2 Support for Political Decision-Making *Portfolio: Support for political decision making*	Lack of information and clarity about the benefits of hydrogen (notably, CO_2 and pollution reduction) among stakeholders and decision-makers whose influence is needed and useful for R&D, planning, demonstration, and deployment		Analysis task: briefs and position papers to address all issues, including CO_2 reduction, climate change
	Theme: H_2 Awareness, understanding & acceptance H_2 Awareness, Understanding, and Acceptance Information Dissemination *Information dissemination — Portfolio*	Broader and deeper information dissemination needed along with targeted dissemination	An element of the Outreach Program managed by Secretariat, it features: IEA HIA web site Annual report Conference strategy Communication/promotion materials Public relations and media Contributions to IEA events and publications	Increase information dissemination by continuing current activities and augmenting with: enhanced conference strategy that features HIA seminars/workshops <200; and possible sponsorship of larger conferences; webinars, possible podcasts, and other IT vehicles Dissemination of safety products Dissemination of analysis products Cooperation with ETDE to expand market for information

(continued overleaf)

Table 39.2 (continued).

Theme and portfolio	Key issues	Approach in place 2009	Approach proposed/potential
H$_2$ Awareness, Understanding, and Acceptance Safety — Portfolio Safety	All aspects of hydrogen safety and consumer comfort with hydrogen	Target audiences include: IEA member countries and IEA family Gleneagles "+5" and potential H$_2$ members hydrogen community Task 19	Expand target audiences: Non-IEA member countries, developing world, and greater energy community Task 19 concludes in 2010 High potential for multiple follow-up tasks in key areas. Definition of successor tasks to occur early in term Task 19 to complete information products and distribute with Secretariat support
H$_2$ Awareness, Understanding, and Acceptance Outreach	Inform and engage	An element of the Outreach Program directed by ExCo and managed by the Secretariat	Inform and engage through information dissemination and targeted networking; build participation; influence stakeholders and decision makers Potential participants: Members, experts, sponsors

Table 39.2 (continued).

Theme and portfolio	Key issues	Approach in place 2009	Approach proposed/potential
			Stakeholders and decision-makers: IEA, IEA member countries; IEA non-member countries; Government and Industry; Hydrogen Community; Renewable Community

Table 39.3 highlights the accomplishments of each task that was active during the previous term 2004–2009. Where applicable, the table indicates the number and name of any successor task currently under way.

Table 39.4 captures metrics on overall IEA HIA accomplishments during the completed 5 year term 2004–2009.

39.5
The 2009–2015 IEA HIA Work Program Timeline

Getting back to the present and the IEA HIA Strategic Plan 2009–2105, Table 39.5 sets forth a timeline for the Agreement's tasks and activities through 2015. As of December 2009, nine tasks were active. Five milestones associated with the 2009–2015 Strategic Plan's Collaborative RD&D theme have already been met:

- Task 18 – Integrated Systems Analysis has closed.
- Task 21 – BioHydrogen, part of the H_2 Production Portfolio theme, was extended for a 5 year period (3 years firm with an additional 2 year option).
- Task 22 – Fundamental and Applied Hydrogen Storage Materials, part of the Storage Portfolio, was extended 3 years as planned.
- Task 29 – Distributed and Community Hydrogen (DISCO H_2) was approved as part of the Integrated H_2 Systems Portfolio.
- Task 28 – Large-Scale Hydrogen Delivery Infrastructure, was approved as part of the Integration in Existing Infrastructure Portfolio.

With respect to milestones associated with the Analysis theme, Task 30 – Global Analysis of Hydrogen Systems was approved.

Table 39.3 Task accomplishments, benefits, and success stories in completed term 2004–2009.

Advancement of Science and Technology via pre-commercial collaborative Production RD&D programs	
Task	Accomplishments, benefits, and success stories in completed term
Task 15 Photobiological Production (completed 2004; succeeded by Task 21 BioHydrogen	R&D progress toward development of H_2 production by microalgae A novel, sustainable photobiological production of molecular hydrogen upon a reversible inactivation of the oxygen evolution in the green alga *Chlamydomonas reinhardtii* (Subtask A) Identification of accessory genes and gene products necessary for the photoproduction of H_2 in *Chlamydomonas reinhardtii*. Finding that STA7 and starch metabolism play an important role in *C. reinhardtii* H_2 photoproduction. (Subtask A) Identification and characterization of *tla1*, a novel gene involved in the regulation of the Chl antenna size in photosynthesis in *C. reinhardtii* (Subtask B) The generation of 11.6 mol of H_2 per mole of glucose-6-phosphate using enzymes of the oxidative pentose phosphate cycle coupled to a hydrogenase purified from *Pyrococcus furiosus* (Subtask C) The development of both smaller and larger photobioreactors (Subtask D)
Task 16 Hydrogen from Carbon-Containing Materials (completed 2005; succeeded by Task 27 – Near-Term Market Routes to H_2 by Co-Utilization of Biomass as RE with Fossil)	State-of-the-art reports for all three Task 16 subtasks: **Subtask A** on the potential for cost reduction of large-scale processing from natural gas with precombustion decarbonization of fossil energy; **Subtask B on prospects for H_2 from biomass from an industry perspective**; and **Subtask C on small-scale reformer technology** for distributed near to medium term H_2 supply **Substantial industry participation** on a challenging scope of work was an HIA first that serves as a benchmark for future industry participation
Task 20 Hydrogen from Water Photolysis (completed 2007; succeeded by Task 26 – WaterPhotolysis)	Development, acceptance, and operation of two multi-year R&D PEC programs, one at the US DOE and the other, called "NanoPEC" under the EU 7th Framework Program

Table 39.3 (continued).

	Pioneered Fe_2O_3 (hematite) as very promising, abundant, low-cost, and environmentally benign photoanode material Maturing PEC water-splitting tandem concepts Photoelectrochemical (PEC) work on tungsten trioxide led to the development of novel, highly sensitive, reliable, and low-cost **pollution control sensors for auto industry**
Task 21 BioHydrogen	Better genomic understanding of hydrogen-producing strict anaerobes New assessment method for overall analysis of biohydrogen (Subtask D) has been screened
Task 23 Small-Scale Reformers for On-Site Hydrogen Supply (SSR for Hydrogen)	Contributing to development of norms for small-scale reformers to harmonize industrialization. This effort, which includes carbon capture, is crucial to the development of the hydrogen infrastructure and future distributed generation capability Subtask 3 market studies stand to materially facilitate HIA analysis efforts Fast-tracking the deployment process of market introduction and penetration of small-scale reformers for on-site hydrogen supply from multiple feedstocks, fossil, and renewable
Task 24 Wind Energy and Hydrogen Integration	Setting the stage for large-scale use of renewable wind energy for hydrogen production in the near future by addressing the entire wind to hydrogen production chain from technical, economical, social, environmental, market, and legal perspectives Exploring in detail possible applications for hydrogen, especially full wind and hydrogen integration by means of hydrogen storage and electrical conversion that balances the original wind energy production, allowing an approach to demand that closes the gap with conventional energies
Task 25 High-Temperature Production of Hydrogen	Poised to elaborate world-wide knowledge on specific high temperature ($>500\,°C$) processes (solar and nuclear) that will support production of massive quantities of zero-emission hydrogen Producing **summary sheets** on high-temperature processes in **general and detailed** versions

(continued overleaf)

Table 39.3 (continued).

Task 26 NEW	On track to create database on advanced materials for water photolysis
Task 17 Solid- and Liquid-State Storage (completed 2006; succeeded by Task 22 – Storage)	Evolved into largest global R&D collaboration on hydrogen storage materials of its time, contributing to R&D, information dissemination, and transfer of technology Huge contribution to the literature with 900+ publications and presentations plus 17 patents
Task 22 Fundamental and Applied Hydrogen Storage Materials	World s largest collaboration to-date on hydrogen storage materials R&D Biannual week-long Task 22 meetings serve as the ultimate global forum for expert cooperation on hydrogen storage R&D (emphasizing materials and the transportation sector), the grand challenge in hydrogen. As of December 2008 it had produced 450+ publications/articles, 450+ presentations, and 16 patents
Assessment of market environment including nonenergy sector; and analysis, safety, and economics	
Task 18 Integrated Systems Evaluation (completed 2009; succeeded by Task 28 – Infra; Task 29 – DISCO H_2; and Task 30 – Analysis)	World s best address for worldwide information and analysis on hydrogen and integrated systems Database with 200+ national documents National organizations database National projects database State-of-the-art analysis entitled *Demonstration Project Evaluations* Used technical simulations that may be applied to other projects to replicate results General conclusions in critical areas of system evaluations, data monitoring, modeling tools, system design, control systems, and cost–benefit analysis Synthesis, lessons learned, and trend analysis relate to permitting, funding, and technology performance More than a dozen relevant case studies

Table 39.3 (continued).

Task 19 Hydrogen Safety	Contribution to global understanding of H_2 safety through studies and databases laying the foundation for codes and standards Phase 1 laid theoretical groundwork for Phase 2 testing program to evaluate the effects of equipment or system failures under a range of real-life scenarios, environments, and mitigation measures Subtask A Activity 1 produced a *Survey of Hydrogen Risk Methods* in Phase 1 Subtask A Activity 2 produced *Comparative Risk Assessment Studies of Hydrogen and Hydrocarbon Fueling Stations* Subtask A Activity 3 produced a *Knowledge Gaps White Paper*

Table 39.4 Overall IEA HIA accomplishments during the completed term 2004–2009.

Criterion	Achievements/targets	Actual 2004–2009
Membership	Number of members at end of term	22 + three pending
Tasks	Number of R&D tasks active during period	13 + 1 in definition
Level of effort	Number of person years	~712
Expert meetings		88
Publications/articles	HIA summary publications	22
	Expert publications/articles	1153
Presentations	HIA ExCo/Secretariat – Internal to IEA	12
	HIA ExCo/Secretariat – External to IEA	37
	Expert	1015
Support	Direct member support for operating agents	~US$2 million
HIA budget	Cumulative operating budget	US$0.85 million

Under the H_2 Awareness, Understanding, and Acceptance theme, Task 31 – Hydrogen Safety became the direct successor of Task 19 in the Safety Portfolio.

Formulation of other tasks – new and successor tasks – will continue apace during this period, as described in this chapter.

39.6
Conclusion and Final Remarks

The IEA HIA is in a period of growth and is well positioned for continued success in its core business of collaborative RD&D, in addition to analysis for technical optimization and market preparation, and hydrogen community outreach.

Table 39.5 IEA HIA work program timeline, current, and proposed/future tasks.

	Mid 2009	2010	2011	2012	2013	Mid 2014
Task 18: Integrated systems evaluation						
Successor: Remote community modeling						
Task 19: Hydrogen safety						
Successor safety: became Task 31						
Task 21: Biohydrogen						
Successor: Biohydrogen						
Task 22: Fundamental and Applied Hydrogen storage materials						
New task: Applied storage						
Task 23: Small-scale reformers for on-site hydrogen supply (SSR)						
Successor: Market studies for SSR						
Task 24: Wind energy & H_2 integration						
Successor: Componentry & Low temp electrolysis						
Task 25: High temperature H_2 production						
Successor: High temp electrolysis						
Task 26: Advanced materials for water photolysis						
New task: PEC Devices						
Task 27: Near-term market routes to H_2 by Co-utilization of biomass as RE with fossil						
Successor: Gasification & Gas cleanup						
Task 28: Infrastructure and mass storage						
New task: Global H2 systems analysis						
New task: Catalyst research						
New task: Production componentry						

Key to work plan above:
- Solid black line = current task
- Short-dash broken line = task extension
- Arrow at end = task is expected to continue after end of 2015 term
- Solid gray line = successor task
- Long-dash gray broken line = new task

The Agreement is the world's premier global resource for technical expertise in hydrogen RD&D, offering its members a *collaboration-based value proposition* that facilitates world-class research and increases their impact. Members benefit from the Agreement's:

- **Neutral international profile**: The Agreement is a reliable, unbiased forum through which Members may access technical experts and achieve global reach to engage governments, academia, and industry.

- **Leveraged global resources**: Members may engage an established network of researchers and shared resources in the areas of science and technology, market analyses, and outreach. The IEA HIA Portfolio includes shorter term and long-term, pre-competitive activities. In all cases, intellectual property (IP) is treated with care.
- **Track record of success**: For 30+ years, IEA HIA Tasks have facilitated collaborative research, and its membership continues to grow.

In conclusion, the IEA HIA goes forward in pursuit of its mission to accelerate hydrogen implementation and widespread utilization, while securing its reputation as the world reference for hydrogen energy RD&D.

References

1. IEA HIA (2009) Hydrogen Implementing Agreement. End of Term Report 2004–2009 and Strategic Plan: 2009–2015, IEA HIA, Bethesda, MD, http://ieahia.org/pdfs/StrategicPlan2009-2015.pdf.
2. Hydrogen Implementing Agreement and IEA – Hydrogen Coordination Group (2006) *Hydrogen Production and Storage: R&D Priorities and Gaps*, OECD/IEA, Paris.
3. Hydrogen Implementing Agreement and IEA – Hydrogen Coordination Group (2006) *Hydrogen Production and Storage: R&D Priorities and Gaps*, OECD/IEA, Paris, p. 5.
4. McKinsey & Company (2010) A Portfolio of Power-Trains for Europe: A Fact-Based Analysis – the Role of Battery Electric Vehicles, Plug-In Hybrids and Fuel Cell Electric Vehicles, Commission of the European Union, Brussels.
5. Op. Cit, Hydrogen Implementing Agreement and IEA – Hydrogen Coordination Group (2006) *Hydrogen Production and Storage: R&D Priorities and Gaps*, OECD/IEA, Paris, pp. 31–32.
6. International Energy Agency Hydrogen Implementing Agreement (2010) *2009 Annual Report on the IEA Agreement on the Production and Utilization of Hydrogen*, IEA HIA, Bethesda, MD.

Further Reading

Energy Strategies Party, Ltd. (2004) *Pursuit of the Future: 25 Years of IEA Research Towards Realisation of Hydrogen Energy Systems*, National Library of Australia: International Energy Agency Hydrogen Implementing Agreement.

International Energy Agency (2005) *Prospects for Hydrogen and Fuel Cells*, OECD/IEA, Paris.

40
Overview of Fuel Cell and Hydrogen Organizations and Initiatives Worldwide
Bernd Emonts

40.1
Introduction

In the future, fuel cells and hydrogen will become more important in terms of an efficient and clean energy supply and also in terms of sustainability. However, before these technologies can be brought to market, considerable efforts are required in the areas of research and development (R&D), demonstration, market introduction, information, education and training, and the qualification of specialists. Throughout the world, numerous organizations and initiatives have been set up over the last 20 years. They collectively represent the interests of commercial enterprises and politics, and also research and educational institutions – all of which want to bring fuel cells and hydrogen to market maturity and implement them on a widespread basis – and they lend these groups more impact. Using different approaches, and to a greater or less extent, they aim to establish fuel cells and hydrogen in various sectors of future energy technology applications. In so doing, they often invest significant sums of money. The organizations and initiatives operate on different levels, ranging from the global, European, and national areas right down to the regional level and individual fields of action. Going beyond this, partnerships, networks, and initiatives also offer support and an opportunity for cooperation in the targeted implementation of specific applications. Against this background, this chapter surveys a selection of fuel cells and hydrogen organizations and initiatives, providing relevant contact information (current at the time of writing).

40.2
International Level

Cooperative hydrogen and fuel-cell technology R&D will play a central role in advancing the transition to a global hydrogen economy. Two organizations, the International Partnership for Hydrogen and Fuel Cells in the Economy (IPHE) and the International Energy Agency (IEA), combine and coordinate activities on this level in the areas of research, development, and market introduction. They aim to facilitate the transition to a competitive global hydrogen economy.

40.2.1
International Partnership for Hydrogen and Fuel Cells in the Economy

IPHE was established in 2003 as an international institution to accelerate the transition to a hydrogen economy (Table 40.1). It provides a mechanism for partners to organize, coordinate, and implement effective, efficient, and focused international research, development, demonstration, and commercial utilization activities related to hydrogen and fuel-cell technologies. Each of the 18 IPHE partner countries has committed to accelerate the development of hydrogen and fuel-cell technologies to improve the security of their energy supply, environment, and economy. IPHE provides a forum for advancing policies and uniform codes and standards that can accelerate the cost-effective transition to a hydrogen economy, and it educates and informs stakeholders and the general public on the benefits of, and challenges to, establishing a hydrogen economy [1].

IPHE recommends that the following actions be taken by public and private sectors to further capitalize on the benefits of energy security and emission reductions offered by the widespread adoption of hydrogen and fuel-cell technologies [2]:

- stimulating early markets through government procurement and early deployment and making available incentive programs for early adopters
- continuing to improve product performance and cost-effectiveness through sustained investment in research, development, and demonstration (RD&D) and infrastructure
- expanding the use of variable renewable energy production through the integration of hydrogen and fuel-cell technologies
- motivating private and public sector financial investments in hydrogen and fuel-cell technology developers and manufacturers

Table 40.1 Contact information for the International Partnership for Hydrogen and Fuel Cells in the Economy (IPHE).

Homepage [1]	
www.iphe.net	IPHE International Partnership for Hydrogen and Fuel Cells in the Economy

Secretariat	IPHE Secretariat, NOW GmbH, Fasanenstrasse 5, 10623 Berlin, Germany
	secretariat@iphe.net
	Steffi Rode *steffi.rode@now-gmbh.de*
Partner countries	Australia, Brazil, Canada, China, European Union, France, Germany, Iceland, India, Japan, Republic of Korea, New Zealand, Norway, Russian Federation, South Africa, UK, USA

- improving education/public outreach, skills and training, codes, standards, and regulations.

IPHE is a forum for member governments to share information and policy experiences with the goal of integrating hydrogen and fuel-cell technologies into the economy. The members collectively account for over 85% of global gross domestic product (GDP), over 75% of the global electricity, and more than 65% of global greenhouse gas emissions. IPHE members take advantage of their global leadership position and national expertise to collaborate extensively to advance hydrogen and fuel-cell technologies. Continuing to work collectively is the only way to ensure that hydrogen and fuel cells play their key role in benefiting our future energy portfolios. IPHE has recently established a new set of strategic priorities and is structuring the organization's future activities based around the following four areas of focus:

- accelerating the market penetration and early adoption of hydrogen and fuel-cell technologies and their supporting infrastructure
- policy and regulatory actions to support widespread deployment
- raising the profile with policy-makers and the public
- monitoring hydrogen, fuel-cell, and complementary technology developments.

Since IPHE was formed, the organization has succeeded in establishing an effective operational structure to facilitate international collaboration and networking both among researchers and at the highest levels of government. Through IPHE, nearly all partner countries have either completed or initiated roadmaps or national strategies for hydrogen and fuel-cell R&D.

40.2.2
International Energy Agency

The IEA is an autonomous organization which works to ensure reliable, affordable, and clean energy for its 28 member countries and beyond. Founded in response to the 1973/1974 oil crisis, the IEA's initial role was to help countries coordinate a collective response to major disruptions in oil supply through the release of emergency oil stocks to the markets. While this continues to be a key aspect of its work, IEA has evolved and expanded. It is at the heart of global dialogue on energy, providing authoritative and unbiased research, statistics, analysis, and recommendations. Today, the IEA's four main areas of focus are as follows [3]:

- **Energy security**: promoting diversity, efficiency, and flexibility within all energy sectors.
- **Economic development**: ensuring the stable supply of energy to IEA member countries and promoting free markets to foster economic growth and eliminate energy poverty.
- **Environmental awareness**: enhancing international knowledge of options for tackling climate change.

- **Engagement worldwide**: working closely with non-member countries, especially major producers and consumers, to find solutions to shared energy and environmental concerns.

Ensuring energy security and addressing climate change cost-effectively are key global challenges. Tackling these issues will require efforts from stakeholders worldwide. To find solutions, the public and private sectors must work together, sharing burdens of resources, while at the same time multiplying results and outcomes.

Through its broad range of more than 40 multilateral technology initiatives (Implementing Agreements), IEA enables member and non-member countries, businesses, industries, international organizations, and non-government organizations to share research on breakthrough technologies, to fill existing research gaps, to build pilot plants, and to carry out deployment or demonstration programs. In short, their work can comprise any technology-related activity that supports energy security, economic growth, environmental protection, and engagement worldwide. A new initiative may be created at any time, provided that at least two IEA member countries agree to work on it together. At the core of a network of senior energy technology experts, these initiatives are also a fundamental building block for facilitating the entry of new and improved energy technologies in the marketplace.

40.2.2.1 Implementing Agreement on Advanced Fuel Cells

The Implementing Agreement for a program of RD&D on advanced fuel cells was signed by seven countries in Paris on 2 April 1990.

Since then, further countries have signed the Implementing Agreement and the program has grown from two to 27 annexes in total. The participating countries are Australia, Austria, Belgium, Canada, Denmark, Finland, France, Germany, Italy, Japan, Korea, Mexico, The Netherlands, Norway, Sweden, Switzerland, Turkey, and the USA [4]. Table 40.2 contains contact information for the IEA Advanced Fuel Cells Executive Committee and Executive Committee Secretary [5].

The aim of the IEA Advanced Fuel Cells program is to advance the state of understanding of all contracting parties in the field of advanced fuel cells. It achieves this through a coordinated program of research, technology development, and systems analysis on molten carbonate fuel cells (MCFCs), solid oxide fuel cells (SOFCs), and polymer electrolyte fuel cells (PEFCs). There is a strong emphasis on information exchange through task meetings, workshops, and reports. The work is undertaken on a task-sharing basis with each participating country providing an agreed level of effort over the period of the task. The current phase of the IEA Implementing Agreement runs until February 2015 and comprises the six annexes shown in Table 40.3.

The IEA's Committee on Energy Research and Technology (CERT) approved a 5 year extension to the Advanced Fuel Cells Implementing Agreement in February 2009. The extension is under way and will run until February 2014. The Implementing Agreement covers fuel-cell technology and its potential applications in stationary power generation, portable power applications, and transport.

Table 40.2 Contact information for the IEA Advanced Fuel Cells Implementing Agreement.

Homepage [4]
www.ieafuelcell.com

Function	ExCo Chairman	ExCo Vice-Chairmen		Secretary	
Name	Prof. Detlef Stolten	Mrs Nancy Garland	Dr. Angelo Moreno	Mrs. Heather Haydock	Dr. Louise Evans
Organization	Forschungszentrum Jülich	US Department of Energy	ENEA	AEA	AEA
Country	Germany	USA	Italy	UK	UK
E-mail	d.stolten@ fz-juelich.de	Nancy.Garland@ ee.doe.gov	angelo.moreno@ casaccia.enea.it	heather.haydock@ aeat.co,uk	louise.evans@ aeat.co.uk

40.2.2.2 Hydrogen Implementing Agreement

The vision of the IEA Implementing Agreement on Hydrogen Production and Utilization is one of clean sustainable energy supply of global proportions that plays a key role in all sectors of the economy. To achieve this vision, the work of the Agreement is directed towards the development of advanced technologies, including direct solar production systems and low-temperature metal hydrides and room-temperature carbon nanostructures for storage [3]. The IEA Hydrogen Implementing Agreement (HIA) was established in 1977 to pursue collaborative hydrogen R&D and information exchange among its member countries. Through the creation and conduct of some 30 annexes or tasks, HIA has facilitated and managed a comprehensive range of hydrogen R&D and analysis activities. It envisions a hydrogen future on a clean, sustainable energy supply that plays a key role in all sectors of the global economy. With more than 30 years of operation, 23 members including the European Union (EU), and significant accomplishments to its credit, HIA is the premier global resource for technical expertise in hydrogen R&D [6]. Contact information for the Chairman, Vice-Chairmen, and the Secretary of the HIA Executive Committee is given in Table 40.4.

Interest in hydrogen technologies has heightened over the past decade, driven by the challenge of climate change coupled with the relationship between energy and economic development. These concerns are reflected in the IEA HIA 2009–2014 Strategic Plan, which devised a Strategic Framework encompassing its vision, mission, strategy, themes, and portfolios, together with a Program of Work elaborating tasks and activities. For the period 2009–2014, HIA has identified three major themes that stem from its mission and vision:

Table 40.3 Descriptions of current annexes [4].

No.	Annex	Objectives	Duration
22	Polymer electrolyte fuel cells	To reduce the cost and improve the performance of PEFCs, direct methanol fuel cells (DMFCs), and corresponding fuel-cell systems	2009–2014
23	Molten carbonate fuel cells	To assist the commercialization of MCFC systems through collaborative research and development	2009–2014
24	Solid oxide fuel cells	To assist, through international cooperation, the development of SOFC technologies	2009–2014
25	Fuel cells for stationary applications	To understand better how stationary fuel-cell systems may be deployed in energy systems	2009–2014
26	Fuel cells for transportation	To understand better how fuel cells may be deployed in transportation applications	2009–2014
27	Fuel cells for portable applications	To assist, through international cooperation, with the development of portable fuel cells towards commercialization	2009–2014
28	Systems analysis (in preparation)	To assist the development of fuel cells through analysis work to enable a better interpretation of the current status, and the future potential, of the technology. This work will provide a competent and factual information base for technical and economic studies	2011–2014

Table 40.4 Contact information for the IEA Hydrogen Implementing Agreement.

Homepage [6]
www.ieahia.org

HYDROGEN IMPLEMENTING AGREEMENT

Function	ExCo Chairman	ExCo Vice-Chairmen		Secretary
Name	Antonio Garcia-Conde	Dr. Steven Pearce	Jan K. Jensen	Mary-Rose de Valladares
Organization	INTA	Solid Energy New Zealand Ltd.	DGC	IEA-HIA
Country	Spain	New Zealand	Denmark	USA
E-mail	glezgca@inta.es	steven.pearce@solidenergy.co.nz	jkj@dgc.dk	mvalladares@ieahia.org

- collaborative RD&D advancing hydrogen science and technology
- an analysis positioning hydrogen for technical progress, and optimizing hydrogen for market preparation and deployment in order to support the political decision-making process
- hydrogen awareness, understanding, and acceptance fostering technology diffusion and commercialization.

These themes are both goals and priorities. Each theme is associated with a set of portfolios containing the tasks and activities shown in Table 40.5. In summary, all current tasks will continue in the new term 2009–2014. Of these tasks, nine are expected to continue past mid-2014, the end of the new term. If the work program is realized as planned, the End-of-Term Report for 2009–2014 will report on progress in a total of 21 tasks.

40.3
European Level

40.3.1
Fuel Cells and Hydrogen Joint Undertaking

The Fuel Cells and Hydrogen Joint Undertaking (FCH JU) is a unique public–private partnership supporting research, technological development, and demonstration activities in fuel-cell and hydrogen energy technologies in Europe [8]. Its aim is to accelerate the market introduction of these technologies, realizing their potential as an instrument in achieving a carbon-lean energy system. Fuel cells, as an efficient conversion technology, and hydrogen, as a clean energy carrier, have great potential to help fight carbon dioxide emissions, to reduce the dependence on hydrocarbons and to contribute to economic growth. The objective of FCH JU is to bring these benefits to Europeans through a concentrated effort from all sectors. The three members of the FCH JU are the European Commission (EC), the fuel-cell, and hydrogen industries, represented by the NEW Industry Grouping (NEW IG), and the research community, represented by Research Grouping N.ERGHY (N.ERGHY).

FCH JU is the result of long-standing cooperation between representatives of industry, the scientific community, public authorities, technology users, and society at large in the context of the European Hydrogen and Fuel Cell Technology Platform. The Platform was launched under the EU's Sixth Framework Programme for Research (FP6) as a grouping of stakeholders, led by companies representing the entire supply chain for fuel-cell and hydrogen energy technologies. The Platform concluded that fuel-cell and hydrogen technologies could play a significant role in a new, cleaner energy system for Europe. However, if these were to make a significant market penetration in transport and power generation, there would need to be research, development, and deployment strategies in which all the stakeholders are committed to common objectives. The European Hydrogen and Fuel Cell Technology Platform has indicated the way forward in a number of strategy

Table 40.5 Descriptions of current and proposed tasks [6, 7].

Task	Annex	Objectives	Duration
21	Biohydrogen	To create a better understanding of the natural processes and genetic regulation involved in reactions in order to use hydrogenases (enzymes) and genetically engineered organisms	2005–2010
22	Fundamental and applied hydrogen storage materials development	To develop reversible or regenerative hydrogen storage materials and systems for use in stationary applications	2006–2009
23	Small-scale reformers for on-site hydrogen supply	To develop technologies for on-site hydrogen production from hydrocarbons (fossil and renewables)	2006–2009
24	Wind energy and hydrogen integration	To explore to the possibilities of hydrogen production using electrolysis with wind energy, and also applications for this hydrogen with special emphasis on H_2 storing and power balancing	2006–2009
25	High-temperature production of hydrogen	To identify and classify high-temperature processors (HTPs) and establish different and coherent criteria for each family of HTPs	2007–2009
26	Water photolysis	To improve the expertise in areas of materials theory, synthesis, and characterization, and also data and database management in order to advance water splitting related to photoelectrode materials science	2008–2011
27	Near-term market routes to hydrogen by co-utilization	To identify and evaluate the most attractive process pathways towards a large-scale demonstration of biomass co-gasification with fossil fuels and for renewable-based H_2	2008–2011
28	Large-scale hydrogen delivery infrastructure	To be defined with regard to geological storage, pipelines, and "mass" storage	2010
29	Distributed and community hydrogen	To further the optimization and replication of green hydrogen within distributed and community energy systems	2010
30	Global hydrogen system analysis	To perform analysis to enable informed decisions that will lead to sustainable clean energy systems	2010
31	Hydrogen safety	To deal with safety through analysis, testing, and the development of target information products	2010

papers, most importantly in the Strategic Research Agenda [9], the Deployment Strategy [10], and the Implementation Plan – Status 2006 [11]. Based on this shared vision, FCH JU was established by a European Council Regulation on 30 May 2008 as a public–private partnership between the EC, European industry, and research organizations to accelerate the development and deployment of fuel-cell and hydrogen technologies.

It is estimated that the activities of FCH JU will help to reduce time to market for hydrogen and fuel-cell technologies by between 2 and 5 years and will therefore have a faster impact on improving energy efficiency and security of supply and reducing greenhouse gases and pollution. In addition, FCH JU is to deliver robust hydrogen supply and fuel-cell technologies developed to the point of commercial lift-off. For the automotive sector, the aim is to achieve breakthroughs in bottleneck technologies and to enable industry to take the large-scale commercialization decisions necessary to achieve mass market growth in the time frame 2015–2020. For stationary fuel-cell systems (domestic and commercial) and portable applications, FCH JU will provide the technology base to initiate market growth from 2010 to 2015. FCH JU supports

- long-term and breakthrough-orientated research
- research and technological development
- demonstration and
- support actions, including pre-normative research.

The first call for proposals was organized in 2008 and new calls will be organized annually until 2013. Ongoing projects will be brought to conclusion by 2017. The research agenda outlining the activities that will be supported by FCH JU is set out in the Multi-Annual Implementation Plan (MAIP). MAIP is translated into annual research priorities each year in an Annual Implementation Plan (AIP) containing the specific topics for the calls for proposals. The FCH JU MAIP is divided into four main application areas with cross-cutting issues such as socio-economic studies (see Figure 40.1). The agenda comprises the entire cycle of research, technological development, and demonstration, including support actions.

40.3.1.1 FCH JU Members

The EC and NEW IG are founding members of FCH JU. Fuel-cell and hydrogen technologies have a great potential to contribute to a number of the EC's key policy goals, including CO_2 reduction, energy security, cleaner transport, and economic competitiveness. Accordingly, the EC has financed European fuel-cell and hydrogen research since 1986. Since 2008, under the 7th Framework Programme for Research, the EC's funding has been channeled through the Joint Undertaking. It has committed a budget of €467 million for the 6 years of the program [12].

NEW IG is an association of European-based companies with activities related to fuel cells and hydrogen. Currently, it has over 60 companies as members, representing a major share of Europe's hydrogen and fuel-cell industries. These companies are developing products for application in transportation, stationary applications, hydrogen production, components, and portable fuel cells. NEW

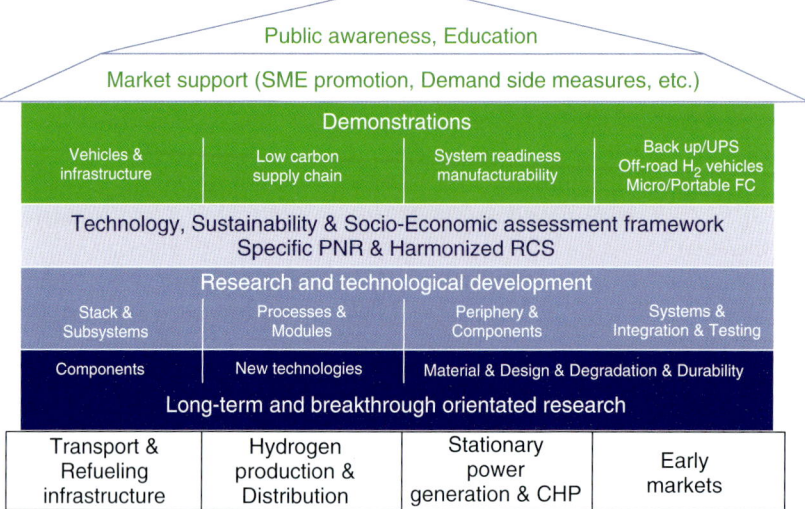

Figure 40.1 FCH JU working plan.

IG has its own secretariat in Brussels, which deals with membership issues, coordinates the members' activities, and provides a link to the FCH JU Programme Office [13].

Member companies lead the planning process for the R&D agenda of FCH JU, thus ensuring the commercial relevance of the projects. Through FCH JU, companies also provide expertise to policy-makers and receive information on policy initiatives in Europe and throughout the world. NEW IG holds six seats in the FCH JU Governing Board. It matches the EC's funding to FCH JU, with a contribution of at least €470 million. The majority of this funding comes in the form of in-kind contributions from companies participating in research projects.

N.ERGHY is the third member of FCH JU. It represents European research institutes and universities working in the area of fuel cells and hydrogen, and currently has over 50 research organizations as members. Member organizations participate directly in the planning process for the R&D agenda of FCH JU. Through FCH JU, research organizations also provide expertise to policy-makers and receive information on policy initiatives in Europe and throughout the world. Like NEW IG, N.ERGHY also has its own secretariat in Brussels. It also participates in the funding of FCH JU as a minority partner and has one seat in its Governing Board [14].

40.3.1.2 Governance Structure

In legal terms, FCH JU partners have come together to support R&D activities. It is responsible for implementing the Fuel Cells and Hydrogen Joint Technology Initiative (FCH JTI), which is the political initiative proposing this public–private partnership in fuel-cell and hydrogen technologies. To coordinate the inputs of all

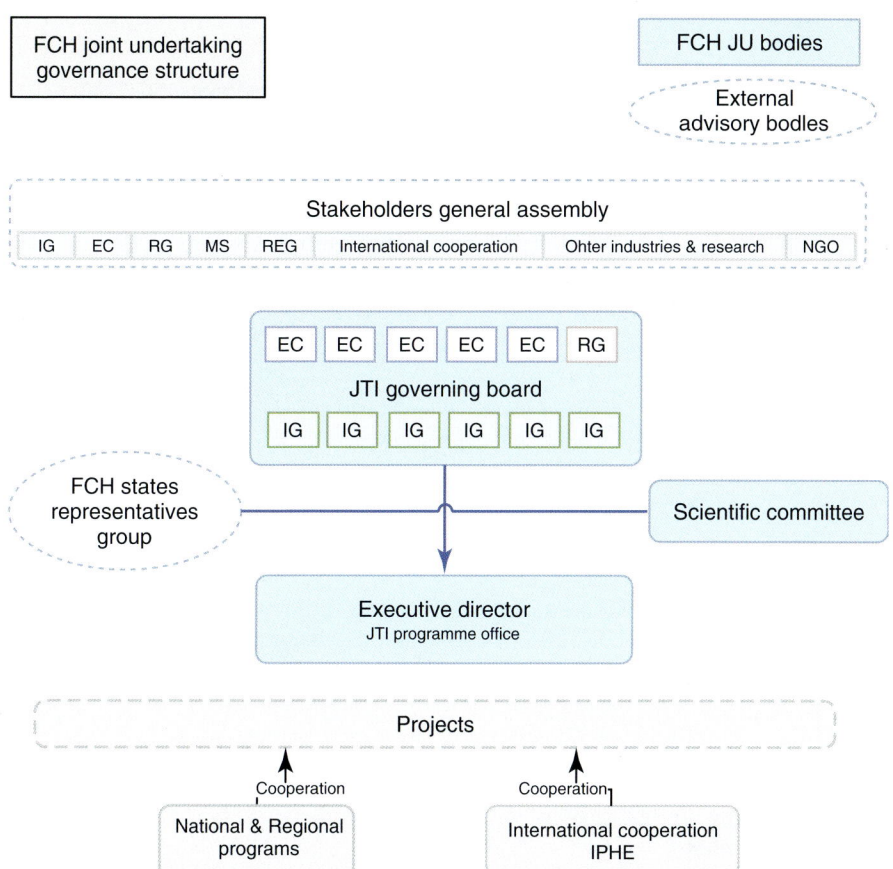

Figure 40.2 Governance structure of FCH JU.

the members and manage the activities, the FCH JU governance structure shown in Figure 40.2 comprises two executive bodies:

- the Governing Board
- the Executive Director assisted by the Programme Office

and three advisory bodies:

- the Scientific Committee
- the States Representatives Group
- the Stakeholders' General Assembly.

The Executive Director and the Programme Office are in charge of the day-to-day management of FCH JU (Table 40.6). The Executive Director is the legal representative of FCH JU. He/she is the chief executive responsible for implementing the Joint Undertaking in accordance with the decisions of the Governing Board.

Table 40.6 Contact information for the Fuel Cells and Hydrogen Joint Undertaking.

Homepage [8] www.fch-ju.eu	FUEL CELLS AND HYDROGEN JOINT UNDERTAKING
Location	Toison d'Or 56–60, 1049 Brussels, Belgium
Governing Board	Gijs van Breda, Shell (Chair)
	Rudolf Strohmeier, DG RTD (Deputy Chair)
Executive Director	Bert de Colvenaer fch-ju@fch.europa.eu
States Representatives Group	Bernard Frois (Chair)
	Georg Menzen (Vice-Chair)
	Aksel Mortensgaard (Vice-Chair)
Scientific Committee	Seven members, nomination process of two further members is ongoing

The Executive Director is supported by the team at the Programme Office. Under the responsibility of the Executive Director, the Programme Office executes all activities from project management to financial matters and communication. The key responsibilities of the Programme Office are as follows:

- organizing the calls for proposals, selection of projects, and management of funding
- managing the R&D agenda of FCH JU in coordination with members and other stakeholders
- communication on FCH JU and fuel-cell and hydrogen technologies.

The Governing Board is the main decision-making body of FCH JU. All three members of FCH JU are represented on the Governing Board: NEW IG has six seats, the EC has five seats, and N.ERGHY has one seat. The Governing Board has overall responsibility for the operations of FCH JU, including the following:

- implementation of activities
- approval of the AIP
- budget, accounts, and the balance sheet
- approval of the list of selected project proposals.

FCH JU supports long-term and breakthrough-orientated research, research and technological development, and also demonstration and support actions, including pre-normative research, following open and competitive calls for project proposals, independent evaluation, and the conclusion of a Consortium Agreement and a Grant Agreement. Project topics in each call are classified by application area:

- transport and refueling infrastructure
- hydrogen production and distribution

- stationary power production and combined heat and power
- early markets
- cross-cutting issues.

40.3.2
European Hydrogen Association

In 2000, five national European hydrogen organizations established the European Hydrogen Association (EHA) and started a close collaboration to promote the use of hydrogen as an energy vector in Europe. In 2004, major European industries active in the development of hydrogen and fuel-cell technologies joined EHA and enforced this effort to create a commercial market for stationary and transport applications and take on a role as market leader for the European hydrogen and fuel-cell sector. EHA promotes the use of hydrogen as a clean energy carrier to power transport and stationary applications. With an office in Brussels, EHA represents 19 national hydrogen and fuel-cell associations in Europe, and also the main European companies involved in hydrogen infrastructure development. Together, they cover more than 300 companies active in hydrogen and fuel-cell application development in Europe [15].

Its unique membership structure has enabled EHA to have up-close insight in local developments and to communicate important issues regarding industrial and regulatory needs to key decision-makers at the EU level. Since 2008, EHA has hosted the European Regions and Municipalities Partnership for Hydrogen and Fuel Cells (HyRaMP). In line with their mission, the activities of EHA focus on the areas of information, representation, expertise, and education and training [15].

- **Information**:
  - Collect relevant information and circulate it between members. Encourage exchanges and joint actions between members.
  - Identify and advertise hydrogen-related capabilities existing in and inform the different relevant segments of society through adapted means. Promote European capabilities with a view to fostering European participation in national and international projects and European programs, in particular by organizing specialized workshops and events. Promote the views of EHA on a national level in Europe by directly contacting relevant national authorities, and on a European level by contacting EU organizations, including the European Parliament and the EC.
- **Representation**:
  - Interface with national and international organizations and with public bodies in Europe in the area of laws, standards, and regulations.
  - Promote cooperation between its members and with similar associations in the rest of the world.
  - Actively promote the creation of national hydrogen associations in European countries where such associations do not exist. Assist potential entrepreneurs

in drawing up statutes and promoting their projects with local authorities, universities, and industrial organizations.
- Actively seek good coordination of hydrogen-related events in Europe.
- Appoint representatives to become members of advisory bodies, scientific institutions, and so on.

- **Expertise**:
  - Identify and maintain a list of experts belonging to member companies and/or organizations that EHA will advertise and propose to appropriate organizations in Europe and in the rest of the world.
  - Encourage R&D initiatives that have innovative potential, in particular by suggesting topics for the EC's R&D program.
  - Effect studies for the benefit of EHA, either directly or by contracting outside specialists. Effect specific socio-economic studies at the European level in order to show the economic–societal impact of the use of hydrogen. Stress the economic and health aspects, and the use of hydrogen as a way toward replacing expendable energies by renewable energies. Encourage the setting-up of technical projects that can facilitate the use of hydrogen or solve critical issues. Propose or participate in the creation of European or international standards, or any other technical documents relating to hydrogen, and advertise such standards or documents. Identify relevant expertise outside Europe and in topics indirectly related to hydrogen, such as renewable energies, and so on.

- **Education and training**:
  - Promote the training of young research students, technicians, and engineers in areas relevant to EHA.
  - Promote the teaching of hydrogen-related techniques and technology, and related socio-economic aspects, and also the introduction of hydrogen-related modules in courses on related topics.

The most important contact information for EHA is given in Table 40.7.

EHA members comprise 20 national organizations. Table 40.8 provides a list of these organizations together with their Internet addresses.

Table 40.7 Contact information for the European Hydrogen Association.

Homepage [15] www.h2euro.org		
Executive Director	Marieke Reijalt	info@h2euro.org
Office Manager	Tanya Carre	info@h2euro.org

Table 40.8   EHA members.

Association	Internet address	Logo
Bulgarian Hydrogen Society	bgh2society.org/en	
Czech Hydrogen Technology Platform	www.hytep.cz	
Danish Partnership for Hydrogen and Fuel Cells	www.hydrogennet.dk	
Dutch Hydrogen and Fuel Cell Association	www.waterstofvereniging.nl	
Flemish Hydrogen and Fuel Cell Association	www.vswb.be	
French Hydrogen Association	www.afh2.org	
German Hydrogen and Fuel Cell Association	www.dwv-info.de	
Hungarian Hydrogen Association	www.e-met.hu	
Irish Hydrogen Association	In preparation	
Italian Hydrogen and Fuel Cell Association	www.h2it.org	
Latvian Hydrogen Association	http://lathydrogen.lv	
Macedonian Association for Hydrogen	www.tmf.ukim.edu.mk	
Norwegian Hydrogen Forum	www.hydrogen.no	
Polish Hydrogen Association	www.hydrogen.edu.pl	

(continued overleaf)

Table 40.8 (continued).

Association	Internet address	Logo
Portuguese Hydrogen Association	www.ap2h2.pt	
Slovene Hydrogen Association	www.dcht.eu	
	www.conot.si	
Spanish Hydrogen Association	www.aeh2.org	
Hydrogen Sweden	www.vatgas.se	
Swiss Hydrogen Association	www.hydropole.ch	
United Kingdom Hydrogen and Fuel Cell Association	www.ukha.org	

40.4
National Level

40.4.1
US Fuel Cell and Hydrogen Energy Association

The Fuel Cell and Hydrogen Energy Association (FCHEA) is an advocacy organization dedicated to the commercialization of fuel cells and hydrogen energy technologies in the USA. FCHEA and its members strive to transform the energy network, fundamentally altering the way in which energy is generated and used. With more than 80 facilities, FCHEA's members represent the full spectrum of the supply chain in the USA including universities, government laboratories and agencies, trade associations, fuel-cell materials, components and systems manufacturers, hydrogen producers and fuel distributors, utilities, and other end users. As the combined team of industry leaders in fuel cells and hydrogen, FCHEA provides members access to a powerful network of contacts.

FCHEA was formed when the US Fuel Cell Council (USFCC) merged with the National Hydrogen Association in October 2010. USFCC had been dedicated to fostering the commercialization of fuel cells in the USA. It aimed to provide technical advice, collect information and issue reports on the industry, raise public awareness of fuel cells and their potential, conduct fuel-cell education programs, provide networking opportunities for developers, suppliers, and potential customers, and establish links to comparable activities in the USA and worldwide. An extensive range of technical, educational, and outreach activities were initiated by USFCC in support of the fuel-cell industry.

40.4.1.1 Working Groups and Committees

FCHEA members lead activities in nine working groups and several sub-group task forces (see Table 40.9). Through these groups, members have an opportunity to develop policies, interact with colleagues and opinion leaders, facilitate codes and standard development, and conduct education and outreach activities on behalf of the fuel-cell and hydrogen energy industries. The formation of new groups is anticipated, given the increased potential for collaboration in FCHEA.

FCHEA is governed by its Board of Directors (see Table 40.10) and its guiding committees. The Executive Committee is composed of FCHEA's officers, who are elected by members. The Market Strategy and Outreach Committee is composed of marketing, communications, and business experts from member companies in collaboration with FCHEA leadership. The Regulatory Affairs Steering Committee is open to participation from Tier 1 and 2 member organizations.

40.4.1.2 Resources

FCHEA has a vast array of resources available to the public, including the following:

- a link to Fuel Cells 2000s State of the States Report
- monthly newsletter entitled "Fuel Cell and Hydrogen Energy Connection"
- fuel-cell industry market reports
- a wide range of fact sheets covering fuel-cell applications and hydrogen energy
- an FCHEA fuel-cell product catalog
- the Hydrogen and Fuel Cell Safety Report
- informative brochures
- the FCHE 2011 Conference Proceedings
- a wide range of technical products.

Table 40.9 Current FCHEA working groups and task forces.

Working group	Task force	
Government affairs	–	
Codes and standards	–	
Materials and components	Gaskets	
Portable power	Transportation regulations	
Power generation	–	
Transportation	Hydrogen quality	Fork lift task force
Aviation	–	
Nuclear	–	
Solid oxide fuel cells	–	
Renewable hydrogen	–	

Table 40.10 FCHEA contact information.

Homepage [16] www.fchea.org		
Location	1211 Connecticut Avenue NW, Suite 600, Washington, DC 20036, USA	
Executive Director	Morry Markowitz	mmarkowitz@fchea.org
Director of Policy	James Warner	jwarner@fchea.org
Technical Director	Robert Wichert	rwichert@fchea.org
Communications Manager	Connor Dolan	cdolan@fchea.org
Senior Advisor	Robert Rose	brose@fuelcells.org

Table 40.11 FCHEA information portal.

Fuel cell and hydrogen energy		
Fuel cell basics	Fuel cell overview	How fuel cells work
	What is a fuel cell?	Types of fuel cells
Hydrogen energy	Hydrogen energy overview	Hydrogen production
	Hydrogen benefits	Hydrogen safety
Applications	Applications overview	Stationary
	Renewables	Portable
	Transportation	Combined heat and power (CHP)
	Military	Fuel cell end users forum

In addition, the FCHEA web site provides popular science information on fuel-cell and hydrogen energy. Table 40.11 summarizes the topics covered.

40.4.2
Canadian Hydrogen and Fuel Cell Association

The Canadian Hydrogen and Fuel Cell Association (CHFCA) works to raise awareness of the economic, environmental, and social benefits of hydrogen and fuel cells. CHFCA is a national, non-profit association providing services and support to Canadian corporations, governments, and educational institutions promoting, developing, demonstrating, and deploying hydrogen and fuel cell products and services in Canada. Nearly 80 members cover most types of hydrogen and fuel-cell technologies, components, systems supply and integration, fueling systems, fuel storage, and engineering and financial services. CHFCA was formed in January 2009 as a result of a merger between the Canadian Hydrogen Association (CHA) and Hydrogen and Fuel Cells Canada (H2FCC). The merger united the members of these two associations to create a vibrant, influential association that represents the majority of the stakeholders in Canada's hydrogen and fuel-cell sector. The aims of CHFCA are as follows:

- raising the profile of the Canadian hydrogen and fuel-cell sector in Canada and around the globe
- enhancing the sector's profile with Canadian governments
- promoting the economic, environmental, and social benefits of hydrogen and fuel-cell technologies and products
- facilitating demonstration projects, such as the Hydrogen Highway, Hydrogen Village, and the Vancouver Fuel Cell Vehicles Program, which allow hydrogen and fuel cells companies to test and perfect their technologies in real-world environments
- providing communications, information sharing, and networking between member organizations through conferences, events, annual meetings, and planning sessions
- supporting the development of regulations, codes, and standards that support the safe and widespread application of hydrogen and fuel-cell products.

More detailed information on CHFCA is available on their well-structured web site. Contact information for CHFCA is also summarized in Table 40.12.

The "Hydrogen and Fuel Cells" section of the CHFCA web site provides compact information and also answers to frequently asked questions. The following areas are covered:

- the need for clean energy
- about hydrogen
- about fuel cells
- FAQ – get the facts
- install hydrogen and fuel cells.

Table 40.12 CHFCA contact information.

Homepage [17] *www.chfca.ca*	 Canadian Hydrogen and Fuel Cell Association		
Location		900 – 1188 West Georgia Street, Vancouver, BC V6E 4A2, Canada	
President and CEO	Eric Denhoff	*edenhoff@chfca.ca*	
Vice-President	Terry Kimmel	*tkimmel@chfca.ca*	
Industry marketing and communications	Javis Lui	*jlui@chfca.ca*	
Finance and administration	Doug Burden	*dburden@chfca.ca*	
Conferences and workshops	Sarah Richards	*srichards@chfca.ca*	

Addressing the last topic, CHFCA provides access to an 85-page guide, published by Air Liquide Canada in 2008, entitled "Permitting Hydrogen and Fuel Cell Installations in Canada." The guide covers codes, standards and regulations, authorities, and approval processes [18]. The "Resources" section of the web site contains a list of CHFCA publications on the Canadian hydrogen and fuel-cell sector with statistics from 2010, a capabilities guide for 2011, and a fuel-cell commercialization roadmap for 2003 and 2008.

40.4.3
German National Organization for Hydrogen and Fuel Cell Technology

NOW GmbH (National Organization for Hydrogen and Fuel Cell Technology) was founded in Germany in 2008. Its mission is to coordinate and manage market preparation programs for products and applications based on hydrogen, fuel cells, and battery electric powertrain technology. Specifically, NOW is responsible for the implementation of the following programs:

- National Innovation Programme for Hydrogen and Fuel Cell Technology (NIP) [19] with its National Development Plan [20]
- Model Regions Electric Mobility.

The German federal government adopted NIP as part of its "high-tech strategy" in spring 2006. The program aims to contribute to maintaining and expanding Germany's good starting position for developing and implementing hydrogen and fuel-cell technology in the marketplace. The Innovation Program includes RD&D of mobile and stationary applications of fuel cells and hydrogen. Complementing fuel cells, battery technology will be further developed as a key technology. The new €500 million program for e-mobility is based on the 2009 economic stimulus package. It ran from 2009 to 2011. Public funding from the Federal Ministry of Transport, Building and Urban Development (BMVBS) and the Federal Ministry of Economics and Technology (BMWi) will amount to €1.4 billion for the period from 2006 to 2016 (NIP). This guarantees substantial institutional funding for basic research within the Max Planck Society (MGP), Helmholtz Association (HGF), and Fraunhofer Gesellschaft (FhG), and also for RD&D projects carried out by industry. Half of the budget is provided by the German government, and industry participants provide the other half. R&D projects are funded through BMWi, which contributes €200 million, and demonstration projects are funded through BMVBS, which contributes €500 million.

NOW is responsible for the coordination and management of the NIP and the Model Regions Electric Mobility Programme of BMVBS. Both programs are designed to prepare the market for these technologies. BMWi funding for R&D in the field of fuel-cell and hydrogen technologies accounted for €25–30 million between 2006 and 2009.

The primary task of NOW is to initiate projects. It evaluates and groups projects together as appropriate, for example, according to geographic and/or thematic

Table 40.13 NOW contact information.

Homepage [21] www.now-gmbh.de	NOW Nationale Organisation Wasserstoff- und Brennstoffzellentechnologie	
Location	Fasanenstrasse 5, 10623 Berlin, Germany	
Supervisory Board	Hilde Trebesch, BMVBS (Chair)	
Advisory Board	Representatives of 18 stakeholder groups	
Management Spokesman	Dr. Klaus Bonhoff	klaus.bonhoff@now-gmbh.de
Secretariat	kontakt@now-gmbh.de	
	Sara Kassem	sarah.kassem@now-gmbh.de

criteria, thus exploiting the benefits of synergy. Furthermore, NOW recognizes the importance of so-called crossover tasks. Production technologies, education, and further training, and also communication at the intersection of politics, industry, science, and public relations, all contribute to raising the general profile of these technologies and their products.

NOW is a state organization. It is wholly owned by the German federal government, represented by the BMVBS. Table 40.13 contains contact information for NOW. The mission and objectives of NOW in terms of implementing NIP are as follows [21]:

- to accelerate development of the market by means of targeted support for and promotion of the hydrogen and fuel-cell sectors in the mobile, stationary, and portable fields
- to establish value chains and value shares in Germany
- to safeguard the technological lead and implement the technology in Germany.

The maximum possible number of industrial companies, small- and medium-sized enterprises, users, and research establishments must therefore be involved in the overall development process, so that Germany can offer competitive cutting-edge technologies in the global market with a large value-added component. The NIP activities safeguard jobs in Germany as an innovative location, and contribute to ensuring efficient, low-emission energy supply and energy use.

40.5
Regional Level

40.5.1
European Regions and Municipalities Partnership for Hydrogen and Fuel Cells

The European Regions and HyRaMP (Table 40.14) was founded in Brussels on 14 April 2008 [22]. HyRaMP's mission is to provide the European Regions and Municipalities with a representative body that is coherent, distinguishable, and influential, particularly within the context of the European FCH JTI. HyRaMP

Table 40.14 Contact information for HYRaMP.

Homepage [22] www.hy-ramp.eu	HyRaMP
Location	Palais des Academies, Rue Ducale 1, 1000 Brussels, Belgium
Secretariat manager	Marieke Reijalt secretariat@hy-ramp.eu
Office assistant	Celine Van Lierde
Members	Belgium, Finland, France, Germany, Italy, The Netherlands, Poland, Portugal, Scandinavia, Slovakia, Slovenia, Spain, UK

allows its members – more than 30 regions and cities in Europe – to play a key role in the implementation of strategies aiming for the uptake of FCH technologies and particularly that of FCH JTI. HyRaMP also aims to accelerate the uptake of electromobility in Europe. Electromobility is considered a prime candidate for reducing the oil dependence of road transport. For European industry, "electromobility" comprises battery electric and hydrogen/fuel-cell vehicles, and also the relevant supply infrastructure. These propulsion technologies offer complementary transport solutions and have many elements in common. HyRaMP will ensure a sufficient critical mass within industry for the deployment of both battery electric and fuel-cell vehicles, and it will help coordinate the creation of the relevant infrastructure across the EU.

40.5.2
Hydrogen and Fuel-Cell Activities in Germany's Federal States

The German federal states have reinforced their activities in the area of hydrogen and fuel cells. Most of them have founded expertise networks, bringing together partners from industry and the research community and initiating research, development, demonstration, and market implementation projects related to the hydrogen and fuel-cell technology.

The objectives of their activities are to sensitize the companies based in the respective federal states to the hydrogen and fuel-cell market, and thus enable them to benefit from the future creation of value. Moreover, the federal states aim to improve the conditions for industry and to safeguard employment. Taken together, the three federal states of NRW, Bavaria, and Baden-Württemberg allocated grants in excess of €50 million for hydrogen and fuel-cell projects in 2004–2005. The total budget of these projects is likely to exceed €100 million per year. Table 40.15 lists the most important initiatives of the German federal states.

Most federal states, such as NRW and Hamburg, have initiatives and funding programs. For example, NRW has funded 85 projects since 2000. The annual funding amount averaged ~€10 million. Owing to a new large-scale project known as the "*Hydrogen Highway*," the annual funding amount for the coming 3 years

Table 40.15 Initiatives of the German federal states.

Initiative	Internet address
Die Brennstoffzellen- und Batterie-Allianz Baden-Württemberg	www.bba-bw.de
Koordinationsstelle der Wasserstoff-Initiative Bayern	www.wiba.de
Koordinierungsstelle Wasserstoff für Berlin	www.element-1.org
HyCologne	www.hycologne.de
hySOLUTIONS Innovative Antriebe für Hamburg	www.hysolutions-hamburg.de
H2 BZ Initiative Hessen	www.h2bz-hessen.de
Wasserstofftechnologie-Initiative Mecklenburg-Vorpommern	www.wti-mv.de
Landesinitiative Brennstoffzelle Niedersachsen	www.brennstoffzelle-nds.de
Netzwerk Brennstoffzelle und Wasserstoff Nordrhein-Westfalen	www.brennstoffzelle-nrw.de
Saarländische Arbeitsgruppe Dezentrale Energieversorgung	www.izes.de
Brennstoffzellen Initiative Sachsen	www.bz-sachsen.de
Brennstoffzellenverband Sachsen-Anhalt	www.brennstoffzelle-sa.de
Wasserstoffinitiative Schleswig-Holstein	www.s-h2.de

will be €15 million. Hamburg funded six major projects between 2003 and 2008 totaling €15.2 million. Between 2009 and 2013, Hamburg will spend €18.4 million on hydrogen and fuel-cell projects.

40.6
Partnerships, Initiatives, and Networks with a Specific Agenda

40.6.1
The California Fuel Cell Partnership

In January 1999, two US government agencies, the California Air Resources Board and California Energy Commission, joined forces with six private sector companies, Ballard Power Systems, DaimlerChrysler, Ford Motor Company, BP, Shell Hydrogen, and ChevronTexaco, to form the California Fuel Cell Partnership (CaFCP) (Table 40.16). The goal was to demonstrate and promote the potential for fuel-cell vehicles as a clean, safe, and practical alternative to vehicles powered by internal combustion engines. Within a very short time, other government agencies and private businesses became members. As a collaboration of 30 organizations, including auto manufacturers, energy providers, government agencies, and fuel-cell technology companies, CaFCP members work together to promote the commercialization of hydrogen fuel cell vehicles with a new technology of the twenty-first century [23].

CaFCP is committed to promoting fuel-cell vehicle commercialization as a means of moving towards a sustainable energy future, increasing energy efficiency and reducing or eliminating air pollution and greenhouse gas emissions. CaFCP members collaborate on activities that advance the technology. For example, California was the first state to designate hydrogen as a transportation fuel. This led to a need to set standards and regulations. CaFCP members provided the US

Table 40.16 Contact information for the CaFCP.

Homepage [23] www.cafcp.org	
Location	3300 Industrial Boulevard, Suite 1000, West Sacramento, CA 95691, USA
Executive Director	Catherine Dunwoody info@cafcp.org

Department of Measurement Standards (DMS) with joint input about hydrogen quality regulations.

Some projects, such as public outreach and first responder education, are ongoing. Other projects arise around one issue, for example, training drivers on how to use the stations. When the issue has been resolved, the project team disbands. The people, however, are all engaged on a day-to-day basis to move fuel-cell vehicles closer to market. Automotive members provide fuel-cell passenger vehicles that are placed in demonstration programs, where they are tested in real-world driving conditions. Energy members work to build hydrogen stations within an infrastructure that is safe, convenient, and fits into the community. Fuel-cell technology members provide fuel cells for passenger vehicles and transit buses. Government members lay the groundwork for demonstration programs by facilitating steps to creating a hydrogen fueling infrastructure.

The main aim of CaFCP is to reach California's goals for cleaner air and reduced greenhouse gases despite the need for full-function cars, pickups, vans, and sports utility vehicles (SUVs) that people want to drive and transit buses in which they want to ride. These vehicles must be comparable to or better than the vehicles driven today, and be better for the environment. Hydrogen-powered fuel-cell vehicles are an option for fulfilling these requirements.

Since 1999, the members of the CaFCP have made steady progress towards a commercial market. By the end of 2008, 250 fuel-cell vehicles had been placed on California's roads, refueling at 26 active stations in the state. California is the first state to recognize hydrogen as a motor vehicle fuel and begin the regulatory process that enables retail fuel sales.

By 2017, automakers expect to place tens of thousands of fuel-cell vehicles in the hands of California consumers. Today, about 300 fuel-cell vehicles have been placed on California's roads and fill at only 22 hydrogen stations in the state. As the number of fuel-cell vehicles in California increases over the next 5–10 years, it is crucial that hydrogen be easily available to the drivers. The experience of the first fuel-cell vehicle drivers must be good if early success is to be made possible. In February 2009, CaFCP's members published an "action plan" that details a strategy for deploying hydrogen fueling stations and fuel-cell vehicles in California. The CaFCP's action plan has three focus areas:

- developing early "hydrogen communities" for passenger vehicles with clusters of retail hydrogen stations in four southern California communities
- expanding the transit program in the San Francisco Bay area with new mixed-use stations that provide fuel for passenger vehicles and transit buses, and also dedicated retail hydrogen stations for passenger vehicles
- developing codes, standards, and regulations with a state-of-the-art hydrogen station in the Sacramento area that will enable regulatory agencies to validate new test procedures and also provide fuel for passenger vehicles in the Sacramento area.

40.6.2
UK Hydrogen Energy Network

The UK Hydrogen Energy Network (H2NET) was established in April 2000 as a joint collaboration between UK industry and academia interested in the development of hydrogen as an energy vector (Table 40.17). Its principal aim is to promote research and discussion on issues connected with the development of the hydrogen energy economy. The formation of H2NET was prompted by wider international developments in the technologies underpinning a hydrogen energy economy. Its objectives are to enhance the current profile of hydrogen energy research in the UK by [24]:

- identifying research requirements and facilitating the development of academic/industrial collaborations in the UK
- providing a forum for the discussion of research, development, and implementation issues related to hydrogen energy exploitation
- disseminating information relating to state-of-the-art research in the hydrogen energy economy.

These objectives are addressed through a regular series of workshops and seminars, backed up by a web site and email discussion group.

H2NET is guided by a steering committee drawn from industrial groups and university departments. It is currently supported by the UK Department of Trade and Industry (DTI) under contract No. 14034676.

40.6.3
Initiative Brennstoffzelle

The Initiative Brennstoffzelle (IBZ) is a German fuel-cell initiative concentrating on fuel-cell heating appliances for domestic energy supply. Together, 13 members including leading companies in the energy industry, well-known appliance manufacturers, the German Energy Agency, and NOW actively support the innovative technology. The partners are involved on various levels:

- in the development of equipment and components
- in field trials and demonstration objects

Table 40.17 Contact information for H$_2$NET.

Homepage [24] www.h2net.org.uk	H$_2$NET
Director	Dr. Geoff Dutton a.g.dutton@rl.ac.uk

Table 40.18 Contact information for IBZ.

Homepage [25] *www.ibz-info.de*		
PR (german: Herausgeber)	EWE AG, Donnerschweer Street 22–26, 26123 Oldenburg, Germany	
Person in charge	Volker Diebels	*presse@ibz-info.de*
Members	BAXI-INNOTECH GmbH, Bosch, dena, EnBW AG, E.ON Ruhrgas AG, erdgas, EWE AG, Hexis AG, MVV Energie AG, NOW GmbH, Vaillant, Verbundnetz Gas AG, Viessmann	

Abbreviation: PR, Publisher.

Table 40.19 Information published by IBZ.

Media	Newsletter	Reports on current events
	Press information	Texts and graphics for download
	IBZ news	Online news
	Brochures	Online brochures
	Animation	Film on the journey inside a fuel cell
Equipment	Areas of application	Fuel cells at home
	Technology	Fuel-cell heating appliances
	Models	Models from three manufacturers
Projects	Callux	Practical trials with fuel cells for homes
	Interactive project map	Location of appliances being tested
Know-how	Online ABC	Glossary articles
	FAQ architects/planners	Answers to 33 specific questions
	FAQ skilled trades	Answers to 17 specific questions

- in the drafting of norms and standards
- in the training of skilled trades.

IBZ also has a clearly structured web site, the address of which is given in Table 40.18.

One of the central tasks of IBZ is to provide factual information. Anyone interested in the topic, particularly schools and training facilities, have access through IBZ to current media [25]. Table 40.19 summarizes the various forms of information covered by IBZ. It does not just aim at the educational sector, but goes beyond this to serve the needs of skilled trades, architects, and planners.

40.7
Conclusion

The specialist community is convinced that fuel cells and hydrogen will play a key role in the energy provision of the twenty-first century. However, before either technology can be put to use in widespread applications, much R&D work is still required. The institutions and enterprises concerned with R&D work will receive support and assistance from the organizations, initiatives, and networks interested in the progress of the two technologies. On an international level, the IEA cooperates with scientific and technical working groups and encourages the exchange of information between circles of experts, while IPHE initiates and coordinates project activities to ensure the widespread dissemination of existing knowledge and the collaborative creation of new information.

In the European context, FCH JU sets the standards by defining the strategy, program, and a very focused budget for potential fields of application. A stringent management structure at FCH JU ensures that the EC funds for the current program totaling €470 million are put to use as intended. With over 19 national fuel-cell and hydrogen organizations and over 300 commercial enterprises working in this field, EHA is the largest and most important umbrella organization in Europe. The main objectives of EHA are collecting and disseminating relevant information, supporting and initiating cooperation, identifying and utilizing expertise, and supporting education and training.

Large national organizations have also emerged in the USA (FCHEA), Canada (CHFCA), and Germany (NOW, NKJ). Whereas the US and Canadian organizations, with some 80 member institutions each, concentrate on providing information material, organizing conferences, and setting up working groups, NOW and NKJ focus on using funding from BMVBS worth €1 billion and funding from BMWi totaling €200 million by 2016 to coordinate the market introduction of products and applications based on hydrogen technology and also fuel-cell and battery drive technology.

On a regional level, more than 30 European regions and cities aim to achieve sufficient critical mass in the industrial sector involved in the development and market introduction of battery and fuel-cell electric vehicles. At the same time, 13 initiatives and competency networks in Germany have joined forces in order to bring partners from industry and research together and thus drive fuel-cell and hydrogen technology forward with coordinated research, development, demonstration, and market introduction.

Organizations, such as CaFCP in California, USA, H2NET in the UK, and IBZ in Germany, have also emerged with a focus on specific applications and specialized markets. In the form of a cooperation between 30 organizations, CaFCP aims to commercialize hydrogen vehicles together with the required infrastructure. H2NET is a cooperation between British industry, universities, and research institutions aiming to establish hydrogen as an energy carrier. In Germany, a consortium of 13 members is working towards the introduction of fuel-cell heating appliances in domestic energy supply. All of the organizations and initiatives worldwide are motivated by the conviction that RD&D, the dissemination of well-founded information, education and training, and financial support are essential if the widespread market introduction of fuel-cell and hydrogen technology is to succeed.

References

1. IPHE (2011) An International Vision for Hydrogen and Fuel Cells, http://www.iphe.net (last accessed 20 January 2012).
2. IPHE (2009) Communiqué Hydrogen and Fuel Cells – A Global Opportunity, IPHE, Berlin.
3. IEA (2012) International Energy Agency, http://www.iea.org (last accessed 20 January 2012).
4. IEA (2012) IEA Advanced Fuel Cells Implementing Agreement, http://www.ieafuelcell.com (last accessed 20 January 2012).
5. IEA AFC (2009) IEA AFC Annual Report, IEA United Kingdom.
6. IEA HIA (2012) Hydrogen Implementing Agreement, http://www.ieahia.org (last accessed 20 January 2012).
7. IEA HIA (2009) IEA HIA Annual Report. HIA Secretariat, Bethesda, MD.
8. FCH JU (2011) Fuel Cells and Hydrogen Joint Undertaking, http://www.fch-ju.eu (last accessed 20 January 2012).
9. European Hydrogen and Fuel Cell Technology Platform (2005) Strategic Research Agenda, HFP, Brussels.
10. European Hydrogen and Fuel Cell Technology Platform (2005) Deployment Strategy, HFP, Brussels.
11. European Hydrogen and Fuel Cell Technology Platform (2007) Implementation Plan – Status 2006, Implementation Panel, HFP, Brussels.
12. EC (2012) European Commission, http://ec.europa.eu (last accessed 20 January 2012).
13. NEW-IG (2012) New Energy World Industry Grouping Fuel Cell and Hydrogen for Sustainability, http://www.fchindustry-jti.eu (last accessed 20 January 2012).
14. N.ERGHY (2011) New European Research Grouping on Fuel Cells and Hydrogen, http://www.nerghy.eu (last accessed 20 January 2012).
15. EHA (2012) European Hydrogen Association, http://www.h2euro.org (last accessed 20 January 2012).
16. FCHEA (2012) Fuel Cell and Hydrogen Energy Association. Transforming the Energy Network, http://www.fchea.org (last accessed 20 January 2012).
17. CHFCA (2012) Canadian Hydrogen and Fuel Cell Association, http://www.chfca.ca (last accessed 20 January 2012).
18. Dubé, J., Hay, D.R., Katz, S., and Oh, S. (2008) *Permitting Hydrogen and Fuel Cell Installations in Canada. A Guide to Codes, Standards and Regulations, Authorities, and Approval Processes*, Air Liquide Canada, Montreal.
19. BMVBS, BMBF, BMWi (2006) Nationales Innovationsprogramm Wasserstoff- und Brennstoffzellentechnologie, Bundesministerium für Verkehr, Bau und Stadtentwicklung, Berlin.
20. Strategierat Wasserstoff Brennstoffzellen (2007) National Development Plan–Version 2.1, Strategierat Wasserstoff Brennstoffzellen, Berlin.
21. NOW (2012) Nationale Organisation Wasserstoff- und Brennstoffzellentechnologie, http://www.now-gmbh.de (last accessed 20 January 2012).
22. HyER (2012) Hydrogen Fuel Cells and Electro-mobility in European Regions, http://www.hy-ramp.eu (last accessed 20 January 2012).
23. CaFCP (2012) California Fuel Cell Partnership, http://www.cafcp.org (last accessed 20 January 2012).
24. H2NET (2008) The UK Hydrogen Energy Network, http://www.h2net.org.uk (last accessed 20 January 2012).
25. IBZ (2010) Initiative Brennstoffzelle, http://www.ibz-info.de (last accessed 20 January 2012).

41
Contributions for Education and Public Awareness

Thorsteinn I. Sigfusson and Bernd Emonts

41.1
Introduction

Around the turn of the millennium there was a strongly felt world movement towards the utilization of hydrogen energy [1, 2]. One of the effects of this was the establishment of IPHE (International Partnership for the Hydrogen Economy), in Washington, DC, in late 2003 [3]. The meeting in Washington was organized by the Department of Energy and was run as a "ministerial" with participants from industry, academia, institutions, and more than two dozen governmental representatives from all over the globe [4].

It was clear from the onset that delegates thought that one of the main tasks to be addressed worldwide was public awareness and education, which soon became one of the focal points of the IPHE organization.

Shortly thereafter, the first author gave a public lecture in the United Nations (UN) headquarters in New York, where a special day was devoted to hydrogen energy. Following a lecture on the origins and pathways for hydrogen energy, there was a slot in the program dedicated to questions and discussions. One of the questions asked at the UN meeting was raised by a national representative and concerned "the links between a hydrogen energy economy and the hydrogen bomb." There was a short silence before the question was answered. For the expert, this was an unexpected and complete misunderstanding of the role of the element as a fuel in a modern society.

The answer was detailed and attempted to detach hydrogen fuel completely from the use of hydrogen atoms in a fusion explosive – where, in fact, there are no links whatsoever. On the other hand, the question made the expert team aware of the great work ahead of it in terms of creating more public awareness of the possibilities and opportunities of the lightest of elements in our society.

We realized that not only were the public subjected to incorrect interpretations of historical events, such as the fate of airships through the demise of the Hindenburg, but there also seemed to be a need for a thorough educational effort to bring the real facts to the table and present an emerging new energy age to the public that ultimately will enjoy its benefits.

Fuel Cell Science and Engineering: Materials, Processes, Systems and Technology,
First Edition. Edited by Detlef Stolten and Bernd Emonts.
© 2012 Wiley-VCH Verlag GmbH & Co. KGaA. Published 2012 by Wiley-VCH Verlag GmbH & Co. KGaA.

In the first years of the millennium, education was seen as crucial to the emerging global transition, providing political leaders, technical specialists, and laypeople with the knowledge necessary to play their own appropriate roles in the transition to a hydrogen-based energy infrastructure.

Education in hydrogen energy was seen as a cross-cutting issue, touching upon almost all levels and individuals in society. The need to inform and educate all government officials, ranging from national to local level, was seen as crucial. Other groups targeted included safety and code officials, university and college students, primary and secondary teachers and students, and the general public.

The identified audience groups have distinct hydrogen educational needs, ranging from general to technical, from broad based to narrowly targeted. Certain audiences require special attention in the near term to facilitate, for example, research, development, and demonstration efforts, whereas the needs of others will grow over time as the transition to a hydrogen economy progresses. In some of the existing demonstration projects in Europe, North America, and Japan, the opportunity to educate and train human capital in the field of hydrogen has been exploited with clear success in building a knowledge base. In some countries and areas, the effect of these demonstration projects will result in public outreach and education of enormous value.

In this chapter, the authors intend to review briefly some important examples of experience with public awareness building worldwide throughout the past decade or so. The review is by no means complete but should give the reader some insight into the subject.

41.2
Information for Interested Laypeople

The provision of a range of information is one of the fundamental prerequisites for the successful implementation of a new technology. This is particularly true in the case of hydrogen and fuel cells as they are substitution products competing with conventional applications. In such a situation, interested laypeople become incredibly important as potential users of the new technology who must be convinced and won over. In addition to a wide range of events such as seminars, exhibitions, and trade fairs, the Internet offers a variety of different information platforms. A selection of these, including organizations, networks, initiatives, and knowledge-based portals, are presented in Table 41.1.

Furthermore, specialized publishers offer books and journals containing information ranging from the basics right up to in-depth expert information. Two examples are

- Christiani: *http://www.christiani.de*
- Hydrogeit: *http://www.hydrogeit-verlag.de*.

Table 41.1 Overview of general information sources available on the Internet.

Type	Name	URL
Organizations	German Hydrogen and Fuel Cell Association (DWV)	http://www.dwv-info.de/
	Ludwig Bölkow Systemtechnik (LBST)	http://www.lbst.de/
	European Hydrogen Association (EHA)	http://www.h2euro.org/
	US Department of Energy (DOE)	http://www.hydrogen.energy.gov
		http://www.eere.energy.gov/hydrogenandfuelcells/
Networks and initiatives	North Rhine-Westphalia	http://fuelcell-nrw.de/
	Hessen	http://www.h2bz-hessen.de/
	Baden-Württemberg	http://www.brennstoffzellen-initiative.de/
	Initiative Brennstoffzelle (Fuel Cell Initiative)	http://www.initiative-brennstoffzelle.de/
	Hydrogen Energy Center	http://www.hydrogenenergycenter.org/
	Nationale Organisation für Wasserstoff- und Brennstoffzellentechnologien (NOW; National Organization for Hydrogen and Fuel Cell Technologies)	http://www.now-gmbh.de/
	Fuel Cell Today	http://www.fuelcelltoday.com/
Knowledge-based portals	Wikipedia	http://en.wikipedia.org/wiki/Fuel_cell
		http://en.wikipedia.org/wiki/Hydrogen
	HyWeb Gazette	http://www.netinform.net/h2/Aktuelles.aspx
	Fuel Cells 2000	http://www.fuelcells.org
	H$_2$DROGEIT	http://www.hydrogeit.de/wasserstoff-und-brennstoffzellen

41.3
Education for School Students and University Students

Future energy use, in terms of a sustainable energy supply, will experience a move towards renewable energy sources and also cleaner and more efficient conversion technologies. School and university curricula are already addressing this issue in the form of supplementary courses. Energy and environmental courses dealing with the topic of hydrogen and fuel-cell technologies are thus becoming a set feature at universities. In the context of the current climate and energy debate, even secondary schools are taking up these pioneering topics by incorporating specialized courses and talks into physics, chemistry, and technology classes. In addition, subject teachers working in other institutions have created environments where school students can conduct their own experiments and get to grips with modern science and engineering, mathematics, and information technology. A good example of such a scheme is the Schools Laboratory in Forschungszentrum Jülich [5].

An example of how the topics of hydrogen and fuel cells are incorporated into the curricula of universities of technology can be taken from the contributions of staff at Forschungszentrum Jülich to the courses taught at RWTH Aachen, the University of Ulm, and Aachen University of Applied Sciences. Three professors and three lecturers run a total of nine courses [6]. The number of students participating in the individual courses each semester varies between 10 and 50. This example emphasizes the synergy, which is also exploited by other partners throughout the world, between independent research institutions and universities in responding to the need to train qualified young scientists.

In some universities, a strong emphasis has been laid on hydrogen and a resulting "center of excellence" created. Many of today's corporations possess enormous knowledge on hydrogen and benefit from workforces of hundreds of knowledgeable employees. One excellent example is the work of a group of scientists at the University of California at Davis [7, 8].

In Japan, a significant effort was made in parallel with the Japanese Hydrogen and Fuel Cell Program in order to educate young students in many aspects of hydrogen energy. This has been done within the Hydrogen and Fuel Cell Demonstration Park, intended for junior high school students and college students, but also focusing on much younger students and their parents. In many ways, this is a brilliant method of introducing a future technology to the world.

In 2007, a new school was initiated in Akureyri in northern Iceland, The Renewable Energy School. This institution, headed by Dr. Bjorn Gunnarsson, has run programs in the area of renewable energy, including hydrogen and fuel cells. One example involves a graduate course intended for full-time study over a 12 month period at master's level. Course modules are of 1–3 weeks' duration and carry two to six ECTS (European Credit Transfer and Accumulation System) credits. They involve 3–4 h of lectures each day with additional laboratory/exercise sessions. Students take an examination following the completion of each module and a final examination at the end of each trimester. Before the completion of the second trimester, each student has to complete, submit, and await approval of a thesis proposal. The thesis proposal must be written with the guidance of a faculty advisor who is appointed by Renewable Energy Science School (RES).

The students generally complete the third trimester in association with universities, corporations, or research institutions worldwide. To date, about 25 students from seven countries have completed the course in 10 locations in Europe, the USA, and Iceland. The professors have been leading academics, innovators, or policy makers/officials from a range of IPHE countries. The course was divided into a few main elements:

- Electrochemistry of Fuel Cells
- Fuel Cell Systems and Technologies
- Hydrogen Production Technology

- Hydrogen Storage
- Fuel Cells in Transportation
- Stationary and Mobile Fuel-Cell Systems
- Policies and Future R&D of Fuel-Cell Technology.

The policies course includes public awareness and public acceptance (Sigfusson, Dvorak, and Gunnarsson, WHEC Essen 2010).

Further, Professor V. Molkov has been organizing and running a series of International Short Courses at the University of Ulster in Jordanstown, Belfast, Northern Ireland. The first course in the series started in 2008 and was entitled Hydrogen and Fuel-Cell Technologies: Safety Issues.

41.4
Electrolyzers and Fuel Cells in Education and Training

Practical experiments and fast experimental setups are also a must in the education sector. A range of specialized vendors offer teaching and demonstration models, in addition to teaching material on hydrogen and fuel-cell technologies. Two established companies with a wide range of products for schools, universities, institutes, and industrial users are Heliocentris and h-tec. The products they offer range from teaching and experimental systems through course materials to system programs for practical applications and tailor-made energy solutions with hydrogen and fuel cells. Depending on the objective and scope of the project involved, they create flexible experiments from a variety of products, including halogen lamps, photovoltaic modules, polymer electrolyte membrane (PEM) electrolyzers, gas storage tanks, PEM fuel cells, consumer electronics, and power electronics. In order to ensure correct implementation, scientific evaluation, and easy-to-follow documentation, products are accompanied by examples and usage instructions.

Both companies are actively involved in information events and also further training and education. They organize competitions for young people, allowing them to explore their interest in new energy technologies and to be creative in implementing their own concepts. For example, Energy Agency North Rhine-Westphalia (NRW) in cooperation with h-tec Hydrogen Energy Systems organized the fifth competition for school students on the topic of "Hydrogen and Fuel Cells" for the 2009–2010 school year [9]. More than 100 senior secondary school students from throughout NRW take part in the competition, comprising two rounds. Projects are worked on by groups with a maximum of three students. Each group is supervised by a teacher. The 20 best teams are then selected to complete the practical part of the competition. These teams are provided with a kit known as the *"fuel-cell box."* How the teams have interpreted and implemented the task assigned to them (see Figure 41.1) is judged at the end of the competition and the winning team is selected by a jury.

Since 2009, Heliocentris Fuel Cells has provided a team of two technicians and a product manager to support participants of the European Shell Eco-Marathon (see Figure 41.2) [10]. Depending on the requirements, the technicians ensure that the

Figure 41.1 Winning team of the Fuel Cell Box competition 2008/2009 during the practical demonstration.

Figure 41.2 A vehicle under way on the EuroSpeedway Lausitz during the Shell Eco-Marathon 2009.

vehicles meet the strict technical regulations. Furthermore, replacement parts are supplied and technical support is offered in the teams' tents.

41.5
Training and Qualification for Trade and Industry

In some cases, regional authorities have joined forces to make an impact in fuel cell and hydrogen education. An outstanding example is in Ulm, Germany, where a Fuel Cell Education and Training Center (*Weiterbildungszentrum Brennstoffzelle Ulm*, WBZU; see Figure 41.3) was inaugurated in 2004. It offers education and training, showcases practical demonstrations, and is involved in public relations activities for industry, trade, schools, universities, policy makers, and the general public [11].

Figure 41.3 Fuel Cell Education and Training Center Ulm.

WBZU's core objective is to monitor new energy technologies, such as fuel cells, batteries, and mini combined heat and power stations, when they are implemented in practice, and to train and educate the various occupational groups at an early stage. In addition, school classes and the interested general public can gain hands-on experience with these new energy innovations at WBZU. WBZU also contributes to opinion forming in the general public, and among decision-makers and politicians, in the form of public debates and discussions.

The center enjoyed 5 years of support from the state of Baden-Württemberg and the Federal Ministry of Economy and Labor. This initial support facilitated the construction of a building devoted to education and training, and the initial provision of fuel-cell hardware. Experimental systems were provided by the German Aerospace Center (Deutsches Zentrum für Luft- und Raumfahrt, DLR), the Fraunhofer Institute for Solar Energy Systems (FhG-ISE) and the Solar and Hydrogen Energy Research Center (Zentrum für Sonnenenergie- und Wasserstoff-Forschung, ZSW). The technical content of the education and training programs will be jointly developed with Forschungszentrum Jülich (FZJ) and representatives of trade and industry. This is truly a remarkable public–private achievement taking place in Germany. The Hydrogen and Fuel Cell Network in NRW is the most powerful of all the German networks. Interesting background materials can be found elsewhere [12, 13].

Within the realms of the UN, hydrogen has also been attracting attention as, for example, has happened with the establishment of the United Nations Industrial Development Organization (UNIDO) International Centre for Hydrogen Energy Technologies (ICHET) in October 2003. The center is the result of an agreement between UNIDO and the Turkish Ministry of Energy and Natural Resources. Under the terms of this agreement, UNIDO-ICHET will act as a conduit for knowledge and technology flow between the developed and developing nations by providing support, facilities, and expertise concerning all aspects of energy conversion technologies involving hydrogen.

The center, initiated by Professor Nejat Veziroglu, aims to be actively involved in communicating the latest technologies to groups and organizations, particularly those in developing countries, that can benefit most from the tangible advantages associated with the use of clean and renewable energy sources. From its base in Istanbul, UNIDO-ICHET provides educational, training, and applied research programs to visiting scientists and engineers, and expertise and support for the establishment of industrial-scale pilot studies at selected locations worldwide [14]. The center is in close contact with UNIDO and enjoys the generous patronage of the Turkish Ministry of Energy.

41.6
Education and Training in the Scientific Arena

In addition to the established programs offered at colleges and universities in the field of hydrogen and fuel-cell research, special tutorials and summer schools are organized by conference organizers and training agencies, supported by well-respected experts in the field. Now in its 10th year, the 10th European SOFC Forum 26–29 June 2012 [15] 10th SOFC Forum (2012) Fuel Cell Tutorial, http://www.efcf.com/tutorial/ will open its next conference in Lucerne in June 2012 with its full-day Fuel-Cell Tutorial for those already working in the field and newcomers who want to learn more about the topic of fuel cells [15]. The tutorial will provide participants with a basic understanding of the technical, chemical, and physical principles that make fuel cells work simply, reliably, and affordably. It will discuss requirements for fuel-cell applications and important advances in research and development in the areas of electrochemistry, materials science, and process engineering. As tutors, Dr. Scherer and Dr. Van Herle have been sharing their expertise with other interested professionals.

Within the framework of further training and education programs in ongoing solid oxide fuel cell (SOFC) research and development projects, the European Union (EU) organizes 1 week summer schools with a different focus each year [16]. The courses are aimed at new academic professionals, students with relevant knowledge, and young scientists. Separate introductory courses aim to fill any gaps in basic knowledge in order to prepare undergraduates and PhD students for participation. The courses help to provide both young scientists and experienced technicians and researchers with a coherent overview of SOFC technology and to enhance their existing knowledge. The aim of the course is to present a comprehensive introduction to SOFC technology starting from the basics and working through modeling to selecting and designing suitable components. Technical and economic aspects are both taken into account.

Further support measures for students, lecturers, and other university employees and staff are covered by diverse training programs offered by the EU [17]. The Erasmus exchange program, named after the Dutch humanist Erasmus von Rotterdam (1465–1536), enables students and lecturers to apply for exchanges in universities throughout the world. In keeping with the humanist tradition associated with Erasmus, the program is rooted in the principle that mankind

Table 41.2 Overview of the European educational program.

Program	Period	Third country	Information
TEMPUS IV	2007–2013	Western Balkans, Eastern Europe, North Africa, Near East, Central Asia, EU Member States	http://eu.daad.de/eu/tempus/05236.html
ERASMUS Mundus II	2007–2013	Worldwide	http://eu.daad.de/eu/erasmus/05332.html
ALFA III	2007–2013	Latin America	http://eu.daad.de/eu/drittlandkooperationen/lateinamerika/alfa/05277.html
ATLANTIS	2006–2013	USA	http://eu.daad.de/eu/drittlandkooperationen/nordamerika/eu-usa/05289.html
TEP	2006–2013	Canada	http://eu.daad.de/eu/drittlandkooperationen/nordamerika/eu-kanada/05294.html
EDULINK	2007–2013	African, Caribbean, and Pacific States	http://eu.daad.de/eu/drittlandkooperationen/akp/edulink/05260.html
ACP Science and Technology	2007–2013	African, Caribbean, and Pacific States	http://eu.daad.de/eu/drittlandkooperationen/akp/09234.html
ICI ECP	2007–2013	Australia New Zealand, Japan, South Korea	http://eu.daad.de/eu/drittlandkooperationen/asien-australien/eu-australien/05271.html
ASEM DUO	2001–	Asia	http://www.asemduo.org/

continuously strives to improve its existence. For students who are interested, the EU program opens up the doors of numerous renowned educational institutions involved in the successful exchange system. Table 41.2 provides an overview of the EU program for educational cooperation with third countries.

41.7
Clarification Assistance in the Political Arena

The political arena is important in that it provides impulses and makes decisions that pave the way and prepare the market for innovations prior to profitability. In order to perform this task, politicians rely on the expertise and neutrality of acknowledged experts. Discussions with experts, workshops, and status and strategy reports are invaluable in providing the political arena with information and evaluation findings, which are essential for appropriate and sustainable political initiatives. The four UN Assessment Reports of the Intergovernmental Panel for Climate Change (IPCC) are a current example of this. A summary of the working group reports for the fourth and last assessment report was published on 17 November 2007 [18].

41.8
Analysis of Public Awareness

Iceland was a founding member of IPHE and the hydrogen energy economy has been tested in Iceland mainly under the auspices of Icelandic New Energy. This company is owned by the Icelandic energy companies and the three international companies, Shell Hydrogen, Norsk Hydro, and Daimler. In an EU project named ECTOS (Ecological TranspOrt System), Icelandic New Energy [19–22] ran three hydrogen-powered fuel-cell buses in Reykjavik from 2003 to 2007. A hydrogen refueling station was built in Iceland and inaugurated in 2003 [4].

Public awareness was assessed rigorously before and during the ECTOS project and the findings have been published. The first public survey in Iceland [23] served as a standard reference for later tests on social acceptance. An initial public survey was conducted in December 2001 before the inauguration of the fueling station and the fuel-cell buses. The Institute for Applied Sociology of the University of Iceland performed a telephone survey and asked 1154 people about issues related to hydrogen. For example, the following question was asked: "Do you have a positive or a negative attitude towards the option of using hydrogen as the main fuel for buses, ships, and cars?" The respondents could indicate a very positive, positive, neutral, negative, or very negative stand. The results indicated that 93% of the respondents took a positive or very positive stand, and only 3% stated that they were against the idea. Finally, the last question posed was: "Do you feel that there is a need for more information about the hydrogen technology?" Some 22% of the respondents stated that they would like to learn more about hydrogen technology, especially young people and women. Gender or age did not play a role in the attitude towards using hydrogen as a fuel.

This outcome was more positive than expected, as the public often tend to take a rather negative stand toward innovations that do not have a known beneficial function. A plausible explanation might be that hydrogen as a fuel has been in academic and public discourse on and off since the 1970s.

Within ECTOS, when the buses were already running on the streets of Reykjavik, a second survey was conducted on board in March 2004. Passengers and other commuters in Reykjavik were engaged in more detailed questionnaires on hydrogen and energy issues. The survey conducted amongst the passengers of the hydrogen buses and more conventional buses had very similar results concerning the attitude towards using hydrogen as the main fuel (86% replied positively and very positively to the same question as posed in December 2001).

Passengers, people living along bus routes, and people in the street were asked about what they initially connect with the concept "hydrogen." Most of the respondents connected hydrogen with water, clean fuel, and a clean environment, even though they were given the option of ticking for explosions, burning Zeppelins, and expensive technology. When asked for their willingness to pay a higher charge for hydrogen during its introduction, people showed understanding, and a majority said that they would be willing to pay 10–20% higher prices for hydrogen during the introductory phase. The questionnaires were designed to profile the aspects relevant in this context, and will be used mostly to benchmark specific aspects during the first stages of the introduction of hydrogen.

Another case study investigated the determinants of knowledge and acceptability of hydrogen vehicles among London residents [24]. Data were collected via a socio-economic survey of over 400 residents. The results indicated that, at present, less than half of respondents have heard of hydrogen as a fuel for transport, and just over one-third are clearly in favor of the introduction of hydrogen vehicles. The key determinant of acceptability was having prior awareness of hydrogen technologies, as identified via logit regression analysis. Hydrogen awareness in turn was found to be related to gender, age, education, and environmental knowledge. These results suggest that there is an opportunity to raise awareness of hydrogen technologies among the remaining three-fifths of the London population, although this is likely to require a differential approach to information provision.

41.9
Conclusion

We have shown how education and the public awareness of hydrogen as an energy carrier and fuel cells as an energy converter have progressed considerably during the first decade of the new millennium. Knowledge has been accumulated, and hydrogen and fuel-cell applications are in use all over the world. These include stationary fuel cells in homes and buildings and fuel-cell transport vehicles. No matter how the economic development of the world evolves over the next decade, the need for education and public awareness in the area will increase and put pressure on corporations, institutions, and technical societies. The world has come a long way towards utilizing the lightest of elements for sustainable fuel [25].

References

1. European Commission (2003) Hydrogen Energy and Fuel Cells, Final Report of the High Level Group, EUR 20719 EN, European Commission, Brussels.
2. Larminie, J. and Dicks, A. (2002) *Fuel Cell Systems Explained*, John Wiley & Sons, Ltd., Chichester.
3. IPHE (2005) Scoping Papers ILC-037-05, International Partnership for the Hydrogen Economy, Washington, DC.
4. Sigfusson, T.I. (2008) *Planet Hydrogen – The Taming of the Proton*, Coxmoor Publishing, Oxford.
5. Forschungszentrum Jülich (2006) Schülerlabor (JULAB), http://www.fz-juelich.de/projects/index.php?index=373 (last accessed 16 December 2011).
6. Forschungszentrum Jülich (2009) IEF-3 Report 2009; Basic Research

7. Ogden, J. (2004) Where will the hydrogen come from? Systems considerations and hydrogen supply, in *The Hydrogen Energy Transition* (eds. D. Sperling and J. Cannon), Elsevier, Amsterdam.
8. Ogden, J., Yang, C.H., Johnson, N., Ni, J., and Lin, Z. (2005) Technical and Economic Assessment of Transition Strategies Toward Widespread Use of Hydrogen as an Energy Carrier Hydrogen Pathways. Report to the United States Department of Energy, UCD-ITS-RR-05-13, University of California Davis, Davis, CA.
9. EnergieAgentur.NRW (2011) 5. Fuel Cell Box Schülerwettbewerb 2009/2010, http://www.fuelcellbox-nrw.de/index.php?id=460 (last accessed 16 December 2011).
10. http://www.heliocentris.com/kunden/systeme/information/kundenprojekte/shell-eco-marathon.html (last accessed 1 December 2009).
11. WBZU (2009) Weiterbildungszentrum Brennstoffzelle Ulm, http://www.wbzu.de/ (last accessed 16 December 2011).
12. Nitsch, J. and Winter, C.-F. (1988) *Hydrogen as an Energy Carrier, Technologies, Systems and Economy*, Springer, Berlin.
13. Winter, C.-F. (ed.) (2000) *The Energies of Change – The Hydrogen Solution*, Gerling Akademie Verlag, Munich.
14. ICHET (2011) ICHET-UNIDO, http://www.unido-ichet.org/ichet.org/activities/conferences_and_workshops/conferences_and_workshops.html (last accessed 16 December 2011).
15. European Fuel Cell Forum (2011) Fuel Cell Tuorial, http://www.efcf.com/tutorial/ (last accessed 16 December 2011).
16. International Solid Oxide Fuel Cell Summer School (2009) 16th International Solid Oxide Fuel Cell Summer School, Ancona, Italy, 2009, http://www.vtt.fi/files/projects/largesofc/brochure_sschportonovo2009_1st.pdf (last accessed 16 December 2011).
17. Europäische Union (2011) ERASMUS im Programm für Lebenslanges Lernen (LLP), http://eu.daad.de/eu/index.html (last accessed 16 December 2011).
18. Pachauri, R.K. and Reisinger, A (eds.) (2007) *Climate Change 2007: Synthesis Report; Contribution of Working Groups I, II and III to the Fourth Assessment Report of the Intergovernment Panel on Climat Change*, IPCC, Geneva, p. 104.
19. Andersen, P.D., Holst Joergensen, B., Eerola, A., Kojonen, T., Loikkanen, T., and Eriksson, E.A. (2005) Building the Nordic Research and Innovation Area in Hydrogen. Nordic Energy Research Summary Report, Nordic Energy H_2 Foresight.
20. Sigfusson, T.I. (2005) in *Renewable Energy 2005* (ed. F. David) Official Publication of the World Renewable Energy Network and UNESCO, Sovereign publications, London, pp. 95–97.
21. Sigfusson, T.I. (2005) L'ile de Jules Verne, in *Decouverte. Revue du Palais de la Decouverte*, Paris, pp. 64–73.
22. Árnason, B. and Sigfusson, T.I. (2000) Iceland – a future hydrogen economy. *Int. J. Hydrogen Energy*, **25** (5), 389–394.
23. Maack, M.H. and Skulason, J.B. (2006) Implementing the hydrogen economy. *J. Cleaner Prod.*, **14** (1), 52–64.
24. O'Garra, T., Mourato, S., and Pearson, P. (2005) Analysing awareness and acceptability of hydrogen vehicles: a London case study. *Int. J. Hydrogen Energy*, **30**, 649–659.
25. Sigfusson, T.I. (2007) Pathways to hydrogen as an energy carrier. *Philos. Trans. R. Soc. London, Ser. A*, **365** (1853), 1025–1042.

Index

a

Aalborg University, Denmark 12
ABAQUS 768, 769, 770
active and passive hybrids 1078–1080
active braze alloys (ABAs) 324
adsorption 1017–1018, 1024–1026
Advent Technologies 11
AFCo 75, 87
agglomerate models 809–810
air brazing 324–325
alkaline direct ethanol fuel cells, assembled with non-platinum catalyst 104–105
alkaline electrolytes, electrodes for 226–227
– alkaline electrolysis 227–228
– alkaline fuel cells (AFCs) 227
– alkaline URFCs 228–229
alkaline fuel cells (AFCs) 222, 227, 418
– carbon dioxide behavior 123–126
– degradation
– – gas diffusion electrodes with Raney nickel catalysts 114–121
– – gas diffusion electrodes with silver catalysts 121–123
– design concepts 99–100
– – alkaline direct ethanol fuel cells assembled with non-platinum catalyst 104–105
– – bipolar stack concept by DLR 101–102
– – cathode gas diffusion electrodes production 113
– – double-skeleton electrodes 106–111
– – Eloflux cell design 100–101
– – falling film cell 101
– – Hydrocell concept 102–103
– – NiO reduction 111–112
– – Ovonics concept 103
– – PTFE-bonded gas diffusion electrodes 105
– – stack design with anion-exchange membranes 104
– – traditional stacks 100
– electrolytes and separators 113–114
– historical introduction and principle 97–99
analysis of variance (ANOVA) 610
analytical modeling, of fuel cells 648
– catalyst layer performance modeling
– – basic equations 648–650
– – Cr poisoning model SOFC cathode 654–657
– – critical current density 652–653
– – feed molecules ideal transport 650
– – optimal catalyst loading 657–658
– – polarization curve 651–652
– – x-shapes 653–654
– PEMFCs and HT-PEMFC polarization curve 658–659
– – high-current regime 661–662
– – low-current regime 660–661
– – one-dimensional cell polarization curve 662
– – oxygen consumption in channel and quasi-two-dimensional polarization curve 663–665
– – oxygen transport in GDL 659–660
anion-exchange membranes (AEMs) 104, 151
anisotropy parameter 676
anode catalyst. *See* low-temperature fuel cells, catalysis in
anode catalyst layer (ACL) 132
anode-supported single cells (ASCs) 452
ANSYS software 769, 771
Apollo program 222, 420

Army Power Division (ATO) 1116
Arrhenius equation 718
Asahi Chemical 388
associative mechanism 393
atomic force microscopy (AFM) 488
attenuation coefficient 496
autothermal reforming (ATR) 428–429, 706, 711, 721, *722*, *723*, *724*, 726–728
auxiliary power units (APUs) 22, 1011
– SOFC-based
– – for aircraft 994–996
– – for automobiles and trucks 993–994

b

balance equations 791, 794, 804, 808
balance of plant (BOP) 73–74, 953
Ballard 26, 29, 30, 1132
barium strontium cobalt ferrite (BSCF) 49
BASF 11
battery chargers 27–29
battery electric vehicles (BEVs) 20, 21–22, 1057–1058
Baxi Group 25
Baxi Innotech 25
Bella Colla 1110
Bernoulli filling 673
bifunctional mechanism 419
bilinear elastic–plastic constitutive model 770
bioanodes 157–158, 163
bioelectrochemical systems (BESs) 147–148, *148*
– applications and concept proofs 164
– – biosensors and environmental monitoring 172–173
– – caustic soda and hydrogen peroxide production 170
– – desalination 169–170
– – electro-assisted anaerobic digestion 168–169
– – greenhouse gas mitigation 171–172
– – heavy metal recovery and removal 172
– – organic alcohols and acids 170–171
– – recalcitrant compounds 171
– – sediments, plants, and photosynthesis 168
– – wastewater treatment 164–168
– internal resistances in *154*
– materials and methods 149
– – biological measurements 152
– – configurations and design 151–152
– – electrochemical measurements 152–156

– – electrode materials 149–151
– – membrane 151
– – performance reporting 156–157
– microbial catalysts 157
– – anode reactions 157–160
– – biological limitations 163
– – cathode reactions 160–162
– – photosynthetic biocatalysts 163
– – pure cultures and mixed microbial communities 162–163
– modeling 173
biofilm 150, 152, 156, 158–159, *159*, 162, 163
biological fuel cells (BFCs) 147
biological oxygen demand (BOD) 172
biological processes and selective oxidation 1029–1030
biosensors and environmental monitoring 172–173
bipolar plate (BP) 132–133, 134, 136, 139
– metallic 363
– – bare plates 363–366
– – coated stainless-steel plates 368–370
– – light alloys 366–367
bipolar stack concept, by DLR 101–102
block diagram. *See* flow diagram
Bloom Energy 6, 8
bonded compliant seals 323–324
– bonded foil seal concept 327–328
– brazing 324–327
bonded foil seal concept 327–328
Bosanquet equation 738
b+w Electronics 29
brazing 210, 324–327
Brillouin light scattering (BLS) 535
Butler–Volmer equation 385, 742, 743, 807, 830, 846
Butler–Volmer–Monod model 173
Butler–Volmer-type equation 823, *824*, *825*, *826*
button cells 251

c

California Air Resources Board 1132
California Energy Commission 1132
California Fuel Cell Partnership (CaFCP) 1132–1133, 1204–1205
Callux program 1142
Canada 1136–1137
Canadian Hydrogen and Fuel Cell Association (CHFCA) 1198–1200
carbonaceous energy carriers 920, 957
carbon dioxide behavior 123–126
carbon nanotubes (CNTs) 416, 422
Case Western Reserve University 12

catalysis in 420–421
catalyst islanding 545
catalyst material 221, 227, 228, 230, 231
catalyst poisoning 409, 417, 418, 419
catalyst poisoning, by sulfur-containing substances in crude oil fractions 1015–1016
catalyst supports, for electron conduction 397–402
catalytic partial oxidation (CPO) 427–428
catalytic steam reforming (SR) 426
– carbon formation 427
– desulfurization 426–427
cathode catalyst. *See* low-temperature fuel cells, catalysis in
cathode catalyst layer (CCL) 132, 133
cathode current-collector layer (CCCL) 253, 254–255
cathode functional layer (CFL) 254, 255
caustic soda and hydrogen peroxide production 170
Celanese 11
Cellera Technologies 392
cell-level modeling 755–758, 821–825
– cold start operation 893–898
– dimensionality 882–884
– flow maladjustment 900
– large-scale fuel-cell simulation 898–899
– model validation 903, 905
– nonisothermal modeling 888, 890–891
– single-phase flow 901
– transient operation 884–888
– two-phase flow 891–893, 902–903
cell reversal 554
cell-to-edge design 304
cell-to-frame design 304–305
cell voltage 546
Center for Solar Energy and Hydrogen Research [Zentrum für Sonnenenergie -und Wasserstoff-Forschung (ZSW)] 15
Centre for Solar Energy and Hydrogen Research Baden-Württemberg 12
Ceramic Fuel Cell Limited (CFCL) 25
ceramic seals 318–319
CeramTec 8
chemical fuel cells (CFCs) 147
chemical oxygen demand (COD) 165, 169
chemisorption 1018
– hydrofining 1021–1022
– H_2S removal 1018–1020
– SO_2 removal 1020–1021
– S-Zorb process 1020
ChevronTexaco 1132

China 1137
– Beijing Olympics and Shanghai World Expo 1138
– GEF/UNDP-China FCB demonstration 1138
classical random graph models 690–691
Clausius-Clapeyron equation 341
Claus process 1022
clean energy partnership (CEP) (Germany) 1122–1124
– stationary power and early market applications 1142–1143
– – regional activities 1143
CO_2 evolution, visualized by means of synchrotron X-ray radiography 508–509
coefficient of thermal expansion (CTE) 309, 310–311, 312–313, 319
color intrusion method 476
combined heat, hydrogen, and power (CHHP) 89
combined heat and power (CHP) 86, 89, 90
combustion reactions 68
combustor 984
complex nonlinear least-squares (CNLS) fit 447–449, 452, 458, 464
compressive seals 319–320
– hybrid mica seals 321–323
– metal gaskets 320
– mica-based seals 320–321, *322*
computational fluid dynamic (CFD) 703–705
– analysis 779–781
– based design 728–730
– high-performance computing, for fuel cells 705–711
– high-performance computing-based modeling for fuel-cell systems 711–712
– – CFD principles 712–715
– – core component modeling of HT-PEFC auxiliary power unit 721–728
– – heat transfer 717
– – mixtures and reactions 717–719
– – multiphase flows 719–720
– – porous media 720–721
– – turbulence 715–716
computer-aided design (CAD) 947
COMSOL software 772
confocal microscopy 488
confounding 610
ConocoPhillips 1020
conventional external reforming MCFC 71–72
coplanar electrodes, SC-SOFCs with 52
– cell performance 52–53
– – cell stacks 55

coplanar electrodes, SC-SOFCs with (*contd.*)
– – electrode shape 54–55
– – electrode thickness 54
– – electrode width 53
– – electrolyte surface roughness 54
– – electrolyte thickness 54
– – fuel choice and lower operating temperatures 55–56
– – gas flow direction 55
– – gas mixture composition (fuel-to-oxygen ratio) 55
– – inter-electrode gap 53
– limitations and challenges 57–58
– – cell efficiency 59
– – chemical interaction between coplanar electrodes 58
– – current collection 59
– – gas intermixing 58
– – microscale electrodes 59
– – miniaturization limits 58
– – nickel instability 59
– miniaturization 56–57
core-shell nanostructure 395
corrosion 551, 554, 555, 557, 558, 559, 560, 562, 563, 564
Cr poisoning model SOFC cathode 654–657
crude oil 1011–1012
– catalyst poisoning by sulfur-containing substances in fractions of 1015–1016
– chemical classes of sulfur-containing substances in 1013–1015
– routes for inserting sulfur into molecules 1012–1013
Cummins 23
current density distribution 548–549
current interrupt method 156
cyclic voltammetry (CV) 112, 156, 549–550

d

DaimlerChrysler 15, 1132
Dais Analytic Corporation (DAC) 25
Dalian Institute of Chemical Physics 12
Danish Power Systems 11
Dantherm 29
Dapozol. *See* membrane electrode assembly (MEAs)
Darcy–Forchheimer law 721
d-band theory 387
degradation, caused by dynamic operation and starvation conditions 543–546
– measurement techniques 546
– – current density distribution 548–549
– – cyclic voltammetry 549–550

– – dynamic operation at standard conditions 550–553
– – reference electrode 546–548
– starvation conditions 553
– – hydrogen starvation during start-up/shut-down 555–558
– – local hydrogen starvation 558–561
– – materials and designs 563
– – mitigation 562
– – operation strategies 563–565
– – overall hydrogen starvation 553–555
– – oxygen starvation 561–562
Delphi and PNNL 7
demonstration projects 1119–1120
– stationary power and early market applications 1139–1140
– – Denmark 1142
– – European Union 1143–1144
– – Germany 1142–1143
– – Japan 1140–1141
– – South Korea 1145
– – United States 1144–1145
– transportation demonstrations 1120–1122
– – automaker demonstration programs 1138–1139
– – Canada 1136–1137
– – China 1137–1138
– – European Union 1133–1136
– – Germany 1122–1125
– – Japan 1126–1129
– – South Korea 1137
– – United States 1130–1133
Denmark 1142
density functional theory (DFT) 386
Department of Defense (DOD), US 1108–1109
Department of Energy (DOE) hydrogen–air fuel cell system 381
desalination 169–170
design of experiments (DOE) 598–599
– 2^2 factorial design 599–601
– 2^3 factorial design 604–609
– 2^{n-k} fractional factorial designs 609–610
– 3^2 factorial design 601–604
Desulfovibrio microorganisms 1012
desulfurization, for fuel-cell systems 1011
– application 1034–1038
– in gas phase 1016–1017
– – absorption 1017
– – adsorption 1017–1018
– – chemisorption 1018–1022
– – sulfur recovery 1022
– in liquid phase
– – adsorption 1024–1026

– – desulfurization with overcritical fluids 1030–1031
– – distillation   1031–1032
– – hydrodesulfurization with presaturator 1022–1024
– – ionic liquids   1026–1028
– – processes with nonporous membranes 1032–1034
– – processes with porous membranes 1032
– – selective oxidation   1028–1030
– sulfur-containing molecules
– – catalyst poisoning by sulfur-containing substances in crude oil fractions 1015–1016
– – chemical classes of sulfur-containing substances in crude oil   1013–1015
– – crude oil   1011–1012
– – routes for inserting sulfur into molecules in crude oil   1012–1013
diffusion bonding   210
direct carbon fuel cells (DCFCs)   89
direct ethanol fuel cells (DEFCs)   136–137
direct fuel cells   417–418
direct internal reforming (DIR)   73
direct methanol fuel cells (DMFCs)   12–13, 131, 134, 135–136, 507–508
– battery chargers   27–29
– CO_2 evolution visualized by means of synchrotron X-ray radiography   508–509
– combined approach of neutron radiography and local current density measurements 509–510
– for light traction   14–17
– miniaturized   141
– passively operating   142
– for portable applications   13–14
direct numerical simulations (DNSs) 847, 848
direct oxidation   979
dissociative mechanism   393
distribution coefficient   1027
distribution function of relaxation times (DRT)   449, 450–452
diverse energy   1112, 1144
DLR   8
Doosan Heavy Industries and Construction (DHI)   75
double-skeleton electrodes   106
– preparation and electrode materials 106–108
– PTFE-bonded gas diffusion electrodes dry preparation   108–111
DuPont   388

durability   220, 221, 223, 226, 230, 231, 233, 236, 240
dusty gas model (DGM)   757
dynamic hydrogen electrode (DHE)   547, 548
dynamic load cycling aging test
– data post processing   586–587
– setting of test conditions (test inputs)   585
– test output measurement   585–586
dynamic mechanical analysis (DMA)   534
dynario system   27

e
ECTOS (Ecological TranspOrt System) 1220
edges, stochastic modeling of   688
– MCMC simulation for edge rearrangement 689–690
education and public awareness   1211–1212
– clarification assistance in political arena 1219–1220
– education and training in the scientific area 1218–1219
– education for school students and university students   1213–1215
– electrolyzers and fuel cells in education and training   1215–1216
– information for interested laypeople 1212, *1213*
– public awareness analysis   1220–1221
– training and qualification for trade and industry   1216–1218
efficiency   220, 221, 224–225, 226, *228*, 229, *230*, *231*, 236, 239, 240
elastoplasticity   770, 771, 777, 782
Elcomax   11
ElectraGen systems   29
electro-assisted anaerobic digestion   168–169
electrocatalysis, in fuel cells   408–410
– alkaline fuel cells   418
– catalysis in direct fuel cells   420–421
– direct fuel cells   417–418
– electrocatalyst degradation   421
– hydrogen oxidation and CO poisoning 418–419
– oxygen reduction, in PEMFCs   410–417
electrocatalyst
– degradation   421
– for oxygen reduction   393–397
electrochemical impedance spectroscopy (EIS) 156, 441, 534, 820
– principles of   443–445
– – frequency response analyzers   445–446
electrochemically active platinum surface area (ECA)   552

electrochemistry 740–741
– contimuum-level approach 741–742
– mesoscale approach 742–744
electrode models 804–806
– agglomerate models 809–810
– spatially lumped models 806–808
– thin-film models 808–809
– volume-averaged models 810–811
electrodes, for hydrogen oxidation 384–388
electrolyte types *221*
electrolyzer efficiency 224
electron beam welding 209
electron-transfer mechanisms 159
– direct 160, 162
– indirect 160, 162
electroosmotic drag 389
Eloflux cell design 100–101
embossing 208
EnBW 25
ENE-FARM program 1140
energy storage *220, 226, 236, 240*
– seasonal 237–239
energy supply 1154
Engineering Advancement Association (ENAA), Japan 1126
environmental engineering 157, 159
E.ON Ruhrgas 25
E Plus 1112
equivalent circuit model (ECM) 447, 449, 452, 453–455
– definition and validation 458–465
– – anodic water partial pressure dependence 461–462
– – cathodic oxygen partial pressure dependence 460–461
– – thermal activation 462–464
Erasmus exchange program 1218–1219
etching techniques 207
ethanol 420
Euler–Euler approach 719
European Fuel Cell GmbH (EFC) 25
European Hydrogen Association (EHA) 1193
– members *1195–1196*
European Regions and Municipalities Partnership for Hydrogen and Fuel Cells (HyRaMP) 1201–1202, 1136
European Union 1133
– fuel cell bus projects 1134–1135
– H2moves Scandinavia 1135–1136
– Joint Undertaking on Fuel Cells and Hydrogen (FCH JU) 1133
– regional and member country activities 1136
– stationary power and early market applications 1143–1144
– Strategic Energy Technology (SET) 1133
EWW 25
exchange programs 1218–1219
exergy analysis 930–932
external reformer 979, 980
extracellular electron transfer (EET) 157

f

2^2 factorial design 599–601
2^3 factorial design 604–609
3^2 factorial design 601–604
failure mode and effect analysis (FMEA) 946
falling film cell 101
Faraday's law 756, 808
fast marching method 693
Federal Ministry of Education and Research [Bundesministerium für Bildung und Forschung (BMBF)], Germany 13
Fick's law 737
finite element method (FEM) 769
flooding 389, 851
flow diagram *931*, 939, 940, 941, 947
flow rule 777
FLUENT package 713, 714, 717, 719, 725, 727
fluorescent *in situ* hybridization (FISH) 152
focused ion beam (FIB) 488
Ford 1132
Ford Focus FCV Hybrid 1059
Forschungszentrum Jülich (FZJ) 7, 8, 12, 15, 249, 270, 469
Fourier transform infrared (FTIR) spectroscopy 625
Fraunhofer IKTS 6, 6
Fraunhofer Institute for Solar Energy Systems (FhG-ISE) 12, 13
Fuel Cell And Hydrogen Energy Association (FCHEA), USA 1196–1197, *1198*
– resources 1197–1198
– working groups and committees 1197
fuel cell and hydrogen organizations and worldwide initiatives 1181
– European level
– – European Hydrogen Association (EHA) 1193–1196
– – Fuel Cells and Hydrogen Joint Undertaking (FCH JU) 1187, 1189–1193
– international level 1181–1182
– – International Energy Agency (IEA) 1183–1187

– – International Partnership for Hydrogen
    and Fuel Cells in the Economy (IPHE)
    1182–1183
– national level
– – Canadian Hydrogen and Fuel Cell
    Association (CHFCA)   1198–1200
– – National Organization for Hydrogen and
    Fuel Cell Technology (NOW)
    1200–1201
– – US Fuel Cell and Hydrogen Energy
    Association (FCHEA)   1196–1198
– partnerships, initiatives, and networks with
    specific agenda
– – California Fuel Cell Partnership (CaFCP)
    1204–1205
– – Initiative Brennstoffzelle (IBZ)
    1206–1207
– – UK Hydrogen Energy Network (H_2NET)
    1206
– regional level
– – European Regions and Municipalities
    Partnership for Hydrogen and Fuel
    Cells   1201–1202
– – hydrogen and fuel-cell activities, in
    Germany's federal states   1202–1204
Fuel Cell Energy (FCE)   8, 75
fuel-cell hybrid electric vehicles (FCHVs)
    18, 20, 21
Fuel Cell Institute of Shanghai Jiaotong
    University   75
fuel cell powertrains   1045
– buses   1092
– components
– – electrical storage   1065–1067
– – electric machines   1067–1069
– – fuel cell systems for automotive
    applications   1063–1065
– – hydrogen storage   1061–1063
– – vehicle drivetrain cost comparison
    1070–1072
– passenger cars   1089
– vehicle powertrains potentials and
    challenges
– – hybrid electric fuel cell vehicles
    1058–1059
– – non-hybrid fuel cell powertrains   1058
– – powertrain comparison   1055–1058
– – propulsion systems   1049–1050,
    1052–1054
– – triple-hybrid fuel cell vehicles
    1060–1061
– vehicle requirements
– – driving resistance   1045–1046, *1047*,
    1048

– – energy conversion and driving cycles
    1046–1049
fuel cell process engineering methodologies
    597
– analysis methods   628
– – nonlinear systems of equations model
    evaluation   637–639
– – pinch-point analysis   639–641
– – predictive method for vapor–liquid and
    liquid–liquid equilibria determination
    630–637
– – system analysis via statistical methods
    628–630
– parameter optimization   604, 628, 638
– verification methods   597–598
– – conversion determination in reforming
    processes   616–628
– – design of experiments (DOE)   598–610
– – measurement uncertainty evaluation
    610–616
Fuel Cell Research Center, Duisburg   12
Fuel Cell Systems Testing, Safety and Quality
    Assurance (FCTES^{QA})   79, 573
Fuel Cell Tutorial   1218
fuel cell vehicles (FCVs)   1121, 1125, 1127,
    1132
fuel processing   706, 709, 714, 718, 720
fuel processor   706–708
fuel utilization   45
Fumatech   11, 12
functional layer   654
FutureE Fuel Cell Solutions   29

g
gadolinium-doped ceria (GDC)   280, 287, 291
gadolinium-doped cerium oxide (CGO)   259,
    260–261, 265
gas diffusion electrodes (GDEs)   100, 101,
    105, 106, *109*, 422
– with Raney nickel catalysts   114–121
– with silver catalysts   121–123
gas diffusion layer (GDL)   132, 135, 381, 493,
    503, 505, 507, 508, 509–510, 515, 523, 524,
    659–660, 669, 678, 691, 850–859, 890
– and gas flow channel interface   864–868
– multi-layer model for paper-type   670
– – binder modeling   672–673
– – fiber modeling   671–672
– – model parameter fitting   674–675
– – results   675–676
– – time-series models for non-woven GDLs
    676–677
– optimization   138–139
– – carbon dioxide discharge   141–142

gas diffusion layer (GDL) (*contd.*)
– – flow-field design 139–141
– – miniaturized DMFC 141
– – passively operating DMFC 142
– structural characterization of porous 692
– – connectivity 695–696
– – graph model validation 698
– – multi-layer model validation 696–698
– – pore size distributions 694–695
– – tortuosity 692–693
gas phase, desulfurization in 1016–1017
– absorption 1017
– adsorption 1017–1018
– chemisorption 1018–1022
– sulfur recovery 1022
gas turbine (GT)–MCFC system integration 73–74, *74*
GDOS (glow discharge optical emission spectroscopy) 488
Gemini program 222, 380, 383, 420
General Motors (GM) Research Laboratories 364
Geobacter spp. 158, 160
Geobacter sulfurreducens 160, 172
geometric tortuosity 692
Germany
– clean energy partnership 1122–1124
– H$_2$ Mobility 1125
– North Rhine-Westphalia activities 1124–1125
– stationary power and early market applications 1142–1143
Gibbs energy function 927–928
glass and glass–ceramic sealants 309–318, *312, 316*
Gore 388
greenhouse gases (GHGs) 3, 21
– mitigation 171–172
Griffith crack model 308–309
Grotthus mechanism 344, 389
GSMA 1111, 1112
Guide to the Expression of Uncertainty in Measurement (GUM) 610–611

h

Hakimi–Havel algorithm 688, 689
hardening rule 777
H.C. Starck 8
heat exchanger 985–986
heat management analytical methods 926
– exergy analysis 930–932
– Gibbs energy function 927–928
– Pinch point analysis 928–930
– process optimization 932–940
– system set-up 926–927
heavy metal recovery and removal 172
Heliocentris Fuel Cells 1215
Hexis 25
high-performance computing
– based modeling for fuel-cell systems 711–712
– – CFD principles 712–715
– – core component modeling of HT-PEFC auxiliary power unit 721–728
– – heat transfer 717
– – mixtures and reactions 717–719
– – multiphase flows 719–720
– – porous media 720–721
– – turbulence 715–716
– for fuel cells 705–711
high-resolution transmission electron microscopy (HRTEM) 398
high-temperature polymer electrolyte fuel cell (HT-PEFC) 11, 511–512, 658–665, 706, 819
– actors and development areas 11–12
– cell-level modeling 821–825
– phosphoric acid as electrolyte 827–829
– polarization curve modeling 829–830
– – activation overpotential 830–831
– – mass transport 833–834
– – ohmic resistance 831–833
– stack-level modeling 825–827
hopping mechanism 843
Hückel's law 1014
humidification sensitivity test
– data post processing 578
– setting of test conditions (test inputs) 574–577
– test output measurement 577–578
hybrid electric fuel cell vehicles 1058–1059
hybridization, for fuel cells 1075–1076
– components 1081
– – control unit 1084–1085
– – energy storage 1082–1083
– – fuel cell 1082
– – power electronics 1083–1084
– concepts
– – overview 1085
– – possible concepts 1087–1088
– – types 1085, 1087, *1088*
– hybridization degree 1081
– reasons 1076–1077
– systems analysis 1096–1097
– technical overview 1088
– – applications 1096

– – fuel-cell powertrains   1088–1089, 1091–1092
– – light traction applications   1092–1096
– types
– – active and passive hybrids   1078–1080
– – series and parallel hybrids   1077–1078
hybrid mica seals   321–323
Hydrocell concept   102–103
Hydrogen-Assisted Renewable Power System (HARP) project   1110
hydrogen evolution reaction (HER)   384, 385–386
hydrogen-fed micro fuel cells   134–135
Hydrogen HyWay Program, North Rhine-Westphalia   1124
Hydrogenics   29, 30
H_2 Mobility   1125
hydrogen oxidation and CO poisoning   418–419
hydrogen oxidation reaction (HOR)   384–387, 845, 846, 850
hydrogen starvation   553
– during start-up/shut-down   555–558
Hydrogen Supply/Utilization Technology (HySUT)   1129
hydromechanics model, for mica-based seals   323
Hy-Life project   1129

i
IdaTech   30
IMM (Germany)   23
impedance spectroscopy   441–443
– data analysis
– – complex nonlinear least-squares (CNLS) fit   447–449
– – data quality analysis   446–447
– – distribution function of relaxation times (DRT)   450–452
– electrochemical   443–446
– equivalent circuit model definition and validation   458–465
– – anodic water partial pressure dependence   461–462
– – cathodic oxygen partial pressure dependence   460–461
– – thermal activation   462–464
– process identification   453–458
– – anodic water partial pressure variation   455
– – cathodic oxygen partial pressure variation   456–457
– – temperature variation   454
impregnation. *See* infiltration

indirect internal reforming (IIR)   73
infiltration   275
– applications   282–284
– – anodes produced by   284–290
– – cathodes produced by   290–295
– motivation for   281–282
information dissemination portfolio   1162–1163
infrastructure portfolio integration   1160–1161
Initiative Brennstoffzelle (IBZ)   1206–1207
in situ imaging, at large-scale facilities   493
– applications
– – neutron tomography   513–514
– – synchrotron X-ray tomography   514–517
– DMFCs   507–508
– – CO_2 evolution visualized by means of synchrotron X-ray radiography   508–509
– – combined approach of neutron radiography and local current density measurements   509–510
– HT-PEFCs   511–512
– PEFCs   500
– – neutron radiography   504–507
– – X-rays   500–504
– X-rays and neutrons
– – complementarity   494–495
– – radiography and tomography principles   496–499
Institute of Chemical Process Engineering [Institut für Chemische Verfahrenstechnik(ICVT)]   15
integrated gasification fuel cell (IGFC)   1001, 1003–1006
integrated gasifier combined cycle (IGCC)   90
Intelligent Energy   389, 1033
– schematic diagram for desulfurization process from   *1034*
internal reforming   979
– MCFC   72–73

international collaboration. *See* International Energy Agency (IEA) Hydrogen Implementing Agreement (IEA HIA)
International Electrotechnical Commission 27
International Energy Agency (IEA) 1183–1184
– Committee on Energy Research and Technology (CERT) 1184
– Hydrogen Implementing Agreement 1185–1187, *1188*
– Implementing Agreement on Advanced Fuel Cells 1184–1185
International Energy Agency (IEA) Hydrogen Implementing Agreement (IEA HIA) 1153–1154
– collaborative RD&D 1155–1161
– – integrated systems portfolio 1159–1160
– – integration in existing infrastructure portfolio 1160–1161
– – production portfolio 1157–1158
– – storage portfolio 1158–1159
– hydrogen, analysis positioning 1161
– – market portfolio 1162
– – political decision-making portfolio support 1162
– – technical portfolio 1161–1162
– hydrogen awareness, understanding, and acceptance 1162
– – information dissemination portfolio 1162–1163
– – outreach portfolio 1164–1166
– – safety portfolio 1163–1164
– past as prolog 1166, 1173, *1174–1177*
– 2009–2015 work program timeline 1173, *1178*
– work program 1166, *1167–1173*
International Partnership for Hydrogen and Fuel Cells in the Economy (IPHE) 1181
International Society of Automation 947
Ion America 8
IRD Fuel Cells 14, 16
Ishikawajima-Harima Heavy Industries (IHI) 75
isotropic hardening 777

j

Japan 1126
– Fukuoka prefecture activities 1129
– Hydrogen Highway Project 1129
– Japan Hydrogen and Fuel Cell Demonstration Project (JHFC) 1126–1129
– stationary power and early market applications 1140–1141
– – regional activities 1141
Japanese Automobile Research Institute (JARI) 1126
Japanese Hydrogen and Fuel Cell Program 1214
Japan Gas Association (JGA) 1126
Japan Petroleum Energy Center (JPEC) 1126
Johnson Matthey 14
Joint Committee for Guides in Metrology (JCGM) 598
Joint Undertaking on Fuel Cells and Hydrogen (FCH JU), European Union 1133, 1187, 1189
– governance structure 1190–1193
– members 1189–1190
Jülich substrate concept 252

k

Kansai Electric Power Company (KEPCO) 9
Karlsruhe Research Center (KTI) 210
Knudsen diffusion 737, 738
Korea Institute of Energy Research (KIER) 15
Korea Institute of Science and Technology 12
Korean Electric Power Research Institute (KEPRI) 87
Korean Institute of Energy Research 12
Kramers–Kronig validation 447
Kyocera 6, 9

l

lanthanum strontium chromite manganite (LSCM) 289
lanthanum strontium cobalt ferrite (LSCF) 49, 293, 295, 484
lanthanum strontium cobaltite (LSC) 49, 268
lanthanum strontium cobaltite ferrite (LSC(F)) 249
lanthanum strontium manganite (LSM) 49, 249, 254, *257*, 290–292
lanthanum strontium scandium manganite (LSSM) 49
lanthanum strontium titanate (LST) 286
large eddy simulations (LESs) 716, 726
laser ablation 208
laser welding 209
Lattice Boltzmann methods (LBM) 851, 854
lead acid batteries 237
Leverett function 855
LIBrary for Process Flowsheeting (LIBPF) technology 78

LIBS (laser-induced breakdown spectroscopy)   488
light traction applications   1092
– commercial vehicles   1093–1094
– forklift trucks   1094–1096
– scooters, wheelchairs, and electromobiles   1093
linear sweep voltammetry (LSV). *See* cyclic voltammetry (CV)
liquid phase, desulfurization in
– adsorption   1024–1026
– desulfurization with overcritical fluids   1030–1031
– distillation   1031–1032
– hydrodesulfurization with presaturator   1022–1024
– ionic liquids   1026–1028
– processes with nonporous membranes   1032–1034
– processes with porous membranes   1032
– selective oxidation   1028–1030
lithium ion technology   20
load cycling   588–592
local hydrogen starvation   558–561
long-term durability test
– data post processing   585
– setting of test conditions (test inputs)   583–584
– test output measurement   584
Los Alamos National Laboratory (LANL)   364
low-temperature fuel cells, catalysis in   407–408
– catalysis in hydrogen production   424–425
– – carbon monoxide removal   429–430
– – hydrogen production catalysis from biomass   430
– – from methanol to heavy hydrocarbons   425–429
– catalyst development, characterization, and *in situ* studies   423–424
– electrocatalysis   408–410
– – alkaline fuel cells   418
– – catalysis in direct fuel cells   420–421
– – direct fuel cells   417–418
– – electrocatalyst degradation   421
– – hydrogen oxidation and CO poisoning   418–419
– – oxygen reduction in PEMFCs   410–417
– novel support materials   422–423
low-temperature fuel cells, physical properties analytics of   521–524
– caloric properties   527–529
– gravimetric properties   524–527
– mechanical properties   531–535
– porosity   530–531

m

3M   388
market portfolio   1162
Markov chain Monte Carlo (MCMC) simulation, for edge rearrangement   689–690
measurement uncertainty evaluation   610–611
– Monte Carlo method   612–616
– procedure summary to evaluate and express uncertainty   611–612
mechanical testing   532, 534
membrane electrode assembly (MEAs)   11, 132, 140, 151, 381, 522
– aging phenomena analysis   587–588
membranes, for ion transportation   388–392
Mercedes-Benz   1059
mercury intrusion porosimetry (MIP)   530
metal/N/C catalysts   415–417
metal gaskets   320
metal oxides   402
methane steam reforming   68
methanol   29, 141, 142, 189. *See also* direct methanol fuel cells (DMFCs)
method of standard contact porosimetry (MSCP)   531
mica-based seals   320–321, *322*
microbial catalysts   157
– anode reactions   157–158
– – biocatalysis   158–159
– – electron donors   158
– – electron-transfer mechanisms   159–160
– cathode reactions   160–161
– – biocatalysts   161
– – electron acceptors   162
– – electron-transfer mechanisms   161–162
microbial desalination cell (MDC)   149
microbial electrolysis cell (MEC)   148, 150, 165, 169
microbial electrosynthesis   170
microbial fuel cells. *See* bioelectrochemical systems (BESs)
microelectrodischarge machining (μEDM)   207
microelectromechanical system (MEMS)   135, 201
micro fuel cells   131–132
– GDL optimization   138–139
– – carbon dioxide discharge   141–142
– – flow-field design   139–141
– – miniaturized DMFC   141

micro fuel cells (*contd.*)
– – passively operating DMFC   142
– materials and manufacturing
– – miniaturization   137–138
– PEMFC physical principles   132–134
– types   134
– – direct ethanol fuel cells (DEFCs)   136–137
– – DMFC   135–136
– – hydrogen-fed   134–135
– – micro-reformed hydrogen fuel cell   135
microporous layers (MPLs)   858
micro-reactors   185
– heat and mass transfer in   185–188
– microchannel fuel processor examples   201–205
– microchannel plate heat-exchanger reactors fabrication   206
– – catalyst coating techniques   210–211
– – construction material choice   206–207
– – micromachining techniques   207–209
– – reactor–heat exchanger assembly   210
– – sealing techniques   209–210
– microchannel plate heat-exchanger reactors heat management   190–191
– – carbon monoxide preferential oxidation   197–200
– – carbon monoxide selective methanation   200–201
– – reforming   191–195
– – water gas shift reaction   195–197
– microchannel plate heat-exchanger reactors, specific features from catalyst formulations for   188–190
micro-reformed hydrogen fuel cell   135
microtechnical devices   208
middle distillates   1011, 1012, 1016, 1020, 1024, 1030, 1031, 1033
military applications, for prime power
– portable generators   1116–1117
– soldier power   1115
minimum spanning tree (MST)   695
Mitsubishi Materials Corporation (MMC)   6, 9
mixed reactant   44
Mobion   27, 28
molten carbonate fuel cells (MCFCs)   26
– balance of plant (BOP)   73–74
– CO_2 energy cycle in   88
– conventional and innovative applications   86
– – applications   90
– – carbon capture, storage, and transportation   87–89
– – distributed generation   86–87
– – hydrogen co-generation   89
– – renewable fuels   89
– geometry and materials   70–71
– operating conditions   69–70
– operating principle   67–69
– reforming   71–73, 72
– state of the art   75–76
– technology analysis   76
– – approach   76–79, 77
– – scientific knowledge   81–86
– – technology optimization   79–81
– vendors   75
molten carbonate fuel cells (MCFCs) modeling   791–794
– electrode models   804–806
– – agglomerate models   809–810
– – spatially lumped models   806–808
– – thin-film models   808–809
– – volume-averaged models   810–811
– spatially distributed model   792–793
– – anode gas channels   795–798
– – cathode gas channels   798–799
– – general assumptions   794–795
– – interface to electrode models   804
– – potential field model   800–804
– – solid phase   799–800
Monte Carlo method   612–616
morphology, of material   669, 670. *See also* stochastic modeling, of fuel-cell components
moving-average model, for dependent marking   685–687
MTI Micro Fuel Cells   14
MTU CFC Solutions   75, 87
multi-acid side chain (MASC) ionomers   391
MVV   25

n

N.ERGHY   1190
Nafion membrane   141
nanoindenter   534
nanostructured materials, for fuel cells   379–380
– catalyst supports, for electron conduction   397–402
– electrocatalysts, for oxygen reduction   393–397
– electrodes, for hydrogen oxidation   384–388
– fuel cell and system   380–382
– membranes, for ion transportation   388–392
– triple phase boundary   382–384

nanotechnology 379–380, 382, 396, 402
National Aeronautics and Space
    Administration (NASA) 222
National Hydrogen Association 1196
National Innovation Program (NIP) 1122
National Learning Demonstration 1130,
    1131–1132
National Organization for Hydrogen and Fuel
    Cell Technology (NOW) 1122, 1200–1201
National Renewable Energy Laboratory
    (NREL) 369, 401
natural gas combined cycle (NGCC) 88
Natural Resources Canada
– ecoENERGY Technology Initiative 1136
Navier–Stokes equations 713, 719, 720, 738,
    851
Nernst equation 46
neutron radiography 504–507. *See also*
    X-rays and neutrons
– and local current density measurements,
    combined approach of 509–510
New European Driving Cycle [or Motor
    Vehicle Emissions Group (MVEG)] 20
NEW Industry Grouping (NEW IG) 1189
Nippon Telegraph and Telephone (NNT) 6
Nokia Siemens 1112
non-hybrid fuel cell powertrains 1058
non-linear least-squares (NLS) fitting
    algorithm 447
non-platinum catalysts 415
North Rhine-Westphalia (NRW)
    1124–1125, 1215
NRW Fuel Cell and Hydrogen Network
    1124
Nuvera 30

o

off-grid power 1107, 1109
– base stations 1111–1113
– homes 1109–1111
Ohmic law 843
on/off aging test
– data post processing 580–581
– setting of test conditions (test inputs)
    578
– test output measurement 578–580
one-dimensional cell polarization curve
    662
Oorja Protonics 16, 30
open-cell voltage (OCV) 544, 546,
    551, 563, 582
operando techniques 424
optical microscopy 477–481
organic alcohols and acids 170–171

Ostwald ripening 552
outreach portfolio 1164–1166
overall hydrogen starvation 553–555
Ovonics concept 103
oxide catalysts 107
oxygen consumption, in channel and
    quasi-two-dimensional polarization curve
    663–665
oxygen reduction, in PEMFCs 410–417
– metal/N/C catalysts 415–417
– non-platinum catalysts 415
– platinum-based catalysts 411–415
– platinum-free noble metal catalysts 415
– transition metal chalcogenides 417
oxygen reduction reaction (ORR) 387, 410,
    411, 414–415, 418, 654, 845, 846, 849, 885
oxygen starvation 553
oxygen starvation 561–562

p

parallel hybrid vehicles 1056
Pemeas 11
perfluorinated sulfonic acid (PFSA)
    membranes 388
performance test and standardization
    measurements
– data post processing 583
– setting of test conditions (test inputs)
    581–582
– test output measurement 582–583
peroxides, oxidation with 1029
pervaporation 1033–1034
PHOEBUS specifications *238*
phosphoric acid 11
– as electrolyte 827–829
phosphoric acid fuel 335, *336*
– basic properties and formal considerations
– – acidity and protolytic equilibria 337–338
– – anhydride and condensation reactions
    337
– – composition specifications and
    condensation equilibria 338–339
– equilibria between the polyphosphoric acid
    species and concentrated phosphoric acid
    composition 353–354
– – evaluated literature data 354–356
– proton conductivity, as function of
    composition and temperature
– – electrical conductivity mechanisms 344
– – enthalpy of activation for ionic transport
    350–352
– – evaluated data for dynamic viscosity of
    aqueous phosphoric acid 352–353
– – evaluated literature data 344–346

phosphoric acid fuel (*contd.*)
– – non-arrhenius behavior for ionic transport 346–350
– vapor pressure of water, as function of composition and temperature
– – evaluated literature data 340–343
– – number of independent variables and Gibb's phase rule 339–340
phosphoric acid fuel cell (PAFC) 819–820
photo-oxidation 1028–1029
phyllosilicates 320
physical vapor deposition (PVD) 265
Pinch point analysis 639–641, 928–930
piping and instrumentation (P&I) diagram 946, 947
planar fuel cell 137, 139
planar stacks (pSOFCs) 302, 303
– seal functional requirements 304–306, 306, 319
plasma desulfurization 1028
plasticity 776, 777
platinum-based catalysts 411–415
platinum-free noble metal catalysts 415
plug-in hybrid electric vehicles (PHEVs) 18
Plug Power 12, 30, 389
Poisson line tessellation (PLT) 671
polarization curve modeling 829–830
– activation overpotential 830–831
– mass transport 833–834
– ohmic resistance 831–833
political decision-making portfolio, support for 1162
polybenzimidazole (PBI) 11
polymer electrolyte fuel cells (PEFCs) 493, 500, 587, 628, 705. *See also* polymer electrolyte membrane fuel cells (PEMFCs)
– neutron radiography 504–507
– X-rays 500–504
polymer electrolyte membrane (PEM) 229–230, 380, 382, 383, 388–392, 399, 529, 842–845
– electrolyzers 230–231
– fuel cell 230, 232
– PEMFCs 231
– URFC 231–233
polymer electrolyte membrane fuel cells (PEMFCs) 131, 135, 138, 140, 222, 231, 361, 380, 543, 545, 546, 549, *550*, 551, *555*, *556*, 562, 658–665, 839–842, 879–881. *See also* polymer electrolyte fuel cells (PEFCs)
– bipolar plates 868–869
– catalyst layers 845–850
– cell-level modeling and simulation 881–882

– – cold start operation 893–898
– – dimensionality 882–884
– – flow maladjustment 900
– – large-scale fuel-cell simulation 898–899
– – model validation 903, 905
– – nonisothermal modeling 888, 890–891
– – single-phase flow 901
– – transient operation 884–888
– – two-phase flow 891–893, 902–903
– coolant flow 869
– gas flow channels 859–864
– GDLs
– – and gas flow channel interface 864–868
– – and microporous layers 850–859
– macroscopic approach 842, 848, 851, 855, 858, 872, 881, 911
– mesoscopic approach 842, 851, 872
– metallic bipolar plates 363
– – bare plates 363–366
– – coated stainless-steel plates 368–370
– – light alloys 366–367
– model validation 869–871
– oxygen reduction in 410–411
– – metal/N/C catalysts 415–417
– – non-platinum catalysts 415
– – platinum-based catalysts 411–415
– – platinum-free noble metal catalysts 415
– – transition metal chalcogenides 417
– perspectives 370–371
– – coatings and surface modifications 372–374
– – substrate selection 371–372
– physical principles 132–134
– polymer electrolyte membrane 842–845
– schematic of *840*, *880*
– stack-level modeling and simulation
– – fuel-cell stacks modeling and simulation 907–910
– – model validation 910–11
– – need for 906–907
polytetrafluoroethylene (PTFE) 100, 106, 108, 112, 115, 117, 122, 388
– -bonded gas diffusion electrodes dry preparation 108–111
pore phase, stochastic network model for 677–678
– graph modification 679
– pores detection 678–679
– stochastic modeling of edges 688–690
– vertex model validation 684–685
– vertices marking 685–688
– vertices stochastic modeling 680–684
porous nickel 107
portable generators, military 1116–1117

portable power 1107. *See also* premium power
POSCO Power 75, 86, 87
post-test characterization, of SOFC stacks 469, *490*
– post-test analysis methods 471–472
– reasons 470
– stack dissection 472–473
– – characterization techniques 488
– – lessons 488–489
– – optical microscopy 477–481
– – photography and distance measurements 475–477
– – scanning electron microscopy (SEM) and energy-dispersive x-ray (EDX) analysis 482–484
– – stack embedding 474–475
– – thermography 473
– – topography 482
– – wet chemical analysis 486–488
– – X-ray diffraction (XRD) 484–486
potentiodynamic controlled experiment 153
potentiostatic control 153
power applications. See sealing techniques
power-split hybrid configuration 1056–1057
predictive method for vapor–liquid and liquid–liquid equilibria determination 630–632
– model fuel evaporation 634–637
– residual hydrocarbons in reformer product gas 632–634
premium power 1107. *See also* off-grid power
– market review 1107–1109
– portable applications 1113
– – military applications for prime power 1115
– – remote monitoring and remote sensing 1113–1114
printed circuit board (PCB) 549
process flow sheet 941
production portfolio 1157
– longer term production options 1158
– near-mid-term production options 1157–1158
– near-term production options 1157
proton conductivity, as function of composition and temperature
– electrical conductivity mechanisms 344
– enthalpy of activation for ionic transport 350–352
– evaluated data for dynamic viscosity of aqueous phosphoric acid 352–353

– evaluated literature data 344–346
– non-arrhenius behavior for ionic transport 346–350
Proton Energy Systems 239
proton pump 562
proton transport 843
PROX 429
pulsed laser deposition 277
punching techniques 207–208

q

quality assurance 249, 259, 268, 270, 271, 272, 573
– degradation and lifetime investigations
– – load cycling 588–592
– – membrane electrode assemblies (MEA) aging phenomena analysis 587–588
– experiment design in fuel-cell research 592–593
– standardized test cells 587
– test procedures and standardization measurements
– – dynamic load cycling aging test 585–587
– – fuel cell preconditioning 574
– – humidification sensitivity test 574–578
– – long-term durability test 583–585
– – on/off aging test 578–581
– – performance test 581–583
quasi-*in situ* approach 513

r

Raman spectroscopy 105
Ramberg–Osgood equation 777
random error 611
Raney nickel catalysts 115–121
recalcitrant compounds 171
reference electrode 546–548
reforming 71–73, *72*, 191–195
regenerative fuel cells 219–220, *239*
– applications and systems 236
– – stationary systems for seasonal energy storage 237–239
– for aviation applications 239–240
– electrodes 226
– – for alkaline electrolytes 226–229
– – polymer electrolyte membrane (PEM) 229–233
– flow chart of *223*
– history 222–223
– principles 220–221
– solid oxide electrolyte (SOE) 233–234, *233*
– system design and components 234–236
– thermodynamics 223–226
– time line of developments *222*

relative humidity (RH) 574, 575, *576*, *577*
relaxation time distribution. *See* distribution function of relaxation times (DRT)
renewable electricity certificates system (RECS) 1123
research and development, technical advancement of 3, 4
– application and demonstration in transportation
– – fuel cells and batteries for propulsion 17–22
– – on-board power supply with fuel cells 22–24
– direct methanol fuel cells (DMFCs) 12–13
– – for light traction 14–17
– – for portable applications 13–14
– high-temperature polymer electrolyte fuel cell (HT-PEFC) 11
– – actors and development areas 11–12
– marketable development results
– – DMFC battery chargers 27–29
– – light traction 30
– – submarine 27
– – uninterruptable power supply/backup power 29–30
– solid oxide fuel cells (SOFCs) 9
– – actors and development areas 8–9
– – planar designs 6–8
– – state of cell and stack developments 10–11
– – tubular concepts 4–6
– special markets for fuel cells 26–27
– stationary applications in building technology 24–6
– stationary industrial applications 26
reverse current decay (RCD) mechanism 556, 560–561
reverse fuel cells. *See* regenerative fuel cells
reverse hydrogen electrode (RHE) 547
reversible heat release 890
rigid bonded seals 308–309
– ceramic seals 318–319
– glass and glass–ceramic sealants 309–318
Risø National Laboratory 8
Rittal 29, 30
Rolls Royce 8
round-trip efficiency 224

s

safety portfolio 1163–1164
samarium-doped ceria (SDC) 280, 291
Samsung Advanced Institutes of Technology 11

S-Brane process 1033
scale-free networks 691
scanning electron microscopy (SEM) 152
– and energy-dispersive x-ray (EDX) analysis 482–484
Schröder's paradox 527
Schunk 30
screen-printing method 211, 277, 278
sealing techniques 306–308, *307–308*
– bonded compliant seals 323–328
– compressive seals 319–323
– rigid bonded seals 308–319
sediments, plants, and photosynthesis 168
selective CO methanation (SELMETH) 429, 430
Serenergy 12
serial hybrid vehicles 1056
series and parallel hybrids 1077–1078
SFC Energy 16, 29, 1114
Shell Hydrogen 1132
Shewanella oneidensis 160
Shewanella spp. 158
Siemens 14, 27
silicate-based glass–ceramic sealants, common compositional modifiers for *312*
silver catalysts 107, 121–123
SIMS (secondary ion mass spectrometry) 488
single-chamber fuel cells (SC-FCs) 43–44
– single-chamber solid oxide fuel cells (SC-SOFCs) 44
– – anode materials 48–49
– – anode-supported 51–52
– – applications 60–61
– – basic principles of operation 44–46
– – catalysis in 46–47
– – cathode materials 49
– – with coplanar electrodes 52–59
– – current collection 48
– – electrode and electrolyte materials 48
– – electrolyte materials 50
– – electrolyte-supported 50–51
– – fully porous 59–60
– – heat production and real cell temperature 47–48
– – tubular 60
single-chamber solid oxide fuel cells (SC-SOFCs). *See* single-chamber fuel cells (SC-FCs)
small-angle X-ray scattering (SAXS) 398
SNMS (secondary neutron mass spectrometry) 488
sol–gel methods 211

solid oxide fuel cell (SOFCs)   9,233, 249–250, 275, 301–303, 444, 445, 449, 452, 465, 488
– actors and development areas   8–9
– advances in testing of   268–269
– – dwell time between individual current–voltage points   271
– – reduction temperature   270–271
– – specifications   270
– – testing housing   269–270
– component requirements   279
– control subsystem   989–991
– and electrochemical fundamentals   275–276
– electrode materials   278–280
– – anode materials   280
– – cathode materials   281
– electrodes current status and fabrication methods   276
– – coating electrode material methods   276–278, 278
– fuel delivery   982
– fuel processing subsystem   979–982
– infiltration   282–284
– – anodes produced by   284–290
– – cathodes produced by   290–295
– – motivation for   281–282
– with LSC(F) cathode   259
– – 2000–2006   259–266
– – 2006–2010   266–268
– with LSM cathode   250
– – 1995–1998   250–252
– – 1998–2002   252–254
– – 2002–2005   254–258
– – 2005–2010   259
– oxidant delivery   982–983
– planar designs   6–8
– power conditioning subsystem   987–989
– for power generation   963–965
– power generation subsystem   970–971
– – equipments   977–979
– – fuel utilization   974–975
– – number of stacks and stack arrangement   971–973
– – stack cell voltage and current density   973–974
– – stack operating pressure   976–977
– – stack pressure drop   975
– – stack temperature gradient   975–976
– power systems
– – design   969–970
– – general   965–968
– – portable systems   991–993
– – stationary systems   997–1006
– – transportation systems   993–996
– – types   968–969
– pSOFC seal functional requirements   304–306, 306, 319
– sealing techniques   306–308, 307–308
– – bonded compliant seals   323–328
– – compressive seals   319–323
– – rigid bonded seals   308–319
– state of cell and stack developments   10–11
– system configuration   969–970, 971, 987, 990
– thermal management subsystem   983–986
– tubular concepts   4–6
– water delivery   983
solid oxide fuel cells (SOFC) modeling   733–735
– governing equations   735–736
– – chemical reactions   745–746
– – electrochemistry   740–744
– – energy conservation   739–740
– – mass conservation   736–738
– – momentum conservation   738–739
– macroscale   747
– – cell-level modeling   755–758
– – stack-level modeling   750–755
– – system-level modeling   747–750
– mesoscale   758–760
– nanoscale   761
Solid State Energy Conversion Alliance (SECA)   8
Solvay Solexis   388
sorption-enhanced steam methane reforming (SE-SMR) process   71
South Korea   1137
– Domestic Fleet Program   1137
– Low Carbon, Green Growth program   1137
– stationary power and early market applications   1145
spatially lumped models   806–808
spontaneous galvanic displacement (SGD)   401
spray coating   211
sputtering   277
stack-level modeling   750–755, 825–827
– fuel-cell stacks modeling and simulation   907–910
– model validation   910–11
– need for   906–907
standardization   249, 268, 270, 271, 272
state of the art cells   257
static test inputs   575, 579, 582
stationary systems   997
– integrated gasification fuel cell (IGFC)   1001, 1003–1006
– simple cycle SOFC systems   997–998

stationary systems (*contd.*)
- SOFC/GT hybrid systems   998–1001
steam generator   984
Stefan–Maxwell equation   737, 756, 757
stochastic modeling, of fuel-cell components   669–670
- classical random graph models   690–691
- multi-layer model for paper-type GDLs   670
- - binder modeling   672–673
- - fiber modeling   671–672
- - model parameter fitting   674–675
- - results   675–676
- - time-series models for non-woven GDLs   676–677
- stochastic network model for pore phase   677–678
- - graph modification   679
- - pores detection   678–679
- - stochastic modeling of edges   688–690
- - vertex model validation   684–685
- - vertices marking   685–688
- - vertices stochastic modeling   680–684
- structural characterization of porous GDL   692
- - connectivity   695–696
- - graph model validation   698
- - multi-layer model validation   696–698
- - pore size distributions   694–695
- - tortuosity   692–693
- transport simulations along graph edges   691
Stokes–Einstein equation   348
storage portfolio   1158–1159
strain   532
stress   532
structural diffusion   344, 389
submarine   27
substrate manufacturing technologies   277
sulfonated fluoropolymer membranes   388
sulfur-containing substances in crude oil, chemical classes of   1013–1015
sulfur poisoning   428, 429
summer schools   1218
supercomputers   703
Supplemental Federal Test Procedure (SFTP)   1049
surface plasma modification   373–374
sustainable technology   150, 161, 164, 172, 173
Swiss Paul Scherer Institute   1059
synchrotron X-ray radiography   508–509
Synechocystis sp.   163
syntrophic acetate oxidizers (SAOs)   168
syntrophic acetogenic bacteria (SAB)   168

systematic error   611
system-level modeling   747–750
systems engineering principles   919–920
- chemical equilibrium   923–926
- - heat management analytical methods   926–940
- - process analysis and design   940–945
- construction   956–957
- detailed engineering   945–948
- - drawings and piping   954–955
- - FEMA   950–953
- - P&I diagram   948–950
- - peripheral component selection   953–954
- general considerations   920–923
- procurement   956
systems integration portfolio   1159–1160

t
tactical quiet generators (TQGs)   1116–1117
Tafel equation   830
Tafel-type equation   *824*
tape casting   278
technical portfolio   1161–1162
technical stress   532
Technical University of Denmark   12
technology readiness level (TRL)   919
Technology Validation Program. *See* National Learning Demonstration
thermal management subsystem   983–986
thermography   473
thermogravimetric analysis (TGA)   526
thermomechanically induced stress numerical modeling, in SOFCs   767–768
- chronological overview   768–773, 775
- geometric design effect on stress distribution   778–779
- - CFD analysis   779–781
- - thermomechanically induced stress analysis   782–788
- stress and strain mathematical formulation   773
- - cell, sealant, and wire mesh components   773–774, 776
- - metallic components   776–778
thin-film deposition techniques   279
thin-film models   808–809
Thiobacillus   1013
Toho Gas   6
Tokuyama   392
Tokyo Gas   9
tomography   488
Topsøe Fuel Cells   7
tortuosity   692–693

Toshiba 13, 27
total organic carbon (TOC) 624–625
TOTO 5, 9
transient 1D model 552
transition metal chalcogenides 417
transmission electron microscopy (TEM) 488
transportation systems 993
– SOFC-based APUs
– – for aircraft 994–996
– – for automobiles and trucks 993–994
transport simulations along graph edges 691
triple-hybrid fuel cell vehicles 1060–1061
triple phase boundary 382–384
true stress 532
Truma 23
TU Bergakademie Freiberg 25
TUBITAK Marmara Research Center, Fuel Cell Laboratory 286
tubular SC-SOFCS 60
tubular stacks 302
tutorials 1218

u
UK Hydrogen Energy Network (H$_2$NET) 1206
under-relaxation factor 714
UNIFAC method 632, 633, 634, 638, 639
UNIGEN RFC 239
United Nations Industrial Development Organization (UNIDO)
– International Centre for Hydrogen Energy Technologies (ICHET) 1217, 1218
United States
– DOE Technology Validation Program 1130–1132
– state activities 1132
– – California 1132–1133
– stationary power and early market applications 1144–1145
United Technologies 389
unitized regenerative fuel cell (URFC) 220, 221, 223
– alkaline 228–229
– polymer electrolyte membrane (PEM) 231–233
University of South Carolina 12
upflow anaerobic sludge bed (UASB) 168
UPS Systems 1114
US DOE technical targets 362
user-defined functions (UDFs) 709–710
US Fuel Cell Council (USFCC) 1196
UTC Power and Delphi 8

v
Vaillant 25
vapor pressure of water
– as function of composition and temperature
– – evaluated literature data 340–343
– – number of independent variables and Gibb's phase rule 339–340
variable test inputs 575, 579, 582, 584
vehicular diffusion 344, 389, 842
Versa Power Systems (VPS) 7, 8
vertices
– marking 685
– – degress 687–688
– – moving-average model for dependent marking 685–687
– stochastic modeling 680
– – multilayer representation 680
– – point process model construction and fitting 680–684
virtual material design 699
VNG 25
Vogel–Tammann–Fulcher equation 348, 352
Volkswagen 12
Volkswagen Bora Hy.Power 1059
volume-averaged models 810–811
volume hourly space velocity (VHSV) 189, 190
volume of fluids (VOFs) 719, 720, 854, 867

w
Walden product 348
wastewater treatment 164–168
water gas shift (WGS) reaction 68, 195–197, 204, 425, 639, 640, 706, 923
water management 389
water transport channels (WTCs) 139
wavelength-dispersive X-ray (WDX) analysis 484
Webasto 23
Weiterbildungszentrum Brennstoffzelle Ulm (WBZU) 1216, 1217
wet chemical analysis 486–488

x
X-ray diffraction (XRD) 484–486
X-ray photoelectron spectroscopy (XPS) 111, 112, 124, 125
X-rays 500–502
– dynamic radiographic studies of water transport 503–504
X-rays and neutrons
– complementarity 494–495
– radiography and tomography principles

X-rays and neutrons (*contd.*)
– – artifacts   498–499
– – image normalization procedure   499
– – synchrotron X-ray sources and X-ray tubes   496–497
– – tomography and tomographic reconstruction   497–498
– – transmission and attenuation   496
x-shapes   653–654

y
Yamaha Motor   16
Young's modulus   532
yttria-stabilized zirconia (YSZ)   47, 50, 233, 280, 283, 284–287, 289–295

z
ZSW   26